I Physical Constants of Materials

Material	Modulus of Elasticity E		Modulus of Rigidity G		Poisson's Ratio ν	Unit Weight w		
	Mpsi	GPa	Mpsi	GPa		lbf/in³	lbf/ft³	kN/m³
Aluminum (all alloys)	10.4	71.7	3.9	26.9	0.333	0.098	169	26.6
Beryllium copper	18.0	124.0	7.0	48.3	0.285	0.297	513	80.6
Brass	15.4	106.0	5.82	40.1	0.324	0.309	534	83.8
Carbon steel	30.0	207.0	11.5	79.3	0.292	0.282	487	76.5
Cast iron (gray)	14.5	100.0	6.0	41.4	0.211	0.260	450	70.6
Copper	17.2	119.0	6.49	44.7	0.326	0.322	556	87.3
Douglas fir	1.6	11.0	0.6	4.1	0.33	0.016	28	4.3
Glass	6.7	46.2	2.7	18.6	0.245	0.094	162	25.4
Inconel	31.0	214.0	11.0	75.8	0.290	0.307	530	83.3
Lead	5.3	36.5	1.9	13.1	0.425	0.411	710	111.5
Magnesium	6.5	44.8	2.4	16.5	0.350	0.065	112	17.6
Molybdenum	48.0	331.0	17.0	117.0	0.307	0.368	636	100.0
Monel metal	26.0	179.0	9.5	65.5	0.320	0.319	551	86.6
Nickel silver	18.5	127.0	7.0	48.3	0.322	0.316	546	85.8
Nickel steel	30.0	207.0	11.5	79.3	0.291	0.280	484	76.0
Phosphor bronze	16.1	111.0	6.0	41.4	0.349	0.295	510	80.1
Stainless steel (18-8)	27.6	190.0	10.6	73.1	0.305	0.280	484	76.0
Titanium alloys	16.5	114.0	6.2	42.4	0.340	0.160	276	43.4

McGraw-Hill Series in Mechanical Engineering

CONSULTING EDITORS

Jack P. Holman, *Southern Methodist University*
John R. Lloyd, *Michigan State University*

Mechanical Engineering Design

Seventh Edition

Joseph E. Shigley
Late Professor of the University of Michigan

Charles R. Mischke
Professor Emeritus of Mechanical Engineering, Iowa State University

Richard G. Budynas
Professor of Mechanical Engineering, Rochester Institute of Technology

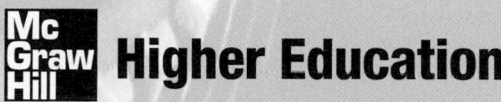
Higher Education

Boston   Burr Ridge, IL   Dubuque, IA   Madison, WI   New York   San Francisco   St. Louis
Bangkok   Bogotá   Caracas   Kuala Lumpur   Lisbon   London   Madrid   Mexico City
Milan   Montreal   New Delhi   Santiago   Seoul   Singapore   Sydney   Taipei   Toronto

The McGraw-Hill Companies

MECHANICAL ENGINEERING DESIGN, SEVENTH EDITION
International Edition 2004

Exclusive rights by McGraw-Hill Education (Asia), for manufacture and export. This book cannot be re-exported from the country to which it is sold by McGraw-Hill. The International Edition is not available in North America.

Published by McGraw-Hill, a business unit of The McGraw-Hill Companies, Inc., 1221 Avenue of the Americas, New York, NY 10020. Copyright © 2004, 2001, 1989, 1983, 1977, 1972, 1963 by The McGraw-Hill Companies, Inc. All rights reserved. No part of this publication may be reproduced or distributed in any form or by any means, or stored in a database or retrieval system, without the prior written consent of The McGraw-Hill Companies, Inc., including, but not limited to, in any network or other electronic storage or transmission, or broadcast for distance learning.
Some ancillaries, including electronic and print components, may not be available to customers outside the United States.

10 09 08 07
20 09 08 07 06
CTF · MPM

Library of Congress Cataloging-in-Publication Data

Shigley, Joseph Edward.
   Mechanical engineering design / Joseph E. Shigley, Charles R. Mischke, Richard G. Budynas.—7th ed.
         p.   cm.
  Includes index.
  ISBN 007-252036-1
  1. Machine design. I. Mischke, Charles R. II. Budynas, Richard G. (Richard Godon). III. Title.
TJ230.S5 2004
621.8'15—dc21   2003048763
                                           CIP

When ordering this title, use ISBN 007-123270-2

Printed in Singapore

www.mhhe.com

Dedication

To the many people who have encouraged me to grow professionally and spiritually over the years, especially my wife, Joanne, and my late mother, Inez. Included also are a great many fellow educators, students, and designers that I am indebted to.

Richard G. Budynas

To the over 3500 of my students who asked many good questions of me, learned to ask them of themselves, to answer them, and then went on to make their Alma Maters proud.

Charles R. Mischke

About the Authors

Joseph E. Shigley (deceased May 1994)
He was Professor Emeritus at the University of Michigan, Fellow in the American Society of Mechanical Engineers, received the Worcester Reed Warner medal in 1977, and their Machine Design Award in 1985. He was the author of eight books, including *Theory of Machines and Mechanisms* (with John J. Uicker, Jr.), and *Applied Mechanics of Materials*. He was Coeditor-in-Chief of the *Standard Handbook of Machine Design*. He began *Machine Design* as sole author in 1956, and it evolved into *Mechanical Engineering Design,* setting the model for such textbooks.

He was awarded the B.S.M.E. and B.S.E.E. degrees of Purdue University, and the M.S. of the University of Michigan.

Joseph Edward Shigley indeed made a difference.

Charles R. Mischke has held faculty positions at the University of Kansas, Pratt Institute, New York, and Iowa State University. He has received many awards for teaching and scholarship, including the Machine Design Award of ASME and Life Fellow of ASME. In addition to numerous research papers, he has authored Elements of Mechanical Analysis, 1963, Introduction to Computer-Aided Design 1968, Mathematical Model Building, 1980, coauthored Mechanical Engineering Design, 1986, 2001, and was Co-Editor-in-Chief of the award-winning Standard Handbook of Machine Design, 1986, 1993.

He was awarded the BSME and MME of Cornell University, and the Ph.D. of the University of Wisconsin. He is a licensed Professional Engineer in Iowa, and Kansas, and lectures in various forums on topics of mutual interest.

Richard G. Budynas is a Professor of Mechanical Engineering at Rochester Institute of Technology. He has over 30 years experience in teaching and practicing mechanical engineering design. He is author of a newly revised McGraw-Hill textbook, *Advanced Strength and Applied Stress Analysis,* Second Edition; and co-author of a newly revised McGraw-Hill reference book, *Roark's Formulas for Stress and Strain,* Seventh Edition.

He was awarded the BME of Union College, MSME of the University of Rochester, and the Ph.D. of the University of Massachusetts. He is a licensed Professional Engineer in the state of New York.

Contents in Brief

Appendixes

Contents

Objectives

This text is intended for students beginning the study of mechanical engineering design. It can also serve as a basic reference book for practicing engineers. The objectives of the text are to:

- Cover the basics of machine design, including the design process, engineering mechanics and materials, failure prevention under static and variable loading, and characteristics of the principal types of mechanical elements.
- Offer a practical approach to the subject through a wide range of real-world applications and examples.
- Encourage readers to link design and analysis.

General Approach: A Return to the Basics

The seventh edition has retained and enhanced the most popular features of earlier editions, while removing many nonessential elements. Coverage focuses on mainstream machine design, with only moderate discussion of statistical methods.

A practical approach is emphasized throughout the text. Concise design problems and examples illustrating the decision-making process are presented.

The writing style is accessible and direct, with many chapters containing reduced discussion of advanced topics.

New to This Edition

The goal has been to make this seventh edition easier for readers to "navigate" in terms of key concepts and procedures in mechanical design. The text contains several new and enhanced features. Among the highlights are:

- *Practical coverage of basic machine design.* Coverage has been streamlined, focusing on major topics that better fit the typical engineering curriculum. Readers can better identify major concept development.
- *Improved coverage of engineering mechanics and failure prevention.* Enhanced discussion of key topics such as fracture and fatigue is provided. Topics are presented in a logical order and coverage is condensed.
- *Enhanced coverage of machine component design.* This edition improves upon the already practical and authoritative earlier coverage of design considerations for major machine components.
- *Revised and expanded problem sets.* Problems have been updated, and new ones have been added. Problems are arranged to progress from basic to challenging in a consistent way. Statistically oriented problems have been drastically reduced, emphasizing deterministic solution of fundamental design problems.

Content Changes and Reorganization

Throughout the text, key chapters have been rewritten and topics have been reorganized to provide greater clarity and logical progression. The writing style is straightforward for easy comprehension. Some highlights are noted below:

Part 1: Basics

Part I provides a logical and unified introduction to machine design. The chapters have been shortened, and they focus sharply on key concepts, allowing students to quickly gain a basic grounding in the subject matter.

- *Chapter 1, Introduction.* Updated with contemporary examples, coverage of design tools and resources, and new information on using the Internet.

- *Chapter 2, Statistical Considerations.* A self-contained overview of statistics that returns to the approach seen in earlier editions (statistics-based examples and problems have been drastically reduced in subsequent chapters).

- *Chapter 3, Materials.* Treated as review material and moved forward in the book. Advanced topics are relocated to chapters where specific materials are applied in design. Coverage of composites has been added.

- *Chapter 4, Load and Stress Analysis.* The chapter has been rewritten, with improved flow and clarity. Additional coverage on loading in general, as well as on equilibrium, free-body diagrams, shear force, bending moments, singularity functions, and curved beams.

- *Chapter 5, Deflection and Stiffness.* Simplified with clear coverage, and added practical examples on superposition, singularity functions, and Castigliano's method.

Part 2: Failure Prevention

Some of the most significant improvements to this edition can be found in Part II, which now features condensed, clear, and concise coverage.

- *Chapter 6, Failures Resulting from Static Loading.* Better organization throughout the chapter, and discussion of fracture mechanics, a key topic.

- *Chapter 7, Fatigue Failure Resulting from Variable Loading.* Significant improvements have been made throughout this crucial chapter. It now moves logically, using a deterministic approach, from general concepts in variable loading failure to discussion of the stress life, strain life, and fracture mechanics methods. New illustrations have been added, and statistical methods have been moved to the end of the chapter.

Part 3: Design of Mechanical Elements

Part 3 covers the design of specific machine components. Throughout this section the presentation of material has been enhanced for better flow. New, more appropriate examples and problems have been added. Coverage has been streamlined and simplified throughout, with most statistics-based examples and problems removed. The topics covered are:

- *Chapter 8, Screws, Fasteners, and the Design of Nonpermanent Joints*
- *Chapter 9, Welding, Bonding, and the Design of Permanent Joints*
- *Chapter 10, Mechanical Springs*
- *Chapter 11, Rolling-Contact Bearings*

- *Chapter 12, Lubrication and Journal Bearings*
- *Chapter 13, Gearing—General*
- *Chapter 14, Spur and Helical Gears*
- *Chapter 15, Bevel and Worm Gears*
- *Chapter 16, Clutches, Brakes, Couplings, and Flywheels*
- *Chapter 17, Flexible Mechanical Elements*
- *Chapter 18, Shafts and Axles*

Learning Tools

This edition features a wealth of elements that can help students quickly understand and use machine design concepts. Key learning tools are outlined below.

Chapter introductions. Each chapter opens with an introduction that explains the chapter's objectives and the importance of topics covered.
Numerous worked-out examples. Each chapter contains a large number of worked-out examples that clarify the material and illustrate basic concepts.
Numerous end-of-chapter problems. Each chapter ends with an array of real-world problems that help reinforce students' understanding of the material.
Extensive use of graphics. Line drawings and photographs are used extensively throughout the book to illustrate material and help students visualize concepts and problems.

Supplements

Online Learning Center Tools

Our new web-based learning center offers an array of practical computer-based tools to help students and instructors explore and master machine design concepts. Key elements include:

Student Supplements

- *Tutorials—Presentation of 10 major concepts, with visuals.* Among the topics covered are pressure vessel design, press and shrink fits, contact stresses, and design for static failure.

- *MATLAB® for machine design.* Includes visual simulations and accompanying source code. The simulations are linked to examples and problems in the text, and demonstrate the ways computational software can be used in mechanical design and analysis.

- *FEPC.* A two-dimensional, finite element analysis software program that provides readers with an easy-to-use tool for understanding the use of FEA in mechanical design. An accompanying Finite Element Primer explains use of the program and application to mechanical design problems. Additional FEM tutorials are also included to give a more detailed look at FEA applications to machine design situations.

- *Fundamentals of engineering (FE) exam questions for machine design.* Engineering test relating to mechanical design, with interactive problems and solutions. These multiple-choice questions serve as effective, self-testing problems as well as for FE exam preparation.

Instructor Supplements (under password protection)

- *Solutions manual.* An instructor's manual that contains solutions to most end-of-chapter nondesign problems. Also available in print format.
- *PowerPoint® slides.* Approximately 200 slides of important figures and tables from the text are provided in PowerPoint format for use in lectures.

Acknowledgments

This edition has benefited from the comments, suggestions, and assistance of numerous colleagues. We would like to acknowledge the following persons:

Leonard L. Bashford, *University of Nebraska*
David Beale, *Auburn University*
P. Thomas Blotter, *Utah State University*
Stephen Boedo, *Rochester Institute of Technology*
Lawrence Carlson, *University of Colorado, Boulder*
Robert Corey, *United States Naval Academy*
Jerry Fuh, *National University of Singapore*
Vladimir Glozman, *California State Polytechnic University, Pomona*
Itzhak Green, *Georgia Institute of Technology*
A. Henry Hagedoorn, *University of Central Florida*
K.R. Halliday, *Ohio University*
Antony Hodgson, *University of British Columbia*
Cecil Huey, Jr., *Clemson University*
E. William Jones, *Mississippi State University*
Tim Lee, *McGill University*
Robert LeMaster, *University of Tennessee*
Liwei Lin, *University of California–Berkeley*
Noah D. Manring, *University of Missouri–Columbia*
David M. McStravick, *Rice University*
Alan H. Nye, *Rochester Institute of Technology*
Keith Nisbitt, *University of Missouri–Rolla*
Marcia O'Malley, *Rice University*
Gordon Pennock, *Purdue University*
R.E. Rowlands, *University of Wisconsin*
Kazuhiro Saitou, *University of Michigan*
Albert J. Shih, *North Carolina State University*
Brian S. Thompson, *Michigan State University*
Ng Heong Wah, *Nanyang Technological University, Singapore*

Special thanks to Om Prakash Agrawal, Southern Illinois University and Vladimir Glozman, California State Polytechnic University, Pomona, for their help in accuracy checking of the text; and to Robert West, Reginald Mitchiner, the late Charles Knight, Shih-Liang (Sid) Wang, Elizabeth Kenyon, and Edward Anderson for their help with the text's Online Learning Center website resources.

List of Symbols

This is a list of common symbols used in machine design and in this book. Specialized use in a subject-matter area often attracts fore and post subscripts and superscripts. To make the table brief enough to be useful the symbol kernels are listed. See Table 14–1, pp. 723–724 for spur and helical gearing symbols, and Table 15–1, pp. 777–778 for bevel-gear symbols.

A	Area, coefficient
$\mathbf{A}$	Area variate
a	Distance, regression constant
$\hat{a}$	Regression constant estimate
$\mathbf{a}$	Distance variate
B	Coefficient
Bhn	Brinell hardness
$\mathbf{B}$	Variate
b	Distance, Weibull shape parameter, range number, regression constant, width
$\hat{b}$	Regression constant estimate
$\mathbf{b}$	Distance variate
C	Basic load rating, bolted-joint constant, center distance, coefficient of variation, column end condition, correction factor, specific heat capacity, spring index
c	Distance, viscous damping, velocity coefficient
CDF	Cumulative distribution function
COV	Coefficient of variation
$\mathbf{c}$	Distance variate
D	Helix diameter
d	Diameter, distance
E	Modulus of elasticity, energy, error
e	Distance, eccentricity, efficiency, Naperian logarithmic base
F	Force, fundamental dimension force
f	Coefficient of friction, frequency, function
fom	Figure of merit
G	Torsional modulus of elasticity
g	Acceleration due to gravity, function
H	Heat, power
H_B	Brinell hardness
HRC	Rockwell C-scale hardness
h	Distance, film thickness
$\hbar_{CR}$	Combined overall coefficient of convection and radiation heat transfer
I	Integral, linear impulse, mass moment of inertia, second moment of area
i	Index
$\mathbf{i}$	Unit vector in x-direction

J	Mechanical equivalent of heat, polar second moment of area, geometry factor
$\mathbf{j}$	Unit vector in the y-direction
K	Service factor, stress-concentration factor, stress-augmentation factor, torque coefficient
k	Marin endurance limit modifying factor, spring rate
$\mathbf{k}$	k variate, unit vector in the z-direction
L	Length, life, fundamental dimension length
$\mathbf{LN}$	Lognormal distribution
l	Length
M	Fundamental dimension mass, moment
$\mathbf{M}$	Moment vector, moment variate
m	Mass, slope, strain-strengthening exponent
N	Normal force, number, rotational speed
$\mathbf{N}$	Normal distribution
n	Load factor, rotational speed, safety factor
n_d	Design factor
P	Force, pressure, diametral pitch
PDF	Probability density function
p	Pitch, pressure, probability
Q	First moment of area, imaginary force, volume
q	Distributed load, notch sensitivity
R	Radius, reaction force, reliability, Rockwell hardness, stress ratio
$\mathbf{R}$	Vector reaction force
r	Correlation coefficient, radius
$\mathbf{r}$	Distance vector
S	Sommerfeld number, strength
$\mathbf{S}$	S variate
s	Distance, sample standard deviation, stress
T	Temperature, tolerance, torque, fundamental dimension time
$\mathbf{T}$	Torque vector, torque variate
t	Distance, Student's t-statistic, time, tolerance
U	Strain energy
$\mathbf{U}$	Uniform distribution
u	Strain energy per unit volume
V	Linear velocity, shear force
v	Linear velocity
W	Cold-work factor, load, weight
$\mathbf{W}$	Weibull distribution
w	Distance, gap, load intensity
$\mathbf{w}$	Vector distance
X	Coordinate, truncated number
x	Coordinate, true value of a number, Weibull parameter
$\mathbf{x}$	x variate
Y	Coordinate
y	Coordinate, deflection
$\mathbf{y}$	y variate
Z	Coordinate, section modulus, viscosity
z	Standard deviation of the unit normal distribution
$\mathbf{z}$	Variate of z

α	Coefficient, coefficient of linear thermal expansion, end-condition for springs, thread angle
β	Bearing angle, coefficient
Δ	Change, deflection
δ	Deviation, elongation
ϵ	Eccentricity ratio, engineering (normal) strain
$\boldsymbol{\epsilon}$	Normal distribution with a mean of 0 and a standard deviation of s
ε	True or logarithmic strain
Γ	Gamma function
γ	Pitch angle, shear strain, specific weight
λ	Slenderness ratio for springs
$\boldsymbol{\lambda}$	Unit lognormal with a mean of 1 and a standard deviation equal to COV
μ	Absolute viscosity, population mean
ν	Poisson ratio
ω	Angular velocity, circular frequency
ϕ	Angle, wave length
ψ	Slope integral
ρ	Radius of curvature
σ	Normal stress
σ'	Von Mises stress
υ	Normal stress variate
σ	Standard deviation
τ	Shear stress
$\boldsymbol{\tau}$	Shear stress variate
θ	Angle, Weibull characteristic parameter
¢	Cost per unit weight
$	Cost

Mechanical Engineering Design

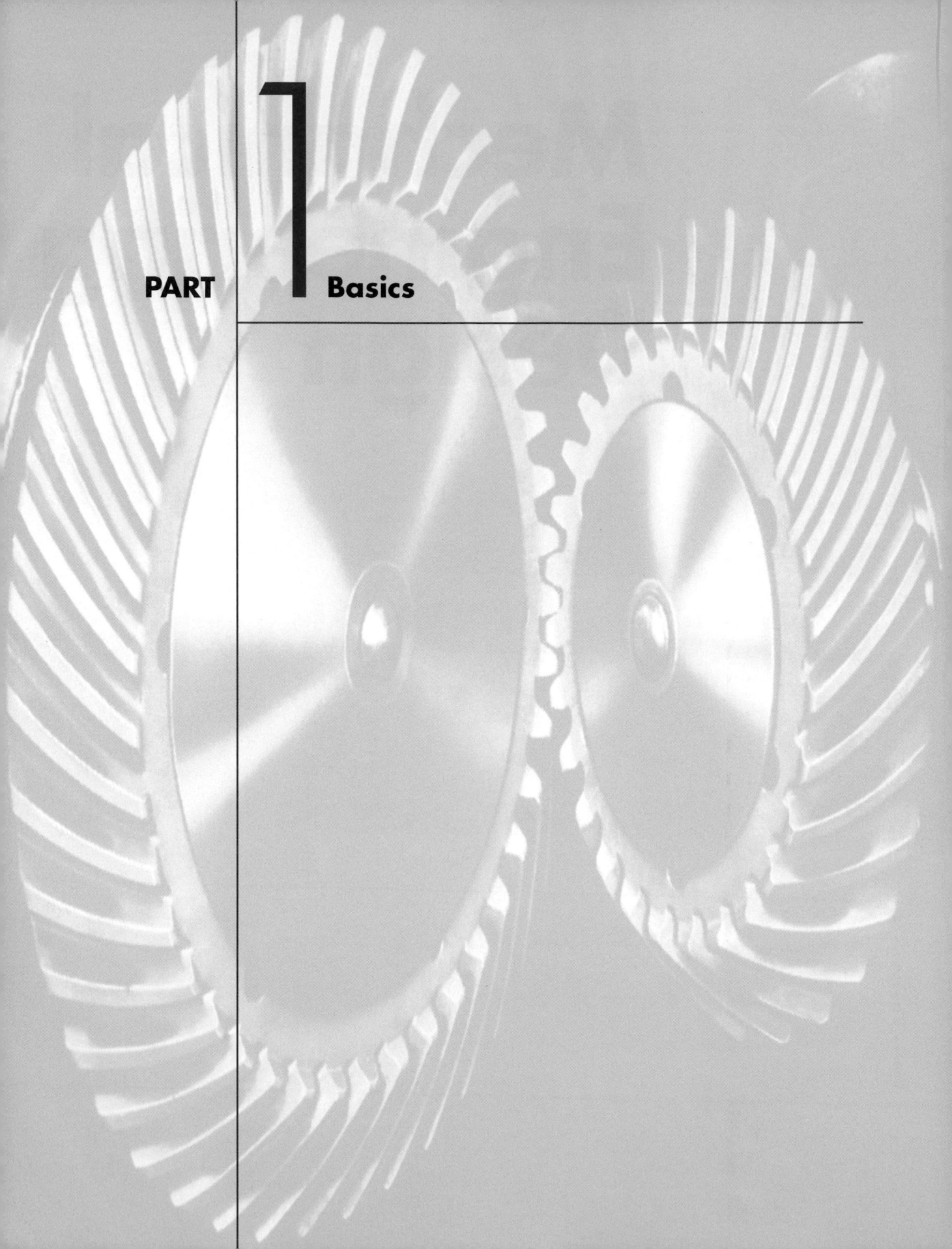

PART 1 Basics

1

Introduction

Chapter Outline

This chapter introduces a number of ideas, many of which are new to you either in context or substance. Mechanical design is a complex undertaking, requiring many skills. A vocabulary that allows large relationships to be subdivided into a series of simple tasks is needed. The complexity of the subject suggests a sequence in which ideas are introduced and revisited. Toward this end, we consider a number of topics briefly in this chapter to serve as an orientation. Later these topics are developed in detail, their bones fleshed out, and they will become part of you.

We first address the nature of design in general, then mechanical engineering design in particular. Design is an iterative process with many interactive phases. Learning and mastering is an ongoing process as you finish your formal education and throughout your career. Many resources exist to support the designer, including sources of information such as the Internet and an abundance of computational design tools. There are roles to be played by codes and standards, ever-present economics, safety, and considerations of product liability.

Then the focus of the chapter narrows, and the primal skill of the designer—adequacy assessment—is considered. Matters of uncertainty have been ever-present in engineering design, and methods have been evolved that draw on growing knowledge. Attention is given to stress and strength and to the distinction between design factor and factor of safety. Reliability is considered, as are units, preferred units, calculations, and significant figures.

Decision-making processes are common to all disciplines in the field of engineering design—not just to mechanical engineering design. But, since our subject *is* mechanical engineering design, we will use mechanical engineering as the vehicle for understanding these decision-making processes and for applying them to practical situations.

The book consists of three parts. Part 1 begins by explaining some differences between design and analysis and introducing some fundamental notions and approaches to design. It continues with a chapter on statistical methods (Chap. 2). A knowledge of statistical analysis is not necessary to study the balance of the material in the book. However, you will find some useful ideas in Chap. 2, as well as methods to aid your understanding and to perfect your design. Next there is a review of material properties, stress analysis, and an introduction to stiffness and deflection analysis.

Part 2, on failure prevention, consists of two chapters on the prevention of failure of mechanical parts. Why machine parts fail and how they can be designed to prevent failure are difficult questions, and so we take two chapters to answer them, one on preventing failure due to static loads, and the other on preventing failure due to fatigue and dynamic loads.

In Part 3, the material of Parts 1 and 2 is applied to the analysis, selection, and design of specific mechanical elements, such as fasteners, weldments, springs, rolling contact bearings, film bearings, gears, belts, chains, and wire ropes. The design of shafts ties together the influences of several of the preceding elements.

1–1 Design

To design is either to formulate a plan for the satisfaction of a specified need or to solve a problem. If the plan results in the creation of something having a physical reality, then the product must be functional, safe, reliable, competitive, usable, manufacturable, and marketable. These terms are defined as follows:

- *Functional:* The product must perform to fill its intended need and customer expectation.

- *Safe:* The product is not hazardous to the user, bystanders, or surrounding property. Hazards that cannot be "designed out" are eliminated by guarding (a protective enclosure); if that is not possible, appropriate directions or warnings are provided.
- *Reliable:* Reliability is the conditional probability, at a given confidence level, that the product will perform its intended function satisfactorily or without failure at a given age.
- *Competitive:* The product is a contender in its market.
- *Usable:* The product is "user friendly," accommodating to human size, strength, posture, reach, force, power, and control.
- *Manufacturable:* The product has been reduced to a "minimum" number of parts, suited to mass production, with dimensions, distortion, and strength under control.
- *Marketable:* The product can be bought, and service (repair) is available.

To remind us that designs are constrained, and have to exhibit qualities known at the outset, a *design imperative* can be expressed as follows:

Design
(subject to certain problem-solving constraints)
a component, system, or process
that will perform a specified task
(subject to certain solution constraints)
optimally.

The parenthetical expressions refer to qualifications placed on the design. The solution methodology is constrained by what the designer knows, or can do; the solution, in addition to being functional, safe, reliable, competitive, usable, manufacturable, and marketable, must also be legal and conform to applicable codes and standards.

It is important that the designer begin by identifying exactly how he or she will recognize a satisfactory alternative, and how to distinguish between two satisfactory alternatives in order to identify the better. From this kernel, optimization strategies can be formed or selected. Then, the following tasks unfold:

- Invent alternative solutions.
- Establish key performance metrics.
- Through analysis and test, simulate and predict the performance of each alternative, retain satisfactory alternatives, and discard unsatisfactory ones.
- Choose the best satisfactory alternative discovered as an approximation to optimality.
- Implement the design.

The characterization of a design task as a design *problem* can introduce the idea that, as a problem, it has a solution. This may not be so. The design space may be empty. Some situations may simply have to be endured. To relieve the absence of solutions, some constraint(s) may have to be renegotiated in order to admit solutions. Then again, even when solutions are possible, the designer may not be creative enough, inventive enough, to conceive of them. This admits to the design problem the necessity of individual talent or skill in this area.

There is usually more than one solution, and distinguishing among them to choose the best may require the ability to handle a large number of solutions without being overwhelmed. Solutions, if they exist, can be characterized as satisfactory, some better than others, some clearly good, and one, the best by some criterion. Solutions can have a time dependency, for what is acceptable today may not be so tomorrow, and vice versa.

Design is an innovative and highly iterative process. It is also a decision-making process. Decisions sometimes have to be made with too little information, occasionally with just the right amount of information, or with a surfeit of partially contradictory information. A man with a watch knows what time it is; with two watches, he is never sure. Decisions are sometimes made tentatively, reserving the right to adjust as more becomes known. The point is that the engineering designer has to be personally comfortable with a decision-making, problem-solving role. It should be a satisfying and welcomed activity. If it is not, there can be personal ramifications (such as stress) that can interfere, even threatening the designer's health.

Design is a communication-intensive activity in which both words and pictures are used, and written and oral forms are employed. Engineers have to communicate effectively and work with people of many disciplines who know more than they do, or less than they do. These are important skills, and an engineer's success depends on them.

A designer's personal resources of creativeness, communicative ability, and problem-solving skill are intertwined with knowledge of technology and first principles. Engineering tools (such as mathematics, statistics, computer, graphics, and language) are combined to produce a plan, which, when carried out, produces a product that is functional, safe, reliable, competitive, usable, manufacturable, and marketable, regardless of who builds it or who uses it.

1–2   Mechanical Engineering Design

Mechanical engineers are associated with the production and processing of energy and with providing the means of production, the tools of transportation, and the techniques of automation. The skill and knowledge base are extensive. Among the disciplinary bases are mechanics of solids and fluids, mass and momentum transport, manufacturing processes, and electrical and information theory. Mechanical engineering design involves all the disciplines of mechanical engineering.

Problems resist compartmentation. A simple journal bearing involves fluid flow, heat transfer, friction, energy transport, material selection, thermomechanical treatments, statistical descriptions, and so on. A building is environmentally controlled. The heating, ventilation, and air-conditioning considerations are sufficiently specialized that some speak of *heating, ventilating, and air-conditioning design* as if it is separate and distinct from mechanical engineering design. Similarly, *internal-combustion engine design, turbomachinery design,* and *jet-engine design* are sometimes considered discrete entities. The leading adjectival string of words preceding the word *design* is merely a product-descriptive aid to the communication process. There are phrases such as *machine design, machine-element design, machine-component design, systems design, and fluid-power design.* All of these phrases are somewhat more focused examples of mechanical engineering design. They all draw on the same bodies of knowledge, are similarly organized, and require similar skills.

In the academic world, with its clustering of knowledge into efficient learning groups, we encounter subjects, courses, disciplines, and fields. Curricula consist of sequences of courses. The arrangement of courses present the opportunity to study machine elements and machines earlier than the last semester. Thus machine design often represents the student's first serious design experience with a substantial knowledge base. Some, but not many machine elements can be understood without a complete thermofluid base, but before you know it, we are into mechanical engineering design.

Science explains what *is;* engineering creates what *never was.* Mathematics is neither science nor engineering. Physics and chemistry are science, but not engineering.

Figure 1–1

The name(s) of the game(s). Note distinctions between analysis, science, and engineering and the significant skills involved.

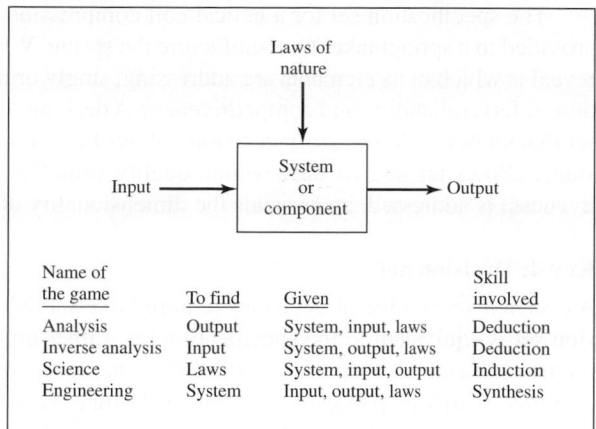

As suggested in Fig. 1–1, it takes one kind of talent to be a scientist and a different talent to create what never was. Engineers and scientists know something of each other's work, but only in rare cases are both talents developed in an individual. It takes talent and ability to create and innovate, talent to be a consistently successful problem solver and decision maker, and talent to be an effective communicator. Preparation, you see, is the developing and polishing of talent, whatever the endeavor.

Rational Decision Making

Designers have to make decisions, few or many, some a priori, some in concert. Rational design making is a systematic design process involving the following key elements.

 Key 1: Suitability, Feasibility, Acceptability

When the military establishment recognized the importance of clear thinking and rational decision making for its general officers, it sought engineers' advice on how to proceed. The military reasoned that engineers commit enormous resources to large projects, with no possibility of starting over. The advice offered can be distilled to the following:

- A contemplated action is *suitable* if its adoption will indeed accomplish the intended purpose.

- A contemplated action is *feasible* if the action can be carried out with the knowledge, personnel, money, and material at hand, or if it can be assembled in time.

- A contemplated action is *acceptable* if the probable results are worth the anticipated costs.

 Key 2: Satisfactory Alternative

If a contemplated action is suitable, feasible, and acceptable, it becomes a *satisfactory alternative,* and it is set aside to be compared with other satisfactory alternatives. If one can compare two satisfactory alternatives and choose the better, an *optimization strategy* can be crafted or selected to deal with a large number of satisfactory alternatives.

 Key 3: Specification Set

A *specification set* is the ensemble of drawings, text, bills of materials, and directions that constitutes the decision record in a form that enables the builder or user to realize function safely, reliably, competitively, and usably, the product having been fabricated and serviced to the customer's satisfaction.

The specification set for a helical coil compression spring for static service can be provided to a springmaker to manufacture the spring. What the specification set does not reveal is which of its elements are addressing, singly or in combination, matters of function, safety, reliability, and competitiveness. A designer needs an alternative (equivalent) set that shows each necessary decision, allows bifurcation into a priori and design decisions, allows tagging to show which quality (function, safety, reliability, or competitiveness) is addressed, and reveals the dimensionality of the problem.

 Key 4: Decision Set

A *decision set* is a list of decisions required to establish the specification set. The decision set is equivalent to the specification set. Either may be deduced from the other on the basis of convenience to clear thinking. The decision set is expressed in terms of the designer's thinking parameters, and it easily focuses on function, safety, reliability, and so on. For example, the *specification set* for a helical coil spring for static service can be displayed as follows:

- Material and condition
- End treatment
- Coil inner or outer diameter and tolerance
- Total turns and tolerance
- Free length and tolerance
- Wire size and tolerance

Note that these items are required by the springmaker to replicate the unique spring ordered. It is not clear how, or if, a coil diameter specification addresses function, safety, and so on, nor does the springmaker care. However, the designer does and therefore organizes the equivalent *decision set* as follows:

- Material and condition
- End treatment
- Force F_1 and end contraction y_1 or F_1 and length L_1 (function)
- Work over a rod: d_{rod} (function)
- Fractional overrun to coil closure ξ: $\xi = 0.15$ [safety, reliability, and spring linearity (robustness of the mathematical model)]
- Wire diameter d (competitiveness through optimality)

Note that these items are used by the designer to identify decisions, what they address, and the dimensionality of the problem. Tolerances can be expressed as a function of median values and can be quantified in a subsequent step. In Chap. 10 you will learn how to prove the equivalence between the decision set and the specification set for a helical coil spring.

Composing a decision set to be properly revealing and useful is a skill developed through knowledge and practice. There is some duplication between entries in the preceding specification set and decision set, but observe the explicit appearance of thinking parameters. The load F_1 that this spring must exhibit at end contraction y_1 (or at spring length L_1) addresses function. Working over a rod is necessary for the spring to function. The fractional overrun-to-closure ξ set at 0.15 addresses safety and reliability, protecting the spring closed solid, intentionally or inadvertently, and preserving the linearity of the spring so the mathematical model remains congruent to nature.

Key 5: A Priori Decisions versus Design Variables

The first five decisions in the previous decision set example can be made a priori (they are called *a priori decisions*). The last decision, that of wire size *d*, is called a *design variable* before we make the decision and the *design decision* after we have made the decision. It is through this variable that the designer attends to issues of preserving function, safety, and reliability, specifically using it to address competitiveness through optimality. In this case, knowledge that there is one independent variable influences selection of the methodology used to establish the wire size *d*.

Key 6: The Adequacy Assessment (Skill 1)

An *adequacy assessment* consists of the cerebral, empirical, and related mathematical modeling steps the designer takes to ensure that a given specification set is satisfactory (suitable, feasible, and acceptable, remember?). An adequacy assessment is the primal skill of the designer. It is how he or she recognizes a satisfactory specification set or alternatively, a corresponding decision set. It is so important that it is called *skill* 1. Much of a first course in design of machine elements focuses on building and refining this skill in many applications. Its centrality is seen in Fig. 1–2.

Key 7: Figure of Merit

If, in the coil spring example, the designer finds several wire sizes that pass the adequacy assessment, he or she uses a *figure of merit* to help identify the best. It is not much of an exaggeration to say that springs "sell by the pound." The volume of material used to form the spring is an index to cost. It is a robust figure of merit, abbreviated f.o.m. or, simply fom. Quantitatively, it can be expressed in the case of the helical coil compression spring as

$$\text{fom} = -\frac{\pi d^2 N_t D}{4}$$

where *d* is the wire diameter, N_t is the total turns, and *D* is the mean coil diameter. The minus sign allows the figure of merit to increase with decreasing volume. A figure of merit is a number whose magnitude is a monotonic index to the merit, or desirability, of the spring. A figure of merit permits rapid choice among several satisfactory designs. Should a large number of satisfactory designs exist, an optimization strategy is used to identify the best without exhaustive examination.

Key 8: A Skill of Synthesis (Skill 2)

The skill of synthesis involves an optimization strategy, a figure of merit, and skill 1. A flowchart of skills 1 and 2 is seen in Fig. 1–2. Note that skill 1 is embedded in skill 2, and it needs to be mastered first.

There is an irony here that should not escape us. In a way we have been building a checklist, but a checklist with an unusual objective. The objective of most checklists is to remind people to perform tasks without overlooking a thinking situation and thus render a credible performance. The checklist here is for the purpose of stimulating and facilitating original and rational thinking, thus enabling engineers with a design task to teach themselves what they need to know about the structure and connectivity of the elements in the task at hand.

Figure 1–2

A logic flowchart of designers' skills 1 and 2. Note that the analysis skill 1 is embedded in the synthesis skill 2. Depending on how the optimization strategy is formulated, skill 2 can amount to antianalysis, much in the same sense that the integral in calculus can be viewed as an antiderivative.

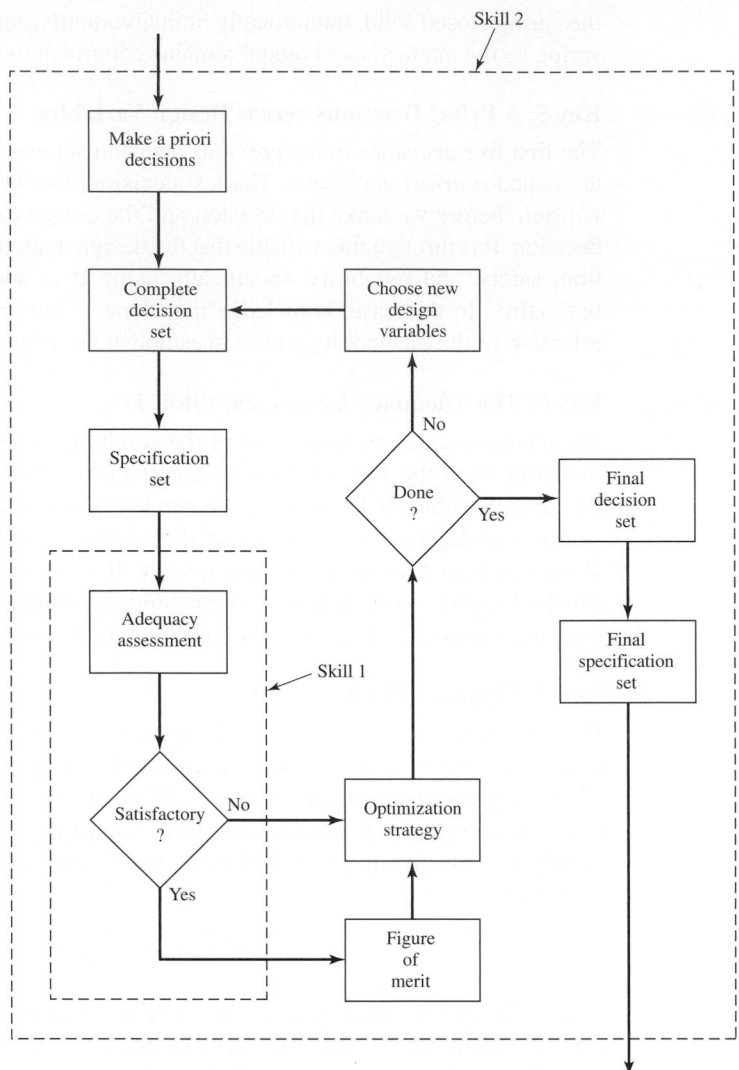

1–3 Interaction between Design Process Elements

The total design process is of interest to us in this chapter. How does it begin? Does the engineer simply sit down at a desk with a blank sheet of paper and jot down some ideas? What happens next? What factors influence or control the decisions that have to be made? Finally, how does this design process end?

The complete design process, from start to finish, is often outlined as in Fig. 1–3. The process begins with a recognition of a need and a decision to do something about it. After many iterations, the process ends with the presentation of the plans for satisfying the need. Depending on the nature of the design task, several design phases may be repeated throughout the life of the product, from inception to termination. In the next several sections, we shall examine these steps in the design process in detail.

Recognition and Identification

Sometimes, but not always, design begins when someone recognizes a need and decides to do something about it. *Recognition of the need* and phrasing the need often constitute

Figure 1–3

The phases in design, acknowledging the many feedbacks and iterations.

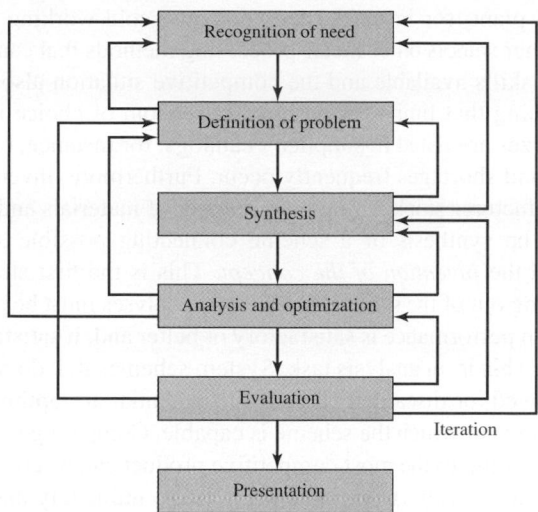

a highly creative act, because the need may be only a vague discontent, a feeling of uneasiness, or a sensing that something is not right. The need is often not evident at all; recognition is usually triggered by a particular adverse circumstance or a set of random circumstances that arise almost simultaneously. For example, the need to do something about a food-packaging machine may be indicated by the noise level, by the variation in package weight, and by slight but perceptible variations in the quality of the packaging or wrap.

It is evident that a sensitive person, one who is easily disturbed by things, is more likely to recognize a need—and also more likely to do something about it. And for this reason sensitive people are generally more creative. A need is easily recognized after someone else has stated it. Thus the need in many countries for cleaner air and water, for more parking facilities in the cities, for better public transportation systems, and for faster traffic flow has become quite evident.

There is a distinct difference between the statement of the need and the identification of the problem that follows this statement (Fig. 1–3). The problem is more specific. If the need is for cleaner air, the problem might be that of reducing the dust discharge from power-plant stacks, of reducing the quantity of irritants from automotive exhausts, or of quickly extinguishing forest fires.

Definition of the problem must include all the specifications for the object that is to be designed. The specifications are the input and output quantities, the characteristics and dimensions of the space the object must occupy, and all the limitations on these quantities. We can regard the object to be designed as something in a black box. In this case we must specify the inputs and outputs of the box, together with their characteristics and limitations. The specifications define the cost, the number to be manufactured, the expected life, the range, the operating temperature, and the reliability. Specified characteristics can include the speeds, feeds, temperature limitations, maximum range, expected variations in the variables, and dimensional and weight limitations.

There are many implied specifications that result either from the designer's particular environment or from the nature of the problem itself. The manufacturing processes that are available, together with the facilities of a certain plant, constitute restrictions on a designer's freedom, and hence are a part of the implied specifications. It may be that a small plant, for instance, does not own cold-working machinery. Knowing this, the designer selects other metal-processing methods that can be performed in the plant. The labor skills available and the competitive situation also constitute implied constraints. Anything that limits the designer's freedom of choice is a constraint. Many materials and sizes are listed in supplier's catalogs, for instance, but these are not all easily available and shortages frequently occur. Furthermore, inventory economics requires that a manufacturer stock a minimum number of materials and sizes.

The synthesis of a scheme connecting possible system elements is sometimes called the *invention of the concept.* This is the first step in the synthesis task. As the fleshing out of the scheme progresses, analyses must be performed to assess whether the system performance is satisfactory or better and, if satisfactory, just how well it will perform. This is an analysis task. System schemes that do not survive analysis are revised, improved, or discarded. Those with potential are optimized to determine the best performance of which the scheme is capable. Competing schemes are compared so that the path leading to the most competitive product can be chosen. Figure 1–3 shows that *synthesis* and *analysis and optimization* are intimately and iteratively related. Synthesis draws heavily on talent. In this iteration the specification set is formed.

We have noted, and we shall do so again and again, that design is an iterative process in which we proceed through several steps, evaluate the results, and then return to an earlier phase of the procedure. Thus we may synthesize several components of a system, analyze and optimize them, and return to synthesis to see what effect this has on the remaining parts of the system. Both analysis and optimization require that we construct or devise abstract models of the system that will admit some form of mathematical analysis. We call these models *mathematical models.* In creating them it is our hope that we can find one that will simulate the real physical system very well.

As indicated in Fig. 1–3, *evaluation* is a significant phase of the total design process. Evaluation is the final proof of a successful design and usually involves the testing of a prototype in the laboratory. Here we wish to discover if the design really satisfies the need or needs. Is it reliable? Will it compete successfully with similar products? Is it economical to manufacture and to use? Is it easily maintained and adjusted? Can a profit be made from its sale or use? How likely is it to result in product-liability lawsuits? And is insurance easily and cheaply obtained? Is it likely that recalls will be needed to replace defective parts or systems?

Communicating the design to others is the final, vital presentation step in the design process. Undoubtedly, many great designs, inventions, and creative works have been lost to posterity simply because the originators were unable or unwilling to explain their accomplishments to others. Presentation is a selling job. The engineer, when presenting a new solution to administrative, management, or supervisory persons, is attempting to sell or to prove to them that this solution is a better one. Unless this can be done successfully, the time and effort spent on obtaining the solution have been largely wasted. When designers sell a new idea, they also sell themselves. If they are repeatedly successful in selling ideas, designs, and new solutions to management, they begin to receive salary increases and promotions; in fact, this is how anyone succeeds in his or her profession.

Design Considerations

Sometimes the strength required of an element in a system is an important factor in the determination of the geometry and the dimensions of the element. In such a situation we say that *strength* is an important design consideration. When we use the expression *design consideration,* we are referring to some characteristic that influences the design of the element or, perhaps, the entire system. Usually quite a number of such characteristics must be considered in a given design situation. Many of the important ones are as follows (not necessarily in order of importance):

1	Functionality	**14**	Noise
2	Strength/stress	**15**	Styling
3	Distortion/deflection/stiffness	**16**	Shape
4	Wear	**17**	Size
5	Corrosion	**18**	Control
6	Safety	**19**	Thermal properties
7	Reliability	**20**	Surface
8	Manufacturability	**21**	Lubrication
9	Utility	**22**	Marketability
10	Cost	**23**	Maintenance
11	Friction	**24**	Volume
12	Weight	**25**	Liability
13	Life	**26**	Remanufacturing/resource recovery

Some of these have to do directly with the dimensions, the material, the processing, and the joining of the elements of the system. Other considerations affect the configuration of the total system. We shall be giving our attention to these factors and other considerations throughout the book.

In this book you will be faced with a great many design situations in which engineering fundamentals must be applied, usually in a mathematical approach, to resolve the problem or problems. This is completely correct and appropriate in an academic atmosphere, where the need is actually to utilize these fundamentals in the resolution of professional problems. To keep the correct perspective, however, it should be observed that in many design situations the important design considerations are such that no calculations or experiments are necessary in order to define an element or system. Students, especially, are often confounded when they run into situations in which it is virtually impossible to make a single calculation and yet an important design decision must be made. These are not extraordinary occurrences at all; they happen every day. Suppose that it is desirable from a sales standpoint—for example, in medical laboratory machinery—to create an impression of great strength and durability. Thicker parts assembled with larger-than-usual oversize bolts can be used to create a rugged-looking machine. Sometimes machines and their parts are designed purely from the standpoint of styling and nothing else. These points are made here so that you will not be misled into believing that there is a rational mathematical approach to every design decision.

1–4 Design Tools and Resources

Today, the engineer has a great variety of tools and resources available to assist in the solution of design problems. Inexpensive microcomputers and robust computer software packages provide tools of immense capability for the design, analysis, and simulation of mechanical components. In addition to these tools, the engineer always needs technical information, either in the form of basic science/engineering behavior or the characteristics of specific off-the-shelf components. Here, the resources can range from science/engineering textbooks to manufacturers' brochures or catalogs. Here too, the computer can play a major role in gathering information.[1]

---

1. An excellent and comprehensive discussion of the process of "gathering information" can be found in Chap. 4, George E. Dieter, *Engineering Design, A Materials and Processing Approach,* 3rd ed., McGraw-Hill, New York, 2000.

Computational Tools

Computer-aided design (CAD) software allows the development of three-dimensional (3-D) designs from which conventional two-dimensional orthographic views with automatic dimensioning can be produced. Manufacturing tool paths can be generated from the 3-D models, and in some cases, parts can be created directly from a 3-D database by using a rapid prototyping and manufacturing method (stereolithography)—*paperless manufacturing!* Another advantage of a 3-D database is that it allows rapid and accurate calculations of mass properties such as mass, location of the center of gravity, and mass moments of inertia. Other geometric properties such as areas and distances between points are likewise easily obtained. There are a great many CAD software packages available, such as Aries, AutoCAD, CadKey, I-Deas/Unigraphics, and ProEngineer, to name a few.

The term *computer-aided engineering* (CAE) generally applies to *all* computer-related engineering applications. With this definition, CAD can be considered as a subset of CAE. Some computer software packages perform specific engineering analysis and/or simulation tasks that assist the designer, but they are not considered tools for the *creation* of the design CAD is. Such software fits into two categories: engineering-based and non-engineering-specific. Some examples of engineering-based software for mechanical engineering applications—software that might also be integrated within a CAD system—include: finite-element analysis (FEA) programs for analysis of stress and deflection, vibration, and heat transfer (e.g., Algor, ANSYS, and MSC/NASTRAN); computational fluid dynamics (CFD) programs for fluid-flow analysis and simulation (e.g., CFD++, FIDAP, and Fluent); and programs for simulation of dynamic force and motion in mechanisms (e.g., ADAMS, DADS, and Working Model).

Examples of non-engineering-specific computer-aided applications include: software for word processing; spreadsheet software (e.g., Excel, Lotus, and Quattro-Pro); and mathematical solvers (e.g., Maple, MathCad, Matlab, Mathematica, and TKsolver).

Your instructor is the best source of information about programs that may be available to you and can recommend those that are useful for specific tasks. *One caution,* however: Computer software is no substitute for the human thought process. *You* are the driver here, the computer is the vehicle to assist you on your journey to a solution. Numbers generated by a computer can be far from the truth if you entered incorrect input, if you misinterpreted the application or the output of the program, if the program contained bugs, etc. It is your responsibility to assure the validity of the results, so be careful to check the application and results carefully, perform benchmark testing by submitting problems with known solutions, and monitor the software company and user-group newsletters.

Acquiring Technical Information

We currently live in what is referred to as the *information age,* when information is generated at an astounding pace. It is difficult, but extremely important, to keep abreast of past and current developments in one's field of study and occupation. The reference in Footnote 1 provides an excellent description of the informational resources available and is highly recommended reading for the serious design engineer. Some sources of information are:

- *Libraries (community, university, and private).* Engineering dictionaries and encyclopedias, textbooks, monographs, handbooks, indexing and abstract services, journals, translations, technical reports, patents, and business sources/brochures/catalogs.

- *Government sources.* Departments of Defense, Commerce, Energy and Transportation; NASA; Government Printing Office; U.S. Patent and Trademark Office; National Technical Information Service; and National Institute for Standards and Technology.

- *Professional societies.* American Society of Mechanical Engineering, Society of Manufacturing Engineers, Society of Automotive Engineers, American Society for Testing and Materials, and American Welding Society.
- *Commercial vendors.* Catalogs, technical literature, test data, samples, and cost information.
- *Internet.* The computer network gateway to websites associated with most of the categories listed above.

This list is not complete. The reader is urged to explore the various sources of information on a regular basis and keep records of the knowledge gained.

1–5 The Design Engineer's Professional Responsibilities

In general, the design engineer is required to satisfy the needs of customers (management, clients, consumers, etc.) and is expected to do so in a competent, responsible, ethical, and professional manner. Much of engineering course work and practical experience focuses on competence, but when does one begin to develop engineering responsibility and professionalism? To start on the road to success, you should start to develop these characteristics early in your educational program. You need to cultivate your professional work ethic and process skills before graduation, so that when you do begin your formal engineering career, you will be prepared to meet the challenges.

It is not obvious to some students, but communication skills play a large role here, and it is the wise student who continuously works to improve these skills—*even if it is not a direct requirement of a course assignment!* Success in engineering (achievements, promotions, raises, etc.) may in large part be due to competence but if you cannot communicate your ideas clearly and concisely, your technical proficiency may be compromised.

You can start to develop your communication skills by keeping a neat and clear journal/logbook of your activities, entering dated entries frequently. (Many companies require their engineers to keep a journal for patent and liability concerns.) Separate journals should be used for each design project (or course subject). When starting a project or problem, in the definition stage, make journal entries quite frequently. Others, as well as yourself, may later question why you made certain decisions. Good chronological records will make it easier to explain your decisions at a later date.

Many engineering students see themselves after graduation as practicing engineers designing, developing, and analyzing products and processes and consider the need of good communication skills, either oral or writing, as secondary. This is far from the truth. Most practicing engineers spend a good deal of time communicating with others, writing proposals and technical reports, and giving presentations and interacting with engineering and nonengineering support personnel. You have the time now to sharpen your communication skills. When given an assignment to write or make any presentation, technical *or* nontechnical, accept it enthusiastically, and work on improving your communication skills. It will be time well spent to learn the skills now rather than on the job.

When you are working on a design problem it is important to develop a systematic approach. Careful attention to the following action steps will help you to organize your solution processing technique.

- *Understand the problem.* Problem definition is probably the most significant step in the engineering design process. Carefully read, understand, and refine the problem statement.
- *Identify the known.* From the refined problem statement describe concisely what information is known and relevant.

- *Identify the unknown and formulate the solution strategy.* State what must be determined, in what order, so as to arrive at a solution to the problem. Sketch the component or system under investigation, identifying known and unknown parameters. Create a flowchart of the steps necessary to reach the final solution. The steps may require the use of free-body diagrams; material properties from tables; equations from first principles, textbooks, or handbooks relating the known and unknown parameters; experimentally or numerically based charts; specific computational tools as discussed in Sec. 1–4; etc.

- *State all assumptions and decisions.* Real design problems generally do not have unique, ideal, closed-form solutions. Selections, such as choice of materials, heat treatments, etc., require decisions. Analyses require assumptions related to the modeling of the real component or system. All assumptions and decisions should be identified and recorded.

- *Analyze the problem.* Using your solution strategy in conjunction with your decisions and assumptions, execute the analysis of the problem. Reference the sources of all equations, tables, charts, software results, etc. Check the credibility of your results. Check the order of magnitude, dimensionality, trends, signs, etc.

- *Evaluate your solution.* Evaluate each step in the solution, noting how changes in strategy, decisions, assumptions, and execution might change the results, in positive or negative ways. If possible, incorporate the positive changes in your final solution.

- *Present your solution.* Here is where your presentation skills are important. At this point, you are selling yourself and your technical abilities. If you cannot skillfully explain what you have done, some or all of your work may be misunderstood and unaccepted. Know your audience.

As stated earlier, all design processes are interactive and iterative. Thus, it may be necessary to repeat some or all of the above steps more than once if less than satisfactory results are obtained.

In order to be effective, all professionals must keep current in their fields of endeavor. The design engineer can satisfy this in a number of ways by: being an active member of a professional society such as the American Society of Mechanical Engineers (ASME), the Society of Automotive Engineers (SAE), the Society of Manufacturing Engineers (SME), etc.; attending meetings, conferences, and seminars of societies, manufacturers, universities, etc.; taking specific graduate courses or programs at universities; regularly reading technical and professional journals; etc. An engineer's education does not end at graduation.

The design engineer's professional obligations include conducting activities in an ethical manner. Reproduced here is the *Engineers' Creed* from the National Society of Professional Engineers (NSPE):

> *As a Professional Engineer, I dedicate my professional knowledge and skill to the advancement and betterment of human welfare.*
> *I pledge:*
> > *To give the utmost of performance;*
> > *To participate in none but honest enterprise;*
> > *To live and work according to the laws of man and the highest standards of professional conduct;*
> > *To place service before profit, the honor and standing of the profession before personal advantage, and the public welfare above all other considerations.*
> *In humility and with need for Divine Guidance, I make this pledge.*
>
> Adopted by the National Society of Professional Engineers, June 1954.
> "The Engineer's Creed" Reprinted by permission of the National Society of Professional Engineers.

1–6 **Codes and Standards**

There was a time when there were no standards for bolts, nuts, and screw threads. One manufacturer would produce, say, $\frac{1}{2}$-inch bolts with 9 threads per inch; another used 12. Some fasteners had left-handed threads and sometimes the thread profiles differed. It wasn't unusual in the early days of the automobile to see a mechanic lay out the fasteners in a line as they were disassembled in order to avoid mixing them during reassembly. This lack of standards and uniformity was costly and inefficient for a great variety of reasons. It is no wonder that people, disgusted and unable to find a replacement for a damaged fastener, sometimes used baling wire to fasten parts together.

A *standard* is a set of specifications for parts, materials, or processes intended to achieve uniformity, efficiency, and a specified quality. One of the important purposes of a standard is to place a limit on the number of items in the specifications so as to provide a reasonable inventory of tooling, sizes, shapes, and varieties.

A *code* is a set of specifications for the analysis, design, manufacture, and construction of something. The purpose of a code is to achieve a specified degree of safety, efficiency, and performance or quality. It is important to observe that safety codes *do not* imply *absolute safety*. In fact, absolute safety is impossible to obtain. Sometimes the unexpected event really does happen. Designing a building to withstand a 120 mi/h wind does not mean that the designers think a 140 mi/h wind is impossible; it simply means that they think it is highly improbable.

All of the organizations and societies listed below have established specifications for standards and safety or design codes. The name of the organization provides a clue to the nature of the standard or code. Some of the standards and codes, as well as addresses, can be obtained in most technical libraries. The organizations of interest to mechanical engineers are

> Aluminum Association (AA)
> American Gear Manufacturers Association (AGMA)
> American Institute of Steel Construction (AISC)
> American Iron and Steel Institute (AISI)
> American National Standards Institute (ANSI)[2]
> American Society for Metals (ASM)
> American Society of Mechanical Engineers (ASME)
> American Society of Testing and Materials (ASTM)
> American Welding Society (AWS)
> American Bearing Manufacturers Association (ABMA)[3]
> British Standards Institution (BSI)
> Industrial Fasteners Institute (IFI)
> Institution of Mechanical Engineers (I. Mech. E.)
> International Bureau of Weights and Measures (BIPM)
> International Standards Organization (ISO)

---

2. In 1966 the American Standards Association (ASA) changed its name to the United States of America Standards Institute (USAS). Then, in 1969, the name was again changed, to American National Standards Institute, as shown above and as it is today. This means that you may occasionally find ANSI standards designated as ASA or USAS.

3. In 1993 the Anti-Friction Bearing Manufacturers Association (AFBMA) changed its name to the American Bearing Manufacturers Association (ABMA).

National Institute for Standards and Technology (NIST)[4]
Society of Automotive Engineers (SAE)

1-7 Economics

The consideration of cost plays such an important role in the design decision process that we could easily spend as much time in studying the cost factor as in the study of the entire subject of design. Here we introduce only a few general approaches and simple rules.

First, observe that nothing can be said in an absolute sense concerning costs. Materials and labor usually show an increasing cost from year to year. But the costs of processing the materials can be expected to exhibit a decreasing trend because of the use of automated machine tools and robots. The cost of manufacturing a single product will vary from city to city and from one plant to another because of overhead, labor, taxes, and freight differentials and the inevitable slight manufacturing variations.

Standard Sizes

The use of standard or stock sizes is a first principle of cost reduction. An engineer who specifies an AISI 1020 bar of hot-rolled steel 53 mm square, called a hot-rolled square, has added cost to the product, provided a bar 50 or 60 mm square, both of which are preferred sizes, would do equally well. The 53-mm size can be obtained by special order or by rolling or machining a 60-mm square, but these approaches add cost to the product. To ensure that standard or preferred sizes are specified, designers must have access to stock lists of the materials they employ. These are available in libraries or can be obtained directly from the suppliers.

A further word of caution regarding the selection of preferred sizes is necessary. Although a great many sizes are usually listed in catalogs, they are not all readily available. Some sizes are used so infrequently that they are not stocked. A rush order for such sizes may mean more expense and delay. Thus you should also have access to a list such as those in Table A–17 for preferred inch and millimeter sizes.

There are many purchased parts, such as motors, pumps, bearings, and fasteners that are specified by designers. In the case of these, too, you should make a special effort to specify parts that are readily available. Parts that are made and sold in large quantities usually cost somewhat less than the odd sizes. The cost of rolling bearings, for example, depends more upon the quantity of production by the bearing manufacturer than upon the size of the bearing.

Large Tolerances

Among the effects of design specifications on costs, those of tolerances are perhaps most significant. Tolerances in design influence the producibility of the end product in many ways; close tolerances may necessitate additional steps in processing or even render a part completely impractical to produce economically. Tolerances cover dimensional variation and surface-roughness range and also the variation in mechanical properties resulting from heat treatment and other processing operations.

Since parts having large tolerances can often be produced by machines with higher production rates, labor costs will be smaller than if skilled workers were required. Also,

---

4. Former National Bureau of Standards (NBS).

Figure 1–4

A breakeven point.

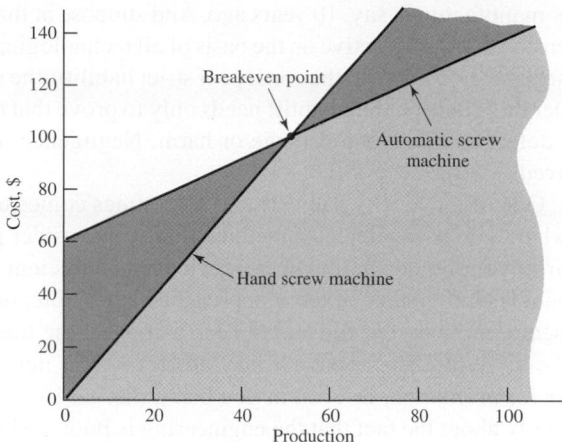

fewer such parts will be rejected in the inspection process, and they are usually easier to assemble.

Breakeven Points

Sometimes it happens that, when two or more design approaches are compared for cost, the choice between the two depends upon a set of conditions such as the quantity of production, the speed of the assembly lines, or some other condition. There then occurs a point corresponding to equal cost, which is called the *breakeven point*.

As an example, consider a situation in which a certain part can be manufactured at the rate of 25 parts per hour on an automatic screw machine or 10 parts per hour on a hand screw machine. Let us suppose, too, that the setup time for the automatic is 3 h and that the labor cost for either machine is $20 per hour, including overhead. Figure 1–4 is a graph of cost versus production by the two methods. The breakeven point corresponds to 50 parts. If the desired production is greater than 50 parts, the automatic machine should be used.

Cost Estimates

There are many ways of obtaining relative cost figures so that two or more designs can be roughly compared. A certain amount of judgment may be required in some instances. For example, we can compare the relative value of two automobiles by comparing the dollar cost per pound of weight. Another way to compare the cost of one design with another is simply to count the number of parts. The design having the smaller number of parts is likely to cost less. Many other cost estimators can be used, depending upon the application, such as area, volume, horsepower, torque, capacity, speed, and various performance ratios.

1–8 Safety and Product Liability

The *strict liability* concept of product liability generally prevails in the United States. This concept states that the manufacturer of an article is liable for any damage or harm that results because of a defect. And it doesn't matter whether the manufacturer knew about the defect, or even could have known about it. For example, suppose an article was manufactured, say, 10 years ago. And suppose at that time the article could not have been considered defective on the basis of all technological knowledge then available. Ten years later, according to the concept of strict liability, the manufacturer is still liable. Thus, under this concept, the plaintiff needs only to prove that the article was defective and that the defect caused some damage or harm. Negligence of the manufacturer need not be proved.

One of the touchy subjects that sometimes comes up in the practice of engineering is what to do if you detect something that you consider poor engineering. If possible, of course, you should attempt to correct it or run sufficient tests to prove that your fears are groundless. If neither of these approaches is feasible, then another approach is to place a memo in the design file and to keep a copy of the memo at home in case the original is "lost." While this approach may protect you, it may also result in you being passed over for promotion, or even in you being discharged. In the long run, if you really feel strongly about the fact that the engineering is poor, and you cannot live with it, then you should transfer or seek a new position.

The best approaches to the prevention of product liability are good engineering in both analysis and design, quality control, and comprehensive testing procedures. Advertising managers often make glowing promises in the warranties and sales literature for a product. These statements should be reviewed carefully by the engineering staff to eliminate excessive promises and to insert adequate warnings and instructions for use.

Several books, all by one author,[5] can be helpful at this point and you may want to consult them from time to time later.

1–9 The Adequacy Assessment

An adequacy assessment was defined in Sec. 1–2, together with other concepts useful in mechanical engineering design. An adequacy assessment is such a useful and important idea that it is also referred to as *skill* 1, the primal skill of the designer. An adequacy assessment begins with the specification set of a simple design or with the entire design description (assembly drawings, detail drawings, and the bill of materials) when the design is complex.

When confronted with a finished design (all decisions made), the designer must ask whether all constraints have been satisfied in a manner that either equals or exceeds expectations or in a less-than expected fashion. Additionally, some criterion is used to examine for merit, and optimal merit is often sought or approximated. An adequacy assessment is a judgment as to whether constraints have been satisfied and optimality realized.

An adequacy assessment is a process by which information is gathered and used in deciding whether constraints and optimality received appropriate attention. It can vary in some detail(s) from case to case, so that it is always familiar, yet somewhat different. It consists of various empirical and theoretical steps and mathematical modeling, ideas that an engineer uses to decide if an existing design (and its parts) is functional, safe, reliable, and so on. An adequacy assessment is composed of simple ideas. One becomes familiar with the process by performing some parts of it and analyzing increasingly complex situations, thus gaining proficiency.

5. L. J. Kamm, *Successful Engineering: A Guide to Achieving Your Career Goals,* McGraw-Hill, New York, 1989; *Real World Engineering: A Guide to Achieving Career Success,* IEEE Press, New York, 1991; *Understanding Electro-Mechanical Engineering: An Introduction to Mechatronics,* IEEE Press, New York, 1996; and *Designing Cost-Efficient Mechanisms,* Society of Automotive Engineers, Warrentown, Pa., 1993.

Begin by asking if each of the 26 design considerations listed in Sec. 1–3 is adequately met, presuming all 26 apply. For example, in considering item 2, strength/stress, the following questions are useful:

- Which strength applies (ultimate, tensile, yield, fatigue, creep, and so on)?
- Which stress applies (tensile, shear, principal, von Mises, and so on)?
- Which pair of stress and strength is most revealing of functional integrity, and which location or locations are critical?
- How accurately are strength and stress known?
- How much of a difference between stress and strength is "enough"?

Similarly, the remaining items in the list of considerations then are addressed.

An effective adequacy assessment identifies the questions to be asked, elicits quantitative estimates of the answers, includes mathematical modeling, and considers compatibility with nature and the marketplace. As with any intellectual skill, proficiency is the result of beginning simply with examples, drill problems in applications, then exercises that demand specific assessments in situations new to the student. The ultimate goal is to cultivate one's knowledge, judgment, proficiency, and consistency.

The adequacy assessment draws from the analysis portions of prior course work, with no respect for the compartmentalization of courses. What you choose to use depends on your goals. And the course that this book supports is no exception. It builds on many disciplinary roots in the reader's prior educational experience. Early problems are simpler, and it is easy to see what is important and to draft a pencil-and-paper adequacy assessment. As you proceed through the book and gain experience, it is useful to revisit earlier problems to see if you would still do them the same way. Your instructor is key to handling the many questions and details that surround skill 1. Consult your instructor as you learn to use skill 1 and apply it to situations developed in this book, or if you are heading in a different direction as necessitated by the course objectives.

1–10 Uncertainty

Uncertainties in machinery design abound. Examples of uncertainties concerning stress and strength include

- Composition of material and the effect of variation on properties.
- Variations in properties from place to place within a bar of stock.
- Effect of processing locally, or nearby, on properties.
- Effect of nearby assemblies such as weldments and shrink fits on stress conditions.
- Effect of thermomechanical treatment on properties.
- Intensity and distribution of loading.
- Validity of mathematical models used to represent reality.
- Intensity of stress concentrations.
- Influence of time on strength and geometry.
- Effect of corrosion.
- Effect of wear.
- Uncertainty as to the length of any list of uncertainties.

Engineers must accommodate uncertainty. Uncertainty always accompanies change. Material properties, load variability, fabrication fidelity, and validity of mathematical models are among concerns to designers.

There are mathematical methods to address uncertainties. The primary techniques are the deterministic and stochastic methods. The deterministic method establishes a *design factor (factor of safety)* based on the absolute uncertainties of a loss-of-function parameter (which could be the load that causes failure) and a maximum allowable parameter (which could be maximum allowable load). Here the design factor, n_d, is defined as

$$n_d = \frac{\text{loss-of-function parameter}}{\text{maximum allowable parameter}} \tag{1-1}$$

If the parameter is load, then the *maximum allowable load* can be found from

$$\text{Maximum allowable load} = \frac{\text{loss-of-function load}}{n_d} \tag{1-2}$$

EXAMPLE 1–1

Consider that the maximum load on a structure is known with an uncertainty of ± 20 percent, and the load causing failure is known with an uncertainty of ± 15 percent. If the load causing failure is nominally 2000 lbf, determine the design factor and the maximum allowable load that will offset the absolute uncertainties.

Solution

To account for its uncertainty the loss-of-function load must increase to $1/0.85$, whereas the maximum allowable load must decrease to $1/1.2$. Thus to offset the absolute uncertainties the design factor should be

Answer

$$n_d = \frac{1/0.85}{1/1.2} = 1.4$$

From Eq. (1–2), the maximum allowable load is found to be

Answer

$$\text{maximum allowable load} = \frac{2000}{1.4} = 1400 \text{ lbf}$$

Stochastic methods are based on the statistical nature of the design parameters and focus on the probability of survival of the design's function (that is, on reliability). Chapter 2 contains background material on applying statistics in design.

1–11 Stress and Strength

The survival of many products depends on how the designer adjusts the maximum stresses in a component to be less than the component's strength at specific locations of interest. The designer must allow the maximum stress to be less than the strength by a sufficient margin so that despite the uncertainties, failure is rare.

In focusing on the stress–strength comparison at a critical (controlling) location, we often look for "strength in the geometry and condition of use." Strengths are stresses at which something of interest occurs, such as the proportional limit, 0.2 percent-offset yielding, or fracture. In many cases, such events represent the stress level at which loss of function occurs.

Strength is a *property* of a material or of a mechanical element. The strength of an element depends upon the choice, the treatment, and the processing of the material.

Consider, for example, a shipment of springs. We can associate a strength with a specific spring. When this spring is incorporated into a machine, external forces are applied that result in load induced stresses in the spring, the magnitudes of which depend upon its geometry and are independent of the material and its processing. If the spring is removed from the machine unharmed, the stress due to the external forces will return to zero. But the strength remains as one of the properties of the spring. Remember, then, that *strength is an inherent property of a part,* a property built into the part because of the use of a particular material and process.

Various metalworking and heat-treating processes, such as forging, rolling, and cold forming, cause variations in the strength from point to point throughout a part. The spring cited above is quite likely to have a strength on the outside of the coils different from its strength on the inside because the spring has been formed by a cold winding process, and the two sides may not have been deformed by the same amount. Remember, too, therefore, that a strength value given for a part may apply to only a particular point or set of points on the part.

In this book we shall use the capital letter S to denote strength, with appropriate markings and subscripts to denote the kind of strength. Thus, S_s is a shear strength, S_y a yield strength, S_u an ultimate strength, and $\bar{S}$ a mean strength obtained by sampling test data.

In accordance with accepted engineering practice, we shall employ the Greek letters σ (sigma) and τ (tau) to designate normal and shear stresses, respectively. Again, various markings and subscripts will indicate some special characteristic. For example, σ_1 is a principal stress, σ_y a stress component in the y direction, and σ_r a stress component in the radial direction.

Stress is a *state property* of a body, which is a function of load, geometry, temperature, and manufacturing processing. In an elementary course in mechanics of materials, stress related to load and geometry is emphasized with some discussion of thermal stresses. However, stresses due to heat-treatments, molding, assembly, etc. are also important and sometimes neglected.

One of the basic problems in dealing with stress and strength is how to relate the two in order to develop a safe, economical, and efficient design. In studying this problem it will be useful to examine this relationship as it applies in a particular branch of engineering design. The American Institute of Steel Construction (AISC), founded in 1921, is a nonprofit society whose objectives are to improve and advance the use of fabricated structural steel. To accomplish these objectives, the AISC publishes manuals, textbooks, specifications, and technical booklets. The best known and most widely used is the *Manual of Steel Construction Allowable Stress Design (ASD),*[6] which holds a highly respected position in engineering literature. Important standards included in this manual are the Specification for the Design, Fabrication, and Erection of Structural Steel for Buildings and the Code of Standard Practice for Steel Buildings and Bridges.

Table 1–1 lists minimum yield strengths S_y and minimum tensile strengths S_u for certain ASTM steels as given in the AISC specifications. In this table, S_y represents either the minimum yield point, if the material has one, or the minimum yield strength.

The ANSI-ASTM standard B483-78 defines *minimum strength* as follows:

A2.1 Standard mechanical property limits for the respective size ranges are based on an analysis of data from standard production material and are established at a level [at] which at least 99 percent of the population of values obtained from all standard material in the size range meets the established value.

---

6. Now in its eighth edition; new editions are published periodically.

Steel Type	ASTM No.	S_y, kpsi	S_u, kpsi	Size, in, up to
Carbon	A36	36	58	8
Carbon	A529	42	60	$\frac{1}{2}$
Low alloy	A572	42	60	6
Low alloy	A572	50	65	2
Stainless	A588	50	70	4
Alloy Q&T	A514	100	110	$2\frac{1}{2}$

Many readers may prefer to qualify *minimum* by always using the adjective *ASTM,* as in *ASTM minimum strength.* The unqualified word *minimum* may be misleading, since there is a chance that up to 1 percent of the materials involved may have a strength smaller than the ASTM minimum.

To ensure that materials having these minimum strengths are actually used in the construction, the AISC specification states that:

> Certified mill test reports or certified reports of tests made by the fabricator or a testing laboratory in accordance with ASTM A6 or A568, as applicable, and the governing specification shall constitute sufficient evidence of conformity with one of the ASTM standards. Additionally, the fabricator shall, if requested, provide an affidavit stating that the structural steel furnished meets the requirements of the grade specified.

The AISC is developing the permissible-stress method and the notion of design factor will not be involved. Note also that the variability in S is not described in any way, and reliability is not mentioned.

Finding reduced values constitutes the first part of the AISC procedure. Following AISC vocabulary, let us designate allowable normal stress σ_{all} and allowable shear stress as τ_{all}. Then the relationship between allowable stresses and specified minimum strengths using the AISC code is specified as:

$$
\begin{aligned}
&\text{TENSION} && 0.45 S_y \leq \sigma_{\text{all}} \leq 0.60 S_y \\
&\text{SHEAR} && \tau_{\text{all}} = 0.40 S_y \\
&\text{BENDING} && 0.60 S_y \leq \sigma_{\text{all}} \leq 0.75 S_y \\
&\text{BEARING} && \sigma_{\text{all}} = 0.90 S_y
\end{aligned}
\tag{1-3}
$$

The next part of the AISC code deals with determining the loads or forces that are used to obtain the stresses. The procedure can be presented succinctly by the equation

$$
F = \sum W_d + \sum W_l + \sum KF_l + F_w + \sum F_{\text{misc}}
\tag{1-4}
$$

where F is the force to be used in the appropriate stress equation; see Chap. 4, for example. The components of this force are defined as follows.

The term $\sum W_d$ is the sum of the *dead loads*. These consist of the weight of the steelwork, the materials fastened to it, and the parts supported by it.

The term $\sum W_l$ is the sum of all the stationary or static *live loads*. This includes the weight of equipment, occupants, fixtures, and the snow load if specified by an applicable code.

Table 1–2

AISC Service Factors for Use in Eq. (1–4)

For supports of elevators	$K = 2$
For cab-operated traveling-crane support girders and their connections	$K = 1.25$
For pendant-operated traveling-crane support girders and their connections	$K = 1.10$
For supports of light machinery, shaft- or motor-driven	$K \geq 1.20$
For supports of reciprocating machinery or power-driven units	$K \geq 1.50$
For hangers supporting floors and balconies	$K = 1.33$

The force or the resultant of forces due to equipment that may cause impact or dynamic loading is also considered to be a live load and is represented by the term F_l. This factor is to be multiplied by a *service factor K* obtained from Table 1–2.

The term F_w in Eq. (1–4) is the wind load on the structure. Appropriate guidelines for this may be specified by local or regional codes.

The term $\sum F_{\text{misc}}$ must be included in some localities to account for the effects of earthquakes, hurricanes, or other extraordinary regional conditions.

The final step in the AISC procedure is to select dimensions of the member to be sized such that the design stress computed from the force F does not exceed the allowable stress as given by Eq. (1–3), in other words, to select dimensions or geometry such that

$$\sigma \leq \sigma_{\text{all}} \quad \text{or} \quad \tau \leq \tau_{\text{all}} \tag{1–5}$$

where σ and τ may be called the design values of the normal and shear stresses, respectively. This presumes a linear proportionality between load and stress.

In summary, Eq. (1–3) defines allowable stresses as the specified minimum strengths reduced by multiplication factors varying from 40 to 90 percent to ensure safety. Equation (1–4) accounts for all possible loads, and the service factors in Table 1–2 provide an additional degree of safety for dynamic loading. This is an implementation of the method of permissible stress with the strength characterized at the 99-percentile level.

1–12   Design Factor and Factor of Safety

The AISC method for relating stress and strength is also used in some other specialized design areas. However, it is not a general approach, since it addresses only specific materials and loadings. In mechanical engineering design there are categories of situations confronting the engineer.

- The product is made in large quantities, is valuable, or is dangerous, justifying elaborate testing of materials, components, and prototypes in the field.

- The product is made in sufficient quantities to justify a modest material test program, perhaps as small as ultimate tensile tests.

- The product is made in such small quantities that no testing of materials is performed at all.

The latter two situations are the more challenging. There is also the faulty product inquiry: Why? How can it be remedied?

A general approach to the allowable-load—loss-of-function load problem is the deterministic *design factor method,* and sometimes called the classical method of design.

The fundamental equation is Eq. (1–2):

$$\text{Allowable load} = \frac{\text{loss-of-function load}}{n_d}$$

where n_d is called the *design factor*.[7] For a given loss-of-function load, doubling the design factor halves the allowable load. To be most useful, this property is made to persist regardless of the linearity, or lack thereof, of stress with load. The adequacy assessment in the design-factor methods consists in part of estimating the *factor of safety* of the completed design. Factor of safety n has the same definition as the design factor, but it differs numerically due to the rounding (usually up) caused by using standard sizes and off-the-shelf components.

In assessing the factor of safety of an element, say a gear tooth that can fail in bending fatigue or surface fatigue, we note the tooth has a factor of safety guarding against bending fatigue and another factor of safety guarding against surface fatigue. If these factors of safety are numerically 1.5 and 1.3, respectively, then surface fatigue will occur before bending fatigue.

When Hooke discovered elastic behavior in metal springs, the insightful concept of stress was still 162 years away. The concept of stress presented the opportunity to express design factor in terms of stress. With the invention of testing machines,[8] where particular modes of failure could be induced, design factors came to be expressed in terms of a stress and a relevant strength, and expressions such as

$$n_d = \frac{S(\text{loss-of-function})}{\sigma(\text{allowable})} = \frac{\text{strength}}{\text{stress}} \tag{1–6}$$

were possible. For contact stress problems[9] where stresses are not linearly proportional to loads, the form changes to

$$n_d = \left(\frac{\text{strength}}{\text{stress}}\right)^3 \qquad \text{for spheres in contact} \tag{1–7}$$

$$n_d = \left(\frac{\text{strength}}{\text{stress}}\right)^2 \qquad \text{for cylinders in contact} \tag{1–8}$$

reflecting the nonlinear relationship between stress and load. In Eqs. (1–6) through (1–8) strengths can be minimums, yields, tensiles, fatigues, and shears, as well as some others. Of course the stress used must correspond in type and units to the strength. Also, the stress and strength must be at the same location.

Equation (1–3) was used to account for any uncertainties regarding the actual strength of a member. Equation (1–4) was used to account for any uncertainties as to actual loads. The design factor in Eq. (1–6) is used to account for *both* uncertainties.

---

7. See D. W. Lapedes, *Dictionary of Scientific and Technical Terms,* McGraw-Hill, New York, 1978, under *factor of safety.*

8. Musschenbrock (1729), Perronet (1768), then Gauthey, Rondelet, Girard, and Williams contributed. The first to employ the hydraulic press and the balanced beam was Lagerhjelm in 1827. The feature that is important is the controlling of the mode of failure during measurement and the introduction of the notion of material strength, verified by test, and using the concept of stress due to Cauchy (1822).

9. See Sec. 4–21.

1–13   **Reliability**

In these days of greatly increasing numbers of liability lawsuits and the need to conform to regulations issued by governmental agencies such as EPA and OSHA, it is very important for the designer and the manufacturer to know the reliability of their product. The *reliability method* of design is one in which we obtain the distribution of stresses and the distribution of strengths and then relate these two in order to achieve an acceptable success rate.

The statistical measure of the probability that a mechanical element will not fail in use is called the *reliability* of that element. The reliability R can be expressed by a number having the range

$$0 \leq R < 1 \tag{1–9}$$

A reliability of $R = 0.90$ means that there is a 90 percent chance that the part will perform its proper function without failure. The failure of 6 parts out of every 1000 manufactured might be considered an acceptable failure rate for a certain class of products. This represents a reliability of

$$R = 1 - \frac{6}{1000} = 0.994$$

or 99.4 percent.

In the *reliability method of design,* the designer's task is to make a judicious selection of materials, processes, and geometry (size) so as to achieve a reliability goal. Thus, if the objective reliability is to be 99.4 percent, as above, what combination of materials, processing, and dimensions is needed to meet this goal?

Analyses that lead to an assessment of reliability address uncertainties, or their estimates, in parameters that describe the situation. Stochastic variables such as stress, strength, load, or size are described in terms of their means, standard deviations, and distributions. If bearing balls are produced by a manufacturing process in which a diameter distribution is created, we can say upon choosing a ball that there is uncertainty as to size. If we wish to consider weight or moment of inertia in rolling, this size uncertainty can be considered to be *propagated* to our knowledge of weight or inertia. There are ways of estimating the statistical parameters describing weight and inertia from those describing size and density. These methods are variously called *propagation of error, propagation of uncertainty,* or *propagation of dispersion*. These methods are integral parts of analysis or synthesis tasks when probability of failure is involved.

The stochastic design factor method and the stochastic method approach to design are relatively new, considering the scope of machine design history. We have records of the historical behavior of ensembles of materials and their statistical descriptions. Using them, the designer can proceed toward a specified reliability goal. When this is done, the geometric sizes of parts will be larger than in the case where the designer has specific stochastic information on the material finally selected.

The statistical approach offers the opportunity to consider reliability goals and other ideas that the methods of antiquity could not. If the loading to which a tractor transmission will be subjected in a field is uncertain, then instrumented drawbars in field studies yield information that is better understood in stochastic rather than deterministic interpretation. Statistics allows the quantification and explanation of variation. Nature is full of individual variability with long-term regularity. Stability is not uniqueness in magnitude, but in a pattern of variation, a mix of systematic and random effects. Statistics allows these to be separated; it is about variation and its measurement, the ways models describe data and data criticize models, and about the sensitive use of data to illuminate the dark places.

1–14 Units and Preferred Units

Units

In the symbolic units equation for Newton's second law, $F = ma$,

$$F = MLT^{-2} \tag{1–10}$$

F stands for force, M for mass, L for length, and T for time. Units chosen for *any* three of these quantities are called *base units*. The first three having been chosen, the fourth unit is called a *derived unit*. When force, length, and time are chosen as base units, the mass is the derived unit and the system that results is called a *gravitational system of units*. When mass, length, and time are chosen as base units, force is the derived unit and the system that results is called an *absolute system of units*.

In some English-speaking countries, the *U.S. customary foot-pound-second system* (fps) and the *inch-pound-second system* (ips) are the two standard gravitational systems most used by engineers. In the fps system the unit of mass is

$$M = \frac{FT^2}{L} = \frac{\text{(pound-force)(second)}^2}{\text{foot}} = \text{lbf} \cdot \text{s}^2/\text{ft} = \text{slug} \tag{1–11}$$

Thus, length, time, and force are the three base units in the fps gravitational system.

The unit of time in the fps system is the second, abbreviated s. The unit of force in the fps system is the pound, more properly the *pound-force*. We shall seldom abbreviate this unit as lbf; the abbreviation lb is permissible, since we shall be dealing only with the U.S. customary gravitational system. In some branches of engineering it is useful to represent 1000 lbf as a kilopound and to abbreviate it as kip. Many writers add the letter "s" to kip to obtain the plural, but to be consistent with the practice of using only singular forms for units we shall not do so here. Thus, 1 kip and 3 kip are used to designate 1000 and 3000 lbf, respectively. Finally, we note in Eq. (1–11) that the derived unit of mass in the fps gravitational system is the lbf · s²/ft, called a *slug;* there is no abbreviation for slug.

The unit of mass in the ips gravitational system is

$$M = \frac{FT^2}{L} = \frac{\text{(pound-force)(second)}^2}{\text{inch}} = \text{lbf} \cdot \text{s}^2/\text{in} \tag{1–12}$$

The units lbf · s²/in have no official name.

The International System of Units (SI) is an absolute system. The base units are the meter, the kilogram (for mass), and the second. The unit of force is derived by using Newton's second law and is called the *newton*. The units constituting the newton (N) are

$$F = \frac{ML}{T^2} = \frac{\text{(kilogram)(meter)}}{\text{(second)}^2} = \text{kg} \cdot \text{m/s}^2 = \text{N} \tag{1–13}$$

The weight of an object is the force exerted upon it by gravity. Designating the weight as W and the acceleration due to gravity as g, we have

$$W = mg \tag{1–14}$$

In the fps system, standard gravity is $g = 32.1740$ ft/s². For most cases this is rounded off to 32.2. Thus the weight of a mass of 1 slug in the fps system is

$$W = mg = (1 \text{ slug})(32.2 \text{ ft/s}^2) = 32.2 \text{ lbf}$$

In the ips system, standard gravity is 386.088 or about 386 in/s². Thus, in this system, a unit mass weighs

$$W = (1 \text{ lbf} \cdot s^2/\text{in})(386 \text{ in/s}^2) = 386 \text{ lbf}$$

With SI units, standard gravity is 9.806 or about 9.81 m/s². Thus, the weight of a 1-kg mass is

$$W = (1 \text{ kg})(9.81 \text{ m/s}^2) = 9.81 \text{ N}$$

A series of names and symbols to form multiples and submultiples of SI units has been established to provide an alternative to the writing of powers of 10. Table A–1 includes these prefixes and symbols.

Rules for Use of SI Units

The International Bureau of Weights and Measures (BIPM), the international standardizing agency for SI, has established certain rules and recommendations for the use of SI. These are intended to eliminate differences that occur among various countries of the world in scientific and technical practices.

Number Groups

Numbers having four or more digits are placed in groups of three and separated by a space instead of a comma. However, the space may be omitted for the special case of numbers having four digits. A period is used as a decimal point. These recommendations avoid the confusion caused by certain European countries in which a comma is used as a decimal point, and by the English use of a centered period. Examples of correct and incorrect usage are as follows:

> 1924 or 1 924 but not 1,924
> 0.1924 or 0.192 4 but not 0.192,4
> 192 423.618 50 but not 192,423.61850

The decimal point should always be preceded by a zero for numbers less than unity.

Use of Prefixes

The multiple and submultiple prefixes in steps of 1000 only are recommended (Table A–1). This means that length is expressed in mm, m, or km, but not in cm, unless a valid reason exists.

When SI units are raised to a power, the prefixes are also raised to the same power. This means that the km² is defined thus:

$$1 \text{ km}^2 = (1000 \text{ m})^2 = (1000)^2 \text{ m}^2 = 10^6 \text{ m}^2$$

Similarly, the mm² is defined as follows:

$$(0.001 \text{ m})^2 = (0.001)^2 \text{ m}^2 = 10^{-6} \text{ m}^2$$

In raising prefixed units to a power, it is permissible, though not often convenient, to use the nonpreferred prefixes such as cm² or dm³.

Except for the kilogram, which is a base unit, prefixes should not be used in the denominators of derived units. Thus the meganewton per square meter, MN/m² (MPa), is satisfactory, but the newton per square millimeter, N/mm², is not to be used.[10] Note that this recommendation avoids a proliferation of derived units.

Double prefixes should not be used. Thus, instead of millimillimeters (mmm), use micrometers (μm).

---

10. When using computer software, such as Finite Element Analysis (FEA), using SI units design analysts may employ newtons and millimeters in an analysis. The stress output will be in units of N/mm², which can readily be converted directly to MPa.

Preferred Units

As a general rule, it is both convenient and good practice to select prefixes such that the number strings will contain no more than four digits to the left of the decimal point. By applying this rule, we find that some of the preferred units are kilopounds per square inch (kpsi) and megapascals (MPa) for stress, pounds per square inch (psi) and kilopascals (kPa) for air or hydraulic pressure, pounds (lbf) and kilonewtons (kN) for force, and inches to the fourth power (in^4) and centimeters to the fourth power (cm^4) for second moment of area.

Table A–1 is of particular value when you are using units with mixed prefixes in an equation. Suppose we wish to solve the deflection equation of an end-loaded cantilever beam of diameter d (see Chap. 5)

$$y = \frac{64Fl^3}{3\pi d^4 E}$$

where $F = 1.30$ kN, $l = 300$ mm, $d = 2.5$ cm, and $E = 207$ GPa. It is convenient to show the solution in two parts, the first containing the numbers, and the second containing the prefixes. Thus

$$y = \frac{64(1.30)(300)^3}{3\pi(2.5)^4(207)} \frac{(kilo)(milli)^3}{(centi)^4(giga)}$$

Now compute the numerical value of the first part and substitute the prefix values in the second. This gives

$$y = [29.48(10)^3]\left[\frac{10^3(10^{-3})^3}{(10^{-2})^4(10^9)}\right] = 29.48(10)^{-4} \text{ m}$$

$$= 2.948 \text{ mm}$$

In many cases, similar equations are used so often that we remember the prefixes to use in order to obtain a result having a convenient prefix. Thus, in the above equation, if we use F in kN, l and d in mm, and E in GPa, the result y is in mm. Try it. Tables A–3 and A–4 will also be of help to you.

1–15 Calculations and Significant Figures

The discussion in this section applies to real numbers, not integers. The accuracy of a real number depends on the number of significant figures describing the number. Usually, but not always, three or four significant figures are necessary for engineering accuracy. Unless otherwise stated, *no less* than three significant figures should be used in your calculations. The number of significant figures is usually inferred by the number of figures given (except for leading zeros). For example, 706, 3.14, and 0.002 19 are assumed to be numbers with three significant figures. For trailing zeros, a little more clarification is necessary. To display 706 to four significant figures insert a trailing zero and display either 706.0, 7.060×10^2, or 0.7060×10^3. Also, consider a number such as 91 600. Scientific notation should be used to clarify the accuracy. For three significant figures express the number as 91.6×10^3. For four significant figures express it as 91.60×10^3.

Computers and calculators display calculations to many significant figures. However, you should never report a number of significant figures of a calculation any greater than the smallest number of significant figures of the numbers used for the calculation. Of course, you should use the greatest accuracy possible when performing a calculation. For example, determine the circumference of a solid shaft with a diameter of $d = 0.40$ in.

The circumference is given by $C = \pi d$. Since d is given with two significant figures, C should be reported with only two significant figures. Now if we used only two significant figures for π our calculator would give $C = 3.1\,(0.40) = 1.24$ in. This rounds off to two significant figures as $C = 1.2$ in. However, using $\pi = 3.141\,592\,654$ as programmed in the calculator, $C = 3.141\,592\,654\,(0.40) = 1.256\,637\,061$ in. This rounds off to $C = 1.3$ in, which is 8.3% higher than the first calculation. Note, however, since d is given with two significant figures, it is implied that the range of d is 0.40 ± 0.005. This means that the calculation of C is only accurate to within $\pm 0.005/0.40 = \pm 0.0125 = \pm 1.25\%$. The calculation could also be one in a series of calculations, and rounding each calculation separately may lead to an accumulation of greater inaccuracy. Thus, it is considered good engineering practice to make all calculations to the greatest accuracy possible and report the results within the accuracy of the given input.

PROBLEMS

The problems in this section touch upon some of the ideas presented in this chapter together with some "thought provoking" exercises in mechanics requiring free-body diagrams.

RESEARCH **1–1** Select a mechanical component from Part 3 of this book (roller bearings, springs, etc.), go to your university's library, and using the *Thomas Register of American Manufacturers,* report on the information obtained on five manufacturers or suppliers.

RESEARCH **1–2** Select a mechanical component from Part 3 of this book (roller bearings, springs, etc.), go to the Internet, and using a search engine, report on the information obtained on five manufacturers or suppliers.

RESEARCH **1–3** Select an organization listed in Sec. 1–6, go to the Internet, and list what information is available on the organization.

RESEARCH **1–4** Go to the Internet and connect to the NSPE website (www.nspe.org). Read the full version of the NSPE Code of Ethics for Engineers and briefly discuss your reading.

ANALYSIS **1–5** A "loose" pin in a hole can be used as a sliding kinematic pair. If a force F is applied to the pin as shown in the figure at an angle to the axis of the pin (and the potential direction of motion), will the pin move? If not, why not? If so, what is the force causing motion? Does reversing the sense of the force F change any of the answers?

Problem 1–5

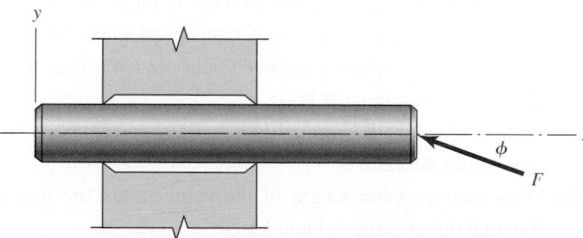

ANALYSIS **1–6** A detent (a lessening of tension, specifically a mechanism that initiates, allows, or locks movement) is a device to index and hold movement between parts. Some move in response to a force, others in response to a torque. A jaw-clutch detent, shown in the figure, has straightflanked teeth like those in a crown gear. The effective radius is r, and the retaining force is F, as shown in the figure. The flank angle θ is measured between the horizontal and the normal force vector, and f is the coefficient of friction. The free body depicts a contact particle on top of the jaw.

Show that

$$F = N(\sin\theta - f\cos\theta)$$

$$T = Fr\frac{1 + f\tan\theta}{\tan\theta - f} = KFr$$

Is the detent self-locking?

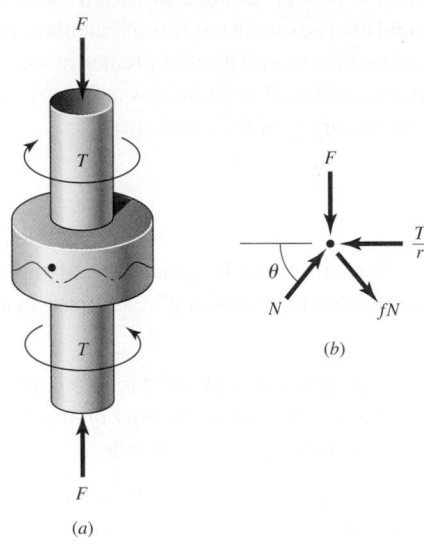

Problem 1–6
A jaw-clutch detent.

(b)

(a)

ANALYSIS

1–7   In Prob. 1–6 the axial force F is supplied by a helical compression spring. The compressive force is F_0 when the jaws are fully engaged. If x is the location of a particle from its lowest possible position, and k is the constant spring rate, then $F = F_0 + kx$.
(a) At what torque T_1 does axial motion begin?
(b) What torque T_2 allows indexing to occur?

ANALYSIS

1–8   A jaw-clutch detent as in Prob. 1–6 has a spring force $F = 10 + 2.5x$ lbf, an effective jaw circle radius of $r = 2$ in, a tooth height of 0.2 in, a tooth flank angle $\theta = 60°$, and a coefficient of friction $f = 0.25$. Estimate the initial $(x = 0)$ and final $(x = 0.2$ in$)$ values of spring force F, normal force N, constant K, and the torque T.

ANALYSIS

1–9   Highway tunnel traffic (two parallel lanes in the same direction) experience indicates the average spacing between vehicles increases with speed. Data from a New York tunnel show that between 15 and 35 mi/h, the space x between vehicles (in miles) is $x = 0.324/(42.1 - v)$ where v is the vehicle's speed in miles per hour.
(a) Ignoring the length of individual vehicles, what speed will give the tunnel the largest volume in vehicles per hour?
(b) Does including the length of the vehicles cut the tunnel capacity prediction significantly? Assume the average vehicle length is 10 ft.
(c) For part (b), does the optimal speed change much?

DESIGN

1–10   The engineering designer must create (invent) the concept and connectivity of the elements that constitute the design, and not lose sight of the need to develop the idea(s) with optimality in mind. A useful figure of merit can be cost, which can be related to the amount of material used (volume or weight). When you think about it, the weight is a function of the geometry and density. When the design is fleshed out, finding the weight is a straightforward, sometimes tedious task. The figure depicts a simple bracket frame that has supports that project from a wall column. The bracket supports a chain-fall hoist. Pinned joints are used to avoid bending. The cost of a link can be *approximated* by $\$ = \cent Al\gamma$, where $\cent$ is the cost of the link per unit weight, A is the cross-sectional area of the prismatic link, l is the pin-to-pin link length, and γ is the specific weight. To be sure, this is approximate because no decisions have been made concerning the geometric form of the links or their fittings. By investigating cost now in this approximate way, one can detect whether a particular set of proportions of the bracket (indexed by angle θ) is advantageous. Is there a preferable angle θ? Show that the figure of merit can be expressed as

$$\text{fom} = -\frac{\gamma \cent W l_2}{S} \left(\frac{1 + \cos^2 \theta}{\sin \theta \cos \theta} \right)$$

where W is the weight of the hoist and load, and S is the allowable tensile or compressive stress in the link material (assume $S = |F/A|$ and no column action). What is the desirable angle θ corresponding to least cost?

Problem 1–10

(a) A chain-hoist bracket frame.
(b) Free body of pin.

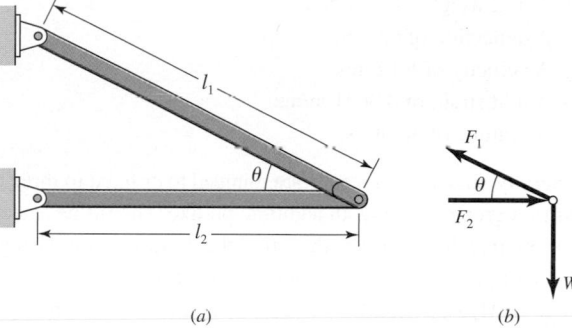

(a)   (b)

ANALYSIS   **1–11**   When one knows the true values x_1 and x_2 and has approximations X_1 and X_2 at hand, one can see where errors may arise. By viewing error as something to be added to an approximation to attain a true value, it follows that the error e_i is related to X_i and x_i as $x_i = X_i + e_i$.

(a) Show that the error in a sum $X_1 + X_2$ is

$$(x_1 + x_2) - (X_1 + X_2) = e_1 + e_2$$

(b) Show that the error in a difference $X_1 - X_2$ is

$$(x_1 - x_2) - (X_1 - X_2) = e_1 - e_2$$

(c) Show that the error in a product $X_1 X_2$ is

$$x_1 x_2 - X_1 X_2 = X_1 X_2 \left(\frac{e_1}{X_1} + \frac{e_2}{X_2} \right)$$

(d) Show that in a quotient X_1/X_2 the error is

$$\frac{x_1}{x_2} - \frac{X_1}{X_2} = \frac{X_1}{X_2} \left(\frac{e_1}{X_1} - \frac{e_2}{X_2} \right)$$

ANALYSIS   **1–12**   Use the true values $x_1 = \sqrt{5}$ and $x_2 = \sqrt{6}$.
(a) Demonstrate the correctness of the error equation from Prob. 1–11 for addition if three correct digits are used for X_1 and X_2.
(b) Demonstrate the correctness of the error equation for addition using three-digit significant numbers for X_1 and X_2.

ANALYSIS

1–13 Convert the following to appropriate SI units:
(a) A stress of 20 000 psi
(b) A force of 350 lbf
(c) A moment of 1200 lbf · in
(d) An area of 2.4 in²
(e) A second moment of area of 17.4 in⁴
(f) An area of 3.6 mi²
(g) A modulus of elasticity of 21 Mpsi
(h) A speed of 45 mi/h
(i) A volume of 60 in³

ANALYSIS

1–14 Convert the following to appropriate ips units:
(a) A length of 1.5 m.
(b) A stress of 600 MPa.
(c) A pressure of 160 kPa.
(d) A section modulus of 1.84 (10⁵) mm³.
(e) A unit weight of 38.1 N/m.
(f) A deflection of 0.05 mm.
(g) A velocity of 6.12 m/s.
(h) A unit strain of 0.0021 m/m.
(i) A volume of 30 liters.

ANALYSIS

1–15 Generally, final design results are rounded to or fixed to three digits because the given data cannot justify a greater display. In addition, prefixes should be selected so as to limit number strings to no more than four digits to the left of the decimal point. Using these rules, as well as those for the choice of prefixes, solve the following relations:
(a) $\sigma = M/Z$, where $M = 200$ N · m and $Z = 15.3$ cm³.
(b) $\sigma = F/A$, where $F = 42$ kN and $A = 6$ cm².
(c) $y = Fl^3/3EI$, where $F = 1200$ N, $l = 800$ mm, $E = 207$ GPa, and $I = 6.4$ cm⁴.
(d) $\theta = Tl/GJ$, where $J = \pi d^4/32$, $T = 1100$ N · m, $l = 250$ mm, $G = 79.3$ GPa, and $d = 25$ mm. Convert results to degrees of angle.

ANALYSIS

1–16 Repeat Prob. 1–15 for the following:
(a) $\sigma = F/wt$, where $F = 600$ N, $w = 20$ mm, and $t = 6$ mm
(b) $I = bh^3/12$, where $b = 8$ mm and $h = 24$ mm
(c) $I = \pi d^4/64$, where $d = 32$ mm
(d) $\tau = 16T/\pi d^3$, where $T = 16$ N · m and $d = 25$ mm

ANALYSIS

1–17 Repeat Prob. 1–15 for:
(a) $\tau = F/A$, where $A = \pi d^2/4$, $F = 120$ kN, and $d = 20$ mm
(b) $\sigma = 32Fa/\pi d^3$, where $F = 800$ N, $a = 800$ mm, and $d = 32$ mm
(c) $Z = (\pi/32d)(d^4 - d_i^4)$ for $d = 36$ mm and $d_i = 26$ mm
(d) $k = (d^4G)/(8D^3N)$, where $d = 1.6$ mm, $G = 79.3$ GPa, $D = 19.2$ mm, and $N = 32$ (a dimensionless number).

2 Statistical Considerations

Statistics in mechanical design provides a method of dealing with characteristics whose values are variable. Products manufactured in large quantities—automobiles, watches, lawnmowers, washing machines, for example—have a life that is variable. One automobile may have so many defects that it must be repaired repeatedly during the first few months of operation while another may operate satisfactorily for years, requiring only minor maintenance.

Methods of quality control are deeply rooted in the use of statistics, and engineering designers need a knowledge of statistics to conform to quality-control standards. The variability inherent in limits and fits, in stress and strength, in bearing clearances, and in a multitude of other characteristics must be described numerically for proper control. It is not satisfactory to say that a product is expected to have a long and troublefree life. We must express such things as product life and product reliability in numerical form in order to achieve a specific quality goal. As noted in Sec. 1–10, uncertainties abound and require quantitative treatment. The algebra of real numbers, by itself, is not well suited to describing the presence of variation.

It is clear that consistencies in nature are stable, not in magnitude, but in the *pattern of variation*. Evidence gathered from nature by measurement is a mixture of systematic and random effects. It is the role of statistics to separate these, and, through the sensitive use of data, illuminate the obscure.

Some students will start this book after completing a formal course in statistics while others may have had brief encounters with statistics in their engineering courses. This contrast in background, together with space and time constraints, makes it very difficult to present an extensive integration of statistics with mechanical engineering design at this stage. Beyond first courses in mechanical design and engineering statistics, the student can begin to meaningfully integrate the two in a second course in design.

The intent of this chapter is to introduce some statistical concepts associated with basic reliability goals.

2–1 Random Variables

Consider an experiment to measure strength in a collection of 20 tensile-test specimens that have been machined from a like number of samples selected at random from a carload shipment of, say, UNS G10200 cold-drawn steel. It is reasonable to expect that there will be differences in the ultimate tensile strengths S_{ut} of each of the individual test specimens. Such differences may occur because of differences in the sizes of the specimens, in the strength of the material itself, or both. Such an experiment is called a *random experiment,* because the specimens are selected at random. The strength S_{ut} determined by this experiment is called a *random,* or a *stochastic, variable.* So a random variable is a variable quantity, such as strength, size, or weight, whose value depends on the outcome of a random experiment.

Let us define a random variable x as the sum of the numbers obtained when two dice are tossed. Either die can display any number from 1 to 6. Figure 2–1 displays all possible outcomes in what is called the *sample space.* Note that x has a specific value for

Figure 2–1

Sample space showing all possible outcomes of the toss of two dice.

1,1	1,2	1,3	1,4	1,5	1,6
2,1	2,2	2,3	2,4	2,5	2,6
3,1	3,2	3,3	3,4	3,5	3,6
4,1	4,2	4,3	4,4	4,5	4,6
5,1	5,2	5,3	5,4	5,5	5,6
6,1	6,2	6,3	6,4	6,5	6,6

Table 2–1

A Probability Distribution

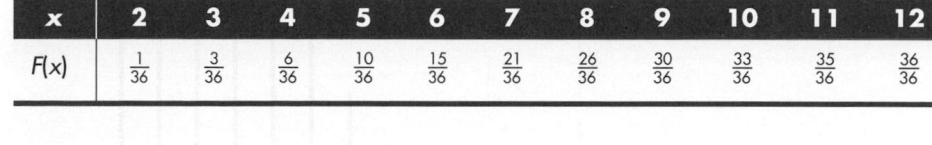

x	2	3	4	5	6	7	8	9	10	11	12
$f(x)$	$\frac{1}{36}$	$\frac{2}{36}$	$\frac{3}{36}$	$\frac{4}{36}$	$\frac{5}{36}$	$\frac{6}{36}$	$\frac{5}{36}$	$\frac{4}{36}$	$\frac{3}{36}$	$\frac{2}{36}$	$\frac{1}{36}$

Table 2–2

A Cumulative Probability Distribution

x	2	3	4	5	6	7	8	9	10	11	12
$F(x)$	$\frac{1}{36}$	$\frac{3}{36}$	$\frac{6}{36}$	$\frac{10}{36}$	$\frac{15}{36}$	$\frac{21}{36}$	$\frac{26}{36}$	$\frac{30}{36}$	$\frac{33}{36}$	$\frac{35}{36}$	$\frac{36}{36}$

Figure 2–2

Frequency distribution.

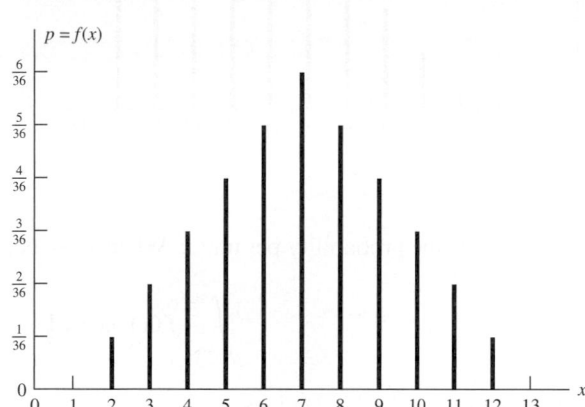

each possible outcome—for example, the event 5, 4; $x = 5 + 4 = 9$. It is useful to form a table showing the values of x and the corresponding values of the probability of x, called $p = f(x)$. This is easily done from Fig. 2–1 merely by adding each outcome, determining how many times a specific value of x arises, and dividing by the total number of possible outcomes. The results are shown in Table 2–1. Any table like this, listing all possible values of a random variable and with the corresponding probabilities, is called a *probability distribution*.

The values of Table 2–1 are plotted in graphical form in Fig. 2–2. Here it is clear that the probability is a function of x. This *probability function* $p = f(x)$ is often called the *frequency function* or, sometimes, the *probability density function* (PDF). The probability that x is less than or equal to a certain value x_i can be obtained from the probability function by summing the probability of all x's up to and including x_i . If we do this with Table 2–1, letting x_i equal 2, then 3, and so on, up to 12, we get Table 2–2, which is called a *cumulative probability distribution*. The *function $F(x)$* in Table 2–2 is called a *cumulative density function* (CDF) of x. In terms of $f(x)$ it may be expressed mathematically in the general form

$$F(x_i) = \sum_{x_j \leq x_i} f(x_j) \tag{2-1}$$

The cumulative distribution may also be plotted as a graph (Fig. 2–3).

The variable x of this example is called a *discrete random variable,* because x has only discrete values. A *continuous random variable* is one that can take on any value in a specified interval; for such variables, graphs like Figs. 2–2 and 2–3 would be plotted as continuous curves. For a continuous probability density function $F(x)$, the probability of obtaining an observation equal to or less than x is given by

$$F(x) = \int_{-\infty}^{x} f(x)\, dx \tag{2-2}$$

Figure 2–3

Cumulative frequency distribution.

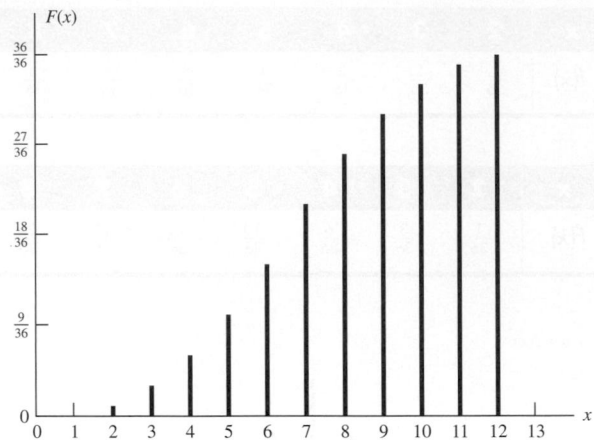

where $f(x)$ is the probability per unit x. When $x \to \infty$, then

$$\int_{-\infty}^{\infty} f(x)\, dx = 1 \qquad (2\text{--}3)$$

Differentiation of Eq. (2–2) gives

$$\frac{dF(x)}{dx} = f(x) \qquad (2\text{--}4)$$

2–2 Arithmetic Mean, Variance, and Standard Deviation

In studying the variations in the mechanical properties and characteristics of mechanical elements, we shall generally be dealing with a finite number of elements. The total number of elements, called the *population,* may in some cases be quite large. In such cases it is usually impractical to measure the characteristics of each member of the population, because this involves destructive testing in some cases, and so we select a small part of the group, called a *sample,* for these determinations. Thus the *population* is the entire group, and the *sample* is a part of the population.

The arithmetic mean of a sample, called the *sample mean,* consisting of N elements, is defined by the equation

$$\bar{x} = \frac{x_1 + x_2 + x_3 + \cdots + x_N}{N} = \frac{1}{N} \sum_{i=1}^{N} x_i \qquad (2\text{--}5)$$

Besides the arithmetic mean, it is useful to have another kind of measure that will tell us something about the spread, or dispersion, of the distribution. For any random variable x, the deviation of the ith observation from the mean is $x_i - \bar{x}$. But since the sum of the deviations so defined is always zero, we square them, and define *sample variance* as

$$s_x^2 = \frac{(x_1 - \bar{x})^2 + (x_2 - \bar{x})^2 + \cdots + (x_N - \bar{x})^2}{N - 1} = \frac{1}{N - 1} \sum_{i=1}^{N} (x_i - \bar{x})^2 \qquad (2\text{--}6)$$

The *sample standard deviation,* defined as the square root of the sample variance, is

$$s_x = \sqrt{\frac{1}{N-1} \sum_{i=1}^{N} (x_i - \bar{x})^2} \qquad (2\text{--}7)$$

Equation (2–7) is not well suited for use in a calculator. For such purposes, use the alternative form

$$s_x = \sqrt{\frac{\sum_{i=1}^{N} x_i^2 - \left(\sum_{i=1}^{N} x_i\right)^2 / N}{N-1}} = \sqrt{\frac{\sum_{i=1}^{N} x_i^2 - N\bar{x}^2}{N-1}} \qquad (2\text{--}8)$$

for the standard deviation.

It should be observed that some authors define the variance and the standard deviation by using N instead of $N - 1$ in the denominator. For large values of N, there is very little difference. For small values, the denominator $N - 1$ actually gives a better estimate of the variance of the population from which the sample is taken.

Equations (2–5) to (2–8) apply specifically to the *sample* of a population. When an entire population is considered, the same equations apply, but $\bar{x}$ and s_x are replaced with the symbols μ_x and, $\hat{\sigma}_x$ respectively. The circumflex accent mark ^, or "hat," is used to avoid confusion with normal stress. For the population variance and standard deviation, N weighting is used in the denominators instead of $N - 1$.

Sometimes we are going to be dealing with the standard deviation of the strength of an element. So you must be careful not to be confused by the notation. Note that we are using the *capital letter S* for *strength* and the *lowercase letter s* for *standard deviation* as shown in the caption of the histogram in Fig. 2–4.

Figure 2–4 is called a *discrete frequency histogram,* which gives the number of occurrences, or class frequency f_i, within a given range. If the data are grouped in this fashion, then the mean and standard deviation are given by

$$\bar{x} = \frac{1}{N} \sum_{i=1}^{k} f_i x_i \qquad (2\text{--}9)$$

and

$$s_x = \sqrt{\frac{\sum_{i=1}^{k} f_i x_i^2 - \left[\left(\sum_{i=1}^{k} f_i x_i\right)^2 / N\right]}{N-1}} = \sqrt{\frac{\sum_{i=1}^{k} f_i x_i^2 - N\bar{x}^2}{N-1}} \qquad (2\text{--}10)$$

Here x_i, f_i, and k are class midpoint, frequency of occurrences within the range of the class, and the total number of classes, respectively. Also, the cumulative density function that gives the probability of an occurrence at class mark of x_i or less is

$$F_i = \frac{f_i w_i}{2} + \sum_{j=1}^{i-1} f_j w_j \qquad (2\text{--}11)$$

where w_i represents the class width at x_i. For Fig. 2–4a, $k = 21$ and the class width is constant at $w = 1$ kpsi.

Figure 2–4

Distribution of tensile properties of hot-rolled UNS G10350 steel, as rolled. These tests were made from round bars varying in diameter from 1 to 9 in. (a) Tensile-strength distributions from 930 heats; $\bar{S}_u = 86.0$ kpsi, $s_{su} = 4.94$ kpsi. (b) Yield-strength distribution from 899 heats; $\bar{S}_y = 49.5$ kpsi, $s_{sy} = 5.36$ kpsi. *(From Metals Handbook, vol. 1, 8th ed., American Society for Metals, Materials Park, OH 44073-0002, fig. 22, p. 64. Reprinted by permission of ASM International®, www.asminternational.org.)*

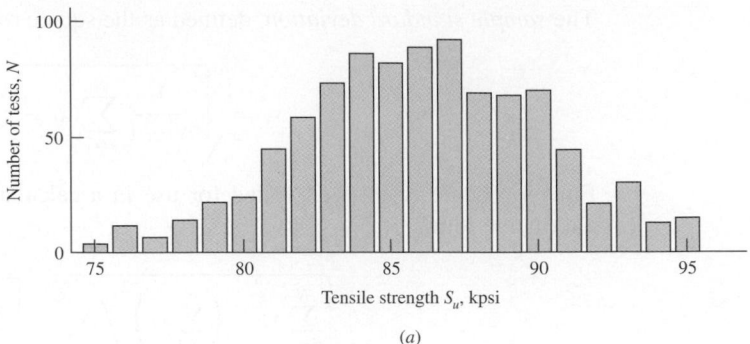

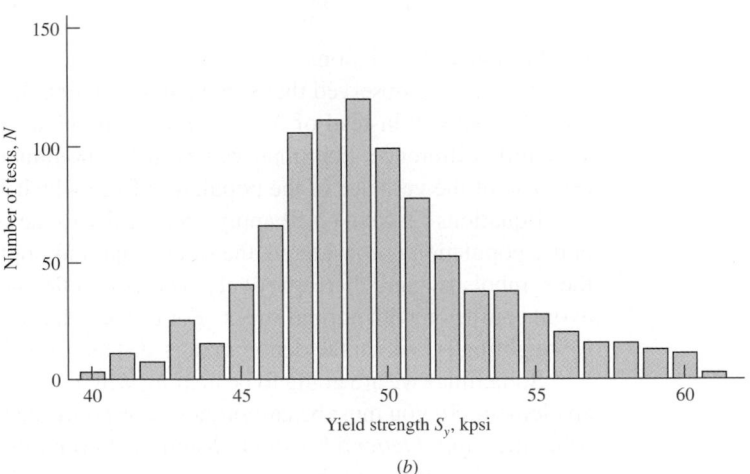

Notation

In this book, we follow the convention of designating vectors by boldface characters, indicative of the fact that two or three quantities, such as direction and magnitude, are necessary to describe them. The same convention is widely used for random variables that can be characterized by specifying a mean and a standard deviation. We shall therefore use boldface characters to designate random variables as well as vectors. No confusion between the two is likely to arise.

The terms *stochastic variable* and *variate* are also used to mean a random variable. A *deterministic quantity* is something that has a single specific value. The mean value of a population is a deterministic quantity, and so is its standard deviation. A stochastic variable can be partially described by the mean and the standard deviation, or by the mean and the *coefficient of variation* defined by

$$C_x = \frac{s_x}{\bar{x}} \tag{2–12}$$

Thus the variate $\mathbf{x}$ for the sample can be expressed in the following two ways:

$$\mathbf{x} = \mathbf{X}(\bar{x}, s_x) = \bar{x}\,\mathbf{X}(1, C_x) \tag{2–13}$$

where $\mathbf{X}$ represents a variate probability distribution function. Note that the deterministic quantities $\bar{x}$, s_x, and C_x are all in normal italic font.

EXAMPLE 2–1 Five tons of 2-in round rod of 1030 hot-rolled steel has been received for workpiece stock. Nine standard-geometry tensile test specimens have been machined from random locations in various rods. In the test report, the ultimate tensile strength was given in kpsi. In ascending order (not necessary), these are displayed in Table 2–3. Find the mean $\bar{x}$, the standard deviation s_x, and the coefficient of variation C_x from the sample, such that these are best estimates of the parent population (the stock your plant will convert to product).

Table 2–3

Data Worksheet from Nine Tensile Test Specimens Taken from a Shipment of 1030 Hot-Rolled Steel Barstock

S_{ut}, kpsi x	x^2
62.8	3 943.84
64.4	4 147.36
65.8	4 329.64
66.3	4 395.69
68.1	4 637.61
69.1	4 774.81
69.8	4 872.04
71.5	5 112.25
74.0	5 476.00
$\sum$ 611.8	41 689.24

Solution From Eqs. (2–5) and (2–8),

$$\bar{x} = \frac{1}{N} \sum_{i=1}^{9} x_i$$

and

$$s_x = \sqrt{\frac{\sum x_i^2 - \left(\sum x_i\right)^2 / N}{N - 1}}$$

It is computationally efficient to generate $\sum x$ and $\sum x^2$ before evaluating $\bar{x}$ and s_x. This has been done in Table 2–3.

$$\bar{x} = \frac{1}{9}(611.8) = 67.98 \text{ kpsi}$$

$$s_x = \sqrt{\frac{41\,689.24 - 611.8^2/9}{9 - 1}} = 3.543 \text{ kpsi}$$

From Eq. (2–12),

$$C_x = \frac{s_x}{\bar{x}} = \frac{3.543}{67.98} = 0.0521$$

All three statistics are estimates of the parent population statistical parameters. Note that these results are independent of the distribution.

Multiple data entries may be identical or may be grouped in histographic form to suggest a distributional shape. If the original data are lost to the designer, the grouped data can still be reduced, although with some loss in computational precision.

EXAMPLE 2–2

The data in Ex. 2–1 have come to the designer in the histographic form of the first two columns of Table 2–4. Using the data in this form, find the mean $\bar{x}$, standard deviation s_x, and the coefficient of variation C_x.

Table 2–4

Grouped Data of Ultimate Tensile Strength from Nine Tensile Test Specimens from a Shipment of 1030 Hot-Rolled Steel Barstock

Class Midpoint x, kpsi	Class Frequency f	Extension	
		fx	fx²
63.5	2	127	8 064.50
66.5	2	133	8 844.50
69.5	3	208.5	14 480.75
72.5	2	145	10 513.50
Σ 9		613.5	41 912.25

The data in Table 2–4 have been extended to provide $\sum f_i x_i$ and $\sum f_i x_i^2$.

Solution

From Eq. (2–9),

$$\bar{x} = \frac{1}{N} \sum_{i=1}^{4} f_i x_i = \frac{1}{9}(613.5) = 68.17 \text{ kpsi}$$

From Eq. (2–10),

$$s_x = \sqrt{\frac{41\,912.25 - 613.5^2/9}{9 - 1}} = 3.391 \text{ kpsi}$$

From Eq. (2–12),

$$C_x = \frac{s_x}{\bar{x}} = \frac{3.391}{68.17} = 0.0497$$

Note the small changes in $\bar{x}$, s_x, and C_x due to small changes in the summation terms.

The descriptive statistics developed, whether from ungrouped or grouped data, describe the ultimate tensile strength $\mathbf{S}_{ut}$ of the material from which we will form parts. Such description is not possible with a single number. In fact, sometimes two or three numbers plus identification or, at least, a robust approximation of the distribution are needed. As you look at the data in Ex. 2–1, consider the answers to these questions:

• Can we characterize the ultimate tensile strength by the mean, $\bar{S}_{ut}$?

• Can we take the lowest ultimate tensile strength of 62.8 kpsi as a minimum? If we do, we will encounter some lesser ultimate strengths, because some of 100 specimens will be lower.

• Can we find the distribution of the ultimate tensile strength of the 1030 stock in Ex. 2–1? Yes, but it will take more specimens and require plotting on coordinates that rectify the data string.

2–3 Probability Distributions

There are a number of standard discrete and continuous probability distributions that are commonly applicable to engineering problems. In this section, we will discuss four important continuous probability distributions; the *Gaussian,* or *normal, distribution;* the *lognormal distribution;* the *uniform distribution;* and the *Weibull distribution.*

The Gaussian (Normal) Distribution

When Gauss asked the question, What distribution is the most likely parent to a set of data?, the answer was the distribution that bears his name. The *Gaussian,* or *normal, distribution* is an important one whose probability density function is expressed in terms of its mean μ_x and its standard deviation $\hat{\sigma}_x$ as

$$f(x) = \frac{1}{\hat{\sigma}_x \sqrt{2\pi}} \exp\left[-\frac{1}{2}\left(\frac{x - \mu_x}{\hat{\sigma}_x}\right)^2\right] \tag{2–14}$$

With the notation described in Sec. 2–2, the normally distributed variate x can be expressed as

$$\mathbf{x} = \mathbf{N}(\mu_x, \hat{\sigma}_x) = \mu_x \mathbf{N}(1, C_x) \tag{2–15}$$

where **N** represents the normal distribution function given by Eq. (2–14).

Since Eq. (2–14) is a probability density function, the area under it, as required, is unity. Plots of Eq. (2–14) are shown in Fig. 2–5 for small and large standard deviations. The bell-shaped curve is taller and narrower for small values of $\hat{\sigma}$ and shorter and broader for large values of $\hat{\sigma}$. Integration of Eq. (2–14) to find the cumulative density function $F(x)$ is not possible in closed form, but must be accomplished numerically. To avoid the need for many tables for different values of μ and $\hat{\sigma}$, the deviation from the mean is expressed in units of standard deviation by the transform

$$\mathbf{z} = \frac{\mathbf{x} - \mu_x}{\hat{\sigma}_x} \tag{2–16}$$

The integral of the transform is tabulated in Table A–10 and sketched in Fig. 2–6. The value of the normal cumulative density function is used so often, and manipulated in so

Figure 2–5

The shape of the normal distribution curve: (a) small $\hat{\sigma}$; (b) large $\hat{\sigma}$.

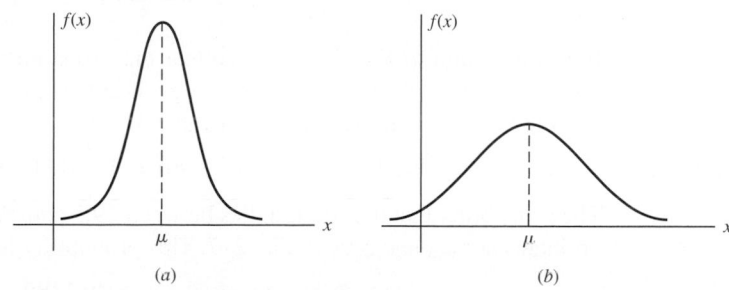

Figure 2–6

The standard normal distribution.

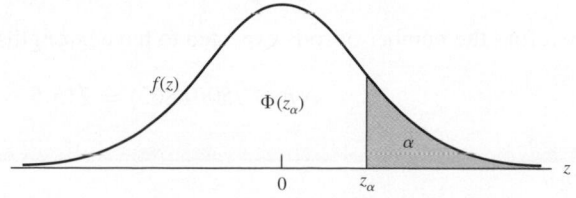

many equations, that it has its own particular symbol, $\Phi(z)$. The transformation variate $\mathbf{z}$ is normally distributed, with a mean of zero and a standard deviation and variance of unity. That is, $\mathbf{z} = \mathbf{N}(0, 1)$. The probability of an observation less than z is $\Phi(z)$ for negative values of z and $1 - \Phi(z)$ for positive values of z in Table A–10.

EXAMPLE 2–3   In a shipment of 250 connecting rods, the mean tensile strength is found to be 45 kpsi and the standard deviation 5 kpsi.
(*a*) Assuming a normal distribution, how many rods can be expected to have a strength less than 39.5 kpsi?
(*b*) How many are expected to have a strength between 39.5 and 59.5 kpsi?

Solution   (*a*) Substituting in Eq. (2–16) gives the standardized z variable as

$$z_{39.5} = \frac{x - \mu_x}{\hat{\sigma}_x} = \frac{S - \bar{S}}{\hat{\sigma}_S} = \frac{39.5 - 45}{5} = -1.10$$

The probability that the strength is less than 39.5 kpsi can be designated as $F(z) = \Phi(-1.10)$. Using Table A–10, and referring to Fig. 2–7, we find $\Phi(z_{39.5}) = 0.1357$. So the number of rods having a strength less than 39.5 kpsi is,

| Figure 2–7

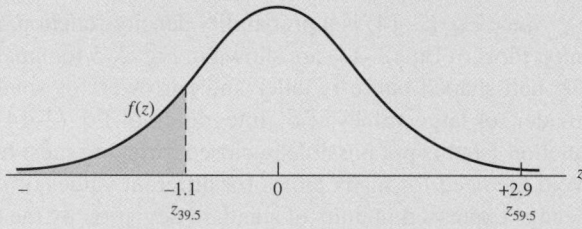

Answer $$N\Phi(z_{39.55}) = 250(0.1357) = 33.9 \approx 34$$

because $\Phi(z_{39.5})$ represents the proportion of the population N having a strength less than 39.5 kpsi.
(*b*) Corresponding to $S = 59.5$ kpsi, we have

$$z_{59.5} = \frac{59.5 - 45}{5} = 2.90$$

Referring again to Fig. 2–7, we see that the probability that the strength is less than 59.5 kpsi is $F(z) = \Phi(z_{59.5})$. Since the z variable is positive, we need to find the value complementary to unity. Thus, from Table A–10,

$$\Phi(2.90) = 1 - \Phi(-2.90) = 1 - 0.001\,87 = 0.998\,13$$

The probability that the strength lies between 39.5 and 59.5 kpsi is the area between the ordinates at $z_{39.5}$ and $z_{59.5}$ in Fig. 2–7. This probability is found to be

$$p = \Phi(z_{59.5}) - \Phi(z_{39.5}) = \Phi(2.90) - \Phi(-1.10)$$
$$= 0.998\,13 - 0.1357 = 0.862\,43$$

Therefore the number of rods expected to have strengths between 39.5 and 59.5 kpsi is

Answer $$Np = 250(0.862) = 215.5 \approx 216$$

The Lognormal Distribution

Sometimes random variables have the following two characteristics:

- The distribution is asymmetrical about the mean.
- The variables have only positive values.

Such characteristics rule out the use of the normal distribution. There are several other distributions that are potentially useful in such situations, one of them being the lognormal (written as a single word) distribution. Especially when life is involved, such as fatigue life under stress or the wear life of rolling bearings, the lognormal distribution may be a very appropriate one to use.

The *lognormal distribution* is one in which the logarithms of the variate have a normal distribution. Thus the variate itself is said to be lognormally distributed. Let this variate be expressed as

$$\mathbf{x} = \mathbf{LN}(\mu_x, \hat{\sigma}_x) \tag{a}$$

Equation (*a*) states that the random variable x is distributed lognormally (*not a logarithm*) and that its mean value is μ_x and its standard deviation is $\hat{\sigma}_x$.

Now use the transformation

$$\mathbf{y} = \ln \mathbf{x} \tag{b}$$

Since, by definition, **y** has a normal distribution, we can write

$$\mathbf{y} = \mathbf{N}(\mu_y, \hat{\sigma}_y) \tag{c}$$

This equation states that the random variable **y** is normally distributed, its mean value is μ_y, and its standard deviation is $\hat{\sigma}_y$.

It is convenient to think of Eq. (*a*) as designating the *parent,* or *principal, distribution* while Eq. (*b*) represents the *companion,* or *subsidiary,* distribution.

The probability density function (PDF) for **x** can be derived from that for **y**; see Eq. (2–14), and substitute y for x in that equation. Thus the PDF for the companion distribution is found to be

$$f(x) = \begin{cases} \dfrac{1}{x\hat{\sigma}_y\sqrt{2\pi}} \exp\left[-\dfrac{1}{2}\left(\dfrac{\ln x - \mu_y}{\hat{\sigma}_y}\right)^2\right] & \text{for } x > 0 \\ 0 & \text{for } x \le 0 \end{cases} \tag{2–17}$$

The companion mean μ_y and standard deviation $\hat{\sigma}_y$ in Eq. (2–17) are obtained from

$$\mu_y = \ln \mu_x - \ln\sqrt{1 + C_x^2} \approx \ln \mu_x - \frac{1}{2}C_x^2 \tag{2–18}$$

$$\hat{\sigma}_y = \sqrt{\ln\left(1 + C_x^2\right)} \approx C_x \tag{2–19}$$

These equations make it possible to use Table A–10 for statistical computations and eliminate the need for a special table for the lognormal distribution.

EXAMPLE 2–4   One thousand specimens of 1020 steel were tested to rupture and the ultimate tensile strengths were reported as grouped data in Table 2–5. From Eq. (2–9),

$$\bar{x} = \frac{63\,625}{1000} = 63.625 \text{ kpsi}$$

Table 2–5

Worksheet for Ex. 2–4

Class Midpoint, kpsi	Frequency f_i	Extension		Observed PDF $f_i/(Nw)$*	Normal Density $f(x)$	Lognormal Density $g(x)$
		$x_i f_i$	$x_i^2 f_i$			
56.5	2	113.0	6 384.5	0.002	0.0035	0.0026
57.5	18	1 035.0	59 512.5	0.018	0.0095	0.0082
58.5	23	1 345.5	78 711.75	0.023	0.0218	0.0209
59.5	31	1 844.5	109 747.75	0.031	0.0434	0.0440
60.5	83	5 021.5	303 800.75	0.083	0.0744	0.0773
61.5	109	6 703.5	412 265.25	0.109	0.110	0.1143
62.5	138	8 625.0	539 062.5	0.138	0.140	0.1434
63.5	151	9 588.5	608 869.75	0.151	0.1536	0.1539
64.5	139	8 965.5	578 274.75	0.139	0.1453	0.1424
65.5	130	8 515.0	577 732.5	0.130	0.1184	0.1142
66.5	82	5 453.0	362 624.5	0.082	0.0832	0.0800
67.5	49	3 307.5	223 256.25	0.049	0.0504	0.0493
68.5	28	1 918.0	131 382.0	0.028	0.0263	0.0268
69.5	11	764.5	53 132.75	0.011	0.0118	0.0129
70.5	4	282.0	19 881.0	0.004	0.0046	0.0056
71.5	2	143.0	10 224.5	0.002	0.0015	0.0022
$\sum$ 1 000		63 625	4 054 864	1.000		

* To compare discrete frequency data with continuous probability density functions f_i must be divided by Nw. Here, N = sample size = 1000; w = width of class interval = 1 kpsi.

From Eq. (2–10),

$$s_x = \sqrt{\frac{4\,054\,864 - 63\,625^2/1000}{1000 - 1}} = 2.594\,245 = 2.594 \text{ kpsi}$$

$$C_x = \frac{s_x}{\bar{x}} = \frac{2.594\,245}{63.625} = 0.040\,773 = 0.0408$$

From Eq. (2–14) the probability density function for a normal distribution with a mean of 63.625 and a standard deviation of 2.594 245 is

$$f(x) = \frac{1}{2.594\,245\sqrt{2\pi}} \exp\left[-\frac{1}{2}\left(\frac{x - 63.625}{2.594\,245}\right)^2\right]$$

For example, $f(63.625) = 0.1538$. The probability density $f(x)$ is evaluated at class midpoints to form the column of normal density in Table 2–5.

EXAMPLE 2–5   Continue Ex. 2–4, but fit a lognormal density function.

Solution   From Eqs. (2–18) and (2–19),

$$\mu_y = \ln \mu_x - \ln \sqrt{1 + C_x^2} = \ln 63.625 - \tfrac{1}{2} \ln(1 + 0.040\,773^2) = 4.1522$$

$$\hat{\sigma}_y = \sqrt{\ln\left(1 + C_x^2\right)} = \sqrt{\ln(1 + 0.040\,773^2)} = 0.0408$$

The probability density of a lognormal distribution is given in Eq. (2–17) as

$$g(x) = \frac{1}{x\,(0.0408)\,\sqrt{2\pi}} \exp\left[-\frac{1}{2} \left(\frac{\ln x - 4.1522}{0.0408} \right)^2 \right] \qquad \text{for } x > 0$$

For example, $g(63.625) = 0.1537$. This lognormal density has been added to Table 2–5. Plot the lognormal PDF superposed on the histogram of Ex. 2–4 along with the normal density. As seen in Fig. 2–8, both normal and lognormal densities fit well.

Figure 2–8

Histogram for Ex. 2–4 and Ex. 2–5 with normal and lognormal probability density functions superposed.

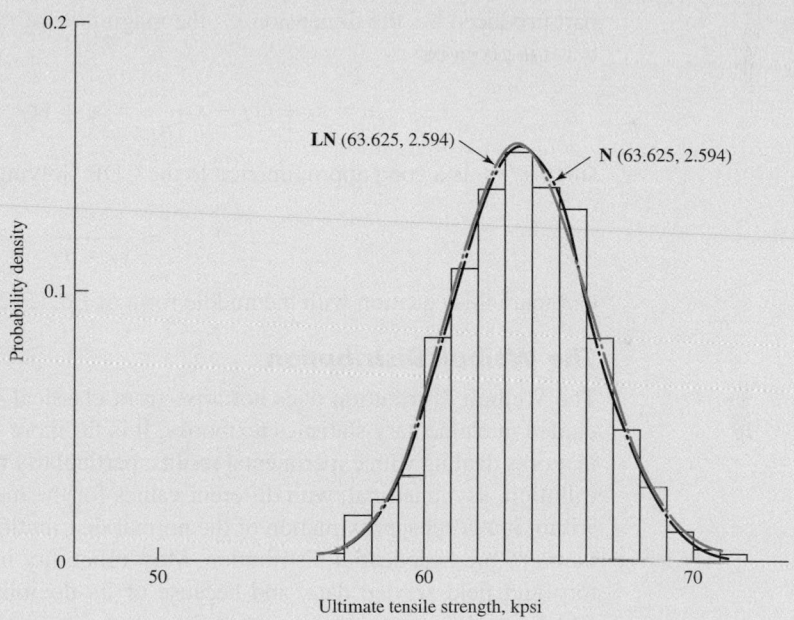

The Uniform Distribution

The uniform distribution is a closed-interval distribution that arises when the chance of an observation is the same as the chance for any other observation. If a is the lower bound and b is the upper bound, then the probability density function (PDF) for the uniform distribution is

$$f(x) = \begin{cases} 1/(b-a) & a \le x \le b \\ 0 & a > x > b \end{cases} \tag{2–20}$$

The cumulative density function (CDF), the integral of $f(x)$, is thus linear in the range $a \leq x \leq b$ given by

$$F(x) = \begin{cases} 0 & x < a \\ (x-a)/(b-a) & a \leq x \leq b \\ 1 & x > b \end{cases} \tag{2-21}$$

The mean and standard deviation are given by

$$\mu_x = \frac{a+b}{2} \tag{2-22}$$

$$\hat{\sigma}_x = \frac{b-a}{2\sqrt{3}} \tag{2-23}$$

The uniform distribution arises, among other places in manufacturing, where a part is mass-produced in an automatic operation and the dimension gradually changes through tool wear and increased tool forces between setups. If n is the part sequence or processing number, and n_f is the sequence number of the final-produced part before another setup, then the dimension x graphs linearly when plotted against the sequence number n. If the last proof part made during the setup has a dimension x_i, and the final part produced has the dimension x_f, the magnitude of the dimension at sequence number n is given by

$$x = x_i + (x_f - x_i)\frac{n}{n_f} = x_i + (x_f - x_i)F(x) \tag{a}$$

since n/n_f is a good approximation to the CDF. Solving Eq. (a) for $F(x)$ gives

$$F(x) = \frac{x - x_i}{x_f - x_i} \tag{b}$$

Compare this equation with the middle form of Eq. (2–21).

The Weibull Distribution

The Weibull distribution does not arise from classical statistics and is usually not included in elementary statistics textbooks. It is far more likely to be discussed and used in works dealing with experimental results, particularly reliability. It is a chameleon distribution, asymmetrical, with different values for the mean and the median. It contains within it a good approximation of the normal distribution as well as an exact representation of the exponential distribution. Most reliability information comes from laboratory and field service data, and because of its flexibility, the Weibull distribution is widely used.

The expression for reliability is the value of the cumulative density function complementary to unity. For the Weibull this value is both explicit and simple. The reliability given by the *three-parameter Weibull distribution* is

$$R(x) = \exp\left[-\left(\frac{x - x_0}{\theta - x_0}\right)^b\right] \qquad x \geq x_0 \geq 0 \tag{2-24}$$

where the three parameters are

$$x_0 = \text{minimum, guaranteed, value of } x$$

$$\theta = \text{a characteristic or scale value } (\theta \geq x_0)$$

$$b = \text{a shape parameter } (b > 0)$$

Figure 2–9

The density function of the Weibull distribution showing the effect of skewness of the shape parameter b.

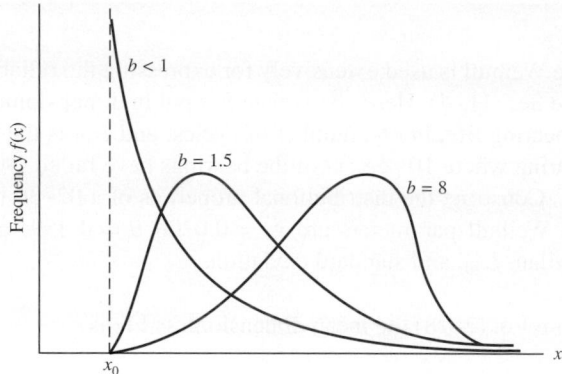

For the special case when $x_0 = 0$, Eq. (2–24) becomes the two-parameter Weibull

$$R(x) = \exp\left[-\left(\frac{x}{\theta}\right)^b\right] \qquad x \geq 0 \qquad (2\text{–}25)$$

The characteristic variate θ serves a role similar to the mean and represents a value of x below which lie 63.2 percent of the observations.

The shape parameter b controls the skewness of the distribution. Figure 2–9 shows that large b's skew the distribution to the right and small b's skew it to the left. In the range $3.3 < b < 3.5$, approximate symmetry is obtained along with a good approximation to the normal distribution. When $b = 1$, the distribution is exponential.

Given a specific required reliability, solving Eq. (2–24) for x yields

$$x = x_0 + (\theta - x_0)\left(\ln\frac{1}{R}\right)^{1/b} \qquad (2\text{–}26)$$

To find the probability function, we note that

$$F(x) = 1 - R(x) \qquad (a)$$

$$f(x) = \frac{dF(x)}{dx} = -\frac{dR(x)}{dx} \qquad (b)$$

Thus, for the Weibull,

$$f(x) = \begin{cases} \dfrac{b}{\theta - x_0}\left(\dfrac{x - x_0}{\theta - x_0}\right)^{b-1}\exp\left[-\left(\dfrac{x - x_0}{\theta - x_0}\right)^b\right] & x \geq x_0 \geq 0 \\ 0 & x \leq x_0 \end{cases} \qquad (2\text{–}27)$$

The mean and standard deviation are given by

$$\mu_x = x_0 + (\theta - x_0)\,\Gamma(1 + 1/b) \qquad (2\text{–}28)$$

$$\hat{\sigma}_x = (\theta - x_0)\,\sqrt{\Gamma(1 + 2/b) - \Gamma^2(1 + 1/b)} \qquad (2\text{–}29)$$

where Γ is the gamma function and may be found tabulated in Table A–34. The notation for a Weibull distribution is[1]

$$\mathbf{x} = \mathbf{W}(x_0, \theta, b) \qquad (2\text{–}30)$$

---

1. To estimate the Weibull parameters from data, see J. E. Shigley and C. R. Mischke, *Mechanical Engineering Design*, 5th ed., 1989, McGraw-Hill, New York, Sec. 4–12. The Weibull parameters are determined for the data given in Ex. 2–4.

EXAMPLE 2–6

The Weibull is used extensively for expressing the reliability of rolling-contact bearings (see Sec. 11–4). Here, the variate x is put in dimensionless form as $x = L/L_{10}$ where L is bearing life, in say, number of cycles; and L_{10} is the manufacturer's rated life of the bearing where 10 percent of the bearings have failed (90 percent reliability).

Construct the distributional properties of a 02–30 mm deep-groove ball bearing if the Weibull parameters are $x_0 = 0.0200$, $\theta = 4.459$, and $b = 1.483$. Find the mean, median, L_{90}, and standard deviation.

Solution

From Eq. (2–28) the mean dimensionless life is

Answer

$$\mu_x = x_0 + (\theta - x_0)\Gamma(1 + 1/b)$$

$$= 0.0200 + (4.459 - 0.0200)\Gamma(1 + 1/1.483) = 4.033$$

This says that the average bearing life is 4.033 L_{10}. The median dimensionless life corresponds to $R = 0.5$, or L_{50}, and from Eq. (2–26) is

Answer

$$x_{0.5} = x_0 + (\theta - x_0)\left(\ln\frac{1}{0.5}\right)^{1/b} = 0.0200 + (4.459 - 0.0200)\left(\ln\frac{1}{0.5}\right)^{1/1.483}$$

$$= 3.487$$

For L_{90}, $R = 0.1$, the dimensionless life x is

Answer

$$x_{0.90} = 0.0200 + (4.459 - 0.0200)\left(\ln\frac{1}{0.1}\right)^{1/1.483} = 7.810$$

The standard deviation of the dimensionless life is given by Eq. (2–29):

Answer

$$\hat{\sigma}_x = (\theta - x_0)\sqrt{\Gamma(1 + 2/b) - \Gamma^2(1 + 1/b)}$$

$$= (4.459 - 0.0200)\sqrt{\Gamma(1 + 2/1.483) - \Gamma^2(1 + 1/1.483)} = 2.753$$

2–4 Propagation of Error

In the equation for axial stress

$$\sigma = \frac{F}{A} \tag{a}$$

suppose both the force F and the area A are random variables. Then Eq. (a) is written as

$$\mathbf{\sigma} = \frac{\mathbf{F}}{\mathbf{A}} \tag{b}$$

and we see that the stress $\mathbf{\sigma}$ is also a random variable. When Eq. (b) is solved, the errors inherent in $\mathbf{F}$ and in $\mathbf{A}$ are said to be *propagated* to the stress variate $\mathbf{\sigma}$. It is not hard to think of many other relations where this will occur.

Suppose we wish to add the two variates $\mathbf{x}$ and $\mathbf{y}$ to form a third variate $\mathbf{z}$. This is written as

$$\mathbf{z} = \mathbf{x} + \mathbf{y} \tag{c}$$

The mean is given as

$$\mu_z = \mu_x + \mu_y \tag{d}$$

Table 2–6

Means and Standard Deviations for Simple Algebraic Operations on Independent (Uncorrelated) Random Variables

Function	Mean (μ)	Standard Deviation ($\hat{\sigma}$)
a	a	0
x	μ_x	$\hat{\sigma}_x$
$\mathbf{x} + a$	$\mu_x + a$	$\hat{\sigma}_x$
$a\mathbf{x}$	$a\mu_x$	$a\hat{\sigma}_x$
$x + y$	$\mu_x + \mu_y$	$\left(\hat{\sigma}_x^2 + \hat{\sigma}_y^2\right)^{1/2}$
$x - y$	$\mu_x - \mu_y$	$\left(\hat{\sigma}_x^2 + \hat{\sigma}_y^2\right)^{1/2}$
xy	$\mu_x\mu_y$	$\mu_x\mu_y\left(C_x^2 + C_y^2 + C_x^2 C_y^2\right)^{1/2}$
x/y	μ_x/μ_y	$\mu_x/\mu_y\left[\left(C_x^2 + C_y^2\right) \Big/ \left(1 + C_y^2\right)\right]^{1/2}$
$\mathbf{x}^n$	$\mu_x^n\left[1 + \dfrac{n(n-1)}{2}C_x^2\right]$	$\lvert n\rvert\mu_x^n C_x\left[1 + \dfrac{(n-1)^2}{4}C_x^2\right]$
$1/\mathbf{x}$	$\dfrac{1}{\mu_x}\left(1 + C_x^2\right)$	$\dfrac{C_x}{\mu_x}\left(1 + C_x^2\right)$
$1/\mathbf{x}^2$	$\dfrac{1}{\mu_x^2}\left(1 + 3C_x^2\right)$	$\dfrac{2C_x}{\mu_x^2}\left(1 + \dfrac{9}{4}C_x^2\right)$
$1/\mathbf{x}^3$	$\dfrac{1}{\mu_x^3}\left(1 + 6C_x^2\right)$	$\dfrac{3C_x}{\mu_x^3}\left(1 + 4C_x^2\right)$
$1/\mathbf{x}^4$	$\dfrac{1}{\mu_x^4}\left(1 + 10C_x^2\right)$	$\dfrac{4C_x}{\mu_x^4}\left(1 + \dfrac{25}{4}C_x^2\right)$
$\sqrt{\mathbf{x}}$	$\sqrt{\mu_x}\left(1 - \dfrac{1}{8}C_x^2\right)$	$\dfrac{\sqrt{\mu_x}}{2}C_x\left(1 + \dfrac{1}{16}C_x^2\right)$
$\mathbf{x}^2$	$\mu_x^2\left(1 + C_x^2\right)$	$2\mu_x^2 C_x\left(1 + \dfrac{1}{4}C_x^2\right)$
$\mathbf{x}^3$	$\mu_x^3\left(1 + 3C_x^2\right)$	$3\mu_x^3 C_x\left(1 + C_x^2\right)$
$\mathbf{x}^4$	$\mu_x^4\left(1 + 6C_x^2\right)$	$4\mu_x^4 C_x\left(1 + \dfrac{9}{4}C_x^2\right)$

Note: The coefficient of variation of variate $\mathbf{x}$ is $C_x = \hat{\sigma}_x/\mu_x$. For small COVs their square is small compared to unity, so the first term in the powers of $\mathbf{x}$ expressions are excellent approximations. For correlated products and quotients see Charles R. Mischke, *Mathematical Model Building,* 2nd rev. ed., Iowa State University Press, Ames, 1980, App. C.

The standard deviation follows the Pythagorean theorem. Thus the standard deviation for *both* addition and subtraction of independent variables is

$$\hat{\sigma}_z = \sqrt{\hat{\sigma}_x^2 + \hat{\sigma}_y^2} \tag{e}$$

Similar relations have been worked out for a variety of functions and are displayed in Table 2–6. The results shown can easily be combined to form other functions.

An unanswered question here is what is the distribution that results from the various operations? For answers to this question, statisticians use closure theorems and the central limit theorem.[2]

---

2. See E. B. Haugen, *Probabilistic Mechanical Design,* Wiley, New York, 1980, pp. 49–54.

EXAMPLE 2–7   A round bar subject to a bending load has a diameter $\mathbf{d} = \mathbf{LN}(2.000, 0.002)$ in. This equivalency states that the mean diameter is $\mu_d = 2.000$ in and the standard deviation is $\hat{\sigma}_d = 0.002$ in. Find the mean and the standard deviation of the second moment of area.

Solution   The second moment of area is given by the equation

$$\mathbf{I} = \frac{\pi \mathbf{d}^4}{64}$$

The coefficient of variation of the diameter is

$$C_d = \frac{\hat{\sigma}_d}{\mu_d} = \frac{0.002}{2} = 0.001$$

Using Table 2–6, we find

Answer $$\mu_I = (\pi/64)\mu_d^4\left(1 + 6C_d^2\right) = (\pi/64)(2.000)^4[1 + 6(0.001)^2] = 0.785 \text{ in}^4$$

Answer $$\hat{\sigma}_I = 4\mu_d^4 C_d\left[1 + (9/4)C_d^2\right] = 4(2.000)^4(0.001)[1 + (9/4)(0.001)^2] = 0.064 \text{ in}^4$$

These results can be expressed in the form

$$\mathbf{I} = (0.785, 0.064) = 0.785\mathbf{LN}(1, 0.0815) \text{ in}^4$$

2–5   Linear Regression

Statisticians use a process of analysis called *regression* to obtain a curve that best fits a set of data points. The process is called *linear regression* when the best-fitting straight line is to be found. The meaning of the word *best* is open to argument, because there can be many meanings. The usual method, and the one employed here, regards a line as "best" if it minimizes the squares of the deviations of the data points from the line.

Figure 2–10 shows a set of data points approximated by the line *AB*. The standard equation of a straight line is

$$y = mx + b \tag{2–31}$$

Figure 2–10

Set of data points approximated by regression line *AB*.

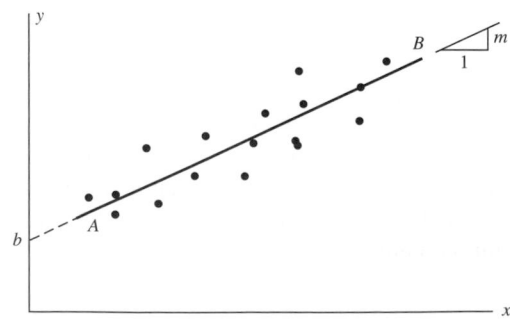

where m is the slope and b is the y intercept. Consider a set of N data points (x_i, y_i). In general, the best-fit line will not intersect a data point. Thus, we can write

$$y_i = mx_i + b + \epsilon_i \qquad (b)$$

where $\epsilon_i = y_i - y$ is the deviation between the data point and the line. The sum of the squares of the deviations is given by[3]

$$\mathcal{E} = \sum \epsilon_i^2 = \sum (y_i - mx_i - b)^2 \qquad (c)$$

Minimizing $\mathcal{E}$, the sum of the squared errors, expecting a stationary point minimum, requires $\partial \mathcal{E} / \partial m = 0$ and $\partial \mathcal{E} / \partial b = 0$. This results in two simultaneous equations for the slope and y intercept denoted as $\hat{m}$ and $\hat{b}$, respectively. Solving these equations results in

$$\hat{m} = \frac{N \sum x_i y_i - \sum x_i \sum y_i}{N \sum x_i^2 - \left(\sum x_i \right)^2} = \frac{\sum x_i y_i - N \bar{x} \bar{y}}{\sum x_i^2 - N \bar{x}^2} \qquad (2\text{--}32)$$

$$\hat{b} = \frac{\sum y_i - \hat{m} \sum x_i}{N} = \bar{y} - \hat{m} \bar{x} \qquad (2\text{--}33)$$

Once you have established a slope and an intercept, the next point of interest is to discover how well x and y correlate with each other. If the data points are scattered all over the xy plane, there is obviously no correlation. But if all the data points coincide with the regression line, then there is perfect correlation. Most statistical data will be in between these extremes. A *correlation coefficient* r, having the range $-1 \leq r \leq +1$, has been devised to answer this question. The formula is

$$r = \hat{m} \frac{s_x}{s_y} \qquad (2\text{--}34)$$

where s_x and s_y are the standard deviations of the x coordinates and y coordinates of the data, respectively. If $r = 0$, there is no correlation; if $r = \pm 1$, there is perfect correlation. A positive or negative r indicates that the regression line has a positive or negative slope, respectively.

The standard deviations for $\hat{m}$ and $\hat{b}$ are given by

$$s_{\hat{m}} = \frac{s_{y \cdot x}}{\sqrt{\sum (x_i - \bar{x})^2}} \qquad (2\text{--}35)$$

$$s_{\hat{b}} = s_{y \cdot x} \sqrt{\frac{1}{N} + \frac{\bar{x}^2}{\sum (x_i - \bar{x})^2}} \qquad (2\text{--}36)$$

where

$$s_{y \cdot x} = \sqrt{\frac{\sum y_i^2 - \hat{b} \sum y_i - \hat{m} \sum x_i y_i}{N - 2}} \qquad (2\text{--}37)$$

is the standard deviation of the scatter of the data from the regression line.

---

3. From this point on, for economy of notation, the limits of the summation of $i(1, N)$ will not be displayed.

EXAMPLE 2–8

A specimen of a medium carbon steel was tested in tension. With an extensometer in place, the specimen was loaded then unloaded, to see if the extensometer reading returned to the no-load reading, then the next higher load was applied. The loads and extensometer elongations were reduced to stress σ and strain ϵ, producing the following data:

σ, psi	5033	10 068	15 104	20 143	35 267
ϵ	0.000 20	0.000 30	0.000 50	0.000 65	0.001 15

Find the mean Young's modulus $\bar{E}$ and its standard deviation. Since the extensometer seems to have an initial reading at no load, use a $y = mx + b$ regression.

Solution

From Table 2–7, $\bar{x} = 0.002\,80/5 = 0.000\,56$, $\bar{y} = 85\,615/5 = 17\,123$. Note that a regression line always contains the data centroid. From Eq. (2–32)

Answer

$$\hat{m} = \frac{5(65.229) - 0.0028(85\,615)}{5(0.000\,002\,125) - 0.0028^2} = 31.03(10^6)\,\text{psi} = \bar{E}$$

From Eq. (2–33)

$$\hat{b} = \frac{0.000\,002\,125(85\,615) - 0.002\,80(65.229)}{5(0.000\,002\,125) - 0.0028^2} = -254.69\,\text{psi}$$

From Eq. (2–34), obtaining s_x and s_y from a statistics calculator routine,

$$\hat{r} = \frac{\hat{m}s_x}{s_y} = \frac{31\,031\,597.85(3\,162\,163\,10^{-4})}{11\,601.11} = 0.998$$

From Eq. (2–37), the scatter about the regression line is measured by the standard deviation $s_{y\cdot x}$ and is equal to

$$s_{y\cdot x} = \sqrt{\frac{\sum y^2 - \hat{b}\sum y - \hat{m}\sum xy}{N - 2}}$$

$$= \sqrt{\frac{2\,004\,328\,267 - (-254.69)85\,615 - 31.03(10^6)(65.229)}{5 - 2}}$$

$$= 811.1\,\text{psi}$$

Table 2–7

Worksheet for Ex. 2–6

y σ, psi	x ϵ	x^2	xy	y^2	$(x - \bar{x})^2$
5 033	0.000 20	0.000 000 040	1.006 600	25 330 089	0.000 000 130
10 068	0.000 30	0.000 000 090	3.020 400	101 364 624	0.000 000 069
15 104	0.000 50	0.000 000 250	7.552 000	228 130 816	0.000 000 004
20 143	0.000 65	0.000 000 423	13.092 950	405 740 449	0.000 000 008
35 261	0.001 15	0.000 001 323	40.557 050	1 243 761 289	0.000 000 348
$\sum$ 85 615	0.002 80	0.000 002 125	65.229 000	2 004 328 267	0.000 000 556

Note: $\bar{y} = 85\,615/5 = 17\,123$ psi, $\bar{x} = 0.002\,80/5 = 0.000\,56$.

From Eq. (2–35), the standard deviation of $\hat{m}$ is

Answer
$$s_{\hat{m}} = \frac{s_{y \cdot x}}{\sqrt{\sum (x - \bar{x})^2}} = \frac{811.1}{\sqrt{0.000\ 000\ 558}} = 1.086(10^6)\ \text{psi} = s_E$$

See Fig. 2–11 for the regression plot.

Figure 2–11

The data from Ex. 2–8 are plotted. The regression line passes through the data centroid and among the data points, minimizing the squared deviations.

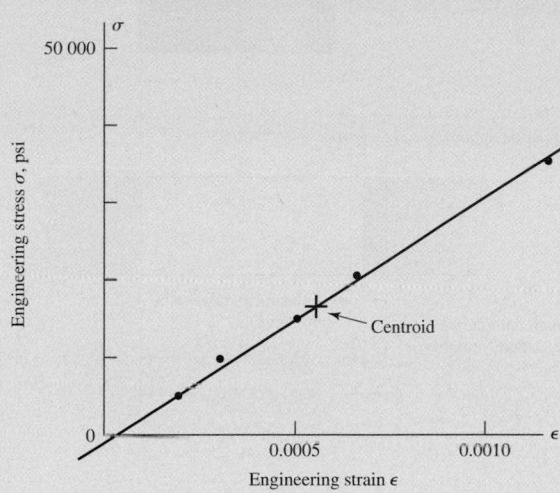

2–6 **Limits and Fits**

The subject of limits and fits really deserves a chapter of its own. The subject is included here because the variability inherent in many of the fit classes is so useful in demonstrating the practical application of the statistical ideas presented.

The designer is free to adopt any geometry of fit for shafts and holes that will ensure the intended function. There is sufficient accumulated experience with commonly recurring situations to make standards useful. There are two standards for limits and fits in the United States, one based on inch units and the other based on metric units.[4] These differ in nomenclature, definitions, and organization. No point would be served by separately studying each of the two systems. The metric version is the newer of the two and is well organized, and so here we present only the metric version but include a set of inch conversions to enable the same system to be used with either system of units.

In using the standard, *capital letters always refer to the hole; lowercase letters are used for the shaft.*

The definitions illustrated in Fig. 2–12 are explained as follows:

- *Basic size* is the size to which limits or deviations are assigned and is the same for both members of the fit.
- *Deviation* is the algebraic difference between a size and the corresponding basic size.
- *Upper deviation* is the algebraic difference between the maximum limit and the corresponding basic size.

---

4. *Preferred Metric Limits and Fits for Cylindrical Parts,* ANSI B4.1-1967. *Preferred Metric Limits and Fits,* ANSI B4.2-1978.

Figure 2–12

Definitions applied to a cylindrical fit.

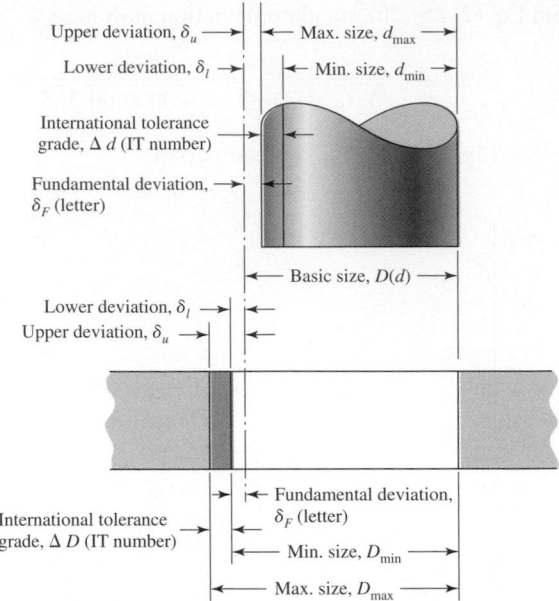

- *Lower deviation* is the algebraic difference between the minimum limit and the corresponding basic size.

- *Fundamental deviation* is either the upper or the lower deviation, depending on which is closer to the basic size.

- *Tolerance* is the difference between the maximum and minimum size limits of a part.

- *International tolerance grade* numbers (IT) designate groups of tolerances such that the tolerances for a particular IT number have the same relative level of accuracy but vary depending on the basic size.

- *Hole basis* represents a system of fits corresponding to a basic hole size. The fundamental deviation is H.

- *Shaft basis* represents a system of fits corresponding to a basic shaft size. The fundamental deviation is h. The shaft-basis system is not included here.

The magnitude of the tolerance zone is the variation in part size and is the same for both the internal and the external dimensions. The tolerance zones are specified in international tolerance grade numbers, called IT numbers. The smaller grade numbers specify a smaller tolerance zone. These range from IT0 to IT16, but only grades IT6 to IT11 are needed for the preferred fits. These are listed in Tables A–11 to A–13 for basic sizes up to 16 in or 400 mm.

The standard uses *tolerance position letters,* with capital letters for internal dimensions (holes) and lowercase letters for external dimensions (shafts). As shown in Fig. 2–12, the fundamental deviation locates the tolerance zone relative to the basic size.

Table 2–8 shows how the letters are combined with the tolerance grades to establish a preferred fit. The ISO symbol for the hole for a sliding fit with a basic size of 32 mm is 32H7. Inch units are not a part of the standard. However, the designation $(1\frac{3}{8}$ in)H7 includes the same information and is recommended for use here. In both cases, the capital letter H establishes the fundamental deviation and the number 7 defines a tolerance grade of IT7.

Table 2-8

Descriptions of Preferred
Fits Using the Basic
Hole System

*Source: Preferred Metric Limits
and Fits*, ANSI B4.2-1978.
See also BS 4500.

Type of Fit	Description	Symbol
Clearance	*Loose running* fit: for wide commercial tolerances or allowances on external members	H11/c11
	Free running fit: not for use where accuracy is essential, but good for large temperature variations, high running speeds, or heavy journal pressures	H9/d9
	Close running fit: for running on accurate machines and for accurate location at moderate speeds and journal pressures	H8/f7
	Sliding fit: where parts are not intended to run freely, but must move and turn freely and locate accurately	H7/g6
	Locational clearance fit: provides snug fit for location of stationary parts, but can be freely assembled and disassembled	H7/h6
Transition	*Locational transition* fit for accurate location, a compromise between clearance and interference	H7/k6
	Locational transition fit for more accurate location where greater interference is permissible	H7/n6
Interference	*Locational interference* fit: for parts requiring rigidity and alignment with prime accuracy of location but without special bore pressure requirements	H7/p6
	Medium drive fit: for ordinary steel parts or shrink fits on light sections, the tightest fit usable with cast iron	H7/s6
	Force fit: suitable for parts that can be highly stressed or for shrink fits where the heavy pressing forces required are impractical	H7/u6

For the sliding fit, the corresponding shaft dimensions are defined by the symbol 32g6 [$(1\frac{3}{8}$ in)g6].

The fundamental deviations for shafts are given in Tables A–11 and A–13. For letter codes c, d, f, g, and h,

Upper deviation = fundamental deviation
Lower deviation = upper deviation − tolerance grade

For letter codes k, n, p, s, and u, the deviations for shafts are

Lower deviation = fundamental deviation
Upper deviation = lower deviation + tolerance grade

The lower deviation H (for holes) is zero. For these, the upper deviation equals the tolerance grade.

As shown in Fig. 2–12, we use the following notation:

$$D = \text{basic size of hole}$$
$$d = \text{basic size of shaft}$$
$$\delta_u = \text{upper deviation}$$
$$\delta_l = \text{lower deviation}$$
$$\delta_F = \text{fundamental deviation}$$
$$\Delta D = \text{tolerance grade for hole}$$
$$\Delta d = \text{tolerance grade for shaft}$$

Note that these quantities are all deterministic. Thus, for the hole,

$$D_{max} = D + \Delta D \qquad D_{min} = D \tag{2-38}$$

For shafts with clearance fits c, d, f, g, and h,

$$d_{max} = d + \delta_F \qquad d_{min} = d + \delta_F - \Delta d \tag{2-39}$$

For shafts with interference fits k, n, p, s, and u,

$$d_{min} = d + \delta_F \qquad d_{max} = d + \delta_F + \Delta d \tag{2-40}$$

EXAMPLE 2–9

Find the shaft and hole dimensions for a loose running fit with a 34-mm basic size.

Solution

From Table 2–8, the ISO symbol is 34H11/c11. From Table A–11, we find that tolerance grade IT11 is 0.160 mm. The symbol 34H11/c11 therefore says that $\Delta D = \Delta d = 0.160$ mm. Using Eq. (2–38) for the hole, we get

Answer

$$D_{max} = D + \Delta D = 34 + 0.160 = 34.160 \text{ mm}$$

Answer

$$D_{min} = D = 34.000 \text{ mm}$$

The shaft is designated as a 34c11 shaft. From Table A–12, the fundamental deviation is $\delta_F = -0.120$ mm. Using Eq. (2–39), we get for the shaft dimensions

Answer

$$d_{max} = d + \delta_F = 34 + (-0.120) = 33.880 \text{ mm}$$

Answer

$$d_{min} = d + \delta_F - \Delta d = 34 + (-0.120) - 0.160 = 33.720 \text{ mm}$$

EXAMPLE 2–10

Find the hole and shaft limits for a medium drive fit using a basic hole size of 2 in.

Solution

The symbol for the fit, from Table 2–8, in inch units is (2 in)H7/s6. For the hole, we use Table A–13 and find the IT7 grade to be $\Delta D = 0.0010$ in. Thus, from Eq. (2–38),

Answer

$$D_{max} = D + \Delta D = 2 + 0.0010 = 2.0010 \text{ in}$$

Answer

$$D_{min} = D = 2.0000 \text{ in}$$

The IT6 tolerance for the shaft is $\Delta d = 0.0006$ in. Also, from Table A–14, the fundamental deviation is $\delta_F = 0.0017$ in. Using Eq. (2–40), we get for the shaft that

Answer

$$d_{min} = d + \delta_F = 2 + 0.0017 = 2.0017 \text{ in}$$

Answer

$$d_{max} = d + \delta_F + \Delta d = 2 + 0.0017 + 0.0006 = 2.0023 \text{ in}$$

2–7 Dimensions and Tolerances

The following terms are used generally in dimensioning:

- *Nominal size.* The size we use in speaking of an element. For example, we may specify a $\frac{1}{2}$-in bolt or a $1\frac{1}{2}$-in pipe. Either the theoretical size or the actual measured size may be quite different. The bolt, say, may actually measure 0.492 in. And the theoretical size of a $1\frac{1}{2}$-in pipe is 1.900 in for the outside diameter.

- *Basic size.* The exact theoretical size. Limiting dimensions in either the plus or the minus direction begin from the basic dimension.

- *Limits.* The stated maximum and minimum dimensions.

- *Tolerance.* The difference between the two limits.

- *Bilateral tolerance.* The variation in both directions from the basic dimension. That is, the basic size is between the two limits, for example, 1.005 ± 0.002 in. The two parts of the tolerance need not be equal.

- *Unilateral tolerance.* The basic dimension is taken as one of the limits, and variation is permitted in only one direction, for example,

$$1.005 \begin{array}{c} + 0.004 \\ - 0.000 \end{array} \text{ in}$$

- *Natural tolerance.* A tolerance equal to plus and minus three standard deviations from the mean. For a normal distribution, this ensures that 99.73 percent of production is within the tolerance limits.

- *Clearance.* A general term that refers to the mating of cylindrical parts such as a bolt and a hole. The word *clearance* is used only when the internal member is smaller than the external member. The *diametral clearance* is the measured difference in the two diameters. The *radial clearance* is the difference in the two radii.

- *Interference.* The opposite of clearance, for mating cylindrical parts in which the internal member is larger than the external member.

- *Allowance.* The minimum stated clearance or the maximum stated interference for mating parts.

Variations in dimensions of parts in a production process may occur purely by chance as well as for specific reasons. For example, the operation temperature of a machine tool changes during start-up, and this may have an effect on the dimensions of parts produced during the first 30 min or so of running time. When all such variations in part dimensions have been eliminated, the production process is said to be in *statistical control.* Under these conditions all variations in dimensions occur at random.

When several parts are assembled, the gap (or interference) depends on the dimensions and tolerances of the individual parts. Consider the array of parallel vectors in Fig. 2–13. The x's are directed to the right, and the y's are directed to the left. As parallel vectors they add and subtract algebraically. The bilateral tolerance is t_i on the mean $\bar{x}_i$, and the bilateral tolerance is t_j on the mean $\bar{y}_j$, all positive numbers. The gap that is not closed is called w. From Fig. 2–13,

$$(x_1 + x_3 + \cdots) - (y_2 + y_4 + \cdots) - w = 0$$

which can be succinctly written as

$$w = \sum x_i - \sum y_j \qquad (2\text{–}41)$$

Figure 2–13

An assembly considered as a group of displacement vectors. (a) Five L-shaped blocks A, B, C, D, E, and one rectangular block F are assembled as shown. The dimensions a, b, c, d, e, and f are toleranced, making them random variables. (b) Corresponding vector diagram, showing right-tending displacement as x and left-tending displacement as y and w.

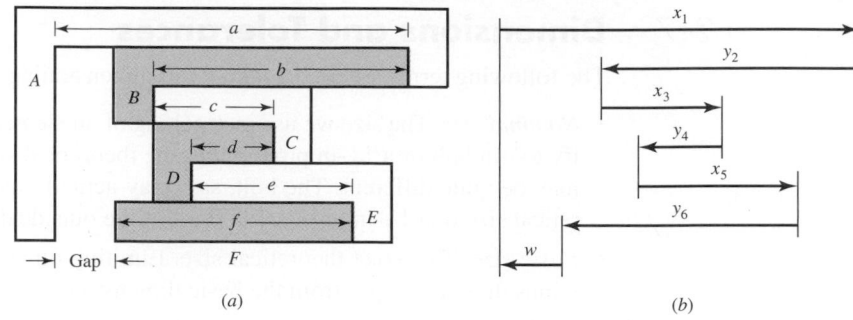

(a)   (b)

The largest gap w_{max} occurs when the right-tending vectors are the largest possible and the left-tending vectors are the smallest possible. We write

$$w_{max} = \sum(x_i + t_i) - \sum(y_j - t_j) = \sum \bar{x}_i - \sum \bar{y}_j + \sum_{all} t \qquad \text{(2–42)}$$

Similarly for the smallest gap, w_{min}, we can write

$$w_{min} = \sum(x_i - t_i) - \sum(y_j + t_j) = \sum \bar{x}_i - \sum \bar{y}_j - \sum_{all} t \qquad \text{(2–43)}$$

The midrange (mean) value $\bar{w} = (w_{max} + w_{min})/2$. Substituting Eqs. (2–42) and (2–43) into the expression for w and simplifying, we obtain

$$\bar{w} = \sum \bar{x}_i - \sum \bar{y}_j \qquad \text{(2–44)}$$

The bilateral tolerance on the gap is $t_w = (w_{max} - w_{min})/2$. Substituting Eqs. (2–42) and (2–43) into the expression for t_w and simplifying, we obtain

$$t_w = \sum_{all} t \qquad \text{(2–45)}$$

Equation (2–45) is the basis for the expression "the stacking of tolerances" used in describing the condition of the gap. All constituent tolerances *add* to give the tolerance on the gap. If the gap instance w is negative, there is interference. Equations (2–44) and (2–45) are the iron rules of tolerance. Since *all* observations of w fall between $\bar{w} - t_w$ and $\bar{w} + t_w$ and none are outside these range numbers, Eqs. (2–44) and (2–45) represent an *absolute tolerance system*.

Gap dimensions near the gap limits are rare events, requiring simultaneous observations near the bounds of individual dimensions, all on a particular side; this suggests a small-risk statistical tolerance system. The operative equation is (2–41) written for random variables:

$$\mathbf{w} = \sum \mathbf{x}_i - \sum \mathbf{y}_j \qquad \text{(2–46)}$$

The mean gap $\bar{w}$ is the algebraic sum of the constituent means:

$$\bar{w} = \sum \bar{x}_i - \sum \bar{y}_j \qquad \text{(2–47)}$$

which agrees with Eq. (2–44). The variance in **w** is the positive sum of all the (uncorrelated) constituent variances:

$$\hat{\sigma}_w^2 = \sum \hat{\sigma}_i^2 + \sum \hat{\sigma}_j^2 = \sum_{\text{all}} \hat{\sigma}^2$$

It follows that the standard deviation of the gap **w** is

$$\hat{\sigma}_w = \sqrt{\sum_{\text{all}} \hat{\sigma}^2} \qquad\qquad (2\text{–}48)$$

Equations (2–47) and (2–48) are the operative equations for the *statistical tolerance system*. A common formulation of Eq. (2–48) is

$$t_w = \sqrt{\sum_{\text{all}} t^2}$$

which requires a presumption of normality in the individual dimensions, a rare occurrence. To find the distribution of **w** and/or the probability of observing values of $w > w_{\text{crit}}$ or, alternatively, $w < w_{\text{crit}}$ requires a computer simulation in most cases. *Monte Carlo* computer simulations are used to determine the distribution of **w** by the following approach:

1   Generate an instance i for each random variable x_i or y_i in the problem by selecting the value of x_i or y_i based on the probability distribution.
2   Calculate w_i using the values of x_i and y_i obtained in step 1.
3   Repeat steps 1 and 2 N times to generate the distribution of w. As the number of trials increases, the reliability of the distribution increases.

EXAMPLE 2–11

A shouldered screw contains three hollow right circular cylindrical parts on the screw before a nut is tightened against the shoulder. To sustain the function, the gap w must equal or exceed 0.003 in. The parts in the assembly depicted in Fig. 2–14 have dimensions and tolerances as follows:

$$a = 1.750 \pm 0.003 \text{ in} \qquad b = 0.750 \pm 0.001 \text{ in}$$
$$c = 0.120 \pm 0.005 \text{ in} \qquad d = 0.875 \pm 0.001 \text{ in}$$

Figure 2–14

An assembly of three cylindrical sleeves of lengths a, b, and c on a shoulder bolt shank of length a. The gap **w** is of interest.

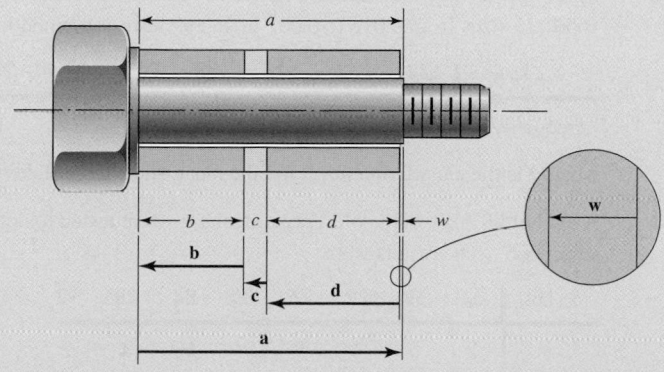

All parts except the part with the dimension d are supplied by vendors. The part containing the dimension d is made in-house.

(a) Estimate the mean and tolerance on the gap **w**.

(b) What value of $\bar{d}$ will assure $w > 0.003$ in?

Solution

(a) We use the preceding equations to find

Answer

$$\bar{w} = 1.750 - 0.750 - 0.120 - 0.875 = 0.005 \text{ in} \qquad (2\text{-}44)$$

$$t_w = 0.003 + 0.001 + 0.005 + 0.001 = 0.010 \text{ in} \qquad (2\text{-}45)$$

Then

$$w_{\max} = \bar{w} + t_w = 0.005 + 0.010 = 0.015 \text{ in} \qquad (2\text{-}42)$$

$$w_{\min} = \bar{w} - t_w = 0.005 - 0.010 = -0.005 \text{ in} \qquad (2\text{-}43)$$

There is both clearance and interference.

(b) If $w_{\min}$ is to be 0.003 in, then

$$\bar{w} = w_{\min} + t_w = 0.003 + 0.010 = 0.013 \text{ in}$$

Solve Eq. (2–47) for $\bar{d}$:

Answer

$$\bar{d} = \bar{a} - \bar{b} - \bar{c} - \bar{w} = 1.750 - 0.750 - 0.120 - 0.013 = 0.867 \text{ in}$$

PROBLEMS

All problems are analysis problems.

2-1 At a constant amplitude, completely reversed bending stress level, the cycles-to-failure experience with 69 specimens of 5160H steel from 1.25-in hexagonal bar stock was as follows:

L	60	70	80	90	100	110	120	130	140	150	160	170	180	190	200	210
f	2	1	3	5	8	12	6	10	8	5	2	3	2	1	0	1

where L is the life in thousands of cycles, and f is the class frequency of failures.

(a) Construct a histogram with class frequency f as ordinate.

(b) Estimate the mean and standard deviation of the life for the population from which the sample was drawn.

2-2 Determinations of the ultimate tensile strength S_{ut} of stainless steel sheet (17-7PH, condition TH 1050), in sizes from 0.016 to 0.062 in, in 197 tests combined into seven classes were

S_{ut}, kpsi	174	182	190	198	206	214	222
Frequency, f	6	9	44	67	53	12	6

where f is the class frequency. Find the mean and standard deviation.

2-3 A total of 58 AISI 1018 cold-drawn steel bars were tested to determine the 0.2 percent offset yield strength S_y. The results were

S_y, kpsi	64	68	72	76	80	84	88	92
f	2	6	6	9	19	10	4	2

where S_y is the class midpoint and f is the class frequency. Estimate the mean and standard deviation of S_y and its PDF assuming a normal distribution.

2–4 The base 10 logarithm of 55 cycles-to-failure observations on specimens subjected to a constant stress level in fatigue have been classified as follows:

y	5.625	5.875	6.125	6.375	6.625	6.875	7.125	7.375	7.625	7.875	8.125
f	1	0	0	3	3	6	14	15	10	2	1

Here y is the class midpoint and f is the class frequency.
(a) Estimate the mean and standard deviation of the population from which the sample was taken and establish the normal PDF.
(b) Plot the histogram and superpose the predicted class frequency from the normal fit.

2–5 A $\frac{1}{2}$-in nominal diameter round is formed in an automatic screw machine operation that is initially set to produce a 0.5000-in diameter and is reset when tool wear produces diameters in excess of 0.5008 in. The stream of parts is thoroughly mixed and produces a uniform distribution of diameters.
(a) Estimate the mean and standard deviation of the large batch of parts from setup to reset.
(b) Find the expressions for the PDF and CDF of the population.
(c) If, by inspection, the diameters less than 0.5002 in are removed, what are the new PDF and CDF as well as the mean and standard deviation of the diameters of the survivors of the inspection?

2–6 The only detail drawing of a machine part has a dimension smudged beyond legibility. The round in question was created in an automatic screw machine and 1000 parts are in stock. A random sample of 50 parts gave a mean dimension of $\bar{d} = 0.6241$ in and a standard deviation of $s = 0.000\,581$ in. Toleranced dimensions elsewhere are given in integral thousandths of an inch. Estimate the missing information on the drawing.

2–7 (a) The CDF of the variate $\mathbf{x}$ is $F(x) = 0.555x - 33$, where x is in millimeters. Find the PDF, the mean, the standard deviation, and the range numbers of the distribution.
(b) In the expression $\sigma = \mathbf{F}/\mathbf{A}$, the force $\mathbf{F} = \mathbf{LN}(3600, 300)$ lbf and the area is $\mathbf{A} = \mathbf{LN}(0.112, 0.001)$ in^2. Estimate the mean, standard deviation, coefficient of variation, and distribution of σ.

2–8 A regression model of the form $y = a_1 x + a_2 x^2$ is desired. From the normal equations

$$\sum y = a_1 \sum x + a_2 \sum x^2$$

$$\sum xy = a_1 \sum x^2 + a_2 \sum x^3$$

show that

$$a_1 = \frac{\sum y \sum x^3 - \sum xy \sum x^2}{\sum x \sum x^3 - \left(\sum x^2\right)^2} \quad \text{and} \quad a_2 = \frac{\sum x \sum xy - \sum y \sum x^2}{\sum x \sum x^3 - \left(\sum x^2\right)^2}$$

For the data set

x	0.0	0.2	0.4	0.6	0.8	1.0
y	0.01	0.15	0.25	0.25	0.17	−0.01

find the regression equation and plot the data with the regression model.

2-9   R. W. Landgraf reported the following axial (push–pull) endurance strengths for steels of differing ultimate strengths:

S_u	S'_e	S_u	S'_e	S_u	S'_e
65	29.5	325	114	280	96
60	30	238	109	295	99
82	45	130	67	120	48
64	48	207	87	180	84
101	51	205	96	213	75
119	50	225	99	242	106
195	78	325	117	134	60
210	87	355	122	145	64
230	105	225	87	227	116
265	105				

(a) Plot the data with S'_e as ordinate and S_u as abscissa.

(b) Using the $y = mx + b$ linear regression model, find the regression line and plot.

2-10   In fatigue studies a parabola of the Gerber type

$$\frac{\sigma_a}{S_e} + \left(\frac{\sigma_m}{S_{ut}}\right)^2 = 1$$

is useful (see Sec. 7–12). Solved for σ_a the preceding equation becomes

$$\sigma_a = S_e - \frac{S_e}{S_{ut}^2}\sigma_m^2$$

This implies a regression model of the form $y = a_0 + a_2 x^2$. Show that the normal equations are

$$\sum y = na_0 + a_2 \sum x^2$$
$$\sum xy = a_0 \sum x + a_2 \sum x^3$$

and that

$$a_0 = \frac{\sum x^3 \sum y - \sum x^2 \sum xy}{n \sum x^3 - \sum x \sum x^2} \quad \text{and} \quad a_2 = \frac{n \sum xy - \sum x \sum y}{n \sum x^3 - \sum x \sum x^2}$$

Plot the data

x	20	40	60	80
y	19	17	13	7

superposed on a plot of the regression line.

2-11   Consider the following data collected on a single helical coil extension spring with an initial extension F_i and a spring rate k suspected of being related by the equation $F = F_i + kx$ where x is the deflection beyond initial. The data are

x, in	0.2	0.4	0.6	0.8	1.0	2.0
F, lbf	7.1	10.3	12.1	13.8	16.2	25.2

(a) Estimate the mean and standard deviation of the initial tension F_i.

(b) Estimate the mean and standard deviation of the spring rate k.

2–12   In the expression for uniaxial strain $\epsilon = \delta/l$, the elongation is specified as $\delta \sim (0.0015, 0.000\,092)$ in and the length as $l \sim (2.0000, 0.0081)$ in. What are the mean, the standard deviation, and the coefficient of variation of the corresponding strain ϵ.

2–13   In Hooke's law for uniaxial stress, $\sigma = \epsilon E$, the strain is given as $\epsilon \sim (0.0005, 0.000\,034)$ and Young's modulus as $E \sim (29.5, 0.885)$ Mpsi. Find the mean, the standard deviation, and the coefficient of variation of the corresponding stress σ in psi.

2–14   The stretch of a uniform rod in tension is given by the formula $\delta = Fl/AE$. Suppose the terms in this equation are random variables and have parameters as follows:

$$\mathbf{F} \sim (14.7, 1.3) \text{ kip} \qquad \mathbf{A} \sim (0.226, 0.003) \text{ in}^2$$
$$\mathbf{l} \sim (1.5, 0.004) \text{ in} \qquad \mathbf{E} \sim (29.5, 0.885) \text{ Mpsi}$$

Estimate the mean, the standard deviation, and the coefficient of variation of the corresponding elongation δ in inches.

2–15   The maximum bending stress in a round bar in flexure occurs in the outer surface and is given by the equation $\sigma = 32M/\pi d^3$. If the moment is specified as $\mathbf{M} \sim (15\,000, 1350)$ lbf $\cdot$ in and the diameter is $\mathbf{d} \sim (2.00, 0.005)$ in, find the mean, the standard deviation, and the coefficient of variation of the corresponding stress σ in psi.

2–16   When a production process is wider than the tolerance interval, inspection rejects a low-end scrap fraction α with $x < x_1$ and an upper-end scrap fraction β with dimensions $x > x_2$. The surviving population has a new density function $g(x)$ related to the original $f(x)$ by a multiplier a. This is because any two observations x_i and x_j will have the same relative probability of occurrence as before. Show that

$$a = \frac{1}{F(x_2) - F(x_1)} = \frac{1}{1 - (\alpha + \beta)}$$

and

$$g(x) = \begin{cases} \dfrac{f(x)}{F(x_2) - F(x_1)} = \dfrac{f(x)}{1 - (\alpha + \beta)} & x_1 \le x \le x_2 \\ 0 & \text{otherwise} \end{cases}$$

2–17   An automatic screw machine produces a run of parts with a uniform distribution $\mathbf{d} = U[0.748, 0.751]$ in because it was not reset when the diameters reached 0.750 in. The square brackets contain range numbers.
   (a) Estimate the mean, standard deviation, and PDF of the original production run if the parts are thoroughly mixed.
   (b) Using the results of Prob. 2–16, find the new mean, standard deviation, and PDF. Superpose the PDF plots and compare.

2–18   A springmaker is supplying helical coil springs meeting the requirement for a spring rate k of 10 ± 1 lbf/in. The test program of the springmaker shows that the distribution of spring rate is well approximated by a normal distribution. The experience with inspection has shown that 8.1 percent are scrapped with $k < 9$ and 5.5 percent are scrapped with $k > 11$. Estimate the probability density function.

2–19   The lives of parts are often expressed as the number of cycles of operation that a specified percentage of a population will exceed before experiencing failure. The symbol L is used to designate this definition of life. Thus we can speak of L10 life as the number of cycles to failure exceeded by 90 percent of a population of parts. Using the mean and standard deviation for the data of Prob. 2–1, a normal distribution model, estimate the corresponding L10 life.

2-20 Fit a normal distribution to the histogram of Prob. 2–1. Superpose the probability density function on the $f/(Nw)$ histographic plot.

2-21 For Prob. 2–2, plot the histogram with $f/(Nw)$ as ordinate and superpose a normal distribution density function on the histographic plot.

2-22 For Prob. 2–3, plot the histogram with $f/(Nw)$ as ordinate and superpose a normal distribution probability density function on the histographic plot.

2-23 A 1018 cold-drawn steel has a 0.2 percent tensile yield strength $\mathbf{S}_y = \mathbf{N}(78.4, 5.90)$ kpsi. A round rod in tension is subjected to a load $\mathbf{P} = \mathbf{N}(40, 8.5)$ kip. If rod diameter d is 1.000 in, what is the probability that a random static tensile load P from $\mathbf{P}$ imposed on the shank with a 0.2 percent tensile load S_y from $\mathbf{S}_y$ will not yield?

2-24 A hot-rolled 1035 steel has a 0.2 percent tensile yield strength $\mathbf{S}_y = \mathbf{LN}(49.6, 3.81)$ kpsi. A round rod in tension is subjected to a load $\mathbf{P} = \mathbf{LN}(30, 5.1)$ kip. If the rod diameter d is 1.000 in, what is the probability that a random static tensile load P from $\mathbf{P}$ on a shank with a 0.2 percent yield strength S_y from $\mathbf{S}_y$ will not yield?

2-25 Three blocks A, B, and C and a grooved block D have dimensions a, b, c, and d as follows:

$$a = 1.000 \pm 0.001 \text{ in} \qquad b = 2.000 \pm 0.003 \text{ in}$$
$$c = 3.000 \pm 0.005 \text{ in} \qquad d = 6.020 \pm 0.006 \text{ in}$$

The blocks are assembled as shown in the figure.

Problem 2–25

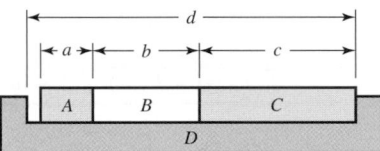

(a) Using the absolute tolerance system, determine the nominal gap $\bar{w}$ and its bilateral tolerance amplitude t_w.

(b) Using the statistical tolerance system, determine the nominal gap $\bar{w}$ and its bilateral tolerance amplitude t_w.

2-26 If $x = a \pm \Delta a$, $y = b \pm \Delta b$, and $z = c \pm \Delta c$, show that for the volume V of a rectangular parallelepiped, $V = xyz$,

$$\frac{\Delta V}{\bar{V}} = \frac{\Delta a}{\bar{a}} + \frac{\Delta b}{\bar{b}} + \frac{\Delta c}{\bar{c}}$$

Use this result to place range numbers (absolute tolerance bounds) on the volume of a rectangular parallelepiped with the dimensions

$$a = 1.250 \pm 0.001 \text{ in} \qquad b = 1.875 \pm 0.002 \text{ in} \qquad c = 2.750 \pm 0.003 \text{ in}$$

2-27 A pivot in a linkage has a pin as depicted in the figure whose dimension $a \pm t_a$ is to be established. The thickness of the link clevis is 1.000 ± 0.002 in. The designer has concluded that a gap of between 0.004 and 0.14 in will satisfactorily sustain the function of the linkage pivot.

(a) Determine the dimension a and its tolerance using the absolute tolerance method.

(b) Establish the dimension a and its tolerance using the statistical tolerance method.

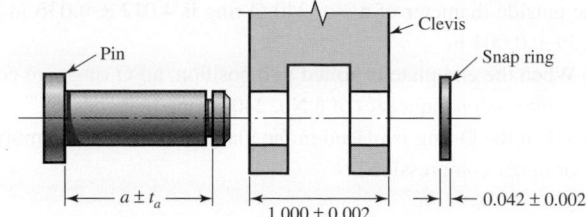

Problem 2–27
Dimensions in inches.

Pin

Clevis

Snap ring

$a \pm t_a$

1.000 ± 0.002

0.042 ± 0.002

2–28 A guide pin is required to align the assembly of a two-part fixture. The nominal size of the pin is 15 mm. Make the dimensional decisions for a 15-mm basic size locational clearance fit.

2–29 An interference fit of a cast-iron hub of a gear on a steel shaft is required. Make the dimensional decisions for a 45-mm basic size medium drive fit.

2–30 A pin is required for forming a linkage pivot. Find the dimensions required for a 50-mm basic size pin and clevis with a sliding fit.

2–31 A journal bearing and bushing need to be described. The nominal size is 1 in. What dimensions are needed for a 1-in basic size with a close running fit if this is a lightly loaded journal and bushing assembly?

2–32 A circular cross section O ring has the dimensions shown in the figure. In particular, a No. 240 O ring has an inside diameter D_i and a cross-section diameter W of

$$D_i = 3.734 \pm 0.028 \text{ in} \qquad W = 0.139 \pm 0.004 \text{ in}$$

Using the absolute tolerance system, estimate the mean outside diameter $\bar{D}_o$ and its bilateral tolerance.

Problem 2–32

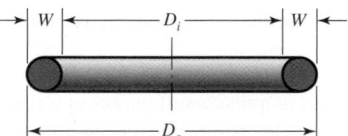

W D_i W

D_o

2–33 A No. 370 O ring has the dimensions

$$D_i = 208.92 \pm 1.30 \text{ mm} \qquad W = 5.33 \pm 0.13 \text{ mm}$$

Using the absolute tolerance method, estimate the mean outside diameter $\bar{D}_o$ and its bilateral tolerance.

2–34 Estimate the mean outside diameter $\bar{D}_o$ and its bilateral tolerance of the No. 240 O ring of Prob. 2–32 if **W** is independent of **D_i** and the statistical tolerancing method is used.

2–35 Find the outside diameter $\bar{D}_o$ and its bilateral tolerance for the No. 370 O ring of Prob. 2–33 if **W** is independent of **D_i** and the statistical tolerancing method is used.

2–36 The gland for an O ring under internal pressure is shown in the figure on the next page. For a No. 240 O ring, the recommended gland dimensions are

$$G = 0.185 \pm 0.005 \text{ in} \qquad F = 0.106 \pm 0.003 \text{ in}$$

One manufacturer of O rings recommends the gland outer diameter

$$Y_{\max} = D_o \qquad Y_{\min} = \max[0.99\bar{D}_o, \bar{D}_o - 0.060] \text{ in}$$

The outside diameter of a No. 240 O ring is 4.012 ± 0.036 in and the cross-section diameter is 0.139 ± 0.004 in.

(a) When the end plate is bolted into position, all O rings are compressed. What is the minimum compression (squeeze) of a No. 240 O ring in inches?

(b) When the O ring is placed in the gland prior to the assembly of the end plate, is the ring free or under compression?

Problems 2–36 and 2–37

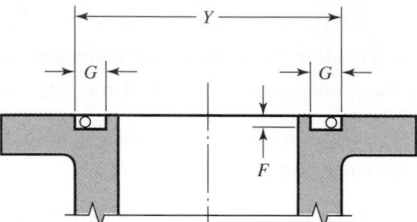

2–37   The gland for an O ring under internal pressure is shown in the figure. For a No. 370 O ring, the recommended gland dimensions are

$$F = 4.32 \pm 0.13 \text{ mm} \qquad G = 7.24 \pm 0.13 \text{ mm}$$

One manufacturer of O rings recommends that the gland's largest diameter be

$$Y_{\max} = D_o \qquad Y_{\min} = \max[0.99D_o, D_o - 1.52] \text{ mm}$$

The outside diameter of a No. 370 O ring is 219.58 ± 0.34 mm, and the cross-sectional diameter is 5.33 ± 0.13 mm.

(a) When the end plate is bolted in position, the O ring is compressed. What is the minimum compression (squeeze) of a No. 370 O ring in millimeters?

(b) When an O ring is placed in the gland prior to assembly of the end plate, is the ring free or under compression?

2–38   Two flanges of a bolted joint compress a soft gasket in a controlled manner so that the "squeeze" is between 0.020 in and 0.040 in. Too little squeeze results in leakage and too much in gasket "flow." As shown in the figure, a shouldered cap screw is used to control the gasket compression,

Problem 2–38

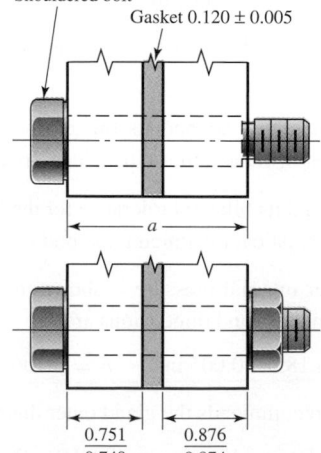

and the critical factor in sustaining the function of the joint is the length a of the cap screw. If the dimensions of the members are as shown and gasket production thickness is 0.120 ± 0.005 in, find the length $\bar{a}$ and the tolerance t_a using the absolute tolerance system and the statistical tolerance system.

2-39 The tensile 0.2 percent offset yield strength of AISI 1137 cold-drawn steel rounds up to 1 inch in diameter from 2 mills and 25 heats is reported histographically as follows:

S_y	93	95	97	99	101	103	105	107	109	111
f	19	25	38	17	12	10	5	4	4	2

where S_y is the class midpoint in kpsi and f is the number in each class. Presuming the distribution is normal, what is the yield strength exceeded by 99 percent of the population?

2-40 Repeat Prob. 2–39, presuming the distribution is lognormal. What is the yield strength exceeded by 99 percent of the population? Compare the normal fit of Prob. 2–39 with the lognormal fit by superposing the PDFs and the histographic PDF.

2-41 A 1046 steel, water-quenched and tempered for 2 h at 1210°F, has a mean tensile strength of 105 kpsi and a yield mean strength of 82 kpsi. Test data from endurance strength testing at 10^4-cycle life give $(\mathbf{S}'_{fe})_{10^4} = \mathbf{W}[79, \ 86.2, \ 2.60]$ kpsi. What are the mean, standard deviation, and coefficient of variation of $(\mathbf{S}'_{fe})_{10^4}$?

2-42 An ASTM grade 40 cast iron has the following result from testing for ultimate tensile strength: $\mathbf{S}_{ut} = \mathbf{W}[27.7, 46.2, 4.38]$ kpsi. Find the mean and standard deviation of $\mathbf{S}_{ut}$, and estimate the chance that the ultimate strength is less than 40 kpsi.

2-43 A cold-drawn 301SS stainless steel has an ultimate tensile strength given by $\mathbf{S}_{ut} = \mathbf{W}[151.9, 193.6, 8.00]$ kpsi. Find the mean and standard deviation.

2-44 A 100-70-04 nodular iron has tensile and yield strengths described by

$$\mathbf{S}_{ut} = \mathbf{W}[47.6, 125.6, 11.84] \text{ kpsi}$$
$$\mathbf{S}_y = \mathbf{W}[64.1, 81.0, 3.77] \text{ kpsi}$$

What is the chance that S_{ut} is less than 100 kpsi? What is the chance that S_y is less than 70 kpsi?

2-45 A 1038 heat-treated steel bolt in finished form provided the material from which a tensile test specimen was made. The testing of many such bolts led to the description $\mathbf{S}_{ut} = \mathbf{W}[122.3, 134.6, 3.64]$ kpsi. What is the probability that the bolts meet the SAE grade 5 requirement of a minimum tensile strength of 120 kpsi? What is the probability that the bolts meet the SAE grade 7 requirement of a minimum tensile strength of 133 kpsi?

2-46 A 5160H steel was tested in fatigue and the distribution of cycles to failure at constant stress level was found to be $\mathbf{n} = \mathbf{W}[36.9, 133.6, 2.66]$ in 10^3 cycles. Plot the PDF of n and the PDF of the lognormal distribution having the same mean and standard deviation. What is the L10 life (see Prob. 2–19) predicted by both distributions?

2-47 A material was tested at steady fully reversed loading to determine the number of cycles to failure using 100 specimens. The results were

$(10^5)L$	3.05	3.55	4.05	4.55	5.05	5.55	6.05	6.55	7.05	7.55	8.05	8.55	9.05	9.55	10.05
f	3	7	11	16	21	13	13	6	2	0	4	3	0	0	1

where L is the life in cycles and f is the number in each class. Assuming a lognormal distribution, plot the theoretical PDF and the histographic PDF for comparison.

2–48 The ultimate tensile strength of an AISI 1117 cold-drawn steel is Weibullian, with $\mathbf{S}_u = \mathbf{W}[70.3, 84.4, 2.01]$. What are the mean, the standard deviation, and the coefficient of variation?

2–49 A 60-45-15 nodular iron has a 0.2 percent yield strength S_y with a mean of 49.0 kpsi, a standard deviation of 4.2 kpsi, and a guaranteed yield strength of 33.8 kpsi. What are the Weibull parameters θ and b?

2–50 A 35018 malleable iron has a 0.2 percent offset yield strength given by the Weibull distribution $\mathbf{S}_y = \mathbf{W}[34.7, 39.0, 2.93]$ kpsi. What are the mean, the standard deviation, and the coefficient of variation?

2–51 The histographic results of steady load tests on 237 rolling-contact bearings are:

L	1	2	3	4	5	6	7	8	9	10	11	12
f	11	22	38	57	31	19	15	12	11	9	7	5

where L is the life in millions of revolutions and f is the number of failures. Fit a lognormal distribution to these data and plot the PDF with the histographic PDF superposed. From the lognormal distribution, estimate the life at which 10 percent of the bearings under this steady loading will have failed.

3

Materials

Chapter Outline

The selection of a material for a machine part or a structural member is one of the decisions the designer is called on to make. The decision is usually made before the dimensions of the part are established. After choosing the process of creating the desired geometry and the material (the two cannot be divorced), the designer can proportion the member so that loss of function can be avoided or the chance of loss of function can be held to an acceptable risk.

In Chaps. 4 and 5, methods for estimating stresses and deflections of machine members are presented. These estimates are based on the properties of the material from which the member will be made. For deflections and stability evaluations, for example, the elastic (stiffness) properties of the material are required, and evaluations of stress at a critical location in a machine member require a comparison with the strength of the material at that location in the geometry and condition of use. This strength is a material property found by testing and is adjusted to the geometry and condition of use as necessary.

As important as stress and deflection are in the design of mechanical parts, the selection of a material is not always based on these factors. Many parts carry no loads on them whatever. Parts may be designed merely to fill up space or for aesthetic qualities. Members must frequently be designed to also resist corrosion. Sometimes temperature effects are more important in design than stress and strain. So many other factors besides stress and strain may govern the design of parts that the designer must have the versatility that comes only with a broad background in materials and processes.

3–1 Material Strength and Stiffness

The standard tensile test is used to obtain a variety of material characteristics and strengths that are used in design. Figure 3–1 illustrates a typical tension-test specimen and its characteristic dimensions.[1] The original diameter d_0 and the gauge length l_0, used to measure the deflections, are recorded before the test is begun. The specimen is then mounted in the test machine and slowly loaded in tension while the load P and deflection are observed. The load is converted to stress by the calculation

$$\sigma = \frac{P}{A_0} \tag{3–1}$$

where $A_0 = \frac{1}{4}\pi d_0^2$ is the original area of the specimen.

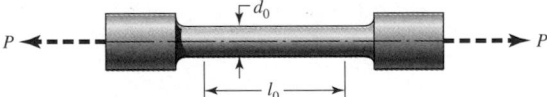

Figure 3–1

A typical tension-test specimen. Some of the standard dimensions used for d_0 are 2.5, 6.25, and 12.5 mm and 0.505 in, but other sections and sizes are in use. Common gauge lengths l_0 used are 10, 25, and 50 mm and 1 and 2 in.

---

1. See ASTM standards E8 and E-8 m for standard dimensions.

Figure 3–2

Stress-strain diagram obtained from the standard tensile test (a) Ductile material; (b) brittle material.
pl marks the proportional limit; el, the elastic limit; y, the offset-yield strength as defined by offset strain OA; u, the maximum or ultimate strength; and f, the fracture strength.

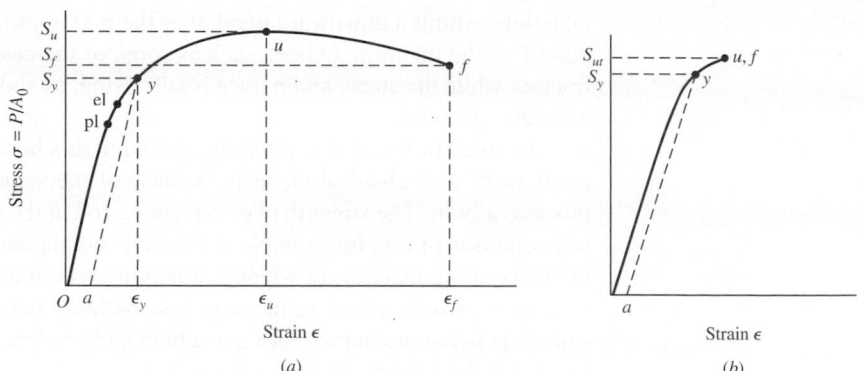

(a)

(b)

The deflection, or extension of the gage length, is given by $l - l_0$ where l is the gauge length corresponding to the load P. The strain is calculated from

$$\epsilon = \frac{l - l_0}{l_0} \tag{3-2}$$

At the conclusion of, or during, the test, the results are plotted as a *stress-strain diagram*. Figure 3–2 depicts typical stress-strain diagrams for ductile and brittle materials. Ductile materials deform much more than brittle materials.

Point *pl* in Fig. 3–2a is called the *proportional limit*. This is the point at which the curve first begins to deviate from a straight line. No permanent set will be observable in the specimen if the load is removed at this point. In the linear range, the uniaxial stress-strain relation is given by *Hooke's law* as

$$\sigma = E\epsilon \tag{3-3}$$

where the constant of proportionality E, the slope of the linear part of the stress-strain curve, is called *Young's modulus* or the *modulus of elasticity*. E is a measure of the stiffness of a material, and since strain is dimensionless, the units of E are the same as stress. Steel, for example, has a modulus of elasticity of about 30 Mpsi (207 GPa) *regardless of heat treatment, carbon content, or alloying*. Stainless steel is about 27.5 Mpsi (190 GPa).

Point el in Fig. 3–2 is called the *elastic limit*. If the specimen is loaded beyond this point, the deformation is said to be plastic and the material will take on a permanent set when the load is removed. Between pl and el the diagram is not a perfectly straight line, even though the specimen is elastic.

During the tension test, many materials reach a point at which the strain begins to increase very rapidly without a corresponding increase in stress. This point is called the *yield point*. Not all materials have an obvious yield point, especially for brittle materials. For this reason, *yield strength* S_y is often defined by an *offset method* as shown in Fig. 3–2, where line *ay* is drawn at slope E. Point *a* corresponds to a definite or stated amount of permanent set, usually 0.2 percent of the original gauge length ($\epsilon = 0.002$), although 0.01, 0.1, and 0.5 percent are sometimes used.

The *ultimate*, or *tensile*, *strength* S_u or S_{ut} corresponds to point *u* in Fig. 3–2 and is the maximum stress reached on the stress-strain diagram.[2] As shown in Fig. 3–2a, some

2. Usage varies. For a long time engineers used the term *ultimate strength*, hence the subscript *u* in S_u or S_{ut}. However, in material science and metallurgy the term *tensile strength* is used.

materials exhibit a downward trend after the maximum stress is reached and fracture at point f on the diagram. Others, such as some of the cast irons and high-strength steels, fracture while the stress-strain trace is still rising, as shown in Fig. 3–2b, where points u and f are identical.

As noted in Sec. 1–11, *strength*, as used in this book, is a built-in property of a material, or of a mechanical element, because of the selection of a particular material or process or both. The strength of a connecting rod at the critical location in the geometry and condition of use, for example, is the same no matter whether it is already an element in an operating machine or whether it is lying on a workbench awaiting assembly with other parts. On the other hand, *stress* is something that occurs in a part, usually as a result of its being assembled into a machine and loaded. However, stresses may be built into a part by processing or handling. For example, shot peening produces a compressive *stress* in the outer surface of a part, and also improves the fatigue strength of the part. Thus, in this book we will be very careful in distinguishing between *strength*, designated by S, and *stress*, designated by σ or τ.

The diagrams in Fig. 3–2 are called *engineering* stress-strain diagrams because the stresses and strains calculated in Eqs. (3–1) and (3–2) are not *true* values. The stress calculated in Eq. (3–1) is based on the original area *before* the load is applied. In reality, as the load is applied the area reduces so that the *actual* or *true stress* is larger than the *engineering stress*. To obtain the true stress for the diagram the load and the cross-sectional area must be measured simultaneously during the test. Figure 3–2a represents a ductile material where the stress appears to decrease from points u to f. Typically, beyond point u the specimen begins to "neck" at a location of weakness where the area reduces dramatically, as shown in Fig. 3–3. For this reason, the true stress is much higher than the engineering stress at the necked section.

The engineering strain given by Eq. (3–2) is based on net change in length from the *original* length. In plotting the *true stress-strain diagram*, it is customary to use a term called *true strain* or, sometimes, *logarithmic strain*. True strain is the sum of the incremental elongations divided by the *current* gauge length at load P, or

$$\varepsilon = \int_{l_0}^{l} \frac{dl}{l} = \ln \frac{l}{l_0} \tag{3–4}$$

where the symbol ε is used to represent true strain. The most important characteristic of a true stress-strain diagram (Fig. 3–4) is that the true stress continually increases all the way to fracture. Thus, as shown in Fig. 3–4, the true fracture stress σ_f is greater than the true ultimate stress σ_u. Contrast this with Fig. 3–2, where the engineering fracture strength S_f is less than the engineering ultimate strength S_u.

Compression tests are more difficult to conduct, and the geometry of the test specimens differs from the geometry of those used in tension tests. The reason for this is that the specimen may buckle during testing or it may be difficult to distribute the stresses evenly. Other difficulties occur because ductile materials will bulge after yielding. However, the results can be plotted on a stress-strain diagram also, and the same strength definitions can be applied as used in tensile testing. For most materials the compressive strengths are about the same as the tensile strengths. When substantial differences occur between tensile and compressive strengths, however, as is the case with the cast irons,

Figure 3–3

Tension specimen after necking.

Figure 3–4

True stress-strain diagram plotted in Cartesian coordinates.

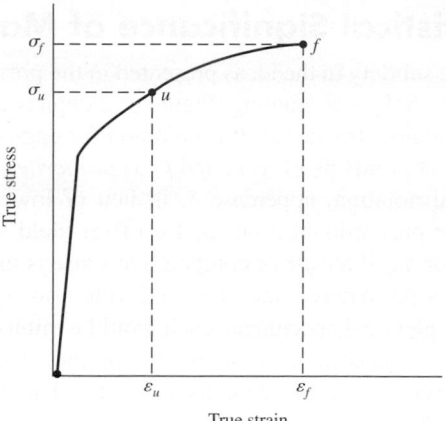

the tensile and compressive strengths should be stated separately, S_{ut}, S_{uc}, where S_{uc} is reported as a *positive* quantity.

Torsional strengths are found by twisting solid circular bars and recording the torque and the twist angle. The results are then plotted as a *torque-twist diagram.* The shear stresses in the specimen are linear with respect to radial location, being zero at the center of the specimen and maximum at the outer radius r (see Chap. 4). The maximum shear stress $\tau_{\max}$ is related to the angle of twist θ by

$$\tau_{\max} = \frac{Gr}{l_0}\theta \qquad (3\text{–}5)$$

where θ is in radians, r is the radius of the specimen, l_0 is the gauge length, and G is the material stiffness property called the *shear modulus* or the *modulus of rigidity*. The maximum shear stress is also related to the applied torque T as

$$\tau_{\max} = \frac{Tr}{J} \qquad (3\text{–}6)$$

where $J = \frac{1}{2}\pi r^4$ is the polar second moment of area of the cross section.

The torque-twist diagram will be similar to Fig. 3–2, and, using Eqs. (3–5) and (3–6), the modulus of rigidity can be found as well as the elastic limit and the *torsional yield strength* S_{sy}. The maximum point on a torque-twist diagram, corresponding to point u on Fig. 3–2, is T_u. The equation

$$S_{su} = \frac{T_u r}{J} \qquad (3\text{–}7)$$

defines the *modulus of rupture* for the torsion test. Note that it is incorrect to call S_{su} the ultimate torsional strength, as the outermost region of the bar is in a plastic state at the torque T_u and the stress distribution is no longer linear.

All of the stresses and strengths defined by the stress-strain diagram of Fig. 3–2 and similar diagrams are specifically known as *engineering stresses* and *strengths* or *nominal stresses* and *strengths*. These are the values normally used in all engineering design calculations. The adjectives *engineering* and *nominal* are used here to emphasize that the stresses are computed by using the *original* or *unstressed cross-sectional area* of the specimen. In this book we shall use these modifiers only when we specifically wish to call attention to this distinction.

3–2 The Statistical Significance of Material Properties

There is some subtlety in the ideas presented in the previous section that should be pondered carefully before continuing. Figure 3–2 depicts the result of a *single* tension test (*one* specimen, now fractured). It is common for engineers to consider these important *stress* values (at points pl, el, *y*, *u*, and *f*) as *properties* and to denote them as *strengths* with a special notation, uppercase S, in lieu of lowercase sigma σ, with subscripts added: S_{pl} for proportional limit, S_y for offset yield strength, S_u for ultimate tensile strength (S_{ut} or S_{uc} if tensile or compressive sense is important).

All this is premature, since Fig. 3–2 refers to *a single specimen.* If there were 10 nominally identical specimens, each would exhibit *different* strengths. What one is observing is a consistent pattern of distribution, and, as observed in Chap. 2, the description has to be stochastic (statistical). The idea of strength, a property, is distributional. To see this, consider the following table, which is a histographic report containing the maximum stresses of 1000 tensile tests on a 1020 steel from a single heat. Here we are seeking the ultimate tensile strength S_{ut}. Recall that these data were analyzed in Ex. 2–4.

Class Frequency f_i	2	18	23	31	83	109	138	151	139	130	82	49	28	11	4	2
Class Midpoint x_i, kpsi	56.5	57.5	58.5	59.5	60.5	61.5	62.5	63.5	64.5	65.5	66.5	67.5	68.5	69.5	70.5	71.5

From Ex. 2–3, the mean of the sample is $\bar{x} = 63.62$ kpsi, the standard deviation is $s_x = 2.594$ kpsi, and the coefficient of variation is $C_x = 0.040\,77$. Assuming a normal distribution, the probability density function is

$$f(x) = \frac{1}{2.594\sqrt{2\pi}} \exp\left[-\frac{1}{2}\left(\frac{x - 63.62}{2.594}\right)^2\right]$$

The description of the strength S_{ut} is then expressed in terms of its statistical parameters and its distribution type. In this case $\mathbf{S}_{ut} = \mathbf{N}\,(63.62, 2.594)$ kpsi. A plot of this distribution is given in Fig. 2–8 together with the histogram and the data fit to a lognormal distribution.

Note that the test program has described 1020 property $\mathbf{S}_{ut}$ for only one heat of one supplier. Testing is an involved and expensive process. Tables of properties are often prepared to be helpful to other persons. A stochastic quantity is described by its mean, standard deviation, and distribution type. Many tables display a single number, which is often the mean, or some percentile, such as the 99th percentile. *Always* read the footnotes to the table. If no qualification is made in a single-entry table, the table is subject to serious doubt.

Since it is no surprise that useful descriptions of a property are stochastic in nature, engineers, when ordering property tests, should couch the instructions so the data generated are enough for them to observe the statistical parameters and to identify the distributional characteristic. The tensile test program on 1000 specimens of 1020 steel is a large one. If you were faced with putting something in a table of ultimate tensile strengths and constrained to a single number, what would it be and just how would your footnote read?

3–3   **Strength and Cold Work**

Cold working is the process of plastic straining below the recrystallization temperature in the plastic region of the stress-strain diagram. Materials can be deformed plastically by the application of heat, as in blacksmithing or hot rolling, but the resulting mechanical properties are quite different from those obtained by cold working. The purpose of this section is to explain what happens to the significant mechanical properties of a material when that material is cold-worked.

Consider the stress-strain diagram of Fig. 3–5a. Here a material has been stressed beyond the yield strength at y to some point i, in the plastic region, and then the load removed. At this point the material has a permanent plastic deformation ϵ_p. If the load corresponding to point i is now reapplied, the material will be elastically deformed by the amount ϵ_e. Thus at point i the total unit strain consists of the two components ϵ_p and ϵ_e and is given by the equation

$$\epsilon = \epsilon_p + \epsilon_e \qquad (a)$$

This material can be unloaded and reloaded any number of times from and to point i, and it is found that the action always occurs along the straight line that is approximately parallel to the initial elastic line Oy. Thus

$$\epsilon_e = \frac{\sigma_i}{E} \qquad (b)$$

The material now has a higher yield point, is less ductile as a result of a reduction in strain capacity, and is said to be *strain-hardened*. If the process is continued, increasing ϵ_p, the material can become brittle and exhibit sudden fracture.

It is possible to construct a similar diagram, as in Fig. 3–5b, where the abscissa is the area deformation and the ordinate is the applied load. The *reduction in area* corresponding to the load P_f, at fracture, is defined as

$$R = \frac{A_0 - A_f}{A_0} = 1 - \frac{A_f}{A_0} \qquad (3\text{–}8)$$

Figure 3–5

(a) Stress-strain diagram showing unloading and reloading at point *I* in the plastic region; (b) analogous load-deformation diagram.

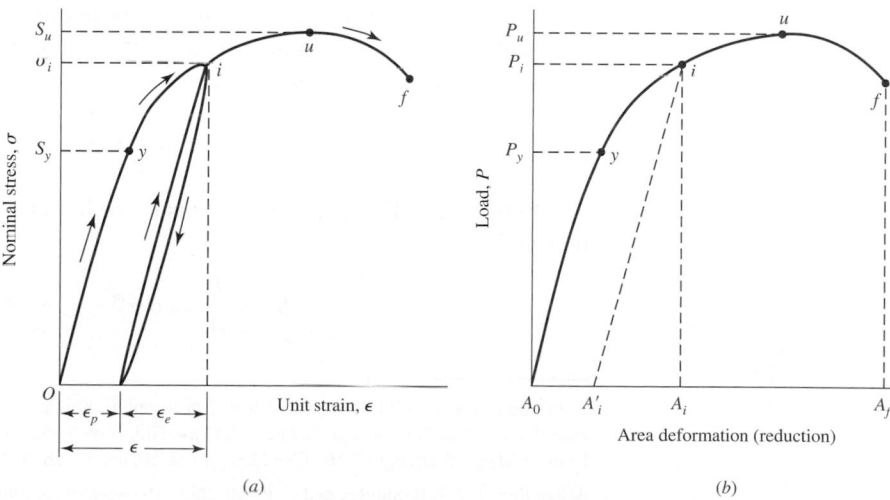

(a)

(b)

where A_0 is the original area. The quantity R in Eq. (3–8) is usually expressed in percent and tabulated in lists of mechanical properties as a measure of *ductility*. See Appendix Table A–20, for example. Ductility is an important property because it measures the ability of a material to absorb overloads and to be cold-worked. Thus such operations as bending, drawing, heading, and stretch forming are metal-processing operations that require ductile materials.

Figure 3–5b can also be used to define the quantity of cold work. The *cold-work factor W* is defined as

$$W = \frac{A_0 - A_i'}{A_0} \approx \frac{A_0 - A_i}{A_0} \tag{3–9}$$

where A_i' corresponds to the area after the load P_i has been released. The approximation in Eq. (3–9) results because of the difficulty of measuring the small diametral changes in the elastic region. If the amount of cold-work is known, then Eq. (3–9) can be solved for the area A_i'. The result is

$$A_i' = A_0(1 - W) \tag{3–10}$$

Cold working a material produces a new set of values for the strengths, as can be seen from stress-strain diagrams. Datsko[3] describes the plastic region of the true stress–true strain diagram by the equation

$$\sigma = \sigma_0 \varepsilon^m \tag{3–11}$$

where
σ = true stress

σ_0 = a strength coefficient, or strain-strengthening coefficient

ε = true plastic strain

m = strain-strengthening exponent

It can be shown[4] that

$$m = \varepsilon_u \tag{3–12}$$

provided that the load-deformation curve exhibits a stationary point (a place of zero slope).

Difficulties arise when using the gauge length to evaluate the true strain in the plastic range, since necking causes the strain to be nonuniform. A more satisfactory relation can be obtained by using the area at the neck. Assuming that the change in volume of the material is small, $Al = A_0 l_0$. Thus, $l/l_0 = A_0/A$, and the true strain is given by

$$\varepsilon = \ln \frac{l}{l_0} = \ln \frac{A_0}{A} \tag{3–13}$$

Returning to Fig. 3–5b, if point i is to the left of point u, that is, $P_i < P_u$, then the new yield strength is

$$S_y' = \frac{P_i}{A_i'} = \sigma_0 \varepsilon_i^m \qquad P_i \le P_u \tag{3–14}$$

3. Joseph Datsko, "Solid Materials," Chap. 7 in Joseph E. Shigley and Charles R. Mischke (eds.), *Standard Handbook of Machine Design,* 2nd ed., McGraw-Hill, New York, 1996. See also Joseph Datsko, "New Look at Material Strength," *Machine Design,* vol. 58, no. 3, Feb. 6, 1986, pp. 81–85.

4. See Sec. 5–2, J. E. Shigley and C. R. Mischke, *Mechanical Engineering Design,* 6th ed., McGraw-Hill, New York, 2001.

Because of the reduced area, that is, because $A_i' < A_0$, the ultimate strength also changes, and is

$$S_u' = \frac{P_u}{A_i'} \qquad \text{(c)}$$

Since $P_u = S_u A_0$, we find, with Eq. (3–10), that

$$S_u' = \frac{S_u A_0}{A_0(1 - W)} = \frac{S_u}{1 - W} \qquad \varepsilon_i \leq \varepsilon_u \qquad \text{(3–15)}$$

which is valid only when point i is to the left of point u.

For points to the right of u, the yield strength is approaching the ultimate strength, and, with small loss in accuracy,

$$S_u' \approx S_y' \approx \sigma_0 \varepsilon_i^m \qquad \varepsilon_i \leq \varepsilon_u \qquad \text{(3–16)}$$

A little thought will reveal that a bar will have the same ultimate load in tension after being strain-strengthened in tension as it had before. The new strength is of interest to us not because the static ultimate load increases, but—since fatigue strengths are correlated with the local ultimate strengths—because the fatigue strength improves. Also the yield strength increases, giving a larger range of sustainable *elastic* loading.

EXAMPLE 3–1

An annealed AISI 1018 steel (see Table A–22) has $S_y = 32.0$ kpsi, $S_u = 49.5$ kpsi, $\sigma_f = 91.1$ kpsi, $\sigma_0 = 90$ kpsi, $m = 0.25$, and $\varepsilon_f = 1.05$ in/in. Find the new values of the strengths if the material is given 15 percent cold work.

Solution

From Eq. (3–12), we find the true strain corresponding to the ultimate strength to be

$$\varepsilon_u = m = 0.25$$

The ratio A_0/A_i is, from Eq. (3–9),

$$\frac{A_0}{A_i} = \frac{1}{1 - W} = \frac{1}{1 - 0.15} = 1.176$$

The true strain corresponding to 15 percent cold work is obtained from Eq. (3–13). Thus

$$\varepsilon_i = \ln \frac{A_0}{A_i} = \ln 1.176 = 0.1625$$

Since $\varepsilon_i < \varepsilon_u$, Eqs. (3–14) and (3–15) apply. Therefore,

Answer

$$S_y' = \sigma_0 \varepsilon_i^m = 90(0.1625)^{0.25} = 57.1 \text{ kpsi}$$

Answer

$$S_u' = \frac{S_u}{1 - W} = \frac{49.5}{1 - 0.15} = 58.2 \text{ kpsi}$$

3–4 Hardness

The resistance of a material to penetration by a pointed tool is called *hardness*. Though there are many hardness-measuring systems, we shall consider here only the two in greatest use.

Rockwell hardness tests are described by ASTM standard hardness method E–18 and measurements are quickly and easily made, they have good reproducibility, and the test machine for them is easy to use. In fact, the hardness number is read directly from a dial. Rockwell hardness scales are designated as $A, B, C, \ldots$, etc. The indenters are described as a diamond, a $\frac{1}{16}$-in-diameter ball, and a diamond for scales A, B, and C, respectively, where the load applied is either 60, 100, or 150 kg. Thus the Rockwell B scale, designated R_B, uses a 100-kg load and a No. 2 indenter, which is a $\frac{1}{16}$-in-diameter ball. The Rockwell C scale R_C uses a diamond cone, which is the No. 1 indenter, and a load of 150 kg. Hardness numbers so obtained are relative. Therefore a hardness $R_C = 50$ has meaning only in relation to another hardness number using the same scale.

The *Brinell hardness* is another test in very general use. In testing, the indenting tool through which force is applied is a ball and the hardness number H_B is found as a number equal to the applied load divided by the spherical surface area of the indentation. Thus the units of H_B are the same as those of stress, though they are seldom used. Brinell hardness testing takes more time, since H_B must be computed from the test data. The primary advantage of both methods is that they are nondestructive in most cases. Both are empirically and directly related to the ultimate strength of the material tested. This means that the strength of parts could, if desired, be tested part by part during manufacture.

For *steels,* the relationship between the minimum ultimate strength and the Brinell hardness number for $200 \leq H_B \leq 450$ is found to be

$$S_u = \begin{cases} 0.495 H_B & \text{kpsi} \\ 3.41 H_B & \text{MPa} \end{cases} \tag{3–17}$$

which may also be called the ASTM minimum (see Sec. 1–11).

Similar relationships for *cast iron* can be derived from data supplied by Krause.[5] Data from 72 tests of gray iron produced by one foundry and poured in two sizes of test bars are reported in graph form. The minimum strength, as defined by the ASTM, is found from these data to be

$$S_u = \begin{cases} 0.23 H_B - 12.5 \text{ kpsi} \\ 1.58 H_B - 86 \text{ MPa} \end{cases} \tag{3–18}$$

Walton[6] shows a chart from which the SAE minimum strength can be obtained. The result is

$$S_u = 0.2375 H_B - 16 \text{ kpsi} \tag{3–19}$$

which is even more conservative than the values obtained from Eq. (3–18).

5. D. E. Krause, "Gray Iron—A Unique Engineering Material," ASTM Special Publication 455, 1969, pp. 3–29, as reported in Charles F. Walton (ed.), *Iron Castings Handbook,* Iron Founders Society, Inc., Cleveland, 1971, pp. 204, 205.

6. Ibid.

EXAMPLE 3–2   It is necessary to ensure that a certain part supplied by a foundry always meets or exceeds ASTM No. 20 specifications for cast iron (see Table A–24). What hardness should be specified?

Solution   From Eq. (3–18), with $(S_u)_{min} = 20$ kpsi, we have

Answer
$$H_B = \frac{S_u + 12.5}{0.23} = \frac{20 + 12.5}{0.23} = 141$$

If the foundry can control the hardness within 20 points, routinely, then specify $145 < H_B < 165$. This imposes no hardship on the foundry and assures the designer that ASTM grade 20 will always be supplied at a predictable cost.

Stochastic Results

For 111 data pairs of carbon and low-alloy wrought steels, $200 \le H_B \le 450$,

$$S_{ut} = \begin{cases} 0.495(1, 0.041)H_B \text{ kpsi} \\ 3.41(1, 0.041)H_B \text{ MPa} \end{cases} \tag{3–20}$$

When hardness is a random variable, then, corresponding to Eqs. (3–20),

$$\mathbf{S}_{ut} = \begin{cases} 0.495(1, 0.041)\mathbf{H}_B \text{ kpsi} \\ 3.41(1, 0.041)\mathbf{H}_B \text{ MPa} \end{cases} \tag{3–21}$$

For cast irons the equations corresponding to Eqs. (3–18) become

$$\mathbf{S}_{ut} = \begin{cases} 0.23\mathbf{H}_B - 9 + (0, 1.5) \text{ kpsi} \\ 1.58\mathbf{H}_B - 62 + (0, 10) \text{ MPa} \end{cases} \tag{3–22}$$

EXAMPLE 3–3   Brinell hardness tests were made on a random sample of five steel parts during processing. The results were H_B values of 248, 253, 247, 244, and 246.
(a) Estimate the mean and the standard deviation of the ultimate strength in SI units.
(b) The ASTM minimum is established at a level that 99 percent of the population can meet or exceed. On the basis of this definition, what minimum ultimate strength corresponds to this sample testing?

Solution   (a) Using the equations of the normal distribution, we first obtain the mean and standard deviation of the hardness as

$$\bar{H}_B = 247.6 \text{ MPa} \qquad \hat{\sigma}_{HB} = 3.36 \text{ MPa} \qquad C_{HB} = 0.0136$$

The ultimate strength is the product of two stochastic quantities. Thus, from Eq. (3–21), we have

$$\mathbf{S}_u = 3.41(1, 0.041)\mathbf{H}_B = 3.41(1, 0.041)247.6(1, 0.0136) \text{ MPa}$$

From Table 2–6 we find the mean of the product of two stochastic quantities to be $\mu_{xy} = \mu_x\mu_y$. Therefore the mean value of $\mathbf{S}_u$ is

Answer

$$\bar{S}_u = 3.41(247.6) = 844.3 \text{ MPa}$$

The coefficient of variation is found from the same table, and is

$$C_{Su} = \left(C_x^2 + C_y^2\right)^{1/2} = [0.041^2 + 0.0136^2]^{1/2} = 0.0432$$

Therefore

Answer

$$\hat{\sigma}_{Su} = \bar{S}_u C_{Su} = 844.3(0.0432) = 36.5 \text{ MPa}$$

(b) For $\Phi(z) = 0.01$, from Table A–10, we find $z = -2.326$. Thus the minimum ultimate strength (deterministic) is

Answer

$$S_u = \bar{S}_u(1 + z_{0.99}C_{Su}) = 844.3[1 + (-2.326)(0.0432)] = 760 \text{ MPa}$$

3–5 Impact Properties

An external force applied to a structure or part is called an *impact load* if the time of application is less than one-third the lowest natural period of vibration of the part or structure. Otherwise it is called simply a *static load*.

The *Charpy* (commonly used) and *Izod* (rarely used) *notched-bar tests* utilize bars of specified geometries to determine brittleness and impact strength. These tests are helpful in comparing several materials and in the determination of low-temperature brittleness. In both tests the specimen is struck by a pendulum released from a fixed height, and the energy absorbed by the specimen, called the *impact value*, can be computed from the height of swing after fracture, but is read from a dial that essentially "computes" the result.

The effect of temperature on impact values is shown in Fig. 3–6 for a material showing a ductile-brittle transition. Not all materials show this transition. Notice the narrow region of critical temperatures where the impact value increases very rapidly. In the low-temperature region the fracture appears as a brittle, shattering type, whereas the appearance is a tough, tearing type above the critical-temperature region. The critical

Figure 3–6

A mean trace shows the effect of temperature on impact values. The result of interest is the brittle-ductile transition temperature, often defined as the temperature at which the mean trace passes through the 15 ft · lbf level. The critical temperature *is* dependent on the geometry of the notch, which is why the Charpy V notch is closely defined.

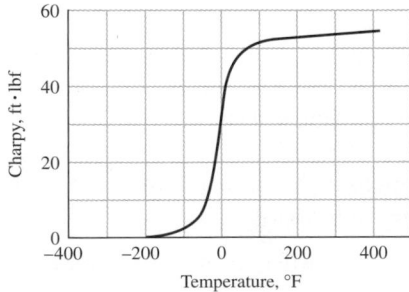

Figure 3–7

Influence of strain rate on tensile properties.

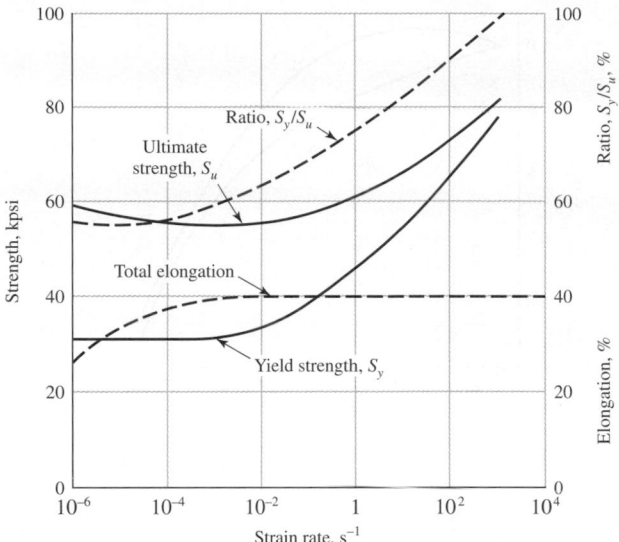

temperature seems to be dependent on both the material and the geometry of the notch. For this reason designers should not rely too heavily on the results of notched-bar tests.

The average strain rate used in obtaining the stress-strain diagram is about 0.001 in/(in · s) or less. When the strain rate is increased, as it is under impact conditions, the strengths increase, as shown in Fig. 3–7. In fact, at very high strain rates the yield strength seems to approach the ultimate strength as a limit. But note that the curves show little change in the elongation. This means that the ductility remains about the same. Also, in view of the sharp increase in yield strength, a mild steel could be expected to behave elastically throughout practically its entire strength range under impact conditions.

The Charpy and Izod tests really provide toughness data under dynamic, rather than static, conditions. It may well be that impact data obtained from these tests are as dependent on the notch geometry as they are on the strain rate. For these reasons it may be better to use the concepts of notch sensitivity, fracture toughness, and fracture mechanics, discussed in Chaps. 6 and 7, to assess the possibility of cracking or fracture.

3–6   Temperature Effects

Strength and ductility, or brittleness, are properties affected by the temperature of the operating environment.

The effect of temperature on the static properties of steels is typified by the strength versus temperature chart of Fig. 3–8. Note that the tensile strength changes only a small amount until a certain temperature is reached. At that point it falls off rapidly. The yield strength, however, decreases continuously as the environmental temperature is increased. There is a substantial increase in ductility, as might be expected, at the higher temperatures.

Many tests have been made of ferrous metals subjected to constant loads for long periods of time at elevated temperatures. The specimens were found to be permanently deformed during the tests, even though at times the actual stresses were less than the yield strength of the material obtained from short-time tests made at the same temperature. This continuous deformation under load is called *creep*.

Figure 3–8

A plot of the results of 145 tests of 21 carbon and alloy steels showing the effect of operating temperature on the yield strength S_y and the ultimate strength S_{ut}. The ordinate is the ratio of the strength at the operating temperature to the strength at room temperature. The standard deviations were $0.0442 \leq \hat{\sigma}_{Sy} \leq 0.152$ for S_y and $0.099 \leq \hat{\sigma}_{Sut} \leq 0.11$ for S_{ut}. (*Data source: E. A. Brandes (ed.), Smithells Metal Reference Book, 6th ed., Butterworth, London, 1983 pp. 22–128 to 22–131.*)

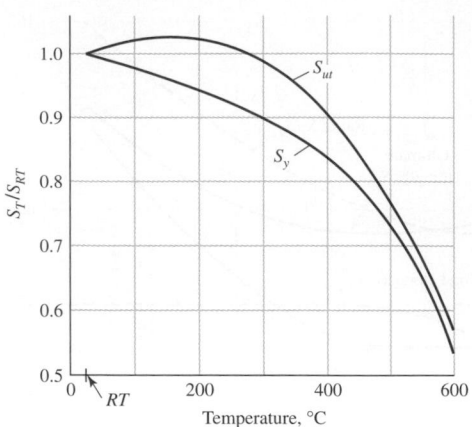

Figure 3–9

Creep-time curve.

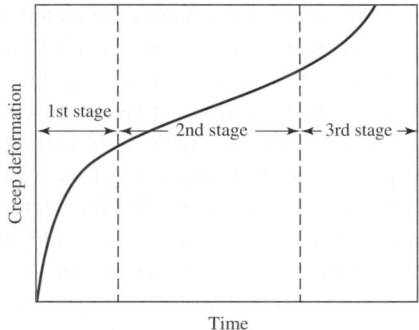

One of the most useful tests to have been devised is the long-time creep test under constant load. Figure 3–9 illustrates a curve that is typical of this kind of test. The curve is obtained at a constant stated temperature. A number of tests are usually run simultaneously at different stress intensities. The curve exhibits three distinct regions. In the first stage are included both the elastic and the plastic deformation. This stage shows a decreasing creep rate, which is due to the strain hardening. The second stage shows a constant minimum creep rate caused by the annealing effect. In the third stage the specimen shows a considerable reduction in area, the true stress is increased, and a higher creep eventually leads to fracture.

When the operating temperatures are lower than the transition temperature (Fig. 3–6), the possibility arises that a part could fail by a brittle fracture. This subject will be discussed in Chap. 6.

Of course, heat treatment, as will be shown, is used to make substantial changes in the mechanical properties of a material.

Heating due to electric and gas welding also changes the mechanical properties. Such changes may be due to clamping during the welding process, as well as heating; the resulting stresses then remain when the parts have cooled and the clamps have been removed. Hardness tests can be used to learn whether the strength has been changed by welding, but such tests will not reveal the presence of residual stresses.

3–7   Numbering Systems

The Society of Automotive Engineers (SAE) was the first to recognize the need, and to adopt a system, for the numbering of steels. Later the American Iron and Steel Institute (AISI) adopted a similar system. In 1975 the SAE published the Unified Numbering System for Metals and Alloys (UNS); this system also contains cross-reference numbers for other material specifications.[7] The UNS uses a letter prefix to designate the material, as, for example, G for the carbon and alloy steels, A for the aluminum alloys, C for the copper-base alloys, and S for the stainless or corrosion-resistant steels. For some materials, not enough agreement has as yet developed in the industry to warrant the establishment of a designation.

For the steels, the first two numbers following the letter prefix indicate the composition, excluding the carbon content. The various compositions used are as follows:

G10	Plain carbon	G46	Nickel-molybdenum
G11	Free-cutting carbon steel with more sulfur or phosphorus	G48	Nickel-molybdenum
		G50	Chromium
G13	Manganese	G51	Chromium
G23	Nickel	G52	Chromium
G25	Nickel	G61	Chromium-vanadium
G31	Nickel-chromium	G86	Chromium-nickel-molybdenum
G33	Nickel-chromium	G87	Chromium-nickel-molybdenum
G40	Molybdenum	G92	Manganese-silicon
G41	Chromium-molybdenum	G94	Nickel-chromium-molybdenum
G43	Nickel-chromium-molybdenum		

The second number pair refers to the approximate carbon content. Thus, G10400 is a plain carbon steel with a nominal carbon content of 0.40 percent (0.37 to 0.44 percent). The fifth number following the prefix is used for special situations. For example, the old designation AISI 52100 represents a chromium alloy with about 100 points of carbon. The UNS designation is G52986.

The UNS designations for the stainless steels, prefix S, utilize the older AISI designations for the first three numbers following the prefix. The next two numbers are reserved for special purposes. The first number of the group indicates the approximate composition. Thus 2 is a chromium-nickel-manganese steel, 3 is a chromium-nickel steel, and 4 is a chromium alloy steel. Sometimes stainless steels are referred to by their alloy content. Thus S30200 is often called an 18-8 stainless steel, meaning 18 percent chromium and 8 percent nickel.

The prefix for the aluminum group is the letter A. The first number following the prefix indicates the processing. For example, A9 is a wrought aluminum, while A0 is a casting alloy. The second number designates the main alloy group as shown in Table 3–1.

---

7. Many of the materials discussed in the balance of this chapter are listed in the Appendix tables. Be sure to review these.

Table 3–1

Aluminum Alloy Designations

Aluminum 99.00% pure and greater	Ax1xxx
Copper alloys	Ax2xxx
Manganese alloys	Ax3xxx
Silicon alloys	Ax4xxx
Magnesium alloys	Ax5xxx
Magnesium-silicon alloys	Ax6xxx
Zinc alloys	Ax7xxx

The third number in the group is used to modify the original alloy or to designate the impurity limits. The last two numbers refer to other alloys used with the basic group.

The American Society for Testing and Materials (ASTM) numbering system for cast iron is in widespread use. This system is based on the tensile strength. Thus ASTM A18 speaks of classes; e.g., 30 cast iron has a minimum tensile strength of 30 kpsi. Note from Appendix A-24, however, that the *typical* tensile strength is 31 kpsi. You should be careful to designate which of the two values is used in design and problem work because of the significance of factor of safety.

3–8   **Sand Casting**

Sand casting is a basic low-cost process, and it lends itself to economical production in large quantities with practically no limit to the size, shape, or complexity of the part produced.

In sand casting, the casting is made by pouring molten metal into sand molds. A pattern, constructed of metal or wood, is used to form the cavity into which the molten metal is poured. Recesses or holes in the casting are produced by sand cores introduced into the mold. The designer should make an effort to visualize the pattern and casting in the mold. In this way the problems of core setting, pattern removal, draft, and solidification can be studied. Castings to be used as test bars of cast iron are cast separately and properties may vary.

Steel castings are the most difficult of all to produce, because steel has the highest melting temperature of all materials normally used for casting. This high temperature aggravates all casting problems.

The following rules will be found quite useful in the design of any sand casting:

1   All sections should be designed with a uniform thickness.
2   The casting should be designed so as to produce a gradual change from section to section where this is necessary.
3   Adjoining sections should be designed with generous fillets or radii.
4   A complicated part should be designed as two or more simple castings to be assembled by fasteners or by welding.

Steel, gray iron, brass, bronze, and aluminum are most often used in castings. The minimum wall thickness for any of these materials is about 5 mm, though with particular care, thinner sections can be obtained with some materials.

3–9   **Shell Molding**

The shell-molding process employs a heated metal pattern, usually made of cast iron, aluminum, or brass, which is placed in a shell-molding machine containing a mixture of dry sand and thermosetting resin. The hot pattern melts the plastic, which, together with the sand, forms a shell about 5 to 10 mm thick around the pattern. The shell is then baked at from 400 to 700°F for a short time while still on the pattern. It is then stripped from the pattern and placed in storage for use in casting.

In the next step the shells are assembled by clamping, bolting, or pasting; they are placed in a backup material, such as steel shot; and the molten metal is poured into the cavity. The thin shell permits the heat to be conducted away so that solidification takes place rapidly. As solidification takes place, the plastic bond is burned and the mold collapses. The permeability of the backup material allows the gases to escape and the casting to air-cool. All this aids in obtaining a fine-grain, stress-free casting.

Shell-mold castings feature a smooth surface, a draft that is quite small, and close tolerances. In general, the rules governing sand casting also apply to shell-mold casting.

3–10 Investment Casting

Investment casting uses a pattern that may be made from wax, plastic, or other material. After the mold is made, the pattern is melted out. Thus a mechanized method of casting a great many patterns is necessary. The mold material is dependent upon the melting point of the cast metal. Thus a plaster mold can be used for some materials while others would require a ceramic mold. After the pattern is melted out, the mold is baked or fired; when firing is completed, the molten metal may be poured into the hot mold and allowed to cool.

If a number of castings are to be made, then metal or permanent molds may be suitable. Such molds have the advantage that the surfaces are smooth, bright, and accurate, so that little, if any, machining is required. *Metal-mold castings* are also known as *die castings* and *centrifugal castings*.

3–11 Powder-Metallurgy Process

The powder-metallurgy process is a quantity-production process that uses powders from a single metal, several metals, or a mixture of metals and nonmetals. It consists essentially of mechanically mixing the powders, compacting them in dies at high pressures, and heating the compacted part at a temperature less than the melting point of the major ingredient. The particles are united into a single strong part similar to what would be obtained by melting the same ingredients together. The advantages are (1) the elimination of scrap or waste material, (2) the elimination of machining operations, (3) the low unit cost when mass-produced, and (4) the exact control of composition. Some of the disadvantages are (1) the high cost of dies, (2) the lower physical properties, (3) the higher cost of materials, (4) the limitations on the design, and (5) the limited range of materials that can be used. Parts commonly made by this process are oil-impregnated bearings, incandescent lamp filaments, cemented-carbide tips for tools, and permanent magnets. Some products can be made only by powder metallurgy: surgical implants, for example. The structure is different from what can be obtained by melting the same ingredients.

3–12 Hot-Working Processes

By *hot working* are meant such processes as rolling, forging, hot extrusion, and hot pressing, in which the metal is heated above its recrystallation temperature.

Hot rolling is usually used to create a bar of material of a particular shape and dimension. Figure 3–10 shows some of the various shapes that are commonly produced by the hot-rolling process. All of them are available in many different sizes as well as in

Figure 3–10

Common shapes available through hot rolling.

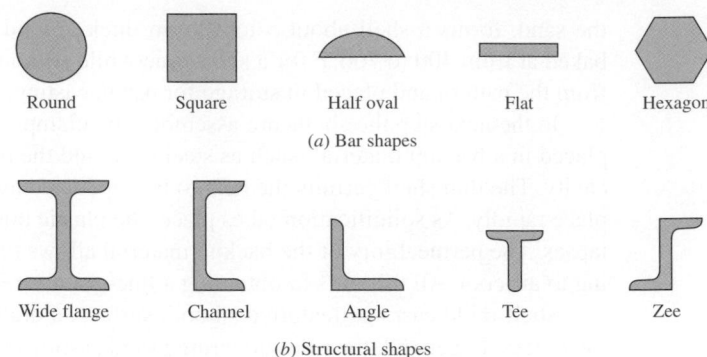

(a) Bar shapes

Round   Square   Half oval   Flat   Hexagon

Wide flange   Channel   Angle   Tee   Zee

(b) Structural shapes

different materials. The materials most available in the hot-rolled bar sizes are steel, aluminum, magnesium, and copper alloys.

Tubing can be manufactured by hot-rolling strip or plate. The edges of the strip are rolled together, creating seams that are either butt-welded or lap-welded. Seamless tubing is manufactured by roll-piercing a solid heated rod with a piercing mandrel.

Extrusion is the process by which great pressure is applied to a heated metal billet or blank, causing it to flow through a restricted orifice. This process is more common with materials of low melting point, such as aluminum, copper, magnesium, lead, tin, and zinc. Stainless steel extrusions are available on a more limited basis.

Forging is the hot working of metal by hammers, presses, or forging machines. In common with other hot-working processes, forging produces a refined grain structure that results in increased strength and ductility. Compared with castings, forgings have greater strength for the same weight. In addition, drop forgings can be made smoother and more accurate than sand castings, so that less machining is necessary. However, the initial cost of the forging dies is usually greater than the cost of patterns for castings, although the greater unit strength rather than the cost is usually the deciding factor between these two processes.

3–13   Cold-Working Processes

By *cold working* is meant the forming of the metal while at a low temperature (usually room temperature). In contrast to parts produced by hot working, cold-worked parts have a bright new finish, are more accurate, and require less machining.

Cold-finished bars and shafts are produced by rolling, drawing, turning, grinding, and polishing. Of these methods, by far the largest percentage of products are made by the cold-rolling and cold-drawing processes. Cold rolling is now used mostly for the production of wide flats and sheets. Practically all cold-finished bars are made by cold drawing but even so are sometimes mistakenly called "cold-rolled bars." In the drawing process, the hot-rolled bars are first cleaned of scale and then drawn by pulling them through a die that reduces the size about $\frac{1}{32}$ to $\frac{1}{16}$ in. This process does not remove material from the bar but reduces, or "draws" down, the size. Many different shapes of hot-rolled bars may be used for cold drawing.

Cold rolling and cold drawing have the same effect upon the mechanical properties. The cold-working process does not change the grain size but merely distorts it. Cold working results in a large increase in yield strength, an increase in ultimate strength and hardness, and a decrease in ductility. In Fig. 3–11 the properties of a cold-drawn bar are compared with those of a hot-rolled bar of the same material.

Figure 3–11

Stress-strain diagram for hot-rolled and cold-drawn UNS G10350 steel.

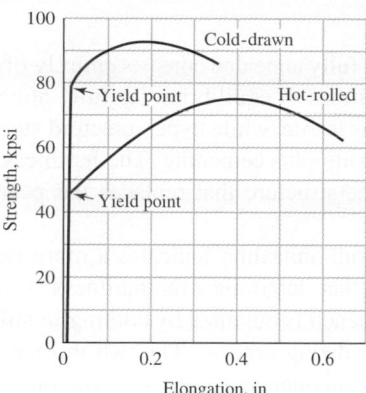

Heading is a cold-working process in which the metal is gathered, or upset. This operation is commonly used to make screw and rivet heads and is capable of producing a wide variety of shapes. *Roll threading* is the process of rolling threads by squeezing and rolling a blank between two serrated dies. *Spinning* is the operation of working sheet material around a rotating form into a circular shape. *Stamping* is the term used to describe punch-press operations such as *blanking, coining, forming,* and *shallow drawing*.

3–14   The Heat Treatment of Steel

Heat treatment of steel refers to time- and temperature-controlled processes that relieve residual stresses and/or modifies material properties such as hardness (strength), ductility, and toughness. Other mechanical or chemical operations are sometimes grouped under the heading of heat treatment. The common heat-treating operations are annealing, quenching, tempering, and case hardening.

Annealing

When a material is cold- or hot-worked, residual stresses are built in, and, in addition, the material usually has a higher hardness as a result of these working operations. These operations change the structure of the material so that it is no longer represented by the equilibrium diagram. Full annealing and normalizing is a heating operation that permits the material to transform according to the equilibrium diagram. The material to be annealed is heated to a temperature that is approximately 100°F above the critical temperature. It is held at this temperature for a time that is sufficient for the carbon to become dissolved and diffused through the material. The object being treated is then allowed to cool slowly, usually in the furnace in which it was treated. If the transformation is complete, then it is said to have a full anneal. Annealing is used to soften a material and make it more ductile, to relieve residual stresses, and to refine the grain structure.

The term *annealing* includes the process called *normalizing*. Parts to be normalized may be heated to a slightly higher temperature than in full annealing. This produces a coarser grain structure, which is more easily machined if the material is a low-carbon steel. In the normalizing process the part is cooled in still air at room temperature. Since this cooling is more rapid than the slow cooling used in full annealing, less time is available for equilibrium, and the material is harder than fully annealed steel. Normalizing is often used as the final treating operation for steel. The cooling in still air amounts to a slow quench.

Quenching

Eutectoid steel that is fully annealed consists entirely of pearlite, which is obtained from austenite under conditions of equilibrium. A fully annealed hypoeutectoid steel would consist of pearlite plus ferrite, while hypereutectoid steel in the fully annealed condition would consist of pearlite plus cementite. The hardness of steel of a given carbon content depends upon the structure that replaces the pearlite when full annealing is not carried out.

The absence of full annealing indicates a more rapid rate of cooling. The rate of cooling is the factor that determines the hardness. A controlled cooling rate is called *quenching*. A mild quench is obtained by cooling in still air, which, as we have seen, is obtained by the normalizing process. The two most widely used media for quenching are water and oil. The oil quench is quite slow but prevents quenching cracks caused by rapid expansion of the object being treated. Quenching in water is used for carbon steels and for medium-carbon, low-alloy steels.

The effectiveness of quenching depends upon the fact that when austenite is cooled it does not transform into pearlite instantaneously but requires time to initiate and complete the process. Since the transformation ceases at about 800°F, it can be prevented by rapidly cooling the material to a lower temperature. When the material is cooled rapidly to 400°F or less, the austenite is transformed into a structure called *martensite*. Martensite is a supersaturated solid solution of carbon in ferrite and is the hardest and strongest form of steel.

If steel is rapidly cooled to a temperature between 400 and 800°F and held there for a sufficient length of time, the austenite is transformed into a material that is generally called *bainite*. Bainite is a structure intermediate between pearlite and martensite. Although there are several structures that can be identified between the temperatures given, depending upon the temperature used, they are collectively known as bainite. By the choice of this transformation temperature, almost any variation of structure may be obtained. These range all the way from coarse pearlite to fine martensite.

Tempering

When a steel specimen has been fully hardened, it is very hard and brittle and has high residual stresses. The steel is unstable and tends to contract on aging. This tendency is increased when the specimen is subjected to externally applied loads, because the resultant stresses contribute still more to the instability. These internal stresses can be relieved by a modest heating process called *stress relieving,* or a combination of stress relieving and softening called *tempering* or *drawing*. After the specimen has been fully hardened by being quenched from above the critical temperature, it is reheated to some temperature below the critical temperature for a certain period of time and then allowed to cool in still air. The temperature to which it is reheated depends upon the composition and the degree of hardness or toughness desired.[8] This reheating operation releases the carbon held in the martensite, forming carbide crystals. The structure obtained is called *tempered martensite*. It is now essentially a superfine dispersion of iron carbide(s) in fine-grained ferrite.

The effect of heat-treating operations upon the various mechanical properties of a low alloy steel is shown graphically in Fig. 3–12.

---

8. For the quantitative aspects of tempering in plain carbon and low-alloy steels, see Charles R. Mischke, "The Strength of Cold-Worked and Heat-Treated Steels," Chap. 8 in Joseph E. Shigley and Charles R. Mischke (eds.), *Standard Handbook of Machine Design,* 2nd ed., McGraw-Hill, New York, 1996.

Figure 3–12

The effect of thermal-mechanical history on the mechanical properties of AISI 4340 steel. *(Prepared by the International Nickel Company.)*

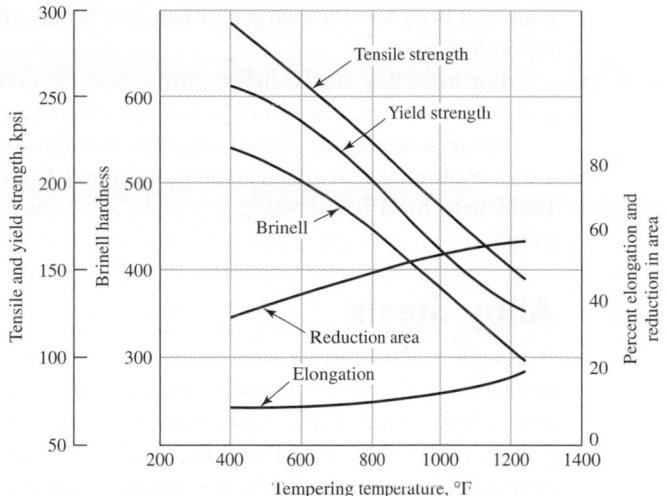

Condition	Tensile strength, kpsi	Yield strength, kpsi	Reduction in area, %	Elongation in 2 in, %	Brinell hardness, Bhn
Normalized	200	147	20	10	410
As rolled	190	144	18	9	380
Annealed	120	99	43	18	228

Case Hardening

The purpose of case hardening is to produce a hard outer surface on a specimen of low-carbon steel while at the same time retaining the ductility and toughness in the core. This is done by increasing the carbon content at the surface. Either solid, liquid, or gaseous carburizing materials may be used. The process consists of introducing the part to be carburized into the carburizing material for a stated time and at a stated temperature, depending upon the depth of case desired and the composition of the part. The part may then be quenched directly from the carburization temperature and tempered, or in some cases it must undergo a double heat treatment in order to ensure that both the core and the case are in proper condition. Some of the more useful case-hardening processes are pack carburizing, gas carburizing, nitriding, cyaniding, induction hardening, and flame hardening. In the last two cases carbon is not added to the steel in question, generally a medium carbon steel, for example SAE/AISI 1144.

Quantitative Estimation of Properties of Heat-Treated Steels

Courses in metallurgy (or material science) for mechanical engineers usually present the addition method of Crafts and Lamont for the prediction of heat-treated properties from the Jominy test for plain carbon steels.[9] If this is not been in your prerequisite experience, then refer to the *Standard Handbook of Machine Design,* where the addition method is covered with examples.[10] If this book is a textbook for a machine elements

9. W. Crafts and J. L. Lamont, *Hardenability and Steel Selection,* Pitman and Sons, London, 1949.

10. Charles R. Mischke, Chap. 8 in Joseph E. Shigley and Charles R. Mischke (eds.), *Standard Handbook of Machine Design,* 2nd ed., McGraw-Hill, New York, 1996, pp. 8-10–8-32.

course, it is a good class project (many hands make light work) to study the method and report to the class.

For low-alloy steels, the multiplication method of Grossman[11] and Field[12] is explained in the *Standard Handbook of Machine Design* (pp. 13-27–13-30).

Modern Steels and Their Properties Handbook explains how to predict the Jominy curve by the method of Grossman and Field from a ladle analysis and grain size.[13] Bethlehem Steel has developed a circular plastic slide rule that is convenient to the purpose.

3–15 Alloy Steels

Although a plain carbon steel is an alloy of iron and carbon with small amounts of manganese, silicon, sulfur, and phosphorus, the term *alloy steel* is applied when one or more elements other than carbon are introduced in sufficient quantities to modify its properties substantially. The alloy steels not only possess more desirable physical properties but also permit a greater latitude in the heat-treating process.

Chromium

The addition of chromium results in the formation of various carbides of chromium that are very hard, yet the resulting steel is more ductile than a steel of the same hardness produced by a simple increase in carbon content. Chromium also refines the grain structure so that these two combined effects result in both increased toughness and increased hardness. The addition of chromium increases the critical range of temperatures and moves the eutectoid point to the left. Chromium is thus a very useful alloying element.

Nickel

The addition of nickel to steel also causes the eutectoid point to move to the left and increases the critical range of temperatures. Nickel is soluble in ferrite and does not form carbides or oxides. This increases the strength without decreasing the ductility. Case hardening of nickel steels results in a better core than can be obtained with plain carbon steels. Chromium is frequently used in combination with nickel to obtain the toughness and ductility provided by the nickel and the wear resistance and hardness contributed by the chromium.

Manganese

Manganese is added to all steels as a deoxidizing and desulfurizing agent, but if the sulfur content is low and the manganese content is over 1 percent, the steel is classified as a manganese alloy. Manganese dissolves in the ferrite and also forms carbides. It causes the eutectoid point to move to the left and lowers the critical range of temperatures. It increases the time required for transformation so that oil quenching becomes practicable.

Silicon

Silicon is added to all steels as a deoxidizing agent. When added to very-low-carbon steels, it produces a brittle material with a low hysteresis loss and a high magnetic permeability. The principal use of silicon is with other alloying elements, such as manganese, chromium, and vanadium, to stabilize the carbides.

---

11. M. A. Grossman, *AIME,* February 1942.

12. J. Field, *Metals Progress,* March 1943.

13. *Modern Steels and Their Properties,* 7th ed., Handbook 2757, Bethlehem Steel, 1972, pp. 46–50.

Molybdenum

While molybdenum is used alone in a few steels, it finds its greatest use when combined with other alloying elements, such as nickel, chromium, or both. Molybdenum forms carbides and also dissolves in ferrite to some extent, so that it adds both hardness and toughness. Molybdenum increases the critical range of temperatures and substantially lowers the transformation point. Because of this lowering of the transformation point, molybdenum is most effective in producing desirable oil-hardening and air-hardening properties. Except for carbon, it has the greatest hardening effect, and because it also contributes to a fine grain size, this results in the retention of a great deal of toughness.

Vanadium

Vanadium has a very strong tendency to form carbides; hence it is used only in small amounts. It is a strong deoxidizing agent and promotes a fine grain size. Since some vanadium is dissolved in the ferrite, it also toughens the steel. Vanadium gives a wide hardening range to steel, and the alloy can be hardened from a higher temperature. It is very difficult to soften vanadium steel by tempering; hence, it is widely used in tool steels.

Tungsten

Tungsten is widely used in tool steels because the tool will maintain its hardness even at red heat. Tungsten produces a fine, dense structure and adds both toughness and hardness. Its effect is similar to that of molybdenum, except that it must be added in greater quantities.

3–16   Corrosion-Resistant Steels

Iron-base alloys containing at least 12 percent chromium are called *stainless steels*. The most important characteristic of these steels is their resistance to many, but not all, corrosive conditions. The four types available are the ferritic chromium steels, the austenitic chromium-nickel steels, and the martensitic and precipitation-hardenable stainless steels.

The ferritic chromium steels have a chromium content ranging from 12 to 27 percent. Their corrosion resistance is a function of the chromium content, so that alloys containing less than 12 percent still exhibit some corrosion resistance, although they may rust. The quench-hardenability of these steels is a function of both the chromium and the carbon content. The very high carbon steels have good quench hardenability up to about 18 percent chromium, while in the lower carbon ranges it ceases at about 13 percent. If a little nickel is added, these steels retain some degree of hardenability up to 20 percent chromium. If the chromium content exceeds 18 percent, they become difficult to weld, and at the very high chromium levels the hardness becomes so great that very careful attention must be paid to the service conditions. Since chromium is expensive, the designer will choose the lowest chromium content consistent with the corrosive conditions.

The chromium-nickel stainless steels retain the austenitic structure at room temperature; hence, they are not amenable to heat treatment. The strength of these steels can be greatly improved by cold working. They are not magnetic unless cold-worked. Their work hardenability properties also cause them to be difficult to machine. All the chromium-nickel steels may be welded. They have greater corrosion-resistant properties than the plain chromium steels. When more chromium is added for greater corrosion resistance, more nickel must also be added if the austenitic properties are to be retained.

3–17 Casting Materials

Gray Cast Iron

Of all the cast materials, gray cast iron is the most widely used. This is because it has a very low cost, is easily cast in large quantities, and is easy to machine. The principal objections to the use of gray cast iron are that it is brittle and that it is weak in tension. In addition to a high carbon content (over 1.7 percent and usually greater than 2 percent), cast iron also has a high silicon content, with low percentages of sulfur, manganese, and phosphorus. The resultant alloy is composed of pearlite, ferrite, and graphite, and under certain conditions the pearlite may decompose into graphite and ferrite. The resulting product then contains all ferrite and graphite. The graphite, in the form of thin flakes distributed evenly throughout the structure, darkens it; hence, the name *gray cast iron*.

Gray cast iron is not readily welded, because it may crack, but this tendency may be reduced if the part is carefully preheated. Although the castings are generally used in the as-cast condition, a mild anneal reduces cooling stresses and improves the machinability. The tensile strength of gray cast iron varies from 100 to 400 MPa (15 to 60 kpsi), and the compressive strengths are 3 to 4 times the tensile strengths. The modulus of elasticity varies widely, with values extending all the way from 75 to 150 GPa (11 to 22 Mpsi).

Ductile and Nodular Cast Iron

Because of the lengthy heat treatment required to produce malleable cast iron, engineers have long desired a cast iron that would combine the ductile properties of malleable iron with the ease of casting and machining of gray iron and at the same time would possess these properties in the as-cast conditions. A process for producing such a material using magnesium-containing material seems to fulfill these requirements.

Ductile cast iron, or *nodular cast iron,* as it is sometimes called, is essentially the same as malleable cast iron, because both contain graphite in the form of spheroids. However, ductile cast iron in the as-cast condition exhibits properties very close to those of malleable iron, and if a simple 1-h anneal is given and is followed by a slow cool, it exhibits even more ductility than the malleable product. Ductile iron is made by adding MgFeSi to the melt; since magnesium boils at this temperature, it is necessary to alloy it with other elements before it is introduced.

Ductile iron has a high modulus of elasticity (172 GPa or 25 Mpsi) as compared with gray cast iron, and it is elastic in the sense that a portion of the stress-strain curve is a straight line. Gray cast iron, on the other hand, does not obey Hooke's law, because the modulus of elasticity steadily decreases with increase in stress. Like gray cast iron, however, nodular iron has a compressive strength that is higher than the tensile strength, although the difference is not as great. In 40 years it has become extensively used.

White Cast Iron

If all the carbon in cast iron is in the form of cementite and pearlite, with no graphite present, the resulting structure is white and is known as *white cast iron*. This may be produced in two ways. The composition may be adjusted by keeping the carbon and silicon content low, or the gray-cast-iron composition may be cast against chills in order to promote rapid cooling. By either method, a casting with large amounts of cementite is produced, and as a result the product is very brittle and hard to machine but also very resistant to wear. A chill is usually used in the production of gray-iron castings in order to provide a very hard surface within a particular area of the casting, while at the same time retaining the more desirable gray structure within the remaining portion. This produces a relatively tough casting with a wear-resistant area.

Malleable Cast Iron

If white cast iron within a certain composition range is annealed, a product called *malleable cast iron* is formed. The annealing process frees the carbon so that it is present as graphite, just as in gray cast iron but in a different form. In gray cast iron the graphite is present in a thin flake form, while in malleable cast iron it has a nodular form and is known as *temper carbon*. A good grade of malleable cast iron may have a tensile strength of over 350 MPa (50 kpsi), with an elongation of as much as 18 percent. The percentage elongation of a gray cast iron, on the other hand, is seldom over 1 percent. Because of the time required for annealing (up to 6 days for large and heavy castings), malleable iron is necessarily somewhat more expensive than gray cast iron.

Alloy Cast Irons

Nickel, chromium, and molybdenum are the most common alloying elements used in cast iron. Nickel is a general-purpose alloying element, usually added in amounts up to 5 percent. Nickel increases the strength and density, improves the wearing qualities, and raises the machinability. If the nickel content is raised to 10 to 18 percent, an austenitic structure with valuable heat- and corrosion-resistant properties results. Chromium increases the hardness and wear resistance and, when used with a chill, increases the tendency to form white iron. When chromium and nickel are both added, the hardness and strength are improved without a reduction in the machinability rating. Molybdenum added in quantities up to 1.25 percent increases the stiffness, hardness, tensile strength, and impact resistance. It is a widely used alloying element.

Cast Steels

The advantage of the casting process is that parts having complex shapes can be manufactured at costs less than fabrication by other means, such as welding. Thus the choice of steel castings is logical when the part is complex and when it must also have a high strength. The higher melting temperatures for steels do aggravate the casting problems and require closer attention to such details as core design, section thicknesses, fillets, and the progress of cooling. The same alloying elements used for the wrought steels can be used for cast steels to improve the strength and other mechanical properties. Cast-steel parts can also be heat-treated to alter the mechanical properties, and, unlike the cast irons, they can be welded.

3–18   Nonferrous Metals

Aluminum

The outstanding characteristics of aluminum and its alloys are their strength-weight ratio, their resistance to corrosion, and their high thermal and electrical conductivity. The density of aluminum is about 2770 kg/m^3 (0.10 lbf/in^3), compared with 7750 kg/m^3 (0.28 lbf/in^3) for steel. Pure aluminum has a tensile strength of about 90 MPa (13 kpsi), but this can be improved considerably by cold working and also by alloying with other materials. The modulus of elasticity of aluminum, as well as of its alloys, is 71.7 GPa (10.4 Mpsi), which means that it has about one-third the stiffness of steel.

Considering the cost and strength of aluminum and its alloys, they are among the most versatile materials from the standpoint of fabrication. Aluminum can be processed by sand casting, die casting, hot or cold working, or extruding. Its alloys can be machined, press-worked, soldered, brazed, or welded. Pure aluminum melts at 660°C (1215°F), which makes it very desirable for the production of either permanent or sand-mold castings. It is commercially available in the form of plate, bar, sheet, foil, rod, and tube and in structural and extruded shapes. Certain precautions must be taken in joining aluminum by soldering, brazing, or welding; these joining methods are not recommended for all alloys.

The corrosion resistance of the aluminum alloys depends upon the formation of a thin oxide coating. This film forms spontaneously because aluminum is inherently very reactive. Constant erosion or abrasion removes this film and allows corrosion to take place. An extra-heavy oxide film may be produced by the process called *anodizing*. In this process the specimen is made to become the anode in an electrolyte, which may be chromic acid, oxalic acid, or sulfuric acid. It is possible in this process to control the color of the resulting film very accurately.

The most useful alloying elements for aluminum are copper, silicon, manganese, magnesium, and zinc. Aluminum alloys are classified as *casting alloys* or *wrought alloys*. The casting alloys have greater percentages of alloying elements to facilitate casting, but this makes cold working difficult. Many of the casting alloys, and some of the wrought alloys, cannot be hardened by heat treatment. The alloys that are heat-treatable use an alloying element that dissolves in the aluminum. The heat treatment consists of heating the specimen to a temperature that permits the alloying element to pass into solution, then quenching so rapidly that the alloying element is not precipitated. The aging process may be accelerated by heating slightly, which results in even greater hardness and strength. One of the better-known heat-treatable alloys is duraluminum, or 2017 (4 percent Cu, 0.5 percent Mg, 0.5 percent Mn). This alloy hardens in 4 days at room temperature. Because of this rapid aging, the alloy must be stored under refrigeration after quenching and before forming, or it must be formed immediately after quenching. Other alloys (such as 5053) have been developed that age-harden much more slowly, so that only mild refrigeration is required before forming. After forming, they are artificially aged in a furnace and possess approximately the same strength and hardness as the 2024 alloys. Those alloys of aluminum that cannot be heat-treated can be hardened only by cold working. Both work hardening and the hardening produced by heat treatment may be removed by an annealing process.

Magnesium

The density of magnesium is about 1800 kg/m^3 (0.065 lb/in^3), which is two-thirds that of aluminum and one-fourth that of steel. Since it is the lightest of all commercial metals, its greatest use is in the aircraft and automotive industries, but other uses are now being found for it. Although the magnesium alloys do not have great strength, because of their light weight the strength-weight ratio compares favorably with the stronger aluminum and steel alloys. Even so, magnesium alloys find their greatest use in applications where strength is not an important consideration. Magnesium will not withstand elevated temperatures; the yield point is definitely reduced when the temperature is raised to that of boiling water.

Magnesium and its alloys have a modulus of elasticity of 45 GPa (6.5 Mpsi) in tension and in compression, although some alloys are not as strong in compression as in tension. Curiously enough, cold working reduces the modulus of elasticity. A range of cast magnesium alloys are also available.

Titanium

Titanium and its alloys are similar in strength to moderate-strength steel but weigh half as much as steel. The material exhibits very good resistence to corrosion, has low thermal conductivity, is nonmagnetic, and has high-temperature strength. Its modulus of elasticity is between those of steel and aluminum at 16.5 Mpsi (114 GPa). Because of its many advantages over steel and aluminum, applications include: aerospace and military aircraft structures and components, marine hardware, chemical tanks and processing equipment, fluid handling systems, and human internal replacement devices. The disadvantages of titanium are its high cost compared to steel and aluminum and the difficulty of machining it.

Copper-Base Alloys

When copper is alloyed with zinc, it is usually called *brass*. If it is alloyed with another element, it is often called *bronze*. Sometimes the other element is specified too, as, for example, *tin bronze* or *phosphor bronze*. There are hundreds of variations in each category.

Brass with 5 to 15 Percent Zinc

The low-zinc brasses are easy to cold work, especially those with the higher zinc content. They are ductile but often hard to machine. The corrosion resistance is good. Alloys included in this group are *gilding brass* (5 percent Zn), *commercial bronze* (10 percent Zn), and *red brass* (15 percent Zn). Gilding brass is used mostly for jewelry and articles to be gold plated; it has the same ductility as copper but greater strength, accompanied by poor machining characteristics. Commercial bronze is used for jewelry and for forgings and stampings, because of its ductility. Its machining properties are poor, but it has excellent cold-working properties. Red brass has good corrosion resistance as well as high-temperature strength. Because of this it is used a great deal in the form of tubing or piping to carry hot water in such applications as radiators or condensers.

Brass with 20 to 36 Percent Zinc

Included in the intermediate-zinc group are *low brass* (20 percent Zn), *cartridge brass* (30 percent Zn), and *yellow brass* (35 percent Zn). Since zinc is cheaper than copper, these alloys cost less than those with more copper and less zinc. They also have better machinability and slightly greater strength; this is offset, however, by poor corrosion resistance and the possibility of cracking at points of residual stresses. Low brass is very similar to red brass and is used for articles requiring deep-drawing operations. Of the copper-zinc alloys, cartridge brass has the best combination of ductility and strength. Cartridge cases were originally manufactured entirely by cold working; the process consisted of a series of deep draws, each draw being followed by an anneal to place the material in condition for the next draw, hence the name cartridge brass. Although the hot-working ability of yellow brass is poor, it can be used in practically any other fabricating process and is therefore employed in a large variety of products.

When small amounts of lead are added to the brasses, their machinability is greatly improved and there is some improvement in their abilities to be hot-worked. The addition of lead impairs both the cold-working and welding properties. In this group are *low-leaded brass* ($32\frac{1}{2}$ percent Zn, $\frac{1}{2}$ percent Pb), *high-leaded brass* (34 percent Zn, 2 percent Pb), and *free-cutting brass* ($35\frac{1}{2}$ percent Zn, 3 percent Pb). The low-leaded brass is not only easy to machine but has good cold-working properties. It is used for various screw-machine parts. High-leaded brass, sometimes called *engraver's brass,* is used for instrument, lock, and watch parts. Free-cutting brass is also used for screw-machine parts and has good corrosion resistance with excellent mechanical properties.

Admiralty metal (28 percent Zn) contains 1 percent tin, which imparts excellent corrosion resistance, especially to saltwater. It has good strength and ductility but only fair machining and working characteristics. Because of its corrosion resistance it is used in power-plant and chemical equipment. *Aluminum brass* (22 percent Zn) contains 2 percent aluminum and is used for the same purposes as admiralty metal, because it has nearly the same properties and characteristics. In the form of tubing or piping, it is favored over admiralty metal, because it has better resistance to erosion caused by high-velocity water.

Brass with 36 to 40 Percent Zinc

Brasses with more than 38 percent zinc are less ductile than cartridge brass and cannot be cold-worked as severely. They are frequently hot-worked and extruded. *Muntz metal* (40 percent Zn) is low in cost and mildly corrosion-resistant. *Naval brass* has the same composition as Muntz metal except for the addition of 0.75 percent tin, which contributes to the corrosion resistance.

Bronze

Silicon bronze, containing 3 percent silicon and 1 percent manganese in addition to the copper, has mechanical properties equal to those of mild steel, as well as good corrosion resistance. It can be hot- or cold-worked, machined, or welded. It is useful wherever corrosion resistance combined with strength is required.

Phosphor bronze, made with up to 11 percent tin and containing small amounts of phosphorus, is especially resistant to fatigue and corrosion. It has a high tensile strength and a high capacity to absorb energy, and it is also resistant to wear. These properties make it very useful as a spring material.

Aluminum bronze is a heat-treatable alloy containing up to 12 percent aluminum. This alloy has strength and corrosion-resistance properties that are better than those of brass, and in addition, its properties may be varied over a wide range by cold working, heat treating, or changing the composition. When iron is added in amounts up to 4 percent, the alloy has a high endurance limit, a high shock resistance, and excellent wear resistance.

Beryllium bronze is another heat-treatable alloy, containing about 2 percent beryllium. This alloy is very corrosion resistant and has high strength, hardness, and resistance to wear. Although it is expensive, it is used for springs and other parts subjected to fatigue loading where corrosion resistance is required.

With slight modification most copper-based alloys are available in cast form.

3–19 Plastics

The term *thermoplastics* is used to mean any plastic that flows or is moldable when heat is applied to it; the term is sometimes applied to plastics moldable under pressure. Such plastics can be remolded when heated.

A *thermoset* is a plastic for which the polymerization process is finished in a hot molding press where the plastic is liquefied under pressure. Thermoset plastics cannot be remolded.

Table 3–2 lists some of the most widely used thermoplastics, together with some of their characteristics and the range of their properties. Table 3–3, listing some of the thermosets, is similar. These tables are presented for information only and should not be used to make a final design decision. The range of properties and characteristics that can be obtained with plastics is very great. The influence of many factors, such as cost, moldability, coefficient of friction, weathering, impact strength, and the effect of fillers and reinforcements, must be considered. Manufacturers' catalogs will be found quite helpful in making possible selections.

Table 3–2

The Thermoplastics *Source:* These data have been obtained from the *Machine Design Materials Reference Issue*, published by Penton/IPC, Cleveland. These reference issues are published about every 2 years and constitute an excellent source of data on a great variety of materials.

Name	S_u, kpsi	E, Mpsi	Hardness Rockwell	Elongation %	Dimensional Stability	Heat Resistance	Chemical Resistance	Processing
ABS group	2–8	0.10–0.37	60–110R	3–50	Good	*	Fair	EMST
Acetal group	8–10	0.41–0.52	80–94M	40–60	Excellent	Good	High	M
Acrylic	5–10	0.20–0.47	92–110M	3–75	High	*	Fair	EMS
Fluoroplastic group	0.50–7	· · ·	50–80D	100–300	High	Excellent	Excellent	MPR[†]
Nylon	8–14	0.18–0.45	112–120R	10–200	Poor	Poor	Good	CEM
Phenylene oxide	7–18	0.35–0.92	115R, 106L	5–60	Excellent	Good	Fair	EFM
Polycarbonate	8–16	0.34–0.86	62–91M	10–125	Excellent	Excellent	Fair	EMS
Polyester	8–18	0.28–1.6	65–90M	1–300	Excellent	Poor	Excellent	CLMR
Polyimide	6–50	· · ·	88–120M	Very low	Excellent	Excellent	Excellent[†]	CLMP
Polyphenylene sulfide	14–19	0.11	122R	1.0	Good	Excellent	Excellent	M
Polystyrene group	1.5–12	0.14–0.60	10–90M	0.5–60	· · ·	Poor	Poor	EM
Polysulfone	10	0.36	120R	50–100	Excellent	Excellent	Excellent[†]	EFM
Polyvinyl chloride	1.5–7.5	0.35–0.60	65–85D	40–450	· · ·	Poor	Poor	EFM

*Heat-resistant grades available.
[†]With exceptions.
C Coatings L Laminates R Resins E Extrusions M Moldings S Sheet F Foams P Press and sinter methods T Tubing

Table 3–3

The Thermosets *Source:* These data have been obtained from the *Machine Design Materials Reference Issue*, published by Penton/IPC, Cleveland. These reference issues are published about every 2 years and constitute an excellent source of data on a great variety of materials.

Name	S_u, kpsi	E, Mpsi	Hardness Rockwell	Elongation %	Dimensional Stability	Heat Resistance	Chemical Resistance	Processing
Alkyd	3–9	0.05–0.30	99M*	· · ·	Excellent	Good	Fair	M
Allylic	4–10	· · ·	105–120M	· · ·	Excellent	Excellent	Excellent	CM
Amino group	5–8	0.13–0.24	110–120M	0.30–0.90	Good	Excellent*	Excellent*	LR
Epoxy	5–20	0.03–0.30*	80–120M	1–10	Excellent	Excellent	Excellent	CMR
Phenolics	5–9	0.10–0.25	70–95E	· · ·	Excellent	Excellent	Good	EMR
Silicones	5–6	· · ·	80–90M	· · ·	· · ·	Excellent	Excellent	CL MR

*With exceptions.
C Coatings L Laminates R Resins E Extrusions M Moldings S Sheet F Foams P Press and sinter methods T Tubing

3–20 **Composite Materials**[14]

Composite materials are formed from two or more dissimilar materials, each of which contributes to the final properties. Unlike metallic alloys, the materials in a composite remain distinct from each other at the macroscopic level.

Most engineering composites consist of two materials: a reinforcement called a *filler* and a *matrix*. The filler provides stiffness and strength; the matrix holds the material together and serves to transfer load among the discontinuous reinforcements. The most common reinforcements, illustrated in Fig. 3–13, are continuous fibers, either straight or woven, short chopped fibers, and particulates. The most common matrices are various plastic resins although other materials including metals are used.

Metals and other traditional engineering materials are uniform, or isotropic, in nature. This means that material properties, such as strength, stiffness, and thermal conductivity, are independent of both position within the material and the choice of coordinate system. The discontinuous nature of composite reinforcements, though, means that material properties can vary with both position and direction. For example, an epoxy resin reinforced with continuous graphite fibers will have very high strength and stiffness in the direction of the fibers, but very low properties normal or transverse to the fibers. For this reason, structures of composite materials are normally constructed of multiple plies (laminates) where each ply is oriented to achieve optimal structural stiffness and strength performance.

High strength-to-weight ratios, up to 5 times greater than those of high-strength steels, can be achieved. High stiffness-to-weight ratios can also be obtained, as much as 8 times greater than those of structural metals. For this reason, composite materials are becoming very popular in automotive, aircraft, and spacecraft applications where weight is a premium.

The directionality of properties of composite materials increases the complexity of structural analyses. Isotropic materials are fully defined by two engineering constants: Young's modulus E and Poisson's ratio ν. A single ply of a composite material, however, requires four constants, defined with respect to the ply coordinate system. The constants are two Young's moduli (the longitudinal modulus in the direction of the fibers, E_1, and the transverse modulus normal to the fibers, E_2), one Poisson's ratio (ν_{12}, called the major Poisson's ratio), and one shear modulus (G_{12}). A fifth constant, the minor Poisson's ratio, ν_{21}, is determined through the reciprocity relation, $\nu_{21}/E_2 = \nu_{12}/E_1$. Combining this with multiple plies oriented at different angles makes structural analysis of complex structures unapproachable by manual techniques. For this reason,

Figure 3–13

Composites categorized by type of reinforcement.

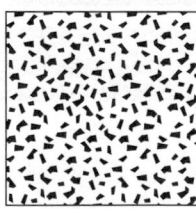 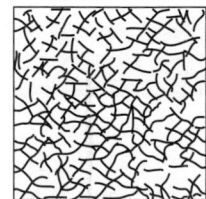 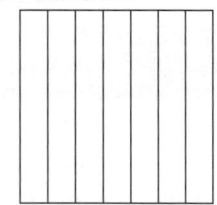 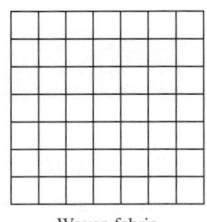

Particulate composite　　Randomly oriented short fiber composite　　Unidirectional continuous fiber composite　　Woven fabric composite

14. For references see I. M. Daniel and O. Ishai, *Engineering Mechanics of Composite Materials,* Oxford University Press, 1994, and *ASM Engineered Materials Handbook: Composites,* ASM International, Materials Park, OH, 1988.

computer software is available to calculate the properties of a laminated composite construction.[15]

PROBLEMS

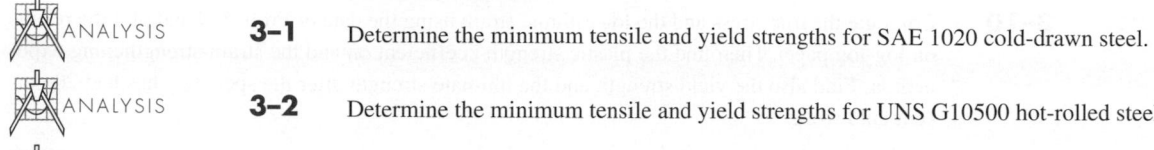

ANALYSIS **3-1** Determine the minimum tensile and yield strengths for SAE 1020 cold-drawn steel.

ANALYSIS **3-2** Determine the minimum tensile and yield strengths for UNS G10500 hot-rolled steel.

ANALYSIS **3-3** For the materials in Probs. 3–1 and 3–2, compare the following properties: minimum tensile and yield strengths, ductility, and stiffness.

ANALYSIS **3-4** Assuming you were specifying an AISI 1040 steel for an application where you desired to maximize the yield strength, how would you specify it?

ANALYSIS **3-5** Assuming you were specifying an AISI 1040 steel for an application where you desired to maximize the ductility, how would you specify it?

ANALYSIS **3-6** Determine the yield strength-to-weight density ratios (called *specific strength*) in units of inches for UNS G10350 hot-rolled steel, 2024-T4 aluminum, Ti-6A1-4V titanium alloy, and ASTM No. 30 gray cast iron.

ANALYSIS **3-7** Determine the stiffness-to-weight density ratios (called *specific modulus*) in units of inches for UNS G10350 hot-rolled steel, 2024-T4 aluminum, Ti-6A1-4V titanium alloy, and ASTM No. 30 gray cast iron.

ANALYSIS **3-8** *Poisson's ratio ν* is a material property and is the ratio of the lateral strain and the longitudinal strain for a member in tension. For a homogeneous, isotropic material, the modulus of rigidity G is related to Young's modulus as

$$G = \frac{E}{2(1 + \nu)}$$

Using the tabulated values of G and E, determine Poisson's ratio for steel, aluminum, beryllium copper, and gray cast iron.

ANALYSIS **3-9** A specimen of medium-carbon steel having an initial diameter of 0.503 in was tested in tension using a gauge length of 2 in. The following data were obtained for the elastic and plastic states:

Elastic State		Plastic State	
Load P, lbf	Elongation, in	Load P, lbf	Area A_i, in²
1 000	0.0004	8 800	0.1984
2 000	0.0006	9 200	0.1978
3 000	0.0010	9 100	0.1963
4 000	0.0013	13 200	0.1924
7 000	0.0023	15 200	0.1875
8 400	0.0028	17 000	0.1563
8 800	0.0036	16 400	0.1307
9 200	0.0089	14 800	0.1077

15. About Composite Materials Software listing, http://composite.about.com/cs/software/index.htm.

Note that there is some overlap in the data. Plot the engineering or nominal stress-strain diagram using two scales for the unit strain ϵ, one from zero to about 0.02 in/in and the other from zero to maximum strain. From this diagram find the modulus of elasticity, the 0.2 percent offset yield strength, the ultimate strength, and the percent reduction in area.

 ANALYSIS   **3-10**   Compute the true stress and the logarithmic strain using the data of Prob. 3–9 and plot the results on log-log paper. Then find the plastic strength coefficient σ_0 and the strain-strengthening exponent m. Find also the yield strength and the ultimate strength after the specimen has had 20 percent cold work.

ANALYSIS   **3-11**   The stress-strain data from a tensile test on a cast-iron specimen are

Engineering stress, kpsi	5	10	16	19	26	32	40	46	49	54
Engineering strain, $\epsilon \cdot 10^{-3}$ in/in	0.20	0.44	0.80	1.0	1.5	2.0	2.8	3.4	4.0	5.0

Plot the stress-strain locus and find the 0.1 percent offset yield strength, and the tangent modulus of elasticity at zero stress and at 20 kpsi.

ANALYSIS   **3-12**   Having completed the task of Prob. 3–11, an engineer is unsatisfied with the accuracy of the result, depending as it does on drawing tangents to unknown curves. Noting that the stress-strain locus on Cartesian coordinates looks smooth and parabolic, the engineer writes

$$\epsilon = a_1\sigma + a_2\sigma^2$$

Find the constants a_1 and a_2 by regression (see Prob. 2–8). From this regression equation, determine the tangent modulus at zero load and at 20 kpsi, as well as the 0.1 percent offset yield strength. How do these determinations compare with the technique of Prob. 3–11? What can you say about the error?

ANALYSIS   **3-13**   A straight bar of arbitrary cross section and thickness h is cold-formed to an inner radius R about an anvil as shown in the figure. Some surface at distance N having an original length L_{AB} will remain unchanged in length after bending. This length is

$$L_{AB} = L_{AB'} = \frac{\pi(R+N)}{2}$$

The lengths of the outer and inner surfaces, after bending, are

$$L_o = \frac{\pi}{2}(R+h) \qquad L_i = \frac{\pi}{2}R$$

Using Eq. (5–8), we then find the true strains to be

$$\varepsilon_o = \ln\frac{R+h}{R+N} \qquad \varepsilon_i = \ln\frac{R}{R+N}$$

Tests show that $|\varepsilon_o| = |\varepsilon_i|$. Show that

$$N = R\left[\left(1+\frac{h}{R}\right)^{1/2} - 1\right]$$

and

$$\varepsilon_o = \ln\left(1+\frac{h}{R}\right)^{1/2}$$

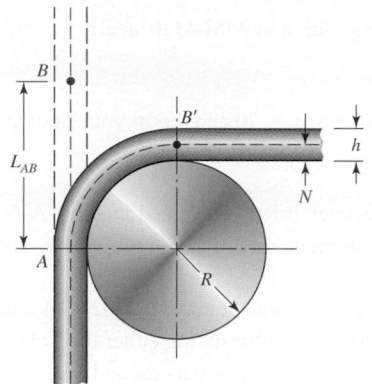

Problem 3–13

ANALYSIS **3–14** The tabulated data for a torque-twist test of a solid torsion specimen of diameter 12.5 mm with a gauge length of 350 mm is given below. Plot the data and, using a linear regression, determine the shear modulus for the material. Also estimate the shear yield strength from this set of data.

$T(N \cdot m)$	0	7.7	15.3	23.0	30.7	38.3	46.0	53.7	61.4	69.0	76.7	80.0	85.0	90.0	95.0	100.0
θ (deg.)	0	0.38	0.80	1.24	1.64	2.01	2.40	2.85	3.25	3.80	4.50	5.10	6.48	8.01	9.58	11.18

ANALYSIS **3–15** The following data for the offset yield strength were taken for 500 samples of a material batch. Determine the mean value, the standard deviation, and coefficient of variation. Assuming a log-normal distribution, express the variate $\mathbf{S}_y$, the probability distribution of S_y, and plot the continuous and discrete probability distributions together.

Class frequency f_i	2	9	30	65	101	112	90	54	25	9	2	1
Class midpoint x_i, kpsi	38.5	39.5	40.5	41.5	42.5	43.5	44.5	45.5	46.5	47.5	48.5	49.5

ANALYSIS **3–16** A hot-rolled AISI 1212 steel is given 20 percent cold work. Determine the new values of the yield and ultimate strengths.

ANALYSIS **3–17** A steel member has a Brinell of $H_B = 250$. Estimate the ultimate strength of the steel in MPa.

ANALYSIS **3–18** Brinell hardness tests were made on a random sample of 10 steel parts during processing. The results were H_B values of 252 (2), 260, 254, 257 (2), 249 (3), and 251. Estimate the mean and standard deviation of the ultimate strength in kpsi.

ANALYSIS **3–19** Repeat Prob. 3–18 assuming the material to be cast iron.

ANALYSIS **3–20** *Toughness* is a term that relates to both strength and ductility. The fracture toughness, for example, is defined as the total area under the stress-strain curve to fracture, $u_T = \int_0^{\epsilon_f} \sigma \, d\epsilon$. This area, called the *modulus of toughness,* is the strain energy per unit volume required to cause the material to fracture. A similar term, but defined within the elastic limit of the material, is called the *modulus of resilience,* $u_R = \int_0^{\epsilon_y} \sigma d\epsilon$, where ϵ_y is the strain at yield. If the stress-strain is linear to $\sigma = S_y$, then it can be shown that $u_R = S_y^2/2E$.

For the material in Prob. 3–9: (*a*) Determine the modulus of resilience, and (*b*) Estimate the modulus of toughness, assuming that the last data point corresponds to fracture.

ANALYSIS **3–21** What is the material composition of AISI 4340 steel?

RESEARCH **3–22** Search the website noted in Sec. 3–20 and report your findings.

RESEARCH **3–23** Research the material Inconel, briefly described in Table A–5. Compare it to various carbon and alloy steels in stiffness, strength, ductility, and toughness. What makes this material so special?

RESEARCH **3–24** Pick a specific material given in the tables (e.g., 2024-T4 aluminum, SAE 1040 steel), and consult a local or regional distributor (consulting either the Yellow Pages or the Thomas Register) to obtain as much information as you can about cost and availability of the material and in what form (bar, plate, etc.).

4

Load and Stress Analysis

Chapter Outline

One of the main objectives of this book is to describe how specific machine components function and how to design or specify them so that they function safely without failing structurally. Although earlier discussion has described structural strength in terms of load or stress versus strength, failure of function for structural reasons may arise from other factors such as excessive deformations or deflections.

Here it is assumed that the reader has completed basic courses in statics of rigid bodies and mechanics of materials and is quite familiar with the analysis of loads, and the stresses and deformations associated with the basic load states of simple prismatic elements. In this chapter and Chap. 5 we will review and extend these topics briefly. Complete derivations will not be presented here, and the reader is urged to return to basic textbooks and notes on these subjects.

This chapter begins with a review of equilibrium and free-body diagrams associated with load-carrying components. One must understand the nature of forces before attempting to perform an extensive stress or deflection analysis of a mechanical component. An extremely useful tool in handling discontinuous loading of structures employs *Macaulay* or *singularity functions*. Singularity functions are described in Sec. 4–3 as applied to the shear forces and bending moments in beams. In Chap. 5, the use of singularity functions will be expanded to show their real power in handling deflections of complex geometry and statically indeterminate problems.

Machine components transmit forces and motion from one point to another. The transmission of force can be envisioned as a flow or force distribution that can be further visualized by isolating internal surfaces within the component. Force distributed over a surface leads to the concept of stress, stress components, and stress transformations (Mohr's circle) for all possible surfaces at a point.

The remainder of the chapter is devoted to the stresses associated with the basic loading of prismatic elements, such as uniform loading, bending, and torsion, and topics with major design ramifications such as stress concentrations, thin- and thick-walled pressurized cylinders, rotating rings, press and shrink fits, thermal stresses, curved beams, and contact stresses.

4–1   Equilibrium and Free-Body Diagrams

Equilibrium

The word *system* will be used to denote any *isolated* part or portion of a machine or structure—including all of it if desired—that we wish to study. A system, under this definition, may consist of a particle, several particles, a part of a rigid body, an entire rigid body, or even several rigid bodies.

If we assume that the system to be studied is motionless or, at most, has constant velocity, then the system has zero acceleration. Under this condition the system is said to be in *equilibrium*. The phrase *static equilibrium* is also used to imply that the system is *at rest*. For equilibrium, the forces and moments acting on the system balance such that

$$\sum \mathbf{F} = 0 \tag{4–1}$$

$$\sum \mathbf{M} = 0 \tag{4–2}$$

which states that *the sum of all force* and the *sum of all moment vectors* acting upon a system in equilibrium is zero.

Free-Body Diagrams

In examining the problem of analyzing the behavior, performance, or efficiency of a complex structure or device, such as an automobile, photocopier, or mechanical drive

Figure 4–1

The isolation of a subsystem.

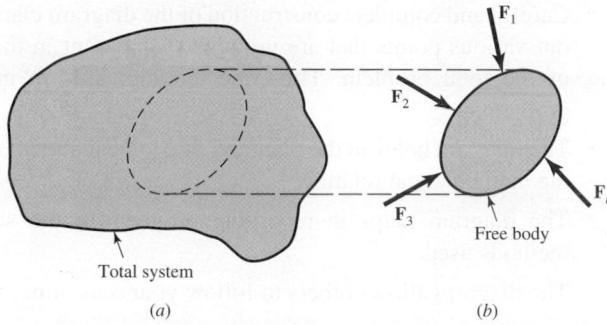

Total system

(a)

Free body

(b)

train, the beginner is faced with a bewildering array of complicated parts and geometries. Fortunately, the problem is not as difficult as it seems. One of the most powerful analytical techniques of mechanics is that of isolating or freeing a portion of the total system in our imagination in order to study the behavior of one of its segments. When the segment is isolated, the original effect of the system on the segment is replaced by the interacting forces and moments. Figure 4–1 is a symbolic illustration of the process. Let Fig. 4–1a be a total system, such as a mechanical drive train. Then we might decide to analyze just one part or segment of the system, such as a shaft, or even several members joined together. We remove this segment, as in Fig. 4–1b, and replace the effect of the whole system on the segment by various forces and moments that would necessarily act at the interfaces of the segment and the system. Although these forces may be internal effects upon the whole system, they are external effects when applied to the segment. These interface forces are represented symbolically by the force vectors F_1, F_2, F_3, and F_i in Fig. 4–1b. The isolated subsystem that results, together with all forces and moments due to any external effects and the reactions with the main system, is called a *free-body diagram*.[1]

We can greatly simplify the analysis of a very complex structure or machine by successively isolating each element and studying and analyzing it by the use of free-body diagrams. When all the members have been treated in this manner, the knowledge can be assembled to yield information concerning the behavior of the total system. Thus, free-body diagramming is essentially a means of breaking a complicated problem into manageable segments, analyzing these simple problems, and then, usually, putting the information together again.

Using free-body diagrams for force analysis serves the following important purposes:

- The diagram establishes the directions of reference axes, provides a place to record the dimensions of the subsystem and the magnitudes and directions of the known forces, and helps in assuming the directions of unknown forces.

- The diagram simplifies your thinking because it provides a place to store one thought while proceeding to the next.

- The diagram provides a means of communicating your thoughts clearly and unambiguously to other people.

---

[1]Professor Gaetano Lanza of MIT in 1885 in *Applied Mechanics* used the "method of sections" in trusses, but the "free-body diagram" as we now know it was championed by Professor Irving Church of Cornell in *Mechanics of Engineering* in 1887. It was late in the development of mechanics that Professor Church made it possible for engineers to teach themselves what they needed to know, rigorously. "Draw the free-body diagram" is an admonition that is key to your becoming an engineer.

- Careful and complete construction of the diagram clarifies fuzzy thinking by bringing out various points that are not always apparent in the statement or in the geometry of the total problem. Thus, the diagram aids in understanding all facets of the problem.

- The diagram helps in the planning of a logical attack on the problem and in setting up the mathematical relations.

- The diagram helps in recording progress in the solution and in illustrating the methods used.

- The diagram allows others to follow your reasoning, showing *all* forces.

EXAMPLE 4–1   Figure 4–2*a* shows a simplified rendition of a gear reducer where the input and output shafts AB and CD are rotating at constant speeds ω_i and ω_o, respectively. The input and output torques (torsional moments) are $T_i = 240$ lbf · in and T_o, respectively. The shafts are supported in the housing by bearings at A, B, C, and D. The pitch radii of gears G_1 and G_2 are $r_1 = 0.75$ in and $r_2 = 1.5$ in, respectively. Draw the free-body diagrams of each member and determine the net reaction forces and moments at all points.

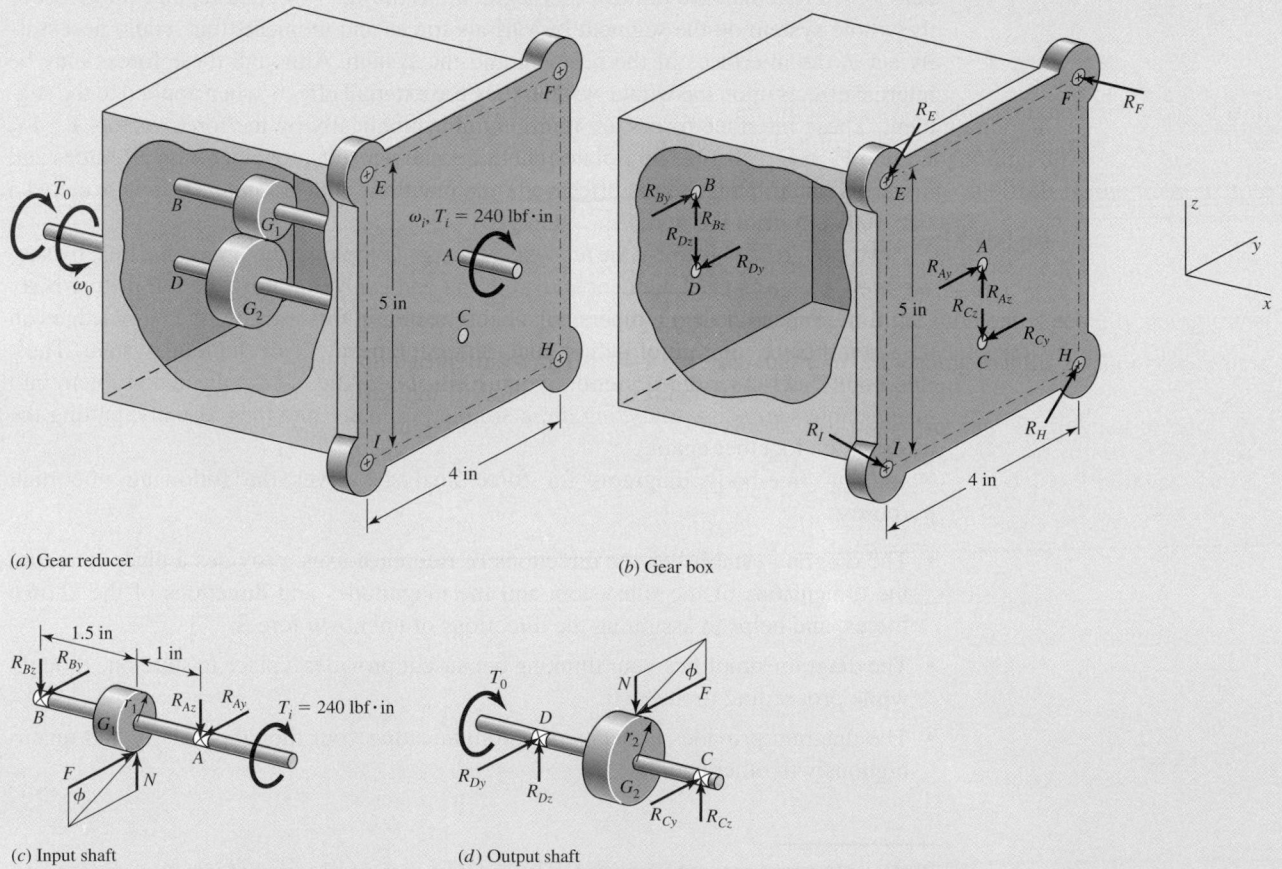

(*a*) Gear reducer

(*b*) Gear box

(*c*) Input shaft

(*d*) Output shaft

Figure 4–2

(*a*) Gear reducer; (*b–d*) free-body diagrams. Diagrams are not drawn to scale.

Solution First, we will list all simplifying assumptions.

1 Gears G_1 and G_2 are simple spur gears with a standard pressure angle $\phi = 20°$ (see Sec. 13–5).
2 The bearings are self-aligning and the shafts can be considered to be simply supported.
3 The weight of each member is negligible.
4 Friction is negligible.
5 The mounting bolts at E, F, H, and I are the same size.

The separate free-body diagrams of the members are shown in Figs. 4–2b–d. Note that Newton's third law, called *the law of action and reaction*, is used extensively where each member mates. The force transmitted between the spur gears is not tangential but at the pressure angle ϕ. Thus, $N = F \tan \phi$.

Summing moments about the x axis of shaft AB in Fig. 4–9d gives

$$\sum M_x = F(0.75) - 240 = 0$$

$$F = 320 \text{ lbf}$$

The normal force is $N = 320 \tan 20° = 116.5$ lbf.

Using the equilibrium equations for Figs. 4–2c and d, the reader should verify that: $R_{Ay} = 192$ lbf, $R_{Az} = 69.9$ lbf, $R_{By} = 128$ lbf, $R_{Bz} = 46.6$ lbf, $R_{Cy} = 192$ lbf, $R_{Cz} = 69.9$ lbf, $R_{Dy} = 128$ lbf, $R_{Dz} = 46.6$ lbf, and $T_o = 480$ lbf · in. The direction of the output torque T_o is opposite $\omega_o +$ because it is the resistive load on the system opposing the motion ω_o.

Note in Fig. 4–2b the net force from the bearing reactions is zero whereas the net moment about the x axis is $2.25 (192) + 2.25 (128) = 720$ lbf · in. This value is the same as $T_i + T_o = 240 + 480 = 720$ lbf · in, as shown in Fig. 4–2a. The reaction forces R_E, R_F, R_H, and R_I, from the mounting bolts cannot be determined from the equilibrium equations as there are too many unknowns. Only three equations are available, $\sum F_y = \sum F_z = \sum M_x = 0$. In case you were wondering about assumption 5, here is where we will use it (see Sec. 8–12). The gear box tends to rotate about the x axis because of a pure torsional moment of 720 lbf · in. The bolt forces must provide an equal but opposite torsional moment. The center of rotation relative to the bolts lies at the centroid of the bolt cross-sectional areas. Thus if the bolt areas are equal: the center of rotation is at the center of the four bolts, a distance of $\sqrt{(4/2)^2 + (5/2)^2} = 3.202$ in from each bolt; the bolt forces are equal ($R_E = R_F = R_H = R_I = R$), and each bolt force is perpendicular to the line from the bolt to the center of rotation. This gives a net torque from the four bolts of $4R(3.202) = 720$. Thus, $R_E = R_F = R_H = R_I = 56.22$ lbf.

4–2 Shear Force and Bending Moments in Beams

Figure 4–3a shows a beam supported by reactions R_1 and R_2 and loaded by the concentrated forces F_1, F_2, and F_3. If the beam is cut at some section located at $x = x_1$ and the left-hand portion is removed as a free body, an *internal shear force* V and *bending moment* M must act on the cut surface to ensure equilibrium. The shear force is obtained by summing the forces on the isolated section. The bending moment is the sum of the moments of the forces to the left of the section taken about an axis through the isolated section. The sign conventions used for bending moment and shear force in this book are

Figure 4–3

Free-body diagram of simply-supported beam with V and M shown in positive directions.

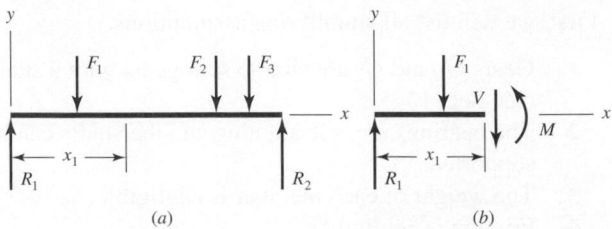

(a)   (b)

Figure 4–4

Sign conventions for bending and shear.

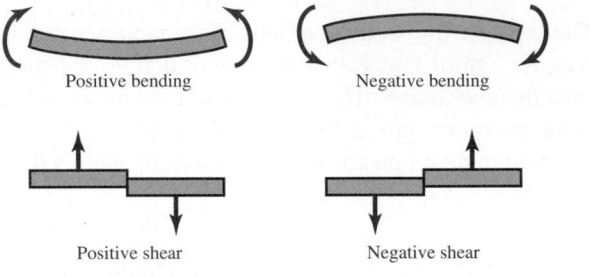

Positive bending   Negative bending

Positive shear   Negative shear

Figure 4–5

Distributed load on beam.

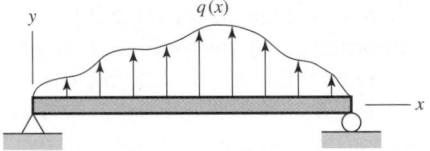

shown in Fig. 4–4. Shear force and bending moment are related by the equation

$$V = \frac{dM}{dx} \tag{4–3}$$

Sometimes the bending is caused by a distributed load $q(x)$, as shown in Fig. 4–5; $q(x)$ is called the *load intensity* with units of force per unit length and is positive in the positive y direction. It can be shown that differentiating Eq. (4–3) results in

$$\frac{dV}{dx} = \frac{d^2M}{dx^2} = q \tag{4–4}$$

Normally the applied distributed load is directed downward and labeled w (e.g., see Fig. 4–7). In this case, $w = -q$.

Equations (4–3) and (4–4) reveal additional relations if they are integrated. Thus, if we integrate between, say, x_A and x_B, we obtain

$$\int_{V_A}^{V_B} dV = \int_{x_A}^{x_B} q \, dx = V_B - V_A \tag{4–5}$$

which states that *the change in shear force from A to B is equal to the area of the loading diagram between x_A and x_B.*

In a similar manner,

$$\int_{M_A}^{M_B} dM = \int_{x_A}^{x_B} V \, dx = M_B - M_A \tag{4–6}$$

which states that *the change in moment from A to B is equal to the area of the shear-force diagram between x_A and x_B.*

4–3 **Singularity Functions**

The five singularity functions defined in Table 4–1 constitute a useful and easy means of integrating across discontinuities. By their use, general expressions for shear force and bending moment in beams can be written when the beam is loaded by concentrated moments or forces. As shown in the table, the concentrated moment and force functions are zero for all values of x not equal to a. The functions are undefined for values of $x = a$. Note that the unit step, ramp, and parabolic functions are zero only for values of x that are less than a. The integration properties shown in the table constitute a part of the mathematical definition too. The first two integrations of $q(x)$ for $V(x)$ and $M(x)$ do not require constants of integration provided *all* loads on the beam are accounted for in $q(x)$. The examples that follow show how these functions are used.

Table 4–1

Singularity (Macaulay[†]) Functions

Function	Graph of $f_n(x)$	Meaning
Concentrated moment (unit doublet)	$\langle x-a\rangle^{-2}$	$\langle x-a\rangle^{-2} = 0 \quad x \neq a$ $\langle x-a\rangle^{-2} = \pm\infty \quad x = a$ $\int \langle x-a\rangle^{-2}\, dx = \langle x-a\rangle^{-1}$
Concentrated force (unit impulse)	$\langle x-a\rangle^{-1}$	$\langle x-a\rangle^{-1} = 0 \quad x \neq a$ $\langle x-a\rangle^{-1} = +\infty \quad x = a$ $\int \langle x-a\rangle^{-1}\, dx = \langle x-a\rangle^{0}$
Unit step	$\langle x-a\rangle^{0}$	$\langle x-a\rangle^{0} = \begin{cases} 0 & x < a \\ 1 & x \geq a \end{cases}$ $\int \langle x-a\rangle^{0}\, dx = \langle x-a\rangle^{1}$
Ramp	$\langle x-a\rangle^{1}$	$\langle x-a\rangle^{1} = \begin{cases} 0 & x < a \\ x-a & x \geq a \end{cases}$ $\int \langle x-a\rangle^{1}\, dx = \dfrac{\langle x-a\rangle^{2}}{2}$
Parabolic	$\langle x-a\rangle^{2}$	$\langle x-a\rangle^{2} = \begin{cases} 0 & x < a \\ (x-a)^2 & x \geq a \end{cases}$ $\int \langle x-a\rangle^{2}\, dx = \dfrac{\langle x-a\rangle^{3}}{3}$

[†]W. H. Macaulay, "Note on the deflection of beams," *Messenger of Mathematics,* vol. 48, pp. 129–130, 1919.

EXAMPLE 4–2   Derive expressions for the loading, shear-force, and bending-moment diagrams for the beam of Fig. 4–6.

| **Figure 4–6**

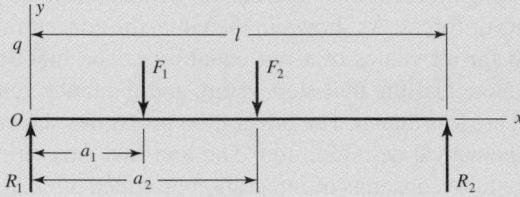

Solution   Using Table 4–1 and $q(x)$ for the loading function, we find

$$q = R_1 \langle x \rangle^{-1} - F_1 \langle x - a_1 \rangle^{-1} - F_2 \langle x - a_2 \rangle^{-1} + R_2 \langle x - l \rangle^{-1} \qquad (1)$$

Next, we use Eq. (4–5) to get the shear force.

$$V = \int q \, dx = R_1 \langle x \rangle^0 - F_1 \langle x - a_1 \rangle^0 - F_2 \langle x - a_2 \rangle^0 + R_2 \langle x - l \rangle^0 \qquad (2)$$

Note that $V = 0$ at $x = 0$.

A second integration, in accordance with Eq. (4–6), yields

$$M = \int V \, dx = R_1 \langle x \rangle^1 - F_1 \langle x - a_1 \rangle^1 - F_2 \langle x - a_2 \rangle^1 + R_2 \langle x - l \rangle^1 \qquad (3)$$

The reactions R_1 and R_2 can be found by taking a summation of moments and forces as usual, *or* they can be found by noting that the shear force and bending moment must be zero everywhere except in the region $0 \le x \le l$. This means that Eq. (2) should give $V = 0$ at x slightly larger than l. Thus

$$R_1 - F_1 - F_2 + R_2 = 0 \qquad (4)$$

Since the bending moment should also be zero in the same region, we have, from Eq. (3),

$$R_1 l - F_1(l - a_1) - F_2(l - a_2) = 0 \qquad (5)$$

Equations (4) and (5) can now be solved for the reactions R_1 and R_2.

EXAMPLE 4–3   Figure 4–7a shows the loading diagram for a beam cantilevered at A with a uniform load of 20 lbf/in acting on the portion 3 in $\le x \le 7$ in, and a concentrated counterclockwise moment of 240 lbf · in at $x = 10$ in. Derive the shear-force and bending-moment relations, and the support reactions M_1 and R_1.

Solution   Following the procedure of Example 4–2, we find the load intensity function to be

$$q = -M_1 \langle x \rangle^{-2} + R_1 \langle x \rangle^{-1} - 20 \langle x - 3 \rangle^0 + 20 \langle x - 7 \rangle^0 - 240 \langle x - 10 \rangle^{-2} \qquad (1)$$

Note that the $20 \langle x - 7 \rangle^0$ term was necessary to "turn off" the uniform load at C. Integrating successively gives

$$V = -M_1 \langle x \rangle^{-1} + R_1 \langle x \rangle^0 - 20 \langle x - 3 \rangle^1 + 20 \langle x - 7 \rangle^1 - 240 \langle x - 10 \rangle^{-1} \qquad (2)$$

$$M = -M_1 \langle x \rangle^0 + R_1 \langle x \rangle^1 - 10 \langle x - 3 \rangle^2 + 10 \langle x - 7 \rangle^2 - 240 \langle x - 10 \rangle^0 \qquad (3)$$

Figure 4–7

(a) Loading diagram for a beam cantilevered at A.
(b) Shear-force diagram.
(c) Bending-moment diagram.

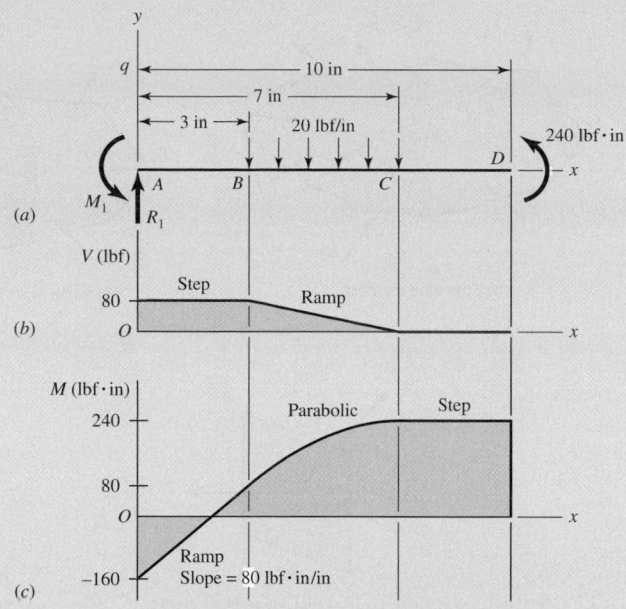

The reactions are found by making x slightly larger than 10 in, where both V and M are zero in this region. Equation (2) will then give

$$-M_1(0) + R_1(1) - 20(10 - 3) + 20(10 - 7) - 240(0) = 0$$

which yields $R_1 = 80$ lbf.

From Eq. (3) we get

$$-M_1(1) + 80(10) - 10(10 - 3)^2 + 10(10 - 7)^2 - 240(1) = 0$$

which yields $M_1 = 160$ lbf · in.

Figures 4–7b and c show the shear-force and bending-moment diagrams. Note that the impulse terms in Eq. (2), $-M_1\langle x\rangle^{-1}$ and $-240\langle x - 10\rangle^{-1}$, are physically not forces and are not shown in the V diagram. Also note that both the M_1 and 240 lbf · in moments are counterclockwise and negative singularity functions; however, by the convention shown in Fig. 4–3 the M_1 and 240 lbf · in are negative and positive bending moments, respectively, which is reflected in Fig. 4–7c.

4–4   **Stress**

Consider a general solid body loaded as shown in Fig. 4–8a. P_i are applied forces and R_i are possible support forces. To determine the state of stress at point Q in the body, it is necessary to expose a surface containing point Q. This is done by making a planar slice, or break, through the body intersecting Q. The orientation of this slice is arbitrary, but it is generally made in a convenient plane where the state of stress can be determined easily or where certain geometric relations can be utilized. The slice, illustrated in Fig. 4–8b, is arbitrarily oriented by the surface normal x. This establishes the yz plane. The external forces on the portion of the remaining body are shown, as well as the internal force distribution across the exposed surface. The units of the force distribution are force per unit area. In general, the force distribution will not be uniform across the surface, and will be neither normal nor tangential to the surface at a given point.

Figure 4–8

Internal force (stress) distributions.

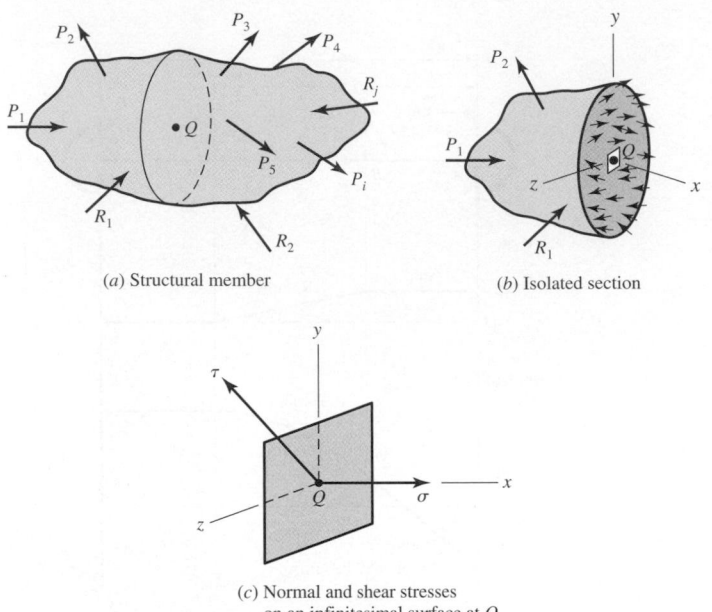

(a) Structural member

(b) Isolated section

(c) Normal and shear stresses
on an infinitesimal surface at Q

However, the force distribution at a point will have components in the normal and tangential directions giving rise to a *normal stress* and a *tangential shear stress,* respectively. Normal and shear stresses are labeled by the Greek symbols σ and τ, respectively, as shown in Fig. 4–8c. If the direction of σ is outward from the surface it is considered a *tensile stress* and is a positive normal stress. If σ is into the surface it is a *compressive stress* and commonly considered to be a negative quantity.

4–5 Cartesian Stress Components

The slice in Fig. 4–8b establishes the normal direction to the plane of the slice, the x direction, and from equilibrium the normal and shear stress distributions are established. Figure 4–9 shows the normal and net shear stresses on an infinitesimal surface area at point Q. The normal stress is labeled as σ_x. The symbol σ indicates a normal stress and the subscript x indicates the direction of the surface normal. For Cartesian components, it is necessary to establish the y direction, which again is arbitrary. Once this is established, the z direction follows directly for a right-handed rectangular Cartesian coordinate system. The net shear stress $(\tau_x)_{\text{net}}$ can then be resolved into components in the y and z directions labeled as τ_{xy} and τ_{xz}, respectively (see Fig. 4–9). Note that double subscripts are necessary for the shear. The first subscript indicates the direction of the surface normal whereas the second subscript is the direction of the stress.

To completely describe the state of stress at a point it would be necessary to examine all surfaces by making different planar slices through the point. Since different planar slices would necessitate different coordinates and different free-body diagrams, the stresses on each slice would be, in general, quite different. Then to understand the complete state of stress at a point, every possible surface intersecting the point should be examined. However, this would require an infinite number of slices intersecting the point. That is, if the point were to be completely isolated from the body, it would be described by an infinitesimal sphere. Naturally, this would be impossible to do, and it is also unnecessary since there is a simple method for accomplishing the same result, called

Figure 4-9

Stress components on surface normal to x direction.

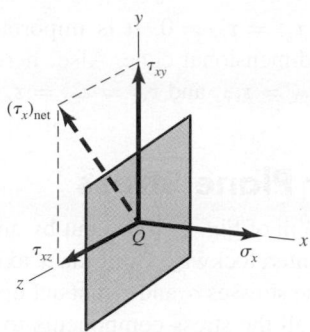

Figure 4-10

(a) General three-dimensional stress. (b) Plane stress with "cross-shears" equal.

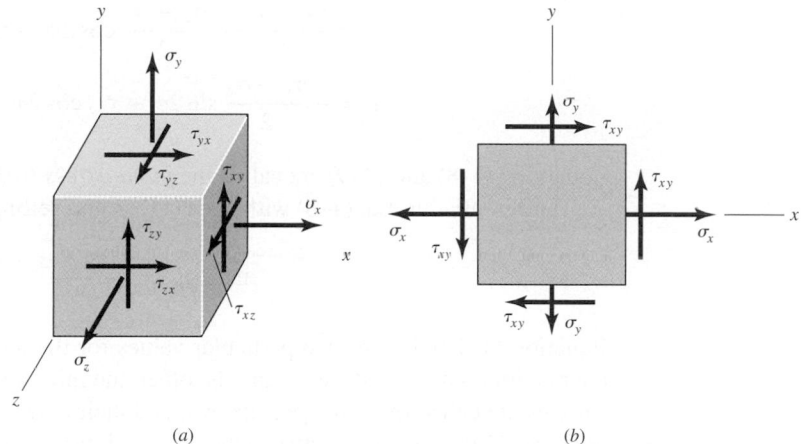

(a)   (b)

coordinate transformation or *Mohr's circles*. The method is described in the next two sections.

Returning to the general body of Fig. 4–8a, a planar slice can be made through point Q perpendicular to the y direction. The procedure for describing the state of stress on the infinitesimal area normal to the y direction is identical to the technique used on Fig. 4–9. However, the stresses on the surface whose normal is in the y direction will be designated as σ_y, τ_{yx}, and τ_{yz} (see Fig. 4–10a). A third orthogonal slice is made perpendicular to the z direction, and the resulting stresses are σ_z, τ_{zx}, and τ_{zy}. Thus, the state of stress at a point described by three mutually perpendicular surfaces is shown in Fig. 4–10a. It can be shown through coordinate transformation that this is sufficient to determine the state of stress on *any* surface intersecting the point. As the dimensions of the cube in Fig. 4–10a approach zero, the stresses on the hidden faces become equal and opposite to those on the opposing visible faces.

In general, a complete state of stress is defined by nine stress components, σ_x, σ_y, σ_z, τ_{xy}, τ_{xz}, τ_{yx}, τ_{yz}, τ_{zx}, and τ_{zy}. For equilibrium, "cross-shears" are equal, hence

$$\tau_{yx} = \tau_{xy} \qquad \tau_{zy} = \tau_{yz} \qquad \tau_{xz} = \tau_{zx} \tag{4-7}$$

This reduces the number of stress components for most three-dimensional states of stress from nine to six quantities, σ_x, σ_y, σ_z, τ_{xy}, τ_{yz}, and τ_{zx}.

A very common state of stress occurs when the stresses on one surface are zero. When this occurs the state of stress is called *plane stress*. Figure 4–10b shows a state of plane stress, arbitrarily assuming that the normal for the stress-free surface is the

z direction such that $\sigma_z = \tau_{zx} = \tau_{zy} = 0$. It is important to note that the element in Fig. 4–10b is still a three-dimensional cube. Also, here it is assumed that the cross-shears are equal such that $\tau_{yx} = \tau_{xy}$, and $\tau_{yz} = \tau_{zy} = \tau_{xz} = \tau_{zx} = 0$.

4–6   Mohr's Circle for Plane Stress

Suppose the $dx\ dy\ dz$ element of Fig. 4–10b is cut by an oblique plane with a normal n at an arbitrary angle ϕ counterclockwise from the x axis as shown in Fig. 4–11. This section is concerned with the stresses σ and τ that act upon this oblique plane. By summing the forces caused by all the stress components to zero, the stresses σ and τ are found to be

$$\sigma = \frac{\sigma_x + \sigma_y}{2} + \frac{\sigma_x - \sigma_y}{2} \cos 2\phi + \tau_{xy} \sin 2\phi \qquad (4\text{–}8)$$

$$\tau = -\frac{\sigma_x - \sigma_y}{2} \sin 2\phi + \tau_{xy} \cos 2\phi \qquad (4\text{–}9)$$

Equations (4–8) and (4–9) are called the *plane-stress transformation equations.*

Differentiating Eq. (4–8) with respect to ϕ and setting the result equal to zero gives

$$\tan 2\phi_p = \frac{2\tau_{xy}}{\sigma_x - \sigma_y} \qquad (4\text{–}10)$$

Equation (4–10) defines two particular values for the angle $2\phi_p$, one of which defines the maximum normal stress σ_1 and the other, the minimum normal stress σ_2. These two stresses are called the *principal stresses,* and their corresponding directions, the *principal directions.* The angle between the principal directions is 90°. It is important to note that Eq. (4–10) can be written in the form

$$\frac{\sigma_x - \sigma_y}{2} \sin 2\phi_p - \tau_{xy} \cos 2\phi_p = 0 \qquad (a)$$

Comparing this with Eq. (4–9), we see that $\tau = 0$, meaning that the *surfaces containing principal stresses have zero shear stresses.*

In a similar manner, we differentiate Eq. (4–9), set the result equal to zero, and obtain

$$\tan 2\phi_s = -\frac{\sigma_x - \sigma_y}{2\tau_{xy}} \qquad (4\text{–}11)$$

| Figure 4–11

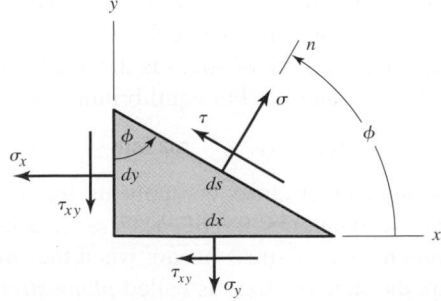

Equation (4–11) defines the two values of $2\phi_s$ at which the shear stress τ reaches an extreme value. The angle between the surfaces containing the maximum shear stresses is 90°. Equation (4–11) can also be written as

$$\frac{\sigma_x - \sigma_y}{2} \cos 2\phi_p + \tau_{xy} \sin 2\phi_p = 0 \qquad (b)$$

Substituting this into Eq. (4–8) yields

$$\sigma = \frac{\sigma_x + \sigma_y}{2} \qquad (4\text{--}12)$$

Equation (4–12) tells us that the two surfaces containing the maximum shear stresses also contain equal normal stresses of $(\sigma_x + \sigma_y)/2$.

Comparing Eqs. (4–10) and (4–11), we see that $\tan 2\phi_s$ is the negative reciprocal of $\tan 2\phi_p$. This means that $2\phi_s$ and $2\phi_p$ are angles 90° apart, and thus the angles between the surfaces containing the maximum shear stresses and the surfaces containing the principal stresses are $\pm 45°$.

Formulas for the two principal stresses can be obtained by substituting the angle $2\phi_p$ from Eq. (4–10) in Eq. (4–8). The result is

$$\sigma_1, \sigma_2 = \frac{\sigma_x + \sigma_y}{2} \pm \sqrt{\left(\frac{\sigma_x - \sigma_y}{2}\right)^2 + \tau_{xy}^2} \qquad (4\text{--}13)$$

In a similar manner the two extreme-value shear stresses are found to be

$$\tau_1, \tau_2 = \pm \sqrt{\left(\frac{\sigma_x - \sigma_y}{2}\right)^2 + \tau_{xy}^2} \qquad (4\text{--}14)$$

Your particular attention is called to the fact that an extreme value of the shear stress *may not be the same as the maximum value*. See Sec. 4–7.

It is important to note that the equations given to this point are quite sufficient for performing any plane stress transformation. However, extreme care must be exercised when applying them. For example, say you are attempting to determine the principal state of stress for a problem where $\sigma_x = 14$ MPa, $\sigma_y = -10$ MPa, and $\tau_{xy} = -16$ MPa. Equation (4–10) yields $\phi_p = -26.57°$ and $63.43°$ to locate the principal stress surfaces, whereas, Eq. (4–13) gives $\sigma_1 = 22$ MPa and $\sigma_2 = -18$ MPa for the principal stresses. If all we wanted was the principal stresses, we would be finished. However, what if we wanted to draw the element containing the principal stresses properly oriented relative to the x, y axes? Well, we have two values of ϕ_p and two values for the principal stresses. How do we know which value of ϕ_p corresponds to which value of the principal stress? To clear this up we would need to substitute the value of ϕ_p into Eq. (4–8) to determine the normal stress corresponding to that angle.

A graphical method for expressing the relations developed in this section, called *Mohr's circle diagram,* is a very effective means of visualizing the stress state at a point and keeping track of the directions of the various components associated with plane stress. Equations (4–8) and (4–9) can be shown to be a set of parametric equations for σ and τ, where the parameter is 2ϕ. The relationship between σ and τ is that of a circle plotted in the σ, τ plane, where the center of the circle is located at $C = (\sigma, \tau) = [(\sigma_x + \sigma_y)/2, 0]$ and has a radius of $R = \sqrt{[(\sigma_x - \sigma_y)/2]^2 + \tau_{xy}^2}$. A problem arises in the sign of the shear stress. The transformation equations are based on a positive ϕ being counterclockwise, as shown in Fig. 4–11. If a positive τ were plotted above the σ axis, points would rotate clockwise on the circle 2ϕ in the opposite direction of rotation on the element. It would be convenient if the rotations were in the same direction. One could solve the problem easily by plotting positive τ below the axis. However, the classical approach to Mohr's circle uses a different convention for the shear stress.

Mohr's Circle Shear Convention

This convention is followed in drawing Mohr's circle:

- Shear stresses tending to rotate the element clockwise (cw) are plotted *above* the σ axis.
- Shear stresses tending to rotate the element counterclockwise (ccw) are plotted *below* the σ axis.

For example, consider the right face of the element in Fig. 4–10b. By Mohr's circle convention the shear stress shown is plotted *below* the σ axis because it tends to rotate the element counterclockwise. The shear stress on the top face of the element is plotted *above* the σ axis because it tends to rotate the element clockwise.

In Fig. 4–12 we create a coordinate system with normal stresses plotted along the abscissa and shear stresses plotted as the ordinates. On the abscissa, tensile (positive) normal stresses are plotted to the right of the origin O and compressive (negative) normal stresses to the left. On the ordinate, clockwise (cw) shear stresses are plotted up; counterclockwise (ccw) shear stresses are plotted down.

Using the stress state of Fig. 4–10b, we plot Mohr's circle, Fig. 4–12, by first looking at the right surface of the element containing σ_x to establish the sign of σ_x and the cw or ccw direction of the shear stress. The right face is called the *x face* where $\phi = 0°$. If σ_x is positive and the shear stress τ_{xy} is ccw as shown in Fig. 4–10b, we can establish point A with coordinates $(\sigma_x, \tau_{xy}^{ccw})$ in Fig. 4–12. Next, we look at the top y *face,* where

Figure 4–12

Mohr's circle diagram.

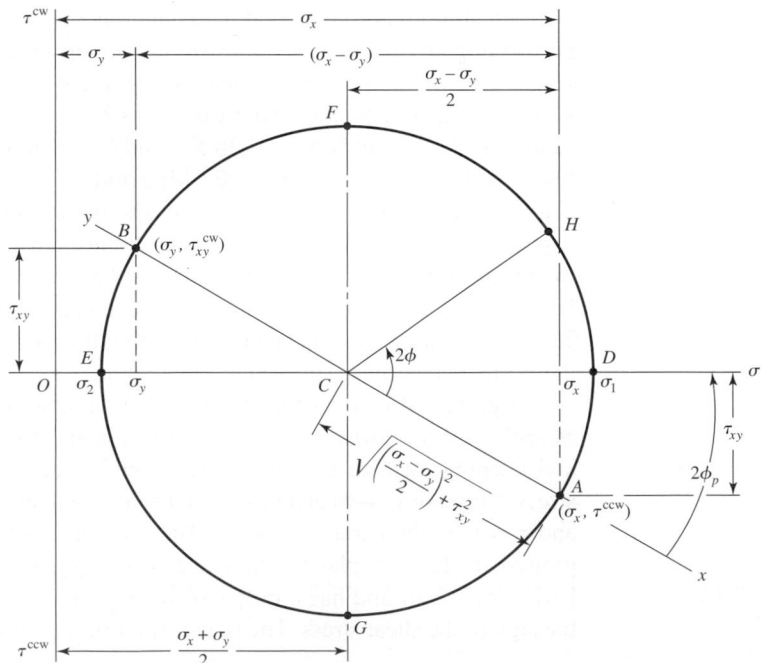

$\phi = 90°$, which contains σ_y, and repeat the process to obtain point B with coordinates $(\sigma_y, \tau_{xy}^{cw})$ as shown in Fig. 4–12. The two states of stress for the element are $\Delta\phi = 90°$ from each other on the element so they will be $2\Delta\phi = 180°$ from each other on Mohr's circle. Points A and B are the same vertical distance from the σ axis. Thus, AB must be on the diameter of the circle, and the center of the circle C is where AB intersects the σ axis. With points A and B on the circle, and center C, the complete circle can then be drawn. Note that the extended ends of line AB are labeled x and y as references to the normals to the surfaces for which points A and B represent the stresses.

The entire Mohr's circle represents the state of stress at a *single* point in a structure. Each point on the circle represents the stress state for a *specific* surface intersecting the point in the structure. Each pair of points on the circle 180° apart represent the state of stress on an element whose surfaces are 90° apart. Once the circle is drawn, the states of stress can be visualized for various surfaces intersecting the point being analyzed. For example, the principal stresses σ_1 and σ_2 are points D and E, respectively, and their values obviously agree with Eq. (4–13). We also see that the shear stresses are zero on the surfaces containing σ_1 and σ_2. The two extreme-value shear stresses, one clockwise and one counterclockwise, occur at F and G with magnitudes equal to the radius of the circle. The surfaces at F and G each also contain normal stresses of $(\sigma_x + \sigma_y)/2$ as noted earlier in Eq. (4–12). Finally, the state of stress on an arbitrary surface located at an angle ϕ counterclockwise from the x face is point H.

At one time, Mohr's circle was used graphically where it was drawn to scale very accurately and values were measured by using a scale and protractor. Here, we are strictly using Mohr's circle as a visualization aid and will use a semigraphical approach, calculating values from the properties of the circle. This is illustrated by the following example.

EXAMPLE 4–4   A stress element has $\sigma_x = 80$ MPa and $\tau_{xy} = -50$ MPa cw, as shown in Fig. 4–13a.

(*a*) Using Mohr's circle, find the principal stresses and directions, and show these on a stress element correctly aligned with respect to the xy coordinates. Draw another stress element to show τ_1 and τ_2, find the corresponding normal stresses, and label the drawing completely.

(*b*) Repeat part *a* using an algebraic approach.

Solution   (*a*) In the semigraphical approach used here, we first make an approximate freehand sketch of Mohr's circle and then use the geometry of the figure to obtain the desired information.

Draw the σ and τ axes first (Fig. 4–13b) and from the x face locate $\sigma_x = 80$ MPa along the σ axis. On the x face of the element, we see that the shear stress is 50 MPa in the cw direction. Thus, for the x face, this establishes point A (80, 50^{cw}) MPa. Corresponding to the y face, the stress is $\sigma = 0$ and $\tau = 50$ MPa in the ccw direction. This locates point B (0, 50^{ccw}) MPa. The line AB forms the diameter of the required circle, which can now be drawn. The intersection of the circle with the σ axis defines σ_1 and σ_2 as shown. Now, noting the triangle ACD, indicate on the sketch the length of the legs AD and CD as 50 and 40 MPa, respectively. The length of the hypotenuse AC is

$$\tau_1 = \sqrt{(50)^2 + (40)^2} = 64.0 \text{ MPa}$$

and this should be labeled on the sketch too. Since intersection C is 40 MPa from the origin, the principal stresses are now found to be

$$\sigma_1 = 40 + 64 = 104 \text{ MPa} \qquad \text{and} \qquad \sigma_2 = 40 - 64 = -24 \text{ MPa}$$

Figure 4–13

All stresses in MPa.

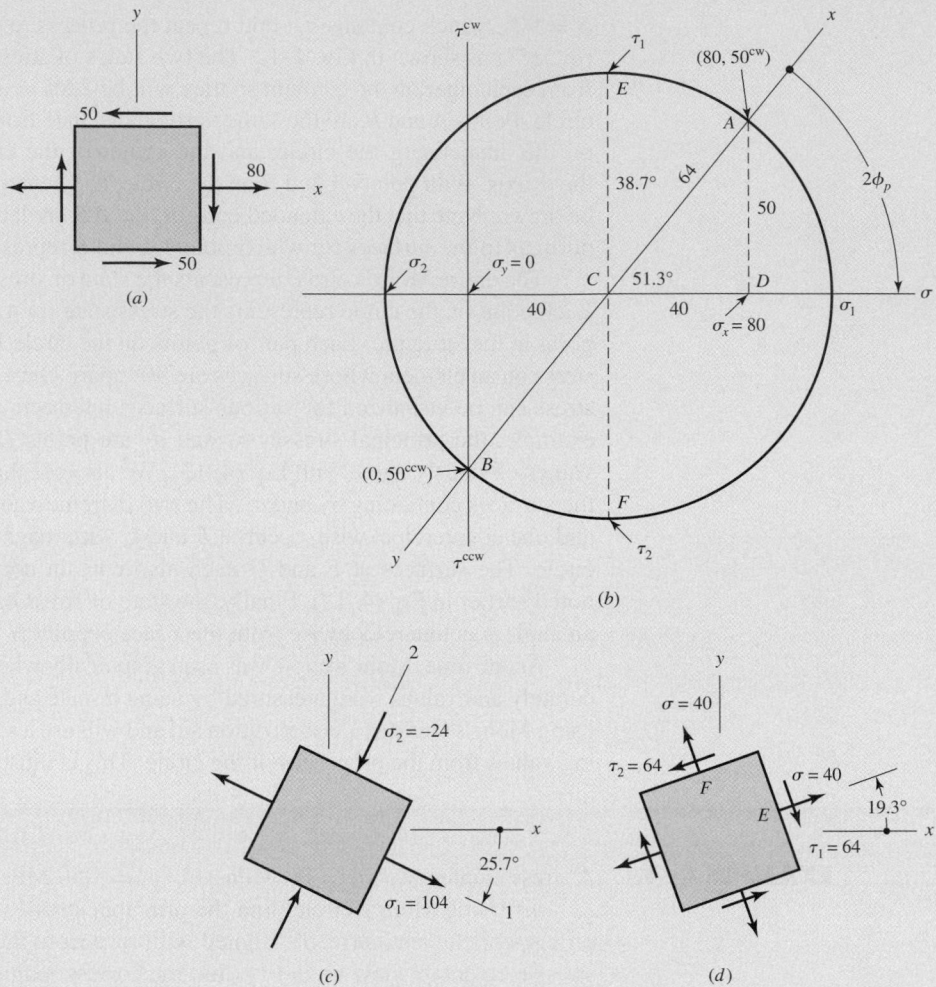

(a)

(b)

(c)

(d)

The angle 2ϕ from the x axis cw to σ_1 is

$$2\phi_p = \tan^{-1} \tfrac{50}{40} = 51.3°$$

To draw the principal stress element (Fig. 4–13c), sketch the x and y axes parallel to the original axes. The angle ϕ_p on the stress element must be measured in the *same* direction as is the angle $2\phi_p$ on the Mohr circle. Thus, from x measure 25.7° (half of 51.3°) clockwise to locate the σ_1 axis. The σ_2 axis is 90° from the σ_1 axis and the stress element can now be completed and labeled as shown. Note that there are *no* shear stresses on this element.

The two maximum shear stresses occur at points E and F in Fig. 4–13b. The two normal stresses corresponding to these shear stresses are each 40 MPa, as indicated. Point E is 38.7° ccw from point A on Mohr's circle. Therefore, in Fig. 4–13d, draw a stress element oriented 19.3° (half of 38.7°) ccw from x. The element should then be labeled with magnitudes and directions as shown.

In constructing these stress elements it is important to indicate the x and y directions of the original reference system. This completes the link between the original machine element and the orientation of its principal stresses.

(b) An algebraic approach is programmable. From Eq. (4–10),

$$\phi_p = \frac{1}{2}\tan^{-1}\left(\frac{2\tau_{xy}}{\sigma_x - \sigma_y}\right) = \frac{1}{2}\tan^{-1}\left(\frac{2(-50)}{80}\right) = -25.7°, 64.3°$$

From Eq. (4–8), for the first angle $\phi_p = -25.7°$,

$$\sigma = \frac{80+0}{2} + \frac{80-0}{2}\cos[2(-25.7)] + (-50)\sin[2(-25.7)] = 104.03 \text{ MPa}$$

The shear on this surface is obtained from Eq. (4–9) as

$$\tau = -\frac{80-0}{2}\sin[2(-25.7)] + (-50)\cos[2(-25.7)] = 0 \text{ MPa}$$

which confirms that 104.03 MPa is a principal stress. From Eq. (4–8), for $\phi_p = 64.3°$,

$$\sigma = \frac{80+0}{2} + \frac{80-0}{2}\cos[2(64.3)] + (-50)\sin[2(64.3)] = -24.03 \text{ MPa}$$

Substituting $\phi_p = 64.3°$ into Eq. (4–9) again yields $\tau = 0$, indicating that -24.03 MPa is also a principal stress. Once the principal stresses are calculated they can be ordered such that $\sigma_1 \geq \sigma_2$. Thus, $\sigma_1 = 104.03$ MPa and $\sigma_2 = -24.03$ MPa.

Since for $\sigma_1 = 104.03$ MPa, $\phi_p = -25.7°$, and since ϕ is defined positive ccw in the transformation equations, we rotate *clockwise* 25.7° for the surface containing σ_1. We see in Fig. 4–13c that this totally agrees with the semigraphical method.

To determine τ_1 and τ_2, we first use Eq. (4–11) to calculate ϕ_s:

$$\phi_s = \frac{1}{2}\tan^{-1}\left(-\frac{\sigma_x - \sigma_y}{2\tau_{xy}}\right) = \frac{1}{2}\tan^{-1}\left(-\frac{80}{2(-50)}\right) = 19.3°, 109.3°$$

For $\phi_s = 19.3°$, Eqs. (4–8) and (4–9) yield

$$\sigma = \frac{80+0}{2} + \frac{80-0}{2}\cos[2(19.3)] + (-50)\sin[2(19.3)] = 40.0 \text{ MPa}$$

$$\tau = -\frac{80-0}{2}\sin[2(19.3)] + (-50)\cos[2(19.3)] = -64.0 \text{ MPa}$$

Remember that Eqs. (4–8) and (4–9) are *coordinate* transformation equations. Imagine that we are rotating the x, y axes 19.3° counterclockwise and y will now point up and to the left. So a negative shear stress will point down and to the right as shown in Fig. 4–13d. Thus again, results agree with the semigraphical method.

For $\phi_s = 109.3°$, Eqs. (4–8) and (4–9) give $\sigma = 40.0$ MPa and $\tau = +64.0$ MPa. Using the same logic for the coordinate transformation we find that results again agree with Fig. 4–13d.

4–7 General Three-Dimensional Stress

As in the case of plane stress, a particular orientation of a stress element occurs in space for which all shear-stress components are zero. When an element has this particular orientation, the normals to the faces correspond to the principal directions, and the normal

Figure 4–14

Mohr's circles for three-dimensional stress.

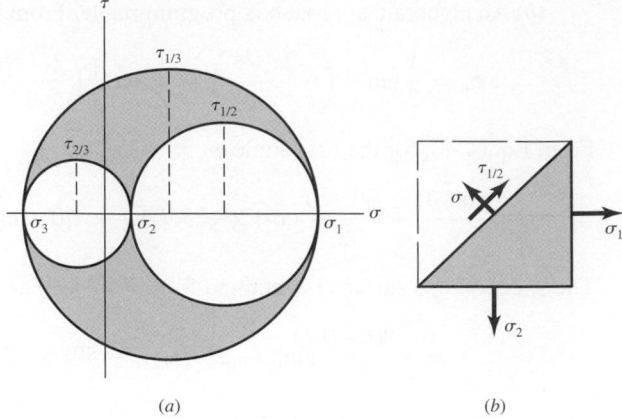

stresses associated with these faces are the principal stresses. Since there are six faces, there are three principal directions and three principal stresses σ_1, σ_2, and σ_3.

In our studies of plane stress we were able to specify any stress state σ_x, σ_y, and τ_{xy} and find the principal stresses and principal directions. But six components of stress are required to specify a general state of stress in three dimensions, and the problem of determining the principal stresses and directions is more difficult. In design, three-dimensional transformations are rarely performed since most maximum stress states occur under plane stress conditions. One notable exception is contact stress, which is not a case of plane stress, where the three principal stresses are given in Sec. 4–20. In fact, *all* states of stress are truly three-dimensional, where they might be described one- or two-dimensionally with respect to *specific* coordinate axes. Here it is most important to understand the relationship amongst the *three* principal stresses. The process in finding the three principal stresses from the six stress components σ_x, σ_y, σ_z, τ_{xy}, τ_{yz}, and τ_{zx}, involves finding the roots of the cubic equation[2]

$$\sigma^3 - (\sigma_x + \sigma_y + \sigma_z)\sigma^2 + (\sigma_x\sigma_y + \sigma_x\sigma_z + \sigma_y\sigma_z - \tau_{xy}^2 - \tau_{yz}^2 - \tau_{zx}^2)\sigma$$
$$- (\sigma_x\sigma_y\sigma_z + 2\tau_{xy}\tau_{yz}\tau_{zx} - \sigma_x\tau_{yz}^2 - \sigma_y\tau_{zx}^2 - \sigma_z\tau_{xy}^2) = 0 \qquad (4\text{--}15)$$

In plotting Mohr's circles for three-dimensional stress, the principal normal stresses are ordered so that $\sigma_1 \geq \sigma_2 \geq \sigma_3$. Then the result appears as in Fig. 4–14a. The stress coordinates σ, τ for any arbitrarily located plane will always lie on the boundaries or within the shaded area.

Figure 4–14a also shows the three *principal shear stresses* $\tau_{1/2}$, $\tau_{2/3}$, and $\tau_{1/3}$.[3] Each of these occurs on the two planes, one of which is shown in Fig. 4–14b. The figure shows that the principal shear stresses are given by the equations

$$\tau_{1/2} = \frac{\sigma_1 - \sigma_2}{2} \qquad \tau_{2/3} = \frac{\sigma_2 - \sigma_3}{2} \qquad \tau_{1/3} = \frac{\sigma_1 - \sigma_3}{2} \qquad (4\text{--}16)$$

---

[2]For development of this equation and further elaboration of three-dimensional stress transformations see: Richard G. Budynas, *Advanced Strength and Applied Stress Analysis,* 2nd ed., McGraw-Hill, New York, 1999, pp. 46–78.

[3]Note the difference between this notation and that for a shear stress, say, τ_{xy}. The use of the shilling mark is not accepted practice, but it is used here to emphasize the distinction.

Of course, $\tau_{max} = \tau_{1/3}$ when the normal principal stresses are ordered ($\sigma_1 > \sigma_2 > \sigma_3$), so always order your principal stresses. Do this in any computer code you generate and you'll always generate τ_{max}.

4–8 Elastic Strain

Normal strain ϵ is defined and discussed in Sec. 3–1 for the tensile specimen. The elongation per unit length of the bar is called *unit elongation or strain,* and is given by

$$\epsilon = \frac{\delta}{l} \tag{4–17}$$

where δ is the total elongation of the bar within the length l. Hooke's law for the tensile specimen is given by Eq. (3–3) as

$$\sigma = E\epsilon \tag{4–18}$$

where the constant E is called *Young's modulus* or the *modulus of elasticity.*

By substituting $\sigma = F/A$ and $\epsilon = \delta/l$ into Eq. (4–18) and rearranging, we obtain the equation for the total elongation of a bar loaded axially in tension as

$$\delta = \frac{Fl}{AE} \tag{a}$$

This equation also applies to the contraction due to axial compression where F is negative.

Experiments demonstrate that when a material is placed in tension, there exists not only an axial strain, but also a negative lateral strain (contraction). Poisson demonstrated that these two strains are proportional to each other within the range of Hooke's law. This constant is expressed as

$$\nu = -\frac{\text{lateral strain}}{\text{axial strain}} \tag{4–19}$$

and is known as *Poisson's ratio*, which is about 0.3 for most structural metals. This relation also applies for axial compression, except that a lateral expansion takes place instead. See Table A–5 for values of Poisson's ratio ν for common materials.

Shear strain γ is the change in a right angle of a stress element when subjected to pure shear stress, and Hooke's law for shear is given by

$$\tau = G\gamma \tag{4–20}$$

where the constant G is the *shear modulus of elasticity* or *modulus of rigidity.*

It can be shown for a linear, isotropic, homogeneous material, the three elastic constants are related to each other as follows by

$$E = 2G(1 + \nu) \tag{4–21}$$

There are many experimental techniques that can be used to measure strain. Thus, if the relationship between stress and strain is known, the stress state at a point can be calculated after the state of strain has been measured. We define the principal strains as the normal strains in the direction of the principal normal stresses. It is true that the shear strains are zero, just as the shear stresses are zero, on an element aligned in the principal directions. Table 4–2 contains the relations for all three types of normal stress states.

Table 4–2

Elastic Stress-Strain Relations

Type of Stress	Principal Strains	Principal Stresses
Uniaxial	$\epsilon_1 = \dfrac{\sigma_1}{E}$	$\sigma_1 = E\epsilon_1$
	$\epsilon_2 = -\nu\epsilon_1$	$\sigma_2 = 0$
	$\epsilon_3 = -\nu\epsilon_1$	$\sigma_3 = 0$
Biaxial	$\epsilon_1 = \dfrac{\sigma_1}{E} - \dfrac{\nu\sigma_2}{E}$	$\sigma_1 = \dfrac{E(\epsilon_1 + \nu\epsilon_2)}{1 - \nu^2}$
	$\epsilon_2 = \dfrac{\sigma_2}{E} - \dfrac{\nu\sigma_1}{E}$	$\sigma_2 = \dfrac{E(\epsilon_2 + \nu\epsilon_1)}{1 - \nu^2}$
	$\epsilon_3 = -\dfrac{\nu\sigma_1}{E} - \dfrac{\nu\sigma_2}{E}$	$\sigma_3 = 0$
Triaxial	$\epsilon_1 = \dfrac{\sigma_1}{E} - \dfrac{\nu\sigma_2}{E} - \dfrac{\nu\sigma_3}{E}$	$\sigma_1 = \dfrac{E\epsilon_1(1 - \nu) + \nu E(\epsilon_2 + \epsilon_3)}{1 - \nu - 2\nu^2}$
	$\epsilon_2 = \dfrac{\sigma_2}{E} - \dfrac{\nu\sigma_1}{E} - \dfrac{\nu\sigma_3}{E}$	$\sigma_2 = \dfrac{E\epsilon_2(1 - \nu) + \nu E(\epsilon_1 + \epsilon_3)}{1 - \nu - 2\nu^2}$
	$\epsilon_3 = \dfrac{\sigma_3}{E} - \dfrac{\nu\sigma_1}{E} - \dfrac{\nu\sigma_2}{E}$	$\sigma_3 = \dfrac{E\epsilon_3(1 - \nu) + \nu E(\epsilon_1 + \epsilon_2)}{1 - \nu - 2\nu^2}$

Note 1: Plane stress situation requires one principal stress to be zero. Plane strain situation requires one principal strain associated with one principal stress to be zero.

Note 2: *Uniaxial, Biaxial,* and *Triaxial,* are terminology for special cases of one-dimensional, two-dimensional, and three-dimensional states of stress, respectively, where the orthogonal faces of an element contain normal stresses only and no shear stresses.

4–9 Uniformly Distributed Stresses

The assumption of a uniform distribution of stress is frequently made in design. The result is then often called *pure tension, pure compression,* or *pure shear,* depending upon how the external load is applied to the body under study. The word *simple* is sometimes used instead of *pure* to indicate that there are no other complicating effects. The tension rod is typical. Here a tension load F is applied through pins at the ends of the bar. The assumption of uniform stress means that if we cut the bar at a section remote from the ends and remove one piece, we can replace its effect by applying a uniformly distributed force of magnitude σA to the cut end. So the stress σ is said to be uniformly distributed. It is calculated from the equation

$$\sigma = \frac{F}{A} \tag{4–22}$$

This assumption of uniform stress distribution requires that:

- The bar be straight and of a homogeneous material
- The line of action of the force contains the centroid of the section
- The section be taken remote from the ends and from any discontinuity or abrupt change in cross section

For simple compression, Eq. (4–22) is applicable with F normally being considered a negative quantity. Also, a slender bar in compression may fail by buckling, and this possibility must be eliminated from consideration before Eq. (4–22) is used.[4]

Use of the equation

$$\tau = \frac{F}{A} \tag{4–23}$$

for a body, say, a bolt, in shear assumes a uniform stress distribution too. It is very difficult in practice to obtain a uniform distribution of shear stress. The equation is included because occasions do arise in which this assumption is utilized.

4–10 Normal Stresses for Beams in Bending

In deriving the relations for the normal bending stresses in straight beams, the following assumptions are made:

1 The beam is subjected to pure bending; this means that the shear force is zero, and that no torsion or axial loads are present.
2 The material is isotropic and homogeneous.
3 The material obeys Hooke's law.
4 The beam is initially straight with a cross section that is constant throughout the beam length.
5 The beam has an axis of symmetry in the plane of bending.
6 The proportions of the beam are such that it would fail by bending rather than by crushing, wrinkling, or sidewise buckling.
7 Plane cross sections of the beam remain plane during bending.

In Fig. 4–15a we visualize a portion of a straight beam acted upon by the positive bending moment M. The y axis is the axis of symmetry. The x axis is coincident with the *neutral axis* of the section, and the xz plane, which contains the neutral axes of all cross sections, is called the *neutral plane*. Elements of the beam coincident with this plane have zero strain. The location of the neutral axis with respect to the cross section has not yet been defined.

Figure 4–15

Straight beam in positive bending.

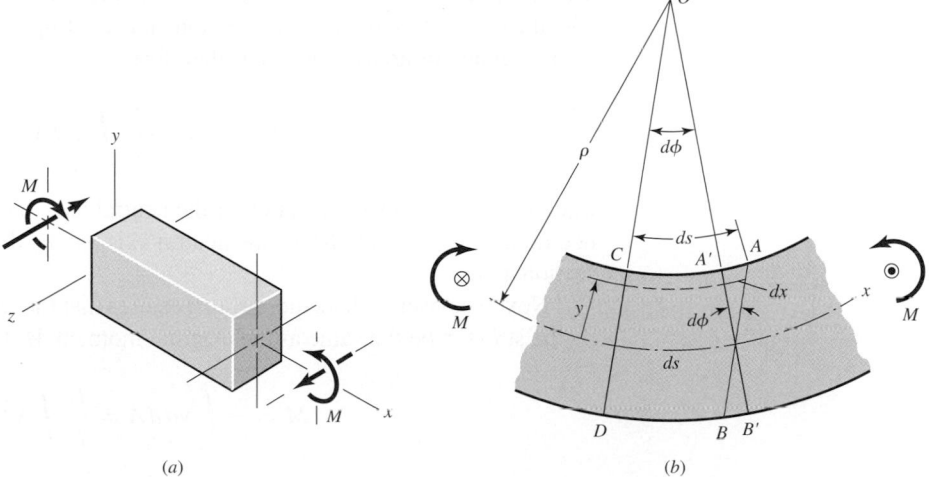

(a) (b)

---

[4]See Sec. 5–11.

Application of the positive moment will cause the upper surface of the beam to bend downward, and the neutral axis will then be curved, as shown in Fig. 4–15b. Because of the curvature, a section AB originally parallel to CD, since the beam was straight, will rotate through the angle $d\phi$ to $A'B'$. Since AB and $A'B'$ are both straight lines, we have utilized the assumption that plane sections remain plane during bending. If we now specify the radius of curvature of the neutral axis as ρ, the length of a differential element of the neutral axis as ds, and the angle subtended by the two adjacent sides CD and $A'B'$ as $d\phi$, then, from the definition of curvature, we have

$$\frac{1}{\rho} = \frac{d\phi}{ds} \tag{a}$$

As shown in Fig. 4–15b, the *elongation* of a "fiber" at distance y from the neutral axis is

$$dx = -y\, d\phi \tag{b}$$

The strain is the elongation divided by the original length, or

$$\epsilon = \frac{dx}{ds} \tag{c}$$

Solving Eqs. (a), (b), and (c) simultaneously gives

$$\epsilon = -\frac{y}{\rho} \tag{d}$$

Thus the strain is proportional to the distance y from the neutral axis. Now, since $\sigma = E\epsilon$, we have for the stress

$$\sigma = -\frac{Ey}{\rho} \tag{e}$$

We are now dealing with pure bending, which means that there are no axial forces acting on the beam. We can state this in mathematical form by summing all the horizontal forces acting on the cross section and equating this sum to zero. The force acting on an element of area dA is $\sigma\, dA$; therefore,

$$\int \sigma\, dA = -\frac{E}{\rho} \int y\, dA = 0 \tag{f}$$

Equation (f) defines the location of the neutral axis. The moment of the area about the neutral axis is zero, and hence the neutral axis passes through the centroid of the cross-sectional area.

Next we observe that equilibrium requires that the internal bending moment created by the stress σ be the same as the external moment M. In other words,

$$M = -\int y\sigma\, dA = \frac{E}{\rho} \int y^2\, dA \tag{g}$$

The second integral in Eq. (g) is the *second moment of area* about the z axis. This is

$$I = \int y^2\, dA \tag{4–24}$$

Figure 4–16

Bending stresses according to
Eq. (4–26).

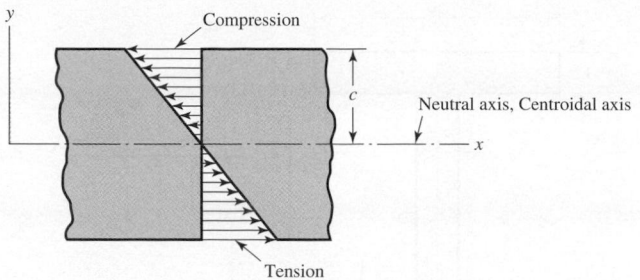

If we next solve Eqs. (*g*) and (4–24) and rearrange them, we have

$$\frac{1}{\rho} = \frac{M}{EI} \tag{4–25}$$

This is an important equation in the determination of the deflection of beams, and we shall employ it in Chap. 5. Finally, we eliminate ρ from Eqs. (*e*) and (4–25) and obtain

$$\sigma = -\frac{My}{I} \tag{4–26}$$

Equation (4–26) states that the bending stress σ is directly proportional to the distance y from the neutral axis and the bending moment M, as shown in Fig. 4–16. It is customary to designate $c = y_{max}$, to omit the negative sign, and to write

$$\sigma = \frac{Mc}{I} \tag{4–27}$$

where it is understood that Eq. (4–27) gives the maximum stress. Then tensile or compressive maximum stresses are determined by inspection when the sense of the moment is known.

Equation (4–27) is often written in the two alternative forms

$$\sigma = \frac{M}{I/c} = \frac{M}{Z} \tag{4–28}$$

where $Z = I/c$ is called the *section modulus*.

EXAMPLE 4–5

A beam having a T section with the dimensions shown in Fig. 4–17 is subjected to a bending moment of 1600 N · m that causes tension at the top surface. Locate the neutral axis and find the maximum tensile and compressive bending stresses.

Solution

The area of the composite section is $A = 1956 \text{ mm}^2$. Now divide the T section into two rectangles, numbered 1 and 2, and sum the moments of these areas about the top edge. We then have

$$1956c_1 = 12(75)(6) + 12(88)(56)$$

and hence $c_1 = 33.0$ mm. Therefore $c_2 = 100 - 33.0 = 67.0$ mm.

Figure 4–17

Dimensions in millimeters.

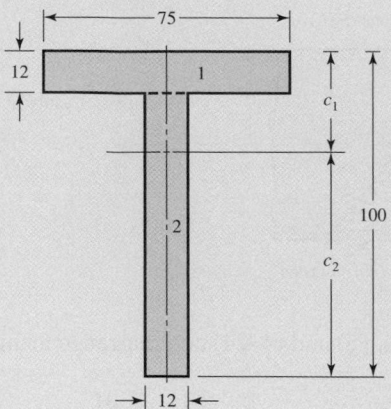

Next we calculate the second moment of area of each rectangle about its own centroidal axis. Using Table A–18, we find for the top rectangle

$$I_1 = \frac{bh^3}{12} = \frac{7.5(1.2)^3}{12} = 1.08 \text{ cm}^4$$

As indicated in Sec. 1–14, it is permissible to use nonpreferred prefixes when raising prefixed units to a power. In this case the use of cm in this equation instead of mm or m yields the shortest number string and, hence, is preferred. For the bottom rectangle, we have

$$I_2 = \frac{bh^3}{12} = \frac{1.2(8.8)^3}{12} = 68.15 \text{ cm}^4$$

We now employ the parallel-axis theorem to obtain the second moment of area of the composite figure about its own centroidal axis. This theorem states

$$I_x = I_{cg} + Ad^2$$

where I_{cg} is the second moment of area about its own centroidal axis and I_x is the second moment of area about any parallel axis a distance d removed. For the top rectangle, the distance is

$$d_1 = 33.0 - 6 = 27.0 \text{ mm}$$

and for the bottom rectangle,

$$d_2 = 67.0 - 44 = 23.0 \text{ mm}$$

Using the parallel-axis theorem for both rectangles, we now find that

$$I = [1.08 + 9(2.70)^2] + [68.15 + 10.56(2.30)^2]$$
$$= 190.7 \text{ cm}^4$$

Finally, the maximum tensile stress, which occurs at the top surface, is found to be

$$\sigma = \frac{Mc_1}{I} = \frac{1600(3.30)}{190.7} = 27.69 \text{ MPa}$$

Note that we have used Table A–3 in solving this equation. Similarly, the maximum compressive stress at the lower surface is found to be

Answer

$$\sigma = -\frac{Mc_2}{I} = -\frac{1600(6.70)}{190.7} = -56.21 \text{ MPa}$$

EXAMPLE 4–6   Determine the diameter for the solid round shaft, 18 in long, shown in Fig. 4–18a. The shaft is supported by self-aligning bearings at the ends. Mounted upon the shaft are a V-belt sheave, which contributes a radial load of 400 lbf to the shaft, and a gear, which contributes a radial load of 150 lbf. The two loads are in the same plane and have the same directions. The bending stress is not to exceed 10 kpsi.

Figure 4–18

(a) Shaft drawing, dimensions in inches; (b) loading diagram; (c) shear-force diagram; (d) bending-moment diagram.

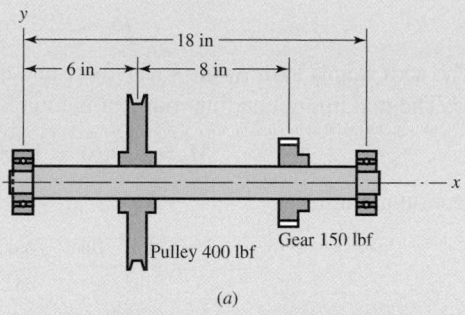

(a)

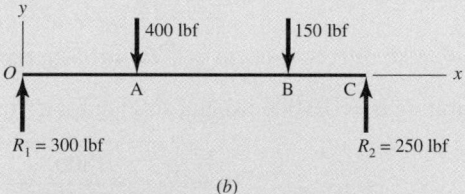

(b)

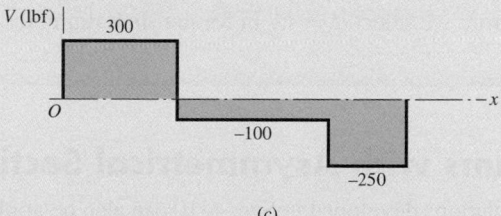

(c)

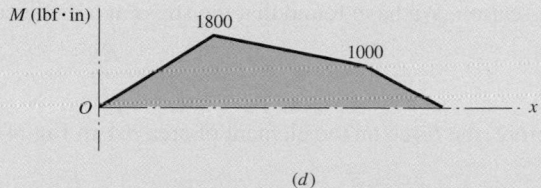

(d)

Solution
As indicated in Chap. 1, certain assumptions are necessary. In this problem we decide on the following conditions:

- The weight of the shaft is neglected.
- Since the bearings are self-aligning, the shaft is assumed to be simply supported and the loads and bearing reactions to be concentrated.
- The normal bending stress is assumed to govern the design at the location of the maximum bending moment.

The assumed loading diagram is shown in Fig. 4–18b. Summing moments about C gives

$$\sum M_C = -18R_1 + 12(400) + 4(150) = 0$$
$$R_1 = 300 \text{ lbf}$$

To get R_2, we can sum forces, or we can sum moments about point O. With either method, we find

$$R_2 = 250 \text{ lbf}$$

The next step is to draw the shear-force and bending-moment diagrams (Fig. 4–18c and d). The maximum bending moment is found to be

$$M = 300(6) = 1800 \text{ lbf} \cdot \text{in}$$

The section modulus is

$$Z = \frac{I}{c} = \frac{\pi d^4/64}{d/2} = \frac{\pi d^3}{32} = 0.0982d^3$$

Then, using Eq. (4–28),

$$\sigma = \frac{M}{Z} = \frac{1800}{0.0982d^3}$$

Substituting $\sigma = 10\,000$ psi and solving for d yields

$$d = \sqrt[3]{\frac{1800}{0.0982(10000)}} = 1.22 \text{ in}$$

Answer
Therefore we select $d = 1\frac{1}{4}$ in for the shaft diameter.

4–11 Beams with Asymmetrical Sections

The relations developed in Sec. 4–10 can also be applied to beams having asymmetrical sections, provided that the plane of bending coincides with one of the two principal axes of the section. We have found that the stress at a distance y from the neutral axis is

$$\sigma = -\frac{Ey}{\rho} \tag{a}$$

Therefore, the force on the element of area dA in Fig. 4–19 is

$$dF = \sigma \, dA = -\frac{Ey}{\rho} \, dA$$

| Figure 4–19

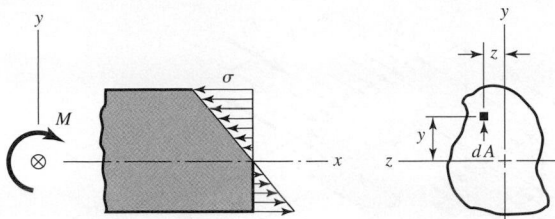

Taking moments of this force about the y axis and integrating across the section gives

$$M_y = \int z \, dF = \int \sigma z \, dA = -\frac{E}{\rho} \int yz \, dA \qquad (b)$$

We recognize that the last integral in Eq. (*b*) is the product of inertia I_{yz}. If the bending moment on the beam is in the plane of one of the principal axes, say the xy plane, then

$$I_{yz} = \int yz \, dA = 0 \qquad (c)$$

With this restriction, the relations developed in Sec. 4–10 hold for any cross-sectional shape. Of course, this means that the designer has a special responsibility to ensure that the bending loads do, in fact, come onto the beam in a principal plane!

4–12 Shear Stresses for Beams in Bending

Most beams have *both* shear forces and bending moments present. It is only occasionally that we encounter beams subjected to pure bending, that is to say, beams having zero shear force. And yet, the flexure formula was developed on the assumption of pure bending. As a matter of fact, the reason for assuming pure bending was simply to eliminate the complicating effects of shear force in the development. For engineering purposes, the flexure formula is valid no matter whether a shear force is present or not. For this reason, we shall utilize the same normal bending-stress distribution [Eqs. (4–26) and (4–27)] when shear forces are present too.

In Fig. 4–20, we show a beam of constant cross section subjected to a shear force V and a bending moment M. The direction of the bending moment is easier to visualize by associating the hollow vector with your right hand. The hollow vector points in the negative z direction. If you will place the thumb of your right hand in the negative z direction, then your fingers, when bent, will indicate the direction of the moment M. By Eq. (4–3), the relationship of V to M is

$$V = \frac{dM}{dx} \qquad (a)$$

At some point along the beam, we cut a transverse section of length dx at a distance y_1 above the neutral axis, as illustrated. We remove this section to study the forces that act. Because a shear force is present, the bending moment is changing as we move along the x axis. Thus, we can designate the bending moment as M on the near side of the section and as $M + dM$ on the far side. The moment M produces a normal stress σ, and the moment $M + dM$, a normal stress $\sigma + d\sigma$, as shown. These normal stresses produce normal forces on the vertical faces of the element, the compressive force on the far side being greater than on the near side. The resultant of these two would cause the section

Figure 4–20

Beam section isolation.
Note: Only forces in the
x direction are shown on
the dx element.

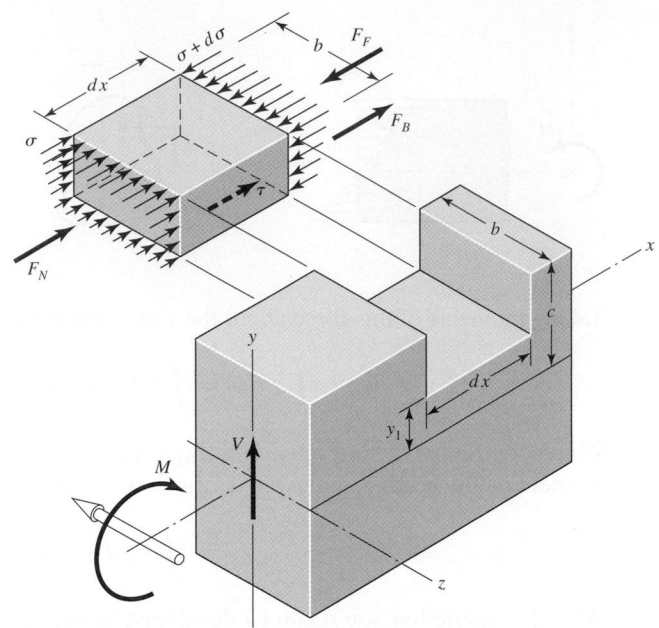

to tend to slide in the $-x$ direction, and so this resultant must be balanced by a shear force acting in the $+x$ direction on the bottom of the section. This shear force results in a shear stress τ, as shown. Thus, there are three resultant forces acting on the element in the x direction: F_N, due to σ, acts on the near face; F_F, due to $\sigma + d\sigma$, acts on the far face; and F_B, due to τ, acts on the bottom face. Let us evaluate these forces.

For the near face, select an element of area dA. The stress acting on this area is σ, and so the force is the stress times the area, or $\sigma\, dA$. The force acting on the entire near face is the sum of all the $\sigma\, dA$'s, or

$$F_N = \int_{y_1}^{c} \sigma\, dA \tag{b}$$

where the limits indicate that we integrate from the bottom $y = y_1$ to the top $y = c$. For $\sigma = My/I$ [Eq. (4–26)], Eq. (b) becomes

$$F_N = \frac{M}{I} \int_{y_1}^{c} y\, dA \tag{c}$$

Note that the negative sign in Eq. (4–26) is omitted, as it is assumed that σ is compression in Fig. 4–19.

The force on the far face is found in a similar manner. It is

$$F_F = \int_{y_1}^{c} (\sigma + d\sigma)\, dA = \frac{M + dM}{I} \int_{y_1}^{c} y\, dA \tag{d}$$

The force on the bottom face is the shear stress τ times the area of the bottom face. Since this area is $b\, dx$, we have

$$F_B = \tau b\, dx \tag{e}$$

Summing these three forces in the x direction gives

$$\sum F_x = +F_N - F_F + F_B = 0 \tag{f}$$

If we substitute Eqs. (c) and (d) for F_N and F_F and solve the result for F_B, we get

$$F_B = F_F - F_N = \frac{M + dM}{I} \int_{y_1}^{c} y \, dA - \frac{M}{I} \int_{y_1}^{c} y \, dA = \frac{dM}{I} \int_{y_1}^{c} y \, dA \qquad (g)$$

Next, using Eq. (e) for F_B and solving for the shear stress gives

$$\tau = \frac{dM}{dx} \frac{1}{Ib} \int_{y_1}^{c} y \, dA \qquad (h)$$

Then, by the use of Eq. (a), we finally get the shear-stress formula as

$$\tau = \frac{V}{Ib} \int_{y_1}^{c} y \, dA \qquad (4\text{–}29)$$

In this equation, the integral is the first moment of the area of the isolated vertical face about the neutral axis. This moment is usually designated Q. Thus,

$$Q = \int_{y_1}^{c} y \, dA = \bar{y}' A' \qquad (4\text{–}30)$$

where, for the isolated area from y_1 to c, $\bar{y}'$ is the distance from the y axis to the area centroid and A' is the area. With this final simplification, Eq. (4–29) may be written as

$$\tau = \frac{VQ}{Ib} \qquad (4\text{–}31)$$

In using this equation, note that b is the width of the section at the particular distance y_1 from the neutral axis. Also, I is the second moment of area of the *entire* section about the neutral axis.

EXAMPLE 4–7

A beam 12 in long is to support a load of 488 lbf acting 3 in from the left support, as shown in Fig. 4–21a. Basing the design only on bending stress, a designer has selected a 3-in aluminum channel with the cross-sectional dimensions shown. If the direct shear is neglected, the stress in the beam may be actually higher than the designer thinks. Determine the principal stresses considering bending and direct shear and compare them with that considering bending only.

Solution

The loading, shear-force, and bending-moment diagrams are shown in Fig. 4–21b. If the direct shear force is included in the analysis, the maximum stresses at the top and bottom of the beam will be the same as if only bending were considered. The maximum stresses are

$$\sigma = \pm \frac{Mc}{I} = \pm \frac{1098(1.5)}{1.66} = \pm 992 \text{ psi}$$

| Figure 4–21

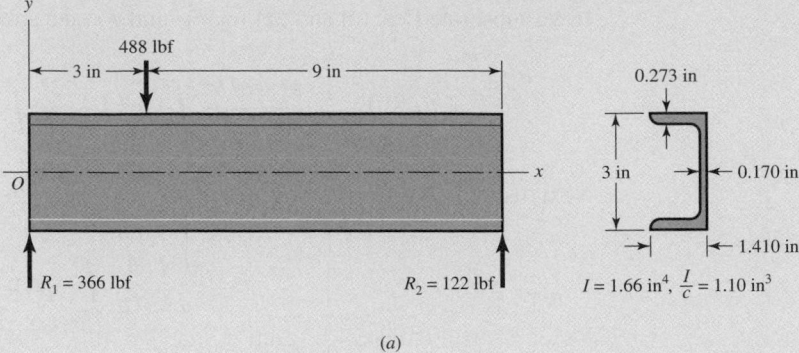

(a)

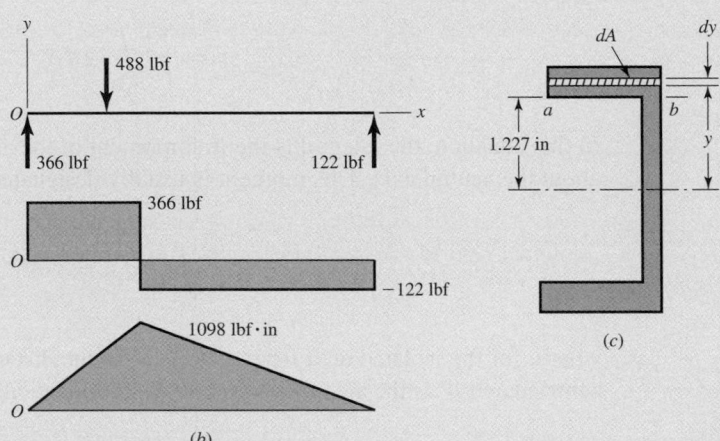

(b)

(c)

However, the maximum stress due to the combined bending and direct shear stresses may be maximum at the point $(3^-, 1.227)$ that is just to the left of the applied load, where the web joins the flange. To simplify the calculations we assume a cross section with square corners (Fig. 4–21c). The normal stress at section ab, with $x = 3$ in, is

$$\sigma = -\frac{My}{I} = -\frac{1098(1.227)}{1.66} = -812 \text{ psi}$$

For the shear stress at section ab, considering the area above ab and using Eq. (4–30) gives

$$Q = \bar{y}'A' = \frac{1}{2}(1.5 + 1.227)(1.410)(0.273) = 0.525 \text{ in}^3$$

Using Eq. (4–33) with $V = 366$ lbf, $I = 1.66$ in^4, $Q = 0.525$ in^3, and $b = 0.170$ in yields

$$\tau_{xy} = -\frac{VQ}{Ib} = -\frac{366(0.525)}{1.66(0.170)} = -681 \text{ psi}$$

The negative sign comes from recognizing that the shear stress is down on an x face of a $dx \, dy$ element at the location being considered.

The principal stresses at the point can now be determined. Using Eq. (4–13), we find that at $x = 3^-$ in, $y = 1.227$ in,

$$\sigma_1, \sigma_2 = \frac{\sigma_x + \sigma_y}{2} \pm \sqrt{\left(\frac{\sigma_x - \sigma_y}{2}\right)^2 + \tau_{xy}^2}$$

$$= \frac{-813 + 0}{2} \pm \sqrt{\left(\frac{-813 - 0}{2}\right)^2 + (-681)^2} = 387, -1200 \text{ psi}$$

For a point at $x = 3^-$ in, $y = -1.227$ in, the principal stresses are $\sigma_1, \sigma_2 = 1200$, -387 psi. Thus we see that the maximum principal stresses are ± 1200 psi, 21 percent higher than thought by the designer.

Shear Stresses in Standard-Section Beams

The shear stress distribution in a beam depends on how Q/b varies as a function of y_1. Here we will show how to determine the shear stress distribution for a beam with a rectangular cross section and provide results of maximum values of shear stress for other standard cross sections. Figure 4–22 shows a portion of a beam with a rectangular cross section, subjected to a shear force V and a bending moment M. As a result of the bending moment, a normal stress σ is developed on a cross section such as A-A, which is in compression above the neutral axis and in tension below. To investigate the shear stress at a distance y_1 above the neutral axis, we select an element of area dA at a distance y above the neutral axis. Then, $dA = b \, dy$, and so Eq. (4–30) becomes

$$Q = \int_{y_1}^{c} y \, dA = b \int_{y_1}^{c} y \, dy = \frac{by^2}{2}\bigg|_{y_1}^{c} = \frac{b}{2}\left(c^2 - y_1^2\right) \qquad (a)$$

Figure 4–22

Shear stresses in a rectangular beam.

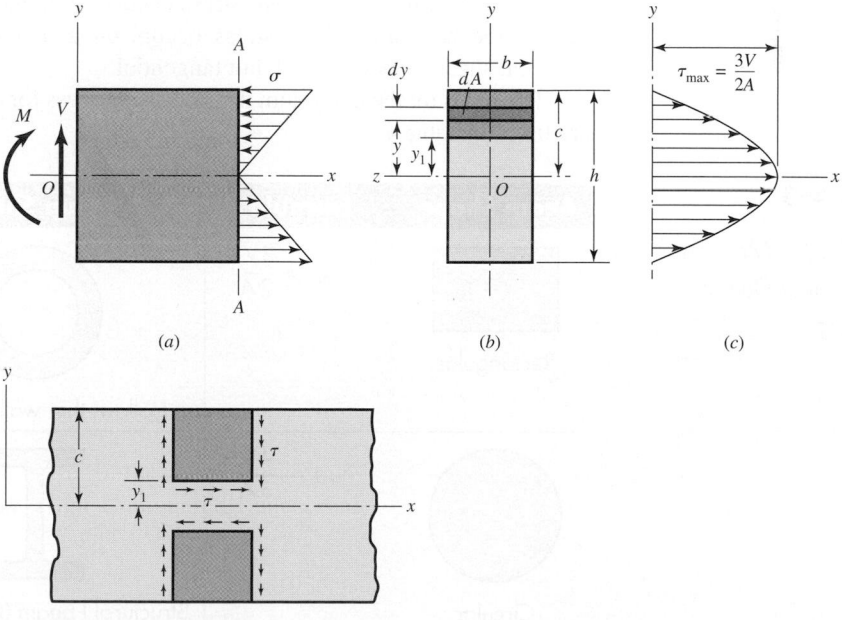

(a)

(b)

(c)

(d)

Substituting this value for Q into Eq. (4–31) gives

$$\tau = \frac{V}{2I}\left(c^2 - y_1^2\right) \tag{4–32}$$

This is the general equation for shear stress in a rectangular beam. To learn something about it, let us make some substitutions. From Table A–18, we learn that the second moment of area for a rectangular section is $I = bh^3/12$; substituting $h = 2c$ and $A = bh = 2bc$ gives

$$I = \frac{Ac^2}{3} \tag{b}$$

If we now use this value of I for Eq. (4–32) and rearrange, we get

$$\tau = \frac{3V}{2A}\left(1 - \frac{y_1^2}{c^2}\right) \tag{4–33}$$

We note that the maximum shear stress exists when $y_1 = 0$, which is at the bending neutral axis. Thus

$$\tau_{max} = \frac{3V}{2A} \tag{4–34}$$

for a rectangular section. As we move away from the neutral axis, the shear stress decreases parabolically until it is zero at the outer surfaces where $y_1 = \pm c$, as shown in Fig. 4–22c. It is particularly interesting and significant here to observe that the shear stress is maximum at the bending neutral axis, where the normal stress due to bending is zero, and that the shear stress is zero at the outer surfaces, where the bending stress is a maximum. Horizontal shear stress is always accompanied by vertical shear stress of the same magnitude, and so the distribution can be diagrammed as shown in Fig. 4–22d. Figure 4–22c shows that the shear τ on the vertical surfaces varies with y. We are almost always interested in the horizontal shear, τ in Fig. 4–22d, which is nearly uniform with constant y. The maximum horizontal shear occurs where the vertical shear is largest. This is usually at the neutral axis but may not be if the width b is smaller somewhere else. Furthermore, if the section is such that b can be minimized on a plane not horizontal, then the horizontal shear stress occurs on an inclined plane. For example, with tubing, the horizontal shear stress occurs on a radial plane and the corresponding "vertical shear" is not vertical, but tangential.

Formulas for the maximum flexural shear stress for the most commonly used shapes are listed in Table 4–3.

Table 4–3

Formulas for Maximum Shear Stress Due to Bending

Beam Shape	Formula	Beam Shape	Formula
Rectangular	$\tau_{max} = \dfrac{3V}{2A}$	Hollow, thin-walled round	$\tau_{max} = \dfrac{2V}{A}$
Circular	$\tau_{max} = \dfrac{4V}{3A}$	Structural I beam (thin-walled) — Web	$\tau_{max} = \dfrac{V}{A_{web}}$

4-13   **Torsion**

Any moment vector that is collinear with an axis of a mechanical element is called a *torque vector*, because the moment causes the element to be twisted about that axis. A bar subjected to such a moment is also said to be in *torsion*.

As shown in Fig. 4–23, the torque T applied to a bar can be designated by drawing arrows on the surface of the bar to indicate direction or by drawing torque-vector arrows along the axes of twist of the bar. Torque vectors are the hollow arrows shown on the x axis in Fig. 4–23. Note that they conform to the right-hand rule for vectors.

The *angle of twist,* in radians, for a solid round bar is

$$\theta = \frac{Tl}{GJ} \qquad (4\text{–}35)$$

where T = torque

l = length

G = modulus of rigidity

J = polar second moment of area

For a solid round bar, the shear stress is zero at the center and maximum at the surface. The distribution is proportional to the radius ρ and is

$$\tau = \frac{T\rho}{J} \qquad (4\text{–}36)$$

Designating r as the radius to the outer surface, we have

$$\tau_{\max} = \frac{Tr}{J} \qquad (4\text{–}37)$$

The assumptions used in the analysis are:

- The bar is acted upon by a pure torque, and the sections under consideration are remote from the point of application of the load and from a change in diameter.
- Adjacent cross sections originally plane and parallel remain plane and parallel after twisting, and any radial line remains straight.
- The material obeys Hooke's law.

| **Figure 4–23**

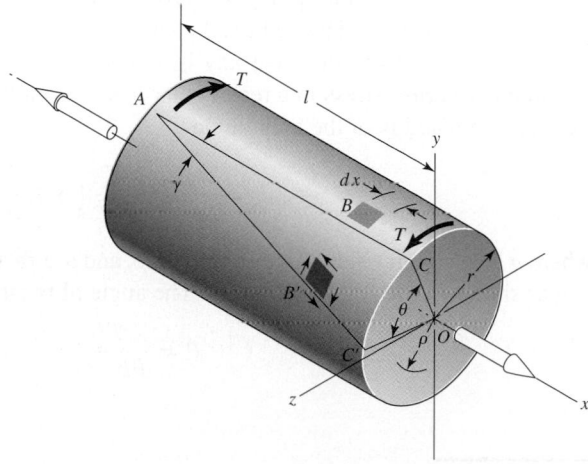

Equation (4–37) applies *only* to circular sections. For a solid round section,

$$J = \frac{\pi d^4}{32} \qquad (4\text{–}38)$$

where d is the diameter of the bar. For a hollow round section,

$$J = \frac{\pi}{32}(d_o^4 - d_i^4) \qquad (4\text{–}39)$$

where the subscripts o and i refer to the outside and inside diameters, respectively.

In using Eq. (4–37) it is often necessary to obtain the torque T from a consideration of the power and speed of a rotating shaft. For convenience when USC units are used, three forms of this relation are

$$H = \frac{FV}{33\,000} = \frac{2\pi T n}{33\,000(12)} = \frac{\pi T n}{198\,000} = \frac{T n}{63\,025} \qquad (4\text{–}40)$$

where H = power, hp

T = torque, lbf · in

n = shaft speed, rev/min

F = force, lbf

V = velocity, ft/min

When SI units are used, the equation is

$$H = T\omega \qquad (4\text{–}41)$$

where H = power, W

T = torque, N · m

ω = angular velocity, rad/s

The torque T corresponding to the power in watts is given approximately by

$$T = 9.55\frac{H}{n} \qquad (4\text{–}42)$$

where n is in revolutions per minute.

There are some applications in machinery for noncircular-cross-section members and shafts where a regular polygonal cross section is useful in transmitting torque to a gear or pulley that can have an axial change in position. Because no key or keyway is needed, the possibility of a lost key is avoided. Saint Venant (1855) showed that the maximum shearing stress in a rectangular $b \times c$ section bar occurs in the middle of the *longest* side b and is of the magnitude

$$\tau_{\max} = \frac{T}{\alpha b c^2} \doteq \frac{T}{bc^2}\left(3 + \frac{1.8}{b/c}\right) \qquad (4\text{–}43)$$

where b is the longer side, c the shorter side, and α a factor that is a function of the ratio b/c as shown in the following table.[5] The angle of twist is given by

$$\theta = \frac{Tl}{\beta bc^3 G} \qquad (4\text{–}44)$$

---

[5]S. Timoshenko, *Strength of Materials*, Part I, 3rd ed., D. Van Nostrand Company, New York, 1955, p. 290.

where β is a function of b/c, as shown in the table.

b/c	1.00	1.50	1.75	2.00	2.50	3.00	4.00	6.00	8.00	10	∞
α	0.208	0.231	0.239	0.246	0.258	0.267	0.282	0.299	0.307	0.313	0.333
β	0.141	0.196	0.214	0.228	0.249	0.263	0.281	0.299	0.307	0.313	0.333

In Eqs. (4–43) and (4–44) b and c are the width (long side) and thickness (short side) of the bar, respectively. They cannot be interchanged. Equation (4–43) is also approximately valid for equal-sided angles; these can be considered as two rectangles, each of which is capable of carrying half the torque.[6]

---

[6]For other sections see W. C. Young and R. G. Budynas, *Roark's Formulas for Stress and Strain,* 7th ed., McGraw-Hill, New York, 2002.

EXAMPLE 4–8

Figure 4–24 shows a crank loaded by a force $F = 300$ lbf that causes twisting and bending of a $\frac{3}{4}$-in-diameter shaft fixed to a support at the origin of the reference system. In actuality, the support may be an inertia that we wish to rotate, but for the purposes of a stress analysis we can consider this a statics problem.

(*a*) Draw separate free-body diagrams of the shaft AB and the arm BC, and compute the values of all forces, moments, and torques that act. Label the directions of the coordinate axes on these diagrams.

(*b*) Compute the maxima of the torsional stress and the bending stress in the arm BC and indicate where these act.

(*c*) Locate a stress element on the top surface of the shaft at A, and calculate all the stress components that act upon this element.

(*d*) Determine the maximum normal and shear stresses at A.

Solution

(*a*) The two free-body diagrams are shown in Fig. 4–25. The results are

AT C: $\qquad$ $\mathbf{F} = -300\mathbf{j}$ lbf, $\mathbf{T} = -450\mathbf{k}$ lbf · in

AT END B OF ARM BC: $\qquad$ $\mathbf{F} = 300\mathbf{j}$ lbf, $\mathbf{M} = 1200\mathbf{i}$ lbf · in, $\mathbf{T} = 450\mathbf{k}$ lbf · in

AT END B OF SHAFT AB: $\qquad$ $\mathbf{F} = -300\mathbf{j}$ lbf, $\mathbf{T} = -1200\mathbf{i}$ lbf · in, $\mathbf{M} = -450\mathbf{k}$ lbf · in

AT A: $\qquad$ $\mathbf{F} = 300\mathbf{j}$ lbf, $\mathbf{M} = 1950\mathbf{k}$ lbf · in, $\mathbf{T} = 1200\mathbf{i}$ lbf · in

| **Figure 4–24**

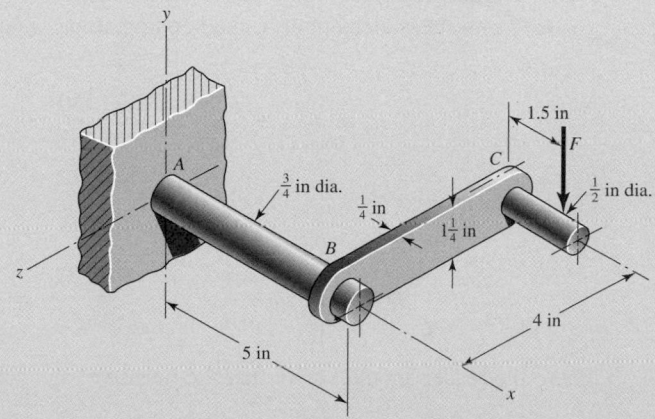

| Figure 4–25

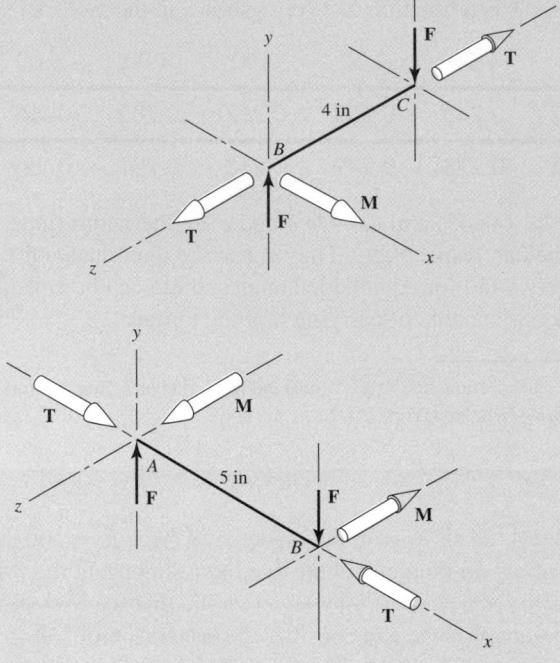

(b) For arm BC, the bending moment will reach a maximum near the shaft at B. If we assume this is 1200 lbf · in, then the bending stress for a rectangular section will be

Answer
$$\sigma = \frac{M}{I/c} = \frac{6M}{bh^2} = \frac{6(1200)}{0.25(1.25)^2} = 18\,400 \text{ psi}$$

Of course, this is not exactly correct, because at B the moment is actually being transferred into the shaft, probably through a weldment.

For the torsional stress, use Eq. (4–43). Thus

Answer
$$\tau_{\max} = \frac{T}{bc^2}\left(3 + \frac{1.8}{b/c}\right) = \frac{450}{1.25(0.25^2)}\left(3 + \frac{1.8}{1.25/0.25}\right) = 19\,400 \text{ psi}$$

This stress occurs at the middle of the $1\frac{1}{4}$-in side.

(c) For a stress element at A, the bending stress is tensile and is

Answer
$$\sigma_x = \frac{M}{I/c} = \frac{32M}{\pi d^3} = \frac{32(1950)}{\pi(0.75)^3} = 47\,100 \text{ psi}$$

The torsional stress is

Answer
$$\tau_{xz} = \frac{-T}{J/c} = \frac{-16T}{\pi d^3} = \frac{-16(1200)}{\pi(0.75)^3} = -14\,500 \text{ psi}$$

where the reader should verify that the negative sign accounts for the direction of τ_{xz}.

(d) Point A is in a state of plane stress where the stresses are in the xz plane. Thus the principal stresses are given by Eq. (4–13) with subscripts corresponding to the x, z axes.

Answer   The maximum normal stress is then given by

$$\sigma_1 = \frac{\sigma_x + \sigma_z}{2} + \sqrt{\left(\frac{\sigma_x - \sigma_z}{2}\right)^2 + \tau_{xz}^2}$$

$$= \frac{47.1 + 0}{2} + \sqrt{\left(\frac{47.1 - 0}{2}\right)^2 + (-14.5)^2} = 51.2 \text{ kpsi}$$

Answer   The maximum shear stress at A occurs on surfaces different than the surfaces containing the principal stresses or the surfaces containing the bending and torsional shear stresses. The maximum shear stress is given by Eq. (4–14), again with modified subscripts, and is given by

$$\tau_1 = \sqrt{\left(\frac{\sigma_x - \sigma_z}{2}\right)^2 + \tau_{xz}^2} = \sqrt{\left(\frac{47.1 - 0}{2}\right)^2 + (-14.5)^2} = 27.7 \text{ kpsi}$$

EXAMPLE 4–9   The 1.5-in-diameter solid steel shaft shown in Fig. 4–26a is simply supported at the ends. Two pulleys are keyed to the shaft where pulley B is of diameter 4.0 in and pulley C is of diameter 8.0 in. Considering bending and torsional stresses only, determine the locations and magnitudes of the greatest tensile, compressive, and shear stresses in the shaft.

Solution   Figure 4–26b shows the net forces, reactions, and torsional moments on the shaft. Although this is a three-dimensional problem and vectors might seem appropriate, we will look at the components of the moment vector by performing a two-plane analysis. Figure 4–26c shows the loading in the xy plane, as viewed down the z axis, where bending moments are actually vectors in the z direction. Thus we label the moment diagram as M_z versus x. For the xz plane, we look down the y axis, and the moment diagram is M_y versus x as shown in Fig. 4–26d.

The net moment on a section is the vector sum of the components. That is,

$$M = \sqrt{M_y^2 + M_z^2} \tag{1}$$

At point B,

$$M_B = \sqrt{2000^2 + 8000^2} = 8246 \text{ lbf} \cdot \text{in}$$

At point C,

$$M_C = \sqrt{4000^2 + 4000^2} = 5657 \text{ lbf} \cdot \text{in}$$

Thus the maximum bending moment is 8246 lbf · in and the maximum bending stress at pulley B is

$$\sigma = \frac{M \, d/2}{\pi d^4/64} = \frac{32M}{\pi d^3} = \frac{32(8246)}{\pi(1.5^3)} = 24 \text{ } 890 \text{ psi}$$

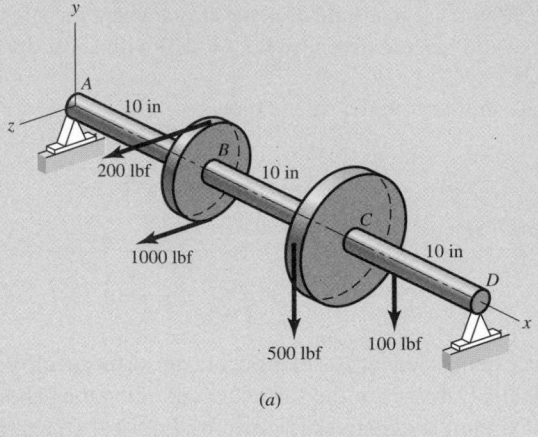

(a)

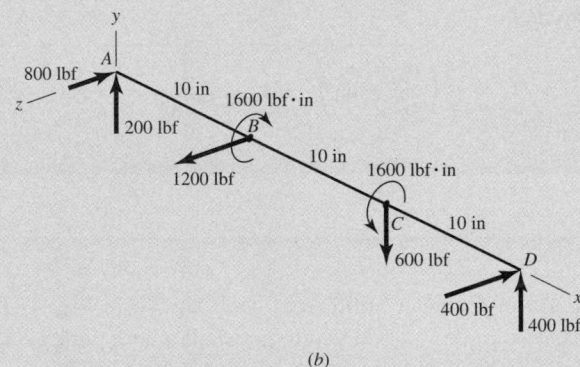

(b)

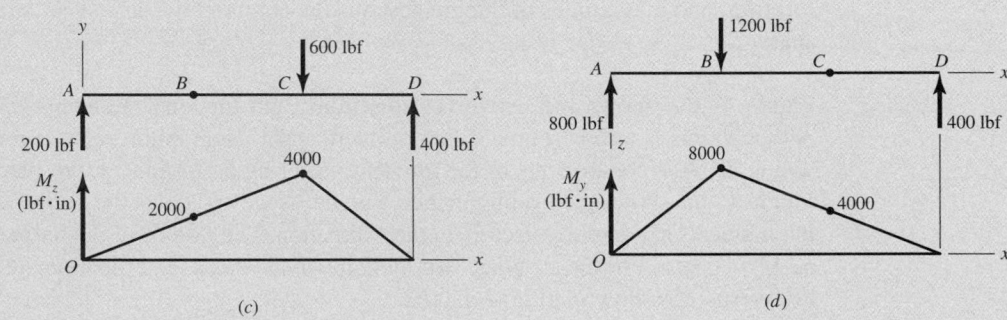

(c)   (d)

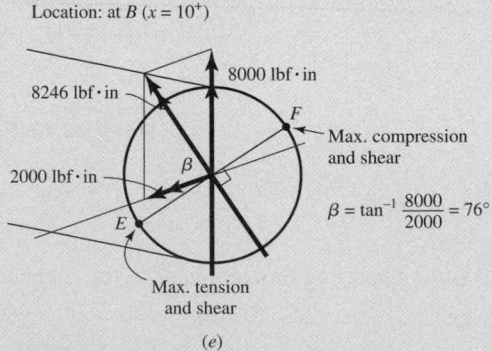

Location: at B ($x = 10^+$)

8246 lbf·in

8000 lbf·in

F

Max. compression
and shear

2000 lbf·in

β

$\beta = \tan^{-1}\dfrac{8000}{2000} = 76°$

E

Max. tension
and shear

(e)

| **Figure 4–26**

The maximum torsional shear stress occurs between B and C and is

$$\tau = \frac{T\,d/2}{\pi d^4/32} = \frac{16T}{\pi d^3} = \frac{16(1600)}{\pi(1.5^3)} = 2414 \text{ psi}$$

The maximum bending and torsional shear stresses occur just to the right of pulley B at points E and F as shown in Fig. 4–26e. At point E, the maximum tensile stress will be σ_1 given by

$$\sigma_1 = \frac{\sigma}{2} + \sqrt{\left(\frac{\sigma}{2}\right)^2 + \tau^2} = \frac{24\,890}{2} + \sqrt{\left(\frac{24\,890}{2}\right)^2 + 2414^2} = 25\,120 \text{ psi}$$

At point F, the maximum compressive stress will be σ_2 given by

$$\sigma_2 = \frac{-\sigma}{2} - \sqrt{\left(\frac{-\sigma}{2}\right)^2 + \tau^2} = \frac{-24\,890}{2} - \sqrt{\left(\frac{-24\,890}{2}\right)^2 + 2414^2} = -25\,120 \text{ psi}$$

The extreme shear stress also occurs at E and F and is

$$\tau_1 = \sqrt{\left(\frac{\pm\sigma}{2}\right)^2 + \tau^2} = \sqrt{\left(\frac{\pm 24\,890}{2}\right)^2 + 2414^2} = 12\,680 \text{ psi}$$

Closed Thin-Walled Tubes ($t \ll r$)[7]

In closed thin-walled tubes, it can be shown that the product of shear stress times thickness of the wall τt is constant, meaning that the shear stress τ is inversely proportional to the wall thickness t. The total torque T on a tube such as depicted in Fig. 4–27 is given by

$$T = \int \tau t r\,ds = (\tau t) \int r\,ds = \tau t(2A_m) = 2A_m t \tau$$

Figure 4–27

The depicted cross section is elliptical, but the section need not be symmetrical nor of constant thickness.

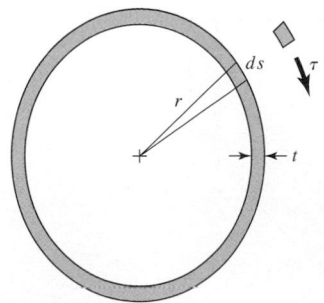

---

[7]See Sec. 3–13, F. P. Beer, E. R. Johnston, and J. T. De Wolf, *Mechanics of Materials,* 3rd ed., McGraw-Hill, New York, 2001.

where A_m is the *area enclosed by the section median line*. Solving for τ gives

$$\tau = \frac{T}{2A_m t} \qquad (4\text{–}45)$$

For constant wall thickness t, the angular twist (radians) per unit of length of the tube θ_1 is given by

$$\theta_1 = \frac{TL_m}{4GA_m^2 t} \qquad (4\text{–}46)$$

where L_m is the *perimeter of the section median line*. These equations presume the buckling of the tube is prevented by ribs, stiffeners, bulkheads, and so on, and that the stresses are below the proportional limit.

EXAMPLE 4–10

A welded steel tube is 40 in long, has a $\frac{1}{8}$-in wall thickness, and a 2.5-in by 3.6-in rectangular cross section as shown in Fig. 4–28. Assume an allowable shear stress of 11 500 psi and a shear modulus of $11.5(10^6)$ psi.

(a) Estimate the allowable torque T.
(b) Estimate the angle of twist due to the torque.

Solution

(a) Within the section median line, the area enclosed is

$$A_m = (2.5 - 0.125)(3.6 - 0.125) = 8.253 \text{ in}^2$$

and the length of the median perimeter is

$$L_m = 2[(2.5 - 0.125) + (3.6 - 0.125)] = 11.70 \text{ in}$$

Answer

From Eq. (4–45) the torque T is

$$T = 2A_m t\tau = 2(8.253)0.125(11\,500) = 23\,728 \text{ lbf} \cdot \text{in}$$

Answer

(b) The angle of twist θ from Eq. (4–46) is

$$\theta = \theta_1 l = \frac{TL_m}{4GA_m^2 t}l = \frac{23\,728(11.70)}{4(11.5 \times 10^6)(8.253^2)(0.125)}(40) = 0.0284 \text{ rad} = 1.63°$$

Figure 4–28

A rectangular steel tube produced by welding.

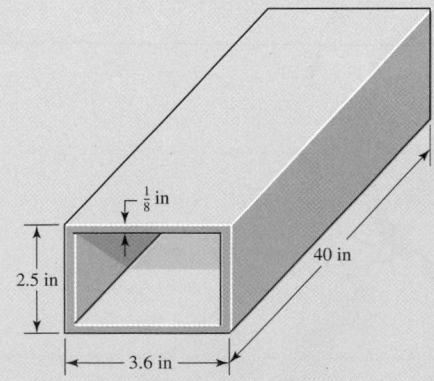

EXAMPLE 4–11   Compare the shear stress on a circular cylindrical tube with an outside diameter of 1 in and an inside diameter of 0.9 in, predicted by Eq. (4–37), to that estimated by Eq. (4–45).

Solution   From Eq. (4–37),

$$\tau_{\max} = \frac{Tr}{J} = \frac{Tr}{(\pi/32)\left(d_o^4 - d_i^4\right)} = \frac{T(0.5)}{(\pi/32)(1^4 - 0.9^4)} = 14.809T$$

From Eq. (4–45),

$$\tau = \frac{T}{2A_m t} = \frac{T}{2(\pi 0.95^2/4)0.05} = 14.108T$$

Taking Eq. (4–37) as correct, the error in the thin-wall estimate is −4.7 percent.

Open Thin-Walled Sections

When the median wall line is not closed, it is said to be *open*. Figure 4–29 presents some examples. Open sections in torsion, where the wall is thin, have relations derived from the membrane analogy theory as follows:

$$\tau = G\theta_1 c = \frac{3T}{Lc^2} \tag{4–47}$$

where τ is the shear stress, G is the shear modulus, θ_1 is the angle of twist per unit length, T is torque, and L is the length of the median line. The wall thickness is designated c (rather than t) to remind you that you are in open sections. By studying the table that follows Eq. (4–44) you will discover that membrane theory presumes $b/c \rightarrow \infty$. Note that open thin-walled sections in torsion should be avoided in design. As indicated in Eq. (4–47), the shear stress and the angle of twist are inversely proportional to c^2 and c^3, respectively. Thus, for small wall thickness, stress and twist can become quite large. For example, consider the thin round tube with a slit in Fig. 4–29. For a ratio of wall thickness of outside diameter of $c/d_o = 0.1$, the open section has greater magnitudes of stress and angle of twist by factors of 12.3 and 61.5, respectively, compared to a closed section of the same dimensions.

Figure 4–29

Some open thin-wall sections.

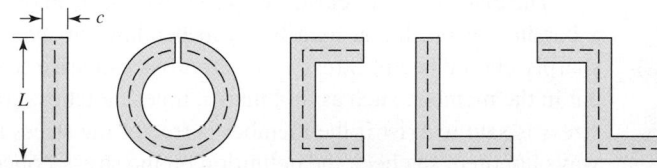

EXAMPLE 4–12   A 12-in-long strip of steel is 1/8 in thick and 1 in wide, as shown in Fig. 4–30. If the allowable shear stress is 11 500 psi and the shear modulus is 11.5(10^6) psi, find the torque corresponding to the allowable shear stress and the angle of twist, in degrees, (*a*) using Eq. (4–47) and (*b*) using Eqs. (4–43) and (4–44).

Solution

(a) The length of the median line is 1 in. From Eq. (4–47),

$$T = \frac{Lc^2\tau}{3} = \frac{(1)(1/8)^2 11\,500}{3} = 59.90 \text{ lbf} \cdot \text{in}$$

$$\theta = \theta_1 l = \frac{3Tl}{Lc^3 G} = \frac{3(59.90)12}{(1)(1/8)^3 11.5(10^6)} = 0.0960 \text{ rad} = 5.5°$$

A torsional spring rate k_t can be expressed as T/θ:

$$k_t = 59.90/0.0960 = 624 \text{ lbf} \cdot \text{in/rad}$$

(b) From Eq. (4–43),

$$T = \frac{\tau_{\max} bc^2}{3 + 1.8/(b/c)} = \frac{11\,500(1)(0.125)^2}{3 + 1.8/(1/0.125)} = 55.72 \text{ lbf} \cdot \text{in}$$

From Eq. (4–44), with $b/c = 1/0.125 = 8$,

$$\theta = \frac{Tl}{\beta bc^3 G} = \frac{55.72(12)}{0.307(1)0.125^3(11.5)10^6} = 0.0970 \text{ rad} = 5.6°$$

$$k_t = 55.72/0.0970 = 574 \text{ lbf} \cdot \text{in/rad}$$

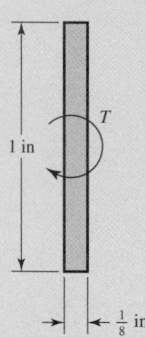

Figure 4–30

The cross-section of a thin strip of steel subjected to a torsional moment T.

1 in

$\frac{1}{8}$ in

T

4–14 Stress Concentration

In the development of the basic stress equations for tension, compression, bending, and torsion, it was assumed that no geometric irregularities occurred in the member under consideration. But it is quite difficult to design a machine without permitting some changes in the cross sections of the members. Rotating shafts must have shoulders designed on them so that the bearings can be properly seated and so that they will take thrust loads; and the shafts must have key slots machined into them for securing pulleys and gears. A bolt has a head on one end and screw threads on the other end, both of which account for abrupt changes in the cross section. Other parts require holes, oil grooves, and notches of various kinds. Any discontinuity in a machine part alters the stress distribution in the neighborhood of the discontinuity so that the elementary stress equations no longer describe the state of stress in the part at these locations. Such discontinuities are called *stress raisers,* and the regions in which they occur are called areas of *stress concentration.*

The distribution of elastic stress across a section of a member may be uniform as in a bar in tension, linear as a beam in bending, or even rapid and curvaceous as in a sharply curved beam. Stress concentrations can arise from some irregularity not inherent in the member, such as tool marks, holes, notches, grooves, or threads. The *nominal stress* is said to exist if the member is free of the stress raiser. This definition is not always honored, so check the definition on the stress-concentration chart or table you are using.

To help in understanding this effect, examine Fig. 4–31. Note that the stress trajectories are uniform everywhere except in the vicinity of the hole. But at the hole these lines of force must bend to get around. Stress concentration is a highly localized effect. The stress on the tension plate is highest at the edge of the hole on plane A-A; this stress drops rapidly as points are examined farther from the hole edge and soon becomes uniform again.

Figure 4–31

Stress distribution near a hole in a plate loaded in tension. The tensile stress on a section B-B, remote from the hole is $\sigma = F/A$, where $A = wt$ and t is the plate thickness. On a section at A-A, through the hole, the area $A_0 = (w - d)t$ and the nominal stress is $\sigma_0 = F/A_0$. Note the difference between the nominal stress and the stress at a section remote from the discontinuity.

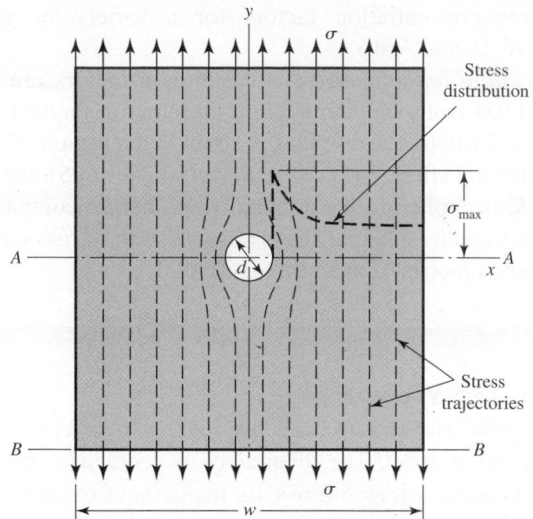

A *theoretical*, or *geometric, stress-concentration factor* K_t or K_{ts} is used to relate the actual maximum stress at the discontinuity to the nominal stress. The factors are defined by the equations

$$K_t = \frac{\sigma_{\max}}{\sigma_0} \qquad K_{ts} = \frac{\tau_{\max}}{\tau_0} \qquad (4\text{–}48)$$

where K_t is used for normal stresses and K_{ts} for shear stresses. The nominal stress σ_0 or τ_0 is more difficult to define. Generally, it is the stress calculated by using the elementary stress equations and the net area, or net cross section. But sometimes the gross cross section is used instead, and so it is always wise to double check your source of K_t or K_{ts} before calculating the maximum stress.

The subscript t in K_t means that this stress-concentration factor depends for its value only on the *geometry* of the part. That is, the particular material used has no effect on the value of K_t. This is why it is called a *theoretical* stress-concentration factor.

The analysis of geometric shapes to determine stress-concentration factors is a difficult problem, and not many solutions can be found. One such solution is that of an infinite plate containing an elliptical hole loaded in uniform tension. The result is

$$K_t = 1 + \frac{2b}{a} \qquad (4\text{–}49)$$

where, after the hole in Fig. 4–31 is replaced with an ellipse, b is the half-width, a is the half-height, and $w \to \infty$. Thus, for a circular hole, $b = a$ and $K_t = 3$ as $w \to \infty$.

Note that Eq. (4–49) can be applied to a transverse crack, where $b \gg a$, or to a longitudinal crack ($b \ll a$). This is useful in fracture mechanics, discussed in Chap. 6.

Most stress-concentration factors are found by using experimental techniques.[8] Though the finite-element method has been used, the fact that the elements are indeed finite prevents finding the true maximum stress. Experimental approaches generally used include photoelasticity, grid methods, brittle-coating methods, and electrical strain-gauge methods. Of course, the grid and strain-gauge methods both suffer from the same drawback as the finite-element method.

---

[8]The best source book is R. E. Peterson, *Stress Concentration Factors*, Wiley, New York, 1974.

Stress-concentration factors for a variety of geometries may be found in Tables A–15 and A–16.

In *static loading,* stress-concentration factors are applied as follows. In ductile ($\epsilon_f \geq 0.05$) materials, the stress-concentration factor *is not* usually applied to predict the critical stress, because plastic strain in the region of the stress is localized and has a strengthening effect. In *brittle materials* ($\epsilon_f < 0.05$), the geometric stress-concentration factor K_t is applied to the nominal stress before comparing it with strength. Gray cast iron has so many inherent stress raisers that the stress raisers introduced by the designer have only a modest (but additive) effect.

EXAMPLE 4–13

Be Alert to Viewpoint

On a "spade" rod end (or lug) a load is transferred through a pin to a rectangular-cross-section rod or strap. The theoretical or geometric stress-concentration factor for this geometry is known as follows, on the basis of the net area $A = (w - d)t$ as shown in Fig. 4–32.

d/w	0.15	0.20	0.25	0.30	0.35	0.40	0.45	0.50
K_t	7.4	5.4	4.6	3.7	3.2	2.8	2.6	2.45

As presented in the table, K_t is a decreasing monotone. This rod end is similar to the square-ended lug depicted in Table A-15-12.

$$\sigma_{\max} = K_t \sigma_0 \qquad (a)$$

$$\sigma_{\max} = \frac{K_t F}{A} = K_t \frac{F}{(w - d)t} \qquad (b)$$

It is insightful to base the stress concentration factor on the *unnotched* area, wt. Let

$$\sigma_{\max} = K_t' \frac{F}{wt} \qquad (c)$$

By equating Eqs. (b) and (c) and solving for K_t' we obtain

$$K_t' = \frac{wt}{F} \frac{F}{(w - d)t} = \frac{K_t}{1 - d/w} \qquad (d)$$

A regression curve-fit for the data in the above table in the form $K_t = a(d/w)^b$ with the aid of a ln ln transform gives the result $a = \exp(0.204\,521\,2) = 1.227, b = -0.935$, and $r = -0.997$. Thus

$$K_t = 1.227 \left(\frac{d}{w}\right)^{-0.935} \qquad (e)$$

which is a decreasing monotone (and unexciting). However, from Eq. (d),

$$K_t' = \frac{1.227}{1 - d/w} \left(\frac{d}{w}\right)^{-0.935} \qquad (f)$$

Form another table from Eq. (f):

d/w	0.15	0.20	0.25	0.30	0.35	0.40	0.45	0.50	0.55	0.60
K_t'	8.507	6.907	5.980	5.403	5.038	4.817	4.707	4.692	4.769	4.946

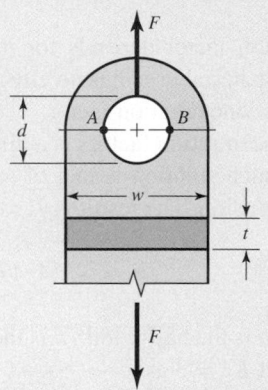

Figure 4–32

A round-ended lug end to a rectangular cross-section rod. The maximum tensile stress in the lug occurs at locations A and B. The net area $A = (w - d)t$ is used in the definition of K_t, but there is an advantage to using the total area wt.

which shows a stationary-point minimum for K_t'. This can be found by differentiating Eq. (f) with respect to d/w and setting it equal to zero:

$$\frac{dK_t'}{d(d/w)} = \frac{(1 - d/w)ab(d/w)^{b-1} + a(d/w)^b}{[1 - (d/w)]^2} = 0$$

where $b = -0.935$, from which

$$\left(\frac{d}{w}\right)^* = \frac{b}{b-1} = \frac{-0.935}{-0.935 - 1} = 0.483$$

with a corresponding K_t' of 4.687. Knowing the section $w \times t$ lets the designer specify the strongest lug immediately by specifying a pin diameter of $0.483w$ (or, as a rule of thumb, of half the width). The theoretical K_t data in the original form, or a plot based on the data using net area, would not suggest this. The right viewpoint can suggest valuable insights.

4–15 Stresses in Pressurized Cylinders

Cylindrical pressure vessels, hydraulic cylinders, gun barrels, and pipes carrying fluids at high pressures develop both radial and tangential stresses with values that depend upon the radius of the element under consideration. In determining the radial stress σ_r and the tangential stress σ_t, we make use of the assumption that the longitudinal elongation is constant around the circumference of the cylinder. In other words, a right section of the cylinder remains plane after stressing.

Referring to Fig. 4–33, we designate the inside radius of the cylinder by r_i, the outside radius by r_o, the internal pressure by p_i, and the external pressure by p_o. Then it can be shown that tangential and radial stresses exist whose magnitudes are

$$\sigma_t = \frac{p_i r_i^2 - p_o r_o^2 - r_i^2 r_o^2 (p_o - p_i)/r^2}{r_o^2 - r_i^2}$$

$$\sigma_r = \frac{p_i r_i^2 - p_o r_o^2 + r_i^2 r_o^2 (p_o - p_i)/r^2}{r_o^2 - r_i^2} \tag{4–50}$$

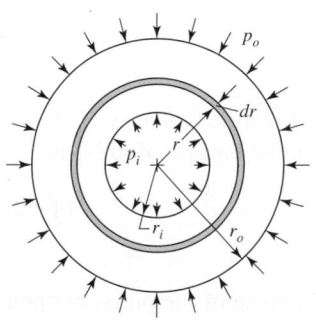

Figure 4–33

A cylinder subjected to both internal and external pressure.

As usual, positive values indicate tension and negative values, compression.

The special case of $p_o = 0$ gives

$$\sigma_t = \frac{r_i^2 p_i}{r_o^2 - r_i^2}\left(1 + \frac{r_o^2}{r^2}\right)$$

$$\sigma_r = \frac{r_i^2 p_i}{r_o^2 - r_i^2}\left(1 - \frac{r_o^2}{r^2}\right) \tag{4–51}$$

The equations of set (4–51) are plotted in Fig. 4–34 to show the distribution of stresses over the wall thickness. It should be realized that longitudinal stresses exist when the end reactions to the internal pressure are taken by the pressure vessel itself. This stress is found to be

$$\sigma_l = \frac{p_i r_i^2}{r_o^2 - r_i^2} \tag{4–52}$$

We further note that Eqs. (4–50), (4–51), and (4–52) apply only to sections taken a significant distance from the ends and away from any areas of stress concentration.

Figure 4–34

Distribution of stresses in a thick-walled cylinder subjected to internal pressure.

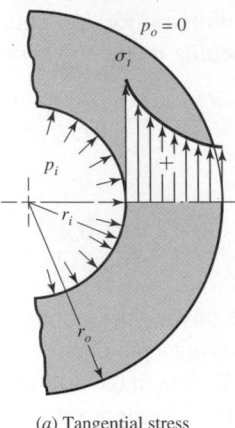

(a) Tangential stress distribution

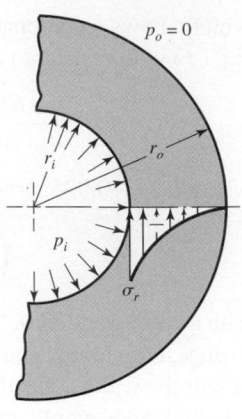

(b) Radial stress distribution

Thin-Walled Vessels

When the wall thickness of a cylindrical pressure vessel is about one-twentieth, or less, of its radius, the radial stress that results from pressurizing the vessel is quite small compared with the tangential stress. Under these conditions the tangential stress can be obtained as follows: Let an internal pressure p be exerted on the wall of a cylinder of thickness t and inside diameter d_i. The force tending to separate two halves of a unit length of the cylinder is pd_i. This force is resisted by the tangential stress, also called the *hoop stress,* acting uniformly over the stressed area. We then have $pd_i = 2t\sigma_t$, or

$$(\sigma_t)_{av} = \frac{pd_i}{2t} \tag{4–53}$$

This equation gives the *average* tangential stress and is valid regardless of the wall thickness. For a thin-walled vessel an approximation to the maximum tangential stress is

$$(\sigma_t)_{max} = \frac{p(d_i + t)}{2t} \tag{4–54}$$

where $d_i + t$ is the average diameter.

In a closed cylinder, the longitudinal stress σ_l exists because of the pressure upon the ends of the vessel. If we assume this stress is also distributed uniformly over the wall thickness, we can easily find it to be

$$\sigma_l = \frac{pd_i}{4t} \tag{4–55}$$

EXAMPLE 4–14

An aluminum-alloy pressure vessel is made of tubing having an outside diameter of 8 in and a wall thickness of $\frac{1}{4}$ in.

(a) What pressure can the cylinder carry if the permissible tangential stress is 12 kpsi and the theory for thin-walled vessels is assumed to apply?

(b) On the basis of the pressure found in part (a), compute all of the stress components using the theory for thick-walled cylinders.

Solution

(a) Here $d_i = 8 - 2(0.25) = 7.5$ in, $r_i = 7.5/2 = 3.75$ in, and $r_o = 8/2 = 4$ in. Then $t/r_i = 0.25/3.75 = 0.067$. Since this ratio is greater than $\frac{1}{20}$, the theory for thin-walled vessels may not yield safe results.

We first solve Eq. (4–54) to obtain the allowable pressure. This gives

Answer
$$p = \frac{2t\,(\sigma_t)_{\max}}{d_i + t} = \frac{2(0.25)(12)(10)^3}{7.5 + 0.25} = 774 \text{ psi}$$

Then, from Eq. (4–55), we find the average longitudinal stress to be

$$\sigma_l = \frac{pd_i}{4t} = \frac{774(7.5)}{4(0.25)} = 5810 \text{ psi}$$

(b) The maximum tangential stress will occur at the inside radius, and so we use $r = r_i$ in the first equation of Eq. (4–51). This gives

Answer
$$(\sigma_t)_{\max} = \frac{r_i^2 p_i}{r_o^2 - r_i^2}\left(1 + \frac{r_o^2}{r_i^2}\right) = p_i \frac{r_o^2 + r_i^2}{r_o^2 - r_i^2} = 774\frac{4^2 + 3.75^2}{4^2 - 3.75^2} = 12\,000 \text{ psi}$$

Similarly, the maximum radial stress is found, from the second equation of Eq. (4–51) to be

Answer
$$\sigma_r = -p_i = -774 \text{ psi}$$

Equation (4–52) gives the longitudinal stress as

Answer
$$\sigma_l = \frac{p_i r_i^2}{r_o^2 - r_i^2} = \frac{774(3.75)^2}{4^2 - 3.75^2} = 5620 \text{ psi}$$

These three stresses, σ_t, σ_r, and σ_l, are principal stresses, since there is no shear on these surfaces. Note that there is no significant difference in the tangential stresses in parts (a) and (b), and so the thin-wall theory can be considered satisfactory.

4–16 **Stresses in Rotating Rings**

Many rotating elements, such as flywheels and blowers, can be simplified to a rotating ring to determine the stresses. When this is done it is found that the same tangential and radial stresses exist as in the theory for thick-walled cylinders except that they are caused by inertial forces acting on all the particles of the ring. The tangential and radial stresses so found are subject to the following restrictions:

- The outside radius of the ring, or disk, is large compared with the thickness $r_o \geq 10t$.
- The thickness of the ring or disk is constant.
- The stresses are constant over the thickness.

The stresses are

$$\sigma_t = \rho\omega^2\left(\frac{3 + \nu}{8}\right)\left(r_i^2 + r_o^2 + \frac{r_i^2 r_o^2}{r^2} - \frac{1 + 3\nu}{3 + \nu}r^2\right)$$

$$\sigma_r = \rho\omega^2\left(\frac{3 + \nu}{8}\right)\left(r_i^2 + r_o^2 - \frac{r_i^2 r_o^2}{r^2} - r^2\right)$$

(4–56)

where r is the radius to the stress element under consideration, ρ is the mass density, and ω is the angular velocity of the ring in radians per second. For a rotating disk, use $r_i = 0$ in these equations.

4-17 Press and Shrink Fits

When two cylindrical parts are assembled by shrinking or press fitting one part upon another, a contact pressure is created between the two parts. The stresses resulting from this pressure may easily be determined with the equations of the preceding sections.

Figure 4–35b shows two cylindrical members that have been assembled with a shrink fit. A contact pressure p exists between the members at the nominal transition radius R, causing radial stresses $\sigma_r = -p$ in each member at the contacting surfaces. From Sec. 4–15, we find the tangential stress at the transition radius of the inner member to be

$$(\sigma_t)_i\bigg|_{r=R} = -p\frac{R^2 + r_i^2}{R^2 - r_i^2} \tag{4–57}$$

In the same manner, the tangential stress at the inner surface of the outer member is found to be

$$(\sigma_t)_o\bigg|_{r=R} = p\frac{r_o^2 + R^2}{r_o^2 - R^2} \tag{4–58}$$

These equations cannot be solved until the contact pressure is known. In obtaining a shrink fit, the transition radius of the inner member is made larger than the radius of the outer member. The difference in these dimensions is called the *radial interference* and is the radial deformation that the two members must experience. Since these dimensions are usually known, the deformation should be introduced in order to evaluate the stresses. As shown in Fig. 4–35a, δ_i and δ_o symbolize the changes in the radii of the inner and outer members, respectively. The total radial interference is, therefore,

$$\delta = |\delta_i| + |\delta_o| \tag{a}$$

The tangential strain at the transition radius of the outer cylinder is measured by the change in circumference, and is

$$(\epsilon_t)_o = \frac{2\pi(R + \delta_o) - 2\pi R}{2\pi R} = \frac{\delta_o}{R} \tag{b}$$

and so $\delta_o = R(\epsilon_t)_o$. Assuming biaxial stress, from Table 4–2, Sec. 4–8,

$$(\epsilon_t)_o = \frac{(\sigma_t)_o}{E_o} - \frac{\nu_o(\sigma_r)_o}{E_o} \tag{c}$$

Figure 4–35

Notation for press and shrink fits. (a) Unassembled parts; (b) after assembly.

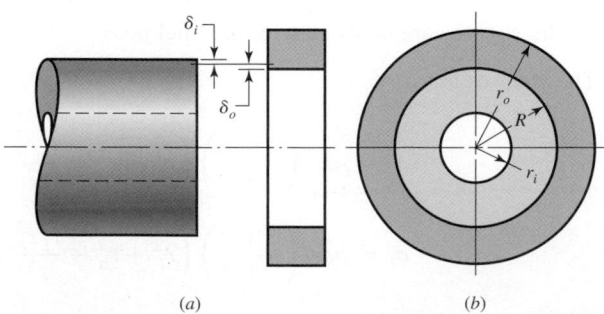

(a)   (b)

Substituting Eqs. (b), (4–58), and $(\sigma_r)_o = -p$ into Eq. (c) gives

$$\delta_o = \frac{pR}{E_o}\left(\frac{r_o^2 + R^2}{r_o^2 - R^2} + \nu_o\right) \qquad (d)$$

This is the change in radius of the outer member at the interface. In a similar manner, the change in radius of the inner member is found to be

$$\delta_i = -\frac{pR}{E_i}\left(\frac{R^2 + r_i^2}{R^2 - r_i^2} - \nu_i\right) \qquad (e)$$

Then, from Eq. (a), we have for the total deformation

$$\delta = \frac{pR}{E_o}\left(\frac{r_o^2 + R^2}{r_o^2 - R^2} + \nu_o\right) + \frac{pR}{E_i}\left(\frac{R^2 + r_i^2}{R^2 - r_i^2} - \nu_i\right) \qquad (4\text{–}59)$$

This equation can be solved for the pressure p when the radial interference δ is given. If the two members are of the same material, $E_o = E_i = E$, $\nu_o = \nu_i$, and the relation simplifies to

$$p = \frac{E\delta}{R}\left[\frac{(r_o^2 - R^2)(R^2 - r_i^2)}{2R^2(r_o^2 - r_i^2)}\right] \qquad (4\text{–}60)$$

The value of the interface pressure p from either Eq. (4–59) or Eq. (4–60) can now be used to obtain the stress state at the specified radius in either cylinder.

Assumptions

In addition to the assumptions both stated and implied by the development, it is necessary to assume that both members have the same length. In the case of a hub that has been press-fitted onto a shaft, this assumption would not be true, and there would be an increased pressure at each end of the hub. It is customary to allow for this condition by employing of a stress-concentration factor. The value of this factor depends upon the contact pressure and the design of the female member, but its theoretical value is seldom greater than 2.

4–18 Temperature Effects

When the temperature of an unrestrained body is uniformly increased, the body expands, and the normal strain is

$$\epsilon_x = \epsilon_y = \epsilon_z = \alpha(\Delta T) \qquad (4\text{–}61)$$

where α is the coefficient of thermal expansion and ΔT is the temperature change, in degrees. In this action the body experiences a simple volume increase with the components of shear strain all zero.

If a straight bar is restrained at the ends so as to prevent lengthwise expansion and then is subjected to a uniform increase in temperature, a compressive stress will develop because of the axial constraint. The stress is

$$\sigma = -\epsilon E = -\alpha(\Delta T)E \qquad (4\text{–}62)$$

In a similar manner, if a uniform flat plate is restrained at the edges and also subjected to a uniform temperature rise, the compressive stress developed is given by the equation

$$\sigma = -\frac{\alpha(\Delta T)E}{1 - \nu} \qquad (4\text{–}63)$$

Table 4–4

Coefficients of Thermal Expansion (Linear Mean Coefficients for the Temperature Range 0–100°C)

Material	Celsius Scale (°C^{-1})	Fahrenheit Scale (°F^{-1})
Aluminum	23.9(10)$^{-6}$	13.3(10)$^{-6}$
Brass, cast	18.7(10)$^{-6}$	10.4(10)$^{-6}$
Carbon steel	10.8(10)$^{-6}$	6.0(10)$^{-6}$
Cast iron	10.6(10)$^{-6}$	5.9(10)$^{-6}$
Magnesium	25.2(10)$^{-6}$	14.0(10)$^{-6}$
Nickel steel	13.1(10)$^{-6}$	7.3(10)$^{-6}$
Stainless steel	17.3(10)$^{-6}$	9.6(10)$^{-6}$
Tungsten	4.3(10)$^{-6}$	2.4(10)$^{-6}$

Figure 4–36

Thermal stresses in an infinite slab during heating and cooling.

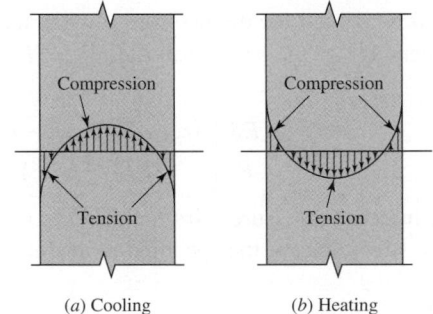

(a) Cooling   (b) Heating

The stresses expressed by Eqs. (4–62) and (4–63) are called *thermal stresses*. They arise because of a temperature change in a clamped or restrained member. Such stresses, for example, occur during welding, since parts to be welded must be clamped before welding. Table 4–4 lists approximate values of the coefficients of thermal expansion.

A *thermal stress* can also arise because of the existence of a *temperature gradient* in a member. Figure 4–36 is an example. Shown are the stress distributions within a slab of infinite dimensions during heating and cooling. During cooling, the maximum stress is the surface tension. During heating, the external surfaces are hot and tend to expand but are restrained by the cooler center. This causes compression in the surface and tension in the center as shown.

4–19 Curved Beams in Bending

The distribution of stress in a curved flexural member is determined by using the following assumptions:

- The cross section has an axis of symmetry in a plane along the length of the beam.
- Plane cross sections remain plane after bending.
- The modulus of elasticity is the same in tension as in compression.

We shall find that the neutral axis and the centroidal axis of a curved beam, unlike the axes of a straight beam, are not coincident and also that the stress does not vary linearly from the neutral axis. The notation shown in Fig. 4–37 is defined as follows:

$$r_o = \text{radius of outer fiber}$$
$$r_i = \text{radius of inner fiber}$$

Figure 4–37

Note that y is positive in the direction toward the center of curvature, point O.

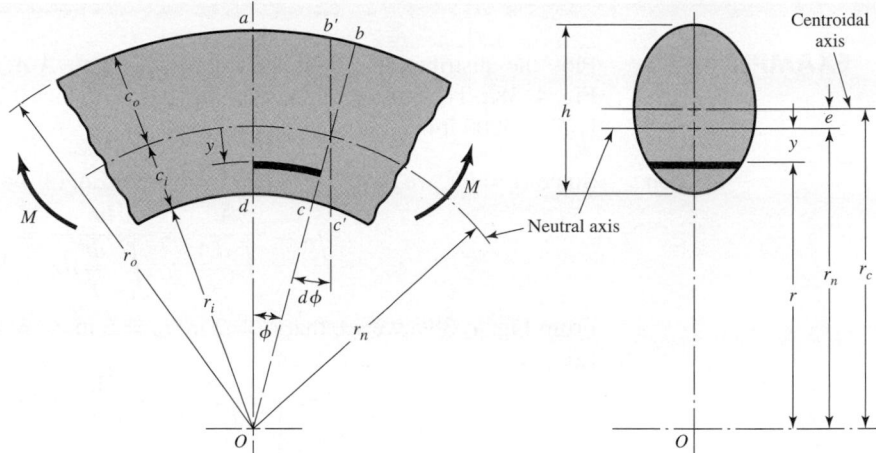

$$h = \text{depth of section}$$
$$c_o = \text{distance from neutral axis to outer fiber}$$
$$c_i = \text{distance from neutral axis to inner fiber}$$
$$r_n = \text{radius of neutral axis}$$
$$r_c = \text{radius of centroidal axis}$$
$$e = \text{distance from centroidal axis to neutral axis}$$
$$M = \text{bending moment; positive } M \text{ decreases curvature}$$

Figure 4–37 shows that the neutral and centroidal axes are not coincident.[9] It turns out that the location of the neutral axis with respect to the center of curvature O is given by the equation

$$r_n = \frac{A}{\int \frac{dA}{r}} \tag{4–64}$$

The stress distribution can be found by balancing the external applied moment against the internal resisting moment. The result is found to be

$$\sigma = \frac{My}{Ae(r_n - y)} \tag{4–65}$$

where M is positive in the direction shown in Fig. 4–37. Equation (4–65) shows that the stress distribution is hyperbolic. The critical stresses occur at the inner and outer surfaces where $y = c_i$ and $y = c_o$, respectively, and are

$$\sigma_i = \frac{Mc_i}{Aer_i} \qquad \sigma_o = -\frac{Mc_o}{Aer_o} \tag{4–66}$$

These equations are valid for pure bending. In the usual and more general case, such as a crane hook, the U frame of a press, or the frame of a clamp, the bending moment is due to forces acting to one side of the cross section under consideration. In this case the bending moment is computed about the *centroidal axis*, not the neutral axis. Also, an additional axial tensile or compressive stress must be added to the bending stresses given by Eqs. (4–65) and (4–66) to obtain the resultant stresses acting on the section.

---

[9]For a complete development of the relations in this section, see Joseph E. Shigley, *Mechanical Engineering Design,* first metric edition, McGraw-Hill, New York, 1986, pp. 72–75.

EXAMPLE 4–15

Plot the distribution of stresses across section *A-A* of the crane hook shown in Fig. 4–38a. The cross section is rectangular, with $b = 0.75$ in and $h = 4$ in, and the load is $F = 5000$ lbf.

Solution

Since $A = bh$, we have $dA = b\,dr$ and, from Eq. (4–64),

$$r_n = \frac{A}{\int \dfrac{dA}{r}} = \frac{bh}{\int_{r_i}^{r_o} \dfrac{b}{r}\,dr} = \frac{h}{\ln \dfrac{r_o}{r_i}} \tag{1}$$

From Fig. 4–38b, we see that $r_i = 2$ in, $r_o = 6$ in, $r_c = 4$ in, and $A = 3$ in². Thus, from Eq. (1),

$$r_n = \frac{h}{\ln(r_o/r_i)} = \frac{4}{\ln \frac{6}{2}} = 3.641 \text{ in}$$

and so the eccentricity is $e = r_c - r_n = 4 - 3.641 = 0.359$ in. The moment M is positive and is $M = Fr_c = 5000(4) = 20\,000$ lbf · in. Adding the axial component of stress to Eq. (4–65) gives

$$\sigma = \frac{F}{A} + \frac{My}{Ae(r_n - y)} = \frac{5000}{3} + \frac{(20\,000)(3.641 - r)}{3(0.359)r} \tag{2}$$

Substituting values of r from 2 to 6 in results in the stress distribution shown in Fig. 4–38c. The stresses at the inner and outer radii are found to be 16.9 and −5.6 kpsi, respectively, as shown.

Figure 4–38

(a) Plan view of crane hook;
(b) cross section and notation;
(c) resulting stress distribution. There is no stress concentration.

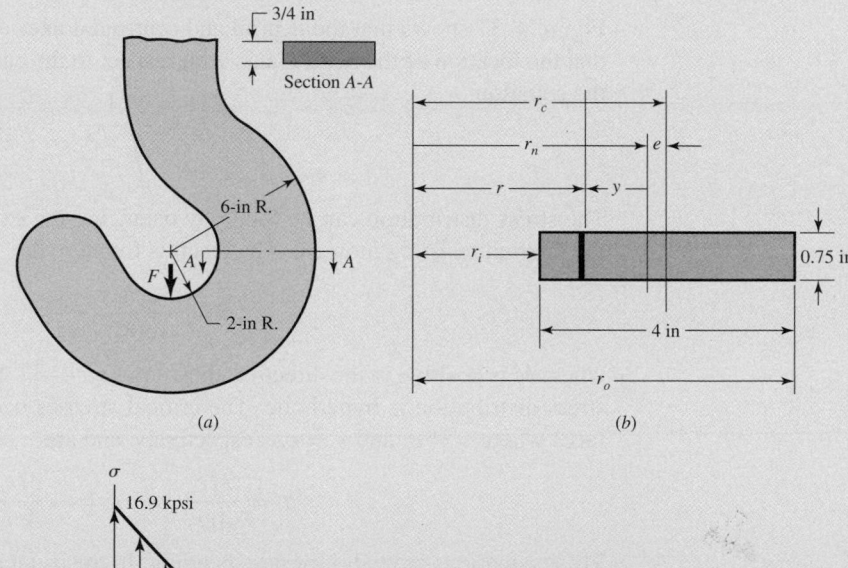

(a)

(b)

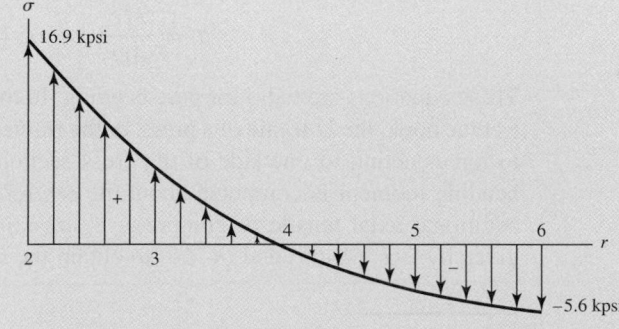

(c)

Note in the hook example, the symmetrical rectangular cross section causes the maximum tensile stress to be 3 times greater than the maximum compressive stress. If we wanted to design the hook to use material more effectively we would use more material at the inner radius and less material at the outer radius. For this reason, trapezoidal, T, or unsymmetric I, cross sections are commonly used. Sections most frequently encountered in the stress analysis of curved beams are shown in Table 4–5.

Approximate Calculations

Calculating r_n and r_c numerically and subtracting the difference can lead to large errors if not done carefully since r_n and r_c are typically large values compared to e. Since e is in the denominator of Eqs. (4–65) and (4–66), a large error in e can lead to an inaccurate stress calculation. Furthermore, if you have a complex section that the tables do not handle, alternative methods for determining r_n, r_c, and e are needed. For a quick and simple approximation of e, let us return to the definition of r_n. Equation (4–64) can be written in the form

$$\int \frac{r_n - r}{r} \, dA = 0 \tag{a}$$

Consider a new coordinate variable s taken from the *centroidal axis*, where $s = r_c - r$. Substituting $r = r_c - s$ and $r_n = r_c - e$ into Eq. (a) results in

$$\int \frac{r_c - e - (r_c - s)}{r_c - s} \, dA = \int \frac{s - e}{r_c - s} \, dA = 0 \tag{b}$$

The second integral can be separated, and solving for e results in

$$e = \frac{\displaystyle\int \frac{s}{r_c - s} \, dA}{\displaystyle\int \frac{dA}{r_c - s}} \tag{4-67}$$

The term in the denominator of both integrals, $(r_c - s)^{-1}$, can be expanded binomially as

$$(r_c - s)^{-1} \doteq r_c^{-1} \left(1 + \frac{s}{r_c}\right) \tag{c}$$

provided $s \ll r_c$, where only the linear term in s/r_c is retained. Thus Eq. (4–67) is

$$e \doteq \frac{\displaystyle\int s\left(1 + \frac{s}{r_c}\right) dA}{\displaystyle\int \left(1 + \frac{s}{r_c}\right) dA} \tag{d}$$

Since s is taken from the centroidal axis, $\int s \, dA = 0$, and therefore $\int s^2 \, dA = I$, the second-moment area about the centroidal axis of the section. Thus the Eq. (d) reduces to

$$e \doteq \frac{I}{r_c A} \tag{4-68}$$

This approximation is good for a large curvature where e is small with $r_n \doteq r_c$ and $s \doteq y$. In this case, Eq. (4–65) can be approximated by $\sigma \doteq Ms/Aer$. Substituting e from Eq. (4–68) yields

$$\sigma \doteq \frac{Ms}{I} \frac{r_c}{r} \tag{4-69}$$

Table 4–5

Formulas for Sections of
Curved Beams

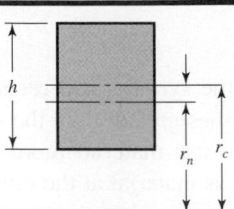

$$r_c = r_i + \frac{h}{2}$$

$$r_n = \frac{h}{\ln(r_o/r_i)}$$

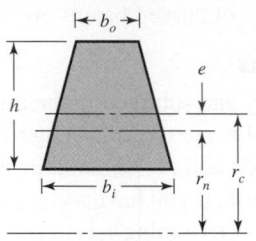

$$r_c = r_i + \frac{h}{3}\frac{b_i + 2b_o}{b_i + b_o}$$

$$r_n = \frac{A}{b_o - b_i + [(b_i r_o - b_o r_i)/h]\ln(r_o/r_i)}$$

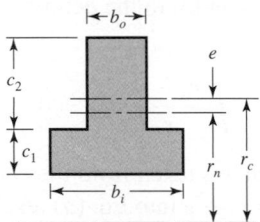

$$r_c = r_i + \frac{b_i c_1^2 + 2b_o c_1 c_2 + b_o c_2^2}{2(b_o c_2 + b_i c_1)}$$

$$r_n = \frac{b_i c_1 + b_o c_2}{b_i \ln[(r_i + c_1)/r_i)] + b_o \ln[r_o/(r_i + c_1)]}$$

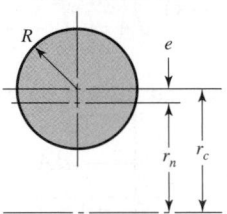

$$r_c = r_i + R$$

$$r_n = \frac{R^2}{2\left(r_c - \sqrt{r_c^2 - R^2}\right)}$$

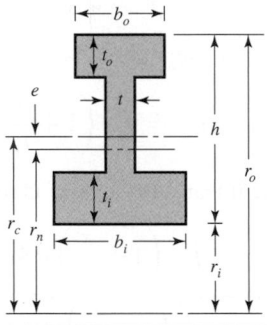

$$r_c = r_i + \frac{\frac{1}{2}h^2 t + \frac{1}{2}t_i^2(b_i - t) + t_o(b_o - t)(h - t_o/2)}{t_i(b_i - t) + t_o(b_o - t) + ht}$$

$$r_n = \frac{t_i(b_i - t) + t_o(b_o - t) + ht_o}{b_i \ln\dfrac{r_i + t}{r_i} + t \ln\dfrac{r_o - t_o}{r_i + t_i} + b_o \ln\dfrac{r_o}{r_o - t_o}}$$

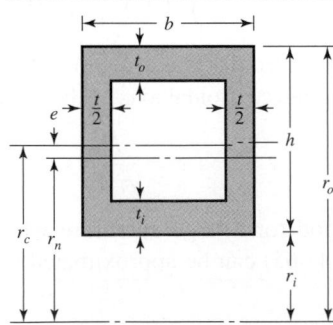

$$r_c = r_i + \frac{\frac{1}{2}h^2 t + \frac{1}{2}t_i^2(b - t) + t_o(b - t)(h - t_o/2)}{ht + (b - t)(t_i + t_o)}$$

$$r_n = \frac{(b - t)(t_i + t_o) + ht}{b\left(\ln\dfrac{r_i + t_i}{r_i} + t \ln\dfrac{r_o}{r_o + t_o}\right) + t \ln\dfrac{r_o - t_o}{r_i + t_i}}$$

Figure 4–39

Discrete representation of a curved beam cross section.

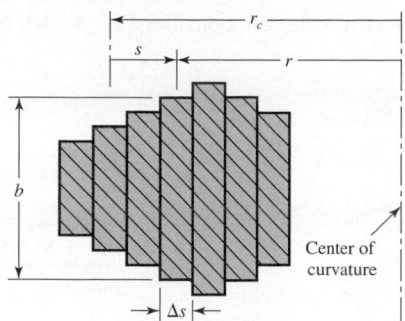

As $r_c \rightarrow \infty$, $r \rightarrow r_c$ and $s \rightarrow y$, we arrive at the straight-beam bending stress equation $\sigma = My/I$. Note that the negative sign is missing because y in Fig. 4–37 is vertically downward, opposite that for the straight-beam equation.

Keep in mind that Eqs. (4–68) and (4–69) are estimates and degrade significantly as r_c decreases relative to the beam depth.

Numerical Calculations

The integrals of Eq. (4–67) can be evaluated numerically by using a spreadsheet; creating a simple program; or using a mathematics program such as Matlab, Mathcad, Maple, or Mathematica. The area can be divided into infinitesimal rectangles of thickness Δs and width b evaluated at s as shown in Fig. 4–39. The resulting equations for r_c and e would then be

$$r_c = \frac{\sum rb \, \Delta s}{\sum b \, \Delta s} \qquad (4\text{–}70)$$

$$e = \frac{\sum \dfrac{s}{r_c - s} b \, \Delta s}{\sum \dfrac{b \, \Delta s}{r_c - s}} \qquad (4\text{–}71)$$

EXAMPLE 4–16

Consider the circular section in Table 4–5 with $r_c = 3$ in and $R = 1$ in. Determine e by using the formula from the table, approximately by using Eq. (4–68), and numerically by using Eq. (4–71). Compare the results of the three solutions.

Solution

Using the formula from Table 4–5 gives

$$r_n = \frac{R^2}{2\left(r_c - \sqrt{r_c^2 - R^2}\right)} = \frac{1^2}{2\left(3 - \sqrt{3^2 - 1}\right)} = 2.91421 \text{ in}$$

This gives an eccentricity of

$$e = r_c - r_n = 3 - 2.91421 = 0.08579 \text{ in}$$

The approximate method, using Eq. (4–68), yields

$$e \doteq \frac{I}{r_c A} = \frac{\pi R^4/4}{r_c(\pi R^2)} = \frac{R^2}{4r_c} = \frac{1^2}{4(3)} = 0.08333 \text{ in}$$

This differs from the exact solution by −2.9 percent.

For the numerical solution, consider Fig. 4–40. For a circle,

$$b = 2\sqrt{R^2 - s^2}$$ (a)

Substituting this into Eq. (4–71) gives

$$e = \frac{\sum \dfrac{s\sqrt{R^2 - s^2}}{r_c - s}\Delta s}{\sum \dfrac{\sqrt{R^2 - s^2}\Delta s}{r_c - s}}$$ (b)

Although this may look complicated, it is rather easy to solve numerically. Figure 4–41a shows a simple Visual Basic program that can be implemented in the spreadsheet program, Excel. In the program, SUM1 and SUM2 are the numerator and

Figure 4–40

Circular cross section of a curved beam.

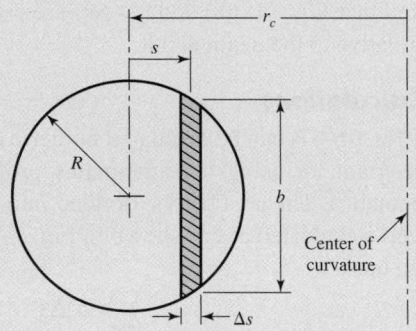

Figure 4–41

(a) Visual Basic program to determine e; (b) spreadsheet entries.

```
Function ecc (rc, R)
DS = 2 * R/1000  'Divide the diameter into 1000 increments
S = −R C DS/2  'Start the summation at s = −R + DS/2
          'for the calculation of the numerator (SUM1)
          'and the denominator (SUM2)
    SUM1 = 0        'Initialize the summations
    SUM2 = 0
    For I = 1 To 1000 Step 1     'Summation loop
        SUM1 = SUM1 + DS * (S * Sqr (R^2 − S^2)) / (rc − S)
        SUM2 = SUM2 + DS * (Sqr (R^2 − S^2)) / (rc − S)
        S = S + DS     'increment S
    Next I
ecc = SUM1/SUM2     'Determine e
End Function
```

(a)

	A	B	C
1	3	1	= ecc(A1, B1)

(b)

denominator of Eq. (*b*), respectively. Figure 4–41*b* shows the entries made into row 1 of the spreadsheet. In the first row, substituting $r_c = 3$ into column A, $R = 1$ into column B, and the expression "= ecc(A1, B1)" into column C will yield

$$e = 0.08579 \text{ in}$$

which is in error by 0.0031 percent.

The table below shows the error in the approximate and numerical solutions as $r_c \rightarrow R$. One can see that for a large radius of curvature the approximate method is quite adequate. However, as $r_c \rightarrow R$ the error becomes appreciable, whereas the error in the numerical method remains very low.

Values of *e* for a Circular Section for Varying r_c / R

r_c	R	Exact *e*	Approximate *e*	% Error	Numerical *e*	% Error
10	1	0.025063	0.025000	−0.25	0.025064	0.0030
8	1	0.031373	0.031250	−0.39	0.031374	0.0030
6	1	0.041960	0.041667	−0.70	0.041961	0.0030
4	1	0.063508	0.062500	−1.59	0.063510	0.0030
3	1	0.085786	0.083333	−2.86	0.085789	0.0031
2	1	0.133975	0.125000	−6.70	0.133979	0.0033
1.5	1	0.190983	0.166667	−12.73	0.190990	0.0038
1.25	1	0.250000	0.200000	−20.00	0.250012	0.0048

4–20 Contact Stresses

When two bodies having curved surfaces are pressed together, point or line contact changes to area contact, and the stresses developed in the two bodies are three-dimensional. Contact-stress problems arise in the contact of a wheel and a rail, in automotive valve cams and tappets, in mating gear teeth, and in the action of rolling bearings. Typical failures are seen as cracks, pits, or flaking in the surface material.

The most general case of contact stress occurs when each contacting body has a double radius of curvature; that is, when the radius in the plane of rolling is different from the radius in a perpendicular plane, both planes taken through the axis of the contacting force. Here we shall consider only the two special cases of contacting spheres and contacting cylinders.[10] The results presented here are due to Hertz and so are frequently known as *Hertzian stresses*.

Spherical Contact

When two solid spheres of diameters d_1 and d_2 are pressed together with a force F, a circular area of contact of radius a is obtained. Specifying E_1, v_1 and E_2, v_2 as the respective elastic constants of the two spheres, the radius a is given by the equation

$$a = \sqrt[3]{\frac{3F}{8} \frac{\left(1 - v_1^2\right)/E_1 + \left(1 - v_2^2\right)/E_2}{1/d_1 + 1/d_2}} \qquad (4\text{–}72)$$

---

[10]A more comprehensive presentation of contact stresses may be found in Arthur P. Boresi and Richard J. Schmidt, *Advanced Mechanics of Materials,* 6th ed., Wiley, New York, 2003 pp. 589–623.

Figure 4–42

(a) Two spheres held in contact by force F; (b) contact stress has a hemispherical distribution across contact zone diameter $2a$.

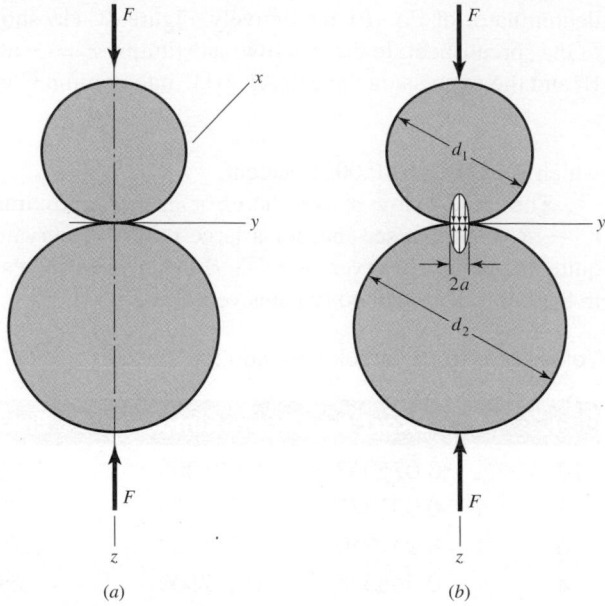

(a) (b)

The pressure distribution within the contact area of each sphere is hemispherical, as shown in Fig. 4–42. The maximum pressure occurs at the center of the contact area and is

$$p_{\max} = \frac{3F}{2\pi a^2} \tag{4–73}$$

Equations (4–72) and (4–73) are perfectly general and also apply to the contact of a sphere and a plane surface or of a sphere and an internal spherical surface. For a plane surface, use $d = \infty$. For an internal surface, the diameter is expressed as a negative quantity.

The maximum stresses occur on the z axis, and these are principal stresses. Their values are

$$\sigma_1 = \sigma_2 = \sigma_x = \sigma_y = -p_{\max}\left[\left(1 - \left|\frac{z}{a}\right|\tan^{-1}\frac{1}{|z/a|}\right)(1+v) - \frac{1}{2\left(1 + \dfrac{z^2}{a^2}\right)}\right]$$

$$\tag{4–74}$$

$$\sigma_3 = \sigma_z = \frac{-p_{\max}}{1 + \dfrac{z^2}{a^2}} \tag{4–75}$$

These equations are valid for either sphere, but the value used for Poisson's ratio must correspond with the sphere under consideration. The equations are even more complicated when stress states off the z axis are to be determined, because here the x and y coordinates must also be included. But these are not required for design purposes, because the maxima occur on the z axis.

Mohr's circles for the stress state described by Eqs. (4–74) and (4–75) are a point and two coincident circles. Since $\sigma_1 = \sigma_2$, we have $\tau_{1/2} = 0$ and

$$\tau_{\max} = \tau_{1/3} = \tau_{2/3} = \frac{\sigma_1 - \sigma_3}{2} = \frac{\sigma_2 - \sigma_3}{2} \tag{4–76}$$

Figure 4–43 is a plot of Eqs. (4–74), (4–75), and (4–76) for a distance to $3a$ below the surface. Note that the shear stress reaches a maximum value slightly below the surface. It is the opinion of many authorities that this maximum shear stress is responsible for the surface fatigue failure of contacting elements. The explanation is that a crack originates at the point of maximum shear stress below the surface and progresses to the surface and that the pressure of the lubricant wedges the chip loose.

Cylindrical Contact

Figure 4–44 illustrates a similar situation in which the contacting elements are two cylinders of length l and diameters d_1 and d_2. As shown in Fig. 4–44b, the area of contact

Figure 4–43

Magnitude of the stress components below the surface as a function of the maximum pressure of contacting spheres. Note that the maximum shear stress is slightly below the surface at $z = 0.48a$ and is approximately $0.3p_{max}$. The chart is based on a Poisson ratio of 0.30. Note that the normal stresses are all compressive stresses.

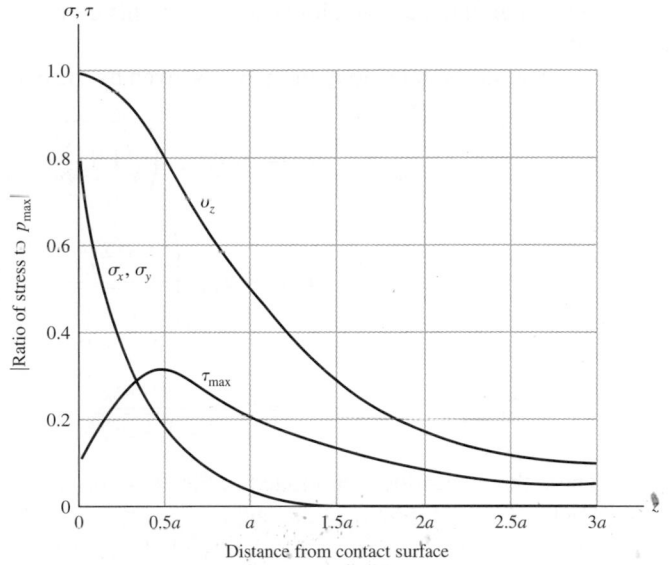

Figure 4–44

(a) Two right circular cylinders held in contact by forces F uniformly distributed along cylinder length l. (b) Contact stress has an elliptical distribution across the contact zone width $2b$.

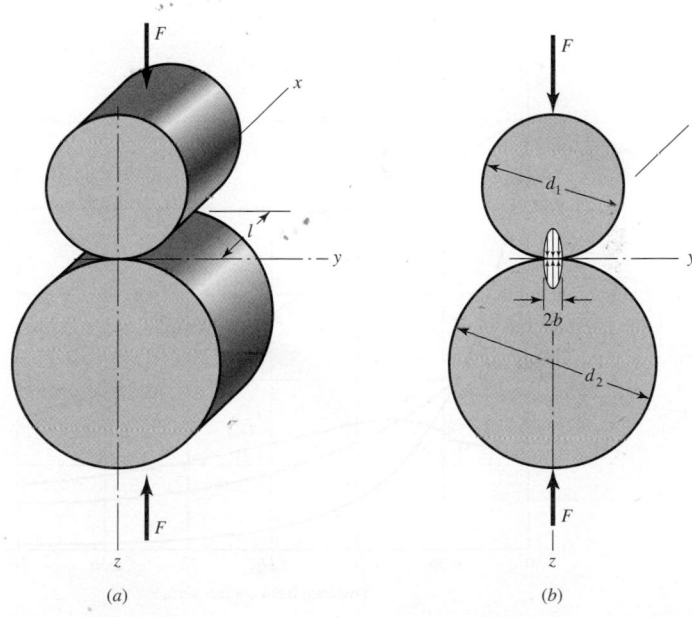

(a) (b)

is a narrow rectangle of width $2b$ and length l, and the pressure distribution is elliptical. The half-width b is given by the equation

$$b = \sqrt{\frac{2F}{\pi l} \frac{\left(1 - v_1^2\right)/E_1 + \left(1 - v_2^2\right)/E_2}{1/d_1 + 1/d_2}} \tag{4–77}$$

The maximum pressure is

$$p_{max} = \frac{2F}{\pi bl} \tag{4–78}$$

Equations (4–77) and (4–78) apply to a cylinder and a plane surface, such as a rail, by making $d = \infty$ for the plane surface. The equations also apply to the contact of a cylinder and an internal cylindrical surface; in this case d is made negative for the internal surface.

The stress state along the z axis is given by the equations

$$\sigma_x = -2v p_{max} \left(\sqrt{1 + \frac{z^2}{b^2}} - \left|\frac{z}{b}\right|\right) \tag{4–79}$$

$$\sigma_y = -p_{max} \left(\frac{1 + 2\dfrac{z^2}{b^2}}{\sqrt{1 + \dfrac{z^2}{b^2}}} - 2\left|\frac{z}{b}\right|\right) \tag{4–80}$$

$$\sigma_3 = \sigma_z = \frac{-p_{max}}{\sqrt{1 + z^2/b^2}} \tag{4–81}$$

These three equations are plotted in Fig. 4–45 up to a distance of $3b$ below the surface. For $0 \le z \le 0.436b$, $\sigma_1 = \sigma_x$, and $\tau_{max} = (\sigma_1 - \sigma_3)/2 = (\sigma_x - \sigma_z)/2$. For $z \ge 0.436b$, $\sigma_1 = \sigma_y$, and $\tau_{max} = (\sigma_y - \sigma_z)/2$. A plot of τ_{max} is also included in Fig. 4–45, where the greatest value occurs at $z/b = 0.786$ with a value of $0.300\, p_{max}$.

Figure 4–45

Magnitude of the stress components below the surface as a function of the maximum pressure for contacting cylinders. Shear stress τ becomes the largest of the three shear stresses at about $z/b = 0.75$. Its maximum value is $0.30 p_{max}$. The chart is based on a Poisson ratio of 0.30. Note that all normal stresses are compressive stresses.

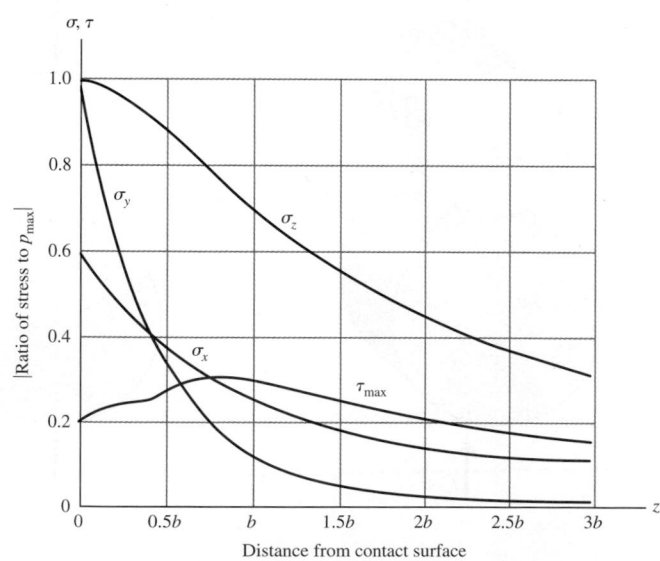

Hertz (1881) provided the preceding mathematical models of the stress field when the contact zone is free of shear stress. Another important contact stress case is *line-of-contact* with friction providing the shearing stress on the contact zone. Such shearing stresses are small with cams and rollers, but in cams with flatfaced followers, wheel-rail contact, and gear teeth, the stresses are elevated above the Hertzian field. Investigations of the effect on the stress field due to normal and shear stresses in the contact zone was begun theoretically by Lundberg (1939), and continued by Mindlin (1949), Smith-Liu (1949), and Poritsky (1949) independently. The Smith-Liu equations[11] for contacting and sliding cylinders are

$$b = \sqrt{\frac{2F}{\pi l} \frac{\left(1 - v_1^2\right)/E_1 + \left(1 - v_2^2\right)/E_2}{1/d_1 + 1/d_2}}$$

$$M = \sqrt{(b + y)^2 + z^2}$$

$$N = \sqrt{(b - y)^2 + z^2}$$

$$\phi_1 = \frac{\pi(M + N)}{MN\sqrt{2MN + 2y^2 + 2z^2 - 2b^2}}$$

$$\phi_2 = \frac{\pi(M - N)}{MN\sqrt{2MN + 2y^2 + 2z^2 - 2b^2}}$$

$$\Delta' = \frac{2}{1/R_1 + 1/R_2}\left(\frac{1 - v_1^2}{E_1} + \frac{1 - v_2^2}{E_2}\right)$$

$$\sigma_x = -\frac{2vb}{\pi\Delta'}\left\{z\left(\frac{b^2 + y^2 + z^2}{b}\phi_1 - \frac{\pi}{b} - 2y\phi_2\right)\right.$$
$$\left. + f\left[(y^2 - b^2 - z^2)\phi_2 + \frac{\pi y}{b} + (b^2 - y^2 - z^2)\frac{y}{b}\phi_1\right]\right\} \qquad (4\text{–}82)$$

$$\sigma_y = -\frac{b}{\pi\Delta'}\left\{z\left(\frac{b^2 + 2z^2 + 2y^2}{b}\phi_1 - \frac{2\pi}{b} - 3y\phi_2\right)\right.$$
$$\left. + f\left[(2y^2 - 2b^2 - 3z^2)\phi_2 + \frac{2\pi y}{b} + 2(b^2 - y^2 - z^2)\frac{y}{b}\phi_1\right]\right\} \qquad (4\text{–}83)$$

$$\sigma_z = -\frac{b}{\pi\Delta'}[z(b\phi_1 - y\phi_2 + fz\phi_2)] \qquad (4\text{–}84)$$

$$\tau_{yz} = -\frac{b}{\pi\Delta'}\left\{z^2\phi_2 + f\left[(b^2 + 2y^2 + 2z^2)\frac{z}{b}\phi_1 - 2\pi\frac{z}{b} - 3yz\phi_2\right]\right\} \qquad (4\text{–}85)$$

where f is the coefficient of sliding friction.

In the Smith-Liu case, axes y and z of Fig. 4–41a are no longer principal axes, and σ_y and σ_z are no longer principal stresses. A glance at these equations will convince the reader that although this was reported in 1953,[12] applications grew slowly as computing

---

[11] A widely held library reference is G. A. Castleberry, "Analyzing Contact Stresses More Accurately," *Machine Design,* April 12, 1984, pp. 92–97.

[12] J. O. Smith and Chang Keng Liu, "Stresses Due to Tangential and Normal Loads on an Elastic Solid with Application to Some Contact Stress Problems," *Journal of Applied Mechanics,* June 1953.

aids improved. A way to study these equations is to program them to create a table running $-1 \leq y/b \leq 1$ across the top and $0 \leq y/b \leq 1$ down. Fill the body of the table with one of σ_x, σ_y, σ_z, τ_{yz}, σ_1, σ_2, σ_3, $\tau_{1/2}$, $\tau_{1/3}$, $\tau_{2/3}$, and τ_{max}, normalized by dividing by the maximum Hertzian contact zone stress, p_{max}. Failure theory investigators examine (1) maximum compressive stress, (2) maximum tensile stress, (3) largest maximum shear stress, and (4) largest alternating shear-stress amplitude. See the problems at the end of the chapter.

4–21   Summary

The ability to quantify the stress condition at a critical location in a machine element is an important skill of the engineer. Why? Whether the member fails or not is assessed by comparing the (damaging) stress at a critical location with the corresponding material strength at this location. This chapter has addressed the description of stress.

Stresses can be estimated with great precision where the geometry is sufficiently simple that theory easily provides the necessary quantitative relationships. In other cases, approximations are used. There are numerical approximations such as finite element analysis (FEA), whose results tend to converge on the true values. There are experimental measurements, strain gauging, for example, allowing *inference* of stresses from the measured strain conditions. Whatever the method(s), the goal is a robust description of the stress condition at a critical location.

The nature of research results and understanding in any field is that the longer we work on it, the more involved things seem to be, and new approaches are sought to help with the complications. As newer schemes are introduced, engineers, hungry for the improvement the new approach *promises,* begin to use the approach. Optimism usually recedes, as further experience adds concerns. Tasks that promised to extend the capabilities of the nonexpert eventually show that expertise is not optional.

In stress analysis, the computer can be helpful if the necessary equations are available. Spreadsheet analysis can quickly reduce complicated calculations for parametric studies, easily handling "what if" questions relating trade-offs (e.g., less of a costly material or more of a cheaper material). It can even give insight into optimization opportunities.

When the necessary equations are not available, then methods such as FEA are attractive, but cautions are in order. Even when you have access to a powerful FEA code, you should be near an expert while you are learning. There are nagging questions of convergence at discontinuities. Elastic analysis is much easier than elastic-plastic analysis. The results are no better than the modeling of reality that was used to formulate the problem.

PROBLEMS

ANALYSIS

4–1   The symbol W is used in the various figure parts to specify the weight of an element. If not given, assume the parts are weightless. For each figure part, sketch a free-body diagram of each element, including the frame. Try to get the forces in the proper directions, but do not compute magnitudes.

ANALYSIS

4–2   Using the figure part selected by your instructor, sketch a free-body diagram of each element in the figure. Compute the magnitude and direction of each force using an algebraic or vector method, as specified.

ANALYSIS

4–3   Find the reactions at the supports and plot the shear-force and bending-moment diagrams for each of the beams shown in the figure on page 168. Label the diagrams properly.

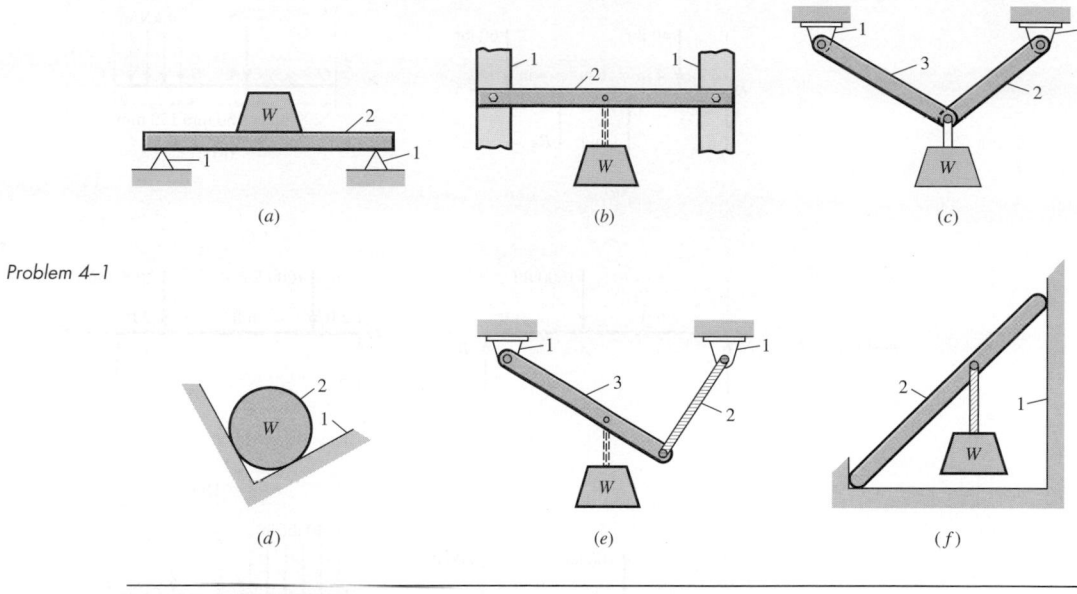

Problem 4–1

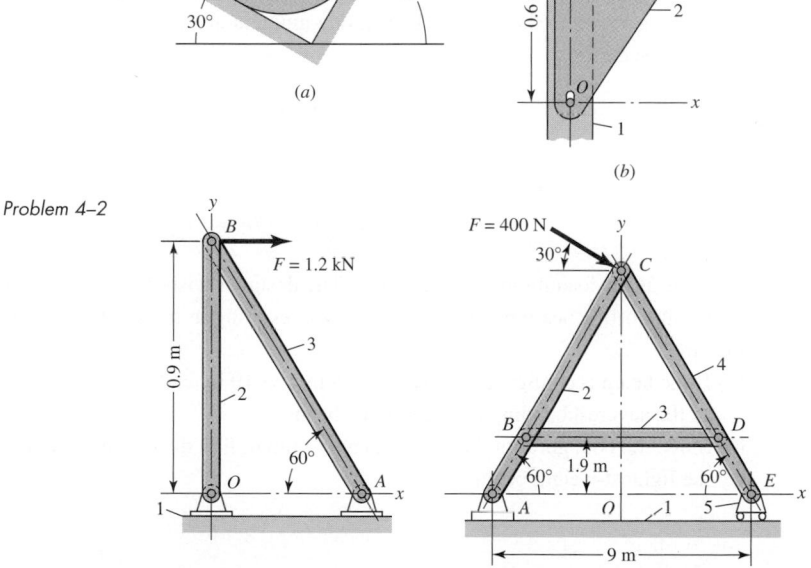

Problem 4–2

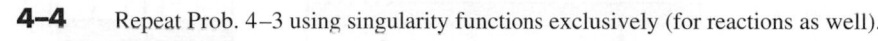

4–4 Repeat Prob. 4–3 using singularity functions exclusively (for reactions as well).

4–5 Select a beam from Table A–9 and find general expressions for the loading, shear-force, bending-moment, and support reactions. Use the method specified by your instructor.

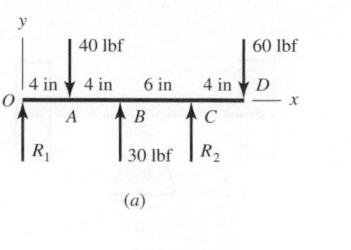

(a)

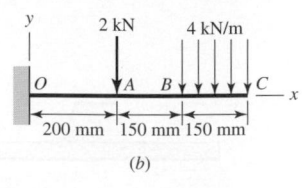

(b)

Problem 4–3

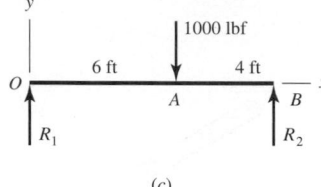

(c)

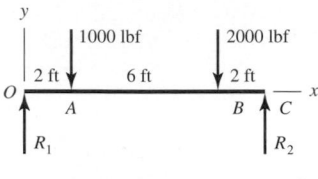

(d)

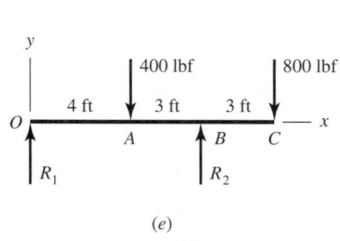

(e)

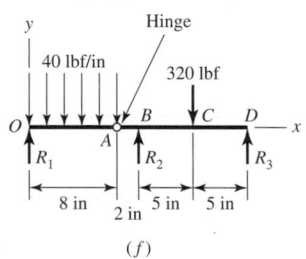

(f)

DESIGN

4–6   A beam carrying a uniform load is simply supported with the supports set back a distance a from the ends as shown in the figure. The bending moment at x can be found from summing moments to zero at section x:

$$\sum M = M + \frac{1}{2}w(a+x)^2 - \frac{1}{2}wlx = 0$$

or

$$M = \frac{w}{2}[lx - (a+x)^2]$$

where w is the loading intensity in lbf/in. The designer wishes to minimize the necessary weight of the supporting beam by choosing a setback resulting in the smallest possible maximum bending stress.

(a) If the beam is configured with $a = 2.25$ in, $l = 10$ in, and $w = 100$ lbf/in, find the magnitude of the severest bending moment in the beam.

(b) Since the configuration in part (a) is not optimal, find the optimal setback a that will result in the lightest-weight beam.

Problem 4–6

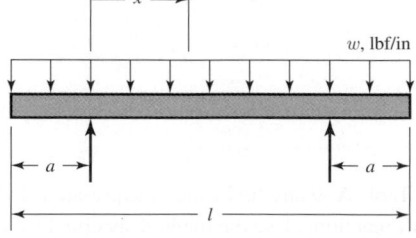
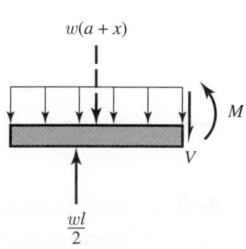

ANALYSIS

4–7

An artist wishes to construct a mobile using pendants, string, and span wire with eyelets as shown in the figure.

(*a*) At what positions w, x, y, and z should the suspension strings be attached to the span wires?

(*b*) Is the mobile stable? If so, justify; if not, suggest a remedy.

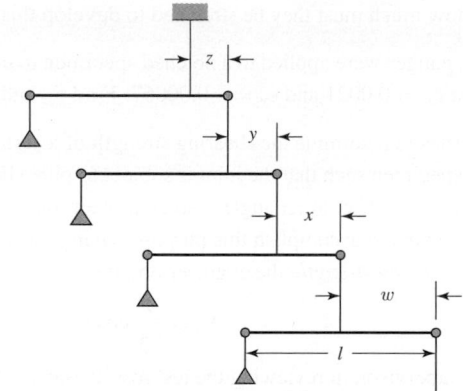

Problem 4–7

ANALYSIS

4–8

For each of the plane stress states listed below, draw a Mohr's circle diagram properly labeled, find the principal normal and shear stresses, and determine the angle from the x axis to σ_1. Draw stress elements as in Fig. 4–13*c* and *d* and label all details.

(*a*) $\sigma_x = 12$, $\sigma_y = 6$, $\tau_{xy} = 4$ cw

(*b*) $\sigma_x = 16$, $\sigma_y = 9$, $\tau_{xy} = 5$ ccw

(*c*) $\sigma_x = 10$, $\sigma_y = 24$, $\tau_{xy} = 6$ ccw

(*d*) $\sigma_x = 9$, $\sigma_y = 19$, $\tau_{xy} = 8$ cw

ANALYSIS

4–9

Repeat Prob. 4–8 for:

(*a*) $\sigma_x = -4$, $\sigma_y = 12$, $\tau_{xy} = 7$ ccw

(*b*) $\sigma_x = 6$, $\sigma_y = -5$, $\tau_{xy} = 8$ ccw

(*c*) $\sigma_x = -8$, $\sigma_y = 7$, $\tau_{xy} = 6$ cw

(*d*) $\sigma_x = 9$, $\sigma_y = -6$, $\tau_{xy} = 3$ cw

ANALYSIS

4–10

Repeat Prob. 4–8 for:

(*a*) $\sigma_x = 20$, $\sigma_y = -10$, $\tau_{xy} = 8$ cw

(*b*) $\sigma_x = 30$, $\sigma_y = -10$, $\tau_{xy} = 10$ ccw

(*c*) $\sigma_x = -10$, $\sigma_y = 18$, $\tau_{xy} = 9$ cw

(*d*) $\sigma_x = -12$, $\sigma_y = 22$, $\tau_{xy} = 12$ cw

ANALYSIS

4–11

For each of the stress states listed below, find all three principal normal and shear stresses. Draw a complete Mohr's three-circle diagram and label all points of interest.

(*a*) $\sigma_x = 10$, $\sigma_y = -4$

(*b*) $\sigma_x = 10$, $\tau_{xy} = 4$ ccw

(*c*) $\sigma_x = -2$, $\sigma_y = -8$, $\tau_{xy} = 4$ cw

(*d*) $\sigma_x = 10$, $\sigma_y = -30$, $\tau_{xy} = 10$ ccw

ANALYSIS

4–12

Repeat Prob. 4–11 for:

(*a*) $\sigma_x = -80$, $\sigma_y = -30$, $\tau_{xy} = 20$ cw

(*b*) $\sigma_x = 30$, $\sigma_y = -60$, $\tau_{xy} = 30$ cw

(*c*) $\sigma_x = 40$, $\sigma_z = -30$, $\tau_{xy} = 20$ ccw

(*d*) $\sigma_x = 50$, $\sigma_z = -20$, $\tau_{xy} = 30$ cw

 ANALYSIS

4–13 A $\frac{1}{2}$-in-diameter steel tension rod is 72 in long and carries a load of 2000 lbf. Find the tensile stress, the total deformation, the unit strains, and the change in the rod diameter.

 ANALYSIS

4–14 Twin diagonal aluminum alloy tension rods 15 mm in diameter are used in a rectangular frame to prevent collapse. The rods can safely support a tensile stress of 135 MPa. If the rods are initially 3 m in length, how much must they be stretched to develop this stress?

 ANALYSIS

4–15 Electrical strain gauges were applied to a notched specimen to determine the stresses in the notch. The results were $\epsilon_x = 0.0021$ and $\epsilon_y = -0.00067$. Find σ_x and σ_y if the material is carbon steel.

 DESIGN

4–16 An engineer wishes to determine the shearing strength of a certain epoxy cement. The problem is to devise a test specimen such that the joint is subject to pure shear. The joint shown in the figure, in which two bars are offset at an angle θ so as to keep the loading force F centroidal with the straight shanks, seems to accomplish this purpose. Using the contact area A and designating S_{su} as the *ultimate shearing strength,* the engineer obtains

$$S_{su} = \frac{F}{A} \cos \theta$$

The engineer's supervisor, in reviewing the test results, says the expression should be

$$S_{su} = \frac{F}{A} \left(1 + \frac{1}{4} \tan^2 \theta \right)^{1/2} \cos \theta$$

Resolve the discrepancy. What is your position?

Problem 4–16

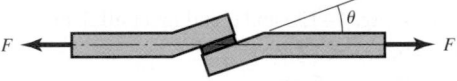

 ANALYSIS

4–17 The state of stress at a point is $\sigma_x = -2$, $\sigma_y = 6$, $\sigma_z = -4$, $\tau_{xy} = 3$, $\tau_{yz} = 2$, and $\tau_{zx} = -5$ kpsi. Determine the principal stresses, draw a complete Mohr's three-circle diagram, labeling all points of interest, and report the maximum shear stress for this case.

 ANALYSIS

4–18 Repeat Prob. 4–17 with $\sigma_x = 10$, $\sigma_y = 0$, $\sigma_z = 10$, $\tau_{xy} = 20$, $\tau_{yz} = -10\sqrt{2}$, and $\tau_{zx} = 0$ MPa.

 ANALYSIS

4–19 Repeat Prob. 4–17 with $\sigma_x = 1$, $\sigma_y = 4$, $\sigma_z = 4$, $\tau_{xy} = 2$, $\tau_{yz} = -4$, and $\tau_{zx} = -2$ kpsi.

 ANALYSIS

4–20 The Roman method for addressing uncertainty in design was to build a copy of a design that was satisfactory and had proven durable. Although the early Romans did not have the intellectual tools to deal with scaling size up or down, you do. Consider a simply supported, rectangular-cross-section beam with a concentrated load F, as depicted in the figure.

Problem 4–20

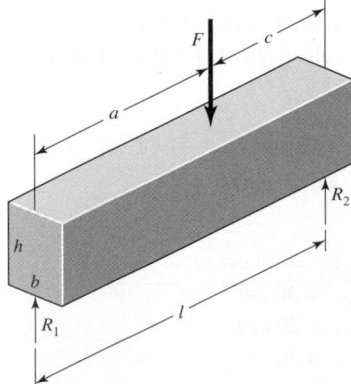

(*a*) Show that the stress-to-load equation is

$$F = \frac{\sigma bh^2 l}{6ac}$$

(*b*) Subscript every parameter with *m* (for model) and divide into the above equation. Introduce a scale factor, $s = a_m/a = b_m/b = c_m/c$ etc. Since the Roman method was to not "lean on" the material any more than the proven design, set $\sigma_m/\sigma = 1$. Express F_m in terms of the scale factors and F, and comment on what you have learned.

ANALYSIS **4–21** Using our experience with concentrated loading on a simple beam, Prob. 4–20, consider a uniformly loaded simple beam (Table A–9–7).

(*a*) Show that the stress-to-load equation for a rectangular-cross-section beam is given by

$$W = \frac{4}{3}\frac{\sigma bh^2}{l}$$

where $W = wl$.

(*b*) Subscript every parameter with *m* (for model) and divide the model equation into the prototype equation. Introduce the scale factor *s* as in Prob. 4–20, setting $\sigma_m/\sigma = 1$. Express W_m and w_m in terms of the scale factor, and comment on what you have learned.

ANALYSIS **4–22** The Chicago North Shore & Milwaukee Railroad was an electric railway running between the cities in its corporate title. It had passenger cars as shown in the figure, which weighed 104.4 kip, had 32-ft, 8-in truck centers, 7-ft-wheelbase trucks, and a coupled length of 55 ft, $3\frac{1}{4}$ in. Consider the case of a single car on a 100-ft-long, simply supported deck plate girder bridge.

(*a*) What was the largest bending moment in the bridge?

(*b*) Where on the bridge was the moment located?

(*c*) What was the position of the car on the bridge?

(*d*) Under which axle is the bending moment?

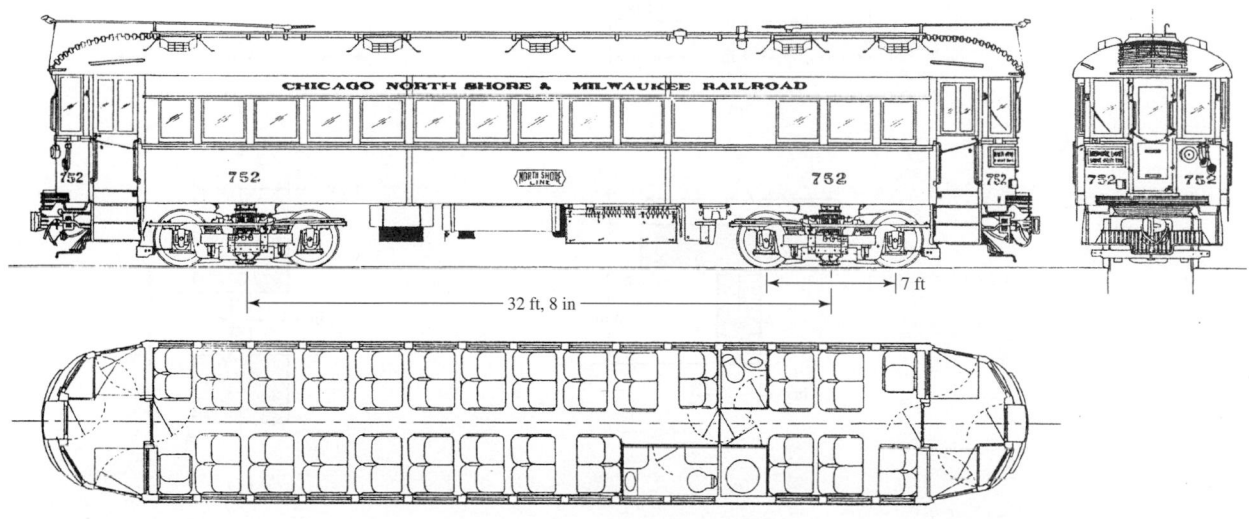

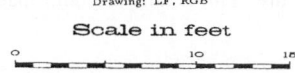

Scale in feet

Coaches 752-776

AS ORIGINALLY BUILT

Problem 4–22
Copyright 1963 by Central Electric Railfans Association, Bull. 107, p. 145, reproduced by permission.

ANALYSIS

4–23 For each section illustrated, find the second moment of area, the location of the neutral axis, and the distances from the neutral axis to the top and bottom surfaces. Suppose a positive bending moment of 10 kip · in is applied; find the resulting stresses at the top and bottom surfaces and at every abrupt change in cross section.

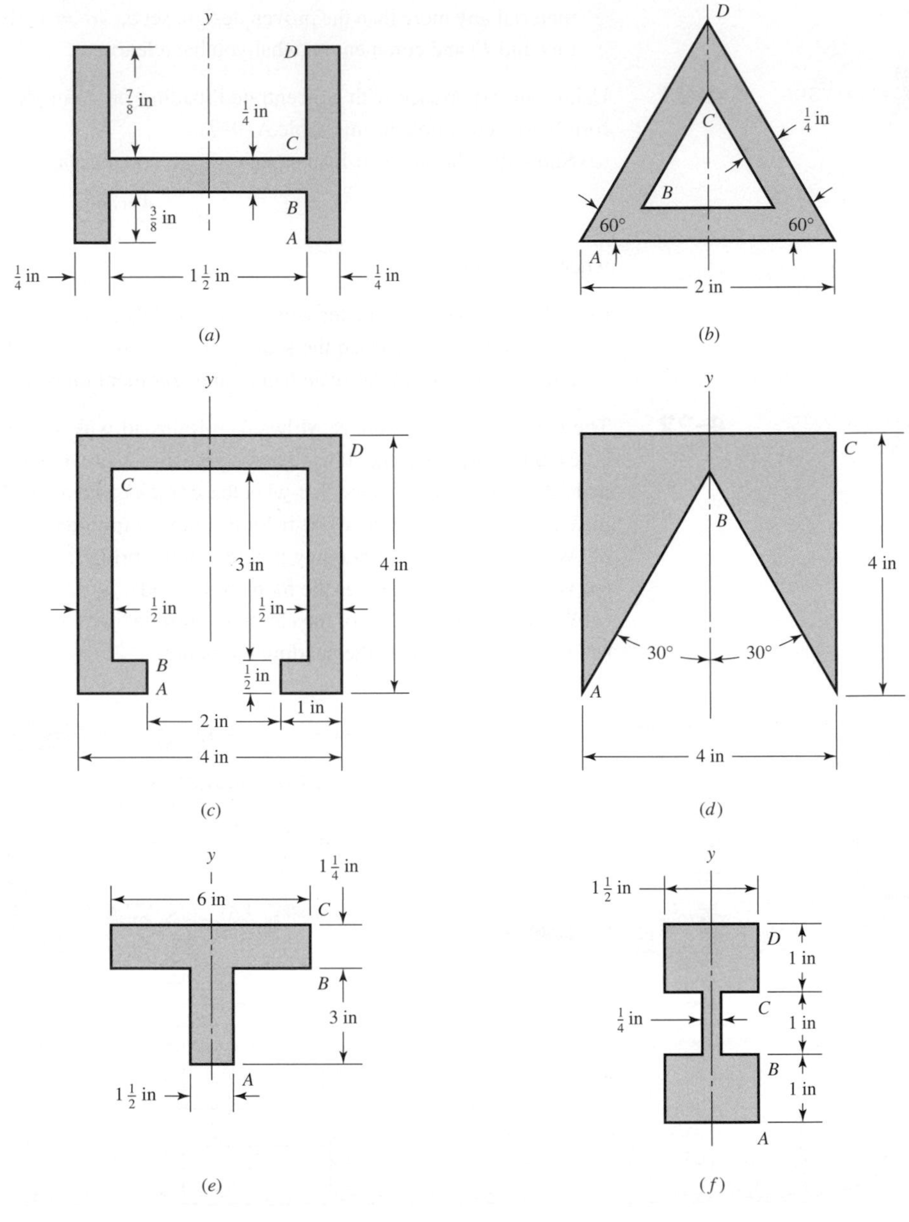

Problem 4–23

(a)

(b)

(c)

(d)

(e)

(f)

ANALYSIS

4–24 Find the x and y coordinates of the center of curvature corresponding to the place where the beam is bent the most, for each beam shown in the figure. The beams are both made of Douglas fir (see Table A–5) and have rectangular sections.

4–25 For each beam illustrated in the figure, find the locations and magnitudes of the maximum tensile bending stress and the maximum shear stress due to V.

Problem 4–24

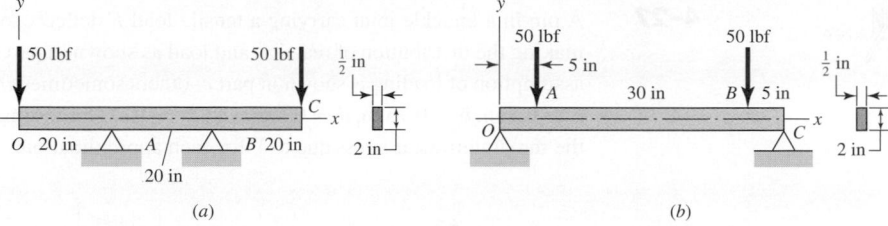

(a)   (b)

Problem 4–25

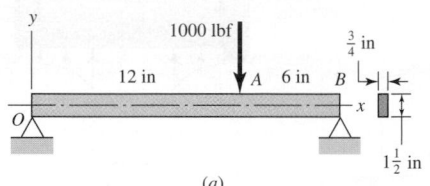   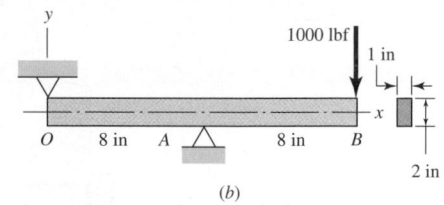

(a)   (b)

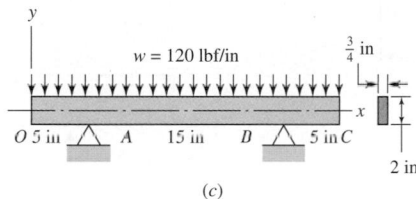   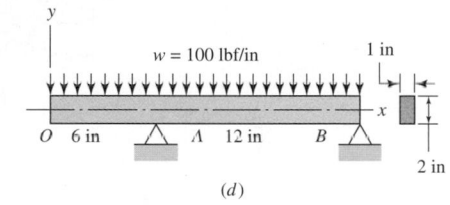

(c)   (d)

ANALYSIS

4–26 The figure illustrates a number of beam sections. Use an allowable stress of 1.2 kpsi for wood and 12 kpsi for steel and find the maximum safe uniformly distributed load that each beam can carry if the given lengths are between simple supports.

(a) Wood joist $1\frac{1}{2}$ by $9\frac{1}{2}$ in and 12 ft long

(b) Steel tube, 2 in OD by $\frac{3}{8}$-in wall thickness, 48 in long

(c) Hollow steel tube 3 by 2 in, outside dimensions, formed from $\frac{3}{16}$-in material and welded, 48 in long

(d) Steel angles $3 \times 3 \times \frac{1}{4}$ in and 72 in long

(e) A 5.4-lb, 4-in steel channel, 72 in long

(f) A 4-in $\times$ 1-in steel bar, 72 in long

Problem 4–26

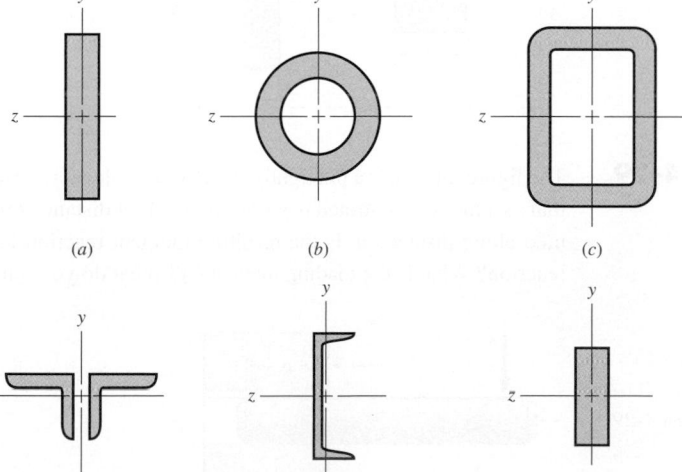

(a)   (b)   (c)

(d)   (e)   (f)

ANALYSIS

4–27   A pin in a knuckle joint carrying a tensile load F deflects somewhat on account of this loading, making the distribution of reaction and load as shown in part b of the figure. The usual designer's assumption of loading is shown in part c; others sometimes choose the loading shown in part d. If $a = 0.5$ in, $b = 0.75$ in, $d = 0.5$ in, and $F = 1000$ lbf, estimate the maximum bending stress and the maximum shear stress due to V for each approximation.

Problem 4–27

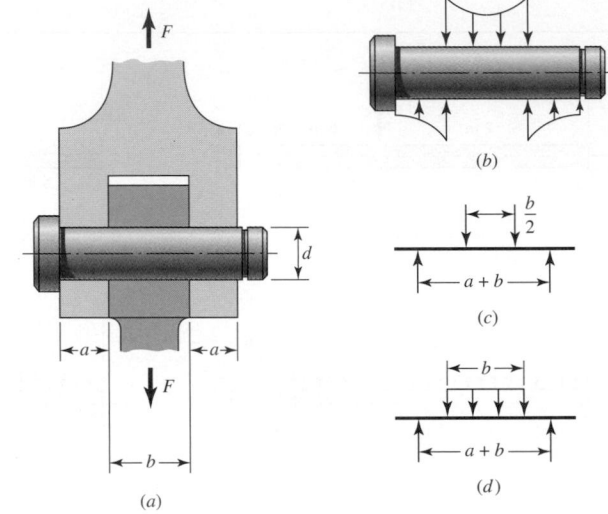

DESIGN

4–28   The figure illustrates two gears that are to be mounted on a shaft on the x axis. The gear loads are $F_1 = 2$ kip and $F_2 = 1.1$ kip. If one bearing is located at A, should the second bearing be located between the gears, or outboard, say, near B? Possible solutions will involve designing for minimum bending moment, or for equal bearing reactions.

Problem 4–28

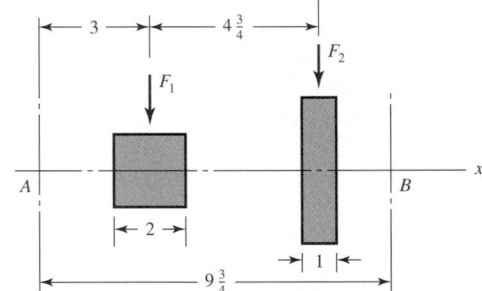

DESIGN

4–29   The figure illustrates a pin tightly fitted into a hole of a substantial member. A usual analysis is one that assumes concentrated reactions R and M at distance l from F. Suppose the reaction is distributed along distance a. Is the resulting moment reaction larger or smaller than the concentrated reaction? What is the loading intensity q? What do you think of using the usual assumption?

Problem 4–29

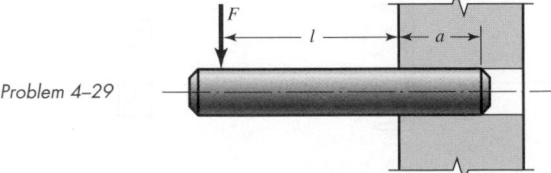

4–30 For a given cross section, a uniformly loaded beam of length l is simply supported as shown in the figure. If the supports are set back a short distance, the largest transverse bending moment is reduced. If the largest transverse bending moment can be minimized, the smallest beam section can be used. Write a computer program to discover the optimal setback a. Submit a listing of the programs you have written, the input and output of the production run, and the analysis on which your programming is based.

Problem 4–30

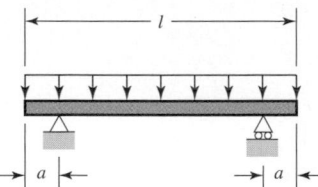

4–31 Consider a simply supported beam of rectangular cross section of constant width b and variable depth h, so proportioned that the maximum stress σ_x at the outer surface due to bending is constant, when subjected to a load F at a distance a from the left support and a distance c from the right support. Show that the depth h at location x is given by

$$h - \sqrt{\frac{6Fcx}{lb\sigma_{max}}} \qquad 0 \le x \le a$$

4–32 Consider a simply supported static beam of circular cross section of diameter d, so proportioned by varying the diameter such that the maximum stress σ_x at the surface due to bending is constant, when subjected to a steady load F located at a distance a from the left support and a distance b from the right support. Show that the diameter d at a location x is given by

$$d = \left(\frac{32Fbx}{\pi l\sigma_{max}}\right)^{1/3} \qquad 0 \le x \le a$$

4–33 Two steel thin-wall tubes of axial length z are to be compared. The first is of square cross section, side length b, and wall thickness t. The second is a round of diameter b and wall thickness t. The largest allowable shear stress is τ_{all} and is to be the same in both cases. How does the angle of twist per unit length compare in each case?

4–34 Begin with a 1-in-square thin-wall steel tube, wall thickness $t = 0.05$ in, length 40 in, then introduce corner radii of inside radii r_i, with allowable shear stress τ_{all} of 11 500 psi, shear modulus of $11.5(10^6)$ psi; now form a table. Use a column of inside corner radii in the range $0 \le r_i \le 0.45$ in. Useful columns include median line radius r_m, periphery of the median line L_m, area enclosed by median curve, torque T, and the angular twist θ. The cross section will vary from square to circular round. A computer program will reduce the calculation effort. Study the table. What have you learned?

Problem 4–34

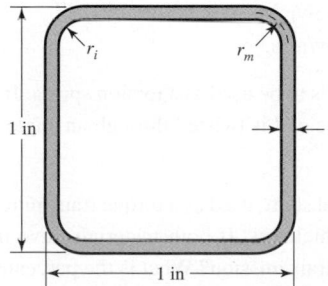

ANALYSIS

4–35 An unequal leg angle shown in the figure carries a torque T. Show that

$$T = \frac{G\theta_1}{3} \sum L_i c_i^3$$

$$\tau_{max} = G\theta_1 c_{max}$$

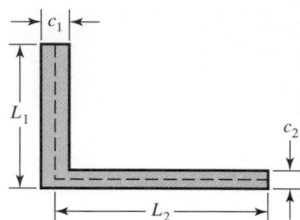

Problem 4–35

ANALYSIS

4–36 In Prob. 4–35 the angle has one leg thickness $\frac{1}{16}$ in and the other $\frac{3}{32}$ in, with both leg lengths $\frac{1}{2}$ in. The allowable shear stress is $\tau_{all} = 11\,500$ psi for this steel angle.
(a) Find the torque carried by each leg, and the largest shear stress therein.
(b) Find the angle of twist per unit length of the section.

ANALYSIS

4–37 Two 12 in long thin rectangular steel strips are placed together as shown. Using a maximum allowable shear stress of 11 500 psi, determine the maximum torque and angular twist, and the torsional spring rate. Compare these with a single strip of cross section 1 in by $\frac{1}{8}$ in.

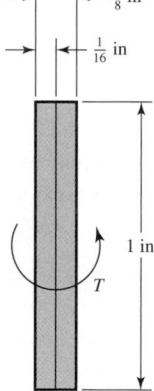

Problem 4–37

ANALYSIS

4–38 Using a maximum allowable shear stress of 8000 psi, find the shaft diameter needed to transmit 50 hp when
(a) The shaft speed is 2000 rev/min.
(b) The shaft speed is 200 rev/min.

ANALYSIS

4–39 A 15-mm-diameter steel bar is to be used as a torsion spring. If the torsional stress in the bar is not to exceed 110 MPa when one end is twisted through an angle of 30°, what must be the length of the bar?

ANALYSIS

4–40 A 70-mm-diameter solid steel shaft, used as a torque transmitter, is replaced with a 70-mm hollow shaft having a 6-mm wall thickness. If both materials have the same strength, what is the percentage reduction in torque transmission? What is the percentage reduction in shaft weight?

ANALYSIS

4–41 A hollow steel shaft is to transmit 5400 N · m of torque and is to be sized so that the torsional stress does not exceed 150 MPa.

(a) If the inside diameter is three-fourths of the outside diameter, what size shaft should be used? Use preferred sizes.

(b) What is the stress on the inside of the shaft when full torque is applied?

DESIGN

4–42 The figure shows an endless-belt conveyor drive roll. The roll has a diameter of 6 in and is driven at 5 rev/min by a geared-motor source rated at 1 hp. Determine a suitable shaft diameter d_C for an allowable torsional stress of 14 kpsi.

(a) What would be the stress in the shaft you have sized if the motor starting torque is twice the running torque?

(b) Is bending stress likely to be a problem? What is the effect of different roll lengths B on bending?

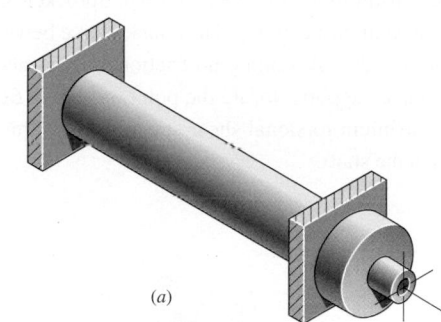

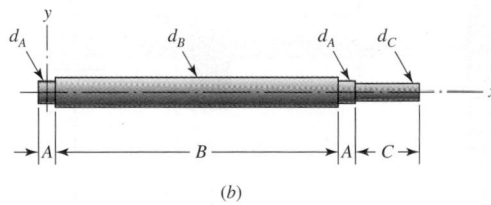

Problem 4–42

(a)

(b)

DESIGN

4–43 The conveyer drive roll in the figure for Prob. 4–42 is 150 mm in diameter and is driven at 8 rev/min by a geared-motor source rated at 1 kW. Find a suitable shaft diameter d_C based on an allowable torsional stress of 75 MPa.

ANALYSIS

4–44 For the same cross-sectional area $A = s^2 = \pi d^2/4$, for a square cross-sectional area shaft and a circular cross-sectional area shaft, in torsion which has the higher maximum shear stress, and by what multiple is it higher?

ANALYSIS

4–45 For the same cross-sectional area $A = s^2 = \pi d^2/4$, for a square cross-sectional area shaft and a circular cross-sectional area shaft, both of length l, in torsion which has the greater angular twist θ, and by what multiple is it greater?

ANALYSIS

4–46 W. C. Young and R. G. Budynas, in *Roark's Formulas for Stress and Strain,* consider torsion of a rectangular cross section $2a$ by $2b$, where $2a$ is the longer side, and the relation for the maximum shear stress in the center of the longest side as a function of b/a is given by

$$\tau_{\max} = \frac{3T}{8ab^2} \left[1 + 0.6095\frac{b}{a} + 0.8865\left(\frac{b}{a}\right)^2 - 1.8023\left(\frac{b}{a}\right)^3 + 0.9100\left(\frac{b}{a}\right)^4 \right]$$

Compare the first two terms of the polynomial expression in b/a with Eq. (4–43). If you find a discrepancy, comment about the reason(s). Equation (4–43) has the form

$$\tau_{max} = \frac{T}{bc^2}\frac{1}{\alpha}$$

From the tabulation of α versus b/c, plot $1/\alpha$ as ordinate and $1/(b/c)$ as abscissa. If the data string is linear or nearly so, find a_0 and a_1 of

$$\frac{1}{\alpha} = a_0 + a_1\frac{1}{b/c}$$

by linear regression. Compare the results with Eq. (4–43).

4–47   A torque of $T = 1000$ lbf · in is applied to the shaft EFG, which is running at constant speed and contains gear F. Gear F transmits torque to shaft $ABCD$ through gear C, which drives the chain sprocket at B, transmitting a force P as shown. Sprocket B, gear C, and gear F have pitch diameters of 6, 10, and 5 in, respectively. The contact force between the gears is transmitted through the pressure angle $\phi = 20°$. Assuming no frictional losses and considering the bearings at A, D, E, and G to be simple supports, locate the point on shaft $ABCD$ that contains the maximum tensile bending and maximum torsional shear stresses. From this, determine the maximum tensile and shear stresses in the shaft.

Problem 4–47

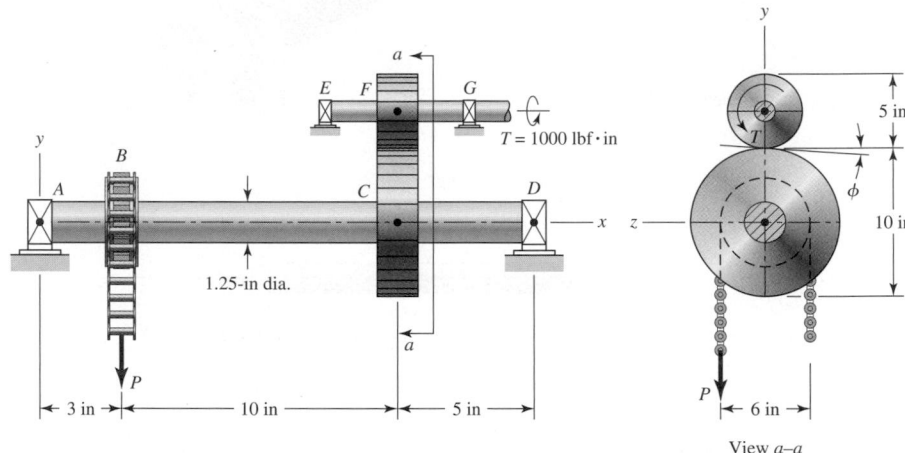

View a–a

ANALYSIS

4–48   If the tension-loaded plate of Fig. 4–31 is infinitely wide, then the stress state anywhere in the plate can be described in polar coordinates as

$$\sigma_r = \frac{\sigma}{2}\left(1 - \frac{d^2}{4r^2}\right) - \frac{\sigma}{2}\left(1 - \frac{d^2}{4r^2}\right)\left(1 - \frac{3d^2}{4r^2}\right)\cos 2\theta$$

$$\sigma_\theta = \frac{\sigma}{2}\left(1 + \frac{d^2}{4r^2}\right) + \frac{\sigma}{2}\left(1 + \frac{3d^4}{16r^4}\right)\cos 2\theta$$

$$\tau_{r\theta} = \frac{\sigma}{2}\left(1 - \frac{d^2}{4r^2}\right)\left(1 + \frac{3d^2}{4r^2}\right)\sin 2\theta$$

for the radial, tangential, and shear components, respectively. Here r is the distance from the center to the point of interest and θ is measured positive counterclockwise from the x axis.

(a) Find the stress components at the top and side of the hole for $r = d/2$.

(b) Plot a graph of the stress distribution for σ_θ, similar to that of Fig. 4–31, out to $r = 20$ mm for $d = 10$ mm.

 ANALYSIS

4–49   Considering the stress concentration at point A in the figure, determine the maximum normal and shear stresses at A if $F = 200$ lbf.

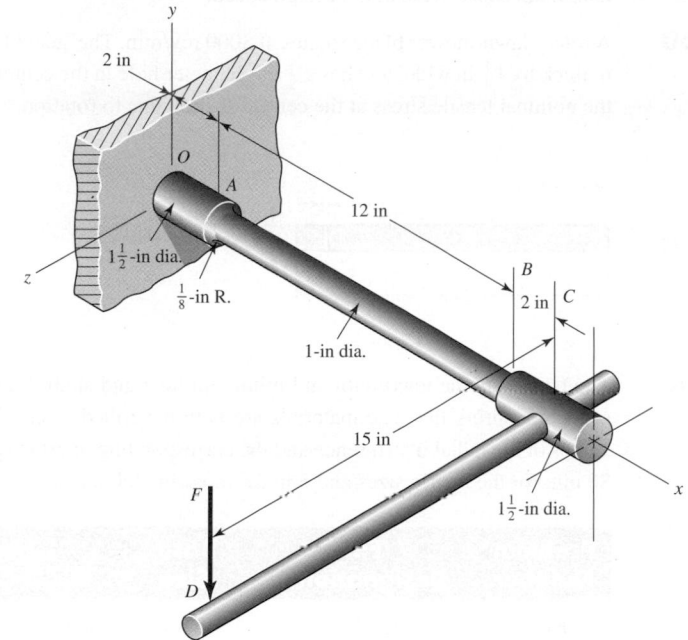

Problem 4–49

 ANALYSIS

4–50   Develop the formulas for the maximum radial and tangential stresses in a thick-walled cylinder due to internal pressure only.

 ANALYSIS

4–51   Repeat Prob. 4–50 where the cylinder is subject to external pressure only. At what radii do the maximum stresses occur?

 ANALYSIS

4–52   Develop the stress relations for a thin-walled spherical pressure vessel.

 ANALYSIS

4–53   A pressure cylinder has a diameter of 6 in and has a $\frac{1}{4}$-in wall thickness. What pressure can this vessel carry if the maximum shear stress is not to exceed 4000 psi?

 ANALYSIS

4–54   A pressure vessel has an outside diameter of 240 mm and a wall thickness of 10 mm. If the internal pressure is 2400 kPa, what is the maximum shear stress in the vessel walls?

 ANALYSIS

4–55   An AISI 1020 cold-drawn steel tube has an ID of $1\frac{1}{4}$ in and an OD of $1\frac{3}{4}$ in. What maximum external pressure can this tube take if the largest principal normal stress is not to exceed 80 percent of the minimum yield strength of the material?

 ANALYSIS

4–56   An AISI 1020 cold-drawn steel tube has an ID of $1\frac{3}{4}$ in and an OD of $2\frac{3}{8}$ in. What maximum internal pressure can this tube take if the largest principal normal stress is not to exceed 80 percent of the minimum yield strength of the material?

 ANALYSIS

4–57   Find the maximum shear stress in a 10-in circular saw if it runs idle at 7200 rev/min. The saw is 14 gauge (0.0747 in) and is used on a $\frac{3}{4}$-in arbor. The thickness is uniform. What is the maximum radial component of stress?

 ANALYSIS

4–58   The maximum recommended speed for a 300-mm-diameter abrasive grinding wheel is 2069 rev/min. Assume that the material is isotropic; use a bore of 25 mm, $v = 0.24$, and a mass density of 3320 kg/m^3; and find the maximum tensile stress at this speed.

ANALYSIS

4–59 An abrasive cutoff wheel has a diameter of 6 in, is $\frac{1}{16}$ in thick, and has a 1-in bore. It weighs 6 oz and is designed to run at 10 000 rev/min. If the material is isotropic and $\nu = 0.20$, find the maximum shear stress at the design speed.

ANALYSIS

4–60 A rotary lawn-mower blade rotates at 3000 rev/min. The steel blade has a uniform cross section $\frac{1}{8}$ in thick by $1\frac{1}{4}$ in wide, and has a $\frac{1}{2}$-in-diameter hole in the center as shown in the figure. Estimate the nominal tensile stress at the central section due to rotation.

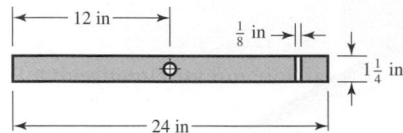

Problem 4–60

ANALYSIS

4–61 to 4–66 The table lists the maximum and minimum hole and shaft dimensions for a variety of standard press and shrink fits. The materials are both hot-rolled steel. Find the maximum and minimum values of the radial interference and the corresponding interface pressure. Use a collar diameter of 80 mm for the metric sizes and 3 in for those in inch units.

Problem Number	Fit Designation*	Basic Size	Hole		Shaft	
			D_{max}	D_{min}	d_{max}	d_{min}
4–61	40H7/p6	40 mm	40.025	40.000	40.042	40.026
4–62	(1.5 in)H7/p6	1.5 in	1.5010	1.5000	1.5016	1.5010
4–63	40H7/s6	40 mm	40.025	40.000	40.059	40.043
4–64	(1.5 in)H7/s6	1.5 in	1.5010	1.5000	1.5023	1.5017
4–65	40H7/u6	40 mm	40.025	40.000	40.076	40.060
4–66	(1.5 in)H7/u6	1.5 in	1.5010	1.5000	1.5030	1.5024

*Notes: See Table 2–8 for description of fits.

ANALYSIS

4–67 to 4–70 The table gives data concerning the shrink fit of two cylinders of differing materials and dimensional specification in inches. Elastic constants for different materials may be found in Table A–5. Identify the radial interference δ, then find the interference pressure p, the tangential normal stress on both sides of the fit surface, and the radial displacements δ_i and δ_0. If dimensional tolerances are given at fit surfaces, repeat the problem for the highest and lowest stress levels.

Problem Number	Inner Cylinder			Outer Cylinder		
	Material	d_i	d_0	Material	D_i	D_0
4–67	Steel	0	1.002	Steel	1.000	2.00
4–68	Steel	0	1.002	Cast iron	1.000	2.00
4–69	Steel	0	1.002/1.003	Steel	1.000/1.001	2.00
4–70	Steel	0	2.005/2.003	Aluminum	2.000/2.002	4.00

ANALYSIS

4–71 Force fits of a shaft and gear are assembled in an air-operated arbor press. An estimate of assembly force and torque capacity of the fit is needed. Assume the coefficient of friction is f, the fit interface pressure is p, the nominal shaft or hole radius is R, and the axial length of the gear bore is l.

(a) Show that the estimate of the axial force is $F_{ax} = 2\pi f \, Rlp$.

(b) Show the estimate of the torque capacity of the fit is $T = 2\pi f \, R^2 lp$.

ANALYSIS

4–72   A utility hook was formed from a 1-in-diameter round rod into the geometry shown in the figure. What are the stresses at the inner and outer surfaces at section *A-A* if the load *F* is 1000 lbf?

Problem 4–72

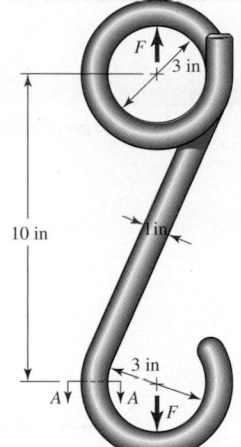

ANALYSIS

4–73   The steel eyebolt shown in the figure is loaded with a force *F* of 100 lbf. The bolt is formed of $\frac{1}{4}$-in-diameter wire to a $\frac{3}{8}$-in radius in the eye and at the shank. Estimate the stresses at the inner and outer surfaces at sections *A-A* and *B-B*.

Problem 4–73

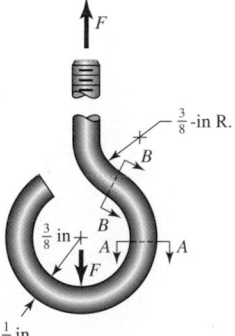

ANALYSIS

4–74   Shown in the figure is a 12-gauge (0.1094-in) by $\frac{3}{4}$-in latching spring that supports a load of $F = 3$ lbf. The inside radius of the bend is $\frac{1}{8}$ in. Estimate the stresses at the inner and outer surfaces at the critical section.

ANALYSIS

4–75   The cast-iron bell-crank lever depicted in the figure is acted upon by forces F_1 of 250 lbf and F_2 of 333 lbf. The section *A-A* at the central pivot has a curved inner surface with a radius of $r_i = 1$ in. Estimate the stresses at the inner and outer surfaces of the curved portion of the lever.

ANALYSIS

4–76   The crane hook depicted in Fig. 4–38 has a 1-in-diameter hole in the center of the critical section. For a load of 5 kip, estimate the bending stresses at the inner and outer surfaces at the critical section.

ANALYSIS

4–77   A 20-kip load is carried by the crane hook shown in the figure. The cross section of the hook uses two concave flanks. The width of the cross section is given by $b = 2/r$, where r is the radius from the center. The inside radius r_i is 2 in, and the outside radius $r_o = 6$ in. Find the stresses at the inner and outer surfaces at the critical section by (a) exact integration and (b) a numerical solution using 1000 increments of Δs. *Note:* r_c must also be evaluated numerically.

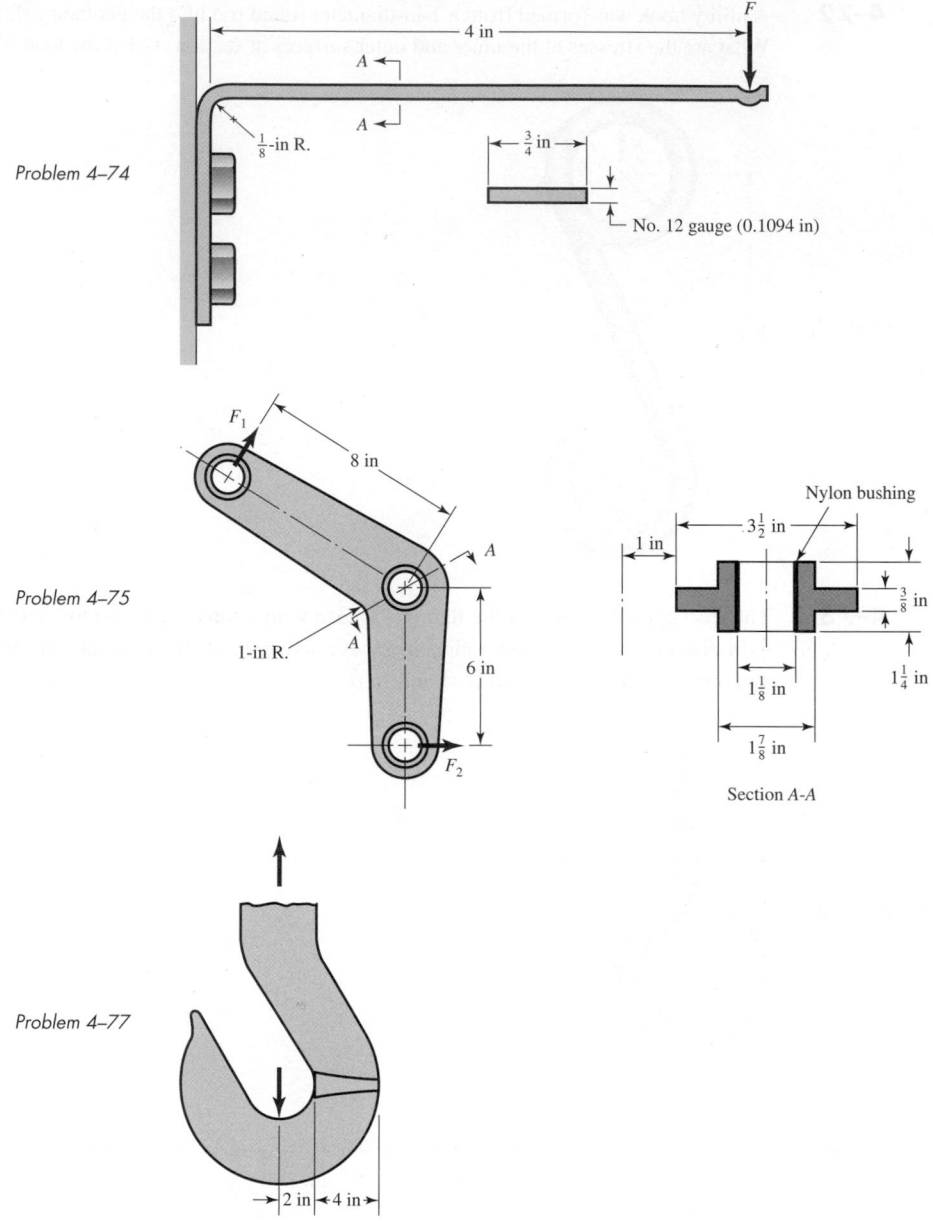

Problem 4–74

Problem 4–75

Section A-A

Problem 4–77

ANALYSIS **4–78** An offset tensile link is shaped to clear an obstruction with a geometry as shown in the figure. The cross section at the critical location is elliptical, with a major axis of 4 in and a minor axis of 2 in. For a load of 20 kip, estimate the stresses at the inner and outer surfaces of the critical section.

ANALYSIS **4–79** A cast-steel C frame as shown in the figure has a rectangular cross section of 1 in by 1.6 in, with a 0.4-in-radius semicircular notch on both sides that forms midflank fluting as shown. Estimate A, r_c, r_n, and e, and for a load of 3000 lbf, estimate the inner and outer surface stresses at the throat C. *Note:* Table 4–5 can be used to determine r_n for this section. From the table, the integral $\int dA/r$ can be evaluated for a rectangle and a circle by evaluating A/r_n for each shape [see Eq. (4–64)]. Subtracting A/r_n of the circle from that of the rectangle yields $\int dA/r$ for the C frame, and r_n can then be evaluated.

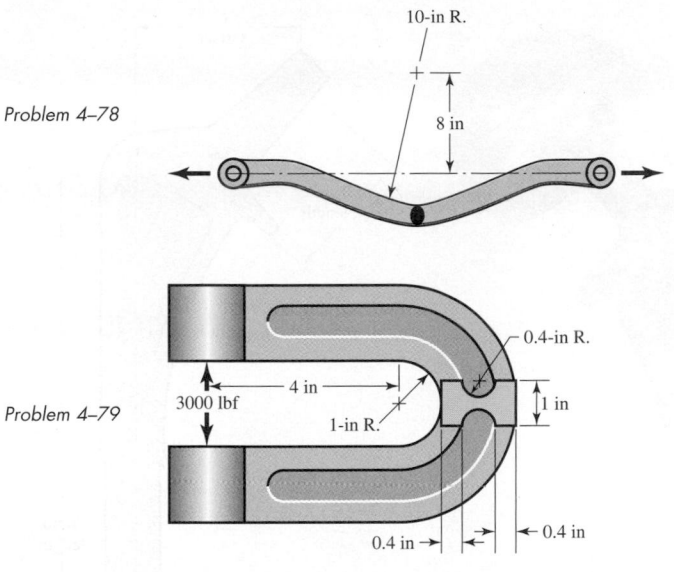

Problem 4–78

Problem 4–79

4–80 Solve Prob. 4–79 numerically, using 100 and 1000 elements. Compare your results with Prob. 4–79, if solved previously.

ANALYSIS **4–81** Two carbon steel balls, each 25 mm in diameter, are pressed together by a force F. In terms of the force F, find the maximum values of the principal stress, and the maximum shear stress, in MPa.

ANALYSIS **4–82** One of the balls in Prob. 4–81 is replaced by a flat carbon steel plate. If $F = 18$ N, at what depth does the maximum shear stress occur?

ANALYSIS **4–83** An aluminum alloy roller with diameter 1 in and length 2 in rolls on the inside of a cast-iron ring having an inside radius of 4 in, which is 2 in thick. Find the maximum contact force F that can be used if the shear stress is not to exceed 4000 psi.

ANALYSIS **4–84** The figure on page 184 shows a hip prosthesis containing a stem that is cemented into a reamed cavity in the femur. The cup is cemented and fastened to the hip with bone screws. Shown are porous layers of titanium into which bone tissue will grow to form a longer-lasting bond than that afforded by cement alone. The bearing surfaces are a plastic cup and a titanium femoral head. The lip shown in the figures bears against the cutoff end of the femur to transfer the load to the leg from the hip. Walking will induce several million stress fluctuations per year for an average person, so there is danger that the prosthesis will loosen the cement bonds or that metal cracks may occur because of the many repetitions of stress. Prostheses like this are made in many different sizes. Typical dimensions are ball diameter 50 mm, stem diameter 15 mm, stem length 155 mm, offset 38 mm, and neck length 39 mm. Develop an outline to follow in making a complete stress analysis of this prosthesis. Describe the material properties needed, the equations required, and how the loading is to be defined.

ANALYSIS **4–85** Simplify Eqs. (4–74), (4–75), and (4–76) by setting $z = 0$ and finding σ_x/p_{max}, σ_y/p_{max}, σ_z/p_{max}, and $\tau_{2/3}/p_{max}$ and, for cast iron, check the ordinate intercepts of the four loci in Fig. 4–43.

ANALYSIS **4–86** A 6-in-diameter cast-iron wheel, 2 in wide, rolls on a flat steel surface carrying a 800-lbf load.
(a) Find the Hertzian stresses σ_x, σ_y, σ_z, and $\tau_{2/3}$.
(b) What happens to the stresses at a point A that is 0.010 in below the wheel rim surface during a revolution?

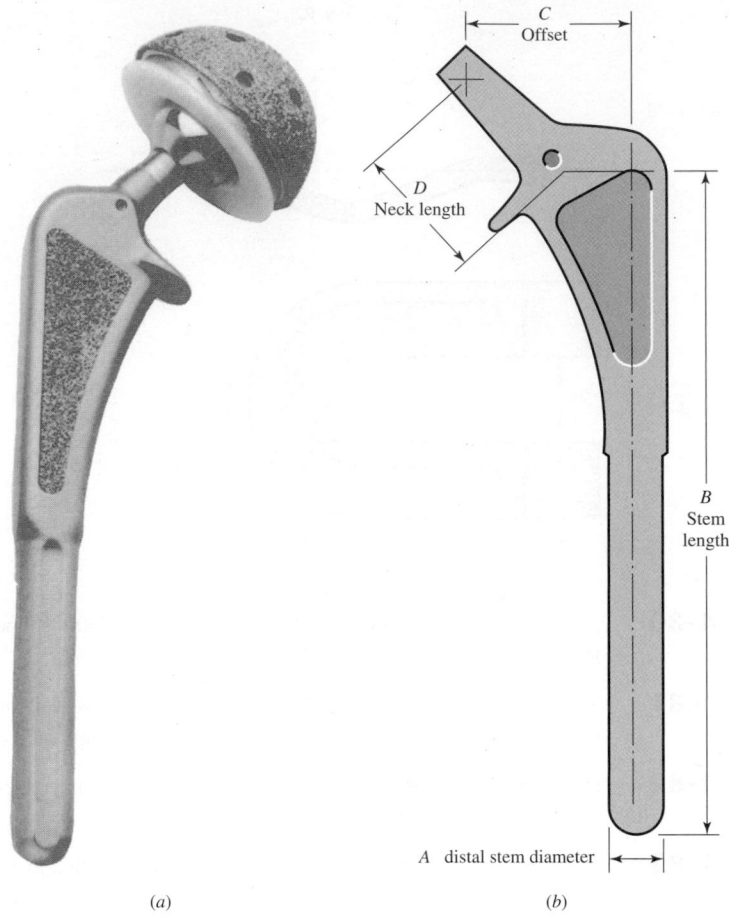

Problem 4–84

Porous hip prosthesis. (Photograph and
drawing courtesy of Zimmer, Inc., Warsaw, Indiana.)

(a)

(b)

4–87   Write an interactive computer program that will evaluate Eqs. (4–82) to (4–85) and fill out a table $0 \leq y/b \leq 1$ across the columns and $0 \leq z/b \leq 1$ for the lines, with the body of the table with any stress of interest: σ_x, σ_y, σ_z, τ_{yz}, σ_1, σ_2, σ_3, and $\tau_{\max}$.

4–88   Use the program of Prob. 4–87 to solve Prob. 4–86.

4–89   Use the program of Prob. 4–87 to solve Prob. 4–86 with a coefficient of friction $f = 0.10$. Notice the changes.

5

Deflection and Stiffness

A body is said to be *rigid* if it exhibits no change in size or shape under the influences of forces or couples. All real bodies deform under load, either elastically or plastically. Classification of a real body as rigid is an idealization. A body can be sufficiently insensitive to deformation that a presumption of rigidity does not affect an analysis enough to warrant a nonrigid treatment. Unlike treatment based on an *assumption* of rigidity, treatment based on the *decision* to ignore deformation represents such a small deviation from actuality that a rigid treatment is robust. If the body deformation later proves to be not negligible, then declaring rigidity was a poor decision, not a poor assumption. Strictly, *flexibility* is the ability of a body to distort by bending—that is, it is but one form of distortion—but the term is often used interchangeably with distortion. Defining the choice as bilateral—rigid *or* flexible—leaves unanswered other possibilities. A wire rope is flexible, but in tension it can be robustly rigid and it distorts enormously under attempts at compressive loading. The same body can be both rigid and nonrigid.

Deflection analysis enters into design situations in many ways. A snap ring, or retaining ring, must be flexible enough to be bent without permanent deformation and assembled with other parts, and then it must be rigid enough to hold the assembled parts together. In a transmission, the gears must be supported by a rigid shaft. If the shaft bends too much, that is, if it is too flexible, the teeth will not mesh properly, and the result will be excessive impact, noise, wear, and early failure. In rolling sheet or strip steel to prescribed thicknesses, the rolls must be crowned, that is, curved, so that the finished product will be of uniform thickness. Thus, to design the rolls it is necessary to know exactly how much they will bend when a sheet of steel is rolled between them. Sometimes mechanical elements must be designed to have a particular force-deflection characteristic. The suspension system of an automobile, for example, must be designed within a very narrow range to achieve an optimum vibration frequency for all conditions of vehicle loading, because the human body is comfortable only within a limited range of frequencies.

The size of a load-bearing component is often determined on deflections, rather than limits on stress.

This chapter considers distortion of single bodies due to geometry (shape) and loading, then, briefly, the behavior of ensembles of bodies.

5–1 Spring Rates

Elasticity is that property of a material that enables it to regain its original configuration after having been deformed. A *spring* is a mechanical element that exerts a force when deformed. Figure 5–1*a* shows a straight beam of length *l* simply supported at the ends

Figure 5–1

(*a*) A linear spring; (*b*) a stiffening spring; (*c*) a softening spring.

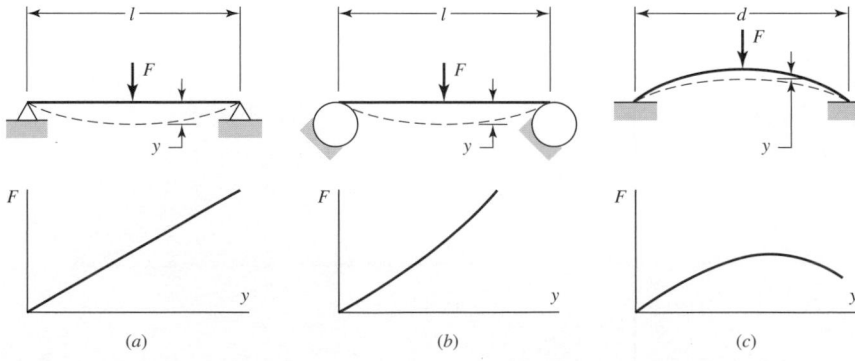

and loaded by the transverse force F. The deflection y is linearly related to the force, as long as the elastic limit of the material is not exceeded, as indicated by the graph. This beam can be described as a *linear spring*.

In Fig. 5–1b a straight beam is supported on two cylinders such that the length between supports decreases as the beam is deflected by the force F. A larger force is required to deflect a short beam than a long one, and hence the more this beam is deflected, the stiffer it becomes. Also, the force is not linearly related to the deflection, and hence this beam can be described as a *nonlinear stiffening spring*.

Figure 5–1c is a dish-shaped round disk. The force necessary to flatten the disk increases at first and then decreases as the disk approaches a flat configuration, as shown by the graph. Any mechanical element having such a characteristic is called a *nonlinear softening spring*.

If we designate the general relationship between force and deflection by the equation

$$F = F(y) \qquad (a)$$

then *spring rate* is defined as

$$k(y) = \lim_{\Delta y \to 0} \frac{\Delta F}{\Delta y} = \frac{dF}{dy} \qquad (5\text{–}1)$$

where y must be measured in the direction of F and at the point of application of F. Most of the force-deflection problems encountered in this book are linear, as in Fig. 5–1a. For these, k is a constant, also called the *spring constant;* consequently Eq. (5–1) is written

$$k = \frac{F}{y} \qquad (5\text{–}2)$$

We might note that Eqs. (5–1) and (5–2) are quite general and apply equally well for torques and moments, provided angular measurements are used for y. For linear displacements, the units of k are often pounds per inch or newtons per meter, and for angular displacements, pound-inches per radian or newton-meters per radian.

5–2   Tension, Compression, and Torsion

The relation for the total extension or deformation of a uniform bar has already been developed in Sec. 4–8, Eq. (a). It is repeated here for convenience:

$$\delta = \frac{Fl}{AE} \qquad (5\text{–}3)$$

This equation does not apply to a long bar loaded in compression if there is a possibility of buckling. Using Eqs. (5–2) and (5–3), we see that the spring constant of an axially loaded bar is

$$k = \frac{AE}{l} \qquad (5\text{–}4)$$

The angular deflection of a uniform round bar subjected to a twisting moment T was given in Eq. (4–35), and is

$$\theta = \frac{Tl}{GJ} \qquad (5\text{–}5)$$

where θ is in radians. If we multiply Eq. (5–5) by $180/\pi$ and substitute $J = \pi d^4/32$ for a solid round bar, we obtain

$$\theta = \frac{583.6Tl}{Gd^4} \qquad (5\text{–}6)$$

where θ is in degrees.

Equation (5–5) can be rearranged to give the torsional spring rate as

$$k = \frac{T}{\theta} = \frac{GJ}{l} \qquad (5\text{–}7)$$

When the word *simple* is used to describe the loading, the meaning is that no other load is present and that no geometric complexities are present. Thus, a bar loaded in *simple tension* is a uniform bar acted upon by a tensile load directed along the centroidal axis, and acted upon by no other loads.

5–3 Deflection Due to Bending

Beams deflect a great deal more than axially loaded members, and the problem of bending probably occurs more often than any other loading problem in design. Shafts, axles, cranks, levers, springs, brackets, and wheels, as well as many other elements, must often be treated as beams in the design and analysis of mechanical structures and systems. The subject of bending, however, is one that you should have studied as preparation for reading this book. It is for this reason that we include here only a brief review to establish the nomenclature and conventions to be used throughout this book.

In Sec. 4–10 we developed the relation for the curvature of a beam subjected to a bending moment M [Eq. (4–25)]. The relation is

$$\frac{1}{\rho} = \frac{M}{EI} \qquad (5\text{–}8)$$

where ρ is the radius of curvature. From studies in mathematics we also learn that the curvature of a plane curve is given by the equation

$$\frac{1}{\rho} = \frac{d^2y/dx^2}{[1 + (dy/dx)^2]^{3/2}} \qquad (5\text{–}9)$$

where the interpretation here is that y is the deflection of the beam at any point x along its length. The slope of the beam at any point x is

$$\theta = \frac{dy}{dx} \qquad (a)$$

For many problems in bending, the slope is very small, and for these the denominator of Eq. (5–9) can be taken as unity. Equation (5–8) can then be written

$$\frac{M}{EI} = \frac{d^2y}{dx^2} \qquad (b)$$

Noting Eqs. (4–3) and (4–4) and successively differentiating Eq. (b) yields

$$\frac{V}{EI} = \frac{d^3y}{dx^3} \qquad (c)$$

$$\frac{q}{EI} = \frac{d^4y}{dx^4} \qquad (d)$$

It is convenient to display these relations in a group as follows:

$$\frac{q}{EI} = \frac{d^4y}{dx^4} \tag{5-10}$$

$$\frac{V}{EI} = \frac{d^3y}{dx^3} \tag{5-11}$$

$$\frac{M}{EI} = \frac{d^2y}{dx^2} \tag{5-12}$$

$$\theta = \frac{dy}{dx} \tag{5-13}$$

$$y = f(x) \tag{5-14}$$

The nomenclature and conventions are illustrated by the beam of Fig. 5–2. Here, a beam of length $l = 20$ in is loaded by the uniform load $w = 80$ lbf per inch of beam length. The x axis is positive to the right, and the y axis positive upward. All quantities— loading, shear, moment, slope, and deflection—have the same sense as y; they are positive if upward, negative if downward.

The values of the quantities at the ends of the beam, where $x = 0$ and $x = l$, are called their *boundary values*. For this reason the beam problem is often called a *boundary-value problem*. The reactions $R_1 = R_2 = +800$ lbf and the shear forces $V_0 = +800$ lbf and $V_l = -800$ lbf are easily computed by using the methods of Chap. 4. The bending moment is zero at each end because the beam is simply supported. For a simply-supported beam, the deflections are zero at each end. The slopes at each end are initially unknown, but from a view of Fig. 5–2e it is obvious that the slope is negative at the left boundary and positive at the right boundary.

| **Figure 5–2**

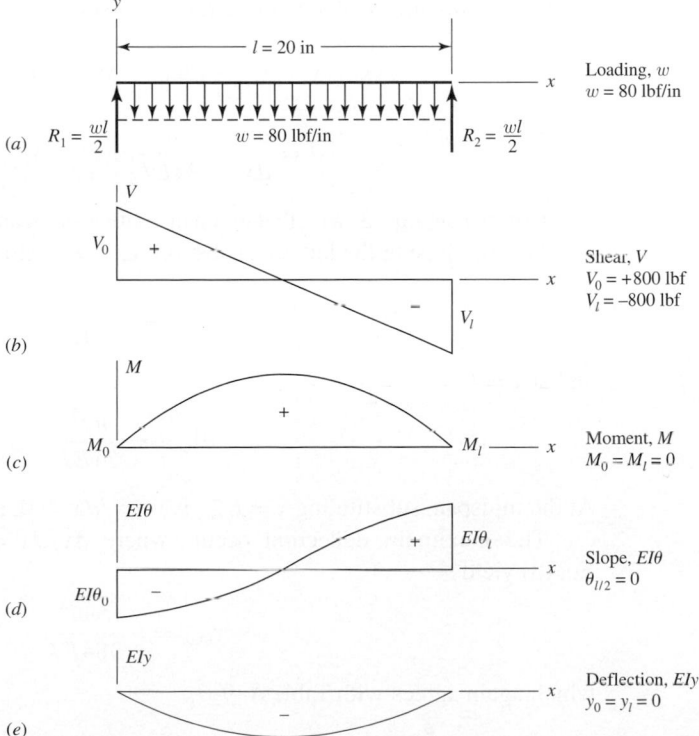

EXAMPLE 5–1   For the beam in Fig. 5–2, the bending moment equation, for $0 \leq x \leq l$, is

$$M = \frac{wl}{2}x - \frac{w}{2}x^2$$

Using Eq. (5–12), determine the equations for the slope and deflection of the beam, the slopes at the ends, and the maximum deflection.

Solution   Integrating Eq. (5–12) as an indefinite integral we have

$$EI\frac{dy}{dx} = \int M\,dx = \frac{wl}{4}x^2 - \frac{w}{6}x^3 + C_1 \tag{1}$$

where C_1 is a constant of integration that is evaluated from geometric boundary conditions. We could impose that the slope is zero at the midspan of the beam, since the beam and loading are symmetric relative to the midspan. However, we will use the given boundary conditions of the problem and verify that the slope is zero at the midspan. Integrating Eq. (1) gives

$$EIy = \iint M\,dx = \frac{wl}{12}x^3 - \frac{w}{24}x^4 + C_1 x + C_2 \tag{2}$$

The boundary conditions for the simply supported beam are $y = 0$ at $x = 0$ and l. Applying the first condition, $y = 0$ at $x = 0$, to Eq. (2) results in $C_2 = 0$. Applying the second condition to Eq. (2) with $C_2 = 0$,

$$EIy(l) = \frac{wl}{12}l^3 - \frac{w}{24}l^4 + C_1 l = 0$$

Solving for C_1 yields $C_1 = -wl^3/24$. Substituting the constants back into Eqs. (1) and (2) and solving for the deflection and slope results in

$$y = \frac{wx}{24EI}(2lx^2 - x^3 - l^3) \tag{3}$$

$$\theta = \frac{dy}{dx} = \frac{w}{24EI}(6lx^2 - 4x^3 - l^3) \tag{4}$$

Comparing Eq. (3) with that given in Table A–9, beam 7, we see complete agreement. For the slope at the left end, substituting $x = 0$ into Eq. (4) yields

$$\theta|_{x=0} = -\frac{wl^3}{24EI}$$

and at $x = l$,

$$\theta|_{x=l} = \frac{wl^3}{24EI}$$

At the midspan, substituting $x = l/2$ gives $dy/dx = 0$, as earlier suspected.
   The maximum deflection occurs where $dy/dx = 0$. Substituting $x = l/2$ into Eq. (3) yields

$$y_{max} = -\frac{5wl^4}{384EI}$$

which again agrees with Table A–9–7.

The approach used in the example is fine for simple beams with continuous loading. However, for beams with discontinuous loading and/or geometry such as a step shaft with multiple gears, flywheels, pulleys, etc., the approach becomes unwieldy. The following section discusses bending deflections in general and the techniques that are provided in this chapter.

5–4 ## Beam Deflection Methods

Equations (5–10) through (5–14) are the basis for relating the intensity of loading q, vertical shear V, bending moment M, slope of the neutral surface θ, and the transverse deflection y. Beams have intensities of loading that range from $q = $ constant (uniform loading), variable intensity $q(x)$, to Dirac delta functions (concentrated loads).

The intensity of loading usually consists of piecewise contiguous zones, the expressions for which are integrated through Eqs. (5–10) to (5–14) with varying degrees of difficulty. Another approach is to represent the deflection $y(x)$ as a Fourier series, which is capable of representing single-valued functions with a finite number of finite discontinuities, then differentiating through Eqs. (5–14) to (5–10), and stopping at some level where the Fourier coefficients can be evaluated. A complication is the piecewise continuous nature of some beams (shafts) that are stepped-diameter bodies.

All of the above constitute, in one form or another, formal integration methods, which, with properly selected problems, result in solutions for q, V, M, θ, and y. These solutions may be

1   Closed-form, or
2   Represented by infinite series, which amount to closed form if the series are rapidly convergent, or
3   Approximations obtained by evaluating the first or the first and second terms.

The series solutions can be made equivalent in convenience to the closed-form solution by the use of a computer. Roark's[1] formulas are committed to CD-ROM and can be used on a personal computer.

There are many techniques employed to solve the integration problem for beam deflection. Some of the popular methods include:

- Superposition (see Sec. 4–5)
- The moment-area method[2]
- Singularity functions (see Sec. 4–6)
- Numerical integration[3]

The methods described in this chapter are easy to implement and can handle a large array of problems.

There are methods that do not deal with Eqs. (5–10) to (5–14) directly. An energy method, based on Castigliano's theorem, is quite powerful for problems not suitable for the methods mentioned earlier and is discussed in Secs. 5–7 to 5–10. Finite element programs are also quite useful for determining beam deflections.

---

1. Warren C. Young and Richard G. Budynas, *Roark's Formulas for Stress and Strain,* 7th ed., McGraw-Hill, New York, 2002.

2. See Chap. 9, F. P. Beer, E. R. Johnston Jr., and J. T. DeWolf, *Mechanics of Materials,* 3rd ed., McGraw-Hill, New York, 2001.

3. See Sec. 4–4, J. E. Shigley and C. R. Mischke, *Mechanical Engineering Design,* 6th ed., McGraw-Hill, New York, 2001.

5–5 Finding Beam Deflections by Superposition

The results of many simple load cases and boundary conditions have been solved and are available. Table A–9 provides a limited number of cases. Roark's[4] provides a much more comprehensive listing. Superposition resolves the effect of combined loading on a structure by determining the effects of each load separately and adding the results algebraically. Superposition may be applied provided: (1) each effect is linearly related to the load that produces it, (2) a load does not create a condition that affects the result of another load, and (3) the deformations resulting from any specific load are not large enough to appreciably alter the geometric relations of the parts of the structural system.

The following examples are illustrations of the use of superposition.

---

4. Warren C. Young and Richard G. Budynas, *Roark's Formulas for Stress and Strain,* 7th ed., McGraw-Hill, New York, 2002.

EXAMPLE 5–2

Consider the uniformly loaded beam with a concentrated force as shown in Fig. 5–3. Using superposition, determine the reactions and the deflection as a function of x.

Solution

Considering each load state separately, we can superpose beams 6 and 7 of Table A–9. For the reactions we find

Answer
$$R_1 = \frac{Fb}{l} + \frac{wl}{2}$$

Answer
$$R_2 = \frac{Fa}{l} + \frac{wl}{2}$$

The loading of beam 6 is discontinuous and separate deflection equations are given for regions AB and BC. Beam 7 loading is not discontinuous so there is only one equation. Superposition yields

Answer
$$y_{AB} = \frac{Fbx}{6EIl}(x^2 + b^2 - l^2) + \frac{wx}{24EI}(2lx^2 - x^3 - l^3)$$

Answer
$$y_{BC} = \frac{Fa(l-x)}{6EIl}(x^2 + a^2 - 2lx) + \frac{wx}{24EI}(2lx^2 - x^3 - l^3)$$

| Figure 5–3

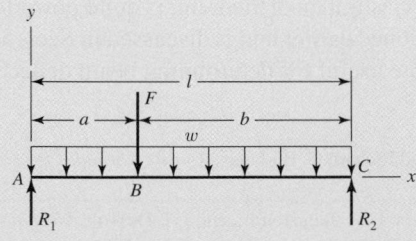

If we wanted to determine the maximum deflection in the previous example, we would set $dy/dx = 0$ and solve for the value of x where the deflection is a maximum. If $a = l/2$, the maximum deflection would obviously occur at $x = l/2$ because of symmetry. However, if $a < l/2$, where would the maximum be? It can be shown that as the force F moves toward the left support, the maximum deflection moves toward the left support also, but not as much as F (see Prob. 5–34). Thus, we would set $dy_{BC}/dx = 0$ and solve for x.

Sometimes it may not be obvious that we can use superposition with the tables at hand, as demonstrated in the next example.

EXAMPLE 5–3

Consider the beam in Fig. 5–4a and determine the deflection equations using superposition.

Solution

For region AB we can superpose beams 7 and 10 of Table A–9 to obtain

Answer

$$y_{AB} = \frac{wx}{24EI}(2lx^2 - x^3 - l^3) + \frac{Fax}{6EIl}(l^2 - x^2)$$

For region BC, how do we represent the uniform load? Considering the uniform load *only*, the beam deflects as shown in Fig. 5–4b. Region BC is straight since there is no bending moment due to w. The slope of the beam at B is θ_B and is obtained by taking the derivative of y given in the table with respect to x and setting $x = l$. Thus,

$$\frac{dy}{dx} = \frac{d}{dx}\left[\frac{wx}{24EI}(2lx^2 - x^3 - l^3)\right] = \frac{w}{24EI}(6lx^2 - 4x^3 - l^3)$$

Substituting $x = l$ gives

$$\theta_B = \frac{w}{24EI}(6ll^2 - 4l^3 - l^3) = \frac{wl^3}{24EI}$$

The deflection in region BC due to w is $\theta_B(x - l)$, and adding this to the deflection due to F yields

Answer

$$y_{BC} = \frac{wl^3}{24EI}(x - l) + \frac{F(x - l)}{6EI}[(x - l)^2 - a(3x - l)]$$

Figure 5–4

(a) Beam with uniformly distributed load; (b) deflections due to uniform load only.

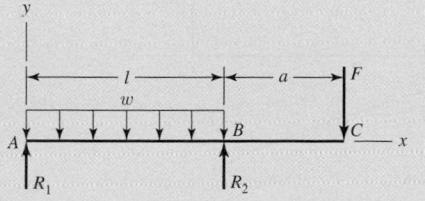

(a)

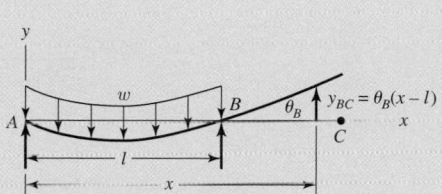

(b) Deflections due to uniform load only

EXAMPLE 5–4

Figure 5–5a shows a cantilever beam with an end load. Normally we model this problem by considering the left support as rigid. After testing the rigidity of the wall it was found that the translational stiffness of the wall was k_t force per unit vertical deflection, and the rotational stiffness was k_r moment per unit angular (radian) deflection (see Fig. 5–5b). Determine the deflection equation for the beam under the load F.

Solution

Here we will superpose the *modes* of deflection. They are: (1) translation due to the compression of spring k_t, (2) rotation of the spring k_r, and (3) the elastic deformation of the beam given by Table A–9–1. The force in spring k_t is $R_1 = F$, giving a deflection from Eq. (5–2) of

$$y_1 = -\frac{F}{k_t} \tag{1}$$

The moment in spring k_r is $M_1 = Fl$. This gives a clockwise rotation of $\theta = Fl/k_r$. Considering this mode of deflection only, the beam rotates rigidly counterclockwise, leading to a deflection equation of

$$y_2 = -\frac{Fl}{k_r}x \tag{2}$$

Finally, the elastic deformation of the beam from Table A–9–1 is

$$y_3 = \frac{Fx^2}{6EI}(x - 3l) \tag{3}$$

Adding the deflections from each mode yields

Answer

$$y = \frac{Fx^2}{6EI}(x - 3l) - \frac{F}{k_t} - \frac{Fl}{k_r}x$$

| **Figure 5–5**

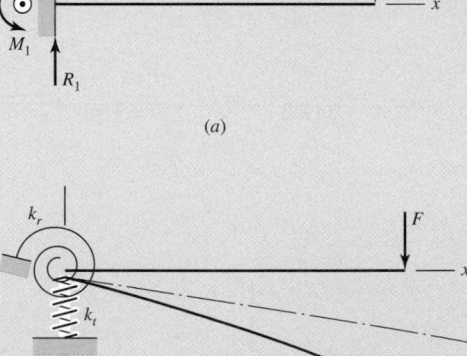

(a)

(b)

5–6   **Beam Deflections by Singularity Functions**

Introduced in Sec. 4–3, singularity functions are excellent for managing discontinuities, and their application is a simple extension of what was presented in the earlier section. They are easy to program, and as will be seen later, they can greatly simplify statically indeterminate problems. The following examples illustrate the use of singularity functions to evaluate deflections of statically determinate beam problems.

EXAMPLE 5–5   Consider the beam of Table A–9–6, which is a simply supported beam having a concentrated load F not in the center. Develop the deflection equations using singularity functions.

Solution   First, write the load intensity equation from the free-body diagram,

$$q = R_1 \langle x \rangle^{-1} - F \langle x - a \rangle^{-1} + R_2 \langle x - l \rangle^{-1} \tag{1}$$

Integrating twice results in

$$V = R_1 \langle x \rangle^{0} - F \langle x - a \rangle^{0} + R_2 \langle x - l \rangle^{0} \tag{2}$$

$$M = R_1 \langle x \rangle^{1} - F \langle x - a \rangle^{1} + R_2 \langle x - l \rangle^{1} \tag{3}$$

Recall that as long as the q equation is complete, integration constants are unnecessary; therefore, they are not included up to this point. From statics, setting $V = M = 0$ for x slightly greater than l yields $R_1 = Fb/l$ and $R_2 = Fa/l$. Thus Eq. (3) becomes

$$M = \frac{Fb}{l} \langle x \rangle^{1} - F \langle x - a \rangle^{1} + \frac{Fa}{l} \langle x - l \rangle^{1}$$

Integrating Eqs. (5–12) and (5–13) as indefinite integrals gives

$$EI\frac{dy}{dx} = \frac{Fb}{2l} \langle x \rangle^{2} - \frac{F}{2} \langle x - a \rangle^{2} + \frac{Fa}{2l} \langle x - l \rangle^{2} + C_1$$

$$EIy = \frac{Fb}{6l} \langle x \rangle^{3} - \frac{F}{6} \langle x - a \rangle^{3} + \frac{Fa}{6l} \langle x - l \rangle^{3} + C_1 x + C_2$$

Note that the first singularity term in both equations always exists, so $\langle x \rangle^{2} = x^2$ and $\langle x \rangle^{3} = x^3$. Also, the last singularity term in both equations does not exist until $x = l$, where it is zero, and since there is no beam for $x > l$ we can drop the last term. Thus

$$EI\frac{dy}{dx} = \frac{Fb}{2l} x^2 - \frac{F}{2} \langle x - a \rangle^{2} + C_1 \tag{4}$$

$$EIy = \frac{Fb}{6l} x^3 - \frac{F}{6} \langle x - a \rangle^{3} + C_1 x + C_2 \tag{5}$$

The constants of integration C_1 and C_2 are evaluated by using the two boundary conditions $y = 0$ at $x = 0$ and $y = 0$ at $x = l$. The first condition, substituted into Eq. (5), gives $C_2 = 0$ (recall that $\langle 0 - a \rangle^{3} = 0$). The second condition, substituted into Eq. (5), yields

$$0 = \frac{Fb}{6l} l^3 - \frac{F}{6} (l - a)^3 + C_1 l = \frac{Fbl^2}{6} - \frac{Fb^3}{6} + C_1 l$$

Solving for C_1,

$$C_1 = -\frac{Fb}{6l}(l^2 - b^2)$$

Finally, substituting C_1 and C_2 in Eq. (5) and simplifying produces

$$y = \frac{F}{6EIl}[bx(x^2 + b^2 - l^2) - l\langle x - a \rangle^3] \tag{6}$$

Comparing Eq. (6) with the two deflection equations in Table A–9–6, we note that the use of singularity functions enables us to express the deflection equation with a single equation.

EXAMPLE 5–6

Determine the deflection equation for the simply supported beam with the load distribution shown in Fig. 5–6.

Solution

This is a good beam to add to our table for later use with superposition. The load intensity equation for the beam is

$$q = R_1 \langle x \rangle^{-1} - w \langle x \rangle^0 + w \langle x - a \rangle^0 + R_2 \langle x - l \rangle^{-1} \tag{1}$$

where the $w \langle x - a \rangle^0$ is necessary to "turn off" the uniform load at $x = a$.

From statics, the reactions are

$$R_1 = \frac{wa}{2l}(2l - a) \qquad R_2 = \frac{wa^2}{2l} \tag{2}$$

For simplicity, we will retain the form of Eq. (1) for integration and substitute the values of the reactions in later.

Two integrations of Eq. (1) reveal

$$V = R_1 \langle x \rangle^0 - w \langle x \rangle^1 + w \langle x - a \rangle^1 + R_2 \langle x - l \rangle^0 \tag{3}$$

$$M = R_1 \langle x \rangle^1 - \frac{w}{2} \langle x \rangle^2 + \frac{w}{2} \langle x - a \rangle^2 + R_2 \langle x - l \rangle^1 \tag{4}$$

As in the previous example, singularity functions of order zero or greater starting at $x = 0$ can be replaced by normal polynomial functions. Also, once the reactions are determined, singularity functions starting at the extreme right end of the beam can be omitted. Thus, Eq. (4) can be rewritten as

$$M = R_1 x - \frac{w}{2} x^2 + \frac{w}{2} \langle x - a \rangle^2 \tag{5}$$

| Figure 5–6

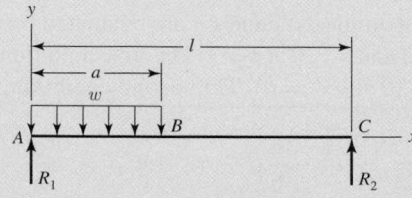

Integrating two more times for slope and deflection gives

$$EI\frac{dy}{dx} = \frac{R_1}{2}x^2 - \frac{w}{6}x^3 + \frac{w}{6}\langle x-a\rangle^3 + C_1 \tag{6}$$

$$EIy = \frac{R_1}{6}x^3 - \frac{w}{24}x^4 + \frac{w}{24}\langle x-a\rangle^4 + C_1x + C_2 \tag{7}$$

The boundary conditions are $y=0$ at $x=0$ and $y=0$ at $x=l$. Substituting the first condition in Eq. (7) shows $C_2 = 0$. For the second condition

$$0 = \frac{R_1}{6}l^3 - \frac{w}{24}l^4 + \frac{w}{24}(l-a)^4 + C_1l$$

Solving for C_1 and substituting into Eq. (7) yields

$$EIy = \frac{R_1}{6}x(x^2 - l^2) - \frac{w}{24}x(x^3 - l^3) - \frac{w}{24l}x(l-a)^4 + \frac{w}{24}\langle x-a\rangle^4$$

Finally, substitution of R_1 from Eq. (2) and simplifying results gives

Answer
$$y = \frac{w}{24EIl}[2ax(2l-a)(x^2-l^2) - xl(x^3-l^3) - x(l-a)^4 + l\langle x-a\rangle^4]$$

EXAMPLE 5–7   Figure 5–7a shows an end-loaded cantilever step shaft. The end reactions have already been determined by using statics. Determine the deflection curve and the deflections at B and C.

Solution   The load-intensity equation is

$$q = F\langle x\rangle^{-1} - Fl\langle x\rangle^{-2} - F\langle x-l\rangle^{-1} \tag{1}$$

| **Figure 5–7**

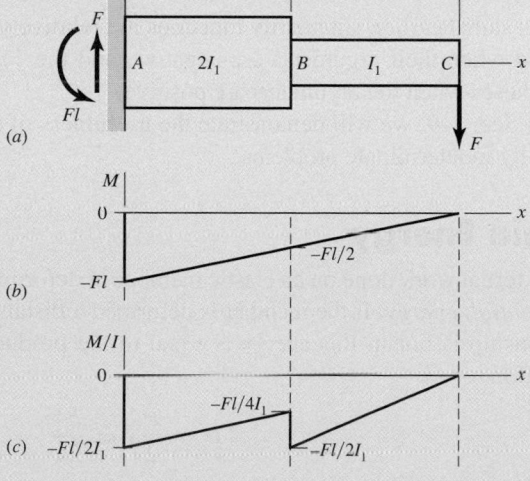

(a)

(b)

(c)

Integrations produce

$$V = F\langle x \rangle^0 - Fl\langle x \rangle^{-1} - F\langle x - l \rangle^0 \tag{2}$$

$$M = F\langle x \rangle^1 - Fl\langle x \rangle^0 - F\langle x - l \rangle^1 = Fx - Fl \tag{3}$$

The resulting M diagram is shown in Fig. 5–7b.

Since the second area moment varies for this beam, we will draw the M/I diagram as shown in Fig. 5–7c. The step drop in the diagram can be accomplished easily by subtracting a step term of magnitude $(Fl)/(4I_1)$ at $x = l/2$ in the singularity equation. Also, the slope of the M/I diagram increases by $F/2I_1$ at $x = l/2$, requiring an additional ramp function at this location. Thus

$$\frac{M}{I} = \frac{F}{2I_1}x - \frac{Fl}{2I_1} - \frac{Fl}{4I_1}\left\langle x - \frac{l}{2} \right\rangle^0 + \frac{F}{2I_1}\left\langle x - \frac{l}{2} \right\rangle^1 \tag{4}$$

Since $Ed^2y/dx^2 = M/I$, two integrations of Eq. (4) yield

$$E\frac{dy}{dx} = \frac{F}{4I_1}x^2 - \frac{Fl}{2I_1}x - \frac{Fl}{4I_1}\left\langle x - \frac{l}{2} \right\rangle^1 + \frac{F}{4I_1}\left\langle x - \frac{l}{2} \right\rangle^2 + C_1 \tag{5}$$

$$Ey = \frac{F}{12I_1}x^3 - \frac{Fl}{4I_1}x^2 - \frac{Fl}{8I_1}\left\langle x - \frac{l}{2} \right\rangle^2 + \frac{F}{12I_1}\left\langle x - \frac{l}{2} \right\rangle^3 + C_1 x + C_2 \tag{6}$$

The boundary conditions for the cantilever beam are $dy/dx = 0$ and $y = 0$ at $x = 0$. Substituting these into Eqs. (5) and (6) yields $C_1 = C_2 = 0$, and Eq. (6) simplifies to

Answer
$$y = \frac{F}{24EI_1}\left(2x^3 - 6lx^2 - 3l\left\langle x - \frac{l}{2} \right\rangle^2 + 2\left\langle x - \frac{l}{2} \right\rangle^3 \right) \tag{7}$$

At $x = l/2$ and l,

Answer
$$y|_{x=l/2} = \frac{F}{24EI_1}\left[2\left(\frac{l}{2}\right)^3 - 6l\left(\frac{l}{2}\right)^2 - 3l(0) + 2(0) \right] = -\frac{5Fl^3}{96EI_1}$$

Answer
$$y|_{x=l} = \frac{F}{24EI_1}\left[2(l)^3 - 6l(l)^2 - 3l\left(l - \frac{l}{2}\right)^2 + 2\left(l - \frac{l}{2}\right)^3 \right] = -\frac{3Fl^3}{16EI_1}$$

As stated earlier, singularity functions are relatively simple to program, as they are omitted when their arguments are negative, and the $\langle \ \rangle$ brackets are replaced with $(\ )$ parentheses when the arguments are positive.

In Sec. 5–9, we will demonstrate the usefulness of singularity functions in solving statically indeterminate problems.

5–7 Strain Energy

The external work done on an elastic member in deforming it is transformed into *strain,* or *potential, energy.* If the member is deformed a distance y, and if the force-deflection relationship is linear, this energy is equal to the product of the average force and the deflection, or

$$U = \frac{F}{2}y = \frac{F^2}{2k} \tag{a}$$

This equation is general in the sense that the force F can also mean torque, or moment, provided, of course, that consistent units are used for k. By substituting appropriate expressions for k, strain-energy formulas for various simple loadings may be obtained. For tension and compression and for torsion, for example, we employ Eqs. (5–4) and (5–7) and obtain

$$U = \frac{F^2 l}{2AE} \qquad \text{tension and compression} \qquad (5\text{–}15)$$

$$U = \frac{T^2 l}{2GJ} \qquad \text{torsion} \qquad (5\text{–}16)$$

To obtain an expression for the strain energy due to direct shear, consider the element with one side fixed in Fig. 5–8a. The force F places the element in pure shear, and the work done is $U = F\delta/2$. Since the shear strain is $\gamma = \delta/l = \tau/G = F/AG$, we have

$$U = \frac{F^2 l}{2AG} \qquad \text{direct shear} \qquad (5\text{–}17)$$

The strain energy stored in a beam or lever by bending may be obtained by referring to Fig. 5–8b. Here AB is a section of the elastic curve of length ds having a radius of curvature ρ. The strain energy stored in this element of the beam is $dU = (M/2)d\theta$. Since $\rho \, d\theta = ds$, we have

$$dU = \frac{M \, ds}{2\rho} \qquad (b)$$

We can eliminate ρ by using Eq. (5–8). Thus

$$dU = \frac{M^2 \, ds}{2EI} \qquad (c)$$

For small deflections, $ds \doteq dx$. Then, for the entire beam

$$U = \int \frac{M^2 \, dx}{2EI} \qquad \text{bending} \qquad (5\text{–}18)$$

Sometimes the strain energy stored in a unit volume u is a useful quantity. By dividing Eqs. (5–15) to (5–17) by the total volume lA, and setting $F/A = \pm\sigma$ for

| **Figure 5–8**

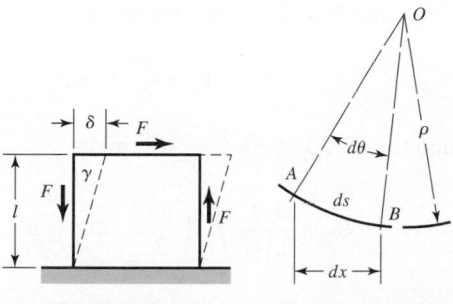

(a) Pure shear element   (b) Beam bending element

tension or compression, $F/A = \tau$ for direct shear, and $Td/(2J) = \tau_{max}$ for torsion, we obtain

$$u = \begin{cases} \dfrac{\sigma^2}{2E} & \text{tension and compression} \\[2mm] \dfrac{\tau^2}{2G} & \text{direct shear} \\[2mm] \dfrac{\tau_{max}^2}{4G} & \text{torsion} \end{cases} \qquad (5\text{--}19)$$

It is interesting to note, from Eq. (5–19), that the development of a high stress in a material with a low modulus of elasticity, or rigidity, will result in the greatest amount of energy storage.

Equation (5–18) is exact only when a beam is subject to pure bending. Even when shear is present, Eq. (5–18) continues to give quite good results, except for very short beams. The strain energy due to shear loading of a beam is a complicated problem. An approximate solution can be obtained by using Eq. (5–17) with a correction factor whose value depends upon the shape of the cross section. If we use C for the correction factor and V for the shear force, then the strain energy due to shear in bending is the integral of Eq. (5–17), or

$$U = \int \frac{CV^2 \, dx}{2AG} \qquad \text{bending shear} \qquad (5\text{--}20)$$

Values of the factor C are listed in Table 5–1.

Table 5–1

Strain-Energy Correction Factors for Shear

Source: Richard G. Budynas, *Advanced Strength and Applied Stress Analysis,* 2nd ed., McGraw-Hill, New York, 1999. Copyright © 1999 The McGraw-Hill Companies.

Beam Cross-Sectional Shape	Factor C
Rectangular	1.2
Circular	1.11
Thin-walled tubular, round	2.00
Box sections[†]	1.00
Structural sections[†]	1.00

[†]Use area of web only.

EXAMPLE 5–8

Find the strain energy due to shear in a rectangular cross-section beam, simply supported, and having a uniformly distributed load.

Solution

Using Appendix Table A–9–7, we find the shear force to be

$$V = \frac{wl}{2} - wx$$

Substituting into Eq. (5–20), with $C = 1.2$, gives

Answer

$$U = \frac{1.2}{2AG} \int_0^l \left(\frac{wl}{2} - wx \right)^2 dx = \frac{w^2 l^3}{20AG}$$

EXAMPLE 5–9   A cantilever has a concentrated load F at the end, as shown in Fig. 5–9. Find the strain energy in the beam by neglecting shear.

| **Figure 5–9**

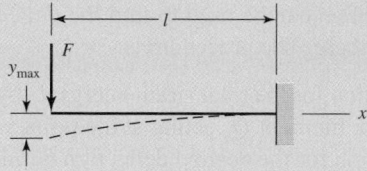

Solution   At any point x along the beam, the moment is $M = -Fx$. Substituting this value of M into Eq. (5–18), we find

Answer

$$U = \int_0^l \frac{F^2 x^2\, dx}{2EI} = \frac{F^2 l^3}{6EI}$$

5–8   Castigliano's Theorem

A most unusual, powerful, and often surprisingly simple approach to deflection analysis is afforded by an energy method called Castigliano's theorem. It is a unique way of analyzing deflections and is even useful for finding the reactions of indeterminate structures. Castigliano's theorem states that *when forces act on elastic systems subject to small displacements, the displacement corresponding to any force, collinear with the force, is equal to the partial derivative of the total strain energy with respect to that force.* The terms *force* and *displacement* in this statement are broadly interpreted to apply equally to moments and angular displacements. Mathematically, the theorem of Castigliano is

$$\delta_i = \frac{\partial U}{\partial F_i} \tag{5–21}$$

where δ_i is the displacement of the point of application of the force F_i in the direction of F_i. For rotational displacement Eq. (5–21) can be written as

$$\theta_i = \frac{\partial U}{\partial M_i} \tag{5–22}$$

where θ_i is the rotational displacement, in radians, of the moment M_i in the direction of M_i.

As an example, apply Castigliano's theorem using Eqs. (5–15) and (5–16) to get the axial and torsional deflections. The results are

$$\delta = \frac{\partial}{\partial F}\left(\frac{F^2 l}{2AE}\right) = \frac{Fl}{AE} \tag{a}$$

$$\theta = \frac{\partial}{\partial T}\left(\frac{T^2 l}{2GJ}\right) = \frac{Tl}{GJ} \tag{b}$$

Compare Eqs. (*a*) and (*b*) with Eqs. (5–3) and (5–5). In Example 5–8, the strain energy for a cantilever having a concentrated end load was found. According to Castigliano's theorem, the deflection at the end of the beam is

$$y = \frac{\partial U}{\partial F} = \frac{\partial}{\partial F}\left(\frac{F^2 l^3}{6EI}\right) = \frac{Fl^3}{3EI}$$

which checks with Table A–9–1.

Castigliano's theorem can be used to find the deflection at a point even though no force or moment acts there. The procedure is:

1. Set up the equation for the total strain energy U by including the energy due to a fictitious force or moment Q_i acting at the point whose deflection is to be found.
2. Find an expression for the desired deflection δ_i, in the direction of Q_i, by taking the derivative of the total strain energy with respect to Q_i.
3. Since Q_i is a fictitious force, solve the expression obtained in step 2 by setting Q_i equal to zero. Thus,

$$\delta_i = \left.\frac{\partial U}{\partial Q_i}\right|_{Q_i=0} \tag{5–23}$$

EXAMPLE 5–10

The cantilever of Ex. 5–9 is a carbon steel bar 10 in long with a 1-in diameter and is loaded by a force $F = 100$ lbf.

(a) Find the maximum deflection using Castigliano's theorem, including that due to shear.

(b) What error is introduced if shear is neglected?

Solution

(a) From Eq. (5–20) and Example 5–9 data, the total strain energy is

$$U = \frac{F^2 l^3}{6EI} + \int_0^l \frac{CV^2\,dx}{2AG} \tag{1}$$

For the cantilever, $V = F$. Also, $C = 1.11$, from Table 5–1. Performing the integration and substituting these values in Eq. (1) gives, for the total strain energy,

$$U = \frac{F^2 l^3}{6EI} + \frac{1.11 F^2 l}{2AG} \tag{2}$$

Then, according to the theorem, the deflection of the end is

$$y = \frac{\partial U}{\partial F} = \frac{Fl^3}{3EI} + \frac{1.11 Fl}{AG} \tag{3}$$

We also find that

$$I = \frac{\pi d^4}{64} = \frac{\pi (1)^4}{64} = 0.0491\ \text{in}^4$$

$$A = \frac{\pi d^2}{4} = \frac{\pi (1)^2}{4} = 0.7854\ \text{in}^2$$

Substituting these values, together with $F = 100$ lbf, $l = 10$ in, $E = 30$ Mpsi, and $G = 11.5$ Mpsi, in Eq. (3) gives

Answer

$$y = 0.022\ 63 + 0.000\ 12 = 0.022\ 75\ \text{in}$$

Note that the result is positive because it is in the *same* direction as the force F.

(b) The error in neglecting shear is found to be about 0.54 percent.

In performing any integrations, it is generally better to take the partial derivative with respect to the load F_i first. This is true especially if the force is a fictitious force Q_i, since it can be set to zero as soon as the derivative is taken. This is demonstrated in the next example. The forms for deflection can then be rewritten. Here we will assume, for axial and torsional loading, that material and cross section properties can vary along the length of the members. From Eqs. (5–15), (5–16), and (5–18),

$$\delta_i = \frac{\partial U}{\partial F_i} = \int \frac{1}{AE} \left(F \frac{\partial F}{\partial F_i} \right) dx \qquad \text{tension and compression} \qquad (5\text{–}24)$$

$$\theta_i = \frac{\partial U}{\partial M_i} = \int \frac{1}{GJ} \left(T \frac{\partial T}{\partial M_i} \right) dx \qquad \text{torsion} \qquad (5\text{–}25)$$

$$\delta_i = \frac{\partial U}{\partial F_i} = \int \frac{1}{EI} \left(M \frac{\partial M}{\partial F_i} \right) dx \qquad \text{bending} \qquad (5\text{–}26)$$

EXAMPLE 5–11

For Ex. 5–7, determine the deflections at points B and C using Castigliano's method.

Solution

With energy methods, coordinate systems are arbitrary, since energy is a scalar quantity. Returning to Fig. 5–7a, imagine x originating at the right end and positive to the left (this eliminates the need for the reactions in the moment equation). Isolation will reveal that the bending moment, for $0 \le x < l$, is

$$M = -Fx \qquad (1)$$

Since F is at C and in the direction of the desired deflection, the deflection at C from Eq. (5–26) is

$$\delta_C = \frac{\partial U}{\partial F} = \int_0^l \frac{1}{EI} \left(M \frac{\partial M}{\partial F} \right) dx \qquad (2)$$

Substituting Eq. (1) into (2), noting that $I = I_1$ for $0 \le x \le l/2$, and $I = 2I_1$ for $l/2 \le x \le l$, we get

Answer

$$\delta_C = \frac{1}{E} \left[\int_0^{l/2} \frac{1}{I_1} (-Fx)(-x)\, dx + \int_{l/2}^l \frac{1}{2I_1} (-Fx)(-x)\, dx \right]$$

$$= \frac{1}{E} \left[\frac{Fl^3}{24I_1} + \frac{7Fl^3}{48I_1} \right] = \frac{3Fl^3}{16EI_1}$$

which is positive, as it is in the direction of F, and agrees with Ex. 5–7.

For B, a fictitious force Q_i is necessary at the point. Assuming Q_i acts down at B, and x is as before, the moment equation is

$$M = -Fx \qquad\qquad 0 \le x \le l/2$$

$$M = -Fx - Q_i \left(x - \frac{l}{2} \right) \qquad l/2 \le x \le l$$

or, in terms of singularity functions,

$$M = -Fx - Q_i \left\langle x - \frac{l}{2} \right\rangle^1 \qquad (3)$$

For Eq. (5–23), we need $\partial U / \partial Q_i$ from Eq. (5–26), for which in turn we need $\partial M / \partial Q_i$. From Eq. (3),

$$\frac{\partial M}{\partial Q_i} = -\left\langle x - \frac{l}{2}\right\rangle^1 \tag{4}$$

Once the derivative is taken, Q_i can be set to zero so from Eq. (3), $M = -Fx$ and

$$\delta_B = \int_0^l \frac{1}{EI}(Fx)\left\langle x - \frac{l}{2}\right\rangle^1 dx$$

But the singularity function is zero for $0 \le x \le l/2$, and $I = 2I_1$ for $l/2 \le x \le l$. Thus,

$$\delta_B = \frac{F}{E(2I_1)} \int_{l/2}^l (x)\left(x - \frac{l}{2}\right) dx = \frac{F}{2EI_1}\left[\frac{x^3}{3} - \frac{lx^2}{4}\right]_{l/2}^l$$

Answer

$$= \frac{5Fl^3}{96EI_1}$$

which again is positive, in the direction of Q_i, and agrees with Ex. 5–7.

EXAMPLE 5–12

For the wire form of diameter d shown in Fig. 5–10a, determine the deflection of point B in the direction of the applied force F (neglect the effect of bending shear).

Solution

It is very important to include the loading effects on all parts of the structure. Coordinate systems are not important, but loads must be consistent with the problem. Thus appropriate use of *free-body diagrams* is essential here. The reader should verify that the reactions as functions of F in elements BC, CD, and GD are as shown in Fig. 5–10b.

The deflection of B in the direction of F is given by

$$\delta_B = \frac{\partial U}{\partial F}$$

so the partial derivatives in Eqs. (5–24) to (5–26) will all be taken with respect to F. Element BC is in bending only so from Eq. (5–26),[5]

$$\frac{\partial U_{BC}}{\partial F} = \frac{1}{EI}\int_0^a (-Fy)(-y)\, dy = \frac{Fa^3}{3EI} \tag{1}$$

Element CD is in bending and in torsion. The torsion is constant so Eq. (5–25) can be written as

$$\frac{\partial U}{\partial F_i} = \left(T\frac{\partial T}{\partial F_i}\right)\frac{l}{GJ}$$

---

[5]. It is very tempting to mix techniques and try to use superposition also, for example. However, some subtle things can occur that you may visually miss. It is highly recommended that if you are using Castigliano's theorem on a problem, you use it for all parts of the problem.

Figure 5-10

(Source: Richard G. Budynas, Advanced Strength and Applied Stress Analysis, 2nd ed., McGraw-Hill, New York, 1999. Copyright © 1999 The McGraw-Hill Companies.)

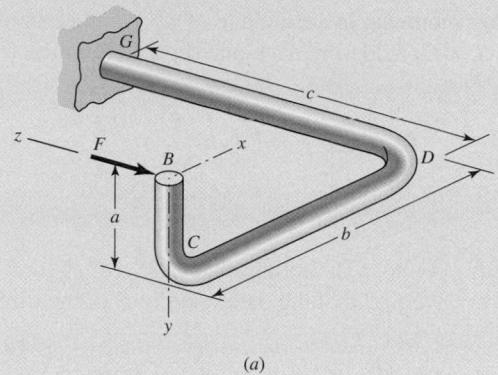

(a)

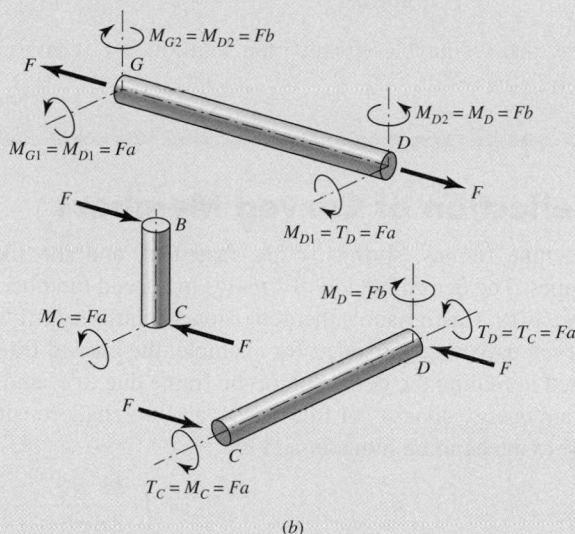

(b)

where l is the length of the member. So for the torsion in member CD, $F_i = F$, $T = Fa$, and $l = b$. Thus,

$$\left(\frac{\partial U_{CD}}{\partial F}\right)_{\text{torsion}} = (Fa)(a)\frac{b}{GJ} = \frac{Fa^2b}{GJ} \tag{2}$$

For the bending in CD,

$$\left(\frac{\partial U_{CD}}{\partial F}\right)_{\text{bending}} = \frac{1}{EI}\int_0^b (-Fx)(-x)\,dx = \frac{Fb^3}{3EI} \tag{3}$$

Member DG is axially loaded and is bending in two planes. The axial loading is constant, so Eq. (5–24) can be written as

$$\frac{\partial U}{\partial F_i} = \left(F\frac{\partial F}{\partial F_i}\right)\frac{l}{AE}$$

where l is the length of the member. Thus, for the axial loading of DG, $F = F_i$, $l = c$, and

$$\left(\frac{\partial U_{DG}}{\partial F}\right)_{\text{axial}} = \frac{Fc}{AE} \tag{4}$$

The bending moments in each plane of DG are constant along the length of $M_y = Fb$ and $M_x = Fa$. Considering each one separately in the form of Eq. (5–26) gives

$$\left(\frac{\partial U_{DG}}{\partial F}\right)_{\text{bending}} = \frac{1}{EI} \int_0^c (Fb)(b) \, dz + \frac{1}{EI} \int_0^c (Fa)(a) \, dz$$

(5)

$$= \frac{Fc(a^2 + b^2)}{EI}$$

Adding Eqs. (1) to (5), noting that $I = \pi d^4/64$, $J = 2I$, $A = \pi d^2/4$, and $G = E/[2(1 + v)]$, we find that the deflection of B in the direction of F is

Answer

$$(\delta_B)_F = \frac{4F}{3\pi E d^4}[16(a^3 + b^3) + 48c(a^2 + b^2) + 48(1 + v)a^2b + 3cd^2]$$

Now that we have completed the solution, see if you can physically account for each term in the result.

5–9 Deflection of Curved Members

Machine frames, springs, clips, fasteners, and the like frequently occur as curved shapes. The determination of stresses in curved members has already been described in Sec. 4–19. Castigliano's theorem is particularly useful for the analysis of deflections in curved parts too. Consider, for example, the curved frame of Fig. 5–11a. We are interested in finding the deflection of the frame due to F and in the direction of F. The total strain energy consists of four terms, and we shall consider each separately. The first is due to the bending moment and is[6]

$$U_1 = \int \frac{M^2 \, d\theta}{2AeE}$$

(5–27)

In this equation, the eccentricity e is

$$e = R - r_n$$

(5–28)

where r_n is the radius of the neutral axis as defined in Sec. 4–19 and shown in Fig. 4–37.

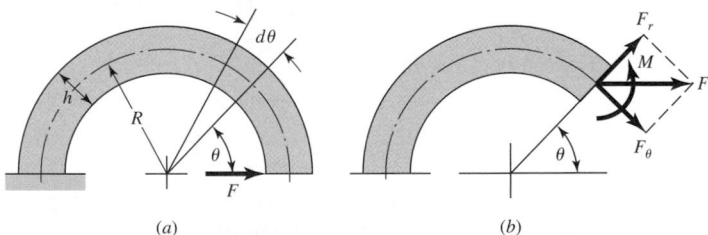

(a)   (b)

Figure 5–11

(a) Curved bar loaded by force F. R = radius to centroidal axis of section; h = section thickness. (b) Diagram showing forces acting on section taken at angle θ. $F_r = V$ = shear component of F; F_θ is component of F normal to section; M is moment caused by force F.

6. See Richard G. Budynas, op. cit., Sec. 6.7.

An approximate result can be obtained by using the equation

$$U_1 \doteq \int \frac{M^2 R \, d\theta}{2EI} \qquad \frac{R}{h} > 10 \tag{5-29}$$

which is obtained directly from Eq. (5–18). Note the limitation on the use of Eq. (5–29).

The strain energy component due to the normal force F_θ consists of two parts, one of which is axial and analogous to Eq. (5–15). This part is

$$U_2 = \int \frac{F_\theta^2 R \, d\theta}{2AE} \tag{5-30}$$

The force F_θ also produces a moment, which opposes the moment M in Fig. 5–11b. The resulting strain energy will be subtractive and is

$$U_3 = -\int \frac{M F_\theta \, d\theta}{AE} \tag{5-31}$$

The negative sign of Eq. (5–31) can be appreciated by referring to both parts of Fig. 5–11. Note that the moment M tends to decrease the angle $d\theta$. On the other hand, the moment due to F_θ tends to increase $d\theta$. Thus U_3 is negative. If F_θ had been acting in the opposite direction, then both M and F_θ would tend to decrease the angle $d\theta$.

The fourth and last term is the shear energy due to F_r. Adapting Eq. (5–20) gives

$$U_4 = \int \frac{C F_r^2 R \, d\theta}{2AG} \tag{5-32}$$

where C is the correction factor of Table 5–1.

Combining the four terms gives for the total strain energy

$$U = \int \frac{M^2 \, d\theta}{2AeE} + \int \frac{F_\theta^2 R \, d\theta}{2AE} - \int \frac{M F_\theta \, d\theta}{AE} + \int \frac{C F_r^2 R \, d\theta}{2AG} \tag{5-33}$$

The deflection produced by the force F can now be found. It is

$$\delta = \frac{\partial U}{\partial F} = \int_0^\pi \frac{M}{AeE} \left(\frac{\partial M}{\partial F} \right) d\theta + \int_0^\pi \frac{F_\theta R}{AE} \left(\frac{\partial F_\theta}{\partial F} \right) d\theta$$

$$- \int_0^\pi \frac{1}{AE} \frac{\partial (M F_\theta)}{\partial F} d\theta + \int_0^\pi \frac{C F_r R}{AG} \left(\frac{\partial F_r}{\partial F} \right) d\theta \tag{5-34}$$

Using Fig. 5–11b, we find

$$M = FR \sin\theta \qquad\qquad \frac{\partial M}{\partial F} = R \sin\theta$$

$$F_\theta = F \sin\theta \qquad\qquad \frac{\partial F_\theta}{\partial F} = \sin\theta$$

$$M F_\theta = F^2 R \sin^2\theta \qquad\qquad \frac{\partial M F_\theta}{\partial F} = 2FR \sin^2\theta$$

$$F_r = F \cos\theta \qquad\qquad \frac{\partial F_r}{\partial F} = \cos\theta$$

Substituting all these into Eq. (5–34) and factoring yields

$$\delta = \frac{FR^2}{AeE} \int_0^\pi \sin^2\theta\, d\theta + \frac{FR}{AE} \int_0^\pi \sin^2\theta\, d\theta - \frac{2FR}{AE} \int_0^\pi \sin^2\theta\, d\theta$$

$$+ \frac{CFR}{AG} \int_0^\pi \cos^2\theta\, d\theta$$

$$= \frac{\pi FR^2}{2AeE} + \frac{\pi FR}{2AE} - \frac{\pi FR}{AE} + \frac{\pi CFR}{2AG} = \frac{\pi FR^2}{2AeE} - \frac{\pi FR}{2AE} + \frac{\pi CFR}{2AG} \qquad (5\text{–}35)$$

Because the first term contains the square of the radius, the second two terms will be small if the frame has a large radius. Also, if $R/h > 10$, Eq. (5–29) can be used. An approximate result then turns out to be

$$\delta \approx \frac{\pi FR^3}{2EI} \qquad (5\text{–}36)$$

The determination of the deflection of a curved member loaded by forces at right angles to the plane of the member is more difficult, but the method is the same.[7] We shall include here only one of the more useful solutions to such a problem, though the methods for all are similar. Figure 5–12 shows a cantilevered ring segment having a span angle ϕ. Assuming $R/h > 10$, the strain energy neglecting direct shear, is obtained from the equation

$$U = \int_0^\phi \frac{M^2 R\, d\theta}{2EI} + \int_0^\phi \frac{T^2 R\, d\theta}{2GJ} \qquad (5\text{–}37)$$

The moments and torques acting on a section at B, due to the force F, are

$$M = FR\sin\theta \qquad T = FR(1 - \cos\theta)$$

The deflection δ of the ring segment at F and in the direction of F is then found to be

$$\delta = \frac{\partial U}{\partial F} = \frac{FR^3}{2}\left(\frac{\alpha}{EI} + \frac{\beta}{GJ}\right) \qquad (5\text{–}38)$$

where the coefficients α and β are dependent on the span angle ϕ and are defined as follows:

$$\alpha = \phi - \sin\phi\cos\phi$$
$$\beta = 3\phi - 4\sin\phi + \sin\phi\cos\phi \qquad (5\text{–}39)$$

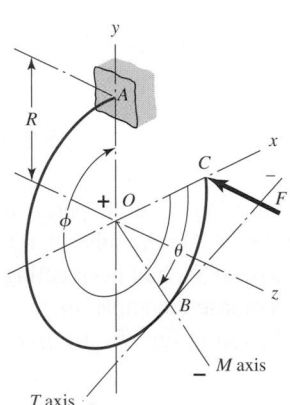

Figure 5–12

Ring ABC in the xy plane subject to force F parallel to the z axis. Corresponding to a ring segment CB at angle θ from the point of application of F, the moment axis is a line BO and the torque axis is a line in the xy plane tangent to the ring at B. Note the positive directions of the T and M axes.

---

7. For more solutions than are included here, see Joseph E. Shigley, "Curved Beams and Rings," Chap. 16 in Joseph E. Shigley and Charles R. Mischke (eds.), *Standard Handbook of Machine Design*, 2nd ed., McGraw-Hill, New York, 1996.

EXAMPLE 5–13 Deflection in a Variable-Cross-Section Punch-Press Frame

The general result expressed in Eq. (5–35),

$$\delta = \frac{\pi FR^2}{2AeE} - \frac{\pi FR}{2AE} + \frac{\pi CFR}{2AG}$$

Figure 5–13

(a) A steel punch press has a C frame with a varying-depth rectangular cross section depicted. The cross section varies sinusoidally from 2 × 2 in at $\theta = 0°$ to 2 × 6 in at $\theta = 90°$, and back to 2 × 2 in at $\theta = 180°$. Of immediate interest to the designer is the deflection in the load axis direction under the load.
(b) Finite element model.
(Source: Richard G. Budynas, Advanced Strength and Applied Stress Analysis, 2nd ed., McGraw-Hill, New York, 1999. Copyright © 1999 The McGraw-Hill Companies.)

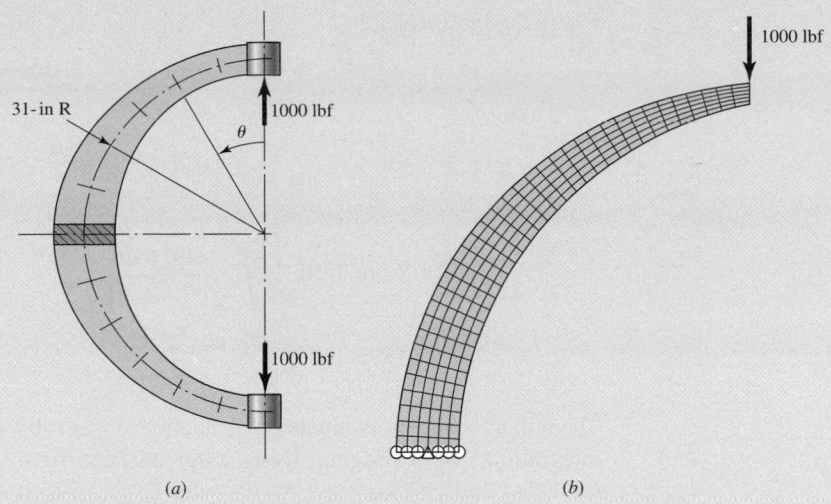

(a)   (b)

is useful in sections that are uniform and in which the centroidal locus is circular. This makes integration tractable and general, and insightful results are obtained. The bending moment is largest where the material is farthest from the load axis. Strengthening requires a larger second area moment I. A variable-depth cross section is attractive, but it makes the integration to a closed form impossible. However, if you are seeking results, numerical integration with computer assistance is satisfactory.

Consider the steel C frame depicted in Fig. 5–13a in which the centroidal radius is 32 in, the cross section at the ends is 2 in × 2 in, and the depth varies sinusoidally with an amplitude of 2 in. The load is 1000 lbf. It follows that $C = 1.2$, $G = 11.5(10^6)$ psi, $E = 30(10^6)$ psi. The outer and inner radii are

$$R_{\text{out}} = 33 + 2\sin\theta$$

$$R_{\text{in}} = 31 - 2\sin\theta$$

$$h = R_{\text{out}} - R_{\text{in}} = 2(1 + 2\sin\theta)$$

$$A = bh = 4(1 + 2\sin\theta)$$

$$r_n = \frac{h}{\ln[(R + h/2)/(R - h/2)]} = \frac{2(1 + 2\sin\theta)}{\ln[(33 + 2\sin\theta)/(31 - 2\sin\theta)]}$$

$$e = R - r_n = 32 - r_n$$

Note that

$$M = FR\sin\theta \qquad \partial M/\partial F = R\sin\theta$$

$$F_\theta = F\sin\theta \qquad \partial F_\theta/\partial F = \sin\theta$$

$$MF_\theta = F^2R\sin^2\theta \qquad \partial MF_\theta/\partial F = 2FR\sin^2\theta$$

$$F_r = F\cos\theta \qquad \partial F_r/\partial F = \cos\theta$$

Substitution of the terms into Eq. (5–34) yields

$$\delta = I_1 + I_2 + I_3 \tag{1}$$

where the integrals are

$$I_1 = 8.5333(10^{-3}) \int_0^\pi \frac{\sin^2 \theta \, d\theta}{(1 + 2\sin\theta)\left[32 - \dfrac{2(1 + 2\sin\theta)}{\ln\left(\dfrac{33 + 2\sin\theta}{31 - 2\sin\theta}\right)}\right]} \tag{2}$$

$$I_2 = -2.6667(10^{-4}) \int_0^\pi \frac{\sin^2 \theta \, d\theta}{1 + 2\sin\theta} \tag{3}$$

$$I_3 = 8.3478(10^{-4}) \int_0^\pi \frac{\cos^2 \theta \, d\theta}{1 + 2\sin\theta} \tag{4}$$

The integrals may be evaluated in a number of ways: by a program using Simpson's rule integration,[8] by a program using a spreadsheet, or by mathematics software. Using MathCad and checking the results with Excel gives the integrals as $I_1 = 0.076\,615$, $I_2 = -0.000\,159$, and $I_3 = 0.000\,773$. Substituting these into Eq. (1) gives

Answer

$$\delta = 0.077\,23 \text{ in}$$

Finite element (FE) programs are also very accessible. Figure 5–13b shows a simple half-model, using symmetry, of the press consisting of 216 plane-stress (2-D) elements. Creating the model and analyzing it to obtain a solution took minutes. Doubling the results from the FE analysis yielded $\delta = 0.07790$ in, a less than 1 percent variation from the results of the numerical integration.

---

8. See Case Study 4, p. 203, J. E. Shigley and C. R. Mischke, *Mechanical Engineering Design*, 6th ed., McGraw-Hill, New York, 2001.

5–10   Statically Indeterminate Problems

A system in which the laws of statics are not sufficient to determine all the unknown forces or moments is said to be *statically indeterminate*. Problems of which this is true are solved by writing the appropriate equations of static equilibrium and additional equations pertaining to the deformation of the part. In all, the number of equations must equal the number of unknowns.

A simple example of a statically indeterminate problem is furnished by the nested helical springs in Fig. 5–14a. When this assembly is loaded by the compressive force F, it deforms through the distance δ. What is the compressive force in each spring?

Only one equation of static equilibrium can be written. It is

$$\sum F = F - F_1 - F_2 = 0 \tag{a}$$

which simply says that the total force F is resisted by a force F_1 in spring 1 plus the force F_2 in spring 2. Since there are two unknowns and only one equation, the system is statically indeterminate.

To write another equation, note the deformation relation in Fig. 5–14b. The two springs have the same deformation. Thus, we obtain the second equation as

$$\delta_1 = \delta_2 = \delta \tag{b}$$

| **Figure 5–14**

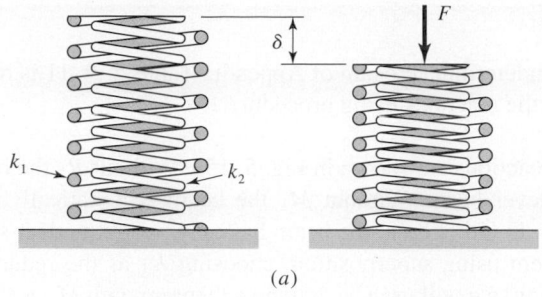

(a)

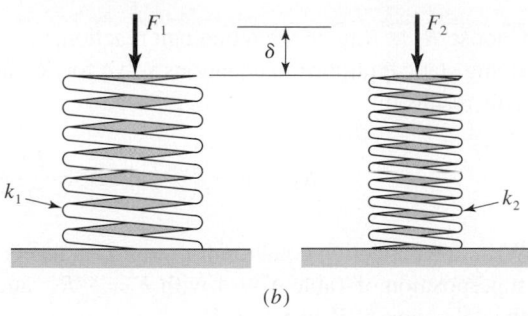

(b)

If we now substitute Eq. (5–2) in Eq. (b), we have

$$\frac{F_1}{k_1} = \frac{F_2}{k_2}$$

(c)

Now we solve Eq. (c) for F_1 and substitute the result in Eq. (a). This gives

$$F - \frac{k_1}{k_2}F_2 - F_2 = 0 \quad \text{or} \quad F_2 = \frac{k_2 F}{k_1 + k_2}$$

(d)

This completes the solution, because with F_2 known, F_1 can be found from Eq. (c).

In the spring example, obtaining the necessary deformation equation was very straightforward. In Sec. 4–17, equations are developed for press and shrink fits. This also is a statically indeterminate problem where the interface pressures between the mating cylinders are equal but unknown through equilibrium conditions. Here, at the interface a deformation equation is necessary and is obtained, as in the nested spring problem. This method normally works fine when unknowns occur at a common interface. However, for other situations, the deformation relations may not be as easy. A more structured approach may be necessary. Here we will show two basic procedures for general statically indeterminate problems.

Procedure 1

1 Choose the redundant reaction(s).

2 Write the equations of static equilibrium for the remaining reactions in terms of the applied loads and the redundant reaction(s) of step 1.

3 Write the deflection equation(s) for the point(s) at the locations of the redundant reaction(s) of step 1 in terms of the applied loads and the redundant reaction(s) of step 1. Normally the deflection(s) is (are) zero. If a redundant reaction is a moment, the corresponding deflection equation is a rotational deflection equation.

4 The equations from steps 2 and 3 can now be solved to determine the reactions.

In step 1 the deflection equations can be solved in any of the standard ways. Here we will demonstrate the use of superposition and Castigliano's theorem on a beam problem.

EXAMPLE 5–14

The indeterminate beam of Appendix Table A–9–11 is reproduced in Fig. 5–15. Determine the reactions using procedure 1.

Solution

The reactions are shown in Fig. 5–15*b*. Without R_2 the beam is a statically determinate cantilever beam. Without M_1 the beam is a statically determinate simply supported beam. In either case, the beam has only *one* redundant support. We will first solve this problem using superposition, choosing R_2 as the redundant reaction. For the second solution, we will use Castigliano's theorem with M_1 as the redundant reaction.

Solution 1

1   Choose R_2 at B to be the redundant reaction.
2   Using static equilibrium equations solve for R_1 and M_1 in terms of F and R_2. This results in

$$R_1 = F - R_2 \qquad M_1 = \frac{Fl}{2} - R_2 l \tag{1}$$

3   Write the deflection equation for point B in terms of F and R_2. Using superposition of Table A–9–1 with $F = -R_2$, and Table A–9–2 with $a = l/2$, the deflection of B, at $x = l$, is

$$\delta_B = -\frac{R_2 l^2}{6EI}(l - 3l) + \frac{F(l/2)^2}{6EI}\left(\frac{l}{2} - 3l\right) = \frac{R_2 l^3}{3EI} - \frac{5Fl^3}{48EI} = 0 \tag{2}$$

4   Equation (2) can be solved for R_2 directly. This yields

Answer

$$R_2 = \frac{5F}{16} \tag{3}$$

Next, substituting R_2 into Eqs. (1) completes the solution, giving

Answer

$$R_1 = \frac{11F}{16} \qquad M_1 = \frac{3Fl}{16} \tag{4}$$

Note that the solution agrees with what is given in Table A–9–11.

Solution 2

1   Choose M_1 at O to be the redundant reaction.
2   Using static equilibrium equations solve for R_1 and R_2 in terms of F and M_1. This results in

$$R_1 = \frac{F}{2} + \frac{M_1}{l} \qquad R_2 = \frac{F}{2} - \frac{M_1}{l} \tag{5}$$

| **Figure 5–15**

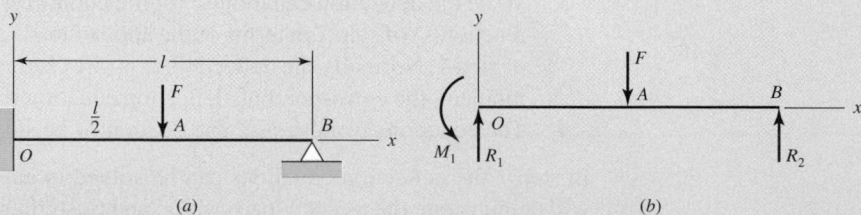

(a)   (b)

3 Since M_1 is the redundant reaction at O, write the equation for the angular deflection at point O. From Castigliano's theorem this is

$$\theta_O = \frac{\partial U}{\partial M_1} \tag{6}$$

We can apply Eq. (5–26), using the variable x as shown in Fig. 5–15b. However, simpler terms can be found by using a variable $\hat{x}$ that starts at B and is positive to the left. With this and the expression for R_2 from Eq. (5) the moment equations are

$$M = \left(\frac{F}{2} - \frac{M_1}{l}\right)\hat{x} \qquad\qquad 0 \le \hat{x} \le \frac{l}{2} \tag{7}$$

$$M = \left(\frac{F}{2} - \frac{M_1}{l}\right)\hat{x} - F\left(\hat{x} - \frac{l}{2}\right) \qquad \frac{l}{2} \le \hat{x} \le l \tag{8}$$

For both equations

$$\frac{\partial M}{\partial M_1} = -\frac{\hat{x}}{l} \tag{9}$$

Substituting Eqs. (7) to (9) in Eq. (6), using the form of Eq. (5–26) where $F_i = M_1$, gives

$$\theta_O = \frac{\partial U}{\partial M_1} = \frac{1}{EI}\left\{\int_0^{l/2}\left(\frac{F}{2} - \frac{M_1}{l}\right)\hat{x}\left(-\frac{\hat{x}}{l}\right)d\hat{x} + \int_{l/2}^{l}\left[\left(\frac{F}{2} - \frac{M_1}{l}\right)\hat{x}\right.\right.$$
$$\left.\left. - F\left(\hat{x} - \frac{l}{2}\right)\right]\left(-\frac{\hat{x}}{l}\right)d\hat{x}\right\} = 0$$

Canceling $1/EIl$, and combining the first two integrals, simplifies this quite readily to

$$\left(\frac{F}{2} - \frac{M_1}{l}\right)\int_0^{l}\hat{x}^2\,d\hat{x} - F\int_{l/2}^{l}\left(\hat{x} - \frac{l}{2}\right)\hat{x}\,d\hat{x} = 0$$

Integrating gives

$$\left(\frac{F}{2} - \frac{M_1}{l}\right)\frac{l^3}{3} - \frac{F}{3}\left[l^3 - \left(\frac{l}{2}\right)^3\right] + \frac{Fl}{4}\left[l^2 - \left(\frac{l}{2}\right)^2\right] = 0$$

which reduces to

$$M_1 = \frac{3Fl}{16} \tag{10}$$

4 Substituting Eq. (10) into (5) results in

$$R_1 = \frac{11F}{16} \qquad R_2 = \frac{5F}{16} \tag{11}$$

which again agrees with Table A–9–11.

For some problems even procedure 1 can be a task. Procedure 2 eliminates some tricky geometric problems that would complicate procedure 1. We will describe the procedure for a beam problem.

Procedure 2

1   Write the equations of static equilibrium for the beam in terms of the applied loads and unknown restraint reactions.
2   Write the deflection equation for the beam in terms of the applied loads and unknown restraint reactions.
3   Apply boundary conditions consistent with the restraints.
4   Solve the equations from steps 1 and 3.

EXAMPLE 5–15

The rods AD and CE shown in Fig. 5–16a each have a diameter of 10 mm. The second-area moment of beam ABC is $I = 62.5(10^3)$ mm^4. The modulus of elasticity of the material used for the rods and beam is $E = 200$ GPa. The threads at the ends of the rods are single-threaded with a pitch of 1.5 mm. The nuts are first snugly fit with bar ABC horizontal. Next the nut at A is tightened one full turn. Determine the resulting tension in each rod and the deflections of points A and C.

Solution

There is a lot going on in this problem; a rod shortens, the rods stretch, and the beam bends. Let's try the procedure!

1   The free-body diagram of the beam is shown in Fig. 5–16b. Summing forces and moments about B gives

$$F_B - F_A - F_C = 0 \tag{1}$$

$$4F_A - 3F_C = 0 \tag{2}$$

2   Using singularity functions, we find the moment equation for the beam is

$$M = -F_A x + F_B \langle x - 0.2 \rangle^1$$

where x is in meters. Integration yields

$$EI\frac{dy}{dx} = -\frac{F_A}{2}x^2 + \frac{F_B}{2}\langle x - 0.2 \rangle^2 + C_1$$

$$EIy = -\frac{F_A}{6}x^3 + \frac{F_B}{6}\langle x - 0.2 \rangle^3 + C_1 x + C_2 \tag{3}$$

The term $EI = 200(10^9)\,62.5(10^{-9}) = 1.25(10^4)$ N · m^2.

| Figure 5–16

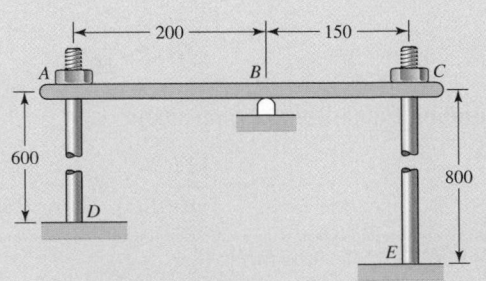

(a)

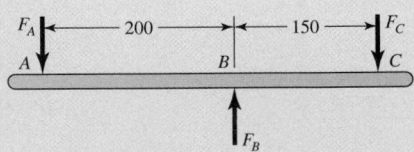

(b) Free-body diagram of beam ABC

3 The upward deflection of point A is $(Fl/AE)_{AD} - Np$, where the first term is the elastic stretch of AD, N is the number of turns of the nut, and p is the pitch of the thread. Thus, the deflection of A is

$$y_A = \frac{F_A(0.6)}{\dfrac{\pi}{4}(0.010)^2(200)(10^9)} - (1)(0.0015) \tag{4}$$

$$= 3.8197(10^{-8})F_A - 1.5(10^{-3})$$

The upward deflection of point C is $(Fl/AE)_{CE}$, or

$$y_C = \frac{F_C(0.8)}{\dfrac{\pi}{4}(0.010)^2(200)(10^9)} = 5.093(10^{-8})F_C \tag{5}$$

Equations (4) and (5) will now serve as the boundary conditions for Eq. (3). At $x = 0$, $y = y_A$. Substituting Eq. (4) into (3) with $x = 0$ and $EI = 1.25(10^4)$, noting that the singularity function is zero for $x = 0$, gives

$$4.7746(10^{-4})F_A + C_2 = -18.75 \tag{6}$$

At $x = 0.2$ m, $y = 0$, and Eq. (3) yields

$$-1.3333(10^{-3})F_A + 0.2C_1 + C_2 = 0 \tag{7}$$

At $x = 0.35$ m, $y = y_C$. Substituting Eq. (5) into (3) with $x = 0.35$ m and $EI = 1.25(10^4)$ gives

$$-7.1458(10^{-3})F_A + 5.625(10^{-4})F_B - 6.3662(10^{-4})F_C + 0.35C_1 + C_2 = 0 \tag{8}$$

Equations (1), (2), (6), (7), and (8) are five equations in F_A, F_B, F_C, C_1, and C_2. Written in matrix form, they are

$$\begin{bmatrix} -1 & 1 & -1 & 0 & 0 \\ 4 & 0 & -3 & 0 & 0 \\ -4.7746(10^{-4}) & 0 & 0 & 0 & 1 \\ -1.3333(10^{-3}) & 0 & 0 & 0.2 & 1 \\ -7.1458(10^{-3}) & 5.625(10^{-4}) & -6.3662(10^{-4}) & 0.35 & 1 \end{bmatrix} \begin{Bmatrix} F_A \\ F_B \\ F_C \\ C_1 \\ C_2 \end{Bmatrix} = \begin{Bmatrix} 0 \\ 0 \\ -18.75 \\ 0 \\ 0 \end{Bmatrix}$$

Solving these equations yields

Answer $\qquad F_A = 2988$ N $\qquad F_B = 6971$ N $\qquad F_C = 3983$ N

$\qquad C_1 = 106.54$ N·m^2 $\qquad C_2 = -17.324$ N·m^3

Equation (3) can be reduced to

$$y = -(39.84x^3 - 92.95\langle x - 0.2\rangle^3 - 8.523x + 1.386)(10^{-3})$$

Answer At $x = 0$, $y = y_A = -1.386(10^{-3})$ m $= -1.386$ mm. At $x = 0.35$ m,

Answer $y = y_C = -[39.84(0.35)^3 - 92.95(0.35 - 0.2)^3 - 8.523(0.35) + 1.386](10^{-3})$

$$= 0.203(10^{-3}) \text{ m} = 0.203 \text{ mm}$$

Note that we could have easily incorporated the stiffness of the support at *B* if we were given a spring constant.

5–11 Compression Members—General

The analysis and design of compression members can differ significantly from that of members loaded in tension or in torsion. If you were to take a long rod or pole, such as a meterstick, and apply gradually increasing compressive forces at each end, nothing would happen at first, but then the stick would bend (buckle), and finally bend so much as to fracture. Try it. The other extreme would occur if you were to saw off, say, a 5-mm length of the meterstick and perform the same experiment on the short piece. You would then observe that the failure exhibits itself as a mashing of the specimen, that is, a simple compressive failure. For these reasons it is convenient to classify compression members according to their length and according to whether the loading is central or eccentric. The term *column* is applied to all such members except those in which failure would be by simple or pure compression. Columns can be categorized then as:

1 Long columns with central loading
2 Intermediate-length columns with central loading
3 Columns with eccentric loading
4 Struts or short columns with eccentric loading

Classifying columns as above makes it possible to develop methods of analysis and design specific to each category. Furthermore, these methods will also reveal whether or not you have selected the category appropriate to your particular problem. The four sections that follow correspond, respectively, to the four categories of columns listed above.

S. Timoshenko[9] offered a telling demonstration of the role of stability of members in compression. Consider a straight *rigid* bar carrying a compressive load *P* as shown in Fig. 5–17. The straight, concentrically loaded bar is held vertical by a pair of springs with a combined spring rate of *k* lbf/in. For a light load any perturbation creating a displacement of the bar to the right or left causes the spring to restore the bar to the vertical position (stable equilibrium). At some high load the spring is unable to correct, and the displacement increases (unstable equilibrium). At some intermediate load equilibrium is

| **Figure 5–17**

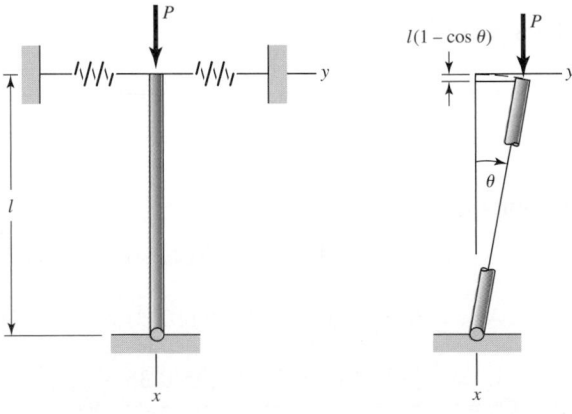

9. S. Timoshenko, *Theory of Elastic Stability*, McGraw-Hill, New York, 1936, pp. 77–78.

neutral. For a small displacement θ the load P is lowered a distance $l(1 - \cos\theta)$. The loss of potential energy of the load, E_p, is

$$E_p = Pl(1 - \cos\theta) = Pl\left[1 - \left(1 - \frac{\theta^2}{2} + \frac{\theta^4}{4!} - \cdots\right)\right] \doteq \frac{Pl\theta^2}{2}$$

for the first two terms of the cosine series. The increase in strain energy of the springs, E_{strain}, is

$$E_{\text{strain}} = \frac{k(l\theta)^2}{2}$$

The strain energy of the bar is zero if rigid and small if not. For neutral equilibrium $E_{\text{strain}} = E_p$, or

$$\frac{k(l\theta)^2}{2} = \frac{Pl\theta^2}{2}$$

from which

$$P_{\text{crit}} = kl \qquad (5\text{–}40)$$

The strength of the column material does not control. The stresses in the column are not involved in Eq. (5–40). The failure to carry the load is due to a condition of elastic instability. The years of experience that engineers accumulate guarding against and monitoring stress-controlled failures dulls alertness to instability. General experience with "heft," proportions, and what works and does not work is not a reliable guide in column action. Such failures can be sudden and catastrophic.

5–12 Long Columns with Central Loading

The relationship between the critical load and the column material and geometry is developed with reference to Fig. 5–18a. We assume a bar of length l loaded by a force P acting along the centroidal axis on rounded or pinned ends. The figure shows that the bar is bent in the positive y direction. This requires a negative moment, and hence

$$M = -Py \qquad (a)$$

Figure 5–18

(a) Both ends rounded or pivoted; (b) both ends fixed; (c) one end free, one end fixed; (d) one end rounded and pivoted, and one end fixed.

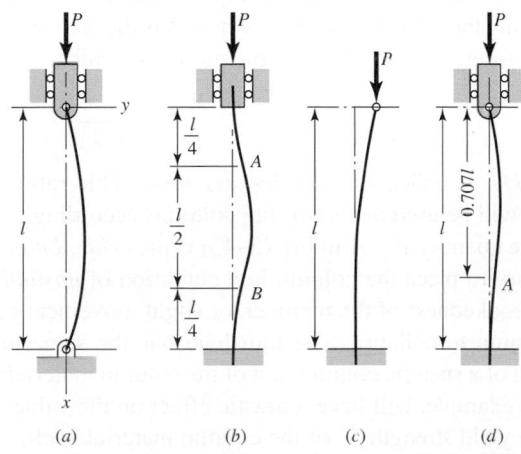

(a) (b) (c) (d)

If the bar should happen to bend in the negative y direction, a positive moment would result, and so $M = -Py$, as before. Using Eq. (5–12), we write

$$\frac{d^2 y}{dx^2} = -\frac{P}{EI} y \qquad (b)$$

or

$$\frac{d^2 y}{dx^2} + \frac{P}{EI} y = 0 \qquad (5\text{–}41)$$

This resembles the well-known differential equation for simple harmonic motion. The solution is

$$y = A \sin \sqrt{\frac{P}{EI}} x + B \cos \sqrt{\frac{P}{EI}} x \qquad (c)$$

where A and B are constants of integration and must be determined from the boundary conditions of the problem. We evaluate them using the conditions that $y = 0$ at $x = 0$ and at $x = l$. This gives $B = 0$, and

$$0 = A \sin \sqrt{\frac{P}{EI}} l \qquad (d)$$

The trivial solution of no buckling occurs with $A = 0$. However, if $A \neq 0$, then

$$\sin \sqrt{\frac{P}{EI}} l = 0 \qquad (e)$$

Equation (e) is satisfied by $\sqrt{P/EI}\, l = n\pi$, where $n = 1, 2, 3, \ldots$. Solving for P when $n = 1$ gives the first critical load

$$P_{cr} = \frac{\pi^2 EI}{l^2} \qquad (5\text{–}42)$$

which is called the *Euler column formula;* it applies only to rounded-end columns. If we substitute these results back in Eq. (c), we get the equation of the deflection curve as

$$y = A \sin \frac{\pi x}{l} \qquad (f)$$

which indicates that the deflection curve is a half-wave sine. We are interested only in the minimum critical load, which occurs with $n = 1$. However, though it is not of any importance here, values of n greater than 1 result in deflection curves that cross the axis at points of inflection and are multiples of half-wave sines.

Using the relation $I = Ak^2$, where A is the area and k the radius of gyration, enables us to rearrange Eq. (5–42) into the more convenient form

$$\frac{P_{cr}}{A} = \frac{\pi^2 E}{(l/k)^2} \qquad (5\text{–}43)$$

where l/k is called the *slenderness ratio*. This ratio, rather than the actual column length, will be used in classifying columns according to length categories.

The quantity P_{cr}/A in Eq. (5–43) is the *critical unit load*. It is the load per unit area necessary to place the column in a condition of *unstable equilibrium*. In this state any small crookedness of the member, or slight movement of the support or load, will cause the column to collapse. The unit load has the same units as strength, but this is the strength of a specific column, not of the column material. Doubling the length of a member, for example, will have a drastic effect on the value of P_{cr}/A but no effect at all on, say, the yield strength S_y of the column material itself.

Equation (5–43) shows that the critical unit load depends only upon the modulus of elasticity and the slenderness ratio. Thus a column obeying the Euler formula made of high-strength alloy steel is *no stronger* than one made of low-carbon steel, since E is the same for both.

The critical loads for columns with different end conditions can be obtained by solving the differential equation or by comparison. Figure 5–18b shows a column with both ends fixed. The inflection points are at A and B, a distance $l/4$ from the ends. The distance AB is the same curve as a rounded-end column. Substituting the length $l/2$ for l in Eq. (5–42), we obtain

$$P_{cr} = \frac{\pi^2 EI}{(l/2)^2} = \frac{4\pi^2 EI}{l^2} \tag{5–44}$$

In Fig. 5–18c is shown a column with one end free and one end fixed. This curve is equivalent to half the curve for columns with rounded ends, so that if a length of $2l$ is substituted in Eq. (5–42), the critical load becomes

$$P_{cr} = \frac{\pi^2 EI}{(2l)^2} = \frac{\pi^2 EI}{4l^2} \tag{5–45}$$

A column with one end fixed and one end rounded, as in Fig. 5–18d, occurs frequently. The inflection point is at A, a distance of $0.707l$ from the rounded end. Therefore

$$P_{cr} = \frac{\pi^2 EI}{(0.707l)^2} = \frac{2\pi^2 EI}{l^2} \tag{5–46}$$

We can account for these various end conditions by writing the Euler equation in the two following forms:

$$P_{cr} = \frac{C\pi^2 EI}{l^2} \qquad \frac{P_{cr}}{A} = \frac{C\pi^2 E}{(l/k)^2} \tag{5–47}$$

Here, the factor C is called the *end-condition constant,* and it may have any one of the theoretical values $\frac{1}{4}$, 1, 2, and 4, depending upon the manner in which the load is applied. In practice it is difficult, if not impossible, to fix the column ends so that the factor $C = 2$ or $C = 4$ would apply. Even if the ends are welded, some deflection will occur. Because of this, some designers never use a value of C greater than unity. However, if liberal factors of safety are employed, and if the column load is accurately known, then a value of C not exceeding 1.2 for both ends fixed, or for one end rounded and one end fixed, is not unreasonable, since it supposes only partial fixation. Of course, the value $C = \frac{1}{4}$ must always be used for a column having one end fixed and one end free. These recommendations are summarized in Table 5–2.

When Eq. (5–47) is solved for various values of the unit load P_{cr}/A in terms of the slenderness ratio l/k, we obtain the curve PQR shown in Fig. 5–19. Since the yield strength of the material has the same units as the unit load, the horizontal line through S_y and Q has been added to the figure. This would appear to make the figure cover the entire range of compression problems from the shortest to the longest compression member. Thus it would appear that any compression member having an l/k value less than $(l/k)_Q$ should be treated as a pure compression member while all others are to be treated as Euler columns. Unfortunately, this is not true.

In the actual design of a member that functions as a column, the designer will be aware of the end conditions shown in Fig. 5–18, and will endeavor to configure the ends, using bolts, welds, or pins, for example, so as to achieve the required ideal end condition.

Table 5–2

End-Condition Constants for Euler Columns [to Be Used with Eq. (5–47)]

Column End Conditions	End-Condition Constant C		
	Theoretical Value	Conservative Value	Recommended Value*
Fixed-free	$\frac{1}{4}$	$\frac{1}{4}$	$\frac{1}{4}$
Rounded-rounded	1	1	1
Fixed-rounded	2	1	1.2
Fixed-fixed	4	1	1.2

*To be used only with liberal factors of safety when the column load is accurately known.

Figure 5–19

Euler curve plotted using Eq. (5–47) with $C = 1$.

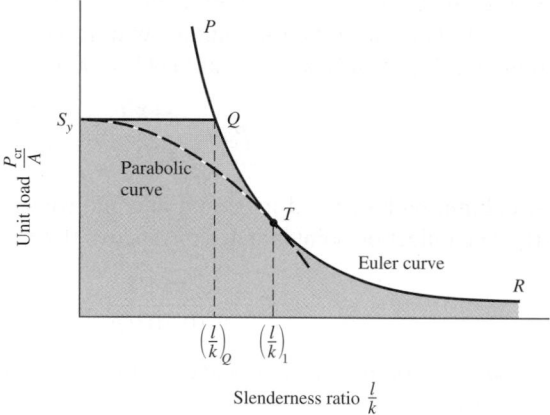

In spite of these precautions, the result, following manufacture, is likely to contain defects such as initial crookedness or load eccentricities. The existence of such defects and the methods of accounting for them will usually involve a factor-of-safety approach or a stochastic analysis. These methods work well for long columns and for simple compression members. However, tests show numerous failures for columns with slenderness ratios below and in the vicinity of point Q, as shown in the shaded area in Fig. 5–19. These have been reported as occurring even when near-perfect geometric specimens were used in the testing procedure.

A column failure is always sudden, total, unexpected, and hence dangerous. There is no advance warning. A beam will bend and give visual warning that it is overloaded, but not so for a column. For this reason neither simple compression methods nor the Euler column equation should be used when the slenderness ratio is near $(l/k)_Q$. Then what should we do? The usual approach is to choose some point T on the Euler curve of Fig. 5–19. If the slenderness ratio is specified as $(l/k)_1$ corresponding to point T, then use the Euler equation only when the actual slenderness ratio is greater than $(l/k)_1$. Otherwise, use one of the methods in the sections that follow. See Examples 5–17 and 5–18.

Most designers select point T such that $P_{cr}/A = S_y/2$. Using Eq. (5–47), we find the corresponding value of $(l/k)_1$ to be

$$\left(\frac{l}{k}\right)_1 = \left(\frac{2\pi^2 C E}{S_y}\right)^{1/2} \tag{5–48}$$

5–13   **Intermediate-Length Columns with Central Loading**

Over the years there have been a number of column formulas proposed and used for the range of l/k values for which the Euler formula is not suitable. Many of these are based on the use of a single material; others, on a so-called safe unit load rather than the critical value. Most of these formulas are based on the use of a linear relationship between the slenderness ratio and the unit load. The *parabolic* or *J. B. Johnson formula* now seems to be the preferred one among designers in the machine, automotive, aircraft, and structural-steel construction fields.

The general form of the parabolic formula is

$$\frac{P_{cr}}{A} = a - b \left(\frac{l}{k}\right)^2 \qquad (5\text{--}49)$$

where a and b are constants that are evaluated by fitting a parabola to the Euler curve of Fig. 5–19 as shown by the dashed line ending at T. If the parabola is begun at S_y, then $a = S_y$. If point T is selected as previously noted, then Eq. (5–48) gives the value of $(l/k)_1$ and the constant b is found to be

$$b = \left(\frac{S_y}{2\pi}\right)^2 \frac{1}{CE} \qquad (a)$$

Upon substituting the known values of a and b into Eq. (5–49), we obtain, for the parabolic equation,

$$\frac{P_{cr}}{A} = S_y - \left(\frac{S_y}{2\pi} \frac{l}{k}\right)^2 \frac{1}{CE} \qquad \frac{l}{k} \le \left(\frac{l}{k}\right)_1 \qquad (5\text{--}50)$$

5–14   **Columns with Eccentric Loading**

We have noted before that deviations from an ideal column, such as load eccentricities or crookedness, are likely to occur during manufacture and assembly. Though these deviations are often quite small, it is still convenient to have a method of dealing with them. Frequently, too, problems occur in which load eccentricities are unavoidable.

Figure 5–20a shows a column in which the line of action of the column forces is separated from the centroidal axis of the column by the eccentricity e. This problem is

Figure 5–20

Notation for an eccentrically loaded column.

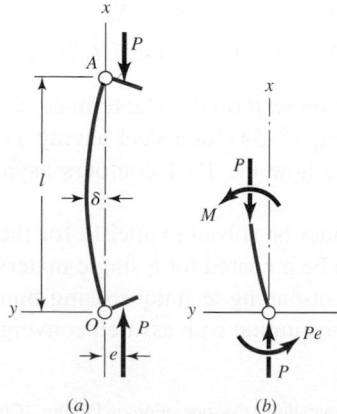

developed by using the free-body diagram of Fig. 5–20b and summing the moments to zero about the origin O. This gives

$$\sum M_O = M + Pe + Py = 0 \tag{a}$$

Using Eq. (5–12) and rearranging produces the differential equation

$$\frac{d^2y}{dx^2} + \frac{P}{EI}y = -\frac{Pe}{EI} \tag{b}$$

Equation (b) is solved in a manner quite similar to others we have solved in this chapter. The boundary conditions are found to be

$$\begin{aligned} x = 0 \qquad y = 0 \\ x = l \qquad y = 0 \end{aligned}$$

By substituting $x = l/2$ in the resulting solution, we find the deflection at midspan and the maximum bending moment to be

$$\delta = e\left[\sec\left(\frac{l}{2}\sqrt{\frac{P}{EI}}\right) - 1\right] \tag{5–51}$$

$$M_{\max} = -P(e + \delta) = -Pe\sec\left(\frac{l}{2}\sqrt{\frac{P}{EI}}\right) \tag{5–52}$$

The maximum *compressive* stress at midspan is found by superposing the axial component and the bending component. This gives

$$\sigma_c = \frac{P}{A} - \frac{Mc}{I} = \frac{P}{A} - \frac{Mc}{Ak^2} \tag{c}$$

Substituting $M_{\max}$ from Eq. (5–52) yields

$$\sigma_c = \frac{P}{A}\left[1 + \frac{ec}{k^2}\sec\left(\frac{l}{2k}\sqrt{\frac{P}{EA}}\right)\right] \tag{5–53}$$

Note that the length l occurs only in the secant function. Since the secant of an acute angle is a number greater than unity, the length amplifies the eccentricity through the secant term. By imposing the compressive yield strength S_{yc} as the maximum value of σ_c, we can write Eq. (5–53) in the form

$$\frac{P}{A} = \frac{S_{yc}}{1 + (ec/k^2)\sec[(l/2k)\sqrt{P/AE}]} \tag{5–54}$$

This is called the *secant column formula*. The term ec/k^2 is called the *eccentricity ratio*. Figure 5–21 is a plot of Eq. (5–54) for a steel having a compressive (and tensile) yield strength of 40 kpsi. Note how the P/A contours asymptotically approach the Euler curve as l/k increases.

Equation (5–54) cannot be solved explicitly for the load P. Design charts, in the fashion of Fig. 5–21, can be prepared for a single material if much column design is to be done. Otherwise, a root-finding technique using numerical methods must be used. With these, successive substitution with assured convergence can be applied.[10]

---

[10]See Charles R. Mischke, "Computational Considerations in Design," Chap. 5 in Joseph E. Shigley and Charles R. Mischke (eds.), *Standard Handbook of Machine Design,* 2 ed., McGraw-Hill, New York, 1996, p. 5.13.

Figure 5–21

Comparison of secant and Euler equations. For steel with $S_y = 40$ kpsi.

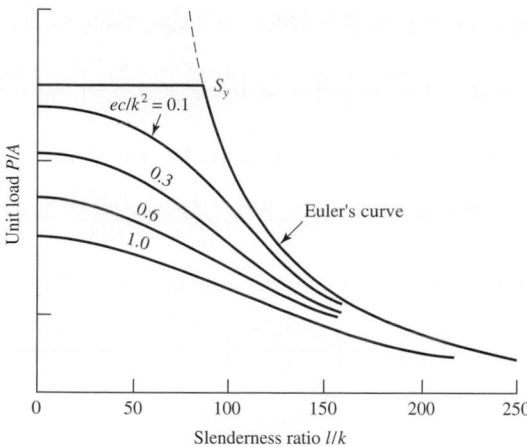

EXAMPLE 5–16 Develop specific Euler equations for the sizes of columns having
(a) Round cross sections
(b) Rectangular cross sections

Solution (a) Using $A = \pi d^2/4$ and $k = \sqrt{I/A} = [(\pi d^4/64)/(\pi d^2/4)]^{1/2} = d/4$ with Eq. (5–47) gives

Answer
$$d = \left(\frac{64 P_{cr} l^2}{\pi^3 C E}\right)^{1/4}$$ (5–55)

(b) For the rectangular column, we specify a height h and a width b with the restriction that $h \le b$. If the end conditions are the *same* for buckling in both directions, then buckling will occur in the direction of the least thickness. Therefore

$$I = \frac{bh^3}{12} \qquad A = bh \qquad k^2 = I/A = \frac{h^2}{12}$$

Substituting these in Eq. (5–47) gives

Answer
$$b = \frac{12 P_{cr} l^2}{\pi^2 C E h^3}$$ (5–56)

Note very particularly, though, that rectangular columns do not generally have the same end conditions in both directions.

EXAMPLE 5–17 Specify the diameter of a round column 1.5 m long that is to carry a maximum load estimated to be 22 kN. Use a design factor $n_d = 4$ and consider the ends as pinned (rounded). The column material selected has a minimum yield strength of 500 MPa and a modulus of elasticity of 207 GPa.

Solution We shall design the column for a critical load of

$$P_{cr} = n_d P = 4(22) = 88 \text{ kN}$$

Then, using Eq. (5–55) with $C = 1$ (see Table 5–2) gives

$$d = \left(\frac{64 P_{cr} l^2}{\pi^3 C E}\right)^{1/4} = \left[\frac{64(88)(1.5)^3}{\pi^3(1)(207)}\right]^{1/4} \left(\frac{10^3}{10^9}\right)^{1/4} (10^3) = 41.48 \text{ mm}$$

Table A–17 shows that the preferred size is 45 mm. The slenderness ratio for this size is

$$\frac{l}{k} = \frac{l}{d/4} = \frac{1.5(10^3)}{45/4} = 133.3$$

To be sure that this is an Euler column, we use Eq. (5–48) and obtain

$$\left(\frac{l}{k}\right)_1 = \left(\frac{2\pi^2 C E}{S_y}\right)^{1/2} = \left[\frac{2\pi^2(1)(207)}{500}\right]^{1/2} \left(\frac{10^9}{10^6}\right)^{1/2} = 90.4$$

which indicates that it is indeed an Euler column. So select

Answer

$$d = 45 \text{ mm}$$

EXAMPLE 5–18 Repeat Ex. 5–16 for J. B. Johnson columns.

Solution (a) For round columns, Eq. (5–50) yields

Answer

$$d = 2\left(\frac{P_{cr}}{\pi S_y} + \frac{S_y l^2}{\pi^2 C E}\right)^{1/2} \qquad (5\text{–}57)$$

(b) For a rectangular section with dimensions $h \le b$, we find

Answer

$$b = \frac{P_{cr}}{h S_y \left(1 - \dfrac{3l^2 S_y}{\pi^2 C E h^2}\right)} \qquad h \le b \qquad (5\text{–}58)$$

EXAMPLE 5–19 Choose a set of dimensions for a rectangular link that is to carry a maximum compressive load of 5000 lbf. The material selected has a minimum yield strength of 75 kpsi and a modulus of elasticity $E = 30$ Mpsi. Use a design factor of 4 and an end condition constant $C = 1$ for buckling in the weakest direction, and design for (a) a length of 15 in, and (b) a length of 8 in with a minimum thickness of $\frac{1}{2}$ in.

Solution (a) Using Eq. (5–48), we find the limiting slenderness ratio to be

$$\left(\frac{l}{k}\right)_1 = \left(\frac{2\pi^2 C E}{S_y}\right)^{1/2} = \left[\frac{2\pi^2(1)(30)(10^6)}{75(10)^3}\right]^{1/2} = 88.9$$

By using $P_{cr} = n_d P = 4(5000) = 20\,000$ lbf, Eqs. (5–56) and (5–58) are solved, using various values of h, to form Table 5–3. The table shows that a cross section of $\frac{5}{8}$ by $\frac{3}{4}$ in, which is marginally suitable, gives the least area.

(b) An approach similar to that in part (a) is used with $l = 8$ in. All trial computations are found to be in the J. B. Johnson region of l/k values. A minimum area occurs when the section is a near square. Thus a cross section of $\frac{1}{2}$ by $\frac{3}{4}$ in is found to be suitable and safe.

Table 5–3

Table Generated to Solve Ex. 5–19, part (a)

h	b	A	l/k	Type	Eq. No.
0.375	3.46	1.298	139	Euler	(5–56)
0.500	1.46	0.730	104	Euler	(5–56)
0.625	0.76	0.475	83	Johnson	(5–58)
0.5625	1.03	0.579	92	Euler	(5–56)

5–15 Struts, or Short Compression Members

A short bar loaded in pure compression by a force P acting along the centroidal axis will shorten in accordance with Hooke's law, until the stress reaches the elastic limit of the material. At this point, permanent set is introduced and usefulness as a machine member may be at an end. If the force P is increased still more, the material either becomes "barrel-like" or fractures. When there is eccentricity in the loading, the elastic limit is encountered at smaller loads.

A *strut* is a *short compression member* such as the one shown in Fig. 5–22. The compressive stress in the x direction at point D in an intermediate section is the sum of a simple component P/A and a flexural component My/I; that is,

$$\sigma_c = \frac{P}{A} + \frac{My}{I} = \frac{P}{A} + \frac{PeyA}{IA} = \frac{P}{A}\left(1 + \frac{ey}{k^2}\right) \qquad (5\text{–}59)$$

where $k = (I/A)^{1/2}$ and is the radius of gyration, y is the coordinate of point D, and e is the eccentricity of loading. The y coordinate of a line parallel to the x axis along which the normal stress is zero is found by setting Eq. (5–59) equal to zero and solving for y. This gives

$$y = -\frac{k^2}{e} \qquad (a)$$

As the eccentricity is increased, the line of zero stress moves toward the cross-section centroid. As e is decreased, the line moves far from the section, and the entire section has a compressive normal stress. The largest compressive stress occurs at point B in Fig. 5–22, where $y = c$. Equation (5–59) then becomes

$$\sigma_c = \frac{P}{A}\left(1 + \frac{ec}{k^2}\right) \qquad (5\text{–}60)$$

Note that the length of the strut does not appear in Eq. (5–60). In order to use the equation for design or analysis, we ought, therefore, to know the range of lengths for which the equation is valid. In other words, how long is a short member?

The difference between the secant formula and Eq. (5–60) is that the secant equation, unlike Eq. (5–60), accounts for an increased bending moment due to bending

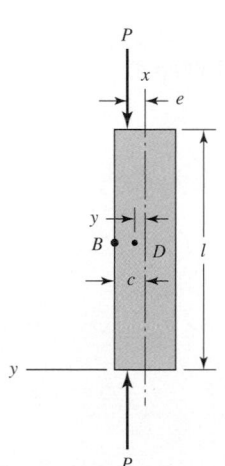

Figure 5–22

Eccentrically loaded strut.

deflection. Thus the secant equation shows the eccentricity to be magnified by the bending deflection. This difference between the two formulas suggests that one way of differentiating between a "secant column" and a strut, or short compression member, is to say that in a strut, the effect of bending deflection must be limited to a certain small percentage of the eccentricity. If we decide that the limiting percentage is to be 1 percent of e, then, from Eq. (5–51), the limiting slenderness ratio turns out to be

$$\left(\frac{l}{k}\right)_2 = 0.282 \left(\frac{AE}{P}\right)^{1/2} \tag{5–61}$$

This equation then gives the limiting slenderness ratio for using Eq. (5–60). If the actual slenderness ratio is greater than $(l/k)_2$, then use the secant formula; otherwise, use Eq. (5–60).

EXAMPLE 5–20

Figure 5–23a shows a workpiece clamped to a milling machine table by a bolt tightened to a tension of 2000 lbf. The clamp contact is offset from the centroidal axis of the strut by a distance $e = 0.10$ in, as shown in part b of the figure. The strut, or block, is steel, 1 in square and 4 in long, as shown. Determine the maximum compressive stress in the block.

Solution

First we find $A = bh = 1(1) = 1$ in^2, $I = bh^3/12 = 1(1)^3/12 = 0.0833$ in^4, $k^2 = I/A = 0.0833/1 = 0.0833$ in^2, and $l/k = 4/(0.0833)^{1/2} = 13.9$. Equation (5–61) gives the limiting slenderness ratio as

$$\left(\frac{l}{k}\right)_2 = 0.282 \left(\frac{AE}{P}\right)^{1/2} = 0.282 \left[\frac{1(30)(10^6)}{1000}\right]^{1/2} = 48.8$$

Thus the block could be as long as

$$l = 48.8k = 48.8(0.0833)^{1/2} = 14.1 \text{ in}$$

before it need be treated by using the secant formula. So Eq. (5–60) applies and the maximum compressive stress is

Answer

$$\sigma_c = \frac{P}{A}\left(1 + \frac{ec}{k^2}\right) = \frac{1000}{1}\left[1 + \frac{0.1(0.5)}{0.0833}\right] = 1600 \text{ psi}$$

Figure 5–23

A strut that is part of a workpiece clamping assembly.

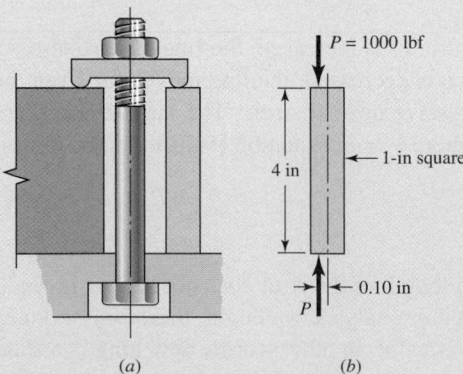

(a) (b)

5–16   **Shock and Impact**

Impact refers to the collision of two masses with initial relative velocity. In some cases it is desirable to achieve a known impact in design; for example, this is the case in the design of coining, stamping, and forming presses. In other cases, impact occurs because of excessive deflections, or because of clearances between parts, and in these cases it is desirable to minimize the effects. The rattling of mating gear teeth in their tooth spaces is an impact problem caused by shaft deflection and the clearance between the teeth. This impact causes gear noise and fatigue failure of the tooth surfaces. The clearance space between a cam and follower or between a journal and its bearing may result in crossover impact and also cause excessive noise and rapid fatigue failure.

Shock is a more general term that is used to describe any suddenly applied force or disturbance. Thus the study of shock includes impact as a special case. There are two general approaches to the study of shock, depending upon whether only statics is used in the analysis or both statics and dynamics are used. Lifetimes have been spent investigating shock and impact phenomena. For reasons of space, the material presented here is intended only to indicate the scope of the subject and to give a basic understanding of what is involved.

Figure 5–24 represents a highly simplified mathematical model of an automobile in collision with a rigid obstruction. Here m_1 is the lumped mass of the engine. The displacement, velocity, and acceleration are described by the coordinate x_1 and its time derivatives. The lumped mass of the vehicle less the engine is denoted by m_2, and its motion by the coordinate x_2 and its derivatives. Springs k_1, k_2, and k_3 represent the linear and nonlinear stiffnesses of the various structural elements that compose the vehicle. Friction can and should be included, but is not shown in this model. The determination of the spring rates for such a complex structure will almost certainly have to be performed experimentally. Once these values—the k's, m's, and frictional coefficients—are obtained, a set of nonlinear differential equations can be written and a computer solution obtained for any impact velocity.

Figure 5–25 is another impact model. Here mass m_1 has an initial velocity v and is just coming into contact with spring k_1. The part or structure to be analyzed is represented by mass m_2 and spring k_2. The problem facing the designer is to find the maximum deflection of m_2 and the maximum force exerted by k_2 against m_2. In the analysis it doesn't matter whether k_1 is fastened to m_1 or to m_2, since we are interested only in a

Figure 5–24

Two-degree-of-freedom mathematical model of an automobile in collision with a rigid obstruction.

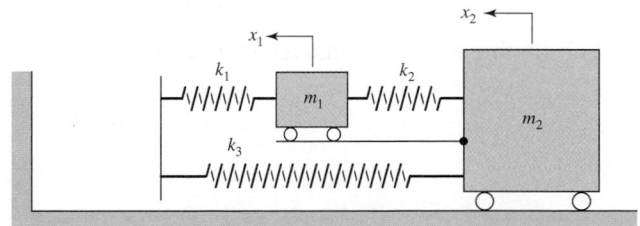

Figure 5–25

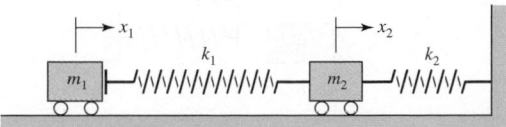

solution up to the point in time for which x_2 reaches a maximum. That is, the solution for the rebound isn't needed. The differential equations are not difficult to derive. They are

$$m_1\ddot{x}_1 + k_1(x_1 - x_2) = 0$$
$$m_2\ddot{x}_2 + k_2x_2 - k_1(x_1 - x_2) = 0$$

(5–62)

Equation pair (5–62) is rather awkward to use in obtaining a general analytical solution. But, if the values of the m's and k's are specified, you can probably find an integration routine in your computer-program library for use in obtaining a solution.

5–17 Suddenly Applied Loading

A simple case of impact is illustrated in Fig. 5–26a. Here a weight W, moving at a constant velocity v on a frictionless surface, strikes a cantilever of stiffness EI and length l. We want to find the maximum deflection and the maximum bending moment in the beam due to the impact.

Figure 5–26b shows an abstract model of the system. Using Table A–9–1, we find the spring rate to be $k = F/y = 3EI/l^3$. We choose to count the time at the instant the weight strikes the spring. Thus at $t = 0$ the motion is started with $y = 0$ and $\dot{y} = v$. Neglecting the beam mass, the differential equation is

$$\frac{W}{g}\ddot{y} = -ky$$

(a)

where the spring force ky is negative because it is opposite to the deflection y. The solution to this equation is well known, and is

$$y = A\cos\omega t + B\sin\omega t$$

(5–63)

where

$$\omega = \sqrt{\frac{kg}{W}}$$

(5–64)

is the circular frequency of vibration. From Eq. (5–63), the velocity is

$$\dot{y} = -A\omega\sin\omega t + B\omega\cos\omega t$$

With the initial conditions at $t = 0$ of $y = 0$ and $\dot{y} = v$, the constants are

$$A = 0 \qquad B = v/\omega$$

Substituting these back into Eq. (5–63) yields

$$y = \frac{v}{\omega}\sin\omega t$$

(5–65)

Figure 5–26

(a) Collision of a weight with a cantilever beam; (b) modeling the cantilever beam as a spring with rate k.

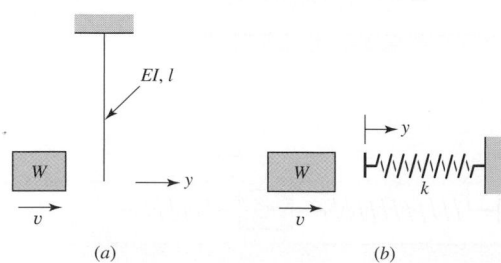

Of course, this solution is valid only as long as the weight remains in contact with the beam. The maximum deflection is

$$y_{max} = \frac{v}{\omega} = \frac{v}{\sqrt{kg/W}} = v\sqrt{\frac{Wl^3}{3EIg}} \qquad (5\text{–}66)$$

Also, the maximum bending moment is the product of the spring force and the beam length. The result is

$$M_{max} = kly_{max} = v\sqrt{\frac{3EIW}{gl}} \qquad (5\text{–}67)$$

As a second example, consider the weight W, in Fig. 5–27a, falling a distance h and impacting some structure or member whose spring rate is k. We choose the origin of the coordinate y corresponding to the position of the weight when time t is zero, as before. Two free-body diagrams, shown in Fig. 5–27b and c, are necessary—one when $y \leq h$, and another when $y > h$—to account for the spring force. For each of these free-body diagrams we can write Newton's law by stating that the inertia force $(W/g)\ddot{y}$ is equal to the sum of the external forces acting on the weight. We then have

$$\frac{W}{g}\ddot{y} = W \qquad\qquad y \leq h$$

$$\frac{W}{g}\ddot{y} = -k(y - h) + W \qquad y > h$$

(b)

We must also include in the mathematical statement of the problem the knowledge that the weight is released with zero initial velocity. Equation pair (b) constitutes a set of *piecewise differential equations*. Each equation is linear, but each applies only for a certain range of y. The solution to the set is valid for all values of t, but, as before, we are interested in values of y only up until the time that the spring or structure reaches its maximum deflection.

The solution to the first equation in the set is

$$y = \frac{gt^2}{2} \qquad y \leq h \qquad (5\text{–}68)$$

and you can verify this by direct substitution. Equation (5–68) is no longer valid after $y = h$; call this time t_1. Then

$$t_1 = \sqrt{2h/g} \qquad (c)$$

Differentiating Eq. (5–68) to get the velocity gives

$$\dot{y} = gt \qquad y \leq h \qquad (d)$$

Figure 5-27

(a) A weight free to fall a distance h to free end of a spring. (b) Free body of weight during fall. (c) Free body of weight during arrest.

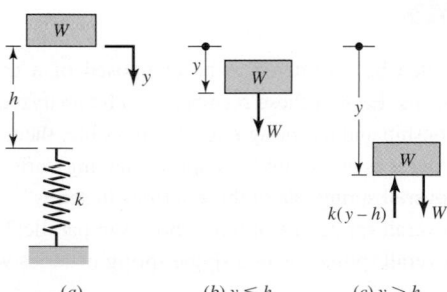

(a)   (b) $y \leq h$ (c) $y > h$

and so the velocity of the weight at $t = t_1$ is

$$\dot{y}_1 = gt_1 = \sqrt{2gh} \tag{e}$$

Having moved from $y = 0$ to $y = h$, we then need to solve the second equation of the set (b). It is convenient to define a new time $t' = t - t_1$. Thus $t' = 0$ at the instant the weight strikes the spring. Applying your knowledge of differential equations, you should find the solution to be

$$y = A \cos \omega t' + B \sin \omega t' + h + \frac{W}{k} \qquad y > h \tag{f}$$

If you cannot derive Eq. (f), you should at least prove to yourself that it is a solution by substituting it and its second derivative in the second equation of set (b) to show that the equality in Eq. (f) is satisfied identically.

The constants A and B are evaluated as shown in the previous solution. When these are substituted back, the result can be transformed to

$$y = \left[\left(\frac{W}{k}\right)^2 + \frac{2Wh}{k} \right]^{1/2} \cos[\omega t' - \phi] + h + \frac{W}{k} \qquad y > h \tag{5-69}$$

where ϕ, though of no interest here, is given by

$$\phi = \frac{\pi}{2} + \tan^{-1} \left(\frac{W}{2kh}\right)^{1/2} \tag{g}$$

The maximum deflection of the spring (structure) occurs when the cosine term in Eq. (5–69) is unity. We designate this as δ and, after rearranging, find it to be

$$\delta = y_{\max} - h = \frac{W}{k} + \frac{W}{k} \left[1 + \left(\frac{2hk}{W}\right) \right]^{1/2} \tag{5-70}$$

The maximum force acting on the spring or structure is now found to be

$$F = k\delta = W + W \left[1 + \left(\frac{2hk}{W}\right) \right]^{1/2} \tag{5-71}$$

Note, in this equation, that if $h = 0$, then $F = 2W$. This says that when the weight is released while in contact with the spring but is not exerting any force on the spring, the largest force is double the weight.

Most systems are not as ideal as those explored here, so be wary about using these relations for nonideal systems.

PROBLEMS

ANALYSIS

5–1 Structures can often be considered to be composed of a combination of tension and torsion members and beams. Each of these members can be analyzed separately to determine its force-deflection relationship and its spring rate. It is possible, then, to obtain the deflection of a structure by considering it as an assembly of springs having various series and parallel relationships.
(*a*) What is the overall spring rate of three springs in series?
(*b*) What is the overall spring rate of three springs in parallel?
(*c*) What is the overall spring rate of a single spring in series with a pair of parallel springs?

ANALYSIS

5-2   The figure shows a torsion bar OA fixed at O, supported at A, and connected to a cantilever AB. The spring rate of the torsion bar is k_T, in newton-meters per radian, and that of the cantilever is k_C, in newtons per meter. What is the overall spring rate based on the deflection y at point B?

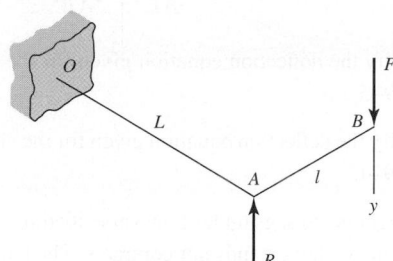

Problem 5-2

ANALYSIS

5-3   A torsion-bar spring consists of a prismatic bar, usually of round cross section, that is twisted at one end and held fast at the other to form a stiff spring. An engineer needs a stiffer one than usual and so considers building in both ends and applying the torque somewhere in the central portion of the span, as shown in the figure. If the bar is uniform in diameter, that is, if $d = d_1 = d_2$, investigate how the allowable angle of twist, the largest torque, and the spring rate depend on the location x at which the torque is applied. *Hint:* Consider two springs in parallel.

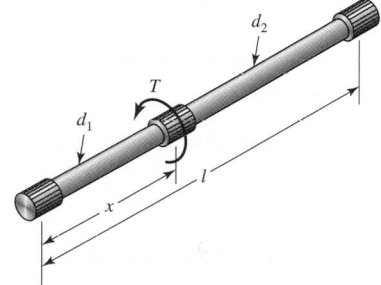

Problem 5-3

ANALYSIS

5-4   An engineer is forced by geometric considerations to apply the torque on the spring of Prob. 5-3 at the location $x = 0.2l$. For a uniform-diameter spring, this would cause the long leg of the span to be underutilized when both legs have the same diameter. If the diameter of the long leg is reduced sufficiently, the shear stress in the two legs can be made equal. How would this change affect the allowable angle of twist, the largest torque, and the spring rate?

ANALYSIS

5-5   A bar in tension has a circular cross section and includes a conical portion of length l, as shown. The task is to find the spring rate of the entire bar. Equation (5-4) is useful for the outer portions of diameters d_1 and d_2, but a new relation must be derived for the tapered section. If α is the apex half-angle, as shown, show that the spring rate of the tapered portion of the shaft is

$$k = \frac{EA_1}{l}\left(1 + \frac{2l}{d_1}\tan\alpha\right)$$

Problem 5-5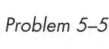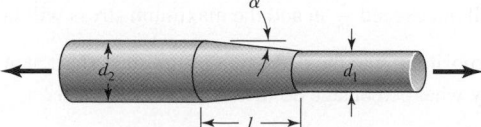

ANALYSIS

5-6 When a hoisting cable is long, the weight of the cable itself contributes to the elongation. If a cable has a weight of w newtons per meter, a length of l meters, and a load P attached to the free end, show that the cable elongation is

$$\delta = \frac{Pl}{AE} + \frac{wl^2}{2AE}$$

ANALYSIS

5-7 Use integration to verify the deflection equation given for the uniformly loaded cantilever beam of appendix Table A–9–3.

ANALYSIS

5-8 Use integration to verify the deflection equation given for the end moment loaded cantilever beam of appendix Table A–9–4.

ANALYSIS

5-9 When an initially straight beam sags under transverse loading, the ends contract because the neutral surface of zero strain neither extends nor contracts. The length of the deflected neutral surface is the same as the original beam length l. Consider a segment of the initially straight beam Δs. After bending, the x-direction component is shorter than Δs, namely, Δx. The contraction is $\Delta s - \Delta x$, and these summed for the entire beam gives the end contraction λ. Show that

$$\lambda \doteq \frac{1}{2} \int_0^l \left(\frac{dy}{dx}\right)^2 dx$$

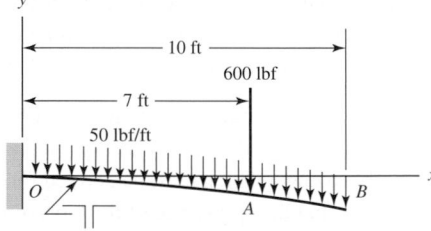ANALYSIS

5-10 A neutral surface of a simply supported beam has the shape

$$y = -\frac{4ax}{l^2}(l - x)$$

where a is the midspan deflection magnitude. Using the result of Prob. 5–9, what is the end contraction from the straight original condition?

ANALYSIS

5-11 The neutral surface of a simply supported beam has the shape

$$y = a \sin \frac{\pi x}{l}$$

Using the result of Prob. 5–9, what is the end contraction from the straight original condition?

ANALYSIS

5-12 The figure shows a cantilever consisting of steel angles size $4 \times 4 \times \frac{1}{2}$ in mounted back to back. Using superposition, find the deflection at B and the maximum stress in the beam.

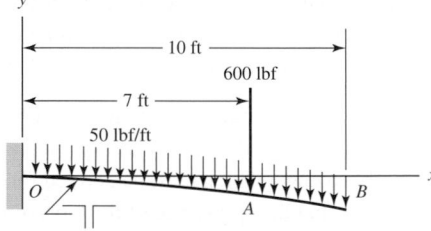

Problem 5–12

ANALYSIS

5-13 A simply supported beam loaded by two forces is shown in the figure. Select a pair of structural steel channels mounted back to back to support the loads in such a way that the deflection at midspan will not exceed $\frac{1}{16}$ in and the maximum stress will not exceed 6 kpsi. Use superposition.

ANALYSIS

5-14 Using superposition, find the deflection of the steel shaft at A in the figure. Find the deflection at midspan. By what percentage do these two values differ?

Problem 5–13

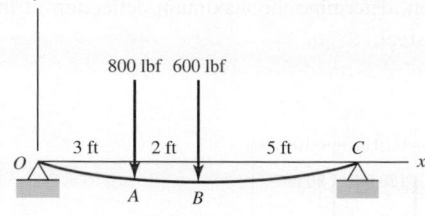

Problem 5–14

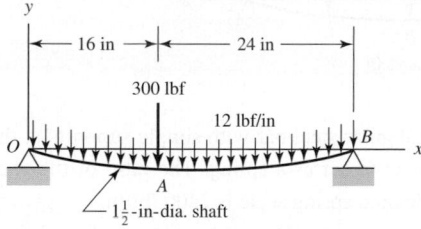

5–15   A rectangular steel bar supports the two overhanging loads shown in the figure. Using superposition, find the deflection at the ends and at the center.

Problem 5–15

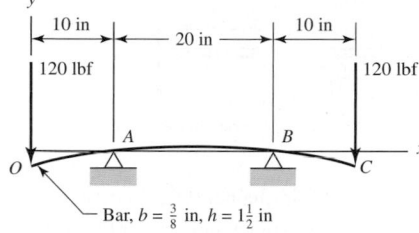

5–16   Using the formulas in Appendix Table A–9 and superposition, find the deflection of the cantilever at B if $I = 13\ \text{in}^4$ and $E = 30$ Mpsi.

Problem 5–16

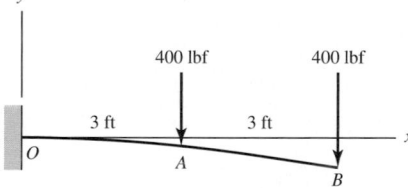

5–17   The cantilever shown in the figure consists of two structural-steel channels size 3 in, 5.0 lbf/ft. Using superposition, find the deflection at A.

Problem 5–17

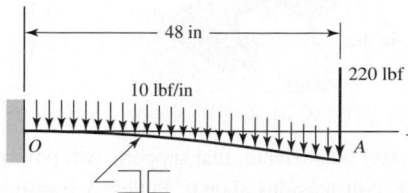

ANALYSIS

5–18 Using superposition, determine the maximum deflection of the beam shown in the figure. The material is carbon steel.

Problem 5–18

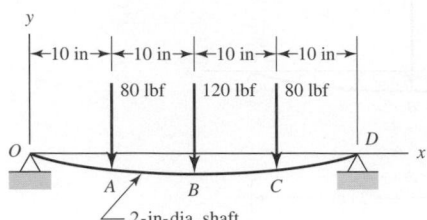

DESIGN

5–19 Illustrated is a rectangular steel bar with simple supports at the ends and loaded by a force F at the middle; the bar is to act as a spring. The ratio of the width to the thickness is to be about $b = 16h$, and the desired spring scale is 2400 lbf/in.

(*a*) Find a set of cross-section dimensions, using preferred sizes.

(*b*) What deflection would cause a permanent set in the spring if this is estimated to occur at a normal stress of 90 kpsi?

Problem 5–19

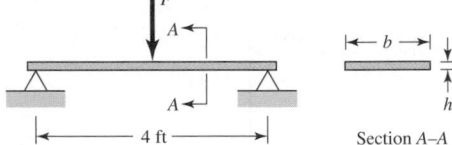

DESIGN

5–20 Illustrated in the figure is a $1\frac{1}{2}$-in-diameter steel countershaft that supports two pulleys. Pulley A delivers power to a machine causing a tension of 600 lbf in the tight side of the belt and 80 lbf in the loose side, as indicated. Pulley B receives power from a motor. The belt tensions on pulley B have the relation $T_1 = 0.125T_2$. Find the deflection of the shaft in the z direction at pulleys A and B. Assume that the bearings constitute simple supports.

Problem 5–20

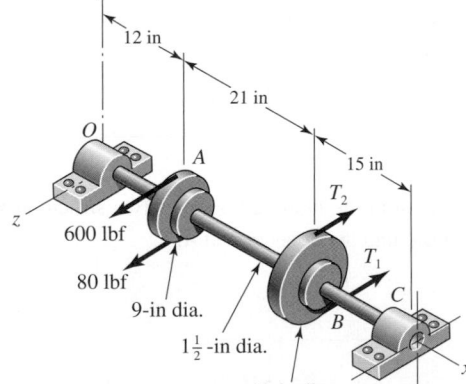

DESIGN

5–21 The figure shows a steel countershaft that supports two pulleys. Pulley C receives power from a motor producing the belt tensions shown. Pulley A transmits this power to another machine through the belt tensions T_1 and T_2 such that $T_1 = 8T_2$.

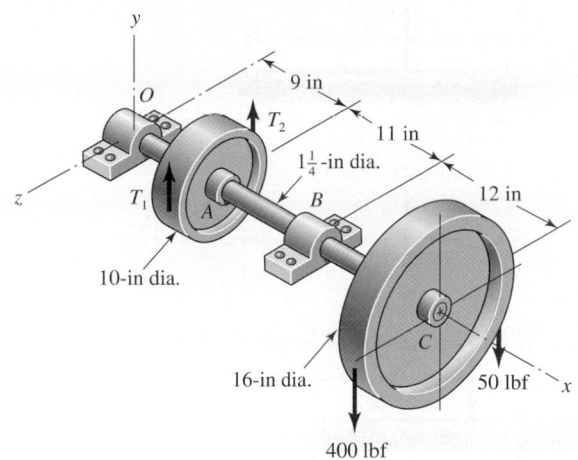

Problem 5–21

(a) Find the deflection of the overhanging end of the shaft, assuming simple supports at the bearings.

(b) If roller bearings are used, the slope of the shaft at the bearings should not exceed 0.06° for good bearing life. What shaft diameter is needed to conform to this requirement? Use $\frac{1}{8}$-in increments in any iteration you may make. What is the deflection at pulley C now?

DESIGN

5–22 The structure of a diesel-electric locomotive is essentially a composite beam supporting a deck. Above the deck are mounted the diesel prime mover, generator or alternator, radiators, switch gear, and auxiliaries. Beneath the deck are found fuel and lubricant tanks, air reservoirs, and small auxiliaries. This assembly is supported at bolsters by the trucks that house the traction motors and brakes. This equipment is distributed as uniformly as possible in the span between the bolsters. In an approximate way, the loading can be viewed as uniform between the bolsters and simply supported. Because the hoods that shield the equipment from the weather have many rectangular access doors, which are mass-produced, it is important that the hood structure be level and plumb and sit on a flat deck. Aesthetics plays a role too. The center sill beam has a second moment of area of $I = 5450$ in^4, the bolsters are 36 ft apart, and the deck loading is 5000 lbf/ft.

(a) What is the camber of the curve to which the deck will be built in order that the service-ready locomotive will have a flat deck?

(b) What equation would you give to locate points on the curve of part (a)?

ANALYSIS

5–23 The designer of a shaft usually has a slope constraint imposed by the bearings used. This limit will be denoted as ξ. If the shaft shown in the figure is to have a uniform diameter d except in the locality of the bearing mounting, it can be approximated as a uniform beam with simple supports. Show that the minimum diameters to meet the slope constraints at the left and right bearings are, respectively,

$$d_L = \left| \frac{32Fb(l^2 - b^2)}{3\pi EI\xi} \right|^{1/4} \qquad d_R = \left| \frac{32Fa(l^2 - a^2)}{3\pi EI\xi} \right|^{1/4}$$

DESIGN

5–24 A shaft is to be designed so that it is supported by roller bearings. The basic geometry is shown in the figure. The allowable slope at the bearings is 0.001 mm/mm without bearing life penalty. For a design factor of 1.28, what uniform-diameter shaft will support the 3.5-kN load 100 mm from the left bearing without penalty? Use $E = 207$ GPa.

ANALYSIS

5–25 Determine the maximum deflection of the shaft of Prob. 5–24.

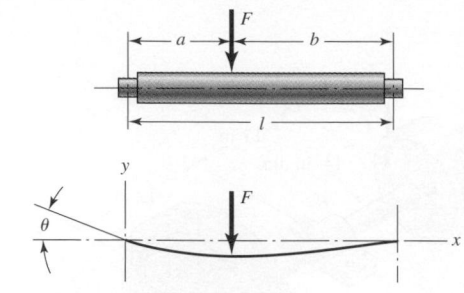

Problem 5–23

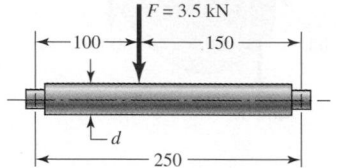

Problem 5–24
Dimensions in millimeters.

DESIGN

5–26 For the shaft shown in the figure, let $a_1 = 4$ in, $b_1 = 12$ in, $a_2 = 10$ in, $F_1 = 100$ lbf, $F_2 = 300$ lbf, and $E = 30$ Mpsi. The shaft is to be sized so that the maximum slope at either bearing A or bearing B does not exceed 0.001 rad. Determine a suitable diameter d.

Problem 5–26

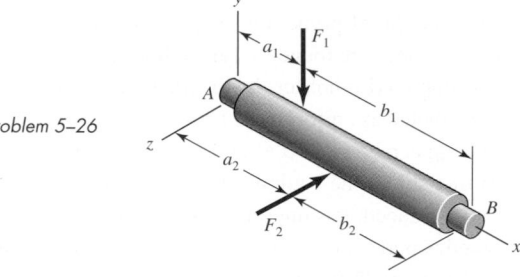

ANALYSIS

5–27 If the diameter of the beam for Prob. 5–26 is 1.375 in, determine the deflection of the beam at $x = 8$ in.

5–28 See Prob. 5–26 and the accompanying figure. The loads and dimensions are $F_1 = 800$ lbf, $F_2 = 600$ lbf, $a_1 = 4$ in, $b_1 = 6$ in, and $a_2 = 7$ in. Find the uniform shaft diameter necessary to limit the slope at the bearings to 0.001 in/in. Use a design factor of $n_d = 1.5$ and $E = 29.8$ Mpsi.

DESIGN

5–29 Shown in the figure is a uniform-diameter shaft with bearing shoulders at the ends; the shaft is subjected to a concentrated moment $M = 1200$ lbf · in. The shaft is of carbon steel and has $a = 5$ in and $l = 9$ in. The slope at the ends must be limited to 0.002 rad. Find a suitable diameter d.

Problem 5–29

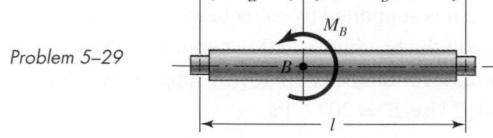

ANALYSIS

5-30   The rectangular member OAB, shown in the figure, is held horizontal by the round hooked bar AC. The modulus of elasticity of both parts is 10 Mpsi. Use superposition to find the deflection at B due to a force $F = 80$ lbf.

Problem 5–30

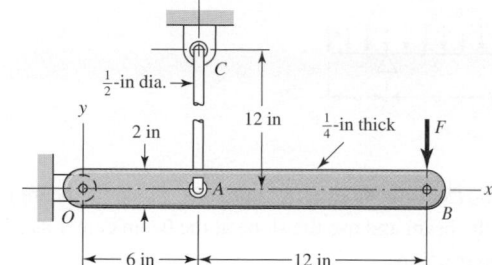

ANALYSIS

5-31   The figure illustrates a torsion-bar spring OA having a diameter $d = 12$ mm. The actuating cantilever AB also has $d = 12$ mm. Both parts are of carbon steel. Use superposition and find the spring rate k corresponding to a force F acting at B.

Problem 5–31

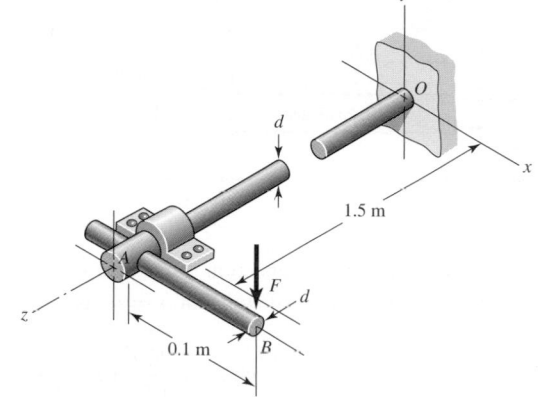

ANALYSIS

5-32   Consider the simply supported beam with an intermediate load in Appendix A–9–6. Determine the deflection equation if the stiffness of the left and right supports are k_1 and k_2, respectively.

ANALYSIS

5-33   Consider the simply supported beam with a uniform load in Appendix A–9–7. Determine the deflection equation if the stiffness of the left and right supports are k_1 and k_2, respectively.

ANALYSIS

5-34   Prove that for a uniform-cross-section beam with simple supports at the ends loaded by a single concentrated load, the location of the maximum deflection will never be outside the range of $0.423l \leq x \leq 0.577l$ regardless of the location of the load along the beam. The importance of this is that you can always get a quick estimate of y_{max} by using $x = l/2$.

ANALYSIS

5-35   Solve Prob. 5–12 using singularity functions. Use statics to determine the reactions.

ANALYSIS

5-36   Solve Prob. 5–13 using singularity functions. Use statics to determine the reactions.

ANALYSIS

5-37   Solve Prob. 5–14 using singularity functions. Use statics to determine the reactions.

5–38 Consider the uniformly loaded simply supported beam with an overhang as shown. Use singularity functions to determine the deflection equation of the beam. Use statics to determine the reactions.

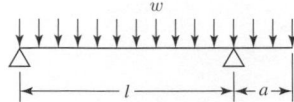

Problem 5–38

5–39 Solve Prob. 5–15 using singularity functions. Since the beam is symmetric, only write the equation for half the beam and use the slope at the beam center as a boundary condition. Use statics to determine the reactions.

5–40 Solve Prob. 5–30 using singularity functions. Use statics to determine the reactions.

5–41 Determine the deflection equation for the beam shown using singularity functions. Since the beam is symmetric, write the equation for only half the beam and use the slope at the beam center as a boundary condition. Use statics to determine the reactions.

Problem 5–41

(From Richard G. Budynas, *Advanced Strength and Applied Stress Analysis*, 2nd ed., McGraw-Hill, New York, 1999. Copyright © 1999 The McGraw-Hill Companies.)

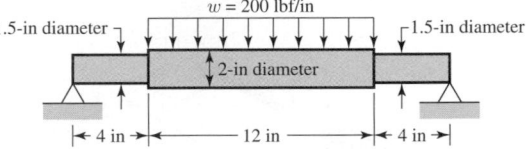

5–42 Examine the expression for the deflection of the cantilever beam, end-loaded, shown in Appendix Table A–9–1 for some intermediate point, $x = a$, as

$$y|_{x=a} = \frac{F_1 a^2}{6EI}(a - 3l)$$

In Table A–9–2, for a cantilever with intermediate load, the deflection at the end is

$$y|_{x=l} = \frac{F_2 a^2}{6EI}(a - 3l)$$

These expressions are remarkably similar and become identical when $F_1 = F_2 = 1$. In other words, the deflection at $x = a$ (station 1) due to a unit load at $x = l$ (station 2) is the same as the deflection at station 2 due to a unit load at station 1. Prove that this is true generally for an elastic body even when the lines of action of the loads are not parallel. This is known as a special case of *Maxwell's reciprocal theorem.* (*Hint:* Consider the potential energy of strain when the body is loaded by two forces in either order of application.)

5–43 A steel shaft of uniform 2-in diameter has a bearing span l of 23 in and an overhang of 7 in on which a coupling is to be mounted. A gear is to be attached 9 in to the right of the left bearing and will carry a radial load of 400 lbf. We require an estimate of the bending deflection at the coupling. Appendix Table A–9–6 is available, but we can't be sure of how to expand the equation to predict the deflection at the coupling.

(*a*) Show how Appendix Table A–9–10 and Maxwell's theorem (see Prob. 5–42) can be used to obtain the needed estimate.

(*b*) Check your work by finding the slope at the right bearing and extending it to the coupling location.

 ANALYSIS   **5–44**   Use Castigliano's theorem to verify the maximum deflection for the uniformly loaded beam of Appendix Table A–9–7. Neglect shear.

ANALYSIS   **5–45**   Solve Prob. 5–17 using Castigliano's theorem. *Hint:* Write the moment equation using a position variable positive to the left starting at the right end of the beam.

ANALYSIS   **5–46**   Solve Prob. 5–30 using Castigliano's theorem.

ANALYSIS   **5–47**   Solve Prob. 5–31 using Castigliano's theorem.

ANALYSIS   **5–48**   Determine the deflection at midspan for the beam of Prob. 5–41 using Castigliano's theorem.

ANALYSIS   **5–49**   Using Castigliano's theorem, determine the deflection of point B in the direction of the force F for the bar shown. The solid bar has a uniform diameter, d. Neglect bending shear.

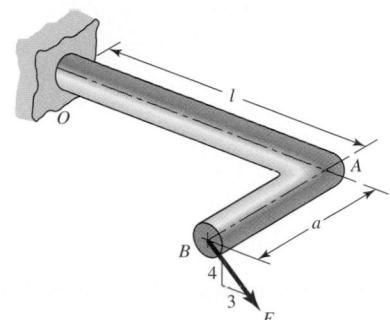

Problem 5–49

ANALYSIS   **5–50**   A cable is made using a 16-gauge (0.0625-in) steel wire and three strands of 12-gauge (0.0801-in) copper wire. Find the stress in each wire if the cable is subjected to a tension of 250 lbf.

ANALYSIS   **5–51**   The figure shows a steel pressure cylinder of diameter 4 in which uses six SAE grade 5 steel bolts having a grip of 12 in. These bolts have a proof strength (see Chap. 8) of 85 kpsi for this size of bolt. Suppose the bolts are tightened to 90 percent of this strength in accordance with some recommendations.

(*a*) Find the tensile stress in the bolts and the compressive stress in the cylinder walls.

(*b*) Repeat part (*a*), but assume now that a fluid under a pressure of 600 psi is introduced into the cylinder.

ANALYSIS   **5–52**   A torsion bar of length L consists of a round core of stiffness $(GJ)_c$ and a shell of stiffness $(GJ)_s$. If a torque T is applied to this composite bar, what percentage of the total torque is carried by the shell?

ANALYSIS   **5–53**   A rectangular aluminum bar $\frac{1}{2}$ in thick and 2 in wide is welded to fixed supports at the ends, and the bar supports a load $W = 800$ lbf, acting through a pin as shown. Find the reactions at the supports.

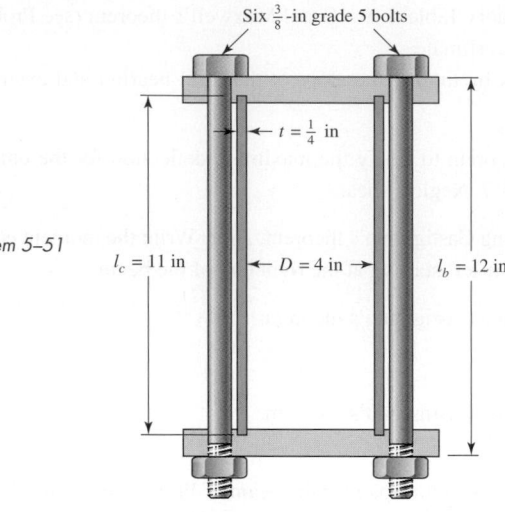

Problem 5–51

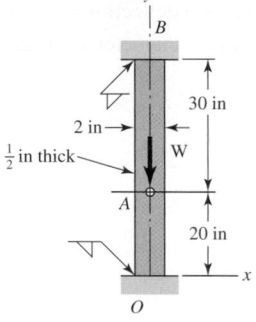

Problem 5–53

5–54 The steel shaft shown in the figure is subjected to a torque T applied at point A. Find the torque reactions at O and B.

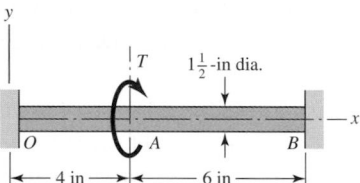

Problem 5–54

5–55 In testing the wear life of gear teeth, the gears are assembled by using a pretorsion. In this way, a large torque can exist even though the power input to the tester is small. The arrangement shown in the figure uses this principle. Note the symbol used to indicate the location of the shaft bearings used in the figure. Gears A, B, and C are assembled first, and then gear C is held fixed. Gear D is assembled and meshed with gear C by twisting it through an angle of $4°$ to provide the pretorsion. Find the maximum shear stress in each shaft resulting from this preload.

5–56 The figure shows a $\frac{3}{8}$- by $1\frac{1}{2}$-in rectangular steel bar welded to fixed supports at each end. The bar is axially loaded by the forces $F_A = 10$ kip and $F_B = 5$ kip acting on pins at A and B. Assuming that the bar will not buckle laterally, find the reactions at the fixed supports. Use procedure 1 from Sec. 5–10.

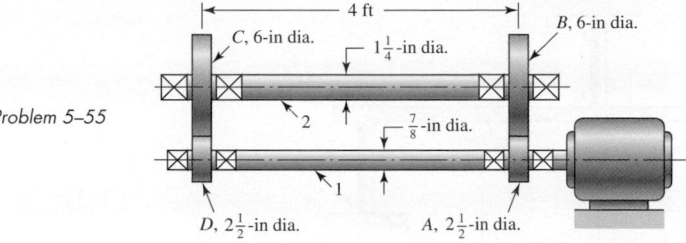

Problem 5–55

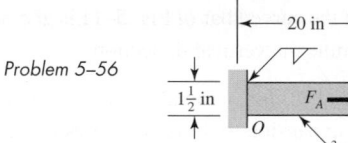

Problem 5–56

5–57   For the beam shown, determine the support reactions using superposition and procedure 1 from Sec. 5–10.

Problem 5–57

(From Richard G. Budynas, *Advanced Strength and Applied Stress Analysis*, 2nd ed., McGraw-Hill, New York, 1999. Copyright © 1999 The McGraw-Hill Companies.)

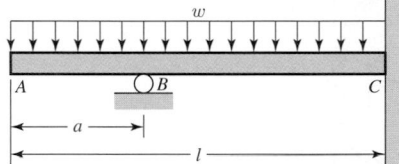

5–58   Solve Prob. 5–57 using Castigliano's theorem and procedure 1 from Sec. 5–10.

5–59   The steel beam $ABCD$ shown is simply supported at A and supported at B and D by steel cables, each having an effective diameter of 12 mm. The second area moment of the beam is $I = 8(10^5)$ mm^4. A force of 20 kN is applied at point C. Using procedure 2 of Sec. 5–10 determine the stresses in the cables and the deflections of B, C, and D. For steel, let $E = 209$ GPa.

Problem 5–59

(From Richard G. Budynas, *Advanced Strength and Applied Stress Analysis*, 2nd ed., McGraw-Hill, New York, 1999. Copyright © 1999 The McGraw-Hill Companies.)

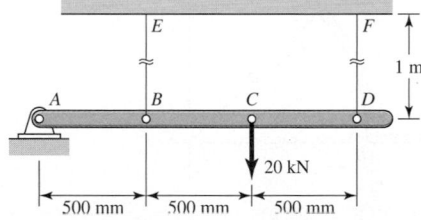

5–60   The steel beam $ABCD$ shown is supported at C as shown and supported at B and D by steel bolts each having a diameter of $\frac{5}{16}$ in. The lengths of BE and DF are 2 and 2.5 in, respectively. The beam has a second area moment of 0.050 in^4. Prior to loading, the nuts are just in contact with the horizontal beam. A force of 500 lbf is then applied at point A. Using procedure 2 of Sec. 5–10, determine the stresses in the bolts and the deflections of points A, B, and D. For steel, let $E = 30$ Mpsi.

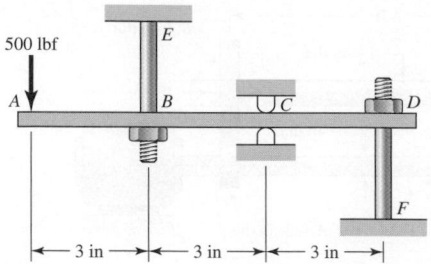

Problem 5–60

(From Richard G. Budynas, *Advanced Strength and Applied Stress Analysis*, 2nd ed., McGraw-Hill, New York, 1999. Copyright © 1999 The McGraw-Hill Companies.)

 ANALYSIS   **5–61**   The horizontal deflection of the right end of the curved bar of Fig. 5–11 is given by Eq. (5–36) for $R/h > 10$. For the same conditions, determine the vertical deflection.

 ANALYSIS   **5–62**   A cast-iron piston ring has a mean diameter of 81 mm, a radial height $h = 6$ mm, and a thickness $b = 4$ mm. The ring is assembled using an expansion tool that separates the split ends a distance δ by applying a force F as shown. Use Castigliano's theorem and determine the deflection δ as a function of F. Use $E = 131$ GPa and assume Eq. (5–29) applies.

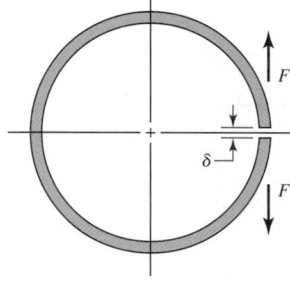

Problem 5–62

 ANALYSIS   **5–63**   For the wire form shown use Castigliano's method to determine the vertical deflection of point A. Consider bending only and assume Eq. (5–29) applies for the curved part.

Problem 5–63

(From Richard G. Budynas, *Advanced Strength and Applied Stress Analysis*, 2nd ed., McGraw-Hill, New York, 1999. Copyright © 1999 The McGraw-Hill Companies.)

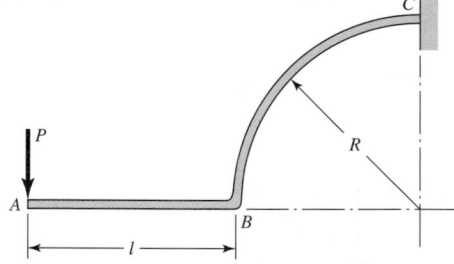

 ANALYSIS   **5–64**   For the wire form shown determine the vertical deflections of points A and B. Consider bending only and assume Eq. (5–29) applies.

ANALYSIS   **5–65**   For the wire form shown, determine the deflection of point A in the y direction. Assume $R/h > 10$ and consider the effects of bending and torsion only. The wire is steel with $E = 200$ GPa, $\nu = 0.29$, and has a diameter of 5 mm. Before application of the 200-N force the wire form is in the xz plane where the radius R is 100 mm.

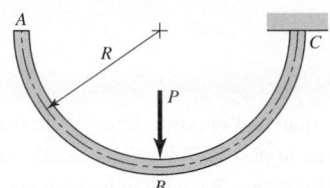

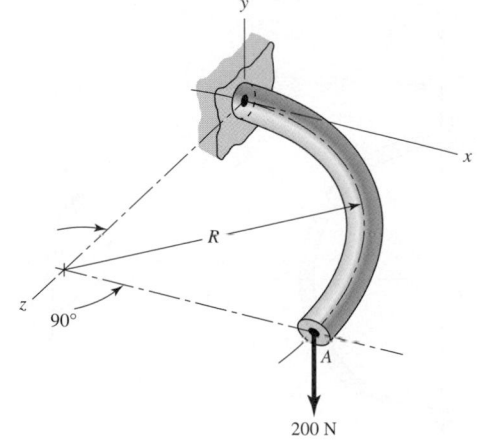

ANALYSIS

5–66 For the wire form shown, determine (*a*) the reactions at points *A* and *B*, (*b*) how the bending moment varies along the wire, and (*c*) the deflection of the load *F*. Assume that the entire energy is described by Eq. (5–29).

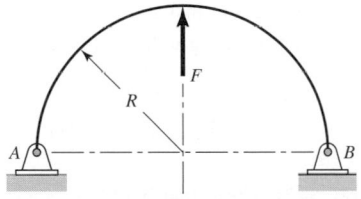

ANALYSIS

5–67 For the curved beam shown, $F = 30$ kN. The material is steel with $E = 207$ GPa and $G = 79$ GPa. Determine the relative deflection of the applied forces.

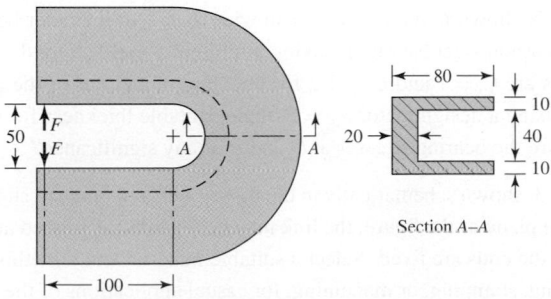

Section *A–A*

(All dimensions in millimeters.)

5–68   Solve Prob. 5–61 using Eq. (5–33).

5–69   A thin ring is loaded by two equal and opposite forces F in part a of the figure. A free-body diagram of one quadrant is shown in part b. This is a statically indeterminate problem, because the moment M_A cannot be found by statics. We wish to find the maximum bending moment in the ring due to the forces F. Assume that the radius of the ring is large so that Eq. (5–29) can be used.

Problem 5–69

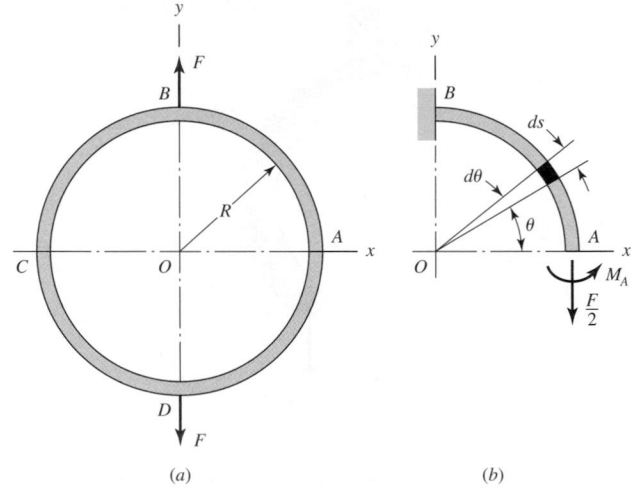

(a) (b)

5–70   Find the increase in the diameter of the ring of Prob. 5–69 due to the forces F and along the y axis.

5–71   A round tubular column has outside and inside diameters of D and d, respectively, and a diametral ratio of $K = d/D$. Show that buckling will occur when the outside diameter is

$$D = \left[\frac{64 P_{cr} l^2}{\pi^3 C E (1 - K^4)} \right]^{1/4}$$

5–72   For the conditions of Prob. 5–71, show that buckling according to the parabolic formula will occur when the outside diameter is

$$D = 2 \left[\frac{P_{cr}}{\pi S_y (1 - K^2)} + \frac{S_y l^2}{\pi^2 C E (1 + K^2)} \right]^{1/2}$$

5–73   Link 2, shown in the figure, is 1 in wide, has $\frac{1}{2}$-in-diameter bearings at the ends, and is cut from low-carbon steel bar stock having a minimum yield strength of 24 kpsi. The end-condition constants are $C = 1$ and $C = 1.2$ for buckling in and out of the plane of the drawing, respectively.
(a) Using a design factor $n_d = 5$, find a suitable thickness for the link.
(b) Are the bearing stresses at O and B of any significance?

5–74   Link 3, shown schematically in the figure, acts as a brace to support the 1.2-kN load. For buckling in the plane of the figure, the link may be regarded as pinned at both ends. For out-of-plane buckling, the ends are fixed. Select a suitable material and a method of manufacture, such as forging, casting, stamping, or machining, for casual applications of the brace in oil-field machinery. Specify the dimensions of the cross section as well as the ends so as to obtain a strong, safe, well-made, and economical brace.

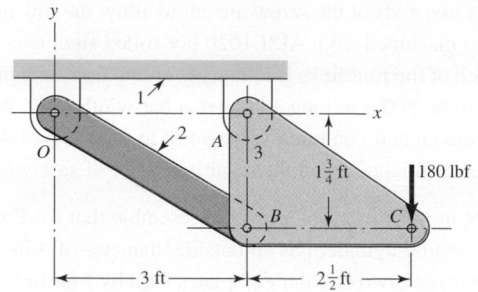

Problem 5–73

$1\frac{1}{4}$ ft

180 lbf

3 ft

$2\frac{1}{2}$ ft

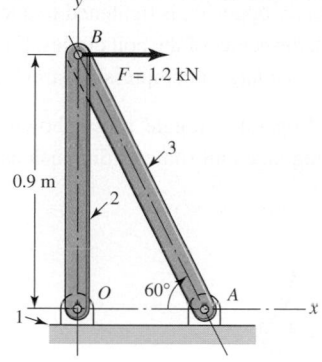

Problem 5–74

B

$F = 1.2$ kN

0.9 m

$60°$

O A

DESIGN

5–75

The hydraulic cylinder shown in the figure has a 3-in bore and is to operate at a pressure of 800 psi. With the clevis mount shown, the piston rod should be sized as a column with both ends rounded for any plane of buckling. The rod is to be made of forged AISI 1030 steel without further heat treatment.

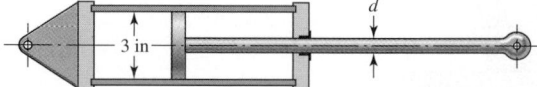

Problem 5–75

d

3 in

(a) Use a design factor $n_d = 3$ and select a preferred size for the rod diameter if the column length is 60 in.

(b) Repeat part (a) but for a column length of 18 in.

(c) What factor of safety actually results for each of the cases above?

DESIGN

5–76

The figure shows a schematic drawing of a vehicular jack that is to be designed to support a maximum mass of 400 kg based on the use of a design factor $n_d = 2.50$. The opposite-handed

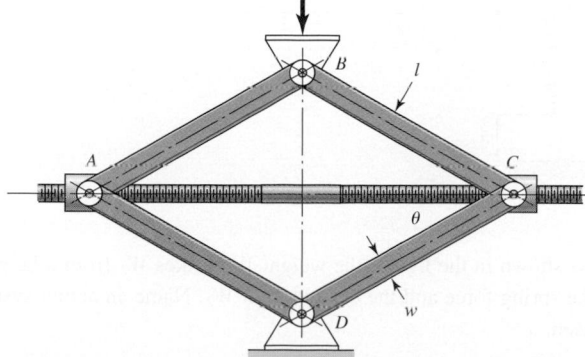

Problem 5–76

W

B

l

A

C

θ

D

w

threads on the two ends of the screw are cut to allow the link angle θ to vary from 15 to 70°. The links are to be machined from AISI 1020 hot-rolled steel bars with a minimum yield strength of 380 MPa. Each of the four links is to consist of two bars, one on each side of the central bearings. The bars are to be 300 mm long and have a bar width of 25 mm. The pinned ends are to be designed to secure an end-condition constant of at least $C = 1.4$ for out-of-plane buckling. Find a suitable preferred thickness and the resulting factor of safety for this thickness.

 ANALYSIS **5-77** If drawn, a figure for this problem would resemble that for Prob. 5–51. A strut that is a standard hollow right circular cylinder has an outside diameter of 4 in and a wall thickness of $\frac{3}{8}$ in and is compressed between two circular end plates held by four bolts equally spaced on a bolt circle of 5.68-in diameter. All four bolts are hand-tightened, and then bolt A is tightened to a tension of 2000 lbf and bolt C, diagonally opposite, is tightened to a tension of 10 000 lbf. The strut axis of symmetry is coincident with the center of the bolt circles. Find the maximum compressive load, the eccentricity of loading, and the largest compressive stress in the strut.

 DESIGN **5-78** Design link CD of the hand-operated toggle press shown in the figure. Specify the cross-section dimensions, the bearing size and rod-end dimensions, the material, and the method of processing.

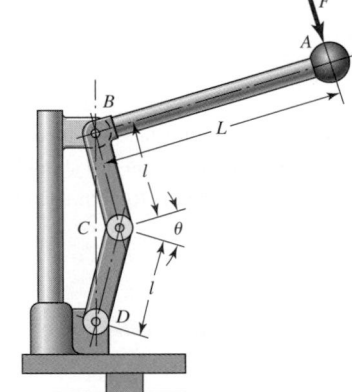

Problem 5–78
$l = 12$ in, $l = 4$ in, $\theta_{min} = 0°$.

 ANALYSIS **5-79** Find expressions for the maximum values of the spring force and deflection y of the impact system shown in the figure. Can you think of a realistic application for this model?

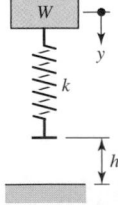

Problem 5–79

 ANALYSIS **5-80** As shown in the figure, the weight W_1 strikes W_2 from a height h. Find the maximum values of the spring force and the deflection of W_2. Name an actual system for which this model might be used.

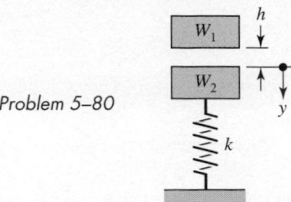

Problem 5–80

5–81 Part *a* of the figure shows a weight W mounted between two springs. If the free end of spring k_1 is suddenly displaced through the distance $x = a$, as shown in part *b*, what would be the maximum displacement y of the weight?

Problem 5–81

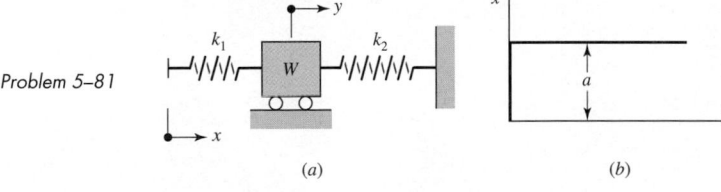

(*a*)   (*b*)

PART 2 **Failure Prevention**

6

Failures Resulting from Static Loading

In Chap. 1 we learned that *strength is a property or characteristic of a mechanical element*. This property results from the material identity, the treatment and processing incidental to creating its geometry, and the loading, and it is at the controlling or critical location. These elements must be identified and considered before we can speak about *the strength* of a part as a useful characteristic of descriptive worth. Tables of properties of engineering materials are *not* about strengths of parts. The strength of a mechanical part does not depend on that part being subjected to its intended load. In fact, this strength property is a characteristic of the element before it is assembled with other elements into a machine or system. The strength of a part is a valuable descriptor, but it exists only if we meet all the qualifications enumerated above.

In addition to considering the strength of a single part, we must be cognizant that the strengths of the mass-produced parts will all be somewhat different from the others in the collection or ensemble because of variations in dimensions, machining, forming, and composition. Descriptors of strength are necessarily statistical in nature, involving parameters such as mean, standard deviations, and distributional identification.

A *static load* is a stationary force or couple applied to a member. To be stationary, the force or couple must be unchanging in magnitude, point or points of application, and direction. A static load can produce axial tension or compression, a shear load, a bending load, a torsional load, or any combination of these. To be considered static, the load cannot change in any manner.

In this chapter we consider the relations between strength and static loading in order to make the decisions concerning material and its treatment, fabrication, and geometry for satisfying the requirements of functionality, safety, reliability, competitiveness, usability, manufacturability, and marketability. How far we go down this list is related to the scope of the examples.

"Failure" is the first word in the chapter title. Failure can mean a part has separated into two or more pieces; has become permanently distorted, thus ruining its geometry; has had its reliability downgraded; or has had its function compromised, whatever the reason. A designer speaking of failure can mean any or all of these possibilities. In this chapter our attention is focused on the predictability of permanent distortion or separation. In strength-sensitive situations the designer must separate mean stress and mean strength at the critical location sufficiently to accomplish his or her purposes.

Figures 6–1 to 6–13 are photographs of several failed parts. The photographs exemplify the need of the designer to be well-versed in failure prevention. Toward this end we shall consider one-, two-, and three-dimensional stress states, with and without stress concentrations, for both ductile and brittle materials.

Figure 6–1

(*a*) Failure of a truck drive-shaft spline due to corrosion fatigue. Note that it was necessary to use clear tape to hold the pieces in place. (*b*) Direct end view of failure.

(*a*)

(*b*)

Figure 6–2

Fatigue failure of an automotive cooling fan due to vibrations caused by a defective water pump.

Figure 6–3

Typical failure of a stamped steel alternator bracket after about 40 000 km. The failure was probably due to residual stresses caused by the cold-forming operation. The high failure rate prompted the manufacturer to redesign the bracket as a die casting. *(Source: Dowling, N. E., Mechanical Behavior of Materials, 2/e, 1999, p. 257. Reprinted by permission of Pearson Education, Inc., Upper Saddle River, New Jersey.)*

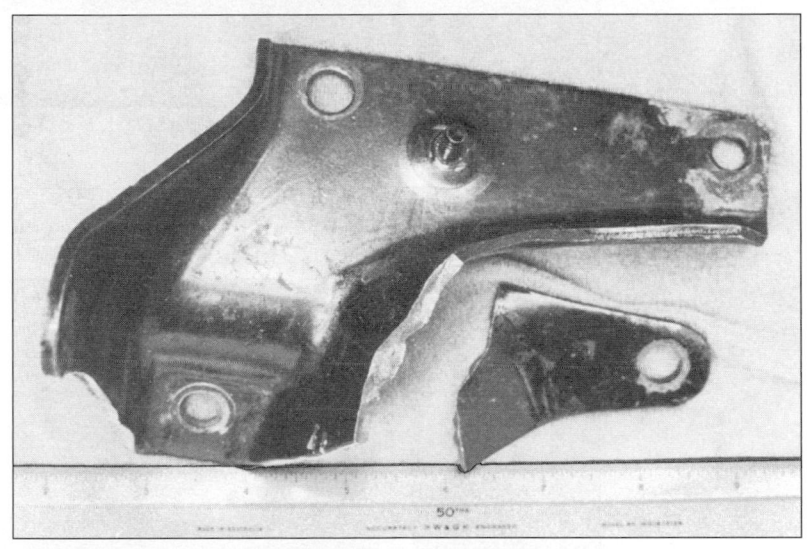

Figure 6–4

Failure of an automotive drag link. The failure occurred after about 225 000 km. Fortunately the car was in park and against a curb. Such a failure results in total disconnect of the steering wheel from the steering mechanism.

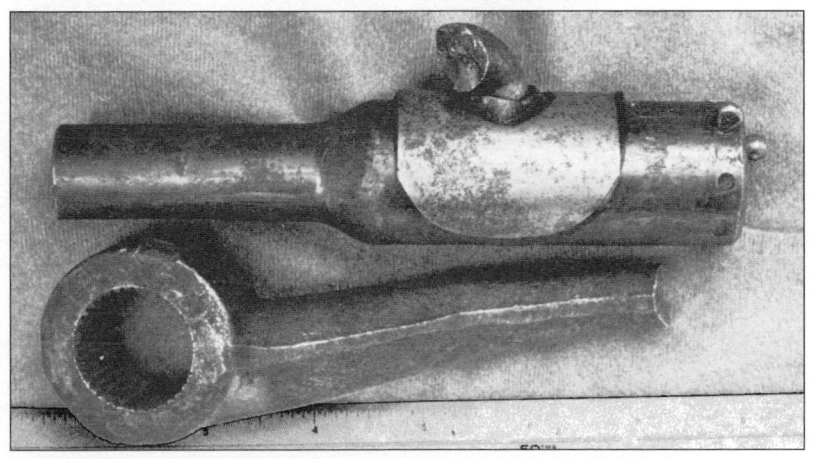

Figure 6–5

Impact failure of a lawn-mower blade driver hub. The blade impacted a surveying pipe marker.

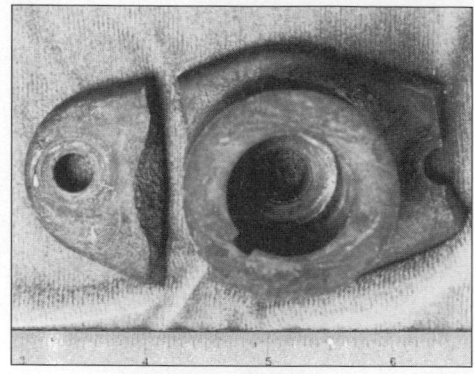

Figure 6–6

Failure of an overhead-pulley retaining bolt on a weightlifting machine. A manufacturing error caused a gap that forced the bolt to take the entire moment load.

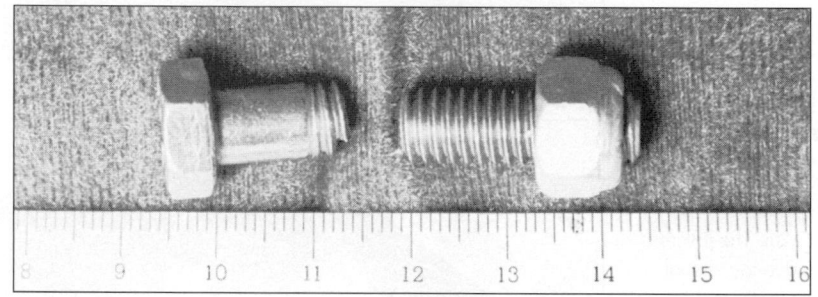

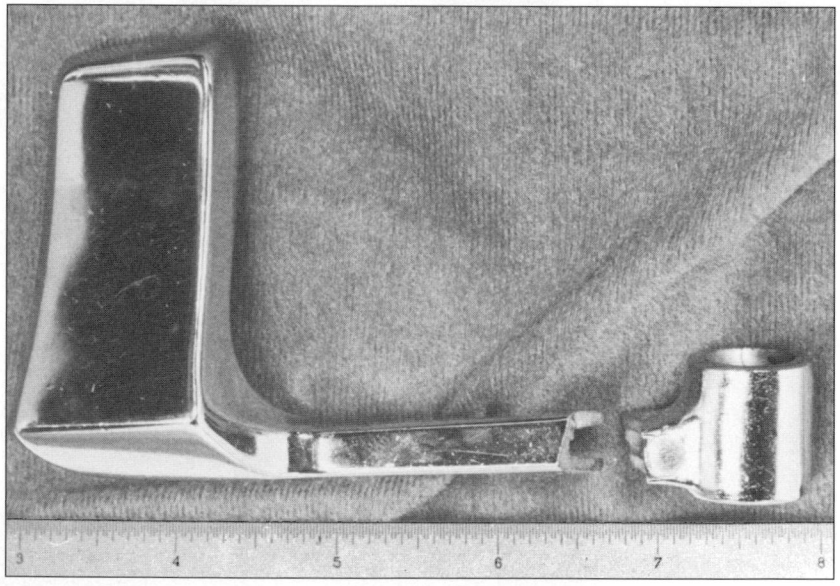

Figure 6–7

Failure of an interior die-cast car-door handle. Failure occurred about every 72 000 km. Probable causes were the electroplating material, stress concentration, the long lever arm required to operate a "sticky" door-release mechanism, and the high actuation forces.

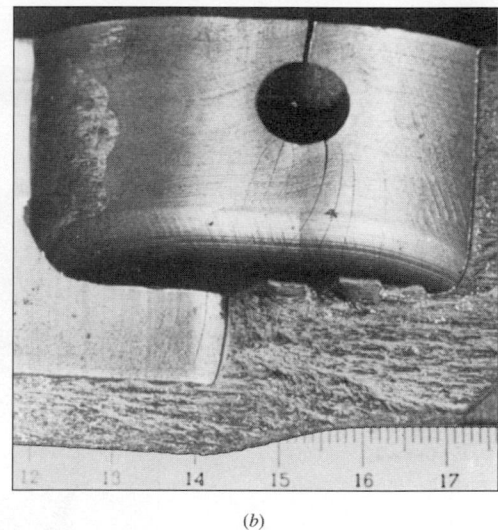

(a)   (b)

Figure 6–8

Chain test fixture that failed in one cycle. To alleviate complaints of excessive wear, the manufacturer decided to case-harden the material. (a) Two halves showing fracture; this is an excellent example of brittle fracture initiated by stress concentration. (b) Enlarged view of one portion to show cracks induced by stress concentration at the support-pin holes.

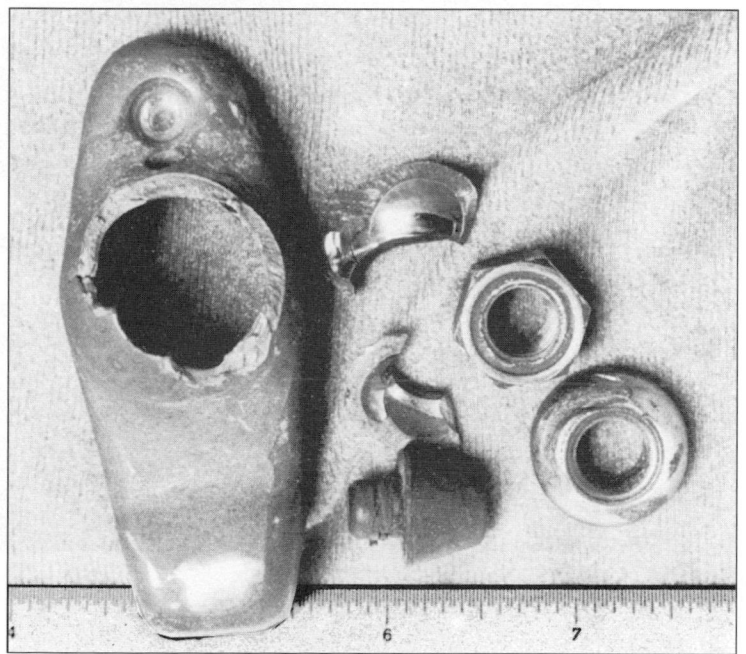

Figure 6–9

Automotive rocker-arm articulation-joint fatigue failure.

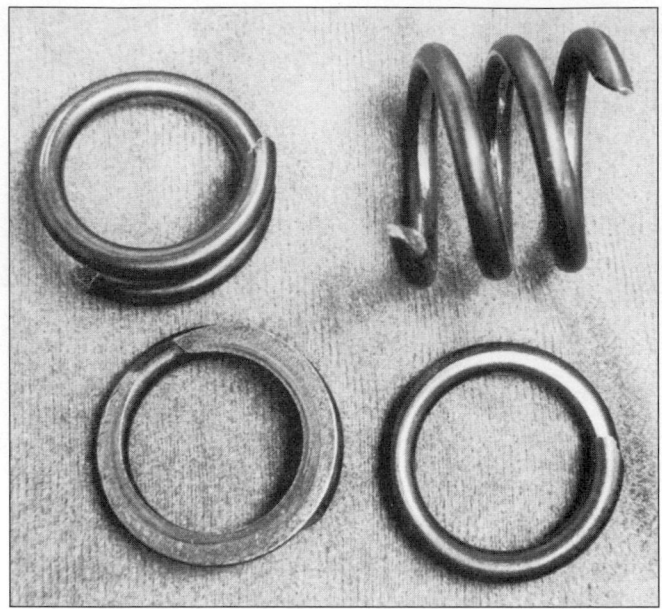

Figure 6–10

Valve-spring failure caused by spring surge in an oversped engine.
The fractures exhibit the classic 45° shear failure.

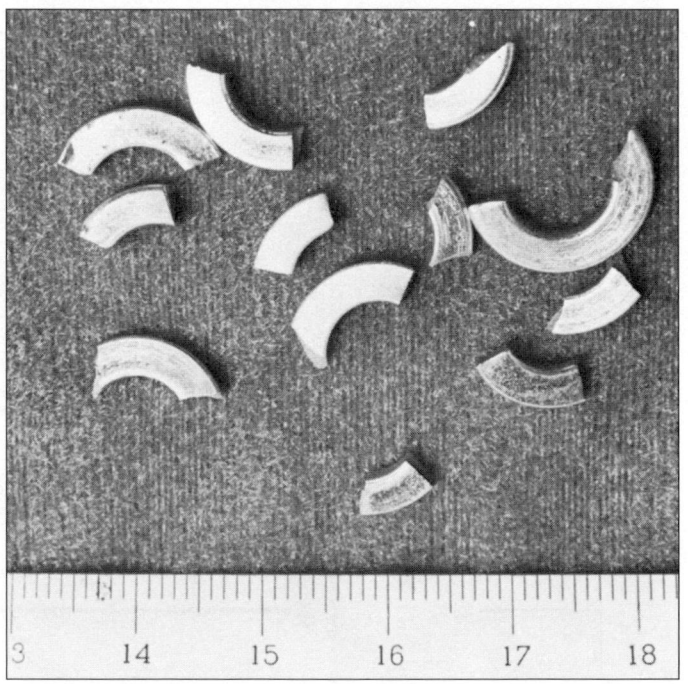

Figure 6–11

Brittle fracture of a lock washer in one-half cycle. The washer failed when
it was installed.

Figure 6–12

Fatigue failure of a die-cast residence door bumper. This bumper is installed on the door hinge to prevent the doorknob from impacting the wall.

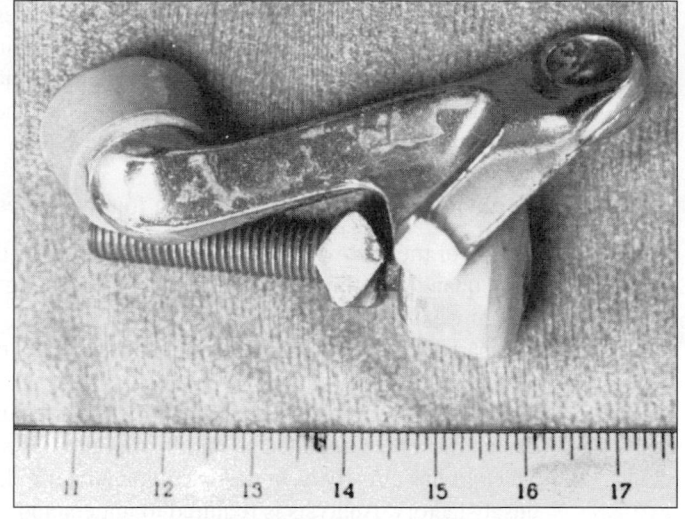

Figure 6–13

A gear failure from a $7\frac{1}{2}$-hp (5.6-kW) American-made outboard motor. The large gear has a $1\frac{7}{8}$-in (47.6-mm) outside diameter and had 21 teeth; 6 are broken. The pinion had 14 teeth; all are broken. Failure occurred when the propeller struck a steel auger placed in the lake bottom as an anchorage. The owner had replaced the shear pin with a substitute pin.

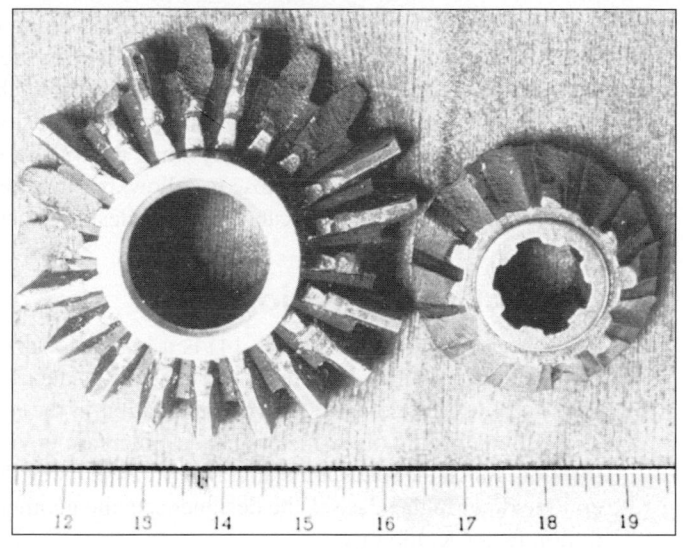

6–1 **Static Strength**

Ideally, in designing any machine element, the engineer should have available the results of a great many strength tests of the particular material chosen. These tests should be made on specimens having the same heat treatment, surface finish, and size as the element the engineer proposes to design; and the tests should be made under exactly the same loading conditions as the part will experience in service. This means that if the part is to experience a bending load, it should be tested with a bending load. If it is to be subjected to combined bending and torsion, it should be tested under combined bending and torsion. If it is made of heat-treated AISI 1040 steel drawn at 500°C with a ground finish, the specimens tested should be of same material prepared in the same manner. Such tests will provide very useful and precise information. Whenever such data are available for design purposes, the engineer can be assured of doing the best possible job of engineering.

The cost of gathering such extensive data prior to design is justified if failure of the part may endanger human life or if the part is manufactured in sufficiently large quantities. Refrigerators and other appliances, for example, have very good reliabilities because the parts are made in such large quantities that they can be thoroughly tested in advance of manufacture. The cost of making these tests is very low when it is divided by the total number of parts manufactured.

You can now appreciate the following four design categories:

1   Failure of the part would endanger human life, or the part is made in extremely large quantities; consequently, an elaborate testing program is justified during design.
2   The part is made in large enough quantities that a moderate series of tests is feasible.
3   The part is made in such small quantities that testing is not justified at all; or the design must be completed so rapidly that there is not enough time for testing.
4   The part has already been designed, manufactured, and tested and found to be unsatisfactory. Analysis is required to understand why the part is unsatisfactory and what to do to improve it.

More often than not it is necessary to design using only published values of yield strength, ultimate strength, percentage reduction in area, and percentage elongation, such as those listed in Appendix A. How can one use such meager data to design against both static and dynamic loads, two- and three-dimensional stress states, high and low temperatures, and very large and very small parts? These and similar questions will be addressed in this chapter and those to follow, but think how much better it would be to have data available that duplicate the actual design situation.

6–2   Stress Concentration

Stress concentration (see Sec. 4–14) is a highly localized effect. In some instances it may be due to a surface scratch. If the material is ductile and the load static, the design load may cause yielding in the critical location in the notch. This yielding can involve strain strengthening of the material and an increase in yield strength at the critical notch location. Since the loads are static, that part can carry them satisfactorily with no general yielding. In these cases the designer sets the geometric (theoretical) stress concentration factor K_t to unity.

The rationale can be expressed as follows. The worst-case scenario is that of an idealized non–strain-strengthening material shown in Fig. 6–14. The stress-strain locus rises linearly to the yield strength S_y, then proceeds at constant stress, which is equal to S_y. Consider a filleted rectangular bar as depicted in Fig. A–15–5, where the cross-section area of the small shank is 1 in². If the material is ductile, with a yield point of 40 kpsi, and the theoretical stress-concentration factor (SCF) K_t is 2,

- A load of 20 kip induces a tensile stress of 20 kpsi in the shank as depicted at point A in Fig. 6–14. At the critical location in the fillet the stress is 40 kpsi, and the SCF is $K = \sigma_{max}/\sigma_{nom} = 40/20 = 2$.
- A load of 30 kip induces a tensile stress of 30 kpsi in the shank at point B. The fillet stress is still 40 kpsi (point D), and the SCF $K = \sigma_{max}/\sigma_{nom} = S_y/\sigma = 40/30 = 1.33$.
- At a load of 40 kip the induced tensile stress (point C) is 40 kpsi in the shank. At the critical location in the fillet the stress (at point E) is 40 kpsi. The SCF $K = \sigma_{max}/\sigma_{nom} = S_y/\sigma = 40/40 = 1$.

Figure 6–14

An idealized stress-strain curve. The dashed line depicts a strain-strengthening material.

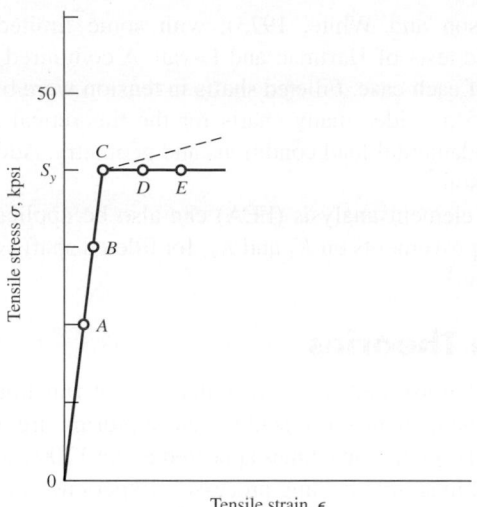

For materials that strain-strengthen, the critical location in the notch has a higher S_y. The shank area is at a stress level a little below 40 kpsi, is carrying load, and is very near its failure-by-general-yielding condition. This is the reason designers do not apply K_t in static loading of a ductile material loaded elastically, instead setting $K_t = 1$.

When using this rule for ductile materials with static loads, be careful to assure yourself that the material is not susceptible to brittle fracture (see Sec. 6–13) in the environment of use. The usual definition of geometric (theoretical) stress-concentration factor for normal stress K_t or shear stress K_{ts} is

$$\sigma_{\max} = K_t \sigma_{\text{nom}} \tag{a}$$

$$\tau_{\max} = K_{ts} \tau_{\text{nom}} \tag{b}$$

Since your attention is on the stress-concentration factor, and the definition of σ_{nom} or τ_{nom} is given in the graph caption or from a computer program, be sure the value of nominal stress is appropriate for the section carrying the load.

Brittle materials do not exhibit a plastic range. A brittle material "feels" the stress concentration factor K_t or K_{ts}, which is applied by using Eq. (a) or (b).

An exception to this rule is a brittle material that inherently contains microdiscontinuity stress concentration, worse than the macrodiscontinuity that the designer has in mind. Sand molding introduces sand particles, air, and water vapor bubbles. The grain structure of cast iron contains graphite flakes (with little strength), which are literally cracks introduced during the solidification process. When a tensile test on a cast iron is performed, the strength reported in the literature *includes* this stress concentration. In such cases K_t or K_{ts} need not be applied.

An important source of stress-concentration factors is R. E. Peterson, who compiled them from his own work and that of others.[1] Peterson developed the style of presentation in which the stress-concentration factor K_t is multiplied by the nominal stress σ_{nom} to estimate the magnitude of the largest stress in the locality. His approximations were based on photoelastic studies of two-dimensional strips (Hartman and Levan,

[1]R. E. Peterson, "Design Factors for Stress Concentration," *Machine Design,* vol. 23, no. 2, February 1951; no. 3, March 1951; no. 5, May 1951; no. 6, June 1951; no. 7, July 1951.

1951; Wilson and White, 1973), with some limited data from three-dimensional photoelastic tests of Hartman and Levan. A contoured graph was included in the presentation of each case. Filleted shafts in tension were based on two-dimensional strips. Table A–15 provides many charts for the theoretical stress-concentration factors for several fundamental load conditions and geometry. Additional charts are also available from Peterson.[2]

Finite element analysis (FEA) can also be applied to obtain stress-concentration factors. Improvements on K_t and K_{ts} for filleted shafts were reported by Tipton, Sorem, and Rolovic.[3]

6–3 Failure Theories

Section 6–1 illustrated some ways that loss of function is manifested. Events such as distortion, permanent set, cracking, and rupturing are among the ways that a machine element fails. Testing machines appeared in the 1700s, and specimens were pulled, bent, and twisted in simple loading processes. Experience in gathering these data was instrumental to the formation of the idea of strain. Cauchy linked stress, strain, and elastic constants in what we today would call theory of elasticity. Always looking for a simple, insightful "mechanism" of failure, humankind persevered. As with all things, postulations lead to predictions, and these predictions are criticized by the use of additional data. Ideas are modified, and the process continues, as does the process of understanding.

If the failure mechanism is simple, then simple tests can give clues. Just what is simple? The tension test is uniaxial (that's simple) and elongations are largest in the axial direction, so strains can be measured and stresses inferred up to "failure." Just what is important: a critical stress, a critical strain, a critical energy? Do growing Mohr's circles, as depicted in Fig. 4–14, expand until they "hit something"? If so, what? If not, what is important? In the next several sections, we shall show failure theories that have helped answer some of these questions.

Unfortunately, there is no universal theory of failure for the general case of material properties and stress state. Instead, over the years several hypotheses have been formulated and tested, leading to today's accepted practices. Being accepted, we will characterize these "practices" as *theories* as most designers do.

Structural metal behavior is typically classified as being ductile or brittle, although under special situations, a material normally considered ductile can fail in a brittle manner (see Sec. 6–13). Ductile materials are normally classified such that $\varepsilon_f \geq 0.05$ and have an identifiable yield strength that is often the same in compression as in tension ($S_{yt} = S_{yc} = S_y$). Brittle materials, $\varepsilon_f < 0.05$, do not exhibit an identifiable yield strength, and are typically classified by ultimate tensile and compressive strengths, S_{ut} and S_{uc}, respectively (where S_{uc} is given as a positive quantity). The generally accepted theories are:

Ductile materials (yield criteria)

- Maximum shear stress (MSS), Sec. 6–4
- Distortion energy (DE), Sec. 6–5
- Ductile Coulomb-Mohr (DCM), Sec. 6–6

[2] R. E. Peterson, *Stress Concentration Factors,* John Wiley & Sons, New York, 1974.

[3] S. M. Tipton, J. R. Sorem Jr., and R. D. Rolovic, "Updated Stress-Concentration Factors for Filleted Shafts in Bending and Tension," *Trans. ASME, Journal of Mechanical Design,* vol. 118, September 1996, pp. 321–327.

Brittle materials (fracture criteria)

- Maximum normal stress (MNS), Sec. 6–8
- Brittle Coulomb-Mohr (BCM) and modifications, Sec. 6–9

It would be inviting if we had one universally accepted theory for each material type, but for one reason or another, they are all used. Later, we will provide rationales for selecting a particular theory. First, we will describe the bases of these theories and apply them to some examples.

6–4   ## Maximum-Shear-Stress Theory for Ductile Materials

The *maximum-shear-stress theory* predicts that *yielding begins whenever the maximum shear stress in any element equals or exceeds the maximum shear stress in a tension-test specimen of the same material when that specimen begins to yield.* The MSS theory is also referred to as the *Tresca* or *Guest theory.*

Many theories are postulated on the basis of the consequences seen from tensile tests. As a strip of a ductile material is subjected to tension, slip lines (called *Lüder lines*) form at approximately 45° with the axis of the strip. These slip lines are the beginning of yield, and when loaded to fracture, fracture lines are also seen at angles approximately 45° with the axis of tension. Since the shear stress is maximum at 45° from the axis of tension, it makes sense to think that this is the mechanism of failure. It will be shown in the next section, that there is a little more going on than this. However, it turns out the MSS theory is an acceptable but conservative predictor of failure; and since engineers are conservative by nature, it is quite often used.

Recall that for simple tensile stress, $\sigma = P/A$, and the maximum shear stress occurs on a surface 45° from the tensile surface with a magnitude of $\tau_{\max} = \sigma/2$. So the maximum shear stress at yield is $\tau_{\max} = S_y/2$. For a general state of stress, three principal stresses can be determined and ordered such that $\sigma_1 \geq \sigma_2 > \sigma_3$. The maximum shear stress is then $\tau_{\max} = (\sigma_1 - \sigma_3)/2$ (see Fig. 4–14). Thus, for a general state of stress, the maximum-shear-stress theory predicts yielding when

$$\tau_{\max} = \frac{\sigma_1 - \sigma_3}{2} \geq \frac{S_y}{2} \qquad \text{or} \qquad \sigma_1 - \sigma_3 \geq S_y \qquad (6\text{–}1)$$

Note that this implies that the yield strength in shear is given by

$$S_{sy} = 0.5 S_y \qquad (6\text{–}2)$$

which, as we will see later is about 15 percent low (conservative).

For design purposes, Eq. (6–1) can be modified to incorporate a factor of safety, n. Thus,

$$\sigma_1 - \sigma_3 = \frac{S_y}{n} \qquad (6\text{–}3)$$

Plane stress problems are very common where one of the principal stresses is zero, and the other two, σ_A and σ_B, are determined from Eq. (4–13). Assuming that $\sigma_A \geq \sigma_B$, there are three cases to consider in using Eq. (6–1) for plane stress:

Case 1: $\sigma_A \geq \sigma_B \geq 0$. For this case, $\sigma_1 = \sigma_A$ and $\sigma_3 = 0$. Equation (6–1) reduces to a yield condition of

$$\sigma_A \geq S_y \qquad (6\text{–}4)$$

Figure 6–15

The maximum-shear-stress (MSS) theory for plane stress, where σ_A and σ_B are the two nonzero principal stresses.

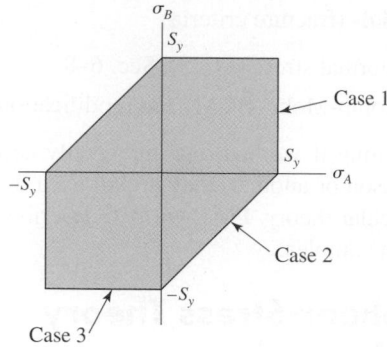

Case 2: $\sigma_A \geq 0 \geq \sigma_B$. Here, $\sigma_1 = \sigma_A$ and $\sigma_3 = \sigma_B$, and Eq. (6–1) becomes

$$\sigma_A - \sigma_B \geq S_y \tag{6–5}$$

Case 3: $0 \geq \sigma_A \geq \sigma_B$. For this case, $\sigma_1 = 0$ and $\sigma_3 = \sigma_B$, and Eq. (6–1) gives

$$\sigma_B \leq -S_y \tag{6–6}$$

Equations (6–4) to (6–6) are represented in Fig. 6–15 by the three lines indicated in the σ_A, σ_B plane. The remaining unmarked lines are cases for $\sigma_B \geq \sigma_A$, which are not normally used. Equations (6–4) to (6–6) can also be converted to design equations by substituting equality for the equal to or greater sign and dividing S_y by n.

6–5 Distortion-Energy Theory for Ductile Materials

The *distortion-energy theory* predicts that *yielding occurs when the distortion strain energy per unit volume reaches or exceeds the distortion strain energy per unit volume for yield in simple tension or compression of the same material.*

The distortion-energy (DE) theory originated from the observation that ductile materials stressed hydrostatically exhibited yield strengths greatly in excess of the values given by the simple tension test. Therefore it was postulated that yielding was not a simple tensile or compressive phenomenon at all, but, rather, that it was related somehow to the angular distortion of the stressed element. To develop the theory, note, in Fig. 6–16a, the unit volume subjected to any three-dimensional stress state designated by the stresses σ_1, σ_2, and σ_3. The stress state shown in Fig. 6–16b is one of hydrostatic tension due to the stresses σ_{av} acting in each of the same principal directions as in Fig. 6–16a. The formula for σ_{av} is simply

$$\sigma_{av} = \frac{\sigma_1 + \sigma_2 + \sigma_3}{3} \tag{a}$$

Thus the element in Fig. 6–16b undergoes pure volume change, that is, no angular distortion. If we regard σ_{av} as a component of σ_1, σ_2, and σ_3, then this component can be subtracted from them, resulting in the stress state shown in Fig. 6–16c. This element is subjected to pure angular distortion, that is, no volume change.

With the help of Table 4–1, we find the strain energy per unit volume subjected to three principal stresses is

$$u = \frac{1}{2}(\epsilon_1\sigma_1 + \epsilon_2\sigma_2 + \epsilon_3\sigma_3)$$

$$= \frac{1}{2E}\left[\sigma_1^2 + \sigma_2^2 + \sigma_3^2 - 2v(\sigma_1\sigma_2 + \sigma_2\sigma_3 + \sigma_3\sigma_1)\right] \tag{b}$$

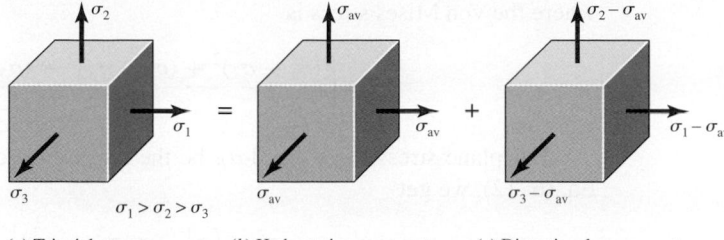

(a) Triaxial stresses (b) Hydrostatic component (c) Distortional component

Figure 6–16

(a) Element with triaxial stresses; this element undergoes both volume change and angular distortion. (b) Element under hydrostatic tension undergoes only volume change. (c) Element has angular distortion without volume change.

The strain energy for producing only volume change u_v can be obtained by substituting σ_{av} for σ_1, σ_2, and σ_3 in Eq. (b). The result is

$$u_v = \frac{3\sigma_{av}^2}{2E}(1 - 2v) \tag{c}$$

If we now substitute the square of Eq. (a) in Eq. (c) and simplify the expression, we get

$$u_v = \frac{1 - 2v}{6E}\left(\sigma_1^2 + \sigma_2^2 + \sigma_3^2 + 2\sigma_1\sigma_2 + 2\sigma_2\sigma_3 + 2\sigma_3\sigma_1\right) \tag{6-7}$$

Then the distortion energy is obtained by subtracting Eq. (6–7) from Eq. (b). This gives

$$u_d = u - u_v = \frac{1 + v}{3E}\left[\frac{(\sigma_1 - \sigma_2)^2 + (\sigma_2 - \sigma_3)^2 + (\sigma_3 - \sigma_1)^2}{2}\right] \tag{6-8}$$

Note that the distortion energy is zero if $\sigma_1 = \sigma_2 = \sigma_3$.

For the simple tensile test, at yield, $\sigma_1 = S_y$ and $\sigma_2 = \sigma_3 = 0$, and from Eq. (6–8) the distortion energy is

$$u_d = \frac{1 + v}{3E}S_y^2 \tag{6-9}$$

So for the general state of stress given by Eq. (6–8), yield is predicted if Eq. (6–8) equals or exceeds Eq. (6–9). This gives

$$\left[\frac{(\sigma_1 - \sigma_2)^2 + (\sigma_2 - \sigma_3)^2 + (\sigma_3 - \sigma_1)^2}{2}\right]^{1/2} \geq S_y \tag{6-10}$$

If we had a simple case of tension σ, then yield would occur when $\sigma \geq S_y$. Thus, the left of Eq. (6–10) can be thought of as a *single, equivalent,* or *effective stress* for the entire general state of stress given by σ_1, σ_2, and σ_3. This effective stress is usually called the *von Mises stress,* σ', named after Dr. R. von Mises, who contributed to the theory. Thus Eq. (6–10), for yield, can be written as

$$\sigma' \geq S_y \tag{6-11}$$

where the von Mises stress is

$$\sigma' = \left[\frac{(\sigma_1 - \sigma_2)^2 + (\sigma_2 - \sigma_3)^2 + (\sigma_3 - \sigma_1)^2}{2} \right]^{1/2} \tag{6-12}$$

For plane stress, let σ_A and σ_B be the two nonzero principal stresses. Then from Eq. (6–12), we get

$$\sigma' = \left(\sigma_A^2 - \sigma_A \sigma_B + \sigma_B^2 \right)^{1/2} \tag{6-13}$$

Equation (6–13) is a rotated ellipse in the σ_A, σ_B plane, as shown in Fig. 6–17 with $\sigma' = S_y$. The dotted lines in the figure represent the MSS theory, which can be seen to be more restrictive, hence, more conservative.[4]

Using *xyz* components of three-dimensional stress, the von Mises stress can be written as

$$\sigma' = \frac{1}{\sqrt{2}} \left[(\sigma_x - \sigma_y)^2 + (\sigma_y - \sigma_z)^2 + (\sigma_z - \sigma_x)^2 + 6\left(\tau_{xy}^2 + \tau_{yz}^2 + \tau_{zx}^2 \right) \right]^{1/2} \tag{6-14}$$

and for plane stress,

$$\sigma' = \left(\sigma_x^2 - \sigma_x \sigma_y + \sigma_y^2 + 3\tau_{xy}^2 \right)^{1/2} \tag{6-15}$$

The distortion-energy theory is also called:

- The von Mises or von Mises–Hencky theory
- The shear-energy theory
- The octahedral-shear-stress theory

Understanding octahedral shear stress will shed some light on why the MSS is conservative. Consider an isolated element in which the normal stresses on each surface are equal to the hydrostatic stress σ_{av}. There are eight surfaces symmetric to the principal directions that contain this stress. This forms an octahedron as shown in Fig. 6–18. The shear stresses on these surfaces are equal and are called the *octahedral shear stresses*

Figure 6–17

The distortion-energy (DE) theory for plane stress states. This is a plot of points obtained from Eq. (6–13) with $\sigma' = S_y$.

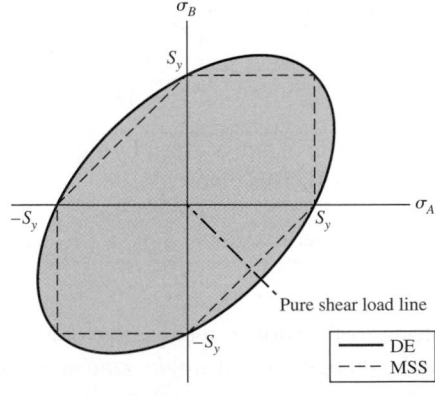

---

[4]The three-dimensional equations for DE and MSS can be plotted relative to three-dimensional $\sigma_1, \sigma_2, \sigma_3$, coordinate axes. The failure surface for DE is a circular cylinder with an axis inclined at 45° from each principal stress axis, whereas the surface for MSS is a hexagon inscribed within the cylinder.

Figure 6–18

Octahedral surfaces.

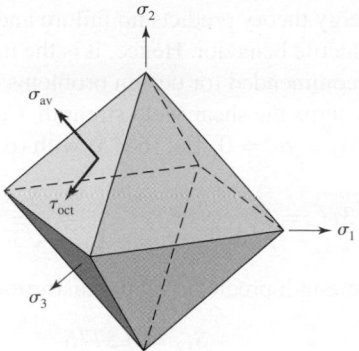

(Fig. 6–18 has only one of the octahedral surfaces labeled). Through coordinate transformations the octahedral shear stress is given by[5]

$$\tau_{\text{oct}} = \frac{1}{3}\left[(\sigma_1 - \sigma_2)^2 + (\sigma_2 - \sigma_3)^2 + (\sigma_3 - \sigma_1)^2\right]^{1/2} \tag{6–16}$$

Under the name of the octahedral-shear-stress theory, *failure is assumed to occur whenever the octahedral shear stress for any stress state equals or exceeds the octahedral shear stress for the simple tension-test specimen at failure.*

As before, on the basis of the tensile test results, yield occurs when $\sigma_1 = S_y$ and $\sigma_2 = \sigma_3 = 0$. From Eq. (6–16) the octahedral shear stress under this condition is

$$\tau_{\text{oct}} = \frac{\sqrt{2}}{3} S_y \tag{6–17}$$

When, for the general stress case, Eq. (6–16) is equal or greater than Eq. (6–17), yield is predicted. This reduces to

$$\left[\frac{(\sigma_1 - \sigma_2)^2 + (\sigma_2 - \sigma_3)^2 + (\sigma_3 - \sigma_1)^2}{2}\right]^{1/2} \geq S_y \tag{6–18}$$

which is identical to Eq. (6–10), verifying that the maximum-octahedral-shear-stress theory is equivalent to the distortion-energy theory.

The model for the MSS theory ignores the contribution of the normal stresses on the 45° surfaces of the tensile specimen. However, these stresses are $P/2A$, and *not* the hydrostatic stresses which are $P/3A$. Herein lies the difference between the MSS and DE theories.

The mathematical manipulation involved in describing the DE theory might tend to obscure the real value and usefulness of the result. The equations given allow the most complicated stress situation to be represented by a single quantity, the von Mises stress, which then can be compared against the yield strength of the material through Eq. (6–11). This equation can be expressed as a design equation by

$$\sigma' = \frac{S_y}{n} \tag{6–19}$$

---

[5]For a derivation, see A. Higdon, E. H. Olsen, W. B. Stiles, J. A. Weese, and W. F. Riley, *Mechanics of Materials,* 4th ed., John Wiley & Sons, New York, 1985, pp. 493–495.

The distortion-energy theory predicts no failure under hydrostatic stress and agrees well with all data for ductile behavior. Hence, it is the most widely used theory for ductile materials and is recommended for design problems unless otherwise specified.

One final note concerns the shear yield strength. Consider a case of pure shear τ_{xy}, where for plane stress $\sigma_x = \sigma_y = 0$; Eq. (6–15) with (6–11) for yield gives

$$\sqrt{3\tau_{xy}^2} = S_y \quad \text{or} \quad \tau_{xy} = \frac{S_y}{\sqrt{3}} = 0.577 S_y \tag{6–20}$$

Thus, the shear yield strength predicted by the distortion-energy theory is

$$S_{sy} = 0.577 S_y \tag{6–21}$$

which as stated earlier, is about 15 percent greater than that predicted by the MSS theory (see the pure shear load line in Fig. 6–17).

EXAMPLE 6–1

A hot-rolled steel has a yield strength of $S_{yt} = S_{yc} = 100$ kpsi and a true strain at fracture of $\varepsilon_f = 0.55$. Estimate the factor of safety for the following principal stress states:

(a) 70, 70, 0 kpsi.
(b) 30, 70, 0 kpsi.
(c) 0, 70, −30 kpsi.
(d) 0, −30, −70 kpsi.
(e) 30, 30, 30 kpsi.

Solution

Since $\varepsilon_f > 0.05$ and S_{yc} and S_{yt} are equal, the material is ductile and the distortion-energy (DE) theory applies. The maximum-shear-stress (MSS) theory will also be applied and compared to the DE results. Note that cases a to d are plane stress states.

(a) The ordered principal stresses are $\sigma_A = \sigma_1 = 70, \sigma_B = \sigma_2 = 70, \sigma_3 = 0$ kpsi.

DE From Eq. (6–13),

$$\sigma' = [70^2 - 70(70) + 70^2]^{1/2} = 70 \text{ kpsi}$$

Answer

$$n = \frac{S_y}{\sigma'} = \frac{100}{70} = 1.43$$

MSS Case 1, using Eq. (6–4),

Answer

$$n = \frac{S_y}{\sigma_A} = \frac{100}{70} = 1.43$$

(b) The ordered principal stresses are $\sigma_A = \sigma_1 = 70, \sigma_B = \sigma_2 = 30, \sigma_3 = 0$ kpsi.

DE

$$\sigma' = [70^2 - 70(30) + 30^2]^{1/2} = 60.8 \text{ kpsi}$$

Answer

$$n = \frac{S_y}{\sigma'} = \frac{100}{60.8} = 1.64$$

MSS Case 1, using Eq. (6–4),

Answer
$$n = \frac{S_y}{\sigma_A} = \frac{100}{70} = 1.43$$

(c) The ordered principal stresses are $\sigma_A = \sigma_1 = 70$, $\sigma_2 = 0$, $\sigma_B = \sigma_3 = -30$ kpsi.

DE
$$\sigma' = [70^2 - 70(-30) + (-30)^2]^{1/2} = 88.9 \text{ kpsi}$$

Answer
$$n = \frac{S_y}{\sigma'} = \frac{100}{88.9} = 1.13$$

MSS   Case 2, using Eq. (6–5),

Answer
$$n = \frac{S_y}{\sigma_A - \sigma_B} = \frac{100}{70 - (-30)} = 1.00$$

(d) The ordered principal stresses are $\sigma_1 = 0$, $\sigma_A = \sigma_2 = -30$, $\sigma_B = \sigma_3 = -70$ kpsi.

DE
$$\sigma' = [(-70)^2 - (-70)(-30) + (-30)^2]^{1/2} = 60.8 \text{ kpsi}$$

Answer
$$n = \frac{S_y}{\sigma'} = \frac{100}{60.8} = 1.64$$

MSS   Case 3, using Eq. (6–6),

Answer
$$n = -\frac{S_y}{\sigma_B} = -\frac{100}{-70} = 1.43$$

(e) The ordered principal stresses are $\sigma_1 = 30$, $\sigma_2 = 30$, $\sigma_3 = 30$ kpsi

DE   From Eq. (6–12),

$$\sigma' = \left[\frac{(30-30)^2 + (30-30)^2 + (30-30)^2}{2} \right]^{1/2} = 0 \text{ kpsi}$$

Answer
$$n = \frac{S_y}{\sigma'} = \frac{100}{0} \rightarrow \infty$$

MSS   From Eq. (6–3),

Answer
$$n = \frac{S_y}{\sigma_1 - \sigma_3} = \frac{100}{30 - 30} \rightarrow \infty$$

A tabular summary of the factors of safety is included for comparisons.

	(a)	(b)	(c)	(d)	(e)
DE	1.43	1.64	1.13	1.64	∞
MSS	1.43	1.43	1.00	1.43	∞

Since the MSS theory is on or within the boundary of the DE theory, it will always predict a factor of safety equal to or less than the DE theory, as can be seen in the table. For each case, except case (e), the coordinates and load lines in the σ_A, σ_B plane are shown in Fig. 6–19. Case (e) is not plane stress. Note that the load line for case (a) is the only

Figure 6–19

Load lines for Ex. 6–1.

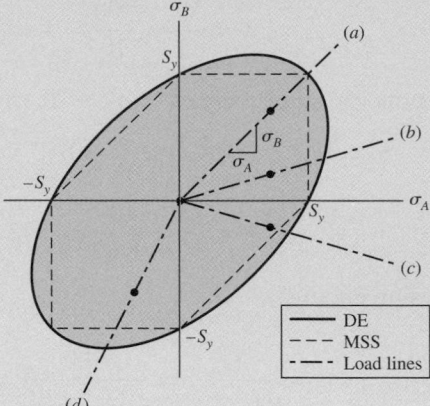

plane stress case given in which the two theories agree, thus giving the same factor of safety.

6–6   Coulomb-Mohr Theory for Ductile Materials

Not all materials have compressive strengths equal to their corresponding tensile values. For example, the yield strength of magnesium alloys in compression may be as little as 50 percent of their yield strength in tension. The ultimate strength of gray cast irons in compression varies from 3 to 4 times greater than the ultimate tensile strength. So, in this section, we are primarily interested in those theories that can be used to predict failure for materials whose strengths in tension and compression are not equal.

Historically, the Mohr theory of failure dates to 1900, a date that is relevant to its presentation. There were no computers, just slide rules, compasses, and French curves. Graphical procedures, common then, are still useful today for visualization. The idea of Mohr is based on three "simple" tests: tension, compression, and shear, to yielding if the material can yield, or to rupture. It is easier to define shear yield strength as S_{sy} than it is to test for it.

The practical difficulties aside, Mohr's hypothesis was to use the results of tensile, compressive, and torsional shear tests to construct the three circles of Fig. 6–20 defining a failure envelope, depicted as line $ABCDE$ in the figure, above the σ axis. The failure envelope need not be straight. The argument amounted to the three Mohr circles describing the stress state in a body (see Fig. 4–14) growing during loading until one of them became tangent to the failure envelope, thereby defining failure. Was the form of the failure envelope straight, circular, or quadratic? A compass or a French curve defined the failure envelope.

A variation of Mohr's theory, called the *Coulomb-Mohr theory* or the *internal-friction theory,* assumes that the boundary BCD in Fig. 6–20 is straight. With this assumption only the tensile and compressive strengths are necessary. Consider the conventional ordering of the principal stresses such that $\sigma_1 \geq \sigma_2 \geq \sigma_3$. The largest circle connects σ_1 and σ_3, as shown in Fig. 6–21. Interpolating the radius of the solid-line circle with the radii of the dashed-line circles with respect to the center distances to point O, we have

$$\frac{\dfrac{\sigma_1 - \sigma_3}{2} - \dfrac{S_t}{2}}{\dfrac{S_c}{2} - \dfrac{S_t}{2}} = \frac{\dfrac{S_t}{2} - \dfrac{\sigma_1 + \sigma_3}{2}}{\dfrac{S_c}{2} + \dfrac{S_t}{2}}$$

Figure 6–20

Three Mohr circles, one for the uniaxial compression test, one for the test in pure shear, and one for the uniaxial tension test, are used to define failure by the Mohr hypothesis. The strengths S_c and S_t are the compressive and tensile strengths, respectively; they can be used for yield or ultimate strength.

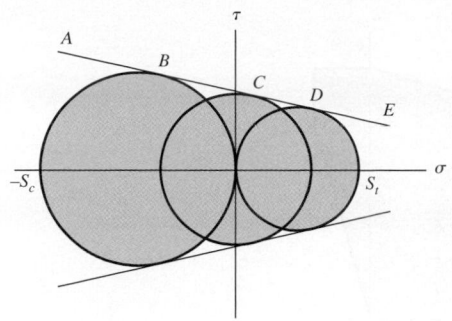

Figure 6–21

(From Richard G. Budynas, *Advanced Strength and Applied Stress Analysis*, 2nd ed., McGraw-Hill, New York, 1999. Copyright © 1999 The McGraw-Hill Companies.)

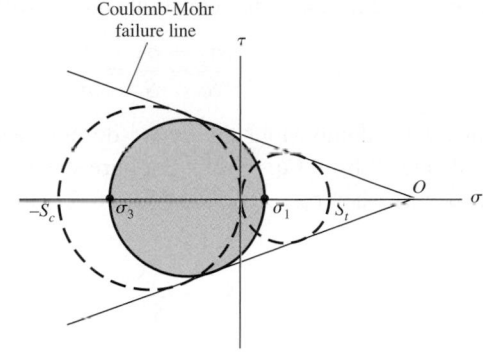

Cross-multiplying and simplifying reduces this equation to

$$\frac{\sigma_1}{S_t} - \frac{\sigma_3}{S_c} = 1 \tag{6–22}$$

where either yield strength or ultimate strength can be used.

For plane stress, when the two nonzero principal stresses are $\sigma_A \geq \sigma_B$, we have a situation similar to the three cases given for the MSS theory, Eqs. (6–4) to (6–6). That is,

Case 1: $\sigma_A \geq \sigma_B \geq 0$. For this case, $\sigma_1 = \sigma_A$ and $\sigma_3 = 0$. Equation (6–22) reduces to a failure condition of

$$\sigma_A \geq S_t \tag{6–23}$$

Case 2: $\sigma_A \geq 0 \geq \sigma_B$. Here, $\sigma_1 = \sigma_A$ and $\sigma_3 = \sigma_B$, and Eq. (6–22) becomes

$$\frac{\sigma_A}{S_t} - \frac{\sigma_B}{S_c} \geq 1 \tag{6–24}$$

Case 3: $0 \geq \sigma_A \geq \sigma_B$. For this case, $\sigma_1 = 0$ and $\sigma_3 = \sigma_B$, and Eq. (6–22) gives

$$\sigma_B \leq -S_c \tag{6–25}$$

A plot of these cases, together with the normally unused cases corresponding to $\sigma_B \geq \sigma_A$, is shown in Fig. 6–22.

Figure 6–22

Plot of the Coulomb-Mohr
theory of failure for plane
stress states.

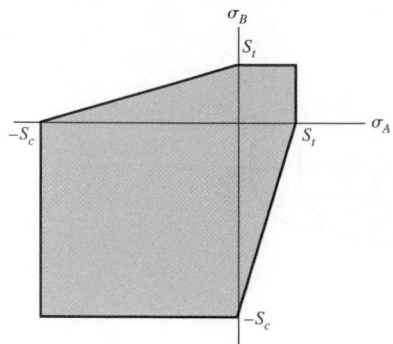

For design equations, incorporating the factor of safety n, divide all strengths by n. For example, Eq. (6–22) as a design equation can be written as

$$\frac{\sigma_1}{S_t} - \frac{\sigma_3}{S_c} = \frac{1}{n} \tag{6–26}$$

Since for the Coulomb-Mohr theory we do not need the torsional shear strength circle we can deduce it from Eq. (6–22). For pure shear τ, $\sigma_1 = -\sigma_3 = \tau$. The torsional yield strength occurs when $\tau_{max} = S_{sy}$, and Eq. (6–22) for yield gives $S_{sy}/S_{yt} - (-S_{sy})/S_{yc} = 1$, or

$$S_{sy} = \frac{S_{yt} S_{yc}}{S_{yt} + S_{yc}} \tag{6–27}$$

EXAMPLE 6–2

A 25-mm-diameter shaft is statically torqued to 230 N · m. It is made of cast 195-T6 aluminum, with a yield strength in tension of 160 MPa and a yield strength in compression of 170 MPa. It is machined to final diameter. Estimate the factor of safety of the shaft.

Solution

Since N · m, cm, and MPa form a consistent set of dimensions, we express the diameter in cm:

$$\tau = \frac{16T}{\pi d^3} = \frac{16(230)}{\pi 2.5^3} = 75 \text{ MPa}$$

The two nonzero principal stresses are 75 and −75 MPa, making the ordered principal stresses $\sigma_1 = 75$, $\sigma_2 = 0$, and $\sigma_3 = -75$ MPa. From Eq. (6–26), for yield,

Answer

$$n = \frac{1}{\sigma_1/S_{yt} - \sigma_3/S_{yc}} = \frac{1}{75/160 - (-75)/170} = 1.10$$

Alternatively, from Eq. (6–27),

$$S_{sy} = \frac{S_{yt} S_{yc}}{S_{yt} + S_{yc}} = \frac{160(170)}{160 + 170} = 82.4 \text{ MPa}$$

and $\tau_{max} = 75$ MPa. Thus,

Answer

$$n = \frac{S_{sy}}{\tau_{max}} = \frac{82.4}{75} = 1.10$$

6–7   **Failure of Ductile Materials Summary**

Having studied some of the various theories of failure, we shall now evaluate them and show how they are applied in design and analysis. In this section we limit our studies to materials and parts that are known to fail in a ductile manner. Materials that fail in a brittle manner will be considered separately because these require different failure theories.

To help decide on appropriate and workable theories of failure, Marin[6] collected data from many sources. Some of the data points used to select failure theories for ductile materials are shown in Fig. 6–23.[7] Mann also collected many data for copper and nickel alloys; if shown, the data points for these would be mingled with those already diagrammed. Figure 6–23 shows that either the maximum-shear-stress theory or the distortion-energy theory is acceptable for design and analysis of materials that would fail in a ductile manner. You may wish to plot other theories using a red or blue pencil on Fig. 6–23 to show why they are not acceptable or are not used.

The selection of one or the other of these two theories is something that you, the engineer, must decide. For design purposes the maximum-shear-stress theory is easy, quick to use, and conservative. If the problem is to learn why a part failed, then the distortion-energy theory may be the best to use; Fig. 6–23 shows that the locus of the distortion-energy theory passes closer to the central area of the data points, and thus is generally a better predictor of failure.

For ductile materials with unequal yield strengths, S_{yt} in tension and S_{yc} in compression, the Mohr theory is the best available. However, the theory requires the results

Figure 6–23

Experimental data superposed on failure theories. *(Reproduced from Fig. 7.11, p. 257, Mechanical Behavior of Materials, 2nd ed., N. E. Dowling, Prentice Hall, Englewood Cliffs, N.J., 1999.)*

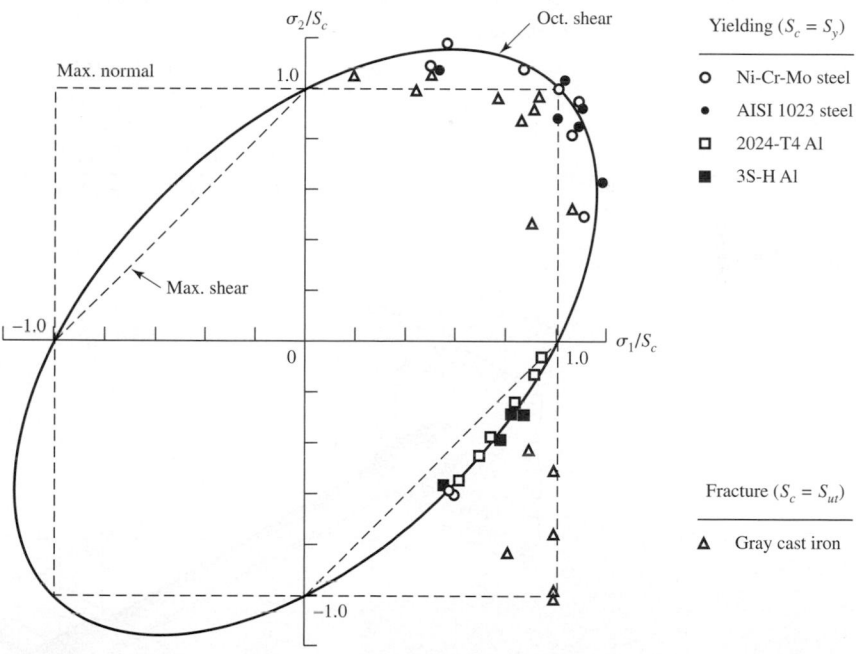

---

[6]Joseph Marin was one of the pioneers in the collection, development, and dissemination of material on the failure of engineering elements. He has published many books and papers on the subject. Here the reference used is Joseph Marin, *Engineering Materials,* Prentice-Hall, Englewood Cliffs, N.J., 1952. (See pp. 156 and 157 for some data points used here.)

[7]Note that some data in Fig. 6–23 are displayed along the top horizontal boundary where $\sigma_B \geq \sigma_A$. This is often done with failure data to thin out congested data points by plotting on the mirror image of the line $\sigma_B = \sigma_A$.

from three separate modes of tests, graphical construction of the failure locus, and fitting the largest Mohr's circle to the failure locus. The alternative to this is to use the Coulomb-Mohr theory, which requires only the tensile and compressive yield strengths and is easily dealt with in equation form.

EXAMPLE 6–3

This example illustrates the use of a failure theory to determine the strength of a mechanical element or component. The example may also clear up any confusion existing between the phrases *strength of a machine part*, *strength of a material*, and *strength of a part at a point*.

A certain force F applied at D near the end of the 15-in lever shown in Fig. 6–24, which is quite similar to a socket wrench, results in certain stresses in the cantilevered bar *OABC*. This bar (*OABC*) is of AISI 1035 steel, forged and heat-treated so that it has a minimum (ASTM) yield strength of 81 kpsi. We presume that this component would be of no value after yielding. Thus the force F required to initiate yielding can be regarded as the strength of the component part. Find this force.

Solution

We will assume that lever *DC* is strong enough and hence not a part of the problem. A 1035 steel, heat-treated, will have a reduction in area of 50 percent or more and hence is a ductile material at normal temperatures. This also means that stress concentration at shoulder A need not be considered. A stress element at A on the top surface will be subjected to a tensile bending stress and a torsional stress. This point, on the 1-in-diameter section, is the weakest section, and governs the strength of the assembly. The two stresses are

$$\sigma_x = \frac{M}{I/c} = \frac{32M}{\pi d^3} = \frac{32(14F)}{\pi(1^3)} = 142.6F$$

$$\tau_{zx} = \frac{Tr}{J} = \frac{16T}{\pi d^3} = \frac{16(15F)}{\pi(1^3)} = 76.4F$$

| **Figure 6–24**

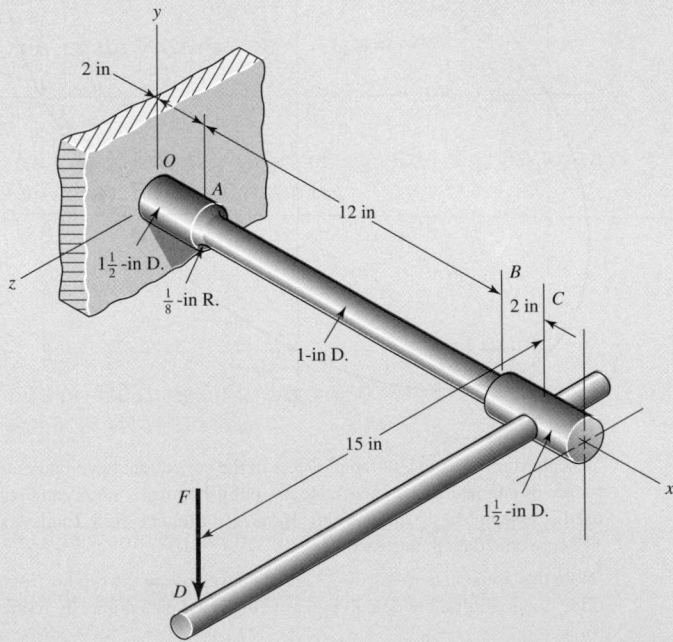

Employing the distortion-energy theory, we find, from Eq. (6–15), that

$$\sigma' = \left(\sigma_x^2 + 3\tau_{zx}^2\right)^{1/2} = \left[(142.6F)^2 + 3(76.4F)^2\right]^{1/2} = 194.5F$$

Equating the von Mises stress to S_y, we solve for F and get

Answer

$$F = \frac{S_y}{194.5} = \frac{81\ 000}{194.5} = 416\ \text{lbf}$$

In this example the strength of the material at point A is $S_y = 81$ kpsi. The strength of the assembly or component is $F = 416$ lbf.

Let us see how to apply the MSS theory. For a point undergoing plane stress with only one non-zero normal stress and one shear stress, the two nonzero principal stresses σ_A and σ_B will have opposite signs and hence fit case 2 for the MSS theory. From Eq. (4–13),

$$\sigma_A - \sigma_B = 2\left[\left(\frac{\sigma_x}{2}\right)^2 + \tau_{zx}^2\right]^{1/2} = \left(\sigma_x^2 + 4\tau_{zx}^2\right)^{1/2}$$

For case 2 of the MSS theory, Eq. (6–5) applies and hence

$$\left(\sigma_x^2 + 4\tau_{zx}^2\right)^{1/2} = S_y$$

$$\left[(142.6F)^2 + 4(76.4F)^2\right]^{1/2} = 209.0F = 81\ 000$$

$$F = 388\ \text{lbf}$$

which is about 7 percent less than found for the DE theory. As stated earlier, the MSS theory is more conservative than the DE theory.

EXAMPLE 6–4

The cantilevered tube shown in Fig. 6–25 is to be made of 2014 aluminum alloy treated to obtain a specified minimum yield strength of 276 MPa. We wish to select a stock-size tube from Table A–8 using a design factor $n_d = 4$. The bending load is $F = 1.75$ kN, the axial tension is $P = 9.0$ kN, and the torsion is $T = 72$ N · m. What is the realized factor of safety?

| **Figure 6–25**

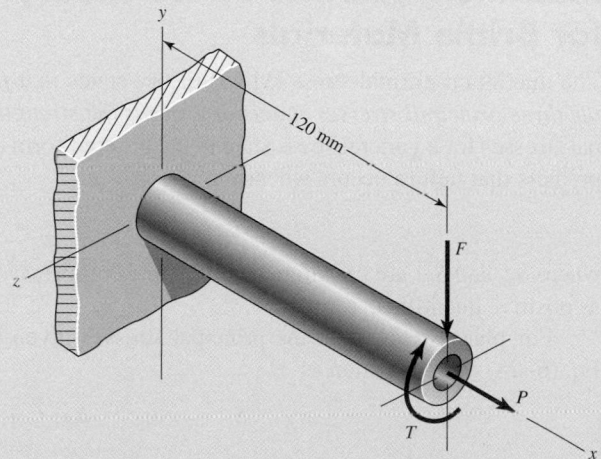

Solution

Since the maximum bending moment is $M = 120F$, the normal stress, for an element on the top surface of the tube at the origin, is

$$\sigma_x = \frac{P}{A} + \frac{Mc}{I} = \frac{9}{A} + \frac{120(1.75)(d_o/2)}{I} = \frac{9}{A} + \frac{105d_o}{I} \qquad (1)$$

where, if millimeters are used for the area properties, the stress is in gigapascals.

The torsional stress at the same point is

$$\tau_{zx} = \frac{Tr}{J} = \frac{72(d_o/2)}{J} = \frac{36d_o}{J} \qquad (2)$$

For accuracy, we choose the distortion-energy theory as the design basis. The von Mises stress, as in the previous example, is

$$\sigma' = \left(\sigma_x^2 + 3\tau_{zx}^2\right)^{1/2} \qquad (3)$$

On the basis of the given design factor, the goal for σ' is

$$\sigma' \leq \frac{S_y}{n_d} = \frac{0.276}{4} = 0.0690 \text{ GPa} \qquad (4)$$

where we have used gigapascals in this relation to agree with Eqs. (1) and (2).

Programming Eqs. (1) to (3) on a spreadsheet and entering metric sizes from Table A–8 reveals that a 42- × 5-mm tube is satisfactory. The von Mises stress is found to be $\sigma' = 0.06043$ GPa for this size. Thus the realized factor of safety is

Answer

$$n = \frac{S_y}{\sigma'} = \frac{0.276}{0.06043} = 4.57$$

For the next size smaller, a 42- × 4-mm tube, $\sigma' = 0.07105$ GPa giving a factor of safety of

$$n = \frac{S_y}{\sigma'} = \frac{0.276}{0.07105} = 3.88$$

6–8 Maximum-Normal-Stress Theory for Brittle Materials

The maximum-normal-stress (MNS) theory states that *failure occurs whenever one of the three principal stresses equals or exceeds the strength.* Again we arrange the principal stresses for a general stress state in the ordered form $\sigma_1 \geq \sigma_2 \geq \sigma_3$. This theory then predicts that failure occurs whenever

$$\sigma_1 \geq S_{ut} \qquad \text{or} \qquad \sigma_3 \leq -S_{uc} \qquad (6\text{–}28)$$

where S_{ut} and S_{uc} are the ultimate tensile and compressive strengths, respectively, given as positive quantities.

For plane stress, with the principal stresses given by Eq. (4–13), with $\sigma_A \geq \sigma_B$, Eq. (6–28) can be written as

$$\sigma_A \geq S_{ut} \qquad \text{or} \qquad \sigma_B \leq -S_{uc} \qquad (6\text{–}29)$$

Figure 6–26

(a) Graph of maximum-normal-stress (MNS) theory of failure for plane stress states. Stress states that plot inside the failure locus are safe. (b) Load line plot.

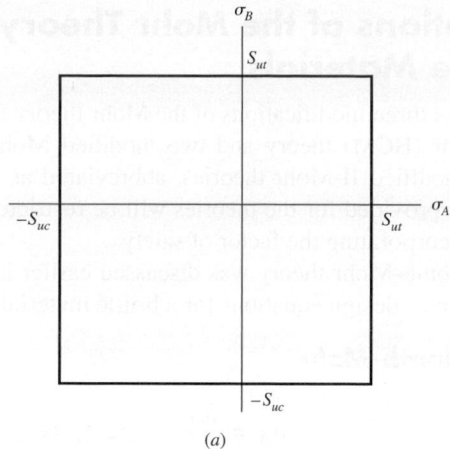

(a)

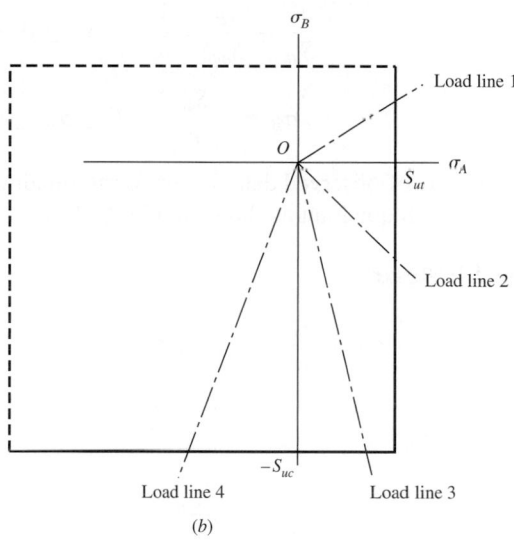

(b)

which is plotted in Fig. 6–26a. As before, the failure criteria equations can be converted to design equations. We can consider two sets of equations for load lines where $\sigma_A \geq \sigma_B$ as

$$\sigma_A = \frac{S_{ut}}{n} \qquad \sigma_A \geq \sigma_B \geq 0 \qquad\qquad\qquad\qquad\qquad \text{Load line 1}$$
$$\sigma_A \geq 0 \geq \sigma_B \quad \text{and} \quad \left| \frac{\sigma_B}{\sigma_A} \right| \leq \frac{S_{uc}}{S_{ut}} \qquad \text{Load line 2} \tag{6–30a}$$

$$\sigma_B = -\frac{S_{uc}}{n} \qquad \sigma_A \geq 0 \geq \sigma_B \quad \text{and} \quad \left| \frac{\sigma_B}{\sigma_A} \right| > \frac{S_{uc}}{S_{ut}} \qquad \text{Load line 3}$$
$$0 \geq \sigma_A \geq \sigma_B \qquad\qquad\qquad\qquad\qquad\qquad \text{Load line 4} \tag{6–30b}$$

where the load lines are shown in Fig. 6–26b.

Before we comment any further on the MNS theory we will explore some modifications to the Mohr theory for brittle materials.

6–9 Modifications of the Mohr Theory for Brittle Materials

We will discuss three modifications of the Mohr theory for brittle materials: the Brittle-Coulomb-Mohr (BCM) theory and two modified Mohr theories called the modified I-Mohr and modified II-Mohr theories, abbreviated as M1M and M2M, respectively. The equations provided for the theories will be restricted to plane stress and be of the design type incorporating the factor of safety.

The Coulomb-Mohr theory was discussed earlier in Sec. 6–6 with Eqs. (6–23) to (6–25). Written as design equations for a brittle material, they are:

Brittle-Coulomb-Mohr

$$\sigma_A = \frac{S_{ut}}{n} \qquad \sigma_A \geq \sigma_B \geq 0 \qquad (6\text{–}31a)$$

$$\frac{\sigma_A}{S_{ut}} - \frac{\sigma_B}{S_{uc}} = \frac{1}{n} \qquad \sigma_A \geq 0 \geq \sigma_B \qquad (6\text{–}31b)$$

$$\sigma_B = -\frac{S_{uc}}{n} \qquad 0 \geq \sigma_A \geq \sigma_B \qquad (6\text{–}31c)$$

On the basis of observed data for the fourth quadrant, the modified I-Mohr theory expands the fourth quadrant as shown in Fig. 6–27.

Modified I-Mohr

$$\sigma_A = \frac{S_{ut}}{n} \qquad \sigma_A \geq \sigma_B \geq 0$$

$$\sigma_A \geq 0 \geq \sigma_B \quad \text{and} \quad \left| \frac{\sigma_B}{\sigma_A} \right| \leq 1 \qquad (6\text{–}32a)$$

Figure 6–27

Biaxial fracture data of gray cast iron compared with various failure criteria. *(Dowling, N. E., Mechanical Behavior of Materials, 2/e, 1999, p. 261. Reprinted by permission of Pearson Education, Inc., Upper Saddle River, New Jersey.)*

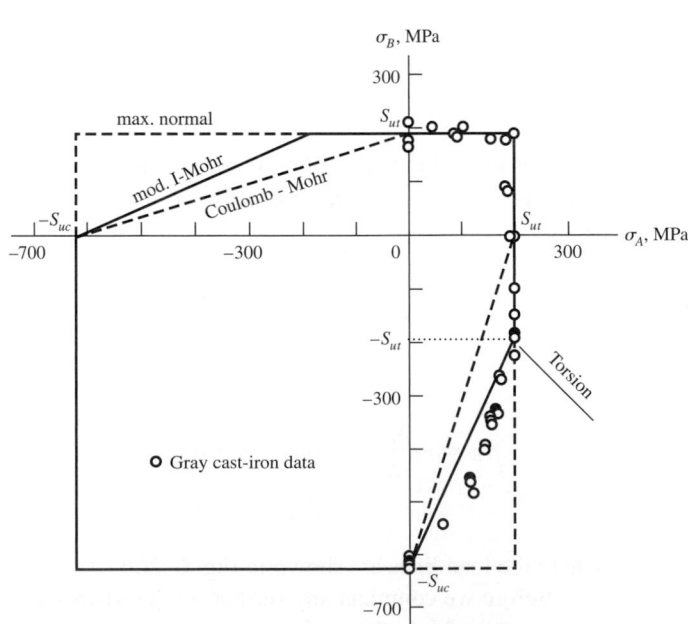

$$\frac{(S_{uc} - S_{ut})\sigma_A}{S_{uc}S_{ut}} - \frac{\sigma_B}{S_{uc}} = \frac{1}{n} \qquad \sigma_A \geq 0 \geq \sigma_B \quad \text{and} \quad \left|\frac{\sigma_B}{\sigma_A}\right| > 1 \qquad (6\text{--}32b)$$

$$\sigma_B = -\frac{S_{uc}}{n} \qquad 0 \geq \sigma_A \geq \sigma_B \qquad (6\text{--}32c)$$

Data are still outside this extended region. The modified II-Mohr theory extends the region in the fourth quadrant where $|\sigma_A/\sigma_B| < 1$ using a parabolic relation. The parabola is in the form $a\sigma_B^2 + b\sigma_B + c\sigma_A = 1$, adjusted such that it is tangent to vertical at $\sigma_A = -\sigma_B = S_{ut}$.

Modified II-Mohr

$$\sigma_A = \frac{S_{ut}}{n} \qquad \sigma_A \geq \sigma_B \geq 0$$

$$\sigma_A \geq 0 \geq \sigma_B \quad \text{and} \quad \left|\frac{\sigma_B}{\sigma_A}\right| \leq 1 \qquad (6\text{--}33a)$$

$$\frac{n\sigma_A}{S_{ut}} + \left(\frac{n\sigma_B + S_{ut}}{S_{ut} - S_{uc}}\right)^2 = 1 \qquad \sigma_A \geq 0 \geq \sigma_B \quad \text{and} \quad \left|\frac{\sigma_B}{\sigma_A}\right| > 1 \qquad (6\text{--}33b)$$

$$\sigma_B = -\frac{S_{uc}}{n} \qquad 0 \geq \sigma_A \geq \sigma_B \qquad (6\text{--}33c)$$

Figure 6.28 shows how the M2M theory conforms to some available data in the fourth quadrant.

Figure 6–28

Fourth quadrant $\sigma_A \sigma_B$ data for a grade 40 cast iron, compared with a modified II-Mohr fracture locus. *(Source of data: Charles F. Walton (ed.), Iron Casting Handbook, Iron Founders' Society, 1971, p. 215, Cleveland, Ohio.)*

σ_A, kpsi	0	0	0	0	18	21	30	32	34	39	40
σ_B, kpsi	−134	−140	−148	−149	−110	−102	−91	−83	−72	−50	−40

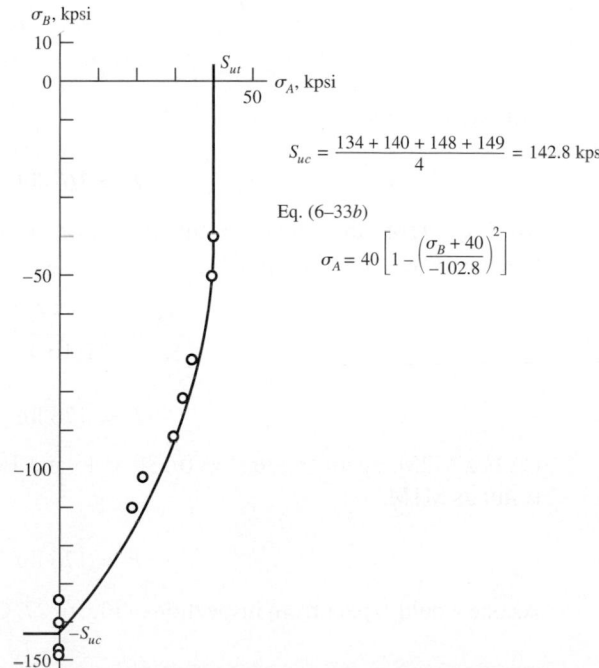

$$S_{uc} = \frac{134 + 140 + 148 + 149}{4} = 142.8 \text{ kpsi}$$

Eq. (6–33b)

$$\sigma_A = 40\left[1 - \left(\frac{\sigma_B + 40}{-102.8}\right)^2\right]$$

EXAMPLE 6–5

Consider the wrench in Ex. 6–3, Fig. 6–24, as made of cast iron, machined to dimension. The force F required to fracture this part can be regarded as the strength of the component part. If the material is ASTM grade 30 cast iron, find the force F with

(a) Coulomb-Mohr failure model.

(b) Mod. I-Mohr failure model.

(c) Mod. II-Mohr failure model.

Solution

We assume that the lever DC is strong enough, and not part of the problem. Since grade 30 cast iron is a brittle material *and* cast iron, the stress-concentration factors K_t and K_{ts} are set to unity. From Table A–24, the tensile ultimate strength is 31 kpsi and the compressive ultimate strength is 109 kpsi. The stress element at A on the top surface will be subjected to a tensile bending stress and a torsional stress. This location, on the 1-in-diameter section fillet, is the weakest location, and it governs the strength of the assembly. The normal stress σ_x and the shear stress at A are given by

$$\sigma_x = K_t \frac{M}{I/c} = K_t \frac{32M}{\pi d^3} = (1)\frac{32(14F)}{\pi(1)^3} = 142.6F$$

$$\tau_{xy} = K_{ts} \frac{Tr}{J} = K_{ts}\frac{16T}{\pi d^3} = (1)\frac{16(15F)}{\pi(1)^3} = 76.4F$$

From Eq. (4–13) the nonzero principal stresses σ_A and σ_B are

$$\sigma_A, \sigma_B = \frac{142.6F + 0}{2} \pm \sqrt{\left(\frac{142.6F - 0}{2}\right)^2 + (76.4F)^2} = 175.8F, -33.2F$$

This puts us in the fourth-quadrant of the σ_A, σ_B plane.

(a) For BCM, Eq. (6–31b) applies with $n = 1$ for failure.

$$\frac{\sigma_A}{S_{ut}} - \frac{\sigma_B}{S_{uc}} = \frac{175.8F}{31(10^3)} - \frac{(-33.2F)}{109(10^3)} = 1$$

Solving for F yields

Answer

$$F = 167 \text{ lbf}$$

(b) For M1M, the slope of the load line is $|\sigma_B/\sigma_A| = 33.2/175.8 = 0.189 < 1$. Obviously, Eq. (6–32a) applies.

$$\frac{\sigma_A}{S_{ut}} = \frac{175.8F}{31(10^3)} = 1$$

Answer

$$F = 176 \text{ lbf}$$

(c) For M2M, again $|\sigma_B/\sigma_A| = 0.189 < 1$, and Eq. (6–33a) applies, giving the same result as M1M:

Answer

$$F = 176 \text{ lbf}$$

As one would expect from inspection of Fig. 6–27, Coulomb-Mohr is more conservative.

Figure 6–29

A plot of experimental data points obtained from tests on cast iron. Shown also are the graphs of three failure theories of possible usefulness for brittle materials. Note points A, B, C, and D. To avoid congestion in the first quadrant, points have been plotted for $\sigma_A > \sigma_B$ as well as for the opposite sense. *(Source of data: Charles F. Walton (ed.),* Iron Castings Handbook, *Iron Founders' Society, 1971, pp. 215, 216, Cleveland, Ohio.)*

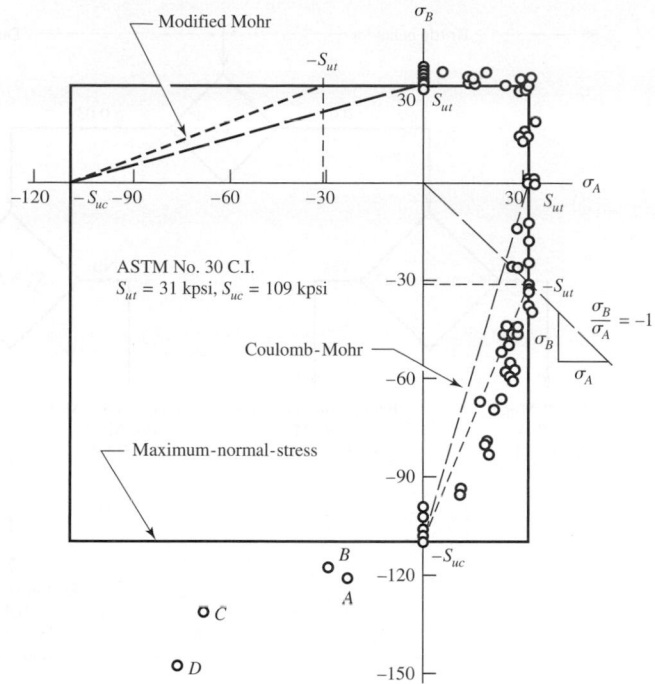

6–10 Failure of Brittle Materials Summary

We have identified failure or strength of brittle materials that conform to the usual meaning of the word *brittle,* relating to those materials whose true strain at fracture is 0.05 or less. We also have to be aware of normally ductile materials that for some reason may develop a brittle fracture or crack if used below the transition temperature. Figure 6–29 shows data for a nominal grade 30 cast iron taken under biaxial stress conditions, with several brittle failure hypotheses shown, superposed. We note the following:

- In the first quadrant the data appear on both sides and along the failure locus of maximum-normal-stress, Coulomb-Mohr, modified I-Mohr, and modified II-Mohr. All failure loci are the same, and data fit well.

- In the fourth quadrant only, the modified II-Mohr hypothesis produces a failure locus that lies in and among the data.

- In the third quadrant the points A, B, C, and D are too few to make any suggestion concerning a fracture locus.

- It is clear that unless the designer is prepared to do considerable testing, he or she should avoid third-quadrant loadings on a contemplated concept. Changing the concept to stay in the first or fourth quadrant is the less costly path.

The pressure to get things done leads engineers to employ loadings that are understood quantitatively. At any point in time, know what we can and cannot do, and act accordingly.

6–11 Selection of Failure Criteria

For ductile behavior the preferred criterion is the distortion-energy theory, although some designers also apply the maximum-shear-stress theory because of its simplicity and conservative nature.

Figure 6–30

Failure theory selection flowchart.

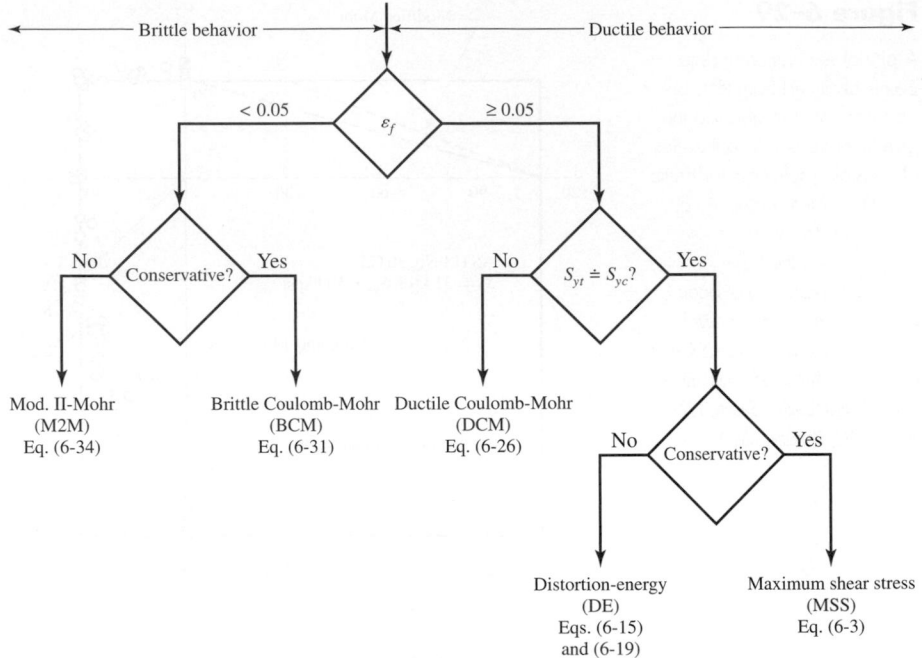

For brittle behavior, the original Mohr hypothesis, constructed with tensile, compression, and torsion tests, with a curved failure locus is the best hypothesis we have. However, the difficulty of applying it without a computer leads engineers to choose modifications, namely, Coulomb Mohr, modified I-Mohr, or modified II-Mohr. Figure 6–30 provides a summary flowchart for the selection of an effective procedure for analyzing or predicting failures from static loading for brittle or ductile behavior.

6–12 Static or Quasi-Static Loading on a Shaft

The fundamental kinematic component of our mechanical universe is the wheel and axle. An essential part of this revolute joint is the shaft. It is a good example of a static, quasi-static, and dynamically loaded body. Inasmuch as this chapter addresses static and quasi-static loading, application of the information developed in this chapter to shafts is useful and necessary.

The stress at an element located on the surface of a solid round shaft of diameter d subjected to bending, axial loading, and twisting is

$$\sigma_x = \frac{32M}{\pi d^3} + \frac{4F}{\pi d^2} \tag{a}$$

$$\tau_{xy} = \frac{16T}{\pi d^3} \tag{b}$$

where the axial component of the normal stress σ_x may be additive or subtractive. We observe that the three loadings M, F, and T occur at a section containing the specific surface element under scrutiny.

By use of Mohr's circle it can be shown that the two *nonzero* principal stresses σ_A and σ_B are

$$\sigma_A, \sigma_B = \frac{\sigma_x}{2} \pm \left[\left(\frac{\sigma_x}{2} \right)^2 + \tau_{xy}^2 \right]^{1/2} \tag{6–34}$$

These principal stresses can be combined to obtain the von Mises stress σ' as

$$\sigma' = \left(\sigma_A^2 - \sigma_A\sigma_B + \sigma_B^2\right)^{1/2} = \left(\sigma_x^2 + 3\tau_{xy}^2\right)^{1/2} \tag{6-35}$$

Should the maximum-shear theory be useful,

$$\tau_{max} = \frac{\sigma_A - \sigma_B}{2} = \frac{1}{2}\left(\sigma_x^2 + 4\tau_{xy}^2\right)^{1/2} \tag{6-36}$$

By substituting Eqs. (*a*) and (*b*) into Eqs. (6–34) and (6–35), we obtain

$$\sigma' = \frac{4}{\pi d^3}[(8M + Fd)^2 + 48T^2]^{1/2} \tag{6-37}$$

for von Mises stress, and using Eq. (6–36) for the maximum-shear-stress approximation, we obtain

$$\tau_{max} = \frac{2}{\pi d^3}[(8M + Fd)^2 + 64T^2]^{1/2} \tag{6-38}$$

Equations (6–37) and (6–38) permit an estimate of σ' or τ_{max} when diameter d is given, or an estimate of d when an allowable value of σ' or τ_{max} is given. For a design factor of n_d the distortion-energy theory of ductile failure gives an allowable stress of

$$\sigma'_{all} = \frac{S_y}{n_d} \tag{6-39}$$

For a design factor of n_d the maximum-shear hypothesis of ductile failure gives an allowable shear stress τ_{all} of

$$\tau_{all} = \frac{S_{sy}}{n_d} = \frac{S_y}{2n_d} \tag{6-40}$$

Static or Quasi-Static Loading of a Shaft—Bending and Torsion

Under many conditions, the axial force F in Eqs. (6–37) and (6–38) is either zero or so small that its effect may be neglected. With $F = 0$, Eqs. (6–37) and (6–38) become

$$\sigma' = \frac{16}{\pi d^3}(4M^2 + 3T^2)^{1/2} \tag{6-41}$$

$$\tau_{max} = \frac{16}{\pi d^3}(M^2 + T^2)^{1/2} \tag{6-42}$$

By substituting the allowable stresses from Eqs. (6–39) and (6–40) we find, for the von Mises theory of failure,

$$d = \left[\frac{16n}{\pi S_y}(4M^2 + 3T^2)^{1/2}\right]^{1/3} \tag{6-43}$$

$$\frac{1}{n} = \frac{16}{\pi d^3 S_y}(4M^2 + 3T^2)^{1/2} \tag{6-44}$$

and for the maximum-shear-stress approximation

$$d = \left[\frac{32n}{\pi S_y}(M^2 + T^2)^{1/2}\right]^{1/3} \tag{6–45}$$

$$\frac{1}{n} = \frac{32}{\pi d^3 S_y}(M^2 + T^2)^{1/2} \tag{6–46}$$

EXAMPLE 6–6

The integral pinion shaft shown in Fig. 6–31a is to be mounted in bearings at the locations shown and is to have a gear (not shown) mounted on the right-hand overhanging end. The loading diagram (Fig. 6–31b) shows the pinion force at A and the gear force at C are in the same plane. Equal and opposite torques T_A and T_B are represented as concentrated at A and C, as are the forces. The bending moment diagram of Fig. 6–31c shows an extreme at A and at B. The smaller diameter at B makes the location B the critical location at the center of the right-hand bearing. Since the shaft is used at intermittent emergencies, its use will not exceed 1000 revolutions at full load, so the problem can be treated as quasi-static. The material is a heat-treated carbon steel with a mean yield strength of 66 kpsi. In this circumstance the project engineer decides to use a design factor of 1.80. What is the smallest right-hand bearing journal diameter determined from distortion-energy theory and by a maximum-shear-stress approximation?

Figure 6–31

Left-hand-bearing diameter is 1.000 in, left-bearing shoulder diameter is 2.000 in, dedendum diameter of gear is 3.43 in; right-hand bearing shoulder diameter is 2.000 in; the overhang shoulder is $1\frac{1}{8}$ in by $\frac{1}{4}$ in long; diameter of overhanging gear seat is 1.000 in.

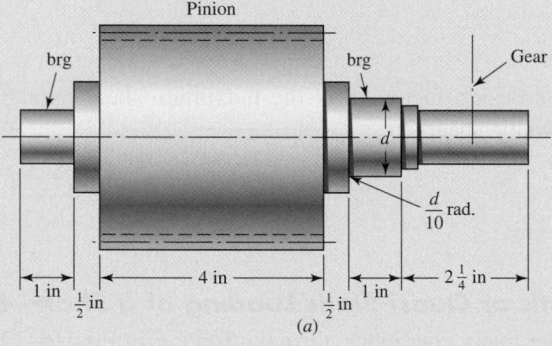

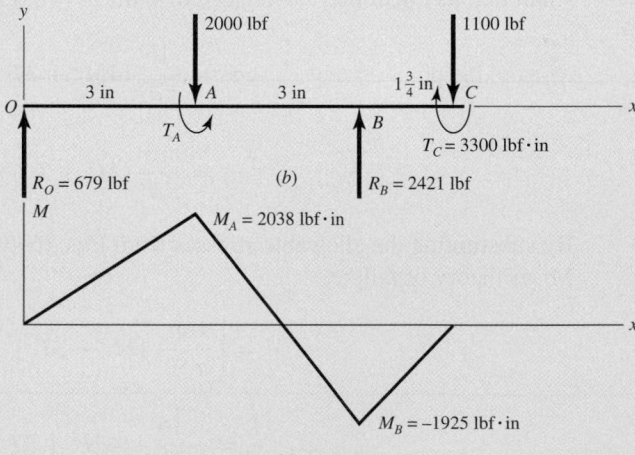

Solution   From Eq. (6–43), for DE,

$$d = \left\{ \frac{16(1.80)}{\pi 66\,000} \left[4(-1925)^2 + 3(3300)^2 \right]^{1/2} \right\}^{1/3} = 0.986 \text{ in}$$

From Eq. (6–45), for MSS,

$$d = \left\{ \frac{32(1.80)}{\pi 66\,000} \left[(-1925)^2 + 3300^2 \right]^{1/2} \right\}^{1/3} = 1.02 \text{ in}$$

6–13   Introduction to Fracture Mechanics

The idea that cracks exist in parts even before service begins, and that cracks can grow during service, has led to the descriptive phrase "damage-tolerant design." The focus of this philosophy is on crack growth until it becomes critical, and the part is removed from service. The analysis tool is *linear elastic fracture mechanics* (LEFM). Inspection and maintenance are essential in the decision to retire parts before cracks reach catastrophic size. Where human safety is concerned, periodic inspections for cracks are mandated by codes and government ordinance.

We shall now briefly examine some of the basic ideas and vocabulary needed for the potential of the approach to be appreciated. The intent here is to make the reader aware of the dangers associated with the sudden brittle fracture of so-called ductile materials. The topic is much too extensive to include in detail here and the reader is urged to read further on this complex subject.[8]

The use of elastic stress-concentration factors provides an indication of the average load required on a part for the onset of plastic deformation, or yielding; these factors are also useful for analysis of the loads on a part that will cause fatigue fracture. However, stress-concentration factors are limited to structures for which all dimensions are precisely known, particularly the radius of curvature in regions of high stress concentration. When there exists a crack, flaw, inclusion, or defect of unknown small radius in a part, the elastic stress-concentration factor approaches infinity as the root radius approaches zero, thus rendering the stress-concentration factor approach useless. Furthermore, even if the radius of curvature of the flaw tip is known, the high local stresses there will lead to local plastic deformation surrounded by a region of elastic deformation. Elastic stress-concentration factors are no longer valid for this situation, so analysis from the point of view of stress-concentration factors does not lead to criteria useful for design when very sharp cracks are present.

By combining analysis of the gross elastic changes in a structure or part that occur as a sharp brittle crack grows with measurements of the energy required to produce new

---

[8]References on brittle fracture include:

H. Tada and P. C. Paris, *The Stress Analysis of Cracks Handbook,* 2nd ed., Paris Productions, St. Louis, 1985.

D. Broek, *Elementary Engineering Fracture Mechanics,* 4th ed., Martinus Nijhoff, London, 1985.

D. Broek, *The Practical Use of Fracture Mechanics,* Kluwar Academic Pub., London, 1988.

David K. Felbeck and Anthony G. Atkins, *Strength and Fracture of Engineering Solids,* Prentice-Hall, Englewood Cliffs, N.J., 1984.

Kåre Hellan, *Introduction to Fracture Mechanics,* McGraw-Hill, New York, 1984.

fracture surfaces, it is possible to calculate the average stress (if no crack were present) that will cause crack growth in a part. Such calculation is possible only for parts with cracks for which the elastic analysis has been completed, and for materials that crack in a relatively brittle manner and for which the fracture energy has been carefully measured. The term *relatively brittle* is rigorously defined in the test procedures,[9] but it means, roughly, *fracture without yielding occurring throughout the fractured cross section*.

Thus glass, hard steels, strong aluminum alloys, and even low-carbon steel below the ductile-to-brittle transition temperature can be analyzed in this way. Fortunately, ductile materials blunt sharp cracks, as we have previously discovered, so that fracture occurs at average stresses of the order of the yield strength, and the designer is prepared for this condition. The middle ground of materials that lie between "relatively brittle" and "ductile" is now being actively analyzed, but exact design criteria for these materials are not yet available.

Quasi-Static Fracture

Many of us have had the experience of observing brittle fracture, whether it is the breaking of a cast-iron specimen in a tensile test or the twist fracture of a piece of blackboard chalk. It happens so rapidly that we think of it as instantaneous, that is, the cross section simply parting. Fewer of us have skated on a frozen pond in the spring, with no one near us, heard a cracking noise, and stopped to observe. The noise is due to cracking. The cracks move slowly enough for us to see them run. The phenomenon is not instantaneous, since some time is necessary to feed the crack energy from the stress field to the crack for propagation. Quantifying these things is important to understanding the phenomenon "in the small." In the large, a static crack may be stable and will not propagate. Some level of loading can render the crack unstable, and the crack propagates to fracture.

The foundation of fracture mechanics was first established by Griffith in 1921 using the stress field calculations for an elliptical flaw in a plate developed by Inglis in 1913. For the infinite plate loaded by an applied uniaxial stress σ in Fig. 6–32, the maximum stress occurs at $(\pm a, 0)$ and is given by

$$(\sigma_y)_{\max} = \left(1 + 2\frac{a}{b}\right)\sigma \tag{6–47}$$

Note that when $a = b$, the ellipse becomes a circle and Eq. (6–47) gives a stress concentration factor of 3. This agrees with the well-known result for an infinite plate with a

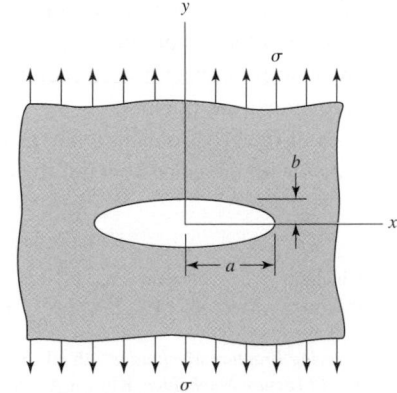

| Figure 6–32

(From Richard G. Budynas, Advanced Strength and Applied Stress Analysis, 2nd ed., McGraw-Hill, New York, 1999. Copyright © 1999 The McGraw-Hill Companies.)

[9]BS 5447:1977 and ASTM E399-78.

circular hole (see Table A–15–1). For a fine crack, $b/a \to 0$, and Eq. (6–47) predicts that $(\sigma_y)_{max} \to \infty$. However, on a microscopic level, an infinitely sharp crack is a hypothetical abstraction that is physically impossible, and when plastic deformation occurs, the stress will be finite at the crack tip.

Griffith showed that the crack growth occurs when the energy release rate from applied loading is greater than the rate of energy for crack growth. Crack growth can be stable or unstable. Unstable crack growth occurs when the *rate* of change of the energy release rate relative to the crack length is equal to or greater than the *rate* of change of the crack growth rate of energy. Griffith's experimental work was restricted to brittle materials, namely glass, which pretty much confirmed his surface energy hypothesis. However, for ductile materials, the energy needed to perform plastic work at the crack tip is found to be much more crucial than surface energy.

Crack Modes and the Stress Intensity Factor

Three distinct modes of crack propagation exist, as shown in Fig. 6–33. A tensile stress field gives rise to mode I, the *opening crack propagation mode,* as shown in Fig. 6–33a. This mode is the most common in practice. Mode II is the *sliding mode,* is due to in-plane shear, and can be seen in Fig. 6–33b. Mode III is the *tearing mode,* which arises from out-of-plane shear, as shown in Fig. 6–33c. Combinations of these modes can also occur. Since mode I is the most common and important mode, the remainder of this section will consider only this mode.

Consider a mode I crack of length $2a$ in the infinite plate of Fig. 6–34. By using complex stress functions, it has been shown that the stress field on a $dx\,dy$ element in the vicinity of the crack tip is given by

$$\sigma_x = \sigma \sqrt{\frac{a}{2r}} \cos \frac{\theta}{2} \left(1 - \sin \frac{\theta}{2} \sin \frac{3\theta}{2} \right) \tag{6–48a}$$

$$\sigma_y = \sigma \sqrt{\frac{a}{2r}} \cos \frac{\theta}{2} \left(1 + \sin \frac{\theta}{2} \sin \frac{3\theta}{2} \right) \tag{6–48b}$$

$$\tau_{xy} = \sigma \sqrt{\frac{a}{2r}} \sin \frac{\theta}{2} \cos \frac{\theta}{2} \cos \frac{3\theta}{2} \tag{6–48c}$$

$$\sigma_z = \begin{cases} 0 & \text{(for plane stress)} \\ v(\sigma_x + \sigma_y) & \text{(for plane strain)} \end{cases} \tag{6–48d}$$

Figure 6–33

Crack propagation modes.

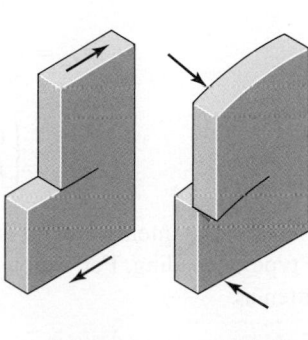

Mode I   Mode II   Mode III

Figure 6–34

Mode I crack model. *(From Richard G. Budynas,* Advanced Strength and Applied Stress Analysis, *2nd ed., McGraw-Hill, New York, 1999. Copyright © 1999 The McGraw-Hill Companies.)*

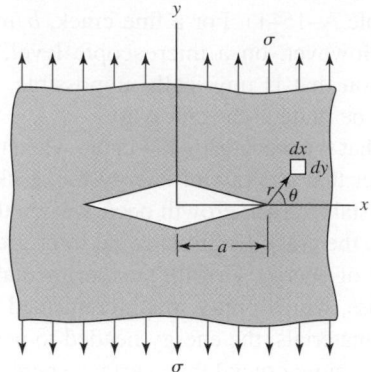

The stress σ_y near the tip, with $\theta = 0$, is

$$\sigma_y|_{\theta=0} = \sigma\sqrt{\frac{a}{2r}} \qquad (a)$$

As with the elliptical crack, we see that $\sigma_y|_{\theta=0} \to \infty$ as $r \to 0$, and again the concept of an infinite stress concentration at the crack tip is inappropriate. The quantity $\sigma_y|_{\theta=0}\sqrt{2r} = \sigma\sqrt{a}$, however, does remain constant as $r \to 0$. It is common practice to define a factor K called the *stress intensity factor* given by

$$K = \sigma\sqrt{\pi a} \qquad (b)$$

where the units are MPa$\sqrt{\text{m}}$ or kpsi$\sqrt{\text{in}}$. Since we are dealing with a mode I crack, Eq. (*b*) is written as

$$K_I = \sigma\sqrt{\pi a} \qquad (6\text{–}49)$$

The stress intensity factor is *not* to be confused with the static stress concentration factors K_t and K_{ts} defined in Secs. 4–14 and 6–2.

Thus Eqs. (6–48) can be rewritten as

$$\sigma_x = \frac{K_I}{\sqrt{2\pi r}}\cos\frac{\theta}{2}\left(1 - \sin\frac{\theta}{2}\sin\frac{3\theta}{2}\right) \qquad (6\text{–}50a)$$

$$\sigma_y = \frac{K_I}{\sqrt{2\pi r}}\cos\frac{\theta}{2}\left(1 + \sin\frac{\theta}{2}\sin\frac{3\theta}{2}\right) \qquad (6\text{–}50b)$$

$$\tau_{xy} = \frac{K_I}{\sqrt{2\pi r}}\sin\frac{\theta}{2}\cos\frac{\theta}{2}\cos\frac{3\theta}{2} \qquad (6\text{–}50c)$$

$$\sigma_z = \begin{cases} 0 & \text{(for plane stress)} \\ \nu(\sigma_x + \sigma_y) & \text{(for plane strain)} \end{cases} \qquad (6\text{–}50d)$$

The stress intensity factor is a function of geometry, size and shape of the crack, and the type of loading. For various load and geometric configurations, Eq. (6–49) can be written as

$$K_I = \beta\sigma\sqrt{\pi a} \qquad (6\text{–}51)$$

where β is the *stress intensity modification factor*. Tables for β are available in the literature for basic configurations.[10] Figures 6–35 to 6–40 present a few examples of β for mode I crack propagation.

Fracture Toughness

When the magnitude of the mode I stress intensity factor reaches a critical value, K_{Ic} crack propagation initiates. The *critical stress intensity factor K_{Ic}* is a material property that depends on the material, crack mode, processing of the material, temperature, loading rate, and the state of stress at the crack site (such as plane stress versus plane strain). The critical stress intensity factor K_{Ic} is also called the *fracture toughness* of the material. The fracture toughness for plane strain is normally lower than that for plane stress. For this reason, the term K_{Ic} is typically defined as the *mode I, plane strain fracture toughness*. Fracture toughness K_{Ic} for engineering metals lies in the range $20 \leq K_{Ic} \leq 200$ MPa $\cdot \sqrt{m}$; for engineering polymers and ceramics, $1 \leq K_{Ic} \leq 5$ MPa $\cdot \sqrt{m}$. For a 4340 steel, where the yield strength due to heat treatment ranges from 800 to 1600 MPa, K_{Ic} *decreases* from 190 to 40 MPa $\cdot \sqrt{m}$.

Figure 6–35

Off-center crack in a plate in longitudinal tension; solid curves are for the crack tip at A; dashed curves are for the tip at B.

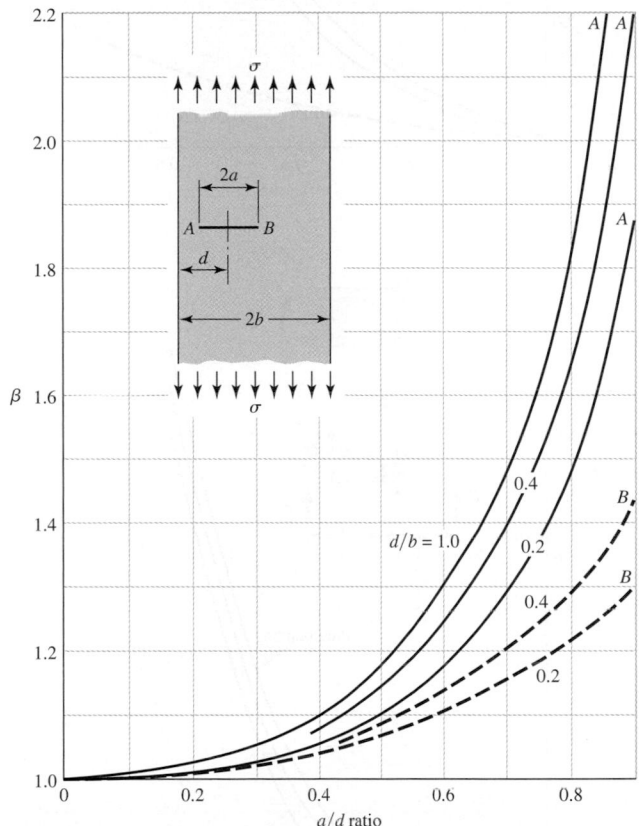

[10]See, for example:

H. Tada and P. C. Paris, *The Stress Analysis of Cracks Handbook,* 2nd ed., Paris Productions, St. Louis, 1985.

G. C. Sib, *Handbook of Stress Intensity Factors for Researchers and Engineers,* Institute of Fracture and Solid Mechanics, Lehigh University, Bethlehem, Pa., 1973.

Y. Murakami, ed., *Stress Intensity Factors Handbook,* Pergamon Press, Oxford, U.K., 1987.

W. D. Pilkey, *Formulas for Stress, Strain, and Structural Matrices,* John Wiley & Sons, New York, 1994.

Figure 6–36

Plate loaded in longitudinal
tension with a crack at the
edge; for the solid curve there
are no constraints to bending;
the dashed curve was
obtained with bending
constraints added.

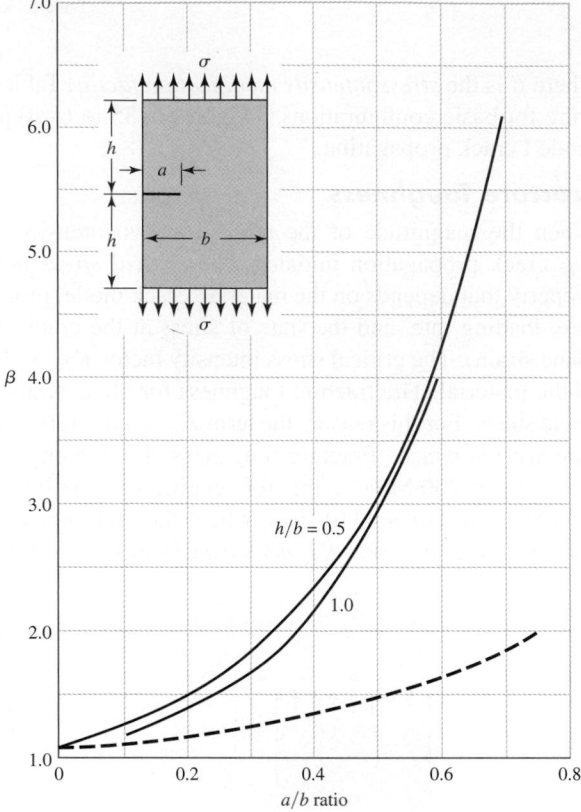

Figure 6–37

Beams of rectangular cross
section having an edge crack.

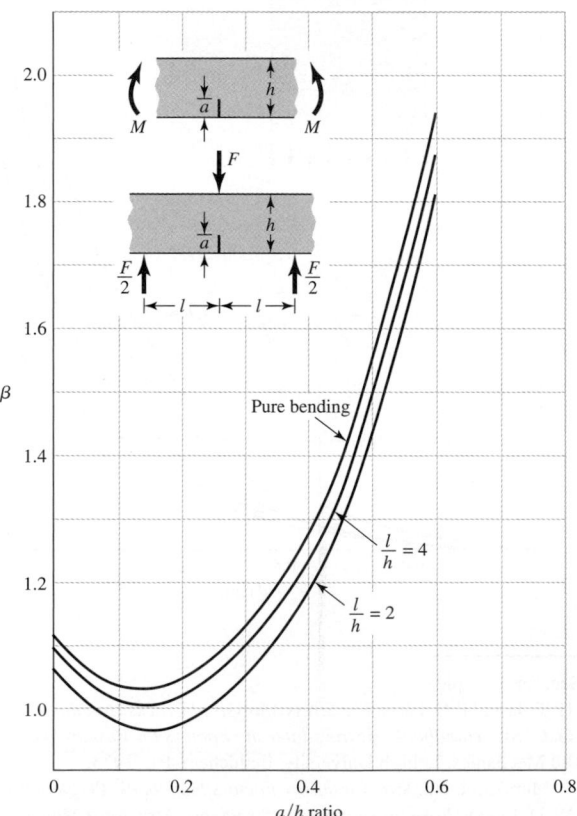

Figure 6–38

Plate in tension containing a circular hole with two cracks.

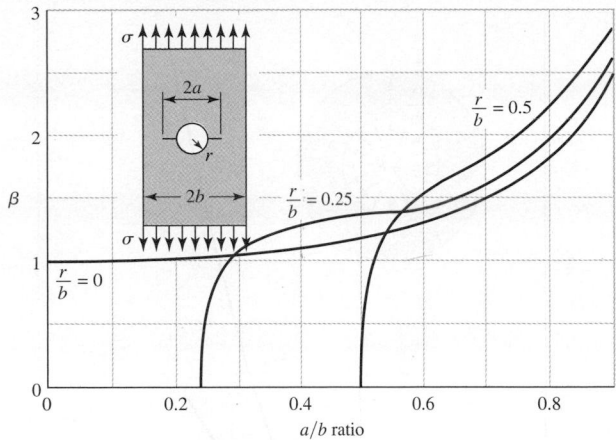

Figure 6–39

A cylinder loading in axial tension having a radial crack of depth a extending completely around the circumference of the cylinder.

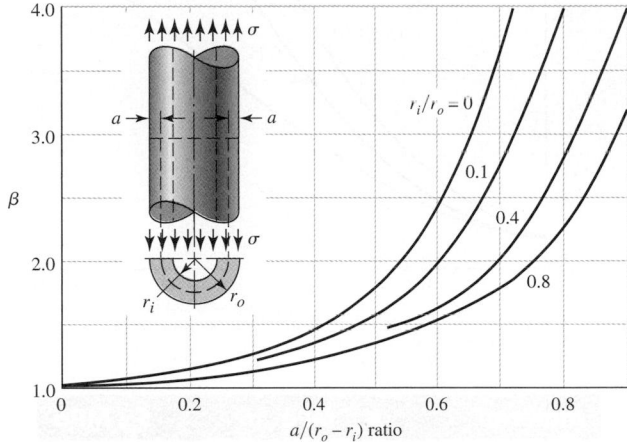

Table 6–1 gives some approximate typical room-temperature values of K_{Ic} for several materials. As previously noted, the fracture toughness depends on many factors and the table is meant only to convey some typical magnitudes of K_{Ic}. For an actual application, it is recommended that the material specified for the application be certified using standard test procedures [see the American Society for Testing and Materials (ASTM) standard E399].

One of the first problems facing the designer is that of deciding whether the conditions exist, or not, for a brittle fracture. Low-temperature operation, that is, operation below room temperature, is a key indicator that brittle fracture is a possible failure mode. Tables of transition temperatures for various materials have not been published, possibly because of the wide variation in values, even for a single material. Thus, in many situations, laboratory testing may give the only clue to the possibility of a brittle fracture. Another key indicator of the possibility of fracture is the ratio of the yield strength to the ultimate strength. A high ratio of S_y/S_u indicates there is only a small ability to absorb energy in the plastic region and hence there is a likelihood of brittle fracture.

The strength-to-stress ratio K_{Ic}/K_I can be used as a factor of safety as

$$n = \frac{K_{Ic}}{K_I}$$

(6–52)

Figure 6–40

Cylinder subjected to internal pressure p, having a radial crack in the longitudinal direction of depth a. Use Eq. (4–51) for the tangential stress at $r = r_0$.

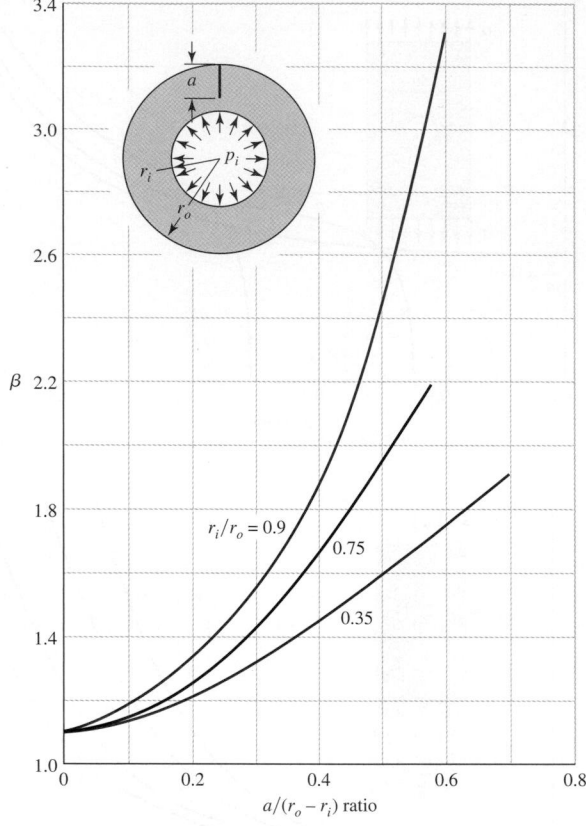

Table 6–1

Values of K_{Ic} for Some Engineering Materials at Room Temperature

Material	K_{Ic}, MPa$\sqrt{m}$	S_y, MPa
Aluminum		
2024	26	455
7075	24	495
7178	33	490
Titanium		
Ti-6AL-4V	115	910
Ti-6AL-4V	55	1035
Steel		
4340	99	860
4340	60	1515
52100	14	2070

EXAMPLE 6–7   A steel ship deck plate is 30 mm thick and 12 m wide. It is loaded with a nominal uni-axial tensile stress of 50 MPa. It is operated below its ductile-to-brittle transition temperature with K_{Ic} equal to 28.3 MPa. If a 65-mm-long central transverse crack is present, estimate the tensile stress at which catastrophic failure will occur. Compare this stress with the yield strength of 240 MPa for this steel.

Solution   For Fig. 6–35, with $d = b$, $2a = 65$ mm and $2b = 12$ m, so that $d/b = 1$ and $a/d = 65/12(10^3) = 0.005$. Since a/d is so small, $\beta = 1$, so that

$$K_I = \sigma \sqrt{\pi a} = 50\sqrt{\pi(32.5 \times 10^{-3})} = 16.0 \text{ MPa } \sqrt{\text{m}}$$

From Eq. (6–52),

$$n = \frac{K_{Ic}}{K_I} = \frac{28.3}{16.0} = 1.77$$

The stress at which catastrophic failure occurs is

Answer

$$\sigma_c = \frac{K_{Ic}}{K_I}\sigma = \frac{28.3}{16.0}(50) = 88.4 \text{ MPa}$$

The yield strength is 240 MPa, and catastrophic failure occurs at $88.4/240 = 0.37$, or at 37 percent of yield. The factor of safety in this circumstance is $K_{Ic}/K_I = 28.3/16 = 1.77$ and *not* $240/50 = 4.8$.

EXAMPLE 6–8   A plate of width 1.4 m and length 2.8 m is required to support a tensile force in the 2.8-m direction of 4.0 MN. Inspection procedures will detect only through-thickness edge cracks larger than 2.7 mm. The two Ti-6AL-4V alloys in Table 6–1 are being considered for this application, for which the safety factor must be 1.3 and minimum weight is important. Which alloy should be used?

Solution   (*a*) We elect first to estimate the thickness required to resist yielding. Since $\sigma = P/wt$, we have $t = P/w\sigma$. But

$$\sigma_{\text{all}} = \frac{S_y}{n} = \frac{910}{1.3} = 700 \text{ MPa}$$

Thus

$$t = \frac{P}{w\sigma_{\text{all}}} = \frac{4.0(10)^3}{1.4(700)} = 4.08 \text{ mm or greater}$$

where we have $S_y = 910$ MPa for the weaker titanium alloy. For the stronger alloy, we have, from Table 6–1,

$$\sigma_{\text{all}} = \frac{1035}{1.3} = 796 \text{ MPa}$$

and so the thickness is

Answer

$$t = \frac{P}{w\sigma_{\text{all}}} = \frac{4.0(10)^3}{1.4(796)} = 3.59 \text{ mm or greater}$$

(*b*) Now let us find the thickness required to prevent crack growth. Using Fig. 6–36, we have

$$\frac{h}{b} = \frac{2.8/2}{1.4} = 1 \qquad \frac{a}{b} = \frac{2.7}{1.4(10^3)} = 0.001\ 93$$

Corresponding to these ratios we find from Fig. 6–36 that $\beta \doteq 1.1$, and $K_I = 1.1\sigma\sqrt{\pi a}$. From Table 6–1, $K_{Ic} = 115$ MPa $\sqrt{\text{m}}$ for the weaker of the two alloys:

$$n = \frac{K_{Ic}}{K_I} = \frac{115\sqrt{10^3}}{1.1\sigma\sqrt{\pi a}}, \qquad \sigma = \frac{K_{Ic}}{n1.1\sqrt{\pi a}}$$

Solving for σ with $n = 1$ gives the fracture stress

$$\sigma = \frac{115}{1.1\sqrt{\pi(2.7 \times 10^{-3})}} = 1135 \text{ MPa}$$

which is greater than the yield strength of 910 MPa, and so yield strength is the basis for the geometry decision. For the stronger alloy $S_y = 1035$ MPa, with $n = 1$ the fracture stress is

$$\sigma = \frac{K_{Ic}}{nK_I} = \frac{55}{1(1.1)\sqrt{\pi(2.7 \times 10^{-3})}} = 542.9 \text{ MPa}$$

which is less than the yield strength. The thickness t is

$$t = \frac{P}{w\sigma_{\text{all}}} = \frac{4.0(10^3)}{1.4(542.9/1.3)} = 6.84 \text{ mm or greater}$$

This example shows that the fracture toughness K_{Ic} limits the geometry when the stronger alloy is used, and so a thickness of 6.84 mm or larger is required. When the weaker alloy is used the geometry is limited by the yield strength, giving a thickness of only 4.08 mm or greater. Thus the weaker alloy leads to a thinner and lighter weight choice since the failure modes differ.

6–14 Stochastic Analysis

Reliability is the probability that machine systems and components will perform their intended function satisfactorily without failure. Up to this point, discussion in this chapter has been restricted to deterministic relations between static stress, strength, and the design factor. Stress and strength, however, are statistical in nature and very much tied to the reliability of the stressed component. Consider the probability density functions for stress and strength, $\boldsymbol{\sigma}$ and $\mathbf{S}$, shown in Fig. 6–41*a*. The mean values of stress and strength are μ_σ and μ_S, respectively. Here, the "average" factor of safety is

$$\bar{n} = \frac{\mu_S}{\mu_\sigma} \qquad (a)$$

The margin of safety for any value of stress σ and strength S is defined as

$$m = S - \sigma \qquad (b)$$

The average part will have a margin of safety of $\bar{m} = \mu_S - \mu_\sigma$. However, for the overlap of the distributions shown by the shaded area in Fig. 6–41*a*, the stress exceeds the strength, the margin of safety is negative, and these parts are expected to fail. This shaded area is called the *interference* of $\boldsymbol{\sigma}$ and $\mathbf{S}$.

Figure 6–41

Plot of density functions showing how the interference of **S** and **σ** is used to obtain the stress margin **m**. (a) Stress and strength distributions. (b) Distribution of interference; the reliability R is the area of the density function for **m** greater than zero; the interference is the area (1 − R).

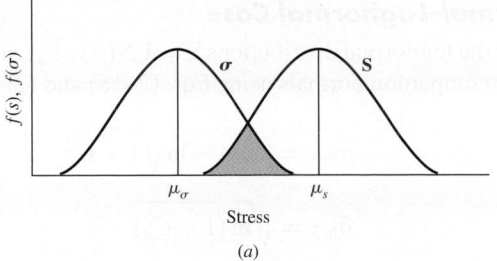

Stress

(a)

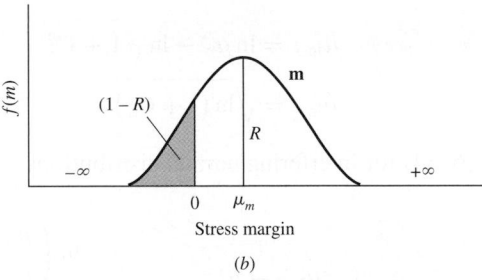

Stress margin

(b)

Figure 6–41b shows the distribution of m, which obviously depends on the distributions of stress and strength. The reliability that a part will perform without failure, R, is the area of the margin of safety distribution for m > 0. The interference is the area 1 − R where parts are expected to fail. We next consider some typical cases involving stress-strength interference.

Normal-Normal Case

Consider the normal distributions, $\mathbf{S} = \mathbf{N}(\mu_S,\ \hat{\sigma}_S)$ and $\sigma = \mathbf{N}(\mu_\sigma,\ \hat{\sigma}_\sigma)$. The stress margin is $\mathbf{m} = \mathbf{S} - \boldsymbol{\sigma}$, and will be normally distributed because the addition or subtraction of normals is normal. Thus $\mathbf{m} = \mathbf{N}(\mu_m, \hat{\sigma}_m)$. Reliability is the probability p that m > 0. That is,

$$R = p(S > \sigma) = p(S - \sigma > 0) = p(m > 0) \tag{6–53}$$

To find the chance that m > 0 we form the z variable of **m** and substitute m = 0. Noting that $\mu_m = \mu_S - \mu_\sigma$ and $\hat{\sigma}_m = (\hat{\sigma}_S^2 + \hat{\sigma}_\sigma^2)^{1/2}$, we write

$$z = \frac{m - \mu_m}{\hat{\sigma}_m} = \frac{0 - \mu_m}{\hat{\sigma}_m} = -\frac{\mu_m}{\hat{\sigma}_m} = -\frac{\mu_S - \mu_\sigma}{\left(\hat{\sigma}_S^2 + \hat{\sigma}_\sigma^2\right)^{1/2}} \tag{6–54}$$

Equation (6–54) is called the *normal coupling equation*. The reliability associated with z is given by

$$R = \int_x^\infty \frac{1}{\sqrt{2\pi}} \exp\left(-\frac{u^2}{2}\right) du = 1 - F = 1 - \Phi(z) \tag{6–55}$$

The body of Table A–10 gives R when z > 0 and (1 − R = F) when z ≤ 0. Noting that $\bar{n} = \mu_S/\mu_\sigma$, square both sides of Eq. (6–54), introduce C_S and C_σ, and solve the resulting quadratic for n to obtain

$$\bar{n} = \frac{1 \pm \sqrt{1 - \left(1 - z^2 C_S^2\right)\left(1 - z^2 C_\sigma^2\right)}}{1 - z^2 C_S^2} \tag{6–56}$$

The plus sign is associated with R > 0.5, and the minus sign with R < 0.5.

Lognormal–Lognormal Case

Consider the lognormal distributions $S = LN(\mu_S, \hat{\sigma}_S)$ and $\sigma = LN(\mu_\sigma, \hat{\sigma}_\sigma)$. If we interfere their companion normals using Eqs. (2–18) and (2–19), we obtain

$$\mu_{\ln S} = \ln \mu_S - \ln \sqrt{1 + C_S^2}$$

$$\hat{\sigma}_{\ln S} = \sqrt{\ln \left(1 + C_S^2\right)} \qquad (strength)$$

and

$$\mu_{\ln \sigma} = \ln \mu_\sigma - \ln \sqrt{1 + C_\sigma^2}$$

$$\hat{\sigma}_{\ln \sigma} = \sqrt{\ln \left(1 + C_\sigma^2\right)} \qquad (stress)$$

Using Eq. (6–54) for interfering normal distributions gives

$$z = -\frac{\mu_{\ln S} - \mu_{\ln \sigma}}{\left(\hat{\sigma}_{\ln S}^2 + \hat{\sigma}_{\ln \sigma}^2\right)^{1/2}} = -\frac{\ln \left(\dfrac{\mu_S}{\mu_\sigma} \sqrt{\dfrac{1 + C_\sigma^2}{1 + C_S^2}}\right)}{\sqrt{\ln \left[\left(1 + C_S^2\right) \left(1 + C_\sigma^2\right)\right]}} \qquad (6\text{–}57)$$

The reliability R is expressed by Eq. (6–55). The design factor $\mathbf{n}$ is the random variable that is the quotient of $\mathbf{S}/\boldsymbol{\sigma}$. The quotient of lognormals is lognormal, so pursuing the z variable of the lognormal $\mathbf{n}$, we note

$$\mu_n = \frac{\mu_S}{\mu_\sigma} \qquad C_n = \sqrt{\frac{C_S^2 + C_\sigma^2}{1 + C_\sigma^2}} \qquad \hat{\sigma}_n = C_n \mu_n$$

The companion normal to $\mathbf{n} = LN(\mu_n, \hat{\sigma}_n)$, from Eqs. (2–18) and (2–19), has a mean and standard deviation of

$$\mu_y = \ln \mu_n - \ln \sqrt{1 + C_n^2} \qquad \hat{\sigma}_y = \sqrt{\ln \left(1 + C_n^2\right)}$$

The z variable for the companion normal y distribution is

$$z = \frac{y - \mu_y}{\hat{\sigma}_y}$$

Failure will occur when the stress is greater than the strength, when $n < 1$, or when $y < 0$.

$$z = \frac{0 - \mu_y}{\hat{\sigma}_y} = -\frac{\mu_y}{\sigma_y} = -\frac{\ln \mu_n - \ln \sqrt{1 + C_n^2}}{\sqrt{\ln \left(1 + C_n^2\right)}} \doteq -\frac{\ln \left(\mu_n / \sqrt{1 + C_n^2}\right)}{\sqrt{\ln \left(1 + C_n^2\right)}} \qquad (6\text{–}58)$$

Solving for μ_n gives

$$\mu_n = \bar{n} = \exp\left[-z\sqrt{\ln \left(1 + C_n^2\right)} + \ln \sqrt{1 + C_n^2}\right] \doteq \exp\left[C_n\left(-z + \frac{C_n}{2}\right)\right]$$

$$(6\text{–}59)$$

Equations (6–56) and (6–59) are remarkable for several reasons:

- They relate design factor $\bar{n}$ to the reliability goal (through z) and the coefficients of variation of strength and stress.
- They are *not* functions of the means of stress and strength.
- They estimate the design factor necessary to achieve the reliability goal before decisions involving means are made. The C_S depends slightly on the particular material. The C_σ has the coefficient of variation (COV) of the load, and that is generally given.

EXAMPLE 6–9

A round cold-drawn 1018 steel rod has an 0.2 percent yield strength $\mathbf{S}_y = \mathbf{N}(78.4, 5.90)$ kpsi and is to be subjected to a static axial load of $\mathbf{P} = \mathbf{N}(50, 4.1)$ kip. What value of the design factor $\bar{n}$ corresponds to a reliability of 0.999 against yielding ($z = -3.09$)? Determine the corresponding diameter of the rod.

Solution $C_S = 5.90/78.4 = 0.0753$, and

$$\sigma = \frac{\mathbf{P}}{A} = \frac{4\mathbf{P}}{\pi d^2}$$

Since the COV of the diameter is an order of magnitude less than the COV of the load or strength, the diameter is treated deterministically:

$$C_\sigma = C_P = \frac{4.1}{50} = 0.082$$

From Eq. (6–56),

$$\bar{n} = \frac{1 + \sqrt{1 - (1 - 3.09^2 0.0753^2)(1 - 3.09^2 0.082^2)}}{1 - 3.09^2 0.0753^2} = 1.4156$$

The diameter is found deterministically:

Answer

$$d = \sqrt{\frac{4\bar{P}}{\pi \bar{S}_y/\bar{n}}} = \sqrt{\frac{4(50\ 000)}{\pi (78\ 400)/1.42}} = 1.074 \text{ in}$$

Check $\mathbf{S}_y = \mathbf{N}(78.4, 5.90)$ kpsi, $\mathbf{P} = \mathbf{N}(50, 4.1)$ kip, and $d = 1.074$ in. Then

$$A = \frac{\pi d^2}{4} = \frac{\pi (1.074^2)}{4} = 0.9059 \text{ in}^2$$

$$\bar{\sigma} = \frac{\bar{P}}{A} = \frac{(50\ 000)}{0.9059} = 55\ 191 \text{ psi}$$

$$C_P = C_\sigma = \frac{4.1}{50} = 0.082$$

$$\hat{\sigma}_\sigma = C_\sigma \bar{\sigma} = 0.082(55\ 191) = 4526 \text{ psi}$$

$$\hat{\sigma}_S = 5.90 \text{ kpsi}$$

From Eq. (6–54)

$$z = -\frac{78.4 - 55.191}{(5.90^2 + 4.526^2)^{1/2}} = -3.12$$

From Appendix Table A–10, $R = \Phi(-3.12) = 0.9991$.

EXAMPLE 6–10   Rework Ex. 6–9 with lognormally distributed stress and strength.

Solution $C_S = 5.90/78.4 = 0.0753$, and $C_\sigma = C_P = 4.1/50 = 0.082$. Then

$$\boldsymbol{\sigma} = \frac{\mathbf{P}}{A} = \frac{4\mathbf{P}}{\pi d^2}$$

$$C_n = \sqrt{\frac{C_S^2 + C_\sigma^2}{1 + C_\sigma^2}} = \sqrt{\frac{0.0753^2 + 0.082^2}{1 + 0.082^2}} = 0.1110$$

From Table A–10, $z = -3.09$. From Eq. (6–59),

$$\bar{n}_d = \exp\left[-(-3.09)\sqrt{\ln(1 + 0.111^2)} + \ln\sqrt{1 + 0.111^2}\right] = 1.416$$

$$d = \sqrt{\frac{4\bar{P}}{\pi \bar{S}_y/\bar{n}}} = \sqrt{\frac{4(50\ 000)}{\pi(78\ 400)/1.416}} = 1.0723 \text{ in}$$

Check $\mathbf{S}_y = \mathbf{LN}(78.4, 5.90)$, $\mathbf{P} = \mathbf{LN}(50, 4.1)$ kip. Then

$$A = \frac{\pi d^2}{4} = \frac{\pi(1.0723^2)}{4} = 0.9031$$

$$\bar{\sigma} = \frac{\bar{P}}{A} = \frac{50\ 000}{0.9031} = 55\ 367 \text{ psi}$$

$$C_\sigma = C_P = \frac{4.1}{50} = 0.082$$

$$\hat{\sigma}_\sigma = C_\sigma \mu_\sigma = 0.082(55\ 367) = 4540 \text{ psi}$$

From Eq. (6–57),

$$z = -\frac{\ln\left(\dfrac{78.4}{55.367}\sqrt{\dfrac{1 + 0.082^2}{1 + 0.0753^2}}\right)}{\sqrt{\ln[(1 + 0.0753^2)(1 + 0.082^2)]}} = -3.1399$$

Appendix Table A–10 gives $R = 0.99914$.

EXAMPLE 6–11   Return to the shaft of Ex. 6–6, and consider that the material has a yield strength $\mathbf{S}_y = \mathbf{LN}(66.0, 5.3)$ kpsi and the applied moment $\mathbf{M} = \mathbf{LN}(1925, 96)$ lbf · in and a correlated torque $\mathbf{T} = \mathbf{LN}(3300, 765)$ lbf · in. If the reliability against yielding at the right-end journal is to be 0.999 995, what is the smallest bearing journal diameter as estimated from distortion energy and maximum shear stress?

Solution   Equation (6–41) can be expressed stochastically as

$$\boldsymbol{\sigma}' = \frac{16}{\pi d^3}(4\mathbf{M}^2 + 3\mathbf{T}^2)^{1/2} = \frac{16\mathbf{M}}{\pi d^3}\left[4 + 3\left(\frac{\mathbf{T}}{\mathbf{M}}\right)^2\right]^{1/2} \tag{1}$$

since there is no stress-concentration factor for static loading. Since $\mathbf{T}$ and $\mathbf{M}$ are correlated, the quotient $\mathbf{T}/\mathbf{M}$ is a deterministic constant, $\bar{T}/\bar{M} = \mu_T/\mu_M$. Thus, from Eq. (1), the variance in stress $\boldsymbol{\sigma}'$ is the same as in $\mathbf{M}$. This results in

$$C_n \doteq \sqrt{C_S^2 + C_\sigma^2} = \left[\left(\frac{5.3}{66} \right)^2 + \left(\frac{96}{1925} \right)^2 \right]^{1/2} = 0.0945$$

For the stated reliability goal, Table A–10 gives $z = -4.417$. Then, from Eq. (6–59),

$$\bar{n} \doteq \exp[C_n(-z + C_n/2)] = \exp[0.0945(4.417 + 0.0945/2)] = 1.52$$

From Eq. (6–43), for DE

$$d = \left\{ \frac{16(1.52)}{\pi 66\,000} [4(-1925)^2 + 3(3300)^2]^{1/2} \right\}^{1/3} = 0.932 \text{ in}$$

and from Eq. (6–45), for MSS

$$d = \left\{ \frac{32(1.52)}{\pi 66\,000} [(-1925)^2 + 3300^2]^{1/2} \right\}^{1/3} = 0.964 \text{ in}$$

The answer utilizes stochastic knowledge of loading ($\mathbf{M}$ and $\mathbf{T}$) and strength ($\mathbf{S}_y$) and responds to a stated reliability goal.

Interference—General

In the previous segments, we employed interference theory to estimate reliability when the distributions are both normal and when they are both lognormal. Sometimes, however, it turns out that the strength has, say, a Weibull distribution while the stress is distributed lognormally. In fact, stresses are quite likely to have a lognormal distribution, because the multiplication of variates that are normally distributed produces a result that approaches lognormal. What all this means is that we must expect to encounter interference problems involving mixed distributions and we need a general method to handle the problem.

It is quite likely that we will use interference theory for problems involving distributions other than strength and stress. For this reason we employ the subscript 1 to designate the strength distribution and the subscript 2 to designate the stress distribution. Figure 6–42 shows these two distributions aligned so that a single cursor x can be used to identify points on both distributions. We can now write

$$\begin{pmatrix} \text{Probability that} \\ \text{stress is less} \\ \text{than strength} \end{pmatrix} = dp(\sigma < x) = dR = F_2(x)\, dF_1(x)$$

By substituting $1 - R_2$ for F_2 and $-dR_1$ for dF_1, we have

$$dR = -[1 - R_2(x)]\, dR_1(x)$$

The reliability for all possible locations of the cursor is obtained by integrating x from $-\infty$ to ∞; but this corresponds to an integration from 1 to 0 on the reliability R_1. Therefore

$$R = -\int_1^0 [1 - R_2(x)]\, dR_1(x)$$

Figure 6–42

(a) PDF of the strength distribution; (b) PDF of the load-induced stress distribution.

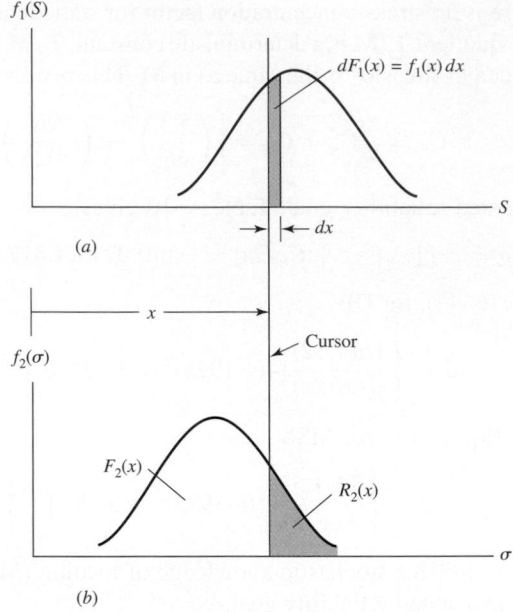

Figure 6–43

Curve shapes of the $R_1 R_2$ plot. In each case the shaded area is equal to $1 - R$ and is obtained by numerical integration. (a) Typical curve for asymptotic distributions; (b) curve shape obtained from lower truncated distributions such as the Weibull.

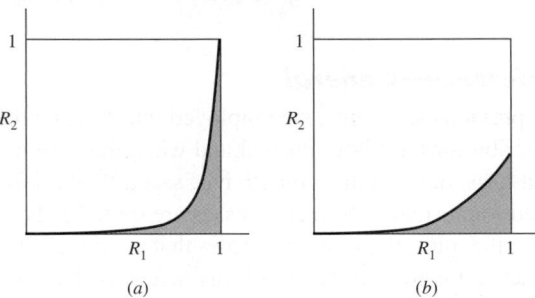

which can be written

$$R = 1 - \int_0^1 R_2 \, dR_1 \tag{6–60}$$

where

$$R_1(x) = \int_x^\infty f_1(S) \, dS \tag{6–61}$$

$$R_2(x) = \int_x^\infty f_2(\sigma) \, d\sigma \tag{6–62}$$

For the usual distributions encountered, plots of R_1 versus R_2 appear as shown in Fig. 6–43. Both of the cases shown are amenable to numerical integration and computer solution. When the reliability is high, the bulk of the integration area is under the right-hand spike of Fig. 6–43a.

PROBLEMS

ANALYSIS

6-1 A ductile hot-rolled steel bar has a minimum yield strength in tension and compression of 50 kpsi. Using the distortion-energy and maximum-shear-stress theories determine the factors of safety for the following plane stress states:

(a) $\sigma_x = 12$ kpsi, $\sigma_y = 6$ kpsi

(b) $\sigma_x = 12$ kpsi, $\tau_{xy} = -8$ kpsi

(c) $\sigma_x = -6$ kpsi, $\sigma_y = -10$ kpsi, $\tau_{xy} = -5$ kpsi

(d) $\sigma_x = 12$ kpsi, $\sigma_y = 4$ kpsi, $\tau_{xy} = 1$ kpsi

ANALYSIS

6-2 Repeat Prob. 6–1 for:

(a) $\sigma_A = 12$ kpsi, $\sigma_B = 12$ kpsi

(b) $\sigma_A = 12$ kpsi, $\sigma_B = 6$ kpsi

(c) $\sigma_A = 12$ kpsi, $\sigma_B = -12$ kpsi

(d) $\sigma_A = -6$ kpsi, $\sigma_B = -12$ kpsi

ANALYSIS

6-3 Repeat Prob. 6–1 for a bar of AISI 1020 cold-drawn steel and:

(a) $\sigma_x = 180$ MPa, $\sigma_y = 100$ MPa

(b) $\sigma_x = 180$ MPa, $\tau_{xy} = 100$ MPa

(c) $\sigma_x = -160$ MPa, $\tau_{xy} = 100$ MPa

(d) $\tau_{xy} = 150$ MPa

ANALYSIS

6-4 Repeat Prob. 6–1 for a bar of AISI 1018 hot-rolled steel and:

(a) $\sigma_A = 100$ MPa, $\sigma_B = 80$ MPa

(b) $\sigma_A = 100$ MPa, $\sigma_B = 10$ MPa

(c) $\sigma_A = 100$ MPa, $\sigma_B = -80$ MPa

(d) $\sigma_A = -80$ MPa, $\sigma_B = -100$ MPa

ANALYSIS

6-5 Repeat Prob. 6–3 by first plotting the failure loci in the σ_A, σ_B plane to scale; then, for each stress state, plot the load line and by graphical measurement estimate the factors of safety.

ANALYSIS

6-6 Repeat Prob. 6–4 by first plotting the failure loci in the σ_A, σ_B plane to scale; then, for each stress state, plot the load line and by graphical measurement estimate the factors of safety.

ANALYSIS

6-7 An ASTM cast iron has minimum ultimate strengths of 30 kpsi in tension and 100 kpsi in compression. Find the factors of safety using the MNS, BCM, M1M, and M2M theories for each of the following stress states. Plot the failure diagrams in the σ_A, σ_B plane to scale and locate the coordinates of each stress state.

(a) $\sigma_x = 20$ kpsi, $\sigma_y = 6$ kpsi

(b) $\sigma_x = 12$ kpsi, $\tau_{xy} = -8$ kpsi

(c) $\sigma_x = -6$ kpsi, $\sigma_y = -10$ kpsi, $\tau_{xy} = -5$ kpsi

(d) $\sigma_x = -12$ kpsi, $\tau_{xy} = 8$ kpsi

ANALYSIS

6-8 For Prob. 6–7, case (d), estimate the factors of safety from the four theories by graphical measurements of the load line.

DECISIONS
IN DESIGN

6-9 Among the decisions a designer must make is selection of the failure locus that is applicable to the material and its static loading. A 1020 hot-rolled steel has the following properties: $S_y = 42$ kpsi, $S_{ut} = 66.2$ kpsi, and true strain at fracture $\varepsilon_f = 0.90$. Plot the failure locus and, for the static stress states at the critical locations listed below, plot the load line and estimate the factor of safety analytically and graphically.

ANALYSIS

(a) $\sigma_x = 9$ kpsi, $\sigma_y = -5$ kpsi.

(b) $\sigma_x = 12$ kpsi, $\tau_{xy} = 3$ kpsi ccw.

(c) $\sigma_x = -4$ kpsi, $\sigma_y = -9$ kpsi, $\tau_{xy} = 5$ kpsi cw.

(d) $\sigma_x = 11$ kpsi, $\sigma_y = 4$ kpsi, $\tau_{xy} = 1$ kpsi cw.

6–10 A 4142 steel Q&T at 80°F exhibits $S_{yt} = 235$ kpsi, $S_{yc} = 275$ kpsi, and $\varepsilon_f = 0.06$. Choose and plot the failure locus and, for the static stresses at the critical locations, which are 10 times those in Prob. 6–9, plot the load lines and estimate the factors of safety analytically and graphically.

6–11 For grade 20 cast iron, Table A–24 gives $S_{ut} = 22$ kpsi, $S_{uc} = 83$ kpsi. Choose and plot the failure locus and, for the static loadings inducing the stresses at the critical locations of Prob. 6–9, plot the load lines and estimate the factors of safety analytically and graphically.

6–12 A cast aluminum 195-T6 has an ultimate strength in tension of $S_{ut} = 36$ kpsi and ultimate strength in compression of $S_{uc} = 35$ kpsi, and it exhibits a true strain at fracture $\varepsilon_f = 0.045$. Choose and plot the failure locus and, for the static loading inducing the stresses at the critical locations of Prob. 6–9, plot the load lines and estimate the factors of safety analytically and graphically.

6–13 An ASTM cast iron, grade 30 (see Table A–24), carries static loading resulting in the stress state listed below at the critical locations. Choose the appropriate failure locus, plot it and the load lines, and estimate the factors of safety analytically and graphically.
(a) $\sigma_A = 20$ kpsi, $\sigma_B = 20$ kpsi.
(b) $\tau_{xy} = 15$ kpsi.
(c) $\sigma_A = \sigma_B = -80$ kpsi.
(d) $\sigma_A = 15$ kpsi, $\sigma_B = -25$ kpsi.

6–14 This problem illustrates that the factor of safety for a machine element depends on the particular point selected for analysis. Here you are to compute factors of safety, based upon the distortion-energy theory, for stress elements at A and B of the member shown in the figure. This bar is made of AISI 1006 cold-drawn steel and is loaded by the forces $F = 0.55$ kN, $P = 8.0$ kN, and $T = 30$ N · m.

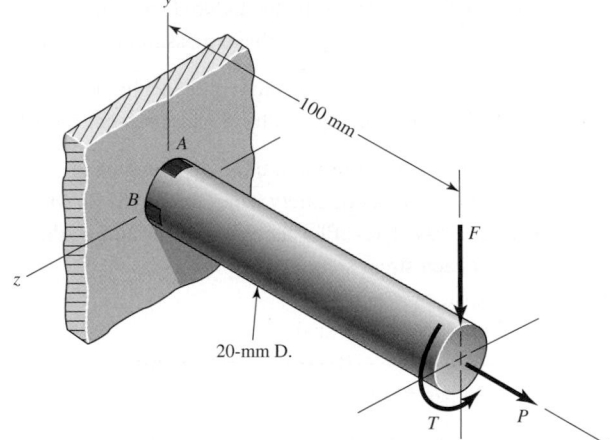

Problem 6–14

6–15* Design the lever arm CD of Fig. 6–24 by specifying a suitable size and material.

6–16 A spherical pressure vessel is formed of 18-gauge (0.05-in) cold-drawn AISI 1018 sheet steel. If the vessel has a diameter of 8 in, estimate the pressure necessary to initiate yielding. What is the estimated bursting pressure?

*The asterisk indicates a problem that may not have a unique result or a particularly challenging problem.

6–17 This problem illustrates that the strength of a machine part can sometimes be measured in units other than those of force or moment. For example, the maximum speed that a flywheel can reach without yielding or fracturing is a measure of its strength. In this problem you have a rotating ring made of hot-forged AISI 1020 steel; the ring has a 6-in inside diameter and a 10-in outside diameter and is 1.5 in thick. What speed in revolutions per minute would cause the ring to yield? At what radius would yielding begin? [*Note:* The maximum radial stress occurs at $r = (r_o r_i)^{1/2}$; see Eq. (4–56).]

6–18 A light pressure vessel is made of 2024-T3 aluminum alloy tubing with suitable end closures. This cylinder has a $3\frac{1}{2}$-in OD, a 0.065-in wall thickness, and $\nu = 0.334$. The purchase order specifies a minimum yield strength of 46 kpsi. What is the factor of safety if the pressure-release valve is set at 500 psi?

6–19 A cold-drawn AISI 1015 steel tube is 300 mm OD by 200 mm ID and is to be subjected to an external pressure caused by a shrink fit. What maximum pressure would cause the material of the tube to yield?

6–20 What speed would cause fracture of the ring of Prob. 6–17 if it were made of grade 30 cast iron?

6–21 The figure shows a shaft mounted in bearings at A and D and having pulleys at B and C. The forces shown acting on the pulley surfaces represent the belt tensions. The shaft is to be made of ASTM grade 25 cast iron using a design factor $n_d = 2.8$. What diameter should be used for the shaft?

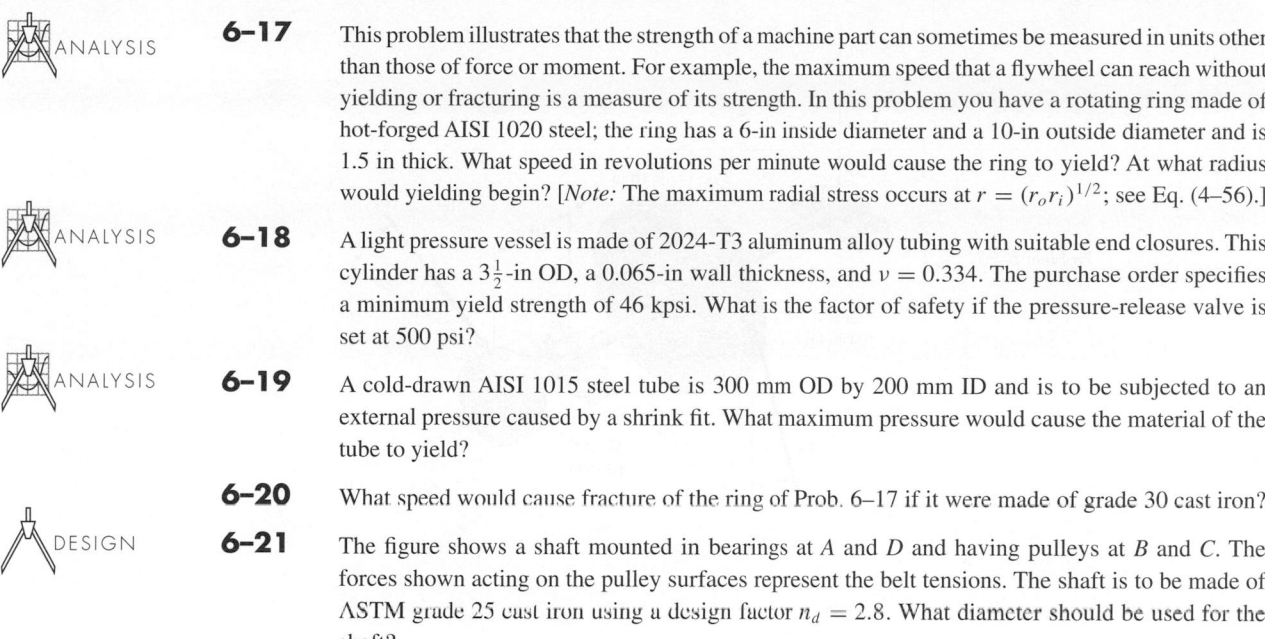

Problem 6–21

6–22 By modern standards, the shaft design of Prob. 6–21 is poor because it is so long. Suppose it is redesigned by halving the length dimensions. Using the same material and design factor as in Prob. 6–21, find the new shaft diameter.

6–23 The gear forces shown act in planes parallel to the yz plane. The force on gear A is 300 lbf. Consider the bearings at O and B to be simple supports. For a static analysis and a factor of safety of 3.5, use distortion energy to determine the minimum safe diameter of the shaft. Consider the material to have a yield strength of 60 kpsi.

6–24 Repeat Prob. 6–23 using maximum-shear-stress.

6–25 The figure is a schematic drawing of a countershaft that supports two V-belt pulleys. For each pulley, the belt tensions are parallel. For pulley A consider the loose belt tension is 15 percent of the tension on the tight side. A cold-drawn UNS G10180 steel shaft of uniform diameter is to be selected for this application. For a static analysis with a factor of safety of 3.0, determine the minimum preferred size diameter. Use the distortion-energy theory.

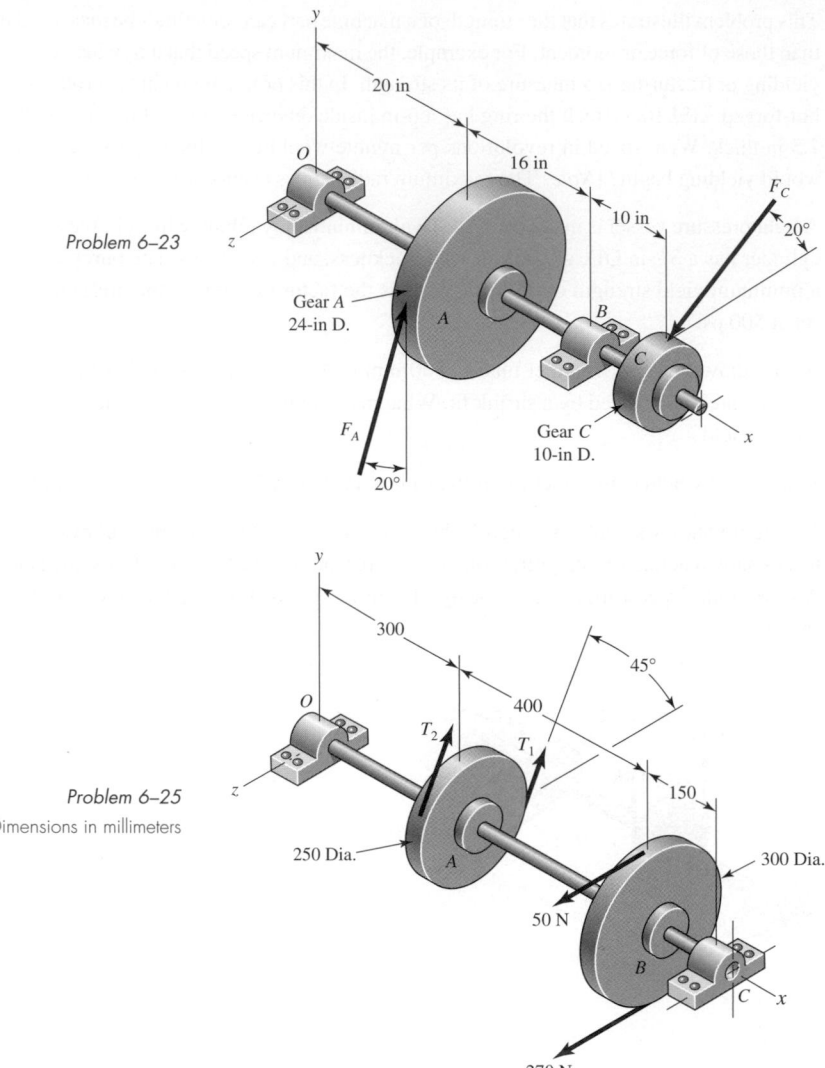

Problem 6–23

Problem 6–25

Dimensions in millimeters

DESIGN

6–26 Repeat Prob. 6–25 using maximum shear stress.

ANALYSIS

6–27 The clevis pin shown in the figure is 12 mm in diameter and has the dimensions $a = 12$ mm and $b = 18$ mm. The pin is machined from AISI 1018 hot-rolled steel (Table A–20) and is to be loaded to no more than 4.4 kN. Determine whether or not the assumed loading of figure c yields a factor of safety any different from that of figure d. Use the maximum-shear-stress theory.

6–28 Repeat Prob. 6–27, but this time use the distortion-energy theory.

ANALYSIS

6–29 A split-ring clamp-type shaft collar is shown in the figure. The collar is 2 in OD by 1 in ID by $\frac{1}{2}$ in wide. The screw is designated as $\frac{1}{4}$-28 UNF. The relation between the screw tightening torque T, the nominal screw diameter d, and the tension in the screw F_i is approximately $T = 0.2\,F_i d$. The shaft is sized to obtain a close running fit. Find the axial holding force F_x of the collar as a function of the coefficient of friction and the screw torque.

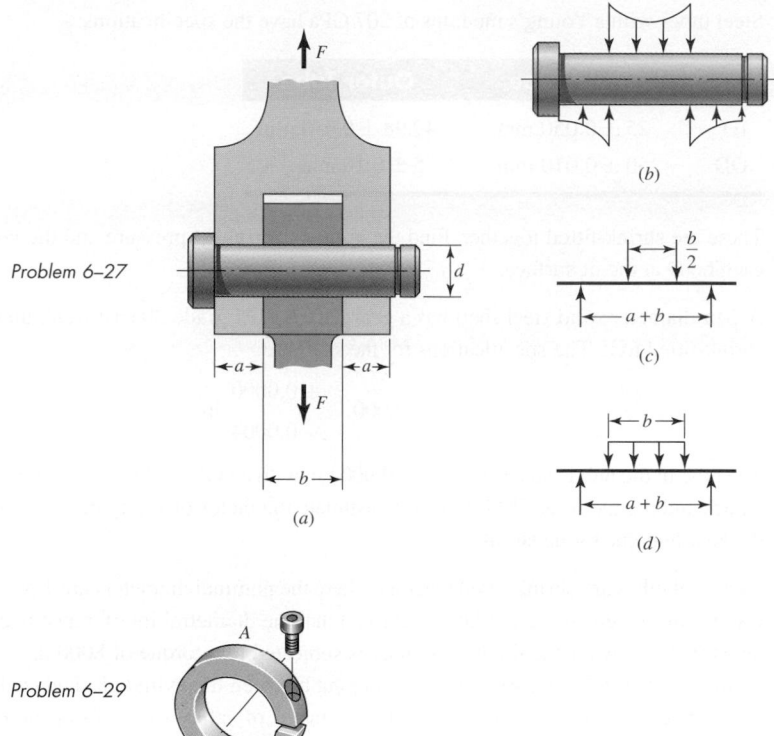

Problem 6–27

(a)

(b)

(c)

(d)

Problem 6–29

ANALYSIS

6–30

Suppose the collar of Prob. 6–29 is tightened by using a screw torque of 190 lbf · in. The collar material is AISI 1040 steel heat-treated to a minimum tensile yield strength of 63 kpsi.

(a) Estimate the tension in the screw.

(b) By relating the tangential stress to the hoop tension, find the internal pressure of the shaft on the ring.

(c) Find the tangential and radial stresses in the ring at the inner surface.

(d) Determine the maximum shear stress and the von Mises stress.

(e) What are the factors of safety based on the maximum-shear-stress hypothesis and the distortion-energy theory?

DESIGN

6–31

In Prob. 6–29, the role of the screw was to induce the hoop tension that produces the clamping. The screw should be placed so that no moment is induced in the ring. Just where should the screw be located?

ANALYSIS

6–32

A tube has another tube shrunk over it. The specifications are:

	Inner Member	**Outer Member**
ID	1.000 ± 0.002 in	1.999 ± 0.0004 in
OD	2.000 ± 0.0004 in	3.000 ± 0.004 in

Both tubes are made of a plain carbon steel.

(a) Find the nominal shrink-fit pressure and the von Mises stresses at the fit surface.

(b) If the inner tube is changed to solid shafting with the same outside dimensions, find the nominal shrink-fit pressure and the von Mises stresses at the fit surface.

6-33 Steel tubes with a Young's modulus of 207 GPa have the specifications:

	Inner Tube	Outer Tube
ID	25 ± 0.050 mm	49.98 ± 0.010 mm
OD	50 ± 0.010 mm	75 ± 0.10 mm

These are shrink-fitted together. Find the nominal shrink-fit pressure and the von Mises stress in each body at the fit surface.

6-34 A 2-in-diameter solid steel shaft has a gear with ASTM grade 20 cast-iron hub ($E = 14.5$ Mpsi) shrink-fitted to it. The specifications for the shaft are

$$2.000 \quad \begin{array}{c} +0.0000 \\ -0.0004 \end{array} \text{ in}$$

The hole in the hub is sized at 1.999 ± 0.0004 in with an OD of $4.00 \pm \frac{1}{32}$ in. Using the midrange values and the modified II Mohr theory, estimate the factor of safety guarding against fracture in the gear hub due to the shrink fit.

6-35 Two steel tubes are shrink-fitted together where the nominal diameters are 1.50, 1.75, and 2.00 in. Careful measurement before fitting revealed that the diametral interference between the tubes to be 0.00246 in. After the fit, the assembly is subjected to a torque of 8000 lbf · in and a bending-moment of 6000 lbf · in. Assuming no slipping between the cylinders, analyze the outer cylinder at the inner and outer radius. Determine the factor of safety using distortion energy with $S_y = 60$ kpsi.

6-36 Repeat Prob. 6–35 for the inner tube.

6-37 For Eqs. (6–50) show that the principal stresses are given by

$$\sigma_1 = \frac{K_I}{\sqrt{2\pi r}} \cos \frac{\theta}{2} \left(1 + \sin \frac{\theta}{2} \right)$$

$$\sigma_2 = \frac{K_I}{\sqrt{2\pi r}} \cos \frac{\theta}{2} \left(1 - \sin \frac{\theta}{2} \right)$$

$$\sigma_3 = \begin{cases} 0 & \text{(plane stress)} \\ \sqrt{\dfrac{2}{\pi r}} \, \nu K_I \cos \dfrac{\theta}{2} & \text{(plane strain)} \end{cases}$$

6-38 Use the results of Prob. 6–37 for plane strain near the tip with $\theta = 0$ and $\nu = \frac{1}{3}$. If the yield strength of the plate is S_y, what is σ_1 when yield occurs?

(a) Use the distortion-energy theory.

(b) Use the maximum-shear-stress theory. Using Mohr's circles, explain your answer.

6-39 A plate 4 in wide, 8 in long, and 0.5 in thick is loaded in tension in the direction of the length. The plate contains a crack as shown in Fig. 6–36 with the crack length of 0.625 in. The material is steel with $K_{Ic} = 70$ kpsi · $\sqrt{\text{in}}$, and $S_y = 160$ kpsi. Determine the maximum possible load that can be applied before the plate (a) yields, and (b) has uncontrollable crack growth.

6-40 A cylinder subjected to internal pressure p_i has an outer diameter of 350 mm and a 25-mm wall thickness. For the cylinder material, $K_{Ic} = 80$ MPa · $\sqrt{\text{m}}$, $S_y = 1200$ MPa, and $S_{ut} = 1350$ MPa. If the cylinder contains a radial crack in the longitudinal direction of depth 12.5 mm determine the pressure that will cause uncontrollable crack growth.

ANALYSIS

6–41 A carbon steel collar of length 1 in is to be machined to inside and outside diameters, respectively, of

$$D_i = 0.750 \pm 0.0004 \text{ in} \qquad D_o = 1.125 \pm 0.002 \text{ in}$$

This collar is to be shrink-fitted to a hollow steel shaft having inside and outside diameters, respectively, of

$$d_i = 0.375 \pm 0.002 \text{ in} \qquad d_o = 0.752 \pm 0.0004 \text{ in}$$

These tolerances are assumed to have a normal distribution, to be centered in the spread interval, and to have a total spread of ± 4 standard deviations. Determine the means and the standard deviations of the tangential stress components for both cylinders at the interface.

6–42 Suppose the collar of Prob. 6–41 has a yield strength of $\mathbf{S}_y = \mathbf{N}(95.5, 6.59)$ kpsi. What is the probability that the material will not yield?

ANALYSIS

6–43 A carbon steel tube has an outside diameter of 1 in and a wall thickness of $\frac{1}{8}$ in. The tube is to carry an internal hydraulic pressure given as $\mathbf{p} = \mathbf{N}(6000, \ 500)$ psi. The material of the tube has a yield strength of $\mathbf{S}_y = \mathbf{N}(50, \ 4.1)$ kpsi. Find the reliability using thin-wall theory.

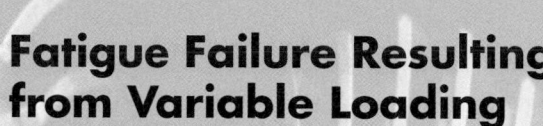

7
Fatigue Failure Resulting from Variable Loading

In Chap. 6 we considered the analysis and design of parts subjected to static loading. The behavior of machine parts is entirely different when they are subjected to time-varying loading. In this chapter we shall examine how parts fail under variable loading and how to proportion them to successfully resist such conditions.

7–1 Introduction to Fatigue in Metals

In most testing of those properties of materials that relate to the stress-strain diagram, the load is applied gradually, to give sufficient time for the strain to fully develop. Furthermore, the specimen is tested to destruction, and so the stresses are applied only once. Testing of this kind is applicable, then, to what are known as *static conditions;* such conditions closely approximate the actual conditions to which many structural and machine members are subjected.

The condition frequently arises, however, in which the stresses vary or they fluctuate between levels. For example, a particular fiber on the surface of a rotating shaft subjected to the action of bending loads undergoes both tension and compression for each revolution of the shaft. If the shaft is part of an electric motor rotating at 1725 rev/min, the fiber is stressed in tension and compression 1725 times each minute. If, in addition, the shaft is also axially loaded (as it would be, for example, by a helical or worm gear), an axial component of stress is superposed upon the bending component. In this case, some stress is always present in any one fiber, but now the *level* of stress is fluctuating. These and other kinds of loading occurring in machine members produce stresses that are called *variable, repeated, alternating,* or *fluctuating* stresses.

Often, machine members are found to have failed under the action of repeated or fluctuating stresses; yet the most careful analysis reveals that the actual maximum stresses were well below the ultimate strength of the material, and quite frequently even below the yield strength. The most distinguishing characteristic of these failures is that the stresses have been repeated a very large number of times. Hence the failure is called a *fatigue failure*.

When machine parts fail statically, they usually develop a very large deflection, because the stress has exceeded the yield strength, and the part is replaced before fracture actually occurs. Thus many static failures give visible warning in advance. But a fatigue failure gives no warning! It is sudden and total, and hence dangerous. It is relatively simple to design against a static failure, because our knowledge is comprehensive. Fatigue is a much more complicated phenomenon, only partially understood, and the engineer seeking competence must acquire as much knowledge of the subject as possible.

A fatigue failure has an appearance similar to a brittle fracture, as the fracture surfaces are flat and perpendicular to the stress axis with the absence of necking. The fracture features of a fatigue failure, however, are quite different from a static brittle fracture arising from three stages of development. *Stage I* is the initiation of one or more microcracks due to cyclic plastic deformation followed by crystallographic propagation extending from two to five grains about the origin. Stage I cracks are not normally discernible to the naked eye. *Stage II* progresses from microcracks to macrocracks forming parallel plateau-like fracture surfaces separated by longitudinal ridges. The plateaus are generally smooth and normal to the direction of maximum tensile stress. These surfaces can be wavy dark and light bands referred to as *beach marks* or *clamshell marks,* as seen in Fig. 7–1. During cyclic loading, these cracked surfaces open and close, rubbing together, and the beach mark appearance depends on the changes in the level or frequency of loading and the corrosive nature of the environment. *Stage III* occurs during the final stress cycle when the remaining material cannot support the loads, resulting in a sudden,

Figure 7-1

Fatigue failure of a bolt due to repeated unidirectional bending. The failure started at the thread root at *A*, propagated across most of the cross section shown by the beach marks at *B*, before final fast fracture at *C*. *(From ASM Handbook, Vol. 12: Fractography, ASM International, Materials Park, OH 44073-0002, fig 50, p. 120. Reprinted by permission of ASM International®, www.asminternational.org.)*

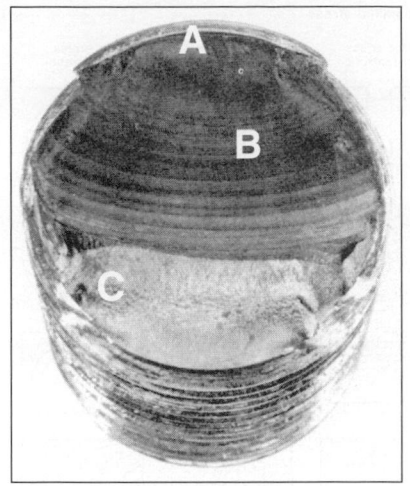

fast fracture. A stage III fracture can be brittle, ductile, or a combination of both. Quite often the beach marks, if they exist, and possible patterns in the stage III fracture called *chevron lines,* point toward the origins of the initial cracks.

There is a good deal to be learned from the fracture patterns of a fatigue failure.[1] Figure 7–2 shows representations of failure surfaces of various part geometries under differing load conditions and levels of stress concentration. Note that, in the case of rotational bending, even the direction of rotation influences the failure pattern.

Fatigue failure is due to crack formation and propagation. A fatigue crack will typically initiate at a discontinuity in the material where the cyclic stress is a maximum. Discontinuities can arise because of:

- Design of rapid changes in cross section, keyways, holes, etc. where stress concentrations occur as discussed in Secs. 4–14 and 6–2.

- Elements that roll and/or slide against each other (bearings, gears, cams, etc.) under high contact pressure, developing concentrated subsurface contact stresses (Sec. 4–20) that can cause surface pitting or spalling after many cycles of the load.

- Carelessness in locations of stamp marks, tool marks, scratches, and burrs; poor joint design; improper assembly; and other fabrication faults.

- Composition of the material itself as processed by rolling, forging, casting, extrusion, drawing, heat treatment, etc. Microscopic and submicroscopic surface and subsurface discontinuities arise, such as inclusions of foreign material, alloy segregation, voids, hard precipitated particles, and crystal discontinuities.

Various conditions that can accelerate crack initiation include residual tensile stresses, elevated temperatures, temperature cycling, a corrosive environment, and high-frequency cycling.

The rate and direction of fatigue crack propagation is primarily controlled by localized stresses and by the structure of the material at the crack. However, as with crack formation, other factors may exert a significant influence, such as environment, temperature, and frequency. As stated earlier, cracks will grow along planes normal to the

---

[1]See the ASM Handbook, *Fractography,* ASM International, Metals Park, Ohio, vol. 12, 9th ed., 1987.

Figure 7–2

Schematics of fatigue fracture surfaces produced in smooth and notched components with round and rectangular cross sections under various loading conditions and nominal stress levels. *(From ASM Handbook, Vol. 11: Failure Analysis and Prevention, ASM International, Materials Park, OH 44073-0002, fig 18, p. 111. Reprinted by permission of ASM International®, www.asminternational.org.)*

maximum tensile stresses. The crack growth process can be explained by fracture mechanics (see Sec. 7–6).

A major reference source in the study of fatigue failure is the 21-volume ASM *Metals Handbook*. Figures 7–1 to 7–8, reproduced with permission from ASM International, are but a minuscule sample of examples of fatigue failures for a great variety of conditions included in the handbook. Comparing Fig. 7–3 with Fig. 7–2, we see that failure occurred by rotating bending stresses, with the direction of rotation being clockwise with respect to the view and with a mild stress concentration and low nominal stress.

Figure 7–3

Fatigue fracture of an AISI 4320 drive shaft. The fatigue failure initiated at the end of the keyway at points *B* and progressed to final rupture at *C*. The final rupture zone is small, indicating that loads were low. *(From* ASM Handbook, Vol. 11: Failure Analysis and Prevention, *ASM International, Materials Park, OH 44073-0002, fig 18, p. 111. Reprinted by permission of ASM International* ®, *www.asminternational.org.)*

Figure 7–4

Fatigue fracture surface of an AISI 8640 pin. Sharp corners of the mismatched grease holes provided stress concentrations that initiated two fatigue cracks indicated by the arrows. *(From* ASM Handbook, Vol. 12: Fractography, *ASM International, Materials Park, OH 44073-0002, fig 520, p. 331. Reprinted by permission of ASM International* ®, *www.asminternational.org.)*

Figure 7–5

Fatigue fracture surface of a forged connecting rod of AISI 8640 steel. The fatigue crack origin is at the left edge, at the flash line of the forging, but no unusual roughness of the flash trim was indicated. The fatigue crack progressed halfway around the oil hole at the left, indicated by the beach marks, before final fast fracture occurred. Note the pronounced shear lip in the final fracture at the right edge. *(From* ASM Handbook, Vol. 12: Fractography, *ASM International, Materials Park, OH 44073-0002, fig 523, p. 332. Reprinted by permission of ASM International®, www.asminternational.org.)*

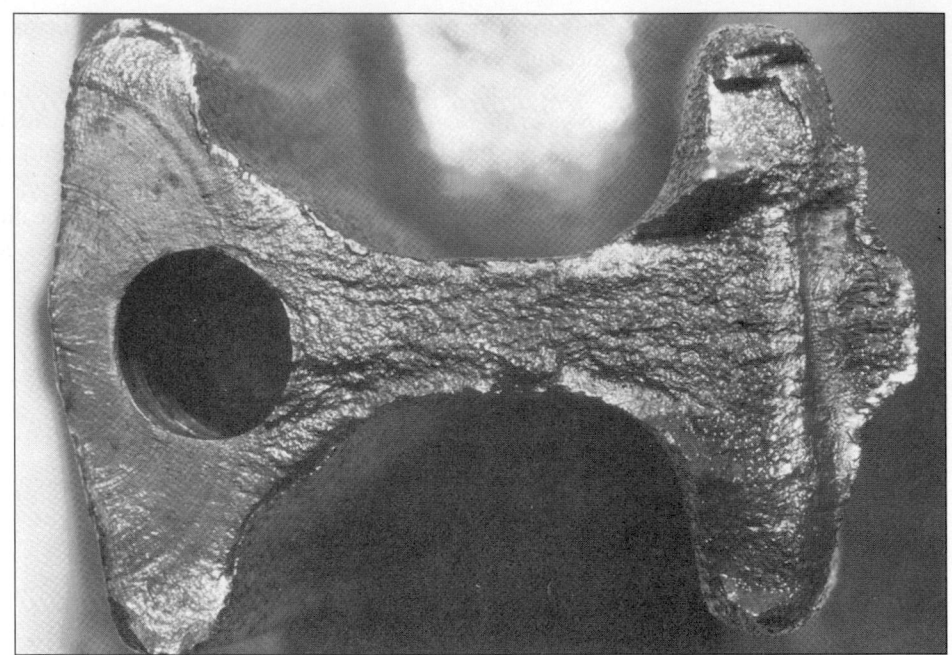

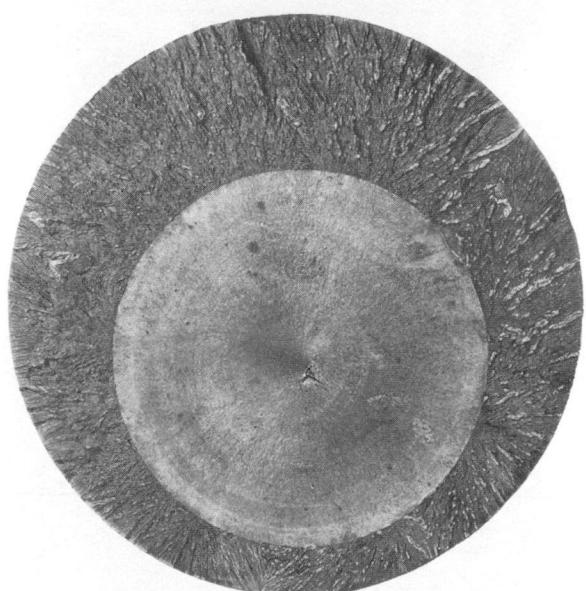

Figure 7–6

Fatigue fracture surface of a 200-mm (8-in) diameter piston rod of an alloy steel steam hammer used for forging. This is an example of a fatigue fracture caused by pure tension where surface stress concentrations are absent and a crack may initiate anywhere in the cross section. In this instance, the initial crack formed at a forging flake slightly below center, grew outward symmetrically, and ultimately produced a brittle fracture without warning. *(From* ASM Handbook, Vol. 12: Fractography, *ASM International, Materials Park, OH 44073-0002, fig 570, p. 342. Reprinted by permission of ASM International®, www.asminternational.org.)*

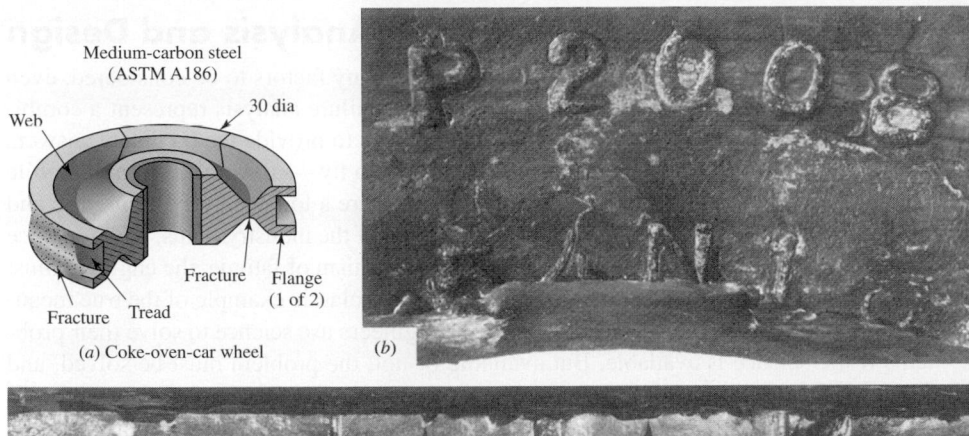

Figure 7–7

Fatigue failure of an ASTM A186 steel double-flange trailer wheel caused by stamp marks. (a) Coke-oven car wheel showing position of stamp marks and fractures in the rib and web. (b) Stamp mark showing heavy impression and fracture extending along the base of the lower row of numbers. (c) Notches, indicated by arrows, created from the heavily indented stamp marks from which cracks initiated along the top at the fracture surface. *(From ASM Handbook, Vol. 11: Failure Analysis and Prevention, ASM International, Materials Park, OH 44073-0002, fig 51, p. 130. Reprinted by permission of ASM International®, www.asminternational.org.)*

Figure 7–8

Aluminum alloy 7075-T73 landing-gear torque-arm assembly redesign to eliminate fatigue fracture at a lubrication hole. (a) Arm configuration, original and improved design (dimensions given in inches). (b) Fracture surface where arrows indicate multiple crack origins. *(From ASM Handbook, Vol. 11: Failure Analysis and Prevention, ASM International, Materials Park, OH 44073-0002, fig 23, p. 114. Reprinted by permission of ASM International®, www.asminternational.org.)*

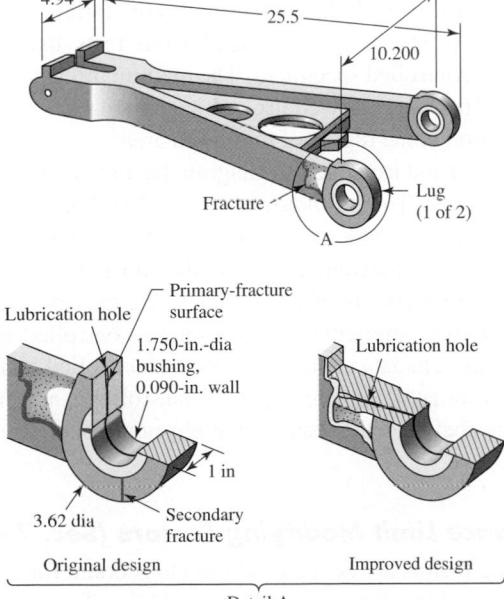

7–2 Approach to Fatigue Failure in Analysis and Design

As noted in the previous section, there are a great many factors to be considered, even for very simple load cases. The methods of fatigue failure analysis represent a combination of engineering and science. Often science fails to provide the complete answers that are needed. But the airplane must still be made to fly—safely. And the automobile must be manufactured with a reliability that will ensure a long and troublefree life and at the same time produce profits for the stockholders of the industry. Thus, while science has not yet completely explained the complete mechanism of fatigue, the engineer must still design things that will not fail. In a sense this is a classic example of the true meaning of engineering as contrasted with science. Engineers use science to solve their problems if the science is available. But available or not, the problem must be solved, and whatever form the solution takes under these conditions is called *engineering*.

In this chapter, we will take a structured approach in the design against fatigue failure. As with static failure, we will attempt to relate to test results performed on simply loaded specimens. However, because of the complex nature of fatigue, there is much more to account for. From this point, we will proceed methodically, and in stages. In an attempt to provide some insight as to what follows in this chapter, a brief description of the remaining sections will be given here.

Fatigue-Life Methods (Secs. 7–3 to 7–6)

Three major approaches used in design and analysis to predict when, if ever, a cyclically loaded machine component will fail in fatigue over a period of time are presented. The premises of each approach are quite different but each adds to our understanding of the mechanisms associated with fatigue. The application, advantages, and disadvantages of each method are indicated. Beyond Sec. 7–6, only one of the methods, the stress-life method, will be pursued for further design applications.

Fatigue Strength and the Endurance Limit (Secs. 7–7 and 7–8)

The strength-life (*S-N*) diagram provides the fatigue strength S_f versus cycle life N of a material. The results are generated from tests using a simple loading of standard laboratory-controlled specimens. The loading often is that of sinusoidally reversing pure bending. The laboratory-controlled specimens are polished without geometric stress concentration at the region of minimum area.

For steel and iron, the *S-N* diagram becomes horizontal at some point. The strength at this point is called the *endurance limit* S'_e and occurs somewhere between 10^6 and 10^7 cycles. The prime mark on S'_e refers to the endurance limit of the *controlled laboratory specimen*. For nonferrous materials that do not exhibit an endurance limit, a fatigue strength at a specific number of cycles, S'_f, may be given, where again, the prime denotes the fatigue strength of the laboratory-controlled specimen.

The strength data are based on many controlled conditions that will not be the same as that for an actual machine part. What follows are practices used to account for the differences between the loading and physical conditions of the specimen and the actual machine part.

Endurance Limit Modifying Factors (Sec. 7–9)

Modifying factors are defined and used to account for differences between the specimen and the actual machine part with regard to surface conditions, size, loading, temperature, reliability, and miscellaneous factors. Loading is still considered to be simple and reversing.

Stress Concentration and Notch Sensitivity (Sec. 7–10)

The actual part may have a geometric stress concentration by which the fatigue behavior depends on the static stress concentration factor and the component material's sensitivity to fatigue damage.

Fluctuating Stresses (Secs. 7–11 to 7–13)

These sections account for simple stress states from fluctuating load conditions that are not purely sinusoidally reversing axial, bending, or torsional stresses.

Combinations of Loading Modes (Sec. 7–14)

Here a procedure based on the distortion-energy theory is presented for analyzing combined fluctuating stress states, such as combined bending and torsion. Here it is assumed that the levels of the fluctuating stresses are not time varying.

Varying, Fluctuating Stresses; Cumulative Fatigue Damage (Sec. 7–15)

The fluctuating stress levels on a machine part may be time varying. Methods are provided to assess the fatigue damage on a cumulative basis.

Remaining Sections

The remaining two sections of the chapter pertain to the special topics of surface fatigue strength and stochastic analysis.

7–3 Fatigue-Life Methods

The three major fatigue life methods used in design and analysis are the *stress-life method*, the *strain-life method*, and the *linear-elastic fracture mechanics method*. These methods attempt to predict the life in number of cycles to failure, N, for a specific level of loading. Life of $1 \leq N \leq 10^3$ cycles is generally classified as *low-cycle fatigue*, whereas *high-cycle fatigue* is considered to be $N > 10^3$ cycles. The stress-life method, based on stress levels only, is the least accurate approach, especially for low-cycle applications. However, it is the most traditional method, since it is the easiest to implement for a wide range of design applications, has ample supporting data, and represents high-cycle applications adequately.

The strain-life method involves more detailed analysis of the plastic deformation at localized regions where the stresses and strains are considered for life estimates. This method is especially good for low-cycle fatigue applications. In applying this method, several idealizations must be compounded, and so some uncertainties will exist in the results. For this reason, it will be discussed only because of its value in adding to the understanding of the nature of fatigue.

The fracture mechanics method assumes a crack is already present and detected. It is then employed to predict crack growth with respect to stress intensity. It is most practical when applied to large structures in conjunction with computer codes and a periodic inspection program.

7–4 The Stress-Life Method

To determine the strength of materials under the action of fatigue loads, specimens are subjected to repeated or varying forces of specified magnitudes while the cycles or stress reversals are counted to destruction. The most widely used fatigue-testing device is the R. R. Moore high-speed rotating-beam machine. This machine subjects the specimen to pure bending (no transverse shear) by means of weights. The specimen, shown in Fig. 7–9, is very carefully machined and polished, with a final polishing in an axial direction to avoid circumferential scratches. Other fatigue-testing machines are available for applying fluctuating or reversed axial stresses, torsional stresses, or combined stresses to the test specimens.

To establish the fatigue strength of a material, quite a number of tests are necessary because of the statistical nature of fatigue. For the rotating-beam test, a constant bending load is applied, and the number of revolutions (stress reversals) of the beam required for failure is recorded. The first test is made at a stress that is somewhat under the ultimate strength of the material. The second test is made at a stress that is less than that used in the first. This process is continued, and the results are plotted as an S-N diagram (Fig. 7–10). This chart may be plotted on semilog paper or on log-log paper. In the case of ferrous metals and alloys, the graph becomes horizontal after the material has been stressed for a certain number of cycles. Plotting on log paper emphasizes the bend in the curve, which might not be apparent if the results were plotted by using Cartesian coordinates.

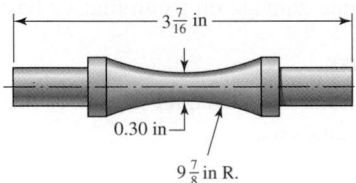

Figure 7–9

Test-specimen geometry for the R. R. Moore rotating-beam machine. The bending moment is uniform over the curved at the highest-stressed portion, a valid test of material, whereas a fracture elsewhere (not at the highest-stress level) is grounds for suspicion of material flaw.

Figure 7–10

An S-N diagram plotted from the results of completely reversed axial fatigue tests. Material: UNS G41300 steel, normalized; $S_{ut} = 116$ kpsi; maximum $S_{ut} = 125$ kpsi. (*Data from NACA Tech. Note 3866, December 1966.*)

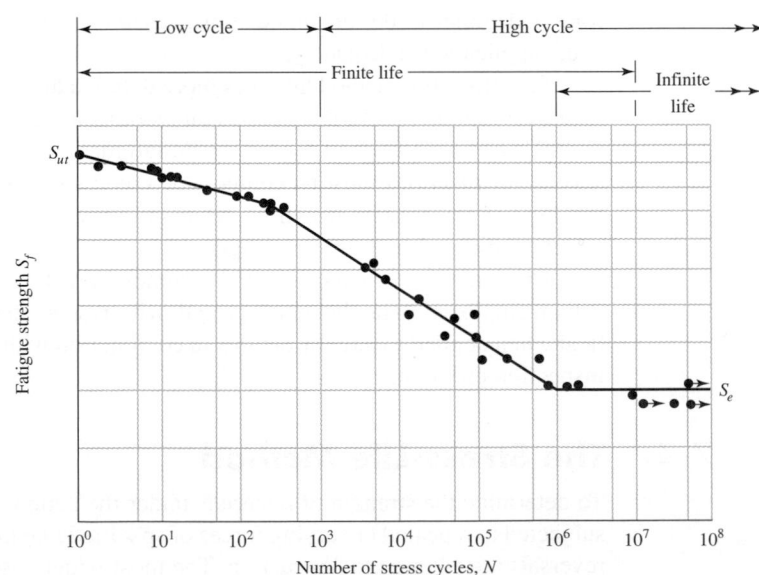

Figure 7–11

S-N bands for representative aluminum alloys, excluding wrought alloys with $S_{ut} < 38$ kpsi. (From R. C. Juvinall, Engineering Considerations of Stress, Strain and Strength. Copyright © 1967 by The McGraw-Hill Companies, Inc. Reprinted by permission.)

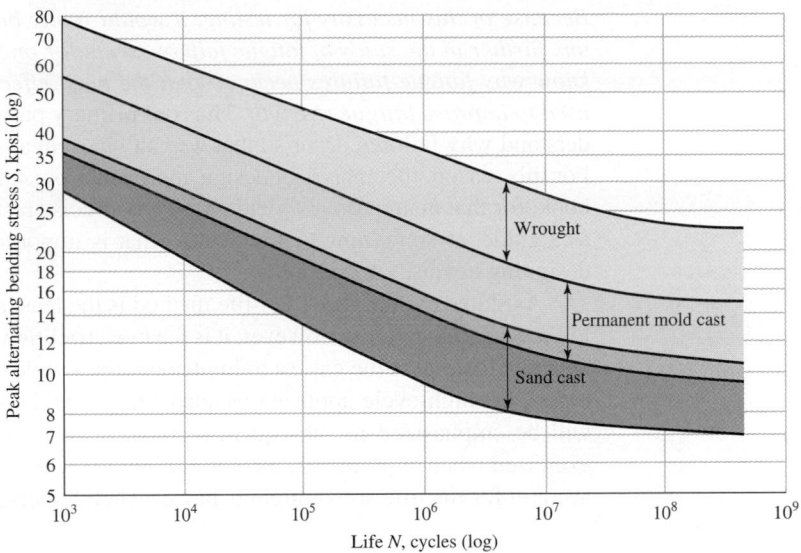

The ordinate of the *S-N* diagram is called the *fatigue strength S_f*; a statement of this strength value must always be accompanied by a statement of the number of cycles *N* to which it corresponds.

Soon we shall learn that *S-N* diagrams can be determined either for a test specimen or for an actual mechanical element. Even when the material of the test specimen and that of the mechanical element are identical, there will be significant differences between the diagrams for the two.

In the case of the steels, a knee occurs in the graph, and beyond this knee failure will not occur, no matter how great the number of cycles. The strength corresponding to the knee is called the *endurance limit S_e*, or the fatigue limit. The graph of Fig. 7–10 never does become horizontal for nonferrous metals and alloys, and hence these materials do not have an endurance limit. Figure 7–11 shows scatter bands indicating the *S-N* curves for most common aluminum alloys excluding wrought alloys having a tensile strength below 38 kpsi. Since aluminum does not have an endurance limit, normally the fatigue strength S_f is reported at a specific number of cycles, normally $N = 5(10^8)$ cycles of reversed stress (see Table A–24).

We note that a stress cycle ($N = 1$) constitutes a single application and removal of a load and then another application and removal of the load in the opposite direction. Thus $N = \frac{1}{2}$ means the load is applied once and then removed, which is the case with the simple tension test.

The body of knowledge available on fatigue failure from $N = 1$ to $N = 1000$ cycles is generally classified as *low-cycle fatigue*, as indicated in Fig. 7–10. *High-cycle fatigue*, then, is concerned with failure corresponding to stress cycles greater than 10^3 cycles.

We also distinguish a *finite-life region* and an *infinite-life region* in Fig. 7–10. The boundary between these regions cannot be clearly defined except for a specific material; but it lies somewhere between 10^6 and 10^7 cycles for steels, as shown in Fig. 7–10.

As noted previously, it is always good engineering practice to conduct a testing program on the materials to be employed in design and manufacture. This, in fact, is a requirement, not an option, in guarding against the possibility of a fatigue failure.

Because of this necessity for testing, it would really be unnecessary for us to proceed any further in the study of fatigue failure except for one important reason: the desire to know why fatigue failures occur so that the most effective method or methods can be used to improve fatigue strength. Thus our primary purpose in studying fatigue is to understand why failures occur so that we can guard against them in an optimum manner. For this reason, the analytical design approaches presented in this book, or in any other book, for that matter, do not yield absolutely precise results. The results should be taken as a guide, as something that indicates what is important and what is not important in designing against fatigue failure.

As stated earlier, the stress-life method is the least accurate approach especially for low-cycle applications. However, it is the most traditional method, with much published data available. It is the easiest to implement for a wide range of design applications and represents high-cycle applications adequately. For these reasons the stress-life method will be emphasized in subsequent sections of this chapter. However, care should be exercised when applying the method for low-cycle applications, as the method does not account for the true stress-strain behavior when localized yielding occurs.

7–5 The Strain-Life Method

The best approach yet advanced to explain the nature of fatigue failure is called by some the *strain-life* method. The approach can be used to estimate fatigue strengths, but when it is so used it is necessary to compound several idealizations, and so some uncertainties will exist in the results. For this reason, the method is presented here only because of its value in explaining the nature of fatigue.

A fatigue failure almost always begins at a local discontinuity such as a notch, crack, or other area of stress concentration. When the stress at the discontinuity exceeds the elastic limit, plastic strain occurs. If a fatigue fracture is to occur, there must exist cyclic plastic strains. Thus we shall need to investigate the behavior of materials subject to cyclic deformation.

In 1910, Bairstow verified by experiment Bauschinger's theory that the elastic limits of iron and steel can be changed, either up or down, by the cyclic variations of stress.[2] In general, the elastic limits of annealed steels are likely to increase when subjected to cycles of stress reversals, while cold-drawn steels exhibit a decreasing elastic limit.

Test specimens subjected to reversed bending are not suitable for strain cycling, because of the difficulty of measuring plastic strains. Consequently, most of the research has been done on axial specimens. By using electrical transducers, it is possible to generate signals that are proportional to the stress and strain, respectively. These signals can then be displayed on an oscilloscope or plotted on an *XY* plotter. R. W. Landgraf has investigated the low-cycle fatigue behavior of a large number of very high-strength steels, and during his research he made many cyclic stress-strain plots.[3] Figure 7–12 has been constructed to show the general appearance of these plots for the first few cycles of controlled cyclic strain. In this case the strength decreases with stress repetitions, as evidenced by the fact that the reversals occur at ever-smaller stress levels. As previously noted, other materials may be strengthened, instead, by cyclic stress reversals.

---

[2]L. Bairstow, "The Elastic Limits of Iron and Steel under Cyclic Variations of Stress," *Philosophical Transactions,* Series A, vol. 210, Royal Society of London, 1910, pp. 35–55.

[3]R. W. Landgraf, *Cyclic Deformation and Fatigue Behavior of Hardened Steels,* Report no. 320, Department of Theoretical and Applied Mechanics, University of Illinois, Urbana, 1968, pp. 84–90.

Figure 7–12

True stress–true strain hysteresis loops showing the first five stress reversals of a cyclic softening material. The graph is slightly exaggerated for clarity. Note that the slope of the line AB is the modulus of elasticity E. The stress range is $\Delta\sigma$, $\Delta\varepsilon_p$ is the plastic-strain range, and $\Delta\varepsilon_e$ is the elastic strain range. The total-strain range is $\Delta\varepsilon = \Delta\varepsilon_p + \Delta\varepsilon_e$.

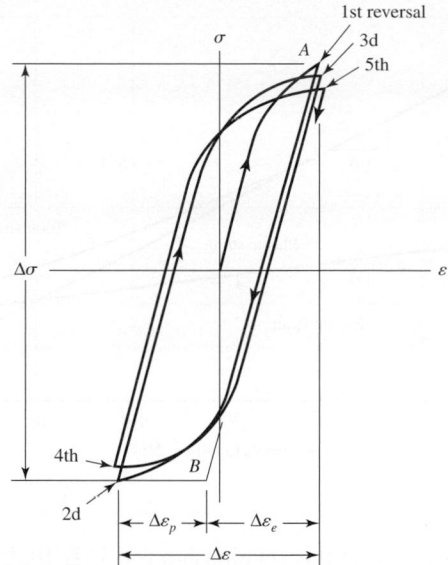

Figure 7–13

Monotonic and cyclic stress-strain results. (a) Ausformed H-11 steel, 660 Brinell; (b) SAE 4142 steel, 400 Brinell.

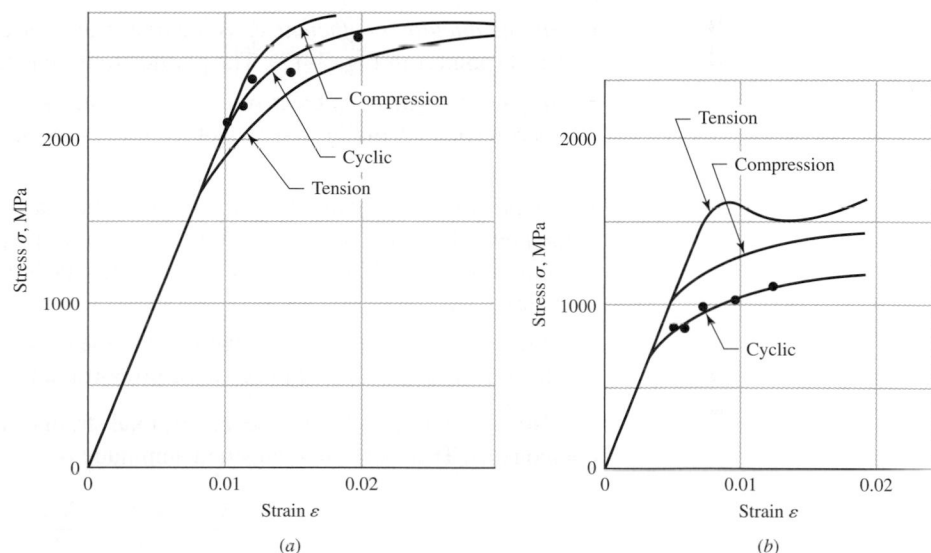

Slightly different results may be obtained if the first reversal occurs in the compressive region; this is probably due to the fatigue-strengthening effect of compression.

Landgraf's paper contains a number of plots that compare the monotonic stress-strain relations in both tension and compression with the cyclic stress-strain curve.[4] Two of these have been redrawn and are shown in Fig. 7–13. The importance of these is that they emphasize the difficulty of attempting to predict the fatigue strength of a material from known values of monotonic yield or ultimate strengths in the low-cycle region.

The SAE Fatigue Design and Evaluation Steering Committee released a report in 1975 in which the life in reversals to failure is related to the strain amplitude $\Delta\varepsilon/2$.[5] The

[4]Ibid., pp. 58–62.

[5]*Technical Report on Fatigue Properties*, SAE J1099, 1975.

Figure 7–14

A log-log plot showing how the fatigue life is related to the true-strain amplitude for hot-rolled SAE 1020 steel. *(Reprinted with permission from SAE J1099_200208 © 2002 SAE International.)*

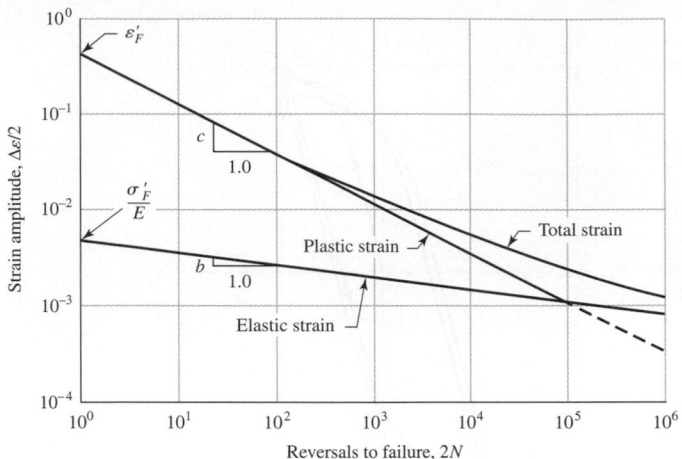

report contains a plot of this relationship for SAE 1020 hot-rolled steel; the graph has been reproduced as Fig. 7–14. To explain the graph, we first define the following terms:

- *Fatigue ductility coefficient* ε'_F is the true strain corresponding to fracture in one reversal (point *A* in Fig. 7–12). The plastic-strain line begins at this point in Fig. 7–14.

- *Fatigue strength coefficient* σ'_F is the true stress corresponding to fracture in one reversal (point *A* in Fig. 7–12). Note in Fig. 7–14 that the elastic-strain line begins at σ'_F / E.

- *Fatigue ductility exponent c* is the slope of the plastic-strain line in Fig. 7–14 and is the power to which the life 2*N* must be raised to be proportional to the true plastic-strain amplitude. If the number of stress reversals is 2*N*, then *N* is the number of cycles.

- *Fatigue strength exponent b* is the slope of the elastic-strain line, and is the power to which the life 2*N* must be raised to be proportional to the true-stress amplitude.

Now, from Fig. 7–12, we see that the total strain is the sum of the elastic and plastic components. Therefore the total strain amplitude is

$$\frac{\Delta\varepsilon}{2} = \frac{\Delta\varepsilon_e}{2} + \frac{\Delta\varepsilon_p}{2} \tag{a}$$

The equation of the plastic-strain line in Fig. 7–14 is

$$\frac{\Delta\varepsilon_p}{2} = \varepsilon'_F (2N)^c \tag{7–1}$$

The equation of the elastic strain line is

$$\frac{\Delta\varepsilon_e}{2} = \frac{\sigma'_F}{E}(2N)^b \tag{7–2}$$

Therefore, from Eq. (*a*), we have for the total-strain amplitude

$$\frac{\Delta\varepsilon}{2} = \frac{\sigma'_F}{E}(2N)^b + \varepsilon'_F (2N)^c \tag{7–3}$$

Table 7–1

Cyclic Properties of Some High-Strength Steels   *Source:* Data from R. W. Landgraf, *Cyclic Deformation and Fatigue Behavior of Hardened Steels,* Report no. 320, Department of Theoretical and Applied Mechanics, University of Illinois, Urbana, 1968.

AISI Number	Processing	Brinell Hardness H_B	Cyclic Yield Strength S'_y, kpsi	Fatigue Strength Coefficient σ'_F, kpsi	Fatigue Ductility Coefficient ε'_F	Fatigue Strength Exponent b	Fatigue Ductility Exponent c	Fatigue Strain-Hardening Exponent m
1045	Q & T 80°F	705	⋯	310	⋯	−0.065	−1.0	0.10
1045	Q & T 360°F	595	250	395	0.07	−0.055	−0.60	0.13
1045	Q & T 500°F	500	185	330	0.25	−0.08	−0.68	0.12
1045	Q & T 600°F	450	140	260	0.35	−0.07	−0.69	0.12
1045	Q & T 720°F	390	110	230	0.45	−0.074	−0.68	0.14
4142	Q & T 80°F	670	300	375	⋯	−0.075	−1.0	0.05
4142	Q & T 400°F	560	250	385	0.07	−0.076	−0.76	0.11
4142	Q & T 600°F	475	195	315	0.09	−0.081	−0.66	0.14
4142	Q & T 700°F	450	155	290	0.40	−0.080	0.73	0.12
4142	Q & T 840°F	380	120	265	0.45	−0.080	−0.75	0.14
4142*	Q & D 550°F	475	160	300	0.20	−0.082	−0.77	0.12
4142	Q & D 650°F	450	155	305	0.60	−0.090	−0.76	0.13
4142	Q & D 800°F	400	130	275	0.50	−0.090	−0.75	0.14

*Deformed 14 percent.

which is the Manson-Coffin relationship between fatigue life and total strain.[6] Some values of the coefficients and exponents are listed in Table 7–1. Many more are included in the SAE J1099 report.

Though Eq. (7–3) is a perfectly legitimate equation for obtaining the fatigue life of a part when the strain and other cyclic characteristics are given, it appears to be of little use to the designer. The question of how to determine the total strain at the bottom of a notch or discontinuity has not been answered. There are no tables or charts of strain concentration factors in the literature. It is possible that strain concentration factors will become available in research literature very soon because of the increase in the use of finite-element analysis. Moreover, finite element analysis can of itself approximate the strains that will occur at all points in the subject structure.[7]

7–6   The Linear-Elastic Fracture Mechanics Method

The first phase of fatigue cracking is designated as stage I fatigue. Crystal slip that extends through several contiguous grains, inclusions, and surface imperfections is presumed to play a role. Since most of this is invisible to the observer, we just say that stage I involves

[6]J. F. Tavernelli and L. F. Coffin, Jr., "Experimental Support for Generalized Equation Predicting Low Cycle Fatigue," and S. S. Manson, discussion, *Trans. ASME, J. Basic Eng.,* vol. 84, no. 4, pp. 533–537.

[7]For further discussion of the strain-life method see N. E. Dowling, *Mechanical Behavior of Materials,* 2nd ed., Prentice-Hall, Englewood Cliffs, N.J., 1999, Chap. 14.

several grains. The second phase, that of crack extension, is called stage II fatigue. The advance of the crack (that is, new crack area is created) does produce evidence that can be observed on micrographs from an electron microscope. The growth of the crack is orderly. Final fracture occurs during stage III fatigue, although fatigue is not involved. When the crack is sufficiently long that $K_I = K_{Ic}$ for the stress amplitude involved, then K_{Ic} is the critical stress intensity for the undamaged metal, and there is sudden, catastrophic failure of the remaining cross section in tensile overload (see Sec. 6–13). Stage III fatigue is associated with rapid acceleration of crack growth then fracture.

Crack Growth

Fatigue cracks nucleate and grow when stresses vary and there is some tension in each stress cycle. Consider the stress to be fluctuating between the limits of σ_{min} and σ_{max}, where the stress range is defined as $\Delta\sigma = \sigma_{max} - \sigma_{min}$. From Eq. (6–51) the stress intensity is given by $K_I = \beta\sigma\sqrt{\pi a}$. Thus, for $\Delta\sigma$, the stress intensity range per cycle is

$$\Delta K_I = \beta(\sigma_{max} - \sigma_{min})\sqrt{\pi a} = \beta\Delta\sigma\sqrt{\pi a} \qquad (7\text{–}4)$$

To develop fatigue strength data, a number of specimens of the same material are tested at various levels of $\Delta\sigma$. Cracks nucleate at or very near a free surface or large discontinuity. Assuming an initial crack length of a_i, crack growth as a function of the number of stress cycles N will depend on $\Delta\sigma$, that is, ΔK_I. For ΔK_I below some threshold value $(\Delta K_I)_{th}$ a crack will not grow. Figure 7–15 represents the crack length a as a function of N for three stress levels $(\Delta\sigma)_3 > (\Delta\sigma)_2 > (\Delta\sigma)_1$, where $(\Delta K_I)_3 > (\Delta K_I)_2 > (\Delta K_I)_1$. Notice the effect of the higher stress range in Fig. 7–15 in the production of longer cracks at a particular cycle count.

When the rate of crack growth per cycle, da/dN in Fig. 7–15, is plotted as shown in Fig. 7–16, the data from all three stress range levels superpose to give a sigmoidal locus. The three stages of crack development are observable, and the stage II data are linear on log-log coordinates, within the domain of linear elastic fracture mechanics (LEFM) validity. A group of similar curves can be generated by changing the stress ratio $R = \sigma_{min}/\sigma_{max}$ of the experiment.

When we present a simplified procedure for estimating the remaining life of a cyclically stressed part after discovery of a crack. This requires the assumption that plane

Figure 7–15

The increase in crack length a from an initial length of a_i as a function of cycle count for three stress ranges, $(\Delta\sigma)_3 > (\Delta\sigma)_2 > (\Delta\sigma)_1$.

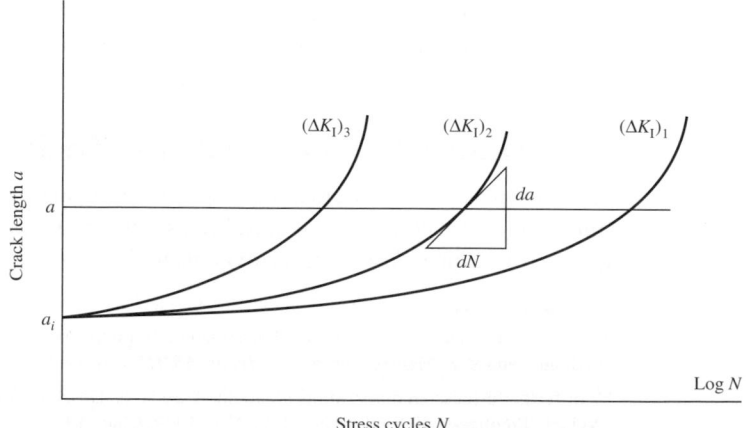

Figure 7–16

When *da/dN* is measured in Fig. 7–15 and plotted on loglog coordinates, the data for different stress ranges *superpose*, giving rise to a sigmoid curve as shown. $(\Delta K_{\mathrm{I}})_{\mathrm{th}}$ is the threshold value of ΔK_{I}, below which a crack does not grow. From threshold to rupture an aluminum alloy will spend 85–90 percent of life in region I, 5–8 percent in region II, and 1–2 percent in region III.

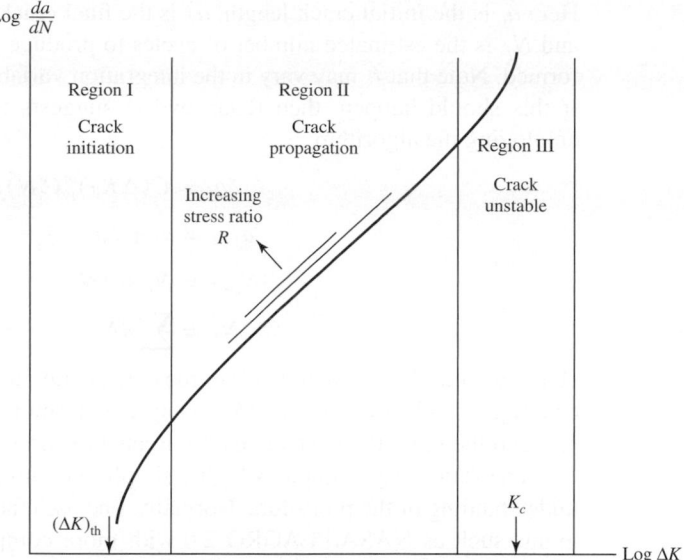

Table 7–2

Conservative Values of Factor *C* and Exponent *m* in Eq. (7–5) for Various Forms of Steel $(R \doteq 0)$

Material	$C, \dfrac{\text{m/cycle}}{\left(\text{MPa}\sqrt{\text{m}}\right)^m}$	$C, \dfrac{\text{in/cycle}}{\left(\text{kpsi}\sqrt{\text{in}}\right)^m}$	m
Ferritic-pearlitic steels	$6.89(10^{-12})$	$3.60(10^{-10})$	3.00
Martensitic steels	$1.36(10^{-10})$	$6.60(10^{-9})$	2.25
Austenitic stainless steels	$5.61(10^{-12})$	$3.00(10^{-10})$	3.25

From J.M. Barsom and S.T. Rolfe, *Fatigue and Fracture Control in Structures,* 2nd ed., Prentice Hall, Upper Saddle River, NJ, 1987, pp. 288–291, Copyright ASTM International. Reprinted with permission.

strain conditions prevail.[8] Assuming a crack is discovered early in stage II, the crack growth in region II of Fig. 7–16 can be approximated by the *Paris equation,* which is of the form

$$\frac{da}{dN} = C(\Delta K_{\mathrm{I}})^m \tag{7–5}$$

where *C* and *m* are empirical constants and ΔK_{I} is given by Eq. (7–4). Representative, but conservative, values of *C* and *m* for various classes of steels are listed in Table 7–2. Substituting Eq. (7–4) and integrating gives

$$\int_0^{N_f} dN = N_f = \frac{1}{C} \int_{a_i}^{a_f} \frac{da}{(\beta \Delta \sigma \sqrt{\pi a})^m} \tag{7–6}$$

[8]Recommended references are: Dowling, op. cit.; J. A. Collins, *Failure of Materials in Mechanical Design,* John Wiley & Sons, New York, 1981; H. O. Fuchs and R. I. Stephens, *Metal Fatigue in Engineering,* John Wiley & Sons, New York, 1980; and Harold S. Reemsnyder, "Constant Amplitude Fatigue Life Assessment Models," *SAE Trans. 820688,* vol. 91, Nov. 1983.

Here a_i is the initial crack length, a_f is the final crack length corresponding to failure, and N_f is the estimated number of cycles to produce a failure *after* the initial crack is formed. Note that β may vary in the integration variable (e.g., see Figs. 6–35 to 6–40). If this should happen, then Reemsnyder[9] suggests the use of numerical integration employing the algorithm

$$\delta a_j = C(\Delta K_I)_j^m (\delta N)_j$$

$$a_{j+1} = a_j + \delta a_j$$

$$N_{j+1} = N_j + \delta N_j \tag{7–7}$$

$$N_f = \sum \delta N_j$$

Here δa_j and δN_j are increments of the crack length and the number of cycles. The procedure is to select a value of δN_j, using a_i determine β and compute ΔK_I, determine δa_j, and then find the next value of a. Repeat the procedure until $a = a_f$.

The following example is highly simplified with β constant in order to give some understanding of the procedure. Normally, one uses fatigue crack growth computer programs such as NASA/FLAGRO 2.0 with more comprehensive theoretical models to solve these problems.

---

[9]Op. cit.

EXAMPLE 7–1

The bar shown in Fig. 7–17 is subjected to a repeated moment $0 \leq M \leq 1200$ lbf · in. The bar is AISI 4430 steel with $S_{ut} = 185$ kpsi, $S_y = 170$ kpsi, and $K_{Ic} = 74$ kpsi$\sqrt{\text{in}}$. Material tests on various specimens of this material with identical heat treatment indicate worst-case constants of $C = 3.8(10^{-11})$(in/cycle)/(kpsi$\sqrt{\text{in}}$)^m and $m = 3.0$. As shown, a nick of size 0.004 in has been discovered on the bottom of the bar. Estimate the number of cycles of life remaining.

Solution

The stress range $\Delta\sigma$ is always computed by using the nominal (uncracked) area. Thus

$$\frac{I}{c} = \frac{bh^2}{6} = \frac{0.25(0.5)^2}{6} = 0.0104 \text{ in}^3$$

Therefore, before the crack initiates, the stress range is

$$\Delta\sigma = \frac{\Delta M}{I/c} = \frac{1200}{0.0104} = 115.4(10^3) \text{ psi} = 115.4 \text{ kpsi}$$

which is below the yield strength. As the crack grows, it will eventually become long enough such that the bar will completely yield or undergo a brittle fracture. For the ratio of S_y/S_{ut} it is highly unlikely that the bar will reach complete yield. For brittle fracture,

| Figure 7–17

designate the crack length as a_f. If $\beta = 1$, then from Eq. (6–51) with $K_I = K_{Ic}$, we approximate a_f as

$$a_f = \frac{1}{\pi}\left(\frac{K_{Ic}}{\beta\sigma_{max}}\right)^2 \doteq \frac{1}{\pi}\left(\frac{73}{115.4}\right)^2 = 0.127 \text{ in}$$

From Fig. 6–37, we compute the ratio a_f/h as

$$\frac{a_f}{h} = \frac{0.127}{0.5} = 0.254$$

Thus a_f/h varies from near zero to approximately 0.254. From Fig. 6–37, for this range β is nearly constant at approximately 1.07. We will assume it to be so, and re-evaluate a_f as

$$a_f = \frac{1}{\pi}\left(\frac{73}{1.07(115.4)}\right)^2 = 0.111 \text{ in}$$

Thus, from Eq. (7–6), the estimated remaining life is

$$N_f = \frac{1}{C}\int_{a_i}^{a_f}\frac{da}{(\beta\Delta\sigma\sqrt{\pi a})^m} = \frac{1}{3.8(10^{-11})}\int_{0.004}^{0.111}\frac{da}{[1.07(115.4)\sqrt{\pi a}]^3}$$

$$= -\frac{5.02(10^3)}{\sqrt{a}}\bigg|_{0.004}^{0.111} = 64.3\,(10^3)\text{ cycles}$$

7–7   The Endurance Limit

The determination of endurance limits by fatigue testing is now routine, though a lengthy procedure. Generally, stress testing is preferred to strain testing for endurance limits.

For preliminary and prototype design and for some failure analysis as well, a quick method of estimating endurance limits is needed. There are great quantities of data in the literature on the results of rotating-beam tests and simple tension tests of specimens taken from the same bar or ingot. By plotting these as in Fig. 7–18, it is possible to see whether there is any correlation between the two sets of results. The graph appears to suggest that the endurance limit ranges from about 40 to 60 percent of the tensile strength for steels up to about 212 kpsi (1460 MPa). Beginning at about $S_{ut} = 212$ kpsi (1460 MPa), the scatter appears to increase, but the trend seems to level off, as suggested by the dashed horizontal line at $S'_e = 107$ kpsi (740 MPa).

Another series of tests, this time for various microstructures, is shown in Table 7–3. In this table the endurance limits vary from about 23 to 63 percent of the tensile strength.[10]

Now, it is important to observe that the dispersion of the endurance limit is *not* due to a dispersion in the tensile strengths of the specimen, but rather that the spread occurs even when the tensile strengths of a large number of specimens remain exactly the same. Keep this in mind when choosing factors of safety.

---

[10]But see H. O. Fuchs and R. I. Stephens, *Metal Fatigue in Engineering*, Wiley, New York, 1980, pp. 69–71, which reports a range of 35 to 60 percent for steels having $S_{ut} < 1400$ MPa and as low as 20 percent for high-strength steels.

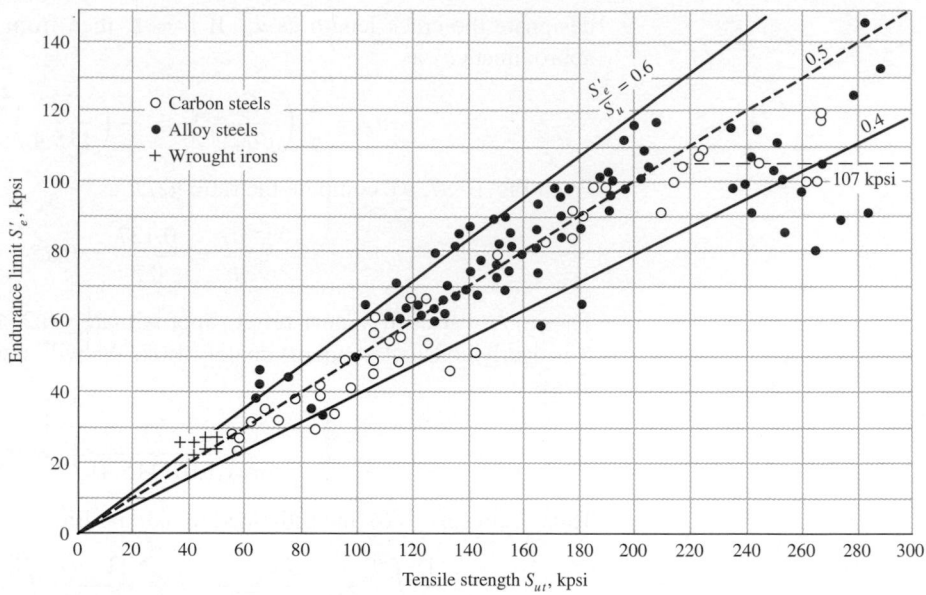

Figure 7–18

Graph of endurance limits versus tensile strengths from actual test results for a large number of wrought irons and steels. Ratios of S'_e/S_{ut} of 0.60, 0.50, and 0.40 are shown by the solid and dashed lines. Note also the horizontal dashed line for $S'_e = 107$ kpsi. Points shown having a tensile strength greater than 214 kpsi have a mean endurance limit of $S'_e = 107$ kpsi and a standard deviation of 13.5 kpsi. *(Collated from data compiled by H. J. Grover, S. A. Gordon, and L. R. Jackson in Fatigue of Metals and Structures, Bureau of Naval Weapons Document NAVWEPS 00-25-534, 1960; and from Fatigue Design Handbook, SAE, 1968, p. 42.)*

Table 7–3

Endurance-Limit Ratio S'_e/S_{ut} for Various Steel Microstructures

Source: Adapted from L. Sors, *Fatigue Design of Machine Components,* Pergamon Press, Oxford, England, 1971.

	Ferrite		Pearlite		Martensite	
	Range	**Average**	**Range**	**Average**	**Range**	**Average**
Carbon steel	0.57–0.63	0.60	0.38–0.41	0.40	$\cdots$	0.25
Alloy steel	$\cdots$	$\cdots$	$\cdots$	$\cdots$	0.23–0.47	0.35

We wish now to present a method for estimating endurance limits. Note that estimates obtained from quantities of data obtained from many sources probably have a large spread and might deviate significantly from the results of actual laboratory tests of the mechanical properties of specimens obtained through strict purchase-order specifications. Since the area of uncertainty is greater, compensation must be made by employing larger design factors than would be used for static design.

Mischke[11] has analyzed a great deal of actual test data from several sources and concluded that endurance limit can, indeed, be related to tensile strength. For steels, the

---

[11]Charles R. Mischke, "Prediction of Stochastic Endurance Strength," *Trans. of ASME, Journal of Vibration, Acoustics, Stress, and Reliability in Design,* vol. 109, no. 1, January 1987, pp. 113–122.

relationship is

$$S'_e = \begin{cases} 0.504 S_{ut} \text{ kpsi or MPa} & S_{ut} \le 212 \text{ k} \\ 107 \text{ kpsi} & S_{ut} > 212 \text{ k} \\ 740 \text{ MPa} & S_{ut} > 1460 \end{cases}$$

where S_{ut} is the *minimum* tensile strength. The prime mark
to the *rotating-beam specimen* itself. We wish to reserve the ~~~~~~~~~~~~~~~~S_e for the
endurance limit of any particular machine element subjected to any kind of loading.
Soon we shall learn that the two strengths may be quite different.

The data of Table 7–3 emphasize the difficulty of attempting to provide a single rule
for deriving the endurance limit from the tensile strength. The table also shows a part of
the cause of this difficulty. Steels treated to give different microstructures have different
S'_e/S_{ut} ratios. It appears that the more ductile microstructures have a higher ratio.
Martensite has a very brittle nature and is highly susceptible to fatigue-induced cracking; thus the ratio is low. When designs include detailed heat-treating specifications to
obtain specific microstructures, it is possible to use an estimate of the endurance limit
based on test data for the particular microstructure; such estimates are much more reliable and indeed should be used.

The endurance limits for various classes of cast irons, polished or machined, are
given in Table A–24. Aluminum alloys do not have an endurance limit. The fatigue
strengths of some aluminum alloys at $5(10^8)$ cycles of reversed stress are given in
Table A–24.

7–8 Fatigue Strength

As shown in Fig. 7–10, a region of low-cycle fatigue extends from $N = 1$ to about 10^3
cycles. In this region the fatigue strength S_f is only slightly smaller than the tensile
strength S_{ut}. An analytical approach has been given by Mischke[12] for both high-cycle
and low-cycle regions, requiring the parameters of the Manson-Coffin equation plus the
strain-strengthening exponent m. Engineers often have to work with less information.

Figure 7–10 indicates that the high-cycle fatigue domain extends from 10^3 cycles
for steels to the endurance limit life N_e, which is about 10^6 to 10^7 cycles. The purpose of
this section is to develop methods of approximation of the *S-N* diagram in the high-cycle
region, when information may be as sparse as the results of a simple tension test. Experience has shown high-cycle fatigue data are rectified by a logarithmic transform to both
stress and cycles-to-failure. Engineers can work with Eq. (7–2) in the following way:

$$(S_f)_{10^3 \text{ cycles}} = \sigma'_F (2 \cdot 10^3)^b = f S_{ut}$$

where f is the fraction of S_{ut} represented by $(S_f)_{10^3 \text{ cycles}}$. Solving for f gives

$$f = \frac{\sigma'_F}{S_{ut}} (2 \cdot 10^3)^b \qquad (7\text{--}9)$$

Now, from Eq. (3–11), $\sigma'_F = \sigma_0 \varepsilon^m$, with $\varepsilon = \varepsilon'_F$. If this true-stress–true-strain equation
is not known, the SAE approximation[13] for steels with $H_B \le 500$ may be used:

$$\sigma'_F = S_{ut} + 50 \text{ kpsi} \qquad \text{or} \qquad \sigma'_F = S_{ut} + 345 \text{ MPa} \qquad (7\text{--}10)$$

[12]J. E. Shigley and C. R. Mischke, *Standard Handbook of Machine Design,* 2nd ed., McGraw-Hill,
New York, 1996, pp. 13.34–13.37.

[13]*Fatigue Design Handbook,* vol. 4, Society of Automotive Engineers, New York, 1958, p. 27.

The exponent b is found from $\sigma_a = S_e = \sigma'_F(2N_e)^b$ as

$$b = -\frac{\log(\sigma'_F/S_e)}{\log(2N_e)} \tag{7-11}$$

Thus the equation $S_f = \sigma'_F(2N)^b$ is known. For example, if $S_{ut} = 105$ kpsi and $S_e = 52.5$ kpsi at 10^6 cycles-to-failure,

$$\sigma'_F = 105 + 50 = 155 \text{ kpsi}$$

$$b = -\frac{\log(155/52.5)}{\log(2 \cdot 10^6)} = -0.0746$$

$$f = \frac{155}{105}(2 \cdot 10^3)^{-0.0746} = 0.837$$

and

$$S_f = 155(2N)^{-0.0746} \tag{a}$$

Empirically, the common curve fit is

$$S_f = aN^b \tag{7-12}$$

where N is cycles to failure and the constants a and b are defined by the points $10^3, (S_f)_{10^3}$ and $10^6, S_e$, with $(S_f)_{10^3} = f S_{ut}$. Substituting these two points in Eq. (7–12) gives

$$a = \frac{(f S_{ut})^2}{S_e} \tag{7-13}$$

$$b = -\frac{1}{3}\log\left(\frac{f S_{ut}}{S_e}\right) \tag{7-14}$$

Continuing the informal example,

$$a = \frac{0.837^2 \, 105^2}{52.5} = 147.1 \text{ kpsi}$$

$$b = -\frac{1}{3}\log\frac{0.837(105)}{52.5} = -0.0746$$

and the resulting equation is

$$S_f = 147.1N^{-0.0746} \tag{b}$$

Note that $2^{-0.0746}(155) = 147.2$, so Eqs. (a) and (b) are really the same. There are popular curve fits with f treated as a constant, normally 0.9, but f varies with S_{ut}. The following table shows the nature of such an approximation:

S_{ut}, kpsi	60	90	120	200
f	0.93	0.86	0.82	0.77

If a completely reversed stress σ_a is given, setting $S_f = \sigma_a$ in Eq. (7–12), the number of cycles-to-failure can be expressed as

$$N = \left(\frac{\sigma_a}{a}\right)^{1/b} \tag{7-15}$$

Low-cycle fatigue is often defined (see Fig. 7–10) as failure that occurs in a range of $1 \le N \le 10^3$ cycles. On a loglog plot such as Fig. 7–10 the failure locus in this range is nearly linear below 10^3 cycles. A straight line between 10^3, $f S_{ut}$ and 1, S_{ut} (transformed) is conservative, and it is given by

$$S_f \ge S_{ut} N^{(\log f)/3} \qquad 1 \le N \le 10^3 \qquad (7\text{–}16)$$

EXAMPLE 7–2

A 1050 HR steel has an ultimate tensile strength of $S_{ut} = 105$ kpsi and a yield strength of 60 kpsi.

(a) Estimate the rotating-beam endurance limit at 10^6 cycles.

(b) Estimate the endurance strength for a polished rotating-beam specimen corresponding to 10^4 cycles to failure.

(c) Estimate the expected life under a completely reversed stress of 55 kpsi.

Solution

(a) From Eq. (7–8),

Answer

$$S'_e = 0.504(105) = 52.9 \text{ kpsi}$$

(b) From Eq. (7–10),

$$\sigma'_F = 105 + 50 = 155 \text{ kpsi}$$

From Eq. (7–11), with $S_e = S'_e$,

$$b = -\frac{\log(155/52.9)}{\log(2 \cdot 10^6)} = -0.0741$$

From Eq. (7–9),

$$f = \frac{155}{105}(2 \cdot 10^3)^{-0.0741} = 0.840$$

From Eq. (7–13),

$$a = \frac{[(0.840)(105)]^2}{52.9} = 147.1 \text{ kpsi}$$

From Eq. (7–14),

$$b = -\frac{1}{3} \log \left[\frac{0.840(105)}{52.9} \right] = -0.0740$$

$$S_f = 147.2 N^{-0.0740}$$

Answer

$$(S_f)_{10^4} = 147.1(10^4)^{-0.0740} = 74.4 \text{ kpsi}$$

(c) From Eq. (7–15),

Answer

$$N = \left(\frac{55}{147.1} \right)^{1/-0.0740} = 593\,810 = 5.9(10^5) \text{ cycles}$$

7–9 Endurance Limit Modifying Factors

We have seen that the rotating-beam specimen used in the laboratory to determine endurance limits is prepared very carefully and tested under closely controlled conditions. It is unrealistic to expect the endurance limit of a mechanical or structural member to match the values obtained in the laboratory. Some differences include

- *Material:* composition, basis of failure, variability
- *Manufacturing:* method, heat treatment, fretting corrosion, surface condition, stress concentration
- *Environment:* corrosion, temperature, stress state, relaxation times
- *Design:* size, shape, life, stress state, stress concentration, speed, fretting, galling

Marin[14] identified factors that quantified the effects of surface condition, size, loading, temperature, and miscellaneous items. The question of whether to adjust the endurance limit by subtractive corrections or multiplicative corrections was resolved by an extensive statistical analysis of a 4340 (electric furnace, aircraft quality) steel, in which a correlation coefficient of 0.85 was found for the multiplicative form and 0.40 for the additive form. A Marin equation is therefore written as

$$S_e = k_a k_b k_c k_d k_e k_f S'_e \qquad (7\text{–}17)$$

where k_a = surface condition modification factor

k_b = size modification factor

k_c = load modification factor

k_d = temperature modification factor

k_e = reliability factor[15]

k_f = miscellaneous-effects modification factor

S'_e = rotary-beam test specimen endurance limit

S_e = endurance limit at the critical location of a machine part in the geometry and condition of use

When endurance tests of parts are not available, estimations are made by applying Marin factors to the endurance limit.

Surface Factor k_a

The surface of a rotating-beam specimen is highly polished, with a final polishing in the axial direction to smooth out any circumferential scratches. The surface modification factor depends on the quality of the finish of the actual part surface and on the tensile strength of the part material. To find quantitative expressions for common finishes of machine parts (ground, machined, or cold-drawn, hot-rolled, and as-forged), the coordinates

---

[14]Joseph Marin, *Mechanical Behavior of Engineering Materials,* Prentice-Hall, Englewood Cliffs, N.J., 1962, p. 224.

[15]Complete stochastic analysis is presented in Sec. 7–17. Until that point the presentation is one of a deterministic nature. However, we must take care of the known scatter in the fatigue data. This means that we will not carry out a true reliability analysis at this time but will attempt to answer the question: What is the probability that a *known* (assumed) stress will exceed the strength of a randomly selected component made from this material population?

of data points were recaptured from a plot of endurance limit versus ultimate tensile strength of data gathered by Lipson and Noll and reproduced by Horger.[16] The result of regression analysis by Mischke was of the form

$$k_a = aS_{ut}^b \tag{7–18}$$

where S_{ut} is the minimum tensile strength and a and b are to be found in Table 7–4.

Table 7–4

Parameters for Marin Surface Modification Factor, Eq. (7–18)

Surface Finish	Factor a S_{ut}, kpsi	Factor a S_{ut}, MPa	Exponent b
Ground	1.34	1.58	−0.085
Machined or cold-drawn	2.70	4.51	−0.265
Hot-rolled	14.4	57.7	−0.718
As-forged	39.9	272.	−0.995

From C.J. Noll and C. Lipson, "Allowable Working Stresses," *Society for Experimental Stress Analysis*, vol. 3, no. 2, 1946 p. 29. Reproduced by O.J. Horger (eds.) *Metals Engineering Design ASME Handbook*, McGraw-Hill, New York. Copyright © 1953 by The McGraw-Hill Companies, Inc. Reprinted by permission.

EXAMPLE 7–3

A steel has a minimum ultimate strength of 520 MPa and a machined surface. Estimate k_a.

Solution

From Table 7–4, $a = 4.51$ and $b = −0.265$. Then, from Eq. (7–18)

$$k_a = 4.51(520)^{-0.265} = 0.860$$

Size Factor k_b

The size factor has been evaluated using 133 sets of data points.[17] The results for bending and torsion may be expressed as

$$k_b = \begin{cases} (d/0.3)^{-0.107} = 0.879d^{-0.107} & 0.11 \leq d \leq 2 \text{ in} \\ 0.91d^{-0.157} & 2 < d \leq 10 \text{ in} \\ (d/7.62)^{-0.107} = 1.24d^{-0.107} & 2.79 \leq d \leq 51 \text{ mm} \\ 1.51d^{-0.157} & 51 < d \leq 254 \text{ mm} \end{cases} \tag{7–19}$$

For axial loading there is no size effect, so

$$k_b = 1 \tag{7–20}$$

but see k_c.

One of the problems that arises in using Eq. (7–19) is what to do when a round bar in bending is not rotating, or when a noncircular cross section is used. For example,

---

[16]C. J. Noll and C. Lipson, "Allowable Working Stresses," *Society for Experimental Stress Analysis,* vol. 3, no. 2, 1946, p. 29. Reproduced by O. J. Horger (ed.), *Metals Engineering Design ASME Handbook,* McGraw-Hill, New York, 1953, p. 102.

[17]Mischke, op. cit., Table 3.

what is the size factor for a bar 6 mm thick and 40 mm wide? The approach to be used here employs an *effective dimension* d_e obtained by equating the volume of material stressed at and above 95 percent of the maximum stress to the same volume in the rotating-beam specimen.[18] It turns out that when these two volumes are equated, the lengths cancel, and so we need only consider the areas. For a rotating round section, the 95 percent stress area is the area in a ring having an outside diameter d and an inside diameter of $0.95d$. So, designating the 95 percent stress area $A_{0.95\sigma}$, we have

$$A_{0.95\sigma} = \frac{\pi}{4}[d^2 - (0.95d)^2] = 0.0766d^2 \tag{7-21}$$

This equation is also valid for a rotating hollow round. For nonrotating solid or hollow rounds, the 95 percent stress area is twice the area outside of two parallel chords having a spacing of $0.95d$, where d is the diameter. Using an exact computation, this is

$$A_{0.95\sigma} = 0.01046d^2 \tag{7-22}$$

with d_e in Eq. (7–21), setting Eqs. (7–21) and (7–22) equal to each other enables us to solve for the effective diameter. This gives

$$d_e = 0.370d \tag{7-23}$$

as the effective size of a round corresponding to a nonrotating solid or hollow round.

A rectangular section of dimensions $h \times b$ has $A_{0.95\sigma} = 0.05hb$. Using the same approach as before,

$$d_e = 0.808(hb)^{1/2} \tag{7-24}$$

Table 7–5 provides $A_{0.95\sigma}$ areas of common structural shapes undergoing non-rotating bending.

---

[18]See R. Kuguel, "A Relation between Theoretical Stress Concentration Factor and Fatigue Notch Factor Deduced from the Concept of Highly Stressed Volume," *Proc. ASTM,* vol. 61, 1961, pp. 732–748.

EXAMPLE 7–4

A steel shaft loaded in bending is 32 mm in diameter, abutting a filleted shoulder 38 mm in diameter. The shaft material has a mean ultimate tensile strength of 690 MPa. Estimate the Marin size factor k_b if the shaft is used in
(*a*) A rotating mode.
(*b*) A nonrotating mode.

Solution

(*a*) From Eq. (7–19),

Answer

$$k_b = \left(\frac{d}{7.62}\right)^{-0.107} = \left(\frac{32}{7.62}\right)^{-0.107} = 0.858$$

(*b*) From Table 7–5,

$$d_e = 0.37d = 0.37(32) = 11.84 \text{ mm}$$

From Eq. (7–19),

Answer

$$k_b = \left(\frac{11.84}{7.62}\right)^{-0.107} = 0.954$$

Table 7-5

$A_{0.95\sigma}$ Areas of Common Nonrotating Structural Shapes

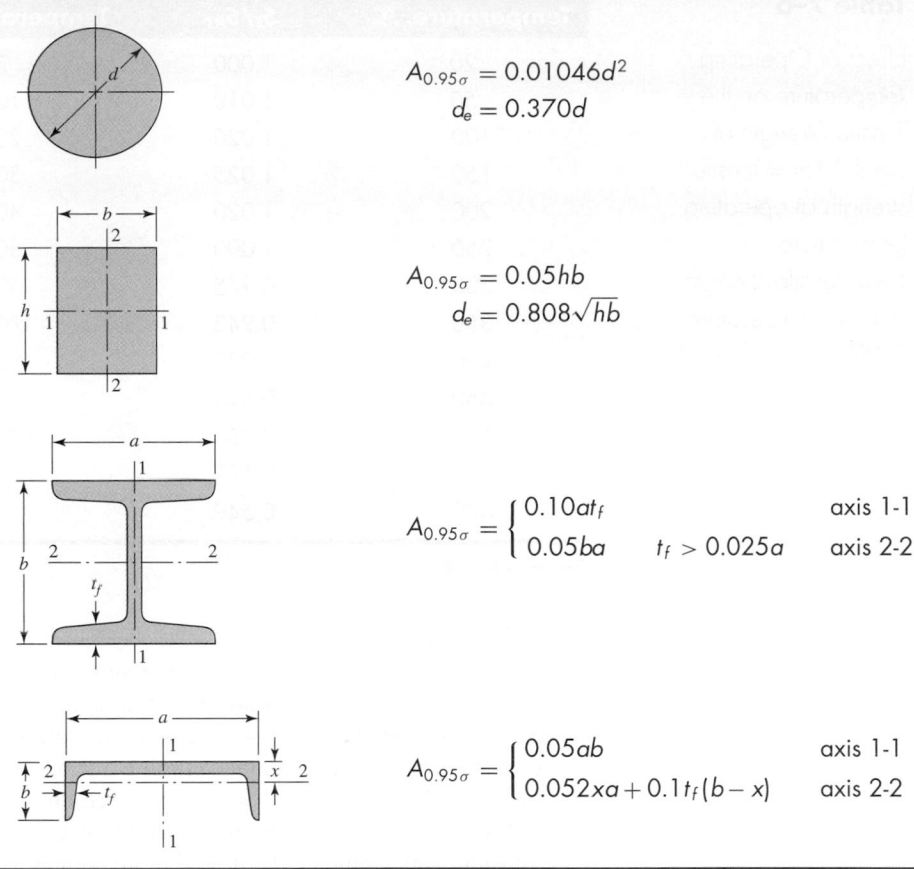

$$A_{0.95\sigma} = 0.01046d^2$$
$$d_e = 0.370d$$

$$A_{0.95\sigma} = 0.05hb$$
$$d_e = 0.808\sqrt{hb}$$

$$A_{0.95\sigma} = \begin{cases} 0.10at_f & \text{axis 1-1} \\ 0.05ba & t_f > 0.025a \quad \text{axis 2-2} \end{cases}$$

$$A_{0.95\sigma} = \begin{cases} 0.05ab & \text{axis 1-1} \\ 0.052xa + 0.1t_f(b-x) & \text{axis 2-2} \end{cases}$$

Loading Factor k_c

When fatigue tests are carried out with rotating bending, axial (push-pull), and torsional loading, the endurance limits differ with S_{ut}. This is discussed further in Sec. 7–17. Here, we will specify average values of the load factor as

$$k_c = \begin{cases} 1 & \text{bending} \\ 0.85 & \text{axial} \\ 0.59 & \text{torsion}^{19} \end{cases} \tag{7-25}$$

Temperature Factor k_d

When operating temperatures are below room temperature, brittle fracture is a strong possibility and should be investigated first. When the operating temperatures are higher than room temperature, yielding should be investigated first because the yield strength drops off so rapidly with temperature; see Fig. 3–8. Any stress will induce creep in a material operating at high temperatures; so this factor must be considered too. Finally, it may be true that there is no fatigue limit for materials operating at high temperatures. Because of the reduced fatigue resistance, the failure process is, to some extent, dependent on time.

---

[19]Use this only for pure torsional fatigue loading. When torsion is combined with other stresses, such as bending, $k_c = 1$ and the combined loading is managed by using the effective von Mises stress as in Sec. 6–5. *Note:* For pure torsion, the distortion energy predicts that $(k_c)_{torsion} = 0.577$.

Table 7–6

Effect of Operating Temperature on the Tensile Strength of Steel.* (S_T = tensile strength at operating temperature; S_{RT} = tensile strength at room temperature; $0.099 \le \hat{\sigma} \le 0.110$)

Temperature, °C	S_T/S_{RT}	Temperature, °F	S_T/S_{RT}
20	1.000	70	1.000
50	1.010	100	1.008
100	1.020	200	1.020
150	1.025	300	1.024
200	1.020	400	1.018
250	1.000	500	0.995
300	0.975	600	0.963
350	0.943	700	0.927
400	0.900	800	0.872
450	0.843	900	0.797
500	0.768	1000	0.698
550	0.672	1100	0.567
600	0.549		

*Data source: Fig. 3–8.

The limited amount of data available show that the endurance limit for steels increases slightly as the temperature rises and then begins to fall off in the 400 to 700°F range, not unlike the behavior of the tensile strength shown in Fig. 3–8. For this reason it is probably true that the endurance limit is related to tensile strength at elevated temperatures in the same manner as at room temperature.[20] It seems quite logical, therefore, to employ the same relations to predict endurance limit at elevated temperatures as are used at room temperature, at least until more comprehensive data become available. At the very least, this practice will provide a useful standard against which the performance of various materials can be compared.

Table 7–6 has been obtained from Fig. 3–8 by using only the tensile-strength data. Note that the table represents 145 tests of 21 different carbon and alloy steels. A fourth-order polynomial curve fit to the data underlying Fig. 3–8 gives

$$k_d = 0.975 + 0.432(10^{-3})T_F - 0.115(10^{-5})T_F^2$$
$$+ 0.104(10^{-8})T_F^3 - 0.595(10^{-12})T_F^4 \qquad (7\text{–}26)$$

where $70 \le T_F \le 1000°F$.

Two types of problems arise when temperature is a consideration. If the rotating-beam endurance limit is known at room temperature, then use

$$k_d = \frac{S_T}{S_{RT}} \qquad (7\text{–}27)$$

from Table 7–6 or Eq. (7–26) and proceed as usual. If the rotating-beam endurance limit is not given, then compute it using Eq. (7–8) and the temperature-corrected tensile strength obtained by using the factor from Table 7–6. Then use $k_d = 1$.

---

[20]For more, see Table 2 of ANSI/ASME B106. 1M-1985 shaft standard, and E. A. Brandes (ed.), *Smithell's Metals Reference Book,* 6th ed., Butterworth, London, 1983, pp. 22–134 to 22–136, where endurance limits from 100 to 650°C are tabulated.

EXAMPLE 7–5   A 1035 steel has a tensile strength of 70 kpsi and is to be used for a part that sees 450°F in service. Estimate the Marin temperature modification factor and $(S_e)_{450°}$ if
(a) The room-temperature endurance limit by test is $(S'_e)_{70°} = 39.0$ kpsi.
(b) Only the tensile strength at room temperature is known.

Solution   (a) First, from Eq. (7–26),

$$k_d = 0.975 + 0.432(10^{-3})(450) - 0.115(10^{-5})(450^2)$$
$$+ 0.104(10^{-8})(450^3) - 0.595(10^{-12})(450^4) = 1.007$$

Thus,

Answer $$(S_e)_{450°} = k_d(S'_e)_{70°} = 1.007(39.0) = 39.3 \text{ kpsi}$$

(b) Interpolating from Table 7–6 gives

$$(S_T/S_{RT})_{450°} = 1.018 + (0.995 - 1.018)\frac{450 - 400}{500 - 400} = 1.007$$

Thus, the tensile strength at 450°F is estimated as

$$(S_{ut})_{450°} = (S_T/S_{RT})_{450°}(S_{ut})_{70°} = 1.007(70) = 70.5 \text{ kpsi}$$

From Eq. (7–8) then,

Answer $$(S_e)_{450°} = 0.504\,(S_{ut})_{450°} = 0.504(70.5) = 35.5 \text{ kpsi}$$

Part *a* gives the better estimate due to actual testing of the particular material.

Reliability Factor k_e

The discussion presented here accounts for the scatter of data such as shown in Fig. 7–18 where the mean endurance limit is shown to be $S'_e/S_{ut} \doteq 0.5$, or as given by Eq. (7–8). Most endurance strength data are reported as mean values. Data presented by Haugen and Wirsching[21] show standard deviations of endurance strengths of less than 8 percent. Thus the reliability modification factor to account for this can be written as

$$k_e = 1 - 0.08\,z_a \tag{7–28}$$

where z_a is defined by Eq. (2–16) and values for any desired reliability can be determined from Table A–10. Table 7–7 gives reliability factors for some standard specified reliabilities.

For a more comprehensive approach to reliability, see Sec. 7–17.

Miscellaneous-Effects Factor k_f

Though the factor k_f is intended to account for the reduction in endurance limit due to all other effects, it is really intended as a reminder that these must be accounted for, because actual values of k_f are not always available.

Residual stresses may either improve the endurance limit or affect it adversely. Generally, if the residual stress in the surface of the part is compression, the endurance

---

[21]E. B. Haugen and P. H. Wirsching, "Probabilistic Design," *Machine Design*, vol. 47, no. 12, 1975, pp. 10–14.

Table 7–7

Reliability Factors k_e Corresponding to 8 Percent Standard Deviation of the Endurance Limit

Reliability, %	Transformation Variate z_a	Reliability Factor k_e
50	0	1.000
90	1.288	0.897
95	1.645	0.868
99	2.326	0.814
99.9	3.091	0.753
99.99	3.719	0.702
99.999	4.265	0.659
99.9999	4.753	0.620

Figure 7–19

The failure of a case-hardened part in bending or torsion. In this example, failure occurs in the core.

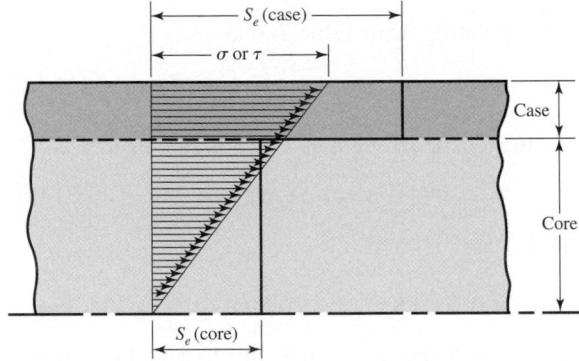

limit is improved. Fatigue failures appear to be tensile failures, or at least to be caused by tensile stress, and so anything that reduces tensile stress will also reduce the possibility of a fatigue failure. Operations such as shot peening, hammering, and cold rolling build compressive stresses into the surface of the part and improve the endurance limit significantly. Of course, the material must not be worked to exhaustion.

The endurance limits of parts that are made from rolled or drawn sheets or bars, as well as parts that are forged, may be affected by the so-called *directional characteristics* of the operation. Rolled or drawn parts, for example, have an endurance limit in the transverse direction that may be 10 to 20 percent less than the endurance limit in the longitudinal direction.

Parts that are case-hardened may fail at the surface or at the maximum core radius, depending upon the stress gradient. Figure 7–19 shows the typical triangular stress distribution of a bar under bending or torsion. Also plotted as a heavy line in this figure are the endurance limits S_e for the case and core. For this example the endurance limit of the core rules the design because the figure shows that the stress σ or τ, whichever applies, at the outer core radius, is appreciably larger than the core endurance limit.

Of course, if stress concentration is also present, the stress gradient is much steeper, and hence failure in the core is unlikely.

Corrosion

It is to be expected that parts that operate in a corrosive atmosphere will have a lowered fatigue resistance. This is, of course, true, and it is due to the roughening or pitting of the surface by the corrosive material. But the problem is not so simple as the one of finding the endurance limit of a specimen that has been corroded. The reason for this is that the corrosion and the stressing occur at the same time. Basically, this means that in time any part will fail when subjected to repeated stressing in a corrosive atmosphere. There is no fatigue limit. Thus the designer's problem is to attempt to minimize the factors that affect the fatigue life; these are:

- Mean or static stress
- Alternating stress
- Electrolyte concentration
- Dissolved oxygen in electrolyte
- Material properties and composition
- Temperature
- Cyclic frequency
- Fluid flow rate around specimen
- Local crevices

Electrolytic Plating

Metallic coatings, such as chromium plating, nickel plating, or cadmium plating, reduce the endurance limit by as much as 50 percent. In some cases the reduction by coatings has been so severe that it has been necessary to eliminate the plating process. Zinc plating does not affect the fatigue strength. Anodic oxidation of light alloys reduces bending endurance limits by as much as 39 percent but has no effect on the torsional endurance limit.

Metal Spraying

Metal spraying results in surface imperfections that can initiate cracks. Limited tests show reductions of 14 percent in the fatigue strength.

Cyclic Frequency

If, for any reason, the fatigue process becomes time-dependent, then it also becomes frequency-dependent. Under normal conditions, fatigue failure is independent of frequency. But when corrosion or high temperatures, or both, are encountered, the cyclic rate becomes important. The slower the frequency and the higher the temperature, the higher the crack propagation rate and the shorter the life at a given stress level.

Frettage Corrosion

The phenomenon of frettage corrosion is the result of microscopic motions of tightly fitting parts or structures. Bolted joints, bearing-race fits, wheel hubs, and any set of tightly fitted parts are examples. The process involves surface discoloration, pitting, and eventual fatigue. The frettage factor k_f depends upon the material of the mating pairs and ranges from 0.24 to 0.90.

7–10 Stress Concentration and Notch Sensitivity

In Sec. 4–14 it was pointed out that the existence of irregularities or discontinuities, such as holes, grooves, or notches, in a part increases the theoretical stresses significantly in the immediate vicinity of the discontinuity. Equation (4–48) defined a stress concentration factor K_t (or K_{ts}), which is used with the nominal stress to obtain the maximum resulting stress due to the irregularity or defect. It turns out that some materials are not fully sensitive to the presence of notches and hence, for these, a reduced value of K_t can be used. For these materials, the maximum stress is, in fact,

$$\sigma_{\max} = K_f \sigma_0 \qquad \text{or} \qquad \tau_{\max} = K_{fs} \tau_0 \tag{7–29}$$

where K_f is a reduced value of K_t and σ_0 is the nominal stress. The factor K_f is commonly called a *fatigue stress-concentration factor,* and hence the subscript f. So it is convenient to think of K_f as a stress-concentration factor reduced from K_t because of lessened sensitivity to notches. The resulting factor is defined by the equation

$$K_f = \frac{\text{maximum stress in notched specimen}}{\text{stress in notch-free specimen}} \tag{a}$$

Notch sensitivity q is defined by the equation

$$q = \frac{K_f - 1}{K_t - 1} \qquad \text{or} \qquad q_{\text{shear}} = \frac{K_{fs} - 1}{K_{ts} - 1} \tag{7–30}$$

where q is usually between zero and unity. Equation (7–30) shows that if $q = 0$, then $K_f = 1$, and the material has no sensitivity to notches at all. On the other hand, if $q = 1$, then $K_f = K_t$, and the material has full notch sensitivity. In analysis or design work, find K_t first, from the geometry of the part. Then specify the material, find q, and solve for K_f from the equation

$$K_f = 1 + q(K_t - 1) \qquad \text{or} \qquad K_{fs} = 1 + q_{\text{shear}}(K_{ts} - 1) \tag{7–31}$$

For steels and 2024 aluminum alloys, use Fig. 7–20 to find q for bending and axial loading. For shear loading, use Fig. 7–21. In using these charts it is well to know that the actual test results from which the curves were derived exhibit a large amount of scatter. Because of this scatter it is always safe to use $K_f = K_t$ if there is any doubt about the true value of q. Also, note that q is not far from unity for large notch radii.

The notch sensitivity of the cast irons is very low, varying from 0 to about 0.20, depending upon the tensile strength. To be on the conservative side, it is recommended that the value $q = 0.20$ be used for all grades of cast iron.

Figure 7–20

Notch-sensitivity charts for steels and UNS A92024-T wrought aluminum alloys subjected to reversed bending or reversed axial loads. For larger notch radii, use the values of q corresponding to the $r = 0.16$-in (4-mm) ordinate. (*From George Sines and J. L. Waisman (eds.), Metal Fatigue, McGraw-Hill, New York. Copyright © 1969 by The McGraw-Hill Companies, Inc. Reprinted by permission.*)

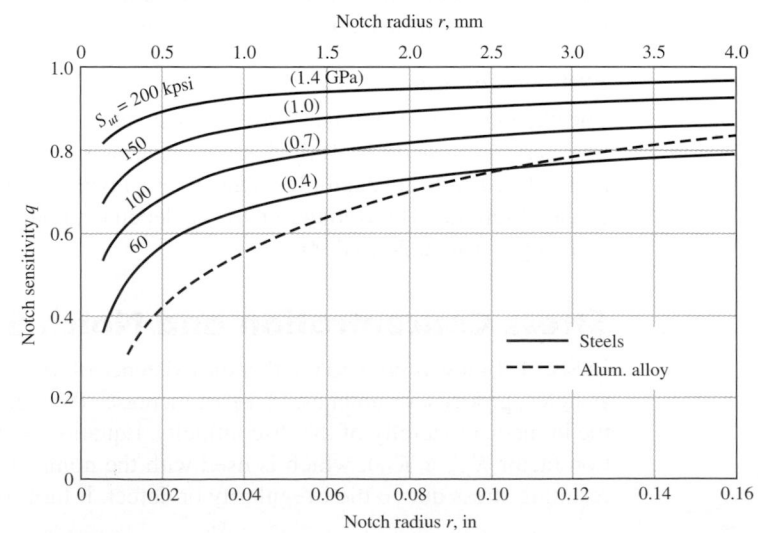

Figure 7–21

Notch-sensitivity curves for materials in reversed torsion. For larger notch radii, use the values of q_{shear} corresponding to $r = 0.16$ in (4 mm).

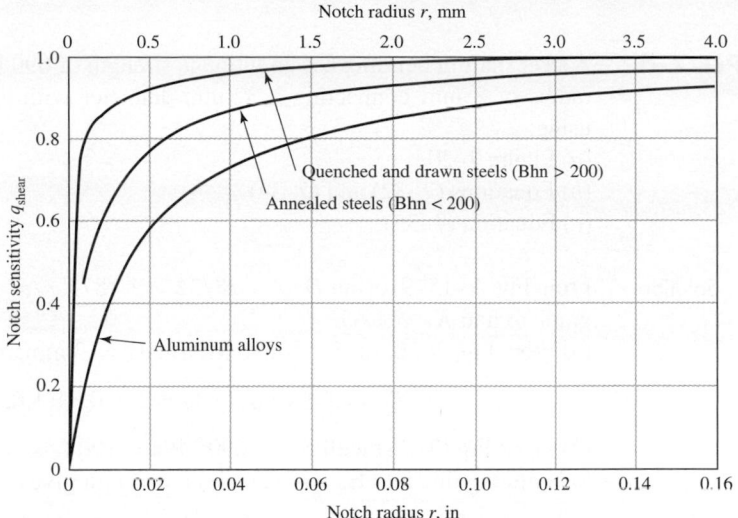

Figure 7–20 has as its basis the *Neuber equation*, which is given by

$$K_f = 1 + \frac{K_t - 1}{1 + \sqrt{a/r}} \tag{7–32}$$

where $\sqrt{a}$ is defined as the *Neuber constant* and is a material constant. Equating Eqs. (7–30) and (7–32) yields the notch sensitivity equation

$$q = \frac{1}{1 + \dfrac{\sqrt{a}}{\sqrt{r}}} \tag{7–33}$$

For steel, with S_{ut} in kpsi, the Neuber constant can be approximated by a third-order polynomial fit of data as

$$\begin{aligned}
\sqrt{a} = {}& 0.245\,799 - 0.307\,794(10^{-2})S_{ut} \\
& + 0.150\,874(10^{-4})S_{ut}^2 - 0.266\,978(10^{-7})S_{ut}^3
\end{aligned} \tag{7–34}$$

To use Eq. (7–32) or (7–33) for torsion for low-alloy steels, increase the ultimate strength by 20 kpsi in Eq. (7–34) and apply this value of $\sqrt{a}$.

A distinction in the configuration of the notch is accounted for in the modified Neuber equation (after Heywood), where the fatigue stress-concentration factor K_f is given as

$$K_f = \frac{K_t}{1 + \dfrac{2(K_t - 1)}{K_t}\dfrac{\sqrt{a}}{\sqrt{r}}} \tag{7–35}$$

where Table 7–8 gives values of $\sqrt{a}$ for steels for transverse holes, shoulders, and grooves.

Table 7–8

Heywood's Parameter $\sqrt{a}$ for Steels

Feature	$\sqrt{a}\left(\sqrt{\text{in}}\right)$, S_{ut} in kpsi	$\sqrt{a}\left(\sqrt{\text{mm}}\right)$, S_{ut} in MPa
Transverse hole	$5/S_{ut}$	$174/S_{ut}$
Shoulder	$4/S_{ut}$	$139/S_{ut}$
Groove	$3/S_{ut}$	$104/S_{ut}$

EXAMPLE 7–6
A steel shaft in bending has an ultimate strength of 690 MPa and a shoulder with a fillet radius of 3 mm connecting a 32-mm diameter with a 38-mm diameter. Estimate K_f using:
(a) Figure 7–20.
(b) Equations (7–32) and (7–34).
(c) Equation (7–35).

Solution
From Fig. A–15–9, using $D/d = 38/32 = 1.1875, r/d = 3/32 = 0.093\ 75$, we read the graph to find $K_t \doteq 1.65$.
(a) From Fig. 7–20, for $S_{ut} = 690$ MPa and $r = 3$ mm, $q \doteq 0.84$. Thus, from Eq. (7–31)

$$K_f = 1 + q(K_t - 1) \doteq 1 + 0.84(1.65 - 1) = 1.55$$

(b) From Eq. (7–34) with $S_{ut} = 690$ MPa $= 100$ kpsi, $\sqrt{a} = 0.062\sqrt{in} = 0.312\sqrt{mm}$. Substituting this into Eq. (7–32) with $r = 3$ mm gives

$$K_f = 1 + \frac{K_t - 1}{1 + \sqrt{a/r}} \doteq 1 + \frac{1.65 - 1}{1 + \dfrac{0.312}{\sqrt{3}}} = 1.55$$

(c) From Table 7–8,

$$\sqrt{a} = \frac{139}{S_{ut}} = \frac{139}{690} = 0.2015\ \sqrt{mm}$$

From Eq. (7–35),

$$K_f = \frac{K_t}{1 + \dfrac{2(K_t - 1)}{K_t}\dfrac{\sqrt{a}}{\sqrt{r}}} = \frac{1.65}{1 + \dfrac{2(1.65 - 1)}{1.65}\dfrac{0.2015}{\sqrt{3}}} = 1.51$$

which is 2.5 percent lower than in parts *a* and *b*.

For simple loading, it is acceptable to reduce the endurance limit by either dividing the unnotched specimen endurance limit by K_f or multiplying the reversing stress by K_f. However, in dealing with combined stress problems that may involve more than one value of fatigue-concentration factor, the stresses are multiplied by K_f.

EXAMPLE 7–7
Consider an unnotched specimen with an endurance limit of 55 kpsi. If the specimen was notched such that $K_f = 1.6$, what would be the factor of safety against failure for $N > 10^6$ cycles at a reversing stress of 30 kpsi?
(a) Solve by reducing S'_e.
(b) Solve by increasing the applied stress.

Solution
(a) The endurance limit of the notched specimen is given by

$$S_e = \frac{S'_e}{K_f} = \frac{55}{1.6} = 34.4 \text{ kpsi}$$

and the factor of safety is

$$n = \frac{S_e}{\sigma_a} = \frac{34.4}{30} = 1.15$$

(b) The maximum stress can be written as

$$(\sigma_a)_{\max} = K_f\sigma_a = 1.6(30) = 48.0 \text{ kpsi}$$

and the factor of safety is

$$n = \frac{S'_e}{K_f\sigma_a} = \frac{55}{48} = 1.15$$

When cycles to failure, N_f, are less than 10^6 there is experimental evidence that the fatigue stress-concentration factor $(K_f)_{N_f}$ is less than K_f. Studies have shown that as N_f approaches 10^3 cycles, $(K_f)_{N_f}$ for high-strength (usually low-ductility) metals approaches K_f, whereas for low-strength (usually ductile) metals $(K_f)_{N_f}$ approaches unity. A conservative approach is to keep K_f constant within the range $10^3 \le N \le 10^6$. To account for a reduction in K_f, Shigley and Mischke[22] suggest defining a notch sensitivity at 10^3 cycles, q_{10^3}, given by

$$q_{10^3} = \frac{(K_f)_{10^3} - 1}{K_f - 1} = -0.18 + 0.43(10^{-2})S_{ut} - 0.45(10^{-5})S_{ut}^2 \qquad (7\text{-}36)$$

where S_{ut} is in kpsi, $S_{ut} < 330$ kpsi, and K_f is the fatigue stress-concentration factor for 10^6 cycles. Solving for $(K_f)_{10^3}$ yields

$$(K_f)_{10^3} = 1 - (K_f - 1)\left[0.18 - 0.43(10^{-2})S_{ut} + 0.45(10^{-5})S_{ut}^2\right] \qquad (7\text{-}37)$$

Assuming a straight-line loglog S-N plot for the notched specimen from $f S_{ut}/(K_f)_{10^3}$, 10^3 to S'_e/K_f, 10^6 gives for $S_f/(K_f)_N$, N

$$(K_f)_N = \frac{(K_f)_{10^3}^2}{K_f} N^{(1/3)\log[K_f/(K_f)_{10^3}]} \qquad (7\text{-}38)$$

---

[22]J. E. Shigley and C. R. Mischke, *Mechanical Engineering Design*, 6th ed., McGraw-Hill, New York, 2001, p. 386–387.

EXAMPLE 7–8

Using the results of Ex. 7–6, find the value of $(K_f)_{10^5}$.

Solution

From Ex. 7–6, $K_f = 1.51$ (part c), $S_{ut} = 690/6.89 = 100$ kpsi. From Eq. (7–37),

$$(K_f)_{10^3} = 1 - (K_f - 1)\left[0.18 - 0.43(10^{-2})S_{ut} + 0.45(10^{-5})S_{ut}^2\right]$$

$$= 1 - (1.51 - 1)[0.18 - 0.43(10^{-2})100 + 0.45(10^{-5})100^2]$$

$$= 1.105$$

From Eq. (7–38),

Answer

$$(K_f)_{10^3} = \frac{1.105^2}{1.51} 10^{5(1/3)\log(1.51/1.105)} = 1.36$$

Up to this point, examples illustrated each factor in Marin's equation and stress concentrations alone. Let us consider a number of factors occurring simultaneously.

EXAMPLE 7–9   A 1015 hot-rolled steel bar has been machined to a diameter of 1 in. It is to be placed in reversed axial loading for 70 000 cycles to failure in an operating environment of 550°F. Using ASTM minimum properties, and a reliability of 99 percent, estimate the endurance limit and fatigue strength at 70 000 cycles.

Solution   From Table A–20, $S_{ut} = 50$ kpsi at 70°F. Since the rotating-beam specimen endurance limit is not known at room temperature, we determine the ultimate strength at the elevated temperature first, using Table 7–6. From Table 7–6,

$$\left(\frac{S_T}{S_{RT}}\right)_{550°} = \frac{0.995 + 0.963}{2} = 0.979$$

The ultimate strength at 550°F is then

$$(S_{ut})_{550°} = (S_T/S_{RT})_{550°}\,(S_{ut})_{70°} = 0.979(50) = 49.0 \text{ kpsi}$$

The rotating-beam specimen endurance limit at 550°F is then estimated from Eq. (7–8) as

$$S_e' = 0.504(49) = 24.7 \text{ kpsi}$$

Next, we determine the Marin factors. For the machined surface, Eq. (7–18) with Table 7–4 gives

$$k_a = aS_{ut}^b = 2.70(49^{-0.265}) = 0.963$$

For axial loading, from Eq. (7–20), the size factor $k_b = 1$, and the loading factor is $k_c = 0.85$, from Eq. (7–25). The temperature factor $k_d = 1$, since we accounted for the temperature in modifying the ultimate strength and consequently the endurance limit. For 99 percent reliability, from Table 7–7, $k_e = 0.814$. Finally, since no other conditions were given, the miscellaneous factor is $k_f = 1$. The endurance limit for the part is estimated by Eq. (7–17) as

$$S_e = k_a k_b k_c k_d k_e k_f S_e'$$

Answer

$$= 0.963(1)(0.85)(1)(0.814)(1)24.7 = 16.5 \text{ kpsi}$$

For the fatigue strength at 70 000 cycles we need to construct the S-N equation. From Sec. 7–8, Eq. (7–10),

$$\sigma_F' = 49 + 50 = 99 \text{ kpsi}$$

From Eq. (7–11),

$$b = -\frac{\log(\sigma_F'/S_e)}{\log(2N_e)} = -\frac{\log(99/16.5)}{\log(2 \cdot 10^6)} = -0.1235$$

Equation (7–9) then gives

$$f = \frac{\sigma_F'}{S_{ut}}(2 \cdot 10^3)^b = \frac{99}{49}(2 \cdot 10^3)^{-0.1235} = 0.790$$

From Eq. (7–13),

$$a = \frac{(f S_{ut})^2}{S_e} = \frac{[0.790(49)]^2}{16.5} = 90.8 \text{ kpsi}$$

Finally, Eq. (7–12) gives

Answer

$$S_f = a N^b = 90.8(70\,000)^{-0.1235} = 22.9 \text{ kpsi}$$

EXAMPLE 7–10

Figure 7–22a shows a rotating axle simply supported in ball bearings at A and D and loaded by a nonrotating force F of 6.8 kN. Using ASTM "minimum" strengths, estimate the life of the part.

Solution

From Fig. 7–22b we learn that failure will probably occur at B rather than at C or at the point of maximum moment. Point B has a smaller cross section, a higher bending moment, and a higher stress-concentration factor than C, and the location of maximum moment has a larger size and no stress-concentration factor.

Figure 7–22

(a) Shaft drawing showing all dimensions in millimeters; all fillets 3-mm radius. The shaft rotates and the load is stationary; material is machined from AISI 1050 cold-drawn steel. (b) Bending-moment diagram.

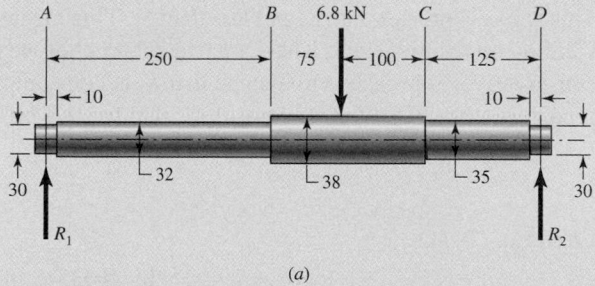

(a)

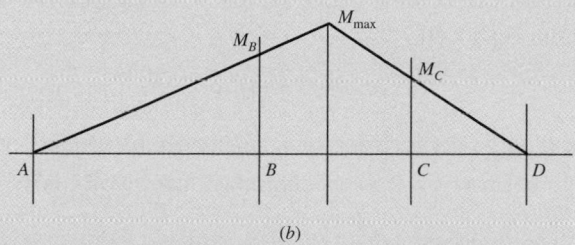

(b)

We shall solve the problem by first estimating the strength at point B, since the strength will be different elsewhere, and comparing this strength with the stress at the same point.

From Table A–20 we find $S_{ut} = 690$ MPa and $S_y = 580$ MPa. The endurance limit S'_e is estimated as

$$S'_e = 0.504(690) = 347.8 \text{ MPa}$$

From Eq. (7–18) and Table 7–4,

$$k_a = 4.51(690)^{-0.265} = 0.798$$

From Eq. (7–19),

$$k_b = (32/7.62)^{-0.107} = 0.858$$

Since $k_c = k_d = k_e = k_f = 1$,

$$S_e = 0.798(0.858)347.8 = 238 \text{ MPa}$$

To find the geometric stress-concentration factor K_t we enter Fig. A–15–9 with $D/d = 38/32 = 1.1875$ and $r/d = 3/32 = 0.093\,75$ and read $K_t \doteq 1.65$. From Table 7–8, $\sqrt{a} = 139/690 = 0.201\ \sqrt{\text{mm}}$. From Eq. (7–35),

$$K_f = \cfrac{K_t}{1 + \cfrac{2(K_t - 1)}{K_t}\cfrac{\sqrt{a}}{\sqrt{r}}} \doteq \cfrac{1.65}{1 + \cfrac{2(1.65 - 1)}{1.65}\cfrac{0.201}{\sqrt{3}}} = 1.51$$

This stress-concentration factor applies to 10^6 cycles or more.

The next step is to estimate the bending stress at point B. The bending moment is

$$M_B = R_1 x = \frac{225F}{550}250 = \frac{225(6.8)}{550}250 = 695 \text{ N} \cdot \text{m}$$

Just to the left of B the section modulus is $I/c = \pi d^3/32 = \pi 3.2^3/32 = 3.22 \text{ cm}^3$. The reversing bending stress is, assuming infinite life,

$$\sigma = K_f \frac{M_B}{I/c} = 1.51\frac{695}{3.22} = 325.9 \text{ MPa}$$

This stress is greater than S_e and less than S_y. This means we have both finite life and no yielding on the first cycle. There are two approaches we may take at this point. The first, conservative, approach is to assume that K_f is constant regardless of N. The second approach is to use Eq. (7–38). We will see that for this problem, approach 1 is far easier.

Approach 1. From Eq. (7–10),

$$\sigma'_F = S_{ut} + 345 = 690 + 345 = 1035 \text{ MPa}$$

From Eq. (7–11),

$$b = -\frac{\log(\sigma'_F/S_e)}{\log(2N_e)} = -\frac{\log(1035/238)}{\log(2 \cdot 10^6)} = -0.1013$$

From Eq. (7–9),

$$f = \frac{\sigma'_F}{S_{ut}}(2 \cdot 10^3)^b = \frac{1035}{690}(2 \cdot 10^3)^{-0.1013} = 0.695$$

From Eq. (7–13),

$$a = \frac{(f S_{ut})^2}{S_e} = \frac{[0.695(690)]^2}{238} = 966.3 \text{ MPa}$$

Answer
(Approach 1)

From Eq. (7–15),

$$N = \left(\frac{\sigma_a}{a}\right)^{1/b} = \left(\frac{325.9}{966.3}\right)^{-1/0.1013} = 45.7(10^3) \text{ cycles}$$

Approach 2. Using Eq. (7–38) requires a value of N, which is unknown. Because we don't know K_N, instead of applying the fatigue stress-concentration factor to stress σ we will reduce the strengths in the *S-N* diagram by the factor. Note that this approach is not advisable for combined stress problems. The nominal stress is

$$\sigma_{\text{nom}} = \frac{M_B}{I/c} = \frac{695}{3.22} = 215.8 \text{ MPa}$$

From Eq. (7–37), with $S_{ut} = 690/6.9 = 100$ kpsi,

$$(K_f)_{10^3} = 1 - (K_f - 1)\left[0.18 - 0.43(10^{-2})S_{ut} + 0.45(10^{-5})S_{ut}^2\right]$$

$$= 1 - (1.51 - 1)[0.18 - 0.43(10^{-2})100 + 0.45(10^{-5})100^2]$$

$$= 1.105$$

The *S-N* equation incorporating the fatigue stress-concentration factor can be written as $N = (\sigma_a/a')^{1/b'}$, where in form similar to that of Eqs. (7–13) and (7–14),

$$a' = \frac{[f\,S_{ut}/(K_f)_{10^3}]^2}{S_e/K_f} = a\frac{K_f}{(K_f)_{10^3}^2} = 966.3\frac{1.51}{1.105^2} = 1195 \text{ MPa}$$

$$b' = -\frac{1}{3}\log\left[\frac{f\,S_{ut}/(K_f)_{10^3}}{S_e/K_f}\right] = b - \frac{1}{3}\log\frac{K_f}{(K_f)_{10^3}}$$

$$= -0.1013 - \frac{1}{3}\log\frac{1.51}{1.105} = -0.147$$

where use of a and b from approach 1 was used. The number of cycles can now be determined from Eq. (7–15) as

Answer
(Approach 2)

$$N = \left(\frac{\sigma_a}{a'}\right)^{1/b'} = \left(\frac{215.8}{1195}\right)^{-1/0.147} = 113.9(10^3) \text{ cycles}$$

We see that approach 1 is indeed more conservative than approach 2, by a factor of 2.5. This is the effect of a logarithmic scale. Approach 2 predicts a fatigue stress-concentration factor according to Eq. (7–38) of

$$(K_f)_{113.9(10^3)} = \frac{(K_f)_{10^3}^2}{K_f} N^{(1/3)\log[K_f/(K_f)_{10^3}]}$$

$$= \frac{1.105^2}{1.51}[113.9(10^3)]^{(1/3)\log(1.51/1.105)} = 1.37$$

Without any further knowledge of the behavior of the material given, one might select the solution of approach 1. However, if it is known that K_f approaches unity for 10^3 cycles for this material, then approach 2 would be appropriate.

7–11 **Characterizing Fluctuating Stresses**

Fluctuating stresses in machinery often take the form of a sinusoidal pattern because of the nature of some rotating machinery. However, other patterns, some quite irregular, do occur. It has been found that in periodic patterns exhibiting a single maximum and a single minimum of force, the shape of the wave is not important, but the peaks on both the high side (maximum) and the low side (minimum) are important. Thus $F_{\max}$ and $F_{\min}$ in a cycle of force can be used to characterize the force pattern. It is also true that ranging above and below some baseline can be equally effective in characterizing the force pattern. If the largest force is $F_{\max}$ and the smallest force is $F_{\min}$, then a steady component and an alternating component can be constructed as follows:

$$F_m = \frac{F_{\max} + F_{\min}}{2} \qquad F_a = \left| \frac{F_{\max} - F_{\min}}{2} \right|$$

where F_m is the midrange component of force, and F_a is the amplitude component of force.

Figure 7–23 illustrates some of the various stress-time traces that occur. The components of stress, some of which are shown in Fig. 7–23d, are

$$\sigma_{\min} = \text{minimum stress} \qquad \sigma_m = \text{midrange component}$$
$$\sigma_{\max} = \text{maximum stress} \qquad \sigma_r = \text{range of stress}$$
$$\sigma_a = \text{amplitude component} \qquad \sigma_s = \text{static or steady stress}$$

Figure 7–23

Some stress-time relations: (a) fluctuating stress with high-frequency ripple; (b and c) nonsinusoidal fluctuating stress; (d) sinusoidal fluctuating stress; (e) repeated stress; (f) completely reversed sinusoidal stress.

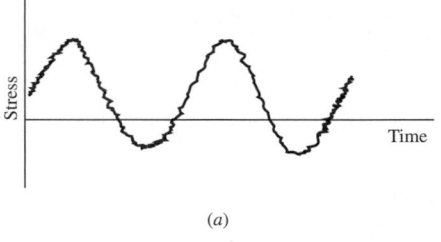

(a)

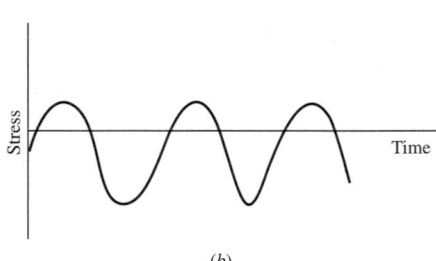

(b)

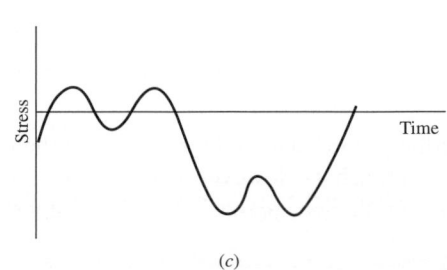

(c)

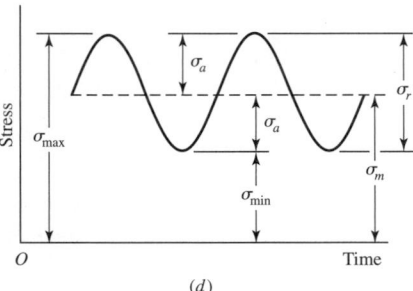

(d)

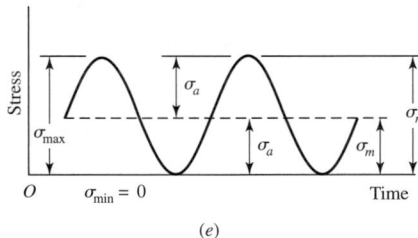

(e)

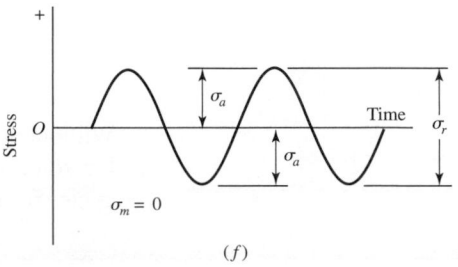

(f)

The steady, or static, stress is *not* the same as the midrange stress; in fact, it may have any value between σ_{min} and σ_{max}. The steady stress exists because of a fixed load or pre-load applied to the part, and it is usually independent of the varying portion of the load. A helical compression spring, for example, is always loaded into a space shorter than the free length of the spring. The stress created by this initial compression is called the steady, or static, component of the stress. It is not the same as the midrange stress.

We shall have occasion to apply the subscripts of these components to shear stresses as well as normal stresses.

The following relations are evident from Fig. 7–23:

$$\sigma_m = \frac{\sigma_{max} + \sigma_{min}}{2}$$

$$\sigma_a = \left| \frac{\sigma_{max} - \sigma_{min}}{2} \right| \tag{7–39}$$

In addition to Eq. (7–39), the *stress ratio*

$$R = \frac{\sigma_{min}}{\sigma_{max}} \tag{7–40}$$

and the stress ratio

$$A = \frac{\sigma_a}{\sigma_m} \tag{7–41}$$

are also defined and used in connection with fluctuating stresses.

Equations (7–39) utilize symbols σ_a and σ_m as the stress components at the location under scrutiny. This means, in the absence of a notch, σ_a and σ_m are equal to the nominal stresses σ_{ao} and σ_{mo} induced by loads F_a and F_m, respectively; in the presence of a notch they are $K_f \sigma_{ao}$ and $K_f \sigma_{mo}$, respectively, as long as the material remains without plastic strain. In other words, the fatigue stress concentration factor K_f is applied to *both* components.

When the steady stress component is high enough to induce localized notch yielding, the designer has a problem. The first-cycle local yielding produces plastic strain and strain-strengthening. This is occurring at the location where fatigue crack nucleation and growth are most likely. The material properties (S_y and S_{ut}) are new and difficult to quantify. The prudent engineer controls the concept, material and condition of use, and geometry so that no plastic strain occurs. There are discussions concerning possible ways of quantifying what is occurring under localized and general yielding in the presence of a notch, referred to as the *nominal mean stress* method, *residual stress* method, and the like.[23] The nominal mean stress method (set $\sigma_a = K_f \sigma_{ao}$ and $\sigma_m = \sigma_{mo}$) gives roughly comparable results to the residual stress method, but both are *approximations*.

There is the method of Dowling[24] for ductile material, which, for materials with a pronounced yield point and approximated by an elastic–perfectly plastic behavior

---

[23]R. C. Juvinall, *Stress, Strain, and Strength,* McGraw-Hill, New York, 1967, articles 14.9–14.12; R. C. Juvinall and K. M. Marshek, *Fundamentals of Machine Component Design,* Wiley, New York, 1991, Sec. 8.11; M. E. Dowling, *Mechanical Behavior of Materials,* 2nd ed., Prentice Hall, Englewood Cliffs, N.J., 1999, Secs. 10.3–10.5.

[24]Dowling, op. cit., p. 437–438.

model, quantitatively expresses the steady stress component stress-concentration factor K_{fm} as

$$K_{fm} = K_f \qquad\qquad K_f|\sigma_{\max,o}| < S_y$$

$$K_{fm} = \frac{S_y - K_f\sigma_{ao}}{|\sigma_{mo}|} \qquad K_f|\sigma_{\max,o}| > S_y \qquad\qquad (7\text{--}42)$$

$$K_{fm} = 0 \qquad\qquad K_f|\sigma_{\max,o} - \sigma_{\min,o}| > 2S_y$$

For the purposes of this book, for ductile materials in fatigue,

- Avoid localized plastic strain at a notch. Set $\sigma_a = K_f\sigma_{a,o}$ and $\sigma_m = K_f\sigma_{mo}$.
- When plastic strain at a notch cannot be avoided, use Eqs. (7–42); or conservatively, set $\sigma_a = K_f\sigma_{ao}$ and use $K_{mf} = 1$, that is, $\sigma_m = \sigma_{mo}$.

7–12 Fatigue Failure Criteria for Fluctuating Stress

Now that we have defined the various components of stress associated with a part subjected to fluctuating stress, we want to vary both the midrange stress and the stress amplitude, or alternating component, to learn something about the fatigue resistance of parts when subjected to such situations. Three methods of plotting the results of such tests are in general use and are shown in Figs. 7–24, 7–25, and 7–26.

The *modified Goodman diagram* of Fig. 7–24 has the midrange stress plotted along the abscissa and all other components of stress plotted on the ordinate, with tension in the positive direction. The endurance limit, fatigue strength, or finite-life strength, whichever applies, is plotted on the ordinate above and below the origin. The midrange-stress line is a 45° line from the origin to the tensile strength of the part. The modified Goodman diagram consists of the lines constructed to S_e (or S_f) above and below the

Figure 7–24

Modified Goodman diagram showing all the strengths and the limiting values of all the stress components for a particular midrange stress.

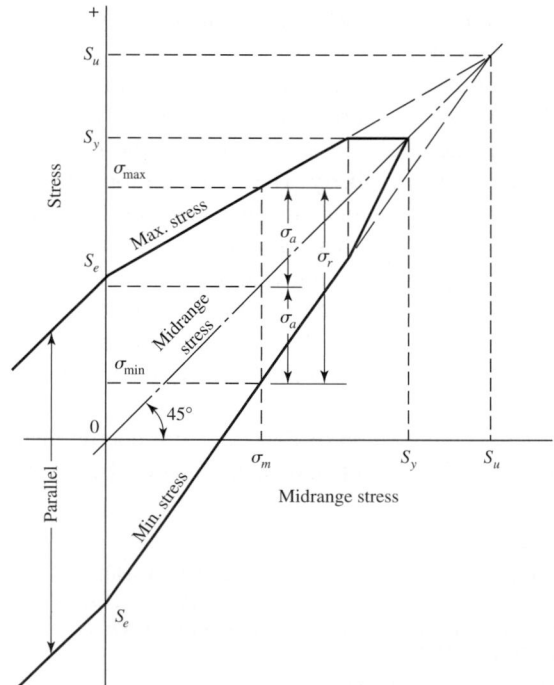

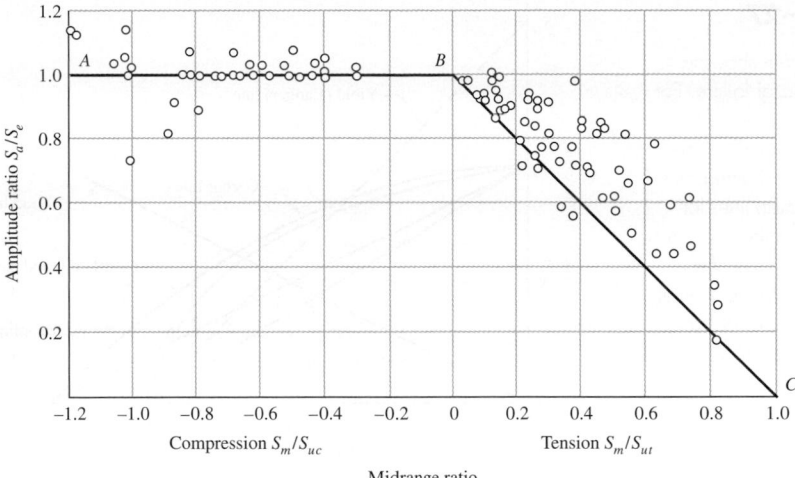

Figure 7–25

Plot of fatigue failures for midrange stresses in both tensile and compressive regions. Normalizing the data by using the ratio of steady strength component to tensile strength S_m/S_{ut}, steady strength component to compressive strength S_m/S_{uc}, and strength amplitude component to endurance limit S_a/S_e enables a plot of experimental results for a variety of steels. [Data source: Thomas J. Dolan, "Stress Range," Sec. 6.2 in O. J. Horger (ed.), ASME Handbook—Metals Engineering Design, McGraw-Hill, New York, 1953.]

Figure 7–26

Master fatigue diagram created for AISI 4340 steel having $S_{ut} = 158$ and $S_y = 147$ kpsi. The stress components at A are $\sigma_{min} = 20$, $\sigma_{max} = 120$, $\sigma_m = 70$, and $\sigma_a = 50$, all in kpsi. (Source: H. J. Grover, Fatigue of Aircraft Structures, U.S. Government Printing Office, Washington, D.C., 1966, pp. 317, 322. See also J. A. Collins, Failure of Materials in Mechanical Design, Wiley, New York, 1981, p. 216.)

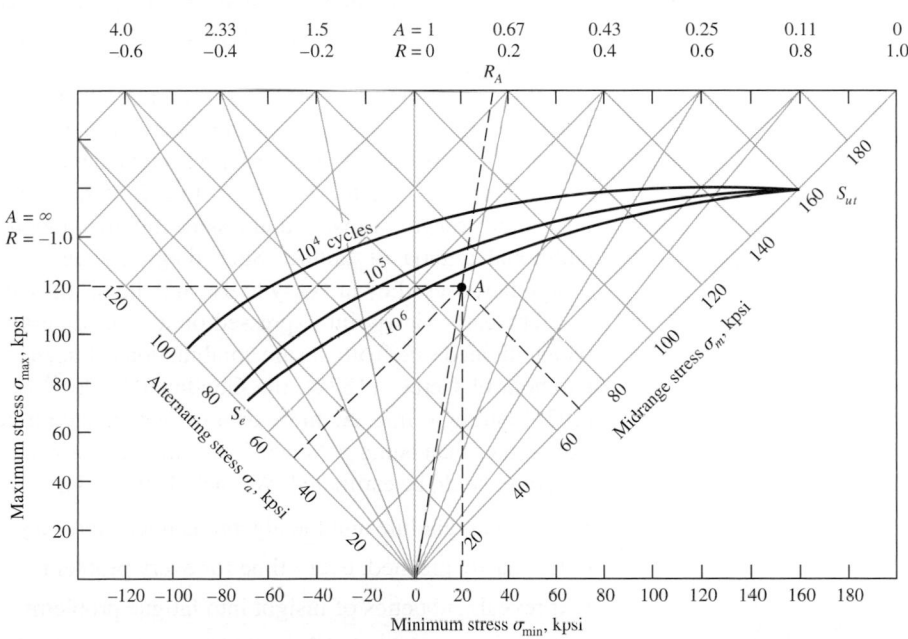

origin. Note that the yield strength is also plotted on both axes, because yielding would be the criterion of failure if σ_{max} exceeded S_y.

Another way to display test results is shown in Fig. 7–25. Here the abscissa represents the ratio of the midrange strength S_m to the ultimate strength, with tension plotted to the right and compression to the left. The ordinate is the ratio of the alternating strength to the endurance limit. The line BC then represents the modified Goodman

Figure 7–27

Fatigue diagram showing various criteria of failure. For each criterion, points on or "above" the respective line indicate failure. Some point A on the Goodman line, for example, gives the strength S_m as the limiting value of σ_m corresponding to the strength S_a, which, paired with σ_m, is the limiting value of σ_a.

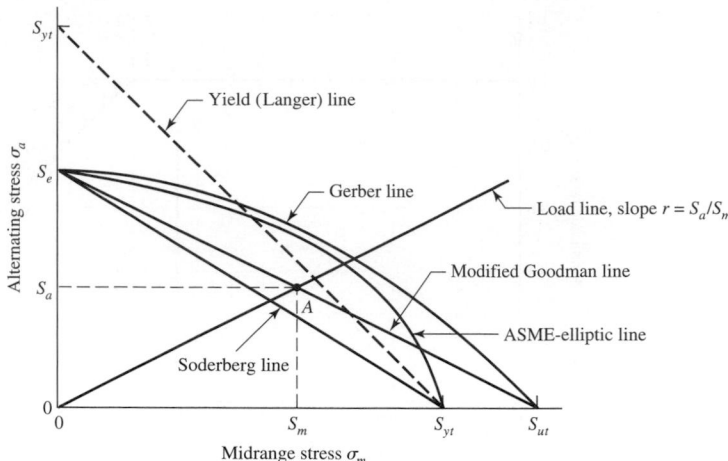

criterion of failure. Note that the existence of midrange stress in the compressive region has little effect on the endurance limit.

The very clever diagram of Fig. 7–26 is unique in that it displays four of the stress components as well as the two stress ratios. A curve representing the endurance limit for values of R beginning at $R = -1$ and ending with $R = 1$ begins at S_e on the σ_a axis and ends at S_{ut} on the σ_m axis. Constant-life curves for $N = 10^5$ and $N = 10^4$ cycles have been drawn too. Any stress state, such as the one at A, can be described by the minimum and maximum components, or by the midrange and alternating components. And safety is indicated whenever the point described by the stress components lies below the constant-life line.

When the midrange stress is compression, failure occurs whenever $\sigma_a = S_e$ or whenever $\sigma_{\max} = S_{yc}$, as indicated by the left-hand side of Fig. 7–25. Neither a fatigue diagram nor any other failure criteria need be developed.

In Fig. 7–27, the tensile side of Fig. 7–25 has been redrawn in terms of strengths, instead of strength ratios, with the same modified Goodman criterion together with four additional criteria of failure. Such diagrams are often constructed for analysis and design purposes; they are easy to use and the results can be scaled off directly.

The early viewpoint expressed on a $\sigma_a\sigma_m$ diagram was that there existed a locus which divided safe from unsafe combinations of σ_a, σ_m. Ensuing proposals included the parabola of Gerber (1874), the Goodman (1890)[25] (straight) line, and the Soderberg (1930) (straight) line. As more data were generated it became clear that a fatigue criterion, rather than being a "fence," was more like a zone or band wherein the probability of failure could be estimated. We include the failure criterion of Goodman because

- It is a straight line and the algebra is linear and easy.
- It is easily graphed, every time for every problem.
- It reveals subtleties of insight into fatigue problems.
- Answers can be scaled from the diagrams as a check on the algebra.

We also caution that it is deterministic and the phenomenon is not. It is biased and we cannot quantify the bias. It is not conservative. It is a stepping-stone to understanding; it is history; and to read the work of other engineers and to have meaningful oral exchanges with them, it is necessary that you understand the Goodman approach should it arise.

---

[25]It is difficult to date Goodman's work because it went through several modifications and was never published.

Either the fatigue limit S_e or the finite-life strength S_f is plotted on the ordinate of Fig. 7–27. These values will have already been corrected using the Marin factors of Eq. (7–17). Note that the yield strength S_{yt} is plotted on the ordinate too. This serves as a reminder that first-cycle yielding rather than fatigue might be the criterion of failure.

The midrange-stress axis of Fig. 7–27 has the yield strength S_{yt} and the tensile strength S_{ut} plotted along it.

Five criteria of failure are diagrammed in Fig. 7–27: the Soderberg, the modified Goodman, the Gerber, the ASME-elliptic, and yielding. The diagram shows that only the Soderberg criterion guards against any yielding, but is biased low.

Considering the modified Goodman line as a criterion, point A represents a limiting point with an alternating strength S_a and midrange strength S_m. The slope of the load line shown is defined as $r = S_a/S_m$.

The criterion equation for the Soderberg line is

$$\frac{S_a}{S_e} + \frac{S_m}{S_{yt}} = 1 \tag{7–43}$$

Similarly, we find the modified Goodman relation to be

$$\frac{S_a}{S_e} + \frac{S_m}{S_{ut}} = 1 \tag{7–44}$$

Examination of Fig. 7–25 shows that both a parabola and an ellipse have a better opportunity to pass among the midrange tension data and to permit quantification of the probability of failure. The Gerber failure criterion is written as

$$\frac{S_a}{S_e} + \left(\frac{S_m}{S_{ut}}\right)^2 = 1 \tag{7–45}$$

and the ASME-elliptic is written as

$$\left(\frac{S_a}{S_e}\right)^2 + \left(\frac{S_m}{S_y}\right)^2 = 1 \tag{7–46}$$

The *Langer* first-cycle-yielding criterion is used in connection with the fatigue locus:

$$\frac{S_a}{S_{yt}} + \frac{S_m}{S_{yt}} = 1 \tag{7–47}$$

The stresses $n\sigma_a$ and $n\sigma_m$ can replace S_a and S_m, where n is the design factor or factor of safety. Then, Eq. (7–43), the Soderberg line, becomes

$$\frac{\sigma_a}{S_e} + \frac{\sigma_m}{S_y} = \frac{1}{n} \tag{7–48}$$

Equation (7–44), the modified Goodman line, becomes

$$\frac{\sigma_a}{S_e} + \frac{\sigma_m}{S_{ut}} = \frac{1}{n} \tag{7–49}$$

Equation (7–45), the Gerber line, becomes

$$\frac{n\sigma_a}{S_e} + \left(\frac{n\sigma_m}{S_{ut}}\right)^2 = 1 \tag{7–50}$$

Equation (7–46), the ASME-elliptic line, becomes

$$\left(\frac{n\sigma_a}{S_e}\right)^2 + \left(\frac{n\sigma_m}{S_y}\right)^2 = 1 \tag{7–51}$$

We will emphasize the Gerber and ASME-elliptic for fatigue failure criterion and the Langer for first-cycle yielding. However, conservative designers often use the modified

Goodman criterion, so we will continue to include it in our discussions. The failure criteria are used in conjunction with a load line, $r = S_a/S_m = \sigma_a/\sigma_m$. Principal intersections are tabulated in Tables 7–9 to 7–11. Formal expressions for fatigue factor of safety are given in the lower panel of Tables 7–9 to 7–11.

Some examples will help solidify the ideas just discussed.

Table 7–9

Amplitude and Steady Coordinates of Strength and Important Intersections in First Quadrant for Modified Goodman and Langer Failure Criteria

Intersecting Equations	Intersection Coordinates
$\dfrac{S_a}{S_e} + \dfrac{S_m}{S_{ut}} = 1$	$S_a = \dfrac{r\,S_e\,S_{ut}}{r\,S_{ut} + S_e}$
Load line $r = \dfrac{S_a}{S_m}$	$S_m = \dfrac{S_a}{r}$
$\dfrac{S_a}{S_y} + \dfrac{S_m}{S_y} = 1$	$S_a = \dfrac{r\,S_y}{1 + r}$
Load line $r = \dfrac{S_a}{S_m}$	$S_m = \dfrac{S_y}{1 + r}$
$\dfrac{S_a}{S_e} + \dfrac{S_m}{S_{ut}} = 1$	$S_m = \dfrac{(S_y - S_e)\,S_{ut}}{S_{ut} - S_e}$
$\dfrac{S_a}{S_y} + \dfrac{S_m}{S_y} = 1$	$S_a = S_y - S_m,\ r_{\text{crit}} = S_a/S_m$

Fatigue factor of safety

$$n_f = \frac{1}{\dfrac{\sigma_a}{S_e} + \dfrac{\sigma_m}{S_{ut}}}$$

Table 7–10

Amplitude and Steady Coordinates of Strength and Important Intersections in First Quadrant for Gerber and Langer Failure Criteria

Intersecting Equations	Intersection Coordinates
$\dfrac{S_a}{S_e} + \left(\dfrac{S_m}{S_{ut}}\right)^2 = 1$	$S_a = \dfrac{r^2 S_{ut}^2}{2 S_e}\left[-1 + \sqrt{1 + \left(\dfrac{2 S_e}{r\,S_{ut}}\right)^2}\right]$
Load line $r = \dfrac{S_a}{S_m}$	$S_m = \dfrac{S_a}{r}$
$\dfrac{S_a}{S_y} + \dfrac{S_m}{S_y} = 1$	$S_a = \dfrac{r\,S_y}{1 + r}$
Load line $r = \dfrac{S_a}{S_m}$	$S_m = \dfrac{S_y}{1 + r}$
$\dfrac{S_a}{S_e} + \left(\dfrac{S_m}{S_{ut}}\right)^2 = 1$	$S_m = \dfrac{S_{ut}^2}{2 S_e}\left[1 - \sqrt{1 + \left(\dfrac{2 S_e}{S_{ut}}\right)^2\left(1 - \dfrac{S_y}{S_e}\right)}\right]$
$\dfrac{S_a}{S_y} + \dfrac{S_m}{S_y} = 1$	$S_a = S_y - S_m,\ r_{\text{crit}} = S_a/S_m$

Fatigue factor of safety

$$n_f = \frac{1}{2}\left(\frac{S_{ut}}{\sigma_m}\right)^2 \frac{\sigma_a}{S_e}\left[-1 + \sqrt{1 + \left(\frac{2\sigma_m S_e}{S_{ut}\sigma_a}\right)^2}\right] \qquad \sigma_m > 0$$

Table 7–11

Amplitude and Steady Coordinates of Strength and Important Intersections in First Quadrant for ASME-Elliptic and Langer Failure Criteria

Intersecting Equations	Intersection Coordinates
$\left(\dfrac{S_a}{S_e}\right)^2 + \left(\dfrac{S_m}{S_y}\right)^2 = 1$	$S_a = \sqrt{\dfrac{r^2 S_e^2 S_y^2}{S_e^2 + r^2 S_y^2}}$
Load line $r = S_a/S_m$	$S_m = \dfrac{S_a}{r}$
$\dfrac{S_a}{S_y} + \dfrac{S_m}{S_y} = 1$	$S_a = \dfrac{r S_y}{1 + r}$
Load line $r = S_a/S_m$	$S_m = \dfrac{S_y}{1 + r}$
$\left(\dfrac{S_a}{S_e}\right)^2 + \left(\dfrac{S_m}{S_y}\right)^2 = 1$	$S_a = 0, \quad \dfrac{2 S_y S_e^2}{S_e^2 + S_y^2}$
$\dfrac{S_a}{S_y} + \dfrac{S_m}{S_y} = 1$	$S_m = S_y - S_a, \quad r_{\text{crit}} = S_a/S_m$
Fatigue factor of safety	

$$n_f = \sqrt{\frac{1}{(\sigma_a/S_e)^2 + (\sigma_m/S_y)^2}}$$

EXAMPLE 7–11

A 1.5-in-diameter bar has been machined from an AISI 1050 cold-drawn bar. This part is to withstand a fluctuating tensile load varying from 0 to 16 kip. Because of the ends, and the fillet radius, a fatigue stress-concentration factor K_f is 1.85 for 10^6 or larger life. Find S_a and S_m and the factor of safety guarding against fatigue and first-cycle yielding, using (a) the Gerber fatigue line and (b) the ASME-elliptic fatigue line.

Solution

We begin with some preliminaries. From Table A–20, $S_{ut} = 100$ kpsi and $S_y = 84$ kpsi. Note that $F_a = F_m = 8$ kip. The Marin factors are, deterministically,

$$k_a = 2.70(100)^{-0.265} = 0.797: \text{ Eq. (7–18), Table 7–4}$$

$$k_b = 1 \text{ (axial loading, see } k_c)$$

$$k_c = 0.85: \text{ Eq. (7–25)}$$

$$k_d = k_e = k_f = 1$$

$$S_e = 0.797(1)0.850(1)(1)(1)0.504(100) = 34.1 \text{ kpsi: Eqs. (7–8), (7–17)}$$

The nominal axial stress components σ_{ao} and σ_{mo} are

$$\sigma_{ao} = \frac{4F_a}{\pi d^2} = \frac{4(8)}{\pi 1.5^2} = 4.53 \text{ kpsi} \qquad \sigma_{mo} = \frac{4F_m}{\pi d^2} = \frac{4(8)}{\pi 1.5^2} = 4.53 \text{ kpsi}$$

Applying K_f to both components σ_{ao} and σ_{mo} constitutes a prescription of no notch yielding:

$$\sigma_a = K_f \sigma_{ao} = 1.85(4.53) = 8.38 \text{ kpsi} = \sigma_m$$

Figure 7–28

Principal points A, B, C, and D on the designer's diagram drawn for Gerber, Langer, and load line.

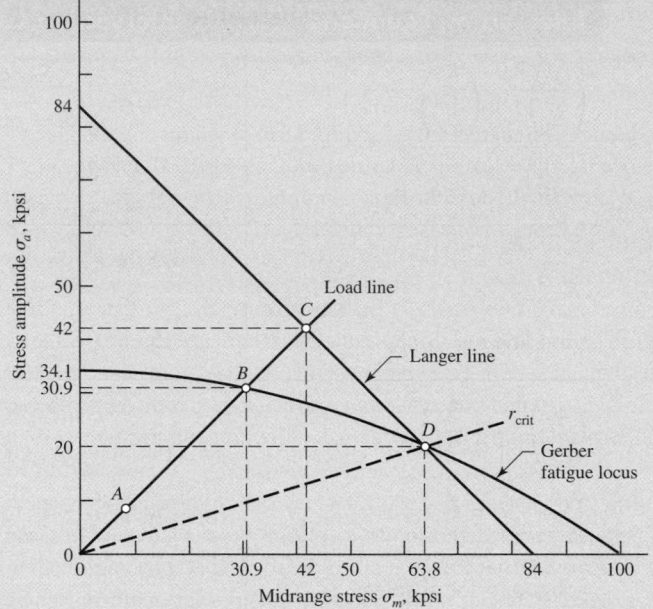

The load line slope is $r = \sigma_a/\sigma_m = 1$.

(a) From the first panel of Table 7–10,

$$S_a = \frac{(1)^2 100^2}{2(34.1)} \left\{ -1 + \sqrt{1 + \left[\frac{2(34.1)}{(1)100}\right]^2} \right\} = 30.9 \text{ kpsi}$$

$$S_m = \frac{S_a}{r} = \frac{30.9}{1} = 30.9 \text{ kpsi}$$

In Fig. 7–28 the intersection of the load line and the Gerber line is point B. Point A on the load line represents the stress components σ_a and σ_m. Point C represents the intersection of the load line and the Langer yield line. From Table 7–10,

$$S_a = \frac{r S_y}{1 + r} = \frac{(1)84}{1 + 1} = 42 \text{ kpsi} \qquad S_m = \frac{S_a}{r} = \frac{42}{1} = 42 \text{ kpsi}$$

The load line is the locus of possible stress states. As loading increases, point A will move toward point B and point C. The first encounter is with point B on the Gerber line, so the threat to the part is from fatigue. The factor of safety in fatigue n_f is

$$n_f = \frac{OB}{OA} = \frac{(S_a)_{\text{Gerber}}}{\sigma_a} = \frac{30.9}{8.38} = 3.69$$

Alternatively, from the fourth panel of Table 7–10,

$$n_f = \frac{1}{2}\left(\frac{100}{8.38}\right)^2 \left(\frac{8.38}{34.1}\right) \left\{ -1 + \sqrt{1 + \left[\frac{2(8.38)34.1}{100(8.38)}\right]^2} \right\} = 3.68$$

indicating rounding error. The factor of safety guarding against first-cycle yielding is

$$n_y = \frac{OC}{OA} = \frac{(S_a)_{\text{Langer}}}{\sigma_a} = \frac{42.0}{8.38} = 5.01$$

This confirms that there is no local yielding at the fillet. Point D represents the changeover from fatigue failure to first-cycle yielding. The coordinates of point D can be found from the simultaneous solution of Eqs. (7–45) and (7–47), as shown in the third panel of Table 7–10:

$$S_m = \frac{100^2}{2(34.1)} \left[1 - \sqrt{1 + \left(\frac{2(34.1)}{100} \right)^2 \left(1 - \frac{84}{34.1} \right)} \right] = 63.8 \text{ kpsi}$$

$$S_a = S_y - S_m = 84 - 63.8 = 20.2 \text{ kpsi}$$

The critical slope of the load line r_{crit} is

$$r_{\text{crit}} = \frac{S_a}{S_m} = \frac{20.2}{63.8} = 0.317$$

The facts that the load line slope is $r = 1$ and $r_{\text{crit}} < r$ confirm that there is a primary threat from fatigue.

(b) From panel 1 of Table 7–11, with $r = 1$, we obtain the coordinates S_a and S_m of point B in Fig. 7–29:

$$S_a = \sqrt{\frac{(1)^2 34.1^2 84^2}{34.1^2 + (1)^2 84^2}} = 31.6 \text{ kpsi} \qquad S_m = \frac{S_a}{r} = \frac{31.6}{1} = 31.6 \text{ kpsi}$$

Figure 7–29

Principal points A, B, C, and D on the designer's diagram drawn for ASME-elliptic, Langer, and load lines.

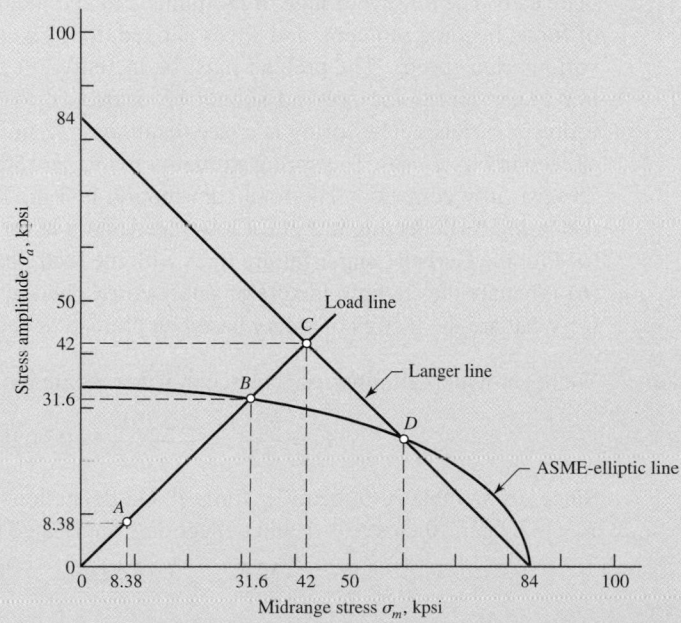

From panel 3 of Table 7–11, the coordinates S_a and S_m of point D in Fig. 7–29 are

$$S_a = \frac{2}{84(1/34.1^2 + 1/84^2)} = 23.8 \text{ kpsi} \qquad S_m = S_y - S_a = 84 - 23.8 = 60.2 \text{ kpsi}$$

$$r_{\text{crit}} = \frac{S_a}{S_m} = \frac{23.8}{60.2} = 0.395$$

Since $r_{\text{crit}} < r$, the primary threat is from fatigue. The factor of safety in fatigue n_f is given by

$$n_f = \frac{S_a}{\sigma_a} = \frac{31.6}{8.38} = 3.77$$

Alternatively, from the fourth panel of Table 7–11,

$$n_f = \sqrt{\frac{1}{(8.38/34.1)^2 + (8.38/84)^2}} = 3.77$$

The factor of safety guarding against first-cycle yielding n_y is

$$n_y = \frac{(S_a)_y}{\sigma_a} = \frac{42}{8.38} = 5.01$$

The Gerber and the ASME-elliptic fatigue failure criteria are very close to each other and are used interchangeably. The ANSI/ASME Standard B106.1M–1985 uses ASME-elliptic for shafting.

EXAMPLE 7–12

A flat-leaf spring is used to retain an oscillating flat-faced follower in contact with a plate cam. The follower range of motion is 2 in and fixed, so the alternating component of force, bending moment, and stress is fixed, too. The spring is preloaded to adjust to various cam speeds. The preload must be increased to prevent follower float or jump. For lower speeds the preload should be decreased to obtain longer life of cam and follower surfaces. The spring is a steel cantilever 32 in long, 2 in wide, and $\frac{1}{4}$ in thick, as seen in Fig. 7–30a. The spring strengths are $S_{ut} = 150$ kpsi, $S_y = 127$ kpsi, and $S_e = 28$ kpsi fully corrected. The total cam motion is 2 in. The designer wishes to preload the spring by deflecting it 2 in for low speed and 5 in for high speed.

(a) Plot the Gerber-Langer failure lines with the load line.

(b) What are the strength factors of safety corresponding to 2 in and 5 in preload?

(c) What are the factors of safety based on preload deflection?

Solution

We begin with preliminaries. The second area moment of the cantilever cross section is

$$I = \frac{bh^3}{12} = \frac{2(0.25)^3}{12} = 0.00260 \text{ in}^4$$

Since, from Table A–9, beam 1, force F and deflection y in a cantilever are related by $F = 3EIy/l^3$, then stress σ and deflection y are related by

$$\sigma = \frac{Mc}{I} = \frac{32Fc}{I} = \frac{32(3EIy)}{l^3}\frac{c}{I} = \frac{96Ecy}{l^3} = Ky$$

Figure 7–30

Cam follower retaining spring. (a) Geometry; (b) designer's fatigue diagram for Ex. 7–12.

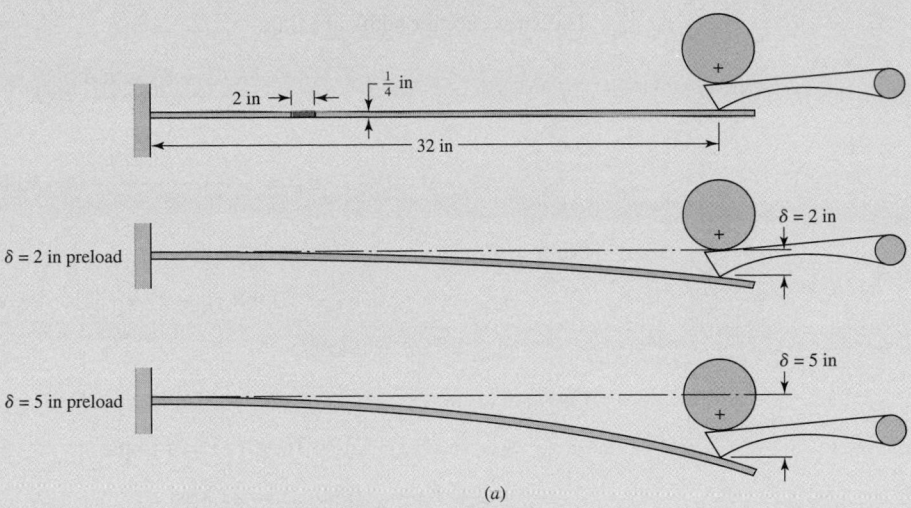

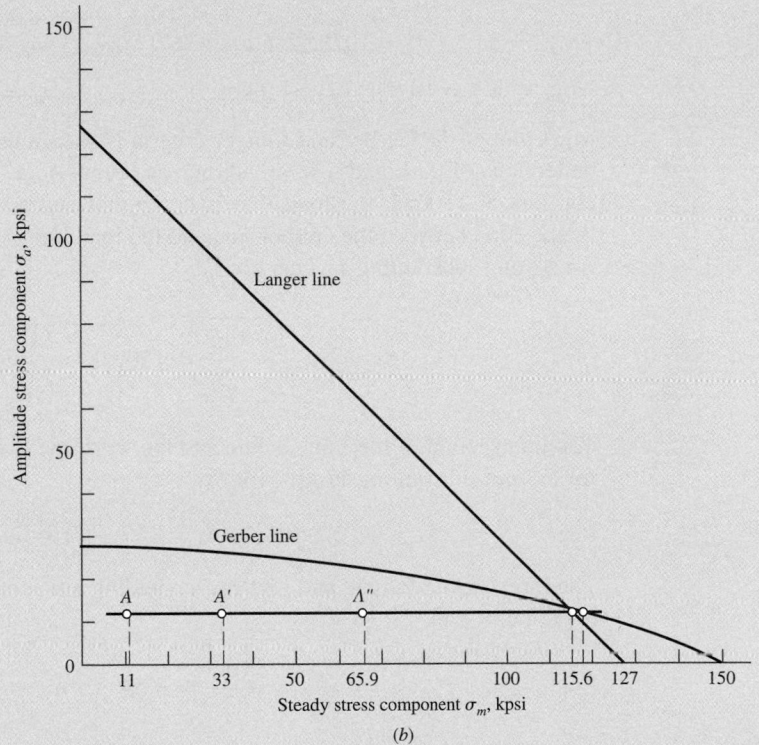

where
$$K = \frac{96(30 \cdot 10^6)0.125}{32^3} = 10.99(10^3) \text{ psi/in} = 10.99 \text{ kpsi/in}$$

Now the minimums and maximums of F, y, and σ can be defined by

$$y_{\min} = \delta \qquad\qquad y_{\max} = 2 + \delta$$
$$F_{\min} = 3EI\delta/l^3 \qquad F_{\max} = 3EI(2 + \delta)/l^3$$
$$\sigma_{\min} = K\delta \qquad\qquad \sigma_{\max} = K(2 + \delta)$$

The stress components are thus

$$\sigma_a = \frac{K(2+\delta) - K\delta}{2} = K$$

$$\sigma_m = \frac{K(2+\delta) + K\delta}{2} = K(1+\delta)$$

For $\delta = 0$,

$$\sigma_{max} = K(2+\delta) = 10.99(2) = 22 \text{ kpsi} \qquad \sigma_a = (22-0)/2 = 11 \text{ kpsi}$$

$$\sigma_{min} = K\delta = 0 \text{ kpsi} \qquad \sigma_m = (22+0)/2 = 11 \text{ kpsi}$$

For $\delta = 2$,

$$\sigma_{max} = K(2+2) = 10.99(4) = 44 \text{ kpsi} \qquad \sigma_a = (44-22)/2 = 11 \text{ kpsi}$$

$$\sigma_{min} = K\delta = 10.99(2) = 22 \text{ kpsi} \qquad \sigma_m = (44+22)/2 = 33 \text{ kpsi}$$

For $\delta = 5$,

$$\sigma_{max} = K(2+\delta) = 10.99(2+5) = 76.9 \text{ kpsi} \qquad \sigma_a = (76.9-54.9)/2 = 11 \text{ kpsi}$$

$$\sigma_{min} = K\delta = 10.99(5) = 54.9 \text{ kpsi} \qquad \sigma_m = (76.9+54.9)/2 = 65.6 \text{ kpsi}$$

(*a*) A plot of the Gerber and Langer criteria is shown in Fig. 7–30*b*. The three preload deflections of 0, 2, and 5 in are shown as points *A*, *A′*, and *A″*. Note that since σ_a is constant at 11 kpsi, the load line is horizontal and does not contain the origin. The intersection between the Gerber line and the load line is found from solving Eq. (7–45) for S_m and substituting 11 kpsi for S_a:

$$S_m = S_{ut}\sqrt{1 - \frac{S_a}{S_e}} = 150\sqrt{1 - \frac{11}{28}} = 116.9 \text{ kpsi}$$

The intersection of the Langer line and the load line is found from solving Eq. (7–47) for S_m and substituting 11 kpsi for S_a:

$$S_m = S_y - S_a = 127 - 11 = 116 \text{ kpsi}$$

The threats from fatigue and first-cycle yielding are approximately equal.
(*b*) For $\delta = 2$ in,

$$n_f = \frac{S_m}{\sigma_m} = \frac{116.9}{33} = 3.54 \qquad n_y = \frac{116}{33} = 3.52$$

and for $\delta = 5$ in,

$$n_f = \frac{116.9}{65.9} = 1.77 \qquad n_y = \frac{116}{65.9} = 1.76$$

(*c*) A factor of safety based on preload deflection involves finding the preload deflection associated with failure. Treating fatigue as the threat, Gerber's relation can be expressed as

$$\frac{11}{28} + \left(\frac{K\delta^*}{150}\right)^2 = 1$$

from which

$$\delta^* = \frac{150}{K}\sqrt{1 - \frac{11}{28}} = \frac{150}{11}\sqrt{1 - \frac{11}{28}} = 10.6 \text{ in}$$

For slower speeds the factor of safety can be defined as

Answer

$$n = \frac{\text{loss-of-function preload deflection}}{\text{working preload deflection}} = \frac{\delta^*}{\delta} = \frac{10.6}{2} = 5.3$$

For higher speeds the same definition applies and the factor of safety is

Answer

$$n = \frac{10.6}{5} = 2.12$$

EXAMPLE 7–13

A steel bar undergoes cyclic loading such that $\sigma_{\max} = 60$ kpsi and $\sigma_{\min} = -20$ kpsi. For the material, $S_{ut} - 80$ kpsi, $S_y = 65$ kpsi, a fully corrected endurance limit of $S_e = 40$ kpsi, and $f = 0.9$. Estimate the number of cycles to a fatigue failure using:
(a) Modified Goodman criterion.
(b) Gerber criterion.

Solution

From the given stresses,

$$\sigma_a = \frac{60 - (-20)}{2} = 40 \text{ kpsi} \qquad \sigma_m = \frac{60 + (-20)}{2} = 20 \text{ kpsi}$$

From the material properties, Eqs. (7–13) to (7–15) give

$$a = \frac{(f\,S_{ut})^2}{S_e} = \frac{[0.9(80)]^2}{40} = 129.6 \text{ kpsi}$$

$$b = -\frac{1}{3}\log\left(\frac{f\,S_{ut}}{S_e}\right) = -\frac{1}{3}\log\left[\frac{0.9(80)}{40}\right] = -0.0851$$

$$N = \left(\frac{S_f}{a}\right)^{1/b} = \left(\frac{S_f}{129.6}\right)^{-1/0.0851} \qquad (1)$$

where S_f replaced σ_a in Eq. (7–15).
(a) For the Goodman criterion with $n = 1$, $S_e = S_f$, Eq. (7–49) gives

$$S_f = \frac{\sigma_a}{1 - \dfrac{\sigma_m}{S_{ut}}} = \frac{40}{1 - \dfrac{20}{80}} = 53.3 \text{ kpsi}$$

Substituting this into Eq. (1) yields

Answer

$$N - \left(\frac{53.3}{129.6}\right)^{-1/0.0851} = 3.40(10^4) \text{ cycles}$$

(b) For Gerber, similar to part (a), from Eq. (7–50),

$$S_f = \frac{\sigma_a}{1 - \left(\dfrac{\sigma_m}{S_{ut}}\right)^2} = \frac{40}{1 - \left(\dfrac{20}{80}\right)^2} = 42.7 \text{ kpsi}$$

Again, from Eq. (1),

Answer

$$N = \left(\frac{42.7}{129.6}\right)^{-1/0.0851} = 4.68(10^5) \text{ cycles}$$

Comparing the answers, we see a large difference in the results. Again, the modified Goodman criterion is conservative as compared to Gerber for which the moderate difference in S_f is then magnified by a logarithmic S, N relationship.

For many *brittle* materials, the first quadrant fatigue failure criteria follows a concave upward Smith-Dolan locus represented by

$$\frac{S_a}{S_e} = \frac{1 - S_m/S_{ut}}{1 + S_m/S_{ut}} \tag{7–52}$$

or as a design equation,

$$\frac{n\sigma_a}{S_e} = \frac{1 - n\sigma_m/S_{ut}}{1 + n\sigma_m/S_{ut}} \tag{7–53}$$

For a radial load line of slope r, we substitute S_a/r for S_m in Eq. (7–52) and solve for S_a, obtaining

$$S_a = \frac{r S_{ut} + S_e}{2}\left[-1 + \sqrt{1 + \frac{4r S_{ut} S_e}{(r S_{ut} + S_e)^2}}\right] \tag{7–54}$$

The fatigue diagram for a brittle material differs markedly from that of a ductile material:

- Yielding is not involved since the material may not have a yield strength.
- Characteristically, the compressive ultimate strength exceeds the ultimate tensile strength severalfold.
- First-quadrant fatigue failure locus is concave-upward (Smith-Dolan), for example, and as flat as Goodman. Brittle materials are more sensitive to midrange stress, being lowered, but compressive midrange stresses are beneficial.
- Not enough work has been done on brittle fatigue to discover insightful generalities, so we stay in the first and a bit of the second quadrant.

The most likely domain of designer use is in the range from $-S_{ut} \le \sigma_m \le S_{ut}$. The locus in the first quadrant is Goodman, Smith-Dolan, or something in between. The portion of the second quadrant that is used is represented by a straight line between the points $-S_{ut}$, S_{ut} and 0, S_e, which has the equation

$$S_a = S_e + \left(\frac{S_e}{S_{ut}} - 1\right) S_m \qquad -S_{ut} \le S_m \le 0 \quad \text{(for cast iron)} \tag{7–55}$$

Table A–24 gives properties of gray cast iron. The endurance limit stated is really $k_a k_b S_e'$ and only corrections k_c, k_d, k_e, and k_f need be made. The average k_c for axial and torsional loading is 0.9.

EXAMPLE 7–14

A grade 30 gray cast iron is subjected to a load F applied to a 1 by $\frac{3}{8}$-in cross-section link with a $\frac{1}{4}$-in-diameter hole drilled in the center as depicted in Fig. 7–31a. The surfaces are machined. In the neighborhood of the hole, what is the factor of safety guarding against failure under the following conditions:

(a) The load $F = 1000$ lbf tensile, steady.

(b) The load is 1000 lbf repeatedly applied.

(c) The load fluctuates between -1000 lbf and 300 lbf without column action.

Use the Smith-Dolan fatigue locus.

Solution

Some preparatory work is needed. From Table A–24, $S_{ut} = 31$ kpsi, $S_{uc} = 109$ kpsi, $k_a k_b S_e' = 14$ kpsi. Since k_c for axial loading is 0.9, then $S_e = (k_a k_b S_e')k_c = 14(0.9) = 12.6$ kpsi. From Table A–15–1, $A = t(w - d) = 0.375(1 - 0.25) = 0.281$ in^2, $d/w = 0.25/1 = 0.25$, and $K_t = 2.45$. The notch sensitivity for cast iron is 0.20 (see p. 336), so

$$K_f = 1 + q(K_t - 1) = 1 + 0.20(2.45 - 1) = 1.29$$

$$(a)\ \sigma_a = \frac{K_f F_a}{A} = \frac{1.29(0)}{0.281} = 0 \qquad \sigma_m = \frac{K_f F_m}{A} = \frac{1.29(1000)}{0.281}(10^{-3}) = 4.59 \text{ kpsi}$$

and

Answer

$$n = \frac{S_{ut}}{\sigma_m} = \frac{31.0}{4.59} = 6.75$$

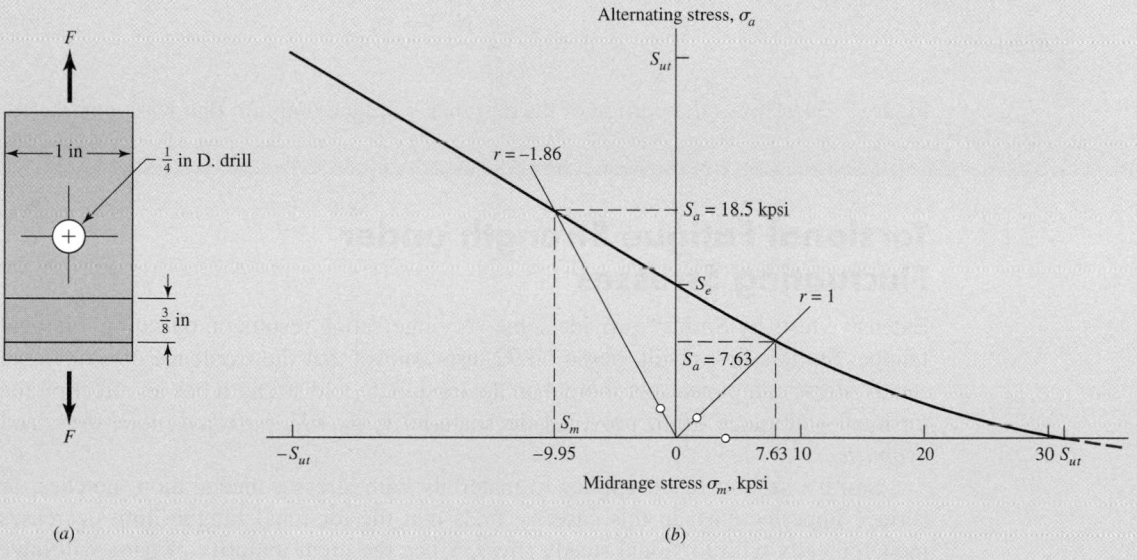

(a)   (b)

Figure 7–31

The grade 30 cast-iron part in axial fatigue with (a) its geometry displayed and (b) its designer's fatigue diagram for the circumstances of Ex. 7–14.

(b)
$$F_a = F_m = \frac{F}{2} = \frac{1000}{2} = 500 \text{ lbf}$$

$$\sigma_a = \sigma_m = \frac{K_f F_a}{A} = \frac{1.29(500)}{0.281}(10^{-3}) = 2.30 \text{ kpsi}$$

$$r = \frac{\sigma_a}{\sigma_m} = 1$$

From Eq. (7–54),

$$S_a = \frac{(1)31 + 12.6}{2}\left[-1 + \sqrt{1 + \frac{4(1)31(12.6)}{[(1)31 + (12.6)]^2}}\right] = 7.63 \text{ kpsi}$$

Answer
$$n = \frac{S_a}{\sigma_a} = \frac{7.63}{2.30} = 3.32$$

(c) $\quad F_a = \frac{1}{2}|300 - (-1000)| = 650 \text{ lbf} \qquad \sigma_a = \frac{1.29(650)}{0.281}(10^{-3}) = 3.0 \text{ kpsi}$

$$F_m = \frac{1}{2}[300 + (-1000)] = -350 \text{ lbf} \qquad \sigma_m = \frac{1.29(-350)}{0.281}(10^{-3}) = -1.61 \text{ kpsi}$$

$$r = \frac{\sigma_a}{\sigma_m} = \frac{3.0}{-1.61} = -1.86$$

From Eq. (7–55), $S_a = S_e + (S_e/S_{ut} - 1)S_m$ and $S_m = S_a/r$. It follows that

$$S_a = \frac{S_e}{1 - \frac{1}{r}\left(\frac{S_e}{S_{ut}} - 1\right)} = \frac{12.6}{1 - \frac{1}{-1.86}\left(\frac{12.6}{31} - 1\right)} = 18.5 \text{ kpsi}$$

Answer
$$n = \frac{S_a}{\sigma_a} = \frac{18.5}{3.0} = 6.17$$

Figure 7–31b shows the portion of the designer's fatigue diagram that was constructed.

7–13 Torsional Fatigue Strength under Fluctuating Stresses

Extensive tests by Smith[26] provide some very interesting results on pulsating torsional fatigue. Smith's first result, based on 72 tests, shows that the existence of a torsional steady-stress component not more than the torsional yield strength has no effect on the torsional endurance limit, provided the material is *ductile, polished, notch-free,* and *cylindrical.*

Smith's second result applies to materials with stress concentration, notches, or surface imperfections. In this case, he finds that the torsional fatigue limit decreases monotonically with torsional steady stress. Since the great majority of parts will have

[26]James O. Smith, "The Effect of Range of Stress on the Fatigue Strength of Metals," *Univ. of Ill. Eng. Exp. Sta. Bull.* 334, 1942.

surfaces that are less than perfect, this result indicates Gerber, ASME-elliptic, and other approximations are useful. Joerres[27] of Associated Spring-Barnes Group, confirms Smith's results and recommends the use of the modified Goodman relation for pulsating torsion. In constructing the Goodman diagram, Joerres uses

$$S_{su} = 0.67S_{ut} \qquad (7\text{–}56)$$

Also, from Chap. 6, $S_{sy} = 0.577S_{yt}$ from distortion-energy theory, and the mean load factor k_c is given by Eq. (7–25), or 0.577. This is discussed further in Chap. 10.

7–14 Combinations of Loading Modes

In Sec. 7–9 we learned that a load factor k_c is used to obtain the endurance limit and hence that the result is dependent on whether the loading is axial, bending, or torsion. In this section we want to answer the question, How do we proceed when the loading is a mixture of, say, axial, bending, and torsional loads? In addition to the complication introduced by the fact that a separate endurance limit is associated with each mode of loading, there may also be multiple stress-concentration factors, one also for each mode of loading. Fortunately, the answer turns out to be rather simple. Assuming that all stress components are completely reversing and are always in time phase with each other,

1 For the strength, use the fully corrected endurance limit for bending, S_e.
2 Apply the appropriate fatigue stress-concentration factors to the torsional stress, the bending stress, and the axial stress components.
3 Multiply any alternating axial stress components by the factor $1/k_{c,\text{ax}}$.
4 Enter the resultant stresses into a Mohr's circle analysis and find the principal stresses.
5 Using the results of step 4, find the von Mises alternating stress σ_a'.
6 Compare σ_a' with S_a to find the factor of safety.

If the stress components are not in phase but have the same frequency, the maxima can be found by expressing each component in trigonometric terms, using phase angles, and then finding the sum. If two or more stress components have differing frequencies, the problem is difficult; one solution is to assume that the two (or more) components often reach an in-phase condition, so that their magnitudes are additive.

   If midrange stresses are also present, then steps 4 and 5 can be repeated for them and the resulting steady von Mises stress component σ_m' used with σ_a' in forming a Gerber or ASME-elliptic solution. Both the steady and amplitude components are augmented by K_f or K_{fs} stress-concentration factor.

---

[27]Robert E. Joerres, "Springs," Chap. 24 in Joseph E. Shigley and Charles R. Mischke, *Standard Handbook of Machine Design*, 2nd ed., McGraw-Hill, New York, 1996.

EXAMPLE 7–15   A rotating shaft is made of 42- × 4-mm AISI 1018 cold-drawn steel tubing and has a 6-mm-diameter hole drilled transversely through it. Estimate the factor of safety guarding against fatigue and static failures using the Gerber and Langer failure criteria for the following loading conditions:
(*a*) The shaft is subjected to a completely reversed torque of 120 N · m in phase with a completely reversed bending moment of 150 N · m.

(b) The shaft is subjected to a pulsating torque fluctuating from 20 to 160 N · m and a steady bending moment of 150 N · m.

Solution

Here we follow the procedure of estimating the strengths and then the stresses, followed by relating the two.

From Table A–20 we find the minimum strengths to be $S_{ut} = 440$ MPa and $S_{yt} = 370$ MPa. The endurance limit of the rotating-beam specimen is $0.504(440) = 222$ MPa. The surface factor, obtained from Eq. (7–18) and Table 7–4, is

$$k_a = 4.51 S_{ut}^{-0.265} = 4.51(440)^{-0.265} = 0.899$$

From Eq. (7–20) the size factor is

$$k_b = \left(\frac{d}{7.62}\right)^{-0.107} = \left(\frac{42}{7.62}\right)^{-0.107} = 0.833$$

The remaining Marin factors are all unity, so the modified endurance strength S_e is

$$S_e = 0.899(0.833)222 = 166 \text{ MPa}$$

(a) Theoretical stress-concentration factors are found from Table A–16. Using $a/D = 6/42 = 0.143$ and $d/D = 34/42 = 0.810$, and using linear interpolation, we obtain $A = 0.798$ and $K_t = 2.366$ for bending; and $A = 0.89$ and $K_{ts} = 1.75$ for torsion. Thus, for bending,

$$Z_{net} = \frac{\pi A}{32D}(D^4 - d^4) = \frac{\pi(0.798)}{32(4.2)}[(4.2)^4 - (3.4)^4] = 3.31 \text{ cm}^3$$

and for torsion

$$J_{net} = \frac{\pi A}{32}(D^4 - d^4) = \frac{\pi(0.89)}{32}[(4.2)^4 - (3.4)^4] = 15.5 \text{ cm}^4$$

Next, using Figs. 7–20 and 7–21 with a notch radius of 3 mm we find the notch sensitivities to be 0.78 for bending and 0.96 for torsion. The two corresponding fatigue stress-concentration factors are obtained from Eq. (7–31) as

$$K_f = 1 + q(K_t - 1) = 1 + 0.78(2.366 - 1) = 2.07$$
$$K_{fs} = 1 + 0.96(1.75 - 1) = 1.72$$

The bending stress is now found to be

$$\sigma_{xa} = K_f \frac{M}{Z_{net}} = 2.07\frac{150}{3.31} = 93.8 \text{ MPa}$$

and the torsional stress is

$$\tau_{xya} = K_{fs} \frac{TD}{2J_{net}} = 1.72\frac{120(4.2)}{2(15.5)} = 28.0 \text{ MPa}$$

The von Mises steady-stress component σ'_m is zero. The amplitude component σ'_a is given by

$$\sigma'_a = \left(\sigma_{xa}^2 + 3\tau_{xya}^2\right)^{1/2} = [93.8^2 + 3(28^2)]^{1/2} = 105.6 \text{ MPa}$$

Figure 7–32

Designer's fatigue diagram for Ex. 7–15.

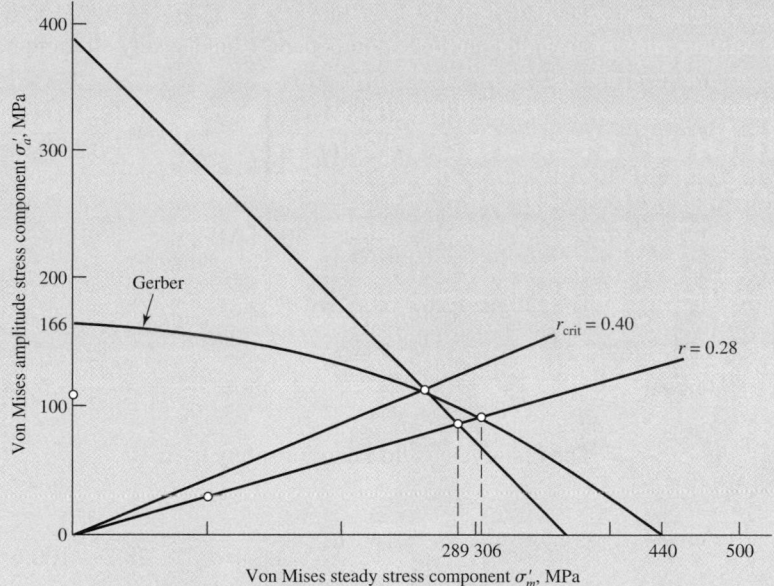

Since $S_e = S_a$, the fatigue factor of safety n_f is

$$n_f = \frac{S_a}{\sigma_a'} = \frac{166}{105.6} = 1.57$$

The first-cycle yield factor of safety is

$$n_y = \frac{S_{yt}}{\sigma_a'} = \frac{370}{105.6} = 3.50$$

There is no localized yielding; the threat is from fatigue. See Fig. 7–32.

(*b*) This part asks us to find the factors of safety when the alternating component is due to pulsating torsion, and a steady component is due to both torsion and bending. We have $T_a = (160 - 20)/2 = 70$ N · m and $T_m = 20 + 70 = 90$ N · m. The corresponding amplitude and steady-stress components are

$$\tau_{xya} = K_{fs}\frac{T_a D}{2J_{\text{net}}} = 1.72\frac{70(4.2)}{2(15.5)} = 16.3 \text{ MPa}$$

$$\tau_{xym} = K_{fs}\frac{T_m D}{2J_{\text{net}}} = 1.72\frac{90(4.2)}{2(15.5)} = 21.0 \text{ MPa}$$

The steady bending stress component σ_{xm} is

$$\sigma_{xm} = K_f\frac{M_m}{Z_{\text{net}}} = 2.07\frac{150}{3.31} = 93.8 \text{ MPa}$$

The von Mises components σ_a' and σ_m' are

$$\sigma_a' = [3(16.3)^2]^{1/2} = 28.2 \text{ MPa}$$

$$\sigma_m' = [93.8^2 + 3(21)^2]^{1/2} = 100.6 \text{ MPa}$$

The slope of the load line is $r = \sigma'_a/\sigma'_m = 28.2/100.6 = 0.28$. From Table 7–10, the strength amplitude component S_a and steady-strength component S_m are

$$S_a = \frac{0.28^2 440^2}{2(166)} \left\{ -1 + \sqrt{1 + \left[\frac{2(166)}{0.28(440)} \right]^2} \right\} = 85.7 \text{ MPa}$$

$$S_m = \frac{85.7}{0.28} = 306.1 \text{ MPa}$$

The fatigue factor of safety n_f is

Answer

$$n_f = \frac{S_a}{\sigma'_a} = \frac{85.7}{28.2} = 3.04$$

The first-cycle yield factor of safety n_y is

Answer

$$n_y = \frac{S_y}{\sigma'_a + \sigma'_m} = \frac{370}{28.2 + 100.6} = 2.87$$

There is no notch yielding. The threat is from first-cycle yielding at the notch. See the plot in Fig. 7–32.

7–15 Varying, Fluctuating Stresses; Cumulative Fatigue Damage

Instead of a single fully reversed stress history block composed of n cycles, suppose a machine part, at a critical location, is subjected to

- A fully reversed stress σ_1 for n_1 cycles, σ_2 for n_2 cycles, ..., or
- A "wiggly" time line of stress exhibiting many and different peaks and valleys.

What stresses are significant, what counts as a cycle, and what is the measure of damage incurred? Consider a fully reversed cycle with stresses varying 60, 80, 40, and 60 kpsi and a second fully reversed cycle −40, −60, −20, and −40 kpsi as depicted in Fig. 7–33a. First, it is clear that to impose the pattern of stress in Fig. 7–33a on a part it is necessary that the time trace look like the solid line plus the dashed line in Fig. 7–33a. Figure 7–33b moves the snapshot to exist beginning with 80 kpsi and ending with 80 kpsi. Acknowledging the existence of a single stress-time trace is to discover a "hidden" cycle shown as the dashed line in Fig. 7–33b. If there are 100 applications of the all-positive stress cycle, then 100 applications of the all-negative stress cycle, the hidden cycle is applied but once. If the all-positive stress cycle is applied alternately with the all-negative stress cycle, the hidden cycle is applied 100 times.

To ensure that the hidden cycle is not lost, begin on the snapshot with the largest (or smallest) stress and add previous history to the right side, as was done in Fig. 7–33b. Characterization of a cycle takes on a max–min–same max (or min–max–same min) form. We identify the hidden cycle first by moving along the dashed-line trace in Fig. 7–33b identifying a cycle with an 80-kpsi max, a 60-kpsi min, and returning to 80 kpsi. Mentally deleting the used part of the trace (the dashed line) leaves a 40, 60,

Figure 7–33

Variable stress diagram prepared for assessing cumulative damage.

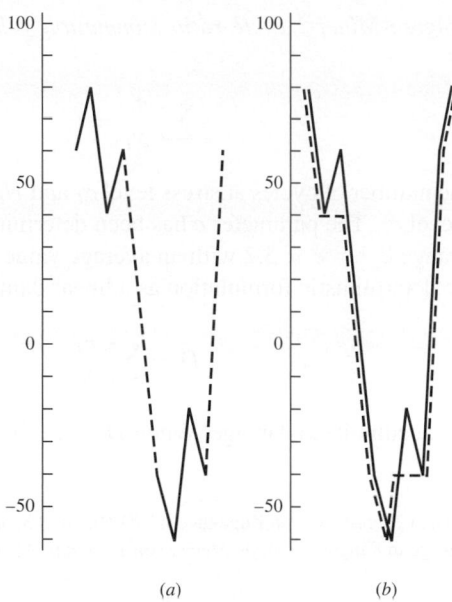

(a)   (b)

40 cycle and a −40, −20, −40 cycle. Since failure loci are expressed in terms of stress amplitude component σ_a and steady component σ_m, we use Eq. (7–39) to construct the table below:

Cycle Number	σ_{max}	σ_{min}	σ_a	σ_m
1	80	−60	70	10
2	60	40	10	50
3	−20	−40	10	−30

The most damaging cycle is number 1. It could have been lost.

Methods for counting cycles include:

- Number of tensile peaks to failure.
- All maxima above the waveform mean, all minima below.
- The global maxima between crossings above the mean and the global minima between crossings below the mean.
- All positive slope crossings of levels above the mean, and all negative slope crossings of levels below the mean.
- A modification of the preceding method with only one count made between successive crossings of a level associated with each counting level.
- Each local maxi-min excursion is counted as a half-cycle, and the associated amplitude is half-range.
- The preceding method plus consideration of the local mean.
- Rain-flow counting technique.

The method used here amounts to a variation of the *rain-flow counting technique*.

The *Palmgren-Miner*[28] *cycle-ratio summation rule,* also called *Miner's rule,* is written

$$\sum \frac{n_i}{N_i} = c \qquad (7\text{--}57)$$

where n_i is the number of cycles at stress level σ_i and N_i is the number of cycles to failure at stress level σ_i. The parameter c has been determined by experiment; it is usually found in the range $0.7 < c < 2.2$ with an average value near unity.

Using the deterministic formulation as a linear damage rule we write

$$D = \sum \frac{n_i}{N_i} \qquad (7\text{--}58)$$

where D is the accumulated damage. When $D = c = 1$, failure ensues.

---

[28]A. Palmgren, "Die Lebensdauer von Kugellagern," *ZVDI,* vol. 68, pp. 339–341, 1924; M. A. Miner, "Cumulative Damage in Fatigue," *J. Appl. Mech.,* vol. 12, *Trans. ASME,* vol. 67, pp. A159–A164, 1945.

EXAMPLE 7–16

For the loading of Fig. 7–33 on a part, the following properties at the critical location exist: $S_{ut} = 151$ kpsi, $\sigma_0 = 210$ kpsi, $\varepsilon_f = 0.45$, $m = 0.09$, $S_e = 67.5$ kpsi. Estimate the number of repetitions of the stress-time block in Fig. 7–33 that can be made before failure.

Solution

$$\sigma_F' = \sigma_0 \varepsilon^m = 210(0.45)^{0.09} = 195.4 \text{ kpsi}$$

From Eq. (7–11),

$$b = -\frac{\log(\sigma_F'/S_e)}{\log(2N_e)} = -\frac{\log(195.4/67.5)}{\log(2 \cdot 10^6)} = -0.07326$$

From Eq. (7–9),

$$f = \frac{\sigma_F'}{S_{ut}}(2 \cdot 10^3)^b = \frac{195.4}{151}(2 \cdot 10^3)^{-0.07326} = 0.7415$$

From Eq. (7–13),

$$a = \frac{(f S_{ut})^2}{S_e} = \frac{[0.7415(151)]^2}{67.5} = 186 \text{ kpsi}$$

so

$$S_f = 186 N^{-0.07326} \qquad N = \left(\frac{\sigma_a}{186}\right)^{-1/0.07326}$$

We prepare to add two columns to the previous table. Using the Gerber fatigue criterion, Eq. (7–45), with $S_a = S_f$, on the designer's fatigue diagram we can write

$$S_f = \begin{cases} \dfrac{\sigma_a}{1 - (\sigma_m/S_{ut})^2} & \sigma_m > 0 \\ S_e & \sigma_m \le 0 \end{cases} \qquad (1)$$

Cycle 1: r = 70/10 = 7, and the strength amplitude from Table 7–9 is

$$S_a = \frac{7^2 151^2}{2(67.5)} \left\{ -1 + \sqrt{1 + \left[\frac{2(67.5)}{7(151)} \right]^2} \right\} = 67.2 \text{ kpsi}$$

Since $\sigma_a > S_a$, that is, 70 > 67.2, life is reduced. From Eq. (1),

$$S_f = \frac{70}{1 - (10/151)^2} = 70.3 \text{ kpsi} \qquad N = \left(\frac{70.3}{186} \right)^{-1/0.07326} = 0.586(10^6) \text{ cycles}$$

Cycle 2: r = 10/50 = 0.2, and the strength amplitude is

$$S_a = \frac{0.2^2 151^2}{2(67.5)} \left\{ -1 + \sqrt{1 + \left[\frac{2(67.5)}{0.2(151)} \right]^2} \right\} = 24.2 \text{ kpsi}$$

Since $\sigma_a < S_a$, that is 10 < 24.2, then $S_f = S_e$ and indefinite life follows.

Cycle 3: r = 10/−30 = −0.333, and since $\sigma_m < 0$, $S_f = S_e$, indefinite life follows.

Cycle Number	S_f, kpsi	N, cycles
1	70.3	0.58(10⁶)
2	67.5	∞
3	67.5	∞

From Eq. (7–62) the damage per block is

$$D = \sum \frac{n_i}{N_i} = N \left[\frac{1}{0.586(10^6)} + \frac{1}{\infty} + \frac{1}{\infty} \right] = \frac{N}{0.586(10^6)}$$

Answer   Setting $D = 1$ yields $N = 0.586(10^6)$ cycles.

    To further illustrate the use of the Miner rule, let us choose a steel having the properties $S_{ut} = 80$ kpsi, $S'_{e,0} = 40$ kpsi, and $f = 0.9$, where we have used the designation $S'_{e,0}$ instead of the more usual S'_e to indicate the endurance limit of the *virgin,* or *undamaged, material.* The log *S*–log *N* diagram for this material is shown in Fig. 7–34 by the heavy solid line. Now apply, say, a reversed stress $\sigma_1 = 60$ kpsi for $n_1 = 3000$ cycles. Since $\sigma_1 > S'_{e,0}$, the endurance limit will be damaged, and we wish to find the new endurance limit $S'_{e,1}$ of the damaged material using the Miner rule. The equation of the virgin material failure line in Fig. 7–34 in the 10^3 to 10^6 cycle range is

$$S_f = aN^b = 129.6N^{-0.085\ 091}$$

The cycles to failure at stress level $\sigma_1 = 60$ kpsi are

$$N_1 = \left(\frac{\sigma_1}{129.6} \right)^{-1/0.085\ 091} = \left(\frac{60}{129.6} \right)^{-1/0.085\ 091} = 8520 \text{ cycles}$$

Figure 7–34 shows that the material has a life $N_1 = 8520$ cycles at 60 kpsi, and consequently, after the application of σ_1 for 3000 cycles, there are $N_1 - n_1 = 5520$ cycles of

Figure 7–34

Use of the Miner rule to predict the endurance limit of a material that has been overstressed for a finite number of cycles.

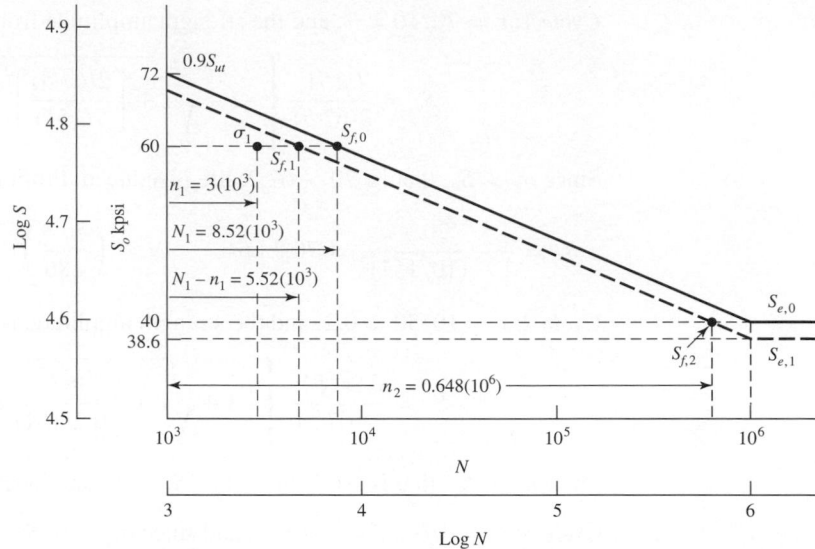

life remaining at σ_1. This locates the finite-life strength $S_{f,1}$ of the damaged material, as shown in Fig. 7–34. To get a second point, we ask the question: With n_1 and N_1 given, how many cycles of stress $\sigma_2 = S'_{e,0}$ can be applied before the damaged material fails? This corresponds to n_2 cycles of stress reversal, and hence, from Eq. (7–57), we have

$$\frac{n_1}{N_1} + \frac{n_2}{N_2} = 1 \tag{a}$$

or

$$n_2 = \left(1 - \frac{n_1}{N_1}\right) N_2 \tag{b}$$

Then

$$n_2 = \left[1 - \frac{3(10)^3}{8.52(10)^3}\right](10^6) = 0.648(10^6) \text{ cycles}$$

This corresponds to the finite-life strength $S_{f,2}$ in Fig. 7–34. A line through $S_{f,1}$ and $S_{f,2}$ is the log S–log N diagram of the damaged material according to the Miner rule. The new endurance limit is $S_{e,1} = 38.6$ kpsi.

We could leave it at this, but a little more investigation can be helpful. We have two points on the new fatigue locus, $N_1 - n_1, \sigma_1$ and n_2, σ_2. It is useful to prove that the slope of the new line is still b. For the equation $S_f = a'N^{b'}$, where the values of a' and b' are established by two points α and β. The equation for b' is

$$b' = \frac{\log \sigma_\alpha/\sigma_\beta}{\log N_\alpha/N_\beta} \tag{c}$$

Examine the denominator of Eq. (c):

$$\log \frac{N_\alpha}{N_\beta} = \log \frac{N_1 - n_1}{n_2} = \log \frac{N_1 - n_1}{(1 - n_1/N_1)N_2} = \log \frac{N_1}{N_2}$$

$$= \log \frac{(\sigma_1/a)^{1/b}}{(\sigma_2/a)^{1/b}} = \log \left(\frac{\sigma_1}{\sigma_2}\right)^{1/b} = \frac{1}{b} \log \left(\frac{\sigma_1}{\sigma_2}\right)$$

Substituting this into Eq. (c) with $\sigma_\alpha/\sigma_\beta = \sigma_1/\sigma_2$ gives

$$b' = \frac{\log(\sigma_1/\sigma_2)}{(1/b) \log(\sigma_1/\sigma_2)} = b$$

Figure 7–35

Use of the Manson method to predict the endurance limit of a material that has been overstressed for a finite number of cycles.

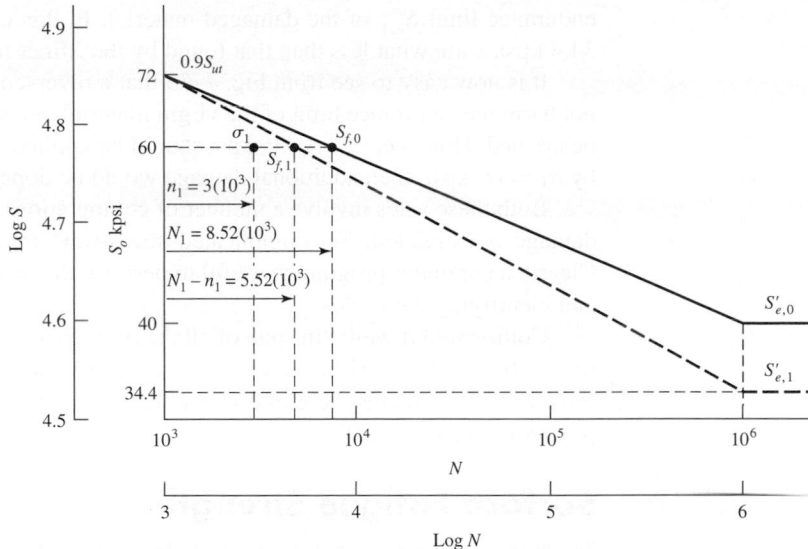

which means the damaged material line has the same slope as the virgin material line; therefore, the lines are parallel. This information can be helpful in writing a computer program for the Palmgren-Miner hypothesis.

Though the Miner rule is quite generally used, it fails in two ways to agree with experiment. First, note that this theory states that the static strength S_{ut} is damaged, that is, decreased, because of the application of σ_1; see Fig. 7–34 at $N = 10^3$ cycles. Experiments fail to verify this prediction.

The Miner rule, as given by Eq. (7–53), does not account for the order in which the stresses are applied, and hence ignores any stresses less than $S'_{e,0}$. But it can be seen in Fig. 7–34 that a stress σ_3 in the range $S'_{e,1} < \sigma_3 < S'_{e,0}$ would cause damage if applied after the endurance limit had been damaged by the application of σ_1.

Manson's[29] approach overcomes both of the deficiencies noted for the Palmgren-Miner method; historically it is a much more recent approach, and it is just as easy to use. Except for a slight change, we shall use and recommend the Manson method in this book. Manson plotted the S–log N diagram instead of a log S–log N plot as is recommended here. Manson also resorted to experiment to find the point of convergence of the S–log N lines corresponding to the static strength, instead of arbitrarily selecting the intersection of $N = 10^3$ cycles with $S = 0.9S_{ut}$ as is done here. Of course, it is always better to use experiment, but our purpose in this book has been to use the simple test data to learn as much as possible about fatigue failure.

The method of Manson, as presented here, consists in having all log S–log N lines, that is, lines for both the damaged and the virgin material, converge to the same point, $0.9S_{ut}$ at 10^3 cycles. In addition, the log S–log N lines must be constructed in the same historical order in which the stresses occur.

The data from the preceding example are used for illustrative purposes. The results are shown in Fig. 7–35. Note that the strength $S_{f,1}$ corresponding to $N_1 - n_1 = 5.52(10^3)$ cycles is found in the same manner as before. Through this point and through $0.9S_{ut}$ at 10^3 cycles, draw the heavy dashed line to meet $N = 10^6$ cycles and define the

[29]S. S. Manson, A. J. Nachtigall, C. R. Ensign, and J. C. Fresche, "Further Investigation of a Relation for Cumulative Fatigue Damage in Bending," *Trans. ASME, J. Eng. Ind.,* ser. B, vol. 87, No. 1, pp. 25–35, February 1965.

endurance limit $S'_{e,1}$ of the damaged material. In this case the new endurance limit is 34.4 kpsi, somewhat less than that found by the Miner method.

It is now easy to see from Fig. 7–35 that a reversed stress $\sigma = 36$ kpsi, say, would not harm the endurance limit of the virgin material, no matter how many cycles it might be applied. However, if $\sigma = 36$ kpsi should be applied *after* the material was damaged by $\sigma_1 = 60$ kpsi, then additional damage would be done.

Both these rules involve a number of computations, which are repeated every time damage is estimated. For complicated stress-time traces, this might be every cycle. Clearly a computer program is useful to perform the tasks, including scanning the trace and identifying the cycles.

Collins said it well: "In spite of all the problems cited, the Palmgren linear damage rule is frequently used because of its simplicity and the experimental fact that other more complex damage theories do not always yield a significant improvement in failure prediction reliability."[30]

7–16 Surface Fatigue Strength

The surface fatigue mechanism is not definitively understood. The contact-affected zone, in the absence of surface shearing tractions, entertains compressive principal stresses. Rotary fatigue has its cracks grown at or near the surface in the presence of tensile stresses that are associated with crack propagation, to catastrophic failure. There are shear stresses in the zone, which are largest just below the surface. Cracks seem to grow from this stratum until small pieces of material are expelled, leaving pits on the surface. Because engineers had to design durable machinery before the surface fatigue phenomenon was understood in detail, they had taken the posture of conducting tests, observing pits on the surface, and declaring failure at an arbitrary projected area of hole, and they related this to the Hertzian contact pressure. This compressive stress did not produce the failure directly, but whatever the failure mechanism, whatever the stress type that was instrumental in the failure, the contact stress was an *index* to its magnitude.

Buckingham[31] conducted a number of tests relating the fatigue at 10^8 cycles to endurance strength (Hertzian contact pressure). While there is evidence of an endurance limit at about $3(10^7)$ cycles for cast materials, hardened steel rollers showed no endurance limit up to $4(10^8)$ cycles. Subsequent testing on hard steel shows no endurance limit. Hardened steel exhibits such high fatigue strengths that its use in resisting surface fatigue is widespread.

Our studies thus far have dealt with the failure of a machine element by yielding, by fracture, and by fatigue. The endurance limit obtained by the rotating-beam test is frequently called the *flexural endurance limit,* because it is a test of a rotating beam. In this section we shall study a property of *mating materials* called the *surface endurance shear.* The design engineer must frequently solve problems in which two machine elements mate with one another by rolling, sliding, or a combination of rolling and sliding contact. Obvious examples of such combinations are the mating teeth of a pair of gears, a cam and follower, a wheel and rail, and a chain and sprocket. A knowledge of the surface strength of materials is necessary if the designer is to create machines having a long and satisfactory life.

When two surfaces roll or roll and slide against one another with sufficient force, a pitting failure will occur after a certain number of cycles of operation. Authorities are not in complete agreement on the exact mechanism of the pitting; although the subject

[30]J. A. Collins, *Failure of Materials in Mechanical Design,* John Wiley & Sons, New York, 1981, p. 243.

[31]Earle Buckingham, *Analytical Mechanics of Gears,* McGraw-Hill, New York, 1949.

is quite complicated, they do agree that the Hertz stresses, the number of cycles, the surface finish, the hardness, the degree of lubrication, and the temperature all influence the strength. In Sec. 4–20 it was learned that, when two surfaces are pressed together, a maximum shear stress is developed slightly below the contacting surface. It is postulated by some authorities that a surface fatigue failure is initiated by this maximum shear stress and then is propagated rapidly to the surface. The lubricant then enters the crack that is formed and, under pressure, eventually wedges the chip loose.

To determine the surface fatigue strength of mating materials, Buckingham designed a simple machine for testing a pair of contacting rolling surfaces in connection with his investigation of the wear of gear teeth. Buckingham and, later, Talbourdet gathered large numbers of data from many tests so that considerable design information is now available. To make the results useful for designers, Buckingham defined a *load-stress factor*, also called a *wear factor*, which is derived from the Hertz equations. Equations (4–77) and (4–78) for contacting cylinders are found to be

$$b = \sqrt{\frac{2F}{\pi l}\frac{\left(1 - v_1^2\right)/E_1 + \left(1 - v_2^2\right)/E_2}{(1/d_1) + (1/d_2)}} \tag{7–59}$$

$$p_{max} = \frac{2F}{\pi bl} \tag{7–60}$$

where b = half width of rectangular contact area

  F = contact force

  l = length of cylinders

  v = Poisson's ratio

  E = modulus of elasticity

  d = cylinder diameter

It is more convenient to use the cylinder radius, so let $2r = d$. If we then designate the length of the cylinders as w (for width of gear, bearing, cam, etc.) instead of l and remove the square root sign, Eq. (7–59) becomes

$$b^2 = \frac{4F}{\pi w}\frac{\left(1 - v_1^2\right)/E_1 + \left(1 - v_2^2\right)/E_2}{1/r_1 + 1/r_2} \tag{7–61}$$

We can define a *surface endurance strength* S_C using

$$p_{max} = \frac{2F}{\pi bw} \tag{7–62}$$

as

$$S_C = \frac{2F}{\pi bw} \tag{7–63}$$

which may also be called *contact strength,* the *contact fatigue strength,* or the *Hertzian endurance strength.* The strength is the contacting pressure which, after a specified number of cycles, will cause failure of the surface. Such failures are often called *wear* because they occur over a very long time. They should not be confused with abrasive wear, however. By substituting the value of b in Eq. (7–61) and substituting the result into Eq. (7–63), we obtain

$$\frac{F}{w}\left(\frac{1}{r_1} + \frac{1}{r_2}\right) = \pi S_C^2 \left[\frac{1 - v_1^2}{E_1} + \frac{1 - v_2^2}{E_2}\right] = K_1 \tag{7–64}$$

The left expression consists of parameters a designer may seek to control independently. The central expression consists of material properties that come with the material and condition specification. The third expression is the parameter K_1, Buckingham's load-stress factor, determined by a test fixture with values F, w, r_1, r_2 and the number of cycles associated with the first tangible evidence of fatigue. In gear studies a similar K factor is used:

$$K_g = \frac{K_1}{4} \sin \phi \qquad (7\text{--}65)$$

where ϕ is the tooth pressure angle, and the term $[(1 - v_1^2)/E_1 + (1 - v_2^2)/E_2]$ is defined as $1/(\pi C_P^2)$, so that

$$S_C = C_P \sqrt{\frac{F}{w} \left(\frac{1}{r_1} + \frac{1}{r_2} \right)} \qquad (7\text{--}66)$$

Buckingham and others reported K_1 for 10^8 cycles and nothing else. This gives only one point on the $S_C N$ curve. For cast metals this may be sufficient, but for wrought steels, heat-treated, some idea of the slope is useful in meeting design goals of other than 10^8 cycles.

Experiments show that K_1 versus N, K_g versus N, and S_C versus N data are rectified by loglog transformation. This suggests that

$$K_1 = \alpha_1 N^{\beta_1} \qquad K_g = a N^b \qquad S_C = \alpha N^\beta$$

The three exponents are given by

$$\beta_1 = \frac{\log(K_1/K_2)}{\log(N_1/N_2)} \qquad b = \frac{\log(K_{g1}/K_{g2})}{\log(N_1/N_2)} \qquad \beta = \frac{\log(S_{C1}/S_{C2})}{\log(N_1/N_2)} \qquad (7\text{--}67)$$

Data on induction-hardened steel on steel give $(S_C)_{10^7} = 271$ kpsi and $(S_C)_{10^8} = 239$ kpsi, so β, from Eq. (7–67), is

$$\beta = \frac{\log(271/239)}{\log(10^7/10^8)} = -0.055$$

It may be of interest that the American Gear Manufacturers Association (AGMA) uses -0.056 between $10^4 < N < 10^{10}$ if the designer has no data to the contrary beyond 10^7 cycles.

A longstanding correlation in steels between S_C and H_B at 10^8 cycles is

$$(S_C)_{10^8} = \begin{cases} 0.4 H_B - 10 \text{ kpsi} \\ 2.76 H_B - 70 \text{ MPa} \end{cases} \qquad (7\text{--}68)$$

AGMA uses

$$_{0.99}(S_C)_{10^7} = 0.327 H_B + 26 \text{ kpsi} \qquad (7\text{--}69)$$

Equation (7–66) can be used in design to find an allowable surface stress by using a design factor. Since this equation is nonlinear in its stress-load transformation, the designer must decide if loss of function denotes inability to carry the load. If so, then to find the allowable stress, one divides the load F by the design factor n_d:

$$\sigma_C = C_P \sqrt{\frac{F}{w n_d} \left(\frac{1}{r_1} + \frac{1}{r_2} \right)} = \frac{C_P}{\sqrt{n_d}} \sqrt{\frac{F}{w} \left(\frac{1}{r_1} + \frac{1}{r_2} \right)} = \frac{S_C}{\sqrt{n_d}}$$

and $n_d = (S_C/\sigma_C)^2$. If the loss of function is focused on stress, then $n_d = S_C/\sigma_C$. It is recommended that an engineer

- Decide whether loss of function is failure to carry load or stress.
- Define the design factor and factor of safety accordingly.

- Announce what he or she is using and why.
- Be prepared to defend his or her position.

In this way everyone who is party to the communication knows what a design factor (or factor of safety) of 2 means and adjusts, if necessary, the judgmental perspective.

7–17 Stochastic Analysis

As already demonstrated in this chapter, there are a great many factors to consider in a fatigue analysis, much more so than in a static analysis. So far, each factor has been treated in a deterministic manner, and if not obvious, these factors are subject to variability and control the overall reliability of the results. When reliability is important, then fatigue testing must certainly be undertaken. There is no other way. Consequently, the methods of stochastic analysis presented here and in other sections of this book constitute guidelines that enable the designer to obtain a good understanding of the various issues involved and help in the development of a safe and reliable design.

In this section, key stochastic modifications to the deterministic features and equations described in earlier sections are provided in the same order of presentation.

Endurance Limit

To begin, a method for estimating endurance limits, the *tensile strength correlation method*, is presented. The ratio $\phi = S'_e / \bar{S}_{ut}$ is called the *fatigue ratio*.[32] For ferrous metals, most of which exhibit an endurance limit, the endurance limit is used as a numerator. For materials that do not show an endurance limit, an endurance strength at a specified number of cycles to failure is used and noted. Gough[33] reported the stochastic nature of the fatigue ratio ϕ for several classes of metals, and this is shown in Fig. 7–36. The first item to note is that the coefficient of variation is of the order 0.10 to 0.15, and the distribution varies for classes of metals. The second item to note is that Gough's data include materials of no interest to engineers. In the absence of testing, engineers use the correlation that ϕ represents to estimate the endurance limit S'_e from the mean ultimate strength $\bar{S}_{ut}$.

Gough's data are for ensembles of metals, some chosen for metallurgical interest, and include materials that are not commonly selected for machine parts. Mischke[34] analyzed data for 133 common steels and treatments in varying diameters in rotating bending,[35] and the result was

$$\phi = 0.445d^{-0.107}\mathbf{LN}(1, 0.138)$$

where d is the specimen diameter in inches and $\mathbf{LN}(1, 0.138)$ is a unit lognormal variate with a mean of 1 and a standard deviation (and coefficient of variation) of 0.138. For the standard R. R. Moore specimen,

$$\phi_{0.30} = 0.445(0.30)^{-0.107}\mathbf{LN}(1, 0.138) = 0.506\mathbf{LN}(1, 0.138)$$

---

[32]From this point, since we will be dealing with statistical distributions in terms of means, standard deviations, etc. A key quantity, the ultimate strength, will here be presented by its mean value, $\bar{S}_{ut}$. This means that certain terms that were defined earlier in terms of the minimum value of S_{ut} will change slightly.

[33]In J. A. Pope, *Metal Fatigue,* Chapman and Hall, London, 1959.

[34]Charles R. Mischke, "Prediction of Stochastic Endurance Strength," *Trans. ASME, Journal of Vibration, Acoustics, Stress, and Reliability in Design,* vol. 109, no. 1, January 1987, pp. 113–122.

[35]Data from H. J. Grover, S. A. Gordon, and L. R. Jackson, *Fatigue of Metals and Structures,* Bureau of Naval Weapons, Document NAVWEPS 00-2500435, 1960.

Figure 7–36

The lognormal probability density PDF of the fatigue ratio ϕ_b of Gough.

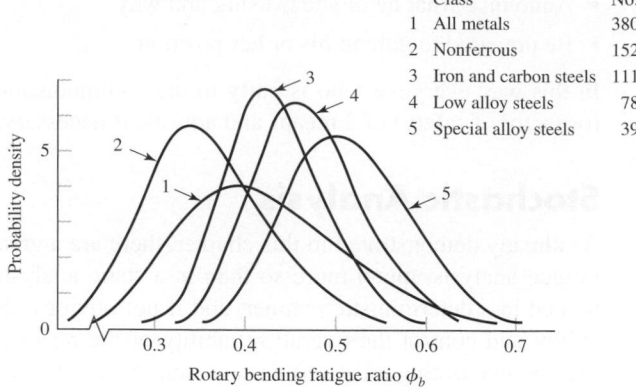

Also, 25 plain carbon and low-alloy steels with $S_{ut} > 212$ kpsi are described by

$$\mathbf{S}'_e = 107\mathbf{LN}(1, 0.139) \text{ kpsi}$$

In summary, for the rotating-beam specimen,

$$\mathbf{S}'_e = \begin{cases} 0.506\bar{S}_{ut}\mathbf{LN}(1, 0.138) \text{ kpsi or MPa} & \bar{S}_{ut} \leq 212 \text{ kpsi (1460 MPa)} \\ 107\mathbf{LN}(1, 0.139) \text{ kpsi} & \bar{S}_{ut} > 212 \text{ kpsi} \\ 740\mathbf{LN}(1, 0.139) \text{ MPa} & \bar{S}_{ut} > 1460 \text{ MPa} \end{cases} \quad (7\text{–}70)$$

where $\bar{S}_{ut}$ is the *mean* ultimate tensile strength.

Equations (7–70) represent the state of information before an engineer has chosen a material. In choosing, the designer has made a random choice from the ensemble of possibilities, and the statistics can give the odds of disappointment. If the testing is limited to finding an estimate of the ultimate tensile strength mean $\bar{S}_{ut}$ with the chosen material, Eqs. (7–70) are directly helpful. If there is to be rotary-beam fatigue testing, then statistical information on the endurance limit is gathered and there is no need for the correlation above.

Table 7–12 compares approximate mean values of the fatigue ratio $\bar{\phi}_{0.30}$ for several classes of ferrous materials.

Endurance Limit Modifying Factors

A Marin equation can be written as

$$\mathbf{S}_e = \mathbf{k}_a k_b \mathbf{k}_c \mathbf{k}_d \mathbf{k}_f \mathbf{S}'_e \quad (7\text{–}71)$$

where the size factor k_b is deterministic and remains unchanged from that given in Sec. 7–9. Also, since we are performing a stochastic analysis, the "reliability factor" k_e is unnecessary here.

The surface factor $\mathbf{k}_a$ cited earlier in deterministic form as Eq. (7–19) is now given in stochastic form by

$$\mathbf{k}_a = a\bar{S}_{ut}^b \mathbf{LN}(1, C) \qquad (\bar{S}_{ut} \text{ in kpsi or MPa}) \quad (7\text{–}72)$$

where Table 7–13 gives values of a, b, and C for various surface conditions.

Table 7–12

Comparison of Approximate Values of Mean Fatigue Ratio for Some Classes of Metals

Material Class	$\bar{\phi}_{0.30}$
Wrought steels	0.50
Cast steels	0.40
Powdered steels	0.38
Gray cast iron	0.35
Malleable cast iron	0.40
Normalized nodular cast iron	0.33

Table 7–13

Parameters in Marin Surface Condition Factor

Surface Finish	$k_a = aS_{ut}^b$ LN(1, C)		b	Coefficient of Variation, C
	a kpsi	a MPa		
Ground*	1.34	1.58	−0.086	0.120
Machined or Cold-rolled	2.67	4.45	−0.265	0.058
Hot-rolled	14.5	58.1	−0.719	0.110
As-forged	39.8	271	−0.995	0.145

*Due to the wide scatter in ground surface data, an alternate function is $k_a = 0.878$LN(1, 0.120). *Note:* S_{ut} in kpsi or MPa.

EXAMPLE 7–17

A steel has a mean ultimate strength of 520 MPa and a machined surface. Estimate k_a.

Solution

From Table 7–13,

$$k_a = 4.45(520)^{-0.265}\text{LN}(1, 0.058)$$

$$\bar{k}_a = 4.45(520)^{-0.265}(1) = 0.848$$

$$\hat{\sigma}_{ka} = C\bar{k}_a = (0.058)4.45(520)^{-0.265} = 0.049$$

Answer

so $k_a = \text{LN}(0.848, 0.049)$.

The load factor k_c for axial and torsional loading is given by

$$(k_c)_{\text{axial}} = 1.23\bar{S}_{ut}^{-0.0778}\text{LN}(1, 0.125) \tag{7–73}$$

$$(k_c)_{\text{torsion}} = 0.328\bar{S}_{ut}^{0.125}\text{LN}(1, 0.125) \tag{7–74}$$

There are fewer data to study for axial fatigue. Equation (7–73) was deduced from the data of Landgraf and of Grover, Gordon, and Jackson (as cited earlier).

Torsional data are sparser, and Eq. (7–74) is deduced from data in Grover et al. Notice the mild sensitivity to strength in the axial and torsional load factor, so k_c in these cases is not constant. Average values are shown in the last column of Table 7–14, and as footnotes to Tables 7–15 and 7–16. Table 7–17 shows the influence of material classes on the load factor k_c. Distortion energy theory predicts $(k_c)_{\text{torsion}} = 0.577$ for materials to which the distortion-energy theory applies. For bending, $k_c = \text{LN}(1, 0)$.

Table 7–14

Parameters in Marin Loading Factor

Mode of Loading	α		β	C	Average k_c
	kpsi	MPa			
Bending	1	1	0	0	1
Axial	1.23	1.43	−0.078	0.125	0.85
Torsion	0.328	0.258	0.125	0.125	0.59

$$k_c = \alpha \bar{S}_{ut}^{\beta} LN(1, C)$$

Table 7–15

Average Marin Loading Factor for Axial Load

$\bar{S}_{ut}$, kpsi	k_c^*
50	0.907
100	0.860
150	0.832
200	0.814

*Average entry 0.85.

Table 7–16

Average Marin Loading Factor for Torsional Load

$\bar{S}_{ut}$, kpsi	k_c^*
50	0.535
100	0.583
150	0.614
200	0.636

*Average entry 0.59.

Table 7–17

Average Marin Torsional Loading Factor k_c for Several Materials

Material	Range	n	$\bar{k}_c$	$\hat{\sigma}_{kc}$
Wrought steels	0.52–0.69	31	0.60	0.03
Wrought Al	0.43–0.74	13	0.55	0.09
Wrought Cu and alloy	0.41–0.67	7	0.56	0.10
Wrought Mg and alloy	0.49–0.60	2	0.54	0.08
Titanium	0.37–0.57	3	0.48	0.12
Cast iron	0.79–1.01	9	0.90	0.07
Cast Al, Mg, and alloy	0.71–0.91	5	0.85	0.09

Source: The table is an extension of P. G. Forrest, *Fatigue of Metals,* Pergamon Press, London, 1962, Table 17, p. 110, with standard deviations estimated from range and sample size using Table A–1 in J. B. Kennedy and A. M. Neville, *Basic Statistical Methods for Engineers and Scientists,* 3rd ed., Harper & Row, New York, 1986, pp. 54–55.

EXAMPLE 7–18

Estimate the Marin loading factor $\mathbf{k}_c$ for a 1–in-diameter bar that is used as follows.
(a) In bending. It is made of steel with $\mathbf{S}_{ut} = 100\mathbf{LN}(1, 0.035)$ kpsi, and the designer intends to use the correlation $\mathbf{S}'_e = \boldsymbol{\phi}_{0.30}\bar{S}_{ut}$ to predict $\mathbf{S}'_e$.
(b) In bending, but endurance testing gave $\mathbf{S}'_e = 55\mathbf{LN}(1, 0.081)$ kpsi.
(c) In push-pull (axial) fatigue, $\mathbf{S}_{ut} = \mathbf{LN}(86.2, 3.92)$ kpsi, and the designer intended to use the correlation $\mathbf{S}'_e = \boldsymbol{\phi}_{0.30}\bar{S}_{ut}$.
(d) In torsional fatigue. The material is cast iron, and $\mathbf{S}'_e$ is known by test.

Solution

(a) Since the bar is in bending,

Answer
$$\mathbf{k}_c = (1, 0)$$

(b) Since the test is in bending and use is in bending,

Answer
$$\mathbf{k}_c = (1, 0)$$

(c) From Eq. (7–73),

Answer
$$(\mathbf{k}_c)_{ax} = 1.23(86.2)^{-0.0778}\mathbf{LN}(1, 0.125)$$
$$\bar{k}_c = 1.23(86.2)^{-0.0778}(1) = 0.870$$
$$\hat{\sigma}_{kc} = C\bar{k}_c = 0.125(0.870) = 0.109$$

(d) From Table 7–17, $\bar{k}_c = 0.90$, $\hat{\sigma}_{kc} = 0.07$, and

Answer
$$C_{kc} = \frac{0.07}{0.90} = 0.08$$

The temperature factor $\mathbf{k}_d$ is

$$\mathbf{k}_d = \bar{k}_d\mathbf{LN}(1, 0.11) \tag{7–75}$$

where $\bar{k}_d = k_d$, given by Eq. (7–26).

Finally, $\mathbf{k}_e$ is, as before, the miscellaneous factor that can come about from a great many considerations, as discussed in Sec. 7–9, where now statistical distributions, possibly from testing, are considered.

Stress Concentration and Notch Sensitivity

Notch sensitivity q was defined by Eq. (2–30). The stochastic equivalent is

$$\mathbf{q} = \frac{\mathbf{K}_f - 1}{K_t - 1} \tag{7–76}$$

where K_t is the theoretical (or geometric) stress-concentration factor, a deterministic quantity. A study of lines 3 and 4 of Table 2–6 will reveal that adding a scalar to (or subtracting one from) a variate $\mathbf{x}$ will affect only the mean. Also, multiplying (or dividing) by a scalar affects both the mean and standard deviation. With this in mind, we can

Table 7–18

Coefficients of Variation C_{Kf} for Steels

Notch Type	Coefficient of Variation C_{Kf}
Transverse hole	0.10
Shoulder	0.11
Groove	0.15

Notes: Heywood's coefficients of variation. Notch sensitivity charts can be avoided using a modified Neuber equation. See Sec. 7–10.

relate the statistical parameters of the fatigue stress-concentration factor $\mathbf{K}_f$ to those of notch sensitivity $\mathbf{q}$. It follows that

$$\mathbf{q} = \mathbf{LN}\left(\frac{\bar{K}_f - 1}{K_t - 1}, \frac{C\bar{K}_f}{K_t - 1}\right)$$

where $C = C_{Kf}$ and

$$\bar{q} = \frac{\bar{K}_f - 1}{K_t - 1}$$

$$\hat{\sigma}_q = \frac{C\bar{K}_f}{K_t - 1} \tag{7–77}$$

$$C_q = \frac{C\bar{K}_f}{\bar{K}_f - 1}$$

The fatigue stress-concentration factor $\mathbf{K}_f$ has been investigated more in England than in the United States. Values of C_{Kf} for transverse holes, shoulders, and grooves are listed in Table 7–18. Once $\mathbf{K}_f$ is described, $\mathbf{q}$ can also be quantified using the set Eqs. (7–77).

The modified Neuber equation (after Heywood) gives the fatigue stress concentration factor as

$$\mathbf{K}_f = \bar{K}_f \mathbf{LN}\left(1, C_{K_f}\right) \tag{7–78}$$

where $\bar{K}_f = K_f$, given by Eq. (7–35).

EXAMPLE 7–19

Estimate $\mathbf{K}_f$ and $\mathbf{q}$ for the steel shaft given in Ex. 7–6.

Solution

From Ex. 7–6 and Eq. (7–35), $K_f = 1.51$. From Table 7–18, $C_{K_f} = 0.11$. Thus, from Eq. (7–78),

Answer

$$\mathbf{K}_f = 1.51\mathbf{LN}(1, 0.11)$$

From Eq. (7–77), with $K_t = 1.65$ from Ex. 7–6,

$$\bar{q} = \frac{1.51 - 1}{1.65 - 1} = 0.785$$

$$C_q = \frac{C_{K_f}\bar{K}_f}{\bar{K}_f - 1} = \frac{0.11(1.51)}{1.51 - 1} = 0.326$$

$$\hat{\sigma}_q = C_q\bar{q} = 0.326(0.785) = 0.256$$

So,

Answer

$$\mathbf{q} = \mathbf{LN}(0.785, 0.256)$$

EXAMPLE 7–20

The bar shown in Fig. 7–37 is machined from a cold-rolled flat having an ultimate strength of $\mathbf{S}_{ut} = \mathbf{LN}(87.6, 5.74)$ kpsi. The axial load shown is completely reversed. The load amplitude is $\mathbf{F}_a = \mathbf{LN}(1000, 120)$ lbf.
(a) Estimate the reliability.
(b) Reestimate the reliability when a rotating bending endurance test shows that $\mathbf{S}'_e = \mathbf{LN}(40, 2)$ kpsi.

Solution

(a) From Eq. (7–70), $\mathbf{S}'_e = 0.506\bar{S}_{ut}\mathbf{LN}(1, 0.138) = 0.506(87.6)\mathbf{LN}(1, 0.138)$

$$= 44.3\mathbf{LN}(1, 0.138) \text{ kpsi}$$

From Eq. (7–72) and Table 7–13,

$$\mathbf{k}_a = 2.67\bar{S}_{ut}^{-0.265}\mathbf{LN}(1, 0.058) = 2.67(87.6)^{-0.265}\mathbf{LN}(1, 0.058)$$

$$= 0.816\mathbf{LN}(1, 0.058)$$

$$k_b = 1 \qquad \text{(axial loading)}$$

From Eq. (7–73),

$$\mathbf{k}_c = 1.23\bar{S}_{ut}^{-0.0778}\mathbf{LN}(1, 0.125) = 1.23(87.6)^{-0.0778}\mathbf{LN}(1, 0.125)$$

$$= 0.869\mathbf{LN}(1, 0.125)$$

$$\mathbf{k}_d = \mathbf{k}_f = (1, 0)$$

The endurance strength, from Eq. (7–71), is

$$\mathbf{S}_e = \mathbf{k}_a k_b \mathbf{k}_c \mathbf{k}_d \mathbf{k}_f \mathbf{S}'_e$$

$$\mathbf{S}_e = 0.816\mathbf{LN}(1, 0.058)(1)0.868\mathbf{LN}(1, 0.125)(1)(1)44.3\mathbf{LN}(1, 0.138)$$

The parameters of $\mathbf{S}_e$ are

$$\bar{S}_e = 0.816(0.868)44.3 = 31.4 \text{ kpsi}$$

$$C_{Se} = (0.058^2 + 0.125^2 + 0.138^2)^{1/2} = 0.195$$

so $\mathbf{S}_e = 31.4\mathbf{LN}(1, 0.195)$ kpsi.

In computing the stress, the section at the hole governs. Using the terminology of Table A–15–1 we find $d/w = 0.50$, therefore $K_t = 2.18$. From Tables 7–8 and 7–18, $\sqrt{a} = 5/S_{ut} = 5/87.6 = 0.057$ and $C_{kf} = 0.10$. From Eqs. (7–35) and (7–78) with $r = 0.375$ in,

$$\mathbf{K}_f = \cfrac{K_t}{1 + \cfrac{2(K_t - 1)}{K_t}\cfrac{\sqrt{a}}{\sqrt{r}}}\mathbf{LN}\left(1, C_{K_f}\right) = \cfrac{2.18}{1 + \cfrac{2(2.18 - 1)}{2.18}\cfrac{0.057}{\sqrt{0.375}}}\mathbf{LN}(1, 0.10)$$

$$= 1.98\mathbf{LN}(1, 0.10)$$

| **Figure 7–37**

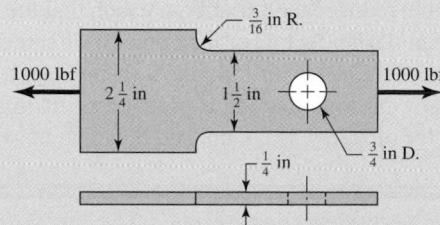

The stress at the hole is

$$\sigma = \mathbf{K}_f \frac{\mathbf{F}}{A} = 1.98\mathbf{LN}(1, 0.10)\frac{1000\mathbf{LN}(1, 0.12)}{0.25(0.75)}$$

$$\bar{\sigma} = 1.98\frac{1000}{0.25(0.75)}10^{-3} = 10.56 \text{ kpsi}$$

$$C_\sigma = (0.10^2 + 0.12^2)^{1/2} = 0.156$$

so stress can be expressed as $\sigma = 10.56\mathbf{LN}(1, 0.156)$ kpsi.

The endurance limit is considerably greater than the load-induced stress, indicating that finite life is not a problem. For interfering lognormal-lognormal distributions, Eq. (6–57) gives

$$z = -\frac{\ln\left(\dfrac{\bar{S}_e}{\bar{\sigma}}\sqrt{\dfrac{1 + C_\sigma^2}{1 + C_{S_e}^2}}\right)}{\sqrt{\ln\left[\left(1 + C_{S_e}^2\right)\left(1 + C_\sigma^2\right)\right]}} = -\frac{\ln\left(\dfrac{31.4}{10.56}\sqrt{\dfrac{1 + 0.156^2}{1 + 0.195^2}}\right)}{\sqrt{\ln[(1 + 0.195^2)(1 + 0.156^2)]}} = -4.36$$

From Table A–10 the probability of failure $p_f = \Phi(-4.36) = .000\,006\,66$, and the reliability is

Answer

$$R = 1 - 0.000\,006\,66 = 0.999\,993\,34$$

(*b*) The rotary endurance tests are described by $\mathbf{S}'_e = 40\mathbf{LN}(1, 0.05)$ kpsi whose mean is *less* than the predicted mean in part *a*. The mean endurance strength $\bar{S}_e$ is

$$\bar{S}_e = 0.816(0.868)40 = 28.3 \text{ kpsi}$$

$$C_{Se} = (0.058^2 + 0.125^2 + 0.05^2)^{1/2} = 0.147$$

so the endurance strength can be expressed as $\mathbf{S}_e = 28.3\mathbf{LN}(1, 0.147)$ kpsi. From Eq. (6–57),

$$z = -\frac{\ln\left(\dfrac{28.3}{10.56}\sqrt{\dfrac{1 + 0.156^2}{1 + 0.147^2}}\right)}{\sqrt{\ln[(1 + 0.147^2)(1 + 0.156^2)]}} = -4.63$$

Using Table A–10, we see the probability of failure $p_f = \Phi(-4.63) = 0.000\,001\,87$, and

$$R = 1 - 0.000\,001\,87 = 0.999\,998\,13$$

an increase! The reduction in the probability of failure is $(0.000\,001\,87 - 0.000\,006\,66)/0.000\,006\,66 = -0.72$, a reduction of 72 percent. We are analyzing an existing design, so in part (*a*) the factor of safety was $\bar{n} = \bar{S}/\bar{\sigma} = 31.3/10.56 = 2.96$. In part (*b*) $\bar{n} = 28.3/10.56 = 2.68$, a *decrease*. This example gives you the opportunity to see the role of the design factor. Given knowledge of $\bar{S}$, C_S, $\bar{\sigma}$, C_σ, and reliability (through z), the mean factor of safety (as a design factor) separates $\bar{S}$ and $\bar{\sigma}$ so that the reliability goal is achieved. Knowing $\bar{n}$ alone *says nothing about the probability of failure*. Looking at $\bar{n} = 2.96$ and $\bar{n} = 2.68$ says nothing about the respective probabilities of failure. The tests did not reduce $\bar{S}_e$ significantly, but reduced the variation C_S such that the reliability was *increased*.

When a mean design factor (or mean factor of safety) defined as $\bar{S}_e/\bar{\sigma}$ is said to be *silent* on matters of frequency of failures, it means that a scalar factor of safety by itself does not offer any information about probability of failure. Nevertheless, some engineers let the factor of safety speak up, and they can be wrong in their conclusions.

As revealing as Ex. 7–20 is concerning the meaning (and lack of meaning) of a design factor or factor of safety, let us remember that the rotary testing associated with part (*b*) changed *nothing* about the part, but only our knowledge about the part. The mean endurance limit was 40 kpsi all the time, and our adequacy assessment had to move with what was known.

Fluctuating Stresses

Deterministic failure loci that lie among the data are candidates for regression models. Included among these are the Gerber, ASME-elliptic, and, for brittle materials, Smith-Dolan models, which use mean values in their presentation. The Gerber parabola is

$$\frac{\bar{S}_a}{\bar{S}_e} + \left(\frac{\bar{S}_m}{\bar{S}_{ut}}\right)^2 = 1 \tag{7–79}$$

Just as the deterministic failure loci are located by endurance strength and ultimate tensile (or yield) strength, so too are stochastic failure loci located by $\mathbf{S}_e$ and by $\mathbf{S}_{ut}$ or $\mathbf{S}_y$. Figure 7–32 shows how the mean locus of Eq. (7–79) fits a parabola to form the Gerber mean locus. We also need to establish a contour located one standard deviation from the mean. Since stochastic loci are most likely to be used with a radial load line, we will develop the equation using the load line slope $r = \bar{S}_a/\bar{S}_m$. Substituting $\bar{S}_m = \bar{S}_a/r$ in Eq. (7–79) and solving for $\bar{S}_a$ gives

$$\bar{S}_a = \frac{r^2 \bar{S}_{ut}^2}{2\bar{S}_e}\left[-1 + \sqrt{1 + \left(\frac{2\bar{S}_e}{r\bar{S}_{ut}}\right)^2}\right] \tag{7–80}$$

Because of the positive correlation between $\mathbf{S}_e$ and $\mathbf{S}_{ut}$, we increment $\bar{S}_e$ by $C_{Se}\bar{S}_e$, $\bar{S}_{ut}$ by $C_{Sut}\bar{S}_{ut}$, and $\bar{S}_a$ by $C_{Sa}\bar{S}_a$, substitute into Eq. (7–80), and solve for C_{Sa} to obtain

$$C_{Sa} = \frac{(1 + C_{Sut})^2}{1 + C_{Se}} \frac{\left\{-1 + \sqrt{1 + \left[\dfrac{2\bar{S}_e(1 + C_{Se})}{r\bar{S}_{ut}(1 + C_{Sut})}\right]^2}\right\}}{\left[-1 + \sqrt{1 + \left(\dfrac{2\bar{S}_e}{r\bar{S}_{ut}}\right)^2}\right]} - 1 \tag{7–81}$$

Equation (7–81) can be viewed as an interpolation formula for C_{Sa}, which falls between C_{Se} and C_{Sut} depending on load line slope r. Note that $\mathbf{S}_a = \bar{S}_a\mathbf{LN}(1, C_{Sa})$.

The ASME-elliptic criterion is expressed in terms of its means as

$$\left(\frac{\bar{S}_a}{\bar{S}_e}\right)^2 + \left(\frac{\bar{S}_m}{\bar{S}_y}\right)^2 = 1 \tag{7–82}$$

Similarly, substituting $\bar{S}_m = \bar{S}_a/r$ into Eq. (7–82) and solving for $\bar{S}_a$ gives

$$\bar{S}_a = \frac{r\bar{S}_y\bar{S}_e}{\sqrt{r^2\bar{S}_y^2 + \bar{S}_e^2}} \tag{7–83}$$

Similarly, we increment $\bar{S}_e$ by $C_{Se}\bar{S}_e$, $\bar{S}_y$ by $C_{Sy}\bar{S}_y$, and $\bar{S}_a$ by $C_{Sa}\bar{S}_a$, substitute into Eq. (7–83), and solve for C_{Sa}:

$$C_{Sa} = (1 + C_{Sy})(1 + C_{Se})\sqrt{\frac{r^2\bar{S}_y^2 + \bar{S}_e^2}{r^2\bar{S}_y^2(1 + C_{Sy})^2 + \bar{S}_e^2(1 + C_{Se})^2} - 1} \qquad (7\text{–}84)$$

Many *brittle* materials follow a Smith-Dolan failure criterion, written deterministically as

$$\frac{n\sigma_a}{S_e} = \frac{1 - n\sigma_m/S_{ut}}{1 + n\sigma_m/S_{ut}} \qquad (7\text{–}85)$$

Expressed in terms of its means,

$$\frac{\bar{S}_a}{\bar{S}_e} = \frac{1 - \bar{S}_m/\bar{S}_{ut}}{1 + \bar{S}_m/\bar{S}_{ut}} \qquad (7\text{–}86)$$

For a radial load line slope of r, we substitute $\bar{S}_a/r$ for $\bar{S}_m$ and solve for $\bar{S}_a$, obtaining

$$\bar{S}_a = \frac{r\bar{S}_{ut} + \bar{S}_e}{2}\left[-1 + \sqrt{1 + \frac{4r\bar{S}_{ut}\bar{S}_e}{(r\bar{S}_{ut} + \bar{S}_e)^2}}\right] \qquad (7\text{–}87)$$

and the expression for C_{Sa} is

$$C_{Sa} = \frac{r\bar{S}_{ut}(1 + C_{Sut}) + \bar{S}_e(1 + C_{Se})}{2\bar{S}_a}$$
$$\cdot\left\{-1 + \sqrt{1 + \frac{4r\bar{S}_{ut}\bar{S}_e(1 + C_{Se})(1 + C_{Sut})}{[r\bar{S}_{ut}(1 + C_{Sut}) + \bar{S}_e(1 + C_{Se})]^2}}\right\} - 1 \qquad (7\text{–}88)$$

EXAMPLE 7–21

A rotating shaft experiences a steady torque $\mathbf{T} = 1360\mathbf{LN}(1, 0.05)$ lbf · in, and at a shoulder with a 1.1-in small diameter, a fatigue stress-concentration factor $\mathbf{K}_f = 1.50\mathbf{LN}(1, 0.11)$, $K_{fs} = 1.28\mathbf{LN}(1, 0.11)$, and at that location a bending moment of $\mathbf{M} = 1260\mathbf{LN}(1, 0.05)$ lbf · in. The material of which the shaft is machined is hot-rolled 1035 with $\mathbf{S}_{ut} = 86.2\mathbf{LN}(1, 0.045)$ kpsi and $\mathbf{S}_y = 56.0\mathbf{LN}(1, 0.077)$ kpsi. Estimate the reliability using a stochastic Gerber failure zone.

Solution

Establish the endurance strength. From Eqs. (7–70) to (7–72) and Eq. (7–19),

$$\mathbf{S}'_e = 0.506(86.2)\mathbf{LN}(1, 0.138) = 43.6\mathbf{LN}(1, 0.138) \text{ kpsi}$$
$$\mathbf{k}_a = 2.67(86.2)^{-0.265}\mathbf{LN}(1, 0.058) = 0.820\mathbf{LN}(1, 0.058)$$
$$k_b = (1.1/0.30)^{-0.107} = 0.870$$
$$\mathbf{k}_c = \mathbf{k}_d = \mathbf{k}_f = \mathbf{LN}(1, 0)$$
$$\mathbf{S}_e = 0.820\mathbf{LN}(1, 0.058)0.870(43.6)\mathbf{LN}(1, 0.138)$$
$$\bar{S}_e = 0.820(0.870)43.6 = 31.1 \text{ kpsi}$$
$$C_{Se} = (0.058^2 + 0.138^2)^{1/2} = 0.150$$

and so $S_e = 31.1\mathbf{LN}(1, 0.150)$ kpsi.

Stress (in kpsi):

$$\sigma_a = \frac{32\mathbf{K}_f\mathbf{M}_a}{\pi d^3} = \frac{32(1.50)\mathbf{LN}(1, 0.11)1.26\mathbf{LN}(1, 0.05)}{\pi(1.1)^3}$$

$$\bar{\sigma}_a = \frac{32(1.50)1.26}{\pi(1.1)^3} = 14.5 \text{ kpsi}$$

$$C_{\sigma a} = (0.11^2 + 0.05^2)^{1/2} = 0.121$$

$$\boldsymbol{\tau}_m = \frac{16\mathbf{K}_{fs}\mathbf{T}_m}{\pi d^3} = \frac{16(1.28)\mathbf{LN}(1, 0.11)1.36\mathbf{LN}(1, 0.05)}{\pi(1.1)^3}$$

$$\bar{\tau}_m = \frac{16(1.28)1.36}{\pi(1.1)^3} = 6.66 \text{ kpsi}$$

$$C_{\tau m} = (0.11^2 + 0.05^2)^{1/2} = 0.121$$

$$\bar{\sigma}'_a = \left(\bar{\sigma}_a^2 + 3\bar{\tau}_a^2\right)^{1/2} = [14.5^2 + 3(0)^2]^{1/2} = 14.5 \text{ kpsi}$$

$$\bar{\sigma}'_m = \left(\bar{\sigma}_m^2 + 3\bar{\tau}_m^2\right)^{1/2} = [0 + 3(6.66)^2]^{1/2} = 11.54 \text{ kpsi}$$

$$r = \frac{\bar{\sigma}'_a}{\bar{\sigma}'_m} = \frac{14.5}{11.54} = 1.26$$

Strength: From Eqs. (7–80) and (7–81),

$$\bar{S}_a = \frac{1.26^2 86.2^2}{2(31.1)} \left\{ -1 + \sqrt{1 + \left[\frac{2(31.1)}{1.26(86.2)}\right]^2} \right\} = 28.9 \text{ kpsi}$$

$$C_{Sa} = \frac{(1 + 0.045)^2}{1 + 0.150} \frac{-1 + \sqrt{1 + \left[\frac{2(31.1)(1 + 0.15)}{1.26(86.2)(1 + 0.045)}\right]^2}}{-1 + \sqrt{1 + \left[\frac{2(31.1)}{1.26(86.2)}\right]^2}} - 1 = 0.134$$

Reliability: Since $\mathbf{S}_a = 28.9\mathbf{LN}(1, 0.134)$ kpsi and $\sigma'_a = 14.5\mathbf{LN}(1, 0.121)$ kpsi, Eq. (6–58) gives

$$z = -\frac{\ln\left(\frac{\bar{S}_a}{\bar{\sigma}}\sqrt{\frac{1 + C_\sigma^2}{1 + C_{S_a}^2}}\right)}{\sqrt{\ln\left[\left(1 + C_{S_a}^2\right)\left(1 + C_\sigma^2\right)\right]}} = -\frac{\ln\left(\frac{28.9}{14.5}\sqrt{\frac{1 + 0.121^2}{1 + 0.134^2}}\right)}{\sqrt{\ln[(1 + 0.134^2)(1 + 0.121^2)]}} = -3.83$$

From Table A–10 the probability of failure is $p_f = 0.000\,065$, and the reliability is, against fatigue,

Answer
$$R = 1 - p_f = 1 - 0.000\,065 = 0.999\,935$$

The chance of first-cycle yielding is estimated by interfering $\mathbf{S}_y$ with σ'_{max}. The quantity σ'_{max} is formed from $\sigma'_a + \sigma'_m$. The mean of σ'_{max} is $\bar{\sigma}'_a + \bar{\sigma}'_m = 14.5 + 11.54 = 26.04$ kpsi. The coefficient of variation of the sum is 0.121, since both COVs are 0.121, thus $C_{\sigma\,max} = 0.121$. We interfere $\mathbf{S}_y = 56\mathbf{LN}(1, 0.077)$ kpsi with

$\sigma'_{max} = 26.04\mathbf{LN}\,(1, 0.121)$ kpsi. The corresponding z variable is

$$z = -\frac{\ln\left(\dfrac{56}{26.04}\sqrt{\dfrac{1+0.121^2}{1+0.077^2}}\right)}{\sqrt{\ln[(1+0.077^2)(1+0.121^2)]}} = -5.39$$

which represents, from Table A–10, a probability of failure of approximately 0.0^7358 [which represents $3.58(10^{-8})$] of first-cycle yield in the fillet.

The probability of observing a fatigue failure exceeds the probability of a yield failure, something a deterministic analysis does not foresee and in fact could lead one to expect a yield failure should a failure occur. Look at the $\sigma'_a\mathbf{S}_a$ interference and the $\sigma'_{max}\mathbf{S}_y$ interference and examine the z expressions. These control the relative probabilities. A deterministic analysis is oblivious to this and can mislead. Check your statistics text for events that are not mutually exclusive, but are independent, to quantify the probability of failure:

$$p_f = p(\text{yield}) + p(\text{fatigue}) - p(\text{yield and fatigue})$$

$$= p(\text{yield}) + p(\text{fatigue}) - p(\text{yield})\,p(\text{fatigue})$$

$$= 0.358(10^{-7}) + 0.65(10^{-4}) - 0.358(10^{-7})0.65(10^{-4}) = 0.650(10^{-4})$$

$$R = 1 - 0.650(10^{-4}) = 0.999\,935$$

against either or both modes of failure.

Examine Fig. 7–38, which depicts the results of Ex. 7–16. The problem distribution of $\mathbf{S}_e$ was compounded of historical experience with $\mathbf{S}'_e$ and the uncertainty manifestations due to features requiring Marin considerations. The Gerber "failure zone" displays

Figure 7–38

Designer's fatigue diagram for Ex. 7–21.

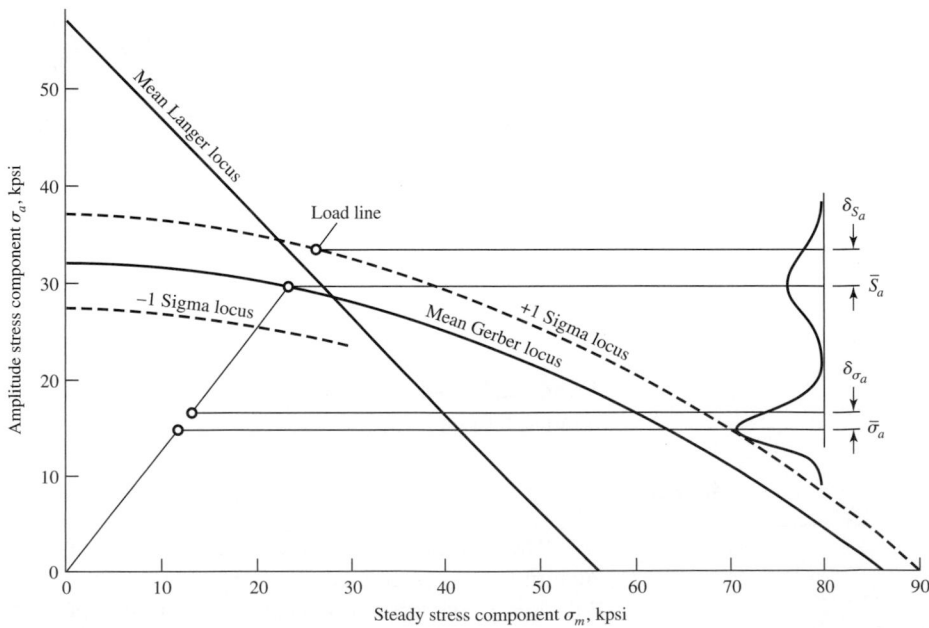

this. The interference with load-induced stress predicts the risk of failure. If additional information is known (R. R. Moore testing, with or without Marin features), the stochastic Gerber can accommodate to the information. Usually, the accommodation to additional test information is movement and contraction of the failure zone. In its own way the stochastic failure model accomplishes more precisely what the deterministic models and conservative postures intend. Additionally, stochastic models can estimate the probability of failure, something a deterministic approach cannot address.

The Design Factor in Fatigue

The designer, in envisioning how to execute the geometry of a part subject to the imposed constraints, can begin making a priori decisions without realizing the impact on the design task. Now is the time to note how these things are related to the reliability goal.

The mean value of the design factor is given by Eq. (6–59),

$$\bar{n} = \exp\left[-z\sqrt{\ln\left(1 + C_n^2\right)} + \ln\sqrt{1 + C_n^2}\right] \doteq \exp[C_n(-z + C_n/2)] \quad (6\text{–}59)$$

in which, from Table 2–6 for the quotient $\mathbf{n} = \mathbf{S}/\boldsymbol{\sigma}$,

$$C_n = \sqrt{\frac{C_S^2 + C_\sigma^2}{1 + C_\sigma^2}}$$

where C_S is the COV of the significant strength and C_σ is the COV of the significant stress at the critical location. Note that $\bar{n}$ is a function of the reliability goal (through z) and the COVs of the strength and stress. There are no means present, just measures of variability. The nature of C_S in a fatigue situation may be C_{Se} for fully reversed loading, or C_{Sa} otherwise. Also, experience shows $C_{Se} > C_{Sa} > C_{Sut}$, so C_{Se} can be used as a conservative estimate of C_{Sa}. If the loading is bending or axial, the form of σ_a' might be

$$\sigma_a' = \mathbf{K}_f \frac{\mathbf{M}_a c}{I} \qquad \text{or} \qquad \sigma_a' = \mathbf{K}_f \frac{\mathbf{F}}{A}$$

respectively. This makes the COV of σ_a', namely $C_{\sigma_a'}$, expressible as

$$C_{\sigma_a'} = \left(C_{Kf}^2 + C_F^2\right)^{1/2}$$

again a function of variabilities. The COV of $\mathbf{S}_e$, namely C_{Se}, is

$$C_{Se} = \left(C_{ka}^2 + C_{kc}^2 + C_{kd}^2 + C_{kf}^2 + C_{Se'}^2\right)^{1/2}$$

again, a function of variabilities. An example will be useful.

EXAMPLE 7–22

A strap to be made from a cold-drawn steel strip workpiece is to carry a fully reversed axial load $\mathbf{F} = \mathbf{LN}(1000, 120)$ lbf as shown in Fig. 7–39. Consideration of adjacent parts established the geometry as shown in the figure, except for the thickness t. Make a decision as to the magnitude of the design factor if the reliability goal is to be 0.999 95, then make a decision as to the workpiece thickness t.

Solution   Let us take each a priori decision and note the consequence:

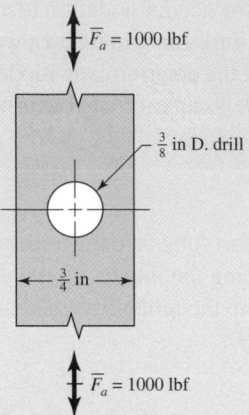

Figure 7–39

A strap with a thickness t is subjected to a fully reversed axial load of 1000 lbf. Example 7–22 considers the thickness necessary to attain a reliability of 0.999 95 against a fatigue failure.

A Priori Decision	**Consequence**
Use 1018 CD steel	$\bar{S}_{ut} = 87.6$ kpsi, $C_{Sut} = 0.0655$
Function:	
Carry axial load	$C_F = 0.12$, $C_{kc} = 0.125$
$R \geq 0.999\ 95$	$z = -3.891$
Machined surfaces	$C_{ka} = 0.058$
Hole critical	$C_{Kf} = 0.10$, $C_{\sigma_a'} = (0.10^2 + 0.12^2)^{1/2} = 0.156$
Ambient temperature	$C_{kd} = 0$
Correlation method	$C_{S_e'} = 0.138$
Hole drilled	$C_{Se} = (0.058^2 + 0.125^2 + 0.138^2)^{1/2} = 0.195$

$$C_n = \sqrt{\frac{C_{Se}^2 + C_{\sigma_a'}^2}{1 + C_{\sigma_a'}^2}} = \sqrt{\frac{0.195^2 + 0.156^2}{1 + 0.156^2}} = 0.2467$$

$$\bar{n} = \exp\left[-(-3.891)\sqrt{\ln(1 + 0.2467^2)} + \ln\sqrt{1 + 0.2467^2} \right]$$
$$= 2.65$$

These eight a priori decisions have quantified the mean design factor as $\bar{n} = 2.65$. Proceeding deterministically hereafter we write

$$\sigma_a' = \frac{\bar{S}_e}{\bar{n}} = \bar{K}_f \frac{\bar{F}}{(w - d)t}$$

from which

$$t = \frac{\bar{K}_f \bar{n} \bar{F}}{(w - d)\bar{S}_e}$$

To evaluate the preceding equation we need $\bar{S}_e$ and $\bar{K}_f$. The Marin factors are

$$\mathbf{k}_a = 2.67\bar{S}_{ut}^{-0.265}\mathbf{LN}(1, 0.058) = 2.67(87.6)^{-0.265}\mathbf{LN}(1, 0.058)$$
$$\bar{k}_a = 0.816$$
$$k_b = 1 \ (\text{see } \mathbf{k}_c)$$
$$\mathbf{k}_c = 1.23\bar{S}_{ut}^{-0.078}\mathbf{LN}(1, 0.125) = 0.868\mathbf{LN}(1, 0.125)$$
$$\bar{k}_c = 0.868$$
$$\bar{k}_d = \bar{k}_f = 1$$

and the endurance strength is

$$\bar{S}_e = 0.816(0.868)(1)(1)(1)0.506(87.6) = 31.4 \text{ kpsi}$$

The hole governs. From Table A–15–1 we find $d/w = 0.50$, therefore $K_t = 2.18$. From Table 7–7 $\sqrt{a} = 5/\bar{S}_{ut} = 5/87.6 = 0.057$, $r = 0.375$ in. From Eq. (7–34) the fatigue stress concentration factor is

$$\bar{K}_f = \frac{2.18}{1 + \dfrac{2}{\sqrt{0.375}}\dfrac{2.18 - 1}{2.18}0.057} = 1.98$$

The thickness t can now be determined from $S_e/\bar{n} = \bar{K}_f \bar{F}/A$,

$$t \geq \frac{\bar{K}_f \bar{n} \bar{F}}{(w-d)S_e} = \frac{1.98(2.65)1000}{(0.75-0.375)31\ 400} = 0.446 \text{ in}$$

Use $\frac{1}{2}$-in-thick strap for the workpiece. The $\frac{1}{2}$-in thickness attains and, in the rounding to available nominal size, exceeds the reliability goal.

The example demonstrates that, for a given reliability goal, the fatigue design factor that facilitates its attainment is decided by the variabilities of the situation. Furthermore, the necessary design factor is not a constant independent of the way the concept unfolds. Rather, it is a function of a number of seemingly unrelated a priori decisions that are made in giving definition to the concept. The involvement of stochastic methodology can be limited to defining the necessary design factor. In particular, in the example, the design factor is not a function of the design variable t; rather, t follows from the design factor.

PROBLEMS

Problems 7–1 to 7–31 are to be solved by deterministic methods. Problems 7–32 to 7–38 are to be solved by stochastic methods. Problems 7–39 to 7–46 are computer problems.

Deterministic Problems

 ANALYSIS

7-1 A $\frac{3}{16}$-in drill rod was heat-treated and ground. The measured hardness was found to be 490 Brinell. Estimate the endurance strength if the rod is used in rotating bending.

ANALYSIS

7-2 Estimate S'_e for the following materials:
(a) AISI 1020 CD steel.
(b) AISI 1080 HR steel.
(c) 2024 T3 aluminum.
(d) AISI 4340 steel heat-treated to a tensile strength of 250 kpsi.

ANALYSIS

7-3 Estimate the fatigue strength of a rotating-beam specimen made of AISI 1020 hot-rolled steel corresponding to a life of 12.5 kilocycles of stress reversal. Also, estimate the life of the specimen corresponding to a stress amplitude of 36 kpsi. The known properties are $S_{ut} = 66.2$ kpsi, $\sigma_0 = 115$ kpsi, $m = 0.22$, and $\varepsilon_f = 0.90$.

ANALYSIS

7-4 Derive Eq. (7–16). For the specimen of Prob. 7–3, estimate the strength corresponding to 500 cycles.

ANALYSIS

7-5 For the interval $10^3 \leq N \leq 10^6$ cycles, develop an expression for the fatigue strength $(S'_f)_{ax}$ for the polished specimens of 4130 used to obtain Fig. 7–10. The ultimate strength is $S_{ut} = 125$ kpsi and the endurance limit is $(S'_e)_{ax} = 49$ kpsi.

7-6 Estimate the endurance strength of a 32-mm-diameter rod of AISI 1035 steel having a machined finish and heat-treated to a tensile strength of 710 MPa.

 ANALYSIS

7-7 Two steels are being considered for manufacture of as-forged connecting rods. One is AISI 4340 Cr-Mo-Ni steel capable of being heat-treated to a tensile strength of 260 kpsi. The other is a plain carbon steel AISI 1040 with an attainable S_{ut} of 113 kpsi. If each rod is to have a size giving an equivalent diameter d_e of 0.75 in, is there any advantage to using the alloy steel for this fatigue application?

 ANALYSIS

7-8 A solid round bar, 25 mm in diameter, has a groove 2.5-mm deep with a 2.5-mm radius machined into it. The bar is made of AISI 1018 CD steel and is subjected to a purely reversing torque of 200 N · m. For the S-N curve of this material, let $f = 0.9$.

 ANALYSIS

(*a*) Estimate the number of cycles to failure.

(*b*) If the bar is also placed in an environment with a temperature of 450°C, estimate the number of cycles to failure.

 **7–9** A solid square rod is cantilevered at one end. The rod is 0.8 m long and supports a completely reversing transverse load at the other end of ±1 kN. The material is AISI 1045 hot-rolled steel. If the rod must support this load for 10^4 cycles with a factor of safety of 1.5, what dimension should the square cross section have? Neglect any stress concentrations at the support end and assume that $f = 0.9$.

 **7–10** A rectangular bar is cut from an AISI 1018 cold-drawn steel flat. The bar is 60 mm wide by 10 mm thick and has a 12-mm hole drilled through the center as depicted in Table A–15–1. The bar is concentrically loaded in push-pull fatigue by axial forces F_a, uniformly distributed across the width. Using a design factor of $n_d = 1.8$, estimate the largest force F_a that can be applied ignoring column action.

 **7–11** Bearing reactions R_1 and R_2 are exerted on the shaft shown in the figure, which rotates at 1150 rev/min and supports a 10-kip bending force. Use a 1095 HR steel. Specify a diameter d using a design factor of $n_d = 1.6$ for a life of 3 min. The surfaces are machined.

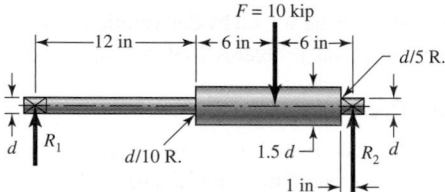

Problem 7–11

7–12 A bar of steel has the minimum properties $S_e = 276$ MPa, $S_y = 413$ MPa, and $S_{ut} = 551$ MPa. The bar is subjected to a steady torsional stress of 103 MPa and an alternating bending stress of 172 MPa. Find the factor of safety guarding against a static failure, and either the factor of safety guarding against a fatigue failure or the expected life of the part. For the fatigue analysis use:

(*a*) Modified Goodman criterion.

(*b*) Gerber criterion.

(*c*) ASME-elliptic criterion.

7–13 Repeat Prob. 7–12 but with a steady torsional stress of 138 MPa and an alternating bending stress of 69 MPa.

7–14 Repeat Prob. 7–12 but with a steady torsional stress of 103 MPa, an alternating torsional stress of 69 MPa, and an alternating bending stress of 83 MPa.

7–15 Repeat Prob. 7–12 but with an alternating torsional stress of 207 MPa.

7–16 Repeat Prob. 7–12 but with an alternating torsional stress of 103 MPa and a steady bending stress of 103 MPa.

7–17 The cold-drawn AISI 1018 steel bar shown in the figure is subjected to a tensile load fluctuating between 800 and 3000 lbf. Estimate the factors of safety n_y and n_f using (*a*) a Gerber fatigue failure criterion as part of the designer's fatigue diagram, and (*b*) a ASME-elliptic fatigue failure criterion as part of the designer's fatigue diagram.

7–18 Repeat Prob. 7–17, with the load fluctuating between −800 and 3000 lbf. Assume no buckling.

7–19 Repeat Prob. 7–17, with the load fluctuating between 800 and −3000 lbf. Assume no buckling.

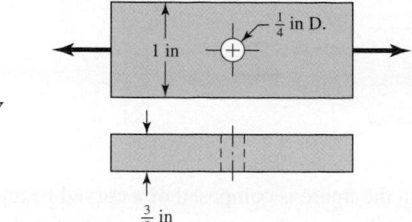

Problem 7–17

ANALYSIS **7–20** The figure shows a formed round-wire cantilever spring subjected to a varying force. The hardness tests made on 25 springs gave a minimum hardness of 380 Brinell. It is apparent from the mounting details that there is no stress concentration. A visual inspection of the springs indicates that the surface finish corresponds closely to a hot-rolled finish. What number of applications is likely to cause failure? Solve using:
(*a*) Modified Goodman criterion.
(*b*) Gerber criterion.

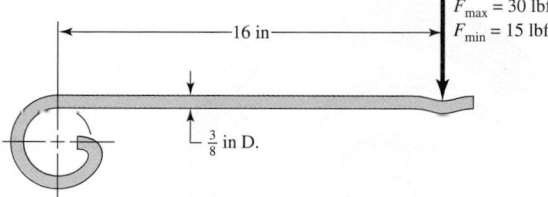

Problem 7–20

ANALYSIS **7–21** The figure is a drawing of a 3- by 18-mm latching spring. A preload is obtained during assembly by shimming under the bolts to obtain an estimated initial deflection of 2 mm. The latching operation itself requires an additional deflection of exactly 4 mm. The material is ground high-carbon steel, bent then hardened and tempered to a minimum hardness of 490 Bhn. The radius of the bend is 3 mm. Estimate the yield strength to be 90 percent of the ultimate strength.
(*a*) Find the maximum and minimum latching forces.
(*b*) Is it likely the spring will fail in fatigue? Use the Gerber criterion.

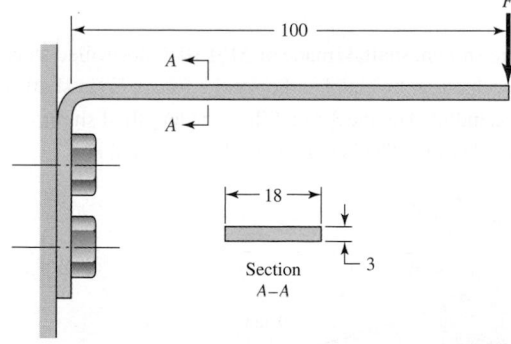

Problem 7–21

Dimensions in millimeters

ANALYSIS **7–22** Repeat Prob. 21, part *b*, using the modified Goodman criterion.

ANALYSIS **7–23** The figure shows the free-body diagram of a connecting-link portion having stress concentration at three sections. The dimensions are $r = 0.25$ in, $d = 0.75$ in, $h = 0.50$ in, $w_1 = 3.75$ in, and $w_2 = 2.5$ in. The forces F fluctuate between a tension of 4 kip and a compression of 16 kip. Neglect column action and find the least factor of safety if the material is cold-drawn AISI 1018 steel.

Problem 7–23

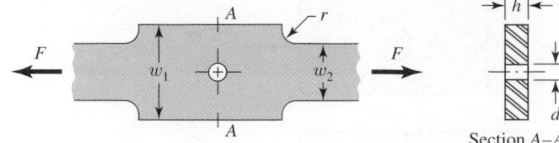

Section A–A

ANALYSIS

7–24 The torsional coupling in the figure is composed of a curved beam of square cross section that is welded to an input shaft and output plate. A torque is applied to the shaft and cycles from zero to T. The cross section of the beam has dimensions of 5 by 5 mm, and the centroidal axis of the beam describes a curve of the form $r = 10\,\theta/\pi$, where r and θ are in mm and radians, respectively $(2\pi \le \theta \le 6\pi)$. The curved beam has a machined surface with yield and ultimate strength values of 420 and 770 MPa, respectively.

(a) Determine the maximum allowable value of T such that the coupling will have an infinite life with a factor of safety, $n = 3$, using the modified Goodman criterion.

(b) Repeat part (a) using the Gerber criterion.

(c) Using T found in part (b), determine the factor of safety guarding against yield.

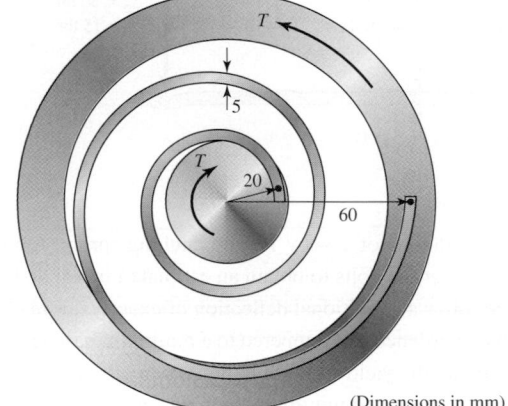

Problem 7–24

(Dimensions in mm)

ANALYSIS

7–25 Repeat Prob. 7–24 ignoring curvature effects on the bending stress.

ANALYSIS

7–26 In the figure shown, shaft A, made of AISI 1010 hot-rolled steel, is welded to a fixed support and is subjected to loading by equal and opposite forces F via shaft B. A theoretical stress concentration K_{ts} of 1.6 is induced by the 3-mm fillet. The length of shaft A from the fixed support to the connection at shaft B is 1 m. The load F cycles from 0.5 to 2 kN.

Problem 7–26

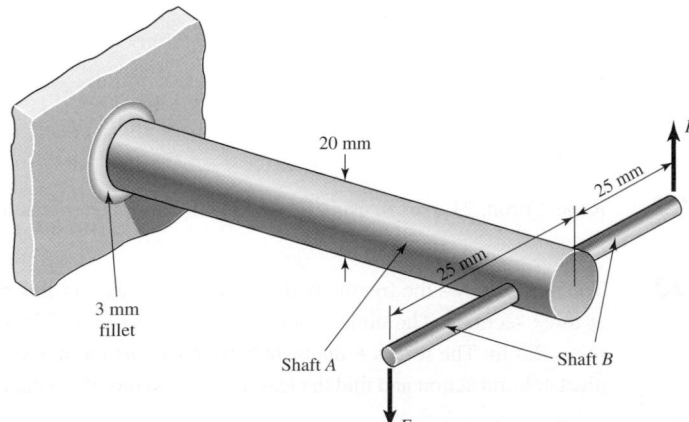

(a) For shaft A, find the factor of safety for infinite life using the modified Goodman fatigue failure criterion.

(b) Repeat part (a) using the Gerber fatigue failure criterion.

7–27 A schematic of a clutch-testing machine is shown. The steel shaft rotates at a constant speed ω. An axial load is applied to the shaft and is cycled from zero to P. The torque T induced by the clutch face onto the shaft is given by

$$T = \frac{fP(D+d)}{4}$$

where D and d are defined in the figure and f is the coefficient of friction of the clutch face. The shaft is machined with $S_y = 800$ MPa and $S_{ut} = 1000$ MPa. The theoretical stress concentration factors for the fillet are 3.0 and 1.8 for the axial and torsional loading, respectively.

(a) Assume the load variation P is synchronous with shaft rotation. With $f = 0.3$, find the maximum allowable load P such that the shaft will survive a minimum of 10^6 cycles with a factor of safety of 3. Use the modified Goodman criterion. Determine the corresponding factor of safety guarding against yielding.

(b) Suppose the shaft is not rotating, but the load P is cycled as shown. With $f = 0.3$, find the maximum allowable load P so that the shaft will survive a minimum of 10^6 cycles with a factor of safety of 3. Use the modified Goodman criterion. Determine the corresponding factor of safety guarding against yielding.

Problem 7–27

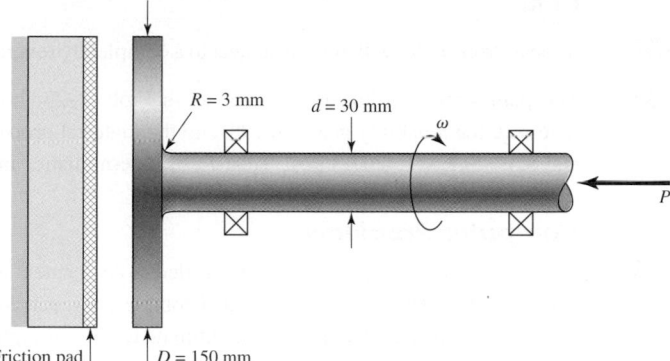

7–28 For the clutch of Prob. 7–27, the external load P is cycled between 20 kN and 80 kN. Assuming that the shaft is rotating synchronous with the external load cycle, estimate the number of cycles to failure. Use the modified Goodman fatigue failure criteria.

7–29 A flat leaf spring has fluctuating stress of $\sigma_{max} = 420$ MPa and $\sigma_{min} = 140$ MPa applied for $5 (10^4)$ cycles. If the load changes to $\sigma_{max} = 350$ MPa and $\sigma_{min} = -200$ MPa, how many cycles should the spring survive? The material is AISI 1040 CD and has a fully corrected endurance strength of $S_e = 200$ MPa. Assume that $f = 0.9$.

(a) Use Miner's method.

(b) Use Manson's method.

7–30 A machine part will be cycled at ± 48 kpsi for $4 (10^3)$ cycles. Then the loading will be changed to ± 38 kpsi for $6 (10^4)$ cycles. Finally, the load will be changed to ± 32 kpsi. How many cycles of operation can be expected at this stress level? For the part, $S_{ut} = 76$ kpsi, $f = 0.9$, and has a fully corrected endurance strength of $S_e = 30$ kpsi.

(a) Use Miner's method.

(b) Use Manson's method.

7–31 A rotating-beam specimen with an endurance limit of 50 kpsi and an ultimate strength of 100 kpsi is cycled 20 percent of the time at 70 kpsi, 50 percent at 55 kpsi, and 30 percent at 40 kpsi. Let $f = 0.9$ and estimate the number of cycles to failure.

Stochastic Problems

ANALYSIS | **7–32** | Solve Prob. 7–1 if the hardness of production pieces is found to be $\mathbf{H}_B = 495\mathbf{LN}(1, 0.03)$.

DESIGN | **7–33** | The situation is similar to that of Prob. 7–10 wherein the imposed completely reversed axial load $F_a = 15\mathbf{LN}(1, 0.20)$ kN is to be carried by the link with a thickness to be specified by you, the designer. Use the 1018 cold-drawn steel of Prob. 7–10 with $\mathbf{S}_{ut} = 440\mathbf{LN}(1, 0.30)$ MPa and $\mathbf{S}_{yt} = 370\mathbf{LN}(1, 0.061)$. The reliability goal must exceed 0.999. Using the correlation method, specify the thickness t.

ANALYSIS | **7–34** | A solid round steel bar is machined to a diameter of 1.25 in. A groove $\frac{1}{8}$ in deep with a radius of $\frac{1}{8}$ in is cut into the bar. The material has a mean tensile strength of 110 kpsi. A completely reversed bending moment $M = 1400$ lbf · in is applied. Estimate the reliability. The size factor should be based on the gross diameter. The bar rotates.

7–35 | Repeat Prob. 7–34, with a completely reversed torsional moment of $T = 1400$ lbf · in applied.

ANALYSIS | **7–36** | A $1\frac{1}{4}$-in-diameter hot-rolled steel bar has a $\frac{1}{8}$-in diameter hole drilled transversely through it. The bar is nonrotating and is subject to a completely reversed bending moment of $M = 1600$ lbf · in in the same plane as the axis of the transverse hole. The material has a mean tensile strength of 58 kpsi. Estimate the reliability. The size factor should be based on the gross size. Use Table A–16 for K_t.

7–37 | Repeat Prob. 7–36, with the bar subject to a completely reversed torsional moment of 2400 lbf · in.

DESIGN | **7–38** | The plan view of a link is the same as in Prob. 7–23; however, the forces F are completely reversed, the reliability goal is 0.998, and the material properties are $\mathbf{S}_{ut} = 64\mathbf{LN}(1, 0.045)$ kpsi and $\mathbf{S}_y = 54\mathbf{LN}(1, 0.077)$ kpsi. Treat F_a as deterministic, and specify the thickness h.

Computer Problems

ANALYSIS | **7–39** | A $\frac{1}{4}$ by $1\frac{1}{2}$-in steel bar has a $\frac{3}{4}$-in drilled hole located in the center, much as is shown in Table A–15–1. The bar is subjected to a completely reversed axial load with a deterministic load of 1200 lbf. The material has a mean ultimate tensile strength of $\bar{S}_{ut} = 80$ kpsi.
(*a*) Estimate the reliability.
(*b*) Conduct a computer simulation to confirm your answer to part *a*.

DESIGN | **7–40** | From your experience with Prob. 7–39 and Ex. 7–20, you observed that for completely reversed axial and bending fatigue, it is possible to

- Observe the COVs associated with a priori design considerations.

- Note the reliability goal.

- Find the mean design factor $\bar{n}_d$ which will permit making a geometric design decision that will attain the goal using deterministic methods in conjunction with $\bar{n}_d$.

Formulate an interactive computer program that will enable the user to find $\bar{n}_d$. While the material properties $\mathbf{S}_{ut}$, $\mathbf{S}_y$, and the load COV must be input by the user, all of the COVs associated with $\Phi_{0.30}$, $\mathbf{k}_a$, $\mathbf{k}_c$, $\mathbf{k}_d$, and $\mathbf{K}_f$ can be internal, and answers to questions will allow C_σ and C_S, as well as C_n and $\bar{n}_d$, to be calculated. Later you can add improvements. Test your program with problems you have already solved.

7–41 | When using the Gerber fatigue failure criterion in a stochastic problem, Eqs. (7–80) and (7–81) are useful. They are also computationally complicated. It is helpful to have a computer subroutine or procedure that performs these calculations. When writing an executive program, and it is appropriate to find S_a and C_{Sa}, a simple call to the subroutine does this with a minimum of effort. Also, once the subroutine is tested, it is always ready to perform. Write and test such a program.

7–42 Repeat Problem. 7–41 for the ASME-elliptic fatigue failure locus, implementing Eqs. (7–83) and (7–84).

7–43 Repeat Prob. 7–41 for the Smith-Dolan fatigue failure locus, implementing Eqs. (7–87) and (7–88).

7–44 Write and test computer subroutines or procedures that will implement
(a) Table 7–4, returning a, b, C, and $\bar{k}_a$.
(b) Equation (7–19) using Table 7–5, returning k_b.
(c) Table 7–14, returning α, β, C, and $\bar{k}_c$.
(d) Equations (7–26) and (7–75), returning $\bar{k}_d$ and C_{kd}.

7–45 Write and test a computer subroutine or procedure that implements Eqs. (7–76) and (7–77), returning $\bar{q}$, $\hat{\sigma}_q$, and C_q.

7–46 Write and test a computer subroutine or procedure that implements Eq. (7–35) and Tables 7–8 and 7–18, returning $\sqrt{a}$, C_{Kf}, and $\bar{K}_f$.

Summary of Parts 1 and 2

The first recommendation is to reread Chap. 1. With the experience you have gathered so far, you will gain from doing it. With the meat you have added to the bare bones of the introductory chapter, it will have a greater meaning. In Sec. 1–3, there are over two dozen design considerations. We have addressed item 2 in detail, the question of the strength/stress relationship in a loss-of-function for ductile and brittle materials, for steady and fatigue loading, and for finite and indefinite life. We have also started on item 7, reliability, as it applies to stress/strength relationships. In investigating the stress/strength relations, the reader should now be prepared to

- Identify the critical location(s), either by inspection, or, if not obvious, by analyzing the several candidates, and identifying the "worst case."
- Identify the significant strength at that location.
- Identify the significant stress at that location.
- Address the question of whether the disparity between stress and strength is sufficient such that function will be preserved in the face of service loading.

This preparation took a long time because an extensive set of ideas and insights had to be identified in and among your prerequisite studies, and placed in a useful context.

The question of stiffness, distortion, and deflection, item 3, and their influence on loss of function has also been addressed. The reader should now be prepared to identify

- The level of distortion that risks loss of function.
- The location(s) at which loss-of-function due to distortion is possible.
- The level of distortion present.
- Whether the difference is sufficient.

Some other considerations will be touched on in Part 3, and those just noted will be further developed for the application at hand. As we proceed into Part 3 our focus becomes more specific as we consider particular machine elements and their applications.

For now, the reader should feel comfortable with a kit of tools from which an adequacy assessment is devised. Skill 1 will take on additional substance as applications unfold. In addition to focus on individual elements, design/synthesis ideas will appear more often, and skill 2 will take form and grow.

PART

3

Design of Mechanical Elements

8

Screws, Fasteners, and the Design of Nonpermanent Joints

Chapter Outline

The helical-thread screw was undoubtably an extremely important mechanical invention. It is the basis of power screws, which change angular motion to linear motion to transmit power or to develop large forces (presses, jacks, etc.), and threaded fasteners, an important element in nonpermanent joints.

This book presupposes a knowledge of the elementary methods of fastening. Typical methods of fastening or joining parts use such devices as bolts, nuts, cap screws, setscrews, rivets, spring retainers, locking devices, pins, keys, welds, and adhesives. Studies in engineering graphics and in metal processes often include instruction on various joining methods, and the curiosity of any person interested in mechanical engineering naturally results in the acquisition of a good background knowledge of fastening methods. Contrary to first impressions, the subject is one of the most interesting in the entire field of mechanical design.

One of the key targets of current design for manufacture is to reduce the number of fasteners. However, there will always be a need for fasteners to facilitate disassembly for whatever purposes. For example, jumbo jets such as Boeing's 747 require as many as 2.5 million fasteners, some of which cost several dollars apiece. To keep costs down, aircraft manufacturers, and their subcontractors, constantly review new fastener designs, installation techniques, and tooling.

The number of innovations in the fastener field over any period you might care to mention has been tremendous. An overwhelming variety of fasteners are available for the designer's selection. Serious designers generally keep specific notebooks on fasteners alone. Methods of joining parts are extremely important in the engineering of a quality design, and it is necessary to have a thorough understanding of the performance of fasteners and joints under all conditions of use and design.

8–1 Thread Standards and Definitions

The terminology of screw threads, illustrated in Fig. 8–1, is explained as follows:

The *pitch* is the distance between adjacent thread forms measured parallel to the thread axis. The pitch in U.S. units is the reciprocal of the number of thread forms per inch N.

The *major diameter d* is the largest diameter of a screw thread.

The *minor diameter d_r or d_1* is the smallest diameter of a screw thread.

The *lead l*, not shown, is the distance the nut moves parallel to the screw axis when the nut is given one turn. For a single thread, as in Fig. 8–1, the lead is the same as the pitch.

A *multiple-threaded* product is one having two or more threads cut beside each other (imagine two or more strings wound side by side around a pencil). Standardized products such as screws, bolts, and nuts all have single threads; a *double-threaded* screw has a lead equal to twice the pitch, a *triple-threaded* screw has a lead equal to 3 times the pitch, and so on.

All threads are made according to the *right-hand rule* unless otherwise noted.

The *American National (Unified)* thread standard has been approved in this country and in Great Britain for use on all standard threaded products. The thread angle is 60° and the crests of the thread may be either flat or rounded.

Figure 8–2 shows the thread geometry of the metric M and MJ profiles. The M profile replaces the inch class and is the basic ISO 68 profile with 60° symmetric threads. The MJ profile has a rounded fillet at the root of the external thread and a larger minor diameter of both the internal and external threads. This profile is especially useful where high fatigue strength is required.

Figure 8–1

Terminology of screw threads. Sharp vee threads shown for clarity; the crests and roots are actually flattened or rounded during the forming operation.

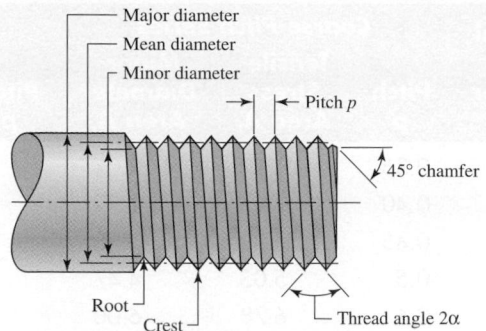

Figure 8–2

Basic thread profile for metric M and MJ threads. $D(d) =$ basic major diameter of internal (external) thread; $D_1(d_1)$ = basic minor diameter of internal (external) thread; $D_2(d_2)$ = basic pitch diameter of internal (external) thread; p — pitch; $H = 0.5(3)^{1/2} p$.

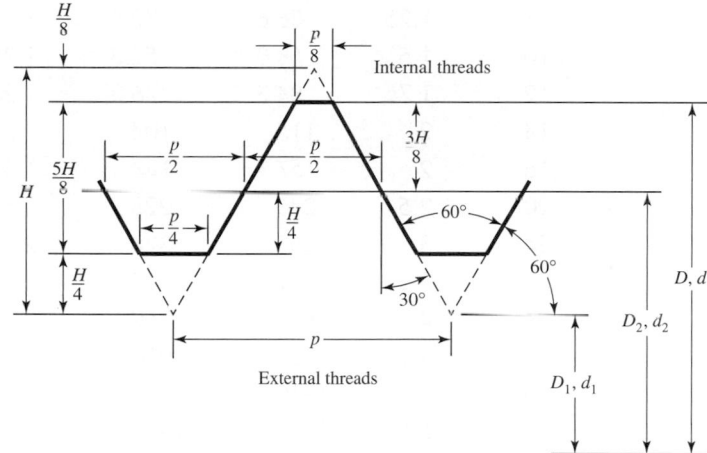

Tables 8–1 and 8–2 will be useful in specifying and designing threaded parts. Note that the thread size is specified by giving the pitch p for metric sizes and by giving the number of threads per inch N for the Unified sizes. The screw sizes in Table 8–2 with diameter under $\frac{1}{4}$ in are numbered or gauge sizes. The second column in Table 8–2 shows that a No. 8 screw has a nominal major diameter of 0.1640 in.

A great many tensile tests of threaded rods have shown that an unthreaded rod having a diameter equal to the mean of the pitch diameter and minor diameter will have the same tensile strength as the threaded rod. The area of this unthreaded rod is called the tensile-stress area A_t of the threaded rod; values of A_t are listed in both tables.

Two major Unified thread series are in common use: UN and UNR. The difference between these is simply that a root radius must be used in the UNR series. Because of reduced thread stress-concentration factors, UNR series threads have improved fatigue strengths. Unified threads are specified by stating the nominal major diameter, the number of threads per inch, and the thread series, for example, $\frac{5}{8}$ in-18 UNRF or 0.625 in-18 UNRF.

Metric threads are specified by writing the diameter and pitch in millimeters, in that order. Thus, M12 × 1.75 is a thread having a nominal major diameter of 12 mm and a pitch of 1.75 mm. Note that the letter M, which precedes the diameter, is the clue to the metric designation.

Table 8–1

Diameters and Areas of Coarse-Pitch and Fine-Pitch Metric Threads. (All Dimensions in Millimeters)*

Nominal Major Diameter d	Coarse-Pitch Series			Fine-Pitch Series		
	Pitch p	Tensile-Stress Area A_t	Minor-Diameter Area A_r	Pitch p	Tensile-Stress Area A_t	Minor-Diameter Area A_r
1.6	0.35	1.27	1.07			
2	0.40	2.07	1.79			
2.5	0.45	3.39	2.98			
3	0.5	5.03	4.47			
3.5	0.6	6.78	6.00			
4	0.7	8.78	7.75			
5	0.8	14.2	12.7			
6	1	20.1	17.9			
8	1.25	36.6	32.8	1	39.2	36.0
10	1.5	58.0	52.3	1.25	61.2	56.3
12	1.75	84.3	76.3	1.25	92.1	86.0
14	2	115	104	1.5	125	116
16	2	157	144	1.5	167	157
20	2.5	245	225	1.5	272	259
24	3	353	324	2	384	365
30	3.5	561	519	2	621	596
36	4	817	759	2	915	884
42	4.5	1120	1050	2	1260	1230
48	5	1470	1380	2	1670	1630
56	5.5	2030	1910	2	2300	2250
64	6	2680	2520	2	3030	2980
72	6	3460	3280	2	3860	3800
80	6	4340	4140	1.5	4850	4800
90	6	5590	5360	2	6100	6020
100	6	6990	6740	2	7560	7470
110				2	9180	9080

*The equations and data used to develop this table have been obtained from ANSI B1.1-1974 and B18.3.1-1978. The minor diameter was found from the equation $d_r = d - 1.226\,869p$, and the pitch diameter from $d_m = d - 0.649\,519p$. The mean of the pitch diameter and the minor diameter was used to compute the tensile-stress area.

Square and Acme threads, shown in Fig. 8–3a and b, respectively, are used on screws when power is to be transmitted. Table 8–3 lists the preferred pitches for inch-series Acme threads. However, other pitches can be and often are used, since the need for a standard for such threads is not great.

Modifications are frequently made to both Acme and square threads. For instance, the square thread is sometimes modified by cutting the space between the teeth so as to have an included thread angle of 10 to 15°. This is not difficult, since these threads are usually cut with a single-point tool anyhow; the modification retains most of the high efficiency inherent in square threads and makes the cutting simpler. Acme threads are

Table 8–2

Diameters and Area of Unified Screw Threads UNC and UNF*

Size Designation	Nominal Major Diameter in	Coarse Series—UNC			Fine Series—UNF		
		Threads per Inch N	Tensile-Stress Area A_t in^2	Minor-Diameter Area A_r in^2	Threads per Inch N	Tensile-Stress Area A_t in^2	Minor-Diameter Area A_r in^2
0	0.0600				80	0.001 80	0.001 51
1	0.0730	64	0.002 63	0.002 18	72	0.002 78	0.002 37
2	0.0860	56	0.003 70	0.003 10	64	0.003 94	0.003 39
3	0.0990	48	0.004 87	0.004 06	56	0.005 23	0.004 51
4	0.1120	40	0.006 04	0.004 96	48	0.006 61	0.005 66
5	0.1250	40	0.007 96	0.006 72	44	0.008 80	0.007 16
6	0.1380	32	0.009 09	0.007 45	40	0.010 15	0.008 74
8	0.1640	32	0.014 0	0.011 96	36	0.014 74	0.012 85
10	0.1900	24	0.017 5	0.014 50	32	0.020 0	0.017 5
12	0.2160	24	0.024 2	0.020 6	28	0.025 8	0.022 6
$\frac{1}{4}$	0.2500	20	0.031 8	0.026 9	28	0.036 4	0.032 6
$\frac{5}{16}$	0.3125	18	0.052 4	0.045 4	24	0.058 0	0.052 4
$\frac{3}{8}$	0.3750	16	0.077 5	0.067 8	24	0.087 8	0.080 9
$\frac{7}{16}$	0.4375	14	0.106 3	0.093 3	20	0.118 7	0.109 0
$\frac{1}{2}$	0.5000	13	0.141 9	0.125 7	20	0.159 9	0.148 6
$\frac{9}{16}$	0.5625	12	0.182	0.162	18	0.203	0.189
$\frac{5}{8}$	0.6250	11	0.226	0.202	18	0.256	0.240
$\frac{3}{4}$	0.7500	10	0.334	0.302	16	0.373	0.351
$\frac{7}{8}$	0.8750	9	0.462	0.419	14	0.509	0.480
1	1.0000	8	0.606	0.551	12	0.663	0.625
$1\frac{1}{4}$	1.2500	7	0.969	0.890	12	1.073	1.024
$1\frac{1}{2}$	1.5000	6	1.405	1.294	12	1.581	1.521

*This table was compiled from ANSI B1.1-1974. The minor diameter was found from the equation $d_r = d - 1.299\,038p$, and the pitch diameter from $d_m = d - 0.649\,519p$. The mean of the pitch diameter and the minor diameter was used to compute the tensile-stress area.

Figure 8–3

(a) Square thread; (b) Acme thread.

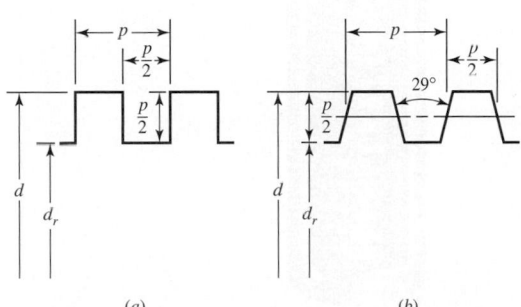

(a) (b)

Table 8–3

Preferred Pitches for
Acme Threads

d, in	$\frac{1}{4}$	$\frac{5}{16}$	$\frac{3}{8}$	$\frac{1}{2}$	$\frac{5}{8}$	$\frac{3}{4}$	$\frac{7}{8}$	1	$1\frac{1}{4}$	$1\frac{1}{2}$	$1\frac{3}{4}$	2	$2\frac{1}{2}$	3
p, in	$\frac{1}{16}$	$\frac{1}{14}$	$\frac{1}{12}$	$\frac{1}{10}$	$\frac{1}{8}$	$\frac{1}{6}$	$\frac{1}{6}$	$\frac{1}{5}$	$\frac{1}{5}$	$\frac{1}{4}$	$\frac{1}{4}$	$\frac{1}{4}$	$\frac{1}{3}$	$\frac{1}{2}$

sometimes modified to a stub form by making the teeth shorter. This results in a larger minor diameter and a somewhat stronger screw.

8–2 The Mechanics of Power Screws

A power screw is a device used in machinery to change angular motion into linear motion, and, usually, to transmit power. Familiar applications include the lead screws of lathes, and the screws for vises, presses, and jacks.

An application of power screws to a power-driven jack is shown in Fig. 8–4. You should be able to identify the worm, the worm gear, the screw, and the nut. Is the worm gear supported by one bearing or two?

Figure 8–4

The Joyce worm-gear screw jack. (*Courtesy Joyce-Dayton Corp., Dayton, Ohio.*)

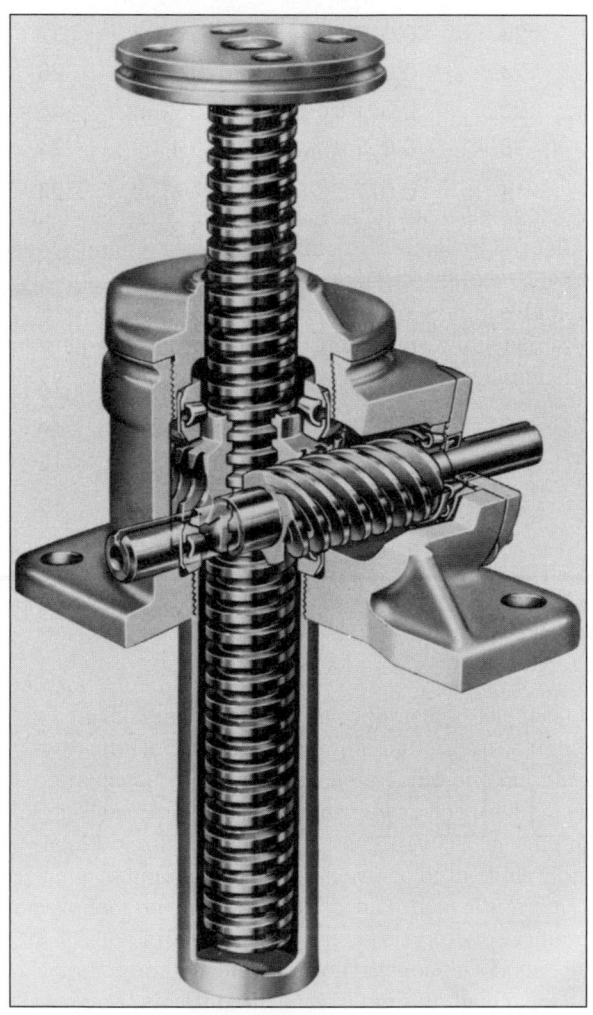

Figure 8–5

Portion of a power screw.

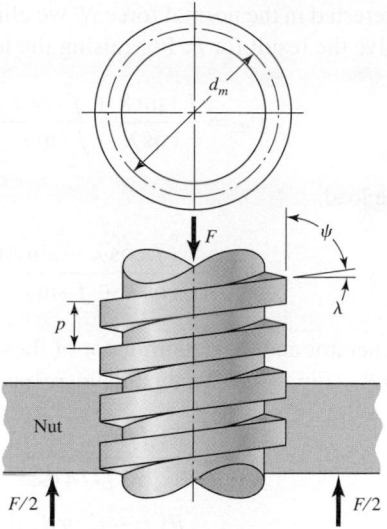

Figure 8–6

Force diagrams: (a) lifting the load; (b) lowering the load.

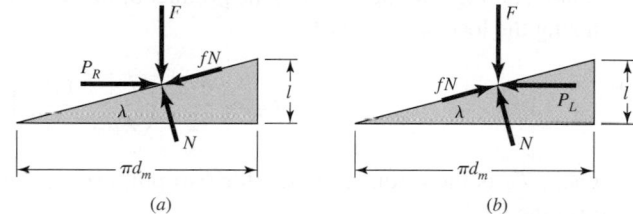

(a)   (b)

In Fig. 8–5 a square-threaded power screw with single thread having a mean diameter d_m, a pitch p, a lead angle λ, and a helix angle ψ is loaded by the axial compressive force F. We wish to find an expression for the torque required to raise this load, and another expression for the torque required to lower the load.

First, imagine that a single thread of the screw is unrolled or developed (Fig. 8–6) for exactly a single turn. Then one edge of the thread will form the hypotenuse of a right triangle whose base is the circumference of the mean-thread-diameter circle and whose height is the lead. The angle λ, in Figs. 8–5 and 8–6, is the lead angle of the thread. We represent the summation of all the unit axial forces acting upon the normal thread area by F. To raise the load, a force P_R acts to the right (Fig. 8–6a), and to lower the load, P_L acts to the left (Fig. 8–6b). The friction force is the product of the coefficient of friction f with the normal force N, and acts to oppose the motion. The system is in equilibrium under the action of these forces, and hence, for raising the load, we have

$$\sum F_H = P_R - N \sin\lambda - fN \cos\lambda = 0$$

$$\sum F_V = F + fN \sin\lambda - N \cos\lambda = 0$$

(a)

In a similar manner, for lowering the load, we have

$$\sum F_H = -P_L - N \sin\lambda + fN \cos\lambda = 0$$

$$\sum F_V = F - fN \sin\lambda - N \cos\lambda = 0$$

(b)

Since we are not interested in the normal force N, we eliminate it from each of these sets of equations and solve the result for P. For raising the load, this gives

$$P_R = \frac{F(\sin\lambda + f\cos\lambda)}{\cos\lambda - f\sin\lambda} \qquad (c)$$

and for lowering the load,

$$P_L = \frac{F(f\cos\lambda - \sin\lambda)}{\cos\lambda + f\sin\lambda} \qquad (d)$$

Next, divide the numerator and the denominator of these equations by $\cos\lambda$ and use the relation $\tan\lambda = l/\pi d_m$ (Fig. 8–6). We then have, respectively,

$$P_R = \frac{F[(l/\pi d_m) + f]}{1 - (fl/\pi d_m)} \qquad (e)$$

$$P_L = \frac{F[f - (l/\pi d_m)]}{1 + (fl/\pi d_m)} \qquad (f)$$

Finally, noting that the torque is the product of the force P and the mean radius $d_m/2$, for raising the load we can write

$$T_R = \frac{F d_m}{2}\left(\frac{l + \pi f d_m}{\pi d_m - fl}\right) \qquad (8\text{--}1)$$

where T_R is the torque required for two purposes: to overcome thread friction and to raise the load.

The torque required to lower the load, from Eq. (f), is found to be

$$T_L = \frac{F d_m}{2}\left(\frac{\pi f d_m - l}{\pi d_m + fl}\right) \qquad (8\text{--}2)$$

This is the torque required to overcome a part of the friction in lowering the load. It may turn out, in specific instances where the lead is large or the friction is low, that the load will lower itself by causing the screw to spin without any external effort. In such cases, the torque T_L from Eq. (8–2) will be negative or zero. When a positive torque is obtained from this equation, the screw is said to be *self-locking*. Thus the condition for self-locking is

$$\pi f d_m > l$$

Now divide both sides of this inequality by πd_m. Recognizing that $l/\pi d_m = \tan\lambda$, we get

$$f > \tan\lambda \qquad (8\text{--}3)$$

This relation states that self-locking is obtained whenever the coefficient of thread friction is equal to or greater than the tangent of the thread lead angle.

An expression for efficiency is also useful in the evaluation of power screws. If we let $f = 0$ in Eq. (8–1), we obtain

$$T_0 = \frac{Fl}{2\pi} \qquad (g)$$

which, since thread friction has been eliminated, is the torque required only to raise the load. The efficiency is therefore

$$e = \frac{T_0}{T_R} = \frac{Fl}{2\pi T_R} \tag{8-4}$$

The preceding equations have been developed for square threads where the normal thread loads are parallel to the axis of the screw. In the case of Acme or other threads, the normal thread load is inclined to the axis because of the thread angle 2α and the lead angle λ. Since lead angles are small, this inclination can be neglected and only the effect of the thread angle (Fig. 8–7a) considered. The effect of the angle α is to increase the frictional force by the wedging action of the threads. Therefore the frictional terms in Eq. (8–1) must be divided by $\cos \alpha$. For raising the load, or for tightening a screw or bolt, this yields

$$T_R = \frac{Fd_m}{2} \left(\frac{l + \pi f d_m \sec \alpha}{\pi d_m - fl \sec \alpha} \right) \tag{8-5}$$

In using Eq. (8–5), remember that it is an approximation because the effect of the lead angle has been neglected.

For power screws, the Acme thread is not as efficient as the square thread, because of the additional friction due to the wedging action, but it is often preferred because it is easier to machine and permits the use of a split nut, which can be adjusted to take up for wear.

Usually a third component of torque must be applied in power-screw applications. When the screw is loaded axially, a thrust or collar bearing must be employed between the rotating and stationary members in order to carry the axial component. Figure 8–7b shows a typical thrust collar in which the load is assumed to be concentrated at the mean collar diameter d_c. If f_c is the coefficient of collar friction, the torque required is

$$T_c = \frac{Ff_cd_c}{2} \tag{8-6}$$

For large collars, the torque should probably be computed in a manner similar to that employed for disk clutches.

Figure 8–7

(a) Normal thread force is increased because of angle α; (b) thrust collar has frictional diameter d_c.

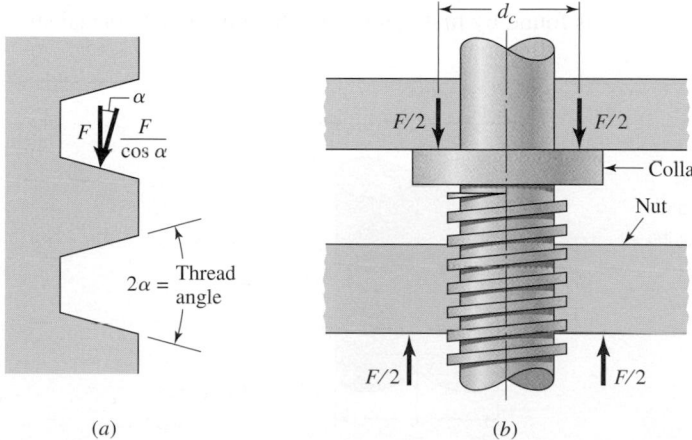

(a)

(b)

Nominal body stresses in power screws can be related to thread parameters as follows. The maximum nominal shear stress τ in torsion of the screw body can be expressed as

$$\tau = \frac{16T}{\pi d_r^3} \tag{8-7}$$

The axial stress σ in the body of the screw due to load F is

$$\sigma = \frac{F}{A} = \frac{4F}{\pi d_r^2} \tag{8-8}$$

in the absence of column action. For a short column the J. B. Johnson equation is

$$\left(\frac{F}{A}\right)_{crit} = S_y - \left(\frac{S_y}{2\pi}\frac{l}{k}\right)^2 \frac{1}{CE} \tag{8-9}$$

after Eq. (5–50).

Nominal thread stresses in power screws can be related to thread parameters as follows. The bearing stress in Fig. 8–8, σ_B, is

$$\sigma_B = -\frac{F}{\pi d_m n_t p/2} = -\frac{2F}{\pi d_m n_t p} \tag{8-10}$$

where n_t is the number of engaged threads. The bending stress at the root of the thread σ_b is found from

$$\frac{I}{c} = \frac{(\pi d_r n_t)(p/2)^2}{6} = \frac{\pi}{24} d_r n_t p^2 \qquad M = \frac{Fp}{4}$$

so

$$\sigma_b = \frac{M}{I/c} = \frac{Fp}{4}\frac{24}{\pi d_r n_t p^2} = \frac{6F}{\pi d_r n_t p} \tag{8-11}$$

The transverse shear stress τ at the center of the root of the thread due to load F is

$$\tau = \frac{3V}{2A} = \frac{3}{2}\frac{F}{\pi d_r n_t p/2} = \frac{3F}{\pi d_r n_t p} \tag{8-12}$$

and at the top of the root it is zero. The von Mises stress σ' at the top of the root "plane" is found by first identifying the orthogonal normal stresses and the shear stresses. From

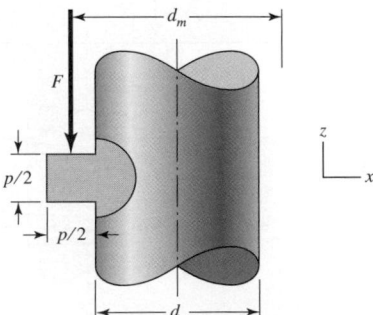

Figure 8–8

Geometry of square thread useful in finding bending and transverse shear stresses at the thread root.

the coordinate system of Fig. 8–8, we note

$$\sigma_x = \frac{6F}{\pi d_r n_t p} \qquad \tau_{xy} = 0$$

$$\sigma_y = 0 \qquad \tau_{yz} = \frac{16T}{\pi d_r^3}$$

$$\sigma_z = -\frac{4F}{\pi d_r^2} \qquad \tau_{zx} = 0$$

then use Eq. (6–14) of Sec. 6–5.

The screw-thread form is complicated from an analysis viewpoint. Remember the origin of the tensile-stress area A_t, which comes from experiment. A power screw lifting a load is in compression and its thread pitch is *shortened* by elastic deformation. Its engaging nut is in tension and its thread pitch is *lengthened*. The engaged threads cannot share the load equally. Some experiments show that the first engaged thread carries 0.38 of the load, the second 0.25, the third 0.18, and the seventh is free of load. In estimating thread stresses by the equations above, substituting $0.38F$ for F and setting n_t to 1 will give the largest level of stresses in the thread-nut combination.

EXAMPLE 8–1

A square-thread power screw has a major diameter of 32 mm and a pitch of 4 mm with double threads, and it is to be used in an application similar to that in Fig. 8–4. The given data include $f = f_c = 0.08$, $d_c = 40$ mm, and $F = 6.4$ kN per screw.
(a) Find the thread depth, thread width, pitch diameter, minor diameter, and lead.
(b) Find the torque required to raise and lower the load.
(c) Find the efficiency during lifting the load.
(d) Find the body stresses, torsional and compressive.
(e) Find the bearing stress.
(f) Find the thread stresses bending at the root, shear at the root, and von Mises stress and maximum shear stress at the same location.

Solution

(a) From Fig. 8–3a the thread depth and width are the same and equal to half the pitch, or 2 mm. Also

$$d_m = d - p/2 = 32 - 4/2 = 30 \text{ mm}$$

Answer

$$d_r = d - p = 32 - 4 = 28 \text{ mm}$$

$$l = np = 2(4) = 8 \text{ mm}$$

(b) Using Eqs. (8–1) and (8–6), the torque required to turn the screw against the load is

$$T_R = \frac{F d_m}{2} \left(\frac{l + \pi f d_m}{\pi d_m - fl} \right) + \frac{F f_c d_c}{2}$$

$$= \frac{6.4(30)}{2} \left[\frac{8 + \pi(0.08)(30)}{\pi(30) - 0.08(8)} \right] + \frac{6.4(0.08)40}{2}$$

Answer

$$= 15.94 + 10.24 = 26.18 \text{ N} \cdot \text{m}$$

Using Eqs. (8–2) and (8–6), the load-lowering torque is

$$T_L = \frac{Fd_m}{2}\left(\frac{\pi f d_m - l}{\pi d_m + fl}\right) + \frac{Ff_c d_c}{2}$$

$$= \frac{6.4(30)}{2}\left[\frac{\pi(0.08)30 - 8}{\pi(30) + 0.08(8)}\right] + \frac{6.4(0.08)(40)}{2}$$

Answer

$$= -0.466 + 10.24 = 9.77 \text{ N} \cdot \text{m}$$

The minus sign in the first term indicates that the screw alone is not self-locking and would rotate under the action of the load except for the fact that the collar friction is present and must be overcome, too. Thus the torque required to rotate the screw "with" the load is less than is necessary to overcome collar friction alone.

(c) The overall efficiency in raising the load is

Answer

$$e = \frac{Fl}{2\pi T_R} = \frac{6.4(8)}{2\pi(26.18)} = 0.311$$

(d) The body shear stress τ due to torsional moment T_R at the outside of the screw body is

Answer

$$\tau = \frac{16T_R}{\pi d_r^3} = \frac{16(26.18)(10^3)}{\pi(28^3)} = 6.07 \text{ MPa}$$

The axial nominal normal stress σ is

Answer

$$\sigma = -\frac{4F}{\pi d_r^2} = -\frac{4(6.4)10^3}{\pi(28^2)} = -10.39 \text{ MPa}$$

(e) The bearing stress σ_B is, with one thread carrying $0.38F$,

Answer

$$\sigma_B = -\frac{2(0.38F)}{\pi d_m(1)p} = -\frac{2(0.38)(6.4)10^3}{\pi(30)(1)(4)} = -12.9 \text{ MPa}$$

(f) The thread-root bending stress σ_b with one thread carrying $0.38F$ is

$$\sigma_b = \frac{6(0.38F)}{\pi d_r(1)p} = \frac{6(0.38)(6.4)10^3}{\pi(28)(1)4} = 41.5 \text{ MPa}$$

The transverse shear at the extreme of the root cross section due to bending is zero. However, there is a circumferential shear stress at the extreme of the root cross section of the thread as shown in part (d) of 6.07 MPa. The three-dimensional stresses, after Fig. 8–8, noting the y coordinate is into the page, are

$$\sigma_x = 41.5 \text{ MPa} \qquad \tau_{xy} = 0$$

$$\sigma_y = 0 \qquad \tau_{yz} = 6.07 \text{ MPa}$$

$$\sigma_z = -10.39 \text{ MPa} \qquad \tau_{zx} = 0$$

Equation (6–14) of Sec. 6–5 can be written as

Answer

$$\sigma' = \frac{1}{\sqrt{2}}\{(41.5 - 0)^2 + [0 - (-10.39)]^2 + (-10.39 - 41.5)^2 + 6(6.07)^2\}^{1/2}$$

$$= 48.7 \text{ MPa}$$

Alternatively, you can determine the principal stresses and then use Eq. (6–12) to find the von Mises stress. This would prove helpful in evaluating τ_{max} as well. The principal stresses can be found from Eq. (4–15); however, sketch the stress element and note that there are no shear stresses on the x face. This means that σ_x is a principal stress. The remaining stresses can be transformed by using the plane stress equation, Eq. (4–13). Thus, the remaining principal stresses are

$$\frac{-10.39}{2} \pm \sqrt{\left(\frac{-10.39}{2}\right)^2 + 6.07^2} = 2.79, -13.18 \text{ MPa}$$

Ordering the principal stresses gives $\sigma_1, \sigma_2, \sigma_3 = 41.5, 2.79, -13.18$ MPa. Substituting these into Eq. (6–12) yields

Answer
$$\sigma' = \left\{ \frac{[41.5 - 2.79]^2 + [2.79 - (-13.18)]^2 + [-13.18 - 41.5]^2}{2} \right\}^{1/2}$$

$$= 48.7 \text{ MPa}$$

The maximum shear stress is given by Eq. (4–16), where $\tau_{max} = \tau_{1/3}$, giving

Answer
$$\tau_{max} = \frac{\sigma_1 - \sigma_3}{2} = \frac{41.5 - (-13.18)}{2} = 27.3 \text{ MPa}$$

Table 8–4

Screw Bearing Pressure p_b

Source: H. A. Rothbart, *Mechanical Design and Systems Handbook*, 2nd ed., McGraw-Hill, New York, 1985.

Screw Material	Nut Material	Safe p_b, psi	Notes
Steel	Bronze	2500–3500	Low speed
Steel	Bronze	1600–2500	10 fpm
	Cast iron	1800–2500	8 fpm
Steel	Bronze	800–1400	20–40 fpm
	Cast iron	600–1000	20–40 fpm
Steel	Bronze	150–240	50 fpm

Ham and Ryan[1] showed that the coefficient of friction in screw threads is independent of axial load, practically independent of speed, decreases with heavier lubricants, shows little variation with combinations of materials, and is best for steel on bronze. Sliding coefficients of friction in power screws are about 0.10–0.15.

Table 8–4 shows safe bearing pressures on threads, to protect the moving surfaces from abnormal wear. Table 8–5 shows the coefficients of sliding friction for common material pairs. Table 8–6 shows coefficients of starting and running friction for common material pairs.

---

[1]Ham and Ryan, *An Experimental Investigation of the Friction of Screw-threads,* Bulletin 247, University of Illinois Experiment Station, Champaign-Urbana, Ill., June 7, 1932.

Table 8–5

Coefficients of Friction f for Threaded Pairs

Source: H. A. Rothbart, *Mechanical Design and Systems Handbook,* 2nd ed., McGraw-Hill, New York, 1985.

Screw Material	Nut Material			
	Steel	**Bronze**	**Brass**	**Cast Iron**
Steel, dry	0.15–0.25	0.15–0.23	0.15–0.19	0.15–0.25
Steel, machine oil	0.11–0.17	0.10–0.16	0.10–0.15	0.11–0.17
Bronze	0.08–0.12	0.04–0.06	—	0.06–0.09

Table 8–6

Thrust-Collar Friction Coefficients

Source: H. A. Rothbart, *Mechanical Design and Systems Handbook,* 2nd ed., McGraw-Hill, New York, 1985.

Combination	Running	Starting
Soft steel on cast iron	0.12	0.17
Hard steel on cast iron	0.09	0.15
Soft steel on bronze	0.08	0.10
Hard steel on bronze	0.06	0.08

8–3 Threaded Fasteners

As you study the sections on threaded fasteners and their use, be alert to the stochastic and deterministic viewpoints. In most cases the threat is from overproof loading of fasteners, and this is best addressed by statistical methods. The threat from fatigue is lower, and deterministic methods can be adequate.

Figure 8–9 is a drawing of a standard hexagon-head bolt. Points of stress concentration are at the fillet, at the start of the threads (runout), and at the thread-root fillet in the plane of the nut when it is present. See Table A–29 for dimensions. The diameter of the washer face is the same as the width across the flats of the hexagon. The thread length of inch-series bolts, where D is the nominal diameter, is

$$L_T = \begin{cases} 2D + \frac{1}{4} \text{ in} & L \leq 6 \text{ in} \\ 2D + \frac{1}{2} \text{ in} & L > 6 \text{ in} \end{cases} \tag{8–13}$$

and for metric bolts is

$$L_T = \begin{cases} 2D + 6 & L \leq 125 \quad D \leq 48 \\ 2D + 12 & 125 < L \leq 200 \\ 2D + 25 & L > 200 \end{cases} \tag{8–14}$$

where the dimensions are in millimeters. The ideal bolt length is one in which only one or two threads project from the nut after it is tightened. Bolt holes may have burrs or sharp edges after drilling. These could bite into the fillet and increase stress concentration. Therefore, washers must always be used under the bolt head to prevent this. They should be of hardened steel and loaded onto the bolt so that the rounded edge of the stamped hole faces the washer face of the bolt. Sometimes it is necessary to use washers under the nut too.

The purpose of a bolt is to clamp two or more parts together. The clamping load stretches or elongates the bolt; the load is obtained by twisting the nut until the bolt has elongated almost to the elastic limit. If the nut does not loosen, this bolt tension remains

Figure 8–9

Hexagon-head bolt; note the washer face, the fillet under the head, the start of threads, and the chamfer on both ends. Bolt lengths are always measured from below the head.

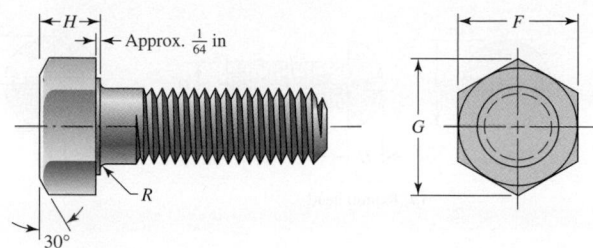

Figure 8–10

Typical cap-screw heads: (a) fillister head; (b) flat head; (c) hexagonal socket head. Cap screws are also manufactured with hexagonal heads similar to the one shown in Fig. 8–9, as well as a variety of other head styles. This illustration uses one of the conventional methods of representing threads.

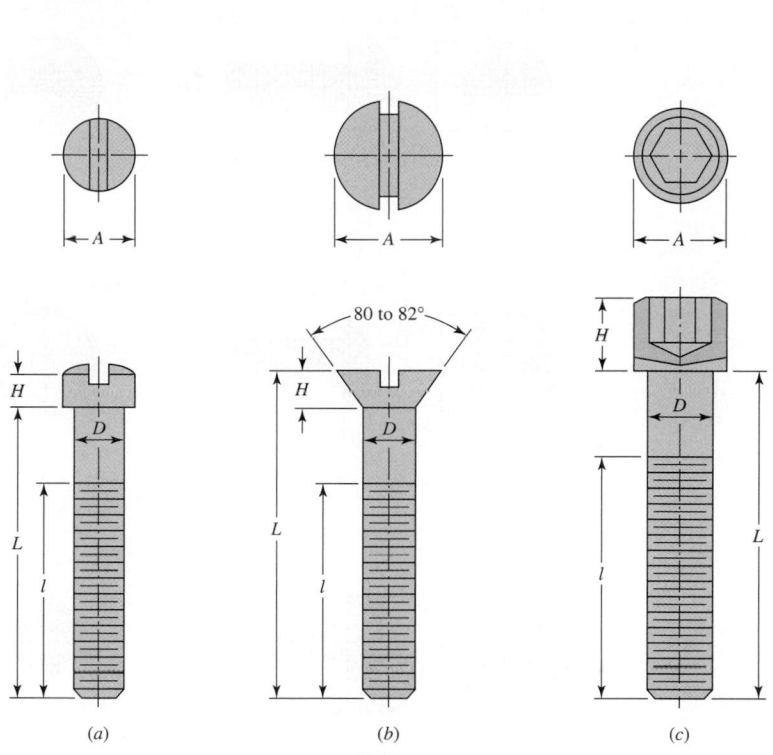

as the preload or clamping force. When tightening, the mechanic should, if possible, hold the bolt head stationary and twist the nut; in this way the bolt shank will not feel the thread-friction torque.

The head of a hexagon-head cap screw is slightly thinner than that of a hexagon-head bolt. Dimensions of hexagon-head cap screws are listed in Table A–30. Hexagon-head cap screws are used in the same applications as bolts and also in applications in which one of the clamped members is threaded. Three other common cap-screw head styles are shown in Fig. 8–10.

A variety of machine-screw head styles are shown in Fig. 8–11. Inch-series machine screws are generally available in sizes from No. 0 to about $\frac{3}{8}$ in.

Several styles of hexagonal nuts are illustrated in Fig. 8–12; their dimensions are given in Table A–31. The material of the nut must be selected carefully to match that of the bolt. During tightening, the first thread of the nut tends to take the entire load; but yielding occurs, with some strengthening due to the cold work that takes place, and the load is eventually divided over about three nut threads. For this reason you should never reuse nuts; in fact, it can be dangerous to do so.

Figure 8–11

Types of heads used on machine screws.

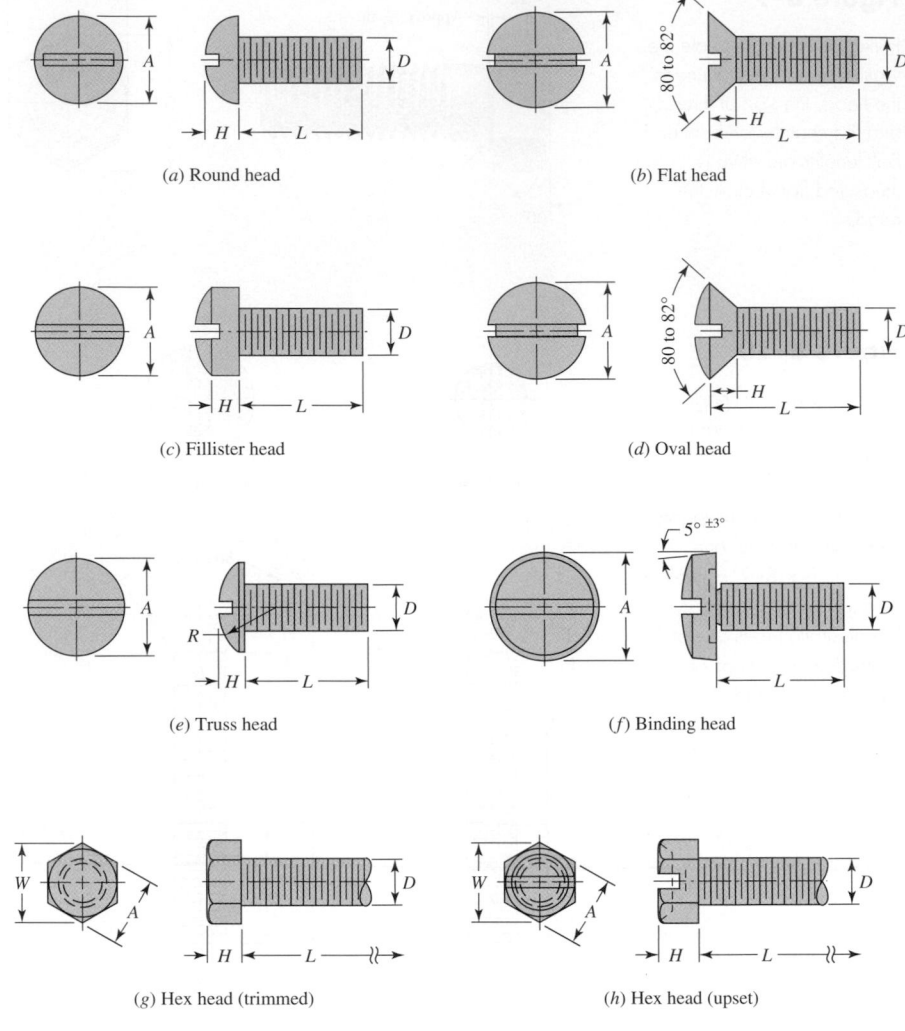

(a) Round head

(b) Flat head

(c) Fillister head

(d) Oval head

(e) Truss head

(f) Binding head

(g) Hex head (trimmed)

(h) Hex head (upset)

Figure 8–12

Hexagonal nuts: (a) end view, general; (b) washer-faced regular nut; (c) regular nut chamfered on both sides; (d) jam nut with washer face; (e) jam nut chamfered on both sides.

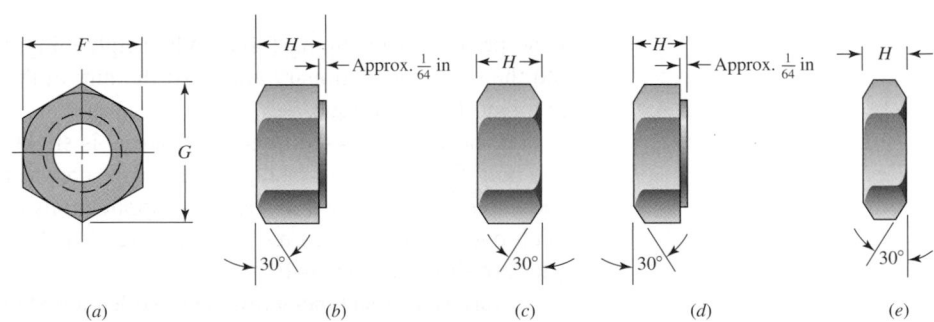

(a) (b) (c) (d) (e)

8–4   **Joints—Fastener Stiffness**

When a connection is desired that can be disassembled without destructive methods and that is strong enough to resist external tensile loads, moment loads, and shear loads, or a combination of these, then the simple bolted joint using hardened-steel washers is a good solution. Such a joint can also be dangerous unless it is properly designed and assembled by a *trained* mechanic.

Figure 8–13

A bolted connection loaded in tension by the forces *P*. Note the use of two washers. Note how the threads extend into the body of the connection. This is usual and is desired. L_G is the grip of the connection.

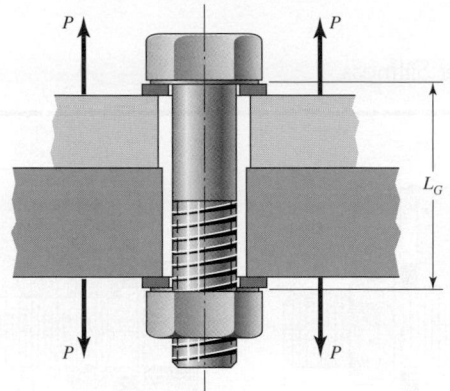

Figure 8–14

Section of cylindrical pressure vessel. Hexagon-head cap screws are used to fasten the cylinder head to the body. Note the use of an O-ring seal. L'_G is the effective grip of the connection (see Table 8–7).

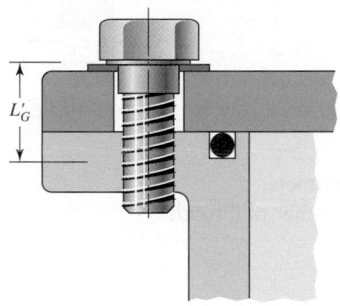

A section through a tension-loaded bolted joint is illustrated in Fig. 8–13. Notice the clearance space provided by the bolt holes. Notice, too, how the bolt threads extend into the body of the connection.

As noted previously, the purpose of the bolt is to clamp the two, or more, parts together. Twisting the nut stretches the bolt to produce the clamping force. This clamping force is called the *pretension* or *bolt preload*. It exists in the connection after the nut has been properly tightened no matter whether the external tensile load *P* is exerted or not.

Of course, since the members are being clamped together, the clamping force that produces tension in the bolt induces compression in the members.

Figure 8–14 shows another tension-loaded connection. This joint uses cap screws threaded into one of the members. An alternative approach to this problem (of not using a nut) would be to use studs. A stud is a rod threaded on both ends. The stud is screwed into the lower member first; then the top member is positioned and fastened down with hardened washers and nuts. The studs are regarded as permanent, and so the joint can be disassembled merely by removing the nut and washer. Thus the threaded part of the lower member is not damaged by reusing the threads.

The *spring rate* is a limit as expressed in Eq. (5–1). For an elastic member such as a bolt, as we learned in Eq. (5–2), it is the ratio between the force applied to the member and the deflection produced by that force. We can use Eq. (5–4) and the results of Prob. 5–1 to find the stiffness constant of a fastener in any bolted connection.

The *grip* L_G of a connection is the total thickness of the clamped material. In Fig. 8–13 the grip is the sum of the thicknesses of both members and both washers. In Fig. 8–14 the effective grip is given in Table 8–7.

The stiffness of the portion of a bolt or screw within the clamped zone will generally consist of two parts, that of the unthreaded shank portion and that of the threaded portion. Thus the stiffness constant of the bolt is equivalent to the stiffnesses of two

Table 8–7

Suggested Procedure for Finding Fastener Stiffness

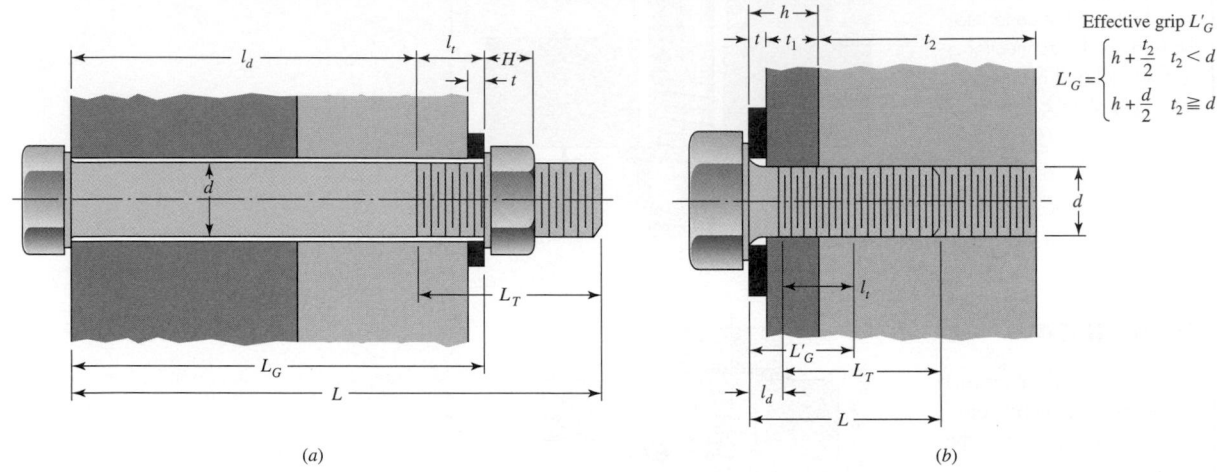

| (a) | (b) |

Given fastener diameter d
and pitch p or number of threads

Grip is thickness L_G

Washer thickness from
Table A–32 or A–33
Threaded length L_T
Inch series:

$$L_T = \begin{cases} 2D + \frac{1}{4} \text{ in,} & L \leq 6 \text{ in} \\ 2D + \frac{1}{2} \text{ in,} & L > 6 \text{ in} \end{cases}$$

Metric series:

$$L_T = \begin{cases} 2D + 6 \text{ mm,} & L \leq 125, D \leq 48 \text{ mm} \\ 2D + 12 \text{ mm,} & 125 < L \leq 200 \text{ mm} \\ 2D + 25 \text{ mm,} & L > 200 \text{ mm} \end{cases}$$

Effective grip

$$L'_G = \begin{cases} h + t_2/2, & t_2 < d \\ h + d/2, & t_2 \geq d \end{cases}$$

Fastener length: $L > L_G + H$

Round up using Table A–17[*]

Fastener length: $L > h + 1.5d$

Length of useful unthreaded
portion: $l_d = L - L_T$
Length of threaded portion:
$l_t = L_G - l_d$

Length of useful unthreaded
portion: $l_d = L - L_T$
Length of useful threaded
portion: $l_t = L'_G - l_d$

Area of unthreaded portion:
$A_d = \pi d^2/4$
Area of threaded portion:
A_t, Table 8–1 or 8–2
Fastener stiffness:
$$k_b = \frac{A_d A_t E}{A_d l_t + A_t l_d}$$

[*]Bolts and cap screws may not be available in all the preferred lengths listed in Table A–17. Large fasteners may not be available in fractional inches or in millimeter lengths ending in a nonzero digit. Check with your bolt supplier for availability.

springs in series. Using the results of Prob. 5–1, we find

$$\frac{1}{k} = \frac{1}{k_1} + \frac{1}{k_2} \qquad \text{or} \qquad k = \frac{k_1 k_2}{k_1 + k_2} \qquad (8\text{–}15)$$

for two springs in series. From Eq. (5–4), the spring rates of the threaded and unthreaded portions of the bolt in the clamped zone are, respectively,

$$k_t = \frac{A_t E}{l_t} \qquad k_d = \frac{A_d E}{l_d} \qquad (8\text{–}16)$$

where A_t = tensile-stress area (Tables 8–1, 8–2)

$\quad\quad\; l_t$ = length of threaded portion of grip

$\quad\quad\, A_d$ = major-diameter area of fastener

$\quad\quad\; l_d$ = length of unthreaded portion in grip

Substituting these stiffnesses in Eq. (8–15) gives

$$k_b = \frac{A_d A_t E}{A_d l_t + A_t l_d} \qquad (8\text{–}17)$$

where k_b is the estimated effective stiffness of the bolt or cap screw in the clamped zone. For short fasteners, the one in Fig. 8–14, for example, the unthreaded area is small and so the first of the expressions in Eq. (8–16) can be used to find k_b. For long fasteners, the threaded area is relatively small, and so the second expression in Eq. (8–16) can be used. Table 8–7 is useful.

8–5   Joints—Member Stiffness

In the previous section, we determined the stiffness of the fastener in the clamped zone. In this section, we wish to study the stiffnesses of the members in the clamped zone. Both of these stiffnesses must be known in order to learn what happens when the assembled connection is subjected to an external tensile loading.

There may be more than two members included in the grip of the fastener. All together these act like compressive springs in series, and hence the total spring rate of the members is

$$\frac{1}{k_m} = \frac{1}{k_1} + \frac{1}{k_2} + \frac{1}{k_3} + \cdots + \frac{1}{k_i} \qquad (8\text{–}18)$$

If one of the members is a soft gasket, its stiffness relative to the other members is usually so small that for all practical purposes the others can be neglected and only the gasket stiffness used.

If there is no gasket, the stiffness of the members is rather difficult to obtain, except by experimentation, because the compression spreads out between the bolt head and the nut and hence the area is not uniform. There are, however, some cases in which this area can be determined.

Ito[2] has used ultrasonic techniques to determine the pressure distribution at the member interface. The results show that the pressure stays high out to about 1.5 bolt radii. The

---

[2]Y. Ito, J. Toyoda, and S. Nagata, "Interface Pressure Distribution in a Bolt-Flange Assembly," ASME paper no. 77-WA/DE-11, 1977.

Figure 8–15

Compression of a member with the equivalent elastic properties represented by a frustum of a hollow cone. Here, l represents the grip length.

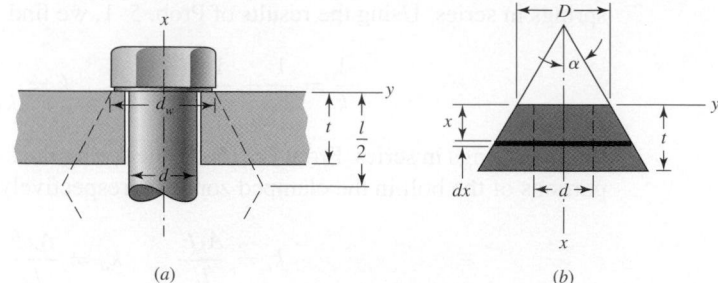

(a)   (b)

pressure, however, falls off farther away from the bolt. Thus Ito suggests the use of Rotscher's pressure-cone method for stiffness calculations with a variable cone angle. This method is quite complicated, and so here we choose to use a simpler approach using a fixed cone angle.

Figure 8–15 illustrates the general cone geometry using a half-apex angle α. An angle $\alpha = 45°$ has been used, but Little[3] reports that this overestimates the clamping stiffness. When loading is restricted to a washer-face annulus (hardened steel, cast iron, or aluminum), the proper apex angle is smaller. Osgood[4] reports a range of $25° \leq \alpha \leq 33°$ for most combinations. In this book we shall use $\alpha = 30°$ except in cases in which the material is insufficient to allow the frusta to exist.

Referring now to Fig. 8–15b, the contraction of an element of the cone of thickness dx subjected to a compressive force P is, from Eq. (5–3),

$$d\delta = \frac{P\,dx}{EA} \qquad (a)$$

The area of the element is

$$A = \pi\left(r_o^2 - r_i^2\right) = \pi\left[\left(x\tan\alpha + \frac{D}{2}\right)^2 - \left(\frac{d}{2}\right)^2\right]$$

$$= \pi\left(x\tan\alpha + \frac{D+d}{2}\right)\left(x\tan\alpha + \frac{D-d}{2}\right) \qquad (b)$$

Substituting this in Eq. (a) and integrating gives a total contraction of

$$\delta = \frac{P}{\pi E}\int_0^t \frac{dx}{[x\tan\alpha + (D+d)/2][x\tan\alpha + (D-d)/2]} \qquad (c)$$

Using a table of integrals, we find the result to be

$$\delta = \frac{P}{\pi Ed\tan\alpha}\ln\frac{(2t\tan\alpha + D - d)(D + d)}{(2t\tan\alpha + D + d)(D - d)} \qquad (d)$$

Thus the spring rate or stiffness of this frustum is

$$k = \frac{P}{\delta} = \frac{\pi Ed\tan\alpha}{\ln\dfrac{(2t\tan\alpha + D - d)(D + d)}{(2t\tan\alpha + D + d)(D - d)}} \qquad (8\text{–}19)$$

[3]R. E. Little, "Bolted Joints: How Much Give?" *Machine Design,* Nov. 9, 1967.

[4]C. C. Osgood, "Saving Weight on Bolted Joints," *Machine Design,* Oct. 25, 1979.

With $\alpha = 30°$, this becomes

$$k = \frac{0.5774\pi \, Ed}{\ln \dfrac{(1.155t + D - d)(D + d)}{(1.155t + D + d)(D - d)}} \tag{8–20}$$

Equation (8–20), or (8–19), must be solved separately for each frustum in the joint. Then individual stiffnesses are assembled to obtain k_m using Eq. (8–18).

If the members of the joint have the same Young's modulus E with symmetrical frusta back to back, then they act as two identical springs in series. From Eq. (8–18) we learn that $k_m = k/2$. Using the grip as $l = 2t$ and d_w as the diameter of the washer face, we find the spring rate of the members to be

$$k_m = \frac{\pi \, Ed \tan\alpha}{2\ln \dfrac{(l\tan\alpha + d_w - d)\,(d_w + d)}{(l\tan\alpha + d_w + d)\,(d_w - d)}} \tag{8–21}$$

The diameter of the washer face is about 50 percent greater than the fastener diameter for standard hexagon-head bolts and cap screws. Thus we can simplify Eq. (8–21) by letting $d_w = 1.5d$. If we also use $\alpha = 30°$, then Eq. (8–21) can be written as

$$k_m = \frac{0.5774\pi \, Ed}{2\ln \left(5\dfrac{0.5774l + 0.5d}{0.5774l + 2.5d}\right)} \tag{8–22}$$

It is easy to program the numbered equations in this section, and you should do so. The time spent in programming will save many hours of formula plugging.

To see how good Eq. (8–21) is, solve it for k_m/Ed:

$$\frac{k_m}{Ed} = \frac{\pi \tan\alpha}{2\ln\left[\dfrac{(l\tan\alpha + d_w - d)\,(d_w + d)}{(l\tan\alpha + d_w + d)\,(d_w - d)}\right]}$$

Earlier in the section use of $\alpha = 30°$ was recommended for hardened steel, cast iron, or aluminum members. Wileman, Choudury, and Green[5] conducted a finite element study of this problem. The results, which are depicted in Fig. 8–16, agree with the $\alpha = 30°$ recommendation, coinciding exactly at the aspect ratio $d/l = 0.4$. Additionally, they offered an exponential curve-fit of the form

$$\frac{k_m}{Ed} = A \exp(Bd/l) \tag{8–23}$$

with constants A and B defined in Table 8–8. For standard washer faces and members of the same material, Eq. (8–23) offers a simple calculation for member stiffness k_m. For departure from these conditions, Eq. (8–20) remains the basis for approaching the problem.

[5]J. Wileman, M. Choudury, and I. Green, "Computation of Member Stiffness in Bolted Connections," *Trans. ASME, J. Mech. Design,* vol. 113, December 1991, pp. 432–437.

Figure 8–16

The dimensionless plot of stiffness versus aspect ratio of the members of a bolted joint, showing the relative accuracy of methods of Rotscher, Mischke, and Motosh, compared to a finite-element analysis (FEA) conducted by Wileman, Choudury, and Green.

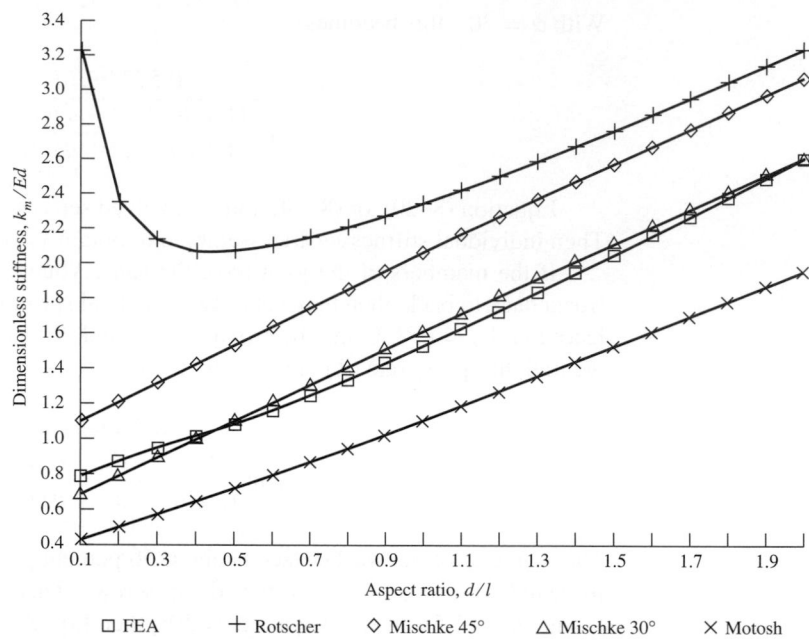

Dimensionless stiffness, k_m/Ed versus Aspect ratio, d/l

□ FEA + Rotscher ◇ Mischke 45° △ Mischke 30° × Motosh

Table 8–8

Stiffness Parameters of Various Member Materials[†]

[†]*Source:* J. Wileman, M. Choudury, and I. Green, "Computation of Member Stiffness in Bolted Connections," *Trans. ASME, J. Mech. Design,* vol. 113, December 1991, pp. 432–437.

Material Used	Poisson Ratio	Elastic GPa	Modulus Mpsi	A	B
Steel	0.291	207	30.0	0.787 15	0.628 73
Aluminum	0.334	71	10.3	0.796 70	0.638 16
Copper	0.326	119	17.3	0.795 68	0.635 53
Gray cast iron	0.211	100	14.5	0.778 71	0.616 16
General expression				0.789 52	0.629 14

EXAMPLE 8–2

Two $\frac{1}{2}$-in-thick steel plates with a modulus of elasticity of $30(10^6)$ psi are clamped by washer-faced $\frac{1}{2}$-in-diameter UNC SAE grade 5 bolts with a 0.095-in-thick washer under the nut. Find the member spring rate k_m using the method of conical frusta, and compare the result with the finite element analysis (FEA) curve-fit method of Wileman et al.

Solution

The grip is $0.5 + 0.5 + 0.095 = 1.095$ in. Using Eq. (8–22) with $l = 1.095$ and $d = 0.5$ in, we write

$$k_m = \frac{0.5774\pi 30(10^6)0.5}{2\ln\left[5\dfrac{0.5774(1.095) + 0.5(0.5)}{0.5774(1.095) + 2.5(0.5)}\right]} = 15.97(10^6) \text{ lbf/in}$$

From Table 8–8, $A = 0.787\ 15$, $B = 0.628\ 73$. Equation (8–23) gives

$$k_m = 30(10^6)(0.5)(0.787\ 15)\exp[0.628\ 73(0.5)/1.095]$$
$$= 15.73(10^6)\ \text{lbf/in}$$

For this case, the difference between the results for Eqs. (8–22) and (8–23) is less than 2 percent.

8–6   Bolt Strength

In the specification standards for bolts, the strength is specified by stating ASTM minimum quantities, the *minimum proof strength,* or *minimum proof load,* and the *minimum tensile strength.*

The *proof load* is the maximum load (force) that a bolt can withstand without acquiring a permanent set. The *proof strength* is the quotient of the proof load and the tensile-stress area. The proof strength thus corresponds roughly to the proportional limit and corresponds to 0.0001 in permanent set in the fastener (first measurable deviation from elastic behavior). The value of the mean proof strength, the mean tensile strength, and the corresponding standard deviations are not part of the specification codes, so it is the designer's responsibility to obtain these values, perhaps by laboratory testing, before designing to a reliability specification. Figure 8–17 shows the distribution of ultimate tensile strength from a bolt production run. If the ASTM minimum strength equals or exceeds 120 kpsi, the bolts can be offered as SAE grade 5. The designer does not see this histogram. Instead, in Table 8–9, the designer sees the entry $S_{ut} = 120$ kpsi under the $\frac{1}{4}$–1-in size in grade 5 bolts. Similarly, minimum strengths are shown in Tables 8–10 and 8–11.

The SAE specifications are found in Table 8–9. The bolt grades are numbered according to the tensile strengths, with decimals used for variations at the same strength level. Bolts and screws are available in all grades listed. Studs are available in grades 1, 2, 4, 5, 8, and 8.1. Grade 8.1 is not listed.

Figure 8–17

Histogram of bolt ultimate tensile strength based on 539 tests displaying a mean ultimate tensile strength $\bar{S}_{ut} = 145.1$ kpsi and a standard deviation of $\hat{\sigma}_{S_{ut}} = 10.3$ kpsi.

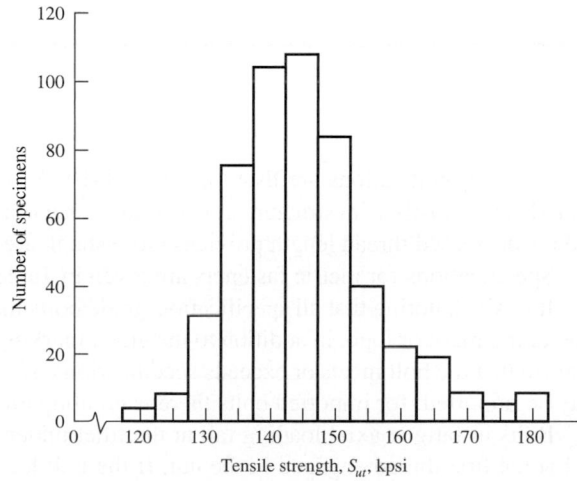

Table 8–9

SAE Specifications for Steel Bolts

SAE Grade No.	Size Range Inclusive, in	Minimum Proof Strength,* kpsi	Minimum Tensile Strength,* kpsi	Minimum Yield Strength,* kpsi	Material	Head Marking
1	$\frac{1}{4}$–$1\frac{1}{2}$	33	60	36	Low or medium carbon	
2	$\frac{1}{4}$–$\frac{3}{4}$	55	74	57	Low or medium carbon	
	$\frac{7}{8}$–$1\frac{1}{2}$	33	60	36		
4	$\frac{1}{4}$–$1\frac{1}{2}$	65	115	100	Medium carbon, cold-drawn	
5	$\frac{1}{4}$–1	85	120	92	Medium carbon, Q&T	
	$1\frac{1}{8}$–$1\frac{1}{2}$	74	105	81		
5.2	$\frac{1}{4}$–1	85	120	92	Low-carbon martensite, Q&T	
7	$\frac{1}{4}$–$1\frac{1}{2}$	105	133	115	Medium-carbon alloy, Q&T	
8	$\frac{1}{4}$–$1\frac{1}{2}$	120	150	130	Medium-carbon alloy, Q&T	
8.2	$\frac{1}{4}$–1	120	150	130	Low-carbon martensite, Q&T	

*Minimum strengths are strengths exceeded by 99 percent of fasteners.

ASTM specifications are listed in Table 8–10. ASTM threads are shorter because ASTM deals mostly with structures; structural connections are generally loaded in shear, and the decreased thread length provides more shank area.

Specifications for metric fasteners are given in Table 8–11.

It is worth noting that all specification-grade bolts made in this country bear a manufacturer's mark or logo, in addition to the grade marking, on the bolt head. Such marks confirm that the bolt meets or exceeds specifications. If such marks are missing, the bolt may be imported; for imported bolts there is no obligation to meet specifications.

Bolts in fatigue axial loading fail at the fillet under the head, at the thread runout, and at the first thread engaged in the nut. If the bolt has a standard shoulder under the

Table 8–10

ASTM Specifications for Steel Bolts

ASTM Desig-nation No.	Size Range, Inclusive, in	Minimum Proof Strength,* kpsi	Minimum Tensile Strength,* kpsi	Minimum Yield Strength,* kpsi	Material	Head Marking
A307	$\frac{1}{4}$–$1\frac{1}{2}$	33	60	36	Low carbon	
A325, type 1	$\frac{1}{2}$–1	85	120	92	Medium carbon, Q&T	
	$1\frac{1}{8}$–$1\frac{1}{2}$	74	105	81		
A325, type 2	$\frac{1}{2}$–1	85	120	92	Low-carbon, martensite, Q&T	
	$1\frac{1}{8}$–$1\frac{1}{2}$	74	105	81		
A325, type 3	$\frac{1}{2}$–1	85	120	92	Weathering steel, Q&T	
	$1\frac{1}{8}$–$1\frac{1}{2}$	74	105	81		
A354, grade BC	$\frac{1}{4}$–$2\frac{1}{2}$	105	125	109	Alloy steel, Q&T	
	$2\frac{3}{4}$–4	95	115	99		
A354, grade BD	$\frac{1}{4}$–4	120	150	130	Alloy steel, Q&T	
A449	$\frac{1}{4}$–1	85	120	92	Medium-carbon, Q&T	
	$1\frac{1}{8}$–$1\frac{1}{2}$	74	105	81		
	$1\frac{3}{4}$–3	55	90	58		
A490, type 1	$\frac{1}{2}$–$1\frac{1}{2}$	120	150	130	Alloy steel, Q&T	
A490, type 3	$\frac{1}{2}$–$1\frac{1}{2}$	120	150	130	Weathering steel, Q&T	

*Minimum strengths are strengths exceeded by 99 percent of fasteners.

Table 8–11

Metric Mechanical-Property Classes for Steel Bolts, Screws, and Studs*

Property Class	Size Range, Inclusive	Minimum Proof Strength,[†] MPa	Minimum Tensile Strength,[†] MPa	Minimum Yield Strength,[†] MPa	Material	Head Marking
4.6	M5–M36	225	400	240	Low or medium carbon	4.6
4.8	M1.6–M16	310	420	340	Low or medium carbon	4.8
5.8	M5–M24	380	520	420	Low or medium carbon	5.8
8.8	M16–M36	600	830	660	Medium carbon, Q&T	8.8
9.8	M1.6–M16	650	900	720	Medium carbon, Q&T	9.8
10.9	M5–M36	830	1040	940	Low-carbon martensite, Q&T	10.9
12.9	M1.6–M36	970	1220	1100	Alloy, Q&T	12.9

*The thread length for bolts and cap screws is

$$L_T = \begin{cases} 2d+6 & L \le 125 \\ 2d+12 & 125 < L \le 200 \\ 2d+25 & L > 200 \end{cases}$$

where L is the bolt length. The thread length for structural bolts is slightly shorter than given above.

[†] Minimum strengths are strength exceeded by 99 percent of fasteners.

head, it has a value of K_f from 2.1 to 2.3, *and* this shoulder fillet is protected from scratching or scoring by a washer. If the thread runout has a 15° or less half-cone angle, the stress is higher at the first engaged thread in the nut. Bolts are sized by examining the loading at the plane of the washer face of the nut. This is the weakest part of the bolt *if and only if* the conditions above are satisfied (washer protection of the shoulder fillet and thread runout $\le 15°$). Inattention to this requirement has led to a record of 15 percent fastener fatigue failure under the head, 20 percent at thread runout, and 65 percent where the designer is focusing attention. It does little good to concentrate on the plane of the nut washer face if it is not the weakest location.

Nuts are graded so that they can be mated with their corresponding grade of bolt. The purpose of the nut is to have its threads deflect to distribute the load of the bolt more evenly to the nut. The nut's properties are controlled in order to accomplish this. The grade of the nut should be the grade of the bolt.

8–7 Tension Joints—The External Load

Let us now consider what happens when an external tensile load P, as in Fig. 8–13, is applied to a bolted connection. It is to be assumed, of course, that the clamping force, which we will call the *preload* F_i, has been correctly applied by tightening the nut *before P* is applied. The nomenclature used is:

F_i = preload

P = external tensile load

P_b = portion of P taken by bolt

P_m = portion of P taken by members

$F_b = P_b + F_i$ = resultant bolt load

$F_m = P_m - F_i$ = resultant load on members

C = fraction of external load P carried by bolt

$1 - C$ = fraction of external load P carried by members

The load P is tension, and it causes the connection to stretch, or elongate, through some distance δ. We can relate this elongation to the stiffnesses by recalling that k is the force divided by the deflection. Thus

$$\delta = \frac{P_b}{k_b} \qquad \text{and} \qquad \delta = \frac{P_m}{k_m} \qquad (a)$$

or

$$P_m = P_b \frac{k_m}{k_b} \qquad (b)$$

Since $P = P_b + P_m$, we have

$$P_b = \frac{k_b P}{k_b + k_m} = CP \qquad (c)$$

and

$$P_m = P - P_b = (1 - C)P \qquad (d)$$

where

$$C = \frac{k_b}{k_b + k_m} \qquad (e)$$

is called the *stiffness constant of the joint*. The resultant bolt load is

$$F_b = P_b + F_i = CP + F_i \qquad F_m < 0 \qquad (8\text{--}24)$$

Table 8–12

Computation of Bolt and Member Stiffnesses. Steel members clamped using a $\frac{1}{2}$ in-13 NC steel bolt. $C = \dfrac{k_b}{k_b + k_m}$

Bolt Grip, in	Stiffnesses, Mlb/in		C	1 − C
	k_b	k_m		
2	2.57	12.69	0.168	0.832
3	1.79	11.33	0.136	0.864
4	1.37	10.63	0.114	0.886

and the resultant load on the connected members is

$$F_m = P_m - F_i = (1 - C)P - F_i \qquad F_m < 0 \qquad (8\text{–}25)$$

Of course, these results are valid only as long as some clamping load remains in the members; this is indicated by the qualifier in the equations.

Table 8–12 is included to provide some information on the relative values of the stiffnesses encountered. The grip contains only two members, both of steel, and no washers. The ratios C and $1 - C$ are the coefficients of P in Eqs. (8–24) and (8–25), respectively. They describe the proportion of the external load taken by the bolt and by the members, respectively. In all cases, the members take over 80 percent of the external load. Think how important this is when fatigue loading is present. Note also that making the grip longer causes the members to take an even greater percentage of the external load.

8–8 Relating Bolt Torque to Bolt Tension

Having learned that a high preload is very desirable in important bolted connections, we must next consider means of ensuring that the preload is actually developed when the parts are assembled.

If the overall length of the bolt can actually be measured with a micrometer when it is assembled, the bolt elongation due to the preload F_i can be computed using the formula $\delta = F_i l / (AE)$. Then the nut is simply tightened until the bolt elongates through the distance δ. This ensures that the desired preload has been attained.

The elongation of a screw cannot usually be measured, because the threaded end is often in a blind hole. It is also impractical in many cases to measure bolt elongation. In such cases the wrench torque required to develop the specified preload must be estimated. Then torque wrenching, pneumatic-impact wrenching, or the turn-of-the-nut method may be used.

The torque wrench has a built-in dial that indicates the proper torque.

With impact wrenching, the air pressure is adjusted so that the wrench stalls when the proper torque is obtained, or in some wrenches, the air automatically shuts off at the desired torque.

The turn-of-the-nut method requires that we first define the meaning of snug-tight. The *snug-tight* condition is the tightness attained by a few impacts of an impact wrench, or the full effort of a person using an ordinary wrench. When the snug-tight condition is attained, all additional turning develops useful tension in the bolt. The turn-of-the-nut method requires that you compute the fractional number of turns necessary to develop the required preload from the snug-tight condition. For example, for heavy hexagonal structural bolts, the turn-of-the-nut specification states that the nut should be turned a minimum of 180° from the snug-tight condition under optimum conditions. Note that

Table 8–13

Distribution of Preload F_i for 20 Tests of Unlubricated Bolts Torqued to 90 N · m

| 23.6, | 27.6, | 28.0, | 29.4, | 30.3, | 30.7, | 32.9, | 33.8, | 33.8, | 33.8, |
| 34.7, | 35.6, | 35.6, | 37.4, | 37.8, | 37.8, | 39.2, | 40.0, | 40.5, | 42.7 |

*Mean value $\bar{F}_i = 34.3$ kN. Standard deviation, $\hat{\sigma} = 4.91$ kN.

this is also about the correct rotation for the wheel nuts of a passenger car. Problems 8–15 to 8–17 illustrate the method further.

Although the coefficients of friction may vary widely, we can obtain a good estimate of the torque required to produce a given preload by combining Eqs. (8–5) and (8–6):

$$T = \frac{F_i d_m}{2} \left(\frac{l + \pi f d_m \sec \alpha}{\pi d_m - f l \sec \alpha} \right) + \frac{F_i f_c d_c}{2} \tag{a}$$

Since $\tan \lambda = l/\pi d_m$, we divide the numerator and denominator of the first term by πd_m and get

$$T = \frac{F_i d_m}{2} \left(\frac{\tan \lambda + f \sec \alpha}{1 - f \tan \lambda \sec \alpha} \right) + \frac{F_i f_c d_c}{2} \tag{b}$$

The diameter of the washer face of a hexagonal nut is the same as the width across flats and equal to $1\frac{1}{2}$ times the nominal size. Therefore the mean collar diameter is $d_c = (d + 1.5d)/2 = 1.25d$. Equation (b) can now be arranged to give

$$T = \left[\left(\frac{d_m}{2d} \right) \left(\frac{\tan \lambda + f \sec \alpha}{1 - f \tan \lambda \sec \alpha} \right) + 0.625 f_c \right] F_i d \tag{c}$$

We now define a *torque coefficient K* as the term in brackets, and so

$$K = \left(\frac{d_m}{2d} \right) \left(\frac{\tan \lambda + f \sec \alpha}{1 - f \tan \lambda \sec \alpha} \right) + 0.625 f_c \tag{8–26}$$

Equation (c) can now be written

$$T = K F_i d \tag{8–27}$$

The coefficient of friction depends upon the surface smoothness, accuracy, and degree of lubrication. On the average, both f and f_c are about 0.15. The interesting fact about Eq. (8–26) is that $K \doteq 0.20$ for $f = f_c = 0.15$ no matter what size bolts are employed and no matter whether the threads are coarse or fine.

Blake and Kurtz have published results of numerous tests of the torquing of bolts.[6] By subjecting their data to a statistical analysis, we can learn something about the distribution of the torque coefficients and the resulting preload. Blake and Kurtz determined the preload in quantities of unlubricated and lubricated bolts of size $\frac{1}{2}$ in-20 UNF when torqued to 800 lbf · in. This corresponds roughly to an M12 × 1.25 bolt torqued to 90 N · m. The statistical analyses of these two groups of bolts, converted to SI units, are displayed in Tables 8–13 and 8–14.

[6]J. C. Blake and H. J. Kurtz, "The Uncertainties of Measuring Fastener Preload," *Machine Design*, vol. 37, Sept. 30, 1965, pp. 128–131.

Table 8–14

Distribution of Preload F_i for 10 Tests of Lubricated Bolts Torqued to 90 N · m

| 30.3, | 32.5, | 32.5, | 32.9, | 32.9, | 33.8, | 34.3, | 34.7, | 37.4, | 40.5 |

*Mean value $\overline{F}_i = 34.18$ kN. Standard deviation, $\hat{\sigma} = 2.88$ kN.

Table 8–15

Torque Factors K for Use with Eq. (8–27)

Bolt Condition	K
Nonplated, black finish	0.30
Zinc-plated	0.20
Lubricated	0.18
Cadmium-plated	0.16
With Bowman Anti-Seize	0.12
With Bowman-Grip nuts	0.09

We first note that both groups have about the same mean preload, 34 kN. The unlubricated bolts have a standard deviation of 4.9 kN and a COV of about 0.15. The lubricated bolts have a standard deviation of 3 kN and a COV of about 0.9.

The means obtained from the two samples are nearly identical, approximately 34 kN; using Eq. (8–27), we find, for both samples, $K = 0.208$.

Bowman Distribution, a large manufacturer of fasteners, recommends the values shown in Table 8–15. In this book we shall use these values and use $K = 0.2$ when the bolt condition is not stated.

EXAMPLE 8–3

A $\frac{3}{4}$ in-16 UNF × $2\frac{1}{2}$ in SAE grade 5 bolt is subjected to a load P of 6 kip in a tension joint. The initial bolt tension is $F_i = 25$ kip. The bolt and joint stiffnesses are $k_b = 6.50$ and $k_m = 13.8$ Mlb/in, respectively.

(*a*) Determine the preload and service load stresses in the bolt. Compare these to the SAE minimum proof strength of the bolt.

(*b*) Specify the torque necessary to develop the preload, using Eq. (8–27).

(*c*) Specify the torque necessary to develop the preload, using Eq. (8–26) with $f = f_c = 0.15$.

Solution

From Table 8–2, $A_t = 0.373$ in^2.

(*a*) The preload stress is

Answer

$$\sigma_i = \frac{F_i}{A_t} = \frac{25}{0.373} = 67.02 \text{ kpsi}$$

The stiffness constant is

$$C = \frac{k_b}{k_b + k_m} = \frac{6.5}{6.5 + 13.8} = 0.320$$

From Eq. (8–24), the stress under the service load is

$$\sigma_b = \frac{F_b}{A_t} = \frac{CP + F_i}{A_t} = C\frac{P}{A_t} + \sigma_i$$

Answer

$$= 0.320\frac{6}{0.373} + 67.02 = 72.17 \text{ kpsi}$$

From Table 8–10, the SAE minimum proof strength of the bolt is $S_p = 85$ kpsi. The preload and service load stresses are respectively 21 and 15 percent less than the proof strength.

(*b*) From Eq. (8–27), the torque necessary to achieve the preload is

Answer

$$T = K F_i d = 0.2(25)(10^3)(0.75) = 3750 \text{ lbf} \cdot \text{in}$$

(*c*) The minor diameter can be determined from the minor area in Table 8–2. Thus $d_r = \sqrt{4A_r/\pi} = \sqrt{4(0.351)/\pi} = 0.6685$ in. Thus, the mean diameter is $d_m = (0.75 + 0.6685)/2 = 0.7093$ in. The lead angle is

$$\lambda = \tan^{-1}\frac{l}{\pi d_m} = \tan^{-1}\frac{1}{\pi d_m N} = \tan^{-1}\frac{1}{\pi(0.7093)(16)} = 1.6066°$$

For $\alpha - 30°$, Eq. (8–26) gives

$$T = \left\{\left[\frac{0.7093}{2(0.75)}\right]\left[\frac{\tan 1.6066° + 0.15(\sec 30°)}{1 - 0.15(\tan 1.6066°)(\sec 30°)}\right] + 0.625(0.15)\right\} 25(10^3)(0.75)$$

$$= 3551 \text{ lbf} \cdot \text{in}$$

which is 5.3 percent less than the value found in part (*b*).

8–9 Statically Loaded Tension Joint with Preload

Equations (8–24) and (8–25) represent the forces in a bolted joint with preload. The tensile stress in the bolt can be found as in Ex. 8–3 as

$$\sigma_b = \frac{CP}{A_t} + \frac{F_i}{A_t} \tag{a}$$

The limiting value of σ_b is the proof strength S_p. Thus, with the introduction of a *load factor n*, Eq. (*a*) becomes

$$\frac{CnP}{A_t} + \frac{F_i}{A_t} = S_p \tag{b}$$

or

$$n = \frac{S_p A_t - F_i}{CP} \tag{8–28}$$

Here we have called *n* a load factor rather than a factor of safety, though the two ideas are somewhat related. Any value of $n > 1$ in Eq. (8–28) ensures that the bolt stress is less than the proof strength.

Another means of ensuring a safe joint is to require that the external load be smaller than that needed to cause the joint to separate. If separation does occur, then the entire external load will be imposed on the bolt. Let P_0 be the value of the external load that would cause joint separation. At separation, $F_m = 0$ in Eq. (8–25), and so

$$(1 - C)P_0 - F_i = 0 \tag{c}$$

Let the factor of safety against joint separation be

$$n_0 = \frac{P_0}{P} \tag{d}$$

Substituting $P_0 = n_0 P$ in Eq. (c), we find

$$n_0 = \frac{F_i}{P(1 - C)} \tag{8–29}$$

as a load factor guarding against joint separation.

Figure 8–18 is the stress-strain diagram of a good-quality bolt material. Notice that there is no clearly defined yield point and that the diagram progresses smoothly up to fracture, which corresponds to the tensile strength. This means that no matter how much preload is given the bolt, it will retain its load-carrying capacity. This is what keeps the bolt tight and determines the joint strength. The pre-tension is the "muscle" of the joint, and its magnitude is determined by the bolt strength. If the full bolt strength is not used in developing the pre-tension, then money is wasted and the joint is weaker.

Good-quality bolts can be preloaded into the plastic range to develop more strength. Some of the bolt torque used in tightening produces torsion, which increases the principal tensile stress. However, this torsion is held only by the friction of the bolt head and nut; in time it relaxes and lowers the bolt tension slightly. Thus, as a rule, a bolt will either fracture during tightening, or not at all.

Above all, do not rely too much on wrench torque; it is not a good indicator of preload. Actual bolt elongation should be used whenever possible—especially with fatigue loading. In fact, if high reliability is a requirement of the design, then preload should always be determined by bolt elongation.

Russell, Burdsall & Ward Inc. (RB&W) recommendations for preload are 60 kpsi for SAE grade 5 bolts for nonpermanent connections, and that A325 bolts (equivalent to SAE grade 5) used in structural applications be tightened to proof load or beyond

Figure 8–18

Typical stress-strain diagram for bolt materials showing proof strength S_p, yield strength S_y, and ultimate tensile strength S_{ut}.

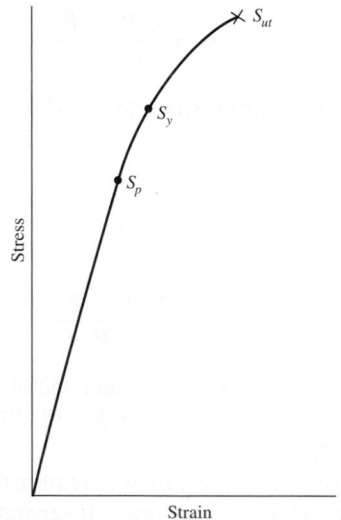

(85 kpsi up to a diameter of 1 in).[7] Bowman[8] recommends a preload of 75 percent of proof load, which is about the same as the RB&W recommendations for reused bolts. In view of these guidelines, it is recommended for both static and fatigue loading that the following be used for preload:

$$F_i = \begin{cases} 0.75 F_p & \text{for nonpermanent connections, reused fasteners} \\ 0.90 F_p & \text{for permanent connections} \end{cases} \tag{8-30}$$

where F_p is the proof load, obtained from the equation

$$F_p = A_t S_p \tag{8-31}$$

Here S_p is the proof strength obtained from Tables 8–9 to 8–11. For other materials, an approximate value is $S_p = 0.85 S_y$. Be very careful not to use a soft material in a threaded fastener. For high-strength steel bolts used as structural steel connectors, if advanced tightening methods are used, tighten to yield.

You can see that the RB&W recommendations on preload are in line with what we have encountered in this chapter. The purposes of development were to give the reader the perspective to appreciate Eqs. (8–30) and a methodology with which to handle cases more specifically than the recommendations.

---

[7]Russell, Burdsall & Ward Inc., *Helpful Hints for Fastener Design and Application,* Mentor, Ohio, 1965, p. 42.

[8]Bowman Distribution–Barnes Group, *Fastener Facts,* Cleveland, 1985, p. 90.

EXAMPLE 8-4

Figure 8–19 is a cross section of a grade 25 cast-iron pressure vessel. A total of N bolts are to be used to resist a separating force of 36 kip.

(*a*) Determine k_b, k_m, and C.

(*b*) Find the number of bolts required for a load factor of 2 where the bolts may be reused when the joint is taken apart.

Solution

(*a*) The grip is $L_G = 1.50$ in. From Table A–31, the nut thickness is $\frac{35}{64}$ in. Adding two threads beyond the nut of $\frac{2}{11}$ in gives a bolt length of

$$L = \frac{35}{64} + 1.50 + \frac{2}{11} = 2.229 \text{ in}$$

| **Figure 8–19**

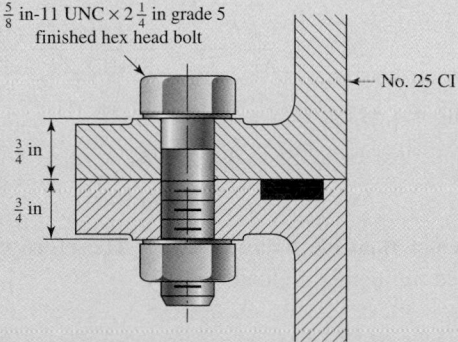

$\frac{5}{8}$ in-11 UNC $\times$ 2 $\frac{1}{4}$ in grade 5 finished hex head bolt

No. 25 CI

$\frac{3}{4}$ in

$\frac{3}{4}$ in

From Table A–17 the next fraction size bolt is $L = 2\frac{1}{4}$ in. From Eq. (8–13), the thread length is $L_T = 2(0.625) + 0.25 = 1.50$ in. Thus the length of the unthreaded portion in the grip is $l_d = 2.25 - 1.50 = 0.75$ in. The threaded length in the grip is $l_t = L_G - l_d = 0.75$ in. From Table 8–2, $A_t = 0.226$ in^2. The major-diameter area is $A_d = \pi(0.625)^2/4 = 0.3068$ in^2. The bolt stiffness is then

Answer

$$k_b = \frac{A_d A_t E}{A_d l_t + A_t l_d} = \frac{0.3068(0.226)(30)}{0.3068(0.75) + 0.226(0.75)}$$

$$= 5.21 \text{ Mlbf/in}$$

From Table A–24, for no. 25 cast iron we will use $E = 14$ Mpsi. The stiffness of the members, from Eq. (8–22), is

Answer

$$k_m = \frac{0.5774\pi E d}{2 \ln\left(5\dfrac{0.5774l + 0.5d}{0.5774l + 2.5d}\right)} = \frac{0.5774\pi(14)(0.625)}{2 \ln\left[5\dfrac{0.5774(1.5) + 0.5(0.625)}{0.5774(1.5) + 2.5(0.625)}\right]}$$

$$= 8.95 \text{ Mlbf/in}$$

If you are using Eq. (8–23), from Table 8–8, $A = 0.778\,71$ and $B = 0.616\,16$, and

$$k_m = EdA \, \exp(Bd/l)$$

$$= 14(0.625)(0.778\,71) \, \exp[0.616\,16(0.625)/1.5]$$

$$= 8.81 \text{ Mlbf/in}$$

which is only 1.6 percent lower than the previous result.

From the first calculation for k_m, the stiffness constant C is

Answer

$$C = \frac{k_b}{k_b + k_m} = \frac{5.21}{5.21 + 8.95} = 0.368$$

(*b*) From Table 8–9, $S_p = 85$ kpsi. Then, using Eqs. (8–30) and (8–31), we find the recommended preload to be

$$F_i = 0.75 A_t S_p = 0.75(0.226)(85) = 14.4 \text{ kip}$$

For *N* bolts, Eq. (8–28) can be written

$$n = \frac{S_p A_t - F_i}{C(P/N)} \tag{1}$$

or

$$N = \frac{CnP}{S_p A_t - F_i} = \frac{0.368(2)(36)}{85(0.226) - 14.4} = 5.52$$

With six bolts, Eq. (1) gives

$$n = \frac{85(0.226) - 14.4}{0.368(36/6)} = 2.18$$

which is greater than the required value. Therefore we choose six bolts and use the recommended tightening preload.

8–10 Gasketed Joints

If a full gasket is present in the joint, the gasket pressure p is found by dividing the force in the member by the gasket area per bolt. Thus, for N bolts,

$$p = -\frac{F_m}{A_g/N} \qquad (a)$$

With a load factor n, Eq. (8–25) can be written as

$$F_m = (1 - C)nP - F_i \qquad (b)$$

Substituting this into Eq. (a) gives the gasket pressure as

$$p = [F_i - nP(1 - C)]\frac{N}{A_g} \qquad (8\text{–}32)$$

In full-gasketed joints uniformity of pressure on the gasket is important. To maintain adequate uniformity of pressure adjacent bolts should not be placed more than six nominal diameters apart on the bolt circle. To maintain wrench clearance, bolts should be placed at least three diameters apart. A rough rule for bolt spacing around a bolt circle is

$$3 \le \frac{\pi D_b}{Nd} \le 6 \qquad (8\text{–}33)$$

where D_b is the diameter of the bolt circle and N is the number of bolts.

8–11 Fatigue Loading of Tension Joints

Tension-loaded bolted joints subjected to fatigue action can be analyzed directly by the methods of Chap. 7. Table 8–16 lists average fatigue stress-concentration factors for the fillet under the bolt head and also at the beginning of the threads on the bolt shank. These are already corrected for notch sensitivity and for surface finish. Designers should be aware that situations may arise in which it would be advisable to investigate these factors more closely, since they are only average values. In fact, Peterson[9] observes that the distribution of typical bolt failures is about 15 percent under the head, 20 percent at the end of the thread, and 65 percent in the thread at the nut face.

Use of rolled threads is the predominant method of thread-forming in screw fasteners, where Table 8–16 applies. In thread-rolling the amount of cold work and strain-strengthening is unknown to the designer; therefore, fully corrected (including K_f) axial endurance strength is reported in Table 8–17. For cut threads, the methods of Chap. 7 are useful. Anticipate that the endurance strengths will be considerably lower.

Most of the time, the type of fatigue loading encountered in the analysis of bolted joints is one in which the externally applied load fluctuates between zero and some

Table 8–16

Fatigue Stress-Concentration Factors K_f for Threaded Elements

SAE Grade	Metric Grade	Rolled Threads	Cut Threads	Fillet
0 to 2	3.6 to 5.8	2.2	2.8	2.1
4 to 8	6.6 to 10.9	3.0	3.8	2.3

[9]R. E. Peterson, *Stress Concentration Factors*, Wiley, New York, 1974, p. 253.

Table 8–17

Fully Corrected Endurance Strengths for Bolts and Screws with Rolled Threads*

Grade or Class	Size Range	Endurance Strength
SAE 5	$\frac{1}{4}$–1 in	18.6 kpsi
	$1\frac{1}{8}$–$1\frac{1}{2}$ in	16.3 kpsi
SAE 7	$\frac{1}{4}$–$1\frac{1}{2}$ in	20.6 kpsi
SAE 8	$\frac{1}{4}$–$1\frac{1}{2}$ in	23.2 kpsi
ISO 8.8	M16–M36	129 MPa
ISO 9.8	M1.6–M16	140 MPa
ISO 10.9	M5–M36	162 MPa
ISO 12.9	M1.6–M36	190 MPa

*Repeatedly-applied, axial loading, fully corrected.

Figure 8–20

Designer's fatigue diagram showing a Goodman failure locus and how a load line is used to define failure and safety in preloaded bolted joints in fatigue. Point B represents nonfailure; point C, failure.

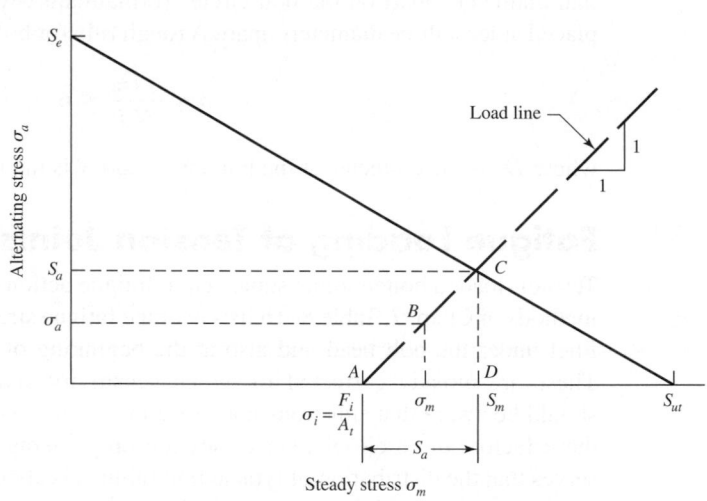

maximum force P. This would be the situation in a pressure cylinder, for example, where a pressure either exists or does not exist. For such cases, $F_{max} = F_b$ and $F_{min} = F_i$ and the alternating component of the force is $F_a = (F_{max} - F_{min})/2 = (F_b - F_i)/2$. Dividing this by A_t yields the alternating component of the bolt stress. Employing the notation from Sec. 8–7 with Eq. (8–24), we obtain

$$\sigma_a = \frac{F_b - F_i}{2A_t} = \frac{(CP + F_i) - F_i}{2A_t} = \frac{CP}{2A_t} \qquad (8\text{–}34)$$

The mean stress is equal to the alternating component plus the minimum stress, $\sigma_i = F_i/A_t$, which results in

$$\sigma_m = \frac{CP}{2A_t} + \frac{F_i}{A_t} \qquad (8\text{–}35)$$

On the designer's fatigue diagram, shown in Fig. 8–20, the load line is

$$\sigma_m = \sigma_a + \sigma_i \qquad (8\text{–}36)$$

The next problem is to find the strength components S_a and S_m of the fatigue failure locus. These depend on the failure criteria:

Goodman:

$$\frac{S_a}{S_e} + \frac{S_m}{S_{ut}} = 1 \tag{8–37}$$

Gerber:

$$\frac{S_a}{S_e} + \left(\frac{S_m}{S_{ut}}\right)^2 = 1 \tag{8–38}$$

ASME-elliptic:

$$\left(\frac{S_a}{S_e}\right)^2 + \left(\frac{S_m}{S_p}\right)^2 = 1 \tag{8–39}$$

For simultaneous solution between Eq. (8–36), as $S_m = S_a + \sigma_i$, and each of Eqs. (8–37) to (8–39) gives

Goodman:

$$S_a = \frac{S_e(S_{ut} - \sigma_i)}{S_{ut} + S_e} \tag{8–40}$$

$$S_m = S_a + \sigma_i \tag{8–41}$$

Gerber:

$$S_a = \frac{1}{2S_e}\left[S_{ut}\sqrt{S_{ut}^2 + 4S_e(S_e + \sigma_i)} - S_{ut}^2 - 2\sigma_i S_e \right] \tag{8–42}$$

$$S_m = S_a + \sigma_i$$

ASME-elliptic:

$$S_a = \frac{S_e}{S_p^2 + S_e^2}\left(S_p\sqrt{S_p^2 + S_e^2 - \sigma_i^2} - \sigma_i S_e \right) \tag{8–43}$$

$$S_m = S_a + \sigma_i$$

When using relations of this section, be sure to use K_f for both σ_a and σ_m. Otherwise, the slope of the load line will not remain 1 to 1.

Examination of Eqs. (8–37) to (8–43) shows parametric equations that relate the coordinates of interest to the form of the criteria. The factor of safety guarding against fatigue is given by

$$n_f = \frac{S_a}{\sigma_a} \tag{8–44}$$

Applying this to the Goodmen criterion, for example, with Eqs. (8–34) and (8–40) and $\sigma_i = F_i/A_t$ gives

$$n_f = \frac{2S_e(S_{ut}A_t - F_i)}{CP(S_{ut} + S_e)} \tag{8–45}$$

when preload F_i is present. With no preload, $C = 1$, $F_i = 0$, and Eq. (8–45) becomes

$$n_{f0} = \frac{2S_e S_{ut} A_t}{P(S_{ut} + S_e)} \tag{8–46}$$

Preload is beneficial for resisting fatigue when n_f/n_{f0} is greater than unity. For Goodman, Eqs. (8–45) and (8–46) with $n_f/n_{f0} \geq 1$ puts an upper bound on the preload F_i of

$$F_i \leq (1 - C)S_{ut} A_t \tag{8–47}$$

If this cannot be achieved, and n_f is unsatisfactory, use the Gerber or ASME-elliptic criterion to obtain a less conservative assessment. If the design is still not satisfactory, additional bolts and/or a different size bolt may be called for. Bolts loosen, as they are friction devices, and cyclic loading and vibration as well as other effects allow the fasteners to lose tension with time. How does one fight loosening? Within strength limitations, the higher the preload the better. A rule of thumb is that preloads of 60 percent of proof load rarely loosen. If more is better, how much more? Well, not enough to create reused fasteners as a future threat. Alternatively, fastener-locking schemes can be employed.

After solving Eq. (8–44), you should also check the possibility of yielding, using the proof strength

$$n_p = \frac{S_p}{\sigma_m + \sigma_a} \tag{8–48}$$

EXAMPLE 8–5

Figure 8–21 shows a connection using cap screws. The joint is subjected to a fluctuating force whose maximum value is 5 kip per screw. The required data are: cap screw, 5/8 in-11 NC, SAE 5; hardened-steel washer, $t_w = \frac{1}{16}$ in thick; steel cover plate, $t_1 = \frac{5}{8}$ in, $E_s = 30$ Mpsi; and cast-iron base, $t_2 = \frac{5}{8}$ in, $E_{ci} = 16$ Mpsi.
(a) Find k_b, k_m, and C using the assumptions given in the caption of Fig. 8–21.
(b) Find all factors of safety and explain what they mean.

Solution

(a) For the symbols of Figs. 8–15 and 8–21, $h = t_1 + t_w = 0.6875$ in, $l = h + d/2 = 1$ in, and $D_2 = 1.5d = 0.9375$ in. The joint is composed of three frusta; the upper two frusta are steel and the lower one is cast iron.

For the upper frustum: $t = l/2 = 0.5$ in, $D = 0.9375$ in, and $E = 30$ Mpsi. Using these values in Eq. (8–20) gives $k_1 = 46.46$ Mlbf/in.

Figure 8–21

Pressure-cone frustum member model for a cap screw. For this model the significant sizes are

$$l = \begin{cases} h + t_2/2 & t_2 < d \\ h + d/2 & t_2 \geq d \end{cases}$$

$D_1 = d_w + l \tan \alpha = $
$1.5d + 0.577l$
$D_2 = d_w = 1.5d$
where l = effective grip. The solutions are for $\alpha = 30°$ and $d_w = 1.5d$.

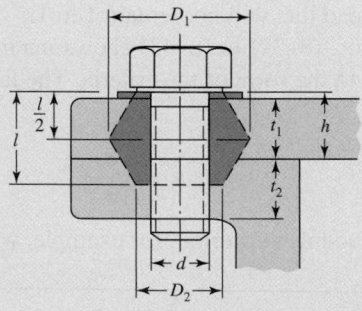

For the middle frustum: $t = h - l/2 = 0.1875$ in and $D = 0.9375 + 2(l - h)$ $\tan 30° = 1.298$ in. With these and $E_s = 30$ Mpsi, Eq. (8–20) gives $k_2 = 197.43$ Mlbf/in.

The lower frustum has $D = 0.9375$ in, $t = l - h = 0.3125$ in, and $E_{ci} = 16$ Mpsi. The same equation yields $k_3 = 32.39$ Mlbf/in.

Substituting these three stiffnesses into Eq. (8–18) gives $k_m = 17.40$ Mlbf/in. The cap screw is short and threaded all the way. Using $l = 1$ in for the grip and $A_t = 0.226$ in^2 from Table 8–2, we find the stiffness to be $k_b = A_t E / l = 6.78$ Mlbf/in. Thus the joint constant is

Answer
$$C = \frac{k_b}{k_b + k_m} = \frac{6.78}{6.78 + 17.40} = 0.280$$

(*b*) Equation (8–30) gives the preload as

$$F_i = 0.75 F_p = 0.75 A_t S_p = 0.75(0.226)(85) = 14.4 \text{ kip}$$

where from Table 8–9, $S_p = 85$ kpsi for an SAE grade 5 cap screw. Using Eq. (8–28), we obtain the load factor as

Answer
$$n = \frac{S_p A_t - F_i}{C P} = \frac{85(0.226) - 14.40}{0.280(5)} = 3.44$$

This factor prevents the bolt stress from becoming equal to the proof strength.

Next, using Eq. (8–29), we have

Answer
$$n_0 = \frac{F_i}{P(1 - C)} = \frac{14.40}{5(1 - 0.280)} = 4.00$$

If the force P gets too large, the joint will separate and the bolt will take the entire load. This factor guards against that event.

For the remaining factors, refer to Fig. 8–22. This diagram contains the modified Goodman line, the Gerber line, the proof-strength line, and the load line. The intersection

Figure 8–22

Designer's fatigue diagram for preloaded bolts, drawn to scale, showing the modified Goodman locus, the Gerber locus, and the Langer proof-strength locus, with an exploded view of the area of interest. The strengths used are $S_p = 85$ kpsi, $S_e = 18.6$ kpsi, and $S_{ut} = 120$ kpsi. The coordinates are A, $\sigma_i = 63.72$ kpsi; B, $\sigma_a = 3.10$ kpsi, $\sigma_m = 66.82$ kpsi; C, $S_a = 7.55$ kpsi, $S_m = 71.29$ kpsi; D, $S_a = 10.64$ kpsi, $S_m = 74.36$ kpsi; E, $S_a = 11.32$ kpsi, $S_m = 75.04$ kpsi.

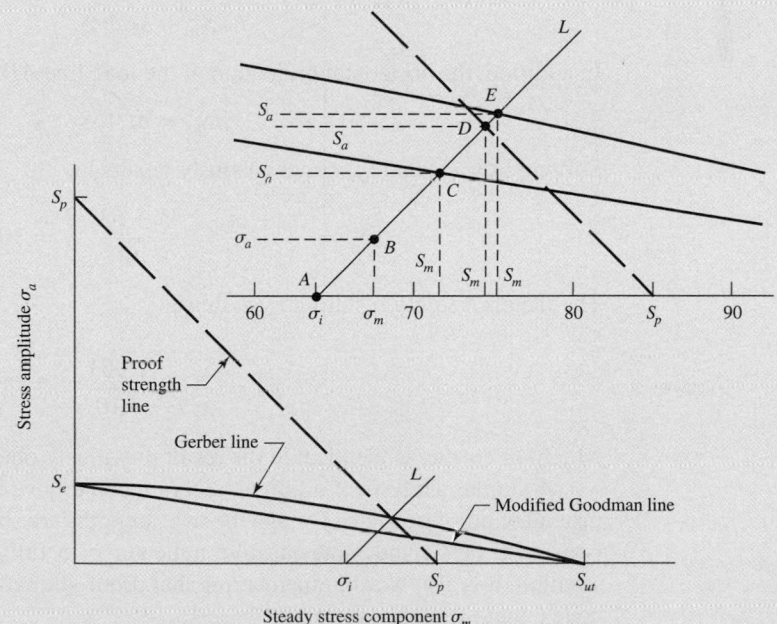

of the load line L with the respective failure lines at points C, D, and E defines a set of strengths S_a and S_m at each intersection. Point B represents the stress state σ_a, σ_m. Point A is the preload stress σ_i. Therefore the load line begins at A and makes an angle having a unit slope. This angle is $45°$ only when both stress axes have the same scale.

The factors of safety are found by dividing the distances AC, AD, and AE by the distance AB. Note that this is the same as dividing S_a for each theory by σ_a.

The quantities shown in the caption of Fig. 8–22 are obtained as follows:

Point A

$$\sigma_i = \frac{F_i}{A_t} = \frac{14.40}{0.226} = 63.72 \text{ kpsi}$$

Point B

$$\sigma_a = \frac{CP}{2A_t} = \frac{0.280(5)}{2(0.226)} = 3.10 \text{ kpsi}$$

$$\sigma_m = \sigma_a + \sigma_i = 3.10 + 63.72 = 66.82 \text{ kpsi}$$

Point C
This is the modified Goodman criteria. From Table 8–17, we find $S_e = 18.6$ kpsi. Then, using Eq. (8–40), we get

$$S_a = \frac{S_e(S_{ut} - \sigma_i)}{S_{ut} + S_e} = \frac{18.6(120 - 63.72)}{120 + 18.6} = 7.55 \text{ kpsi}$$

The factor of safety is found to be

Answer
$$n_f = \frac{S_a}{\sigma_a} = \frac{7.55}{3.10} = 2.44$$

Point D
This is on the proof-strength line where

$$S_m + S_a = S_p \tag{1}$$

In addition, the horizontal projection of the load line AD is

$$S_m = \sigma_i + S_a \tag{2}$$

Solving Eqs. (1) and (2) simultaneously results in

$$S_a = \frac{S_p - \sigma_i}{2} = \frac{85 - 63.72}{2} = 10.64 \text{ kpsi}$$

The factor of safety resulting from this is

Answer
$$n_p = \frac{S_a}{\sigma_a} = \frac{10.64}{3.10} = 3.43$$

which, of course, is identical to the result previously obtained by using Eq. (8–28).

A similar analysis of a fatigue diagram could have been done using yield strength instead of proof strength. Though the two strengths are somewhat related, proof strength is a much better and more positive indicator of a fully loaded bolt than is the yield strength. It is also worth remembering that proof-strength values are specified in design codes; yield strengths are not.

We found $n_f = 2.44$ on the basis of fatigue and the modified Goodman line, and $n_p = 3.43$ on the basis of proof strength. Thus the danger of failure is by fatigue, not by overproof loading. These two factors should always be compared to determine where the greatest danger lies.

Point E

For the Gerber criterion, from Eq. (8–42),

$$S_a = \frac{1}{2S_e}\left[S_{ut}\sqrt{S_{ut}^2 + 4S_e(S_e + \sigma_i)} - S_{ut}^2 - 2\sigma_i S_e\right]$$

$$= \frac{1}{2(18.6)}\left[120\sqrt{120^2 + 4(18.6)(18.6 + 63.72)} - 120^2 - 2(63.72)(18.6)\right]$$

$$= 11.33 \text{ kpsi}$$

Thus for the Gerber criterion the safety factor is

Answer

$$n_f = \frac{S_a}{\sigma_a} = \frac{11.33}{3.10} = 3.65$$

which is greater than $n_p = 3.43$ and contradicts the conclusion earlier that the danger of failure is fatigue. Figure 8–22 clearly shows the conflict where point D lies between points C and E. Again, the conservative nature of the Goodman criterion explains the discrepancy and the designer must form his or her own conclusion.

8–12   Shear Joints

Joints can and should be loaded in shear so that the fasteners see no additional stress beyond the initial tightening. The shear loading is resisted in two principal ways:

- The shear load is carried by friction between the members and ensured by the clamping action of the bolts or cap screws. Should the friction be insufficient, the shear load is carried by only two of the fasteners in the pattern. This occurs because errors in hole size and placement preclude a uniform sharing of the shear load. The analysis problem involves identifying the two fasteners that represent the worst case.

- The shear load is carried by dowel pins in reamed holes, placed in both parts while clamped together to ensure alignment. The dowels will carry the shear load. Pins are often tapered to allow firm setting and easy removal if necessary.

In days when driving hot rivets was a common structural joining method, the driving ensured that the rivets filled every hole completely and, upon cooling, provided a clamping preload. This kind of rivet pattern can share a shear load.

Integral to the analysis of a shear joint is locating the center of relative motion between the two members. In Fig. 8–23 let A_1 to A_5 be the respective cross-sectional areas of a group of five pins, or hot-driven rivets, or tight-fitting shoulder bolts. Under this assumption the rotational pivot point lies at the centroid of the cross-sectional area pattern of the pins, rivets, or bolts. Using statics, we learn that the centroid G is located

Figure 8–23

Centroid of pins, rivets, or bolts.

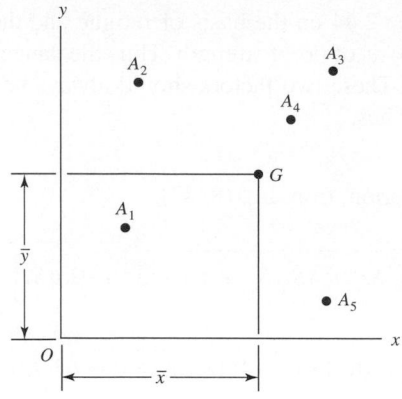

Figure 8–24

(a) Beam bolted at both ends with distributed load; (b) free-body diagram of beam; (c) enlarged view of bolt group centered at O showing primary and secondary resultant shear forces.

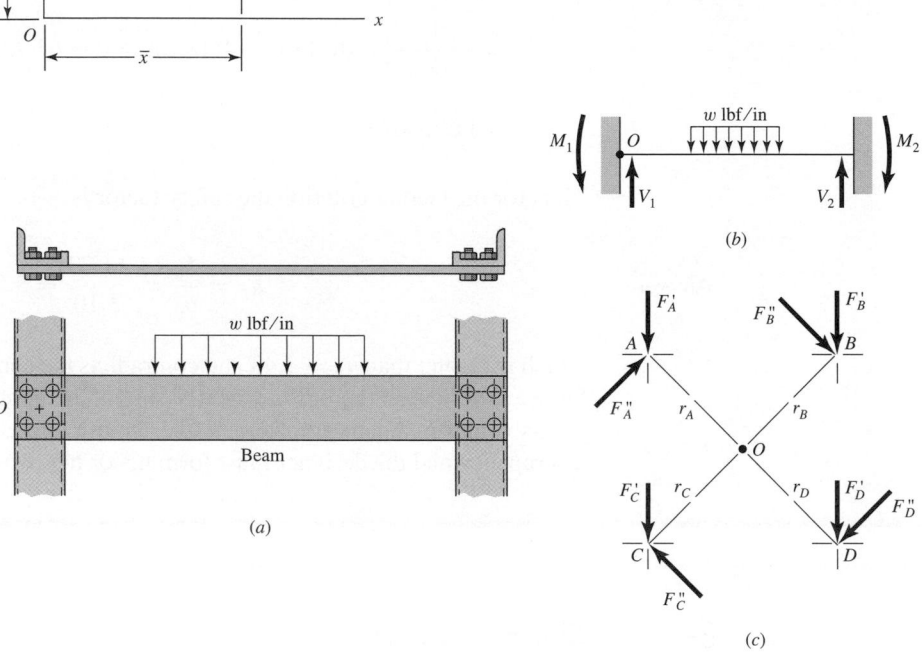

by the coordinates $\bar{x}$ and $\bar{y}$, where x_1 and y_i are the distances to the ith area center:

$$\bar{x} = \frac{A_1 x_1 + A_2 x_2 + A_3 x_3 + A_4 x_4 + A_5 x_5}{A_1 + A_2 + A_3 + A_4 + A_5} = \frac{\sum_1^n A_i x_i}{\sum_1^n A_i}$$

$$\bar{y} = \frac{A_1 y_1 + A_2 y_2 + A_3 y_3 + A_4 y_4 + A_5 y_5}{A_1 + A_2 + A_3 + A_4 + A_5} = \frac{\sum_1^n A_i y_i}{\sum_1^n A_i}$$

(8–49)

In many instances the centroid can be located by symmetry.

An example of eccentric loading of fasteners is shown in Fig. 8–24. This is a portion of a machine frame containing a beam subjected to the action of a bending load. In this case, the beam is fastened to vertical members at the ends with specially prepared load-sharing bolts. You will recognize the schematic representation in Fig. 8–24b as a statically indeterminate beam with both ends fixed and with moment and shear reactions at each end.

For convenience, the centers of the bolts at the left end of the beam are drawn to a larger scale in Fig. 8–24c. Point O represents the centroid of the group, and it is assumed in this example that all the bolts are of the same diameter. Note that the forces shown in Fig. 8–24c are the *resultant* forces acting on the pins with a net force and moment equal and opposite to the *reaction* loads V_1 and M_1 acting at O. The total load taken by each bolt will be calculated in three steps. In the first step the shear V_1 is divided equally among the bolts so that each bolt takes $F' = V_1/n$, where n refers to the number of bolts in the group and the force F' is called the *direct load,* or *primary shear.*

It is noted that an equal distribution of the direct load to the bolts assumes an absolutely rigid member. The arrangement of the bolts or the shape and size of the members sometimes justifies the use of another assumption as to the division of the load. The direct loads F' are shown as vectors on the loading diagram (Fig. 8–24c).

The *moment load,* or *secondary shear,* is the additional load on each bolt due to the moment M_1. If r_A, r_B, r_C, etc., are the radial distances from the centroid to the center of each bolt, the moment and moment loads are related as follows:

$$M_1 = F''_A r_A + F''_B r_B + F''_C r_C + \cdots \tag{a}$$

where the F'' are the moment loads. The force taken by each bolt depends upon its radial distance from the centroid; that is, the bolt farthest from the centroid takes the greatest load, while the nearest bolt takes the smallest. We can therefore write

$$\frac{F''_A}{r_A} = \frac{F''_B}{r_B} = \frac{F''_C}{r_C} \tag{b}$$

where again, the diameters of the bolts are assumed equal. If not, then one replaces F'' in Eq. (b) with the shear stresses $\tau'' = 4F''/\pi d^2$ for each bolt. Solving Eqs. (a) and (b) simultaneously, we obtain

$$F''_n = \frac{M_1 r_n}{r_A^2 + r_B^2 + r_C^2 + \cdots} \tag{8–50}$$

where the subscript n refers to the particular bolt whose load is to be found. These moment loads are also shown as vectors on the loading diagram.

In the third step the direct and moment loads are added vectorially to obtain the resultant load on each bolt. Since all the bolts or rivets are usually the same size, only that bolt having the maximum load need be considered. When the maximum load is found, the strength may be determined by using the various methods already described.

EXAMPLE 8–6

Shown in Fig. 8–25 is a 15- by 200-mm rectangular steel bar cantilevered to a 250-mm steel channel using two pins, located at E and F, and four bolts located at A, B, C, and D.

Figure 8–25

Dimensions in millimeters. O' and $\bar{x}$ are used in part (b) of Ex. 8–6.

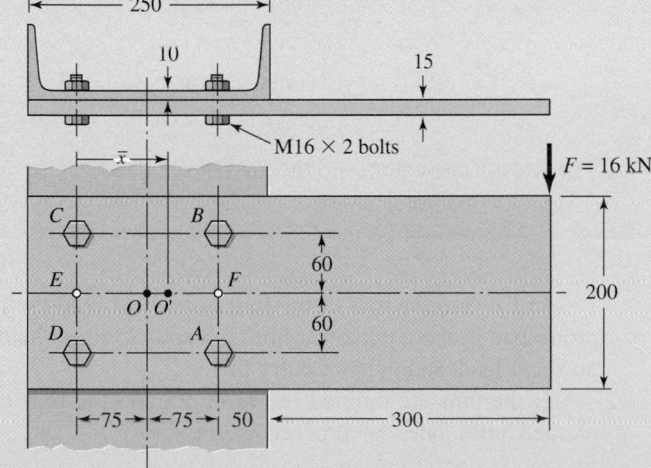

(*a*) On the basis of a steady external load of 16 kN, find the shear loads in the *pins* should the clamping action fail.

(*b*) Investigate increasing the area of pin *F* to obtain equal shear stresses.

Solution

(*a*) The point *O*, the centroid of the pin cross-sectional areas, is found by symmetry. If a free-body diagram of the beam is considered, the shear reaction *V* would pass through the centroid *O*, and the moment reaction would be about *O*. The reactions are $V = 16$ kN, and the moment is $M = 16(425) = 6800$ N · m. The primary shear force F' on each pin is equal to $V/2 = 16/2 = 8$ kN. The secondary shear forces F'' are also equal (and opposite), and from Eq. (8–50),

$$F'' = \frac{Mr}{2r^2} = \frac{M}{2r} = \frac{6800}{2(75)} = 45.33 \text{ kN}$$

making the resultants $F_E = 45.33 - 8 = 37.33$ kN (upward) and $F_F = 45.33 + 8 = 53.33$ kN (downward).

(*b*) Since the pin forces differ, the designer may wish to work the pin material equally and consider increasing the size of A_F to $A_F = A_E 53.33/37.33 = 1.43 A_E$. It is instructive to pursue this as it represents an instance of free bodies teaching us what we need to know. From Eq. (8–49), the distance from *E* to the centroid $\bar{x}$ is

$$\bar{x} = \frac{A_E(0) + A_F(150)}{A_E + A_F} = \frac{0 + 1.43(150)A_E}{A_E + 1.43A_E} = \frac{1.43(150)}{1 + 1.43} = 88.3 \text{ mm}$$

The new moment reaction about the new pivot point O' is $M = 16(61.7 + 350) = 6587$ N · m and the shear reaction is still 16 kN. The total reactions at *E* and *F* can be found by taking moments about O' and summing forces vertically. The equations are

$$88.3 F_E + 61.7 F_F = 6587$$

$$-F_E + F_F = 16$$

from which the simultaneous solution for F_E and F_F gives

$$F_E = 37.33 \text{ kN}$$

$$F_F = 53.33 \text{ kN}$$

The reactions have not changed. (Why?) The shear stresses on the pin cross sections are

$$\tau_E = \frac{F_E}{A_E} = \frac{37.3}{A_E}$$

$$\tau_F = \frac{F_F}{A_F} = \frac{F_F}{1.43 A_E} = \frac{53.33}{1.43 A_E} = \frac{37.3}{A_E}$$

The stresses are equal, and the goal is achieved.

In Ex. 8–6 the role of the four bolts is to hold the joint together. A preload of 0.6 proof load is about the lower limit; about 0.75 proof load would be better. The bolts see no shear load, and the pins carry that.

If the pins are omitted in Fig. 8–25 and the four bolt shanks are tightly fitted in reamed holes, one may proceed as in Ex. 8–7.

EXAMPLE 8-7   The situation is similar to that of Ex. 8–6 except that there are no pins and the bolts are tightly fitted. For a $F = 16$ kN load find
(a) The resultant load on each bolt
(b) The maximum shear stress in each bolt
(c) The maximum bearing stress
(d) The critical bending stress in the bar

Solution   (a) Point O, the centroid of the bolt group in Fig. 8–25, is found by symmetry. If a free-body diagram of the beam were constructed, the shear reaction V would pass through O and the moment reactions M would be about O. These reactions are

$$V = 16 \text{ kN} \qquad M = 16(425) = 6800 \text{ N} \cdot \text{m}$$

In Fig. 8–26, the bolt group has been drawn to a larger scale and the reactions are shown. The distance from the centroid to the center of each bolt is

$$r = \sqrt{(60)^2 + (75)^2} = 96.0 \text{ mm}$$

The primary shear load per bolt is

$$F' = \frac{V}{n} = \frac{16}{4} = 4 \text{ kN}$$

Since the secondary shear forces are equal, Eq. (8–50) becomes

$$F'' = \frac{Mr}{4r^2} = \frac{M}{4r} = \frac{6800}{4(96.0)} = 17.7 \text{ kN}$$

The primary and secondary shear forces are plotted to scale in Fig. 8–26 and the resultants obtained by using the parallelogram rule. The magnitudes are found by measurement

| **Figure 8–26**

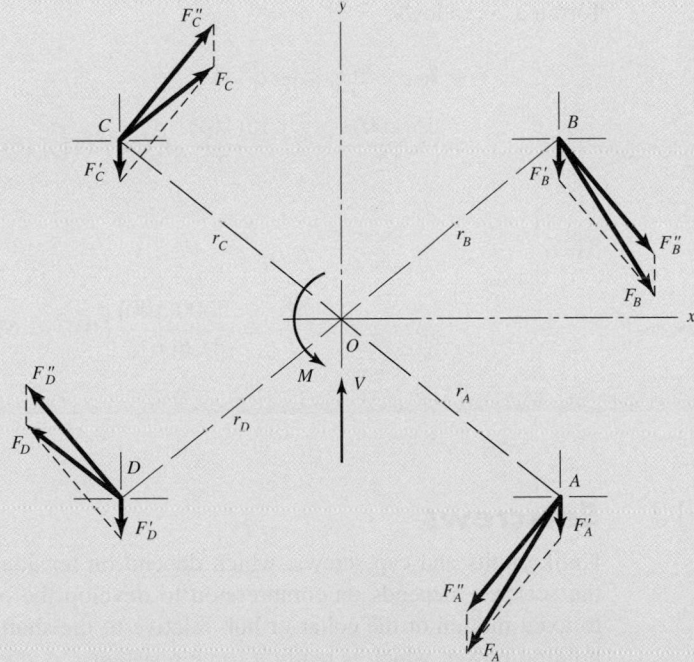

(or analysis) to be

Answer
$$F_A = F_B = 21.0 \text{ kN}$$

Answer
$$F_C = F_D = 14.8 \text{ kN}$$

(b) Bolts A and B are critical because they carry the largest shear load. Does this shear act on the threaded portion of the bolt, or on the unthreaded portion? The bolt length will be 25 mm plus the height of the nut plus about 2 mm for a washer. Table A–31 gives the nut height as 14.8 mm. Including two threads beyond the nut, this adds up to a length of 43.8 mm, and so a bolt 46 mm long will be needed. From Eq. (8–14) we compute the thread length as $L_T = 38$ mm. Thus the unthreaded portion of the bolt is $46 - 38 = 8$ mm long. This is less than the 15 mm for the plate in Fig. 8–25, and so the bolt will tend to shear across its minor diameter. Therefore the shear-stress area is $A_s = 144$ mm^2, and so the shear stress is

Answer
$$\tau = \frac{F}{A_s} = -\frac{21.0(10)^3}{144} = 146 \text{ MPa}$$

(c) The channel is thinner than the bar, and so the largest bearing stress is due to the pressing of the bolt against the channel web. The bearing area is $A_b = td = 10(16) = 160$ mm^2. Thus the bearing stress is

Answer
$$\sigma = -\frac{F}{A_b} = -\frac{21.0(10)^3}{160} = -131 \text{ MPa}$$

(d) The critical bending stress in the bar is assumed to occur in a section parallel to the y axis and through bolts A and B. At this section the bending moment is

$$M = 16(300 + 50) = 5600 \text{ N} \cdot \text{m}$$

The second moment of area through this section is obtained by the use of the transfer formula, as follows:

$$I = I_{bar} - 2(I_{holes} + \bar{d}^2 A)$$

$$= \frac{15(200)^3}{12} - 2\left[\frac{15(16)^3}{12} + (60)^2(15)(16)\right] = 8.26(10)^6 \text{ mm}^4$$

Then

Answer
$$\sigma = \frac{Mc}{I} = \frac{5600(100)}{8.26(10)^6}(10)^3 = 67.8 \text{ MPa}$$

8–13 Setscrews

Unlike bolts and cap screws, which depend on tension to develop a clamping force, the setscrew depends on compression to develop the clamping force. The resistance to axial motion of the collar or hub relative to the shaft is called *holding power*. This holding power, which is really a force resistance, is due to frictional resistance of the contacting portions of the collar and shaft as well as any slight penetration of the setscrew into the shaft.

Figure 8–27 shows the point types available with socket setscrews. These are also manufactured with screwdriver slots and with square heads.

Table 8–18 lists values of the seating torque and the corresponding holding power for inch-series setscrews. The values listed apply to both axial holding power, for resisting

Figure 8–27

Socket setscrews: (a) flat point; (b) cup point; (c) oval point; (d) cone point; (e) half-dog point.

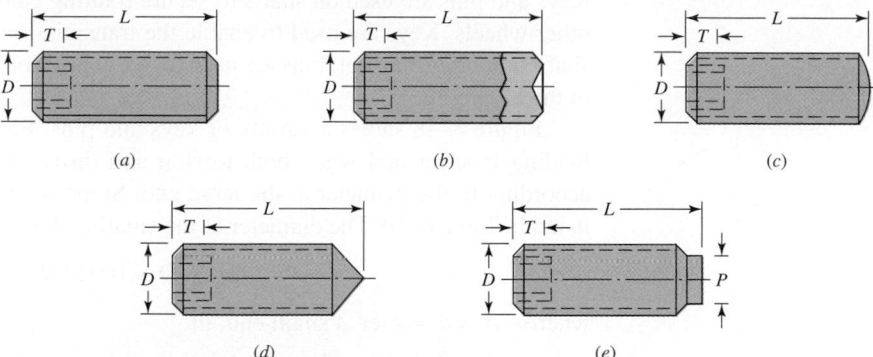

Table 8–18

Typical Holding Power (Force) for Socket Setscrews*

Source: Unbrako Division, SPS Technologies, Jenkintown, Pa.

Size, in	Seating Torque, lbf · in	Holding Power, lbf
#0	1.0	50
#1	1.8	65
#2	1.8	85
#3	5	120
#4	5	160
#5	10	200
#6	10	250
#8	20	385
#10	36	540
$\frac{1}{4}$	87	1000
$\frac{5}{16}$	165	1500
$\frac{3}{8}$	290	2000
$\frac{7}{16}$	430	2500
$\frac{1}{2}$	620	3000
$\frac{9}{16}$	620	3500
$\frac{5}{8}$	1325	4000
$\frac{3}{4}$	2400	5000
$\frac{7}{8}$	5200	6000
1	7200	7000

*Based on allow-steel screw against steel shaft, class 3A coarse or fine threads in class 2B holes, and cup-point socket setscrews.

thrust, and the tangential holding power, for resisting torsion. Typical factors of safety are 1.5 to 2.0 for static loads and 4 to 8 for various dynamic loads.

Setscrews should have a length of about half of the shaft diameter. Note that this practice also provides a rough rule for the radial thickness of a hub or collar.

8–14 Keys and Pins

Keys and pins are used on shafts to secure rotating elements, such as gears, pulleys, or other wheels. Keys are used to enable the transmission of torque from the shaft to the shaft-supported element. Pins are used for axial positioning and for the transfer of torque or thrust or both.

Figure 8–28 shows a variety of keys and pins. Pins are useful when the principal loading is shear and when both torsion and thrust are present. Taper pins are sized according to the diameter at the large end. Some of the most useful sizes of these are listed in Table 8–19. The diameter at the small end is

$$d = D - 0.0208L \tag{8–51}$$

where $d =$ diameter at small end, in

$D =$ diameter at large end, in

$L =$ length, in

Figure 8–28

(a) Square key; (b) round key; (c and d) round pins; (e) taper pin; (f) split tubular spring pin. The pins in parts (e) and (f) are shown longer than necessary, to illustrate the chamfer on the ends, but their lengths should be kept smaller than the hub diameters to prevent injuries due to projections on rotating parts.

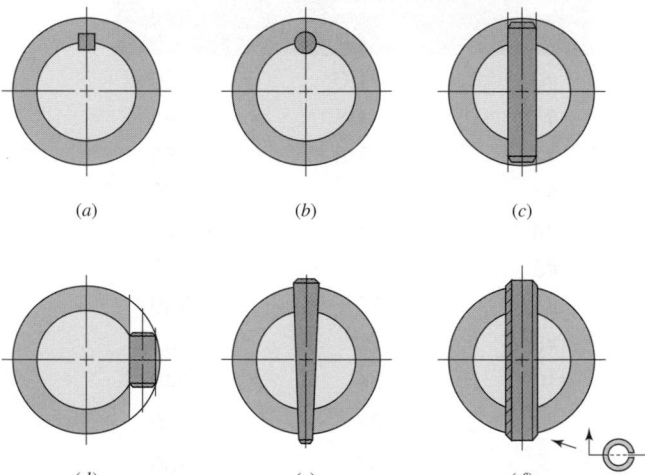

(a)   (b)   (c)

(d)   (e)   (f)

Table 8–19

Dimensions at Large End of Some Standard Taper Pins—Inch Series

Size	Commercial		Precision	
	Maximum	Minimum	Maximum	Minimum
4/0	0.1103	0.1083	0.1100	0.1090
2/0	0.1423	0.1403	0.1420	0.1410
0	0.1573	0.1553	0.1570	0.1560
2	0.1943	0.1923	0.1940	0.1930
4	0.2513	0.2493	0.2510	0.2500
6	0.3423	0.3403	0.3420	0.3410
8	0.4933	0.4913	0.4930	0.4920

Table 8–20

Inch Dimensions for Some Standard Square- and Rectangular-Key Applications

Source: Joseph E. Shigley, "Unthreaded Fasteners," Chap. 22 in Joseph E. Shigley and Charles R. Mischke (eds.), *Standard Handbook of Machine Design,* 2nd ed., McGraw-Hill, New York, 1996.

| Shaft Diameter | | Key Size | | Keyway Depth |
Over	To (Incl.)	w	h	
$\frac{5}{16}$	$\frac{7}{16}$	$\frac{3}{32}$	$\frac{3}{32}$	$\frac{3}{64}$
$\frac{7}{16}$	$\frac{9}{16}$	$\frac{1}{8}$	$\frac{3}{32}$	$\frac{3}{64}$
		$\frac{1}{8}$	$\frac{1}{8}$	$\frac{1}{16}$
$\frac{9}{16}$	$\frac{7}{8}$	$\frac{3}{16}$	$\frac{1}{8}$	$\frac{1}{16}$
		$\frac{3}{16}$	$\frac{3}{16}$	$\frac{3}{32}$
$\frac{7}{8}$	$1\frac{1}{4}$	$\frac{1}{4}$	$\frac{3}{16}$	$\frac{3}{32}$
		$\frac{1}{4}$	$\frac{1}{4}$	$\frac{1}{8}$
$1\frac{1}{4}$	$1\frac{3}{8}$	$\frac{5}{16}$	$\frac{1}{4}$	$\frac{1}{8}$
		$\frac{5}{16}$	$\frac{5}{16}$	$\frac{5}{32}$
$1\frac{3}{8}$	$1\frac{3}{4}$	$\frac{3}{8}$	$\frac{1}{4}$	$\frac{1}{8}$
		$\frac{3}{8}$	$\frac{3}{8}$	$\frac{3}{16}$
$1\frac{3}{4}$	$2\frac{1}{4}$	$\frac{1}{2}$	$\frac{3}{8}$	$\frac{3}{16}$
		$\frac{1}{2}$	$\frac{1}{2}$	$\frac{1}{4}$
$2\frac{1}{4}$	$2\frac{3}{4}$	$\frac{5}{8}$	$\frac{7}{16}$	$\frac{7}{32}$
		$\frac{5}{8}$	$\frac{5}{8}$	$\frac{5}{16}$
$2\frac{3}{4}$	$3\frac{1}{4}$	$\frac{3}{4}$	$\frac{1}{2}$	$\frac{1}{4}$
		$\frac{3}{4}$	$\frac{3}{4}$	$\frac{3}{8}$

For less important applications, a dowel pin or a drive pin can be used. A large variety of these are listed in manufacturers' catalogs.[10]

The square key, shown in Fig. 8–28*a*, is also available in rectangular sizes. Standard sizes of these, together with the range of applicable shaft diameters, are listed in Table 8–20. The length of the key is based on the hub length and the torsional load to be transferred.

The gib-head key, in Fig. 8–29, is tapered so that, when firmly driven, it acts to prevent relative axial motion. This also gives the advantage that the hub position can be adjusted for the best axial location. The head makes removal possible without access to the other end, but the projection may be hazardous.

Also shown in Fig. 8–29*a* is the sled-runner keyway. This has less stress concentration than an end-milled keyway.

The Woodruff key, shown in Fig. 8–29*b*, is of general usefulness, especially when a wheel is to be positioned against a shaft shoulder, since the keyslot need not be machined into the shoulder stress-concentration region. The use of the Woodruff key also yields better concentricity after assembly of the wheel and shaft. This is especially important at high speeds, as, for example, with a turbine wheel and shaft. Dimensions

[10]See also Joseph E. Shigley, "Unthreaded Fasteners," Chap. 22 in Joseph E. Shigley and Charles R. Mischke (eds.), *Standard Handbook of Machine Design,* 2nd ed., McGraw-Hill, New York, 1996.

Figure 8–29

(a) Gib-head key;
(b) Woodruff key.

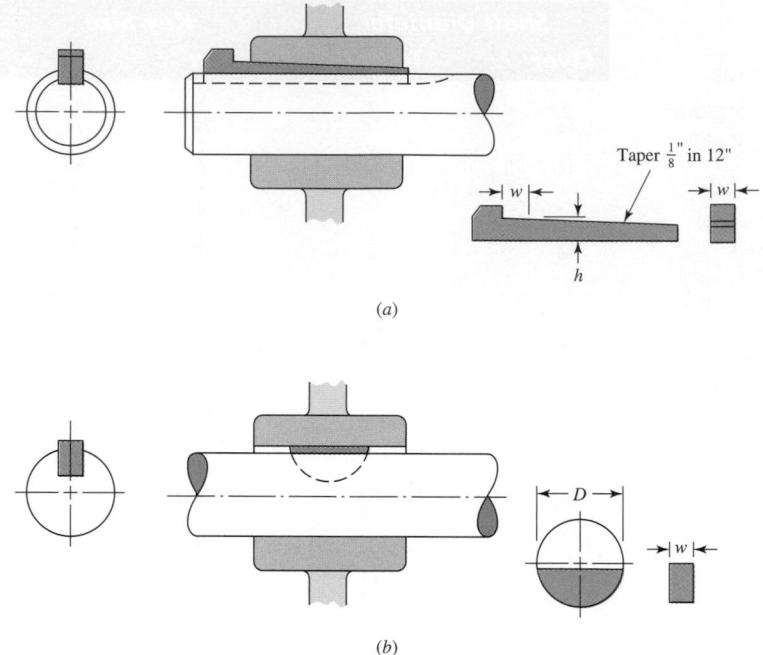

Taper $\frac{1}{8}$" in 12"

(a)

(b)

for some standard Woodruff key sizes can be found in Table 8–21, and Table 8–22 gives the shaft diameters for which the different keyseat widths are suitable.

Stress-concentration factors for keyways depend for their values upon the fillet radius at the bottom and ends of the keyway, according to Peterson.[11] For fillets cut by standard milling-machine cutters, Peterson's charts give $K_t = 2.14$ for bending and $K_{ts} = 2.62$ for torsion. These are for end-milled keyseats, which permit definite key positioning longitudinally. In bending, the sled-runner seat will have less stress concentration. The values of K_t for the sled-runner keyway are based on the full values of I and J for shafting; that is, the second moments are not reduced by the slot dimensions. Other results reported are 1.79 for an end-milled keyway and 1.38 for the sled-runner type.

The use of shrink fits eliminates the use of keys to transmit torque. These were discussed in Sec. 2–6, and corresponding data are listed in Tables A–11 to A–14. Peterson reports $K_f = 2.0$ for bending at each end of a press-fitted collar. Tests seem to indicate that the interface pressure has little effect on K_t. However, fretting corrosion introduces a complication.

A retaining ring is frequently used instead of a shaft shoulder or a sleeve to axially position a component on a shaft or in a housing bore. As shown in Fig. 8–30, a groove is cut in the shaft or bore to receive the spring retainer. The tapered design of both the internal and external rings ensures uniform pressure against the bottom of the groove. For sizes, dimensions, and ratings, the manufacturers' catalogs should be consulted.

---

[11]R. E. Peterson, *Stress Concentration Factors,* Wiley, New York, 1974, pp. 245, 266, 267.

Table 8–21

Dimensions of Woodruff Keys—Inch Series

Key Size w	D	Height b	Offset e	Keyseat Depth Shaft	Hub
$\frac{1}{16}$	$\frac{1}{4}$	0.109	$\frac{1}{64}$	0.0728	0.0372
$\frac{1}{16}$	$\frac{3}{8}$	0.172	$\frac{1}{64}$	0.1358	0.0372
$\frac{3}{32}$	$\frac{3}{8}$	0.172	$\frac{1}{64}$	0.1202	0.0529
$\frac{3}{32}$	$\frac{1}{2}$	0.203	$\frac{3}{64}$	0.1511	0.0529
$\frac{3}{32}$	$\frac{5}{8}$	0.250	$\frac{1}{16}$	0.1981	0.0529
$\frac{1}{8}$	$\frac{1}{2}$	0.203	$\frac{3}{64}$	0.1355	0.0685
$\frac{1}{8}$	$\frac{5}{8}$	0.250	$\frac{1}{16}$	0.1825	0.0685
$\frac{1}{8}$	$\frac{3}{4}$	0.313	$\frac{1}{16}$	0.2455	0.0685
$\frac{5}{32}$	$\frac{5}{8}$	0.250	$\frac{1}{16}$	0.1669	0.0841
$\frac{5}{32}$	$\frac{3}{4}$	0.313	$\frac{1}{16}$	0.2299	0.0841
$\frac{5}{32}$	$\frac{7}{8}$	0.375	$\frac{1}{16}$	0.2919	0.0841
$\frac{3}{16}$	$\frac{3}{4}$	0.313	$\frac{1}{16}$	0.2143	0.0997
$\frac{3}{16}$	$\frac{7}{8}$	0.375	$\frac{1}{16}$	0.2763	0.0997
$\frac{3}{16}$	1	0.438	$\frac{1}{16}$	0.3393	0.0997
$\frac{1}{4}$	$\frac{7}{8}$	0.375	$\frac{1}{16}$	0.2450	0.1310
$\frac{1}{4}$	1	0.438	$\frac{1}{16}$	0.3080	0.1310
$\frac{1}{4}$	$1\frac{1}{4}$	0.547	$\frac{5}{64}$	0.4170	0.1310
$\frac{5}{16}$	1	0.438	$\frac{1}{16}$	0.2768	0.1622
$\frac{5}{16}$	$1\frac{1}{4}$	0.547	$\frac{5}{64}$	0.3858	0.1622
$\frac{5}{16}$	$1\frac{1}{2}$	0.641	$\frac{7}{64}$	0.4798	0.1622
$\frac{3}{8}$	$1\frac{1}{4}$	0.547	$\frac{5}{64}$	0.3545	0.1935
$\frac{3}{8}$	$1\frac{1}{2}$	0.641	$\frac{7}{64}$	0.4485	0.1935

Table 8–22

Sizes of Woodruff Keys Suitable for Various Shaft Diameters

Keyseat Width, in	Shaft Diameter, in From	To (inclusive)
$\frac{1}{16}$	$\frac{5}{16}$	$\frac{1}{2}$
$\frac{3}{32}$	$\frac{3}{8}$	$\frac{7}{8}$
$\frac{1}{8}$	$\frac{3}{8}$	$1\frac{1}{2}$
$\frac{5}{32}$	$\frac{1}{2}$	$1\frac{5}{8}$
$\frac{3}{16}$	$\frac{9}{16}$	2
$\frac{1}{4}$	$\frac{11}{16}$	$2\frac{1}{4}$
$\frac{5}{16}$	$\frac{3}{4}$	$2\frac{3}{8}$
$\frac{3}{8}$	1	$2\frac{5}{8}$

Figure 8–30

Typical uses for retaining rings. (a) External ring and (b) its application; (c) internal ring and (d) its application.

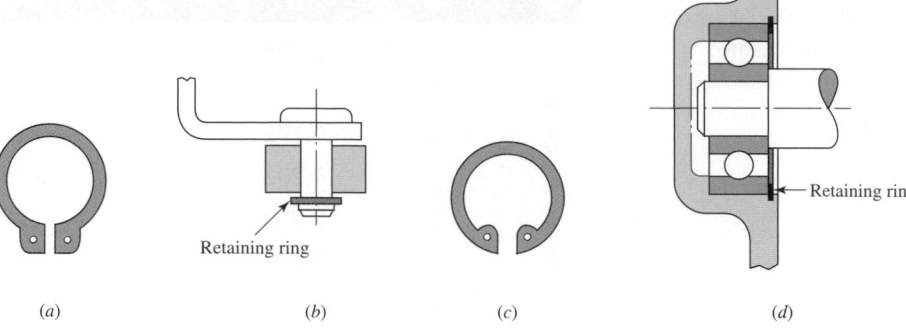

Retaining ring

Retaining ring

(a)　　　　　　　(b)　　　　(c)　　　　(d)

EXAMPLE 8–8

A UNS G10350 steel shaft, heat-treated to a minimum yield strength of 75 kpsi, has a diameter of $1\frac{7}{16}$ in. The shaft rotates at 600 rev/min and transmits 40 hp through a gear. Select an appropriate key for the gear.

Solution

A $\frac{3}{8}$-in square key is selected, UNS G10200 cold-drawn steel being used. The design will be based on a yield strength of 65 kpsi. A factor of safety of 2.80 will be employed in the absence of exact information about the nature of the load.

The torque is obtained from the horsepower equation

$$T = \frac{63\ 025H}{n} = \frac{(63\ 025)(40)}{600} = 4200 \text{ lbf} \cdot \text{in}$$

From Fig. 8–31, the force F at the surface of the shaft is

$$F = \frac{T}{r} = \frac{4200}{1.4375/2} = 5850 \text{ lbf}$$

By the distortion-energy theory, the shear strength is

$$S_{sy} = 0.577S_y = (0.577)(65) = 37.5 \text{ kpsi}$$

Failure by shear across the area ab will create a stress of $\tau = F/tl$. Substituting the strength divided by the factor of safety for τ gives

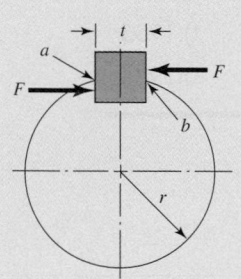

Figure 8–31

$$\frac{S_{sy}}{n} = \frac{F}{tl} \qquad \text{or} \qquad \frac{37.5(10)^3}{2.80} = \frac{5850}{0.375l}$$

or $l = 1.16$ in. To resist crushing, the area of one-half the face of the key is used:

$$\frac{S_y}{n} = \frac{F}{tl/2} \qquad \text{or} \qquad \frac{65(10)^3}{2.80} = \frac{5850}{0.375l/2}$$

and $l = 1.34$ in. The hub length of a gear is usually greater than the shaft diameter, for stability. If the key, in this example, is made equal in length to the hub, it would therefore have ample strength, since it would probably be $1\frac{7}{16}$ in or longer.

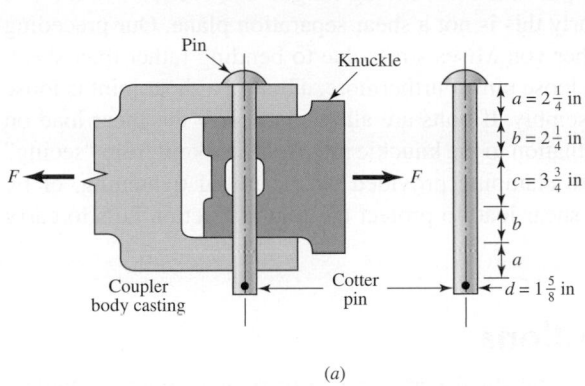

(a)   (b)

Figure 8–32

National AAR E freight car coupler and some geometric detail.

Loose Pins

In some applications a pin is not a tight fit in its holes but is loose enough to allow movement. One such example is the knuckle pin for a railroad car or locomotive coupler as seen in Fig. 8–32. The pin is loose enough to allow unimpeded rotation of the knuckle even in the presence of dirt and rust. Problem 4–27 presented an exercise in the bending and shear stresses present in a loose-fitting clevis pin, suggesting two methods of analysis. For a similar model, the effective beam span length, from Fig. 8–32, is

$$l = \frac{2.25}{2} + 2.25 + 3.75 + 2.25 + \frac{2.25}{2} = 10.5 \text{ in}$$

The distance from the top reaction to the top concentrated flexural loading is $(a+b)/2 = 2.25$ in. The largest bending moment in the beam is estimated as

$$M_{\max} = \frac{F}{2}(2.25) = 1.125F$$

The associated bending stress σ and the von Mises stress σ'_b are

$$\sigma'_b = \sigma = \frac{Mc}{I} = \frac{32M}{\pi d^3} = \frac{32(1.125F)}{\pi(1.625^3)} = 2.67F$$

The shearing stress due to double shear at the shearing separation planes (at the top and bottom surfaces of the knuckle) is

$$\tau_{\max} = \frac{4V}{3A} = \frac{4}{3}\frac{V}{\pi d^2/4} = \frac{4}{3}\frac{F/2}{\pi(1.625^2)/4} = 0.321F$$

The associated von Mises stress is $\sigma'_s = \sqrt{3}\tau_{\max} = 0.556F$. The ratio is

$$\frac{\sigma'_b}{\sigma'_s} = \frac{2.67}{0.556} = 4.8$$

Freight cars, including the coupling system, are designed to tolerate a large load F. The history of the buffeting force F is not known, but if the pin fails in bending fatigue rather than shear fatigue, the location and character of the failure reveal where the stress is the highest. In this service the pin failed in bending fatigue at a section about $3\frac{1}{2}$ in below the head of the pin.[12] Clearly this is not a shear separation plane. Our preceding analysis indicates a 4.8-fold higher von Mises stress due to bending rather than shear. Always investigate bending with loose pins. Furthermore, a bolt in a shear joint is loose in its hole to allow easy joint assembly. If bolts are allowed to carry the shear load on their body, they are in a similar situation to the knuckle pin. Bolts are kept from "seeing" such a loading by friction due to clamping provided by the initial tightening, or by providing tight pins to carry the shear load to protect the joint if friction fails to carry the load.

8–15 Stochastic Considerations

The data available to analyze the statistical performance of bolted joints are widely scattered and in such small quantity that a reliable analysis is difficult to make. The methods of Parts 1 and 2 of this book can be used to perform the analysis when data are available.

One of the random variables needed is the tensile strength. The codes specify only the minimum value. Figure 8–17 illustrates data for one series of tests that might be used as a guide. For these tests the probability that $\mathbf{S}_{ut}$ will equal or exceed the minimum specification works out as 99.2 percent.

Another random variable that would be desirable is the proof strength $\mathbf{S}_p$. The data of Tables 8–13 and 8–14 do give some help on the preload variable $\mathbf{F}_i$. The coefficients of variation turn out to be $C_{Fi} = 0.143$ for unlubricated bolts and 0.084 for lubricated bolts.

The modified Goodman line will be of little value in analyzing the statistical performance of bolted joints, because it is always on the conservative side of failure points. Thus a nonlinear theory, such as the Gerber parabolic theory, is needed. Equation (8–43) reveals that the determination of the mean-strength variate $\mathbf{S}_a$ is quite complex for the Gerber theory. A computer solution might be the way to go.

The following example illustrates the use of statistics for a simple preload investigation.

---

[12]The lower part of the pin dropped to the roadbed, the upper part of the pin bounced out, the knuckle pulled out, and the train parted, the emergency brakes automatically engaged, and the crew substituted a new pin. The two parts of the pin were found by a track-walker within 15 feet of each other. The parts were unseen by the train crew, since the incident occurred at night and the pieces were under the train, to the back.

EXAMPLE 8–9

A $\frac{3}{8}$ in-24 UNF SAE grade 5 bolt is lubricated and then tightened to a torque-wrench specification of 450 lbf · in. This corresponds to a preload of 90 percent of proof strength. [See Eq. (8–30).] Data in the footnote of Table 8–14 show that the coefficient of variation of the preload is $C_{Fi} = 2.88/34.18 = 0.0843$. Assuming normal distributions and $C_{Sp} = 0.08$:

(a) Find the mean and the standard deviation of the bolt tension.

(b) What is the probability that the proof strength will be exceeded?

Solution	From Table 8–2, $A_t = 0.087\ 8$ in^2, from Table 8–15, $K = 0.18$, and from Table 8–9, the minimum proof strength, based on a 99 percent reliability is, $(S_p)_{0.99} = 85$ kpsi.

(*a*) From Eq. (8–27),

$$\bar{F}_i = \frac{\bar{T}}{Kd} = \frac{450}{0.18\,(0.375)} = 6667\ \text{lbf}$$

and

$$\hat{\sigma}_{Fi} = C_{Fi}F_i = 0.084(6667) = 560\ \text{lbf}$$

Answer	Thus, $\mathbf{F}_i = \mathbf{N}\,(6667, 560)$ lbf. The corresponding stress is

$$\boldsymbol{\sigma}_i = \frac{\mathbf{F}_i}{A_t} = \frac{\mathbf{N}(6.667,\ 0.560)}{0.0878} = \mathbf{N}(75.9,\ 6.38)\ \text{kpsi}$$

(*b*) To determine the probability that the proof strength will be exceeded, we need to interfere $\boldsymbol{\sigma}_i$ with $\mathbf{S}_p$. The mean is determined from Eq. (2–16) as

$$\bar{S}_p = \frac{(S_p)_{0.99}}{1 + z_{0.99}C_{Sp}} = \frac{85}{1 + (-2.326)(0.08)} = 104.4\ \text{kpsi}$$

The standard deviation is $\hat{\sigma}_{Sp} = C_{Sp}\bar{S}_p = 0.08(104.4) = 8.35$ kpsi. From Eq. (6–54),

$$z = -\frac{\bar{S}_p - \bar{\sigma}_i}{\left(\hat{\sigma}_{Sp}^2 + \hat{\sigma}_{\sigma i}^2\right)^{1/2}} = -\frac{104.4 - 75.9}{(8.35^2 + 6.38^2)^{1/2}} = -2.71$$

Answer	From Table A–10, $\Phi = 0.00336$. Thus the probability of a single bolt being preloaded over the proof strength is 0.34 percent.

PROBLEMS

8–1 A power screw is 25 mm in diameter and has a thread pitch of 5 mm.
(*a*) Find the thread depth, the thread width, the mean and root diameters, and the lead, provided square threads are used.
(*b*) Repeat part (*a*) for Acme threads.

8–2 Using the information in the footnote of Table 8–1, show that the tensile-stress area is

$$A_t = \frac{\pi}{4}(d - 0.938\ 194p)^2$$

8–3 Show that for zero collar friction the efficiency of a square-thread screw is given by the equation

$$e = \tan\lambda\,\frac{1 - f\tan\lambda}{\tan\lambda + f}$$

Plot a curve of the efficiency for lead angles up to 45°. Use $f = 0.08$.

8–4 A single-threaded 25-mm power screw is 25 mm in diameter with a pitch of 5 mm. A vertical load on the screw reaches a maximum of 6 kN. The coefficients of friction are 0.05 for the collar and 0.08 for the threads. The frictional diameter of the collar is 40 mm. Find the overall efficiency and the torque to "raise" and "lower" the load.

ANALYSIS

8–5 The machine shown in the figure can be used for a tension test but not for a compression test. Why? Can both screws have the same hand?

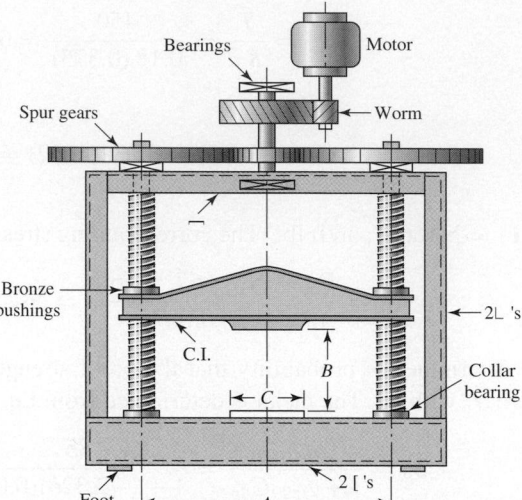

Problem 8–5

ANALYSIS

8–6 The press shown for Prob. 8–5 has a rated load of 5000 lbf. The twin screws have Acme threads, a diameter of 3 in, and a pitch of $\frac{1}{2}$ in. Coefficients of friction are 0.05 for the threads and 0.06 for the collar bearings. Collar diameters are 5 in. The gears have an efficiency of 95 percent and a speed ratio of 75:1. A slip clutch, on the motor shaft, prevents overloading. The full-load motor speed is 1720 rev/min.

(*a*) When the motor is turned on, how fast will the press head move?

(*b*) What should be the horsepower rating of the motor?

ANALYSIS

8–7 A screw clamp similar to the one shown in the figure has a handle with diameter $\frac{3}{16}$ in made of cold-drawn AISI 1006 steel. The overall length is 3 in. The screw is $\frac{7}{16}$ in-14 UNC and is $5\frac{3}{4}$ in long, overall. Distance A is 2 in. The clamp will accommodate parts up to $4\frac{3}{16}$ in high.

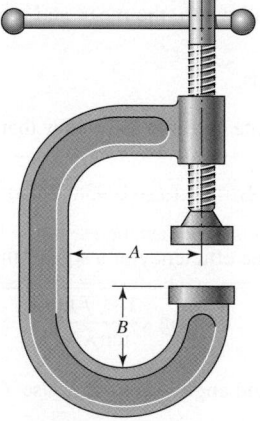

Problem 8–7

(*a*) What screw torque will cause the handle to bend permanently?

(*b*) What clamping force will the answer to part (*a*) cause if the collar friction is neglected and if the thread friction is 0.075?

(*c*) What clamping force will cause the screw to buckle?

(*d*) Are there any other stresses or possible failures to be checked?

 ANALYSIS

8-8 The C clamp shown in the figure for Prob. 8–7 uses a $\frac{5}{8}$ in-6 Acme thread. The frictional coefficients are 0.15 for the threads and for the collar. The collar, which in this case is the anvil striker's swivel joint, has a friction diameter of $\frac{7}{16}$ in. Calculations are to be based on a maximum force of 6 lbf applied to the handle at a radius of $2\frac{3}{4}$ in from the screw centerline. Find the clamping force.

 ANALYSIS

8-9 Find the power required to drive a 40-mm power screw having double square threads with a pitch of 6 mm. The nut is to move at a velocity of 48 mm/s and move a load of $F = 10$ kN. The frictional coefficients are 0.10 for the threads and 0.15 for the collar. The frictional diameter of the collar is 60 mm.

8-10 A single square-thread power screw has an input power of 3 kW at a speed of 1 rev/s. The screw has a diameter of 36 mm and a pitch of 6 mm. The frictional coefficients are 0.14 for the threads and 0.09 for the collar, with a collar friction radius of 45 mm. Find the axial resisting load F and the combined efficiency of the screw and collar.

 ANALYSIS

8-11 A bolted joint is to have a grip consisting of two $\frac{1}{2}$-in steel plates and one wide $\frac{1}{2}$-in American Standard plain washer to fit under the head of the $\frac{1}{2}$ in-13 × 1.75 in UNC hex-head bolt.

(*a*) What is the length of the thread L_T for this diameter inch-series bolt?

(*b*) What is the length of the grip L_G?

(*c*) What is the height H of the nut?

(*d*) Is the bolt long enough? If not, round to the next larger preferred length (Table A–17).

(*e*) What is the length of the shank and threaded portions of the bolt within the grip? These lengths are needed in order to estimate the bolt spring rate k_b.

 ANALYSIS

8-12 A bolted joint is to have a grip consisting of two 14-mm steel plates and one 14R metric plain washer to fit under the head of the M14 × 2 hex-head bolt, 50 mm long.

(*a*) What is the length of the thread L_T for this diameter metric coarse-pitch series bolt?

(*b*) What is the length of the grip L_G?

(*c*) What is the height H of the nut?

(*d*) Is the bolt long enough? If not, round to the next larger preferred length (Table A–17).

(*e*) What is the length of the shank and the threaded portions of the bolt within the grip? These lengths are needed in order to estimate bolt spring rate k_b.

 ANALYSIS

8-13 A blanking disk 0.875 in thick is to be fastened to a spool whose flange is 1 in thick, using eight $\frac{1}{2}$ in-13 × 1.75 in hex-head cap screws.

(*a*) What is the length of threads L_T for this cap screw?

(*b*) What is the effective length of the grip L'_G?

(*c*) Is the length of this cap screw sufficient? If not, round up.

(*d*) Find the shank length l_d and the useful thread length l_t within the grip. These lengths are needed for the estimate of the fastener spring rate k_b.

 ANALYSIS

8-14 A blanking disk is 20 mm thick and is to be fastened to a spool whose flange is 25 mm thick, using eight M12 × 40 hex-head metric cap screws.

(*a*) What is the length of the threads L_T for this fastener?

(*b*) What is the effective grip length L'_G?

(*c*) Is the length of this fastener sufficient? If not, round to the next preferred length.

(*d*) Find the shank length l_d and the useful threaded length in the grip l_t. These lengths are needed in order to estimate the fastener spring rate k_b.

 ANALYSIS

8-15 A $\frac{3}{4}$ in-16 UNF series SAE grade 5 bolt has a $\frac{3}{4}$-in ID tube 13 in long, clamped between washer faces of bolt and nut by turning the nut snug and adding one-third of a turn. The tube OD is the washer-face diameter $d_w = 1.5d = 1.5(0.75) = 1.125$ in = OD.

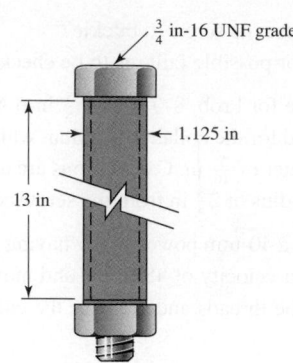

$\frac{3}{4}$ in-16 UNF grade

1.125 in

13 in

Problem 8–15

(*a*) What is the spring rate of the bolt and the tube, if the tube is made of steel? What is the joint constant *C*?

(*b*) When the one-third turn-of-nut is applied, what is the initial tension F_i in the bolt?

(*c*) What is the bolt tension at opening if additional tension is applied to the bolt external to the joint?

ANALYSIS

8–16   From your experience with Prob. 8–15, generalize your solution to develop a turn-of-nut equation

$$N_t = \frac{\theta}{360°} = \left(\frac{k_b + k_m}{k_b k_m} \right) F_i N$$

where N_t = turn of the nut from snug tight

θ = turn of the nut in degrees

N = number of thread/in ($1/p$ where p is pitch)

F_i = initial preload

k_b, k_m = spring rates of the bolt and members, respectively

Use this equation to find the relation between torque-wrench setting *T* and turn-of-nut N_t. ("Snug tight" means the joint has been tightened to perhaps half the intended preload to flatten asperities on the washer faces and the members. Then the nut is loosened and retightened finger tight, and the nut is rotated the number of degrees indicated by the equation. Properly done, the result is competitive with torque wrenching.)

ANALYSIS

8–17   RB&W[13] recommends turn-of-nut from snug fit to preload as follows: 1/3 turn for bolt grips of 1–4 diameters, 1/2 turn for bolt grips 4–8 diameters, and 2/3 turn for grips of 8–12 diameters. These recommendations are for structural steel fabrication (permanent joints), producing preloads of 100 percent of proof strength and beyond. Machinery fabricators with fatigue loadings and possible joint disassembly have much smaller turns-of-nut. The RB&W recommendation enters the nonlinear plastic deformation zone.

Problem 8–17

Turn-of-nut method

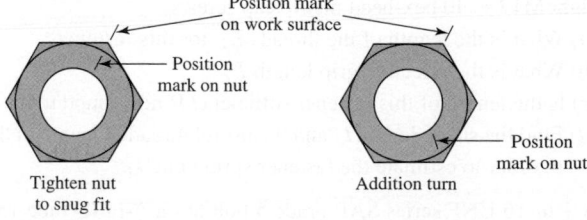

Position mark on work surface

Position mark on nut

Position mark on nut

Tighten nut to snug fit

Addition turn

---

[13]Russell, Burdsall & Ward, Inc., Metal Forming Specialists, Mentor, Ohio.

(*a*) For Ex. 8–4, use Eq. (8–27) with $K = 0.2$ to estimate the torque necessary to establish the desired preload. Then, using the results from Prob. 8–16, determine the turn of the nut in degrees. How does this compare with the RB&W recommendations?

(*b*) Repeat part (*a*) for Ex. 8–5.

8–18 Take Eq. (8–22) and express $k_m/(Ed)$ as a function of l/d, then compare with Eq. (8–23) for $d/l = 0.5$.

8–19 A joint has the same geometry as Ex. 8–4, but the lower member is steel. Use Eq. (8–23) to find the spring rate of the members in the grip. *Hint:* Equation (8–23) applies to the stiffness of two sections of a joint of one material. If each section has the same thickness, then what is the stiffness of one of the sections?

ADEQUACY
ASSESSMENT

8–20 The figure illustrates the connection of a cylinder head to a pressure vessel using 10 bolts and a confined-gasket seal. The effective sealing diameter is 150 mm. Other dimensions are: $A = 100$, $B = 200$, $C = 300$, $D = 20$, and $E = 20$, all in millimeters. The cylinder is used to store gas at a static pressure of 6 MPa. ISO class 8.8 bolts with a diameter of 12 mm have been selected. This provides an acceptable bolt spacing. What load factor n results from this selection?

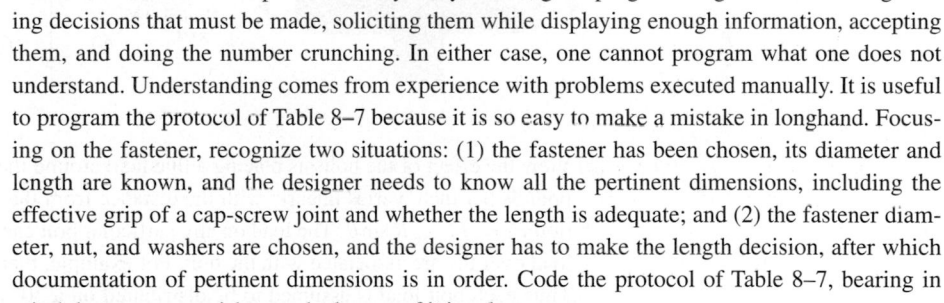

Problem 8–20
Cylinder head is steel; cylinder is
grade 30 cast iron.

8–21 The computer can be very helpful to the engineer. In matters of analysis it can take the drudgery out of calculations and improve accuracy. In synthesis, good programming is a matter of organizing decisions that must be made, soliciting them while displaying enough information, accepting them, and doing the number crunching. In either case, one cannot program what one does not understand. Understanding comes from experience with problems executed manually. It is useful to program the protocol of Table 8–7 because it is so easy to make a mistake in longhand. Focusing on the fastener, recognize two situations: (1) the fastener has been chosen, its diameter and length are known, and the designer needs to know all the pertinent dimensions, including the effective grip of a cap-screw joint and whether the length is adequate; and (2) the fastener diameter, nut, and washers are chosen, and the designer has to make the length decision, after which documentation of pertinent dimensions is in order. Code the protocol of Table 8–7, bearing in mind that you may wish to embed some of it in a larger program.

8–22 Figure P8–20 illustrates the connection of a cylinder head to a pressure vessel using 10 bolts and a confined-gasket seal. The effective sealing diameter is 150 mm. Other dimensions are: $A = 100$, $B = 200$, $C = 300$, $D = 20$, and $E = 25$, all in millimeters. The cylinder is used to store gas at a static pressure of 6 MPa. ISO class 8.8 bolts with a diameter of 12 mm have been selected. This provides an acceptable bolt spacing. What load factor n results from this selection?

8–23 We wish to alter the figure for Prob. 8–22 by decreasing the inside diameter of the seal to the diameter $A = 100$ mm. This makes an effective sealing diameter of 120 mm. Then, by using cap screws instead of bolts, the bolt circle diameter B can be reduced as well as the outside diameter C. If the same bolt spacing and the same edge distance are used, then eight 12-mm cap screws can be used on a bolt circle with $B = 160$ mm and an outside diameter of 260 mm, a substantial savings. With these dimensions and all other data the same as in Prob. 8–22, find the load factor.

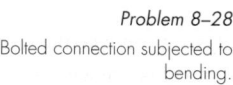 ANALYSIS **8-24** In the figure for Prob. 8–20, the bolts have a diameter of $\frac{1}{2}$ in and the cover plate is steel, with $D = \frac{1}{2}$ in. The cylinder is cast iron, with $E = \frac{5}{8}$ in and a modulus of elasticity of 18 Mpsi. The $\frac{1}{2}$-in SAE washer to be used under the nut has OD $= 1.062$ in and is 0.095 in thick. Find the stiffnesses of the bolt and the members and the joint constant C.

8-25 The same as Prob. 8–24, except that $\frac{1}{2}$-in cap screws are used with washers (see Fig. 8–21).

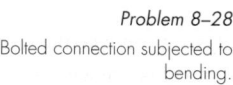 ANALYSIS **8-26** In addition to the data of Prob. 8–24, the dimensions of the cylinder are $A = 3.5$ in and an effective seal diameter of 4.25 in. The internal static pressure is 1500 psi. The outside diameter of the head is $C = 8$ in. The diameter of the bolt circle is 6 in, and so a bolt spacing in the range of 3 to 5 bolt diameters would require from 8 to 13 bolts. Select 10 SAE grade 5 bolts and find the resulting load factor n.

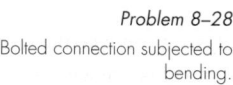 ANALYSIS **8-27** A $\frac{3}{8}$-in class 5 cap screw and steel washer are used to secure a cap to a cast-iron frame of a machine having a blind threaded hole. The washer is 0.065 in thick. The frame has a modulus of elasticity of 14 Mpsi and is $\frac{1}{4}$ in thick. The screw is 1 in long. The material in the frame also has a modulus of elasticity of 14 Mpsi. Find the stiffnesses k_b and k_m of the bolt and members.

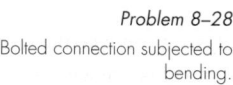 ANALYSIS **8-28** Bolts distributed about a bolt circle are often called upon to resist an external bending moment as shown in the figure. The external moment is 12 kip · in and the bolt circle has a diameter of 8 in. The neutral axis for bending is a diameter of the bolt circle. What needs to be determined is the most severe external load seen by a bolt in the assembly.

Problem 8–28

Bolted connection subjected to bending.

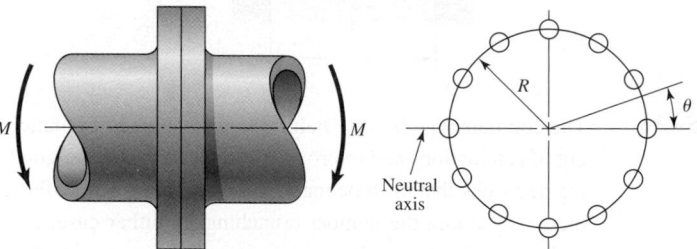

(*a*) View the effect of the bolts as placing a line load around the bolt circle whose intensity F_b', in pounds per inch, varies linearly with the distance from the neutral axis according to the relation $F_b' = F_{b,\max}' R \sin \theta$. The load on any particular bolt can be viewed as the effect of the line load over the arc associated with the bolt. For example, there are 12 bolts shown in the figure. Thus each bolt load is assumed to be distributed on a 30° arc of the bolt circle. Under these conditions, what is the largest bolt load?

(*b*) View the largest load as the intensity $F_{b,\max}'$ multiplied by the arc length associated with each bolt and find the largest bolt load.

(*c*) Express the load on any bolt as $F = F_{\max} \sin \theta$, sum the moments due to all the bolts, and estimate the largest bolt load. Compare the results of these three approaches to decide how to attack such problems in the future.

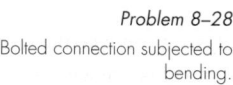 ANALYSIS **8-29** The figure shows a cast-iron bearing block that is to be bolted to a steel ceiling joist and is to support a gravity load. Bolts used are M20 ISO 8.8 with coarse threads and with 3.4-mm-thick steel washers under the bolt head and nut. The joist flanges are 20 mm in thickness, and the dimension A, shown in the figure, is 20 mm. The modulus of elasticity of the bearing block is 135 GPa.

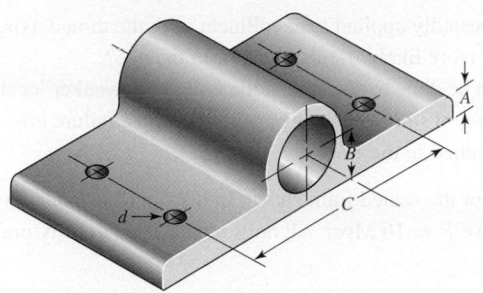

Problem 8–29

(*a*) Find the wrench torque required if the fasteners are lubricated during assembly and the joint is to be permanent.

(*b*) Determine the load factor for the design if the gravity load is 15 kN.

ANALYSIS   **8–30**   The upside-down steel A frame shown in the figure is to be bolted to steel beams on the ceiling of a machine room using ISO grade 8.8 bolts. This frame is to support the 40-kN radial load as illustrated. The total bolt grip is 48 mm, which includes the thickness of the steel beam, the A-frame feet, and the steel washers used. The bolts are size M20 × 2.5.

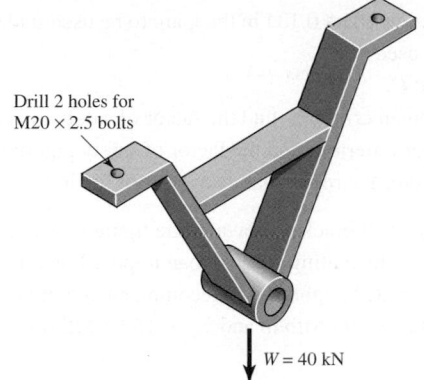

Drill 2 holes for
M20 × 2.5 bolts

Problem 8–30

$W = 40$ kN

(*a*) What tightening torque should be used if the connection is permanent and the fasteners are lubricated?

(*b*) What portion of the external load is taken by the bolts? By the members?

ANALYSIS   **8–31**   If the pressure in Prob. 8–20 is cycling between 0 and 6 MPa, determine the fatigue factor of safety using the:

(*a*) Goodman criterion.

(*b*) Gerber criterion.

(*c*) ASME-elliptic criterion.

ANALYSIS   **8–32**   In the figure for Prob. 8–20, let $A = 0.9$ m, $B = 1$ m, $C = 1.10$ m, $D = 20$ mm, and $E = 25$ mm. The cylinder is made of ASTM No. 35 cast iron ($E = 96$ GPa), and the head, of low-carbon steel. There are thirty-six M10 × 1.5 ISO 10.9 bolts tightened to 75 percent of proof load. During use, the cylinder pressure fluctuates between 0 and 550 kPa. Find the factor of safety guarding against a fatigue failure of a bolt using the:

(*a*) Goodman criterion.

(*b*) Gerber criterion.

(*c*) ASME-elliptic criterion.

ANALYSIS   **8–33**   A 1-in-diameter hot-rolled AISI 1144 steel rod is hot-formed into an eyebolt similar to that shown in the figure for Prob. 4–73, with an inner 2-in-diameter eye. The threads are 1 in-12 UNF and are die-cut.

(a) For a repeatedly applied load collinear with the thread axis, using the Gerber criterion fatigue failure is more likely in the thread or in the eye?

(b) What can be done to strengthen the bolt at the weaker location?

(c) If the factor of safety guarding against a fatigue failure is $n_f = 2$, what repeatedly applied load can be applied to the eye?

 ANALYSIS **8-34** The section of the sealed joint shown in the figure is loaded by a repeated force $P = 6$ kip. The members have $E = 16$ Mpsi. All bolts have been carefully preloaded to $F_i = 25$ kip each.

$\frac{3}{4}$ in-16 UNF $\times$ $2\frac{1}{2}$ in
SAE grade 5

Problem 8–34

$1\frac{1}{2}$ in

No. 40 CI

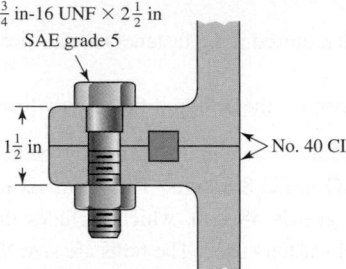

(a) If hardened-steel washers 0.134 in thick are to be used under the head and nut, what length of bolts should be used?

(b) Find k_b, k_m, and C.

(c) Using the Goodman criterion, find the factor of safety guarding against a fatigue failure.

(d) Using the Gerber criterion, find the factor of safety guarding against a fatigue failure.

(e) Find the load factor guarding against overproof loading.

 ANALYSIS **8-35** Suppose the welded steel bracket shown in the figure is bolted underneath a structural-steel ceiling beam to support a fluctuating vertical load imposed on it by a pin and yoke. The bolts are $\frac{1}{2}$ in coarse-thread SAE grade 5, tightened to recommended preload. The stiffnesses have already been computed and are $k_b = 4.94$ Mlb/in and $k_m = 15.97$ Mlb/in.

Problem 8–35

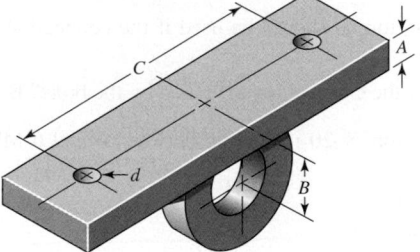

(a) Assuming that the bolts, rather than the welds, govern the strength of this design, determine the safe repeated load P that can be imposed on this assembly using the Goodman criterion and a fatigue design factor of 2.

(b) Repeat part (a) using the Gerber criterion.

(c) Compute the load factors based on the load found in part (b).

 ANALYSIS **8-36** Using the Gerber fatigue criterion and a fatigue-design factor of 2, determine the external repeated load P that a $1\frac{1}{4}$-in SAE grade 5 coarse-thread bolt can take compared with that for a fine-thread bolt. The joint constants are $C = 0.30$ for coarse- and 0.32 for fine-thread bolts.

 ANALYSIS **8-37** An M30 $\times$ 3.5 ISO 8.8 bolt is used in a joint at recommended preload, and the joint is subject to a repeated tensile fatigue load of $P = 80$ kN per bolt. The joint constant is $C = 0.33$. Find the

load factors and the factor of safety guarding against a fatigue failure based on the Gerber fatigue criterion.

 ANALYSIS **8–38** The figure shows a fluid-pressure linear actuator (hydraulic cylinder) in which $D = 4$ in, $t = \frac{3}{8}$ in, $L = 12$ in, and $w = \frac{3}{4}$ in. Both brackets as well as the cylinder are of steel. The actuator has been designed for a working pressure of 2000 psi. Six $\frac{3}{8}$-in SAE grade 5 coarse-thread bolts are used, tightened to 75 percent of proof load.

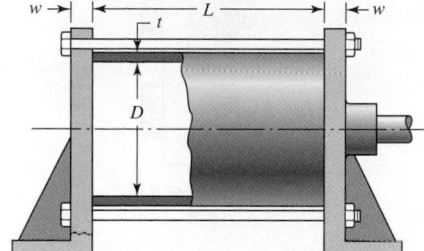

Problem 8–38

(a) Find the stiffnesses of the bolts and members, assuming that the entire cylinder is compressed uniformly and that the end brackets are perfectly rigid.

(b) Using the Goodman fatigue criterion, find the factor of safety guarding against a fatigue failure.

(c) Repeat part (b) using the Gerber fatigue criterion.

(d) What pressure would be required to cause total joint separation?

 ANALYSIS **8–39** The figure shows a bolted lap joint that uses SAE grade 8 bolts. The members are made of cold-drawn AISI 1040 steel. Find the safe tensile shear load F that can be applied to this connection if the following factors of safety are specified: shear of bolts 3, bearing on bolts 2, bearing on members 2.5, and tension of members 3.

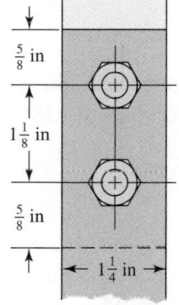 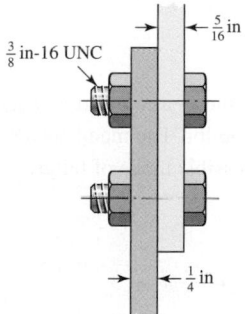

Problem 8–39

 ANALYSIS **8–40** The bolted connection shown in the figure uses SAE grade 5 bolts. The members are hot-rolled AISI 1018 steel. A tensile shear load $F = 4000$ lbf is applied to the connection. Find the factor of safety for all possible modes of failure.

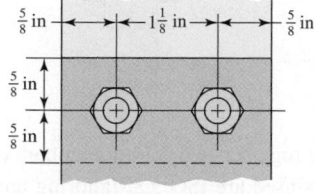 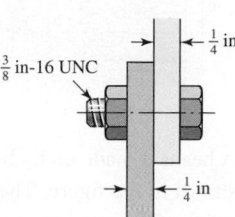

Problem 8–40

ANALYSIS

8–41   A bolted lap joint using SAE grade 5 bolts and members made of cold-drawn SAE 1040 steel is shown in the figure. Find the tensile shear load F that can be applied to this connection if the following factors of safety are specified: shear of bolts 1.8, bearing on bolts 2.2, bearing on members 2.4, and tension of members 2.6.

Problem 8–41

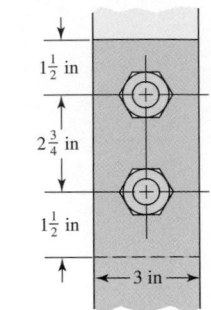

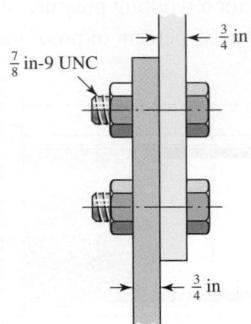

ANALYSIS

8–42   The bolted connection shown in the figure is subjected to a tensile shear load of 20 kip. The bolts are SAE grade 5 and the material is cold-drawn AISI 1015 steel. Find the factor of safety of the connection for all possible modes of failure.

Problem 8–42

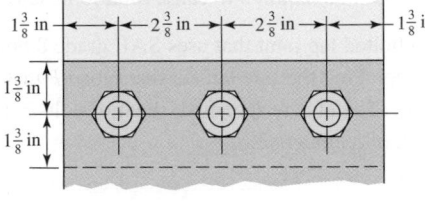

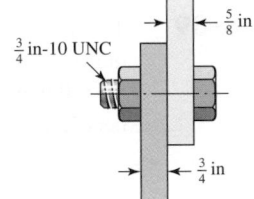

ANALYSIS

8–43   The figure shows a connection that employs three SAE grade 5 bolts. The tensile shear load on the joint is 5400 lbf. The members are cold-drawn bars of AISI 1020 steel. Find the factor of safety for each possible mode of failure.

Problem 8–43

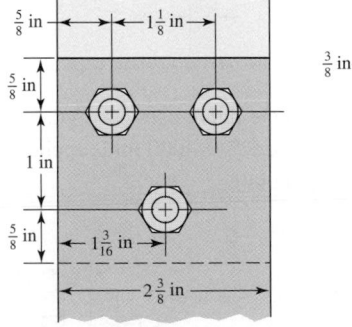

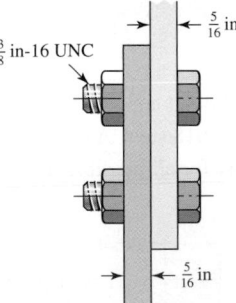

ANALYSIS

8–44   A beam is made up by bolting together two cold-drawn bars of AISI 1018 steel as a lap joint, as shown in the figure. The bolts used are ISO 5.8. Ignoring any twisting, determine the factor of safety of the connection.

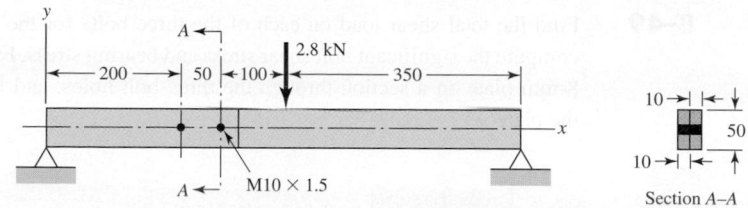

Section A–A

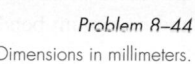ANALYSIS   **8–45**   Standard design practice, as exhibited by the solutions to Probs. 8–39 to 8–43, is to assume that the bolts, or rivets, share the shear equally. For many situations, such an assumption may lead to an unsafe design. Consider the yoke bracket of Prob. 8–35, for example. Suppose this bracket is bolted to a wide-flange *column* with the centerline through the two bolts in the vertical direction. A vertical load through the yoke-pin hole at distance B from the column flange would place a shear load on the bolts as well as a tensile load. The tensile load comes about because the bracket tends to pry itself about the bottom corner, much like a claw hammer, exerting a large tensile load on the upper bolt. In addition, it is almost certain that both the spacing of the bolt holes and their diameters will be slightly different on the column flange from what they are on the yoke bracket. Thus, unless yielding occurs, only one of the bolts will take the shear load. The designer has no way of knowing which bolt this will be.

In this problem the bracket is 8 in long, $A = \frac{1}{2}$ in, $B = 3$ in, $C = 6$ in, and the column flange is $\frac{1}{2}$ in thick. The bolts are $\frac{1}{2}$ in UNC SAE 5. Steel washers 0.095 in thick are used under the nuts. The nuts are tightened to 75 percent of proof load. The vertical yoke-pin load is 3000 lbf. If the upper bolt takes all the shear load as well as the tensile load, how closely does the bolt stress approach the proof strength?

ANALYSIS   **8–46**   The bearing of Prob. 8–29 is bolted to a vertical surface and supports a horizontal shaft. The bolts used have coarse threads and are M20 ISO 5.8. The joint constant is $C = 0.30$, and the dimensions are $A = 20$ mm, $B = 50$ mm, and $C = 160$ mm. The bearing base is 240 mm long. The bearing load is 12 kN. If the bolts are tightened to 75 percent of proof load, will the bolt stress exceed the proof strength? Use worst-case loading, as discussed in Prob. 8–45.

ANALYSIS   **8–47**   A split-ring clamp-type shaft collar such as is described in Prob. 6–29 must resist an axial load of 1000 lbf. Using a design factor of $n = 3$ and a coefficient of friction of 0.12, specify an SAE Grade 5 cap screw using fine threads. What wrench torque should be used if a lubricated screw is used?

ANALYSIS   **8–48**   A vertical channel 152×76 (see Table A–7) has a cantilever bolted to it as shown. The channel is hot-rolled AISI 1006 steel. The bar is of hot-rolled AISI 1015 steel. The bolts are M12 $\times$ 1.75 ISO 5.8. For a design factor of 2.8, find the safe force F that can be applied to the cantilever.

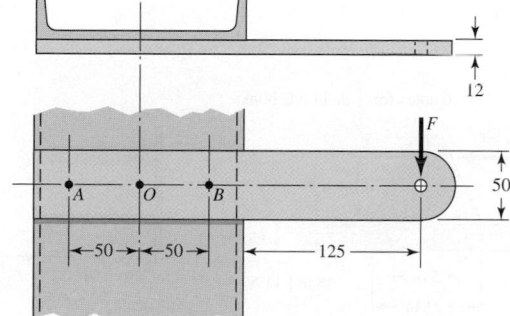

ANALYSIS

8–49 Find the total shear load on each of the three bolts for the connection shown in the figure and compute the significant bolt shear stress and bearing stress. Find the second moment of area of the 8-mm plate on a section through the three bolt holes, and find the maximum bending stress in the plate.

Problem 8–49
Dimensions in millimeters.

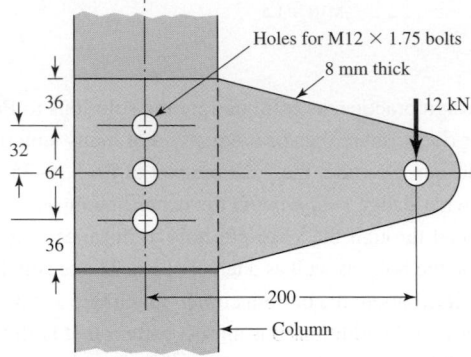

ANALYSIS

8–50 A $\frac{3}{8}$- × 2-in AISI 1018 cold-drawn steel bar is cantilevered to support a static load of 300 lbf as illustrated. The bar is secured to the support using two $\frac{1}{2}$ in-13 UNC SAE 5 bolts. Find the factor of safety for the following modes of failure: shear of bolt, bearing on bolt, bearing on member, and strength of member.

Problem 8–50

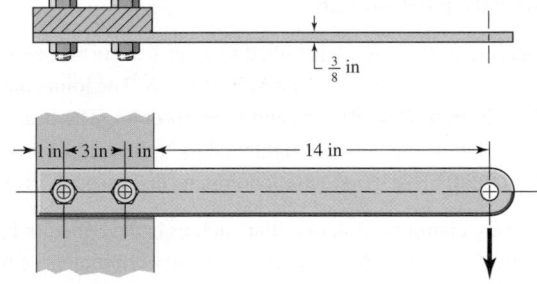

ANALYSIS

8–51 The figure shows a welded fitting which has been tentatively designed to be bolted to a channel so as to transfer the 2500-lbf load into the channel. The channel is made of hot-rolled low-carbon steel having a minimum yield strength of 46 kpsi; the two fitting plates are of hot-rolled stock having a minimum S_y of 45.5 kpsi. The fitting is to be bolted using six standard SAE grade 2 bolts. Check the strength of the design by computing the factor of safety for all possible modes of failure.

Problem 8–51
Dimensions in inches.

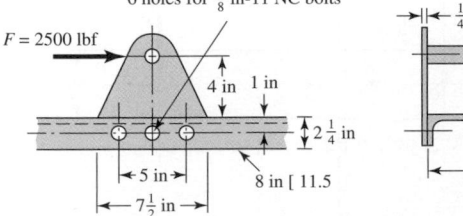

DESIGN

8–52   A cantilever is to be attached to the flat side of a 6-in, 13.0-lbf/in channel used as a column. The cantilever is to carry a load as shown in the figure. To a designer the choice of a bolt array is usually an a priori decision. Such decisions are made from a background of knowledge of the effectiveness of various patterns.

Problem 8–52

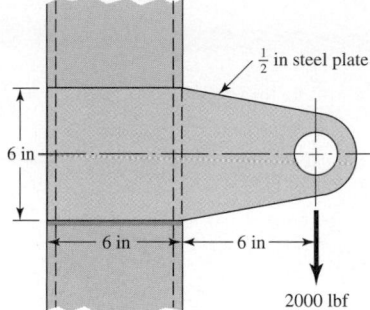

$\frac{1}{2}$ in steel plate

6 in

6 in

6 in

2000 lbf

(a) If two fasteners are used, should the array be arranged vertically, horizontally, or diagonally? How would you decide?

(b) If three fasteners are used, should a linear or triangular array be used? For a triangular array, what should be the orientation of the triangle? How would you decide?

8–53   Using your experience with Prob. 8–52, specify a bolt pattern for Prob. 8–52, and size the bolts.

8–54   Determining the joint stiffness of nonsymmetric joints of two or more different materials using a frustum of a hollow cone can be time-consuming and prone to error. Develop a computer program to determine k_m for a joint composed of two different materials of differing thickness. Test the program to determine k_m for problems such as Ex. 8–5 and Probs. 8–19, 8–20, 8–22, 8–24, and 8–27.

9

Welding, Bonding, and the Design of Permanent Joints

Form can more readily pursue function with the help of joining processes such as welding, brazing, soldering, cementing, and gluing—processes that are used extensively in manufacturing today. Whenever parts have to be assembled or fabricated, there is usually good cause for considering one of these processes in preliminary design work. Particularly when sections to be joined are thin, one of these methods may lead to significant savings. The elimination of individual fasteners, with their holes and assembly costs, is an important factor. Also, some of the methods allow rapid machine assembly, furthering their attractiveness.

Riveted permanent joints were common as the means of fastening rolled steel shapes to one another to form a permanent joint. The childhood fascination of seeing a cherry-red hot rivet thrown with tongs across a building skeleton to be unerringly caught by a person with a conical bucket, to be hammered pneumatically into its final shape, is all but gone. Two developments relegated riveting to lesser prominence. The first was the development of high-strength steel bolts whose preload could be controlled. The second was the improvement of welding, competing both in cost and in latitude of possible form.

9–1 Welding Symbols

A weldment is fabricated by welding together a collection of metal shapes, cut to particular configurations. During welding, the several parts are held securely together, often by clamping or jigging. The welds must be precisely specified on working drawings, and this is done by using the welding symbol, shown in Fig. 9–1, as standardized by the American Welding Society (AWS). The arrow of this symbol points to the joint to be welded. The body of the symbol contains as many of the following elements as are deemed necessary:

- Reference line
- Arrow

Figure 9–1

The AWS standard welding symbol showing the location of the symbol elements.

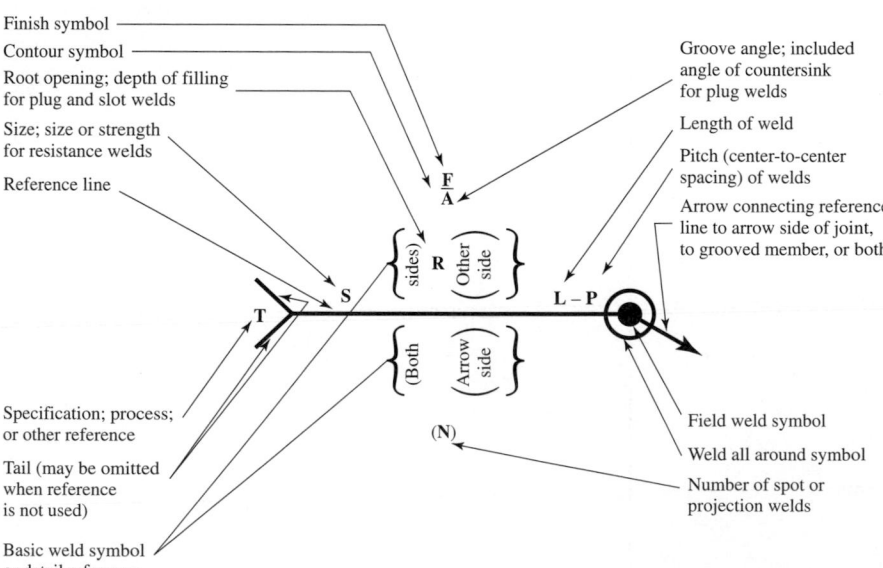

- Basic weld symbols as in Fig. 9–2
- Dimensions and other data
- Supplementary symbols
- Finish symbols
- Tail
- Specification or process

The *arrow side* of a joint is the line, side, area, or near member to which the arrow points. The side opposite the arrow side is the *other side*.

Figures 9–3 to 9–6 illustrate the types of welds used most frequently by designers. For general machine elements most welds are fillet welds, though butt welds are used a great deal in designing pressure vessels. Of course, the parts to be joined must be arranged so that there is sufficient clearance for the welding operation. If unusual joints are required because of insufficient clearance or because of the section shape, the design may be a poor one and the designer should begin again and endeavor to synthesize another solution.

Since heat is used in the welding operation, there are metallurgical changes in the parent metal in the vicinity of the weld. Also, residual stresses may be introduced because of clamping or holding or, sometimes, because of the order of welding. Usually

Figure 9–2

Arc- and gas-weld symbols.

			Type of weld				
		Plug or slot	Square	Groove			
Bead	Fillet			V	Bevel	U	J
⌓	◺	⏢	‖	⋁	⋁	⋃	⋃

Figure 9–3

Fillet welds. (*a*) The number indicates the leg size; the arrow should point only to one weld when both sides are the same. (*b*) The symbol indicates that the welds are intermittent and staggered 60 mm along on 200-mm centers.

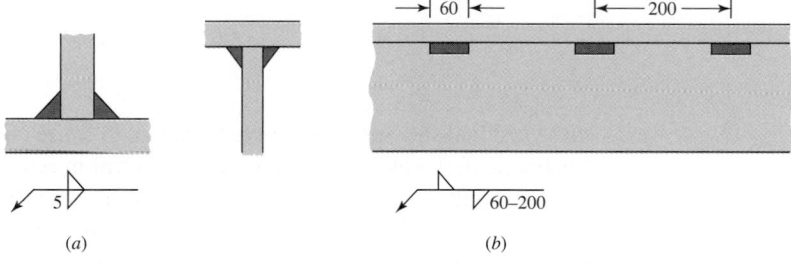

(*a*) (*b*)

Figure 9–4

The circle on the weld symbol indicates that the welding is to go all around.

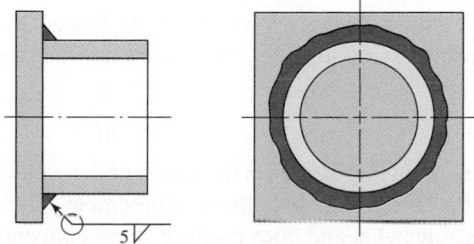

Figure 9–5

Butt or groove welds:
(a) square butt-welded on both sides; (b) single V with 60° bevel and root opening of 2 mm; (c) double V; (d) single bevel.

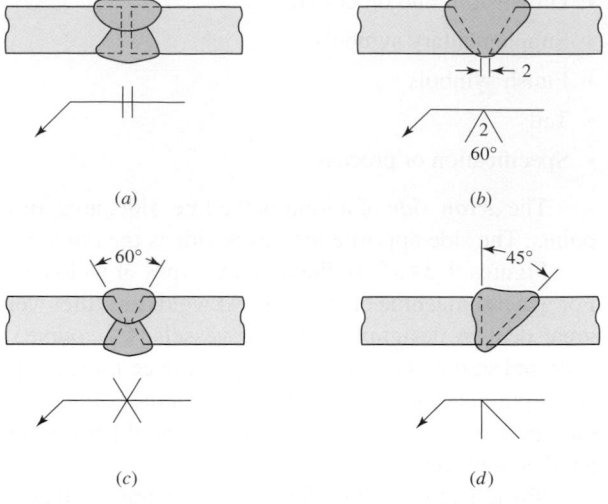

(a)   *(b)*

(c)   *(d)*

Figure 9–6

Special groove welds: (a) T joint for thick plates; (b) U and J welds for thick plates;
(c) corner weld (may also have a bead weld on inside for greater strength but should not be used for heavy loads);
(d) edge weld for sheet metal and light loads.

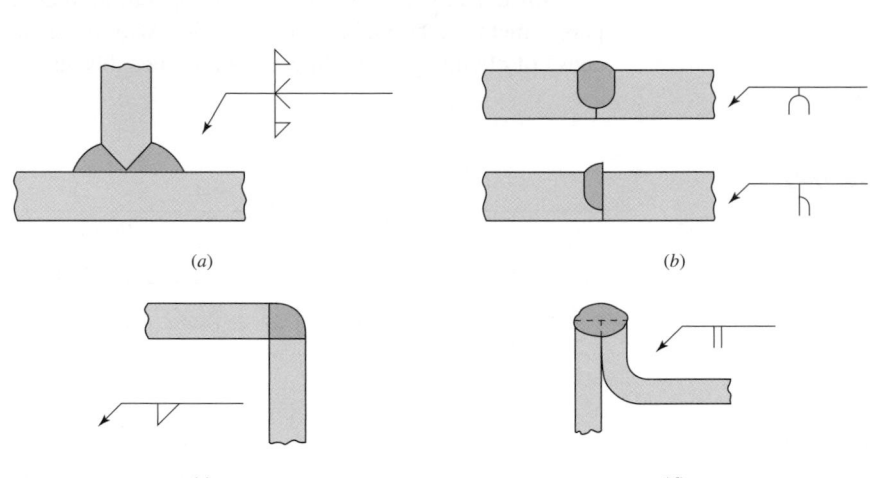

(a)   *(b)*

(c)   *(d)*

these residual stresses are not severe enough to cause concern; in some cases a light heat treatment after welding has been found helpful in relieving them. When the parts to be welded are thick, a preheating will also be of benefit. If the reliability of the component is to be quite high, a testing program should be established to learn what changes or additions to the operations are necessary to ensure the best quality.

9–2 Butt and Fillet Welds

Figure 9–7a shows a single V-groove weld loaded by the tensile force F. For either tension or compression loading, the average normal stress is

$$\sigma = \frac{F}{hl} \tag{9–1}$$

where h is the weld throat and l is the length of the weld, as shown in the figure. Note that the value of h does not include the reinforcement. The reinforcement can be desirable, but it varies somewhat and does produce stress concentration at point A in the figure. If fatigue loads exist, it is good practice to grind or machine off the reinforcement.

Figure 9–7

A typical butt joint.

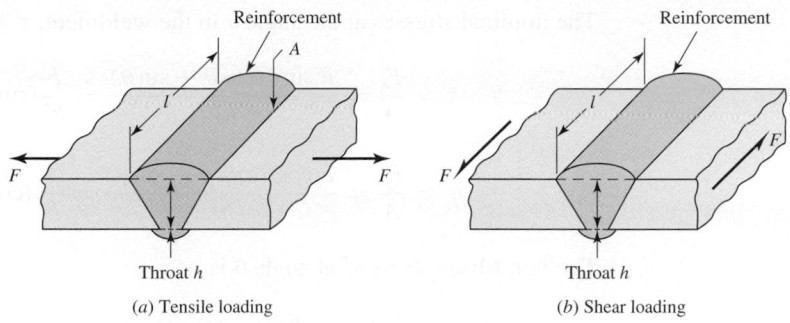

(a) Tensile loading   (b) Shear loading

Figure 9–8

A transverse fillet weld.

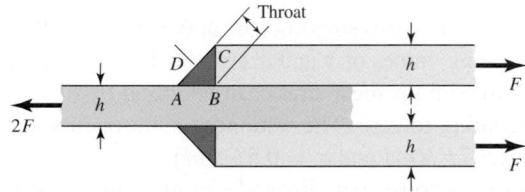

Figure 9–9

Free body from Fig. 9–8.

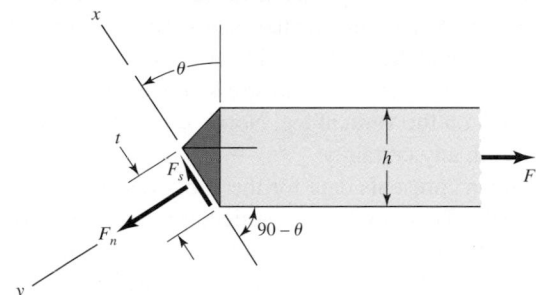

The average stress in a butt weld due to shear loading (Fig. 9–7b) is

$$\tau = \frac{F}{hl} \tag{9–2}$$

Figure 9–8 illustrates a typical transverse fillet weld. In Fig. 9–9 a portion of the welded joint has been isolated from Fig. 9–8 as a free body. At angle θ the forces on each weldment consist of a normal force F_n and a shear force F_s. Summing forces in the x and y directions gives

$$F_s = F \sin \theta \tag{a}$$

$$F_n = F \cos \theta \tag{b}$$

Using the law of sines for the triangle in Fig. 9–9 yields

$$\frac{t}{\sin 45°} = \frac{h}{\sin(90° - \theta + 45°)} = \frac{h}{\sin(135° - \theta)} = \frac{\sqrt{2}h}{\cos \theta + \sin \theta}$$

Solving for the throat length t gives

$$t = \frac{h}{\cos \theta + \sin \theta} \tag{c}$$

The nominal stresses at the angle θ in the weldment, τ and σ, are

$$\tau = \frac{F_s}{A} = \frac{F\sin\theta(\cos\theta + \sin\theta)}{hl} = \frac{F}{hl}(\sin\theta\cos\theta + \sin^2\theta) \qquad (d)$$

$$\sigma = \frac{F_n}{A} = \frac{F\cos\theta(\cos\theta + \sin\theta)}{hl} = \frac{F}{hl}(\cos^2\theta + \sin\theta\cos\theta) \qquad (e)$$

The von Mises stress σ' at angle θ is

$$\sigma' = (\sigma^2 + 3\tau^2)^{1/2} = \frac{F}{hl}[(\cos^2\theta + \sin\theta\cos\theta)^2 + 3(\sin^2\theta + \sin\theta\cos\theta)^2]^{1/2} \qquad (f)$$

The largest von Mises stress occurs at $\theta = 62.5°$ with a value of $\sigma' = 2.16F/(hl)$. The corresponding values of τ and σ are $\tau = 1.196F/(hl)$ and $\sigma = 0.623F/(hl)$.

The maximum shear stress can be found by differentiating Eq. (d) with respect to θ and equating to zero. The stationary point occurs at $\theta = 67.5°$ with a corresponding $\tau_{\max} = 1.207F/(hl)$ and $\sigma = 0.5F/(hl)$.

There are some experimental and analytical results that are helpful in evaluating Eqs. (d) through (f) and consequences. A model of the transverse fillet weld of Fig. 9–8 is easily constructed for photoelastic purposes and has the advantage of a balanced loading condition. Norris constructed such a model and reported the stress distribution along the sides AB and BC of the weld.[1] An approximate graph of the results he obtained is shown as Fig. 9–10a. Note that stress concentration exists at A and B on the horizontal leg and at B on the vertical leg. Norris states that he could not determine the stresses at A and B with any certainty.

Salakian[2] presents data for the stress distribution across the throat of a fillet weld (Fig. 9–10b). This graph is of particular interest because we have just learned that it is the throat stresses that are used in design. Again, the figure shows stress concentration at point B. Note that Fig. 9–10a applies either to the weld metal or to the parent metal, and that Fig. 9–10b applies only to the weld metal.

Equations (a) through (f) and their consequences seem familiar, and we can become comfortable with them. The net result of photoelastic and finite element analysis of transverse fillet weld geometry is more like that shown in Fig. 9–10 than those given by mechanics of materials or elasticity methods. The most important concept here is that we have *no analytical approach that predicts the existing stresses*. The geometry of the fillet is crude by machinery standards, and even if it were ideal, the macrogeometry is too abrupt and complex for our methods. There are also subtle bending stresses due to eccentricities. Still, in the absence of robust analysis, weldments must be specified and the resulting joints must be safe. The approach has been to use a simple *and conservative* model, verified by testing as conservative. The approach has been to

- Consider the external loading to be carried by shear forces on the throat area of the weld. By ignoring the normal stress on the throat, the shearing stresses are inflated sufficiently to render the model conservative.

---

[1]C. H. Norris, "Photoelastic Investigation of Stress Distribution in Transverse Fillet Welds," *Welding J.,* vol. 24, 1945, p. 557s.

[2]A. G. Salakian and G. E. Claussen, "Stress Distribution in Fillet Welds: A Review of the Literature," *Welding J.,* vol. 16, May 1937, pp. 1–24.

Figure 9–10

Stress distribution in fillet welds: (a) stress distribution on the legs as reported by Norris; (b) distribution of principal stresses and maximum shear stress as reported by Salakian.

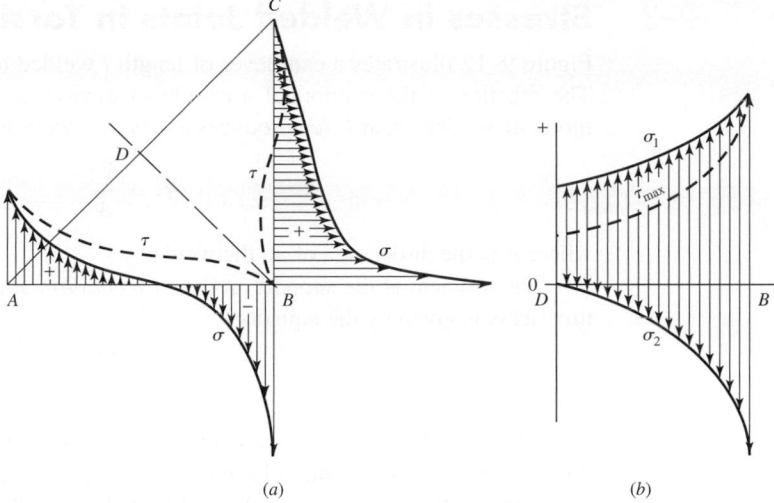

(a)   (b)

Figure 9–11

Parallel fillet welds.

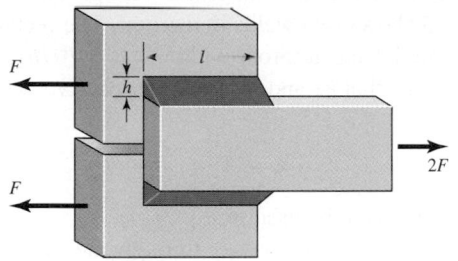

- Use distortion energy for significant stresses.
- Circumscribe typical cases by code.

For this model, the basis for weld analysis or design employs

$$\tau = \frac{F}{0.707hl} = \frac{1.414F}{hl} \qquad (9\text{–}3)$$

which assumes the entire force F is accounted for by a shear stress in the minimum throat area. Note that this inflates the maximum estimated shear stress by a factor of $1.414/1.207 = 1.17$. Further, consider the parallel fillet welds shown in Fig. 9–11 where, as in Fig. 9–8, each weld transmits a force F. However, in the case of Fig. 9–11, the maximum shear stress *is* at the minimum throat area and corresponds to Eq. (9–3).

Under circumstances of combined loading we

- Examine primary shear stresses due to external forces.
- Examine secondary shear stresses due to torsional and bending moments.
- Estimate the strength(s) of the parent metal(s).
- Estimate the strength of deposited weld metal.
- Estimate permissible load(s) for parent metal(s).
- Estimate permissible load for deposited weld metal.

9–3 **Stresses in Welded Joints in Torsion**

Figure 9–12 illustrates a cantilever of length l welded to a column by two fillet welds. The reaction at the support of a cantilever always consists of a shear force V and a moment M. The shear force produces a *primary shear* in the welds of magnitude

$$\tau' = \frac{V}{A} \tag{9–4}$$

where A is the throat area of all the welds.

The moment at the support produces *secondary shear* or *torsion* of the welds, and this stress is given by the equation

$$\tau'' = \frac{Mr}{J} \tag{9–5}$$

where r is the distance from the centroid of the weld group to the point in the weld of interest and J is the second polar moment of area of the weld group about the centroid of the group. When the sizes of the welds are known, these equations can be solved and the results combined to obtain the maximum shear stress. Note that r is usually the farthest distance from the centroid of the weld group.

Figure 9–13 shows two welds in a group. The rectangles represent the throat areas of the welds. Weld 1 has a throat width $b_1 = 0.707h_1$, and weld 2 has a throat width $d_2 = 0.707h_2$. Note that h_1 and h_2 are the respective weld sizes. The throat area of both welds together is

$$A = A_1 + A_2 = b_1 d_1 + b_2 d_2 \tag{a}$$

This is the area that is to be used in Eq. (9–4).

The x axis in Fig. 9–13 passes through the centroid G_1 of weld 1. The second moment of area about this axis is

$$I_x = \frac{b_1 d_1^3}{12}$$

Similarly, the second moment of area about an axis through G_1 parallel to the y axis is

$$I_y = \frac{d_1 b_1^3}{12}$$

Figure 9–12

This is a *moment connection;* such a connection produces *torsion* in the welds.

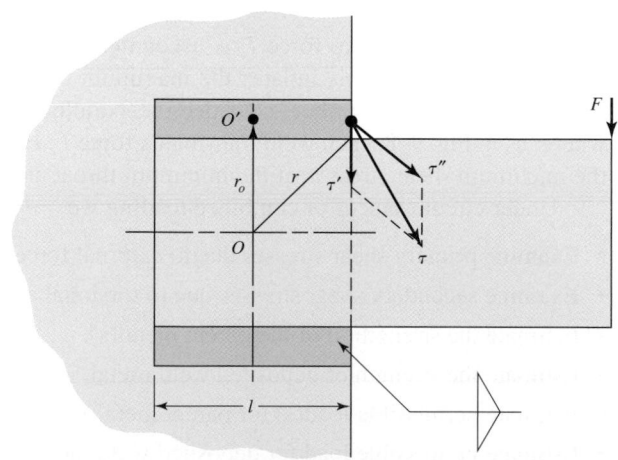

| Figure 9–13

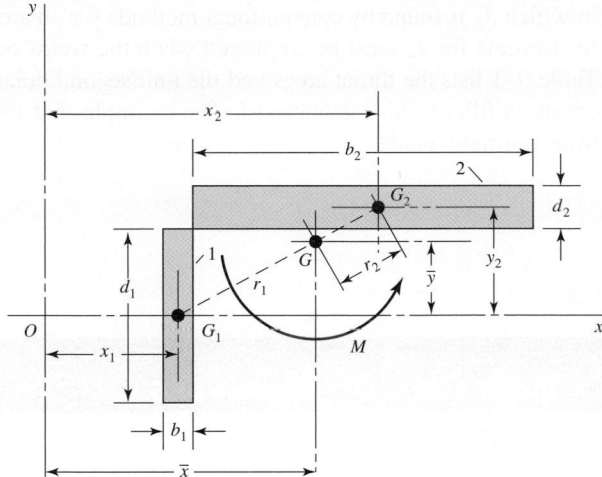

Thus the second polar moment of area of weld 1 about its own centroid is

$$J_{G1} = I_x + I_y = \frac{b_1 d_1^3}{12} + \frac{d_1 b_1^3}{12} \qquad (b)$$

In a similar manner, the second polar moment of area of weld 2 about its centroid is

$$J_{G2} = \frac{b_2 d_2^3}{12} + \frac{d_2 b_2^3}{12} \qquad (c)$$

The centroid G of the weld group is located at

$$\bar{x} = \frac{A_1 x_1 + A_2 x_2}{A} \qquad \bar{y} = \frac{A_1 y_1 + A_2 y_2}{A}$$

Using Fig. 9–13 again, we see that the distances r_1 and r_2 from G_1 and G_2 to G, respectively, are

$$r_1 = [(\bar{x} - x_1)^2 + \bar{y}^2]^{1/2} \qquad r_2 = [(y_2 - \bar{y})^2 + (x_2 - \bar{x})^2]^{1/2}$$

Now, using the parallel-axis theorem, we find the second polar moment of area of the weld group to be

$$J = \left(J_{G1} + A_1 r_1^2 \right) + \left(J_{G2} + A_2 r_2^2 \right) \qquad (d)$$

This is the quantity to be used in Eq. (9–5). The distance r must be measured from G and the moment M computed about G.

The reverse procedure is that in which the allowable shear stress is given and we wish to find the weld size. The usual procedure is to estimate a probable weld size and then to use iteration.

Observe in Eqs. (b) and (c) the quantities b_1^3 and d_2^3, respectively, which are the cubes of the weld widths. These quantities are small and can be neglected. This leaves the terms $b_1 d_1^3/12$ and $d_2 b_2^3/12$, which make J_{G1} and J_{G2} linear in the weld width. Setting the weld widths b_1 and d_2 to unity leads to the idea of treating each fillet weld as a line. The resulting second moment of area is then a *unit second polar moment of area*. The advantage of treating the weld size as a line is that the value of J_u is the same regardless of the weld size. Since the throat width of a fillet weld is $0.707h$, the relationship between J and the unit value is

$$J = 0.707h J_u \qquad (9\text{–}6)$$

in which J_u is found by conventional methods for an area having unit width. The transfer formula for J_u must be employed when the welds occur in groups, as in Fig. 9–12. Table 9–1 lists the throat areas and the unit second polar moments of area for the most common fillet welds encountered. The example that follows is typical of the calculations normally made.

Table 9–1

Torsional Properties of Fillet Welds*

Weld	Throat Area	Location of G	Unit Second Polar Moment of Area
	$A = 0.707hd$	$\bar{x} = 0$ $\bar{y} = d/2$	$J_u = d^3/12$
	$A = 1.414hd$	$\bar{x} = b/2$ $\bar{y} = d/2$	$J_u = \dfrac{d(3b^2 + d^2)}{6}$
	$A = 0.707h(2b + d)$	$\bar{x} = \dfrac{b^2}{2(b + d)}$ $\bar{y} = \dfrac{d^2}{2(b + d)}$	$J_u = \dfrac{(b + d)^4 - 6b^2 d^2}{12(b + d)}$
	$A = 0.707h(2b + d)$	$\bar{x} = \dfrac{b^2}{2b + d}$ $\bar{y} = d/2$	$J_u = \dfrac{8b^3 + 6bd^2 + d^3}{12} - \dfrac{b^4}{2b + d}$
	$A = 1.414h(b + d)$	$\bar{x} = b/2$ $\bar{y} = d/2$	$J_u = \dfrac{(b + d)^3}{6}$
	$A = 1.414\pi hr$		$J_u = 2\pi r^3$

*G is centroid of weld group; h is weld size; plane of torque couple is in the plane of the paper; all welds are of unit width.

EXAMPLE 9–1

A 50-kN load is transferred from a welded fitting into a 200-mm steel channel as illustrated in Fig. 9–14. Estimate the maximum stress in the weld.

Solution[3]

(*a*) Label the ends and corners of each weld by letter. Sometimes it is desirable to label each weld of a set by number. See Fig. 9–15.

(*b*) Estimate the primary shear stress τ'. As shown in Fig. 9–14, each plate is welded to the channel by means of three 6-mm fillet welds. Figure 9–15 shows that we have divided the load in half and are considering only a single plate. From case 4 of Table 9–1 we find the throat area as

$$A = 0.707(6)[2(56) + 190] = 1280 \text{ mm}^2$$

Then the primary shear stress is

$$\tau' = \frac{V}{A} = \frac{25(10)^3}{1280} = 19.5 \text{ MPa}$$

Figure 9–14

Dimensions in millimeters.

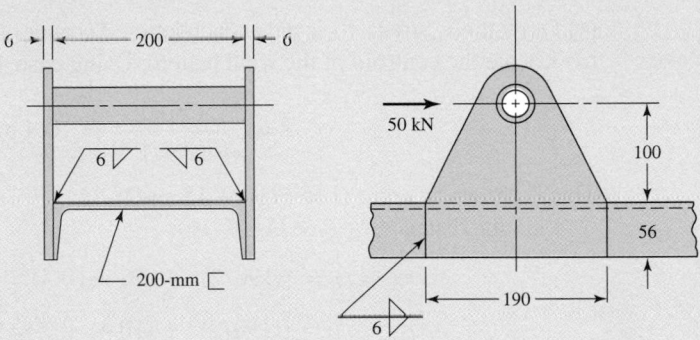

Figure 9–15

Diagram showing the weld geometry; all dimensions in millimeters. Note that *V* and *M* represent loads applied by the welds *to the plate*.

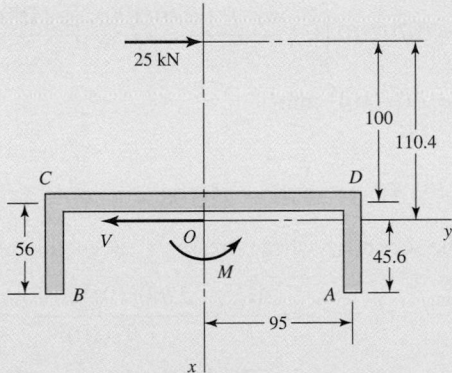

---

[3]We are indebted to Professor George Piotrowski of the University of Florida for the detailed steps, presented here, of his method of weld analysis. J.E.S., C.R.M., R.G.B.

Figure 9–16

Free-body diagram of one of the side plates.

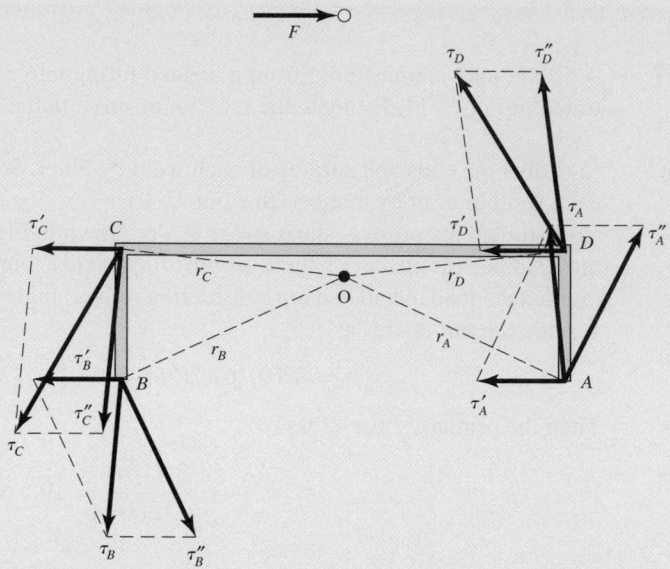

(c) Draw the τ' stress, to scale, at each lettered corner or end. See Fig. 9–16.

(d) Locate the centroid of the weld pattern. Using case 4 of Table 9–1, we find

$$\bar{x} = \frac{(56)^2}{2(56) + 190} = 10.4 \text{ mm}$$

This is shown as point O on Figs. 9–15 and 9–16.

(e) Find the distances r_i (see Fig. 9–16):

$$r_A = r_B = [(190/2)^2 + (56 - 10.4)^2]^{1/2} = 105 \text{ mm}$$

$$r_C = r_D = [(190/2)^2 + (10.4)^2]^{1/2} = 95.6 \text{ mm}$$

These distances can also be scaled from the drawing.

(f) Find J. Using case 4 of Table 9–1 again, we get

$$J = 0.707(6) \left[\frac{8(56)^3 + 6(56)(190)^2 + (190)^3}{12} - \frac{(56)^4}{2(56) + 190} \right]$$

$$= 7.07(10)^6 \text{ mm}^4$$

(g) Find M:

$$M = Fl = 25(100 + 10.4) = 2760 \text{ N} \cdot \text{m}$$

(h) Estimate the secondary shear stresses τ'' at each lettered end or corner:

$$\tau_A'' = \tau_B'' = \frac{Mr}{J} = \frac{2760(10)^3(105)}{7.07(10)^6} = 41.0 \text{ MPa}$$

$$\tau_C'' = \tau_D'' = \frac{2760(10)^3(95.6)}{7.07(10)^6} = 37.3 \text{ MPa}$$

(i) Draw the τ'' stress, to scale, at each corner and end. See Fig. 9–16. Note that this is a free-body diagram of one of the side plates, and therefore the τ' and τ'' stresses represent what the channel is doing to the plate (through the welds) to hold the plate in equilibrium.

(*j*) At each letter, combine the two stress components as vectors. This gives

$$\tau_A = \tau_B = 37 \text{ MPa}$$

$$\tau_C = \tau_D = 44 \text{ MPa}$$

(*k*) Identify the most highly stressed point:

Answer

$$\tau_{\max} = \tau_C = \tau_D = 44 \text{ MPa}$$

9-4 Stresses in Welded Joints in Bending

Figure 9–17*a* shows a cantilever welded to a support by fillet welds at top and bottom. A free-body diagram of the beam would show a shear-force reaction V and a moment reaction M. The shear force produces a primary shear in the welds of magnitude

$$\tau' = \frac{V}{A} \tag{a}$$

where A is the total throat area.

The moment M induces a throat shear stress component of 0.707τ in the welds.[4] Treating the two welds of Fig. 9–17*b* as lines we find the unit second moment of area to be

$$I_u = \frac{bd^2}{2} \tag{b}$$

The second moment of area I, based on weld throat area, is

$$I = 0.707h I_u = 0.707h \frac{bd^2}{2} \tag{c}$$

The nominal throat shear stress is now found to be

$$\tau = \frac{Mc}{I} = \frac{Md/2}{0.707hbd^2/2} = \frac{1.414M}{bdh} \tag{d}$$

The model gives the coefficient of 1.414, in contrast to the predictions of Sec. 9–2 of 1.197 from distortion energy, or 1.207 from maximum shear. The conservatism of the model's 1.414 is not that it is simply larger than either 1.196 or 1.207, but the tests carried out to validate the model show that it is large enough.

Figure 9–17

A rectangular cross-section cantilever welded to a support at the top and bottom edges.

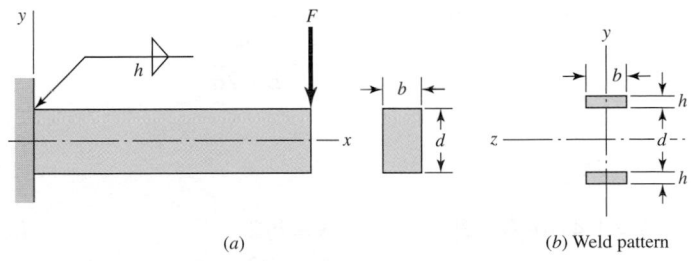

(*a*) (*b*) Weld pattern

---

[4]According to the model described before Eq. (9–3), the moment is carried by components of the shear stress 0.707τ parallel to the *x*-axis of Fig. 9–17. The *y* components cancel.

The second moment of area in Eq. (d) is based on the distance d between the two welds. If this moment is found by treating the two welds as having rectangular footprints, the distance between the weld throat centroids is approximately ($d + h$). This would produce a slightly larger second moment of area, and result in a smaller level of stress. This method of treating welds as a line does not interfere with the conservatism of the model. It also makes Table 9–2 possible with all the conveniences that ensue.

Table 9–2

Bending Properties of Fillet Welds*

Weld	Throat Area	Location of G	Unit Second Moment of Area
	$A = 0.707hd$	$\bar{x} = 0$ $\bar{y} = d/2$	$I_u = \dfrac{d^3}{12}$
	$A = 1.414hd$	$\bar{x} = b/2$ $\bar{y} = d/2$	$I_u = \dfrac{d^3}{6}$
	$A = 1.414hb$	$\bar{x} = b/2$ $\bar{y} = d/2$	$I_u = \dfrac{bd^2}{2}$
	$A = 0.707h(2b + d)$	$\bar{x} = \dfrac{b^2}{2b + d}$ $\bar{y} = d/2$	$I_u = \dfrac{d^2}{12}(6b + d)$
	$A = 0.707h(b + 2d)$	$\bar{x} = b/2$ $\bar{y} = \dfrac{d^2}{b + 2d}$	$I_u = \dfrac{2d^3}{3} - 2d^2\bar{y} + (b + 2d)\bar{y}^2$
	$A = 1.414h(b + d)$	$\bar{x} = b/2$ $\bar{y} = d/2$	$I_u = \dfrac{d^2}{6}(3b + d)$

Table 9–2

Continued

Weld	Throat Area	Location of G	Unit Second Moment of Area
	$A = 0.707h(b + 2d)$	$\bar{x} = b/2$ $\bar{y} = \dfrac{d^2}{b + 2d}$	$I_u = \dfrac{2d^3}{3} - 2d^2\bar{y} + (b + 2d)\bar{y}^2$
	$A = 1.414h(b + d)$	$\bar{x} = b/2$ $\bar{y} = d/2$	$I_u = \dfrac{d^2}{6}(3b + d)$
	$A = 1.414\pi hr$		$I_u = \pi r^3$

*I_u, unit second moment of area, is taken about a horizontal axis through G, the centroid of the weld group, h is weld size; the plane of the bending couple is normal to the plane of the paper and parallel to the y-axis; all welds are of the same size.

9–5   ## The Strength of Welded Joints

The matching of the electrode properties with those of the parent metal is usually not so important as speed, operator appeal, and the appearance of the completed joint. The properties of electrodes vary considerably, but Table 9–3 lists the minimum properties for some electrode classes.

It is preferable, in designing welded components, to select a steel that will result in a fast, economical weld even though this may require a sacrifice of other qualities such as machinability. Under the proper conditions, all steels can be welded, but best results will be obtained if steels having a UNS specification between G10140 and G10230 are chosen. All these steels have a tensile strength in the hot-rolled condition in the range of 60 to 70 kpsi.

The designer can choose factors of safety or permissible working stresses with more confidence if he or she is aware of the values of those used by others. One of the best standards to use is the American Institute of Steel Construction (AISC) code for building construction.[5] The permissible stresses are now based on the yield strength of the material instead of the ultimate strength, and the code permits the use of a variety of ASTM structural steels having yield strengths varying from 33 to 50 kpsi. Provided the loading is the same, the code permits the same stress in the weld metal as in the parent metal. For these ASTM steels, $S_y = 0.5S_u$. Table 9–4 lists the formulas specified by the code for calculating these permissible stresses for various loading conditions. The factors of safety implied by this code are easily calculated. For tension, $n = 1/0.60 = 1.67$. For shear, $n = 0.577/0.40 = 1.44$, using the distortion-energy theory as the criterion of failure.

---

[5]For a copy, either write the AISC, 400 N. Michigan Ave., Chicago, IL 60611, or contact on the Internet at www.aisc.org.

Table 9–3

Minimum Weld-Metal Properties

AWS Electrode Number*	Tensile Strength kpsi (MPa)	Yield Strength, kpsi (MPa)	Percent Elongation
E60xx	62 (427)	50 (345)	17–25
E70xx	70 (482)	57 (393)	22
E80xx	80 (551)	67 (462)	19
E90xx	90 (620)	77 (531)	14–17
E100xx	100 (689)	87 (600)	13–16
E120xx	120 (827)	107 (737)	14

*The American Welding Society (AWS) specification code numbering system for electrodes. This system uses an E prefixed to a four- or five-digit numbering system in which the first two or three digits designate the approximate tensile strength. The last digit includes variables in the welding technique, such as current supply. The next-to-last digit indicates the welding position, as, for example, flat, or vertical, or overhead. The complete set of specifications may be obtained from the AWS upon request.

Table 9–4

Stresses Permitted by the AISC Code for Weld Metal

Type of Loading	Type of Weld	Permissible Stress	n*
Tension	Butt	$0.60S_y$	1.67
Bearing	Butt	$0.90S_y$	1.11
Bending	Butt	$0.60–0.66S_y$	1.52–1.67
Simple compression	Butt	$0.60S_y$	1.67
Shear	Butt or fillet	$0.30S_{ut}^\dagger$	

*The factor of safety n has been computed by using the distortion-energy theory.

†Shear stress on base metal should not exceed $0.40S_y$ of base metal.

It is important to observe that the electrode material is often the strongest material present. If a bar of AISI 1010 steel is welded to one of 1018 steel, the weld metal is actually a mixture of the electrode material and the 1010 and 1018 steels. Furthermore, a welded cold-drawn bar has its cold-drawn properties replaced with the hot-rolled properties in the vicinity of the weld. Finally, remembering that the weld metal is usually the strongest, do check the stresses in the parent metals.

The AISC code, as well as the AWS code, for bridges includes permissible stresses when fatigue loading is present. The designer will have no difficulty in using these codes, but their empirical nature tends to obscure the fact that they have been established by means of the same knowledge of fatigue failure already discussed in Chap. 7. Of course, for structures covered by these codes, the actual stresses *cannot* exceed the permissible stresses; otherwise the designer is legally liable. But in general, codes tend to conceal the actual margin of safety involved.

The fatigue stress-concentration factors listed in Table 9–5 are suggested for use. These factors should be used for the parent metal as well as for the weld metal. Table 9–6 gives steady-load information and minimum fillet sizes.

Table 9–5

Fatigue Stress-Concentration Factors, K_{fs}

Type of Weld	K_{fs}
Reinforced butt weld	1.2
Toe of transverse fillet weld	1.5
End of parallel fillet weld	2.7
T-butt joint with sharp corners	2.0

Table 9-6

Allowable Steady Loads and Minimum Fillet Weld Sizes

Schedule A: Allowable Load for Various Sizes of Fillet Welds

	Strength Level of Weld Metal (EXX)						
	60*	70*	80	90*	100	110*	120
$\tau =$	Allowable shear stress on throat, ksi (1000 psi) of fillet weld or partial penetration groove weld						
	18.0	21.0	24.0	27.0	30.0	33.0	36.0
†f =	Allowable Unit Force on Fillet Weld, kip/linear in						
	12.73h	14.85h	16.97h	19.09h	21.21h	23.33h	25.45h
Leg Size h, in	Allowable Unit Force for Various Sizes of Fillet Welds kip/linear in						
1	12.73	14.85	16.97	19.09	21.21	23.33	25.45
7/8	11.14	12.99	14.85	16.70	18.57	20.41	22.27
3/4	9.55	11.14	12.73	14.32	15.92	17.50	19.09
5/8	7.96	9.28	10.61	11.93	13.27	14.58	15.91
1/2	6.37	7.42	8.48	9.54	10.61	11.67	12.73
7/16	5.57	6.50	7.42	8.35	9.28	10.21	11.14
3/8	4.77	5.57	6.36	7.16	7.95	8.75	9.54
5/16	3.98	4.64	5.30	5.97	6.63	7.29	7.95
1/4	3.18	3.71	4.24	4.77	5.30	5.83	6.36
3/16	2.39	2.78	3.18	3.58	3.98	4.38	4.77
1/8	1.59	1.86	2.12	2.39	2.65	2.92	3.18
1/16	0.795	0.930	1.06	1.19	1.33	1.46	1.59

*Fillet welds actually tested by the joint AISC-AWS Task Committee.

†f = 0.707h τ$_{all}$.

Schedule B: Minimum Fillet Weld Size, h

Material Thickness of Thicker Part Joined, in	Weld Size, in
*To $\frac{1}{4}$ incl.	$\frac{1}{8}$
Over $\frac{1}{4}$ To $\frac{1}{2}$	$\frac{3}{16}$
Over $\frac{1}{2}$ To $\frac{3}{4}$	$\frac{1}{4}$
†Over $\frac{3}{4}$ To $1\frac{1}{2}$	$\frac{5}{16}$
Over $1\frac{1}{2}$ To $2\frac{1}{4}$	$\frac{3}{8}$
Over $2\frac{1}{4}$ To 6	$\frac{1}{2}$
Over 6	$\frac{5}{8}$

Not to exceed the thickness of the thinner part.

* Minimum size for bridge application does not go below $\frac{3}{16}$ in.

†For minimum fillet weld size, schedule does not go above $\frac{5}{16}$ in fillet weld for every $\frac{3}{4}$ in material.

Source: From Omer W. Blodgett (ed.), Stress Allowables Affect Weldment Design, D412, The James F. Lircoln Arc Welding Foundation, Cleveland, May 1991, p. 3. Reprinted by permission of Lincoln Electric Company.

9–6 Static Loading

Some examples of statically loaded joints are useful in comparing and contrasting the conventional method of analysis and the welding code methodology.

EXAMPLE 9–2

A $\frac{1}{2}$-in by 2-in rectangular-cross-section 1015 bar carries a static load of 16.5 kip. It is welded to a gusset plate with a $\frac{3}{8}$-in fillet weld 2 in long on both sides with an E70XX electrode as depicted in Fig. 9–18. Use the welding code method.
(a) Is the weld metal strength satisfactory?
(b) Is the attachment strength satisfactory?

Solution

(a) From Table 9–6, allowable force per unit length for a $\frac{3}{8}$-in E70 electrode metal is 5.57 kip/in of weldment; thus

$$F = 5.57l = 5.57(4) = 22.28 \text{ kip}$$

Since 22.28 > 16.5 kip, weld metal strength is satisfactory.
(b) Check shear in attachment adjacent to the welds. From Table 9–4 and Table A–20, from which $S_y = 27.5$ kpsi, the allowable attachment shear stress is

$$\tau_{\text{all}} = 0.4S_y = 0.4(27.5) = 11 \text{ kpsi}$$

The shear stress τ on the base metal adjacent to the weld is

$$\tau = \frac{F}{2hl} = \frac{16.5}{2(0.375)2} = 11 \text{ kpsi}$$

Since $\tau_{\text{all}} \geq \tau$, the attachment is satisfactory near the weld beads. The tensile stress in the shank of the attachment σ is

$$\sigma = \frac{F}{tl} = \frac{16.5}{(1/2)2} = 16.5 \text{ kpsi}$$

The allowable tensile stress σ_{all}, from Table 9–4, is $0.6S_y$ and, with welding code safety level preserved,

$$\sigma_{\text{all}} = 0.6S_y = 0.6(27.5) = 16.5 \text{ kpsi}$$

Since $\sigma_{\text{all}} \geq \sigma$, the shank tensile stress is satisfactory.

| **Figure 9–18**

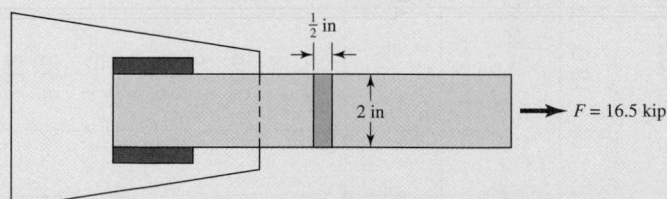

EXAMPLE 9–3   A specially rolled A36 structural steel section for the attachment has a cross section as shown in Fig. 9–19 and has yield and ultimate tensile strengths of 36 and 58 kpsi, respectively. It is statically loaded through the attachment centroid by a load of $F = 24$ kip. Unsymmetrical weld tracks can compensate for eccentricity such that there is no moment to be resisted by the welds. Specify the weld track lengths l_1 and l_2 for a $\frac{5}{16}$-in fillet weld using an E70XX electrode. This is part of a design problem in which the design variables include weld lengths and the fillet leg size.

Solution   The y coordinate of the section centroid of the attachment is

$$\bar{y} = \frac{\sum y_i A_i}{\sum A_i} = \frac{1(0.75)2 + 3(0.375)2}{0.75(2) + 0.375(2)} = 1.67 \text{ in}$$

Summing moments about point B to zero gives

$$\sum M_B = 0 = -F_1 b + F\bar{y} = -F_1(4) + 24(1.67)$$

from which

$$F_1 = 10 \text{ kip}$$

It follows that

$$F_2 = 24 - 10.0 = 14.0 \text{ kip}$$

The weld throat areas have to be in the ratio $14/10 = 1.4$, that is, $l_2 = 1.4l_1$. The weld length design variables are coupled by this relation, so l_1 is the weld length design variable. The other design variable is the fillet weld leg size h, which has been decided by the problem statement. From Table 9–4, the allowable shear stress on the throat τ_{all} is

$$\tau_{\text{all}} = 0.3(70) = 21 \text{ kpsi}$$

The shear stress τ on the 45° throat is

$$\tau = \frac{F}{(0.707)h(l_1 + l_2)} = \frac{F}{(0.707)h(l_1 + 1.4l_1)}$$

$$= \frac{F}{(0.707)h(2.4l_1)} = \tau_{\text{all}} = 21 \text{ kpsi}$$

from which the weld length l_1 is

$$l_1 = \frac{24}{21(0.707)0.3125(2.4)} = 2.16 \text{ in}$$

and

$$l_2 = 1.4l_1 = 1.4(2.16) = 3.02 \text{ in}$$

| **Figure 9–19**

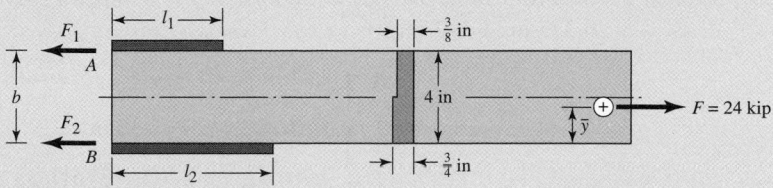

These are the weld-bead lengths required by weld metal strength. The attachment shear stress allowable in the base metal, from Table 9–4, is

$$\tau_{all} = 0.4S_y = 0.4(36) = 14.4 \text{ kpsi}$$

The shear stress τ in the base metal adjacent to the weld is

$$\tau = \frac{F}{h(l_1 + l_2)} = \frac{F}{h(l_1 + 1.4l_1)} = \frac{F}{h(2.4l_1)} = \tau_{all} = 14.4 \text{ kpsi}$$

from which

$$l_1 = \frac{F}{14.4h(2.4)} = \frac{24}{14.4(0.3125)2.4} = 2.22 \text{ in}$$

$$l_2 = 1.4l_1 = 1.4(2.22) = 3.11 \text{ in}$$

These are the weld-bead lengths required by base metal (attachment) strength. The base metal controls the weld lengths. For the allowable tensile stress σ_{all} in the shank of the attachment, the AISC allowable for tension members is $0.6S_y$; therefore,

$$\sigma_{all} = 0.6S_y = 0.6(36) = 21.6 \text{ kpsi}$$

The nominal tensile stress σ is *uniform* across the attachment cross section because of the load application at the centroid. The stress σ is

$$\sigma = \frac{F}{A} = \frac{24}{0.75(2) + 2(0.375)} = 10.7 \text{ kpsi}$$

Since $\sigma_{all} \geq \sigma$, the shank section is satisfactory. With l_1 set to a nominal $2\frac{1}{4}$ in, l_2 should be $1.4(2.25) = 3.15$ in.

Decision   Set $l_1 = 2\frac{1}{4}$ in, $l_2 = 3\frac{1}{4}$ in. The small magnitude of the departure from $l_2/l_1 = 1.4$ is not serious. The joint is essentially moment-free.

EXAMPLE 9–4   Perform an adequacy assessment of the statically loaded welded cantilever carrying 500 lbf depicted in Fig. 9–20. The cantilever is made of AISI 1018 HR steel and welded with a $\frac{3}{8}$-in fillet weld as shown in the figure. An E6010 electrode was used, and the design factor was 3.0.

(a) Use the conventional method for the weld metal.
(b) Use the conventional method for the attachment (cantilever) metal.
(c) Use a welding code for the weld metal.

Solution   (a) From Table 9–3, $S_y = 50$ kpsi, $S_{ut} = 62$ kpsi. From Table 9–2, second pattern, $b = 0.375$ in, $d = 2$ in, so

$$A = 1.414hd = 1.414(0.375)2 = 1.06 \text{ in}^2$$

$$I_u = d^3/6 = 2^3/6 = 1.33 \text{ in}^3$$

$$I = 0.707hI_u = 0.707(0.375)1.33 = 0.353 \text{ in}^4$$

| Figure 9–20

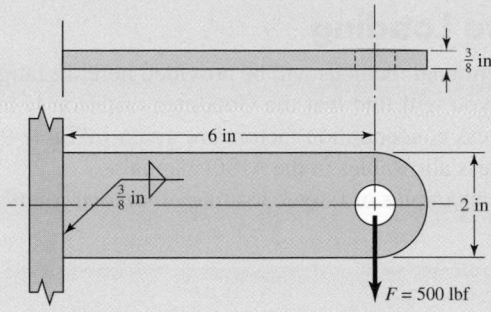

Primary shear:

$$\tau' = \frac{F}{A} = \frac{500(10^{-3})}{1.06} = 0.472 \text{ kpsi}$$

Secondary shear:

$$\tau'' = \frac{Mr}{I} = \frac{500(10^{-3})(6)(1)}{0.353} = 8.50 \text{ kpsi}$$

The shear magnitude τ is the Pythagorean combination

$$\tau = (\tau'^2 + \tau''^2)^{1/2} = (0.472^2 + 8.50^2)^{1/2} = 8.51 \text{ kpsi}$$

The factor of safety based on a minimum strength and the distortion-energy criterion is

Answer
$$n = \frac{S_{sy}}{\tau} = \frac{0.577(50)}{8.51} = 3.39$$

Since $n \geq n_d$, that is, $3.39 \geq 3.0$, the weld metal has satisfactory strength.
(b) From Table A–20, minimum strengths are $S_{ut} = 58$ kpsi and $S_y = 32$ kpsi. Then

$$\sigma = \frac{M}{I/c} = \frac{M}{bd^2/6} = \frac{500(10^{-3})6}{0.375(2^2)/6} = 12 \text{ kpsi}$$

Answer
$$n = \frac{S_y}{\sigma} = \frac{32}{12} = 2.67$$

Since $n < n_d$, that is, $2.67 < 3.0$, the joint is unsatisfactory as to the attachment strength.
(c) From part (a), $\tau = 8.51$ kpsi. For an E6010 electrode Table 9–6 gives the allowable shear stress τ_{all} as 18 kpsi. Since $\tau < \tau_{all}$, the weld is satisfactory. Since the code already has a design factor of $0.577(50)/18 = 1.6$ included at the equality, the corresponding factor of safety to part (a) is

Answer
$$n = 1.6\frac{18}{8.51} = 3.38$$

which is consistent.

9-7 Fatigue Loading

The conventional methods will be provided here. In fatigue, the Gerber criterion is best; however, you will find that the Goodman criterion is in common use. Recall, that the fatigue stress concentration factors are given in Table 9–5. For welding codes, see the fatigue stress allowables in the AISC manual.

Some examples of fatigue loading of welded joints follow.

EXAMPLE 9-5

The 1018 steel strap of Fig. 9–21 has a 1000-lbf, completely reversed load applied. Determine the factor of safety of the weldment for infinite life.

Solution

From Table A–20 for the 1018 attachment metal the strengths are $S_{ut} = 58$ kpsi and $S_y = 32$ kpsi. For the E6010 electrode, $S_{ut} = 62$ kpsi and $S_y = 50$ kpsi. The fatigue stress-concentration factor, from Table 9–5, is $K_{fs} = 2.7$. From Table 7–4, $k_a = 39.9(58)^{-0.995} = 0.702$. The shear area is:

$$A = 2(0.707)0.375(2) = 1.061 \text{ in}^2$$

For a uniform shear stress on the throat, $k_b = 1$.

From Eq. (7–25) for torsion (shear),

$$k_c = 0.59 \qquad k_d = k_e = k_f = 1$$

From Eqs. (7–8) and (7–17),

$$S_{se} = 0.702(1)0.59(1)(1)(1)0.504(58) = 12.1 \text{ kpsi}$$

$$K_{fs} = 2.7 \qquad F_a = 1000 \text{ lbf} \qquad F_m = 0$$

Only primary shear is present:

$$\tau'_a = \frac{K_{fs}F_a}{A} = \frac{2.7(1000)}{1.061} = 2545 \text{ psi} \qquad \tau'_m = 0 \text{ psi}$$

| Figure 9–21

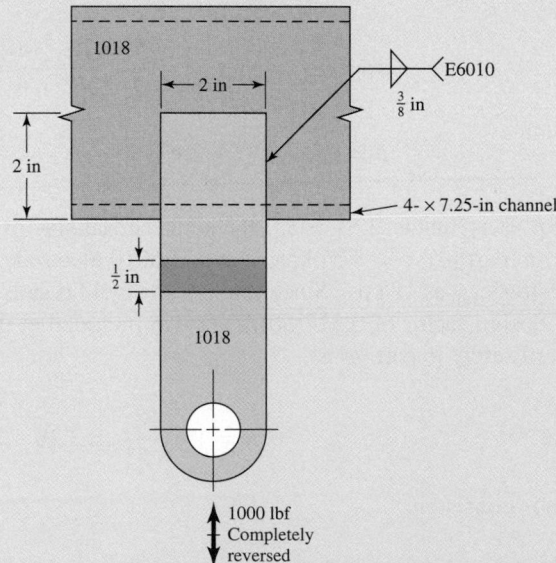

1018

2 in

E6010

$\frac{3}{8}$ in

2 in

4- × 7.25-in channel

$\frac{1}{2}$ in

1018

1000 lbf
Completely
reversed

In the absence of a midrange component, the fatigue factor of safety n_f is given by

Answer
$$n_f = \frac{S_{se}}{\tau_a'} = \frac{12\ 100}{2545} = 4.75$$

EXAMPLE 9–6   The 1018 steel strap of Fig. 9–22 has a repeatedly applied load of 2000 lbf ($F_a = F_m = 1000$ lbf). Determine the fatigue factor of safety fatigue strength of the weldment.

Solution   From Table 7–4, $k_a = 39.9(58)^{-0.995} = 0.702$.

$$A = 2(0.707)0.375(2) = 1.061 \text{ in}^2$$

For uniform shear stress on the throat $k_b = 1$.
From Eq. (7–25), $k_c = 0.59$. From Eqs. (7–8) and (7–17),

$$S_{se} = 0.702(1)0.59(1)(1)(1)0.504(58) = 12.1 \text{ kpsi}$$

From Table 9–5, $K_{fs} = 2$. Only primary shear is present:

$$\tau_a' = \tau_m' = \frac{K_{fs} F_a}{A} = \frac{2(1000)}{1.061} = 1885 \text{ psi}$$

From Eq. (7–56), $S_{su} \doteq 0.67 S_{ut}$. This, together with the Gerber fatigue failure criterion for shear stresses from Table 7–10, gives

$$n_f = \frac{1}{2}\left(\frac{0.67 S_{ut}}{\tau_m}\right)^2 \frac{\tau_a}{S_{se}}\left[-1 + \sqrt{1 + \left(\frac{2\tau_m S_{se}}{0.67 S_{ut}\tau_a}\right)^2}\right]$$

Answer
$$n_f = \frac{1}{2}\left[\frac{0.67(58)}{1.885}\right]^2 \frac{1.885}{12.1}\left\{-1 + \sqrt{1 + \left[\frac{2(1.885)12.1}{0.67(58)1.885}\right]^2}\right\} = 5.9$$

| Figure 9–22

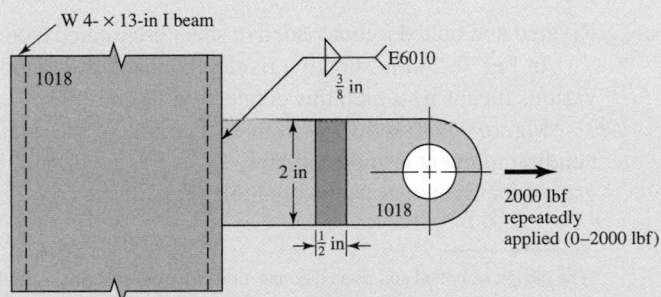

Figure 9–23

(a) Spot welding; (b) seam welding.

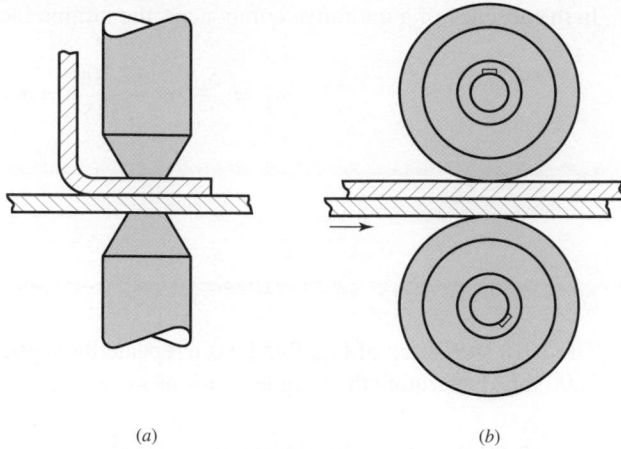

(a)   (b)

9–8 Resistance Welding

The heating and consequent welding that occur when an electric current is passed through several parts that are pressed together is called *resistance welding*. *Spot welding* and *seam welding* are forms of resistance welding most often used. The advantages of resistance welding over other forms are the speed, the accurate regulation of time and heat, the uniformity of the weld, and the mechanical properties that result. In addition the process is easy to automate, and filler metal and fluxes are not needed.

The spot- and seam-welding processes are illustrated schematically in Fig. 9–23. Seam welding is actually a series of overlapping spot welds, since the current is applied in pulses as the work moves between the rotating electrodes.

Failure of a resistance weld occurs either by shearing of the weld or by tearing of the metal around the weld. Because of the possibility of tearing, it is good practice to avoid loading a resistance-welded joint in tension. Thus, for the most part, design so that the spot or seam is loaded in pure shear. The shear stress is then simply the load divided by the area of the spot. Because the thinner sheet of the pair being welded may tear, the strength of spot welds is often specified by stating the load per spot based on the thickness of the thinnest sheet. Such strengths are best obtained by experiment.

Somewhat larger factors of safety should be used when parts are fastened by spot welding rather than by bolts or rivets, to account for the metallurgical changes in the materials due to the welding.

9–9 Bolted and Riveted Joints Loaded in Shear[6]

Riveted and bolted joints loaded in shear are treated exactly alike in design and analysis.

In Fig. 9–24a is shown a riveted connection loaded in shear. Let us now study the various means by which this connection might fail.

Figure 9–24b shows a failure by bending of the rivet or of the riveted members. The bending moment is approximately $M = Ft/2$, where F is the shearing force and t is the grip of the rivet, that is, the total thickness of the connected parts. The bending stress in

---

[6]The design of bolted and riveted connections for boilers, bridges, buildings, and other structures in which danger to human life is involved is strictly governed by various construction codes. When designing these structures, the engineer should refer to the *American Institute of Steel Construction Handbook,* the American Railway Engineering Association specifications, or the Boiler Construction Code of the American Society of Mechanical Engineers.

Figure 9–24

Modes of failure in shear loading of a bolted or riveted connection: (a) shear loading; (b) bending of rivet; (c) shear of rivet; (d) tensile failure of members; (e) bearing of rivet on members or bearing of members on rivet; (f) shear tear-out; (g) tensile tear-out.

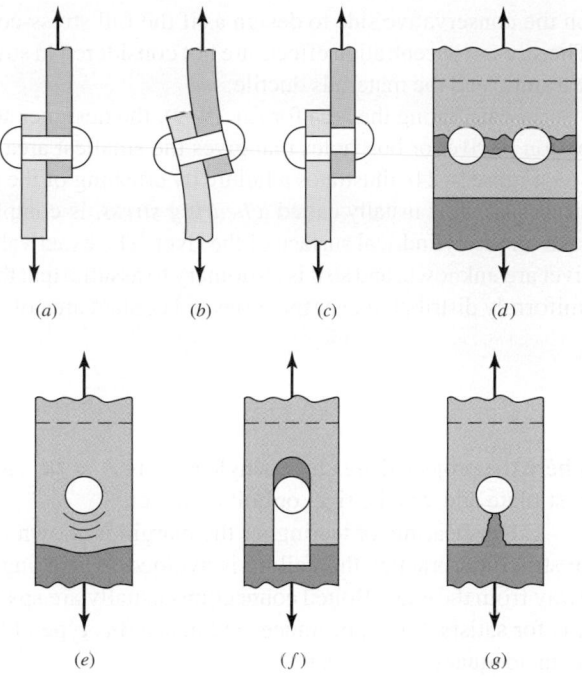

the members or in the rivet is, neglecting stress concentration,

$$\sigma = \frac{M}{I/c} \tag{9–7}$$

where I/c is the section modulus for the weakest member or for the rivet or rivets, depending upon which stress is to be found. The calculation of the bending stress in this manner is an assumption, because we do not know exactly how the load is distributed to the rivet or the relative deformations of the rivet and the members. Although this equation can be used to determine the bending stress, it is seldom used in design; instead its effect is compensated for by an increase in the factor of safety.

In Fig. 9–24c failure of the rivet by pure shear is shown; the stress in the rivet is

$$\tau = \frac{F}{A} \tag{9–8}$$

where A is the cross-sectional area of all the rivets in the group. It may be noted that it is standard practice in structural design to use the nominal diameter of the rivet rather than the diameter of the hole, even though a hot-driven rivet expands and nearly fills up the hole.

Rupture of one of the connected membes or plates by pure tension is illustrated in Fig. 9–24d. The tensile stress is

$$\sigma = \frac{F}{A} \tag{9–9}$$

where A is the net area of the plate, that is, the area reduced by an amount equal to the area of all the rivet holes. For brittle materials and static loads and for either ductile or brittle materials loaded in fatigue, the stress-concentration effects must be included. It is true that the use of a bolt with an initial preload and, sometimes, a rivet will place the area around the hole in compression and thus tend to nullify the effects of stress concentration, but unless definite steps are taken to ensure that the preload does not relax, it is on the conservative side to design as if the full stress-concentration effect were present. The stress-concentration effects are not considered in structural design, because the loads are static and the materials ductile.

In calculating the area for Eq. (9–9), the designer should, of course, use the combination of rivet or bolt holes that gives the smallest area.

Figure 9–24e illustrates a failure by crushing of the rivet or plate. Calculation of this stress, which is usually called a *bearing stress,* is complicated by the distribution of the load on the cylindrical surface of the rivet. The exact values of the forces acting upon the rivet are unknown, and so it is customary to assume that the components of these forces are uniformly distributed over the projected contact area of the rivet. This gives for the stress

$$\sigma = -\frac{F}{A} \tag{9-10}$$

where the projected area for a single rivet is $A = td$. Here, t is the thickness of the thinnest plate and d is the rivet or bolt diameter.

Edge shearing, or tearing, of the margin is shown in Fig. 9–24f and g, respectively. In structural practice this failure is avoided by spacing the rivets at least $1\frac{1}{2}$ diameters away from the edge. Bolted connections usually are spaced an even greater distance than this for satisfactory appearance, and hence this type of failure may usually be neglected in an adequacy assessment.

In structural design it is customary to select in advance the number of rivets, their diameters, and their spacing. The strength is then determined for each method of failure. If the strength is found to be unsatisfactory, a change is made in the diameter, spacing, or the number of rivets used to increase the strength. The mode of operation is to complete the decision set and perform an adequacy assessment; if the design is satisfactory, the next step is to estimate the figure of merit and continue to improve the figure of merit if possible. It is not usual instructional practice to consider the combined effects of the various methods of failure.

In nonpermanent joints using bolts, upon disassembly the fasteners are replaced, but the members, and other elements such as splice plates or lug plates, are reused. They are permanent members of a nonpermanent joint. The focus in Chap. 8 was on the fasteners, not on the members of the joint. In this chapter we have the opportunity to examine the permanent members of the joint.

In a rivet joint, the rivets all share the load in shear, bearing in the rivet, bearing in the member, and shear in the rivet. Other failures are participated in by only some of the joint. In a bolted joint, shear is taken by clamping friction, and bearing does not exist. When bolt preload is lost, one bolt begins to carry the shear and bearing until yielding slowly brings other fasteners in to share the shear and bearing. Finally, all participate, and this is the basis of most bolted-joint analysis if loss of bolt preload is complete. The usual analysis involves

- Bearing in the bolt (all bolts participate)
- Bearing in members (all holes participate)
- Shear of bolt (all bolts participate eventually)
- Distinguishing between thread and shank shear
- Edge shearing and tearing of member (edge bolts participate)
- Tensile yielding of member across bolt holes
- Checking member capacity

EXAMPLE 9–7

Two 1- by 4-in 1018 cold-rolled steel bars are butt-spliced with two $\frac{1}{2}$- by 4-in 1018 cold-rolled splice plates using four $\frac{3}{4}$ in-16 UNF grade 5 bolts as depicted in Fig. 9–25. For a design factor of $n_d = 1.5$ estimate the static load F that can be carried if the bolts lose preload.

Solution

From Table A–20, minimum strengths of $S_y = 54$ kpsi and $S_{ut} = 64$ kpsi are found for the members, and from Table 8–9 minimum strengths of $S_p = 85$ kpsi and $S_{ut} = 120$ kpsi for the bolts are found.

$F/2$ is transmitted by each of the splice plates, but since the areas of the splice plates are half those of the center bars, the stresses associated with the plates are the same. So for stresses associated with the plates, the force and areas used will be those of the center plates.

Bearing in bolts, all bolts loaded:

$$\sigma = \frac{F}{2td} = \frac{S_p}{n_d}$$

$$F = \frac{2td\,S_p}{n_d} = \frac{2(1)\left(\frac{3}{4}\right)85}{1.5} = 85 \text{ kip}$$

Bearing in members, all bolts active:

$$\sigma = \frac{F}{2td} = \frac{(S_y)_{\text{mem}}}{n_d}$$

$$F = \frac{2td\,(S_y)_{\text{mem}}}{n_d} = \frac{2(1)\left(\frac{3}{4}\right)54}{1.5} = 54 \text{ kip}$$

| **Figure 9–25**

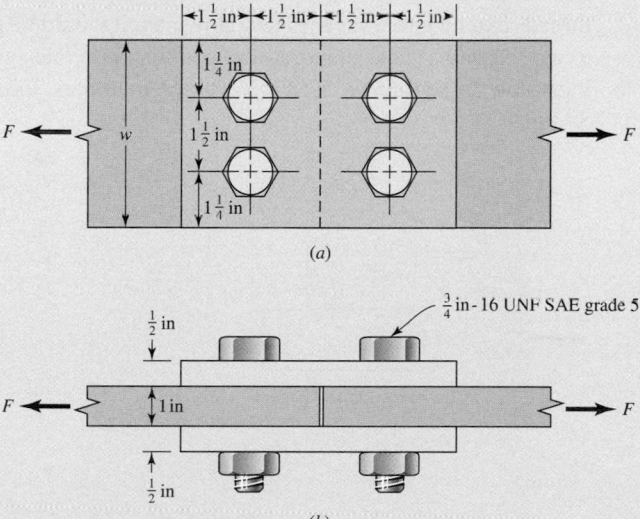

(a)

(b)

Shear of bolt, all bolts active: If the bolt threads do not extend into the shear planes for four shanks:

$$\tau = \frac{F}{4\pi d^2/4} = 0.577\frac{S_p}{n_d}$$

$$F = 0.577\pi d^2 \frac{S_p}{n_d} = 0.577\pi(0.75)^2\frac{85}{1.5} = 57.8 \text{ kip}$$

If the bolt threads extend into a shear plane:

$$\tau = \frac{F}{4A_r} = 0.577\frac{S_p}{n_d}$$

$$F = \frac{0.577(4)A_r S_p}{n_d} = \frac{0.577(4)0.351(85)}{1.5} = 45.9 \text{ kip}$$

Edge shearing of member at two margin bolts: From Fig. 9–26,

$$\tau = \frac{F}{4at} = \frac{0.577(S_y)_{\text{mem}}}{n_d}$$

$$F = \frac{4at\,0.577(S_y)_{\text{mem}}}{n_d} = \frac{4(1.125)(1)0.577(54)}{1.5} = 93.5 \text{ kip}$$

Tensile yielding of members across bolt holes:

$$\sigma = \frac{F}{\left[4 - 2\left(\frac{3}{4}\right)\right]t} = \frac{(S_y)_{\text{mem}}}{n_d}$$

$$F = \frac{\left[4 - 2\left(\frac{3}{4}\right)\right]t(S_y)_{\text{mem}}}{n_d} = \frac{\left[4 - 2\left(\frac{3}{4}\right)\right](1)54}{1.5} = 90 \text{ kip}$$

Member yield:

$$F = \frac{wt(S_y)_{\text{mem}}}{n_d} = \frac{4(1)54}{1.5} = 144 \text{ kip}$$

On the basis of bolt shear, the limiting value of the force is 45.9 kip, assuming the threads extend into a shear plane. However, it would be poor design to allow the threads to extend into a shear plane. So, assuming a *good* design based on bolt shear, the limiting value of the force is 57.8 kip. For the members, the bearing stress limits the load to 54 kip.

Figure 9–26

Edge shearing of member.

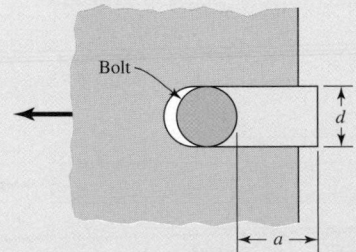

Bolt

d

a

9–10 **Adhesive Bonding[7]**

The use of polymeric adhesives to join components for structural, semistructural, and nonstructural applications has expanded greatly in recent years as a result of the unique advantages adhesives may offer for certain assembly processes and the development of new adhesives with improved robustness and environmental acceptability. The increasing complexity of modern assembled structures and the diverse types of materials used have led to many joining applications that would not be possible with more conventional joining techniques. Adhesives are also being used either in conjunction with or to replace mechanical fasteners and welds. Reduced weight, sealing capabilities, and reduced part count and assembly time, as well as improved fatigue and corrosion resistance, all combine to provide the designer with opportunities for customized assembly. In 1998, for example, adhesives were a $20 billion industry with 24 trillion pounds of adhesives produced and sold. Figure 9–27 illustrates the numerous places where adhesives are used on a modern automobile. Indeed, the fabrication of many modern vehicles, devices, and structures is dependent on adhesives.

In well-designed joints and with proper processing procedures, use of adhesives can result in significant reductions in weight. Eliminating mechanical fasteners eliminates the weight of the fasteners, and also may permit the use of thinner-gauge materials because stress concentrations associated with the holes are eliminated. The capability of

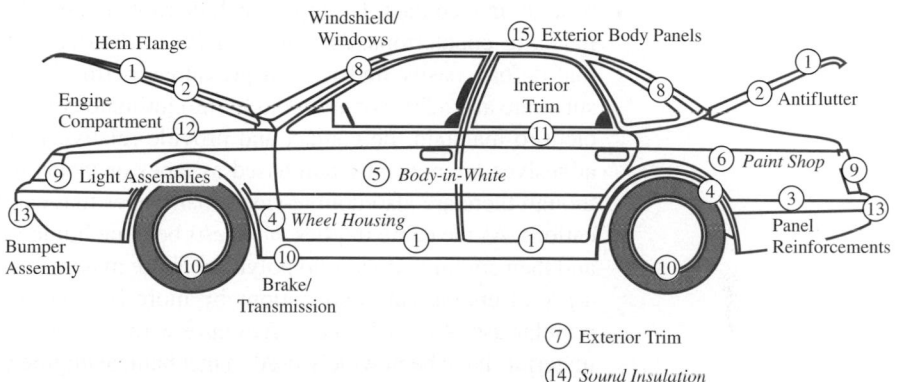

Figure 9–27

Diagram of an automobile body showing at least 15 locations at which adhesives and sealants could be used or are being used. Particular note should be made of the windshield (8), which is considered a load-bearing structure in modern automobiles and is adhesively bonded. Also attention should be paid to hem flange bonding (1), in which adhesives are used to bond and seal. Adhesives are used to bond friction surfaces in brakes and clutches (10). Antiflutter adhesive bonding (2) helps control deformation of hood and trunk lids under wind shear. Thread-sealing adhesives are used in engine applications (12). *(From A. V. Pocius, Adhesion and Adhesives Technology, 2nd edition, Hanser Publishers, Munich, 2002. Reprinted by permission.)*

---

[7]For a more extensive discussion of this topic, see J. E. Shigley and C. R. Mischke, *Mechanical Engineering Design,* 6th ed., McGraw-Hill, New York, 2001, Sec. 9–11. This section was prepared with the assistance of Professor David A. Dillard, Professor of Engineering Science and Mechanics and Director of the Center for Adhesive and Sealant Science, Virginia Polytechnic Institute and State University, Blacksburg, Virginia, and with the encouragement and technical support of the Bonding Systems Division of 3M, Saint Paul, Minnesota.

polymeric adhesives to dissipate energy can significantly reduce noise, vibration, and harshness (NVH), crucial in modern automobile performance. Adhesives can be used to assemble heat-sensitive materials or components that might be damaged by drilling holes for mechanical fasteners. They can be used to join dissimilar materials or thin-gauge stock that cannot be joined through other means.

Types of Adhesive

There are numerous adhesive types for various applications. They may be classified in a variety of ways depending on their chemistry (e.g., epoxies, polyurethanes, poly-imides), their form (e.g., paste, liquid, film, pellets, tape), their type (e.g., hot melt, reactive hot melt, thermosetting, pressure sensitive, contact), or their load-carrying capability (structural, semistructural, or nonstructural).

Structural adhesives are relatively strong adhesives that are normally used well below their glass transition temperature; common examples include epoxies and certain acrylics. Such adhesives can carry significant stresses, and they lend themselves to structural applications. For many engineering applications, semistructural applications (where failure would be less critical) and nonstructural applications (of headliners, etc., for aesthetic purposes) are also of significant interest to the design engineer, providing cost-effective means required for assembly of finished products. These include *contact adhesives,* where a solution or emulsion containing an elastomeric adhesive is coated onto both adherends, the solvent is allowed to evaporate, and then the two adherends are brought into contact. Examples include rubber cement and adhesives used to bond laminates to countertops. *Pressure-sensitive adhesives* are very low modulus elastomers that deform easily under small pressures, permitting them to wet surfaces. When the substrate and adhesive are brought into intimate contact, van der Waals forces are sufficient to maintain the contact and provide relatively durable bonds. Pressure-sensitive adhesives are normally purchased as tapes or labels for nonstructural applications, although there are also double-sided foam tapes that can be used in semistructural applications. As the name implies, *hot melts* become liquid when heated, wetting the surfaces and then cooling into a solid polymer. These materials are increasingly applied in a wide array of engineering applications by more sophisticated versions of the glue guns in popular use. *Anaerobic adhesives* cure within narrow spaces deprived of oxygen; such materials have been widely used in mechanical engineering applications to lock bolts or bearings in place. Cure in other adhesives may be induced by exposure to ultraviolet light or electron beams, or it may be catalyzed by certain materials that are ubiquitous on many surfaces, such as water.

Table 9–7 presents important strength properties of commonly used adhesives.

Stress Distributions

Good design practice normally requires that adhesive joints be constructed in such a manner that the adhesive carries the load in shear rather than tension. Bonds are typically much stronger when loaded in shear rather than in tension across the bond plate. Lap-shear joints represent an important family of joints, both for test specimens to evaluate adhesive properties and for actual incorporation into practical designs. Generic types of lap joints that commonly arise are illustrated in Fig. 9–28.

The simplest analysis of lap joints suggests the applied load is uniformly distributed over the bond area. Lap joint test results, such as those obtained following the ASTM D1002 for single-lap joints, report the "apparent shear strength" as the breaking load divided by the bond area. Although this simple analysis can be adequate for stiff

Table 9–7

Mechanical Performance of Various Types of Adhesives *Source:* From A. V. Pocius, *Adhesion and Adhesives Technology,* Hanser Publishers, Munich, 2002. Reprinted by permission.

Adhesive Chemistry or Type	Room Temperature Lap-Shear Strength, MPa (psi)		Peel Strength Per Unit Width, kN/m (lbf/in)	
Pressure-sensitive	0.01–0.07	(2–10)	0.18–0.88	(1–5)
Starch-based	0.07–0.7	(10–100)	0.18–0.88	(1–5)
Cellosics	0.35–3.5	(50–500)	0.18–1.8	(1–10)
Rubber-based	0.35–3.5	(50–500)	1.8–7	(10–40)
Formulated hot melt	0.35–4.8	(50–700)	0.88–3.5	(5–20)
Synthetically designed hot melt	0.7–6.9	(100–1000)	0.88–3.5	(5–20)
PVAc emulsion (white glue)	1.4–6.9	(200–1000)	0.88–1.8	(5–10)
Cyanoacrylate	6.9–13.8	(1000–2000)	0.18–3.5	(1–20)
Protein-based	6.9–13.8	(1000–2000)	0.18–1.8	(1–10)
Anaerobic acrylic	6.9–13.8	(1000–2000)	0.18–1.8	(1–10)
Urethane	6.9–17.2	(1000–2500)	1.8–8.8	(10–50)
Rubber-modified acrylic	13.8–24.1	(2000–3500)	1.8–8.8	(10–50)
Modified phenolic	13.8–27.6	(2000–4000)	3.6–7	(20–40)
Unmodified epoxy	10.3–27.6	(1500–4000)	0.35–1.8	(2–10)
Bis-maleimide	13.8–27.6	(2000–4000)	0.18–3.5	(1–20)
Polyimide	13.8–27.6	(2000–4000)	0.18–0.88	(1–5)
Rubber-modified epoxy	20.7–41.4	(3000–6000)	4.4–14	(25–80)

adherends bonded with a soft adhesive over a relatively short bond length, significant peaks in shear stress occur except for the most flexible adhesives. In an effort to point out the problems associates with such practice, ASTM D4896 outlines some of the concerns associated with taking this simplistic view of stresses within lap joints.

In 1938, O. Volkersen presented an analysis of the lap joint, known as the *shear-lag model.* It provides valuable insights into the shear-stress distributions in a host of lap joints. Bending induced in the single-lap joint due to eccentricity significantly complicates the analysis, so here we will consider a symmetric double-lap joint to illustrate the principles. The shear-stress distribution for the double lap joint of Fig. 9–29 is given by

$$\tau(x) = \frac{P\omega}{4b\,\sinh(\omega l/2)}\cosh(\omega x) + \left[\frac{P\omega}{4b\,\cosh(\omega l/2)}\left(\frac{2E_o t_o - E_i t_i}{2E_o t_o + E_i t_i}\right)\right.$$

$$\left. + \frac{(\alpha_i - \alpha_o)\,\Delta T\omega}{(1/E_o t_o + 2/E_i t_i)\cosh(\omega l/2)}\right]\sinh(\omega x) \qquad (9\text{–}11)$$

where

$$\omega = \sqrt{\frac{G}{h}\left(\frac{1}{E_o t_o} + \frac{2}{E_i t_i}\right)}$$

Figure 9–28

Common types of lap joints used in mechanical design: (a) single lap; (b) double lap; (c) scarf; (d) bevel; (e) step; (f) butt strap; (g) double butt strap; (h) tubular lap. (*Adapted from R. D. Adams, J. Comyn, and W. C. Wake,* Structural Adhesive Joints in Engineering, *2nd ed., Chapman and Hall, New York, 1997.*)

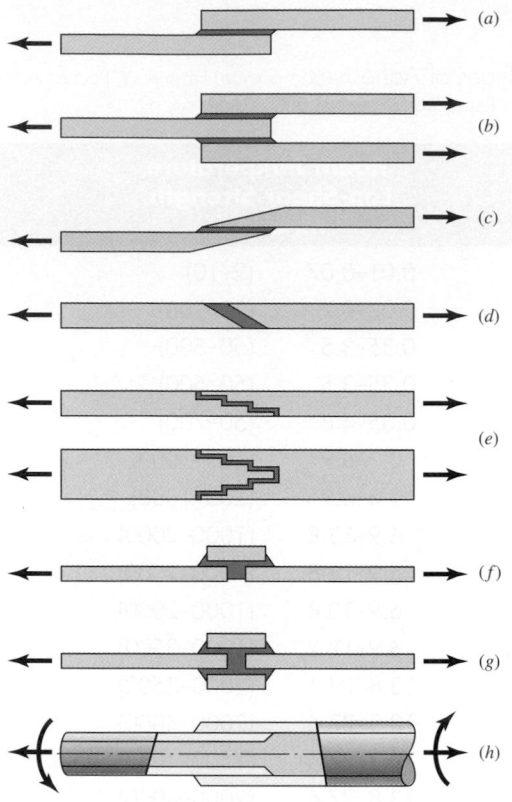

Figure 9–29

Double-lap joint.

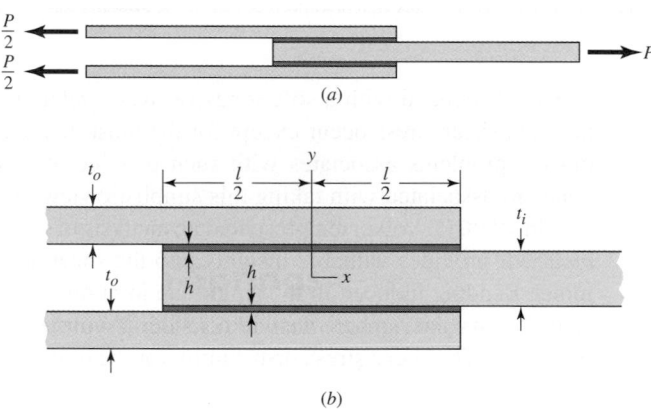

and E_o, t_o, α_o, and E_i, t_i, α_i, are the modulus, thickness, coefficient of thermal expansion for the outer and inner adherend, respectively; G, h, b, and l are the shear modulus, thickness, width, and length of the adhesive, respectively; and ΔT is a change in temperature of the joint. If the adhesive is cured at an elevated temperature such that the stress-free temperature of the joint differs from the service temperature, the mismatch in thermal expansion of the outer and inner adherends induces a thermal shear across the adhesive.

EXAMPLE 9–8

The double-lap joint depicted in Fig. 9–29 consists of aluminum outer adherends and an inner steel adherend. The assembly is cured at 250°F and is stress-free at 200°F. The completed bond is subjected to an axial load of 2000 lbf at a service temperature of 70°F. The width b is 1 in, the length of the bond l is 1 in. Additional information is tabulated below:

	G, psi	E, psi	α, in/(in · °F)	Thickness, in
Adhesive	0.2(10⁶)		55(10⁻⁶)	0.020
Outer adherend		10(10⁶)	13.3(10⁻⁶)	0.150
Inner adherend		30(10⁶)	6.0(10⁻⁶)	0.100

Sketch a plot of the shear stress as a function of the length of the bond due to (*a*) thermal stress, (*b*) load-induced stress, and (*c*) the sum of stresses in *a* and *b*; and (*d*) find where the largest shear stress is maximum.

Solution

In Eq. (9–11) the parameter ω is given by

$$\omega = \sqrt{\frac{G}{h}\left(\frac{1}{E_o t_o} + \frac{2}{E_i t_i}\right)}$$

$$= \sqrt{\frac{0.2(10^6)}{0.020}\left[\frac{1}{10(10^6)0.15} + \frac{2}{30(10^6)0.10}\right]} = 3.65 \text{ in}^{-1}$$

(*a*) For the thermal component, $\alpha_i - \alpha_o = 6(10^{-6}) - 13.3(10^{-6}) = -7.3(10^{-6})$, $\Delta T = 70 - 200 = -130°F$,

$$\tau_{th}(x) = \frac{(\alpha_i - \alpha_o)\Delta T \omega \sinh(\omega x)}{(1/E_o t_o + 2/E_i t_i)\cosh(\omega l/2)}$$

$$\tau_{th}(x) = \frac{-7.3(10^{-6})(-130)3.65 \ \sinh(3.65x)}{\left[\dfrac{1}{10(10^6)0.150} + \dfrac{2}{30(10^6)0.100}\right]\cosh\left[\dfrac{3.65(1)}{2}\right]}$$

$$= 816.4 \ \sinh(3.65x)$$

The thermal stress is plotted in Fig. (9–30) and tabulated at $x = -0.5, 0,$ and 0.5 in the table below.

(*b*) The bond is "balanced" ($E_o t_o = E_i t_i/2$), so the load-induced stress is given by

$$\tau_P(x) = \frac{P\omega \cosh(\omega x)}{4b \sinh(\omega l/2)} = \frac{2000(3.65)\cosh(3.65x)}{4(1)3.0208} = 604.1 \cosh(3.65x) \qquad (1)$$

The load-induced stress is plotted in Fig. (9–30) and tabulated at $x = -0.5, 0,$ and 0.5 in the table below.

(*c*) Total stress table (in psi):

	τ(−0.5)	τ(0)	τ(0.5)
Thermal only	−2466	0	2466
Load-induced only	1922	604	1922
Combined	−544	604	4388

Figure 9–30

Plot for Ex. 9–8.

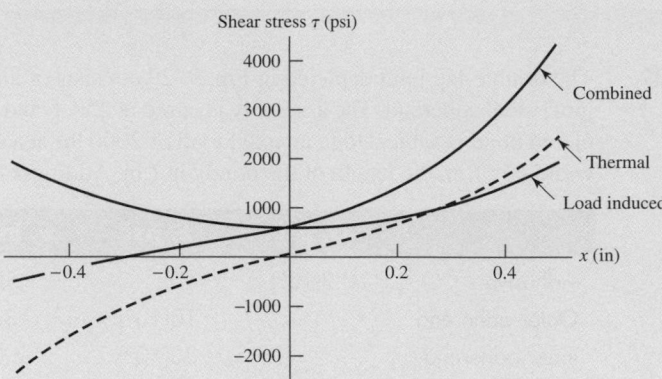

(*d*) The maximum shear stress predicted by the shear-lag model will always occur at the ends. See the plot in Fig. 9–30. Since the residual stresses are always present, significant shear stresses may already exist prior to application of the load. The large stresses present for the combined-load case could result in local yielding of a ductile adhesive or failure of a more brittle one. The significance of the thermal stresses serves as a caution against joining dissimilar adherends when large temperature changes are involved. Note also that the average shear stress due to the load is $\tau_{avg} = P/(2bl) = 1000$ psi. Equation (1) produced a maximum of 1922 psi, almost double the average.

Although design considerations for single-lap joints are beyond the scope of this chapter, one should note that the load eccentricity is an important aspect in the stress state of single-lap joints. Adherend bending can result in shear stresses that may be as much as double those given for the double-lap configuration (for a given total bond area). In addition, peel stresses can be quite large and often account for joint failure. Finally, plastic bending of the adherends can lead to high strains, which less ductile adhesives cannot withstand, leading to bond failure as well. Bending stresses in the adherends at the end of the overlap can be four times greater than the average stress within the adherend; thus, they must be considered in the design. Figure 9–31 shows the shear and peel stresses present in a typical single-lap joint that corresponds to the ASTM D1002 test specimen. Note that the shear stresses are significantly larger than predicted by the Volkersen analysis, a result of the increased adhesive strains associated with adherend bending.

Joint Design

Some basic guidelines that should be used in adhesive joint design include:

- Design to place bondline in shear, not peel. Beware of peel stresses focused at bond terminations. When necessary, reduce peel stresses through tapering the adherend ends, increasing bond area where peel stresses occur, or utilizing rivets at bond terminations where peel stresses can initiate failures.

- Where possible, use adhesives with adequate ductility. The ability of an adhesive to yield reduces the stress concentrations associated with the ends of joints and increases the toughness to resist debond propagation.

- Recognize environmental limitations of adhesives and surface preparation methods. Exposure to water, solvents, and other diluents can significantly degrade adhesive performance in some situations, through displacing the adhesive from the surface or degrading the polymer. Certain adhesives may be susceptible to environmental stress

Figure 9–31

Stresses within a single-lap joint. (*a*) Lap-joint tensile forces have a line of action that is not initially parallel to the adherend sides. (*b*) As the load increases the adherends and bond bend. (*c*) In the locality of the end of an adherend peel and shear stresses appear, and the peel stresses often induce joint failure. (*d*) The seminal Goland and Reissner stress predictions (*J. Appl. Mech.*, vol. 77, 1944) are shown. (*Note that the predicted shear-stress maximum is higher than that predicted by the Volkersen shear-lag model because of adherend bending.*)

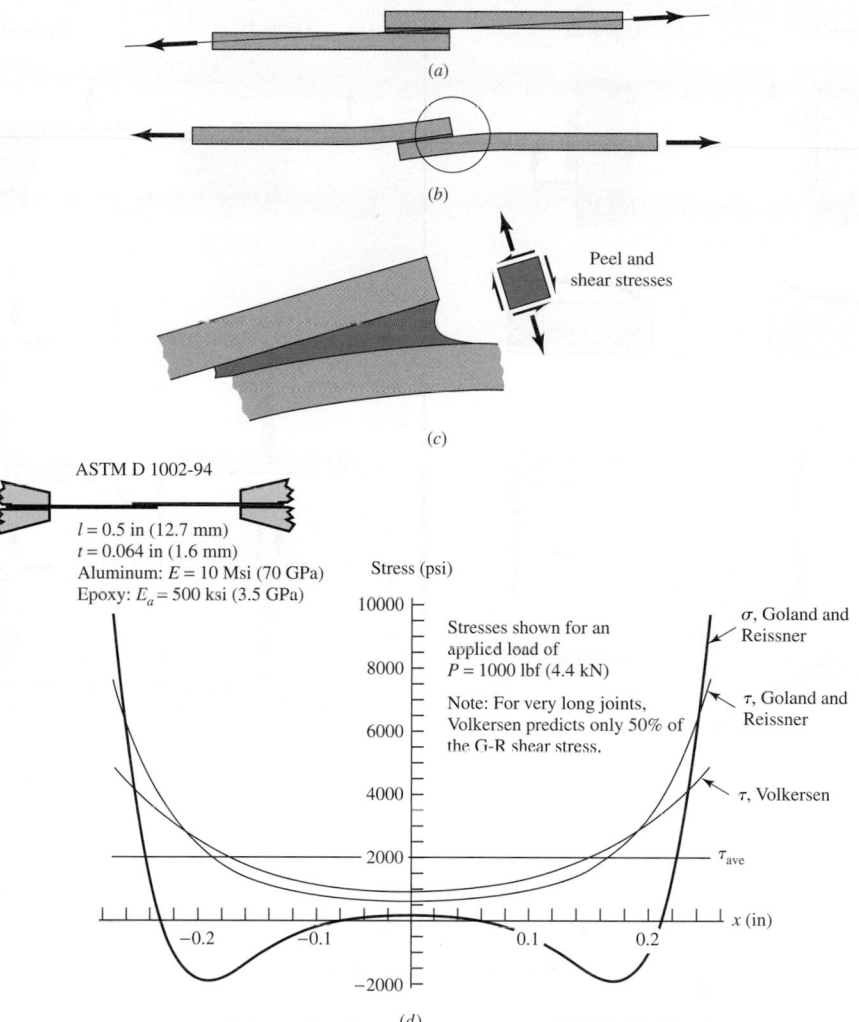

(*a*)

(*b*)

Peel and shear stresses

(*c*)

ASTM D 1002-94

$l = 0.5$ in (12.7 mm)
$t = 0.064$ in (1.6 mm)
Aluminum: $E = 10$ Msi (70 GPa)
Epoxy: $E_a = 500$ ksi (3.5 GPa)

Stress (psi)

Stresses shown for an applied load of $P = 1000$ lbf (4.4 kN)

Note: For very long joints, Volkersen predicts only 50% of the G-R shear stress.

σ, Goland and Reissner

τ, Goland and Reissner

τ, Volkersen

τ_{ave}

x (in)

(*d*)

cracking in the presence of certain solvents. Exposure to ultraviolet light can also degrade adhesives.

- Design in a way that permits or facilitates inspections of bonds where possible. A missing rivet or bolt is often easy to detect, but debonds or unsatisfactory adhesive bonds are not readily apparent.

- Allow for sufficient bond area so that the joint can tolerate some debonding before going critical. This increases the likelihood that debonds can be detected. Having some regions of the overall bond at relatively low stress levels can significantly improve durability and reliability.

- Where possible, bond to multiple surfaces to offer support to loads in any direction. Bonding an attachment to a single surface can place peel stresses on the bond, whereas bonding to several adjacent planes tends to permit arbitrary loads to be carried predominantly in shear.

- Adhesives can be used in conjunction with spot welding. The process is known as *weld bonding*. The spot welds serve to fixture the bond until it is cured.

Figure 9–32 presents examples of improvements in adhesive bonding.

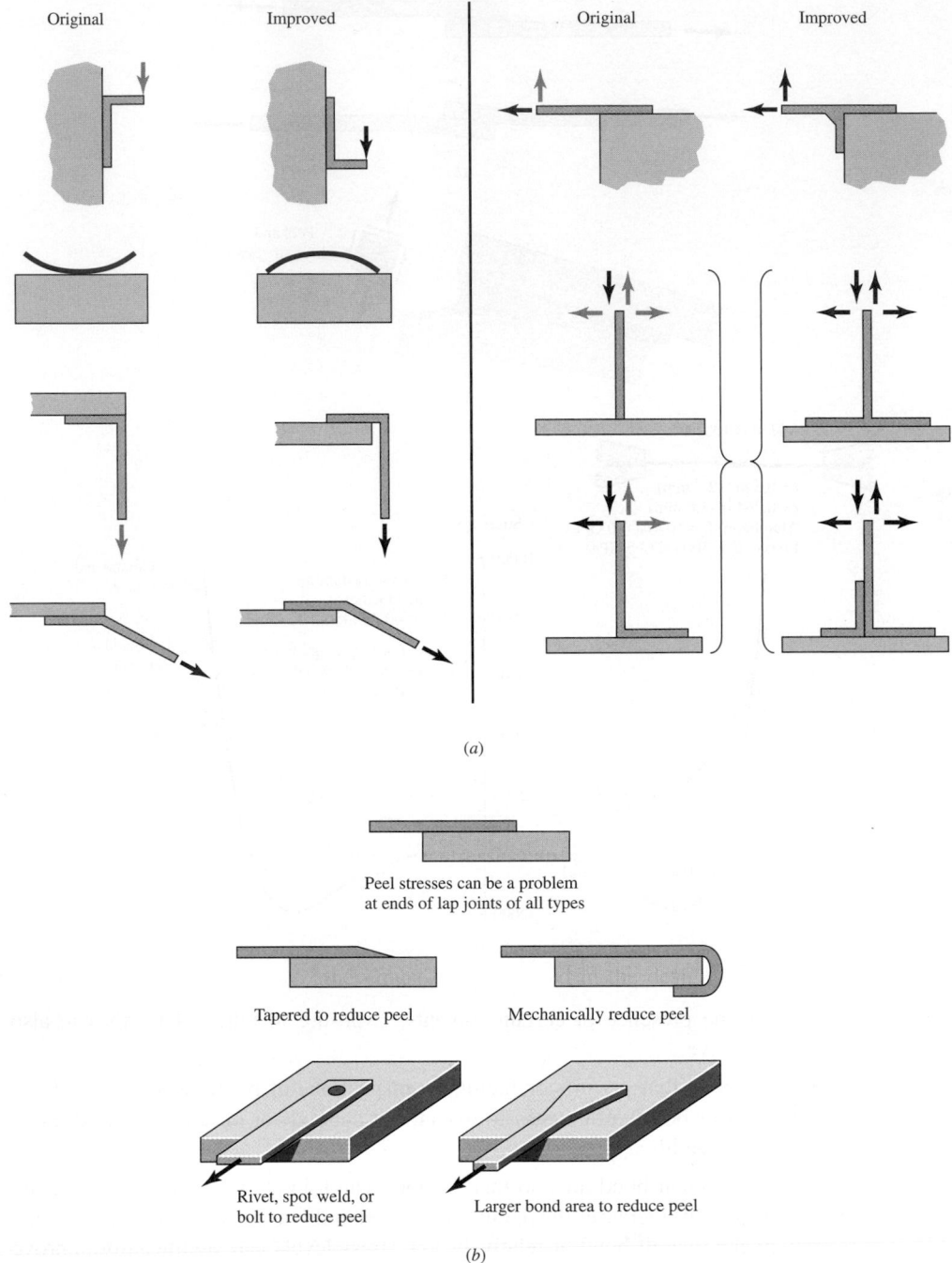

(a)

Peel stresses can be a problem
at ends of lap joints of all types

Tapered to reduce peel   Mechanically reduce peel

Rivet, spot weld, or   Larger bond area to reduce peel
bolt to reduce peel

(b)

Figure 9–32

Design practices that improve adhesive bonding. (a) Gray load vectors are to be avoided as resulting strength is poor. (b) Means to reduce peel stresses in lap-type joints.

References

A number of good references are available for analyzing and designing adhesive bonds, including the following:

G. P. Anderson, S. J. Bennett, and K. L. DeVries, *Analysis and Testing of Adhesive Bonds,* Academic Press, New York, 1977.

R. D. Adams, J. Comyn, and W. C. Wake, *Structural Adhesive Joints in Engineering,* 2nd ed., Chapman and Hall, New York, 1997.

H. F. Brinson (ed.), *Engineered Materials Handbook, vol. 3: Adhesives and Sealants,* ASM International, Metals Park, Ohio, 1990.

A. J. Kinloch, *Adhesion and Adhesives: Science and Technology,* Chapman and Hall, New York, 1987.

A. J. Kinloch (ed.), *Durability of Structural Adhesives,* Applied Science Publishers, New York, 1983.

W. A. Lees, *Adhesives in Engineering Design,* Springer-Verlag, New York, 1984.

F. L. Matthews, *Joining Fibre-Reinforced Plastics,* Elsevier, New York, 1986.

A. V. Pocius, *Adhesion and Adhesives Technology: An Introduction,* Hanser, New York, 1997.

The Internet is also a good source of information. For example, try this website: www.3m.com/adhesives.

PROBLEMS

ANALYSIS

9–1 The figure shows a horizontal steel bar $\frac{3}{8}$ in thick loaded in steady tension and welded to a vertical support. Find the load F that will cause a shear stress of 20 kpsi in the throats of the welds.

Problem 9–1

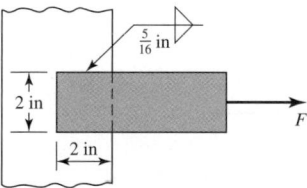

9–2 For the weldment of Prob. 9–1 the electrode specified is E7010. For the electrode metal, what is the allowable load on the weldment?

ANALYSIS

9–3 The members being joined in Prob. 9–1 are cold-rolled 1018 for the bar and hot-rolled 1018 for the vertical support. What load on the weldment is allowable because member metal is incorporated into the welds?

ANALYSIS

9–4 A $\frac{5}{16}$-in steel bar is welded to a vertical support as shown in the figure. What is the shear stress in the throat of the welds if the force F is 32 kip?

Problem 9–4

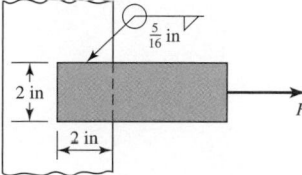

9–5 ANALYSIS A $\frac{3}{4}$-in-thick steel bar, to be used as a beam, is welded to a vertical support by two fillet welds as illustrated.

(*a*) Find the safe bending force F if the permissible shear stress in the welds is 20 kpsi.

(*b*) In part *a* you found a simple expression for F in terms of the allowable shear stress. Find the allowable load if the electrode is E7010, the bar is hot-rolled 1020, and the support is hot-rolled 1015.

Problem 9–5

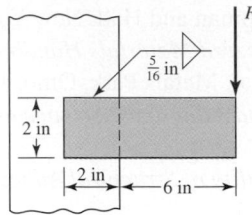

9–6 ANALYSIS The figure shows a weldment just like that of Prob. 9–5 except that there are four welds instead of two. Show that the weldment is twice as strong as that of Prob. 9–5.

Problem 9–6

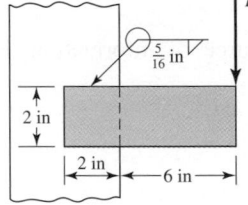

9–7 ANALYSIS The weldment shown in the figure is subjected to an alternating force F. The hot-rolled steel bar is 10 mm thick and is of AISI 1010 steel. The vertical support is likewise of 1010 steel. The electrode is 6010. Estimate the fatigue load F the bar will carry if three 6-mm fillet welds are used.

Problem 9–7

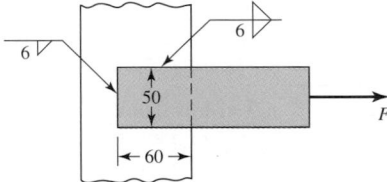

Dimensions in millimeters

9–8 The permissible shear stress for the weldment illustrated is 140 MPa. Estimate the load, F, that will cause this stress in the weldment throat.

Problem 9–8

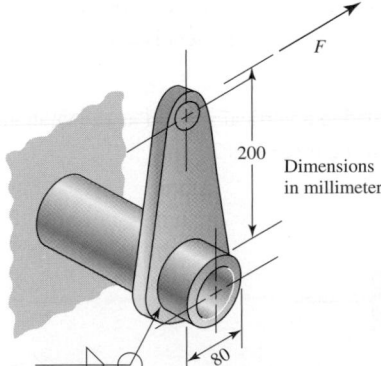

Dimensions in millimeters

DESIGN

9–9

In the design of weldments in torsion it is helpful to have a hierarchical perception of the relative efficiency of common patterns. For example, the weld-bead patterns shown in Table 9–1 can be ranked for desirability. Assume the space available is an $a \times a$ square. Use a formal figure of merit that is directly proportional to J and inversely proportional to the volume of weld metal laid down:

$$\text{fom} = \frac{J}{\text{vol}} = \frac{0.707h J_u}{(h^2/2)l} = 1.414\frac{J_u}{hl}$$

A tactical figure of merit could omit the constant, that is, $\text{fom}' = J_u/(hl)$. Rank the six patterns of Table 9–1 from most to least efficient.

DESIGN

9–10

The space available for a weld-bead pattern subject to bending is $a \times a$. Place the patterns of Table 9–2 in hierarchical order of efficiency of weld metal placement to resist bending. A formal figure of merit can be directly proportion to I and inversely proportional to the volume of weld metal laid down:

$$\text{fom} = \frac{I}{\text{vol}} = \frac{0.707h I_u}{(h^2/2)l} = 1.414\frac{I_u}{hl}$$

The tactical figure of merit can omit the constant 1.414, that is, $\text{fom}' = I_u/(hl)$. Omit the patterns intended for T beams and I beams. Rank the remaining seven.

DESIGN

9–11

Among the possible forms of weldment problems are the following:

• The attachment and the member(s) exist and only the weld specifications need to be decided.

• The members exist, but both the attachment and the weldment must be designed.

• The attachment, member(s), and weldment must be designed.

What follows is a design task of the first category. The attachment shown in the figure is made of 1018 HR steel $\frac{1}{2}$ in thick. The static force is 25 kip. The member is 4 in wide, such as that shown in Prob. 9–4. Specify the weldment (give the pattern, electrode number, type of weld, length of weld, and leg size).

Problem 9–11

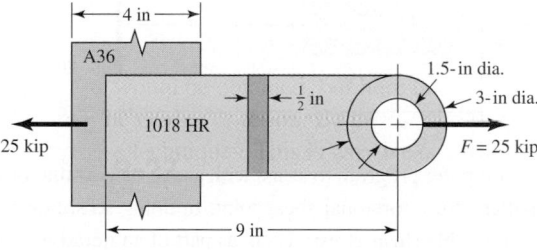

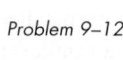
DESIGN

9–12

The attachment shown carries a bending load of 3 kip. The clearance a is to be 6 in. The load is a static 3000 lbf. Specify the weldment (give the pattern, electrode number, type of weld, length of weld, and leg size).

Problem 9–12

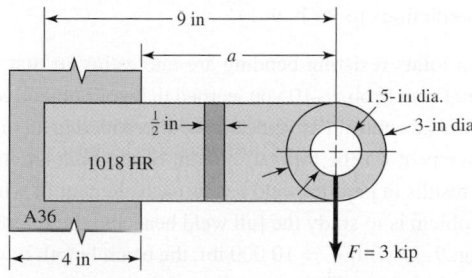

DESIGN

9–13   The attachment in Prob. 9–12 has not had its length determined. The static force is 3 kip; the clearance a is to be 6 in. The member is 4 in wide. Specify the weldment (give the pattern, electrode number, type of weld, length of bead, and leg size). Specify the attachment length.

Problem 9–13

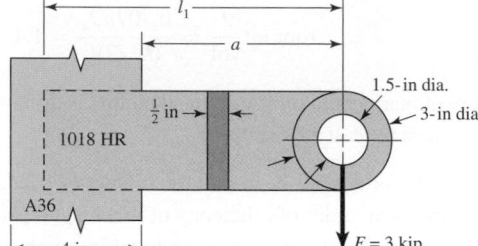

DESIGN

9–14   A vertical column of A36 structural steel ($S_y = 36$ kpsi, $S_{ut} = 58$–80 kpsi) is 10 in wide. An attachment has been designed to the point shown in the figure. The static load of 20 kip is applied, and the clearance a of 6.25 in has to be equaled or exceeded. The attachment is 1018 hot-rolled steel, to be made from $\frac{1}{2}$-in plate with weld-on bosses when all dimensions are known. Specify the weldment (give the pattern, electrode number, type of weld, length of weld bead, and leg size). Specify also the length l_1 for the attachment.

Problem 9–14

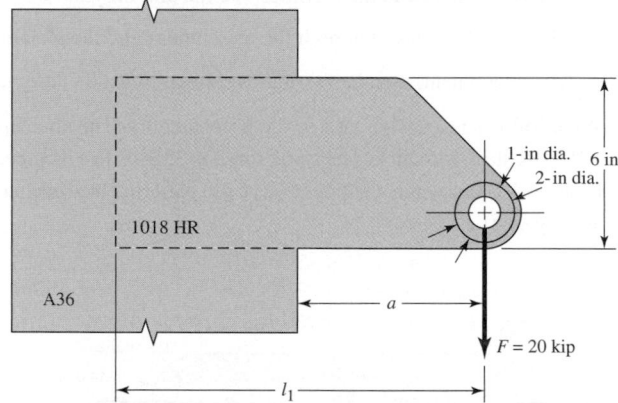

ANALYSIS

9–15   Write a computer program to assist with a task such as that of Prob. 9–14 with a rectangular weld-bead pattern for a torsional shear joint. In doing so solicit the force F, the clearance a, and the largest allowable shear stress. Then, as part of an iterative loop, solicit the dimensions b and d of the rectangle. These can be your design variables. Output all the parameters after the leg size has been determined by computation. In effect this will be your adequacy assessment when you stop iterating. Include the figure of merit $J_u/(hl)$ in the output. The fom and the leg size h with available width will give you a useful insight into the nature of this class of welds. Use your program to verify your solutions to Prob. 9–14.

DESIGN

9–16   Fillet welds in joints resisting bending are interesting in that they can be simpler than those resisting torsion. From Prob. 9–10 you learned that your objective is to place weld metal as far away from the weld-bead centroid as you can, but distributed in an orientation parallel to the x axis. Furthermore, placement on the top and bottom of the built-in end of a cantilever with rectangular cross section results in parallel weld beads, each element of which is in the ideal position. The object of this problem is to study the full weld bead and the interrupted weld-bead pattern. Consider the case of Fig. 9–17 with $F = 10\,000$ lbf, the beam length $a = 10$ in, $b = 8$ in, and $d = 8$ in. For the second case, for the interrupted weld consider a centered gap of $b_1 = 2$ in existing in the top and bottom welds. Study the two cases with $\tau_{all} = 12.8$ kpsi. What do you notice about τ, σ, and τ_{max}? Compare the fom′.

 ANALYSIS

9–17 For a rectangular weld-bead track resisting bending, develop the necessary equations to treat cases of vertical welds, horizontal welds, and weld-all-around patterns with depth d and width b and allowing central gaps in parallel beads of length b_1 and d_1. Do this by superposition of parallel tracks, vertical tracks subtracting out the gaps. Then put the two together for a rectangular weld bead with central gaps of length b_1 and d_1. Show that the results are

$$A = 1.414(b - b_1 + d - d_1)h$$

$$I_u = \frac{(b - b_1)d^2}{2} + \frac{d^3 - d_1^3}{6}$$

$$I = 0.707hI_u$$

$$l = 2(b - b_1) + 2(d - d_1)$$

$$\text{fom} = \frac{I_u}{hl}$$

 ANALYSIS

9–18 Write a computer program based on the Prob. 9–17 protocol. Solicit the largest allowable shear stress, the force F, and the clearance a, as well as the dimensions b and d. Begin an iterative loop by soliciting b_1 and d_1. Either or both of these can be your design variables. Program to find the leg size corresponding to a shear-stress level at the maximum allowable at a corner. Output all your parameters including the figure of merit. Use the program to check any previous problems to which it is applicable. Play with it in a "what if" mode and learn from the trends in your parameters.

 DESIGN

9–19 When comparing two different weldment patterns it is useful to observe the resistance to bending or torsion and the volume of weld metal deposited. Measure of effectiveness, defined as second moment of area divided by weld-metal volume, is useful. If a 6-in by 8-in section of a cantilever carries a static 10 kip bending load 10 in from the weldment plane, with an allowable shear stress of 12 800 psi realized, compare horizontal weldments with vertical weldments. The horizontal beads are to be 6 in long and the vertical beads, 8 in long.

ANALYSIS

9–20 A torque $T = 20(10^3)$ lbf · in is applied to the weldment shown. Estimate the maximum shear stress in the weld throat.

Problem 9–20

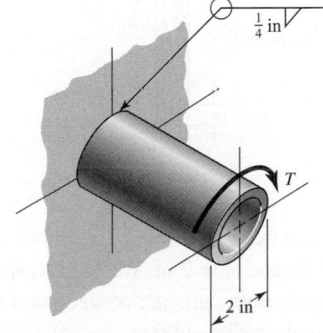

ANALYSIS

9-21 Find the maximum shear stress in the throat of the weld metal in the figure.

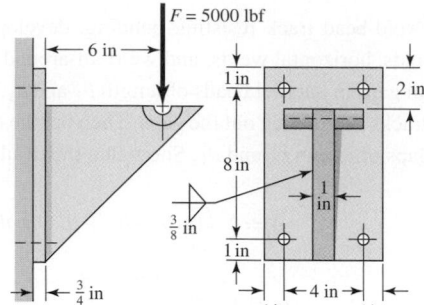

Problem 9–21

ANALYSIS

9-22 The figure shows a welded steel bracket loaded by a static force F. Estimate the factor of safety if the allowable shear stress in the weld throat is 120 MPa.

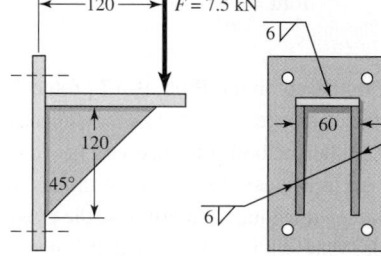

Problem 9–22

Dimensions in millimeters

ANALYSIS

9-23 The figure shows a formed sheet-steel bracket. Instead of securing it to the support with machine screws, welding has been proposed. If the combined stress in the weld metal is limited to 900 psi, estimate the total load W the bracket will support. The dimensions of the top flange are the same as the mounting flange.

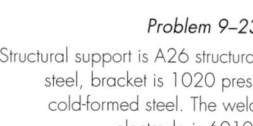

Problem 9–23

Structural support is A26 structural steel, bracket is 1020 press cold-formed steel. The weld electrode is 6010.

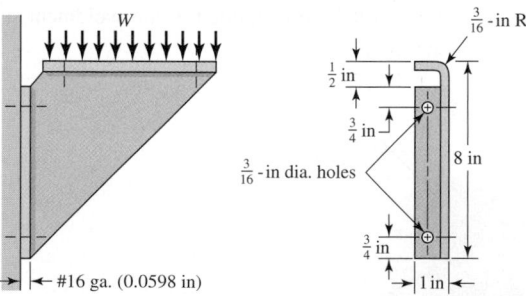

ANALYSIS

9-24 Without bracing, a machinist can exert only about 100 lbf on a wrench or tool handle. The lever shown in the figure has $t = \frac{1}{2}$ in and $w = 2$ in. We wish to specify the fillet-weld size to secure the lever to the tubular part at A. Both parts are of steel, and the shear stress in the weld throat should not exceed 3000 psi. Find a safe weld size.

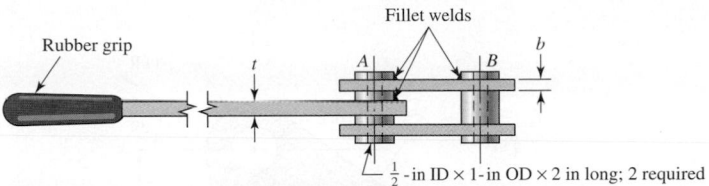

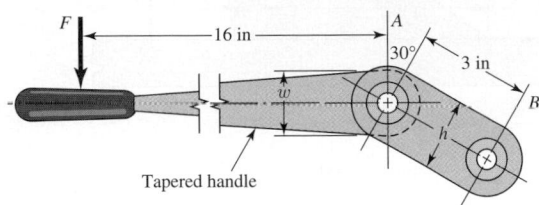

Problem 9–24

ANALYSIS

9–25   Estimate the safe static load F for the weldment shown in the figure if an E6010 electrode is used and the design factor is to be 2. Use conventional analysis.

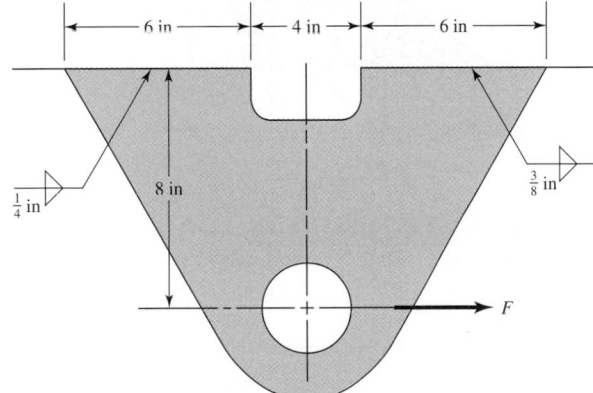

Problem 9–25

ANALYSIS

9–26   Brackets, such as the one shown, are used in mooring small watercraft. Failure of such brackets is usually caused by bearing pressure of the mooring clip against the side of the hole. Our purpose here is to get an idea of the static and dynamic margins of safety involved. We use a bracket 1/4 in thick made of hot-rolled 1018 steel. We then assume wave action on the boat will create force F no greater than 1200 lbf.

(a) Identify the moment M that produces a shear stress on the throat resisting bending action with a "tension" at A and "compression" at C.

(b) Find the force component F_y that produces a shear stress at the throat resisting a "tension" throughout the weld.

(c) Find the force component F_x that produces an in-line shear throughout the weld.

(d) Find A, I_u, and I using Table 9–2, in part.

(e) Find the shear stress τ_1 at A due to F_y and M, the shear stress τ_2 due to F_x, and combine to find τ.

(f) Find the factor of safety guarding against shear yielding in the weldment.

(g) Find the factor of safety guarding against a static failure in the parent metal at the weld.

(h) Find the factor of safety guarding against a fatigue failure in the weld metal using a Gerber failure criterion.

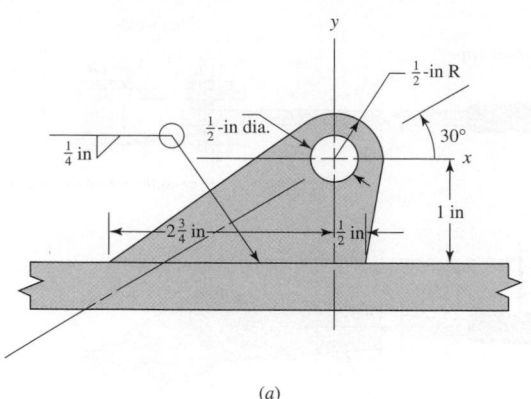

(a)

Problem 9–26
Small watercraft mooring bracket.

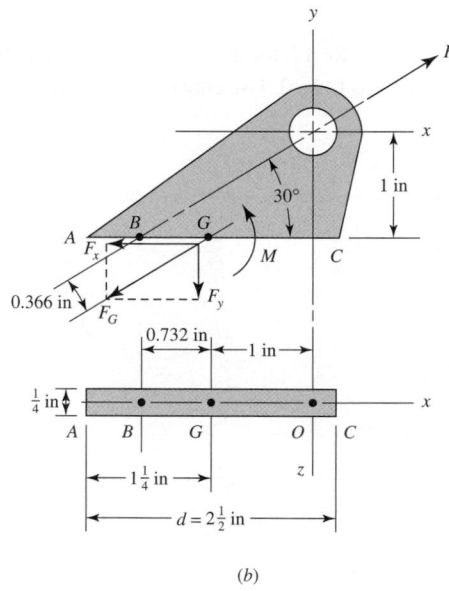

(b)

ANALYSIS

9–27   For the sake of perspective it is always useful to look at the matter of scale. Double all dimensions in Prob. 9–5 and find the allowable load. By what factor has it increased? First make a guess, then carry out the computation. Would you expect the same ratio if the load had been variable?

ANALYSIS

9–28   Hardware stores often sell plastic hooks that can be mounted on walls with pressure-sensitive adhesive foam tape. Two designs are shown in (a) and (b) of the figure. Indicate which one you would buy and why.

ANALYSIS

9–29   For a balanced double-lap joint cured at room temperature, Volkersen's equation simplifies to

$$\tau(x) = \frac{P\omega\cosh(\omega x)}{4b\sinh(\omega l/2)} = A_1\cosh(\omega x)$$

(a) Show that the average stress $\bar{\tau}$ is $P/(2bl)$.
(b) Show that the largest shear stress is $P\omega/[4b\tanh(\omega l/2)]$.
(c) Define a stress-augmentation factor K such that

$$\tau(l/2) = K\bar{\tau}$$

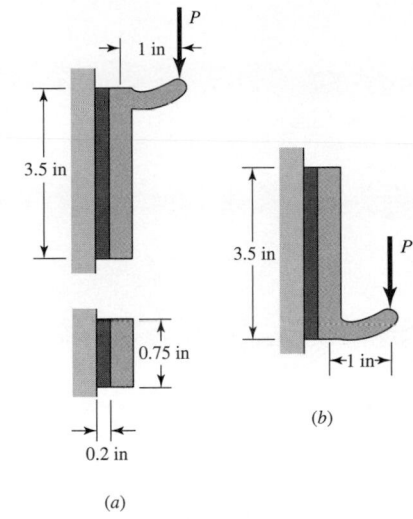

Problem 9–28

(a)

(b)

and it follows that

$$K = \frac{P\omega}{4b\tanh(\omega l/2)} \frac{2bl}{P} = \frac{\omega l/2}{\tanh(\omega l/2)} = \frac{\omega l}{2} \frac{\exp(\omega l/2) + \exp(-\omega l/2)}{\exp(\omega l/2) - \exp(-\omega l/2)}$$

9–30 Program the shear-lag solution for the shear-stress state into your computer using Eq. (9–12). Determine the maximum shear stress for each of the following scenarios:

Part	E_a, psi	t_o, in	t_i, in	E_o, psi	E_i, psi	h, in
a	$0.2(10^6)$	0.125	0.250	$30(10^6)$	$30(10^6)$	0.005
b	$0.2(10^6)$	0.125	0.250	$30(10^6)$	$30(10^6)$	0.015
c	$0.2(10^6)$	0.125	0.125	$30(10^6)$	$30(10^6)$	0.005
d	$0.2(10^6)$	0.125	0.250	$30(10^6)$	$10(10^6)$	0.005

Provide plots of the actual stress distributions predicted by this analysis. You may omit thermal stresses from the calculations, assuming that the service temperature is similar to the stress-free temperature. If the allowable shear stress is 800 psi and the load to be carried is 300 lbf, estimate the respective factors of safety for each geometry. Let $l = 1.25$ in and $b = 1$ in.

10

Mechanical Springs

When a designer wants rigidity, negligible deflection is an acceptable approximation as long as it does not compromise function. Flexibility is sometimes needed and is often provided by metal bodies with cleverly controlled geometry. These bodies can exhibit flexibility to the degree the designer seeks. Such flexibility can be linear or nonlinear in relating deflection to load. These devices allow controlled application of force or torque; the storing and release of energy can be another purpose. Flexibility allows temporary distortion for access and the immediate restoration of function. Because of machinery's value to designers, springs have been intensively studied; moreover, they are mass-produced (and therefore low cost), and ingenious configurations have been found for a variety of desired applications. In this chapter we will discuss the more frequently used types of springs, their necessary parametric relationships, and their design.

In general, springs may be classified as wire springs, flat springs, or special-shaped springs, and there are variations within these divisions. Wire springs include helical springs of round or square wire, made to resist and deflect under tensile, compressive, and torsional loads. Flat springs include cantilever and elliptical types, wound motor- or clock-type power springs, and flat spring washers, usually called Belleville springs.

10–1 Stresses in Helical Springs

Figure 10–1a shows a round-wire helical compression spring loaded by the axial force F. We designate D as the *mean coil diameter* and d as the *wire diameter*. Now imagine that the spring is cut at some point (Fig. 10–1b), a portion of it removed, and the effect of the removed portion replaced by the net internal reactions. Then, as shown in the figure, from equilibrium the cut portion would contain a direct shear force F and a torsion $T = FD/2$.

To visualize the torsion, picture a coiled garden hose. Now pull one end of the hose in a straight line perpendicular to the plane of the coil. As each turn of hose is pulled off the coil, the hose twists or turns about its own axis. The flexing of a helical spring creates a torsion in the wire in a similar manner.

The maximum stress in the wire may be computed by superposition of the direct shear stress given by Eq. (4–23) and the torsional shear stress given by Eq. (4–37). The result is

$$\tau_{\max} = \frac{Tr}{J} + \frac{F}{A} \qquad (a)$$

Figure 10–1

(a) Axially loaded helical spring; (b) free-body diagram showing that the wire is subjected to a direct shear and a torsional shear.

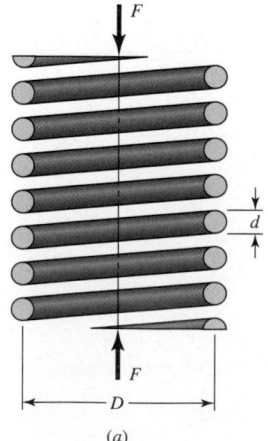

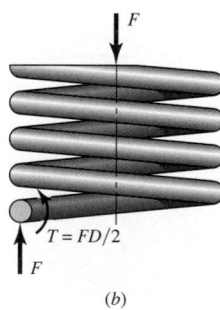

(a)

(b)

at the *inside* fiber of the spring. Substitution of $\tau_{max} = \tau$, $T = FD/2$, $r = d/2$, $J = \pi d^4/32$, and $A = \pi d^2/4$ gives

$$\tau = \frac{8FD}{\pi d^3} + \frac{4F}{\pi d^2} \qquad (10-1)$$

Now we define the *spring index*

$$C = \frac{D}{d} \qquad (10-2)$$

which is a measure of coil curvature. With this relation, Eq. (10–1) can be rearranged to give

$$\tau = K_s \frac{8FD}{\pi d^3} \qquad (10-3)$$

where K_s is a *shear-stress correction factor* and is defined by the equation

$$K_s = \frac{2C + 1}{2C} \qquad (10-4)$$

For most springs, C ranges from about 6 to 12. Equation (10–3) is quite general and applies for both static and dynamic loads.

The use of square or rectangular wire is not recommended for springs unless space limitations make it necessary. Springs of special wire shapes are not made in large quantities, unlike those of round wire; they have not had the benefit of refining development and hence may not be as strong as springs made from round wire. When space is severely limited, the use of nested round-wire springs should always be considered. They may have an economical advantage over the special-section springs, as well as a strength advantage.

10–2 The Curvature Effect

Equation (10–1) is based on the wire being straight. However, the curvature of the wire increases the stress on the inside of the spring but decreases it only slightly on the outside. This curvature stress is primarily important in fatigue because the loads are lower and there is no opportunity for localized yielding. For static loading, these stresses can normally be neglected because of strain-strengthening with the first application of load.

Unfortunately, it is necessary to find the curvature factor in a roundabout way. The reason for this is that the published equations also include the effect of the direct shear stress. Suppose K_s in Eq. (10–3) is replaced by another K factor, which corrects for both curvature and direct shear. Then this factor is given by either of the equations

$$K_W = \frac{4C - 1}{4C - 4} + \frac{0.615}{C} \qquad (10-5)$$

$$K_B = \frac{4C + 2}{4C - 3} \qquad (10-6)$$

The first of these is called the *Wahl factor,* and the second, the *Bergsträsser factor.*[1] Since the results of these two equations differ by less than 1 percent, Eq. (10–6) is preferred.

---

[1]Cyril Samónov, "Some Aspects of Design of Helical Compression Springs," *Int. Symp. Design and Synthesis,* Tokyo, 1984.

The curvature correction factor can now be obtained by canceling out the effect of the direct shear. Thus, using Eq. (10–6) with Eq. (10–4), the curvature correction factor is found to be

$$K_c = \frac{K_B}{K_s} = \frac{2C(4C + 2)}{(4C - 3)(2C + 1)} \tag{10–7}$$

Now, K_s, K_B or K_W, and K_c are simply stress correction factors applied multiplicatively to Tr/J at the critical location to estimate a particular stress. There is *no* stress concentration factor. In this book we will use $\tau = K_B(8FD)/(\pi d^3)$ to predict the largest shear stress.

10–3   Deflection of Helical Springs

The deflection-force relations are quite easily obtained by using Castigliano's theorem. The total strain energy for a helical spring is composed of a torsional component and a shear component. From Eqs. (5–16) and (5–17), the strain energy is

$$U = \frac{T^2 l}{2GJ} + \frac{F^2 l}{2AG} \tag{a}$$

Substituting $T = FD/2$, $l = \pi DN$, $J = \pi d^4/32$, and $A = \pi d^2/4$ results in

$$U = \frac{4F^2 D^3 N}{d^4 G} + \frac{2F^2 DN}{d^2 G} \tag{b}$$

where $N = N_a$ = number of active coils. Then using Castigliano's theorem, Eq. (5–21), to find total deflection y gives

$$y = \frac{\partial U}{\partial F} = \frac{8FD^3 N}{d^4 G} + \frac{4FDN}{d^2 G} \tag{c}$$

Since $C = D/d$, Eq. (c) can be rearranged to yield

$$y = \frac{8FD^3 N}{d^4 G}\left(1 + \frac{1}{2C^2}\right) \doteq \frac{8FD^3 N}{d^4 G} \tag{10–8}$$

The spring rate, also called the *scale* of the spring, is $k = F/y$, and so

$$k \doteq \frac{d^4 G}{8D^3 N} \tag{10–9}$$

10–4   Compression Springs

The four types of ends generally used for compression springs are illustrated in Fig. 10–2. A spring with *plain ends* has a noninterrupted helicoid; the ends are the same as if a long spring had been cut into sections. A spring with plain ends that are *squared* or *closed* is obtained by deforming the ends to a zero-degree helix angle. Springs should always be both squared and ground for important applications, because a better transfer of the load is obtained.

Table 10–1 shows how the type of end used affects the number of coils and the spring length.[2] Note that the digits 0, 1, 2, and 3 appearing in Table 10–1 are often used without

---

[2]For a thorough discussion and development of these relations, see Cyril Samónov, "Computer-Aided Design of Helical Compression Springs," ASME paper No. 80-DET-69, 1980.

Figure 10–2

Types of ends for compression springs: (a) both ends plain; (b) both ends squared; (c) both ends squared and ground; (d) both ends plain and ground.

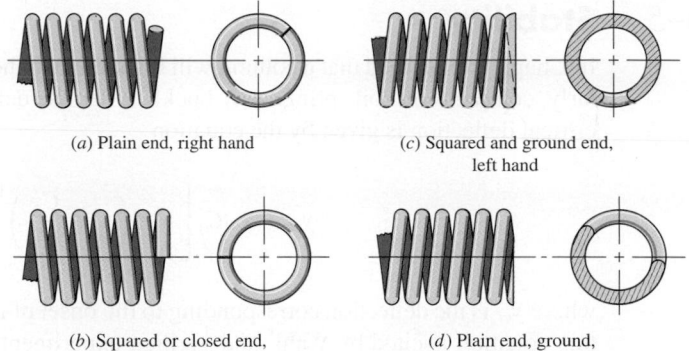

(a) Plain end, right hand

(c) Squared and ground end, left hand

(b) Squared or closed end, right hand

(d) Plain end, ground, left hand

Table 10–1

Formulas for Compression-Spring Dimensions. ($N_a =$ Number of Active Coils)

Source: From *Design Handbook*, 1987, p. 32. Courtesy of Associated Spring.

		Type of Spring Ends		
Term	**Plain**	**Plain and Ground**	**Squared or Closed**	**Squared and Ground**
End coils, N_e	0	1	2	2
Total coils, N_t	N_a	$N_a + 1$	$N_a + 2$	$N_a + 2$
Free length, L_0	$pN_a + d$	$p(N_a + 1)$	$pN_a + 3d$	$pN_a + 2d$
Solid length, L_s	$d(N_t + 1)$	dN_t	$d(N_t + 1)$	dN_t
Pitch, p	$(L_0 - d)/N_a$	$L_0/(N_a + 1)$	$(L_0 - 3d)/N_a$	$(L_0 - 2d)/N_a$

question. *Some of these need closer scrutiny as they may not be integers.* This depends on how a springmaker forms the ends. Forys[3] pointed out that squared and ground ends give a solid length L_s of

$$L_s = (N_t - a)d$$

where a varies, with an average of 0.75, so the entry dN_t in Table 10–1 may be overstated. The way to check these variations is to take springs from a particular springmaker, close them solid, and measure the solid height. Another way is to look at the spring and count the wire diameters in the solid stack.

Set removal or *presetting* is a process used in the manufacture of compression springs to induce useful residual stresses. It is done by making the spring longer than needed and then compressing it to its solid height. This operation *sets* the spring to the required final free length and, since the torsional yield strength has been exceeded, induces residual stresses opposite in direction to those induced in service. Springs to be preset should be designed so that 10 to 30 percent of the initial free length is removed during the operation. If the stress at the solid height is greater than 1.3 times the torsional yield strength, distortion may occur. If this stress is much less than 1.1 times, it is difficult to control the resulting free length.

Set removal increases the strength of the spring and so is especially useful when the spring is used for energy-storage purposes. However, set removal should not be used when springs are subject to fatigue.

---

[3]Edward L. Forys, "Accurate Spring Heights," *Machine Design,* vol. 56, no. 2, January 26, 1984.

10–5 Stability

In Chap. 4 we learned that a column will buckle when the load becomes too large. Similarly, compression coil springs may buckle when the deflection becomes too large. The critical deflection is given by the equation

$$y_{cr} = L_0 C_1' \left[1 - \left(1 - \frac{C_2'}{\lambda_{eff}^2} \right)^{1/2} \right] \tag{10-10}$$

where y_{cr} is the deflection corresponding to the onset of instability. Samónov[4] states that this equation is cited by Wahl[5] and verified experimentally by Haringx.[6] The quantity λ_{eff} in Eq. (10–10) is the *effective slenderness ratio* and is given by the equation

$$\lambda_{eff} = \frac{\alpha L_0}{D} \tag{10-11}$$

C_1' and C_2' are elastic constants defined by the equations

$$C_1' = \frac{E}{2(E - G)}$$

$$C_2' = \frac{2\pi^2(E - G)}{2G + E}$$

Equation (10–11) contains the *end-condition constant* α. This depends upon how the ends of the spring are supported. Table 10–2 gives values of α for usual end conditions. Note how closely these resemble the end conditions for columns.

Absolute stability occurs when, in Eq. (10–10), the term C_2'/λ_{eff}^2 is greater than unity. This means that the condition for absolute stability is that

$$L_0 < \frac{\pi D}{\alpha} \left[\frac{2(E - G)}{2G + E} \right]^{1/2} \tag{10-12}$$

Table 10–2

End-Condition Constants α for Helical Compression Springs*

End Condition	Constant α
Spring supported between flat parallel surfaces (fixed ends)	0.5
One end supported by flat surface perpendicular to spring axis (fixed); other end pivoted (hinged)	0.707
Both ends pivoted (hinged)	1
One end clamped; other end free	2

*Ends supported by flat surfaces must be squared and ground.

---

[4]"Computer-Aided Design."

[5]A. M. Wahl, *Mechanical Springs,* 2d ed., McGraw-Hill, New York, 1963.

[6]J. A. Haringx, "On Highly Compressible Helical Springs and Rubber Rods and Their Application for Vibration-Free Mountings," I and II, *Philips Res. Rep.,* vol. 3, December 1948, pp. 401–449, and vol. 4, February 1949, pp. 49–80.

For steels, this turns out to be

$$L_0 < 2.63\frac{D}{\alpha}$$ (10–13)

For squared and ground ends $\alpha = 0.5$ and $L_0 < 5.26D$.

10–6 Spring Materials

Springs are manufactured either by hot- or cold-working processes, depending upon the size of the material, the spring index, and the properties desired. In general, prehardened wire should not be used if $D/d < 4$ or if $d > \frac{1}{4}$ in. Winding of the spring induces residual stresses through bending, but these are normal to the direction of the torsional working stresses in a coil spring. Quite frequently in spring manufacture, they are relieved, after winding, by a mild thermal treatment.

A great variety of spring materials are available to the designer, including plain carbon steels, alloy steels, and corrosion-resisting steels, as well as nonferrous materials such as phosphor bronze, spring brass, beryllium copper, and various nickel alloys. Descriptions of the most commonly used steels will be found in Table 10–3. The UNS steels listed in Appendix A should be used in designing hot-worked, heavy-coil springs, as well as flat springs, leaf springs, and torsion bars.

Spring materials may be compared by an examination of their tensile strengths; these vary so much with wire size that they cannot be specified until the wire size is known. The material and its processing also, of course, have an effect on tensile strength. It turns out that the graph of tensile strength versus wire diameter is almost a straight line for some materials when plotted on log-log paper. Writing the equation of this line as

$$S_{ut} = \frac{A}{d^m}$$ (10–14)

furnishes a good means of estimating minimum tensile strengths when the intercept A and the slope m of the line are known. Values of these constants have been worked out from recent data and are given for strengths in units of kpsi and MPa in Table 10–4. In Eq. (10–14) when d is measured in millimeters, then A is in MPa · mm^m and when d is measured in inches, then A is in kpsi · in^m.

Although the torsional yield strength is needed to design the spring and to analyze the performance, spring materials customarily are tested only for tensile strength—perhaps because it is such an easy and economical test to make. A very rough estimate of the torsional yield strength can be obtained by assuming that the tensile yield strength is between 60 and 90 percent of the tensile strength. Then the distortion-energy theory can be employed to obtain the torsional yield strength ($S_{ys} = 0.577S_y$). This approach results in the range

$$0.35S_{ut} \le S_{sy} \le 0.52S_{ut}$$ (10–15)

for steels.

For wires listed in Table 10–5, the maximum allowable shear stress in a spring can be seen in column 3. Music wire and hard-drawn steel spring wire have a low end of range $S_{sy} = 0.45S_{ut}$. Valve spring wire, Cr-Va, Cr-Si, and other (not shown) hardened and tempered carbon and low-alloy steel wires as a group have $S_{sy} \ge 0.50S_{ut}$. Many nonferrous materials (not shown) as a group have $S_{sy} \ge 0.35S_{ut}$. In view of this,

Table 10–3

High-Carbon and Alloy Spring Steels

Source: From Harold C. R. Carlson, "Selection and Application of Spring Materials," *Mechanical Engineering,* vol. 78, 1956, pp. 331–334.

Name of Material	Similar Specifications	Description
Music wire, 0.80–0.95C	UNS G10850 AISI 1085 ASTM A228-51	This is the best, toughest, and most widely used of all spring materials for small springs. It has the highest tensile strength and can withstand higher stresses under repeated loading than any other spring material. Available in diameters 0.12 to 3 mm (0.005 to 0.125 in). Do not use above 120°C (250°F) or at subzero temperatures.
Oil-tempered wire, 0.60–0.70C	UNS G10650 AISI 1065 ASTM 229-41	This general-purpose spring steel is used for many types of coil springs where the cost of music wire is prohibitive and in sizes larger than available in music wire. Not for shock or impact loading. Available in diameters 3 to 12 mm (0.125 to 0.5000 in), but larger and smaller sizes may be obtained. Not for use above 180°C (350°F) or at subzero temperatures.
Hard-drawn wire, 0.60–0.70C	UNS G10660 AISI 1066 ASTM A227-47	This is the cheapest general-purpose spring steel and should be used only where life, accuracy, and deflection are not too important. Available in diameters 0.8 to 12 mm (0.031 to 0.500 in). Not for use above 120°C (250°F) or at subzero temperatures.
Chrome-vanadium	UNS G61500 AISI 6150 ASTM 231-41	This is the most popular alloy spring steel for conditions involving higher stresses than can be used with the high-carbon steels and for use where fatigue resistance and long endurance are needed. Also good for shock and impact loads. Widely used for aircraft-engine valve springs and for temperatures to 220°C (425°F). Available in annealed or pretempered sizes 0.8 to 12 mm (0.031 to 0.500 in) in diameter.
Chrome-silicon	UNS G92540 AISI 9254	This alloy is an excellent material for highly stressed springs that require long life and are subjected to shock loading. Rockwell hardnesses of C50 to C53 are quite common, and the material may be used up to 250°C (475°F). Available from 0.8 to 12 mm (0.031 to 0.500 in) in diameter.

Table 10–4

Constants A and m of $S_{ut} = A/d^m$ for Estimating Minimum Tensile Strength of Common Spring Wires
Source: From *Design Handbook*, 1987, p. 19. Courtesy of Associated Spring.

Material	ASTM No.	Exponent m	Diameter, in	A, kpsi·in^m	Diameter, mm	A, MPa·mm^m	Relative Cost of wire
Music wire*	A228	0.145	0.004–0.256	201	0.10–6.5	2211	2.6
OQ&T wire[†]	A229	0.187	0.020–0.500	147	0.5–12.7	1855	1.3
Hard-drawn wire[‡]	A227	0.190	0.028–0.500	140	0.7–12.7	1783	1.0
Chrome-vanadium wire[§]	A232	0.168	0.032–0.437	169	0.8–11.1	2005	3.1
Chrome-silicon wire[‖]	A401	0.108	0.063–0.375	202	1.6–9.5	1974	4.0
302 Stainless wire[#]	A313	0.146	0.013–0.10	169	0.3–2.5	1867	7.6–11
		0.263	0.10–0.20	128	2.5–5	2065	
		0.478	0.20–0.40	90	5–10	2911	
Phosphor-bronze wire**	B159	0	0.004–0.022	145	0.1–0.6	1000	8.0
		0.028	0.022–0.075	121	0.6–2	913	
		0.064	0.075–0.30	110	2–7.5	932	

*Surface is smooth, free of defects, and has a bright, lustrous finish.
[†] Has a slight heat-treating scale which must be removed before plating.
[‡] Surface is smooth and bright with no visible marks.
[§] Aircraft-quality tempered wire, can also be obtained annealed.
[‖] Tempered to Rockwell C49, but may be obtained untempered.
[#] Type 302 stainless steel.
**Temper CA510.

Joerres[7] uses the maximum allowable torsional stress for static application shown in Table 10–6. For specific materials for which you have torsional yield information use this table as a guide. Joerres provides set-removal information in Table 10–6, that $S_{sy} \geq 0.65 S_{ut}$ increases strength through cold work, but at the cost of an additional operation by the springmaker. Sometimes the additional operation can be done by the manufacturer during assembly. Some correlations with carbon steel springs show that the tensile yield strength of spring wire in torsion can be estimated from $0.75 S_{ut}$. The corresponding estimate of the yield strength in shear based on distortion energy theory is $S_{sy} = 0.577(0.75) S_{ut} = 0.433 S_{ut} \doteq 0.45 S_{ut}$. Samónov discusses the problem of allowable stress and shows that

$$S_{sy} = \tau_{\text{all}} = 0.56 S_{ut} \tag{10–16}$$

for high-tensile spring steels, which is close to the value given by Joerres for hardened alloy steels. He points out that this value of allowable stress is specified by Draft Standard 2089 of the German Federal Republic when Eq. (10–3) is used without stress-correction factor.

---

[7] Robert E. Joerres, "Springs," Chap. 24 in Joseph E. Shigley and Charles R. Mischke (eds.), *Standard Handbook of Machine Design,* 2nd ed., McGraw-Hill, New York, 1996.

Table 10–5

Mechanical Properties of Some Spring Wires

Material	Elastic Limit, Percent of S_{ut} Tension	Torsion	Diameter d, in	E Mpsi	GPa	G Mpsi	GPa
Music wire A228	65–75	45–60	<0.032	29.5	203.4	12.0	82.7
			0.033–0.063	29.0	200	11.85	81.7
			0.064–0.125	28.5	196.5	11.75	81.0
			>0.125	28.0	193	11.6	80.0
HD spring A227	60–70	45–55	<0.032	28.8	198.6	11.7	80.7
			0.033–0.063	28.7	197.9	11.6	80.0
			0.064–0.125	28.6	197.2	11.5	79.3
			>0.125	28.5	196.5	11.4	78.6
Oil tempered A239	85–90	45–50		28.5	196.5	11.2	77.2
Valve spring A230	85–90	50–60		29.5	203.4	11.2	77.2
Chrome-vanadium A231	88–93	65–75		29.5	203.4	11.2	77.2
A232	88–93			29.5	203.4	11.2	77.2
Chrome-silicon A401	85–93	65–75		29.5	203.4	11.2	77.2
Stainless steel							
A313*	65–75	45–55		28	193	10	69.0
17-7PH	75–80	55–60		29.5	208.4	11	75.8
414	65–70	42–55		29	200	11.2	77.2
420	65–75	45–55		29	200	11.2	77.2
431	72–76	50–55		30	206	11.5	79.3
Phosphor-bronze B159	75–80	45–50		15	103.4	6	41.4
Beryllium-copper B197	70	50		17	117.2	6.5	44.8
	75	50–55		19	131	7.3	50.3
Inconel alloy X-750	65–70	40–45		31	213.7	11.2	77.2

*Also includes 302, 304, and 316.

Note: See Table 10–6 for allowable torsional stress design values.

Table 10–6

Maximum Allowable Torsional Stresses for Helical Compression Springs in Static Applications

Source: Robert E. Joerres, "Springs," Chap. 24 in Joseph E. Shigley and Charles R. Mischke (eds.), *Standard Handbook of Machine Design,* 2nd ed., McGraw-Hill, New York, 1996.

Material	Maximum Percent of Tensile Strength Before Set Removed (includes K_W or K_B)	After Set Removed (includes K_s)
Music wire and cold-drawn carbon steel	45	60–70
Hardened and tempered carbon and low-alloy steel	50	65–75
Austenitic stainless steels	35	55–65
Nonferrous alloys	35	55–65

EXAMPLE 10–1

A helical compression spring is made of no. 16 music wire. The outside diameter of the spring is $\frac{7}{16}$ in. The ends are squared and there are $12\frac{1}{2}$ total turns.

(a) Estimate the torsional yield strength of the wire.

(b) Estimate the static load corresponding to the yield strength.

(c) Estimate the scale of the spring.

(d) Estimate the deflection that would be caused by the load in part (b).

(e) Estimate the solid length of the spring.

(f) What length should the spring be to ensure that when it is compressed solid and then released, there will be no permanent change in the free length?

(g) Given the length found in part (f), is buckling a possibility?

(h) What is the pitch of the body coil?

Solution

(a) From Table A–28, the wire diameter is $d = 0.037$ in. From Table 10–4, we find $A = 201$ kpsi $\cdot$ in^m and $m = 0.145$. Therefore, from Eq. (10–14)

$$S_{ut} = \frac{A}{d^m} = \frac{201}{0.037^{0.145}} = 324 \text{ kpsi}$$

Then, from Table 10–6,

Answer

$$S_{sy} = 0.45 S_{ut} = 0.45(324) = 146 \text{ kpsi}$$

(b) The mean spring coil diameter is $D = \frac{7}{16} - 0.037 = 0.400$ in, and so the spring index is $C = 0.400/0.037 = 10.8$. Then, from Eq. (10–6),

$$K_B = \frac{4C + 2}{4C - 3} = \frac{4(10.8) + 2}{4(10.8) - 3} = 1.124$$

Now rearrange Eq. (10–3) replacing K_s and τ with K_B and S_{ys}, respectively, and solve for F:

Answer

$$F = \frac{\pi d^3 S_{sy}}{8 K_B D} = \frac{\pi (0.037^3) 146(10^3)}{8(1.124)\,0.400} = 6.46 \text{ lbf}$$

(c) From Table 10–1, $N_a = 12.5 - 2 = 10.5$ turns. In Table 10–5, $G = 11.85$ Mpsi, and the scale of the spring is found to be, from Eq. (10–9),

Answer

$$k = \frac{d^4 G}{8 D^3 N_a} = \frac{0.037^4 \,(11.85) 10^6}{8(0.400^3) 10.5} = 4.13 \text{ lbf/in}$$

Answer

(d)

$$y = \frac{F}{k} = \frac{6.46}{4.13} = 1.56 \text{ in}$$

(e) From Table 10–1,

Answer

$$L_s = (N_t + 1)d = (12.5 + 1)0.037 = 0.500 \text{ in}$$

Answer

(f)

$$L_0 = y + L_s = 1.56 + 0.500 = 2.06 \text{ in.}$$

(g) To avoid buckling, Eq. (10–13) and Table 10–2 give

$$L_0 < 2.63 \frac{D}{\alpha} = 2.63 \frac{0.400}{0.5} = 2.10 \text{ in}$$

Mathematically, a free length of 2.06 in is less than 2.10 in, and buckling is unlikely. However, the forming of the ends will control how close α is to 0.5. This has to be investigated and an inside rod or exterior tube or hole may be needed.

(*h*) Finally, from Table 10–1, the pitch of the body coil is

Answer

$$p = \frac{L_0 - 3d}{N_a} = \frac{2.06 - 3(0.037)}{10.5} = 0.186 \text{ in}$$

10–7 Helical Compression Spring Design for Static Service

The preferred range of spring index is $4 \leq C \leq 12$, with the lower indexes being more difficult to form (because of the danger of surface cracking) and springs with higher indexes tending to tangle often enough to require individual packing. This can be the first item of the design assessment. The recommended range of active turns is $3 \leq N_a \leq 15$. To maintain linearity when a spring is about to close, it is necessary to avoid the gradual touching of coils (due to nonperfect pitch). A helical coil spring force-deflection characteristic is ideally linear. Practically, it is nearly so, but not at each end of the force-deflection curve. The spring force is not reproducible for very small deflections, and near closure, nonlinear behavior begins as the number of active turns diminishes as coils begin to touch. The designer confines the spring's operating point to the central 75 percent of the curve between no load, $F = 0$, and closure, $F = F_s$. Thus, the maximum operating force should be limited to $F_{\max} \leq \frac{7}{8} F_s$. Defining the fractional overrun to closure as ξ, where

$$F_s = (1 + \xi) F_{\max} \tag{10–17}$$

it follows that

$$F_s = (1 + \xi) F_{\max} = (1 + \xi) \left(\frac{7}{8}\right) F_s$$

From the outer equality $\xi = 1/7 = 0.143 \doteq 0.15$. Thus, it is recommended that $\xi \geq 0.15$.

In addition to the relationships and material properties for springs, we now have some recommended design conditions to follow, namely:

$$4 \leq C \leq 12 \tag{10–18}$$

$$3 \leq N_a \leq 15 \tag{10–19}$$

$$\xi \geq 0.15 \tag{10–20}$$

$$n_s \geq 1.2 \tag{10–21}$$

When considering designing a spring for high volume production, the figure of merit can be the cost of the wire from which the spring is wound. The fom would be proportional to the relative material cost, weight density, and volume:

$$\text{fom} = -(\text{relative material cost}) \frac{\gamma \pi^2 d^2 N_t D}{4} \tag{10–22}$$

For comparisons between steels, the specific weight γ can be omitted.

Spring design is an open-ended process. There are many decisions to be made, and many possible solution paths as well as solutions. In the past, charts, nomographs, and "spring design slide rules" were used by many to simplify the spring design problem. Today, the computer enables the designer to create programs in many different

Figure 10–3

Helical coil compression spring design flowchart for static loading.

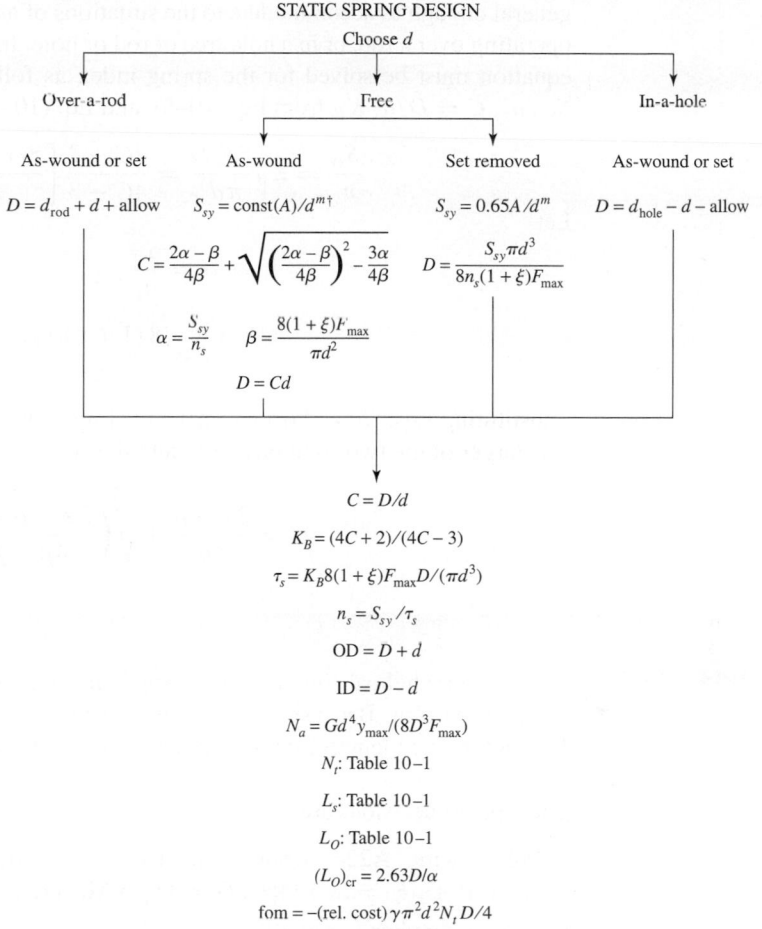

STATIC SPRING DESIGN

Choose d

Over-a-rod | Free | In-a-hole

As-wound or set | As-wound | Set removed | As-wound or set

$D = d_{\text{rod}} + d + \text{allow}$

$S_{sy} = \text{const}(A)/d^{m\dagger}$

$S_{sy} = 0.65A/d^m$

$D = d_{\text{hole}} - d - \text{allow}$

$$C = \frac{2\alpha - \beta}{4\beta} + \sqrt{\left(\frac{2\alpha - \beta}{4\beta}\right)^2 - \frac{3\alpha}{4\beta}}$$

$$D = \frac{S_{sy}\pi d^3}{8n_s(1 + \xi)F_{\max}}$$

$$\alpha = \frac{S_{sy}}{n_s} \qquad \beta = \frac{8(1 + \xi)F_{\max}}{\pi d^2}$$

$$D = Cd$$

$C = D/d$

$K_B = (4C + 2)/(4C - 3)$

$\tau_s = K_B 8(1 + \xi)F_{\max}D/(\pi d^3)$

$n_s = S_{sy}/\tau_s$

$\text{OD} = D + d$

$\text{ID} = D - d$

$N_a = Gd^4 y_{\max}/(8D^3 F_{\max})$

N_t: Table 10–1

L_s: Table 10–1

L_0: Table 10–1

$(L_0)_{\text{cr}} = 2.63D/\alpha$

$\text{fom} = -(\text{rel. cost})\gamma\pi^2 d^2 N_t D/4$

Print or display: d, D, C, OD, ID, N_a, N_t, L_s, L_0, $(L_0)_{\text{cr}}$, n_s, fom

Build a table, conduct design assessment by inspection

Eliminate infeasible designs by showing active constraints

Choose among satisfactory designs using the figure of merit

† const is found from Table 10–6

formats—direct programming, spreadsheet, MATLAB, etc. Commercial programs are also available.[8] There are almost as many ways to create a spring-design program as there are programmers. Here, we will suggest one possible design approach.

Design Strategy

Make the a priori decisions, with hard-drawn steel wire the first choice (relative material cost is 1.0). Choose a wire size d. With all decisions made, generate a column of parameters: d, D, C, OD or ID, N_a, L_s, L_0, $(L_0)_{\text{cr}}$, n_s, and fom. By incrementing wire sizes available, we can scan the table of parameters and apply the design recommendations by inspection. After wire sizes are eliminated, choose the spring design with the highest figure of merit. This will give the optimal design despite the presence of a discrete design variable d and aggregation of equality and inequality constraints. The column vector of information can be generated by using the flowchart displayed in Fig. 10–3. It is

[8]For example, see *Advanced Spring Design,* a program developed jointly between the Spring Manufacturers Institute (SMI), www.smihq.org, and Universal Technical Systems, Inc. (UTS), www.uts.com.

general enough to accommodate to the situations of as-wound and set-removed springs, operating over a rod, or in a hole free of rod or hole. In as-wound springs the controlling equation must be solved for the spring index as follows. From Eq. (10–3) with $\tau = S_{sy}/n_s$, $C = D/d$, K_B from Eq. (10–6), and Eq. (10–17),

$$\frac{S_{sy}}{n_s} = K_B \frac{8 F_s D}{\pi d^3} = \frac{4C + 2}{4C - 3} \left[\frac{8(1 + \xi) F_{max} C}{\pi d^2} \right] \tag{a}$$

Let

$$\alpha = \frac{S_{sy}}{n_s} \tag{b}$$

$$\beta = \frac{8(1 + \xi) F_{max}}{\pi d^2} \tag{c}$$

Substituting Eqs. (b) and (c) into (a) and simplifying yields a quadratic equation in C. The larger of the two solutions will yield the spring index

$$C = \frac{2\alpha - \beta}{4\beta} + \sqrt{\left(\frac{2\alpha - \beta}{4\beta} \right)^2 - \frac{3\alpha}{4\beta}} \tag{10–23}$$

EXAMPLE 10–2

A music wire helical compression spring is needed to support a 20-lbf load after being compressed 2 in. Because of assembly considerations the solid height cannot exceed 1 in and the free length cannot be more than 4 in. Design the spring.

Solution

The a priori decisions are

- Music wire, A228; from Table 10–4, $A = 201\,000$ psi-in^m; $m = 0.145$; from Table 10–5, $E = 28.5$ Mpsi, $G = 11.75$ Mpsi (expecting $d > 0.064$ in)
- Ends squared and ground
- Function: $F_{max} = 20$ lbf, $y_{max} = 2$ in
- Safety: use design factor at solid height of $(n_s)_d = 1.2$
- Robust linearity: $\xi = 0.15$
- Use as-wound spring (cheaper), $S_{sy} = 0.45 S_{ut}$ from Table 10–6
- Decision variable: $d = 0.080$ in, music wire gage #30, Table A–25. From Fig. 10–3 and Table 10–6,

$$S_{sy} = 0.45 \frac{201\,000}{0.080^{0.145}} = 130\,455 \text{ psi}$$

From Fig. 10–3 or Eq. (10–23)

$$\alpha = \frac{S_{sy}}{n_s} = \frac{130\,455}{1.2} = 108\,713 \text{ psi}$$

$$\beta = \frac{8(1 + \xi) F_{max}}{\pi d^2} = \frac{8(1 + 0.15)20}{\pi (0.080^2)} = 9151.4 \text{ psi}$$

$$C = \frac{2(108\,713) - 9151.4}{4(9151.4)} + \sqrt{\left[\frac{2(108\,713) - 9151.4}{4(9151.4)} \right]^2 - \frac{3(108\,713)}{4(9151.4)}} = 10.53$$

Continuing with Fig. 10–3:

$$D = Cd = 10.53(0.080) = 0.8424$$

$$K_B = \frac{4(10.53) + 2}{4(10.53) - 3} = 1.128$$

$$\tau_s = 1.128\frac{8(1 + 0.15)20(0.8424)}{\pi(0.080)^3} = 108\,700 \text{ psi}$$

$$n_s = \frac{130\,445}{108\,700} = 1.2$$

$$OD = 0.843 + 0.080 = 0.923 \text{ in}$$

$$N_a = \frac{0.080^4(11.75)10^6(2)}{8(0.843)^3 20} = 10.05 \text{ turns}$$

$$N_t = 10.05 + 2 = 12.05 \text{ total turns}$$

$$L_s = 0.080(12.05) = 0.964 \text{ in}$$

$$L_0 = 0.964 + (1 + 0.15)2 = 3.264 \text{ in}$$

$$(L)_{cr} = 2.63(0.843/0.5) = 4.43 \text{ in}$$

$$\text{fom} = -2.6\pi^2(0.080)^2 12.05(0.843)/4 = -0.417$$

Repeat the above for other wire diameters and form a table (easily accomplished with a spreadsheet program):

d:	0.063	0.067	0.071	0.075	0.080	0.085	0.090	0.095
D	0.391	0.479	0.578	0.688	0.843	1.017	1.211	1.427
C	6.205	7.153	8.143	9.178	10.53	11.96	13.46	15.02
OD	0.454	0.546	0.649	0.763	0.923	1.102	1.301	1.522
N_a	39.1	26.9	19.3	14.2	10.1	7.3	5.4	4.1
L_s	2.587	1.936	1.513	1.219	0.964	0.790	0.668	0.581
L_0	4.887	4.236	3.813	3.519	3.264	3.090	2.968	2.881
$(L_0)_{cr}$	2.06	2.52	3.04	3.62	4.43	5.35	6.37	7.51
n_s	1.2	1.2	1.2	1.2	1.2	1.2	1.2	1.2
fom	−0.409	−0.399	−0.398	−0.404	−0.417	−0.438	−0.467	−0.505

Now examine the table and perform the adequacy assessment. The constraint $3 \leq N_a \leq 15$ rules out wire diameters less than 0.075 in. The spring index constraint $4 \leq C \leq 12$ rules out diameters larger than 0.085 in. The $L_s > 1$ constraint rules out diameters less than 0.080 in. The $L_0 > 4$ constraint rules out diameters less than 0.071 in. The buckling criterion rules out free lengths longer than $(L_0)_{cr}$, which rules out diameters less than

0.075 in. The factor of safety n_s is exactly 1.20 because the mathematics forced it. Had the spring been in a hole or over a rod, the helix diameter would be chosen without reference to $(n_s)_d$. The result is that there are only two springs in the feasible domain, one with a wire diameter of 0.080 in and the other with a wire diameter of 0.085. The figure of merit decides and the decision is the design with 0.080 in wire diameter.

Having designed a spring, will we have it made to our specifications? Not necessarily. There are vendors who stock literally thousands of music wire compression springs. By browsing their catalogs, we will usually find several that are close. Maximum deflection and maximum load are listed in the display of characteristics. Check to see if this allows soliding without damage. Often it does not. Spring rates may only be close. At the very least this situation allows a small number of springs to be ordered "off the shelf" for testing. The decision often hinges on the economics of special order versus the acceptability of a close match.

EXAMPLE 10–3

Indexing is used in machine operations when a circular part being manufactured must be divided into a certain number of segments. Figure 10–4 shows a portion of an indexing fixture used to successively position a part for the operation. When the knob is momentarily pulled up, part 6, which holds the workpiece, is rotated about a vertical axis to the next position and locked in place by releasing the index pin. In this example we wish to design the spring to exert a force of about 3 lbf and to fit in the space defined in the figure caption.

Solution

Since the fixture is not a high-production item, a stock spring will be selected. These are available in music wire. In one catalog there are 76 stock springs available having an outside diameter of 0.480 in and designed to work in a $\frac{1}{2}$-in hole. These are made in seven different wire sizes, ranging from 0.038 up to 0.063 in, and in free lengths from $\frac{1}{2}$ to $2\frac{1}{2}$ in, depending upon the wire size.

Figure 10–4

Part 1, pull knob; part 2, tapered retaining pin; part 3, hardened bushing with press fit; part 4, body of fixture; part 5, indexing pin; part 6, workpiece holder. Space of the spring is $\frac{5}{8}$ in OD, $\frac{1}{4}$ in ID, and $1\frac{3}{8}$ in long, with the pin down as shown. The pull knob must be raised $\frac{3}{4}$ in to permit indexing.

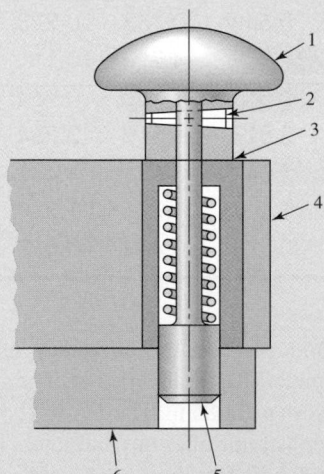

Since the pull knob must be raised $\frac{3}{4}$ in for indexing and the space for the spring is $1\frac{3}{8}$ in long when the pin is down, the solid length cannot be more than $\frac{5}{8}$ in.

Let us begin by selecting a spring having an outside diameter of 0.480 in, a wire size of 0.051 in, a free length of $1\frac{3}{4}$ in, $11\frac{1}{2}$ total turns, and plain ends. Then $m = 0.145$ and $A = 201$ kpsi · in^m for music wire. Then

$$S_{sy} = 0.45\frac{A}{d^m} = 0.45\frac{201}{0.051^{0.145}} = 139.3 \text{ kpsi}$$

With plain ends, from Table 10–1, the number of active turns is

$$N_a = N_t = 11.5 \text{ turns}$$

The mean coil diameter is $D = \text{OD} - d = 0.480 - 0.051 = 0.429$ in. From Eq. (10–9) the spring rate is, for $G = 11.85(10^6)$ psi from Table 10–5,

$$k = \frac{d^4 G}{8D^3 N_a} = \frac{0.051^4(11.85)10^6}{8(0.429)^3 11.5} = 11.0 \text{ lbf/in}$$

From Table 10–1, the solid height L_s is

$$L_s = d(N_t + 1) = 0.051(11.5 + 1) = 0.638 \text{ in}$$

The spring force when the pin is down, F_{min}, is

$$F_{min} = k y_{min} = 11.0(1.75 - 1.375) = 4.13 \text{ lbf}$$

When the spring is compressed solid, the spring force F_s is

$$F_s = k y_s = k(L_0 - L_s) = 11.0(1.75 - 0.638) = 12.2 \text{ lbf}$$

Since the spring index is $C = D/d = 0.429/0.051 = 8.41$,

$$K_B = \frac{4C + 2}{4C - 3} = \frac{4(8.41) + 2}{4(8.41) - 3} = 1.163$$

and for the as-wound spring, the shear stress when compressed solid is

$$\tau_s = K_B \frac{8F_s D}{\pi d^3} = 1.163\frac{8(12.2)0.429}{\pi(0.051)^3} = 116\,850 \text{ psi}$$

The factor of safety when the spring is compressed solid is

$$n_s - \frac{S_{sy}}{\tau_s} = \frac{139.3}{116.9} = 1.19$$

Since n_s is marginally adequate and L_s is larger than $\frac{5}{8}$ in, we must investigate other springs with a smaller wire size. After several investigations another spring has possibilities. It is as-wound music wire, $d = 0.045$ in, 20 gauge (see Table A–25) $\text{OD} = 0.480$ in, $N_t = 11.5$ turns, $L_0 = 1.75$ in. S_{sy} is still 139.3 kpsi, and

$$D = \text{OD} - d = 0.480 - 0.045 = 0.435 \text{ in}$$

$$N_a = N_t = 11.5 \text{ turns}$$

$$k = \frac{0.045^4(11.85)10^6}{8(0.435)^3 11.5} = 6.42 \text{ lbf/in}$$

$$L_s = d(N_t + 1) = 0.045(11.5 + 1) = 0.563 \text{ in}$$

$$F_{\min} = ky_{\min} = 6.42(1.75 - 1.375) = 2.41 \text{ lbf}$$

$$F_s = 6.42(1.75 - 0.563) = 7.62 \text{ lbf}$$

$$C = \frac{D}{d} = \frac{0.435}{0.045} = 9.67$$

$$K_B = \frac{4(9.67) + 2}{4(9.67) - 3} = 1.140$$

$$\tau_s = 1.140 \frac{8(7.62)0.435}{\pi(0.045)^3} = 105\,600 \text{ psi}$$

$$n_s = \frac{S_{sy}}{\tau_s} = \frac{139.3}{105.6} = 1.32$$

Now $n_s > 1.2$, buckling is not possible as the coils are guarded by the hole surface, and the solid length is less than $\frac{5}{8}$ in, so this spring is selected. By using a stock spring, we take advantage of economy of scale.

10–8 Critical Frequency of Helical Springs

If a wave is created by a disturbance at one end of a swimming pool, this wave will travel down the length of the pool, be reflected back at the far end, and continue in this back-and-forth motion until it is finally damped out. The same effect occurs in helical springs, and it is called *spring surge*. If one end of a compression spring is held against a flat surface and the other end is disturbed, a compression wave is created that travels back and forth from one end to the other exactly like the swimming-pool wave.

Spring manufacturers have taken slow-motion movies of automotive valve-spring surge. These pictures show a very violent surging, with the spring actually jumping out of contact with the end plates. Figure 10–5 is a photograph of a failure caused by such surging.

When helical springs are used in applications requiring a rapid reciprocating motion, the designer must be certain that the physical dimensions of the spring are not such as to create a natural vibratory frequency close to the frequency of the applied force; otherwise, resonance may occur, resulting in damaging stresses, since the internal damping of spring materials is quite low.

The governing equation for the translational vibration of a spring is the wave equation

$$\frac{\partial^2 u}{\partial x^2} = \frac{W}{kgl^2}\frac{\partial^2 u}{\partial t^2} \tag{10–24}$$

where k = spring rate

g = acceleration due to gravity

l = length of spring

W = weight of spring

x = coordinate along length of spring

u = motion of any particle at distance x

Figure 10–5

Valve-spring failure in an overrevved engine. Fracture is along the 45° line of maximum principal stress associated with pure torsional loading.

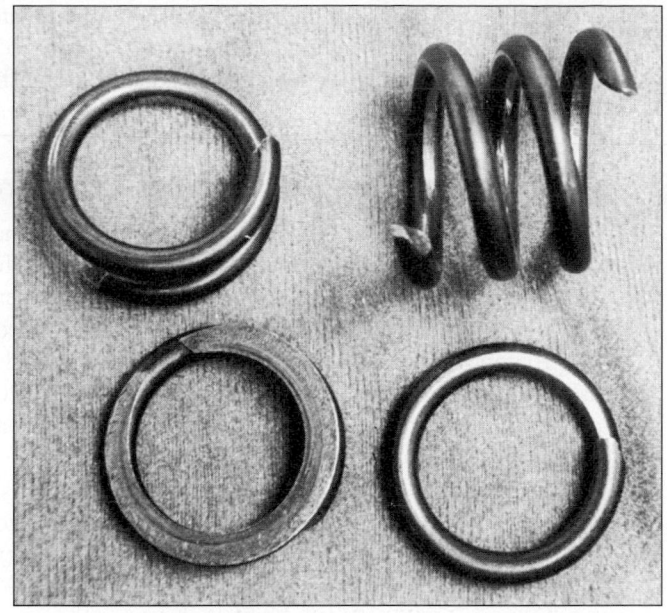

The solution to this equation is harmonic and depends on the given physical properties as well as the end conditions of the spring. The harmonic, *natural,* frequencies for a spring placed between two flat and parallel plates, in radians per second, are

$$\omega = m\pi\sqrt{\frac{kg}{W}} \qquad m = 1, 2, 3, \ldots$$

where the fundamental frequency is found for $m = 1$, the second harmonic for $m = 2$, and so on. We are usually interested in the frequency in cycles per second; since $\omega = 2\pi f$, we have, for the fundamental frequency in hertz,

$$f = \frac{1}{2}\sqrt{\frac{kg}{W}} \tag{10–25}$$

assuming the spring ends are always in contact with the plates.

Wolford and Smith[9] show that the frequency is

$$f = \frac{1}{4}\sqrt{\frac{kg}{W}} \tag{10–26}$$

where the spring has one end against a flat plate and the other end free. They also point out that Eq. (10–25) applies when one end is against a flat plate and the other end is driven with a sine-wave motion.

The weight of the active part of a helical spring is

$$W = AL\gamma = \frac{\pi d^2}{4}(\pi DN_a)(\gamma) = \frac{\pi^2 d^2 DN_a\gamma}{4} \tag{10–27}$$

where γ is the specific weight.

[9]J. C. Wolford and G. M. Smith, "Surge of Helical Springs," *Mech. Eng. News,* vol. 13, no. 1, February 1976, pp. 4–9.

The fundamental critical frequency should be greater than 15 to 20 times the frequency of the force or motion of the spring in order to avoid resonance with the harmonics. If the frequency is not high enough, the spring should be redesigned to increase k or decrease W.

10-9 Fatigue Loading of Helical Compression Springs

Springs are almost always subject to fatigue loading. In many instances the number of cycles of required life may be small, say, several thousand for a padlock spring or a toggle-switch spring. But the valve spring of an automotive engine must sustain millions of cycles of operation without failure; so it must be designed for infinite life.

To improve the fatigue strength of dynamically loaded springs, shot peening can be used. It can increase the torsional fatigue strength by 20 percent or more. Shot size is about $\frac{1}{64}$ in, so spring coil wire diameter and pitch must allow for complete coverage of the spring surface.

The best data on the torsional endurance limits of spring steels are those reported by Zimmerli.[10] He discovered the surprising fact that size, material, and tensile strength have no effect on the endurance limits (infinite life only) of spring steels in sizes under $\frac{3}{8}$ in (10 mm). We have already observed that endurance limits tend to level out at high tensile strengths (Fig. 7–18), but the reason for this is not clear. Zimmerli suggests that it may be because the original surfaces are alike or because plastic flow during testing makes them the same. Unpeened springs were tested from a minimum torsional stress of 20 kpsi to a maximum of 90 kpsi and peened springs in the range 20 kpsi to 135 kpsi. The corresponding endurance strength components for infinite life were found to be

Unpeened:

$$S_{sa} = 35 \text{ kpsi (241 MPa)} \qquad S_{sm} = 55 \text{ kpsi (379 MPa)} \qquad (10\text{--}28)$$

Peened:

$$S_{sa} = 57.5 \text{ kpsi (398 MPa)} \qquad S_{sm} = 77.5 \text{ kpsi (534 MPa)} \qquad (10\text{--}29)$$

For example, given an unpeened spring with $S_{su} = 211.5$ kpsi, the Gerber ordinate intercept, from Eq. (7–45), is

$$S_{se} = \frac{S_{sa}}{1 - \left(\dfrac{S_{sm}}{S_{su}}\right)^2} = \frac{35}{1 - \left(\dfrac{55}{211.5}\right)^2} = 37.5 \text{ kpsi}$$

For the Goodman failure criterion, the intercept would be 47.3 kpsi. Each possible wire size would change these numbers, since S_{su} would change.

An extended study[11] of available literature regarding torsional fatigue found that for polished, notch-free, cylindrical specimens subjected to torsional shear stress, the maximum alternating stress that may be imposed without causing failure is *constant* and independent of the mean stress in the cycle provided that the maximum stress range does not equal or exceed the torsional yield strength of the metal. With notches and abrupt section changes this consistency is not found. Springs are free of notches and surfaces are often very smooth. This failure criterion is known as the *Sines failure criterion* in torsional fatigue.

---

[10]F. P. Zimmerli, "Human Failures in Spring Applications," *The Mainspring,* no. 17, Associated Spring Corporation, Bristol, Conn., August–September 1957.

[11]Oscar J. Horger (ed.), *Metals Engineering: Design Handbook,* McGraw-Hill, New York, 1953, p. 84.

In constructing certain failure criteria on the designers' torsional fatigue diagram, the torsional modulus of rupture S_{su} is needed. We shall continue to employ Eq. (7–56), which is

$$S_{su} = 0.67 S_{ut} \tag{10–30}$$

In the case of shafts and many other machine members, fatigue loading in the form of completely reversed stresses is quite ordinary. Helical springs, on the other hand, are never used as both compression and extension springs. In fact, they are usually assembled with a preload so that the working load is additional. Thus the stress-time diagram of Fig. 7–23d expresses the usual condition for helical springs. The worst condition, then, would occur when there is no preload, that is, when $\tau_{\min} = 0$.

Now, we define

$$F_a = \frac{F_{\max} - F_{\min}}{2} \tag{10–31a}$$

$$F_m = \frac{F_{\max} + F_{\min}}{2} \tag{10–31b}$$

where the subscripts have the same meaning as those of Fig. 7–23d when applied to the axial spring force F. Then the shear stress amplitude is

$$\tau_a = K_B \frac{8 F_a D}{\pi d^3} \tag{10–32}$$

where K_B is the Bergsträsser factor, obtained from Eq. (10–6), and corrects for both direct shear and the curvature effect. As noted in Sec. 10–2, the Wahl factor K_W can be used instead, if desired.

The midrange shear stress is given by the equation

$$\tau_m = K_B \frac{8 F_m D}{\pi d^3} \tag{10–33}$$

EXAMPLE 10–4

An as-wound helical compression spring, made of music wire, has a wire size of 0.092 in, an outside coil diameter of $\frac{9}{16}$ in, a free length of $4\frac{3}{8}$ in, 21 active coils, and both ends squared and ground. The spring is unpeened. This spring is to be assembled with a preload of 5 lbf and will operate with a maximum load of 35 lbf during use.

(a) Estimate the factor of safety guarding against fatigue failure using a torsional Gerber fatigue failure criterion with Zimmerli data.

(b) Repeat part (a) using the Sines torsional fatigue criterion (steady stress component has no effect), with Zimmerli data.

(c) Repeat using a torsional Goodman failure criterion with Zimmerli data.

(d) Estimate the critical frequency of the spring.

Solution

The mean coil diameter is $D = 0.5625 - 0.092 = 0.4705$ in. The spring index is $C = D/d = 0.4705/0.092 = 5.11$. Then

$$K_B = \frac{4C + 2}{4C - 3} = \frac{4(5.11) + 2}{4(5.11) - 3} = 1.287$$

From Eqs. (10–31),

$$F_a = \frac{35 - 5}{2} = 15 \text{ lbf} \qquad F_m = \frac{35 + 5}{2} = 20 \text{ lbf}$$

The alternating shear-stress component is found from Eq. (10–32) to be

$$\tau_a = K_B \frac{8 F_a D}{\pi d^3} = (1.287) \frac{8(15)0.4705}{\pi (0.092)^3} (10^{-3}) = 29.7 \text{ kpsi}$$

Equation (10–33) gives the midrange shear-stress component

$$\tau_m = K_B \frac{8 F_m D}{\pi d^3} = 1.287 \frac{8(20)0.4705}{\pi (0.092)^3} (10^{-3}) = 39.6 \text{ kpsi}$$

From Table 10–4 we find $A = 201$ kpsi $\cdot$ in^m and $m = 0.145$. The ultimate tensile strength is estimated from Eq. (10–14) as

$$S_{ut} = \frac{A}{d^m} = \frac{201}{0.092^{0.145}} = 284.1 \text{ kpsi}$$

Also the shearing ultimate strength is estimated from

$$S_{su} = 0.67 S_{ut} = 0.67(284.1) = 190.3 \text{ kpsi}$$

The load-line slope $r = \tau_a / \tau_m = 29.7/39.6 = 0.75$.
(a) The Gerber ordinate intercept for the Zimmerli data, Eq. (10–28), is

$$S_{se} = \frac{S_{sa}}{1 - (S_{sm}/S_{su})^2} = \frac{35}{1 - (55/190.3)^2} = 38.2 \text{ kpsi}$$

The amplitude component of strength S_{sa}, from Table 7–10, is

$$S_{sa} = \frac{r^2 S_{su}^2}{2 S_{se}} \left[-1 + \sqrt{1 + \left(\frac{2 S_{se}}{r S_{su}}\right)^2} \right]$$

$$= \frac{0.75^2 190.3^2}{2(38.2)} \left\{ -1 + \sqrt{1 + \left[\frac{2(38.2)}{0.75(190.3)}\right]^2} \right\} = 35.8 \text{ kpsi}$$

and the fatigue factor of safety n_f is given by

Answer

$$n_f = \frac{S_{sa}}{\tau_a} = \frac{35.8}{29.7} = 1.21$$

(b) The Sines failure criterion ignores S_{sm} so that, for the Zimmerli data with $S_{sa} = 35$ kpsi,

Answer

$$n_f = \frac{S_{sa}}{\tau_a} = \frac{35}{29.7} = 1.18$$

(c) The ordinate intercept S_{se} for the Goodman failure criterion with the Zimmerli data is

$$S_{se} = \frac{S_{sa}}{1 - (S_{sm}/S_{su})} = \frac{35}{1 - (55/190.3)} = 49.2 \text{ kpsi}$$

The amplitude component of the strength S_{sa} for the Goodman criterion, from Table 7–8, is

$$S_{sa} = \frac{r S_{se} S_{su}}{r S_{su} + S_{se}} = \frac{0.75(49.2)190.3}{0.75(190.3) + 49.2} = 36.6 \text{ kpsi}$$

The fatigue factor of safety is given by

Answer
$$n_f = \frac{S_{sa}}{\tau_a} = \frac{36.6}{29.7} = 1.23$$

(d) Using Eq. (10–9) and Table 10–5, we estimate the spring rate as

$$k = \frac{d^4 G}{8 D^3 N_a} = \frac{0.092^4 [11.75(10^6)]}{8(0.4705)^3 21} = 48.1 \text{ lbf/in}$$

From Eq. (10–27) we estimate the spring weight as

$$W = \frac{\pi^2 (0.092^2) 0.4705(21) 0.284}{4} = 0.0586 \text{ lbf}$$

and from Eq. (10–25) the frequency of the fundamental wave is

Answer
$$f_n = \frac{1}{2} \left[\frac{48.1(386)}{0.0586} \right]^{1/2} = 281 \text{ Hz}$$

If the operating or exciting frequency is more than $281/20 = 14.1$ Hz, the spring may have to be redesigned.

We used three approaches to estimate the fatigue factor of safety in Ex. 10–5. The results, in order of smallest to largest, were 1.18 (Sines), 1.21 (Gerber), and 1.23 (Goodman). Although the results were very close to one another, using the Zimmerli data as we have, the Sines criterion will always be the most conservative and the Goodman the least. If we perform a fatigue analysis using strength properties as was done in Chap. 7, different results would be obtained, but here the Goodman criterion would be more conservative than the Gerber criterion. Be prepared to see designers or design software using any one of these techniques. This is why we cover them. Which criterion is correct? Remember, we are performing *estimates* and only testing will reveal the truth—*statistically.*

10–10 Helical Compression Spring Design for Fatigue Loading

Let us begin with the statement of a problem. In order to compare a static spring to a dynamic spring, we shall design the spring in Ex. 10–2 for dynamic service.

EXAMPLE 10–5

A music wire helical compression spring with infinite life is needed to resist a dynamic load that varies from 5 to 20 lbf at 5 Hz while the end deflection varies from $\frac{1}{2}$ to 2 in. Because of assembly considerations, the solid height cannot exceed 1 in and the free length cannot be more than 4 in. The springmaker has the following wire sizes in stock: 0.069, 0.071, 0.080, 0.085, 0.090, 0.095, 0.105, and 0.112 in.

Solution The a priori decisions are:

- Material and condition: for music wire, $A = 201$ kpsi $\cdot$ in^m, $m = 0.145$, $G = 11.75(10^6)$ psi; relative cost is 2.6
- Surface treatment: unpeened
- End treatment: squared and ground
- Robust linearity: $\xi = 0.15$
- Set: use in as-wound condition
- Fatigue-safe: $n_f = 1.5$ using the Sines-Zimmerli fatigue-failure criterion
- Function: $F_{min} = 5$ lbf, $F_{max} = 20$ lbf, $y_{min} = 0.5$ in, $y_{max} = 2$ in, spring operates free (no rod or hole)
- Decision variable: wire size d

The figure of merit will be the volume of wire to wind the spring, Eq. (10–22). The design strategy will be to set wire size d, build a table, inspect the table, and choose the satisfactory spring with the highest figure of merit.

Solution Set $d = 0.112$ in. Then

$$F_a = \frac{20 - 5}{2} = 7.5 \text{ lbf} \qquad F_m = \frac{20 + 5}{2} = 12.5 \text{ lbf}$$

$$k = \frac{F_{max}}{y_{max}} = \frac{20}{2} = 10 \text{ lbf/in}$$

$$S_{ut} = \frac{201}{0.112^{0.145}} = 276.1 \text{ kpsi}$$

$$S_{su} = 0.67(276.1) = 185.0 \text{ kpsi}$$

$$S_{sy} = 0.45(276.1) = 124.2 \text{ kpsi}$$

From Eq. (10–28), with the Sines criterion, $S_{se} = S_{sa} = 35$ kpsi. Equation (10–23) can be used to determine C with S_{se}, n_f, and F_a in place of S_{sy}, n_s, and $(1 + \xi)F_{max}$, respectively. Thus,

$$\alpha = \frac{S_{se}}{n_f} = \frac{35\ 000}{1.5} = 23\ 333 \text{ psi}$$

$$\beta = \frac{8F_a}{\pi d^2} = \frac{8(7.5)}{\pi(0.112^2)} = 1522.5 \text{ psi}$$

$$C = \frac{2(23\ 333) - 1522.5}{4(1522.5)} + \sqrt{\left[\frac{2(23\ 333) - 1522.5}{4(1522.5)}\right]^2 - \frac{3(23\ 333)}{4(1522.5)}} = 14.005$$

$$D = Cd = 14.005(0.112) = 1.569 \text{ in}$$

$$F_s = (1 + \xi)F_{max} = (1 + 0.15)20 = 23 \text{ lbf}$$

$$N_a = \frac{d^4 G}{8 D^3 k} = \frac{0.112^4 (11.75)(10^6)}{8(1.569)^3 10} = 5.98 \text{ turns}$$

$$N_t = N_a + 2 = 5.98 + 2 = 7.98 \text{ turns}$$

$$L_s = d N_t = 0.112(7.98) = 0.894 \text{ in}$$

$$L_0 = L_s + \frac{F_s}{k} = 0.894 + \frac{23}{10} = 3.194 \text{ in}$$

$$\text{ID} = 1.569 - 0.112 = 1.457 \text{ in}$$

$$\text{OD} = 1.569 + 0.112 = 1.681 \text{ in}$$

$$y_s = L_0 - L_s = 3.194 - 0.894 = 2.30 \text{ in}$$

$$(L_0)_{\text{cr}} < \frac{2.63 D}{\alpha} = 2.63 \frac{(1.569)}{0.5} = 8.253 \text{ in}$$

$$K_B = \frac{4(14.005) + 2}{4(14.005) - 3} = 1.094$$

$$W = \frac{\pi^2 d^2 D N_a \gamma}{4} = \frac{\pi^2 0.112^2 (1.569) 5.98 (0.284)}{4} = 0.0825 \text{ lbf}$$

$$f_n = 0.5 \sqrt{\frac{386 k}{W}} = 0.5 \sqrt{\frac{386(10)}{0.0825}} = 108 \text{ Hz}$$

$$\tau_a = K_B \frac{8 F_a D}{\pi d^3} = 1.094 \frac{8(7.5) 1.569}{\pi 0.112^3} = 23\,334 \text{ psi}$$

$$\tau_m = \tau_a \frac{F_m}{F_a} = 23\,334 \frac{12.5}{7.5} = 38\,890 \text{ psi}$$

$$\tau_s = \tau_a \frac{F_s}{F_a} = 23\,334 \frac{23}{7.5} = 71\,560 \text{ psi}$$

$$n_f = \frac{S_{sa}}{\tau_a} = \frac{35\,000}{23\,334} = 1.5$$

$$n_s = \frac{S_{sy}}{\tau_s} = \frac{124\,200}{71\,560} = 1.74$$

$$\text{fom} = -(\text{relative material cost}) \pi^2 d^2 N_t D / 4$$

$$= -2.6 \pi^2 (0.112^2)(7.98) 1.569 / 4 = -1.01$$

Inspection of the results shows that all conditions are satisfied and the design is acceptable. Are there better solutions? Repeat the process using the other available wire sizes and develop the following table:

d:	0.069	0.071	0.080	0.085	0.090	0.095	0.105	0.112
D	0.297	0.332	0.512	0.632	0.767	0.919	1.274	1.569
ID	0.228	0.261	0.432	0.547	0.677	0.824	1.169	1.457
OD	0.366	0.403	0.592	0.717	0.857	1.014	1.379	1.681
C	4.33	4.67	6.40	7.44	8.53	9.67	12.14	14.00
N_a	127.2	102.4	44.8	30.5	21.3	15.4	8.63	6.0
L_s	8.916	7.414	3.740	2.750	2.100	1.655	1.116	0.895
L_0	11.216	9.714	6.040	5.050	4.400	3.955	3.416	3.195
$(L_0)_{cr}$	1.562	1.744	2.964	3.325	4.036	4.833	6.703	8.250
n_f	1.50	1.50	1.50	1.50	1.50	1.50	1.50	1.50
n_s	1.86	1.85	1.82	1.81	1.79	1.78	1.75	1.74
f_n	87.5	89.7	96.9	99.7	101.9	103.8	106.6	108
fom	−1.17	−1.12	−0.983	−0.948	−0.930	−0.927	−0.958	−1.01

The problem-specific inequality constraints are

$$L_s \le 1 \text{ in}$$

$$L_0 \le 4 \text{ in}$$

$$f_n \ge 5(20) = 100 \text{ Hz}$$

The general constraints are

$$3 \le N_a \le 15$$

$$4 \le C \le 16$$

$$(L_0)_{cr} > L_0$$

The tight constraint is L_s, which limits the smallest diameter of wire to 0.112 in.

10–11   Extension Springs

Extension springs differ from compression springs in that they carry tensile loading, they require some means of transferring the load from the support to the body of the spring, and the spring body is wound with an initial tension. The load transfer can be done with a threaded plug or a swivel hook; both of these add to the cost of the finished product, and so one of the methods shown in Fig. 10–6 is usually employed.

Stresses in the body of the extension spring are handled the same as compression springs. In designing a spring with a hook end, bending and torsion in the hook must be included in the analysis. In Fig. 10–7a and b a commonly used method of designing the end is shown. The maximum tensile stress at A, due to bending and axial loading, is given by

$$\sigma_A = F \left[(K)_A \frac{16D}{\pi d^3} + \frac{4}{\pi d^2} \right] \tag{10–34}$$

Figure 10–6

Types of ends used on extension springs. *(Courtesy of Associated Spring.)*

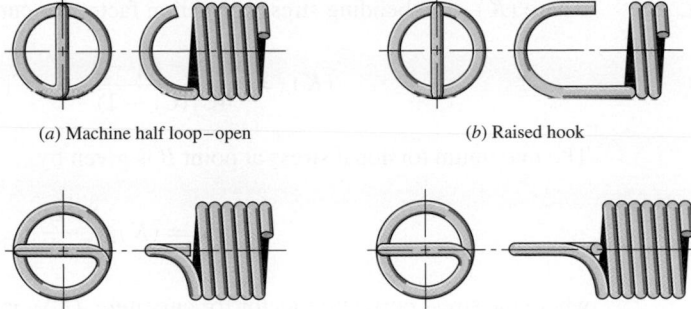

(*a*) Machine half loop–open   (*b*) Raised hook

(*c*) Short twisted loop   (*d*) Full twisted loop

Figure 10–7

Ends for extension springs. (*a*) Usual design; stress at *A* is due to combined axial force and bending moment. (*b*) Side view of part *a*; stress is mostly torsion at *B*. (*c*) Improved design; stress at *A* is due to combined axial force and bending moment. (*d*) Side view of part *c*; stress at *B* is mostly torsion.

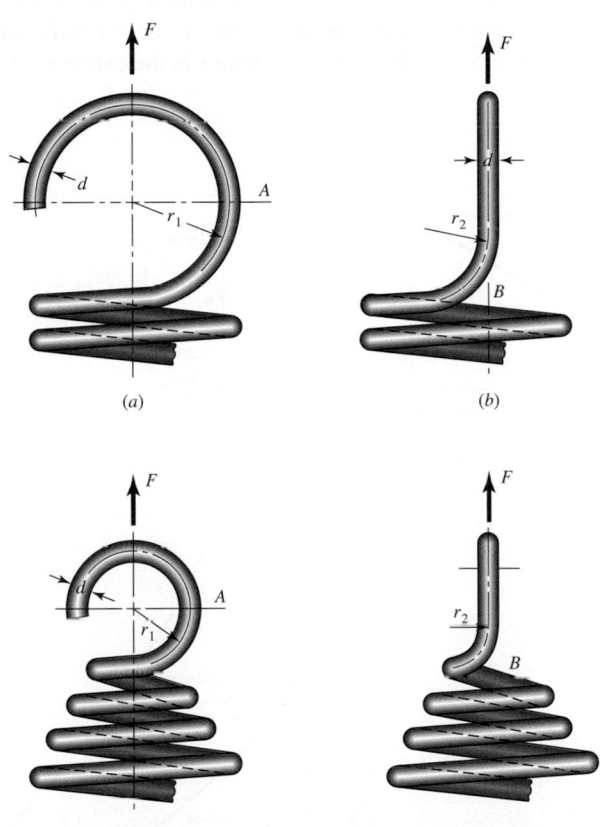

(*a*)   (*b*)

(*c*)   (*d*)

Note: Radius r_1 is in the plane of the end coil for curved beam bending stress. Radius r_2 is at a right angle to the end coil for torsional shear stress.

where $(K)_A$ is a bending stress correction factor for curvature, given by

$$(K)_A = \frac{4C_1^2 - C_1 - 1}{4C_1(C_1 - 1)} \qquad C_1 = \frac{2r_1}{d} \qquad (10\text{--}35)$$

The maximum torsional stress at point B is given by

$$\tau_B = (K)_B \frac{8FD}{\pi d^3} \qquad (10\text{--}36)$$

where the stress correction factor for curvature, $(K)_B$, is

$$(K)_B = \frac{4C_2 - 1}{4C_2 - 4} \qquad C_2 = \frac{2r_2}{d} \qquad (10\text{--}37)$$

Figure 10–7c and d show an improved design due to a reduced coil diameter.

When extension springs are made with coils in contact with one another, they are said to be *close-wound*. Spring manufacturers prefer some initial tension in close-wound springs in order to hold the free length more accurately. The corresponding load-deflection curve is shown in Fig. 10–8a, where y is the extension beyond the free length L_0 and F_i

Figure 10–8

(a) Geometry of the force F and extension y locus of an extension spring; (b) geometry of the extension spring; and (c) torsional stresses due to initial tension as a function of spring index C in helical extension springs.

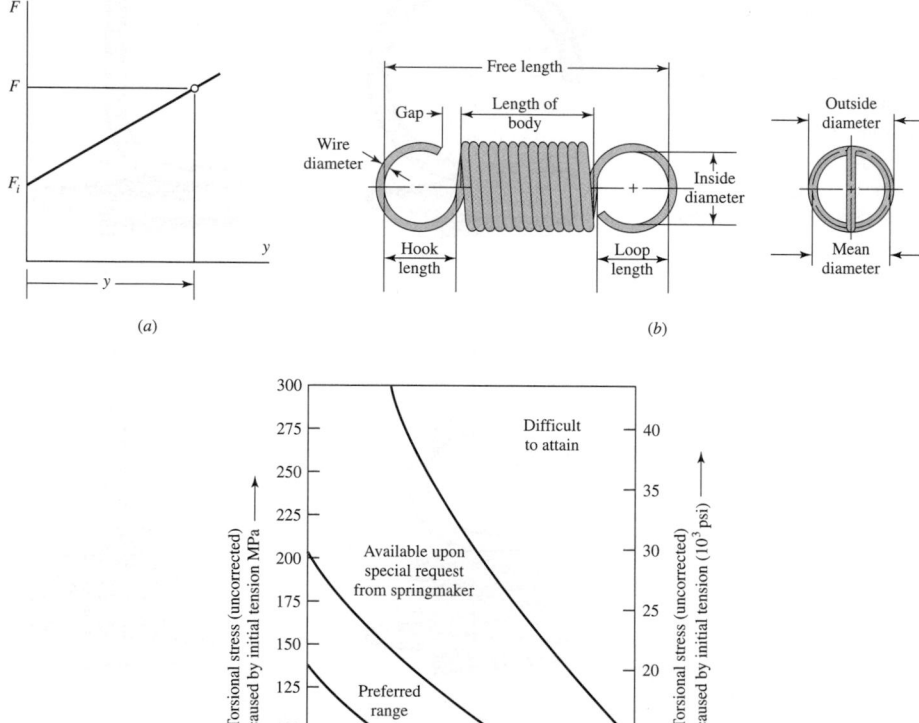

is the initial tension in the spring that must be exceeded before the spring deflects. The load-deflection relation is then

$$F = F_i + ky \tag{10-38}$$

where k is the spring rate. The free length L_0 of a spring measured inside the end loops or hooks as shown in Fig. 10–8b can be expressed as

$$L_0 = 2(D - d) + (N_b + 1)d = (2C - 1 + N_b)d \tag{10-39}$$

where D is the mean coil diameter, N_b is the number of body coils, and C is the spring index. With ordinary twisted end loops as shown in Fig. 10–8b, to account for the deflection of the loops in determining the spring rate k, the equivalent number of active helical turns N_a for use in Eq. (10–9) is

$$N_a = N_b + \frac{G}{E} \tag{10-40}$$

where G and E are the shear and tensile moduli of elasticity, respectively (see Prob. 10–31).

The initial tension in an extension spring is created in the winding process by twisting the wire as it is wound onto the mandrel. When the spring is completed and removed from the mandrel, the initial tension is locked in because the spring cannot get any shorter. The amount of initial tension that a springmaker can routinely incorporate is as shown in Fig. 10–8c. The preferred range can be expressed in terms of the *uncorrected torsional* stress τ_i as

$$\tau_i = \frac{33\,500}{\exp(0.105C)} \pm 1000 \left(4 - \frac{C - 3}{6.5}\right) \text{ psi} \tag{10-41}$$

where C is the spring index.

Guidelines for the maximum allowable corrected stresses for static applications of extension springs are given in Table 10–7.

Table 10–7

Maximum Allowable Stresses (K_W or K_B corrected) for Helical Extension Springs in Static Applications

Source: From *Design Handbook*, 1987, p. 52. Courtesy of Associated Spring.

| | Percent of Tensile Strength | | |
| | In Torsion | | In Bending |
Materials	Body	End	End
Patented, cold-drawn or hardened and tempered carbon and low-alloy steels	45–50	40	75
Austenitic stainless steel and nonferrous alloys	35	30	55

This information is based on the following conditions: set not removed and low temperature heat treatment applied. For springs that require high initial tension, use the same percent of tensile strength as for end.

EXAMPLE 10–6

A hard-drawn steel wire extension spring has a wire diameter of 0.035 in, an outside coil diameter of 0.248 in, hook radii of $r_1 = 0.106$ in and $r_2 = 0.089$ in, and an initial tension of 1.19 lbf. The number of body turns is 12.17. From the given information:
(a) Determine the physical parameters of the spring.
(b) Check the initial preload stress conditions.
(c) Find the factors of safety under a static 5.25-lbf load.

Solution

(a)

$$D = OD - d = 0.248 - 0.035 = 0.213 \text{ in}$$

$$C = \frac{D}{d} = \frac{0.213}{0.035} = 6.086$$

$$K_B = \frac{4C + 2}{4C - 3} = 1.234$$

Eq. (10–40): $N_a = N_b + G/E = 12.17 + 11.5/28.7 = 12.57$ turns

Eq. (10–9): $k = \dfrac{d^4 G}{8 D^3 N_a} = \dfrac{0.035^4 (11.5)10^6}{8(0.213^3)12.57} = 17.76$ lbf/in

Eq. (10–39): $L_0 = (2C - 1 + N_b)d = [2(6.086) - 1 + 12.17]0.035 = 0.817$ in

The deflection under the service load is

$$y_{\max} = \frac{F_{\max} - F_i}{k} = \frac{5.25 - 1.19}{17.76} = 0.229 \text{ in}$$

where the spring length becomes $L = L_0 + y = 0.817 + 0.229 = 1.046$ in.
(b) The uncorrected initial stress is given by Eq. (10–3) without the correction factor. That is,

$$(\tau_i)_{\text{uncorr}} = \frac{8 F_i D}{\pi d^3} = \frac{8(1.19)0.213(10^{-3})}{\pi(0.035^3)} = 15.1 \text{ kpsi}$$

The preferred range is given by Eq. (10–41) and for this case is

$$(\tau_i)_{\text{pref}} = \frac{33\,500}{\exp(0.105C)} \pm 1000 \left(4 - \frac{C - 3}{6.5} \right)$$

$$= \frac{33\,500}{\exp[0.105(6.086)]} \pm 1000 \left(4 - \frac{6.086 - 3}{6.5} \right)$$

$$= 17\,681 \pm 3525 = 21.2, 14.2 \text{ kpsi}$$

Answer

Thus, the initial tension of 15.1 kpsi is in the preferred range.
(c) For hard-drawn wire, Table 10–4 gives $m = 0.190$ and $A = 140$ kpsi $\cdot$ in^m. From Eq. (10–14)

$$S_{ut} = \frac{A}{d^m} = \frac{140}{0.035^{0.190}} = 264.7 \text{ kpsi}$$

For torsional shear in the main body of the spring, from Table 10–7,

$$S_{sy} = 0.45 S_{ut} = 0.45(264.7) = 119.1 \text{ kpsi}$$

The shear stress under the service load is

$$\tau_{max} = \frac{8K_B F_{max} D}{\pi d^3} = \frac{8(1.234)5.25(0.213)}{\pi(0.035^3)}(10^{-3}) = 82.0 \text{ kpsi}$$

Thus, the factor of safety is

Answer
$$n = \frac{S_{sy}}{\tau_{max}} = \frac{119.1}{82.0} = 1.45$$

For the end-hook bending at A,

$$C_1 = 2r_1/d = 2(0.106)/0.0.035 = 6.057$$

From Eq. (10–35)

$$(K)_A = \frac{4C_1^2 - C_1 - 1}{4C_1(C_1 - 1)} = \frac{4(6.057^2) - 6.057 - 1}{4(6.057)(6.057 - 1)} = 1.14$$

From Eq. (10–34)

$$\sigma_A = F_{max}\left[(K)_A \frac{16D}{\pi d^3} + \frac{4}{\pi d^2}\right]$$

$$= 5.25\left[1.14\frac{16(0.213)}{\pi(0.035^3)} + \frac{4}{\pi(0.035^2)}\right](10^{-3}) = 156.9 \text{ kpsi}$$

The yield strength, from Table 10–7, is given by

$$S_y = 0.75S_{ut} = 0.75(264.7) = 198.5 \text{ kpsi}$$

The factor of safety for end-hook bending at A is then

Answer
$$n_A = \frac{S_y}{\sigma_A} = \frac{198.5}{156.9} = 1.27$$

For the end-hook in torsion at B, from Eq. (10–37)

$$C_2 = 2r_2/d = 2(0.089)/0.035 = 5.086$$

$$(K)_B = \frac{4C_2 - 1}{4C_2 - 4} = \frac{4(5.086) - 1}{4(5.086) - 4} = 1.18$$

and the corresponding stress, given by Eq. (10–36), is

$$\tau_B = (K)_B \frac{8F_{max}D}{\pi d^3} = 1.18\frac{8(5.25)0.213}{\pi(0.035^3)}(10^{-3}) = 78.4 \text{ kpsi}$$

Using Table 10–7 for yield strength, the factor of safety for end-hook torsion at B is

Answer
$$n_B = \frac{(S_{sy})_B}{\tau_B} = \frac{0.4(264.7)}{78.4} = 1.35$$

Yield due to bending of the end hook will occur first.

Next, let us consider a fatigue problem.

EXAMPLE 10–7
The helical coil extension spring of Ex. 10–6 is subjected to a dynamic loading from 1.5 to 5 lbf. Estimate the factors of safety using the Gerber failure criterion for (a) coil fatigue, (b) coil yielding, (c) end-hook bending fatigue at point A of Fig. 10–7a, and (d) end-hook torsional fatigue at point B of Fig. 10–7b.

Solution
A number of quantities are the same as in Ex. 10–6: $d = 0.035$ in, $S_{ut} = 264.7$ kpsi, $D = 0.213$ in, $r_1 = 0.106$ in, $C = 6.086$, $K_B = 1.234$, $(K)_A = 1.14$, $(K)_B = 1.18$, $N_b = 12.17$ turns, $L_0 = 0.817$ in, $k = 17.76$ lbf/in, $F_i = 1.19$ lbf, and $(\tau_i)_{\text{uncorr}} = 15.1$ kpsi. Then

$$F_a = (F_{\max} - F_{\min})/2 = (5 - 1.5)/2 = 1.75 \text{ lbf}$$

$$F_m = (F_{\max} + F_{\min})/2 = (5 + 1.5)/2 = 3.25 \text{ lbf}$$

The strengths from Ex. 10–6 include $S_{ut} = 264.7$ kpsi, $S_y = 198.5$ kpsi, and $S_{sy} = 119.1$ kpsi. The ultimate shear strength is estimated from Eq. (10–30) as

$$S_{su} = 0.67 S_{ut} = 0.67(264.7) = 177.3 \text{ kpsi}$$

(a) Body-coil fatigue:

$$\tau_a = \frac{8 K_B F_a D}{\pi d^3} = \frac{8(1.234)1.75(0.213)}{\pi (0.035^3)}(10^{-3}) = 27.3 \text{ kpsi}$$

$$\tau_m = \frac{F_m}{F_a}\tau_a = \frac{3.25}{1.75}27.3 = 50.7 \text{ kpsi}$$

Using the Zimmerli data of Eq. (10–28) gives

$$S_{se} = \frac{S_{sa}}{1 - \left(\dfrac{S_{sm}}{S_{su}}\right)^2} = \frac{35}{1 - \left(\dfrac{55}{177.3}\right)^2} = 38.7 \text{ kpsi}$$

From Table 7–10, the Gerber fatigue criterion for shear is

Answer

$$(n_f)_{\text{body}} = \frac{1}{2}\left(\frac{S_{su}}{\tau_m}\right)^2 \frac{\tau_a}{S_{se}}\left[-1 + \sqrt{1 + \left(2\frac{\tau_m}{S_{su}}\frac{S_{se}}{\tau_a}\right)^2}\right]$$

$$= \frac{1}{2}\left(\frac{177.3}{50.7}\right)^2 \frac{27.3}{38.7}\left[-1 + \sqrt{1 + \left(2\frac{50.7}{177.3}\frac{38.7}{27.3}\right)^2}\right] = 1.24$$

(b) The load-line for the coil body begins at $S_{sm} = \tau_i$ and has a slope $r = \tau_a/(\tau_m - \tau_i)$. It can be shown that the intersection with the yield line is given by $(S_{sa})_y = \frac{r}{r+1}(S_{sy} - \tau_i)$. Consequently, $\tau_i = (F_i/F_a)\tau_a = (1.19/1.75)27.3 = 18.6$ kpsi, $r = 27.3/(50.7 - 18.6) = 0.850$, and

$$(S_{sa})_y = \frac{0.850}{0.850 + 1}(119.1 - 18.6) = 46.2 \text{ kpsi}$$

Thus,

Answer

$$(n_y)_{\text{body}} = \frac{(S_{sa})_y}{\tau_a} = \frac{46.2}{27.3} = 1.69$$

(*c*) End-hook bending fatigue: using Eqs. (10–34) and (10–35) gives

$$\sigma_a = F_a \left[(K)_A \frac{16D}{\pi d^3} + \frac{4}{\pi d^2} \right]$$

$$= 1.75 \left[1.14 \frac{16(0.213)}{\pi (0.035^3)} + \frac{4}{\pi (0.035^2)} \right] (10^{-3}) = 52.3 \text{ kpsi}$$

$$\sigma_m = \frac{F_m}{F_a} \sigma_a = \frac{3.25}{1.75} 52.3 = 97.1 \text{ kpsi}$$

To estimate the tensile endurance limit using the distortion-energy theory,

$$S_e = S_{se}/0.577 = 38.7/0.577 = 67.1 \text{ kpsi}$$

Using the Gerber criterion for tension gives

Answer

$$(n_f)_A = \frac{1}{2} \left(\frac{S_{ut}}{\sigma_m} \right)^2 \frac{\sigma_a}{S_e} \left[-1 + \sqrt{1 + \left(2 \frac{\sigma_m}{S_{ut}} \frac{S_e}{\sigma_a} \right)^2} \right]$$

$$= \frac{1}{2} \left(\frac{264.7}{97.1} \right)^2 \frac{52.3}{67.1} \left[-1 + \sqrt{1 + \left(2 \frac{97.1}{264.7} \frac{67.1}{52.3} \right)^2} \right] = 1.08$$

(*d*) End-hook torsional fatigue: from Eq. (10–36)

$$(\tau_a)_B = (K)_B \frac{8 F_a D}{\pi d^3} = 1.18 \frac{8(1.75)0.213}{\pi (0.035^3)} (10^{-3}) = 26.1 \text{ kpsi}$$

$$(\tau_m)_B = \frac{F_m}{F_a} (\tau_a)_B = \frac{3.25}{1.75} 26.1 = 48.5 \text{ kpsi}$$

Then, again using the Gerber criterion, we obtain

Answer

$$(n_f)_B = \frac{1}{2} \left(\frac{S_{su}}{\tau_m} \right)^2 \frac{\tau_a}{S_{se}} \left[-1 + \sqrt{1 + \left(2 \frac{\tau_m}{S_{su}} \frac{S_{se}}{\tau_a} \right)^2} \right]$$

$$= \frac{1}{2} \left(\frac{177.3}{48.5} \right)^2 \frac{26.1}{38.7} \left[-1 + \sqrt{1 + \left(2 \frac{48.5}{177.3} \frac{38.7}{26.1} \right)^2} \right] = 1.30$$

The analyses in Exs. 10–6 and 10–7 show how extension springs differ from compression springs. The end hooks are usually the weakest part, with bending usually controlling. We should also appreciate that a fatigue failure separates the extension spring under load. Flying fragments, lost load, and machine shutdown are threats to personal safety as well as machine function. For these reasons higher design factors are used in extension-spring design than in the design of compression springs.

In Ex. 10–7 we estimated the endurance limit for the hook in bending using the Zimmerli data, which are based on torsion in compression springs and the distortion theory. An alternative method is to use Table 10–8, which is based on a stress-ratio of $R = \tau_{\min}/\tau_{\max} = 0$. For this case, $\tau_a = \tau_m = \tau_{\max}/2$. Label the strength values of Table 10–8 as S_r

Table 10–8

Maximum Allowable Stresses for ASTM A228 and Type 302 Stainless Steel Helical Extension Springs in Cyclic Applications

Source: From *Design Handbook*, 1987, p. 52. Courtesy of Associated Spring.

Number	Percent of Tensile Strength		
	In Torsion		In Bending
of Cycles	Body	End	End
10^5	36	34	51
10^6	33	30	47
10^7	30	28	45

This information is based on the following conditions: not shot-peened, no surging and ambient environment with a low temperature heat treatment applied. Stress ratio = 0.

for bending or S_{sr} for torsion. Then for torsion, for example, $S_{sa} = S_{sm} = S_{sr}/2$ and the Gerber ordinate intercept, given by Eq. (7–45), is

$$S_{se} = \frac{S_{sa}}{1 - (S_{sm}/S_{su})^2} = \frac{S_{sr}/2}{1 - \left(\dfrac{S_{sr}/2}{S_{su}}\right)^2} \tag{10–42}$$

So in Ex. 10–7 an estimate for the bending endurance limit from Table 10–8 would be

$$S_r = 0.45 S_{ut} = 0.45(264.7) = 119.1 \text{ kpsi}$$

and from Eq. (10–42)

$$S_e = \frac{S_r/2}{1 - [S_r/(2S_{ut})]^2} = \frac{119.1/2}{1 - \left(\dfrac{119.1/2}{264.7}\right)^2} = 62.7 \text{ kpsi}$$

Using this in place of 67.1 kpsi in Ex. 10–7 results in $(n_f)_A = 1.03$, a reduction of 5 percent.

10–12   Helical Coil Torsion Springs

When a helical coil spring is subjected to end torsion, it is called a *torsion spring*. It is usually close-wound, as is a helical coil extension spring, but with negligible initial tension. There are single-bodied and double-bodied types as depicted in Fig. 10–9. As shown in the figure, torsion springs have ends configured to apply torsion to the coil body in a convenient manner, with short hook, hinged straight offset, straight torsion, and special ends. The ends ultimately connect a force at a distance from the coil axis to apply a torque. The most frequently encountered (and least expensive) end is the straight torsion end. If intercoil friction is to be avoided completely, the spring can be wound with a pitch that just separates the body coils. Helical coil torsion springs are usually used with a rod or arbor for reactive support when ends cannot be built in, to maintain alignment, and to provide buckling resistance if necessary.

The wire in a torsion spring is in bending, in contrast to the torsion encountered in helical coil compression and extension springs. The springs are designed to wind tighter in service. As the applied torque increases, the inside diameter of the coil decreases. Care must be taken so that the coils do not interfere with the pin, rod, or arbor. The bending mode in the coil might seem to invite square- or rectangular-cross-section wire, but cost, range of materials, and availability discourage its use.

Figure 10–9

Torsion springs. *(Courtesy of Associated Spring.)*

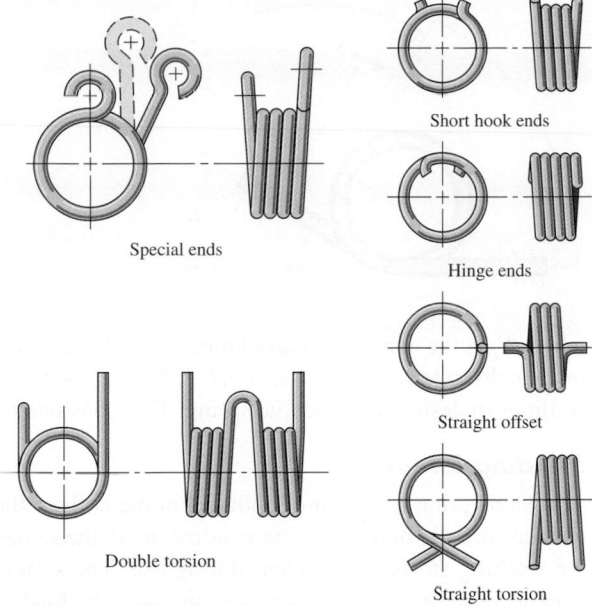

Special ends

Short hook ends

Hinge ends

Double torsion

Straight offset

Straight torsion

Table 10–9

End Position Tolerances for Helical Coil Torsion Springs (for D/d Ratios up to and Including 16)

Source: From *Design Handbook,* 1987, p. 52. Courtesy of Associated Spring.

Total Coils	Tolerance: ± Degrees*
Up to 3	8
Over 3–10	10
Over 10–20	15
Over 20–30	20
Over 30	25

*Closer tolerances available on request.

Torsion springs are familiar in clothespins, window shades, and animal traps, where they may be seen around the house, and out of sight in counterbalance mechanisms, ratchets, and a variety of other machine components. There are many stock springs that can be purchased off-the-shelf from a vendor. This selection can add economy of scale to small projects, avoiding the cost of custom design and small-run manufacture.

Describing the End Location

In specifying a torsion spring, the ends must be located relative to each other. Commercial tolerances on these relative positions are listed in Table 10–9. The simplest scheme for expressing the initial unloaded location of one end with respect to the other is in terms of an angle β defining the partial turn present in the coil body as $N_p = \beta/360°$, as shown in Fig. 10–10. For analysis purposes the nomenclature of Fig. 10–10 can be used. Communication with a springmaker is often in terms of the back-angle α.

The number of body turns N_b is the number of turns in the free spring body by count. The body-turn count is related to the initial position angle β by

$$N_b = \text{integer} + \frac{\beta}{360°} = \text{integer} + N_p$$

Figure 10–10

The free-end location angle is β. The rotational coordinate θ is proportional to the product Fl. Its back angle is α. For all positions of the moving end $\theta + \alpha = \Sigma = $ constant.

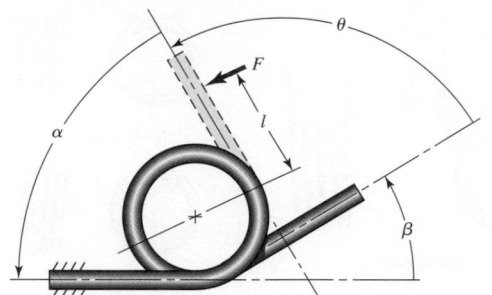

where N_p is the number of partial turns. The above equation means that N_b takes on non-integer, discrete values such as $5.3, 6.3, 7.3, \ldots$, with successive differences of 1 as possibilities in designing a specific spring. This consideration will be discussed later.

Bending Stress

A torsion spring has bending induced in the coils, rather than torsion. This means that residual stresses built in during winding are in the same direction but of opposite sign to the working stresses that occur during use. The strain-strengthening locks in residual stresses opposing working stresses *provided* the load is always applied in the winding sense. Torsion springs can operate at bending stresses exceeding the yield strength of the wire from which it was wound.

The bending stress can be obtained from curved-beam theory expressed in the form

$$\sigma = K \frac{Mc}{I}$$

where K is a stress-correction factor. The value of K depends on the shape of the wire cross section and whether the stress sought is at the inner or outer fiber. Wahl analytically determined the values of K to be, for round wire,

$$K_i = \frac{4C^2 - C - 1}{4C(C - 1)} \qquad K_o = \frac{4C^2 + C - 1}{4C(C + 1)} \tag{10–43}$$

where C is the spring index and the subscripts i and o refer to the inner and outer fibers, respectively. In view of the fact that K_o is always less than unity, we shall use K_i to estimate the stresses. When the bending moment is $M = Fr$ and the section modulus $I/c = d^3/32$, we express the bending equation as

$$\sigma = K_i \frac{32Fr}{\pi d^3} \tag{10–44}$$

which gives the bending stress for a round-wire torsion spring.

Deflection and Spring Rate

For torsion springs, angular deflection can be expressed in radians or revolutions (turns). If a term contains revolution units the term will be expressed with a prime sign. The spring rate k' is expressed in units of torque/revolution (lbf · in/rev or N · mm/rev) and moment is proportional to angle θ' expressed in turns rather than radians. The spring rate, if linear, can be expressed as

$$k' = \frac{M_1}{\theta_1'} = \frac{M_2}{\theta_2'} = \frac{M_2 - M_1}{\theta_2' - \theta_1'} \tag{10–45}$$

where the moment M can be expressed as Fl or Fr.

The angle subtended by the end deflection of a cantilever, when viewed from the built-in ends, is y/l rad. From Table A–9–1

$$\theta_e = \frac{y}{l} = \frac{Fl^2}{3EI} = \frac{Fl^2}{3E(\pi d^4/64)} = \frac{64Ml}{3\pi d^4 E} \tag{10–46}$$

For a straight torsion end spring, end corrections such as Eq. (10–46) must be added to the body-coil deflection. The strain energy in bending is, from Eq. (5–18),

$$U = \int \frac{M^2\, dx}{2EI}$$

For a torsion spring $M = Fl = Fr$, and integration must be accomplished over the length of the body-coil wire. The force F will deflect through a distance $r\theta$ where θ is the angular deflection of the coil body, in radians. Applying Castigliano's theorem gives

$$r\theta = \frac{\partial U}{\partial F} = \int_0^{\pi DN_b} \frac{\partial}{\partial F}\left(\frac{F^2 r^2\, dx}{2EI}\right) = \int_0^{\pi DN_b} \frac{Fr^2\, dx}{EI}$$

Substituting $I = \pi d^4/64$ for round wire and solving for θ gives

$$\theta = \frac{64 FrDN_b}{d^4 E} = \frac{64 MDN_b}{d^4 E}$$

The total angular deflection in radians is obtained by adding Eq. (10–46) for each end of lengths l_1, l_2:

$$\theta_t = \frac{64MDN_b}{d^4 E} + \frac{64Ml_1}{3\pi d^4 E} + \frac{64Ml_2}{3\pi d^4 E} = \frac{64MD}{d^4 E}\left(N_b + \frac{l_1 + l_2}{3\pi D}\right) \tag{10–47}$$

The equivalent number of active turns N_a is expressed as

$$N_a = N_b + \frac{l_1 + l_2}{3\pi D} \tag{10–48}$$

The spring rate k in torque per radian is

$$k = \frac{Fr}{\theta_t} = \frac{M}{\theta_t} = \frac{d^4 E}{64 D N_a} \tag{10–49}$$

The spring rate may also be expressed as torque per turn. The expression for this is obtained by multiplying Eq. (10–49) by 2π rad/turn. Thus spring rate k' (units torque/turn) is

$$k' = \frac{2\pi d^4 E}{64 D N_a} = \frac{d^4 E}{10.2 D N_a} \tag{10–50}$$

Tests show that the effect of friction between the coils and arbor is such that the constant 10.2 should be increased to 10.8. The equation above becomes

$$k' = \frac{d^4 E}{10.8 D N_a} \tag{10–51}$$

(units torque per turn). Equation (10–51) gives better results. Also Eq. (10–47) becomes

$$\theta'_t = \frac{10.8MD}{d^4 E}\left(N_b + \frac{l_1 + l_2}{3\pi D}\right) \tag{10–52}$$

Torsion springs are frequently used over a round bar or pin. When the load is applied to a torsion spring, the spring winds up, causing a decrease in the inside diameter of the coil body. It is necessary to ensure that the inside diameter of the coil never becomes equal to or less than the diameter of the pin, in which case loss of spring function would ensue. The helix diameter of the coil D' becomes

$$D' = \frac{N_b D}{N_b + \theta'_c} \tag{10–53}$$

where θ'_c is the angular deflection of the body of the coil in number of turns, given by

$$\theta'_c = \frac{10.8 M D N_b}{d^4 E} \tag{10–54}$$

The new inside diameter $D'_i = D' - d$ makes the diametral clearance Δ between the body coil and the pin of diameter D_p equal to

$$\Delta = D' - d - D_p = \frac{N_b D}{N_b + \theta'_c} - d - D_p \tag{10–55}$$

Equation (10–55) solved for N_b is

$$N_b = \frac{\theta'_c (\Delta + d + D_p)}{D - \Delta - d - D_p} \tag{10–56}$$

which gives the number of body turns corresponding to a specified diametral clearance of the arbor. This angle may not be in agreement with the necessary partial-turn remainder. Thus the diametral clearance may be exceeded but not equaled.

Static Strength

First column entries in Table 10–6 can be divided by 0.577 (from distortion-energy theory) to give

$$S_y = \begin{cases} 0.78 S_{ut} & \text{Music wire and cold-drawn carbon steels} \\ 0.87 S_{ut} & \text{OQ\&T carbon and low-alloy steels} \\ 0.61 S_{ut} & \text{Austenitic stainless steel and nonferrous alloys} \end{cases} \tag{10–57}$$

Fatigue Strength

Since the spring wire is in bending, the Sines equation is not applicable. The Sines model is in the presence of pure torsion. Since Zimmerli's results were for compression springs (wire in pure torsion), we will use the repeated bending stress ($R = 0$) values provided by Associated Spring in Table 10–10. As in Eq. (10–40) we will use the Gerber fatigue-failure criterion incorporating the Associated Spring $R = 0$ fatigue strength S_r:

$$S_e = \frac{S_r/2}{1 - \left(\dfrac{S_r/2}{S_{ut}}\right)^2} \tag{10–58}$$

The value of S_r (and S_e) has been corrected for size, surface condition, and type of loading, but not for temperature or miscellaneous effects. The Gerber fatigue criterion is now defined. The strength-amplitude component is given by Table 7–10 as

$$S_a = \frac{r^2 S_{ut}^2}{2 S_e} \left[-1 + \sqrt{1 + \left(\frac{2 S_e}{r S_{ut}}\right)^2} \right] \tag{10–59}$$

Table 10–10

Maximum Recommended Bending Stresses (K_B Corrected) for Helical Torsion Springs in Cyclic Applications as Percent of S_{ut}

Source: Courtesy of Associated Spring.

Fatigue Life, cycles	ASTM A228 and Type 302 Stainless Steel		ASTM A230 and A232	
	Not Shot-Peened	Shot-Peened*	Not Shot-Peened	Shot-Peened*
10^5	53	62	55	64
10^6	50	60	53	62

This information is based on the following conditions: no surging, springs are in the "as-stress-relieved" condition.

*Not always possible.

where the slope of the load line is $r = M_a/M_m$. The load line is radial through the origin of the designer's fatigue diagram. The factor of safety guarding against fatigue failure is

$$n_f = \frac{S_a}{\sigma_a} \qquad (10\text{–}60)$$

Alternatively, we can find n_f directly by using Table 7–10:

$$n_f = \frac{1}{2}\frac{\sigma_a}{S_e}\left(\frac{S_{ut}}{\sigma_m}\right)^2\left[-1 + \sqrt{1 + \left(2\frac{\sigma_m}{S_{ut}}\frac{S_e}{\sigma_a}\right)^2}\right] \qquad (10\text{–}61)$$

EXAMPLE 10–8

A stock spring is shown in Fig. 10–11. It is made from 0.072-in-diameter music wire and has $4\frac{1}{4}$ body turns with straight torsion ends. It works over a pin of 0.400 in diameter. The coil outside diameter is $\frac{19}{32}$ in.

(a) Find the maximum operating torque and corresponding rotation for static loading.

(b) Estimate the inside coil diameter and pin diametral clearance when the spring is subjected to the torque in part (a).

Figure 10–11

Angles α, β, and θ are measured between the straight-end centerline translated to the coil axis. Coil OD is 19/32 in.

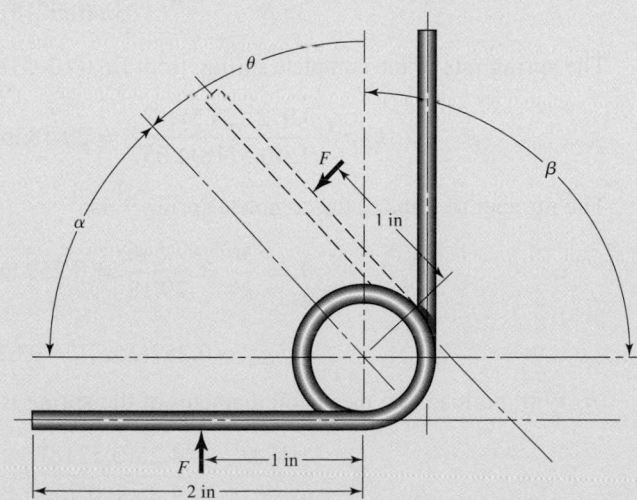

(*c*) Estimate the fatigue factor of safety n_f if the applied moment varies between $M_{min} = 1$ to $M_{max} = 5$ in · lbf.

Solution
(*a*) For music wire, from Table 10–4 we find that $A = 201$ kpsi · in^m and $m = 0.145$. Therefore,

$$S_{ut} = \frac{A}{d^m} = \frac{201}{(0.072)^{0.145}} = 294.4 \text{ kpsi}$$

Using Eq. (10–57) gives

$$S_y = 0.78 S_{ut} = 0.78(294.4) = 229.6 \text{ kpsi}$$

The mean coil diameter is $D = 19/32 - 0.072 = 0.5218$ in. The spring index $C = D/d = 0.5218/0.072 = 7.247$. The bending stress correction factor K_i, from Eq. (10–43), is

$$K_i = \frac{4(7.247)^2 - 7.247 - 1}{4(7.247)(7.247 - 1)} = 1.115$$

Now rearrange Eq. (10–44), substitute S_y for σ, and solve for the maximum torque Fr to obtain

$$M_{max} = (Fr)_{max} = \frac{\pi d^3 S_y}{32 K_i} = \frac{\pi (0.072)^3 229\,600}{32(1.115)} = 7.546 \text{ lbf · in}$$

Note that no factor of safety has been used. Next, from Eq. (10–54), the number of turns of the coil body θ_c' is

$$\theta_c' = \frac{10.8 M D N_b}{d^4 E} = \frac{10.8(7.546)0.5218(4.25)}{0.072^4(28.5)10^6} = 0.236 \text{ turn}$$

Answer
$$(\theta_c')_{deg} = 0.236(360°) = 85.0°$$

The active number of turns N_a, from Eq. (10–48), is

$$N_a = N_b + \frac{l_1 + l_2}{3\pi D} = 4.25 + \frac{1 + 1}{3\pi(0.5218)} = 4.657 \text{ turns}$$

The spring rate of the complete spring, from Eq. (10–51), is

$$k' = \frac{0.072^4(28.5)10^6}{10.8(0.5218)4.657} = 29.18 \text{ lbf · in/turn}$$

The number of turns of the complete spring θ' is

$$\theta' = \frac{M}{k'} = \frac{7.546}{29.18} = 0.259 \text{ turn}$$

Answer
$$(\theta_s')_{deg} = 0.259(360°) = 93.24°$$

(*b*) With no load, the mean coil diameter of the spring is 0.5218 in. From Eq. (10–53),

$$D' = \frac{N_b D}{N_b + \theta_c'} = \frac{4.25(0.5218)}{4.25 + 0.236} = 0.494 \text{ in}$$

The diametral clearance between the inside of the spring coil and the pin at load is

Answer

$$\Delta = D' - d - D_p = 0.494 - 0.072 - 0.400 = 0.022 \text{ in}$$

(*c*) Fatigue:

$$M_a = (M_{\max} - M_{\min})/2 = (5 - 1)/2 = 2 \text{ lbf} \cdot \text{in}$$

$$M_m = (M_{\max} + M_{\min})/2 = (5 + 1)/2 = 3 \text{ lbf} \cdot \text{in}$$

$$r = \frac{M_a}{M_m} = \frac{2}{3}$$

$$\sigma_a = K_i \frac{32M_a}{\pi d^3} = 1.115 \frac{32(2)}{\pi 0.072^3} = 60\ 857 \text{ psi}$$

$$\sigma_m = \frac{M_m}{M_a}\sigma_a = \frac{3}{2}(60\ 857) = 91\ 286 \text{ psi}$$

From Table 10–10, $S_r = 0.50S_{ut} = 0.50(294.4) = 147.2$ kpsi. Then

$$S_e = \frac{147.2/2}{1 - \left(\dfrac{147.2/2}{294.4}\right)^2} = 78.51 \text{ kpsi}$$

The amplitude component of the strength S_a, from Eq. (10–59), is

$$S_a = \frac{(2/3)^2 294.4^2}{2(78.51)}\left[-1 + \sqrt{1 + \left(\frac{2}{2/3}\frac{78.51}{294.4}\right)^2}\right] = 68.85 \text{ kpsi}$$

The fatigue factor of safety is

Answer

$$n_f = \frac{S_a}{\sigma_a} = \frac{68.85}{60.86} = 1.13$$

10–13 Belleville Springs

The inset of Fig. 10–12 shows a coned-disk spring, commonly called a *Belleville spring*. Although the mathematical treatment is beyond the scope of this book, you should at least become familiar with the remarkable characteristics of these springs.

Aside from the obvious advantage that a Belleville spring occupies only a small space, variation in the h/t ratio will produce a wide variety of load-deflection curve shapes, as illustrated in Fig. 10–12. For example, using an h/t ratio of 2.83 or larger gives an S curve that might be useful for snap-acting mechanisms. A reduction of the ratio to a value between 1.41 and 2.1 causes the central portion of the curve to become horizontal, which means that the load is constant over a considerable deflection range.

A higher load for a given deflection may be obtained by nesting, that is, by stacking the springs in parallel. On the other hand, stacking in series provides a larger deflection for the same load, but in this case there is danger of instability.

Figure 10–12

Load-deflection curves for Belleville springs. *(Courtesy of Associated Spring.)*

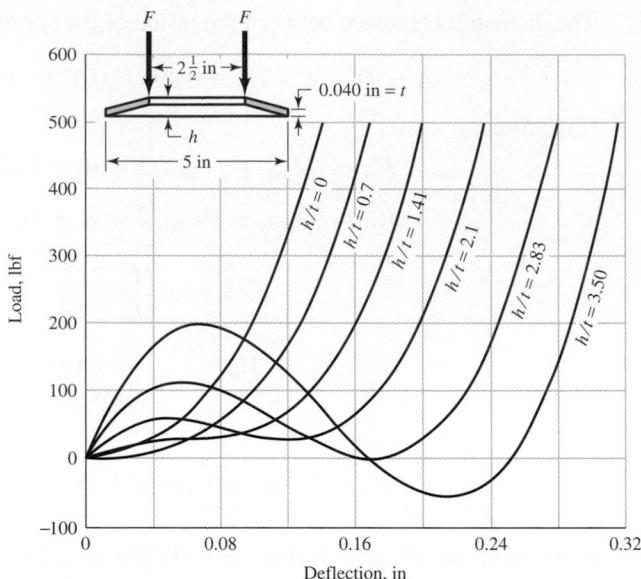

10–14 Miscellaneous Springs

The extension spring shown in Fig. 10–13 is made of slightly curved strip steel, not flat, so that the force required to uncoil it remains constant; thus it is called a *constant-force spring*. This is equivalent to a zero spring rate. Such springs can also be manufactured having either a positive or a negative spring rate.

A *volute spring,* shown in Fig. 10–14a, is a wide, thin strip, or "flat," of material wound on the flat so that the coils fit inside one another. Since the coils do not stack, the solid height of the spring is the width of the strip. A variable-spring scale, in a compression volute spring, is obtained by permitting the coils to contact the support. Thus, as the deflection increases, the number of active coils decreases. The volute spring has another important advantage that cannot be obtained with round-wire springs: if the coils are wound so as to contact or slide on one another during action, the sliding friction will serve to damp out vibrations or other unwanted transient disturbances.

A *conical spring,* as the name implies, is a coil spring wound in the shape of a cone (see Prob. 10–22). Most conical springs are compression springs and are wound with round wire. But a volute spring is a conical spring too. Probably the principal advantage of this type of spring is that it can be wound so that the solid height is only a single wire diameter.

Flat stock is used for a great variety of springs, such as clock springs, power springs, torsion springs, cantilever springs, and hair springs; frequently it is specially shaped to create certain spring actions for fuse clips, relay springs, spring washers, snap rings, and retainers.

In designing many springs of flat stock or strip material, it is often economical and of value to proportion the material so as to obtain a constant stress throughout the spring material. A uniform-section cantilever spring has a stress

$$\sigma = \frac{M}{I/c} = \frac{Fx}{I/c} \qquad (a)$$

Figure 10–13

Constant-force spring. *(Courtesy of Vulcan Spring & Mfg. Co. Telford, PA. www.vulcanspring.com.)*

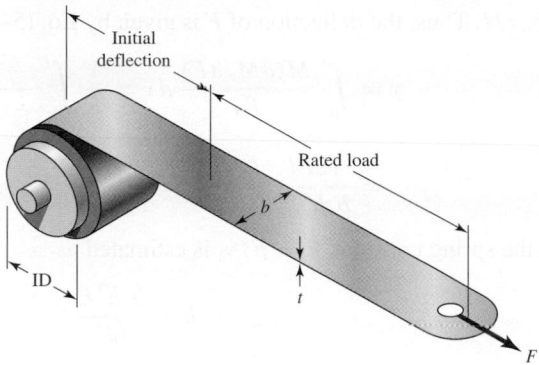

Figure 10–14

(a) A volute spring; (b) a flat triangular spring.

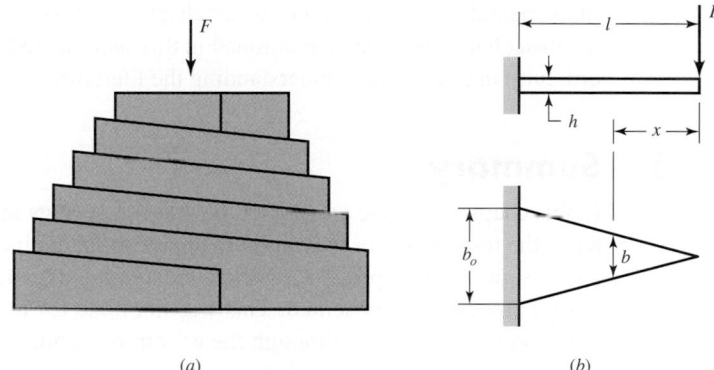

(a)   (b)

which is proportional to the distance x if I/c is a constant. But there is no reason why I/c need be a constant. For example, one might design such a spring as that shown in Fig. 10–14b, in which the thickness h is constant but the width b is permitted to vary. Since, for a rectangular section, $I/c = bh^2/6$, we have, from Eq. (a),

$$\frac{bh^2}{6} = \frac{Fx}{\sigma}$$

or

$$b = \frac{6Fx}{h^2\sigma} \tag{b}$$

Since b is linearly related to x, the width b_o at the base of the spring is

$$b_o = \frac{6Fl}{h^2\sigma} \tag{10–62}$$

Good approximations for deflections can be found easily by using Castigliano's theorem. To demonstrate this, assume that deflection of the triangular flat spring is primarily due to bending and we can neglect the transverse shear force.[12] The bending moment as a function of x is $M = -Fx$ and the beam width at x can be expressed as

---

[12]Note that, because of shear, the width of the beam cannot be zero at $x = 0$. So, there is already some simplification in the design model. All of this can be accounted for in a more sophisticated model.

$b = b_o x / l$. Thus, the deflection of F is given by Eq. (5–26) as

$$y = \int_0^l \frac{M(\partial M / \partial F)}{EI} \, dx = \frac{1}{E} \int_0^l \frac{-Fx(-x)}{\frac{1}{12}(b_o x / l) h^3} \, dx$$

$$= \frac{12Fl}{b_o h^3 E} \int_0^l x \, dx = \frac{6Fl^3}{b_o h^3 E} \tag{10–63}$$

Thus the spring constant, $k = F/y$, is estimated as

$$k = \frac{b_o h^3 E}{6l^3} \tag{10–64}$$

The methods of stress and deflection analysis illustrated in previous sections of this chapter have served to illustrate that springs may be analyzed and designed by using the fundamentals discussed in the earlier chapters of this book. This is also true for most of the miscellaneous springs mentioned in this section, and you should now experience no difficulty in reading and understanding the literature of such springs.

10–15 Summary

In this chapter we have considered helical coil springs in considerable detail in order to show the importance of viewpoint in approaching engineering problems, their analysis, and design. For compression springs undergoing static and fatigue loads, the complete design process was presented. This was not done for extension and torsion springs, as the process is the same, although the governing conditions are not. The governing conditions, however, were provided and extension to the design process from what was provided for the compression spring should be straightforward. Problems are provided at the end of the chapter, and it is hoped that the reader will develop additional, similar, problems to tackle.

Stochastic considerations are notably missing in this chapter. The complexity and nuances of the deterministic approach alone are enough to handle in a first presentation of spring design. Springmakers offer a vast array of information concerning tolerances on springs.[13] This, together with the material in Chaps. 2, 6, and 7, should provide the reader with ample ability to advance and incorporate statistical analyses in their design evaluations.

As spring problems become more computationally involved, programmable calculators and computers must be used. Spreadsheet programming is very popular for repetitive calculations. As mentioned earlier, commercial programs are available. With these programs, backsolving can be performed; that is, when the final objective criteria are entered, the program determines the input values.

PROBLEMS

 ANALYSIS

10–1 Make a two-view drawing or a good freehand sketch of a helical compression spring closed to its solid height and having a wire diameter of $\frac{1}{2}$ in, outside diameter of 4 in, and one active coil. The spring is to have plain ends. Make another drawing of the same spring with ends plain and ground.

---

[13]See, for example, Associated Spring–Barnes Group, *Design Handbook,* Bristol, Conn., 1987.

ANALYSIS

10–2   It is instructive to examine the question of the units of the parameter A of Eq. (10–14). Show that for U.S. customary units the units for A_{uscu} are kpsi · in^m and for SI units are MPa · mm^m for A_{SI}. which make the dimensions of both A_{uscu} and A_{SI} different for every material to which Eq. (10–14) applies. Also show that the conversion from A_{uscu} to A_{SI} is given by

$$A_{SI} = 6.895(25.40)^m A_{uscu}$$

ANALYSIS

10–3   A helical compression spring is wound using 0.105-in-diameter music wire. The spring has an outside diameter of 1.225 in with plain ground ends, and 12 total coils.
(a) What should the free length be to ensure that when the spring is compressed solid the torsional stress does not exceed the yield strength, that is, that it is solid-safe?
(b) What force is needed to compress this spring to closure?
(c) Estimate the spring rate.
(d) Is there a possibility that the spring might buckle in service?

ANALYSIS

10–4   The spring in Prob. 10–3 is to be used with a static load of 30 lbf. Perform a design assessment represented by Eqs. (10–13) and (10–18) through (10–21) if the spring is closed to solid height.

ANALYSIS

10–5   A helical compression spring is made of hard-drawn spring steel wire 2 mm in diameter and has an outside diameter of 22 mm. The ends are plain and ground, and there are $8\frac{1}{2}$ total coils.
(a) The spring is wound to a free length, which is the largest possible with a solid-safe property. Find this free length.
(b) What is the pitch of this spring?
(c) What force is needed to compress the spring to its solid length?
(d) Estimate the spring rate.
(e) Will the spring buckle in service?

ANALYSIS

10–6   The spring of Prob. 10–5 is to be used with a static load of 75 N. Perform a design assessment represented by Eqs. (10–13) and (10–18) through (10–21) if the spring closed to solid height.

ANALYSIS

10–7
to
10–17   Listed below are six springs described in customary units and five springs described in SI units. Investigate these squared-and-ground-ended helical compression springs to see if they are solid-safe. If not, what is the largest free length to which they can be wound using $n_s = 1.2$?

Problem Number	d, in	OD, in	L_0, in	N_t	Material
10–7	0.006	0.036	0.63	40	A228 music wire
10–8	0.012	0.120	0.81	15.1	B159 phosphor-bronze
10–9	0.040	0.240	0.75	10.4	A313 stainless steel
10–10	0.135	2.0	2.94	5.25	A227 hard-drawn steel
10–11	0.144	1.0	3.75	13.0	A229 OQ&T steel
10–12	0.192	3.0	9.0	8.0	A232 chrome-vanadium

	d, mm	OD, mm	L_0, mm	N_t	Material
10–13	0.2	0.91	15.9	40	A313 stainless steel
10–14	1.0	6.10	19.1	10.4	A228 music wire
10–15	3.4	50.8	74.6	5.25	A229 OQ&T spring steel
10–16	3.7	25.4	95.3	13.0	B159 phosphor-bronze
10–17	4.3	76.2	228.6	8.0	A232 chrome-vanadium

DESIGN

10–18   A static service music wire helical compression spring is needed to support a 20-lbf load after being compressed 2 in. The solid height of the spring cannot exceed $1\frac{1}{2}$ in. The free length must not exceed 4 in. The static factor of safety must equal or exceed 1.2. For robust linearity use a fractional overrun to closure ξ of 0.15. There are two springs to be designed.

(a) The spring must operate over a $\frac{3}{4}$-in rod. A 0.050-in diametral clearance allowance should be adequate to avoid interference between the rod and the spring due to out-of-round coils. Design the spring.

(b) The spring must operate in a 1-in-diameter hole. A 0.050-in diametral clearance allowance should be adequate to avoid interference between the spring and the hole due to swelling of the spring diameter as the spring is compressed and out-of-round coils. Design the spring.

ANALYSIS

10–19   Not all springs are made in a conventional way. Consider the special steel spring in the illustration.

(a) Find the pitch, solid height, and number of active turns.

(b) Find the spring rate. Assume the material is A227 HD steel.

(c) Find the force F_s required to close the spring solid.

(d) Find the shear stress in the spring due to the force F_s.

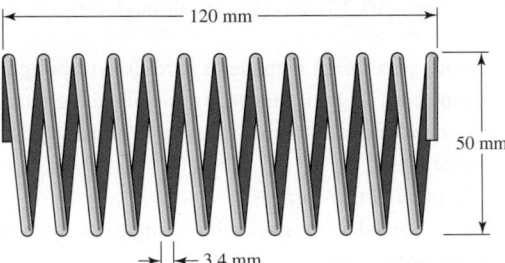

Problem 10–19

DESIGN

10–20   A holding fixture for a workpiece $1\frac{1}{2}$ in thick at clamp locations is being designed. The detail of one of the clamps is shown in the figure. A spring is required to drive the clamp upward while removing or inserting a workpiece. A clamping force of 10 lbf is satisfactory. The base plate is $\frac{5}{8}$ in thick. The clamp screw has a $\frac{7}{16}$ in-20 UNF thread. It is useful to have the free length L_0 short enough so that the clamp screw can compress the spring upon fixture reassembly during inspection and service, say $L_0 \leq 1.5 + \frac{3}{8}$ in. The spring cannot close solid at a length greater than $1\frac{1}{4}$ in. The safety factor when compressed solid should be $n_s \geq 1.2$, and at service load $n_1 \geq 1.5$. Design a suitable helical coil compression spring for this fixture.

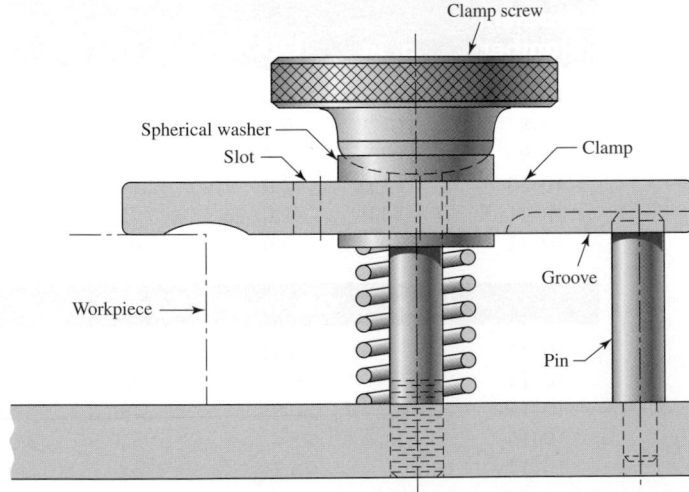

Problem 10–20
Clamping fixture.

10–21 Your instructor will provide you with a stock spring supplier's catalog, or pages reproduced from it. Accomplish the task of Prob. 10–20 by selecting an available stock spring. (This is design by *selection.*)

10–22 The figure shows a conical compression helical coil spring where R_1 and R_2 are the initial and final coil radii, respectively, d is the diameter of the wire, and N_a is the total number of active coils. The wire cross section primarily transmits a torsional moment, which changes with the coil radius. Let the coil radius be given by

$$R = R_1 + \frac{R_2 - R_1}{2\pi N_a}\theta$$

where θ is in radians. Use Castigliano's method to estimate the spring rate as

$$k = \frac{d^4 G}{16N_a(R_2 + R_1)\left(R_2^2 + R_1^2\right)}$$

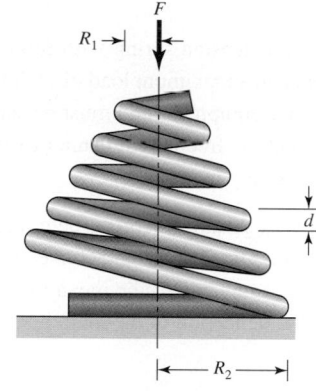

Problem 10–22
(From Richard G. Budynas, *Advanced Strength and Applied Stress Analysis*, 2nd ed., McGraw-Hill, New York, 1999. Copyright © 1999 The McGraw-Hill Companies.)

10–23 A helical coil compression spring is needed for food service machinery. The load varies from a minimum of 4 lbf to a maximum of 18 lbf. The spring deflection varies from a minimum of 1 in to a maximum of 2.5 in. The outside diameter of the spring cannot exceed $2\frac{1}{2}$ in. The springmaker has available suitable dies for drawing 0.080-, 0.0915-, 0.1055-, and 0.1205-in-diameter wire. Using a fatigue design factor n_f of 1.5, and the Gerber-Zimmerli fatigue-failure criterion, design a suitable spring.

10–24 Solve Prob. 10–23 using the Goodman-Zimmerli fatigue-failure criterion.

10–25 Solve Prob. 10–23 using the Sines-Zimmerli fatigue-failure criterion.

10–26 Design the spring of Ex. 10–5 using the Gerber fatigue-failure criterion.

10–27 Solve Prob. 10–26 using the Goodman-Zimmerli fatigue-failure criterion.

10–28 A hard-drawn spring steel extension spring is to be designed to carry a static load of 18 lbf with an extension of $\frac{1}{2}$ in using a design factor of $n_y = 1.5$ in bending. Use full-coil end hooks with the fullest bend radius of $r = D/2$ and $r_2 = 2d$. The free length must be less than 3 in, and the body turns must be fewer than 30. Integer and half-integer body turns allow end hooks to be placed in the same plane. This adds extra cost and is done only when necessary.

ANALYSIS

10–29   The extension spring shown in the figure has full-twisted loop ends. The material is AISI 1065 OQ&T wire. The spring has 84 coils and is close-wound with a preload of 16 lbf.

(a) Find the closed length of the spring.

(b) Find the torsional stress in the spring corresponding to the preload.

(c) Estimate the spring rate.

(d) What load would cause permanent deformation?

(e) What is the spring deflection corresponding to the load found in part d?

Problem 10–29

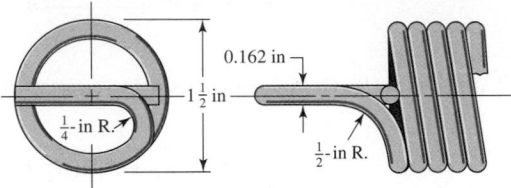

DESIGN

10–30   Design an infinite-life helical coil extension spring with full end loops and generous loop-bend radii for a minimum load of 9 lbf and a maximum load of 18 lbf, with an accompanying stretch of $\frac{1}{4}$ in. The spring is for food-service equipment and must be stainless steel. The outside diameter of the coil cannot exceed 1 in, and the free length cannot exceed $2\frac{1}{2}$ in. Using a fatigue design factor of $n_f = 2$, complete the design.

ANALYSIS

10–31   Prove Eq. (10–40). *Hint:* Using Castigliano's theorem, determine the deflection due to bending of an end hook alone as if the hook were fixed at the end connecting it to the body of the spring. Consider the wire diameter d small as compared to the mean radius of the hook, $R = D/2$. Add the deflections of the end hooks to the deflection of the main body to determine the final spring constant, then equate it to Eq. (10–9).

ANALYSIS

10–32   The figure shows a finger exerciser used by law-enforcement officers and athletes to strengthen their grip. It is formed by winding A227 hard-drawn steel wire around a mandrel to obtain $2\frac{1}{2}$ turns when the grip is in the closed position. After winding, the wire is cut to leave the two legs as handles. The plastic handles are then molded on, the grip is squeezed together, and a wire clip is placed around the legs to obtain initial "tension" and to space the handles for the best initial gripping position. The clip is formed like a figure 8 to prevent it from coming off. When the grip is in the closed position, the stress in the spring should not exceed the permissible stress.

(a) Determine the configuration of the spring before the grip is assembled.

(b) Find the force necessary to close the grip.

Problem 10–32

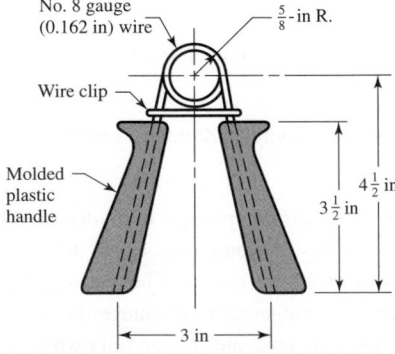

ANALYSIS

10–33   The rat trap shown in the figure uses two opposite-image torsion springs. The wire has a diameter of 0.081 in, and the outside diameter of the spring in the position shown is $\frac{1}{2}$ in. Each spring has 11 turns. Use of a fish scale revealed a force of about 8 lbf is needed to set the trap.
(a) Find the probabable configuration of the spring prior to assembly.
(b) Find the maximum stress in the spring when the trap is set.

Problem 10–33

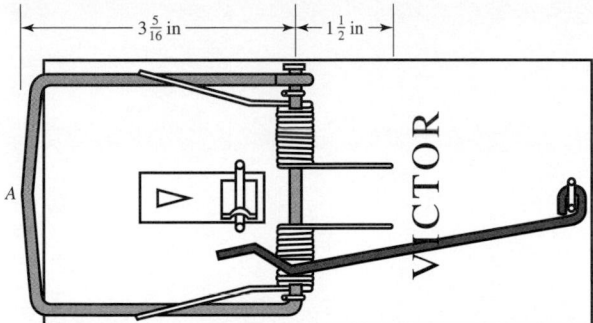

ANALYSIS

10–34   Wire form springs can be made in a variety of shapes. The clip shown operates by applying a force F. The wire diameter is d, the length of the straight section is l, and Young's modulus is E. Consider the effects of bending only, with $d \ll R$, and use Castigliano's theorem to determine the spring constant, k.

Problem 10–34

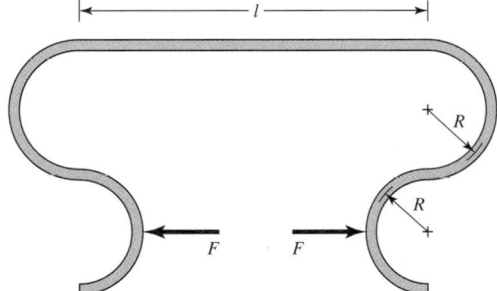

10–35   Using the experience gained with Prob. 10–23, write a computer program that would help in the design of helical coil compression springs.

10–36   Using the experience gained with Prob. 10–30, write a computer program that would help in the design of a helical coil extension spring.

11

Rolling-Contact Bearings

The terms *rolling-contact bearing, antifriction bearing,* and *rolling bearing* are all used to describe that class of bearing in which the main load is transferred through elements in rolling contact rather than in sliding contact. In a rolling bearing the starting friction is about twice the running friction, but still it is negligible in comparison with the starting friction of a sleeve bearing. Load, speed, and the operating viscosity of the lubricant do affect the frictional characteristics of a rolling bearing. It is probably a mistake to describe a rolling bearing as "antifriction," but the term is used generally throughout the industry.

From the mechanical designer's standpoint, the study of antifriction bearings differs in several respects when compared with the study of other topics because the bearings they specify have already been designed. The specialist in antifriction-bearing design is confronted with the problem of designing a group of elements that compose a rolling bearing: these elements must be designed to fit into a space whose dimensions are specified; they must be designed to receive a load having certain characteristics; and finally, these elements must be designed to have a satisfactory life when operated under the specified conditions. Bearing specialists must therefore consider such matters as fatigue loading, friction, heat, corrosion resistance, kinematic problems, material properties, lubrication, machining tolerances, assembly, use, and cost. From a consideration of all these factors, bearing specialists arrive at a compromise that, in their judgment, is a good solution to the problem as stated.

We begin with an overview of bearing types; then we note that bearing life cannot be described in deterministic form. We introduce the invariant, the statistical distribution of life, which is strongly Weibullian. There are some useful deterministic equations addressing load versus life at constant reliability, and we introduce the catalog rating at rating life.

The reliability-life relationship involves Weibullian statistics. The load-life-reliability relationship, combines statistical and deterministic relationships giving the designer a way to move from the desired load and life to the catalog rating in *one* equation.

Ball bearings also resist thrust, and a unit of thrust does different damage per revolution than a unit of radial load, so we must find the equivalent pure radial load that does the same damage as the existing radial and thrust loads. Next, variable loading, stepwise and continuous, is approached, and the equivalent pure radial load doing the same damage is quantified. Oscillatory loading is mentioned.

With this preparation we have the tools to consider the selection of ball and cylindrical roller bearings. The question of misalignment is quantitatively approached.

Tapered roller bearings have some complications, and our experience so far contributes to understanding them.

Having the tools to find the proper catalog ratings, we make decisions (selections), we perform a design assessment, and the bearing reliability is quantified. Lubrication and mounting conclude our introduction. Vendors' manuals should be consulted for specific details relating to bearings of their manufacture.

11–1 Bearing Types

Bearings are manufactured to take pure radial loads, pure thrust loads, or a combination of the two kinds of loads. The nomenclature of a ball bearing is illustrated in Fig. 11–1, which also shows the four essential parts of a bearing. These are the outer ring, the inner ring, the balls or rolling elements, and the separator. In low-priced bearings, the separator is sometimes omitted, but it has the important function of separating the elements so that rubbing contact will not occur.

Figure 11-1

Nomenclature of a ball bearing. *(General Motors Corp. Used with permission, GM Media Archives.)*

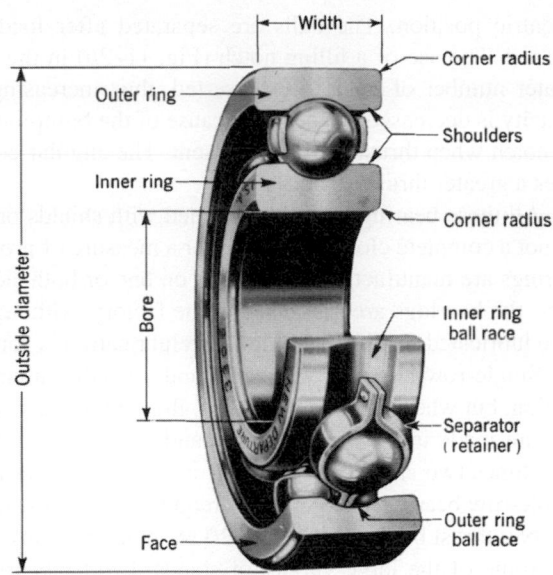

Figure 11-2

Various types of ball bearings.

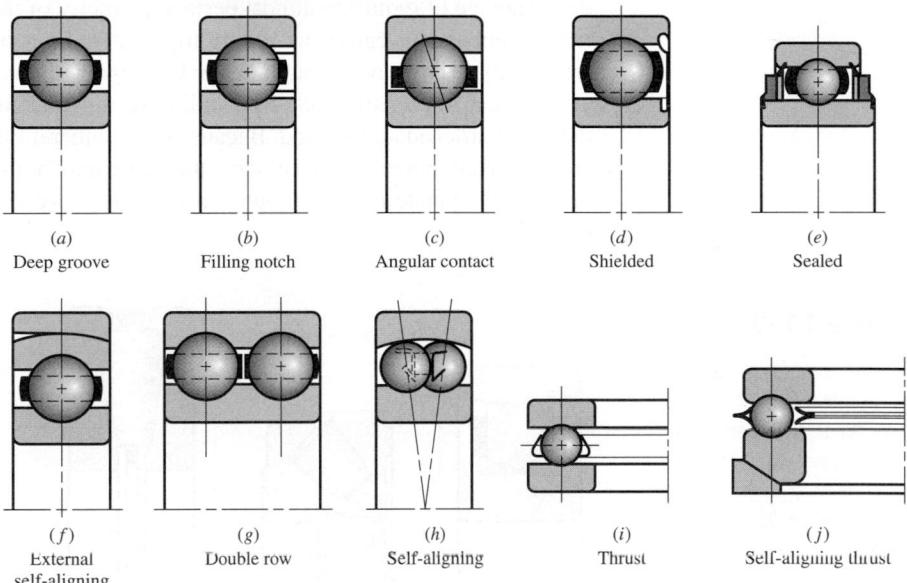

(*a*)	(*b*)	(*c*)	(*d*)	(*e*)
Deep groove	Filling notch	Angular contact	Shielded	Sealed

(*f*)	(*g*)	(*h*)	(*i*)	(*j*)
External self-aligning	Double row	Self-aligning	Thrust	Self-aligning thrust

In this section we include a selection from the many types of standardized bearings that are manufactured. Most bearing manufacturers provide engineering manuals and brochures containing lavish descriptions of the various types available. In the small space available here, only a meager outline of some of the most common types can be given. So you should include a survey of bearing manufacturers' literature in your studies of this section.

Some of the various types of standardized bearings that are manufactured are shown in Fig. 11–2. The single-row deep-groove bearing will take radial load as well as some thrust load. The balls are inserted into the grooves by moving the inner ring to an eccentric position. The balls are separated after loading, and the separator is then inserted. The use of a filling notch (Fig. 11–2b) in the inner and outer rings enables a greater number of balls to be inserted, thus increasing the load capacity. The thrust capacity is decreased, however, because of the bumping of the balls against the edge of the notch when thrust loads are present. The angular-contact bearing (Fig. 11–2c) provides a greater thrust capacity.

All these bearings may be obtained with shields on one or both sides. The shields are not a complete closure but do offer a measure of protection against dirt. A variety of bearings are manufactured with seals on one or both sides. When the seals are on both sides, the bearings are lubricated at the factory. Although a sealed bearing is supposed to be lubricated for life, a method of relubrication is sometimes provided.

Single-row bearings will withstand a small amount of shaft misalignment of deflection, but where this is severe, self-aligning bearings may be used. Double-row bearings are made in a variety of types and sizes to carry heavier radial and thrust loads. Sometimes two single-row bearings are used together for the same reason, although a double-row bearing will generally require fewer parts and occupy less space. The one-way ball thrust bearings (Fig. 11–2i) are made in many types and sizes.

Some of the large variety of standard roller bearings available are illustrated in Fig. 11–3. Straight roller bearings (Fig. 11–3a) will carry a greater radial load than ball bearings of the same size because of the greater contact area. However, they have the disadvantage of requiring almost perfect geometry of the raceways and rollers. A slight misalignment will cause the rollers to skew and get out of line. For this reason, the retainer must be heavy. Straight roller bearings will not, of course, take thrust loads.

Helical rollers are made by winding rectangular material into rollers, after which they are hardened and ground. Because of the inherent flexibility, they will take considerable misalignment. If necessary, the shaft and housing can be used for raceways instead of separate inner and outer races. This is especially important if radial space is limited.

Figure 11–3

Types of roller bearings:
(a) straight roller; (b) spherical roller, thrust; (c) tapered roller, thrust; (d) needle; (e) tapered roller; (f) steep-angle tapered roller. (*Courtesy of The Timken Company.*)

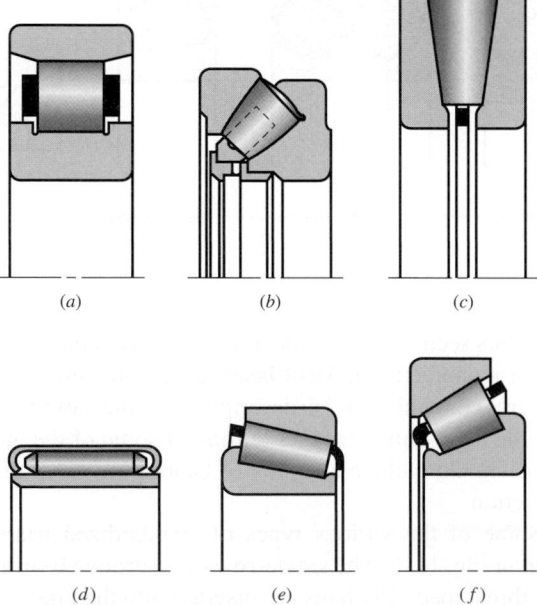

(a)   (b)   (c)

(d)   (e)   (f)

The spherical-roller thrust bearing (Fig. 11–3*b*) is useful where heavy loads and misalignment occur. The spherical elements have the advantage of increasing their contact area as the load is increased.

Needle bearings (Fig. 11–3*d*) are very useful where radial space is limited. They have a high load capacity when separators are used, but may be obtained without separators. They are furnished both with and without races.

Tapered roller bearings (Fig. 11–3*e, f*) combine the advantages of ball and straight roller bearings, since they can take either radial or thrust loads or any combination of the two, and in addition, they have the high load-carrying capacity of straight roller bearings. The tapered roller bearing is designed so that all elements in the roller surface and the raceways intersect at a common point on the bearing axis.

The bearings described here represent only a small portion of the many available for selection. Many special-purpose bearings are manufactured, and bearings are also made for particular classes of machinery. Typical of these are:

- Instrument bearings, which are high-precision and are available in stainless steel and high-temperature materials
- Nonprecision bearings, usually made with no separator and sometimes having split or stamped sheet-metal races
- Ball bushings, which permit either rotation or sliding motion or both
- Bearings with flexible rollers

11–2 Bearing Life

When the ball or roller of rolling-contact bearings rolls, contact stresses occur on the inner ring, the rolling element, and on the outer ring. Because the curvature of the contacting elements in the axial direction is different from that in the radial direction, the equations for these stresses are more involved than in the Hertz or Smith-Liu equations presented in Chapter 4. If a bearing is clean and properly lubricated, is mounted and sealed against the entrance of dust and dirt, is maintained in this condition, and is operated at reasonable temperatures, then metal fatigue will be the only cause of failure. Inasmuch as metal fatigue implies many millions of stress applications successfully endured, we need a quantitative life measure. Common life measures are

- Number of revolutions of the inner ring (outer ring stationary) until the first tangible evidence of fatigue
- Number of hours of use at a standard angular speed until the first tangible evidence of fatigue

The commonly used term is *bearing life,* which is applied to either of the measures just mentioned. It is important to realize, as in all fatigue, life as defined above is a stochastic variable and, as such, has both a distribution and associated statistical parameters. The life measure of an individual bearing is defined as the total number of revolutions (or hours at a constant speed) of bearing operation until the failure criterion is developed. Under ideal conditions, the fatigue failure consists of spalling of the load-carrying surfaces. The American Bearing Manufacturers Association (ABMA) standard states that the failure criterion is the first evidence of fatigue. The fatigue criterion used by the Timken Company laboratories is the spalling or pitting of an area of 0.01 in². Timken also observes that the useful life of the bearing may extend considerably beyond this point. This is an operational definition of fatigue failure in rolling bearings.

The *rating life* is a term sanctioned by the ABMA and used by most manufacturers. The rating life of a group of nominally identical ball or roller bearings is defined as the number of revolutions (or hours at a constant speed) that 90 percent of a group of bearings will achieve or exceed before the failure criterion develops. The terms *minimum life, L_{10} life,* and B_{10} *life* are also used as synonyms for rating life. The rating life is the 10th percentile location of the bearing group's revolutions-to-failure distribution.

Median life is the 50th percentile life of a group of bearings. The term *average life* has been used as a synonym for median life, contributing to confusion. When many groups of bearings are tested, the median life is between 4 and 5 times the L_{10} life.

11–3 Bearing Load Life at Rated Reliability

When nominally identical groups are tested to the life-failure criterion at different loads, the data are plotted on a graph as depicted in Fig. 11–4 using a log-log transformation. To establish a single point, load F_1 and the rating life of group one $(L_{10})_1$ are the coordinates that are logarithmically transformed. The reliability associated with this point, and all other points, is 0.90. Thus we gain a glimpse of the load-life function at 0.90 reliability. Using a regression equation of the form

$$FL^{1/a} = \text{constant} \tag{11-1}$$

the result of many tests for various kinds of bearings result in

- $a = 3$ for ball bearings
- $a = 10/3$ for roller bearings (cylindrical and tapered roller)

A bearing manufacturer may choose a rated cycle value of 10^6 revolutions (or in the case of the Timken Company, $90(10^6)$ revolutions) or otherwise, as declared in the manufacturer's catalog to correspond to a basic load rating in the catalog for each bearing manufactured, as their rating life. We shall call this the *catalog load rating* and display it algebraically as C_{10}, to denote it as the 10th percentile rating life for a particular bearing in the catalog. From Eq. (11–1) we can write

$$F_1 L_1^{1/a} = F_2 L_2^{1/a} \tag{11-2}$$

and associate load F_1 with C_{10}, life measure L_1 with L_{10}, and write

$$C_{10} L_{10}^{1/a} = F L^{1/a}$$

where the units of L are revolutions.

Figure 11–4

Typical bearing load-life log-log curve.

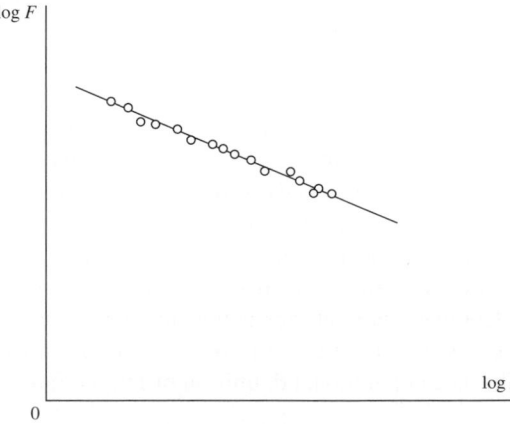

log F

log L

0

Further, we can write

$$C_{10}(L_R n_R 60)^{1/a} = F_D(L_D n_D 60)^{1/a}$$

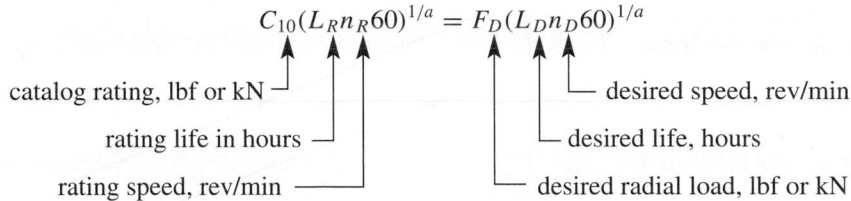

catalog rating, lbf or kN

rating life in hours

rating speed, rev/min

desired speed, rev/min

desired life, hours

desired radial load, lbf or kN

Solving for C_{10} gives

$$C_{10} = F_D \left(\frac{L_D n_D 60}{L_R n_R 60} \right)^{1/a} \tag{11–3}$$

EXAMPLE 11–1

Consider SKF, which rates its bearings for 1 million revolutions, so that L_{10} life is $60 L_R n_R = 10^6$ revolutions. The $L_R n_R 60$ product produces a familiar number. Timken, for example, uses $90(10^6)$ revolutions. If you desire a life of 5000 h at 1725 rev/min with a load of 400 lbf with a reliability of 90 percent, for which catalog rating would you search in an SKF catalog?

Solution

From Eq. (11–3),

$$C_{10} = F_D \left(\frac{L_D n_D 60}{L_R n_R 60} \right)^{1/a} = 400 \left[\frac{5000(1725)60}{10^6} \right]^{1/3} = 3211 \text{ lbf} = 14.3 \text{ kN}$$

If a bearing manufacturer rates bearings at 500 h at $33\frac{1}{3}$ rev/min with a reliability of 0.90, then $L_R n_R 60 = 500(33\frac{1}{3})60 = 10^6$ revolutions. The tendency is to substitute 10^6 for $L_R n_R 60$ in Eq. (11–3). Although it is true that the 60 terms in Eq. (11–3) as displayed cancel algebraically, they are worth keeping, because at some point in your keystroke sequence on your hand-held calculator the manufacturer's magic number (10^6 or some other number) will appear to remind you of what the rating basis is and those manufacturers' catalogs to which you are limited. Of course, if you evaluate the bracketed quantity in Eq. (11–3) by alternating between numerator and denominator entries, the magic number will not appear and you will have lost an opportunity to check.

11–4 Bearing Survival: Reliability versus Life

At constant load, the life measure distribution is right skewed as depicted in Fig. 11–5. Candidates for a distributional curve fit include lognormal and Weibull. The Weibull is by far the most popular, largely because of its ability to adjust to varying amounts of skewness. If the life measure is expressed in dimensionless form as $x = L/L_{10}$, then the reliability can be expressed as [recall Eq. (2–24)]

$$R = \exp \left[-\left(\frac{x - x_0}{\theta - x_0} \right)^b \right] \tag{11–4}$$

where R = reliability

x = life measure dimensionless variate, L/L_{10}

x_0 = guaranteed, or "minimum," value of the variate

Figure 11–5

Constant reliability contours. Point A represents the catalog rating C_{10} at $x = L/L_{10} = 1$. Point B is on the target reliability locus R_D, with a load of C_{10}. Point D is a point on the desired reliability contour exhibiting the design life $x_D = L_D/L_{10}$ at the design load F_D.

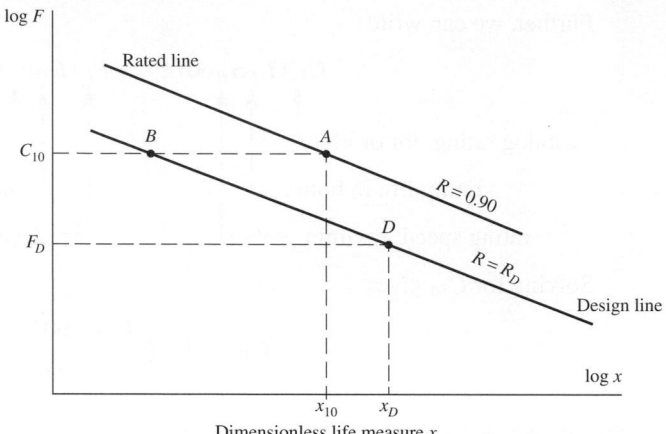

θ = characteristic parameter corresponding to the 63.2121 percentile value of the variate

b = shape parameter that controls the skewness

Because there are three distributional parameters, x_0, θ, and b, the Weibull has a robust ability to conform to a data string. Also, in Eq. (11–4) an explicit expression for the cumulative distribution function is possible:

$$F = 1 - R = 1 - \exp\left[-\left(\frac{x - x_0}{\theta - x_0}\right)^b\right] \qquad (11\text{–}5)$$

EXAMPLE 11–2

Construct the distributional properties of a 02-30 mm deep-groove ball bearing if the Weibull parameters are $x_0 = 0.02$, $(\theta - x_0) = 4.439$, and $b = 1.483$. Find the mean, median, 10th percentile life, standard deviation, and coefficient of variation.

Solution

From Eq. (2–28) the mean dimensionless life μ_x is

Answer

$$\mu_x = x_0 + (\theta - x_0)\Gamma\left(1 + \frac{1}{b}\right) = 0.02 + 4.439\Gamma\left(1 + \frac{1}{1.483}\right) = 4.033$$

The median dimensionless life is, from Eq. (2–26) where $R = 0.5$,

Answer

$$x_{0.50} = x_0 + (\theta - x_0)\left(\ln\frac{1}{R}\right)^{1/b} = 0.02 + 4.439\left(\ln\frac{1}{0.5}\right)^{1/1.483}$$

$$= 3.487$$

The 10th percentile value of the dimensionless life x is

Answer

$$x_{0.10} = 0.02 + 4.439\left(\ln\frac{1}{0.90}\right)^{1/1.483} \doteq 1 \qquad \text{(as it should be)}$$

The standard deviation of the dimensionless life is given by Eq. (2–29):

Answer

$$\hat{\sigma}_x = (\theta - x_0)\left[\Gamma\left(1 + \frac{2}{b}\right) - \Gamma^2\left(1 + \frac{1}{b}\right)\right]^{1/2}$$

$$= 4.439\left[\Gamma\left(1 + \frac{2}{1.483}\right) - \Gamma^2\left(1 + \frac{1}{1.483}\right)\right]^{1/2} = 2.753$$

The coefficient of variation of the dimensionless life is

Answer

$$C_x = \frac{\hat{\sigma}_x}{\mu_x} = \frac{2.753}{4.033} = 0.683$$

11–5   Relating Load, Life, and Reliability

This is the designer's problem. The desired load is not the manufacturer's test load or catalog entry. The desired speed is different from the vendor's test speed, and the reliability expectation is much higher than the 0.90 accompanying the catalog entry. Figure 11–5 shows the situation. The catalog information is plotted as point A, whose coordinates are (the logs of) C_{10} and $x_{10} = L_{10}/L_{10} = 1$, a point on the 0.90 reliability contour. The design point is at D, with the coordinates (the logs of) F_D and x_D, a point that is on the $R = R_D$ reliability contour. The designer must move from point D to point A via point B as follows. Along a constant reliability contour (BD), Eq. (11–2) applies:

$$F_B x_B^{1/a} = F_D x_D^{1/a}$$

from which

$$F_B = F_D \left(\frac{x_D}{x_B}\right)^{1/a} \tag{a}$$

Along a constant load line (AB), Eq. (11–4) applies:

$$R_D = \exp\left[-\left(\frac{x_B - x_0}{\theta - x_0}\right)^b\right]$$

Solving for x_B gives

$$x_B = x_0 + (\theta - x_0)\left(\ln\frac{1}{R_D}\right)^{1/b}$$

Now substitute this in Eq. (a) to obtain

$$F_B = F_D\left(\frac{x_D}{x_B}\right)^{1/a} = F_D\left[\frac{x_D}{x_0 + (\theta - x_0)(\ln 1/R_D)^{1/b}}\right]^{1/a}$$

However, $F_B = C_{10}$, so

$$C_{10} = F_D\left[\frac{x_D}{x_0 + (\theta - x_0)(\ln 1/R_D)^{1/b}}\right]^{1/a} \tag{11–6}$$

As useful as Eq. (11–6) is, one's attention to keystrokes and their sequence on a hand-held calculator strays, and, as a result, the most common error is keying in the inappropriate logarithm. We have the opportunity here to make Eq. (11–6) more error-proof. Note that

$$\ln \frac{1}{R_D} = \ln \frac{1}{1 - p_f} = \ln(1 + p_f + \cdots) \doteq p_f = 1 - R_D$$

where p_f is the probability for failure. Equation (11–6) can be written as

$$C_{10} \doteq F_D \left[\frac{x_D}{x_0 + (\theta - x_0)(1 - R_D)^{1/b}} \right]^{1/a} \qquad R \geq 0.90 \qquad (11\text{–}7)$$

Loads are often nonsteady, so that the desired load is multiplied by an application factor a_f. The steady load $a_f F_D$ does the same damage as the variable load F_D does to the rolling surfaces. This point will be elaborated later.

EXAMPLE 11–3

The design load on a ball bearing is 413 lbf and an application factor of 1.2 is appropriate. The speed of the shaft is to be 300 rev/min, the life to be 30 kh with a reliability of 0.99. What is the C_{10} catalog entry to be sought (or exceeded) when searching for a deep-groove bearing in a manufacturer's catalog on the basis of 10^6 revolutions for rating life? The Weibull parameters are $x_0 = 0.02$, $(\theta - x_0) = 4.439$, and $b = 1.483$.

Solution

$$x_D = \frac{L}{L_{10}} = \frac{60 L_D n_D}{60 L_R n_R} = \frac{60(30\ 000)300}{10^6} = 540$$

Thus, the design life is 540 times the L_{10} life. For a ball bearing, $a = 3$. Then, from Eq. (11–7),

Answer

$$C_{10} = (1.2)(413) \left[\frac{540}{0.02 + 4.439(1 - 0.99)^{1/1.483}} \right]^{1/3} = 6696 \text{ lbf}$$

We have learned to identify the catalog basic load rating corresponding to a steady radial load F_D, a desired life L_D, and a speed n_D.

Shafts generally have two bearings. Often these bearings are different. If the bearing reliability of the shaft with its pair of bearings is to be R, then R is related to the individual bearing reliabilities R_A and R_B by

$$R = R_A R_B$$

First, we observe that if the product $R_A R_B$ equals R, then, in general, R_A and R_B are both greater than R. Since the failure of either or both of the bearings results in the shutdown of the shaft, then A or B or both can create a failure. Second, in sizing bearings one can begin by making R_A and R_B equal to the square root of the reliability goal, $\sqrt{R}$. In Ex. 11–3, if the bearing was one of a pair, the reliability goal would be $\sqrt{0.99}$, or 0.995. The bearings selected are discrete in their reliability property in your problem, so the selection procedure "rounds up," and the overall reliability exceeds the goal R. Third, it may be possible, if $R_A > \sqrt{R}$, to round down on B yet have the product $R_A R_B$ still exceed the goal R.

11–6   **Combined Radial and Thrust Loading**

A ball bearing is capable of resisting radial loading and a thrust loading. Furthermore, these can be combined. Consider F_a and F_r to be the axial thrust and radial loads, respectively, and F_e to be the *equivalent radial load* that does the same damage as the combined radial and thrust loads together. A rotation factor V is defined such that $V = 1$ when the inner ring rotates and $V = 1.2$ when the outer ring rotates. Two dimensionless groups can now be formed: F_e/VF_r and F_a/VF_r. When these two dimensionless groups are plotted as in Fig. 11–6, the data fall in a gentle curve that is well approximated by two straight-line segments. The abscissa e is defined by the intersection of the two lines. The equations for the two lines shown in Fig. 11–6 are

$$\frac{F_e}{VF_r} = 1 \qquad \text{when } \frac{F_a}{VF_r} \le e \tag{11–8a}$$

$$\frac{F_e}{VF_r} = X + Y\frac{F_a}{VF_r} \qquad \text{when } \frac{F_a}{VF_r} > e \tag{11–8b}$$

where, as shown, X is the ordinate intercept and Y is the slope of the line for $F_a/VF_r > e$. It is common to express Eqs. (11–8a) and (11–8b) as a single equation,

$$F_e = X_i VF_r + Y_i F_a \tag{11–9}$$

where $i = 1$ when $F_a/VF_r \le e$ and $i = 2$ when $F_a/VF_r > e$. Table 11–1 lists values of X_1, Y_1, X_2, and Y_2 as a function of e, which in turn is a function of F_a/C_0, where C_0 is the bearing static load catalog rating.

In these equations, the rotation factor V is intended to correct for the rotating-ring conditions. The factor of 1.2 for outer-ring rotation is simply an acknowledgment that the fatigue life is reduced under these conditions. Self-aligning bearings are an exception: they have $V = 1$ for rotation of either ring.

The X and Y factors in Eqs. (11–8a) and (11–8b) depend upon the geometry of the bearing, including the number of balls and the ball diameter. The ABMA recommendations

Figure 11–6

The relationship of dimensionless group $F_e/(VF_r)$ and $F_a/(VF_r)$ and the straight-line segments representing the data.

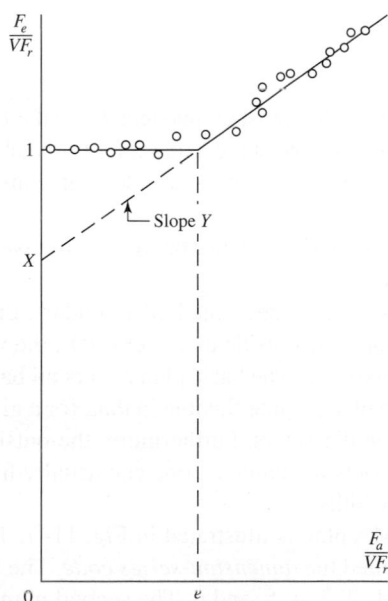

Table 11–1

Equivalent Radial Load Factors for Ball Bearings

| F_a/C_0 | e | $F_a/(VF_r) \le e$ | | $F_a/(VF_r) > e$ | |
		X_1	Y_1	X_2	Y_2
0.014*	0.19	1.00	0	0.56	2.30
0.021	0.21	1.00	0	0.56	2.15
0.028	0.22	1.00	0	0.56	1.99
0.042	0.24	1.00	0	0.56	1.85
0.056	0.26	1.00	0	0.56	1.71
0.070	0.27	1.00	0	0.56	1.63
0.084	0.28	1.00	0	0.56	1.55
0.110	0.30	1.00	0	0.56	1.45
0.17	0.34	1.00	0	0.56	1.31
0.28	0.38	1.00	0	0.56	1.15
0.42	0.42	1.00	0	0.56	1.04
0.56	0.44	1.00	0	0.56	1.00

*Use 0.014 if $F_a/C_0 < 0.014$.

Figure 11–7

The basic ABMA plan for boundary dimensions. These apply to ball bearings, straight roller bearings, and spherical roller bearings, but not to inch-series ball bearings or tapered roller bearings. The contour of the corner is not specified. It may be rounded or chamfered, but it must be small enough to clear the fillet radius specified in the standards.

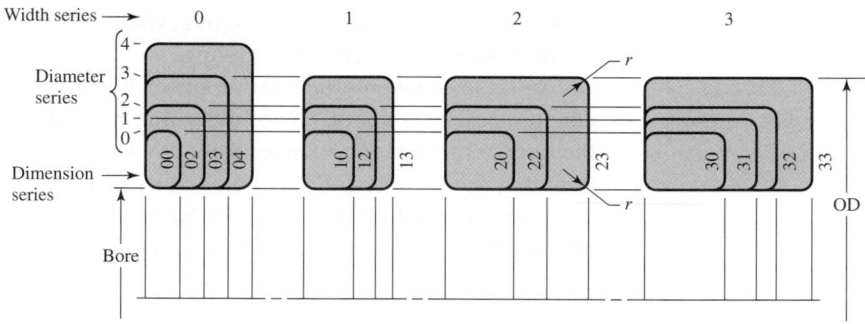

are based on the ratio of the thrust component F_a to the *basic static load rating* C_0 and a variable reference value e. The static load rating C_0 is tabulated, along with the basic dynamic load rating C_{10}, in many of the bearing manufacturers' publications; see Table 11–2, for example.

Since straight or cylindrical roller bearings will take no axial load, or very little, the Y factor is always zero.

The ABMA has established standard boundary dimensions for bearings, which define the bearing bore, the outside diameter (OD), the width, and the fillet sizes on the shaft and housing shoulders. The basic plan covers all ball and straight roller bearings in the metric sizes. The plan is quite flexible in that, for a given bore, there is an assortment of widths and outside diameters. Furthermore, the outside diameters selected are such that, for a particular outside diameter, one can usually find a variety of bearings having different bores and widths.

This basic ABMA plan is illustrated in Fig. 11–7. The bearings are identified by a two-digit number called the *dimension-series code*. The first number in the code is from the *width series*, 0, 1, 2, 3, 4, 5, and 6. The second number is from the *diameter series*

Table 11–2

Dimensions and Load Ratings for Single-Row 02-Series Deep-Groove and Angular-Contact Ball Bearings

Bore, mm	OD, mm	Width, mm	Fillet Radius, mm	Shoulder Diameter, mm d_S	Shoulder Diameter, mm d_H	Load Ratings, kN Deep Groove C_{10}	Load Ratings, kN Deep Groove C_0	Load Ratings, kN Angular Contact C_{10}	Load Ratings, kN Angular Contact C_0
10	30	9	0.6	12.5	27	5.07	2.24	4.94	2.12
12	32	10	0.6	14.5	28	6.89	3.10	7.02	3.05
15	35	11	0.6	17.5	31	7.80	3.55	8.06	3.65
17	40	12	0.6	19.5	34	9.56	4.50	9.95	4.75
20	47	14	1.0	25	41	12.7	6.20	13.3	6.55
25	52	15	1.0	30	47	14.0	6.95	14.8	7.65
30	62	16	1.0	35	55	19.5	10.0	20.3	11.0
35	72	17	1.0	41	65	25.5	13.7	27.0	15.0
40	80	18	1.0	46	72	30.7	16.6	31.9	18.6
45	85	19	1.0	52	77	33.2	18.6	35.8	21.2
50	90	20	1.0	56	82	35.1	19.6	37.7	22.8
55	100	21	1.5	63	90	43.6	25.0	46.2	28.5
60	110	22	1.5	70	99	47.5	28.0	55.9	35.5
65	120	23	1.5	74	109	55.9	34.0	63.7	41.5
70	125	24	1.5	79	114	61.8	37.5	68.9	45.5
75	130	25	1.5	86	119	66.3	40.5	71.5	49.0
80	140	26	2.0	93	127	70.2	45.0	80.6	55.0
85	150	28	2.0	99	136	83.2	53.0	90.4	63.0
90	160	30	2.0	104	146	95.6	62.0	106	73.5
95	170	32	2.0	110	156	108	69.5	121	85.0

(outside), 8, 9, 0, 1, 2, 3, and 4. Figure 11–7 shows the variety of bearings that may be obtained with a particular bore. Since the dimension-series code does not reveal the dimensions directly, it is necessary to resort to tabulations. The 02 series is used here as an example of what is available. See Table 11–2.

The housing and shaft shoulder diameters listed in the tables should be used whenever possible to secure adequate support for the bearing and to resist the maximum thrust loads (Fig. 11–8). Table 11–3 lists the dimensions and load ratings of some straight roller bearings.

To assist the designer in the selection of bearings, most of the manufacturers' handbooks contain data on bearing life for many classes of machinery, as well as information on load-application factors. Such information has been accumulated the hard way, that is, by experience, and the beginner designer should utilize this information until he or she gains enough experience to know when deviations are possible. Table 11–4 contains recommendations on bearing life for some classes of machinery. The load-application factors in Table 11–5 serve the same purpose as factors of safety; use them to increase the equivalent load before selecting a bearing.

Figure 11–8

Shaft and housing shoulder diameters d_S and d_H should be adequate to ensure good bearing support.

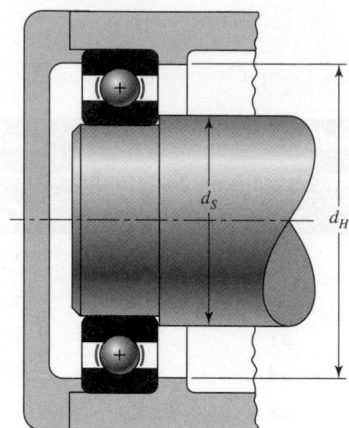

Table 11–3

Dimensions and Basic Load Ratings for Cylindrical Roller Bearings

Bore, mm	02-Series				03-Series			
	OD, mm	Width, mm	Load Rating, kN		OD, mm	Width, mm	Load Rating, kN	
			C_{10}	C_0			C_{10}	C_0
25	52	15	16.8	8.8	62	17	28.6	15.0
30	62	16	22.4	12.0	72	19	36.9	20.0
35	72	17	31.9	17.6	80	21	44.6	27.1
40	80	18	41.8	24.0	90	23	56.1	32.5
45	85	19	44.0	25.5	100	25	72.1	45.4
50	90	20	45.7	27.5	110	27	88.0	52.0
55	100	21	56.1	34.0	120	29	102	67.2
60	110	22	64.4	43.1	130	31	123	76.5
65	120	23	76.5	51.2	140	33	138	85.0
70	125	24	79.2	51.2	150	35	151	102
75	130	25	93.1	63.2	160	37	183	125
80	140	26	106	69.4	170	39	190	125
85	150	28	119	78.3	180	41	212	149
90	160	30	142	100	190	43	242	160
95	170	32	165	112	200	45	264	189
100	180	34	183	125	215	47	303	220
110	200	38	229	167	240	50	391	304
120	215	40	260	183	260	55	457	340
130	230	40	270	193	280	58	539	408
140	250	42	319	240	300	62	682	454
150	270	45	446	260	320	65	781	502

Table 11–4

Bearing-Life
Recommendations
for Various Classes
of Machinery

Type of Application	Life, kh
Instruments and apparatus for infrequent use	Up to 0.5
Aircraft engines	0.5–2
Machines for short or intermittent operation where service interruption is of minor importance	4–8
Machines for intermittent service where reliable operation is of great importance	8–14
Machines for 8-h service that are not always fully utilized	14–20
Machines for 8-h service that are fully utilized	20–30
Machines for continuous 24-h service	50–60
Machines for continuous 24-h service where reliability is of extreme importance	100–200

Table 11–5

Load-Application Factors

Type of Application	Load Factor
Precision gearing	1.0–1.1
Commercial gearing	1.1–1.3
Applications with poor bearing seals	1.2
Machinery with no impact	1.0–1.2
Machinery with light impact	1.2–1.5
Machinery with moderate impact	1.5–3.0

The static load rating is given in bearing catalog tables. It comes from the equations

$$C_0 = M n_b d_b^2 \qquad \text{(ball bearings)}$$

and

$$C_0 = M n_r l_c d \qquad \text{(roller bearings)}$$

where C_0 = bearing static load rating, lbf (kN)

n_b = number of balls

n_r = number of rollers

d_b = diameter of balls, in (mm)

d = diameter of rollers, in (mm)

l_c = length of contact line, in (mm)

and M takes on the values of which the following table is representative:

M	in and lbf	mm and kN
Radial ball	$1.78(10)^3$	$5.11(10)^3$
Ball thrust	$7.10(10)^3$	$20.4(10)^3$
Radial roller	$3.13(10)^3$	$8.99(10)^3$
Roller thrust	$14.2(10)^3$	$40.7(10)^3$

EXAMPLE 11–4   An SKF 6210 angular-contact ball bearing has an axial load F_a of 400 lbf and a radial load F_r of 500 lbf applied with the outer ring stationary. The basic static load rating C_0 is 4450 lbf and the basic load rating C_{10} is 7900 lbf. Estimate the L_{10} life at a speed of 720 rev/min.

Solution $V = 1$ and $F_a/C_0 = 400/4450 = 0.090$. Interpolate for e in Table 11–1:

F_a/C_0	e	
0.084	0.28	
0.090	e	from which $e = 0.285$
0.110	0.30	

$F_a/(V F_r) = 400/[(1)500] = 0.8 > 0.285$. Thus, interpolate for Y_2:

F_a/C_0	Y_2	
0.084	1.55	
0.090	Y_2	from which $Y_2 = 1.527$
0.110	1.45	

From Eq. (11–9),

$$F_e = X_2 V F_r + Y_2 F_a = 0.56(1)500 + 1.527(400) = 890.8 \text{ lbf}$$

With $L_D = L_{10}$ and $F_D = F_e$, solving Eq. (11–3) for L_{10} gives

Answer
$$L_{10} = \frac{60 L_R n_R}{60 n_D} \left(\frac{C_{10}}{F_e} \right)^a = \frac{10^6}{60(720)} \left(\frac{7900}{890.8} \right)^3 = 16\ 150 \text{ h}$$

We now know how to combine a steady radial load and a steady thrust load into an equivalent steady radial load F_e that inflicts the same damage per revolution as the radial–thrust combination.

11–7   Variable Loading

Bearing loads are frequently variable and occur in some identifiable patterns:

- Piecewise constant loading in a cyclic pattern
- Continuously variable loading in a repeatable cyclic pattern
- Random variation

Equation (11–1) can be written as

$$F^a L = \text{constant} = K \tag{a}$$

Note that F may already be an equivalent steady radial load for a radial–thrust load combination. Figure 11–9 is a plot of F^a as ordinate and L as abscissa for Eq. (a). If a load level of F_1 is selected and run to the failure criterion, then the area under the F_1-L_1 trace is numerically equal to K. The same is true for a load level F_2; that is, the area under the F_2-L_2 trace is numerically equal to K. The linear damage theory says that in the case of load level F_1, the area from $L = 0$ to $L = L_A$ does damage measured by $F_1^a L_A = D$.

Figure 11–9

Plot of F^a as ordinate and L as abscissa for $F^a L$ = constant. The linear damage hypothesis says that in the case of load F_1, the area under the curve from $L = 0$ to $L = L_A$ is a measure of the damage $D = F_1^a L_A$. The complete damage to failure is measured by $C_{10}^a L_B$.

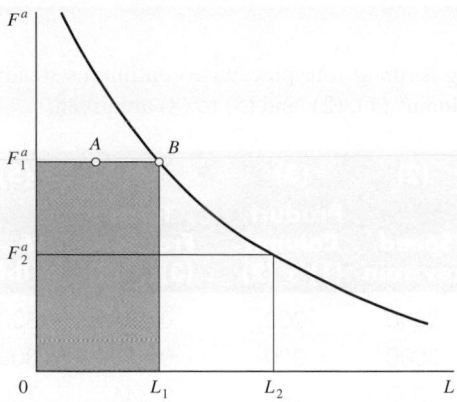

Figure 11–10

A three-part piecewise-continuous periodic loading cycle involving loads F_{e1}, F_{e2}, and F_{e3}. F_{eq} is the equivalent steady load inflicting the same damage when run for $l_1 + l_2 + l_3$ revolutions, doing the same damage D per period.

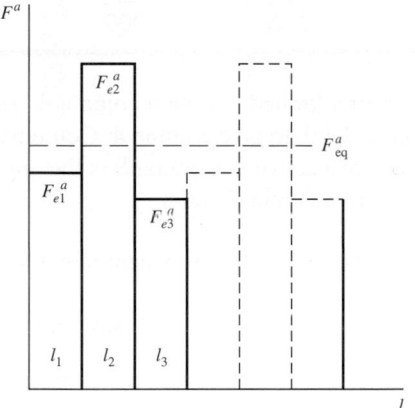

Consider the piecewise continuous cycle depicted in Fig. 11–10. The loads F_{ei} are equivalent steady radial loads for combined radial–thrust loads. The damage done by loads F_{e1}, F_{e2}, and F_{e3} is

$$D = F_{e1}^a l_1 + F_{e2}^a l_2 + F_{e3}^a l_3 \qquad (b)$$

where l_i is the number of revolutions at life L_i. The equivalent steady load F_{eq} when run for $l_1 + l_2 + l_3$ revolutions does the same damage D. Thus

$$D = F_{eq}^a (l_1 + l_2 + l_3) \qquad (c)$$

Equating Eqs. (b) and (c), and solving for F_{eq}, we get

$$F_{eq} = \left[\frac{F_{e1}^a l_1 + F_{e2}^a l_2 + F_{e3}^a l_3}{l_1 + l_2 + l_3} \right]^{1/a} = \left[\sum f_i F_{ei}^a \right]^{1/a} \qquad (11\text{–}10)$$

where f_i is the fraction of revolution run up under load F_{ei}. Since l_i can be expressed as $n_i t_i$, where n_i is the rotational speed at load F_{ei} and t_i is the duration of that speed, then it follows that

$$F_{eq} = \left[\frac{\sum n_i t_i F_{ei}^a}{\sum n_i t_i} \right]^{1/a} \qquad (11\text{–}11)$$

The character of the individual loads can change, so an application factor (a_f) can be prefixed to each F_{ei} as $(a_{fi} F_{ei})^a$; then Eq. (11–10) can be written

$$F_{eq} = \left[\sum f_i (a_{fi} F_{ei})^a \right]^{1/a} \qquad L_{eq} = \frac{K}{F_{eq}^a} \qquad (11\text{–}12)$$

EXAMPLE 11–5

A ball bearing is run at four piecewise continuous steady loads as shown in the following table. Columns (1), (2), and (5) to (8) are given.

(1) Time Fraction	(2) Speed, rev/min	(3) Product, Column (1) × (2)	(4) Turns Fraction, (3)/∑(3)	(5) F_{ri}, lbf	(6) F_{ai}, lbf	(7) F_{ei}, lbf	(8) a_{fi}	(9) $a_{fi}F_{ei}$, lbf
0.1	2000	200	0.077	600	300	794	1.10	873
0.1	3000	300	0.115	300	300	626	1.25	795
0.3	3000	900	0.346	750	300	878	1.10	966
0.5	2400	1200	0.462	375	300	668	1.25	835
		2600	1.000					

Columns 1 and 2 are multiplied to obtain column 3. The column 3 entry is divided by the sum of column 3, 2600, to give column 4. Columns 5, 6, and 7 are the radial, axial, and equivalent loads respectively. Column 8 is the appropriate application factor. Column 9 is the product of columns 7 and 8.

Solution

From Eq. (11–10), with $a = 3$, the equivalent radial load F_e is

Answer

$$F_e = \left[0.077(873)^3 + 0.115(795)^3 + 0.346(966)^3 + 0.462(835)^3\right]^{1/3} = 884 \text{ lbf}$$

Sometimes the question after several levels of loading is: How much life is left if the next level of stress is held until failure? Failure occurs under the linear damage hypothesis when the damage D equals the constant $K = F^a L$. Taking the first form of Eq. (11–10), we write

$$F_{eq}^a L_{eq} = F_{e1}^a l_1 + F_{e2}^a l_2 + F_{e3}^a l_3$$

and note that

$$K = F_{e1}^a L_1 = F_{e2}^a L_2 = F_{e3}^a L_3$$

and K also equals

$$K = F_{e1}^a l_1 + F_{e2}^a l_2 + F_{e3}^a l_3 = \frac{K}{L_1}l_1 + \frac{K}{L_2}l_2 + \frac{K}{L_3}l_3 = K\sum \frac{l_i}{L_i}$$

From the outer parts of the preceding equation we obtain

$$\sum \frac{l_i}{L_i} = 1 \tag{11–13}$$

This equation was advanced by Palmgren in 1924, and again by Miner in 1945. See Eq. (7–58).

The second kind of load variation mentioned is continuous, periodic variation, depicted by Fig. 11–11. The differential damage done by F^a during rotation through the angle $d\theta$ is

$$dD = F^a d\theta$$

Figure 11–11

A continuous load variation of
a cyclic nature whose period
is ϕ.

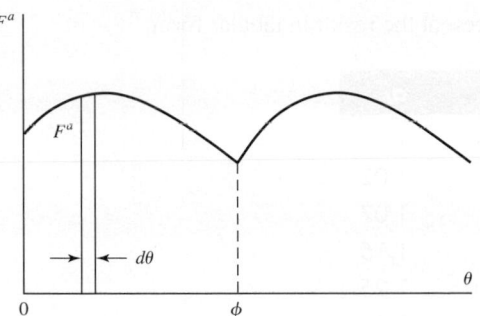

An example of this would be a cam whose bearings rotate with the cam through the
angle $d\theta$. The total damage during a complete cam rotation is given by

$$D = \int dD = \int_0^\phi F^a d\theta = F_{eq}^a \phi$$

from which, solving for the equivalent load, we obtain

$$F_{eq} = \left[\frac{1}{\phi} \int_0^\phi F^a d\theta \right]^{1/a} \qquad L_{eq} = \frac{K}{F_{eq}^a} \qquad (11\text{–}14)$$

The value of ϕ is often 2π, although other values occur. Numerical integration is often
useful to carry out the indicated integration, particularly when a is not an integer and
trigonometric functions are involved. We have now learned how to find the steady
equivalent load that does the same damage as a continuously varying cyclic load.

EXAMPLE 11–6

The operation of a particular rotary pump involves a power demand of $P = \bar{P} + A'
\sin\theta$ where $\bar{P}$ is the average power. The bearings feel the same variation as $F =
\bar{F} + A\sin\theta$. Develop an application factor a_f for this application of ball bearings.

Solution

From Eq. (11–14), with $a = 3$,

$$F_{eq} = \left(\frac{1}{2\pi} \int_0^{2\pi} F^a d\theta \right)^{1/a} = \left(\frac{1}{2\pi} \int_0^{2\pi} (\bar{F} + A\sin\theta)^3 d\theta \right)^{1/3}$$

$$= \left[\frac{1}{2\pi} \left(\int_0^{2\pi} \bar{F}^3 d\theta + 3\bar{F}^2 A \int_0^{2\pi} \sin\theta \, d\theta + 3\bar{F} A^2 \int_0^{2\pi} \sin^2\theta d\theta \right. \right.$$

$$\left. \left. + A^3 \int_0^{2\pi} \sin^3\theta \, d\theta \right) \right]^{1/3}$$

$$F_{eq} = \left[\frac{1}{2\pi} (2\pi \bar{F}^3 + 0 + 3\pi \bar{F} A^2 + 0) \right]^{1/3} = \bar{F} \left[1 + \frac{3}{2} \left(\frac{A}{\bar{F}} \right)^2 \right]^{1/3}$$

In terms of $\bar{F}$, the application factor is

$$a_f = \left[1 + \frac{3}{2} \left(\frac{A}{\bar{F}} \right)^2 \right]^{1/3}$$

We can present the result in tabular form:

A/F̄	a_f
0	1
0.2	1.02
0.4	1.07
0.6	1.15
0.8	1.25
1.0	1.36

11–8 Selection of Ball and Cylindrical Roller Bearings

We have enough information concerning the loading of rolling-contact ball and roller bearings to develop the steady equivalent radial load that will do as much damage to the bearing as the existing loading. Now let's put it to work.

EXAMPLE 11–7

The second shaft on a parallel-shaft 25-hp foundry crane speed reducer contains a helical gear with a pitch diameter of 8.08 in. Helical gears transmit components of force in the tangential, radial, and axial directions (see Chap. 13). The components of the gear force transmitted to the second shaft are shown in Fig. 11–12, at point A. The bearing reactions at C and D, assuming simple-supports, are also shown. A ball bearing is to be selected for location C to accept the thrust, and a cylindrical roller bearing is to be utilized

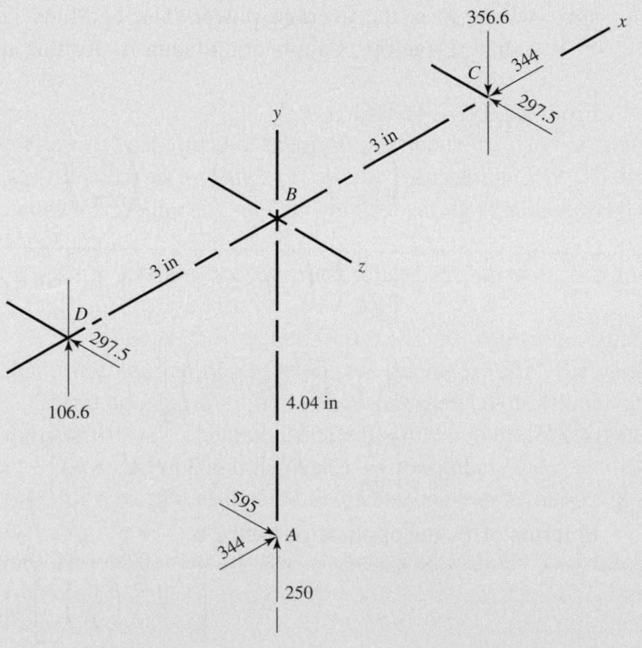

Figure 11–12

Forces in pounds applied to the second shaft of the helical gear speed reducer of Ex. 11–7.

at location D. The life goal of the speed reducer is 10 kh, with a reliability factor for the ensemble of all four bearings (both shafts) to equal or exceed 0.96 for the Weibull parameters of Ex. 11–3. The application factor is to be 1.2.

(a) Select the roller bearing for location D.

(b) Select the ball bearing (angular contact) for location C, assuming the inner ring rotates.

Solution   The torque transmitted is $T = 595(4.04) = 2404$ lbf · in. The speed at the rated horsepower, given by Eq. (4–40) is

$$n_D = \frac{63\,025H}{T} = \frac{63\,025(25)}{2404} = 655.4 \text{ rev/min}$$

The radial load at D is $\sqrt{106.6^2 + 297.5^2} = 316.0$ lbf, and the radial load at C is $\sqrt{356.6^2 + 297.5^2} = 464.4$ lbf. The individual bearing reliabilities, if equal, must be at least $\sqrt[4]{0.96} = 0.98985 \doteq 0.99$. The dimensionless design life for both bearings is

$$x_D = \frac{L}{L_{10}} = \frac{60L_D n_D}{60L_R n_R} = \frac{60(10\,000)655.4}{10^6} = 393.2$$

(a) From Eq. (11–7), the Weibull parameters of Ex. 11–3, an application factor of 1.2, and $a = 10/3$ for the roller bearing at D, the catalog rating should be equal to or greater than

$$C_{10} = a_f F_D \left[\frac{x_D}{x_0 + (\theta - x_0)(1 - R_D)^{1/b}} \right]^{1/a}$$

$$= 1.2(316.0) \left[\frac{393.2}{0.02 + 4.439(1 - 0.99)^{1/1.483}} \right]^{3/10} = 3591 \text{ lbf} = 16.0 \text{ kN}$$

Answer   The absence of a thrust component makes the selection procedure simple. Choose a 02-25 mm series, or a 03-25 mm series cylindrical roller bearing from Table 11–3.

(b) The ball bearing at C involves a thrust component. This selection procedure requires an iterative procedure. Assuming $F_a/(VF_r) > e$,

1   Choose Y_2 from Table 11–1.
2   Find C_{10}.
3   Tentatively identify a suitable bearing from Table 11–2, note C_0.
4   Using F_a/C_0 enter Table 11–1 to obtain a new value of Y_2.
5   Find C_{10}.
6   If the same bearing is obtained, stop.
7   If not, take next bearing and go to step 4.

As a first approximation, take the middle entry from Table 11–1:

$$X_2 = 0.56 \qquad Y_2 = 1.63.$$

From Eq. (11–8b), with $V = 1$,

$$\frac{F_e}{VF_r} = X + \frac{Y}{V}\frac{F_a}{F_r} = 0.56 + 1.63\frac{344}{(1)464.4} = 1.77$$

$$F_e = 1.77VF_r = 1.77(1)464.4 = 822 \text{ lbf} \qquad \text{or} \qquad 3.66 \text{ kN}$$

From Eq. (11–7), with $a = 3$,

$$C_{10} = 1.2(3.66)\left[\frac{393.2}{0.02 + 4.439(1 - 0.99)^{1/1.483}}\right]^{1/3} = 53.4 \text{ kN}$$

From Table 11–2, angular-contact bearing 02-60 mm has $C_{10} = 55.9$ kN. C_0 is 35.5 kN. Step 4 becomes, with F_a in kN,

$$\frac{F_a}{C_0} = \frac{344(4.45)10^{-3}}{35.5} = 0.0431$$

which makes e from Table 11–1 approximately 0.24. Now $F_a/[VF_r] = 344/[(1)464.4] = 0.74$, which is greater than 0.24, so we find Y_2 by interpolation:

F_a/C_0	Y_2	
0.042	1.85	
0.043	Y_2	from which $Y_2 = 1.84$
0.056	1.71	

From Eq. (11–8b),

$$\frac{F_e}{VF_r} = 0.56 + 1.84\frac{344}{464.4} = 1.92$$

$$F_e = 1.92VF_r = 1.92(1)464.4 = 892 \text{ lbf} \quad \text{or} \quad 3.97 \text{ kN}$$

The prior calculation for C_{10} changes only in F_e, so

$$C_{10} = \frac{3.97}{3.66}53.4 = 57.9 \text{ kN}$$

From Table 11–2 an angular contact bearing 02-65 mm has $C_{10} = 63.7$ kN and C_0 of 41.5 kN. Again,

$$\frac{F_a}{C_0} = \frac{344(4.45)10^{-3}}{41.5} = 0.0369$$

making e approximately 0.23. Now from before, $F_a/VF_r = 0.74$, which is greater than 0.23. We find Y_2 again by interpolation:

F_a/C_0	Y_2	
0.028	1.99	
0.0369	Y_2	from which $Y_2 = 1.90$
0.042	1.85	

From Eq. (11–8b),

$$\frac{F_e}{VF_r} = 0.56 + 1.90\frac{344}{464.4} = 1.967$$

$$F_e = 1.967VF_r = 1.967(1)464.4 = 913.5 \text{ lbf} \quad \text{or} \quad 4.065 \text{ kN}$$

The prior calculation for C_{10} changes only in F_e, so

$$C_{10} = \frac{4.07}{3.66} 53.4 = 59.4 \text{ kN}$$

Answer   From Table 11–2 an angular-contact 02-65 mm is still selected, so the iteration is complete.

11–9   ## Selection of Tapered Roller Bearings

Tapered roller bearings have a number of features that make them complicated. As we address the differences between tapered roller and ball and cylindrical roller bearings, note that the underlying fundamentals are the same, but that there are differences in detail. Moreover, bearing and cup combinations are not necessarily priced in proportion to capacity. Any catalog displays a mix of high-production, low-production, and successful special-order designs. Bearing suppliers have computer programs that will take your problem descriptions, give intermediate design assessment information, and list a number of satisfactory cup-and-cone combinations in order of decreasing cost. Company sales offices provide access to comprehensive engineering services to help designers select and apply their bearings. At a large original equipment manufacturer's plant, there may be a resident bearing company representative.

Take a few minutes to go to your department's design library and look at a bearing supplier's engineering catalog, such as The Timken Company's *Bearing Selection Handbook—Revised* (1986). There is a log of engineering information and detail, based on long and successful experience. All we can do here is introduce the vocabulary, show congruence to fundamentals that were learned earlier, offer examples, and develop confidence. Finally, problems should reinforce the learning experience.

Form

The four components of a tapered roller bearing assembly are the

- Cone (inner ring)
- Cup (outer ring)
- Tapered rollers
- Cage (spacer-retainer)

The assembled bearing consists of two separable parts: (1) the cone assembly: the cone, the rollers, and the cage; and (2) the cup. Bearings can be made as single-row, two-row, four-row, and thrust-bearing assemblies. Additionally, auxiliary components such as spacers and closures can be used.

A tapered roller bearing can carry both radial and thrust (axial) loads, or any combination of the two. However, even when an external thrust load is not present, the radial load will induce a thrust reaction within the bearing because of the taper. To avoid the separation of the races and the rollers, this thrust must be resisted by an equal and opposite force. One way of generating this force is to always use at least two tapered roller bearings on a shaft. Two bearings can be mounted with the cone backs facing each other, in a configuration called *direct mounting,* or with the cone fronts facing each other, in what is called *indirect mounting.* Figure 11–13 shows the nomenclature of a tapered roller bearing, and the point G through which radial and axial components of load act.

Figure 11-13

Nomenclature of a tapered roller bearing. Point G is the location of the effective load center; use this point to estimate the radial bearing load. *(Courtesy of The Timken Company.)*

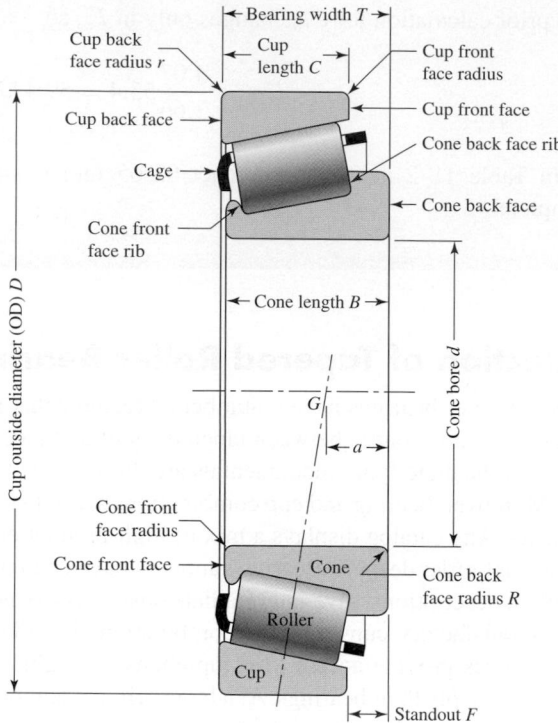

A radial load will induce a thrust reaction. The *load zone* includes about half the rollers and subtends an angle of approximately 180°. Using the symbol $F_{a(180)}$ for the induced thrust load from a radial load with a 180° load zone, Timken provides the equation

$$F_{a(180)} = \frac{0.47 F_r}{K} \tag{11-15}$$

where the K factor is geometry-specific, coming from the relationship

$$K = 0.389 \cot \alpha$$

where α is half the included cup angle. The K factor is the ratio of the radial load rating to the thrust load rating. The K factor can be first approximated with 1.5 for a radial bearing and 0.75 for a steep angle bearing in the preliminary selection process. After a possible bearing is identified, the exact value of K for each bearing can be found in the *Bearing Selection Handbook—Revised* (1986) in the case of Timken bearings.

Notation

The catalog rating C corresponding to 90 percent reliability was denoted C_{10} earlier in the chapter, the subscript 10 denoting 10 percent failure level. Timken denoted its catalog ratings as C_{90}, the subscript 90 standing for "at 90 million revolutions." The failure fraction is still 10 percent (90 percent reliability). This should produce no difficulties since Timken's catalog ratings for radial and thrust loads display neither C_{90} nor $C_{a(90)}$ at the head of the columns. See Fig. 11–15, which is a reproduction of two Timken catalog pages.

Figure 11–14

Comparison of mounting stability between indirect and direct mountings. *(Courtesy of The Timken Company.)*

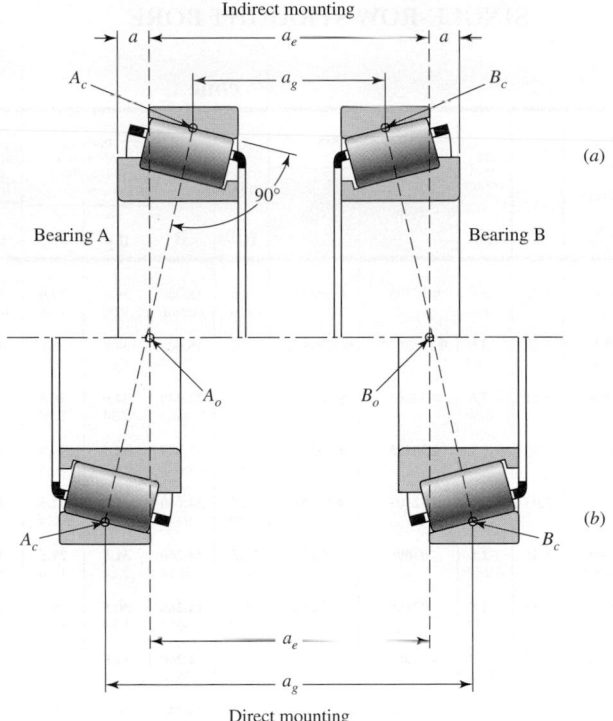

Indirect mounting

Bearing A

Bearing B

(a)

(b)

Direct mounting

Location of Reactions

Figure 11–14 shows a pair of tapered roller bearings mounted directly (*b*) and indirectly (*a*) with the bearing reaction locations A_0 and B_0 shown for the shaft. For the shaft as a beam, the span is a_e, the effective spread. It is through points A_0 and B_0 that the radial loads act perpendicular to the shaft axis, and the thrust loads act along the shaft axis. The geometric spread a_g for the direct mounting is greater than for the indirect mounting. With indirect mounting the bearings are closer together compared to the direct mounting; however, the system stability is the same (a_e is the same in both cases). Thus direct and indirect mounting involve space and compactness needed or desired, but with the same system stability.

Relating Load, Life, and Reliability

Recall Eq. (11–7) for a three-parameter Weibull model,

$$C_{10} = F_D \left[\frac{x_D}{x_0 + (\theta - x_0)\,(1 - R_D)^{1/b}} \right]^{1/a}$$

Solving for x_D gives

$$x_D = x_0 + (\theta - x_0)(1 - R_D)^{1/b} \left(\frac{C_{10}}{F_D} \right)^a$$

SINGLE-ROW STRAIGHT BORE

bore	outside diameter	width	rating at 500 rpm for 3000 hours L₁₀		fac-tor	eff. load center	part numbers		cone				cup			
			one-row radial	thrust			cone	cup	max shaft fillet radius	width	backing shoulder diameters		max housing fillet radius	width	backing shoulder diameters	
d	**D**	**T**	**N lbf**	**N lbf**	**K**	**a②**			**R①**	**B**	**d_b**	**d_a**	**r①**	**C**	**D_b**	**D_a**
25.000 0.9843	52.000 2.0472	16.250 0.6398	8190 1840	5260 1180	1.56	−3.6 −0.14	◆30205	◆30205	1.0 0.04	15.000 0.5906	30.5 1.20	29.0 1.14	1.0 0.04	13.000 0.5118	46.0 1.81	48.5 1.91
25.000 0.9843	52.000 2.0472	19.250 0.7579	9520 2140	9510 2140	1.00	−3.0 −0.12	◆32205-B	◆32205-B	1.0 0.04	18.000 0.7087	34.0 1.34	31.0 1.22	1.0 0.04	15.000 0.5906	43.5 1.71	49.5 1.95
25.000 0.9843	52.000 2.0472	22.000 0.8661	13200 2980	7960 1790	1.66	−7.6 −0.30	◆33205	◆33205	1.0 0.04	22.000 0.8661	34.0 1.34	30.5 1.20	1.0 0.04	18.000 0.7087	44.5 1.75	49.0 1.93
25.000 0.9843	62.000 2.4409	18.250 0.7185	13000 2930	6680 1500	1.95	−5.1 −0.20	◆30305	◆30305	1.5 0.06	17.000 0.6693	32.5 1.28	30.0 1.18	1.5 0.06	15.000 0.5906	55.0 2.17	57.0 2.24
25.000 0.9843	62.000 2.4409	25.250 0.9941	17400 3910	8930 2010	1.95	−9.7 −0.38	◆32305	◆32305	1.5 0.06	24.000 0.9449	35.0 1.38	31.5 1.24	1.5 0.06	20.000 0.7874	54.0 2.13	57.0 2.24
25.159 0.9905	50.005 1.9687	13.495 0.5313	6990 1570	4810 1080	1.45	−2.8 −0.11	07096	07196	1.5 0.06	14.260 0.5614	31.5 1.24	29.5 1.16	1.0 0.04	9.525 0.3750	44.5 1.75	47.0 1.85
25.400 1.0000	50.005 1.9687	13.495 0.5313	6990 1570	4810 1080	1.45	−2.8 −0.11	07100	07196	1.0 0.04	14.260 0.5614	30.5 1.20	29.5 1.16	1.0 0.04	9.525 0.3750	44.5 1.75	47.0 1.85
25.400 1.0000	50.005 1.9687	13.495 0.5313	6990 1570	4810 1080	1.45	−2.8 −0.11	07100-S	07196	1.5 0.06	14.260 0.5614	31.5 1.24	29.5 1.16	1.0 0.04	9.525 0.3750	44.5 1.75	47.0 1.85
25.400 1.0000	50.292 1.9800	14.224 0.5600	7210 1620	4620 1040	1.56	−3.3 −0.13	L44642	L44610	3.5 0.14	14.732 0.5800	36.0 1.42	29.5 1.16	1.3 0.05	10.668 0.4200	44.5 1.75	47.0 1.85
25.400 1.0000	50.292 1.9800	14.224 0.5600	7210 1620	4620 1040	1.56	−3.3 −0.13	L44643	L44610	1.3 0.05	14.732 0.5800	31.5 1.24	29.5 1.16	1.3 0.05	10.668 0.4200	44.5 1.75	47.0 1.85
25.400 1.0000	51.994 2.0470	15.011 0.5910	6990 1570	4810 1080	1.45	−2.8 −0.11	07100	07204	1.0 0.04	14.260 0.5614	30.5 1.20	29.5 1.16	1.3 0.05	12.700 0.5000	45.0 1.77	48.0 1.89
25.400 1.0000	56.896 2.2400	19.368 0.7625	10900 2450	5740 1290	1.90	−6.9 −0.27	1780	1729	0.8 0.03	19.837 0.7810	30.5 1.20	30.0 1.18	1.3 0.05	15.875 0.6250	49.0 1.93	51.0 2.01
25.400 1.0000	57.150 2.2500	19.431 0.7650	11700 2620	10900 2450	1.07	−3.0 −0.12	M84548	M84510	1.5 0.06	19.431 0.7650	36.0 1.42	33.0 1.30	1.5 0.06	14.732 0.5800	48.5 1.91	54.0 2.13
25.400 1.0000	58.738 2.3125	19.050 0.7500	11600 2610	6560 1470	1.77	−5.8 −0.23	1986	1932	1.3 0.05	19.355 0.7620	32.5 1.28	30.5 1.20	1.3 0.05	15.080 0.5937	52.0 2.05	54.0 2.13
25.400 1.0000	59.530 2.3437	23.368 0.9200	13900 3140	13000 2930	1.07	−5.1 −0.20	M84249	M84210	0.8 0.03	23.114 0.9100	36.0 1.42	32.5 1.27	1.5 0.06	18.288 0.7200	49.5 1.95	56.0 2.20
25.400 1.0000	60.325 2.3750	19.842 0.7812	11000 2480	6550 1470	1.69	−5.1 −0.20	15578	15523	1.3 0.05	17.462 0.6875	32.5 1.28	30.5 1.20	1.5 0.06	15.875 0.6250	51.0 2.01	54.0 2.13
25.400 1.0000	61.912 2.4375	19.050 0.7500	12100 2730	7280 1640	1.67	−5.8 −0.23	15101	15243	0.8 0.03	20.638 0.8125	32.5 1.28	31.5 1.24	2.0 0.08	14.288 0.5625	54.0 2.13	58.0 2.28
25.400 1.0000	62.000 2.4409	19.050 0.7500	12100 2730	7280 1640	1.67	−5.8 −0.23	15100	15245	3.5 0.14	20.638 0.8125	38.0 1.50	31.5 1.24	1.3 0.05	14.288 0.5625	55.0 2.17	58.0 2.28
25.400 1.0000	62.000 2.4409	19.050 0.7500	12100 2730	7280 1640	1.67	−5.8 −0.23	15101	15245	0.8 0.03	20.638 0.8125	32.5 1.28	31.5 1.24	1.3 0.05	14.288 0.5625	55.0 2.17	58.0 2.28
25.400 1.0000	62.000 2.4409	19.050 0.7500	12100 2730	7280 1640	1.67	−5.8 −0.23	15102	15245	1.5 0.06	20.638 0.8125	34.0 1.34	31.5 1.24	1.3 0.05	14.288 0.5625	55.0 2.17	58.0 2.28
25.400 1.0000	62.000 2.4409	20.638 0.8125	12100 2730	7280 1640	1.67	−5.8 −0.23	15101	15244	0.8 0.03	20.638 0.8125	32.5 1.28	31.5 1.24	1.3 0.05	15.875 0.6250	55.0 2.17	58.0 2.28
25.400 1.0000	63.500 2.5000	20.638 0.8125	12100 2730	7280 1640	1.67	−5.8 −0.23	15101	15250	0.8 0.03	20.638 0.8125	32.5 1.28	31.5 1.24	1.3 0.05	15.875 0.6250	56.0 2.20	59.0 2.32
25.400 1.0000	63.500 2.5000	20.638 0.8125	12100 2730	7280 1640	1.67	−5.8 −0.23	15101	15250X	0.8 0.03	20.638 0.8125	32.5 1.28	31.5 1.24	1.5 0.06	15.875 0.6250	55.0 2.17	59.0 2.32
25.400 1.0000	64.292 2.5312	21.433 0.8438	14500 3250	13500 3040	1.07	−3.3 −0.13	M86643	M86610	1.5 0.06	21.433 0.8438	38.0 1.50	36.5 1.44	1.5 0.06	16.670 0.6563	54.0 2.13	61.0 2.40

Figure 11–15

Catalog entry of single-row straight-bore Timken roller bearings, in part. (*Courtesy of The Timken Company.*)

SINGLE-ROW STRAIGHT BORE

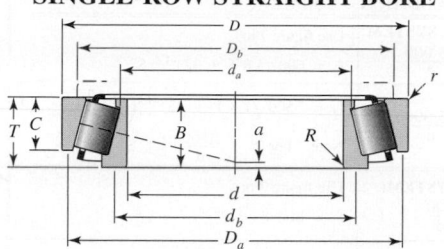

bore d	outside diameter D	width T	rating at 500 rpm for 3000 hours L₁₀ — one-row radial N lbf	thrust N lbf	factor K	eff. load center a②	part numbers cone	cup	cone max shaft fillet radius R①	width B	backing shoulder diameters d_b	d_a	cup max housing fillet radius r①	width C	backing shoulder diameters D_b	D_a
25.400 / 1.0000	65.088 / 2.5625	22.225 / 0.8750	13100 / 2950	16400 / 3690	0.80	−2.3 / −0.09	23100	23256	1.5 / 0.06	21.463 / 0.8450	39.0 / 1.54	34.5 / 1.36	1.5 / 0.06	15.875 / 0.6250	53.0 / 2.09	63.0 / 2.48
25.400 / 1.0000	66.421 / 2.6150	23.812 / 0.9375	18400 / 4140	8000 / 1800	2.30	−9.4 / −0.37	2687	2631	1.3 / 0.05	25.433 / 1.0013	33.5 / 1.32	31.5 / 1.24	1.3 / 0.05	19.050 / 0.7500	58.0 / 2.28	60.0 / 2.36
25.400 / 1.0000	68.262 / 2.6875	22.225 / 0.8750	15300 / 3440	10900 / 2450	1.40	−5.1 / −0.20	02473	02420	0.8 / 0.03	22.225 / 0.8750	34.5 / 1.36	33.5 / 1.32	1.5 / 0.06	17.462 / 0.6875	59.0 / 2.32	63.0 / 2.48
25.400 / 1.0000	72.233 / 2.8438	25.400 / 1.0000	18400 / 4140	17200 / 3870	1.07	−4.6 / −0.18	HM88630	HM88610	0.8 / 0.03	25.400 / 1.0000	39.5 / 1.56	39.5 / 1.56	2.3 / 0.09	19.842 / 0.7812	60.0 / 2.36	69.0 / 2.72
25.400 / 1.0000	72.626 / 2.8593	30.162 / 1.1875	22700 / 5110	13000 / 2910	1.76	−10.2 / −0.40	3189	3120	0.8 / 0.03	29.997 / 1.1810	35.5 / 1.40	35.0 / 1.38	3.3 / 0.13	23.812 / 0.9375	61.0 / 2.40	67.0 / 2.64
26.157 / 1.0298	62.000 / 2.4409	19.050 / 0.7500	12100 / 2730	7280 / 1640	1.67	−5.8 / −0.23	15103	15245	0.8 / 0.03	20.638 / 0.8125	33.0 / 1.30	32.5 / 1.28	1.3 / 0.05	14.288 / 0.5625	55.0 / 2.17	58.0 / 2.28
26.162 / 1.0300	63.100 / 2.4843	23.812 / 0.9375	18400 / 4140	8000 / 1800	2.30	−9.4 / −0.37	2682	2630	1.5 / 0.06	25.433 / 1.0013	34.5 / 1.36	32.0 / 1.26	0.8 / 0.03	19.050 / 0.7500	57.0 / 2.24	59.0 / 2.32
26.162 / 1.0300	66.421 / 2.6150	23.812 / 0.9375	18400 / 4140	8000 / 1800	2.30	−9.4 / −0.37	2682	2631	1.5 / 0.06	25.433 / 1.0013	34.5 / 1.36	32.0 / 1.26	1.3 / 0.05	19.050 / 0.7500	58.0 / 2.28	60.0 / 2.36
26.975 / 1.0620	58.738 / 2.3125	19.050 / 0.7500	11600 / 2610	6560 / 1470	1.77	−5.8 / −0.23	1987	1932	0.8 / 0.03	19.355 / 0.7620	32.5 / 1.28	31.5 / 1.24	1.3 / 0.05	15.080 / 0.5937	52.0 / 2.05	54.0 / 2.13
† 26.988 / † 1.0625	50.292 / 1.9800	14.224 / 0.5600	7210 / 1620	4620 / 1040	1.56	−3.3 / −0.13	L44649	L44610	3.5 / 0.14	14.732 / 0.5800	37.5 / 1.48	31.0 / 1.22	1.3 / 0.05	10.668 / 0.4200	44.5 / 1.75	47.0 / 1.85
† 26.988 / † 1.0625	60.325 / 2.3750	19.842 / 0.7812	11000 / 2480	6550 / 1470	1.69	−5.1 / −0.20	15580	15523	3.5 / 0.14	17.462 / 0.6875	38.5 / 1.52	32.0 / 1.26	1.5 / 0.06	15.875 / 0.6250	51.0 / 2.01	54.0 / 2.13
† 26.988 / † 1.0625	62.000 / 2.4409	19.050 / 0.7500	12100 / 2730	7280 / 1640	1.67	−5.8 / −0.23	15106	15245	0.8 / 0.03	20.638 / 0.8125	33.5 / 1.32	33.0 / 1.30	1.3 / 0.05	14.288 / 0.5625	55.0 / 2.17	58.0 / 2.28
† 26.988 / † 1.0625	66.421 / 2.6150	23.812 / 0.9375	18400 / 4140	8000 / 1800	2.30	−9.4 / −0.37	2688	2631	1.5 / 0.06	25.433 / 1.0013	35.0 / 1.38	33.0 / 1.30	1.3 / 0.05	19.050 / 0.7500	58.0 / 2.28	60.0 / 2.36
28.575 / 1.1250	56.896 / 2.2400	19.845 / 0.7813	11600 / 2610	6560 / 1470	1.77	−5.8 / −0.23	1985	1930	0.8 / 0.03	19.355 / 0.7620	34.0 / 1.34	33.5 / 1.32	0.8 / 0.03	15.875 / 0.6250	51.0 / 2.01	54.0 / 2.11
28.575 / 1.1250	57.150 / 2.2500	17.462 / 0.6875	11000 / 2480	6550 / 1470	1.69	−5.1 / −0.20	15590	15520	3.5 / 0.14	17.462 / 0.6875	39.5 / 1.56	33.5 / 1.32	1.5 / 0.06	13.495 / 0.5313	51.0 / 2.01	53.0 / 2.09
28.575 / 1.1250	58.738 / 2.3125	19.050 / 0.7500	11600 / 2610	6560 / 1470	1.77	−5.8 / −0.23	1985	1932	0.8 / 0.03	19.355 / 0.7620	34.0 / 1.34	33.5 / 1.32	1.3 / 0.05	15.080 / 0.5937	52.0 / 2.05	54.0 / 2.13
28.575 / 1.1250	58.738 / 2.3125	19.050 / 0.7500	11600 / 2610	6560 / 1470	1.77	−5.8 / −0.23	1988	1932	3.5 / 0.14	19.355 / 0.7620	39.5 / 1.56	33.5 / 1.32	1.3 / 0.05	15.080 / 0.5937	52.0 / 2.05	54.0 / 2.13
28.575 / 1.1250	60.325 / 2.3750	19.842 / 0.7812	11000 / 2480	6550 / 1470	1.69	−5.1 / −0.20	15590	15523	3.5 / 0.14	17.462 / 0.6875	39.5 / 1.56	33.5 / 1.32	1.5 / 0.06	15.875 / 0.6250	51.0 / 2.01	54.0 / 2.13
28.575 / 1.1250	60.325 / 2.3750	19.845 / 0.7813	11600 / 2610	6560 / 1470	1.77	−5.8 / −0.23	1985	1931	0.5 / 0.03	19.355 / 0.7620	34.0 / 1.34	33.5 / 1.32	1.3 / 0.05	15.875 / 0.6250	52.0 / 2.05	55.0 / 2.17

① These maximum fillet radii will be cleared by the bearing corners.
② Minus value indicates center is inside cone backface.
† For standard class **ONLY**, the maximum metric size is a whole millimetre value.
* For "J" part tolerances—see metric tolerances, page 73, and fitting practice, page 65.
◆ ISO cone and cup combinations are designated with a common part number and should be purchased as an assembly. For ISO bearing tolerances—see metric tolerances, page 73, and fitting practice, page 65.

Figure 11–15

(Continued)

Figure 11–16

Temperature factor f_T as a function of speed and bearing operating temperature. For speed S less than 15 000/d use equation shown in inset when d is bearing bore in millimeters (less than 600/d when bearing bore is in inches). *(Courtesy of The Timken Company.)*

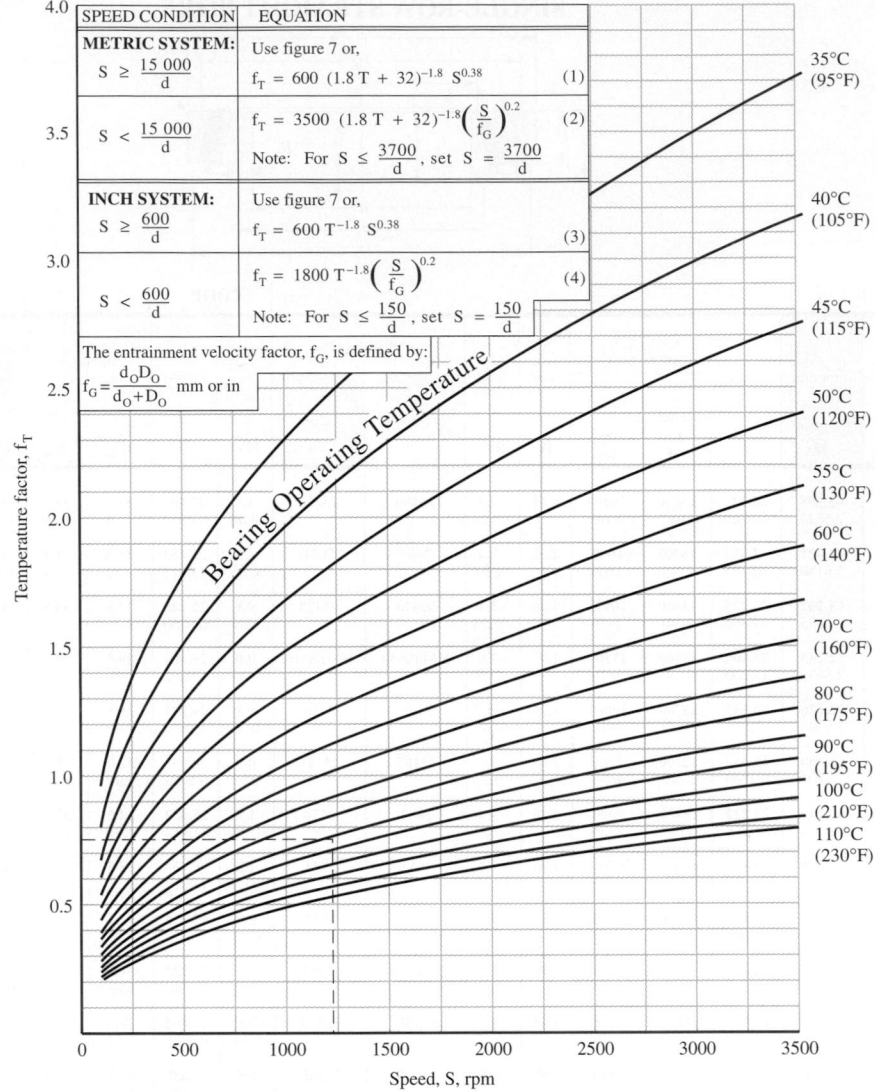

Timken uses a two-parameter Weibull model with $x_0 = 0$, $\theta = 4.48$, and $b = \frac{3}{2}$. So, x_D for a Timken tapered roller bearing, with $a = \frac{10}{3}$, is

$$x_D = 4.48(1 - R_D)^{2/3} \left(\frac{C_{10}}{F_D}\right)^{10/3}$$

Now x_D is the desired life in multiples of rating life. The Timken design life equation is written in terms of revolutions, and for Timken's $L_{10} = 90(10^6)$ revolutions, we can express $x_D = L_D/90(10^6)$. From the equation for x_D above,

$$L_D = 4.48(1 - R_D)^{2/3} \left(\frac{C_{10}}{F_D}\right)^{10/3} 90(10^6)$$

where L_D is in revolutions.

Timken writes this equation as

$$L_D = a_1 a_2 a_3 a_4 \left(\frac{C_{10}}{F_D}\right)^{10/3} 90(10^6) \qquad (11\text{–}16)$$

4.48 $(1 - R_D)^{2/3}$ ———

bearing material ———

$a_4 = 1$ (spall size is 0.01 in²)

$a_3 = a_{3k} a_{3l} a_{3m}$

alignment

lubricant $a_{3l} = f_T f_v$

load zone

Temperature factor f_T can be found in Fig. 11–16, and viscosity factor f_v can be found in Fig. 11–17.

For the usual case, $a_2 = a_{3k} = a_{3m} = 1$, and solving the preceding equation for C_{10} gives

$$C_{10} = a_f P \left[\frac{I_{,D}}{4.48 f_T f_v (1 - R_D)^{2/3} 90(10^6)}\right]^{3/10} \qquad (L_D \text{ in revolutions}) \quad (11\text{–}17)$$

where F_D is replaced by $a_f P$. The load P is the dynamic equivalent load of the combination F_r and F_a of Sec. 11–6. The particular values of X and Y are given in Table 11–6,

---

Figure 11-17

Viscosity factor f_v as a function of oil viscosity. (This graph applies to petroleum oil with a viscosity index of approximately 90.) *(Courtesy of The Timken Company.)*

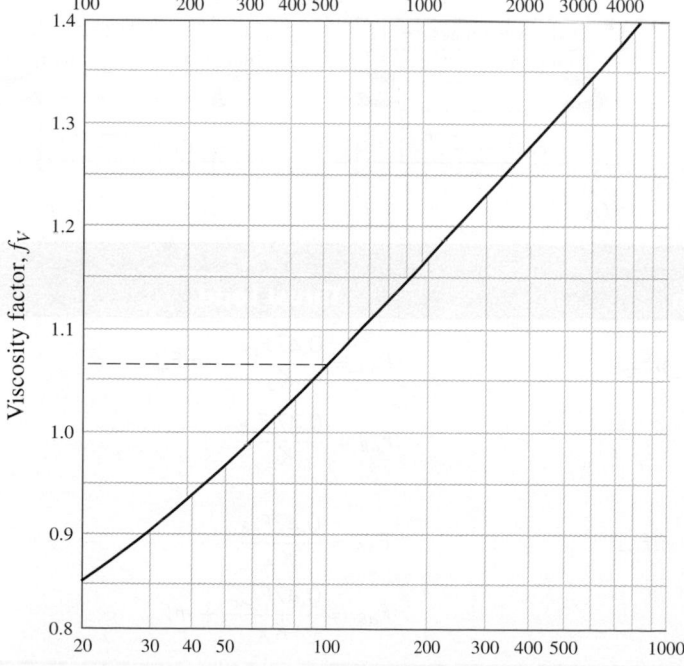

Viscosity, Saybolt universal seconds (SUS) @ 100° F

Viscosity, centistokes (cSt) @ 40° C

in the various expressions for the radial equivalent load P in the right-hand column. In using the table, first determine whether the design is direct mounting ($m = 1$) or indirect mounting ($m = -1$). Next, evaluate the thrust conditions, and depending on which condition is met, apply the appropriate sets of thrust load and/or dynamic equivalent radial load equations. This will be demonstrated in the example that follows.

Table 11–6

Dynamic Equivalent Radial Load Equations for P (*Source:* Courtesy of The Timken Company.)

Two-Row Mounting, Fixed or Floating (with No External Thrust, $F_{ae} = 0$) Similar Bearing Series

For two-row similar bearing series with no external thrust, $F_{ae} = 0$, the dynamic equivalent radial load P equals F_{rAB} or F_{rC}. Since F_{rAB} or F_{rC} is the radial load on the two-row assembly, the two-row basic dynamic radial load rating, $C_{90(2)}$, is to be used to calculate bearing life.

OPTIONAL APPROACH FOR DETERMINING DYNAMIC EQUIVALENT RADIAL LOADS

The following is a general approach to determining the dynamic equivalent radial loads and therefore is more suitable for programmable calculators and computer programming. Here a factor m has to be defined as $+1$ for direct-mounted single-row or two-row bearings or -1 for indirect-mounted bearings. Also a sign convention is necessary for the external thrust F_{ae} as follows:

a. In case of external thrust *applied to the shaft* (typical rotating cone application), F_{ae} to the right is positive, to the left is negative.

b. When external thrust is *applied to the housing* (typical rotating cup application), F_{ae} to the right is negative, to the left is positive.

1. Single-Row Mounting

Design

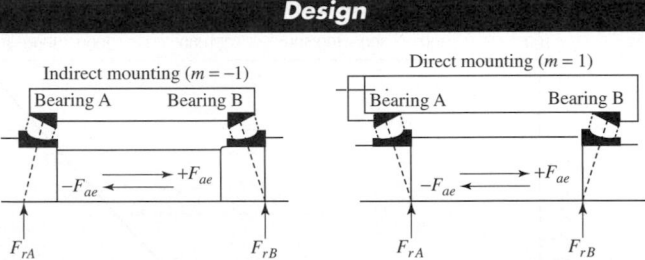

Thrust Condition	Thrust Load	Dynamic Equivalent Radial Load
$\dfrac{0.47F_{rA}}{K_A} \leq \dfrac{0.47F_{rB}}{K_B} - mF_{ae}$	$F_{aA} = \dfrac{0.47F_{rB}}{K_B} - mF_{ae}$	$P_A = 0.4F_{rA} + K_A F_{aA}$
	$F_{aB} = \dfrac{0.47F_{rB}}{K_B}$	$P_B = F_{rB}$
$\dfrac{0.47F_{rA}}{K_A} > \dfrac{0.47F_{rB}}{K_B} - mF_{ae}$	$F_{aA} = \dfrac{0.47F_{rA}}{K_A}$	$P_A = F_{rA}$
	$F_{aB} = \dfrac{0.47F_{rAB}}{K_A} + mF_{ae}$	$P_B = 0.4F_{rB} + K_B F_{aB}$

Note: If $P_A < F_{rA}$, use $P_A = F_{rA}$ or if $P_B < F_{rB}$, use $P_B = F_{rB}$.

Table 11–6

(Continued)

2. Two-Row Mounting—Fixed Bearing with External Thrust, F_{ae}

(Similar or Dissimilar Series)

Design

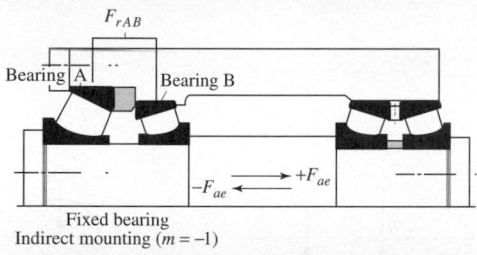

Fixed bearing
Indirect mounting ($m = -1$)

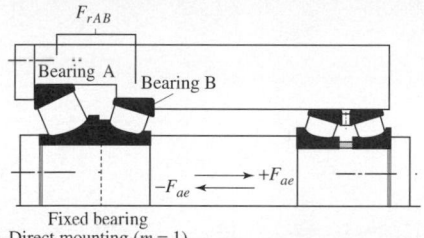

Fixed bearing
Direct mounting ($m = 1$)

Thrust Condition*	Dynamic Equivalent Radial Load
$F_{ae} \leq \dfrac{0.6\,F_{r\,AB}}{K}$	$P_A = \dfrac{K_A}{K_A + K_B}(F_{r\,AB} - 1.67\,m K_B F_{ae})$
	$P_B = \dfrac{K_B}{K_A + K_B}(F_{r\,AB} + 1.67\,m K_A F_{ae})$
$F_{ae} > \dfrac{0.6\,F_{r\,AB}}{K}$	$P_A = 0.4 F_{r\,AB} - m K_A F_{ae}$
	$P_B = 0.4 F_{r\,AB} + m K_B F_{ae}$

* If $m F_{ae}$ is positive, $K = K_B$; If $m F_{ae}$ is negative, $K = K_A$.

Note: F_{rAB} is the radial load on the two-row assembly. The single-row basic dynamic radial load rating, C_{90}, is to be applied in calculating life by the above equations.

EXAMPLE 11–8

The shaft depicted in Fig. 11–18a carries a helical gear with a tangential force of 3980 N, a separating force of 1770 N, and a thrust force of 1690 N at the pitch cylinder with directions shown. The pitch diameter of the gear is 200 mm. The shaft runs at a speed of 1050 rev/min, and the span (effective spread) between the direct-mount bearings is 150 mm. The design life is to be 5000 h and an application factor of 1 is appropriate. The lubricant will be ISO VG 68 (68 cSt at 40°C) oil with an estimated operating temperature of 55°C. If the reliability of the bearing set is to be 0.99, select suitable single-row tapered-roller Timken bearings.

Solution

The reactions in the xz plane from Fig. 11–18b are

$$R_{zA} = \frac{3980(50)}{150} = 1327\,\text{N}$$

$$R_{zB} = \frac{3980(100)}{150} = 2653\,\text{N}$$

Figure 11–18

Essential geometry of helical gear and shaft. Length dimensions in mm, loads in N, couple in N · mm. (a) Sketch (not to scale) showing thrust, radial, and tangential forces. (b) Forces in xz plane. (c) Forces in xy plane.

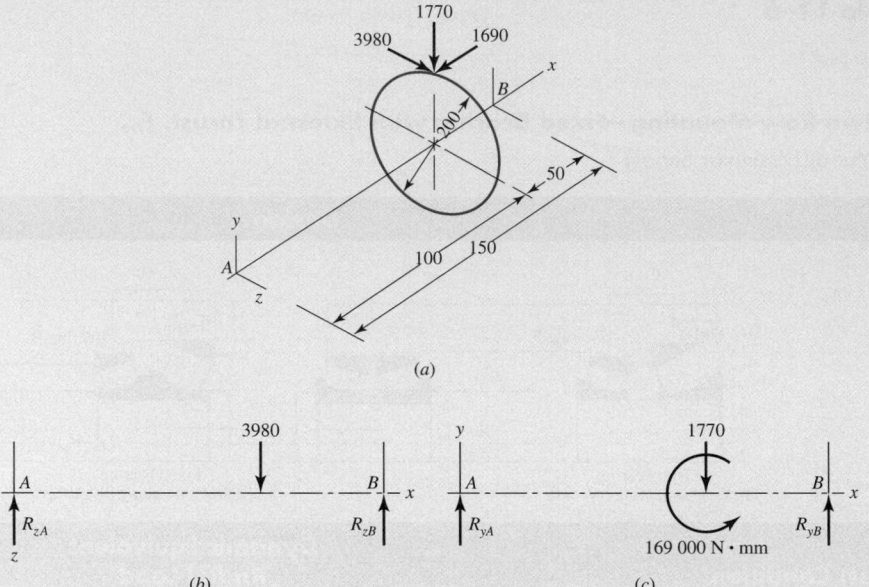

(a)

(b)

(c)

The reactions in the xy plane from Fig. 11–18c are

$$R_{yA} = \frac{1770(50)}{150} + \frac{169\,000}{150} = 1716.7 = 1717 \text{ N}$$

$$R_{yB} = \frac{1770(100)}{150} - \frac{169\,000}{150} = 53.3 \text{ N}$$

The radial loads F_{rA} and F_{rB} are the vector additions of R_{yA} and R_{zA}, and R_{yB} and R_{zB}, respectively:

$$F_{rA} = \left(R_{zA}^2 + R_{yA}^2\right)^{1/2} = (1327^2 + 1717^2)^{1/2} = 2170 \text{ N}$$

$$F_{rB} = \left(R_{zB}^2 + R_{yB}^2\right)^{1/2} = (2653^2 + 53.3^2)^{1/2} = 2654 \text{ N}$$

Trial 1: We will use $K_A = K_B = 1.5$ to start. From Table 11–6, noting that $m = +1$ for direct mounting and F_{ae} to the right is positive, we write

$$\frac{0.47 F_{rA}}{K_A} < ? > \frac{0.47 F_{rB}}{K_B} - m F_{ae}$$

$$\frac{0.47(2170)}{1.5} < ? > \left[\frac{0.47(2654)}{1.5} - (+1)(-1690)\right]$$

$$680 < 2522$$

We use the upper set of equations in Table 11–6 to find the thrust loads:

$$F_{aA} = \frac{0.47 F_{rB}}{K_B} - m F_{ae} = \frac{0.47(2654)}{1.5} - (+1)(-1690) = 2522 \text{ N}$$

$$F_{aB} \frac{0.47 F_{rB}}{K_B} = \frac{0.47(2654)}{1.5} = 832 \text{ N}$$

The dynamic equivalent loads P_A and P_B are

$$P_A = 0.4F_{rA} + K_A F_{aA} = 0.4(2170) + 1.5(2522) = 4651 \text{ N}$$

$$P_B = F_{rB} = 2654 \text{ N}$$

From Fig. 11–16 for 1050 rev/min at 55°C, $f_T = 1.31$. From Fig. 11–17, $f_v = 1.01$. For use in Eq. (11–16), $a_{3l} = f_T f_v = 1.31(1.01) = 1.32$. The catalog basic load rating corresponding to the load–life–reliability goals is given by Eq. (11–17). Estimate R_D as $\sqrt{0.99} = 0.995$ for each bearing. For bearing A, from Eq. (11–17) the catalog entry C_{10} should equal or exceed

$$C_{10} = (1)(4651) \left[\frac{5000(1050)60}{(4.48)1.32(1 - 0.995)^{2/3}90(10^6)} \right]^{3/10} = 11\,466 \text{ N}$$

From Fig. 11–14, tentatively select type TS 15100 cone and 15245 cup, which will work: $K_A = 1.67$, $C_{10} = 12\,100$ N.

For bearing B, from Eq. (11–17), the catalog entry C_{10} should equal or exceed

$$C_{10} = (1)2654 \left[\frac{5000(1050)60}{(4.48)1.32(1 - 0.995)^{2/3}90(10^6)} \right]^{3/10} = 6543 \text{ N}$$

Tentatively select the bearing identical to bearing A, which will work: $K_B = 1.67$, $C_{10} = 12\,100$ N.

Trial 2: Use $K_A = K_B = 1.67$ from tentative bearing selection. The sense of the previous inequality $680 < 2521$ is still the same, so the same equations apply:

$$F_{aA} = \frac{0.47 F_{rB}}{K_B} - m F_{ae} = \frac{0.47(2654)}{1.67} - (+1)(-1690) = 2437 \text{ N}$$

$$F_{aB} = \frac{0.47 F_{rB}}{K_B} = \frac{0.47(2654)}{1.67} = 747 \text{ N}$$

$$P_A = 0.4 F_{rA} + K_A F_{aA} = 0.4(2170) + 1.67(2437) = 4938 \text{ N}$$

$$P_B = F_{rB} = 2654 \text{ N}$$

For bearing A, from Eq. (11–17) the corrected catalog entry C_{10} should equal or exceed

$$C_{10} = (1)(4938) \left[\frac{5000(1050)60}{(4.48)1.32(1 - 0.995)^{2/3}90(10^6)} \right]^{3/10} = 12\,174 \text{ N}$$

Although this catalog entry exceeds slightly the tentative selection for bearing A, we will keep it since the reliability of bearing B exceeds 0.995. In the next section we will quantitatively show that the combined reliability of bearing A and B will exceed the reliability goal of 0.99.

For bearing B, $P_B = F_{rB} = 2654$ N. From Eq. (11–17),

$$C_{10} = (1)2654 \left[\frac{5000(1050)60}{(4.48)1.32(1 - 0.995)^{2/3}90(10^6)} \right]^{3/10} = 6543 \text{ N}$$

Select cone and cup 15100 and 15245, respectively, for both bearing A and B. Note from Fig. 11–14 the effective load center is located at $a = -5.8$ mm, that is, 5.8 mm into the cup from the back. Thus the shoulder-to-shoulder dimension should be $150 - 2(5.8) = 138.4$ mm. Note, also, the calculation for the second bearing C_{10} contains the same bracketed expression as for the first. For example, on the first trial C_{10} for bearing A is 11 466 N. C_{10} for bearing B can be easily calculated by

$$(C_{10})_B = \frac{(C_{10})_A}{P_A} P_B = \frac{11\ 466}{4651} 2654 = 6543 \text{ N}$$

The computational effort can be simplified only after this is understood, and not until then.

11–10 Design Assessment for Selected Rolling-Contact Bearings

In textbooks machine elements typically are treated singly. This can lead the reader to the presumption that a design assessment involves only that element, in this case a rolling-contact bearing. The immediately adjacent elements (the shaft journal and the housing bore) have immediate influence on the performance. Other elements, further removed (gears producing the bearing load), also have influence. Just as some say, "If you pull on something in the environment, you find that it is attached to everything else." This should be intuitively obvious to those involved with machinery. How, then, can one check shaft attributes that aren't mentioned in a problem statement? Possibly, because the bearing hasn't been designed yet (in fine detail). All this points out the necessary iterative nature of designing, say, a speed reducer. If power, speed, and reduction are stipulated, then gear sets can be roughed in, their sizes, geometry, and location estimated, shaft forces and moments identified, bearings tentatively selected, seals identified; the bulk is beginning to make itself evident, the housing and lubricating scheme as well as the cooling considerations become clearer, shaft overhangs and coupling accommodations appear. It is time to iterate, now addressing each element again, knowing much more about all of the others. When you have completed the necessary iterations, you will know what you need for the design assessment for the bearings. In the meantime you do as much of the design assessment as you can, avoiding bad selections, even if tentative. Always keep in mind that you eventually have to do it all in order to pronounce your completed design satisfactory.

An outline of a design assessment for a rolling contact bearing includes, at a minimum,

- Bearing reliability for the load imposed and life expected
- Shouldering on shaft and housing satisfactory
- Journal finish, diameter and tolerance compatible
- Housing finish, diameter and tolerance compatible
- Lubricant type according to manufacturer's recommendations; lubricant paths and volume supplied to keep operating temperature satisfactory
- Preloads, if required, are supplied

Since we are focusing on rolling-contact bearings, we can address bearing reliability quantitatively, as well as shouldering. Other quantitative treatment will have to wait until the materials for shaft and housing, surface quality, and diameters and tolerances are known.

Bearing Reliability

Equation (11–6) can be solved for the reliability R_D in terms of C_{10}, the basic load rating of the selected bearing:

$$R = \exp\left(-\left\{\frac{x_D\left(\dfrac{a_f F_D}{C_{10}}\right)^a - x_0}{\theta - x_0}\right\}^b\right) \tag{11–18}$$

Equation (11–7) can likewise be solved for R_D:

$$R \doteq 1 - \left\{\frac{x_D\left(\dfrac{a_f F_D}{C_{10}}\right)^a - x_0}{\theta - x_0}\right\}^b \qquad R \geq 0.90 \tag{11–19}$$

EXAMPLE 11–9

In Ex. 11–3, the minimum required load rating for 99 percent reliability, at $x_D = L/L_{10} = 540$, is $C_{10} = 6671$ lbf $= 29.7$ kN. From Table 11–2 a 02-40 mm deep-groove ball bearing would satisfy the requirement. If the bore in the application had to be 70 mm or larger (selecting a 02-70 mm deep-groove ball bearing), what is the resulting reliability?

Solution

From Table 11–2, for a 02-70 mm deep-groove ball bearing, $C_{10} = 61.8$ kN $= 13\,888$ lbf. Using Eq. (11–19), recalling from Ex. 11–3 that $a_f = 1.2$, $F_D = 413$ lbf, $x_0 = 0.02$, $(\theta - x_0) = 4.439$, and $b = 1.489$, we can write

$$R \doteq 1 - \left\{\frac{540\left[\dfrac{1.2(413)}{13\,888}\right]^3 - 0.02}{4.439}\right\}^{1.489} = 0.999\,965$$

which, as expected, is much higher than 0.99 from Ex. 11–3.

In tapered roller bearings, or other bearings for a two-parameter Weibull distribution, Eq. (11–18) becomes, for $x_0 = 0$, $\theta = 4.48$, $b = \frac{3}{2}$,

$$R = \exp\left\{-\left[\frac{x_D}{\theta(C_{10}/[a_f F_D])^a}\right]^b\right\}$$

$$= \exp\left\{-\left[\frac{x_D}{4.48\,f_T\,f_v(C_{10}/[a_f F_D])^{10/3}}\right]^{3/2}\right\} \tag{11–20}$$

and Eq. (11–19) becomes

$$R \doteq 1 - \left\{\frac{x_D}{\theta[C_{10}/(a_f F_D)]^a}\right\}^b = 1 - \left\{\frac{x_D}{4.48\,f_T\,f_v[C_{10}/(a_f F_D)]^{10/3}}\right\}^{3/2} \tag{11–21}$$

EXAMPLE 11-10   In Ex. 11–8 bearings A and B (cone 15100 and cup 15245) have $C_{10} = 12\ 100$ N. What is the reliability of the pair of bearings A and B?

Solution   The desired life x_D was $5000(1050)60/[90(10^6)] = 3.5$ rating lives. Using Eq. (11–21) for bearing A, where from Ex. 11–8, $F_D = P_A = 4938$ N, $f_T f_v = 1.32$, and $a_f = 1$, gives

$$R_A \doteq 1 - \left\{ \frac{3.5}{4.48(1.32)\,[12\ 100/\,(1 \times 4938)]^{10/3}} \right\}^{3/2} = 0.994\ 846$$

which is less than 0.995, as expected. Using Eq. (11–21) for bearing B with $F_D = P_B = 2654$ N gives

$$R_B \doteq 1 - \left\{ \frac{3.5}{4.48(1.32)\,[12\ 100/\,(1 \times 2654)]^{10/3}} \right\}^{3/2} = 0.999\ 769$$

Answer   The reliability of the bearing pair is

$$R = R_A R_B = 0.994\ 846(0.999\ 769) = 0.994\ 616$$

which is greater than the overall reliability goal of 0.99. When two bearings are made identical for simplicity, or reducing the number of spares, or other stipulation, and the loading is not the same, both can be made smaller and still meet a reliability goal. If the loading is disparate, then the more heavily loaded bearing can be chosen for a reliability goal just slightly larger than the overall goal.

An additional example is useful to show what happens in cases of pure thrust loading.

EXAMPLE 11-11   Consider a constrained housing as depicted in Fig. 11–19 with two direct-mount tapered roller bearings resisting an external thrust F_{ae} of 8000 N. The shaft speed is 950 rev/min, the desired life is 10 000 h, the expected shaft diameter is approximately 1 in. The lubricant is ISO VG 150 (150 cSt at 40°C) oil with an estimated bearing operating temperature of 80°C. The reliability goal is 0.95. The application factor is appropriately $a_f = 1$.
(a) Choose a suitable tapered roller bearing for A.
(b) Choose a suitable tapered roller bearing for B.
(c) Find the reliabilities R_A, R_B, and R.

Solution   (a) The bearing reactions at A are

$$F_{rA} = F_{rB} = 0$$

$$F_{aA} = F_{ae} = 8000 \text{ N}$$

Since bearing B is unloaded, we will start with $R = R_A = 0.95$. From Table 11–6,

$$\frac{0.47 F_{rA}}{K_A} < ? > \frac{0.47 F_{rB}}{K_B} - m F_{ae}$$

Figure 11-19

The constrained housing of Ex. 11-11.

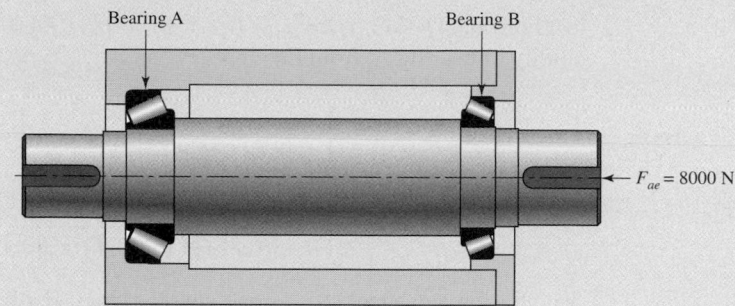

Noting that F_{ae} to the right is positive (Table 11–6), with direct mounting $m = +1$, we can write

$$\frac{0(0)}{K_A} < ? > \left[\frac{0.47(0)}{K_B} - (+1)(-8000)\right]$$

$$0 < 8000 \text{ N}$$

The top set of equations in Table 11–6 applies, so

$$F_{aA} = \frac{0.47(0)}{K_B} - (+1)(-8000) = 8000 \text{ N}$$

$$F_{aB} = \frac{0.47(0)}{K_B} = 0$$

If we set $K_A = 1$, we can find C_{10} in the thrust column and avoid iteration:

$$P_A = 0.4F_{rA} + K_A F_{aA} = 0.4(0) + (1)8000 = 8000 \text{ N}$$

$$P_B = F_{rB} = 0$$

The required life is

$$L_D = 10\,000(950)60 = 570(10^6) \text{ rev}$$

Under the given conditions, $f_T = 0.76$ from Fig. 11–16, and $f_v = 1.12$ from Fig. 11–17. This gives $f_T f_v = 0.76(1.12) = 0.85$. Then, from Eq. (11–17), for bearing A

$$C_{10} = a_f P \left[\frac{L_D}{4.48 f_T f_v (1 - R_D)^{2/3} 90(10^6)}\right]^{3/10}$$

$$= (1)8000 \left[\frac{570(10^6)}{4.48(0.85)(1 - 0.95)^{2/3}\, 90(10^6)}\right]^{3/10} = 16\,970 \text{ N}$$

Answer   Figure 11–14 presents one possibility in the 1-in bore (25.4-mm) size: cone, HM88630, cup HM88610 with a thrust rating $(C_{10})_a = 17\,200$ N.

Answer   (*b*) Bearing *B* experiences no load, and the cheapest bearing of this bore size will do, including a ball or roller bearing.

(c) For Eq. (11–21), $x_D = L_D/L_{10} = 570(10^6)/90(10^6) = 6.333$. Thus the actual reliability of bearing A, from Eq. (11–21), is

Answer

$$R \doteq 1 - \left\{ \frac{x_D}{4.48\, f_T\, f_v\, [C_{10}/(a_f F_D)]^{10/3}} \right\}^{3/2}$$

$$\doteq 1 - \left\{ \frac{6.333}{4.48(0.85)\,[17\ 200/(1 \times 8000)]^{10/3}} \right\}^{3/2} = 0.953$$

which is greater than 0.95, as one would expect. For bearing B,

Answer

$$F_D = P_B = 0$$

$$R_B \doteq 1 - \left[\frac{6.333}{0.85(4.48)(17\ 200/0)^{10/3}} \right]^{3/2} = 1 - 0 = 1$$

as one would expect. The combined reliability of bearings A and B as a pair is

Answer

$$R = R_A R_B = 0.953(1) = 0.953$$

which is greater than the reliability goal of 0.95, as one would expect.

Matters of Fit

Table 11–2 (and Fig. 11–8), which shows the rating of single-row, 02-series, deep-groove and angular-contact ball bearings, includes shoulder diameters recommended for the shaft seat of the inner ring and the shoulder diameter of the outer ring, denoted d_S and d_H, respectively. The shaft shoulder can be greater than d_S but not enough to obstruct the annulus. It is important to maintain concentricity and perpendicularity with the shaft centerline, and to that end the shoulder diameter should equal or exceed d_S. The housing shoulder diameter d_H is to be equal to or less than d_H to maintain concentricity and perpendicularity with the housing bore axis. Neither the shaft shoulder nor the housing shoulder features should allow interference with the free movement of lubricant through the bearing annulus.

In a tapered roller bearing (Fig. 11–14), the cup housing shoulder diameter should be equal to or less than D_b. The shaft shoulder for the cone should be equal to or greater than d_b. Additionally, free lubricant flow is not to be impeded by obstructing any of the annulus. In splash lubrication, common in speed reducers, the lubricant is thrown to the housing cover (ceiling) and is directed in its draining by ribs to a bearing. In direct mounting, a tapered roller bearing pumps oil from outboard to inboard. An oil passageway to the outboard side of the bearing needs to be provided. The oil returns to the sump as a consequence of bearing pump action. With an indirect mount, the oil is directed to the inboard annulus, the bearing pumping it to the outboard side. An oil passage from the outboard side to the sump has to be provided.

11–11   Lubrication

The contacting surfaces in rolling bearings have a relative motion that is both rolling and sliding, and so it is difficult to understand exactly what happens. If the relative velocity of the sliding surfaces is high enough, then the lubricant action is hydrodynamic (see Chap. 12). *Elastohydrodynamic lubrication* (EHD) is the phenomenon that occurs when a lubricant is introduced between surfaces that are in pure rolling contact. The contact of gear teeth and that found in rolling bearings and in cam-and-follower surfaces are typical examples. When a lubricant is trapped between two surfaces in rolling contact, a tremendous increase in the pressure within the lubricant film occurs. But viscosity is exponentially related to pressure, and so a very large increase in viscosity occurs in the lubricant that is trapped between the surfaces. Leibensperger[1] observes that the change in viscosity in and out of contact pressure is equivalent to the difference between cold asphalt and light sewing machine oil.

The purposes of an antifriction-bearing lubricant may be summarized as follows:

1 To provide a film of lubricant between the sliding and rolling surfaces
2 To help distribute and dissipate heat
3 To prevent corrosion of the bearing surfaces
4 To protect the parts from the entrance of foreign matter

Either oil or grease may be employed as a lubricant. The following rules may help in deciding between them.

Use Grease When	Use Oil When
1. The temperature is not over 200°F.	1. Speeds are high.
2. The speed is low.	2. Temperatures are high.
3. Unusual protection is required from the entrance of foreign matter.	3. Oiltight seals are readily employed.
4. Simple bearing enclosures are desired.	4. Bearing type is not suitable for grease lubrication.
5. Operation for long periods without attention is desired.	5. The bearing is lubricated from a central supply which is also used for other machine parts.

11–12 Mounting and Enclosure

There are so many methods of mounting antifriction bearings that each new design is a real challenge to the ingenuity of the designer. The housing bore and shaft outside diameter must be held to very close limits, which of course is expensive. There are usually one or more counterboring operations, several facing operations and drilling, tapping, and threading operations, all of which must be performed on the shaft, housing, or cover plate. Each of these operations contributes to the cost of production, so that the designer, in ferreting out a trouble-free and low-cost mounting, is faced with a difficult and important problem. The various bearing manufacturers' handbooks give many mounting details in almost every design area. In a text of this nature, however, it is possible to give only the barest details.

The most frequently encountered mounting problem is that which requires one bearing at each end of a shaft. Such a design might use one ball bearing at each end, one tapered roller bearing at each end, or a ball bearing at one end and a straight roller bearing at the other. One of the bearings usually has the added function of positioning or

---

[1] R. L. Leibensperger, "When Selecting a Bearing," *Machine Design,* vol. 47, no. 8, April 3, 1975, pp. 142–147.

Figure 11–20

A common bearing mounting.

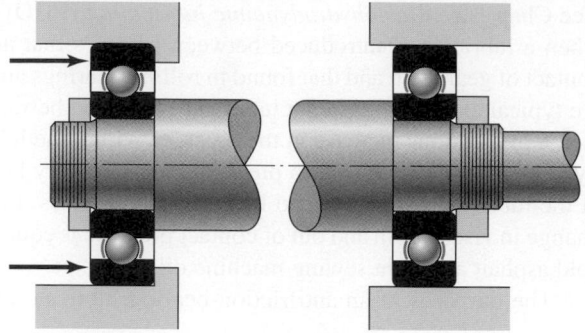

Figure 11–21

An alternative bearing mounting to that in Fig. 11–20.

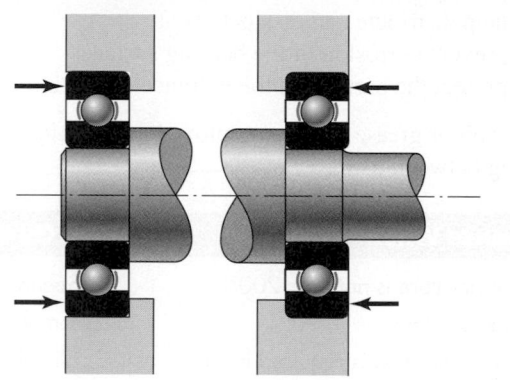

axially locating the shaft. Figure 11–20 shows a very common solution to this problem. The inner rings are backed up against the shaft shoulders and are held in position by round nuts threaded onto the shaft. The outer ring of the left-hand bearing is backed up against a housing shoulder and is held in position by a device that is not shown. The outer ring of the right-hand bearing floats in the housing.

There are many variations possible on the method shown in Fig. 11–20. For example, the function of the shaft shoulder may be performed by retaining rings, by the hub of a gear or pulley, or by spacing tubes or rings. The round nuts may be replaced by retaining rings or by washers locked in position by screws, cotters, or taper pins. The housing shoulder may be replaced by a retaining ring; the outer ring of the bearing may be grooved for a retaining ring, or a flanged outer ring may be used. The force against the outer ring of the left-hand bearing is usually applied by the cover plate, but if no thrust is present, the ring may be held in place by retaining rings.

Figure 11–21 shows an alternative method of mounting in which the inner races are backed up against the shaft shoulders as before but no retaining devices are required. With this method the outer races are completely retained. This eliminates the grooves or threads, which cause stress concentration on the overhanging end, but it requires accurate dimensions in an axial direction or the employment of adjusting means. This method has the disadvantage that if the distance between the bearings is great, the temperature rise during operation may expand the shaft enough to destroy the bearings.

It is frequently necessary to use two or more bearings at one end of a shaft. For example, two bearings could be used to obtain additional rigidity or increased load capacity or to cantilever a shaft. Several two-bearing mountings are shown in Fig. 11–22. These may be used with tapered roller bearings, as shown, or with ball bearings. In either case it should be noted that the effect of the mounting is to preload the bearings in an axial direction.

Figure 11–22

Two-bearing mountings. *(Courtesy of The Timken Company.)*

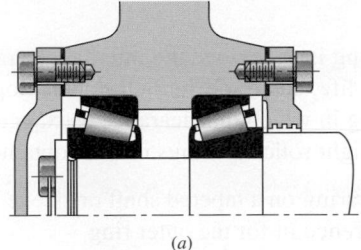 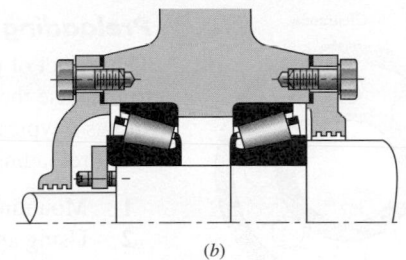

(a)   (b)

Figure 11–23

Mounting for a washing-machine spindle. (Courtesy of The Timken Company.)

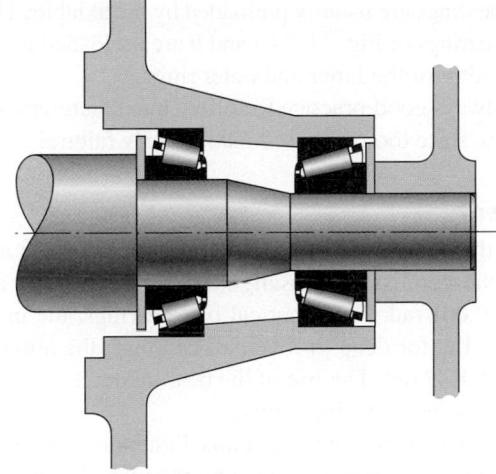

Figure 11–24

Arrangements of angular ball bearings. (a) DF mounting; (b) DB mounting; (c) DT mounting. *(Courtesy of The Timken Company.)*

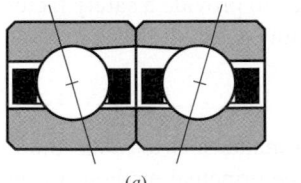 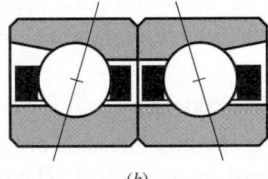 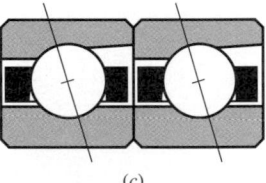

(a)   (b)   (c)

Figure 11–23 shows another two-bearing mounting. Note the use of washers against the cone backs.

When maximum stiffness and resistance to shaft misalignment is desired, pairs of angular-contact ball bearings (Fig. 11–2) are often used in an arrangement called *duplexing*. Bearings manufactured for duplex mounting have their rings ground with an offset, so that when a pair of bearings is tightly clamped together, a preload is automatically established. As shown in Fig. 11–24, three mounting arrangements are used. The face-to-face mounting, called DF, will take heavy radial loads and thrust loads from either direction. The DB mounting (back to back) has the greatest aligning stiffness and is also good for heavy radial loads and thrust loads from either direction. The tandem arrangement, called the DT mounting, is used where the thrust is always in the same direction; since the two bearings have their thrust functions in the same direction, a preload, if required, must be obtained in some other manner.

Bearings are usually mounted with the rotating ring a press fit, whether it be the inner or outer ring. The stationary ring is then mounted with a push fit. This permits the stationary ring to creep in its mounting slightly, bringing new portions of the ring into the load-bearing zone to equalize wear.

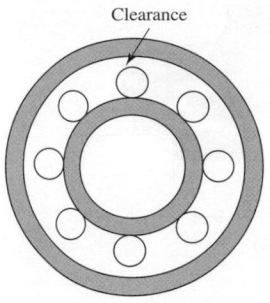

Figure 11–25

Clearance in an off-the-shelf bearing, exaggerated for clarity.

Preloading

The object of preloading is to remove the internal clearance usually found in bearings, to increase the fatigue life, and to decrease the shaft slope at the bearing. Figure 11–25 shows a typical bearing in which the clearance is exaggerated for clarity.

Preloading of straight roller bearings may be obtained by:

1   Mounting the bearing on a tapered shaft or sleeve to expand the inner ring
2   Using an interference fit for the outer ring
3   Purchasing a bearing with the outer ring preshrunk over the rollers

Ball bearings are usually preloaded by the axial load built in during assembly. However, the bearings of Fig. 11–24*a* and *b* are preloaded in assembly because of the differences in widths of the inner and outer rings.

It is always good practice to follow manufacturers' recommendations in determining preload, since too much will lead to early failure.

Alignment

Based on the general experience with rolling bearings as expressed in manufacturers' catalogs, the permissible misalignment in cylindrical and tapered roller bearings is limited to 0.001 rad. For spherical ball bearings, the misalignment should not exceed 0.0087 rad. But for deep-groove ball bearings, the allowable range of misalignment is 0.0035 to 0.0047 rad. The life of the bearing decreases significantly when the misalignment exceeds the allowable limits.

Additional protection against misalignment is obtained by providing the full shoulders (see Fig. 11–8) recommended by the manufacturer. Also, if there is any misalignment at all, it is good practice to provide a safety factor of around 2 to account for possible increases during assembly.

Enclosures

To exclude dirt and foreign matter and to retain the lubricant, the bearing mountings must include a seal. The three principal methods of sealings are the felt seal, the commercial seal, and the labyrinth seal (Fig. 11–26).

Felt seals may be used with grease lubrication when the speeds are low. The rubbing surfaces should have a high polish. Felt seals should be protected from dirt by placing them in machined grooves or by using metal stampings as shields.

The *commercial seal* is an assembly consisting of the rubbing element and, generally, a spring backing, which are retained in a sheet-metal jacket. These seals are usually made by press fitting them into a counterbored hole in the bearing cover. Since they obtain the sealing action by rubbing, they should not be used for high speeds.

The *labyrinth seal* is especially effective for high-speed installations and may be used with either oil or grease. It is sometimes used with flingers. At least three

Figure 11–26

Typical sealing methods.
(General Motors Corp. Used with permission, GM Media Archives.)

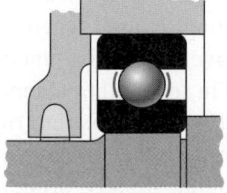
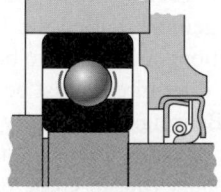
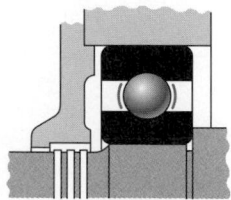

(*a*) Felt seal   (*b*) Commercial seal   (*c*) Labyrinth seal

grooves should be used, and they may be cut on either the bore or the outside diameter. The clearance may vary from 0.010 to 0.040 in, depending upon the speed and temperature.

PROBLEMS

Since each bearing manufacturer makes individual decisions with respect to materials, treatments, and manufacturing processes, manufacturers' experiences with bearing life distribution differ. In solving the following problems, we will use the experience of two manufacturers, tabulated as follows:

Manufacturer	Rating Life, revolutions	Weibull Parameters Rating Lives		
		x_0	θ	b
1	$90(10^6)$	0	4.48	1.5
2	$1(10^6)$	0.02	4.459	1.483

Tables 11–2 and 11–3 are based on manufacturer 2.

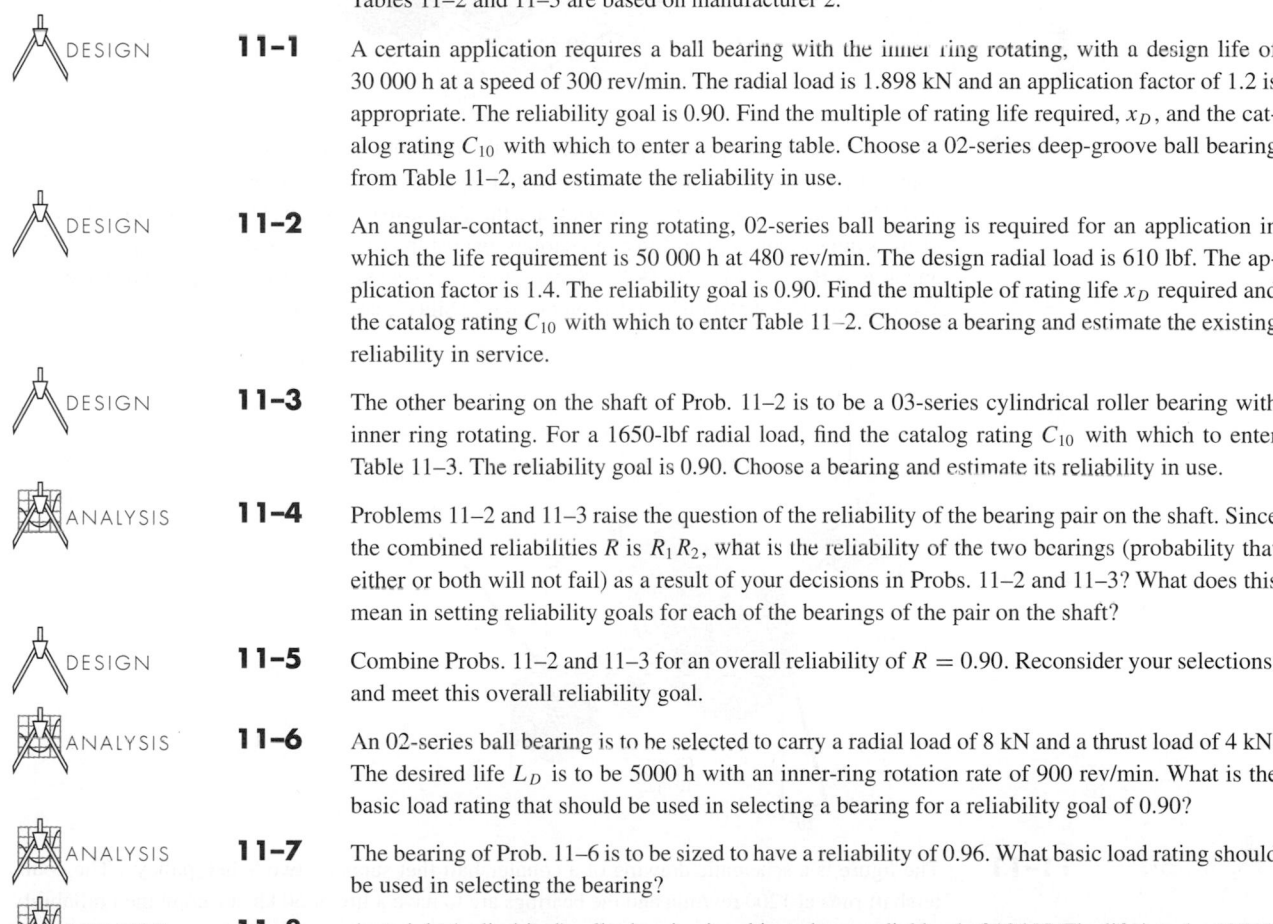

DESIGN   **11-1**   A certain application requires a ball bearing with the inner ring rotating, with a design life of 30 000 h at a speed of 300 rev/min. The radial load is 1.898 kN and an application factor of 1.2 is appropriate. The reliability goal is 0.90. Find the multiple of rating life required, x_D, and the catalog rating C_{10} with which to enter a bearing table. Choose a 02-series deep-groove ball bearing from Table 11–2, and estimate the reliability in use.

DESIGN   **11-2**   An angular-contact, inner ring rotating, 02-series ball bearing is required for an application in which the life requirement is 50 000 h at 480 rev/min. The design radial load is 610 lbf. The application factor is 1.4. The reliability goal is 0.90. Find the multiple of rating life x_D required and the catalog rating C_{10} with which to enter Table 11–2. Choose a bearing and estimate the existing reliability in service.

DESIGN   **11-3**   The other bearing on the shaft of Prob. 11–2 is to be a 03-series cylindrical roller bearing with inner ring rotating. For a 1650-lbf radial load, find the catalog rating C_{10} with which to enter Table 11–3. The reliability goal is 0.90. Choose a bearing and estimate its reliability in use.

ANALYSIS   **11-4**   Problems 11–2 and 11–3 raise the question of the reliability of the bearing pair on the shaft. Since the combined reliabilities R is $R_1 R_2$, what is the reliability of the two bearings (probability that either or both will not fail) as a result of your decisions in Probs. 11–2 and 11–3? What does this mean in setting reliability goals for each of the bearings of the pair on the shaft?

DESIGN   **11-5**   Combine Probs. 11–2 and 11–3 for an overall reliability of $R = 0.90$. Reconsider your selections, and meet this overall reliability goal.

ANALYSIS   **11-6**   An 02-series ball bearing is to be selected to carry a radial load of 8 kN and a thrust load of 4 kN. The desired life L_D is to be 5000 h with an inner-ring rotation rate of 900 rev/min. What is the basic load rating that should be used in selecting a bearing for a reliability goal of 0.90?

ANALYSIS   **11-7**   The bearing of Prob. 11–6 is to be sized to have a reliability of 0.96. What basic load rating should be used in selecting the bearing?

ANALYSIS   **11-8**   A straight (cylindrical) roller bearing is subjected to a radial load of 12 kN. The life is to be 4000 h at a speed of 750 rev/min and exhibit a reliability of 0.90. What basic load rating should be used in selecting the bearing from a catalog of manufacturer 2?

DESIGN

11-9 Shown in the figure is a gear-driven squeeze roll that mates with an idler roll, below. The roll is designed to exert a normal force of 30 lbf/in of roll length and a pull of 24 lbf/in on the material being processed. The roll speed is 300 rev/min, and a design life of 30 000 h is desired. Use an application factor of 1.2, and select a pair of angular-contact 02-series ball bearings from Table 11–2 to be mounted at 0 and A. Use the same size bearings at both locations and a combined reliability of at least 0.92.

Problem 11–9
Idler roll is below powered roll.
Dimensions in inches.

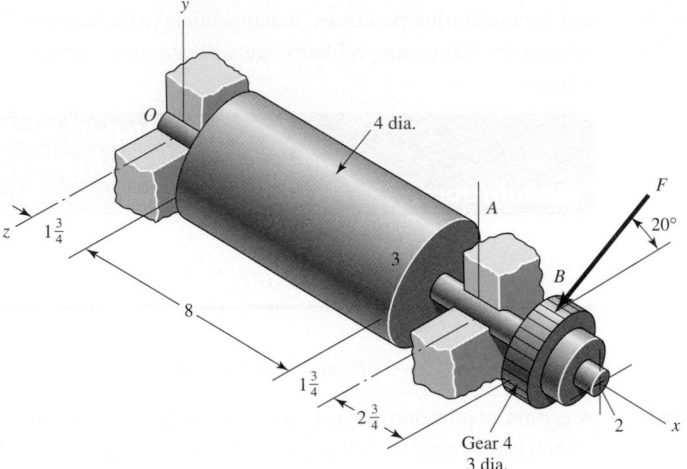

DESIGN

11-10 The figure shown is a geared countershaft with an overhanging pinion at C. Select an angular-contact ball bearing from Table 11–2 for mounting at O and a straight roller bearing for mounting at B. The force on gear A is $F_A = 600$ lbf, and the shaft is to run at a speed of 480 rev/min. Solution of the statics problem gives force of bearings against the shaft at O as $\mathbf{R}_O = -387\mathbf{j} + 467\mathbf{k}$ lbf, and at B as $\mathbf{R}_B = 316\mathbf{j} - 1615\mathbf{k}$ lbf. Specify the bearings required, using an application factor of 1.4, a desired life of 50 000 h, and a combined reliability goal of 0.90.

Problem 11–10
Dimensions in inches.

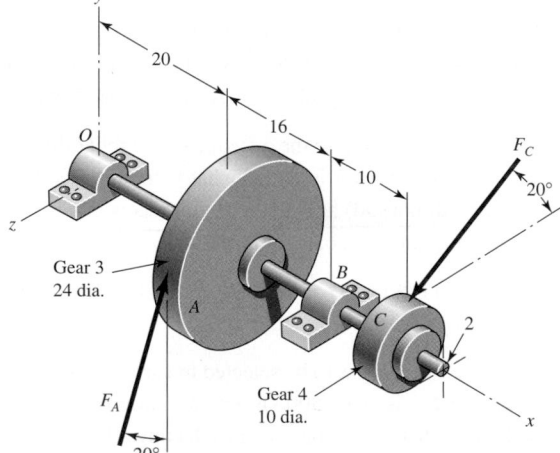

DESIGN

11-11 The figure is a schematic drawing of a countershaft that supports two V-belt pulleys. The countershaft runs at 1200 rev/min and the bearings are to have a life of 60 kh at a combined reliability of 0.999. The belt tension on the loose side of pulley A is 15 percent of the tension on the tight side. Select deep-groove bearings from Table 10–2 for use at O and E, each to have a 25-mm bore, using an application factor of unity.

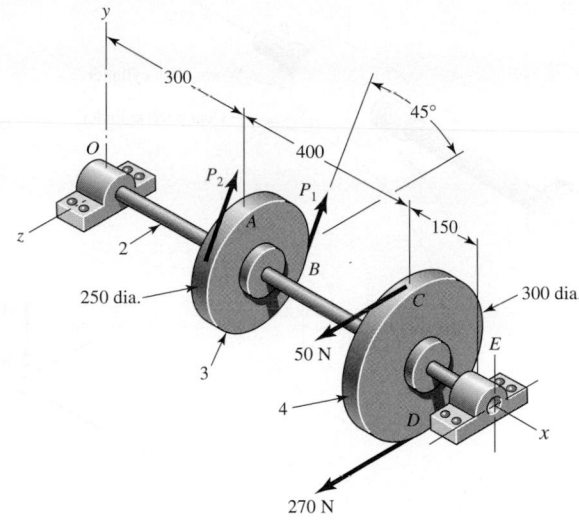

Problem 11–11
Dimensions in millimeters.

ANALYSIS

11–12 The bearing lubricant (513 SUS at 100°F) operating point is 135°F. A countershaft is supported by two tapered roller bearings using an indirect mounting. The radial bearing loads are 560 lbf for the left-hand bearing and 1095 for the right-hand bearing. The shaft rotates at 400 rev/min and is to have a desired life of 40 kh. Use an application factor of 1.4 and a combined reliability goal of 0.90. Using an initial $K = 1.5$, find the required radial rating for each bearing. Select the bearings from Fig. 11–14.

DESIGN

11–13 A gear-reduction unit uses the countershaft depicted in the figure. Find the two bearing reactions. The bearings are to be angular-contact ball bearings, having a desired life of 40 kh when used at 200 rev/min. Use 1.2 for the application factor and a reliability goal for the bearing pair of 0.95. Select the bearings from Table 11–2.

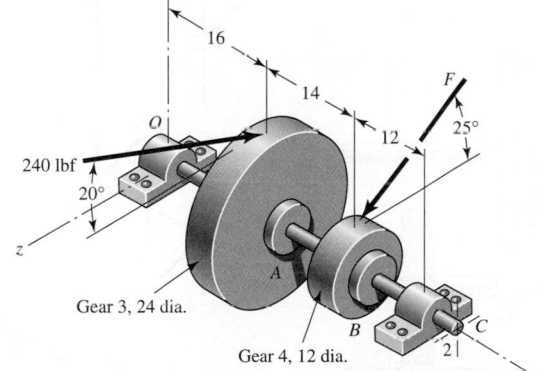

Problem 11–13
Dimensions in inches.

DESIGN

11–14 The worm shaft shown in part *a* of the figure transmits 1.35 hp at 600 rev/min. A static force analysis gave the results shown in part *b* of the figure. Bearing *A* is to be an angular-contact ball bearing mounted to take the 555-lbf thrust load. The bearing at *B* is to take only the radial load, so a straight roller bearing will be employed. Use an application factor of 1.3, a desired life of 25 kh, and a reliability goal, combined, of 0.99. Specify each bearing.

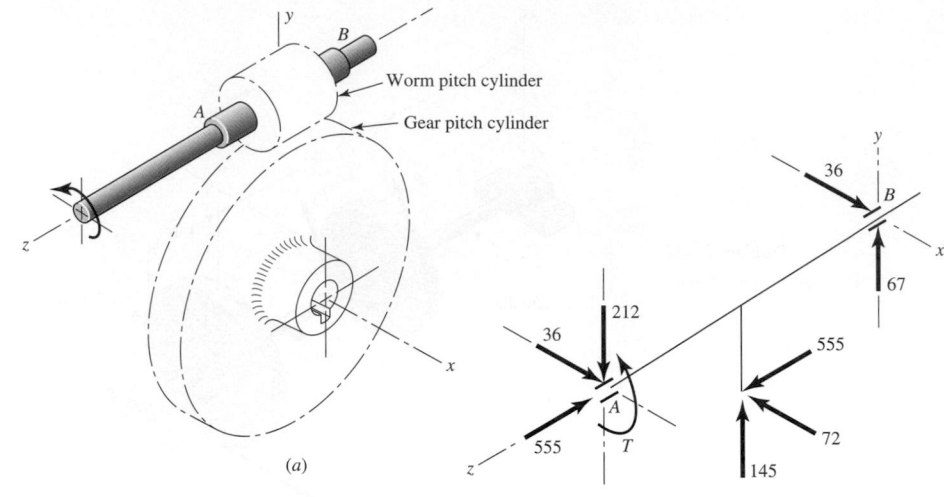

Problem 11–14
(a) Worm and worm gear; (b) force analysis of worm shaft, forces in pounds.

(a)

(b)

ANALYSIS

11–15 In bearings tested at 2000 rev/min with a steady radial load of 18 kN, a set of bearings showed an L_{10} life of 115 h and an L_{80} life of 600 h. The basic load rating of this bearing is 39.6 kN. Estimate the Weibull shape factor b and the characteristic life θ for a two-parameter model. This manufacturer rates ball bearings at 1 million revolutions.

DESIGN

11–16 A 16-tooth pinion drives the double-reduction spur-gear train in the figure. All gears have 25° pressure angles. The pinion rotates ccw at 1200 rev/min and transmits power to the gear train. The

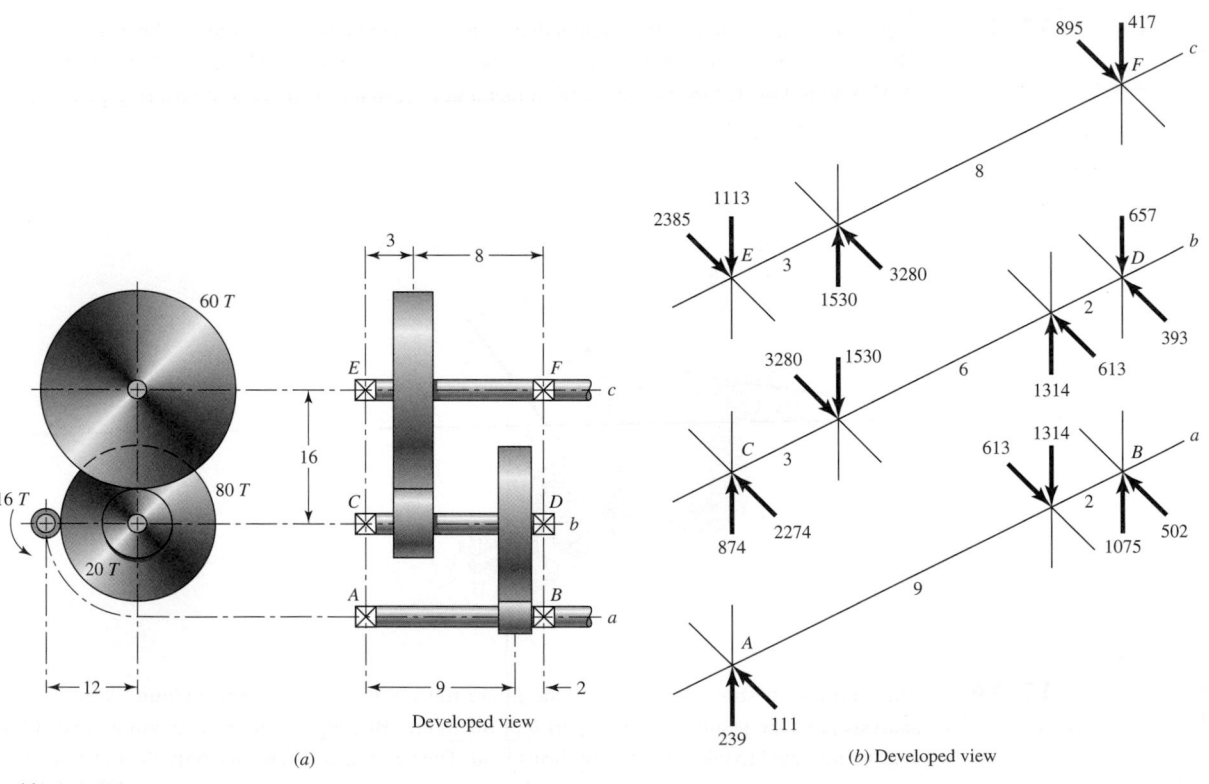

Developed view

(a)

(b) Developed view

Problem 11–16
(a) Drive detail; (b) force analysis on shafts. Forces in pounds; linear dimensions in inches.

shaft has not yet been designed, but the free bodies have been generated. The shaft speeds are 1200 rev/min, 240 rev/min, and 80 rev/min. A bearing study is commencing with a 10-kh life and a gearbox bearing ensemble reliability of 0.99. An application factor of 1.2 is appropriate. Specify the six bearings.

11–17 ANALYSIS

Different bearing metallurgy affects bearing life. A manufacturer reports that a particular heat treatment increases bearing life at least threefold. A bearing identical to that of Prob. 11–15 except for the heat treatment, loaded to 18 kN and run at 2000 rev/min, revealed an L_{10} life of 360 h and an L_{80} life of 2000 h. Do you agree with the manufacturer's assertion concerning increased life?

11–18 ANALYSIS

Estimate the remaining life in revolutions of an 02-30 mm angular-contact ball bearing already subjected to 200 000 revolutions with a radial load of 18 kN, if it is now to be subjected to a change in load to 30 kN.

11–19 ANALYSIS

The same 02-30 angular-contact ball bearing as in Prob. 11–18 is to be subjected to a two-step loading cycle of 4 min with a loading of 18 kN, and one of 6 min with a loading of 30 kN. This cycle is to be repeated until failure. Estimate the total life in revolutions, hours, and loading cycles.

11–20 ANALYSIS

The expression $F^a L = $ constant can be written using $x = L/L_{10}$, and it can be expressed as $F^a x = K$ or $\log F = (1/a) \log K - (1/a) \log x$. This is a straight line on a log-log plot, and it is the basis of Fig. 11–5. For the geometric insight provided, produce Fig. 11–5 to scale using Ex. 11–3, and

> For point D: find $F_D = 1.2(413) = 495.6$ lbf, $\log F_D$, x_D, $\log x_D$, K_D
> For point B: find x_B, $\log x_B$, F_B, $\log F_B$, K_B
> For point A: find $F_A = F_B = C_{10}$, $\log F_A$, K_{10}

and plot to scale. On this plot, also show the line containing C_{10}, the basic load rating, of the *selected* bearing.

12

Lubrication and Journal Bearings

The object of lubrication is to reduce friction, wear, and heating of machine parts that move relative to each other. A lubricant is any substance that, when inserted between the moving surfaces, accomplishes these purposes. In a sleeve bearing, a shaft, or *journal,* rotates or oscillates within a sleeve, or *bushing,* and the relative motion is sliding. In an antifriction bearing, the main relative motion is rolling. A follower may either roll or slide on the cam. Gear teeth mate with each other by a combination of rolling and sliding. Pistons slide within their cylinders. All these applications require lubrication to reduce friction, wear, and heating.

The field of application for journal bearings is immense. The crankshaft and connecting-rod bearings of an automotive engine must operate for thousands of miles at high temperatures and under varying load conditions. The journal bearings used in the steam turbines of power-generating stations are said to have reliabilities approaching 100 percent. At the other extreme there are thousands of applications in which the loads are light and the service relatively unimportant; a simple, easily installed bearing is required, using little or no lubrication. In such cases an antifriction bearing might be a poor answer because of the cost, the elaborate enclosures, the close tolerances, the radial space required, the high speeds, or the increased inertial effects. Instead, a nylon bearing requiring no lubrication, a powder-metallurgy bearing with the lubrication "built in," or a bronze bearing with ring oiling, wick feeding, or solid-lubricant film or grease lubrication might be a very satisfactory solution. Recent metallurgy developments in bearing materials, combined with increased knowledge of the lubrication process, now make it possible to design journal bearings with satisfactory lives and very good reliabilities.

Much of the material we have studied thus far in this book has been based on fundamental engineering studies, such as statics, dynamics, the mechanics of solids, metal processing, mathematics, and metallurgy. In the study of lubrication and journal bearings, additional fundamental studies, such as chemistry, fluid mechanics, thermodynamics, and heat transfer, must be utilized in developing the material. While we shall not utilize all of them in the material to be included here, you can now begin to appreciate better how the study of mechanical engineering design is really an integration of most of your previous studies and a directing of this total background toward the resolution of a single objective.

12–1   Types of Lubrication

Five distinct forms of lubrication may be identified:

1   Hydrodynamic
2   Hydrostatic
3   Elastohydrodynamic
4   Boundary
5   Solid film

Hydrodynamic lubrication means that the load-carrying surfaces of the bearing are separated by a relatively thick film of lubricant, so as to prevent metal-to-metal contact, and that the stability thus obtained can be explained by the laws of fluid mechanics. Hydrodynamic lubrication does not depend upon the introduction of the lubricant under pressure, though that may occur; but it does require the existence of an adequate supply at all times. The film pressure is created by the moving surface itself pulling the lubricant into a wedge-shaped zone at a velocity sufficiently high to create the pressure necessary to separate the surfaces against the load on the bearing. Hydrodynamic lubrication is also called *full-film,* or *fluid, lubrication.*

Hydrostatic lubrication is obtained by introducing the lubricant, which is sometimes air or water, into the load-bearing area at a pressure high enough to separate the surfaces with a relatively thick film of lubricant. So, unlike hydrodynamic lubrication, this kind of lubrication does not require motion of one surface relative to another. We shall not deal with hydrostatic lubrication in this book, but the subject should be considered in designing bearings where the velocities are small or zero and where the frictional resistance is to be an absolute minimum.

Elastohydrodynamic lubrication is the phenomenon that occurs when a lubricant is introduced between surfaces that are in rolling contact, such as mating gears or rolling bearings. The mathematical explanation requires the Hertzian theory of contact stress and fluid mechanics.

Insufficient surface area, a drop in the velocity of the moving surface, a lessening in the quantity of lubricant delivered to a bearing, an increase in the bearing load, or an increase in lubricant temperature resulting in a decrease in viscosity—any one of these— may prevent the buildup of a film thick enough for full-film lubrication. When this happens, the highest asperities may be separated by lubricant films only several molecular dimensions in thickness. This is called *boundary lubrication*. The change from hydrodynamic to boundary lubrication is not at all a sudden or abrupt one. It is probable that a mixed hydrodynamic- and boundary-type lubrication occurs first, and as the surfaces move closer together, the boundary-type lubrication becomes predominant. The viscosity of the lubricant is not of as much importance with boundary lubrication as is the chemical composition.

When bearings must be operated at extreme temperatures, a *solid-film lubricant* such as graphite or molybdenum disulfide must be used because the ordinary mineral oils are not satisfactory. Much research is currently being carried out in an effort, too, to find composite bearing materials with low wear rates as well as small frictional coefficients.

12–2 Viscosity

In Fig. 12–1 let a plate *A* be moving with a velocity *U* on a film of lubricant of thickness *h*. We imagine the film as composed of a series of horizontal layers and the force *F* causing these layers to deform or slide on one another just like a deck of cards. The layers in contact with the moving plate are assumed to have a velocity *U*; those in contact with the stationary surface are assumed to have a zero velocity. Intermediate layers have velocities that depend upon their distances *y* from the stationary surface. Newton's viscous effect states that the shear stress in the fluid is proportional to the rate of change of velocity with respect to *y*. Thus

$$\tau = \frac{F}{A} = \mu \frac{du}{dy} \tag{12–1}$$

| Figure 12–1

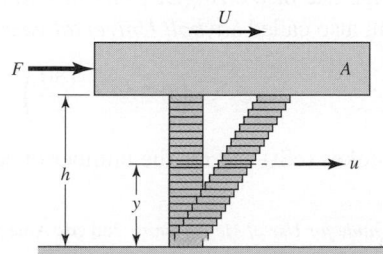

where μ is the constant of proportionality and defines *absolute viscosity*, also called *dynamic viscosity*. The derivative du/dy is the rate of change of velocity with distance and may be called the rate of shear, or the velocity gradient. The viscosity μ is thus a measure of the internal frictional resistance of the fluid. For most lubricating fluids, the rate of shear is constant, and $du/dy = U/h$. Thus, from Eq. (12–1),

$$\tau = \frac{F}{A} = \mu \frac{U}{h} \tag{12–2}$$

Fluids exhibiting this characteristic are said to be *Newtonian fluids*. The unit of viscosity in the ips system is seen to be the pound-force-second per square inch; this is the same as stress or pressure multiplied by time. The ips unit is called the *reyn*, in honor of Sir Osborne Reynolds.

The absolute viscosity is measured by the pascal-second (Pa · s) in SI; this is the same as a Newton-second per square meter. The conversion from ips units to SI is the same as for stress. For example, multiply the absolute viscosity in reyns by 6890 to convert to units of Pa · s.

The American Society of Mechanical Engineers (ASME) has published a list of cgs units that are not to be used in ASME documents.[1] This list results from a recommendation by the International Committee of Weights and Measures (CIPM) that the use of cgs units with special names be discouraged. Included in this list is a unit of force called the *dyne* (dyn), a unit of dynamic viscosity called the *poise* (P), and a unit of kinematic viscosity called the *stoke* (St). All of these units have been, and still are, used extensively in lubrication studies.

The poise is the cgs unit of dynamic or absolute viscosity, and its unit is the dyne-second per square centimeter (dyn · s/cm^2). It has been customary to use the centipoise (cP) in analysis, because its value is more convenient. When the viscosity is expressed in centipoises, it is designated by Z. The conversion from cgs units to SI and ips units is as follows:

$$\mu(\text{Pa} \cdot \text{s}) = (10)^{-3} Z \text{ (cP)}$$

$$\mu(\text{reyn}) = \frac{Z \text{ (cP)}}{6.89(10)^6}$$

$$\mu(\text{mPa} \cdot \text{s}) = 6.89 \, \mu'(\mu\text{reyn})$$

In using ips units, the microreyn (μreyn) is often more convenient. The symbol μ' will be used to designate viscosity in μreyn such that $\mu = \mu'/(10^6)$.

The ASTM standard method for determining viscosity uses an instrument called the Saybolt Universal Viscosimeter. The method consists of measuring the time in seconds for 60 mL of lubricant at a specified temperature to run through a tube 17.6 mm in diameter and 12.25 mm long. The result is called the *kinematic viscosity*, and in the past the unit of the square centimeter per second has been used. One square centimeter per second is defined as a *stoke*. By the use of the *Hagen-Poiseuille law*, the kinematic viscosity based upon seconds Saybolt, also called *Saybolt Universal viscosity* (SUV) in seconds, is

$$Z_k = \left(0.22t - \frac{180}{t} \right) \tag{12–3}$$

where Z_k is in centistokes (cSt) and t is the number of seconds Saybolt.

---

[1]*ASME Orientation and Guide for Use of Metric Units*, 2nd ed., American Society of Mechanical Engineers, 1972, p. 13.

Figure 12–2

A comparison of the viscosities of various fluids.

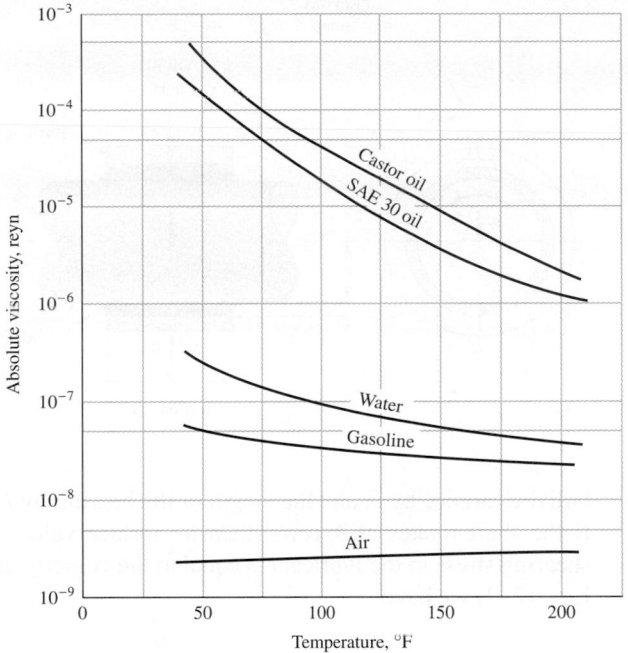

In SI, the kinematic viscosity ν has the unit of the square meter per second (m²/s), and the conversion is

$$\nu(\text{m}^2/\text{s}) = 10^{-6} Z_k \text{ (cSt)}$$

Thus, Eq. (12–3) becomes

$$\nu = \left(0.22t - \frac{180}{t} \right) (10^{-6}) \tag{12-4}$$

To convert to dynamic viscosity, we multiply ν by the density in SI units. Designating the density as ρ with the unit of the kilogram per cubic meter, we have

$$\mu = \rho \left(0.22t - \frac{180}{t} \right) (10^{-6}) \tag{12-5}$$

where μ is in pascal-seconds.

Figure 12–2 shows the absolute viscosity in the ips system of a number of fluids often used for lubrication purposes and their variation with temperature.

12–3   Petroff's Equation

The phenomenon of bearing friction was first explained by Petroff on the assumption that the shaft is concentric. Though we shall seldom make use of Petroff's method of analysis in the material to follow, it is important because it defines groups of dimensionless parameters and because the coefficient of friction predicted by this law turns out to be quite good even when the shaft is not concentric.

Let us now consider a vertical shaft rotating in a guide bearing. It is assumed that the bearing carries a very small load, that the clearance space is completely filled with oil, and that leakage is negligible (Fig. 12–3). We denote the radius of the shaft by r, the

Figure 12–3

Petroff's lightly loaded journal bearing consisting of a shaft journal and a bushing with an axial-groove internal lubricant reservoir. The linear velocity gradient is shown in the end view. The clearance c is several thousandths of an inch and is grossly exaggerated for presentation purposes.

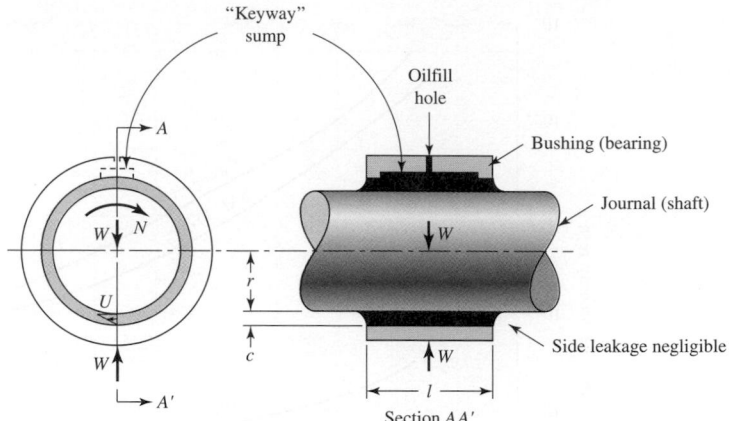

radial clearance by c, and the length of the bearing by l, all dimensions being in inches. If the shaft rotates at N rev/s, then its surface velocity is $U = 2\pi r N$ in/s. Since the shearing stress in the lubricant is equal to the velocity gradient times the viscosity, from Eq. (12–2) we have

$$\tau = \mu \frac{U}{h} = \frac{2\pi r \mu N}{c} \qquad (a)$$

where the radial clearance c has been substituted for the distance h. The force required to shear the film is the stress times the area. The torque is the force times the lever arm r. Thus

$$T = (\tau A)(r) = \left(\frac{2\pi r \mu N}{c}\right)(2\pi r l)(r) = \frac{4\pi^2 r^3 l \mu N}{c} \qquad (b)$$

If we now designate a small force on the bearing by W, in pounds-force, then the pressure P, in pounds-force per square inch of projected area, is $P = W/2rl$. The frictional force is fW, where f is the coefficient of friction, and so the frictional torque is

$$T = fWr = (f)(2rlP)(r) = 2r^2 flP \qquad (c)$$

Substituting the value of the torque from Eq. (c) in Eq. (b) and solving for the coefficient of friction, we find

$$f = 2\pi^2 \frac{\mu N}{P} \frac{r}{c} \qquad (12\text{–}6)$$

Equation (12–6) is called *Petroff's equation* and was first published in 1883. The two quantities $\mu N/P$ and r/c are very important parameters in lubrication. Substitution of the appropriate dimensions in each parameter will show that they are dimensionless.

The *bearing characteristic number,* or the *Sommerfeld number,* is defined by the equation

$$S = \left(\frac{r}{c}\right)^2 \frac{\mu N}{P} \qquad (12\text{–}7)$$

The Sommerfeld number is very important in lubrication analysis because it contains many of the parameters that are specified by the designer. Note that it is also dimensionless. The quantity r/c is called the *radial clearance ratio.* If we multiply both sides of Eq. (12–6) by this ratio, we obtain the interesting relation

$$f\frac{r}{c} = 2\pi^2 \frac{\mu N}{P}\left(\frac{r}{c}\right)^2 = 2\pi^2 S \qquad (12\text{–}8)$$

12–4   Stable Lubrication

The difference between boundary and hydrodynamic lubrication can be explained by reference to Fig. 12–4. This plot of the change in the coefficient of friction versus the bearing characteristic $\mu N/P$ was obtained by the McKee brothers in an actual test of friction.[2] The plot is important because it defines stability of lubrication and helps us to understand hydrodynamic and boundary, or thin-film, lubrication.

Recall Petroff's bearing model in the form of Eq. (12–6) predicts that f is proportional to $\mu N/P$, that is, a straight line from the origin in the first quadrant. On the coordinates of Fig. 12–4 the locus to the right of point C is an example. Petroff's model presumes thick-film lubrication, that is, no metal-to-metal contact, the surfaces being completely separated by a lubricant film.

The McKee abscissa was ZN/P (centipoise × rev/min/psi) and the value of abscissa B in Fig. 12–4 was 30. The corresponding $\mu N/P$ (reyn × rev/s/psi) is $0.33(10^{-6})$. Designers keep $\mu N/P \geq 1.7(10^{-6})$, which corresponds to $ZN/P \geq 150$. A design constraint to keep thick film lubrication is to be sure that

$$\frac{\mu N}{P} \geq 1.7(10^{-6}) \qquad (a)$$

Suppose we are operating to the right of line BA and something happens, say, an increase in lubricant temperature. This results in a lower viscosity and hence a smaller value of $\mu N/P$. The coefficient of friction decreases, not as much heat is generated in shearing the lubricant, and consequently the lubricant temperature drops. Thus the region to the right of line BA defines *stable lubrication* because variations are self-correcting.

To the left of line BA, a decrease in viscosity would increase the friction. A temperature rise would ensue, and the viscosity would be reduced still more. The result would be compounded. Thus the region to the left of line BA represents *unstable lubrication*.

It is also helpful to see that a small viscosity, and hence a small $\mu N/P$, means that the lubricant film is very thin and that there will be a greater possibility of some

Figure 12–4

The variation of the coefficient of friction f with $\mu N/P$.

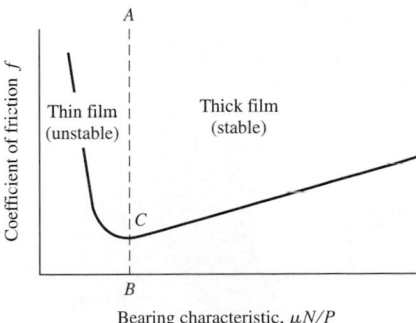

[2]S. A. McKee and T. R. McKee, "Journal Bearing Friction in the Region of Thin Film Lubrication," *SAE J.*, vol. 31, 1932, pp. (T)371–377.

metal-to-metal contact, and hence of more friction. Thus, point C represents what is probably the beginning of metal-to-metal contact as $\mu N/P$ becomes smaller.

12–5   Thick-Film Lubrication

Let us now examine the formation of a lubricant film in a journal bearing. Figure 12–5a shows a journal that is just beginning to rotate in a clockwise direction. Under starting conditions, the bearing will be dry, or at least partly dry, and hence the journal will climb or roll up the right side of the bearing as shown in Fig. 12–5a.

Now suppose a lubricant is introduced into the top of the bearing as shown in Fig. 12–5b. The action of the rotating journal is to pump the lubricant around the bearing in a clockwise direction. The lubricant is pumped into a wedge-shaped space and forces the journal over to the other side. A *minimum film thickness* h_0 occurs, not at the bottom of the journal, but displaced clockwise from the bottom as in Fig. 12–5b. This is explained by the fact that a film pressure in the converging half of the film reaches a maximum somewhere to the left of the bearing center.

Figure 12–5 shows how to decide whether the journal, under hydrodynamic lubrication, is eccentrically located on the right or on the left side of the bearing. Visualize the journal beginning to rotate. Find the side of the bearing upon which the journal tends to roll. Then, if the lubrication is hydrodynamic, mentally place the journal on the opposite side.

The nomenclature of a journal bearing is shown in Fig. 12–6. The dimension c is the *radial clearance* and is the difference in the radii of the bushing and journal. In

Figure 12–5

Formation of a film.

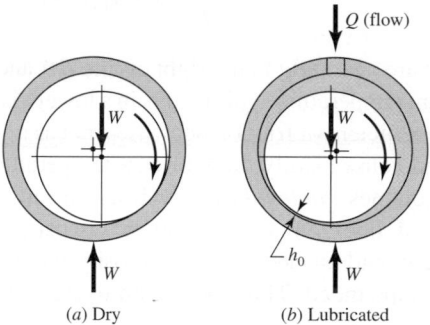

(a) Dry   (b) Lubricated

Figure 12–6

Nomenclature of a partial journal bearing.

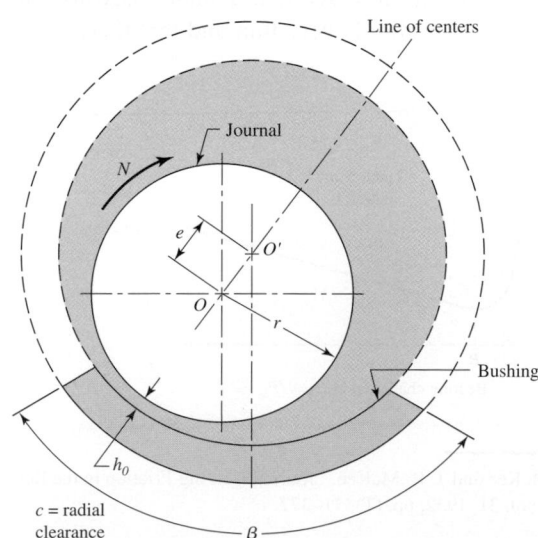

Fig. 12–6 the center of the journal is at O and the center of the bearing at O'. The distance between these centers is the *eccentricity* and is denoted by e. The *minimum film thickness* is designated by h_0, and it occurs at the line of centers. The film thickness at any other point is designated by h. We also define an *eccentricity ratio* ϵ as

$$\epsilon = \frac{e}{c}$$

The bearing shown in the figure is known as a *partial bearing*. If the radius of the bushing is the same as the radius of the journal, it is known as a *fitted bearing*. If the bushing encloses the journal, as indicated by the dashed lines, it becomes a *full bearing*. The angle β describes the angular length of a partial bearing. For example, a 120° partial bearing has the angle β equal to 120°.

12–6 Hydrodynamic Theory

The present theory of hydrodynamic lubrication originated in the laboratory of Beauchamp Tower in the early 1880s in England. Tower had been employed to study the friction in railroad journal bearings and learn the best methods of lubricating them. It was an accident or error, during the course of this investigation, that prompted Tower to look at the problem in more detail and that resulted in a discovery that eventually led to the development of the theory.

Figure 12–7 is a schematic drawing of the journal bearing that Tower investigated. It is a partial bearing, having a diameter of 4 in, a length of 6 in, and a bearing arc of 157°, and having bath-type lubrication, as shown. The coefficients of friction obtained by Tower in his investigations on this bearing were quite low, which is now not surprising. After testing this bearing, Tower later drilled a $\frac{1}{2}$-in-diameter lubricator hole through the top. But when the apparatus was set in motion, oil flowed out of this hole. In an effort to prevent this, a cork stopper was used, but this popped out, and so it was necessary to drive a wooden plug into the hole. When the wooden plug was pushed out too, Tower, at this point, undoubtedly realized that he was on the verge of discovery. A pressure gauge connected to the hole indicated a pressure of more than twice the unit bearing load. Finally, he investigated the bearing film pressures in detail throughout the bearing width and length and reported a distribution similar to that of Fig. 12–8.[3]

The results obtained by Tower had such regularity that Osborne Reynolds concluded that there must be a definite equation relating the friction, the pressure, and the

Figure 12–7

Schematic representation of the partial bearing used by Tower.

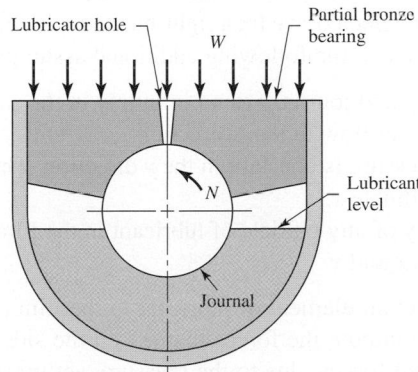

[3]Beauchamp Tower, "First Report on Friction Experiments," *Proc. Inst. Mech. Eng.*, November 1883, pp. 632–666; "Second Report," ibid., 1885, pp. 58–70; "Third Report," ibid., 1888, pp. 173–205; "Fourth Report," ibid., 1891, pp. 111–140.

Figure 12–8

Approximate pressure-
distribution curves obtained
by Tower.

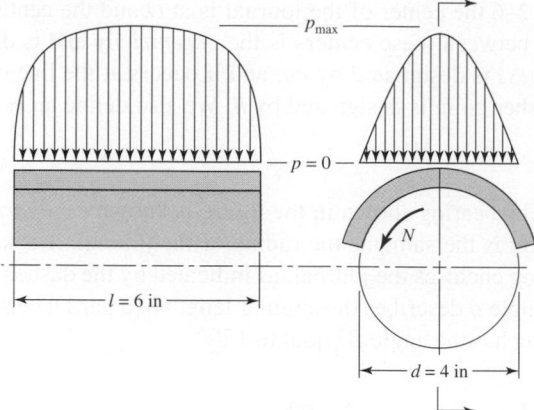

velocity. The present mathematical theory of lubrication is based upon Reynolds' work following the experiment by Tower.[4] The original differential equation, developed by Reynolds, was used by him to explain Tower's results. The solution is a challenging problem that has interested many investigators ever since then, and it is still the starting point for lubrication studies.

Reynolds pictured the lubricant as adhering to both surfaces and being pulled by the moving surface into a narrowing, wedge-shaped space so as to create a fluid or film pressure of sufficient intensity to support the bearing load. One of the important simplifying assumptions resulted from Reynolds' realization that the fluid films were so thin in comparison with the bearing radius that the curvature could be neglected. This enabled him to replace the curved partial bearing with a flat bearing, called a *plane slider bearing*. Other assumptions made were:

1 The lubricant obeys Newton's viscous effect, Eq. (12–1).
2 The forces due to the inertia of the lubricant are neglected.
3 The lubricant is assumed to be incompressible.
4 The viscosity is assumed to be constant throughout the film.
5 The pressure does not vary in the axial direction.

Figure 12–9*a* shows a journal rotating in the clockwise direction supported by a film of lubricant of variable thickness h on a partial bearing, which is fixed. We specify that the journal has a constant surface velocity U. Using Reynolds' assumption that curvature can be neglected, we fix a right-handed xyz reference system to the stationary bearing. We now make the following additional assumptions:

6 The bushing and journal extend infinitely in the z direction; this means there can be no lubricant flow in the z direction.
7 The film pressure is constant in the y direction. Thus the pressure depends only on the coordinate x.
8 The velocity of any particle of lubricant in the film depends only on the coordinates x and y.

We now select an element of lubricant in the film (Fig. 12–9*a*) of dimensions dx, dy, and dz, and compute the forces that act on the sides of this element. As shown in Fig. 12–9*b*, normal forces, due to the pressure, act upon the right and left sides of the

---

[4]Osborne Reynolds, "Theory of Lubrication, Part I," *Phil. Trans. Roy. Soc. London,* 1886.

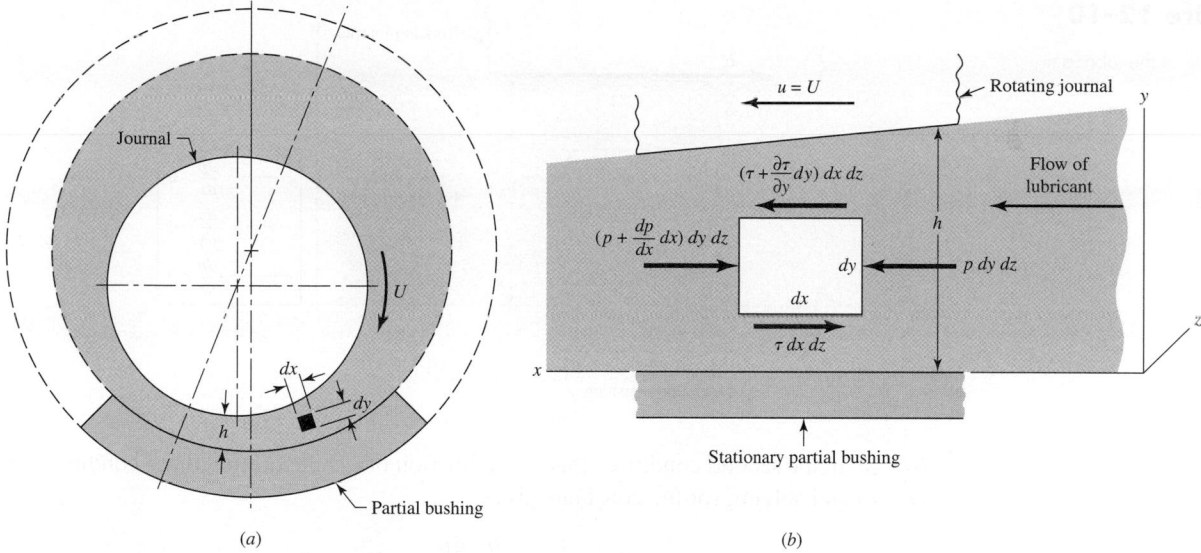

| Figure 12-9

element, and shear forces, due to the viscosity and to the velocity, act upon the top and bottom sides. Summing the forces in the x direction gives

$$\sum F_x = p\,dy\,dz - \left(p + \frac{dp}{dx}dx\right)dy\,dz - \tau\,dx\,dz + \left(\tau + \frac{\partial\tau}{\partial y}dy\right)dx\,dz = 0 \qquad (a)$$

This reduces to

$$\frac{dp}{dx} = \frac{\partial\tau}{\partial y} \qquad (b)$$

From Eq. (12–1), we have

$$\tau = \mu\frac{\partial u}{\partial y} \qquad (c)$$

where the partial derivative is used because the velocity u depends upon both x and y. Substituting Eq. (c) in Eq. (b), we obtain

$$\frac{dp}{dx} = \mu\frac{\partial^2 u}{\partial y^2} \qquad (d)$$

Holding x constant, we now integrate this expression twice with respect to y. This gives

$$\frac{\partial u}{\partial y} = \frac{1}{\mu}\frac{dp}{dx}y + C_1$$

$$u = \frac{1}{2\mu}\frac{dp}{dx}y^2 + C_1 y + C_2 \qquad (e)$$

Note that the act of holding x constant means that C_1 and C_2 can be functions of x. We now assume that there is no slip between the lubricant and the boundary surfaces. This gives two sets of boundary conditions for evaluating the constants C_1 and C_2:

$$\text{At} \quad y = 0, \ u = 0$$
$$\text{At} \quad y = h, \ u = U \qquad (f)$$

Figure 12–10

Velocity of the lubricant.

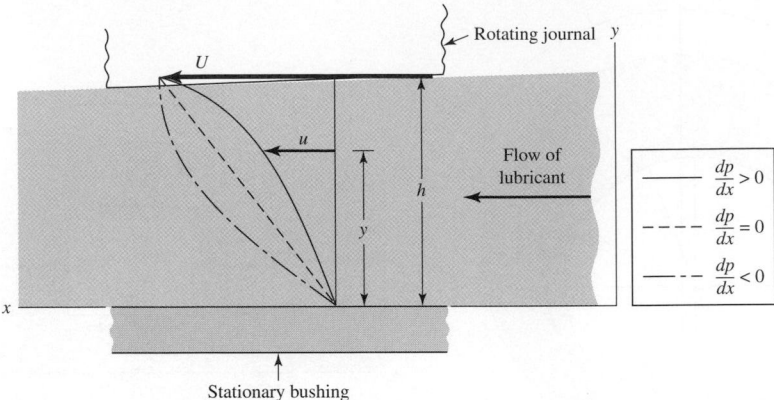

Notice, in the second condition, that h is a function of x. Substituting these conditions in Eq. (e) and solving for the constants gives

$$C_1 = \frac{U}{h} - \frac{h}{2\mu} \frac{dp}{dx} \qquad C_2 = 0$$

or

$$u = \frac{1}{2\mu} \frac{dp}{dx}(y^2 - hy) + \frac{U}{h} y \qquad (12\text{–}9)$$

This equation gives the velocity distribution of the lubricant in the film as a function of the coordinate y and the pressure gradient dp/dx. The equation shows that the velocity distribution across the film (from $y = 0$ to $y = h$) is obtained by superposing a parabolic distribution onto a linear distribution. Figure 12–10 shows the superposition of these distributions to obtain the velocity for particular values of x and dp/dx. In general, the parabolic term may be additive or subtractive to the linear term, depending upon the sign of the pressure gradient. When the pressure is maximum, $dp/dx = 0$ and the velocity is

$$u = \frac{U}{h} y \qquad (g)$$

which is a linear relation.

We next define Q as the volume of lubricant flowing in the x direction per unit time. By using a width of unity in the z direction, the volume may be obtained by the expression

$$Q = \int_0^h u \, dy \qquad (h)$$

Substituting the value of u from Eq. (12–9) and integrating gives

$$Q = \frac{Uh}{2} - \frac{h^3}{12\mu} \frac{dp}{dx} \qquad (i)$$

The next step uses the assumption of an incompressible lubricant and states that the flow is the same for any cross section. Thus

$$\frac{dQ}{dx} = 0$$

From Eq. (*i*),

$$\frac{dQ}{dx} = \frac{U}{2}\frac{dh}{dx} - \frac{d}{dx}\left(\frac{h^3}{12\mu}\frac{dp}{dx}\right) = 0$$

or

$$\frac{d}{dx}\left(\frac{h^3}{\mu}\frac{dp}{dx}\right) = 6U\frac{dh}{dx} \tag{12-10}$$

which is the classical Reynolds equation for one-dimensional flow. It neglects side leakage, that is, flow in the z direction. A similar development is used when side leakage is not neglected. The resulting equation is

$$\frac{\partial}{\partial x}\left(\frac{h^3}{\mu}\frac{\partial p}{\partial x}\right) + \frac{\partial}{\partial z}\left(\frac{h^3}{\mu}\frac{\partial p}{\partial z}\right) = 6U\frac{\partial h}{\partial x} \tag{12-11}$$

There is no general analytical solution to Eq. (12–11); approximate solutions have been obtained by using electrical analogies, mathematical summations, relaxation methods, and numerical and graphical methods. One of the important solutions is due to Sommerfeld[5] and may be expressed in the form

$$\frac{r}{c}f = \phi\left[\left(\frac{r}{c}\right)^2\frac{\mu N}{P}\right] \tag{12-12}$$

where ϕ indicates a functional relationship. Sommerfeld found the functions for half-bearings and full bearings by using the assumption of no side leakage.

12-7 Design Considerations

We may distinguish between two groups of variables in the design of sliding bearings. In the first group are those whose values either are given or are under the control of the designer. These are:

1 The viscosity μ
2 The load per unit of projected bearing area, P
3 The speed N
4 The bearing dimensions r, c, β, and l

Of these four variables, the designer usually has no control over the speed, because it is specified by the overall design of the machine. Sometimes the viscosity is specified in advance, as, for example, when the oil is stored in a sump and is used for lubricating and cooling a variety of bearings. The remaining variables, and sometimes the viscosity, may be controlled by the designer and are therefore the *decisions* the designer makes. In other words, when these four decisions are made, the design is complete.

In the second group are the dependent variables. The designer cannot control these except indirectly by changing one or more of the first group. These are:

1 The coefficient of friction f
2 The temperature rise ΔT
3 The volume flow rate of oil Q
4 The minimum film thickness h_0

---

[5]A. Sommerfeld, "Zur Hydrodynamischen Theorie der Schmiermittel-Reibung" ("On the Hydrodynamic Theory of Lubrication"), *Z. Math. Physik,* vol. 50, 1904, pp. 97–155.

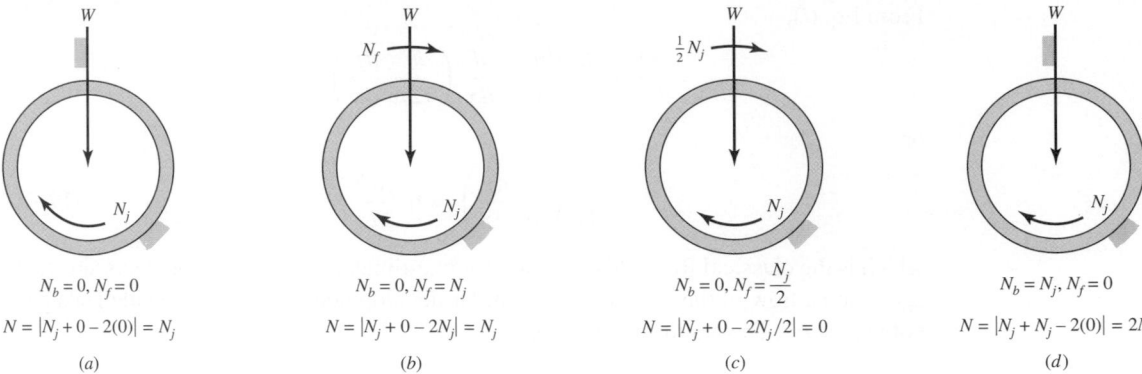

Figure 12–11

How the significant speed varies. (a) Common bearing case. (b) Load vector moves at the same speed as the journal. (c) Load vector moves at half journal speed, no load can be carried. (d) Journal and bushing move at same speed, load vector stationary, capacity halved.

This group of variables tells us how well the bearing is performing, and hence we may regard them as *performance factors*. Certain limitations on their values must be imposed by the designer to ensure satisfactory performance. These limitations are specified by the characteristics of the bearing materials and of the lubricant. The fundamental problem in bearing design, therefore, is to define satisfactory limits for the second group of variables and then to decide upon values for the first group such that these limitations are not exceeded.

Significant Angular Speed

In the next section we will examine several important charts relating key variables to the Sommerfeld number. To this point we have assumed that only the journal rotates and it is the journal rotational speed that is used in the Sommerfeld number. It has been discovered that the angular speed N that is significant to hydrodynamic film bearing performance is[6]

$$N = |N_j + N_b - 2N_f| \tag{12–13}$$

where N_j = journal angular speed, rev/s

N_b = bearing angular speed, rev/s

N_f = load vector angular speed, rev/s

When determining the Sommerfeld number for a general bearing, use Eq. (12–13) when entering N. Figure 12–11 shows several situations for determining N.

Trumpler's Design Criteria for Journal Bearings

Because the bearing assembly creates the lubricant pressure to carry a load, it reacts to loading by changing its eccentricity, which reduces the minimum film thickness h_0 until the load is carried. What is the limit of smallness of h_0? Close examination reveals that the moving adjacent surfaces of the journal and bushing are not smooth but consist of a series of asperities that pass one another, separated by a lubricant film. In starting a

---

[6]Paul Robert Trumpler, *Design of Film Bearings,* Macmillan, New York, 1966, pp. 103–119.

bearing under load from rest there is metal-to-metal contact and surface asperities are broken off, free to move and circulate with the oil. Unless a filter is provided, this debris accumulates. Such particles have to be free to tumble at the section containing the minimum film thickness without snagging in a togglelike configuration, creating additional damage and debris. Trumpler, an accomplished bearing designer, provides a throat of at least 200 μin to pass particles from ground surfaces.[7] He also provides for the influence of size (tolerances tend to increase with size) by stipulating

$$h_0 \geq 0.0002 + 0.000\,04d \text{ in} \tag{a}$$

where d is the journal diameter in inches.

A lubricant is a mixture of hydrocarbons that reacts to increasing temperature by vaporizing the lighter components, leaving behind the heavier. This process (bearings have lots of time) slowly increases the viscosity of the remaining lubricant, which increases heat generation rate and elevates lubricant temperatures. This sets the stage for future failure. For light oils, Trumpler limits the maximum film temperature T_{max} to

$$T_{max} \leq 250°\text{F} \tag{b}$$

Some oils can operate at slightly higher temperatures. Always check with the lubricant manufacturer.

A journal bearing often consists of a ground steel journal working against a softer, usually nonferrous, bushing. In starting under load there is metal-to-metal contact, abrasion, and the generation of wear particles, which, over time, can change the geometry of the bushing. The starting load divided by the projected area is limited to

$$\frac{W_{st}}{lD} \leq 300 \text{ psi} \tag{c}$$

If the load on a journal bearing is suddenly increased, the increase in film temperature in the annulus is immediate. Since ground vibration due to passing trucks, trains, and earth tremors is often present, Trumpler used a design factor of 2 or more on the running load, but not on the starting load of Eq. (c):

$$n_d \geq 2 \tag{d}$$

Many of Trumpler's designs are operating today, long after his consulting career is over; clearly they constitute good advice to the beginning designer.

12-8 The Relations of the Variables

Before proceeding to the problem of design, it is necessary to establish the relationships between the variables. Albert A. Raimondi and John Boyd, of Westinghouse Research Laboratories, used an iteration technique to solve Reynolds' equation on the digital computer.[8] This is the first time such extensive data have been available for use by designers, and consequently we shall employ them in this book.[9]

---

[7]Op. cit., pp. 192–194.

[8]A. A. Raimondi and John Boyd, "A Solution for the Finite Journal Bearing and Its Application to Analysis and Design, Parts I, II, and III," *Trans. ASLE,* vol. 1, no. 1, in *Lubrication Science and Technology,* Pergamon, New York, 1958, pp. 159–209.

[9]See also the earlier companion paper, John Boyd and Albert A. Raimondi, "Applying Bearing Theory to the Analysis and Design of Journal Bearings, Part I and II," *J. Appl. Mechanics,* vol. 73, 1951, pp. 298–316.

The Raimondi and Boyd papers were published in three parts and contain 45 detailed charts and 6 tables of numerical information. In all three parts, charts are used to define the variables for length-diameter (l/d) ratios of 1:4, 1:2, and 1 and for beta angles of 60 to 360°. Under certain conditions the solution to the Reynolds equation gives negative pressures in the diverging portion of the oil film. Since a lubricant cannot usually support a tensile stress, Part III of the Raimondi-Boyd papers assumes that the oil film is ruptured when the film pressure becomes zero. Part III also contains data for the infinitely long bearing; since it has no ends, this means that there is no side leakage. The charts appearing in this book are from Part III of the papers, and are for full journal bearings $(\beta = 360°)$ only. Space does not permit the inclusion of charts for partial bearings. This means that you must refer to the charts in the original papers when beta angles of less than 360° are desired. The notation is very nearly the same as in this book, and so no problems should arise.

Viscosity Charts (Figs. 12–12 to 12–14)

One of the most important assumptions made in the Raimondi-Boyd analysis is that *viscosity of the lubricant is constant as it passes through the bearing*. But since work is done on the lubricant during this flow, the temperature of the oil is higher when it leaves the loading zone than it was on entry. And the viscosity charts clearly indicate that the viscosity drops off significantly with a rise in temperature. Since the analysis is based on a constant viscosity, our problem now is to determine the value of viscosity to be used in the analysis.

Some of the lubricant that enters the bearing emerges as a side flow, which carries away some of the heat. The balance of the lubricant flows through the load-bearing zone and carries away the balance of the heat generated. In determining the viscosity to be used we shall employ a temperature that is the average of the inlet and outlet temperatures, or

$$T_{av} = T_1 + \frac{\Delta T}{2} \tag{12–14}$$

where T_1 is the inlet temperature and ΔT is the temperature rise of the lubricant from inlet to outlet. Of course, the viscosity used in the analysis must correspond to T_{av}.

Viscosity varies considerably with temperature in a nonlinear fashion. The ordinates in Figs. 12–12 to 12–14 are not logarithmic, as the decades are of differing vertical length. These graphs represent the temperature versus viscosity functions for common grades of lubricating oils in both customary engineering and SI units. We have the temperature versus viscosity function only in graphical form, unless curve fits are developed. See Table 12–1.

One of the objectives of lubrication analysis is to determine the oil outlet temperature when the oil and its inlet temperature are specified. This is a trial-and-error type of problem. In an analysis, the temperature rise will first be estimated. This allows for the viscosity to be determined from the chart. With the value of the viscosity, the analysis is performed where the temperature rise is then computed. With this, a new estimate of the temperature rise is established. This process is continued until the estimated and computed temperatures agree.

To illustrate, suppose we have decided to use SAE 30 oil in an application in which the oil inlet temperature is $T_1 = 180°F$. We begin by estimating that the temperature rise

Figure 12–12

Viscosity–temperature chart in U.S. customary units. *(Raimondi and Boyd.)*

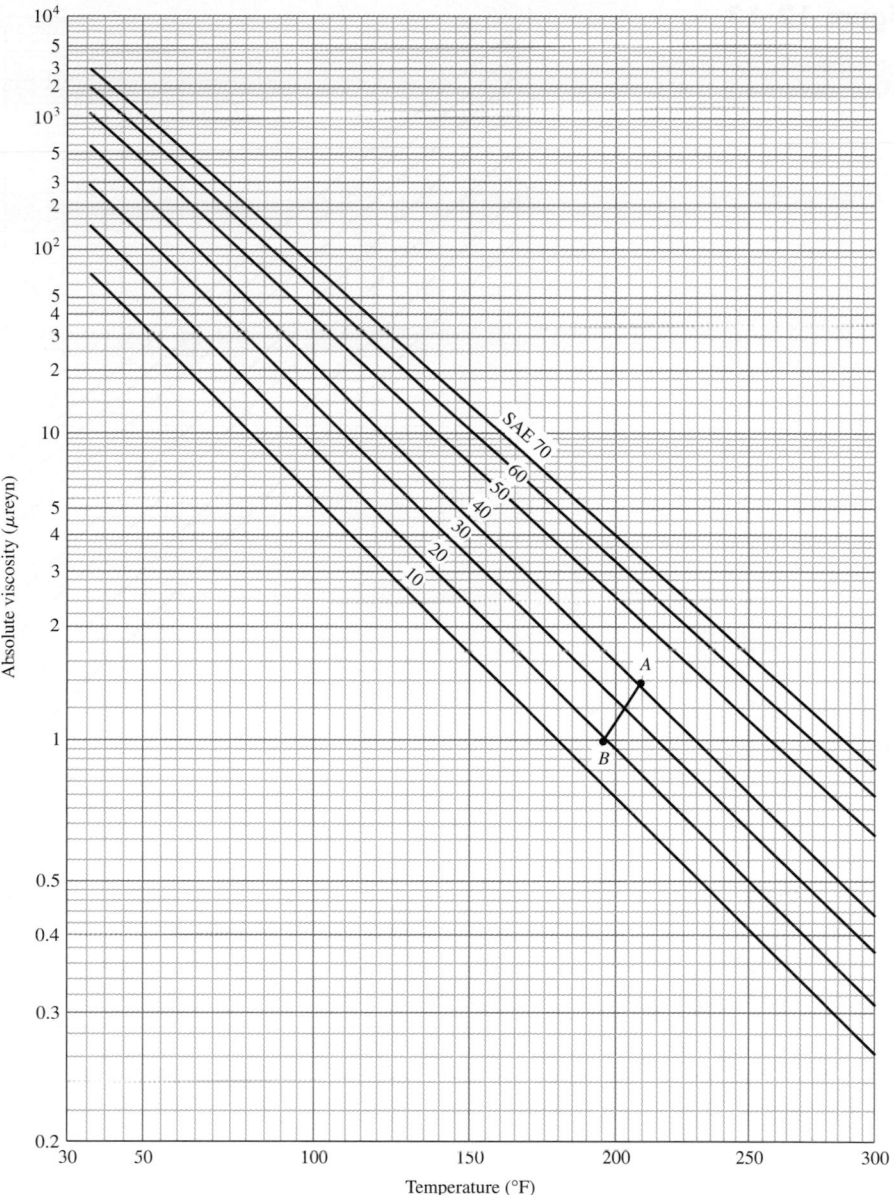

will be $\Delta T = 30°F$. Then, from Eq. (12–14),

$$T_{av} = T_1 + \frac{\Delta T}{2} = 180 + \frac{30}{2} = 195°F$$

From Fig. 12–12 we follow the SAE 30 line and find that $\mu = 1.40$ μreyn at 195°F. So we use this viscosity (in an analysis to be explained in detail later) and find that the temperature rise is actually $\Delta T = 54°F$. Thus Eq. (12–14) gives

$$T_{av} = 180 + \frac{54}{2} = 207°F$$

Figure 12–13

Viscosity–temperature chart in SI units. *(Adapted from Fig. 12–12.)*

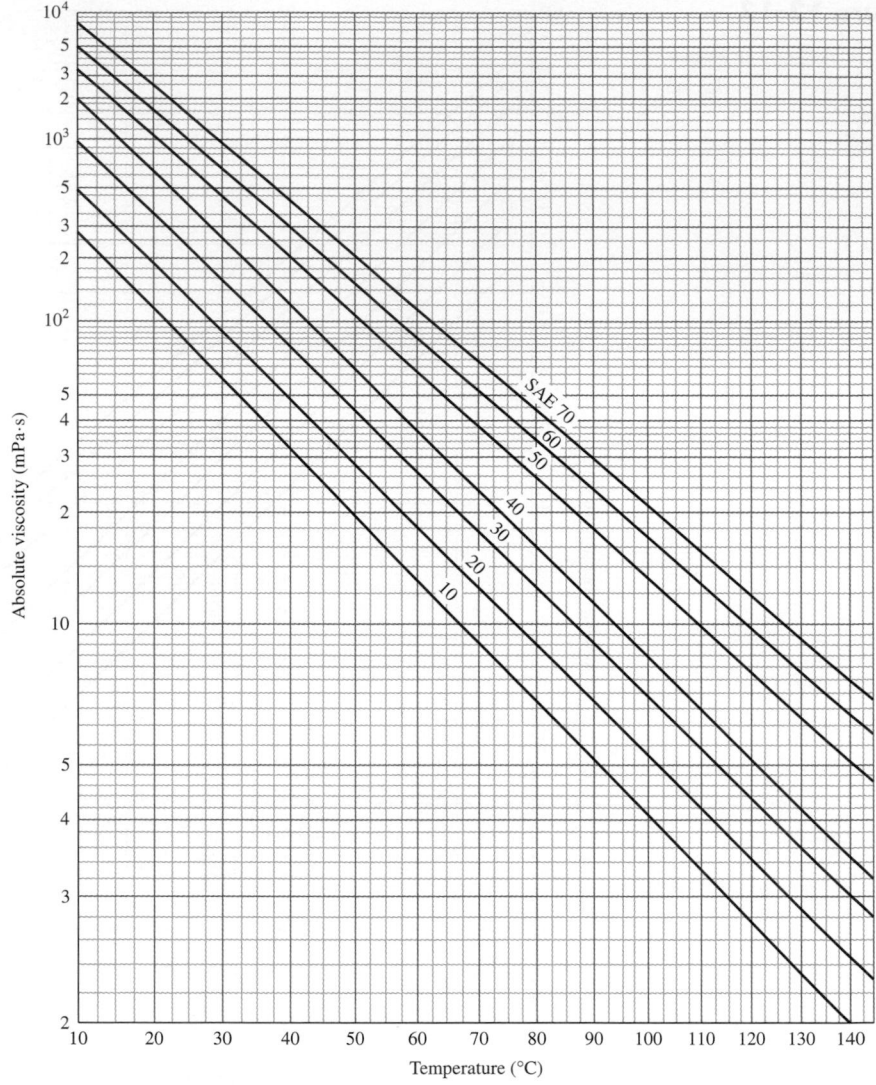

This corresponds to point *A* on Fig. 12–12, which is above the SAE 30 line and indicates that the viscosity used in the analysis was too high.

For a second guess, try $\mu = 1.00\ \mu$reyn. Again we run through an analysis and this time find that $\Delta T = 30°$F. This gives an average temperature of

$$T_{av} = 180 + \frac{30}{2} = 195°\text{F}$$

and locates point *B* on Fig. 12–12.

If points *A* and *B* are fairly close to each other and on opposite sides of the SAE 30 line, a straight line can be drawn between them with the intersection locating the correct values of viscosity and average temperature to be used in the analysis. For this illustration, we see from the viscosity chart that they are $T_{av} = 203°$F and $\mu = 1.20\ \mu$reyn.

Figure 12-14

Chart for multiviscosity lubricants. This chart was derived from known viscosities at two points, 100 and 210°F, and the results are believed to be correct for other temperatures.

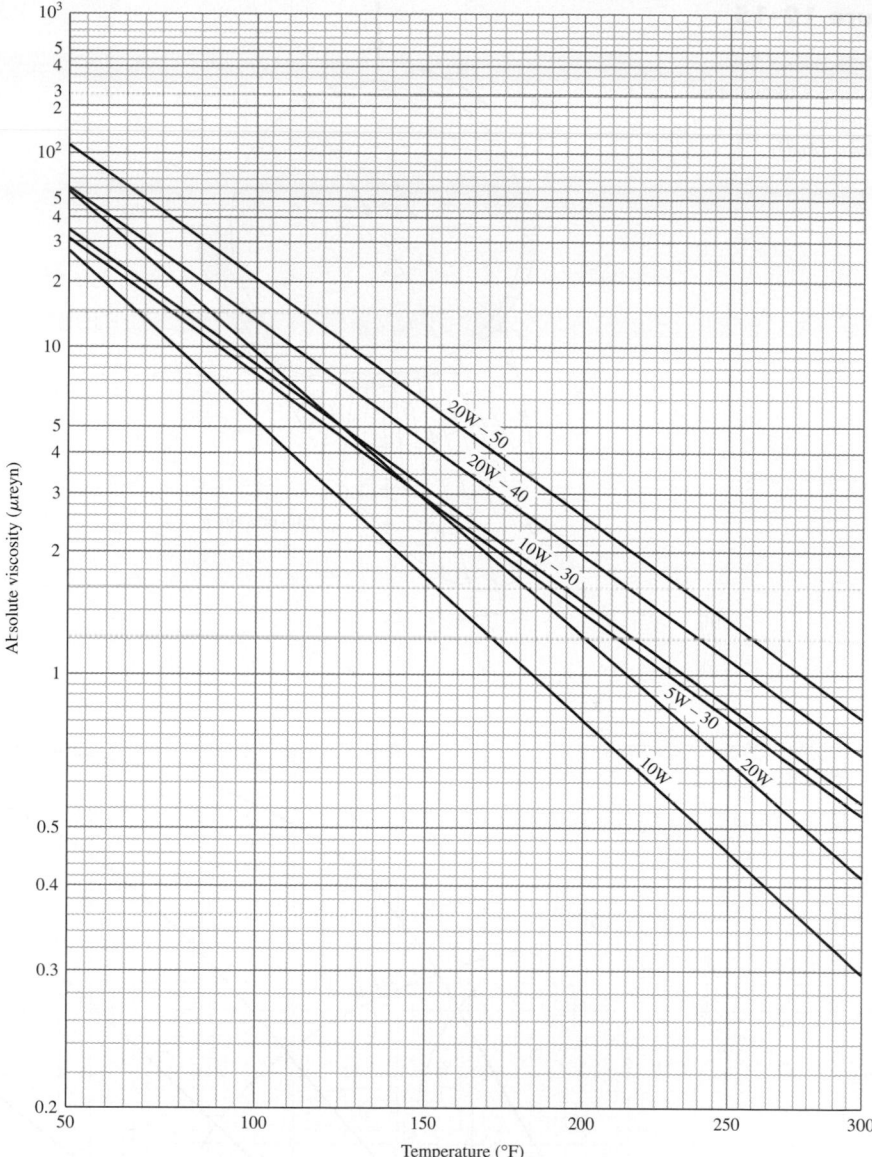

Table 12-1

Curve Fits* to Approximate the Viscosity versus Temperature Functions for SAE Grades 10 to 60

Source: A. S. Seireg and S. Dandage, "Empirical Design Procedure for the Thermodynamic Behavior of Journal Bearings," *J. Lubrication Technology,* vol. 104, April 1982, pp. 135–148.

Oil Grade, SAE	Viscosity μ_0, reyn	Constant b, °F
10	$0.0158(10^{-6})$	1157.5
20	$0.0136(10^{-6})$	1271.6
30	$0.0141(10^{-6})$	1360.0
40	$0.0121(10^{-6})$	1474.4
50	$0.0170(10^{-6})$	1509.6
60	**$0.0187(10^{-6})$**	**1564.0**

*$\mu = \mu_0 \exp [b/(T + 95)]$, T in °F.

Figure 12–15

Polar diagram of the film–pressure distribution showing the notation used. *(Raimondi and Boyd.)*

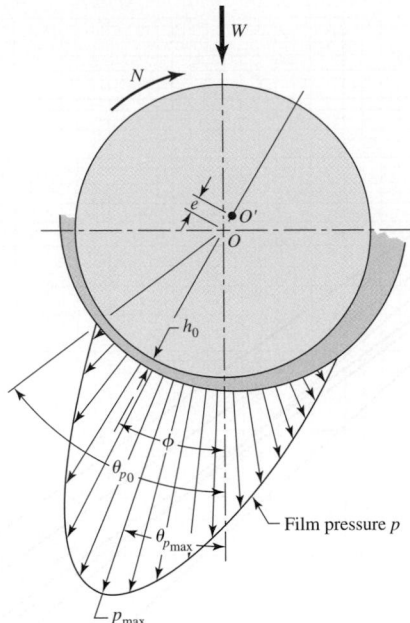

Figure 12–16

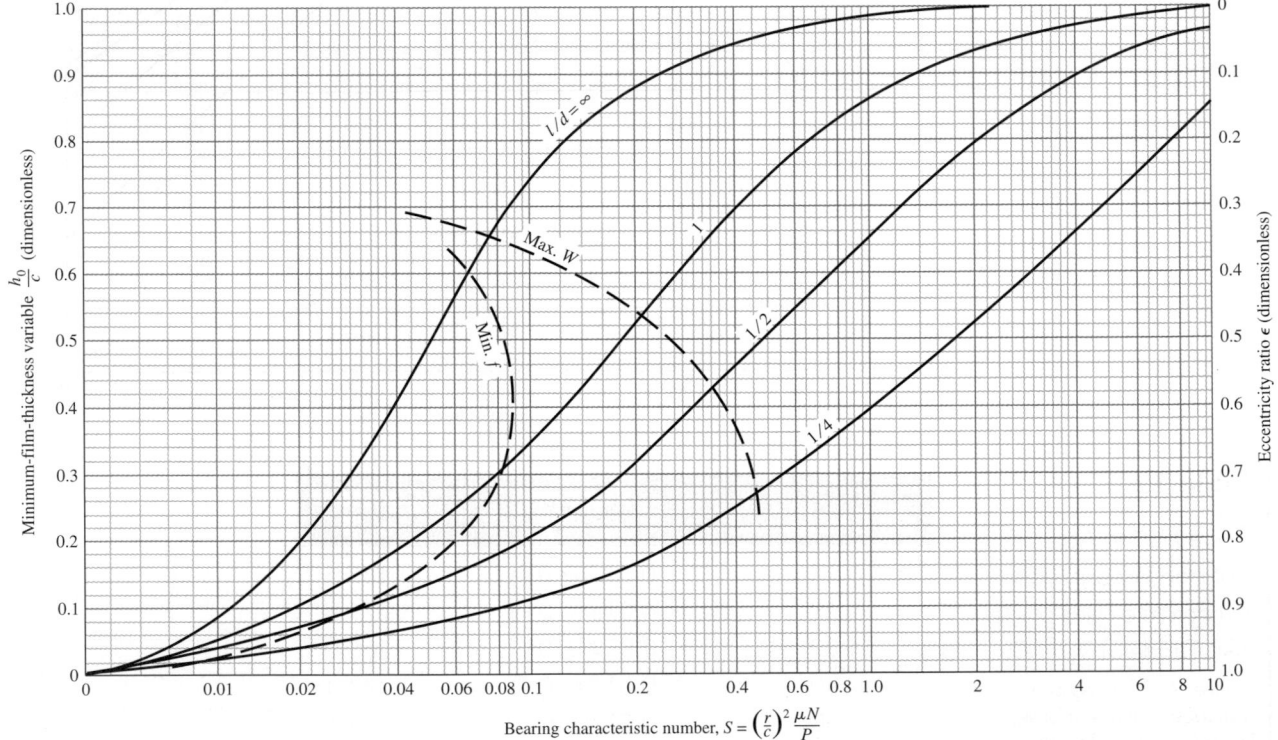

Chart for minimum film-thickness variable and eccentricity ratio. The left boundary of the zone defines the optimal h_0 for minimum friction; the right boundary is optimum h_0 for load. *(Raimondi and Boyd.)*

Figure 12–17

Chart for determining the position of the minimum film thickness h_0. *(Raimondi and Boyd.)*

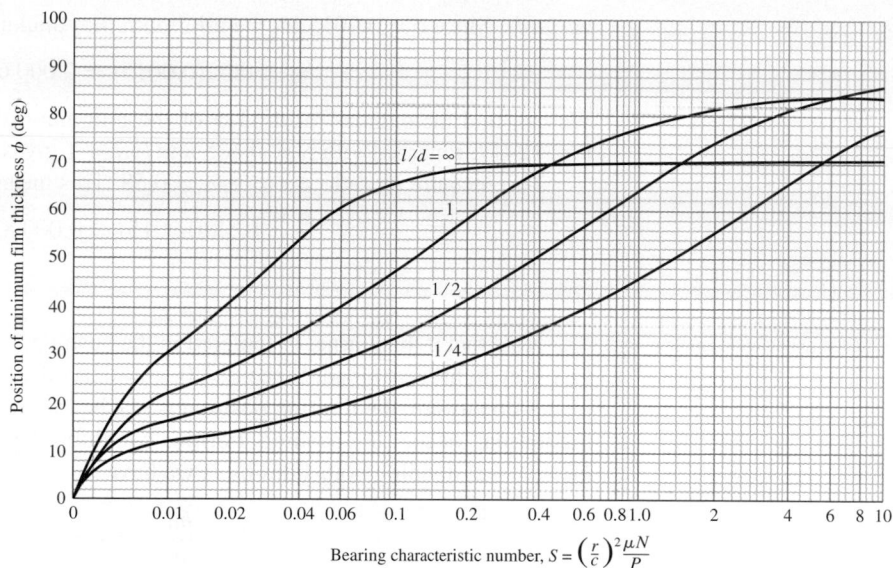

Bearing characteristic number, $S = \left(\dfrac{r}{c}\right)^2 \dfrac{\mu N}{P}$

The remaining charts from Raimondi and Boyd relate several variables to the Sommerfeld number. These variables are

Minimum film thickness (Figs. 12–16 and 12–17)
Coefficient of friction (Fig. 12–18)
Lubricant flow (Figs. 12–19 and 12–20)
Film pressure (Figs. 12–21 and 12–22)

Figure 12–15 shows the notation used for the variables. We will describe the use of these curves in a series of four examples using the same set of given parameters.

Minimum Film Thickness

In Fig. 12–16, *the minimum film-thickness* variable h_0/c and *eccentricity ratio* $\epsilon = e/c$ are plotted against the Sommerfeld number S with contours for various values of l/d. The corresponding angular position of the minimum film thickness is found in Fig. 12–17.

EXAMPLE 12–1

Determine h_0 and e using the following given parameters: $\mu = 4\ \mu$reyn, $N = 30$ rev/s, $W = 500$ lbf (bearing load), $r = 0.75$ in, $c = 0.0015$ in, and $l = 1.5$ in.

Solution

The nominal bearing pressure (in projected area of the journal) is

$$P = \frac{W}{2rl} = \frac{500}{2(0.75)1.5} = 222 \text{ psi}$$

The Sommerfeld number is, from Eq. (12–7), where $N = N_j = 30$ rev/s,

$$S = \left(\frac{r}{c}\right)^2 \left(\frac{\mu N}{P}\right) = \left(\frac{0.75}{0.0015}\right)^2 \left[\frac{4(10^{-6})30}{222}\right] = 0.135$$

Also, $l/d = 1.50/[2(0.75)] = 1$. Entering Fig. 12–16 with $S = 0.135$ and $l/d = 1$ gives $h_0/c = 0.42$ and $\epsilon = 0.58$. The quantity h_0/c is called the *minimum film thickness variable.* Since $c = 0.0015$ in, the minimum film thickness h_0 is

$$h_0 = 0.42(0.0015) = 0.000\ 63 \text{ in}$$

We can find the angular location ϕ of the minimum film thickness from the chart of Fig. 12–17. Entering with $S = 0.135$ and $l/d = 1$ gives $\phi = 53°$.

The eccentricity ratio is $\epsilon = e/c = 0.58$. This means the eccentricity e is

$$e = 0.58(0.0015) = 0.000\ 87 \text{ in}$$

Note that if the journal is centered in the bushing, $e = 0$ and $h_0 = c$, corresponding to a very light (zero) load. Since $e = 0$, $\epsilon = 0$. As the load is increased the journal displaces downward; the limiting position is reached when $h_0 = 0$ and $e = c$, that is, when the journal touches the bushing. For this condition the eccentricity ratio is unity. Since $h_0 = c - e$, dividing both sides by c, we have

$$\frac{h_0}{c} = 1 - \epsilon$$

Design optima are sometimes *maximum load,* which is a load-carrying characteristic of the bearing, and sometimes *minimum parasitic power loss* or *minimum coefficient of friction.* Dashed lines appear on Fig. 12–16 for maximum load and minimum coefficient of friction, so you can easily favor one of maximum load or minimum coefficient of friction, but not both. The zone between the two dashed-line contours might be considered a desirable location for a design point.

Coefficient of Friction

The friction chart, Fig. 12–18, has the *friction variable* $(r/c)f$ plotted against Sommerfeld number S with contours for various values of the l/d ratio.

EXAMPLE 12–2

Using the parameters given in Ex. 12–1, determine the coefficient of friction, the torque to overcome friction, and the power loss to friction.

Solution

We enter Fig. 12–18 with $S = 0.135$ and $l/d = 1$ and find $(r/c)f = 3.50$. The coefficient of friction f is

$$f = 3.50\ c/r = 3.50(0.0015/0.75) = 0.0070$$

The friction torque on the journal is

$$T = fWr = 0.007(500)0.75 = 2.62 \text{ lbf} \cdot \text{in}$$

The power loss in horsepower is

$$(hp)_{\text{loss}} = \frac{TN}{1050} = \frac{2.62(30)}{1050} = 0.075 \text{ hp}$$

or, expressed in Btu/s,

$$H = \frac{2\pi TN}{778(12)} = \frac{2\pi(2.62)30}{778(12)} = 0.0529 \text{ Btu/s}$$

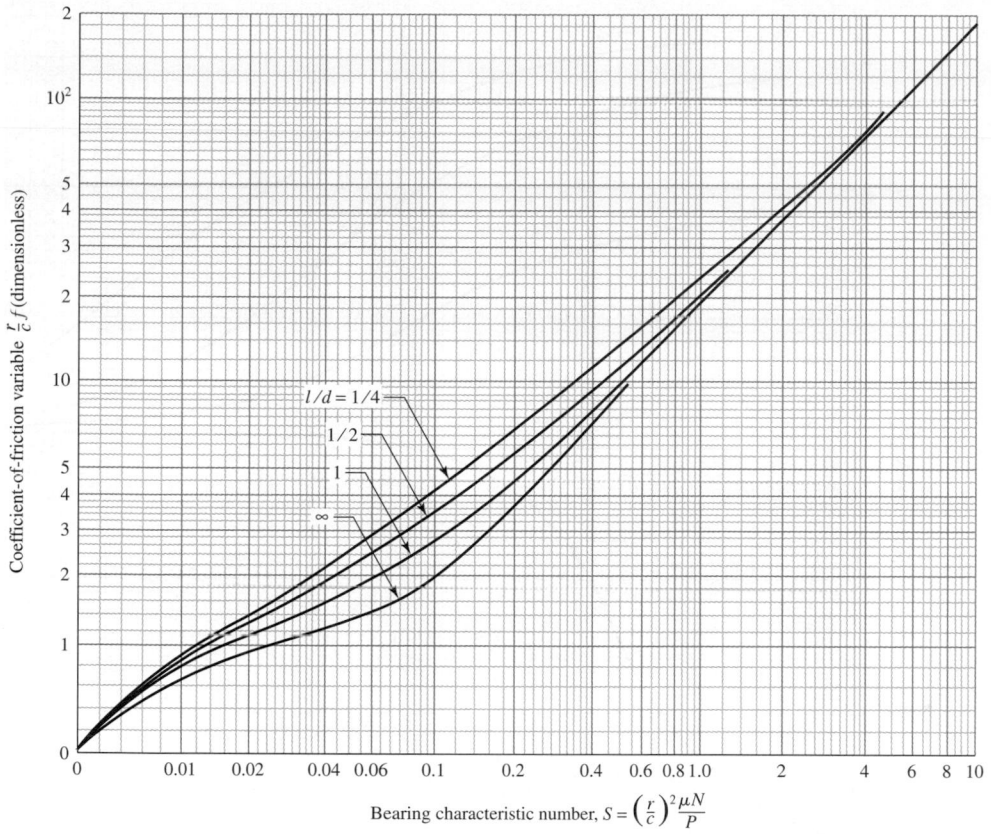

Figure 12–18

Chart for coefficient-of-friction variable; note that Petroff's equation is the asymptote. *(Raimondi and Boyd.)*

Lubricant Flow

Figures 12–19 and 12–20 are used to determine the lubricant flow and side flow.

EXAMPLE 12–3

Continuing with the parameters of Ex. 12–1, determine the total volumetric flow rate Q and the side flow rate Q_s.

Solution

To estimate the lubricant flow, enter Fig. 12–19 with $S = 0.135$ and $l/d = 1$ to obtain $Q/(rcNl) = 4.28$. The total volumetric flow rate is

$$Q = 4.28rcNl = 4.28(0.75)0.0015(30)1.5 = 0.217 \text{ in}^3/\text{s}$$

From Fig. 12–20 we find the *flow ratio* $Q_s/Q = 0.655$ and Q_s is

$$Q_s = 0.655Q = 0.655(0.217) = 0.142 \text{ in}^3/\text{s}$$

Figure 12–19

Chart for flow variable. *Note:* Not for pressure-fed bearings. *(Raimondi and Boyd.)*

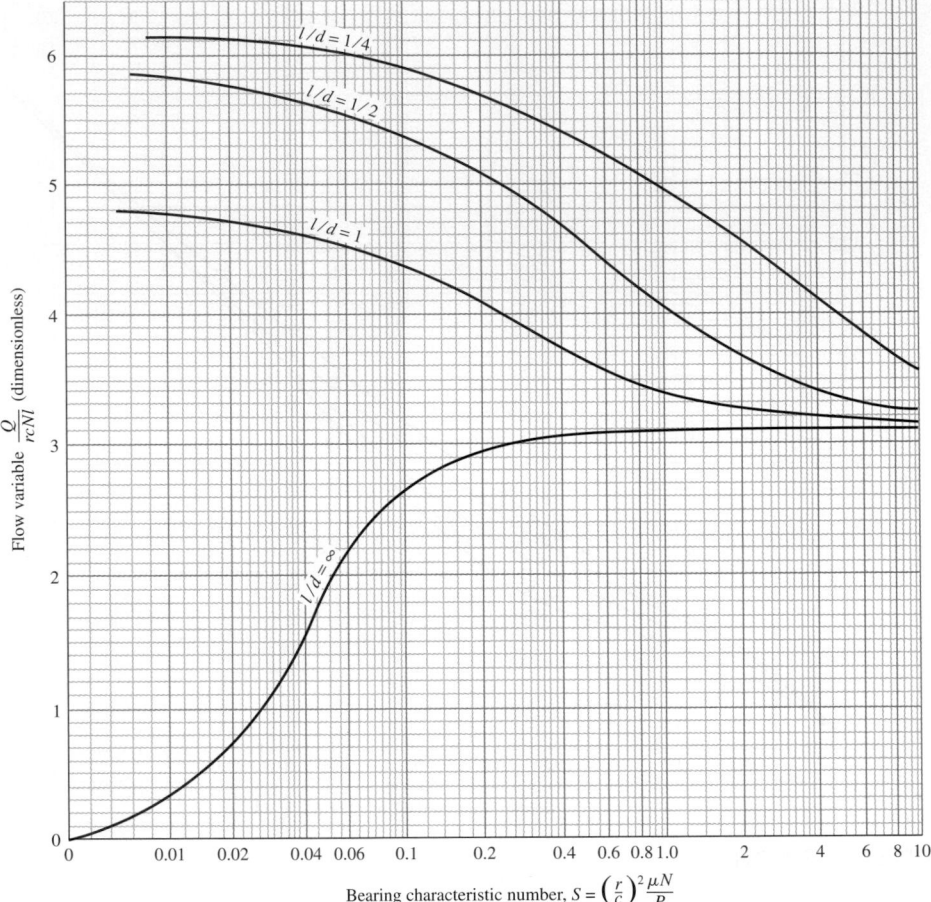

Figure 12–20

Chart for determining the ratio of side flow to total flow. *(Raimondi and Boyd.)*

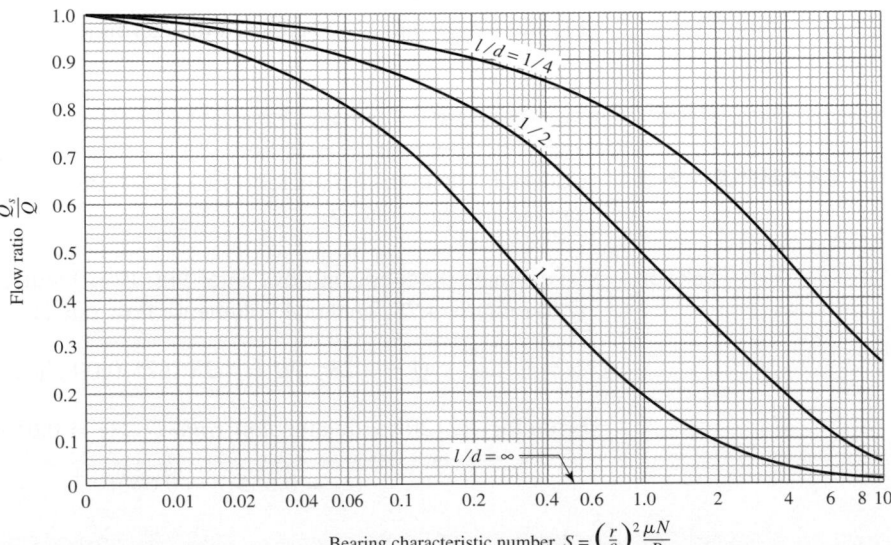

Figure 12–21

Chart for determining the maximum film pressure. *Note: Not for pressure-fed bearings.* *(Raimondi and Boyd.)*

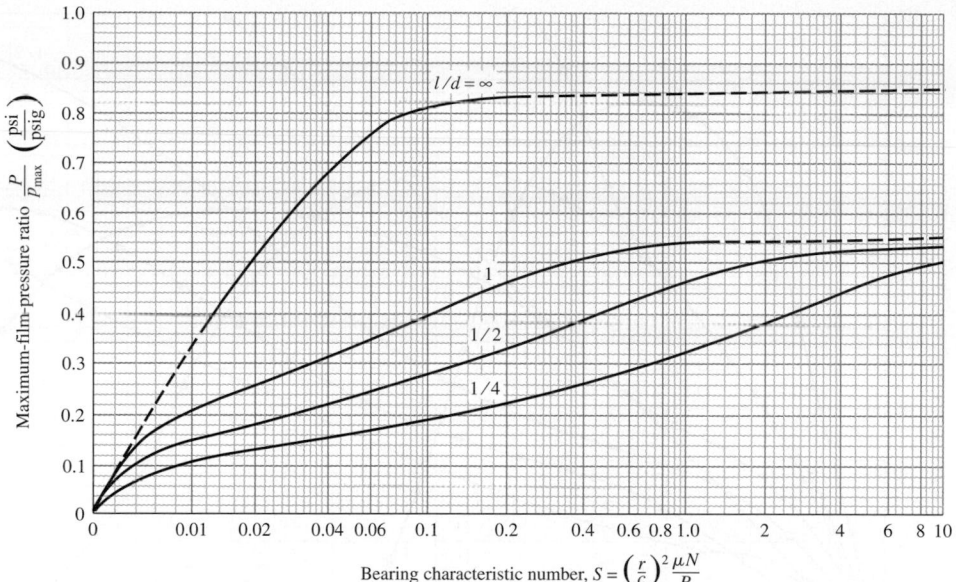

The side leakage Q_s is from the lower part of the bearing, where the internal pressure is above atmospheric pressure. The leakage forms a fillet at the journal-bushing external junction, and it is carried by journal motion to the top of the bushing, where the internal pressure is below atmospheric pressure and the gap is much larger, to be "sucked in" and returned to the lubricant sump. That portion of side leakage that leaks away from the bearing has to be made up by adding oil to the bearing sump periodically by maintenance personnel.

Film Pressure

The maximum pressure developed in the film can be estimated by finding the pressure ratio P/p_{max} from the chart in Fig. 12–21. The locations where the terminating and maximum pressures occur, as defined in Fig 12–15, are determined from Fig. 12–22.

EXAMPLE 12–4

Using the parameters given in Ex. 12–1, determine the maximum film pressure and the locations of the maximum and terminating pressures.

Solution

Entering Fig. 12–21 with $S = 0.135$ and $l/d = 1$, we find $P/p_{max} = 0.42$. The maximum pressure p_{max} is therefore

$$p_{max} = \frac{P}{0.42} = \frac{222}{0.42} = 529 \text{ psi}$$

With $S = 0.135$ and $l/d = 1$, from Fig. 12–22, $\theta_{p_{max}} = 18.5°$ and the terminating position θ_{p_0} is 75°.

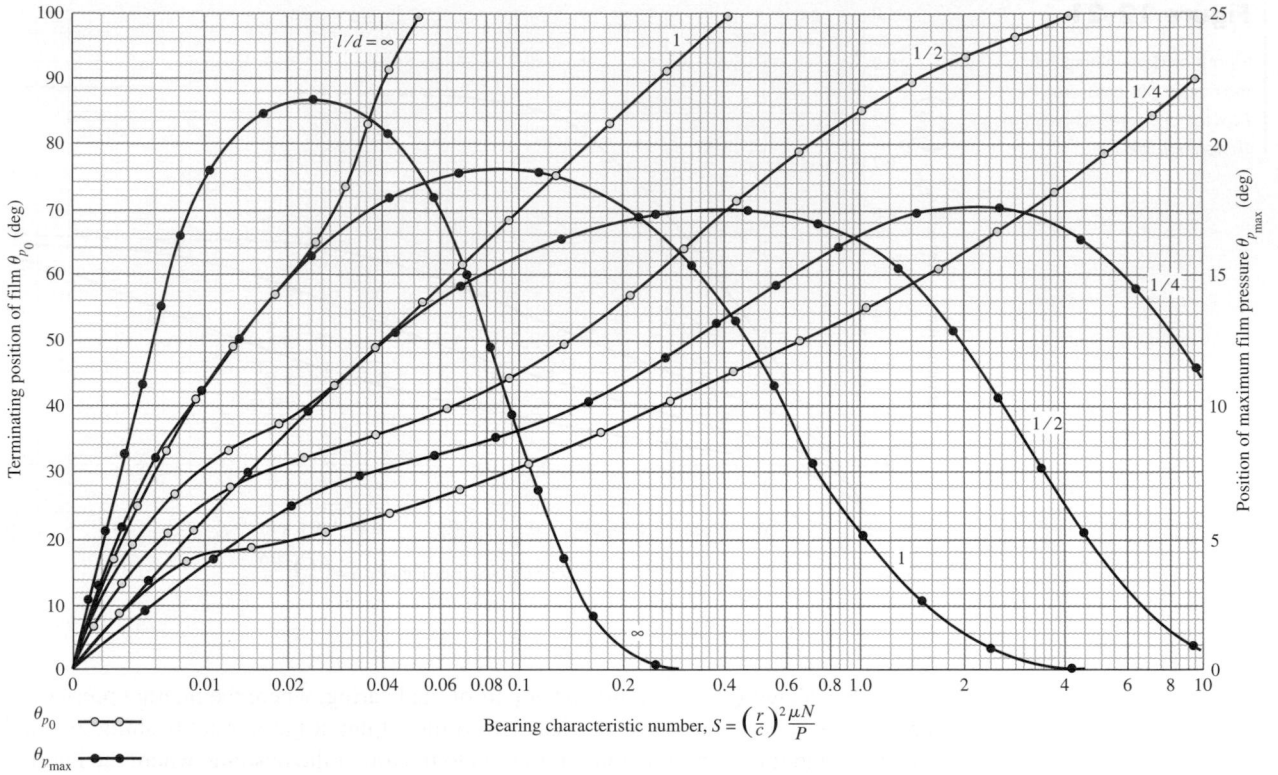

Figure 12–22

Chart for finding the terminating position of the lubricant film and the position of maximum film pressure. *(Raimondi and Boyd.)*

Examples 12–1 to 12–4 demonstrate how the Raimondi and Boyd charts are used. It should be clear that we do not have journal-bearing parametric relations as equations, but in the form of charts. Moreover, the examples were simple because the steady-state equivalent viscosity was given. We will now show how the average film temperature (and the corresponding viscosity) is found from energy considerations.

Lubricant Temperature Rise

The temperature of the lubricant rises until the rate at which work is done by the journal on the film through fluid shear is the same as the rate at which heat is transferred to the greater surroundings. The specific arrangement of the bearing plumbing affects the quantitative relationships. See Fig. 12–23. A lubricant sump (internal or external to the bearing housing) supplies lubricant at sump temperature T_s to the bearing annulus at temperature $T_s = T_1$. The lubricant passes once around the bushing and is delivered at a higher lubricant temperature $T_1 + \Delta T$ to the sump. Some of the lubricant leaks out of the bearing at a mixing-cup temperature of $T_1 + \Delta T/2$ and is returned to the sump. The sump may be a keyway-like groove in the bearing cap or a larger chamber up to half the bearing circumference. It can occupy "all" of the bearing cap of a split bearing. In such a bearing the side leakage occurs from the lower portion and is sucked back in, into the ruptured film arc. The sump could be well removed from the journal-bushing interface.

Figure 12–23

Schematic of a journal bearing with an external sump with cooling; lubricant makes one pass before returning to the sump.

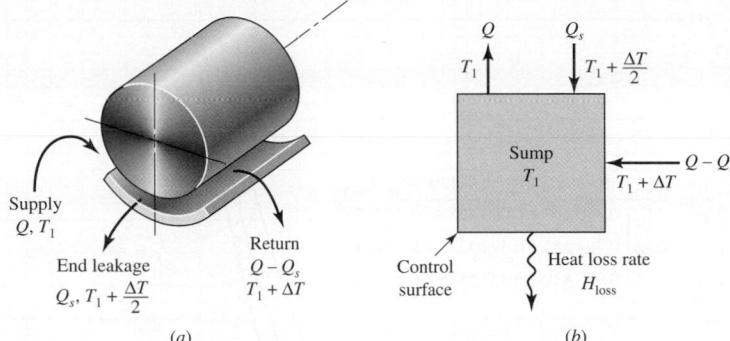

(a) $\qquad\qquad\qquad\qquad\qquad$ (b)

Let

$$Q = \text{volumetric oil-flow rate into the bearing, in}^3\text{/s}$$
$$Q_s = \text{volumetric side-flow leakage rate out of the bearing and to the sump, in}^3\text{/s}$$
$$Q - Q_s = \text{volumetric oil-flow discharge from annulus to sump, in}^3\text{/s}$$
$$T_1 = \text{oil inlet temperature (equal to sump temperature } T_s), \, ^\circ\text{F}$$
$$\Delta T = \text{temperature rise in oil between inlet and outlet,} \, ^\circ\text{F}$$
$$\rho = \text{lubricant density, lbm/in}^3$$
$$C_p = \text{specific heat capacity of lubricant, Btu/(lbm} \cdot {}^\circ\text{F)}$$
$$J = \text{Joulean heat equivalent, in} \cdot \text{lbf/Btu}$$
$$H = \text{heat rate, Btu/s}$$

Using the sump as a control region, we can write an enthalpy balance. Using T_1 as the datum temperature gives

$$H_{\text{loss}} = \rho C_p Q_s \Delta T/2 + \rho C_p (Q - Q_s) \Delta T = \rho C_p Q \Delta T \left(1 - \frac{1}{2}\frac{Q_s}{Q}\right) \qquad (a)$$

The thermal energy loss at steady state H_{loss} is equal to the rate the journal does work on the film is $H_{\text{loss}} = \dot{W} = 2\pi T N/J$. The torque $T = fWr$, the load in terms of pressure is $W = 2Prl$, and multiplying numerator and denominator by the clearance c gives

$$H_{\text{loss}} = \frac{4\pi \, PrlNc}{J}\frac{rf}{c} \qquad (b)$$

Equating Eqs. (a) and (b) and rearranging results in

$$\frac{J\rho C_p \, \Delta T}{4\pi \, P} = \frac{rf/c}{(1 - 0.5 Q_s/Q)\,[Q/(rcNl)]} \qquad (c)$$

For common petroleum lubricants $\rho = 0.0311$ lbm/in^3, $C_p = 0.42$ Btu/(lbm $\cdot$ $^\circ$F), and $J = 778(12) = 9336$ in $\cdot$ lbf/Btu; therefore the left term of Eq. (c) is

$$\frac{J\rho C_p \, \Delta T}{4\pi \, P} = \frac{9336(0.0311)0.42\Delta T_F}{4\pi \, P_{\text{psi}}} = 9.70\frac{\Delta T_F}{P_{\text{psi}}}$$

thus

$$\frac{9.70\Delta T_F}{P_{\text{psi}}} = \frac{rf/c}{\left(1 - \frac{1}{2} Q_s/Q\right)(Q/rcN_jl)} \qquad (12\text{–}15)$$

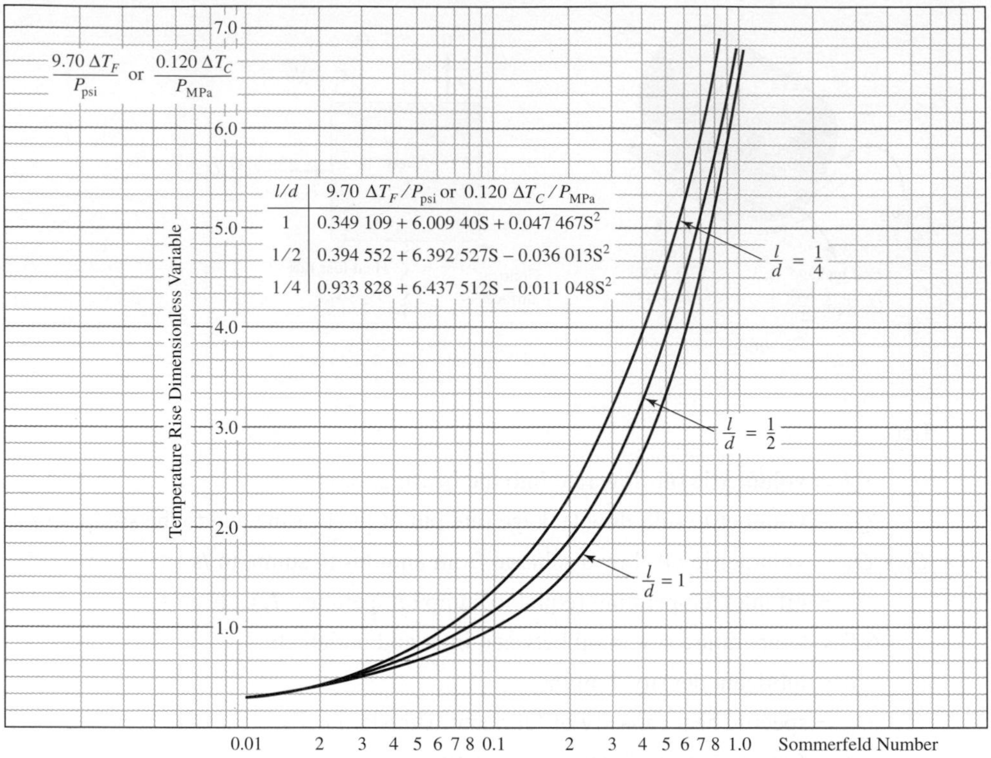

Figure 12–24

Figures 12–18, 12–19, and 12–20 combined to reduce iterative table look-up. *(Source: Chart based on work of Raimondi and Boyd boundary condition (2), i.e., no negative lubricant pressure developed. Chart is for full journal bearing using single lubricant pass, side flow emerges with temperature rise $\Delta T/2$, thru flow emerges with temperature rise ΔT, and entire flow is supplied at datum sump temperature.)*

where ΔT_F is the temperature rise in °F and P_{psi} is the bearing pressure in psi. The right side of Eq. (12–15) can be evaluated from Figs. 12–18, 12–19, and 12–20 for various Sommerfeld numbers and l/d ratios to give Fig. 12–24. It is easy to show that the left side of Eq. (12–15) can be expressed as $0.120\Delta T_C/P_{MPa}$ where ΔT_C is expressed in °C and the pressure P_{MPa} is expressed in MPa. The ordinate in Fig. 12–24 is either $9.70\,\Delta T_F/P_{psi}$ or $0.120\Delta T_C/P_{MPa}$, which is not surprising since both are dimensionless in proper units and *identical in magnitude*. Since solutions to bearing problems involve iteration and reading many graphs can introduce errors, Fig. 12–24 reduces three graphs to one, a step in the proper direction.

Interpolation

According to Raimondi and Boyd, interpolation of the chart data for other l/d ratios can be done by using the equation

$$y = \frac{1}{(l/d)^3}\left[-\frac{1}{8}\left(1-\frac{l}{d}\right)\left(1-2\frac{l}{d}\right)\left(1-4\frac{l}{d}\right)y_\infty + \frac{1}{3}\left(1-2\frac{l}{d}\right)\left(1-4\frac{l}{d}\right)y_1 \right.$$
$$\left. -\frac{1}{4}\left(1-\frac{l}{d}\right)\left(1-4\frac{l}{d}\right)y_{1/2} + \frac{1}{24}\left(1-\frac{l}{d}\right)\left(1-2\frac{l}{d}\right)y_{1/4}\right] \tag{12–16}$$

where y is the desired variable within the interval $\infty > l/d > \frac{1}{4}$ and y_∞, y_1, $y_{1/2}$, and $y_{1/4}$ are the variables corresponding to l/d ratios of ∞, 1, $\frac{1}{2}$, and $\frac{1}{4}$, respectively.

12–9   Steady-State Conditions in Self-Contained Bearings

The case in which the lubricant carries away all of the enthalpy increase from the journal-bushing pair has already been discussed. Bearings in which the warm lubricant stays within the bearing housing will now be addressed. These bearings are called *self-contained* bearings because the lubricant sump is within the bearing housing and the lubricant is cooled within the housing. These bearings are described as *pillow-block* or *pedestal* bearings. They find use on fans, blowers, pumps, and motors, for example. Integral to design considerations for these bearings is dissipating heat from the bearing housing to the surroundings at the same rate that enthalpy is being generated within the fluid film.

In a self-contained bearing the sump can be positioned as a keywaylike cavity in the bushing, the ends of the cavity not penetrating the end planes of the bushing. Film oil exits the annulus at about one-half of the relative peripheral speeds of the journal and bushing and slowly tumbles the sump lubricant, mixing with the sump contents. Since the film in the top "half" of the cap has cavitated, it contributes essentially nothing to the support of the load, but it does contribute friction. Bearing caps are in use in which the "keyway" sump is expanded peripherally to encompass the top half of the bearing. This reduces friction for the same load, but the included angle β of the bearing has been reduced to $180°$. Charts for this case were included in the Raimondi and Boyd paper.

The heat given up by the bearing housing may be estimated from the equation

$$H_{\text{loss}} = \hbar_{\text{CR}} A (T_b - T_\infty) \tag{12–17}$$

where H_{loss} = heat dissipated, Btu/h

$\hbar_{\text{CR}}$ = combined overall coefficient of radiation and convection heat transfer, Btu/(h · ft^2 · °F)

A = surface area of bearing housing, ft^2

T_b = surface temperature of the housing, °F

T_∞ = ambient temperature, °F

The overall coefficient $\hbar_{\text{CR}}$ depends on the material, surface coating, geometry, even the roughness, the temperature difference between the housing and surrounding objects, and air velocity. After Karelitz,[10] and others, in ordinary industrial environments, the overall coefficient $\hbar_{\text{CR}}$ can be treated as a constant. Some representative values are

$$\hbar_{\text{CR}} = \begin{cases} 2 \text{ Btu/(h · ft}^2 \cdot \text{°F)} & \text{for still air} \\ 2.7 \text{ Btu/(h · ft}^2 \cdot \text{°F)} & \text{for shaft-stirred air} \\ 5.9 \text{ Btu/(h · ft}^2 \cdot \text{°F)} & \text{for air moving at 500 ft/min} \end{cases} \tag{12–18}$$

An expression similar to Eq. (12–17) can be written for the temperature difference $T_f - T_b$ between the lubricant film and the housing surface. This is possible because the bushing and housing are metal and very nearly isothermal. If one defines $\bar{T}_f$ as the *average* film temperature (halfway between the lubricant inlet temperature T_s and the

---

[10]G. B. Karelitz, "Heat Dissipation in Self-Contained Bearings," Trans. ASME, Vol. 64, 1942, p. 463; D. C. Lemmon and E. R. Booser, "Bearing Oil-Ring Performance," Trans. ASME, J. Bas. Engin., Vol. 88, 1960, p. 327.

| **Table 12–2**

Lubrication System	Conditions	Range of α
Oil ring	Moving air	1–2
	Still air	$\frac{1}{2}$–1
Oil bath	Moving air	$\frac{1}{2}$–1
	Still air	$\frac{1}{5}$–$\frac{2}{5}$

outlet temperature $T_s + \Delta T$), then the following proportionality has been observed between $\bar{T}_f - T_b$ and the difference between the housing surface temperature and the ambient temperature, $T_b - T_\infty$:

$$\bar{T}_f - T_b = \alpha(T_b - T_\infty) \qquad (a)$$

where $\bar{T}_f$ is the average film temperature and α is a constant depending on the lubrication scheme and the bearing housing geometry. Equation (a) may be used to estimate the bearing housing temperature. Table 12–2 provides some guidance concerning suitable values of α. The work of Karelitz allows the broadening of the application of the charts of Raimondi and Boyd, to be applied to a variety of bearings beyond the natural circulation pillow-block bearing.

Solving Eq. (a) for T_b and substituting into Eq. (12–17) gives the bearing heat loss rate to the surroundings as

$$H_{\text{loss}} = \frac{\hbar_{\text{CR}} A}{1 + \alpha}(\bar{T}_f - T_\infty) \qquad (12\text{–}19a)$$

and rewriting Eq. (a) gives

$$T_b = \frac{\bar{T}_f + \alpha T_\infty}{1 + \alpha} \qquad (12\text{–}19b)$$

In beginning a steady-state analysis the average film temperature is unknown, hence the viscosity of the lubricant in a self-contained bearing is unknown. Finding the equilibrium temperatures is an iterative process wherein a trial average film temperature (and the corresponding viscosity) is used to compare the heat generation rate and the heat loss rate. An adjustment is made to bring these two heat rates into agreement. This can be done on paper with a tabular array to help adjust $\bar{T}_f$ to achieve equality between heat generation and loss rates. A root-finding algorithm can be used. Even a simple one can be programmed for a digital computer.

Because of the shearing action there is a uniformly distributed energy release in the lubricant that heats the lubricant as it works its way around the bearing. The temperature is uniform in the radial direction but increases from the sump temperature T_s by an amount ΔT during the lubricant pass. The exiting lubricant mixes with the sump contents, being cooled to sump temperature. The lubricant in the sump is cooled because the bushing and housing metal are at a nearly uniform lower temperature because of heat losses by convection and radiation to the surroundings at ambient temperature T_∞. In the usual configurations of such bearings, the bushing and housing metal temperature is approximately midway between the average film temperature $\bar{T}_f = T_s + \Delta T/2$ and the ambient temperature T_∞. The heat generation rate H_{gen}, at steady state, is equal to the work rate from the frictional torque T. Expressing this in Btu/h requires the conversion constants 2545 Btu/(hp · h) and 1050 (lbf · in)(rev/s)/hp results in $H_{\text{gen}} = 2545\,TN/1050$. Then from Eq. ($b$), Sec. 12–3, the torque is $T = 4\pi^2 r^3 l\mu/c$, resulting in

$$H_{\text{gen}} = \frac{2545}{1050} \frac{4\pi^2 r^3 l\mu N}{c} N = \frac{95.69 \mu N^2 lr^3}{c} \tag{b}$$

Equating this to Eq. (12–19a) and solving for $\bar{T}_f$ give

$$\bar{T}_f = T_\infty + 95.69(1+\alpha)\frac{\mu N^2 lr^3}{\hbar_{\text{CR}} Ac} \tag{12–20}$$

EXAMPLE 12–5

Consider a pillow-block bearing with a keyway sump, whose journal rotates at 900 rev/min in shaft-stirred air at 70°F with $\alpha = 1$. The lateral area of the bearing is 40 in^2. The lubricant is SAE grade 20 oil. The gravity radial load is 100 lbf and the l/d ratio is unity. The bearing has a journal diameter of $2.000 + 0.000/-0.002$ in, a bushing bore of $2.002 + 0.004/-0.000$ in. For a minimum clearance assembly estimate the steady-state temperatures as well as the minimum film thickness and coefficient of friction.

Solution

The minimum radial clearance, c_{min}, is

$$c_{\text{min}} = \frac{2.002 - 2.000}{2} = 0.001 \text{ in}$$

$$P = \frac{W}{ld} = \frac{100}{(2)2} = 25 \text{ psi}$$

$$S = \left(\frac{r}{c}\right)^2 \frac{\mu N}{P} = \left(\frac{1}{0.001}\right)^2 \frac{\mu'(15)}{10^6(25)} = 0.6\,\mu'$$

where μ' is viscosity in μreyn. The friction horsepower loss, $(\text{hp})_f$, is found as follows:

$$(\text{hp})_f = \frac{fWrN}{1050} = \frac{WNc}{1050}\frac{fr}{c} = \frac{100(900/60)0.001}{1050}\frac{fr}{c} = 0.001\,429\frac{fr}{c} \text{ hp}$$

The heat generation rate H_{gen}, in Btu/h, is

$$H_{\text{gen}} = 2545(\text{hp})_f = 2545(0.001\,429)fr/c = 3.637\,fr/c \text{ Btu/h}$$

From Eq. (12–19a) with $\hbar_{\text{CR}} = 2.7$ Btu/(h · ft^2 · °F), the rate of heat loss to the environment H_{loss} is

$$H_{\text{loss}} = \frac{\hbar_{\text{CR}}A}{\alpha+1}(\bar{T}_f - 70) = \frac{2.7(40/144)}{(1+1)}(\bar{T}_f - 70) = 0.375(\bar{T}_f - 70) \text{ Btu/h}$$

Build a table as follows for trial values of $\bar{T}_f$ of 190 and 195°F:

Trial $\bar{T}_f$	μ'	S	fr/c	H_{gen}	H_{loss}
190	1.15	0.69	13.6	49.5	45.0
195	1.03	0.62	12.2	44.4	46.9

The temperature at which $H_{gen} = H_{loss} = 46.3$ Btu/h is 193.4°F. Rounding $\bar{T}_f$ to 193°F we find $\mu' = 1.08$ μreyn and $S = 0.6(1.08) = 0.65$. From Fig. 12–24, $9.70\Delta T_F/P = 4.25$°F/psi and thus

$$\Delta T_F = 4.25P/9.70 = 4.25(25)/9.70 = 11.0°F$$

$$T_1 = T_s = \bar{T}_f - \Delta T/2 = 193 - 11/2 = 187.5° F$$

$$T_{max} = T_1 + \Delta T_F = 187.5 + 11 = 198.5°F$$

From Eq. (12–19b)

$$T_b = \frac{T_f + \alpha T_\infty}{1 + \alpha} = \frac{193 + (1)70}{1 + 1} = 131.5°F$$

with $S = 0.65$, the minimum film thickness from Fig. 12–16 is

$$h_0 = \frac{h_0}{c}c = 0.79(0.001) = 0.000\ 79 \text{ in}$$

The coefficient of friction from Fig. 12–18 is

$$f = \frac{fr}{c}\frac{c}{r} = 12.8\frac{0.001}{1} = 0.012\ 8$$

The parasitic friction torque T is

$$T = fWr = 0.012\ 8(100)(1) = 1.28 \text{ lbf} \cdot \text{in}$$

12–10 Clearance

In designing a journal bearing for thick-film lubrication, the engineer must select the grade of oil to be used, together with suitable values for P, N, r, c, and l. A poor selection of these or inadequate control of them during manufacture or in use may result in a film that is too thin, so that the oil flow is insufficient, causing the bearing to overheat and, eventually, fail. Furthermore, the radial clearance c is difficult to hold accurate during manufacture, and it may increase because of wear. What is the effect of an entire range of radial clearances, expected in manufacture, and what will happen to the bearing performance if c increases because of wear? Most of these questions can be answered and the design optimized by plotting curves of the performance as functions of the quantities over which the designer has control.

Figure 12–25 shows the results obtained when the performance of a particular bearing is calculated for a whole range of radial clearances and is plotted with clearance as the independent variable. The bearing used for this graph is the one of Examples 12–1 to 12–4 with SAE 20 oil at an inlet temperature of 100°F. The graph shows that if the clearance is too tight, the temperature will be too high and the minimum film thickness too low. High temperatures may cause the bearing to fail by fatigue. If the oil film is too thin, dirt particles may be unable to pass without scoring or may embed themselves in the bearing. In either event, there will be excessive wear and friction, resulting in high temperatures and possible seizing.

To investigate the problem in more detail, Table 12–3 was prepared using the two types of preferred running fits that seem to be most useful for journal-bearing design

Figure 12–25

A plot of some performance characteristics of the bearing of Exs. 12–1 to 12–4 for radial clearances of 0.0005 to 0.003 in. The bearing outlet temperature is designated T_2. New bearings should be designed for the shaded zone, because wear will move the operating point to the right.

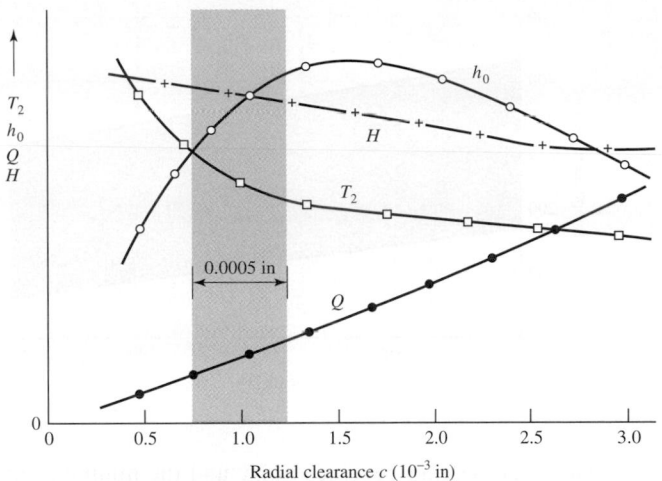

Radial clearance c (10^{-3} in)

Table 12–3

Maximum, Minimum, and Average Clearances for 1.5-in-Diameter Journal Bearings Based on Type of Fit

		Clearance c, in		
Type of Fit	**Symbol**	**Maximum**	**Average**	**Minimum**
Close-running	H8/f7	0.001 75	0.001 125	0.000 5
Free-running	H9/d9	0.003 95	0.002 75	0.001 55

Table 12–4

Performance of 1.5-in-Diameter Journal Bearing with Various Clearances. (SAE 20 Lubricant, $T_1 = 100°F$, $N = 30$ r/s, $W = 500$ lbf, $l = 1.5$ in)

c, in	T_2, °F	h_0, in	f	Q, in³/s,	H, Btu/s
0.000 5	226	0.000 38	0.011 3	0.061	0.086
0.001 125	142	0.000 65	0.009 0	0.153	0.068
0.001 55	133	0.000 77	0.008 7	0.218	0.066
0.001 75	128	0.000 76	0.008 4	0.252	0.064
0.002 75	118	0.000 73	0.007 9	0.419	0.060
0.003 95	113	0.000 69	0.007 7	0.617	0.059

(see Table 2–8). The results shown in Table 12–3 were obtained by using Eqs. (2–38) and (2–39) of Sec. 2–6. Notice that there is a slight overlap, but the range of clearances for the free-running fit is about twice that of the close-running fit.

The six clearances of Table 12–3 were used in a computer program to obtain the numerical results shown in Table 12–4. These conform to the results of Fig. 12–25, too. Both the table and the figure show that a tight clearance results in a high temperature. Figure 12–26 can be used to estimate an upper temperature limit when the characteristics of the application are known.

It would seem that a large clearance will permit the dirt particles to pass through and also will permit a large flow of oil, as indicated in Table 12–4. This lowers the temperature and increases the life of the bearing. However, if the clearance becomes too

Figure 12–26

Temperature limits for mineral oils. The lower limit is for oils containing antioxidants and applies when oxygen supply is unlimited. The upper limit applies when insignificant oxygen is present. The life in the shaded zone depends on the amount of oxygen and catalysts present. *(Source: M. J. Neale (ed.),* Tribology Handbook, Section *B1, Newnes-Butterworth, London, 1975.)*

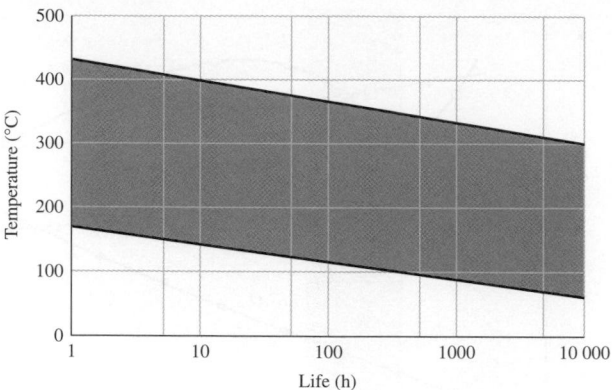

large, the bearing becomes noisy and the minimum film thickness begins to decrease again.

In between these two limitations there exists a rather large range of clearances that will result in satisfactory bearing performance.

When both the production tolerance and the future wear on the bearing are considered, it is seen, from Fig. 12–25, that the best compromise is a clearance range slightly to the left of the top of the minimum-film-thickness curve. In this way, future wear will move the operating point to the right and increase the film thickness and decrease the operating temperature.

12–11 Pressure-Fed Bearings

The load-carrying capacity of self-contained natural-circulating journal bearings is quite restricted. The factor limiting better performance is the heat-dissipation capability of the bearing. A first thought of a way to increase heat dissipation is to cool the sump with an external fluid such as water. The high-temperature problem is in the film where the heat is generated but cooling is not possible in the film until later. This does not protect against exceeding the maximum allowable temperature of the lubricant. A second alternative is to reduce the *temperature rise* in the film by dramatically increasing the rate of lubricant flow. The lubricant itself is reducing the temperature rise. A water-cooled sump may still be in the picture. To increase lubricant flow, an external pump must be used with lubricant supplied at pressures of tens of pounds per square inch gage. Because the lubricant is supplied to the bearing under pressure, such bearings are called *pressure-fed bearings.*

To force a greater flow through the bearing and thus obtain an increased cooling effect, a common practice is to use a circumferential groove at the center of the bearing, with an oil-supply hole located opposite the load-bearing zone. Such a bearing is shown in Fig. 12–27. The effect of the groove is to create two half-bearings, each having a smaller l/d ratio than the original. The groove divides the pressure-distribution curve into two lobes and reduces the minimum film thickness, but it has wide acceptance among lubrication engineers because such bearings carry more load without overheating.

To set up a method of solution for oil flow, we shall assume a groove ample enough that the pressure drop in the groove itself is small. Initially we will neglect eccentricity and then apply a correction factor for this condition. The oil flow, then, is the amount that flows out of the two halves of the bearing in the direction of the concentric shaft. If we neglect the rotation of the shaft, the flow of the lubricant is caused by the supply

Figure 12–27

Centrally located full annular groove. *(Courtesy of the Cleveland Graphite Bronze Company, Division of Clevite Corporation.)*

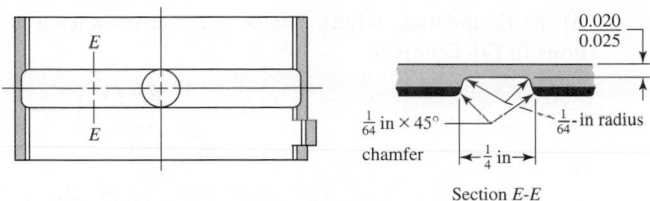

Section E-E

Figure 12–28

Flow of lubricant from a pressure-fed bearing having a central annular groove.

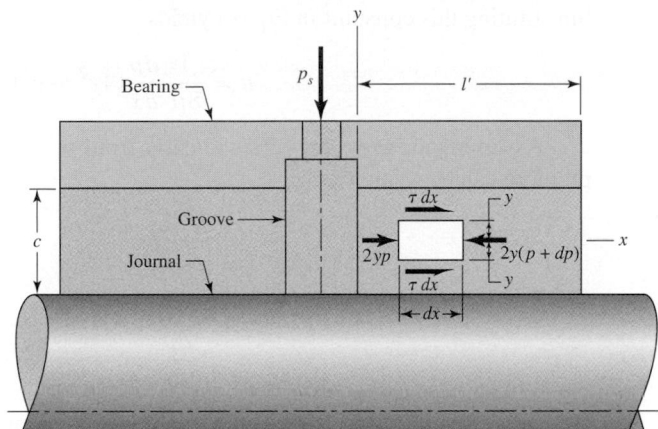

pressure p_s, shown in Fig. 12–28. Laminar flow is assumed, with the pressure varying linearly from $p = p_s$ at $x = 0$, to $p = 0$ at $x = l'$. Consider the static equilibrium of an element of thickness dx, height $2y$, and unit depth. Note particularly that the origin of the reference system has been chosen at the midpoint of the clearance space and symmetry about the x axis is implied with the shear stresses τ being equal on the top and bottom surfaces. The equilibrium equation in the x direction is

$$-2y(p + dp) + 2yp + 2\tau\, dx = 0 \qquad (a)$$

Expanding and canceling terms, we find that

$$\tau = y\frac{dp}{dx} \qquad (b)$$

Newton's equation for viscous flow [Eq. (12–1)] is

$$\tau = \mu\frac{du}{dy} \qquad (c)$$

Now eliminating τ from Eqs. (b) and (c) gives

$$\frac{du}{dy} = \frac{1}{\mu}\frac{dp}{dx}\, y \qquad (d)$$

Treating dp/dx as a constant and integrating with respect to y gives

$$u = \frac{1}{2\mu}\frac{dp}{dx}\, y^2 + C_1 \qquad (e)$$

At the boundaries, where $y = \pm c/2$, the velocity u is zero. Using one of these conditions in Eq. (e) gives

$$0 = \frac{1}{2\mu}\frac{dp}{dx}\left(\frac{c}{2}\right)^2 + C_1$$

or

$$C_1 = -\frac{c^2}{8\mu}\frac{dp}{dx}$$

Substituting this constant in Eq. (e) yields

$$u = \frac{1}{8\mu}\frac{dp}{dx}(4y^2 - c^2) \tag{f}$$

Assuming the pressure varies linearly from p_s to 0 at $x = 0$ to l', respectively, the pressure can be written as

$$p = p_s - \frac{p_s}{l'}x \tag{g}$$

and therefore the pressure gradient is given by

$$\frac{dp}{dx} = -\frac{p_s}{l'} \tag{h}$$

We can now substitute Eq. (h) in Eq. (f) to get the relationship between the oil velocity and the coordinate y:

$$u = \frac{p_s}{8\mu l'}(c^2 - 4y^2) \tag{12–21}$$

Figure 12–29 shows a graph of this relation fitted into the clearance space c so that you can see how the velocity of the lubricant varies from the journal surface to the bearing surface. The distribution is parabolic, as shown, with the maximum velocity occurring at the center, where $y = 0$. The magnitude is, from Eq. (12–21),

$$u_{max} = \frac{p_s c^2}{8\mu l'} \tag{i}$$

To consider eccentricity, as shown in Fig. 12–30, the film thickness is $h = c - e\cos\theta$. Substituting h for c in Eq. (i), with the average ordinate of a parabola being two-thirds the maximum, the average velocity at any angular position θ is

$$u_{av} = \frac{2}{3}\frac{p_s h^2}{8\mu l'} = \frac{p_s}{12\mu l'}(c - e\cos\theta)^2 \tag{j}$$

We still have a little further to go in this analysis; so please be patient. Now that we have an expression for the lubricant velocity, we can compute the amount of lubricant

Figure 12–29

Parabolic distribution of the lubricant velocity.

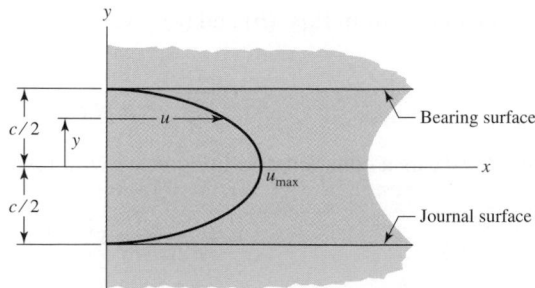

| **Figure 12–30**

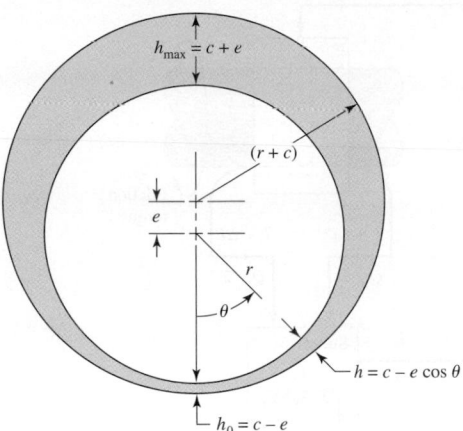

that flows out both ends; the elemental side flow at any position θ (Fig. 12–30) is

$$dQ_s = 2u_{\text{av}}\, dA = 2u_{\text{av}}(rh\, d\theta) \qquad (k)$$

where dA is the elemental area. Substituting u_{av} from Eq. (j) and h from Fig. 12–30 gives

$$dQ_s = \frac{p_s r}{6\mu l'}(c - e\cos\theta)^3\, d\theta \qquad (l)$$

Integrating around the bearing gives the total side flow as

$$Q_s = \int dQ_s = \frac{p_s r}{6\mu l'}\int_0^{2\pi}(c - e\cos\theta)^3\, d\theta = \frac{p_s r}{6\mu l'}(2\pi c^3 + 3\pi\, ce^2)$$

Rearranging, with $\epsilon = e/c$, gives

$$Q_s = \frac{\pi p_s r c^3}{3\mu l'}(1 + 1.5\epsilon^2) \qquad (12\text{--}22)$$

In analyzing the performance of pressure-fed bearings, the bearing length should be taken as l', as defined in Fig. 12–28. The characteristic pressure in each of the two bearings that constitute the pressure-fed bearing assembly P is given by

$$P = \frac{W/2}{2rl'} = \frac{W}{4rl'} \qquad (12\text{--}23)$$

The charts for flow variable and flow ratio (Figs. 12–19 and 12–20) do not apply to pressure-fed bearings. Also, the maximum film pressure given by Fig. 12–21 must be increased by the oil supply pressure p_s to obtain the total film pressure.

Since the oil flow has been increased by forced feed, Eq. (12–14) will give a temperature rise that is too high because the side flow carries away all the heat generated. The plumbing in a pressure-fed bearing is depicted schematically in Fig. 12–31. The oil leaves the sump at the externally maintained temperature T_s at the volumetric rate Q_s. The heat gain of the fluid passing through the bearing is

$$H_{\text{gain}} = 2\,\rho C_p(Q_s/2)\Delta T = \rho C_p Q_s \Delta T \qquad (m)$$

At steady state, the rate at which the journal does frictional work on the fluid film is

$$H_f = \frac{2\pi T N}{J} = \frac{2\pi f\, W r N}{J} = \frac{2\pi\, W N c}{J}\frac{f r}{c} \qquad (n)$$

Figure 12–31

Pressure-fed centrally located full annular-groove journal bearing with external, coiled lubricant sump.

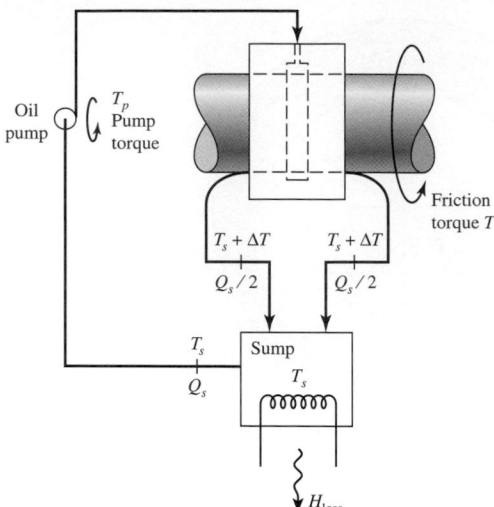

Equating the heat gain to the frictional work and solving for ΔT gives

$$\Delta T = \frac{2\pi W N c}{J \rho C_p Q_s} \frac{fr}{c} \qquad (o)$$

Substituting Eq. (12–22) for Q_s in the equation for ΔT gives

$$\Delta T = \frac{2\pi}{J \rho C_p} W N c \frac{fr}{c} \frac{3\mu l'}{(1 + 1.5\epsilon^2)\pi p_s r c^3}$$

The Sommerfeld number may be expressed as

$$S = \left(\frac{r}{c}\right)^2 \frac{\mu N}{P} = \left(\frac{r}{c}\right)^2 \frac{4rl'\mu N}{W}$$

Solving for $\mu N l'$ in the Sommerfeld expression; substituting in the ΔT expression; and using $J = 9336$ lbf · in/Btu, $\rho = 0.0311$ lbf/in³, and $C_p = 0.42$ Btu/(lbf · °F), we find

$$\Delta T_F = \frac{3(fr/c)SW^2}{2J\rho C_p p_s r^4} \frac{1}{(1 + 1.5\epsilon^2)} = \frac{0.0123(fr/c)SW^2}{(1 + 1.5\epsilon^2) p_s r^4} \qquad (12\text{–}24)$$

where ΔT_F is ΔT in °F. The corresponding equation in SI units uses the bearing load W in kN, lubricant supply pressure p_s in kPa, and the journal radius r in mm:

$$\Delta T_C = \frac{978(10^6)}{1 + 1.5\epsilon^2} \frac{(fr/c)SW^2}{p_s r^4} \qquad (12\text{–}25)$$

An analysis example of a pressure-fed bearing will be useful.

EXAMPLE 12–6   A circumferential-groove pressure-fed bearing is lubricated with SAE grade 20 oil supplied at a gauge pressure of 30 psi. The journal diameter d_j is 1.750 in, with a unilateral tolerance of -0.002 in. The central circumferential bushing has a diameter d_b of 1.753 in, with a unilateral tolerance of $+0.004$ in. The l'/d ratio of the two "half-bearings" that constitute the complete pressure-fed bearing is 1/2. The journal angular speed is

300 rev/min, or 50 rev/s, and the radial steady load is 900 lbf. The external sump is maintained at 120°F as long as the necessary heat transfer does not exceed 800 Btu/h.
(a) Find the steady-state average film temperature.
(b) Compare h_0, T_{max}, and P_{st} with the Trumpler criteria.
(c) Estimate the volumetric side flow Q_s, the heat loss rate H_{loss}, and the parasitic friction torque.

Solution   (a)

$$r = \frac{d_j}{2} = \frac{1.750}{2} = 0.875 \text{ in}$$

$$c_{min} = \frac{(d_b)_{min} - (d_j)_{max}}{2} = \frac{1.753 - 1.750}{2} = 0.0015 \text{ in}$$

Since $l'/d = 1/2$, $l' = d/2 = r = 0.875$ in. Then the pressure due to the load is

$$P = \frac{W}{4rl'} = \frac{900}{4(0.875)0.875} = 294 \text{ psi}$$

The Sommerfeld number S can be expressed as

$$S = \left(\frac{r}{c}\right)^2 \frac{\mu N}{P} = \left(\frac{0.875}{0.0015}\right)^2 \frac{\mu'}{(10^6)} \frac{50}{294} = 0.0579\mu' \tag{1}$$

We will use a tabulation method to find the average film temperature. The first trial average film temperature $\bar{T}_f$ will be 170°F. Using the Seireg curve fit of Table 12–1, we obtain

$$\mu' = 0.0136 \exp[1271.6/(170 + 95)] = 1.650 \ \mu\text{reyn}$$

From Eq. (1)

$$S = 0.0579\mu' = 0.0579(1.650) = 0.0955$$

From Fig. (12–18), $fr/c = 3.3$, and from Fig. (12–16), $\epsilon = 0.80$. From Eq. (12–24),

$$\Delta T_F = \frac{0.0123(3.3)0.0955(900^2)}{[1 + 1.5(0.80)^2]30(0.875^4)} = 91.1°F$$

$$T_{av} = T_s + \frac{\Delta T}{2} = 120 + \frac{91.1}{2} = 165.6°F$$

We form a table, adding a second line with $\bar{T}_f = 168.5°F$:

Trial $\bar{T}_f$	μ'	S	fr/c	ϵ	ΔT_F	T_{av}
170	1.65	0.0955	3.3	0.800	91.1	165.6
168.5	1.693	0.0980	3.39	0.792	97.1	168.5

If the iteration had not closed, one could plot trial $\bar{T}_f$ against resulting T_{av} and draw a straight line between them, the intersection with a $\bar{T}_f = T_{av}$ line defining the new trial $\bar{T}_f$.

The result of this tabulation is $\bar{T}_f = 168.5$, $\Delta T_F = 97.1°F$, and $T_{max} = 120 + 97.1 = 217.1°F$

(b) Since $h_0 = (1 - \epsilon)c$,

$$h_0 = (1 - 0.792)0.0015 = 0.000\,312\ \text{in}$$

The required four Trumpler criteria, from "Significant Angular Speed" in Sec. 12–7 are

$$h_0 \geq 0.0002 + 0.000\,04(1.750) = 0.000\,270\ \text{in} \qquad \text{(OK)}$$

$$T_{max} = T_s + \Delta T = 120 + 97.1 = 217.1°F \qquad \text{(OK)}$$

$$P_{st} = \frac{W_{st}}{4rl'} = \frac{900}{4(0.875)0.875} = 294\ \text{psi} \qquad \text{(OK)}$$

The factor of safety on the load is approximately unity. (Not OK.)

(c) From Eq. (12–22),

$$Q_s = \frac{\pi(30)0.875(0.0015)^3}{3(1.693)10^{-6}(0.875)}[1 + 1.5(0.80)^2] = 0.123\ \text{in}^3/\text{s}$$

$$H_{loss} = \rho C_p Q_s \Delta T = 0.0311(0.42)0.123(97.1) = 0.156\ \text{Btu/s}$$

or 562 Btu/h or 0.221 hp. The parasitic friction torque T is

$$T = fWr = \frac{fr}{c}Wc = 3.39(900)0.0015 = 4.58\ \text{lbf} \cdot \text{in}$$

12–12   Loads and Materials

Some help in choosing unit loads and bearing materials is afforded by Tables 12–5 and 12–6. Since the diameter and length of a bearing depend upon the unit load, these tables will help the designer to establish a starting point in the design.

Table 12–5

Range of Unit Loads in Current Use for Sleeve Bearings

	Unit Load	
Application	**psi**	**MPa**
Diesel engines:		
Main bearings	900–1700	6–12
Crankpin	1150–2300	8–15
Wristpin	2000–2300	14–15
Electric motors	120–250	0.8–1.5
Steam turbines	120–250	0.8–1.5
Gear reducers	120–250	0.8–1.5
Automotive engines:		
Main bearings	600–750	4–5
Crankpin	1700–2300	10–15
Air compressors:		
Main bearings	140–280	1–2
Crankpin	280–500	2–4
Centrifugal pumps	100–180	0.6–1.2

The length-diameter ratio l/d of a bearing depends upon whether it is expected to run under thin-film-lubrication conditions. A long bearing (large l/d ratio) reduces the coefficient of friction and the side flow of oil and therefore is desirable where thin-film or boundary-value lubrication is present. On the other hand, where forced-feed or positive lubrication is present, the l/d ratio should be relatively small. The short bearing length results in a greater flow of oil out of the ends, thus keeping the bearing cooler. Current practice is to use an l/d ratio of about unity, in general, and then to increase this ratio if thin-film lubrication is likely to occur and to decrease it for thick-film lubrication or high temperatures. If shaft deflection is likely to be severe, a short bearing should be used to prevent metal-to-metal contact at the ends of the bearings.

You should always consider the use of a partial bearing if high temperatures are a problem, because relieving the non-load-bearing area of a bearing can very substantially reduce the heat generated.

The two conflicting requirements of a good bearing material are that it must have a satisfactory compressive and fatigue strength to resist the externally applied loads and that it must be soft and have a low melting point and a low modulus of elasticity. The second set of requirements is necessary to permit the material to wear or break in, since the material can then conform to slight irregularities and absorb and release foreign particles. The resistance to wear and the coefficient of friction are also important because all bearings must operate, at least for part of the time, with thin-film or boundary lubrication.

Additional considerations in the selection of a good bearing material are its ability to resist corrosion and, of course, the cost of producing the bearing. Some of the commonly used materials are listed in Table 12–6, together with their composition and characteristics.

Table 12–6

Some Characteristics of Bearing Alloys

Alloy Name	Thickness, in	SAE Number	Clearance Ratio r/c	Load Capacity	Corrosion Resistance
Tin-base babbitt	0.022	12	600–1000	1.0	Excellent
Lead-base babbitt	0.022	15	600–1000	1.2	Very good
Tin-base babbitt	0.004	12	600–1000	1.5	Excellent
Lead-base babbitt	0.004	15	600–1000	1.5	Very good
Leaded bronze	Solid	792	500–1000	3.3	Very good
Copper-lead	0.022	480	500–1000	1.9	Good
Aluminum alloy	Solid		400–500	3.0	Excellent
Silver plus overlay	0.013	17P	600–1000	4.1	Excellent
Cadmium (1.5% Ni)	0.022	18	400–500	1.3	Good
Trimetal 88*				4.1	Excellent
Trimetal 77†				4.1	Very good

*This is a 0.008-in layer of copper-lead on a steel back plus 0.001 in of tin-base babbitt.
†This is a 0.013-in layer of copper-lead on a steel back plus 0.001 in of lead-base babbitt.

Bearing life can be increased very substantially by depositing a layer of babbitt, or other white metal, in thicknesses from 0.001 to 0.014 in over steel backup material. In fact, a copper-lead layer on steel to provide strength, combined with a babbitt overlay to enhance surface conformability and corrosion resistance, makes an excellent bearing.

Small bushings and thrust collars are often expected to run with thin-film or boundary lubrication. When this is the case, improvements over a solid bearing material can be made to add significantly to the life. A powder-metallurgy bushing is porous and permits the oil to penetrate into the bushing material. Sometimes such a bushing may be enclosed by oil-soaked material to provide additional storage space. Bearings are frequently ball-indented to provide small basins for the storage of lubricant while the journal is at rest. This supplies some lubrication during starting. Another method of reducing friction is to indent the bearing wall and to fill the indentations with graphite.

With all these tentative decisions made, a lubricant can be selected and the hydrodynamic analysis made as already presented. The values of the various performance parameters, if plotted as in Fig. 12–25, for example, will then indicate whether a satisfactory design has been achieved or additional iterations are necessary.

12–13 Bearing Types

A bearing may be as simple as a hole machined into a cast-iron machine member. It may still be simple yet require detailed design procedures, as, for example, the two-piece grooved pressure-fed connecting-rod bearing in an automotive engine. Or it may be as elaborate as the large water-cooled, ring-oiled bearings with built-in reservoirs used on heavy machinery.

Figure 12–32 shows two types of bushings. The solid bushing is made by casting, by drawing and machining, or by using a powder-metallurgy process. The lined bushing is usually a split type. In one method of manufacture the molten lining material is cast continuously on thin strip steel. The babbitted strip is then processed through presses, shavers, and broaches, resulting in a lined bushing. Any type of grooving may be cut into the bushings. Bushings are assembled as a press fit and finished by boring, reaming, or burnishing.

Flanged and straight two-piece bearings are shown in Fig. 12–33. These are available in many sizes in both thick- and thin-wall types, with or without lining material. A locking lug positions the bearing and effectively prevents axial or rotational movement of the bearing in the housing.

Some typical groove patterns are shown in Fig. 12–34. In general, the lubricant may be brought in from the end of the bushing, through the shaft, or through the bushing. The flow may be intermittent or continuous. The preferred practice is to bring the

Figure 12–32

Sleeve bushings.

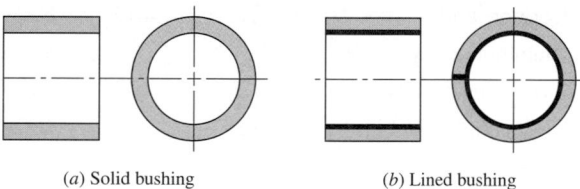

(a) Solid bushing   (b) Lined bushing

Figure 12–33

Two-piece bushings.

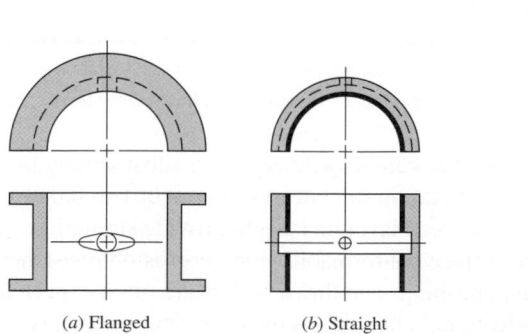

(a) Flanged   (b) Straight

Figure 12–34

Developed views of typical groove patterns. *(Courtesy of the Cleveland Graphite Bronze Company, Division of Clevite Corporation.)*

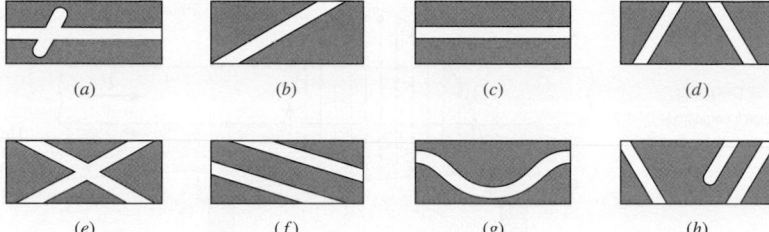

(a)   (b)   (c)   (d)

(e)   (f)   (g)   (h)

Figure 12–35

Fixed-pad thrust bearing. *(Courtesy of Westinghouse Electric Corporation.)*

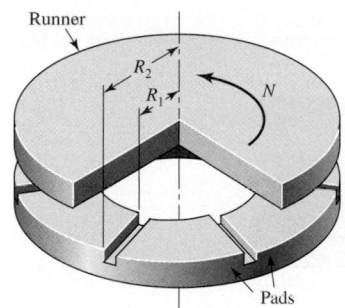

oil in at the center of the bushing so that it will flow out both ends, thus increasing the flow and cooling action.

12–14   Thrust Bearings

This chapter is devoted to the study of the mechanics of lubrication and its application to the design and analysis of journal bearings. The design and analysis of thrust bearings is an important application of lubrication theory, too. A detailed study of thrust bearings is not included here, because it would not contribute anything significantly different and because of space limitations. Having studied this chapter, you should experience no difficulty in reading the literature on thrust bearings and applying that knowledge to actual design situations.[11]

Figure 12–35 shows a fixed-pad thrust bearing consisting essentially of a runner sliding over a fixed pad. The lubricant is brought into the radial grooves and pumped into the wedge-shaped space by the motion of the runner. Full-film, or hydrodynamic, lubrication is obtained if the speed of the runner is continuous and sufficiently high, if the lubricant has the correct viscosity, and if it is supplied in sufficient quantity. Figure 12–36 provides a picture of the pressure distribution under conditions of full-film lubrication.

We should note that bearings are frequently made with a flange, as shown in Fig. 12–37. The flange positions the bearing in the housing and also takes a thrust load. Even when it is grooved, however, and has adequate lubrication, such an arrangement is not theoretically a hydrodynamically lubricated thrust bearing. The reason for this is that the clearance space is not wedge-shaped but has a uniform thickness. Similar reasoning would apply to various designs of thrust washers.

---

[11]Harry C. Rippel, *Cast Bronze Thrust Bearing Design Manual,* International Copper Research Association, Inc., 825 Third Ave., New York, NY 10022, 1967. CBBI, 14600 Detroit Ave., Cleveland, OH, 44107, 1967.

Figure 12–36

Pressure distribution of lubricant in a thrust bearing. *(Courtesy of Copper Research Corporation.)*

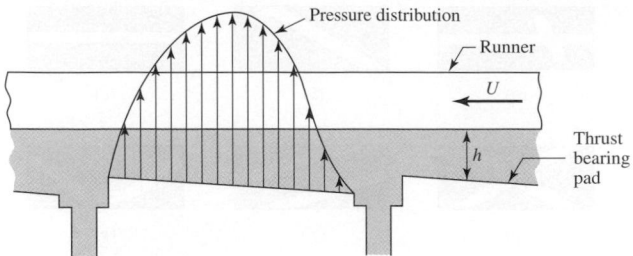

Figure 12–37

Flanged sleeve bearing takes both radial and thrust loads.

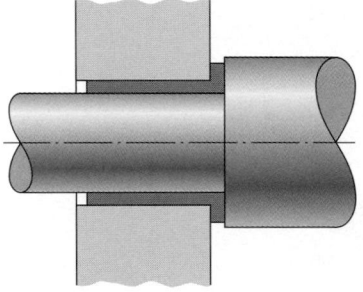

12–15 Boundary-Lubricated Bearings

When two surfaces slide relative to each other with only a partial lubricant film between them, *boundary lubrication* is said to exist. Boundary- or thin-film lubrication occurs in hydrodynamically lubricated bearings when they are starting or stopping, when the load increases, when the supply of lubricant decreases, or whenever other operating changes happen to occur. There are, of course, a very large number of cases in design in which boundary-lubricated bearings must be used because of the type of application or the competitive situation.

The coefficient of friction for boundary-lubricated surfaces may be greatly decreased by the use of animal or vegetable oils mixed with the mineral oil or grease. Fatty acids, such as stearic acid, palmitic acid, or oleic acid, or several of these, which occur in animal and vegetable fats, are called *oiliness agents*. These acids appear to reduce friction, either because of their strong affinity for certain metallic surfaces or because they form a soap film that binds itself to the metallic surfaces by a chemical reaction. Thus the fatty-acid molecules bind themselves to the journal and bearing surfaces with such great strength that the metallic asperities of the rubbing metals do not weld or shear.

Fatty acids will break down at temperatures of 250°F or more, causing increased friction and wear in thin-film-lubricated bearings. In such cases the *extreme-pressure,* or EP, lubricants may be mixed with the fatty-acid lubricant. These are composed of chemicals such as chlorinated esters or tricresyl phosphate, which form an organic film between the rubbing surfaces. Though the EP lubricants make it possible to operate at higher temperatures, there is the added possibility of excessive chemical corrosion of the sliding surfaces.

When a bearing operates partly under hydrodynamic conditions and partly under dry or thin-film conditions, a *mixed-film lubrication* exists. If the lubricant is supplied by hand oiling, by drop or mechanical feed, or by wick feed, for example, the bearing is operating under mixed-film conditions. In addition to occurring with a scarcity of lubricant, mixed-film conditions may be present when

- The viscosity is too low.
- The bearing speed is too low.
- The bearing is overloaded.
- The clearance is too tight.
- Journal and bearing are not properly aligned.

Relative motion between surfaces in contact in the presence of a lubricant is called *boundary lubrication*. This condition is present in hydrodynamic film bearings during starting, stopping, overloading, or lubricant deficiency. Some bearings are boundary lubricated (or dry) at all times. To signal this an adjective is placed before the word "bearing." Commonly applied adjectives (to name a few) are thin-film, boundary friction, Oilite, Oiles, and bushed-pin. The applications include situations in which thick film will not develop and there are low journal speed, oscillating journal, padded slides, light loads, and lifetime lubrication. The characteristics include considerable friction, ability to tolerate expected wear without losing function, and light loading. Such bearings are limited by lubricant temperature, speed, pressure, galling, and cumulative wear. Table 12–7 gives some properties of a range of bushing materials.

Table 12–7

Some Materials for Boundary-Lubricated Bearings and Their Operating Limits

Material	Maximum Load, psi	Maximum Temperature, °F	Maximum Speed, fpm	Maximum PV Value*
Cast bronze	4 500	325	1 500	50 000
Porous bronze	4 500	150	1 500	50 000
Porous iron	8 000	150	800	50 000
Phenolics	6 000	200	2 500	15 000
Nylon	1 000	200	1 000	3 000
Teflon	500	500	100	1 000
Reinforced Teflon	2 500	500	1 000	10 000
Teflon fabric	60 000	500	50	25 000
Delrin	1 000	180	1 000	3 000
Carbon-graphite	600	750	2 500	15 000
Rubber	50	150	4 000	
Wood	2 000	150	2 000	15 000

*P = load, psi; V = speed, fpm.

Linear Sliding Wear

Consider the sliding block depicted in Fig. 12–38, moving along a plate with contact pressure P' acting over area A, in the presence of a coefficient of sliding friction f_s. The linear measure of wear w is expressed in inches or millimeters. The work done by force $f_s P A$ during displacement S is $f_s P A S$ or $f_s P A V t$, where V is the sliding velocity and t is time. The material volume removed due to wear is wA and is proportional to the work done, that is, $wA \propto f_s P A V t$, or

$$wA = KPAVt$$

Figure 12–38

Sliding block subjected to wear.

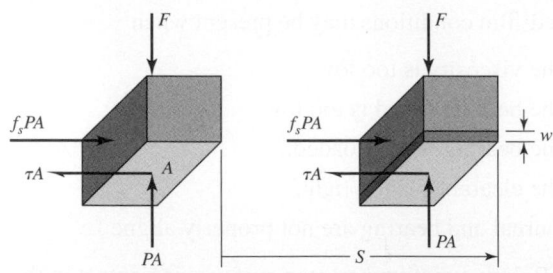

Table 12–8

Wear Factors in U.S. Customary Units*

Source: Oiles America Corp., Plymouth, MI 48170.

Bushing Material	Wear Factor K	Limiting PV
Oiles 800	$3(10^{-10})$	18 000
Oiles 500	$0.6(10^{-10})$	46 700
Polyactal copolymer	$50(10^{-10})$	5 000
Polyactal homopolymer	$60(10^{-10})$	3 000
66 nylon	$200(10^{-10})$	2 000
66 nylon + 15% PTFE	$13(10^{-10})$	7 000
+ 15% PTFE + 30% glass	$16(10^{-10})$	10 000
+ 2.5% MoS$_2$	$200(10^{-10})$	2 000
6 nylon	$200(10^{-10})$	2 000
Polycarbonate + 15% PTFE	$75(10^{-10})$	7 000
Sintered bronze	$102(10^{-10})$	8 500
Phenol + 25% glass fiber	$8(10^{-10})$	11 500

*dim$[K] = \text{in}^3 \cdot \min/(\text{lbf} \cdot \text{ft} \cdot \text{h})$, dim $[PV] = \text{psi} \cdot \text{ft}/\min$.

Table 12–9

Coefficients of Friction

Source: Oiles America Corp., Plymouth, MI 48170.

Type	Bearing	f_s
Placetic	Oiles 80	0.05
Composite	Drymet ST	0.03
	Toughmet	0.05
Met	Cermet M	0.05
	Oiles 2000	0.03
	Oiles 300	0.03
	Oiles 500SP	0.03

where K is the proportionality factor, which includes f_s, and is determined from laboratory testing. The linear wear is then expressed as

$$w = KPVt \qquad (12\text{–}26)$$

In US customary units, P is expressed in psi, V in fpm (i.e., ft/min), and t in hours. This makes the units of K in$^3 \cdot \min/(\text{lbf} \cdot \text{ft} \cdot \text{h})$. Units commonly used for K are cm$^3 \cdot \min/(\text{kgf} \cdot \text{m} \cdot \text{h})$, where 1 kgf = 9.806 N. Tables 12–8 and 12–9 give some wear factors and coefficients of friction from one manufacturer.

It is useful to include a modifying factor f_1 depending on motion type, load, and speed and an environment factor f_2 to account for temperature and cleanliness conditions (see Tables 12–10 and 12–11). These factors account for departures from the laboratory conditions under which K was measured. Equation (12–26) can now be written as

$$w = f_1 f_2 K P V t \qquad (12\text{--}27)$$

Wear, then, is proportional to PV, material property K, operating conditions f_1 and f_2, and time t.

Table 12–10

Motion-Related Factor f_1

Mode of Motion	Characteristic Pressure P, psi		Velocity V, ft/min	f_1*
Rotary	720 or less		3.3 or less	1.0
			3.3–33	1.0–1.3
			33–100	1.3–1.8
	720–3600		3.3 or less	1.5
			3.3–33	1.5–2.0
			33–100	2.0–2.7
Oscillatory	720 or less	>30°	3.3 or less	1.3
			3.3–100	1.3–2.4
		<30°	3.3 or less	2.0
			3.3–100	2.0–3.6
	720–3600	>30°	3.3 or less	2.0
			3.3–100	2.0–3.2
		<30°	3.3 or less	3.0
			3.3–100	3.0–4.8
Reciprocating	720 or less		33 or less	1.5
			33–100	1.5–3.8
	720–3600		33 or less	2.0
			33–100	2.0–7.5

*Values of f_1 based on results over an extended period of time on automotive manufacturing machinery.

Table 12–11

Environmental Factor f_2

Source: Oiles America Corp., Plymouth, MI 48170.

Ambient Temperature, °F	Foreign Matter	f_2
140 or lower	No	1.0
140 or lower	Yes	3.0–6.0
140–210	No	3.0–6.0
140–210	Yes	6.0–12.0

Bushing Wear

Consider a pin of diameter D, rotating at speed N, in a bushing of length L, and supporting a stationary radial load F. The nominal pressure P is given by

$$P = \frac{F}{DL} \qquad (12\text{--}28)$$

Figure 12–39

Pressure distribution on a boundary-lubricated bushing.

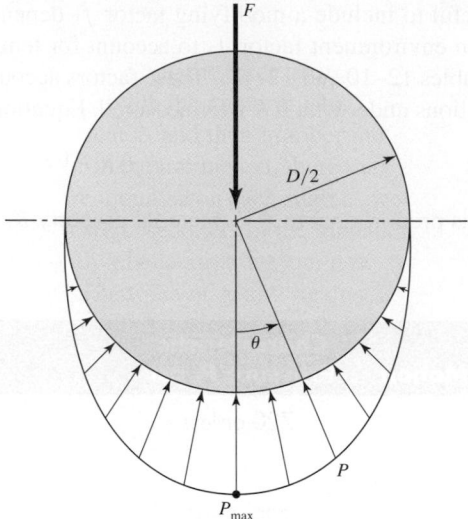

and if N is in rev/min and D is in inches, velocity in ft/min is given by

$$V = \frac{\pi DN}{12} \tag{12–29}$$

Thus PV, in psi · ft/min, is

$$PV = \frac{F}{DL}\frac{\pi DN}{12} = \frac{\pi}{12}\frac{FN}{L} \tag{12–30}$$

Note the independence of PV from the journal diameter D.

A time-wear equation similar to Eq. (12–27) can be written. However, before doing so, it is important to note that Eq. (12–28) provides the nominal value of P. Figure 12–39 provides a more accurate representation of the pressure distribution, which can be written as

$$p = P_{max}\cos\theta \qquad -\frac{\pi}{2} \leq \theta \leq \frac{\pi}{2}$$

The vertical component of $p\,dA$ is $p\,dA\cos\theta = [pL(D/2)\,d\theta]\cos\theta = P_{max}(DL/2)\cos^2\theta\,d\theta$. Integrating this from $\theta = -\pi/2$ to $\pi/2$ yields F. Thus,

$$\int_{-\pi/2}^{\pi/2} P_{max}\left(\frac{DL}{2}\right)\cos^2\theta\,d\theta = \frac{\pi}{4}P_{max}DL = F$$

or

$$P_{max} = \frac{4}{\pi}\frac{F}{DL} \tag{12–31}$$

Substituting V from Eq. (12–29) and P_{max} for P from Eq. (12–31) into Eq. (12–27) gives

$$w = f_1 f_2 K \frac{4}{\pi}\frac{F}{DL}\frac{\pi DNt}{12} = \frac{f_1 f_2 KFNt}{3L} \tag{12–32}$$

In designing a bushing, because of various trade-offs it is recommended that the length/diameter ratio be in the range

$$0.5 \leq L/D \leq 2 \tag{12–33}$$

EXAMPLE 12–7

An Oiles SP 500 alloy brass bushing is 1 in long with a 1-in bore and operates in a clean environment at 70°F. The allowable wear without loss of function is 0.005 in. The radial load is 700 lbf. The peripheral velocity is 33 ft/min. Estimate the number of revolutions for radial wear to be 0.005 in. See Fig. 12–40 and Table 12–12 from the manufacturer.

Solution

From Table 12–8, $K = 0.6(10^{-10})$ in^3 · min/(lbf · ft · h); Tables 12–10 and 12–11, $f_1 = 1.3$, $f_2 = 1$; and Table 12–12, $PV = 46\,700$ psi · ft/min, $P_{max} = 3560$ psi, $V_{max} = 100$ ft/min. From Eqs. (12–31), (12–29), and (12–30),

$$P_{max} = \frac{4}{\pi}\frac{F}{DL} = \frac{4}{\pi}\frac{700}{(1)(1)} = 891\text{ psi} < 3560\text{ psi} \qquad \text{(OK)}$$

$$P = \frac{F}{DL} = \frac{700}{(1)(1)} = 700\text{ psi}$$

$$V = 33\text{ ft/min} < 100\text{ ft/min} \qquad \text{(OK)}$$

$$PV = 700(33) = 23\,100\text{ psi} \cdot \text{ft/min} < 46\,700\text{ psi} \cdot \text{ft/min} \qquad \text{(OK)}$$

Equation (12–32) is

$$w = f_1 f_2 K \frac{4}{\pi}\frac{F}{DL}\frac{\pi D N t}{12} = f_1 f_2 K \frac{4}{\pi}\frac{F}{DL}Vt$$

Figure 12–40

Journal/bushing for Ex. 12–7.

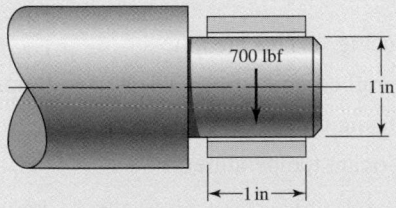

700 lbf

1 in

1 in

Table 12–12

Oiles 500 SP (SPBN · SPWN) Service Range and Properties

Source: Oiles America Corp., Plymouth, MI 48170.

Service Range	Units	Allowable
Characteristic pressure P_{max}	psi	<3560
Velocity V_{max}	ft/min	<100
PV product	(psi)(ft/min)	<46 700
Temperature T	°F	<300

Properties	Test Method, Units	Value
Tensile strength	(ASTM E8) psi	>110 000
Elongation	(ASTM E8) %	>12
Compressive strength	(ASTM E9) psi	49 770
Brinell hardness	(ASTM E10) HB	>210
Coefficient of thermal expansion	(10^{-5}) °C	>1.6
Specific gravity		8.2

Solving for t gives

$$t = \frac{\pi D L w}{4 f_1 f_2 K V F} = \frac{\pi (1)(1) 0.005}{4(1.3)(1) 0.6 (10^{-10}) 33 (700)} = 2180 \text{ h} = 130\ 770 \text{ min}$$

The rotational speed is

$$N = \frac{12V}{\pi D} = \frac{12(33)}{\pi(1)} = 126 \text{ r/min}$$

Answer

$$\text{Cycles} = Nt = 126(130\ 770) = 16.5(10^6) \text{ rev}$$

Temperature Rise

At steady state, the rate at which work is done against bearing friction equals the rate at which heat is transferred from the bearing housing to the surroundings by convection and radiation. The rate of heat generation in Btu/h is given by $f_s F V / J$, or

$$H_{\text{gen}} = \frac{f_s F (\pi D)(60N)}{12J} = \frac{5\pi f_s F D N}{J} \tag{12–34}$$

where N is journal speed in rev/min and $J = 778$ ft · lbf/Btu. The rate at which heat is transferred to the surroundings, in Btu/h, is

$$H_{\text{loss}} = \hbar_{\text{CR}} A \Delta T = \hbar_{\text{CR}} A (T_b - T_\infty) = \frac{\hbar_{\text{CR}} A}{2}(T_f - T_\infty) \tag{12–35}$$

where

A = housing surface area, ft^2

$\hbar_{\text{CR}}$ = overall combined coefficient of heat transfer, Btu/(h · ft^2 · °F)

T_b = housing metal temperature, °F

T_f = lubricant temperature, °F

The empirical observation that T_b is about midway between T_f and T_∞ has been incorporated in Eq. (12–35). Equating Eqs. (12–34) and (12–35) gives

$$T_f = T_\infty + \frac{10\pi f_s F D N}{J \hbar_{\text{CR}} A} \tag{12–36}$$

Although this equation seems to indicate the temperature rise $T_f - T_\infty$ is independent of length L, the housing surface area generally is a function of L. The housing surface area can be initially estimated, and as tuning of the design proceeds, improved results will converge. If the bushing is to be housed in a pillow block, the surface area can be roughly estimated from

$$A \doteq \frac{2\pi D L}{144} \tag{12–37}$$

Substituting Eq. (12–37) into Eq. (12–36) gives

$$T_f \doteq T_\infty + \frac{10\pi f_s F D N}{J \hbar_{\text{CR}} (2\pi D L / 144)} = T_\infty + \frac{720 f_s F N}{J \hbar_{\text{CR}} L} \tag{12–38}$$

EXAMPLE 12–8

Choose an Oiles 500 bushing to give a maximum wear of 0.001 in for 800 h of use with a 300 rev/min journal and 50 lbf radial load. Use $\hbar_{CR} = 2.7$ Btu/(h · ft^2 · °F), $T_{max} = 300$°F, $f_s = 0.03$, and a design factor $n_d = 2$. Table 12–13 lists the available bushing sizes from the manufacturer.

Solution

With a design factor n_d, substitute $n_d F$ for F. First, estimate the bushing length using Eq. (12–32) with $f_1 = f_2 = 1$, and $K = 0.6(10^{-10})$ from Table 12–8:

$$L = \frac{f_1 f_2 K n_d F N t}{3w} = \frac{1(1)0.6(10^{-10})2(50)300(800)}{3(0.001)} = 0.48 \text{ in} \qquad (1)$$

From Eq. (12–38) with $f_s = 0.03$ from Table 12–9, $\hbar_{CR} = 2.7$ Btu/(h · ft^2 · °F), and $n_d F$ for F,

$$L \doteq \frac{720 \, f_s n_d F N}{J \hbar_{CR}(T_f - T_\infty)} = \frac{720(0.03)2(50)300}{778(2.7)(300 - 70)} = 1.34 \text{ in}$$

The two results bracket L such that $0.48 \leq L \leq 1.34$ in. As a start let $L = 1$ in. From Table 12–13, we select $D = 1$ in from the midrange of available bushings.

Table 12–13

Available Bushing Sizes (in inches) of One Manufacturer*

ID	OD	$\frac{1}{2}$	$\frac{5}{8}$	$\frac{3}{4}$	$\frac{7}{8}$	1	$1\frac{1}{4}$	$1\frac{1}{2}$	$1\frac{3}{4}$	2	$2\frac{1}{2}$	3	$3\frac{1}{2}$	4	5
$\frac{1}{2}$	$\frac{3}{4}$	•	•	•	•	•									
$\frac{5}{8}$	$\frac{7}{8}$		•	•		•	•								
$\frac{3}{4}$	$1\frac{1}{8}$		•	•		•	•								
$\frac{7}{8}$	$1\frac{1}{4}$			•		•	•	•							
1	$1\frac{3}{8}$			•		•	•	•		•					
1	$1\frac{1}{2}$			•		•	•	•		•					
$1\frac{1}{4}$	$1\frac{5}{8}$					•	•	•	•	•					
$1\frac{1}{2}$	2					•	•	•	•	•					
$1\frac{3}{4}$	$2\frac{1}{4}$						•	•	•	•		•	•	•	•
2	$2\frac{1}{2}$							•		•	•	•	•		
$2\frac{1}{4}$	$2\frac{3}{4}$							•		•	•	•	•		
$2\frac{1}{2}$	3							•		•	•	•			
$2\frac{3}{4}$	$3\frac{3}{8}$							•		•	•	•	•		
3	$3\frac{5}{8}$									•	•	•	•	•	
$3\frac{1}{2}$	$4\frac{1}{8}$									•	•		•	•	
4	$4\frac{3}{4}$									•	•		•	•	
$4\frac{1}{2}$	$5\frac{3}{8}$										•		•	•	•
5	6										•		•	•	•

*In a display such as this a manufacturer is likely to show catalog numbers where the • appears.

Trial 1: $D = L = 1$ in.

Eq. (12–31): $\quad P_{\max} = \dfrac{4}{\pi} \dfrac{n_d F}{DL} = \dfrac{4}{\pi} \dfrac{2(50)}{1(1)} = 127$ psi < 3560 psi $\quad$ (OK)

$$P = \frac{n_d F}{DL} = \frac{2(50)}{1(1)} = 100 \text{ psi}$$

Eq. (12–29): $\quad V = \dfrac{\pi DN}{12} = \dfrac{\pi(1)300}{12} = 78.5$ ft/min < 100 ft/min $\quad$ (OK)

$$PV = 100(78.5) = 7850 \text{ psi} \cdot \text{ft/min} < 46\,700 \text{ psi} \cdot \text{ft/min} \quad \text{(OK)}$$

From Table 12–9,

V	f_1	
33	1.3	
78.5	f_1	=> $f_1 = 1.64$
100	1.8	

Our second estimate is $L \geq 0.48(1.64) = 0.787$ in. From Table 12–13, there is not much available for $L = \frac{7}{8}$ in. So staying with $L = 1$ in, try $D = \frac{1}{2}$ in.

Trial 2: $D = 0.5$ in, $L = 1$ in.

$$P_{\max} = \frac{4}{\pi} \frac{n_d F}{DL} = \frac{4}{\pi} \frac{2(50)}{0.5(1)} = 255 \text{ psi} < 3560 \text{ psi} \quad \text{(OK)}$$

$$P = \frac{n_d F}{DL} = \frac{2(50)}{0.5(1)} = 200 \text{ psi}$$

$$V = \frac{\pi DN}{12} = \frac{\pi(0.5)300}{12} = 39.3 \text{ ft/min} < 100 \text{ ft/min} \quad \text{(OK)}$$

Note that PV is not a function of D, and since we did not change L, PV will remain the same:

$$PV = 200(39.3) = 7860 \text{ psi} \cdot \text{ft/min} < 46\,700 \text{ psi} \cdot \text{ft/min} \quad \text{(OK)}$$

From Table 12–9, $f_1 = 1.34$, $L \geq 1.34(0.48) = 0.643$ in. There are many $\frac{3}{4}$-in bushings to select from. The smallest diameter in Table 12–13 is $D = \frac{1}{2}$ in. This gives an L/D ratio of 1.5, which is acceptable according to Eq. (12–33).

Trial 3: $D = 0.5$ in, $L = 0.75$ in. From trial 2, $V = 39.3$ ft/min does not change.

$$P_{\max} = \frac{4}{\pi} \frac{n_d F}{DL} = \frac{4}{\pi} \frac{2(50)}{0.5(0.75)} = 340 \text{ psi} < 3560 \text{ psi} \quad \text{(OK)}$$

$$P = \frac{n_d F}{DL} = \frac{2(50)}{0.5(0.75)} = 267 \text{ psi}$$

$$PV = 267(39.3) = 10\,490 \text{ psi} \cdot \text{ft/min} < 46\,700 \text{ psi} \cdot \text{ft/min} \quad \text{(OK)}$$

Answer

Select any of the bushings from the trials, where the optimum, from trial 3, is $D = \frac{1}{2}$ in and $L = \frac{3}{4}$ in. Other factors may enter in the overall design that make the other bushings more appropriate.

PROBLEMS

ANALYSIS

12–1 A full journal bearing has a journal diameter of 1.000 in, with a unilateral tolerance of −0.0015 in. The bushing bore has a diameter of 1.0015 in and a unilateral tolerance of 0.003 in. The l/d ratio is unity. The load is 250 lbf and the journal runs at 1100 rev/min. If the average viscosity is 8 μreyn, find the minimum film thickness, the power loss, and the side flow for the minimum clearance assembly.

ANALYSIS

12–2 A full journal bearing has a journal diameter of 1.250 in, with a unilateral tolerance of −0.001 in. The bushing bore has a diameter of 1.252 in and a unilateral tolerance of 0.003 in. The bearing is 2.5 in long. The journal load is 400 lbf and it runs at a speed of 1150 rev/min. Using an average viscosity of 10 μreyn find the minimum film thickness, the maximum film pressure, and the total oil-flow rate for the minimum clearance assembly.

ANALYSIS

12–3 A journal bearing has a journal diameter of 3.000 in, with a unilateral tolerance of −0.001 in. The bushing bore has a diameter of 3.005 in and a unilateral tolerance of 0.004 in. The bushing is 1.5 in long. The journal speed is 600 rev/min and the load is 800 lbf. For both SAE 10 and SAE 40, lubricants, find the minimum film thickness and the maximum film pressure for an operating temperature of 150°F for the minimum clearance assembly.

ANALYSIS

12–4 A journal bearing has a journal diameter of 3.000 in with a unilateral tolerance of −0.003 in. The bushing bore has a diameter of 3.006 in and a unilateral tolerance of 0.004 in. The bushing is 3 in long and supports a 600-lbf load. The journal speed is 750 rev/min. Find the minimum oil film thickness and the maximum film pressure for both SAE 10 and SAE 20W-40 lubricants, for the tightest assembly if the operating film temperature is 140°F.

ANALYSIS

12–5 A full journal bearing has a journal with a diameter of 2.000 in and a unilateral tolerance of −0.0012 in. The bushing has a bore with a diameter of 2.0024 and a unilateral tolerance of 0.002 in. The bushing is 1 in long and supports a load of 600 lbf at a speed of 800 rev/min. Find the minimum film thickness, the power loss, and the total lubricant flow if the average film temperature is 130°F and SAE 20 lubricant is used. The tightest assembly is to be analyzed.

ANALYSIS

12–6 A full journal bearing has a shaft journal diameter of 25 mm with a unilateral tolerance of −0.01 mm. The bushing bore has a diameter of 25.04 mm with a unilateral tolerance of 0.03 mm. The l/d ratio is unity. The bushing load is 1.25 kN, and the journal rotates at 1200 rev/min. Analyze the minimum clearance assembly if the average viscosity is 50 mPa · s to find the minimum oil film thickness, the power loss, and the percentage of side flow.

ANALYSIS

12–7 A full journal bearing has a shaft journal with a diameter of 30.00 mm and a unilateral tolerance of −0.015 mm. The bushing bore has a diameter of 30.05 mm with a unilateral tolerance of 0.035 mm. The bushing bore is 50 mm in length. The bearing load is 2.75 kN and the journal rotates at 1120 rev/min. Analyze the minimum clearance assembly and find the minimum film thickness, the coefficient of friction, and the total oil flow if the average viscosity is 60 mPa · s.

ANALYSIS

12–8 A journal bearing has a shaft diameter of 75.00 mm with a unilateral tolerance of −0.02 mm. The bushing bore has a diameter of 75.10 mm with a unilateral tolerance of 0.06 mm. The bushing is 36 mm long and supports a load of 2 kN. The journal speed is 720 rev/min. For the minimum clearance assembly find the minimum film thickness, the heat loss rate, and the maximum lubricant pressure for SAE 20 and SAE 40 lubricants operating at an average film temperature of 60°C.

ANALYSIS

12–9 A full journal bearing is 25 mm long. The shaft journal has a diameter of 50 mm with a unilateral tolerance of −0.01 mm. The bushing bore has a diameter of 50.05 mm with a unilateral tolerance of 0.01 mm. The load is 2000 N and the journal speed is 840 rev/min. For the minimum clearance assembly find the minimum oil-film thickness, the power loss, and the side flow if the operating temperature is 55°C and SAE 30 lubricating oil is used.

12–10 A $1\frac{1}{4}$- $\times 1\frac{1}{4}$-in sleeve bearing supports a load of 700 lbf and has a journal speed of 3600 rev/min. An SAE 10 oil is used having an average temperature of 160°F. Using Fig. 12–16, estimate the radial clearance for minimum coefficient of friction f and for maximum load-carrying capacity W. The difference between these two clearances is called the clearance range. Is the resulting range attainable in manufacture?

12–11 A full journal bearing has a shaft diameter of 80.00 mm with a unilateral tolerance of −0.01 mm. The l/d ratio is unity. The bushing has a bore diameter of 80.08 mm with a unilateral tolerance of 0.03 mm. The SAE 30 oil supply is in an axial-groove sump with a steady-state temperature of 60°C. The radial load is 3000 N. Estimate the average film temperature, the minimum film thickness, the heat loss rate, and the lubricant side-flow rate for the minimum clearance assembly, if the journal speed is 8 rev/s.

12–12 A $2\frac{1}{2}$- $\times 2\frac{1}{2}$-in sleeve bearing uses grade 20 lubricant. The axial-groove sump has a steady-state temperature of 110°F. The shaft journal has a diameter of 2.500 in with a unilateral tolerance of −0.001 in. The bushing bore has a diameter of 2.504 in with a unilateral tolerance of 0.001 in. The journal speed is 1120 rev/min and the radial load is 1200 lbf. Estimate
(*a*) The magnitude and location of the minimum oil-film thickness.
(*b*) The eccentricity.
(*c*) The coefficient of friction.
(*d*) The power loss rate.
(*e*) Both the total and side oil-flow rates.
(*f*) The maximum oil-film pressure and its angular location.
(*g*) The terminating position of the oil film.
(*h*) The average temperature of the side flow.
(*i*) The oil temperature at the terminating position of the oil film.

12–13 A set of sleeve bearings has a specification of shaft journal diameter of 1.250 in with a unilateral tolerance of −0.001 in. The bushing bore has a diameter of 1.252 in with a unilateral tolerance of 0.003 in. The bushing is $1\frac{1}{4}$ in long. The radial load is 250 lbf and the shaft rotational speed is 1750 rev/min. The lubricant is SAE 10 oil and the axial-groove sump temperature at steady state T_s is 120°F. For the c_{min}, c_{median}, and c_{max} assemblies analyze the bearings and observe the changes in S, ϵ, f, Q, Q_s, ΔT, T_{max}, $\bar{T}_f$, and hp.

12–14 An interpolation equation was given by Raimondi and Boyd, and it is displayed as Eq. (12–16). This equation is a good candidate for a computer program. Write such a program for interactive use. Once ready for service it can save time and reduce errors. Another version of this program can be used with a subprogram that contains curve fits to Raimondi and Boyd charts for computer use.

12–15 A natural-circulation pillow-block bearing has a journal diameter D of 2.500 in with a unilateral tolerance of −0.001 in. The bushing bore diameter B is 2.504 in with a unilateral tolerance of 0.004 in. The shaft runs at an angular speed of 1120 rev/min; the bearing uses SAE grade 20 oil and carries a steady load of 300 lbf in shaft-stirred air at 70°F. The lateral area of the pillow-block housing is 60 in². Perform a design assessment using minimum radial clearance for a load of 600 lbf and 300 lbf. Use Trumpler's criteria.

12–16 An eight-cylinder diesel engine has a front main bearing with a journal diameter of 3.500 in and a unilateral tolerance of −0.003 in. The bushing bore diameter is 3.505 in with a unilateral tolerance of +0.005 in. The bushing length is 2 in. The pressure-fed bearing has a central annular groove 0.250 in wide. The SAE 30 oil comes from a sump at 120°F using a supply pressure of 50 psig. The sump's heat-dissipation capacity is 5000 Btu/h per bearing. For a minimum radial clearance, a speed of 2000 rev/min, and a radial load of 4600 lbf, find the average film temperature and apply Trumpler's criteria in your design assessment.

DESIGN **12–17** A pressure-fed bearing has a journal diameter of 50.00 mm with a unilateral tolerance of −0.05 mm. The bushing bore diameter is 50.084 mm with a unilateral tolerance of 0.10 mm. The length of the bushing is 55 mm. Its central annular groove is 5 mm wide and is fed by SAE 30 oil is 55°C at 200 kPa supply gauge pressure. The journal speed is 2880 rev/min carrying a load of 10 kN. The sump can dissipate 300 watts per bearing if necessary. For minimum radial clearances, perform a design assessment using Trumpler's criteria.

DESIGN **12–18** Design a central annular-groove pressure-fed bearing with an l'/d ratio of 0.5, using SAE grade 20 oil, the lubricant supplied at 30 psig. The exterior oil cooler can maintain the sump temperature at 120°F for heat dissipation rates up to 1500 Btu/h. The load to be carried is 900 lbf at 3000 rev/min. The groove width is $\frac{1}{4}$ in. Use nominal journal diameter d as one design variable and c as the other. Use Trumpler's criteria for your adequacy assessment.

DESIGN **12–19** Repeat design problem Prob. 12–18 using the nominal bushing bore B as one decision variable and the radial clearance c as the other. Again, Trumpler's criteria to be used.

ANALYSIS **12–20** Table 12–1 gives the Seireg and Dandage curve fit for the absolute viscosity in customary U.S. engineering units. Show that in SI units of mPa · s and a temperature of C degrees Celsius, the viscosity can be expressed as

$$\mu = 6.89(10^6)\mu_0 \exp[(b/(1.8C + 127))]$$

where μ_0 and b are from Table 12–1. If the viscosity μ_0' is expressed in μreyn, then

$$\mu = 6.89\mu_0' \exp[(b/(1.8C + 127))]$$

What is the viscosity of a grade 30 oil at 79°C?

DESIGN **12–21** For Prob. 12–18 a satisfactory design is

$$d = 2.000^{+0}_{-0.001} \text{ in} \qquad b = 2.005^{+0.003}_{-0} \text{ in}$$

Double the size of the bearing dimensions and quadruple the load to 3600 lbf.
(a) Analyze the scaled-up bearing for median assembly.
(b) Compare the results of a similar analysis for the 2-in bearing, median assembly.

ANALYSIS **12–22** An Oiles SP 500 alloy brass bushing is 1 in long with a 1-in bore and operates in a clean environment at 70°F. The allowable wear without loss of function is 0.005 in. The radial load is 500 lbf. The shaft speed is 200 rev/min. Estimate the number of revolutions for radial wear to be 0.005 in.

DESIGN **12–23** Choose an Oiles SP 500 alloy brass bushing to give a maximum wear of 0.002 in for 1000 h of use with a 400 rev/min journal and 100 lbf radial load. Use $\hbar_{CR} = 2.7$ Btu/(h · ft² · °F), $T_{max} = 300$°F, $f_s = 0.03$, and a design factor $n_d = 2$. Table 12–13 lists the bushing sizes available from the manufacturer.

13

Gears—General

Chapter Outline

This chapter addresses gear geometry, the kinematic relations, and the forces transmitted by the four principal types of gears: spur, helical, bevel, and worm gears. The forces transmitted between meshing gears supply torsional moments to shafts for motion and power transmission and create forces and moments that affect the shaft and its bearings. The next two chapters will address stress, strength, safety, and reliability of the four types of gears.

13–1   Types of Gears

Spur gears, illustrated in Fig. 13–1, have teeth parallel to the axis of rotation and are used to transmit motion from one shaft to another, parallel, shaft. Of all types, the spur gear is the simplest and, for this reason, will be used to develop the primary kinematic relationships of the tooth form.

Helical gears, shown in Fig. 13–2, have teeth inclined to the axis of rotation. Helical gears can be used for the same applications as spur gears and, when so used, are not as noisy, because of the more gradual engagement of the teeth during meshing. The inclined tooth also develops thrust loads and bending couples, which are not present with spur gearing. Sometimes helical gears are used to transmit motion between nonparallel shafts.

Bevel gears, shown in Fig. 13–3, have teeth formed on conical surfaces and are used mostly for transmitting motion between intersecting shafts. The figure actually illustrates *straight-tooth bevel gears. Spiral bevel gears* are cut so the tooth is no longer straight, but forms a circular arc. *Hypoid gears* are quite similar to spiral bevel gears except that the shafts are offset and nonintersecting.

Figure 13–1

Spur gears are used to transmit rotary motion between parallel shafts.

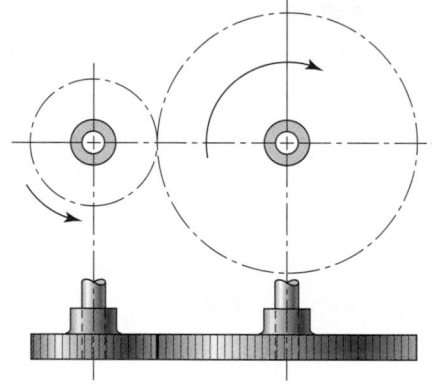

Figure 13–2

Helical gears are used to transmit motion between parallel or nonparallel shafts.

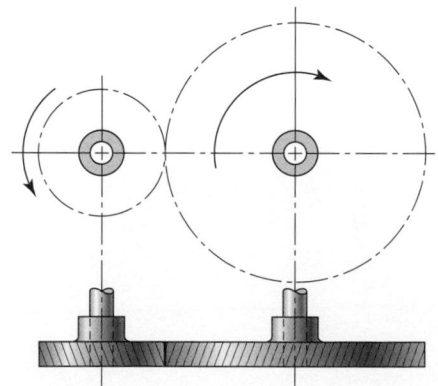

Figure 13–3

Bevel gears are used to transmit rotary motion between intersecting shafts.

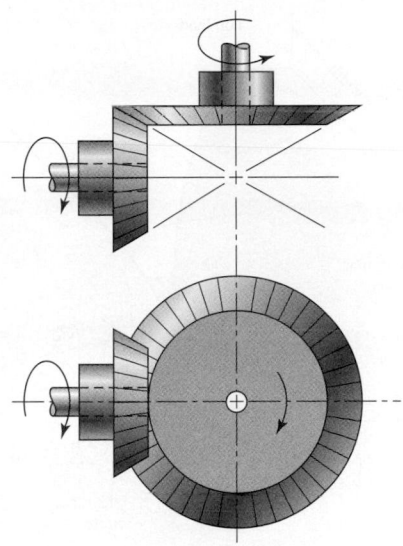

Figure 13–4

Worm gearsets are used to transmit rotary motion between nonparallel and nonintersecting shafts.

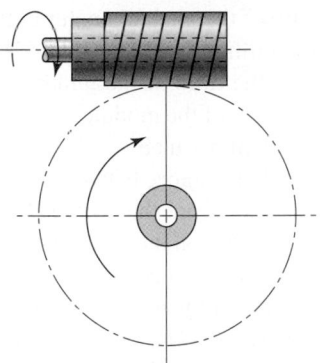

Worms and *worm gears,* shown in Fig. 13–4, represent the fourth basic gear type. As shown, the worm resembles a screw. The direction of rotation of the worm gear, also called the worm wheel, depends upon the direction of rotation of the worm and upon whether the worm teeth are cut right-hand or left-hand. Worm-gear sets are also made so that the teeth of one or both wrap partly around the other. Such sets are called *single-enveloping* and *double-enveloping* worm-gear sets. Worm-gear sets are mostly used when the speed ratios of the two shafts are quite high, say, 3 or more.

13–2   Nomenclature

The terminology of spur-gear teeth is illustrated in Fig. 13–5. The *pitch circle* is a theoretical circle upon which all calculations are usually based; its diameter is the *pitch diameter.* The pitch circles of a pair of mating gears are tangent to each other. A *pinion* is the smaller of two mating gears. The larger is often called the *gear.*

The *circular pitch p* is the distance, measured on the pitch circle, from a point on one tooth to a corresponding point on an adjacent tooth. Thus the circular pitch is equal to the sum of the *tooth thickness* and the *width of space.*

Figure 13–5

Nomenclature of spur-gear teeth.

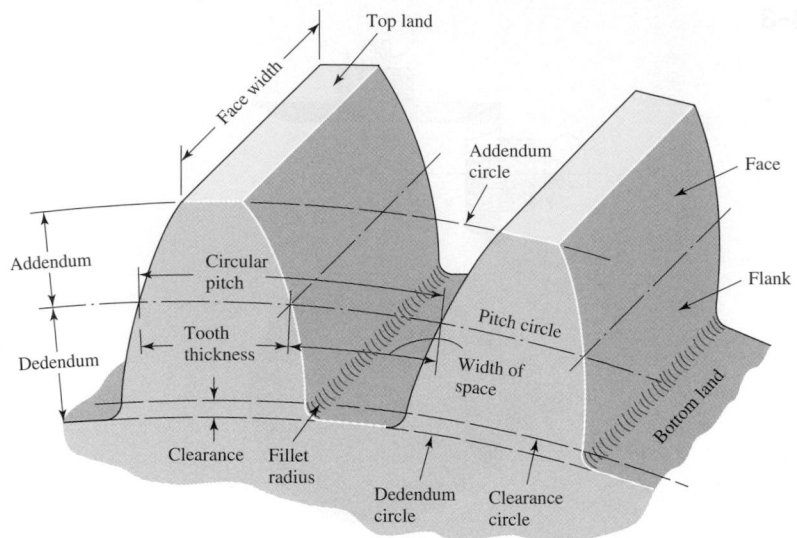

The *module m* is the ratio of the pitch diameter to the number of teeth. The customary unit of length used is the millimeter. The module is the index of tooth size in SI.

The *diametral pitch P* is the ratio of the number of teeth on the gear to the pitch diameter. Thus, it is the reciprocal of the module. Since diametral pitch is used only with U.S. units, it is expressed as teeth per inch.

The *addendum a* is the radial distance between the *top land* and the pitch circle. The *dedendum b* is the radial distance from the *bottom land* to the pitch circle. The *whole depth* h_t is the sum of the addendum and the dedendum.

The *clearance circle* is a circle that is tangent to the addendum circle of the mating gear. The *clearance c* is the amount by which the dedendum in a given gear exceeds the addendum of its mating gear. The *backlash* is the amount by which the width of a tooth space exceeds the thickness of the engaging tooth measured on the pitch circles.

You should prove for yourself the validity of the following useful relations:

$$P = \frac{N}{d} \qquad (13–1)$$

$$m = \frac{d}{N} \qquad (13–2)$$

$$p = \frac{\pi d}{N} = \pi m \qquad (13–3)$$

$$pP = \pi \qquad (13–4)$$

where P = diametral pitch, teeth per inch
   N = number of teeth
   d = pitch diameter, in
   m = module, mm
   d = pitch diameter, mm
   p = circular pitch

13–3   **Conjugate Action**

The following discussion assumes the teeth to be perfectly formed, perfectly smooth, and absolutely rigid. Such an assumption is, of course, unrealistic, because the application of forces will cause deflections.

Mating gear teeth acting against each other to produce rotary motion are similar to cams. When the tooth profiles, or cams, are designed so as to produce a constant angular-velocity ratio during meshing, these are said to have *conjugate action*. In theory, at least, it is possible arbitrarily to select any profile for one tooth and then to find a profile for the meshing tooth that will give conjugate action. One of these solutions is the *involute profile*, which, with few exceptions, is in universal use for gear teeth and is the only one with which we should be concerned.

When one curved surface pushes against another (Fig. 13–6), the point of contact occurs where the two surfaces are tangent to each other (point *c*), and the forces at any instant are directed along the common normal *ab* to the two curves. The line *ab*, representing the direction of action of the forces, is called the *line of action*. The line of action will intersect the line of centers *O-O* at some point *P*. The angular-velocity ratio between the two arms is inversely proportional to their radii to the point *P*. Circles drawn through point *P* from each center are called *pitch circles,* and the radius of each circle is called the *pitch radius*. Point *P* is called the *pitch point.*

Figure 13–6 is useful in making another observation. A pair of gears is really pairs of cams that act through a small arc and, before running off the involute contour, are replaced by another identical pair of cams. The cams can run in either direction and are configured to transmit a constant angular-velocity ratio. If involute curves are used, the gears tolerate changes in center-to-center distance with *no* variation in constant angular-velocity ratio. Furthermore, the rack profiles are straight-flanked, making primary tooling simpler.

To transmit motion at a constant angular-velocity ratio, the pitch point must remain fixed; that is, all the lines of action for every instantaneous point of contact must pass through the same point *P*. In the case of the involute profile, it will be shown that all points of contact occur on the same straight line *ab*, that all normals to the tooth profiles at the point of contact coincide with the line *ab*, and, thus, that these profiles transmit uniform rotary motion.

Figure 13–6

Cam A and follower B in contact. When the contacting surfaces are involute profiles, the ensuing conjugate action produces a constant angular-velocity ratio.

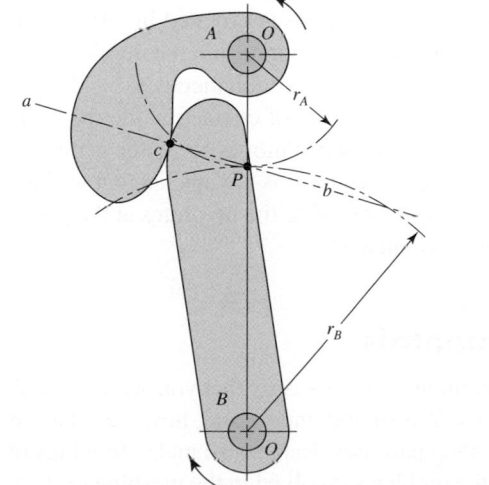

Figure 13–7

(a) Generation of an involute;
(b) involute action.

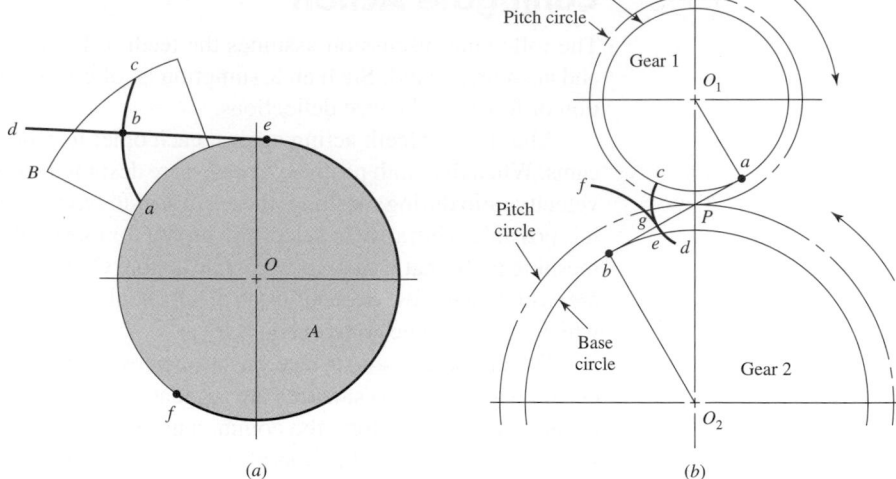

(a)   (b)

13–4   Involute Properties

An involute curve may be generated as shown in Fig. 13–7a. A partial flange B is attached to the cylinder A, around which is wrapped a cord *def,* which is held tight. Point b on the cord represents the tracing point, and as the cord is wrapped and unwrapped about the cylinder, point b will trace out the involute curve ac. The radius of the curvature of the involute varies continuously, being zero at point a and a maximum at point c. At point b the radius is equal to the distance be, since point b is instantaneously rotating about point e. Thus the generating line de is normal to the involute at all points of intersection and, at the same time, is always tangent to the cylinder A. The circle on which the involute is generated is called the *base circle.*

Let us now examine the involute profile to see how it satisfies the requirement for the transmission of uniform motion. In Fig. 13–7b, two gear blanks with fixed centers at O_1 and O_2 are shown having base circles whose respective radii are O_1a and O_2b. We now imagine that a cord is wound clockwise around the base circle of gear 1, pulled tight between points a and b, and wound counterclockwise around the base circle of gear 2. If, now, the base circles are rotated in different directions so as to keep the cord tight, a point g on the cord will trace out the involutes cd on gear 1 and ef on gear 2. The involutes are thus generated simultaneously by the tracing point. The tracing point, therefore, represents the point of contact, while the portion of the cord ab is the generating line. The point of contact moves along the generating line; the generating line does not change position, because it is always tangent to the base circles; and since the generating line is always normal to the involutes at the point of contact, the requirement for uniform motion is satisfied.

13–5   Fundamentals

Among other things, it is necessary that you actually be able to draw the teeth on a pair of meshing gears. You should understand, however, that you are not doing this for manufacturing or shop purposes. Rather, we make drawings of gear teeth to obtain an understanding of the problems involved in the meshing of the mating teeth.

Figure 13–8

Construction of an involute curve.

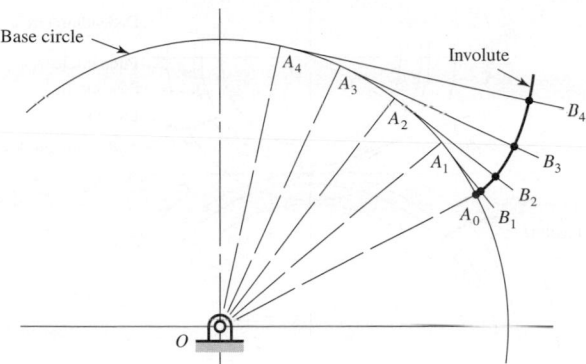

First, it is necessary to learn how to construct an involute curve. As shown in Fig. 13–8, divide the base circle into a number of equal parts, and construct radial lines OA_0, OA_1, OA_2, etc. Beginning at A_1, construct perpendiculars A_1B_1, A_2B_2, A_3B_3, etc. Then along A_1B_1 lay off the distance A_1A_0, along A_2B_2 lay off twice the distance A_1A_0, etc., producing points through which the involute curve can be constructed.

To investigate the fundamentals of tooth action, let us proceed step by step through the process of constructing the teeth on a pair of gears.

When two gears are in mesh, their pitch circles roll on one another without slipping. Designate the pitch radii as r_1 and r_2 and the angular velocities as ω_1 and ω_2, respectively. Then the pitch-line velocity is

$$V = |r_1\omega_1| = |r_2\omega_2|$$

Thus the relation between the radii on the angular velocities is

$$\left|\frac{\omega_1}{\omega_2}\right| = \frac{r_2}{r_1} \tag{13–5}$$

Suppose now we wish to design a speed reducer such that the input speed is 1800 rev/min and the output speed is 1200 rev/min. This is a ratio of 3:2; the gear pitch diameters would be in the same ratio, for example, a 4-in pinion driving a 6-in gear. The various dimensions found in gearing are always based on the pitch circles.

Suppose we specify that an 18-tooth pinion is to mesh with a 30-tooth gear and that the diametral pitch of the gearset is to be 2 teeth per inch. Then, from Eq. (13–1), the pitch diameters of the pinion and gear are, respectively,

$$d_1 = \frac{N_1}{P} = \frac{18}{2} = 9 \text{ in} \qquad d_2 = \frac{N_2}{P} = \frac{30}{2} = 15 \text{ in}$$

The first step in drawing teeth on a pair of mating gears is shown in Fig. 13–9. The center distance is the sum of the pitch radii, in this case 12 in. So locate the pinion and gear centers O_1 and O_2, 12 in apart. Then construct the pitch circles of radii r_1 and r_2. These are tangent at P, the *pitch point*. Next draw line ab, the common tangent, through the pitch point. We now designate gear 1 as the driver, and since it is rotating counterclockwise, we draw a line cd through point P at an angle ϕ to the common tangent ab. The line cd has three names, all of which are in general use. It is called the *pressure line,* the *generating line,* and the *line of action.* It represents the direction in which the resultant force acts between the gears. The angle ϕ is called the *pressure angle,* and it usually has values of 20 or 25°, though $14\frac{1}{2}°$ was once used.

Figure 13–9

Circles of a gear layout.

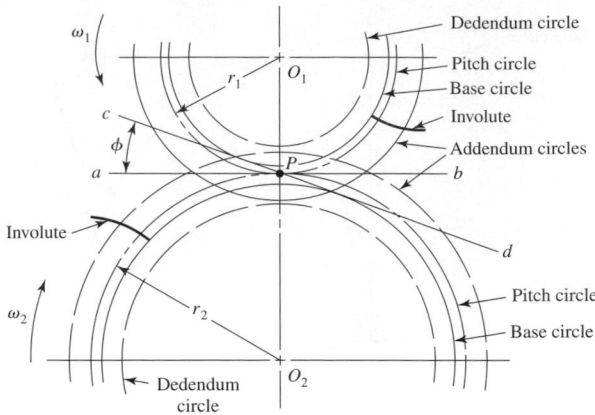

Figure 13–10

Base circle radius can be related to the pressure angle ϕ and the pitch circle radius by $r_b = r \cos \phi$.

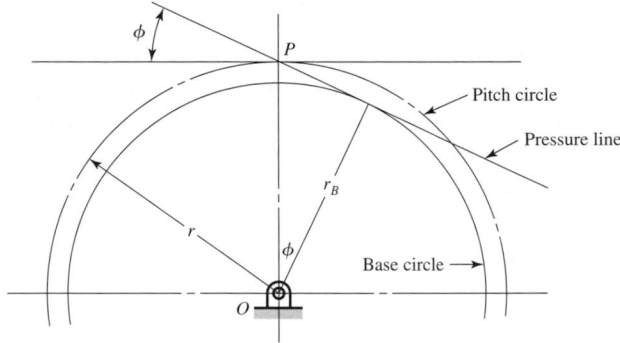

Next, on each gear draw a circle tangent to the pressure line. These circles are the *base circles*. Since they are tangent to the pressure line, the pressure angle determines their size. As shown in Fig. 13–10, the radius of the base circle is

$$r_b = r \cos \phi \tag{13–6}$$

where r is the pitch radius.

Now generate an involute on each base circle as previously described and as shown in Fig. 13–9. This involute is to be used for one side of a gear tooth. It is not necessary to draw another curve in the reverse direction for the other side of the tooth, because we are going to use a template which can be turned over to obtain the other side.

The addendum and dedendum distances for standard interchangeable teeth are, as we shall learn later, $1/P$ and $1.25/P$, respectively. Therefore, for the pair of gears we are constructing,

$$a = \frac{1}{P} = \frac{1}{2} = 0.500 \text{ in} \qquad b = \frac{1.25}{P} = \frac{1.25}{2} = 0.625 \text{ in}$$

Using these distances, draw the addendum and dedendum circles on the pinion and on the gear as shown in Fig. 13–9.

Next, using heavy drawing paper, or preferably, a sheet of 0.015- to 0.020-in clear plastic, cut a template for each involute, being careful to locate the gear centers properly with respect to each involute. Figure 13–11 is a reproduction of the template used to create some of the illustrations for this book. Note that only one side of the tooth profile is formed on the template. To get the other side, turn the template over. For some problems you might wish to construct a template for the entire tooth.

Figure 13–11

A template for drawing gear teeth.

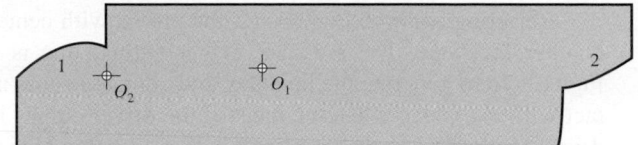

Figure 13–12

Tooth action.

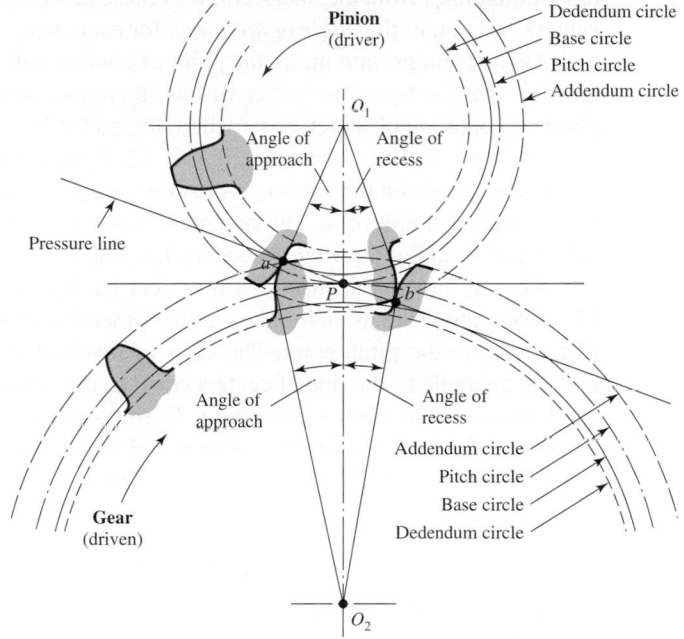

To draw a tooth, we must know the tooth thickness. From Eq. (13–4), the circular pitch is

$$p = \frac{\pi}{P} = \frac{\pi}{2} = 1.57 \text{ in}$$

Therefore, the tooth thickness is

$$t = \frac{p}{2} = \frac{1.57}{2} = 0.785 \text{ in}$$

measured on the pitch circle. Using this distance for the tooth thickness as well as the tooth space, draw as many teeth as desired, using the template, after the points have been marked on the pitch circle. In Fig. 13–12 only one tooth has been drawn on each gear. You may run into trouble in drawing these teeth if one of the base circles happens to be larger than the dedendum circle. The reason for this is that the involute begins at the base circle and is undefined below this circle. So, in drawing gear teeth, we usually draw a radial line for the profile below the base circle. The actual shape, however, will depend upon the kind of machine tool used to form the teeth in manufacture, that is, how the profile is generated.

The portion of the tooth between the clearance circle and the dedendum circle includes the fillet. In this instance the clearance is

$$c = b - a = 0.625 - 0.500 = 0.125 \text{ in}$$

The construction is finished when these fillets have been drawn.

Referring again to Fig. 13–12, the pinion with center at O_1 is the driver and turns counterclockwise. The pressure, or generating, line is the same as the cord used in Fig. 13–7a to generate the involute, and contact occurs along this line. The initial contact will take place when the flank of the driver comes into contact with the tip of the driven tooth. This occurs at point a in Fig. 13–12, where the addendum circle of the driven gear crosses the pressure line. If we now construct tooth profiles through point a and draw radial lines from the intersections of these profiles with the pitch circles to the gear centers, we obtain the *angle of approach* for each gear.

As the teeth go into mesh, the point of contact will slide up the side of the driving tooth so that the tip of the driver will be in contact just before contact ends. The final point of contact will therefore be where the addendum circle of the driver crosses the pressure line. This is point b in Fig. 13–12. By drawing another set of tooth profiles through b, we obtain the *angle of recess* for each gear in a manner similar to that of finding the angles of approach. The sum of the angle of approach and the angle of recess for either gear is called the *angle of action*. The line ab is called *the line of action*.

We may imagine a *rack* as a spur gear having an infinitely large pitch diameter. Therefore, the rack has an infinite number of teeth and a base circle which is an infinite distance from the pitch point. The sides of involute teeth on a rack are straight lines making an angle to the line of centers equal to the pressure angle. Figure 13–13 shows an involute rack in mesh with a pinion. Corresponding sides on involute teeth are parallel curves; the *base pitch* is the constant and fundamental distance between them along a common normal as shown in Fig. 13–13. The base pitch is related to the circular pitch by the equation

$$p_b = p_c \cos \phi \qquad (13\text{–}7)$$

where p_b is the base pitch.

Figure 13–14 shows a pinion in mesh with an *internal*, or *ring, gear*. Note that both of the gears now have their centers of rotation on the same side of the pitch point. Thus the positions of the addendum and dedendum circles with respect to the pitch circle are reversed; the addendum circle of the internal gear lies *inside* the pitch circle. Note, too, from Fig. 13–14, that the base circle of the internal gear lies inside the pitch circle near the addendum circle.

Another interesting observation concerns the fact that the operating diameters of the pitch circles of a pair of meshing gears need not be the same as the respective design pitch diameters of the gears, though this is the way they have been constructed in Fig. 13–12. If we increase the center distance, we create two new operating pitch circles having larger diameters because they must be tangent to each other at the pitch point.

Figure 13–13

Involute-toothed pinion and rack.

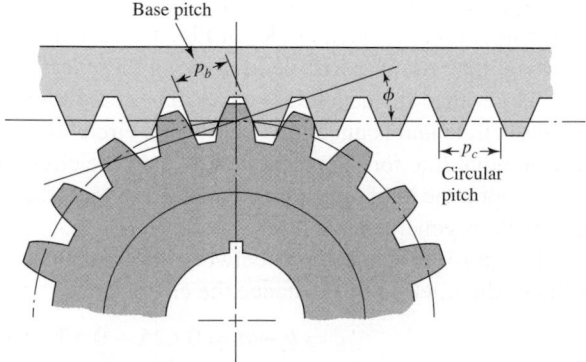

Base pitch

p_b

ϕ

p_c
Circular
pitch

Figure 13–14

Internal gear and pinion.

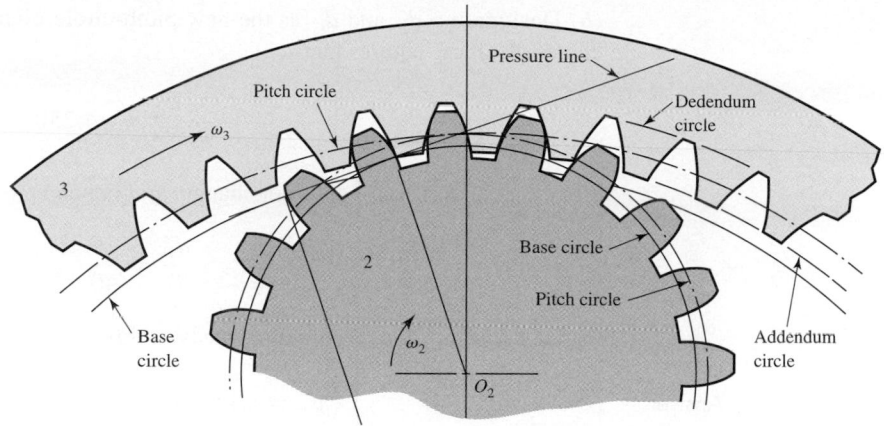

Thus the pitch circles of gears really do not come into existence until a pair of gears are brought into mesh.

Changing the center distance has no effect on the base circles, because these were used to generate the tooth profiles. Thus the base circle is basic to a gear. Increasing the center distance increases the pressure angle and decreases the length of the line of action, but the teeth are still conjugate, the requirement for uniform motion transmission is still satisfied, and the angular-velocity ratio has not changed.

EXAMPLE 13–1

A gearset consists of a 16-tooth pinion driving a 40-tooth gear. The diametral pitch is 2, and the addendum and dedendum are $1/P$ and $1.25/P$, respectively. The gears are cut using a pressure angle of $20°$.

(a) Compute the circular pitch, the center distance, and the radii of the base circles.

(b) In mounting these gears, the center distance was incorrectly made $\frac{1}{4}$ in larger. Compute the new values of the pressure angle and the pitch-circle diameters.

Solution

Answer   (a)
$$p = \frac{\pi}{P} = \frac{\pi}{2} = 1.57 \text{ in}$$

The pitch diameters of the pinion and gear are, respectively,

$$d_P = \frac{16}{2} = 8 \text{ in} \qquad d_G = \frac{40}{2} = 20 \text{ in}$$

Therefore the center distance is

$$\frac{d_P + d_G}{2} = \frac{8 + 20}{2} = 14 \text{ in}$$

Since the teeth were cut on the $20°$ pressure angle, the base-circle radii are found to be, using $r_b = r \cos \phi$,

Answer
$$r_b \text{ (pinion)} = \frac{8}{2} \cos 20° = 3.76 \text{ in}$$

Answer
$$r_b \text{ (gear)} = \frac{20}{2} \cos 20° = 9.40 \text{ in}$$

(b) Designating d'_P and d'_G as the new pitch-circle diameters, the $\frac{1}{4}$-in increase in the center distance requires that

$$\frac{d'_P + d'_G}{2} = 14.250 \tag{1}$$

Also, the velocity ratio does not change, and hence

$$\frac{d'_P}{d'_G} = \frac{16}{40} \tag{2}$$

Solving Eqs. (1) and (2) simultaneously yields

Answer
$$d'_P = 8.143 \text{ in} \qquad d'_G = 20.357 \text{ in}$$

Since $r_b = r \cos\phi$, the new pressure angle is

Answer
$$\phi' = \cos^{-1}\frac{r_b \text{ (pinion)}}{d'_P/2} = \cos^{-1}\frac{3.76}{8.143/2} = 22.56°$$

13–6 Contact Ratio

The zone of action of meshing gear teeth is shown in Fig. 13–15. We recall that tooth contact begins and ends at the intersections of the two addendum circles with the pressure line. In Fig. 13–15 initial contact occurs at a and final contact at b. Tooth profiles drawn through these points intersect the pitch circle at A and B, respectively. As shown, the distance AP is called the *arc of approach* q_a, and the distance PB, the *arc of recess* q_r. The sum of these is the *arc of action* q_t.

Now, consider a situation in which the arc of action is exactly equal to the circular pitch, that is, $q_t = p$. This means that one tooth and its space will occupy the entire arc AB. In other words, when a tooth is just beginning contact at a, the previous tooth is simultaneously ending its contact at b. Therefore, during the tooth action from a to b, there will be exactly one pair of teeth in contact.

Next, consider a situation in which the arc of action is greater than the circular pitch, but not very much greater, say, $q_t \doteq 1.2p$. This means that when one pair of teeth is just entering contact at a, another pair, already in contact, will not yet have reached b. Thus,

Figure 13–15

Definition of contact ratio.

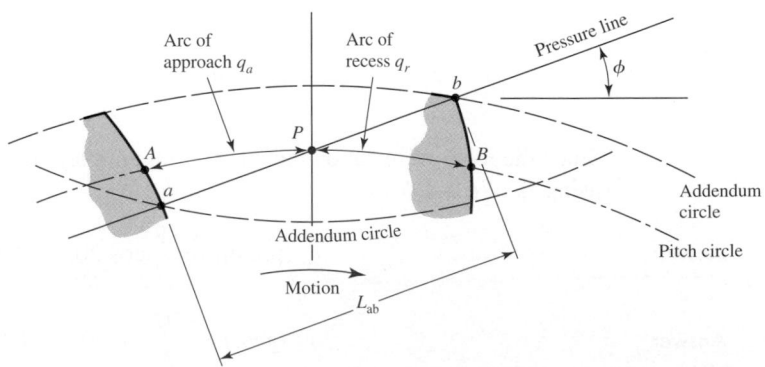

for a short period of time, there will be two teeth in contact, one in the vicinity of A and another near B. As the meshing proceeds, the pair near B must cease contact, leaving only a single pair of contacting teeth, until the procedure repeats itself.

Because of the nature of this tooth action, either one or two pairs of teeth in contact, it is convenient to define the term *contact ratio* m_c as

$$m_c = \frac{q_t}{p} \tag{13–8}$$

a number that indicates the average number of pairs of teeth in contact. Note that this ratio is also equal to the length of the path of contact divided by the base pitch. Gears should not generally be designed having contact ratios less than about 1.20, because inaccuracies in mounting might reduce the contact ratio even more, increasing the possibility of impact between the teeth as well as an increase in the noise level.

An easier way to obtain the contact ratio is to measure the line of action ab instead of the arc distance AB. Since ab in Fig. 13–15 is tangent to the base circle when extended, the base pitch p_b must be used to calculate m_c instead of the circular pitch as in Eq. (13–8). If the length of the line of action is L_{ab}, the contact ratio is

$$m_c = \frac{L_{ab}}{p \cos \phi} \tag{13–9}$$

in which Eq. (13–7) was used for the base pitch.

13–7 Interference

The contact of portions of tooth profiles that are not conjugate is called *interference*. Consider Fig. 13–16. Illustrated are two 16-tooth gears that have been cut to the now obsolete $14\frac{1}{2}°$ pressure angle. The driver, gear 2, turns clockwise. The initial and final points of contact are designated A and B, respectively, and are located on the pressure line. Now notice that the points of tangency of the pressure line with the base circles C and D are located *inside* of points A and B. Interference is present.

The interference is explained as follows. Contact begins when the tip of the driven tooth contacts the flank of the driving tooth. In this case the flank of the driving tooth first makes contact with the driven tooth at point A, and this occurs *before* the involute portion of the driving tooth comes within range. In other words, contact is occurring below the base circle of gear 2 on the *noninvolute* portion of the flank. The actual effect is that the involute tip or face of the driven gear tends to dig out the noninvolute flank of the driver.

In this example the same effect occurs again as the teeth leave contact. Contact should end at point D or before. Since it does not end until point B, the effect is for the tip of the driving tooth to dig out, or interfere with, the flank of the driven tooth.

When gear teeth are produced by a generation process, interference is automatically eliminated because the cutting tool removes the interfering portion of the flank. This effect is called *undercutting;* if undercutting is at all pronounced, the undercut tooth is considerably weakened. Thus the effect of eliminating interference by a generation process is merely to substitute another problem for the original one.

The smallest number of teeth on a spur pinion and gear,[1] one-to-one gear ratio, which can exist without interference is N_P. This number of teeth for spur gears is

---

[1]Robert Lipp, "Avoiding Tooth Interference in Gears," *Machine Design*, Vol. 54, No. 1, 1982, pp. 122, 124.

Figure 13–16

Interference in the action of gear teeth. (This is actually a rather poor figure; J. E. Shigley drew the tooth shape using circular arcs, which is incorrect, to answer a student's question many years ago.)

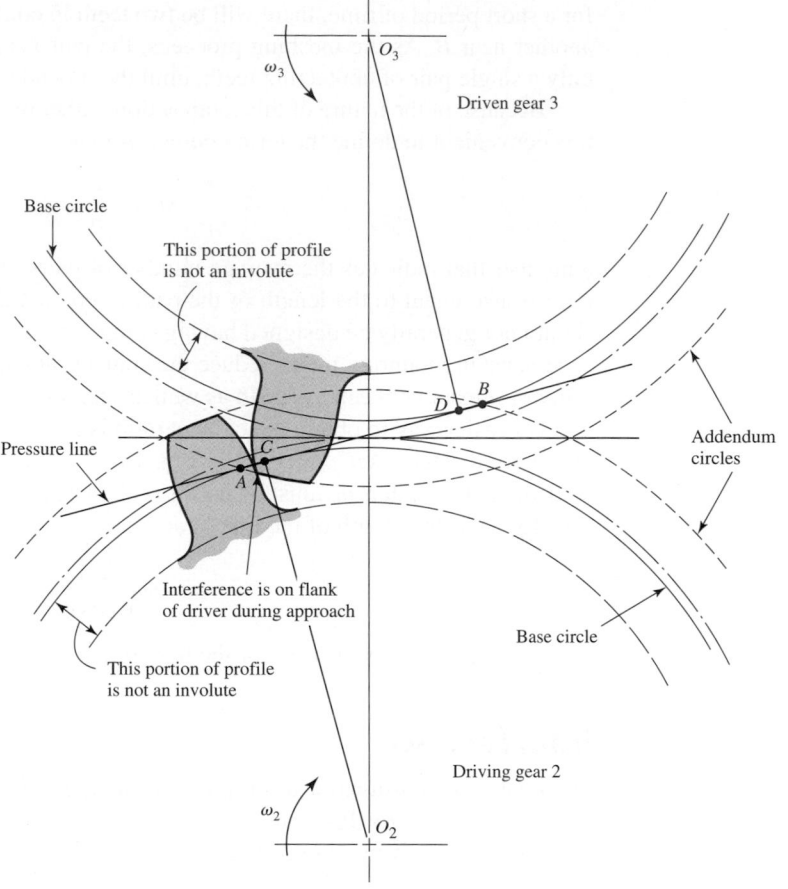

given by

$$N_P = \frac{4k}{6\sin^2\phi}\left(1 + \sqrt{1 + 3\sin^2\phi}\right) \tag{13–10}$$

where $k = 1$ for full-depth teeth, 0.8 for stub teeth and ϕ = pressure angle.

For a 20° pressure angle, with $k = 1$,

$$N_P = \frac{4(1)}{6\sin^2 20°}\left(1 + \sqrt{1 + 3\sin^2 20°}\right) = 12.3 = 13 \text{ teeth}$$

Thus 13 teeth on pinion and gear are interference-free. Realize that 12.3 teeth is possible in meshing arcs, but for fully rotating gears, 13 teeth represents the least number. For a $14\frac{1}{2}°$ pressure angle, $N_P = 23$ teeth, so one can appreciate why few $14\frac{1}{2}°$-tooth systems are used, as the higher pressure angles can produce a smaller pinion with accompanying smaller center-to-center distances.

If the mating gear has more teeth than the pinion, that is, $m_G = N_G/N_P = m$ is more than one, then the smallest number of teeth on the pinion without interference is given by

$$N_P = \frac{2k}{(1 + 2m)\sin^2\phi}\left(m + \sqrt{m^2 + (1 + 2m)\sin^2\phi}\right) \tag{13–11}$$

If $m = 4$, $\phi = 20°$,

$$N_P = \frac{2(1)}{(1 + 2[4]) \sin^2 20°}\left[4 + \sqrt{4^2 + (1 + 2[4]) \sin^2 20°}\right] = 15.4 = 16 \text{ teeth}$$

Thus a 16-tooth pinion will mesh with a 64-tooth gear without interference.

The smallest spur pinion that will operate with a rack without interference is

$$N_P = \frac{4(k)}{2 \sin^2 \phi} \tag{13–12}$$

For a 20° pressure angle full-depth tooth the smallest number of pinion teeth is

$$N_P = \frac{4(1)}{2 \sin^2 20°} = 17.1 = 18 \text{ teeth}$$

The largest gear with a specified pinion that is interference-free is

$$N_G = \frac{N_P^2 \sin^2 \phi - 4k^2}{4k - 2N_P \sin^2 \phi} \tag{13–13}$$

For a 13-tooth pinion with a pressure angle ϕ of 20°,

$$N_G = \frac{13^2 \sin^2 20° - 4(1)^2}{4(1) - 2(13) \sin^2 20°} - 16.45 = 16 \text{ teeth}$$

For a 13-tooth spur pinion, the maximum number of gear teeth possible without interference is 16.

Since gear-shaping tools amount to contact with a rack, and the gear-hobbing process is similar, the minimum number of teeth to prevent interference to prevent undercutting by the hobbing process is equal to the value of N_P when N_G is infinite.

The importance of the problem of teeth that have been weakened by undercutting cannot be overemphasized. Of course, interference can be eliminated by using more teeth on the pinion. However, if the pinion is to transmit a given amount of power, more teeth can be used only by increasing the pitch diameter.

Interference can also be reduced by using a larger pressure angle. This results in a smaller base circle, so that more of the tooth profile becomes involute. The demand for smaller pinions with fewer teeth thus favors the use of a 25° pressure angle even though the frictional forces and bearing loads are increased and the contact ratio decreased.

13–8   The Forming of Gear Teeth

There are a large number of ways of forming the teeth of gears, such as *sand casting, shell molding, investment casting, permanent-mold casting, die casting,* and *centrifugal casting.* Teeth can also be formed by using the *powder-metallurgy process;* or, by using *extrusion,* a single bar of aluminum may be formed and then sliced into gears. Gears that carry large loads in comparison with their size are usually made of steel and are cut with either *form cutters* or *generating cutters.* In form cutting, the tooth space takes the exact form of the cutter. In generating, a tool having a shape different from the tooth profile is moved relative to the gear blank so as to obtain the proper tooth shape. One of the newest and most promising of the methods of forming teeth is called *cold forming,* or *cold rolling,* in which dies are rolled against steel blanks to form the teeth. The mechanical properties of the metal are greatly improved by the rolling process, and a high-quality generated profile is obtained at the same time.

Gear teeth may be machined by milling, shaping, or hobbing. They may be finished by shaving, burnishing, grinding, or lapping.

Gears made of thermoplastics such as nylon, polycarbonate, acetal are quite popular and are easily manufactured by *injection molding*. These gears are of low to moderate precision, low in cost for high production quantities, and capable of light loads, and can run without lubrication.

Milling

Gear teeth may be cut with a form milling cutter shaped to conform to the tooth space. With this method it is theoretically necessary to use a different cutter for each gear, because a gear having 25 teeth, for example, will have a different-shaped tooth space from one having, say, 24 teeth. Actually, the change in space is not too great, and it has been found that eight cutters may be used to cut with reasonable accuracy any gear in the range of 12 teeth to a rack. A separate set of cutters is, of course, required for each pitch.

Shaping

Teeth may be generated with either a pinion cutter or a rack cutter. The pinion cutter (Fig. 13–17) reciprocates along the vertical axis and is slowly fed into the gear blank to the required depth. When the pitch circles are tangent, both the cutter and the blank rotate slightly after each cutting stroke. Since each tooth of the cutter is a cutting tool, the teeth are all cut after the blank has completed one rotation. The sides of an involute rack tooth are straight. For this reason, a rack-generating tool provides an accurate method of cutting gear teeth. This is also a shaping operation and is illustrated by the drawing of Fig. 13–18. In operation, the cutter reciprocates and is first fed into the gear blank until the pitch circles are tangent. Then, after each cutting stroke, the gear blank

Figure 13–17

Generating a spur gear with a pinion cutter. *(Courtesy of Boston Gear Works, Inc.)*

Figure 13–18

Shaping teeth with a rack. (This is a drawing-board figure that J. E. Shigley executed about 35 years ago in response to a question from a student at the University of Michigan.)

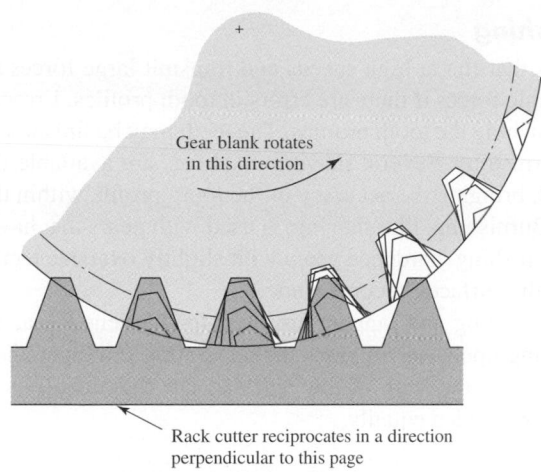

Gear blank rotates in this direction

Rack cutter reciprocates in a direction perpendicular to this page

Figure 13–19

Hobbing a worm gear. *(Courtesy of Boston Gear Works, Inc.)*

and cutter roll slightly on their pitch circles. When the blank and cutter have rolled a distance equal to the circular pitch, the cutter is returned to the starting point, and the process is continued until all the teeth have been cut.

Hobbing

The hobbing process is illustrated in Fig. 13–19. The hob is simply a cutting tool that is shaped like a worm. The teeth have straight sides, as in a rack, but the hob axis must be turned through the lead angle in order to cut spur-gear teeth. For this reason, the teeth generated by a hob have a slightly different shape from those generated by a rack cutter. Both the hob and the blank must be rotated at the proper angular-velocity ratio. The hob is then fed slowly across the face of the blank until all the teeth have been cut.

Finishing

Gears that run at high speeds and transmit large forces may be subjected to additional dynamic forces if there are errors in tooth profiles. Errors may be diminished somewhat by finishing the tooth profiles. The teeth may be finished, after cutting, by either shaving or burnishing. Several shaving machines are available that cut off a minute amount of metal, bringing the accuracy of the tooth profile within the limits of 250 μin.

Burnishing, like shaving, is used with gears that have been cut but not heat-treated. In burnishing, hardened gears with slightly oversize teeth are run in mesh with the gear until the surfaces become smooth.

Grinding and lapping are used for hardened gear teeth after heat treatment. The grinding operation employs the generating principle and produces very accurate teeth. In lapping, the teeth of the gear and lap slide axially so that the whole surface of the teeth is abraded equally.

13–9 Straight Bevel Gears

When gears are used to transmit motion between intersecting shafts, some form of bevel gear is required. A bevel gearset is shown in Fig. 13–20. Although bevel gears are usually made for a shaft angle of 90°, they may be produced for almost any angle. The teeth may be cast, milled, or generated. Only the generated teeth may be classed as accurate.

The terminology of bevel gears is illustrated in Fig. 13–20. The pitch of bevel gears is measured at the large end of the tooth, and both the circular pitch and the pitch diameter are calculated in the same manner as for spur gears. It should be noted that the clearance is uniform. The pitch angles are defined by the pitch cones meeting at the apex, as shown in the figure. They are related to the tooth numbers as follows:

$$\tan \gamma = \frac{N_P}{N_G} \qquad \tan \Gamma = \frac{N_G}{N_P} \tag{13–14}$$

Figure 13–20

Terminology of bevel gears.

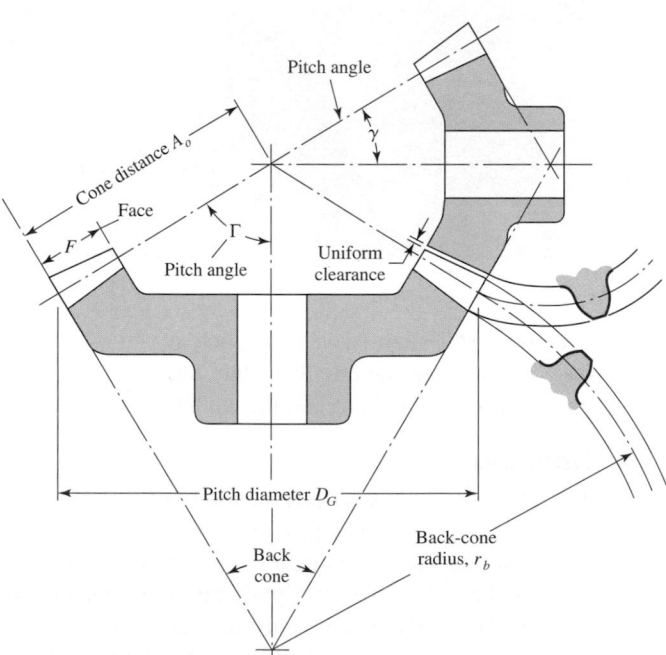

where the subscripts P and G refer to the pinion and gear, respectively, and where γ and Γ are, respectively, the pitch angles of the pinion and gear.

Figure 13–20 shows that the shape of the teeth, when projected on the back cone, is the same as in a spur gear having a radius equal to the back-cone distance r_b. This is called Tredgold's approximation. The number of teeth in this imaginary gear is

$$N' = \frac{2\pi r_b}{p} \tag{13-15}$$

where N' is the *virtual number of teeth* and p is the circular pitch measured at the large end of the teeth. Standard straight-tooth bevel gears are cut by using a 20° pressure angle, unequal addenda and dedenda, and full-depth teeth. This increases the contact ratio, avoids undercut, and increases the strength of the pinion.

13–10 Parallel Helical Gears

Helical gears, used to transmit motion between parallel shafts, are shown in Fig. 13–2. The helix angle is the same on each gear, but one gear must have a right-hand helix and the other a left-hand helix. The shape of the tooth is an involute helicoid and is illustrated in Fig. 13–21. If a piece of paper cut in the shape of a parallelogram is wrapped around a cylinder, the angular edge of the paper becomes a helix. If we unwind this paper, each point on the angular edge generates an involute curve. This surface obtained when every point on the edge generates an involute is called an *involute helicoid*.

The initial contact of spur-gear teeth is a line extending all the way across the face of the tooth. The initial contact of helical-gear teeth is a point that extends into a line as the teeth come into more engagement. In spur gears the line of contact is parallel to the axis of rotation; in helical gears the line is diagonal across the face of the tooth. It is this gradual engagement of the teeth and the smooth transfer of load from one tooth to another that gives helical gears the ability to transmit heavy loads at high speeds. Because of the nature of contact between helical gears, the contact ratio is of only minor importance, and it is the contact area, which is proportional to the face width of the gear, that becomes significant.

Helical gears subject the shaft bearings to both radial and thrust loads. When the thrust loads become high or are objectionable for other reasons, it may be desirable to use double helical gears. A double helical gear (herringbone) is equivalent to two helical gears of opposite hand, mounted side by side on the same shaft. They develop opposite thrust reactions and thus cancel out the thrust load.

When two or more single helical gears are mounted on the same shaft, the hand of the gears should be selected so as to produce the minimum thrust load.

Figure 13–21

An involute helicoid.

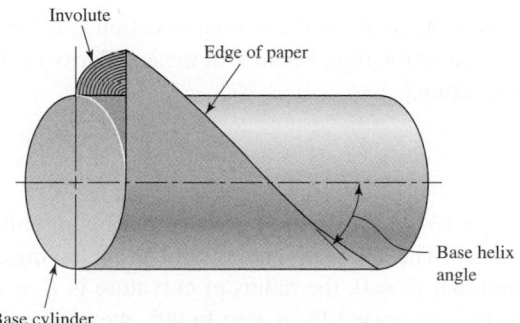

Involute

Edge of paper

Base helix angle

Base cylinder

Figure 13–22

Nomenclature of helical
gears.

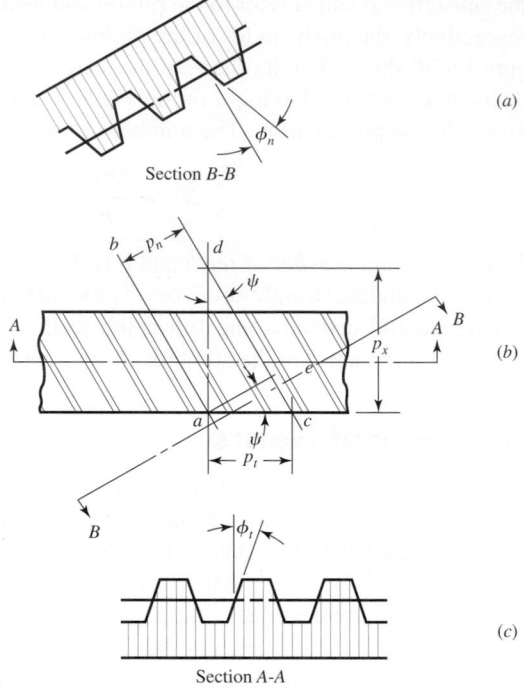

(a)

Section B-B

(b)

(c)

Section A-A

Figure 13–22 represents a portion of the top view of a helical rack. Lines *ab* and *cd* are the centerlines of two adjacent helical teeth taken on the same pitch plane. The angle ψ is the *helix angle*. The distance *ac* is the *transverse circular pitch* p_t in the plane of rotation (usually called the *circular pitch*). The distance *ae* is the *normal circular pitch* p_n and is related to the transverse circular pitch as follows:

$$p_n = p_t \cos \psi \tag{13–16}$$

The distance *ad* is called the *axial pitch* p_x and is related by the expression

$$p_x = \frac{p_t}{\tan \psi} \tag{13–17}$$

Since $p_n P_n = \pi$, the *normal diametral pitch* is

$$P_n = \frac{P_t}{\cos \psi} \tag{13–18}$$

The pressure angle ϕ_n in the normal direction is different from the pressure angle ϕ_t in the direction of rotation, because of the angularity of the teeth. These angles are related by the equation

$$\cos \psi = \frac{\tan \phi_n}{\tan \phi_t} \tag{13–19}$$

Figure 13–23 illustrates a cylinder cut by an oblique plane *ab* at an angle ψ to a right section. The oblique plane cuts out an arc having a radius of curvature of R. For the condition that $\psi = 0$, the radius of curvature is $R = D/2$. If we imagine the angle ψ to be slowly increased from zero to $90°$, we see that R begins at a value of $D/2$ and

Figure 13–23

A cylinder cut by an oblique plane.

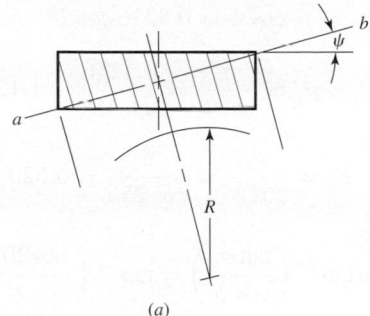

(a)

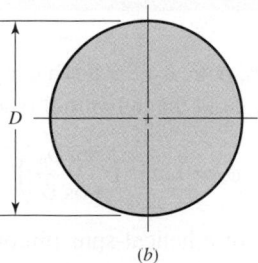

(b)

increases until, when $\psi = 90°$, $R = \infty$. The radius R is the apparent pitch radius of a helical-gear tooth when viewed in the direction of the tooth elements. A gear of the same pitch and with the radius R will have a greater number of teeth, because of the increased radius. In helical-gear terminology this is called the *virtual number of teeth*. It can be shown by analytical geometry that the virtual number of teeth is related to the actual number by the equation

$$N' = \frac{N}{\cos^3 \psi} \qquad (13\text{–}20)$$

where N' is the virtual number of teeth and N is the actual number of teeth. It is necessary to know the virtual number of teeth in design for strength and also, sometimes, in cutting helical teeth. This apparently larger radius of curvature means that few teeth may be used on helical gears, because there will be less undercutting.

EXAMPLE 13–2   A stock helical gear has a normal pressure angle of 20°, a helix angle of 25°, and a transverse diametral pitch of 6 teeth/in, and has 18 teeth. Find:
(*a*) The pitch diameter
(*b*) The transverse, the normal, and the axial pitches
(*c*) The normal diametral pitch
(*d*) The transverse pressure angle

Solution

Answer   (*a*)
$$d = \frac{N}{P_t} = \frac{18}{6} = 3 \text{ in}$$

Answer   (*b*)
$$p_t = \frac{\pi}{P_t} = \frac{\pi}{6} = 0.5236 \text{ in}$$

Answer
$$p_n = p_t \cos \psi = 0.5236 \cos 25° = 0.4745 \text{ in}$$

Answer
$$p_x = \frac{p_t}{\tan \psi} = \frac{0.5236}{\tan 45°} = 1.123 \text{ in}$$

Answer   (c)
$$P_n = \frac{P_t}{\cos \psi} = \frac{6}{\cos 25°} = 6.620 \text{ teeth/in}$$

Answer   (d)
$$\phi_t = \tan^{-1}\left(\frac{\tan \phi_n}{\cos \psi}\right) = \tan^{-1}\left(\frac{\tan 20°}{\cos 25°}\right) = 21.88°$$

Just like teeth on spur gears, helical-gear teeth can interfere. Equation (13–19) can be solved for the pressure angle ϕ_t in the tangential (rotation) direction to give

$$\phi_t = \tan^{-1}\left(\frac{\tan \phi_n}{\cos \psi}\right)$$

The smallest tooth number N_P of a helical-spur pinion that will run without interference[2] with a gear with the same number of teeth is

$$N_P = \frac{4k \cos \psi}{6 \sin^2 \phi_t}\left(1 + \sqrt{1 + 3 \sin^2 \phi_t}\right) \tag{13–21}$$

If the normal pressure angle ϕ_n is 20°, the helix angle ψ is 30°, then ϕ_t is

$$\phi_t = \tan^{-1}\left(\frac{\tan 20°}{\cos 30°}\right) = 22.80°$$

$$N_P = \frac{4(1) \cos 30°}{6 \sin^2 22.80°}\left(1 + \sqrt{1 + 3 \sin^2 22.80°}\right) = 8.48 = 9 \text{ teeth}$$

For a given gear ratio $m_G = N_G/N_P = m$, the smallest pinion tooth count is

$$N_P = \frac{2k \cos \psi}{(1 + 2m) \sin^2 \phi_t}\left[m + \sqrt{m^2 + (1 + 2m) \sin^2 \phi_t}\right] \tag{13–22}$$

The smallest pinion that can be run with a rack is

$$N_P = \frac{4k \cos \psi}{2 \sin^2 \phi_t} \tag{13–23}$$

For a normal pressure angle ϕ_n of 20° and a helix angle ψ of 30°, and $\phi_t = 22.80°$,

$$N_P = \frac{4(1) \cos 30°}{2 \sin^2 22.80°} = 11.5 = 12 \text{ teeth}$$

The largest gear with a specified pinion is given by

$$N_G = \frac{N_P^2 \sin^2 \phi_t - 4k^2 \cos^2 \psi}{4k \cos \psi - 2N_P \sin^2 \phi_t} \tag{13–24}$$

---

[2]Op. cit., Robert Lipp, *Machine Design*, pp. 122, 124.

For a nine-tooth pinion with a pressure angle ϕ_n of 20°, a helix angle ψ of 30°, and recalling that the tangential pressure angle ϕ_t is 22.80°,

$$N_G = \frac{9^2 \sin^2 22.80° - 4(1)^2 \cos^2 30°}{4(1) \cos 30° - 2(9) \sin^2 22.80°} = 12.02 = 12$$

For helical-gear teeth the number of teeth in mesh across the width of the gear will be greater than unity and a term called *face-contact ratio* is used to describe it. This increase of contact ratio, and the gradual sliding engagement of each tooth, results in quieter gears.

13-11   Worm Gears

The nomenclature of a worm gear is shown in Fig. 13–24. The worm and worm gear of a set have the same hand of helix as for crossed helical gears, but the helix angles are usually quite different. The helix angle on the worm is generally quite large, and that on the gear very small. Because of this, it is usual to specify the lead angle λ on the worm and helix angle ψ_G on the gear; the two angles are equal for a 90° shaft angle. The worm lead angle is the complement of the worm helix angle, as shown in Fig. 13–24.

In specifying the pitch of worm gearsets, it is customary to state the *axial pitch p_x* of the worm and the *transverse circular pitch p_t*, often simply called the circular pitch, of the mating gear. These are equal if the shaft angle is 90°. The pitch diameter of the gear is the diameter measured on a plane containing the worm axis, as shown in Fig. 13–24; it is the same as for spur gears and is

$$d_G = \frac{N_G p_t}{\pi} \tag{13-25}$$

Figure 13–24

Nomenclature of a single-enveloping worm gearset.

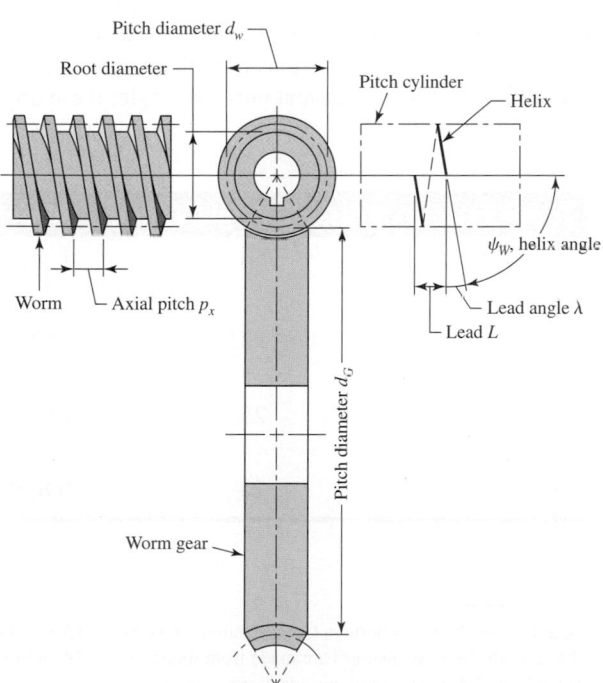

Since it is not related to the number of teeth, the worm may have any pitch diameter; this diameter should, however, be the same as the pitch diameter of the hob used to cut the worm-gear teeth. Generally, the pitch diameter of the worm should be selected so as to fall into the range

$$\frac{C^{0.875}}{3.0} \leq d_W \leq \frac{C^{0.875}}{1.7} \tag{13–26}$$

where C is the center distance. These proportions appear to result in optimum horse-power capacity of the gearset.

The *lead L* and the *lead angle* λ of the worm have the following relations:

$$L = p_x N_W \tag{13–27}$$

$$\tan \lambda = \frac{L}{\pi d_W} \tag{13–28}$$

13–12 Tooth Systems[3]

A *tooth system* is a standard that specifies the relationships involving addendum, dedendum, working depth, tooth thickness, and pressure angle. The standards were originally planned to attain interchangeability of gears of all tooth numbers, but of the same pressure angle and pitch.

Table 13–1 contains the standards most used for spur gears. A $14\frac{1}{2}^\circ$ pressure angle was once used for these but is now obsolete; the resulting gears had to be comparatively larger to avoid interference problems.

Table 13–2 is particularly useful in selecting the pitch or module of a gear. Cutters are generally available for the sizes shown in this table.

Table 13–3 lists the standard tooth proportions for straight bevel gears. These sizes apply to the large end of the teeth. The nomenclature is defined in Fig. 13–20.

Standard tooth proportions for helical gears are listed in Table 13–4. Tooth proportions are based on the normal pressure angle; these angles are standardized the same as

Table 13–1

Standard and Commonly Used Tooth Systems for Spur Gears

Tooth System	Pressure Angle ϕ, deg	Addendum a	Dedendum b
Full depth	20	$1/P_d$ or $1m$	$1.25/P_d$ or $1.25m$
			$1.35/P_d$ or $1.35m$
	$22\frac{1}{2}$	$1/P_d$ or $1m$	$1.25/P_d$ or $1.25m$
			$1.35/P_d$ or $1.35m$
	25	$1/P_d$ or $1m$	$1.25/P_d$ or $1.25m$
			$1.35/P_d$ or $1.35m$
Stub	20	$0.8/P_d$ or $0.8m$	$1/P_d$ or $1m$

---

[3]Standardized by the American Gear Manufacturers Association (AGMA). Write AGMA for a complete list of standards, because changes are made from time to time. The address is: 1500 King Street, Suite 201, Alexandria, VA 22314; or, www.agma.org.

Table 13–2

Tooth Sizes in General Uses

Diametral Pitch	
Coarse	2, $2\frac{1}{4}$, $2\frac{1}{2}$, 3, 4, 6, 8, 10, 12, 16
Fine	20, 24, 32, 40, 48, 64, 80, 96, 120, 150, 200

Modules	
Preferred	1, 1.25, 1.5, 2, 2.5, 3, 4, 5, 6, 8, 10, 12, 16, 20, 25, 32, 40, 50
Next Choice	1.125, 1.375, 1.75, 2.25, 2.75, 3.5, 4.5, 5.5, 7, 9, 11, 14, 18, 22, 28, 36, 45

Table 13–3

Tooth Proportions for 20° Straight Bevel-Gear Teeth

Item	Formula
Working depth	$h_k = 2.0/P$
Clearance	$c = (0.188/P) + 0.002$ in
Addendum of gear	$a_G = \dfrac{0.54}{P} + \dfrac{0.460}{P(m_{90})^2}$
Gear ratio	$m_G = N_G/N_P$
Equivalent 90° ratio	$m_{90} = m_G$ when $\Gamma = 90°$
	$m_{90} = \sqrt{m_G \dfrac{\cos \gamma}{\cos \Gamma}}$ when $\Gamma \neq 90°$
Face width	$F = 0.3 A_0$ or $F = \dfrac{10}{P}$, whichever is smaller
Minimum number of teeth	Pinion: 16, 15, 14, 13
	Gear: 16, 17, 20, 30

Table 13–4

Standard Tooth Proportions for Helical Gears

Quantity*	Formula	Quantity*	Formula
Addendum	$\dfrac{1.00}{P_n}$	External gears:	
Dedendum	$\dfrac{1.25}{P_n}$	Standard center distance	$\dfrac{D + d}{2}$
Pinion pitch diameter	$\dfrac{N_P}{P_n \cos \psi}$	Gear outside diameter	$D + 2a$
Gear pitch diameter	$\dfrac{N_G}{P_n \cos \psi}$	Pinion outside diameter	$d + 2a$
Normal arc tooth thickness[†]	$\dfrac{\pi}{P_n} - \dfrac{B_n}{2}$	Gear root diameter	$D - 2b$
Pinion base diameter	$d \cos \phi_t$	Pinion root diameter	$d - 2b$
		Internal gears:	
Gear base diameter	$D \cos \phi_t$	Center distance	$\dfrac{D - d}{2}$
Base helix angle	$\tan^{-1}(\tan \psi \cos \phi_t)$	Inside diameter	$D - 2a$
		Root diameter	$D + 2b$

*All dimensions are in inches, and angles are in degrees.
[†]B_n is the normal backlash.

Table 13–5

Recommended Pressure Angles and Tooth Depths for Worm Gearing

Lead Angle λ, deg	Pressure Angle ϕ_n, deg	Addendum a	Dedendum b_G
0–15	$14\frac{1}{2}$	$0.3683p_x$	$0.3683p_x$
15–30	20	$0.3683p_x$	$0.3683p_x$
30–35	25	$0.2865p_x$	$0.3314p_x$
35–40	25	$0.2546p_x$	$0.2947p_x$
40–45	30	$0.2228p_x$	$0.2578p_x$

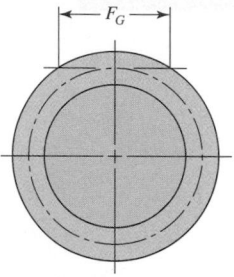

Figure 13–25

A graphical depiction of the face width of the worm of a worm gearset.

for spur gears. Though there will be exceptions, the face width of helical gears should be at least 2 times the axial pitch to obtain good helical-gear action.

Tooth forms for worm gearing have not been highly standardized, perhaps because there has been less need for it. The pressure angles used depend upon the lead angles and must be large enough to avoid undercutting of the worm-gear tooth on the side at which contact ends. A satisfactory tooth depth, which remains in about the right proportion to the lead angle, may be obtained by making the depth a proportion of the axial circular pitch. Table 13–5 summarizes what may be regarded as good practice for pressure angle and tooth depth.

The *face width* F_G of the worm gear should be made equal to the length of a tangent to the worm pitch circle between its points of intersection with the addendum circle, as shown in Fig. 13–25.

13–13 Gear Trains

Consider a pinion 2 driving a gear 3. The speed of the driven gear is

$$n_3 = \left| \frac{N_2}{N_3} n_2 \right| = \left| \frac{d_2}{d_3} n_2 \right|$$

(13–29)

where n = revolutions or rev/min

       N = number of teeth

       d = pitch diameter

Equation (13–29) applies to any gearset no matter whether the gears are spur, helical, bevel, or worm. The absolute-value signs are used to permit complete freedom in choosing positive and negative directions. In the case of spur and parallel helical gears, the directions ordinarily correspond to the right-hand rule and are positive for counterclockwise rotation.

Rotational directions are somewhat more difficult to deduce for worm and crossed helical gearsets. Figure 13–26 will be of help in these situations.

The gear train shown in Fig. 13–27 is made up of five gears. The speed of gear 6 is

$$n_6 = -\frac{N_2}{N_3} \frac{N_3}{N_4} \frac{N_5}{N_6} n_2$$

(a)

Hence we notice that gear 3 is an idler, that its tooth numbers cancel in Eq. (a), and hence that it affects only the direction of rotation of gear 6. We notice, furthermore, that

Figure 13–26

Thrust, rotation, and hand relations for crossed helical gears. Note that each pair of drawings refers to a single gearset. These relations also apply to worm gearsets.
(Reproduced by permission, Boston Gear Division, Colfax Corp.)

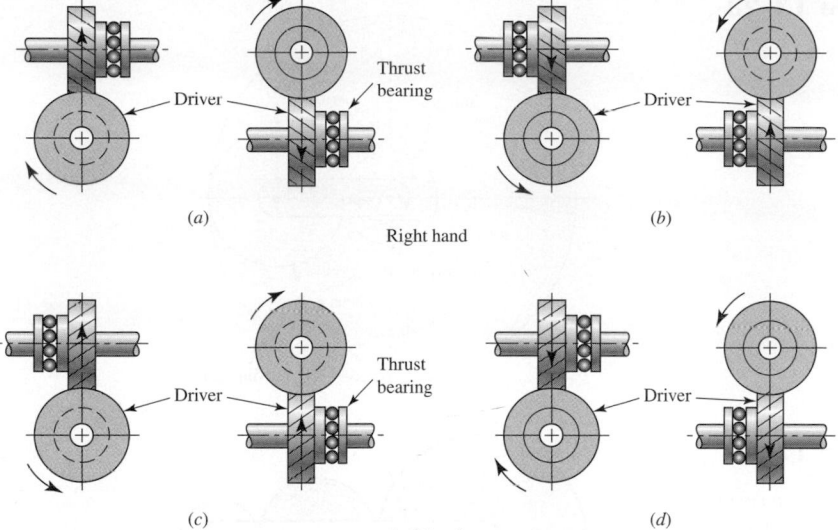

(a)

Right hand

(b)

(c)

Left hand

(d)

Figure 13–27

A gear train.

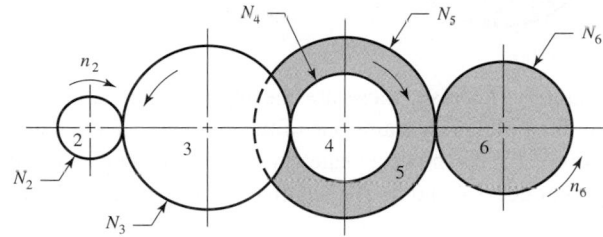

gears 2, 3, and 5 are drivers, while 3, 4, and 6 are driven members. We define the *train value e* as

$$e = \frac{\text{product of driving tooth numbers}}{\text{product of driven tooth numbers}} \tag{13–30}$$

Note that pitch diameters can be used in Eq. (13–30) as well. When Eq. (13–30) is used for spur gears, e is positive if the last gear rotates in the same sense as the first, and negative if the last rotates in the opposite sense.

Now we can write

$$n_L = e n_F \tag{13–31}$$

where n_L is the speed of the last gear in the train and n_F is the speed of the first.

Unusual effects can be obtained in a gear train by permitting some of the gear axes to rotate about others. Such trains are called *planetary*, or *epicyclic, gear trains*. Planetary trains always include a *sun gear*, a *planet carrier* or *arm*, and one or more *planet gears*, as shown in Fig. 13–28. Planetary gear trains are unusual mechanisms because they have two degrees of freedom; that is, for constrained motion, a planetary train must have two inputs. For example, in Fig. 13–28 these two inputs could be the motion of any two of the elements of the train. We might, in Fig. 13–28, say, specify that the sun gear rotates at 100 rev/min clockwise and that the ring gear rotates at 50 rev/min

Figure 13–28

A planetary gear train.

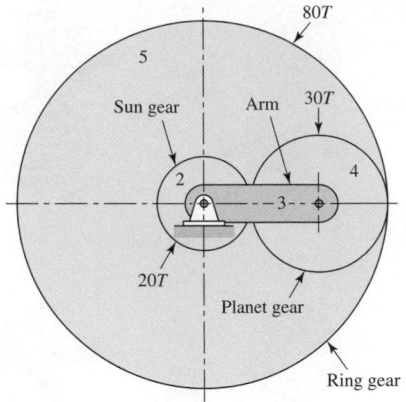

Figure 13–29

A gear train on the arm of a planetary gear train.

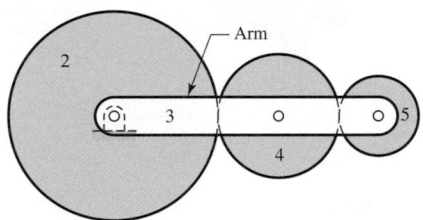

counterclockwise; these are the inputs. The output would be the motion of the arm. In most planetary trains one of the elements is attached to the frame and has no motion. Figure 13–29 shows a planetary train composed of a sun gear 2, an arm or carrier 3, and planet gears 4 and 5. The angular velocity of gear 2 relative to the arm in rev/min is

$$n_{23} = n_2 - n_3 \qquad (b)$$

Also, the velocity of gear 5 relative to the arm is

$$n_{53} = n_5 - n_3 \qquad (c)$$

Dividing Eq. (c) by Eq. (b) gives

$$\frac{n_{53}}{n_{23}} = \frac{n_5 - n_3}{n_2 - n_3} \qquad (d)$$

Equation (d) expresses the ratio of gear 5 to that of gear 2, and both velocities are taken relative to the arm. Now this ratio is the same and is proportional to the tooth numbers, whether the arm is rotating or not. It is the train value. Therefore, we may write

$$e = \frac{n_5 - n_3}{n_2 - n_3} \qquad (e)$$

This equation can be used to solve for the output motion of any planetary train. It is more conveniently written in the form

$$e = \frac{n_L - n_A}{n_F - n_A} \qquad (13\text{–}32)$$

where n_F = rev/min of first gear in planetary train

n_L = rev/min of last gear in planetary train

n_A = rev/min of arm

EXAMPLE 13–3
In Fig. 13–28 the sun gear is the input, and it is driven clockwise at 100 rev/min. The ring gear is held stationary by being fastened to the frame. Find the rev/min and direction of rotation of the arm and gear 4.

Solution
Designate $n_F = n_2 = -100$ rev/min, and $n_L = n_5 = 0$. Unlocking gear 5 and holding the arm stationary, in our imagination, we find

$$e = -\left(\frac{20}{30}\right)\left(\frac{30}{80}\right) = -0.25$$

Substituting this value in Eq. (13–32) gives

$$-0.25 = \frac{0 - n_A}{(-100) - n_A}$$

or

Answer
$$n_A = -20 \text{ rev/min}$$

To obtain the speed of gear 4, we follow the procedure outlined by Eqs. (b), (c), and (d). Thus

$$n_{43} = n_4 - n_3 \qquad n_{23} = n_2 - n_3$$

and so

$$\frac{n_{43}}{n_{23}} = \frac{n_4 - n_3}{n_2 - n_3} \tag{1}$$

But

$$\frac{n_{43}}{n_{23}} = -\frac{20}{30} = -\frac{2}{3} \tag{2}$$

Substituting the known values in Eq. (1) gives

$$-\frac{2}{3} = \frac{n_4 - (-20)}{(-100) - (-20)}$$

Solving gives

Answer
$$n_4 = 33\frac{1}{3} \text{ rev/min}$$

An alternative method, useful for planetary gear trains, employs relative *instantaneous centers*. When two gears mesh, at any instant each gear has the same velocity at the point of mesh, which is a relative instantaneous center of the two gears. If we can determine the velocity of two points on a body, the instantaneous center of rotation of the body can be found by drawing perpendiculars to the two velocity vectors. The center is located at the intersection of the perpendiculars and the angular velocity of the body is the velocity of a point on the body divided by the distance between the point and the instantaneous center. If the perpendiculars of the two velocity vectors are collinear, there is no intersection, but the angular velocity of the body is the difference in the velocities divided by the distance between the two points. The instantaneous center is found by drawing a line between the tips of the velocity vectors, drawn to scale, and the intersection of this line with the collinear perpendiculars locates the center.

EXAMPLE 13–4 Using relative instantaneous centers:
(a) Repeat Ex. 13–3, except let the ring gear rotate counterclockwise at 40 rev/min.
(b) Repeat Ex. 13–3, as stated, with the ring gear being stationary.

Solution (a) Planet gear 4 meshes with two gears, the sun gear 2 and ring gear 5. Figure 13–30 shows gear 4. Point A is the mesh point with gear 2 and the velocity of point A is given by $v_A = r_2 n_2$ where r_2 is the pitch radius of gear 2 (do not worry about units). Point B meshes with gear 5, and the velocity of point B is given by $v_B = r_5 n_5$. The two velocities of gear 4 are shown in Fig. 13–30, where the directions of the velocities agree with the rotations of gears 2 and 5. Since the perpendiculars to v_A and v_B are collinear, from Fig. 13–30 the angular velocity of gear 4 is

$$n_4 = \frac{v_B - v_A}{2r_4} = \frac{r_5 n_5 - (-r_2 n_2)}{2r_4} = \frac{N_5 n_5 + N_2 n_2}{2N_4}$$

$$= \frac{80(40) + 20(100)}{2(30)} = 86\tfrac{2}{3} \text{ rev/min}$$

(1)

where the pitch radii are proportional to the number of teeth. The instantaneous center is at I and it is obvious that the rotation direction of gear 4 is counterclockwise and thus positive.

The arm is rotating about the center of gear 2; the velocity of C is the average of v_A and v_B. The radius of the arm is the sum of the radii of gears 2 and 4. Thus the arm angular velocity is

$$n_A = \frac{v_C}{r_2 + r_4} = \frac{(v_A + v_B)/2}{r_2 + r_4} = \frac{(-N_2 n_2 + N_5 n_5)/2}{N_2 + N_4}$$

$$= \frac{[-20(100) + 80(40)]/2}{20 + 30} = 12 \text{ rev/min}$$

(2)

and obviously counterclockwise, so it is positive.

Figure 13–30

Planet gear for Ex. 13–4a.

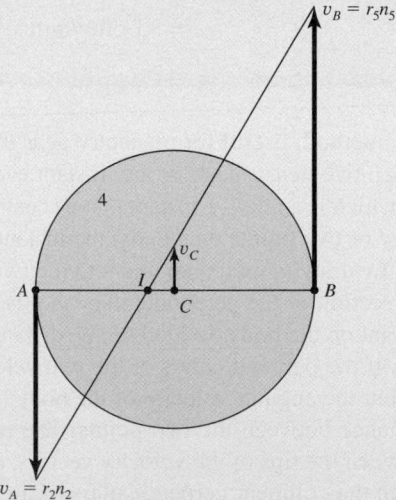

(*b*) With the ring gear stationary, $n_5 = 0$, the instantaneous center of gear 4 moves to *B*, and from Eq. (1),

$$n_4 = \frac{N_2 n_2}{2 N_4} = \frac{20(100)}{2(30)} = 33\frac{1}{3} \text{ rev/min}$$

and is counterclockwise, thus positive. From Eq. (2), with $n_5 = 0$,

$$n_A = \frac{-N_2 n_2 / 2}{N_2 + N_4} = \frac{-20(100)/2}{20 + 30} = -20 \text{ rev/min}$$

Thus, the arm is rotating clockwise.

EXAMPLE 13–5

Figure 13–31 shows a gear train consisting of a pair of *miter gears* (same-size bevel gears) having 16 teeth each, a 4-tooth right-hand worm, and a 40-tooth worm gear. The speed of gear 2 is given as $n_2 = +200$ rev/min, which corresponds to counterclockwise about the *y* axis. What is the speed and direction of rotation of the worm gear?

Solution

Answer

$$n_5 = -\left(\frac{16}{16}\right)\left(\frac{4}{40}\right)(200) = -20 \text{ rev/min}$$

Gear 5 rotates clockwise (negative) 20 rev/min about the *z* axis in a right-handed coordinate system.

Figure 13–31

A miter gearset and a worm gearset in train.

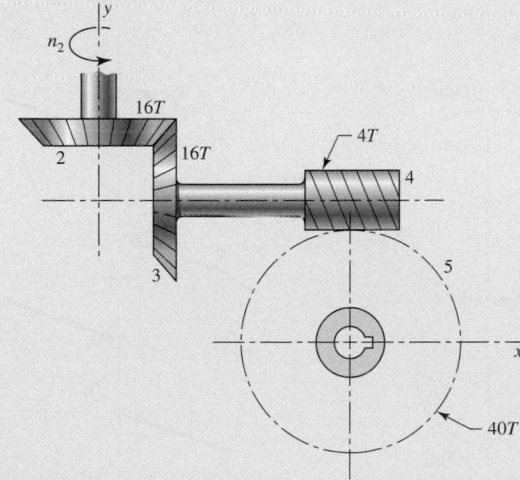

13–14   **Force Analysis—Spur Gearing**

Before beginning the force analysis of gear trains, let us agree on the notation to be used. Beginning with the numeral 1 for the frame of the machine, we shall designate the input gear as gear 2, and then number the gears successively 3, 4, etc., until we arrive at the last gear in the train. Next, there may be several shafts involved, and usually one or two gears are mounted on each shaft as well as other elements. We shall designate the shafts, using lowercase letters of the alphabet, a, b, c, etc.

With this notation we can now speak of the force exerted by gear 2 against gear 3 as F_{23}. The force of gear 2 against a shaft a is F_{2a}. We can also write F_{a2} to mean the force of a shaft a against gear 2. Unfortunately, it is also necessary to use superscripts to indicate directions. The coordinate directions will usually be indicated by the x, y, and z coordinates, and the radial and tangential directions by superscripts r and t. With this notation, F_{43}^t is the tangential component of the force of gear 4 acting against gear 3.

Figure 13–32a shows a pinion mounted on shaft a rotating clockwise at n_2 rev/min and driving a gear on shaft b at n_3 rev/min. The reactions between the mating teeth occur along the pressure line. In Fig. 13–32b the pinion has been separated from the gear and the shaft, and their effects have been replaced by forces. F_{a2} and T_{a2} are the force and torque, respectively, exerted by shaft a against pinion 2. F_{32} is the force exerted by gear 3 against the pinion. Using a similar approach, we obtain the free-body diagram of the gear shown in Fig. 13–32c.

In Fig. 13–33, the free-body diagram of the pinion has been redrawn and the forces have been resolved into tangential and radial components. We now define

$$W_t = F_{32}^t \qquad (a)$$

as the *transmitted load*. This tangential load is really the useful component, because the radial component F_{32}^r serves no useful purpose. It does not transmit power. The applied torque and the transmitted load are seen to be related by the equation

$$T = \frac{d}{2} W_t \qquad (13\text{–}33)$$

where we have used $T = T_{a2}$ and $d = d_2$ to obtain a general relation.

Figure 13–32

Free-body diagrams of the forces and moments acting upon two gears of a simple gear train.

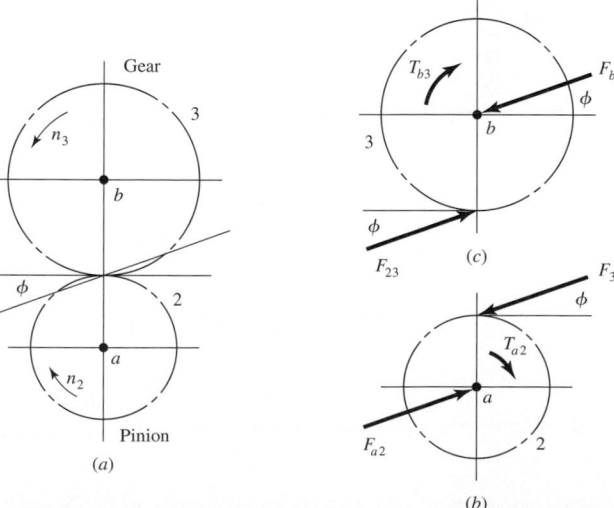

Figure 13–33

Resolution of gear forces.

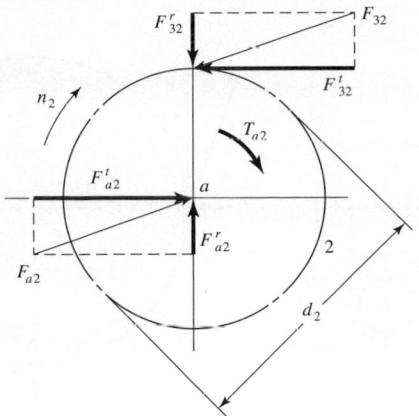

If next we designate the pitch-line velocity to be V, where $V = \pi\,dn/12$ and is in feet per minute, the power H may be obtained from the equation

$$H = \frac{W_t V}{33\,000} \tag{13-34}$$

The corresponding equation in SI is

$$W_t = \frac{60(10)^3 H}{\pi dn} \tag{13-35}$$

where W_t = transmitted load, kN

H = power, kW

d = gear diameter, mm

n = speed, rev/min

EXAMPLE 13–6

Pinion 2 in Fig. 13–34a runs at 1750 rev/min and transmits 2.5 kW to idler gear 3. The teeth are cut on the 20° full-depth system and have a module of $m = 2.5$ mm. Draw a free body diagram of gear 3 and show all the forces that act upon it.

Solution

The pitch diameters of gears 2 and 3 are

$$d_2 = N_2 m = 20(2.5) = 50 \text{ mm}$$

$$d_3 = N_3 m = 50(2.5) = 125 \text{ mm}$$

From Eq. (13–35) we find the transmitted load to be

$$W_t = \frac{60(10)^3 H}{\pi d_2 n} = \frac{60(10)^3(2.5)}{\pi(50)(1750)} = 0.546 \text{ kN}$$

Thus, the tangential force of gear 2 on gear 3 is $F_{23}^t = 0.546$ kN, as shown in Fig. 13–34b. Therefore

$$F_{23}^r = F_{23}^t \tan 20° = (0.546) \tan 20° = 0.199 \text{ kN}$$

Figure 13–34

A gear train containing an idler gear. (a) The gear train. (b) Free-body of the idler gear.

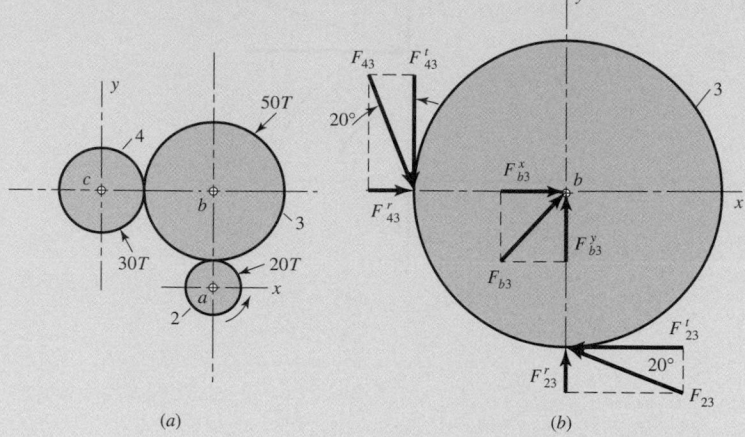

(a)   (b)

and so

$$F_{23} = \frac{F_{23}^t}{\cos 20°} = \frac{0.546}{\cos 20°} = 0.581 \text{ kN}$$

Since gear 3 is an idler, it transmits no power (torque) to its shaft, and so the tangential reaction of gear 4 on gear 3 is also equal to W_t. Therefore

$$F_{43}^t = 0.546 \text{ kN} \qquad F_{43}^r = 0.199 \text{ kN} \qquad F_{43} = 0.581 \text{ kN}$$

and the directions are shown in Fig. 13–34b.

The shaft reactions in the x and y directions are

$$F_{b3}^x = -(F_{23}^t + F_{43}^r) = -(-0.546 + 0.199) = 0.347 \text{ kN}$$

$$F_{b3}^y = -(F_{23}^r + F_{43}^t) = -(0.199 - 0.546) = 0.347 \text{ kN}$$

The resultant shaft reaction is

$$F_{b3} = \sqrt{(0.347)^2 + (0.347)^2} = 0.491 \text{ kN}$$

These are shown on the figure.

13–15   Force Analysis—Bevel Gearing

In determining shaft and bearing loads for bevel-gear applications, the usual practice is to use the tangential or transmitted load that would occur if all the forces were concentrated at the midpoint of the tooth. While the actual resultant occurs somewhere between the midpoint and the large end of the tooth, there is only as mall error in making this assumption. For the transmitted load, this gives

$$W_t = \frac{T}{r_{av}} \qquad\qquad (13\text{–}36)$$

where T is the torque and r_{av} is the pitch radius at the midpoint of the tooth for the gear under consideration.

Figure 13–35

Bevel-gear tooth forces.

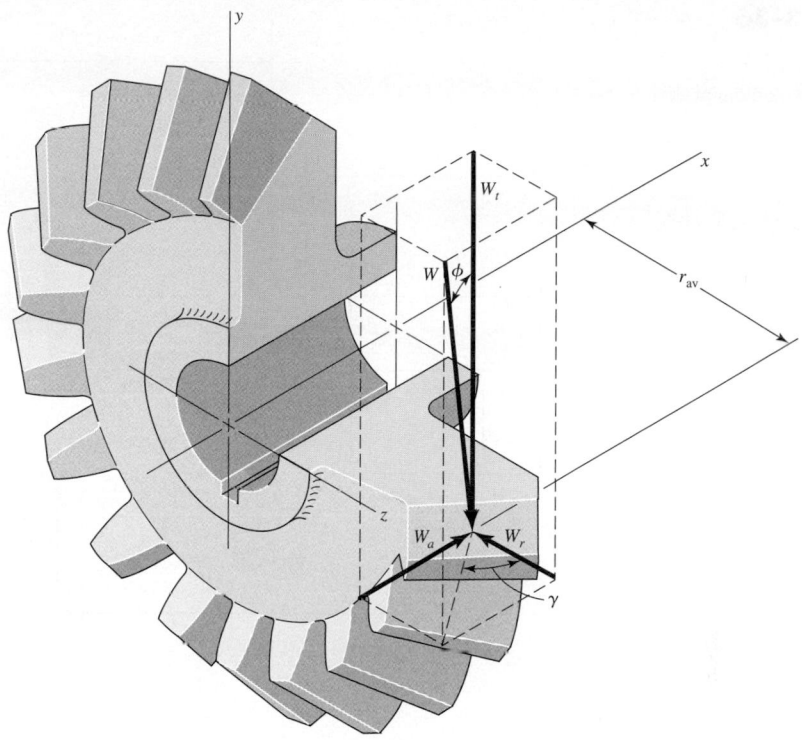

The forces acting at the center of the tooth are shown in Fig. 13–35. The resultant force W has three components: a tangential force W_t, a radial force W_r, and an axial force W_a. From the trigonometry of the figure,

$$W_r = W_t \tan \phi \cos \gamma \qquad (13\text{–}37)$$

$$W_a = W_t \tan \phi \sin \gamma \qquad (13\text{–}38)$$

The three forces W_t, W_r, and W_a are at right angles to each other and can be used to determine the bearing loads by using the methods of statics.

EXAMPLE 13–7   The bevel pinion in Fig. 13–36a rotates at 600 rev/min in the direction shown and transmits 5 hp to the gear. The mounting distances, the location of all bearings, and the average pitch radii of the pinion and gear are shown in the figure. For simplicity, the teeth have been replaced by pitch cones. Bearings A and C should take the thrust loads. Find the bearing forces on the gearshaft.

Solution   The pitch angles are

$$\gamma = \tan^{-1}\left(\frac{3}{9}\right) = 18.4° \qquad \Gamma = \tan^{-1}\left(\frac{9}{3}\right) = 71.6°$$

The pitch-line velocity corresponding to the average pitch radius is

$$V = \frac{2\pi r_P n}{12} = \frac{2\pi(1.293)(600)}{12} = 406 \text{ ft/min}$$

Figure 13–36

(a) Bevel-gear set of Ex. 13–7.
(b) Free body diagram of shaft
CD. Dimensions in inches.

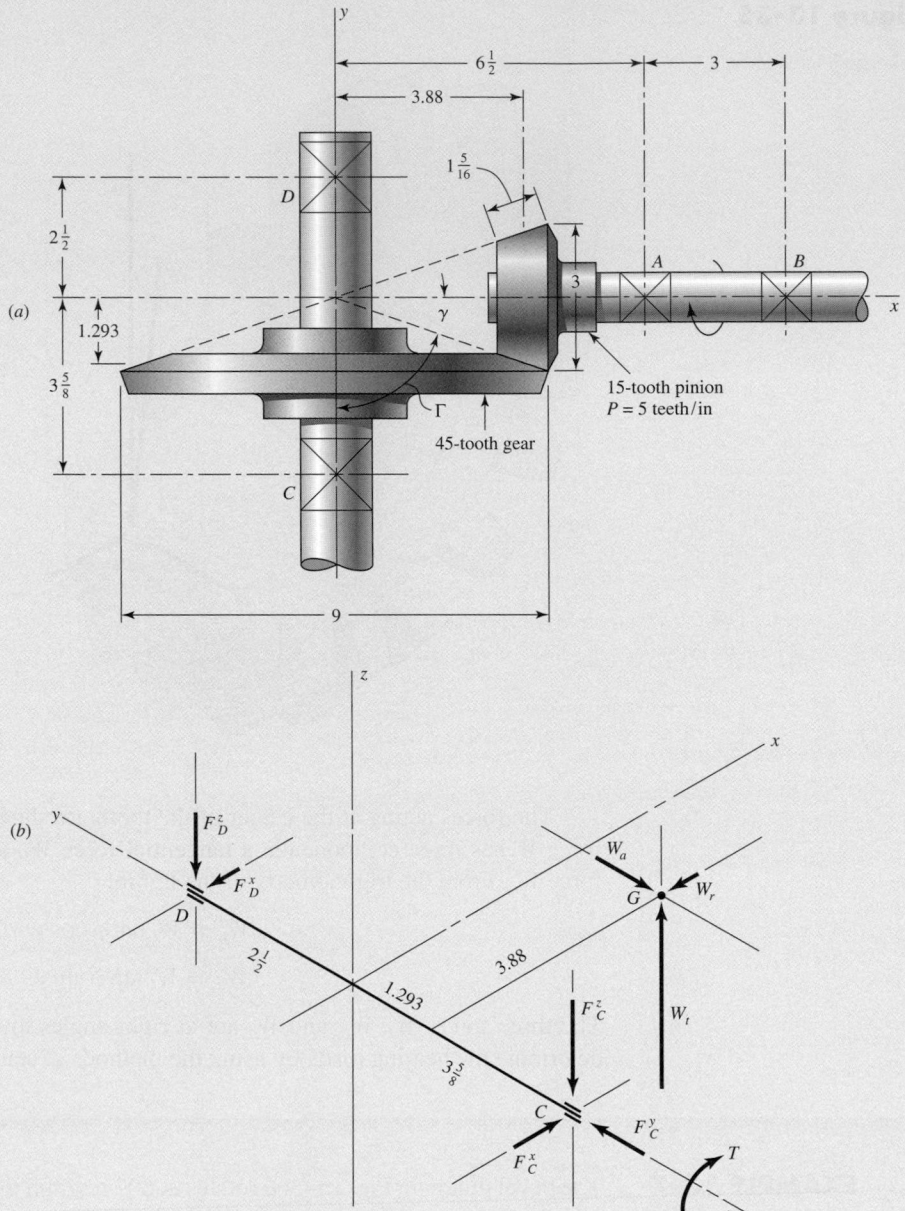

Therefore the transmitted load is

$$W_t = \frac{33\,000H}{V} = \frac{(33\,000)(5)}{406} = 406\,\text{lbf}$$

which acts in the positive z direction, as shown in Fig. 13–36b. We next have

$$W_r = W_t \tan\phi \cos\Gamma = 406 \tan 20° \cos 71.6° = 46.6\,\text{lbf}$$

$$W_a = W_t \tan\phi \sin\Gamma = 406 \tan 20° \sin 71.6° = 140\,\text{lbf}$$

where W_r is in the $-x$ direction and W_a is in the $-y$ direction, as illustrated in the isometric sketch of Fig. 13–36b.

In preparing to take a sum of the moments about bearing D, define the position vector from D to G as

$$\mathbf{R}_G = 3.88\mathbf{i} - (2.5 + 1.293)\mathbf{j} = 3.88\mathbf{i} - 3.793\mathbf{j}$$

We shall also require a vector from D to C:

$$\mathbf{R}_C = -(2.5 + 3.625)\mathbf{j} = -6.125\mathbf{j}$$

Then, summing moments about D gives

$$\mathbf{R}_G \times \mathbf{W} + \mathbf{R}_C \times \mathbf{F}_C + \mathbf{T} = 0 \tag{1}$$

When we place the details in Eq. (1), we get

$$\begin{aligned} &(3.88\mathbf{i} - 3.793\mathbf{j}) \times (-46.6\mathbf{i} - 140\mathbf{j} + 406\mathbf{k}) \\ &+ (-6.125\mathbf{j}) \times \left(F_C^x\mathbf{i} + F_C^y\mathbf{j} + F_C^z\mathbf{k}\right) + T\mathbf{j} = 0 \end{aligned} \tag{2}$$

After the two cross products are taken, the equation becomes

$$(-1540\mathbf{i} - 1575\mathbf{j} - 720\mathbf{k}) + \left(-6.125F_C^z\mathbf{i} + 6.125F_C^x\mathbf{k}\right) + T\mathbf{j} = 0$$

from which

$$\mathbf{T} = 1575\mathbf{j} \text{ lbf} \cdot \text{in} \qquad F_C^x = 118 \text{ lbf} \qquad F_C^z = -251 \text{ lbf} \tag{3}$$

Now sum the forces to zero. Thus

$$\mathbf{F}_D + \mathbf{F}_C + \mathbf{W} = 0 \tag{4}$$

When the details are inserted, Eq. (4) becomes

$$\left(F_D^x\mathbf{i} + F_D^z\mathbf{k}\right) + \left(118\mathbf{i} + F_C^y\mathbf{j} - 251\mathbf{k}\right) + (-46.6\mathbf{i} - 140\mathbf{j} + 406\mathbf{k}) = 0 \tag{5}$$

First we see that $F_C^y = 140$ lbf, and so

Answer
$$\mathbf{F}_C = 118\mathbf{i} + 140\mathbf{j} - 251\mathbf{k} \text{ lbf}$$

Then, from Eq. (5),

Answer
$$\mathbf{F}_D = -71.4\mathbf{i} - 155\mathbf{k} \text{ lbf}$$

These are all shown in Fig. 13–36b in the proper directions. The analysis for the pinion shaft is quite similar.

13–16 Force Analysis—Helical Gearing

Figure 13–37 is a three-dimensional view of the forces acting against a helical-gear tooth. The point of application of the forces is in the pitch plane and in the center of the gear face. From the geometry of the figure, the three components of the total (normal) tooth force W are

$$\begin{aligned} W_r &= W \sin \phi_n \\ W_t &= W \cos \phi_n \cos \psi \\ W_a &= W \cos \phi_n \sin \psi \end{aligned} \tag{13–39}$$

Figure 13–37

Tooth forces acting on a right-hand helical gear.

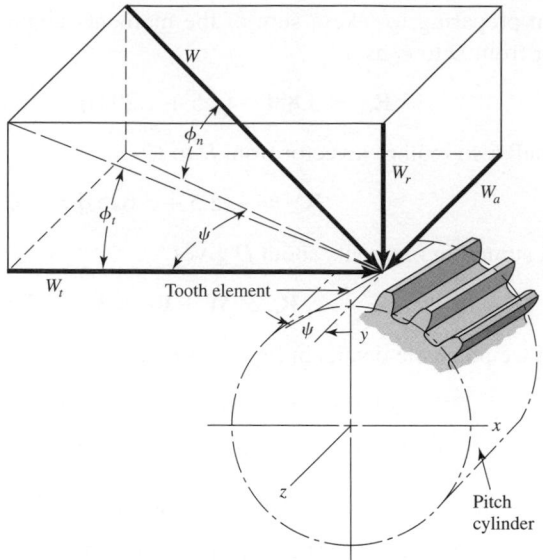

where W = total force

W_r = radial component

W_t = tangential component, also called transmitted load

W_a = axial component, also called thrust load

Usually W_t is given and the other forces are desired. In this case, it is not difficult to discover that

$$W_r = W_t \tan \phi_t$$

$$W_a = W_t \tan \psi$$

$$W = \frac{W_t}{\cos \phi_n \cos \psi}$$

(13–40)

EXAMPLE 13–8

In Fig. 13–38 a 1-hp electric motor runs at 1800 rev/min in the clockwise direction, as viewed from the positive x axis. Keyed to the motor shaft is an 18-tooth helical pinion having a normal pressure angle of 20°, a helix angle of 30°, and a normal diametral pitch

Figure 13–38

The motor and gear train of Ex. 13–8.

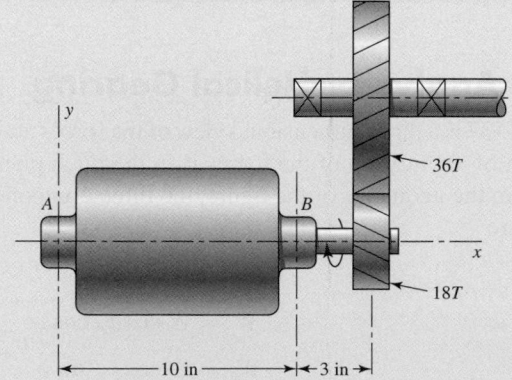

of 12 teeth/in. The hand of the helix is shown in the figure. Make a three-dimensional sketch of the motor shaft and pinion, and show the forces acting on the pinion and the bearing reactions at A and B. The thrust should be taken out at A.

Solution From Eq. (13–19) we find

$$\phi_t = \tan^{-1} \frac{\tan \phi_n}{\cos \psi} = \tan^{-1} \frac{\tan 20°}{\cos 30°} = 22.8°$$

Also, $P_t = P_n \cos \psi = 12 \cos 30° = 10.39$ teeth/in. Therefore the pitch diameter of the pinion is $d_p = 18/10.39 = 1.732$ in. The pitch-line velocity is

$$V = \frac{\pi d n}{12} = \frac{\pi (1.732)(1800)}{12} = 816 \text{ ft/min}$$

The transmitted load is

$$W_t = \frac{33\,000 H}{V} = \frac{(33\,000)(1)}{816} = 40.4 \text{ lbf}$$

From Eq. (13–40) we find

$$W_r = W_t \tan \phi_t = (40.4) \tan 22.8° = 17.0 \text{ lbf}$$

$$W_a = W_t \tan \psi = (40.4) \tan 30° = 23.3 \text{ lbf}$$

$$W = \frac{W_t}{\cos \phi_n \cos \psi} = \frac{40.4}{\cos 20° \cos 30°} = 49.6 \text{ lbf}$$

These three forces, W_r in the $-y$ direction, W_a in the $-x$ direction, and W_t in the $+z$ direction, are shown acting at point C in Fig. 13–39. We assume bearing reactions at A and B as shown. Then $F_A^x = W_a = 23.3$ lbf. Taking moments about the z axis,

$$-(17.0)(13) + (23.3)\left(\frac{1.732}{2}\right) + 10 F_B^y = 0$$

or $F_B^y = 20.1$ lbf. Summing forces in the y direction then gives $F_A^y = 3.1$ lbf. Taking moments about the y axis, next

$$10 F_B^z - (40.4)(13) = 0$$

or $F_B^z = 52.5$ lbf. Summing forces in the z direction and solving gives $F_A^z = 12.1$ lbf. Also, the torque is $T = W_t d_p/2 = (40.4)(1.732/2) = 35$ lbf · in.

Figure 13–39

Free body diagram of motor shaft of Ex. 13–8.

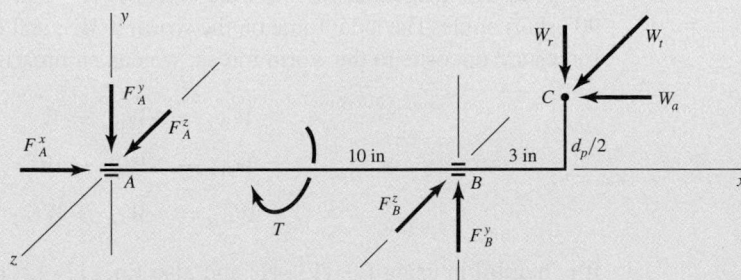

EXAMPLE 13–9   Solve Example 13–8 using vectors.

Solution   The force at C is

$$\mathbf{W} = -23.3\mathbf{i} - 17.0\mathbf{j} + 40.4\mathbf{k} \text{ lbf}$$

Position vectors to B and C from origin A are

$$\mathbf{R}_B = 10\mathbf{i} \qquad \mathbf{R}_C = 13\mathbf{i} + 0.866\mathbf{j}$$

Taking moments about A, we have

$$\mathbf{R}_B \times \mathbf{F}_B + \mathbf{T} + \mathbf{R}_C \times \mathbf{W} = 0$$

Using the directions assumed in Fig. 13–39 and substituting values gives

$$10\mathbf{i} \times \left(F_B^y\mathbf{j} - F_B^z\mathbf{k}\right) - T\mathbf{i} + (13\mathbf{i} + 0.866\mathbf{j}) \times (-23.3\mathbf{i} - 17.0\mathbf{j} + 40.4\mathbf{k}) = 0$$

When the cross products are formed, we get

$$\left(10F_B^y\mathbf{k} + 10F_B^z\mathbf{j}\right) - T\mathbf{i} + (35\mathbf{i} - 525\mathbf{j} - 201\mathbf{k}) = 0$$

whence $T = 35$ lbf · in, $F_B^y = 20.1$ lbf, and $F_B^z = 52.5$ lbf.
    Next,

$$\mathbf{F}_A = -\mathbf{F}_B - \mathbf{W}, \text{ and so } \mathbf{F}_A = 23.3\mathbf{i} - 3.1\mathbf{j} + 12.1\mathbf{k} \text{ lbf.}$$

13–17   Force Analysis—Worm Gearing

If friction is neglected, then the only force exerted by the gear will be the force W, shown in Fig. 13–40, having the three orthogonal components W^x, W^y, and W^z. From the geometry of the figure, we see that

$$\begin{aligned}
W^x &= W \cos\phi_n \sin\lambda \\
W^y &= W \sin\phi_n \\
W^z &= W \cos\phi_n \cos\lambda
\end{aligned} \qquad (13\text{–}41)$$

We now use the subscripts W and G to indicate forces acting against the worm and gear, respectively. We note that W^y is the separating, or radial, force for both the worm and the gear. The tangential force on the worm is W^x and is W^z on the gear, assuming a 90° shaft angle. The axial force on the worm is W^z, and on the gear, W^x. Since the gear forces are opposite to the worm forces, we can summarize these relations by writing

$$\begin{aligned}
W_{Wt} &= -W_{Ga} = W^x \\
W_{Wr} &= -W_{Gr} = W^y \\
W_{Wa} &= -W_{Gt} = W^z
\end{aligned} \qquad (13\text{–}42)$$

It is helpful in using Eq. (13–41) and also Eq. (13–42) to observe that *the gear axis is parallel to the x direction and the worm axis is parallel to the z direction* and that we are employing a right-handed coordinate system.

Figure 13–40

Drawing of the pitch cylinder of a worm, showing the forces exerted upon it by the worm gear.

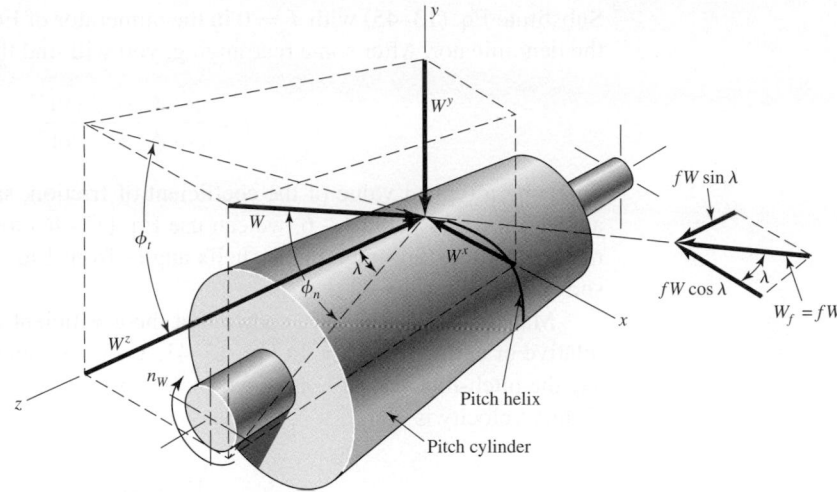

In our study of spur-gear teeth we have learned that the motion of one tooth relative to the mating tooth is primarily a rolling motion; in fact, when contact occurs at the pitch point, the motion is pure rolling. In contrast, the relative motion between worm and worm-gear teeth is pure sliding, and so we must expect that friction plays an important role in the performance of worm gearing. By introducing a coefficient of friction f, we can develop another set of relations similar to those of Eq. (13–41). In Fig. 13–40 we see that the force W acting normal to the worm-tooth profile produces a frictional force $W_f = fW$, having a component $fW \cos \lambda$ in the negative x direction and another component $fW \sin \lambda$ in the positive z direction. Equation (13–41) therefore becomes

$$W^x = W(\cos \phi_n \sin \lambda + f \cos \lambda)$$
$$W^y = W \sin \phi_n \qquad\qquad\qquad (13\text{–}43)$$
$$W^z = W(\cos \phi_n \cos \lambda - f \sin \lambda)$$

Equation (13–42), of course, still applies.

Inserting $-W_{Gt}$ from Eq. (13–42) for W^z in Eq. (13–43) and multiplying both sides by f, we find the frictional force W_f to be

$$W_f = fW = \frac{f W_{Gt}}{f \sin \lambda - \cos \phi_n \cos \lambda} \qquad (13\text{–}44)$$

A useful relation between the two tangential forces, W_{Wt} and W_{Gt}, can be obtained by equating the first and third parts of Eqs. (13–42) and (13–43) and eliminating W. The result is

$$W_{Wt} = W_{Gt} \frac{\cos \phi_n \sin \lambda + f \cos \lambda}{f \sin \lambda - \cos \phi_n \cos \lambda} \qquad (13\text{–}45)$$

Efficiency η can be defined by using the equation

$$\eta = \frac{W_{Wt}(\text{without friction})}{W_{Wt}(\text{with friction})} \qquad (a)$$

Substitute Eq. (13–45) with $f = 0$ in the numerator of Eq. (a) and the same equation in the denominator. After some rearranging, you will find the efficiency to be

$$\eta = \frac{\cos \phi_n - f \tan \lambda}{\cos \phi_n + f \cot \lambda} \qquad (13\text{–}46)$$

Selecting a typical value of the coefficient of friction, say $f = 0.05$, and the pressure angles shown in Table 13–6, we can use Eq. (13–46) to get some useful design information. Solving this equation for helix angles from 1 to 30° gives the interesting results shown in Table 13–6.

Many experiments have shown that the coefficient of friction is dependent on the relative or sliding velocity. In Fig. 13–41, V_G is the pitch-line velocity of the gear and V_W the pitch-line velocity of the worm. Vectorially, $\mathbf{V}_W = \mathbf{V}_G + \mathbf{V}_S$; consequently, the sliding velocity is

$$V_S = \frac{V_W}{\cos \lambda} \qquad (13\text{–}47)$$

Published values of the coefficient of friction vary as much as 20 percent, undoubtedly because of the differences in surface finish, materials, and lubrication. The values on the chart of Fig. 13–42 are representative and indicate the general trend.

Table 13–6

Efficiency of Worm Gearsets for $f = 0.05$

Helix Angle ψ, deg	Efficiency η, %
1.0	25.2
2.5	45.7
5.0	62.0
7.5	71.3
10.0	76.6
15.0	82.7
20.0	85.9
30.0	89.1

Figure 13–41

Velocity components in worm gearing.

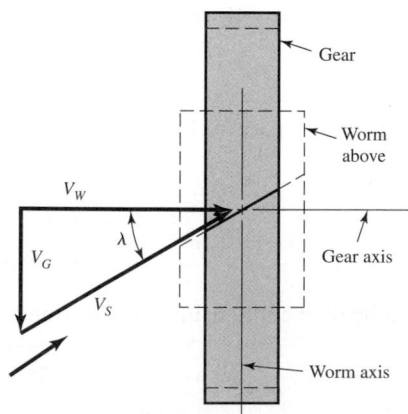

Figure 13–42

Representative values of the coefficient of friction for worm gearing. These values are based on good lubrication. Use curve *B* for high-quality materials, such as a case-hardened steel worm mating with a phosphor-bronze gear. Use curve *A* when more friction is expected, as with a cast-iron worm mating with a cast-iron worm gear.

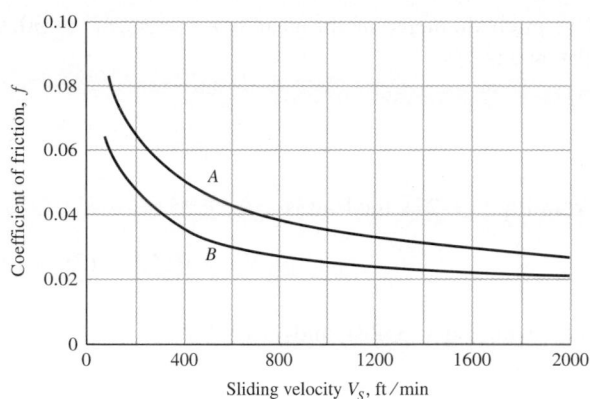

EXAMPLE 13–10

A 2-tooth right-hand worm transmits 1 hp at 1200 rev/min to a 30-tooth worm gear. The gear has a transverse diametral pitch of 6 teeth/in and a face width of 1 in. The worm has a pitch diameter of 2 in and a face width of $2\frac{1}{2}$ in. The normal pressure angle is $14\frac{1}{2}°$. The materials and quality of work needed are such that curve *B* of Fig. 13–42 should be used to obtain the coefficient of friction.

(*a*) Find the axial pitch, the center distance, the lead, and the lead angle.

(*b*) Figure 13–43 is a drawing of the worm gear oriented with respect to the coordinate system described earlier in this section; the gear is supported by bearings *A* and *B*. Find the forces exerted by the bearings against the worm-gear shaft, and the output torque.

Solution

(*a*) The axial pitch is the same as the transverse circular pitch of the gear, which is

Answer

$$p_x = p_t = \frac{\pi}{P} = \frac{\pi}{6} = 0.5236 \text{ in}$$

Figure 13–43

The pitch cylinders of the worm gear train of Ex. 13–10.

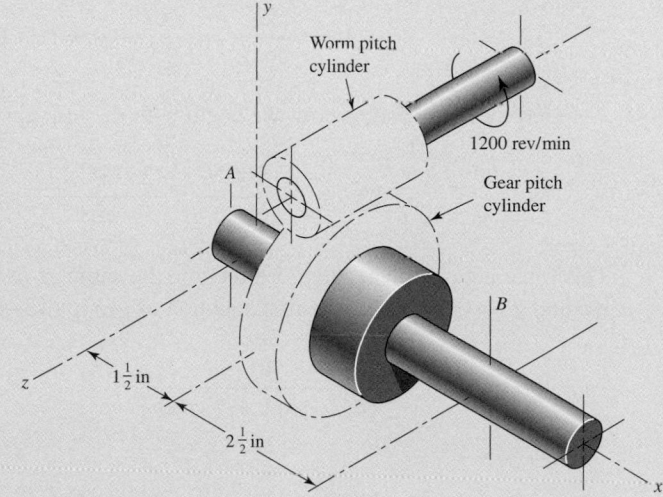

The pitch diameter of the gear is $d_G = N_G/P = 30/6 = 5$ in. Therefore, the center distance is

Answer
$$C = \frac{d_W + d_G}{2} = \frac{2 + 5}{2} = 3.5 \text{ in}$$

From Eq. (13–27), the lead is

$$L = p_x N_W = (0.5236)(2) = 1.0472 \text{ in}$$

Answer   Also using Eq. (13–28), find

Answer
$$\lambda = \tan^{-1} \frac{L}{\pi d_W} = \tan^{-1} \frac{1.0472}{\pi(2)} = 9.46°$$

(b) Using the right-hand rule for the rotation of the worm, you will see that your thumb points in the positive z direction. Now use the bolt-and-nut analogy (the worm is right-handed, as is the screw thread of a bolt), and turn the bolt clockwise with the right hand while preventing nut rotation with the left. The nut will move axially along the bolt toward your right hand. Therefore the surface of the gear (Fig. 13–43) in contact with the worm will move in the negative z direction. Thus, the gear rotates clockwise about x, with your right thumb pointing in the negative x direction.

The pitch-line velocity of the worm is

$$V_W = \frac{\pi d_W n_W}{12} = \frac{\pi(2)(1200)}{12} = 628 \text{ ft/min}$$

The speed of the gear is $n_G = (\frac{2}{30})(1200) = 80$ rev/min. Therefore the pitch-line velocity of the gear is

$$V_G = \frac{\pi d_G n_G}{12} = \frac{\pi(5)(80)}{12} = 105 \text{ ft/min}$$

Then, from Eq. (13–47), the sliding velocity V_S is found to be

$$V_S = \frac{V_W}{\cos \lambda} = \frac{628}{\cos 9.46°} = 637 \text{ ft/min}$$

Getting to the forces now, we begin with the horsepower formula

$$W_{Wt} = \frac{33\,000H}{V_W} = \frac{(33\,000)(1)}{628} = 52.5 \text{ lbf}$$

This force acts in the negative x direction, the same as in Fig. 13–40. Using Fig. 13–42, we find $f = 0.03$. Then, the first equation of group (13–42) and (13–43) gives

$$W = \frac{W^x}{\cos \phi_n \sin \lambda + f \cos \lambda}$$

$$= \frac{52.5}{\cos 14.5° \sin 9.46° + 0.03 \cos 9.46°} = 278 \text{ lbf}$$

Also, from Eq. (13–43),

$$W^y = W \sin \phi_n = 278 \sin 14.5° = 69.6 \text{ lbf}$$

$$W^z = W(\cos \phi_n \cos \lambda - f \sin \lambda)$$

$$= 278(\cos 14.5° \cos 9.46° - 0.03 \sin 9.46°) = 264 \text{ lbf}$$

We now identify the components acting on the gear as

$$W_{Ga} = -W^x = 52.5 \text{ lbf}$$

$$W_{Gr} = -W^y = -69.6 \text{ lbf}$$

$$W_{Gt} = -W^z = -264 \text{ lbf}$$

At this point a three-dimensional line drawing should be made in order to simplify the work to follow. An isometric sketch, such as the one of Fig. 13–44, is easy to make and will help you to avoid errors.

We shall make B a thrust bearing in order to place the gearshaft in compression. Thus, summing forces in the x direction gives

Answer

$$F_B^x = -52.5 \text{ lbf}$$

Taking moments about the z axis, we have

Answer

$$-(52.5)(2.5) - (69.6)(1.5) + 4F_B^y = 0 \qquad F_B^y = 58.9 \text{ lbf}$$

Taking moments about the y axis,

Answer

$$(264)(1.5) - 4F_B^z = 0 \qquad F_B^z = 99 \text{ lbf}$$

Figure 13–44

An isometric sketch used in Ex. 13–10.

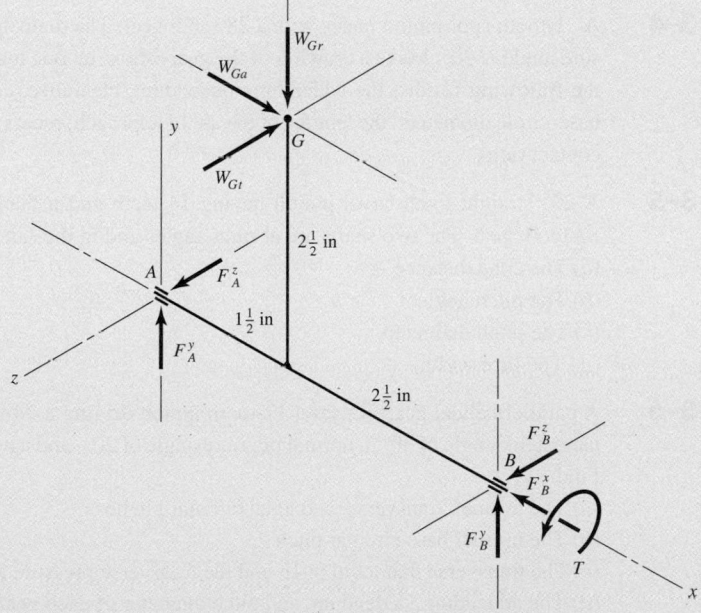

These three components are now inserted on the sketch as shown at B in Fig. 13–44. Summing forces in the y direction,

Answer
$$-69.6 + 58.9 + F_A^y = 0 \qquad F_A^y = 10.7 \text{ lbf}$$

Similarly, summing forces in the z direction,

Answer
$$-264 + 99 + F_A^z = 0 \qquad F_A^z = 165 \text{ lbf}$$

These two components can now be placed at A on the sketch. We still have one more equation to write. Summing moments about x,

Answer
$$-(264)(2.5) + T = 0 \qquad T = 660 \text{ lbf} \cdot \text{in}$$

It is because of the frictional loss that this output torque is less than the product of the gear ratio and the input torque.

PROBLEMS

ANALYSIS

13–1 A 17-tooth spur pinion has a diametral pitch of 8, runs at 1120 rev/min, and drives a gear at a speed of 544 rev/min. Find the number of teeth on the gear and the theoretical center-to-center distance.

ANALYSIS

13–2 A 15-tooth spur pinion has a module of 3 mm and runs at a speed of 1600 rev/min. The driven gear has 60 teeth. Find the speed of the driven gear, the circular pitch, and the theoretical center-to-center distance.

ANALYSIS

13–3 A spur gearset has a module of 4 mm and a velocity ratio of 2.80. The pinion has 20 teeth. Find the number of teeth on the driven gear, the pitch diameters, and the theoretical center-to-center distance.

ANALYSIS

13–4 A 21-tooth spur pinion mates with a 28-tooth gear. The diametral pitch is 3 teeth/in and the pressure angle is 20°. Make a drawing of the gears showing one tooth on each gear. Find and tabulate the following results: the addendum, dedendum, clearance, circular pitch, tooth thickness, and base-circle diameters; the lengths of the arc of approach, recess, and action; and the base pitch and contact ratio.

ANALYSIS

13–5 A 20° straight-tooth bevel pinion having 14 teeth and a diametral pitch of 6 teeth/in drives a 32-tooth gear. The two shafts are at right angles and in the same plane. Find:
(a) The cone distance
(b) The pitch angles
(c) The pitch diameters
(d) The face width

ANALYSIS

13–6 A parallel helical gearset uses a 17-tooth pinion driving a 34-tooth gear. The pinion has a right-hand helix angle of 30°, a normal pressure angle of 20°, and a normal diametral pitch of 5 teeth/in. Find:
(a) The normal, transverse, and axial circular pitches
(b) The normal base circular pitch
(c) The transverse diametral pitch and the transverse pressure angle
(d) The addendum, dedendum, and pitch diameter of each gear

ANALYSIS	**13-7**	A parallel helical gearset consists of a 19-tooth pinion driving a 57-tooth gear. The pinion has a left-hand helix angle of 20°, a normal pressure angle of $14\frac{1}{2}°$, and a normal diametral pitch of 10 teeth/in. Find: (*a*) The normal, transverse, and axial circular pitches (*b*) The transverse diametral pitch and the transverse pressure angle (*c*) The addendum, dedendum, and pitch diameter of each gear
ANALYSIS	**13-8**	In Ex. 13–1 find: (*a*) The smallest pinion tooth count that will run with itself (*b*) The smallest pinion tooth count at a ratio $m_G = 2.5$ (*c*) The smallest pinion that will run with a rack (*d*) The largest gear tooth count possible with this pinion
ANALYSIS	**13-9**	For Ex. 13–2 find: (*a*) The smallest pinion tooth count that will run with itself (*b*) The smallest pinion tooth count that will run with a rack (*c*) The largest gear tooth count possible with this pinion
ANALYSIS	**13-10**	The decision has been made to use $\phi_n = 20°$, $P_t = 6$ teeth/in, and $\psi = 30°$ for a 2:1 reduction. Choose a suitable pinion and gear tooth count to avoid interference.
ANALYSIS	**13-11**	Repeat Problem 13–10 with a 6:1 reduction.
	13-12	By employing a pressure angle larger than standard, it is possible to use fewer pinion teeth, and hence obtain smaller gears without undercutting during machining. If the gears are spur gears, what is the smallest possible pressure angle ϕ_t that can be obtained without undercutting for a 9-tooth pinion to mesh with a rack?
ANALYSIS	**13-13**	A parallel-shaft gearset consists of an 18-tooth helical pinion driving a 32-tooth gear. The pinion has a left-hand helix angle of 25°, a normal pressure angle of 20°, and a normal module of 3 mm. Find: (*a*) The normal, transverse, and axial circular pitches (*b*) The transverse module and the transverse pressure angle (*c*) The pitch diameters of the two gears
ANALYSIS	**13-14**	The double-reduction helical gearset shown in the figure is driven through shaft *a* at a speed of 900 rev/min. Gears 2 and 3 have a normal diametral pitch of 10 teeth/in, a 30° helix angle, and a

Problem 13–14
Dimensions in inches.

normal pressure angle of 20°. The second pair of gears in the train, gears 4 and 5, have a normal diametral pitch of 6 teeth/in, a 25° helix angle, and a normal pressure angle of 20°. The tooth numbers are: $N_2 = 14$, $N_3 = 54$, $N_4 = 16$, $N_5 = 36$. Find:

(a) The directions of the thrust force exerted by each gear upon its shaft

(b) The speed and direction of shaft c

(c) The center distance between shafts

 ANALYSIS **13–15** Shaft a in the figure rotates at 600 rev/min in the direction shown. Find the speed and direction of rotation of shaft d.

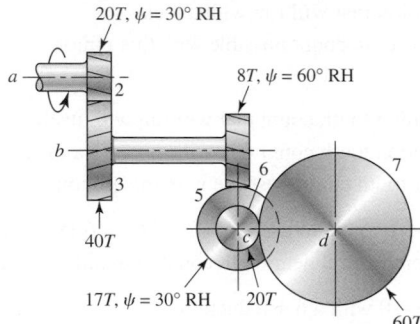

Problem 13–15

 ANALYSIS **13–16** The mechanism train shown consists of an assortment of gears and pulleys to drive gear 9. Pulley 2 rotates at 1200 rev/min in the direction shown. Determine the speed and direction of rotation of gear 9.

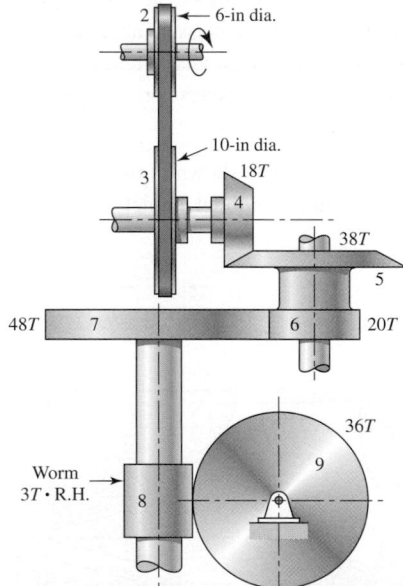

Problem 13–16

 ANALYSIS **13–17** The figure shows a gear train consisting of a pair of helical gears and a pair of miter gears. The helical gears have a $17\frac{1}{2}°$ normal pressure angle and a helix angle as shown. Find:

(a) The speed of shaft c

(b) The distance between shafts a and b

(c) The diameter of the miter gears

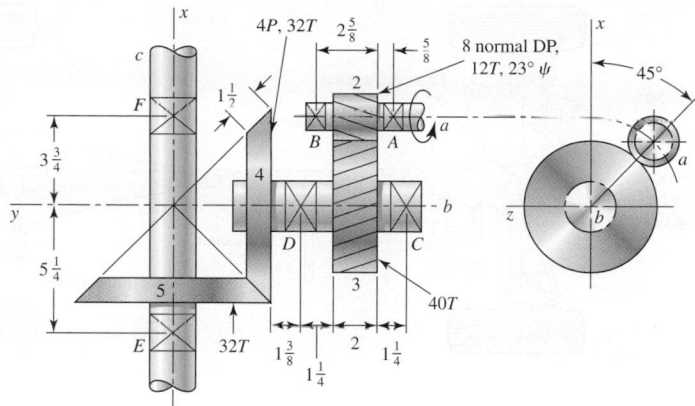

Problem 13–17
Dimensions in inches.

ANALYSIS   **13–18**   The tooth numbers for the automotive differential shown in the figure are $N_2 = 17$, $N_3 = 54$, $N_4 = 11$, $N_5 = N_6 = 16$. The drive shaft turns at 1200 rev/min.

(*a*) What are the wheel speeds if the car is traveling in a straight line on a good road surface?

(*b*) Suppose the right wheel is jacked up and the left wheel resting on a good road surface. What is the speed of the right wheel?

(*c*) Suppose, with a rear-wheel drive vehicle, the auto is parked with the right wheel resting on a wet icy surface. Does the answer to part (*b*) give you any hint as to what would happen if you started the car and attempted to drive on?

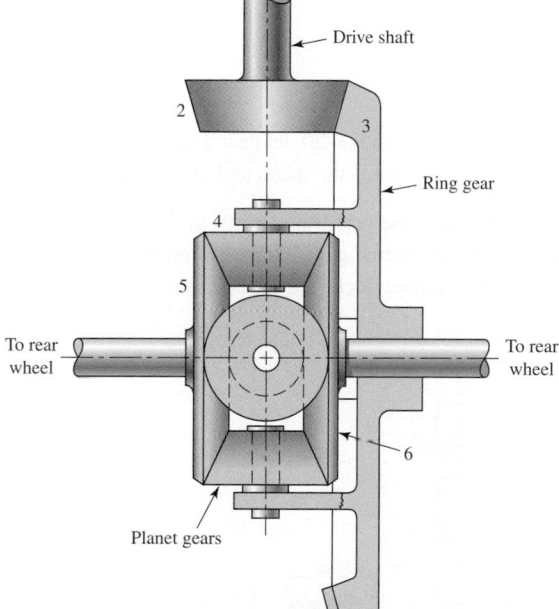

Problem 13–18

ANALYSIS   **13–19**   The figure illustrates an all-wheel drive concept using three differentials, one for the front axle, another for the rear, and the third connected to the drive shaft.

(*a*) Explain why this concept may allow greater acceleration.

(*b*) Suppose either the center of the rear differential, or both, can be locked for certain road conditions. Would either or both of these actions provide greater traction? Why?

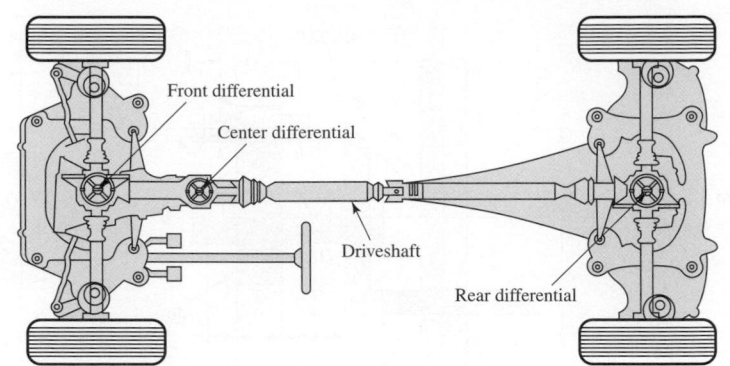

Problem 13–19
The Audi "Quattro concept,"
showing the three differentials that
provide permanent all-wheel
drive. *(Reprinted by permission of
Audi of America, Inc.)*

ANALYSIS

13–20   In the reverted planetary train illustrated, find the speed and direction of rotation of the arm if gear 2 is unable to rotate and gear 6 is driven at 12 rev/min in the clockwise direction.

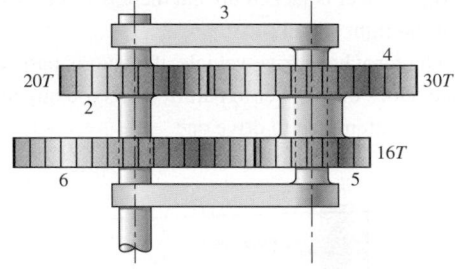

Problem 13–20

ANALYSIS

13–21   In the gear train of Prob. 13–20, let gear 2 be driven at 180 rev/min counterclockwise while gear 6 is held stationary. What is the speed and direction of rotation of the arm?

ANALYSIS

13–22   Tooth numbers for the gear train shown in the figure are $N_2 = 12$, $N_3 = 16$, and $N_4 = 12$. How many teeth must internal gear 5 have? Suppose gear 5 is fixed. What is the speed of the arm if shaft a rotates counterclockwise at 320 rev/min?

Problem 13–22

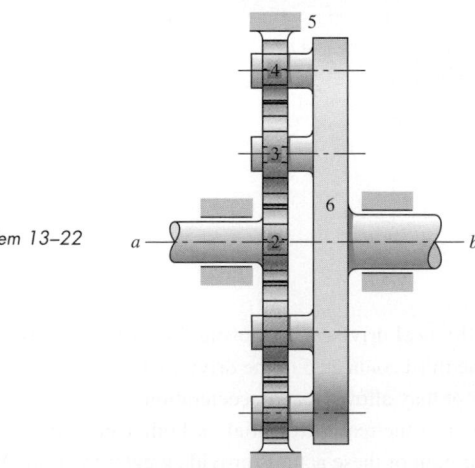

 ANALYSIS **13–23** The tooth numbers for the gear train illustrated are $N_2 = 24$, $N_3 = 18$, $N_4 = 30$, $N_6 = 36$, and $N_7 = 54$. Gear 7 is fixed. If shaft b is turned through 5 revolutions, how many turns will shaft a make?

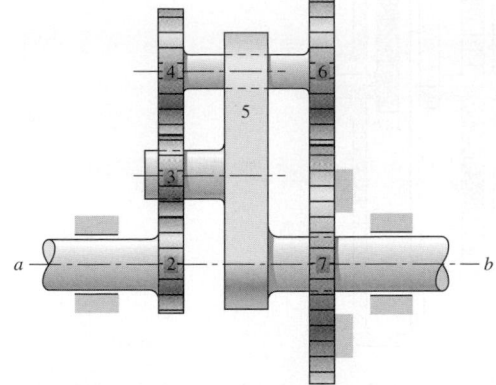

Problem 13–23

 ANALYSIS **13–24** By using nonstandard gears, it is possible to mate a $99T$ with a $100T$ gear at the same distance between centers as would be required to mate a $100T$ gear with a $101T$ gear. The planetary train shown in the figure is based on this idea.

(*a*) Find the ratio of the speed of the output shaft to the speed of the input shaft.

(*b*) The housing for this planetary train is cylindrical, with the axis of the cylinder coincident with the axes of the input and output shafts. If the pitch of gears 4 and 5 is 10 teeth/in of pitch diameter, and if these gears have standard addenda, what should be the inside diameter of this housing?

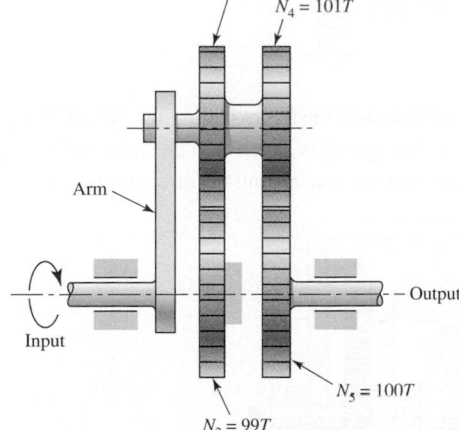

Problem 13–24

 ANALYSIS **13–25** The epicyclic train shown in the figure has the arm attached to shaft a, and sun gear 2 to shaft b. Gear 5, with 111 teeth, is an internal gear and is part of the frame. The two planets, gears 3 and 4, are both fixed to the same planet shaft. If this train is used as an in-line speed reducer, which is the input shaft, a or b? Will both shafts then rotate in the same or in the opposite directions?

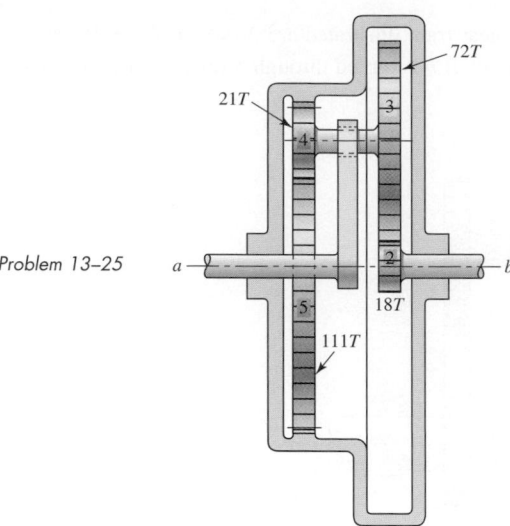

Problem 13–25

ANALYSIS

13–26 The figure shows a speed reducer in which the input shaft a is in line with output shaft b. The tooth numbers are $N_2 = 24$, $N_3 = 18$, $N_5 = 22$, and $N_6 = 64$. Find the ratio of the output speed to the input speed. Will both shafts rotate in the same direction? Note that gear 6 is a fixed internal gear.

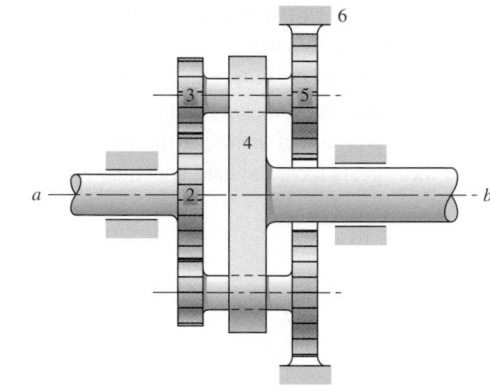

Problem 13–26

ANALYSIS

13–27 The speed reducer shown in the figure has pinion 2 fixed. The planets are gears 3 and 4, both fixed to the planet shaft. Sun gear 5 is attached to the output shaft. Input shaft a drives the arm. Find the overall speed ratio of this reducer, and the direction of rotation of the output shaft.

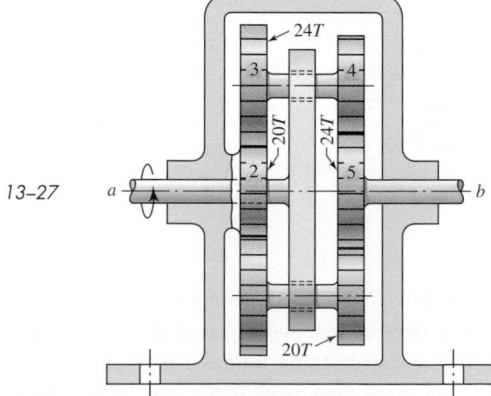

Problem 13–27

ANALYSIS

13–28 Shaft *a* in the figure has a power input of 75 kW at a speed of 1000 rev/min in the counterclockwise direction. The gears have a module of 5 mm and a 20° pressure angle. Gear 3 is an idler.
(*a*) Find the force F_{3b} that gear 3 exerts against shaft *b*.
(*b*) Find the torque T_{4c} that gear 4 exerts on shaft *c*.

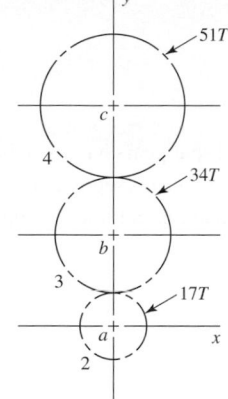

Problem 13–28

ANALYSIS

13–29 The 24*T* 6-pitch 20° pinion 2 shown in the figure rotates clockwise at 1000 rev/min and is driven at a power of 25 hp. Gears 4, 5, and 6 have 24, 36, and 144 teeth, respectively. What torque can arm 3 deliver to its output shaft? Draw free-body diagrams of the arm and of each gear and show all forces that act upon them.

Problem 13–29

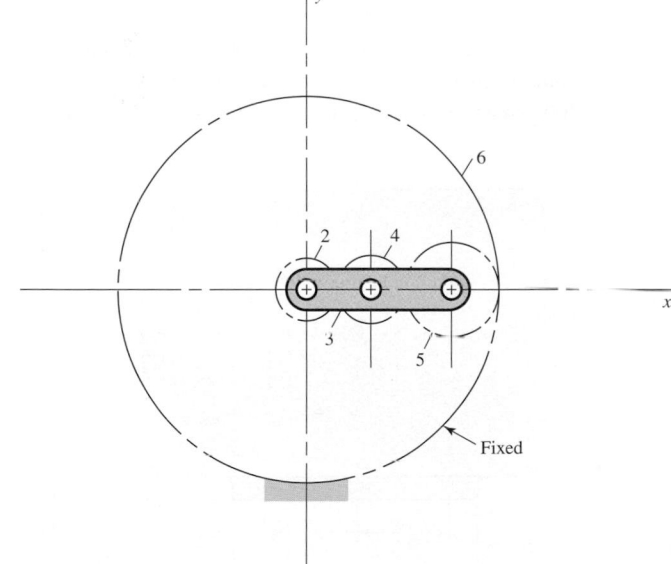

ANALYSIS

13–30 The gears shown in the figure have a diametral pitch of 2 teeth per inch and a 20° pressure angle. The pinion rotates at 1800 rev/min clockwise and transmits 200 hp through the idler pair to gear 5 on shaft *c*. What forces do gears 3 and 4 transmit to the idler shaft?

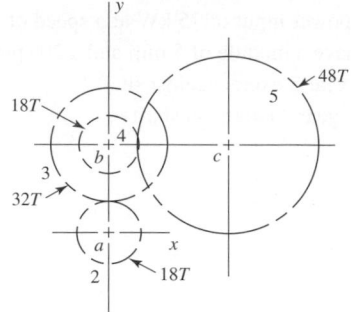

Problem 13–30

13–31 The figure shows a pair of shaft-mounted spur gears having a diametral pitch of 5 teeth/in with an 18-tooth 20° pinion driving a 45-tooth gear. The horsepower input is 32 maximum at 1800 rev/min. Find the direction and magnitude of the maximum forces acting on bearings A, B, C, and D.

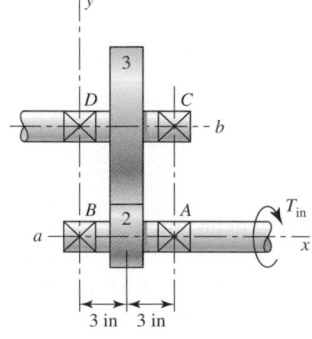

Problem 13–31

13–32 The figure shows the electric-motor frame dimensions for a 30-hp 900 rev/min motor. The frame is bolted to its support using four $\frac{3}{4}$-in bolts spaced $11\frac{1}{4}$ in apart in the view shown and 14 in apart when viewed from the end of the motor. A 4 diametral pitch 20° spur pinion having 20 teeth and a face width of 2 in is keyed to the motor shaft. This pinion drives another gear whose axis is in the same xz plane. Determine the maximum shear and tensile forces on the mounting bolts based on 200 percent overload torque. Does the direction of rotation matter?

Problem 13–32

NEMA No. 364 frame; dimensions in inches. The z axis is directed out of the paper.

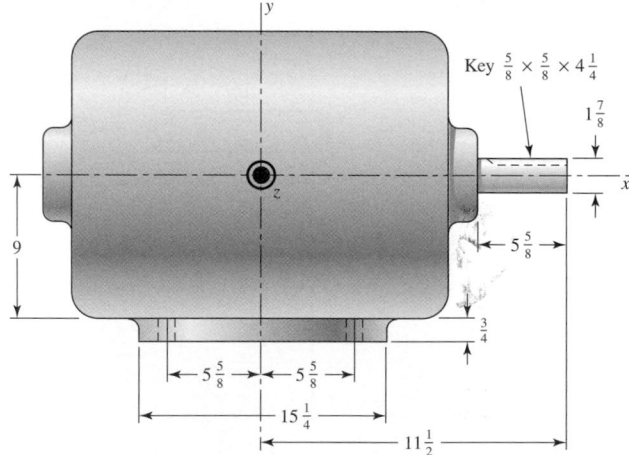

13–33 The figure shows a 16T 20° straight bevel pinion driving a 32T gear, and the location of the bearing centerlines. Pinion shaft a receives 2.5 hp at 240 rev/min. Determine the bearing reactions at A and B if A is to take both radial and thrust loads.

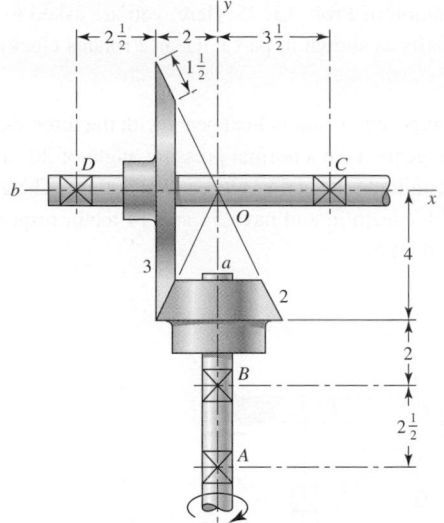

Problem 13–33
Dimensions in inches.

ANALYSIS **13–34** The figure shows a 10 diametral pitch 15-tooth 20° straight bevel pinion driving a 25-tooth gear. The transmitted load is 30 lbf. Find the bearing reactions at C and D on the output shaft if D is to take both radial and thrust loads.

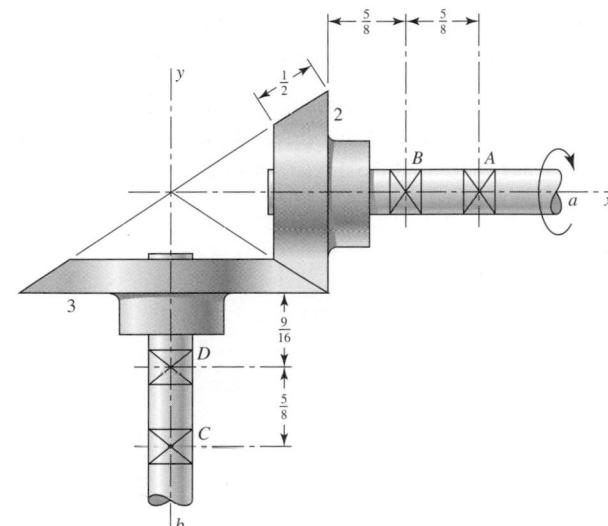

Problem 13–34
Dimensions in inches.

ANALYSIS **13–35** The gears in the two trains shown in the figure have a normal diametral pitch of 4, a normal pressure angle of 20°, and a 30° helix angle. For both gear trains the transmitted load is 800 lbf. In part a the pinion rotates counterclockwise about the y axis. Find the force exerted by each gear in part a on its shaft.

Problem 13–35

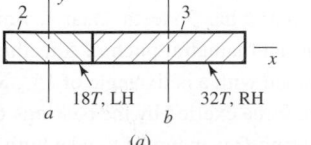

(a)

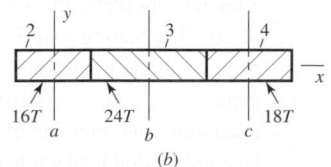

(b)

ANALYSIS

13–36   This is a continuation of Prob. 13–35. Here, you are asked to find the forces exerted by gears 2 and 3 on their shafts as shown in part *b*. Gear 2 rotates clockwise about the *y* axis. Gear 3 is an idler.

13–37   A gear train is composed of four helical gears with the three shaft axes in a single plane, as shown in the figure. The gears have a normal pressure angle of 20° and a 30° helix angle. Shaft *b* is an idler and the transmitted load acting on gear 3 is 500 lbf. The gears on shaft *b* both have a normal diametral pitch of 7 teeth/in and have 54 and 14 teeth, respectively. Find the forces exerted by gears 3 and 4 on shaft *b*.

Problem 13–37

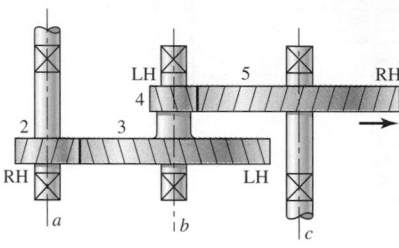

ANALYSIS

13–38   In the figure for Prob. 13–31, pinion 2 is to be a right-hand helical gear having a helix angle of 30°, a normal pressure angle of 20°, 16 teeth, and a normal diametral pitch of 6 teeth/in. A 25-hp motor drives shaft *a* at a speed of 1720 rev/min clockwise about the *x* axis. Gear 3 has 42 teeth. Find the reaction exerted by bearings *C* and *D* on shaft *b*. One of these bearings is to take both radial and thrust loads. This bearing should be selected so as to place the shaft in compression.

ANALYSIS

13–39   Gear 2, in the figure, has 16 teeth, a 20° transverse angle, a 15° helix angle, and a normal diametral pitch of 8 teeth/in. Gear 2 drives the idler on shaft *b*, which has 36 teeth. The driven gear on shaft *c* has 28 teeth. If the driver rotates at 1720 rev/min and transmits $7\frac{1}{2}$ hp, find the radial and thrust load on each shaft.

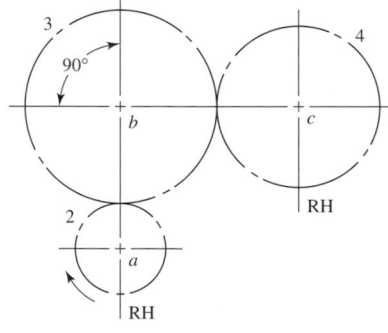

Problem 13–39

ANALYSIS

13–40   The figure shows a double-reduction helical gearset. Pinion 2 is the driver, and it receives a torque of 1200 lbf · in from its shaft in the direction shown. Pinion 2 has a normal diametral pitch of 8 teeth/in, 14 teeth, and a normal pressure angle of 20° and is cut right-handed with a helix angle of 30°. The mating gear 3 on shaft *b* has 36 teeth. Gear 4, which is the driver for the second pair of gears in the train, has a normal diametral pitch of 5 teeth/in, 15 teeth, and a normal pressure angle of 20° and is cut left-handed with a helix angle of 15°. Mating gear 5 has 45 teeth. Find the magnitude and direction of the force exerted by the bearings *C* and *D* on shaft *b* if bearing *C* can take only radial load while bearing *D* is mounted to take both radial and thrust load.

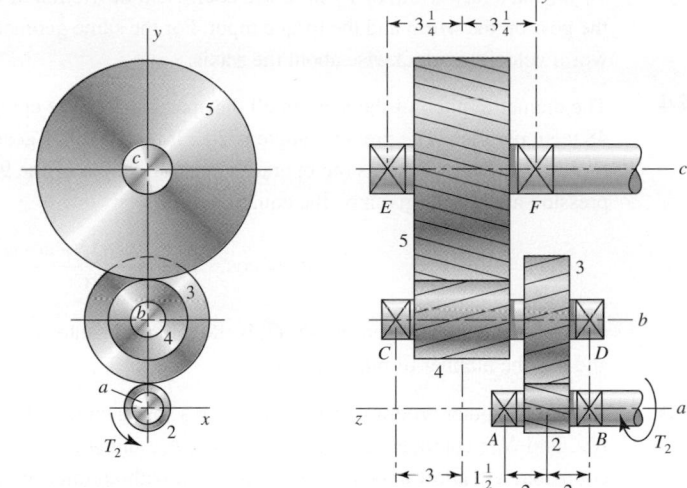

Problem 13–40
Dimensions in inches.

 ANALYSIS **13–41** A right-hand single-tooth hardened-steel (hardness not specified) worm has a catalog rating of 2000 W at 600 rev/min when meshed with a 48-tooth cast-iron gear. The axial pitch of the worm is 25 mm, the normal pressure angle is $14\frac{1}{2}^{\circ}$, the pitch diameter of the worm is 100 mm, and the face widths of the worm and gear are, respectively, 100 mm and 50 mm. The figure shows bearings A and B on the worm shaft symmetrically located with respect to the worm and 200 mm apart. Determine which should be the thrust bearing, and find the magnitudes and directions of the forces exerted by both bearings.

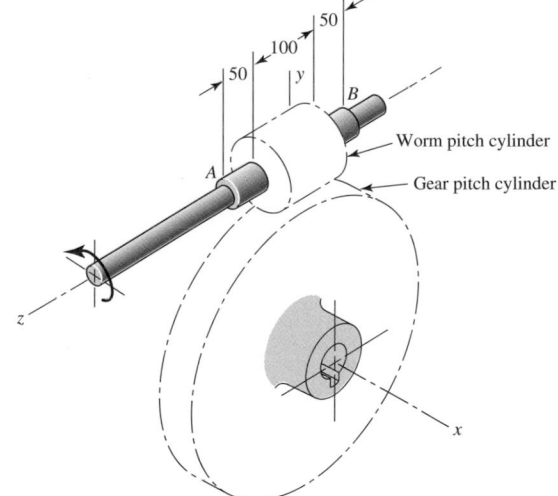

Problem 13–41
Dimensions in millimeters.

 ANALYSIS **13–42** The hub diameter and projection for the gear of Prob. 13–41 are 100 and 37.5 mm, respectively. The face width of the gear is 50 mm. Locate bearings C and D on opposite sides, spacing C 10 mm from the gear on the hidden face (see figure) and D 10 mm from the hub face. Find the output torque and the magnitudes and directions of the forces exerted by the bearings on the gearshaft.

ANALYSIS **13–43** A 2-tooth left-hand worm transmits $\frac{1}{2}$ hp at 900 rev/min to a 36-tooth gear having a transverse diametral pitch of 10 teeth/in. The worm has a normal pressure angle of $14\frac{1}{2}^{\circ}$, a pitch diameter of

$1\frac{1}{2}$ in, and a face width of $1\frac{1}{2}$ in. Use a coefficient of friction of 0.05 and find the force exerted by the gear on the worm and the torque input. For the same geometry as shown for Prob. 13–41, the worm velocity is clockwise about the z axis.

13–44   The diametral pitch of the teeth in all the spur gears of the epicyclic gear train of Prob. 13–24 is 48 teeth per inch. The pressure angle is 20°. If the 100:101 gear mesh is set to the proper center-to-center distance, what is the operative pressure angle of the 99:100 gear mesh? Show that the pressure angle ϕ' is given by the equation

$$\phi' = \cos^{-1}\left[\frac{(1+m_G)N_P \cos\phi}{2PC'}\right]$$

where for the 99:100 gear mesh, C' is the imposed center-to-center distance, m_G the gear ratio, and N_P the number of pinion teeth.

13–45   Write a computer program that will analyze a spur gear or helical-mesh gear, accepting ϕ_n, ψ, P_t, N_P, and N_G; compute m_G, d_P, d_G, p_t, p_n, p_x, and ϕ_t; and give advice as to the smallest tooth count that will allow a pinion to run with itself without interference, run with its gear, and run with a rack. Also have it give the largest tooth count possible with the intended pinion.

14

Spur and Helical Gears

Chapter Outline

This chapter is devoted primarily to analysis and design of spur and helical gears to resist bending failure of the teeth as well as pitting failure of tooth surfaces. Failure by bending will occur when the significant tooth stress equals or exceeds either the yield strength or the bending endurance strength. A surface failure occurs when the significant contact stress equals or exceeds the surface endurance strength. The first two sections present a little of the history of the analyses from which current methodology developed.

The American Gear Manufacturers Association[1] (AGMA) has for many years been the responsible authority for the dissemination of knowledge pertaining to the design and analysis of gearing. The methods this organization presents are in general use in the United States when strength and wear are primary considerations. In view of this fact it is important that the AGMA approach to the subject be presented here.

The general AGMA approach requires a great many charts and graphs—too many for a single chapter in this book. We have omitted many of these here by choosing a single pressure angle and by using only full-depth teeth. This simplification reduces the complexity but does not prevent the development of a basic understanding of the approach. Furthermore, the simplification makes possible a better development of the fundamentals and hence should constitute an ideal introduction to the use of the general AGMA method.[2] Sections 14–1 and 14–2 are elementary and serve as an examination of the foundations of the AGMA method. Table 14–1 is largely AGMA nomenclature.

14–1 The Lewis Bending Equation

Wilfred Lewis introduced an equation for estimating the bending stress in gear teeth in which the tooth form entered into the formulation. The equation, announced in 1892, still remains the basis for most gear design today.

To derive the basic Lewis equation, refer to Fig. 14–1a, which shows a cantilever of cross-sectional dimensions F and t, having a length l and a load W^t, uniformly distributed across the face width F. The section modulus I/c is $Ft^2/6$, and therefore the bending stress is

$$\sigma = \frac{M}{I/c} = \frac{6W^t l}{Ft^2} \tag{a}$$

Gear designers denote the components of gear-tooth forces as W_t, W_r, W_a or W^t, W^r, W^a interchangeably. The latter notation leaves room for post-subscripts essential to free-body diagrams. For instance, for gears 2 and 3 in mesh, W^t_{23} is the transmitted force of

---

[1] 500 Montgomery Street, Suite 350, Alexandria, VA 22314-1560.

[2] The standards ANSI/AGMA 2001-C95 (revised AGMA 2001-B88) and ANSI/AGMA 2101-C95 (metric edition of ANSI/AGMA 2001-C95), *Fundamental Rating Factors and Calculation Methods for Involute Spur and Helical Gear Teeth,* are used in this chapter. The use of American National Standards is completely voluntary; their existence does not in any respect preclude people, whether they have approved the standards or not, from manufacturing, marketing, purchasing, or using products, processes, or procedures not conforming to the standards.

The American National Standards Institute does not develop standards and will in no circumstances give an interpretation of any American National Standard. Requests for interpretation of these standards should be addressed to the American Gear Manufacturers Association. [Tables or other self-supporting sections may be quoted or extracted in their entirety. Credit line should read: "Extracted from ANSI/AGMA Standard 2001-C95 or 2101-C95 *Fundamental Rating Factors and Calculation Methods for Involute Spur and Helical Gear Teeth*" with the permission of the publisher, American Gear Manufacturers Association, 500 Montgomery Street, Suite 350, Alexandria, Virginia 22314-1560.] The foregoing is adapted in part from the ANSI foreword to these standards.

Table 14–1

Symbols, Their Names, and Locations*

Symbol	Name	Where Found
b	Net width of face of narrowest member	Eq. (14–16)
C_e	Mesh alignment correction factor	Eq. (14–35)
C_f	Surface condition factor	Eq. (14–16)
C_H	Hardness-ratio factor	Eq. (14–18)
C_{ma}	Mesh alignment factor	Eq. (14–34)
C_{mc}	Load correction factor	Eq. (14–31)
C_{mf}	Face load-distribution factor	Eq. (14–30)
C_p	Elastic coefficient	Eq. (14–13)
C_{pf}	Pinion proportion factor	Eq. (14–32)
C_{pm}	Pinion proportion modifier	Eq. (14–33)
d	Operating pitch diameter of pinion	Eq. (14–1)
d_P	Pitch diameter, pinion	Eq. (14–22)
d_G	Pitch diameter, gear	Eq. (14–22)
E	Modulus of elasticity	Eq. (14–10)
F	Net face width of narrowest member	Eq. (14–15)
f_P	Pinion surface finish	Fig. 14–13
H	Power	Fig. 14–17
H_B	Brinell hardness	Ex. 14–3
H_{BG}	Brinell hardness of gear	Sec. 14–12
H_{BP}	Brinell hardness of pinion	Sec. 14–12
hp	Horsepower	Ex. 14–1
h_t	Gear-tooth whole depth	Sec. 14–16
I	Geometry factor of pitting resistance	Eq. (14–16)
J	Geometry factor for bending strength	Eq. (14–15)
K	Contact load factor for pitting resistance	Eq. (7–72)
K_B	Rim-thickness factor	Eq. (14–40)
K_f	Fatigue stress-concentration factor	Eq. (14–9)
K_m	Load-distribution factor	Eq. (14–30)
K_o	Overload factor	Eq. (14–15)
K_R	Reliability factor	Eq. (14–17)
K_s	Size factor	Sec. 14–10
K_T	Temperature factor	Eq. (14–17)
K_v	Dynamic factor	Eq. (14–17)
m	Metric module	Eq. (14–15)
m_B	Backup ratio	Eq. (14–34)
m_G	Gear ratio (never less than 1)	Eq. (14–22)
m_N	Load-sharing ratio	Eq. (14–21)
N	Number of stress cycles	Fig. 14–14
N_G	Number of teeth on gear	Eq. (14–22)
N_P	Number of teeth on pinion	Eq. (14–22)
n	Speed	Ex. 14–1

(Continued)

Table 14–1

Symbols, Their Names, and Locations* (*Continued*)

Symbol	Name	Where Found
n_P	Pinion speed	Ex. 14–4
P	Diametral pitch	Eq. (14–2)
P_d	Diametral pitch of pinion	Eq. (14–15)
p_N	Normal base pitch	Eq. (14–24)
P_n	Normal circular pitch	Eq. (14–24)
P_x	Axial pitch	Eq. (14–19)
Q_v	Transmission accuracy level number	Eq. (14–29)
R	Reliability	Eq. (14–38)
R_a	Root-mean-squared roughness	Fig. 14–13
r_f	Tooth fillet radius	Fig. 14–1
r_G	Pitch-circle radius, gear	In standard
r_P	Pitch-circle radius, pinion	In standard
r_{bP}	Pinion base-circle radius	Eq. (14–25)
r_{bG}	Gear base-circle radius	Eq. (14–25)
S_C	Buckingham surface endurance strength	Ex. 14–3
S_c	AGMA surface endurance strength	Eq. (14–18)
S_t	AGMA bending strength	Eq. (14–17)
S	Bearing span	Fig. 14–10
S_1	Pinion offset from center span	Fig. 14–10
S_F	Safety factor—bending	Eq. (14–41)
S_H	Safety factor—pitting	Eq. (14–42)
W^t or W_t^t	Transmitted load	Fig. 14–1
Y_N	Stress cycle factor for bending strength	Fig. 14–14
Z_N	Stress cycle factor for pitting resistance	Fig. 14–15
β	Exponent	Eq. (14–44)
σ	Bending stress	Eq. (14–2)
σ_C	Contact stress from Hertzian relationships	Eq. (14–14)
σ_c	Contact stress from AGMA relationships	Eq. (14–16)
σ_{all}	Allowable bending stress	Eq. (14–17)
$\sigma_{c,all}$	Allowable contact stress, AGMA	Eq. (14–18)
ϕ	Pressure angle	Eq. (14–12)
ϕ_t	Transverse pressure angle	Eq. (14–23)
ψ	Helix angle at standard pitch diameter	Ex. 14–2

*Because ANSI/AGMA 2001-C95 introduced a significant amount of new nomenclature, this summary and references is provided for use until the reader's vocabulary has grown.

†See preference rationale following Eq. (*a*), Sec. 14–1.

body 2 on body 3, and W_{32}^t is the transmitted force of body 3 on body 2. When working with double- or triple-reduction speed reducers, this notation is compact and essential to clear thinking. Since gear-force components rarely take exponents, this causes no complication. Pythagorean combinations, if necessary, can be treated with parentheses or avoided by expressing the relations trigonometrically.

| **Figure 14-1**

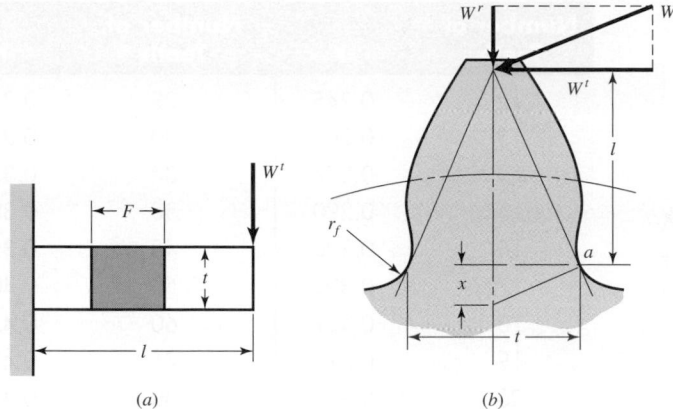

(a)   (b)

Referring now to Fig. 14–1*b*, we assume that the maximum stress in a gear tooth occurs at point *a*. By similar triangles, you can write

$$\frac{t/2}{x} = \frac{l}{t/2} \qquad \text{or} \qquad x = \frac{t^2}{4l} \tag{b}$$

By rearranging Eq. (*a*),

$$\sigma = \frac{6W^t l}{Ft^2} = \frac{W^t}{F} \frac{1}{t^2/6l} = \frac{W^t}{F} \frac{1}{t^2/4l} \frac{1}{\frac{4}{6}} \tag{c}$$

If we now substitute the value of *x* from Eq. (*b*) in Eq. (*c*) and multiply the numerator and denominator by the circular pitch *p*, we find

$$\sigma = \frac{W^t p}{F\left(\frac{2}{3}\right) xp} \tag{d}$$

Letting $y = 2x/3p$, we have

$$\sigma = \frac{W^t}{Fpy} \tag{14-1}$$

This completes the development of the original Lewis equation. The factor *y* is called the *Lewis form factor,* and it may be obtained by a graphical layout of the gear tooth or by digital computation.

In using this equation, most engineers prefer to employ the diametral pitch in determining the stresses. This is done by substituting $P = \pi/p$ and $Y = \pi y$ in Eq. (14–1). This gives

$$\sigma = \frac{W^t P}{FY} \tag{14-2}$$

where

$$Y = \frac{2xP}{3} \tag{14-3}$$

The use of this equation for *Y* means that only the bending of the tooth is considered and that the compression due to the radial component of the force is neglected. Values of *Y* obtained from this equation are tabulated in Table 14–2.

Table 14–2

Values of the Lewis Form Factor Y (These Values Are for a Normal Pressure Angle of 20°, Full-Depth Teeth, and a Diametral Pitch of Unity in the Plane of Rotation)

Number of Teeth	Y	Number of Teeth	Y
12	0.245	28	0.353
13	0.261	30	0.359
14	0.277	34	0.371
15	0.290	38	0.384
16	0.296	43	0.397
17	0.303	50	0.409
18	0.309	60	0.422
19	0.314	75	0.435
20	0.322	100	0.447
21	0.328	150	0.460
22	0.331	300	0.472
24	0.337	400	0.480
26	0.346	Rack	0.485

The use of Eq. (14–3) also implies that the teeth do not share the load and that the greatest force is exerted at the tip of the tooth. But we have already learned that the contact ratio should be somewhat greater than unity, say about 1.5, to achieve a quality gearset. If, in fact, the gears are cut with sufficient accuracy, the tip-load condition is not the worst, because another pair of teeth will be in contact when this condition occurs. Examination of run-in teeth will show that the heaviest loads occur near the middle of the tooth. Therefore the maximum stress probably occurs while a single pair of teeth is carrying the full load, at a point where another pair of teeth is just on the verge of coming into contact.

Dynamic Effects

When a pair of gears is driven at moderate or high speed and noise is generated, it is certain that dynamic effects are present. One of the earliest efforts to account for an increase in the load due to velocity employed a number of gears of the same size, material, and strength. Several of these gears were tested to destruction by meshing and loading them at zero velocity. The remaining gears were tested to destruction at various pitch-line velocities. For example, if a pair of gears failed at 500 lbf tangential load at zero velocity and at 250 lbf at velocity V_1, then a *velocity factor,* designated K_v, of 2 was specified for the gears at velocity V_1. Then another, identical, pair of gears running at a pitch-line velocity V_1 could be assumed to have a load equal to twice the tangential or transmitted load.

Note that the definition of dynamic factor K_v has been altered. AGMA standards ANSI/AGMA 2110-C95 and 2101-C95 contain this caution:

> Dynamic factor K_v has been redefined as the reciprocal of that used in previous standards. It is now greater than 1.0. In earlier AGMA standards it was less than 1.0.

Care must be taken in referring to work done prior to this change in the standards.

In the nineteenth century, Carl G. Barth first expressed the velocity factor, and in terms of the current AGMA standards, they are represented as

$$K_v = \frac{600 + V}{600} \qquad \text{(cast iron, cast profile)} \qquad (14\text{--}4a)$$

$$K_v = \frac{1200 + V}{1200} \qquad \text{(cut or milled profile)} \qquad (14\text{--}4b)$$

where V is the pitch-line velocity in feet per minute. It is also quite probable, because of the date that the tests were made, that the tests were conducted on teeth having a cycloidal profile instead of an involute profile. Cycloidal teeth were in general use in the nineteenth century because they were easier to cast than involute teeth. Equation (14–4a) is called the *Barth equation*. The Barth equation is often modified into Eq. (14–4b), for cut or milled teeth. Later AGMA added

$$K_v = \frac{50 + \sqrt{V}}{50} \qquad \text{(hobbed or shaped profile)} \qquad (14\text{--}5a)$$

$$K_v = \sqrt{\frac{78 + \sqrt{V}}{78}} \qquad \text{(shaved or ground profile)} \qquad (14\text{--}5b)$$

In SI units, Eqs. (14–4a) through (14–5b) become

$$K_v = \frac{3.05 + V}{3.05} \qquad \text{(cast iron, cast profile)} \qquad (14\text{--}6a)$$

$$K_v = \frac{6.1 + V}{6.1} \qquad \text{(cut or milled profile)} \qquad (14\text{--}6b)$$

$$K_v = \frac{3.56 + \sqrt{V}}{3.56} \qquad \text{(hobbed or shaped profile)} \qquad (14\text{--}6c)$$

$$K_v = \sqrt{\frac{5.56 + \sqrt{V}}{5.56}} \qquad \text{(shaved or ground profile)} \qquad (14\text{--}6d)$$

where V is in meters per second (m/s).

Introducing the velocity factor into Eq. (14–2) gives

$$\sigma = \frac{K_v W^t P}{F Y} \qquad (14\text{--}7)$$

The metric version of this equation is

$$\sigma = \frac{K_v W^t}{F m Y} \qquad (14\text{--}8)$$

where the face width F and the module m are both in millimeters (mm). Expressing the tangential component of load W^t in newtons (N) then results in stress units of megapascals (MPa).

As a general rule, spur gears should have a face width F from 3 to 5 times the circular pitch p.

Equations (14–7) and (14–8) are important because they form the basis for the AGMA approach to the bending strength of gear teeth. They are in general use for estimating the capacity of gear drives when life and reliability are not important considerations. The equations can be useful in obtaining a preliminary estimate of gear sizes needed for various applications.

EXAMPLE 14–1

A stock spur gear is available having a diametral pitch of 8 teeth/in, a $1\frac{1}{2}$-in face, 16 teeth, and a pressure angle of 20° with full-depth teeth. The material is AISI 1020 steel in as-rolled condition. Use a design factor of $n_d = 3$ to rate the horsepower output of the gear corresponding to a speed of 1200 rev/m and moderate applications.

Solution

The term *moderate applications* seems to imply that the gear can be rated by using the yield strength as a criterion of failure. From Table A–20, we find $S_{ut} = 55$ kpsi and $S_y = 30$ kpsi. A design factor of 3 means that the allowable bending stress is $30/3 = 10$ kpsi. The pitch diameter is $N/P = 16/8 = 2$ in, so the pitch-line velocity is

$$V = \frac{\pi dn}{12} = \frac{\pi(2)1200}{12} = 628 \text{ ft/min}$$

The velocity factor from Eq. (14–4b) is found to be

$$K_v = \frac{1200 + V}{1200} = \frac{1200 + 628}{1200} = 1.52$$

Table 14–2 gives the form factor as $Y = 0.296$ for 16 teeth. We now arrange and substitute in Eq. (14–7) as follows:

$$W^t = \frac{FY\sigma_{\text{all}}}{K_v P} = \frac{1.5(0.296)10\,000}{1.52(8)} = 365 \text{ lbf}$$

The horsepower that can be transmitted is

Answer

$$hp = \frac{W^t V}{33\,000} = \frac{365(628)}{33\,000} = 6.95 \text{ hp}$$

It is important to emphasize that this is a rough estimate, and that this approach must not be used for important applications. The example is intended to help you understand some of the fundamentals that will be involved in the AGMA approach.

EXAMPLE 14–2

Estimate the horsepower rating of the gear in the previous example based on obtaining an infinite life in bending.

Solution

The rotating-beam endurance limit is estimated from Eq. (7–8)

$$S'_e = 0.504 S_{ut} = 0.504(55) = 27.7 \text{ kpsi}$$

To obtain the surface finish Marin factor k_a we refer to Table 7–4 for machined surface, finding $a = 2.70$ and $b = -0.265$. Then Eq. (7–18) gives the surface finish Marin factor k_a as

$$k_a = a S_{ut}^b = 2.70(55)^{-0.265} = 0.934$$

The next step is to estimate the size factor k_b. From Table 13–1, the sum of the addendum and dedendum is

$$l = \frac{1}{P} + \frac{1.25}{P} = \frac{1}{8} + \frac{1.25}{8} = 0.281 \text{ in}$$

The tooth thickness t in Fig. 14–1b is given in Sec. 14–1 [Eq. (b)] as $t = (4lx)^{1/2}$ when $x = 3Y/(2P)$ from Eq. (14–3). Therefore, since from Ex. 14–1 $Y = 0.296$ and $P = 8$,

$$x = \frac{3Y}{2P} = \frac{3(0.296)}{2(8)} = 0.0555 \text{ in}$$

then

$$t = (4lx)^{1/2} = [4(0.281)0.0555]^{1/2} = 0.250 \text{ in}$$

We have recognized the tooth as a cantilever beam of rectangular cross section, so the equivalent rotating-beam diameter must be obtained from Eq. (7–24):

$$d_e = 0.808(hb)^{1/2} = 0.808(Ft)^{1/2} = 0.808[1.5(0.250)]^{1/2} = 0.495 \text{ in}$$

Then, Eq. (7–19) gives k_b as

$$k_b = \left(\frac{d_e}{0.30}\right)^{-0.107} = \left(\frac{0.495}{0.30}\right)^{-0.107} = 0.948$$

The load factor k_c from Eq. (7–25) is unity. With no information given concerning temperature and reliability we will set $k_d = k_e = 1$.

Two effects are used to evaluate the miscellaneous-effects Marin factor k_f. The first of these is the effect of one-way bending. In general, a gear tooth is subjected only to one-way bending. Exceptions include idler gears and gears used in reversing mechanisms.

For one-way bending the steady and alternating stress components are $\sigma_a = \sigma_m = \sigma/2$ where σ is the largest repeatedly applied bending stress as given in Eq. (14–7). If a material exhibited a Goodman failure locus,

$$\frac{S_a}{S_e'} + \frac{S_m}{S_{ut}} = 1$$

Since S_a and S_m are equal for one-way bending, we substitute S_a for S_m and solve the preceding equation for S_a, giving

$$S_a = \frac{S_e' S_{ut}}{S_e' + S_{ut}}$$

Now replace S_a with $\sigma/2$, and in the denominator replace S_e' with $0.504S_{ut}$ to obtain

$$\sigma = \frac{2S_e' S_{ut}}{0.504S_{ut} + S_{ut}} = \frac{2S_e'}{0.504 + 1} = 1.33S_e'$$

Now $k_f = \sigma/S_e' = 1.33S_e'/S_e' = 1.33$. However, a Gerber fatigue locus gives mean values of

$$\frac{S_a}{S_e'} + \left(\frac{S_m}{S_{ut}}\right)^2 = 1$$

Setting $S_a = S_m$ and solving the quadratic in S_a gives

$$S_a = \frac{S_{ut}^2}{2S_e'}\left(-1 + \sqrt{1 + \frac{4S_e'^2}{S_{ut}^2}}\right)$$

Setting $S_a = \sigma/2$, $S_{ut} = S_e'/0.504$ gives

$$\sigma = \frac{S_e'}{0.504^2}\left[-1 + \sqrt{1 + 4(0.504)^2}\right] = 1.65 S_e'$$

and $k_e = \sigma/S_e' = 1.65$. Since a Gerber locus runs in and among fatigue data and Goodman does not, we will use $k_e = 1.65$.

The second effect to be accounted for in using the miscellaneous-effects Marin factor k_f is stress concentration, for which we will use our fundamentals from Chap. 7. For a 20° full-depth tooth the radius of the root fillet is denoted r_f, where

$$r_f = \frac{0.300}{P} = \frac{0.300}{8} = 0.0375 \text{ in}$$

From Fig. A–15–6

$$\frac{r}{d} = \frac{r_f}{t} = \frac{0.0375}{0.250} = 0.15$$

Since $D/d = \infty$, we approximate with $D/d = 3$, giving $K_t = 1.68$. From Eq. (7–35) and Table 7–8 for a shoulder,

$$K_f = \frac{K_t}{1 + \dfrac{2}{\sqrt{r}}\dfrac{K_t - 1}{K_t}\sqrt{a}} = \frac{1.68}{1 + \dfrac{2}{\sqrt{0.0375}}\dfrac{1.68 - 1}{1.68}\dfrac{4}{55}} = 1.29$$

The miscellaneous-effects Marin factor for stress concentration can be expressed as

$$k_f = \frac{1}{K_f} = \frac{1}{1.29} = 0.775$$

The final value of k_f is the product of the two k_f factors, that is, $1.65(0.775) = 1.28$. The Marin equation for the fully corrected endurance strength is

$$S_e = k_a k_b k_c k_d k_e k_f S_e'$$
$$= 0.934(0.948)(1)(1)(1)1.28(27.7) = 31.4 \text{ kpsi}$$

For a design factor of $n_d = 3$, as used in Ex. 14–1, applied to the load or strength, the allowable bending stress is

$$\sigma_{\text{all}} = \frac{S_e}{n_d} = \frac{31.4}{3} = 10.47 \text{ kpsi}$$

The transmitted load W^t is

$$W^t = \frac{FY\sigma_{\text{all}}}{K_v P} = \frac{1.5(0.296)10\,470}{1.52(8)} = 382 \text{ lbf}$$

and the power is, with $V = 628$ ft/min from Ex. 14–1,

$$hp = \frac{W^t V}{33\,000} = \frac{382(628)}{33\,000} = 7.3 \text{ hp}$$

Again, it should be emphasized that these results should be accepted only as preliminary estimates to alert you to the nature of bending in gear teeth. Note that the repeatedly-applied allowable stress ($\sigma_{\text{all}} = 10\,470$ psi) is slightly higher than the allowable stress in Ex. 14–1 ($\sigma_{\text{allow}} = 10$ kpsi). This condition could be corrected by using a design factor somewhat larger than three.

In Ex. 14–2 our resources (Fig. A–15–6) did not directly address stress concentration in gear teeth. A photoelastic investigation by Dolan and Broghamer reported in 1942 constitutes a primary source of information on stress concentration.[3] Mitchiner and Mabie[4] interpret the results in term of fatigue stress-concentration factor K_f as

$$K_f = H + \left(\frac{t}{r}\right)^L \left(\frac{t}{l}\right)^M \qquad (14\text{–}9)$$

where $H = 0.34 - 0.458\ 366\ 2\phi$

$L = 0.316 - 0.458\ 366\ 2\phi$

$M = 0.290 + 0.458\ 366\ 2\phi$

$r = \dfrac{(b - r_f)^2}{(d/2) + b - r_f}$

In these equations l and t are from the layout in Fig. 14–1, ϕ is the pressure angle, r_f is the fillet radius, b is the dedendum, and d is the pitch diameter. It is left as an exercise for the reader to compare K_f from Eq. (14–9) with the results of using the approximation of Fig. A–15–6 in Ex. 14–2.

14–2 Surface Durability

In this section we are interested in the failure of the surfaces of gear teeth, which is generally called *wear*. *Pitting,* as explained in Sec. 7–16, is a surface fatigue failure due to many repetitions of high contact stresses. Other surface failures are *scoring,* which is a lubrication failure, and *abrasion,* which is wear due to the presence of foreign material.

To obtain an expression for the surface-contact stress, we shall employ the Hertz theory. In Eq. (3–78) it was shown that the contact stress between two cylinders may be computed from the equation

$$p_{\max} = \frac{2F}{\pi b l} \qquad (a)$$

where $p_{\max}$ = largest surface pressure

F = force pressing the two cylinders together

l = length of cylinders

and half-width b is obtained from Eq. (4–77):

$$b = \left\{ \frac{2F}{\pi l} \frac{\left[(1 - \nu_1^2)/E_1\right] + \left[(1 - \nu_2^2)/E_2\right]}{(1/d_1) + (1/d_2)} \right\}^{1/2} \qquad (14\text{–}10)$$

where ν_1, ν_2, E_1, and E_2 are the elastic constants and d_1 and d_2 are the diameters, respectively, of the two contacting cylinders.

To adapt these relations to the notation used in gearing, we replace F by $W^t/\cos\phi$, d by $2r$, and l by the face width F. With these changes, we can substitute the value of b

---

[3]T. J. Dolan and E. I. Broghamer, *A Photoelastic Study of the Stresses in Gear Tooth Fillets,* Bulletin 335, Univ. Ill. Exp. Sta., March 1942, See also R. E. Peterson, *Stress Concentration Factors,* Wiley, New York, 1974, pp. 250, 251.

[4]R. G. Mitchiner and H. H. Mabie, "Determination of the Lewis Form Factor and the AGMA Geometry Factor J of External Spur Gear Teeth," *J. Mech. Des.,* Vol. 104, No. 1, Jan. 1982, pp. 148–158.

as given by Eq. (14–10) in Eq. (a). Replacing p_{max} by σ_C, the *surface compressive stress (Hertzian stress)* is found from the equation

$$\sigma_C^2 = \frac{W^t}{\pi F \cos\phi} \frac{(1/r_1) + (1/r_2)}{\left[(1 - v_1^2)/E_1\right] + \left[(1 - v_2^2)/E_2\right]} \tag{14–11}$$

where r_1 and r_2 are the instantaneous values of the radii of curvature on the pinion- and gear-tooth profiles, respectively, at the point of contact. By accounting for load sharing in the value of W^t used, Eq. (14–11) can be solved for the Hertzian stress for any or all points from the beginning to the end of tooth contact. Of course, pure rolling exists only at the pitch point. Elsewhere the motion is a mixture of rolling and sliding. Equation (14–11) does not account for any sliding action in the evaluation of stress.

We note that AGMA uses μ for Poisson's ratio instead of v as is used here.

We have already noted that the first evidence of wear occurs near the pitch line. The radii of curvature of the tooth profiles at the pitch point are

$$r_1 = \frac{d_P \sin\phi}{2} \qquad r_2 = \frac{d_G \sin\phi}{2} \tag{14–12}$$

where ϕ is the pressure angle and d_P and d_G are the pitch diameters of the pinion and gear, respectively.

Note, in Eq. (14–11), that the denominator of the second group of terms contains four elastic constants, two for the pinion and two for the gear. As a simple means of combining and tabulating the results for various combinations of pinion and gear materials, AGMA defines an *elastic coefficient* C_p by the equation

$$C_p = \left[\frac{1}{\pi \left(\dfrac{1 - v_P^2}{E_P} + \dfrac{1 - v_G^2}{E_G} \right)} \right]^{1/2} \tag{14–13}$$

With this simplification, and the addition of a velocity factor K_v, Eq. (14–11) can be written as

$$\sigma_C = -C_p \left[\frac{K_v W^t}{F \cos\phi} \left(\frac{1}{r_1} + \frac{1}{r_2} \right) \right]^{1/2} \tag{14–14}$$

where the sign is negative because σ_C is a compressive stress.

EXAMPLE 14–3

The pinion of Examples 14–1 and 14–2 is to be mated with a 50-tooth gear manufactured of ASTM No. 50 cast iron. Using the tangential load of 382 lbf, estimate the factor of safety of the drive based on the possibility of a surface fatigue failure.

Solution

From Table A–5 we find the elastic constants to be $E_P = 30$ Mpsi, $v_P = 0.292$, $E_G = 14.5$ Mpsi, $v_G = 0.211$. We substitute these in Eq. (14–13) to get the elastic coefficient as

$$C_p = \left\{ \frac{1}{\pi \left[\dfrac{1 - (0.292)^2}{30(10^6)} + \dfrac{1 - (0.211)^2}{14.5(10^6)} \right]} \right\}^{1/2} = 1817$$

From Example 14–1, the pinion pitch diameter is $d_P = 2$ in. The value for the gear is $d_G = 50/8 = 6.25$ in. Then Eq. (14–12) is used to obtain the radii of curvature at the pitch points. Thus

$$r_1 = \frac{2 \sin 20°}{2} = 0.342 \text{ in} \qquad r_2 = \frac{6.25 \sin 20°}{2} = 1.069 \text{ in}$$

The face width is given as $F = 1.5$ in. Use $K_v = 1.52$ from Example 4–1. Substituting all these values in Eq. (14–14) with $\phi = 20°$ gives the contact stress as

$$\sigma_C = -1817 \left[\frac{1.52(380)}{1.5 \cos 20°} \left(\frac{1}{0.342} + \frac{1}{1.069} \right) \right]^{1/2} = -72\,400 \text{ psi}$$

The surface endurance strength of cast iron can be estimated from

$$S_C = 0.32 H_B \text{ kpsi}$$

for 10^8 cycles, where S_C is in kpsi. Table A–24 gives $H_B = 262$ for ASTM No. 50 cast iron. Therefore $S_C = 0.32(262) = 83.8$ kpsi. Contact stress is not linear with transmitted load [see Eq. (14–14)]. If the factor of safety is defined as the loss-of-function load divided by the imposed load, then the ratio of loads is the ratio of stresses squared. In other words,

$$n = \frac{\text{loss-of-function load}}{\text{imposed load}} = \frac{S_C^2}{\sigma_C^2} = \left(\frac{83.8}{72.4} \right)^2 = 1.34$$

One is free to define factor of safety as S_C/σ_C. Awkwardness comes when one compares the factor of safety in bending fatigue with the factor of safety in surface fatigue for a particular gear. Suppose the factor of safety of this gear in bending fatigue is 1.20 and the factor of safety in surface fatigue is 1.34 as above. The threat, since 1.34 is greater than 1.20, is in bending fatigue since both numbers are based on load ratios. If the factor of safety in surface fatigue is based on $S_C/\sigma_C = \sqrt{1.34} = 1.16$, then 1.20 is greater than 1.16, but the threat is not from surface fatigue. The surface fatigue factor of safety can be defined either way. One way has the burden of requiring a squared number before numbers that instinctively seem comparable can be compared.

In addition to the dynamic factor K_v already introduced, there are transmitted load excursions, nonuniform distribution of the transmitted load over the tooth contact, and the influence of rim thickness on bending stress. Tabulated strength values can be means, ASTM minimums, or of unknown heritage. In surface fatigue there are no endurance limits. Endurance strengths have to be qualified as to corresponding cycle count, and the slope of the *S-N* curve needs to be known. In bending fatigue there is a definite change in slope of the *S-N* curve near 10^6 cycles, but some evidence indicates that an endurance limit does not exist. Gearing experience leads to cycle counts of 10^{11} or more. Evidence of diminishing endurance strengths in bending have been included in AGMA methodology.

14–3   AGMA Stress Equations

Two fundamental stress equations are used in the AGMA methodology, one for bending stress and another for pitting resistance (contact stress). In AGMA terminology, these are called *stress numbers* and are designated by a lowercase letter *s* instead of the Greek lower case σ we have used in this book (and shall continue to use). The fundamental equations are

$$
\sigma = \begin{cases}
W^t K_o K_v K_s \dfrac{P_d}{F} \dfrac{K_m K_B}{J} & \text{(U.S. customary units)} \\[2ex]
W^t K_o K_v K_s \dfrac{1}{b m_t} \dfrac{K_H K_B}{Y_J} & \text{(SI units)}
\end{cases}
\tag{14–15}
$$

where for U.S. customary units (SI units),

W^t is the tangential transmitted load, lbf (N)
K_o is the overload factor
K_v is the dynamic factor
K_s is the size factor
P_d is the transverse diameteral pitch
F (b) is the face width of the narrower member, in (mm)
K_m (K_H) is the load-distribution factor
K_B is the rim-thickness factor
J (Y_J) is the geometry factor for bending strength (which includes root fillet stress-concentration factor K_f)
(m_t) is the transverse metric module

Before you try to digest the meaning of all these terms in Eq. (14–15), view it as advice from AGMA concerning items the designer should consider *whether he or she follows the voluntary standard or not*. These items include issues such as

- Transmitted load magnitude
- Overload
- Dynamic augmentation of transmitted load
- Size
- Geometry: pitch and face width
- Distribution of load across the teeth
- Rim support of the tooth
- Lewis form factor and root fillet stress concentration

The fundamental equation for pitting resistance (contact stress) is

$$
\sigma_c = \begin{cases}
C_p \sqrt{W^t K_o K_v K_s \dfrac{K_m}{d_P F} \dfrac{C_f}{I}} & \text{(U.S. customary units)} \\[2ex]
Z_E \sqrt{W^t K_o K_v K_s \dfrac{K_H}{d_{w1} b} \dfrac{Z_R}{Z_I}} & \text{(SI units)}
\end{cases}
\tag{14–16}
$$

where W^t, K_o, K_v, K_s, K_m, F, and b are the same terms as defined for Eq. (14–15). For U.S. customary units (SI units), the additional terms are

C_p (Z_E) is an elastic coefficient, $\sqrt{\text{lbf/in}^2}$ ($\sqrt{\text{N/mm}^2}$)
C_f (Z_R) is the surface condition factor
d_P (d_{w1}) is the pitch diameter of the *pinion,* in (mm)
I (Z_I) is the geometry factor for pitting resistance

The evaluation of all these factors is explained in the sections that follow. The development of Eq. (14–16) is clarified in the second part of Sec. 14–5.

14–4 AGMA Strength Equations

Instead of using the term *strength*, AGMA uses data termed *allowable stress numbers* and designates these by the symbol S_a. It will be less confusing here if we continue the practice in this book of using the uppercase letter S to designate strength and the lowercase Greek letters σ and τ for stress. To make it perfectly clear we shall use the term *AGMA strength* as a replacement for the phrase *allowable stress numbers* as used by AGMA.

Following this convention, values for *AGMA bending strength*, designated here as S_t, are to be found in Figs. 14–2, 14–3, and 14–4, and in Tables 14–3 and 14–4. Since AGMA strengths are not identified with other strengths such as S_{ut}, S_e, or S_y as used elsewhere in this book, their use should be restricted to gear problems.

In the AGMA approach the strengths are modified by various factors that produce limiting values of the bending stress and the contact stress. Using similar notation to that

Figure 14–2

Allowable bending stress number for through-hardened steels. The SI equations are $\sigma_{FP} = 0.533 H_B$ $+ 88.3$ MPa, grade 1, and $\sigma_{FP} = 0.703 H_B + 113$ MPa, grade 2. *(Source: ANSI/AGMA 2001-C95 and 2101-C95.)*

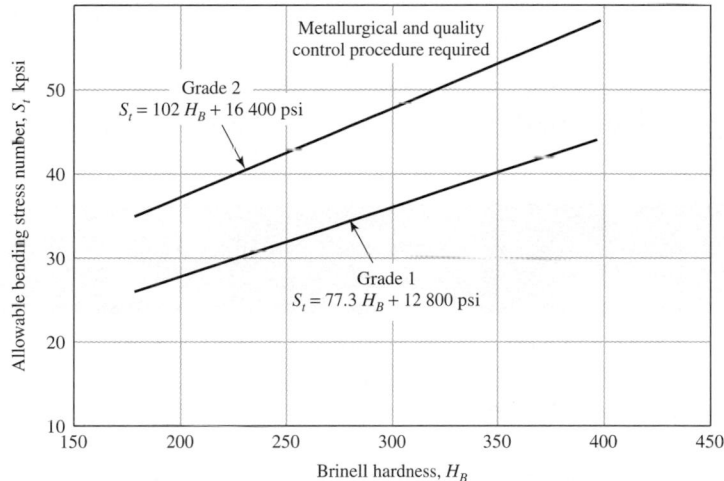

Figure 14–3

Allowable bending stress number for nitrided through-hardened steel gears (i.e., AISI 4140, 4340), S_t. The SI equations are $\sigma_{FP} = 0.568 H_B + 83.8$ MPa, grade 1, and $\sigma_{FP} = 0.749 H_B + 110$ MPa, grade 2. *(Source: ANSI/AGMA 2001-C95 and 2101-C95.)*

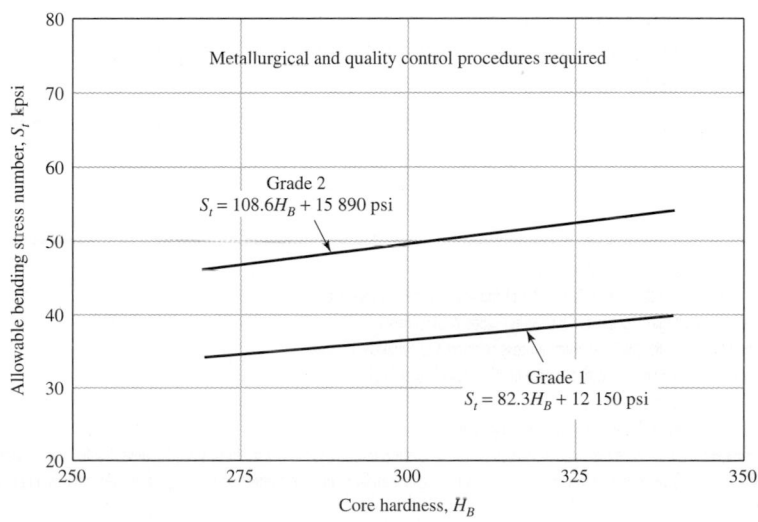

Figure 14–4

Allowable bending stress numbers for nitriding steel gears S_t. The SI equations are $\sigma_{FP} = 0.594H_B + 87.76$ MPa Nitralloy grade 1
$\sigma_{FP} = 0.784H_B + 114.81$ MPa Nitralloy grade 2
$\sigma_{FP} = 0.7255H_B + 63.89$ MPa 2.5% chrome, grade 1
$\sigma_{FP} = 0.7255H_B + 153.63$ MPa 2.5% chrome, grade 2
$\sigma_{FP} = 0.7255H_B + 201.91$ MPa 2.5% chrome, grade 3
(Source: ANSI/AGMA 2001-C95, 2101-C95.)

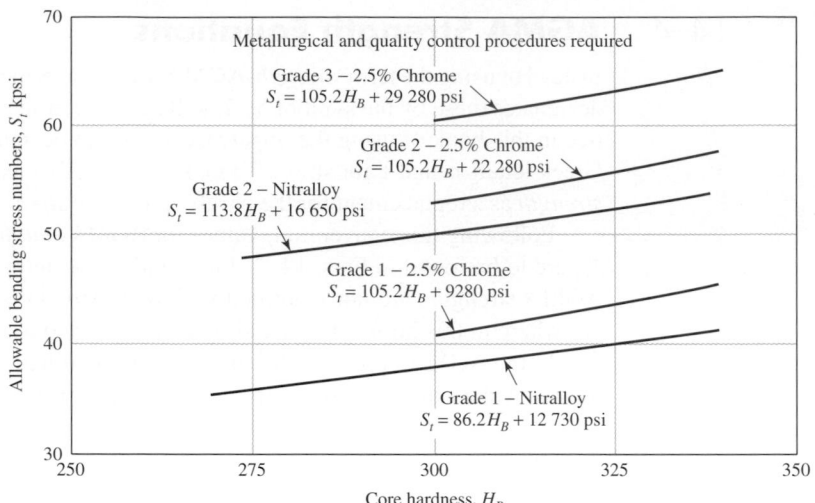

Table 14–3

Repeatedly Applied Bending Strength S_t at 10^7 Cycles and 0.99 Reliability for Steel Gears
Source: ANSI/AGMA 2001-C95.

Material Designation	Heat Treatment	Minimum Surface Hardness[1]	Allowable Bending Stress Number S_t,[2] psi		
			Grade 1	**Grade 2**	**Grade 3**
Steel[3]	Through-hardened	See Fig. 14–2	See Fig. 14–2	See Fig. 14–2	—
	Flame[4] or induction hardened[4] with type A pattern[5]	See Table 8*	45 000	55 000	—
	Flame[4] or induction hardened[4] with type B pattern[5]	See Table 8*	22 000	22 000	—
	Carburized and hardened	See Table 9*	55 000	65 000 or 70 000[6]	75 000
	Nitrided[4,7] (through-hardened steels)	83.5 HR15N	See Fig. 14–3	See Fig. 14–3	—
Nitralloy 135M, Nitralloy N, and 2.5% chrome (no aluminum)	Nitrided[4,7]	87.5 HR15N	See Fig. 14–4	See Fig. 14–4	See Fig. 14–4

Notes: See ANSI/AGMA 2001-C95 for references cited in notes 1–7.

[1] Hardness to be equivalent to that at the root diameter in the center of the tooth space and face width.

[2] See tables 7 through 10 for major metallurgical factors for each stress grade of steel gears.

[3] The steel selected must be compatible with the heat treatment process selected and hardness required.

[4] The allowable stress numbers indicated may be used with the case depths prescribed in 16.1.

[5] See figure 12 for type A and type B hardness patterns.

[6] If bainite and microcracks are limited to grade 3 levels, 70,000 psi may be used.

[7] The overload capacity of nitrided gears is low. Since the shape of the effective S-N curve is flat, the sensitivity to shock should be investigated before proceeding with the design. [7]

*Tables 8 and 9 of ANSI/AGMA 2001-C95 are comprehensive tabulations of the major metallurgical factors affecting S_t and S_c of flame-hardened and induction-hardened (Table 8) and carburized and hardened (Table 9) steel gears.

Table 14–4

Repeatedly Applied Bending Strength S_t for Iron and Bronze Gears at 10^7 Cycles and 0.99 Reliability
Source: ANSI/AGMA 2001-C95.

Material	Material Designation[1]	Heat Treatment	Typical Minimum Surface Hardness[2]	Allowable Bending Stress Number, S_t,[3] psi
ASTM A48 gray cast iron	Class 20	As cast	—	5000
	Class 30	As cast	174 HB	8500
	Class 40	As cast	201 HB	13 000
ASTM A536 ductile (nodular) Iron	Grade 60–40–18	Annealed	140 HB	22 000–33 000
	Grade 80–55–06	Quenched and tempered	179 HB	22 000–33 000
	Grade 100–70–03	Quenched and tempered	229 HB	27 000–40 000
	Grade 120–90–02	Quenched and tempered	269 HB	31 000–44 000
Bronze		Sand cast	Minimum tensile strength 40 000 psi	5700
	ASTM B–148 Alloy 954	Heat treated	Minimum tensile strength 90 000 psi	23 600

Notes:
[1]See ANSI/AGMA 2004-B89, *Gear Materials and Heat Treatment Manual.*
[2]Measured hardness to be equivalent to that which would be measured at the root diameter in the center of the tooth space and face width.
[3]The lower values should be used for general design purposes. The upper values may be used when:
   High quality material is used.
   Section size and design allow maximum response to heat treatment.
   Proper quality control is effected by adequate inspection.
   Operating experience justifies their use.

in Eq. (1–6) we shall term the resulting modifications the allowable bending stress σ_{all} and the allowable contact stress $\sigma_{c,\text{all}}$. The equation for the allowable bending stress is

$$\sigma_{\text{all}} = \begin{cases} \dfrac{S_t}{S_F}\dfrac{Y_N}{K_T K_R} & \text{(U.S. customary units)} \\[2ex] \dfrac{\sigma_{FP}}{S_F}\dfrac{Y_N}{Y_\theta Y_Z} & \text{(SI units)} \end{cases} \qquad (14\text{--}17)$$

where for U.S. customary units (SI units),

    S_t (σ_{FP}) are allowable bending stress, lbf/in^2 (N/mm^2)
    Y_N is the stress cycle factor for bending stress
    K_T (Y_θ) are the temperature factors
    K_R (Y_Z) are the reliability factors
    S_F is the AGMA factor of safety, a stress ratio

The equation for the allowable contact stress $\sigma_{c,\text{all}}$ is

$$\sigma_{c,\text{all}} = \begin{cases} \dfrac{S_c}{S_H} \dfrac{Z_N C_H}{K_T K_R} & \text{(U.S. customary units)} \\[2ex] \dfrac{\sigma_{HP}}{S_H} \dfrac{Z_N Z_W}{Y_\theta Y_Z} & \text{(SI units)} \end{cases} \qquad (14\text{--}18)$$

where the upper equation is in U.S. customary units and the lower equation is in SI units, Also,

S_c (σ_{HP}) are the allowable contact stress, lbf/in^2 (N/mm^2)
Z_N is the stress cycle life factor
C_H (Z_W) are the hardness ratio factors for pitting resistance
K_T (Y_θ) are the temperature factors
K_R (Y_Z) are the reliability factors
S_H is the AGMA factor of safety, a stress ratio

The values for the AGMA allowable contact stress, designated here as S_c, are to be found in Fig. 14–5 and Tables 14–5, 14–6, and 14–7.

AGMA allowable stress numbers (strengths) for bending and contact stress are for

- Unidirectional loading
- 10 million stress cycles
- 99 percent reliability

Figure 14–5

Contact-fatigue strength S_c at 10^7 cycles and 0.99 reliability for through-hardened steel gears. The SI equations are $\sigma_{HP} = 2.22H_B + 200$ MPa, grade 1, and $\sigma_{HP} = 2.41H_B + 237$ MPa, grade 2. *(Source: ANSI/AGMA 2001-C95 and 2101-C95.)*

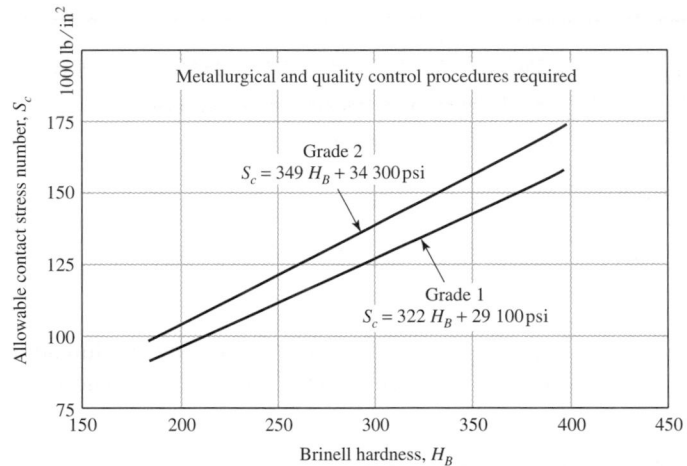

Table 14–5

Nominal Temperature Used in Nitriding and Hardnesses Obtained

Source: Darle W. Dudley, *Handbook of Practical Gear Design*, rev. ed., McGraw-Hill, New York, 1984.

Steel	Temperature before nitriding, °F	Nitriding, °F	Hardness, Rockwell C Scale Case	Core
Nitralloy 135*	1150	975	62–65	30–35
Nitralloy 135M	1150	975	62–65	32–36
Nitralloy N	1000	975	62–65	40–44
AISI 4340	1100	975	48–53	27–35
AISI 4140	1100	975	49–54	27–35
31 Cr Mo V 9	1100	975	58–62	27–33

*Nitralloy is a trademark of the Nitralloy Corp., New York.

Table 14–6

Repeatedly Applied Contact Strength S_c at 10^7 Cycles and 0.99 Reliability for Steel Gears
Source: ANSI/AGMA 2001-C95.

Material Designation	Heat Treatment	Minimum Surface Hardness[1]	Allowable Contact Stress Number,[2] S_c, psi		
			Grade 1	Grade 2	Grade 3
Steel[3]	Through hardened[4]	See Fig. 14–5	See Fig. 14–5	See Fig. 14–5	—
	Flame[5] or induction hardened[5]	50 HRC	170 000	190 000	—
		54 HRC	175 000	195 000	—
	Carburized and hardened[5]	See Table 9*	180 000	225 000	275 000
	Nitrided[5] (through hardened steels)	83.5 HR15N	150 000	163 000	175 000
		84.5 HR15N	155 000	168 000	180 000
2.5% chrome (no aluminum)	Nitrided[5]	87.5 HR15N	155 000	172 000	189 000
Nitralloy 135M	Nitrided[5]	90.0 HR15N	170 000	183 000	195 000
Nitralloy N	Nitrided[5]	90.0 HR15N	172 000	188 000	205 000
2.5% chrome (no aluminum)	Nitrided[5]	90.0 HR15N	176 000	196 000	216 000

Notes: See ANSI/AGMA 2001-C95 for references cited in notes 1–5.
[1] Hardness to be equivalent to that at the start of active profile in the center of the face width.
[2] See Tables 7 through 10 for major metallurgical factors for each stress grade of steel gears.
[3] The steel selected must be compatible with the heat treatment process selected and hardness required.
[4] These materials must be annealed or normalized as a minimum.
[5] The allowable stress numbers indicated may be used with the case depths prescribed in 16.1.
*Table 9 of ANSI/AGMA 2001 C95 is a comprehensive tabulation of the major metallurgical factors affecting S_t and S_c of carburized and hardened steel gears.

The factors in this section, too, will be evaluated in subsequent sections.

When two-way (reversed) loading occurs, as with idler gears, AGMA recommends using 70 percent of S_t values. This is equivalent to $1/0.70 = 1.43$ as a value of k_e in Ex. 14–2. The recommendation falls between the value of $k_e = 1.33$ for a Goodman failure locus and $k_e = 1.65$ for a Gerber failure locus.

14–5 Geometry Factors *I* and *J* (Z_I and Y_J)

We have seen how the factor Y is used in the Lewis equation to introduce the effect of tooth form into the stress equation. The AGMA factors[5] I and J are intended to accomplish the same purpose in a more involved manner.

The determination of I and J depends upon the *face-contact ratio* m_F. This is defined as

$$m_F = \frac{F}{p_x} \tag{14-19}$$

where p_x is the axial pitch and F is the face width. For spur gears, $m_F = 0$.

---

[5] A useful reference is AGMA 908-B89, *Geometry Factors for Determining Pitting Resistance and Bending Strength of Spur, Helical and Herringbone Gear Teeth.*

Table 14–7

Repeatedly Applied Contact Strength S_c 10^7 Cycles and 0.99 Reliability for Iron and Bronze Gears
Source: ANSI/AGMA 2001-C95.

Material	Material Designation[1]	Heat Treatment	Typical Minimum Surface Hardness[2]	Allowable Contact Stress Number,[3] S_{cr} psi
ASTM A48 gray cast iron	Class 20	As cast	—	50 000–60 000
	Class 30	As cast	174 HB	65 000–75 000
	Class 40	As cast	201 HB	75 000–85 000
ASTM A536 ductile (nodular) iron	Grade 60–40–18	Annealed	140 HB	77 000–92 000
	Grade 80–55–06	Quenched and tempered	179 HB	77 000–92 000
	Grade 100–70–03	Quenched and tempered	229 HB	92 000–112 000
	Grade 120–90–02	Quenched and tempered	269 HB	103 000–126 000
Bronze	—	Sand cast	Minimum tensile strength 40 000 psi	30 000
	ASTM B-148 Alloy 954	Heat treated	Minimum tensile strength 90 000 psi	65 000

Notes:

[1]See ANSI/AGMA 2004-B89, *Gear Materials and Heat Treatment Manual.*

[2]Hardness to be equivalent to that at the start of active profile in the center of the face width.

[3]The lower values should be used for general design purposes. The upper values may be used when:
   High-quality material is used.
   Section size and design allow maximum response to heat treatment.
   Proper quality control is effected by adequate inspection.
   Operating experience justifies their use.

Low-contact-ratio (LCR) helical gears having a small helix angle or a thin face width, or both, have face-contact ratios less than unity ($m_F \leq 1$), and will not be considered here. Such gears have a noise level not too different from that for spur gears. Consequently we shall consider here only spur gears with $m_F = 0$ and conventional helical gears with $m_F > 1$.

Bending-Strength Geometry Factor J (Y_J)

The AGMA factor J employs a modified value of the Lewis form factor, also denoted by Y; a *fatigue stress-concentration factor K_f*; and a tooth *load-sharing ratio m_N*. The resulting equation for J is

$$J = \frac{Y}{K_f m_N} \tag{14-20}$$

It is important to note that the form factor Y in Eq. (14–20) is *not* the Lewis factor at all. The value of Y here is obtained from a generated layout of the tooth profile in the normal plane and is based on the highest point of single-tooth contact.

The factor K_f in Eq. (14–20) is called a *stress correction factor* by AGMA. It is based on a formula deduced from a photoelastic investigation of stress concentration in gear teeth over 50 years ago.

The load-sharing ratio m_N is equal to the face width divided by the minimum total length of the lines of contact. This factor depends on the transverse contact ratio m_p, the face-contact ratio m_F, the effects of any profile modifications, and the tooth deflection. For spur gears, $m_N = 1.0$. For helical gears having a face-contact ratio $m_F > 2.0$, a conservative approximation is given by the equation

$$m_N = \frac{p_N}{0.95Z} \tag{14–21}$$

where p_N is the normal base pitch and Z is the length of the line of action in the transverse plane (distance L_{ab} in Fig. 13–15).

Use Fig. 14–6 to obtain the geometry factor J for spur gears having a 20° pressure angle and full-depth teeth. Use Figs. 14–7 and 14–8 for helical gears having a 20° normal pressure angle and face-contact ratios of $m_F = 2$ or greater. For other gears, consult the AGMA standard.

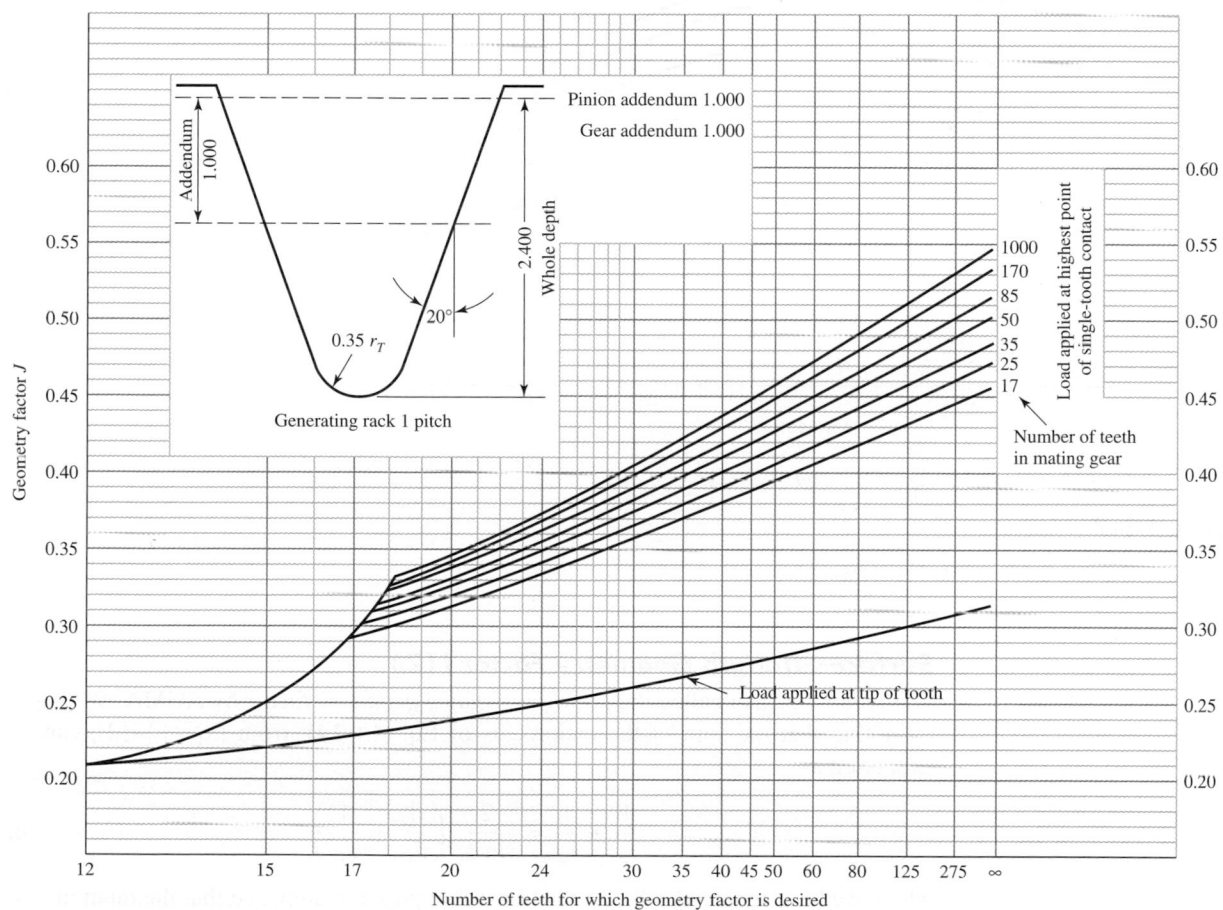

Figure 14–6

Spur-gear geometry factors J. *(Source: ANSI/AGMA 218.01.)*

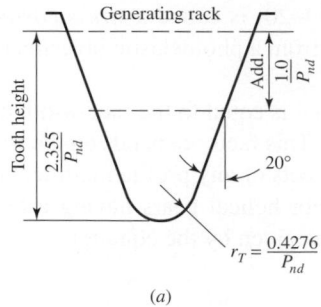

(a)

$$m_N = \frac{p_N}{0.95Z}$$

Value for Z is for an element of indicated numbers of teeth and a 75-tooth mate

Normal tooth thickness of pinion and gear tooth each reduced 0.024 in to provide 0.048 in total backlash for one normal diametral pitch

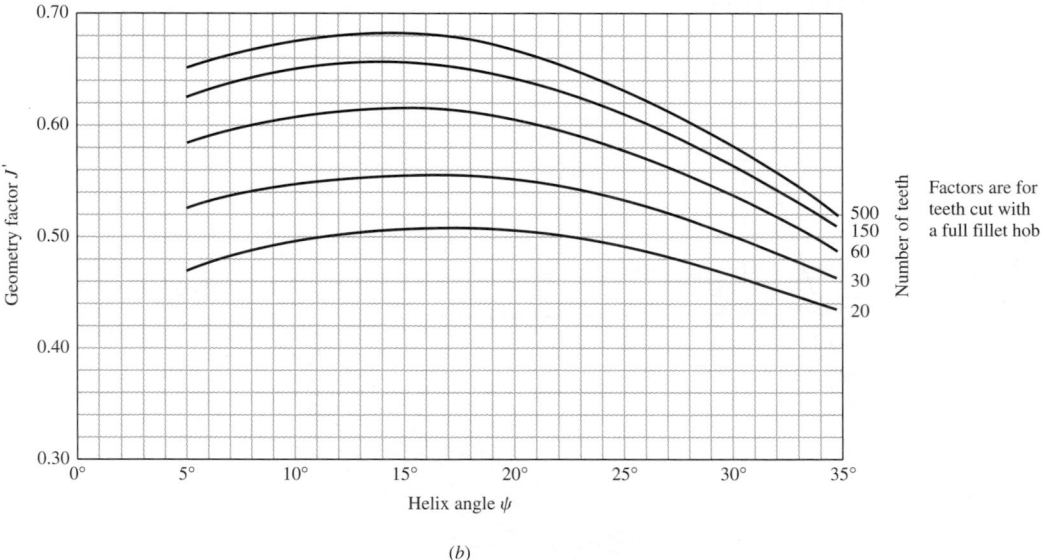

(b)

Figure 14–7

Helical-gear geometry factors J'. *(ANSI/AGMA 218.01.)*

Surface-Strength Geometry Factor I (Z_I)

The factor I is also called the *pitting-resistance geometry factor* by AGMA. We begin by noting that the sum of the reciprocals of Eq. (14–14), from Eq. (14–12), can be expressed as

$$\frac{1}{r_1} + \frac{1}{r_2} = \frac{2}{\sin \phi_t} \left(\frac{1}{d_P} + \frac{1}{d_G} \right) \qquad (a)$$

where we have replaced ϕ by ϕ_t, the transverse pressure angle, so that the relation will apply to helical gears too. Now define *speed ratio* m_G as

$$m_G = \frac{N_G}{N_P} = \frac{d_G}{d_P} \qquad (14\text{–}22)$$

Figure 14–8

J'-factor multipliers for use with Fig. 14–7 to find J.
(ANSI/AGMA 218.01.)

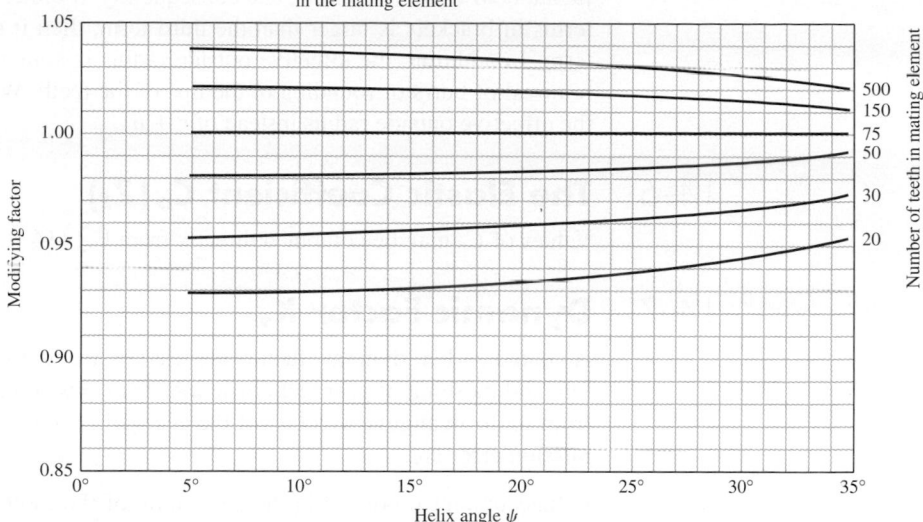

The modifying factor can be applied to the J factor when other than 75 teeth are used in the mating element

Equation (a) can now be written

$$\frac{1}{r_1} + \frac{1}{r_2} = \frac{2}{d_P \sin \phi_t} \frac{m_G + 1}{m_G} \tag{b}$$

Now substitute Eq. (b) for the sum of the reciprocals in Eq. (14–14). The result is found to be

$$\sigma_c = -\sigma_C = C_p \left[\frac{K_V W^t}{d_P F} \frac{1}{\dfrac{\cos \phi_t \sin \phi_t}{2} \dfrac{m_G}{m_G + 1}} \right]^{1/2} \tag{c}$$

The geometry factor I for external spur and helical gears is the denominator of the second term in the brackets in Eq. (c). By adding the load-sharing ratio m_N, we obtain a factor valid for both spur and helical gears. The equation is then written as

$$I = \begin{cases} \dfrac{\cos \phi_t \sin \phi_t}{2m_N} \dfrac{m_G}{m_G + 1} & \text{external gears} \\[3mm] \dfrac{\cos \phi_t \sin \phi_t}{2m_N} \dfrac{m_G}{m_G - 1} & \text{internal gears} \end{cases} \tag{14–23}$$

where $m_N = 1$ for spur gears. In solving Eq. (14–21) for m_N, note that

$$p_N = p_n \cos \phi_n \tag{14–24}$$

where p_n is the normal circular pitch. If a layout of the gears is not made, the quantity Z, for use in Eq. (14–21), can be obtained from the equation

$$Z = \left[(r_P + a)^2 - r_{bP}^2 \right]^{1/2} + \left[(r_G + a)^2 - r_{bG}^2 \right]^{1/2} - (r_P + r_G) \sin \phi_t \tag{14–25}$$

where r_P and r_G are the pitch radii and r_{bP} and r_{bG} the base-circle radii of the pinion and gear, respectively.[6] Recall from Eq. (13–6), the radius of the base circle is

$$r_b = r \cos \phi_t \tag{14–26}$$

---

[6]For a development, see Joseph E. Shigley and John J. Uicker Jr., *Theory of Machines and Mechanisms,* McGraw-Hill, New York, 1980, p. 262.

Certain precautions must be taken in using Eq. (14–25). The tooth profiles are not conjugate below the base circle, and consequently, if either one or the other of the first two terms in brackets is larger than the third term, then it should be replaced by the third term. In addition, the effective outside radius is sometimes less than $r + a$, owing to removal of burrs or rounding of the tips of the teeth. When this is the case, always use the effective outside radius instead of $r + a$.

14–6 The Elastic Coefficient $C_p (Z_E)$

Values of C_p may be computed directly from Eq. (14–13) or obtained from Table 14–8.

14–7 Dynamic Factor K_v

As noted earlier, dynamic factors are used to account for inaccuracies in the manufacture and meshing of gear teeth in action. *Transmission error* is defined as the departure from uniform angular velocity of the gear pair. Some of the effects that produce transmission error are:

- Inaccuracies produced in the generation of the tooth profile; these include errors in tooth spacing, profile lead, and runout
- Vibration of the tooth during meshing due to the tooth stiffness
- Magnitude of the pitch-line velocity
- Dynamic unbalance of the rotating members
- Wear and permanent deformation of contacting portions of the teeth
- Gearshaft misalignment and the linear and angular deflection of the shaft
- Tooth friction

In an attempt to gain some control over these effects, AGMA has defined a set of *quality-control numbers*.[7] These numbers define the tolerances for gears of various sizes manufactured to a specified quality class. Classes 3 to 7 will include most commercial-quality gears. Classes 8 to 12 are of precision quality. The AGMA *transmission accuracy-level number* Q_v can be taken as the same as the quality number. The following equations for the dynamic factor are based on these Q_v numbers:

$$K_v = \begin{cases} \left(\dfrac{A + \sqrt{V}}{A} \right)^B & V \text{ in ft/min} \\[3ex] \left(\dfrac{A + \sqrt{200V}}{A} \right)^B & V \text{ in m/s} \end{cases} \quad (14\text{–}27)$$

where

$$A = 50 + 56(1 - B)$$
$$B = 0.25(12 - Q_v)^{2/3} \quad (14\text{–}28)$$

and the maximum velocity, representing the end point of the Q_v curve, is given by

$$(V_t)_{\max} = \begin{cases} [A + (Q_v - 3)]^2 & \text{ft/min} \\[2ex] \dfrac{[A + (Q_v - 3)]^2}{200} & \text{m/s} \end{cases} \quad (14\text{–}29)$$

[7]AGMA 390.01.

Table 14-8

Elastic Coefficient C_p (Z_E), $\sqrt{psi}$ ($\sqrt{MPa}$)   Source: AGMA 218.01

Pinion Material	Pinion Modulus of Elasticity E_p psi (MPa)*	Gear Material and Modulus of Elasticity E_G, lbf/in² (MPa)*					
		Steel 30×10^6 (2×10^5)	Malleable Iron 25×10^6 (1.7×10^5)	Nodular Iron 24×10^6 (1.7×10^5)	Cast Iron 22×10^6 (1.5×10^5)	Aluminum Bronze 17.5×10^6 (1.2×10^5)	Tin Bronze 16×10^6 (1.1×10^5)
Steel	30×10^6 (2×10^5)	2300 (191)	2180 (181)	2160 (179)	2100 (174)	1950 (162)	1900 (158)
Malleable iron	25×10^6 (1.7×10^5)	2180 (181)	2090 (174)	2070 (172)	2020 (168)	1900 (158)	1850 (154)
Nodular iron	24×10^6 (1.7×10^5)	2160 (179)	2070 (172)	2050 (170)	2000 (166)	1880 (156)	1830 (152)
Cast iron	22×10^6 (1.5×10^5)	2100 (174)	2020 (168)	2000 (166)	1960 (163)	1850 (154)	1800 (149)
Aluminum bronze	17.5×10^6 (1.2×10^5)	1950 (162)	1900 (158)	1880 (156)	1850 (154)	1750 (145)	1700 (141)
Tin bronze	16×10^6 (1.1×10^5)	1900 (158)	1850 (154)	1830 (152)	1800 (149)	1700 (141)	1650 (137)

Poisson's ratio = 0.30.

*When more exact values for modulus of elasticity are obtained from roller contact tests, they may be used.

Figure 14–9

Dynamic factor K_v. The equations to these curves are given by Eq. (14–27) and the end points by Eq. (14–29). *(ANSI/AGMA 2001-C95.)*

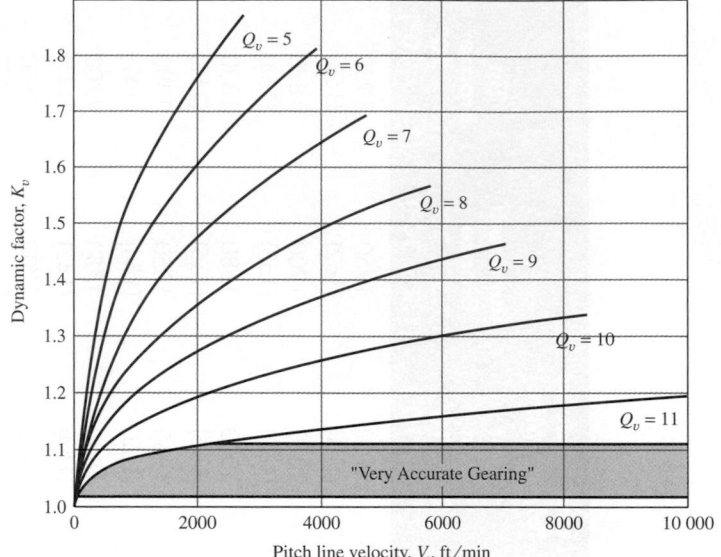

Figure 14–9 is a graph of K_v, the dynamic factor, as a function of pitch-line speed for graphical estimates of K_v.

14–8 Overload Factor K_o

The overload factor K_o is intended to make allowance for all externally applied loads in excess of the nominal tangential load W^t in a particular application. Examples include variations in torque from the mean value due to firing of cylinders in an internal combustion engine or reaction to torque variations in a piston pump drive. Others call a similar factor an application factor or a service factor. These are established after considerable field experience in a particular application.[8]

14–9 Surface Condition Factor C_f (Z_R)

The surface condition factor C_f or Z_R is used only in the pitting resistance equation, Eq. (14–16). It depends on

- Surface finish as affected by, but not limited to, cutting, shaving, lapping, grinding, shotpeening
- Residual stress
- Plastic effects (work hardening)

Standard surface conditions for gear teeth have not yet been established. When a detrimental surface finish effect is known to exist, AGMA suggests a value of C_f greater than unity.

---

[8]An extensive list of service factors appears in Howard B. Schwerdlin, "Couplings," Chap. 29 in Joseph E. Shigley and Charles R. Mischke (eds.), *Standard Handbook of Machine Design,* 2nd ed., McGraw-Hill, New York, 1996.

14–10   **Size Factor K_s**

The size factor reflects nonuniformity of material properties due to size. It depends upon

- Tooth size
- Diameter of part
- Ratio of tooth size to diameter of part
- Face width
- Area of stress pattern
- Ratio of case depth to tooth size
- Hardenability and heat treatment

Standard size factors for gear teeth have not yet been established for cases where there is a detrimental size effect. In such cases AGMA recommends a size factor greater than unity. If there is no detrimental size effect, use unity.

AGMA has identified and provided a location symbol for size factor. Also, AGMA suggests $K_s = 1$, which makes K_s a placeholder in Eqs. (14–15) and (14–16) until more information is gathered. Following the standard in this manner is a failure to apply all of your knowledge. From Table 13–1, $l = a + b = 2.25/P$. The tooth thickness t in Fig. 14–6 is given in Sec. 14–1, Eq. (b), as $t = \sqrt{4lx}$ where $x = 3Y/(2P)$ from Eq. (14–3). From Eq. (7–24) the equivalent diameter d_e of a rectangular section in bending is $d_e = 0.808\sqrt{Ft}$. From Eq. (7–19) $k_b = (d_e/0.3)^{-0.107}$. Noting that AGMA K_s is the reciprocal of k_b, we find the result of all the algebraic substitution is

$$K_s = \frac{1}{k_b} = 1.192 \left(\frac{F\sqrt{Y}}{P} \right)^{0.0535} \qquad (a)$$

AGMA K_s can be viewed as Lewis's geometry incorporated into the Marin size factor in fatigue. You may set AGMA $K_s = 1$, or you may elect to use the preceding Eq. (a). This is a point to discuss with your instructor. We will use Eq. (a) to remind you that you have a choice. If K_s in Eq. (a) is less than 1, use $K_s = 1$.

14–11   **Load-Distribution Factor K_m (K_H)**

The load-distribution factor modified the stress equations to reflect nonuniform distribution of load across the line of contact. The ideal is to locate the gear "midspan" between two bearings at the zero slope place when the load is applied. However, this is not always possible. The following procedure is applicable to

- Net face width to pinion pitch diameter ratio $F/d \leq 2$
- Gear elements mounted between the bearings
- Face widths up to 40 in
- Contact, when loaded, across the full width of the narrowest member

The load-distribution factor under these conditions is given by

$$K_m = 1 + C_{mc}(C_{pf}C_{pm} + C_{ma}C_e) \qquad (14–30)$$

where

$$C_{mc} = \begin{cases} 1 & \text{for uncrowned teeth} \\ 0.8 & \text{for crowned teeth} \end{cases} \tag{14-31}$$

$$C_{pf} = \begin{cases} \dfrac{F}{10d} - 0.025 & F \le 1 \text{ in} \\[2mm] \dfrac{F}{10d} - 0.0375 + 0.0125F & 1 < F \le 17 \text{ in} \\[2mm] \dfrac{F}{10d} - 0.1109 + 0.0207F - 0.000\,228F^2 & 17 < F \le 40 \text{ in} \end{cases} \tag{14-32}$$

Note that for values of $F/(10d) < 0.05$, $F/(10d) = 0.05$ is used.

$$C_{pm} = \begin{cases} 1 & \text{for straddle-mounted pinion with } S_1/S < 0.175 \\ 1.1 & \text{for straddle-mounted pinion with } S_1/S \ge 0.175 \end{cases} \tag{14-33}$$

$$C_{ma} = A + BF + CF^2 \qquad \text{(see Table 14-9 for values of } A, B, \text{ and } C) \tag{14-34}$$

$$C_e = \begin{cases} 0.8 & \text{for gearing adjusted at assembly, or compatibility} \\ & \text{is improved by lapping, or both} \\ 1 & \text{for all other conditions} \end{cases} \tag{14-35}$$

See Fig. 14-10 for definitions of S and S_1 for use with Eq. (14-33), and see Fig. 14-11 for graph of C_{ma}.

Table 14-9

Empirical Constants A, B, and C for Eq. (14-34), Face Width F in Inches*

Source: ANSI/AGMA 2001-C95.

Condition	A	B	C
Open gearing	0.247	0.0167	$-0.765(10^{-4})$
Commercial, enclosed units	0.127	0.0158	$-0.930(10^{-4})$
Precision, enclosed units	0.0675	0.0128	$-0.926(10^{-4})$
Extraprecision enclosed gear units	0.00360	0.0102	$-0.822(10^{-4})$

*See ANSI/AGMA 2101-C95, pp. 21–22, for SI formulation.

Figure 14-10

Definition of distances S and S_1 used in evaluating C_{pm}, Eq. (14-33). *(ANSI/AGMA 2001-C95.)*

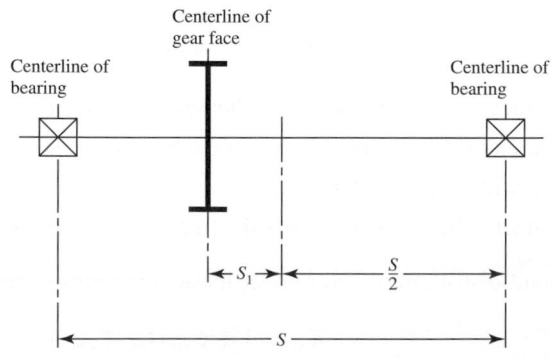

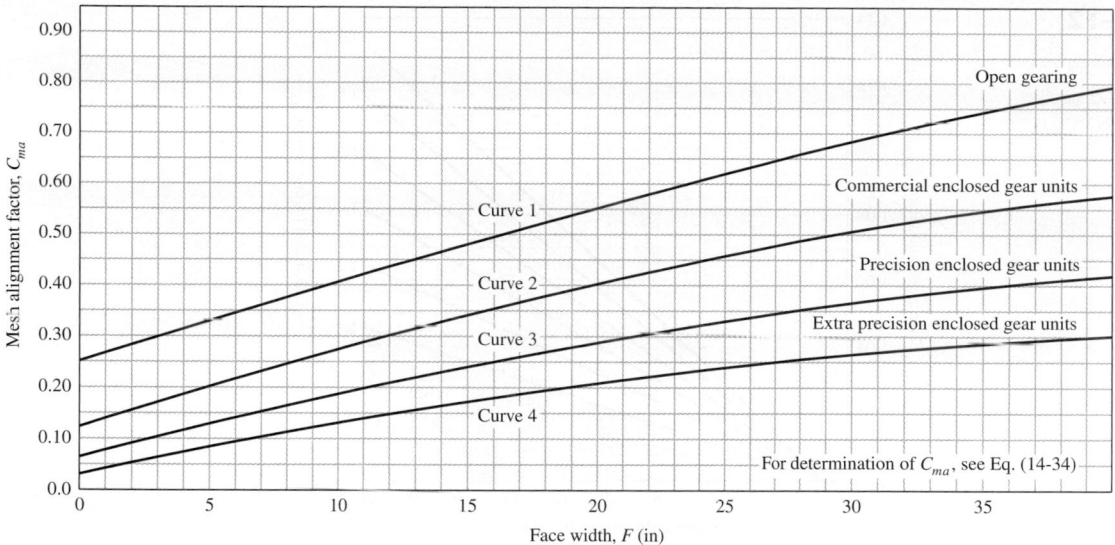

Figure 14–11

Mesh alignment factor C_{ma}. Curve-fit equations in Table 14–9. *(ANSI/AGMA 2001-C95.)*

14–12 Hardness-Ratio Factor C_H

The pinion generally has a smaller number of teeth than the gear and consequently is subjected to more cycles of contact stress. If both the pinion and the gear are through-hardened, then a uniform surface strength can be obtained by making the pinion harder than the gear. A similar effect can be obtained when a surface-hardened pinion is mated with a through-hardened gear. The hardness-ratio factor C_H is used *only for the gear*. Its purpose is to adjust the surface strengths for this effect. The values of C_H are obtained from the equation

$$C_H = 1.0 + A'(m_G - 1.0) \tag{14–36}$$

where

$$A' = 8.98(10^{-3}) \left(\frac{H_{BP}}{H_{BG}} \right) - 8.29(10^{-3}) \quad 1.2 \leq \frac{H_{BP}}{H_{BG}} \leq 1.7$$

The terms H_{BP} and H_{BG} are the Brinell hardness (10-mm ball at 3000-kg load) of the pinion and gear, respectively. The term m_G is the speed ratio and is given by Eq. (14–22). See Fig. 14–12 for a graph of Eq. (14–36). For

$$\frac{H_{BP}}{H_{BG}} < 1.2, \quad A' = 0$$

$$\frac{H_{BP}}{H_{BG}} > 1.7, \quad A' = 0.006\,98$$

When surface-hardened pinions with hardnesses of 48 Rockwell C scale (Rockwell C48) or harder are run with through-hardened gears (180–400 Brinell), a work hardening occurs. The C_H factor is a function of pinion surface finish f_P and the mating gear hardness. Figure 14–13 displays the relationships:

$$C_H = 1 + B'(450 - H_{BG}) \tag{14–37}$$

Figure 14–12

Hardness ratio factor C_H (through-hardened steel). *(ANSI/AGMA 2001-C95.)*

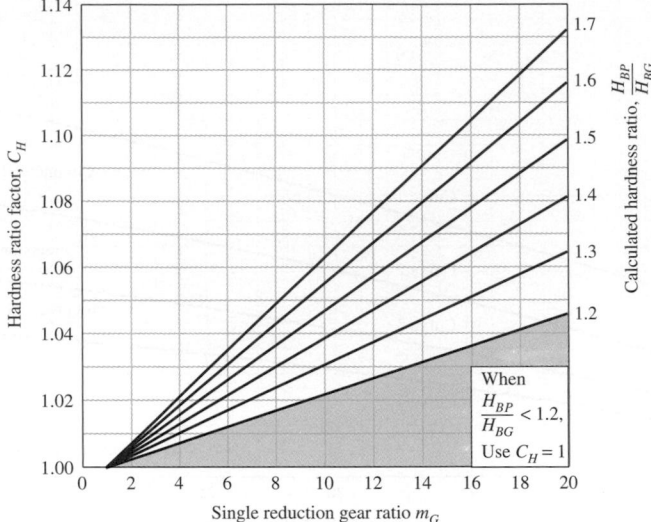

Figure 14–13

Hardness ratio factor C_H (surface-hardened steel pinion). *(ANSI/AGMA 2001-C95.)*

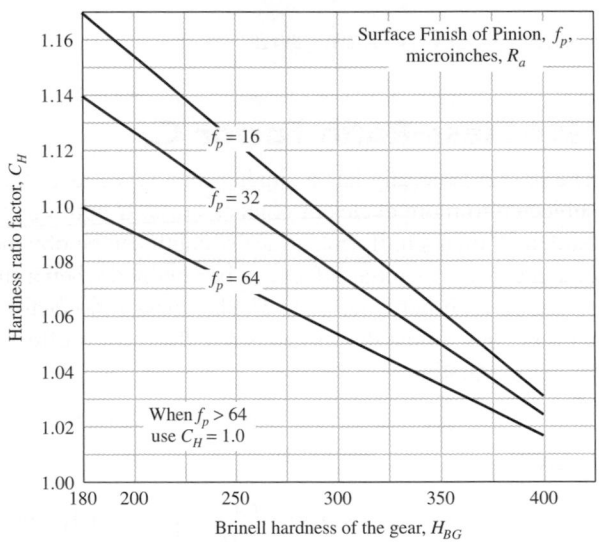

where $B' = 0.000\ 75\ \exp[-0.0112 f_P]$ and f_P is the surface finish of the pinion expressed as root-mean-square roughness R_a in μ in.

14–13 Stress Cycle Life Factors Y_N and Z_N

The AGMA strengths as given in Figs. 14–2 through 14–4, in Tables 14–3 and 14–4 for bending fatigue, and in Fig. 14–5 and Tables 14–5 and 14–6 for contact-stress fatigue are based on 10^7 load cycles repeatedly applied. The purpose of the load cycle factors Y_N and Z_N is to modify the AGMA strength for lives other than 10^7 cycles. Values for these factors are given in Figs. 14–14 and 14–15. Note that for 10^7 cycles $Y_N = Z_N = 1$ on each graph. Note also that the equations for Y_N and Z_N change on either side of 10^7 cycles. For life goals slightly higher than 10^7 cycles, the mating gear may be experiencing fewer than 10^7 cycles and the equations for $(Y_N)_P$ and $(Y_N)_G$ can be different. The same comment applies to $(Z_N)_P$ and $(Z_N)_G$.

Figure 14–14

Repeatedly applied bending strength stress-cycle factor Y_N. *(ANSI/AGMA 2001-C95.)*

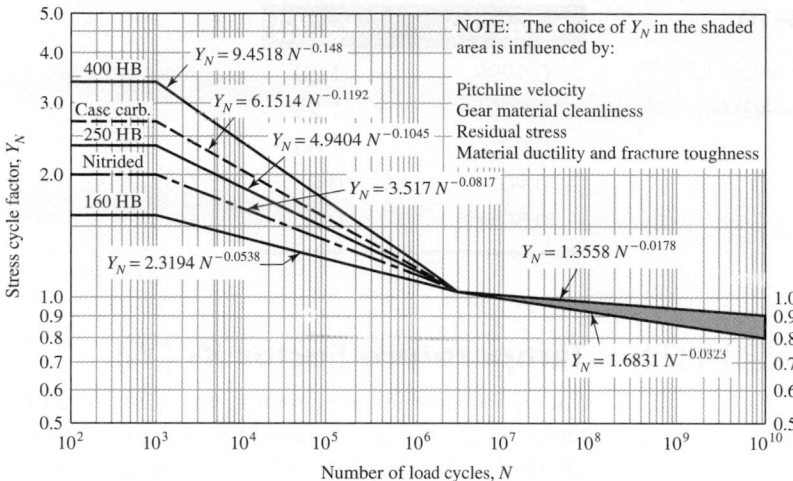

Figure 14–15

Pitting resistance stress cycle factor Z_N. *(ANSI/AGMA 2001-C95.)*

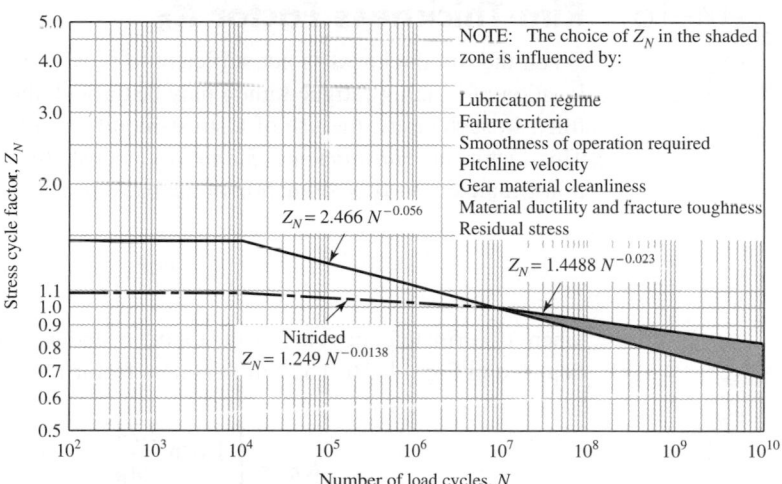

14–14   Reliability Factor K_R (Y_Z)

The reliability factor accounts for the effect of the statistical distributions of material fatigue failures. Load variation is not addressed here. The AGMA strengths S_t and S_c are based on a reliability of 99 percent. Table 14–10 is based on data developed by the U.S. Navy for bending and contact-stress fatigue failures.

The functional relationship between K_R and reliability is highly nonlinear. When interpolation is required, linear interpolation is too crude. A log transformation to each quantity produces a linear string. A least-squares regression fit is

$$K_R = \begin{cases} 0.658 - 0.0759 \ln(1 - R) & 0.5 < R < 0.99 \\ 0.50 - 0.109 \ln(1 - R) & 0.99 \leq R \leq 0.9999 \end{cases} \tag{14–38}$$

For cardinal values of R, take K_R from the table. Otherwise use the logarithmic interpolation afforded by Eqs. (14–38).

Table 14–10

Reliability Factors KR (YZ)

Source: ANSI/AGMA
2001-C95.

Reliability	K_R (Y_Z)
0.9999	1.50
0.999	1.25
0.99	1.00
0.90	0.85
0.50	0.70

14–15 Temperature Factor K_T (Y_θ)

For oil or gear-blank temperatures up to 250°F (120°C), use $K_T = Y_\theta = 1.0$. For higher temperatures, the factor should be greater than unity. Heat exchangers may be used to ensure that operating temperatures are considerably below this value, as is desirable for the lubricant.

14–16 Rim-Thickness Factor K_B

When the rim thickness is not sufficient to provide full support for the tooth root, the location of bending fatigue failure may be through the gear rim rather than at the tooth fillet. In such cases, the use of a stress-modifying factor K_B or (t_R) is recommended. This factor, the *rim-thickness factor K_B*, adjusts the estimated bending stress for the thin-rimmed gear. It is a function of the backup ratio m_B,

$$m_B = \frac{t_R}{h_t} \tag{14–39}$$

where t_R = rim thickness below the tooth, in, and h_t = the tooth height. The geometry is depicted in Fig. 14–16. The rim-thickness factor K_B is given by

$$K_B = \begin{cases} 1.6\ln\dfrac{2.242}{m_B} & m_B < 1.2 \\ 1 & m_B \geq 1.2 \end{cases} \tag{14–40}$$

Figure 14–16

Rim thickness factor K_B.
(ANSI/AGMA 2001-C95.)

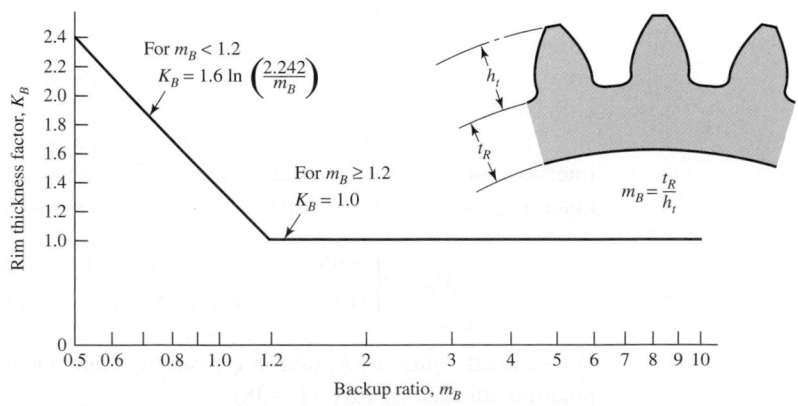

Figure 14–16 also gives the value of K_B graphically. The rim-thickness factor K_B is applied in addition to the 0.70 reverse-loading factor when applicable.

14–17 Safety Factors S_F and S_H

The ANSI/AGMA standards 2001-C95 and 2101-C95 have reintroduced safety factor S_F guarding against bending fatigue failure and safety factor S_H guarding against pitting failure.

The definition of S_F, from Eq. (14–17), is

$$S_F = \frac{S_t Y_N/(K_T K_R)}{\sigma} = \frac{\text{fully corrected bending strength}}{\text{bending stress}} \tag{14–41}$$

where σ is estimated from Eq. (14–15). It is a strength-over-stress definition in a case where the stress is linear with the transmitted load.

The definition of S_H, from Eq. (14–18), is

$$S_H = \frac{S_c Z_N C_H/(K_T K_R)}{\sigma_c} = \frac{\text{fully corrected contact strength}}{\text{contact stress}} \tag{14–42}$$

when σ_c is estimated from Eq. (14–16). This, too, is a strength-over-stress definition but in a case where the stress is *not* linear with the transmitted load W^t.

While the definition of S_H does not interfere with its intended function, a caution is required when comparing S_F with S_H in an analysis in order to ascertain the nature and severity of the threat to loss of function. To render S_H linear with the transmitted load, W^t it could have been defined as

$$S_H = \left(\frac{\text{fully corrected contact strength}}{\text{contact stress imposed}}\right)^2 \tag{14–43}$$

with the exponent 2 for linear or helical contact, or an exponent of 3 for crowned teeth (spherical contact). With the AGMA definition, Eq. (14–42), compare S_F with S_H^2 (or S_H^3 for crowned teeth) when trying to identify the threat to loss of function with confidence.

The role of the overload factor K_o is to include predictable excursions of load beyond W^t based on experience. A safety factor is intended to account for unquantifiable elements in addition to K_o. When designing a gear mesh, the quantity S_F becomes a design factor $(S_F)_d$ within the meanings used in this book. The quantity S_F evaluated as part of a design assessment is a factor of safety. This applies equally well to the quantity S_H.

14–18 Analysis

Description of the AGMA procedure is highly detailed. The best review is a "road map" for bending fatigue and contact-stress fatigue. Figure 14–17 identifies the AGMA bending stress equation, the endurance strength in bending equation, and the factor of safety S_F. Figure 14–18 displays the contact-stress equation, the contact fatigue endurance strength equation, and the factor of safety S_H. When analyzing a gear problem, this figure is a useful reference.

The following example of a gear mesh analysis is intended to make all the details presented concerning the AGMA method more familiar.

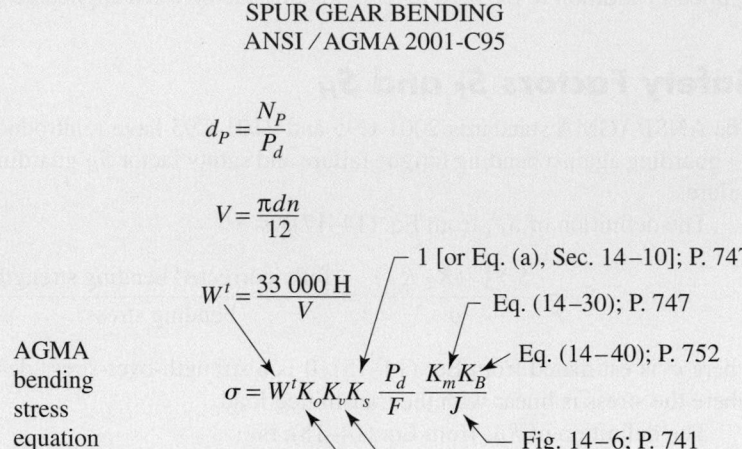

SPUR GEAR BENDING
ANSI / AGMA 2001-C95

$$d_P = \frac{N_P}{P_d}$$

$$V = \frac{\pi d n}{12}$$

$$W^t = \frac{33\,000\,H}{V}$$

AGMA bending stress equation Eq. (14–15)

$$\sigma = W^t K_o K_v K_s \frac{P_d}{F} \frac{K_m K_B}{J}$$

- 1 [or Eq. (a), Sec. 14–10]; P. 747
- Eq. (14–30); P. 747
- Eq. (14–40); P. 752
- Fig. 14–6; P. 741
- Eq. (14–27); P. 744
- Table below

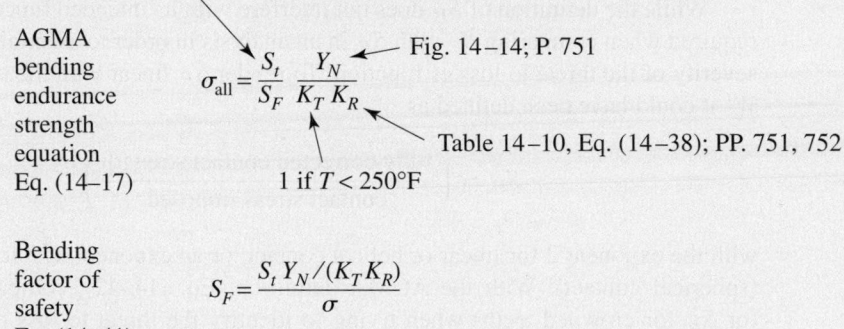

AGMA bending endurance strength equation Eq. (14–17)

$$\sigma_{\text{all}} = \frac{S_t}{S_F} \frac{Y_N}{K_T K_R}$$

- $0.99(S_t)_{10^7}$ Tables 14–3, 14–4; PP. 736, 737
- Fig. 14–14; P. 751
- Table 14–10, Eq. (14–38); PP. 751, 752
- 1 if $T < 250°F$

Bending factor of safety Eq. (14–41)

$$S_F = \frac{S_t Y_N / (K_T K_R)}{\sigma}$$

Remember to compare S_F with S_H^2 when deciding whether bending or wear is the threat to function. For crowned gears compare S_F with S_H^3.

Table of Overload Factors, K_o

	Driven Machine		
Power source	Uniform	Moderate shock	Heavy shock
Uniform	1.00	1.25	1.75
Light shock	1.25	1.50	2.00
Medium shock	1.50	1.75	2.25

Figure 14–17

Roadmap of AGMA bending equations. *(ANSI/AGMA 2001-C95.)*

SPUR GEAR WEAR
ANSI / AGMA 2001-C95

$$d_P = \frac{N_P}{P_d}$$

$$V = \frac{\pi d n}{12}$$

$$W^t = \frac{33\,000\,H}{V}$$

AGMA
contact
stress
equation
Eq. (14–16)

$$\sigma_c = C_p \left(W^t K_o K_v K_s \frac{K_m}{d_P F} \frac{C_f}{I} \right)^{1/2}$$

1 [or Eq. (a), Sec. 14–10]; P. 747
Eq. (14–30); P. 747
1
Eq. (14–23); P. 743
Eq. (14–27); P. 744
Table below

Eq. (14–13), Table 14–8; PP. 732, 745

AGMA
contact
endurance
strength
Eq. (14–18)

$$\sigma_{c,\text{all}} = \frac{S_c Z_N C_H}{S_H K_T K_R}$$

$0.99 (S_c)_{10^7}$ Tables, 14–6, 14–7; PP. 739, 740
Fig. 14–15; P. 751
Section 14–12, gear only; PP. 749, 750
Table 14–10, Eqs. (14–38); PP. 751, 752
1 if $T < 250°$ F

Wear
factor of
safety
Eq. (14–42)

$$S_H = \frac{S_c Z_N C_H / (K_T K_R)}{\sigma_c}$$

Gear only

Remember to compare S_F with S_H^2 when deciding whether bending
or wear is the threat to function. For crowned gears compare S_F with S_H^3.

Table of Overload Factors, K_o

| | Driven Machine | | |
Power source	Uniform	Moderate shock	Heavy shock
Uniform	1.00	1.25	1.75
Light shock	1.25	1.50	2.00
Medium shock	1.50	1.75	2.25

Figure 14–18

Roadmap of AGMA wear equations. *(ANSI/AGMA 2001-C95.)*

EXAMPLE 14–4

A 17-tooth 20° pressure angle spur pinion rotates at 1800 rev/min and transmits 4 hp to a 52-tooth disk gear. The diametral pitch is 10 teeth/in, the face width 1.5 in, and the quality standard is No. 6. The gears are straddle-mounted with bearings immediately adjacent. The pinion is a grade 1 steel with a hardness of 240 Brinell tooth surface and through-hardened core. The gear is steel, through-hardened also, grade 1 material, with a Brinell hardness of 200, tooth surface and core. Poisson's ratio is 0.30, $J_P = 0.30$, $J_G = 0.40$, and Young's modulus is $30(10^6)$ psi. The loading is smooth because of motor and load. Assume a pinion life of 10^8 cycles and a reliability of 0.90, and use $Y_N = 1.3558N^{-0.0178}$, $Z_N = 1.4488N^{-0.023}$. The tooth profile is uncrowned. This is a commercial enclosed gear unit.

(a) Find the factor of safety of the gears in bending.
(b) Find the factor of safety of the gears in wear.
(c) By examining the factors of safety, identify the threat to each gear and to the mesh.

Solution

There will be many terms to obtain so use Figs. 14–17 and 14–18 as guides to what is needed.

$$d_P = N_P/P_d = 17/10 = 1.7 \text{ in} \qquad d_G = 52/10 = 5.2 \text{ in}$$

$$V = \frac{\pi d_P n_P}{12} = \frac{\pi (1.7) 1800}{12} = 801.1 \text{ ft/min}$$

$$W^t = \frac{33\,000\,H}{V} = \frac{33\,000(4)}{801.1} = 164.8 \text{ lbf}$$

Assuming uniform loading, $K_o = 1$. To evaluate K_v, from Eq. (14–28) with a quality number $Q_v = 6$,

$$B = 0.25(12 - 6)^{2/3} = 0.8255$$

$$A = 50 + 56(1 - 0.8255) = 59.77$$

Then from Eq. (14–27) the dynamic factor is

$$K_v = \left(\frac{59.77 + \sqrt{801.1}}{59.77} \right)^{0.8255} = 1.377$$

To determine the size factor, K_s, the Lewis form factor is needed. From Table 14–2, with $N_P = 17$ teeth, $Y_P = 0.303$. Interpolation for the gear with $N_G = 52$ teeth yields $Y_G = 0.412$. Thus from Eq. (a) of Sec. 14–10, with $F = 1.5$ in,

$$(K_s)_P = 1.192 \left(\frac{1.5\sqrt{0.303}}{10} \right)^{0.0535} = 1.043$$

$$(K_s)_G = 1.192 \left(\frac{1.5\sqrt{0.412}}{10} \right)^{0.0535} = 1.052$$

The load distribution factor K_m is determined from Eq. (14–30), where five terms are needed. They are, where $F = 1.5$ in when needed:

Uncrowned, Eq. (14–30): $C_{mc} = 1$,
Eq. (14–32): $C_{pf} = 1.5/[10(1.7)] - 0.0375 + 0.0125(1.5) = 0.0695$
Bearings immediately adjacent, Eq. (14–33): $C_{pm} = 1$
Commercial enclosed gear units (Fig. 14–11): $C_{ma} = 0.15$
Eq. (14–35): $C_e = 1$

Thus,

$$K_m = 1 + C_{mc}(C_{pf}C_{pm} + C_{ma}C_e) = 1 + (1)[0.0695(1) + 0.15(1)] = 1.22$$

Assuming constant thickness gears, the rim-thickness factor $K_B = 1$. The speed ratio is $m_G = N_G/N_P = 52/17 = 3.059$. The load cycle factors given in the problem statement, with $N(\text{pinion}) = 10^8$ cycles and $N(\text{gear}) = 10^8/m_G = 10^8/3.059$ cycles, are

$$(Y_N)_P = 1.3558(10^8)^{-0.0178} = 0.977$$

$$(Y_N)_G = 1.3558(10^8/3.059)^{-0.0178} = 0.996$$

From Table 14.10, with a reliability of 0.9, $K_R = 0.85$. From Fig. 14–18, the temperature and surface condition factors are $K_T = 1$ and $C_f = 1$. From Eq. (14–23), with $m_N = 1$ for spur gears,

$$I = \frac{\cos 20° \sin 20°}{2} \frac{3.059}{3.059 + 1} = 0.121$$

From Table 14–8, $C_p = 2300\sqrt{\text{psi}}$.

Next, we need the terms for the AGMA endurance strength equations. From Table 14–3, for grade 1 steel with $H_{BP} = 240$ and $H_{BG} = 200$, we use Fig. 14–2, which gives

$$(S_t)_P = 77.3(240) + 12\ 800 = 31\ 350 \text{ psi}$$

$$(S_t)_G = 77.3(200) + 12\ 800 = 28\ 260 \text{ psi}$$

Similarly, from Table 14–6, we use Fig. 14–5, which gives

$$(S_c)_P = 322(240) + 29\ 100 = 106\ 400 \text{ psi}$$

$$(S_c)_G = 322(200) + 29\ 100 = 93\ 500 \text{ psi}$$

From Fig. 14–15,

$$(Z_N)_P = 1.4488(10^8)^{-0.023} = 0.948$$

$$(Z_N)_G = 1.4488(10^8/3.059)^{-0.023} = 0.973$$

For the hardness ratio factor C_H, the hardness ratio is $H_{BP}/H_{BG} = 240/200 = 1.2$. Then, from Sec. 14–12,

$$A' = 8.98(10^{-3})(H_{BP}/H_{BG}) - 8.29(10^{-3})$$

$$= 8.98(10^{-3})(1.2) - 8.29(10^{-3}) = 0.002\ 49$$

Thus, from Eq. (14–36),

$$C_H = 1 + 0.002\ 49(3.059 - 1) = 1.005$$

(*a*) **Pinion tooth bending.** Substituting the appropriate terms for the pinion into Eq. (14–15) gives

$$(\sigma)_P = \left(W^t K_o K_v K_s \frac{P_d}{F} \frac{K_m K_B}{J}\right)_P = 164.8(1)1.377(1.043)\frac{10}{1.5}\frac{1.22\,(1)}{0.30}$$

$$= 6417 \text{ psi}$$

Substituting the appropriate terms for the pinion into Eq. (14–41) gives

Answer
$$(S_F)_P = \left(\frac{S_t Y_N/(K_T K_R)}{\sigma}\right)_P = \frac{31\,350(0.977)/[1(0.85)]}{6417} = 5.62$$

Gear tooth bending. Substituting the appropriate terms for the gear into Eq. (14–15) gives

$$(\sigma)_G = 164.8(1)1.377(1.052)\frac{10}{1.5}\frac{1.22(1)}{0.40} = 4854 \text{ psi}$$

Substituting the appropriate terms for the gear into Eq. (14–41) gives

Answer
$$(S_F)_G = \frac{28\,260(0.996)/[1(0.85)]}{4854} = 6.82$$

(*b*) **Pinion tooth wear.** Substituting the appropriate terms for the pinion into Eq. (14–16) gives

$$(\sigma_c)_P = C_p\left(W^t K_o K_v K_s \frac{K_m}{d_p F}\frac{C_f}{I}\right)^{1/2}_P$$

$$= 2300\left[164.8(1)1.377(1.043)\frac{1.22}{1.7(1.5)}\frac{1}{0.121}\right]^{1/2} = 70\,360 \text{ psi}$$

Substituting the appropriate terms for the pinion into Eq. (14–42) gives

Answer
$$(S_H)_P = \left[\frac{S_c Z_N/(K_T K_R)}{\sigma_c}\right]_P = \frac{106\,400(0.948)/[1(0.85)]}{70\,360} = 1.69$$

Gear tooth wear. The only term in Eq. (14–16) that changes for the gear is K_s. Thus,

$$(\sigma_c)_G = \left[\frac{(K_s)_G}{(K_s)_P}\right]^{1/2}(\sigma_c)_P = \left(\frac{1.052}{1.043}\right)^{1/2}70\,360 = 70\,660 \text{ psi}$$

Substituting the appropriate terms for the gear into Eq. (14–42) with $C_H = 1.005$ gives

Answer
$$(S_H)_G = \frac{93\,500(0.973)1.005/[1(0.85)]}{70\,660} = 1.52$$

(*c*) For the pinion, we compare $(S_F)_P$ with $(S_H)^2_P$, or 5.73 with $1.69^2 = 2.86$, so the threat in the pinion is from wear. For the gear, we compare $(S_F)_G$ with $(S_H)^2_G$, or 6.96 with $1.52^2 = 2.31$, so the threat in the gear is also from wear.

There are perspectives to be gained from Ex. 14–4. First, the pinion is overly strong in bending compared to wear. The performance in wear can be improved by surface-hardening techniques, such as flame or induction hardening, nitriding, or carburizing and case hardening, as well as shot peening. This in turn permits the gearset to be made smaller. Second, in bending, the gear is stronger than the pinion, indicating that both the gear core hardness and tooth size could be reduced; that is, we may increase P and reduce diameter of the gears, or perhaps allow a cheaper material. Third, in wear, surface strength equations have the ratio $(Z_N)/K_R$. The values of $(Z_N)_P$ and $(Z_N)_G$ are affected by gear ratio m_G. The designer can control strength by specifying surface hardness. This point will be elaborated later.

Having followed a spur-gear analysis in detail in Ex. 14–4, it is timely to analyze a helical gearset under similar circumstances to observe similarities and differences.

EXAMPLE 14–5

A 17-tooth $20°$ normal pitch-angle helical pinion with a right-hand helix angle of $30°$ rotates at 1800 rev/min when transmitting 4 hp to a 52-tooth helical gear. The normal diametral pitch is 10 teeth/in, the face width is 1.5 in, and the set has a quality number of 6. The gears are straddle-mounted with bearings immediately adjacent. The pinion and gear are made from a through-hardened steel with surface and core hardnesses of 240 Brinell on the pinion an surface and core hardnesses of 200 Brinell on the gear. The transmission is smooth, connecting an electric motor and a centrifugal pump. Assume a pinion life of 10^8 cycles and a reliability of 0.9 and use upper curves in Figs. 14–14 and 14–15.

(a) Find the factors of safety of the gears in bending.
(b) Find the factors of safety of the gears in wear.
(c) By examining the factors of safety identify the threat to each gear and to the mesh.

Solution

All of the parameters in this example are the same as in Ex. 14–4 with the exception that we are using helical gears. Thus, several terms will be the same as Ex. 14–4. The reader should verify that the following terms remain unchanged: $K_o = 1$, $Y_P = 0.303$, $Y_G = 0.412$, $m_G = 3.059$, $(K_s)_P = 1.043$, $(K_s)_G = 1.052$, $(Y_N)_P = 0.977$, $(Y_N)_G = 0.996$, $K_R = 0.85$, $K_T = 1$, $C_f = 1$, $C_p = 2300 \sqrt{\text{psi}}$, $(S_t)_P = 31\,350$ psi, $(S_t)_G = 28\,260$ psi, $(S_c)_P = 106\,380$ psi, $(S_c)_G = 93\,500$ psi, $(Z_N)_P = 0.948$, $(Z_N)_G = 0.973$, and $C_H = 1.005$.

For helical gears, the transverse diametral pitch, given by Eq. (13–18), is

$$P_t = P_n \cos \psi = 10 \cos 30° = 8.660 \text{ teeth/in}$$

Thus, the pitch diameters are $d_P = N_P/P_t = 17/8.660 = 1.963$ in and $d_G = 52/8.660 = 6.005$ in. The pitch-line velocity and transmitted force are

$$V = \frac{\pi d_P n_P}{12} = \frac{\pi (1.963) 1800}{12} = 925 \text{ ft/min}$$

$$W^t = \frac{33\,000 H}{V} = \frac{33\,000(4)}{925} = 142.7 \text{ lbf}$$

As in Ex. 14–4, for the dynamic factor, $B = 0.8255$ and $A = 59.77$. Thus, Eq. (14–27) gives

$$K_v = \left(\frac{59.77 + \sqrt{925}}{59.77} \right)^{0.8255} = 1.404$$

The geometry factor I for helical gears requires a little work. First, the transverse pressure angle is given by Eq. (13–19)

$$\phi_t = \tan^{-1}\left(\frac{\tan \phi_n}{\cos \psi}\right) = \tan^{-1}\left(\frac{\tan 20°}{\cos 30°}\right) = 22.80°$$

The radii of the pinion and gear are $r_P = 1.963/2 = 0.9815$ in and $r_G = 6.004/2 = 3.002$ in, respectively. The addendum is $a = 1/P_n = 1/10 = 0.1$, and the base-circle radii of the pinion and gear are given by Eq. (13–6) with $\phi = \phi_t$:

$$(r_b)_P = r_P \cos \phi_t = 0.9815 \cos 22.80° = 0.9048 \text{ in}$$

$$(r_b)_G = 3.002 \cos 22.80° = 2.767 \text{ in}$$

From Eq. (14–25), the surface strength geometry factor

$$Z = \sqrt{(0.9815 + 0.1)^2 - 0.9048^2} + \sqrt{(3.004 + 0.1)^2 - 2.769^2}$$

$$- (0.9815 + 3.004) \sin 22.80°$$

$$= 0.5924 + 1.4027 - 1.544\,4 = 0.4507 \text{ in}$$

Since the first two terms are less than 1.544 4, the equation for Z stands. From Eq. (14–24) the normal circular pitch p_N is

$$p_N = p_n \cos \phi_n = \frac{\pi}{P_n} \cos 20° = \frac{\pi}{10} \cos 20° = 0.2952 \text{ in}$$

From Eq. (14–21), the load sharing ratio

$$m_N = \frac{p_N}{0.95Z} = \frac{0.2952}{0.95(0.4507)} = 0.6895$$

Substituting in Eq. (14–23), the geometry factor I is

$$I = \frac{\sin 22.80° \cos 22.80°}{2(0.6895)} \frac{3.06}{3.06 + 1} = 0.195$$

From Fig. 14–7, geometry factors $J_P' = 0.45$ and $J_G' = 0.54$. Also from Fig. 14–8 the J-factor multipliers are 0.94 and 0.98, correcting J_P' and J_G' to

$$J_P = 0.45(0.94) = 0.423$$

$$J_G = 0.54(0.98) = 0.529$$

The load-distribution factor K_m is estimated from Eq. (14–32):

$$C_{pf} = \frac{1.5}{10(1.963)} - 0.0375 + 0.0125(1.5) = 0.0577$$

with $C_{mc} = 1$, $C_{pm} = 1$, $C_{ma} = 0.15$ from Fig. 14–11, and $C_e = 1$. Therefore, from Eq. (14–30),

$$K_m = 1 + (1)[0.0577(1) + 0.15(1)] = 1.208$$

(a) **Pinion tooth bending.** Substituting the appropriate terms into Eq. (14–15) using P_t gives

$$(\sigma)_P = \left(W^t K_o K_v K_s \frac{P_t}{F} \frac{K_m K_B}{J}\right)_P = 142.7(1)1.404(1.043)\frac{8.66}{1.5}\frac{1.208(1)}{0.423}$$

$$= 3445 \text{ psi}$$

Substituting the appropriate terms for the pinion into Eq. (14–41) gives

Answer
$$(S_F)_P = \left(\frac{S_t Y_N/(K_T K_R)}{\sigma}\right)_P = \frac{31\,350(0.977)/[1(0.85)]}{3445} = 10.5$$

Gear tooth bending. Substituting the appropriate terms for the gear into Eq. (14–15) gives

$$(\sigma)_G = 142.7(1)1.404(1.052)\frac{8.66}{1.5}\frac{1.208(1)}{0.529} = 2779 \text{ psi}$$

Substituting the appropriate terms for the gear into Eq. (14–41) gives

Answer
$$(S_F)_G = \frac{28\,260(0.996)/[1(0.85)]}{2779} = 11.9$$

(*b*) **Pinion tooth wear.** Substituting the appropriate terms for the pinion into Eq. (14–16) gives

$$(\sigma_c)_P = C_p \left(W^t K_o K_v K_s \frac{K_m}{d_P F}\frac{C_f}{I}\right)^{1/2}_P$$

$$= 2300\left[142.7(1)1.404(1.043)\frac{1.208}{1.963(1.5)}\frac{1}{0.195}\right]^{1/2} = 48\,230 \text{ psi}$$

Substituting the appropriate terms for the pinion into Eq. (14–42) gives

Answer
$$(S_H)_P = \left(\frac{S_c Z_N/(K_T K_R)}{\sigma_c}\right)_P = \frac{106\,400(0.948)/[1(0.85)]}{48\,230} = 2.46$$

Gear tooth wear. The only term in Eq. (14–16) that changes for the gear is K_s. Thus,

$$(\sigma_c)_G = \left[\frac{(K_s)_G}{(K_s)_P}\right]^{1/2}(\sigma_c)_P = \left(\frac{1.052}{1.043}\right)^{1/2}48\,230 = 48\,440 \text{ psi}$$

Substituting the appropriate terms for the gear into Eq. (14–42) with $C_H = 1.005$ gives

Answer
$$(S_H)_G = \frac{93\,500(0.973)1.005/[1(0.85)]}{48\,440} = 2.22$$

(*c*) For the pinion we compare S_F with S_H^2, or 10.5 with $2.46^2 = 6.05$, so the threat in the pinion is from wear. For the gear we compare S_F with S_H^2, or 11.9 with $2.22^2 = 4.93$, so the threat is also from wear in the gear. For the meshing gearset wear controls.

It is worthwhile to compare Ex. 14–4 with Ex. 14–5. The spur and helical gearsets were placed in nearly identical circumstances. The helical gear teeth are of greater length because of the helix and identical face widths. The pitch diameters of the helical gears are larger. The *J* factors and the *I* factor are larger, thereby reducing stresses. The result is larger factors of safety. In the design phase the gearsets in Ex. 14–4 and Ex. 14–5 can be made smaller with control of materials and relative hardnesses.

Now that examples have given the AGMA parameters substance, it is time to examine some desirable (and necessary) relationships between material properties of spur gears in mesh. In bending, the AGMA equations are displayed side by side:

$$\sigma_P = \left(W^t K_o K_v K_s \frac{P_d}{F} \frac{K_m K_B}{J}\right)_P \qquad \sigma_G = \left(W^t K_o K_v K_s \frac{P_d}{F} \frac{K_m K_B}{J}\right)_G$$

$$(S_F)_P = \left(\frac{S_t Y_N/(K_T K_R)}{\sigma}\right)_P \qquad (S_F)_G = \left(\frac{S_t Y_N/(K_T K_R)}{\sigma}\right)_G$$

Equating the factors of safety, substituting for stress and strength, canceling identical terms (K_s virtually equal or exactly equal), and solving for $(S_t)_G$ gives

$$(S_t)_G = (S_t)_P \frac{(Y_N)_P}{(Y_N)_G} \frac{J_P}{J_G} \tag{a}$$

The stress-cycle factor Y_N comes from Fig.14–14, where for a particular hardness, $Y_N = \alpha N^\beta$. For the pinion, $(Y_N)_P = \alpha N_P^\beta$, and for the gear, $(Y_N)_G = \alpha(N_P/m_G)^\beta$. Substituting these into Eq. (a) and simplifying gives

$$(S_t)_G = (S_t)_P m_G^\beta \frac{J_P}{J_G} \tag{14–44}$$

Normally, $m_G > 1$ and $J_G > J_P$, so equation (14–44) shows that the gear can be less strong (lower Brinell hardness) than the pinion for the same safety factor.

EXAMPLE 14–6

In a set of spur gears, a 300-Brinell 18-tooth 16-pitch 20° full-depth pinion meshes with a 64-tooth gear. Both gear and pinion are of grade 1 through-hardened steel. Using $\beta = -0.023$, what hardness can the gear have for the same factor of safety?

Solution

For through-hardened grade 1 steel the pinion strength $(S_t)_P$ is given in Fig. 14–2:

$$(S_t)_P = 77.3(300) + 12\,800 = 35\,990 \text{ psi}$$

From Fig. 14–6 the form factors are $J_P = 0.32$ and $J_G = 0.41$. Equation (14–44) gives

$$(S_t)_G = 35\,990 \left(\frac{64}{18}\right)^{-0.023} \frac{0.32}{0.41} = 27\,280 \text{ psi}$$

Use the equation in Fig. 14–2 again.

Answer

$$(H_B)_G = \frac{27\,280 - 12\,800}{77.3} = 187 \text{ Brinell}$$

The AGMA contact-stress equations also are displayed side by side:

$$(\sigma_c)_P = C_p \left(W^t K_o K_v K_s \frac{K_m}{d_P F} \frac{C_f}{I}\right)_P^{1/2} \qquad (\sigma_c)_G = C_p \left(W^t K_o K_v K_s \frac{K_m}{d_P F} \frac{C_f}{I}\right)_G^{1/2}$$

$$(S_H)_P = \left(\frac{S_c Z_N/(K_T K_R)}{\sigma_c}\right)_P \qquad (S_H)_G = \left(\frac{S_c Z_N C_H/(K_T K_R)}{\sigma_c}\right)_G$$

Equating the factors of safety, substituting the stress relations, and canceling identical terms including K_s gives, after solving for $(S_c)_G$,

$$(S_c)_G = (S_c)_P \frac{(Z_N)_P}{(Z_N)_G} \left(\frac{1}{C_H}\right)_G = (S_C)_P m_G^\beta \left(\frac{1}{C_H}\right)_G$$

where, as in the development of Eq. (14–44), $(Z_N)_P/(Z_N)_G = m_G^\beta$ and the value of β for wear comes from Fig. 14–15. Since C_H is so close to unity, it is usually neglected; therefore

$$(S_c)_G = (S_c)_P m_G^\beta \tag{14–45}$$

EXAMPLE 14–7

For $\beta = -0.056$ for a through-hardened steel, grade 1, continue Ex. 14–6 for wear.

Solution

From Fig. 14–5,

$$(S_c)_P = 322(300) + 29\,100 = 125\,700 \text{ psi}$$

From Eq. (14–45),

$$(S_c)_G = (S_c)_P \left(\frac{64}{18}\right)^{-0.056} = 125\,700 \left(\frac{64}{18}\right)^{-0.056} = 117\,100 \text{ psi}$$

Answer

$$(H_B)_G = \frac{117\,100 - 29\,200}{322} = 273 \text{ Brinell}$$

which is slightly less than the pinion hardness of 300 Brinell.

Equations (14–44) and (14–45) apply as well to helical gears.

14–19 Design of a Gear Mesh

A useful decision set for spur and helical gears includes

- Function: load, speed, reliability, life, K_o
- Unquantifiable risk: design factor n_d
- Tooth system: ϕ, ψ, addendum, dedendum, root fillet radius
- Gear ratio m_G, N_p, N_G
- Quality number Q_v

} a priori decisions

- Diametral pitch P_d
- Face width F
- Pinion material, core hardness, case hardness
- Gear material, core hardness, case hardness

} design decisions

The first item to notice is the dimensionality of the decision set. There are four design decision categories, eight different decisions if you count them separately. This is a larger number than we have encountered before. It is important to use a design strategy that is convenient in either longhand execution or computer implementation. The design decisions have been placed in order of importance (impact on the amount of work to be redone in iterations). The steps are, after the a priori decisions have been made,

- Choose a diametral pitch.
- Examine implications on face width, pitch diameters, and material properties. If not satisfactory, return to pitch decision for change.
- Choose a pinion material and examine core and case hardness requirements. If not satisfactory, return to pitch decision and iterate until no decisions are changed.
- Choose a gear material and examine core and case hardness requirements. If not satisfactory, return to pitch decision and iterate until no decisions are changed.

With these plan steps in mind, we can consider them in more detail.

First select a trial diametral pitch.

Pinion bending:

- Select a median face width for this pitch, $4\pi/P$
- Find the range of necessary ultimate strengths
- Choose a material and a core hardness
- Find face width to meet factor of safety in bending
- Choose face width
- Check factor of safety in bending

Gear bending:

- Find necessary companion core hardness
- Choose a material and core hardness
- Check factor of safety in bending

Pinion wear:

- Find necessary S_c and attendant case hardness
- Choose a case hardness
- Check factor of safety in wear

Gear wear:

- Find companion case hardness
- Choose a case hardness
- Check factor of safety in wear

Completing this set of steps will yield a satisfactory design. Additional designs with diametral pitches adjacent to the first satisfactory design will produce several among which to choose. A figure of merit is necessary in order to choose the best. Unfortunately, a figure of merit in gear design is complex in an academic environment because material and processing cost vary. The possibility of using a process depends on the manufacturing facility if gears are made in house.

After examining Ex. 14–4 and Ex. 14–5 and seeing the wide range of factors of safety, one might entertain the notion of setting all factors of safety equal.[9] In steel gears,

---

[9] In designing gears it makes sense to define the factor of safety in wear as $(S)_H^2$ for uncrowned teeth, so that there is no mix-up. ANSI, in the preface to ANSI/AGMA 2001-C95 and 2101-C95, states "the use is completely voluntary. . . does not preclude anyone from using . . . procedures . . . not conforming to the standards."

wear is usually controlling and $(S_H)_P$ and $(S_H)_G$ can be brought close to equality. The use of softer cores can bring down $(S_F)_P$ and $(S_F)_G$, but there is value in keeping them higher. A tooth broken by bending fatigue not only can destroy the gear set, but can bend shafts, damage bearings, and produce inertial stresses up- and downstream in the power train, causing damage elsewhere if the gear box locks.

EXAMPLE 14-8

Design a 4:1 spur-gear reduction for a 100-hp, three-phase squirrel-cage induction motor running at 1120 rev/min. The load is smooth, providing a reliability of 0.95 at 10^9 revolutions of the pinion. Gearing space is meager. Use Nitralloy 135M, grade 1 material to keep the gear size small. The gears are heat-treated first then nitrided.

Solution

Make the a priori decisions:

- Function: 100 hp, 1120 rev/min, $R = 0.95$, $N = 10^9$ cycles, $K_o = 1$
- Design factor for unquantifiable exingencies: $n_d = 2$
- Tooth system: $\phi_n = 20°$
- Tooth count: $N_P = 18$ teeth, $N_G = 72$ teeth (no interference)
- Quality number: $Q_v = 6$, use grade 1 material
- Assume $m_B \geq 1.2$ in Eq. (14–40), $K_B = 1$

Pitch: Select a trial diametral pitch of $P_d = 4$ teeth/in. Thus, $d_P = 18/4 = 4.5$ in and $d_G = 72/4 = 18$ in. From Table 14–2, $Y_P = 0.309$, $Y_G = 0.4324$ (interpolated). From Fig. 14–6, $J_P = 0.32$, $J_G = 0.415$.

$$V = \frac{\pi d_P n_P}{12} = \frac{\pi(4.5)1120}{12} = 1319 \text{ ft/min}$$

$$W^t = \frac{33\,000H}{V} = \frac{33\,000(100)}{1319} = 2502 \text{ lbf}$$

From Eqs. (14–28) and (14–27),

$$B = 0.25(12 - Q_v)^{2/3} = 0.25(12 - 6)^{2/3} = 0.8255$$

$$A = 50 + 56(1 - 0.8255) = 59.77$$

$$K_v = \left(\frac{59.77 + \sqrt{1319}}{59.77}\right)^{0.8255} = 1.480$$

From Eq. (14–38), $K_R = 0.658 - 0.0759 \ln(1 - 0.95) = 0.885$. From Fig. 14–14,

$$(Y_N)_P = 1.3558(10^9)^{-0.0178} = 0.938$$

$$(Y_N)_G = 1.3558(10^9/4)^{-0.0178} = 0.961$$

From Fig. 14–15,

$$(Z_N)_P = 1.4488(10^9)^{-0.023} = 0.900$$

$$(Z_N)_G = 1.4488(10^9/4)^{-0.023} = 0.929$$

From the recommendation after Eq. (14–8), $3p \leq F \leq 5p$. Try $F = 4p = 4\pi/P = 4\pi/4 = 3.14$ in. From Eq. (a), Sec. 14–10,

$$K_s = 1.192 \left(\frac{F\sqrt{Y}}{P} \right)^{0.0535} = 1.192 \left(\frac{3.14\sqrt{0.309}}{4} \right)^{0.0535} = 1.140$$

From Eqs. (14–31), (14–33), (14–35), $C_{mc} = C_{pm} = C_e = 1$. From Fig. 14–11, $C_{ma} = 0.175$ for commercial enclosed gear units. From Eq. (14–32), $F/(10d_P) = 3.14/[10(4.5)] = 0.0698$. Thus,

$$C_{pf} = 0.0698 - 0.0375 + 0.0125(3.14) = 0.0715$$

From Eq. (14–30),

$$K_m = 1 + (1)[0.0715(1) + 0.175(1)] = 1.247$$

From Table 14–8, for steel gears, $C_p = 2300\sqrt{\text{psi}}$. From Eq. (14–23), with $m_G = 4$ and $m_N = 1$,

$$I = \frac{\cos 20° \sin 20°}{2} \frac{4}{4+1} = 0.1286$$

Pinion tooth bending. With the above estimates of K_s and K_m from the trial dimetral pitch, we check to see if the mesh width F is controlled by bending or wear considerations. Equating Eqs. (14–15) and (14–17), substituting $n_d W^t$ for W^t, and solving for the face width $(F)_{\text{bend}}$ necessary to resist bending fatigue, we obtain

$$(F)_{\text{bend}} = n_d W^t K_o K_v K_s P_d \frac{K_m K_B}{J_P} \frac{K_T K_R}{S_t Y_N} \tag{1}$$

Equating Eqs. (14–16) and (14–18), substituting $n_d W^t$ for W^t, and solving for the face width $(F)_{\text{wear}}$ necessary to resist wear fatigue, we obtain

$$(F)_{\text{wear}} = \left(\frac{C_p Z_N}{S_c K_T K_R} \right)^2 n_d W^t K_o K_v K_s \frac{K_m C_f}{d_P I} \tag{2}$$

From Table 14–5 the hardness range of Nitralloy 135M is Rockwell C32–36 (302–335 Brinell). Choosing a midrange hardness as attainable, using 320 Brinell. From Fig. 14–4,

$$S_t = 86.2(320) + 12\,730 = 40\,310 \text{ psi}$$

Inserting the numerical value of S_t in Eq. (1) to estimate the face width gives

$$(F)_{\text{bend}} = 2(2502)(1)1.48(1.14)4 \frac{1.247(1)(1)0.885}{0.32(40\,310)0.938} = 3.08 \text{ in}$$

From Table 14–6 for Nitralloy 135M, $S_c = 170\,000$ psi. Inserting this in Eq. (2), we find

$$(F)_{\text{wear}} = \left(\frac{2300(0.900)}{170\,000(1)0.885} \right)^2 2(2502)1(1.48)1.14 \frac{1.247(1)}{4.5(0.1286)} = 3.44 \text{ in}$$

Decision   Make face width 3.50 in. Correct K_s and K_m:

$$K_s = 1.192 \left(\frac{3.50\sqrt{0.309}}{4} \right)^{0.0535} = 1.147$$

$$\frac{F}{10 d_P} = \frac{3.50}{10(4.5)} = 0.0778$$

$$C_{pf} = 0.0778 - 0.0375 + 0.0125(3.50) = 0.0841$$

$$K_m = 1 + (1)[0.0841(1) + 0.175(1)] = 1.259$$

The bending stress induced by W^t in bending, from Eq. (14–15), is

$$(\sigma)_P = 2502(1)1.48(1.147) \frac{4}{3.50} \frac{1.259(1)}{0.32} = 19\,100 \text{ psi}$$

The factor of safety in bending of the pinion, from Eq. (14–41), is

$$(S_F)_P = \frac{40\,310(0.938)/[1(0.885)]}{19\,100} = 2.24$$

Decision   **Gear tooth bending.** Use cast gear blank because of the 18-in pitch diameter. Use the same material, heat treatment, and nitriding. The load-induced bending stress is in the ratio of J_P / J_G. Then

$$(\sigma)_G = 19\,100 \frac{0.32}{0.415} = 14\,730 \text{ psi}$$

The factor of safety of the gear in bending is

$$(S_F)_G = \frac{40\,310(0.961)/[1(0.885)]}{14\,730} = 2.97$$

Pinion tooth wear. The contact stress, given by Eq. (14–16), is

$$(\sigma_c)_P = 2300 \left[2502(1)1.48(1.147) \frac{1.259}{4.5(3.5)} \frac{1}{0.129} \right]^{1/2} = 118\,000 \text{ psi}$$

The factor of safety from Eq. (14–42), is

$$(S_H)_P = \frac{170\,000(0.900)/[1(0.885)]}{118\,000} = 1.465$$

By our definition of factor of safety, pinion bending is $(S_F)_P = 2.24$, and wear is $(S_H)_P^2 = (1.465)^2 = 2.15$.

Gear tooth wear. The hardness of the gear and pinion are the same. Thus, from Fig. 14–12, $C_H = 1$, the contact stress on the gear is the same as the pinion, $(\sigma_c)_G = 118\,000$ psi. The wear strength is also the same, $S_c = 170\,000$ psi. The factor of safety of the gear in wear is

$$(S_H)_G = \frac{170\,000(0.929)/[1(0.885)]}{118\,000} = 1.51$$

So, for the gear in bending, $(S_F)_G = 2.97$, and wear $(S_H)_G^2 = (1.51)^2 = 2.29$.

Rim. Keep $m_B \geq 1.2$. The whole depth is $h_t =$ addendum $+$ dedendum $= 1/P_d + 1.25/P_d = 2.25/P_d = 2.25/4 = 0.5625$ in. The rim thickness t_R is

$$t_R \geq m_B h_t = 1.2(0.5625) = 0.675 \text{ in}$$

In the design of the gear blank, be sure the rim thickness exceeds 0.675 in; if it does not, review and modify this mesh design.

This design example showed a satisfactory design for a four-pitch spur-gear mesh. Material could be changed, as could pitch. There are a number of other satisfactory designs, thus a figure of merit is needed to identify the best.

One can appreciate that gear design was one of the early applications of the digital computer to mechanical engineering. A design program should be interactive, presenting results of calculations, pausing for a decision by the designer, and showing the consequences of the decision, with a loop back to change a decision for the better. The program can be structured in totem-pole fashion, with the most influential decision at the top, then tumbling down, decision after decision, ending with the ability to change the current decision or to begin again. Such a program would make a fine class project. Troubleshooting the coding will reinforce your knowledge, adding flexibility as well as bells and whistles in subsequent terms.

Standard gears may not be the most economical design that meets the functional requirements, because no application is standard in all respects.[10] Methods of designing custom gears are well-understood and frequently used in mobile equipment to provide good weight-to-performance index. The required calculations including optimizations are within the capability of a personal computer.

PROBLEMS

Because gearing problems can be difficult, the problems are presented by section.

Section 14–1

14–1 ANALYSIS

A steel spur pinion has a pitch of 6 teeth/in, 22 full-depth teeth, and a 20° pressure angle. The pinion runs at a speed of 1200 rev/min and transmits 15 hp to a 60-tooth gear. If the face width is 2 in, estimate the bending stress.

14–2 ANALYSIS

A steel spur pinion has a diametral pitch of 12 teeth/in, 16 teeth cut full-depth with a 20° pressure angle, and a face width of $\frac{3}{4}$ in. This pinion is expected to transmit 1.5 hp at a speed of 700 rev/min. Determine the bending stress.

14–3 ANALYSIS

A steel spur pinion has a module of 1.25 mm, 18 teeth cut on the 20° full-depth system, and a face width of 12 mm. At a speed of 1800 rev/min, this pinion is expected to carry a steady load of 0.5 kW. Determine the resulting bending stress.

14–4 ANALYSIS

A steel spur pinion has 15 teeth cut on the 20° full-depth system with a module of 5 mm and a face width of 60 mm. The pinion rotates at 200 rev/min and transmits 5 kW to the mating steel gear. What is the resulting bending stress?

---

[10]See H. W. Van Gerpen, C. K. Reece, and J. K. Jensen, *Computer Aided Design of Custom Gears,* Van Gerpen–Reece Engineering, Cedar Falls, Iowa, 1996.

ANALYSIS **14–5** A steel spur pinion has a module of 1 mm and 16 teeth cut on the 20° full-depth system and is to carry 0.15 kW at 400 rev/min. Determine a suitable face width based on an allowable bending stress of 150 MPa.

ANALYSIS **14–6** A 20° full-depth steel spur pinion has 17 teeth and a module of 1.5 mm and is to transmit 0.25 kW at a speed of 400 rev/min. Find an appropriate face width if the bending stress is not to exceed 75 MPa.

ANALYSIS **14–7** A 20° full-depth steel spur pinion has a diametral pitch of 5 teeth/in and 24 teeth and transmits 6 hp at a speed of 50 rev/min. Find an appropriate face width if the allowable bending stress is 20 kpsi.

ANALYSIS **14–8** A steel spur pinion is to transmit 15 hp at a speed of 600 rev/min. The pinion is cut on the 20° full-depth system and has a diametral pitch of 5 teeth/in and 16 teeth. Find a suitable face width based on an allowable stress of 10 kpsi.

DESIGN **14–9** A 20° full-depth steel spur pinion with 18 teeth is to transmit 2.5 hp at a speed of 600 rev/min. Determine appropriate values for the face width and diametral pitch based on an allowable bending stress of 10 kpsi.

DESIGN **14–10** A 20° full-depth steel spur pinion is to transmit 1.5 kW hp at a speed of 900 rev/min. If the pinion has 18 teeth, determine suitable values for the module and face width. The bending stress should not exceed 75 MPa.

Section 14–2

ANALYSIS **14–11** A speed reducer has 20° full-depth teeth and consists of a 22-tooth steel spur pinion driving a 60-tooth cast-iron gear. The horsepower transmitted is 15 at a pinion speed of 1200 rev/min. For a diametral pitch of 6 teeth/in and a face width of 2 in, find the contact stress.

ANALYSIS **14–12** A gear drive consists of a 16-tooth 20° steel spur pinion and a 48-tooth cast-iron gear having a pitch of 12 teeth/in. For a power input of 1.5 hp at a pinion speed of 700 rev/min, select a face width based on an allowable contact stress of 100 kpsi.

ANALYSIS **14–13** A gearset has a diametral pitch of 5 teeth/in, a 20° pressure angle, and a 24-tooth cast-iron spur pinion driving a 48-tooth cast-iron gear. The pinion is to rotate at 50 rev/min. What horsepower input can be used with this gearset if the contact stress is limited to 100 kpsi? and $F = 2.5$ in?

ANALYSIS **14–14** A 20° 20-tooth cast-iron spur pinion having a module of 4 mm drives a 32-tooth cast-iron gear. Find the contact stress if the pinion speed is 1000 rev/min, the face width is 50 mm, and 10 kW of power is transmitted.

ANALYSIS **14–15** A steel spur pinion and gear have a diametral pitch of 12 teeth/in, milled teeth, 17 and 30 teeth, respectively, a 20° pressure angle, and a pinion speed of 525 rev/min. The tooth properties are $S_{ut} = 76$ kpsi, $S_y = 42$ kpsi and the Brinell hardness is 149. For a design factor of 2.25, a face width of $\frac{7}{8}$ in, what is the power rating of the gearset?

ANALYSIS **14–16** A milled-teeth steel pinion and gear pair have $S_{ut} = 113$ kpsi, $S_y = 86$ kpsi and a hardness at the involute surface of 262 Brinell. The diametral pitch is 3 teeth/in, the face width is 2.5 in, and the pinion speed is 870 rev/min. The tooth counts are 20 and 100. For a design factor of 1.5, rate the gearset for power considering both bending and wear.

ANALYSIS **14–17** A 20° full-depth steel spur pinion rotates at 1145 rev/min. It has a module of 6 mm, a face width of 75 mm, and 16 milled teeth. The ultimate tensile strength at the involute is 900 MPa exhibiting a Brinell hardness of 260. The gear is steel with 30 teeth and has identical material strengths. For a design factor of 1.3 find the power rating of the gearset based on the pinion and the gear resisting bending and wear fatigue.

 ANALYSIS

14–18 A steel spur pinion has a pitch of 6 teeth/in, 17 full-depth milled teeth, and a pressure angle of 20°. The pinion has an ultimate tensile strength at the involute surface of 116 kpsi, a Brinell hardness of 232, and a yield strength of 90 kpsi. Its shaft speed is 1120 rev/min, its face width is 2 in, and its mating gear has 51 teeth. Rate the pinion for power transmission if the design factor is 2.

(*a*) Pinion bending fatigue imposes what power limitation?

(*b*) Pinion surface fatigue imposes what power limitation? The gear has identical strengths to the pinion with regard to material properties.

(*c*) Consider power limitations due to gear bending and wear.

(*d*) Rate the gearset.

Sections 14–3 to 14–19

 ANALYSIS

14–19 A commercial enclosed gear drive consists of a 20° spur pinion having 16 teeth driving a 48-tooth gear. The pinion speed is 300 rev/min, the face width 2 in, and the diametral pitch 6 teeth/in. The gears are grade 1 steel, through-hardened at 200 Brinell, made to No. 6 quality standards, uncrowned, and are to be accurately and rigidly mounted. Assume a pinion life of 10^8 cycles and a reliability of 0.90. Determine the AGMA bending and contact stresses and the corresponding factors of safety if 5 hp is to be transmitted.

14–20 A 20° spur pinion with 20 teeth and a module of 2.5 mm transmits 120 W to a 36-tooth gear. The pinion speed is 100 rev/min, and the gears are grade 1, 18 mm face width, through-hardened steel at 200 Brinell, uncrowned, manufactured to a No. 6 quality standard, and considered to be of open gearing quality installation. Find the AGMA bending and contact stresses and the corresponding factors of safety for a pinion life of 10^8 cycles and a reliability of 0.95.

 ANALYSIS

14–21 Repeat Prob. 14–19 using helical gears each with a 20° normal pitch angle and a helix angle of 30° and a normal diametral pitch of 6 teeth/in.

 ANALYSIS

14–22 A spur gearset has 17 teeth on the pinion and 51 teeth on the gear. The pressure angle is 20° and the overload factor $K_o = 1$. The diametral pitch is 6 teeth/in and the face width is 2 in. The pinion speed is 1120 rev/min and its cycle life is to be 10^8 revolutions at a reliability $R = 0.99$. The quality number is 5. The material is a through-hardened steel, grade 1, with Brinell hardnesses of 232 core and case of both gears. For a design factor of 2, rate the gearset for these conditions using the AGMA method.

 ANALYSIS

14–23 In Sec. 14–10, Eq. (*a*) is given for K_s based on the procedure in Ex. 14–2. Derive this equation.

14–24 A speed-reducer has 20° full-depth teeth, and the single-reduction spur-gear gearset has 22 and 60 teeth. The diametral pitch is 4 teeth/in and the face width is $3\frac{1}{4}$ in. The pinion shaft speed is 1145 rev/min. The life goal of 5-year 24-hour-per-day service is about $3(10^9)$ pinion revolutions. The absolute value of the pitch variation is such that the transmission accuracy level number is 6. The materials are 4340 through-hardened grade 1 steels, heat-treated to 250 Brinell, core and case, both gears. The load is moderate shock and the power is smooth. For a reliability of 0.99, rate the speed reducer for power.

 ANALYSIS

14–25 The speed reducer of Prob. 14–24 is to be used for an application requiring 40 hp at 1145 rev/min. Estimate the stresses of pinion bending, gear bending, pinion wear, and gear wear and the attendant AGMA factors of safety $(S_F)_P$, $(S_F)_G$, $(S_H)_P$, and $(S_H)_G$. For the reducer, what is the factor of safety for unquantifiable exingencies in W^t? What mode of failure is the most threatening?

 ANALYSIS

14–26 The gearset of Prob. 14–24 needs improvement of wear capacity. Toward this end the gears are nitrided so that the grade 1 materials have hardnesses as follows: The pinion core is 250 and the pinion case hardness is 390 Brinell, and the gear core hardness is 250 core and 390 case. Estimate the power rating for the new gearset.

14–27

ANALYSIS

The gearset of Prob. 14–24 has had its gear specification changed to 9310 for carburizing and surface hardening with the result that the pinion Brinell hardnesses are 285 core and 580–600 case, and the gear hardnesses are 285 core and 580–600 case. Estimate the power rating for the new gearset.

14–28

ANALYSIS

The gearset of Prob. 14–27 is going to be upgraded in material to a quality of grade 2 9310 steel. Estimate the power rating for the new gearset.

14–29

Matters of scale always improve insight and perspective. Reduce the physical size of the gearset in Prob. 14–24 by one-half and note the result on the estimates of transmitted load W^t and power.

14–30

ANALYSIS

AGMA procedures with cast-iron gear pairs differ from those with steels because life predictions are difficult; consequently $(Y_N)_P$, $(Y_N)_G$, $(Z_N)_P$, and $(Z_N)_G$ are set to unity. The consequence of this is that the fatigue strengths of the pinion and gear materials are the same. The reliability is 0.99 and the life is 10^7 revolution of the pinion ($K_R = 1$). For longer lives the reducer is derated in power. For the pinion and gear set of Prob. 14–24, use grade 40 cast iron for both gears ($H_B = 201$ Brinell). Rate the reducer for power with S_F and S_H equal to unity.

14–31

ANALYSIS

Spur-gear teeth have rolling and slipping contact (often about 8 percent slip). Spur gears tested to wear failure are reported at 10^8 cycles as Buckingham's surface fatigue load-stress factor K. This factor is related to Hertzian contact strength S_C by

$$S_C = \sqrt{\frac{1.4K}{(1/E_1 + 1/E_2)\sin\phi}}$$

where ϕ is the normal pressure angle. Cast iron grade 20 gears with $\phi = 14\frac{1}{2}°$ and 20° pressure angle exhibit a minimum K of 81 and 112 psi, respectively. How does this compare with $S_C = 0.32H_B$ kpsi?

14–32

You've probably noticed that although the AGMA method is based on two equations, the details of assembling all the factors is computationally intensive. To reduce error and omissions, a computer program would be useful. Write a program to perform a power rating of an existing gearset, then use Prob. 14–24, 14–26, 14–27, 14–28, and 14–29 to test your program by comparing the results to your longhand solutions.

14–33

ANALYSIS

In Ex. 14–5 use nitrided grade 1 steel (4140) which produces Brinell hardnesses of 250 core and 500 at the surface (case). Use the upper fatigue curves on Figs. 14–14 and 14–15. Estimate the power capacity of the mesh with factors of safety of $S_F = S_H = 1$.

14–34

ANALYSIS

In Ex. 14–5 use carburized and case-hardened gears of grade 1. Carburizing and case-hardening can produce a 550 Brinell case. The core hardnesses are 200 Brinell. Estimate the power capacity of the mesh with factors of safety of $S_F = S_H = 1$, using the lower fatigue curves in Figs. 14–14 and 14–15.

14–35

ANALYSIS

In Ex. 14–5, use carburized and case-hardened gears of grade 2 steel. The core hardnesses are 200, and surface hardnesses are 600 Brinell. Use the lower fatigue curves of Figs. 14–14 and 14–15. Estimate the power capacity of the mesh using $S_F = S_H = 1$. Compare the power capacity with the results of Prob. 14–34.

15

Bevel and Worm Gears

The American Gear Manufacturers Association (AGMA) has established standards for the analysis and design of the various kinds of bevel and worm gears. Chapter 14 was an introduction to the AGMA methods for spur and helical gears. AGMA has established similar methods for other types of gearing, which all follow the same general approach.

15–1 Bevel Gearing—General

Bevel gears may be classified as follows:

- Straight bevel gears
- Spiral bevel gears
- Zerol bevel gears
- Hypoid gears
- Spiroid gears

A straight bevel gear was illustrated in Fig. 13–35. These gears are usually used for pitch-line velocities up to 1000 ft/min (5 m/s) when the noise level is not an important consideration. They are available in many stock sizes and are less expensive to produce than other bevel gears, especially in small quantities.

A *spiral bevel* gear is shown in Fig. 15–1; the definition of the *spiral angle* is illustrated in Fig. 15–2. These gears are recommended for higher speeds and where the noise level is an important consideration. Spiral bevel gears are the bevel counterpart of the helical gear; it can be seen in Fig. 15–1 that the pitch surfaces and the nature of contact are the same as for straight bevel gears except for the differences brought about by the spiral-shaped teeth.

The *Zerol bevel gear* is a patented gear having curved teeth but with a zero spiral angle. The axial thrust loads permissible for Zerol bevel gears are not as large as those for the spiral bevel gear, and so they are often used instead of straight bevel gears. The Zerol bevel gear is generated by the same tool used for regular spiral bevel gears. For design purposes, use the same procedure as for straight bevel gears and then simply substitute a Zerol bevel gear.

Figure 15–1

Spiral bevel gears. *(Courtesy of Gleason Works, Rochester, N.Y.)*

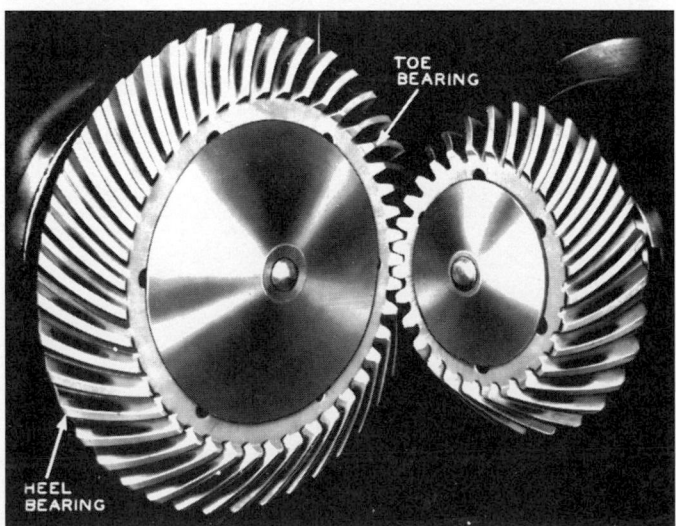

Figure 15–2

Cutting spiral-gear teeth on the basic crown rack.

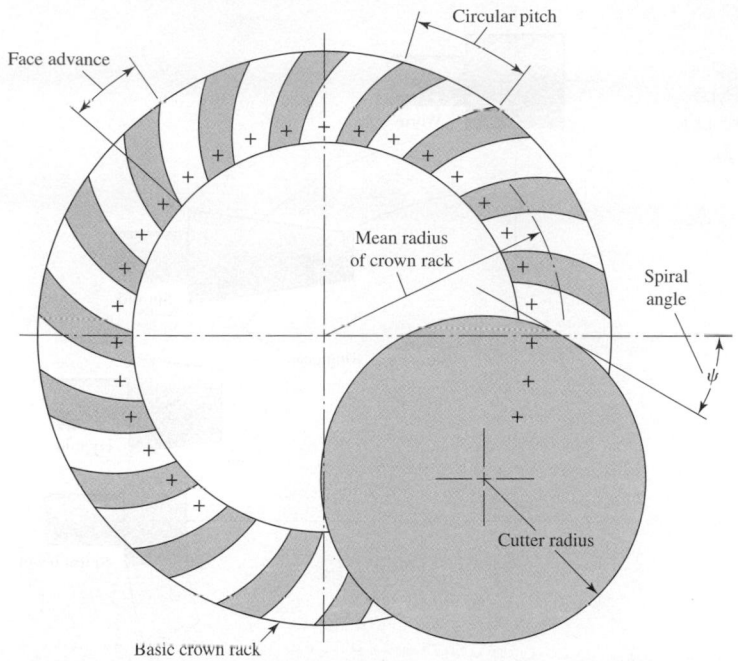

Figure 15–3

Hypoid gears. *(Courtesy of Gleason Works, Rochester, N.Y.)*

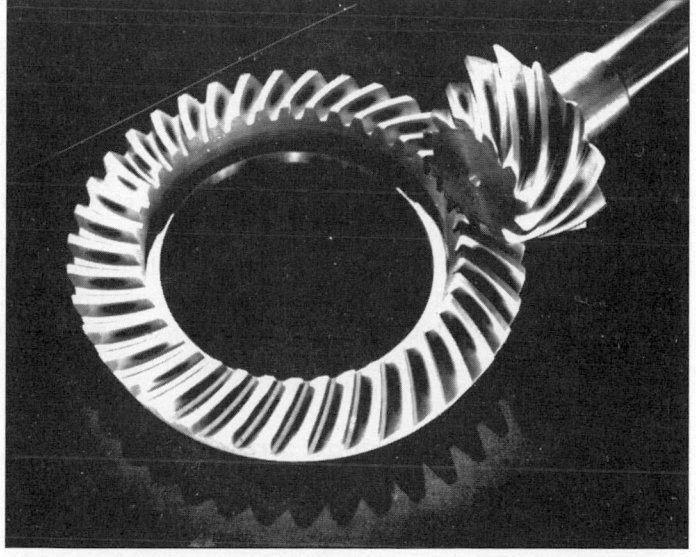

It is frequently desirable, as in the case of automotive differential applications, to have gearing similar to bevel gears but with the shafts offset. Such gears are called *hypoid gears,* because their pitch surfaces are hyperboloids of revolution. The tooth action between such gears is a combination of rolling and sliding along a straight line and has much in common with that of worm gears. Figure 15–3 shows a pair of hypoid gears in mesh.

Figure 15–4 is included to assist in the classification of spiral bevel gearing. It is seen that the hypoid gear has a relatively small shaft offset. For larger offsets, the pinion begins to resemble a tapered worm and the set is then called *spiroid gearing*.

Figure 15–4

Comparison of intersecting- and offset-shaft bevel-type gearings. *(From Gear Handbook by Darle W. Dudley, 1962, p. 2–24.)*

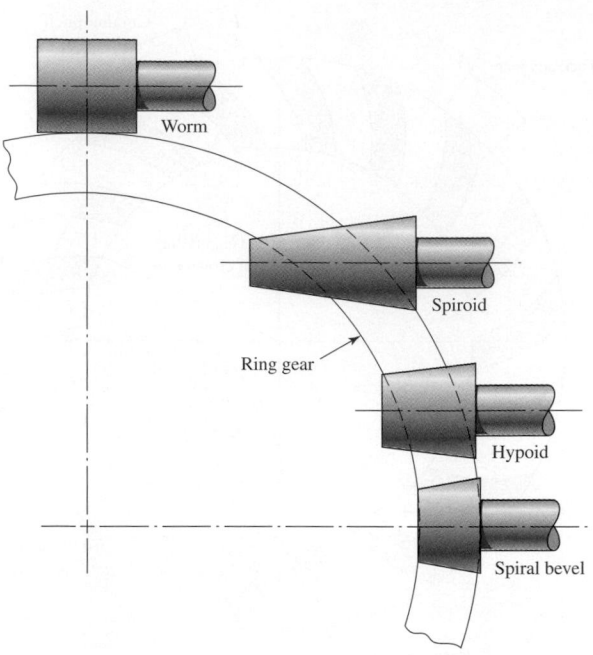

Worm

Spiroid

Ring gear

Hypoid

Spiral bevel

15–2 ## Bevel-Gear Stresses and Strengths

In a typical bevel-gear mounting, Fig. 13–36, for example, one of the gears is often mounted outboard of the bearings. This means that the shaft deflections can be more pronounced and can have a greater effect on the nature of the tooth contact. Another difficulty that occurs in predicting the stress in bevel-gear teeth is the fact that the teeth are tapered. Thus, to achieve perfect line contact passing through the cone center, the teeth ought to bend more at the large end than at the small end. To obtain this condition requires that the load be proportionately greater at the large end. Because of this varying load across the face of the tooth, it is desirable to have a fairly short face width.

Because of the complexity of bevel, spiral bevel, Zerol bevel, hypoid, and spiroid gears, as well as the limitations of space, only a portion of the applicable standards that refer to straight-bevel gears is presented here.[1] Table 15–1 gives the symbols used in ANSI/AGMA 2003-B97.

Fundamental Contact Stress Equation

$$s_c = \sigma_c = C_p \left(\frac{W^t}{F d_P I} K_o K_v K_m C_s C_{xc} \right)^{1/2} \qquad \text{(U.S. customary units)}$$

$$\sigma_H = Z_E \left(\frac{1000 W^t}{b d Z_1} K_A K_v K_{H\beta} Z_x Z_{xc} \right)^{1/2} \qquad \text{(SI units)}$$

(15–1)

We will use σ_c and σ_H for contact stress to remain consistent with our notation.

---

[1]Figures 15–5 to 15–13 and Tables 15–1 to 15–7 have been extracted from ANSI/AGMA 2003-B97, *Rating the Pitting Resistance and Bending Strength of Generated Straight Bevel, Zerol Bevel and Spiral Bevel Gear Teeth* with the permission of the publisher, the American Gear Manufacturers Association, 500 Montgomery Street, Suite 350, Alexandria, VA, 22314-1560.

Table 15–1

Symbols Used in Gear Rating Equations, ANSI/AGMA 2003-B97 Standard   *Source:* ANSI/AGMA 2003-B97.

AGMA Symbol	ISO Symbol	Description	Units
A_m	R_m	Mean cone distance	in (mm)
A_0	R_e	Outer cone distance	in (mm)
C_H	Z_W	Hardness ratio factor for pitting resistance	
C_i	Z_i	Inertia factor for pitting resistance	
C_L	Z_{NT}	Stress cycle factor for pitting resistance	
C_p	Z_E	Elastic coefficient	$[\text{lbf/in}^2]^{0.5}$ $([\text{N/mm}^2]^{0.5})$
C_R	Z_Z	Reliability factor for pitting	
C_{SF}		Service factor for pitting resistance	
C_S	Z_x	Size factor for pitting resistance	
C_{xc}	Z_{xc}	Crowning factor for pitting resistance	
D, d	d_{e2}, d_{e1}	Outer pitch diameters of gear and pinion, respectively	in (mm)
E_G, E_P	E_2, E_1	Young's modulus of elasticity for materials of gear and pinion, respectively	lbf/in^2 (N/mm^2)
e	e	Base of natural (Napierian) logarithms	
F	b	Net face width	in (mm)
F_{eG}, F_{eP}	b'_2, b'_1	Effective face widths of gear and pinion, respectively	in (mm)
f_P	R_{a1}	Pinion surface roughness	$\mu\text{in} (\mu\text{m})$
H_{BG}	H_{B2}	Minimum Brinell hardness number for gear material	HB
H_{BP}	H_{B1}	Minimum Brinell hardness number for pinion material	HB
h_c	$E_{ht\,min}$	Minimum total case depth at tooth middepth	in (mm)
h_e	h'_c	Minimum effective case depth	in (mm)
$h_{e\,lim}$	$h'_{c\,lim}$	Suggested maximum effective case depth limit at tooth middepth	in (mm)
I	Z_I	Geometry factor for pitting resistance	
J	Y_J	Geometry factor for bending strength	
J_G, J_P	Y_{J2}, Y_{J1}	Geometry factor for bending strength for gear and pinion, respectively	
K_F	Y_F	Stress correction and concentration factor	
K_i	Y_i	Inertia factor for bending strength	
K_L	Y_{NT}	Stress cycle factor for bending strength	
K_m	$K_{H\beta}$	Load distribution factor	
K_o	K_A	Overload factor	
K_R	Y_z	Reliability factor for bending strength	
K_S	Y_X	Size factor for bending strength	
K_{SF}		Service factor for bending strength	
K_T	K_θ	Temperature factor	
K_v	K_v	Dynamic factor	
K_x	Y_β	Lengthwise curvature factor for bending strength	
	m_{et}	Outer transverse module	(mm)
	m_{mt}	Mean transverse module	(mm)
	m_{mn}	Mean normal module	(mm)
m_{NI}	ε_{NI}	Load sharing ratio, pitting	
m_{NJ}	ε_{NJ}	Load sharing ratio, bending	
N	z_2	Number of gear teeth	
N_L	n_L	Number of load cycles	
n	z_1	Number of pinion teeth	
n_P	n_1	Pinion speed	rev/min

(Continued)

Table 15–1

Symbols Used in Gear Rating Equations, ANSI/AGMA 2003-B97 Standard (*Continued*)

AGMA Symbol	ISO Symbol	Description	Units
P	P	Design power through gear pair	hp (kW)
P_a	P_a	Allowable transmitted power	hp (kW)
P_{ac}	P_{az}	Allowable transmitted power for pitting resistance	hp (kW)
P_{acu}	P_{azu}	Allowable transmitted power for pitting resistance at unity service factor	hp (kW)
P_{at}	P_{ay}	Allowable transmitted power for bending strength	hp (kW)
P_{atu}	P_{ayu}	Allowable transmitted power for bending strength at unity service factor	hp (kW)
P_d		Outer transverse diametral pitch	in^{-1}
P_m		Mean transverse diametral pitch	in^{-1}
P_{mn}		Mean normal diametral pitch	in^{-1}
Q_v	Q_v	Transmission accuracy number	
q	q	Exponent used in formula for lengthwise curvature factor	
R, r	r_{mpt2}, r_{mpt1}	Mean transverse pitch radii for gear and pinion, respectively	in (mm)
R_t, r_t	r_{myo2}, r_{myo1}	Mean transverse radii to point of load application for gear and pinion, respectively	in (mm)
r_c	r_{c0}	Cutter radius used for producing Zerol bevel and spiral bevel gears	in (mm)
s	g_c	Length of the instantaneous line of contact between mating tooth surfaces	in (mm)
s_{ac}	$\sigma_{H\,lim}$	Allowable contact stress number	lbf/in^2 (N/mm^2)
s_{at}	$\sigma_{F\,lim}$	Bending stress number (allowable)	lbf/in^2 (N/mm^2)
s_c	σ_H	Calculated contact stress number	lbf/in^2 (N/mm^2)
s_F	s_F	Bending safety factor	
s_H	s_H	Contact safety factor	
s_t	σ_F	Calculated bending stress number	lbf/in^2 (N/mm^2)
s_{wc}	σ_{HP}	Permissible contact stress number	lbf/in^2 (N/mm^2)
s_{wt}	σ_{FP}	Permissible bending stress number	lbf/in^2 (N/mm^2)
T_P	T_1	Operating pinion torque	lbf in (Nm)
T_T	θ_T	Operating gear blank temperature	°F (°C)
t_0	s_{ai}	Normal tooth top land thickness at narrowest point	in (mm)
U_c	U_c	Core hardness coefficient for nitrided gear	lbf/in^2 (N/mm^2)
U_H	U_H	Hardening process factor for steel	lbf/in^2 (N/mm^2)
v_t	v_{et}	Pitch-line velocity at outer pitch circle	ft/min (m/s)
Y_{KG}, Y_{KP}	Y_{K2}, Y_{K1}	Tooth form factors including stress-concentration factor for gear and pinion, respectively	
μ_G, μ_P	v_2, v_1	Poisson's ratio for materials of gear and pinion, respectively	
ρ_o	ρ_{yo}	Relative radius of profile curvature at point of maximum contact stress between mating tooth surfaces	in (mm)
ϕ	α_n	Normal pressure angle at pitch surface	
ϕ_t	α_{wt}	Transverse pressure angle at pitch point	
ψ	β_m	Mean spiral angle at pitch surface	
ψ_b	β_{mb}	Mean base spiral angle	

Permissible Contact Stress Number (Strength) Equation

$$s_{wc} = (\sigma_c)_{\text{all}} = \frac{s_{ac}C_L C_H}{S_H K_T C_R} \qquad \text{(U.S. customary units)}$$

$$\sigma_{HP} = \frac{\sigma_{H\lim}Z_{NT}Z_W}{S_H K_\theta Z_Z} \qquad \text{(SI units)}$$

(15–2)

Bending Stress

$$s_t = \frac{W^t}{F}P_d K_o K_v \frac{K_s K_m}{K_x J} \qquad \text{(U.S. customary units)}$$

$$\sigma_F = \frac{1000W^t}{b}\frac{K_A K_v}{m_{et}}\frac{Y_x K_{H\beta}}{Y_\beta Y_J} \qquad \text{(SI units)}$$

(15–3)

Permissible Bending Stress Equation

$$s_{wt} = \frac{s_{at}K_L}{S_F K_T K_R} \qquad \text{(U.S. customary units)}$$

$$\sigma_{FP} = \frac{\sigma_{F\lim}Y_{NT}}{S_F K_\theta Y_z} \qquad \text{(SI units)}$$

(15–4)

15–3 AGMA Equation Factors

Overload Factor K_o (K_A)

The overload factor makes allowance for any externally applied loads in excess of the nominal transmitted load. Table 15–2, from Appendix A of 2003-B97, is included for your guidance.

Safety Factors S_H and S_F

The factors of safety S_H and S_F as defined in 2003-B97 are adjustments to strength, not load, and consequently cannot be used as is to assess (by comparison) whether the threat is from wear fatigue or bending fatigue. Since W^t is the same for the pinion and gear, the comparison of $\sqrt{S_H}$ to S_F allows direct comparison.

Dynamic Factor K_v

In 2003-C87 AGMA changed the definition of K_v to its reciprocal but used the same symbol. Other standards have yet to follow this move. The dynamic factor K_v makes

Table 15–2

Overload Factors K_o (K_A)

Source: ANSI/AGMA 2003-B97.

Character of	Character of Load on Driven Machine			
Prime Mover	**Uniform**	**Light Shock**	**Medium Shock**	**Heavy Shock**
Uniform	1.00	1.25	1.50	1.75 or higher
Light shock	1.10	1.35	1.60	1.85 or higher
Medium shock	1.25	1.50	1.75	2.00 or higher
Heavy shock	1.50	1.75	2.00	2.25 or higher

Note: This table is for speed-decreasing drives. For speed-increasing drives, add $0.01(N/n)^2$ or $0.01(z_2/z_1)^2$ to the above factors.

Figure 15–5

Dynamic factor K_v.
(Source: ANSI/AGMA 2003-B97.)

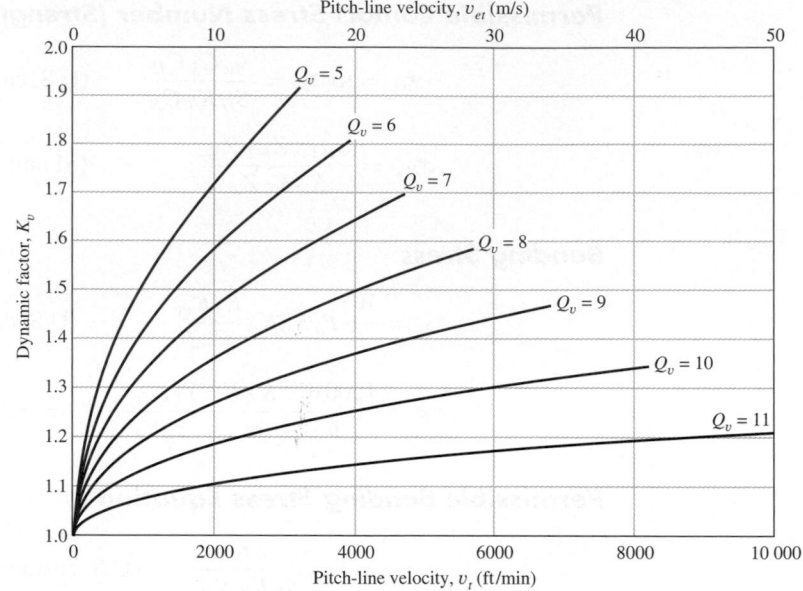

allowance for the effect of gear-tooth quality related to speed and load, and the increase in stress that follows. AGMA uses a *transmission accuracy number* Q_v to describe the precision with which tooth profiles are spaced along the pitch circle. Figure 15–5 shows graphically how pitch-line velocity and transmission accuracy number are related to the dynamic factor K_v. Curve fits are

$$K_v = \left(\frac{A + \sqrt{v_t}}{A}\right)^B \qquad \text{(U.S. customary units)}$$

$$K_v = \left(\frac{A + \sqrt{200v_{et}}}{A}\right)^B \qquad \text{(SI units)}$$

(15–5)

where

$$A = 50 + 56(1 - B)$$
$$B = 0.25(12 - Q_v)^{2/3}$$

(15–6)

and $v_t(v_{et})$ is the pitch-line velocity at outside pitch diameter, expressed in ft/min (m/s):

$$v_t = \pi d_P n_P / 12 \qquad \text{(U.S. customary units)}$$
$$v_{et} = 5.236(10^{-5})d_1 n_1 \qquad \text{(SI units)}$$

(15–7)

The maximum recommended pitch-line velocity is associated with the abscissa of the terminal points of the curve in Fig. 15–5:

$$v_{t\,\text{max}} = [A + (Q_v - 3)]^2 \qquad \text{(U.S. customary units)}$$
$$v_{te\,\text{max}} = \frac{[A + (Q_v - 3)]^2}{200} \qquad \text{(SI units)}$$

(15–8)

where $v_{t\,\text{max}}$ and $v_{et\,\text{max}}$ are in ft/min and m/s, respectively.

Size Factor for Pitting Resistance C_s (Z_x)

$$C_s = \begin{cases} 0.5 & F < 0.5 \text{ in} \\ 0.125F + 0.4375 & 0.5 \leq F \leq 4.5 \text{ in} \quad \text{(U.S. customary units)} \\ 1 & F > 4.5 \text{ in} \end{cases} \quad (15\text{--}9)$$

$$Z_x = \begin{cases} 0.5 & b < 12.7 \text{ mm} \\ 0.004\,92b + 0.4375 & 12.7 \leq b \leq 114.3 \text{ mm} \quad \text{(SI units)} \\ 1 & b > 114.3 \text{ mm} \end{cases}$$

Size Factor for Bending K_s (Y_x)

$$K_s = \begin{cases} 0.4867 + 0.2132/P_d & 0.5 \leq P_d \leq 16 \text{ in}^{-1} \\ 0.5 & P_d > 16 \text{ in}^{-1} \end{cases} \quad \text{(U.S. customary units)}$$

$$\quad (15\text{--}10)$$

$$Y_x = \begin{cases} 0.5 & m_{et} < 1.6 \text{ mm} \\ 0.4867 + 0.008\,339m_{et} & 1.6 \leq m_{et} \leq 50 \text{ mm} \end{cases} \quad \text{(SI units)}$$

Load-Distribution Factor K_m ($K_{H\beta}$)

$$K_m = K_{mb} + 0.0036F^2 \qquad \text{(U.S. customary units)}$$
$$K_{H\beta} = K_{mb} + 5.6(10^{-6})b^2 \qquad \text{(SI units)} \qquad (15\text{--}11)$$

where

$$K_{mb} = \begin{cases} 1.00 & \text{both members straddle-mounted} \\ 1.10 & \text{one member straddle-mounted} \\ 1.25 & \text{neither member straddle-mounted} \end{cases}$$

Crowning Factor for Pitting C_{xc} (Z_{xc})

The teeth of most bevel gears are crowned in the lengthwise direction during manufacture to accommodate to the deflection of the mountings.

$$C_{xc} = Z_{xc} = \begin{cases} 1.5 & \text{properly crowned teeth} \\ 2.0 & \text{or larger uncrowned teeth} \end{cases} \qquad (15\text{--}12)$$

Lengthwise Curvature Factor for Bending Strength K_x (Y_β)

For straight-bevel gears,

$$K_x = Y_\beta = 1 \qquad (15\text{--}13)$$

Pitting Resistance Geometry Factor I (Z_I)

Figure 15–6 shows the geometry factor I (Z_I) for straight-bevel gears with a 20° pressure angle and 90° shaft angle. Enter the figure ordinate with the number of pinion teeth, move to the number of gear-teeth contour, and read from the abscissa.

Bending Strength Geometry Factor J (Y_J)

Figure 15–7 shows the geometry factor J for straight-bevel gears with a 20° pressure angle and 90° shaft angle.

Figure 15–6

Contact geometry factor $I\,(Z_I)$ for coniflex straight-bevel gears with a 20° normal pressure angle and a 90° shaft angle. *(Source: ANSI/AGMA 2003-B97.)*

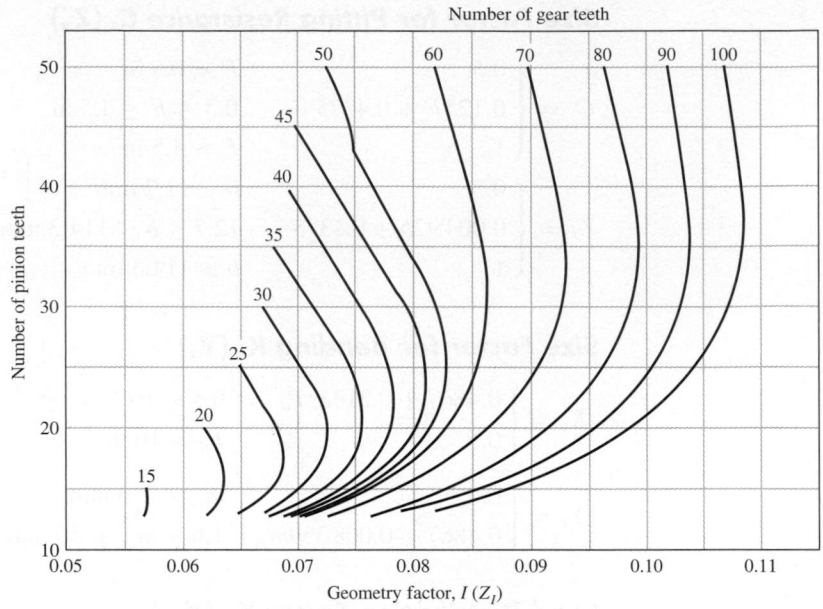

Figure 15–7

Bending factor $J\,(Y_J)$ for coniflex straight-bevel gears with a 20° normal pressure angle and 90° shaft angle. *(Source: ANSI/AGMA 2003-B97.)*

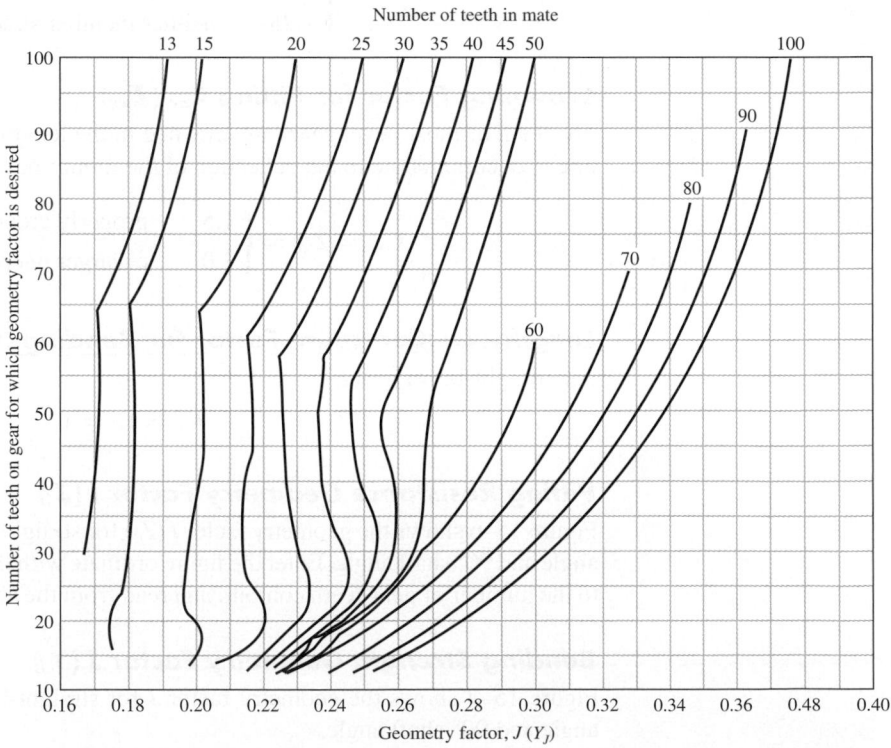

Stress-Cycle Factor for Pitting Resistance C_L (Z_{NT})

$$C_L = \begin{cases} 2 & 10^3 \leq N_L < 10^4 \\ 3.4822 N_L^{-0.0602} & 10^4 \leq N_L \leq 10^{10} \end{cases}$$

$$Z_{NT} = \begin{cases} 2 & 10^3 \leq n_L < 10^4 \\ 3.4822 n_L^{-0.0602} & 10^4 \leq n_L \leq 10^{10} \end{cases}$$

(15–14)

See Fig. 15–8 for a graphical presentation of Eqs. (15–14).

Stress-Cycle Factor for Bending Strength K_L (Y_{NT})

$$K_L = \begin{cases} 2.7 & 10^2 \leq N_L < 10^3 \\ 6.1514 N_L^{-0.1182} & 10^3 \leq N_L < 3(10^6) \\ 1.6831 N_L^{-0.0323} & 3(10^6) \leq N_L \leq 10^{10} \quad \text{general} \\ 1.3558 N_L^{-0.0178} & 3(10^6) \leq N_L \leq 10^{10} \quad \text{critical} \end{cases}$$

$$Y_{NT} = \begin{cases} 2.7 & 10^2 \leq n_L < 10^3 \\ 6.1514 n_L^{-0.1182} & 10^3 \leq n_L < 3(10^6) \\ 1.6831 n_L^{-0.0323} & 3(10^6) \leq n_L \leq 10^{10} \quad \text{general} \\ 1.3558 n_L^{-0.0323} & 3(10^6) \leq n_L \leq 10^{10} \quad \text{critical} \end{cases}$$

(15–15)

See Fig. 15–9 for a plot of Eqs. (15–15).

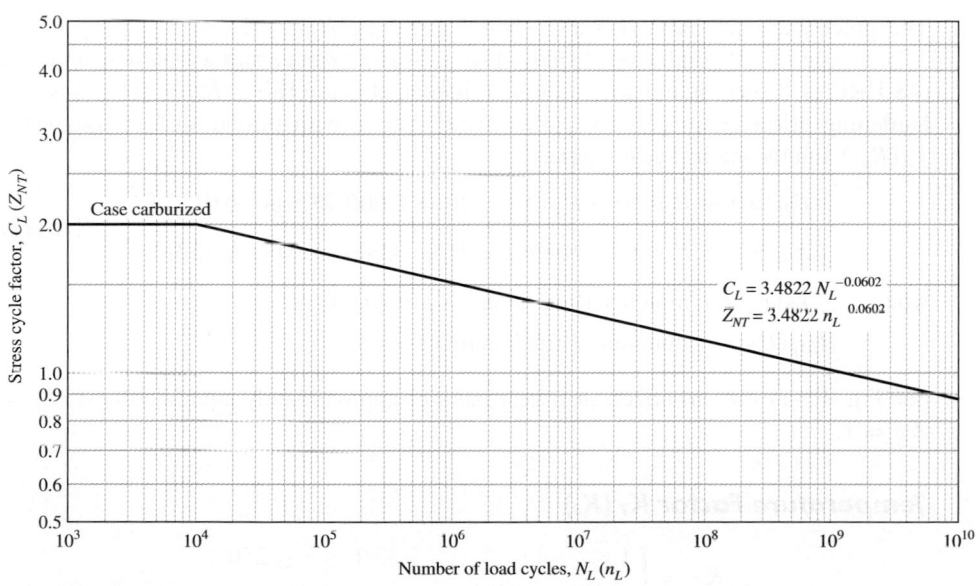

Figure 15–8

Contact stress cycle factor for pitting resistance C_l (Z_{NT}) for carburized case-hardened steel bevel gears. *(Source: ANSI/AGMA 2003-B97.)*

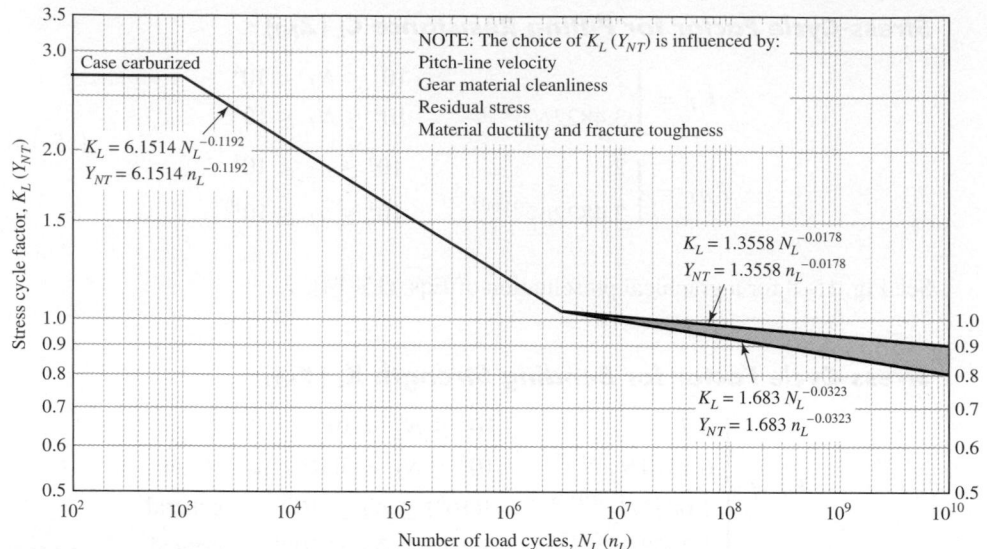

Figure 15–9

Stress cycle factor for bending strength K_L (Y_{NT}) for carburized case-hardened steel bevel gears.
(Source: ANSI/AGMA 2003-B97.)

Hardness-Ratio Factor C_H (Z_W)

$$C_H = 1 + B_1(N/n - 1) \qquad B_1 = 0.008\,98(H_{BP}/H_{BG}) - 0.008\,29$$

$$Z_W = 1 + B_1(z_1/z_2 - 1) \qquad B_1 = 0.008\,98(H_{B1}/H_{B2}) - 0.008\,29 \tag{15–16}$$

The preceding equations are valid when $1.2 \le H_{BP}/H_{BG} \le 1.7$ ($1.2 \le H_{B1}/H_{B2} \le 1.7$). Figure 15–10 graphically displays Eqs. (15–16). When a surface-hardened pinion (48 HRC or harder) is run with a through-hardened gear ($180 \le H_B \le 400$), a work-hardening effect occurs. The C_H(Z_W) factor varies with pinion surface roughness f_P(R_{a1}) and the mating-gear hardness:

$$C_H = 1 + B_2(450 - H_{BG}) \qquad B_2 = 0.000\,75 \exp(-0.0122 f_P)$$

$$Z_W = 1 + B_2(450 - H_{B2}) \qquad B_2 = 0.000\,75 \exp(-0.52 f_P) \tag{15–17}$$

where $\quad f_P(R_{a1}) =$ pinion surface hardness μin (μm)

$\qquad H_{BG}(H_{B2}) =$ minimum Brinell hardness

See Fig. 15–11 for carburized steel gear pairs of approximately equal hardness $C_H = Z_W = 1$.

Temperature Factor K_T (K_θ)

$$K_T = \begin{cases} 1 & 32°\text{F} \le t \le 250°\text{F} \\ (460 + t)/710 & t > 250°\text{F} \end{cases}$$

$$K_\theta = \begin{cases} 1 & 0°\text{C} \le \theta \le 120°\text{C} \\ (273 + \theta)/393 & \theta > 120°\text{C} \end{cases} \tag{15–18}$$

Figure 15–10

Hardness-ratio factor C_H (Z_W) for through-hardened pinion and gear.
(*Source: ANSI/AGMA 2003-B97.*)

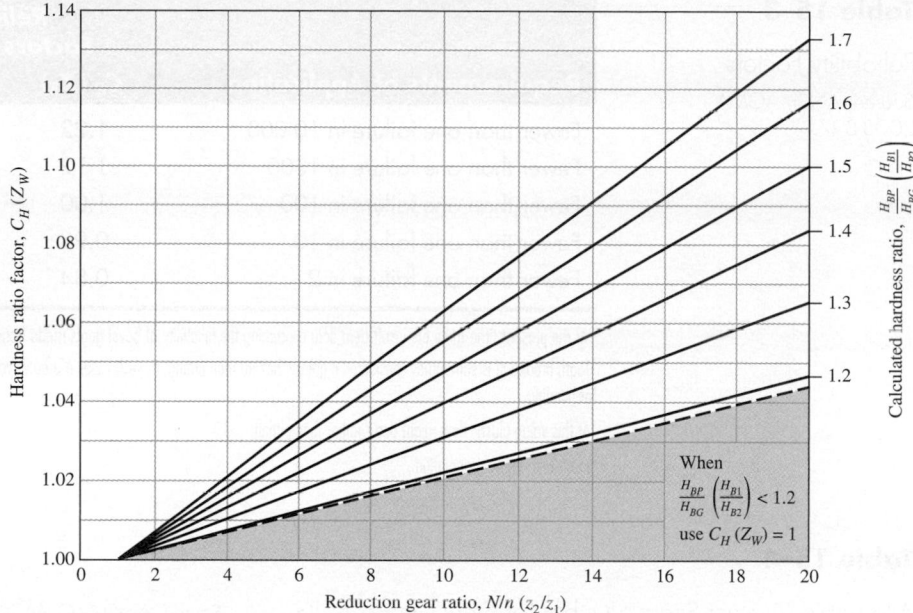

Figure 15–11

Hardness-ratio factor C_H (Z_W) for surface-hardened pinions.
(*Source: ANSI/AGMA 2003-B97.*)

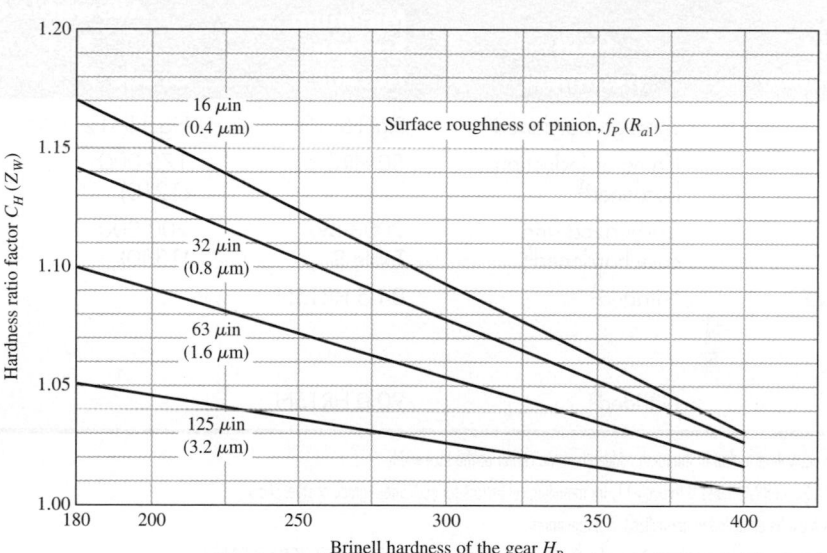

Reliability Factors C_R (Z_Z) and K_R (Y_Z)

Table 15–3 displays the reliability factors. Note that $C_R = \sqrt{K_R}$ and $Z_Z = \sqrt{Y_Z}$. Logarithmic interpolation equations are

$$Y_Z = K_R = \begin{cases} 0.50 - 0.25 \log(1 - R) & 0.99 \leq R \leq 0.999 & (15\text{–}19) \\ 0.70 - 0.15 \log(1 - R) & 0.90 \leq R < 0.99 & (15\text{–}20) \end{cases}$$

The reliability of the stress (fatigue) numbers allowable in Tables 15–4, 15–5, 15–6, and 15–7 is 0.99.

Table 15–3

Reliability Factors

Source: ANSI/AGMA 2003-B97.

Requirements of Application	Reliability Factors for Steel*	
	C_R (Z_Z)	K_R (Y_Z)[†]
Fewer than one failure in 10 000	1.22	1.50
Fewer than one failure in 1000	1.12	1.25
Fewer than one failure in 100	1.00	1.00
Fewer than one failure in 10	0.92	0.85[‡]
Fewer than one failure in 2	0.84	0.70[§]

*At the present time there are insufficient data concerning the reliability of bevel gears made from other materials.

[†]Tooth breakage is sometimes considered a greater hazard than pitting. In such cases a greater value of K_R (Y_Z) is selected for bending.

[‡]At this value plastic flow might occur rather than pitting.

[§]From test data extrapolation.

Table 15–4

Allowable Contact Stress Number for Steel Gears, s_{ac} ($\sigma_{H\,lim}$)   *Source:* ANSI/AGMA 2003-B97.

Material Designation	Heat Treatment	Minimum Surface* Hardness	Allowable Contact Stress Number, s_{ac} ($\sigma_{H\,lim}$) lbf/in² (N/mm²)		
			Grade 1[†]	Grade 2[†]	Grade 3[†]
Steel	Through-hardened[‡]	Fig.15–12	Fig.15–12	Fig.15–12	
	Flame or induction hardened[§]	50 HRC	175 000 (1210)	190 000 (1310)	
	Carburized and case hardened[§]	2003-B97 Table 8	200 000 (1380)	225 000 (1550)	250 000 (1720)
AISI 4140	Nitrided[§]	84.5 HR15N		145 000 (1000)	
Nitralloy 135M	Nitrided[§]	90.0 HR15N		160 000 (1100)	

*Hardness to be equivalent to that at the tooth middepth in the center of the face width.

[†]See ANSI/AGMA 2003-B97, Tables 8 through 11, for metallurgical factors for each stress grade of steel gears.

[‡]These materials must be annealed or normalized as a minumum.

[§]The allowable stress numbers indicated may be used with the case depths prescribed in 21.1, ANSI/AGMA 2003-B97.

Elastic Coefficient for Pitting Resistance C_p (Z_E)

$$C_p = \sqrt{\frac{1}{\pi\left[(1 - \nu_P^2)/E_P + (1 - \nu_G^2)/E_G\right]}}$$

$$Z_E = \sqrt{\frac{1}{\pi\left[(1 - \nu_1^2)/E_1 + (1 - \nu_2^2)/E_2\right]}}$$

(15–21)

Table 15–5

Allowable Contact Stress Number for Iron Gears, s_{ac} ($\sigma_{H\,\text{lim}}$)   *Source:* ANSI/AGMA 2003-B97.

Material	Material Designation ASTM	Material Designation ISO	Heat Treatment	Typical Minimum Surface Hardness	Allowable Contact Stress Number, s_{ac} ($\sigma_{H\,\text{lim}}$) lbf/in^2 (N/mm^2)
Cast iron	ASTM A48 Class 30 Class 40	ISO/DR 185 Grade 200 Grade 300	As cast As cast	175 HB 200 HB	50 000 (345) 65 000 (450)
Ductile (nodular) iron	ASTM A536 Grade 80-55-06 Grade 120-90-02	ISO/DIS 1083 Grade 600-370-03 Grade 800-480-02	Quenched and tempered	180 HB 300 HB	94 000 (650) 135 000 (930)

Table 15–6

Allowable Bending Stress Numbers for Steel Gears, s_{at} ($\sigma_{F\,\text{lim}}$)   *Source:* ANSI/AGMA 2003-B97.

Material Designation	Heat Treatment	Minimum Surface Hardness	Bending Stress Number (Allowable), s_{at} ($\sigma_{F\,\text{lim}}$) lbf/in^2 (N/mm^2) Grade 1*	Grade 2*	Grade 3*
Steel	Through-hardened	Fig. 15–13	Fig. 15–13	Fig. 15–13	
	Flame or induction hardened Unhardened roots Hardened roots	50 HRC	15 000 (85) 22 500 (154)	13 500 (95)	
	Carburized and case hardened†	2003-B97 Table 8	30 000 (205)	35 000 (240)	40 000 (275)
AISI 4140	Nitrided†,‡	84.5 HR15N		22 000 (150)	
Nitralloy 135M	Nitrided†,‡	90.0 HR15N		24 000 (165)	

*See ANSI/AGMA 2003-B97, Tables 8–11, for metallurgical factors for each stress grade of steel gears.

†The allowable stress numbers indicated may be used with the case depths prescribed in 21.1, ANSI/AGMA 2003-B97.

‡The overload capacity of nitrided gears is low. Since the shape of the effective S-N curve is flat, the sensitivity to shock should be investigated before proceeding with the design.

where

C_p = elastic coefficient, 2290 $\sqrt{\text{psi}}$ for steel

Z_E = elastic coefficient, 190 $\sqrt{\text{N/mm}^2}$ for steel

E_P and E_G = Young's moduli for pinion and gear respectively, psi

E_1 and E_2 = Young's moduli for pinion and gear respectively, N/mm^2

Allowable Contact Stress

Tables 15–4 and 15–5 provide values of $s_{ac}(\sigma_H)$ for steel gears and for iron gears, respectively. Figure 15–12 graphically displays allowable stress for grade 1 and 2 materials.

Table 15-7

Allowable Bending Stress Number for Iron Gears, s_{at} ($\sigma_{F\,\text{lim}}$)   Source: ANSI/AGMA 2003-B97.

| Material | Material Designation | | Heat Treatment | Typical Minimum Surface Hardness | Bending Stress Number (Allowable), s_{at} ($\sigma_{F\,\text{lim}}$) lbf/in^2 (N/mm^2) |
	ASTM	ISO			
Cast iron	ASTM A48	ISO/DR 185			
	Class 30	Grade 200	As cast	175 HB	4500 (30)
	Class 40	Grade 300	As cast	200 HB	6500 (45)
Ductile (nodular) iron	ASTM A536 Grade 80-55-06 Grade 120-90-02	ISO/DIS 1083 Grade 600-370-03 Grade 800-480-02	Quenched and tempered	180 HB 300 HB	10 000 (70) 13 500 (95)

Figure 15-12

Allowable contact stress number for through-hardened steel gears, s_{ac} ($\sigma_{H\,\text{lim}}$).
(Source: ANSI/AGMA 2003-B97.)

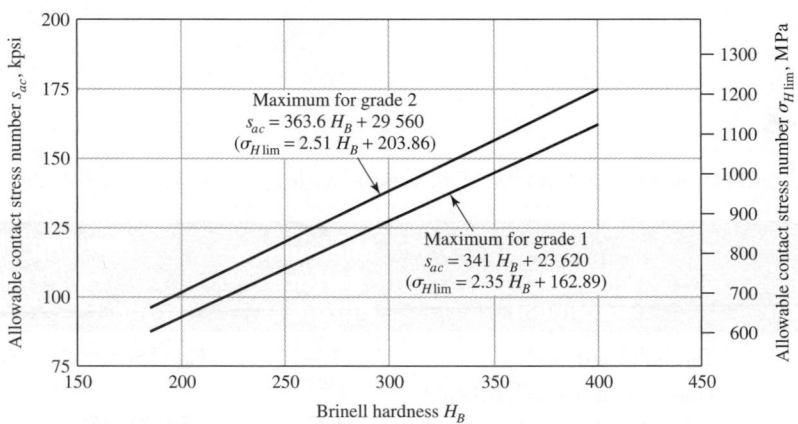

The equations are

$$s_{ac} = 341H_B + 23\,620 \text{ psi} \qquad \text{grade 1}$$
$$\sigma_{H\,\text{lim}} = 2.35H_B + 162.89 \text{ MPa} \qquad \text{grade 1}$$
$$s_{ac} = 363.6H_B + 29\,560 \text{ psi} \qquad \text{grade 2}$$
$$\sigma_{H\,\text{lim}} = 2.51H_B + 203.86 \text{ MPa} \qquad \text{grade 2}$$

$$(15\text{-}22)$$

Allowable Bending Stress Numbers

Tables 15-6 and 15-7 provide s_{at} ($\sigma_{F\,\text{lim}}$) for steel gears and for iron gears, respectively. Figure 15-13 shows graphically allowable bending stress s_{at} ($\sigma_{H\,\text{lim}}$) for through-hardened steels. The equations are

$$s_{at} = 44H_B + 2100 \text{ psi} \qquad \text{grade 1}$$
$$\sigma_{F\,\text{lim}} = 0.30H_B + 14.48 \text{ MPa} \qquad \text{grade 1}$$
$$s_{at} = 48H_B + 5980 \text{ psi} \qquad \text{grade 2}$$
$$\sigma_{H\,\text{lim}} = 0.33H_B + 41.24 \text{ MPa} \qquad \text{grade 2}$$

$$(15\text{-}23)$$

Reversed Loading

AGMA recommends use of 70 percent of allowable strength in cases where tooth load is completely reversed, as in idler gears and reversing mechanisms.

Summary

Figure 15-14 is a "roadmap" for straight-bevel gear wear relations using 2003-B97. Figure 15-15 is a similar guide for straight-bevel gear bending using 2003-B97.

Figure 15–13

Allowable bending stress number for through-hardened steel gears, $s_{at}(\sigma_{F\,\text{lim}})$.
(Source: ANSI/AGMA 2003-B97.)

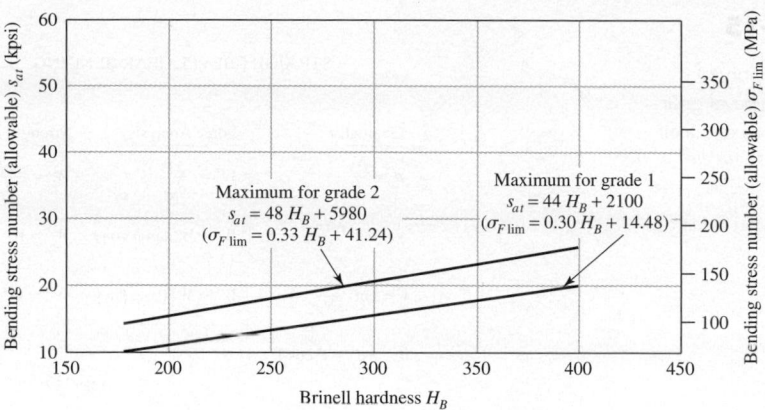

Maximum for grade 2
$s_{at} = 48\,H_B + 5980$
$(\sigma_{F\,\text{lim}} = 0.33\,H_B + 41.24)$

Maximum for grade 1
$s_{at} = 44\,H_B + 2100$
$(\sigma_{F\,\text{lim}} = 0.30\,H_B + 14.48)$

Figure 15–14

"Roadmap" summary of principal straight bevel gear wear equations and their parameters.

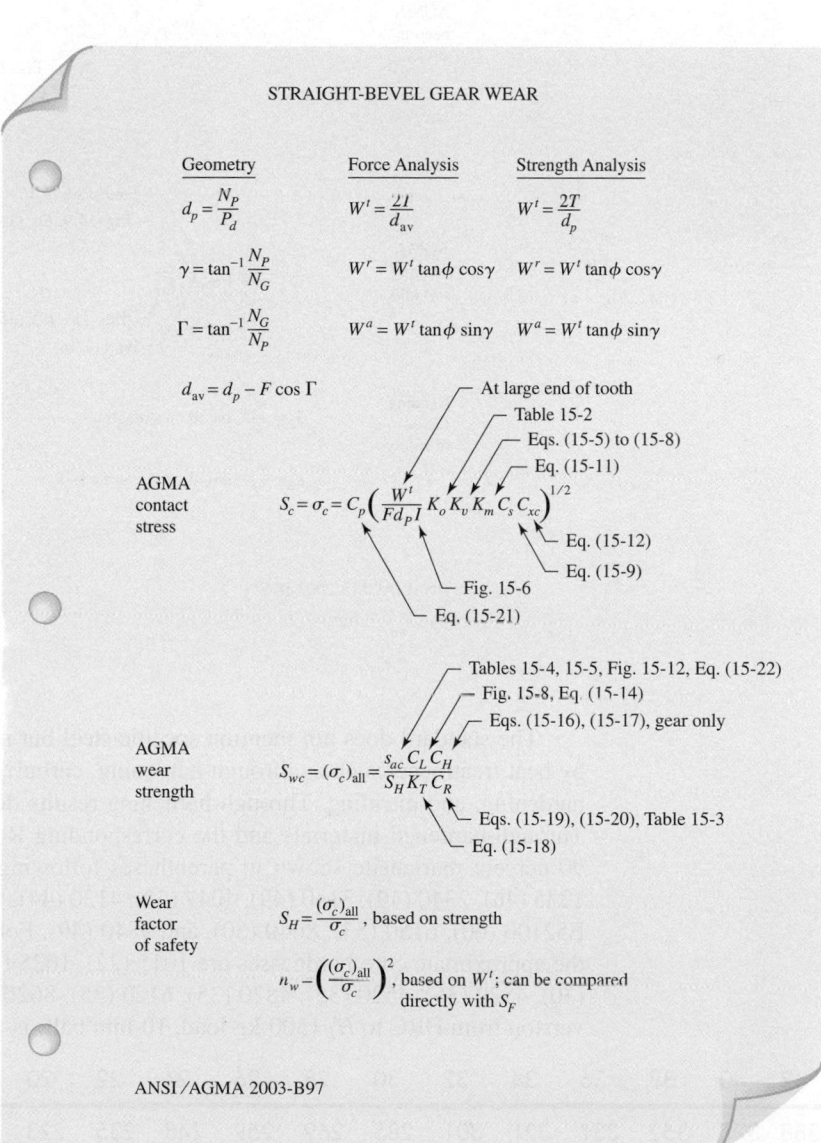

STRAIGHT-BEVEL GEAR WEAR

Geometry	Force Analysis	Strength Analysis
$d_p = \dfrac{N_P}{P_d}$	$W^t = \dfrac{2T}{d_{av}}$	$W^t = \dfrac{2T}{d_p}$
$\gamma = \tan^{-1}\dfrac{N_P}{N_G}$	$W^r = W^t \tan\phi \cos\gamma$	$W^r = W^t \tan\phi \cos\gamma$
$\Gamma = \tan^{-1}\dfrac{N_G}{N_P}$	$W^a = W^t \tan\phi \sin\gamma$	$W^a = W^t \tan\phi \sin\gamma$

$d_{av} = d_p - F\cos\Gamma$

AGMA contact stress

$$S_c = \sigma_c = C_p\left(\frac{W^t}{Fd_pI}K_oK_vK_mC_sC_{xc}\right)^{1/2}$$

— At large end of tooth
— Table 15-2
— Eqs. (15-5) to (15-8)
— Eq. (15-11)
— Eq. (15-12)
— Eq. (15-9)
— Fig. 15-6
— Eq. (15-21)

AGMA wear strength

$$S_{wc} = (\sigma_c)_{\text{all}} = \frac{s_{ac}\,C_L\,C_H}{S_H\,K_T\,C_R}$$

— Tables 15-4, 15-5, Fig. 15-12, Eq. (15-22)
— Fig. 15-8, Eq. (15-14)
— Eqs. (15-16), (15-17), gear only
— Eqs. (15-19), (15-20), Table 15-3
— Eq. (15-18)

Wear factor of safety

$$S_H = \frac{(\sigma_c)_{\text{all}}}{\sigma_c}, \text{ based on strength}$$

$$n_w = \left(\frac{(\sigma_c)_{\text{all}}}{\sigma_c}\right)^2, \text{ based on } W^t; \text{ can be compared directly with } S_F$$

ANSI/AGMA 2003-B97

Figure 15–15

"Roadmap" summary of principal straight-bevel gear bending equations and their parameters.

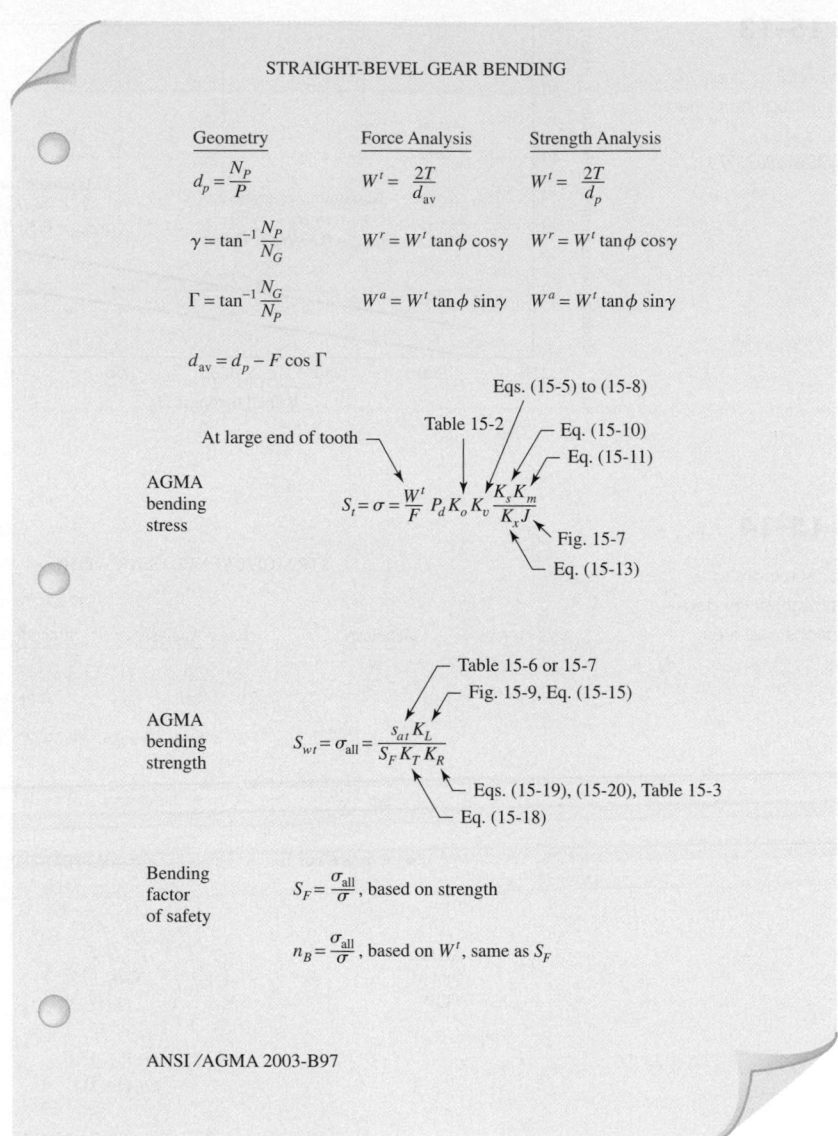

The standard does not mention specific steel but mentions the hardness attainable by heat treatments such as through-hardening, carburizing and case-hardening, flame-hardening, and nitriding. Through-hardening results depend on size (diametral pitch). Through-hardened materials and the corresponding Rockwell C-scale hardness at the 90 percent martensite shown in parentheses following include 1045 (50), 1060 (54), 1335 (46), 2340 (49), 3140 (49), 4047 (52), 4130 (44), 4140 (49), 4340 (49), 5145 (51), E52100 (60), 6150 (53), 8640 (50), and 9840 (49). For carburized case-hard materials the approximate core hardnesses are 1015 (22), 1025 (37), 1118 (33), 1320 (35), 2317 (30), 4320 (35), 4620 (35), 4820 (35), 6120 (35), 8620 (35), and E9310 (30). The conversion from HRC to H_B (300-kg load, 10-mm ball) is

HRC	42	40	38	36	34	32	30	28	26	24	22	20	18	16	14	12	10
H_B	388	375	352	331	321	301	285	269	259	248	235	223	217	207	199	192	187

Most bevel-gear sets are made from carburized case-hardened steel, and the factors incorporated in 2003-B97 largely address these high-performance gears. For through-hardened gears, 2003-B97 is silent on K_L and C_L, and Figs. 15–8 and 15–9 should prudently be considered as approximate.

15–4   Straight-Bevel Gear Analysis

EXAMPLE 15–1   A pair of identical straight-tooth miter gears listed in a catalog has a diametral pitch of 5 at the large end, 25 teeth, a 1.10-in face width, and a 20° normal pressure angle; the gears are grade 1 steel through-hardened with a core and case hardness of 180 Brinell. The gears are uncrowned and intended for general industrial use. They have a quality number of $Q_v = 7$. It is likely that the application intended will require outboard mounting of the gears. Use a safety factor of 1, a 10^7 cycle life, and a 0.99 reliability.

(a) For a speed of 600 rev/min find the power rating of this gearset based on AGMA bending strength.

(b) For the same conditions as in part (a) find the power rating of this gearset based on AGMA wear strength.

(c) For a reliability of 0.995, a gear life of 10^9 revolutions, and a safety factor of $S_F = S_H = 1.5$, find the power rating for this gearset using AGMA strengths.

Solution   From Figs. 15–14 and 15–15,

$$d_P = N_P/P = 25/5 = 5.000 \text{ in}$$

$$v_t = \pi d_P n_P/12 = \pi(5)600/12 = 785.4 \text{ ft/min}$$

Overload factor: uniform-uniform loading, Table 15–2, $K_o = 1.00$.
Safety factor: $S_F = 1$, $S_H = 1$.
Dynamic factor K_v: from Eq. (15–6),

$$B = 0.25(12 - 7)^{2/3} = 0.731$$

$$A = 50 + 56(1 - 0.731) = 65.06$$

$$K_v = \left(\frac{65.06 + \sqrt{785.4}}{65.06}\right)^{0.731} = 1.299$$

From Eq. (15–8),

$$v_{t\,\text{max}} = [65.06 + (7 - 3)]^2 = 4769 \text{ ft/min}$$

$v_t < v_{t\,\text{max}}$, that is, $785.4 < 4769$ ft/min, therefore K_v is valid. From Eq. (15–10),

$$K_s = 0.4867 + 0.2132/5 = 0.529$$

From Eq. (15–11),

$$K_{mb} = 1.25 \quad \text{and} \quad K_m = 1.25 + 0.0036(1.10)^2 = 1.254$$

From Eq. (15–13), $K_x = 1$. From Fig. 15–6, $I = 0.065$; from Fig. 15–7, $J_P = 0.216$, $J_G = 0.216$. From Eq. (15–15),

$$K_L = 1.683(10^7)^{-0.0323} = 0.999\,96 \doteq 1$$

From Eq. (15–14),

$$C_L = 3.4822(10^7)^{-0.0602} = 1.32$$

Since $H_{BP}/H_{BG} = 1$, then from Fig. 15–10, $C_H = 1$. From Eqs. (15–13) and (15–18), $K_x = 1$ and $K_T = 1$, respectively. From Eq. (15–20),

$$K_R = 0.70 - 0.15 \log(1 - 0.99) = 1, \qquad C_R = \sqrt{K_R} = \sqrt{1} = 1$$

(a) *Bending:* From Eq. (15–23),

$$s_{at} = 44(180) + 2100 = 10\,020 \text{ psi}$$

From Eq. (15–3),

$$s_t = \sigma = \frac{W^t}{F} P_d K_o K_v \frac{K_s K_m}{K_x J} = \frac{W^t}{1.10}(5)(1)1.299 \frac{0.529(1.254)}{(1)0.216}$$

$$= 18.13 W^t$$

From Eq. (15–4),

$$s_{wt} = \frac{s_{at} K_L}{S_F K_T K_R} = \frac{10\,020(1)}{(1)(1)(1)} = 10\,020 \text{ psi}$$

Equating s_t and s_{wt},

$$18.13 W^t = 10\,020 \qquad W^t = 552.6 \text{ lbf}$$

Answer

$$H = \frac{W^t v_t}{33\,000} = \frac{552.6(785.4)}{33\,000} = 13.2 \text{ hp}$$

(b) *Wear:* From Fig. 15–12,

$$s_{ac} = 341(180) + 23\,620 = 85\,000 \text{ psi}$$

From Eq. (15–2),

$$\sigma_{c,\text{all}} = \frac{s_{ac} C_L C_H}{S_H K_T C_R} = \frac{85\,000(1.32)(1)}{(1)(1)(1)} = 112\,200 \text{ psi}$$

Now $C_p = 2290\sqrt{\text{psi}}$ from definitions following Eq. (15–21). From Eq. (15–9),

$$C_s = 0.125(1.1) + 0.4375 = 0.575$$

From Eq. (15–12), $C_{xc} = 2$. Substituting in Eq. (15–1) gives

$$\sigma_c = C_p \left(\frac{W^t}{F d_P I} K_o K_v K_m C_s C_{xc} \right)^{1/2}$$

$$= 2290 \left[\frac{W^t}{1.10(5)0.065}(1)1.299(1.254)0.575(2) \right]^{1/2} = 5242\sqrt{W^t}$$

Equating σ_c and $\sigma_{c,\text{all}}$ gives

$$5242\sqrt{W^t} = 112\,200, \qquad W^t = 458.1 \text{ lbf}$$

$$H = \frac{458.1(785.4)}{33\,000} = 10.9 \text{ hp}$$

Rated power for the gearset is

Answer

$$H = \min(12.9, 10.9) = 10.9 \text{ hp}$$

(c) Life goal 10^9 cycles, $R = 0.995$, $S_F = S_H = 1.5$, and from Eq. (15–15),

$$K_L = 1.683(10^9)^{-0.0323} = 0.8618$$

From Eq. (15–19),

$$K_R = 0.50 - 0.25 \log(1 - 0.995) = 1.075, \qquad C_R = \sqrt{K_R} = \sqrt{1.075} = 1.037$$

From Eq. (15–14),

$$C_L = 3.4822(10^9)^{-0.0602} = 1$$

Bending: From Eq. (15–23) and part (a), $s_{at} = 10\,020$ psi. From Eq. (15–3),

$$s_t = \sigma = \frac{W^t}{1.10}5(1)1.299\frac{0.529(1.254)}{(1)0.216} = 18.13W^t$$

From Eq. (15–4),

$$s_{wt} = \frac{s_{at}K_L}{S_F K_I K_R} = \frac{10\,020(0.8618)}{1.5(1)1.075} = 5355 \text{ psi}$$

Equating s_t to s_{wt} gives

$$18.13W^t = 5355 \qquad W^t = 295.4 \text{ lbf}$$

$$H = \frac{295.4(785.4)}{33\,000} = 7.0 \text{ hp}$$

Wear: From Eq. (15–22), and part (b), $s_{ac} = 85\,000$ psi.
Substituting into Eq. (15–2) gives

$$\sigma_{c,\text{all}} = \frac{s_{ac}C_L C_H}{S_H K_I C_R} = \frac{85\,000(1)(1)}{1.5(1)1.037} = 54\,640 \text{ psi}$$

Substituting into Eq. (15–1) gives, from part (b), $\sigma_c = 5242\sqrt{W^t}$.
Equating σ_c to $\sigma_{c,\text{all}}$ gives

$$\sigma_c = \sigma_{c,\text{all}} = 54\,640 = 5242\sqrt{W^t} \qquad W^t = 108.6 \text{ lbf}$$

The wear power is

$$H = \frac{108.6(785.4)}{33\,000} = 2.58 \text{ hp}$$

Answer

The mesh rated power is $H = \min(7.0, 2.58) = 2.6$ hp.

15–5 **Design of a Straight-Bevel Gear Mesh**

A useful decision set for straight-bevel gear design is

- Function
- Design factor
- Tooth system
- Tooth count

} A priori decisions

- Pitch and face width
- Quality number
- Gear material, core and case hardness
- Pinion material, core and case hardness

} Design variables

In bevel gears the quality number is linked to the wear strength. The J factor for the gear can be smaller than for the pinion. Bending strength is not linear with face width, because added material is placed at the small end of the teeth. Consequently, face width is roughly prescribed as

$$F = \min(0.3A_0, 10/P_d) \tag{15–24}$$

where A_0 is the cone distance (see Fig. 13–20), given by

$$A_0 = \frac{d_P}{2 \sin \gamma} = \frac{d_G}{2 \sin \Gamma} \tag{15–25}$$

EXAMPLE 15–2

Design a straight-bevel gear mesh for shaft centerlines that intersect perpendicularly, to deliver 6.85 hp at 900 rev/min with a gear ratio of 3:1, temperature of 300°F, normal pressure angle of 20°, using a design factor of 2. The load is uniform-uniform. Although the minimum number of teeth on the pinion is 13, which will mesh with 31 or more teeth without interference, use a pinion of 20 teeth. The material is to be AGMA grade 1 and the teeth are to be crowned. The reliability goal is 0.995 with a pinion life of 10^9 revolutions.

Solution

First we list the a priori decisions and their immediate consequences.

Function: 6.85 hp at 900 rev/min, gear ratio $m_G = 3$, 300°F environment, neither gear straddle-mounted, $K_{mb} = 1.25$ [Eq. (15–11)], $R = 0.995$ at 10^9 revolutions of the pinion,

Eq. (15–14): $\quad (C_L)_G = 3.4822(10^9/3)^{-0.0602} = 1.068$

$\quad (C_L)_P = 3.4822(10^9)^{-0.0602} = 1$

Eq. (15–15): $\quad (K_L)_G = 1.683(10^9/3)^{-0.0323} = 0.8929$

$\quad (K_L)_P = 1.683(10^9)^{-0.0323} = 0.8618$

Eq. (15–19): $\quad K_R = 0.50 - 0.25 \log(1 - 0.995) = 1.075$

$\quad C_R = \sqrt{K_R} = \sqrt{1.075} = 1.037$

Eq. (15–18): $\quad K_T = C_T = (460 + 300)/710 = 1.070$

Design factor: $n_d = 2$, $S_F = 2$, $S_H = \sqrt{2} = 1.414$.

Tooth system: crowned, straight-bevel gears, normal pressure angle 20°,

Eq. (15–13): $\qquad\qquad\qquad\qquad\qquad K_x = 1$

Eq. (15–12): $\qquad\qquad\qquad\qquad\qquad C_{xc} = 1.5.$

With $N_P = 20$ teeth, $N_G = (3)20 = 60$ teeth and from Fig. 15–14,

$$\gamma = \tan^{-1}(N_P/N_G) = \tan^{-1}(20/60) = 18.43° \qquad \Gamma = \tan^{-1}(60/20) = 71.57°$$

From Figs. 15–6 and 15–7, $I = 0.0825$, $J_P = 0.248$, and $J_G = 0.202$. Note that $J_P > J_G$.

Decision 1: Trial diametral pitch, $P_d = 8$ teeth/in.

Eq. (15–10): $\qquad K_s = 0.4867 + 0.2132/8 = 0.5134$

$$d_P = N_P/P_d = 20/8 = 2.5 \text{ in}$$

$$d_G = 2.5(3) = 7.5 \text{ in}$$

$$v_t = \pi d_P n_P/12 = \pi(2.5)900/12 = 589.0 \text{ ft/min}$$

$$W^t = 33\,000 \text{ hp}/v_t = 33\,000(6.85)/589.0 = 383.8 \text{ lbf}$$

Eq. (15–25): $\qquad A_0 = d_P/(2\sin\gamma) = 2.5/(2\sin 18.43°) = 3.954 \text{ in}$

Eq. (15–24):

$$F = \min(0.3A_0, 10/P_d) = \min[0.3(3.954), 10/8] = \min(1.186, 1.25) = 1.186 \text{ in}$$

Decision 2: Let $F = 1.25$ in. Then,

Eq. (15–9): $\qquad\qquad\qquad C_s = 0.125(1.25) + 0.4375 = 0.5937$

Eq. (15–11): $\qquad\qquad\qquad K_m = 1.25 + 0.0036(1.25)^2 = 1.256$

Decision 3: Let the transmission accuracy number be 6. Then, from Eq. (15–6),

$$B = 0.25(12 - 6)^{2/3} = 0.8255$$

$$A = 50 + 56(1 - 0.8255) = 59.77$$

Eq. (15–5): $\qquad\qquad K_v = \left(\frac{59.77 + \sqrt{589.0}}{59.77}\right)^{0.8255} = 1.325$

Decision 4: Pinion and gear material and treatment. Carburize and case-harden grade ASTM 1320 to

Core 21 HRC (H_B is 229 Brinell)
Case 55-64 HRC (H_B is 515 Brinell)

From Table 15–4, $s_{ac} = 200\,000$ psi and from Table 15–6, $s_{at} = 30\,000$ psi.

Gear bending: From Eq. (15–3), the bending stress is

$$(s_t)_G = \frac{W^t}{F} P_d K_o K_v \frac{K_s K_m}{K_x J_G} = \frac{383.8}{1.25} 8(1)1.325 \frac{0.5134(1.256)}{(1)0.202}$$

$$= 10\,390 \text{ psi}$$

The bending strength, from Eq. (15–4), is given by

$$(s_{wt})_G = \left(\frac{s_{at}K_L}{S_F K_T K_R}\right)_G = \frac{30\,000(0.8929)}{2(1.070)1.075} = 11\,640 \text{ psi}$$

The strength exceeds the stress by a factor of $11640/10390 = 1.12$, giving an actual factor of safety of $(S_F)_G = 2(1.12) = 2.24$.

Pinion bending: The bending stress can be found from

$$(s_t)_P = (s_t)_G \frac{J_G}{J_P} = 10\,390\frac{0.202}{0.248} = 8463 \text{ psi}$$

The bending strength, again from Eq. (15–4), is given by

$$(s_{wt})_P = \left(\frac{s_{at}K_L}{S_F K_T K_R}\right)_P = \frac{30\,000\,(0.8618)}{2(1.070)1.075} = 11\,240 \text{ psi}$$

The strength exceeds the stress by a factor of $11\,240/8463 = 1.33$, giving an actual factor of safety of $(S_F)_P = 2(1.33) = 2.66$.

Gear wear: The load-induced contact stress for the pinion and gear, from Eq. (15–1), is

$$s_c = C_p \left(\frac{W^t}{Fd_P I}K_o K_v K_m C_s C_{xc}\right)^{1/2}$$

$$= 2290\left[\frac{383.8}{1.25(2.5)0.0825}(1)1.325(1.256)0.5937(1.5)\right]^{1/2}$$

$$= 107\,560 \text{ psi}$$

From Eq. (15–2) the contact strength of the gear is

$$(s_{wc})_G = \left(\frac{s_{ac}C_L C_H}{S_H K_T C_R}\right)_G = \frac{200\,000(1.068)(1)}{\sqrt{2}(1.070)1.037} = 136\,120 \text{ psi}$$

The strength exceeds the stress by a factor of $136\,120/107\,560 = 1.266$, giving an actual factor of safety of $(S_H)_G^2 = 1.266^2(2) = 3.21$.

Pinion wear: From Eq. (15–2) the contact strength of the pinion is

$$(s_{wc})_P = \left(\frac{s_{ac}C_L C_H}{S_H K_T C_R}\right)_P = \frac{200\,000(1)(1)}{\sqrt{2}(1.070)1.037} = 127\,450 \text{ psi}$$

The strength exceeds the stress by a factor of $136\,120/127\,450 = 1.068$, giving an actual factor of safety of $(S_H)_P^2 = 1.068^2(2) = 2.28$.

The actual factors of safety are 2.24, 2.66, 3.21, and 2.28. Making a direct comparison of the factors, we note that the threat from gear bending and pinion wear are practically equal. We also note that three of the ratios are comparable. Our goal would be to make changes in the design decisions that drive the factors closer to 2. The next step would be to adjust the design variables. It is obvious that an iterative process is involved. We need a figure of merit to order the designs. A computer program clearly is desirable.

15–6 ## **Worm Gearing—AGMA Equation**

Since they are essentially nonenveloping worm gears, the *crossed helical* gears, shown in Fig. 15–16, can be considered with other worm gearing. Because the teeth of worm gears have *point contact* changing to *line contact* as the gears are used, worm gears are said to "wear in," whereas other types "wear out."

Crossed helical gears, and worm gears too, usually have a 90° shaft angle, though this need not be so. The relation between the shaft and helix angles is

$$\sum = \psi_P \pm \psi_G \tag{15–26}$$

where $\sum$ is the shaft angle. The plus sign is used when both helix angles are of the same hand, and the minus sign when they are of opposite hand. The subscript P in Eq. (15–26) refers to the pinion (worm); the subscript W is used for this same purpose. The subscript G refers to the gear, also called *gear wheel, worm wheel,* or simply the *wheel*. Table 15–8 gives cylindrical worm dimensions common to worm and gear.

Section 13–11 introduced worm gears, and Sec. 13–17 developed the force analysis and efficiency of worm gearing to which we will refer. Here our interest is in strength and durability. Good proportions indicate the mean worm diameter d_m falls in the range

$$\frac{C^{0.875}}{3} \le d_m \le \frac{C^{0.875}}{1.6} \tag{15–27}$$

Figure 15–16

View of the pitch cylinders of a pair of crossed helical gears.

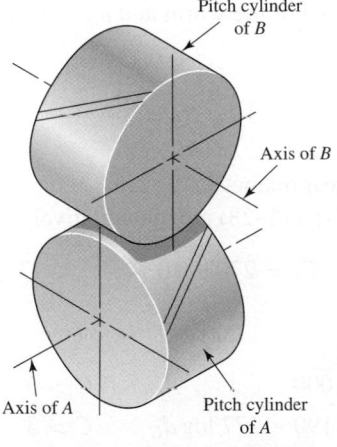

Pitch cylinder of B

Axis of B

Axis of A

Pitch cylinder of A

Table 15–8

Cylindrical Worm Dimensions Common to Both Worm and Gear*

Quantity	Symbol	ϕ_n 14.5° $N_W \le 2$	20° $N_W \le 2$	25° $N_W > 2$
Addendum	a	$0.3183p_x$	$0.3183p_x$	$0.286p_x$
Dedendum	b	$0.3683p_x$	$0.3683p_x$	$0.349p_x$
Whole depth	h_t	$0.6866p_x$	$0.6866p_x$	$0.635p_x$

*The table entries are for a tangential diametral pitch of the gear of $P_t = 1$.

where C is the center-to-center distance.[2] AGMA relates the allowable tangential force on the worm-gear tooth $(W^t)_{all}$ to other parameters by

$$(W^t)_{all} = C_s D_m^{0.8}(F_e)_G C_m C_v \tag{15-28}$$

where C_s = materials factor

D_m = mean gear diameter, in (mm)

F_e = effective face width of the gear (actual face width, but not to exceed $0.67d_m$, the mean worm diameter), in (mm)

C_m = ratio correction factor

C_v = velocity factor

The friction force W_f is given by

$$W_f = \frac{f W^t}{\cos \lambda \cos \phi_n} \tag{15-29}$$

where f = coefficient of friction

λ = lead angle at mean worm diameter

ϕ_n = normal pressure angle

The sliding velocity V_s is

$$V_s = \frac{\pi n_W d_m}{12 \cos \lambda} \tag{15-30}$$

where n_W = rotative speed of the worm and d_m = mean worm diameter. The torque at the worm gear is

$$T_G = \frac{W^t D_m}{2} \tag{15-31}$$

where D_m is the mean gear diameter.

The parameters in Eq. (15–28) are, quantitatively,

$$C_s = 270 + 10.37C^3 \qquad C \le 3 \text{ in} \tag{15-32}$$

For sand-cast gears,

$$C_s = \begin{cases} 1000 & C > 3 & d_G \le 2.5 \text{ in} \\ 1190 - 477 \log d_G & C > 3 & d_G > 2.5 \text{ in} \end{cases} \tag{15-33}$$

For chilled-cast gears,

$$C_s = \begin{cases} 1000 & C > 3 & d_G \le 8 \text{ in} \\ 1412 - 456 \log d_G & C > 3 & d_G > 8 \text{ in} \end{cases} \tag{15-34}$$

---

[2]ANSI/AGMA 6034-A87, March 1988, *Practice for Enclosed Cylindrical Wormgear Speed-Reducers and Gear Motors;* ANSI/AGMA 6030-C87, March 1988, *Design of Industrial Double-Enveloping Wormgears;* and ANSI/AGMA 6022-C93, Dec. 1993, *Design Manual for Cylindrical Wormgearing,* and useful references.

For centrifugally cast gears,

$$
C_s = \begin{cases} 1000 & C > 3 & d_G \leq 25 \text{ in} \\ 1251 - 180 \log d_G & C > 3 & d_G > 25 \text{ in} \end{cases} \tag{15-35}
$$

The ratio correction factor C_m is given by

$$
C_m = \begin{cases} 0.02\sqrt{-m_G^2 + 40 m_G - 76} + 0.46 & 3 < m_G \leq 20 \\ 0.0107\sqrt{-m_G^2 + 56 m_G + 5145} & 20 < m_G \leq 76 \\ 1.1483 - 0.006\,58 m_G & m_G > 76 \end{cases} \tag{15-36}
$$

The velocity factor C_v is given by

$$
C_v = \begin{cases} 0.659 \exp(-0.0011 V_s) & V_s < 700 \text{ ft/min} \\ 13.31 V_s^{-0.571} & 700 \leq V_s < 3000 \text{ ft/min} \\ 65.52 V_s^{-0.774} & V_s > 3000 \text{ ft/min} \end{cases} \tag{15-37}
$$

AGMA reports the coefficient of friction f as

$$
f = \begin{cases} 0.15 & V_s = 0 \\ 0.124 \exp(-0.074 V_s^{0.645}) & 0 < V_s \leq 10 \text{ ft/min} \\ 0.103 \exp(-0.110 V_s^{0.450}) + 0.012 & V_s > 10 \text{ ft/min} \end{cases} \tag{15-38}
$$

Now we examine some worm-gear mesh geometry. The addendum a and dedendum b are

$$
a = \frac{p_x}{\pi} = 0.3183 p_x \tag{15-39}
$$

$$
b = \frac{1.157 p_x}{\pi} = 0.3683 p_x \tag{15-40}
$$

The full depth h_t is

$$
h_t = \begin{cases} \dfrac{2.157 p_x}{\pi} = 0.6866 p_x & p_x \geq 0.16 \text{ in} \\[2ex] \dfrac{2.200 p_x}{\pi} + 0.002 = 0.7003 p_x + 0.002 & p_x < 0.16 \text{ in} \end{cases} \tag{15-41}
$$

The worm outside diameter d_0 is

$$
d_0 = d + 2a \tag{15-42}
$$

The worm root diameter d_r is

$$
d_r = d - 2b \tag{15-43}
$$

The worm-gear throat diameter D_t is

$$
D_t = D + 2a \tag{15-44}
$$

The worm-gear root diameter D_r is

$$D_r = D - 2b \tag{15-45}$$

The clearance c is

$$c = b - a \tag{15-46}$$

The worm face width (maximum) $(F_W)_{\max}$ is

$$(F_W)_{\max} = 2\sqrt{\left(\frac{D_t}{2}\right)^2 - \left(\frac{D}{2} - a\right)^2} \tag{15-47}$$

The worm-gear face width F_G is

$$F_G = \begin{cases} 2d_m/3 & p_x > 0.16 \text{ in} \\ 1.125\sqrt{(d_0 + 2c)^2 - (d_0 - 4a)^2} & p_x \le 0.16 \text{ in} \end{cases} \tag{15-48}$$

The heat loss rate H_{loss} from the worm-gear case in ft · lbf/min is

$$H_{\text{loss}} = 33\,000(1 - e)H_{\text{in}} \tag{15-49}$$

where e is efficiency, given by Eq. (13–46), and H_{in} is the input horsepower from the worm. The overall coefficient $\hbar_{\text{CR}}$ for combined convective and radiative heat transfer from the worm-gear case in ft · lbf/(min · in^2 · °F) is

$$\hbar_{\text{CR}} = \begin{cases} \dfrac{n_W}{6494} + 0.13 & \text{no fan on worm shaft} \\ \dfrac{n_W}{3939} + 0.13 & \text{fan on worm shaft} \end{cases} \tag{15-50}$$

When the case lateral area A is expressed in in^2, the temperature of the oil sump t_s is given by

$$t_s = t_a + \frac{H_{\text{loss}}}{\hbar_{\text{CR}} A} = \frac{33\,000(1 - e)(H)_{\text{in}}}{\hbar_{\text{CR}} A} + t_a \tag{15-51}$$

Bypassing Eqs. (15–49), (15–50), and (15–51) one can apply the AGMA recommendation for minimum lateral area $A_{\min}$ in in^2 using

$$A_{\min} = 43.20 C^{1.7} \tag{15-52}$$

Because worm teeth are inherently much stronger than worm-gear teeth, they are not considered. The teeth in worm gears are short and thick on the edges of the face; midplane they are thinner as well as curved. Buckingham[3] adapted the Lewis equation for this case:

$$\sigma_a = \frac{W_G^t}{p_n(F_e)_G y} \tag{15-53}$$

where $p_n = p_x \cos \lambda$ and y is the Lewis form factor related to circular pitch. For $\phi_n = 14.5°$, $y = 0.100$; $\phi_n = 20°$, $y = 0.125$; $\phi_n = 25°$, $y = 0.150$; $\phi_n = 30°$, $y = 0.175$.

---

[3]Earle Buckingham, *Analytical Mechanics of Gears,* McGraw-Hill, New York, 1949, p. 495.

15–7   **Worm-Gear Analysis**

Compared to other gearing systems worm-gear meshes have a much lower mechanical efficiency. Cooling, for the benefit of the lubricant, becomes a design constraint sometimes resulting in what appears to be an oversize gear case in light of its contents. If the heat can be dissipated by natural cooling, or simply with a fan on the wormshaft, simplicity persists. Water coils within the gear case or lubricant outpumping to an external cooler is the next level of complexity. For this reason, gear-case area is a design decision.

To reduce cooling load, use multiple-thread worms. Also keep the worm pitch diameter as small as possible.

Multiple-thread worms can remove the self-locking feature of many worm-gear drives. When the worm drives the gearset, the mechanical efficiency e_W is given by

$$e_W = \frac{\cos\phi_n - f\tan\lambda}{\cos\phi_n + f\cot\lambda} \tag{15–54}$$

With the gear driving the gearset, the mechanical efficiency e_G is given by

$$e_G = \frac{\cos\phi_n - f\cot\lambda}{\cos\phi_n + f\tan\lambda} \tag{15–55}$$

To ensure that the worm gear will drive the worm,

$$f_{\text{stat}} < \cos\phi_n \tan\lambda \tag{15–56}$$

where values of f_{stat} can be found in ANSI/AGMA 6034-B92. To prevent the worm gear from driving the worm, refer to clause 9 of 6034-B92 for a discussion of self-locking in the static condition.

It is important to have a way to relate the tangential component of the gear force W_G^t to the tangential component of the worm force W_W^t, which includes the role of friction and the angularities of ϕ_n and λ. Refer to Eq. (13–45) solved for W_W^t:

$$W_W^t = W_G^t \frac{\cos\phi_n \sin\lambda + f\cos\lambda}{\cos\phi_n \cos\lambda - f\sin\lambda} \tag{15–57}$$

In the absence of friction

$$W_W^t = W_G^t \tan\lambda$$

The mechanical efficiency of most gearing is very high, which allows power in and power out to be used almost interchangeably. Worm gearsets have such poor efficiencies that we work with, and speak of, output power. The magnitude of the gear transmitted force W_G^t can be related to the output horsepower H_0, the application factor K_a, the efficiency e, and design factor n_d by

$$W_G^t = \frac{33\,000 n_d H_0 K_a}{V_G e} \tag{15–58}$$

We use Eq. (15–57) to obtain the corresponding worm force W_W^t. It follows that

$$H_W = \frac{W_W^t V_W}{33\,000} = \frac{\pi d_W n_W W_W^t}{12(33\,000)} \text{ hp} \tag{15–59}$$

$$H_G = \frac{W_G^t V_G}{33\,000} = \frac{\pi d_G n_G W_G^t}{12(33\,000)} \text{ hp} \tag{15–60}$$

Table 15–9

Largest Lead Angle Associated with a Normal Pressure Angle ϕ_n for Worm Gearing

ϕ_n	Maximum Lead Angle λ_{max}
14.5°	16°
20°	25°
25°	35°
30°	45°

From Eq. (13–44),

$$W_f = \frac{f W_G^t}{f \sin\lambda - \cos\phi_n \cos\lambda} \tag{15–61}$$

The sliding velocity of the worm at the pitch cylinder V_s is

$$V_s = \frac{\pi d_W n_W}{12 \cos\lambda} \tag{15–62}$$

and the friction power H_f is given by

$$H_f = \frac{|W_f| V_s}{33\,000} \text{ hp} \tag{15–63}$$

Table 15–9 gives the largest lead angle λ_{max} associated with normal pressure angle ϕ_n.

EXAMPLE 15–3

A single-thread steel worm rotates at 1800 rev/min, meshing with a 24-tooth worm gear transmitting 3 hp to the output shaft. The worm diameter is 3 in and the tangential diametral pitch of the gear is 4 teeth/in. The normal pressure angle is 14.5°. The ambient temperature is 70°F. The application factor is 1.25 and the design factor is 1; gear face width is 2 in, lateral case area 600 in², and the gear is chill-cast bronze.
(a) Find the gear geometry.
(b) Find the transmitted gear forces and the mesh efficiency.
(c) Is the mesh sufficient to handle the loading?
(d) Estimate the lubricant sump temperature.

Solution

(a) $m_G = N_G/N_W = 24/1 = 24$, $d_G = N_G/P_t = 24/4 = 6.000$ in, $d_P = 3.000$ in. The axial circular pitch p_x is $p_x = \pi/P_t = \pi/4 = 0.7854$ in. $C = (3 + 6)/2 = 4.5$ in.

Eq. (15–39): $a = p_x/\pi = 0.7854/\pi = 0.250$ in

Eq. (15–40): $h = 0.3683 p_x = 0.3683(0.7854) = 0.289$ in

Eq. (15–41): $h_t = 0.6866 p_x = 0.6866(0.7854) = 0.539$ in

Eq. (15–42): $d_0 = 3 + 2(0.250) = 3.500$ in

Eq. (15–43): $d_r = 3 - 2(0.289) = 2.422$ in

Eq. (15–44): $D_t = 6 + 2(0.250) = 6.500$ in

Eq. (15–45): $D_r = 6 - 2(0.289) = 5.422$ in

Eq. (15–46): $c = 0.289 - 0.250 = 0.039$ in

Eq. (15–47): $(F_W)_{max} = 2\sqrt{\left(\frac{6.5}{2}\right)^2 - \left(\frac{6}{2} - 0.250\right)^2} = 3.464$ in

The tangential speeds of the worm, V_W, and gear, V_G, are, respectively,

$$V_W = \pi(3)1800/12 = 1414 \text{ ft/min} \qquad V_G = \frac{\pi(6)1800/24}{12} = 117.8 \text{ ft/min}$$

The lead of the worm, from Eq. (13–27), is $p_x N_W = 0.7854(1) = 0.7854$ in. The lead angle λ, from Eq. (13–28), is

$$\lambda = \tan^{-1} \frac{L}{\pi d_W} = \tan^{-1} \frac{0.7854}{\pi(3)} = 4.764°$$

The normal diametral pitch for a worm gear is the same as for a helical gear, which from Eq. (13–18) with $\psi = \lambda$ is

$$P_n = \frac{P_t}{\cos \lambda} = \frac{4}{\cos 4.764°} = 4.014$$

$$p_n = \frac{\pi}{P_n} = \frac{\pi}{4.014} = 0.7827 \text{ in}$$

The sliding velocity, from Eq. (15–62), is

$$V_s = \frac{\pi d_W n_W}{12 \cos \lambda} = \frac{\pi(3)1800}{12 \cos 4.764°} = 1419 \text{ ft/min}$$

(*b*) The coefficient of friction, from Eq. (15–38), is

$$f = 0.103 \exp[-0.110(1419)^{0.450}] + 0.012 = 0.0178$$

The efficiency e, from Eq. (13–46), is

Answer

$$e = \frac{\cos \phi_n - f \tan \lambda}{\cos \phi_n + f \cot \lambda} = \frac{\cos 14.5° - 0.0178 \tan 4.764°}{\cos 14.5° + 0.0178 \cot 4.764°} = 0.818$$

The designer used $n_d = 1$, $K_a = 1.25$ and an output horsepower of $H_0 = 3$ hp. The gear tangential force component W_G^t, from Eq. (15–58), is

Answer

$$W_G^t = \frac{33\,000 n_d H_0 K_a}{V_G e} = \frac{33\,000(1)3(1.25)}{117.8(0.818)} = 1284 \text{ lbf}$$

Answer

The tangential force on the worm is given by Eq. (15–57):

$$W_W^t = W_G^t \frac{\cos \phi_n \sin \lambda + f \cos \lambda}{\cos \phi_n \cos \lambda - f \sin \lambda}$$

$$= 1284 \frac{\cos 14.5° \sin 4.764° + 0.0178 \cos 4.764°}{\cos 14.5° \cos 4.764° - 0.0178 \sin 4.764°} = 131 \text{ lbf}$$

(*c*)

Eq. (15–34): $\qquad C_s = 1000$

Eq. (15–36): $\qquad C_m = 0.0107\sqrt{-24^2 + 56(24) + 5145} = 0.823$

Eq. (15–37): $\qquad C_v = 13.31(1419)^{-0.571} = 0.211$

Eq. (15–38): $(W^t)_{\text{all}} = C_s d_G^{0.8}(F_e)_G C_m C_v$

$$= 1000(6)^{0.8}(2)0.823(0.211) = 1456 \text{ lbf}$$

Since $W_G^t < (W^t)_{\text{all}}$, the mesh will survive at least 25 000 h. The friction force W_f is given by Eq. (15–61):

$$W_f = \frac{fW_G^t}{f \sin \lambda - \cos \phi_n \cos \lambda} = \frac{0.0178(1284)}{0.0178 \sin 4.764° - \cos 14.5° \cos 4.764°}$$

$$= -23.7 \text{ lbf}$$

The power dissipated in frictional work H_f is given by Eq. (15–63):

$$H_f = \frac{|W_f|V_s}{33\,000} = \frac{|-23.7|1419}{33\,000} = 1.02 \text{ hp}$$

The worm and gear powers, H_W and H_G, are given by

$$H_W = \frac{W_W^t V_W}{33\,000} = \frac{131(1414)}{33\,000} = 5.61 \text{ hp} \qquad H_G = \frac{W_G^t V_G}{33\,000} = \frac{1284(117.8)}{33\,000} = 4.58 \text{ hp}$$

Answer   Gear power is satisfactory. Now,

$$P_n = P_t/\cos \lambda = 4/\cos 4.764° = 4.014$$

$$p_n = \pi/P_n = \pi/4.014 = 0.7827 \text{ in}$$

The bending stress in a gear tooth is given by Buckingham's adaptation of the Lewis equation, Eq. (15–53), as

$$(\sigma)_G = \frac{W_G^t}{p_n F_G y} = \frac{1284}{0.7827(2)(0.1)} = 8200 \text{ psi}$$

Answer   Stress in gear satisfactory.
(d)

Eq. (15–52): $A_{\min} = 43.2C^{1.7} = 43.2(4.5)^{1.7} = 557 \text{ in}^2$

The gear case has a lateral area of 600 in^2.

Eq. (15–49): $H_{\text{loss}} = 33\,000(1 - e)H_{\text{in}} = 33\,000(1 - 0.818)5.61$

$$= 33\,690 \text{ ft} \cdot \text{lbf/min}$$

Eq. (15–50): $\hbar_{\text{CR}} = \dfrac{n_W}{3939} + 0.13 = \dfrac{1800}{3939} + 0.13 = 0.587 \text{ ft} \cdot \text{lbf/(min} \cdot \text{in}^2 \cdot °\text{F)}$

Answer   Eq. (15–51): $t_s = t_a + \dfrac{H_{\text{loss}}}{\hbar_{\text{CR}} A} = 70 + \dfrac{33\,690}{0.587(600)} = 166°\text{F}$

15–8 Designing a Worm-Gear Mesh

A usable decision set for a worm-gear mesh includes

- Function: power, speed, m_G, K_a
- Design factor: n_d
- Tooth system
- Materials and processes
- Number of threads on the worm: N_W

  } A priori decisions

- Axial pitch of worm: p_x
- Pitch diameter of the worm: d_W
- Face width of gear: F_G
- Lateral area of case: A

  } Design variables

Reliability information for worm gearing is not well developed at this time. The use of AGMA Eq. (15–28) together with the factors C_s, C_m, and C_v, with an alloy steel case-hardened worm together with customary nonferrous worm-wheel materials, will result in lives in excess of 25 000 h. The worm-gear materials in the experience base are principally bronzes.

- Tin- and nickel-bronzes (chilled-casting produces hardest surfaces)
- Lead-bronze (high-speed applications)
- Aluminum- and silicon-bronze (heavy load, slow-speed application)

The factor C_s for bronze in the spectrum sand-cast, chilled-cast, and centrifugally cast increases in the same order.

Standardization of tooth systems is not as far along as it is in other types of gearing. For the designer this represents freedom of action, but acquisition of tooling for tooth-forming is more of a problem for in-house manufacturing. When using a subcontractor the designer must be aware of what the supplier is capable of providing with on-hand tooling.

Axial pitches for the worm are usually integers, and quotients of integers are common. Typical pitches are $\frac{1}{4}$, $\frac{5}{16}$, $\frac{3}{8}$, $\frac{1}{2}$, $\frac{3}{4}$, 1, $\frac{5}{4}$, $\frac{6}{4}$, $\frac{7}{4}$, and 2, but others are possible. Table 15–8 shows dimensions common to both worm gear and cylindrical worm for proportions often used. Teeth frequently are stubbed when lead angles are 30° or larger.

Worm-gear design is constrained by available tooling, space restrictions, shaft center-to-center distances, gear ratios needed, and the designer's experience. ANSI/AGMA 6022-C93, *Design Manual for Cylindrical Wormgearing* offers the following guidance. Normal pressure angles are chosen from 14.5°, 17.5°, 20°, 22.5°, 25°, 27.5°, and 30°. The recommended minimum number of gear teeth is given in Table 15–10. The normal range of the number of threads on the worm is 1 through 10. Mean worm pitch diameter is usually chosen in the range given by Eq. (15–27).

A design decision is the axial pitch of the worm. Since acceptable proportions are couched in terms of the center-to-center distance, which is not yet known, one chooses a trial axial pitch p_x. Having N_W and a trial worm diameter d_W,

$$N_G = m_G N_W \qquad P_t = \frac{\pi}{p_x} \qquad d_G = \frac{N_G}{P_t}$$

Table 15–10

Minimum Number of
Gear Teeth for Normal
Pressure Angle ϕ_n

ϕ_n	$(N_G)_{min}$
14.5	40
17.5	27
20	21
22.5	17
25	14
27.5	12
30	10

Then

$$(d_W)_{lo} = C^{0.875}/3 \qquad (d_W)_{hi} = C^{0.875}/1.6$$

Examine $(d_W)_{lo} \le d_W \le (d_W)_{hi}$, and refine the selection of mean worm-pitch diameter to d_{W1} if necessary. Recompute the center-to-center distance as $C = (d_{W1} + d_G)/2$. There is even an opportunity to make C a round number. Choose C and set

$$d_{W2} = 2C - d_G$$

Equations (15–39) through (15–48) apply to one usual set of proportions.

EXAMPLE 15–4

Design a 10-hp 11:1 worm-gear speed-reducer mesh for a lumber mill planer feed drive for 3- to 10-h daily use. A 1720-rev/min squirrel-cage induction motor drives the planer feed ($K_a = 1.25$), and the ambient temperature is 70°F.

Solution

Function: $H_0 = 10$ hp, $m_G = 11$, $n_W = 1720$ rev/min.
Design factor: $n_d = 1.2$.
Materials and processes: case-hardened alloy steel worm, sand-cast bronze gear.
Worm threads: double, $N_W = 2$, $N_G = m_G N_W = 11(2) = 22$ gear teeth acceptable for $\phi_n = 20°$, according to Table 15–10.
Decision 1: Choose an axial pitch of worm $p_x = 1.5$ in. Then,

$$P_t = \pi/p_x = \pi/1.5 = 2.0944$$

$$d_G = N_G/P_t = 22/2.0944 = 10.504 \text{ in}$$

Eq. (15–39): $\quad a = 0.3183 p_x = 0.3183(1.5) = 0.4775$ in (addendum)

Eq. (15–40): $\quad b = 0.3683(1.5) = 0.5525$ in (dedendum)

Eq. (15–41): $\quad h_t = 0.6866(1.5) = 1.030$ in

Decision 2: Choose a mean worm diameter $d_W = 2.000$ in. Then

$$C = (d_W + d_G)/2 = (2.000 + 10.504)/2 = 6.252 \text{ in}$$

$$(d_W)_{lo} = 6.252^{0.875}/3 = 1.657 \text{ in}$$

$$(d_W)_{hi} = 6.252^{0.875}/1.6 = 3.107 \text{ in}$$

The range, given by Eq. (15–27), is $1.657 \leq d_W \leq 3.107$ in, which is satisfactory. Try $d_W = 2.500$ in. Recompute C:

$$C = (2.5 + 10.504)/2 = 6.502 \text{ in}$$

The range is now $1.715 \leq d_W \leq 3.216$ in, which is still satisfactory. Decision: $d_W = 2.500$ in. Then

Eq. (13–27): $\qquad\qquad\qquad L = p_x N_W = 1.5(2) = 3.000 \text{ in}$

Eq. (13–28):

$$\lambda = \tan^{-1}[L/(\pi d_W)] = \tan^{-1}[3/(\pi 2.5)] = 20.905° \quad \text{(from Table 15–9 lead angle OK)}$$

Eq. (15–62): $\quad V_s = \dfrac{\pi d_W n_W}{12 \cos \lambda} = \dfrac{\pi(2.5)1720}{12 \cos 20.905°} = 1205.1 \text{ ft/min}$

$$V_W = \frac{\pi d_W n_W}{12} = \frac{\pi(2.5)1720}{12} = 1125.7 \text{ ft/min}$$

$$V_G = \frac{\pi d_G n_G}{12} = \frac{\pi(10.504)1720/11}{12} = 430.0 \text{ ft/min}$$

Eq. (15–33): $\quad C_s = 1190 - 477 \log 10.504 = 702.8$

Eq. (15–36): $\quad C_m = 0.02\sqrt{-11^2 + 40(11) - 76} + 0.46 = 0.772$

Eq. (15–37): $\quad C_v = 13.31(1205.1)^{-0.571} = 0.232$

Eq. (15–38): $\quad f = 0.103 \exp[-0.11(1205.1)^{0.45}] + 0.012 = 0.0191$

Eq. (15–54): $\quad e_W = \dfrac{\cos 20° - 0.0191 \tan 20.905°}{\cos 20° + 0.0191 \cot 20.905°} = 0.942$

(If the worm gear drives, $e_G = 0.939$.) To ensure nominal 10-hp output, with adjustments for K_a, n_d, and e,

Eq. (15–57): $\quad W_W^t = 1222 \dfrac{\cos 20° \sin 20.905° + 0.0191 \cos 20.905°}{\cos 20° \cos 20.905° - 0.0191 \sin 20.905°} = 495.4 \text{ lbf}$

Eq. (15–58): $\quad W_G^t = \dfrac{33\,000(1.2)10(1.25)}{430(0.942)} = 1222 \text{ lbf}$

Eq. (15–59): $\quad H_W = \dfrac{\pi(2.5)1720(495.4)}{12(33\,000)} = 16.9 \text{ hp}$

Eq. (15–60): $\quad H_G = \dfrac{\pi(10.504)1720/11(1222)}{12(33\,000)} = 15.92 \text{ hp}$

Eq. (15–61): $\quad W_f = \dfrac{0.0191(1222)}{0.0191 \sin 20.905° - \cos 20° \cos 20.905°} = -26.8 \text{ lbf}$

Eq. (15–63): $\quad H_f = \dfrac{|-26.8|1205.1}{33\,000} = 0.979 \text{ hp}$

With $C_s = 702.8$, $C_m = 0.772$, and $C_v = 0.232$,

$$(F_e)_{\text{req}} = \frac{W_G^t}{C_s d_G^{0.8} C_m C_v} = \frac{1222}{702.8(10.504)^{0.8}0.772(0.232)} = 1.479 \text{ in}$$

Decision 3: The available range of $(F_e)_G$ is $1.479 \le (F_e)_G \le 2d_W/3$ or $1.479 \le (F_e)_G \le 1.667$ in. Set $(F_e)_G = 1.5$ in.

Eq. (15–28): $W_{all}^t = 702.8(10.504)^{0.8}1.5(0.772)0.232 = 1239$ lbf

This is greater than 1222 lbf. There is a little excess capacity. The force analysis stands.

Decision 4:

Eq. (15–50): $\hbar_{CR} = \dfrac{n_W}{6494} + 0.13 = \dfrac{1720}{6494} + 0.13 = 0.395$ ft $\cdot$ lbf/(min $\cdot$ in$^2 \cdot$ °F)

Eq. (15–49): $H_{loss} = 33\,000(1 - e)H_W = 33\,000(1 - 0.942)16.9 = 32\,347$ ft $\cdot$ lbf/min

The AGMA area, from Eq. (15–52), is $A_{min} = 43.2C^{1.7} = 43.2(6.502)^{1.7} = 1041.5$ in^2. A rough estimate of the lateral area for 6-in clearances:

Vertical:	$d_W + d_G + 6 = 2.5 + 10.5 + 6 = 19$ in
Width:	$d_G + 6 = 10.5 + 6 = 16.5$ in
Thickness:	$d_W + 6 = 2.5 + 6 = 8.5$ in
Area:	$2(19)16.5 + 2(8.5)19 + 16.5(8.5) \doteq 1090$ in^2

Expect an area of 1100 in^2. Choose: Air-cooled, no fan on worm, with an ambient temperature of 70°F.

$$t_s = t_a + \frac{H_{loss}}{\hbar_{CR} A} = 70 + \frac{32\,350}{0.395(1100)} = 70 + 74.5 = 144.5°\text{F}$$

Lubricant is safe with some margin for smaller area.

Eq. (13–18): $P_n = \dfrac{P_t}{\cos \lambda} = \dfrac{2.094}{\cos 20.905°} = 2.242$

$$p_n = \frac{\pi}{P_n} = \frac{\pi}{2.242} = 1.401 \text{ in}$$

Gear bending stress, for reference, is

Eq. (15–53): $\sigma = \dfrac{W_G^t}{p_n(F_e)_G y} = \dfrac{1222}{1.401(1.5)0.125} = 4652$ psi

The risk is from wear, which is addressed by the AGMA method that provides $(W_G^t)_{all}$.

15–9   Buckingham Wear Load

A precursor to the AGMA method was the method of Buckingham, which identified an allowable wear load in worm gearing. Buckingham showed that the allowable gear-tooth loading for wear can be estimated from

$$\left(W_G^t\right)_{all} = K_w d_G F_e \tag{15–64}$$

where K_w = worm-gear load factor
  d_G = gear-pitch diameter
  F_e = worm-gear effective face width

Table 15–11

Wear Factor K_w for Worm Gearing

Source: Earle Buckingham, *Design of Worm and Spiral Gears,* Industrial Press, New York, 1981.

Material		Thread Angle ϕ_n			
Worm	**Gear**	**$14\frac{1}{2}°$**	**20°**	**25°**	**30°**
Hardened steel*	Chilled bronze	90	125	150	180
Hardened steel*	Bronze	60	80	100	120
Steel, 250 BHN (min.)	Bronze	36	50	60	72
High-test cast iron	Bronze	80	115	140	165
Gray iron†	Aluminum	10	12	15	18
High-test cast iron	Gray iron	90	125	150	180
High-test cast iron	Cast steel	22	31	37	45
High-test cast iron	High-test cast iron	135	185	225	270
Steel 250 BHN (min.)	Laminated phenolic	47	64	80	95
Gray iron	Laminated phenolic	70	96	120	140

*Over 500 BHN surface.
†For steel worms, multiply given values by 0.6.

Table 15–11 gives values for K_w for worm gearsets as a function of the material pairing and the normal pressure angle.

EXAMPLE 15–5

Estimate the allowable gear wear load $(W_G^t)_{\text{all}}$ for the gearset of Ex. 15–4 using Buckingham's wear equation.

Solution

From Table 15–11 for a hardened steel worm and a bronze bear, K_w is given as 80 for $\phi_n = 20°$. Equation (15–64) gives

$$(W_G^t)_{\text{all}} = 80(10.504)1.5 = 1260 \text{ lbf}$$

which is larger than the 1239 lbf of the AGMA method. The method of Buckingham does not have refinements of the AGMA method. [Is $(W_G^t)_{\text{all}}$ linear with gear diameter?]

For material combinations not addressed by AGMA, Buckingham's method allows quantitative treatment.

PROBLEMS

ANALYSIS

15–1

An uncrowned straight-bevel pinion has 20 teeth, a diametral pitch of 6 teeth/in, and a transmission accuracy number of 6. Both the pinion and gear are made of through-hardened steel with a Brinell hardness of 300. The driven gear has 60 teeth. The gearset has a life goal of 10^9 revolutions of the pinion with a reliability of 0.999. The shaft angle is 90°; the pinion speed is 900 rev/min. The face width is 1.25 in, and the normal pressure angle is 20°. The pinion is mounted outboard of its bearings, and the gear is straddle-mounted. Based on the AGMA bending strength, what is the power rating of the gearset? Use $K_0 = 1$, $S_F = 1$, and $S_H = 1$.

15–2 For the gearset and conditions of Prob. 15–1, find the power rating based on the AGMA surface durability.

15–3 An uncrowned straight-bevel pinion has 30 teeth, a diametral pitch of 6, and a transmission accuracy number of 6. The driven gear has 60 teeth. Both are made of No. 30 cast iron. The shaft angle is 90°. The face width is 1.25 in, the pinion speed is 900 rev/min, and the normal pressure angle is 20°. The pinion is mounted outboard of its bearings; the bearings of the gear straddle it. What is the power rating based on AGMA bending strength? (For cast iron gearsets reliability information has not yet been developed. We say the life is greater than 10^7 revolutions; set $K_L = 1$, $C_L = 1$, $C_R = 1$, $K_R = 1$; and apply a factor of safety. Use $S_F = 2$ and $S_H = \sqrt{2}$.)

15–4 For the gearset and conditions of Prob. 15–3, find the power rating based on AGMA surface durability. For the solutions to Probs. 15–3 and 15–4, what is the power rating of the gearset?

15–5 An uncrowned straight-bevel pinion has 22 teeth, a module of 4 mm, and a transmission accuracy number of 5. The pinion and the gear are made of through-hardened steel, both having core and case hardnesses of 180 Brinell. The pinion drives the 24-tooth bevel gear. The shaft angle is 90°, the pinion speed is 1800 rev/min, the face width is 25 mm, and the normal pressure angle is 20°. Both gears have an outboard mounting. Find the power rating based on AGMA pitting resistance if the life goal is 10^9 revolutions of the pinion at 0.999 reliability.

15–6 For the gearset and conditions of Prob. 15–5, find the power rating for AGMA bending strength.

15–7 In straight-bevel gearing, there are some analogs to Eqs. (14–44) and (14–45). If we have a pinion core with a hardness of $(H_B)_{11}$ and we try equal power ratings, the transmitted load W^t can be made equal in all four cases. It is possible to find these relations:

	Core	**Case**
Pinion	$(H_B)_{11}$	$(H_B)_{12}$
Gear	$(H_B)_{21}$	$(H_B)_{22}$

(*a*) For carburized case-hardened gear steel with core AGMA bending strength $(s_{at})_G$ and pinion core strength $(s_{at})_P$, show that the relationship is

$$(s_{at})_G = (s_{at})_P \frac{J_P}{J_G} m_G^{-0.0323}$$

This allows $(H_B)_{21}$ to be related to $(H_B)_{11}$.

(*b*) Show that the AGMA contact strength of the gear case $(s_{ac})_G$ can be related to the AGMA core bending strength of the pinion core $(s_{at})_P$ by

$$(s_{ac})_G = \frac{C_p}{(C_L)_G C_H} \sqrt{\frac{S_H^2}{S_F} \frac{(s_{at})_P (K_L)_P K_x J_P K_T C_s C_{xc}}{N_P I K_s}}$$

If factors of safety are applied to the transmitted load W_t, then $S_H = \sqrt{S_F}$ and S_H^2/S_F is unity. The result allows $(H_B)_{22}$ to be related to $(H_B)_{11}$.

(*c*) Show that the AGMA contact strength of the gear $(s_{ac})_G$ is related to the contact strength of the pinion $(s_{ac})_P$ by

$$(s_{ac})_P = (s_{ac})_G m_G^{0.0602} C_H$$

15-8 Refer to your solution to Probs. 15–1 and 15–2, which is to have a pinion core hardness of 300 Brinell. Use the relations from Prob. 15–7 to establish the hardness of the gear core and the case hardnesses of both gears.

15-9 Repeat Probs. 15–1 and 15–2 with the hardness protocol

	Core	Case
Pinion	300	372
Gear	352	344

which can be established by relations in Prob. 15–7, and see if the result matches transmitted loads W^t in all four cases.

15-10 A catalog of stock bevel gears lists a power rating of 5.2 hp at 1200 rev/min pinion speed for a straight-bevel gearset consisting of a 20-tooth pinion driving a 40-tooth gear. This gear pair has a 20° normal pressure angle, a face width of 0.71 in, and a diametral pitch of 10 teeth/in and is through-hardened to 300 BHN. Assume the gears are for general industrial use, are generated to a transmission accuracy number of 5, and are uncrowned. Given these data, what do you think about the stated catalog power rating?

15-11 Apply the relations of Prob. 15–7 to Ex. 15–1 and find the Brinell case hardness of the gears for equal allowable load W^t in bending and wear. Check your work by reworking Ex. 15–1 to see if you are correct. How would you go about the heat treatment of the gears?

15-12 Your experience with Ex. 15–1 and problems based on it will enable you to write an interactive computer program for power rating of through-hardened steel gears. Test your understanding of bevel-gear analysis by noting the ease with which the coding develops. The hardness protocol developed in Prob. 15–7 can be incorporated at the end of your code, first to display it, then as an option to loop back and see the consequences of it.

15-13 Use your experience with Prob. 15–11 and Ex. 15–2 to design an interactive computer-aided design program for straight-steel bevel gears, implementing the ANSI/AGMA 2003-B97 standard. It will be helpful to follow the decision set in Sec. 15–5, allowing the return to earlier decisions for revision as the consequences of earlier decisions develop.

15-14 A single-threaded steel worm rotates at 1725 rev/min, meshing with a 56-tooth worm gear transmitting 1 hp to the output shaft. The pitch diameter of the worm is 1.50. The tangential diametral pitch of the gear is 8 teeth per inch and the normal pressure angle is 20°. The ambient temperature is 70°F, the application factor is 1.25, the design factor is 1, the gear face is 0.5 in, the lateral case area is 850 in², and the gear is sand-cast bronze.
(*a*) Determine and evaluate the geometric properties of the gears.
(*b*) Determine the transmitted gear forces and the mesh efficiency.
(*c*) Is the mesh sufficient to handle the loading?
(*d*) Estimate the lubricant sump temperature.

15-15 to 15-22 As in Ex. 15–4, design a cylindrical worm-gear mesh to connect a squirrel-cage induction motor to a liquid agitator. The motor speed is 1125 rev/min, and the velocity ratio is to be 10:1. The output power requirement is 25 hp. The shaft axes are 90° to each other. An overload factor K_o (see Table 15–2) makes allowance for external dynamic excursions of load from the nominal or average load W^t. For this service $K_o = 1.25$ is appropriate. Additionally, a design factor n_d of 1.1 is to be included to address other unquantifiable risks. For Probs. 15–15 to 15–17 use the AGMA method for $(W^t_G)_{\text{all}}$ whereas for Probs. 15–18 to 15–22, use the Buckingham method. See Table 15–12.

Table 15–12

Table Supporting Problems 15–15 to 15–22

Problem No.	Method	Materials	
		Worm	**Gear**
15–15	AGMA	Steel, HRC 58	Sand-cast bronze
15–16	AGMA	Steel, HRC 58	Chilled-cast bronze
15–17	AGMA	Steel, HRC 58	Centrifugal-cast bronze
15–18	Buckingham	Steel, 500 Bhn	Chilled-cast bronze
15–19	Buckingham	Steel, 500 Bhn	Cast bronze
15–20	Buckingham	Steel, 250 Bhn	Cast bronze
15–21	Buckingham	High-test cast iron	Cast bronze
15–22	Buckingham	High-test cast iron	High-test cast iron

16

Clutches, Brakes, Couplings, and Flywheels

Chapter Outline

This chapter is concerned with a group of elements usually associated with rotation that have in common the function of storing and/or transferring rotating energy. Because of this similarity of function, clutches, brakes, couplings, and flywheels are treated together in this book.

A simplified dynamic representation of a friction clutch or brake is shown in Fig. 16–1a. Two inertias, I_1 and I_2, traveling at the respective angular velocities ω_1 and ω_2, one of which may be zero in the case of brakes, are to be brought to the same speed by engaging the clutch or brake. Slippage occurs because the two elements are running at different speeds and energy is dissipated during actuation, resulting in a temperature rise. In analyzing the performance of these devices we shall be interested in:

1. The actuating force
2. The torque transmitted
3. The energy loss
4. The temperature rise

The torque transmitted is related to the actuating force, the coefficient of friction, and the geometry of the clutch or brake. This is a problem in statics, which will have to be studied separately for each geometric configuration. However, temperature rise is related to energy loss and can be studied without regard to the type of brake or clutch, because the geometry of interest is that of the heat-dissipating surfaces.

The various types of devices to be studied may be classified as follows:

1. Rim types with internal expanding shoes
2. Rim types with external contracting shoes
3. Band types
4. Disk or axial types
5. Cone types
6. Miscellaneous types

A flywheel is an inertial energy-storage device. It absorbs mechanical energy by increasing its angular velocity and delivers energy by decreasing its velocity. Figure 16–1b is a mathematical representation of a flywheel. An input torque T_i, corresponding to a coordinate θ_i, will cause the flywheel speed to increase. And a load or output torque T_o, with coordinate θ_o, will absorb energy from the flywheel and cause it to slow down. We shall be interested in designing flywheels so as to obtain a specified amount of speed regulation.

Figure 16–1

(a) Dynamic representation of a clutch or brake; (b) mathematical representation of a flywheel.

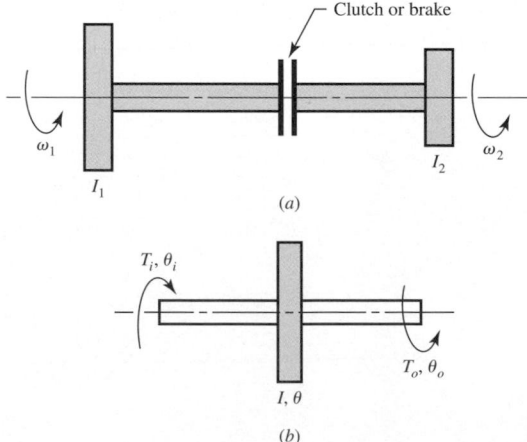

16–1 **Static Analysis of Clutches and Brakes**

Many types of clutches and brakes can be analyzed by following a general procedure. The procedure entails the following tasks:

- Estimate, model, or measure the pressure distribution on the friction surfaces.
- Find a relationship between the largest pressure and the pressure at any point.
- Use the conditions of static equilibrium to find the braking force or torque and the support reactions.

Let us apply these tasks to the doorstop depicted in Fig. 16–2a. The stop is hinged at pin A. A normal pressure distribution $p(u)$ is shown under the friction pad as a function of position u, taken from the right edge of the pad. A similar distribution of shearing frictional traction is on the surface, of intensity $fp(u)$, in the direction of the motion of the floor relative to the pad, where f is the coefficient of friction. The width of the pad into the page is w_2. Thes net force in the y direction and moment about C from the pressure are respectively,

$$N = w_2 \int_0^{w_1} p(u)\,du = p_{\text{av}} w_1 w_2 \tag{a}$$

$$\int_0^{w_1} p(u)u\,du = \bar{u} \int_0^{w_1} p(u)\,du = p_{\text{av}} w_1 \bar{u} \tag{b}$$

We sum the forces in the x-direction to obtain

$$\sum F_x = R_x \mp \int_0^{w_1} f w_2 p(u)\,du = 0$$

where $-$ or $+$ is for rightward or leftward relative motion of the floor, respectively. Assuming f and w_2 constant, solving for R_x gives

$$R_x = \pm \int_0^{w_1} f w_2 p(u)\,du = \pm f w_1 w_2 p_{\text{av}} \tag{c}$$

Summing the forces in the y direction gives

$$\sum F_y = -F + \int_0^{w_1} p(u) w_2\,du + R_y = 0$$

from which

$$R_y = F - w_2 \int_0^{w_1} p(u)\,du = F - p_{\text{av}} w_1 w_2 \tag{d}$$

for either direction. Summing moments about the pin located at A we have

$$\sum M_A = Fb - \int_0^{w_1} w_2 p(u)(c+u)\,du \mp a f w_2 \int_0^{w_1} p(u)\,du = 0$$

A brake shoe is *self-energizing* if its moment sense helps set the brake, *self-deenergizing* if the moment resists setting the brake. Continuing,

$$F = \frac{w_2}{b} \left[\int_0^{w_1} p(u)(c+u)\,du \mp a f \int_0^{w_1} p(u)\,du \right] \tag{e}$$

Figure 16–2

A common doorstop.
(*a*) Free body of the doorstop.
(*b*) Trapezoidal pressure distribution on the foot pad based on linear deformation of pad. (*c*) Free-body diagram for leftward movement of the floor, uniform pressure, Ex. 16–1. (*d*) Free-body diagram for rightward movement of the floor, uniform pressure, Ex. 16–1. (*e*) Free-body diagram for leftward movement of the floor, trapezoidal pressure, Ex. 16–1.

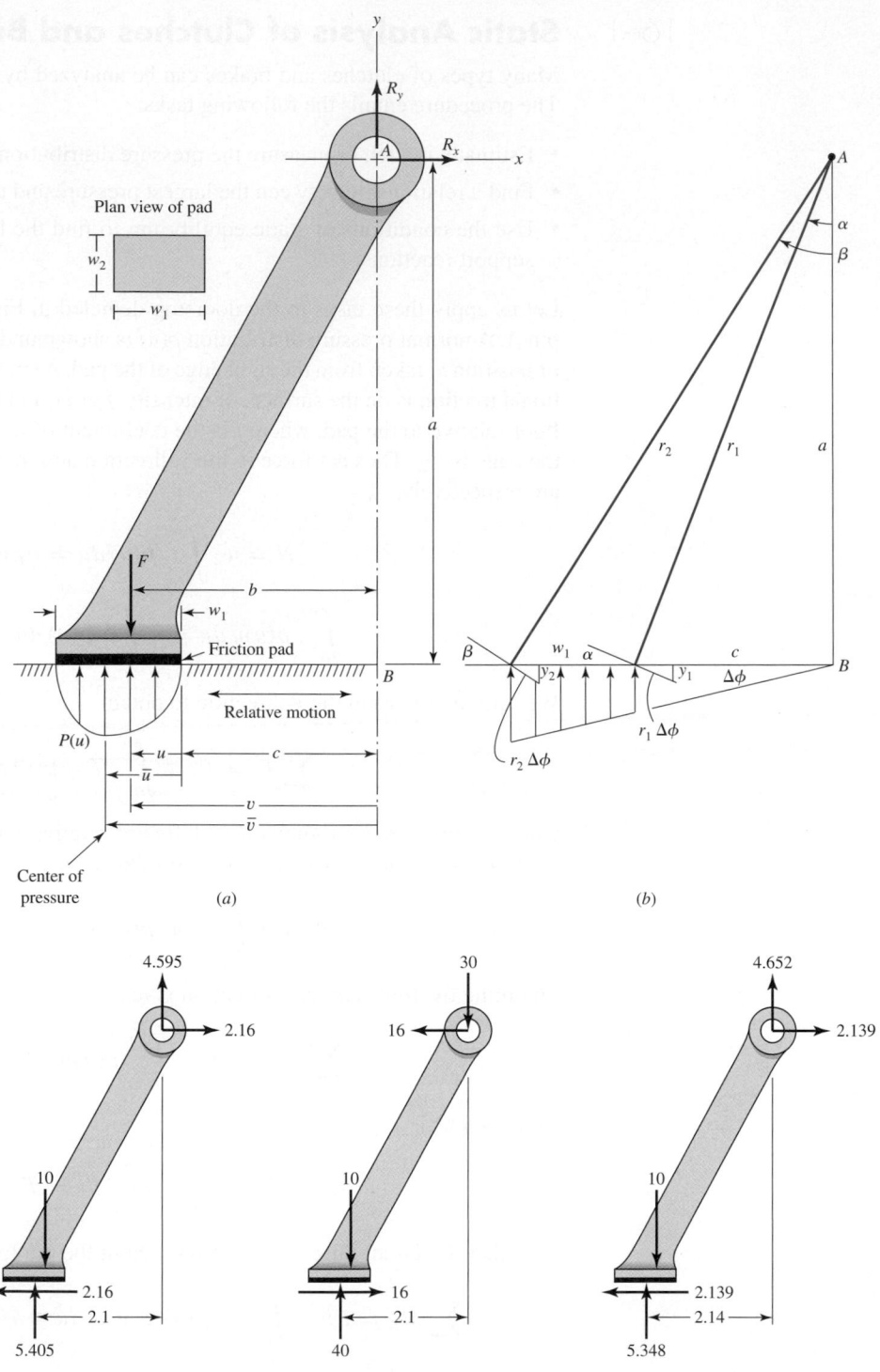

Can F be equal to or less than zero? Only during rightward motion of the floor when the expression in brackets in Eq. (e) is equal to or less than zero. We set the brackets to zero or less:

$$\int_0^{w_1} p(u)(c+u)\,du - af \int_0^{w_1} p(u)\,du \leq 0$$

from which

$$f_{cr} \geq \frac{1}{a} \frac{\int_0^{w_1} p(u)(c+u)\,du}{\int_0^{w_1} p(u)\,du} = \frac{1}{a} \frac{c\int_0^{w_1} p(u)\,du + \int_0^{w_1} p(u)u\,du}{\int_0^{w_1} p(u)\,du}$$

$$f_{cr} \geq \frac{c+\bar{u}}{a} \tag{f}$$

where $\bar{u}$ is the distance of the center of pressure from the right edge of the pad. The conclusion that a *self-acting* or *self-locking* phenomenon is present is independent of our knowledge of the normal pressure distribution $p(u)$. Our ability to *find* the critical value of the coefficient of friction f_{cr} is dependent on our knowledge of $p(u)$, from which we derive $\bar{u}$.

EXAMPLE 16–1   The doorstop depicted in Fig. 16–2a has the following dimensions: $a = 4$ in, $b = 2$ in, $c = 1.6$ in, $w_1 = 1$ in, $w_2 = 0.75$ in, where w_2 is the depth of the pad into the plane of the paper.
(a) For a leftward relative movement of the floor, an actuating force F of 10 lbf, a coefficient of friction of 0.4, use a uniform pressure distribution p_{av}, find R_x, R_y, p_{av}, and the largest pressure p_a.
(b) Repeat part a for rightward relative movement of the floor.
(c) Model the normal pressure to be the "crush" of the pad, much as if it were composed of many small helical coil springs. Find R_x, R_y, p_{av}, and p_a for leftward relative movement of the floor and other conditions as in part a.
(d) For rightward relative movement of the floor, is the doorstop a self-acting brake?

Solution   (a)

Eq. (c): $R_x = f p_{av} w_1 w_2 = 0.4(1)(0.75) p_{av} = 0.3 p_{av}$

Eq. (d): $R_y = F - p_{av} w_1 w_2 = 10 - p_{av}(1)(0.75) = 10 - 0.75 p_{av}$

Eq. (e): $F = \frac{w_2}{b}\left[\int_0^1 p_{av}(c+u)\,du + af \int_0^1 p_{av}\,du\right]$

$$= \frac{w_2}{b}\left(p_{av}c\int_0^1 du + p_{av}\int_0^1 u\,du + af p_{av}\int_0^1 du\right)$$

$$= \frac{w_2 p_{av}}{b}(c + 0.5 + af) = \frac{0.75}{2}[1.6 + 0.5 + 4(0.4)]p_{av}$$

$$= 1.3875 p_{av}$$

Solving for p_{av} gives

$$p_{av} = \frac{F}{1.3875} = \frac{10}{1.3875} = 7.207 \text{ psi}$$

We evaluate R_x and R_y as

Answer

$$R_x = 0.3(7.207) = 2.162 \text{ lbf}$$

Answer

$$R_y = 10 - 0.75(7.207) = 4.595 \text{ lbf}$$

The normal force N on the pad is $F - R_y = 10 - 4.595 = 5.405$ lbf, upward. The line of action is through the center of pressure, which is at the center of the pad. The friction force is $fN = 0.4(5.405) = 2.162$ lbf directed to the left. A check of the moments about A gives

$$\sum M_A = Fb - fNa - N(w_1/2 + c)$$

$$= 10(2) - 0.4(5.405)4 - 5.405(1/2 + 1.6) \doteq 0$$

Answer

The maximum pressure $p_a = p_{av} = 7.207$ psi.

(b)

Eq. (c): $R_x = -fp_{av}w_1w_2 = -0.4(1)(0.75)p_{av} = -0.3p_{av}$

Eq. (d): $R_y = F - p_{av}w_1w_2 = 10 - p_{av}(1)(0.75) = 10 - 0.75p_{av}$

Eq. (e): $F = \frac{w_2}{b}\left[\int_0^1 p_{av}(c+u)\,du + af\int_0^1 p_{av}\,du\right]$

$$= \frac{w_2}{b}\left(p_{av}c\int_0^1 du + p_{av}\int_0^1 u\,du + afp_{av}\int_0^1 du\right)$$

$$= \frac{0.75}{2}p_{av}[1.6 + 0.5 - 4(0.4)] = 0.1875p_{av}$$

from which

$$p_{av} = \frac{F}{0.1875} = \frac{10}{0.1875} = 53.33 \text{ psi}$$

which makes

Answer

$$R_x = -0.3(53.33) = -16 \text{ lbf}$$

Answer

$$R_y = 10 - 0.75(53.33) = -30 \text{ lbf}$$

The normal force N on the pad is $10 + 30 = 40$ lbf upward. The friction shearing force is $fN = 0.4(40) = 16$ lbf to the right. We now check the moments about A:

$$M_A = fNa + Fb - N(c + 0.5) = 16(4) + 10(2) - 40(1.6 + 0.5) = 0$$

Note the change in average pressure from 7.207 psi in part a to 53.3 psi. Also note how directions of forces have changed. The maximum pressure p_a is the same as p_{av}, which has changed from 7.207 psi to 53.3 psi.

(c) We will model the deformation of the pad as follows. If the doorstop rotates $\Delta\phi$ counterclockwise, the right and left edges of the pad will deform down y_1 and y_2, respectively (Fig. 16–2b). From similar triangles, $y_1/(r_1\,\Delta\phi) = c/r_1$ and $y_2/(r_2\,\Delta\phi) = (c + w_1)/r_2$. Thus, $y_1 = c\,\Delta\phi$ and $y_2 = (c + w_1)\,\Delta\phi$. This means that y is directly proportional to the horizontal distance from the pivot point A; that is, $y = C_1 v$, where C_1 is a constant (see Fig. 16–2b). Assuming the pressure is directly proportional to deformation, then $p(v) = C_2 v$, where C_2 is a constant. In terms of u, the pressure is $p(u) = C_2(c + u) = C_2(1.6 + u)$.

Eq. (e):

$$F = \frac{w_2}{b} \left[\int_0^{w_1} p(u)c \, du + \int_0^{w_1} p(u)u \, du + af \int_0^{w_1} p(u) \, du \right]$$

$$= \frac{0.75}{2} \left[\int_0^1 C_2(1.6 + u)1.6 \, du + \int_0^1 C_2(1.6 + u) u \, du + af \int_0^1 C_2(1.6 + u) \, du \right]$$

$$= 0.375C_2[(1.6 + 0.5)1.6 + (0.8 + 0.3333) + 4(0.4)(1.6 + 0.5)] = 2.945C_2$$

Since $F = 10$ lbf, then $C_2 = 10/2.945 = 3.396$ psi/in, and $p(u) = 3.396(1.6 + u)$. The average pressure is given by

Answer $$p_{av} = \frac{1}{w_1} \int_0^{w_1} p(u) \, du = \frac{1}{1} \int_0^1 3.396(1.6 + u) \, du = 3.396(1.6 + 0.5) = 7.132 \text{ psi}$$

The maximum pressure occurs at $u = 1$ in, and is

Answer $$p_a = 3.396(1.6 + 1) = 8.83 \text{ psi}$$

Equations (c) and (d) of Sec. 16–1 are still valid. Thus,

Answer $$R_x = 0.3p_{av} = 0.3(7.131) = 2.139 \text{ lbf}$$

$$R_y = 10 - 0.75p_{av} = 10 - 0.75(7.131) = 4.652 \text{ lbf}$$

The average pressure is $p_{av} = 7.13$ psi and the maximum pressure is $p_a = 8.83$ psi, which is approximately 24 percent higher than the average pressure. The presumption that the pressure was uniform in part a (because the pad was small, or because the arithmetic would be easier?) underestimated the peak pressure. Modeling the pad as a one-dimensional springset is better, but the pad is really a three-dimensional continuum. A theory of elasticity approach or a finite element modeling may be overkill, given uncertainties inherent in this problem, but it still represents better modeling.
(d) To evaluate $\bar{u}$ we need to evaluate two integrations

$$\int_0^c p(u)u \, du = \int_0^1 3.396(1.6 + u)u \, du = 3.396(0.8 + 0.3333) = 3.849$$

$$\int_0^c p(u) \, du = \int_0^1 3.396(1.6 + u) \, du = 3.396(1.6 + 0.5) = 7.132$$

Thus $\bar{u} = 3.849/7.132 = 0.5397$ in. Then, from Eq. (f) of Sec. 16–1, the critical coefficient of friction is

Answer $$f_{cr} \geq \frac{c + \bar{u}}{a} = \frac{1.6 + 0.5397}{4} = 0.535$$

The doorstop friction pad does not have a high enough coefficient of friction to make the doorstop a self-acting brake. The configuration must change and/or the pad material specification must be changed to sustain the function of a doorstop.

16–2 **Internal Expanding Rim Clutches and Brakes**

The internal-shoe rim clutch shown in Fig. 16–3 consists essentially of three elements: the mating frictional surface, the means of transmitting the torque to and from the surfaces, and the actuating mechanism. Depending upon the operating mechanism, such clutches are further classified as *expanding-ring, centrifugal, magnetic, hydraulic,* and *pneumatic.*

The expanding-ring clutch is often used in textile machinery, excavators, and machine tools where the clutch may be located within the driving pulley. Expanding-ring clutches benefit from centrifugal effects; transmit high torque, even at low speeds; and require both positive engagement and ample release force.

The centrifugal clutch is used mostly for automatic operation. If no spring is used, the torque transmitted is proportional to the square of the speed. This is particularly useful for electric-motor drives where, during starting, the driven machine comes up to speed without shock. Springs can also be used to prevent engagement until a certain motor speed is reached, but some shock may occur.

Magnetic clutches are particularly useful for automatic and remote-control systems. Such clutches are also useful in drives subject to complex load cycles (see Sec. 11–7).

Hydraulic and pneumatic clutches are also useful in drives having complex loading cycles and in automatic machinery, or in robots. Here the fluid flow can be controlled remotely using solenoid valves. These clutches are also available as disk, cone, and multiple-plate clutches.

In braking systems, the *internal-shoe* or *drum* brake is used mostly for automotive applications.

To analyze an internal-shoe device, refer to Fig. 16–4, which shows a shoe pivoted at point A, with the actuating force acting at the other end of the shoe. Since the shoe is long, we cannot make the assumption that the distribution of normal forces is uniform. The mechanical arrangement permits no pressure to be applied at the heel, and we will therefore assume the pressure at this point to be zero.

It is the usual practice to omit the friction material for a short distance away from the heel (point A). This eliminates interference, and the material would contribute little to the performance anyway, as will be shown. In some designs the hinge pin is made movable to provide additional heel pressure. This gives the effect of a floating shoe.

Figure 16–3

An internal expanding centrifugal-acting rim clutch. *(Courtesy of the Hilliard Corporation.)*

Figure 16–4

Internal friction shoe geometry.

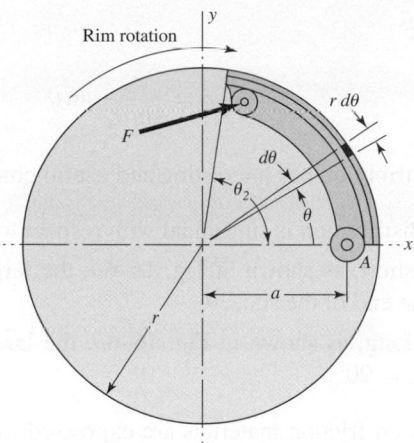

Figure 16–5

The geometry associated with an arbitrary point on the shoe.

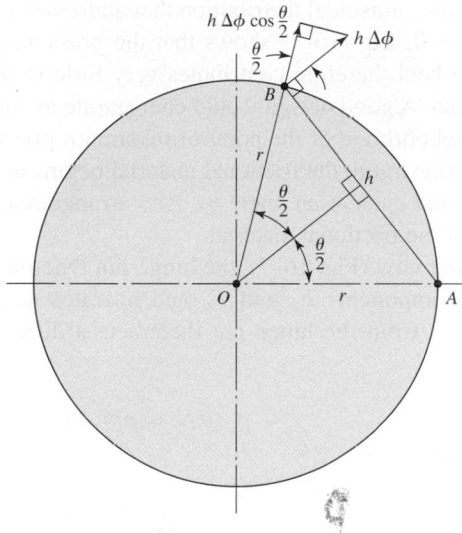

(Floating shoes will not be treated in this book, although their design follows the same general principles.)

Let us consider the pressure p acting upon an element of area of the frictional material located at an angle θ from the hinge pin (Fig. 16–4). We designate the maximum pressure p_a located at an angle θ_a from the hinge pin. To find the pressure distribution on the periphery of the internal shoe, consider point B on the shoe (Fig. 16–5). As in Ex. 16–1, if the shoe deforms by an infinitesimal rotation $\Delta\phi$ about the pivot point A, deformation perpendicular to AB is $h\,\Delta\phi$. From the isosceles triangle AOB, $h = 2r\sin(\theta/2)$, so

$$h\,\Delta\phi = 2r\,\Delta\phi\sin(\theta/2)$$

The deformation perpendicular to the rim is $h\,\Delta\phi\cos(\theta/2)$, which is

$$h\,\Delta\phi\cos(\theta/2) = 2r\,\Delta\phi\sin(\theta/2)\cos(\theta/2) = r\,\Delta\phi\sin\theta$$

Thus, the deformation, and consequently the pressure, is proportional to $\sin\theta$. In terms of the pressure at B and where the pressure is a maximum, this means

$$\frac{p}{\sin\theta} = \frac{p_a}{\sin\theta_a} \tag{a}$$

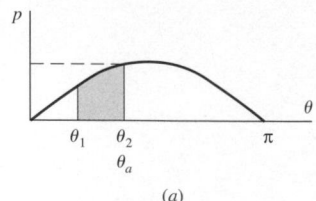

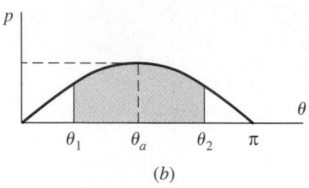

Figure 16–6

Defining the angle θ_a at which the maximum pressure p_a occurs when (a) shoe exists in zone $\theta_1 \leq \theta_2 \leq \pi/2$ and (b) shoe exists in zone $\theta_1 \leq \pi/2 \leq \theta_2$.

Rearranging gives

$$p = \frac{p_a}{\sin \theta_a} \sin \theta \qquad (16\text{–}1)$$

This pressure distribution has interesting and useful characteristics:

- The pressure distribution is sinusoidal with respect to the angle θ.
- If the shoe is short, as shown in Fig. 16–6a, the largest pressure *on the shoe* is p_a occurring at the end of the shoe, θ_2.
- If the shoe is long, as shown in Fig. 16–6b, the largest pressure on the shoe is p_a occurring at $\theta_a = 90°$.

Since limitations on friction materials are expressed in terms of the largest allowable pressure on the lining, the designer wants to think in terms of p_a and not about the amplitude of the sinusoidal distribution that addresses locations off the shoe.

When $\theta = 0$, Eq. (16–1) shows that the pressure is zero. The frictional material located at the heel therefore contributes very little to the braking action and might as well be omitted. A good design would concentrate as much frictional material as possible in the neighborhood of the point of maximum pressure. Such a design is shown in Fig. 16–7. In this figure the frictional material begins at an angle θ_1, measured from the hinge pin A, and ends at an angle θ_2. Any arrangement such as this will give a good distribution of the frictional material.

Proceeding now (Fig. 16–7), the hinge-pin reactions are R_x and R_y. The actuating force F has components F_x and F_y and operates at distance c from the hinge pin. At any angle θ from the hinge pin there acts a differential normal force dN whose magnitude is

$$dN = pbr\, d\theta \qquad (b)$$

Figure 16–7

Forces on the shoe.

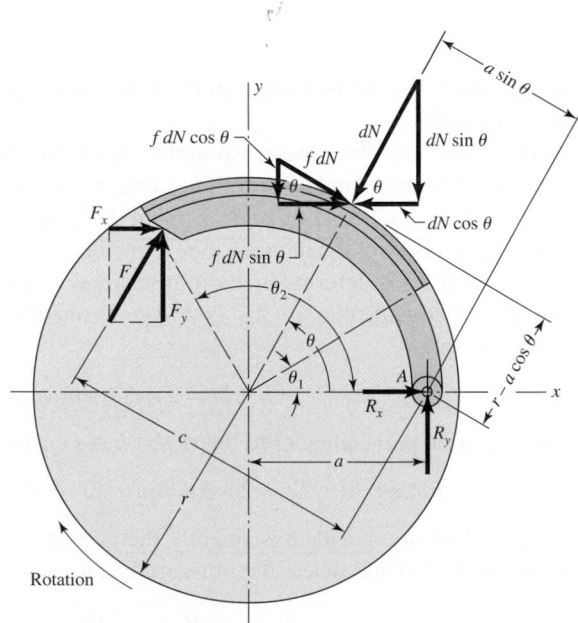

where b is the face width (perpendicular to the paper) of the friction material. Substituting the value of the pressure from Eq. (16–1), the normal force is

$$dN = \frac{p_a b r \sin\theta\, d\theta}{\sin\theta_a} \qquad (c)$$

The normal force dN has horizontal and vertical components $dN\cos\theta$ and $dN\sin\theta$, as shown in the figure. The frictional force $f\,dN$ has horizontal and vertical components whose magnitudes are $f\,dN\sin\theta$ and $f\,dN\cos\theta$, respectively. By applying the conditions of static equilibrium, we may find the actuating force F, the torque T, and the pin reactions R_x and R_y.

We shall find the actuating force F, using the condition that the summation of the moments about the hinge pin is zero. The frictional forces have a moment arm about the pin of $r - a\cos\theta$. The moment M_f of these frictional forces is

$$M_f = \int f\,dN(r - a\cos\theta) = \frac{f p_a b r}{\sin\theta_a}\int_{\theta_1}^{\theta_2}\sin\theta(r - a\cos\theta)\,d\theta \qquad (16\text{–}2)$$

which is obtained by substituting the value of dN from Eq. (c). It is convenient to integrate Eq. (16–2) for each problem, and we shall therefore retain it in this form. The moment arm of the normal force dN about the pin is $a\sin\theta$. Designating the moment of the normal forces by M_N and summing these about the hinge pin give

$$M_N = \int dN(a\sin\theta) = \frac{p_a b r a}{\sin\theta_a}\int_{\theta_1}^{\theta_2}\sin^2\theta\,d\theta \qquad (16\text{–}3)$$

The actuating force F must balance these moments. Thus

$$F = \frac{M_N - M_f}{c} \qquad (16\text{–}4)$$

We see here that a condition for zero actuating force exists. In other words, if we make $M_N = M_f$, self-locking is obtained, and no actuating force is required. This furnishes us with a method for obtaining the dimensions for some self-energizing action. Thus the dimension a in Fig. 16–7 must be such that

$$M_N > M_f \qquad (16\text{–}5)$$

The torque T applied to the drum by the brake shoe is the sum of the frictional forces $f\,dN$ times the radius of the drum:

$$T = \int f r\,dN = \frac{f p_a b r^2}{\sin\theta_a}\int_{\theta_1}^{\theta_2}\sin\theta\,d\theta$$

$$= \frac{f p_a b r^2(\cos\theta_1 - \cos\theta_2)}{\sin\theta_a} \qquad (16\text{–}6)$$

The hinge-pin reactions are found by taking a summation of the horizontal and vertical forces. Thus, for R_x, we have

$$R_x = \int dN\cos\theta - \int f\,dN\sin\theta - F_x$$

$$= \frac{p_a b r}{\sin\theta_a}\left(\int_{\theta_1}^{\theta_2}\sin\theta\cos\theta\,d\theta - f\int_{\theta_1}^{\theta_2}\sin^2\theta\,d\theta\right) - F_x \qquad (d)$$

The vertical reaction is found in the same way:

$$R_y = \int dN \sin\theta + \int f \, dN \cos\theta - F_y$$

$$= \frac{p_a b r}{\sin\theta_a} \left(\int_{\theta_1}^{\theta_2} \sin^2\theta \, d\theta + f \int_{\theta_1}^{\theta_2} \sin\theta \cos\theta \, d\theta \right) - F_y \qquad (e)$$

The direction of the frictional forces is reversed if the rotation is reversed. Thus, for counterclockwise rotation the actuating force is

$$F = \frac{M_N + M_f}{c} \qquad (16\text{–}7)$$

and since both moments have the same sense, the self-energizing effect is lost. Also, for counterclockwise rotation the signs of the frictional terms in the equations for the pin reactions change, and Eqs. (d) and (e) become

$$R_x = \frac{p_a b r}{\sin\theta_a} \left(\int_{\theta_1}^{\theta_2} \sin\theta \cos\theta \, d\theta + f \int_{\theta_1}^{\theta_2} \sin^2\theta \, d\theta \right) - F_x \qquad (f)$$

$$R_y = \frac{p_a b r}{\sin\theta_a} \left(\int_{\theta_1}^{\theta_2} \sin^2\theta \, d\theta - f \int_{\theta_1}^{\theta_2} \sin\theta \cos\theta \, d\theta \right) - F_y \qquad (g)$$

Equations (d), (e), (f), and (g) can be simplified to ease computations. Thus, let

$$A = \int_{\theta_1}^{\theta_2} \sin\theta \cos\theta \, d\theta = \left(\frac{1}{2} \sin^2\theta \right)_{\theta_1}^{\theta_2}$$

$$B = \int_{\theta_1}^{\theta_2} \sin^2\theta \, d\theta = \left(\frac{\theta}{2} - \frac{1}{4} \sin 2\theta \right)_{\theta_1}^{\theta_2} \qquad (16\text{–}8)$$

Then, for clockwise rotation as shown in Fig. 16–7, the hinge-pin reactions are

$$R_x = \frac{p_a b r}{\sin\theta_a}(A - fB) - F_x$$

$$R_y = \frac{p_a b r}{\sin\theta_a}(B + fA) - F_y \qquad (16\text{–}9)$$

For counterclockwise rotation, Eqs. (f) and (g) become

$$R_x = \frac{p_a b r}{\sin\theta_a}(A + fB) - F_x$$

$$R_y = \frac{p_a b r}{\sin\theta_a}(B - fA) - F_y \qquad (16\text{–}10)$$

In using these equations, the reference system always has its origin at the center of the drum. The positive x axis is taken through the hinge pin. The positive y axis is always in the direction of the shoe, even if this should result in a left-handed system.

The following assumptions are implied by the preceding analysis:

1   The pressure at any point on the shoe is assumed to be proportional to the distance from the hinge pin, being zero at the heel. This should be considered from the standpoint that pressures specified by manufacturers are averages rather than maxima.

2 The effect of centrifugal force has been neglected. In the case of brakes, the shoes are not rotating, and no centrifugal force exists. In clutch design, the effect of this force must be considered in writing the equations of static equilibrium.

3 The shoe is assumed to be rigid. Since this cannot be true, some deflection will occur, depending upon the load, pressure, and stiffness of the shoe. The resulting pressure distribution may be different from that which has been assumed.

4 The entire analysis has been based upon a coefficient of friction that does not vary with pressure. Actually, the coefficient may vary with a number of conditions, including temperature, wear, and environment.

EXAMPLE 16–2

The brake shown in Fig. 16–8 is 300 mm in diameter and is actuated by a mechanism that exerts the same force F on each shoe. The shoes are identical and have a face width of 32 mm. The lining is a molded asbestos having a coefficient of friction of 0.32 and a pressure limitation of 1000 kPa. Estimate the maximum

    (*a*) Actuating force F.

    (*b*) Braking capacity.

    (*c*) Hinge-pin reactions.

Solution

(*a*) The right-hand shoe is self-energizing, and so the force F is found on the basis that the maximum pressure will occur on this shoe. Here $\theta_1 = 0°$, $\theta_2 = 126°$, $\theta_a = 90°$, and $\sin \theta_a = 1$. Also,

$$a = \sqrt{(112)^2 + (50)^2} = 122.7 \text{ mm}$$

Integrating Eq. (16–2) from 0 to θ_2 yields

$$M_f = \frac{f p_a b r}{\sin \theta_a} \left[\left(-r \cos \theta \right)_0^{\theta_2} - a \left(\frac{1}{2} \sin^2 \theta \right)_0^{\theta_2} \right]$$

$$= \frac{f p_a b r}{\sin \theta_a} \left(r - r \cos \theta_2 - \frac{a}{2} \sin^2 \theta_2 \right)$$

Figure 16–8

Brake with internal expanding shoes; dimensions in millimeters.

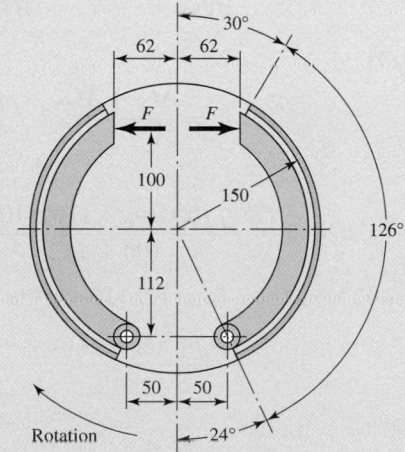

Changing all lengths to meters, we have

$$M_f = (0.32)[1000(10)^3](0.032)(0.150)$$

$$\times \left[0.150 - 0.150 \cos 126° - \left(\frac{0.1227}{2} \right) \sin^2 126° \right]$$

$$= 304 \text{ N} \cdot \text{m}$$

The moment of the normal forces is obtained from Eq. (16–3). Integrating from 0 to θ_2 gives

$$M_N = \frac{p_a b r a}{\sin \theta_a} \left(\frac{\theta}{2} - \frac{1}{4} \sin 2\theta \right)_0^{\theta_2}$$

$$= \frac{p_a b r a}{\sin \theta_a} \left(\frac{\theta_2}{2} - \frac{1}{4} \sin 2\theta_2 \right)$$

$$= [1000(10)^3](0.032)(0.150)(0.1227) \left\{ \frac{\pi}{2} \frac{126}{180} - \frac{1}{4} \sin[(2)(126°)] \right\}$$

$$= 788 \text{ N} \cdot \text{m}$$

From Eq. (16–4), the actuating force is

Answer

$$F = \frac{M_N - M_f}{c} = \frac{788 - 304}{100 + 112} = 2.28 \text{ kN}$$

(b) From Eq. (16–6), the torque applied by the right-hand shoe is

$$T_R = \frac{f p_a b r^2 (\cos \theta_1 - \cos \theta_2)}{\sin \theta_a}$$

$$= \frac{0.32[1000(10)^3](0.032)(0.150)^2(\cos 0° - \cos 126°)}{\sin 90°} = 366 \text{ N} \cdot \text{m}$$

The torque contributed by the left-hand shoe cannot be obtained until we learn its maximum operating pressure. Equations (16–2) and (16–3) indicate that the frictional and normal moments are proportional to this pressure. Thus, for the left-hand shoe,

$$M_N = \frac{788 p_a}{1000} \qquad M_f = \frac{304 p_a}{1000}$$

Then, from Eq. (16–7),

$$F = \frac{M_N + M_f}{c}$$

or

$$2.28 = \frac{(788/1000)p_a + (304/1000)p_a}{100 + 112}$$

Solving gives $p_a = 443$ kPa. Then, from Eq. (16–6), the torque on the left-hand shoe is

$$T_L = \frac{f p_a b r^2 (\cos \theta_1 - \cos \theta_2)}{\sin \theta_a}$$

Since $\sin \theta_a = \sin 90° = 1$, we have

$$T_L = 0.32[443(10)^3](0.032)(0.150)^2(\cos 0° - \cos 126°) = 162 \text{ N} \cdot \text{m}$$

The braking capacity is the total torque:

$$T = T_R + T_L = 366 + 162 = 528 \text{ N} \cdot \text{m}$$

(c) In order to find the hinge-pin reactions, we note that $\sin \theta_a = 1$ and $\theta_1 = 0$. Then Eq. (16–8) gives

$$A = \frac{1}{2} \sin^2 \theta_2 = \frac{1}{2} \sin^2 126° = 0.3273$$

$$B = \frac{\theta_2}{2} - \frac{1}{4} \sin 2\theta_2 = \frac{\pi(126)}{2(180)} - \frac{1}{4} \sin[(2)(126°)] = 1.3373$$

Also, let

$$D = \frac{p_a b r}{\sin \theta_a} = \frac{1000(0.032)(0.150)}{1} = 4.8 \text{ kN}$$

where $p_a = 1000$ kPa for the right-hand shoe. Then, using Eq. (16–9), we have

$$R_x = D(A - fB) - F_x = 4.8[0.3273 - 0.32(1.3373)] - 2.28 \sin 24°$$

$$= -1.410 \text{ kN}$$

$$R_y = D(B + fA) - F_y = 4.8[1.3373 + 0.32(0.3273)] - 2.28 \cos 24°$$

$$= 4.839 \text{ kN}$$

The resultant on this hinge pin is

$$R = \sqrt{(-1.410)^2 + (4.839)^2} = 5.04 \text{ kN}$$

The reactions at the hinge pin of the left-hand shoe are found using Eqs. (16–10) for a pressure of 443 kPa. They are found to be $R_x = 0.678$ kN and $R_y = 0.538$ kN. The resultant is

$$R = \sqrt{(0.678)^2 + (0.538)^2} = 0.866 \text{ kN}$$

The reactions for both hinge pins, together with their directions, are shown in Fig. 16–9.

| **Figure 16–9**

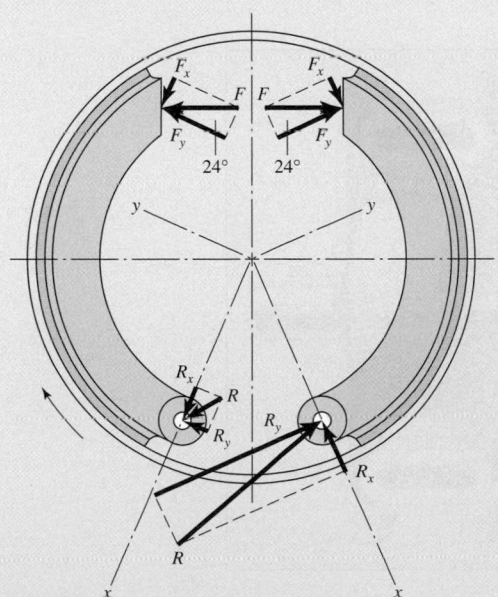

This example dramatically shows the benefit to be gained by arranging the shoes to be self-energizing. If the left-hand shoe were turned over so as to place the hinge pin at the top, it could apply the same torque as the right-hand shoe. This would make the capacity of the brake $(2)(366) = 732$ N · m instead of the present 528 N · m, a 30 percent improvement. In addition, some of the friction material at the heel could be eliminated without seriously affecting the capacity, because of the low pressure in this area. This change might actually improve the overall design because the additional rim exposure would improve the heat-dissipation capacity.

16–3 External Contracting Rim Clutches and Brakes

The patented clutch-brake of Fig. 16–10 has external contracting friction elements, but the actuating mechanism is pneumatic. Here we shall study only pivoted external shoe brakes and clutches, though the methods presented can easily be adapted to the clutch-brake of Fig. 16–10.

Operating mechanisms can be classified as:

1 Solenoids
2 Levers, linkages, or toggle devices
3 Linkages with spring loading
4 Hydraulic and pneumatic devices

The static analysis required for these devices has already been covered in Sec. 4–1. The methods there apply to any mechanism system, including all those used in brakes and clutches. It is not necessary to repeat the material in Chap. 4 that applies directly to such mechanisms. Omitting the operating mechanisms from consideration allows us to concentrate on brake and clutch performance without the extraneous influences introduced by the need to analyze the statics of the control mechanisms.

The notation for external contracting shoes is shown in Fig. 16–11. The moments of the frictional and normal forces about the hinge pin are the same as for the internal

Figure 16–10

An external contracting clutch-brake that is engaged by expanding the flexible tube with compressed air. *(Courtesy of Twin Disc Clutch Company.)*

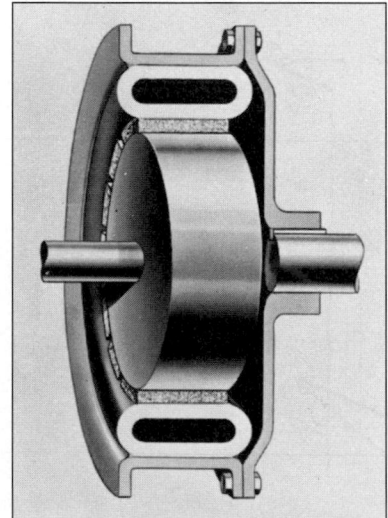

Figure 16–11

Notation of external contacting shoes.

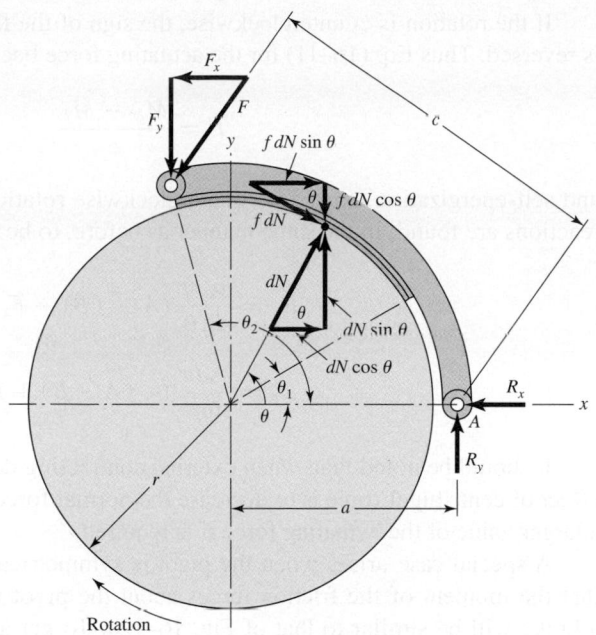

expanding shoes. Equations (16–2) and (16–3) apply and are repeated here for convenience:

$$M_f = \frac{f p_a b r}{\sin \theta_a} \int_{\theta_1}^{\theta_2} \sin \theta (r - a \cos \theta)\, d\theta \tag{16–2}$$

$$M_N = \frac{p_a b r a}{\sin \theta_a} \int_{\theta_1}^{\theta_2} \sin^2 \theta\, d\theta \tag{16–3}$$

Both these equations give positive values for clockwise moments (Fig. 16–11) when used for external contracting shoes. The actuating force must be large enough to balance both moments:

$$F = \frac{M_N + M_f}{c} \tag{16–11}$$

The horizontal and vertical reactions at the hinge pin are found in the same manner as for internal expanding shoes. They are

$$R_x = \int dN \cos \theta + \int f\, dN \sin \theta - F_x \tag{a}$$

$$R_y = \int f\, dN \cos \theta - \int dN \sin \theta + F_y \tag{b}$$

By using Eq. (16–8) and Eq. (c) from Sec. 16–2, we have

$$R_x = \frac{p_a b r}{\sin \theta_a}(A + f B) - F_x$$

$$R_y = \frac{p_a b r}{\sin \theta_a}(f A - B) + F_y \tag{16–12}$$

If the rotation is counterclockwise, the sign of the frictional term in each equation is reversed. Thus Eq. (16–11) for the actuating force becomes

$$F = \frac{M_N - M_f}{c} \tag{16–13}$$

and self-energization exists for counterclockwise rotation. The horizontal and vertical reactions are found, in the same manner as before, to be

$$R_x = \frac{p_a b r}{\sin \theta_a}(A - fB) - F_x$$

$$R_y = \frac{p_a b r}{\sin \theta_a}(-fA - B) + F_y \tag{16–14}$$

It should be noted that, when external contracting designs are used as clutches, the effect of centrifugal force is to decrease the normal force. Thus, as the speed increases, a larger value of the actuating force F is required.

A special case arises when the pivot is symmetrically located and also placed so that the moment of the friction forces about the pivot is zero. The geometry of such a brake will be similar to that of Fig. 16–12a. To get a pressure-distribution relation, we note that lining wear is such as to retain the cylindrical shape, much as a milling machine cutter feeding in the x direction would do to the shoe held in a vise. See Fig. 16–12b. This means the abscissa component of wear is w_0 for all positions θ. If wear in the radial direction is expressed as $w(\theta)$, then

$$w(\theta) = w_0 \cos \theta$$

Using Eq. (12–26) to express radial wear $w(\theta)$ as

$$w(\theta) = KPVt$$

Figure 16–12

(a) Brake with symmetrical pivoted shoe; (b) wear of brake lining.

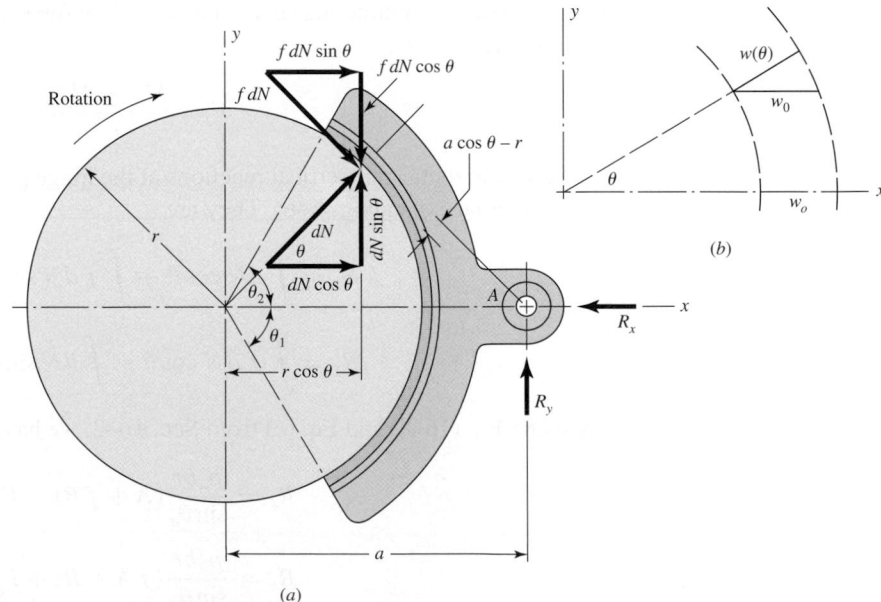

where K is a material constant, P is pressure, V is rim velocity, and t is time. Then, denoting P as $p(\theta)$ above and solving for $p(\theta)$ gives

$$p(\theta) = \frac{w(\theta)}{KVt} = \frac{w_0 \cos\theta}{KVt}$$

Since all elemental surface areas of the friction material see the same rubbing speed for the same duration, $w_0/(KVt)$ is a constant and

$$p(\theta) = (\text{constant})\cos\theta = p_a \cos\theta \tag{c}$$

where p_a is the maximum value of $p(\theta)$.

Proceeding to the force analysis, we observe from Fig. 16–12a that

$$dN = pbr\, d\theta \tag{d}$$

or

$$dN = p_a br \cos\theta\, d\theta \tag{e}$$

The distance a to the pivot is chosen by finding where the moment of the frictional forces M_f is zero. First, this ensures that reaction R_y is at the correct location to establish symmetrical wear. Second, a cosinusoidal pressure distribution is sustained, preserving our predictive ability. Symmetry means $\theta_1 = \theta_2$, so

$$M_f = 2 \int_0^{\theta_2} (f\, dN)(a\cos\theta - r) = 0$$

Substituting Eq. (e) gives

$$2 f p_a br \int_0^{\theta_2} (a\cos^2\theta - r\cos\theta)\, d\theta = 0$$

from which

$$a = \frac{4r\sin\theta_2}{2\theta_2 + \sin 2\theta_2} \tag{16–15}$$

The distance a depends on the pressure distribution. Mislocating the pivot makes M_f zero about a different location, so the brake lining adjusts its local contact pressure, through wear, to compensate. The result is unsymmetrical wear, retiring the shoe lining, hence the shoe, sooner.

With the pivot located according to Eq. (16–15), the moment about the pin is zero, and the horizontal and vertical reactions are

$$R_x = 2 \int_0^{\theta_2} dN \cos\theta = \frac{p_a br}{2}(2\theta_2 + \sin 2\theta_2) \tag{16–16}$$

where, because of symmetry,

$$\int f\, dN \sin\theta = 0$$

Also,

$$R_y = 2 \int_0^{\theta_2} f\, dN \cos\theta = \frac{p_a brf}{2}(2\theta_2 + \sin 2\theta_2) \tag{16–17}$$

where

$$\int dN \sin\theta = 0$$

also because of symmetry. Note, too, that $R_x = -N$ and $R_y = -fN$, as might be expected for the particular choice of the dimension a. Therefore the torque is

$$T = afN \qquad (16–18)$$

16–4 Band-Type Clutches and Brakes

Flexible clutch and brake bands are used in power excavators and in hoisting and other machinery. The analysis follows the notation of Fig. 16–13.

Because of friction and the rotation of the drum, the actuating force P_2 is less than the pin reaction P_1. Any element of the band, of angular length $d\theta$, will be in equilibrium under the action of the forces shown in the figure. Summing these forces in the vertical direction, we have

$$(P + dP)\sin\frac{d\theta}{2} + P\sin\frac{d\theta}{2} - dN = 0 \qquad (a)$$

$$dN = P\,d\theta \qquad (b)$$

since for small angles $\sin d\theta/2 = d\theta/2$. Summing the forces in the horizontal direction gives

$$(P + dP)\cos\frac{d\theta}{2} - P\cos\frac{d\theta}{2} - f\,dN = 0 \qquad (c)$$

$$dP - f\,dN = 0 \qquad (d)$$

Since for small angles, $\cos(d\theta/2) \doteq 1$. Substituting the value of dN from Eq. (b) in (d) and integrating give

$$\int_{P_2}^{P_1} \frac{dP}{P} = f\int_0^\phi d\theta \qquad \text{or} \qquad \ln\frac{P_1}{P_2} = f\phi$$

Figure 16–13

Forces on a brake band.

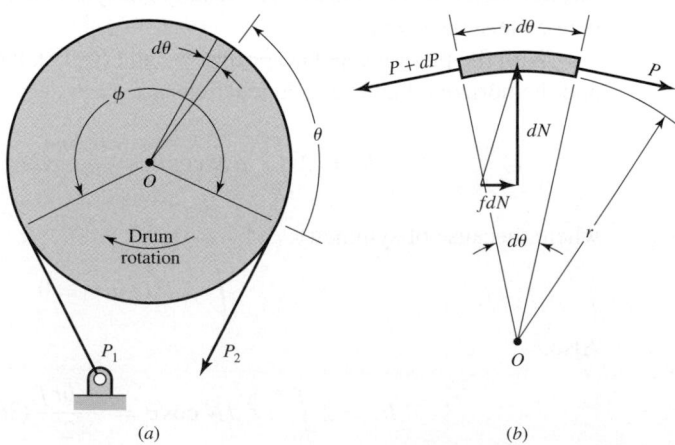

(a) (b)

and

$$\frac{P_1}{P_2} = e^{f\phi} \qquad (16\text{–}19)$$

The torque may be obtained from the equation

$$T = (P_1 - P_2)\frac{D}{2} \qquad (16\text{–}20)$$

The normal force dN acting on an element of area of width b and length $r\,d\theta$ is

$$dN = pbr\,d\theta \qquad (e)$$

where p is the pressure. Substitution of the value of dN from Eq. (b) gives

$$P\,d\theta = pbr\,d\theta$$

Therefore

$$p = \frac{P}{br} = \frac{2P}{bD} \qquad (16\text{–}21)$$

The pressure is therefore proportional to the tension in the band. The maximum pressure p_a will occur at the toe and has the value

$$p_a = \frac{2P_1}{bD} \qquad (16\text{–}22)$$

16–5 **Frictional-Contact Axial Clutches**

An axial clutch is one in which the mating frictional members are moved in a direction parallel to the shaft. One of the earliest of these is the cone clutch, which is simple in construction and quite powerful. However, except for relatively simple installations, it has been largely displaced by the disk clutch employing one or more disks as the operating members. Advantages of the disk clutch include the freedom from centrifugal effects, the large frictional area that can be installed in a small space, the more effective heat-dissipation surfaces, and the favorable pressure distribution. Figure 16–14 shows a

Figure 16–14

Cross-sectional view of a single-plate clutch; A, driver; B, driven plate (keyed to driven shaft); C, actuator.

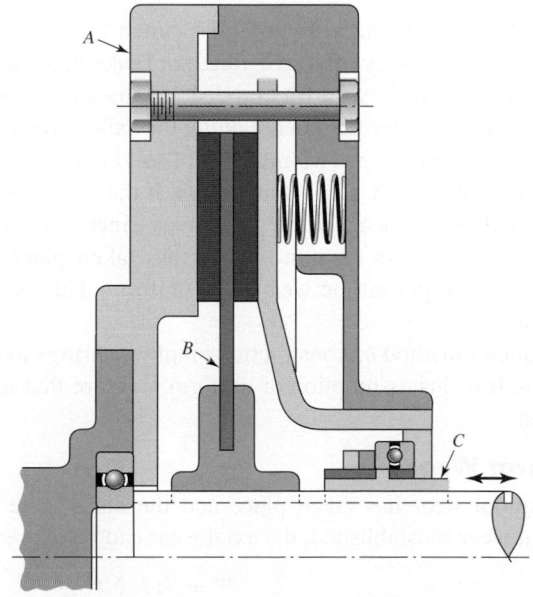

Figure 16–15

An oil-actuated multiple-disk clutch-brake for operation in an oil bath or spray. It is especially useful for rapid cycling. *(Courtesy of Twin Disc Clutch Company.)*

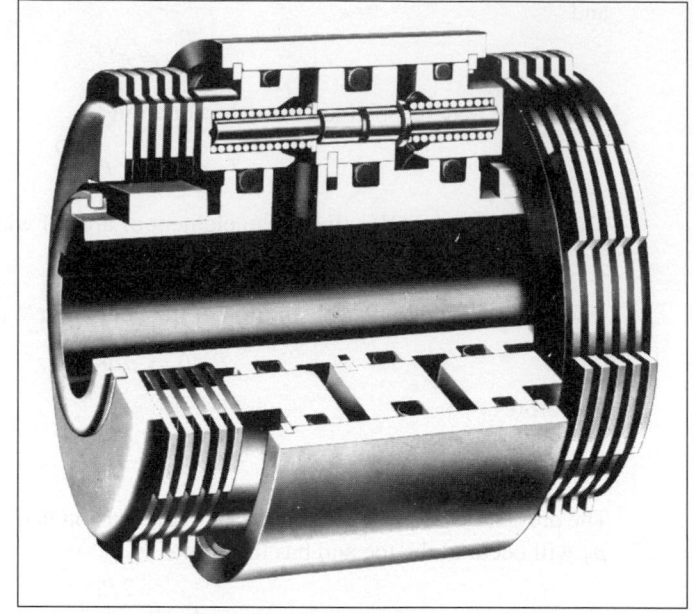

Figure 16–16

Disk friction member.

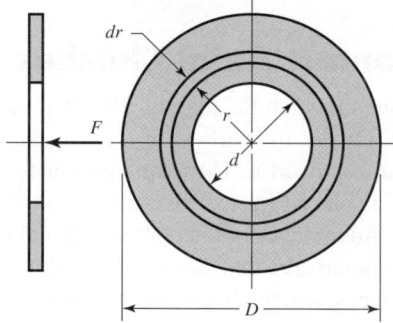

single-plate disk clutch; a multiple-disk clutch-brake is shown in Fig. 16–15. Let us now determine the capacity of such a clutch or brake in terms of the material and geometry.

Figure 16–16 shows a friction disk having an outside diameter D and an inside diameter d. We are interested in obtaining the axial force F necessary to produce a certain torque T and pressure p. Two methods of solving the problem, depending upon the construction of the clutch, are in general use. If the disks are rigid, then the greatest amount of wear will at first occur in the outer areas, since the work of friction is greater in those areas. After a certain amount of wear has taken place, the pressure distribution will change so as to permit the wear to be uniform. This is the basis of the first method of solution.

Another method of construction employs springs to obtain a uniform pressure over the area. It is this assumption of uniform pressure that is used in the second method of solution.

Uniform Wear

After initial wear has taken place and the disks have worn down to a point where uniform wear is established, the axial wear can be expressed by Eq. (12–27) as

$$w = f_1 f_2 K P V t$$

in which only P and V vary from place to place in the rubbing surfaces. By definition uniform wear is constant from place to place; therefore,

$$PV = (\text{constant}) = C_1$$

$$pr\omega = C_2$$

$$pr = C_3 = p_{\max}r_i = p_a r_i = p_a \frac{d}{2} \qquad (a)$$

We can take an expression from Eq. (a), which is the condition for having the same amount of work done at radius r as is done at radius $d/2$. Referring to Fig. 16–16, we have an element of area of radius r and thickness dr. The area of this element is $2\pi r\, dr$, so that the normal force acting upon this element is $dF = 2\pi pr\, dr$. We can find the total normal force by letting r vary from $d/2$ to $D/2$ and integrating. Thus, with pr constant,

$$F = \int_{d/2}^{D/2} 2\pi pr\, dr = \pi p_a d \int_{d/2}^{D/2} dr = \frac{\pi p_a d}{2}(D - d) \qquad (16\text{–}23)$$

The torque is found by integrating the product of the frictional force and the radius:

$$T = \int_{d/2}^{D/2} 2\pi f pr^2\, dr = \pi f p_a d \int_{d/2}^{D/2} r\, dr = \frac{\pi f p_a d}{8}(D^2 - d^2) \qquad (16\text{–}24)$$

By substituting the value of F from Eq. (16–23) we may obtain a more convenient expression for the torque. Thus

$$T = \frac{Ff}{4}(D + d) \qquad (16\text{–}25)$$

In use, Eq. (16–23) gives the actuating force for the selected maximum pressure p_a. This equation holds for any number of friction pairs or surfaces. Equation (16–25), however, gives the torque capacity for only a single friction surface.

Uniform Pressure

When uniform pressure can be assumed over the area of the disk, the actuating force F is simply the product of the pressure and the area. This gives

$$F = \frac{\pi p_a}{4}(D^2 - d^2) \qquad (16\text{–}26)$$

As before, the torque is found by integrating the product of the frictional force and the radius:

$$T = 2\pi f p \int_{d/2}^{D/2} r^2\, dr = \frac{\pi f p}{12}(D^3 - d^3) \qquad (16\text{–}27)$$

Since $p = p_a$, from Eq. (16–26) we can rewrite Eq. (16–27) as

$$T = \frac{Ff}{3} \frac{D^3 - d^3}{D^2 - d^2} \qquad (16\text{–}28)$$

It should be noted for both equations that the torque is for a single pair of mating surfaces. This value must therefore be multiplied by the number of pairs of surfaces in contact.

Let us express Eq. (16–25) for torque during uniform wear as

$$\frac{T}{f F D} = \frac{1 + d/D}{4} \qquad (b)$$

Figure 16–17

Dimensionless plot of Eqs. (b) and (c).

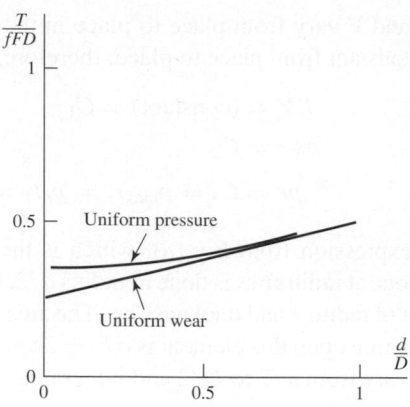

and Eq. (16–28) for torque during uniform pressure (new clutch) as

$$\frac{T}{fFD} = \frac{1}{3}\frac{1 - (d/D)^3}{1 - (d/D)^2}$$

(c)

and plot these in Fig. 16–17. What we see is a dimensionless[1] presentation of Eqs. (b) and (c) which reduces the number of variables from five (T, f, F, D, and d) to three (T/FD, f, and d/D) which are dimensionless. This is the method of Buckingham. The dimensionless groups (called pi terms) are

$$\pi_1 = \frac{T}{FD} \qquad \pi_2 = f \qquad \pi_3 = \frac{d}{D}$$

This allows a five-dimensional space to be reduced to a three-dimensional space. Further, because of the "multiplicative" relation between f and T in Eqs. (b) and (c), it is possible to plot π_1/π_2 versus π_3 in a two-dimensional space (the plane of a sheet of paper) to view all cases over the domain of existence of Eqs. (b) and (c) and to compare, without risk of oversight! By examining Fig. 16–17 we can conclude that a new clutch, Eq. (b), always transmits more torque than an old clutch, Eq. (c). Furthermore, since clutches of this type are proportioned to make the diameter ratio d/D fall in the range $0.6 \le d/D \le 1$, the largest discrepancy between Eq. (b) and Eq. (c) will be

$$\frac{T}{fFD} = \frac{1 + 0.6}{4} = 0.400 \qquad \text{(old clutch, uniform wear)}$$

$$\frac{T}{fFD} = \frac{1}{3}\frac{1 - 0.6^3}{1 - 0.6^2} = 0.4083 \qquad \text{(new clutch, uniform pressure)}$$

so the proportional error is $(0.4083 - 0.400)/0.400 = 0.021$, or about 2 percent. Given the uncertainties in the actual coefficient of friction and the certainty that new clutches get old, there is little reason to use anything but Eqs. (16–23), (16–24), and (16–25).

---

[1]Charles R. Mischke, "Minimizing Engineering Effort," Chap. 11 in Joseph E. Shigley and Charles R. Mischke (co–editors-in-chief), *Standard Handbook of Machine Design,* 2nd ed., McGraw-Hill, New York, 1996, pp. 11.1–11.9, or Charles R. Mischke, *Mathematical Model Building,* 2nd rev. ed., Iowa State University Press, Ames, 1980, pp. 139–164.

16–6   **Disk Brakes**

As indicated in Fig. 16–16, there is no fundamental difference between a disk clutch and a disk brake. The analysis of the preceding section applies to disk brakes too.

We have seen that rim or drum brakes can be designed for self-energization. While this feature is important in reducing the braking effort required, it also has a disadvantage. When drum brakes are used as vehicle brakes, only a slight change in the coefficient of friction will cause a large change in the pedal force required for braking. A not unusual 30 percent reduction in the coefficient of friction due to a temperature change or moisture, for example, can result in a 50 percent change in the pedal force required to obtain the same braking torque obtainable prior to the change. The disk brake has no self-energization, and hence is not so susceptible to changes in the coefficient of friction.

Another type of disk brake is the *floating caliper brake,* shown in Fig. 16–18. The caliper supports a single floating piston actuated by hydraulic pressure. The action is much like that of a screw clamp, with the piston replacing the function of the screw. The floating action also compensates for wear and ensures a fairly constant pressure over the area of the friction pads. The seal and boot of Fig. 16–18 are designed to obtain clearance by backing off from the piston when the piston is released.

Caliper brakes (named for the nature of the actuating linkage) and disk brakes (named for the shape of the unlined surface) press friction material against the face(s) of

Figure 16–18

An automotive disk brake.
(Courtesy DaimlerChrysler Corporation.)

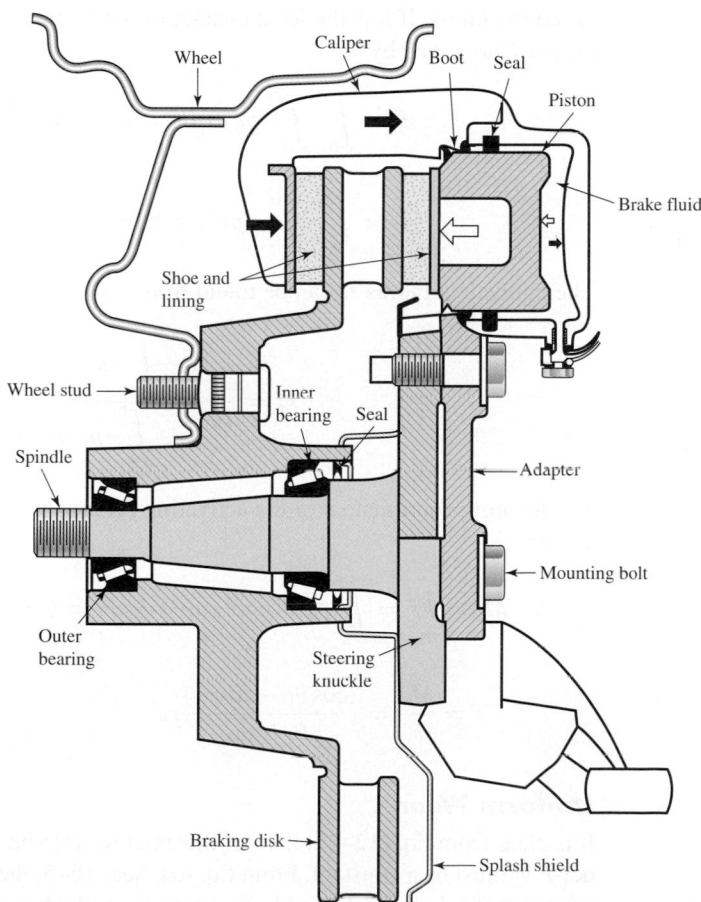

Figure 16–19

Geometry of contact area of an annular-pad segment of a caliper brake.

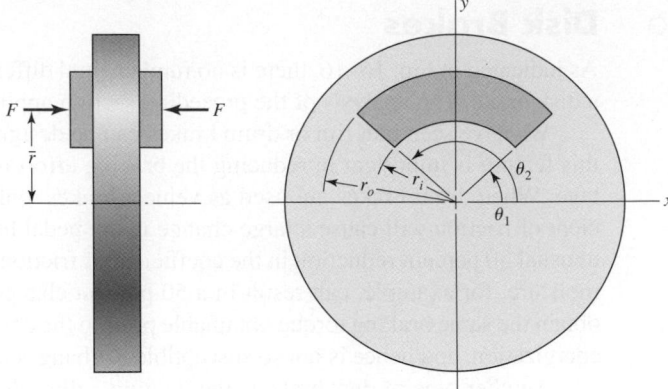

a rotating disk. Depicted in Fig. 16–19 is the geometry of an annular-pad brake contact area. The governing axial wear equation is Eq. (12–27):

$$w = f_1 f_2 K P V t$$

The coordinate $\bar{r}$ locates the line of action of force F that intersects the y axis. Of interest also is the effective radius r_e, which is the radius of an equivalent shoe of infinitesimal radial thickness. If p is the local contact pressure, the actuating force F and the friction torque T are given by

$$F = \int_{\theta_1}^{\theta_2} \int_{r_i}^{r_o} pr \, dr \, d\theta = (\theta_2 - \theta_1) \int_{r_i}^{r_o} pr \, dr \qquad (16\text{–}29)$$

$$T = \int_{\theta_1}^{\theta_2} \int_{r_i}^{r_o} fpr^2 \, dr \, d\theta = (\theta_2 - \theta_1) f \int_{r_i}^{r_o} pr^2 \, dr \qquad (16\text{–}30)$$

The equivalent radius r_e can be found from $f F r_e = T$, or

$$r_e = \frac{T}{fF} = \frac{\displaystyle\int_{r_i}^{r_o} pr^2 \, dr}{\displaystyle\int_{r_i}^{r_o} pr \, dr} \qquad (16\text{–}31)$$

The locating coordinate $\bar{r}$ of the activating force is found by taking moments about the x axis:

$$M_x = F\bar{r} = \int_{\theta_1}^{\theta_2} \int_{r_i}^{r_o} pr(r \sin \theta) \, dr \, d\theta = (\cos \theta_1 - \cos \theta_2) \int_{r_i}^{r_o} pr^2 \, dr$$

$$\bar{r} = \frac{M_x}{F} = \frac{(\cos \theta_1 - \cos \theta_2)}{\theta_2 - \theta_1} r_e \qquad (16\text{–}32)$$

Uniform Wear

It is clear from Eq. (12–27) that for the axial wear to be the same everywhere, the product PV must be a constant. From Eq. (a), Sec. 16–5, the pressure p can be expressed in terms of the largest allowable pressure p_a (which occurs at the inner radius r_i) as

$p = p_a r_i / r$. Equation (16–29) becomes

$$F = (\theta_2 - \theta_1) p_a r_i (r_o - r_i) \qquad (16\text{–}33)$$

Equation (16–30) becomes

$$T = (\theta_2 - \theta_1) f p_a r_i \int_{r_i}^{r_o} r\, dr = \frac{1}{2}(\theta_2 - \theta_1) f p_a r_i \left(r_o^2 - r_i^2\right) \qquad (16\text{–}34)$$

Equation (16–31) becomes

$$r_e = \frac{p_a r_i \displaystyle\int_{r_i}^{r_o} r\, dr}{p_a r_i \displaystyle\int_{r_i}^{r_o} dr} = \frac{r_o^2 - r_i^2}{2}\frac{1}{r_o - r_i} = \frac{r_o + r_i}{2} \qquad (16\text{–}35)$$

Equation (16–32) becomes

$$\bar{r} = \frac{\cos\theta_1 - \cos\theta_2}{\theta_2 - \theta_1}\frac{r_o + r_i}{2} \qquad (16\text{–}36)$$

Uniform Pressure

In this situation, approximated by a new brake, $p = p_a$. Equation (16–29) becomes

$$F = (\theta_2 - \theta_1) p_a \int_{r_i}^{r_o} r\, dr = \frac{1}{2}(\theta_2 - \theta_1) p_a \left(r_o^2 - r_i^2\right) \qquad (16\text{–}37)$$

Equation (16–30) becomes

$$T = (\theta_2 - \theta_1) f p_a \int_{r_i}^{r_o} r^2\, dr = \frac{1}{3}(\theta_2 - \theta_1) f p_a \left(r_o^3 - r_i^3\right) \qquad (16\text{–}38)$$

Equation (16–31) becomes

$$r_e = \frac{p_a \displaystyle\int_{r_i}^{r_o} r^2\, dr}{p_a \displaystyle\int_{r_i}^{r_o} r\, dr} = \frac{r_o^3 - r_i^3}{3}\frac{2}{r_o^2 - r_i^2} = \frac{2}{3}\frac{r_o^3 - r_i^3}{r_o^2 - r_i^2} \qquad (16\text{–}39)$$

Equation (16–32) becomes

$$\bar{r} = \frac{\cos\theta_1 - \cos\theta_2}{\theta_2 - \theta_1}\frac{2}{3}\frac{r_o^3 - r_i^3}{r_o^2 - r_i^2} = \frac{2}{3}\frac{r_o^3 - r_i^3}{r_o^2 - r_i^2}\frac{\cos\theta_1 - \cos\theta_2}{\theta_2 - \theta_1} \qquad (16\text{–}40)$$

EXAMPLE 16–3 Two annular pads, $r_i = 3.875$ in, $r_o = 5.50$ in, subtend an angle of $108°$, have a co-efficient of friction of 0.37, and are actuated by a pair of hydraulic cylinders 1.5 in in diameter. The torque requirement is 13 000 lbf · in. For uniform wear
(a) Find the largest normal pressure p_a.
(b) Estimate the actuating force F.
(c) Find the equivalent radius r_e and force location $\bar{r}$.
(d) Estimate the required hydraulic pressure.

Solution   (a) From Eq. (16–34), with $T = 13\,000/2 = 6500$ lbf · in for each pad,

Answer
$$p_a = \frac{2T}{(\theta_2 - \theta_1)\,f r_i \left(r_o^2 - r_i^2\right)}$$

$$= \frac{2(6500)}{(144° - 36°)(\pi/180)0.37(3.875)(5.5^2 - 3.875^2)} = 315.8 \text{ psi}$$

(b) From Eq. (16–33),

Answer
$$F = (\theta_2 - \theta_1)\,p_a r_i (r_o - r_i) = (144° - 36°)(\pi/180)315.8(3.875)(5.5 - 3.875)$$

$$= 3748 \text{ lbf}$$

(c) From Eq. (16–35),

Answer
$$r_e = \frac{r_o + r_i}{2} = \frac{5.50 + 3.875}{2} = 4.688 \text{ in}$$

From Eq. (16–36),

Answer
$$\bar{r} = \frac{\cos\theta_1 - \cos\theta_2}{\theta_2 - \theta_1}\frac{r_o + r_i}{2} = \frac{\cos 36° - \cos 144°}{(144° - 36°)(\pi/180)}\frac{5.50 + 3.875}{2}$$

$$= 4.024 \text{ in}$$

(d) Each cylinder supplies the actuating force, 3748 lbf.

Answer
$$p_{\text{hydraulic}} = \frac{F}{A_P} = \frac{3748}{\pi(1.5^2/4)} = 2121 \text{ psi}$$

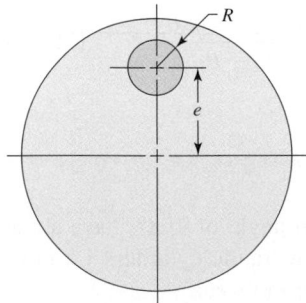

Circular (Button or Puck) Pad Caliper Brake

Figure 16–20 displays the pad geometry. Numerical integration is necessary to analyze this brake since the boundaries are difficult to handle in closed form. Table 16–1 gives the parameters for this brake as determined by Fazekas. The effective radius is given by

$$r_e = \delta e \qquad (16\text{–}41)$$

The actuating force is given by

$$F = \pi R^2 p_{\text{av}} \qquad (16\text{–}42)$$

Figure 16–20

Geometry of circular pad of a caliper brake.

and the torque is given by

$$T = f F r_e \qquad (16\text{–}43)$$

Table 16–1

Parameters for a Circular-Pad Caliper Brake

Source: G. A. Fazekas, "On Circular Spot Brakes," *Trans. ASME, J. Engineering for Industry*, vol. 94, Series B, No. 3, August 1972, pp. 859–863.

$\dfrac{R}{e}$	$\delta = \dfrac{r_e}{e}$	$\dfrac{p_{max}}{p_{av}}$
0.0	1.000	1.000
0.1	0.983	1.093
0.2	0.969	1.212
0.3	0.957	1.367
0.4	0.947	1.578
0.5	0.938	1.875

EXAMPLE 16–4

A button-pad disk brake uses dry sintered metal pads. The pad radius is $\frac{1}{2}$ in, and its center is 2 in from the axis of rotation of the $3\frac{1}{2}$-in-diameter disk. Using half of the largest allowable pressure, $p_{max} = 350$ psi, find the actuating force and the brake torque. The coefficient of friction is 0.31.

Solution

Since the pad radius $R = 0.5$ in and eccentricity $e = 2$ in,

$$\frac{R}{e} = \frac{0.5}{2} = 0.25$$

From Table 16–1, by interpolation, $\delta = 0.963$ and $p_{max}/p_{av} = 1.290$. It follows that the effective radius e is found from Eq. (16–41):

$$r_e = \delta e = 0.963(2) = 1.926 \text{ in}$$

and the average pressure is

$$p_{av} = \frac{p_{max}/2}{1.290} = \frac{350/2}{1.290} = 135.7 \text{ psi}$$

The actuating force F is found from Eq. (16–42) to be

Answer

$$F = \pi R^2 p_{av} = \pi (0.5)^2 135.7 = 106.6 \text{ lbf} \qquad \text{(one side)}$$

The brake torque T is

Answer

$$T = f F r_e = 0.31(106.6)1.926 = 63.65 \text{ lbf} \cdot \text{in} \qquad \text{(one side)}$$

16–7 Cone Clutches and Brakes

The drawing of a *cone clutch* in Fig. 16–21 shows that it consists of a *cup* keyed or splined to one of the shafts, a *cone* that must slide axially on splines or keys on the mating shaft, and a helical *spring* to hold the clutch in engagement. The clutch is disengaged by means of a fork that fits into the shifting groove on the friction cone. The *cone angle* α and the diameter and face width of the cone are the important geometric design parameters.

Figure 16–21

Cross section of a cone clutch.

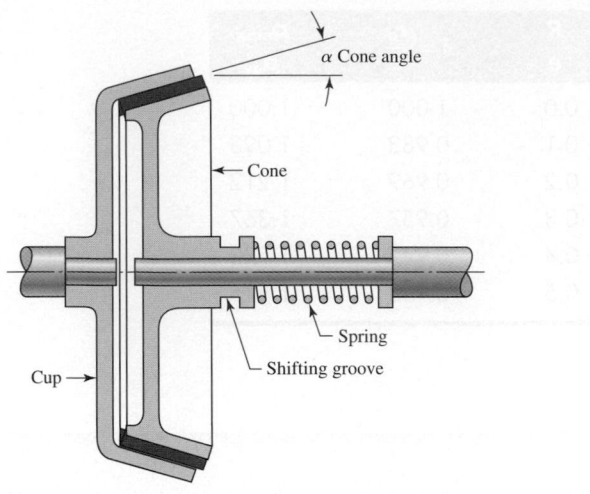

Figure 16–22

Contact area of a cone clutch.

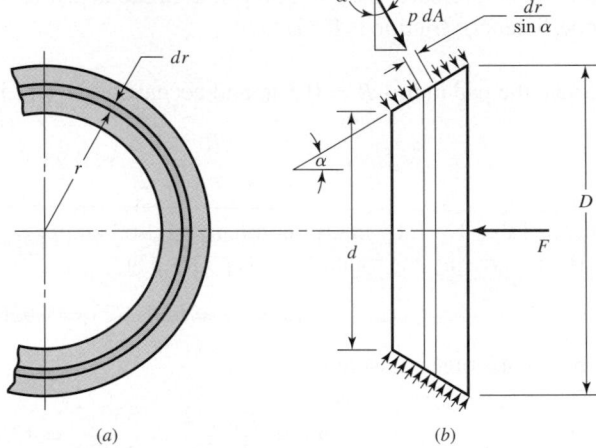

If the cone angle is too small, say, less than about 8°, then the force required to disengage the clutch may be quite large. And the wedging effect lessens rapidly when larger cone angles are used. Depending upon the characteristics of the friction materials, a good compromise can usually be found using cone angles between 10 and 15°.

To find a relation between the operating force F and the torque transmitted, designate the dimensions of the friction cone as shown in Figure 16–22. As in the case of the axial clutch, we can obtain one set of relations for a uniform-wear and another set for a uniform-pressure assumption.

Uniform Wear

The pressure relation is the same as for the axial clutch:

$$p = p_a \frac{d}{2r}$$ (a)

Next, referring to Fig. 16–22, we see that we have an element of area dA of radius r and width $dr/\sin\alpha$. Thus $dA = (2\pi r\,dr)/\sin\alpha$. As shown in Fig. 16–22, the operating force will be the integral of the axial component of the differential force $p\,dA$. Thus

$$F = \int p\,dA \sin\alpha = \int_{d/2}^{D/2} \left(p_a \frac{d}{2r}\right)\left(\frac{2\pi r\,dr}{\sin\alpha}\right)(\sin\alpha)$$

$$= \pi p_a d \int_{d/2}^{D/2} dr = \frac{\pi p_a d}{2}(D-d) \tag{16–44}$$

which is the same result as in Eq. (16–23).

The differential friction force is $fp\,dA$, and the torque is the integral of the product of this force with the radius. Thus

$$T = \int rfp\,dA = \int_{d/2}^{D/2}(rf)\left(p_a \frac{d}{2r}\right)\left(\frac{2\pi r\,dr}{\sin\alpha}\right)$$

$$= \frac{\pi fp_a d}{\sin\alpha}\int_{d/2}^{D/2} r\,dr = \frac{\pi fp_a d}{8\sin\alpha}(D^2-d^2) \tag{16–45}$$

Note that Eq. (16–24) is a special case of Eq. (16–45), with $\alpha = 90°$. Using Eq. (16–44), we find that the torque can also be written

$$T = \frac{Ff}{4\sin\alpha}(D+d) \tag{16–46}$$

Uniform Pressure

Using $p = p_a$, the actuating force is found to be

$$F = \int p_a\,dA \sin\alpha = \int_{d/2}^{D/2}(p_a)\left(\frac{2\pi r\,dr}{\sin\alpha}\right)(\sin\alpha) = \frac{\pi p_a}{4}(D^2-d^2) \tag{16–47}$$

The torque is

$$T = \int rfp_a\,dA = \int_{d/2}^{D/2}(rfp_a)\left(\frac{2\pi r\,dr}{\sin\alpha}\right) = \frac{\pi fp_a}{12\sin\alpha}(D^3-d^3) \tag{16–48}$$

Using Eq. (16–47) in Eq. (16–48) gives

$$T = \frac{Ff}{3\sin\alpha}\frac{D^3-d^3}{D^2-d^2} \tag{16–49}$$

As in the case of the axial clutch, we can write Eq. (16–46) dimensionlessly as

$$\frac{T\sin\alpha}{fFd} = \frac{1+d/D}{4} \tag{b}$$

and write Eq. (16–49) as

$$\frac{T\sin\alpha}{fFd} = \frac{1}{3}\frac{1-(d/D)^3}{1-(d/D)^2} \tag{c}$$

This time there are six (T, α, f, F, D, and d) parameters and four pi terms:

$$\pi_1 = \frac{T}{FD} \qquad \pi_2 = f \qquad \pi_3 = \sin\alpha \qquad \pi_4 = \frac{d}{D}$$

As in Fig. 16–17, we plot $T\,\sin\alpha/(fFD)$ as ordinate and d/D as abscissa. The plots and conclusions are the same. There is little reason for using equations other than Eqs. (16–44), (16–45), and (16–46).

16–8   **Energy Considerations**

When the rotating members of a machine are caused to stop by means of a brake, the kinetic energy of rotation must be absorbed by the brake. This energy appears in the brake in the form of heat. In the same way, when the members of a machine that are initially at rest are brought up to speed, slipping must occur in the clutch until the driven members have the same speed as the driver. Kinetic energy is absorbed during slippage of either a clutch or a brake, and this energy appears as heat.

We have seen how the torque capacity of a clutch or brake depends upon the coefficient of friction of the material and upon a safe normal pressure. However, the character of the load may be such that, if this torque value is permitted, the clutch or brake may be destroyed by its own generated heat. The capacity of a clutch is therefore limited by two factors, the characteristics of the material and the ability of the clutch to dissipate heat. In this section we shall consider the amount of heat generated by a clutching or braking operation. If the heat is generated faster than it is dissipated, we have a temperature-rise problem; that is the subject of the next section.

To get a clear picture of what happens during a simple clutching or braking operation, refer to Fig. 16–1a, which is a mathematical model of a two-inertia system connected by a clutch. As shown, inertias I_1 and I_2 have initial angular velocities of ω_1 and ω_2, respectively. During the clutch operation both angular velocities change and eventually become equal. We assume that the two shafts are rigid and that the clutch torque is constant.

Writing the equation of motion for inertia 1 gives

$$I_1\ddot{\theta}_1 = -T \tag{a}$$

where $\ddot{\theta}_1$ is the angular acceleration of I_1 and T is the clutch torque. A similar equation for I_2 is

$$I_2\ddot{\theta}_2 = T \tag{b}$$

We can determine the instantaneous angular velocities $\dot{\theta}_1$ and $\dot{\theta}_2$ of I_1 and I_2 after any period of time t has elapsed by integrating Eqs. (a) and (b). The results are

$$\dot{\theta}_1 = -\frac{T}{I_1}t + \omega_1 \tag{c}$$

$$\dot{\theta}_2 = \frac{T}{I_2}t + \omega_2 \tag{d}$$

where $\dot{\theta}_1 = \omega_1$ and $\dot{\theta}_2 = \omega_2$ at $t = 0$. The difference in the velocities, sometimes called the relative velocity, is

$$\dot{\theta} = \dot{\theta}_1 - \dot{\theta}_2 = -\frac{T}{I_1}t + \omega_1 - \left(\frac{T}{I_2}t + \omega_2\right)$$
$$= \omega_1 - \omega_2 - T\left(\frac{I_1 + I_2}{I_1 I_2}\right)t \tag{16–50}$$

The clutching operation is completed at the instant in which the two angular velocities $\dot{\theta}_1$ and $\dot{\theta}_2$ become equal. Let the time required for the entire operation be t_1. Then $\dot{\theta} = 0$ when $\dot{\theta}_1 = \dot{\theta}_2$, and so Eq. (16–50) gives the time as

$$t_1 = \frac{I_1 I_2(\omega_1 - \omega_2)}{T(I_1 + I_2)} \tag{16–51}$$

This equation shows that the time required for the engagement operation is directly proportional to the velocity difference and inversely proportional to the torque.

We have assumed the clutch torque to be constant. Therefore, using Eq. (16–50), we find the rate of energy-dissipation during the clutching operation to be

$$u = T\dot{\theta} = T\left[\omega_1 - \omega_2 - T\left(\frac{I_1 + I_2}{I_1 I_2}\right)t\right] \qquad (e)$$

This equation shows that the energy-dissipation rate is greatest at the start, when $t = 0$.

The total energy dissipated during the clutching operation or braking cycle is obtained by integrating Eq. (e) from $t = 0$ to $t = t_1$. The result is found to be

$$\begin{aligned}
E &= \int_0^{t_1} u\, dt = T\int_0^{t_1}\left[\omega_1 - \omega_2 - T\left(\frac{I_1 + I_2}{I_1 I_2}\right)t\right]dt \\
&= \frac{I_1 I_2(\omega_1 - \omega_2)^2}{2(I_1 + I_2)}
\end{aligned} \qquad (16\text{–}52)$$

where Eq. (16–51) was employed. Note that the energy dissipated is proportional to the velocity difference squared and is independent of the clutch torque.

Note that E in Eq. (16–52) is the energy lost or dissipated; this is the energy that is absorbed by the clutch or brake. If the inertias are expressed in U.S. customary units (lbf · in · s²), then the energy absorbed by the clutch assembly is in in · lbf. Using these units, the heat generated in Btu is

$$H = \frac{E}{9336} \qquad (16\text{–}53)$$

In SI, the inertias are expressed in kilogram-meter² units, and the energy dissipated is expressed in joules.

16–9 Temperature Rise

The temperature rise of the clutch or brake assembly can be approximated by the classic expression

$$\Delta T = \frac{H}{C_p W} \qquad (16\text{–}54)$$

where ΔT = temperature rise, °F

C_p = specific heat capacity, Btu/(lb$_m$ · °F); use 0.12 for steel or cast iron

W = mass of clutch or brake parts, lbm

A similar equation can be written for SI units. It is

$$\Delta T = \frac{E}{C_p m} \qquad (16\text{–}55)$$

where ΔT = temperature rise, °C

C_p = specific heat capacity; use 500 J/kg · °C for steel or cast iron

m = mass of clutch or brake parts, kg

The temperature-rise equations above can be used to explain what happens when a clutch or brake is operated. However, there are so many variables involved that it would be most unlikely that such an analysis would even approximate experimental results. For this reason such analyses are most useful, for repetitive cycling, in pinpointing those design parameters that have the greatest effect on performance.

If an object is at initial temperature T_1 in an environment of temperature T_∞, then Newton's cooling model is expressed as

$$\frac{T - T_\infty}{T_1 - T_\infty} = \exp\left(-\frac{\hbar_{CR} A}{W C_p} t\right) \tag{16–56}$$

where
T = temperature at time t, °F
T_1 = initial temperature, °F
T_∞ = environmental temperature, °F
$\hbar_{CR}$ = overall coefficient of heat transfer, Btu/(in$^2 \cdot$ s $\cdot$ °F)
A = lateral surface area, in^2
W = mass of the object, lbm
C_p = specific heat capacity of the object, Btu/(lbm $\cdot$ °F)

Figure 16–23 shows an application of Eq. (16–56). The curve ABC is the exponential decline of temperature given by Eq. (16–56). At time t_B a second application of the brake occurs. The temperature quickly rises to temperature T_2, and a new cooling curve is started. For repetitive brake applications, subsequent temperature peaks $T_3, T_4, \ldots,$ occur until the brake is able to dissipate by cooling between operations an amount of heat equal to the energy absorbed in the application. If this is a production situation with brake applications every t_1 seconds, then a steady state develops in which all the peaks T_{max} and all the valleys T_{min} are repetitive.

The heat-dissipation capacity of disk brakes has to be planned to avoid reaching the temperatures of disk and pad that are detrimental to the parts. When a disk brake has a rhythm such as discussed above, then the rate of heat transfer is described by another Newtonian equation:

$$H_{loss} = \hbar_{CR} A(T - T_\infty) = (h_r + f_v h_c) A(T - T_\infty) \tag{16–57}$$

Figure 16–23

The effect of clutching or braking operations on temperature. T_∞ is the ambient temperature. Note that the temperature rise ΔT may be different for each operation.

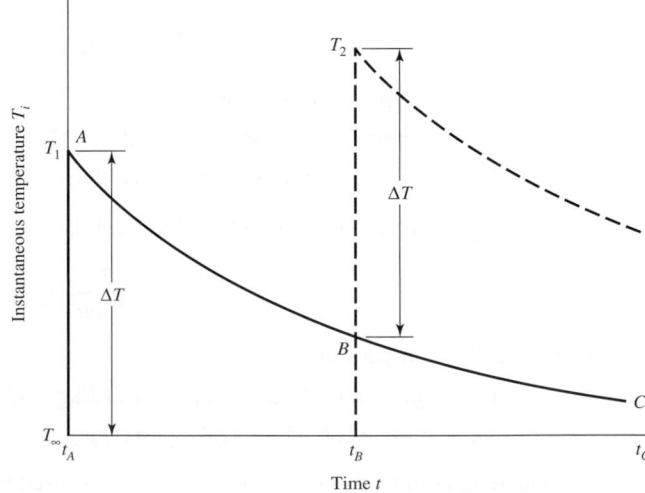

where H_{loss} = rate of energy loss, Btu/s

$\hbar_{CR}$ = overall coefficient of heat transfer, Btu/(in² · s · °F)

h_r = radiation component of $\hbar_{CR}$, Btu/(in² · s · °F), Fig. 16–24a

h_c = convective component of $\hbar_{CR}$, Btu/(in² · s · °F), Fig. 16–24a

f_v = ventilation factor, Fig. 16–24b

T = disk temperature, °F

T_∞ = ambient temperature, °F

The energy E absorbed by the brake stopping an equivalent rotary inertia I in terms of original and final angular velocities ω_o and ω_f is given by Eq. (16–53) with $I_1 = I$

Figure 16–24

(a) Heat-transfer coefficient in still air. (b) Ventilation factors. (*Courtesy of Tolo-o-matic.*)

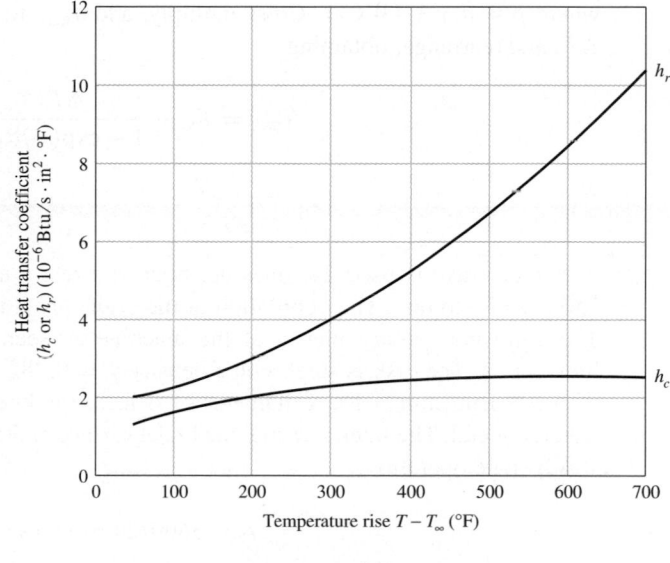

(a)

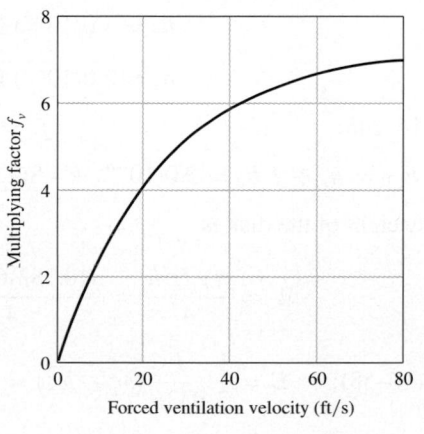

(b)

and $I_2 = 0$,

$$E = \frac{1}{2} \frac{I}{9336} \left(\omega_o^2 - \omega_f^2 \right) \tag{16-58}$$

in Btu. The temperature rise ΔT due to a single stop is

$$\Delta T = \frac{E}{WC} \tag{16-59}$$

T_{max} has to be high enough to transfer E Btu in t_1 seconds. For steady state, rearrange Eq. (16–56) as

$$\frac{T_{min} - T_\infty}{T_{max} - T_\infty} = \exp(-\beta t_1)$$

where $\beta = \hbar_{CR} A / (WC_p)$. Cross-multiply, add T_{max} to both sides, set $T_{max} - T_{min} = \Delta T$, and rearrange, obtaining

$$T_{max} = T_\infty + \frac{\Delta T}{1 - \exp(-\beta t_1)} \tag{16-60}$$

EXAMPLE 16–5

A caliper brake is used 24 times per hour to arrest a machine shaft from a speed of 250 rev/min to rest. The ventilation of the brake provides a mean air speed of 25 ft/s. The equivalent rotary inertia of the machine as seen from the brake shaft is 289 lbm · in · s. The disk is steel with a density $\gamma = 0.282$ lb/in³, a specific heat capacity of 0.108 Btu/(lbm · °F), a diameter of 6 in, a thickness of $\frac{1}{4}$ in. The pads are dry sintered metal. The lateral area of the brake surface is 50 in². Find T_{max} and T_{min} for the steady-state operation.

Solution

$$t_1 = 3600/24 = 150 \text{ s}$$

Assuming a temperature rise of $T_{max} - T_\infty = 200°\text{F}$, from Fig. 16–24a,

$$h_r = 3.0(10^{-6}) \text{ Btu/(in}^2 \cdot \text{s} \cdot °\text{F)}$$

$$h_c = 2.0(10^{-6}) \text{ Btu/(in}^2 \cdot \text{s} \cdot °\text{F)}$$

Fig.16–24b: $$f_v = 4.8$$

$$\hbar_{CR} = h_r + f_v h_c = 3.0(10^{-6}) + 4.8(2.0)10^{-6} = 12.6(10^{-6}) \text{ Btu/(in}^2 \cdot \text{s} \cdot °\text{F)}$$

The weight of the disk is

$$W = \frac{\pi \gamma D^2 h}{4} = \frac{\pi (0.282) 6^2 (0.25)}{4} = 1.99 \text{ lb}$$

Eq. (16–58): $$E = \frac{1}{2} \frac{I}{9336} \left(\omega_o^2 - \omega_f^2 \right) = \frac{289}{2(9336)} \left(\frac{2\pi}{60} 250 \right)^2 = 10.6 \text{ Btu}$$

$$\beta = \frac{\hbar_{CR} A}{WC_p} = \frac{12.6(10^{-6})50}{1.99(0.108)} = 2.93(10^{-3}) \text{ s}^{-1}$$

Eq. (16–59): $$\Delta T = \frac{E}{WC_p} = \frac{10.6}{1.99(0.108)} = 49.3°F$$

Answer   Eq. (16–60): $$T_{max} = 70 + \frac{49.3}{1 - \exp[-2.93(10^{-3})150]} = 209°F$$

Answer $$T_{min} = 209 - 49.3 = 160°F$$

The predicted temperature rise here is $T_{max} - T_\infty = 139°F$. Iterating with revised values of h_r and h_c from Fig. 16–24a, we can make the solution converge to $T_{max} = 220°F$ and $T_{min} = 171°F$.

Table 16–3 for dry sintered metal pads gives a continuous operating maximum temperature of 570–660°F. There is no danger of overheating.

16–10   Friction Materials

A brake or friction clutch should have the following lining material characteristics to a degree that is dependent on the severity of service:

- High and reproducible coefficient of friction
- Imperviousness to environmental conditions, such as moisture
- The ability to withstand high temperatures, together with good thermal conductivity and diffusivity, as well as high specific heat capacity
- Good resiliency
- High resistance to wear, scoring, and galling
- Compatible with the environment
- Flexibility

Table 16–2 gives area of friction surface required for several braking powers. Table 16–3 gives important characteristics of some friction materials for brakes and clutches.

Table 16–2

Area of Friction Material Required for a Given Average Braking Power   *Sources:* M. J. Neale, *The Tribology Handbook*, Butterworth, London, 1973; *Friction Materials for Engineers*, Ferodo Ltd., Chapel-en-le-frith, England, 1968.

Duty Cycle	Typical Applications	Ratio of Area to Average Braking Power, in²/(Btu/s)		
		Band and Drum Brakes	Plate Disk Brakes	Caliper Disk Brakes
Infrequent	Emergency brakes	0.85	2.8	0.28
Intermittent	Elevators, cranes, and winches	2.8	7.1	0.70
Heavy-duty	Excavators, presses	5.6–6.9	13.6	1.41

Table 16–3

Characteristics of Friction Materials for Brakes and Clutches   *Sources:* Ferodo Ltd., Chapel-en-le-frith, England; Scan-pac, Mequon, Wisc.; Raybestos, New York, N.Y. and Stratford, Conn.; Gatke Corp., Chicago, Ill.; General Metals Powder Co., Akron, Ohio; D. A. B. Industries, Troy, Mich.; Friction Products Co., Medina, Ohio.

Material	Friction Coefficient f	Maximum Pressure P_{max}, psi	Maximum Temperature Instantaneous, °F	Maximum Temperature Continuous, °F	Maximum Velocity V_{max}, ft/min	Applications
Cermet	0.32	150	1500	750		Brakes and clutches
Sintered metal (dry)	0.29–0.33	300–400	930–1020	570–660	3600	Clutches and caliper disk brakes
Sintered metal (wet)	0.06–0.08	500	930	570	3600	Clutches
Rigid molded asbestos (dry)	0.35–0.41	100	660–750	350	3600	Drum brakes and clutches
Rigid molded asbestos (wet)	0.06	300	660	350	3600	Industrial clutches
Rigid molded asbestos pads	0.31–0.49	750	930–1380	440–660	4800	Disk brakes
Rigid molded nonasbestos	0.33–0.63	100–150		500–750	4800–7500	Clutches and brakes
Semirigid molded asbestos	0.37–0.41	100	660	300	3600	Clutches and brakes
Flexible molded asbestos	0.39–0.45	100	660–750	300–350	3600	Clutches and brakes
Wound asbestos yarn and wire	0.38	100	660	300	3600	Vehicle clutches
Woven asbestos yarn and wire	0.38	100	500	260	3600	Industrial clutches and brakes
Woven cotton	0.47	100	230	170	3600	Industrial clutches and brakes
Resilient paper (wet)	0.09–0.15	400	300		$PV < 500\,000$ psi·ft/min	Clutches and transmission bands

The manufacture of friction materials is a highly specialized process, and it is advisable to consult manufacturers' catalogs and handbooks, as well as manufacturers directly, in selecting friction materials for specific applications. Selection involves a consideration of the many characteristics as well as the standard sizes available.

The *woven-cotton lining* is produced as a fabric belt that is impregnated with resins and polymerized. It is used mostly in heavy machinery and is usually supplied in rolls up to 50 ft in length. Thicknesses available range from $\frac{1}{8}$ to 1 in, in widths up to about 12 in.

A *woven-asbestos lining* is made in a similar manner to the cotton lining and may also contain metal particles. It is not quite as flexible as the cotton lining and comes in a smaller range of sizes. Along with the cotton lining, the asbestos lining was widely used as a brake material in heavy machinery.

Molded-asbestos linings contain asbestos fiber and friction modifiers; a thermoset polymer is used, with heat, to form a rigid or semirigid molding. The principal use was in drum brakes.

Molded-asbestos pads are similar to molded linings but have no flexibility; they were used for both clutches and brakes.

Sintered-metal pads are made of a mixture of copper and/or iron particles with friction modifiers, molded under high pressure and then heated to a high temperature to fuse the material. These pads are used in both brakes and clutches for heavy-duty applications.

Cermet pads are similar to the sintered-metal pads and have a substantial ceramic content.

Table 16–4 lists properties of typical brake linings. The linings may consist of a mixture of fibers to provide strength and ability to withstand high temperatures, various friction particles to obtain a degree of wear resistance as well as a higher coefficient of friction, and bonding materials.

Table 16–5 includes a wider variety of clutch friction materials, together with some of their properties. Some of these materials may be run wet by allowing them to dip in oil or to be sprayed by oil. This reduces the coefficient of friction somewhat but carries away more heat and permits higher pressures to be used.

Table 16–4

Some Properties of Brake Linings

	Woven Lining	Molded Lining	Rigid Block
Compressive strength, kpsi	10–15	10–18	10–15
Compressive strength, MPa	70–100	70–125	70–100
Tensile strength, kpsi	2.5–3	4–5	3–4
Tensile strength, MPa	17–21	27–35	21–27
Max. temperature, °F	400–500	500	750
Max. temperature, °C	200–260	260	400
Max. speed, ft/min	7500	5000	7500
Max. speed, m/s	38	25	38
Max. pressure, psi	50–100	100	150
Max. pressure, kPa	340–690	690	1000
Frictional coefficient, mean	0.45	0.47	0.40–45

Table 16–5

Friction Materials for Clutches

Material	Friction Coefficient		Max. Temperature		Max. Pressure	
	Wet	Dry	°F	°C	psi	kPa
Cast iron on cast iron	0.05	0.15–0.20	600	320	150–250	1000–1750
Powdered metal* on cast iron	0.05–0.1	0.1–0.4	1000	540	150	1000
Powdered metal* on hard steel	0.05–0.1	0.1–0.3	1000	540	300	2100
Wood on steel or cast iron	0.16	0.2–0.35	300	150	60–90	400–620
Leather on steel or cast iron	0.12	0.3–0.5	200	100	10–40	70–280
Cork on steel or cast iron	0.15–0.25	0.3–0.5	200	100	8–14	50–100
Felt on steel or cast iron	0.18	0.22	280	140	5–10	35–70
Woven asbestos* on steel or cast iron	0.1–0.2	0.3–0.6	350–500	175–260	50–100	350–700
Molded asbestos* on steel or cast iron	0.08–0.12	0.2–0.5	500	260	50–150	350–1000
Impregnated asbestos* on steel or cast iron	0.12	0.32	500–750	260–400	150	1000
Carbon graphite on steel	0.05–0.1	0.25	700–1000	370–540	300	2100

*The friction coefficient can be maintained with ±5 percent for specific materials in this group.

16–11 Miscellaneous Clutches and Couplings

The square-jaw clutch shown in Fig. 16–25a is one form of positive-contact clutch. These clutches have the following characteristics:

1 They do not slip.
2 No heat is generated.
3 They cannot be engaged at high speeds.
4 Sometimes they cannot be engaged when both shafts are at rest.
5 Engagement at any speed is accompanied by shock.

The greatest differences among the various types of positive clutches are concerned with the design of the jaws. To provide a longer period of time for shift action during engagement, the jaws may be ratchet-shaped, spiral-shaped, or gear-tooth-shaped. Sometimes a great many teeth or jaws are used, and they may be cut either circumferentially, so that they engage by cylindrical mating, or on the faces of the mating elements.

Although positive clutches are not used to the extent of the frictional-contact types, they do have important applications where synchronous operation is required, as, for example, in power presses or rolling-mill screw-downs.

Devices such as linear drives or motor-operated screwdrivers must run to a definite limit and then come to a stop. An overload-release type of clutch is required for these applications. Figure 16–25b is a schematic drawing illustrating the principle of operation of such a clutch. These clutches are usually spring-loaded so as to release at a

Figure 16–25

(a) Square-jaw clutch;
(b) overload release clutch
using a detent.

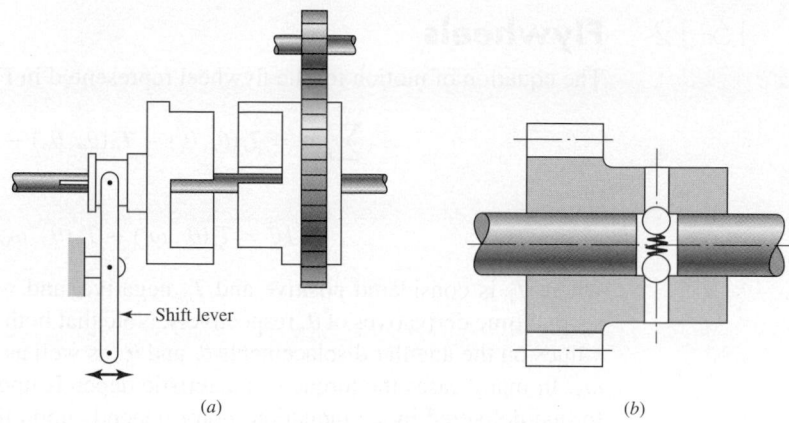

← Shift lever

(a)

(b)

Figure 16–26

Shaft couplings. (a) Plain.
(b) Light-duty toothed coupling.
(c) BOST-FLEX® through-bore
design having elastomer
Insert to transmit torque by
compression; insert permits 1°
misalignment. (d) Three-jaw
coupling available with
bronze, rubber, or
polyurethane insert to minimize
vibration. *(Reproduced by permission,
Boston Gear Division, Colfax Corp.)*

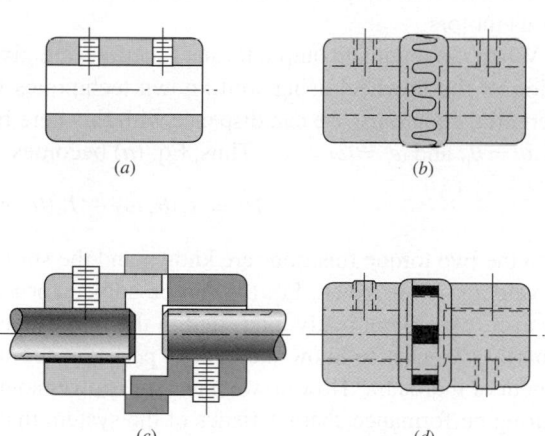

(a)

(b)

(c)

(d)

predetermined torque. The clicking sound which is heard when the overload point is reached is considered to be a desirable signal.

Both fatigue and shock loads must be considered in obtaining the stresses and deflections of the various portions of positive clutches. In addition, wear must generally be considered. The application of the fundamentals discussed in Parts 1 and 2 is usually sufficient for the complete design of these devices.

An overrunning clutch or coupling permits the driven member of a machine to "freewheel" or "overrun" because the driver is stopped or because another source of power increases the speed of the driven mechanism. The construction uses rollers or balls mounted between an outer sleeve and an inner member having cam flats machined around the periphery. Driving action is obtained by wedging the rollers between the sleeve and the cam flats. This clutch is therefore equivalent to a pawl and ratchet with an infinite number of teeth.

There are many varieties of overrunning clutches available, and they are built in capacities up to hundreds of horsepower. Since no slippage is involved, the only power loss is that due to bearing friction and windage.

The shaft couplings shown in Fig. 16–26 are representative of the selection available in catalogs.

16–12 Flywheels

The equation of motion for the flywheel represented in Fig. 16–1b is

$$\sum M = T_i(\theta_i, \dot{\theta}_i) - T_o(\theta_o, \dot{\theta}_o) - I\ddot{\theta} = 0$$

or

$$I\ddot{\theta} = T_i(\theta_i, \omega_i) - T_o(\theta_o, \omega_o) \tag{a}$$

where T_i is considered positive and T_o negative, and where $\dot{\theta}$ and $\ddot{\theta}$ are the first and second time derivatives of θ, respectively. Note that both T_i and T_o may depend for their values on the angular displacements θ_i and θ_o as well as their angular velocities ω_i and ω_o. In many cases the torque characteristic depends upon only one of these. Thus, the torque delivered by an induction motor depends upon the speed of the motor. In fact, motor manufacturers publish charts detailing the torque-speed characteristics of their various motors.

When the input and output torque functions are given, Eq. (a) can be solved for the motion of the flywheel using well-known techniques for solving linear and nonlinear differential equations. We can dispense with this here by assuming a rigid shaft, giving $\theta_i = \theta = \theta_o$ and $\omega_i = \omega = \omega_o$. Thus, Eq. ($a$) becomes

$$I\ddot{\theta} = T_i(\theta, \omega) - T_o(\theta, \omega) \tag{b}$$

When the two torque functions are known and the starting values of the displacement θ and velocity ω are given, Eq. (b) can be solved for θ, ω, and $\ddot{\theta}$ as functions of time. However, we are not really interested in the instantaneous values of these terms at all. Primarily we want to know the overall performance of the flywheel. What should its moment of inertia be? How do we match the power source to the load? And what are the resulting performance characteristics of the system that we have selected?

To gain insight into the problem, a hypothetical situation is diagrammed in Fig. 16–27. An input power source subjects a flywheel to a constant torque T_i while the shaft rotates from θ_1 to θ_2. This is a positive torque and is plotted upward. Equation (b) indicates that a positive acceleration $\ddot{\theta}$ will be the result, and so the shaft velocity increases from ω_1 to ω_2. As shown, the shaft now rotates from θ_2 to θ_3 with zero torque and hence, from Eq. (b), with zero acceleration. Therefore $\omega_3 = \omega_2$. From θ_3 to θ_4 a load, or output torque, of constant magnitude is applied, causing the shaft to slow down from ω_3 to ω_4. Note that the output torque is plotted in the negative direction in accordance with Eq. (b).

The work input to the flywheel is the area of the rectangle between θ_1 and θ_2, or

$$U_i = T_i(\theta_2 - \theta_1) \tag{c}$$

| **Figure 16–27**

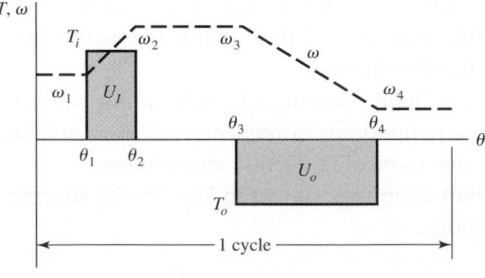

The work output of the flywheel is the area of the rectangle from θ_3 to θ_4, or

$$U_o = T_o(\theta_4 - \theta_3) \tag{d}$$

If U_o is greater than U_i, the load uses more energy than has been delivered to the flywheel and so ω_4 will be less than ω_1. If $U_o = U_i$, ω_4 will be equal to ω_1 because the gains and losses are equal; we are assuming no friction losses. And finally, ω_4 will be greater than ω_1 if $U_i > U_o$.

We can also write these relations in terms of kinetic energy. At $\theta = \theta_1$ the flywheel has a velocity of ω_1 rad/s, and so its kinetic energy is

$$E_1 = \frac{1}{2}I\omega_1^2 \tag{e}$$

At $\theta = \theta_2$ the velocity is ω_2, and so

$$E_2 = \frac{1}{2}I\omega_2^2 \tag{f}$$

Thus the change in kinetic energy is

$$E_2 - E_1 - \frac{1}{2}I\left(\omega_2^2 - \omega_1^2\right) \tag{16-61}$$

Many of the torque displacement functions encountered in practical engineering situations are so complicated that they must be integrated by numerical methods. Figure 16–28, for example, is a typical plot of the engine torque for one cycle of motion of a single-cylinder internal combustion engine. Since a part of the torque curve is negative, the flywheel must return part of the energy back to the engine. Integrating this curve from $\theta = 0$ to 4π and dividing the result by 4π yields the mean torque T_m available to drive a load during the cycle.

It is convenient to define a *coefficient of speed fluctuation* as

$$C_s = \frac{\omega_2 - \omega_1}{\omega} \tag{16-62}$$

Figure 16–28

Relation between torque and crank angle for a one-cylinder, four-stroke–cycle internal combustion engine.

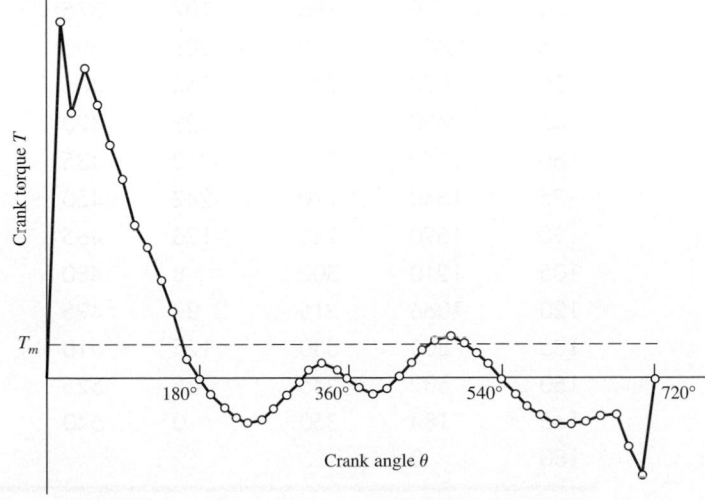

where ω is the nominal angular velocity, given by

$$\omega = \frac{\omega_2 + \omega_1}{2} \qquad (16\text{--}63)$$

Equation (16–61) can be factored to give

$$E_2 - E_1 = \frac{I}{2}(\omega_2 - \omega_1)(\omega_2 + \omega_1)$$

Since $\omega_2 - \omega_1 = C_s\omega$ and $\omega_2 + \omega_1 = 2\omega$, we have

$$E_2 - E_1 = C_s I \omega^2 \qquad (16\text{--}64)$$

Equation (16–64) can be used to obtain an appropriate flywheel inertia corresponding to the energy change $E_2 - E_1$.

EXAMPLE 16–6

Table 16–6 lists values of the torque used to plot Fig. 16–28. The nominal speed of the engine is to be 250 rad/s.

(a) Integrate the torque-displacement function for one cycle and find the energy that can be delivered to a load during the cycle.

(b) Determine the mean torque T_m (see Fig. 16–28).

(c) The greatest energy fluctuation is approximately between $\theta = 15°$ and $\theta = 150°$ on the torque diagram; see Fig. 16–28 and note that $T_o = -T_m$. Using a coefficient of speed fluctuation $C_s = 0.1$, find a suitable value for the flywheel inertia.

(d) Find ω_2 and ω_1.

Solution

(a) Using $n = 48$ intervals of $\Delta\theta = 4\pi/48$, numerical integration of the data of Table 16–6 yields $E = 3368$ in · lbf. This is the energy that can be delivered to the load.

Table 16–6

Plotting Data for Fig. 16–29

θ, deg	T, lbf · in	θ, deg	T, lbf · in	θ, deg	T, lbf · in	θ, deg	T, lbf · in
0	0	195	−107	375	−85	555	−107
15	2800	210	−206	390	−125	570	−206
30	2090	225	−260	405	−89	585	−292
45	2430	240	−323	420	8	600	−355
60	2160	255	−310	435	126	615	−371
75	1840	270	−242	450	242	630	−362
90	1590	285	−126	465	310	645	−312
105	1210	300	−8	480	323	660	−272
120	1066	315	89	495	280	675	−274
135	803	330	125	510	206	690	−548
150	532	345	85	525	107	705	−760
165	184	360	0	540	0	720	0
180	0						

Answer   (b) $$T_m = \frac{3368}{4\pi} = 268 \text{ lbf} \cdot \text{in}$$

(c) The largest positive loop on the torque-displacement diagram occurs between $\theta = 0°$ and $\theta = 180°$. We select this loop as yielding the largest speed change. Subtracting 268 lbf · in from the values in Table 16–6 for this loop gives, respectively, $-268, 2532, 1822, 2162, 1892, 1572, 1322, 942, 798, 535, 264, -84,$ and -268 lbf · in. Numerically integrating $T - T_m$ with respect to θ yields $E_2 - E_1 = 3531$ lbf · in. We now solve Eq. (16–64) for I. This gives

Answer $$I = \frac{E_2 - E_1}{C_s \omega^2} = \frac{3531}{0.1(250)^2} = 0.565 \text{ lbf} \cdot \text{s}^2 \text{ in}$$

(d) Equations (16–62) and (16–63) can be solved simultaneously for ω_2 and ω_1. Substituting appropriate values in these two equations yields

Answer $$\omega_2 = \frac{\omega}{2}(2 + C_s) = \frac{250}{2}(2 + 0.1) = 262.5 \text{ rad/s}$$

Answer $$\omega_1 = 2\omega - \omega_2 = 2(250) - 262.5 = 237.5 \text{ rad/s}$$

These two speeds occur at $\theta = 180°$ and $\theta = 0°$, respectively.

Punch-press torque demand often takes the form of a severe impulse and the running friction of the drive train. The motor overcomes the minor task of overcoming friction while attending to the major task of restoring the flywheel's angular speed. The situation can be idealized as shown in Fig. 16–29. Neglecting the running friction, Euler's equation can be written as

$$T(t_1 - 0) = \frac{1}{2}I(\omega_1^2 - \omega_2^2) = E_2 - E_1$$

where the only significant inertia is that of the flywheel. Punch presses can have the motor and flywheel on one shaft, then, through a gear reduction, drive a slider-crank mechanism that carries the punching tool. The motor can be connected to the punch

Figure 16–29

(a) Punch-press torque demand during punching. (b) Squirrel-cage electric motor torque-speed characteristic.

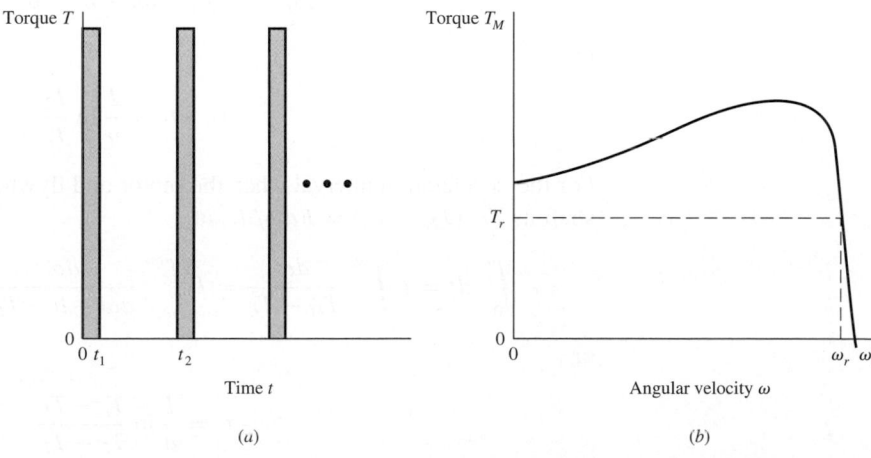

continuously, creating a punching rhythm, or it can be connected on command through a clutch that allows one punch and a disconnect. The motor and flywheel must be sized for the most demanding service, which is steady punching. The work done is given by

$$W = \int_{\theta_1}^{\theta_2} [T(\theta) - T] \, d\theta = \frac{1}{2} I \left(\omega_{max}^2 - \omega_{min}^2 \right)$$

This equation can be arranged to include the coefficient of speed fluctuation C_s as follows:

$$W = \frac{1}{2} I \left(\omega_{max}^2 - \omega_{min}^2 \right) = \frac{I}{2} (\omega_{max} - \omega_{min})(\omega_{max} + \omega_{min})$$

$$= \frac{I}{2} (C_s \bar{\omega})(2\omega_0) = I C_s \bar{\omega} \omega_0$$

When the speed fluctuation is low, $\omega_0 \doteq \bar{\omega}$, and

$$I = \frac{W}{C_s \bar{\omega}^2}$$

An induction motor has a linear torque characteristic $T = a\omega + b$ in the range of operation. The constants a and b can be found from the nameplate speed ω_r and the synchronous speed ω_s:

$$a = \frac{T_r - T_s}{\omega_r - \omega_s} = \frac{T_r}{\omega_r - \omega_s} = -\frac{T_r}{\omega_s - \omega_r}$$

$$b = \frac{T_r \omega_s - T_s \omega_r}{\omega_s - \omega_r} = \frac{T_r \omega_s}{\omega_s - \omega_r}$$

(16–65)

For example, a 3-hp three-phase squirrel-cage ac motor rated at 1125 rev/min has a torque of $63\,025(3)/1125 = 168.1$ lbf · in. The rated angular velocity is $\omega_r = 2\pi n_r/60 = 2\pi(1125)/60 = 117.81$ rad/s, and the synchronous angular velocity $\omega_s = 2\pi(1200)/60 = 125.66$ rad/s. Thus $a = -21.41$ lbf · in · s/rad, and $b = 2690.9$ lbf · in, and we can express $T(\omega)$ as $a\omega + b$. During the interval from t_1 to t_2 the motor accelerates the flywheel according to $I\ddot{\theta} = T_M$ (i.e., $T \, d\omega/dt = T_M$). Separating the equation $T_M = I \, d\omega/dt$ we have

$$\int_{t_1}^{t_2} dt = \int_{\omega_r}^{\omega_2} \frac{I \, d\omega}{T_M} = I \int_{\omega_r}^{\omega_2} \frac{d\omega}{a\omega + b} = \frac{I}{a} \ln \frac{a\omega_2 + b}{a\omega_r + b} = \frac{I}{a} \ln \frac{T_2}{T_r}$$

or

$$t_2 - t_1 = \frac{I}{a} \ln \frac{T_2}{T_r}$$

(16–66)

For the deceleration interval when the motor and flywheel feel the punch torque on the shaft as T_L, $(T_M - T_L) = I \, d\omega/dt$, or

$$\int_0^{t_1} dt = I \int_{\omega_2}^{\omega_r} \frac{d\omega}{T_M - T_L} = I \int_{\omega_2}^{\omega_r} \frac{d\omega}{a\omega + b - T_L} = \frac{I}{a} \ln \frac{a\omega_r + b - T_L}{a\omega_2 + b - T_L}$$

or

$$t_1 = \frac{I}{a} \ln \frac{T_r - T_L}{T_2 - T_L}$$

(16–67)

We can divide Eq. (16–66) by Eq. (16–67) to obtain

$$\frac{T_2}{T_r} = \left(\frac{T_L - T_r}{T_L - T_2}\right)^{(t_2 - t_1)/t_1} \tag{16–68}$$

Equation (16–68) can be solved for T_2 numerically. Having T_2 the flywheel inertia is, from Eq. (16–66),

$$I = \frac{a(t_2 - t_1)}{\ln(T_2/T_r)} \tag{16–69}$$

It is important that a be in units of lbf · in · s/rad so that I has proper units. The constant a should not be in lbf · in per rev/min or lbf · in per rev/s.

PROBLEMS

16–1   The figure shows an internal rim-type brake having an inside rim diameter of 12 in and a dimension $R = 5$ in. The shoes have a face width of $1\frac{1}{2}$ in and are both actuated by a force of 500 lbf. The mean coefficient of friction is 0.28.

(a) Find the maximum pressure and indicate the shoe on which it occurs.

(b) Estimate the braking torque effected by each shoe, and find the total braking torque.

(c) Estimate the resulting hinge-pin reactions.

Problem 16–1

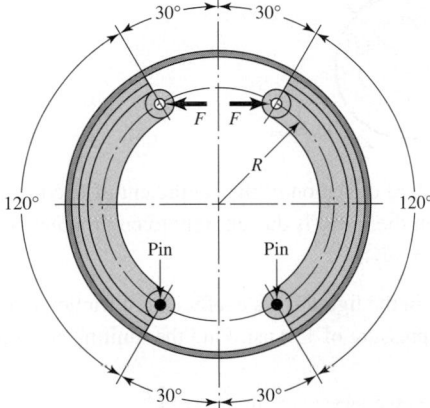

16–2   For the brake in Prob. 16–1, consider the pin and actuator locations to be the same. However, instead of 120°, let the friction surface of the brake shoes be 90° and centrally located. Find the maximum pressure and the total braking torque.

ANALYSIS

16–3   In the figure for Prob. 16–1, the inside rim diameter is 280 mm and the dimension R is 90 mm. The shoes have a face width of 30 mm. Find the braking torque and the maximum pressure for each shoe if the actuating force is 1000 N, the drum rotation is counterclockwise, and $f = 0.30$.

ANALYSIS

16–4   The figure shows a 400-mm-diameter brake drum with four internally expanding shoes. Each of the hinge pins A and B supports a pair of shoes. The actuating mechanism is to be arranged to produce the same force F on each shoe. The face width of the shoes is 75 mm. The material used permits a coefficient of friction of 0.24 and a maximum pressure of 1000 kPa.

(a) Determine the actuating force.

(b) Estimate the brake capacity.

(c) Noting that rotation may be in either direction, estimate the hinge-pin reactions.

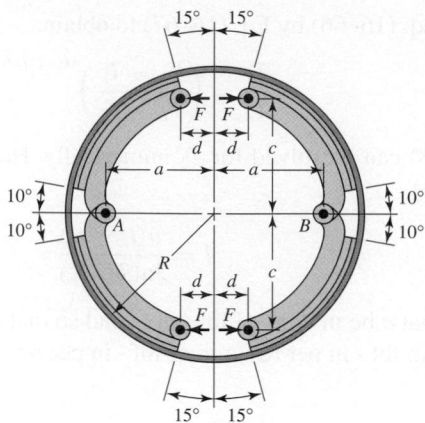

Problem 16–4

The dimensions in millimeters are
$a = 150$, $c = 165$, $R = 200$,
and $d = 50$.

ANALYSIS   **16–5**   The block-type hand brake shown in the figure has a face width of 30 mm and a mean coefficient of friction of 0.25. For an estimated actuating force of 400 N, find the maximum pressure on the shoe and find the braking torque.

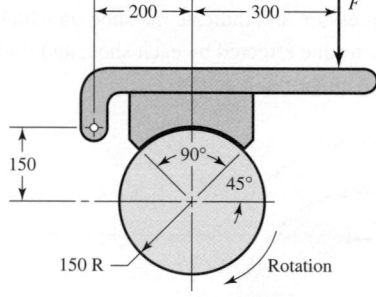

Problem 16–5

Dimensions in millimeters.

ANALYSIS   **16–6**   Suppose the standard deviation of the coefficient of friction in Prob. 16–5 is $\hat{\sigma}_f = 0.025$, where the deviation from the mean is due entirely to environmental conditions. Find the brake torques corresponding to $\pm 3\hat{\sigma}_f$.

ANALYSIS   **16–7**   The brake shown in the figure has a coefficient of friction of 0.30, a face width of 2 in, and a limiting shoe lining pressure of 150 psi. Find the limiting actuating force F and the torque capacity.

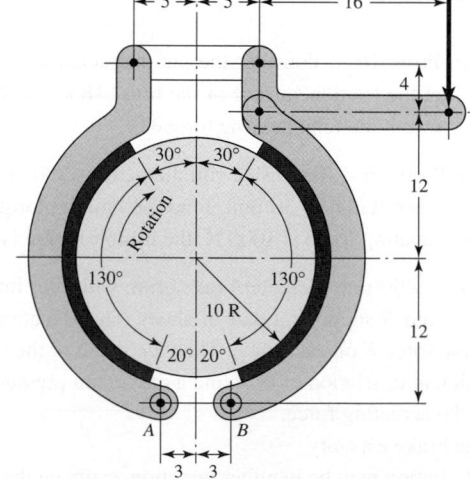

Problem 16–7

Dimensions in inches.

ANALYSIS

16–8 Refer to the symmetrical pivoted external brake shoe of Fig. 16–12 and Eq. (16–15). Suppose the pressure distribution was uniform, that is, the pressure p is independent of θ. What would the pivot distance a' be? If $\theta_1 = \theta_2 = 60°$, compare a with a'.

ANALYSIS

16–9 The shoes on the brake depicted in the figure subtend a 90° arc on the drum of this external pivoted-shoe brake. The actuation force P is applied to the lever. The rotation direction of the drum is counterclockwise, and the coefficient of friction is 0.30.
(a) What should the dimension e be?
(b) Draw the free-body diagrams of the handle lever and both shoe levers, with forces expressed in terms of the actuation force P.
(c) Does the direction of rotation of the drum affect the braking torque?

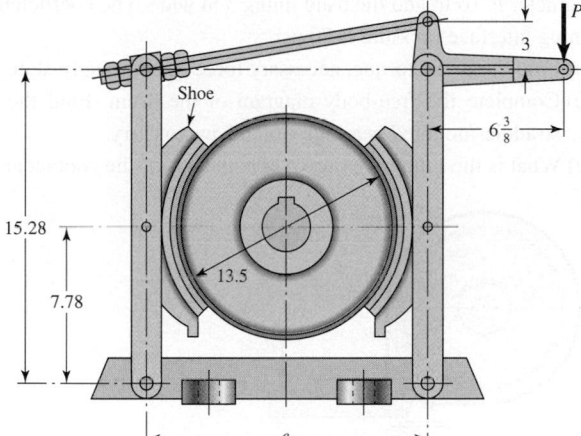

Problem 16–9
Dimensions in inches.

ANALYSIS

16–10 Problem 16–9 is preliminary to analyzing the brake. A molded lining is used dry in the brake of Prob. 16–9 on a cast iron drum. The shoes are 7.5 in wide and subtend a 90° arc. Find the actuation force and the braking torque.

ANALYSIS

16–11 The maximum band interface pressure on the brake shown in the figure is 90 psi. Use a 14-in-diameter drum, a band width of 4 in, a coefficient of friction of 0.25, and an angle-of-wrap of 270°. Find the band tensions and the torque capacity.

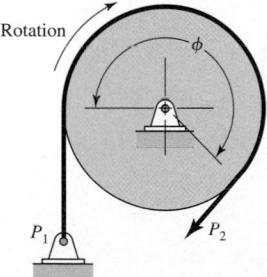

Problem 16–11

ANALYSIS

16–12 The drum for the band brake in Prob. 16–11 is 300 mm in diameter. The band selected has a mean coefficient of friction of 0.28 and a width of 80 mm. It can safely support a tension of 7.6 kN. If the angle of wrap is 270°, find the lining pressure and the torque capacity.

ANALYSIS

16–13 The brake shown in the figure has a coefficient of friction of 0.30 and is to operate using a maximum force F of 400 N. If the band width is 50 mm, find the band tensions and the braking torque.

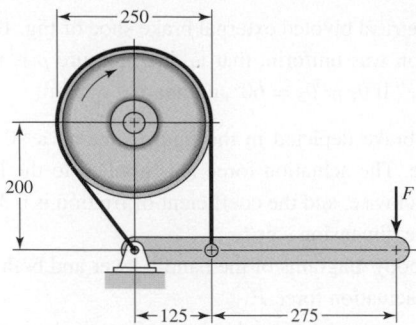

 **ANALYSIS**

16–14   The figure depicts a band brake whose drum rotates counterclockwise at 200 rev/min. The drum diameter is 16 in and the band lining 3 in wide. The coefficient of friction is 0.20. The maximum lining interface pressure is 70 psi.

(*a*) Find the brake torque, necessary force P, and steady-state power.

(*b*) Complete the free-body diagram of the drum. Find the bearing radial load that a pair of straddle-mounted bearings would have to carry.

(*c*) What is the lining pressure p at both ends of the contact arc?

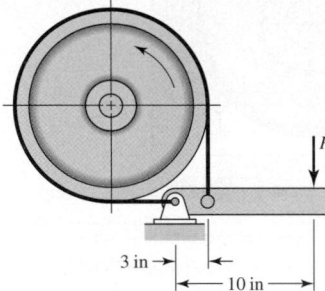

Problem 16–14

 **ANALYSIS**

16–15   The figure shows a band brake designed to prevent "backward" rotation of the shaft. The angle of wrap is $270°$, the band width is $2\frac{1}{8}$ in, and the coefficient of friction is 0.20. The torque to be resisted by the brake is 150 lbf · ft. The diameter of the pulley is $8\frac{1}{4}$ in.

Problem 16–15

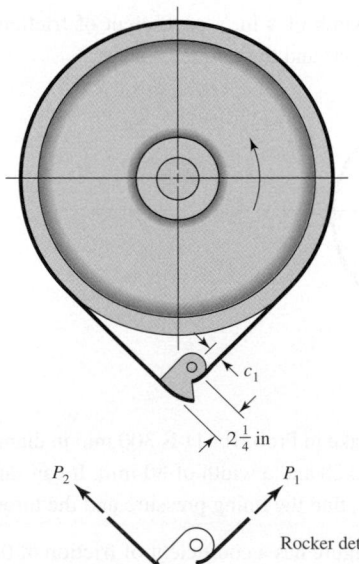

(*a*) What dimension c_1 will just prevent backward motion?

(*b*) If the rocker was designed with $c_1 = 1$ in, what is the maximum pressure between the band and drum at 150 lbf · ft back torque?

(*c*) If the back-torque demand is 100 lbf · in, what is the largest pressure between the band and drum?

 ANALYSIS **16–16** A plate clutch has a single pair of mating friction surfaces 300 mm OD by 225 mm ID. The mean value of the coefficient of friction is 0.25, and the actuating force is 5 kN.

(*a*) Find the maximum pressure and the torque capacity using the uniform-wear model.

(*b*) Find the maximum pressure and the torque capacity using the uniform-pressure model.

 ANALYSIS **16–17** A hydraulically operated multidisk plate clutch has an effective disk outer diameter of 6.5 in and an inner diameter of 4 in. The coefficient of friction is 0.24, and the limiting pressure is 120 psi. There are six planes of sliding present.

(*a*) Using the uniform wear model, estimate the axial force F and the torque T.

(*b*) Let the inner diameter of the friction pairs d be a variable. Complete the following table:

d, in	2	3	4	5	6
T, lbf · in					

(*c*) What does the table show?

DESIGN **16–18** Look again at Prob. 16–17.

(*a*) Show how the optimal diameter d^* is related to the outside diameter D.

(*b*) What is the optimal inner diameter?

(*c*) What does the tabulation show about maxima?

(*d*) Common proportions for such plate clutches lie in the range $0.45 \leq d/D \leq 0.80$. Is the result in part *a* useful?

 ANALYSIS **16–19** A cone clutch has $D = 330$ mm, $d = 306$ mm, a cone length of 60 mm, and a coefficient of friction of 0.26. A torque of 200 N · m is to be transmitted. For this requirement, estimate the actuating force and pressure by both models.

 ANALYSIS **16–20** Show that for the caliper brake the $T/(fFD)$ versus d/D plots are the same as Eqs. (*b*) and (*c*) of Sec. 16–5.

 ANALYSIS **16–21** A two-jaw clutch has the dimensions shown in the figure and is made of ductile steel. The clutch has been designed to transmit 2 kW at 500 rev/min. Find the bearing and shear stresses in the key and the jaws.

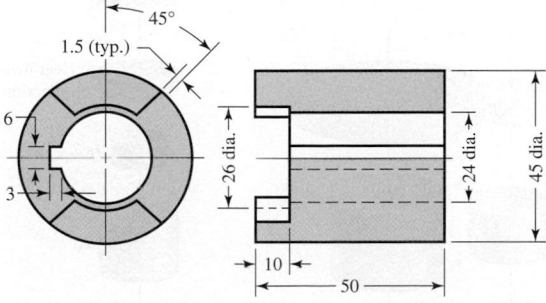

Problem 16–21
Dimensions in millimeters.

 ANALYSIS **16–22** A brake has a normal braking torque of 320 N · m and heat-dissipating surfaces whose mass is 18 kg. Suppose a load is brought to rest in 8.3 s from an initial angular speed of 1800 rev/min using the normal braking torque; estimate the temperature rise of the heat-dissipating surfaces.

16–23 A cast-iron flywheel has a rim whose OD is 60 in and whose ID is 56 in. The flywheel weight is to be such that an energy fluctuation of 5000 ft · lbf will cause the angular speed to vary no more than 240 to 260 rev/min. Estimate the coefficient of speed fluctuation. If the weight of the spokes is neglected, what should be the width of the rim?

16–24 A single-geared blanking press has a stroke of 8 in and a rated capacity of 35 tons. A cam-driven ram is assumed to be capable of delivering the full press load at constant force during the last 15 percent of a constant-velocity stroke. The camshaft has an average speed of 90 rev/min and is geared to the flywheel shaft at a 6:1 ratio. The total work done is to include an allowance of 16 percent for friction.
(a) Estimate the maximum energy fluctuation.
(b) Find the rim weight for an effective diameter of 48 in and a coefficient of speed fluctuation of 0.10.

16–25 Using the data of Table 16–6, find the mean output torque and flywheel inertia required for a three-cylinder in-line engine corresponding to a nominal speed of 2400 rev/min. Use $C_s = 0.30$.

16–26 When a motor armature inertia, a pinion inertia, and a motor torque reside on a motor shaft, and a gear inertia, a load inertia, and a load torque exist on a second shaft, it is useful to reflect all the torques and inertias to one shaft, say, the armature shaft. We need some rules to make such reflection easy. Consider the pinion and gear as disks of pitch radius.

- A torque on a second shaft is reflected to the motor shaft as the load torque divided by the negative of the stepdown ratio.
- An inertia on a second shaft is reflected to the motor shaft as its inertia divided by the stepdown ratio squared.
- The inertia of a disk gear on a second shaft in mesh with a disk pinion on the motor shaft is reflected to the pinion shaft as the *pinion* inertia multiplied by the stepdown ratio squared.

(a) Verify the three rules.
(b) Using the rules, reduce the two-shaft system in the figure to a motor-shaft shish-kebob equivalent. Correctly done, the dynamic response of the shish kebab and the real system are identical.
(c) For a stepdown ratio of $n = 10$ compare the shish-kebab inertias.

Problem 16–26
Dimensions in millimeters.

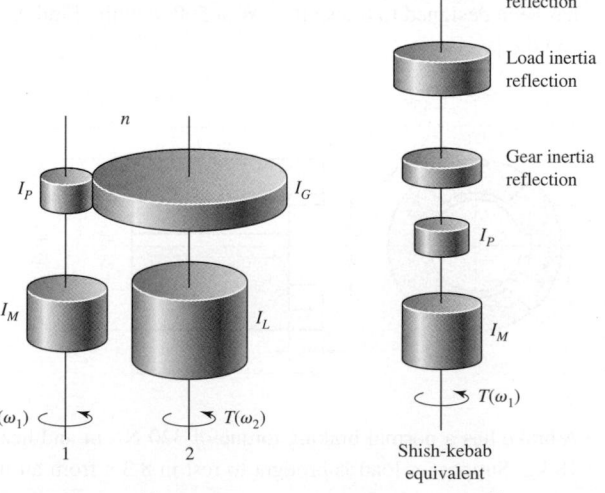

16–27 Apply the rules of Prob. 16–26 to the three-shaft system shown in the figure to create a motor shaft shish kebab.

(*a*) Show that the equivalent inertia I_e is given by

$$I_e = I_M + I_P + n^2 I_P + \frac{I_P}{n^2} + \frac{m^2 I_P}{n^2} + \frac{I_L}{m^2 n^2}$$

(*b*) If the overall gear reduction R is a constant nm, show that the equivalent inertia becomes

$$I_e = I_M + I_P + n^2 I_P + \frac{I_P}{n^2} + \frac{R^2 I_P}{n^4} + \frac{I_L}{R^2}$$

(*c*) If the problem is to minimize the gear-train inertia, find the ratios n and m for the values of $I_P = 1$, $I_M = 10$, $I_L = 100$, and $R = 10$.

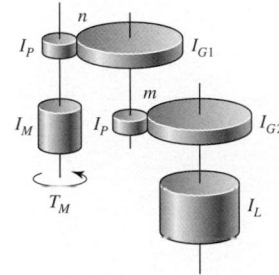

Problem 16–27

16–28  For the conditions of Prob. 16–27, make a plot of the equivalent inertia I_e as ordinate and the stepdown ratio n as abscissa in the range $1 \le n \le 10$. How does the minimum inertia compare to the single-step inertia?

16–29  A punch-press geared 10:1 is to make six punches per minute under circumstances where the torque on the crankshaft is 1300 lbf · ft for $\frac{1}{2}$ s. The motor's nameplate reads 3 bhp at 1125 rev/min for continuous duty. Design a satisfactory flywheel for use on the motor shaft to the extent of specifying material and rim inside and outside diameters as well as its width. As you prepare your specifications, note $\omega_{\max}$, $\omega_{\min}$, the coefficient of speed fluctuation C_s, energy transfer, and peak power that the flywheel transmits to the punch-press. Note power and shock conditions imposed on the gear train because the flywheel is on the motor shaft.

16–30  The punch-press of Prob. 16–29 needs a flywheel for service on the crankshaft of the punch-press. Design a satisfactory flywheel to the extent of specifying material, rim inside and outside diameters, and width. Note $\omega_{\max}$, $\omega_{\min}$, C_s, energy transfer, and peak power the flywheel transmits to the punch. What is the peak power seen in the gear train? What power and shock conditions must the gear-train transmit?

16–31  Compare the designs resulting from the tasks assigned in Probs. 16–29 and 16–30. What have you learned? What recommendations do you have?

17

Flexible Mechanical Elements

Belts, ropes, chains, and other similar elastic or flexible machine elements are used in conveying systems and in the transmission of power over comparatively long distances. It often happens that these elements can be used as a replacement for gears, shafts, bearings, and other relatively rigid power-transmission devices. In many cases their use simplifies the design of a machine and substantially reduces the cost.

In addition, since these elements are elastic and usually quite long, they play an important part in absorbing shock loads and in damping out and isolating the effects of vibration. This is an important advantage as far as machine life is concerned.

Most flexible elements do not have an infinite life. When they are used, it is important to establish an inspection schedule to guard against wear, aging, and loss of elasticity. The elements should be replaced at the first sign of deterioration.

17–1   Belts

The four principal types of belts are shown, with some of their characteristics, in Table 17–1. *Crowned pulleys* are used for flat belts, and *grooved pulleys,* or *sheaves,* for round and V belts. Timing belts require *toothed wheels,* or *sprockets.* In all cases, the pulley axes must be separated by a certain minimum distance, depending upon the belt type and size, to operate properly. Other characteristics of belts are:

• They may be used for long center distances.
• Except for timing belts, there is some slip and creep, and so the angular-velocity ratio between the driving and driven shafts is neither constant nor exactly equal to the ratio of the pulley diameters.
• In some cases an idler or tension pulley can be used to avoid adjustments in center distance that are ordinarily necessitated by age or the installation of new belts.

Figure 17–1 illustrates the geometry of open and closed flat-belt drives. For a flat belt with this drive the belt tension is such that the sag or droop is visible in Fig. 7–2a, when the belt is running. Although the top is preferred for the loose side of the belt, for other belt types either the top or the bottom may be used, because their installed tension is usually greater.

Two types of reversing drives are shown in Fig. 17–2 Notice that both sides of the belt contact the pulleys in Figs. 17–2b and 17–2c, and so these drives cannot be used with V belts or timing belts.

Table 17–1

Characteristics of Some Common Belt Types. Figures are Cross Sections except for the Timing Belt, which is a Side View

Belt Type	Figure	Joint	Size Range	Center Distance
Flat		Yes	$t = \begin{cases} 0.03 \text{ to } 0.20 \text{ in} \\ 0.75 \text{ to } 5 \text{ mm} \end{cases}$	No upper limit
Round		Yes	$d = \frac{1}{8} \text{ to } \frac{3}{4} \text{ in}$	No upper limit
V		None	$b = \begin{cases} 0.31 \text{ to } 0.91 \text{ in} \\ 8 \text{ to } 19 \text{ mm} \end{cases}$	Limited
Timing		None	$p = 2 \text{ mm and up}$	Limited

Figure 17-1

Flat-belt geometry. (a) Open belt. (b) Crossed belt.

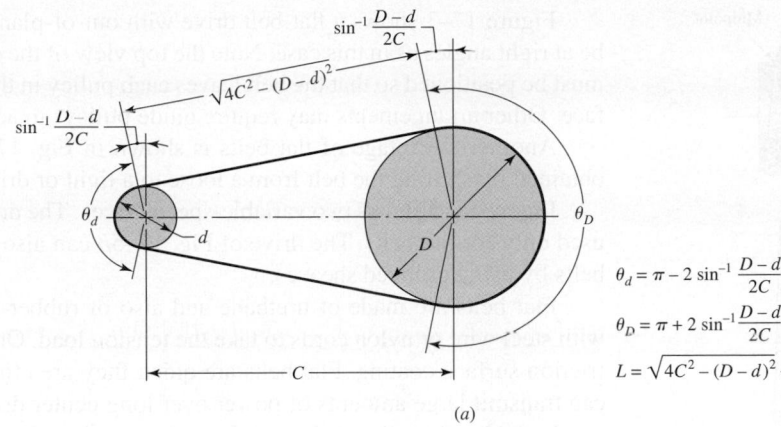

$$\theta_d = \pi - 2 \sin^{-1} \frac{D-d}{2C}$$

$$\theta_D = \pi + 2 \sin^{-1} \frac{D-d}{2C}$$

$$L = \sqrt{4C^2 - (D-d)^2} + \frac{1}{2}(D\theta_D + d\theta_d)$$

(a)

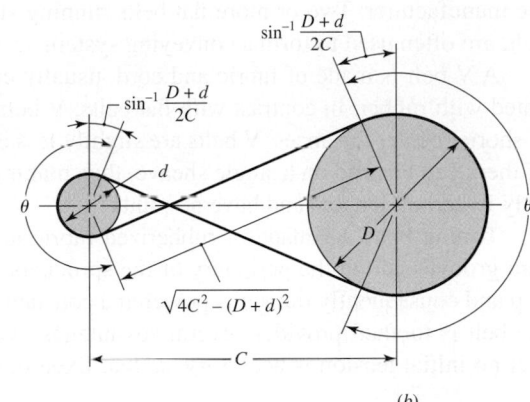

$$\theta = \pi + 2 \sin^{-1} \frac{D+d}{2C}$$

$$L = \sqrt{4C^2 - (D+d)^2} + \frac{1}{2}(D+d)\theta$$

(b)

Figure 17-2

Nonreversing and reversing belt drives. (a) Nonreversing open belt. (b) Reversing crossed belt. Crossed belts must be separated to prevent rubbing if high-friction materials are used.
(c) Reversing open-belt drive.

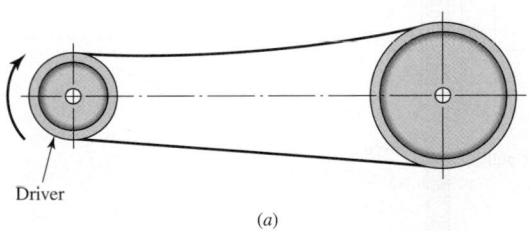

Driver

(a)

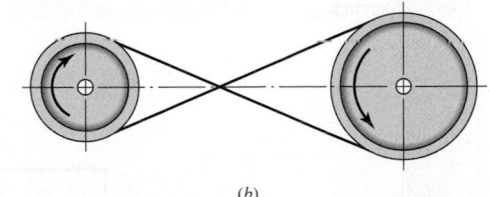

(b)

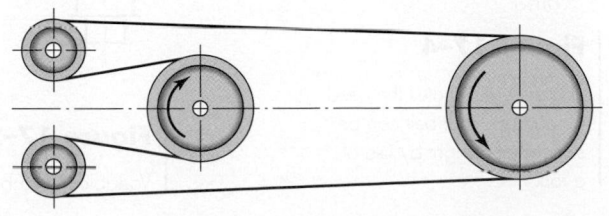

(c)

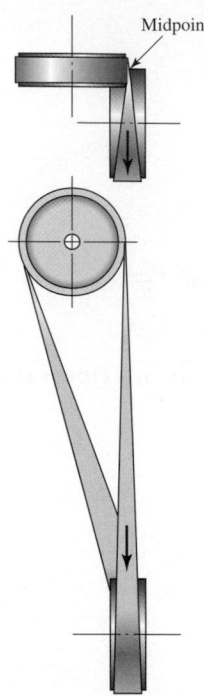

Midpoint

Figure 17–3

Quarter-twist belt drive;
an idler guide pulley must be
used if motion is to be in both
directions.

Figure 17–3 shows a flat-belt drive with out-of-plane pulleys. The shafts need not be at right angles as in this case. Note the top view of the drive in Fig. 17–3. The pulleys must be positioned so that the belt leaves each pulley in the midplane of the other pulley face. Other arrangements may require guide pulleys to achieve this condition.

Another advantage of flat belts is shown in Fig. 17–4, where clutching action is obtained by shifting the belt from a loose to a tight or driven pulley.

Figure 17–5 shows two variable-speed drives. The drive in Fig. 17–5a is commonly used only for flat belts. The drive of Fig. 17–5b can also be used for V belts and round belts by using grooved sheaves.

Flat belts are made of urethane and also of rubber-impregnated fabric reinforced with steel wire or nylon cords to take the tension load. One or both surfaces may have a friction surface coating. Flat belts are quiet, they are efficient at high speeds, and they can transmit large amounts of power over long center distances. Usually, flat belting is purchased by the roll and cut and the ends are joined by using special kits furnished by the manufacturer. Two or more flat belts running side by side, instead of a single wide belt, are often used to form a conveying system.

A V belt is made of fabric and cord, usually cotton, rayon, or nylon, and impregnated with rubber. In contrast with flat belts, V belts are used with similar sheaves and at shorter center distances. V belts are slightly less efficient than flat belts, but a number of them can be used on a single sheave, thus making a multiple drive. V belts are made only in certain lengths and have no joints.

Timing belts are made of rubberized fabric and steel wire and have teeth that fit into grooves cut on the periphery of the sprockets. The timing belt does not stretch or slip and consequently transmits power at a constant angular-velocity ratio. The fact that the belt is toothed provides several advantages over ordinary belting. One of these is that no initial tension is necessary, so that fixed-center drives may be used. Another is

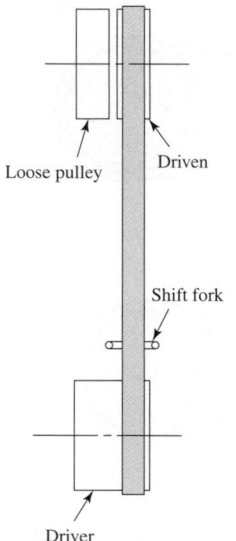

Loose pulley

Driven

Shift fork

Driver

Figure 17–4

This drive eliminates the need for a clutch. Flat belt can be shifted left or right by use of a fork.

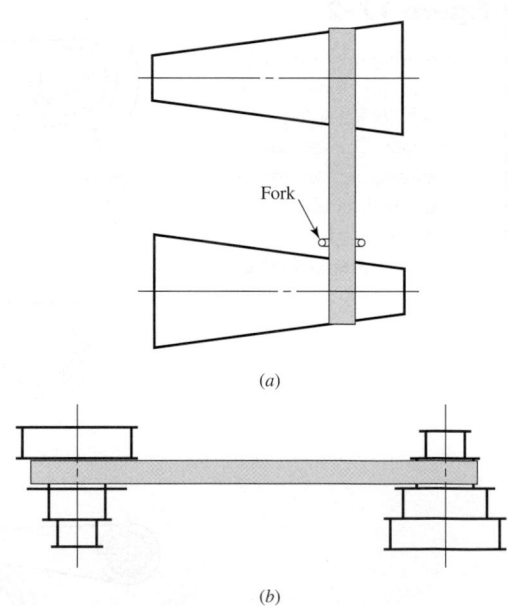

Fork

(a)

(b)

Figure 17–5

Variable-speed belt drives.

the elimination of the restriction on speeds; the teeth make it possible to run at nearly any speed, slow or fast. Disadvantages are the first cost of the belt, the necessity of grooving the sprockets, and the attendant dynamic fluctuations caused at the belt-tooth meshing frequency.

17–2   Flat- and Round-Belt Drives

Modern flat-belt drives consist of a strong elastic core surrounded by an elastomer; these drives have distinct advantages over gear drives or V-belt drives. A flat-belt drive has an efficiency of about 98 percent, which is about the same as for a gear drive. On the other hand, the efficiency of a V-belt drive ranges from about 70 to 96 percent.[1] Flat-belt drives produce very little noise and absorb more torsional vibration from the system than either V-belt or gear drives.

When an open-belt drive (Fig. 17–1a) is used, the contact angles are found to be

$$\theta_d = \pi - 2\sin^{-1}\frac{D-d}{2C}$$

$$\theta_D = \pi + 2\sin^{-1}\frac{D-d}{2C}$$

(17–1)

where D = diameter of large pulley

d = diameter of small pulley

C = center distance

θ = angle of contact

The length of the belt is found by summing the two arc lengths with twice the distance between the beginning and end of contact. The result is

$$L = [4C^2 - (D-d)^2]^{1/2} + \frac{1}{2}(D\theta_D + d\theta_d)$$

(17–2)

A similar set of equations can be derived for the crossed belt of Fig. 17–2b. For this belt, the angle of wrap is the same for both pulleys and is

$$\theta = \pi + 2\sin^{-1}\frac{D+d}{2C}$$

(17–3)

The belt length for crossed belts is found to be

$$L = [4C^2 - (D+d)^2]^{1/2} + \frac{1}{2}(D+d)\theta$$

(17–4)

Firbank[2] explains flat-belt-drive theory in the following way. A change in belt tension due to friction forces between the belt and pulley will cause the belt to elongate or contract and move relative to the surface of the pulley. This motion is caused by *elastic creep* and is associated with sliding friction as opposed to static friction. The action at the driving pulley, through that portion of the angle of contact that is actually transmitting power, is such that the belt moves more slowly than the surface speed of the pulley because of the elastic creep. The angle of contact is made up of the *effective arc,* through

---

[1]A. W. Wallin, "Efficiency of Synchronous Belts and V-Belts," *Proc. Nat. Conf. Power Transmission,* vol. 5, Illinois Institute of Technology, Chicago, Nov. 7–9, 1978, pp. 265–271.

[2]T. C. Firbank, *Mechanics of the Flat Belt Drive,* ASME paper no. 72-PTG-21.

which power is transmitted, and the *idle arc*. For the driving pulley the belt first contacts the pulley with a *tight-side tension* F_1 and a velocity V_1, which is the same as the surface velocity of the pulley. The belt then passes through the idle arc with no change in F_1 or V_1. Then creep or sliding contact begins, and the belt tension changes in accordance with the friction forces. At the end of the effective arc the belt leaves the pulley with a *loose-side tension* F_2 and a reduced speed V_2.

Firbank has used this theory to express the mechanics of flat-belt drives in mathematical form and has verified the results by experiment. His observations include the finding that substantially more power is transmitted by static friction than sliding friction. He also found that the coefficient of friction for a belt having a nylon core and leather surface was typically 0.7, but that it could be raised to 0.9 by employing special surface finishes.

Our model will assume that the friction force on the belt is proportional to the normal pressure along the arc of contact. We seek first a relationship between the tight side tension and slack side tension, similar to that of band brakes but incorporating the consequences of movement, that is, centrifugal tension in the belt. In Fig. 17–6 we see a free body of a small segment of the belt. The differential force dS is due to centrifugal force, dN is the normal force between the belt and pulley, and $f\,dN$ is the shearing traction due to friction at the point of slip. The belt width is b and the thickness is t. The belt mass per unit length is m. The centrifugal force dS can be expressed as

$$dS = (mr\,d\theta)r\omega^2 = mr^2\omega^2\,d\theta = mV^2\,d\theta = F_c\,d\theta \tag{a}$$

where V is the belt speed. Summing forces radially gives

$$\sum F_r = -(F + dF)\frac{d\theta}{2} - F\frac{d\theta}{2} + dN + dS = 0$$

Ignoring the higher-order term, we have

$$dN = F\,d\theta - dS \tag{b}$$

Summing forces tangentially gives

$$\sum F_t = -f\,dN - F + (F + dF) = 0$$

from which, incorporating Eqs. (*a*) and (*b*), we obtain

$$dF = f\,dN = fF\,d\theta - f\,dS = fF\,d\theta - fmr^2\omega^2\,d\theta$$

or

$$\frac{dF}{d\theta} - fF = -fmr^2\omega^2 \tag{c}$$

The solution to this nonhomogeneous first-order linear differential equation is

$$F = A\exp(f\theta) + mr^2\omega^2 \tag{d}$$

where A is an arbitrary constant. Assuming θ starts at the loose side, the boundary condition that F at $\theta = 0$ equals F_2 gives $A = F_2 - mr^2\omega^2$. The solution is

$$F = (F_2 - mr^2\omega^2)\exp(f\theta) + mr^2\omega^2 \tag{17–5}$$

At the end of the angle of wrap ϕ, the tight side,

$$F|_{\theta=\phi} = F_1 = (F_2 - mr^2\omega^2)\exp(f\phi) + mr^2\omega^2 \tag{17–6}$$

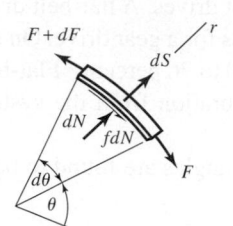

Figure 17–6

Free body of an infinitesimal element of a flat belt in contact with a pulley.

Now we can write

$$\frac{F_1 - mr^2\omega^2}{F_2 - mr^2\omega^2} = \frac{F_1 - F_c}{F_2 - F_c} = \exp(f\phi) \tag{17-7}$$

where, from Eq. (a), $F_c = mr^2\omega^2$. It is also useful that Eq. (17–7) can be written as

$$F_1 - F_2 = (F_1 - F_c)\frac{\exp(f\phi) - 1}{\exp(f\phi)} \tag{17-8}$$

Now F_c is found as follows: with n being the rotational speed, in rev/min, of the pulley of diameter d, the belt speed is

$$V = \pi\, dn/12 \qquad \text{ft/min}$$

The weight w of a foot of belt is given in terms of the weight density γ in lbf/in^3 as $w = 12\gamma bt$ lbf/ft where b and t are in inches. F_c is written as

$$F_c = \frac{w}{g}\left(\frac{V}{60}\right)^2 = \frac{w}{32.17}\left(\frac{V}{60}\right)^2 \tag{e}$$

Figure 17–7 shows a free body of a pulley and part of the belt. The tight side tension F_1 and the loose side tension F_2 have the following additive components:

$$F_1 = F_i + F_c + \Delta F' = F_i + F_c + T/D \tag{f}$$

$$F_2 = F_i + F_c - \Delta F' = F_i + F_c - T/D \tag{g}$$

where
F_i = initial tension
F_c = hoop tension due to centrifugal force
$\Delta F'$ = tension due to the transmitted torque T
D = diameter of the pulley

The difference between F_1 and F_2 is related to the pulley torque. Subtracting Eq. (g) from Eq. (f) gives

$$F_1 - F_2 = \frac{2T}{D} = \frac{T}{D/2} \tag{h}$$

Adding Eqs. (f) and (g) gives

$$F_1 + F_2 = 2F_i + 2F_c$$

Figure 17–7

Forces and torques on a pulley.

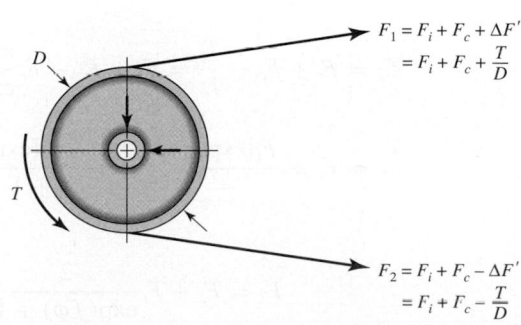

$$F_1 = F_i + F_c + \Delta F'$$
$$= F_i + F_c + \frac{T}{D}$$

$$F_2 = F_i + F_c - \Delta F'$$
$$= F_i + F_c - \frac{T}{D}$$

from which

$$F_i = \frac{F_1 + F_2}{2} - F_c \qquad (i)$$

Dividing Eq. (*i*) by Eq. (*h*), manipulating, and using Eq. (17–7) gives

$$\frac{F_i}{T/D} = \frac{(F_1 + F_2)/2 - F_c}{(F_1 - F_2)/2} = \frac{F_1 + F_2 - 2F_c}{F_1 - F_2} = \frac{(F_1 - F_c) + (F_2 - F_c)}{(F_1 - F_c) - (F_2 - F_c)}$$

$$= \frac{(F_1 - F_c)/(F_2 - F_c) + 1}{(F_1 - F_c)/(F_2 - F_c) - 1} = \frac{\exp(f\phi) + 1}{\exp(f\phi) - 1}$$

from which

$$F_i = \frac{T}{D} \frac{\exp(f\phi) + 1}{\exp(f\phi) - 1} \qquad (17\text{–}9)$$

Equation (17–9) give us a fundamental insight into flat belting. If F_i equals zero, then T equals zero: no initial tension, no torque transmitted. The torque is in proportion to the initial tension. This means that if there is to be a satisfactory flat-belt drive, the initial tension must be (1) provided, (2) sustained, (3) in the proper amount, and (4) maintained by routine inspection.

From Eq. (*f*), incorporating Eq. (17–9) gives

$$F_1 = F_i + F_c + \frac{T}{D} = F_c + F_i + F_i \frac{\exp(f\phi) - 1}{\exp(f\phi) + 1}$$

$$= F_c + \frac{F_i[\exp(f\phi) + 1] + F_i[\exp(f\phi) - 1]}{\exp(f\phi) + 1}$$

$$F_1 = F_c + F_i \frac{2\exp(f\phi)}{\exp(f\phi) + 1} \qquad (17\text{–}10)$$

From Eq. (*g*), incorporating Eq. (17–9) gives

$$F_2 = F_i + F_c - \frac{T}{D} = F_c + F_i - F_i \frac{\exp(f\phi) - 1}{\exp(f\phi) + 1}$$

$$= F_c + \frac{F_i[\exp(f\phi) + 1] - F_i[\exp(f\phi) - 1]}{\exp(f\phi) + 1}$$

$$F_2 = F_c + F_i \frac{2}{\exp(f\phi) + 1} \qquad (17\text{–}11)$$

Equation (17–7) is called the *belting equation,* but Eqs. (17–9), (17–10), and (17–11) reveal how belting works. We plot Eqs. (17–10) and (17–11) as shown in Fig. 17–8 against F_i as abscissa. The initial tension needs to be sufficient so that the difference between the F_1 and F_2 curve is $2T/D$. With no torque transmitted, the least possible belt tension is $F_1 = F_2 = F_c$.

The transmitted horsepower is given by

$$H = \frac{(F_1 - F_2)V}{33\,000} \qquad (j)$$

Manufacturers provide specifications for their belts that include allowable tension F_a (or stress σ_{all}), the tension being expressed in units of force per unit width. Belt life is usually several years. The severity of flexing at the pulley and its effect on life is reflected in a pulley correction factor C_p. Speed in excess of 600 ft/min and its effect on life is reflected in a velocity correction factor C_v. For polyamide and urethane belts use $C_v = 1$. For leather belts see Fig. 17–9. A service factor K_s is used for excursions of load from nominal, applied to the nominal power as $H_d = H_{\text{nom}} K_s n_d$, where n_d is the

Figure 17–8

Plot of initial tension F_i against belt tension F_1 or F_2, showing the intercept F_c, the equations of the loci, and where $2T/D$ is to be found.

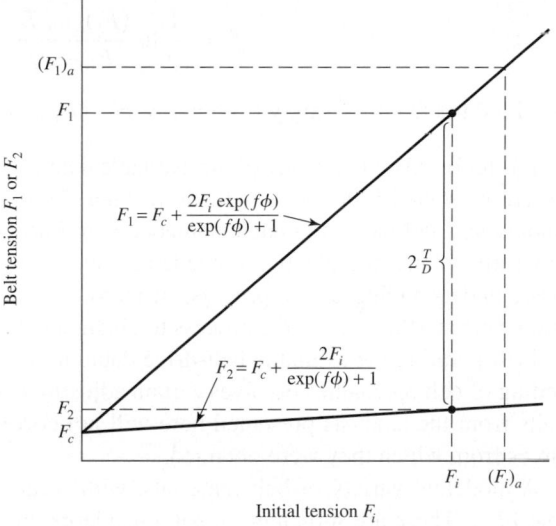

Figure 17–9

Velocity correction factor C_v for leather belts. (*Data source:* Machinery's Handbook, 20th ed., *Industrial Press, New York, 1976, p. 1047.*)

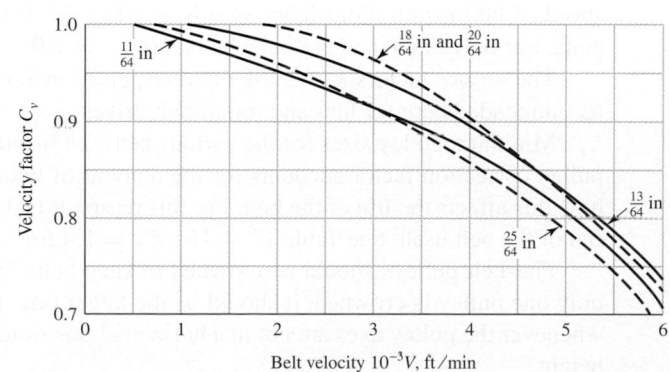

design factor for exigencies. These effects are incorporated as follows:

$$(F_1)_a = bF_aC_pC_v \tag{17-12}$$

where $(F_1)_a$ = allowable largest tension, lbf

$\quad\quad\quad b$ = belt width, in

$\quad\quad\quad F_a$ = manufacturer's allowed tension, lbf/in

$\quad\quad\quad C_p$ = pulley correction factor (Table 17–4)

$\quad\quad\quad C_v$ = velocity correction factor

The steps in analyzing a flat-belt drive can include

1 Find $\exp(f\phi)$ from belt-drive geometry and friction
2 From belt geometry and speed find F_c
3 From $T = 63\,025H_{\text{nom}}K_sn_d/n$ find necessary torque
4 From torque T find the necessary $(F_1)_a - F_2 = 2T/D$
5 Find F_2 from $(F_1)_a - [(F_1)_a - F_2]$
6 From Eq. (i) find the necessary initial tension F_i
7 Check the friction development, $f' < f$. Use Eq. (17–7) solved for f':

$$f' = \frac{1}{\phi}\ln\frac{(F_1)_a - F_c}{F_2 - F_c}$$

8 Find the factor of safety from $n_{fs} = H_a/(H_{\text{nom}}K_s)$

It is unfortunate that many of the available data on belting are from sources in which they are presented in a very simplistic manner. These sources use a variety of charts, nomographs, and tables to enable someone who knows nothing about belting to apply them. Little, if any, computation is needed for such a person to obtain valid results. Since a basic understanding of the process, in many cases, is lacking, there is no way this person can vary the steps in the process to obtain a better design.

Incorporating the available belt-drive data into a form that provides a good understanding of belt mechanics involves certain adjustments in the data. Because of this, the results from the analysis presented here will not correspond exactly with those of the sources from which they were obtained.

A moderate variety of belt materials, with some of their properties, are listed in Table 17–2. These are sufficient for solving a large variety of design and analysis problems. The design equation to be used is Eq. (j).

The values given in Table 17–2 for the allowable belt tension are based on a belt speed of 600 ft/min. For higher speeds, use Fig. 17–9 to obtain C_v values for leather belts. For polyamide and urethane belts, use $C_v = 1.0$.

The service factors K_s for V-belt drives, given in Table 17–15 in Sec. 17–3, are also recommended here for flat- and round-belt drives.

Minimum pulley sizes for the various belts are listed in Tables 17–2 and 17–3. The pulley correction factor accounts for the amount of bending or flexing of the belt and how this affects the life of the belt. For this reason it is dependent on the size and material of the belt used. See Table 17–4. Use $C_p = 1.0$ for urethane belts.

Flat-belt pulleys should be crowned to keep belts from running off the pulleys. If only one pulley is crowned, it should be the larger one. Both pulleys must be crowned whenever the pulley axes are not in a horizontal position. Use Table 17–5 for the crown height.

Table 17-2

Properties of Some Flat- and Round-Belt Materials. (Diameter = d, thickness = t, width = w)

Material	Specification	Size, in	Minimum Pulley Diameter, in	Allowable Tension per Unit Width at 600 ft/min, lbf/in	Specific Weight, lbf/in³	Coefficient of Friction
Leather	1 ply	$t = \frac{11}{64}$	3	30	0.035–0.045	0.4
		$t = \frac{13}{64}$	$3\frac{1}{2}$	33	0.035–0.045	0.4
	2 ply	$t = \frac{18}{64}$	$4\frac{1}{2}$	41	0.035–0.045	0.4
		$t = \frac{20}{64}$	6^a	50	0.035–0.045	0.4
		$t = \frac{23}{64}$	9^a	60	0.035–0.045	0.4
Polyamide[b]	F–0[c]	$t = 0.03$	0.60	10	0.035	0.5
	F–1[c]	$t = 0.05$	1.0	35	0.035	0.5
	F–2[c]	$t = 0.07$	2.4	60	0.051	0.5
	A–2[c]	$t = 0.11$	2.4	60	0.037	0.8
	A–3[c]	$t = 0.13$	4.3	100	0.042	0.8
	A–4[c]	$t = 0.20$	9.5	175	0.039	0.8
	A–5[c]	$t = 0.25$	13.5	275	0.039	0.8
Urethane[d]	$w = 0.50$	$t = 0.062$	See	5.2^e	0.038–0.045	0.7
	$w = 0.75$	$t = 0.078$	Table	9.8^e	0.038–0.045	0.7
	$w = 1.25$	$t = 0.090$	17–3	18.9^e	0.038–0.045	0.7
	Round	$d = \frac{1}{4}$	See	8.3^e	0.038–0.045	0.7
		$d = \frac{3}{8}$	Table	18.6^e	0.038–0.045	0.7
		$d = \frac{1}{2}$	17–3	33.0^e	0.038–0.045	0.7
		$d = \frac{3}{4}$		74.3^e	0.038–0.045	0.7

[a]Add 2 in to pulley size for belts 8 in wide or more.

[b]*Source: Habasit Engineering Manual,* Habasit Belting, Inc., Chamblee (Atlanta), Ga.

[c]Friction cover of acrylonitrile-butadiene rubber on both sides.

[d]*Source:* Eagle Belting Co., Des Plaines, Ill.

[e]At 6% elongation; 12% is maximum allowable value.

Table 17-3

Minimum Pulley Sizes for Flat and Round Urethane Belts. (Listed are the Pulley Diameters in Inches)

Source: Eagle Belting Co., Des Plaines, Ill.

Belt Style	Belt Size, in	Ratio of Pulley Speed to Belt Length, rev/(ft · min)		
		Up to 250	250 to 499	500 to 1000
Flat	0.50 × 0.062	0.38	0.44	0.50
	0.75 × 0.078	0.50	0.63	0.75
	1.25 × 0.090	0.50	0.63	0.75
Round	$\frac{1}{4}$	1.50	1.75	2.00
	$\frac{3}{8}$	2.25	2.62	3.00
	$\frac{1}{2}$	3.00	3.50	4.00
	$\frac{3}{4}$	5.00	6.00	7.00

Table 17–4

Pulley Correction Factor C_P for Flat Belts*

Material	Small-Pulley Diameter, in					
	1.6 to 4	4.5 to 8	9 to 12.5	14, 16	18 to 31.5	Over 31.5
Leather	0.5	0.6	0.7	0.8	0.9	1.0
Polyamide, F–0	0.95	1.0	1.0	1.0	1.0	1.0
F–1	0.70	0.92	0.95	1.0	1.0	1.0
F–2	0.73	0.86	0.96	1.0	1.0	1.0
A–2	0.73	0.86	0.96	1.0	1.0	1.0
A–3	—	0.70	0.87	0.94	0.96	1.0
A–4	—	—	0.71	0.80	0.85	0.92
A–5	—	—	—	0.72	0.77	0.91

*Average values of C_P for the given ranges were approximated from curves in the *Habasit Engineering Manual*, Habasit Belting, Inc., Chamblee (Atlanta), Ga.

Table 17–5

Crown Height and ISO Pulley Diameters for Flat Belts*

ISO Pulley Diameter, in	Crown Height, in	ISO Pulley Diameter, in	Crown Height, in	
			$w \leq 10$ in	$w > 10$ in
1.6, 2, 2.5	0.012	12.5, 14	0.03	0.03
2.8, 3.15	0.012	12.5, 14	0.04	0.04
3.55, 4, 4.5	0.012	22.4, 25, 28	0.05	0.05
5, 5.6	0.016	31.5, 35.5	0.05	0.06
6.3, 7.1	0.020	40	0.05	0.06
8, 9	0.024	45, 50, 56	0.06	0.08
10, 11.2	0.030	63, 71, 80	0.07	0.10

*Crown should be rounded, not angled; maximum roughness is $R_a = $ AA 63 μin.

EXAMPLE 17–1

A polyamide A-3 flat belt 6 in wide is used to transmit 15 hp under light shock conditions where $K_s = 1.25$, and a factor of safety equal to or greater than 1.1 is appropriate. The pulley rotational axes are parallel and in the horizontal plane. The shafts are 8 ft apart. The 6-in driving pulley rotates at 1750 rev/min in such a way that the loose side is on top. The driven pulley is 18 in in diameter. See Fig. 17–10. The factor of safety is for unquantifiable exigencies.
(*a*) Estimate the centrifugal tension F_c and the torque T.
(*b*) Estimate the allowable F_1, F_2, F_i and allowable power H_a.
(*c*) Estimate the factor of safety. Is it satisfactory?

Figure 17–10

The flat-belt drive of Ex. 17–1.

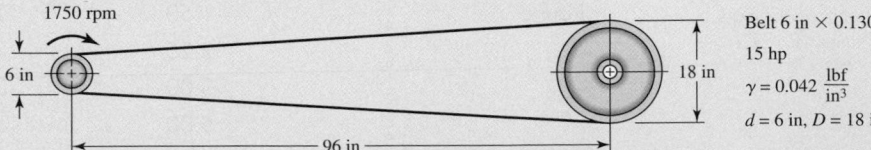

Solution   (a) Eq. (17–1): $\phi = \theta_d = \pi - 2\sin^{-1}\left[\dfrac{18-6}{2(8)12}\right] = 3.0165$ rad

$$\exp(f\phi) = \exp[0.8(3.0165)] = 11.17$$

$$V = \pi(6)1750/12 = 2749 \text{ ft/min}$$

Table 17–2: $w = 12\gamma bt = 12(0.042)6(0.130) = 0.393$ lbf/ft

Answer   Eq. (e): $F_c = \dfrac{w}{g}\left(\dfrac{V}{60}\right)^2 = \dfrac{0.393}{32.17}\left(\dfrac{2749}{60}\right)^2 = 25.6$ lbf

$$T = \frac{63\,025 H_{\text{nom}} K_s n_d}{n} = \frac{63\,025(15)1.25(1.1)}{1750}$$

Answer $= 742.8$ lbf $\cdot$ in

(b) The necessary $(F_1)_a - F_2$ to transmit the torque T, from Eq. (h), is

$$(F_1)_a - F_2 = \frac{2T}{d} = \frac{2(742.8)}{6} = 247.6 \text{ lbf}$$

From Table 17–2 $F_a = 100$ lbf. For polyamide belts $C_v = 1$, and from Table 17–4 $C_p = 0.70$. From Eq. (17–12) the allowable largest belt tension $(F_1)_a$ is

Answer $(F_1)_a = bF_aC_pC_v = 6(100)0.70(1) = 420$ lbf

then

Answer $F_2 = (F_1)_a - [(F_1)_a - F_2] = 420 - 247.6 = 172.4$ lbf

and from Eq. (i)

$$F_i = \frac{(F_1)_a + F_2}{2} - F_c = \frac{420 + 172.4}{2} - 25.6 = 270.6 \text{ lbf}$$

Answer   The combination $(F_1)_a$, F_2, and F_i will transmit the design power of $15(1.25)(1.1) = 20.6$ hp and protect the belt. We check the friction development by solving Eq. (17–7) for f':

$$f' = \frac{1}{\phi}\ln\frac{(F_1)_a - F_c}{F_2 - F_c} = \frac{1}{3.0165}\ln\frac{420 - 25.6}{172.4 - 25.6} = 0.328$$

From Table 17–2, $f = 0.8$. Since $f' < f$, that is, $0.328 < 0.80$, there is no danger of slipping.

(c)

Answer $n_{fs} = \dfrac{H}{H_{\text{nom}}K_s} = \dfrac{20.6}{15(1.25)} = 1.1$ (as expected)

Answer   The belt is satisfactory and the maximum allowable belt tension exists. If the initial tension is maintained, the capacity is the design power of 20.6 hp.

Initial tension is the key to the functioning of the flat belt as intended. There are ways of controlling initial tension. One way is to place the motor and drive pulley on a pivoted mounting plate so that the weight of the motor, pulley, and mounting plate and a share of the belt weight induces the correct initial tension and maintains it. A second way is use of a spring-loaded idler pulley, adjusted to the same task. Both of these methods accommodate to temporary or permanent belt stretch. See Fig. 17–11.

Because flat belts were used for long center-to-center distances, the weight of the belt itself can provide the initial tension. The static belt deflects to an approximate catenary curve, and the dip from a straight belt can be measured against a stretched music wire. This provides a way of measuring and adjusting the dip. From catenary theory the dip is related to the initial tension by

$$d = \frac{12L^2w}{8F_i} = \frac{3L^2w}{2F_i} \tag{17–13}$$

where d = dip, in

L = center-to-center distance, ft

w = weight per foot of the belt, lbf/ft

F_i = initial tension, lbf

In Ex. 17–1 the dip corresponding to a 270.6-lb initial tension is

$$d = \frac{3(8^2)0.393}{2(270.6)} = 0.14 \text{ in}$$

Figure 17–11

Belt-tensioning schemes.
(a) Weighted idler pulley.
(b) Pivoted motor mount.
(c) Catenary-induced tension.

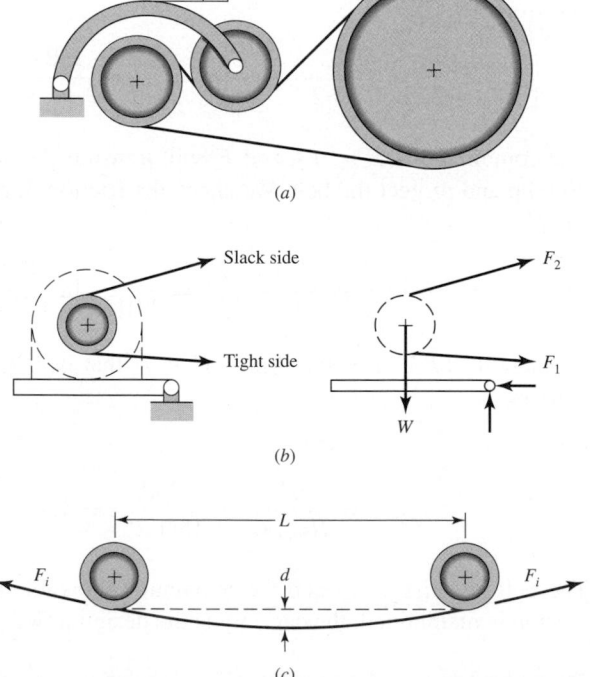

A decision set for a flat belt can be

- Function: power, speed, durability, reduction, service factor, C
- Design factor: n_d
- Initial tension maintenance
- Belt material
- Drive geometry, d, D
- Belt thickness: t
- Belt width: b

Depending on the problem, some or all of the last four could be design variables. Belt cross-sectional area is really the design decision, but available belt thicknesses and widths are discrete choices. Available dimensions are found in suppliers' catalogs.

EXAMPLE 17–2

Design a flat-belt drive to connect horizontal shafts on 16-ft centers. The velocity ratio is to be 2.25:1. The angular speed of the small driving pulley is 860 rev/min, and the nominal power transmission is to be 60 hp under very light shock.

Solution

- Function: $H_{\text{nom}} = 60$ hp, 860 rev/min, 2.25:1 ratio, $K_s = 1.15$, $C = 16$ ft
- Design factor: $n_d = 1.05$
- Initial tension maintenance: catenary
- Belt material: polyamide
- Drive geometry, d, D
- Belt thickness: t
- Belt width: b

The last four could be design variables. Let's make a few more a priori decisions.

Decision

$d = 16$ in, $D = 2.25d = 2.25(16) = 36$ in.

Decision

Use polyamide A-3 belt; therefore $t = 0.13$ in and $C_v = 1$.
Now there is one design decision remaining to be made, the belt width b.

Table 17–2: $\gamma = 0.042$ lbf/in^3 $\qquad f = 0.8 \qquad F_a = 100$ lbf/in at 600 rev/min

Table 17–4: $C_p = 0.94$

Eq. (17–12): $F_{1a} = b(100)0.94(1) = 94.0b$ lbf $\qquad\qquad\qquad$ (1)

$$H_d = H_{\text{nom}} K_s n_d = 60(1.15)1.05 = 72.5 \text{ hp}$$

$$T = \frac{63\ 025 H_d}{n} = \frac{63\ 025(72.5)}{860} = 5310 \text{ lbf} \cdot \text{in}$$

Estimate $\exp(f\phi)$ for full friction development:

Eq. (17–1): $\qquad\qquad \phi = \theta_d = \pi - 2\sin^{-1}\frac{36 - 16}{2(16)12} = 3.037$ rad

$$\exp(f\phi) = \exp[0.80(3.037)] = 11.35$$

Estimate centrifugal tension F_c in terms of belt width b:

$$w = 12\gamma bt = 12(0.042)b(0.13) = 0.0655b \text{ lbf/ft}$$

$$V = \pi dn/12 = \pi(16)860/12 = 3602 \text{ ft/min}$$

Eq. (e):
$$F_c = \frac{w}{g}\left(\frac{V}{60}\right)^2 = \frac{0.0655b}{32.17}\left(\frac{3602}{60}\right)^2 = 7.34b \text{ lbf} \qquad (2)$$

For design conditions, that is, at H_d power level, using Eq. (h) gives

$$(F_1)_a - F_2 = 2T/d = 2(5310)/16 = 664 \text{ lbf} \qquad (3)$$

$$F_2 = (F_1)_a - [(F_1)_a - F_2] = 94.0b - 664 \text{ lbf} \qquad (4)$$

Using Eq. (i) gives

$$F_i = \frac{(F_1)_a + F_2}{2} - F_c = \frac{94.0b + 94.0b - 664}{2} - 7.34b = 86.7b - 332 \text{ lbf} \qquad (5)$$

Place friction development at its highest level, using Eq. (17–7):

$$f\phi = \ln\frac{(F_1)_a - F_c}{F_2 - F_c} = \ln\frac{94.0b - 7.34b}{94.0b - 664 - 7.34b} = \ln\frac{86.7b}{86.7b - 664}$$

Solving the preceding equation for belt width b at which friction is fully developed gives

$$b = \frac{664}{86.7}\frac{\exp(f\phi)}{\exp(f\phi) - 1} = \frac{664}{86.7}\frac{11.38}{11.38 - 1} = 8.40 \text{ in}$$

A belt width greater than 8.40 in will develop friction less than $f = 0.80$. The manufacturer's data indicate that the next available larger width is 10-in.

Decision Use 10-in-wide belt.
It follows that for a 10-in-wide belt

Eq. (2): $\qquad\qquad F_c = 7.34(10) = 73.4 \text{ lbf}$

Eq. (1): $\qquad\qquad (F_1)_a = 94(10) = 940 \text{ lbf}$

Eq. (4): $\qquad\qquad F_2 = 94(10) - 664 = 276 \text{ lbf}$

Eq. (5): $\qquad\qquad F_i = 86.7(10) - 332 = 535 \text{ lbf}$

The transmitted power, from Eq. (3), is

$$H_t = \frac{[(F_1)_a - F_2]V}{33\,000} = \frac{664(3602)}{33\,000} = 72.5 \text{ hp}$$

and the level of friction development f', from Eq. (17–7) is

$$f' = \frac{1}{\phi}\ln\frac{(F_1)_a - F_c}{F_2 - F_c} = \frac{1}{3.037}\ln\frac{940 - 73.4}{276 - 73.4} = 0.479$$

which is less than $f = 0.8$, and thus is satisfactory. Had a 9-in belt width been available, the analysis would show $(F_1)_a = 846$ lbf, $F_2 = 182$ lbf, $F_i = 448$ lbf, and $f' = 0.63$. With a figure of merit available reflecting cost, thicker belts (A-4 or A-5) could be examined to ascertain which of the satisfactory alternatives is best. From Eq. (17–13) the catenary dip is

$$d = \frac{3L^2w}{2F_i} = \frac{3(15^2)0.0655(10)}{2(535)} = 0.413 \text{ in}$$

Figure 17–12

Flat-belt tensions.

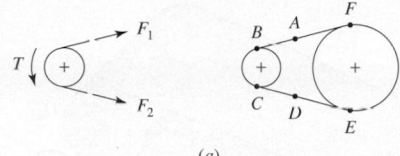

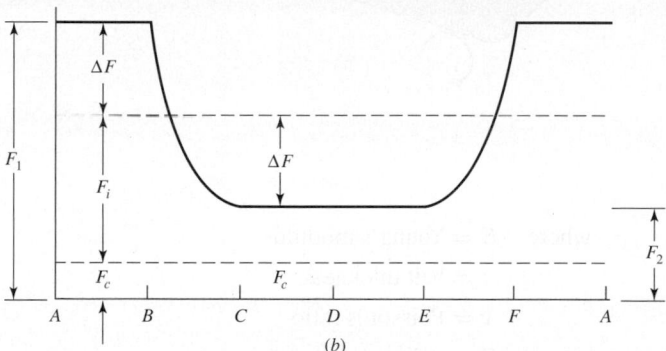

(b)

Figure 17–12 illustrates the variation of flexible flat-belt tensions at some cardinal points during a belt pass.

Flat Metal Belts

Thin flat metal belts with their attendant strength and geometric stability could not be fabricated until laser welding and thin rolling technology made possible belts as thin as 0.002 in and as narrow as 0.026 in. The introduction of perforations allows no-slip applications. Thin metal belts exhibit

- High strength-to-weight ratio
- Dimensional stability
- Accurate timing
- Usefulness to temperatures up to 700°F
- Good electrical and thermal conduction properties

In addition, stainless steel alloys offer "inert," nonabsorbent belts suitable to hostile (corrosive) environments, and can be made sterile for food and pharmaceutical applications.

Thin metal belts can be classified as friction drives, timing or positioning drives, or tape drives. Among friction drives are plain, metal-coated, and perforated belts. Crowned pulleys are used to compensate for tracking errors.

Figure 17–13 shows a thin flat metal belt with the tight tension F_1 and the slack side tension F_2 revealed. The relationship between F_1 and F_2 and the driving torque T is the same as in Eq. (h). Equations (17–9), (17–10), and (17–11) also apply. The largest allowable tension, as in Eq. (17–12), is posed in terms of stress in metal belts. A bending stress is created by making the belt conform to the pulley, and its tensile magnitude σ_b is given by

$$\sigma_b = \frac{Et}{(1 - v^2)D} = \frac{E}{(1 - v^2)(D/t)} \tag{17–14}$$

Figure 17–13

Metal-belt tensions and torques.

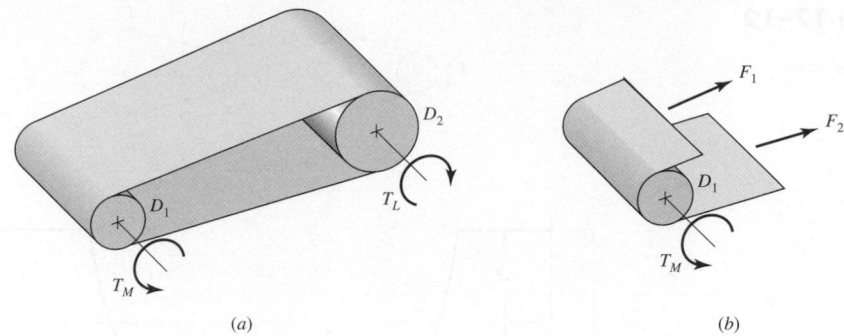

(a)   (b)

where E = Young's modulus

t = belt thickness

ν = Poisson's ratio

D = pulley diameter

The tensile stresses $(\sigma)_1$ and $(\sigma)_2$ imposed by the belt tensions F_1 and F_2 are

$$(\sigma)_1 = F_1/(bt) \qquad \text{and} \qquad (\sigma)_2 = F_2/(bt)$$

The largest tensile stress is $(\sigma_b)_1 + F_1/(bt)$ and the smallest is $(\sigma_b)_2 + F_2/(bt)$. During a belt pass both levels of stress appear.

Although the belts are of simple geometry, the method of Marin is not used because the condition of the butt weldment (to form the loop) is not accurately known, and the testing of coupons is difficult. The belts are run to failure on two equal-sized pulleys. Information concerning fatigue life, as shown in Table 17–6, is obtainable. Tables 17–7 and 17–8 give additional information.

Table 17–6 shows metal belt life expectancies for a stainless steel belt. From Eq. (17–14) with $E = 28$ Mpsi and $\nu = 0.29$, the bending stresses corresponding to the four entries of the table are 48 914, 76 428, 91 805, and 152 855 psi. Using a natural log transformation on stress and passes shows that the regression line ($r = -0.96$) is

$$\sigma = 14\ 169\ 982 N^{-0.407} = 14.17(10^6)N_p^{-0.407} \qquad (17\text{–}15)$$

where N_p is the number of belt passes.

Table 17–6

Belt Life for Stainless Steel Friction Drives*

$\dfrac{D}{t}$	Belt Passes
625	$\geq 10^6$
400	$0.500 \cdot 10^6$
333	$0.165 \cdot 10^6$
200	$0.085 \cdot 10^6$

*Data courtesy of Belt Technologies, Agawam, Mass.

Table 17–7

Minimum Pulley Diameter*

Belt Thickness, in	Minimum Pulley Diameter, in
0.002	1.2
0.003	1.8
0.005	3.0
0.008	5.0
0.010	6.0
0.015	10
0.020	12.5
0.040	25.0

*Data courtesy of Belt Technologies, Agawam, Mass.

Table 17–8

Typical Material Properties, Metal Belts*

Alloy	Yield Strength, kpsi	Young's Modulus, Mpsi	Poisson's Ratio
301 or 302 stainless steel	175	28	0.285
BeCu	170	17	0.220
1075 or 1095 carbon steel	230	30	0.287
Titanium	150	15	—
Inconel	160	30	0.284

*Data courtesy of Belt Technologies, Agawam, Mass.

The selection of a metal flat belt can consist of the following steps:

1 Find $\exp(f\phi)$ from geometry and friction
2 Find endurance strength

$$S_f = 14.17(10^6)N_p^{-0.407} \qquad 301, 302 \text{ stainless}$$

$$S_f = S_y/3 \qquad\qquad\qquad \text{others}$$

3 Allowable tension

$$F_{1a} = \left[S_f - \frac{Et}{(1-v^2)D} \right] tb = ab$$

4 $\Delta F = 2T/D$

5 $F_2 = F_{1a} - \Delta F = ab - \Delta F$

6 $F_i = \dfrac{F_{1a} + F_2}{2} = \dfrac{ab + ab - \Delta F}{2} = ab - \dfrac{\Delta F}{2}$

7 $b_{\min} = \dfrac{\Delta F}{a} \dfrac{\exp(f\phi)}{\exp(f\phi) - 1}$

8 Choose $b > b_{\min}$, $F_1 = ab$, $F_2 = ab - \Delta F$, $F_i = ab - \Delta F/2$, $T = \Delta FD/2$

9 Check frictional development f':

$$f' = \frac{1}{\phi} \ln \frac{F_1}{F_2} \qquad f' < f$$

EXAMPLE 17–3

A friction-drive stainless steel metal belt runs over two 4-in metal pulleys ($f = 0.35$). The belt thickness is to be 0.003 in. For a life exceeding 10^6 belt passes with smooth torque ($K_s = 1$), (a) select the belt if the torque is to be 30 lbf · in, and (b) find the initial tension F_i.

Solution

(a) From step 1, $\phi = \theta_d = \pi$, therefore $\exp(0.35\pi) = 3.00$. From step 2,

$$(S_f)_{10^6} = 14.17(10^6)(10^6)^{-0.407} = 51\,210 \text{ psi}$$

From steps 3, 4, 5, and 6,

$$F_{1a} = \left[51\,210 - \frac{28(10^6)0.003}{(1 - 0.285^2)4} \right] 0.003b = 85.1b \text{ lbf} \tag{1}$$

$$\Delta F = 2T/D = 2(30)/4 = 15 \text{ lbf}$$

$$F_2 = F_{1a} - \Delta F = 85.1b - 15 \text{ lbf} \tag{2}$$

$$F_i = \frac{F_{1a} + F_2}{2} = \frac{85.1b + 15}{2} \text{ lbf} \tag{3}$$

From step 7,

$$b_{\min} = \frac{\Delta F}{a} \frac{\exp(f\phi)}{\exp(f\phi) - 1} = \frac{15}{85.1} \frac{3.00}{3.00 - 1} = 0.264 \text{ in}$$

Decision

Select an available 0.75-in-wide belt 0.003 in thick.

Eq. (1): $\qquad\qquad F_1 = 85.1(0.75) = 63.8 \text{ lbf}$

Eq. (2): $\qquad\qquad F_2 = 85.1(0.75) - 15 = 48.8 \text{ lbf}$

Eq. (3): $\qquad\qquad F_i = (63.8 + 48.8)/2 = 56.3 \text{ lbf}$

$$f' = \frac{1}{\phi} \ln \frac{F_1}{F_2} = \frac{1}{\pi} \ln \frac{63.8}{48.8} = 0.0853$$

Note $f' < f$, that is, $0.0853 < 0.35$.

17–3 V Belts

The cross-sectional dimensions of V belts have been standardized by manufacturers, with each section designated by a letter of the alphabet for sizes in inch dimensions. Metric sizes are designated in numbers. Though these have not been included here, the procedure for analyzing and designing them is the same as presented here. Dimensions, minimum sheave diameters, and the horsepower range for each of the lettered sections are listed in Table 17–9.

Table 17–9

Standard V-Belt Sections

Belt Section	Width a, in	Thickness b, in	Minimum Sheave Diameter, in	hp Range, One or More Belts
A	$\frac{1}{2}$	$\frac{11}{32}$	3.0	$\frac{1}{4}$–10
B	$\frac{21}{32}$	$\frac{7}{16}$	5.4	1–25
C	$\frac{7}{8}$	$\frac{17}{32}$	9.0	15–100
D	$1\frac{1}{4}$	$\frac{3}{4}$	13.0	50–250
E	$1\frac{1}{2}$	1	21.6	100 and up

Table 17–10

Inside Circumferences of Standard V Belts

Section	Circumference, in
A	26, 31, 33, 35, 38, 42, 46, 48, 51, 53, 55, 57, 60, 62, 64, 66, 68, 71, 75, 78, 80, 85, 90, 96, 105, 112, 120, 128
B	35, 38, 42, 46, 48, 51, 53, 55, 57, 60, 62, 64, 65, 66, 68, 71, 75, 78, 79, 81, 83, 85, 90, 93, 97, 100, 103, 105, 112, 120, 128, 131, 136, 144, 158, 173, 180, 195, 210, 240, 270, 300
C	51, 60, 68, 75, 81, 85, 90, 96, 105, 112, 120, 128, 136, 144, 158, 162, 173, 180, 195, 210, 240, 270, 300, 330, 360, 390, 420
D	120, 128, 144, 158, 162, 173, 180, 195, 210, 240, 270, 300, 330, 360, 390, 420, 480, 540, 600, 660
E	180, 195, 210, 240, 270, 300, 330, 360, 390, 420, 480, 540, 600, 660

Table 17–11

Length Conversion Dimensions (Add the Listed Quantity to the Inside Circumference to Obtain the Pitch Length in Inches)

Belt section	A	B	C	D	E
Quantity to be added	1.3	1.8	2.9	3.3	4.5

To specify a V belt, give the belt-section letter, followed by the inside circumference in inches (standard circumferences are listed in Table 17–10). For example, B75 is a B-section belt having an inside circumference of 75 in.

Calculations involving the belt length are usually based on the pitch length. For any given belt section, the pitch length is obtained by adding a quantity to the inside circumference (Tables 17–10 and 17–11). For example, a B75 belt has a pitch length of 76.8 in. Similarly, calculations of the velocity ratios are made using the pitch diameters of the sheaves, and for this reason the stated diameters are usually understood to be the pitch diameters even though they are not always so specified.

The groove angle of a sheave is made somewhat smaller than the belt-section angle. This causes the belt to wedge itself into the groove, thus increasing friction. The exact value of this angle depends on the belt section, the sheave diameter, and the angle of contact. If it is made too much smaller than the belt, the force required to pull the belt out of the groove as the belt leaves the pulley will be excessive. Optimum values are given in the commercial literature.

The minimum sheave diameters have been listed in Table 17–9. For best results, a V belt should be run quite fast: 4000 ft/min is a good speed. Trouble may be encountered if the belt runs much faster than 5000 ft/min or much slower than 1000 ft/min.

The *pitch length* L_p and the center-to-center distance C are

$$L_p = 2C + \pi(D + d)/2 + (D - d)^2/(4C) \qquad (17\text{–}16a)$$

$$C = 0.25 \left\{ \left[L_p - \frac{\pi}{2}(D + d) \right] + \sqrt{\left[L_p - \frac{\pi}{2}(D + d) \right]^2 - 2(D - d)^2} \right\} \qquad (17\text{–}16b)$$

where D = pitch diameter of the large sheave and d = pitch diameter of the small sheave.

In the case of flat belts, there is virtually no limit to the center-to-center distance. Long center-to-center distances are not recommended for V belts because the excessive vibration of the slack side will shorten the belt life materially. In general, the center-to-center distance should be no greater than 3 times the sum of the sheave diameters and no less than the diameter of the larger sheave. Link-type V belts have less vibration, because of better balance, and hence may be used with longer center-to-center distances.

The basis for power ratings of V belts depends somewhat on the manufacturer; it is not often mentioned quantitatively in vendors' literature but is available from vendors. The basis may be a number of hours, 24 000, for example, or a life of 10^8 or 10^9 belt passes. Since the number of belts must be an integer, an undersized belt set that is augmented by one belt can be substantially oversized. Table 17–12 gives power ratings of standard V belts.

The rating, whether in terms of hours or belt passes, is for a belt running on equal-diameter sheaves (180° of wrap), of moderate length, and transmitting a steady load. Deviations from these laboratory test conditions are acknowledged by multiplicative adjustments. If the tabulated power of a belt for a C-section belt is 9.46 hp for a 12-in-diameter sheave at a peripheral speed of 3000 ft/min (Table 17–12), then, when the belt is used under other conditions, the tabulated value H_{tab} is adjusted as follows:

$$H_a = K_1 K_2 H_{\text{tab}} \qquad (17\text{–}17)$$

where H_a = allowable power, per belt, Table 17–12

   K_1 = angle-of-wrap correction factor, Table 17–13

   K_2 = belt length correction factor, Table 17–14

The allowable power can be near to H_{tab}, depending upon circumstances.

In a V belt the effective coefficient of friction f' is $f/\sin(\phi/2)$, which amounts to an augmentation by a factor of about 3 due to the grooves. The effective coefficient of friction f' is sometimes tabulated against *sheave* groove angles of 30°, 34°, and 38°, the tabulated values being 0.50, 0.45, and 0.40, respectively, revealing a belt material-on-metal coefficient of friction of 0.13 for each case. The Gates Rubber Company declares its effective coefficient of friction to be 0.5123 for grooves. Thus

$$\frac{F_1 - F_c}{F_2 - F_c} = \exp(0.5123\phi) \qquad (17\text{–}18)$$

The design power is given by

$$H_d = H_{\text{nom}} K_s n_d \qquad (17\text{–}19)$$

where H_{nom} is the nominal power, K_s is the service factor given in Table 17–15, and n_d is the design factor. The number of belts, N_b, is usually the next higher integer to H_d/H_a.

Table 17–12

Horsepower Ratings of
Standard V Belts

Belt Section	Sheave Pitch Diameter, in	Belt Speed, ft/min				
		1000	2000	3000	4000	5000
A	2.6	0.47	0.62	0.53	0.15	
	3.0	0.66	1.01	1.12	0.93	0.38
	3.4	0.81	1.31	1.57	1.53	1.12
	3.8	0.93	1.55	1.92	2.00	1.71
	4.2	1.03	1.74	2.20	2.38	2.19
	4.6	1.11	1.89	2.44	2.69	2.58
	5.0 and up	1.17	2.03	2.64	2.96	2.89
B	4.2	1.07	1.58	1.68	1.26	0.22
	4.6	1.27	1.99	2.29	2.08	1.24
	5.0	1.44	2.33	2.80	2.76	2.10
	5.4	1.59	2.62	3.24	3.34	2.82
	5.8	1.72	2.87	3.61	3.85	3.45
	6.2	1.82	3.09	3.94	4.28	4.00
	6.6	1.92	3.29	4.23	4.67	4.48
	7.0 and up	2.01	3.46	4.49	5.01	4.90
C	6.0	1.84	2.66	2.72	1.87	
	7.0	2.48	3.94	4.64	4.44	3.12
	8.0	2.96	4.90	6.09	6.36	5.52
	9.0	3.34	5.65	7.21	7.86	7.39
	10.0	3.64	6.25	8.11	9.06	8.89
	11.0	3.88	6.74	8.84	10.0	10.1
	12.0 and up	4.09	7.15	9.46	10.9	11.1
D	10.0	4.14	6.13	6.55	5.09	1.35
	11.0	5.00	7.83	9.11	8.50	5.62
	12.0	5.71	9.26	11.2	11.4	9.18
	13.0	6.31	10.5	13.0	13.8	12.2
	14.0	6.82	11.5	14.6	15.8	14.8
	15.0	7.27	12.4	15.9	17.6	17.0
	16.0	7.66	13.2	17.1	19.2	19.0
	17.0 and up	8.01	13.9	18.1	20.6	20.7
E	16.0	8.68	14.0	17.5	18.1	15.3
	18.0	9.92	16.7	21.2	23.0	21.5
	20.0	10.9	18.7	24.2	26.9	26.4
	22.0	11.7	20.3	26.6	30.2	30.5
	24.0	12.4	21.6	28.6	32.9	33.8
	26.0	13.0	22.8	30.3	35.1	36.7
	28.0 and up	13.4	23.7	31.8	37.1	39.1

That is,

$$N_b \geq \frac{H_d}{H_a} \qquad N_b = 1, 2, 3, \ldots \qquad (17\text{–}20)$$

Designers work on a per-belt basis.

The flat-belt tensions shown in Fig. 17–12 ignored the tension induced by bending the belt about the pulleys. This is more pronounced with V belts, as shown in Fig. 17–14.

The centrifugal tension F_c is given by

$$F_c = K_c \left(\frac{V}{1000}\right)^2 \qquad (17\text{–}21)$$

where K_c is from Table 17–16.

Table 17–13

Angle of Contact Correction Factor K_1 for VV* and V-Flat Drives

$\dfrac{D-d}{C}$	θ, deg	VV	K_1 V Flat
0.00	180	1.00	0.75
0.10	174.3	0.99	0.76
0.20	166.5	0.97	0.78
0.30	162.7	0.96	0.79
0.40	156.9	0.94	0.80
0.50	151.0	0.93	0.81
0.60	145.1	0.91	0.83
0.70	139.0	0.89	0.84
0.80	132.8	0.87	0.85
0.90	126.5	0.85	0.85
1.00	120.0	0.82	0.82
1.10	113.3	0.80	0.80
1.20	106.3	0.77	0.77
1.30	98.9	0.73	0.73
1.40	91.1	0.70	0.70
1.50	82.8	0.65	0.65

*A curvefit for the VV column in terms of θ is
$K_1 = 0.143\,543 + 0.007\,46\,8\,\theta - 0.000\,015\,052\,\theta^2$
in the range $90° \le \theta \le 180°$.

Table 17–14

Belt-Length Correction Factor K_2*

Length Factor	A Belts	Nominal Belt Length, in B Belts	C Belts	D Belts	E Belts
0.85	Up to 35	Up to 46	Up to 75	Up to 128	
0.90	38–46	48–60	81–96	144–162	Up to 195
0.95	48–55	62–75	105–120	173–210	210–240
1.00	60–75	78–97	128–158	240	270–300
1.05	78–90	105–120	162–195	270–330	330–390
1.10	96–112	128–144	210–240	360–420	420–480
1.15	120 and up	158–180	270–300	480	540–600
1.20		195 and up	330 and up	540 and up	660

*Multiply the rated horsepower per belt by this factor to obtain the corrected horsepower.

Table 17–15

Suggested Service Factors K_S for V-Belt Drives

Driven Machinery	Source of Power Normal Torque Characteristic	High or Nonuniform Torque
Uniform	1.0 to 1.2	1.1 to 1.3
Light shock	1.1 to 1.3	1.2 to 1.4
Medium shock	1.2 to 1.4	1.4 to 1.6
Heavy shock	1.3 to 1.5	1.5 to 1.8

Figure 17–14

V-belt tensions.

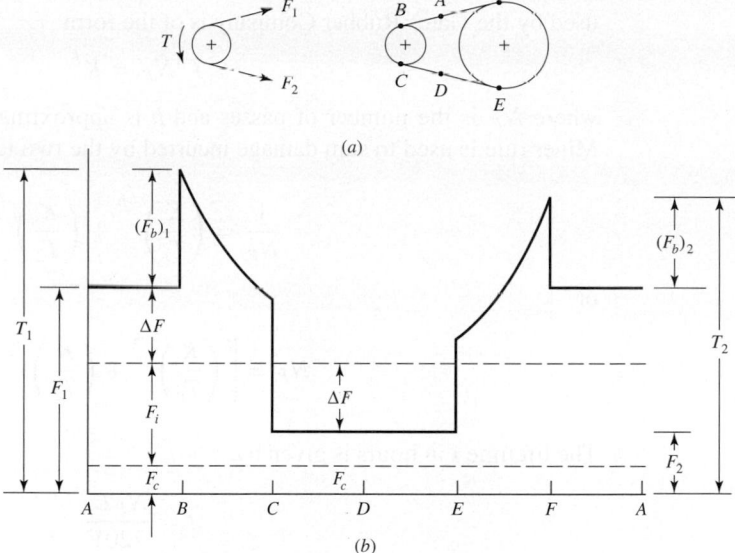

(a)

(b)

The power that is transmitted per belt is based on $\Delta F = F_1 - F_2$, where

$$\Delta F = \frac{63\,025 H_d/N_b}{n(d/2)} \tag{17–22}$$

then from Eq. (17–8) the largest tension F_1 is given by

$$F_1 = F_c + \frac{\Delta F \exp(f\phi)}{\exp(f\phi) - 1} \tag{17–23}$$

From the definition of ΔF, the least tension F_2 is

$$F_2 = F_1 - \Delta F \tag{17–24}$$

From Eq. (j) in Sec. 17–2

$$F_i = \frac{F_1 + F_2}{2} - F_c \tag{17–25}$$

The factor of safety is

$$n_{fs} = \frac{H_a N_b}{H_{nom} K_s} \tag{17–26}$$

Durability (life) correlations are complicated by the fact that the bending induces flexural stresses in the belt; the corresponding belt tension that induces the same maximum tensile stress is F_{b1} at the driving sheave and F_{b2} at the driven pulley. These equivalent tensions are added to F_1 as

$$T_1 = F_1 + (F_b)_1 = F_1 + \frac{K_b}{d}$$

$$T_2 = F_1 + (F_b)_2 = F_1 + \frac{K_b}{D}$$

where K_b is given in Table 17–16. The equation for the tension versus pass trade-off used by the Gates Rubber Company is of the form

$$T^b N_P = K^b$$

where N_P is the number of passes and b is approximately 11. See Table 17–17. The Miner rule is used to sum damage incurred by the two tension peaks:

$$\frac{1}{N_P} = \left(\frac{K}{T_1}\right)^{-b} + \left(\frac{K}{T_2}\right)^{-b}$$

or

$$N_P = \left[\left(\frac{K}{T_1}\right)^{-b} + \left(\frac{K}{T_2}\right)^{-b}\right]^{-1} \tag{17–27}$$

The lifetime t in hours is given by

$$t = \frac{N_P L_p}{720V} \tag{17–28}$$

Table 17–16

Some V-Belt Parameters*

Belt Section	K_b	K_c
A	220	0.561
B	576	0.965
C	1 600	1.716
D	5 680	3.498
E	10 850	5.041
3V	230	0.425
5V	1098	1.217
8V	4830	3.288

*Data courtesy of Gates Rubber Co., Denver, Colo.

Table 17–17

Durability Parameters for Some V-Belt Sections

Source: M. E. Spotts, Design of Machine Elements, 6th ed. Prentice Hall, Englewood Cliffs, N.J., 1985.

Belt Section	10^8 to 10^9 Force Peaks		10^9 to 10^{10} Force Peaks		Minimum Sheave Diameter, in
	K	b	K	b	
A	674	11.089			3.0
B	1193	10.926			5.0
C	2038	11.173			8.5
D	4208	11.105			13.0
E	6061	11.100			21.6
3V	728	12.464	1062	10.153	2.65
5V	1654	12.593	2394	10.283	7.1
8V	3638	12.629	5253	10.319	12.5

The constants K and b have their ranges of validity. If $N_P > 10^9$, report that $N_P = 10^9$ and $t > N_P L_p/(720V)$ without placing confidence in numerical values beyond the validity interval. See the statement about N_P and t near the conclusion of Ex. 17–4.

The analysis of a V-belt drive can consist of the following steps:

- Find V, L_p, C, ϕ, and $\exp(0.5123\phi)$
- Find H_d, H_a, and N_b from H_d/H_a and round up
- Find F_c, ΔF, F_1, F_2, and F_i, and n_{fs}
- Find belt life in number of passes, or hours, if possible

EXAMPLE 17–4

A 10-hp split-phase motor running at 1750 rev/min is used to drive a rotary pump, which operates 24 hours per day. An engineer has specified a 7.4-in small sheave, an 11-in large sheave, and three B112 belts. The service factor of 1.2 was augmented by 0.1 because of the continuous-duty requirement. Analyze the drive and estimate the belt life in passes and hours.

Solution

The peripheral speed V of the belt is

$$V = \pi\, dn/12 = \pi(7.4)1750/12 = 3390 \text{ ft/min}$$

Table 17–11: $L_p = L + L_c = 112 + 1.8 = 113.8$ in

Eq. (17–16b):
$$C = \frac{1}{4}\left\{-\left[\frac{\pi}{2}(11 + 7.4) - 113.8\right]\right.$$
$$\left. + \sqrt{\left[\frac{\pi}{2}(11 + 7.4) - 113.8\right]^2 - 2(11 - 7.4)^2}\right\}$$
$$= 42.4 \text{ in}$$

Eq. (17–1): $\phi = \theta_d = \pi - 2\sin^{-1}(11 - 7.4)/[2(42.4)] = 3.057$ rad

$$\exp[0.5123(3.057)] = 4.788$$

Interpolating in Table 17–12 for $V = 3390$ ft/min gives $H_{\text{tab}} = 4.693$ hp. The wrap angle in degrees is $3.057(180)/\pi = 175°$. From Table 17–13, $K_1 = 0.99$. From Table 17–14, $K_2 = 1.05$. Thus, from Eq. (17–17),

$$H_a = K_1 K_2 H_{\text{tab}} = 0.99(1.05)4.693 = 4.878 \text{ hp}$$

Eq. (17–19): $H_d H_{\text{nom}} K_s n_d = 10(1.2 + 0.1)(1) = 13$ hp

Eq. (17–20): $N_b \geq H_d/H_a = 13/4.878 = 2.67 \rightarrow 3$

From Table 17–16, $K_c = 0.965$. Thus, from Eq. (17–21),

$$F_c = 0.965(3390/1000)^2 = 11.1 \text{ lbf}$$

Eq.(17–22): $$\Delta F = \frac{63\,025(13)/3}{1750(7.4/2)} = 42.2 \text{ lbf}$$

Eq. (17–23): $$F_1 = 11.1 + \frac{42.2(4.788)}{4.788 - 1} = 64.4 \text{ lbf}$$

Eq. (17–24): $\quad F_2 = F_1 - \Delta F = 64.4 - 42.2 = 22.2 \text{ lbf}$

Eq. (17–25): $\quad F_i = \dfrac{64.4 + 22.2}{2} - 11.1 = 32.2 \text{ lbf}$

Eq. (17–26): $\quad n_{fs} = \dfrac{H_a N_b}{H_{\text{nom}} K_s} = \dfrac{4.878(3)}{10(1.3)} = 1.13$

Life: From Table 17–16, $K_b = 576$.

$$F_{b1} = \frac{K_b}{d} = \frac{576}{7.4} = 77.8 \text{ lbf}$$

$$F_{b2} = \frac{576}{11} = 52.4 \text{ lbf}$$

$$T_1 = F_1 + F_{b1} = 64.4 + 77.8 = 142.2 \text{ lbf}$$

$$T_2 = F_1 + F_{b2} = 64.4 + 52.4 = 116.8 \text{ lbf}$$

From Table 17–17, $K = 1193$ and $b = 10.926$.

Eq. (17–27): $\quad N_P = \left[\left(\dfrac{1193}{142.2} \right)^{-10.926} + \left(\dfrac{1193}{116.8} \right)^{-10.926} \right]^{-1} = 11(10^9) \text{ passes}$

Answer Since N_P is out of the validity range of Eq. (17–27), life is reported as greater than 10^9 passes. Then

Answer Eq. (17–28): $\quad t > \dfrac{10^9(113.8)}{720(3390)} = 46\,600 \text{ h}$

17–4 Timing Belts

A timing belt is made of a rubberized fabric coated with a nylon fabric, and has steel wire within to take the tension load. It has teeth that fit into grooves cut on the periphery of the pulleys (Fig. 17–15). A timing belt does not stretch appreciably or slip and consequently transmits power at a constant angular-velocity ratio. No initial tension is needed.

Figure 17–15

Timing-belt drive showing portions of the pulley and belt. Note that the pitch diameter of the pulley is greater than the diametral distance across the top lands of the teeth.

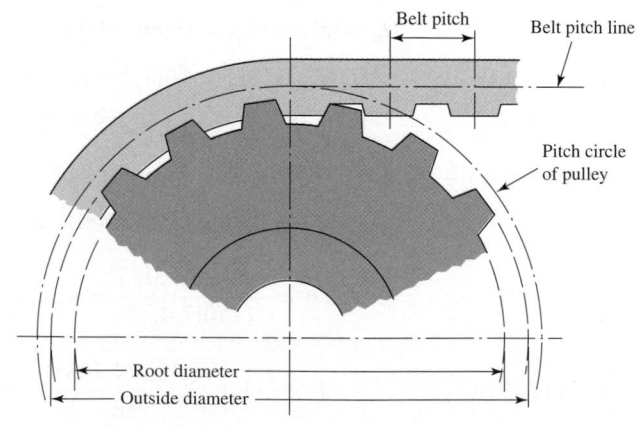

Belt pitch

Belt pitch line

Pitch circle of pulley

Root diameter

Outside diameter

Table 17–18

Standard Pitches of Timing Belts

Service	Designation	Pitch p, in
Extra light	XL	$\frac{1}{5}$
Light	L	$\frac{3}{8}$
Heavy	H	$\frac{1}{2}$
Extra heavy	XH	$\frac{7}{8}$
Double extra heavy	XXH	$1\frac{1}{4}$

Such belts can operate over a very wide range of speeds, have efficiencies in the range of 97 to 99 percent, require no lubrication, and are quieter than chain drives. There is no chordal-speed variation, as in chain drives (see Sec. 17–5), and so they are an attractive solution for precision-drive requirements.

The steel wire, the tension member of a timing belt, is located at the belt pitch line (Fig. 17–15). Thus the pitch length is the same regardless of the thickness of the backing.

The five standard inch-series pitches available are listed in Table 17–18 with their letter designations. Standard pitch lengths are available in sizes from 6 to 180 in. Pulleys come in sizes from 0.60 in pitch diameter up to 35.8 in and with groove numbers from 10 to 120.

The design and selection process for timing belts is so similar to that for V belts that the process will not be presented here. As in the case of other belt drives, the manufacturers will provide an ample supply of information and details on sizes and strengths.

17–5 Roller Chain

Basic features of chain drives include a constant ratio, since no slippage or creep is involved; long life; and the ability to drive a number of shafts from a single source of power.

Roller chains have been standardized as to sizes by the ANSI. Figure 17–16 shows the nomenclature. The pitch is the linear distance between the centers of the rollers. The width is the space between the inner link plates. These chains are manufactured in single, double, triple, and quadruple strands. The dimensions of standard sizes are listed in Table 17–19.

Figure 17–16

Portion of a double-strand roller chain.

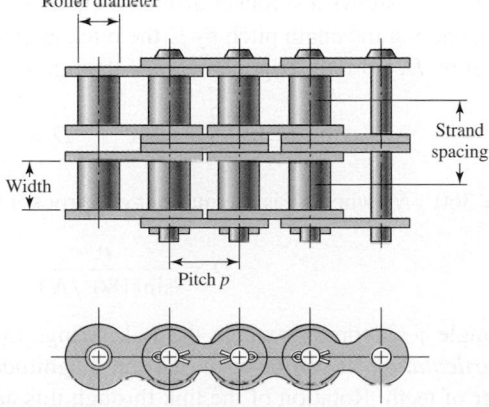

Table 17–19

Dimensions of American Standard Roller Chains—Single Strand

Source: Compiled from ANSI B29.1-1975.

ANSI Chain Number	Pitch, in (mm)	Width, in (mm)	Minimum Tensile Strength, lbf (N)	Average Weight, lbf/ft (N/m)	Roller Diameter, in (mm)	Multiple-Strand Spacing, in (mm)
25	0.250 (6.35)	0.125 (3.18)	780 (3 470)	0.09 (1.31)	0.130 (3.30)	0.252 (6.40)
35	0.375 (9.52)	0.188 (4.76)	1 760 (7 830)	0.21 (3.06)	0.200 (5.08)	0.399 (10.13)
41	0.500 (12.70)	0.25 (6.35)	1 500 (6 670)	0.25 (3.65)	0.306 (7.77)	— —
40	0.500 (12.70)	0.312 (7.94)	3 130 (13 920)	0.42 (6.13)	0.312 (7.92)	0.566 (14.38)
50	0.625 (15.88)	0.375 (9.52)	4 880 (21 700)	0.69 (10.1)	0.400 (10.16)	0.713 (18.11)
60	0.750 (19.05)	0.500 (12.7)	7 030 (31 300)	1.00 (14.6)	0.469 (11.91)	0.897 (22.78)
80	1.000 (25.40)	0.625 (15.88)	12 500 (55 600)	1.71 (25.0)	0.625 (15.87)	1.153 (29.29)
100	1.250 (31.75)	0.750 (19.05)	19 500 (86 700)	2.58 (37.7)	0.750 (19.05)	1.409 (35.76)
120	1.500 (38.10)	1.000 (25.40)	28 000 (124 500)	3.87 (56.5)	0.875 (22.22)	1.789 (45.44)
140	1.750 (44.45)	1.000 (25.40)	38 000 (169 000)	4.95 (72.2)	1.000 (25.40)	1.924 (48.87)
160	2.000 (50.80)	1.250 (31.75)	50 000 (222 000)	6.61 (96.5)	1.125 (28.57)	2.305 (58.55)
180	2.250 (57.15)	1.406 (35.71)	63 000 (280 000)	9.06 (132.2)	1.406 (35.71)	2.592 (65.84)
200	2.500 (63.50)	1.500 (38.10)	78 000 (347 000)	10.96 (159.9)	1.562 (39.67)	.2.817 (71.55)
240	3.00 (76.70)	1.875 (47.63)	112 000 (498 000)	16.4 (239)	1.875 (47.62)	3.458 (87.83)

Figure 17–17 shows a sprocket driving a chain and rotating in a counterclockwise direction. Denoting the chain pitch by p, the pitch angle by γ, and the pitch diameter of the sprocket by D, from the trigonometry of the figure we see

$$\sin\frac{\gamma}{2} = \frac{p/2}{D/2} \qquad \text{or} \qquad D = \frac{p}{\sin(\gamma/2)} \qquad (a)$$

Since $\gamma = 360°/N$, where N is the number of sprocket teeth, Eq. (a) can be written

$$D = \frac{p}{\sin(180°/N)} \qquad (17\text{–}29)$$

The angle $\gamma/2$, through which the link swings as it enters contact, is called the *angle of articulation*. It can be seen that the magnitude of this angle is a function of the number of teeth. Rotation of the link through this angle causes impact between the

Figure 17–17

Engagement of a chain and sprocket.

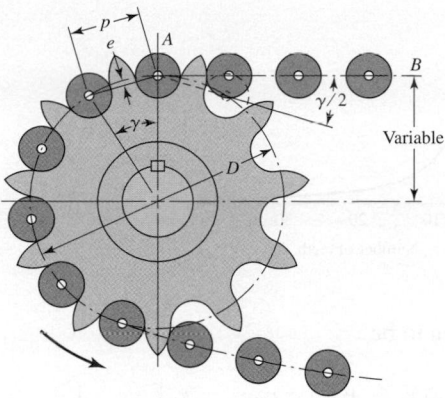

rollers and the sprocket teeth and also wear in the chain joint. Since the life of a properly selected drive is a function of the wear and the surface fatigue strength of the rollers, it is important to reduce the angle of articulation as much as possible.

The number of sprocket teeth also affects the velocity ratio during the rotation through the pitch angle γ. At the position shown in Fig. 17–17, the chain AB is tangent to the pitch circle of the sprocket. However, when the sprocket has turned an angle of $\gamma/2$, the chain line AB moves closer to the center of rotation of the sprocket. This means that the chain line AB is moving up and down, and that the lever arm varies with rotation through the pitch angle, all resulting in an uneven chain exit velocity. You can think of the sprocket as a polygon in which the exit velocity of the chain depends upon whether the exit is from a corner, or from a flat of the polygon. Of course, the same effect occurs when the chain first enters into engagement with the sprocket.

The chain velocity V is defined as the number of feet coming off the sprocket per unit time. Thus the chain velocity in feet per minute is

$$V = \frac{Npn}{12}$$ (17–30)

where N = number of sprocket teeth

         p = chain pitch, in

         n = sprocket speed, rev/min

The maximum exit velocity of the chain is

$$v_{\max} = \frac{\pi Dn}{12} = \frac{\pi np}{12 \sin(\gamma/2)}$$ (b)

where Eq. (a) has been substituted for the pitch diameter D. The minimum exit velocity occurs at a diameter d, smaller than D. Using the geometry of Fig. 17–17, we find

$$d = D \cos \frac{\gamma}{2}$$ (c)

Thus the minimum exit velocity is

$$v_{\min} = \frac{\pi dn}{12} = \frac{\pi np}{12} \frac{\cos(\gamma/2)}{\sin(\gamma/2)}$$ (d)

Now substituting $\gamma/2 = 180°/N$ and employing Eqs. (17–30), (b), and (d), we find the

| **Figure 17–18**

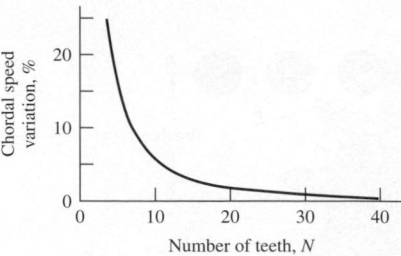

speed variation to be

$$
\frac{\Delta V}{V} = \frac{v_{\max} - v_{\min}}{V} = \frac{\pi}{N} \left[\frac{1}{\sin(180°/N)} - \frac{1}{\tan(180°/N)} \right] \qquad (17\text{–}31)
$$

This is called the *chordal speed variation* and is plotted in Fig. 17–18. When chain drives are used to synchronize precision components or processes, due consideration must be given to these variations. For example, if a chain drive synchronized the cutting of photographic film with the forward drive of the film, the lengths of the cut sheets of film might vary too much because of this chordal speed variation. Such variations can also cause vibrations within the system.

Although a large number of teeth is considered desirable for the driving sprocket, in the usual case it is advantageous to obtain as small a sprocket as possible, and this requires one with a small number of teeth. For smooth operation at moderate and high speeds it is considered good practice to use a driving sprocket with at least 17 teeth; 19 or 21 will, of course, give a better life expectancy with less chain noise. Where space limitations are severe or for very slow speeds, smaller tooth numbers may be used by sacrificing the life expectancy of the chain.

Driven sprockets are not made in standard sizes over 120 teeth, because the pitch elongation will eventually cause the chain to "ride" high long before the chain is worn out. The most successful drives have velocity ratios up to 6:1, but higher ratios may be used at the sacrifice of chain life.

Roller chains seldom fail because they lack tensile strength; they more often fail because they have been subjected to a great many hours of service. Actual failure may be due either to wear of the rollers on the pins or to fatigue of the surfaces of the rollers. Roller-chain manufacturers have compiled tables that give the horsepower capacity corresponding to a life expectancy of 15 kh for various sprocket speeds. These capacities are tabulated in Table 17–20 for 17–tooth sprockets. Table 17–21 displays available tooth counts on sprockets of one supplier. Table 17–22 lists the tooth correction factors for other than 17 teeth. Table 17–23 shows the multiple-strand factors K_2.

The capacities of chains are based on the following:

- 15 000 h at full load
- Single strand
- ANSI proportions
- Service factor of unity
- 100 pitches in length
- Recommended lubrication
- Elongation maximum of 3 percent

- Horizontal shafts
- Two 17-tooth sprockets

The fatigue strength of link plates governs capacity at lower speeds. The American Chain Association (ACA) publication *Chains for Power Transmission and Materials Handling* (1982) gives, for single-strand chain, the nominal power H_1, link-plate limited, as

$$H_1 = 0.004 N_1^{1.08} n_1^{0.9} p^{(3-0.07p)} \qquad \text{hp} \qquad (17\text{–}32)$$

and the nominal power H_2, roller-limited, as

$$H_2 = \frac{1000 K_r N_1^{1.5} p^{0.8}}{n_1^{1.5}} \qquad \text{hp} \qquad (17\text{–}33)$$

where N_1 = number of teeth in the smaller sprocket

         n_1 = sprocket speed, rev/min

         p = pitch of the chain, in

         K_r = 29 for chain numbers 25, 35; 3.4 for chain 41; and 17 for chains 40–240

Table 17–20

Rated Horsepower Capacity of Single-Strand Single-Pitch Roller Chain for a 17-Tooth Sprocket

Source: Compiled from ANSI B29.1-1975 information only section, and from B29.9-1958.

Sprocket Speed, rev/min	ANSI Chain Number					
	25	**35**	**40**	**41**	**50**	**60**
50	0.05	0.16	0.37	0.20	0.72	1.24
100	0.09	0.29	0.69	0.38	1.34	2.31
150	0.13*	0.41*	0.99*	0.55*	1.92*	3.32
200	0.16*	0.54*	1.29	0.71	2.50	4.30
300	0.23	0.78	1.85	1.02	3.61	6.20
400	0.30*	1.01*	2.40	1.32	4.67	8.03
500	0.37	1.24	2.93	1.61	5.71	9.81
600	0.44*	1.46*	3.45*	1.90*	6.72*	11.6
700	0.50	1.68	3.97	2.18	7.73	13.3
800	0.56*	1.89*	4.48*	2.46*	8.71*	15.0
900	0.62	2.10	4.98	2.74	9.69	16.7
1000	0.68*	2.31*	5.48	3.01	10.7	18.3
1200	0.81	2.73	6.45	3.29	12.6	21.6
1400	0.93*	3.13*	7.41	2.61	14.4	18.1
1600	1.05*	3.53*	8.36	2.14	12.8	14.8
1800	1.16	3.93	8.96	1.79	10.7	12.4
2000	1.27*	4.32*	7.72*	1.52*	9.23*	10.6
2500	1.56	5.28	5.51*	1.10*	6.58*	7.57
3000	1.84	5.64	4.17	0.83	4.98	5.76
Type A		**Type B**			**Type C**	

*Estimated from ANSI tables by linear interpolation.

Note: Type A—manual or drip lubrication; type B—bath or disk lubrication; type C—oil-stream lubrication.

(Continued)

Table 17–20

Rated Horsepower Capacity of Single-Strand Single-Pitch Roller Chain for a 17-Tooth Sprocket (*Continued*)

Sprocket Speed, rev/min		ANSI Chain Number							
		80	100	120	140	160	180	200	240
50	Type A	2.88	5.52	9.33	14.4	20.9	28.9	38.4	61.8
100		5.38	10.3	17.4	26.9	39.1	54.0	71.6	115
150		7.75	14.8	25.1	38.8	56.3	77.7	103	166
200		10.0	19.2	32.5	50.3	72.9	101	134	215
300		14.5	27.7	46.8	72.4	105	145	193	310
400		18.7	35.9	60.6	93.8	136	188	249	359
500	Type B	22.9	43.9	74.1	115	166	204	222	0
600		27.0	51.7	87.3	127	141	155	169	
700		31.0	59.4	89.0	101	112	123	0	
800		35.0	63.0	72.8	82.4	91.7	101		
900		39.9	52.8	61.0	69.1	76.8	84.4		
1000		37.7	45.0	52.1	59.0	65.6	72.1		
1200		28.7	34.3	39.6	44.9	49.9	0		
1400		22.7	27.2	31.5	35.6	0			
1600		18.6	22.3	25.8	0				
1800		15.6	18.7	21.6					
2000		13.3	15.9	0					
2500		9.56	0.40						
3000		7.25	0						

Type C		Type C′	

Note: Type A—manual or drip lubrication; type B—bath or disk lubrication; type C—oil-stream lubrication; type C′—type C, but this is a galling region; submit design to manufacturer for evaluation.

Table 17–21

Single-Strand Sprocket Tooth Counts Available from One Supplier*

No.	Available Sprocket Tooth Counts
25	8-30, 32, 34, 35, 36, 40, 42, 45, 48, 54, 60, 64, 65, 70, 72, 76, 80, 84, 90, 95, 96, 102, 112, 120
35	4-45, 48, 52, 54, 60, 64, 65, 68, 70, 72, 76, 80, 84, 90, 95, 96, 102, 112, 120
41	6-60, 64, 65, 68, 70, 72, 76, 80, 84, 90, 95, 96, 102, 112, 120
40	8-60, 64, 65, 68, 70, 72, 76, 80, 84, 90, 95, 96, 102, 112, 120
50	8-60, 64, 65, 68, 70, 72, 76, 80, 84, 90, 95, 96, 102, 112, 120
60	8-60, 62, 63, 64, 65, 66, 67, 68, 70, 72, 76, 80, 84, 90, 95, 96, 102, 112, 120
80	8-60, 64, 65, 68, 70, 72, 76, 78, 80, 84, 90, 95, 96, 102, 112, 120
100	8-60, 64, 65, 67, 68, 70, 72, 74, 76, 80, 84, 90, 95, 96, 102, 112, 120
120	9-45, 46, 48, 50, 52, 54, 55, 57, 60, 64, 65, 67, 68, 70, 72, 76, 80, 84, 90, 96, 102, 112, 120
140	9-28, 30, 31, 32, 33, 34, 35, 36, 37, 39, 40, 42, 43, 45, 48, 54, 60, 64, 65, 68, 70, 72, 76, 80, 84, 96
160	8-30, 32–36, 38, 40, 45, 46, 50, 52, 53, 54, 56, 57, 60, 62, 63, 64, 65, 66, 68, 70, 72, 73, 80, 84, 96
180	13-25, 28, 35, 39, 40, 45, 54, 60
200	9-30, 32, 33, 35, 36, 39, 40, 42, 44, 45, 48, 50, 51, 54, 56, 58, 59, 60, 63, 64, 65, 68, 70, 72
240	9-30, 32, 35, 36, 40, 44, 45, 48, 52, 54, 60

*Morse Chain Company, Ithaca, NY, Type B hub sprockets.

Table 17–22

Tooth Correction Factors, K_1

Number of Teeth on Driving Sprocket	K_1 Pre-extreme Horsepower	K_1 Post-extreme Horsepower
11	0.62	0.52
12	0.69	0.59
13	0.75	0.67
14	0.81	0.75
15	0.87	0.83
16	0.94	0.91
17	1.00	1.00
18	1.06	1.09
19	1.13	1.18
20	1.19	1.28
N	$(N_1/17)^{1.08}$	$(N_1/17)^{1.5}$

Table 17–23

Multiple-Strand Factors K_2

Number of Strands	K_2
1	1.0
2	1.7
3	2.5
4	3.3
5	3.9
6	4.6
8	6.0

The constant 0.004 becomes 0.0022 for no. 41 lightweight chain. The nominal horsepower in Table 17–20 is $H_{nom} = \min(H_1, H_2)$. For example, for $N_1 = 17$, $n_1 = 1000$ rev/min, no. 40 chain with $p = 0.5$ in, from Eq. (17–32),

$$H_1 = 0.004(17)^{1.08}1000^{0.9}0.5^{[3-0.07(0.5)]} = 5.48 \text{ hp}$$

From Eq. (17 33),

$$H_2 = \frac{1000(17)17^{1.5}(0.5^{0.8})}{1000^{1.5}} = 21.64 \text{ hp}$$

The tabulated value in Table 17–20 is $H_{tab} = \min(5.48, 21.64) = 5.48$ hp.

It is preferable to have an odd number of teeth on the driving sprocket (17, 19, ...) and an even number of pitches in the chain to avoid a special link. The approximate length of the chain L in pitches is

$$\frac{L}{p} \doteq \frac{2C}{p} + \frac{N_1 + N_2}{2} + \frac{(N_2 - N_1)^2}{4\pi^2 C/p} \qquad (17\text{–}34)$$

The center-to-center distance C is given by

$$C = \frac{p}{4}\left[-A + \sqrt{A^2 - 8\left(\frac{N_2 - N_1}{2\pi}\right)^2}\right] \qquad (17\text{–}35)$$

where

$$A = \frac{N_1 + N_2}{2} - \frac{L}{p} \tag{17–36}$$

The allowable power H_a is given by

$$H_a = K_1 K_2 H_{\text{tab}} \tag{17–37}$$

where K_1 = correction factor for tooth number other than 17 (Table 17–22)

   K_2 = strand correction (Table 17–23)

The horsepower that must be transmitted H_d is given by

$$H_d = H_{\text{nom}} K_s n_d \tag{17–38}$$

Equation (17–32) is the basis of the pre-extreme power entries (vertical entries) of Table 17–20, and the chain power is limited by link-plate fatigue. Equation (17–33) is the basis for the post-extreme power entries of these tables, and the chain power performance is limited by impact fatigue. The entries are for chains of 100 pitch length and 17-tooth sprocket. For a deviation from this

$$H_2 = 1000 \left[K_r \left(\frac{N_1}{n_1} \right)^{1.5} p^{0.8} \left(\frac{L_p}{100} \right)^{0.4} \left(\frac{15\,000}{h} \right)^{0.4} \right] \tag{17–39}$$

where L_p is the chain length in pitches and h is the chain life in hours. Viewed from a deviation viewpoint, Eq. (17–39) can be written as a trade-off equation in the following form:

$$\frac{H_2^{2.5} h}{N_1^{3.75} L_p} = \text{constant} \tag{17–40}$$

If tooth-correction factor K_1 is used, then omit the term $N_1^{3.75}$. Note that $(N_1^{1.5})^{2.5} = N_1^{3.75}$.

In Eq. (17–40) one would expect the h/L_p term because doubling the hours can require doubling the chain length, other conditions constant, for the same number of cycles. Our experience with contact stresses leads us to expect a load (tension) life relation of the form $F^a L = \text{constant}$. In the more complex circumstance of roller-bushing impact, the Diamond Chain Company has identified $a = 2.5$.

The maximum speed (rev/min) for a chain drive is limited by galling between the pin and the bushing. Tests suggest

$$n_1 \leq 1000 \left[\frac{82.5}{7.95^p (1.0278)^{N_1} (1.323)^{F/1000}} \right]^{1/(1.59 \log p + 1.873)} \quad \text{rev/min}$$

where F is the chain tension in pounds.

EXAMPLE 17–5

Select drive components for a 2:1 reduction, 90-hp input at 300 rev/min, moderate shock, an abnormally long 18-hour day, poor lubrication, cold temperatures, dirty surroundings, short drive $C/p = 25$.

Solution

Function: $H_{\text{nom}} = 90$ hp, $n_1 = 300$ rev/min, $C/p = 25$, $K_s = 1.3$
Design factor: $n_d = 1.5$

Sprocket teeth: $N_1 = 17$ teeth, $N_2 = 34$ teeth, $K_1 = 1$, $K_2 = 1, 1.7, 2.5, 3.3$
Chain number of strands:

$$H_{\text{tab}} = \frac{n_d K_s H_{\text{nom}}}{K_1 K_2} = \frac{1.5(1.3)90}{(1)K_2} = \frac{176}{K_2}$$

Form a table:

Number of Strands	176/K2 (Table 17–23)	Chain Number (Table 17–19)	Lubrication Type
1	176/1 = 176	200	C'
2	176/1.7 = 104	160	C
3	176/2.5 = 70.4	140	B
4	176/3.3 = 53.3	140	B

Decision   3 strands of number 140 chain (H_{tab} is 72.4 hp).
Number of pitches in the chain:

$$\frac{L}{p} = \frac{2C}{p} + \frac{N_1 + N_2}{2} + \frac{(N_2 - N_1)^2}{4\pi^2 C/p}$$

$$= 2(25) + \frac{17 + 34}{2} + \frac{(34 - 17)^2}{4\pi^2(25)} = 75.79 \text{ pitches}$$

Decision   Use 76 pitches. Then $L/p = 76$.
Identify the center-to-center distance: From Eqs. (17–35) and (17–36),

$$A = \frac{N_1 + N_2}{2} - \frac{L}{p} = \frac{17 + 34}{2} - 76 = -50.5$$

$$C = \frac{p}{4}\left[-A + \sqrt{A^2 - 8\left(\frac{N_2 - N_1}{2\pi}\right)^2} \right]$$

$$= \frac{p}{4}\left[50.5 + \sqrt{50.5^2 - 8\left(\frac{34 - 17}{2\pi}\right)^2} \right] = 25.104p$$

For a 140 chain, $p = 1.75$ in. Thus,

$$C = 25.104p = 25.104(1.75) = 43.93 \text{ in}$$

Lubrication: Type B
Comment: This is operating on the pre-extreme portion of the power, so durability estimates other than 15 000 h are not available. Given the poor operating conditions, life will be much shorter.

Lubrication of roller chains is essential in order to obtain a long and trouble-free life. Either a drip feed or a shallow bath in the lubricant is satisfactory. A medium or light mineral oil, without additives, should be used. Except for unusual conditions, heavy oils and greases are not recommended, because they are too viscous to enter the small clearances in the chain parts.

17–6 Wire Rope

Wire rope is made with two types of winding, as shown in Fig. 17–19. The *regular lay,* which is the accepted standard, has the wire twisted in one direction to form the strands, and the strands twisted in the opposite direction to form the rope. In the completed rope the visible wires are approximately parallel to the axis of the rope. Regular-lay ropes do not kink or untwist and are easy to handle.

Lang-lay ropes have the wires in the strand and the strands in the rope twisted in the same direction, and hence the outer wires run diagonally across the axis of the rope. Lang-lay ropes are more resistant to abrasive wear and failure due to fatigue than are regular-lay ropes, but they are more likely to kink and untwist.

Standard ropes are made with a hemp core, which supports and lubricates the strands. When the rope is subjected to heat, either a steel center or a wire-strand center must be used.

Wire rope is designated as, for example, a $1\frac{1}{8}$-in 6×7 haulage rope. The first figure is the diameter of the rope (Fig. 17–19c). The second and third figures are the number of strands and the number of wires in each strand, respectively. Table 17–24 lists some of the various ropes that are available, together with their characteristics and properties. The area of the metal in standard hoisting and haulage rope is $A_m = 0.38d^2$.

When a wire rope passes around a sheave, there is a certain amount of readjustment of the elements. Each of the wires and strands must slide on several others, and presumably some individual bending takes place. It is probable that in this complex action there exists some stress concentration. The stress in one of the wires of a rope passing around a sheave may be calculated as follows. From solid mechanics, we have

$$M = \frac{EI}{\rho} \qquad \text{and} \qquad M = \frac{\sigma I}{c} \qquad (a)$$

where the quantities have their usual meaning. Eliminating M and solving for the stress gives

$$\sigma = \frac{Ec}{\rho} \qquad (b)$$

For the radius of curvature ρ, we can substitute the sheave radius $D/2$. Also, $c = d_w/2$, where d_w is the wire diameter. These substitutions give

$$\sigma = E_r \frac{d_w}{D} \qquad (c)$$

To understand this equation, observe that the individual wire makes a corkscrew figure in space and if you pull on it to determine E it will stretch or give more than its native E would suggest. Therefore E is still the modulus of elasticity of the *wire,* but in its peculiar

Figure 17–19

Types of wire rope; both lays are available in either right or left hand.

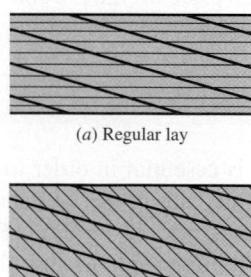

(a) Regular lay

(b) Lang lay

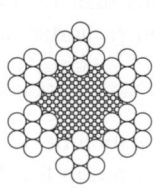

(c) Section of 6×7 rope

Table 17–24

Wire-Rope Data *Source: Compiled from American Steel and Wire Company Handbook.*

Rope	Weight per Foot, lbf	Minimum Sheave Diameter, in	Standard Sizes d, in	Material	Size of Outer Wires	Modulus of Elasticity,* Mpsi	Strength,† kpsi
6 × 7 haulage	$1.50d^2$	$42d$	$\frac{1}{4}$–$1\frac{1}{2}$	Monitor steel	$d/9$	14	100
				Plow steel	$d/9$	14	88
				Mild plow steel	$d/9$	14	76
6 × 19 standard hoisting	$1.60d^2$	$26d$–$34d$	$\frac{1}{4}$–$2\frac{3}{4}$	Monitor steel	$d/13$–$d/16$	12	106
				Plow steel	$d/13$–$d/16$	12	93
				Mild plow steel	$d/13$–$d/16$	12	80
6 × 37 special flexible	$1.55d^2$	$18d$	$\frac{1}{4}$–$3\frac{1}{2}$	Monitor steel	$d/22$	11	100
				Plow steel	$d/22$	11	88
8 × 19 extra flexible	$1.45d^2$	$21d$–$26d$	$\frac{1}{4}$–$1\frac{1}{2}$	Monitor steel	$d/15$–$d/19$	10	92
				Plow steel	$d/15$–$d/19$	10	80
7 × 7 aircraft	$1.70d^2$	—	$\frac{1}{16}$–$\frac{3}{8}$	Corrosion-resistant steel	—	—	124
				Carbon steel	—	—	124
7 × 9 aircraft	$1.75d^2$	—	$\frac{1}{8}$–$1\frac{3}{8}$	Corrosion-resistant steel	—	—	135
				Carbon steel	—	—	143
19-wire aircraft	$2.15d^2$	—	$\frac{1}{32}$–$\frac{5}{16}$	Corrosion-resistant steel	—	—	165
				Carbon steel	—	—	165

*The modulus of elasticity is only approximate; it is affected by the loads on the rope and, in general, increases with the life of the rope.

†The strength is based on the nominal area of the rope. The figures given are only approximate and are based on 1-in rope sizes and $\frac{1}{4}$-in aircraft-cable sizes.

configuration as part of the rope. A value for E equal to the modulus of elasticity of the *rope* gives an approximately correct value for the stress σ. For this reason we say that E_r in Eq. (c) is the modulus of elasticity of the rope, not the wire, recognizing that one can quibble over the name used.

Equation (c) gives the tensile stress σ in the outer wires. The sheave diameter is represented by D. This equation reveals the importance of using a large-diameter sheave. The suggested minimum sheave diameters in Table 17–24 are based on a D/d_w ratio of 400. If possible, the sheaves should be designed for a larger ratio. For elevators and mine hoists, D/d_w is usually taken from 800 to 1000. If the ratio is less than 200, heavy loads will often cause a permanent set in the rope.

A wire rope tension giving the same tensile stress as the sheave bending is called the *equivalent bending load* F_b, given by

$$F_b = \sigma A_m = \frac{E_r d A_m}{D} \tag{17–41}$$

A wire rope may fail because the static load exceeds the ultimate strength of the rope. Failure of this nature is generally not the fault of the designer, but rather that of the operator in permitting the rope to be subjected to loads for which it was not designed.

The first consideration in selecting a wire rope is to determine the static load. This load is composed of the following items:

- The known or dead weight
- Additional loads caused by sudden stops or starts
- Shock loads
- Sheave-bearing friction

When these loads are summed, the total can be compared with the ultimate strength of the rope to find a factor of safety. However, the ultimate strength used in this determination must be reduced by the strength loss that occurs when the rope passes over a curved surface such as a stationary sheave or a pin; see Fig. 17–20.

For an average operation, use a factor of safety of 5. Factors of safety up to 8 or 9 are used if there is danger to human life and for very critical situations. Table 17–25 lists

Figure 17–20

Percent strength loss due to different D/d ratios; derived from standard test data for 6 × 19 and 6 × 17 class ropes. *(Materials provided by the Wire Rope Technical Board (WRTB), Wire Rope Users Manual Third Edition, Second printing. Reprinted by permission.)*

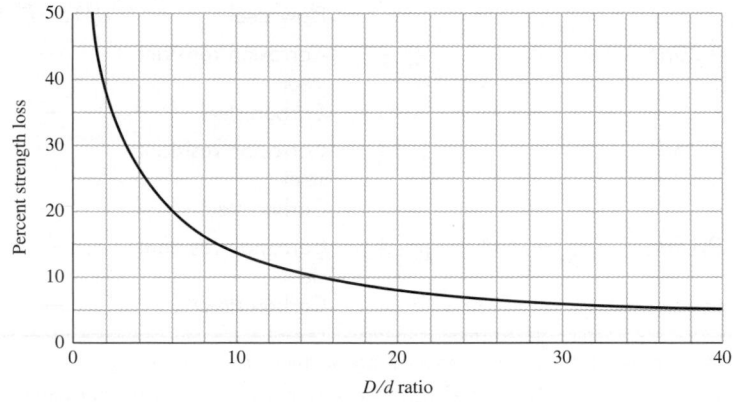

Table 17–25

Minimum Factors of Safety for Wire Rope*

Source: Compiled from a variety of sources, including ANSI A17.1-1978.

Track cables	3.2	Passenger elevators, ft/min:	
Guys	3.5	50	7.60
		300	9.20
Mine shafts, ft:		800	11.25
Up to 500	8.0	1200	11.80
1000–2000	7.0	1500	11.90
2000–3000	6.0		
Over 3000	5.0	Freight elevators, ft/min:	
		50	6.65
Hoisting	5.0	300	8.20
Haulage	6.0	800	10.00
Cranes and derricks	6.0	1200	10.50
Electric hoists	7.0	1500	10.55
Hand elevators	5.0	Powered dumbwaiters, ft/min:	
Private elevators	7.5	50	4.8
Hand dumbwaiter	4.5	300	6.6
Grain elevators	7.5	500	8.0

*Use of these factors does not preclude a fatigue failure.

minimum factors of safety for a variety of design situations. Here, the factor of safety is defined as

$$n = \frac{F_u}{F_t}$$

where F_u is the ultimate wire load and F_t is the largest working tension.

Once you have made a tentative selection of a rope based upon static strength, the next consideration is to ensure that the wear life of the rope and the sheave or sheaves meets certain requirements. When a loaded rope is bent over a sheave, the rope stretches like a spring, rubs against the sheave, and causes wear of both the rope and the sheave. The amount of wear that occurs depends upon the pressure of the rope in the sheave groove. This pressure is called the *bearing pressure;* a good estimate of its magnitude is given by

$$p = \frac{2F}{dD} \qquad (17\text{--}42)$$

where F = tensile force on rope

$\quad\quad\quad d$ = rope diameter

$\quad\quad\quad D$ = sheave diameter

The allowable pressures given in Table 17–26 are to be used only as a rough guide; they may not prevent a fatigue failure or severe wear. They are presented here because they represent past practice and furnish a starting point in design.

A fatigue diagram not unlike an *S-N* diagram can be obtained for wire rope. Such a diagram is shown in Fig. 17–21. Here the ordinate is the pressure-strength ratio p/S_u, and S_u is the ultimate tensile strength of the *wire*. The abscissa is the number of bends that occur in the total life of the rope. The curve implies that a wire rope has a fatigue limit; but this is not true at all. A wire rope that is used over sheaves will eventually fail

Table 17–26

Maximum Allowable Bearing Pressures of Ropes on Sheaves (in psi)

Source: Wire Rope Users Manual, AISI, 1979.

Rope	Wood[a]	Cast Iron[b]	Cast Steel[c]	Chilled Cast Irons[d]	Manganese Steel[e]
Sheave Material					
Regular lay:					
6 × 7	150	300	550	650	1470
6 × 19	250	480	900	1100	2400
6 × 37	300	585	1075	1325	3000
8 × 19	350	680	1260	1550	3500
Lang lay:					
6 × 7	165	350	600	715	1650
6 × 19	275	550	1000	1210	2750
6 × 37	330	660	1180	1450	3300

[a]On end grain of beech, hickory, or gum.

[b]For H_B(min.) = 125.

[c]30–40 carbon; H_B(min.) = 160.

[d]Use only with uniform surface hardness.

[e]For high speeds with balanced sheaves having ground surfaces.

Figure 17–21

Experimentally determined relation between the fatigue life of wire rope and the sheave pressure.

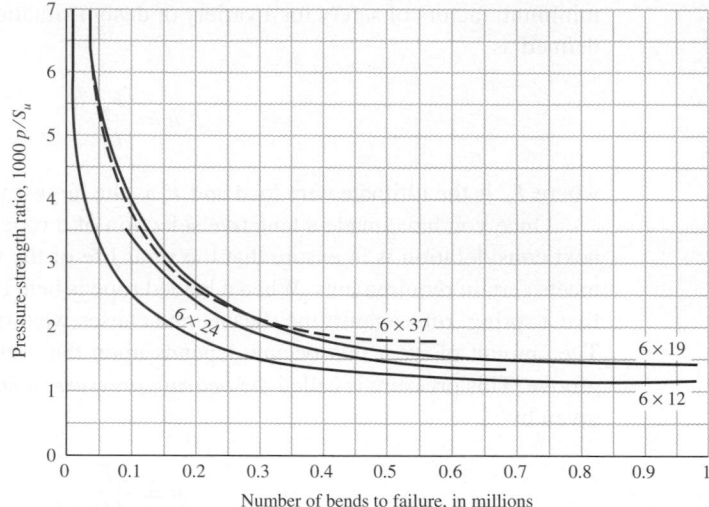

Number of bends to failure, in millions

in fatigue or in wear. However, the graph does show that the rope will have a long life if the ratio p/S_u is less than 0.001. Substitution of this ratio in Eq. (17–42) gives

$$S_u = \frac{2000F}{dD} \tag{17–43}$$

where S_u is the ultimate strength of the *wire*, not the rope, and the units of S_u are related to the units of F. This interesting equation contains the wire strength, the load, the rope diameter, and the sheave diameter—all four variables in a single equation! Dividing both sides of Eq. (17–42) by the ultimate strength of the wires S_u and solving for F gives

$$F_f = \frac{(p/S_u)S_u dD}{2} \tag{17–44}$$

where F_f is interpreted as the allowable fatigue tension as the wire is flexed a number of times corresponding to p/S_u selected from Fig. 17–21 for a particular rope and life expectancy. The factor of safety can be defined in fatigue as

$$n_f = \frac{F_f - F_b}{F_t} \tag{17–45}$$

where F_f is the rope tension strength under flexing and F_t is the tension at the place where the rope is flexing. Unfortunately, the designer often has vendor information that tabulates ultimate rope tension and gives no ultimate-strength S_u information concerning the wires from which the rope is made. Some guidance in strength of individual wires is

Improved plow steel (monitor)	$240 < S_u < 280$ kpsi
Plow steel	$210 < S_u < 240$ kpsi
Mild plow steel	$180 < S_u < 210$ kpsi

In wire-rope usage, the factor of safety has been defined in static loading as $n = F_u/F_t$ or $n = (F_u - F_b)/F_t$, where F_b is the rope tension that would induce the same outer-wire stress as that given by Eq. (c). The factor of safety in fatigue loading can be defined as in Eq. (17–45), or by using a static analysis and compensating with a large factor of safety applicable to static loading, as in Table 17–25. When using factors of safety expressed in codes, standards, corporate design manuals, or wire-rope manufacturers' recommendations or from the literature, be sure to ascertain upon which basis the factor of safety is to be evaluated, and proceed accordingly.

If the rope is made of plow steel, the wires are probably hard-drawn AISI 1070 or 1080 carbon steel. Referring to Table 10–3, we see that this lies somewhere between hard-drawn spring wire and music wire. But the constants m and A needed to solve Eq. (10–14) for S_u are lacking.

Practicing engineers who desire to solve Eq. (17–43) should determine the wire strength S_u for the rope under consideration by unraveling enough wire to test for the Brinell hardness. Then S_u can be found using Eq. (3–17). Fatigue failure in wire rope is not sudden, as in solid bodies, but progressive, and shows as the breaking of an outside wire. This means that the beginning of fatigue can be detected by periodic routine inspection.

Figure 17–22 is another graph showing the gain in life to be obtained by using large D/d ratios. In view of the fact that the life of wire rope used over sheaves is only finite, it is extremely important that the designer specify and insist that periodic inspection, lubrication, and maintenance procedures be carried out during the life of the rope. Table 17–27 gives useful properties of some wire ropes.

For a mine-hoist problem we can develop working equations from the preceding presentation. The wire rope tension F_t due to load and acceleration/deceleration is

$$F_t = \left(\frac{W}{m} + wl \right) \left(1 + \frac{a}{g} \right) \qquad (17\text{–}46)$$

Figure 17–22

Service-life curve based on bending and tensile stresses only. This curve shows that the life corresponding to $D/d = 48$ is twice that of $D/d = 33$. *(Materials provided by the Wire Rope Technical Board (WRTB), Wire Rope Users Manual Third Edition, Second printing. Reprinted by permission.)*

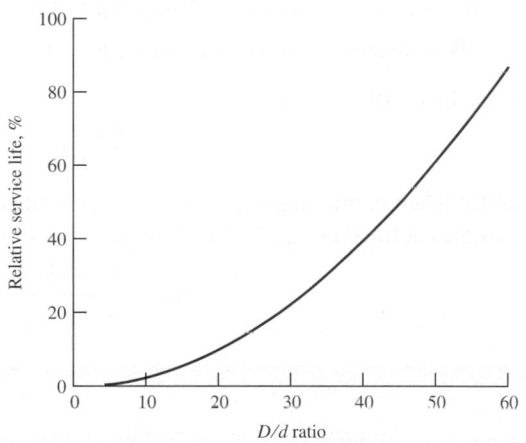

Table 17–27

Some Useful Properties of 6 × 7, 6 × 19, and 6 × 37 Wire Ropes

Wire Rope	Weight per Foot w, lbf/ft	Weight per Foot Including Core w, lbf/ft	Minimum Sheave Diameter D, in	Better Sheave Diameter D, in	Diameter of Wires d_w, in	Area of Metal A_m, in²	Rope Young's Modulus E_r, psi
6 × 7	$1.52d^2$		$42d$	$72d$	$0.111d$	$0.38d^2$	13×10^6
6 × 19	$1.60d^2$	$1.76d^2$	$30d$	$45d$	$0.067d$	$0.40d^2$	12×10^6
6 × 37	$1.55d^2$	$1.71d^2$	$18d$	$27d$	$0.048d$	$0.40d^2$	12×10^6

where W = weight at the end of the rope (cage and load), lbf

m = number of wire ropes supporting the load

w = weight/foot of the wire rope, lbf/ft

l = suspended length of rope, ft

a = maximum acceleration/deceleration experienced, ft/s^2

g = acceleration of gravity, ft/s^2

The fatigue tensile strength in pounds for a specified life F_f is

$$F_f = \frac{(p/S_u)S_u Dd}{2} \tag{17–47}$$

where (p/S_u) = specified life, from Fig. 17–21

S_u = ultimate tensile strength of the wires, psi

D = sheave or winch drum diameter, in

d = nominal wire rope size, in

The *equivalent bending load* F_b is

$$F_b = \frac{E_r d_w A_m}{D} \tag{17–48}$$

where E_r = Young's modulus for the wire rope, Table 17–24 or 17–28, psi

d_w = diameter of the wires, in

A_m = metal cross-sectional area, Table 17–24 or 17–28, in^2

D = sheave or winch drum diameter, in

The static factor of safety n_s is

$$n_s = \frac{F_u - F_b}{F_t} \tag{17–49}$$

Be careful when comparing recommended static factors of safety to Eq. (17–49), as n_s is sometimes defined as F_u/F_t. The fatigue factor of safety n_f is

$$n_f = \frac{F_f - F_b}{F_t} \tag{17–50}$$

EXAMPLE 17–6   Given a 6 × 19 monitor steel (S_u = 240 kpsi) wire rope.

(a) Develop the expressions for rope tension F_t, fatigue tension F_f, equivalent bending tensions F_b, and fatigue factor of safety n_f for a 531.5-ft, 1-ton cage-and-load mine hoist with a starting acceleration of 2 ft/s^2 as depicted in Fig. 17–23. The sheave diameter is 72 in.

(b) Using the expressions developed in part (a), examine the variation in factor of safety n_f for various wire rope diameters d and number of supporting ropes m.

Solution   (a) Rope tension F_t from Eq. (17–46) is given by

Answer
$$F_t = \left(\frac{W}{m} + wl\right)\left(1 + \frac{a}{g}\right) = \left[\frac{2000}{m} + 1.60d^2(531.5)\right]\left(1 + \frac{2}{32.2}\right)$$

$$= \frac{2124}{m} + 903d^2 \text{ lbf}$$

Figure 17–23

Geometry of the mine hoist of Ex. 17–6.

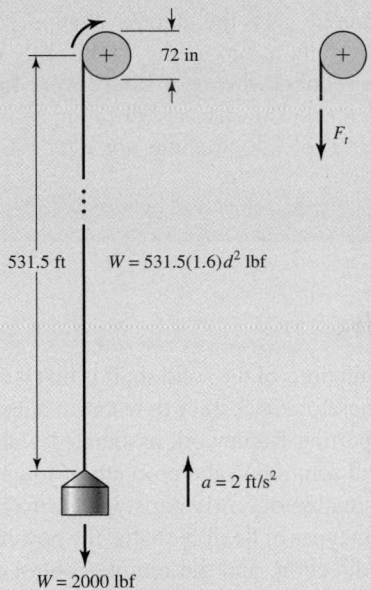

72 in

F_t

531.5 ft $W = 531.5(1.6)d^2$ lbf

$a = 2$ ft/s^2

$W = 2000$ lbf

From Fig. 17–21, use $p/S_u = 0.0014$. Fatigue tension F_f from Eq. (17–47) is given by

Answer

$$F_f = \frac{(p/S_u)S_u Dd}{2} = \frac{0.0014(240\,000)72d}{2} = 12\,096d \text{ lbf}$$

Equivalent bending tension F_b from Eq. (17–48) and Table 17–27 is given by

Answer

$$F_b = \frac{E_r d_w A_m}{D} = \frac{12(10^6)0.067d(0.40d^2)}{72} = 4467d^3 \text{ lbf}$$

Factor of safety n_f in fatigue from Eq. (17–50) is given by

Answer

$$n_f = \frac{F_f - F_b}{F_t} = \frac{12\,096d - 4467d^3}{2124/m + 903d^2}$$

(*b*) Form a table as follows:

| | n_f | | | |
d	m = 1	m = 2	m = 3	m = 4
0.25	1.355	2.641	3.865	5.029
0.375	1.910	3.617	5.150	6.536
0.500	2.336	4.263	5.879	7.254
0.625	2.612	4.573	6.099	7.331
0.750	2.731	4.578	5.911	6.918
0.875	2.696	4.330	5.425	6.210
1.000	2.520	3.882	4.736	5.320

Wire rope sizes are discrete, as is the number of supporting ropes. Note that for each m the factor of safety exhibits a maximum. Predictably the largest factor of safety increases with m. If the required factor of safety were to be 6, only three or four ropes could meet the requirement. The sizes are different: $\frac{5}{8}$-in ropes with three ropes or $\frac{3}{8}$-in ropes with four ropes. The costs include not only the wires, but the grooved winch drums.

17–7 Flexible Shafts

One of the greatest limitations of the solid shaft is that it cannot transmit motion or power around corners. It is therefore necessary to resort to belts, chains, or gears, together with bearings and the supporting framework associated with them. The flexible shaft may often be an economical solution to the problem of transmitting motion around corners. In addition to the elimination of costly parts, its use may reduce noise considerably.

There are two main types of flexible shafts: the power-drive shaft for the transmission of power in a single direction, and the remote-control or manual-control shaft for the transmission of motion in either direction.

The construction of a flexible shaft is shown in Fig. 17–24. The cable is made by winding several layers of wire around a central core. For the power-drive shaft, rotation should be in a direction such that the outer layer is wound up. Remote-control cables

Figure 17–24

Flexible shaft: (a) construction details; (b) a variety of configurations. *(Courtesy of S. S. White Technologies, Inc.)*

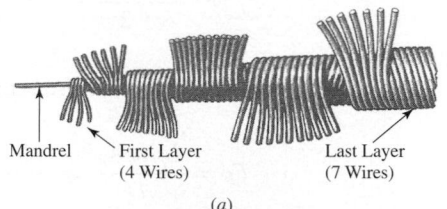

Mandrel   First Layer   Last Layer
           (4 Wires)   (7 Wires)

(a)

(b)

have a different lay of the wires forming the cable, with more wires in each layer, so that the torsional deflection is approximately the same for either direction of rotation.

Flexible shafts are rated by specifying the torque corresponding to various radii of curvature of the casing. A 15-in radius of curvature, for example, will give from 2 to 5 times more torque capacity than a 7-in radius. When flexible shafts are used in a drive in which gears are also used, the gears should be placed so that the flexible shaft runs at as high a speed as possible. This permits the transmission of the maximum amount of horsepower.

PROBLEMS

ANALYSIS

17–1   A 6-in-wide polyamide F-1 flat belt is used to connect a 2-in-diameter pulley to drive a larger pulley with an angular velocity ratio of 0.5. The center-to-center distance is 9 ft. The angular speed of the small pulley is 1750 rev/min as it delivers 2 hp. The service is such that a service factor K_s of 1.25 is appropriate.
(a) Find F_c, F_i, F_{1a}, and F_2.
(b) Find H_a, n_{fs}, and belt length.
(c) Find the dip.

ANALYSIS

17–2   Perspective and insight can be gained by doubling all geometric dimensions and observing the effect on problem parameters. Take the drive of Prob. 17–1, double the dimensions, and compare.

DESIGN

17–3   A flat-belt drive is to consist of two 4-ft-diameter cast-iron pulleys spaced 16 ft apart. Select a belt type to transmit 60 hp at a pulley speed of 380 rev/min. Use a service factor of 1.1 and a design factor of 1.0.

DESIGN

17–4   In solving problems and examining examples, you probably have noticed some recurring forms:

$$w = 12\gamma bt = (12\gamma t)b = a_1 b,$$

$$(F_1)_a = F_a bC_p C_v = (F_a C_p C_v)b = a_0 b$$

$$F_c = \frac{wV^2}{g} = \frac{a_1 b}{32.174}\left(\frac{V}{60}\right)^2 = a_2 b$$

$$(F_1)_a - F_2 = 2T/d = 33\,000 H_d / V = 33\,000 H_{\text{nom}} K_s n_d / V$$

$$F_2 = (F_1)_a - [(F_1)_a - F_2] = a_0 b - 2T/d$$

$$f\phi = \ln\frac{(F_1)_a - F_c}{F_2 - F_c} = \ln\frac{(a_0 - a_2)b}{(a_0 - a_2)b - 2T/d}$$

Show that

$$b = \frac{1}{a_0 - a_2}\frac{33\,000 H_d}{V}\frac{\exp(f\phi)}{\exp(f\phi) - 1}$$

ANALYSIS

17–5   Return to Ex. 17–1 and complete the following.
(a) Find the torque capacity that would put the drive as built at the point of slip, as well as the initial tension F_i.
(b) Find the belt width b that exhibits $n_{fs} = n_d = 1.1$.
(c) For part b find the corresponding F_{1a}, F_c, F_i, F_2, power, and n_{fs}.
(d) What have you learned?

ANALYSIS

17–6   Take the drive of Prob. 17–5 and double the belt width. Compare F_c, F_i, F_{1a}, F_2, H_a, n_{fs}, and dip.

ANALYSIS   **17-7**   Belted pulleys place loads on shafts, inducing bending and loading bearings. Examine Fig. 17–7 and develop an expression for the load the belt places on the pulley, and then apply it to Ex. 17–2.

DESIGN   **17-8**   Example 17–2 resulted in selection of a 10-in-wide A-3 polyamide flat belt. Show that the value of F_1 restoring f to 0.80 is

$$F_1 = \frac{(\Delta F + F_c) \exp f\phi - F_c}{\exp f\phi - 1}$$

and compare the initial tensions.

DESIGN   **17-9**   The line shaft illustrated in the figure is used to transmit power from an electric motor by means of flat-belt drives to various machines. Pulley A is driven by a vertical belt from the motor pulley. A belt from pulley B drives a machine tool at an angle of 70° from the vertical and at a center-to-center distance of 9 ft. Another belt from pulley C drives a grinder at a center-to-center distance of 11 ft. Pulley C has a double width to permit belt shifting as shown in Fig. 17–4. The belt from pulley D drives a dust-extractor fan whose axis is located horizontally 8 ft from the axis of the lineshaft. Additional data are

Machine	Speed, rev/min	Power, hp	Lineshaft Pulley	Diameter, in
Machine tool	400	12.5	B	16
Grinder	300	4.5	C	14
Dust extractor	500	8.0	D	18

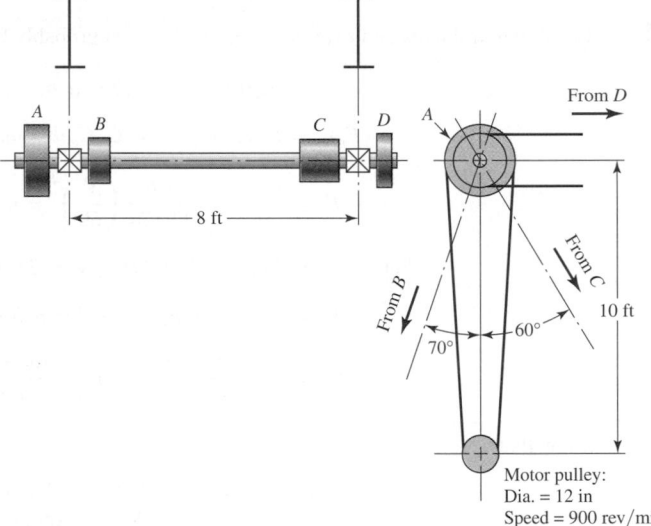

Problem 17–9
(Courtesy of Dr. Ahmed F. Abdel Azim,
Zagazig University, Cairo.)

The power requirements, listed above, account for the overall efficiencies of the equipment. The two line-shaft bearings are mounted on hangers suspended from two overhead wide-flange beams. Select the belt types and sizes for each of the four drives. Make provision for replacing belts from time to time because of wear or permanent stretch.

DESIGN   **17-10**   Two shafts 20 ft apart, with axes in the same horizontal plane, are to be connected with a flat belt in which the driving pulley, powered by a six-pole squirrel-cage induction motor with a 100 brake hp rating at 1140 rev/min, drives the second shaft at half its angular speed. The driven shaft drives light-shock machinery loads. Select a flat belt.

DESIGN **17–11** The mechanical efficiency of a flat-belt drive is approximately 98 percent. Because of its high value, the efficiency is often neglected. If a designer should choose to include it, where would he or she insert it in the flat-belt protocol?

DESIGN **17–12** In metal belts, the centrifugal tension F_c is ignored as negligible. Convince yourself that this is a reasonable problem simplification.

DESIGN **17–13** A designer has to select a metal-belt drive to transmit a power of H_{nom} under circumstances where a service factor of K_s and a design factor of n_d are appropriate. The design goal becomes $H_d = H_{\text{nom}} K_s n_d$. Use Eq. (17–8) to show that the minimum belt width is given by

$$b_{\min} = \frac{1}{a}\left(\frac{33\,000 H_d}{V}\right)\frac{\exp f\theta}{\exp f\theta - 1}$$

where a is the constant from $F_{1a} = ab$.

DESIGN **17–14** Design a friction metal flat-belt drive to connect a 1-hp, four-pole squirrel-cage motor turning at 1750 rev/min to a shaft 15 in away, running at half speed. The circumstances are such that a service factor of 1.2 and a design factor of 1.05 are appropriate. The life goal is 10^6 belt passes, $f = 0.35$, and the environmental considerations require a stainless steel belt.

DESIGN **17–15** A beryllium-copper metal flat belt with $S_f = 56670$ psi is to transmit 5 hp at 1125 rev/min with a life goal of 10^6 belt passes between two shafts 20 in apart whose centerlines are in a horizontal plane. The coefficient of friction between belt and pulley is 0.32. The conditions are such that a service factor of 1.25 and a design factor of 1.1 are appropriate. The driven shaft rotates at one-third the motor-pulley speed. Specify your belt, pulley sizes, and initial tension at installation.

DESIGN **17–16** For the conditions of Prob. 17–15 use a 1095 plain carbon-steel heat-treated belt. Conditions at the driving pulley hub require a pulley outside diameter of 3 in or more. Specify your belt, pulley sizes, and initial tension at installation.

DESIGN **17–17** A single V belt is to be selected to deliver engine power to the wheel-drive transmission of a riding tractor. A 5-hp single-cylinder engine is used. At most, 60 percent of this power is transmitted to the belt. The driving sheave has a diameter of 6.2 in, the driven, 12.0 in. The belt selected should be as close to a 92-in pitch length as possible. The engine speed is governor-controlled to a maximum of 3100 rev/min. Select a satisfactory belt and assess the factor of safety and the belt life in passes.

ANALYSIS **17–18** Two B85 V belts are used in a drive composed of a 5.4-in driving sheave, rotating at 1200 rev/min, and a 16-in driven sheave. Find the power capacity of the drive based on a service factor of 1.25, and find the center-to-center distance.

DESIGN **17–19** A 60-hp four-cylinder internal combustion engine is used to drive a brick-making machine under a schedule of two shifts per day. The drive consists of two 26-in sheaves spaced about 12 ft apart, with a sheave speed of 400 rev/min. Select a V-belt arrangement. Find the factor of safety, and estimate the life in passes and hours.

DESIGN **17–20** A reciprocating air compressor has a 5-ft-diameter flywheel 14 in wide, and it operates at 170 rev/min. An eight-pole squirrel-cage induction motor has nameplate data 50 bhp at 875 rev/min.
(a) Design a V-belt drive.
(b) Can cutting the V-belt grooves in the flywheel be avoided by using a V-flat drive?

ANALYSIS **17–21** The geometric implications of a V-flat drive are interesting.
(a) If the earth's equator was an inextensible string, snug to the spherical earth, and you spliced 6 ft of string into the equatorial cord and arranged it to be concentric to the equator, how far off the ground is the string?

(b) Using the solution to part a, formulate the modifications to the expressions for m_G, θ_d and θ_D, L_p, and C.

(c) As a result of this exercise, how would you revise your solution to part b of Prob. 17–20?

17–22 A 2-hp electric motor running at 1720 rev/min is to drive a blower at a speed of 240 rev/min. Select a V-belt drive for this application and specify standard V belts, sheave sizes, and the resulting center-to-center distance. The motor size limits the center distance to at least 22 in.

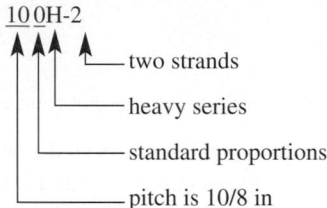

17–23 The standard roller-chain number indicates the chain pitch in inches, construction proportions, series, and number of strands as follows:

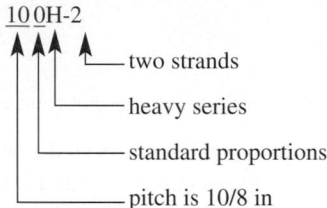

This convention makes the pitch directly readable from the chain number. In Ex. 17–5 ascertain the pitch from the selected chain number and confirm from Table 17–19.

17–24 Equate Eqs. (17–32) and (17–33) to find the rotating speed n_1 at which the power equates and marks the division between the premaximum and the postmaximum power domains.

(a) Show that

$$n_1 = \left[\frac{0.25(10^6) K_r N_1^{0.42}}{p^{(2.2-0.07p)}} \right]^{1/2.4}$$

(b) Find the speed n_1 for a no. 60 chain, $p = 0.75$ in, $N_1 = 17$, $K_r = 17$, and confirm from Table 17–20.

(c) At which speeds is Eq. (17–40) applicable?

17–25 A double-strand no. 60 roller chain is used to transmit power between a 13-tooth driving sprocket rotating at 300 rev/min and a 52-tooth driven sprocket.

(a) What is the allowable horsepower of this drive?

(b) Estimate the center-to-center distance if the chain length is 82 pitches.

(c) Estimate the torque and bending force on the driving shaft by the chain if the actual horsepower transmitted is 30 percent less than the corrected (allowable) power.

17–26 A four-strand no. 40 roller chain transmits power from a 21-tooth driving sprocket to an 84-tooth driven sprocket. The angular speed of the driving sprocket is 2000 rev/min.

(a) Estimate the chain length if the center-to-center distance has to be about 20 in.

(b) Estimate the tabulated horsepower entry H'_{tab} for a 20 000-h life goal.

(c) Estimate the rated (allowable) horsepower that would appear in Table 17–20 for a 20 000-h life.

(d) Estimate the tension in the chain at the allowable power.

17–27 A 700 rev/min 25-hp squirrel-cage induction motor is to drive a two-cylinder reciprocating pump, out-of-doors under a shed. A service factor K_s of 1.5 and a design factor of 1.1 are appropriate. The pump speed is 140 rev/min. Select a suitable chain and sprocket sizes.

17–28 A centrifugal pump is driven by a 50-hp synchronous motor at a speed of 1800 rev/min. The pump is to operate at 900 rev/min. Despite the speed, the load is smooth ($K_s = 1.2$). For a design factor of 1.1 specify a chain and sprockets that will realize a 50 000-h life goal. Let the sprockets be 19T and 38T.

ANALYSIS

17–29 A mine hoist uses a 2-in 6 × 19 monitor-steel wire rope. The rope is used to haul loads of 4 tons from the shaft 480 ft deep. The drum has a diameter of 6 ft, the sheaves are of good-quality cast steel, and the smallest is 3 ft in diameter.

(a) Using a maximum hoisting speed of 1200 ft/min and a maximum acceleration of 2 ft/s², estimate the stresses in the rope.

(b) Estimate the various factors of safety.

DESIGN

17–30 A temporary construction elevator is to be designed to carry workers and materials to a height of 90 ft. The maximum estimated load to be hoisted is 5000 lbf at a velocity not to exceed 2 ft/s. For minimum sheave diameters and acceleration of 4 ft/s², specify the number of ropes required if the 1-in plow-steel 6 × 19 hoisting strand is used.

DESIGN

17–31 A 2000-ft mine hoist operates with a 72-in drum using 6 × 19 monitor-steel wire rope. The cage and load weigh 8000 lbf, and the cage is subjected to an acceleration of 2 ft/s² when starting.

(a) For a single-strand hoist how does the factor of safety $n = F_f/F_t$ vary with the choice of rope diameter?

(b) For four supporting strands of wire rope attached to the cage, how does the factor of safety vary with the choice of rope diameter?

ANALYSIS

17–32 Generalize the results of Prob. 17–31 by representing the factor of safety n as

$$n = \frac{ad}{(b/m) + cd^2}$$

where m is the number of ropes supporting the cage, and a, b, and c are constants. Show that the optimal diameter is $d^* = [b/(mc)]^{1/2}$ and the corresponding maximum attainable factor of safety is $n^* = a[m/(bc)]^{1/2}/2$.

ANALYSIS

17–33 From your results in Prob. 17–32, show that to meet a fatigue factor of safety n_1 the optimal solution is

$$m = \frac{4bcn_1}{a^2} \text{ ropes}$$

having a diameter of

$$d = \frac{a}{2cn_1}$$

Solve Prob. 17–31 if a factor of safety of 2 is required. Show what to do in order to accommodate to the necessary discreteness in the rope diameter d and the number of ropes m.

ANALYSIS

17–34 For Prob. 17–29 estimate the elongation of the rope if a 9000-lbf loaded mine cart is placed on the cage. The results of Prob. 5–6 may be useful.

Computer Programs

In approaching the ensuing computer problems, the following suggestions may be helpful:

• Decide whether an analysis program or a design program would be more useful. In problems as simple as these, you will find the programs similar. For maximum instructional benefit, try the design problem.

• Creating a design program without a figure of merit precludes ranking alternative designs but does not hinder the attainment of satisfactory designs. Your instructor can provide the class design library with commercial catalogs, which not only have price information but define available sizes.

- Quantitative understanding and logic of interrelations are required for programming. Difficulty in programming is a signal to you and your instructor to increase your understanding. The following programs can be accomplished in 100 to 500 lines of code.

- Make programs interactive and user-friendly.

- Let the computer do what it can do best; the user should do what a human can do best.

- Assume the user has a copy of the text and can respond to prompts for information.

- If interpolating in a table is in order, solicit table entries in the neighborhood, and let the computer crunch the numbers.

- In decision steps, allow the user to make the necessary decision, even if it is undesirable. This allows learning of consequences and the use of the program for analysis.

- Display a lot of information in the summary. Show the decision set used up-front for user perspective.

- When a summary is complete, adequacy assessment can be accomplished with ease, so consider adding this feature.

DESIGN

17–35 Your experience with Probs. 17–1 through 17–11 has placed you in a position to write an interactive computer program to design/select flat-belt drive components. A possible decision set is

A Priori Decisions

- Function: H_{nom}, rev/min, velocity ratio, approximate C

- Design factor: n_d

- Initial tension maintenance: catenary

- Belt material: t, d_{min}, allowable tension, density, f

- Drive geometry: d, D

- Belt thickness: t (in material decision)

Design Decisions

- Belt width: b

DESIGN

17–36 Problems 17–12 through 17–16 have given you some experience with flat metal friction belts, indicating that a computer program could be helpful in the design/selection process. A possible decision set is

A Priori Decisions

- Function: H_{nom}, rev/min, velocity ratio approximate C

- Design factor: n_d

- Belt material: S_y, E, ν, d_{min}

- Drive geometry: d, D

- Belt thickness: t

Design Decisions

- Belt width: b

- Length of belt (often standard loop periphery)

DESIGN

17-37 Problems 17–17 through 17–22 have given you enough experience with V belts to convince you that a computer program would be helpful in the design/selection of V-belt drive components. Write such a program.

17-38 Experience with Probs. 17–23 through 17–28 can suggest an interactive computer program to help in the design/selection process of roller-chain elements. A possible decision set is

A Priori Decisions

- Function: power, speed, space, K_s, life goal
- Design factor: n_d
- Sprocket tooth counts: N_1, N_2, K_1, K_2

Design Decisions

- Chain number
- Strand count
- Lubrication system
- Chain length in pitches

(center-to-center distance for reference)

18

Shafts and Axles

Chapter Outline

18–1   Introduction

A *shaft* is a rotating member, usually of circular cross section, used to transmit power or motion. It provides the axis of rotation, or oscillation, of elements such as gears, pulleys, flywheels, cranks, sprockets, and the like and controls the geometry of their motion. An *axle* is a nonrotating member that carries no torque and is used to support rotating wheels, pulleys, and the like. The automotive axle is not a true axle; the term is a carry-over from the horse-and-buggy era, when the wheels rotated on nonrotating members. A *spindle* is a short shaft. Terms such as *lineshaft, headshaft, stub shaft, transmission shaft, countershaft,* and *flexible shaft* are names associated with special usage.

A shaft design really begins after much preliminary work. The design of the machine itself will dictate that certain gears, pulleys, bearings, and other elements will have at least been partially analyzed and their size and spacing tentatively determined. At this stage the design must be studied from the following points of view:

1   Deflection and rigidity
  (*a*) Bending deflection
  (*b*) Torsional deflection
  (*c*) Slope at bearings and shaft-supported elements
  (*d*) Shear deflection due to transverse loading of short shafts
2   Stress and strength
  (*a*) Static strength
  (*b*) Fatigue strength
  (*c*) Reliability

The geometry of a shaft is generally that of a stepped cylinder. Gears, bearings, and pulleys must always be accurately positioned and provision made to accept thrust loads. The use of shaft shoulders is an excellent means of axially locating the shaft elements; these shoulders can be used to preload rolling bearings and to provide the necessary thrust reactions to the rotating elements. For these reasons our analysis will usually involve stepped shafts.

Figure 18–1 shows the stepped shaft supporting the gear of a worm-gear speed reducer. You should be able to determine the reason for each step of the shaft and for the sleeve. Note also how the lubricant is supplied and drained. Can this reducer be used for counterclockwise as well as clockwise rotation? Why?

In deciding on an approach to design, it is necessary to realize that a stress analysis at a specific point on a shaft can be made using only the shaft geometry in the vicinity of that point. Thus the geometry of the entire shaft is not needed. In design it is usually possible to locate the critical areas, size these to meet the strength requirements, and then size the rest of the shaft to meet the requirements of the shaft-supported elements.

Note that the deflection and slope analyses cannot be made until the geometry of the entire shaft has been defined. Thus deflection is a function of the geometry *everywhere*, whereas the stress at a section of interest is a function of *local geometry and moments*. For this reason shaft design allows a consideration of stress and strength first. Then, after tentative values for the shaft dimensions have been established, the determination of the deflection and slope can be made.

The geometric configuration of the shaft to be designed is usually determined from past experience and, most often, is simply a revision of existing models in which a limited number of changes must be made. These changes can result for a variety of reasons, such as the use of a newly designed seal or coupling, a change in the power or speed, bearings of a different size, or the use of newly designed rotating components. Such changes are easy for the designer and need no further explanation.

Figure 18–1

A vertical worm-gear speed reducer. *(Courtesy of the Cleveland Gear Company.)*

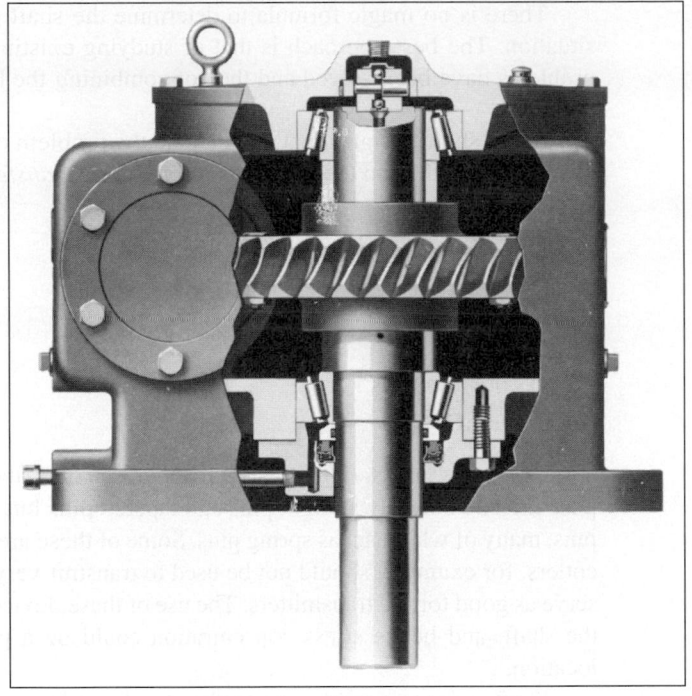

Figure 18–2

(*a*) Choose a shaft configuration to support and locate the two gears and two bearings. (*b*) Solution uses an integral pinion, three shaft shoulders, key and keyway, and sleeve. The housing locates the bearings on their outer rings and receives the thrust loads. (*c*) Choose fan-shaft configuration. (*d*) Solution uses sleeve bearings, a straight-through shaft, locating collars, and setscrews for collars, fan pulley, and fan itself. The fan housing supports the sleeve bearings.

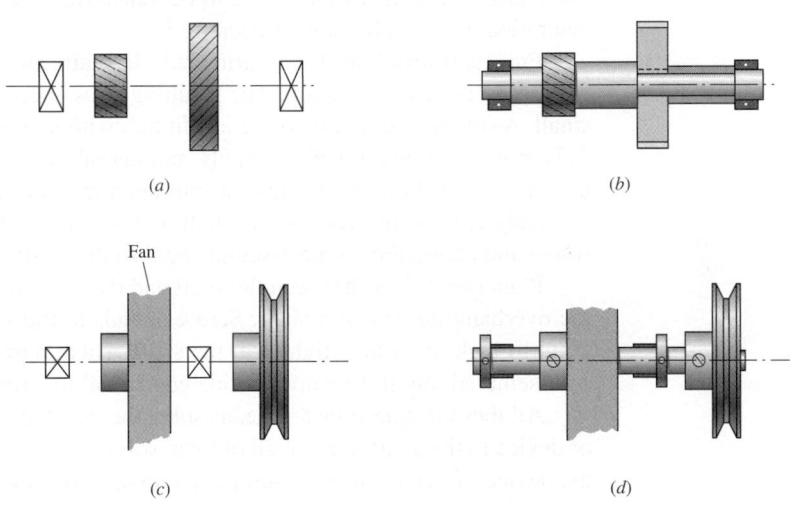

If there is no existing design to use as a starter, then the determination of the shaft geometry may have many solutions. This problem is illustrated by the two examples of Fig. 18–2. In Fig. 18–2*a* a geared countershaft is to be supported by two bearings. In Fig. 18–2*c* a fanshaft is to be configured. A variety of solutions should occur to you for each of these problems. The solutions shown are not necessarily the best ones, but they do illustrate how the shaft-mounted devices are fixed and located in the axial direction, and how provision is made for torque transfer from one element to another. Note that the shaft in Fig. 18–2*a* is subject to bending, torsional, and axial loads. The shaft in Fig. 18–2*c* is subject only to bending and torsion.

There is no magic formula to determine the shaft geometry for any given design situation. The best approach is that of studying existing designs to learn how similar problems have been solved and then of combining the best of these to solve your own problem.

Many shaft-design situations include the problem of transmitting torque from one element to another on the shaft. Common *torque-transfer elements* are:

- Keys
- Splines
- Setscrews
- Pins
- Press or shrink fits
- Tapered fits

Keys, pins, and *setscrews* have already been described in Chap. 8. Pins for this purpose include not only straight pins and tapered pins but also a wide variety of patented pins, many of which act as spring pins. Some of these are for locational purposes only—cotters, for example, should not be used to transmit very much torque—but others will serve as good torque transmitters. The use of these devices requires radial holes through the shaft, and hence stress concentration could be a problem, depending upon their location.

Shaft splines resemble gear teeth cut or forged into the shaft surface. They are used when large amounts of torque are to be transferred. When splines are used, stress concentration is generally quite moderate.

Press and shrink fits for securing hubs to shafts are used both for torque transfer and for preserving axial location. The resulting stress-concentration factor is usually quite small. A similar method is to use a split hub with screws to clamp the hub to the shaft. This method allows for disassembly and lateral adjustments. Another similar method uses a two-part hub consisting of a split inner member that fits into a tapered hole. The assembly is then tightened to the shaft with screws, which force the inner part into the wheel and clamp the whole assembly against the shaft.

Plain tapered fits between the shaft and the shaft-mounted device are often used on the overhanging end of a shaft. Screw threads at the shaft end then permit the use of a nut to lock the wheel tightly to the shaft. This approach is useful because it can be disassembled, but it does not provide good axial location of the wheel on the shaft.

All these torque-transfer means solve the problem of securely anchoring the wheel or device to the shaft, but not all of them solve the problem of accurate axial location of the device. Some of the most-used *locational devices* include:

- Cotter and washer
- Nut and washer
- Sleeve
- Shaft shoulder
- Ring and groove
- Setscrew
- Split hub or tapered two-piece hub
- Collar and screw
- Pins

Figure 18–3

Tapered roller bearings used in a mowing machine spindle. This design represents good practice for the situation in which one or more torque-transfer elements must be mounted outboard. *(Source: Redrawn from material furnished by The Timken Company.)*

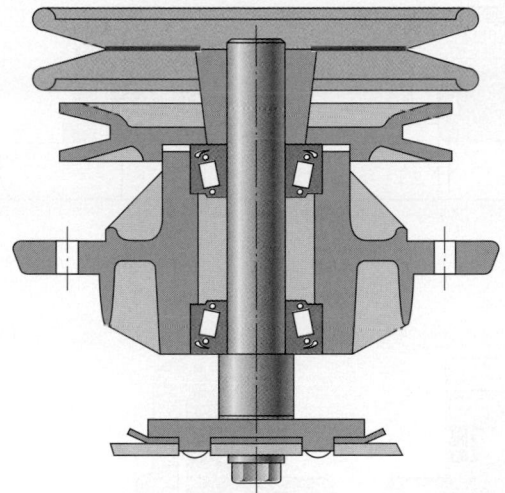

Figure 18–4

A bevel-gear drive in which both pinion and gear are straddle-mounted. *(Source: Redrawn from material furnished by Gleason Machine Division.)*

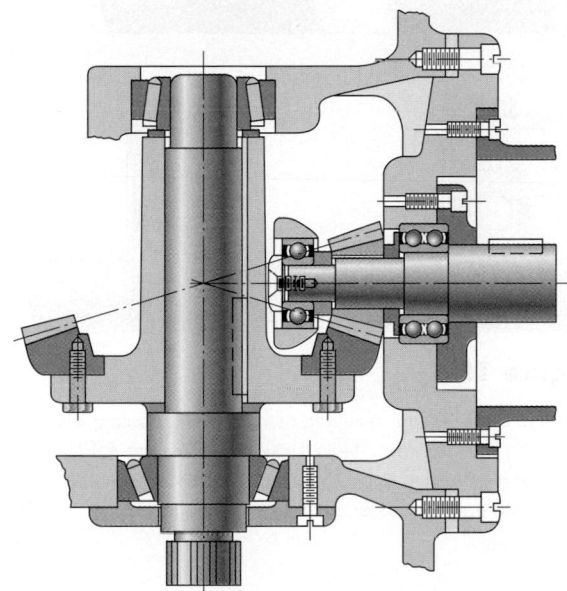

We have already discussed some of these items. The use of a ring fitted into a shaft groove is an economical solution to some problems. The grooves are quite shallow; many of the ring styles available do exert a spring force against the device to be anchored; and sometimes the grooves can be located where the effect of the stress-concentration factor is small or unimportant.

Figure 18–2*d* shows the use of a collar in which holding is obtained by tightening the setscrews against the shaft. An alternative—and better—arrangement uses a split collar and a cap screw to clamp the collar in place.

Figures 18–3 to 18–8 are intended to generate ideas illustrating how shaft features reflect attachments.[1] These successful designs, developed and refined over a period of

---

[1]Figures 18–3 to 18–8 were redrawn from material furnished by New Departure-Hyatt Division, General Motors Corporation.

Figure 18–5

Arrangement showing bearing inner rings press-fitted to shaft while outer rings float in the housing. The axial clearance should be sufficient only to allow for machinery vibrations. Note the labyrinth seal on the right.

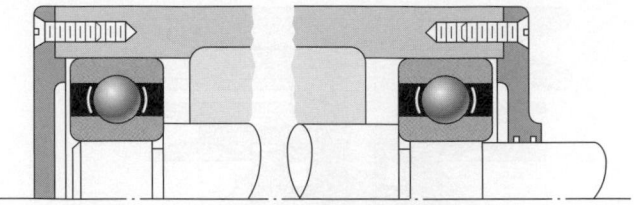

Figure 18–6

Similar to the arrangement of Fig. 18–5 except that the outer bearing rings are preloaded. Note the use of adjusting shims under the end cap.

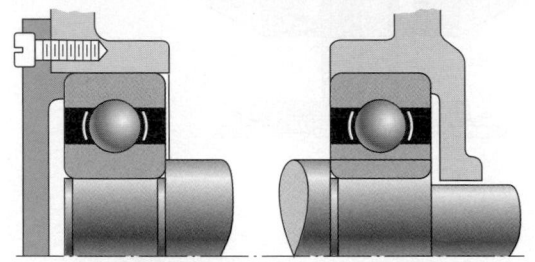

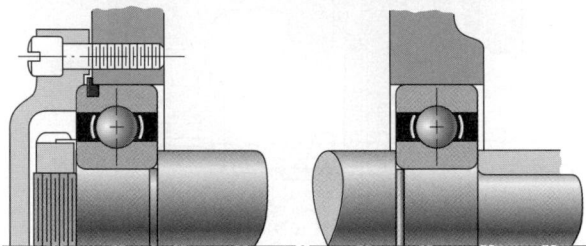

Figure 18–7

In this arrangement the inner ring of the left-hand bearing is locked to the shaft between a nut and a shaft shoulder. The locknut and washer are AFBMA standard. The snap ring in the outer race is used to positively locate the shaft assembly in the axial direction. Note the floating right-hand bearing and the grinding runout grooves in the shaft.

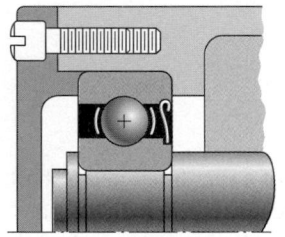

Figure 18–8

This arrangement is similar to Fig. 18–7 in that the left-hand bearing positions the entire shaft assembly. In this case the inner ring is secured to the shaft using a snap ring. Note the use of a shield to prevent dirt generated from within the machine from entering the bearing.

many years, represent very good design practice. Some of the figures in Chap. 11 will also help in the development of good shaft design. Catalog and commercial literature include illustrations and recommendations leading to good design practice.

A Plan for Attacking the Shaft Design Problem

The decision sets we have encountered so far have had one or two decision variables, and the constraints upon them have often been active (tight) when we were finished. A shaft is a complicated geometric form, as you can appreciate after reading this section. A shaft may have a half-dozen different diameters and a similar number of shoulder-to-shoulder lengths. Some of the positions of elements attached to the shaft may have fixed or variable relationships to others, and the bearing locations may be decision variables too. We can quickly accumulate a dozen decision variables.

To help reduce the dimensionality from a dozen independent variables to one or two, we can recognize that if the tight constraints are located in advance, the problem is greatly simplified. Such recognition can be based on experience or on procedures that help when experience is lacking. One method of approaching the optimum while reducing the dimensionality of the problem deals with the functional constraints associated with deformation and stress. We will start with geometric constraints related to deformation, and then go on to strength constraints associated with critical stresses.

18–2   Geometric Constraints

Consider the geometric constraints for a transmission shaft design task. The first step is to size the gears and pulleys for the specified speed and power. The root diameter of the gear teeth or pulley groove plus the necessary radial space for a keyway fixes one constraint on shaft size. Having the gear and pulley size fixes the forces on the system. The second step is to select bearings to provide adequate life under these forces and speeds. After the bearings are selected, the distance between the bearings will be set. Also, for roller bearings, the bearing bore places a limit on shaft size. The third step is to consider shaft deflection and stress as outlined below.

In material bodies distortion is unavoidable under load. We seek to control it to avoid compromising functionality. The slope of a shaft centerline with respect to a rolling-bearing outer ring centerline ought to be less than 0.001 rad for cylindrical and 0.0005 rad for tapered roller bearings. Similarly, it should be less than 0.004 rad for deep-groove roller bearings and, typically, less than 0.0087 rad for spherical ball bearings. At a gear mesh the allowable relative slope of two spur gears with uncrowned teeth should be held to less than 0.0005 rad. It is apparent that there are several stringent distortion limits on a power transmission shaft. The designer has a choice of designing for strength and checking distortion or designing for distortion and checking for strength. Most power transmission shafts have an active distortion constraint, so designing for distortion and checking for strength is appealing.

For perspective, the first move is to find a uniform-diameter shaft that satisfies all the distortion constraints. Going with the odds, since bearing slope constraints are often limiting, we first consider the slope at the bearings of the simply supported shaft in Fig. 18–9. We can use Tables A–9–6 and A–9–8 to develop expressions for the slope at the bearings due to the loads F_i and M_i. Then, using superposition, we can sum the results from all of the loads on the shaft. For the left-bearing, the deflections from the tables of Appendix A are

$$y_{AB} = \frac{F_i b_i x}{6EIl}\left(x^2 + b_i^2 - l^2\right) + \frac{M_i x}{6EIl}\left(x^2 + 3a_i^2 - 6a_i l + 2l^2\right)$$

Differentiating to obtain the slope and setting $x = 0$ for the left bearing results in

$$\theta_A = \frac{1}{6EIl}\left[F_i b_i\left(b_i^2 - l^2\right) + M_i\left(3a_i^2 - 6a_i l + 2l^2\right)\right]$$

Figure 18–9

Simply supported shaft with force F_i and couple M_i applied.

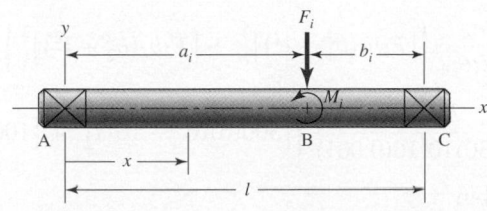

If we have a number of concentrated forces and moments in the xy plane, then using superposition we sum the results obtaining

$$\theta_A = \frac{1}{6EIl} \sum \left[F_i b_i (b_i^2 - l^2) + M_i (3a_i^2 - 6a_i l + 2l^2) \right]$$

Considering the xy plane to be the vertical plane V and the xz plane to be the horizontal plane H, for loading in both planes, the results can be added as vectors to give

$$\theta_A = \frac{1}{6EIl} \left\{ \left[\sum F_i b_i (b_i^2 - l^2) + \sum M_i (3a_i^2 - 6a_i l + 2l^2) \right]_H^2 \right.$$
$$\left. + \left[\sum F_i b_i (b_i^2 - l^2) + \sum M_i (3a_i^2 - 6a_i l + 2l^2) \right]_V^2 \right\}^{1/2}$$

For a solid circular shaft of diameter d, $I = \pi d^4/64$. Let $\theta_A = \theta_{\text{all}}$ be the absolute value of the allowable slope at the bearing, and n_d be the design factor. Then, for the left bearing slope constraint,

$$d = \left| \frac{32 n_d}{3\pi E l \theta_{\text{all}}} \left\{ \left[\sum F_i b_i (b_i^2 - l^2) + \sum M_i (3a_i^2 - 6a_i l + 2l^2) \right]_H^2 \right. \right.$$
$$\left. \left. + \left[\sum F_i b_i (b_i^2 - l^2) + \sum M_i (3a_i^2 - 6a_i l + 2l^2) \right]_V^2 \right\}^{1/2} \right|^{1/4} \quad (18\text{-}1)$$

In a similar fashion, for the right-bearing slope constraint,

$$d = \left| \frac{32 n_d}{3\pi E l \theta_{\text{all}}} \left\{ \left[\sum F_i a_i (l^2 - a_i^2) + \sum M_i (3a_i^2 - l^2) \right]_H^2 \right. \right.$$
$$\left. \left. + \left[\sum F_i a_i (l^2 - a_i^2) + \sum M_i (3a_i^2 - l^2) \right]_V^2 \right\}^{1/2} \right|^{1/4} \quad (18\text{-}2)$$

These equations are worth programming.

EXAMPLE 18–1

The steel shaft depicted in Fig. 18–10 carries two spur gears and has loadings as shown. The bearings located at A and B will be cylindrical roller bearings. The spatial centerline slope at the bearings is limited to 0.001 rad with a design factor of 1.5. Estimate the diameter of a uniform shaft that meets the slope constraints imposed by the bearings.

Solution

From Eq. (18–1),

$$d = \left| \frac{32 n_d}{3\pi E l \theta_{\text{all}}} \left\{ \left[F_1 b_1 (b_1^2 - l^2) \right]_H^2 + \left[F_2 b_2 (b_2^2 - l^2) \right]_V^2 \right\}^{1/2} \right|^{1/4}$$

$$= \left| \frac{32(1.5)}{3\pi (30) 10^6 16 (0.001)} \left\{ \left[300(6)(6^2 - 16^2) \right]^2 + \left[1000(12)(12^2 - 16^2) \right]^2 \right\}^{1/2} \right|^{1/4}$$

Answer

$= 1.964$ in

Figure 18–10

A contemplated shaft is to carry two spur gears between bearings A and B. Reactions and gear loadings are shown in a plane. The y axis is vertical, the other two are horizontal.

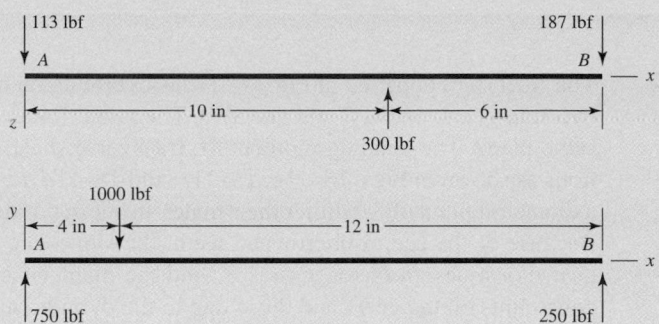

for the left-bearing slope constraint. For the right-bearing slope constraint, from Eq. (18–2),

$$d = \left| \frac{32n_d}{3\pi\, EI\theta_{\text{all}}} \left\{ \left[F_1 a_1 (l^2 - a_1^2) \right]_H^2 + \left[F_2 a_2 (l^2 - a_2^2) \right]_V^2 \right\}^{1/2} \right|^{1/4}$$

$$= \left| \frac{32(1.5)}{3\pi\,(30)(10)^6 16(0.001)} \left\{ \left[300(10)(16^2 - 10^2) \right]^2 + \left[1000(4)(16^2 - 4^2) \right]^2 \right\}^{1/2} \right|^{1/4}$$

Answer $= 1.835$ in

This initial calculation tells a designer that a uniform shaft of diameter 1.964 in will meet the bearing-slope constraints. For any decrease in diameter at the bearing journals and shoulders, diametral increase elsewhere should be made for gear seats and shoulders. The designer has an idea of the "heft" of the shaft, a useful perspective.

In Sec. 5–4 several beam deflection methods were described. For stepped shafts, where the deflections may be sought at a number of different points, integration using either singularity functions or numerical integration is practical. Finite element programs are also very useful here.

Given the bending-moment diagram and the shaft geometry, the deflection and slope at various points can be found. If, in examining the deflections, any value is larger than the allowable deflection y_{all}, a new diameter can be found from

$$d_{\text{new}} = d_{\text{old}} \left| \frac{n_d y_{\text{old}}}{y_{\text{all}}} \right|^{1/4} \qquad (18\text{–}3)$$

where y_{all} is the allowable deflection at that station and n_d is the design factor. Similarly, if any slope is larger than the allowable slope θ_{all}, a new diameter can be found from

$$d_{\text{new}} = d_{\text{old}} \left| \frac{n_d (dy/dx)_{\text{old}}}{(\text{slope})_{\text{all}}} \right|^{1/4} \qquad (18\text{–}4)$$

where $(\text{slope})_{\text{all}}$ is the allowable slope. As a result of these calculations note the largest $d_{\text{new}}/d_{\text{old}}$ ratio, then multiply *all* diameters by this ratio. The tight constraint will be just tight, and all others will be loose. Don't be too concerned about end journal sizes, as their influence is usually negligible. The beauty of the method is that the deflections need to be completed just once and constraints can be rendered loose but for one, diameters all identified without reworking every deflection. Furthermore, the deflections and slopes are (usually) exact. The method lends itself to computer implementation.

EXAMPLE 18-2   The steel shaft depicted in Fig. 18–11a is overhung on the right-bearing side. It carries two spur gears, one at B and one at D. The radial forces of 2000 and 1100 lbf lie in the same plane. The bending moment M, transverse shear V, and torque T have distributions as shown in Figs. 18–11b, 18–11c, and 18–11d, respectively. The gears both have a diametral pitch of 8; neither their mates-in-mesh nor detail about their sizes is shown. Because of the use of uncrowned teeth, the slopes are limited to 0.0005 rad on each gear. For a design factor $n_d = 1.5$, find the diameter of a *uniform* shaft to avoid the constraints on the gears and those due to the cylindrical roller bearings expected to be used.

Solution   For a uniform diameter of 1 in,

$$6EIl = 6(30)10^6 \frac{\pi(1^4)}{64} 6 = 5.3014(10^7) \text{ lbf} \cdot \text{in}^3$$

Figure 18-11

(a) A contemplated shaft carrying two gears and transmitting torque between the gears. The forces lie in a plane. (b) The bending-moment diagram. (c) The transverse-shear diagram. (d) The torque diagram.

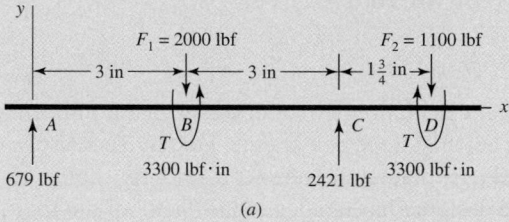

(a)

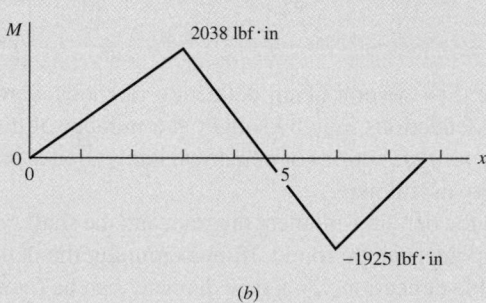

(b)

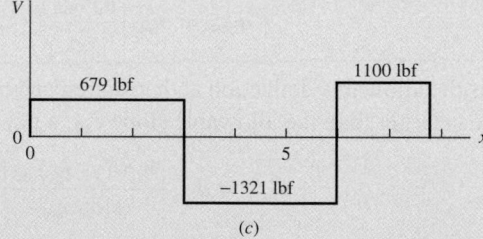

(c)

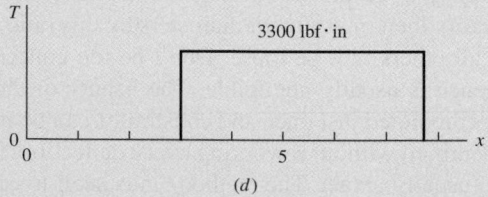

(d)

As in the development of Eqs. (18–1) and (18–2), except that Tables A–9–6 and A–9–10 are used,

$$y_{AB} = \frac{2000(3)x}{5.3014(10^7)}(x^2 + 3^2 - 6^2) + \frac{1100(1.75)x}{5.3014(10^7)}(6^2 - x^2)$$

$$= 7.6866(10^{-5})x^3 - 1.7486(10^{-3})x \text{ in}$$

$$\theta_{AB} = \frac{2000(3)}{5.3014(10^7)}(3x^2 + 3^2 - 6^2) + \frac{1100(1.75)}{5.3014(10^7)}(6^2 - 3x^2)$$

$$= 2.306(10^{-4})x^2 - 1.7486(10^{-3}) \text{ rad}$$

$$y_{BC} = \frac{2000(3)(6 - x)}{5.3014(10^7)}[x^2 + 3^2 - 2(6)x] + \frac{1100(1.75)x}{5.3014(10^7)}(6^2 - x^2)$$

$$= -1.4949(10^{-4})x^3 + 2.0372(10^{-3})x^2 - 7.8601(10^{-3})x + 6.1115(10^{-3}) \text{ in}$$

$$\theta_{BC} = \frac{2000(3)}{5.3014(10^7)}[-3x^2 + 6(6)x - 2(6^2) - 3^2] + \frac{1100(1.75)}{5.3014(10^7)}(6^2 - 3x^2)$$

$$= -4.4846(10^{-4})x^2 + 4.0744(10^{-3})x - 7.8601(10^{-3}) \text{ rad}$$

The deflection of beam A–9–6 for $x \geq l$ is the slope of that beam at l multiplied by $(x - l)$. Thus,

$$y_{CD} = \frac{2000(3)}{5.3014(10^7)}[-3(6^2) + 6(6)6 - 2(6^2) - 3^2](x - 6)$$

$$+ \frac{1100(x - 6)6}{5.3014(10^7)}[(x - 6)^2 - 1.75(3x - 6)]$$

$$= 1.24495(10^{-4})x^3 - 2.8945(10^{-3})x^2 + 2.173(10^{-2})x - 5.3069(10^{-2}) \text{ in}$$

$$\theta_{CD} = \frac{2000(3)}{5.3014(10^7)}[-3(6^2) + 6(6)6 - 2(6^2) - 3^2]$$

$$+ \frac{1100(6)}{5.3014(10^7)}\{3(x - 6)^2 - 1.75[6x - 4(6)]\}$$

$$= 3.7349(10^{-4})x^2 - 5.789(10^{-3})x + 2.173(10^{-2}) \text{ rad}$$

The displacements and slopes at A, B, C, and D are tabulated below.

Position, in	y, in	θ, rad
A (x = 0)	0	−0.0017486
B (x = 3)	−0.003170	0.0003268
C (x = 6)	0	0.0004414
D (x = 7.75)	−0.0005620	−0.0007024

Bearing slopes: From Eq. (18–4), with the subscripts of *d* being position locations,

$$d_A = 1 \left| \frac{1.5(-0.001\ 749)}{0.001} \right|^{1/4} = 1.273 \text{ in}$$

$$d_C = 1 \left| \frac{1.5(0.000\ 4414)}{0.001} \right|^{1/4} = 0.902 \text{ in}$$

Gear-mesh slopes: From Eq. (18–4),

$$d_B = 1 \left| \frac{1.5(0.000\ 3268)}{0.0005} \right|^{1/4} = 0.995 \text{ in}$$

$$d_D = 1 \left| \frac{1.5(-0.000\ 7024)}{0.0005} \right|^{1/4} = 1.205 \text{ in}$$

Center-to-center distance: For 8-diametral-pitch gears the center-to-center expansion distance on commercial-quality gears is 0.010 in. We will divide this between meshing gears, making the allowable transverse deflection 0.005 in. From Eq. (18–3),

$$d_B = 1 \left| \frac{1.5(-0.003\ 170)}{0.005} \right|^{1/4} = 0.988 \text{ in}$$

$$d_D = 1 \left| \frac{1.5(-0.0005620)}{0.005} \right|^{1/4} = 0.641 \text{ in}$$

The shaft diameter, keeping all of these constraints loose but one, is 1.273 in. We also know that the tight constraint (with a margin of 1.5) is the bearing misalignment at the left (bearing *A*).

The transverse shear *V* at a section of a beam in flexure imposes a shearing deflection, which is superposed on the bending deflection. Usually such shearing deflection is less than 1 percent of the transverse bending deflection, and it is seldom evaluated. However, when a shaft's length-to-diameter ratio is less than 10, the shear component of transverse deflection merits attention. There are many short shafts. A tabular method[2] is explained in detail, with examples elsewhere. The method lends itself to computer implementation.

For right-circular cylindrical shafts in torsion the angular deflection θ is given in Eq. (5–5). For a stepped shaft with individual cylinder length l_i and torque T_i, the angular deflection can be estimated from

$$\theta = \sum \theta_i = \sum \frac{T_i l_i}{G_i J_i} \tag{18–5}$$

[2]C.R. Mischke, "Tabular Method for Transverse Shear Deflection," Sec. 37.3 in Joseph E. Shigley and Charles R. Mischke (eds.-in-chief), *Standard Handbook of Machine Design,* 2nd ed., McGraw-Hill, New York, 1966.

or, for a constant torque throughout homogeneous material, from

$$\theta = \frac{T}{G} \sum \frac{l_i}{J_i} \tag{18–6}$$

If torsional stiffness is defined as $k_i = T_i/\theta_i$ and, since $\theta_i = T_i/k_i$ and $\theta = \sum \theta_i = \sum(T_i/k_i)$, for constant torque $\theta = T \sum(1/k_i)$, it follows that the stiffness of the shaft k in terms of segment stiffnesses is

$$\frac{1}{k} = \sum \frac{1}{k_i} \tag{18–7}$$

Note that Eq. (18–5) is not precise, since experimental evidence shows that θ is larger than given by Eq. (18–5).[3]

18–3 Strength Constraints

For static or quasi-static loading of a shaft, the reader should refer to Sec. 6–12.

In this section, we will examine shaft fatigue strength using distortion-energy-Gerber and distortion-energy-elliptic failure models. These two failure models fit the data best. Other methods, still used in practice, are discussed in the next section. The analyses will be presented under deterministic conditions (for stochastic analysis, the reader is urged to review Sec. 7–17).

Any rotating shaft loaded by stationary bending and torsional moments will be stressed in completely reversed bending because of shaft rotation, but the torsional stress will be steady. By using the subscripts a for alternating stress amplitude σ_a and m for midrange or steady stress as in τ_{xym}, we can express the components of stress as

$$\sigma_a = \frac{32M_a}{\pi d^3} \tag{18–8}$$

$$\tau_{xym} = \frac{16T_m}{\pi d^3} \tag{18–9}$$

Expressing the midrange component as a von Mises normal stress gives

$$\sigma_a' = \sigma_{xa} \tag{18–10}$$

$$\sigma_m' = \sqrt{3}\tau_{xym} \tag{18–11}$$

After the curve fit to be used is determined, the fatigue failure criterion is plotted on the designer's fatigue diagram, without the data, as depicted in Fig. 18–12. The curve partitions the plane into a "safe" area (beneath the curve) and an unsafe area (above the curve).

The Gerber fatigue-failure curve of passes among fatigue test points and can be successfully used as a regression line. Statistical chances of being below the curve can be assessed from the data, and the chance of a fatigue failure can be quantified. The Gerber curve is used in conjunction with the Langer curve (first-cycle yielding), and the dog-legged function is called the Gerber-Langer failure curve. The ASME-elliptic fatigue-failure curve likewise passes among the data and can also be used as a regression line. The statistical chances of being below the line can be assessed from the data, and the chance of a fatigue can be quantified. The ANSI/ASME B106.1M-1985 standard,

---

[3]R. Bruce Hopkins, *Design Analysis of Shafts and Beams,* McGraw-Hill, New York, 1970, pp. 93–99.

Figure 18–12

The designer's fatigue-failure diagram for Ex. 18–3. The fatigue-failure locus is DE-elliptic. The stress condition components at the shoulder are shown with the load line.

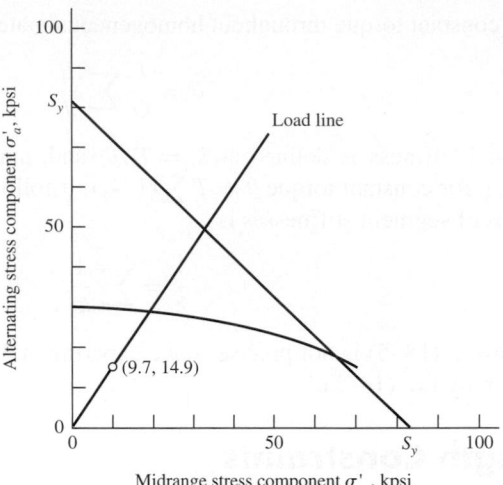

Table 18–1

Fatigue-Strength Criteria for Shafts

Failure Criterion	Basic Formula
Soderberg	$\dfrac{n\sigma_a}{S_e} + \dfrac{n\sigma_m}{S_y} = 1$
Goodman	$\dfrac{n\sigma_a}{S_e} + \dfrac{n\sigma_m}{S_{ut}} = 1$
Gerber	$\dfrac{n\sigma_a}{S_e} + \left(\dfrac{n\sigma_m}{S_{ut}}\right)^2 = 1$
ASME elliptic	$\left(\dfrac{n\sigma_a}{S_e}\right)^2 + \left(\dfrac{n\sigma_m}{S_y}\right)^2 = 1$
Yielding (Langer)	$\dfrac{n}{S_y}(\sigma_a + \sigma_m) = 1$

Note: Substituting S_a for $n\sigma_a$ and S_m for $n\sigma_m$ gives a locus entirely in strength terms.

second printing, is the ASME standard for shafting. The Gerber-Langer and ASME-elliptic-Langer curves offer the best predictors of fatigue failure.

Table 18–1 displays relationships among stress, strength, and design factor (factor of safety) for a number of models. When S_a is substituted for $n\sigma_a$ and S_m for $n\sigma_m$, the equation for the failure curve is obtained for display on the designer's fatigue diagram.

The stress condition at the critical location described by σ_a and σ_m can be plotted, along with the load line upon which it is constrained to move (if it moves). This point helps complete the visual picture.

Literature (textbooks, handbooks, technical papers, and industrial literature) concerning shafts presents a large array of equations. In an engineering analysis, a *failure criterion* is used. A name is associated with each. A displayed equation uses the name in an adjectival fashion, for example, a Gerber failure criterion or a Gerber-Langer failure criterion. Additionally, if the stress components are von Mises components, they are identified with von Mises (or Henckey) or simply called distortion energy (easily abbreviated DE) as in DE-Gerber. Only by attention to vocabulary can we communicate essential information with precision. The short title identifies the significant stress (the failure theory), then a hyphen, followed by a failure locus name (the fatigue data model). Equations cannot usually speak for themselves.

Shaft Diameter Equation for the DE-Gerber Criterion

The von Mises stress-amplitude component σ_a' and the midrange stress component σ_m' are given by

$$\sigma_a' = \left(\sigma_{xa}^2 + 3\tau_{xya}^2\right)^{1/2} = \frac{16}{\pi d^3}\sqrt{4(K_f M_a)^2 + 3(K_{fs} T_a)^2} = \frac{16A}{\pi d^3}$$

(18–12)

$$\sigma_m' = \left(\sigma_{xm}^2 + 3\tau_{xym}^2\right)^{1/2} = \frac{16}{\pi d^3}\sqrt{4(K_f M_m)^2 + 3(K_{fs} T_m)^2} = \frac{16B}{\pi d^3}$$

where A and B are defined by the radicals in Eqs. (18–12). The Gerber fatigue-failure criterion is defined by

$$\frac{S_a}{S_e} + \left(\frac{S_m}{S_{ut}}\right)^2 = \frac{n\sigma_a'}{S_e} + \left(\frac{n\sigma_m'}{S_{ut}}\right)^2 = \frac{16nA}{\pi d^3 S_e} + \left(\frac{16nB}{\pi d^3 S_{ut}}\right)^2 = 1$$

Solving for diameter d gives

$$d = \left(\frac{8nA}{\pi S_e}\left\{1 + \left[1 + \left(\frac{2BS_e}{AS_{ut}}\right)^2\right]^{1/2}\right\}\right)^{1/3}$$

(18–13)

or, solving for $1/n$,

$$\frac{1}{n} = \frac{8A}{\pi d^3 S_e}\left\{1 + \left[1 + \left(\frac{2BS_e}{AS_{ut}}\right)^2\right]^{1/2}\right\}$$

(18–14)

where

$$A = \sqrt{4(K_f M_a)^2 + 3(K_{fs} T_a)^2} \qquad B = \sqrt{4(K_f M_m)^2 + 3(K_{fs} T_m)^2}$$

(18–15)

$$r = \sigma_a'/\sigma_m' = A/B$$

Expressions for d or $1/n$ are constructed for the problem at hand by evaluating A and B and then substituting into Eqs. (18–13) and (18–14). For example, with $M_m = 0$ and $T_a = 0$,

$$A = 2K_f M_a \qquad B = \sqrt{3}K_{fs} T_m$$

and the preceding equations become

$$d = \left(\frac{16nK_f M_a}{\pi S_e}\left\{1 + \left[1 + 3\left(\frac{K_{fs} T_m S_e}{K_f M_a S_{ut}}\right)^2\right]^{1/2}\right\}\right)^{1/3}$$

(18–16)

$$\frac{1}{n} = \frac{16K_f M_a}{\pi d^3 S_e}\left\{1 + \left[1 + 3\left(\frac{K_{fs} T_m S_e}{K_f M_a S_{ut}}\right)^2\right]^{1/2}\right\}$$

(18–17)

$$r = \frac{\sigma_a'}{\sigma_m'} = \frac{2K_f M_a}{\sqrt{3}K_{fs} T_m}$$

(18–18)

Shaft Diameter Equation for the DE-Elliptic Criterion

The previous definitions for σ_a', σ_m', A, and B apply. The elliptic fatigue-failure criterion is defined by

$$\left(\frac{S_a}{S_e}\right)^2 + \left(\frac{S_m}{S_y}\right)^2 = \left(\frac{n\sigma_a'}{S_e}\right)^2 + \left(\frac{n\sigma_m'}{S_y}\right)^2 = \left(\frac{16nA}{\pi d^3 S_e}\right)^2 + \left(\frac{16nB}{\pi d^3 S_y}\right)^2 = 1$$

Solving for the diameter d gives

$$d = \left(\frac{16n}{\pi}\sqrt{\frac{A^2}{S_e^2} + \frac{B^2}{S_y^2}}\right)^{1/3}$$

Substituting for A and B gives expressions for d, $1/n$, and r:

$$d = \left\{\frac{16n}{\pi}\left[4\left(\frac{K_f M_a}{S_e}\right)^2 + 3\left(\frac{K_{fs} T_a}{S_e}\right)^2 + 4\left(\frac{K_f M_m}{S_y}\right)^2 + 3\left(\frac{K_{fs} T_m}{S_y}\right)^2\right]^{1/2}\right\}^{1/3}$$

$$\text{(18-19)}$$

$$\frac{1}{n} = \frac{16}{\pi d^3}\left[4\left(\frac{K_f M_a}{S_e}\right)^2 + 3\left(\frac{K_{fs} T_a}{S_e}\right)^2 + 4\left(\frac{K_f M_m}{S_y}\right)^2 + 3\left(\frac{K_{fs} T_m}{S_y}\right)^2\right]^{1/2} \quad \text{(18-20)}$$

$$r = \frac{\sigma_a'}{\sigma_m'} = \frac{A}{B} = \sqrt{\frac{4(K_f M_a)^2 + 3(K_{fs} T_a)^2}{4(K_f M_m)^2 + 3(K_{fs} T_m)^2}}$$

For the case when $M_m = 0$ and $T_a = 0$, the preceding relationships become

$$d = \left\{\frac{16n}{\pi}\left[4\left(\frac{K_f M_a}{S_e}\right)^2 + 3\left(\frac{K_{fs} T_m}{S_y}\right)^2\right]^{1/2}\right\}^{1/3} \quad \text{(18-21)}$$

$$\frac{1}{n} = \frac{16}{\pi d^3}\left[4\left(\frac{K_f M_a}{S_e}\right)^2 + 3\left(\frac{K_{fs} T_m}{S_y}\right)^2\right]^{1/2} \quad \text{(18-22)}$$

$$r = \frac{\sigma_a'}{\sigma_m'} = \frac{2K_f M_a}{\sqrt{3}K_{fs} T_m}$$

At a shoulder Figs. A–15–8 and A–15–9 provide information about K_t and K_{ts}. For a hole in a solid shaft, Figs. A–15–10 and A–15–11 provide K_t and K_{ts} information. For a hole in a solid shaft, use Table A–16. For grooves use Figs. A–15–14 and A–15–15.

The value of slope at which the load line intersects the junction of the failure curves is designated r_{crit}. It tells whether the threat is from fatigue or first-cycle yielding. If $r > r_{\text{crit}}$, the threat is from fatigue; if $r < r_{\text{crit}}$, the threat is from first-cycle yielding. For the Gerber-Langer intersection the strength components S_a and S_m are given in Table 7–10 as

$$S_m = \frac{S_{ut}^2}{2S_e}\left[1 - \sqrt{1 + \left(\frac{2S_e}{S_{ut}}\right)^2\left(1 - \frac{S_y}{S_e}\right)}\right]$$

$$S_a = S_y - S_m \quad \text{(18-23)}$$

$$r_{\text{crit}} = \frac{S_a}{S_m} = \frac{S_y - S_m}{S_m}$$

For the DE-elliptic-Langer intersection

$$S_a = \frac{2S_y S_e^2}{S_e^2 + S_y^2}$$

$$S_m = S_y - S_a$$

$$r_{\text{crit}} = \frac{S_a}{S_m} = \frac{S_a}{S_y - S_a}$$

(18–24)

EXAMPLE 18–3   At a machined shaft shoulder the small diameter d is 1.100 in, the large diameter D is 1.65 in, and the fillet radius is 0.11 in. The bending moment is 1260 lbf · in and the steady torsion moment is 1100 lbf · in. The heat-treated steel shaft has an ultimate strength of $S_{ut} = 105$ kpsi and a yield strength of $S_y = 82$ kpsi. The reliability goal is 0.99.
(*a*) Determine the fatigue factor of safety of the design using the DE-elliptic criterion.
(*b*) Determine the fatigue factor of safety of the design using the DE-Gerber criterion.
(*c*) Plot the designer's fatigue diagram for part (*a*).

Solution   (*a*) $D/d = 1.65/1.100 - 1.50, r/d = 0.11/1.100 = 0.10, K_t = 1.68$ (Fig. A–15–9), $K_{ts} = 1.42$ (Fig. A–15–8). Then

from Eq. (7–35),

$$K_f = \frac{1.68}{1 + \dfrac{2(1.68 - 1)}{1.68}\dfrac{4/105}{\sqrt{0.11}}} = 1.54$$

$$K_{fs} = \frac{1.42}{1 + \dfrac{2(1.42 - 1)}{1.42}\dfrac{4/105}{\sqrt{0.11}}} = 1.33$$

Eq. (7–8): $S_e' = 0.504(105) = 52.92$ kpsi

Eq. (7–18): $k_a = 2.70(105)^{-0.265} = 0.787$

Eq. (7–19): $k_b = \left(\dfrac{1.100}{0.30}\right)^{-0.107} = 0.870$

$k_c = k_d = k_f = 1$

Table 7–7: $k_e = 0.814$

$$S_e = 0.787(0.870)0.814(52.92) = 29.49 \text{ kpsi}$$

$$r = \frac{2}{\sqrt{3}}\frac{1.54}{1.33}\frac{1260}{1100} = 1.53$$

Answer   Eq. (18–22): $\dfrac{1}{n} = \dfrac{16}{\pi(1.1)^3}\left\{4\left[\dfrac{1.54(1260)}{29\,490}\right]^2 + 3\left[\dfrac{1.33(1100)}{82\,000}\right]^2\right\}^{1/2} = 0.517$

$n = 1.93$

(b) From Eq. (18–17),

$$\frac{1}{n} = \frac{16(1.54)1260}{\pi(1.1^3)29\,490}\left\{1 + \sqrt{1 + 3\left[\frac{1.33(1100)29\,490}{1.54(1260)105\,000}\right]}\right\} = 0.574$$

Answer

$$n = 1.74$$

(c)

$$\sigma_a' = \frac{32(1.54)1260}{\pi(1.1)^3} = 14\,850 \text{ psi}$$

$$\sigma_m' = \frac{16\sqrt{3}(1.33)1100}{\pi(1.1)^3} = 9696 \text{ psi}$$

Eqs. (18–24):

$$S_a = \frac{2(82\,000)29\,490^2}{29\,490^2 + 82\,000^2} = 18\,780 \text{ psi}$$

$$S_m = 82\,000 - 18\,780 = 63\,220 \text{ psi}$$

$$r_{\text{crit}} = \frac{18\,780}{63\,220} = 0.297$$

The designer's fatigue diagram is seen in Fig. 18–12. The fatigue locus is plotted from $(S_a/29.5)^2 + (S_m/82)^2 = 1$.

As seen in Ex. 18–3, the DE-ASME-elliptic has a complication that involves the use of S_y rather than S_{ut}. A direct comparison with other methods is not possible without knowledge of S_y/S_{ut}. In the domain of most power transmission shafts, the ASME-elliptic uses a fatigue-failure curve that is practically coincident with the Gerber curve on the left portion of the curve where most power transmission shafts have their operating point on the designer's fatigue diagram. In this domain one can use either the DE-Gerber or the DE-ASME-elliptic with about the same results.

If one writes the Gerber failure criterion as

$$\frac{S_a}{S_e} + \left(\frac{S_m}{S_{ut}}\right)^2 = 1$$

and the ASME-elliptic failure criterion as

$$\left(\frac{S_a}{S_e}\right)^2 + \left(\frac{S_m}{S_{ut}}\frac{S_{ut}}{S_y}\right)^2 = 1$$

and then equates S_a/S_e from both expressions, the resulting equation is

$$\left(\frac{S_m}{S_{ut}}\right)^4 + \left(\frac{S_m}{S_{ut}}\right)^2\left[\left(\frac{S_{ut}}{S_y}\right)^2 - 2\right] = 0$$

from which the S_m coordinates of the intersecting points of the two curves are

$$\frac{S_m}{S_{ut}} = 0 \quad \text{and} \quad \frac{S_m}{S_{ut}} = \sqrt{2 - \left(\frac{S_{ut}}{S_y}\right)^2}$$

The first intersection occurs at $S_a = S_e$ and $S_m = 0$. The second occurs if $S_{ut}/S_y < \sqrt{2}$. In other words, if S_{ut}/S_m is less than $\sqrt{2}$, the ASME locus is above the Gerber, then it crosses at the given nonzero abscissa. If S_{ut}/S_y is $> \sqrt{2}$, then there is no (real) crossing and the ASME-elliptic is below the Gerber for $S_m > 0$. For Ex. 18–3, $S_{ut}/S_y = 105/82 = 1.28$, which is less than $\sqrt{2}$. Thus, the ASME curve is above the Gerber, curve explaining why the factor of safety was larger for the ASME than the Gerber in the example.

Combined Variable Bending and Variable Torsion

When tests are carried out with variable bending stress σ_a and variable torsional stress τ_a, in phase, the results plot as shown in Fig. 18–13b. For the normalized coordinates of the figure the data fall near a circular arc, which would be an elliptical arc on a $\sigma_a \tau_a$ diagram. The test failure ellipse is expressed as

$$\left(\frac{\sigma_a}{S_e}\right)^2 + \left(\frac{\tau_a}{S_{se}}\right)^2 = \frac{1}{n^2}$$

Substituting $\sigma_a = 32M_a/(\pi d^3)$, $\tau_a = 16T_a/(\pi d^3)$, $S_{se} = S_e/\sqrt{3}$ and including the fatigue stress-concentration factors K_f and K_{fs} multiplicatively associated with their

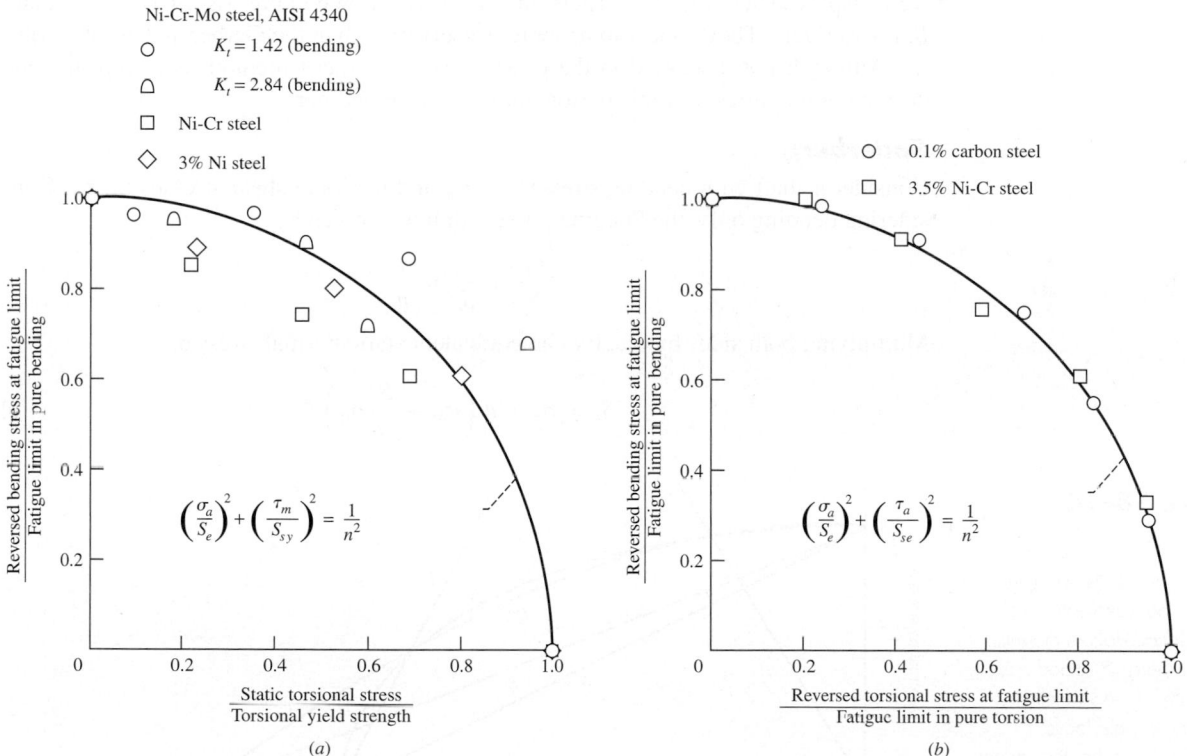

Figure 18–13

Normalized designer's fatigue-failure diagrams with test data for (a) reversed bending and steady torsion (DE-elliptic model) and (b) reversed bending and in-phase reversed torsion. (*Source: ANSI/ASME Standard B106.1.M-1985, second printing, "Design of Transmission Shafting."*)

nominal stresses gives diameter d as

$$d = \left\{ \frac{16n}{\pi S_e} \left[4(K_f M_a)^2 + 3(K_{fs} T_a)^2 \right]^{1/2} \right\}^{1/3} \tag{18-25}$$

Also, the solution for $1/n$ gives

$$\frac{1}{n} = \frac{16}{\pi d^3 S_e} \left[4(K_f M_a)^2 + 3(K_{fs} T_a)^2 \right]^{1/2} \tag{18-26}$$

These same equations can be obtained by simplifying DE-elliptic Eqs. (18–19) and (18–20), as well as by simplifying DE-Gerber Eqs. (18–13) and (18–14) with $B = 0$. Both fatigue equations agree exactly because $r \to \infty$. Equations (18–13) and (18–14) for DE-Gerber and Eqs. (18–19) and (18–20) for DE-elliptic are general cases that can be simplified for a particular application.

18-4 Strength Constraints—Additional Methods

In the previous section, the DE-Gerber and the ASME-elliptic fatigue failure criteria were featured only because they are regarded to be the most accurate in predicting failure. Figure 18–14 shows a number of fatigue-failure curves plotted to scale for the same material. The curves labeled C and D are the DE-Gerber and DE-elliptic fatigue-failure curves. Note that the Gerber curve crosses the Langer curve once, and the elliptic curve crosses the Langer curve twice. The equations for the Soderberg and Goodman were displayed in Table 18–1. Their curves are also shown in Fig. 18–14 as curves A and B, respectively. They appear to be more conservative than the Gerber and elliptic criteria. Although not discussed in the previous section, these methods are used in some design communities. For this reason, they are presented here.

Soderberg

Consider a shaft with bending stresses σ_a, σ_m and torsional shear stresses τ_a, τ_m. Considering bending only, the Soderberg strength line is given by

$$\frac{\sigma_m}{S_y} + \frac{\sigma_a}{S_e} = \frac{1}{n}$$

Multiplying both sides by S_y gives an equivalent static normal stress σ_e,

$$S_y = \sigma_e = n \left(\sigma_m + \frac{S_y}{S_e} \sigma_a \right) \tag{a}$$

Figure 18–14

A designer's fatigue diagram plotted to scale showing the relations between some well-known fatigue-failure models. A, Soderberg; B, Goodman; C, Gerber; D, ASME-elliptic; E, first-cycle yield locus; F, load line. Note that the Langer locus, curve E, makes a 45° angle with the abscissa *only* when the ordinate and abscissa scales are identical.

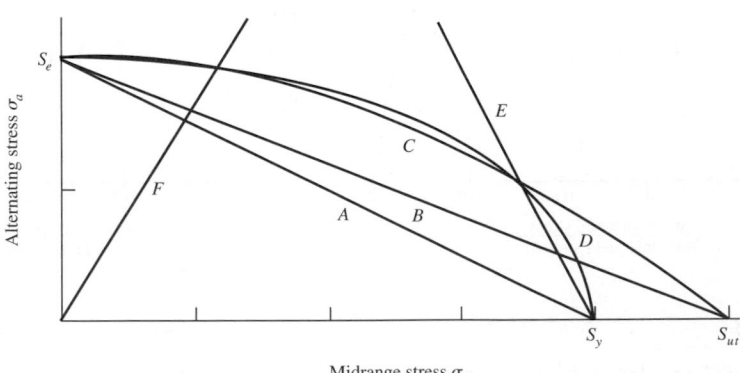

Repeating this considering torsion only gives

$$\tau_e = n \left(\tau_m + \frac{S_y}{S_e} \tau_a \right) \tag{b}$$

where it is assumed that $S_{sy}/S_{se} = S_y/S_e$. The maximum shear stress for an element containing a normal and shear stress is given by

$$\tau_{\max} = \sqrt{\left(\frac{\sigma_e}{2} \right)^2 + \tau_e^2} = \frac{1}{2} \sqrt{\sigma_e^2 + 4\tau_e^2} \tag{c}$$

Failure, according to the maximum-shear-stress theory (MSS), occurs when $\tau_{\max} = S_y/2$. Putting this together with Eqs. (a), (b), and (c) and rearranging gives

$$\left[\left(\sigma_m + \frac{S_y}{S_e} \sigma_a \right)^2 + 4 \left(\tau_m + \frac{S_y}{S_e} \tau_a \right)^2 \right]^{1/2} = \frac{S_y}{n} \tag{d}$$

For a solid circular shaft in bending and torsion, $\sigma = 32 K_f M/(\pi d^3)$, and $\tau = 16 K_{fs} T/(\pi d^3)$. Using these relations and dividing Eq. (d) by S_y gives

$$\frac{1}{n} = \frac{32}{\pi d^3} \left[K_f^2 \left(\frac{M_m}{S_y} + \frac{M_a}{S_e} \right)^2 + K_{fs}^2 \left(\frac{T_m}{S_y} + \frac{T_a}{S_e} \right)^2 \right]^{1/2} \tag{18-27}$$

or, if solving for d,

$$d = \left\{ \frac{32n}{\pi} \left[K_f^2 \left(\frac{M_m}{S_y} + \frac{M_a}{S_e} \right)^2 + K_{fs}^2 \left(\frac{T_m}{S_y} + \frac{T_a}{S_e} \right)^2 \right]^{1/2} \right\}^{1/3} \tag{18-28}$$

Equation (18–28) is sometimes referred to (perhaps in a less complete form) as the *Westinghouse code formula*. Again, a common problem occurs when $M_m = T_a = 0$, and Eqs. (18–27) and (18–28) become

$$\frac{1}{n} = \frac{32}{\pi d^3} \left[\left(K_f \frac{M_a}{S_e} \right)^2 + \left(K_{fs} \frac{T_m}{S_y} \right)^2 \right]^{1/2} \tag{18-29}$$

$$d = \left\{ \frac{32n}{\pi} \left[\left(K_f \frac{M_a}{S_e} \right)^2 + \left(K_{fs} \frac{T_m}{S_y} \right)^2 \right]^{1/2} \right\}^{1/3} \tag{18-30}$$

Equations (18–27) to (18–30) are based on a maximum-shear-stress failure approach. If the distortion-energy theory (DE) is used in place of Eq. (c), Eqs. (18–27) to (18–30) are modified by changing the 32 to 48.

DE-Goodman

The alternating and mean von Mises stresses are given by

$$\sigma_a' = (\sigma_a^2 + 3\tau_a^2)^{1/2} = \left[\left(\frac{32 K_f M_a}{\pi d^3} \right)^2 + 3 \left(\frac{16 K_{fs} T_a}{\pi d^3} \right)^2 \right]^{1/2} \tag{e}$$

$$\sigma_m' = (\sigma_m^2 + 3\tau_m^2)^{1/2} = \left[\left(\frac{32 K_f M_m}{\pi d^3} \right)^2 + 3 \left(\frac{16 K_{fs} T_m}{\pi d^3} \right)^2 \right]^{1/2} \tag{f}$$

The Goodman criterion is given by

$$\frac{\sigma_a'}{S_e} + \frac{\sigma_m'}{S_{ut}} = \frac{1}{n} \tag{g}$$

Substitution of Eqs. (e) and (f) results in

$$\frac{1}{n} = \frac{16}{\pi d^3} \left\{ \frac{1}{S_e} \left[4(K_f M_a)^2 + 3(K_{fs} T_a)^2 \right]^{1/2} \right. $$
$$\left. + \frac{1}{S_{ut}} \left[4(K_f M_m)^2 + 3(K_{fs} T_m)^2 \right]^{1/2} \right\} \tag{18-31}$$

or for d,

$$d = \left(\frac{16n}{\pi} \left\{ \frac{1}{S_e} \left[4(K_f M_a)^2 + 3(K_{fs} T_a)^2 \right]^{1/2} \right. \right.$$
$$\left. \left. + \frac{1}{S_{ut}} \left[4(K_f M_m)^2 + 3(K_{fs} T_m)^2 \right]^{1/2} \right\} \right)^{1/3} \tag{18-32}$$

For the case $M_m = T_a = 0$, Eqs. (18–31) and (18–32) become

$$\frac{1}{n} = \frac{16}{\pi d^3} \left(2 \frac{K_f M_a}{S_e} + \sqrt{3} \frac{K_{fs} T_m}{S_{ut}} \right) \tag{18-33}$$

$$d = \left[\frac{16n}{\pi} \left(2 \frac{K_f M_a}{S_e} + \sqrt{3} \frac{K_{fs} T_m}{S_{ut}} \right) \right]^{1/3} \tag{18-34}$$

EXAMPLE 18–4

Given $M_m = T_a = 0$, $M_a = 1260$ lbf · in, $T_m = 1100$ lbf · in, $S_{ut} = 105$ kpsi, $S_y = 82$ kpsi, and a fully corrected $S_e = 36$ kpsi, $n = 2$, $K_f = 1.73$, and $K_{fs} = 1.31$, determine d using the following fatigue-failure criteria:
(a) MSS-Soderberg.
(b) DE-Goodman.
(c) ASME-elliptic.
(d) DE-Gerber.

Solution

(a) From Eq. (18–31),

Answer

$$d = \left\{ \frac{32n}{\pi} \left[\left(K_f \frac{M_a}{S_e} \right)^2 + \left(K_{fs} \frac{T_m}{S_y} \right)^2 \right]^{1/2} \right\}^{1/3}$$

$$= \left\{ \frac{32(2)}{\pi} \left[\left(1.73 \frac{1260}{36\,000} \right)^2 + \left(1.31 \frac{1100}{82\,000} \right)^2 \right]^{1/2} \right\}^{1/3} = 1.087 \text{ in}$$

(b) From Eq. (18–34),

Answer

$$d = \left[\frac{16n}{\pi} \left(2 \frac{K_f M_a}{S_e} + \sqrt{3} \frac{K_{fs} T_m}{S_{ut}} \right) \right]^{1/3}$$

$$= \left\{ \frac{16(2)}{\pi} \left[2 \frac{1.73(1260)}{36\,000} + \sqrt{3} \frac{1.31(1100)}{105\,000} \right] \right\}^{1/3} = 1.138 \text{ in}$$

(c) From Eq. (18–21),

$$d = \left\{ \frac{16n}{\pi} \left[4 \left(K_f \frac{M_a}{S_e} \right)^2 + 3 \left(K_{fs} \frac{T_m}{S_y} \right)^2 \right]^{1/2} \right\}^{1/3}$$

$$= \left\{ \frac{16(2)}{\pi} \left[4 \left(1.73 \frac{1260}{36\,000} \right)^2 + 3 \left(1.31 \frac{1100}{82\,000} \right)^2 \right]^{1/2} \right\}^{1/3} = 1.083 \text{ in}$$

(d) From Eq. (18–16),

Answer
$$d = \left(\frac{16n K_f M_a}{\pi S_e} \left\{ 1 + \left[1 + 3 \left(\frac{K_{fs} T_m S_e}{K_f M_a S_{ut}} \right)^2 \right]^{1/2} \right\} \right)^{1/3}$$

$$= \left\{ \frac{16(2)1.73(1260)}{\pi(36\,000)} \left(1 + \left[1 + 3 \left(\frac{1.31(1100)36}{1.73(1260)105} \right)^2 \right]^{1/2} \right) \right\}^{1/3} = 1.086 \text{ in}$$

Tabulated, the results are

Criterion	d, in
MSS-Soderberg	1.087
DE-Goodman	1.138
ASME-elliptic	1.083
DE-Gerber	1.086

From the results in Ex. 18–4, we see that the MSS-Soderberg, ASME-elliptic, and DE-Gerber criteria are virtually identical, whereas the DE-Goodman criterion is quite conservative. With the exception of the MSS-Soderberg, the results agree with Fig. 18–14. Curve B is quite low compared to Gerber (C) and ASME-elliptic (D). Hence, the conservative result is from Goodman. The reason that MSS-Soderberg is not more conservative than Goodman, as Fig. 18–14 appears to indicate, is that Fig. 18–14 is based on the distortion-energy theory. Using Eq. (18–30) and replacing the 32 with 48 yields $d = 1.244$ in for the DE-Soderberg criterion, which *is* more conservative than the DE-Goodman result. At first glance, one would think that the MSS-Soderberg criterion would be more conservative than the DE-Soderberg criterion since the MSS theory is more conservative than the DE theory, and it is in the way the equivalent stresses were formulated separately. Returning to Eq. (d), with the special case $\sigma_m = \tau_a = 0$, we find

$$\left[\left(\frac{S_y}{S_e} \sigma_a \right)^2 + 4\tau_m^2 \right]^{1/2} = \frac{S_y}{n} \quad \text{or} \quad \left(\frac{\sigma_a}{S_e} \right)^2 + \left(\frac{2\tau_m}{S_y} \right)^2 = \frac{1}{n^2} \quad (h)$$

Using the maximum-shear-stress theory we can approximate $2\tau_m \doteq \sigma_{em}$, where σ_{em} is an equivalent mean normal stress. Thus, we see that Eq. (h) reverts to a form similar to the ASME-elliptic equation,

$$\left(\frac{\sigma_a}{S_e}\right)^2 + \left(\frac{\sigma_{em}}{S_y}\right)^2 \doteq \frac{1}{n^2}$$

This is why, in Ex. 18–4, the two results were almost identical and the MSS-Soderberg method is an acceptable criterion to use.

18–5 Shaft Materials

Various steels have comparable Young moduli. For that reason, rigidity cannot be controlled by material decisions, but only by geometric decisions. Necessary strength to resist loading stresses affects the choice of materials and their treatments. ANSI 1020–1050 steels are a common choice, as is a 11xx free-machining steel. Heat treating increases expense; if used, include 1340-50, 3140-50, 4140, 4340, 5140, and 8650. It should be used only if there is no other way to obtain necessary strengths. Carburizing grades of 1020, 4320, 4820, and 8620 are chosen when surface hardness is important.

Cold-rolled sections are available to about $3\frac{1}{2}$ in in diameter and hot-rolled sections up to 6 in can be obtained. Larger sizes require forging before machining.

In approaching material selection, the amount to be produced is a salient factor. For low production, turning is the usual primary shaping process. An economic viewpoint may require removing the least material. High production may permit a volume-conservative shaping method (hot or cold forming, casting) and minimum material in the shaft can become a design goal. Properties of the shaft locally depend on its history: cold work, cold forming, rolling of fillet features, heat treatment, including quenching medium, agitation, and tempering regimen.[4]

18–6 Hollow Shafts

A hollow shaft can reduce weight and allow fluids to circulate to moving parts for cooling or lubrication. Thick-walled seamless tubing is available for simpler, smaller shafts.

- Thick-walled tubing may not have sufficient wall material for shouldering. Split-ring shaft collars, tapered hubs, and other less common methods of locating and fastening may have to be used.

- In hollow shafting, whether the geometry of the tube was created by a drawing process, drilling, or boring, it will have to be checked for balance and corrected if necessary.

The equations developed in this chapter can be conveniently modified to apply to the hollow shaft. If the outer diameter is d_o and the inner diameter is d_i, define $K = d_i/d_o$. For torsional and bending loading the quantity $d_o(1 - K^4)^{1/3}$ may be substituted for the diameter d in Eqs. (18–12) through (18–21), for example, as applicable. The algebra of the equations allows explicit solutions for outside diameter d_o when K is known. However, when d_i is fixed, then K is not known, and iterative procedures are necessary to find the outside diameter d_o.

---

[4]See Joseph E. Shigley and Charles R. Mischke (eds.-in-chief), *Standard Handbook of Machine Design,* 2nd ed., McGraw-Hill, New York, 1996. For cold-worked property prediction see Chap. 8, and for heat-treated property prediction see Chaps. 8 and 13.

18–7   **Critical Speeds**

When a shaft is turning, eccentricity causes a centrifugal force deflection, which is resisted by the shaft's flexural rigidity EI. As long as deflections are small, no harm is done. Another potential problem, however, is called *critical speeds:* at certain speeds the shaft is unstable, with deflections increasing without upper bound. It is fortunate that although the dynamic deflection shape is unknown, using a static deflection curve gives an excellent estimate of the lowest critical speed. Such a curve meets the boundary condition of the differential equation (zero moment and deflection at both bearings) and the shaft energy is not particularly sensitive to the exact shape of the deflection curve. Designers seek first critical speeds at least twice the operating speed.

The shaft, because of its own mass, has a critical speed. The ensemble of attachments to a shaft likewise has a critical speed that is much lower than the shaft's intrinsic critical speed. Estimating these critical speeds (and harmonics) is a task of the designer. When geometry is simple, as in a shaft of uniform diameter, simply supported, the task is easy. It can be expressed[5] as

$$\omega_1 = \left(\frac{\pi}{l}\right)^2 \sqrt{\frac{EI}{m}} = \left(\frac{\pi}{l}\right)^2 \sqrt{\frac{gEI}{A\gamma}} \tag{18–35}$$

where m is the mass per unit length, A the cross-sectional area, and γ the specific weight. For an ensemble of attachments, Rayleigh's method for lumped masses gives[6]

$$\omega_1 = \sqrt{\frac{g\sum w_i y_i}{\sum w_i y_i^2}} \tag{18–36}$$

where w_i is the weight of the ith location and y_i is the deflection at the ith body location. It is possible to use Eq. (18–36) for the case of Eq. (18–35) by partitioning the shaft into segments and placing its weight force at the segment centroid as seen in Fig. 18–15. Computer assistance is often used to lessen the difficulty in finding transverse deflections of a stepped shaft. Rayleigh's equation overestimates the critical speed.

Figure 18–15

(a) A uniform-diameter shaft for Eq. (18–39). (b) A segmented uniform-diameter shaft for Eq. (18–40).

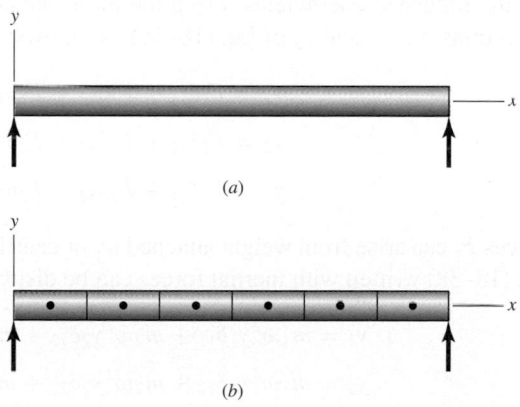

(a)

(b)

---

[5]William T. Thomson and Marie Dillon Dahleh, *Theory of Vibration with Applications,* Prentice Hall, 5th ed., 1998, p. 273.

[6]Thomson, op. cit., p. 357.

Figure 18–16

The influence coefficient δ_{ij} is the deflection at i due to a unit load at j.

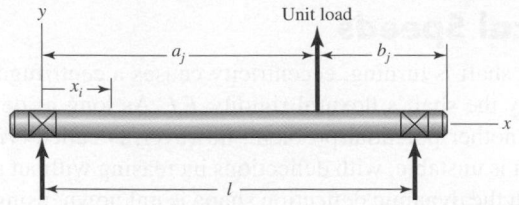

To counter the increasing complexity of detail, we adopt a useful viewpoint. Inasmuch as the shaft is an elastic body, we can use *influence coefficients*. An influence coefficient is the transverse deflection at location i on a shaft due to a unit load at location j on the shaft. From Table A–9–6 we obtain, for a simply supported beam with a single unit load as shown in Fig. 18–16,

$$\delta_{ij} = \begin{cases} \dfrac{b_j x_i}{6EIl}(l^2 - b_j^2 - x_i^2) & x_i \le a_i \\[3mm] \dfrac{a_j(l - x_i)}{6EIl}(2lx_i - a_j^2 - x_i^2) & x_i > a_i \end{cases} \tag{18–37}$$

For three loads the influence coefficients may be displayed as

i	i		
	1	**2**	**3**
1	δ_{11}	δ_{12}	δ_{13}
2	δ_{21}	δ_{22}	δ_{23}
3	δ_{31}	δ_{32}	δ_{33}

Maxwell's reciprocity theorem[7] states that there is a symmetry about the main diagonal, composed of δ_{11}, δ_{22}, and δ_{33}, of the form $\delta_{ij} = \delta_{ji}$. This relation reduces the work of finding the influence coefficients. From the influence coefficients above, one can find the deflections y_1, y_2, and y_3 of Eq. (18–36) as follows:

$$\begin{aligned} y_1 &= F_1\delta_{11} + F_2\delta_{12} + F_3\delta_{13} \\ y_2 &= F_1\delta_{21} + F_2\delta_{22} + F_3\delta_{23} \\ y_3 &= F_1\delta_{31} + F_2\delta_{32} + F_3\delta_{33} \end{aligned} \tag{18–38}$$

The forces F_i can arise from weight attached w_i or centrifugal forces $m_i\omega^2 y_i$. The equation set (18–38) written with inertial forces can be displayed as

$$\begin{aligned} y_1 &= m_1\omega^2 y_1\delta_{11} + m_2\omega^2 y_2\delta_{12} + m_3\omega^2 y_3\delta_{13} \\ y_2 &= m_1\omega^2 y_1\delta_{21} + m_2\omega^2 y_2\delta_{22} + m_3\omega^2 y_3\delta_{23} \\ y_3 &= m_1\omega^2 y_1\delta_{31} + m_2\omega^2 y_2\delta_{32} + m_3\omega^2 y_3\delta_{33} \end{aligned}$$

---

[7]Thomson, op. cit., p. 167.

which can be rewritten as

$$(m_1\delta_{11} - 1/\omega^2)y_1 + (m_2\delta_{12})y_2 + (m_3\delta_{13})y_3 = 0$$

$$(m_1\delta_{21})y_1 + (m_2\delta_{22} - 1/\omega^2)y_2 + (m_3\delta_{23})y_3 = 0 \qquad (a)$$

$$(m_1\delta_{31})y_1 + (m_2\delta_{32})y_2 + (m_3\delta_{33} - 1/\omega^2)y_3 = 0$$

Equation set (a) is three simultaneous equations in terms of y_1, y_2, and y_3. To avoid the trivial solution $y_1 = y_2 = y_3 = 0$, the determinant of the coefficients of y_1, y_2, and y_3 must be zero (eigenvalue problem). Thus,

$$\begin{vmatrix} (m_1\delta_{11} - 1/\omega^2) & m_2\delta_{12} & m_3\delta_{13} \\ m_1\delta_{21} & (m_2\delta_{22} - 1/\omega^2) & m_3\delta_{23} \\ m_1\delta_{31} & m_2\delta_{32} & (m_3\delta_{33} - 1/\omega^2) \end{vmatrix} = 0 \qquad (18\text{–}39)$$

which says that a deflection other than zero exists only at three distinct values of ω, the critical speeds. Expanding the determinant, we obtain

$$\left(\frac{1}{\omega^2}\right)^3 - (m_1\delta_{11} + m_2\delta_{22} + m_3\delta_{33})\left(\frac{1}{\omega^2}\right)^2 + \cdots = 0 \qquad (18\text{–}40)$$

The three roots of Eq. (18–40) can be expressed as $1/\omega_1^2$, $1/\omega_2^2$, and $1/\omega_3^2$. Thus Eq. (18–40) can be written in the form

$$\left(\frac{1}{\omega^2} - \frac{1}{\omega_1^2}\right)\left(\frac{1}{\omega^2} - \frac{1}{\omega_2^2}\right)\left(\frac{1}{\omega^2} - \frac{1}{\omega_3^2}\right) = 0$$

or

$$\left(\frac{1}{\omega^2}\right)^3 - \left(\frac{1}{\omega_1^2} + \frac{1}{\omega_2^2} + \frac{1}{\omega_3^2}\right)\left(\frac{1}{\omega^2}\right)^2 + \cdots = 0 \qquad (18\text{–}41)$$

Comparing Eqs. (18–40) and (18–41) we see that

$$\frac{1}{\omega_1^2} + \frac{1}{\omega_2^2} + \frac{1}{\omega_3^2} = m_1\delta_{11} + m_2\delta_{22} + m_3\delta_{33} \qquad (18\text{–}42)$$

If we had only a single mass m_1 alone, the critical speed would be given by $1/\omega^2 = m_1\delta_{11}$. Denote this critical speed as ω_{11} (which considers only m_1 acting alone). Likewise for m_2 or m_3 acting alone, we similarly define the terms $1/\omega_{22}^2 = m_2\delta_{22}$ or $1/\omega_{33}^2 = m_3\delta_{33}$, respectively. Thus, Eq. (18–42) can be rewritten as

$$\frac{1}{\omega_1^2} + \frac{1}{\omega_2^2} + \frac{1}{\omega_3^2} = \frac{1}{\omega_{11}^2} + \frac{1}{\omega_{22}^2} + \frac{1}{\omega_{33}^2} \qquad (18\text{–}43)$$

If we order the critical speeds such that $\omega_1 < \omega_2 < \omega_3$, then $1/\omega_1^2 \gg 1/\omega_2^2$, and $1/\omega_3^2$. So the first, or fundamental, critical speed ω_1 can be approximated by

$$\frac{1}{\omega_1^2} \doteq \frac{1}{\omega_{11}^2} + \frac{1}{\omega_{22}^2} + \frac{1}{\omega_{33}^2} \qquad (18\text{–}44)$$

This idea can be extended to an n-body shaft:

$$\frac{1}{\omega_1^2} \doteq \sum_{1=1}^{n} \frac{1}{\omega_{ii}^2} \qquad (18\text{–}45)$$

This is called *Dunkerley's equation*. By ignoring the higher mode term(s), the first critical speed estimate is *lower* than actually is the case.

Since Eq. (18–45) has no loads appearing in the equation, it follows that if each load could be placed at some convenient location transformed into an equivalent load, then the critical speed of an array of loads could be found by summing the equivalent loads, all placed at a single convenient location. For the load at station 1, placed at the center of span, denoted with the subscript c, the equivalent load is found from

$$\omega_{11}^2 = \frac{g}{w_1 \delta_{11}} = \frac{g}{w_{1c} \delta_{cc}}$$

or

$$w_{1c} = w_1 \frac{\delta_{11}}{\delta_{cc}} \tag{18–46}$$

EXAMPLE 18–5

Consider a simply supported steel shaft as depicted in Fig. 18–17, with 1 in diameter and a 31-in span between bearings, carrying two gears weighing 35 and 55 lbf.
(*a*) Find the influence coefficients.
(*b*) Find $\sum wy$ and $\sum wy^2$ and the first critical speed using Rayleigh's equation, Eq. (18–36).
(*c*) From the influence coefficients, find ω_{11} and ω_{22}.
(*d*) Using Dunkerley's equation, Eq. (18–45), estimate the first critical speed.
(*e*) Use superposition to estimate the first critical speed.
(*f*) Estimate the shaft's intrinsic critical speed. Suggest a modification to Dunkerley's equation to include the effect of the shaft's mass on the first critical speed of the attachments.

Solution   (*a*)

$$I = \frac{\pi d^4}{64} = \frac{\pi (1)^4}{64} = 0.049\ 09\ \text{in}^4$$

$$6EIl = 6(30)10^6(0.049\ 09)31 = 0.2739(10^9)\ \text{lbf} \cdot \text{in}^3$$

Figure 18–17

(*a*) A 1-in uniform-diameter shaft for Ex. 18–5.
(*b*) Superposing of equivalent loads at the center of the shaft for the purpose of finding the first critical speed.

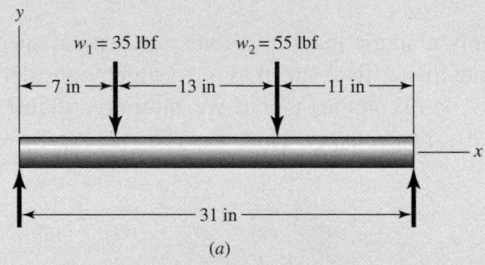

$w_1 = 35$ lbf $w_2 = 55$ lbf

7 in — 13 in — 11 in

31 in

(*a*)

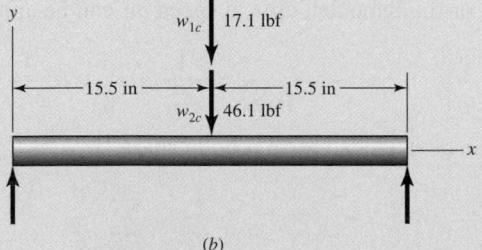

w_{1c} 17.1 lbf

15.5 in — 15.5 in

w_{2c} 46.1 lbf

(*b*)

From Eq. set (18–37),

$$\delta_{11} = \frac{24(7)(31^2 - 24^2 - 7^2)}{0.2739(10^9)} = 2.061(10^{-4}) \text{ in/lbf}$$

$$\delta_{22} = \frac{11(20)(31^2 - 11^2 - 20^2)}{0.2739(10^9)} = 3.534(10^{-4}) \text{ in/lbf}$$

$$\delta_{12} = \delta_{21} = \frac{11(7)(31^2 - 11^2 - 7^2)}{0.2739(10^9)} = 2.224(10^{-4}) \text{ in/lbf}$$

Answer

i	i	
	1	**2**
1	2.061(10⁻⁴)	2.234(10⁻⁴)
2	2.234(10⁻⁴)	3.534(10⁻⁴)

$$y_1 = w_1\delta_{11} + w_2\delta_{12} = 35(2.061)10^{-4} + 55(2.234)10^{-4} = 0.01950 \text{ in}$$

$$y_2 = w_1\delta_{21} + w_2\delta_{22} = 35(2.234)10^{-4} + 55(3.534)10^{-4} = 0.02726 \text{ in}$$

(b)

$$\sum w_i y_i = 35(0.01950) + 55(0.02726) = 2.181 \text{ lbf} \cdot \text{in}$$

Answer

$$\sum w_i y_i^2 = 35(0.01950)^2 + 55(0.02726)^2 = 0.05418 \text{ lbf} \cdot \text{in}^2$$

Answer

$$\omega = \sqrt{\frac{386.1(2.181)}{0.05418}} = 124.7 \text{ rad/s, or 1191 rev/min}$$

(c)

Answer

$$\frac{1}{\omega_{11}^2} = \frac{w_1}{g}\delta_{11}$$

$$\omega_{11} = \sqrt{\frac{g}{w_1\delta_{11}}} = \sqrt{\frac{386.1}{35(2.061)10^{-4}}} = 231.4 \text{ rad/s, or 2210 rev/min}$$

Answer

$$\omega_{22} = \sqrt{\frac{g}{w_2\delta_{22}}} = \sqrt{\frac{386.1}{55(3.534)10^{-4}}} = 140.9 \text{ rad/s, or 1346 rev/min}$$

(d)

$$\frac{1}{\omega_1^2} \doteq \sum \frac{1}{\omega_{ii}^2} = \frac{1}{231.4^2} + \frac{1}{140.9^2} = 6.905(10^{-5}) \tag{1}$$

Answer

$$\omega_1 \doteq \sqrt{\frac{1}{6.905(10^{-5})}} = 120.3 \text{ rad/s, or 1149 rev/min}$$

which is less than part b, as expected.
(e) From Eq. (18–37),

$$\delta_{cc} = \frac{b_{cc}x_{cc}(l^2 - b_{cc}^2 - x_{cc}^2)}{6EIl} = \frac{15.5(15.5)(31^2 - 15.5^2 - 15.5^2)}{0.2739(10^9)}$$

$$= 4.215(10^{-4}) \text{ in/lbf}$$

From Eq. (18–46),

$$w_{1c} = w_1 \frac{\delta_{11}}{\delta_{cc}} = 35 \frac{2.061(10^{-4})}{4.215(10^{-4})} = 17.11 \text{ lbf}$$

$$w_{2c} = w_2 \frac{\delta_{22}}{\delta_{cc}} = 55 \frac{3.534(10^{-4})}{4.213(10^{-4})} = 46.11 \text{ lbf}$$

Answer $$\omega = \sqrt{\frac{g}{\delta_{cc} \sum w_{ic}}} = \sqrt{\frac{386.1}{4.215(10^{-4})(17.11 + 46.11)}} = 120.4 \text{ rad/s, or } 1150 \text{ rev/min}$$

which, except for rounding, agrees with part d, as expected.

(f) For the shaft, $E = 30(10^6)$ psi, $\gamma = 0.282$ lbf/in^3, and $A = \pi(1^2)/4 = 0.7854$ in^2. Considering the shaft alone, the critical speed, from Eq. (18–35), is

Answer $$\omega_s = \left(\frac{\pi}{l}\right)^2 \sqrt{\frac{gEI}{A\gamma}} = \left(\frac{\pi}{31}\right)^2 \sqrt{\frac{386.1(30)10^6(0.049\ 09)}{0.7854(0.282)}}$$

$$= 520.4 \text{ rad/s, or } 4970 \text{ rev/min}$$

We can simply add $1/\omega_s^2$ to the right side of Dunkerley's equation, Eq. (1), to include the shaft's contribution,

Answer $$\frac{1}{\omega_1^2} \doteq \frac{1}{520.4^2} + 6.905(10^{-5}) = 7.274(10^{-5})$$

$$\omega_1 \doteq 117.3 \text{ rad/s, or } 1120 \text{ rev/min}$$

which is slightly less than part d, as expected.

The shaft's first critical speed ω_s is just one more single effect to add to Dunkerley's equation. Since it does not fit into the summation, it is usually written up front.

Answer $$\frac{1}{\omega_1^2} \doteq \frac{1}{\omega_s^2} + \sum_{i=1}^{n} \frac{1}{\omega_{ii}^2} \qquad (18\text{–}47)$$

Common shafts are complicated by the stepped-cylinder geometry, which makes the influence-coefficient determination part of a numerical solution.

18–8 Shaft Design

Many shaft designs are modifications to existing shafts. The modification results from minor changes in bulk geometry of speed reducers, small changes in gear ratio, substitution of helical gears for spur gears or herringbone gears for helical gears. The point is that the designer has a head start. He or she knows the previous shaft's geometry and has a feel for the general size of the new design. The shaft to be designed is nebulous from the beginning. Moreover, other shaft designs are first-time efforts and no approximation is available to help in the visualization.

A shaft embodies a large number of decisions. To view all these decisions as "open" is to create a design space that cannot be easily visualized. As a result, a designer's most valuable tool—geometric thinking—is severely hampered.

An approach for the designer is to come closer to the position of the engineer who has to modify an existing design and has an approximate picture of the result at the outset. Since the active constraint is likely to be from deformation, consider deformation first.

- Find the diameter of the uniform-diameter shaft that meets the deflection and slopes at the bearings and at the power transmission elements. Equations (18–1) and (18–2) will be helpful. Finding that diameter will give you an average size for the shaft. Bearing seats and shoulders will create geometric features at the bearing locations. If the bearing locations are "outboard" of the power transmission features, the bearing journal diameter will have little influence on deflection and slope; thus either it can be ignored, or the central diameter can be increased slightly.

- Consider the power transmission features, shoulders, and hub bores, and make some tentative decisions on step geometry (diameters and length). Now the geometry is becoming less opaque.

- Take your approximate idea of the shaft geometry and perform a deflection and slope analysis. Use Eqs. (18–3) and (18–4) to find the largest d_{new}/d_{old} ratio, then multiply *all* diameters by this ratio. See Ex. 18–2. You will have a stepped shaft that meets all the deflection and slope constraints. Check the new shoulders (especially with rolling-contact bearings) to see if shoulder height still is within manufacturers' recommended range. If adjustments are necessary, construct a new table.

- Begin a strength analysis using DE-Gerber or DE-elliptic theory. Use a desirable shaft material that will not need heat treatment. Heat treatment may increase strength, but the cost of heat treatment will increase the cost of the shaft severalfold. Examine, feature by feature, left bearing shoulder, left gear shoulder, gear keyway, right gear shoulder, right gear keyway, right bearing shoulder, shaft collar locations, snap-ring locations, for example. Examine these features for adequate diameter to see if material strength or diameter needs improvement. This feature-by-feature examination will tell the designer where the critical location is. If the critical feature is of sufficient strength, the designer has a clearer picture of the final design. The engineer may now have a satisfactory shaft. If some material can be pared off the shaft at other features (still keeping relevant constraints loose), consider this. In smaller production runs where volume-conservative initial forming methods are not used, the cost of additional turning to remove material (make chips) may mitigate against any such size reduction.

PROBLEMS

 DESIGN

18–1

The 16-tooth pinion in the figure drives a double-reduction gear train as shown. All gears have 25° pressure angle. The pinion rotates counterclockwise at 1200 rev/min and transmits 50 hp to the gear train. Our focus is on the pinion shaft. None of the shafts have been designed.

(a) For the pinion shaft develop the moment diagram.

(b) For cylindrical roller bearings the shaft slope at the journal should be less than 0.001 rad. For a design factor of $n_d = 2$ estimate the uniform shaft diameter that would meet the deflection constraints.

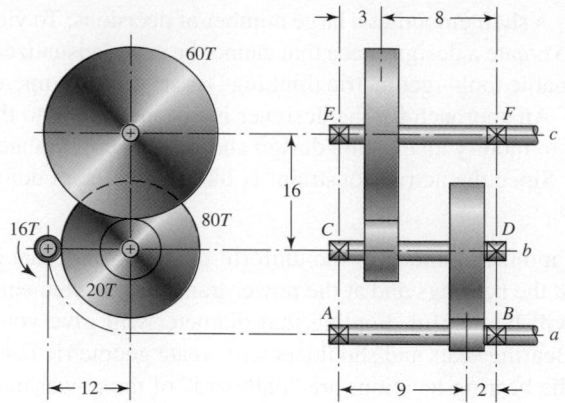

18–2 Investigate the result of Prob. 18–1 for fatigue strength in a preliminary way, whether strength or deflection controls. Use a 1030 hot-rolled steel for this purpose. The likely reliability goal for this shaft is 0.999. Assume a keyway in shaft with K_t and K_{ts} given on page 444 and a fillet radius of 0.02 in.
(a) Use the DE-elliptic fatigue-failure criterion for your factor of safety estimate.
(b) Use the DE-Gerber fatigue-failure criterion in your estimate.

18–3 Having found the pinion shaft of Probs. 18–1 and 18–2 has a tight deflection constraint, design as much of the shaft as you can. The overhang to mount a coupling upon can presume a 6-in extension of the shaft beyond the bearing. Draw your shaft, showing all dimensional decisions.

18–4 All the designs completed in response to Prob. 18–3 will differ. For the sake of discussion, consider the design shown in the figure. The designer specified 02–40-mm ball bearing on the left and an 03–40-mm cylindrical roller bearing on the right. Check this design for adequacy with respect to deformation.

1030 HR Bearing shoulder radii 0.030. Keyway fillet radii radii 0.10.

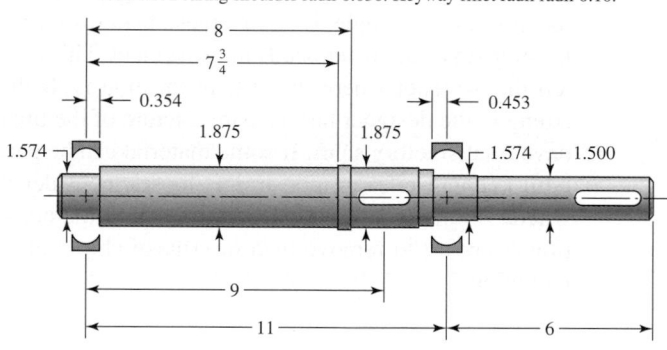

18–5 Check the design in Prob. 18–4 for adequacy in fatigue strength using the DE-elliptic fatigue-failure criterion.

18–6 Now you have the opportunity to see how your design (Prob. 18–3) compares. Using your design completed in response to Prob. 18–3, perform the adequacy checks suggested in Probs. 18–4 and 18–5.

18–7 The experience you have gained with Probs. 18–1 through 18–6 will make the design of shaft *c* of Prob. 18–1 easier. Design shaft *c* of Prob. 18–1 using a design factor of 2 with a reliability goal of 0.995.

 DESIGN **18–8** The design of the intermediate shaft *b* of the speed reducer of Prob. 18–1 is more difficult. The moment diagram does not lie in a plane. It is necessary to use moment diagrams in orthogonal planes. Design shaft *b* of Prob. 18–1 for $R = 0.995$ and a design factor of 2.

 ANALYSIS **18–9** Whether you have had the time to accomplish the design of the shafts *a*, *b*, and *c* of Prob. 18–1 or not, it is useful to consider the interrelationship to the geometry of the shafts, which the rolling-contact bearings will have. Avoid engaging in an iterative design situation unnecessarily. There will be a bearing reliability goal for the six bearings of the reducer. Plan a strategy for identifying the individual bearing reliabilities. The design work you will save will be your own.

DESIGN **18–10** A geared industrial roll shown in the figure is driven at 300 rev/min by a force *F* acting on a 3-in-diameter pitch circle as shown. The roll exerts a normal force of 30 lbf/in of roll length on the material being pulled through. The material passes under the roll. The coefficient of friction is 0.40. Develop the moment and shear diagrams for the shaft modeling the roll force as (*a*) a concentrated force at the center of the roll, and (*b*) a uniformly distributed force along the roll. These diagrams will appear on two orthogonal planes.

Problem 18–10
Material moves under the roll.
Dimensions in inches.

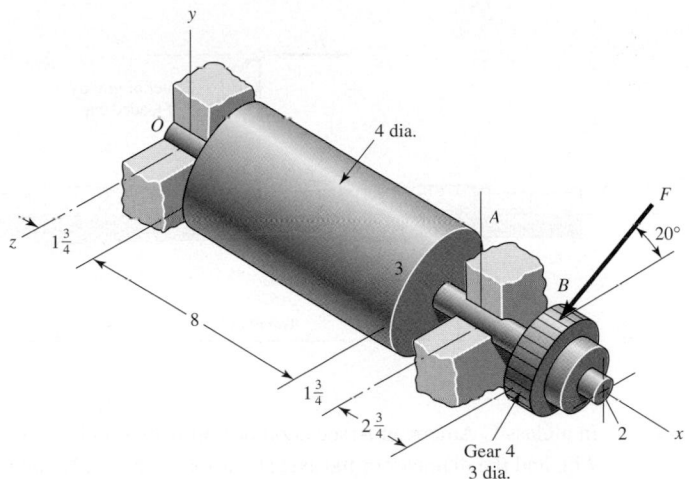

DESIGN **18–11** Using a 1035 hot-rolled steel, estimate the necessary diameters at the locations of peak bending moment using a design factor of 2. There are likely to be fillets at both ends of the right-hand bearing seat, where the bending moment is slightly less than the local extreme. Estimating the fatigue stress-concentration factor as 2, and using a design factor of 2, what is the approximate necessary diameter of the bearing seat using the DE-elliptic fatigue-failure criterion in Prob. 18–10?

DESIGN **18–12** For the situation in Prob. 18–10, what diameter uniform shaft will meet a deflection bearing slope of (*a*) 0.001 rad and (*b*) 0.0005 rad?

DESIGN **18–13** For Prob. 18–10, find the diameter of a uniform shaft to meet the slope limitation of 0.0005 rad at the gear mesh. What is the diameter to meet the slope with a factor of safety of 2?

DESIGN **18–14** Design a shaft for the situation of the industrial roll of Prob. 18–10 with a design factor of 2 and a reliability goal of 0.999 against fatigue failure. Plan for a ball bearing on the left and a cylindrical roller on the right. For deformation use a factor of safety of 2.

ANALYSIS **18–15** The figure shows a proposed design for the industrial roll shaft of Prob. 18–10. Hydrodynamic film bearings are to be used. All surfaces are machined except the journals, which are ground and polished. The material is 1035 HR steel. Perform a design assessment. Is the design satisfactory?

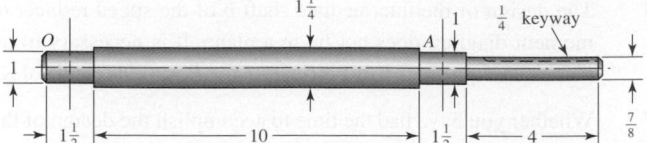

Problem 18–15
Bearing shoulder fillets 0.030 in, others $\frac{1}{16}$ in. Sled-runner keyway is $3\frac{1}{2}$ in long. Dimensions in inches.

ANALYSIS **18–16**

As shown in the figure, the axle of a railroad freight car is tapered, with the least diameter between the rails. This problem will give some insight as to why this is so. A freight-car axle is force-fitted to its wheels with the journals outboard of the wheels, and with journal centers about 80 in apart. The track gauge is $56\frac{1}{2}$ in between the rails, and so the span between the rail centers is about $59\frac{1}{2}$ in. Freight-car wheels are usually of diameter 33 in for cars up to 70-ton capacity. Consider the worst-case location of the center of mass of the car and load to be 72 in above the rails. The worst-case vertical load is to be 42 750 lbf per axle. The worst-case horizontal load due to crosswinds and track curvature is 17 100 lbf per axle, through the center of mass. Construct the bending-moment diagram for the axle.

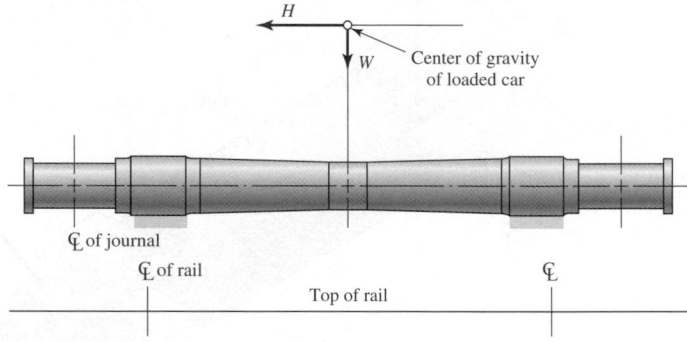

Problem 18–16
A railroad freight-car axle.

ANALYSIS **18–17**

In a class C American Association of Railroads standard axle, the diameter of the wheel seat is 7 in, and the diameter of the axle at midspan is $5\frac{3}{8}$ in. Using the results of Prob. 18–16, compare the bending stress level at the center of the wheel seat and at the axle midspan. Is this difference to be expected?

DESIGN **18–18**

The section of shaft shown in the figure is to be designed to approximate relative sizes of $d = 0.75D$ and $r = D/20$ with diameter d conforming to that of standard metric rolling-bearing bore sizes. The shaft is to be made of SAE 2340 steel, heat-treated to obtain minimum strengths in the shoulder area of 1226-MPa ultimate tensile strength and 1130-MPa yield strength with a Brinell hardness not less than 368. At the shoulder the shaft is subjected to a completely reversed bending moment of 70 N · m, accompanied by a steady torsion of 45 N · m. Use a design factor of 2.5 and size the shaft for an infinite life. The results should be based on the DE-elliptic fatigue-failure criterion.

Problem 18–18
Section of a shaft containing a grinding-relief groove. Unless otherwise specified, the diameter at the root of the groove $d_r = d - 2r$, and though the section of diameter d is ground, the root of the groove is still a machined surface.

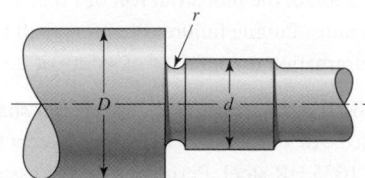

 ANALYSIS

18-19   Repeat Prob. 18–18 using the MSS-Soderberg criterion.

 ANALYSIS

18-20   The rotating solid steel shaft is simply supported by bearings at points B and C and is driven by a gear (not shown) which meshes with the spur gear at D, which has a 6-in pitch diameter. The force F from the drive gear acts at a pressure angle of 20°. The shaft transmits a torque to point A of $T_A = 3000$ lbf · in. The shaft is machined from steel with $S_y = 60$ kpsi and $S_{ut} = 80$ kpsi. Using a factor of safety of 2.5, determine the minimum allowable diameter of the shaft based on (a) a static yield analysis using the distortion energy theory and (b) a fatigue-failure analysis using the four criteria given in Secs. 18–3 and 18–4. Use fatigue stress concentration factors of $K_f = 1.8$ and $K_{fs} = 1.3$.

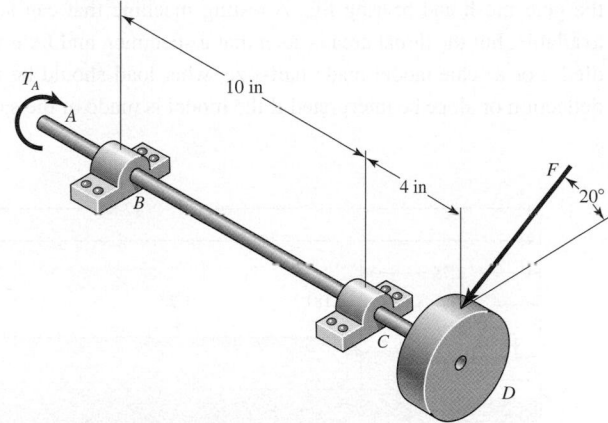

Problem 18–20

 ANALYSIS

18-21   A shaft is loaded in bending and torsion such that $M_a = 600$ lbf · in, $T_a = 400$ lbf · in, $M_m = 500$ lbf · in, and $T_m = 300$ lbf · in. For the shaft, $S_u = 100$ kpsi and $S_y = 80$ kpsi, and a fully corrected endurance limit of $S_e = 30$ kpsi is assumed. Let $K_f = 2.2$ and $K_{fs} = 1.8$. With a design factor of 2.0 determine the minimum acceptable diameter of the shaft using the
(a) DE-Gerber criterion.
(b) DE-elliptic criterion.
(c) MSS-Soderberg criterion.
(d) DE-Goodman criterion.
Discuss and compare the results.

 ANALYSIS

18-22   A transverse drilled and reamed hole can be used in a solid shaft to hold a pin that locates and holds a mechanical element, such as the hub of a gear, in axial position, and allows for the transmission of torque. Since a small-diameter hole introduces high stress concentration, and a larger-diameter hole erodes the area resisting bending and torsion, investigate the existance of a pin diameter with minimum adverse affect on the shaft. Then formulate a design rule. (*Hint:* Use Table A–16.)

 DESIGN

18-23   A slow-speed spur gear with a 1.75-in bore and a $1\frac{1}{2}$-in-long hub has a pitch diameter of 8 in and involute teeth of 20° pressure angle, and it transmits 4.5 hp at 112 rev/min. The shaft is to have a bearing span of 10 in, with gear placed 3 in from the right-hand bearing. Deep-groove ball bearings are to be used. A 2-in overhang for a coupling having a 1-in diameter seat is to be provided at the left of the left-hand bearing. For a design factor of 2, the bearing life is to be 10 kh at a reliability of 0.995 for the bearing pair. Some a priori decisions are shown in the figure. Select appropriate bearings, then dimension the shaft in the neighborhood of the bearings.

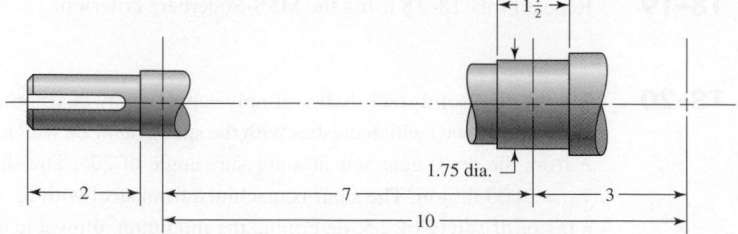

Problem 18–23
Dimensions in inches.

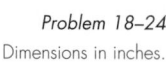

DESIGN   **18-24**   The design of Prob. 18–23 has been completed, resulting in a shaft that is $12\frac{7}{8}$ in long overall. It is important to have assurance that the deflection is within limits necessary to the functioning of the gear mesh and bearing life. A testing machine that can accurately apply a bending load is available, but the throat area is such that a specimen and fixture only up to 8 in long can be handled. For a scale model made half-size, what load should be applied? How should a measured deflection or slope be interpreted if the model is made of the same material as the prototype?

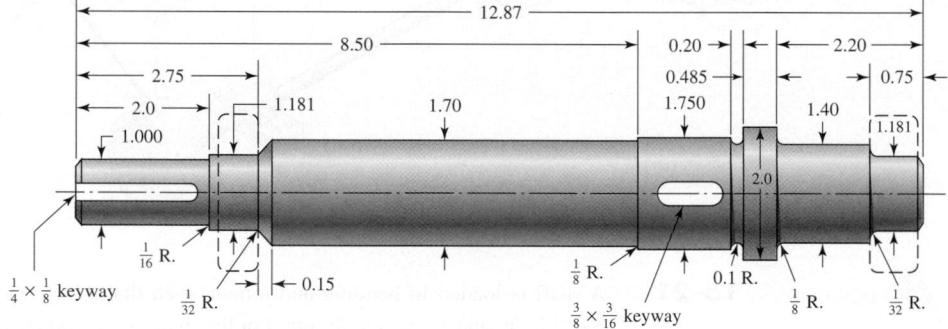

Problem 18–24
Dimensions in inches.

DESIGN
ASSESSMENT   **18-25**   An AISI 1020 cold-drawn steel shaft with the geometry shown in the figure carries a transverse load of 7 kN and a torque of 107 N · m. Examine the shaft for strength and deflection. If the largest allowable slope at the bearings is 0.001 rad and at the gear mesh is 0.0005 rad, what is the factor of safety guarding against damaging distortion? What is the factor of safety guarding against a fatigue failure? If the shaft turns out to be unsatisfactory, what would you recommend to correct the problem?

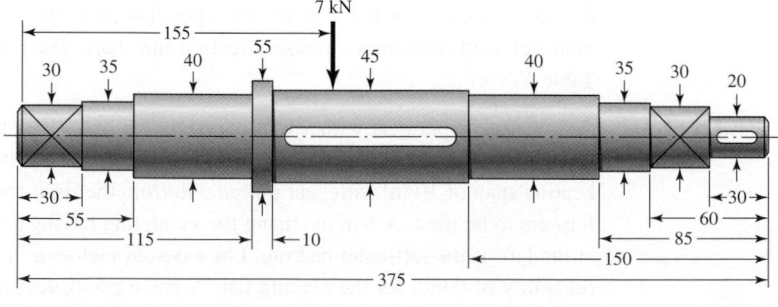

Problem 18–25
Dimensions in millimeters.

All fillets 2 mm

18-26 A 1-in-diameter uniform steel shaft is 24 in long between bearings.

(a) Find the lowest critical speed of the shaft.

(b) If the goal is to double the critical speed, find the new diameter.

(c) A half-size model of the original shaft has what critical speed?

18-27 Demonstrate how rapidly Rayleigh's method converges for the uniform-diameter solid shaft of Prob. 18-26, by partitioning the shaft into first one, then two, and finally three elements.

ANALYSIS **18-28** Compare Eq. (18–40) for the angular frequency of a two-disk shaft with Eq. (18–41), and note that the constants in the two equations are equal.

(a) Develop an expression for the *second* critical speed.

(b) Estimate the second critical speed of the shaft addressed in Ex. 18–5, parts *a* and *b*.

ANALYSIS **18-29** For a uniform-diameter shaft, does hollowing the shaft increase or decrease the critical speed?

18-30 The shaft shown in the figure carries a 20-lbf gear on the left and a 35-lbf gear on the right. Estimate the first critical speed due to the loads, the shaft's critical speed without the loads, and the critical speed of the combination.

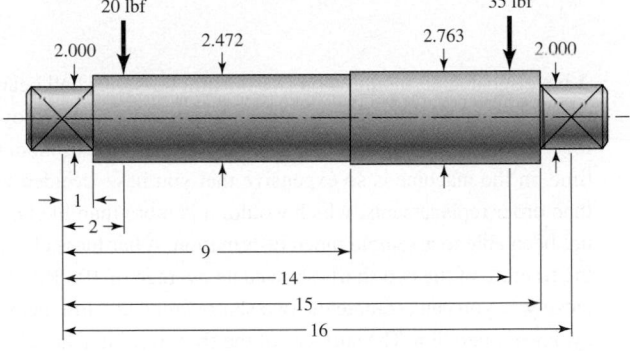

Problem 18–30

Dimensions in inches.

DESIGN **18-31** A shaft is to be designed to support the spur pinion and helical gear shown in the figure on two bearings spaced 28 in center-to-center. Bearing *A* is a cylindrical roller and is to take only radial load; bearing *B* is to take the thrust load of 220 lbf produced by the helical gear and its share of the radial load. The bearing at *B* can be a ball bearing. The radial loads of both gears are in the same plane, and are 660 lbf for the pinion and 220 lbf for the gear. The bearings are to have a life of 2000 h at a combined reliability of 0.995. The shaft speed is 1150 rev/min. Select the bearings and design the shaft. Make a sketch to scale of the shaft showing all fillet sizes, keyways, shoulders, and diameters. Specify the material and its heat treatment.

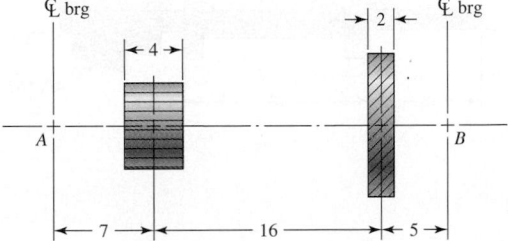

Problem 18–31

Dimensions in inches.

DESIGN **18-32** A heat-treated steel shaft is to be designed to support the spur gear and the overhanging worm shown in the figure. A bearing at *A* takes pure radial load. The bearing at *B* takes the worm-thrust load for either direction of rotation. The dimensions and the loading are shown in the figure; note that the radial loads are in the same plane. Make a complete design of the shaft, including a sketch of the shaft showing all dimensions. Identify the material and its heat treatment (if necessary). Provide an assessment of your final design. The shaft speed 310 rev/min.

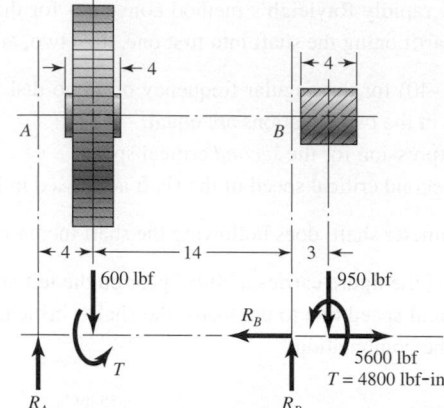

Problem 18–32
Dimensions in inches.

REDESIGN   **18–33**

A bevel-gear shaft mounted on two 40-mm 02-series ball bearings is driven at 1720 rev/min by a motor connected through a flexible coupling. The figure shows the shaft, the gear, and the bearings. The shaft has been giving trouble—in fact, two of them have already failed—and the down time on the machine is so expensive that you have decided to redesign the shaft yourself rather than order replacements, which would, it is more than likely, also fail in a few months. You have not been able to assemble much information. A hardness check of the two shafts in the vicinity of the fracture of the two shafts showed an average of 198 Bhn for one and 204 Bhn of the other. As closely as you can estimate the two shafts failed at a life measure between 600 000 and 1 200 000 cycles of operation. The surfaces of the shaft were machined, but not ground. The fillet sizes were not measured, but they correspond with the recommendations for the ball bearings used. You know that the load is a pulsating or shock-type load, but you have no idea of the magnitude, because the shaft drives an indexing mechanism, and the forces are inertial. The keyways are $\frac{3}{8}$ in wide by $\frac{3}{16}$ in deep. The straight-toothed bevel pinion drives a 48-tooth bevel gear. Specify a new shaft in sufficient detail to ensure a long and trouble-free life.

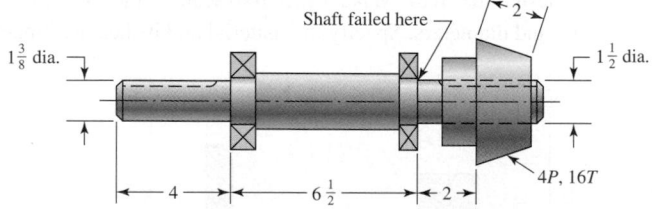

Problem 18–33
Dimensions in inches.

Table A–1

Standard SI Prefixes*†

Name	Symbol	Factor
exa	E	$1\ 000\ 000\ 000\ 000\ 000\ 000 = 10^{18}$
peta	P	$1\ 000\ 000\ 000\ 000\ 000 = 10^{15}$
tera	T	$1\ 000\ 000\ 000\ 000 = 10^{12}$
giga	G	$1\ 000\ 000\ 000 = 10^{9}$
mega	M	$1\ 000\ 000 = 10^{6}$
kilo	k	$1\ 000 = 10^{3}$
hecto‡	h	$100 = 10^{2}$
deka‡	da	$10 = 10^{1}$
deci‡	d	$0.1 = 10^{-1}$
centi‡	c	$0.01 = 10^{-2}$
milli	m	$0.001 = 10^{-3}$
micro	μ	$0.000\ 001 = 10^{-6}$
nano	n	$0.000\ 000\ 001 = 10^{-9}$
pico	p	$0.000\ 000\ 000\ 001 = 10^{-12}$
femto	f	$0.000\ 000\ 000\ 000\ 001 = 10^{-15}$
atto	a	$0.000\ 000\ 000\ 000\ 000\ 001 = 10^{-18}$

*If possible use multiple and submultiple prefixes in steps of 1000.

†Spaces are used in SI instead of commas to group numbers to avoid confusion with the practice in some European countries of using commas for decimal points.

‡Not recommended but sometimes encountered.

Table A–2

Conversion Factors A to Convert Input X to Output Y Using the Formula $Y = AX$*

Multiply Input X	By Factor A	To Get Output Y	Multiply Input X	By Factor A	To Get Output Y
British thermal unit, Btu	1055	joule, J	mile/hour, mi/h	1.61	kilometer/hour, km/h
Btu/second, Btu/s	1.05	kilowatt, kW	mile/hour, mi/h	0.447	meter/second, m/s
calorie	4.19	joule, J	moment of inertia, lbm ·ft^2	0.0421	kilogram-meter^2, kg · m^2
centimeter of mercury (0°C)	1.333	kilopascal, kPa	moment of inertia, lbm · in^2	293	kilogram-millimeter^2, kg · mm^2
centipoise, cP	0.001	pascal-second, Pa · s	moment of section (second moment of area), in^4	41.6	centimeter^4, cm^4
degree (angle)	0.0174	radian, rad	ounce-force, oz	0.278	newton, N
foot, ft	0.305	meter, m	ounce-mass	0.0311	kilogram, kg
foot^2, ft^2	0.0929	meter^2, m^2	pound, lbf†	4.45	newton, N
foot/minute, ft/min	0.0051	meter/second, m/s	pound-foot, lbf · ft	1.36	newton-meter, N · m
foot-pound, ft · lbf	1.35	joule, J	pound/foot^2, lbf/ft^2	47.9	pascal, Pa
foot-pound/ second, ft · lbf/s	1.35	watt, W	pound-inch, lbf · in	0.113	joule, J
foot/second, ft/s	0.305	meter/second, m/s	pound-inch, lbf · in	0.113	newton-meter, N · m
gallon (U.S.), gal	3.785	liter, L	pound/inch, lbf/in	175	newton/meter, N/m
horsepower, hp	0.746	kilowatt, kW	pound/inch^2, psi (lbf/in^2)	6.89	kilopascal, kPa
inch, in	0.0254	meter, m			
inch, in	25.4	millimeter, mm	pound-mass, lbm	0.454	kilogram, kg
inch^2, in^2	645	millimeter^2, mm^2	pound-mass/ second, lbm/s	0.454	kilogram/second, kg/s
inch of mercury (32°F)	3.386	kilopascal, kPa	quart (U.S. liquid), qt	946	milliliter, mL
kilopound, kip	4.45	kilonewton, kN	section modulus, in^3	16.4	centimeter^3, cm^3
kilopound/inch^2, kpsi (ksi)	6.89	megapascal, MPa (N/mm^2)	slug	14.6	kilogram, kg
mass, lbf · s^2/in	175	kilogram, kg	ton (short 2000 lbm)	907	kilogram, kg
mile, mi	1.610	kilometer, km	yard, yd	0.914	meter, m

*Approximate.

† The U.S. Customary system unit of the pound-force is often abbreviated as lbf to distinguish it from the pound-mass, which is abbreviated as lbm.

Table A–3

Optional SI Units for Bending Stress $\sigma = Mc/I$, Torsion Stress $\tau = Tr/J$, Axial Stress $\sigma = F/A$, and Direct Shear Stress $\tau = F/A$

	Bending and Torsion				Axial and Direct Shear		
M, T	I, J	c, r	σ, τ		F	A	σ, τ
N · m*	m⁴	m	Pa		N*	m²	Pa
N · m	cm⁴	cm	MPa (N/mm²)		N†	mm²	MPa (N/mm²)
N · m†	mm⁴	mm	GPa		kN	m²	kPa
kN · m	cm⁴	cm	GPa		kN†	mm²	GPa
N · mm†	mm⁴	mm	MPa (N/mm²)				

* Basic relation.
† Often preferred.

Table A–4

Optional SI Units for Bending Deflection $y = f(Fl^3/EI)$ or $y = f(wl^4/EI)$ and Torsional Deflection $\theta = Tl/GJ$

	Bending Deflection					Torsional Deflection			
F, wl	l	I	E	y	T	l	J	G	θ
N*	m	m⁴	Pa	m	N · m*	m	m⁴	Pa	rad
kN†	mm	mm⁴	GPa	mm	N · m†	mm	mm⁴	GPa	rad
kN	m	m⁴	GPa	μm	N · mm	mm	mm⁴	MPa (N/mm²)	rad
N	mm	mm⁴	kPa	m	N · m	cm	cm⁴	MPa (N/mm²)	rad

* Basic relation.
† Often preferred.

Table A–5

Physical Constants of Materials

Material	Modulus of Elasticity E		Modulus of Rigidity G		Poisson's Ratio ν	Unit Weight w		
	Mpsi	GPa	Mpsi	GPa		lbf/in³	lbf/ft³	kN/m³
Aluminum (all alloys)	10.4	71.7	3.9	26.9	0.333	0.098	169	26.6
Beryllium copper	18.0	124.0	7.0	48.3	0.285	0.297	513	80.6
Brass	15.4	106.0	5.82	40.1	0.324	0.309	534	83.8
Carbon steel	30.0	207.0	11.5	79.3	0.292	0.282	487	76.5
Cast iron (gray)	14.5	100.0	6.0	41.4	0.211	0.260	450	70.6
Copper	17.2	119.0	6.49	44.7	0.326	0.322	556	87.3
Douglas fir	1.6	11.0	0.6	4.1	0.33	0.016	28	4.3
Glass	6.7	46.2	2.7	18.6	0.245	0.094	162	25.4
Inconel	31.0	214.0	11.0	75.8	0.290	0.307	530	83.3
Lead	5.3	36.5	1.9	13.1	0.425	0.411	710	111.5
Magnesium	6.5	44.8	2.4	16.5	0.350	0.065	112	17.6
Molybdenum	48.0	331.0	17.0	117.0	0.307	0.368	636	100.0
Monel metal	26.0	179.0	9.5	65.5	0.320	0.319	551	86.6
Nickel silver	18.5	127.0	7.0	48.3	0.322	0.316	546	85.8
Nickel steel	30.0	207.0	11.5	79.3	0.291	0.280	484	76.0
Phosphor bronze	16.1	111.0	6.0	41.4	0.349	0.295	510	80.1
Stainless steel (18-8)	27.6	190.0	10.6	73.1	0.305	0.280	484	76.0
Titanium alloys	16.5	114.0	6.2	42.4	0.340	0.160	276	43.4

Table A–6

Properties of Structural-Steel Angles*†

w = weight per foot, lbf/ft
m = mass per meter, kg/m
A = area, in^2 (cm^2)
I = second moment of area, in^4 (cm^4)
k = radius of gyration, in (cm)
y = centroidal distance, in (cm)
Z = section modulus, in^3, (cm^3)

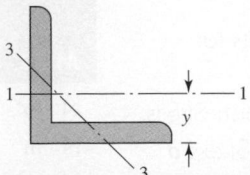

Size, in	w	A	I_{1-1}	k_{1-1}	Z_{1-1}	y	k_{3-3}
$1 \times 1 \times \frac{1}{8}$	0.80	0.234	0.021	0.298	0.029	0.290	0.191
$\times \frac{1}{4}$	1.49	0.437	0.036	0.287	0.054	0.336	0.193
$1\frac{1}{2} \times 1\frac{1}{2} \times \frac{1}{8}$	1.23	0.36	0.074	0.45	0.068	0.41	0.29
$\times \frac{1}{4}$	2.34	0.69	0.135	0.44	0.130	0.46	0.29
$2 \times 2 \times \frac{1}{8}$	1.65	0.484	0.190	0.626	0.131	0.546	0.398
$\times \frac{1}{4}$	3.19	0.938	0.348	0.609	0.247	0.592	0.391
$\times \frac{3}{8}$	4.7	1.36	0.479	0.594	0.351	0.636	0.389
$2\frac{1}{2} \times 2\frac{1}{2} \times \frac{1}{4}$	4.1	1.19	0.703	0.769	0.394	0.717	0.491
$\times \frac{3}{8}$	5.9	1.73	0.984	0.753	0.566	0.762	0.487
$3 \times 3 \times \frac{1}{4}$	4.9	1.44	1.24	0.930	0.577	0.842	0.592
$\times \frac{3}{8}$	7.2	2.11	1.76	0.913	0.833	0.888	0.587
$\times \frac{1}{2}$	9.4	2.75	2.22	0.898	1.07	0.932	0.584
$3\frac{1}{2} \times 3\frac{1}{2} \times \frac{1}{4}$	5.8	1.69	2.01	1.09	0.794	0.968	0.694
$\times \frac{3}{8}$	8.5	2.48	2.87	1.07	1.15	1.01	0.687
$\times \frac{1}{2}$	11.1	3.25	3.64	1.06	1.49	1.06	0.683
$4 \times 4 \times \frac{1}{4}$	6.6	1.94	3.04	1.25	1.05	1.09	0.795
$\times \frac{3}{8}$	9.8	2.86	4.36	1.23	1.52	1.14	0.788
$\times \frac{1}{2}$	12.8	3.75	5.56	1.22	1.97	1.18	0.782
$\times \frac{5}{8}$	15.7	4.61	6.66	1.20	2.40	1.23	0.779
$6 \times 6 \times \frac{3}{8}$	14.9	4.36	15.4	1.88	3.53	1.64	1.19
$\times \frac{1}{2}$	19.6	5.75	19.9	1.86	4.61	1.68	1.18
$\times \frac{5}{8}$	24.2	7.11	24.2	1.84	5.66	1.73	1.18
$\times \frac{3}{4}$	28.7	8.44	28.2	1.83	6.66	1.78	1.17

Table A–6

Properties of Structural-
Steel Angles*†
(Continued)

Size, mm	m	A	I_{1-1}	k_{1-1}	Z_{1-1}	y	k_{3-3}
25 × 25 × 3	1.11	1.42	0.80	0.75	0.45	0.72	0.48
× 4	1.45	1.85	1.01	0.74	0.58	0.76	0.48
× 5	1.77	2.26	1.20	0.73	0.71	0.80	0.48
40 × 40 × 4	2.42	3.08	4.47	1.21	1.55	1.12	0.78
× 5	2.97	3.79	5.43	1.20	1.91	1.16	0.77
× 6	3.52	4.48	6.31	1.19	2.26	1.20	0.77
50 × 50 × 5	3.77	4.80	11.0	1.51	3.05	1.40	0.97
× 6	4.47	5.59	12.8	1.50	3.61	1.45	0.97
× 8	5.82	7.41	16.3	1.48	4.68	1.52	0.96
60 × 60 × 5	4.57	5.82	19.4	1.82	4.45	1.64	1.17
× 6	5.42	6.91	22.8	1.82	5.29	1.69	1.17
× 8	7.09	9.03	29.2	1.80	6.89	1.77	1.16
× 10	8.69	11.1	34.9	1.78	8.41	1.85	1.16
80 × 80 × 6	7.34	9.35	55.8	2.44	9.57	2.17	1.57
× 8	9.63	12.3	72.2	2.43	12.6	2.26	1.56
× 10	11.9	15.1	87.5	2.41	15.4	2.34	1.55
100 × 100 × 8	12.2	15.5	145	3.06	19.9	2.74	1.96
× 12	17.8	22.7	207	3.02	29.1	2.90	1.94
× 15	21.9	27.9	249	2.98	35.6	3.02	1.93
150 × 150 × 10	23.0	29.3	624	4.62	56.9	4.03	2.97
× 12	27.3	34.8	737	4.60	67.7	4.12	2.95
× 15	33.8	43.0	898	4.57	83.5	4.25	2.93
× 18	40.1	51.0	1050	4.54	98.7	4.37	2.92

*Metric sizes also available in sizes of 45, 70, 90, 120, and 200 mm.
†These sizes are also available in aluminum alloy.

Table A–7

Properties of Structural-Steel Channels*

a, b = size, in (mm)
w = weight per foot, lbf/ft
m = mass per meter, kg/m
t = web thickness, in (mm)
A = area, in^2 (cm^2)
I = second moment of area, in^4 (cm^4)
k = radius of gyration, in (cm)
x = centroidal distance, in (cm)
Z = section modulus, in^3 (cm^3)

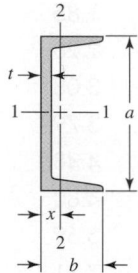

a, in	b, in	t	A	w	I_{1-1}	k_{1-1}	Z_{1-1}	I_{2-2}	k_{2-2}	Z_{2-2}	x
3	1.410	0.170	1.21	4.1	1.66	1.17	1.10	0.197	0.404	0.202	0.436
3	1.498	0.258	1.47	5.0	1.85	1.12	1.24	0.247	0.410	0.233	0.438
3	1.596	0.356	1.76	6.0	2.07	1.08	1.38	0.305	0.416	0.268	0.455
4	1.580	0.180	1.57	5.4	3.85	1.56	1.93	0.319	0.449	0.283	0.457
4	1.720	0.321	2.13	7.25	4.59	1.47	2.29	0.433	0.450	0.343	0.459
5	1.750	0.190	1.97	6.7	7.49	1.95	3.00	0.479	0.493	0.378	0.484
5	1.885	0.325	2.64	9.0	8.90	1.83	3.56	0.632	0.489	0.450	0.478
6	1.920	0.200	2.40	8.2	13.1	2.34	4.38	0.693	0.537	0.492	0.511
6	2.034	0.314	3.09	10.5	15.2	2.22	5.06	0.866	0.529	0.564	0.499
6	2.157	0.437	3.83	13.0	17.4	2.13	5.80	1.05	0.525	0.642	0.514
7	2.090	0.210	2.87	9.8	21.3	2.72	6.08	0.968	0.581	0.625	0.540
7	2.194	0.314	3.60	12.25	24.2	2.60	6.93	1.17	0.571	0.703	0.525
7	2.299	0.419	4.33	14.75	27.2	2.51	7.78	1.38	0.564	0.779	0.532
8	2.260	0.220	3.36	11.5	32.3	3.10	8.10	1.30	0.625	0.781	0.571
8	2.343	0.303	4.04	13.75	36.2	2.99	9.03	1.53	0.615	0.854	0.553
8	2.527	0.487	5.51	18.75	44.0	2.82	11.0	1.98	0.599	1.01	0.565
9	2.430	0.230	3.91	13.4	47.7	3.49	10.6	1.75	0.669	0.962	0.601
9	2.485	0.285	4.41	15.0	51.0	3.40	11.3	1.93	0.661	1.01	0.586
9	2.648	0.448	5.88	20.0	60.9	3.22	13.5	2.42	0.647	1.17	0.583
10	2.600	0.240	4.49	15.3	67.4	3.87	13.5	2.28	0.713	1.16	0.634
10	2.739	0.379	5.88	20.0	78.9	3.66	15.8	2.81	0.693	1.32	0.606
10	2.886	0.526	7.35	25.0	91.2	3.52	18.2	3.36	0.676	1.48	0.617
10	3.033	0.673	8.82	30.0	103	3.43	20.7	3.95	0.669	1.66	0.649
12	3.047	0.387	7.35	25.0	144	4.43	24.1	4.47	0.780	1.89	0.674
12	3.170	0.510	8.82	30.0	162	4.29	27.0	5.14	0.763	2.06	0.674

Table A–7

Properties of Structural-Steel Channels *(Continued)*

a × b, mm	m	t	A	I_{1-1}	k_{1-1}	Z_{1-1}	I_{2-2}	k_{2-2}	Z_{2-2}	x
76 × 38	6.70	5.1	8.53	74.14	2.95	19.46	10.66	1.12	4.07	1.19
102 × 51	10.42	6.1	13.28	207.7	3.95	40.89	29.10	1.48	8.16	1.51
127 × 64	14.90	6.4	18.98	482.5	5.04	75.99	67.23	1.88	15.25	1.94
152 × 76	17.88	6.4	22.77	851.5	6.12	111.8	113.8	2.24	21.05	2.21
152 × 89	23.84	7.1	30.36	1166	6.20	153.0	215.1	2.66	35.70	2.86
178 × 76	20.84	6.6	26.54	1337	7.10	150.4	134.0	2.25	24.72	2.20
178 × 89	26.81	7.6	34.15	1753	7.16	197.2	241.0	2.66	39.29	2.76
203 × 76	23.82	7.1	30.34	1950	8.02	192.0	151.3	2.23	27.59	2.13
203 × 89	29.78	8.1	37.94	2491	8.10	245.2	264.4	2.64	42.34	2.65
229 × 76	26.06	7.6	33.20	2610	8.87	228.3	158.7	2.19	28.22	2.00
229 × 89	32.76	8.6	41.73	3387	9.01	296.4	285.0	2.61	44.82	2.53
254 × 76	28.29	8.1	36.03	3367	9.67	265.1	162.6	2.12	28.21	1.86
254 × 89	35.74	9.1	45.42	4448	9.88	350.2	302.4	2.58	46.70	2.42
305 × 89	41.69	10.2	53.11	7061	11.5	463.3	325.4	2.48	48.49	2.18
305 × 102	46.18	10.2	58.83	8214	11.8	539.0	499.5	2.91	66.59	2.66

*These sizes are also available in aluminum alloy.

Table A–8

Properties of Round
Tubing

w_a = unit weight of aluminum tubing, lbf/ft
w_s = unit weight of steel tubing, lbf/ft
m = unit mass, kg/m
A = area, in^2 (cm^2)
I = second moment of area, in^4 (cm^4)
J = second polar moment of area, in^4 (cm^4)
k = radius of gyration, in (cm)
Z = section modulus, in^3 (cm^3)
d, t = size (OD) and thickness, in (mm)

Size, in	w_a	w_s	A	I	k	Z	J
$1 \times \frac{1}{8}$	0.416	1.128	0.344	0.034	0.313	0.067	0.067
$1 \times \frac{1}{4}$	0.713	2.003	0.589	0.046	0.280	0.092	0.092
$1\frac{1}{2} \times \frac{1}{8}$	0.653	1.769	0.540	0.129	0.488	0.172	0.257
$1\frac{1}{2} \times \frac{1}{4}$	1.188	3.338	0.982	0.199	0.451	0.266	0.399
$2 \times \frac{1}{8}$	0.891	2.670	0.736	0.325	0.664	0.325	0.650
$2 \times \frac{1}{4}$	1.663	4.673	1.374	0.537	0.625	0.537	1.074
$2\frac{1}{2} \times \frac{1}{8}$	1.129	3.050	0.933	0.660	0.841	0.528	1.319
$2\frac{1}{2} \times \frac{1}{4}$	2.138	6.008	1.767	1.132	0.800	0.906	2.276
$3 \times \frac{1}{4}$	2.614	7.343	2.160	2.059	0.976	1.373	4.117
$3 \times \frac{3}{8}$	3.742	10.51	3.093	2.718	0.938	1.812	5.436
$4 \times \frac{3}{16}$	2.717	7.654	2.246	4.090	1.350	2.045	8.180
$4 \times \frac{3}{8}$	5.167	14.52	4.271	7.090	1.289	3.544	14.180

Size, mm	m	A	I	k	Z	J
12×2	0.490	0.628	0.082	0.361	0.136	0.163
16×2	0.687	0.879	0.220	0.500	0.275	0.440
16×3	0.956	1.225	0.273	0.472	0.341	0.545
20×4	1.569	2.010	0.684	0.583	0.684	1.367
25×4	2.060	2.638	1.508	0.756	1.206	3.015
25×5	2.452	3.140	1.669	0.729	1.336	3.338
30×4	2.550	3.266	2.827	0.930	1.885	5.652
30×5	3.065	3.925	3.192	0.901	2.128	6.381
42×4	3.727	4.773	8.717	1.351	4.151	17.430
42×5	4.536	5.809	10.130	1.320	4.825	20.255
50×4	4.512	5.778	15.409	1.632	6.164	30.810
50×5	5.517	7.065	18.118	1.601	7.247	36.226

Table A–9

Shear, Moment, and Deflection of Beams (*Note: Force and moment reactions are positive in the directions shown; equations for shear force V and bending moment M follow the sign conventions given in Sec. 4–2.*)

1 Cantilever—end load

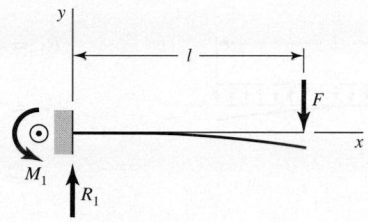

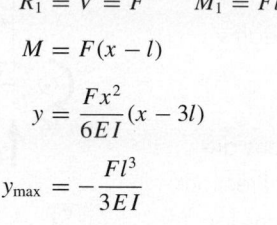

$$R_1 = V = F \qquad M_1 = Fl$$

$$M = F(x - l)$$

$$y = \frac{Fx^2}{6EI}(x - 3l)$$

$$y_{\max} = -\frac{Fl^3}{3EI}$$

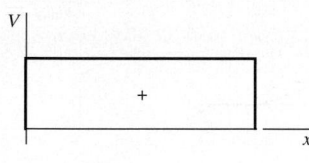

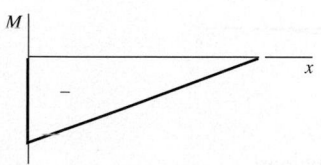

2 Cantilever—intermediate load

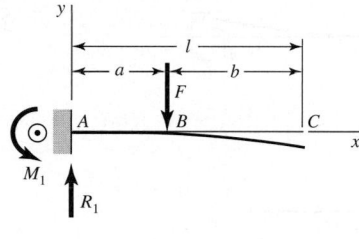

$$R_1 = V = F \qquad M_1 = Fa$$

$$M_{AB} = F(x - a) \qquad M_{BC} = 0$$

$$y_{AB} = \frac{Fx^2}{6EI}(x - 3a)$$

$$y_{BC} = \frac{Fa^2}{6EI}(a - 3x)$$

$$y_{\max} = \frac{Fa^2}{6EI}(a - 3l)$$

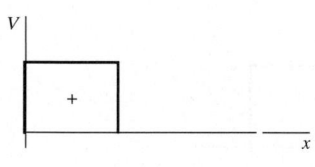

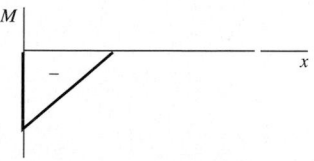

(continued)

Table A–9

Shear, Moment, and Deflection of Beams *(Continued)* *(Note:* Force and moment reactions are positive in the directions shown; equations for shear force V and bending moment M follow the sign conventions given in Sec. 4–2.)

3 Cantilever—uniform load

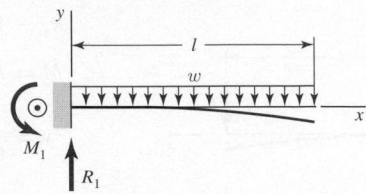

$$R_1 = wl \qquad M_1 = \frac{wl^2}{2}$$

$$V = w(l - x) \qquad M = -\frac{w}{2}(l - x)^2$$

$$y = \frac{wx^2}{24EI}(4lx - x^2 - 6l^2)$$

$$y_{max} = -\frac{wl^4}{8EI}$$

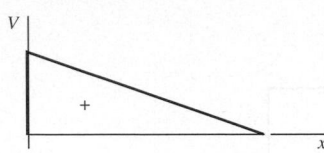

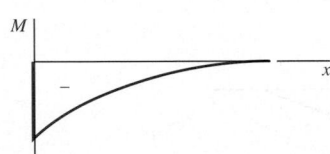

4 Cantilever—moment load

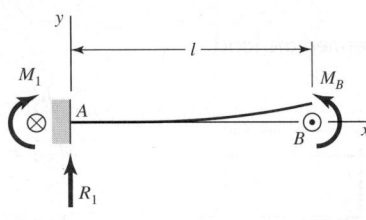

$$R_1 = 0 \qquad M_1 = M_B \qquad M = M_B$$

$$y = \frac{M_B x^2}{2EI} \qquad y_{max} = \frac{M_B l^2}{2EI}$$

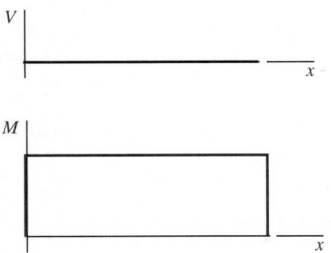

Table A–9

Shear, Moment, and Deflection of Beams *(Continued)* *(Note:* Force and moment reactions are positive in the directions shown; equations for shear force *V* and bending moment *M* follow the sign conventions given in Sec. 4–2.)

5 Simple supports—center load

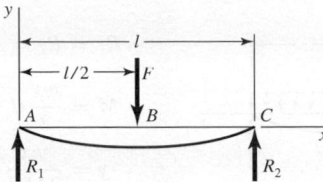

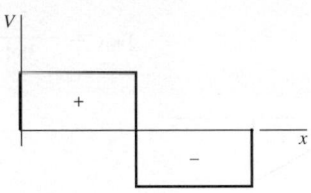

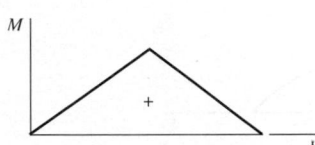

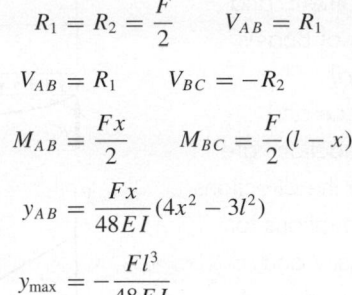

$$R_1 = R_2 = \frac{F}{2} \qquad V_{AB} = R_1$$

$$V_{AB} = R_1 \qquad V_{BC} = -R_2$$

$$M_{AB} = \frac{Fx}{2} \qquad M_{BC} = \frac{F}{2}(l - x)$$

$$y_{AB} = \frac{Fx}{48EI}(4x^2 - 3l^2)$$

$$y_{\max} = -\frac{Fl^3}{48EI}$$

6 Simple supports—intermediate load

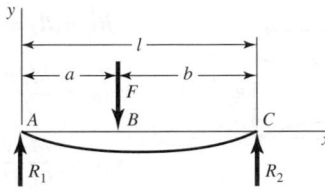

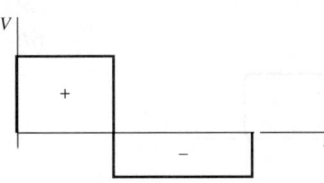

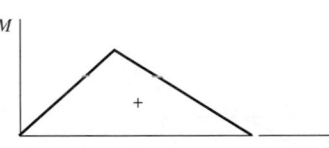

$$R_1 = \frac{Fb}{l} \qquad R_2 = \frac{Fa}{l}$$

$$V_{AB} = R_1 \qquad V_{BC} = -R_2$$

$$M_{AB} = \frac{Fbx}{l} \qquad M_{BC} = \frac{Fa}{l}(l - x)$$

$$y_{AB} = \frac{Fbx}{6EIl}(x^2 + b^2 - l^2)$$

$$y_{BC} = \frac{Fa(l - x)}{6EIl}(x^2 + a^2 - 2lx)$$

(continued)

Table A–9

Shear, Moment, and
Deflection of Beams
(Continued)
(Note: Force and
moment reactions are
positive in the directions
shown; equations for
shear force V and
bending moment M
follow the sign
conventions given in
Sec. 4–2.)

7 Simple supports—uniform load

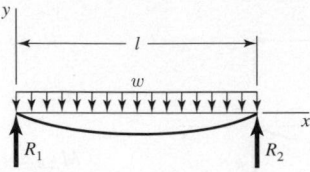

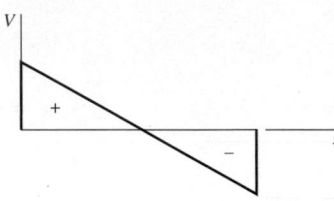

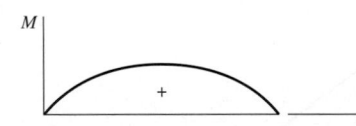

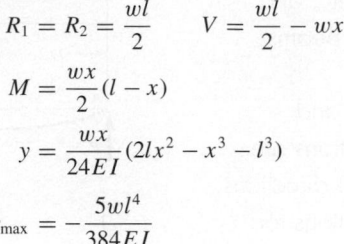

$$R_1 = R_2 = \frac{wl}{2} \qquad V = \frac{wl}{2} - wx$$

$$M = \frac{wx}{2}(l - x)$$

$$y = \frac{wx}{24EI}(2lx^2 - x^3 - l^3)$$

$$y_{max} = -\frac{5wl^4}{384EI}$$

8 Simple supports—moment load

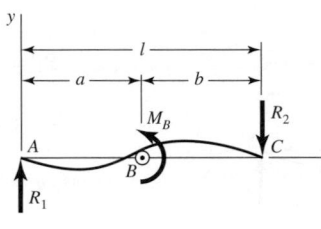

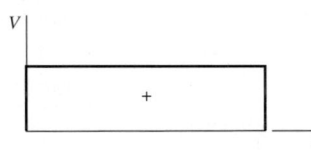

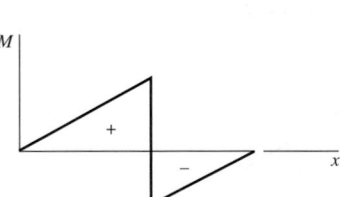

$$R_1 = R_2 = \frac{M_B}{l} \qquad V = \frac{M_B}{l}$$

$$M_{AB} = \frac{M_B x}{l} \qquad M_{BC} = \frac{M_B}{l}(x - l)$$

$$y_{AB} = \frac{M_B x}{6EIl}(x^2 + 3a^2 - 6al + 2l^2)$$

$$y_{BC} = \frac{M_B}{6EIl}[x^3 - 3lx^2 + x(2l^2 + 3a^2) - 3a^2 l]$$

Table A–9

Shear, Moment, and Deflection of Beams *(Continued)* (*Note:* Force and moment reactions are positive in the directions shown; equations for shear force V and bending moment M follow the sign conventions given in Sec. 4–2.)

9 Simple supports—twin loads

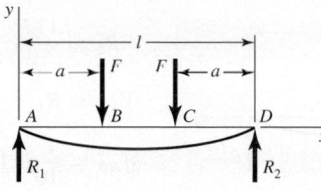

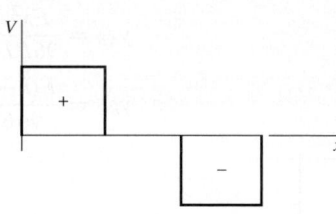

$$R_1 = R_2 = F \qquad V_{AB} = F \qquad V_{BC} = 0$$

$$V_{CD} = -F$$

$$M_{AB} = Fx \qquad M_{BC} = Fa \qquad M_{CD} = F(l-x)$$

$$y_{AB} = \frac{Fx}{6EI}(x^2 + 3a^2 - 3la)$$

$$y_{BC} = \frac{Fa}{6EI}(3x^2 + a^2 - 3lx)$$

$$y_{max} = \frac{Fa}{24EI}(4a^2 - 3l^2)$$

10 Simple supports—overhanging load

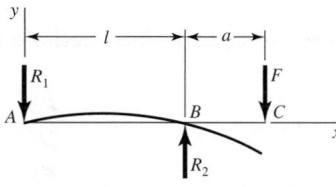

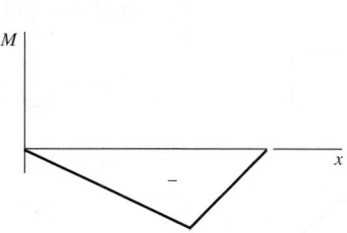

$$R_1 = \frac{Fa}{l} \qquad R_2 = \frac{F}{l}(l+a)$$

$$V_{AB} = -\frac{Fa}{l} \qquad V_{BC} = F$$

$$M_{AB} = -\frac{Fax}{l} \qquad M_{BC} = F(x-l-a)$$

$$y_{AB} = \frac{Fax}{6EIl}(l^2 - x^2)$$

$$y_{BC} = \frac{F(x-l)}{6EI}[(x-l)^2 - a(3x-l)]$$

$$y_c = -\frac{Fa^2}{3EI}(l+a)$$

(continued)

Table A–9

Shear, Moment, and Deflection of Beams *(Continued)* *(Note:* Force and moment reactions are positive in the directions shown; equations for shear force V and bending moment M follow the sign conventions given in Sec. 4–2.)

11 One fixed and one simple support—center load

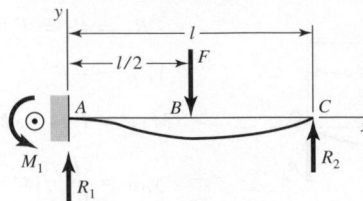

$$R_1 = \frac{11F}{16} \qquad R_2 = \frac{5F}{16} \qquad M_1 = \frac{3Fl}{16}$$

$$V_{AB} = R_1 \qquad V_{BC} = -R_2$$

$$M_{AB} = \frac{F}{16}(11x - 3l) \qquad M_{BC} = \frac{5F}{16}(l - x)$$

$$y_{AB} = \frac{Fx^2}{96EI}(11x - 9l)$$

$$y_{BC} = \frac{F(l-x)}{96EI}(5x^2 + 2l^2 - 10lx)$$

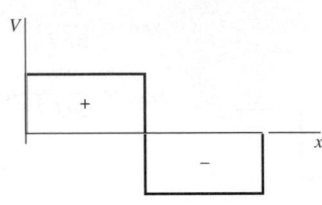

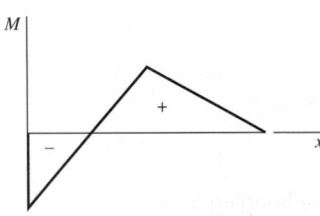

12 One fixed and one simple support—intermediate load

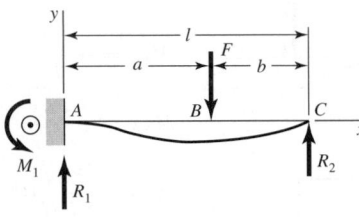

$$R_1 = \frac{Fb}{2l^3}(3l^2 - b^2) \qquad R_2 = \frac{Fa^2}{2l^3}(3l - a)$$

$$M_1 = \frac{Fb}{2l^2}(l^2 - b^2)$$

$$V_{AB} = R_1 \qquad V_{BC} = -R_2$$

$$M_{AB} = \frac{Fb}{2l^3}[b^2l - l^3 + x(3l^2 - b^2)]$$

$$M_{BC} = \frac{Fa^2}{2l^3}(3l^2 - 3lx - al + ax)$$

$$y_{AB} = \frac{Fbx^2}{12EIl^3}[3l(b^2 - l^2) + x(3l^2 - b^2)]$$

$$y_{BC} = y_{AB} - \frac{F(x-a)^3}{6EI}$$

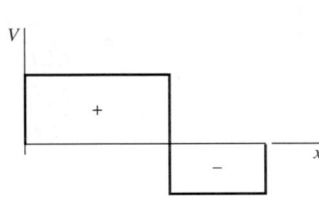

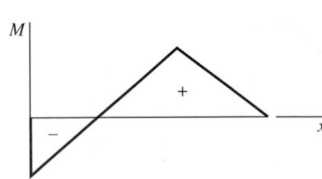

Table A–9

Shear, Moment, and
Deflection of Beams
(Continued)
(*Note:* Force and
moment reactions are
positive in the directions
shown; equations for
shear force V and
bending moment M
follow the sign
conventions given in
Sec. 4–2.)

13 One fixed and one simple support—uniform load

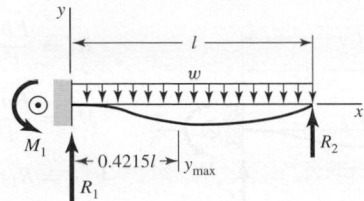

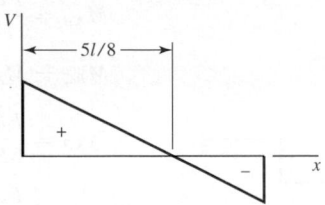

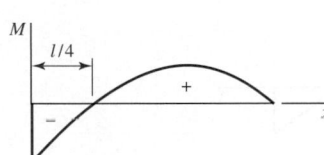

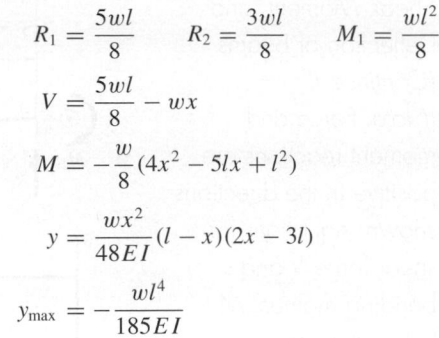

$$R_1 = \frac{5wl}{8} \qquad R_2 = \frac{3wl}{8} \qquad M_1 = \frac{wl^2}{8}$$

$$V = \frac{5wl}{8} - wx$$

$$M = -\frac{w}{8}(4x^2 - 5lx + l^2)$$

$$y = \frac{wx^2}{48EI}(l - x)(2x - 3l)$$

$$y_{max} = -\frac{wl^4}{185EI}$$

14 Fixed supports—center load

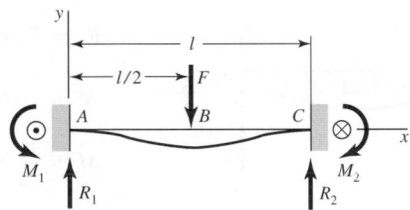

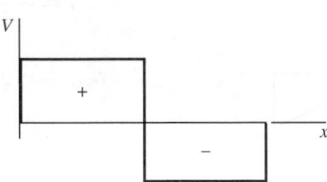

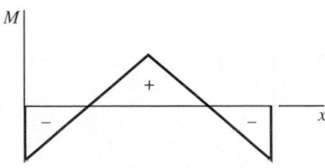

$$R_1 = R_2 = \frac{F}{2} \qquad M_1 = M_2 = \frac{Fl}{8}$$

$$V_{AB} = -V_{BC} = \frac{F}{2}$$

$$M_{AB} = \frac{F}{8}(4x - l) \qquad M_{BC} = \frac{F}{8}(3l - 4x)$$

$$y_{AB} = \frac{Fx^2}{48EI}(4x - 3l)$$

$$y_{max} = -\frac{Fl^3}{192EI}$$

(continued)

Table A–9

Shear, Moment, and Deflection of Beams *(Continued)* *(Note:* Force and moment reactions are positive in the directions shown; equations for shear force V and bending moment M follow the sign conventions given in Sec. 4–2.)

15 Fixed supports—intermediate load

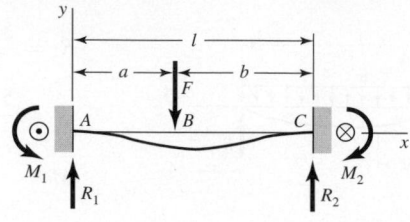

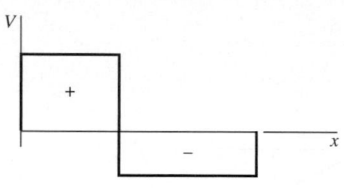

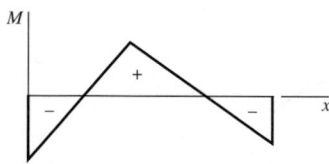

$$R_1 = \frac{Fb^2}{l^3}(3a + b) \qquad R_2 = \frac{Fa^2}{l^3}(3b + a)$$

$$M_1 = \frac{Fab^2}{l^2} \qquad M_2 = \frac{Fa^2 b}{l^2}$$

$$V_{AB} = R_1 \qquad V_{BC} = -R_2$$

$$M_{AB} = \frac{Fb^2}{l^3}[x(3a + b) - al]$$

$$M_{BC} = M_{AB} - F(x - a)$$

$$y_{AB} = \frac{Fb^2 x^2}{6EIl^3}[x(3a + b) - 3al]$$

$$y_{BC} = \frac{Fa^2(l - x)^2}{6EIl^3}[(l - x)(3b + a) - 3bl]$$

16 Fixed supports—uniform load

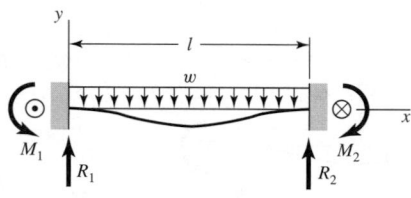

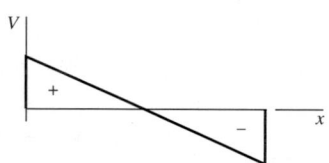

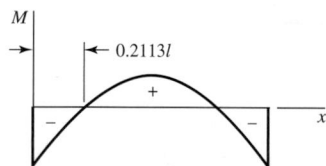

$$R_1 = R_2 = \frac{wl}{2} \qquad M_1 = M_2 = \frac{wl^2}{12}$$

$$V = \frac{w}{2}(l - 2x)$$

$$M = \frac{w}{12}(6lx - 6x^2 - l^2)$$

$$y = -\frac{wx^2}{24EI}(l - x)^2$$

$$y_{\max} = -\frac{wl^4}{384EI}$$

Table A-10

Cumulative Distribution Function of Normal (Gaussian) Distribution

$$\Phi(z_\alpha) = \int_{-\infty}^{z_\alpha} \frac{1}{\sqrt{2\pi}} \exp\left(-\frac{u^2}{2}\right) du$$

$$= \begin{cases} \alpha & z_\alpha \leq 0 \\ 1 - \alpha & z_\alpha > 0 \end{cases}$$

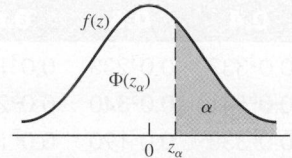

Z_α	0.00	0.01	0.02	0.03	0.04	0.05	0.06	0.07	0.08	0.09
0.0	0.5000	0.4960	0.4920	0.4880	0.4840	0.4801	0.4761	0.4721	0.4681	0.4641
0.1	0.4602	0.4562	0.4522	0.4483	0.4443	0.4404	0.4364	0.4325	0.4286	0.4247
0.2	0.4207	0.4168	0.4129	0.4090	0.4052	0.4013	0.3974	0.3936	0.3897	0.3859
0.3	0.3821	0.3783	0.3745	0.3707	0.3669	0.3632	0.3594	0.3557	0.3520	0.3483
0.4	0.3446	0.3409	0.3372	0.3336	0.3300	0.3264	0.3238	0.3192	0.3156	0.3121
0.5	0.3085	0.3050	0.3015	0.2981	0.2946	0.2912	0.2877	0.2843	0.2810	0.2776
0.6	0.2743	0.2709	0.2676	0.2643	0.2611	0.2578	0.2546	0.2514	0.2483	0.2451
0.7	0.2420	0.2389	0.2358	0.2327	0.2296	0.2266	0.2236	0.2206	0.2177	0.2148
0.8	0.2119	0.2090	0.2061	0.2033	0.2005	0.1977	0.1949	0.1922	0.1894	0.1867
0.9	0.1841	0.1814	0.1788	0.1762	0.1736	0.1711	0.1685	0.1660	0.1635	0.1611
1.0	0.1587	0.1562	0.1539	0.1515	0.1492	0.1469	0.1446	0.1423	0.1401	0.1379
1.1	0.1357	0.1335	0.1314	0.1292	0.1271	0.1251	0.1230	0.1210	0.1190	0.1170
1.2	0.1151	0.1131	0.1112	0.1093	0.1075	0.1056	0.1038	0.1020	0.1003	0.0985
1.3	0.0968	0.0951	0.0934	0.0918	0.0901	0.0885	0.0869	0.0853	0.0838	0.0823
1.4	0.0808	0.0793	0.0778	0.0764	0.0749	0.0735	0.0721	0.0708	0.0694	0.0681
1.5	0.0668	0.0655	0.0643	0.0630	0.0618	0.0606	0.0594	0.0582	0.0571	0.0559
1.6	0.0548	0.0537	0.0526	0.0516	0.0505	0.0495	0.0485	0.0475	0.0465	0.0455
1.7	0.0446	0.0436	0.0427	0.0418	0.0409	0.0401	0.0392	0.0384	0.0375	0.0367
1.8	0.0359	0.0351	0.0344	0.0336	0.0329	0.0322	0.0314	0.0307	0.0301	0.0294
1.9	0.0287	0.0281	0.0274	0.0268	0.0262	0.0256	0.0250	0.0244	0.0239	0.0233
2.0	0.0228	0.0222	0.0217	0.0212	0.0207	0.0202	0.0197	0.0192	0.0188	0.0183
2.1	0.0179	0.0174	0.0170	0.0166	0.0162	0.0158	0.0154	0.0150	0.0146	0.0143
2.2	0.0139	0.0136	0.0132	0.0129	0.0125	0.0122	0.0119	0.0116	0.0113	0.0110
2.3	0.0107	0.0104	0.0102	0.00990	0.00964	0.00939	0.00914	0.00889	0.00866	0.00842
2.4	0.00820	0.00798	0.00776	0.00755	0.00734	0.00714	0.00695	0.00676	0.00657	0.00639
2.5	0.00621	0.00604	0.00587	0.00570	0.00554	0.00539	0.00523	0.00508	0.00494	0.00480
2.6	0.00466	0.00453	0.00440	0.00427	0.00415	0.00402	0.00391	0.00379	0.00368	0.00357
2.7	0.00347	0.00336	0.00326	0.00317	0.00307	0.00298	0.00289	0.00280	0.00272	0.00264
2.8	0.00256	0.00248	0.00240	0.00233	0.00226	0.00219	0.00212	0.00205	0.00199	0.00193
2.9	0.00187	0.00181	0.00175	0.00169	0.00164	0.00159	0.00154	0.00149	0.00144	0.00139

(continued)

Table A–10

Cumulative Distribution Function of Normal (Gaussian) Distribution *(Continued)*

z_α	0.0	0.1	0.2	0.3	0.4	0.5	0.6	0.7	0.8	0.9
3	0.00135	0.0^3968	0.0^3687	0.0^3483	0.0^3337	0.0^3233	0.0^3159	0.0^3108	0.0^4723	0.0^4481
4	0.0^4317	0.0^4207	0.0^4133	0.0^5854	0.0^5541	0.0^5340	0.0^5211	0.0^5130	0.0^6793	0.0^6479
5	0.0^6287	0.0^6170	0.0^7996	0.0^7579	0.0^7333	0.0^7190	0.0^7107	0.0^8599	0.0^8332	0.0^8182
6	0.0^9987	0.0^9530	0.0^9282	0.0^9149	$0.0^{10}777$	$0.0^{10}402$	$0.0^{10}206$	$0.0^{10}104$	$0.0^{11}523$	$0.0^{11}260$

z_α	−1.282	−1.643	−1.960	−2.326	−2.576	−3.090	−3.291	−3.891	−4.417
$F(z_\alpha)$	0.10	0.05	0.025	0.010	0.005	0.001	0.005	0.00005	0.000005
$R(z_\alpha)$	0.90	0.95	0.975	0.999	0.995	0.999	0.9995	0.9999	0.999995

Table A–11

A Selection of International Tolerance Grades—Metric Series (Size Ranges Are for *Over* the Lower Limit and *Including* the Upper Limit. All Values Are in Millimeters)

Source: *Preferred Metric Limits and Fits,* ANSI B4.2-1978. See also BSI 4500.

Basic Sizes	Tolerance Grades					
	IT6	IT7	IT8	IT9	IT10	IT11
0–3	0.006	0.010	0.014	0.025	0.040	0.060
3–6	0.008	0.012	0.018	0.030	0.048	0.075
6–10	0.009	0.015	0.022	0.036	0.058	0.090
10–18	0.011	0.018	0.027	0.043	0.070	0.110
18–30	0.013	0.021	0.033	0.052	0.084	0.130
30–50	0.016	0.025	0.039	0.062	0.100	0.160
50–80	0.019	0.030	0.046	0.074	0.120	0.190
80–120	0.022	0.035	0.054	0.087	0.140	0.220
120–180	0.025	0.040	0.063	0.100	0.160	0.250
180–250	0.029	0.046	0.072	0.115	0.185	0.290
250–315	0.032	0.052	0.081	0.130	0.210	0.320
315–400	0.036	0.057	0.089	0.140	0.230	0.360

Table A-12

Fundamental Deviations for Shafts—Metric Series
(Size Ranges Are for *Over* the Lower Limit and *Including* the Upper Limit. All Values Are in Millimeters)
Source: Preferred Metric Limits and Fits, ANSI B4.2-1978. See also BSI 4500.

Basic Sizes	Upper-Deviation Letter						Lower-Deviation Letter				
	c	d	f	g	h	k	n	p	s	u
0–3	−0.060	−0.020	−0.006	−0.002	0	0	+0.004	+0.006	+0.014	+0.018
3–6	−0.070	−0.030	−0.010	−0.004	0	+0.001	+0.008	+0.012	+0.019	+0.023
6–10	−0.080	−0.040	−0.013	−0.005	0	+0.001	+0.010	+0.015	+0.023	+0.028
10–14	−0.095	−0.050	−0.016	−0.006	0	+0.001	+0.012	+0.018	+0.028	+0.033
14–18	−0.095	−0.050	−0.016	−0.006	0	+0.001	+0.012	+0.018	+0.028	+0.033
18–24	−0.110	−0.065	−0.020	−0.007	0	+0.002	+0.015	+0.022	+0.035	+0.041
24–30	−0.110	−0.065	−0.020	−0.007	0	+0.002	+0.015	+0.022	+0.035	+0.048
30–40	−0.120	−0.080	−0.025	−0.009	0	+0.002	+0.017	+0.026	+0.043	+0.060
40–50	−0.130	−0.080	0.025	−0.009	0	+0.002	+0.017	+0.026	+0.043	+0.070
50–65	−0.140	−0.100	−0.030	−0.010	0	+0.002	+0.020	+0.032	+0.053	+0.087
65–80	−0.150	−0.100	−0.030	−0.010	0	+0.002	+0.020	+0.032	+0.059	+0.102
80–100	−0.170	−0.120	−0.036	−0.012	0	+0.003	+0.023	+0.037	+0.071	+0.124
100–120	−0.180	−0.120	−0.036	−0.012	0	+0.003	+0.023	+0.037	+0.079	+0.144
120–140	−0.200	−0.145	−0.043	−0.014	0	+0.003	+0.027	+0.043	+0.092	+0.170
140–160	−0.210	−0.145	−0.043	−0.014	0	+0.003	+0.027	+0.043	+0.100	+0.190
160–180	−0.230	−0.145	−0.043	−0.014	0	+0.003	+0.027	+0.043	+0.108	+0.210
180–200	−0.240	−0.170	−0.050	−0.015	0	+0.004	+0.031	+0.050	+0.122	+0.236
200–225	−0.260	−0.170	−0.050	−0.015	0	+0.004	+0.031	+0.050	+0.130	+0.258
225–250	−0.280	−0.170	−0.050	−0.015	0	+0.004	+0.031	+0.050	+0.140	+0.284
250–280	−0.300	−0.190	−0.056	−0.017	0	+0.004	+0.034	+0.056	+0.158	+0.315
280–315	−0.330	−0.190	−0.056	−0.017	0	+0.004	+0.034	+0.056	+0.170	+0.350
315–355	−0.360	−0.210	−0.062	−0.018	0	+0.004	+0.037	+0.062	+0.190	+0.390
355–400	−0.400	−0.210	−0.062	−0.018	0	+0.004	+0.037	+0.062	+0.208	+0.435

Table A–13

A Selection of International Tolerance Grades—Inch Series (Size Ranges Are for *Over* the Lower Limit and *Including* the Upper Limit. All Values Are in Inches, Converted from Table A–11)

Basic Sizes	Tolerance Grades					
	IT6	IT7	IT8	IT9	IT10	IT11
0–0.12	0.0002	0.0004	0.0006	0.0010	0.0016	0.0024
0.12–0.24	0.0003	0.0005	0.0007	0.0012	0.0019	0.0030
0.24–0.40	0.0004	0.0006	0.0009	0.0014	0.0023	0.0035
0.40–0.72	0.0004	0.0007	0.0011	0.0017	0.0028	0.0043
0.72–1.20	0.0005	0.0008	0.0013	0.0020	0.0033	0.0051
1.20–2.00	0.0006	0.0010	0.0015	0.0024	0.0039	0.0063
2.00–3.20	0.0007	0.0012	0.0018	0.0029	0.0047	0.0075
3.20–4.80	0.0009	0.0014	0.0021	0.0034	0.0055	0.0087
4.80–7.20	0.0010	0.0016	0.0025	0.0039	0.0063	0.0098
7.20–10.00	0.0011	0.0018	0.0028	0.0045	0.0073	0.0114
10.00–12.60	0.0013	0.0020	0.0032	0.0051	0.0083	0.0126
12.60–16.00	0.0014	0.0022	0.0035	0.0055	0.0091	0.0142

Table A-14

Fundamental Deviations for Shafts—Inch Series (Size Ranges Are for *Over* the Lower Limit and *Including* the Upper Limit. All Values Are in Inches, Converted from Table A–12)

Basic Sizes	Upper-Deviation Letter						Lower-Deviation Letter			
	c	d	f	g	h	k	n	p	s	u
0–0.12	−0.0024	−0.0008	−0.0002	−0.0001	0	0	+0.0002	+0.0002	+0.0006	+0.0007
0.12–0.24	−0.0028	−0.0012	−0.0004	−0.0002	0	0	+0.0003	+0.0005	+0.0007	+0.0009
0.24–0.40	−0.0031	−0.0016	−0.0005	−0.0002	0	0	+0.0004	+0.0006	+0.0009	−0.0011
0.40–0.72	−0.0037	−0.0020	−0.0006	−0.0002	0	0	+0.0005	+0.0007	+0.0011	−0.0013
0.72–0.96	−0.0043	−0.0026	−0.0008	−0.0003	0	+0.0001	+0.0006	+0.0009	+0.0014	+0.0016
0.96–1.20	−0.0043	−0.0026	−0.0008	−0.0003	0	+0.0001	+0.0006	+0.0009	+0.0014	+0.0019
1.20–1.60	−0.0047	−0.0031	−0.0010	−0.0004	0	+0.0001	+0.0007	+0.0010	+0.0017	+0.0024
1.60–2.00	−0.0051	−0.0031	−0.0010	−0.0004	0	+0.0001	+0.0007	+0.0010	+0.0017	+0.0028
2.00–2.60	−0.0055	−0.0039	−0.0012	−0.0004	0	+0.0001	+0.0008	+0.0013	+0.0021	+0.0034
2.60–3.20	−0.0059	−0.0039	−0.0012	−0.0004	0	+0.0001	+0.0008	+0.0013	+0.0023	+0.0040
3.20–4.00	−0.0067	−0.0047	−0.0014	−0.0005	0	+0.0001	+0.0009	+0.0015	+0.0028	+0.0049
4.00–4.80	−0.0071	−0.0047	−0.0014	−0.0005	0	+0.0001	+0.0009	+0.0015	+0.0031	+0.0057
4.80–5.60	−0.0079	−0.0057	−0.0017	−0.0006	0	+0.0001	+0.0011	+0.0017	+0.0036	+0.0067
5.60–6.40	−0.0083	−0.0057	−0.0017	−0.0006	0	+0.0001	+0.0011	+0.0017	+0.0039	+0.0075
6.40–7.20	−0.0091	−0.0057	−0.0017	−0.0006	0	+0.0001	+0.0011	+0.0017	+0.0043	+0.0083
7.20–8.00	−0.0094	−0.0067	−0.0020	−0.0006	0	+0.0002	+0.0012	+0.0020	+0.0048	+0.0093
8.00–9.00	−0.0102	−0.0067	−0.0020	−0.0006	0	+0.0002	+0.0012	+0.0020	+0.0051	+0.0102
9.00–10.00	−0.0110	−0.0067	−0.0020	−0.0006	0	+0.0002	+0.0012	+0.0020	+0.0055	+0.0112
10.00–11.20	−0.0118	−0.0075	−0.0022	−0.0007	0	+0.0002	+0.0013	+0.0022	+0.0062	+0.0124
11.20–12.60	−0.0130	−0.0075	−0.0022	−0.0007	0	+0.0002	+0.0013	+0.0022	+0.0067	+0.0130
12.60–14.20	−0.0142	−0.0083	−0.0024	−0.0007	0	+0.0002	+0.0015	+0.0024	+0.0075	+0.0154
14.20–16.00	−0.0157	−0.0083	−0.0024	−0.0007	0	+0.0002	+0.0015	+0.0024	+0.0082	+0.0171

Table A–15

Charts of Theoretical Stress-Concentration Factors K_t^*

Figure A–15–1

Bar in tension or simple compression with a transverse hole. $\sigma_0 = F/A$, where $A = (w - d)t$ and t is the thickness.

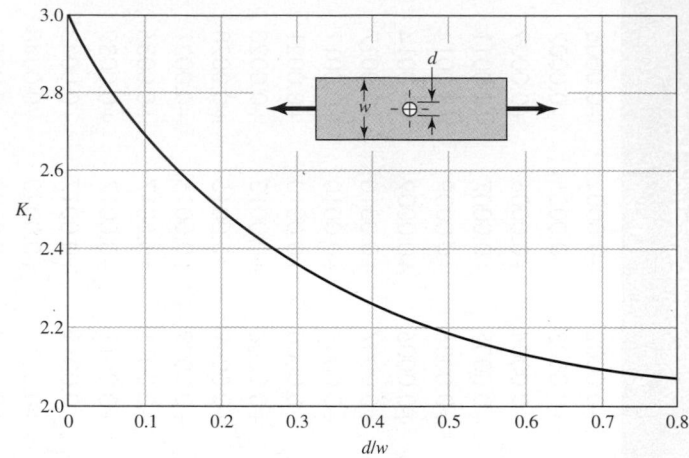

Figure A–15–2

Rectangular bar with a transverse hole in bending. $\sigma_0 = Mc/I$, where $I = (w - d)h^3/12$.

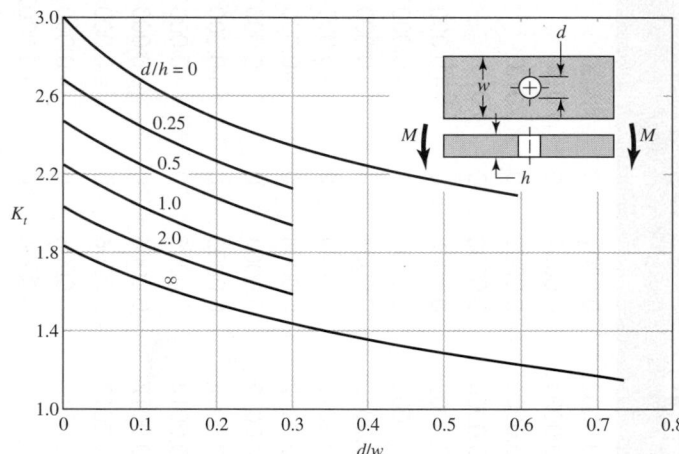

Figure A–15–3

Notched rectangular bar in tension or simple compression. $\sigma_0 = F/A$, where $A = dt$ and t is the thickness.

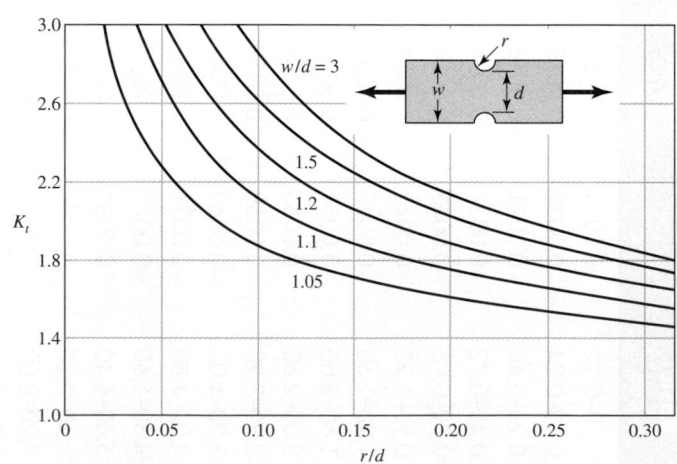

Table A-15

Charts of Theoretical Stress-Concentration Factors K_t^* (Continued)

Figure A-15-4

Notched rectangular bar in bending. $\sigma_0 = Mc/I$, where $c = d/2$, $I = td^3/12$, and t is the thickness.

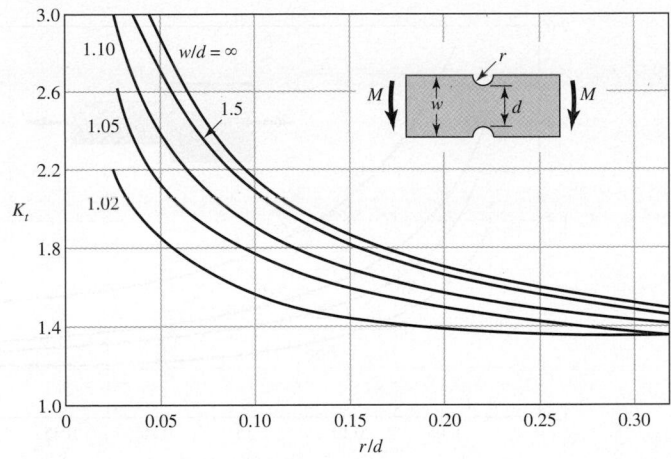

Figure A-15-5

Rectangular filleted bar in tension or simple compression. $\sigma_0 = F/A$, where $A = dt$ and t is the thickness.

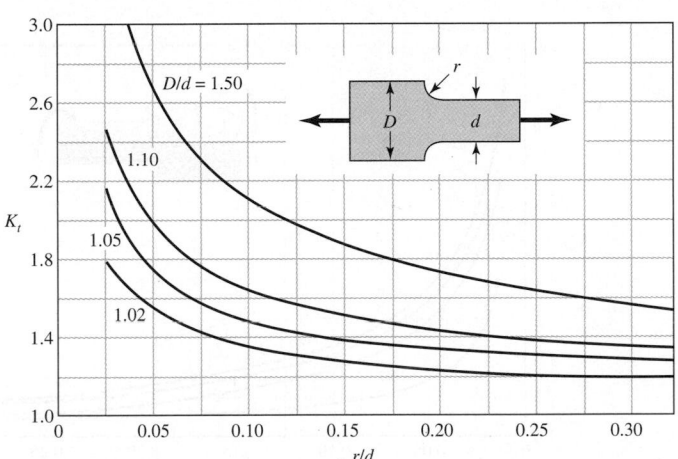

Figure A-15-6

Rectangular filleted bar in bending. $\sigma_0 = Mc/I$, where $c = d/2$, $I = td^3/12$, t is the thickness.

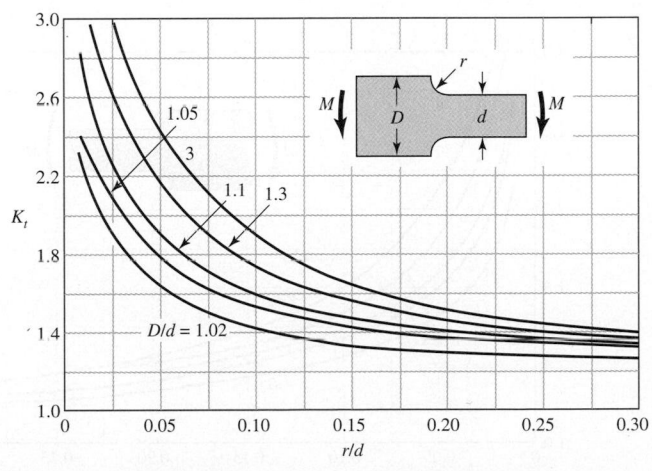

(continued)

*Factors from R. E. Peterson, "Design Factors for Stress Concentration," *Machine Design,* vol. 23, no. 2, February 1951, p. 169; no. 3, March 1951, p. 161, no. 5, May 1951, p. 159; no. 6, June 1951, p. 173; no. 7, July 1951, p. 155. Reprinted with permission from *Machine Design,* a Penton Media Inc. publication.

Table A–15

Charts of Theoretical Stress-Concentration Factors K_t^* (Continued)

Figure A–15–7

Round shaft with shoulder fillet in tension. $\sigma_0 = F/A$, where $A = \pi d^2/4$.

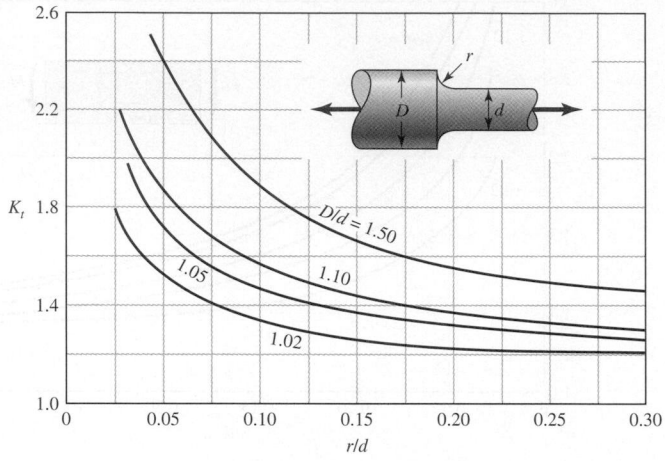

Figure A–15–8

Round shaft with shoulder fillet in torsion. $\tau_0 = Tc/J$, where $c = d/2$ and $J = \pi d^4/32$.

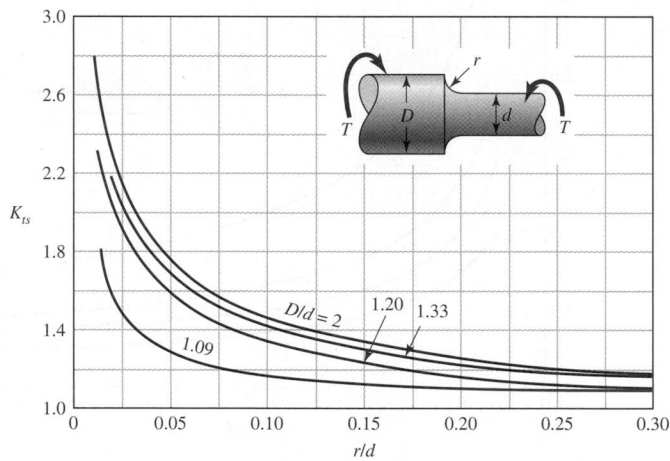

Figure A–15–9

Round shaft with shoulder fillet in bending. $\sigma_0 = Mc/I$, where $c = d/2$ and $I = \pi d^4/64$.

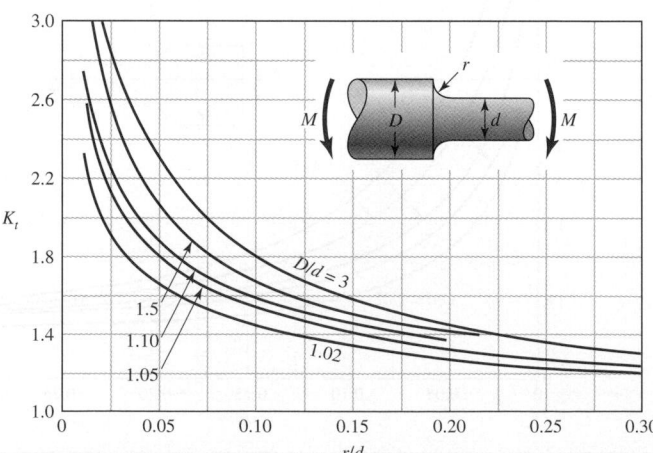

Table A–15

Charts of Theoretical Stress-Concentration Factors K_t^* (Continued)

Figure A–15–10

Round shaft in torsion with transverse hole.

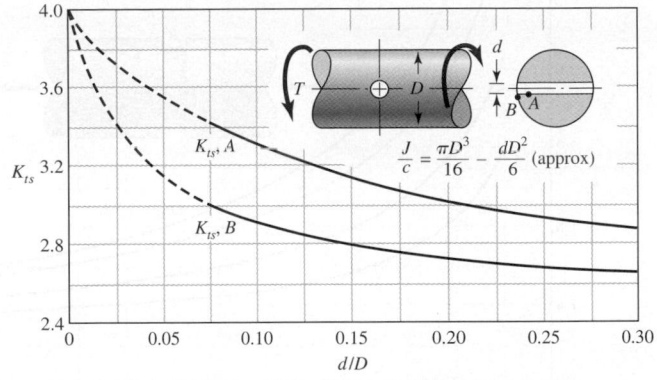

$$\frac{J}{c} = \frac{\pi D^3}{16} - \frac{dD^2}{6} \text{ (approx)}$$

Figure A–15–11

Round shaft in bending with a transverse hole. $\sigma_0 = M/[(\pi D^3/32) - (dD^2/6)]$, approximately.

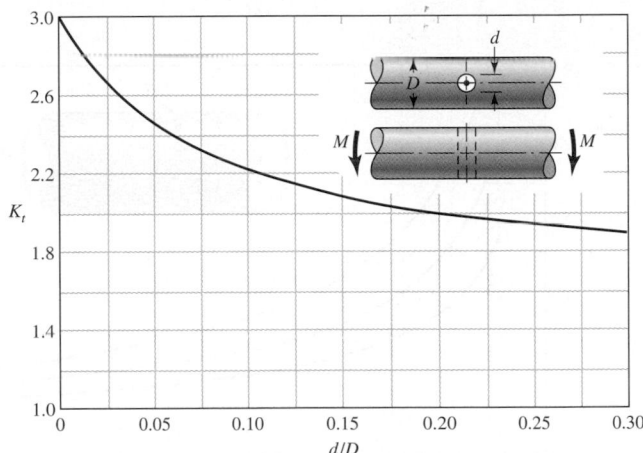

Figure A–15–12

Plate loaded in tension by a pin through a hole. $\sigma_0 = F/A$, where $A = (w - d)t$. When clearance exists, increase K_t 35 to 50 percent. (M. M. Frocht and H. N. Hill, "Stress Concentration Factors around a Central Circular Hole in a Plate Loaded through a Pin in Hole," J. Appl. Mechanics, vol. 7, no. 1, March 1940, p. A-5.)

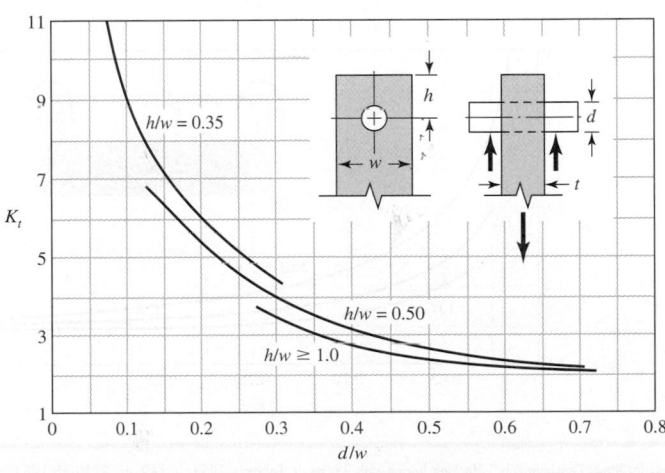

(continued)

*Factors from R. E. Peterson, "Design Factors for Stress Concentration," *Machine Design,* vol. 23, no. 2, February 1951, p. 169; no. 3, March 1951, p. 161, no. 5, May 1951, p. 159; no. 6, June 1951, p. 173; no. 7, July 1951, p. 155. Reprinted with permission from *Machine Design,* a Penton Media Inc. publication.

Table A–15

Charts of Theoretical Stress-Concentration Factors K_t^* (Continued)

Figure A–15–13

Grooved round bar in tension. $\sigma_0 = F/A$, where $A = \pi d^2/4$.

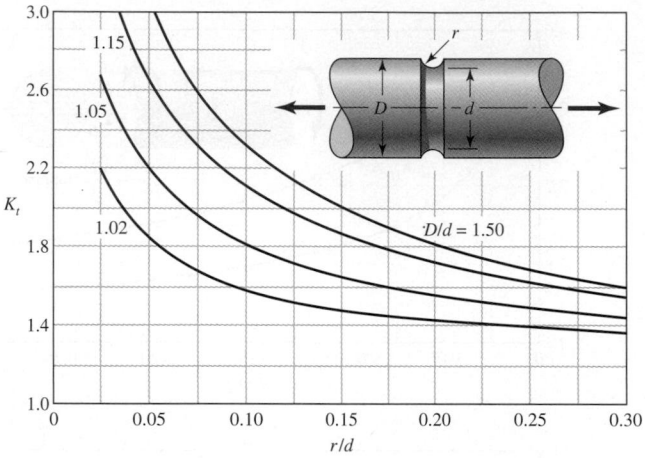

Figure A–15–14

Grooved round bar in bending. $\sigma_0 = Mc/I$, where $c = d/2$ and $I = \pi d^4/64$.

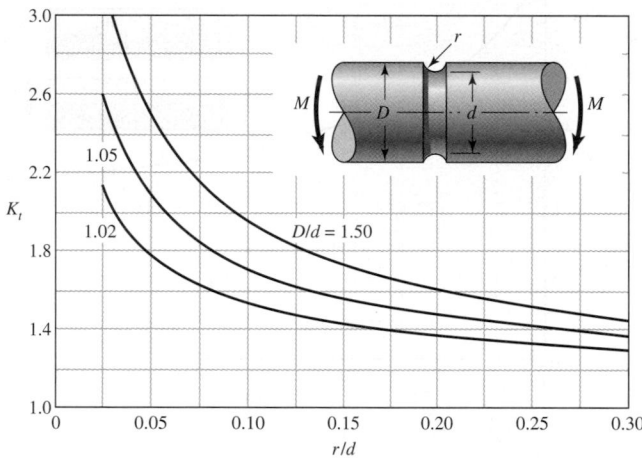

Figure A–15–15

Grooved round bar in torsion. $\tau_0 = Tc/J$, where $c = d/2$ and $J = \pi d^4/32$.

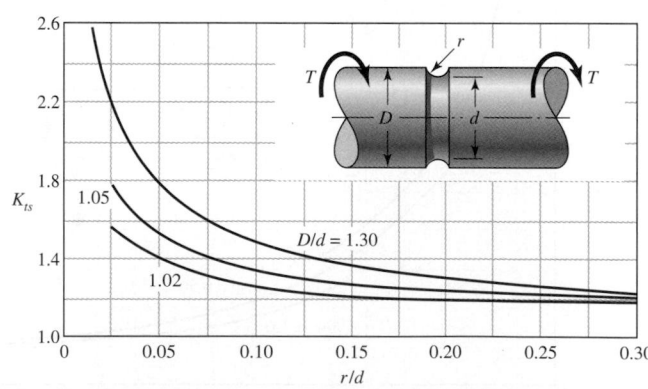

*Factors from R. E. Peterson, "Design Factors for Stress Concentration," *Machine Design*, vol. 23, no. 2, February 1951, p. 169; no. 3, March 1951, p. 161, no. 5, May 1951, p. 159; no. 6, June 1951, p. 173; no. 7, July 1951, p. 155. Reprinted with permission from *Machine Design*, a Penton Media Inc. publication.

Table A–16

Approximate Stress-Concentration Factor K_t for Bending of a Round Bar or Tube with a Transverse Round Hole

Source: R. E. Peterson, *Stress Concentration Factors,* Wiley, New York, 1974, pp. 146, 235.

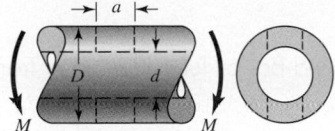

The nominal bending stress is $\sigma_0 = M/Z_{net}$ where Z_{net} is a reduced value of the section modulus and is defined by

$$Z_{net} = \frac{\pi A}{32D}(D^4 - d^4)$$

Values of A are listed in the table. Use $d = 0$ for a solid bar

| | d/D | | | | | |
| | 0.9 | | 0.6 | | 0 | |
a/D	A	K_t	A	K_t	A	K_t
0.050	0.92	2.63	0.91	2.55	0.88	2.42
0.075	0.89	2.55	0.88	2.43	0.86	2.35
0.10	0.86	2.49	0.85	2.36	0.83	2.27
0.125	0.82	2.41	0.82	2.32	0.80	2.20
0.15	0.79	2.39	0.79	2.29	0.76	2.15
0.175	0.76	2.38	0.75	2.26	0.72	2.10
0.20	0.73	2.39	0.72	2.23	0.68	2.07
0.225	0.69	2.40	0.68	2.21	0.65	2.04
0.25	0.67	2.42	0.64	2.18	0.61	2.00
0.275	0.66	2.48	0.61	2.16	0.58	1.97
0.30	0.64	2.52	0.58	2.14	0.54	1.94

(continued)

Table A–16 (Continued)

Approximate Stress-Concentration Factors K_{ts} for a Round Bar or Tube Having a Transverse Round Hole and Loaded in Torsion   Source: R. E. Peterson, *Stress Concentration Factors,* Wiley, New York, 1974, pp. 148, 244.

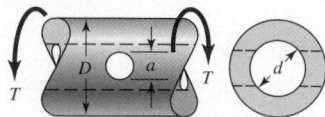

The maximum stress occurs on the inside of the hole, slightly below the shaft surface. The nominal shear stress is $\tau_0 = TD/2J_{net}$, where J_{net} is a reduced value of the second polar moment of area and is defined by

$$J_{net} = \frac{\pi A(D^4 - d^4)}{32}$$

Values of A are listed in the table. Use $d = 0$ for a solid bar.

										d/D	
	0.9			**0.8**			**0.6**			**0.4**			**0**	
a/D	**A**	**K_{ts}**	**A**	**K_{ts}**	**A**	**K_{ts}**	**A**	**K_{ts}**	**A**	**K_{ts}**
0.05	0.96	1.78							0.95	1.77
0.075	0.95	1.82							0.93	1.71
0.10	0.94	1.76	0.93	1.74	0.92	1.72	0.92	1.70	0.92	1.68
0.125	0.91	1.76	0.91	1.74	0.90	1.70	0.90	1.67	0.89	1.64
0.15	0.90	1.77	0.89	1.75	0.87	1.69	0.87	1.65	0.87	1.62
0.175	0.89	1.81	0.88	1.76	0.87	1.69	0.86	1.64	0.85	1.60
0.20	0.88	1.96	0.86	1.79	0.85	1.70	0.84	1.63	0.83	1.58
0.25	0.87	2.00	0.82	1.86	0.81	1.72	0.80	1.63	0.79	1.54
0.30	0.80	2.18	0.78	1.97	0.77	1.76	0.75	1.63	0.74	1.51
0.35	0.77	2.41	0.75	2.09	0.72	1.81	0.69	1.63	0.68	1.47
0.40	0.72	2.67	0.71	2.25	0.68	1.89	0.64	1.63	0.63	1.44

Table A-17

Preferred Sizes and Renard (R-Series) Numbers
(When a choice can be made, use one of these sizes; however, not all parts or items are available in all the sizes shown in the table.)

Fraction of Inches

$\frac{1}{64}$, $\frac{1}{32}$, $\frac{1}{16}$, $\frac{3}{32}$, $\frac{1}{8}$, $\frac{5}{32}$, $\frac{3}{16}$, $\frac{1}{4}$, $\frac{5}{16}$, $\frac{3}{8}$, $\frac{7}{16}$, $\frac{1}{2}$, $\frac{9}{16}$, $\frac{5}{8}$, $\frac{11}{16}$, $\frac{3}{4}$, $\frac{7}{8}$, 1, $1\frac{1}{4}$, $1\frac{1}{2}$, $1\frac{3}{4}$, 2, $2\frac{1}{4}$, $2\frac{1}{2}$, $2\frac{3}{4}$, 3, $3\frac{1}{4}$, $3\frac{1}{2}$, $3\frac{3}{4}$, 4, $4\frac{1}{4}$, $4\frac{1}{2}$, $4\frac{3}{4}$, 5, $5\frac{1}{4}$, $5\frac{1}{2}$, $5\frac{3}{4}$, 6, $6\frac{1}{2}$, 7, $7\frac{1}{2}$, 8, $8\frac{1}{2}$, 9, $9\frac{1}{2}$, 10, $10\frac{1}{2}$, 11, $11\frac{1}{2}$, 12, $12\frac{1}{2}$, 13, $13\frac{1}{2}$, 14, $14\frac{1}{2}$, 15, $15\frac{1}{2}$, 16, $16\frac{1}{2}$, 17, $17\frac{1}{2}$, 18, $18\frac{1}{2}$, 19, $19\frac{1}{2}$, 20

Decimal Inches

0.010, 0.012, 0.016, 0.020, 0.025, 0.032, 0.040, 0.05, 0.06, 0.08, 0.10, 0.12, 0.16, 0.20, 0.24, 0.30, 0.40, 0.50, 0.60, 0.80, 1.00, 1.20, 1.40, 1.60, 1.80, 2.0, 2.4, 2.6, 2.8, 3.0, 3.2, 3.4, 3.6, 3.8, 4.0, 4.2, 4.4, 4.6, 4.8, 5.0, 5.2, 5.4, 5.6, 5.8, 6.0, 7.0, 7.5, 8.5, 9.0, 9.5, 10.0, 10.5, 11.0, 11.5, 12.0, 12.5, 13.0, 13.5, 14.0, 14.5, 15.0, 15.5, 16.0, 16.5, 17.0, 17.5, 18.0, 18.5, 19.0, 19.5, 20

Millimeters

0.05, 0.06, 0.08, 0.10, 0.12, 0.16, 0.20, 0.25, 0.30, 0.40, 0.50, 0.60, 0.70, 0.80, 0.90, 1.0, 1.1, 1.2, 1.4, 1.5, 1.6, 1.8, 2.0, 2.2, 2.5, 2.8, 3.0, 3.5, 4.0, 4.5, 5.0, 5.5, 6.0, 6.5, 7.0, 8.0, 9.0, 10, 11, 12, 14, 16, 18, 20, 22, 25, 28, 30, 32, 35, 40, 45, 50, 60, 80, 100, 120, 140, 160, 180, 200, 250, 300

Renard Numbers*

1st choice, R5: 1, 1.6, 2.5, 4, 6.3, 10
2d choice, R10: 1.25, 2, 3.15, 5, 8
3d choice, R20: 1.12, 1.4, 1.8, 2.24, 2.8, 3.55, 4.5, 5.6, 7.1, 9
4th choice, R40: 1.06, 1.18, 1.32, 1.5, 1.7, 1.9, 2.12, 2.36, 2.65, 3, 3.35, 3.75, 4.25, 4.75, 5.3, 6, 6.7, 7.5, 8.5, 9.5

*May be multiplied or divided by powers of 10.

Table A–18

Geometric Properties

Part 1 Properties of Sections

A = area

G = location of centroid

$I_x = \int x^2 \, dA$ = second moment of area about x axis

$I_{xy} = \int xy \, dA$ = mixed moment of area about x and y axes

$J_G = \int r^2 \, dA = \int (x^2 + y^2) \, dA = I_x + I_y$

= second polar moment of area about axis through G

$k_x^2 = I_x/A$ = squared radius of gyration about x axis

Rectangle

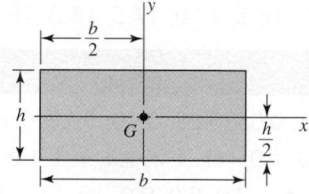

$A = bh \qquad I_x = \dfrac{bh^3}{12} \qquad I_y = \dfrac{b^3 h}{12} \qquad I_{xy} = 0$

Circle

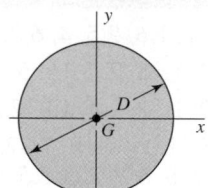

$A = \dfrac{\pi D^2}{4} \qquad I_x = I_y = \dfrac{\pi D^4}{64} \qquad I_{xy} = 0$

Hollow circle

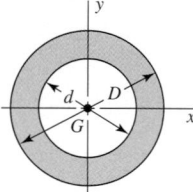

$A = \dfrac{\pi}{4}(D^2 - d^2) \qquad I_x = I_y = \dfrac{\pi}{64}(D^4 - d^4) \qquad I_{xy} = 0$

Table A–18

Geometric Properties
(Continued)

Right triangles

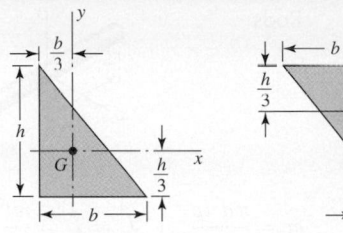

$$A = \frac{bh}{2} \qquad I_x = \frac{bh^3}{36} \qquad I_y = \frac{b^3 h}{36} \qquad I_{xy} = \frac{-b^2 h^2}{72}$$

Right triangles

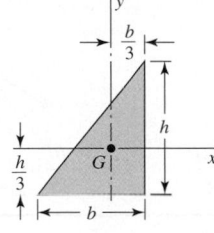 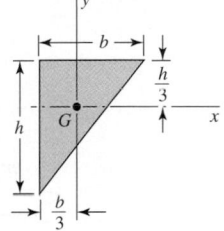

$$A = \frac{bh}{2} \qquad I_x = \frac{bh^3}{36} \qquad I_y = \frac{b^3 h}{36} \qquad I_{xy} = \frac{b^2 h^2}{72}$$

Quarter-circles

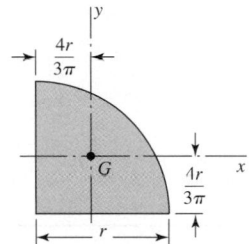 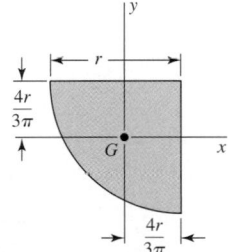

$$A = \frac{\pi r^2}{4} \qquad I_x = I_y = r^4\left(\frac{\pi}{16} - \frac{4}{9\pi}\right) \qquad I_{xy} = r^4\left(\frac{1}{8} - \frac{4}{9\pi}\right)$$

Quarter-circles

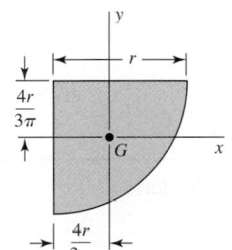 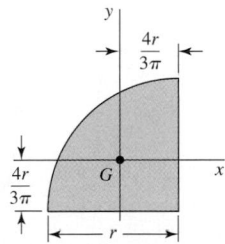

$$A = \frac{\pi r^2}{4} \qquad I_x = I_y = r^4\left(\frac{\pi}{16} - \frac{4}{9\pi}\right) \qquad I_{xy} = r^4\left(\frac{4}{9\pi} - \frac{1}{8}\right)$$

(continued)

Table A–18

Geometric Properties
(Continued)

Part 2 Properties of Solids (ρ = Density, Weight per Unit Volume)

Rods

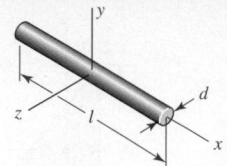

$$m = \frac{\pi d^2 l \rho}{4g} \qquad I_y = I_z = \frac{ml^2}{12}$$

Round disks

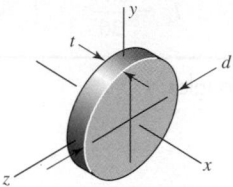

$$m = \frac{\pi d^2 t \rho}{4g} \qquad I_x = \frac{md^2}{8} \qquad I_y = I_z = \frac{md^2}{16}$$

Rectangular prisms

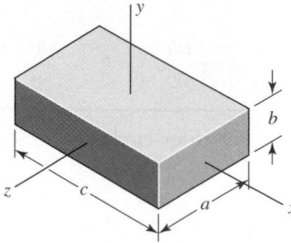

$$m = \frac{abc\rho}{g} \qquad I_x = \frac{m}{12}(a^2 + b^2) \qquad I_y = \frac{m}{12}(a^2 + c^2) \quad I_z = \frac{m}{12}(b^2 + c^2)$$

Cylinders

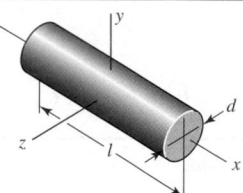

$$m = \frac{\pi d^2 l \rho}{4g} \qquad I_x = \frac{md^2}{8} \qquad I_y = I_z = \frac{m}{48}(3d^2 + 4l^2)$$

Hollow cylinders

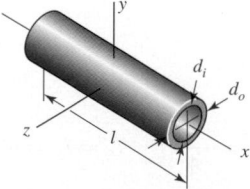

$$m = \frac{\pi \left(d_o^2 - d_i^2\right) l \rho}{4g} \qquad I_x = \frac{m}{8}\left(d_o^2 + d_i^2\right) \qquad I_y = I_z = \frac{m}{48}\left(3d_o^2 + 3d_i^2 + 4l^2\right)$$

Table A–19

American Standard Pipe

Nominal Size, in	Outside Diameter, in	Threads per inch	Wall Thickness, in		
			Standard No. 40	Extra Strong No. 80	Double Extra Strong
$\frac{1}{8}$	0.405	27	0.070	0.098	
$\frac{1}{4}$	0.540	18	0.090	0.122	
$\frac{3}{8}$	0.675	18	0.093	0.129	
$\frac{1}{2}$	0.840	14	0.111	0.151	0.307
$\frac{3}{4}$	1.050	14	0.115	0.157	0.318
1	1.315	$11\frac{1}{2}$	0.136	0.183	0.369
$1\frac{1}{4}$	1.660	$11\frac{1}{2}$	0.143	0.195	0.393
$1\frac{1}{2}$	1.900	$11\frac{1}{2}$	0.148	0.204	0.411
2	2.375	$11\frac{1}{2}$	0.158	0.223	0.447
$2\frac{1}{2}$	2.875	8	0.208	0.282	0.565
3	3.500	8	0.221	0.306	0.615
$3\frac{1}{2}$	4.000	8	0.231	0.325	
4	4.500	8	0.242	0.344	0.690
5	5.563	8	0.263	0.383	0.768
6	6.625	8	0.286	0.441	0.884
8	8.625	8	0.329	0.510	0.895

Table A–20

Deterministic ASTM Minimum Tensile and Yield Strengths for Some Hot-Rolled (HR) and Cold-Drawn (CD) Steels [The strengths listed are estimated ASTM minimum values in the size range 18 to 32 mm ($\frac{3}{4}$ to $1\frac{1}{4}$ in). These strengths are suitable for use with the design factor defined in Sec. 1–10, provided the materials conform to ASTM A6 or A568 requirements or are required in the purchase specifications. Remember that a numbering system is not a specification. See Table 1–1 for certain ASTM steels.]   *Source:* 1986 SAE Handbook, p. 2.15.

1 UNS No.	2 SAE and/or AISI No.	3 Processing	4 Tensile Strength, MPa (kpsi)	5 Yield Strength, MPa (kpsi)	6 Elongation in 2 in, %	7 Reduction in Area, %	8 Brinell Hardness
G10060	1006	HR	300 (43)	170 (24)	30	55	86
		CD	330 (48)	280 (41)	20	45	95
G10100	1010	HR	320 (47)	180 (26)	28	50	95
		CD	370 (53)	300 (44)	20	40	105
G10150	1015	HR	340 (50)	190 (27.5)	28	50	101
		CD	390 (56)	320 (47)	18	40	111
G10180	1018	HR	400 (58)	220 (32)	25	50	116
		CD	440 (64)	370 (54)	15	40	126
G10200	1020	HR	380 (55)	210 (30)	25	50	111
		CD	470 (68)	390 (57)	15	40	131
G10300	1030	HR	470 (68)	260 (37.5)	20	42	137
		CD	520 (76)	440 (64)	12	35	149
G10350	1035	HR	500 (72)	270 (39.5)	18	40	143
		CD	550 (80)	460 (67)	12	35	163
G10400	1040	HR	520 (76)	290 (42)	18	40	149
		CD	590 (85)	490 (71)	12	35	170
G10450	1045	HR	570 (82)	310 (45)	16	40	163
		CD	630 (91)	530 (77)	12	35	179
G10500	1050	HR	620 (90)	340 (49.5)	15	35	179
		CD	690 (100)	580 (84)	10	30	197
G10600	1060	HR	680 (98)	370 (54)	12	30	201
G10800	1080	HR	770 (112)	420 (61.5)	10	25	229
G10950	1095	HR	830 (120)	460 (66)	10	25	248

Table A–21

Mean Mechanical Properties of Some Heat-Treated Steels
[These are typical properties for materials normalized and annealed. The properties for quenched and tempered (Q&T) steels are from a single heat. Because of the many variables, the properties listed are global averages. In all cases, data were obtained from specimens of diameter 0.505 in, machined from 1-in rounds, and of gauge length 2 in. unless noted, all specimens were oil-quenched.]   Source: ASM Metals Reference Book, 2d ed., American Society for Metals, Metals Park, Ohio, 1983.

1 AISI No.	2 Treatment	3 Temperature °C (°F)	4 Tensile Strength MPa (kpsi)	5 Yield Strength, MPa (kpsi)	6 Elongation, %	7 Reduction in Area, %	8 Brinell Hardness
1030	Q&T*	205 (400)	848 (123)	648 (94)	17	47	495
	Q&T*	315 (600)	800 (116)	621 (90)	19	53	401
	Q&T*	425 (800)	731 (106)	579 (84)	23	60	302
	Q&T*	540 (1000)	669 (97)	517 (75)	28	65	255
	Q&T*	650 (1200)	586 (85)	441 (64)	32	70	207
	Normalized	925 (1700)	521 (75)	345 (50)	32	61	149
	Annealed	870 (1600)	430 (62)	317 (46)	35	64	137
1040	Q&T	205 (400)	779 (113)	593 (86)	19	48	262
	Q&T	425 (800)	758 (110)	552 (80)	21	54	241
	Q&T	650 (1200)	634 (92)	434 (63)	29	65	192
	Normalized	900 (1650)	590 (86)	374 (54)	28	55	170
	Annealed	790 (1450)	519 (75)	353 (51)	30	57	149
1050	Q&T*	205 (400)	1120 (163)	807 (117)	9	27	514
	Q&T*	425 (800)	1090 (158)	793 (115)	13	36	444
	Q&T*	650 (1200)	717 (104)	538 (78)	28	65	235
	Normalized	900 (1650)	748 (108)	427 (62)	20	39	217
	Annealed	790 (1450)	636 (92)	365 (53)	24	40	187
1060	Q&T	425 (800)	1080 (156)	765 (111)	14	41	311
	Q&T	540 (1000)	965 (140)	669 (97)	17	45	277
	Q&T	650 (1200)	800 (116)	524 (76)	23	54	229
	Normalized	900 (1650)	776 (112)	421 (61)	18	37	229
	Annealed	790 (1450)	626 (91)	372 (54)	22	38	179
1095	Q&T	315 (600)	1260 (183)	813 (118)	10	30	375
	Q&T	425 (800)	1210 (176)	772 (112)	12	32	363
	Q&T	540 (1000)	1090 (158)	676 (98)	15	37	321
	Q&T	650 (1200)	896 (130)	552 (80)	21	47	269
	Normalized	900 (1650)	1010 (147)	500 (72)	9	13	293
	Annealed	790 (1450)	658 (95)	380 (55)	13	21	192
1141	Q&T	315 (600)	1460 (212)	1280 (186)	9	32	415
	Q&T	540 (1000)	896 (130)	765 (111)	18	57	262

(continued)

Table A–21 *(Continued)*

Mean Mechanical Properties of Some Heat-Treated Steels
[These are typical properties for materials normalized and annealed. The properties for quenched and tempered (Q&T) steels are from a single heat. Because of the many variables, the properties listed are global averages. In all cases, data were obtained from specimens of diameter 0.505 in, machined from 1-in rounds, and of gauge length 2 in. Unless noted, all specimens were oil-quenched.] *Source: ASM Metals Reference Book,* 2d ed., American Society for Metals, Metals Park, Ohio, 1983.

1	2	3	4	5	6	7	8
			Tensile	Yield			
		Temperature	Strength	Strength,	Elongation,	Reduction	Brinell
AISI No.	Treatment	°C (°F)	MPa (kpsi)	MPa (kpsi)	%	in Area, %	Hardness
4130	Q&T*	205 (400)	1630 (236)	1460 (212)	10	41	467
	Q&T*	315 (600)	1500 (217)	1380 (200)	11	43	435
	Q&T*	425 (800)	1280 (186)	1190 (173)	13	49	380
	Q&T*	540 (1000)	1030 (150)	910 (132)	17	57	315
	Q&T*	650 (1200)	814 (118)	703 (102)	22	64	245
	Normalized	870 (1600)	670 (97)	436 (63)	25	59	197
	Annealed	865 (1585)	560 (81)	361 (52)	28	56	156
4140	Q&T	205 (400)	1770 (257)	1640 (238)	8	38	510
	Q&T	315 (600)	1550 (225)	1430 (208)	9	43	445
	Q&T	425 (800)	1250 (181)	1140 (165)	13	49	370
	Q&T	540 (1000)	951 (138)	834 (121)	18	58	285
	Q&T	650 (1200)	758 (110)	655 (95)	22	63	230
	Normalized	870 (1600)	1020 (148)	655 (95)	18	47	302
	Annealed	815 (1500)	655 (95)	417 (61)	26	57	197
4340	Q&T	315 (600)	1720 (250)	1590 (230)	10	40	486
	Q&T	425 (800)	1470 (213)	1360 (198)	10	44	430
	Q&T	540 (1000)	1170 (170)	1080 (156)	13	51	360
	Q&T	650 (1200)	965 (140)	855 (124)	19	60	280

*Water-quenched

Table A–22

Results of Tensile Tests of Some Metals*   Source: J. Datsko, "Solid Materials," chap. 7 in Joseph E. Shigley and Charles R. Mischke (eds.-in-chief), *Standard Handbook of Machine Design*, 2nd ed., McGraw-Hill, New York, 1996, pp. 7.47–7.50.

Number	Material	Condition	Yield S_y, MPa (kpsi)	Ultimate S_u, MPa (kpsi)	Strength (Tensile) Fracture, σ_f, MPa (kpsi)	Coefficient σ_0, MPa (kpsi)	Strain Strength, Exponent m	Fracture Strain ε_f
1018	Steel	Annealed	220 (32.0)	341 (49.5)	628 (91.1)[†]	620 (90.0)	0.25	1.05
1144	Steel	Annealed	358 (52.0)	646 (93.7)	898 (130)[†]	992 (144)	0.14	0.49
1212	Steel	HR	193 (28.0)	424 (61.5)	729 (106)[†]	758 (110)	0.24	0.85
1045	Steel	Q&T 600°F	1520 (220)	1580 (230)	2380 (345)	1880 (273)[†]	0.041	0.81
4142	Steel	Q&T 600°F	1720 (250)	1930 (210)	2340 (340)	1760 (255)[†]	0.048	0.43
303	Stainless steel	Annealed	241 (35.0)	601 (87.3)	1520 (221)[†]	1410 (205)	0.51	1.16
304	Stainless steel	Annealed	276 (40.0)	568 (82.4)	160C (233)[†]	1270 (185)	0.45	1.67
2011	Aluminum alloy	T6	169 (24.5)	324 (47.0)	325 (47.2)[†]	620 (90)	0.28	0.10
2024	Aluminum alloy	T4	296 (43.0)	446 (64.8)	533 (77.3)[†]	689 (100)	0.15	0.18
7075	Aluminum alloy	T6	542 (78.6)	593 (86.0)	706 (102)[†]	882 (128)	0.13	0.18

*Values from one or two heats and believed to be attainable using proper purchase specifications. The fracture strain may vary as much as 100 percent.

[†]Derived value.

Table A-23

Mean Monotonic and Cyclic Stress-Strain Properties of Selected Steels   Source: ASM Metals Reference Book, 2nd ed., American Society for Metals, Metals Park, Ohio, 1983, p. 217.

Grade (a)	Orientation (e)	Description (f)	Hardness HB	Tensile Strength S_{ut} MPa	ksi	Reduction in Area %	True Strain at Fracture ε_f	Modulus of Elasticity E GPa	10^4 psi	Fatigue Strength Coefficient σ'_f MPa	ksi	Fatigue Strength Exponent b	Fatigue Ductility Coefficient ε'_f	Fatigue Ductility Exponent c
A538A (b)	L	STA	405	1515	220	67	1.10	185	27	1655	240	−0.065	0.30	−0.62
A538B (b)	L	STA	460	1860	270	56	0.82	185	27	2135	310	−0.071	0.80	−0.71
A538C (b)	L	STA	480	2000	290	55	0.81	180	26	2240	325	−0.07	0.60	−0.75
AM-350 (c)	L	HR, A		1315	191	52	0.74	195	28	2800	406	−0.14	0.33	−0.84
AM-350 (c)	L	CD	496	1905	276	20	0.23	180	26	2690	390	−0.102	0.10	−0.42
Gainex (c)	LT	HR sheet		530	77	58	0.86	200	29.2	805	117	−0.07	0.86	−0.65
Gainex (c)	L	HR sheet		510	74	64	1.02	200	29.2	805	117	−0.071	0.86	−0.68
H-11	L	Ausformed	660	2585	375	33	0.40	205	30	3170	460	−0.077	0.08	−0.74
RQC-100 (c)	LT	HR plate	290	940	136	43	0.56	205	30	1240	180	−0.07	0.66	−0.69
RQC-100 (c)	L	HR plate	290	930	135	67	1.02	205	30	1240	180	−0.07	0.66	−0.69
10B62	L	Q&T	430	1640	238	38	0.89	195	28	1780	258	−0.067	0.32	−0.56
1005-1009	LT	HR sheet	90	360	52	73	1.3	205	30	580	84	−0.09	0.15	−0.43
1005-1009	LT	CD sheet	125	470	68	66	1.09	205	30	515	75	−0.059	0.30	−0.51
1005-1009	L	CD sheet	125	415	60	64	1.02	200	29	540	78	−0.073	0.11	−0.41
1005-1009	L	HR sheet	90	345	50	80	1.6	200	29	640	93	−0.109	0.10	−0.39
1015	L	Normalized	80	415	60	68	1.14	205	30	825	120	−0.11	0.95	−0.64
1020	L	HR plate	108	440	64	62	0.96	205	29.5	895	130	−0.12	0.41	−0.51
1040	L	As forged	225	620	90	60	0.93	200	29	1540	223	−0.14	0.61	−0.57
1045	L	Q&T	225	725	105	65	1.04	200	29	1225	178	−0.095	1.00	−0.66
1045	L	Q&T	410	1450	210	51	0.72	200	29	1860	270	−0.073	0.60	−0.70
1045	L	Q&T	390	1345	195	59	0.89	205	30	1585	230	−0.074	0.45	−0.68
1045	L	Q&T	450	1585	230	55	0.81	205	30	1795	260	−0.07	0.35	−0.69
1045	L	Q&T	500	1825	265	51	0.71	205	30	2275	330	−0.08	0.25	−0.68
1045	L	Q&T	595	2240	325	41	0.52	205	30	2725	395	−0.081	0.07	−0.60
1144	L	CDSR	265	930	135	33	0.51	195	28.5	1000	145	−0.08	0.32	−0.58

Grade	Orient.	Condition												
1144	L	DAT	305	1035	150	25	0.29	200	28.8	1585	230	−0.09	0.27	−0.53
1541F	L	Q&T forging	290	950	138	49	0.68	205	29.9	1275	185	−0.076	0.68	−0.65
1541F	L	Q&T forging	260	890	129	60	0.93	205	29.9	1275	185	−0.071	0.93	−0.65
4130	L	Q&T	258	895	130	67	1.12	220	32	1275	185	−0.083	0.92	−0.63
4130	L	Q&T	365	1425	207	55	0.79	200	29	1695	246	−0.081	0.89	−0.69
4140	L	Q&T, DAT	310	1075	156	60	0.69	200	29.2	1825	265	−0.08	1.2	−0.59
4142	L	DAT	310	1060	154	29	0.35	200	29	1450	210	−0.10	0.22	−0.51
4142	L	DAT	335	1250	181	28	0.34	200	28.9	1250	181	−0.08	0.06	−0.62
4142	L	Q&T	380	1415	205	48	0.66	205	30	1825	265	−0.08	0.45	−0.75
4142	L	Q&T and deformed	400	1550	225	47	0.63	200	29	1895	275	−0.09	0.50	−0.75
4142	L	Q&T	450	1760	255	42	0.54	205	30	2000	290	−0.08	0.40	−0.73
4142	L	Q&T and deformed	475	2035	295	20	0.22	200	29	2070	300	−0.082	0.20	−0.77
4142	L	Q&T and deformed	450	1930	280	37	0.46	200	29	2105	305	−0.09	0.60	−0.76
4142	L	Q&T	475	1930	280	35	0.43	205	30	2170	315	−0.081	0.09	−0.61
4142	L	Q&T	560	2240	325	27	0.31	205	30	2655	385	−0.089	0.07	−0.76
4340	L	HR, A	243	825	120	43	0.57	195	28	1200	174	−0.095	0.45	−0.54
4340	L	Q&T	409	1470	213	38	0.48	200	29	2000	290	−0.091	0.48	−0.60
4340	L	Q&T	350	1240	180	57	0.84	195	28	1655	240	−0.076	0.73	−0.62
5160	L	Q&T	430	1670	242	42	0.87	195	28	1930	280	−0.071	0.40	−0.57
52100	L	SH, Q&T	518	2015	292	11	0.12	205	30	2585	375	−0.09	0.18	−0.56
9262	L	A	260	925	134	14	0.16	205	30	1040	151	−0.071	0.16	−0.47
9262	L	Q&T	280	1000	145	33	0.41	195	28	1220	177	−0.073	0.41	−0.60
9262	L	Q&T	410	565	227	32	0.38	200	29	1855	269	−0.057	0.38	−0.65
950C (d)	LT	HR plate	159	565	82	64	1.03	205	29.6	1170	170	−0.12	0.95	−0.61
950C (d)	L	HR bar	150	565	82	69	1.19	205	30	970	141	−0.11	0.85	−0.59
950X (d)	L	Plate channel	150	440	64	65	1.06	205	30	625	91	−0.075	0.35	−0.54
950X (d)	L	HR plate	156	530	77	72	1.24	205	29.5	1005	146	−0.10	0.85	−0.61
950X (d)	L	Plate channel	225	695	101	68	1.15	195	28.2	1055	153	−0.08	0.21	−0.53

Notes: (a) AISI/SAE grade, unless otherwise indicated. (b) ASTM designation. (c) Proprietary designation. (d) SAE HSLA grade. (e) Orientation of axis of specimen, relative to rolling direction; L is longitudinal (parallel to rolling direction); LT is long transverse (perpendicular to rolling direction). (f) STA, solution treated and aged; HR, hot rolled; CD, cold drawn; Q&T, quenched and tempered; CDSR, cold drawn strain relieved; DAT, drawn at temperature; A, annealed.

From *ASM Metals Reference Book, 2nd edition*, 1983; ASM International, Materials Park, OH 44073-0002; table 217. Reprinted by permission of ASM International®, www.asminternational.org.

Table A–24

Mechanical Properties of Three Non-Steel Metals

(a) Typical Properties of Gray Cast Iron

[The American Society for Testing and Materials (ASTM) numbering system for gray cast iron is such that the numbers correspond to the minimum tensile strength in kpsi. Thus an ASTM No. 20 cast iron has a minimum tensile strength of 20 kpsi. Note particularly that the tabulations are *typical* of several heats.]

ASTM Number	Tensile Strength S_{ut}, kpsi	Compressive Strength S_{uc}, kpsi	Shear Modulus of Rupture S_{su}, kpsi	Modulus of Elasticity, Mpsi[†]		Endurance Limit* S_e, kpsi	Brinell Hardness H_B	Fatigue Stress-Concentration Factor K_f
				Tension	Torsion			
20	22	83	26	9.6–14	3.9–5.6	10	156	1.00
25	26	97	32	11.5–14.8	4.6–6.0	11.5	174	1.05
30	31	109	40	13–16.4	5.2–6.6	14	201	1.10
35	36.5	124	48.5	14.5–17.2	5.8–6.9	16	212	1.15
40	42.5	140	57	16–20	6.4–7.8	18.5	235	1.25
50	52.5	164	73	18.8–22.8	7.2–8.0	21.5	262	1.35
60	62.5	187.5	88.5	20.4–23.5	7.8–8.5	24.5	302	1.50

*Polished or machined specimens.

[†]The modulus of elasticity of cast iron in compression corresponds closely to the upper value in the range given for tension and is a more constant value than that for tension.

Table A–24

Mechanical Properties of Three Non-Steel Metals *(Continued)*

(b) Mechanical Properties of Some Aluminum Alloys

[These *are typical* properties for sizes of about $\frac{1}{2}$ in; similar properties can be obtained by using proper purchase specifications. The values given for fatigue strength correspond to $50(10^7)$ cycles of completely reversed stress. Alluminum alloys do not have an endurance limit. Yield strengths were obtained by the 0.2 percent offset method.]

Aluminum Association Number	Temper	Yield, S_y, MPa (kpsi)	Strength Tensile, S_u, MPa (kpsi)	Fatigue, S_f, MPa (kpsi)	Elongation in 2 in, %	Brinell Hardness H_B
Wrought:						
2017	O	70 (10)	179 (26)	90 (13)	22	45
2024	O	76 (11)	186 (27)	90 (13)	22	47
	T3	345 (50)	482 (70)	138 (20)	16	120
3003	H12	117 (17)	131 (19)	55 (8)	20	35
	H16	165 (24)	179 (26)	65 (9.5)	14	47
3004	H34	186 (27)	234 (34)	103 (15)	12	63
	H38	234 (34)	276 (40)	110 (16)	6	77
5052	H32	186 (27)	234 (34)	117 (17)	18	62
	H36	234 (34)	269 (39)	124 (18)	10	74
Cast:						
319.0*	T6	165 (24)	248 (36)	69 (10)	2.0	80
333.0†	T5	172 (25)	234 (34)	83 (12)	1.0	100
	T6	207 (30)	289 (42)	103 (15)	1.5	105
335.0*	T6	172 (25)	241 (35)	62 (9)	3.0	80
	T7	248 (36)	262 (38)	62 (9)	0.5	85

*Sand casting.

†Permanent-mold casting.

(c) Mechanical Properties of Some Titanium Alloys

Titanium Alloy	Condition	Yield, S_y (0.2% offset) MPa (kpsi)	Strength Tensile, S_{ut} MPa (kpsi)	Elongation in 2 in, %	Hardness (Brinell or Rockwell)
Ti-35A†	Annealed	210 (30)	275 (40)	30	135 HB
Ti-50A†	Annealed	310 (45)	380 (55)	25	215 HB
Ti-0.2 Pd	Annealed	280 (40)	340 (50)	28	200 HB
Ti-5 Al-2.5 Sn	Annealed	760 (110)	790 (115)	16	36 HRC
Ti-8 Al-1 Mo-1 V	Annealed	900 (130)	965 (140)	15	39 HRC
Ti-6 Al-6 V-2 Sn	Annealed	970 (140)	1030 (150)	14	38 HRC
Ti-6Al-4V	Annealed	900 (130)	830 (120)	14	36 HRC
Ti-13 V-11 Cr-3 Al	Sol. + aging	1207 (175)	1276 (185)	8	40 HRC

†Commercially pure alpha titanium

Table A-25

Stochastic Yield and Ultimate Strengths for Selected Materials   *Source*: Data compiled from "Some Property Data and Corresponding Weibull Parameters for Stochastic Mechanical Design," *Trans. ASME Journal of Mechanical Design*, vol. 114 (March 1992), pp. 29–34.

Material		μ_{Sut}	σ_{Sut}	x_0	θ	b	μ_{Sy}	σ_{Sy}	x_0	θ	b	C_{Sut}	C_{Sy}
1018	CD	87.6	5.74	30.8	90.1	12	78.4	5.90	56	80.6	4.29	0.0655	0.0753
1035	HR	86.2	3.92	72.6	87.5	3.86	49.6	3.81	39.5	50.8	2.88	0.0455	0.0768
1045	CD	117.7	7.13	90.2	120.5	4.38	95.5	6.59	82.1	97.2	2.14	0.0606	0.0690
1117	CD	83.1	5.25	73.0	84.4	2.01	81.4	4.71	72.4	82.6	2.00	0.0632	0.0579
1137	CD	106.5	6.15	96.2	107.7	1.72	98.1	4.24	92.2	98.7	1.41	0.0577	0.0432
12L14	CD	79.6	6.92	70.3	80.4	1.36	78.1	8.27	64.3	78.8	1.72	0.0869	0.1059
1038	HT bolts	133.4	3.38	122.3	134.6	3.64						0.0253	
ASTM40		44.5	4.34	27.7	46.2	4.38						0.0975	
35018	Malleable	53.3	1.59	48.7	53.8	3.18	38.5	1.42	34.7	39.0	2.93	0.0298	0.0369
32510	Malleable	53.4	2.68	44.7	54.3	3.61	34.9	1.47	30.1	35.5	3.67	0.0502	0.0421
Malleable	Pearlitic	93.9	3.83	80.1	95.3	4.04	60.2	2.78	50.2	61.2	4.02	0.0408	0.0462
604515	Nodular	64.8	3.77	53.7	66.1	3.23	49.0	4.20	33.8	50.5	4.06	0.0582	0.0857
100-70-04	Nodular	122.2	7.65	47.6	125.6	11.84	79.3	4.51	64.1	81.0	3.77	0.0626	0.0569
201SS	CD	195.9	7.76	180.7	197.9	2.06						0.0396	
301SS	CD	191.2	5.82	151.9	193.6	8.00	166.8	9.37	139.7	170.0	3.17	0.0304	0.0562
301SS	A	105.0	5.68	92.3	106.6	2.38	46.8	4.70	26.3	48.7	4.99	0.0541	0.1004
304SS	A	85.0	4.14	66.6	86.6	5.11	37.9	3.76	30.2	38.9	2.17	0.0487	0.0992
310SS	A	84.8	4.23	71.6	86.3	3.45						0.0499	
403SS		105.3	3.09	95.7	106.4	3.44	78.5	3.91	64.8	79.9	3.93	0.0293	0.0498
17-7PSS		198.8	9.51	163.3	202.3	4.21	189.4	11.49	144.0	193.8	4.48	0.0478	0.0607
AM350SS	A	149.1	8.29	101.8	152.4	6.68	63.0	5.05	38.0	65.0	5.73	0.0556	0.0802
Ti-6AL-4V		175.4	7.91	141.8	178.5	4.85	163.7	9.03	101.5	167.4	8.18	0.0451	0.0552
2024	0	28.1	1.73	24.2	28.7	2.43						0.0616	
2024	T4	64.9	1.64	60.2	65.5	3.16	40.8	1.83	38.4	41.0	1.32	0.0253	0.0449
7075	T6	67.5	1.50	55.9	68.1	9.26	53.4	1.17	51.2	53.6	1.91	0.0222	0.0219
7075	T6 .025"	75.5	2.10	68.8	76.2	3.53	63.7	1.98	58.9	64.3	2.63	0.0278	0.0311

Table A–26

Stochastic Parameters for Finite Life Fatigue Tests in Selected Metals   *Source:* E. B. Haugen, *Probabilistic Mechanical Design*, Wiley, New York, 1980, Appendix 10–B.

1 Number	2 Condition	3 TS MPa (kpsi)	4 YS MPa (kpsi)	5 Distribution		6 Stress Cycles to Failure 10^4	7 10^5	8 10^6	9 10^7
1046	WQ&T 1210°F	723 (105)	565 (82)	W	x_0	544 (79)	462 (67)	391 (56.7)	
					θ	594 (86.2)	503 (73.0)	425 (61.7)	
					b	2.60	2.75	2.85	
2340	OQ&T 1200°F	799 (116)	661 (96)	W	x_0	579 (84)	510 (74)	420 (61)	
					θ	699 (101.5)	588 (85.4)	496 (72.0)	
					b	4.3	3.4	4.1	
3140	OQ&T 1300°F	744 (108)	599 (87)	W	x_0	510 (74)	455 (66)	393 (57)	
					θ	604 (87.7)	528 (76.7)	463 (67.2)	
					b	5.2	5.0	5.5	
2024 Aluminum	T-4	489 (71)	365 (53)	N	σ	26.3 (3.82)	21.4 (3.11)	17.4 (2.53)	14.0 (2.03)
					μ	143 (20.7)	116 (16.9)	95 (13.8)	77 (11.2)
Ti-6Al-4V	HT-46	1040 (151)	992 (144)	N	σ	39.6 (5.75)	38.1 (5.53)	36.6 (5.31)	35.1 (5.10)
					μ	712 (108)	684 (99.3)	657 (95.4)	493 (71.6)

Statistical parameters from a large number of fatigue tests are listed. Weibull distribution is denoted *W* and the parameters are x_0, "guaranteed" fatigue strength; θ, characteristic fatigue strength; and *b*, shape factor. Normal distribution is denoted *N* and the parameters are μ, mean fatigue strength; and σ, standard deviation of the fatigue strength. The life is in stress-cycles-to-failure. TS = tensile strength, YS = yield strength. All testing by rotating-beam specimen.

Table A–27

Finite Life Fatigue Strengths of Selected Plain Carbon Steels *Source:* Compiled from Table 4 in H. J. Grover, S. A. Gordon, and L. R. Jackson, *Fatigue of Metals and Structures,* Bureau of Naval Weapons Document NAVWEPS 00-25-534, 1960.

Material	Condition	BHN*	Tensile Strength kpsi	Yield Strength kpsi	RA*	Stress Cycles to Failure							
						10^4	$4(10^4)$	10^5	$4(10^5)$	10^6	$4(10^6)$	10^7	10^8
1020	Furnace cooled		58	30	0.63			37	34	30	28	25	
1030	Air-cooled	135	80	45	0.62		51	47	42	38	38	38	
1035	Normal	132	72	35	0.54			44	40	37	34	33	33
	WQT	209	103	87	0.65		80	72	65	60	57	57	57
1040	Forged	195	92	53	0.23			40	40	47	33	33	
1045	HR, N		107	63	0.49	80	70	56	47	47	47	47	
1050	N, AC	164	92	47	0.40	50	48	46	40	38	34	34	
	WQT 1200	196	97	70	0.58		60	57	52	50	50	50	50
.56 MN	N	193	98	47	0.42	61	55	51	47	43	41	41	41
	WQT 1200	277	111	84	0.57	94	81	73	62	57	55	55	55
1060	As Rec.	67 Rb	134	65	0.20	65	60	55	50	48	48	48	
1095		162	84	33	0.37	50	43	40	34	31	30	30	30
	OQT 1200	227	115	65	0.40	77	68	64	57	56	56	56	56
10120		224	117	59	0.12		60	56	51	50	50	50	
	OQT 860	369	180	130	0.15		102	95	91	91	91	91	

*BHN = Brinell hardness number; RA = fractional reduction in area.

Decimal Equivalents of Wire and Sheet-Metal Gauges* (All Sizes Are Given in Inches)

Name of Gauge: Principal Use:	American or Brown & Sharpe Nonferrous Sheet, Wire, and Rod	Birmingham or Stubs Iron Wire Tubing, Ferrous Strip, Flat Wire, and Spring Steel	United States Standard† Ferrous Sheet and Plate, 480 lbf/ft³	Manufacturers Standard Ferrous Sheet	Steel Wire or Washburn & Moen Ferrous Wire Except Music Wire	Music Wire Music Wire	Stubs Steel Wire Steel Drill Rod	Twist Drill Twist Drills and Drill Steel
7/0			0.500		0.490			
6/0	0.580 0		0.468 75		0.461 5	0.004		
5/0	0.516 5		0.437 5		0.430 5	0.005		
4/0	0.460 0	0.454	0.406 25		0.393 8	0.006		
3/0	0.409 6	0.425	0.375		0.362 5	0.007		
2/0	0.364 8	0.380	0.343 75		0.331 0	0.008		
0	0.324 9	0.340	0.312 5		0.306 5	0.009		
1	0.289 3	0.300	0.281 25		0.283 0	0.010	0.227	0.228 0
2	0.257 6	0.284	0.265 625		0.262 5	0.011	0.219	0.221 0
3	0.229 4	0.259	0.25	0.239 1	0.243 7	0.012	0.212	0.213 0
4	0.204 3	0.238	0.234 375	0.224 2	0.225 3	0.013	0.207	0.209 0
5	0.181 9	0.220	0.218 75	0.209 2	0.207 0	0.014	0.204	0.205 5
6	0.162 0	0.203	0.203 125	0.194 3	0.192 0	0.016	0.201	0.204 0
7	0.144 3	0.180	0.187 5	0.179 3	0.177 0	0.018	0.199	0.201 0
8	0.128 5	0.165	0.171 875	0.164 4	0.162 0	0.020	0.197	0.199 0
9	0.114 4	0.148	0.156 25	0.149 5	0.148 3	0.022	0.194	0.196 0
10	0.101 9	0.134	0.140 625	0.134 5	0.135 0	0.024	0.191	0.193 5
11	0.090 74	0.120	0.125	0.119 6	0.120 5	0.026	0.188	0.191 0
12	0.080 81	0.109	0.109 357	0.104 6	0.105 5	0.029	0.185	0.189 0
13	0.071 96	0.095	0.093 75	0.089 7	0.091 5	0.031	0.182	0.185 0
14	0.064 08	0.083	0.078 125	0.074 7	0.080 0	0.033	0.180	0.182 0
15	0.057 07	0.072	0.070 312 5	0.067 3	0.072 0	0.035	0.178	0.180 0
16	0.050 82	0.065	0.062 5	0.059 8	0.062 5	0.037	0.175	0.177 0
17	0.045 26	0.058	0.056 25	0.053 8	0.054 0	0.039	0.172	0.173 0

(continued)

Name of Gauge:	American or Brown & Sharpe	Birmingham or Stubs Iron Wire	United States Standard†	Manufacturers Standard	Steel Wire or Washburn & Moen	Music Wire	Stubs Steel Wire	Twist Drill
Principal Use:	Nonferrous Sheet, Wire, and Rod	Tubing, Ferrous Strip, Flat Wire, and Spring Steel	Ferrous Sheet and Plate, 480 lbf/ft³	Ferrous Sheet	Ferrous Wire Except Music Wire	Music Wire	Steel Drill Rod	Twist Drills and Drill Steel
18	0.040 30	0.049	0.05	0.047 8	0.047 5	0.041	0.168	0.169 5
19	0.035 89	0.042	0.043 75	0.041 8	0.041 0	0.043	0.164	0.166 0
20	0.031 96	0.035	0.037 5	0.035 9	0.034 8	0.045	0.161	0.161 0
21	0.028 46	0.032	0.034 375	0.032 9	0.031 7	0.047	0.157	0.159 0
22	0.025 35	0.028	0.031 25	0.029 9	0.028 6	0.049	0.155	0.157 0
23	0.022 57	0.025	0.028 125	0.026 9	0.025 8	0.051	0.153	0.154 0
24	0.020 10	0.022	0.025	0.023 9	0.023 0	0.055	0.151	0.152 0
25	0.017 90	0.020	0.021 875	0.020 9	0.020 4	0.059	0.148	0.149 5
26	0.015 94	0.018	0.018 75	0.017 9	0.018 1	0.063	0.146	0.147 0
27	0.014 20	0.016	0.017 187 5	0.016 4	0.017 3	0.067	0.143	0.144 0
28	0.012 64	0.014	0.015 625	0.014 9	0.016 2	0.071	0.139	0.140 5
29	0.011 26	0.013	0.014 062 5	0.013 5	0.015 0	0.075	0.134	0.136 0
30	0.010 03	0.012	0.012 5	0.012 0	0.014 0	0.080	0.127	0.128 5
31	0.008 928	0.010	0.010 937 5	0.010 5	0.013 2	0.085	0.120	0.120 0
32	0.007 950	0.009	0.010 156 25	0.009 7	0.012 8	0.090	0.115	0.116 0
33	0.007 080	0.008	0.009 375	0.009 0	0.011 8	0.095	0.112	0.113 0
34	0.006 305	0.007	0.008 593 75	0.008 2	0.010 4		0.110	0.111 0
35	0.005 615	0.005	0.007 812 5	0.007 5	0.009 5		0.108	0.110 0
36	0.005 000	0.004	0.007 031 25	0.006 7	0.009 0		0.106	0.106 5
37	0.004 453		0.006 640 625	0.006 4	0.008 5		0.103	0.104 0
38	0.003 965		0.006 25	0.006 0	0.008 0		0.101	0.101 5
39	0.003 531				0.007 5		0.099	0.099 5
40	0.003 145				0.007 0		0.097	0.098 0

*Specify sheet, wire, and plate by stating the gauge number, the gauge name, and the decimal equivalent in parentheses.

†Reflects present overage and weights of sheet steel.

Table A–29

Dimensions of Square and Hexagonal Bolts

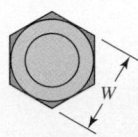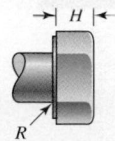

Nominal Size, in	Square		Regular Hexagonal			Head Type Heavy Hexagonal			Structural Hexagonal		
	W	H	W	H	R_{min}	W	H	R_{min}	W	H	R_{min}
$\frac{1}{4}$	$\frac{3}{8}$	$\frac{11}{64}$	$\frac{7}{16}$	$\frac{11}{64}$	0.01						
$\frac{5}{16}$	$\frac{1}{2}$	$\frac{13}{64}$	$\frac{1}{2}$	$\frac{7}{32}$	0.01						
$\frac{3}{8}$	$\frac{9}{16}$	$\frac{1}{4}$	$\frac{9}{16}$	$\frac{1}{4}$	0.01						
$\frac{7}{16}$	$\frac{5}{8}$	$\frac{19}{64}$	$\frac{5}{8}$	$\frac{19}{64}$	0.01						
$\frac{1}{2}$	$\frac{3}{4}$	$\frac{21}{64}$	$\frac{3}{4}$	$\frac{11}{32}$	0.01	$\frac{7}{8}$	$\frac{11}{32}$	0.01	$\frac{7}{8}$	$\frac{5}{16}$	0.009
$\frac{5}{8}$	$\frac{15}{16}$	$\frac{27}{64}$	$\frac{15}{16}$	$\frac{27}{64}$	0.02	$1\frac{1}{16}$	$\frac{27}{64}$	0.02	$1\frac{1}{16}$	$\frac{25}{64}$	0.021
$\frac{3}{4}$	$1\frac{1}{8}$	$\frac{1}{2}$	$1\frac{1}{8}$	$\frac{1}{2}$	0.02	$1\frac{1}{4}$	$\frac{1}{2}$	0.02	$1\frac{1}{4}$	$\frac{15}{32}$	0.021
1	$1\frac{1}{2}$	$\frac{21}{32}$	$1\frac{1}{2}$	$\frac{43}{64}$	0.03	$1\frac{5}{8}$	$\frac{43}{64}$	0.03	$1\frac{5}{8}$	$\frac{39}{64}$	0.062
$1\frac{1}{8}$	$1\frac{11}{16}$	$\frac{3}{4}$	$1\frac{11}{16}$	$\frac{3}{4}$	0.03	$1\frac{13}{16}$	$\frac{3}{4}$	0.03	$1\frac{13}{16}$	$\frac{11}{16}$	0.062
$1\frac{1}{4}$	$1\frac{7}{8}$	$\frac{27}{32}$	$1\frac{7}{8}$	$\frac{27}{32}$	0.03	2	$\frac{27}{32}$	0.03	2	$\frac{25}{32}$	0.062
$1\frac{3}{8}$	$2\frac{1}{16}$	$\frac{29}{32}$	$2\frac{1}{16}$	$\frac{29}{32}$	0.03	$2\frac{3}{16}$	$\frac{29}{32}$	0.03	$2\frac{3}{16}$	$\frac{27}{32}$	0.062
$1\frac{1}{2}$	$2\frac{1}{4}$	1	$2\frac{1}{4}$	1	0.03	$2\frac{3}{8}$	1	0.03	$2\frac{3}{8}$	$\frac{15}{16}$	0.062
Nominal Size, mm											
M5	8	3.58	8	3.58	0.2						
M6			10	4.38	0.3						
M8			13	5.68	0.4						
M10			16	6.85	0.4						
M12			18	7.95	0.6	21	7.95	0.6			
M14			21	9.25	0.6	24	9.25	0.6			
M16			24	10.75	0.6	27	10.75	0.6	27	10.75	0.6
M20			30	13.40	0.8	34	13.40	0.8	34	13.40	0.8
M24			36	15.90	0.8	41	15.90	0.8	41	15.90	1.0
M30			46	19.75	1.0	50	19.75	1.0	50	19.75	1.2
M36			55	23.55	1.0	60	23.55	1.0	60	23.55	1.5

Table A–30

Dimensions of Hexagonal Cap Screws and Heavy Hexagonal Screws (W = Width across Flats; H = Height of Head; See Figure in Table A–29)

Nominal Size, in	Minimum Fillet Radius	Type of Screw Cap W	Heavy W	Height H
$\frac{1}{4}$	0.015	$\frac{7}{16}$		$\frac{5}{32}$
$\frac{5}{16}$	0.015	$\frac{1}{2}$		$\frac{13}{64}$
$\frac{3}{8}$	0.015	$\frac{9}{16}$		$\frac{15}{64}$
$\frac{7}{16}$	0.015	$\frac{5}{8}$		$\frac{9}{32}$
$\frac{1}{2}$	0.015	$\frac{3}{4}$	$\frac{7}{8}$	$\frac{5}{16}$
$\frac{5}{8}$	0.020	$\frac{15}{16}$	$1\frac{1}{16}$	$\frac{25}{64}$
$\frac{3}{4}$	0.020	$1\frac{1}{8}$	$1\frac{1}{4}$	$\frac{15}{32}$
$\frac{7}{8}$	0.040	$1\frac{5}{16}$	$1\frac{7}{16}$	$\frac{35}{64}$
1	0.060	$1\frac{1}{2}$	$1\frac{1}{8}$	$\frac{39}{64}$
$1\frac{1}{4}$	0.060	$1\frac{7}{8}$	2	$\frac{25}{32}$
$1\frac{3}{8}$	0.060	$2\frac{1}{16}$	$2\frac{3}{16}$	$\frac{27}{32}$
$1\frac{1}{2}$	0.060	$2\frac{1}{4}$	$2\frac{3}{8}$	$\frac{15}{16}$

Nominal Size, mm				
M5	0.2	8		3.65
M6	0.3	10		4.15
M8	0.4	13		5.50
M10	0.4	16		6.63
M12	0.6	18	21	7.76
M14	0.6	21	24	9.09
M16	0.6	24	27	10.32
M20	0.8	30	34	12.88
M24	0.8	36	41	15.44
M30	1.0	46	50	19.48
M36	1.0	55	60	23.38

Table A–31

Dimensions of
Hexagonal Nuts

Nominal Size, in	Width W	Height H Regular Hexagonal	Height H Thick or Slotted	Height H JAM
$\frac{1}{4}$	$\frac{7}{16}$	$\frac{7}{32}$	$\frac{9}{32}$	$\frac{5}{32}$
$\frac{5}{16}$	$\frac{1}{2}$	$\frac{17}{64}$	$\frac{21}{64}$	$\frac{3}{16}$
$\frac{3}{8}$	$\frac{9}{16}$	$\frac{21}{64}$	$\frac{13}{32}$	$\frac{7}{32}$
$\frac{7}{16}$	$\frac{11}{16}$	$\frac{3}{8}$	$\frac{29}{64}$	$\frac{1}{4}$
$\frac{1}{2}$	$\frac{3}{4}$	$\frac{7}{16}$	$\frac{9}{16}$	$\frac{5}{16}$
$\frac{9}{16}$	$\frac{7}{8}$	$\frac{31}{64}$	$\frac{39}{64}$	$\frac{5}{16}$
$\frac{5}{8}$	$\frac{15}{16}$	$\frac{35}{64}$	$\frac{23}{32}$	$\frac{3}{8}$
$\frac{3}{4}$	$1\frac{1}{8}$	$\frac{41}{64}$	$\frac{13}{16}$	$\frac{27}{64}$
$\frac{7}{8}$	$1\frac{5}{16}$	$\frac{3}{4}$	$\frac{29}{32}$	$\frac{31}{64}$
1	$1\frac{1}{2}$	$\frac{55}{64}$	1	$\frac{35}{64}$
$1\frac{1}{8}$	$1\frac{11}{16}$	$\frac{31}{32}$	$1\frac{5}{32}$	$\frac{39}{64}$
$1\frac{1}{4}$	$1\frac{7}{8}$	$1\frac{1}{16}$	$1\frac{1}{4}$	$\frac{23}{32}$
$1\frac{3}{8}$	$2\frac{1}{16}$	$1\frac{11}{64}$	$1\frac{3}{8}$	$\frac{25}{32}$
$1\frac{1}{2}$	$2\frac{1}{4}$	$1\frac{9}{32}$	$1\frac{1}{2}$	$\frac{27}{32}$

Nominal Size, mm				
M5	8	4.7	5.1	2.7
M6	10	5.2	5.7	3.2
M8	13	6.8	7.5	4.0
M10	16	8.4	9.3	5.0
M12	18	10.8	12.0	6.0
M14	21	12.8	14.1	7.0
M16	24	14.8	16.4	8.0
M20	30	18.0	20.3	10.0
M24	36	21.5	23.9	12.0
M30	46	25.6	28.6	15.0
M36	55	31.0	34.7	18.0

Table A–32

Basic Dimensions of American Standard Plain Washers (All Dimensions in Inches)

Fastener Size	Washer Size	Diameter ID	Diameter OD	Thickness
#6	0.138	0.156	0.375	0.049
#8	0.164	0.188	0.438	0.049
#10	0.190	0.219	0.500	0.049
$\frac{3}{16}$	0.188	0.250	0.562	0.049
#12	0.216	0.250	0.562	0.065
$\frac{1}{4}$ N	0.250	0.281	0.625	0.065
$\frac{1}{4}$ W	0.250	0.312	0.734	0.065
$\frac{5}{16}$ N	0.312	0.344	0.688	0.065
$\frac{5}{16}$ W	0.312	0.375	0.875	0.083
$\frac{3}{8}$ N	0.375	0.406	0.812	0.065
$\frac{3}{8}$ W	0.375	0.438	1.000	0.083
$\frac{7}{16}$ N	0.438	0.469	0.922	0.065
$\frac{7}{16}$ W	0.438	0.500	1.250	0.083
$\frac{1}{2}$ N	0.500	0.531	1.062	0.095
$\frac{1}{2}$ W	0.500	0.562	1.375	0.109
$\frac{9}{16}$ N	0.562	0.594	1.156	0.095
$\frac{9}{16}$ W	0.562	0.625	1.469	0.109
$\frac{5}{8}$ N	0.625	0.656	1.312	0.095
$\frac{5}{8}$ W	0.625	0.688	1.750	0.134
$\frac{3}{4}$ N	0.750	0.812	1.469	0.134
$\frac{3}{4}$ W	0.750	0.812	2.000	0.148
$\frac{7}{8}$ N	0.875	0.938	1.750	0.134
$\frac{7}{8}$ W	0.875	0.938	2.250	0.165
1 N	1.000	1.062	2.000	0.134
1 W	1.000	1.062	2.500	0.165
$1\frac{1}{8}$ N	1.125	1.250	2.250	0.134
$1\frac{1}{8}$ W	1.125	1.250	2.750	0.165
$1\frac{1}{4}$ N	1.250	1.375	2.500	0.165
$1\frac{1}{4}$ W	1.250	1.375	3.000	0.165
$1\frac{3}{8}$ N	1.375	1.500	2.750	0.165
$1\frac{3}{8}$ W	1.375	1.500	3.250	0.180
$1\frac{1}{2}$ N	1.500	1.625	3.000	0.165
$1\frac{1}{2}$ W	1.500	1.625	3.500	0.180
$1\frac{5}{8}$	1.625	1.750	3.750	0.180
$1\frac{3}{4}$	1.750	1.875	4.000	0.180
$1\frac{7}{8}$	1.875	2.000	4.250	0.180
2	2.000	2.125	4.500	0.180
$2\frac{1}{4}$	2.250	2.375	4.750	0.220
$2\frac{1}{2}$	2.500	2.625	5.000	0.238
$2\frac{3}{4}$	2.750	2.875	5.250	0.259
3	3.000	3.125	5.500	0.284

N = narrow; W = wide; use W when not specified.

Table A–33

Dimensions of Metric Plain Washers (All Dimensions in Millimeters)

Washer Size*	Minimum ID	Maximum OD	Maximum Thickness	Washer Size*	Minimum ID	Maximum OD	Maximum Thickness
1.6 N	1.95	4.00	0.70	10 N	10.85	20.00	2.30
1.6 R	1.95	5.00	0.70	10 R	10.85	28.00	2.80
1.6 W	1.95	6.00	0.90	10 W	10.85	39.00	3.50
2 N	2.50	5.00	0.90	12 N	13.30	25.40	2.80
2 R	2.50	6.00	0.90	12 R	13.30	34.00	3.50
2 W	2.50	8.00	0.90	12 W	13.30	44.00	3.50
2.5 N	3.00	6.00	0.90	14 N	15.25	28.00	2.80
2.5 R	3.00	8.00	0.90	14 R	15.25	39.00	3.50
2.5 W	3.00	10.00	1.20	14 W	15.25	50.00	4.00
3 N	3.50	7.00	0.90	16 N	17.25	32.00	3.50
3 R	3.50	10.00	1.20	16 R	17.25	44.00	4.00
3 W	3.50	12.00	1.40	16 W	17.25	56.00	4.60
3.5 N	4.00	9.00	1.20	20 N	21.80	39.00	4.00
3.5 R	4.00	10.00	1.40	20 R	21.80	50.00	4.60
3.5 W	4.00	15.00	1.75	20 W	21.80	66.00	5.10
4 N	4.70	10.00	1.20	24 N	25.60	44.00	4.60
4 R	4.70	12.00	1.40	24 R	25.60	56.00	5.10
4 W	4.70	16.00	2.30	24 W	25.60	72.00	5.60
5 N	5.50	11.00	1.40	30 N	32.40	56.00	5.10
5 R	5.50	15.00	1.75	30 R	32.40	72.00	5.60
5 W	5.50	20.00	2.30	30 W	32.40	90.00	6.40
6 N	6.65	13.00	1.75	36 N	38.30	66.00	5.60
6 R	6.65	18.80	1.75	36 R	38.30	90.00	6.40
6 W	6.65	25.40	2.30	36 W	38.30	110.00	8.50
8 N	8.90	18.80	2.30				
8 R	8.90	25.40	2.30				
8 W	8.90	32.00	2.80				

N = narrow; R = regular; W = wide.
*Same as screw or bolt size.

Table A–34

Gamma Function*

Source: Reprinted with permission from William H. Beyer (ed.), *Handbook of Tables for Probability and Statistics,* 2nd ed., 1966. Copyright CRC Press, Boca Raton, Florida.

Values of $\Gamma(n) = \int_0^\infty e^{-x} x^{n-1} dx$; $\Gamma(n+1) = n\Gamma(n)$

n	$\Gamma(n)$	n	$\Gamma(n)$	n	$\Gamma(n)$	n	$\Gamma(n)$
1.00	1.000 00	1.25	.906 40	1.50	.886 23	1.75	.919 06
1.01	.994 33	1.26	.904 40	1.51	.886 59	1.76	.921 37
1.02	.988 84	1.27	.902 50	1.52	.887 04	1.77	.923 76
1.03	.983 55	1.28	.900 72	1.53	.887 57	1.78	.926 23
1.04	.978 44	1.29	.899 04	1.54	.888 18	1.79	.928 77
1.05	.973 50	1.30	.897 47	1.55	.888 87	1.80	.931 38
1.06	.968 74	1.31	.896 00	1.56	.889 64	1.81	.934 08
1.07	.964 15	1.32	.894 64	1.57	.890 49	1.82	.936 85
1.08	.959 73	1.33	.893 38	1.58	.891 42	1.83	.939 69
1.09	.955 46	1.34	.892 22	1.59	.892 43	1.84	.942 61
1.10	.951 35	1.35	.891 15	1.60	.893 52	1.85	.945 61
1.11	.947 39	1.36	.890 18	1.61	.894 68	1.86	.948 69
1.12	.943 59	1.37	.889 31	1.62	.895 92	1.87	.951 84
1.13	.939 93	1.38	.888 54	1.63	.897 24	1.88	.955 07
1.14	.936 42	1.39	.887 85	1.64	.898 64	1.89	.958 38
1.15	.933 04	1.40	.887 26	1.65	.900 12	1.90	.961 77
1.16	.929 80	1.41	.886 76	1.66	.901 67	1.91	.965 23
1.17	.936 70	1.42	.886 36	1.67	.903 30	1.92	.968 78
1.18	.923 73	1.43	.886 04	1.68	.905 00	1.93	.972 40
1.19	.920 88	1.44	.885 80	1.69	.906 78	1.94	.976 10
1.20	.918 17	1.45	.885 65	1.70	.908 64	1.95	.979 88
1.21	.915 58	1.46	.885 60	1.71	.910 57	1.96	.983 74
1.22	.913 11	1.47	.885 63	1.72	.912 58	1.97	.987 68
1.23	.910 75	1.48	.885 75	1.73	.914 66	1.98	.991 71
1.24	.908 52	1.49	.885 95	1.74	.916 83	1.99	.995 81
						2.00	1.000 00

*For large positive values of x, $\Gamma(x)$ approximates the asymptotic series

$$x^x e^{-x} \sqrt{\frac{2x}{x}} \left[1 + \frac{1}{12x} + \frac{1}{288x^2} - \frac{139}{51\,840x^3} - \frac{571}{2\,488\,320x^4} + \cdots \right]$$

Appendix B

B–1 Chapter 1

1–8 $10 \le F \le 10.5$ lbf, $13.5 \le N \le 14.2$ lbf, $K = 0.967$, $19.3 \le T \le 20.3$ lbf $\cdot$ in

1–12 (a) $e_1 = 0.006\ 067\ 977$, $e_2 = 0.009\ 489\ 743$, $e = 0.015\ 557\ 720$, (b) $e_1 = -0.003\ 932\ 023$, $e_2 = -0.000\ 510\ 257$, $e = -0.004\ 442\ 280$

1–15 (a) $\sigma = 13.1$ MPa, (b) $\sigma = 70$ MPa, (c) $y = 15.5$ mm, (d) $\theta = 5.18°$

B–2 Chapter 2

2–1 $\bar{x} = 122.9$ kilocycles, $s_x = 30.3$ kilocycles

2–2 $\bar{x} = 198.55$ kpsi, $s_x = 9.55$ kpsi

2–3 $\bar{x} = 78.4$ kpsi, $s_x = 6.57$ kpsi

2–11 (a) $\bar{F}_i = 5.979$ lbf, $s_{Fi} = 0.396$ lbf; (b) $\bar{k} = 9.766$ lbf/in, $s_k = 0.390$ lbf/in

2–19 $L_{10} = 84.1$ kcycles

2–23 $R = 0.987$

2–25 (a) $w = 0.020 \pm 0.015$ in, (b) $w = 0.020 \pm 0.005$ in

2–32 $\bar{D}_0 = 4.012$ in, $t_{D_0} = 0.036$ in; $D_0 = 4.012 \pm 0.036$ in

2–39 $\bar{x} = 98.26$ kpsi, $s_x = 4.30$ kpsi

2–46 $\mu_n = 122.9$ kcycles, $\hat{\sigma}_n = 34.8$ kcycles

B–3 Chapter 3

3–9 $E = 30$ Mpsi, $S_y = 45.5$ kpsi, $S_{ut} = 85.5$ kpsi, area reduction $= 45.8$ percent

3–11 $(S_y)_{0.001} \doteq 35$ kpsi, $E_{\sigma=0} = 25$ Mpsi, $E_{\sigma=20\ kpsi} = 14$ Mpsi

3–14 $G = 77.3$ GPa, $S_{ys} \doteq 200$ MPa

3–18 $\bar{S}_{ut} = 125.2$ kpsi, $\hat{\sigma}_{S_{ut}} = 1.9$ kpsi

3–20 (a) $u_R \doteq 34.5$ in $\cdot$ lbf/in^3, (b) $u_T \doteq 66.7(10^3)$ in $\cdot$ lbf/in^3

B–4 Chapter 4

4–4 (a) $V(x) = -1.43 - 40\langle x - 4\rangle^0 + 30\langle x - 8\rangle^0 + 71.43\langle x - 14\rangle^0 - 60\langle x - 18\rangle^0$ lbf
$M(x) = -1.43x - 40\langle x - 4\rangle^1 + 30\langle x - 8\rangle^1 + 71.43\langle x - 14\rangle^1 - 60\langle x - 18\rangle^1$ lbf $\cdot$ in

4–6 (a) $M_{max} = 253$ lbf $\cdot$ in, (b) $(a/l)^* = 0.207$, $M^* = 214$ lbf $\cdot$ in

4–8 (a) $\sigma_1 = 14$, $\sigma_2 = 4$, $\sigma_3 = 0$, $2\theta = 53.1°$ cw; (b) $\sigma_1 = 18.6$, $\sigma_2 = 6.4$, $\sigma_3 = 0$, $2\theta = 55°$ ccw; (c) $\sigma_1 = 26.2$, $\sigma_2 = 7.78$, $\sigma_3 = 0$, $2\theta = 139.7°$ cw; (d) $\sigma_1 = 23.4$, $\sigma_2 = 4.57$, $\sigma_3 = 0$, $2\theta = 122°$ cw

4–13 $\sigma = 10.2$ kpsi, $\delta = 0.0245$ in, $\epsilon_1 = 0.000\ 340$, $\nu = 0.292$, $\epsilon_2 = -0.000\ 099\ 1$, $\Delta d = -0.000\ 049\ 6$ in

4–18 $\sigma_1 = 30$ MPa, $\sigma_2 = 10$ MPa, $\sigma_3 = -20$ MPa, $\tau_{max} = 25$ MPa

4–22 (a) $M_{max} = 21\ 600$ kip $\cdot$ in, (b) $x_{max} = 523$ in from left or right supports

4–23 (a) $\sigma_A = 42$ kpsi, $\sigma_B = 18.5$ kpsi, $\sigma_C = 2.7$ kpsi, $\sigma_D = -52.7$ kpsi

4–27 $M_{max} = 219$ lbf $\cdot$ in, $\sigma = 17.8$ kpsi, $\tau_{max} = 3.4$ kpsi, both models

4–33 The same

4–37 Two $\frac{1}{16}$-in-thick strips: $T_{max} = 29.95$ lbf $\cdot$ in, $\theta = 0.192$ rad, $k_t = 156$ lbf $\cdot$ in/rad. One $\frac{1}{8}$-in-thick strip: $T_{max} = 59.90$ lbf $\cdot$ in, $\theta = 0.0960$ rad, $k_t = 624$ lbf $\cdot$ in/rad

4–43 $d_C = 45$ mm

4–47 $\sigma_{max} = 11.79$ kpsi, $\tau_{max} = 7.05$ kpsi

4–53 $p_i = 639$ psi

4–57 $(\sigma_r)_{max} = 3656$ psi

4–65 $\delta_{max} = 0.038$ mm, $\delta_{min} = 0.0175$ mm, $p_{max} = 147.5$ MPa, $p_{min} = 67.9$ MPa

4–69 For δ_{max}, $p = 33.75$ kpsi, $(\sigma_t)_o = 56.25$ kpsi, $(\sigma_t)_i = -33.75$ kpsi, $\delta_o = 0.001\ 11$ in, $\delta_i = -0.000\ 398$ in

4–72 $\sigma_i = 26.3$ kpsi, $\sigma_o = -15.8$ kpsi

4–77 $\sigma_i = 71.3$ kpsi, $\sigma_o = -34.2$ kpsi

4–81 $p_{max} = 399F^{1/3}$ MPa, $\sigma_{max} = 399F^{1/3}$ MPa, $\tau_{max} = 120F^{1/3}$ MPa

B–5 Chapter 5

5–1 (a) $k = (1/k_1 + 1/k_2 + 1/k_3)$, (b) $k = k_1 + k_2 + k_3$, (c) $k = [1/k_1 + 1/(k_2 + k_3)]^{-1}$

5–10 $\lambda = 8a^2/(3l)$

5–12 $\sigma_{max} = -20.4$ kpsi, $y = -0.908$ in

5–15 $y_{left} = -0.0506$ in, $y_{right} = -0.0506$ in, $y_{midspan} = 0.0190$ in

5–18 $y_{max} = -0.0130$ in

5–20 $z_A = 0.0368$ in, $z_B = 0.00430$ in

5–26 Use $d = 1\frac{3}{8}$ in

5–30 $y_B = 0.0459$ in

5–37 $y_A = -0.101$ in, $y_{x=20 \text{ in}} = -0.104$ in

5–45 $y_A = -0.133$ in

5–48 $y_{x=10 \text{ in}} = -0.0167$ in

5–51 (a) $\sigma_b = 76.5$ kpsi, $\sigma_c = -15.2$ kpsi, (b) $\sigma_b = 78.4$ kpsi, $\sigma_c = -13.3$ kpsi

5–56 $R_O = 3.89$ kip, $R_C = 1.11$ kip, both in same direction

5–59 $\sigma_{BE} = 140$ MPa, $\sigma_{DF} = 71.2$ MPa, $y_B = -0.670$ mm, $y_C = -2.27$ mm, $y_D = -0.341$ mm

5–64 $\delta_A = (\pi + 4)PR^3/(4EI)$, $\delta_B = \pi PR^3/(4EI)$

5–67 $\delta = 0.476$ mm

5–73 (a) $t = 0.5$ in, (b) No

5–81 $y_{max} = 2k_1a/(k_1 + k_2)$

B–6 Chapter 6

6–2 (a) MSS: $n = 4.17$, DE: $n = 4.17$, (b) MSS: $n = 4.17$, DE: $n = 4.81$, (c) MSS: $n = 2.08$, DE: $n = 2.41$, (c) MSS: $n = 4.17$, DE: $n = 4.81$

6–3 (a) MSS: $n = 2.17$, DE: $n = 2.50$, (b) MSS: $n = 1.45$, DE: $n = 1.56$, (c) MSS: $n = 1.52$, DE: $n = 1.65$, (c) MSS: $n = 1.27$, DE: $n = 1.50$

6–9 (a) DE: $\sigma' = 12.29$ kpsi, $n = 3.42$

6–10 (a) DCM: $\sigma_1 = 90$ kpsi, $\sigma_2 = 0$, $\sigma_3 = -50$ kpsi, $r = -0.56$, $n = 1.77$

6–12 (a) M2M: $n = 3.89$

6–13 (a) $\sigma_A = \sigma_B = 20$ kpsi, $r = 1$, $n = 1.5$

6–18 $(\sigma_t)_{max}$ 13.21 kpsi, $\sigma_l = 6.42$ kpsi, $\sigma_r = -500$ psi, $\sigma' = 11.9$ kpsi, $n = 3.87$

6–21 Using BCM, select $d = 1\frac{3}{8}$ in

6–25 $d = 18$ mm

6–32 (a) $\delta = 0.0005$ in, $p = 3516$ psi, $(\sigma_t)_i = -5860$ psi, $(\sigma_r)_i = -3516$ psi, $(\sigma_t)_o = -9142$ psi, $(\sigma_r)_o = -3516$ psi

6–35 $n_o = 2.81$, $n_i = 2.41$

6–40 $p = 29.2$ MPa

B–7 Chapter 7

7–1 $S_e = 94.4$ kpsi

7–3 $S_e' = 33.4$ kpsi, $\sigma_F' = 112.4$ kpsi, $b = -0.0836$, $f = 0.899$, $a = 106.1$ kpsi, $S_f = 48.2$ kpsi, $N = 409\,530$ cycles

7–5 $(S_f)_{ax} = 162N^{-0.0851}$ kpsi, $10^3 \leq N \leq 10^6$ cycles

7–6 $S_e = 243$ MPa

7–10 $S_e' = 221.8$ MPa, $k_a = 0.899$, $k_b = 1$, $k_c = 0.85$, $S_e = 169.5$ MPa, $K_t = 2.5$, $K_f = 2.09$, $F_a = 21.6$ kN, $F_y = 98.7$ kN

7–12 Yield: $n_y = 1.18$. Fatigue: (a) $n_f = 1.06$, (b) $n_f = 1.31$, (c) $n_f = 1.32$

7–17 $n_y = 5.06$, (a) $n_f = 2.44$, (b) $n_f = 2.55$

7–23 At the fillet $n_f = 1.70$

7–24 (a) $T = 3.42$ N · m, (b) $T = 4.21$ N · m, (c) $n_y = 1.91$

7–27 (a) $P_{all} = 16.1$ kN, $n_y = 5.69$, (b) $P_{all} = 51.4$ kN, $n_y = 3.87$

7–29 (a) 24 900 cycles, (b) 27 900 cycles

7–34 Rotation presumed. $S_e' = 55.7$ **LN**$(1, 0.138)$ kpsi, $k_a = 0.768$ **LN**$(1, 0.058)$, $k_b = 0.879$, $S_e = 37.6$ **LN**$(1, 0.150)$ kpsi, $K_f = 1.598$ **LN**$(1, 0.15)$, $\sigma = 22.8$ **LN**$(1, 0.15)$ kpsi, $z = -2.373$, $R = 0.991$

B–8 Chapter 8

8–1 (a) Thread depth 2.5 mm, thread width 2.5 mm, $d_m = 22.5$ mm, $d_r = 20$ mm, $l = p = 5$ mm

8–4 $T_R = 16.23$ N · m, $T_L = 6.62$ N · m, $e = 0.294$

8–8 $T = 16.5$ lbf · in, $d_m = 0.5417$ in, $l = 0.1667$ in, sec $\alpha = 1.033$, $T = 0.0696F$, $T_c = 0.0328F$, $T_{total} = 0.1024F$, $F = 161$ lbf

8–11 $L_T = 1.25$ in, $L_G = 1.109$ in, $H = 0.4375$ in, $L_G + H = 1.5465$ in, use 1.75 in, $l_d = 0.500$ in, $l_t = 0.609$ in

8–13 $L_T = 1.25$ in, $L > h + 1.5d = 1.625$ in, use 1.75 in, $l_d = 0.500$ in, $l_t = 0.625$ in

8–15 (a) $A_d = 0.442$ in^2, $A_{\text{tube}} = 0.552$ in^2, $k_b = 1.02(10^6)$ lbf/in, $k_m = 1.27(10^6)$ lbf/in, $C = 0.445$, (b) $F_i = 11\,810$ lbf

8–18 Frusta to Wileman ratio is 1.11/1.08

8–22 $n = 4.73$

8–23 $n = 5.84$

8–27 $k_b = 4.63$ Mlbf/in, $k_m = 7.99$ Mlbf/in using frustums

8–34 (a) $L = 2.5$ in, (b) $k_b = 6.78$ Mlbf/in, $k_m = 14.41$ Mlbf/in, $C = 0.320$

8–37 Load: $n = 3.19$. Separation: $n = 4.71$. Fatigue: $n = 3.27$

8–43 Bolt shear: $n = 3.26$. Bolt bearing: $n = 5.99$. Member bearing: $n = 3.71$. Member tension: $n = 5.36$

8–48 $F = 2.22$ kN

8–50 Bearing on bolt, $n = 9.58$; shear of bolt, $n = 5.79$; bearing on members, $n = 5.63$; bending of members, $n = 2.95$

B–9 Chapter 9

9–1 $F = 17.7$ kip

9–3 $F = 11.3$ kip

9–5 (a) $\tau' = 1.13F$ kpsi, $\tau_x'' = \tau_y'' = 5.93F$ kpsi, $\tau_{\max} = 9.22F$ kpsi, $F = 2.17$ kip; (b) $\tau_{\text{all}} = 11$ kpsi, $F_{\text{all}} = 1.19$ kip

9–8 $\tau' = 0$ (why?), $F = 49.2$ kN

9–9 A two-way tie for first, vertical parallel beads, and square beads

9–10 First: horizontal parallel beads. Second: square beads

9–11 Decisions: Pattern; all-around square
Electrode: E60XX
Type: two parallel fillets, two transverse fillets
Length of beads: 12 in
Leg: $\frac{1}{4}$ in

9–20 $\tau_{\max} = 18$ kpsi

9–22 $n = 3.57$

B–10 Chapter 10

10–3 (a) $L_0 = 5.17$ in, (b) $F_{S_{sy}} = 45.2$ lbf, (c) $k = 11.55$ lbf/in, (d) $(L_0)_{\text{cr}} = 5.89$ in, guide spring

10–5 (a) $L_0 = 47.7$ mm, (b) $p = 5.61$ mm, (c) $F_s = 81.1$ N, (d) $k = 2643$ N/m, (e) $(L_0)_{\text{cr}} = 105.2$ mm, needs guidance

10–9 Not solid safe, $L_0 \le 0.577$ in

10–15 Not solid safe, $L_0 \le 66.6$ mm

10–19 (a) $p = 10$ mm, $L_s = 44.2$ mm, $N_a = 12$ turns, (b) $k = 1080$ N/m, (c) $F_s = 81.9$ N, (d) $\tau_s = 271$ MPa

10–29 (a) $L_0 = 16.12$ in, (b) $\tau_i = 14.95$ kpsi, (c) $k = 4.855$ lbf/in, (d) $F = 85.8$ lbf, (e) $y = 14.4$ in

10–33 (a) $k' = 24.7$ lbf $\cdot$ in/turn each, (b) 297 kpsi

10–34 $k = 2EI/[R^2(19\pi R + 18l)]$

B–11 Chapter 11

11–1 $x_D = 540$, $F_D = 2.278$ kN, $C_{10} = 18.59$ kN, 02–30 mm deep-groove ball bearing, $R = 0.919$

11–4 $R = R_1 R_2 = 0.927(0.942) = 0.873$, goal not met

11–8 $x_D = 180$, $C_{10} = 57.0$ kN

11–11 $C_{10} = 8.88$ kN

11–13 $R_O = 195$ N, $R_E = 196$ N, deep-groove 02–25 mm at O and C

11–18 $l_2 = 0.267(10^6)$ rev

B–12 Chapter 12

12–1 $c_{\min} = 0.000\,75$ in, $r = 0.500$ in, $r/c = 667$, $N_j = 18.3$ r/s, $S = 0.261$, $h_0/c = 0.595$, $rf/c = 5.8$, $Q/(rcNl) = 3.98$, $Q_s/Q = 0.5$, $h_0 = 0.000\,446$ in, $H = 0.0134$ Btu/s, $Q = 0.0274$ in^3/s, $Q_s = 0.0137$ in^3/s

12–3 SAE 10: $h_0 = 0.000\,275$ in, $p_{\max} = 847$ psi, $c_{\min} = 0.0025$ in

12–7 $h_0 = 0.0165$ mm, $f = 0.007\,65$, $Q = 1263$ mm^3/s

12–9 $h_0 = 0.010$ mm, $H = 34.3$ W, $Q = 1072$ mm^3/s, $Q_s = 793$ mm^3/s

12–11 $T_{\text{av}} = 65°C$, $h_0 = 0.0272$ mm, $H = 45.2$ W, $Q_s = 1712$ mm^3/s

12–20 15.2 mPa $\cdot$ s

B–13 Chapter 13

13–1 35 teeth, 3.25 in

13–2 400 rev/min, $p = 3\pi$ mm, $C = 112.5$ mm

13–4 $a = 0.3333$ in, $b = 0.4167$ in, $c = 0.0834$ in, $p = 1.047$ in, $t = 0.523$ in, $d_1 = 7$ in, $d_{1b} = 6.578$ in, $d_2 = 9.333$ in, $d_{2b} = 8.77$ in, $p_b = 0.984$ in

13–5 $d_P = 2.333$ in, $d_G = 5.333$ in, $\gamma = 23.63°$, $\Gamma = 66.37°$, $A_0 = 2.911$ in, $F = 0.873$ in

13–8 (a) 13, (b) 15, (c) 18, (d) 16

13–10 10:20 and higher

13–13 (a) $p_n = 3\pi$ mm, $p_t = 10.40$ mm, $p_x = 22.30$ mm, (b) $m_t = 3.310$ mm, $\phi_t = 21.88°$, (c) $d_p = 59.58$ mm, $d_G = 105.92$ mm

13–15 $e = 4/51$, $n_d = 47.06$ rev/min cw

13–22 $n_A = 68.57$ rev/min cw

13–27 $n_b/n_a = 11/36$ same sense

13–33 $\mathbf{F}_A = 71.5\,\mathbf{i} + 53.4\,\mathbf{j} + 350.5\,\mathbf{k}$ lbf, $\mathbf{F}_B = -149.5\,\mathbf{i} - 590.4\,\mathbf{k}$ lbf

13–40 $\mathbf{F}_C = 1565\,\mathbf{i} + 672\,\mathbf{j}$ lbf; $\mathbf{F}_D = 1610\,\mathbf{i} - 425\,\mathbf{j} + 154\,\mathbf{k}$ lbf

B–14 Chapter 14

14–1 $\sigma = 7.63$ kpsi

14–4 $\sigma = 82.6$ MPa

14–7 $F = 2.5$ in

14–10 $m = 2$ mm, $F = 25$ mm

14–14 $\sigma_c = -617$ MPa

14–17 $W^t = 16\,890$ N, $H = 97.2$ kW (pinion bending); $W^t = 3433$ N, $H = 19.8$ kW (pinion and gear wear)

14–18 $W^t = 1356$ lbf, $H = 34.1$ hp (pinion bending); $W^t = 1720$ lbf, $H = 43.3$ hp (gear bending), $W^t = 265$ lbf; $H = 6.67$ hp (pinion and gear wear)

14–22 $W^t = 775$ lbf, $H = 19.5$ hp (pinion bending); $W^t = 300$ lbf, $H = 7.55$ hp (pinion wear) AGMA method accounts for more conditions

14–24 Rating power = min(157.5, 192.9, 53.0, 59.0) = 53 hp

14–28 Rating power = min(270, 335, 240, 267) = 240 hp

14–34 $H = 69.7$ hp

B–15 Chapter 15

15–1 $W_P^t = 690$ lbf, $H_1 = 16.4$ hp, $W_G^t = 620$ lbf, $H_2 = 14.8$ hp

15–2 $W_P^t = 464$ lbf, $H_3 = 11.0$ hp, $W_G^t = 531$ lbf, $H_4 = 12.6$ hp

15–8 Pinion core 300 Bhn, case, 373 Bhn; gear core 339 Bhn, case, 345 Bhn

15–9 All four $W^t = 690$ lbf

15–11 Pinion core 180 Bhn, case, 266 Bhn; gear core, 180 Bhn, case, 266 Bhn

B–16 Chapter 16

16–1 (a) Right shoe: $p_a = 111.4$ psi cw rotation, (b) Right shoe: $T = 2530$ lbf · in; left shoe: 1310 lbf · in; total $T = 3840$ lbf · in, (c) RH shoe: $R^x = -229$ lbf, $R^y = 940$ lbf, $R = 967$ lbf; LH shoe: $R^x = 130$ lbf, $R^y = 171$ lbf, $R = 215$ lbf

16–3 LH shoe: $T = 161.4$ N · m, $p_a = 610$ kPa; RH shoe: $T = 59.0$ N · m, $p_a = 222.8$ kPa, $T_{\text{total}} = 220.4$ N · m

16–5 $p_a = 203$ kN, $T = 38.76$ N · m

16–8 $a' = 1.209r$, $a = 1.170r$

16–10 $P = 1560$ lbf, $T = 29\,980$ lbf · in

16–14 (a) $T = 8200$ lbf · in, $P = 504$ lbf, $H = 26$ hp; (b) $R = 901$ lbf; (c) $p|_{\theta=0} = 70$ psi, $p|_{\theta=270°} = 27.3$ psi

16–17 (a) $F = 1885$ lbf, $T = 7125$ lbf · in; (c) torque capacity exhibits a stationary point maximum

16–18 (a) $d^* = D/\sqrt{3}$; (b) $d^* = 3.75$ in, $T^* = 7173$ lbf · in; (c) $(d/D)^* = 1/\sqrt{3} = 0.577$

16–19 (a) Uniform wear: $p_a = 82.2$ kPa, $F = 949$ N; (b) Uniform pressure: $p_a = 79.1$ kPa, $F = 948$ N

16–23 $C_s = 0.08$, $t = 5.30$ in

16–26 (b) $I_e = I_M + I_P + n^2 I_P + I_L/n^2$; (c) $I_e = 10 + 1 + 10^2(1) + 100/10^2 = 112$

16–27 (c) $n^* = 2.430$, $m^* = 4.115$, which are independent of I_L

B–17 Chapter 17

17–1 (a) $F_c = 0.913$ lbf, $F_i = 101.1$ lbf, $F_{1a} = 147$ lbf, $F_2 = 57$ lbf; (b) $H_a = 2.5$ hp, $n_{fs} = 1.0$; (c) 0.151 in

17–3 A-3 polyamide belt, $b = 6$ in, $F_c = 77.4$ lbf, $T - 10\,946$ lbf · in, $F_1 = 573.7$ lbf, $F_2 = 117.6$ lbf, $F_i = 268.3$ lbf, dip $= 0.562$ in

17–5 (a) $T = 742.8$ lbf · in, $F_i = 148.1$ lbf; (b) $b = 4.13$ in; (c) $F_1 = 293.4$ lbf, $F_c = 17.7$ lbf, $F_i = 147.6$ lbf, $F_2 = 41.5$ lbf, $H = 20.6$ hp, $n_{fs} = 1.1$

17–7 $R^x = (F_1 + F_2)\{1 - 0.5[(D - d)/(2C)]^2\}$, $R^y = (F_1 - F_2)(D - d)/(2C)$. From Ex. 17–2, $R^y = 1214.4$ lbf, $R^x = 34.6$ lbf

17–14 With $d = 2$ in, $D = 4$ in, life of 10^6 passes, $b = 4.5$ in, $n_{fs} = 1.05$

17–17 Select one B90 belt

17–20 Select nine C270 belts, life $> 10^9$ passes, life $>$ 150 000 h

17–24 (b) $n_1 = 1227$ rev/min. Table 17–20 confirms this point occurs in the range 1200 ± 200 rev/min, (c) Eq. (17–40) applicable at speeds exceeding 1227 rev/min for No. 60 chain

17–25 (a) $H_a = 7.91$ hp; (b) $C = 18$ in; (c) $T = 1164$ lbf · in, $F = 744$ lbf

17–27 Four-strand No. 60 chain, $N_1 = 17$ teeth, $N_2 = 84$ teeth, rounded $L/p = 134$, $n_{fs} = 1.17$, life 15 000 h (pre-extreme)

B–18 Chapter 18

18–1 (a) Maximum bending moment 2371 lbf · in, (b) $d_A = 1.625$ in, $d_B = 1.810$ in, average diameter ≥ 1.810 in

18–2 (a) $d = 1.725$ in, (b) $d = 1.687$ in

18–13 $d = 1.371$ in; $d = 1.630$ in for $n_d = 2$

18–18 $d = 24$ mm, $D = 32$ mm, $r = 1.6$ mm

18–20 (a) Static: $d = 1.526$ in; (b) DE-Gerber: $d = 1.929$ in; ASME-elliptic: $d = 1.927$ in; MSS-Soderberg: $d = 1.932$ in; DE-Goodman: $d = 2.008$ in

18–24 Slope: $y_m' = y'$; deflection: $y_m = sy = y/2$; moment: $M_m = s^3 M = M/8$; Force: $F_m = s^2 F = F/4$; same material, same stress levels

18–26 (a) $\omega = 868$ rad/s; (b) $d = 2$ in; (c) $\omega = 1736$ rad/s (doubles)

18–28 (b) $\omega = 466$ rad/s $= 4450$ rev/min

Index

Part 1 Properties of Sections

A = area

G = location of centroid

$I_x = \int x^2 \, dA$ = second moment of area about x axis

$I_{xy} = \int xy \, dA$ = mixed moment of area about x and y axes

$J_G = \int r^2 \, dA = \int (x^2 + y^2) \, dA = I_x + I_y$

    = second polar moment of area about axis through G

$k_x^2 = I_x / A$ = squared radius of gyration about x axis

Rectangle

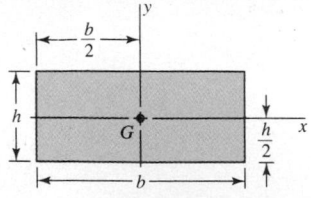

$A = bh$ $\qquad I_x = \dfrac{bh^3}{12}$ $\qquad I_y = \dfrac{b^3 h}{12}$ $\qquad I_{xy} = 0$

Circle

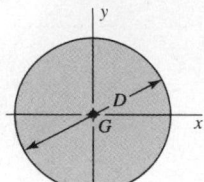

$A = \dfrac{\pi D^2}{4}$ $\qquad I_x = I_y = \dfrac{\pi D^4}{64}$ $\qquad I_{xy} = 0$

Hollow circle

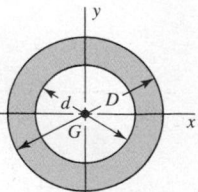

$A = \dfrac{\pi}{4}(D^2 - d^2)$ $\qquad I_x = I_y = \dfrac{\pi}{64}(D^4 - d^4)$ $\qquad I_{xy} = 0$

A MODERN METHOD FOR GUITAR

william leavitt

volume **1**

 **Berklee**
Press

1140 Boylston Street
Boston, MA 02215-3693 USA
(617) 747-2146

Visit Berklee Press Online at
www.berkleepress.com

EXCLUSIVELY DISTRIBUTED BY

HAL•LEONARD®

7777 W. BLUEMOUND RD. P.O. BOX 13819
MILWAUKEE, WISCONSIN 53213

Visit Hal Leonard Online at
www.halleonard.com

Introduction

This book has been specifically designed to accomplish two things:

1. To teach the student to *read* music.

Reading "crutches" have been eliminated as much as possible. Fingering and counting indications have been kept at what I consider a sensible minimum.

2. For the gradual development of dexterity in *both* hands.

This is the physical part of learning to play the guitar and as such cannot be rushed. Practice all material slowly enough to maintain an even tempo. Do not skip or "slight" anything, and also do not attempt to "completely perfect" any one lesson before going on. Playing technique is an accumulative process and you will find that each time you review material already studied it will seem easier to play. (Slow, steady practice and constant review will eventually lead to speed and accuracy.)

I should like to mention at this point that all music presented for study on these pages is original and has been created especially for the guitar. *Each* composition has been designed to advance the student's musical knowledge and playing ability, and yet be as musical as possible. There is no student/teacher division in the duets; both guitar parts are written to be studied by the pupil, and almost all parts will musically stand alone.

I have not included any "old favorites," as guitar arrangements of these songs are available in many existing publications. (Also, you do not learn to *read* music by playing melodies that are familiar to you.)

I have not tried to make this book into a music dictionary by cramming it with pages filled with nothing but musical terms and markings, as it is considerably more important to give the student as much music to play as possible. (The most common and necessary terms and markings are, of course, used and explained. If further information is desired, some very excellent music dictionaries in soft cover editions can be obtained at a small cost.)

I do feel, however, that with this method (as with all others), you must search out additional material to practice, as your ultimate ability depends entirely on how much reading and playing you do.

So good luck, and have fun.

Wm. G. Leavitt

Contents

It is important that the following material be covered in consecutive order. The index on page 126 is for reference purposes only and will prove valuable for review or concentration on specific techniques.

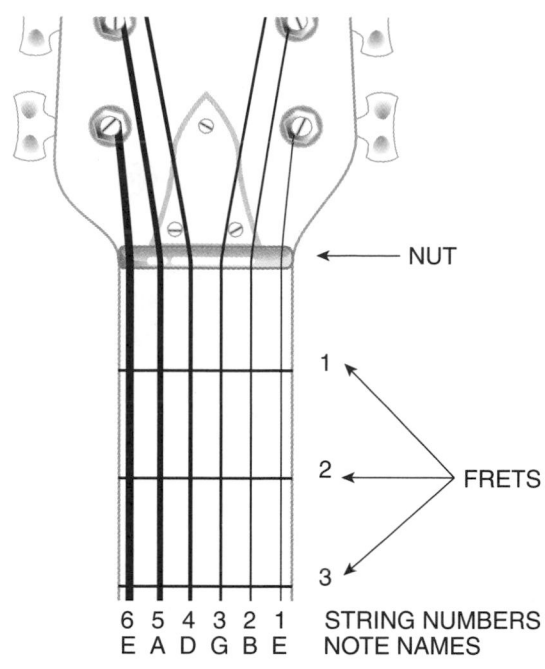

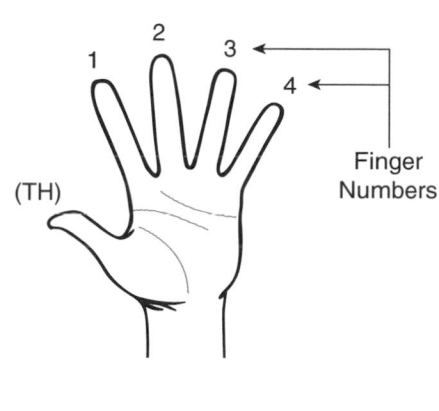

TO TUNE THE GUITAR: (using piano or pitch pipe)

1. Tune the open 1st string to the first E above middle C.
2. Press the 2nd string down at the fifth fret and tune (2nd string) until it sounds exactly the same as the open 1st string.
3. Press the 3rd string down at the fourth fret and tune (3rd string) until it sounds exactly the same as the open 2nd string.
4. Press the 4th string at the fifth fret and tune to the open 3rd string.
5. Press the 5th string at the fifth fret and tune to the open 4th string
6. Press the 6th string at the fifth fret and tune to the open 5th string

THE STAFF: consists of five lines and four spaces, and is divided into *measures* by *bar lines.*

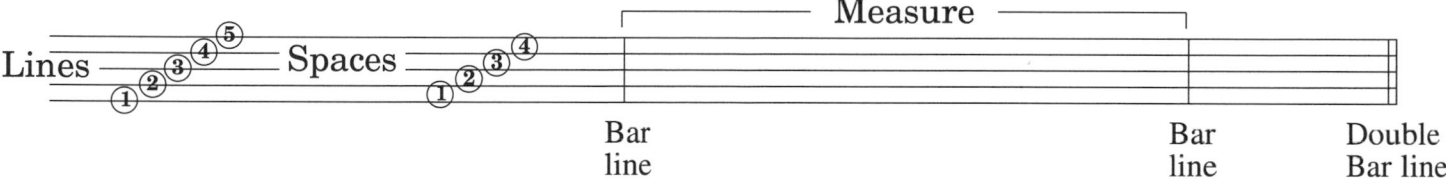

CLEF SIGN: Guitar music is written in the *treble* (or "G") *clef.* The number of sharps (♯) or flats (♭) found next to the clef sign indicate the key signature (to be explained more fully later).

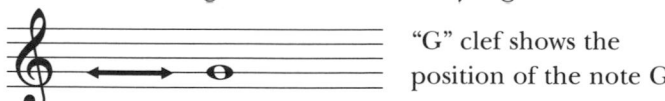

"G" clef shows the position of the note G

COMMON TIME VALUE OF THE NOTES:

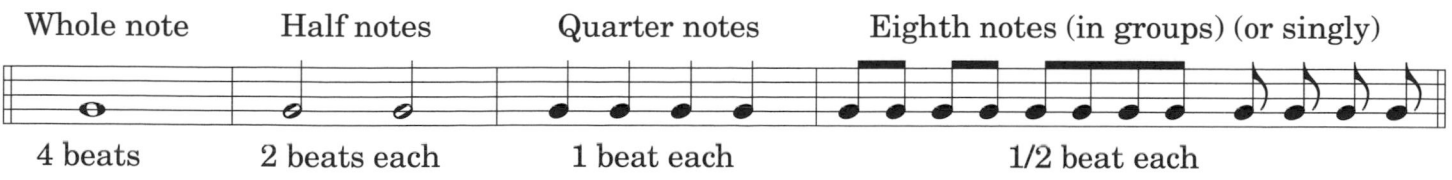

(Continued on next page)

TIME SIGNATURES: Next to the clef sign (at the beginning of a composition), locate the two numbers (like a fraction) or a symbol that represents these numbers. The top number tells how many beats (or counts) in a measure, and the bottom number indicates what kind of note gets one beat.

EXAMPLE: **4/4** means four quarters, or four beats per measure, with a quarter note receiving one beat or *count*. The symbol is **C** .

Notes in the First Position
No sharps or flats—Key of C major.

Order of the notes going up the scale:
A B C D E F G, A B C D E F G, A B...
Start at any point and read left to right.

■ **EXERCISE 1**

count 1 2 3 4

4

▌ Read the *notes*, not the fingering. Fingering numbers will eventually be omitted.

■ EXERCISE 2

■ EXERCISE 3

■ EXERCISE 4

Sea to Sea (duet)

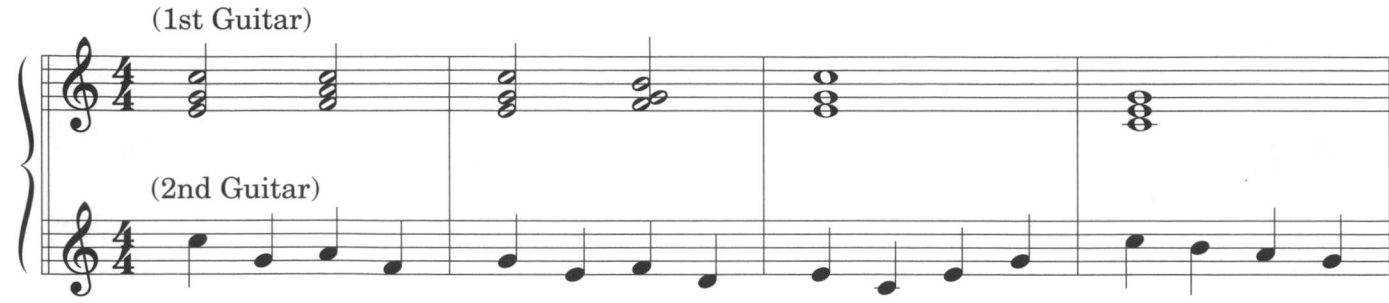

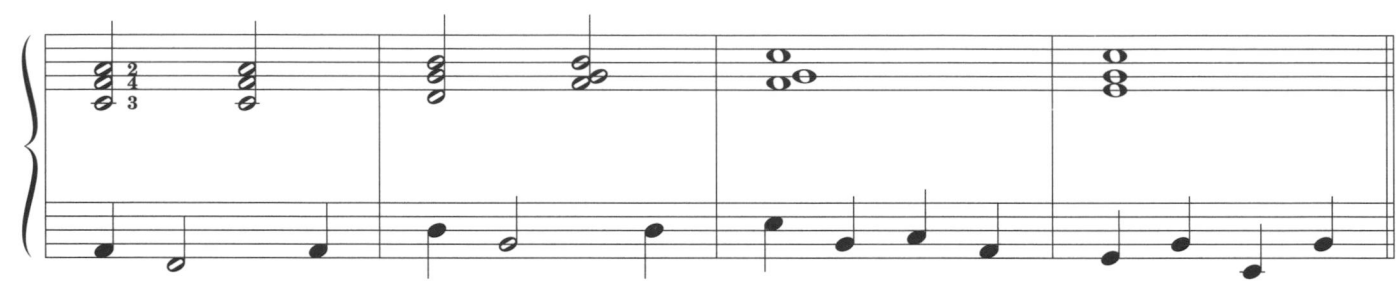

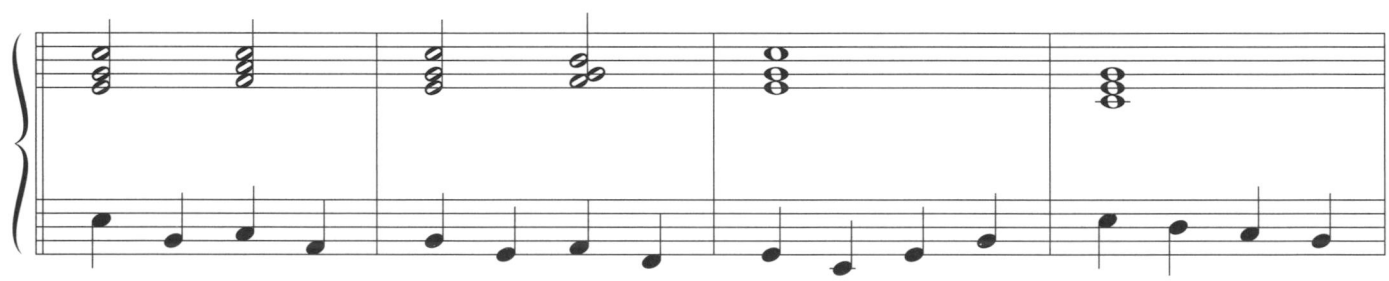

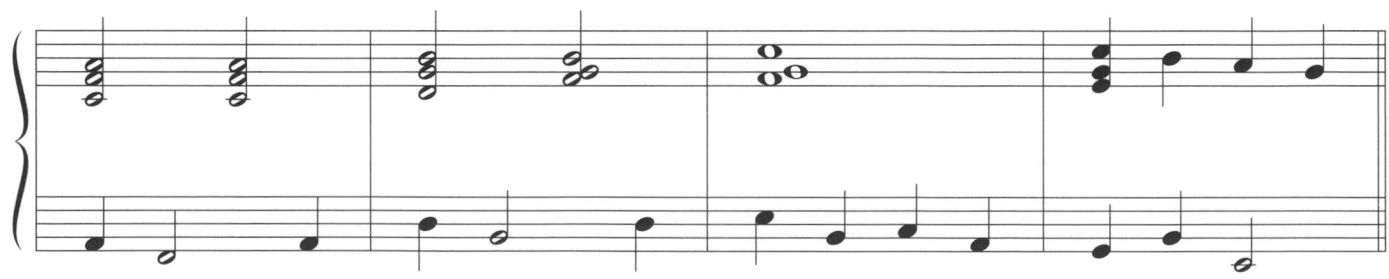

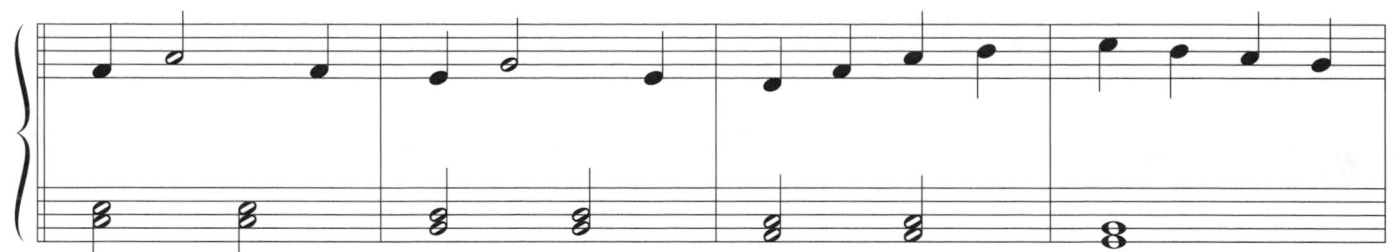

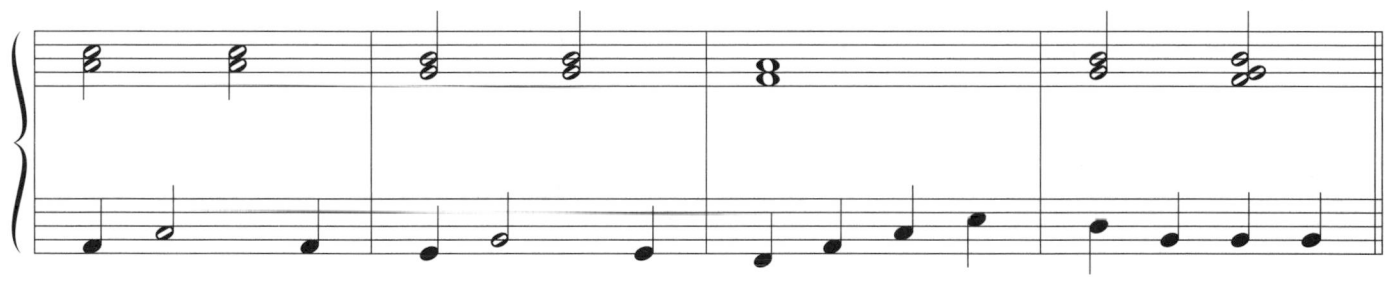

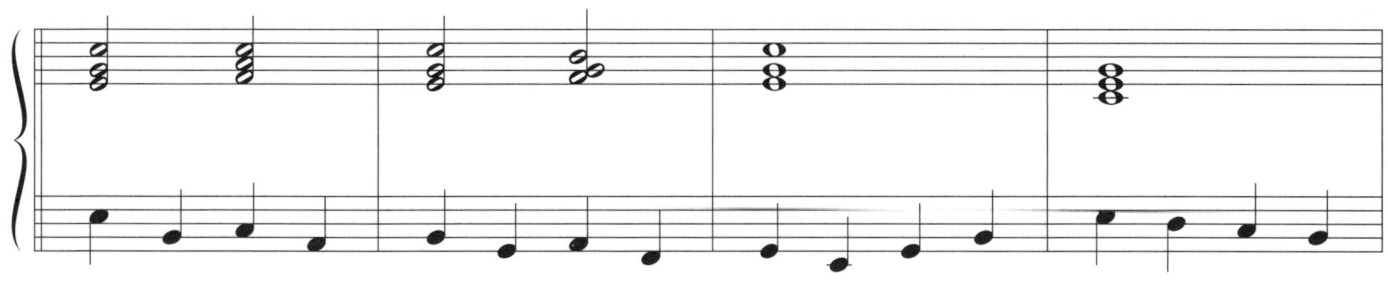

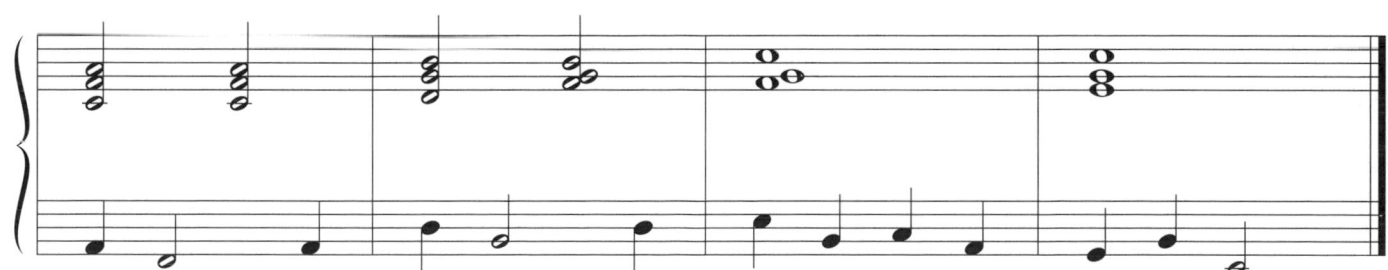

Starting on C one *octave* higher than C found on the 5th string, we complete the *upper register* of the first position.

■ **EXERCISE 5**

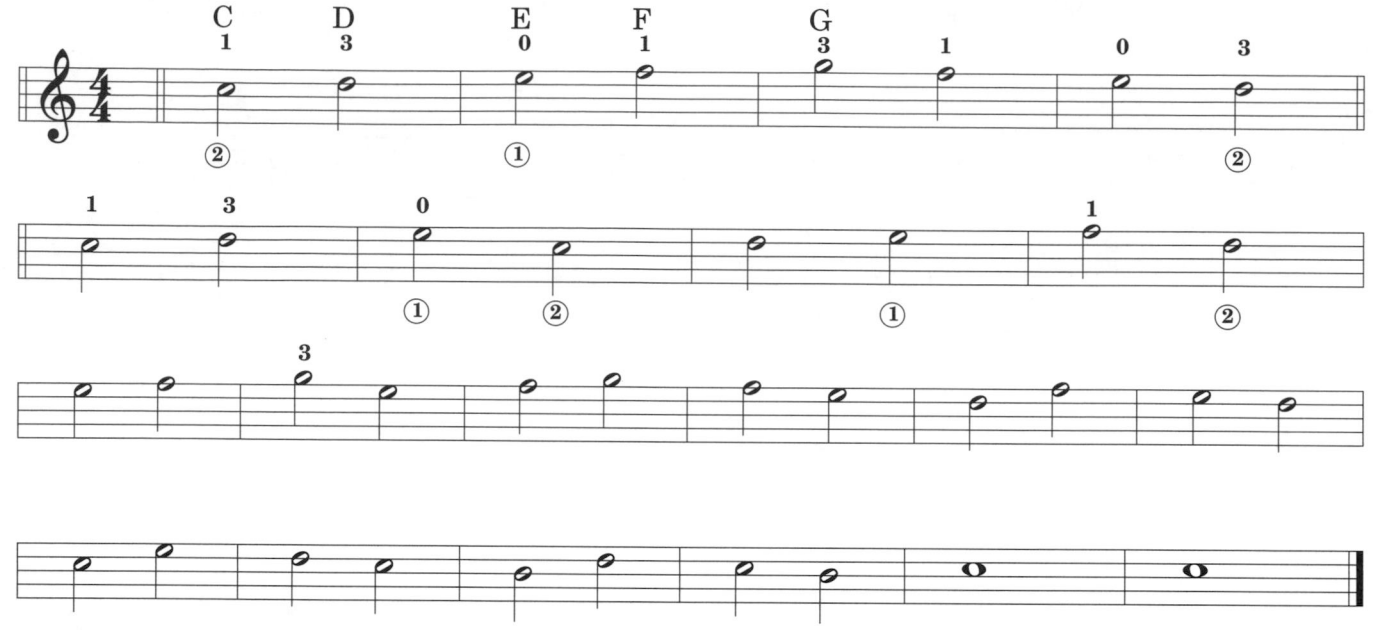

■ **EXERCISE 6**

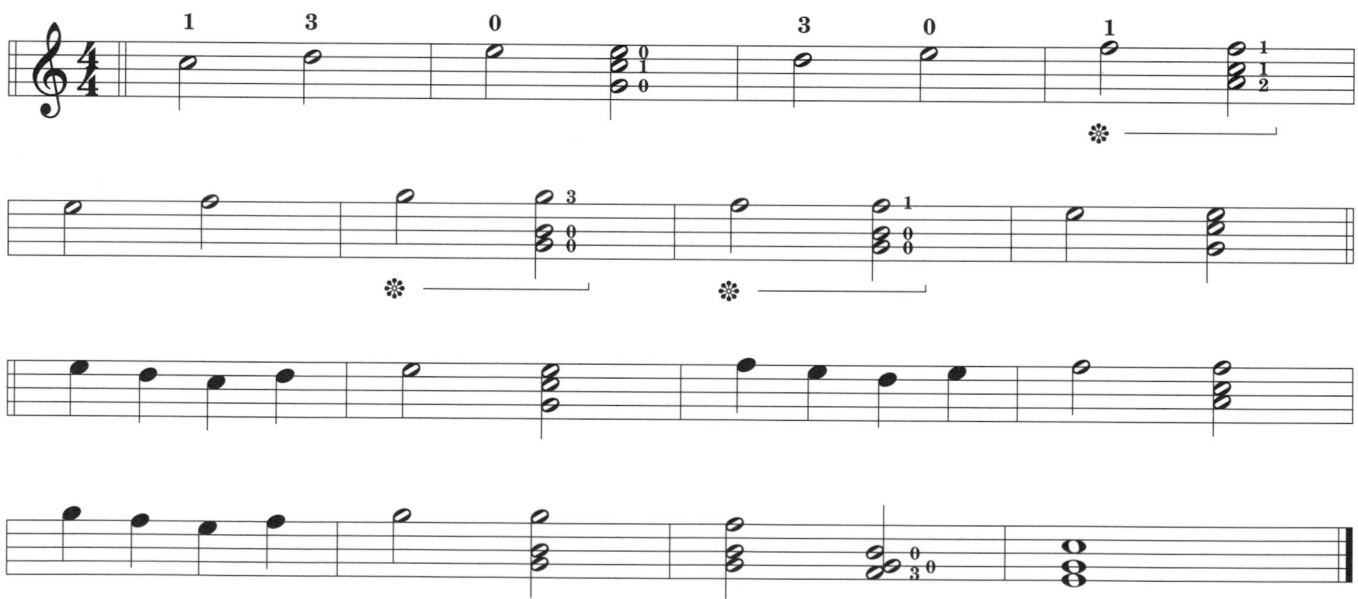

Note and Chord Review

■ **EXERCISE 7**

■ **EXERCISE 8**

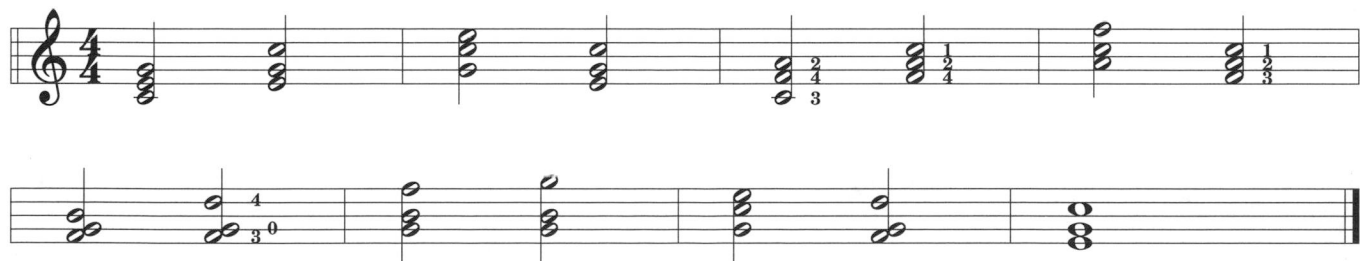

Regular review of all material is a must!

One, Two, Three, Four (duet)

Tempo (Speed): Moderate 4

(1st Guitar)

(2nd Guitar)

Rhythm Accompaniment

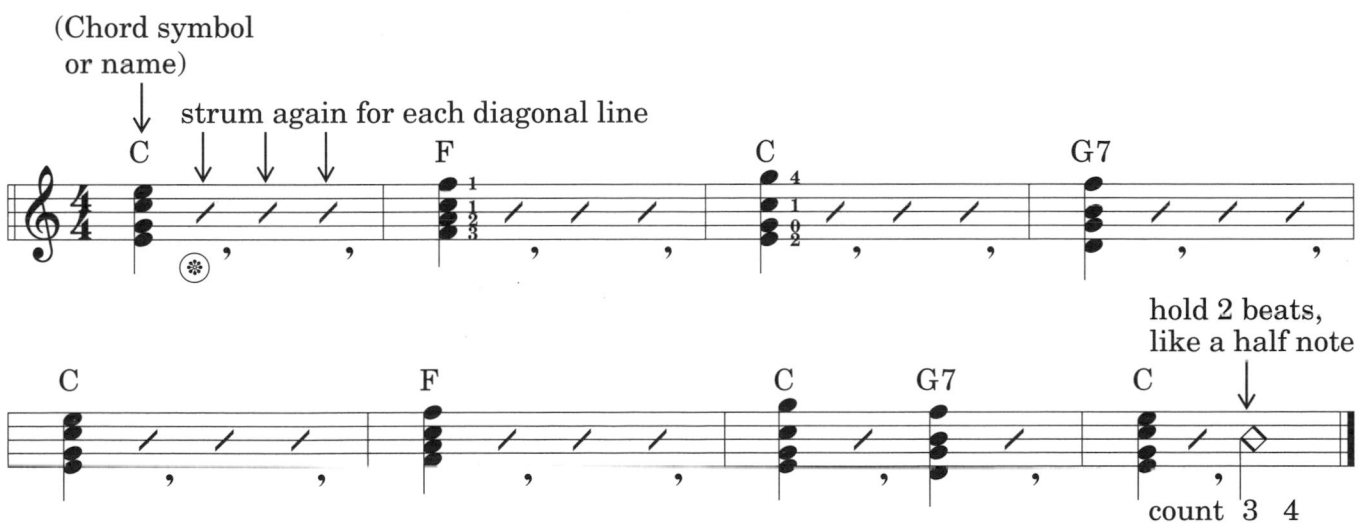

✳A better rhythmic pulse is produced if you relax left-hand pressure at these points (').
However, do not remove fingers from strings. Also, if open strings are involved, mute them with
the side of the right hand at the same instant that you relax left-hand pressure.

■ *Ledger lines* are added below or above the staff for notes too low or too high to appear on the staff.

■ **EXERCISE 9**

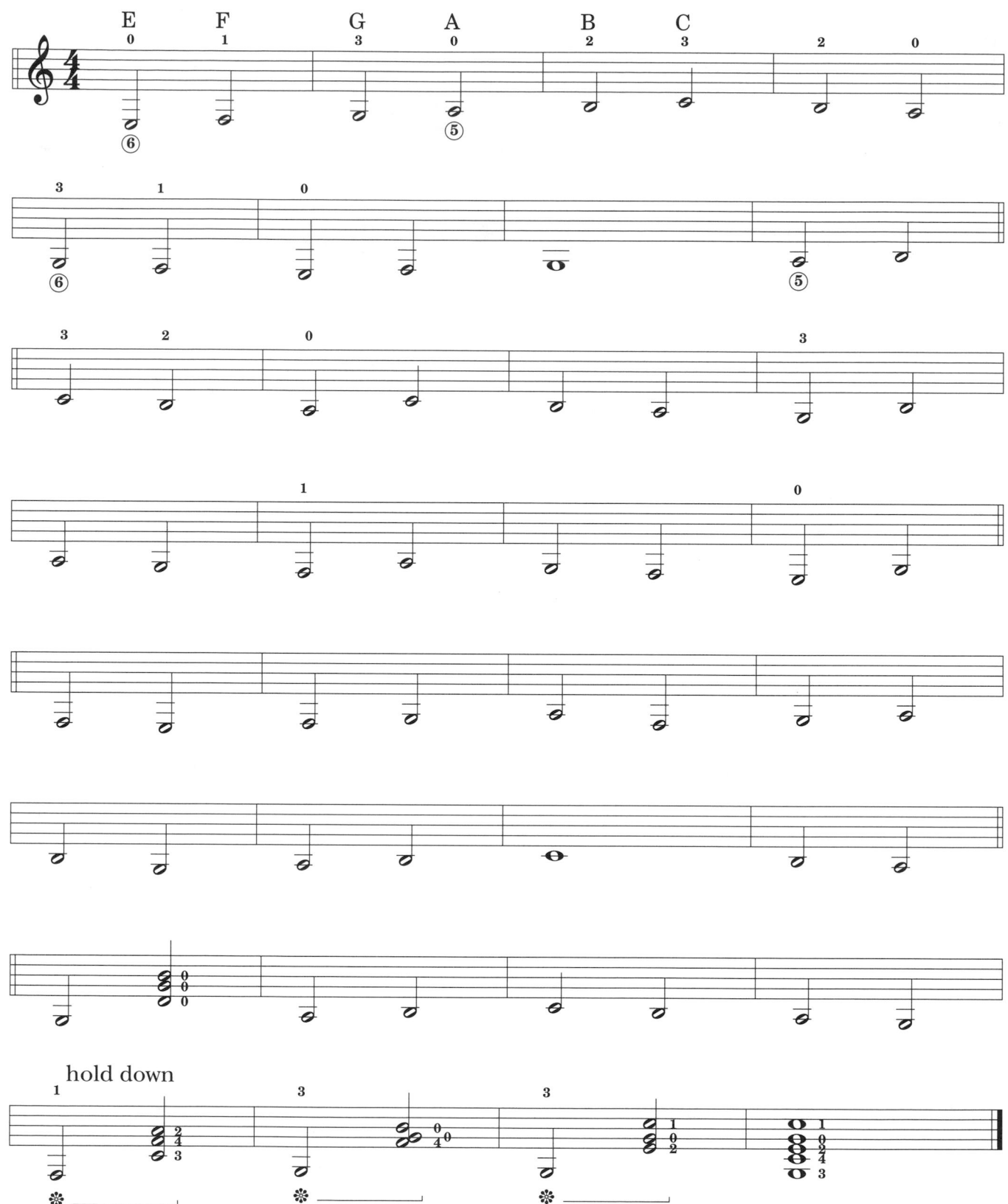

Review

Complete first position—Key of C major.

Imitation Duet

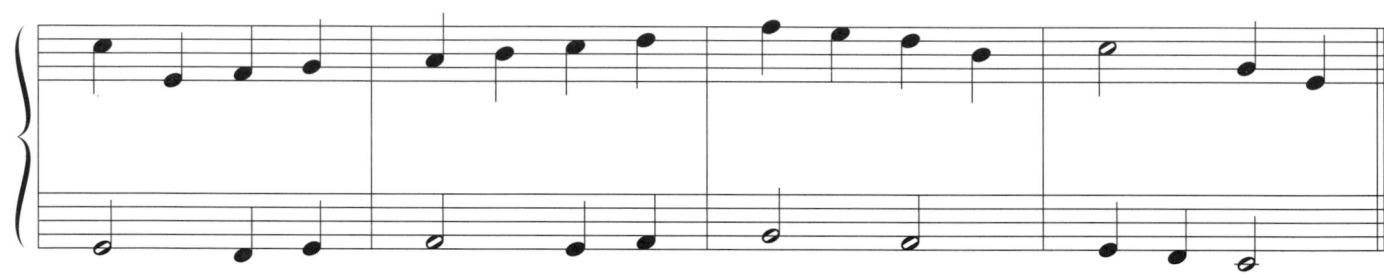

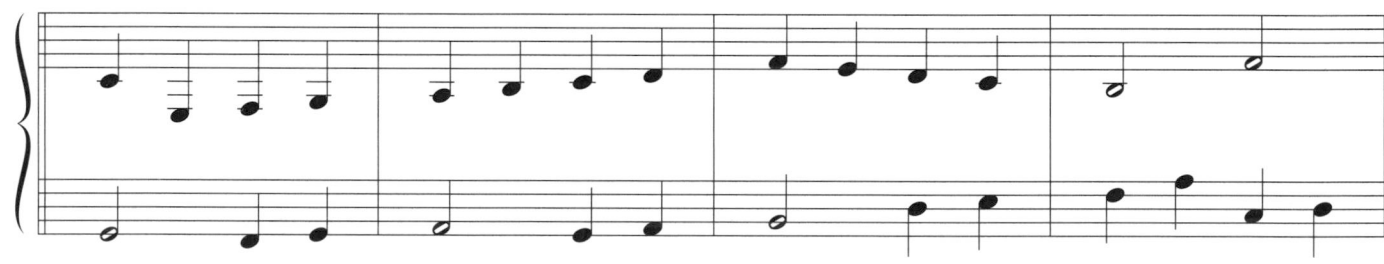

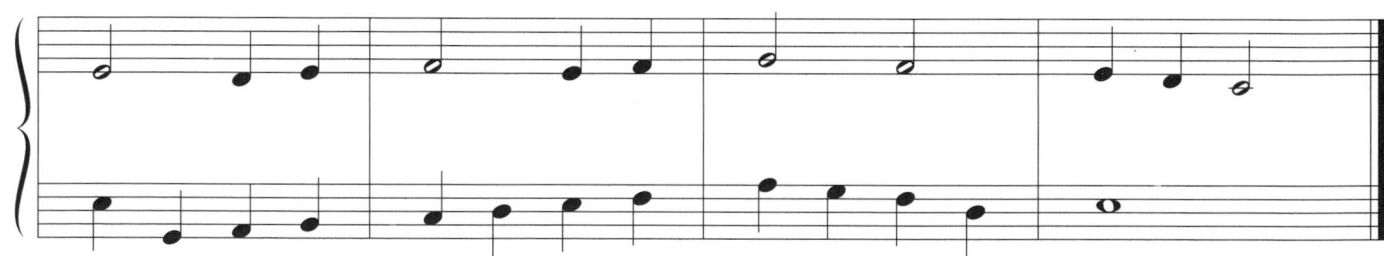

Sharps and Flats

A note that is not altered by a sharp or flat is called *natural*.

▌ A *sharp* (♯) raises the note a half tone (one fret). A *flat* (♭) lowers the note a half tone (one fret).

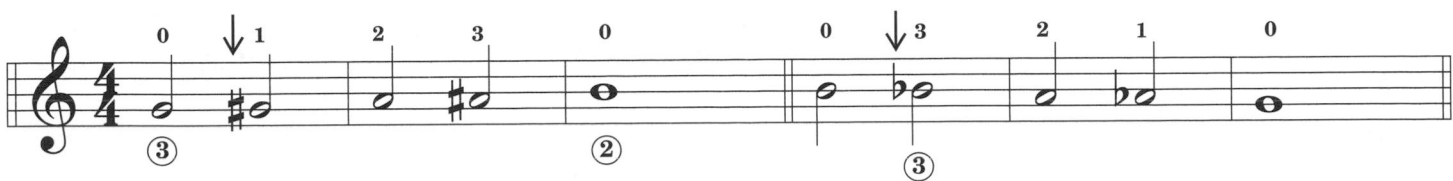

▌ When a sharp or flat appears in the *key signature* (between the clef sign and the time signature), it is used throughout the entire piece.

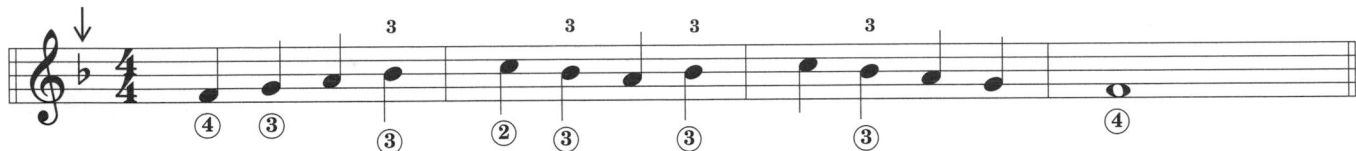

▌ When a sharp or flat that is not in the key signature appears in the piece, it is called an *accidental* and is used only for the remainder of that measure. The next bar line cancels it out.

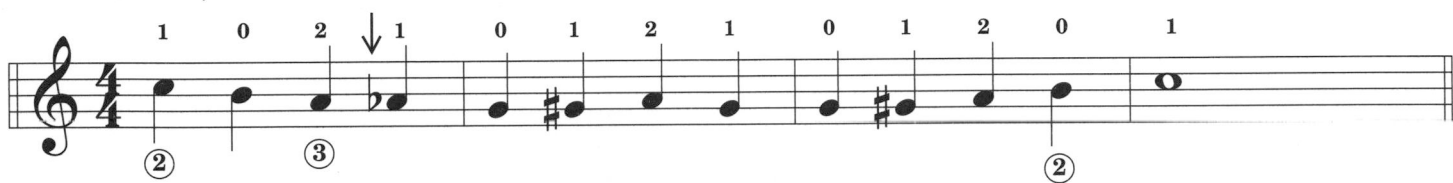

▌ The natural sign (♮) is used to cancel out accidentals within the same measure. It is also used as a reminder that the bar line has cancelled the accidental.

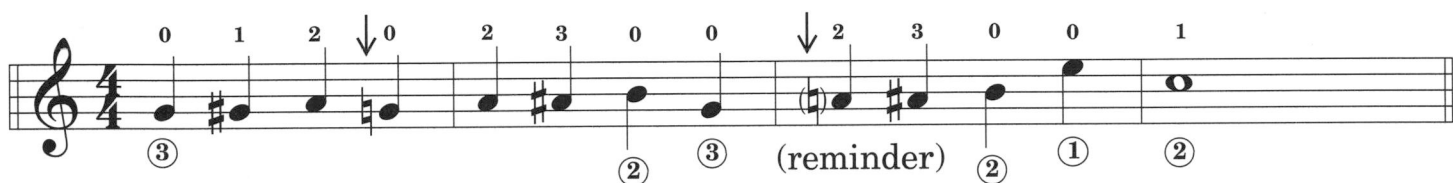

▌ When the natural sign is used to cancel a sharp or flat found in the key signature, cancellation is good only for the remainder of the measure.

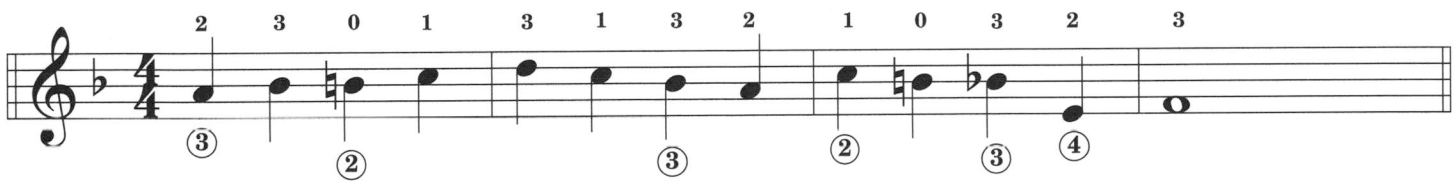

■ **EXERCISE**

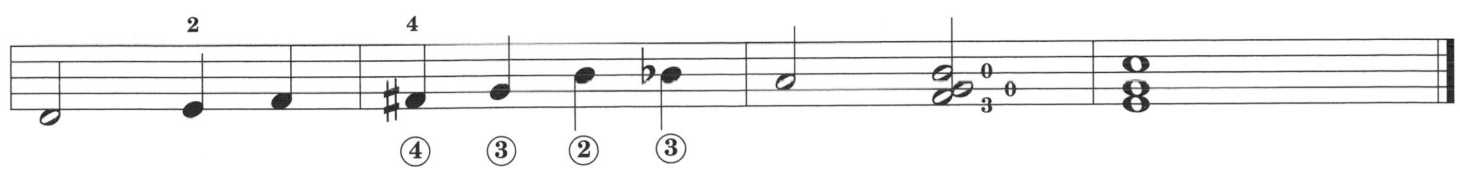

15

Here We Go Again (duet)

❋ *Mute,* or deaden, the 5th string by lightly touching it with the side of the third finger so it will not sound.

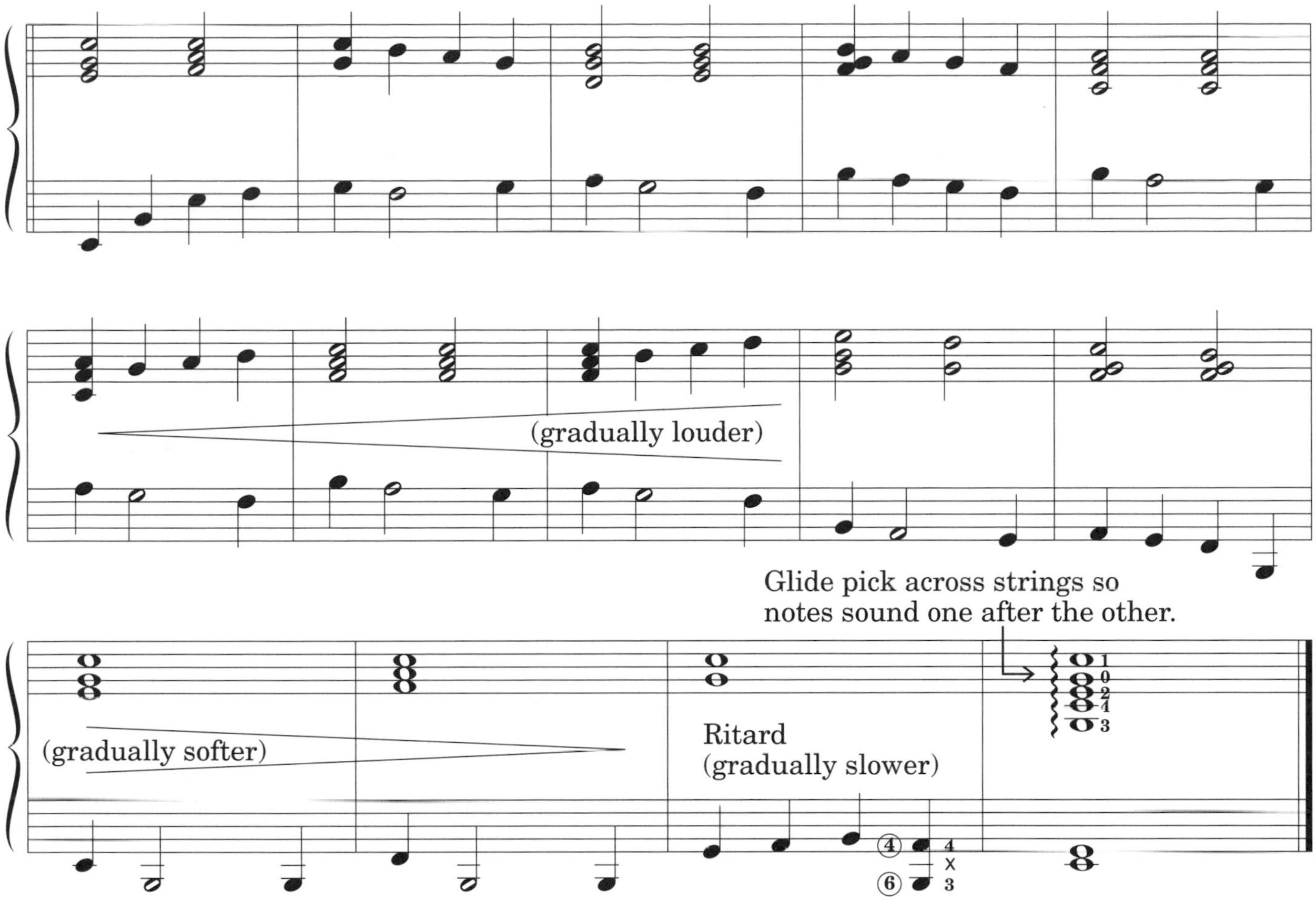

(gradually louder)
Glide pick across strings so
notes sound one after the other.
(gradually softer)
Ritard
(gradually slower)

Rhythm Accompaniment

Bass Notes and Chords

All chord symbols (names) appearing as only a letter are assumed to be *major* chords. A letter followed by the number "7" represents *dominant* 7 chords. A letter followed by a small "m" indicates *minor.*

C
Em
F
C
hold
G7
C
Am
Dm
Em
G7
C
F
C

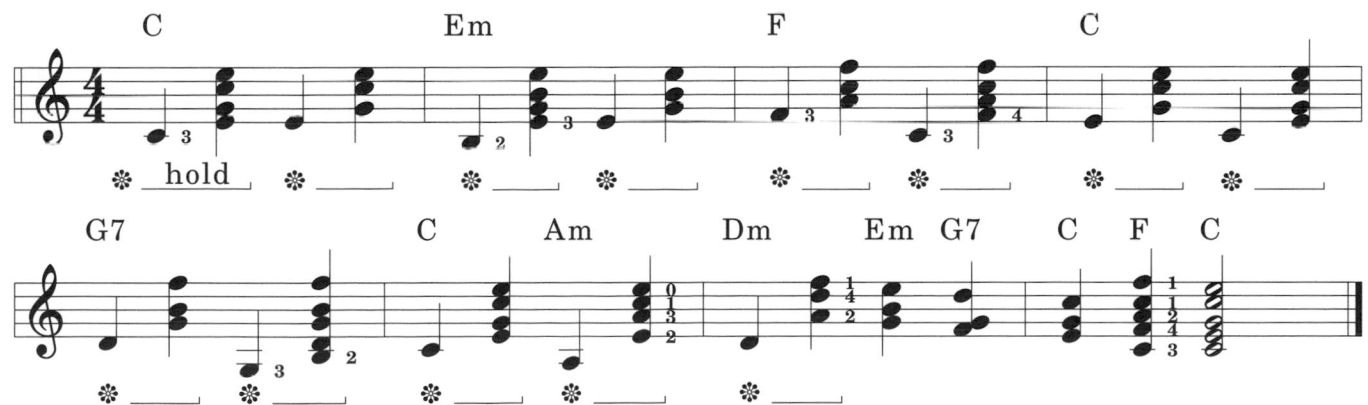

Do not skip or "slight" any lesson material.

Eighth Notes—Counting and Picking

⊓ means "pick downward" ∨ means "pick upward"

■ **EXERCISE 1**

count   1   &   2   &   3   &   4   &

count   1   2   &   3   4   &

■ **EXERCISE 2**

❋ *Fermata* means "hold."

⌢ is a fermata. It indicates to hold the note.

Review of all material is a must.

■ EXERCISE 3

Etude No. 1 (duet)

Fine
(the end)

Rests, Tied Notes, Dotted Notes

COMMON TIME VALUES OF RESTS (periods of silence):

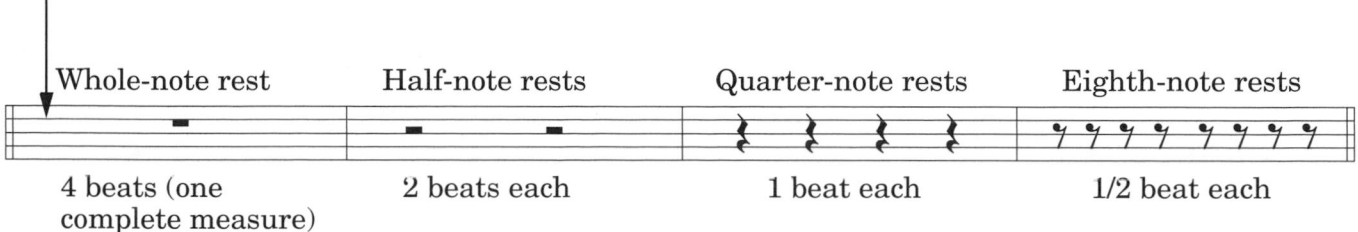

Whole-note rest — 4 beats (one complete measure)

Half-note rests — 2 beats each

Quarter-note rests — 1 beat each

Eighth-note rests — 1/2 beat each

TIED NOTES: When two notes are "tied" together with a curved line, only the first note is picked. The second note is merely held and counted.

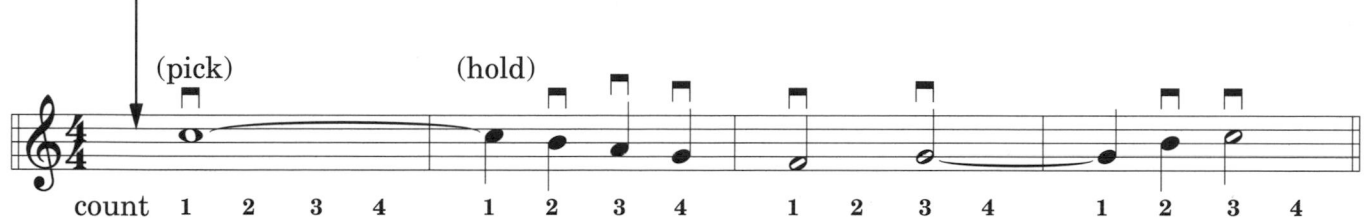

DOTTED NOTES: A "dot" placed after any note increases the time value of the note by one-half. Or you may say a "dot" found next to any note receives half the time value of the note itself.

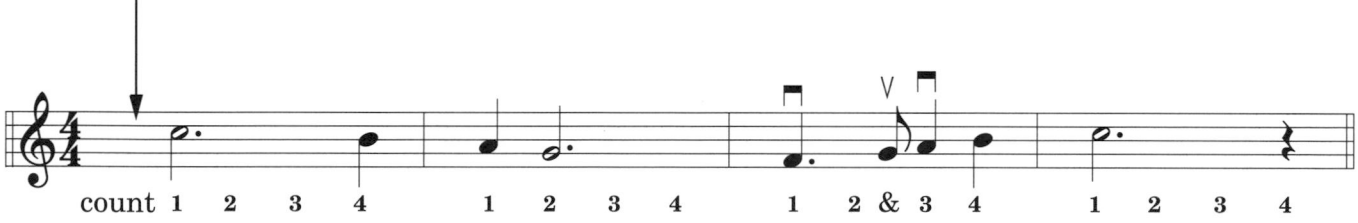

This is the same example as shown above but using "tied" notes. . . .

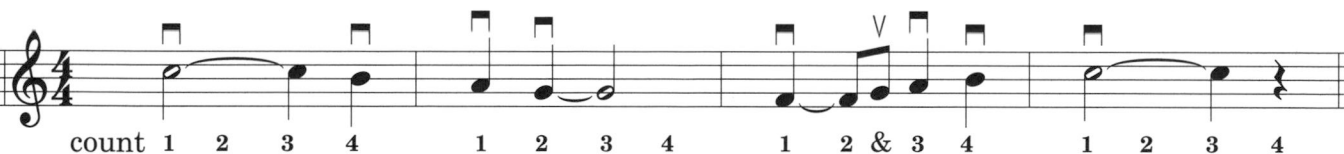

■ EXERCISE

▌ Count aloud as you play.

Etude No. 2 (duet)

First Solo

Solo arrangement: with melody and accompaniment.

Accompaniment chord is
played on the 2nd beat

Melody note is picked on the 1st beat
and held while chord is played

Fine

Be sure to hold all notes for their full time values.

Rhythm Accompaniment

CHORD DIAGRAMS

1. Vertical lines represent strings.
2. Horizontal lines represent frets.
   (See illustration, pg. 3.)
3. Dots represent finger placement.
4. Numbers indicate fingers to be used.
5. Zero means open string.
6. × means muted string.

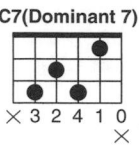
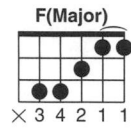
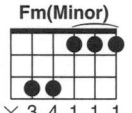
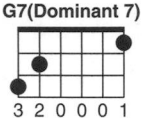

■ EXERCISE 1

▌ Use only the chord forms shown above.

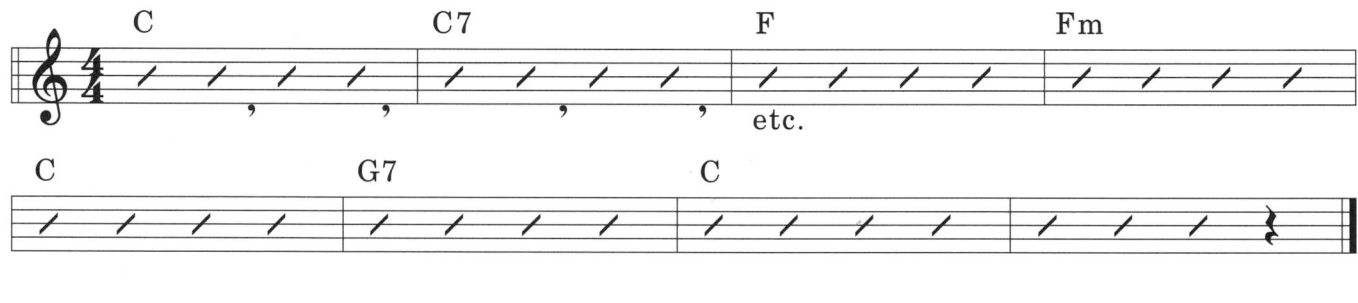

If no number, zero, or × is found below the diagram, do not allow the pick to strike the string

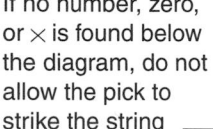
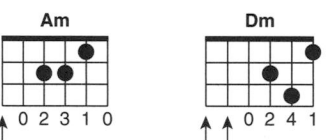

Optional fingered note or open string

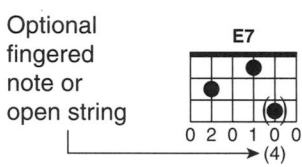

■ EXERCISE 2

■ EXERCISE 3

▌ This exercise combines all forms shown above, and should not be attempted until the preceding chord sequences are mastered at least partially.

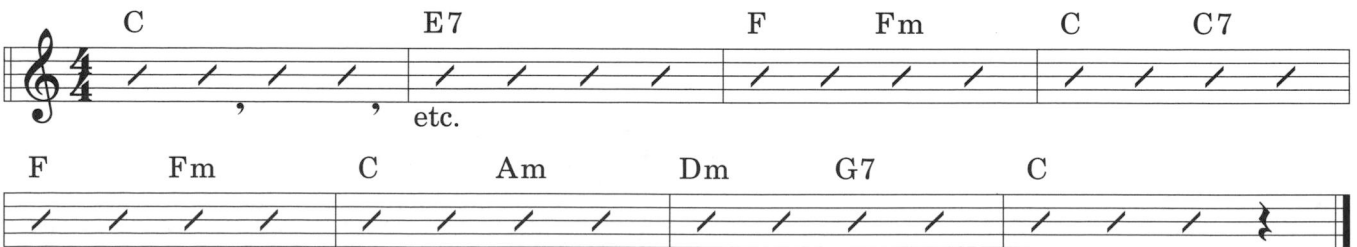

All chord forms must be memorized.

Second Solo

Solo arrangement with melody *above* (and *below*) the chord accompaniment.

Hold all notes for their full value.

Etude No. 3 (duet)

count   1   2   3   4   &

count   1   2   3   &   4   &

1   2   &   3   4

Ritard   Fine

Review everything regularly.

Picking Etude No. 1

For development of the right hand.

PREPARATION

Ritard

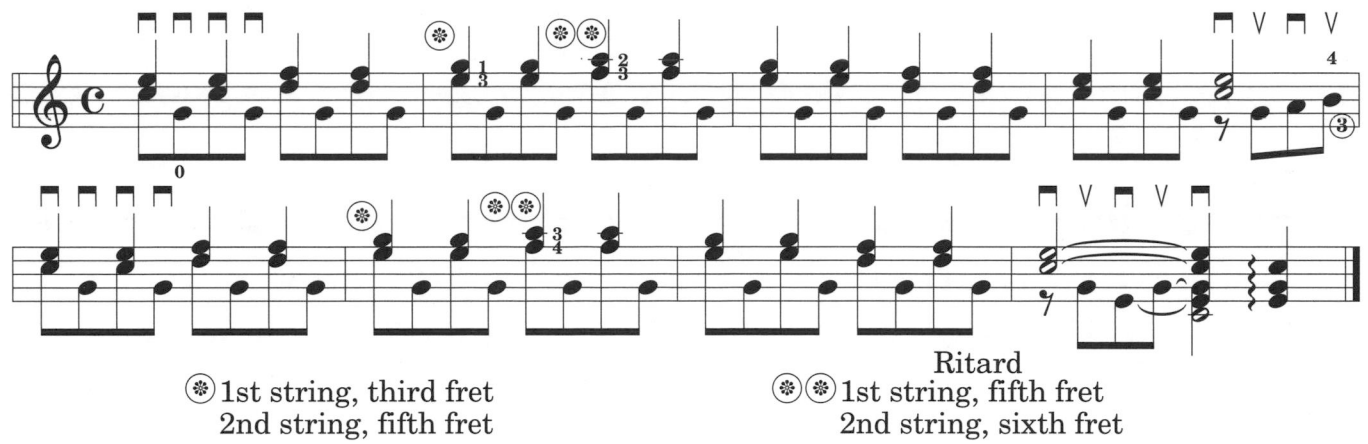

⊛ 1st string, third fret
2nd string, fifth fret

⊛⊛ 1st string, fifth fret
2nd string, sixth fret

Etude

Tempo: Moderately Slow 4

Ritard
A tempo (back to original tempo)
Ritard
Fine

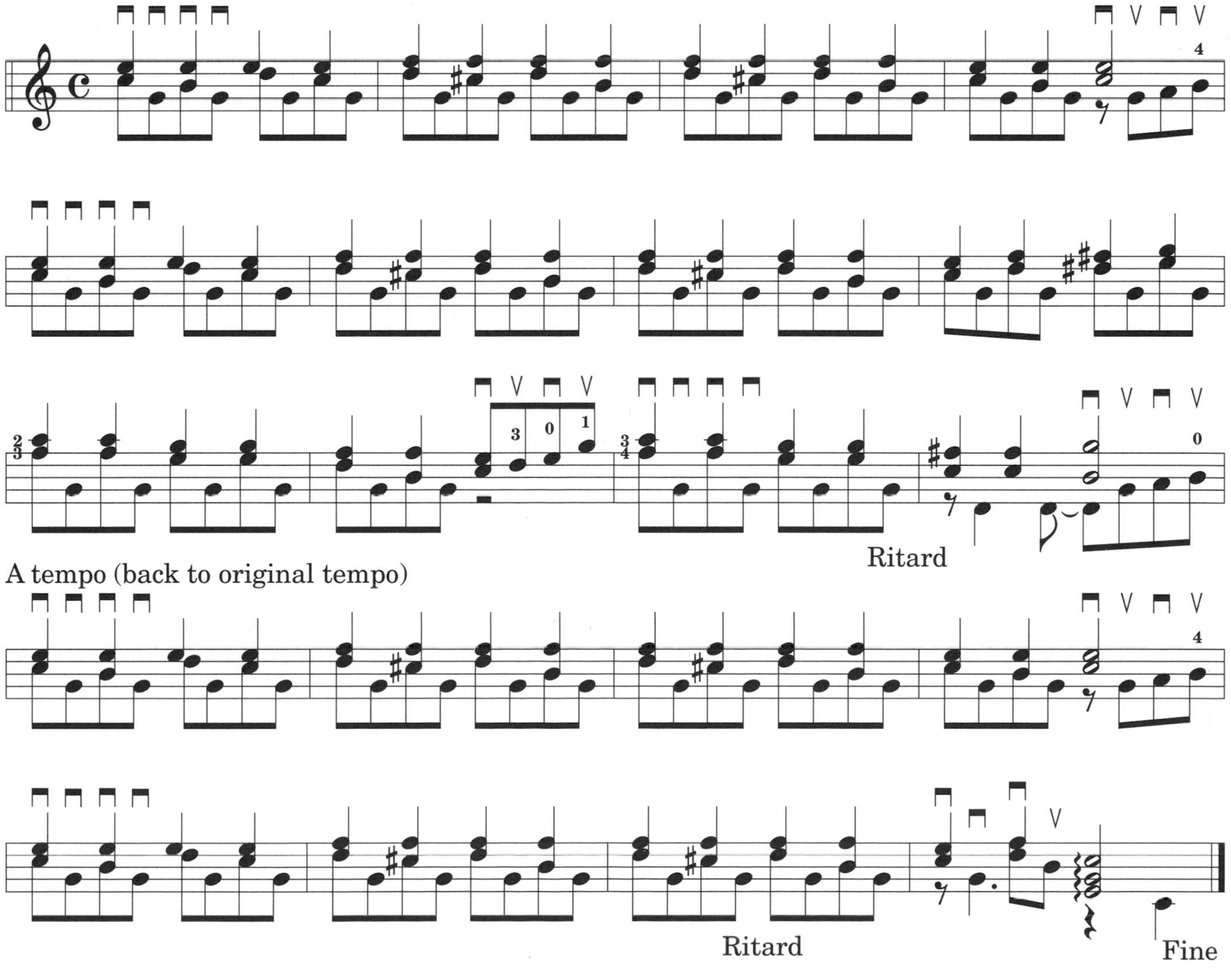

Two, Two (duet)

¢ is often misused to represent 4/4 in popular music.

Play again from
the (𝄋) sign to the
al coda. Then skip
to the coda (𝄌).

Key of G Major (First Position)

(All Fs are sharped.)

Rhythm Accompaniment

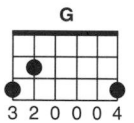

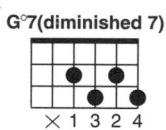

This chord structure is also indicated by the abbreviation "dim." Even though the numeral 7 is often omitted from the symbol, diminished 7 is intended.

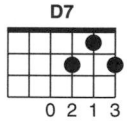

■ **EXERCISE 1**

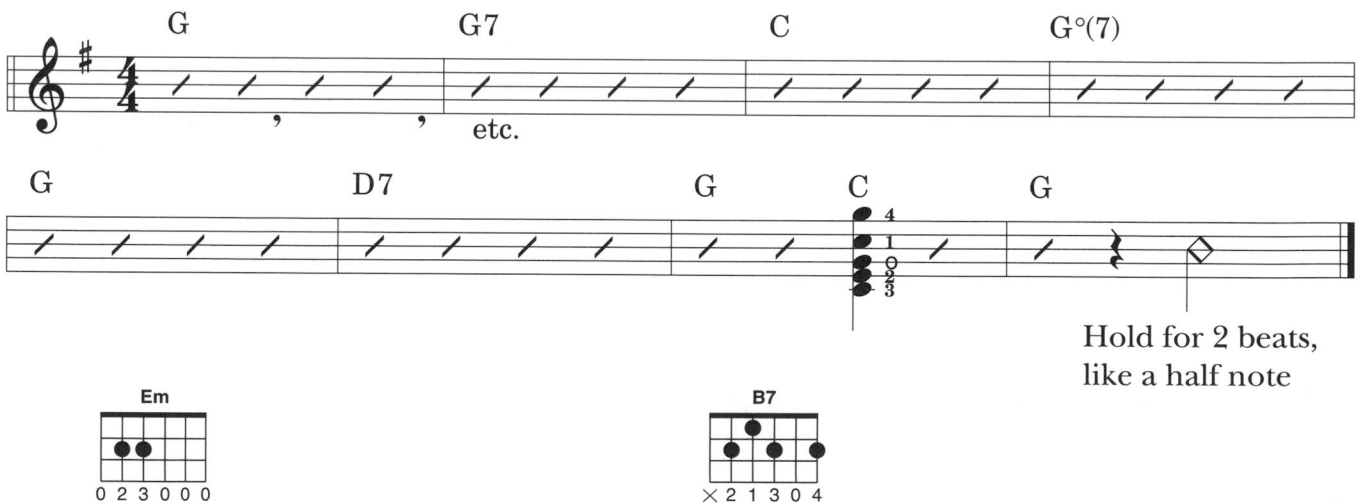

Hold for 2 beats, like a half note

■ **EXERCISE 2**

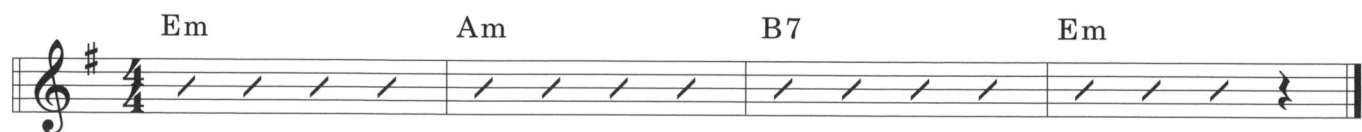

30

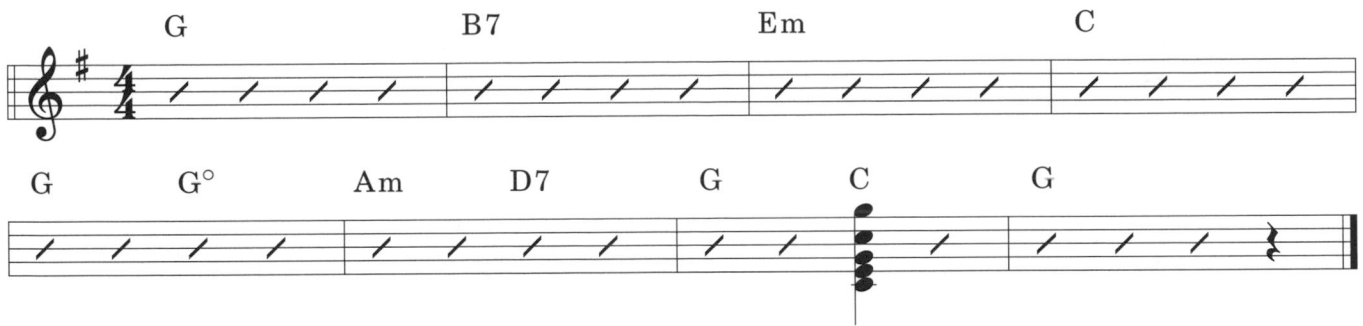

Sixteenth Notes

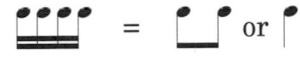

Duet in G

Fine

Picking Etude No. 2

For alternate picking while skipping strings.

❚ Pay very strict attention to "down" and "up" pickings on all eighth-note passages.

Another Duet in G

count   1 & 2 & 3 & 4 &

Ritard   Fine

Key of F Major (First Position)

(All Bs are flatted.)

Rhythm Accompaniment

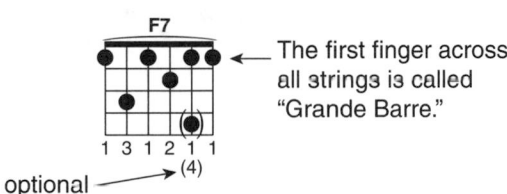

← The first finger across
all strings is called
"Grande Barre."

optional →

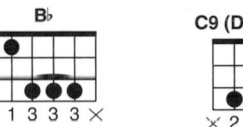

This C9 chord has
the same function as
C7 and is often
substituted for it.

■ **EXERCISE**

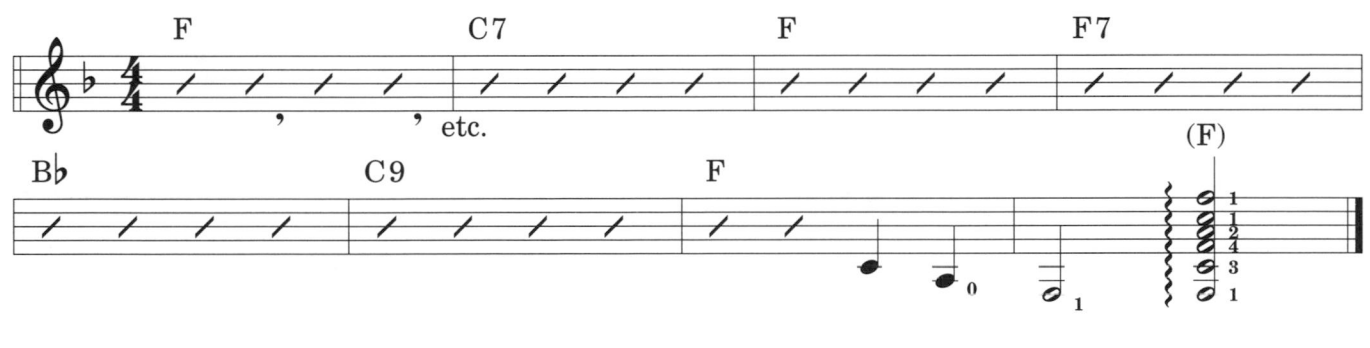

Note the slight
difference between
this D minor fingering
and the one on pg. 24.

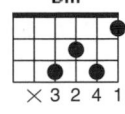

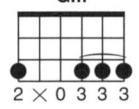

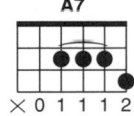

■ **EXERCISE**

■ Several of the forms presented above will take some time to play clearly. Be patient and keep at them.

Duet in F

The Triplet

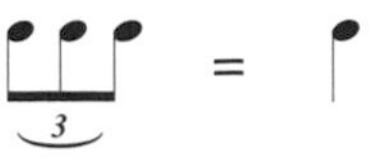

There are two ways to pick consccutive sets of triplets. Practice the entire exercise thoroughly, using first the picking marked Type 1. Then practice using Type 2.

■ EXERCISE

Fine

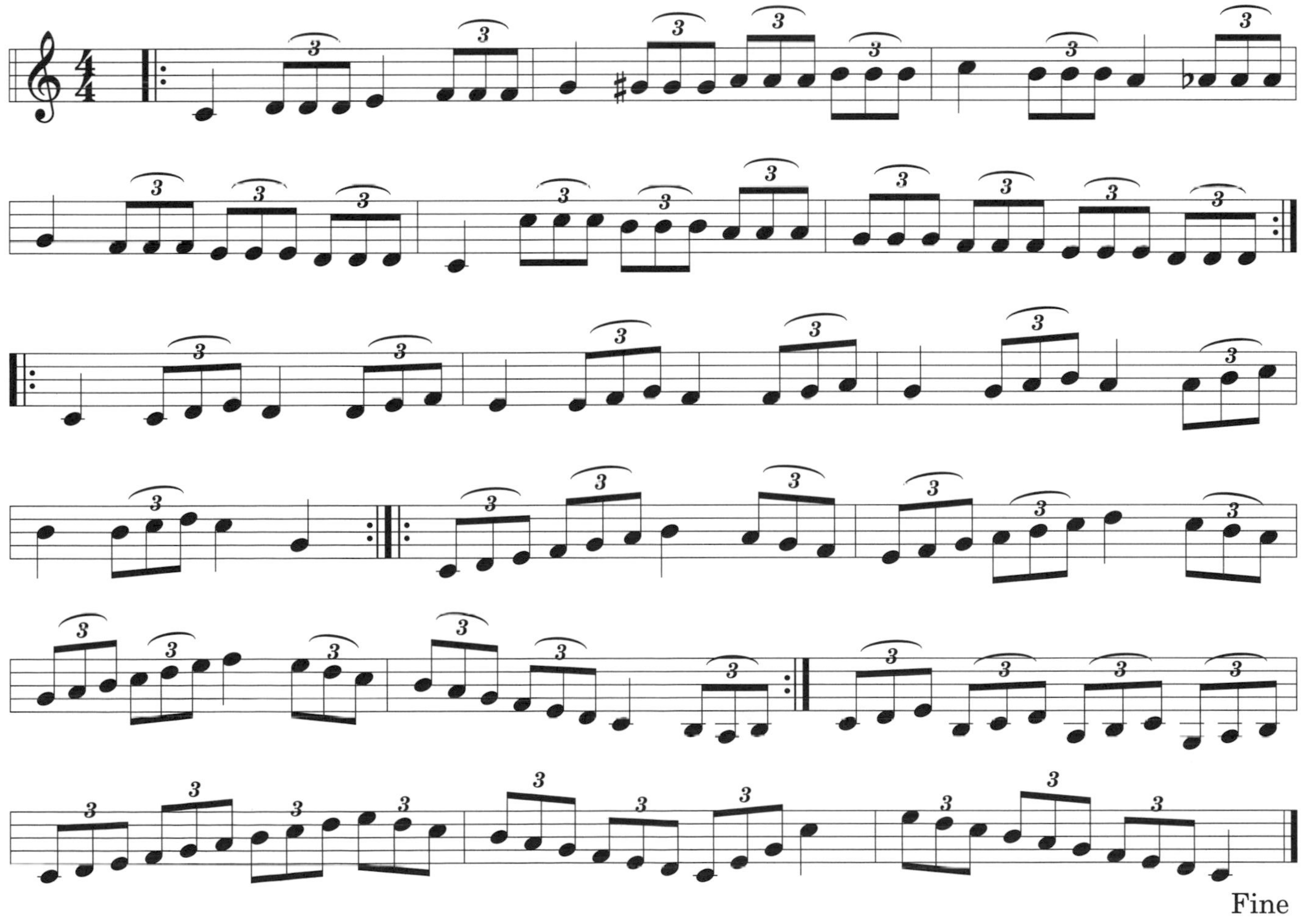

Review all material.

Waltz in F (solo)

A waltz has three beats per measure.

count 2   3   1   2   3 etc.

Rall. ("Rallentando," or slow down)

A tempo
(back in tempo)

Ritard poco a poco (little by little)

Fine

Key of A Minor (First Position)

Relative to C major.

∎ The sixth degree (note) of any major scale is the tonic (first note) of its *relative minor key*. The major and relative minor key signatures are the same. There are three different scales in each minor key.

∎ A Natural Minor: All notes exactly the same as its relative major, C major.

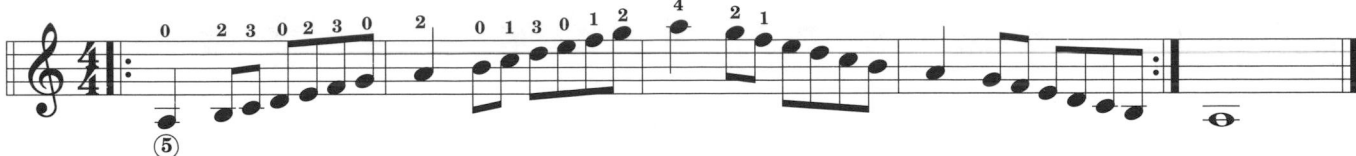

∎ A Harmonic Minor: The 7th degree, counting up from A, is raised a half step.

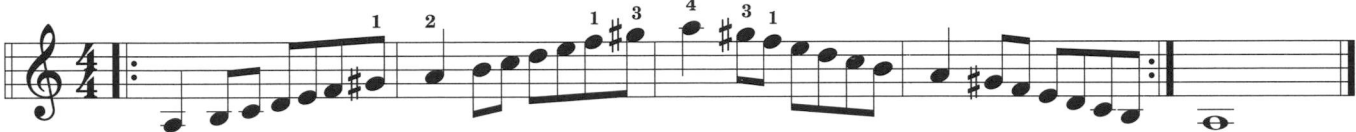

∎ A Melodic Minor: The 6th and 7th degrees are raised *ascending* but return to normal descending.

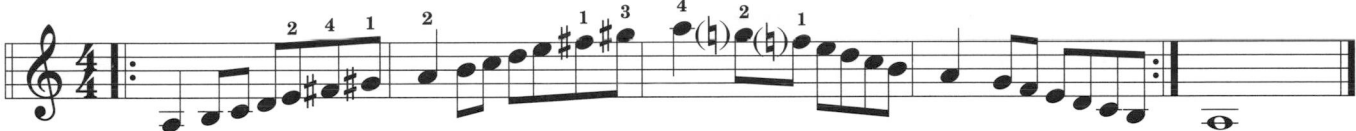

Rhythm Accompaniment

∎ We now begin to observe that many chords have more than one fingering. The choice of which one to use generally depends upon the chord fingerings that immediately precede and/or follow. In the following exercise use the large diagrams *or* the smaller optional fingerings in sequence. Do not mix them!

∎ **EXERCISE**

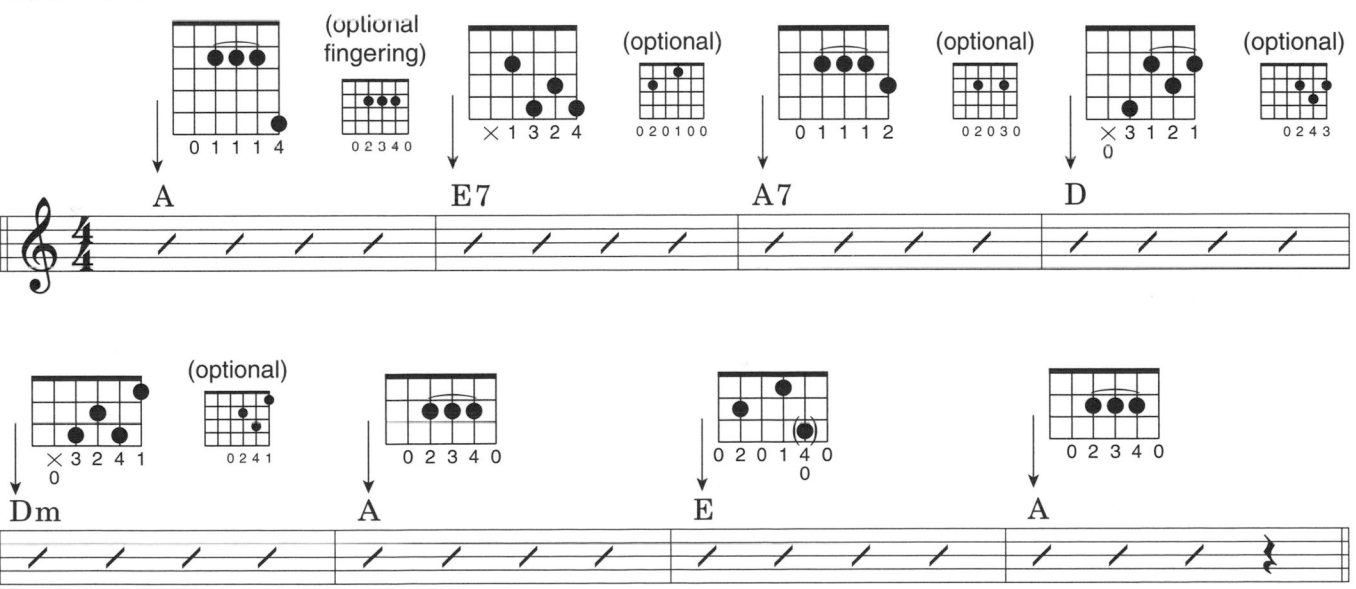

Smooth, melodic accompaniment depends
on the number of chord forms mastered.

Pretty Pickin' (duet)

For alternate picking while skipping strings.

CHORD PREPARATION

Slowly

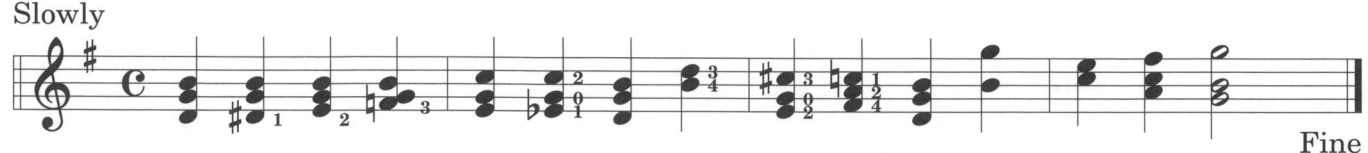

Fine

Duet

Moderate Waltz Tempo

(All notes under the curved line must be kept ringing.)

Crescendo
(get louder)

Diminuendo
(or get softer)

※ (Repeat from the beginning to the coda)

Dotted Eighth and Sixteenth Notes

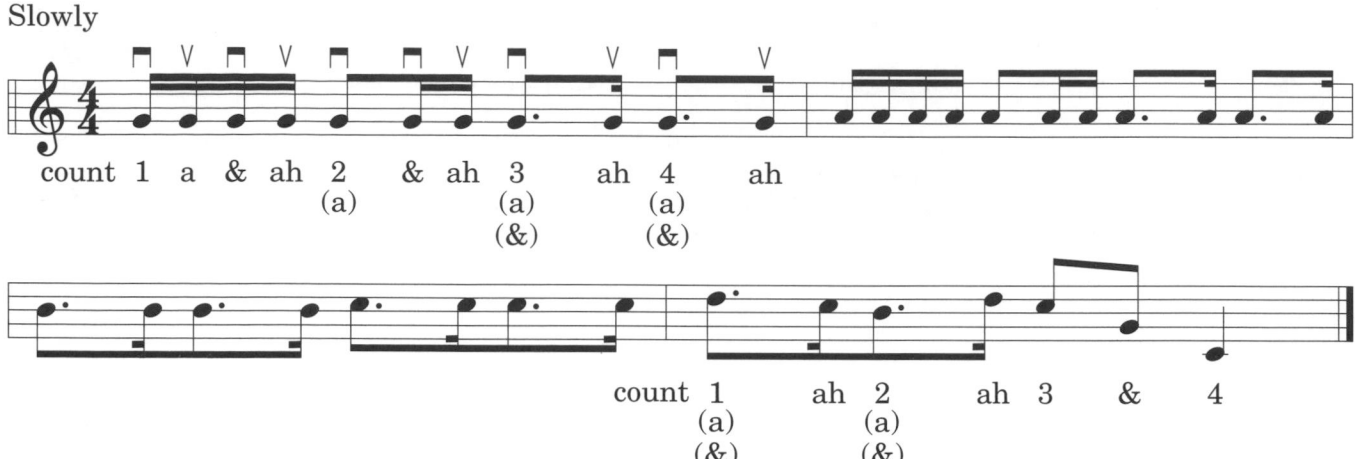

■ EXERCISE 1

■ EXERCISE 2

Note that the above "strict" (or "legitimate") interpretation of dotted eighth and sixteenth notes produces a rather jerky rhythm. In pop music and jazz they are played more *legato* (smoothly, in a flowing manner). This is done by treating them as triplets.

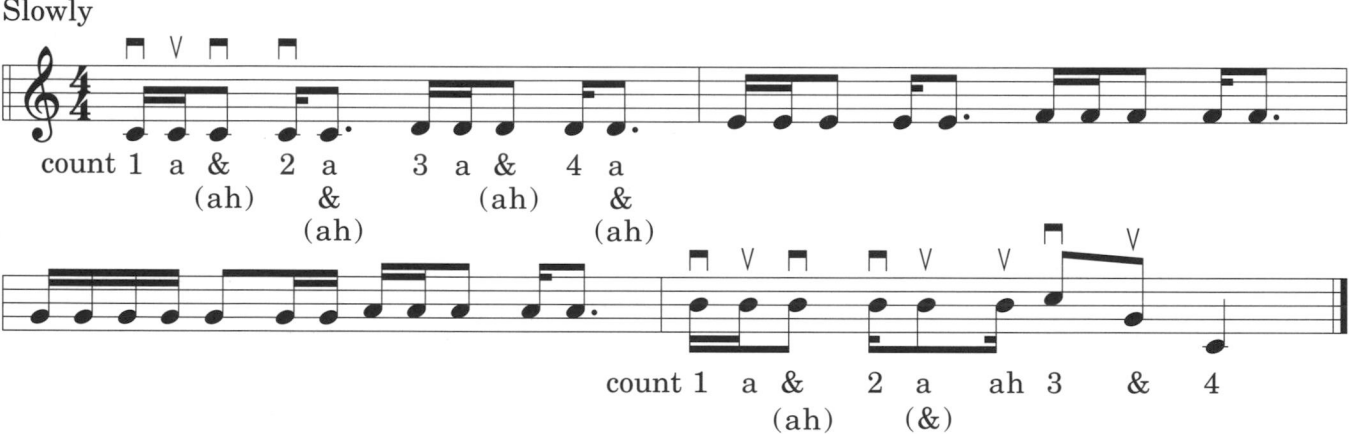

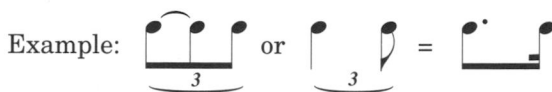

■ EXERCISE 3

Key of E Minor (Scales in First Position)

Relative to G major.

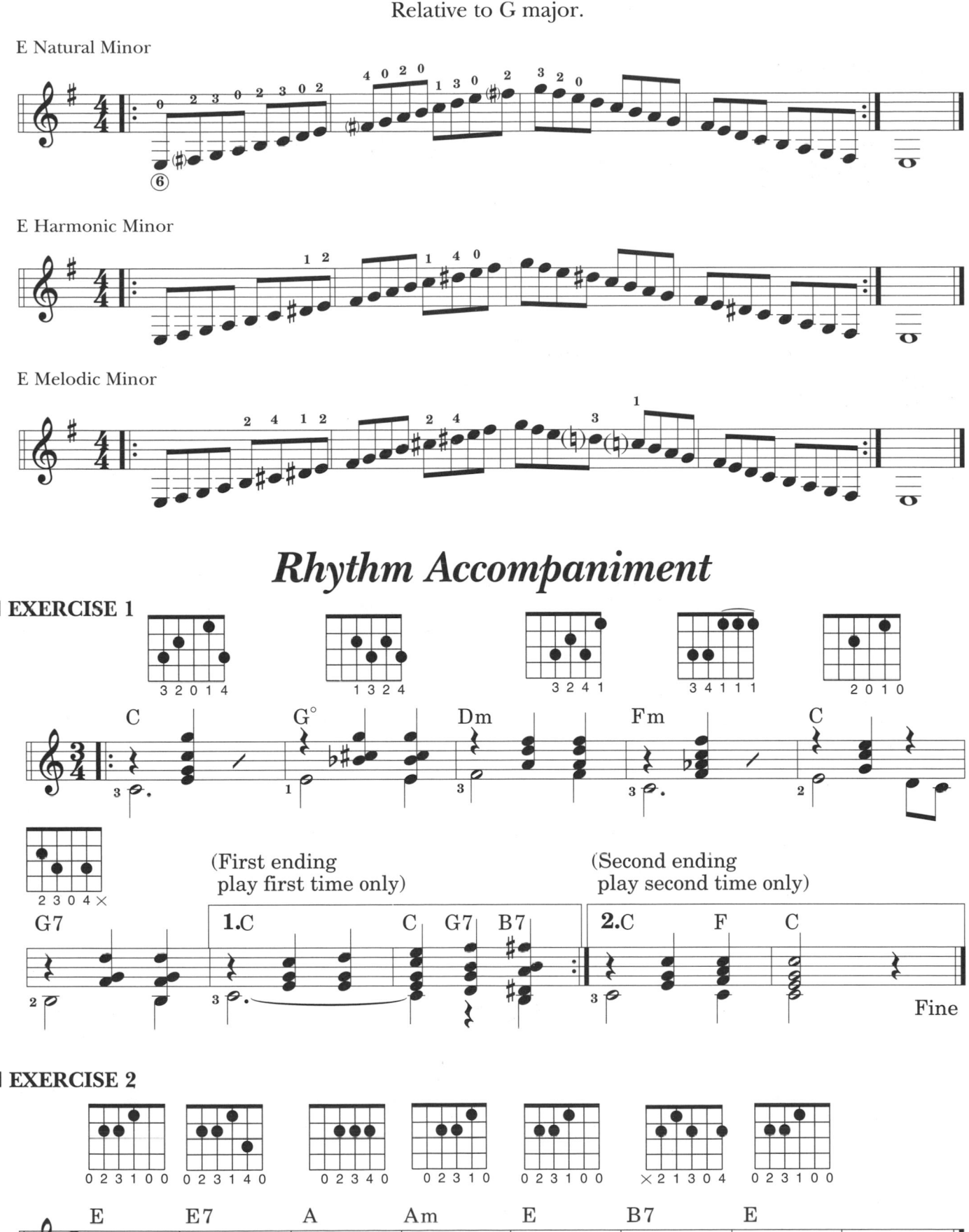

Rhythm Accompaniment

■ **EXERCISE 1**

■ **EXERCISE 2**

In waltz time, chords are muted immediately after the second and third beats.

Take Your Pick (duet)

For alternate picking while skipping strings.

CHORD PREPARATION

Duet

Rhythm Accompaniment

The principle of movable chord forms.

❚ Moving up the fingerboard in pitch, *natural* notes are two frets apart, except E to F, and B to C. They are one fret apart.

EXAMPLE (1st or 6th string)

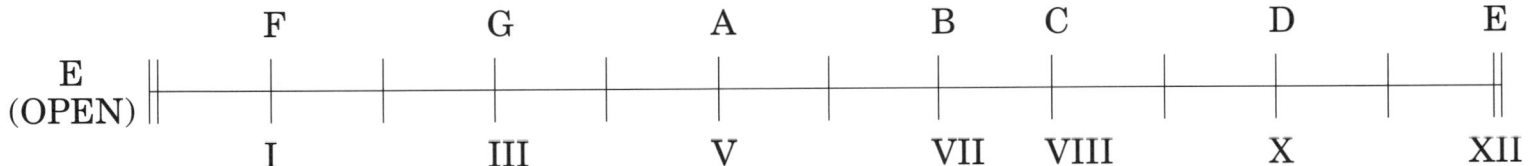

This fact applies to chord playing as follows:

1. If you play F major, F minor, and F7 on the first fret, then (using the same fingering) G major, G minor, and G7 will be on the third fret (two frets above F). Moving still higher, A major, A minor, and A7 will be on the fifth fret, B major, B minor, and B7 on the seventh fret, and C major, C minor, and C7 on the eighth—one fret up from B.
2. All *movable* forms will have *no open strings*.
3. Sharps and flats alter chord positions by one fret, the same as single notes.

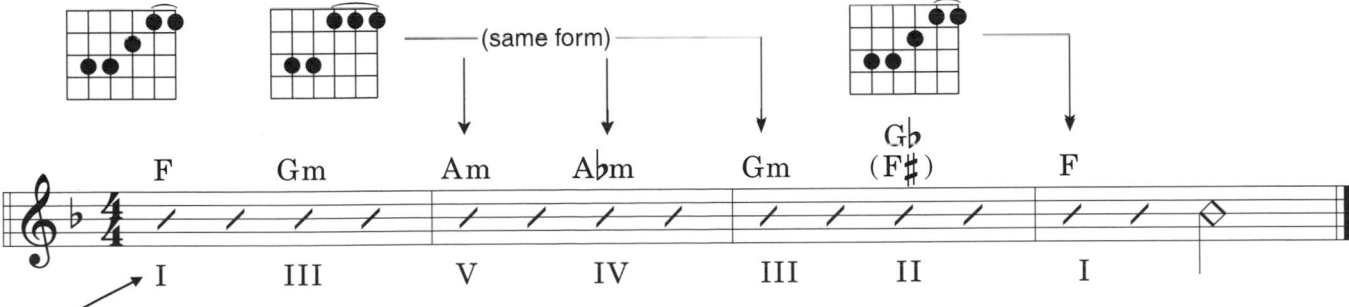

❚ The Roman numerals (called "Position Marks") indicate the frets on which the first finger plays.

❚ On the following pages, all new chord forms will be movable.

Chromatic Scale (First Position)

The *chromatic scale* is made up of semi-tones (half steps).

Speed Studies

Play the following eighth-note patterns at an even speed, slowly at first. Very gradually (over a period of time), increase the tempo. *Memorize the patterns*, and practice each one in all keys. Always start on the *tonic* (first note) of each scale and *transpose* the rest of the notes by the following pattern. (Write them out if necessary.)

PATTERN 1

PATTERN 2

PATTERN 3

First-position F and G scales can be played in two octaves. Play all patterns in *both* octaves.

Key of D Minor (Scales in First Position)

Relative to F major.

D Natural Minor

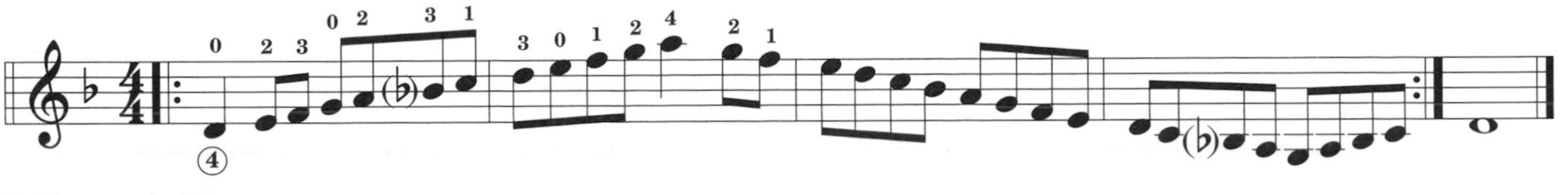

D Harmonic Minor

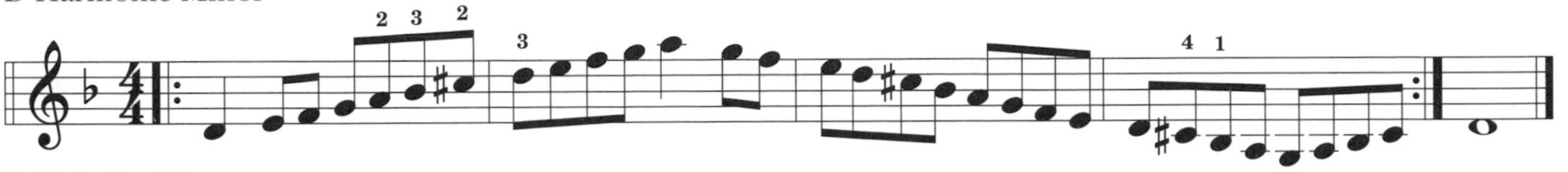

D Melodic Minor

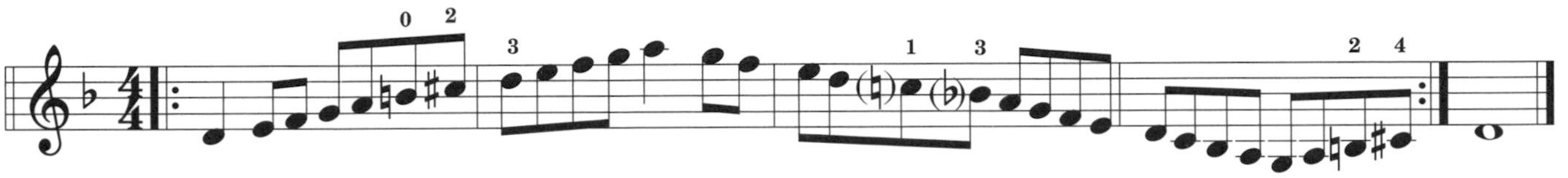

Rhythm Accompaniment

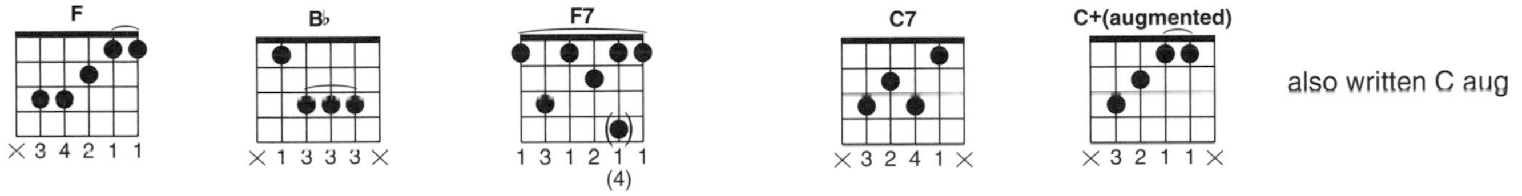

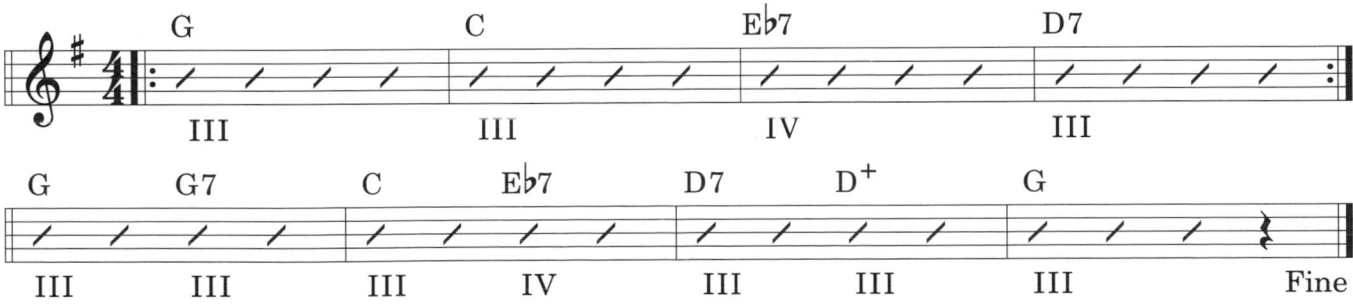

F
×3 4 2 1 1
B♭
×1 3 3 3×
F7
1 3 1 2 1 1
(4)
C7
×3 2 4 1×
C+(augmented)
×3 2 1 1×
also written C aug

■ EXERCISE 1

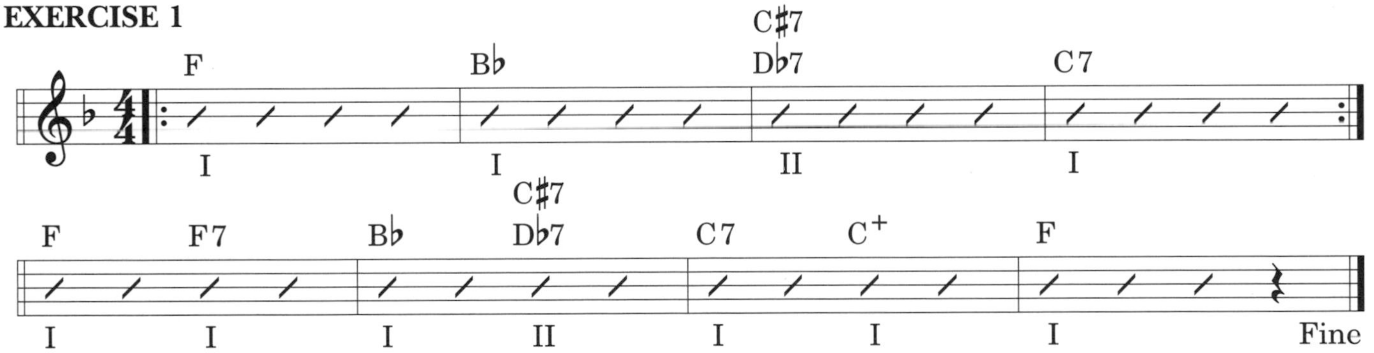

F B♭ C♯7 D♭7 C7
I I II I
F F7 B♭ C♯7 D♭7 C7 C+ F
I I I II I I I Fine

▌ This is the same chord sequence but *transposed* to a different key. Watch the position marks.

■ EXERCISE 2

G C E♭7 D7
III III IV III
G G7 C E♭7 D7 D+ G
III III III IV III III III Fine

The augmented chord can actually be named from any note within the form (C+ = E+ = G♯+ or A♭+). Augmented chords repeat themselves every fifth fret.

Endurance Etude
Picking Etude #3
Hold fourth finger down throughout.

Be sure to observe tempo changes. Also, vary the *dynamics* (degrees of volume, loud and soft) to make the music more interesting to listen to.

Key of B♭ Major (Scales in First Position)

(All Bs and Es are flatted.)

When a key signature has two or more flats, the name of the next-to-last flat is the name of the key.

Rhythm Accompaniment

Fm   × 3 4 1 1 1

B♭m   × 1 3 4 2 1

Mute the 5th string with the tip of your first finger. Mute the 6th by touching it with your thumb. ⟶

G°   × × 1 3 2 4

also written as G dim (see p.30)

■ EXERCISE 1

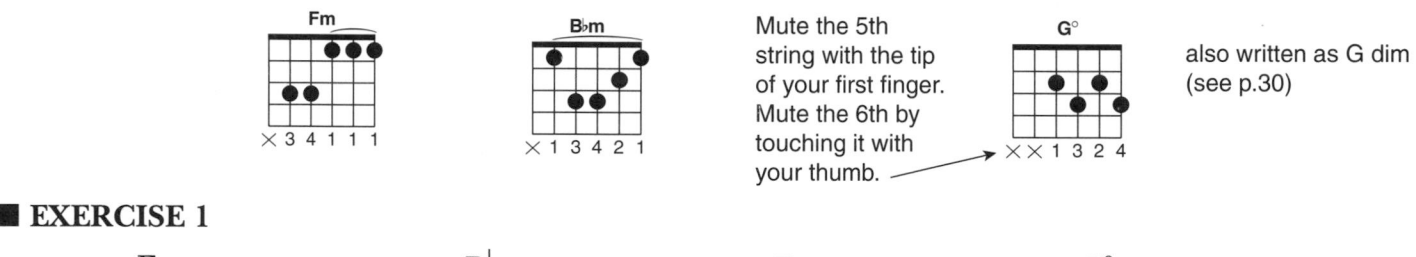

■ EXERCISE 2

This is the same chord sequence but transposed to a different key. Watch the position marks.

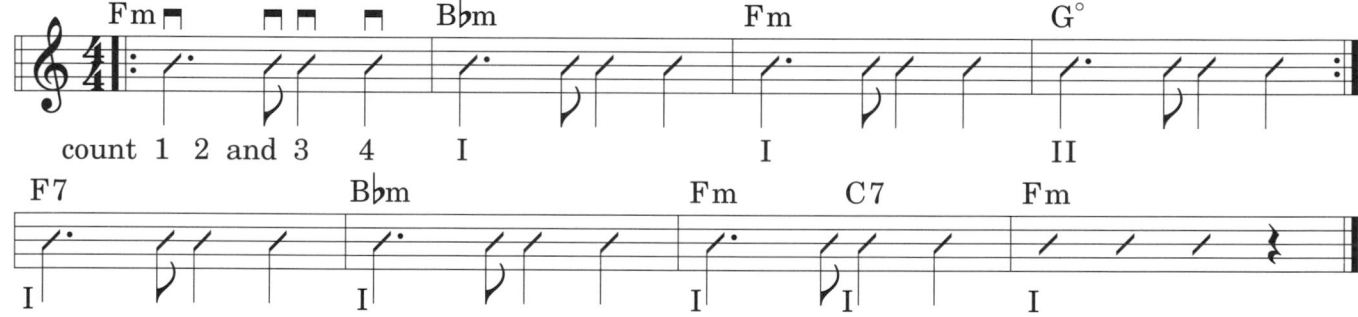

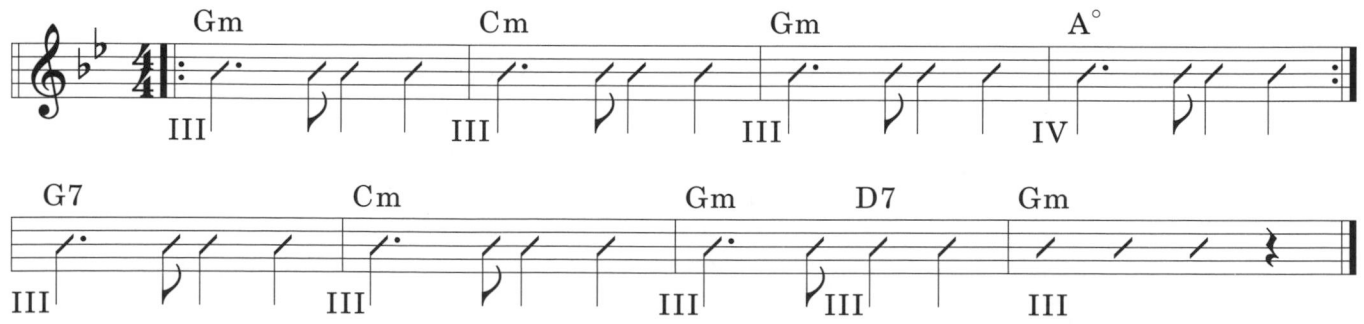

The diminished chord can actually be named from any note in the form (G° = B♭° = C♯° or D♭° = E°).
Diminished chords repeat themselves every fourth fret.

Duet in B♭

(Finger complete chord form.
Do not strum top string.)

Reverse Alternate Picking Study

■ Pay very strict attention to picking as indicated.

(hold down bottom note)

Review all material.

Key of D Major (Scale in First Position)

(All Fs and Cs are sharped.)

In any sharp signature, the first note above
the last sharp is the name of the key.

Duet in D

count (1   2)   3   4   &   1 & 2 & 3 (4)   1) & 2 & 3 (4)

Dot over a note
means staccato

Play like this.   II Fine

Dynamic Etude (duet)
Etude #4

■ Be sure to hold all notes for their full value.

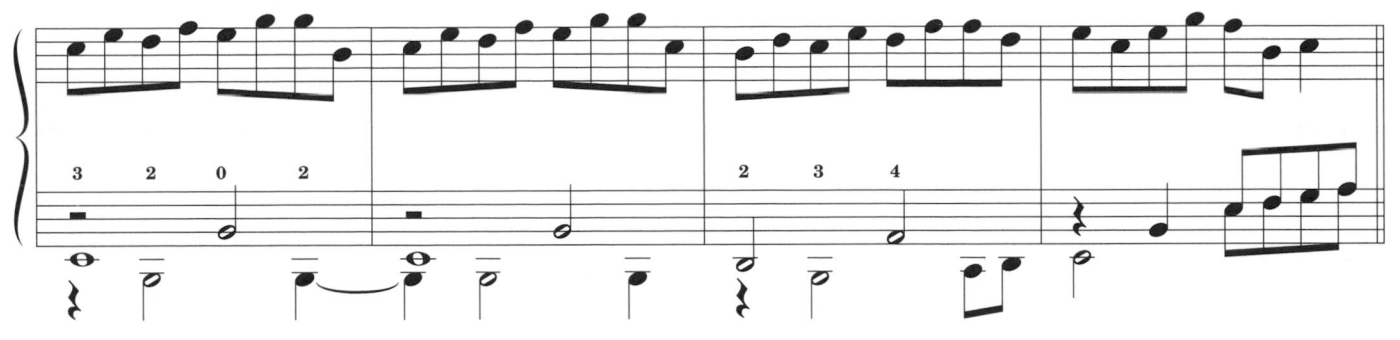

(soft)

(Repeat previous measure)

(loud)

count 1 2 3 4

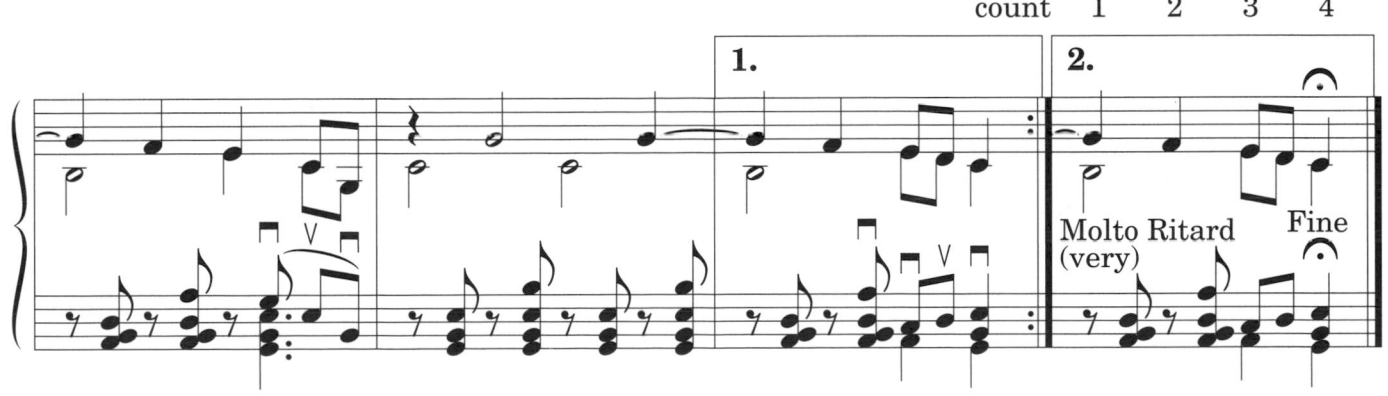

Molto Ritard
(very)

Fine

Key of A Major (First Position)
Duet in A

(All Fs, Cs, and Gs are sharped.)

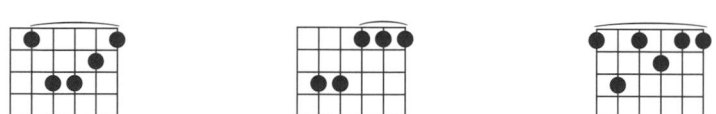

Rhythm Accompaniment

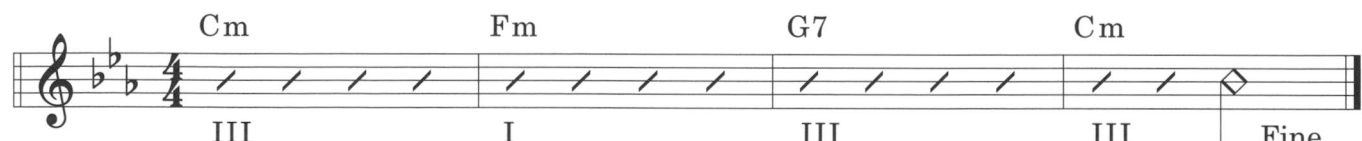

■ **EXERCISE 1**

Cm	Fm	G7	Cm
III	I	III	III

Fine

■ **EXERCISE 2**

Dm	Gm	A7	Dm
V	III	V	V

Fine

Key of E♭ Major (Scale in First Position)

(All Bs, Es, and As are flatted.)

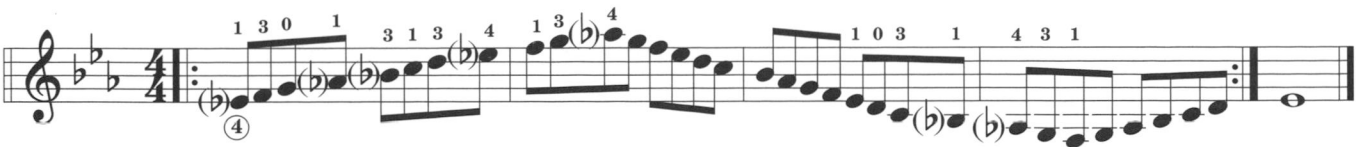

Duet in E♭

▌ Remember the flats. Count time carefully.

Moderate 4 (swing feeling)

NEW FORM → E♭ x 4 3 1 2 1 (x) III

Cm III

Fm I

NEW FORM B♭7 x 1 3 1 4 1 (x) I

E♭ III

Cm

Fm I

B♭7

C-flat–same as B-natural. Open 2nd string or fourth finger, 3rd string.

E♭ III

E♭+ IV

A♭ IV

A♭m IV

E♭ III

A♭ IV

E♭ III   B♭7 I

E♭ III

Fine

Movable Chord Forms

A compilation of movable forms presented in Section I.

Related Fingerings

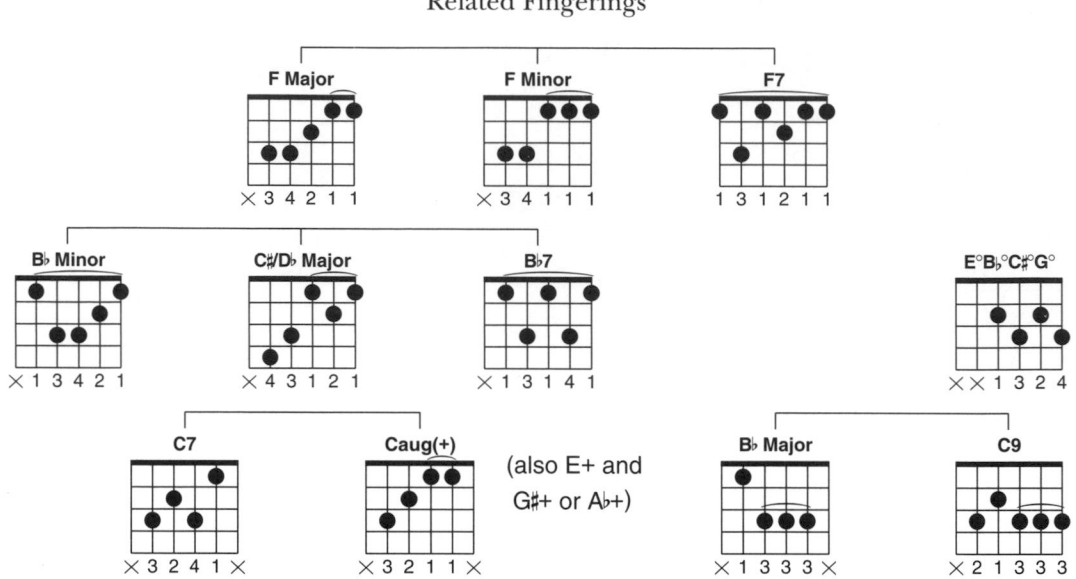

▮ With these eleven forms, you can play the accompaniment to any song in any key, providing:
1. That you understand the principle of movable chord forms discussed on pg. 45, and
2. That you observe the following chart.

Chord Simplification and Substitution Chart

TYPE	WRITTEN					SIMPLIFIED SUBSTITUTION		
Major	C6	Cmaj7	Cmaj9	C_6^9	$Cmaj_7^9$	Use:	C major	
Dominant 7th	C9	C13	C9(11+)	C11+	———	Use:–	C7	
Dom 7-Altered 9th	C7(-9)	C7(♭9)	C13(-9)	C13(♭9)	———	—	C7 or G dim	{build dim chord
	C7(+9)	C7(♯9)	C13(+9)	etc.	———	—	C7 (or G°)	on 5th note above C}
Dom 7-Altered 5th	C7+	C7(+5)	Caug7	C9+	C9(+5) C+9	Use:	C+	{build substitute
	C7(-5)	C7(♭5)	C9(-5)	etc. . .	$C7_{+5}^{-5}$ ——	—	C+ or G♭+	chord on flatted
Dom 7-Altered 5, 9	$C7_{+5}^{-9}$	$C7_{+5}^{+9}$	$C7_{-5}^{+9}$	$C7_{-5}^{-9}$	————	—	C+ or G♭7	5th above C}
Dom 7-Sus 4	C7(sus4)	C7(susF)	C9(sus4)	C9(susF)	C11	Use:	G minor	5th note above C
Minor	Cm6	Cm_6^9				Use:	C minor	
Minor 7th	Cm7	Cm9	Cm11			Use:	Cm	
Min-with Maj7	Cm(♮7)	Cm(♯7)	Cm(maj7)			Use:	G+(5th above C) or Cm	
Min 7-Altered 5th	Cm7(-5)	Cm7(♭5)				Use:	E♭m	{built on minor, (or lowered) 3rd above C}

▮ Of course, having only eleven chord forms at your command will cause you to move up and down the fingerboard much more than is desired for good rhythm playing. The more forms you know, the less distance you have to travel and the more melodic your rhythm playing can become.

Picking—A Different Technique
The principle is to attack each new string with a downstroke.

This technique is older than alternate picking, and less emphasis is placed on it today. However it is one more step in right-hand control and, when mastered, it is very fast in ascending passages.

An example of this technique in use can be found on pg. 48, measure 20 of the "Endurance Etude." This type of picking will be suggested on the following pages from time to time *but only in certain situations* (arpeggios, whole tone scales, etc.) and only in addition to alternate picking. It will be up to you to gradually master and (whenever practical) add this style to your overall right-hand technique. However: the most concentrated effort must still be placed on alternate picking.

✲ (>) Accent mark: strike more sharply

SECTION TWO
Position Playing

Position is determined by the fret on which the first finger plays. It is indicated by a Roman numeral. Strictly speaking, a position on the fingerboard occupies four adjacent frets. Some scales have one or more notes that fall outside this four-fret area, and these notes are to be played by reaching out with the first or fourth finger without shifting the entire hand, i.e. finger stretch or "FS." When the out-of-position note is a scale tone, the finger stretch is determined by the *fingering type* (Type I = first finger stretch, Type IV = fourth finger stretch). When the out-of-position note is not a scale tone and moving upward, use finger stretch 1, and moving downward finger stretch 4—regardless of fingering type. (All scale fingerings introduced from this point on will not use any open strings, and therefore they are movable in the same manner as the chord forms presented earlier. See pg. 45.)

Major Scales
C Major (Fingering Type 1)
Second Position

60

✻ When an out-of-position note is immediately preceded or followed by a note played with the same finger that would normally make the stretch, reverse the usual finger-stretch procedure. Always move back into a position from a finger stretch, never away from it.

EIGHTH-NOTE STUDY

ARPEGGIO STUDY: BROKEN CHORDS

❚ Practice picking as indicated and also with alternate ⊓∨ picking.

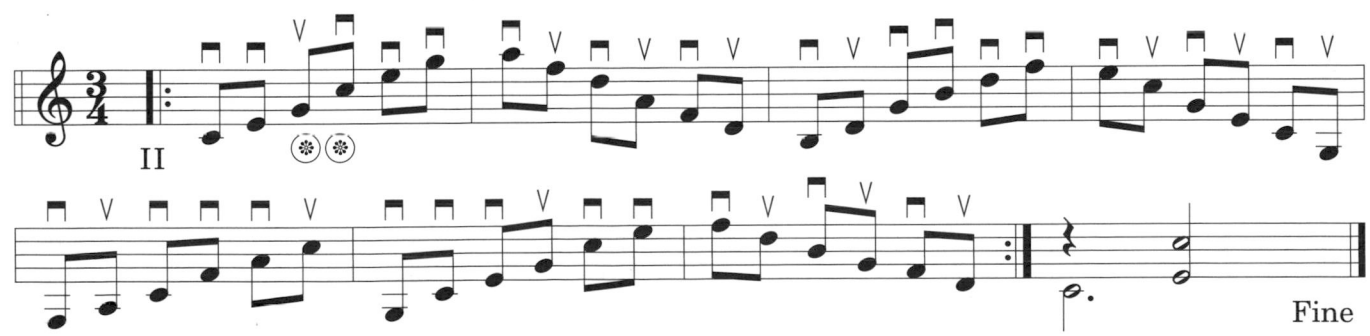

✻ ✻ When two consecutive notes are played with the same finger on adjacent strings, "roll" the finger tip from one string to the next. Do not lift the finger from the string.

Chord Etude No. 1

Practice slowly and evenly, connecting the chords so they flow from one to the next with no silences between them. Observe fingering and position marks!

Etude No. 5 (duet)

Remember, all natural notes on the guitar are two frets apart except E to F and B to C.

Reading Studies

Do not *practice* these two pages. Just read them, but not more than twice through during any single practice session. Do not play them on two consecutive days. Do not go back over any particular section because of a wrong note. Do keep an even tempo and play the proper time values.

By obeying these rules, you will never memorize the reading studies, so they will always be good reading practice. A little later on, it is recommended that you use this procedure with a variety of material, as this is the only way for a guitarist to achieve and maintain any proficiency in reading. (Even when working steadily, we are not reading every day, so "scare yourself" in the privacy of your practice sessions.)

C MAJOR 1 (FINGERING TYPE 1)

C MAJOR 2 (FINGERING TYPE 1)

Fine

If you encounter unusual difficulty reading these pages, go back to pg. 60 and start again.

Ballad (duet)

❋ A position mark in parentheses represents placement of second finger, as first finger is not used.

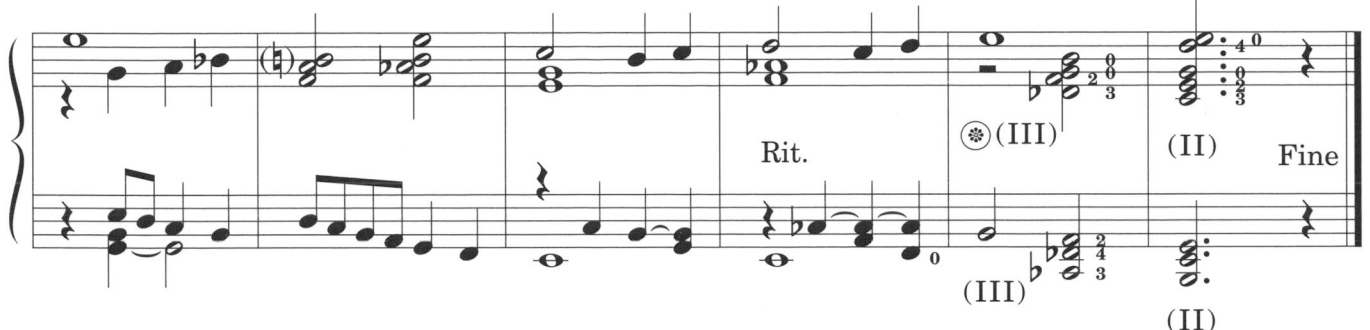

Rit.

※(III)   (II)   Fine

(III)   (II)

Movable Chord Forms
Rhythm Accompaniment, Part Two

The most difficult part of learning to play chords on the guitar is getting the fingers to fall instantly and without conscious effort in the proper arrangement on the fingerboard. This is mainly a physical problem, and a certain amount of practice time seems to be the only solution.

However, I have found that learning new chord forms in a certain order (a sequence of related fingerings) seems to lessen the time normally required to perform them.

Therefore, the following chord forms are presented in a particular order. We will use three of the previously learned fingerings as basic forms. We will alter these forms by moving, or removing, one or more fingers. In this way each new fingering is directly related to the one(s) preceding it.

So, each of the basic forms and each derivative is a preparation for another new chord form.

No specific letter names are given—only the chord type and the string on which the root is found.

Memorize the fingerings for all chord structures in their order of appearance. Do not skip around. Do not change the fingering of any form, even if you already play it in a different way. It will appear later on with "your" fingering, but related to a new set of forms. Practice all chord forms chromatically up and down the fingerboard, observing root (chord) names.

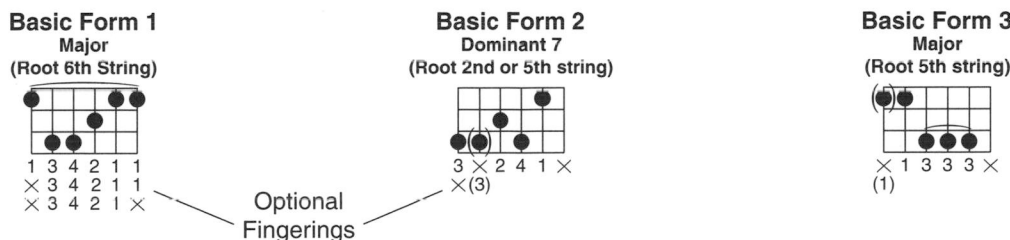

A dot in parentheses (•) means that although the note belongs to the chord, it need not sound and, in many cases, sounds better without it.

Chord Forms

Shown below is Basic Form 1 and seven derivative fingerings. When the basic form has been mastered, the performance of the derivatives is relatively easy to accomplish. Memorize the type of chord (major, minor, etc.) each form produces and the string on which the root (or name) is found. All optional fingerings should eventually be learned, but first concentrate on the one appearing directly below the diagram. It is the preferred one.

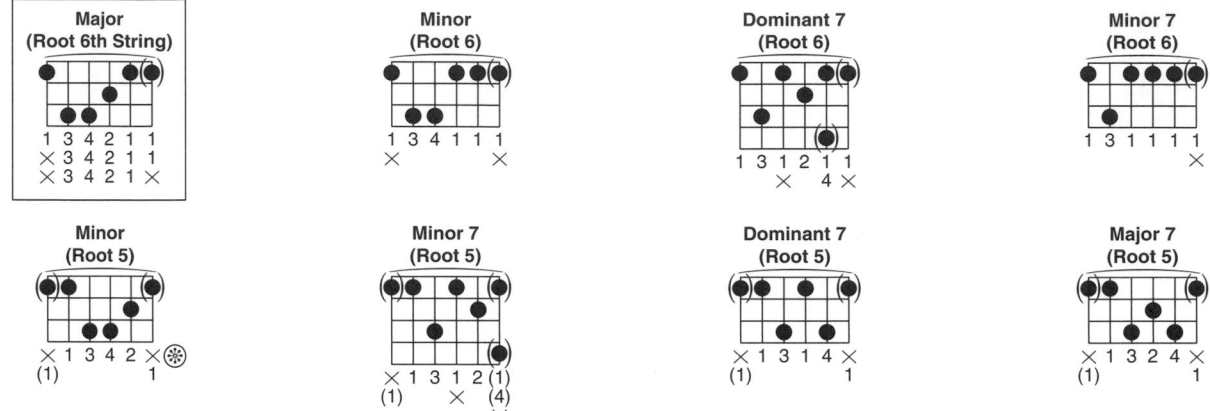

A word about notation:

1. When a chord is indicated by just a letter, it is major.
2. When it is a letter followed by a 7, it is a dominant 7 chord.
3. Minor is indicated by min, m, or a dash (—).
4. Major 7 is indicated by maj7, ma7, or sometimes M7.

■ **EXERCISE**

Use only the forms shown above. Watch the position marks!

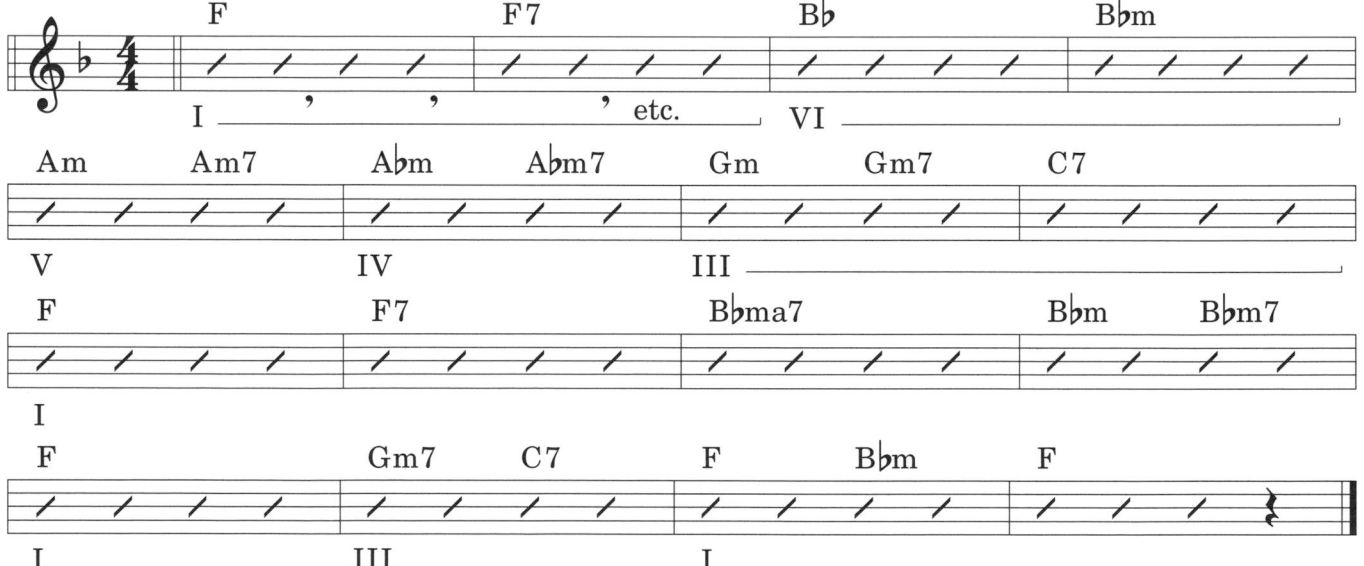

✸ The first string is not very effective in rhythm playing, and even when it is pressed down with a "barre" it is usually best to omit it by making the pick travel in an arc across the strings, passing above it. (⌒·····)

Rhythm Accompaniment—Right-Hand Technique

To most beginners, strumming chords (pushing the pick across the strings so they sound one after the other) is easy and natural.

However, striking the chords so that the sound fits with a modern rhythm section is quite another thing, and requires considerable practice and know-how.

First, by using a combination rotary forearm and loose-wrist motion (snap the wrist as if flicking something from the back of your hand), you produce an explosive attack in which all notes seem to sound simultaneously.

Secondly, the placement of "pressure release points" (𝄐) and accents determines the types of beat produced. (Much more about all this later.)

Picking Etude No. 4

Observe fingering.

Hold third finger down throughout.

✵ *Grace note* to be played slightly before the top note G on the fourth beat.

F Major Scale
(Fingering Type 1A, Second Position)

The F major scale shown above is in the second position even though the first finger plays the first fret on three strings. This is because the three scale tones require stretches by the first finger. The basic four-fret position is never numbered from a stretch.

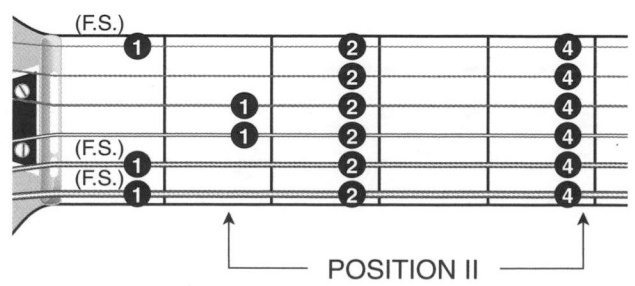

POSITION II

EIGHTH-NOTE STUDY

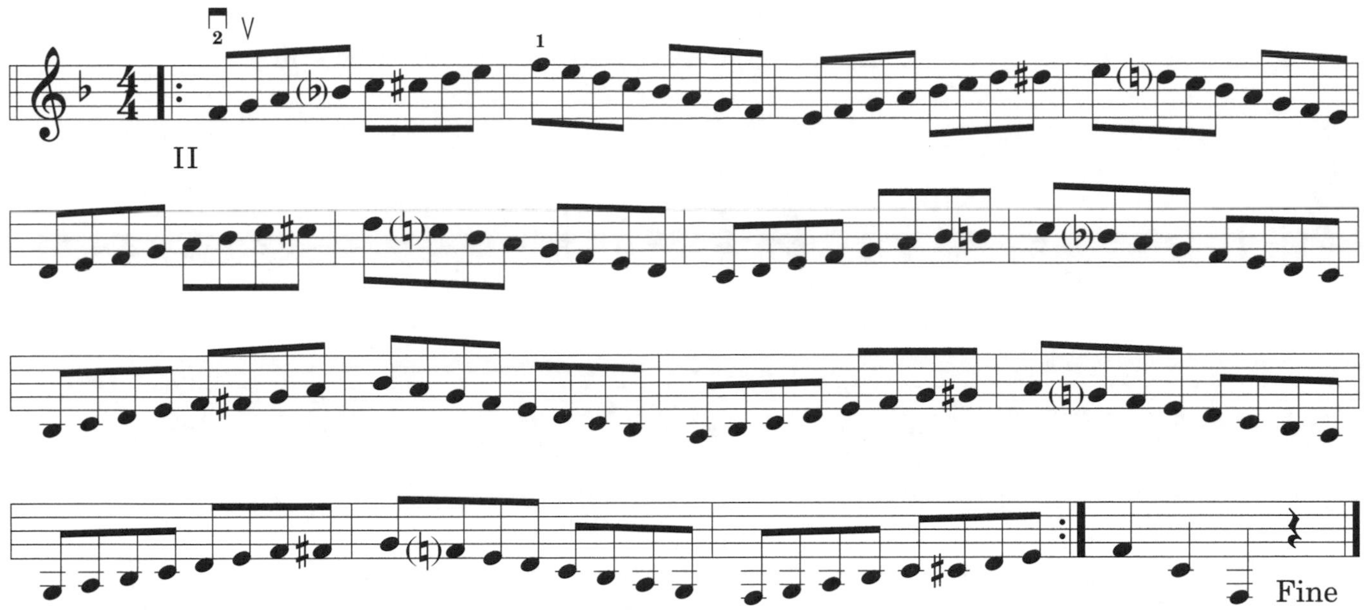

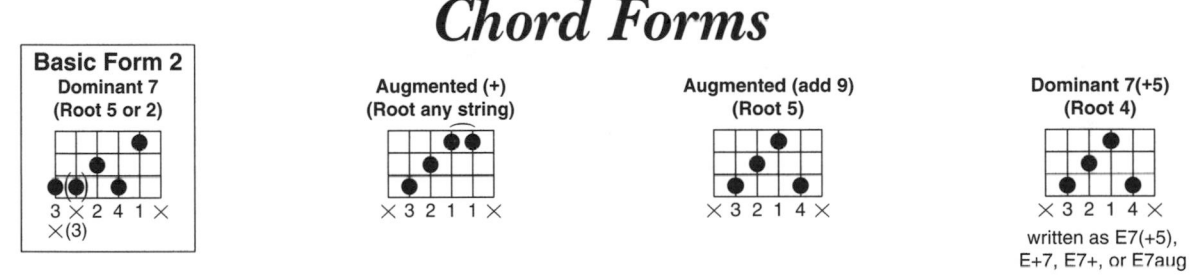

II
Fine

ARPEGGIO STUDY

II
Fine

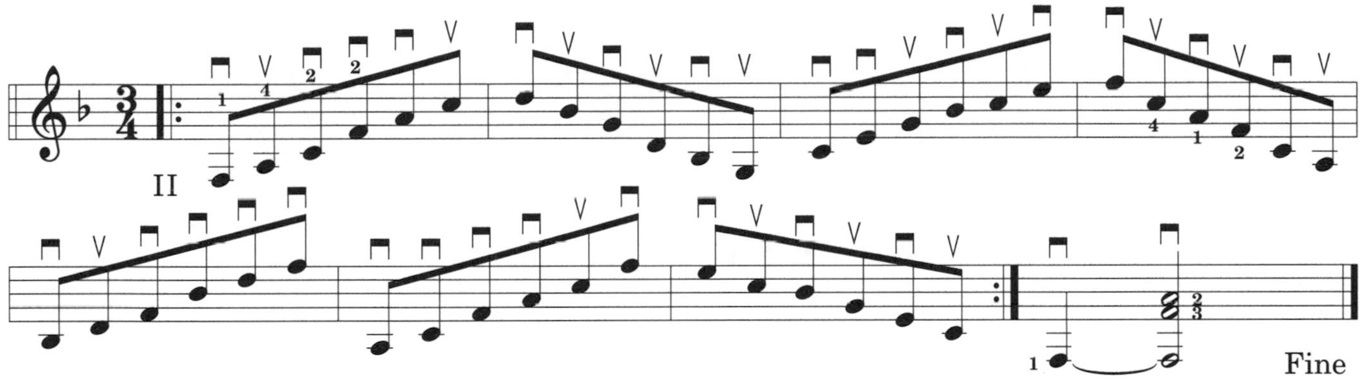

Also practice arpeggios with alternate ⊓ V picking, which is generally the most practical.

Chord Forms

Basic Form 2
Dominant 7
(Root 5 or 2)
3 × 2 4 1 ×
×(3)
Augmented (+)
(Root any string)
× 3 2 1 1 ×
Augmented (add 9)
(Root 5)
× 3 2 1 4 ×
Dominant 7(+5)
(Root 4)
× 3 2 1 4 ×
written as E7(+5),
E+7, E7+, or E7aug

■ EXERCISE

Use the above forms plus some of the preceding ones.

F
E7
E♭7
D7
D+ D+(add9)
I
V
IV
III
G7
G+ G7(+5)
C7
F
B♭7
F
III
IV
III
I

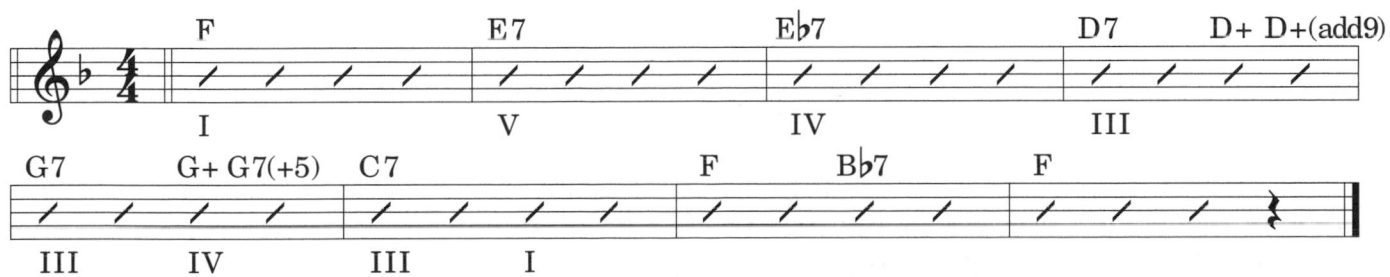

Transpose, write out, and practice all rhythm exercises in one or more higher keys.

Chord Etude No. 2

Rubato: Freedom of tempo. Accelerate and ritard as you wish.

These chord exercises are very important and should be reviewed *regularly,* as they serve many pur poses, such as physical development of the left hand, fingering relationship between chord struc tures, and eventual "chord picture" recognition.

Another Duet in F

II (Hold down all notes under slurs)

I

(also with alternating ⊓V)

Fine

Regular review is a must!

Reading Studies

Just play these Reading Studies: Do not *practice* them, and do not play them on two consecutive days. (See pg. 64.)

F MAJOR 1 (FINGERING TYPE 1A)

F MAJOR 2 (FINGERING TYPE 1A)

F.S.

Fine

FS: Finger Stretch. Stretch the finger; do not move the entire hand.

75

Play It Pretty (duet)

⊛ A temporary change to position III at this point will simplify the fingering this passage, and eliminate the necessity of the open E (preceding the high B♭).

Chord Forms

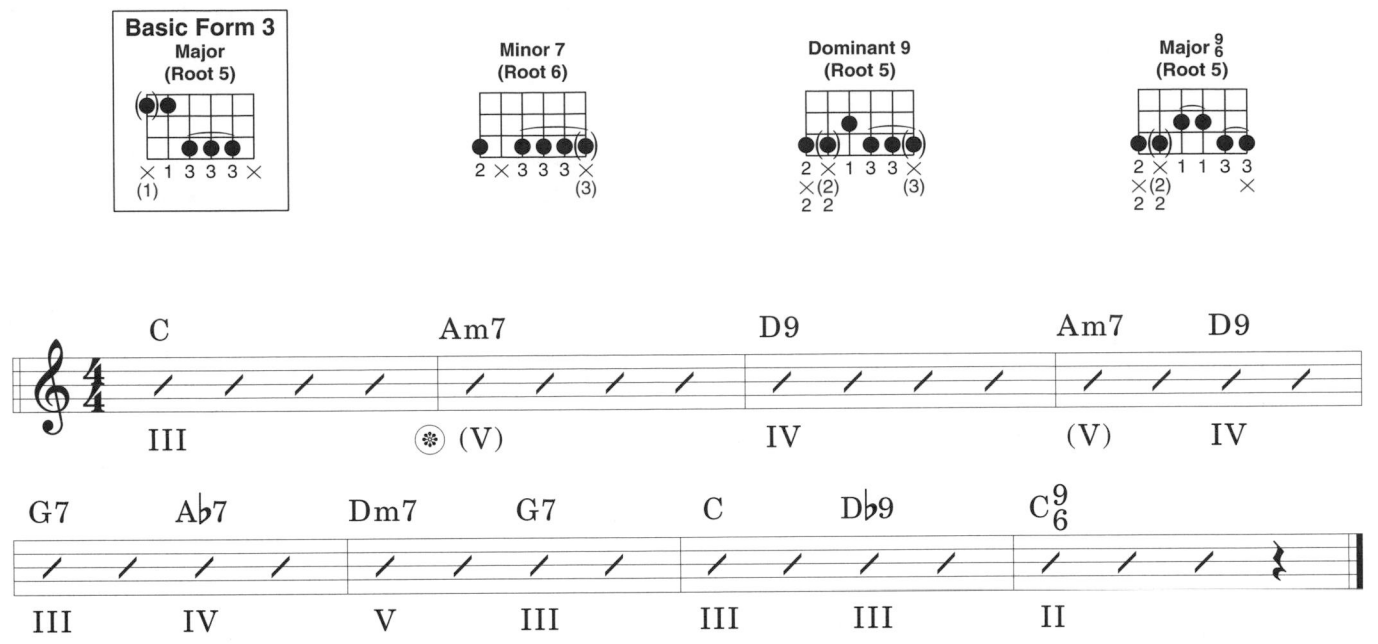

⊛ A position mark in parentheses means that the first finger is omitted from the form. The position number is determined by the lowest fret used.

Triplet Study

Practice using both patterns of picking. See pg. 37.

Speed Study—Fingering Type 1

Maintain an even tempo. Play no faster than perfect coordination in both hands will allow. An increase in speed will come gradually.

Speed Study—Fingering Type 1A

▌ Practice all speed studies as written and with a ♪♫ rhythm. Also play with and without repeats.

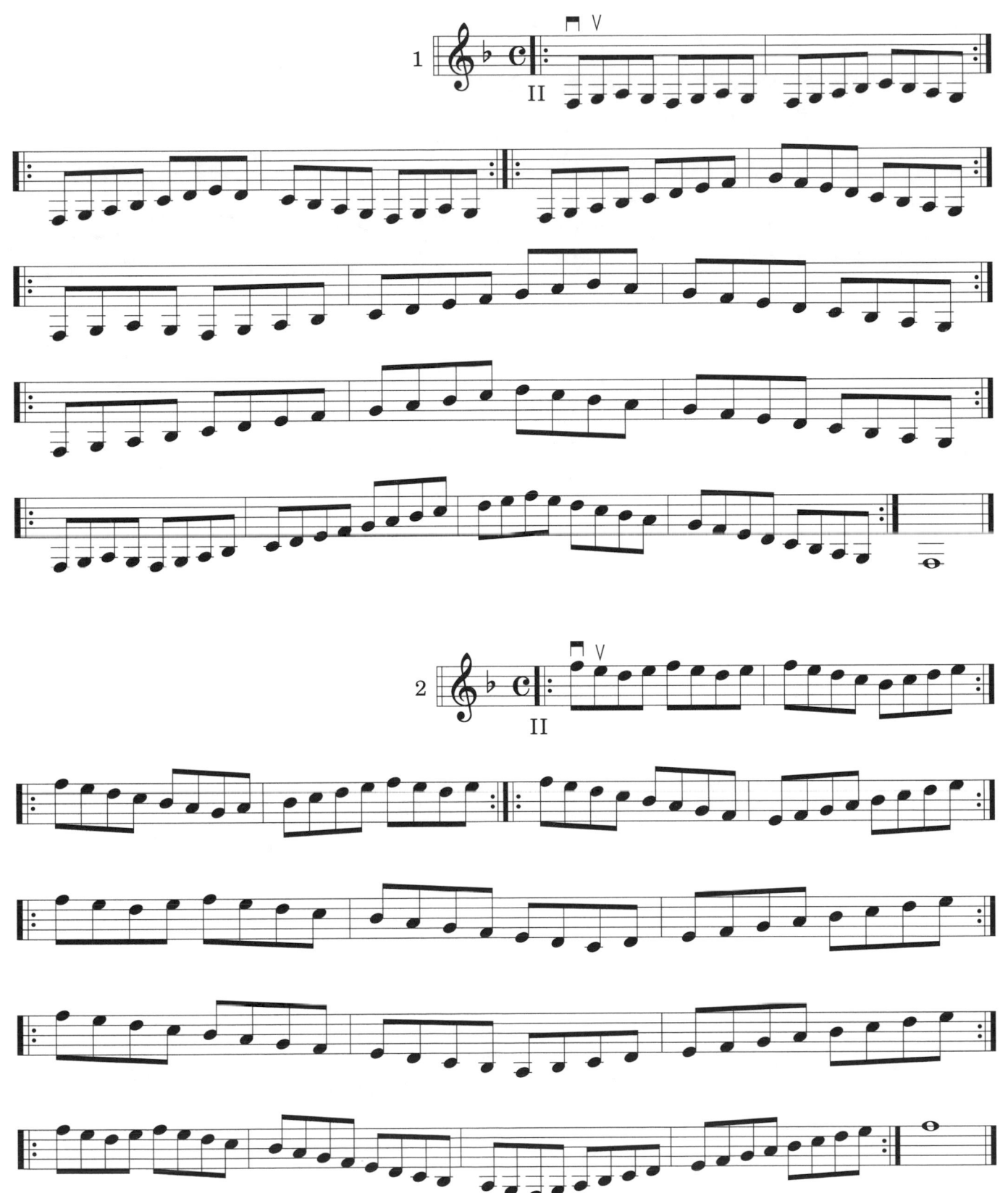

For additional technique-building patterns, see pg. 46.

G Major Scale
(Fingering Type 2, Second Position)

EIGHTH-NOTE STUDY

ARPEGGIO STUDY

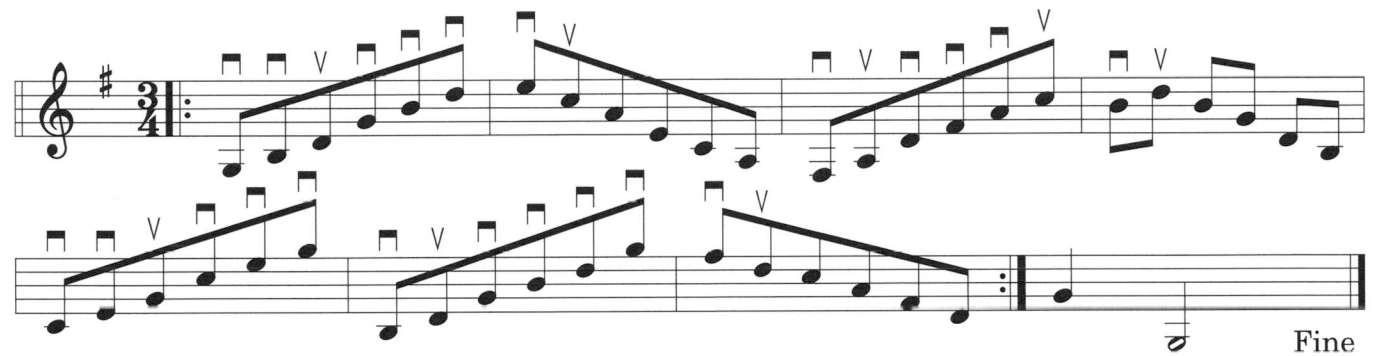

Also practice arpeggios with an alternate ⊓ V picking.

Dotted Eighth and Sixteenth Study

Practice as legitimate ⊓ and as ♩♪ rhythms. See pg. 42.

When two consecutive notes on adjacent strings require the same finger, roll your finger; don't lift it.

Waltz for Two (duet)

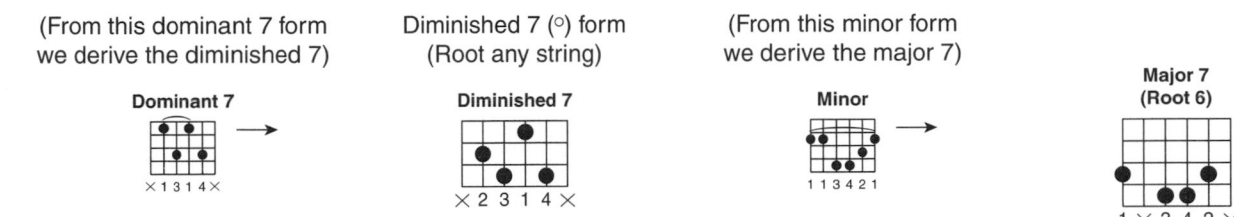
Chord Forms

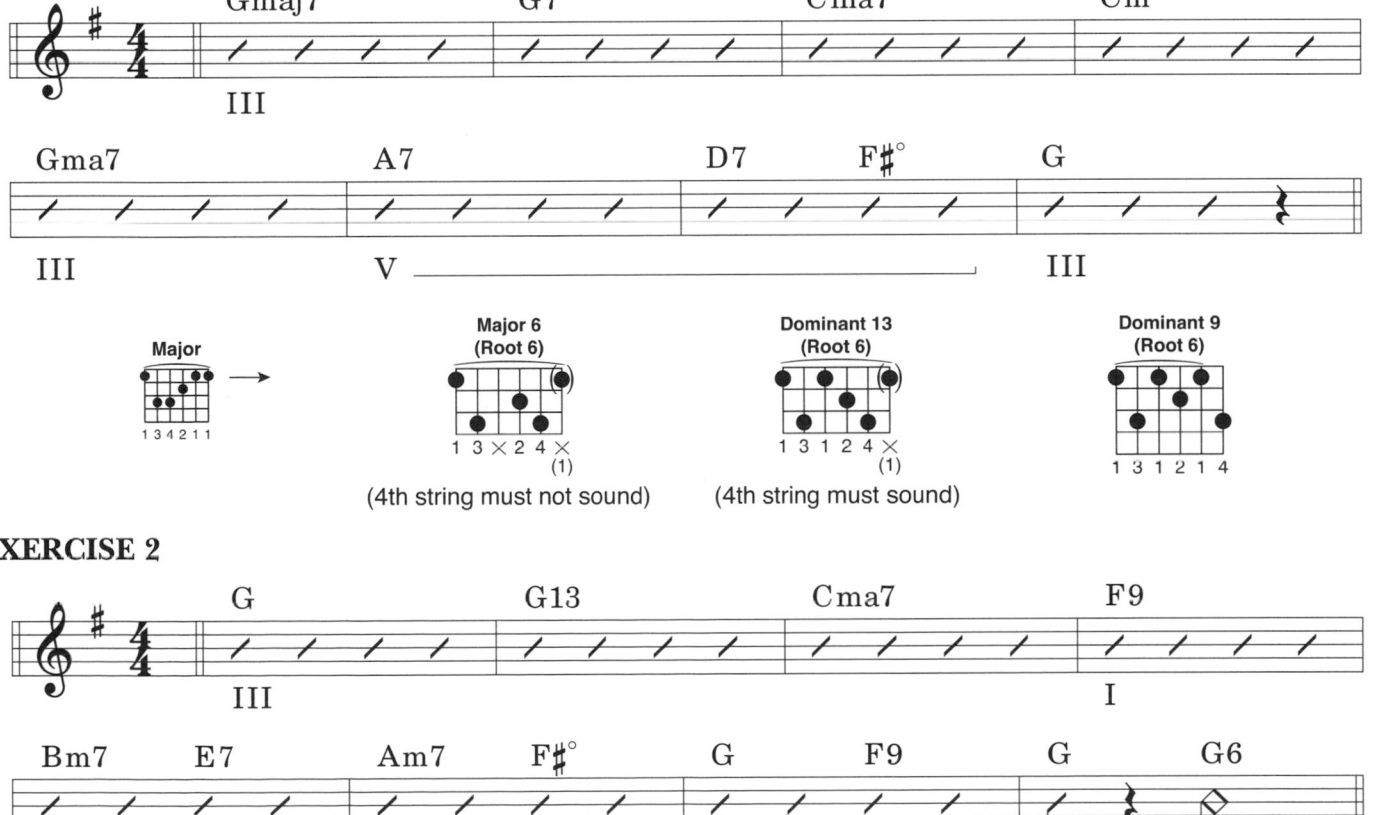

(From this dominant 7 form we derive the diminished 7)

Diminished 7 (°) form (Root any string)

(From this minor form we derive the major 7)

Dominant 7

Diminished 7

Minor

Major 7 (Root 6)

▌ Diminished 7 chords are indicated by Gdim or G°. (The 7th is assumed.)

■ EXERCISE 1

Gmaj7   G7   Cma7   Cm
III

Gma7   A7   D7   F#°   G
III   V   III

Major

Major 6 (Root 6)

Dominant 13 (Root 6)

Dominant 9 (Root 6)

(4th string must not sound)   (4th string must sound)

■ EXERCISE 2

G   G13   Cma7   F9
III   I

Bm7   E7   Am7   F#°   G   F9   G   G6
II   V   III   I   III

(hold chord for 2 beats)

▌ You may substitute 6 and major 7 chords for major chords, and dominant 9 and 13 chords for dominant 7 chords.

83

Reading Studies

▮ Do not "practice" Reading Studies. Just read them.

G MAJOR 1 (FINGERING TYPE 2)

G MAJOR 2 (FINGERING TYPE 2)

▮ Continue on, without stopping, at the same tempo but in waltz time.

Speed not coming? Left-hand accuracy not consistent? Play any scale very slowly.
Watch your left hand. Force your fingers to remain poised over the fingerboard always in readiness.
Don't let them move too far away from the strings when not in use. Concentrate on this.

Blues in G (duet)

The 1st Guitar part of this duet is often played using the "muffled effect." This sound is produced by laying the right hand lightly along the top of the bridge. All strings being played must be kept covered. As this somewhat inhibits picking, the part should first be thoroughly practiced without the muffled effect (or "open").

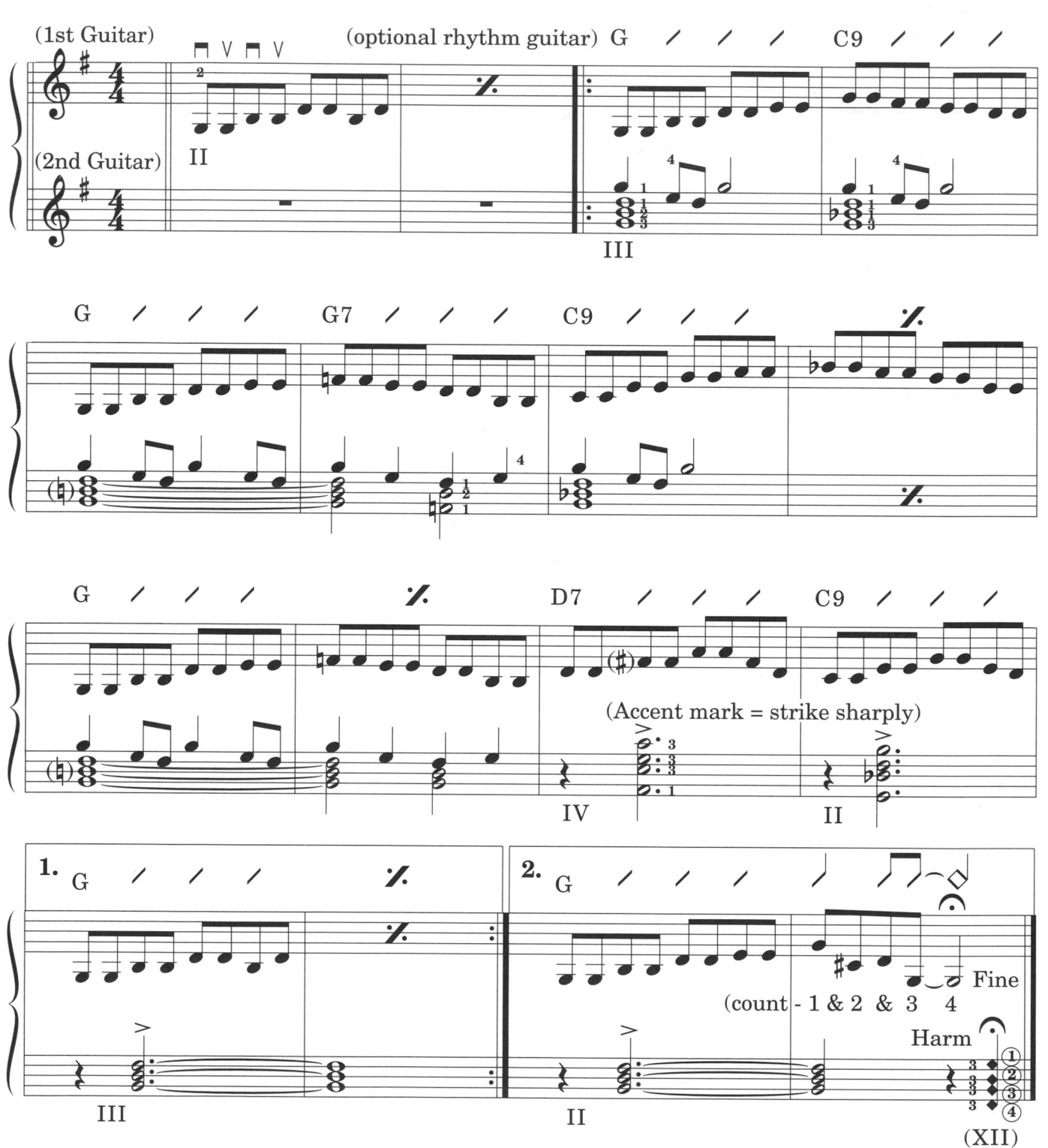

For a different rhythm feel, play all consecutive eighth notes as a rhythm.

Chord Etude No. 3

▌ Observe position marks and fingerings, as they will make possible a smooth performance.

When moving from chord to chord, the best fingering is usually the one
that involves the least motion in the left hand.
Leaving one finger free for possible melodic additions is also an important factor.

Rhythm Accompaniment—Right-Hand Technique

Memorize these symbols.

⊓	Downstroke
V	Upstroke
(❜)	Release finger pressure (of left hand) immediately *after* chord sounds. Do not remove from strings.
×|	Strike deadened strings (fingers in formation on strings, but no pressure).
>	Accent (strike sharply) with more force.

A basic Latin beat, which will work with the cha-cha, beguine, samba, and others.

or

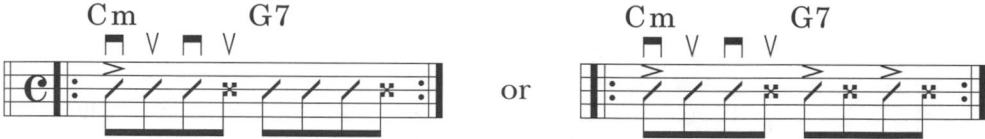

Picking Etude No. 5

Hold down fourth finger throughout.

Review! Review!

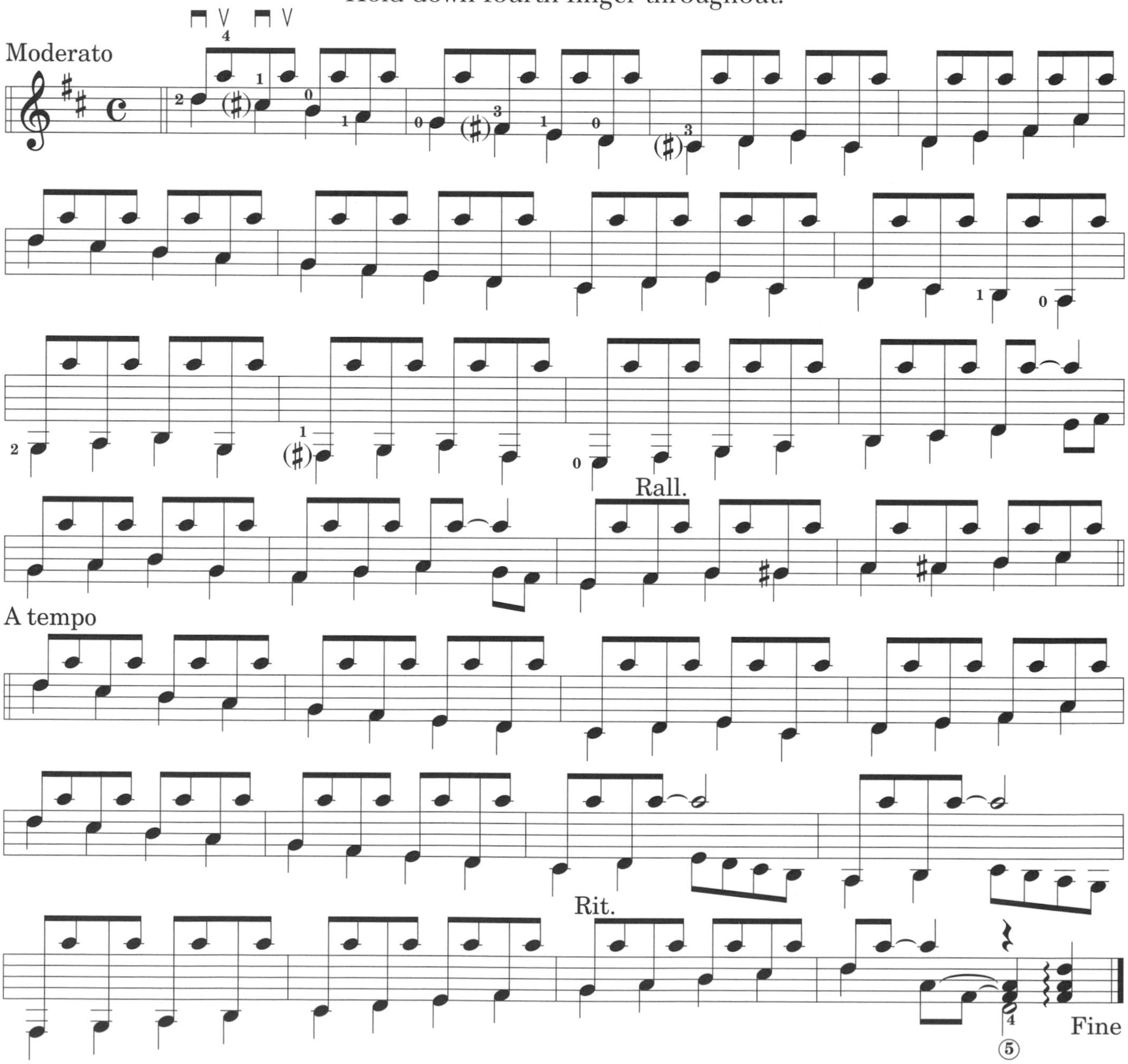

Short and Sweet (duet)

D Major Scale
(Fingering Type 3, Second Position)

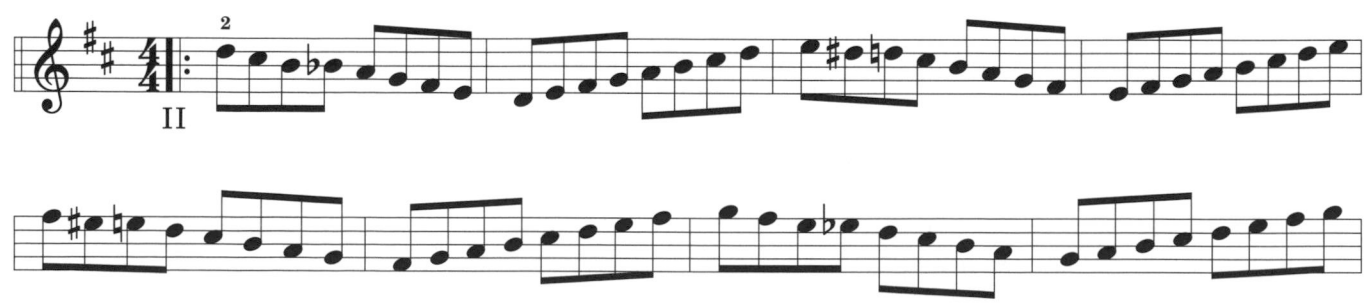

EIGHTH-NOTE STUDY

ARPEGGIO STUDY

Also practice with alternate ⊓ ∨ picking.

Chord Forms

written: Cm7 ♭5
Cm7 -5
Cm7 5♭

■ EXERCISE

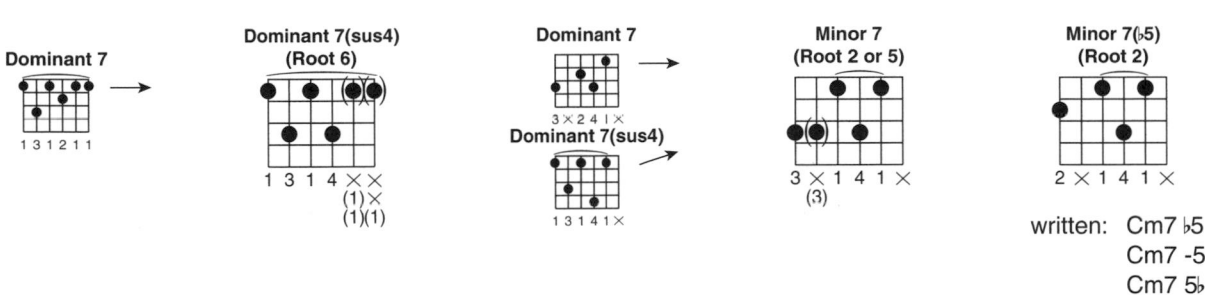

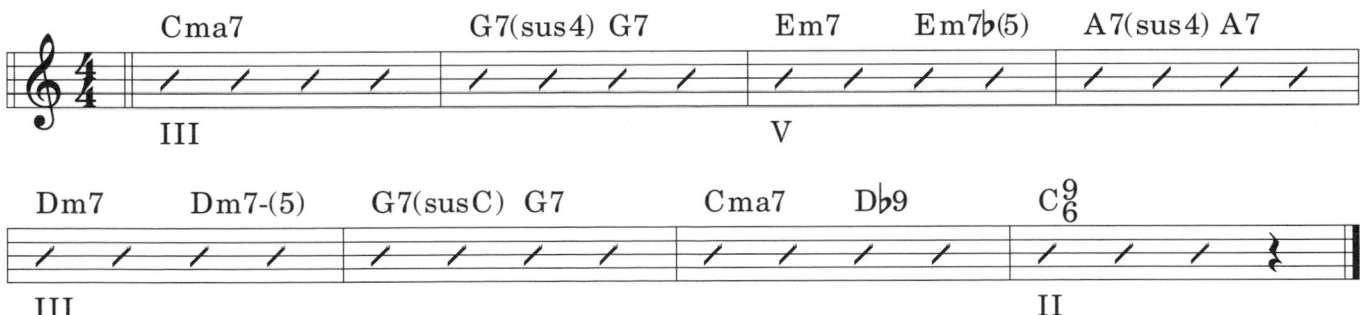

The sus4 refers to the 4th scale degree of the chord so named. The note name (for the 4th) is also used, e.g., G7susC. Sus4 may also be called (natural) 11. The root is on the same string as the sus4 form. For example, in the above exercise you may substitute symbols "G11" and "A11" for sus4.

Melodic Rhythm Study No. 1
Optional Duet With Rhythm Guitar

Be sure to count the rhythm until you can "feel" the phrase. Eventually you will be able to recognize (and feel) entire groups of syncopated notes. In the beginning you should pick *down* for notes falling on the beat, and *up* for those counted "and." This is a definite aid in learning to read these "off beat" rhythms. Later on, when syncopation is no longer a problem, you will vary your picking for the purpose of phrasing and accents.

❋ Rhythm Guitar: Use Latin beat.

❋ Rhythm Guitar ⎣♩♩♩♩⎦ ⎣♩♩♩⎦ or ⎣♩♩♩♩⎦ ⎣♩♩⎦. Remember the substitutions possible on the dominant 7 and major chords.

Chord Etude No. 4

Be sure to hold all notes for their full value.

Staccato, Legato

A dot (.) above or below a note means "staccato" or short.

A line (−) above or below a note means "legato" or long.

Reading Studies

For reading only.

D MAJOR 1 (FINGERING TYPE 3)

D MAJOR 2 (FINGERING TYPE 3)

Fine

Reading music is a combination of instant note (and finger) recognition and playing the "sound" that you "see" in music (along with the relative time durations of the notes, of course). Try this: Play the tonic chord of these reading studies to get your ear in the proper key. Then try to sing the music to yourself as you play it. If your fingers have been over the fingering type enough times, they will automatically play whatever notes (sound patterns) you mentally "hear" on the page. This will take a great deal of time to master, but keep after it. It's worth it!

Dee-Oo-Ett (duet)

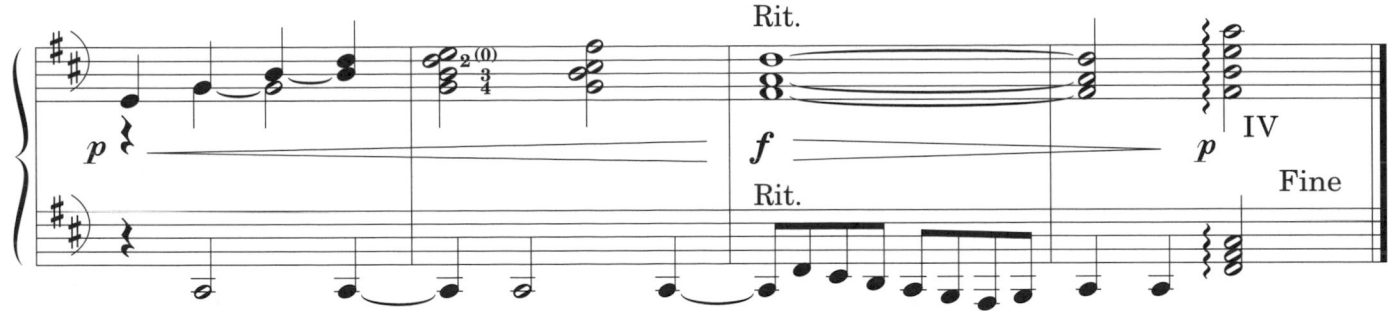

Chord Forms

Dominant 9 → Dominant 9 (Root 5) — Diminished 7 (Root any string) — Major 6 (Root 6)

■ **EXERCISE**

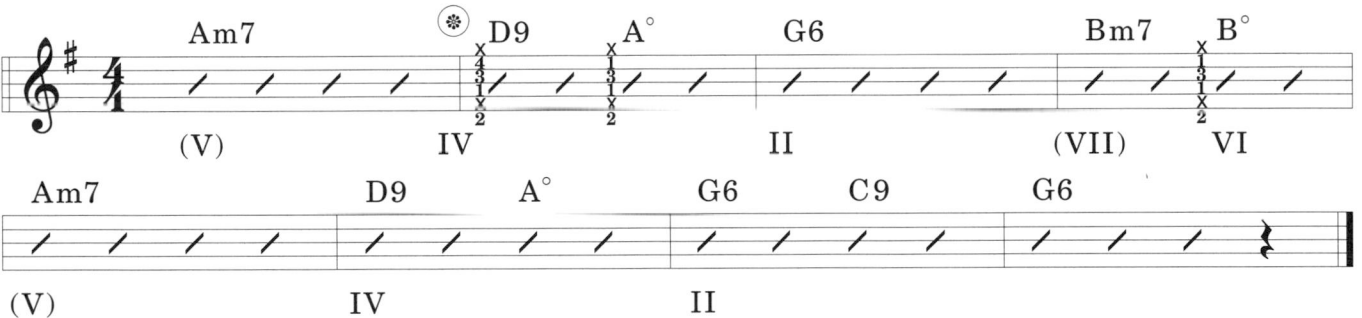

✳ The fingering will be given as shown here whenever two forms are possible in the same position (and also as an occasional reminder).

(Major 7) → Minor 6 (Root 2) — Dominant 9 → Dominant 7(+5) (Root 6)

■ **EXERCISE**

▌ Latin beat—Be sure to release pressure where indicated (⤬).

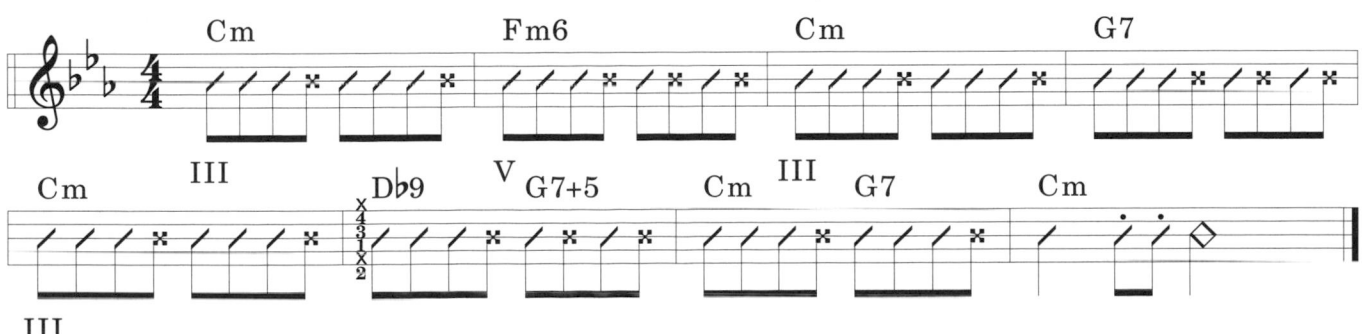

▌ The minor 6 form shown above may also be called a minor 7(♭5) (root 5th string).

Speed Study—Fingering Type 2

Maintain an *even tempo*. Play no faster than perfect coordination in both hands will allow. An increase in speed will come gradually.

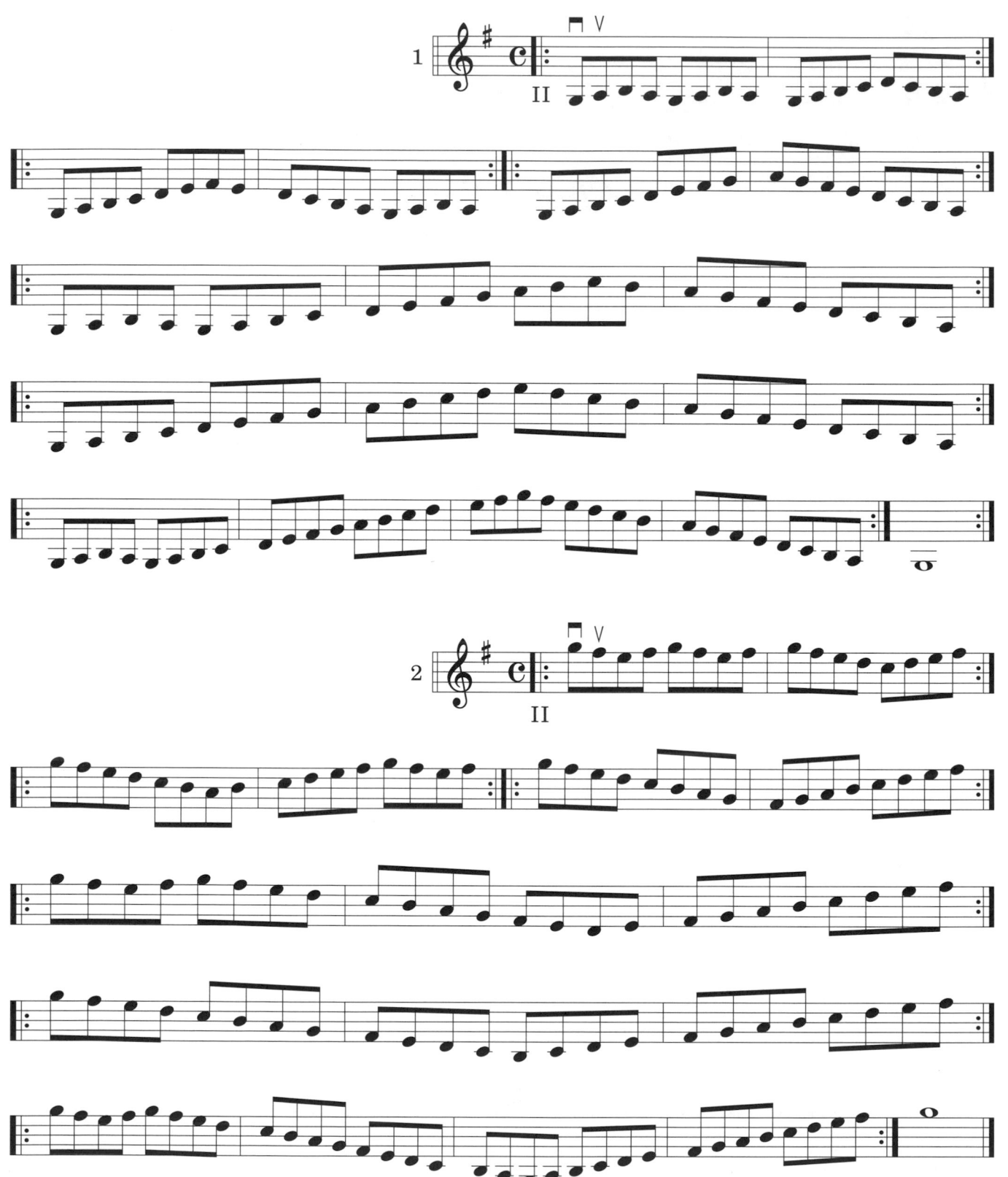

Speed Study—Fingering Type 3

Practice all speed studies as written and with ♩♩ rhythms. Also play them with and without repeats.

For additional technique-building patterns, see pg. 46.

A Major Scale

(Fingering Type 4, Second Position)

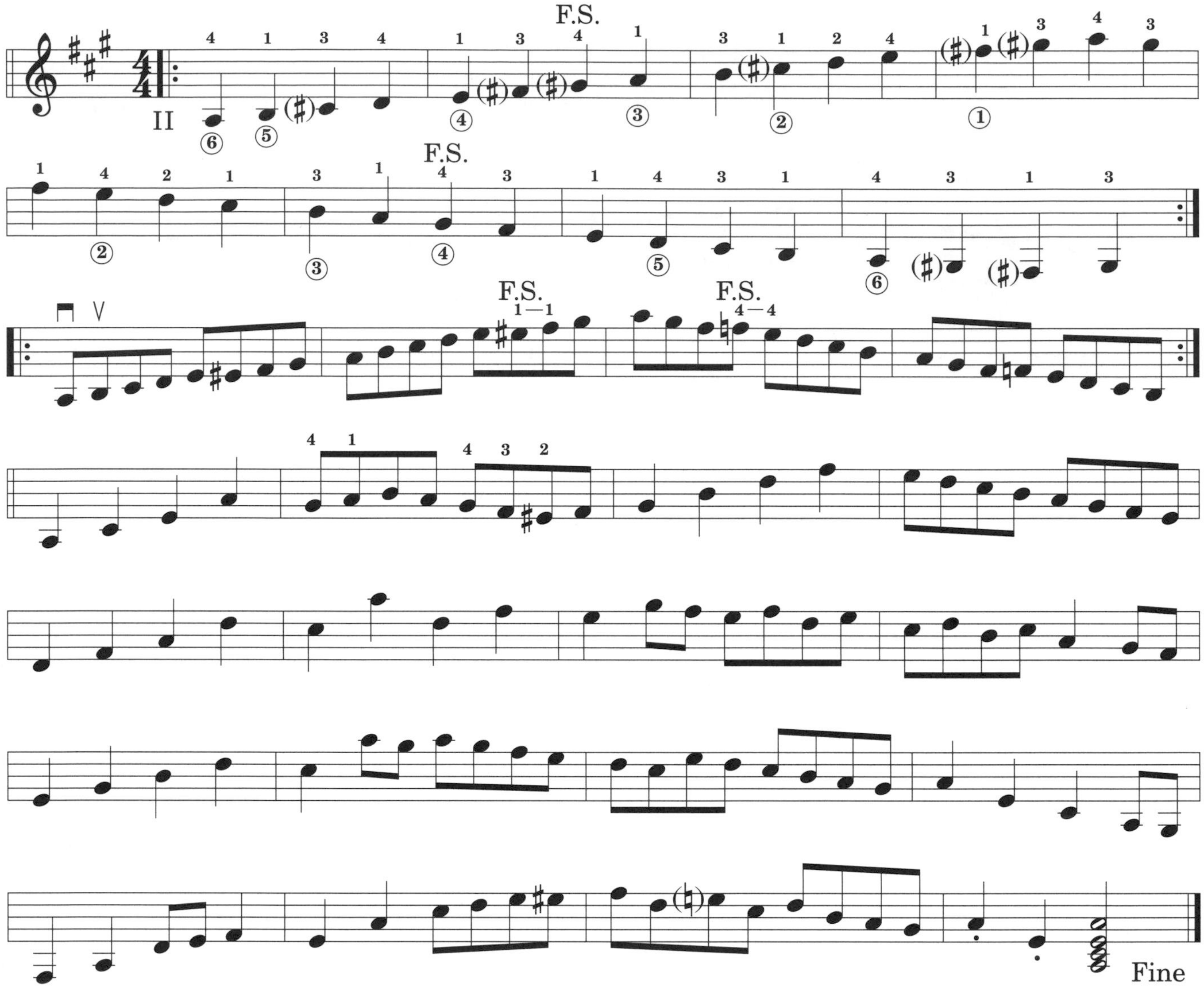

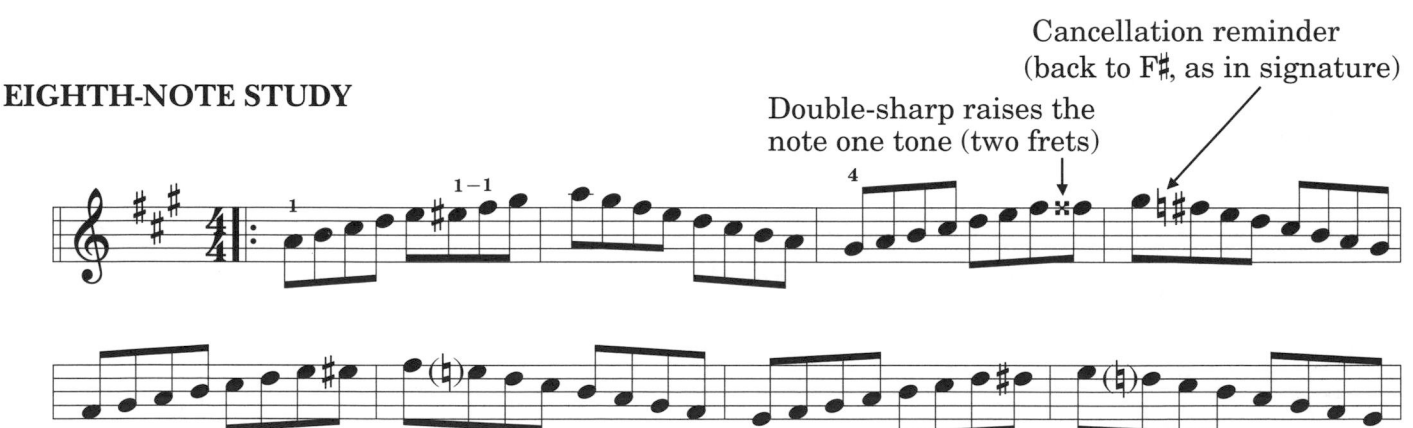

F.S.
II
F.S.
F.S.
F.S.
Fine

EIGHTH-NOTE STUDY

Cancellation reminder
(back to F♯, as in signature)
Double-sharp raises the
note one tone (two frets)

ARPEGGIO STUDY

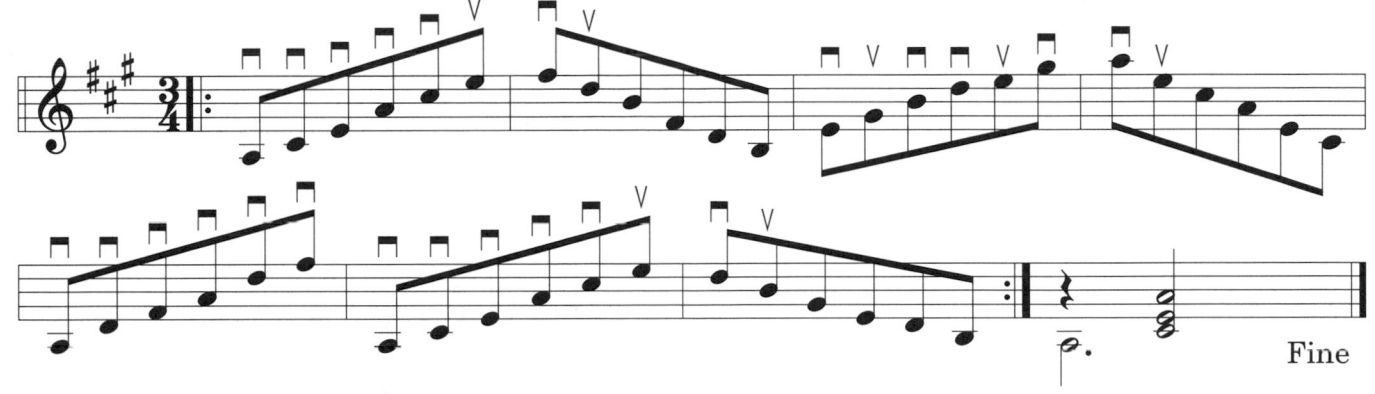

Fine

Also practice arpeggios with alternate ⊓ V picking.

Chord Etude No. 5

Fine

Reading Studies

▋ For reading only.

A MAJOR 1 (FINGERING TYPE 4)

A MAJOR 2 (FINGERING TYPE 4)

A MAJOR 3 (FINGERING TYPE 4)

Tres Sharp (duet)

Sixteenth Note Study

Count carefully. See pg. 31.

count 1 a & ah 2a & ah

1 a&ah 2a & ah 3 a &ah

Chord Forms

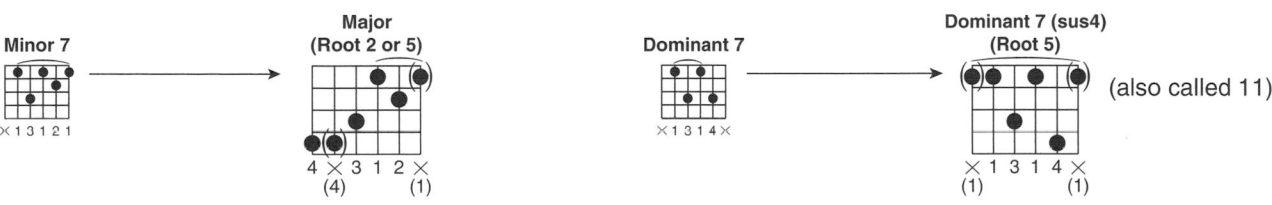

■ **EXERCISE**

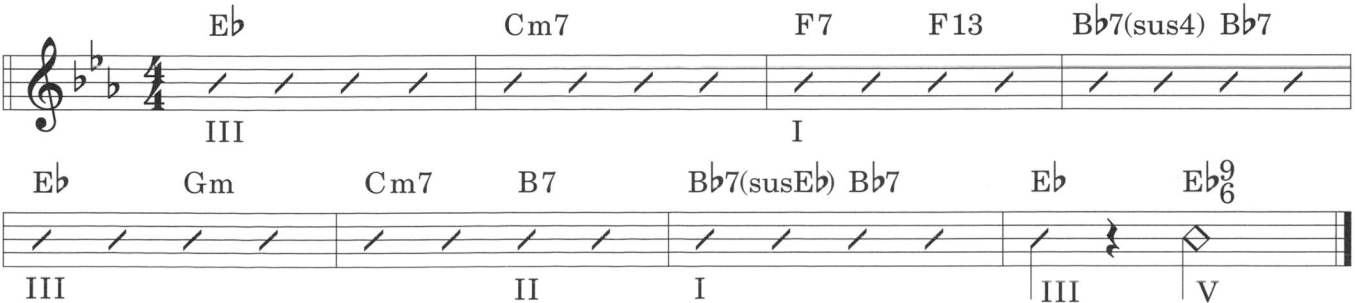

Speed Study—Fingering Type 4, Second Position

▌ As before, keep an even tempo. Play as written and with a ♪♩ rhythm, with and without repeats.

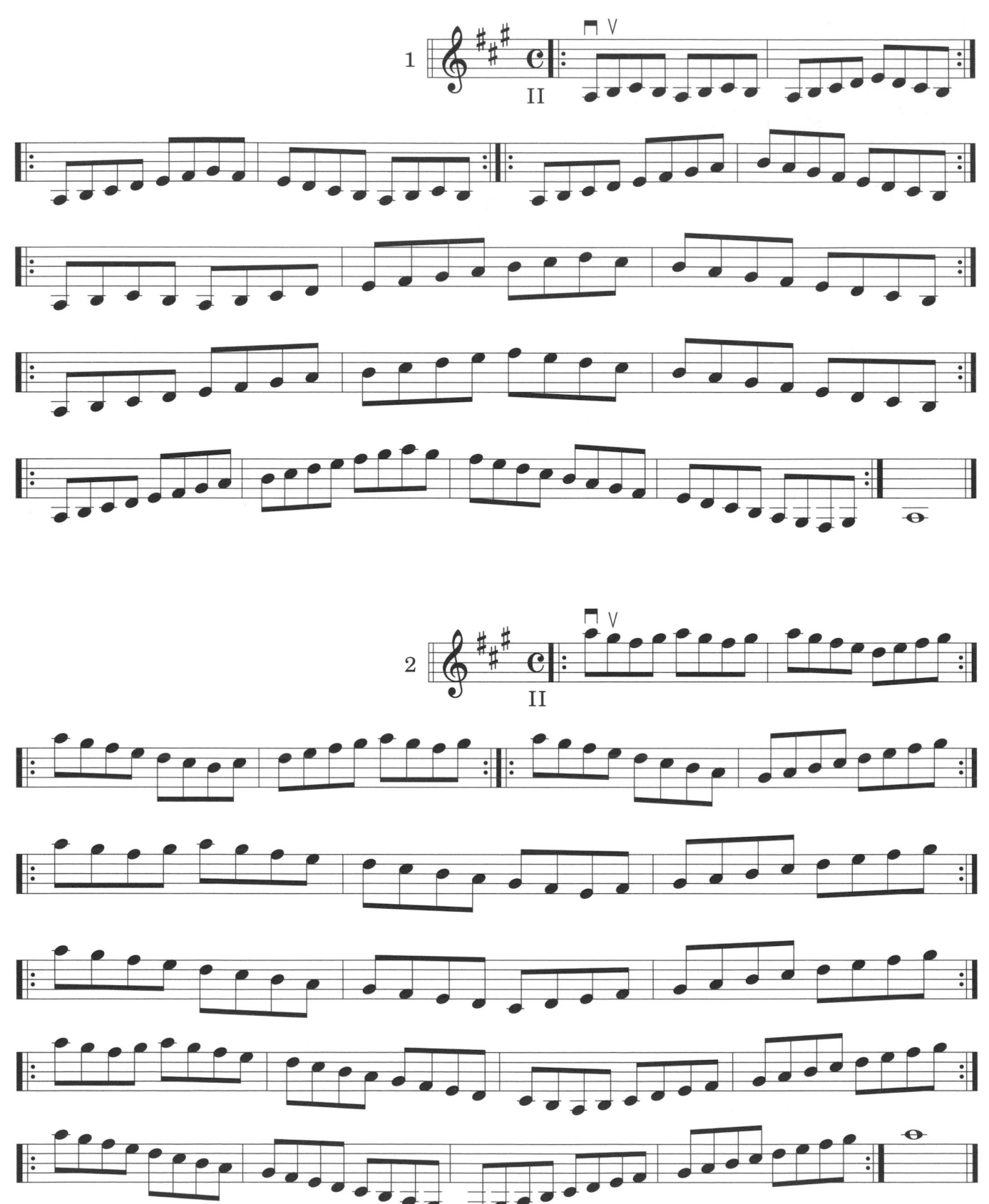

For additional technique-building patterns, see pg. 46.

Chord Forms

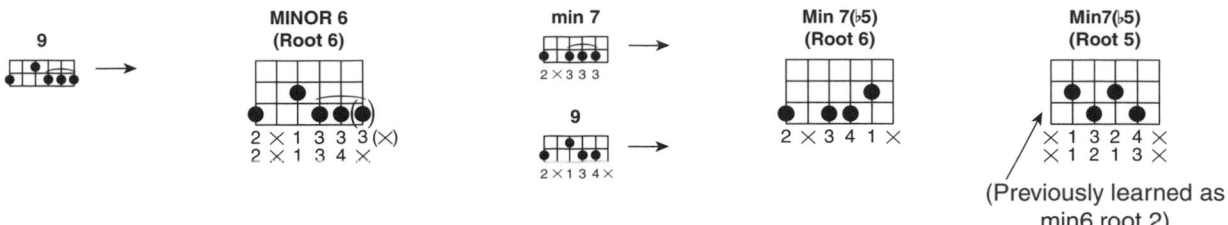

■ EXERCISE 1

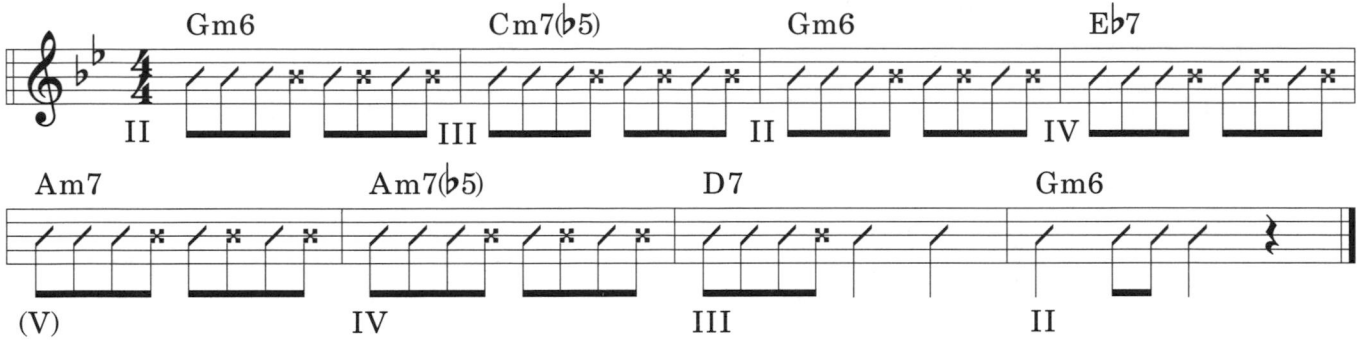

As the minor 6 and minor 7(♭5) forms tend to get confusing, study the following exercises, paying careful attention to the position marks. Play rhythm straight 4 (as written) and also practice using Latin beat. Experiment with various "pressure release" points to vary the accents.

■ EXERCISE 2

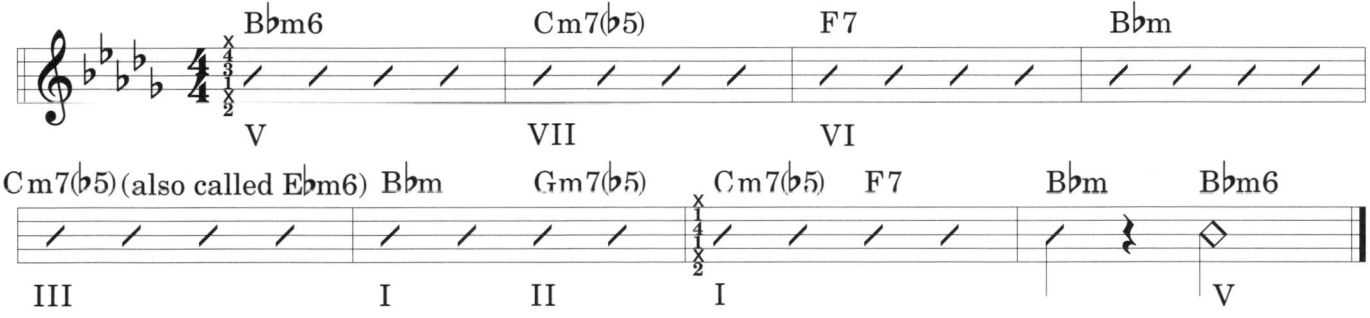

■ EXERCISE 3

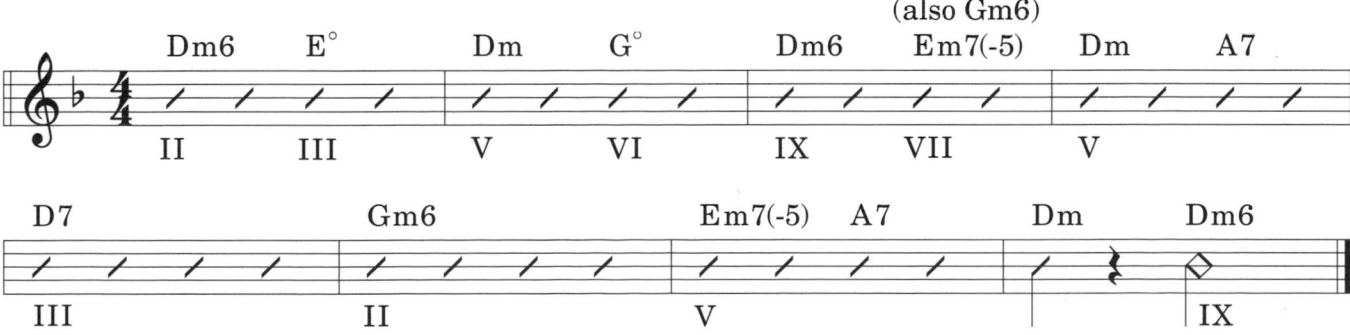

Transpose, write out, and practice all rhythm exercises one or more keys higher.

Second Position Review

Employing the five preceding major scales in position II.

When played as a duet:
- Melody guitar play as written; rhythm guitar play Latin beat.
- Melody guitar play consecutive eighth notes as ♩♪ ; rhythm guitar play straight 4.

FINGERING TYPE 1

Chord Forms

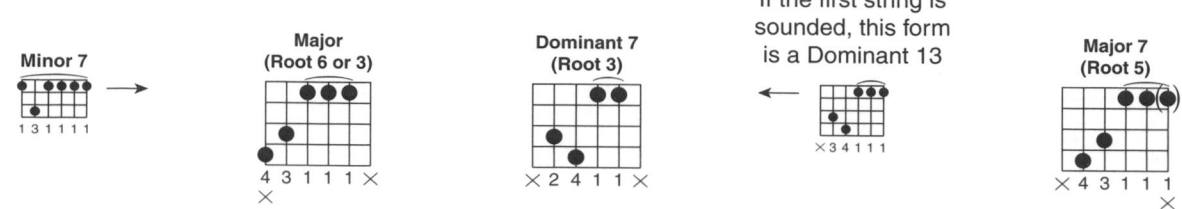

■ EXERCISE 1

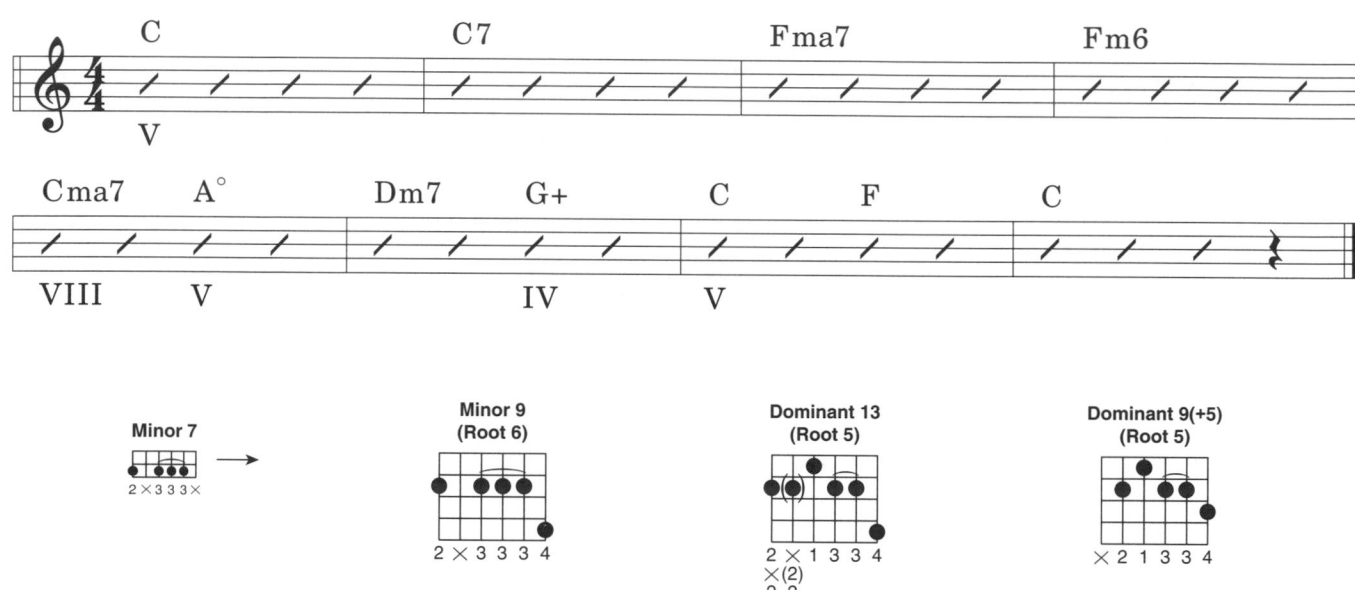

■ EXERCISE 2

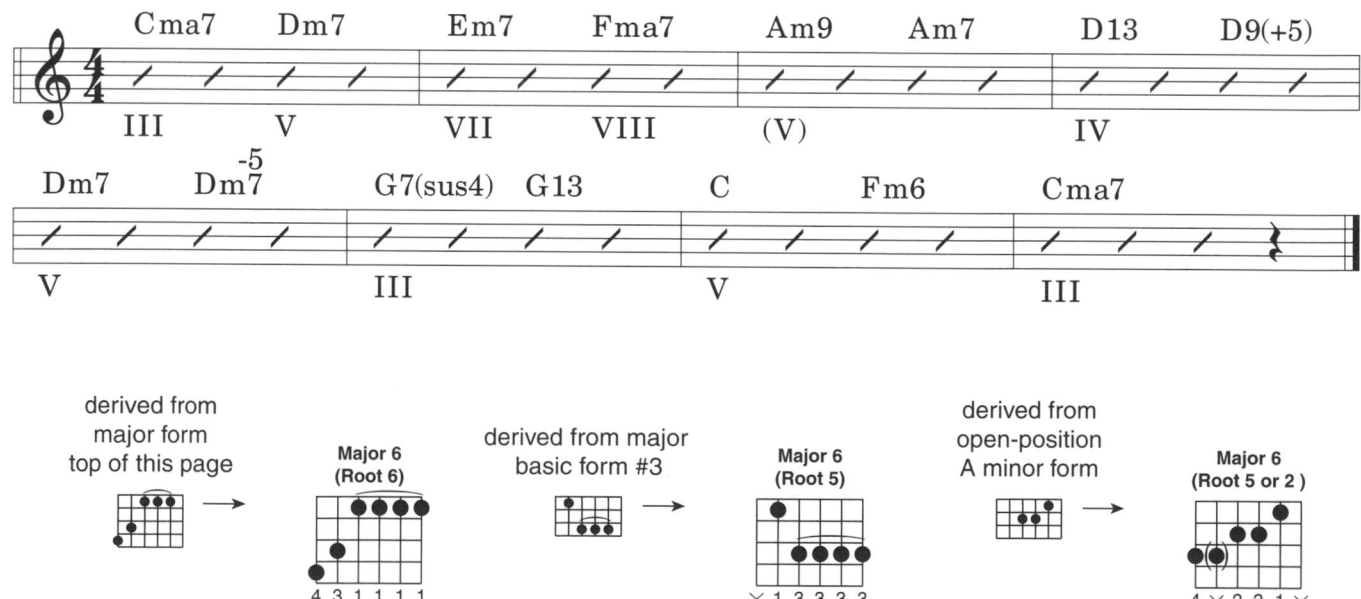

The third major 6 form shown here is, by far, the most valuable,
as it does not use the first string, and therefore has a better rhythm sound.

Quarter-Note Triplets

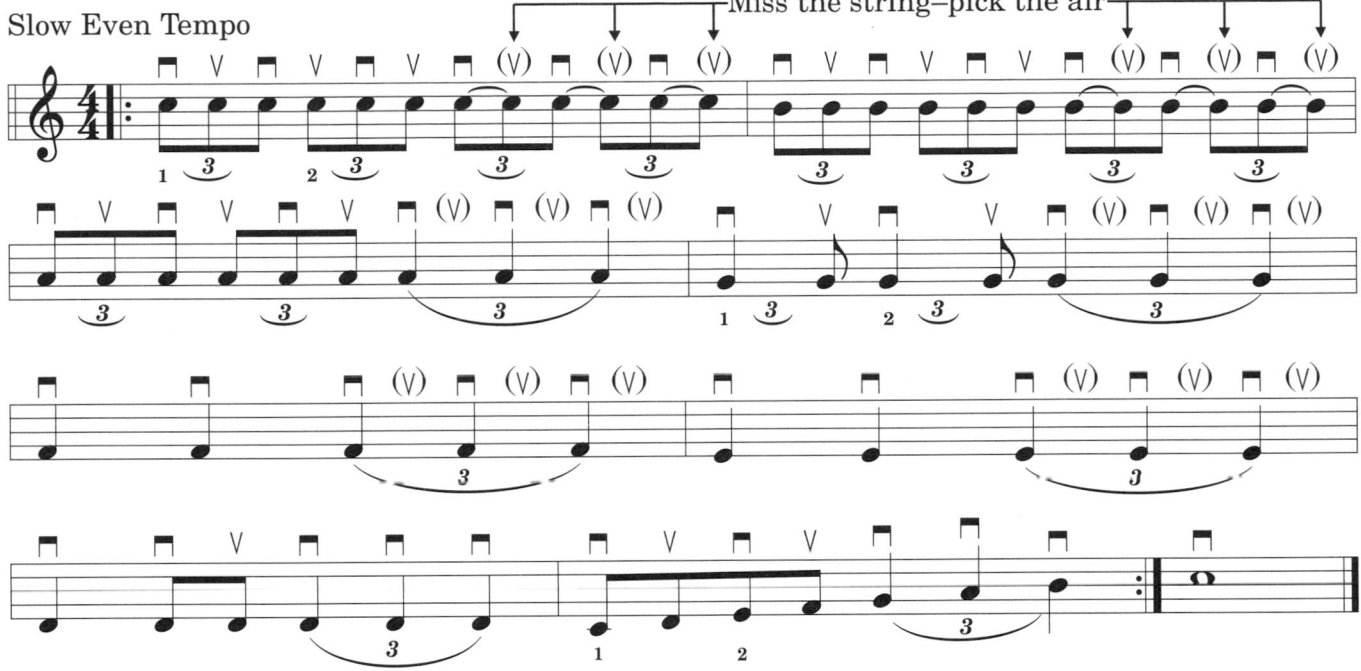

Quarter-note triplets are very difficult to count. The most practical approach is to learn to "feel" them. This can be accomplished (as shown below) by playing two sets of eighth-note triplets using alternate picking and then two more sets of the same—but miss the string with the upstrokes of the pick.

Tap your foot while playing this exercise. Keep at it until you can "feel" the rhythm.

You are now able to read and play in five major keys in the second position. Actually, you can now play in five (major) keys in any position by using these same fingerings (Types 1, 1A, 2, 3, 4) on the higher frets.

Example: Position II major keys C, F, G, D, A

Position III C♯/D♭, F♯/G♭, A♭, E♭, B♭

Of course, you may not yet be able to read in these higher positions, as you have not seen the notes that correspond to these fingering patterns in any area of the fingerboard except the second position.

The following pages show the most used keys in the third position, first position (closed fingering, no open strings), and fourth position. You will be able to concentrate more on the notes, as by now, your fingers should know the patterns.

Major Scales in Third Position
(Most Used)

B-FLAT MAJOR (FINGERING TYPE 4)

E-FLAT MAJOR (FINGERING TYPE 3)

Fine

A-FLAT MAJOR (FINGERING TYPE 2)

Fine

D-FLAT MAJOR (FINGERING TYPE 1)

Double-flat lowers
note one tone

Cancellation
reminder–back to
B♭ as in signature

Fine

Third Position Review
Optional Duet With Rhythm Guitar
Employing the four preceding major scales in position III.

▌ When played as a duet:
- Melody guitar as written; rhythm guitar with optional Latin beat.
- Melody guitar play consecutive eighth notes as a ♩₃♪ rhythm; rhythm guitar play straight 4.

TYPE 2

TYPE 1

Chord Forms

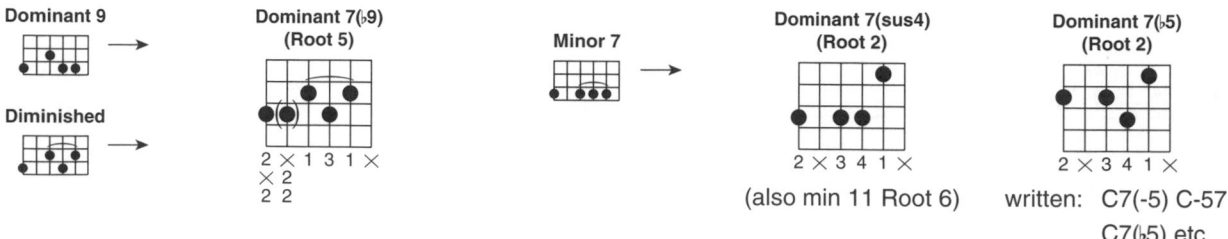

■ EXERCISE 1

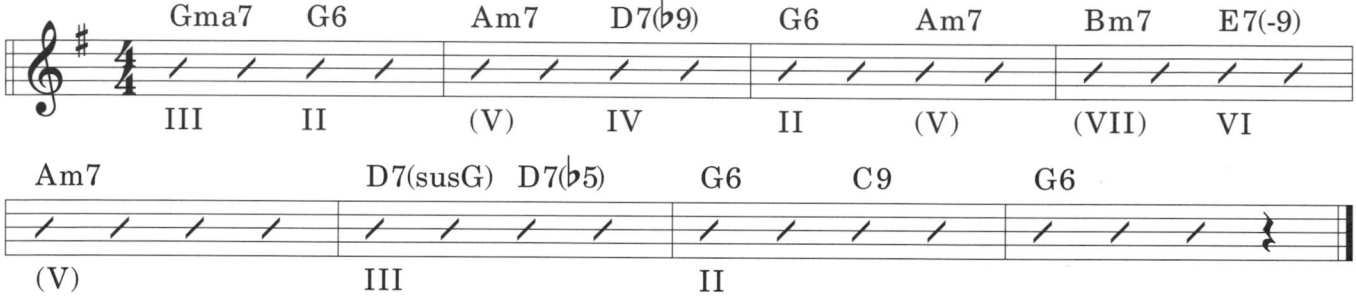

■ (The dominant 7(♭5) form shown above may also be named from the 6th string.)

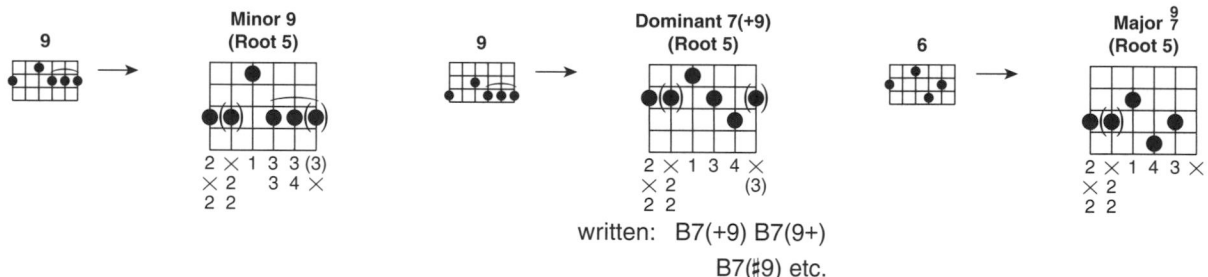

■ EXERCISE 2

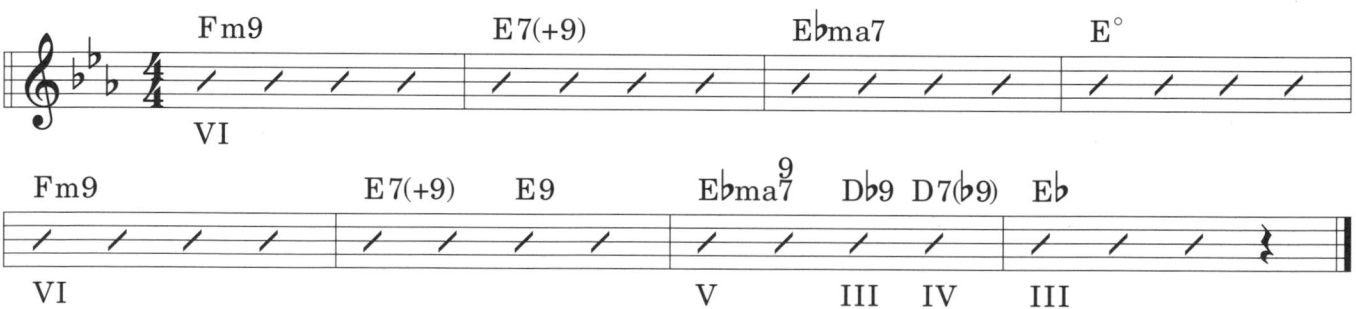

■ The E7(+9) chord used above would be called: E7(♯9), E7 raise 9, or E7 augmented 9. This explicit reference to the altered degree is important.

Major Scales in First Position
(Most Used)
No open strings.

A-FLAT MAJOR (FINGERING TYPE 4)

Fine

D-FLAT MAJOR (FINGERING TYPE 3)

Fine

First Position Review
Optional Duet With Rhythm Guitar
Employing the two preceding major scales in position I.

Melody guitar plays consecutive eighth notes as written and as ♪ rhythms. Rhythm guitar plays a waltz beat for both types of eighth-note rhythms.

Major Scales in Fourth Position
(Most Used)

G MAJOR (FINGERING TYPE 1A)

D MAJOR (FINGERING TYPE 1)

Fine

A MAJOR (FINGERING TYPE 2)

E MAJOR (FINGERING TYPE 3)

Chord Forms

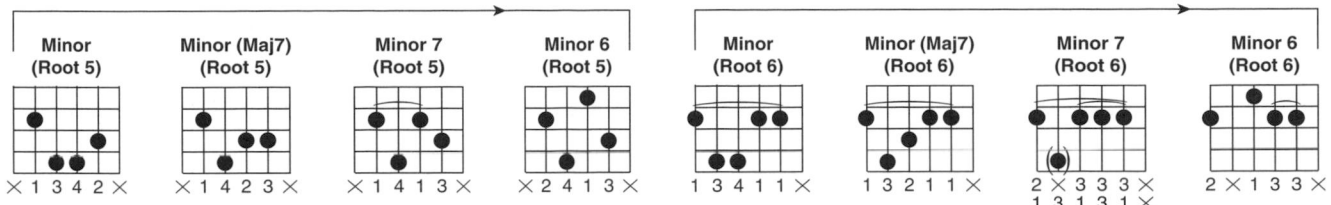

■ EXERCISE 1

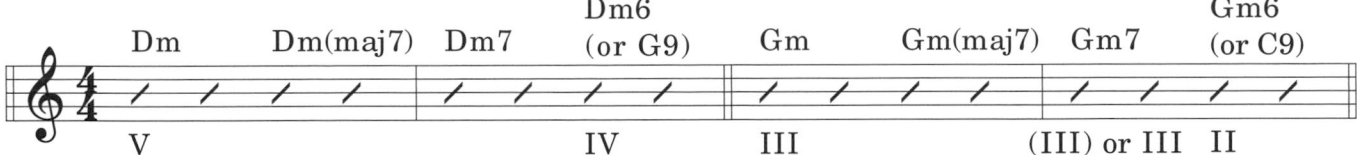

These same minor chord sequences are often found written like this:

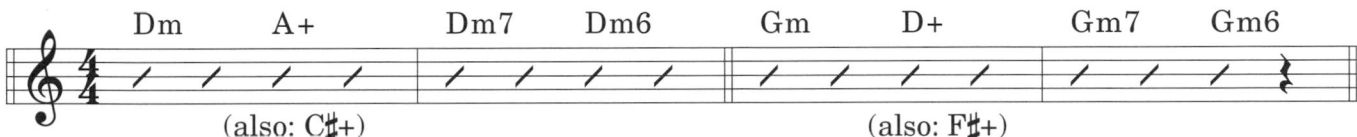

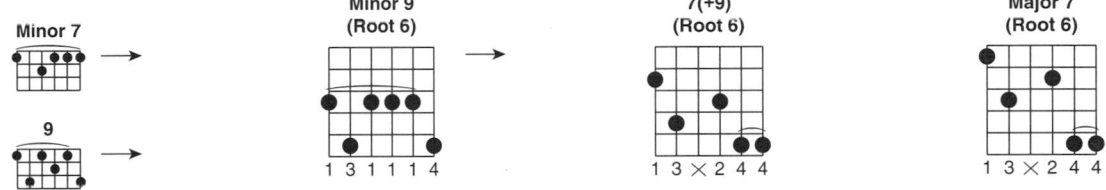

■ EXERCISE 2

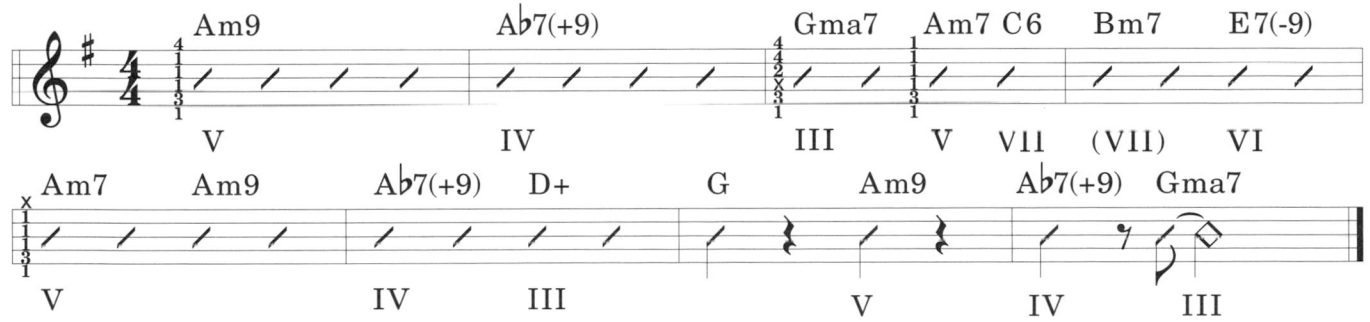

Substitution Tip: ♭5 and ♯5 forms are almost always interchangeable (also +9 and ♭9).

Fourth Position Review
Optional Duet With Rhythm Guitar
Employing the two preceding major scales in position IV.

Melody guitar plays consecutive eighth notes as written and as ♩♪ rhythm. Rhythm guitar plays a waltz beat ♩ ♪ ♪ for both types of eighth-note rhythms.

Find additional reading material. *Be sure it is easy to execute.* Then read five or more pages every day. Play each page *not more* than twice through. Do not practice, do not memorize, and do not use the same pages on consecutive days. Vary the material, and READ, READ, READ, READ.

Chord Forms

The root of this form is one fret below any fingered note. It has four possible names, like the diminished 7 chord.

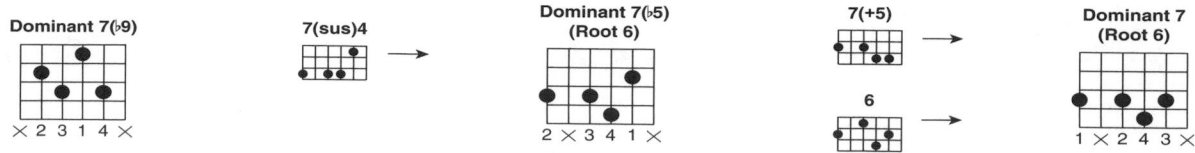

■ EXERCISE

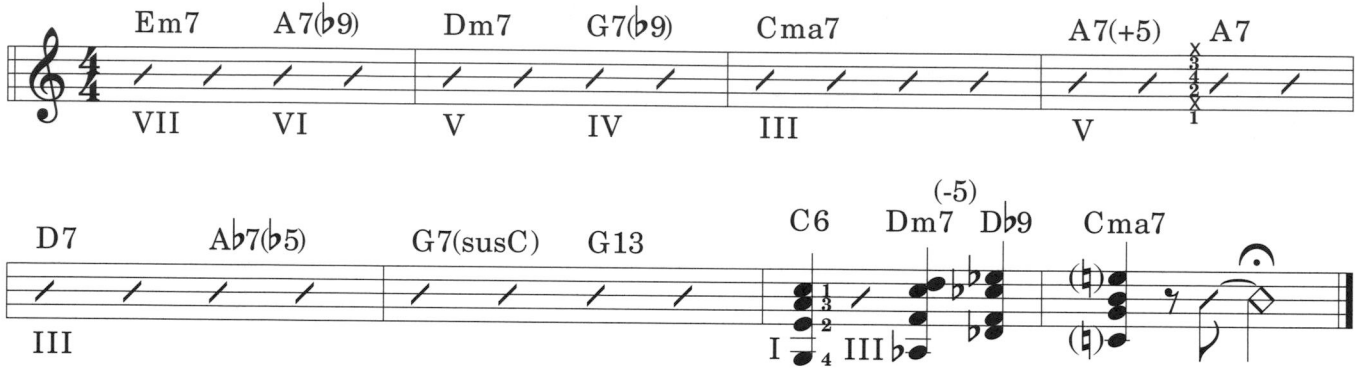

Author's Notes

All forms presented in this book that employ the 6th string (and therefore sound in part in the real bass register) have the root (first) or fifth chordal degrees sounding on the bottom. These are the "strongest" chord tones and *always sound right.*

You have probably seen some of these same forms elsewhere with different chord names indicated. Theoretically, these other names are also correct. However, the bass notes are "weak" chordal degrees and require special handling. This will be discussed thoroughly in a later section. Until then be careful of any forms that use the 6th string and do not have the root or fifth in the bass as they do not always sound right.

In an orchestral rhythm part, the chord symbols used generally indicate the total or complete harmonic structures, and it is not expected (nor is it possible) that you play all degrees at all times. Of course, you should try to play as close as possible to the written sequences. Actually, simplification by omitting some of the chordal degrees is the "norm." (It is best, for now, to omit the higher degrees.)

Examples: For C7+5($\flat$9) you may play: C7(+5)(omit the $\flat$9) or C+
              For G7 ($\flat$9,13) you may play: G7($\flat$9)(omit the 13) or G7
              For F9(sus4) you may play: F7(sus4) (omit the 9)

Be very careful of substitutions, as they must be completely compatible with the chord(s) indicated. (More about this in later volumes.)

Now, in addition to the five major keys in the second position, you should be somewhat familiar with the most used major scales in positions I, III, and IV. You will have to do a great deal of reading in these areas, however, to really know them.

I cannot overemphasize the importance of learning the four major scale fingering types well, as they are the foundation for other kinds of scales. We will gradually add more (major) fingering patterns until, ultimately, we have twelve—one for each key in each position. At the same time, we will learn how to *convert previously practiced* major forms onto jazz minor, harmonic minor, etc.

Our next project (*Modern Method for Guitar, Volume II*) will be to learn the notes on the entire fingerboard by using all fingering types *in the same key.* This will require moving from position to position as we go through the patterns. The sequence of patterns will vary, depending upon the key signature. You will have a definite advantage in learning the fingerboard in this manner, as your fingers "know" the patterns and you can concentrate on the notes.

Remember, learning to play the guitar is an accumulative process. Regular, complete review is absolutely necessary for the gradual improvement and perfection of the techniques.

Index

A MODERN
METHOD
FOR GUITAR

william leavitt

volume

 **Berklee**
Press

DISTRIBUTED BY

HAL•LEONARD®

7777 W. BLUEMOUND RD. P.O. BOX 13819
MILWAUKEE, WISCONSIN 53213

1140 Boylston Street
Boston, MA 02215-3693 USA
(617) 747-2146

Visit Berklee Press Online at
www.berkleepress.com

Visit Hal Leonard Online at
www.halleonard.com

Introduction

This book is a continuation of *Modern Method for Guitar, Volume I.* Most of the terms and techniques are directly evolved from material presented there. For example, the entire fingerboard is covered at once in the five-position C major scale study. This is accomplished by connecting the four basic fingering patterns (Types 1, 2, 3, 4) and one derivative (Type 1A) that hopefully were mastered from the first book. (The sequence of fingering types will vary from position to position up the neck, depending upon the key.)

Study all the material in sequence, as I have tried to relate all new techniques and theoretical concepts to something already learned.

As in the previous volume, all music is original and has been created especially for the presentation and perfection of the lesson material.

Please be advised that the pages devoted to theory are not intended to replace the serious study of this subject with a competent teacher, but rather to intrigue the more inquisitive student, and maybe shed some light on the mysterious workings of music for guitar players in general.

Good luck and have fun.

Wm. G. Leavitt

It is important to learn the following material in consecutive order. The index on pg. 117 is for reference only and will prove valuable for review or concentration on specific techniques.

Outline
Section One

Section Two

SECTION ONE

Four Basic Major-Scale Fingering Patterns

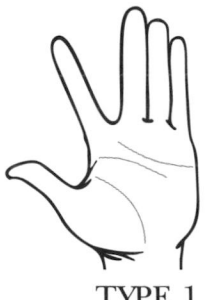 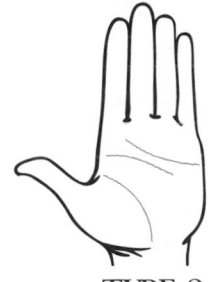 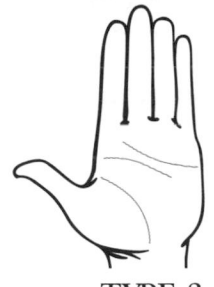 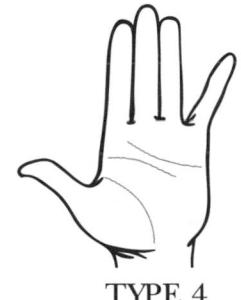

TYPE 1   TYPE 2   TYPE 3   TYPE 4

▪ *(s)*: Finger Stretch

Remember: Do not move entire hand.

TYPE 1   All out-of-position scale tones played with first-finger stretches (See *Vol. I*, pg. 60.)

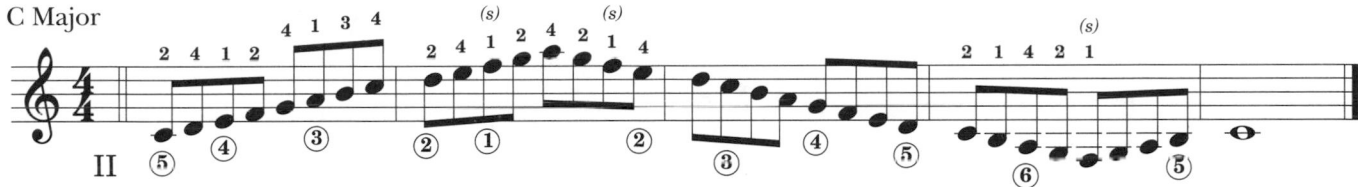

TYPE 2   No finger stretches necessary for scale tones

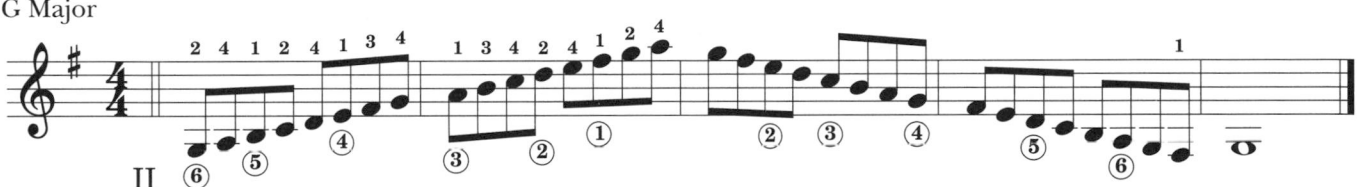

TYPE 3   No stretches

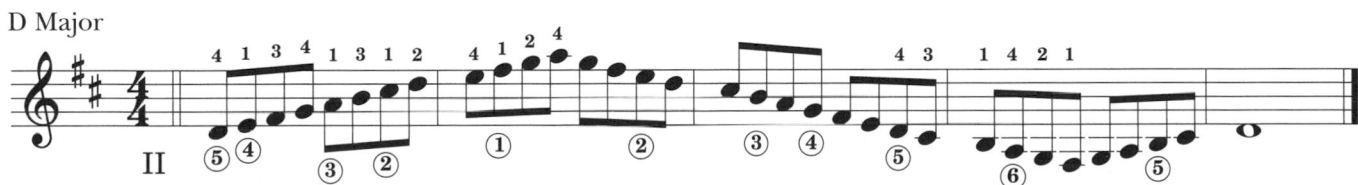

TYPE 4   All out-of-position scale tones played with fourth-finger stretches

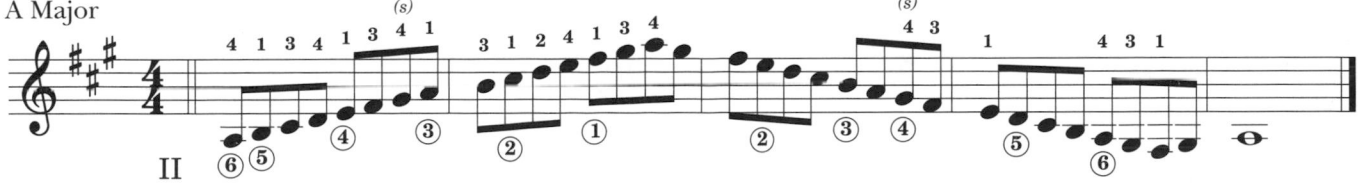

▪ All scales (major and minor, etc.) will be derived from these four basic major-scale patterns. Ultimately, five major keys will be possible in each position with Type 1 and its four derivative fingering patterns: 1A, 1B, 1C, and 1D. This also applies to Type 4 and its derivatives: 4A, 4B, 4C, and 4D. Fingering Types 2 and 3 have no derivative major fingering patterns.

C Major—Ascending (Five Positions)

FINGERING TYPE 1

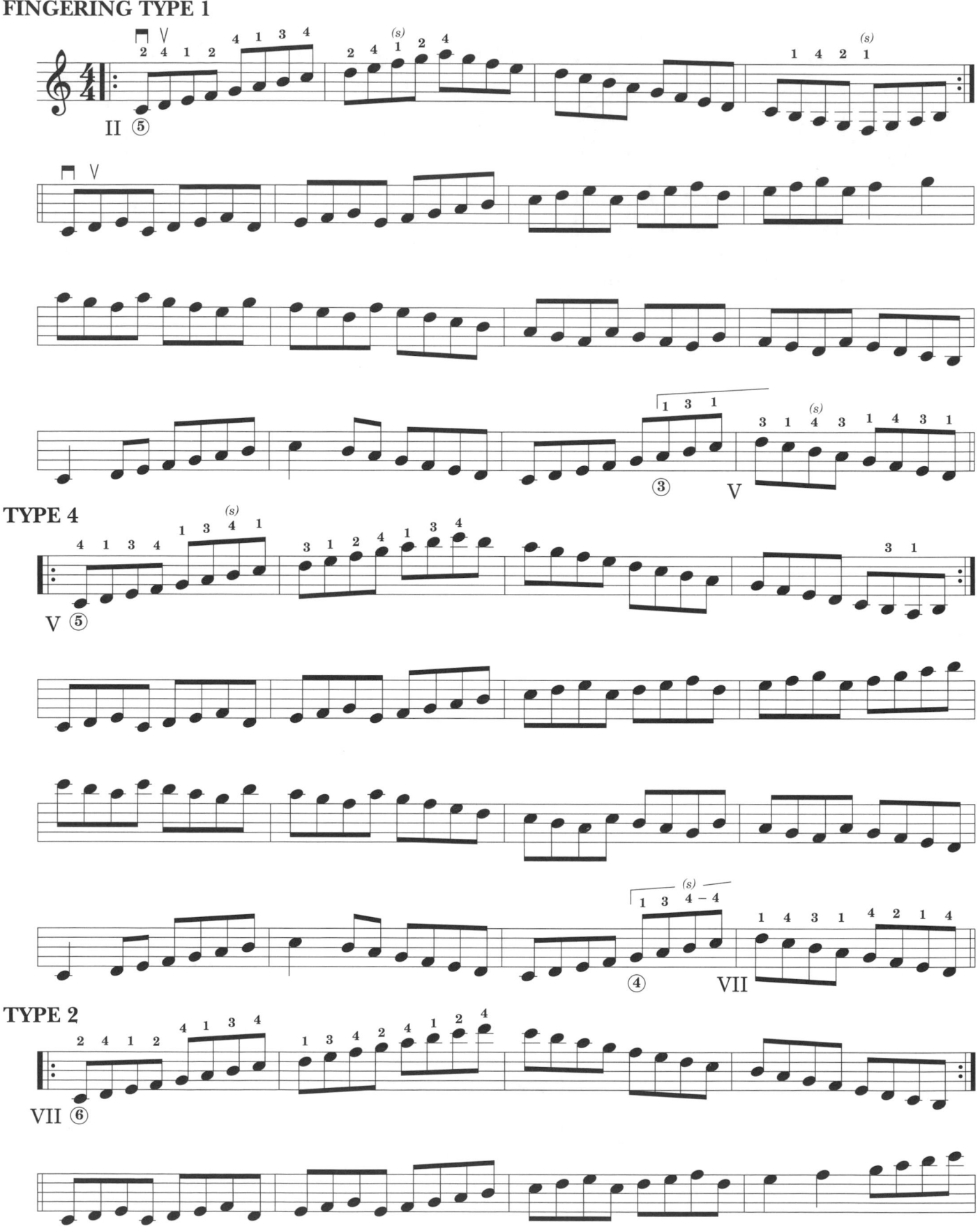

TYPE 4

TYPE 2

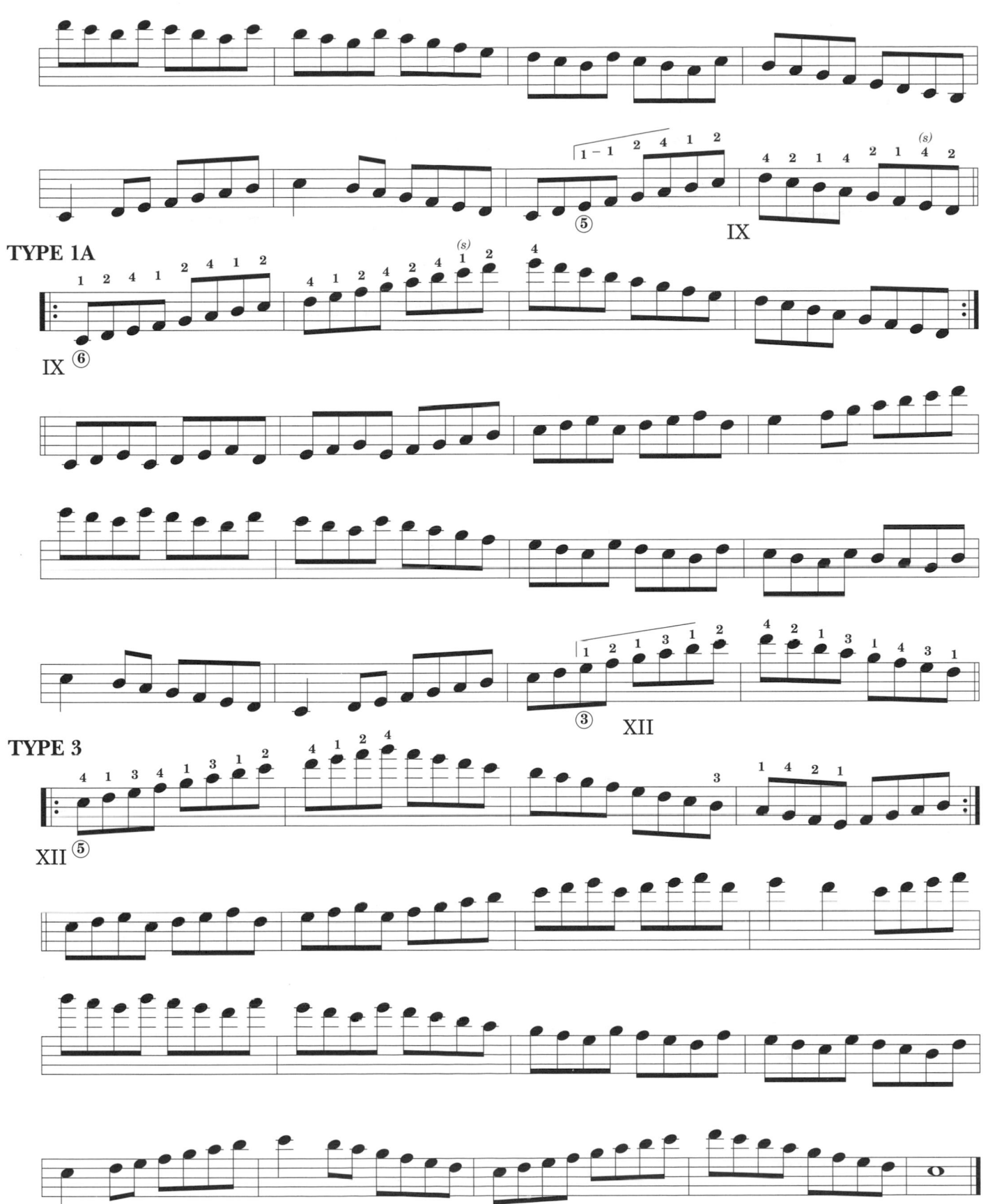

TYPE 1A

TYPE 3

(s): **FINGER STRETCH**

C Major—Descending (Five Positions)

FINGERING TYPE 3

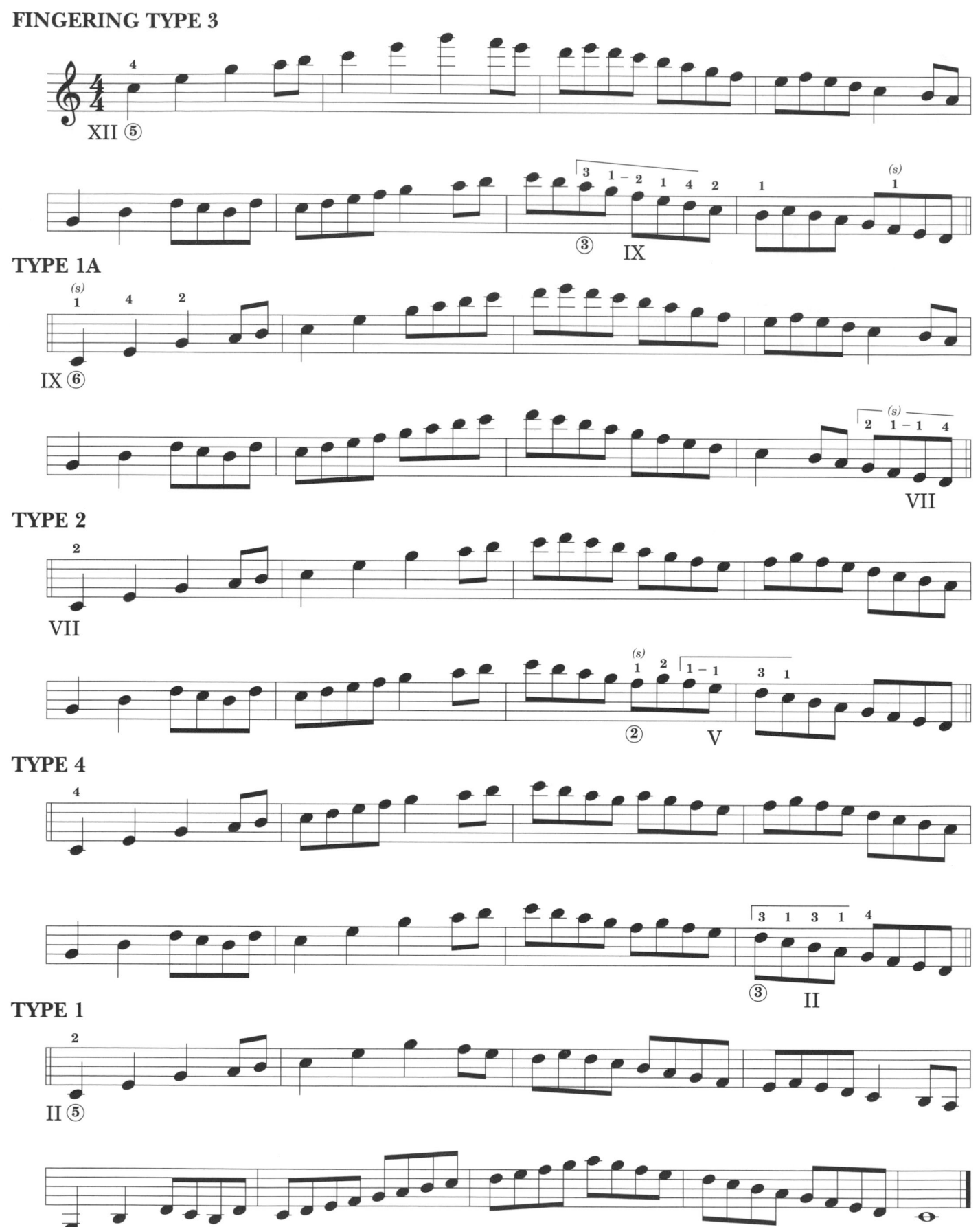

TYPE 1A

TYPE 2

TYPE 4

TYPE 1

Getting Up There (duet)

Chord Etude No. 6

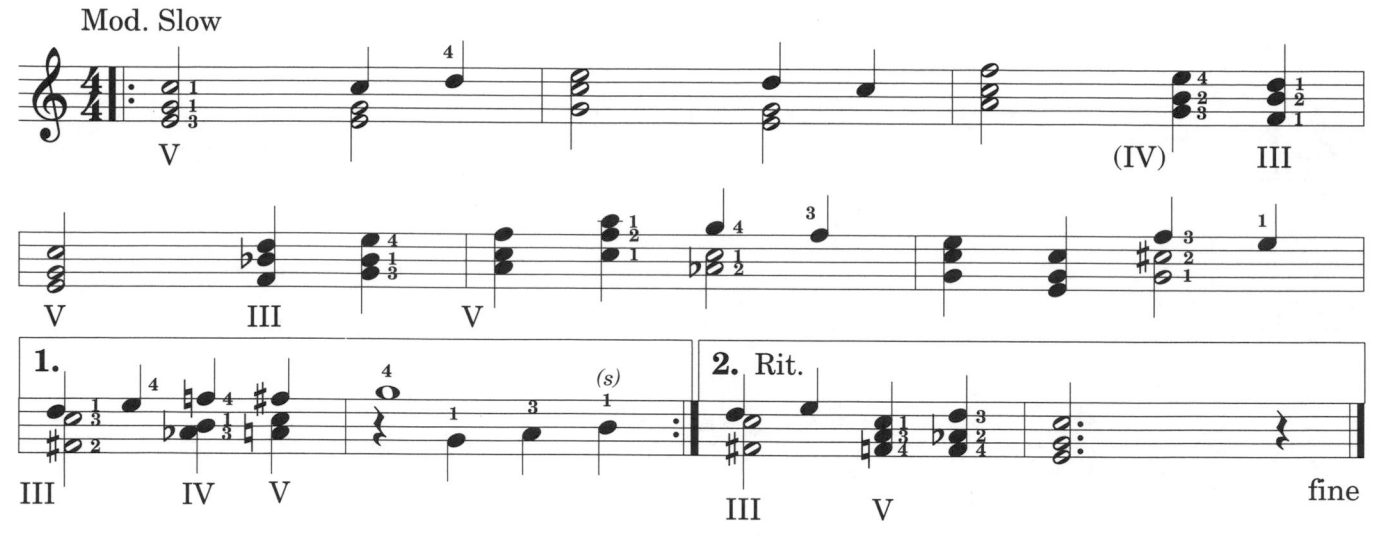

Observe fingering carefully.

Melodic Rhythm Study No. 2

This is a notation comparison, not a duet.

¢ is referred to as "Alla Breve," "Cut Time," or "In Two."

¢ or $\frac{2}{2}$: Half note (♩) gets one beat

$\frac{2}{4}$: Quarter (♩) note gets one beat

Triads (Three-Note Chords)

CONSTRUCTION FROM MAJOR SCALES

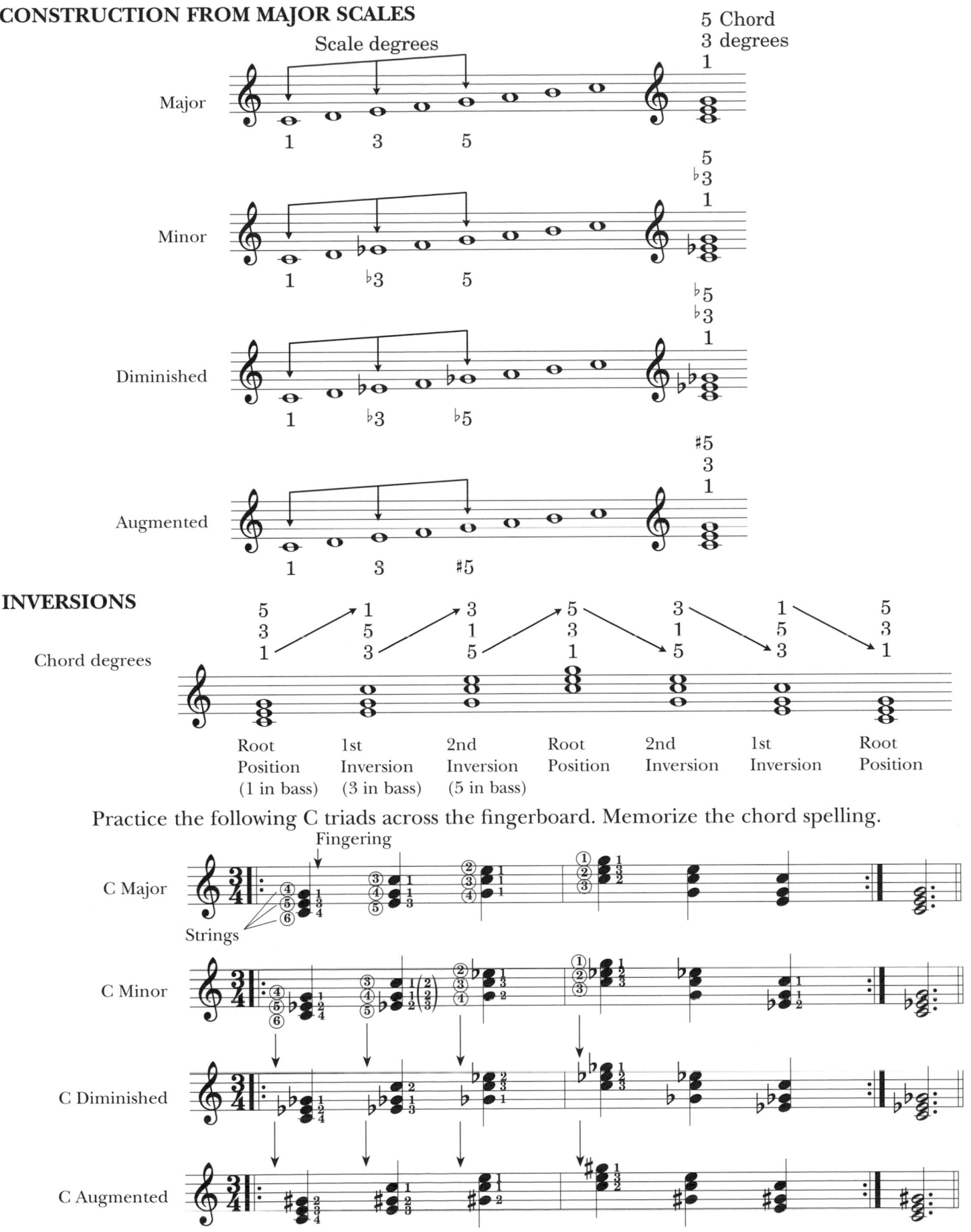

INVERSIONS

Practice the following C triads across the fingerboard. Memorize the chord spelling.

Note common finger and string relationships between most forms.

F Major—Ascending (Five Positions)

FINGERING TYPE 1A

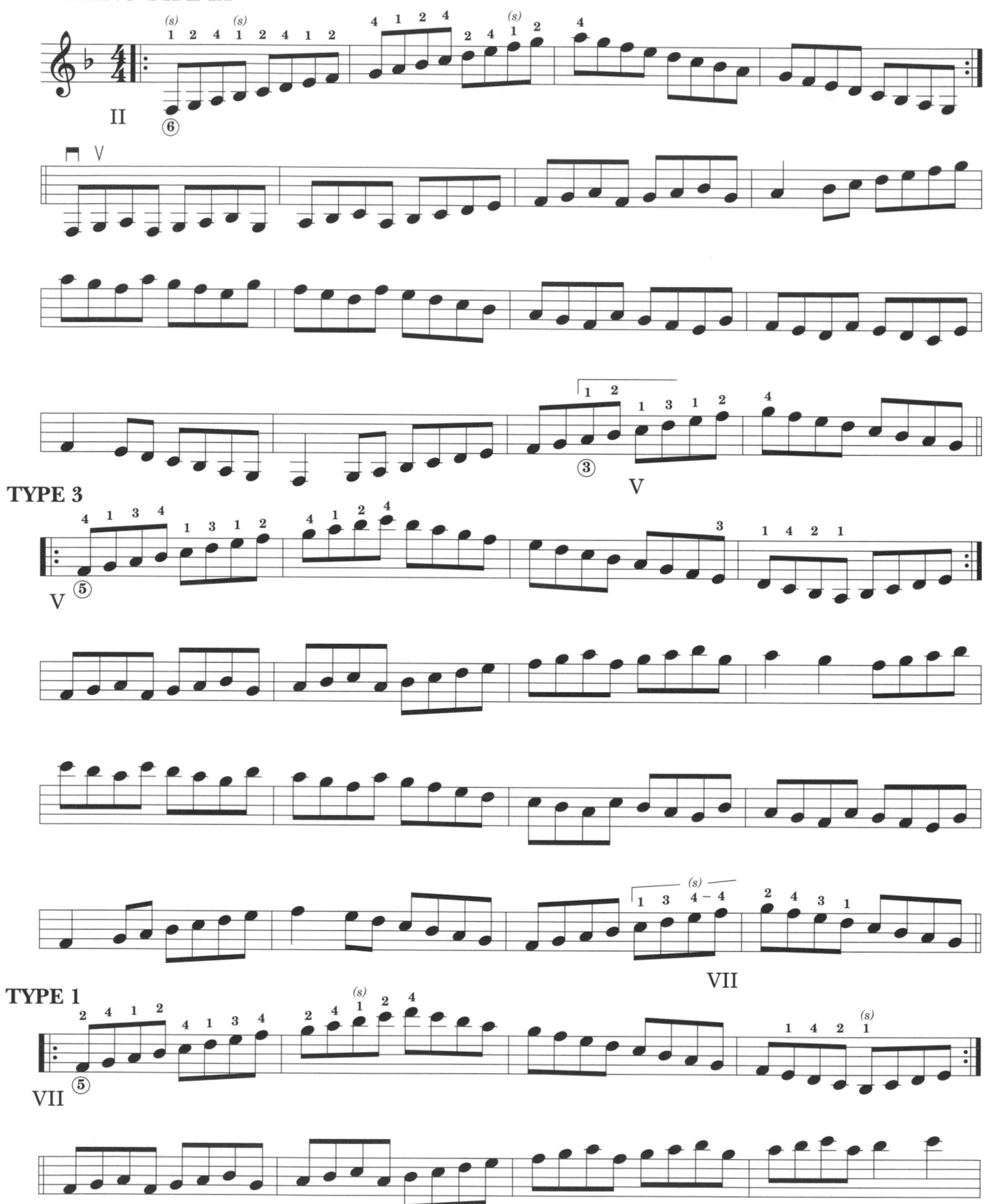

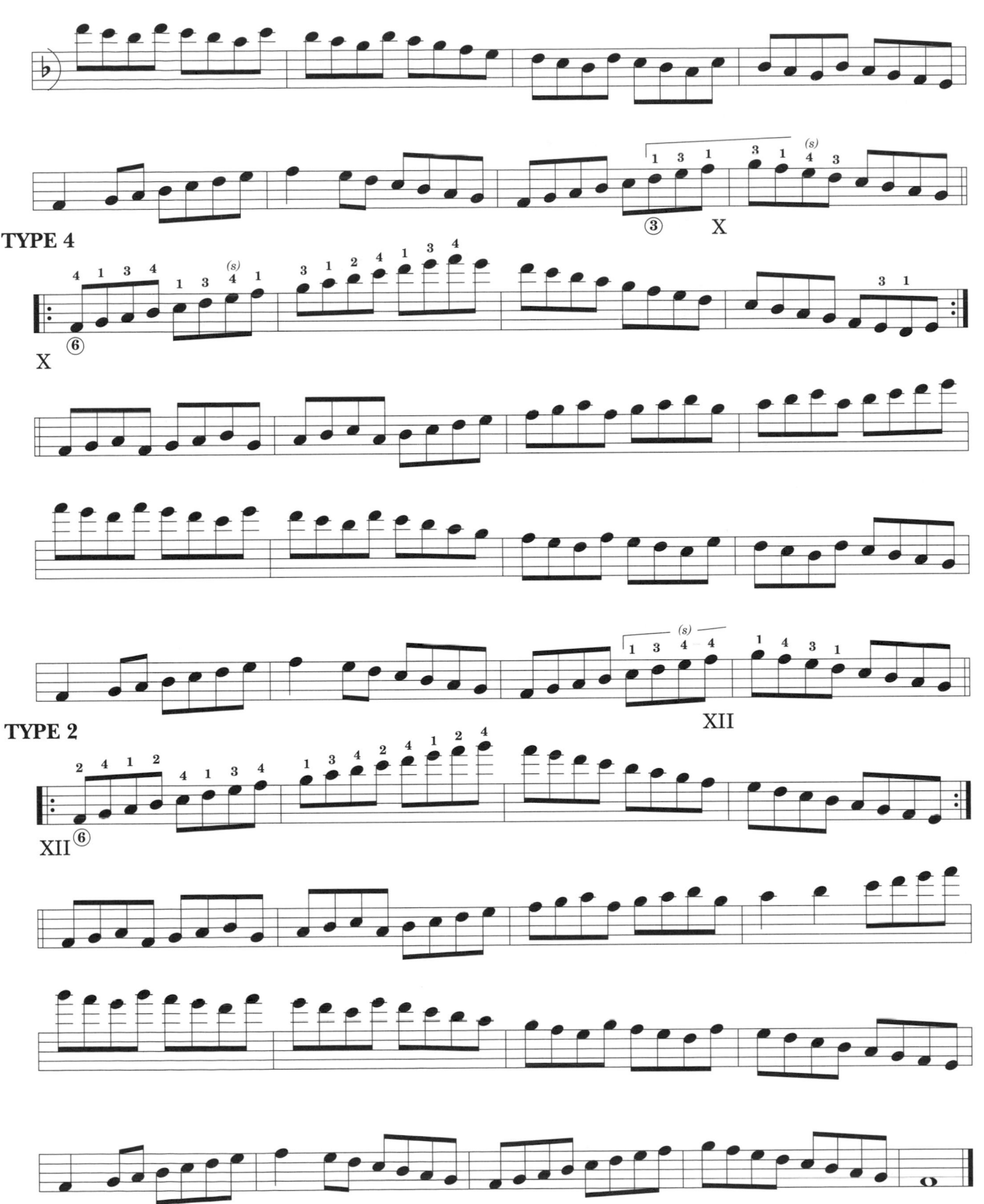

F Major—Descending (Five Positions)

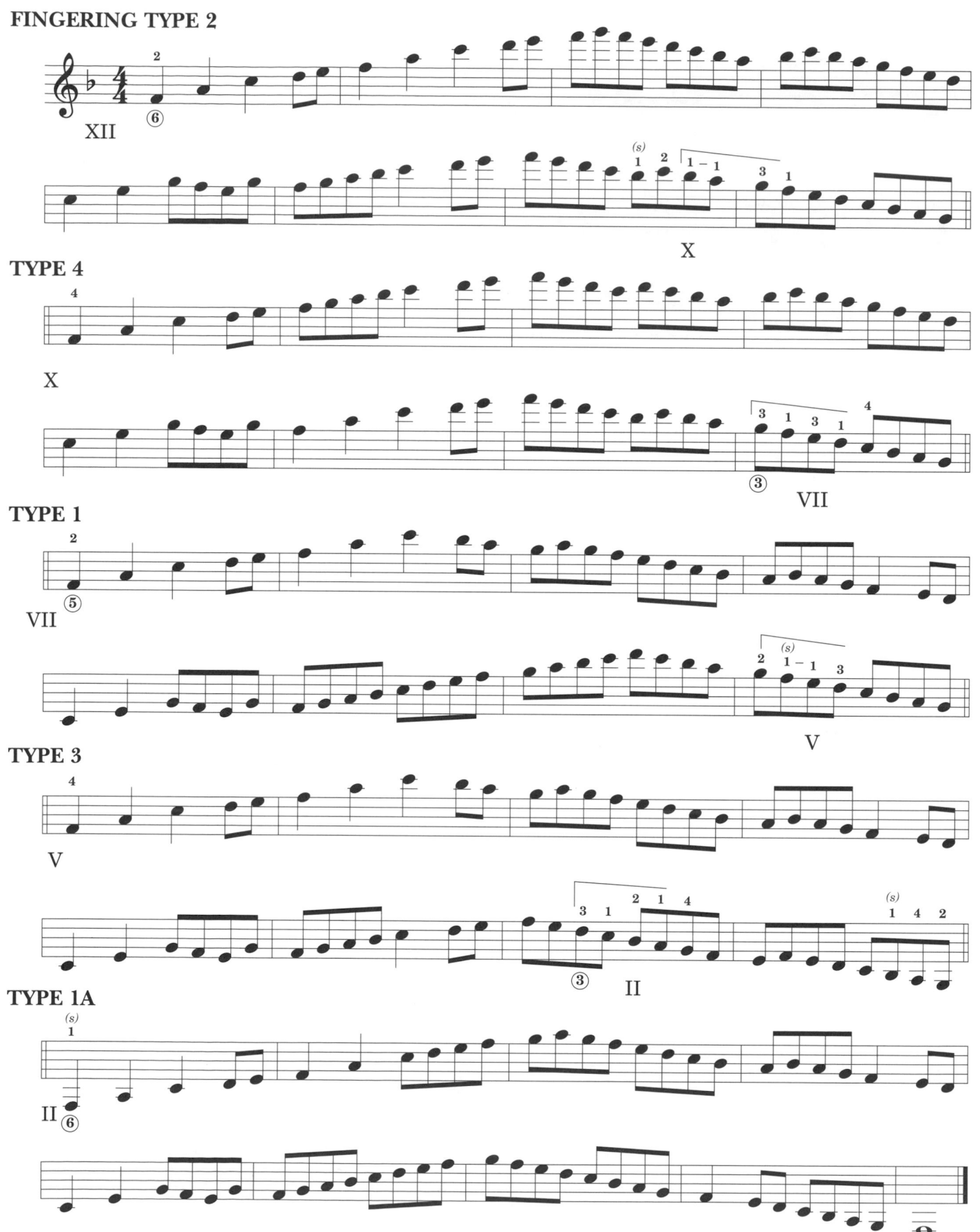

Another Waltz for Two (duet)

Chord Forms

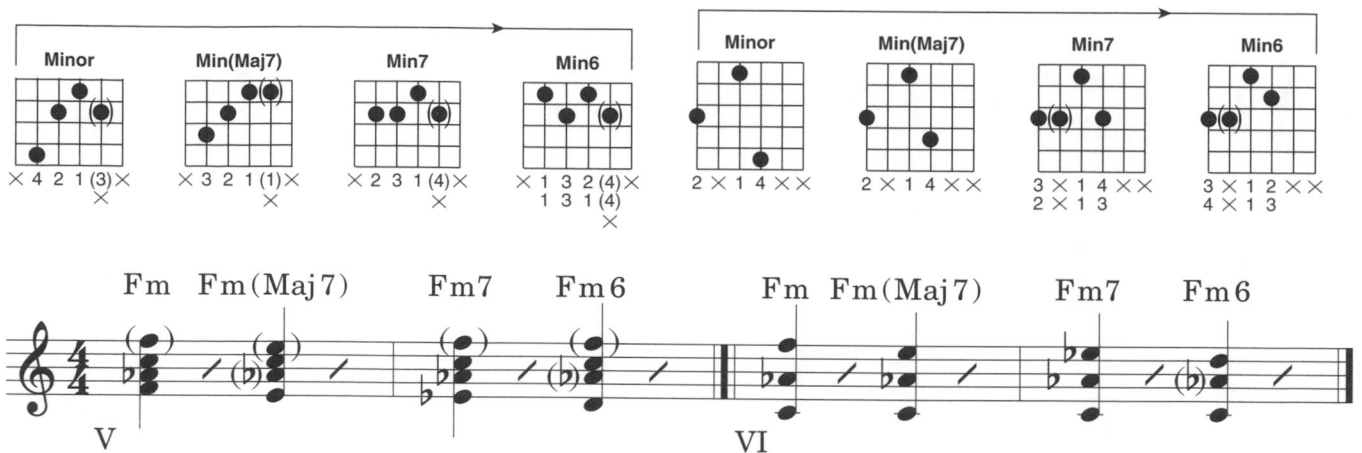

*Also see *Vol. I,* pg. 121.

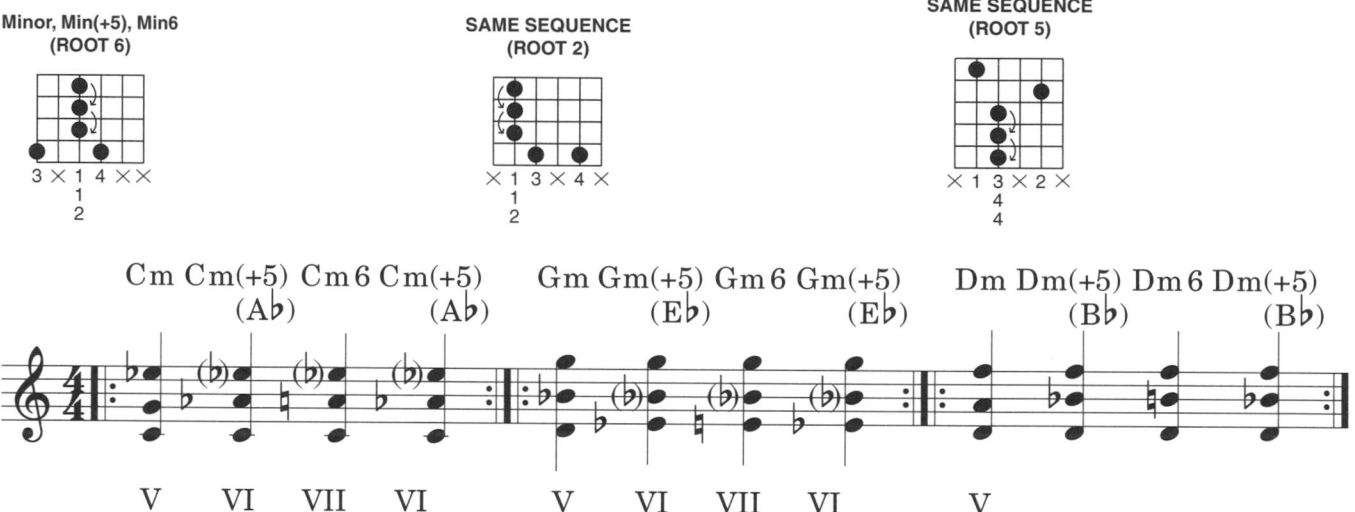

Speed Study

Keep tempo constant throughout.

To practice other fingering patterns, play "Speed Study" as written, but change the key signature to A, D, G, and C.

Triads

Scale (Chord) Degrees

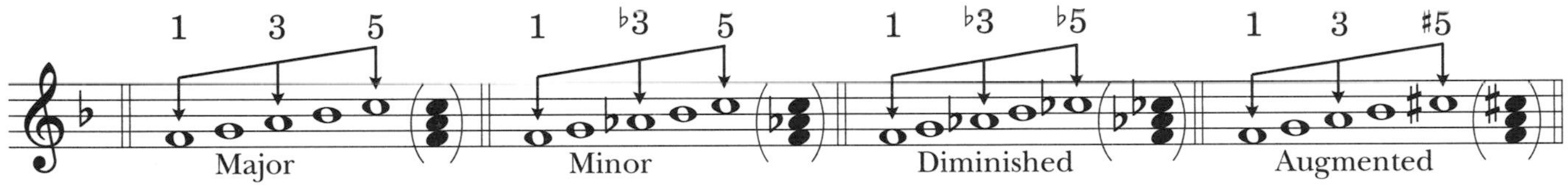

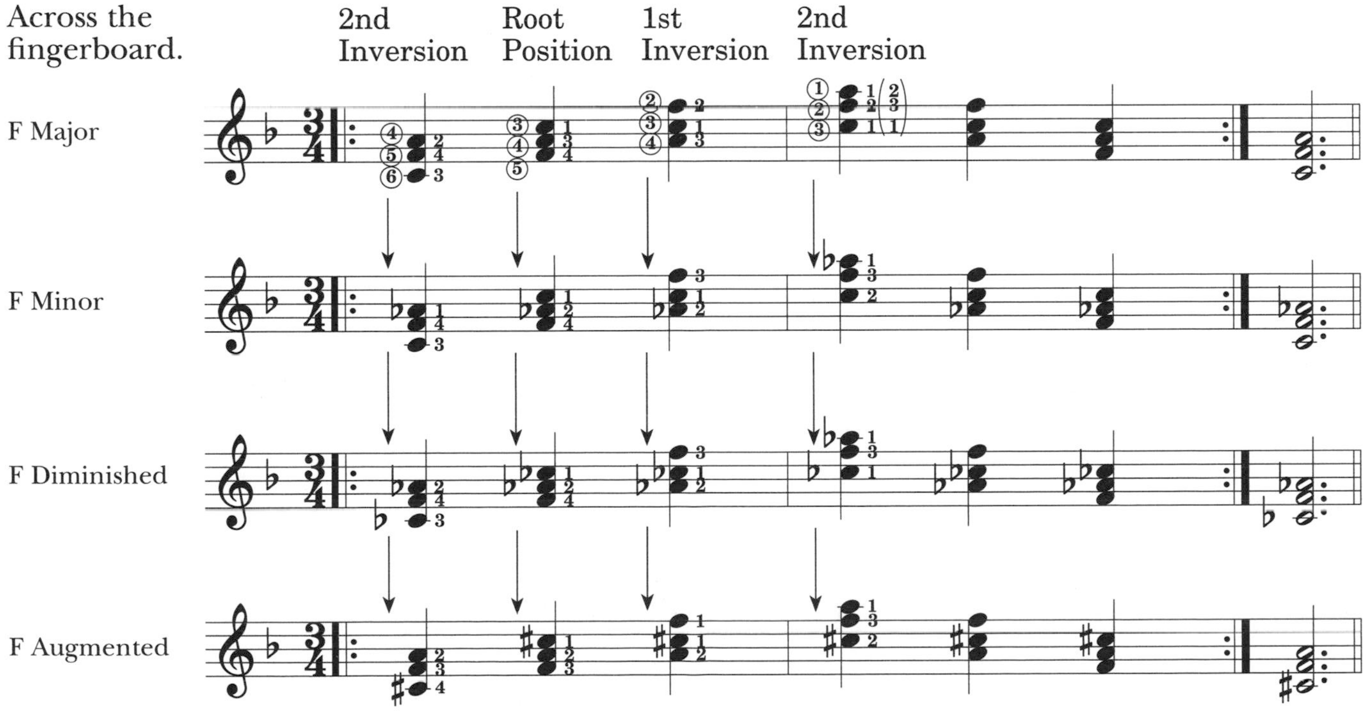

Across the fingerboard.

Observe the fingering: Note common finger(s) between most forms.

Rhythm Guitar—The Right Hand ($\frac{4}{4}$ and 2 Beat)

For a good rhythm section blend, all notes of a chord must seem to explode into sound at the same instant. This can be accomplished by a combination of downward, rotary forearm and loose wrist motion, as if flecking something from the back of your hand. The pick must travel very quickly across the strings to match the sound of the production of a pizzicato note on the bass violin.

NOTATION: ⊓ Downstroke

ⱽ Upstroke

× Strike muffled strings; fingers in formation

, Release pressure immediately after chord sounds

All strokes labeled "Basic" are usually best when used with a guitar alone or with an incomplete rhythm section.

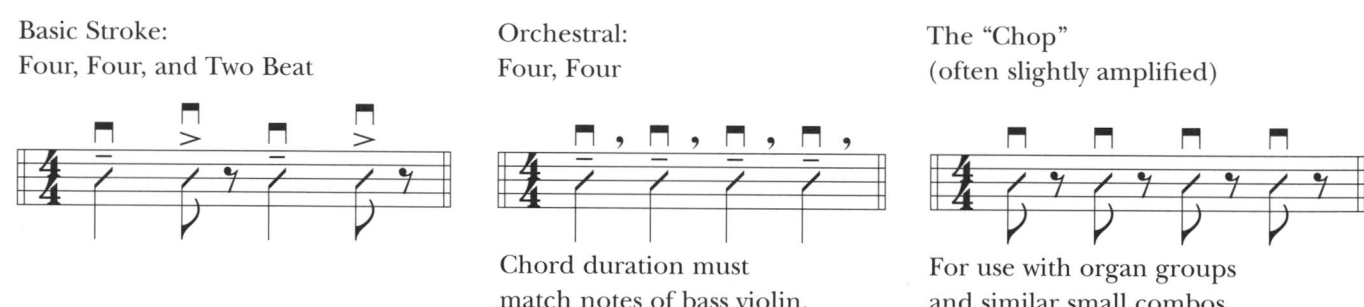

Basic Stroke:
Four, Four, and Two Beat

Orchestral:
Four, Four

Chord duration must
match notes of bass violin.

The "Chop"
(often slightly amplified)

For use with organ groups
and similar small combos.

■ **EXERCISE**

Practice in all three styles, with emphasis on the orchestral stroke.

The principal difficulty in the above orchestral stroke is in producing the sharp, explosive attack while keeping the chord duration long.

Orchestral
"Two Beat"

❋ It is sometimes advisable in practice (and in use) to lightly hit the (muffled) top strings on the returning upstroke where rests are indicated.

■ **EXERCISE**

Be sure to practice at slow, medium, and fast tempos. When learning this style of rhythm playing, it is necessary to tap the foot. First tap on beats 1 and 3, and later tap on 1, 2, 3, and 4.

Orchestral Fast to
Very Fast "Four"

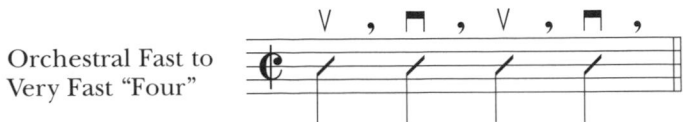

Tap the foot "in two" (i.e. on beats 1 and 3)

Make the upstroke sound as much like the downstroke as possible by favoring the lower strings with the returning upstroke of the pick. There will be a slight natural accent on beats 2 and 4, because the downstroke hits the heavy string first. This is good, as it is comparable to the drummer's use of the hi-hat cymbal on these beats.

■ **EXERCISE**

This right-hand technique is difficult to master but is extremely valuable. It allows you to maintain very bright tempos (steady as a rock) with very little tightening up.

Chord Etude No. 7

$\frac{3}{8}$: Eighth note (♪) gets one beat

All notes connected by a curved line must be kept ringing.

Moderately Fast Waltz

17

G Major—Ascending (Five Positions)

FINGERING TYPE 2

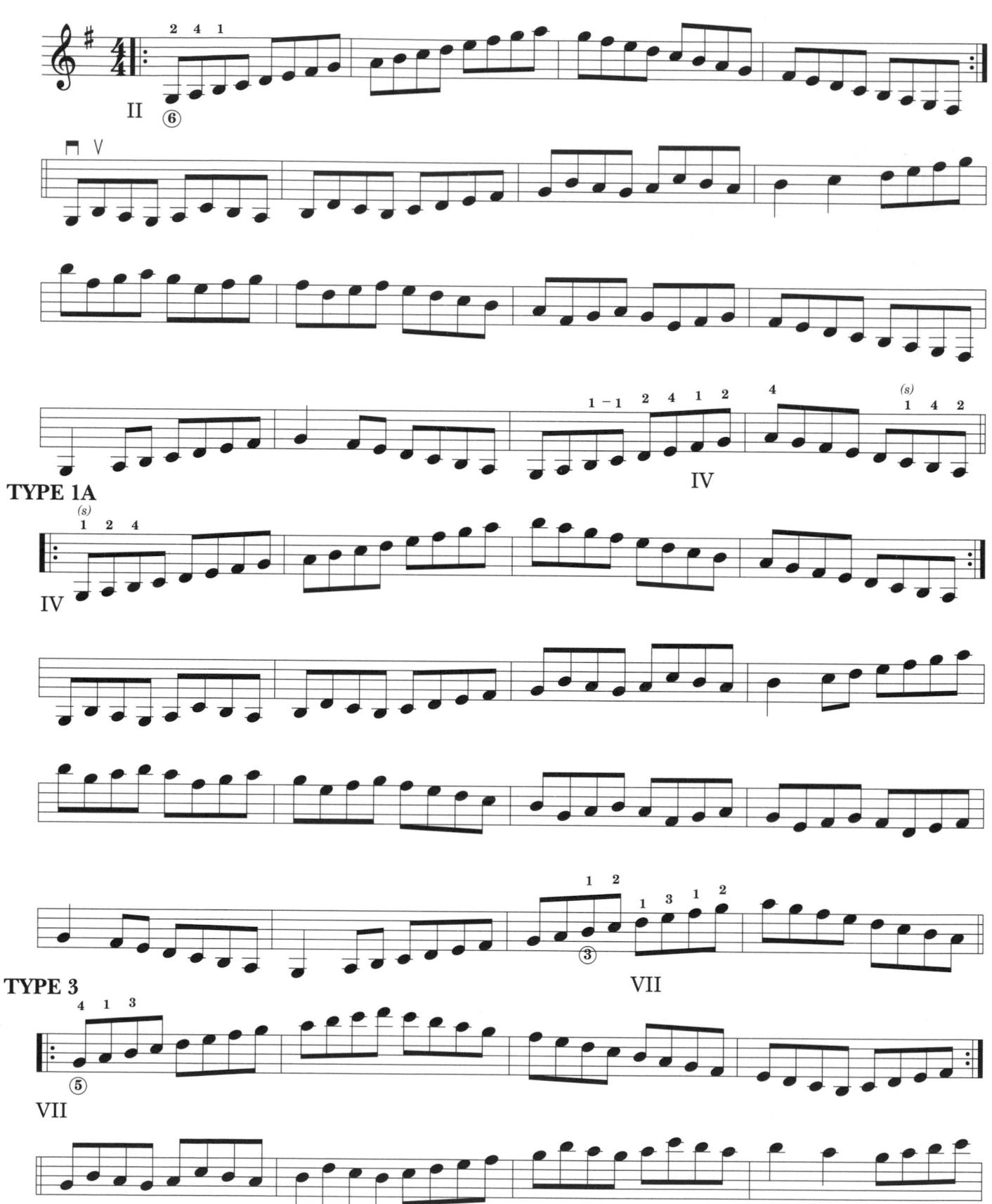

TYPE 1A

TYPE 3

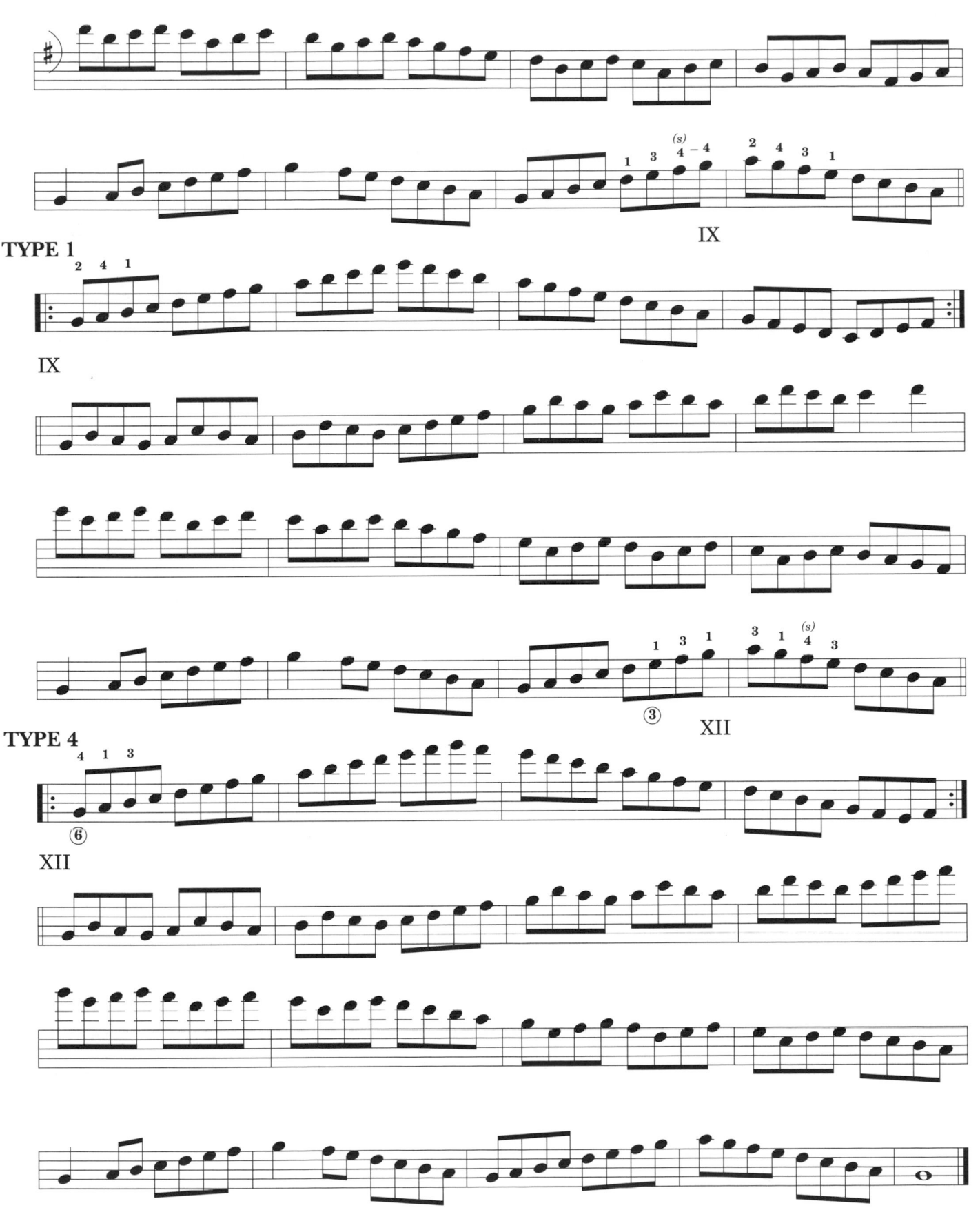

TYPE 1

IX

TYPE 4

XII

G Major—Descending (Five Positions)

FINGERING TYPE 4

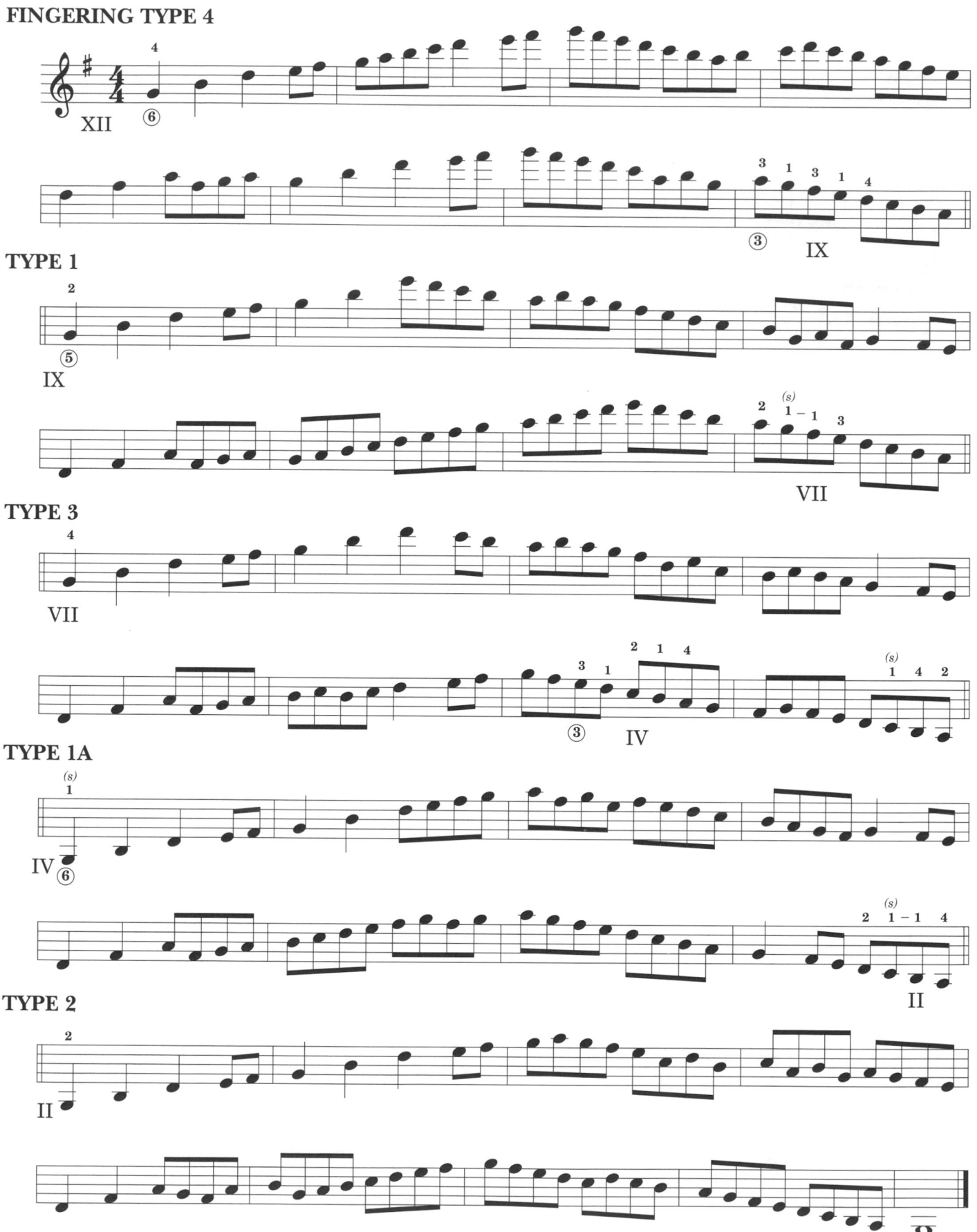

TYPE 1

TYPE 3

TYPE 1A

TYPE 2

Sea-See-Si (duet)

Note durations are relative to tempo. Sixteenth notes will not sound fast if the tempo is slow.

Chord Forms

A.R. = Assumed Root

Most of the chord form pages from here on are highly concentrated. I recommend that you practice one line at a time while going on with the note studies on the following pages. Keep coming back periodically until all forms and sequences are mastered.

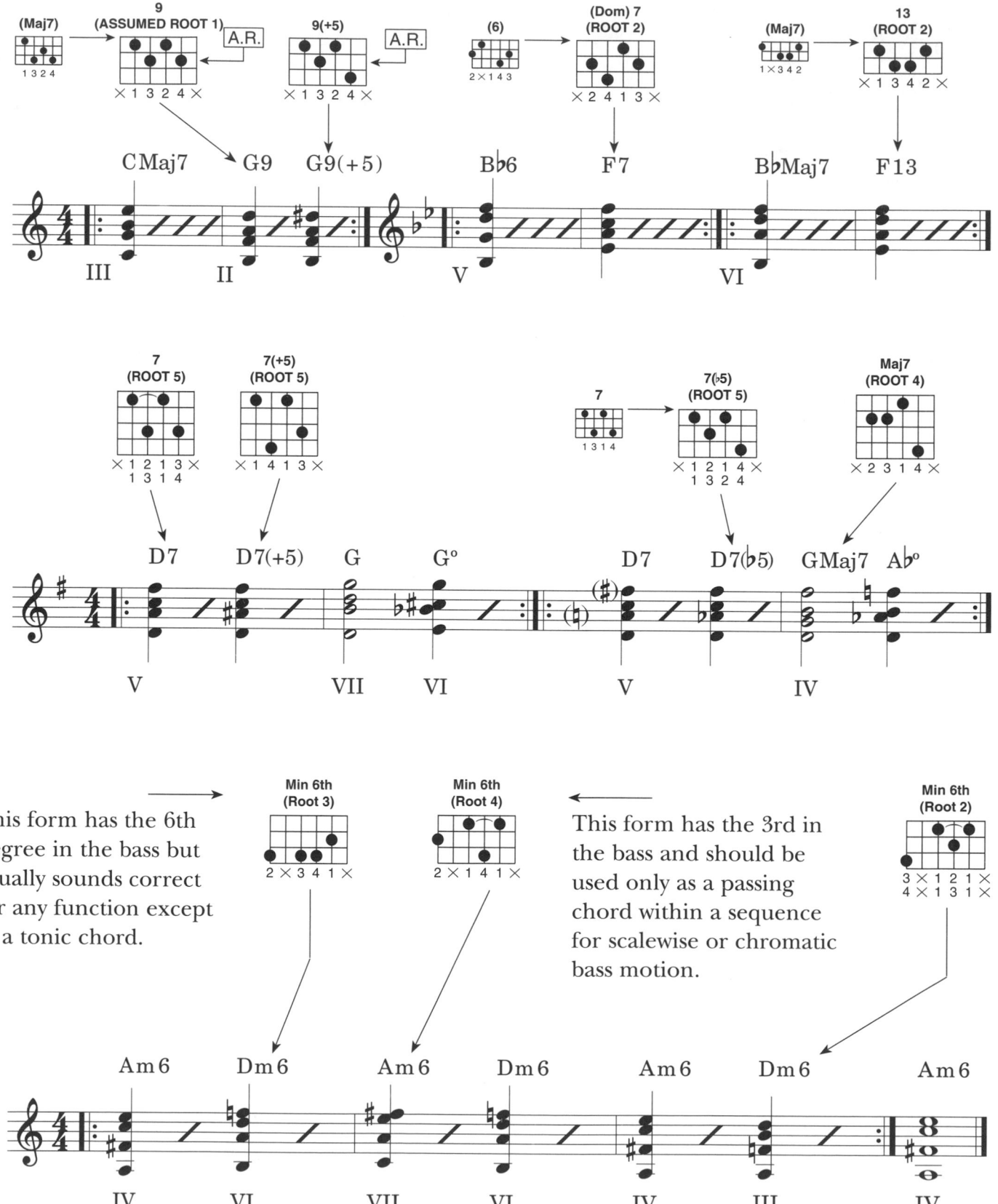

This form has the 6th degree in the bass but usually sounds correct for any function except as a tonic chord.

This form has the 3rd in the bass and should be used only as a passing chord within a sequence for scalewise or chromatic bass motion.

22

Triads

Scale (Chord) Degrees

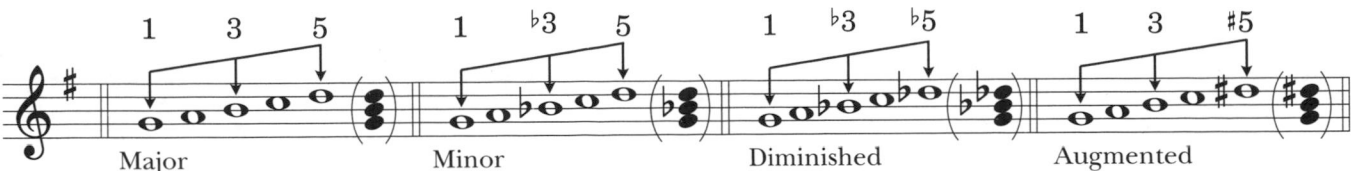

Major Minor Diminished Augmented

Across the fingerboard.

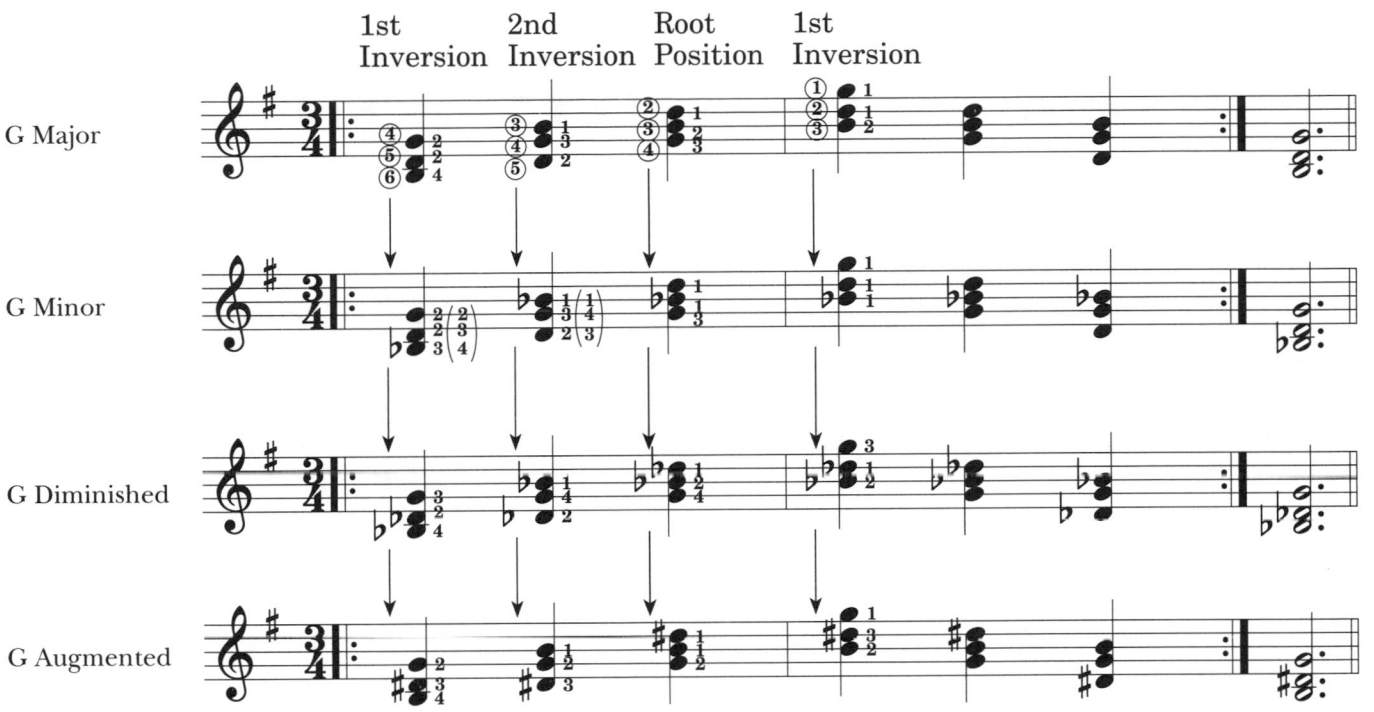

Finger Stretching Exercises

D Major—Ascending (Five Positions)

FINGERING TYPE 3

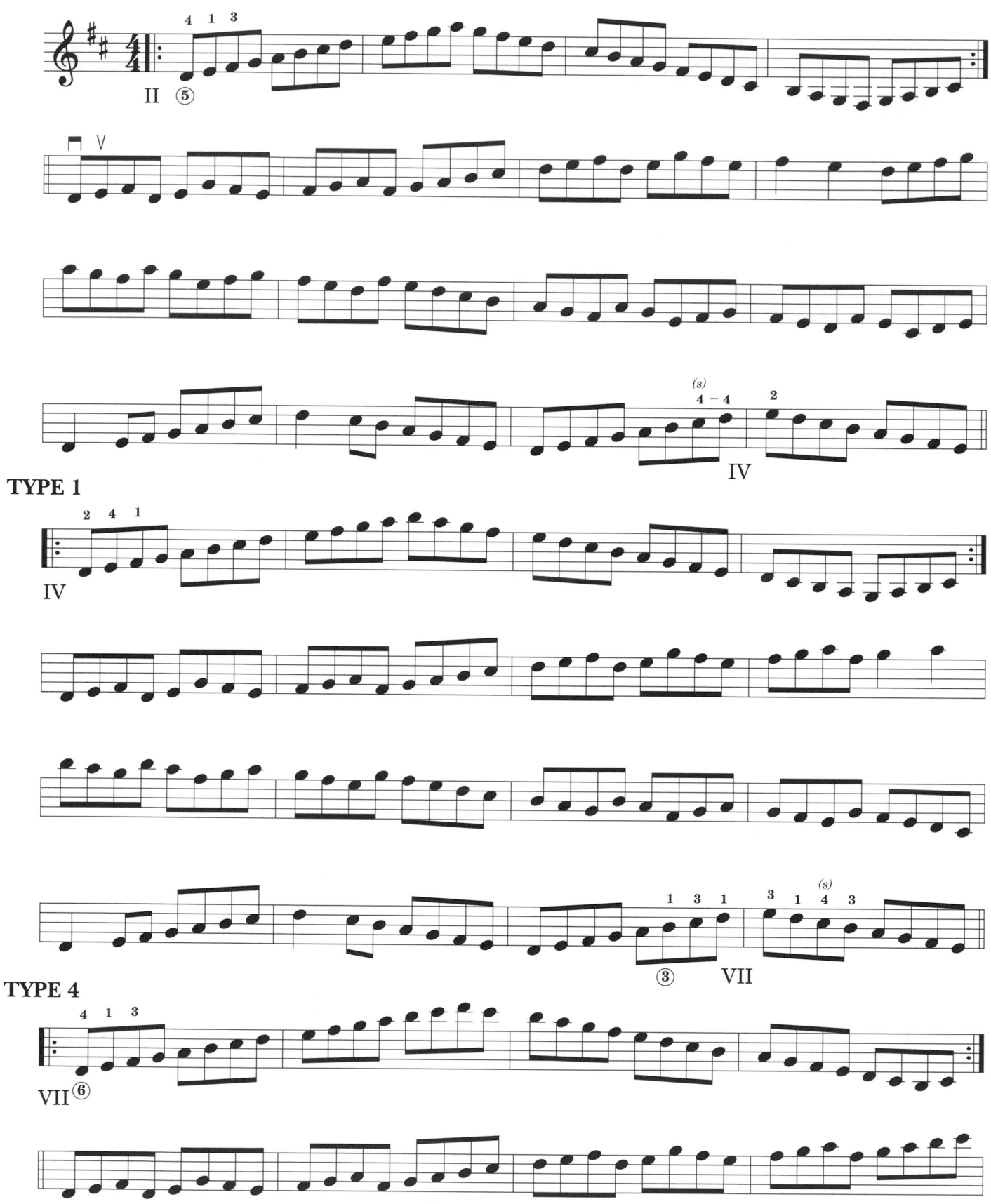

TYPE 1

TYPE 4

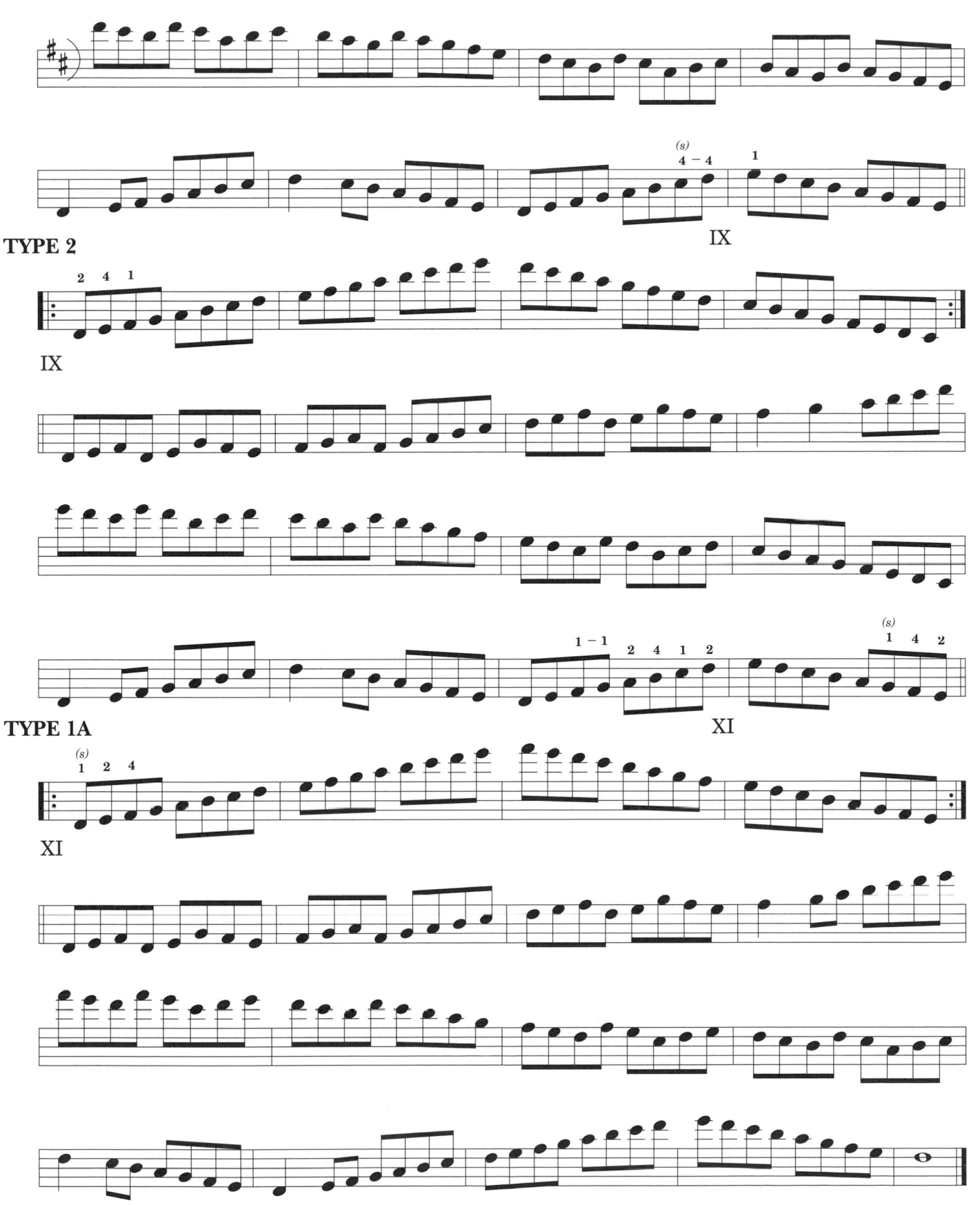

TYPE 2

TYPE 1A

D Major—Descending (Five Positions)

FINGERING TYPE 1A

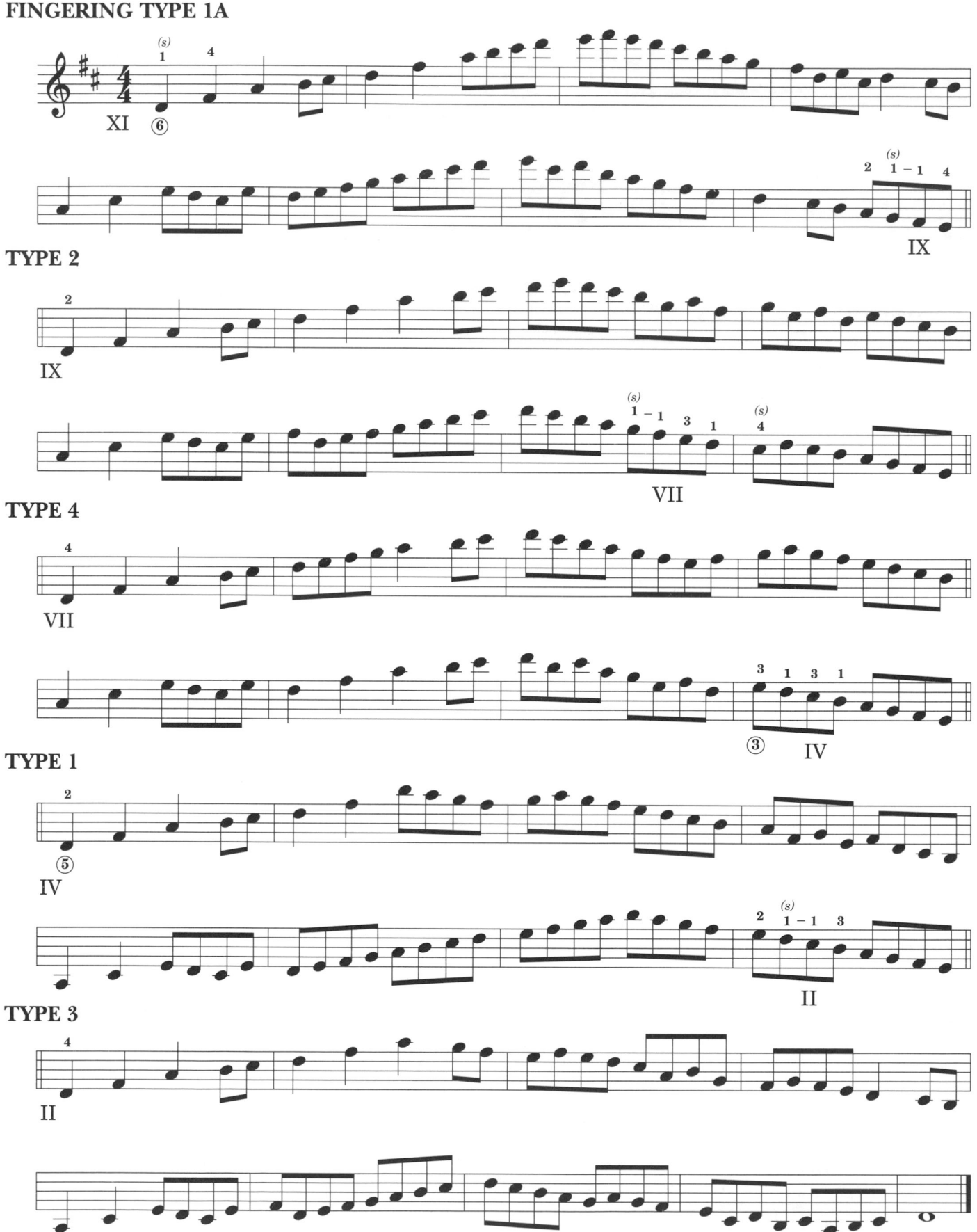

TYPE 2

TYPE 4

TYPE 1

TYPE 3

Melodic Rhythm Study No. 3 (duet)

Intervals

Interval: The number of whole and half steps between two notes.

Simple Intervals

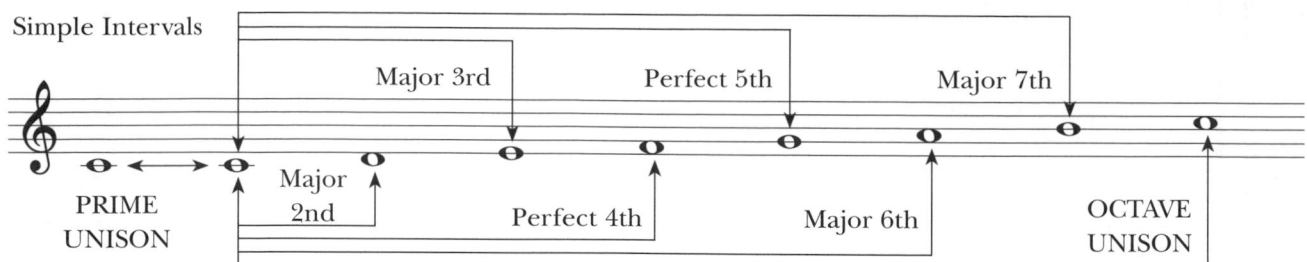

The above numbers represent the scale and chord degrees, as well as the interval from the tonic.

- If the top note is a member of the major scale of the bottom note, the interval is called: major 2nd, major 3rd, major 6th, major 7th, or perfect 4th, perfect 5th, perfect octave.

- Intervals one half step smaller than major are called *minor*. Intervals one half step smaller than *perfect* or a whole step smaller than major are called *diminished*. Any major or perfect interval expanded by one half step is called *augmented*.

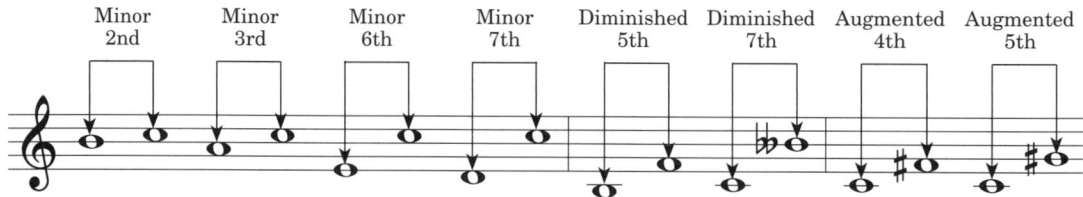

When only the numerical term (3rd, 4th, etc.) is used, major and perfect intervals are intended. Minor, diminished, and augmented intervals must be specifically named.

Compound intervals (larger than one octave) are described by the same terms as simple intervals (one octave or less) from which they are derived. (Example: Major and minor 2nd plus an octave = major and minor 9th.)

Compound Intervals

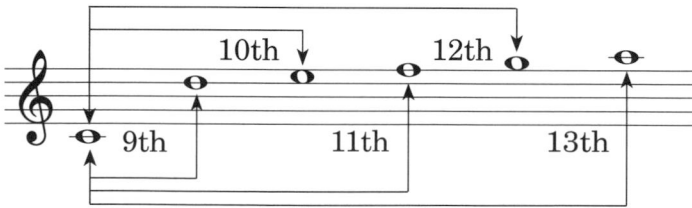

Triads

Scale (Chord) Degrees

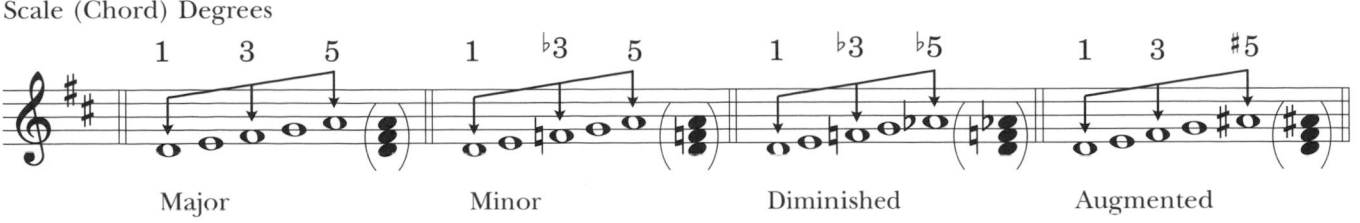

| Major | Minor | Diminished | Augmented |

As you move across and up the fingerboard, carefully observe fingerings and strings.

D Major

Root 1st 2nd Root
Pos. Inv. Inv. Pos.

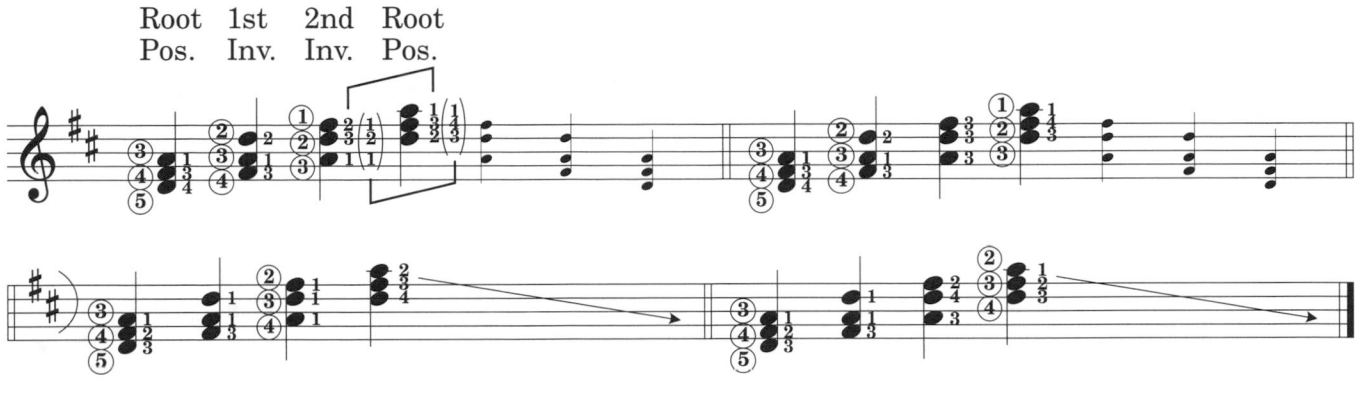

D Minor

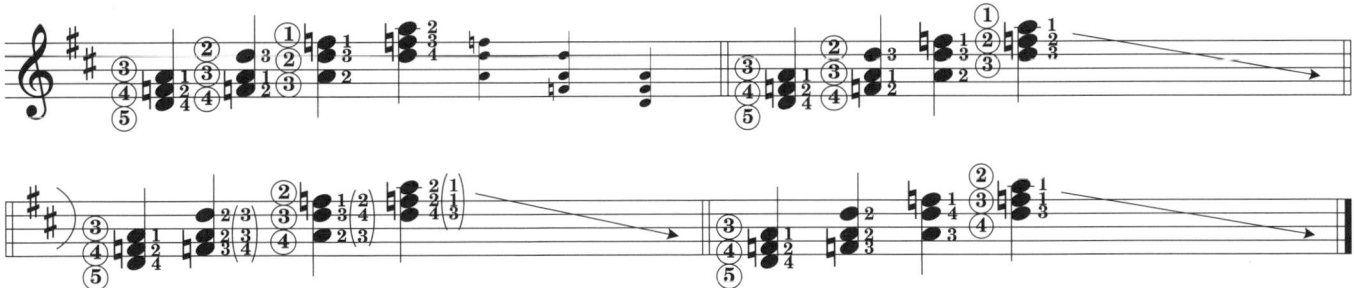

D Diminished

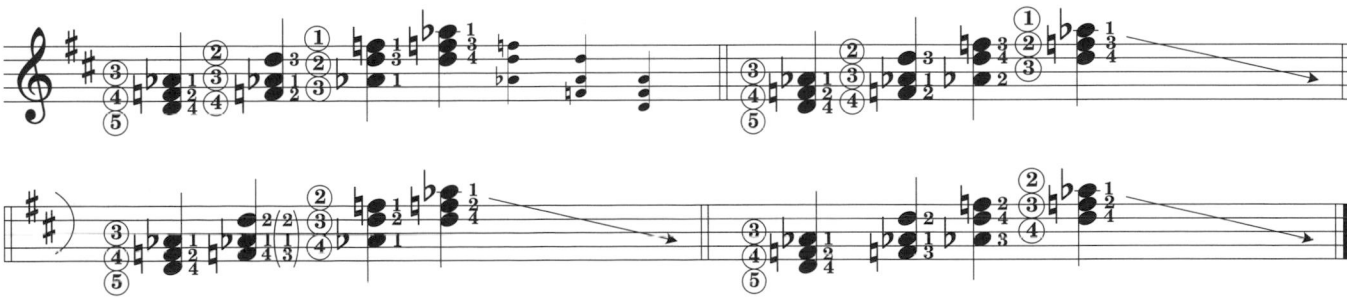

D Augmented

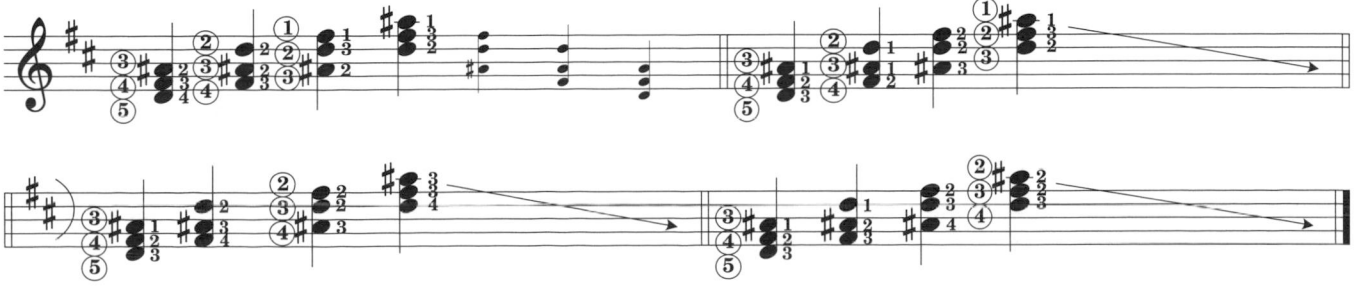

A Major—Ascending (Five Positions)

FINGERING TYPE 4

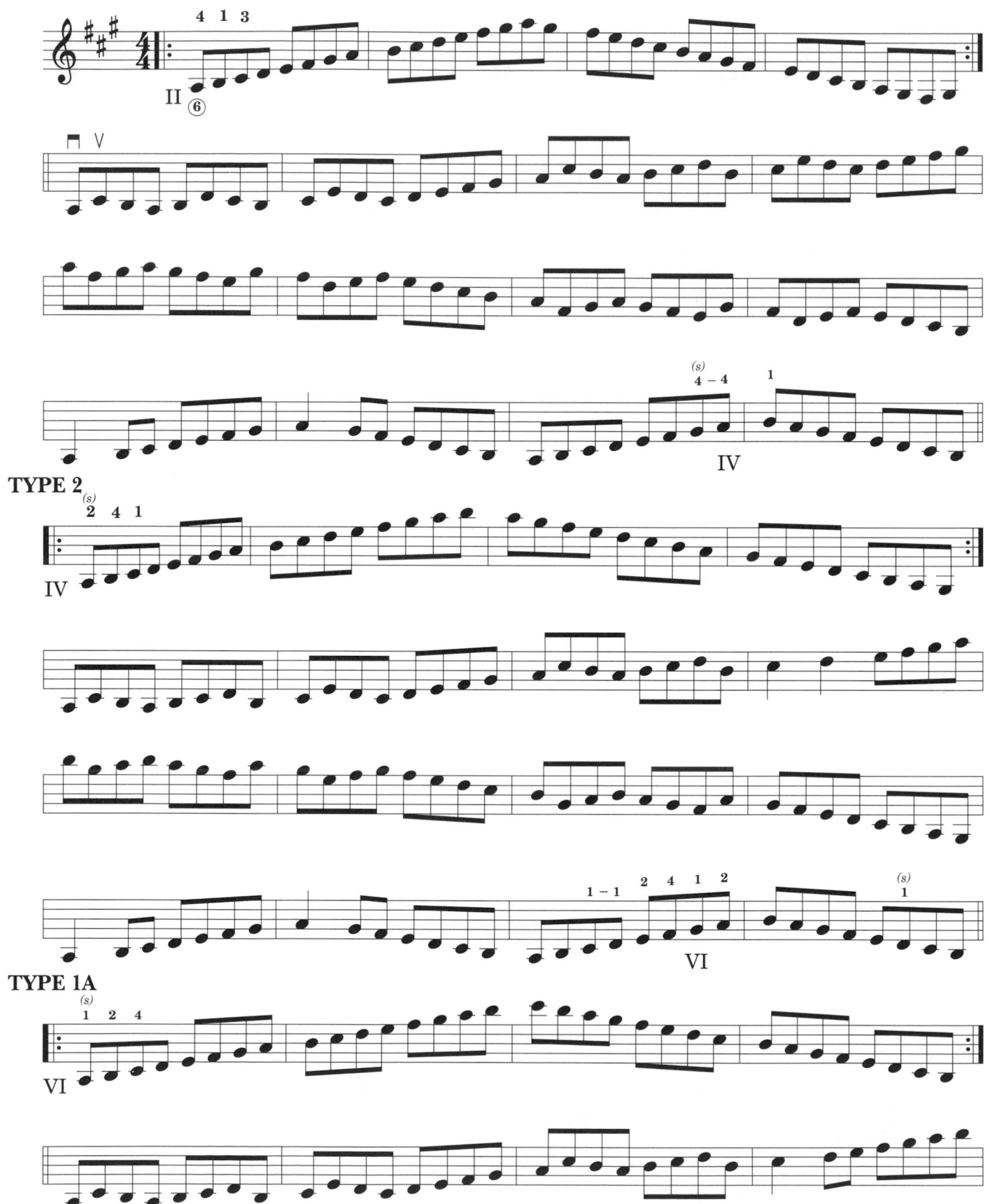

TYPE 2

TYPE 1A

TYPE 3

TYPE 1

A Major—Descending (Five Positions)

FINGERING TYPE 1

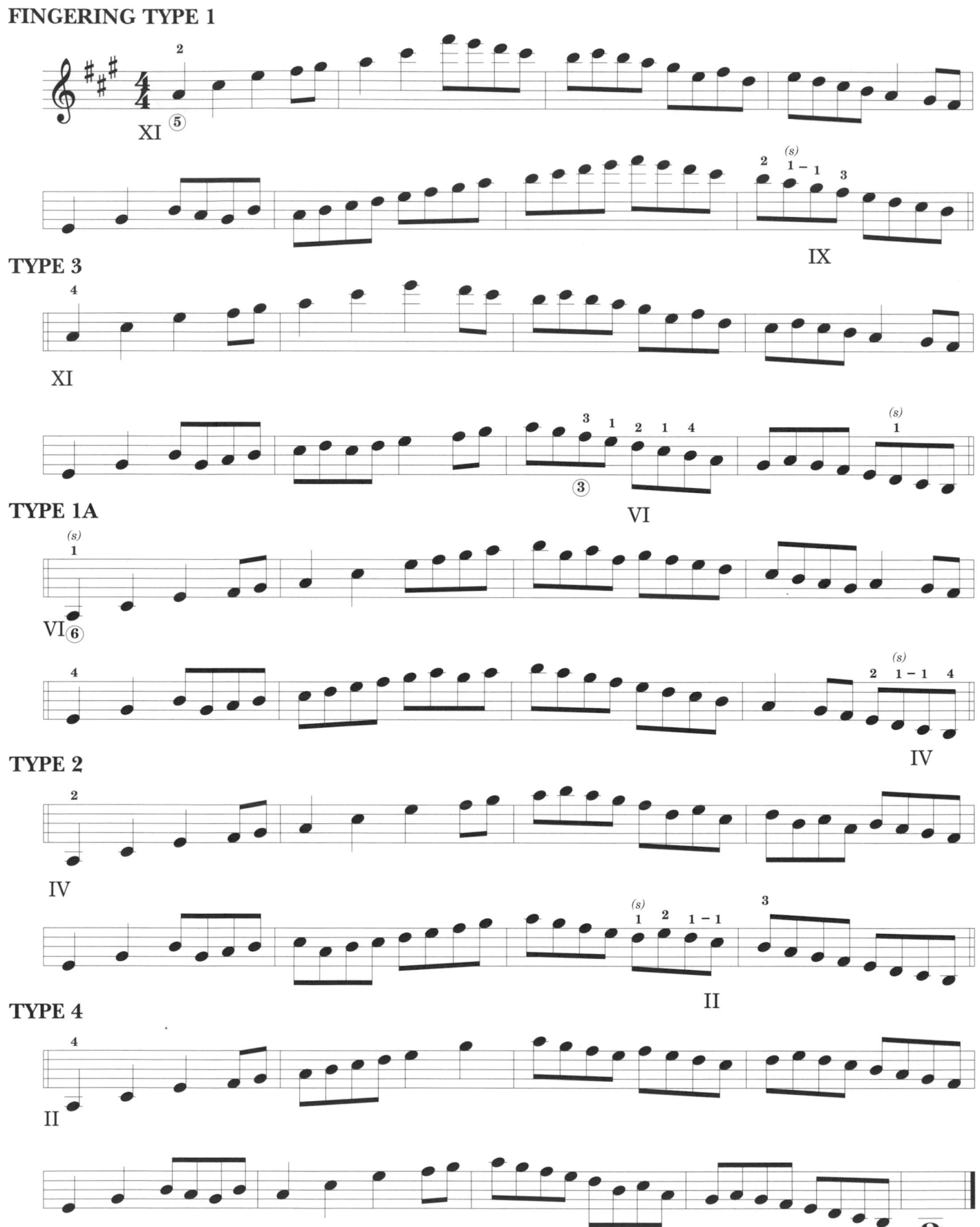

TYPE 3

TYPE 1A

TYPE 2

TYPE 4

Chord Etude No. 8

Rhythm Guitar—Right Hand (Rock-Style Ballad)

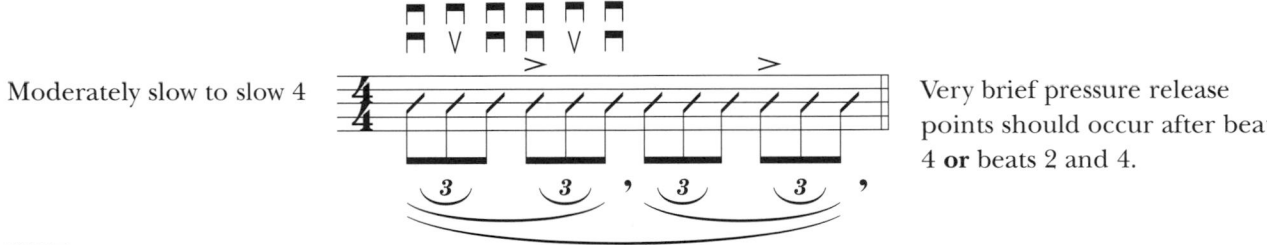

Moderately slow to slow 4

Very brief pressure release points should occur after beat 4 **or** beats 2 and 4.

■ **EXERCISE**

Observe notation.

simile (continue in similar manner)

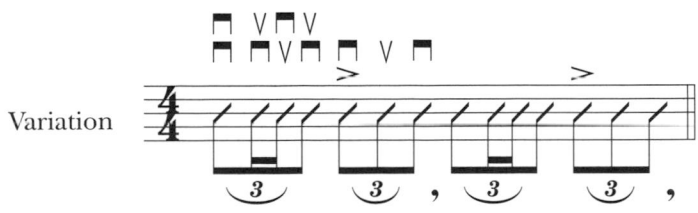

Variation

■ **EXERCISE**

Observe notation.

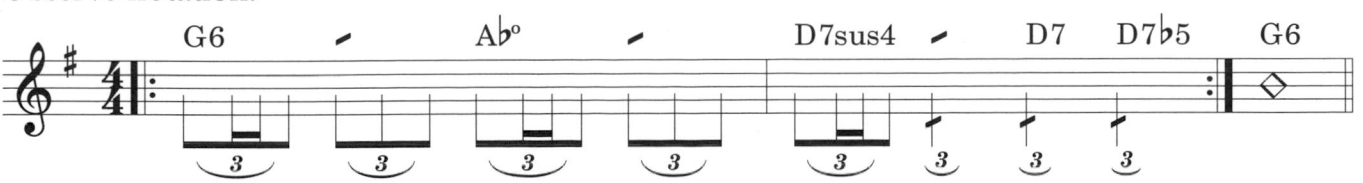

These strokes are used with regular acoustic and amplified (high-register) rhythm playing. Observe notation.

Chord Forms

A.R.: Assumed Root

The low register limit for all dominant 7 forms on this page is E♭ (possibly D.)

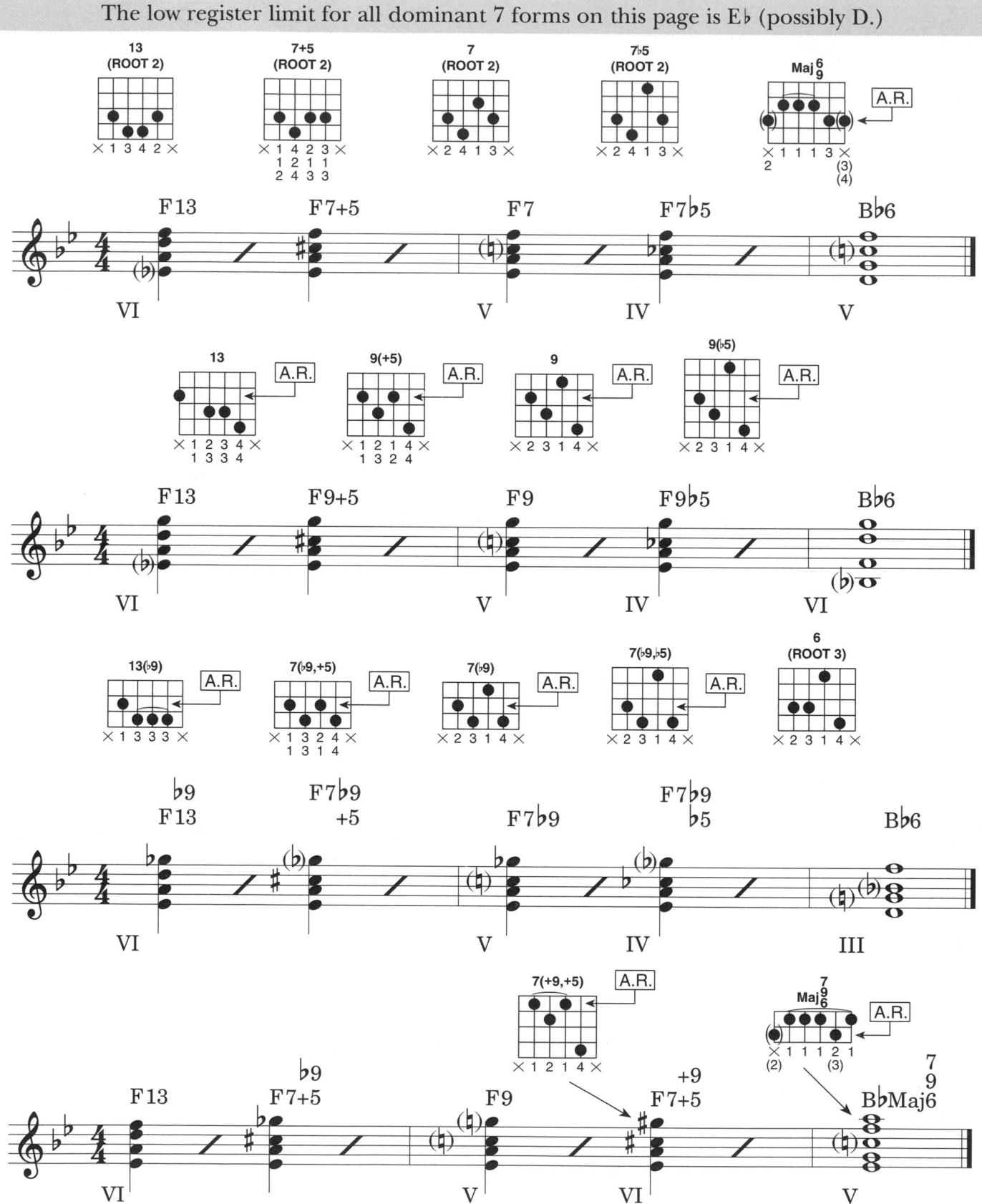

Do not be concerned with the theoretical explanation of the more complex chords. They will be covered in a later section. Most important for now is the physical ability to perform them and eventually memorize all forms, chord types, and root locations.

These are the same forms as those shown on the opposite page. The roots are different and the sequence is reversed. Considerable time will be required to really learn them.

The low register limit for all dominant 7 forms on this page is A♭ (possibly G). Also, all ♭5s on this and the preceding page may be considered Augmented 11 (+11).

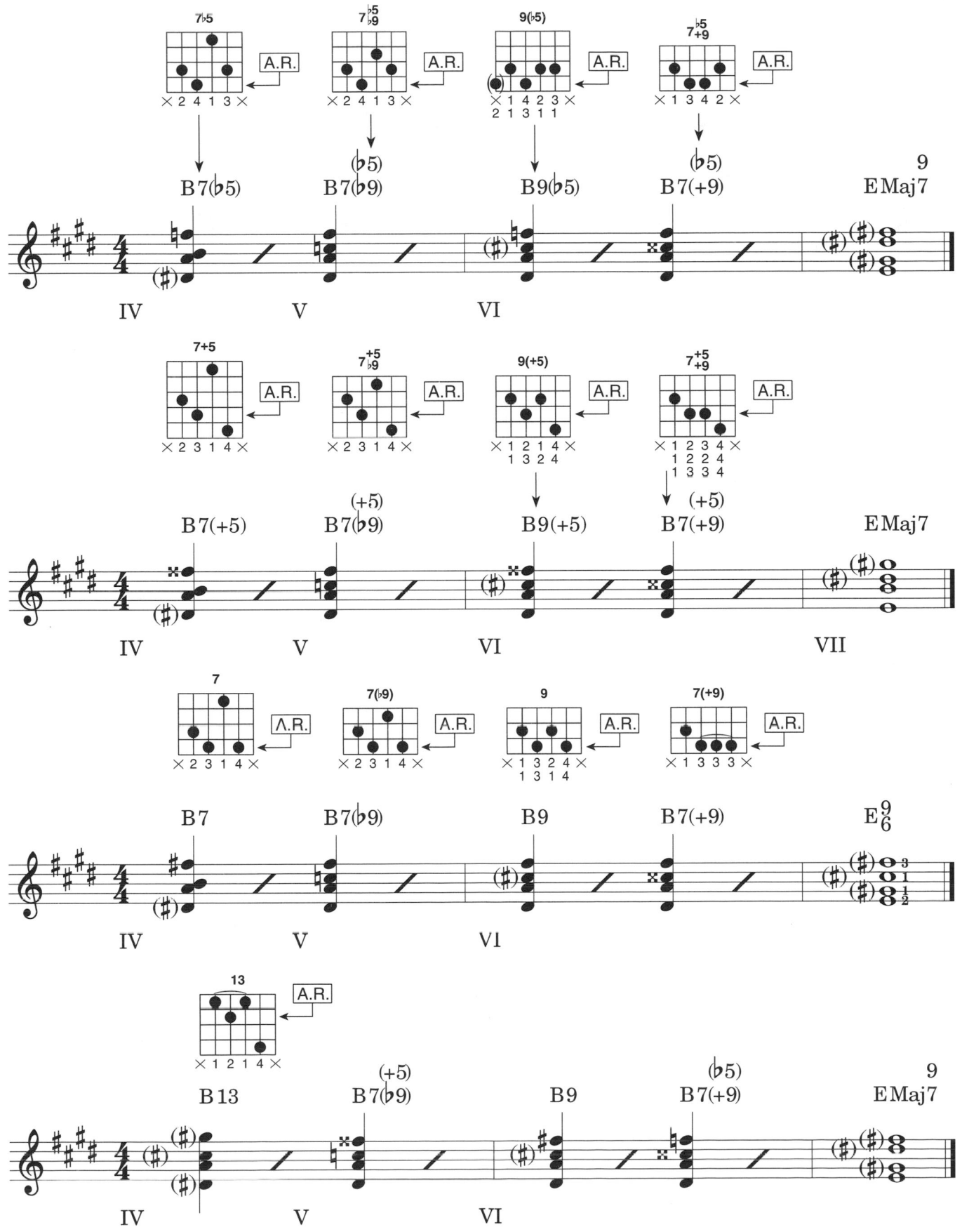

Tranquility (duet)

Sustain all notes full value.

Triads

Scale (Chord) Degrees

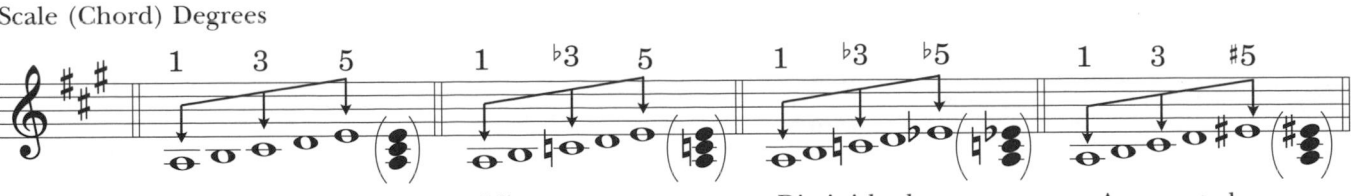

As you move across and up the fingerboard, carefully observe fingerings and strings.

A Major

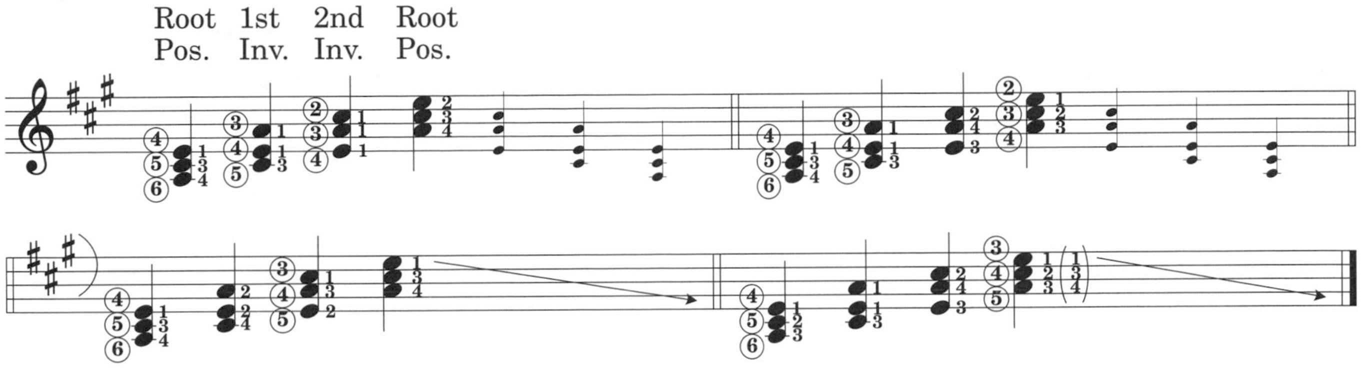

A Minor

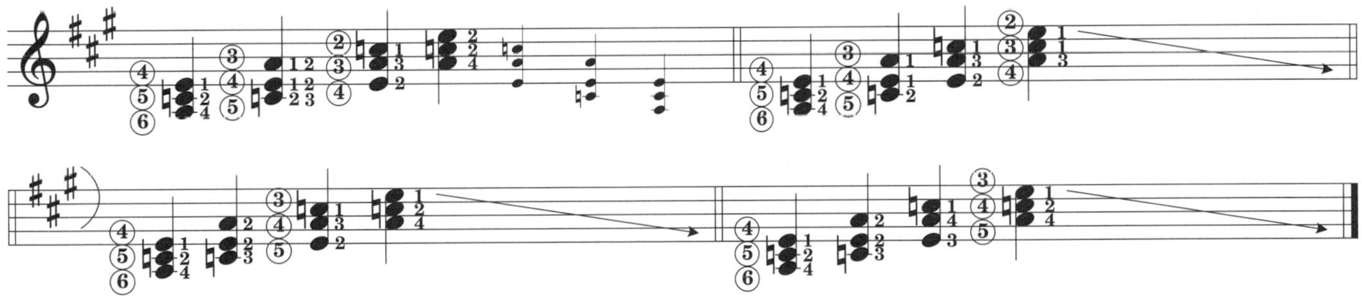

A Diminished

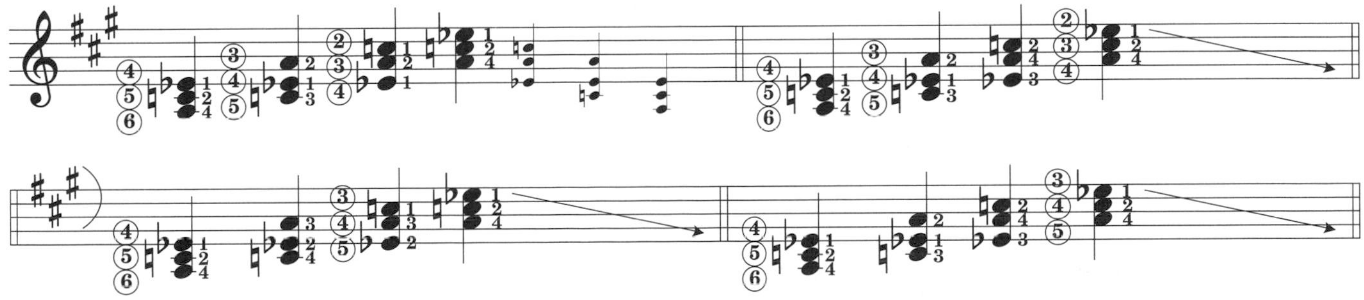

A Augmented

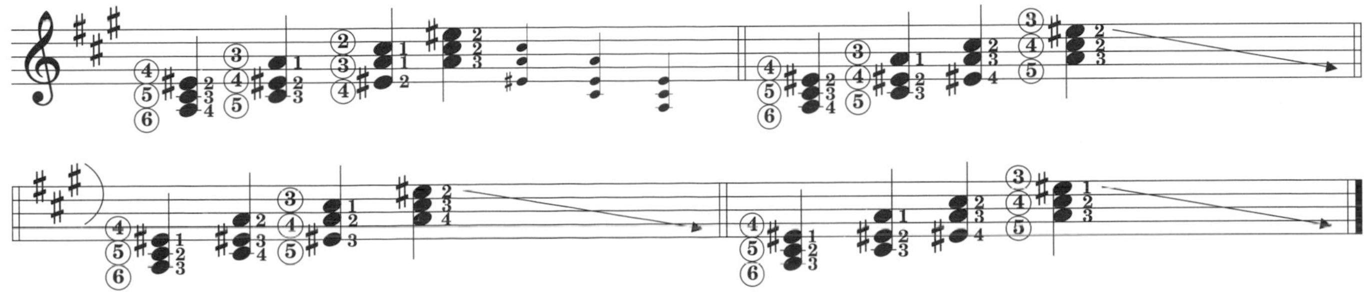

B♭ Major—Ascending (Five Positions)

FINGERING TYPE 4

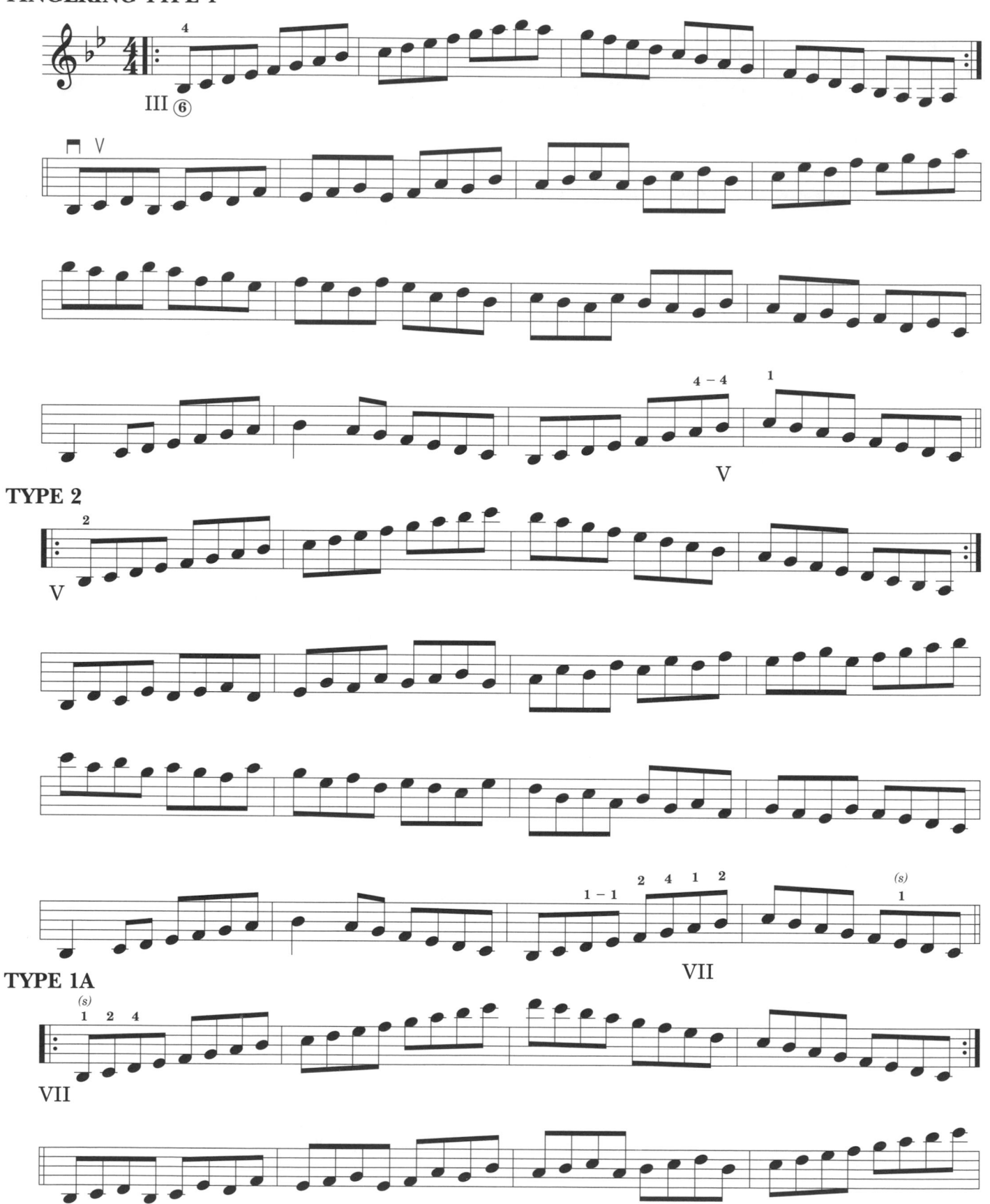

TYPE 2

TYPE 1A

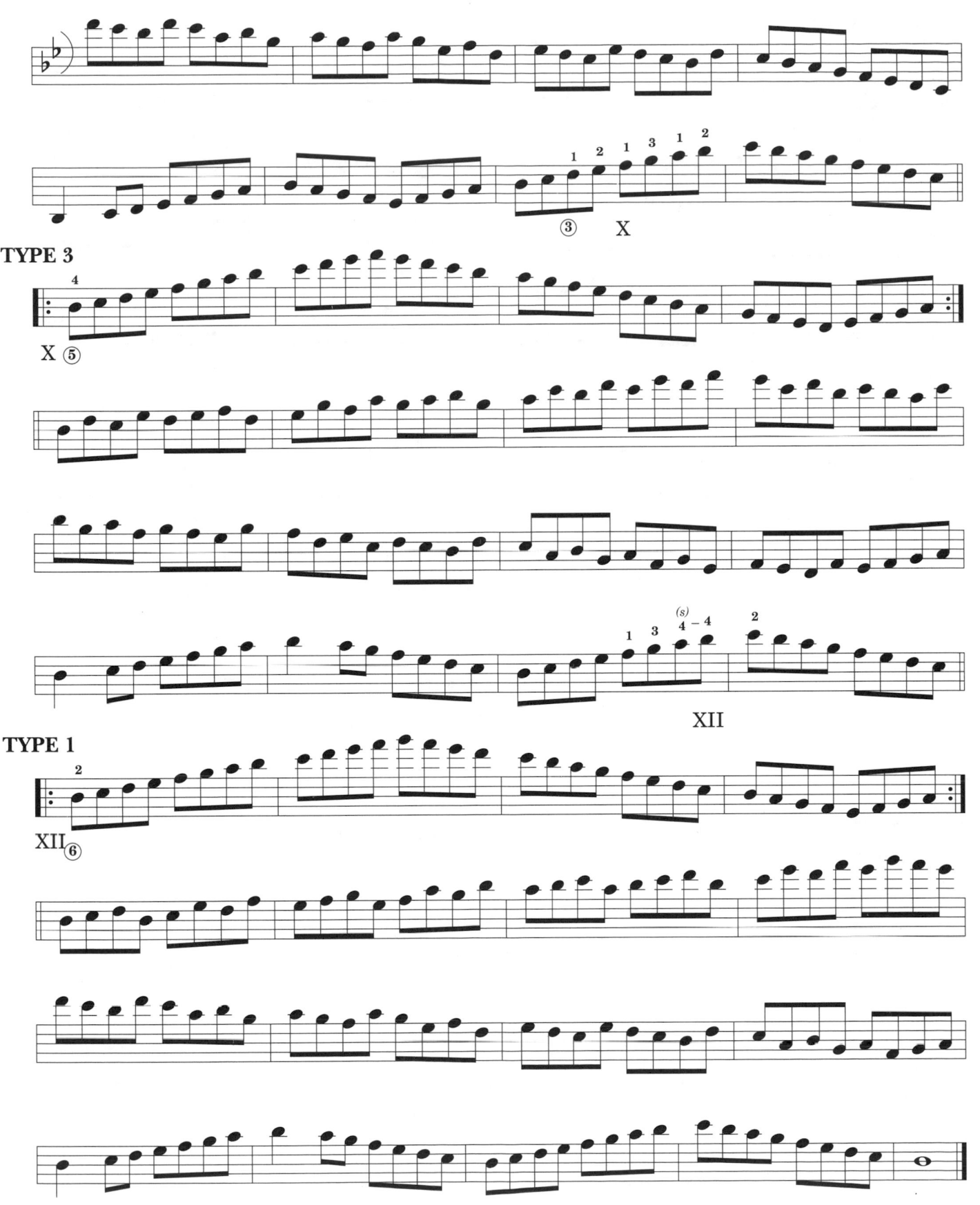

TYPE 3

TYPE 1

B♭ Major—Descending (Five Positions)

FINGERING TYPE 1

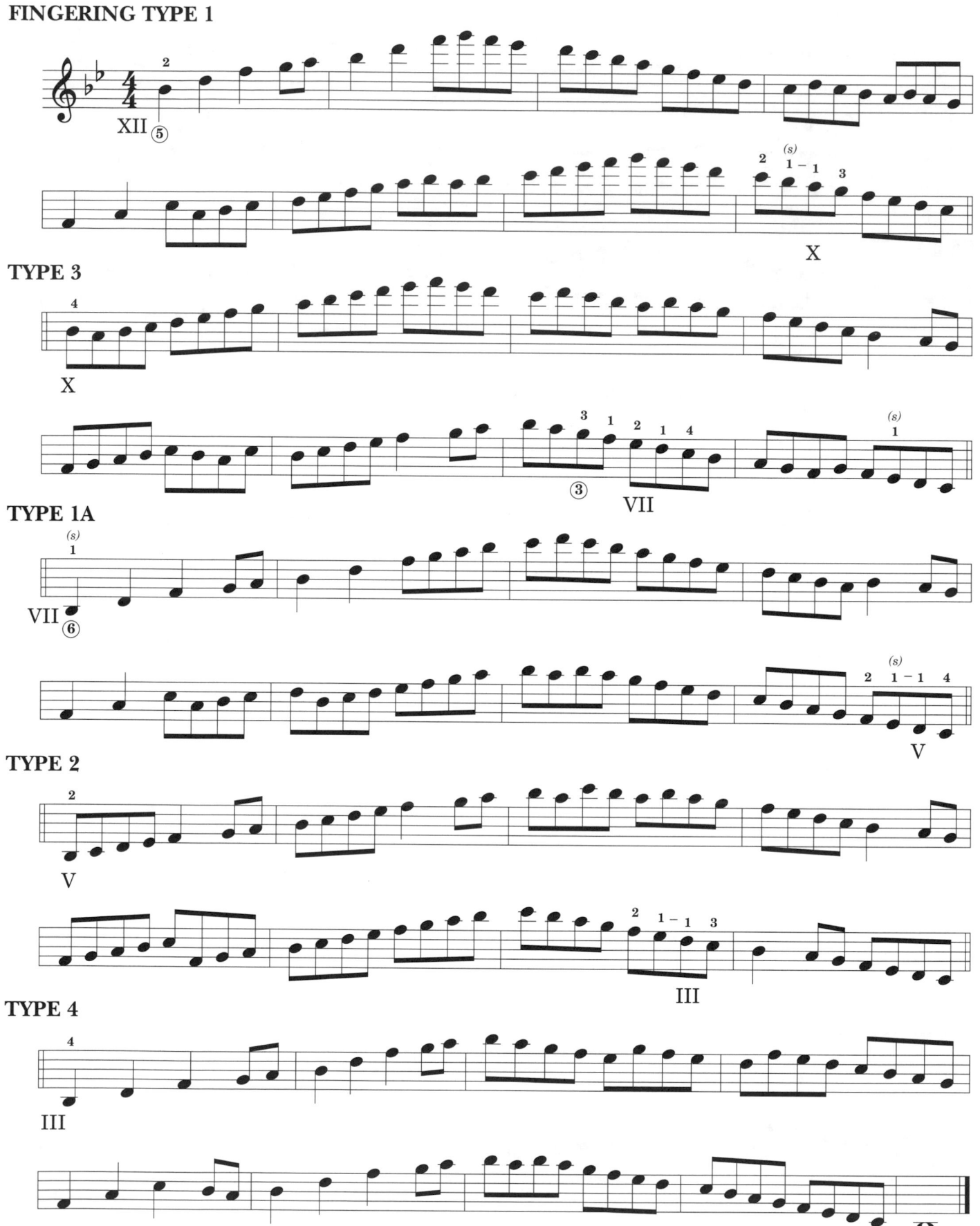

Waltz in B♭ (duet)

Moderate Waltz Tempo

41

Melodic Rhythm Study No. 4

This is a notation comparison, not a duet.

A fast waltz is often best counted "in 1." Beats 2 and 3 are merely felt. 6/8 is usually counted "in 2," each measure being divided in half (like two fast waltz measures). However, a slower 6/8 is counted 1-2-3-4-5-6. Each eighth note gets one full beat.

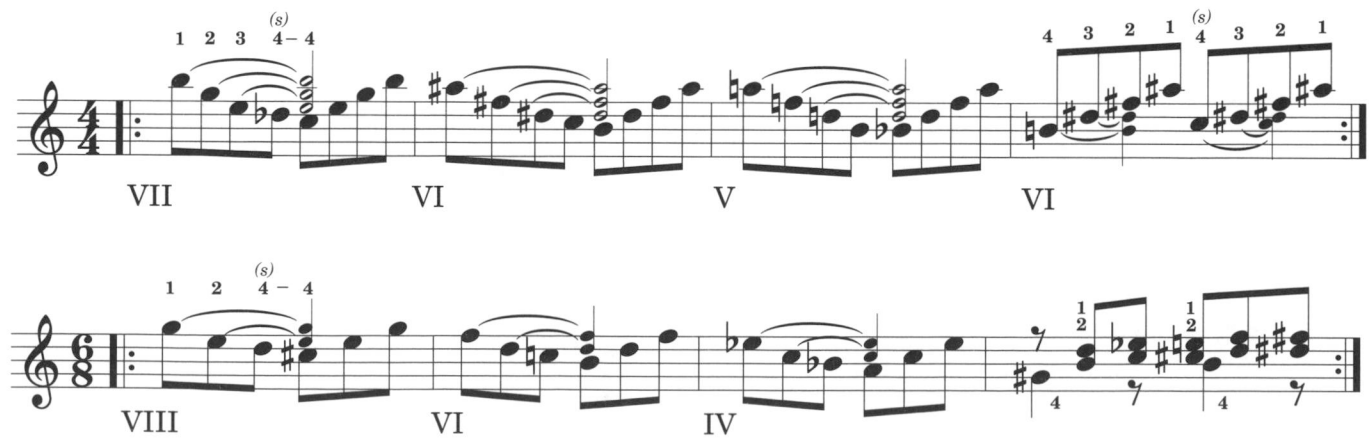

Note durations are relative to tempo and time signatures.

Finger Stretching Exercises

Triads

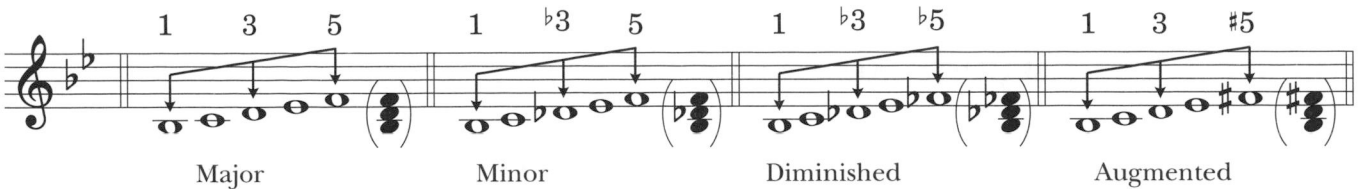

Major   Minor   Diminished   Augmented

As you move across and up the fingerboard, carefully observe fingerings and strings.

B♭ Major

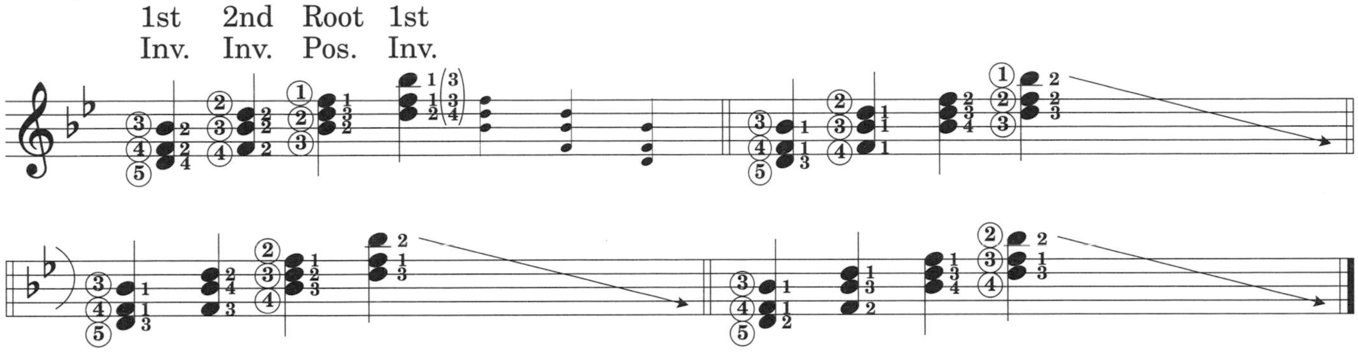

B♭ Minor

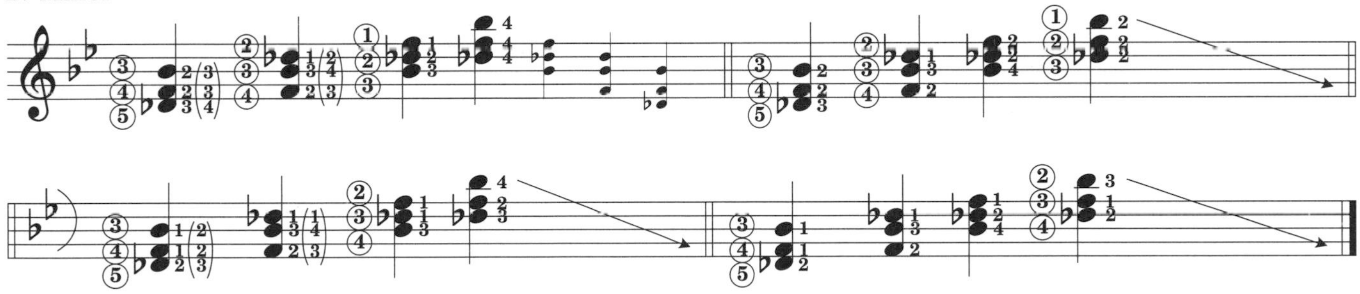

B♭ Diminished

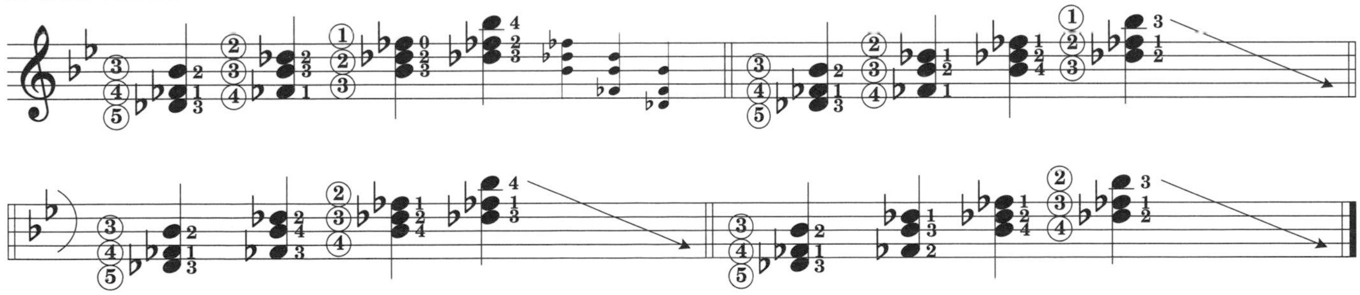

B♭ Augmented

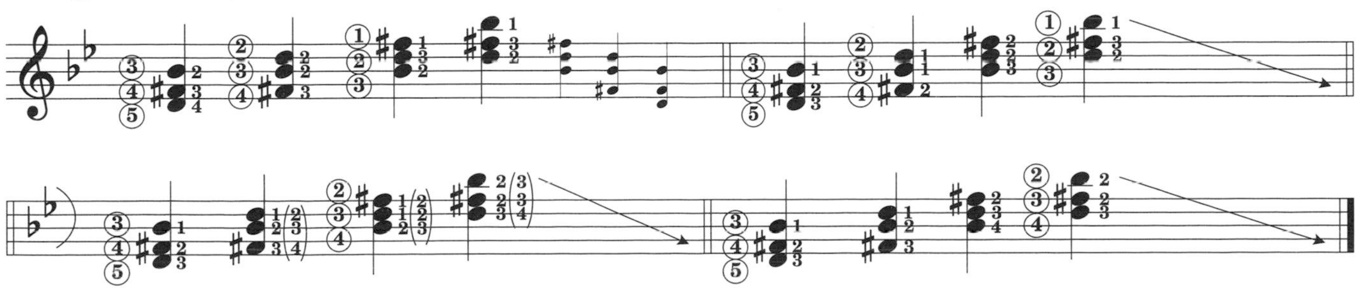

Pentatonic (Five-Note) Scales

A good preparation for arpeggio studies.

MAJOR (1, 2, 3, 5, 6 OF MAJOR SCALE)

Tremolo Study

Tremolo: Quick repetition of the same note.

At first, practice very slowly and evenly. Later, gradually increase the tempo, but keep it steady throughout. Practice all "Loco" (in the same octave as written) and also 8va (one octave higher than written).

Observe picking.

■ **EXERCISE 1**

Abbreviated triplet notation

■ **EXERCISE 2**

Abbreviated eighth note notation

Abbreviated 16th note notation

Abbreviated 32nd note notation

E♭ Major—Ascending (Five Positions)

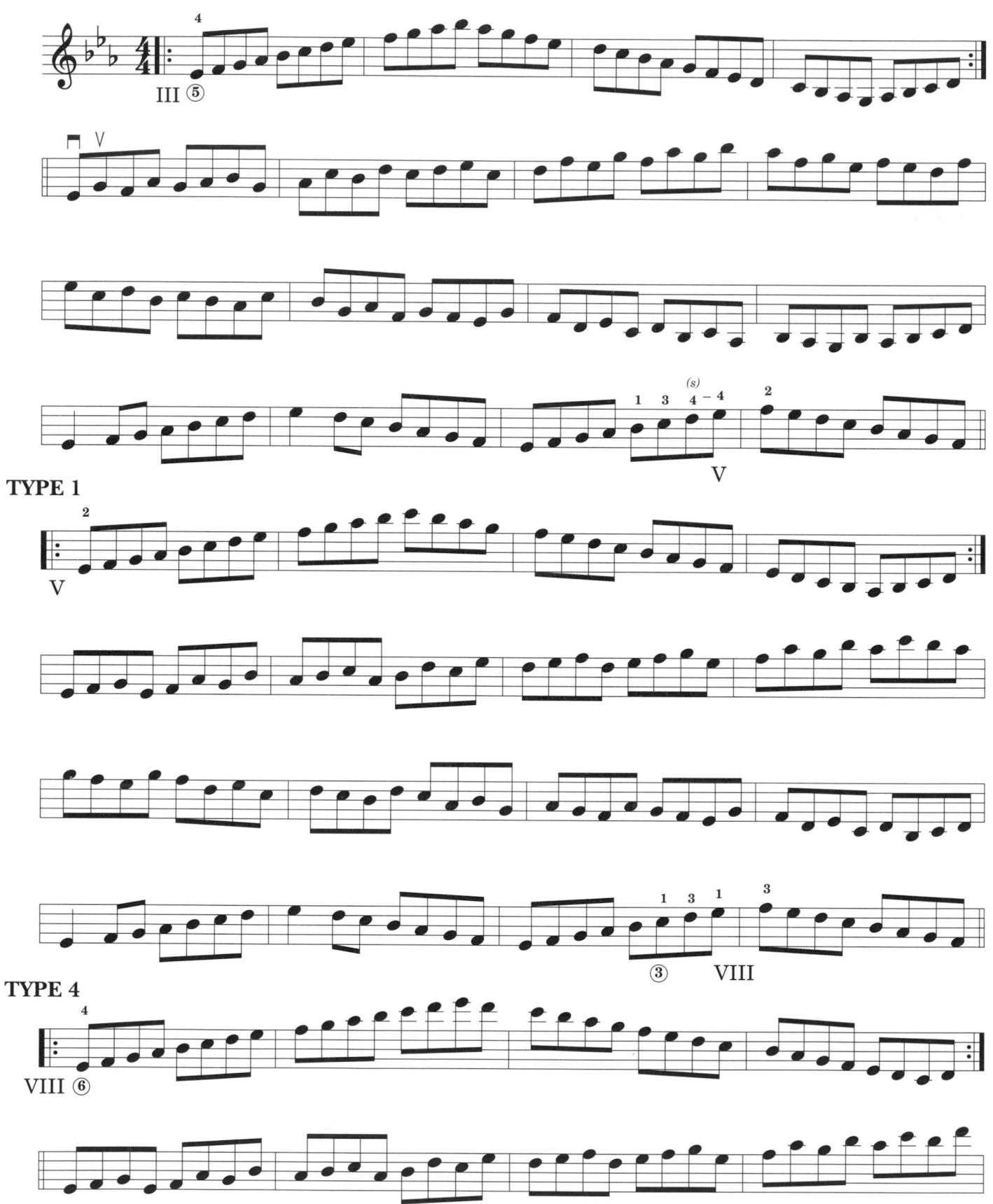

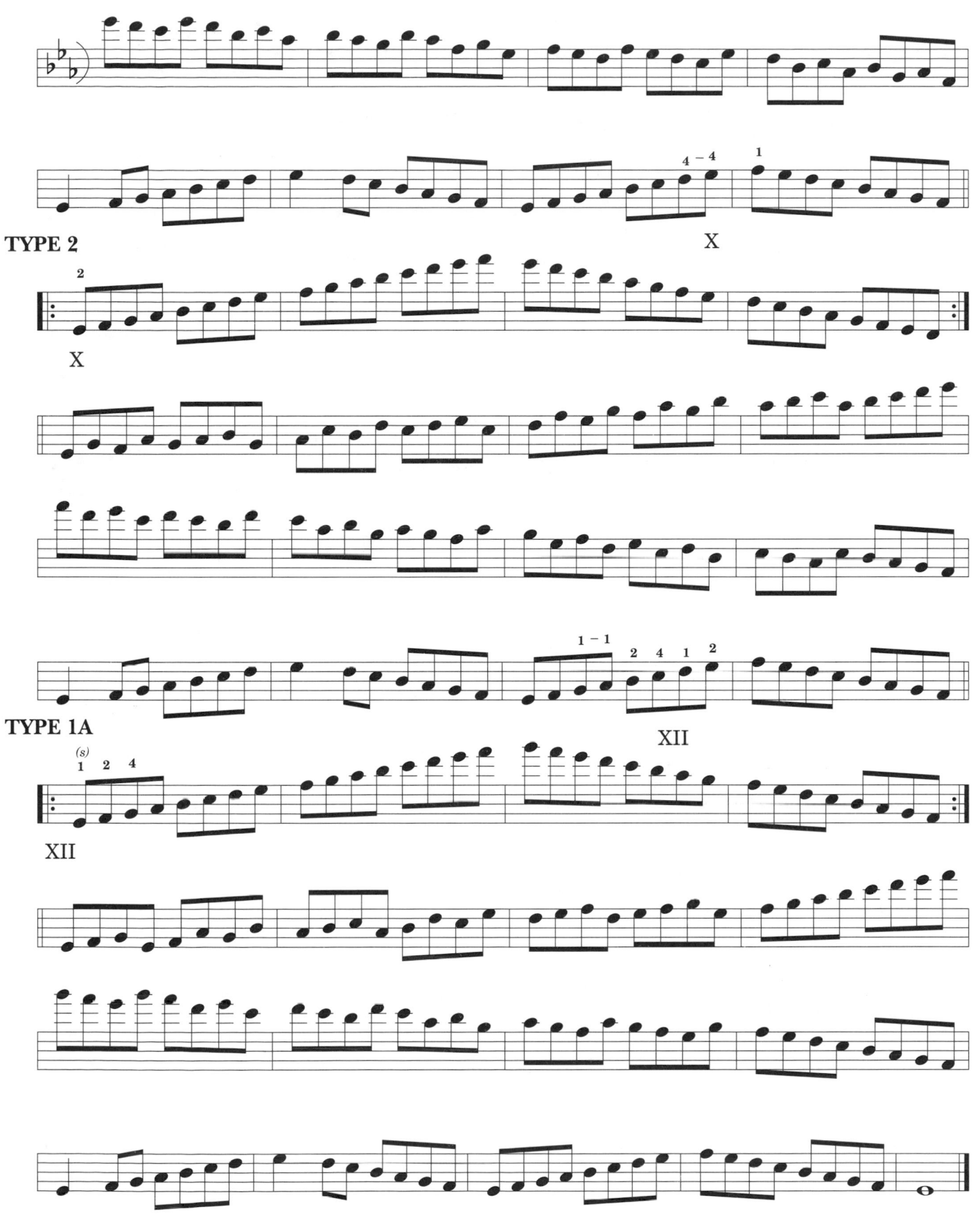

E♭ Major—Descending (Five Positions)

FINGERING TYPE 1A

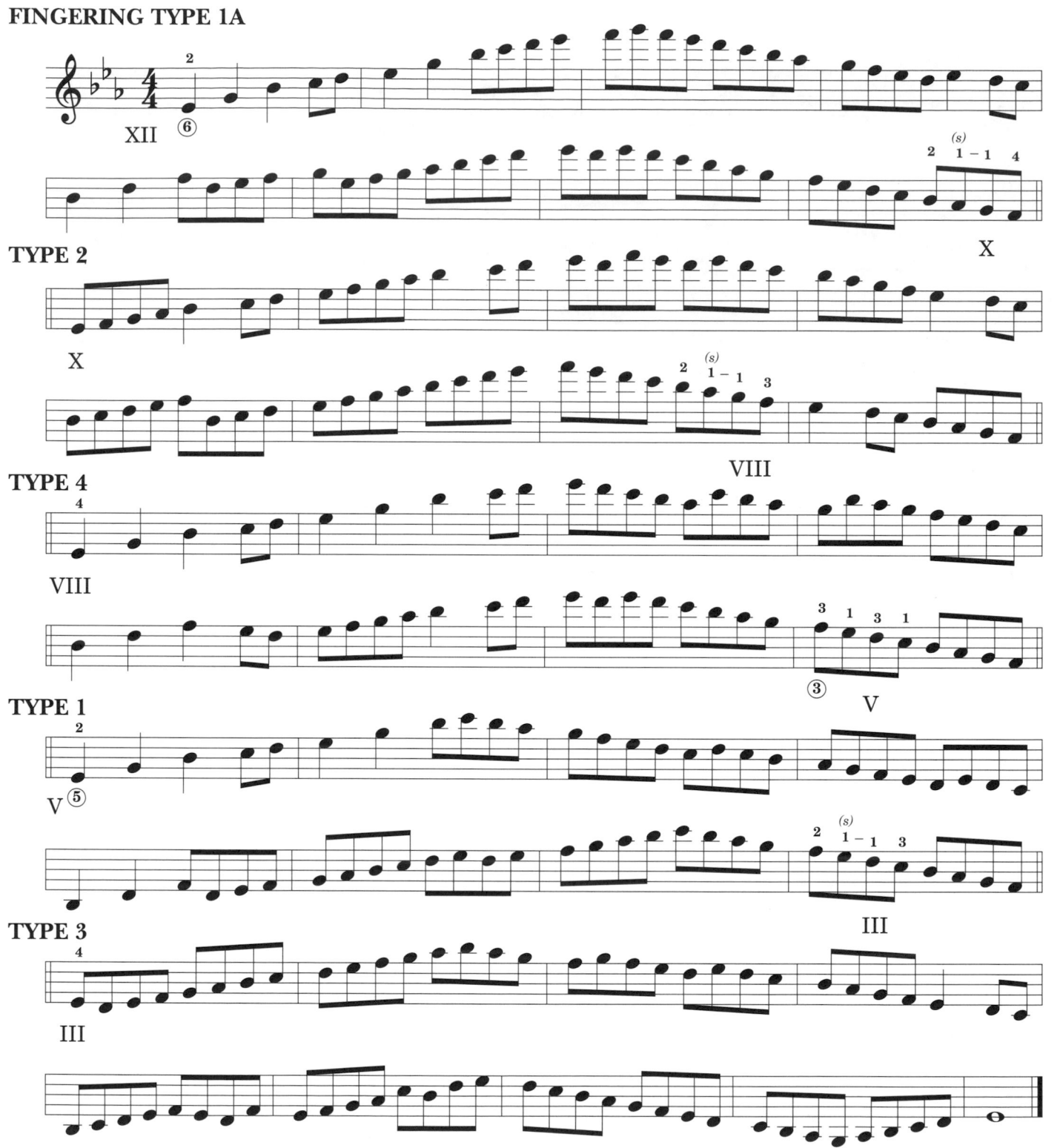

By transposing the preceding five-position major scale studies up or down one half step (one fret or one position), all major scales are now possible.

EXAMPLES: D major position II to D♭ major position I; E♭ major position III to E major position IV.

These same seven (five-position) studies can be used for practice if you merely change the key signatures and position marks. As before, additional reading material must be used to learn these new keys.

Chord Forms

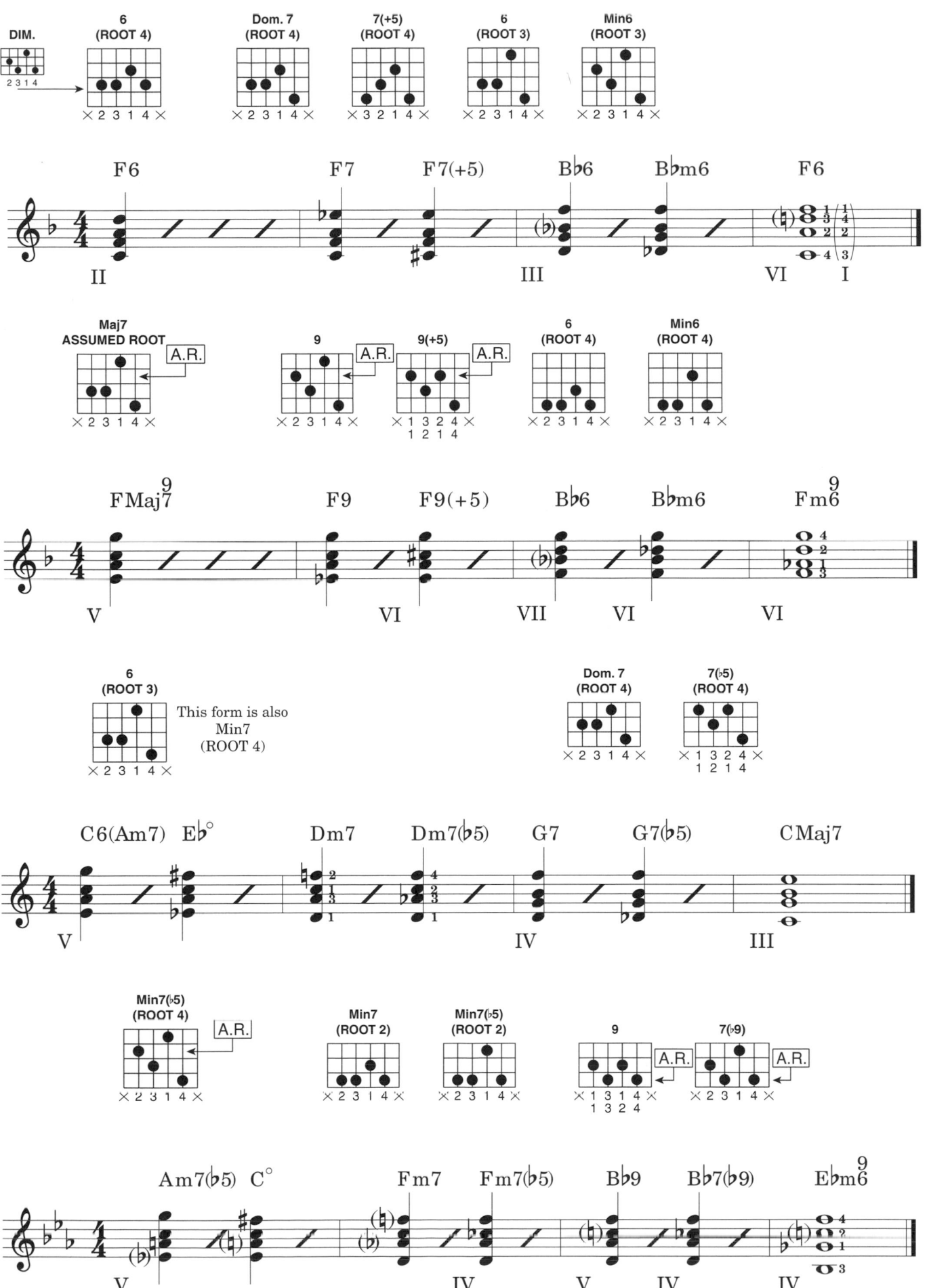

49

Major Scale Review—Positions II, III, V

✸ v.s. quickly

✸ v.s. means turn page.

The construction of a major scale (upwards) from any note is accomplished by using the following series of whole and half step intervals.
(1 = half step, 2 = whole step):

2 2 1 2 2 2 1
C D E F G A B C

2 2 1 2 2 2 1
F G A B♭ C D E F

2 2 1 2 2 2 1
G A B C D E F♯ G

Observe the half steps between the 3rd and 4th, 7th and 1st scale degrees in all major scales. These interval relationships account for the presence of flats or sharps in the various keys.

Triads

Scale (Chord) Degrees

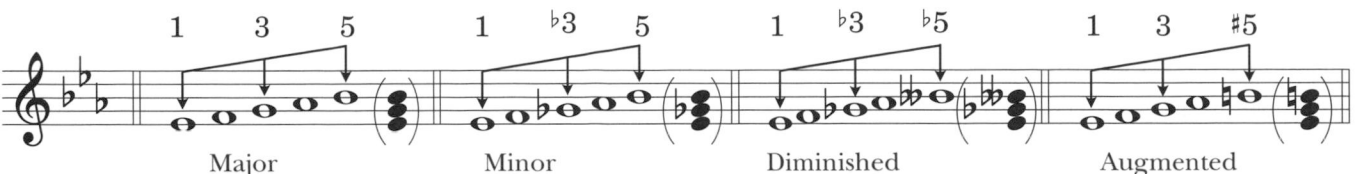

| Major | Minor | Diminished | Augmented |

As you move up and across the fingerboard, carefully observe fingerings and strings.

E♭ Major

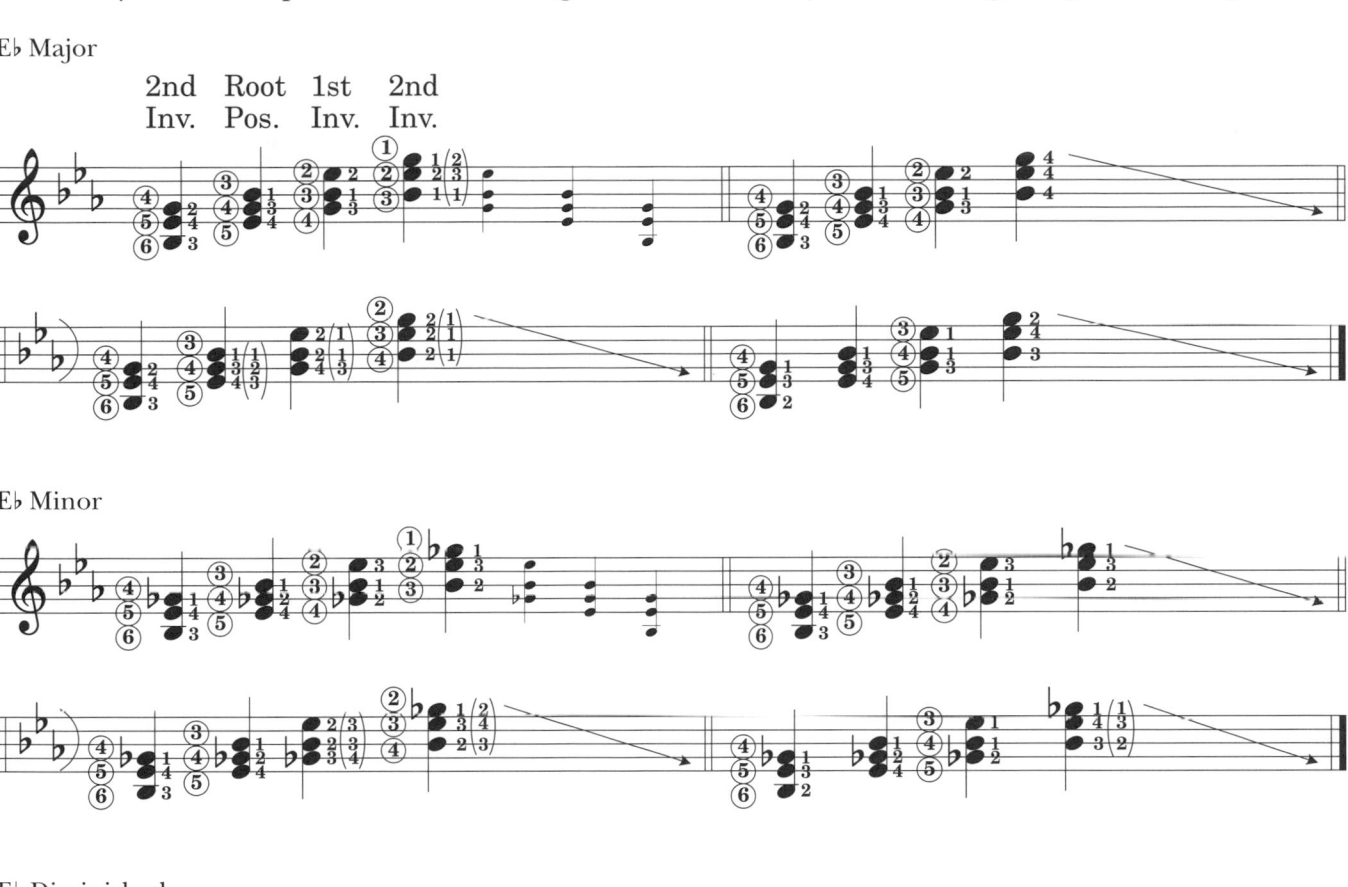

E♭ Minor

E♭ Diminished

E♭ Augmented

Theory: Diatonic Triads (Major Keys)

Diatonic: All notes belong to the key signature.

1.) There are seven notes in every major scale and seven chords common to each key. These diatonic chords are constructed upwards in thirds from each scale tone. The structures (major, minor, diminished, resulting from the scale) will be as follows in all major keys:

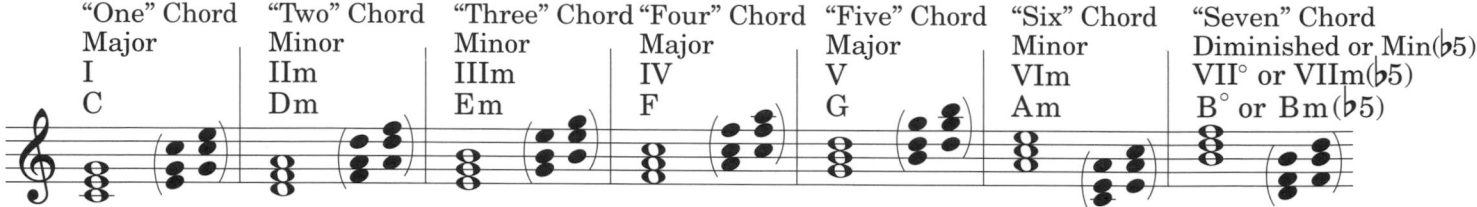

Roman numerals are used to represent these chord structures. (Be careful not to confuse them with position marks.) You must memorize the names and structures in all major keys.

2.) The principal chords and cadences (chord sequences) in major keys are:
   • I-V-I, called **Authentic Cadence**, or C-G-C in the key of C.
   • I-IV-I, called **Plagal Cadence**, or C-F-C in the key of C.
   • Combined I-IV-V-I, **Authentic Cadence**, or C-F-G-C in the key of C.

❋ The II minor chord sometimes replaces IV in the preceding combined authentic cadence: I-IIm-V-I, or C-Dm-G-C in the key of C.

3.) There are three basic chordal sounds in every major key that are represented by these diatonic chord structures, and the following specific terms are used to name them:
   • **Tonic:** I chord
   • **Subdominant:** IV chord
   • **Dominant:** V chord

There are also names for the chords built on all other scale degrees, but we will not discuss them here as they have no direct bearing on the three basic sounds, and they are usually referred to by number, i.e., the "two" (II) chord, the "three" (III) chord, the "six" (VI) chord, etc.

4.) The seven chords in a major key are related to each other with regard to the three basic chordal sounds. The I, IIIm, and VIm all produce a tonic sound. The IIm and IV chords produce a subdominant sound, and the V and VIIm(♭5) produce a dominant sound. These facts will be very important later on for chord substitutions and scale relationships in improvisation.

Memorize chord names and diatonic structures in all major keys.

Diatonic Triads
Key of G Major
Arpeggios and Scales

FINGERING TYPE 2

■ (Also play in position IV, FINGERING TYPE 1A.)

Key of F Major

FINGERING TYPE 3

When two consecutive notes are played by the same finger on adjacent strings, "roll" the fingertip from one string to the next. Do not lift the finger from the string.

Key of B♭ Major

FINGERING TYPE 4

Key of E♭ Major

FINGERING TYPE 1

5th Position Study (duet)

(Play ♪♪ as ♪♪)

■ EXERCISE (MAJOR TRIADS)

This exercise moves up and down the fingerboard, through all inversions on the same three strings (including all four sets of three adjacent strings).

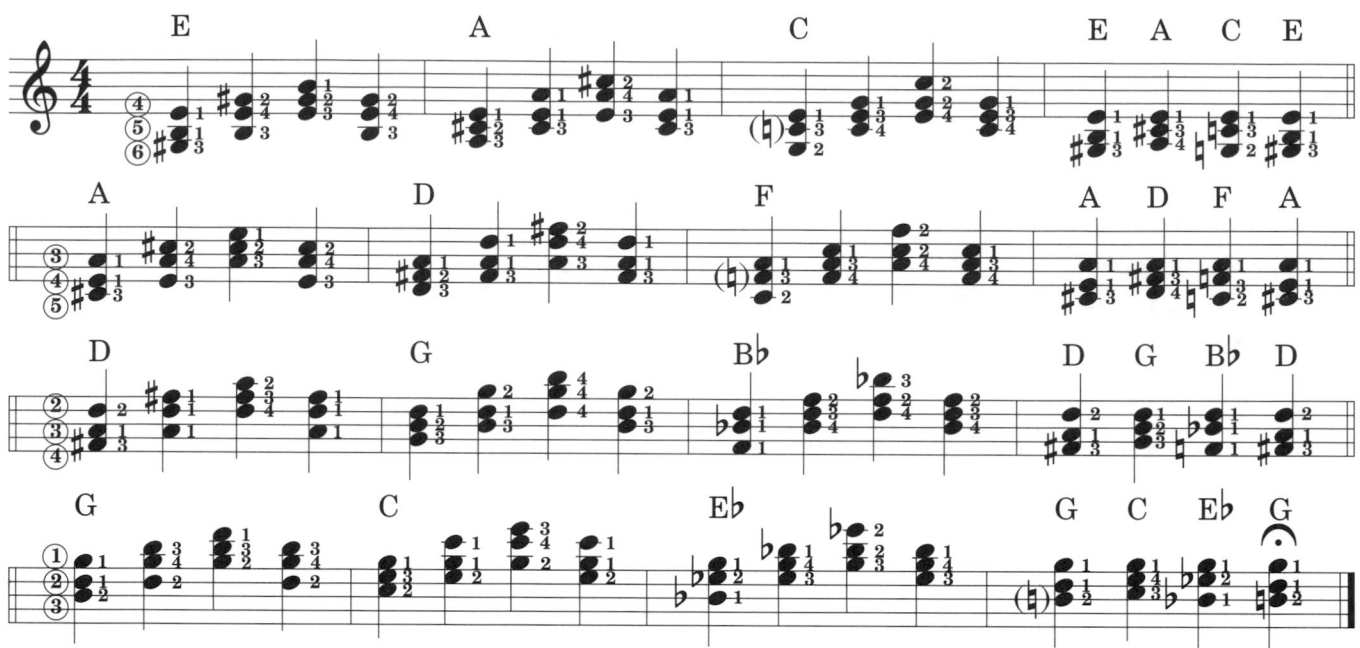

■ EXERCISE (MINOR TRIADS)

This exercise moves up and down the fingerboard, through all inversions on the same three strings (including all four sets of three adjacent strings).

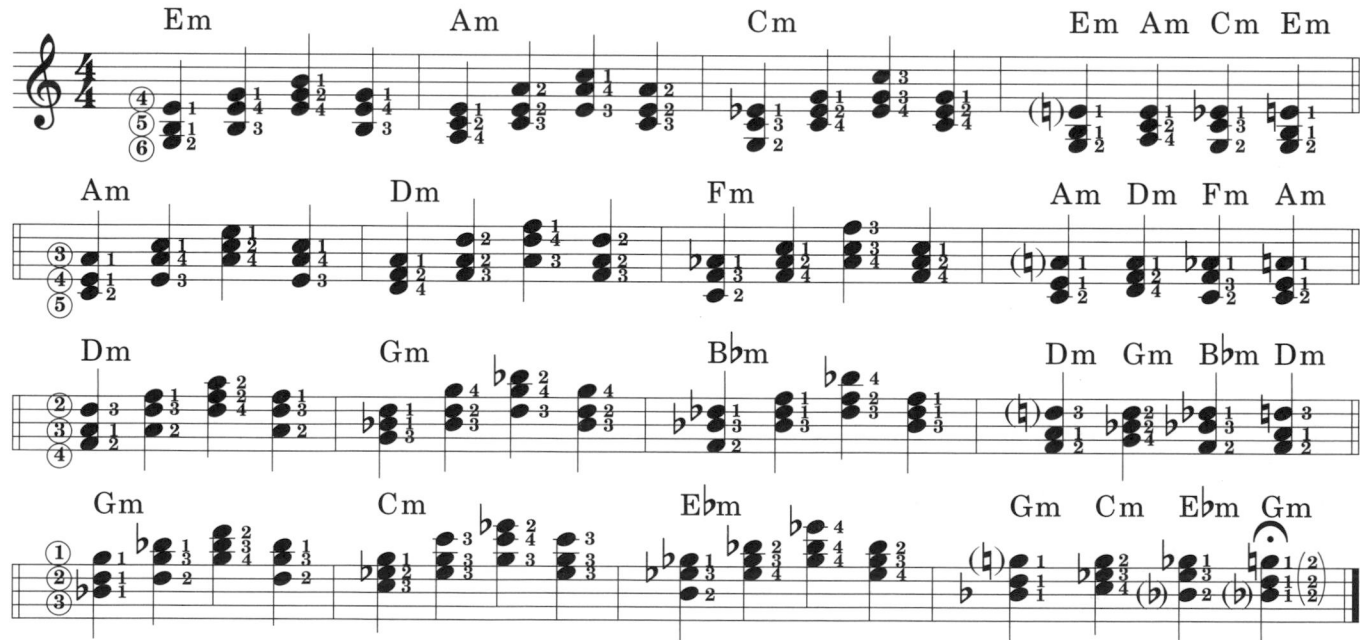

SECTION TWO

One-Octave Arpeggios—Triads
Fingering derived from scales—Across the fingerboard.

Transpose on the guitar by moving up the fingerboard (do not write out), and practice the preceding arpeggios in the following keys: A, B♭, C, D, and E♭. Thoroughly memorize all chord spellings.

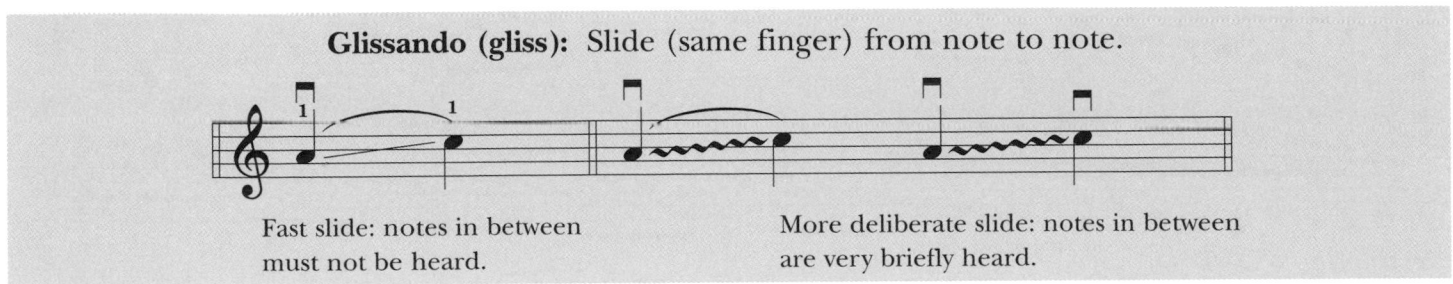

Glissando (gliss): Slide (same finger) from note to note.

Fast slide: notes in between must not be heard.

More deliberate slide: notes in between are very briefly heard.

59

Real Melodic (or Jazz) Minor Scale

The real melodic minor scale is derived from the preceding major scale forms by merely lowering the 3rd degree (note) one half step (one fret). This is a tonic major-to-minor relationship. All notes in this melodic minor scale remain the same—ascending and descending.

In the real melodic (or jazz) minor studies on the following pages, tonic major key signatures are used to simplify the conversion from major to minor. All playing positions are exactly the same.

You must practice these minor scales carefully, as at first they are difficult to "hear." They are worth putting considerable effort into, as they play a very large part in improvisation. (Application will be discussed later.)

C Real Melodic Minor (Five Positions)

FINGERING DERIVED FROM TYPE 1

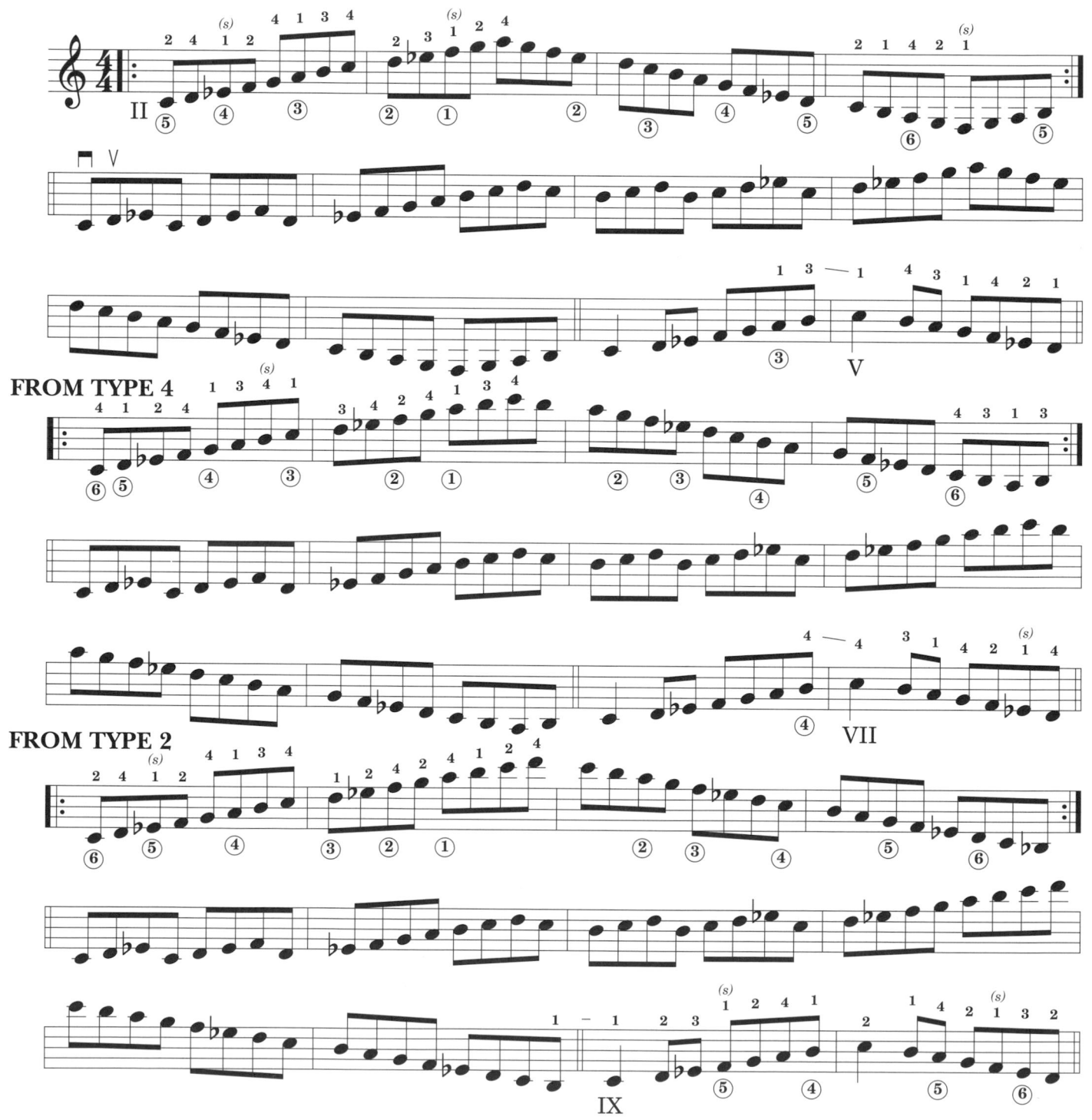

FROM TYPE 4

FROM TYPE 2

FROM TYPE 1A

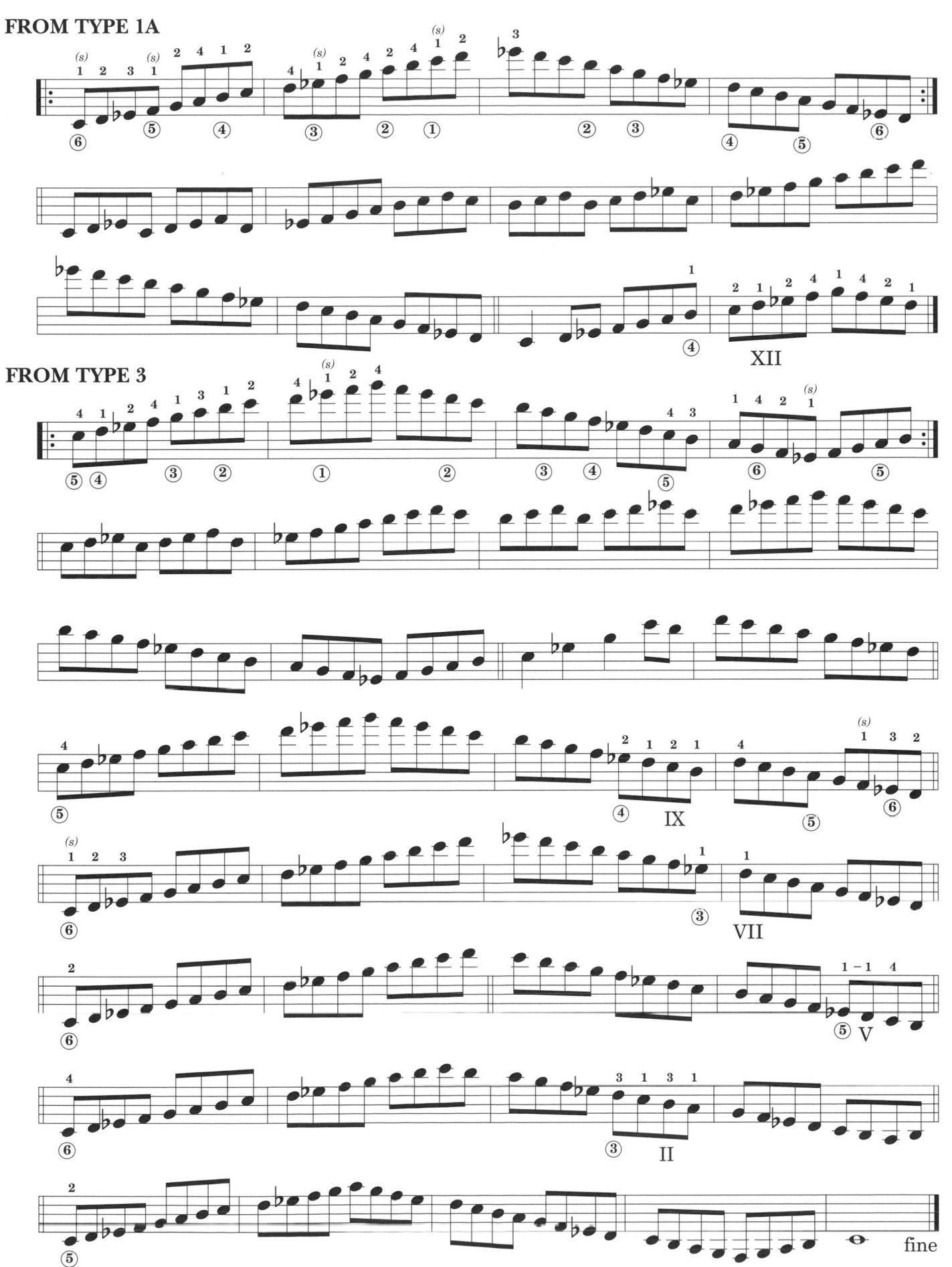

FROM TYPE 3

61

Rhythm Guitar—The Right Hand (Shuffle)

Basic Stroke

A very stable beat, but most practical with an incomplete rhythm section, as all accents fall "on the beat."

■ **EXERCISE**
▌ Observe notation.

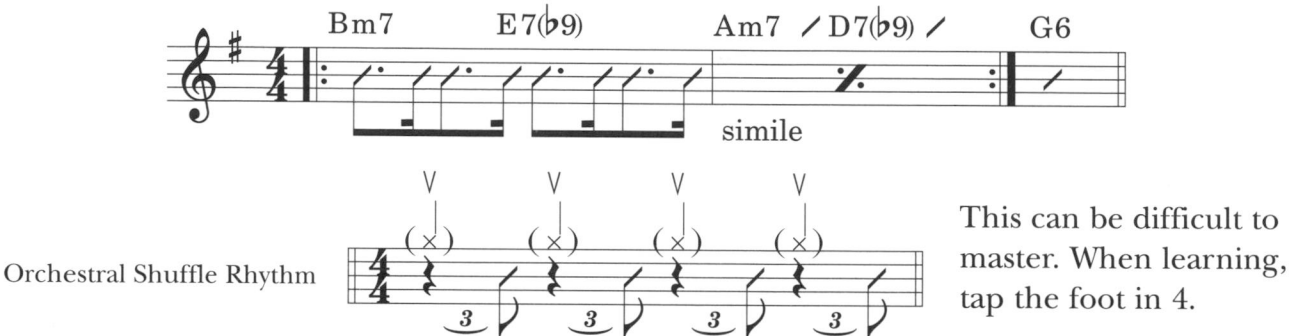

Orchestral Shuffle Rhythm

This can be difficult to master. When learning, tap the foot in 4.

▌ This stroke accents the off-beats and therefore adds a great deal more to a rhythm section.

■ **EXERCISE**
▌ Observe notation.

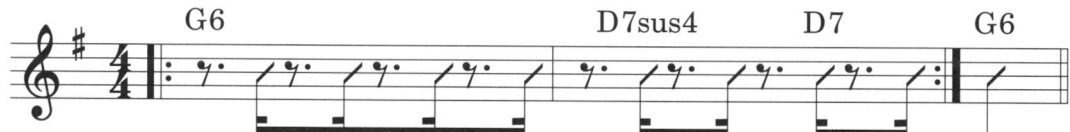

▌ The preceding shuffle rhythm strokes also apply to rhythm parts in 6/8.

Observe notation.

Speed Study
Tempo must be constant throughout.

For practice with other fingerings, change the signature to C, F, D, and A.

Chord Forms

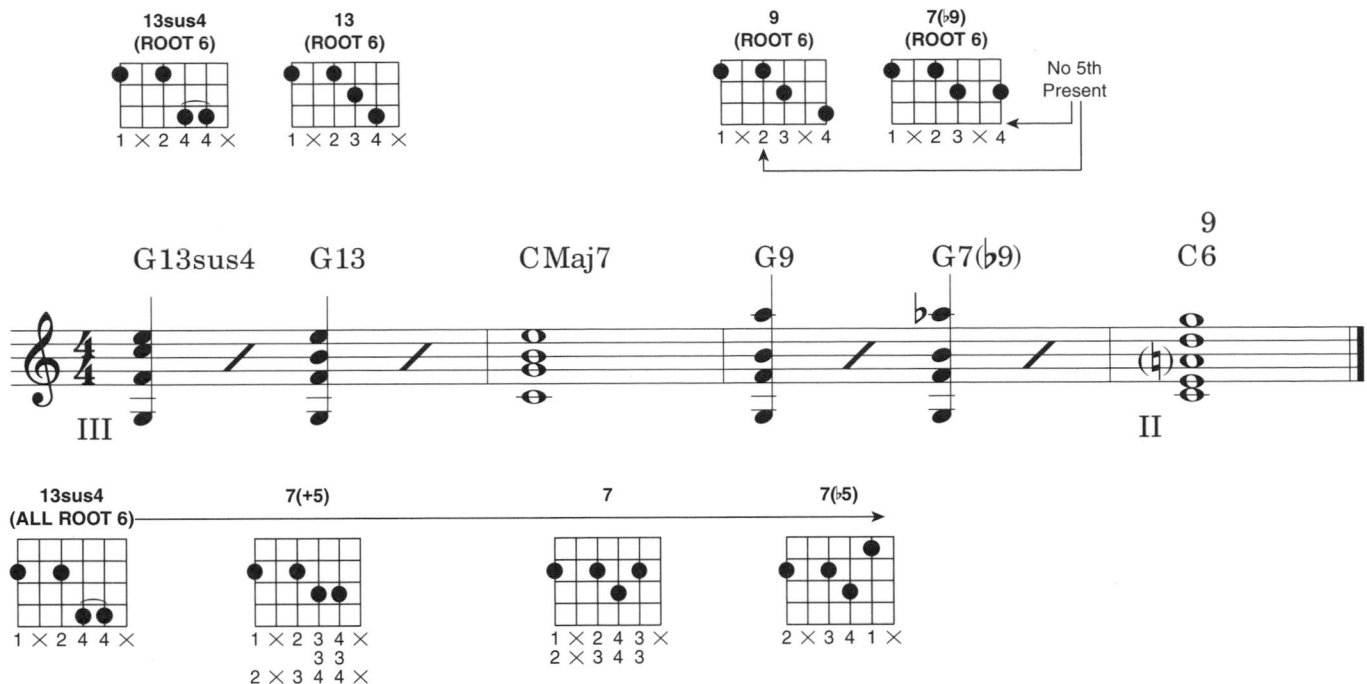

When the ♭5 of any dominant 7 form falls on the 1st, 2nd, or 3rd strings, you may consider it an augmented 11.

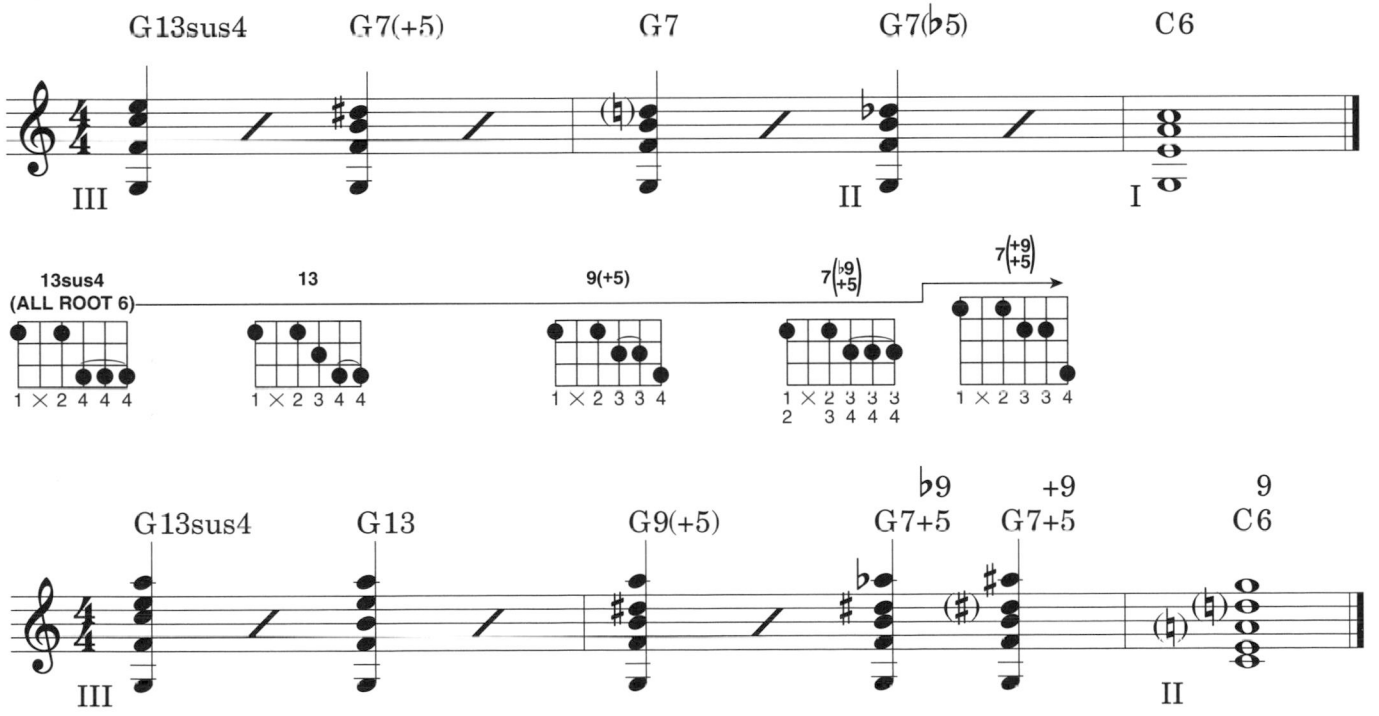

■ EXERCISE

Play through this exercise using some of the above forms. Carefully observe the fingerings (and their relationships to each other).

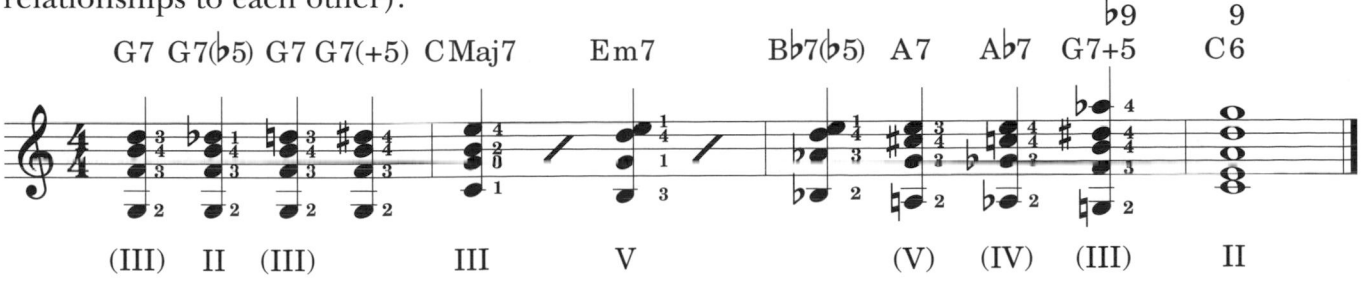

Melodic Rhythm Study No. 5 (duet)

One-Octave Arpeggios—Triads

Fingering derived from scales—Across the fingerboard.

■ Transpose and play in the keys of E♭, F, G, A, and B♭.

■ Transpose and play in keys of D, E♭, F, G, and A.

F Real Melodic Minor (Five Positions)

FINGERING DERIVED FROM TYPE 1A

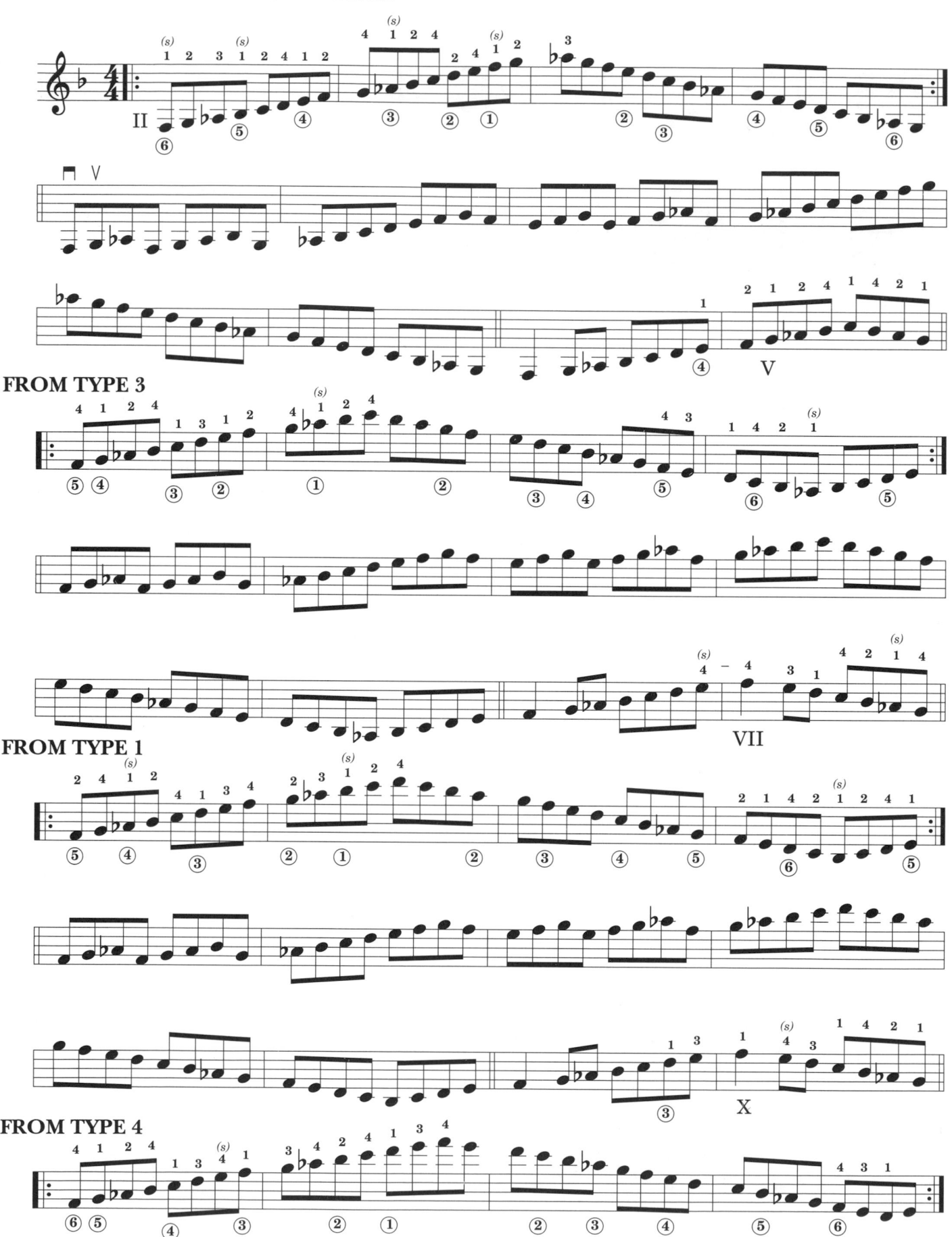

FROM TYPE 3

FROM TYPE 1

FROM TYPE 4

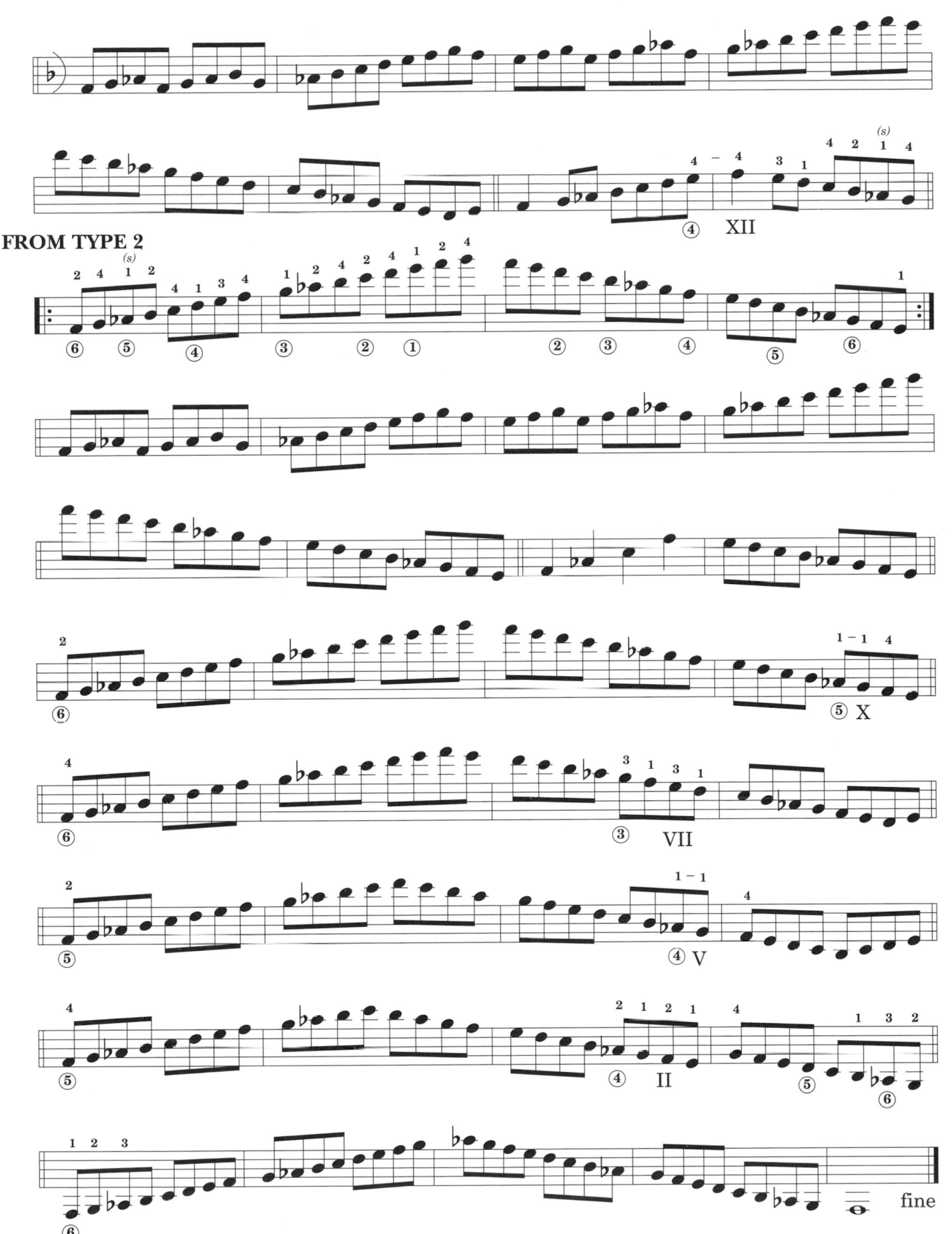

For additional practice on real melodic minor scales, refer to *Volume I*. Play reading and speed studies with lowered 3rd scale degree.

5th Position Study No. 2 (duet)

One-Octave Arpeggios—Triads

Fingerings derived from scales—Across the fingerboard.

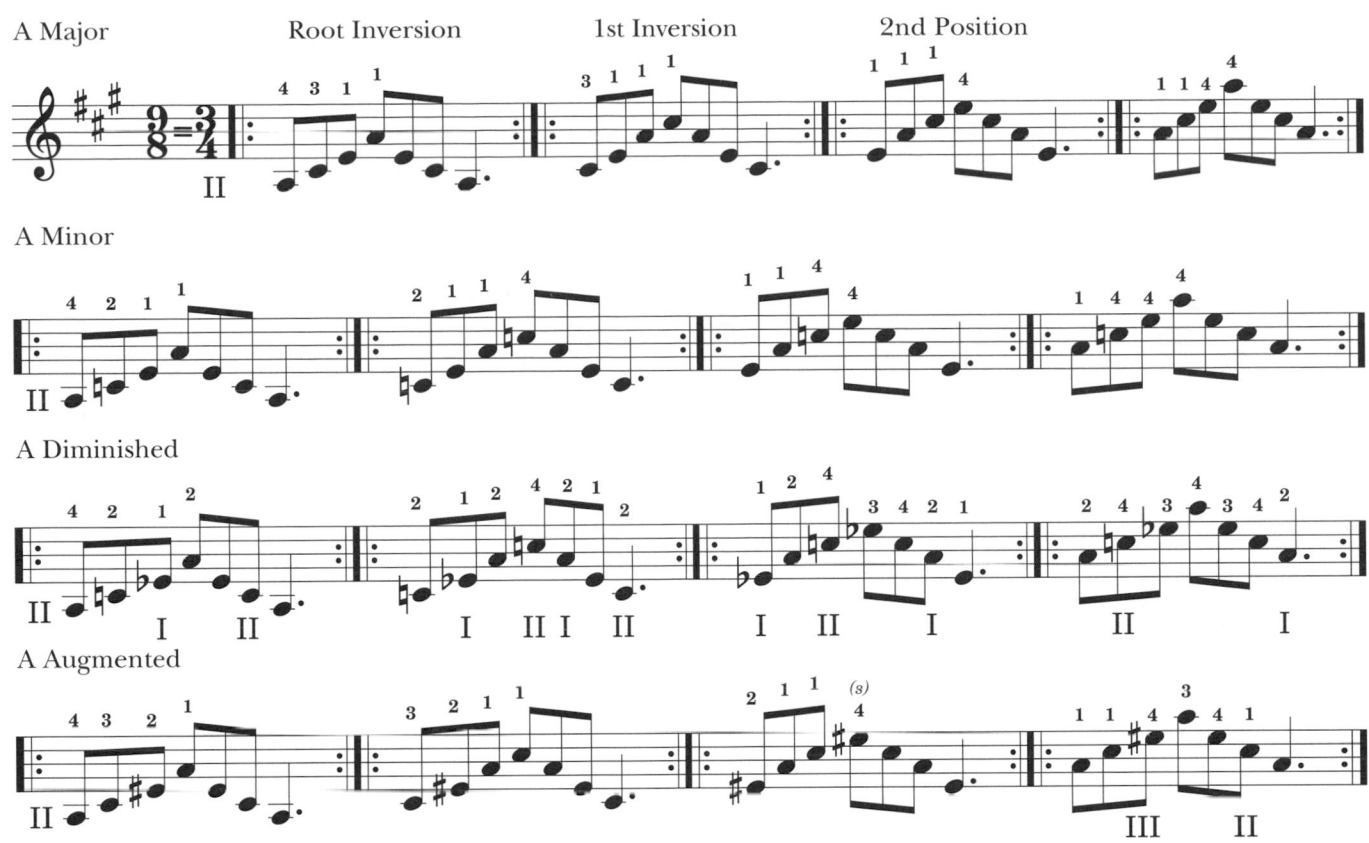

Transpose and play in the keys of B♭, C, D, E♭, and F.

Fingering derived only partly from scales—Across and up the fingerboard.

Transpose and play in the keys of G, A, B♭, C, and D.

Chord Forms

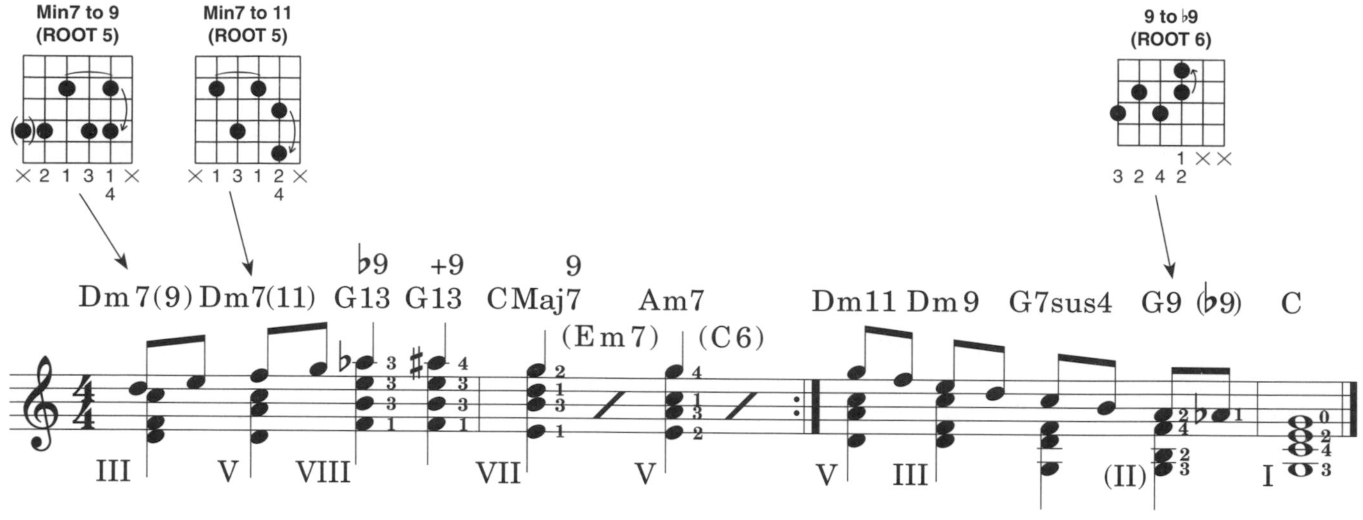

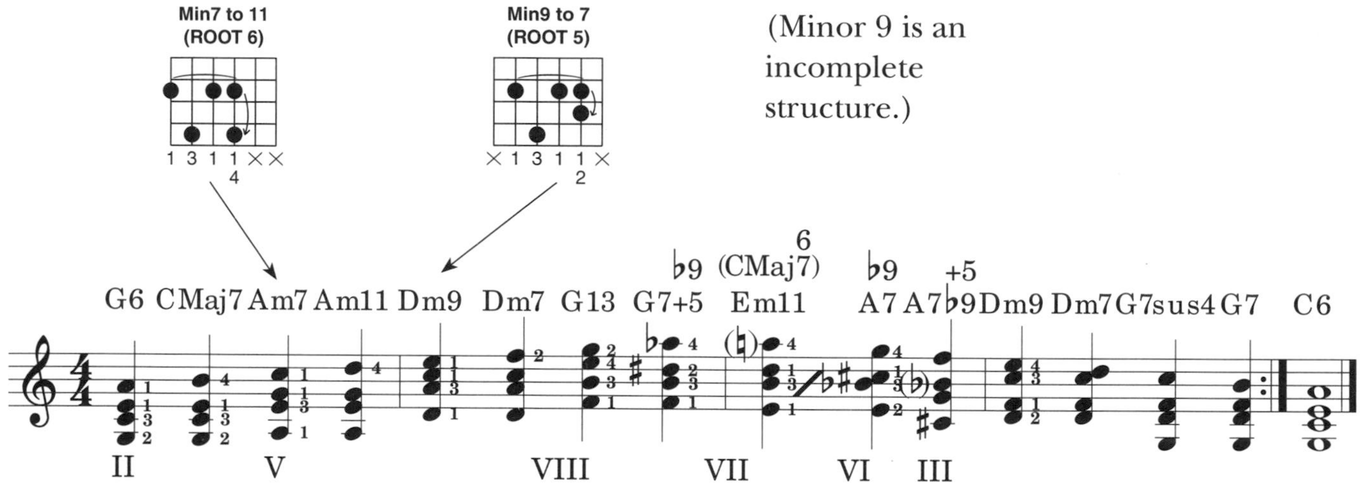

Slur

Ascending: Indicated by a curved line over two or more notes. Pick only the first note and drop the finger(s) of the left hand sharply on the string to produce the remaining note(s).

Descending: Prepare the entire group of notes with the fingers of the left hand in place. Pick only the first note with the right hand. Remove the left hand fingers from the remaining notes of the slur, drawing them toward the palm so as to actually pick the string again.

When blending electric guitar with horns, it is usually best to gliss from note to note when a slur is indicated. This produces no attack whatsoever on the second note and therefore is more "horn-like." (Be careful not to mistake a phrasing mark for a slur. A phrasing mark generally encompasses a large group of notes and indicates a legato or smooth performance of them.) You can also expect the horn player to break the phrase or breathe at the end of a phrasing mark. For a perfect blend you must perform accordingly. Commas are also used to indicate where to break a phrase or "breathe."

Trill

A trill is the effect created by rapidly alternating a note with the next diatonic note above it. Pick only the principal note, drop the finger for the next note sharply on the same string. Then, draw it off toward the palm, actually picking with the left-hand finger, to keep the string vibrating.

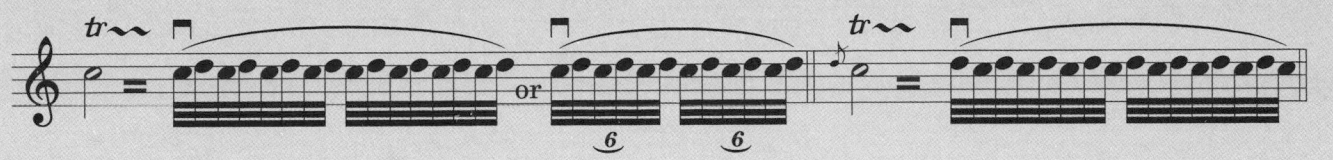

Theory: Diatonic 7th Chords (Major Keys)

▌ All diatonic chords within a key are built upwards in thirds.

1.) By adding another note a third above the diatonic triads, we construct all four-part chords common to a major key. (See diatonic triads, pg. 54.)

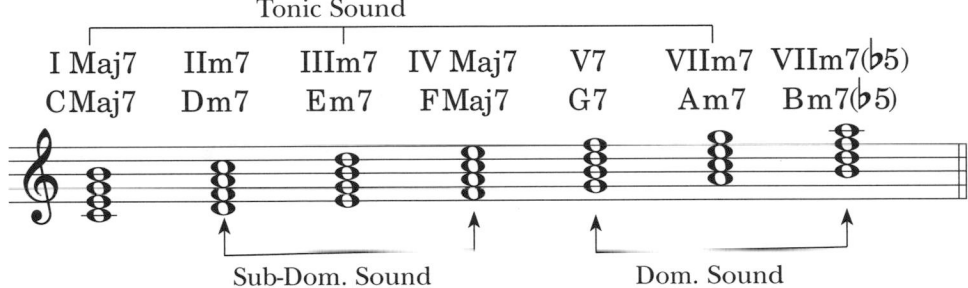

The VIIm7(♭5) is sometimes called "half diminished" (ø). Observe the chord relationships producing the tonic, subdominant, and dominant sounds.

Also: The IIIm7 is often found as an intermediate chord in a subdominant sequence.

EXAMPLES

IV	IIIm7	IIm7	(V7 I)...	IIm7	IIIm7	IV	(V7 I)
FMaj7	Em7	Dm7	(G7CMaj7)..	Dm7	Em7	FMaj7	(G7 CMaj7)

2.) Because of a conflict with the root in the melody, the four-part structures used on the I and IV are often 6th chords, built from major scale degrees 1, 3, 5, and 6. You might say this is a result of the substitution of VIm7 over the root of the I chord and IIm7 over the root of the IV chord: Am7 = C6; Dm7 = F6.

3.) Substitution of IIIm7 or VIm7 for I, IIm7 for I, IIm7 for IV, and VIIm7(♭5) for V7 are especially valuable when creating moving bass lines with strong chordal degrees (I and V) supporting the harmonic structures.

EXAMPLE:

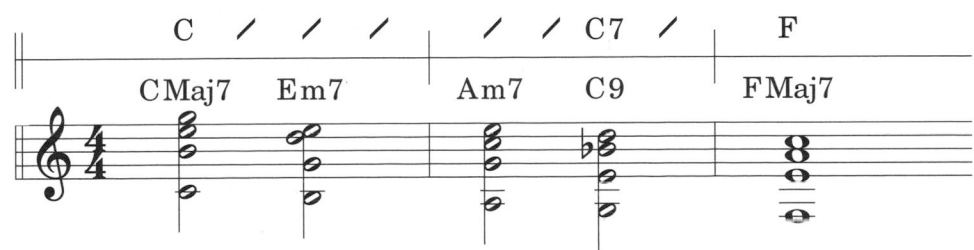

All diatonic chords (names and structures) must be memorized, in all keys.

Arpeggios—Diatonic Sevenths

All four-part chords, all inversions—Key of G major.

FINGERING TYPE 1A

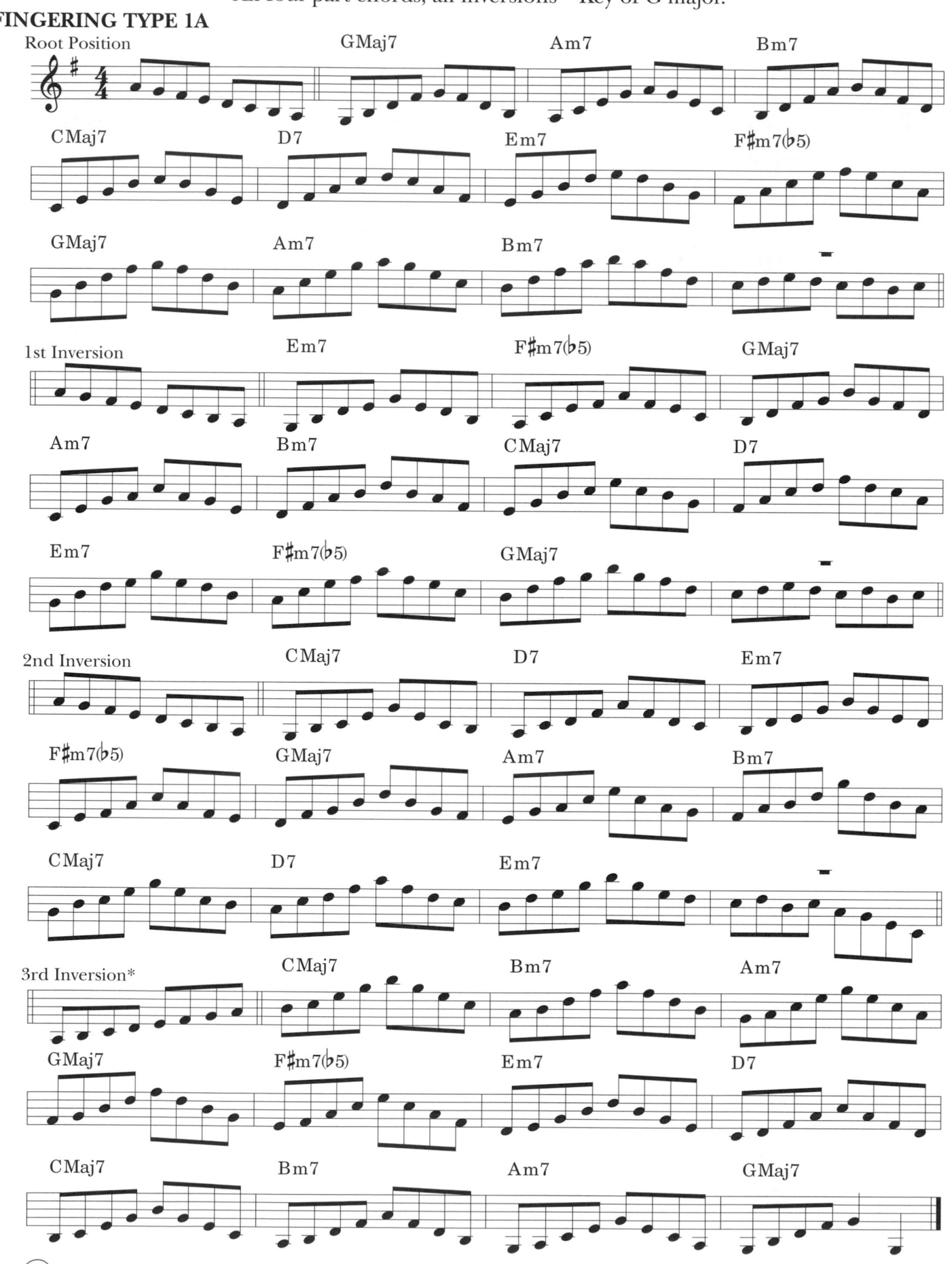

✱ 3rd Inversion: 7th in the bass

72

Arpeggios—Diatonic Sevenths

All four-part chords, all inversions—Key of C major.

FINGERING TYPE 4

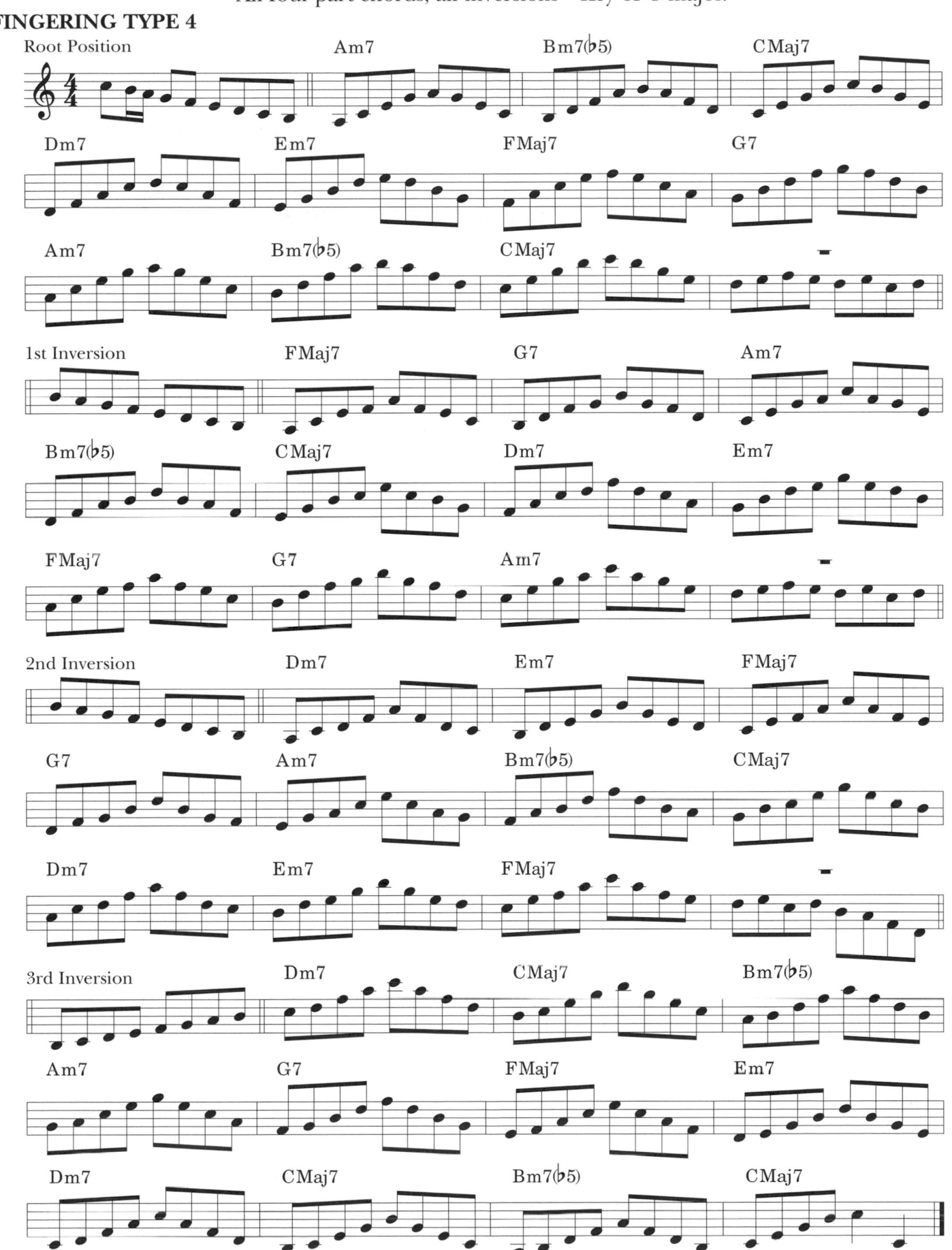

■ (See bottom of pg. 55.)

G Real Melodic Minor (Five Positions)

FINGERING DERIVED FROM TYPE 2

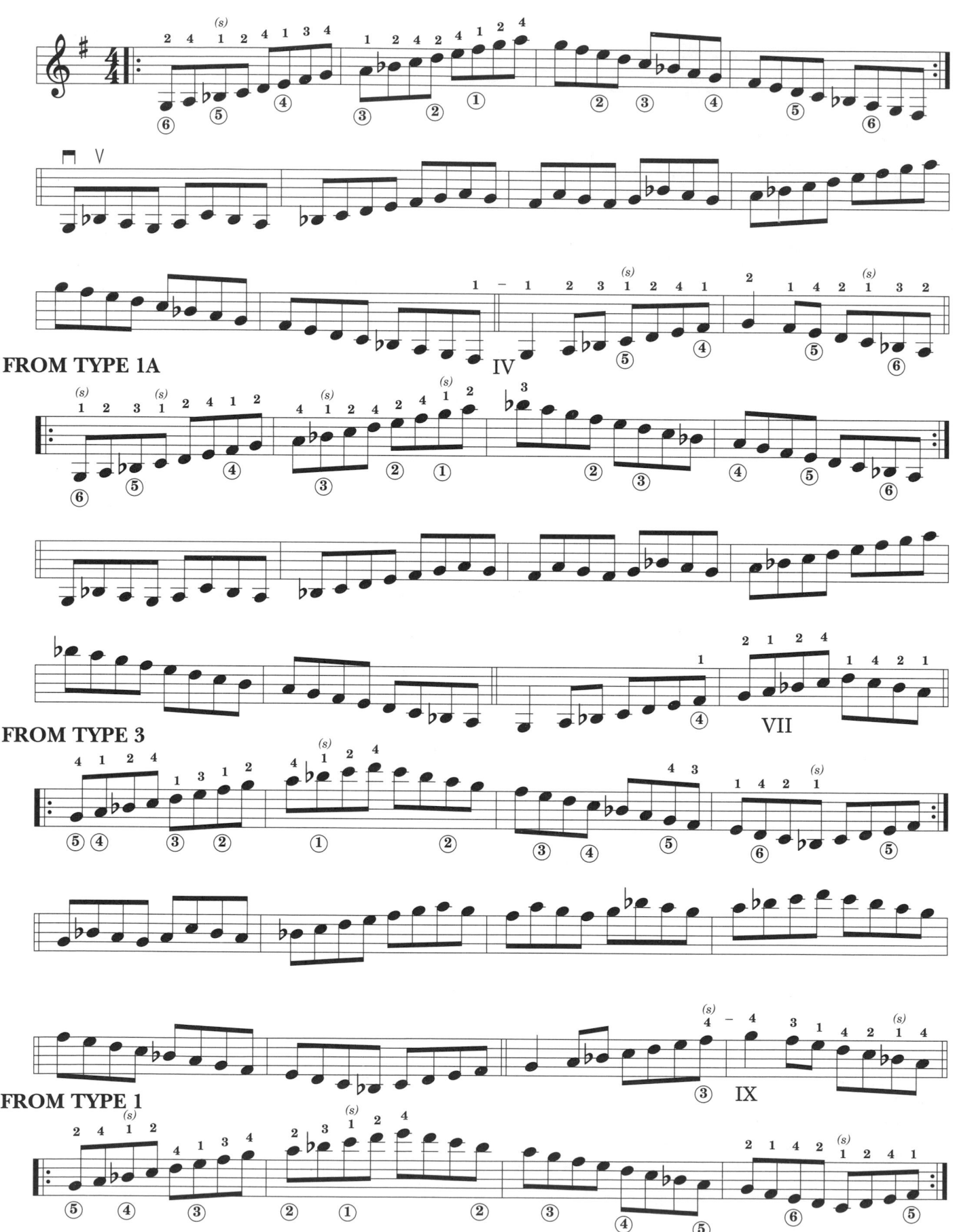

FROM TYPE 1A

FROM TYPE 3

FROM TYPE 1

FROM TYPE 4

Chord Forms

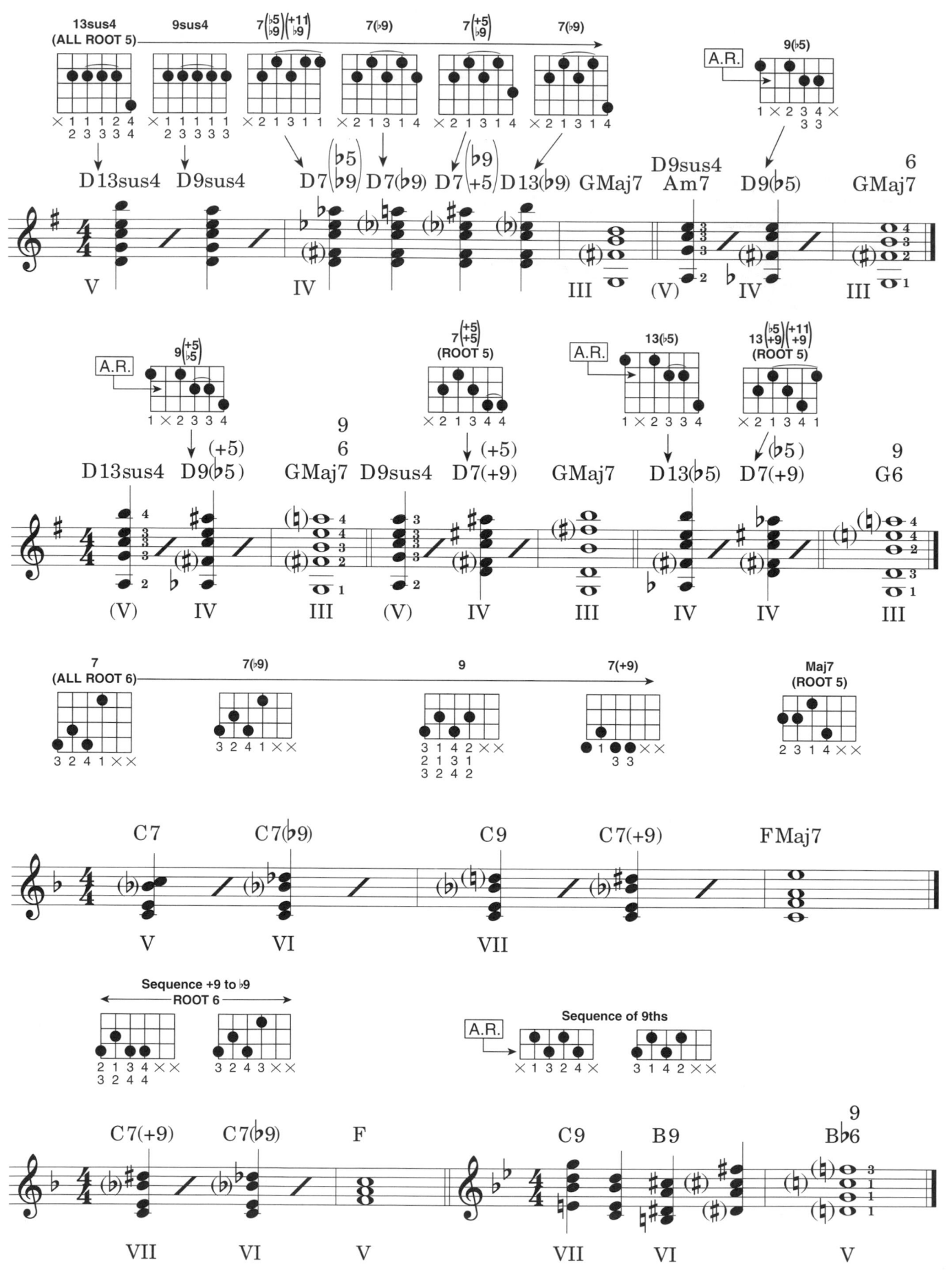

Two-Octave Arpeggios
C Major Triad from the Root
Fingering derived from scales and chords—Across and up the fingerboard.

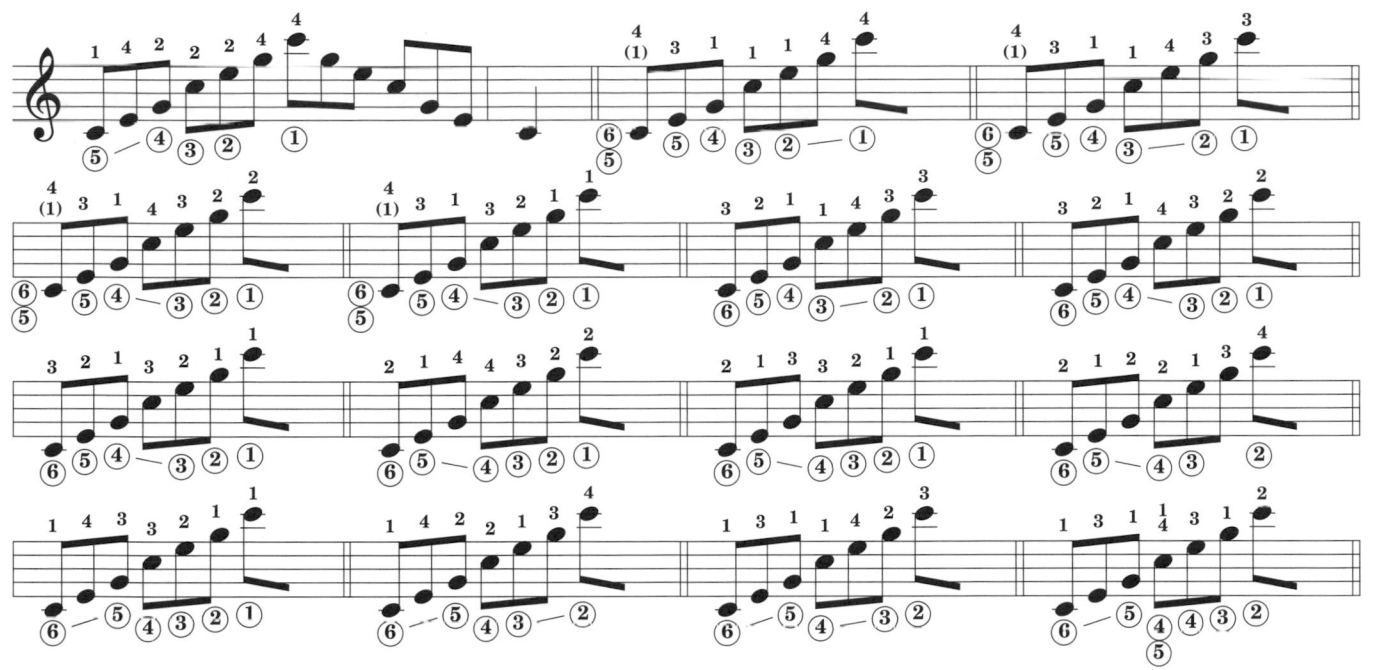

Practice all forms in all possible keys.

Chord Etude No. 9

77

The Wanderer (duet)

Rhythm Guitar—The Right Hand (Waltz)

Arpeggios—Diatonic Sevenths

All four-part chords, all inversions—Key of F major.

FINGERING TYPE 3

Theory: Chord-Scale Relationships

For improvisation.

WITH DIATONIC CHORD STRUCTURES

All the notes of a major scale may be used melodically over the seven chord structures contained in that key. However, any scale tone one half step above a chord tone (1, 3, 5, 7 in diatonic harmony) must be of short duration and used only in passing to a chord tone next to it.

EXAMPLE:

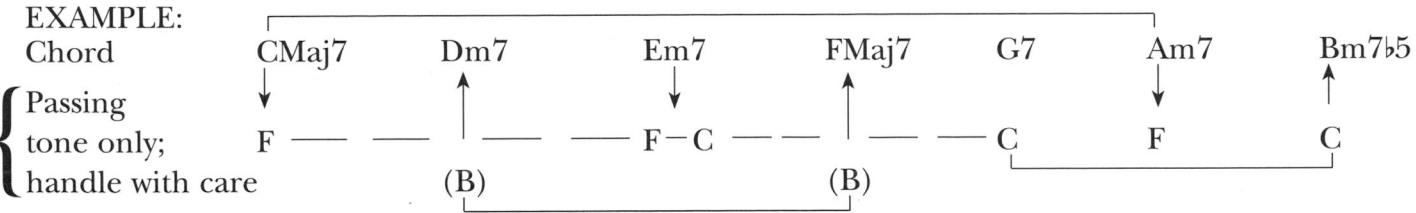

Melodic ideas may be created from scale tones in any order, providing you do not start with, or "lean on," the passing tones discussed above.

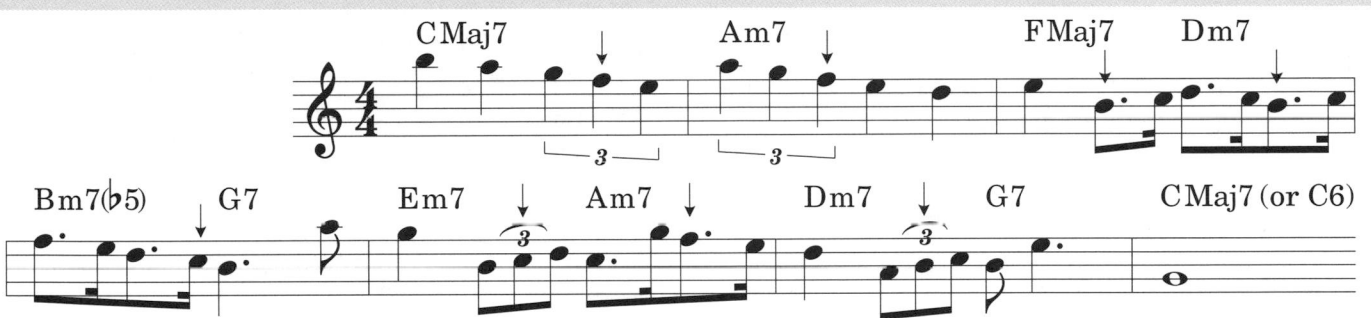

Improvisation: The spontaneous creation of music while playing, usually within the confines of the harmonic content of a song. (All available notes are drawn from chord tones and related scales.)

Before all-out (no holds barred) improvisation is attempted on the chords to a song, it is best if you stay close to the melody and fill in only during notes of long duration.

EXAMPLE:
Straight Melody

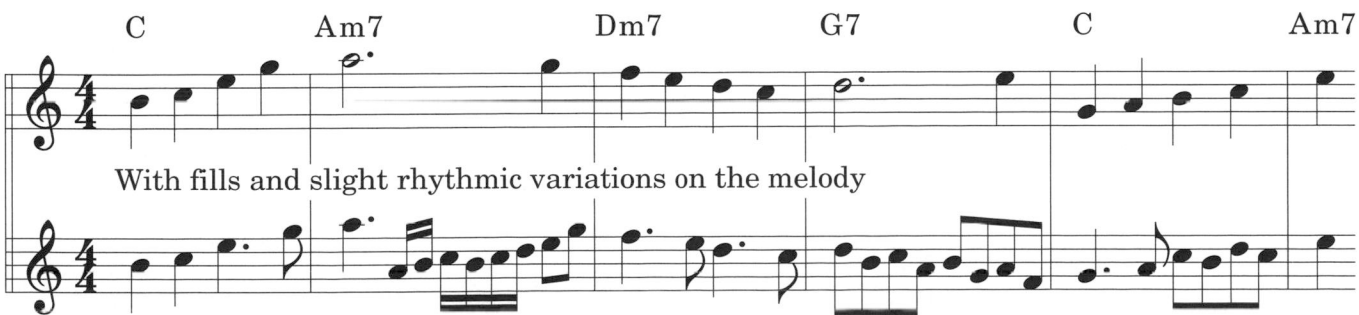

With fills and slight rhythmic variations on the melody

D Real Melodic Minor (Five Positions)

FINGERING DERIVED FROM TYPE 3

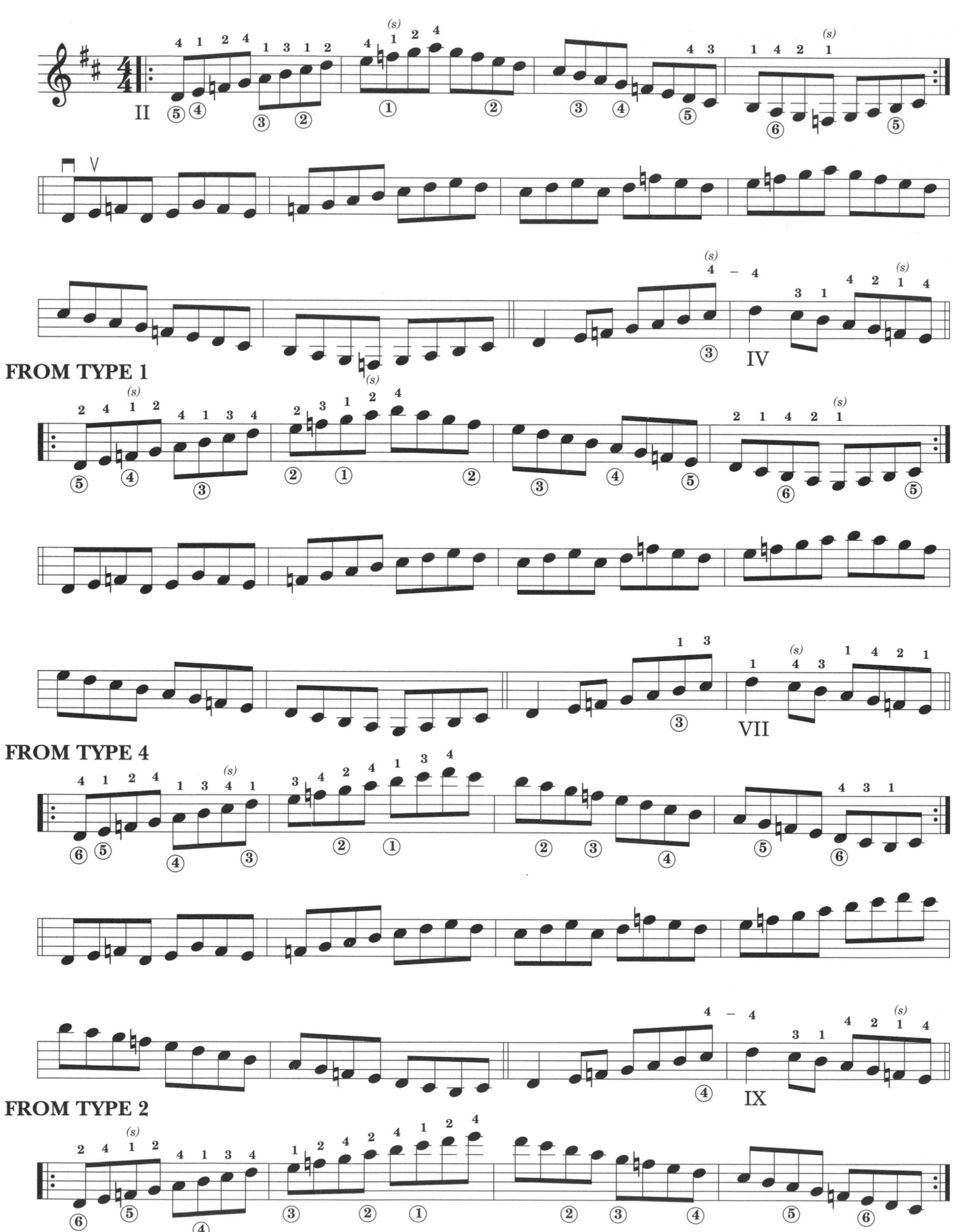

FROM TYPE 1

FROM TYPE 4

FROM TYPE 2

FROM TYPE 1A

Chord Forms—3rd in the Bass

I define the real bass (sounding) range as any note lower in pitch from C, 5th string (3rd fret) or C, 6th string (8th fret).

Any chord voiced with the 3rd degree in the bass has a weak chordal sound, and should be used only when leaping to a new inversion of the same chord, or as a passing chord to produce scalewise or chromatic bass motion.

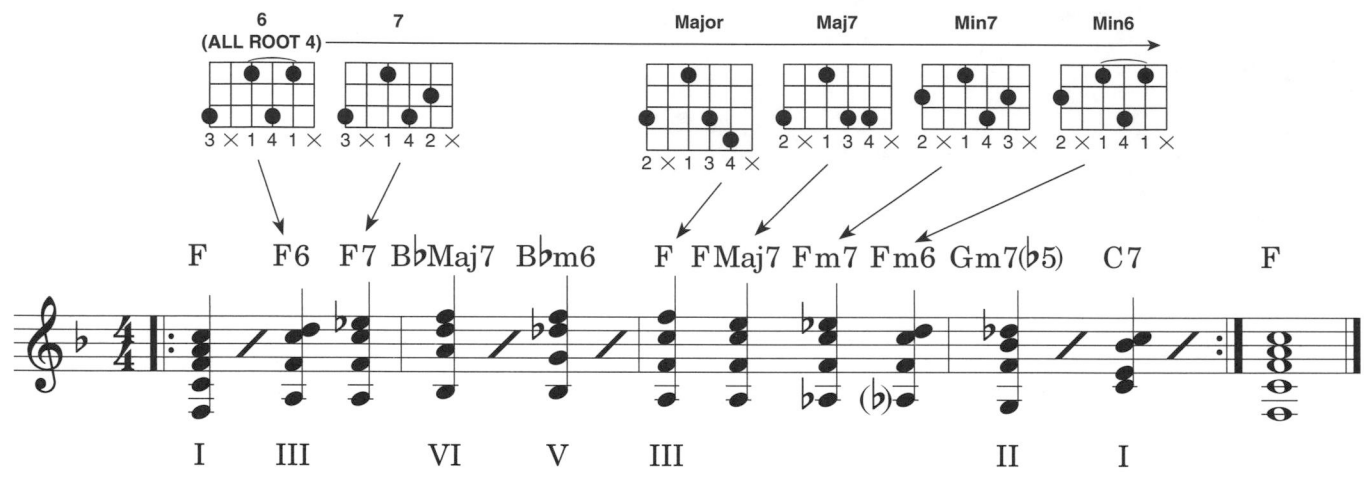

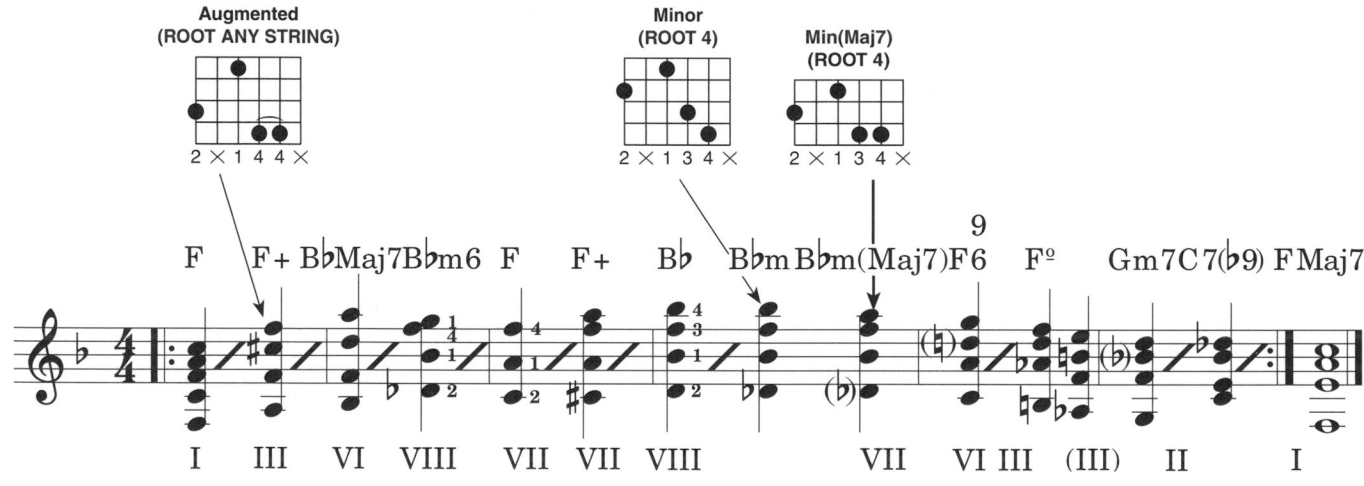

Chord Etude No. 10

Two-Octave Arpeggios

G Major Triad from the 3rd

Fingering derived from scales and chords—Across and up the fingerboard.

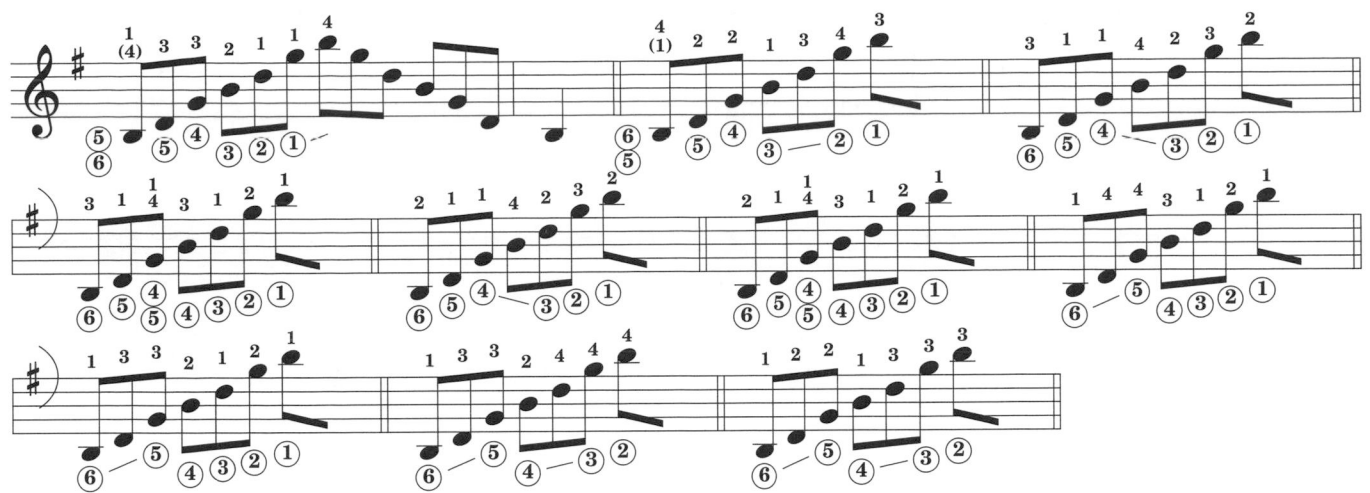

F Major Triad from the 5th

Across and up the fingerboard.

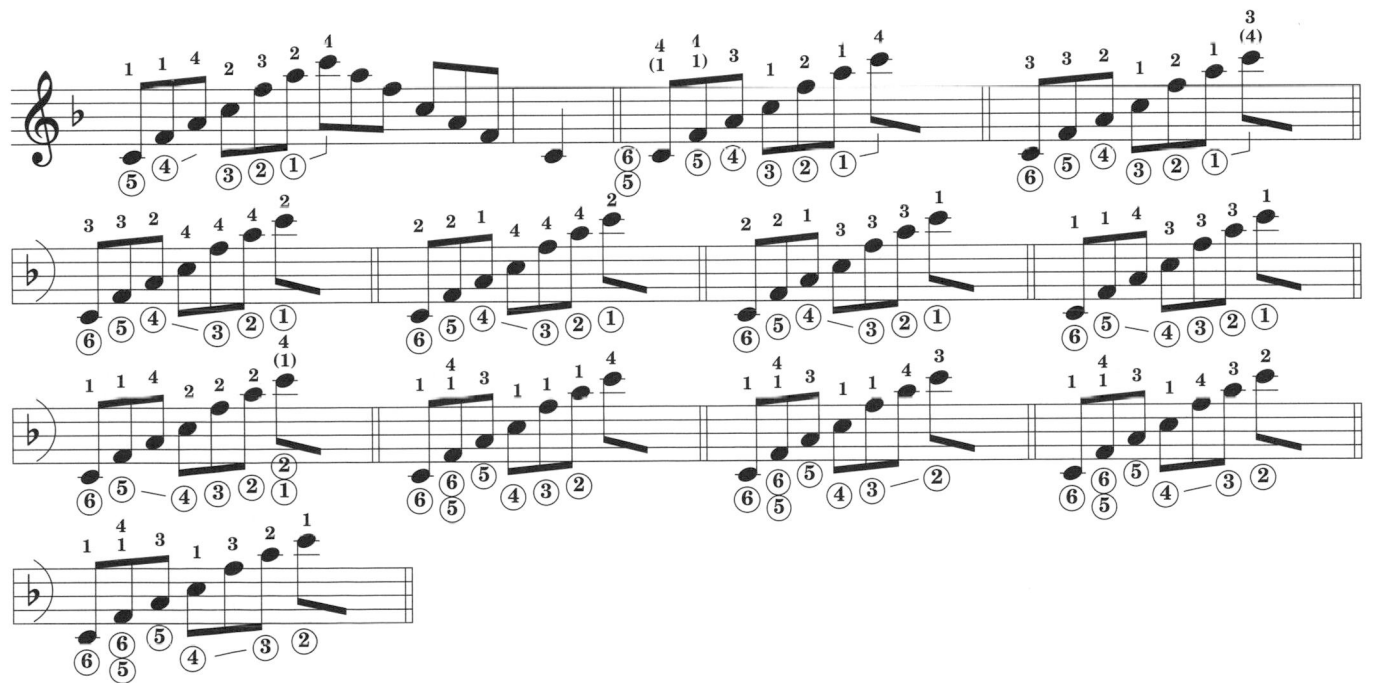

Practice all forms in all possible keys.

Rhythm Guitar—The Right Hand (Jazz Waltz)

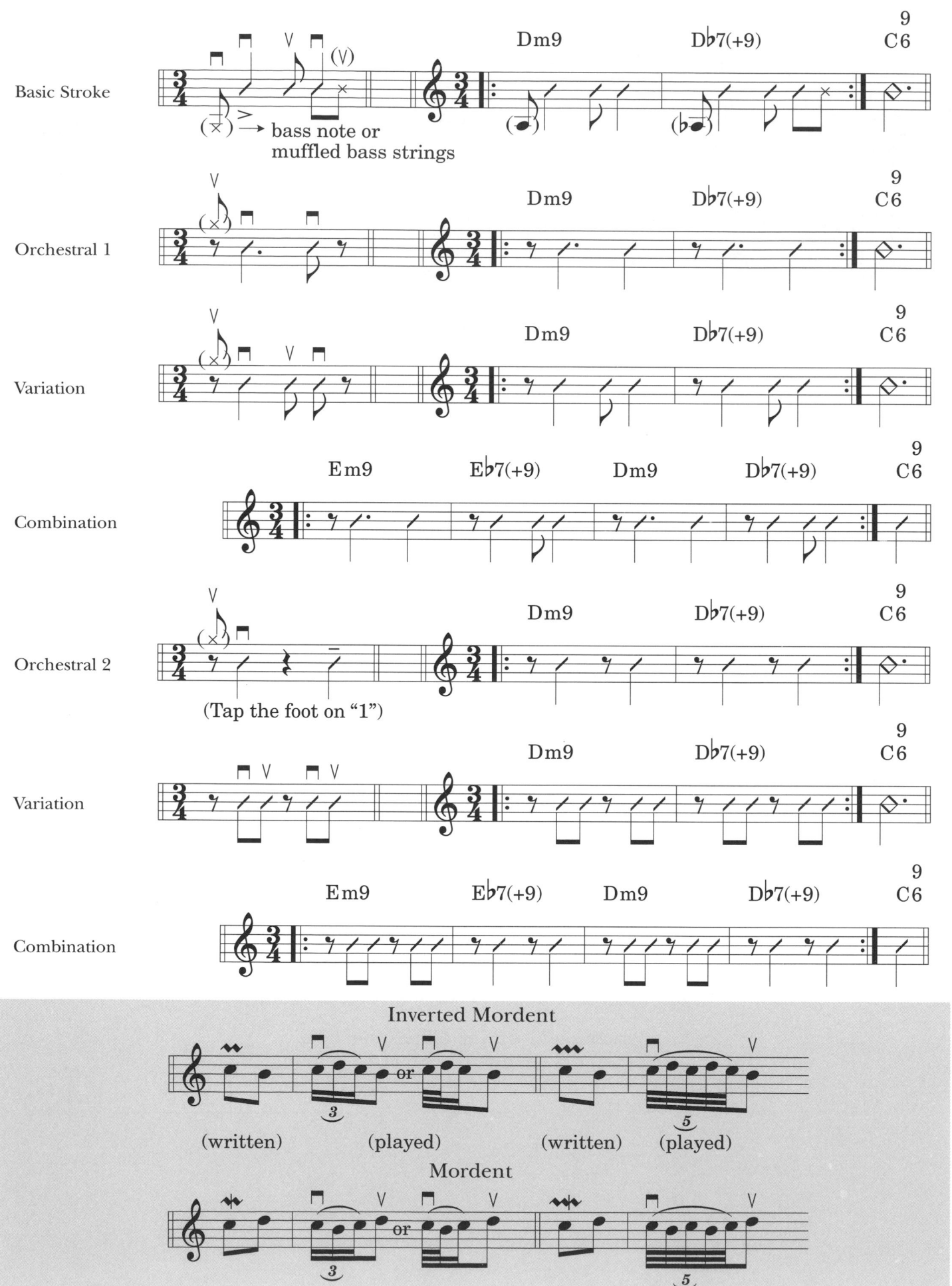

Arpeggios—Diatonic Sevenths

All four-part chords, all inversions—Key of B♭ major.

FINGERING TYPE 2

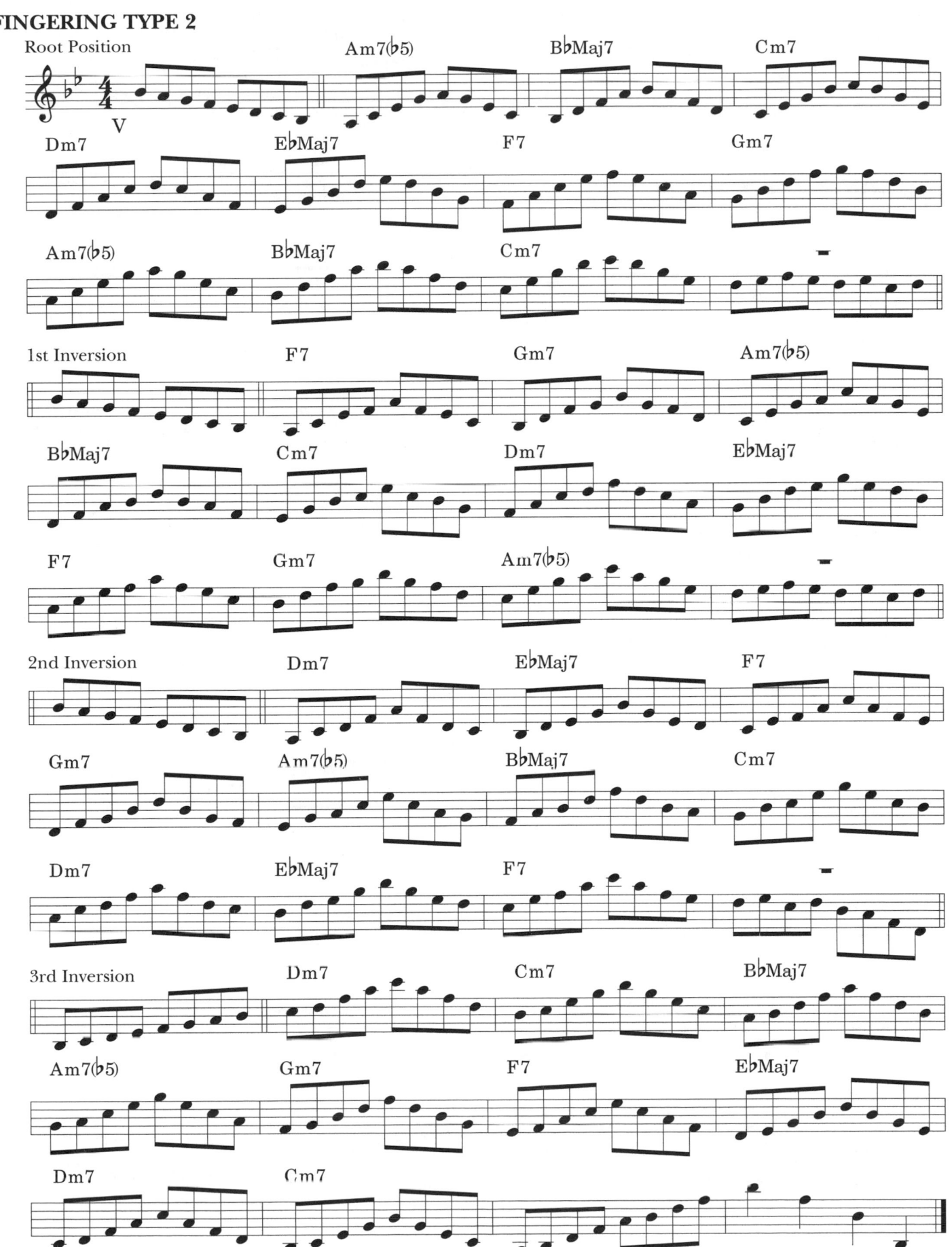

A Real Melodic Minor (Five Positions)

FINGERING DERIVED FROM TYPE 4

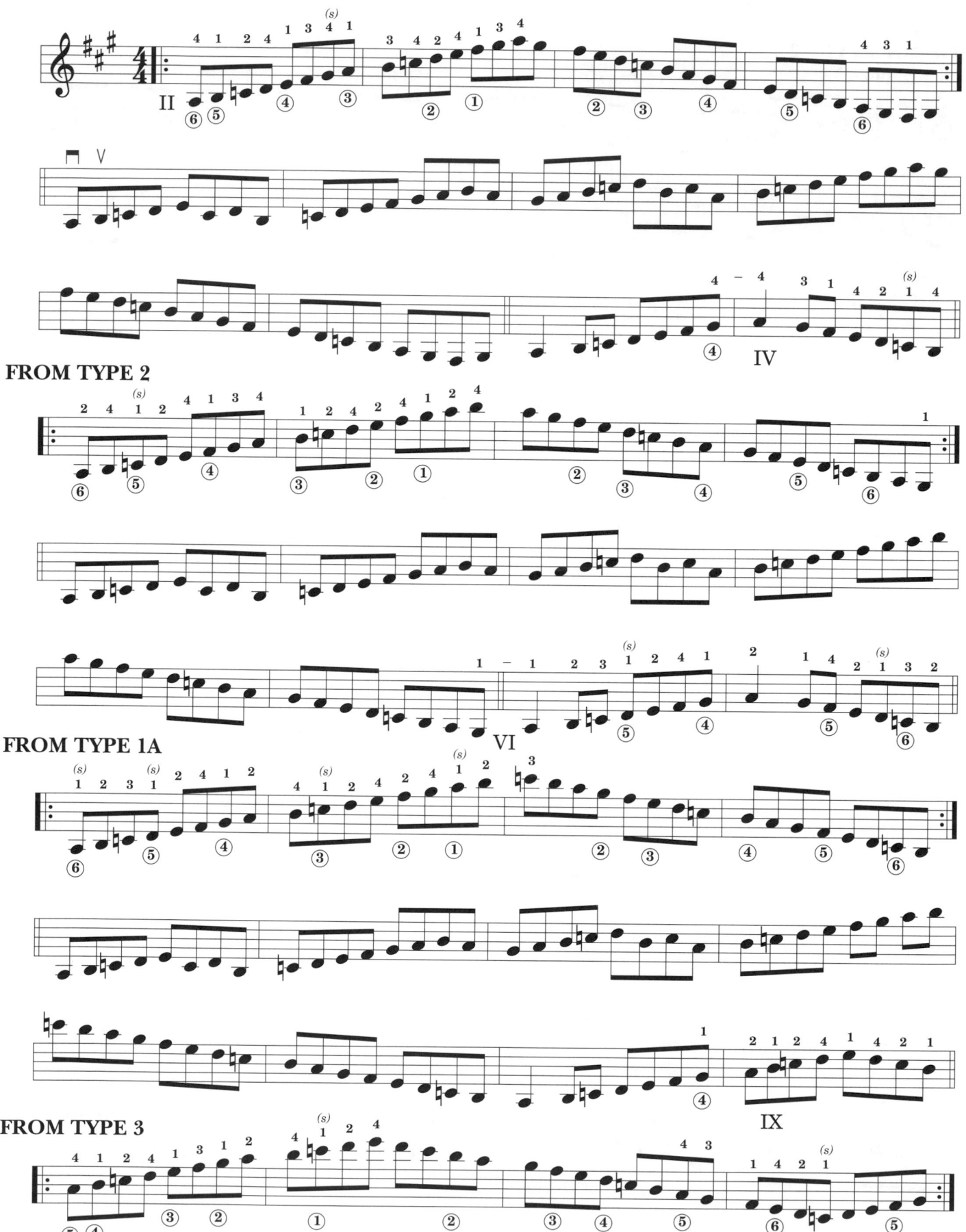

FROM TYPE 2

FROM TYPE 1A

FROM TYPE 3

FROM TYPE 1

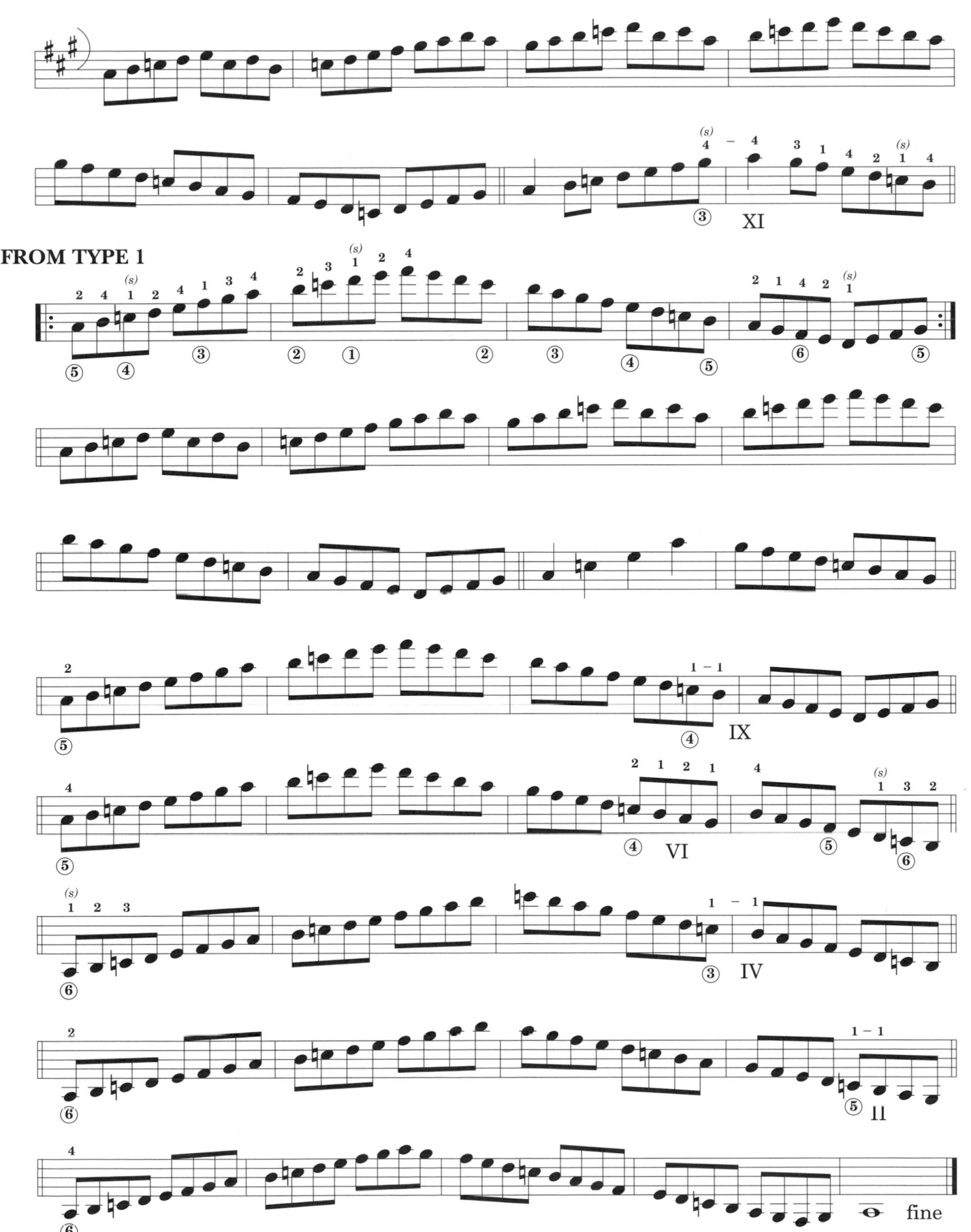

89

Chord Forms

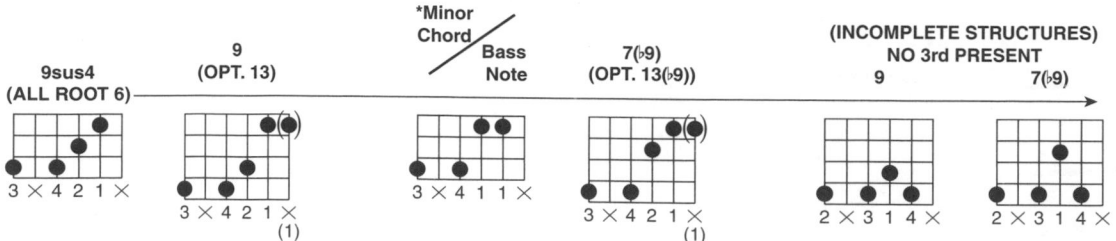

⊛ This is a way to notate symbols for chord structures that might be difficult to name any other way. The basic chord sound is represented above the diagonal line; the bass note it is to be played over is indicated below it.

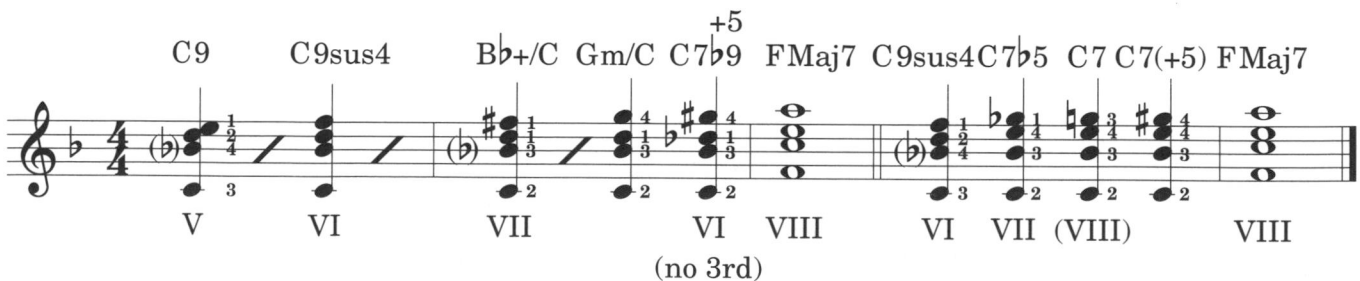

Two-Octave Arpeggios

C Minor Triad from the Root

Fingering derived from scales and chords—Across and up the fingerboard.

Practice all forms in all possible keys.

Appoggiatura (Grace Notes)

The unaccented appoggiatura takes its duration from the preceding beat.

(written) (played) (written) (played)

The accented appoggiatura (usually shown with no slash through the hook) falls directly on the beat.

It is best written out in full.

(written) (played) (written) (played)

Melodic Rhythm Study No. 6 (duet)

Latin beat optional.

Rhythm Guitar—The Right Hand (Cha-cha and Beguine)

Cha-cha: Basic and Orchestral

Orchestral

(Tap the foot in "2")

Variations
Basic and Orchestral

Beguine: Basic Stroke

Variation

Beguine: Orchestral

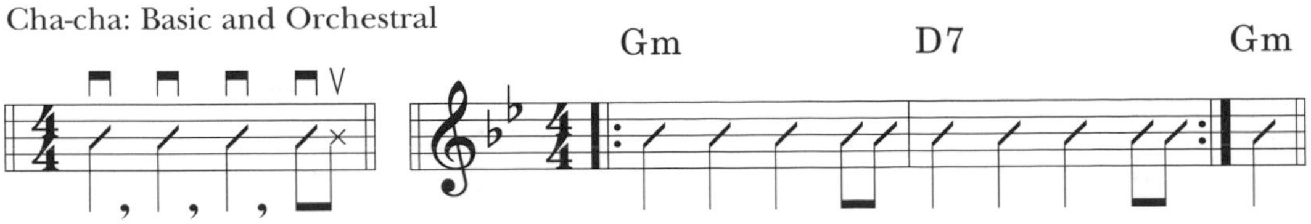

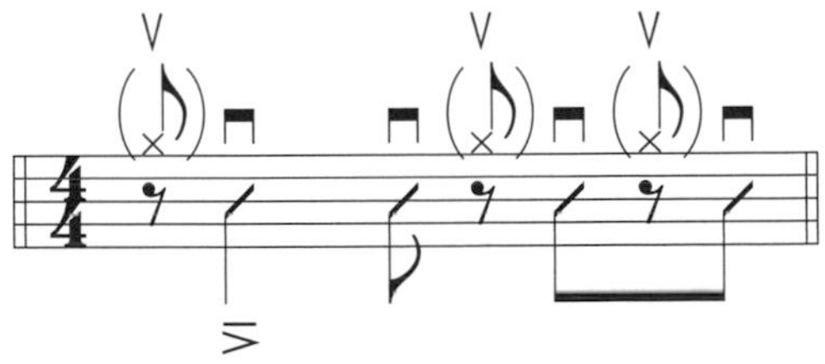

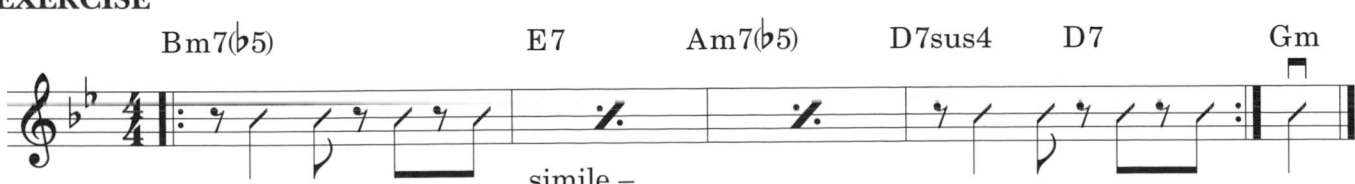

This stroke is difficult to master, but it is very important to right-hand development. When learning, tap the foot on beats 1, 3, and 4, or 1, 2, 3, and 4.

■ **EXERCISE**

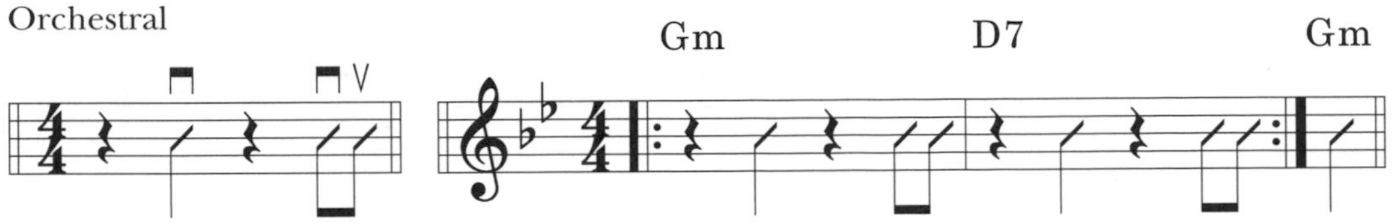

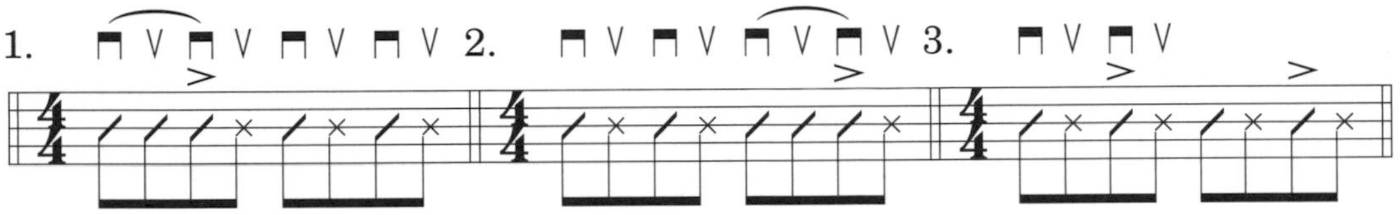

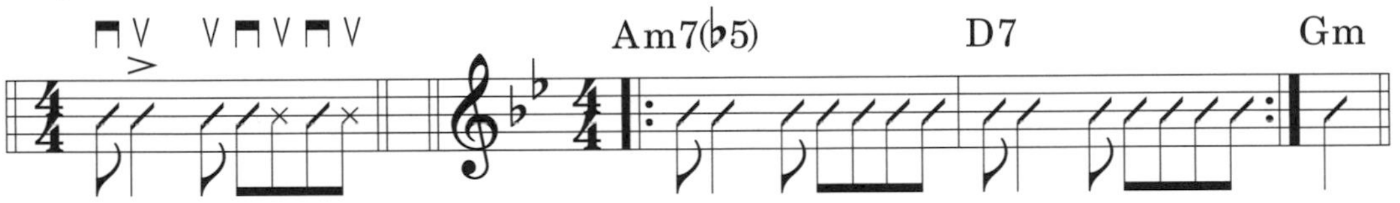

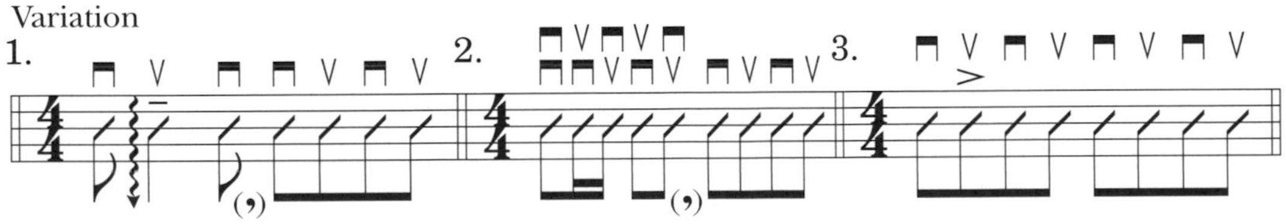

Arpeggios—Diatonic Sevenths

All four-part chords, all inversions—Key of E♭ major.

FINGERING TYPE 1

Root Position

Theory: Chord-Scale Relationships

For improvisation.

- It is very rare when a song remains completely diatonic harmonically, from beginning to end.
- Any chord that does not conform to the diatonic structures actually is a modulation to another key (or scale) for its duration.
- Sometimes a series of nondiatonic chords completely changes the key for a period of time. (This is why there will be references to "the key of the moment" in some of the following discussions on chord-scale relationships.)
- Remember this: The ear has memory but no eyes. Therefore the sound of what has gone before has a definite influence on which scales belong to certain chords in particular situations (but what is yet to sound has no bearing whatsoever).

> **Modulation:** The change of key within a composition or arrangement.

Nondiatonic Minor 7 and Major Chords

1.) Any minor 7 chord not in the key of the moment usually tends to sound like a IIm7 of whatever key it is the 2nd diatonic structure of. (A nondiatonic minor 7 chord actually performs the function of modulation more thoroughly than dominant 7 chords.) Use the major scale from one whole step below the chord name for nondiatonic minor 7.

EXAMPLE:

Chord	C	Cm7	Dm7	G7	E♭m7	A♭7	A♭m7	D♭7	C
Scale	CMaj	B♭Maj	CMaj	→	D♭Maj	→	G♭Maj	→	CMaj

2.) Any major chord that is not in the key (of the moment), not preceded by modulation, and with a nondiatonic root, tends to sound like a IV chord of whatever key it is the 4th diatonic structure of. Use the major scale from the 5th chordal degree of the major chord with a nondiatonic root.

EXAMPLE:

Chord	C	E♭(Maj7)	Dm7	G7	A♭(Maj7)	D♭(Maj7)	C
Scale	CMaj	B♭Maj	CMaj	→	E♭Maj	A♭Maj	Cmaj

3.) Any major chord not in the key (of the moment), not preceded by modulation, and with a scale tone root, tends to sound like a I (tonic) chord. Use the major scale from the chord name of the nondiatonic major chord with a scale-tone root.

EXAMPLE:

Chord	C	E(Maj7)	G9susC	G7	C
Scale	CMaj	EMaj	CMaj	→	→

> The major scale constructed from the 5th chordal degree may be used with any major chord at any time, but the chord-scale relationship on those with diatonic roots will be less perfect and will sound farther "out."
>
> Also: Minor 7 chords are occasionally tonic chords in disguise, so don't overlook the possibility of nondiatonic minor 7 chords actually being a IIIm7 or VIm7 for I. (See pg. 71.)

7th Position Study (duet)

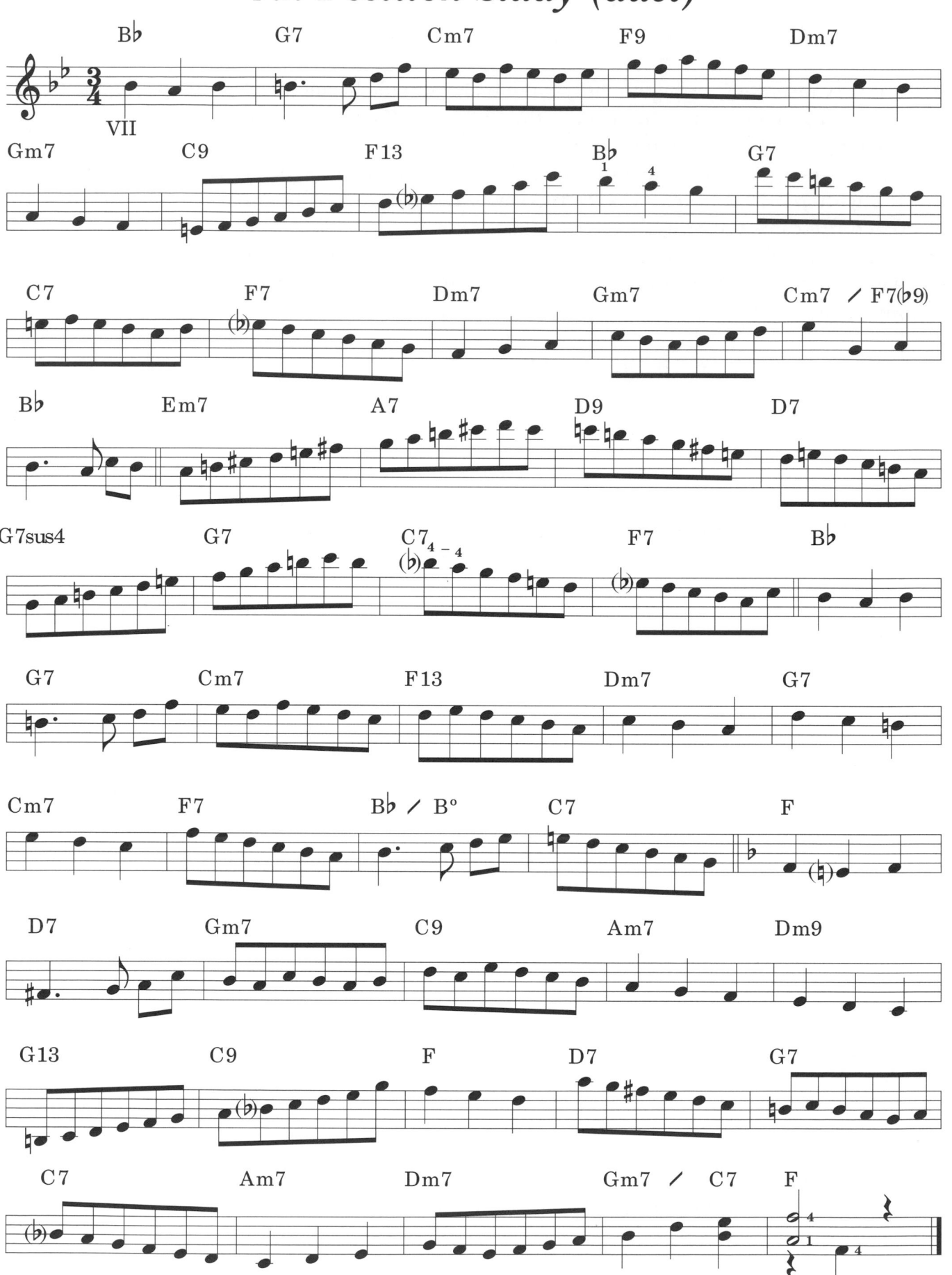

Solo in G

B♭ Real Melodic Minor (Five Positions)

FINGERING DERIVED FROM TYPE 4

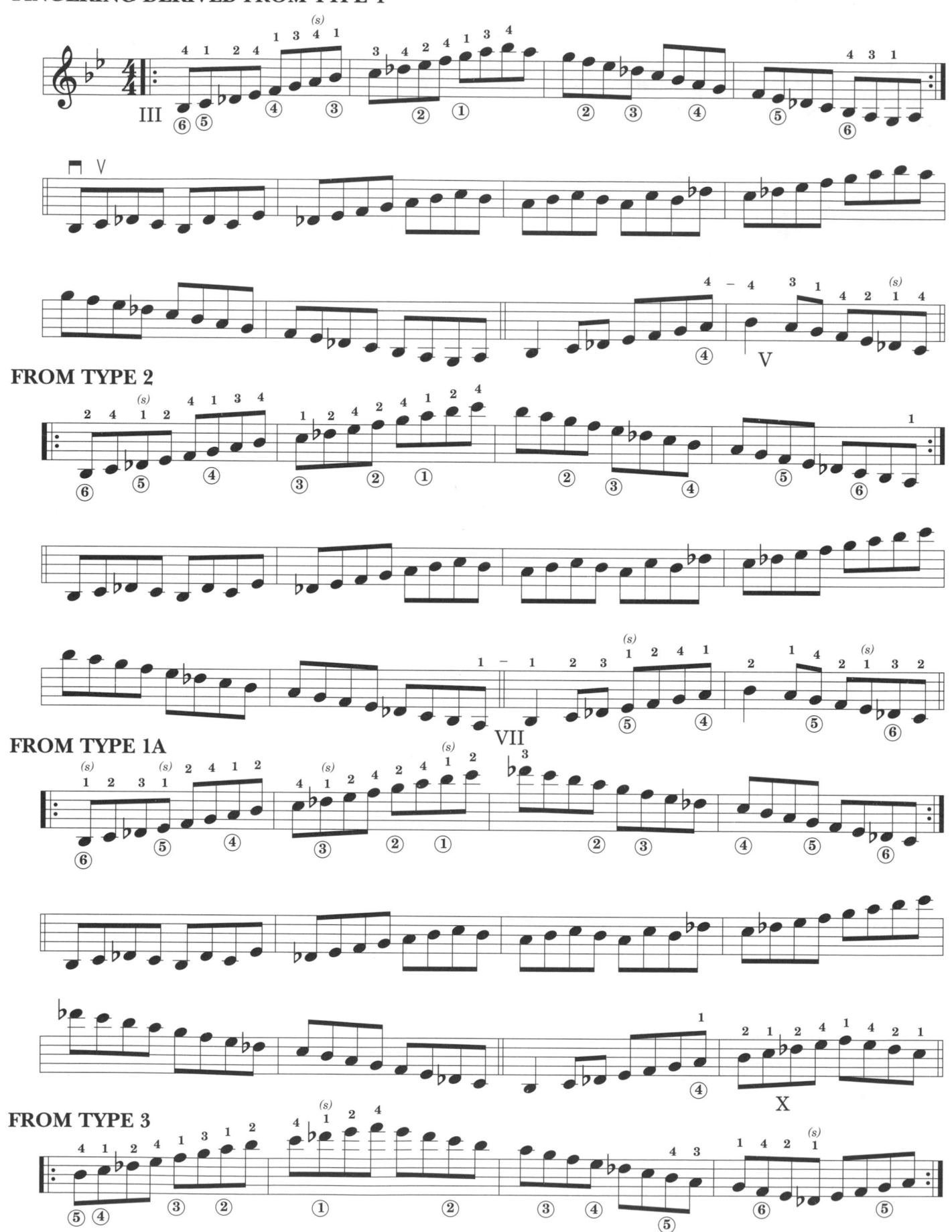

FROM TYPE 2

FROM TYPE 1A

FROM TYPE 3

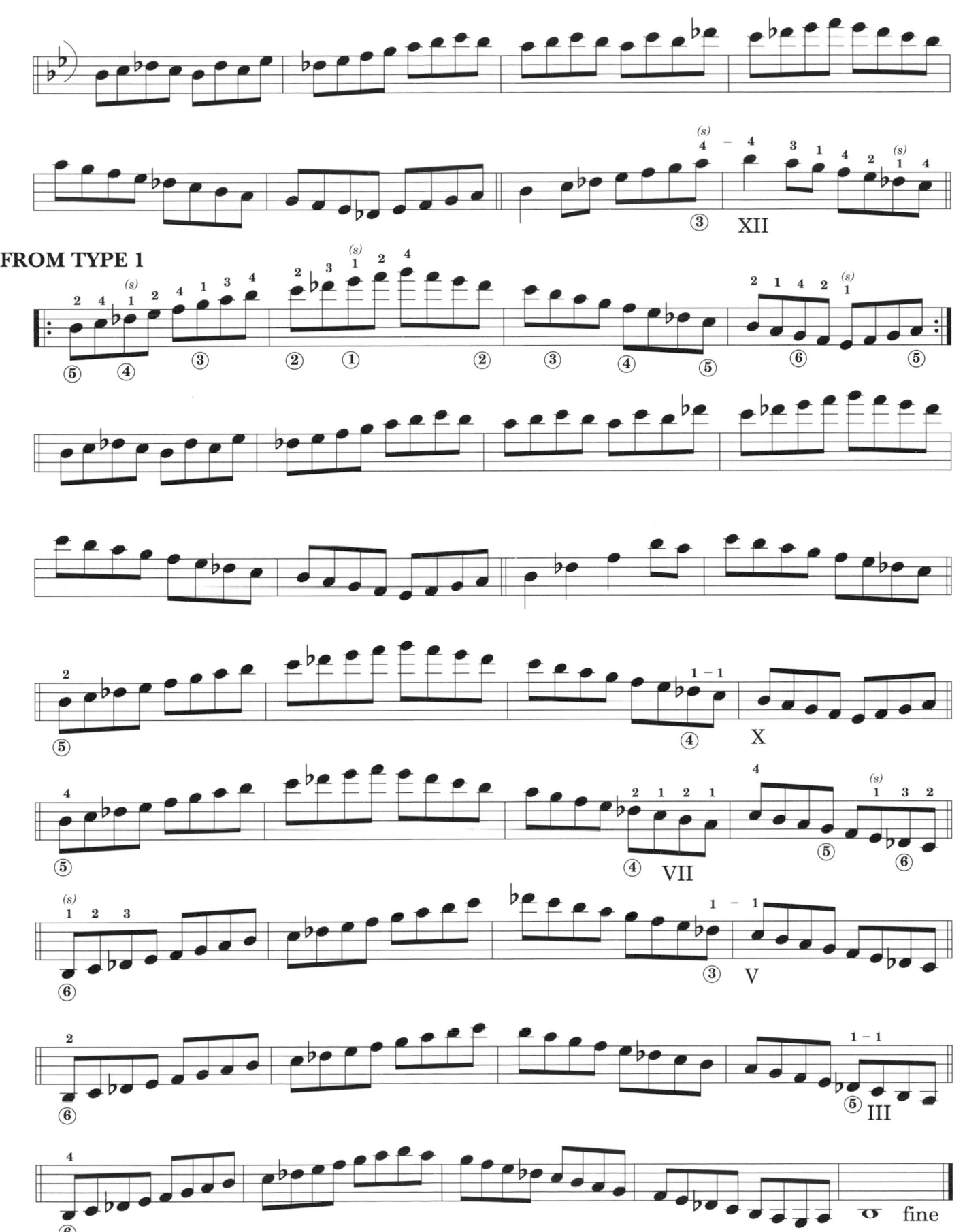

FROM TYPE 1

99

Chord Forms—7th in the Bass

Bass (sounding) range: From approximately C (5th or 6th strings) on down in pitch.

Chord voicings with the 7th degree in the bass have very weak chordal sounds. These forms (like those with the 3rd in the bass) may be used for inversion leaps or as passing chords, but their use must be well justified—such as in a strong descending bass line—or they will sound "wrong."

Chord Etude No. 11

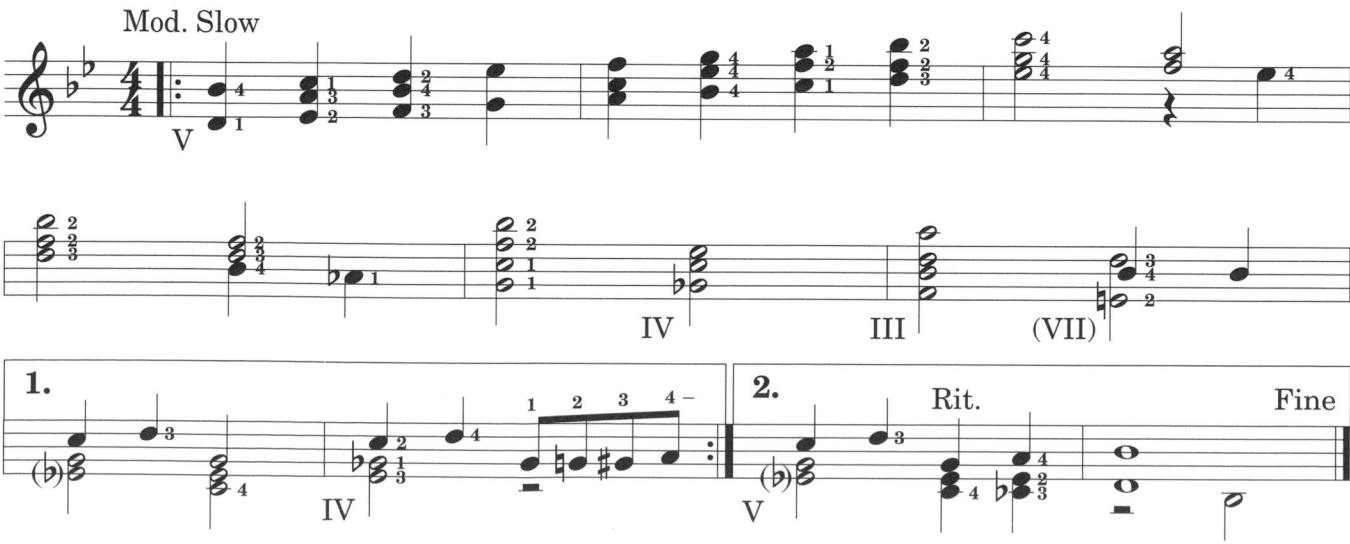

Two-Octave Arpeggios

G Minor Triad from the 3rd

Fingering derived from scales and chords—Across and up the fingerboard.

F Minor Triad from the 5th

Across and up the fingerboard.

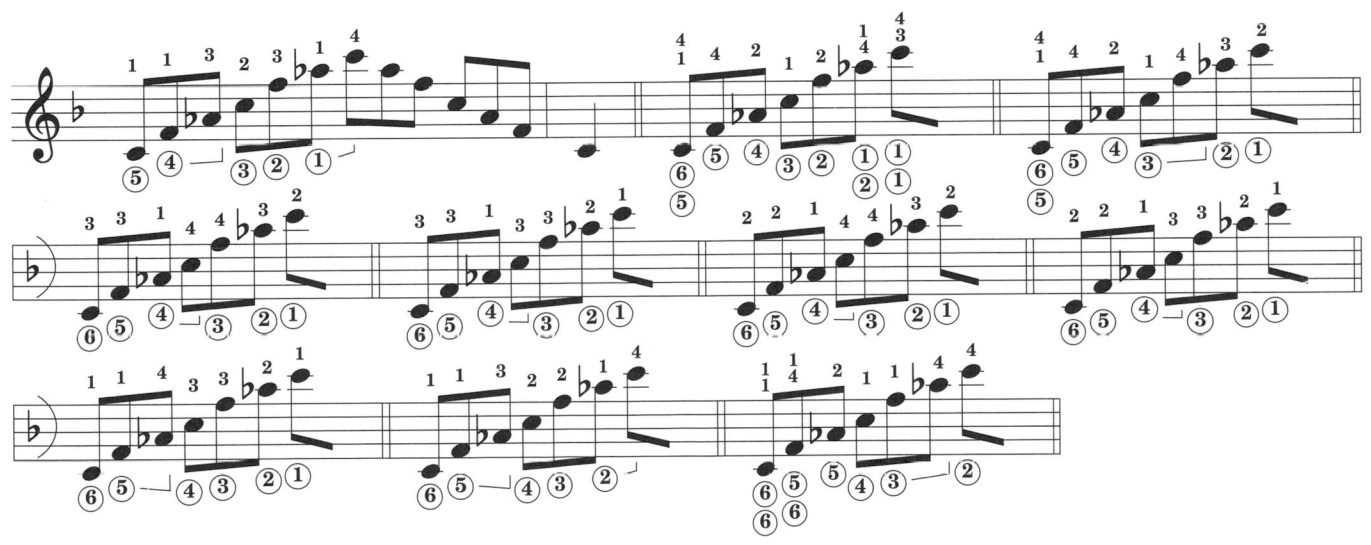

Practice all forms in all possible keys.

Key Signatures: The order of appearance of flats and sharps.

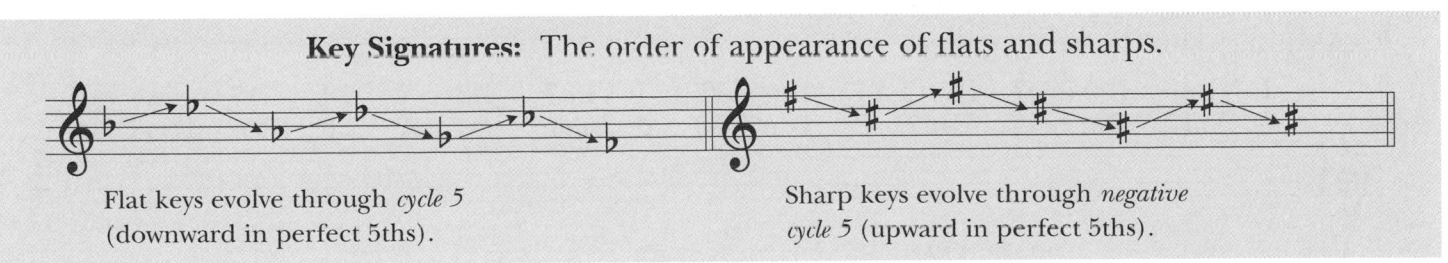

Flat keys evolve through *cycle 5* (downward in perfect 5ths).

Sharp keys evolve through *negative cycle 5* (upward in perfect 5ths).

Theory: Chord-to-Chord Motion
A Brief Discussion

DESCRIPTIONS AND TERMS

1.) Chord sequences (cadences) are represented by numerical terms or numbers that indicate the chords and their structures in the key of the moment. If a single number is used to represent a chord, the structure is assumed to be diatonic (in the indicated key).

EXAMPLE

II-V-I in C = Dm7-G7-C

II-V-I in F = Gm7-C7-F

2.) Nondiatonic structures are represented by two numbers and, if necessary, a descriptive term or symbol.

EXAMPLE

	"one"	"six-seven"	"two-seven"	"five-seven"	"one"
(Key of C)	I	VI7	II7	V7	I
	C	A7	D7	G7	C

	"one"	"one sharp diminished"	"two"	"flat two-seven"	"one"
(Key of C)	I	I#°	IIm7	♭II7	I
	C	C#°	Dm7	D♭7	C

3.) Chord sequences are also described in another way. The word "cycle" followed by a number indicates the interval (distance) from chord root to chord root. In the most common chord progressions (cycle 5, cycle 3, cycle 7), the interval is figured downward. Notice in the following examples that, in use, the direction of bass notes is optional, but the chords have been constructed from the notes a 5th, 3rd, or 7th below.

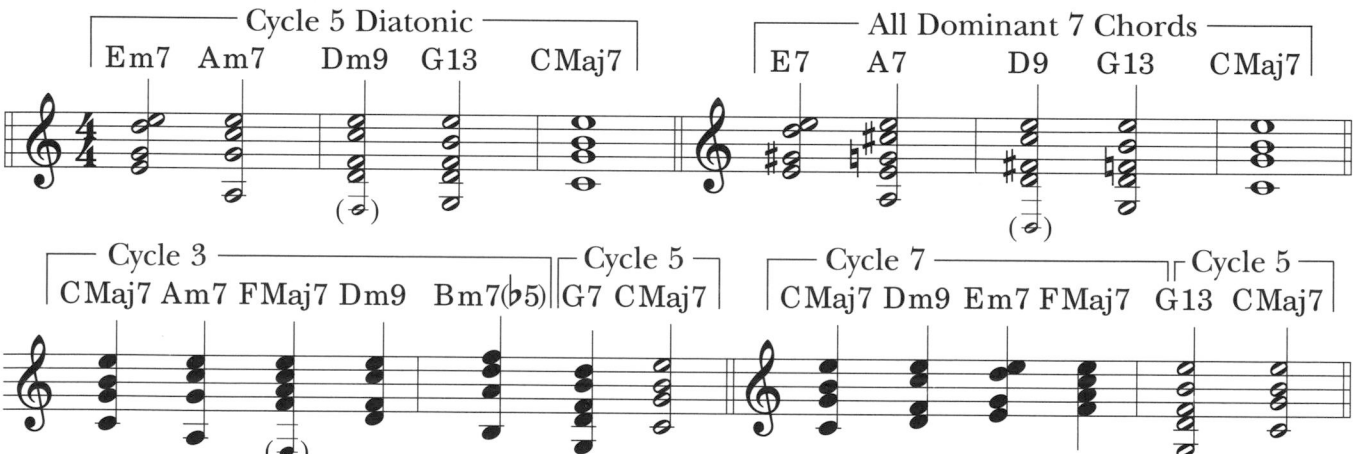

4.) When chord root motion goes up a 3rd, 5th, or 7th, it is called a negative cycle 3, negative cycle 5, or negative cycle 7. (One sequence of two chords is common; further extension of negative cycles is less common.)

Both of the above methods of indicating chord motion are extremely valuable, especially in memorizing and transposing the chords to songs.

EXAMPLE: (First 16 bars)

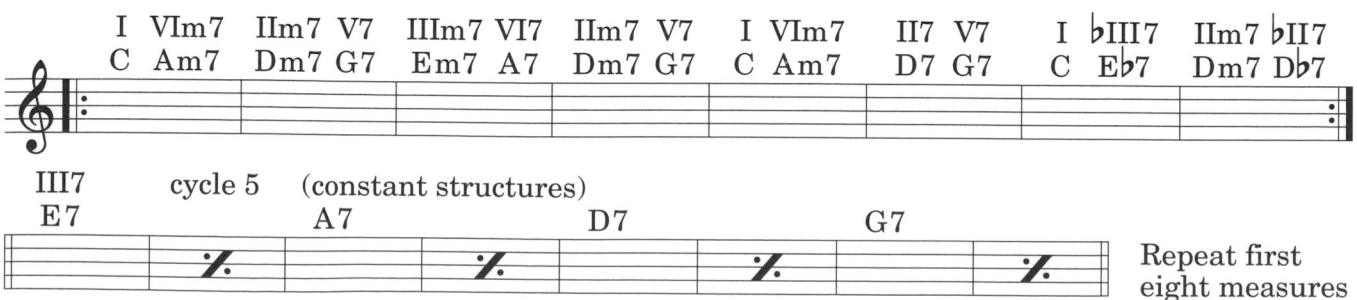

Chromatic Scale

The chromatic scale consists of twelve notes, one half step apart.

FINGERING PATTERN 1

Across fingerboard—no position change.

Examples of application shows use of chromatic scales over augmented and diminished (optional dominant 7(♭9)) chords. Observe the way that sixteenth notes and triplets are used to ensure that the first attack of each beat is a chord tone.

FINGERING PATTERN 2

Across fingerboard—with position changes.
This is less practical than the fingering shown above, as the use of this pattern must be pre-set in order to come out in the proper position.

103

Eb Real Melodic Minor (Five Positions)

FINGERING DERIVED FROM TYPE 3

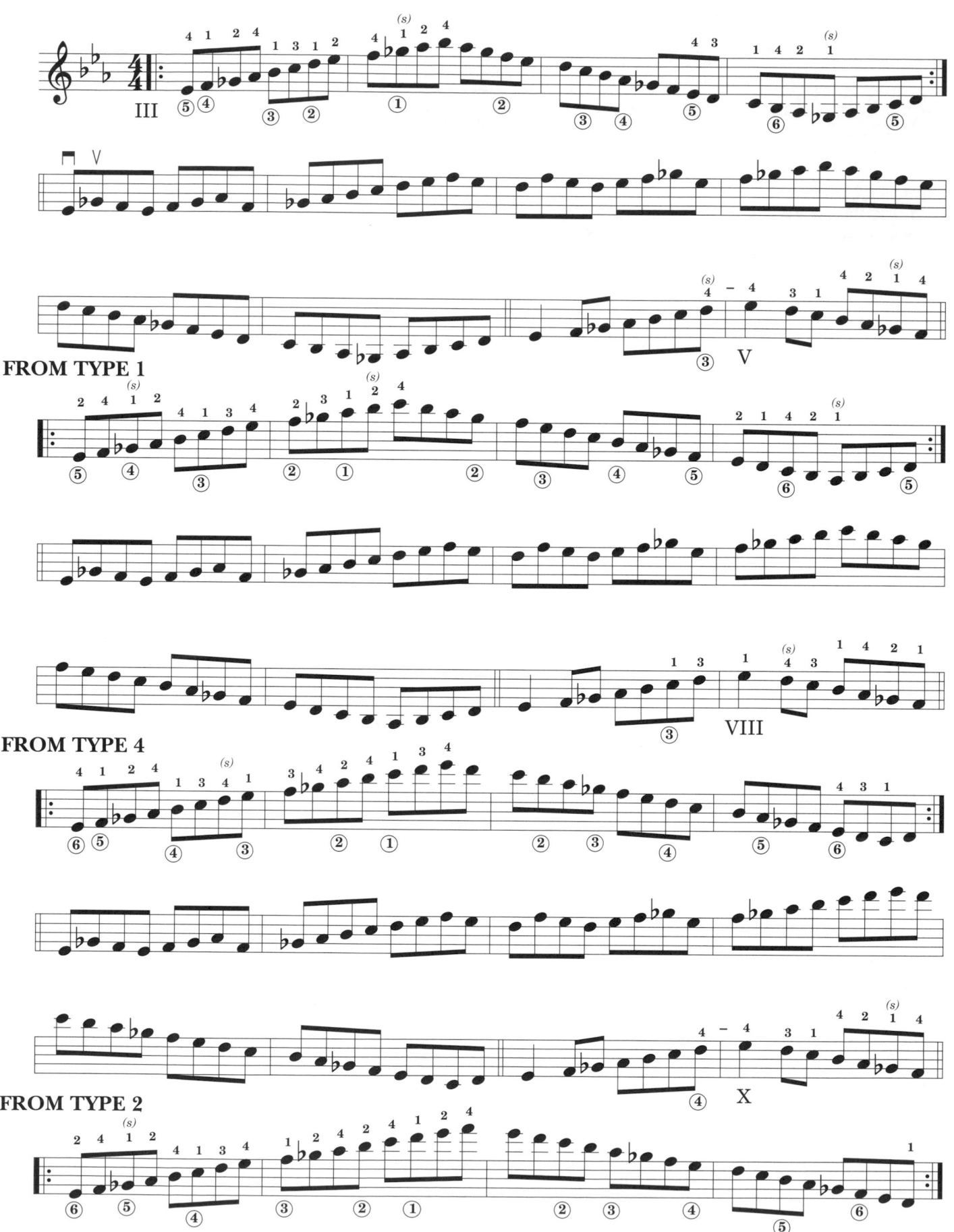

FROM TYPE 1

FROM TYPE 4

FROM TYPE 2

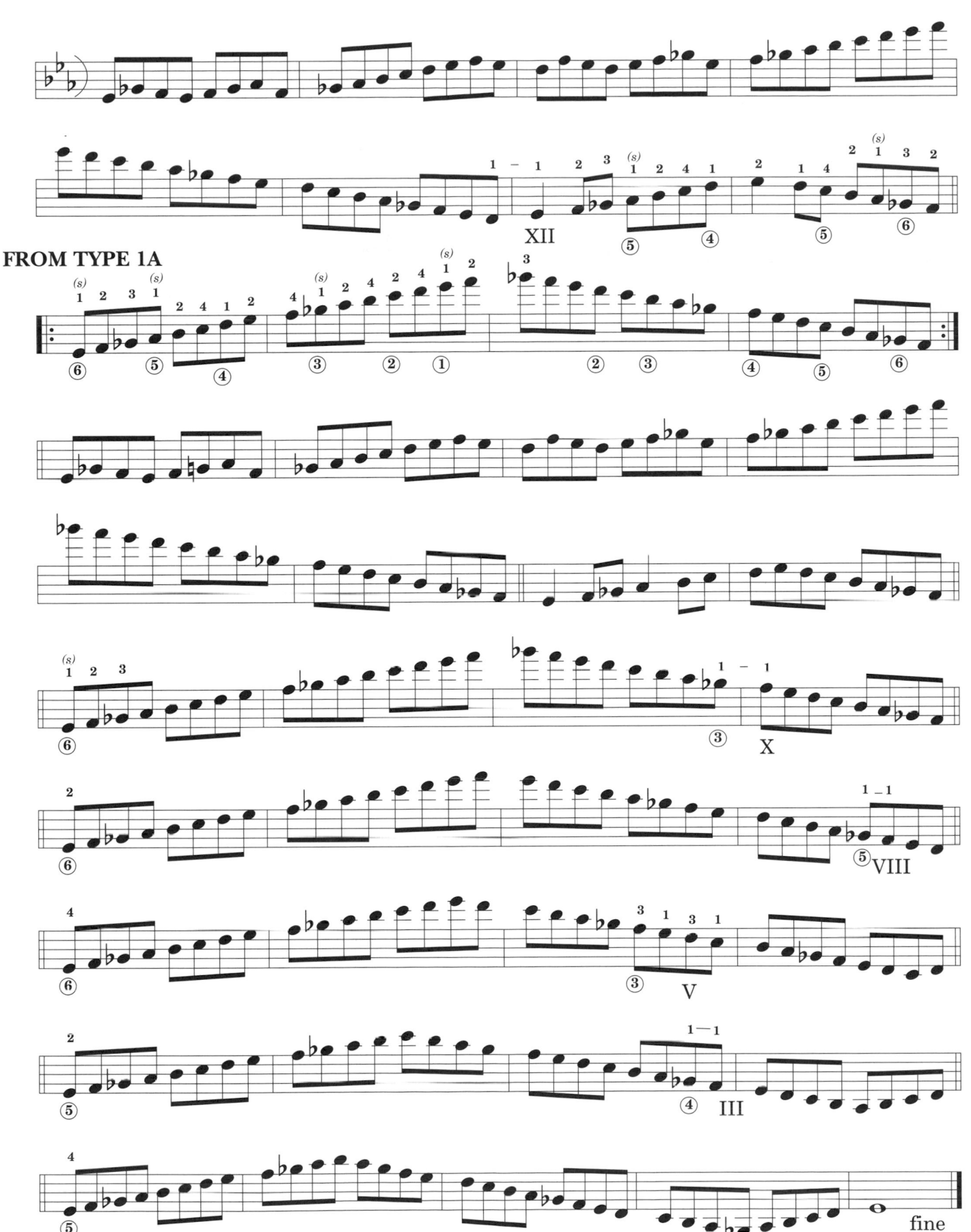

FROM TYPE 1A

Chord Forms

Diminished 7
(with added high degrees)

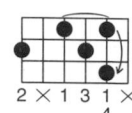

Diminished 7 chords may be named from any chord tone. High degrees (two frets above any diminished chord tone) give you the name of the four dominant 7(♭9) chords with the same sound.

(Gdim = A7(♭9) = B♭dim = C7(♭9) = D♭dim = E♭7(♭9) = Edim = F♯7(♭9))

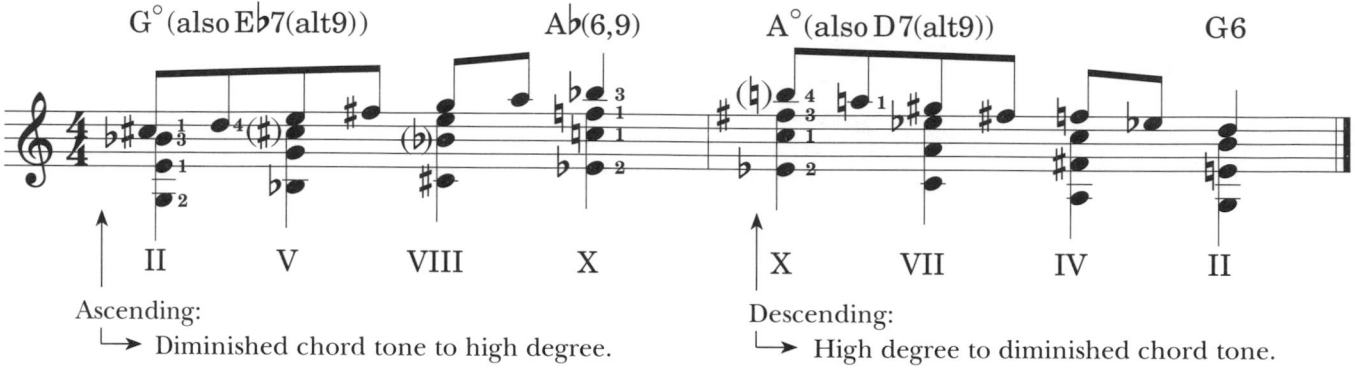

Ascending:
→ Diminished chord tone to high degree.

Descending:
→ High degree to diminished chord tone.

Diminished 7
(with added high degrees)

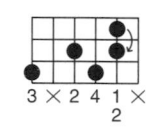

High degrees on diminished 7 chords may also be thought of as the note one fret below any diminished chord tone. (They will be the same four notes as those found two frets above.)

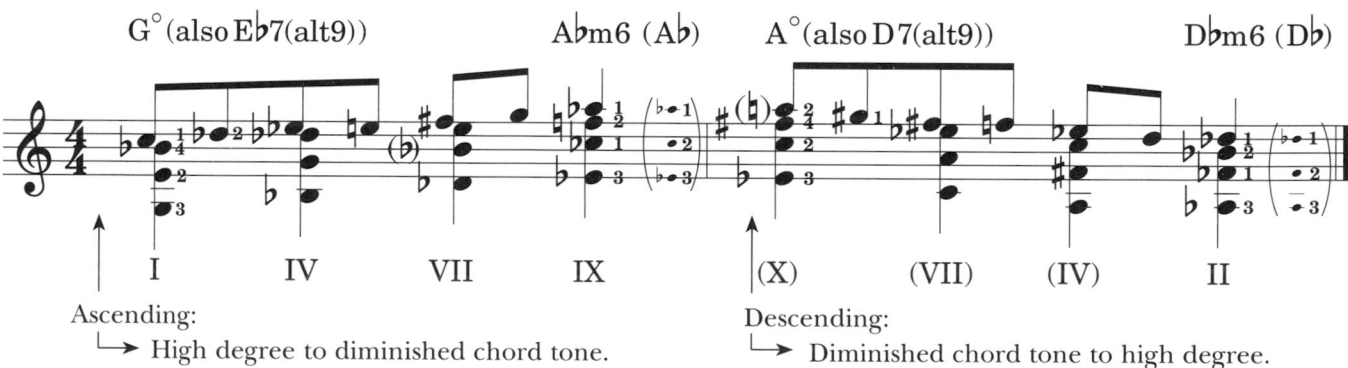

Ascending:
→ High degree to diminished chord tone.

Descending:
→ Diminished chord tone to high degree.

Chord Etude No. 12

106

Speed Study

Tempo must be constant throughout.

For practice with other fingerings, change the key signature to C, F, D, and A. Also use speed studies for real melodic minor scales. Practice all suggested keys with ♭13.

Two-Octave Arpeggios

C Diminished Triad from the Root

Across and up the fingerboard.

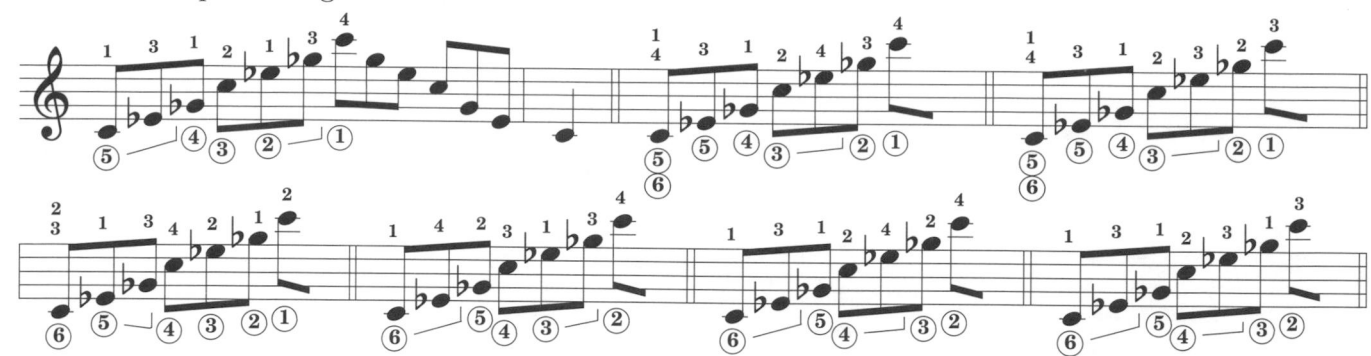

G Diminished Triad from the 3rd

F Diminished Triad from the 5th

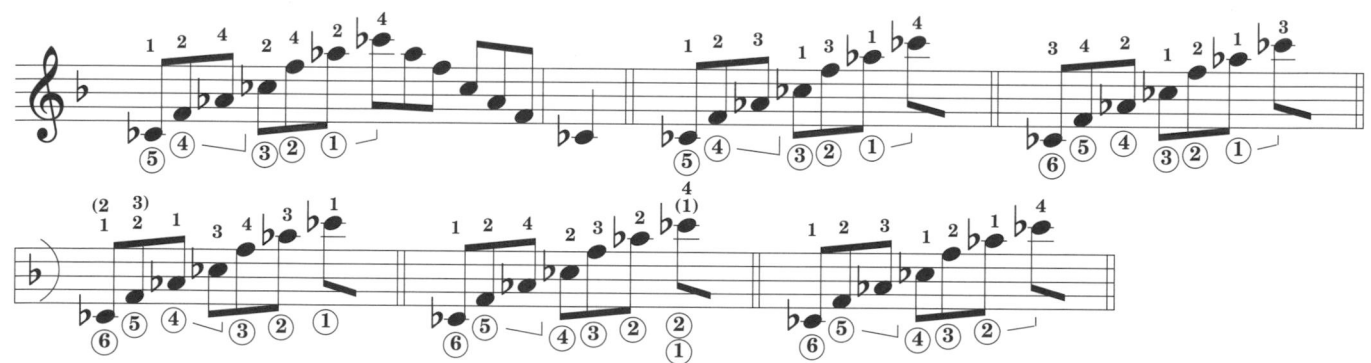

Practice all forms in all possible keys.

Melodic Rhythm No. 7 (duet)

Whole Tone Scales

The whole tone scale consists of six notes, a whole step apart. Each scale tone can be considered the tonic. Therefore only two whole tone scales exist.

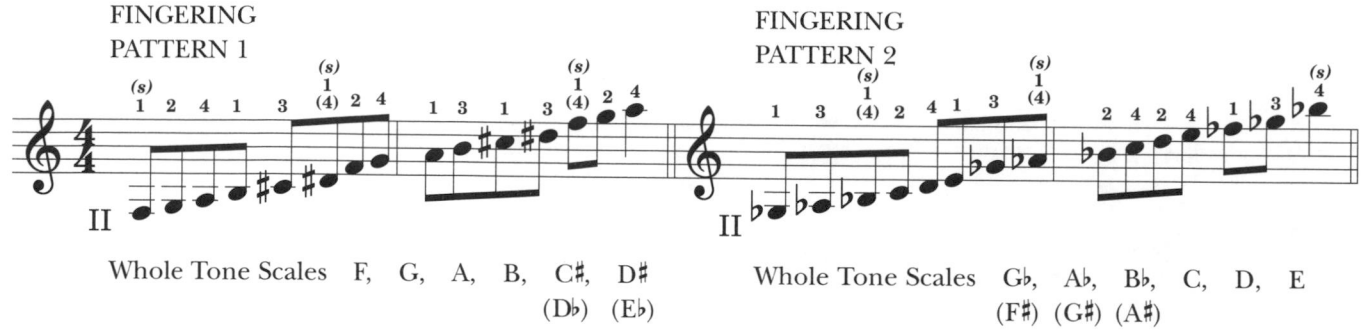

FINGERING PATTERN 1

FINGERING PATTERN 2

Whole Tone Scales F, G, A, B, C♯, D♯
(D♭) (E♭)

Whole Tone Scales G♭, A♭, B♭, C, D, E
(F♯) (G♯) (A♯)

Practice ascending and descending from each finger. (First-finger stretches are the most practical, but eventually include all possibilities.)

PATTERN 1

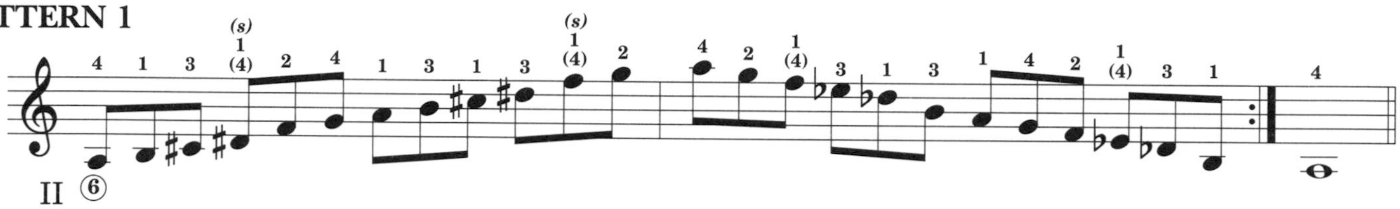

PATTERN 2

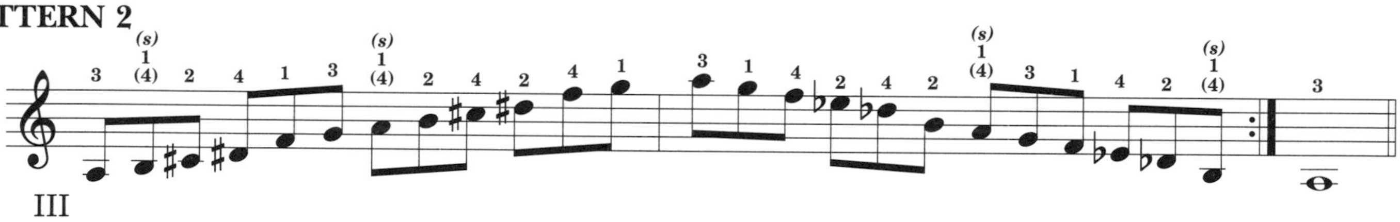

PATTERN 1

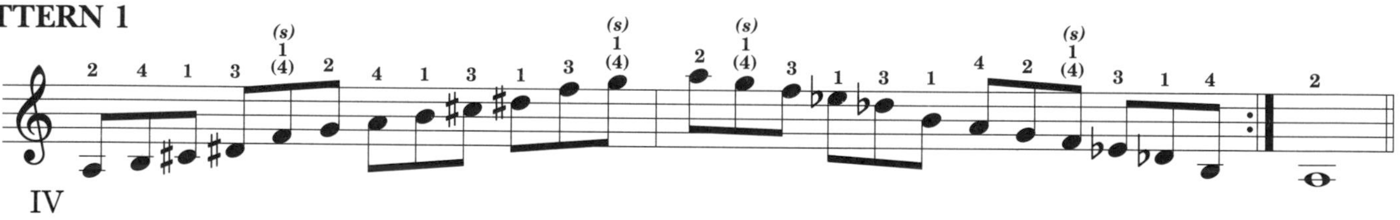

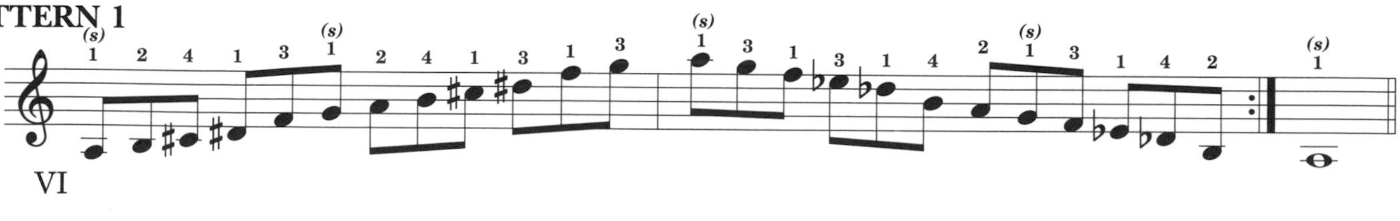

PATTERN 2

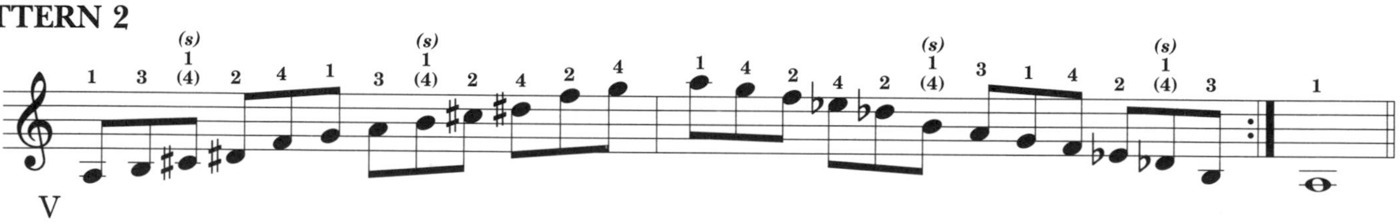

PATTERN 1

Memorize the fingering patterns. Practice both whole tone scales, in all positions.

The principal use of whole tone scales in improvisation is over augmented triads, and (augmented) dominant 7 chords (where the ninth is unaltered, or can be assumed to be unaltered).

EXAMPLE

The following example employs both whole tone scales (same position).

Additional Whole Tone Scale Fingerings with Position Changes:

1.) Across and down the fingerboard as scale ascends (two octaves).

2.) Constant fingering-position change every string (three octaves).

✸ These additional fingerings are less practical for general use.

111

Rhythm Guitar—The Right Hand (Bossa Nova)

Basic Stroke 1

Bass note or
muffled bass strings

Orchestral

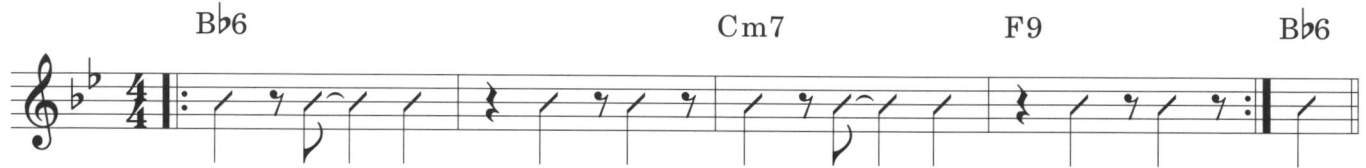

■ **EXERCISE**

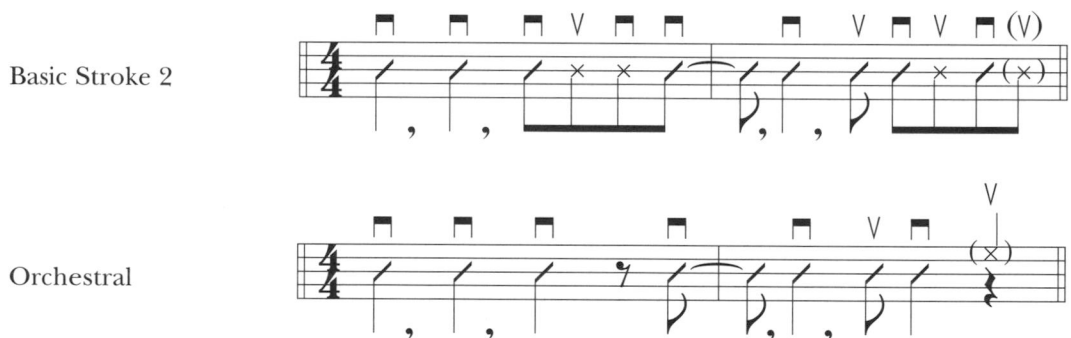

Observe notation. Practice with basic and orchestral strokes.

Basic Stroke 2

Orchestral

■ **EXERCISE**

Practice with each Basic Stroke 2.

Variation 1
Orchestral

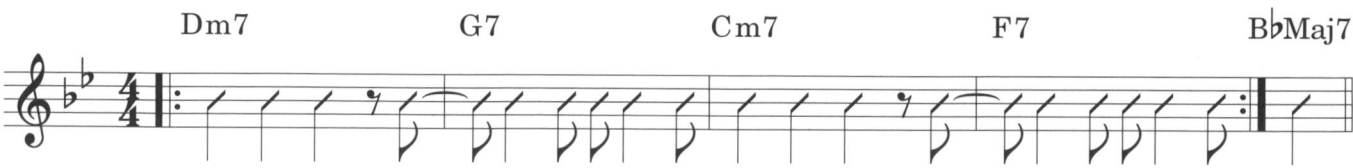

❚ Tap the foot in "2."

■ **EXERCISE**

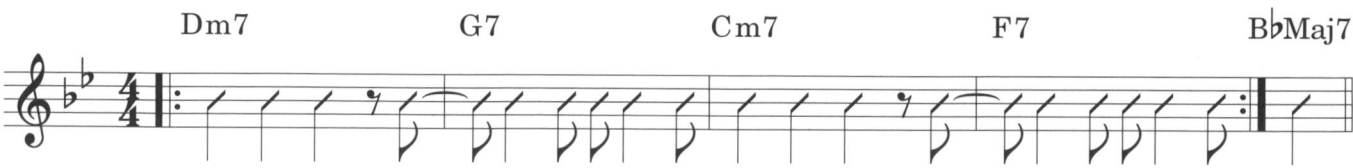

■ Variation 2
Orchestral

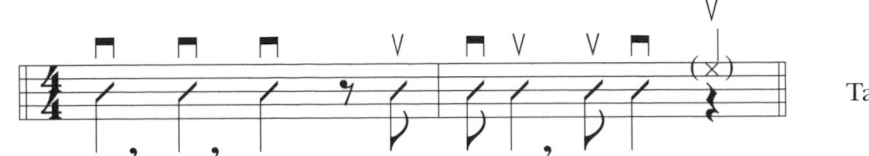

Tap the foot in "2."

■ **EXERCISE**

Cm7   F7   B♭Maj7

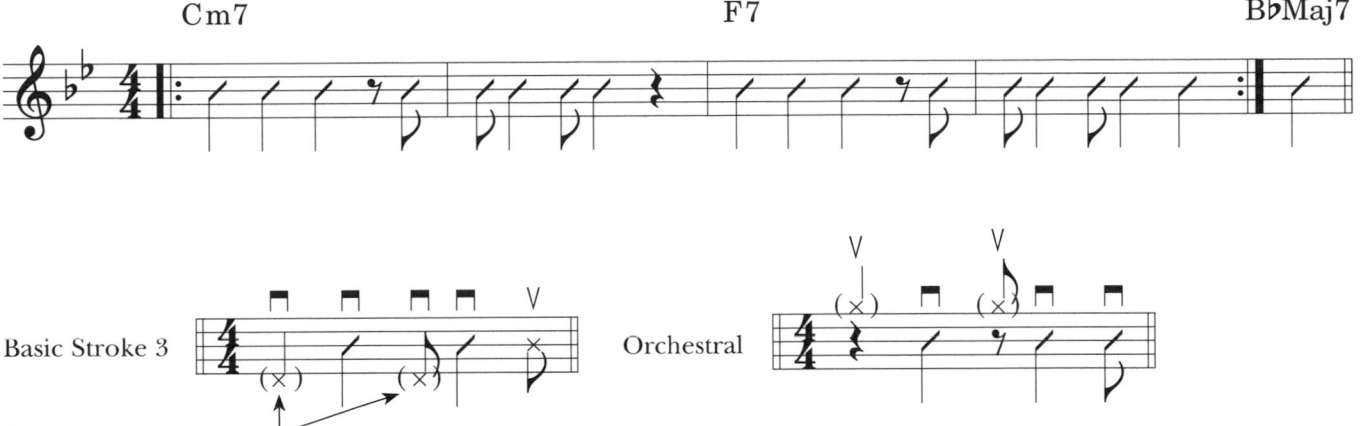

Basic Stroke 3   Orchestral

(Bass note or muffled bass strings)

■ **EXERCISE**

B♭Maj7   Gm7   Cm7   F7   B♭Maj7

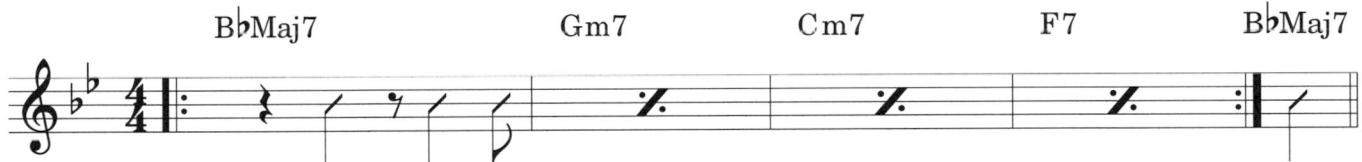

■ Practice this exercise with each Bossa Nova basic stroke #3.

Two-Octave Arpeggios
B♭, F♯, and D Augmented Triads
From the root, 3rd, and augmented 5th—Across and up the fingerboard.

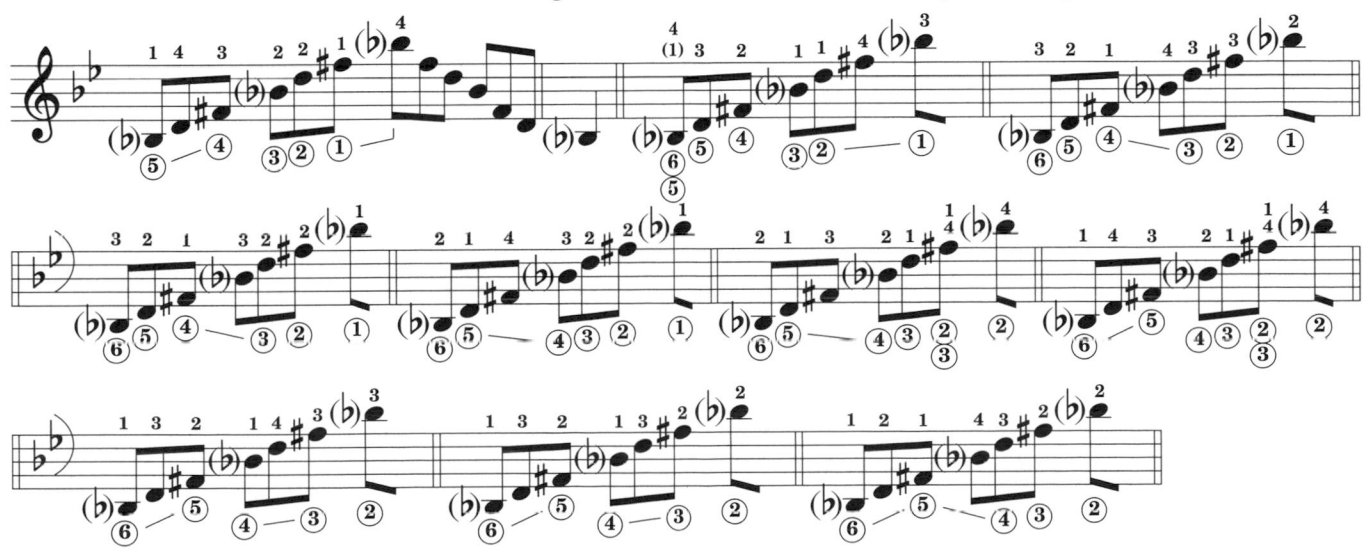

Using the preceding forms, practice and learn augmented triad arpeggios from all notes possible.

113

Theory: Chord-Scale Relationships

For improvisation.

Nondiatonic Minor 6 and (Unaltered*) Dominant 7 Chords

1.) The tonic and subdominant (I and IV) chords in a major key are often found temporarily altered to minor 6 structures (Im6 and IVm6). Use the real melodic minor scale built from the chord name for Im6 and IVm6 in major keys. Be careful of minor 6 chords. Be sure they are actually functioning as Im6 or IVm6 before employing the above. They are often misnamed minor 7(♭5) chords (the diatonic VIIm7(♭5) of a major key) or ninth chords (V7(9) renamed to indicate bass motion).

2.) These same Im6 and IVm6 chords will also appear (harmonically extended) as dominant 7 chords on the 4th and lowered 7th scale degrees (IV7 and ♭VII7). These dominant 7 structures include chord degrees 9, 11+, and 13. Use the real melodic minor scale from the 5th chordal degree of IV7 and ♭VII7.

EXAMPLE

Chord	C	Cm6	C	C7	F	Fm6	C G7	C
Scale	CMaj	C Real Mel. Min.	CMaj	FMaj	CMaj or FMaj	F Real Mel. Min.	CMaj →	

Chord	C	F9	C	C7	F	B♭9	C G7	C
Scale	CMaj	C Real Mel. Min.	CMaj	FMaj	CMaj or FMaj	F Real Mel. Min.	CMaj →	

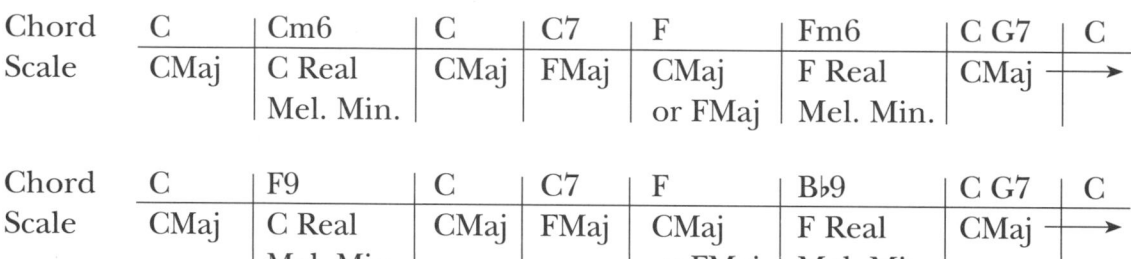

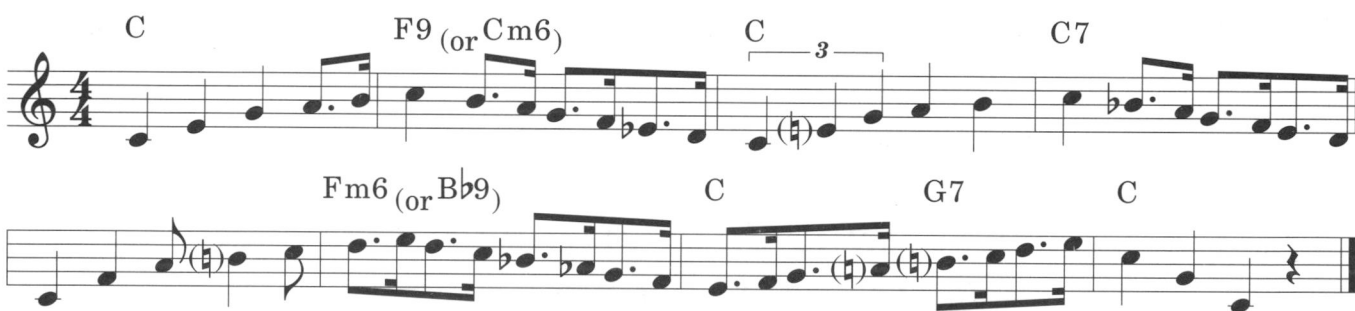

3.) Any unaltered dominant 7 chord with a nondiatonic root (not preceded by a modulating IIm7) tends to sound like ♭VIIm7 of whatever key it is the lowered 7th degree of. All dominant 7 chords with nondiatonic roots include chord degrees 9, 11+, and 13. Use the real melodic minor scale from the 5th chordal degree of the dominant 7 with nondiatonic root.

EXAMPLE

Chord	C	E♭9	Am7	A♭13	Dm9	D♭9	C	B♭9	C
Scale	CMaj	B♭ Real Mel. Min.	CMaj	E♭ Real Mel. Min.	CMaj	A♭ Real Mel. Min.	CMaj	F Real Mel. Min.	CMaj

�֎ Unaltered in this instance means no ♭9, +9, ♭5, or +5.

114

Dm9   Db9   C   Bb9   C

4.) We do not (as yet) have the necessary "scale tools" to properly handle all dominant 7 chords with diatonic roots. Therefore I suggest that, for now, you use the major or real melodic minor scale derived from the **intended tonic*** chord for all dominant 7 chords with scale tone roots, except IV7. (See previous page.)

*Intended tonic: Where the chord would normally resolve to, e.g., B7 to E, E7 to A, A7 to D, etc.

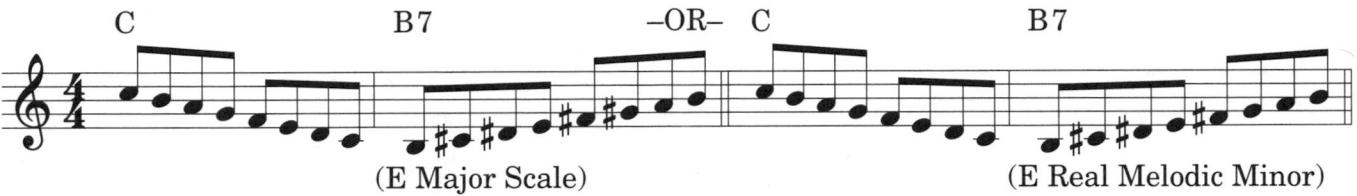

C   B7   –OR– C   B7

(E Major Scale)   (E Real Melodic Minor)

The real melodic minor constructed from the 5th chordal degree may be used on any (unaltered) dominant 7 chord at any time. But, because most dominant 7 chords with scale tone roots have 9 and/or 13 altered by the surrounding key sound, this chord-scale relationship is imperfect.
I recommend that you avoid this for now.

* * * * * * * * * *

You must hear the sound of related scales with chords. Have someone play the changes for you (or use a tape recorder) and experiment with them. Much depends upon your command of the scales, mentally and physically, and upon correct chord names.

It is a very long process to learn (never mind to use!) the chord-scale relationships covering all harmonic situations. Only diligence, perseverance, and considerable experimentation (including thinking, playing, and listening) will make it possible.

I have only scratched the surface of chord and scale relationships in this book. We will pursue this much further in *Volume III*.

Solo in D

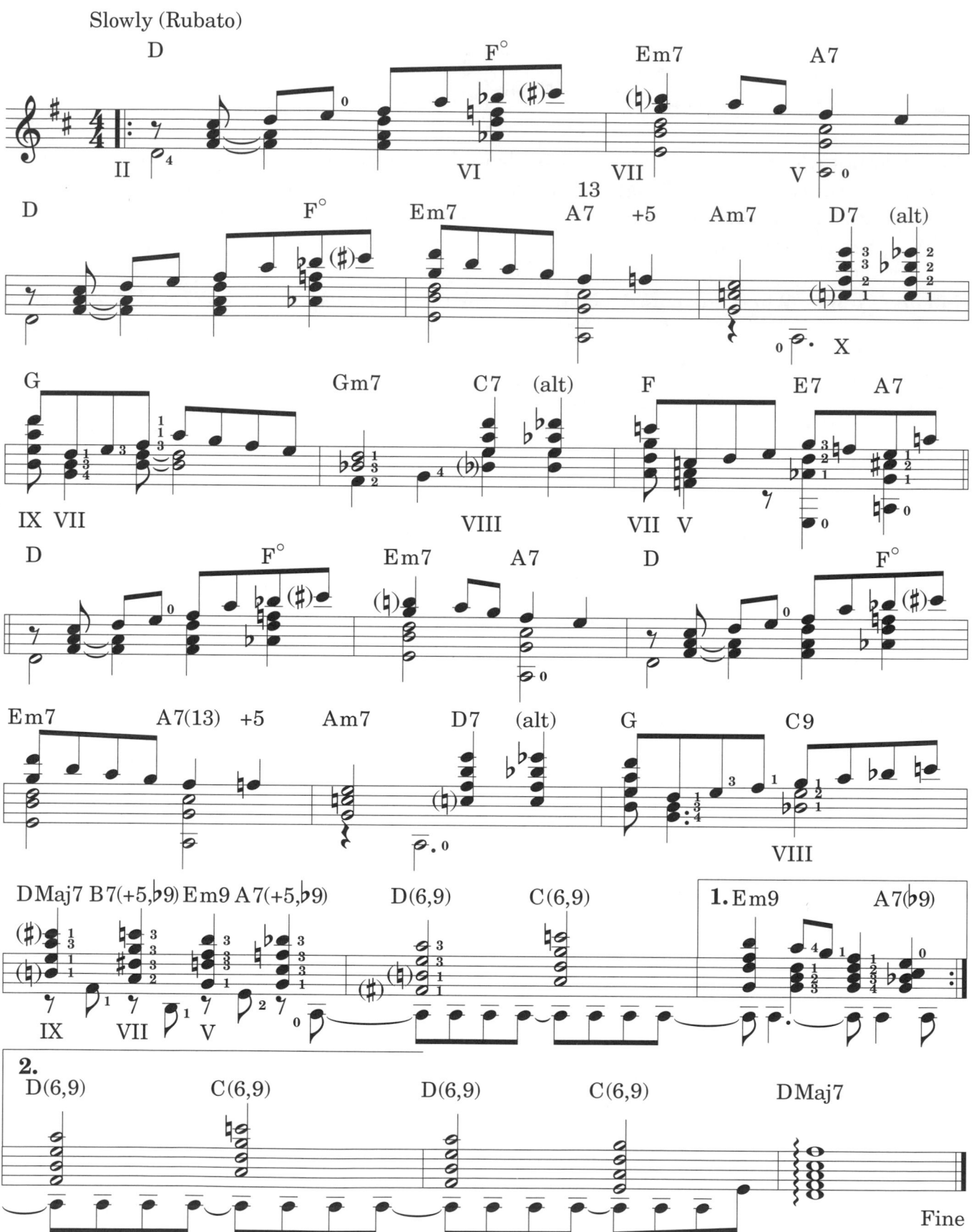

Index

A MODERN METHOD FOR GUITAR

william leavitt

volume 3

Berklee
Press

1140 Boylston Street
Boston, MA 02215-3693 USA
(617) 747-2146

Visit Berklee Press Online at
www.berkleepress.com

DISTRIBUTED BY

HAL•LEONARD®
7777 W. BLUEMOUND RD. P.O. BOX 13819
MILWAUKEE, WISCONSIN 53213

Visit Hal Leonard Online at
www.halleonard.com

Introduction

This book is a continuation of *A Modern Method for Guitar, Volumes I and II.* Most of the terms and techniques are directly evolved from material presented in those books. Fingerings for two-octave scales and arpeggios are developed to the ultimate, in that any other patterns that you may discover will consist of nothing more than combinations of two or more of those presented here. Three-octave patterns will be shown in later volumes, but many can be worked out with the aid of the position-to-position fingerings on pp. 76 and 77.

With regard to chords and harmony, diagrams are totally dispensed of, and everything is worked out from a knowledge of chord spelling and the construction of voicings. There will be further development later in this area of study.

Mastery of the "Right-Hand Rhythms" pages should enable you to perform any rhythmic combinations that may confront you at any time, assuming, of course, that you have the ability to swing. (If this property is lacking, then perhaps you had better throw the pick away!)

Should you be fortunate enough to possess a creative soul, I'm sure that the pages devoted to chord and scale relationships will be a rather large help. In any event this knowledge can certainly keep you out of trouble when you have some on-the-spot filling to do.

As in the preceding volumes, all music is original and has been created especially for the presentation and perfection of the lesson material.

Once again, all the best and good luck.

Wm. G. Leavitt

It is important that you cover the following material in consecutive order. The index on pg. 158 is for reference purposes only and will prove valuable for review or concentration on specific techniques.

Outline

Evolution of Major Scale Fingering Patterns

Type I fingerings evolve through cycle 5 (down a 5th). Using the second position as a sample, we start with C major, fingering Type 1, then proceed to F major (Type 1A), B♭ major (Type 1B), E♭ major (Type 1C), and A♭ major (Type 1D). Observe that each new key requires additional first-finger stretches. Also note the optional fourth-finger stretch shown on the 2nd string of Type 1D. This will occasionally be necessary for certain melodic patterns, such as 3rds.

Type I

TYPE 1

TYPE 1A

TYPE 1B

TYPE 1C

TYPE 1D

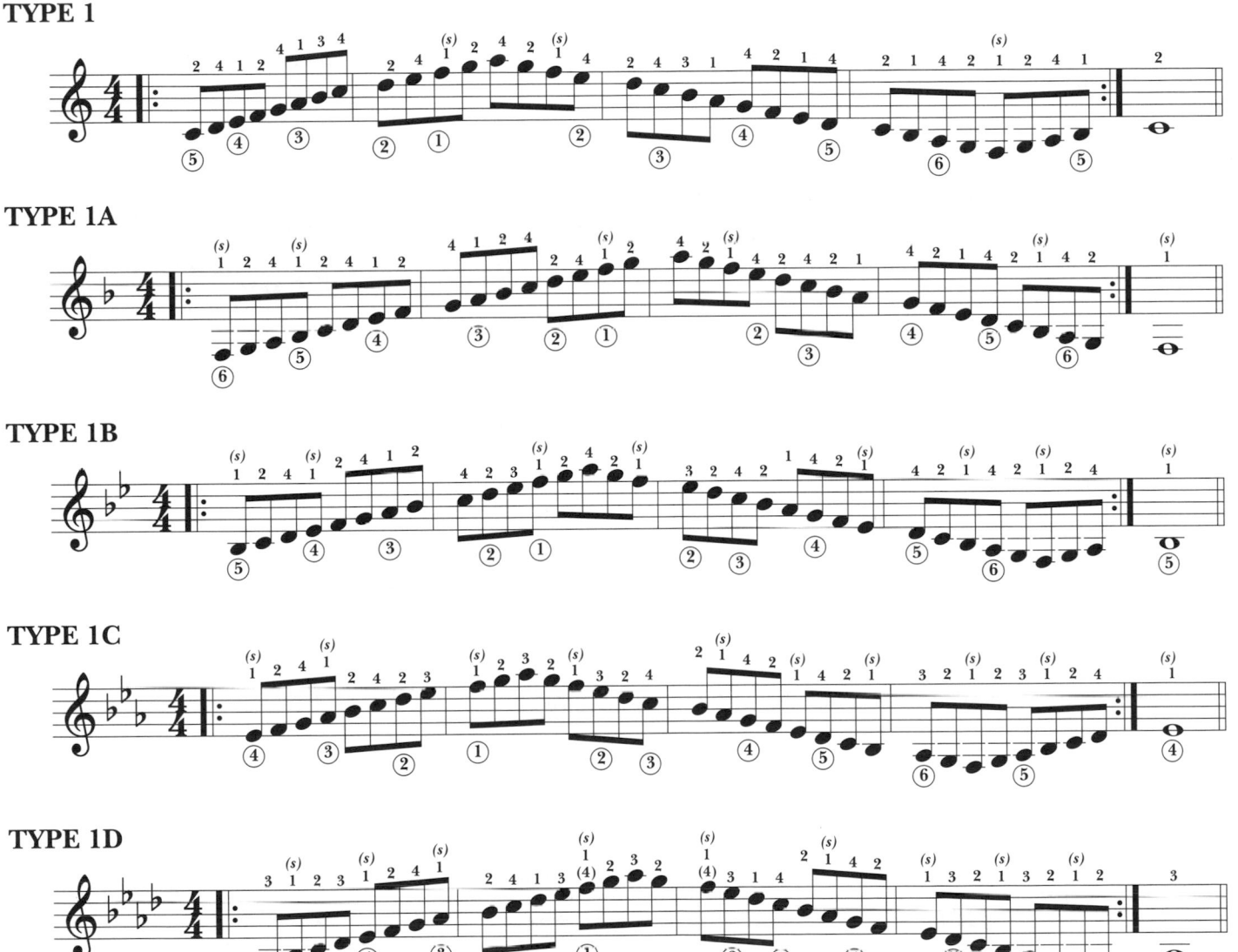

1

Type II

TYPE 2

▌ No derivative fingerings

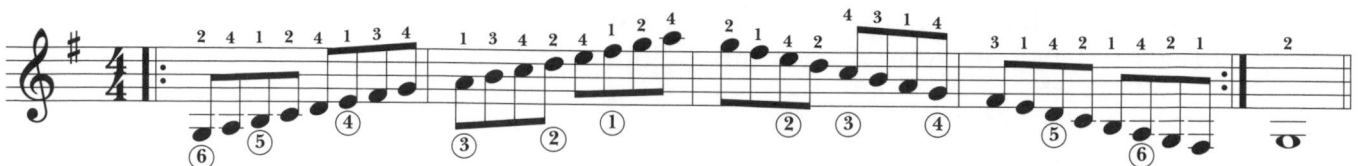

Type III

TYPE 3

▌ No derivative fingerings

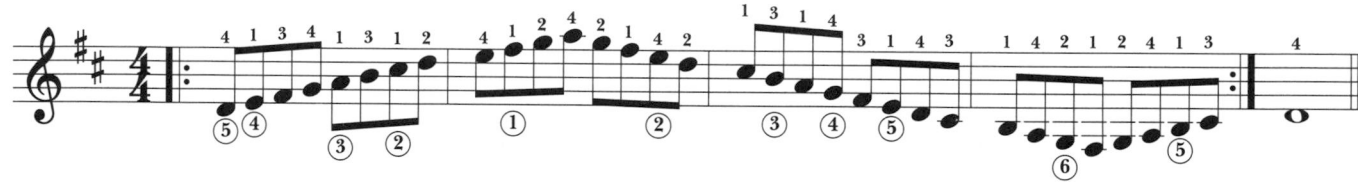

Type IV

TYPE 4

▌ Type 4 fingerings evolve through negative cycle 5 (up a 5th). Using the second position as a sample, we start with A major, fingering Type 4, then proceed to E major (Type 4A), B or C♭ major (Type 4B), F♯ or G♭ major (Type 4C), and C♯ or D♭ major (Type 4D). Observe that each new key requires additional fourth-finger stretches.

▌ Also note that fingering Type 4D is shown with optional first-finger stretches, which actually represent a combination of Types 1 and 4. The combined pattern is usually best.

TYPE 4

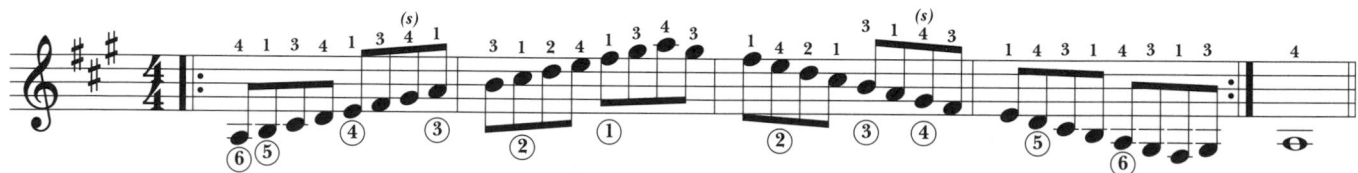

TYPE 4A

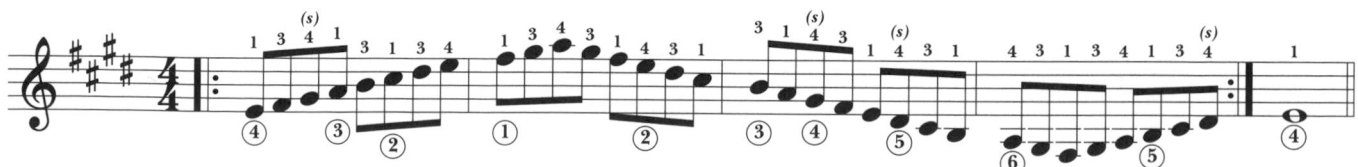

TYPE 4B

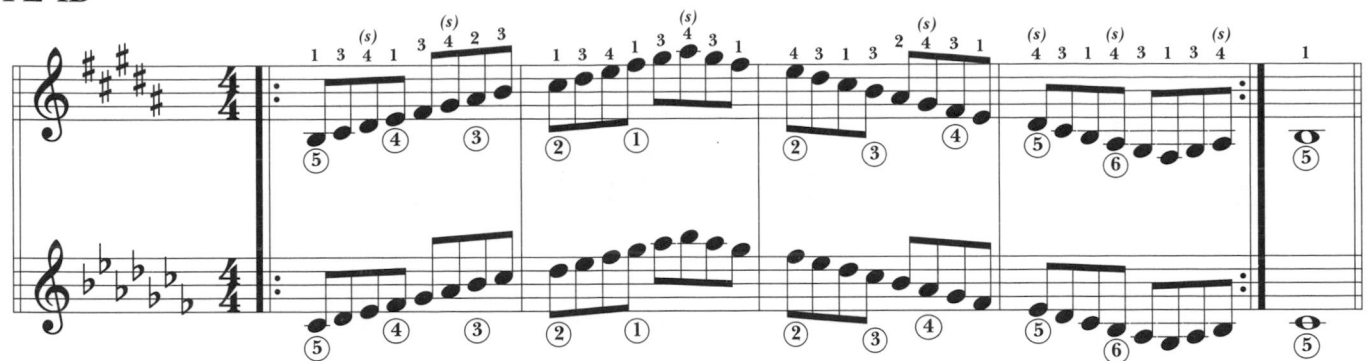

TYPE 4C

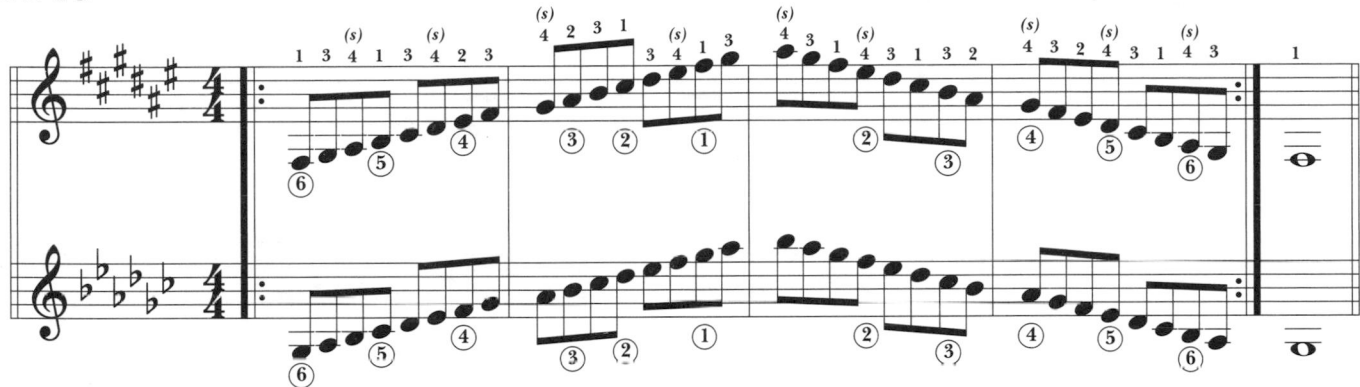

TYPE 4D*

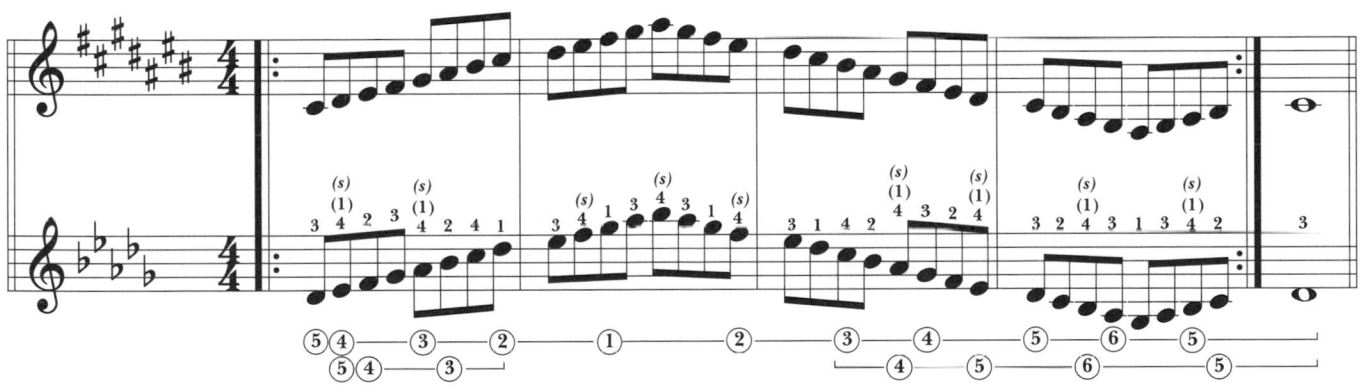

✱ Although this fingering has evolved from Type 4, it is best played in combination with Type 1. On the following pages only this mixed fingering will be shown. (It will be referred to as 1D/4D.)

Familiarity with all twelve major scale fingerings is valuable, especially when reading something for the first time. All forms do not, however, convert to really practical minor scale fingerings. On the following pages, only the nine best minor forms resulting from the conversion of the preceding major patterns will be emphasized. Eventually, all possibilities will be shown.

A Refined Definition of Position

Now that we have encountered many finger stretches with all the fingering possibilities, let's refine the definition of position. Let's now say: *one fret below the placement of the second finger determines the position.*

Speed Study

Tempo must be constant throughout.

Change the signature and practice in other keys in this position. Possible keys include C through all sharps and up to four flats. Later, convert to minor keys.

Solo in B♭

In the following arrangement, strings are indicated by numbers in circles to aid in positioning the chord voicings.

Rhythm Guitar—The Right Hand

Rhumba: Basic and Orchestral

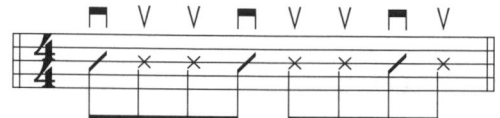

This is difficult but very good for the right hand. It may help to count the eighth notes: 1, 2, 3–1, 2, 3–1, 2 while learning.

■ **EXERCISE**

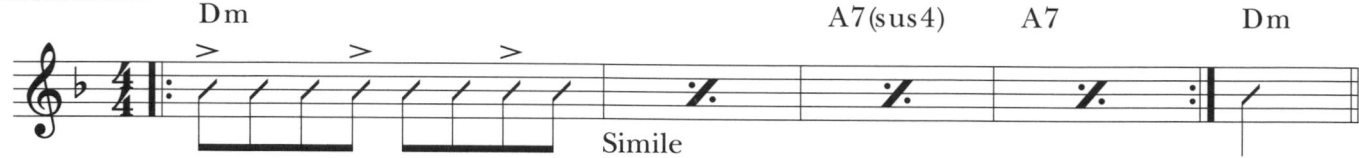

Simile

Variations
Practice with
above exercise.

Optional Orchestral

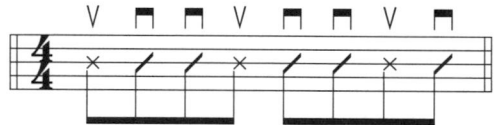

This is the exact opposite of the preceding basic stroke, and it produces complementary accents.

■ **EXERCISE**

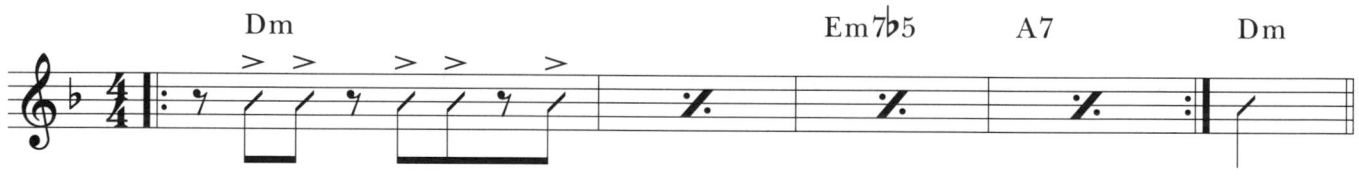

Variations
Practice with
above exercise.

■ (Also see "Orchestral Beguine," *Vol. II*, pg. 93.)

Each note in a chord is called a "voice." These voices are numbered from the top down. The top note is always called the first voice. The note immediately below it is the second voice. The next note down is the third voice, and so on, depending on the number of notes in the chord. This is always the same, regardless of whether the chord appears in close or open harmony.

Triad Studies: Chords in C Major

The following triad studies are primarily to train the fingers to move from chord to chord, with emphasis on related (or economical) finger movement. *Pay strict attention to fingerings.*

CLOSE VOICINGS

* These brackets represent related fingerings. Do not mix them.

OPEN VOICINGS

In the preceding open-voiced triads, the chords on the first stave have the 5th degree on the bottom. Chords on the second stave have the root on the bottom. These are the strongest chord degrees and therefore are the best "bass" notes. The open voicings on the third stave have the 3rd degree on the bottom, but because they do not (and cannot) sound in the "real bass" range, special handling is not necessary. (See *Vol. II*, pg. 84.)

Adjacent Strings—Common Finger Exercises

"Roll" the fingertip from string to string so the notes flow from one to the next without running into each other.

In the following exercise, roll the finger from the tip to the first joint. Do not let the notes ring together as a chord.

Major Scales—Position II

Twelve keys—ascending chromatically.

FINGERING TYPE

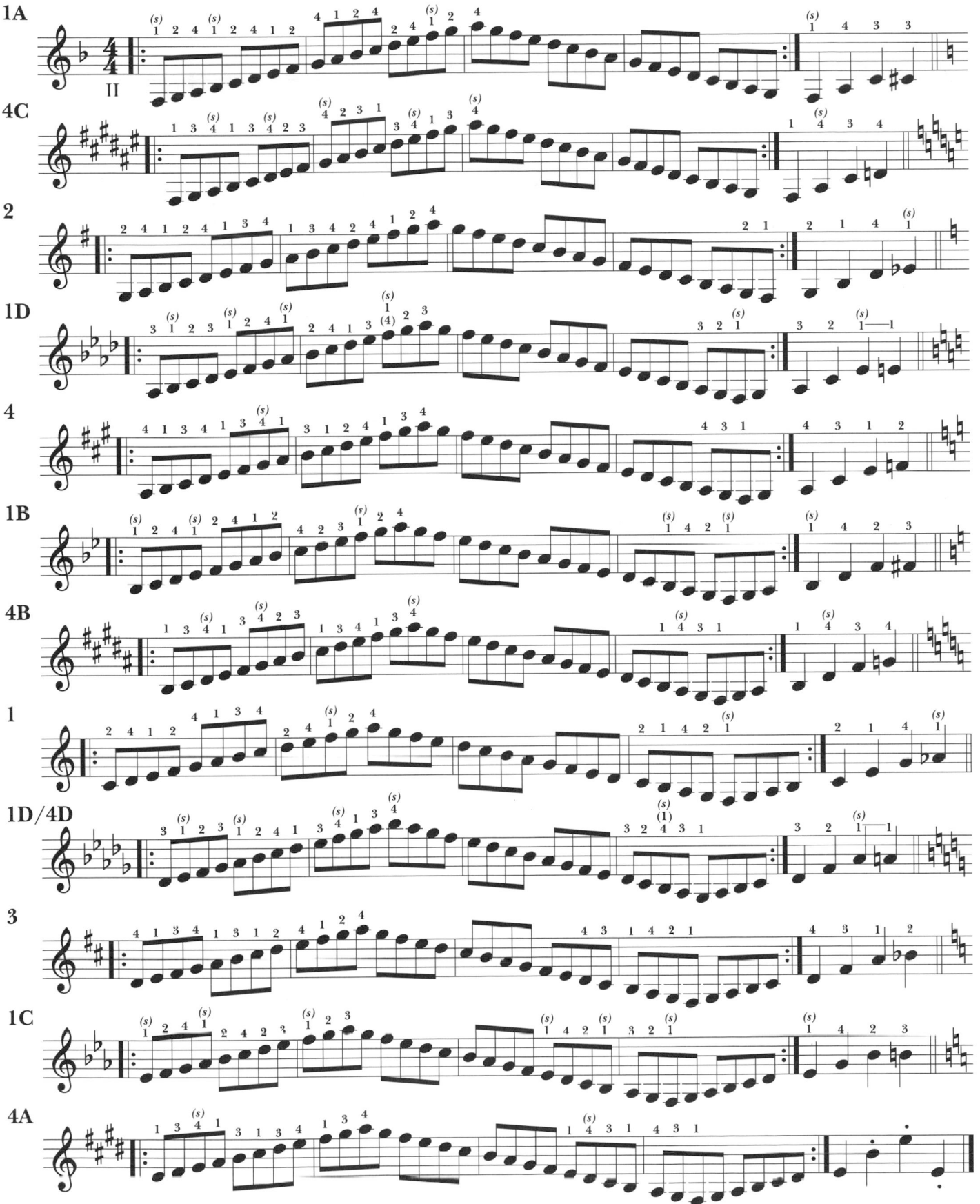

Principal Real Melodic Minor Scales*—Position II
Nine Practical Fingerings

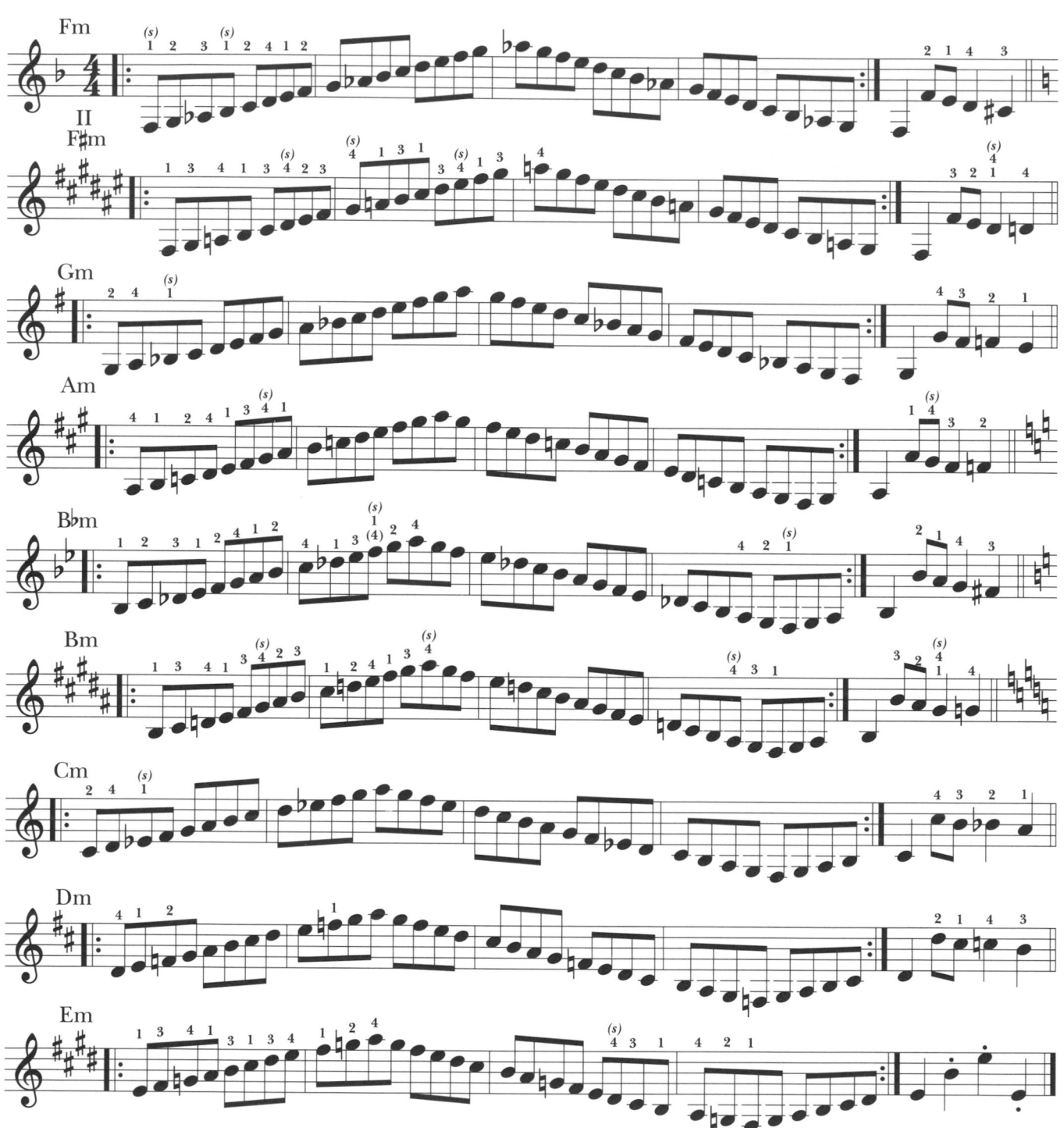

✱ Real melodic minor scale is derived from tonic major scale with ♭3.

Triad Studies—Chords in G Major

Pay strict attention to fingerings.

CLOSE VOICINGS

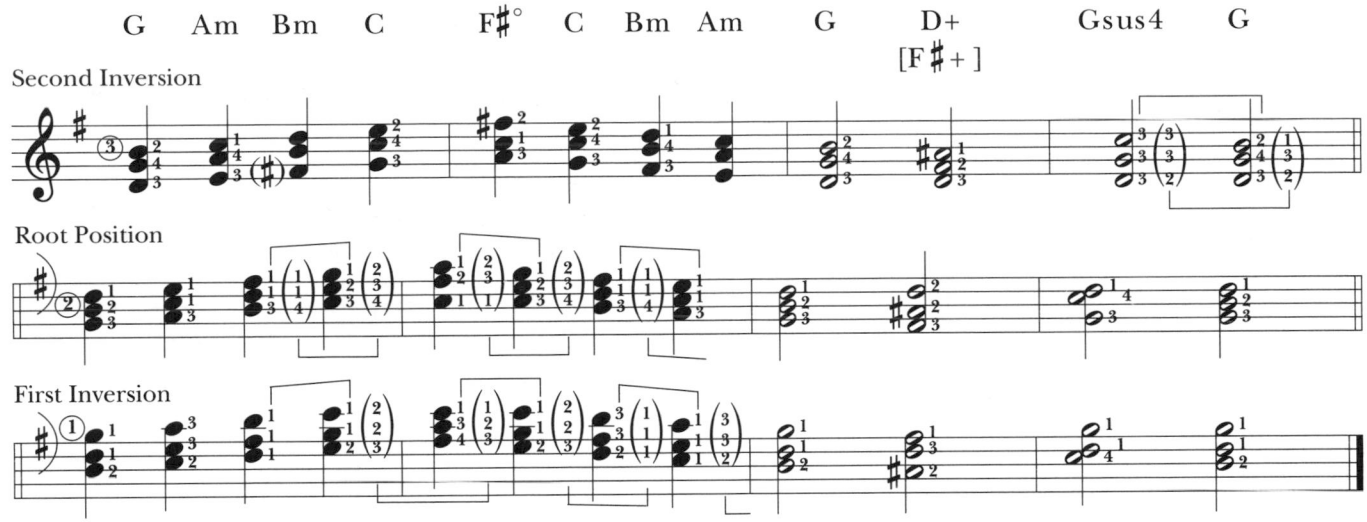

OPEN VOICINGS

❋ The augmented 5th is a weak bass note unless used in passing. Treat #5 the same as the 3rd in the bass. (See *Vol. II*, pg. 84.)

Arpeggios—Three-Note Chords

All major triads (Position V) presented chromatically.

12

Second Inversion

13

Arpeggios—Three-Note Chords

All minor triads (Position V) presented chromatically.

Second Inversion

About Chord Symbols

Chord symbols are a form of musical shorthand to indicate chord structures. They can sometimes be so explicit as to indicate both the harmonic content as well as the voicing and melodic potential. The following facts may help clear up some of the discrepancies that exist in their interpretation.

Any chord symbol involving the number 7 or higher (9, 11, 13) and containing no descriptive term or special mark (maj, min, —, dim, °, etc.) always represents a dominant 7 structure.

The abbreviation "alt" (for altered) means to play the indicated chord degree chromatically altered up and/or down. This term is used exclusively with the 5th degree of major chords and minor 7 chords, and with the 5th and 9th degrees of dominant 7 chords. When the term alt appears with no specific chord degree indicated—and this only happens with dominant 7 chords—then chromatically alter both the 5th and 9th degrees (in either or both directions) in the same structure.

REFERENCE CHART FOR MAJOR SCALE FINGERING TYPES

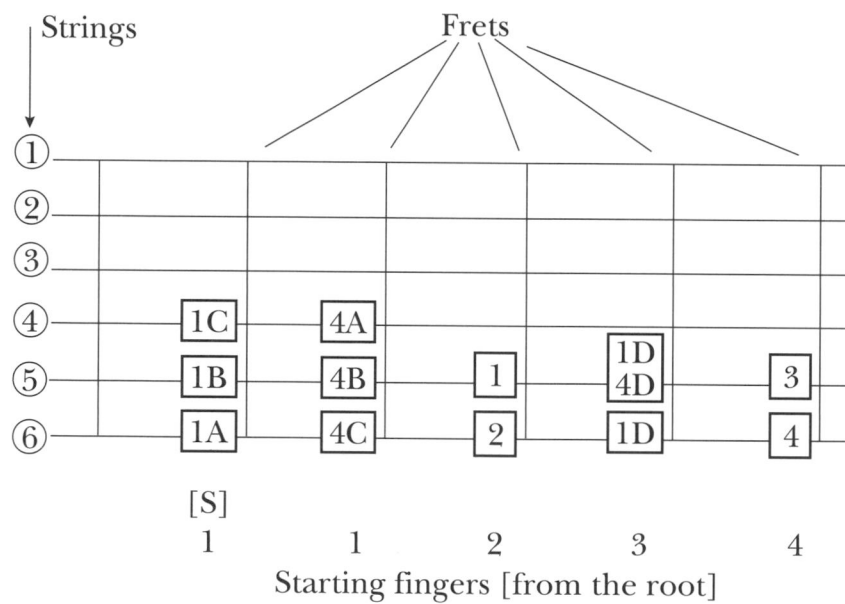

EXAMPLE: Notated in Position V (All notes are roots.)

Major Scales—Position III

Twelve keys—descending chromatically.

FINGERING TYPE

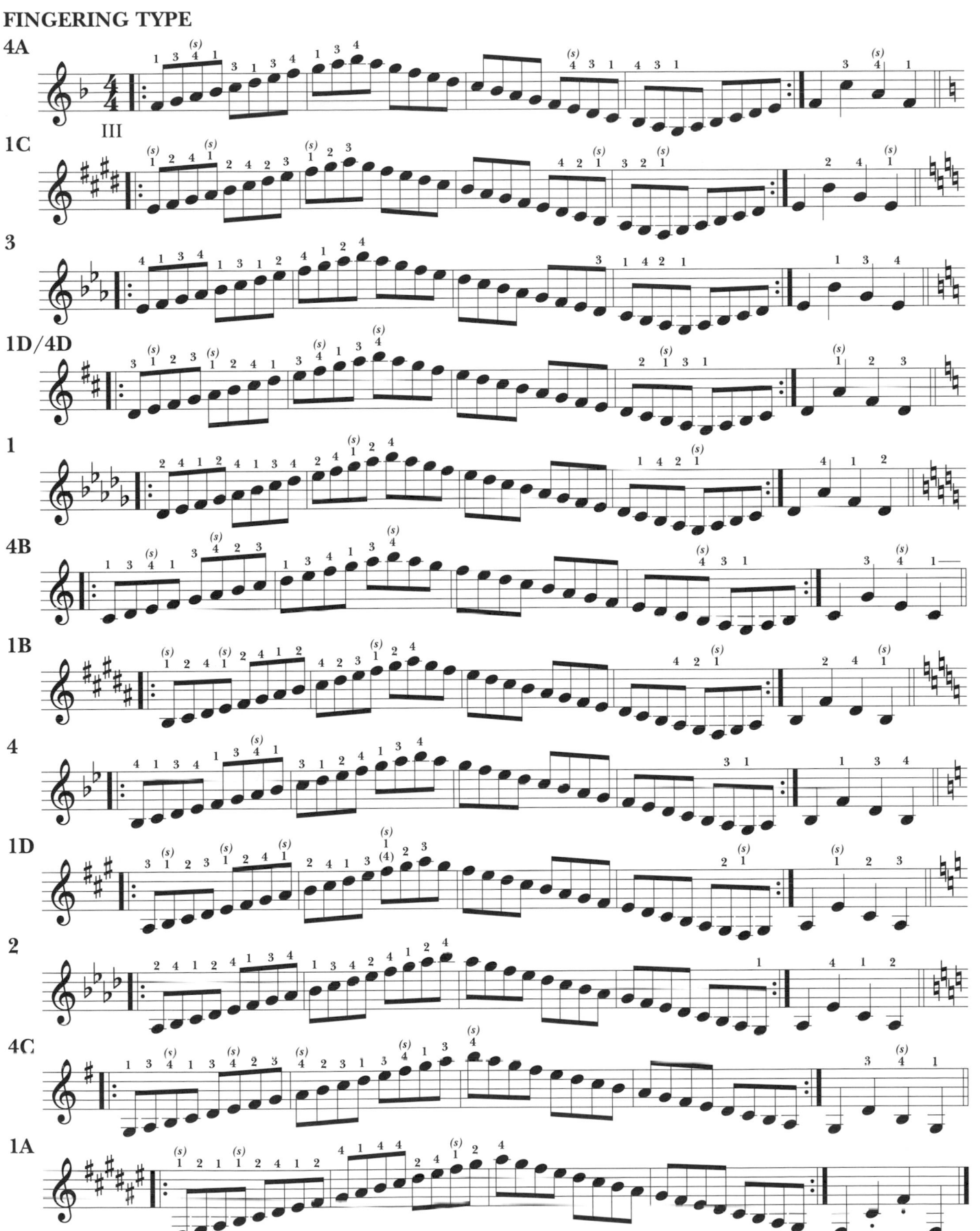

Melodic Rhythm Study No. 8 (duet)

Easy Swing Tempo

About Practicing

Because the guitar is a percussive instrument, it is easy and natural to play staccato phrasing. Therefore emphasis should be placed on legato practice of all studies—a smooth performance of connected notes (with absolutely minimal silences between attacks). This type of phrasing is considerably more difficult and consequently more beneficial. A slow, strict tempo is best for legato practice, as the slightest inaccuracy is far more apparent.

The amount of time involved in practice varies with the individual, as concentration spans vary from person to person. For most students, I suggest that instead of one long session, the maximum benefit is derived from two or three shorter periods of daily practice.

Triad Studies—Chords in F Major

Pay strict attention to fingerings.

CLOSE VOICINGS

OPEN VOICINGS

✱ All voicings in this sequence have the 3rd in the bass. (See *Vol. II*, pg. 84.)

Technical Study

Practice with all possible fingerings, picking each note and also picking on the first note of each triplet group and slurring the rest.

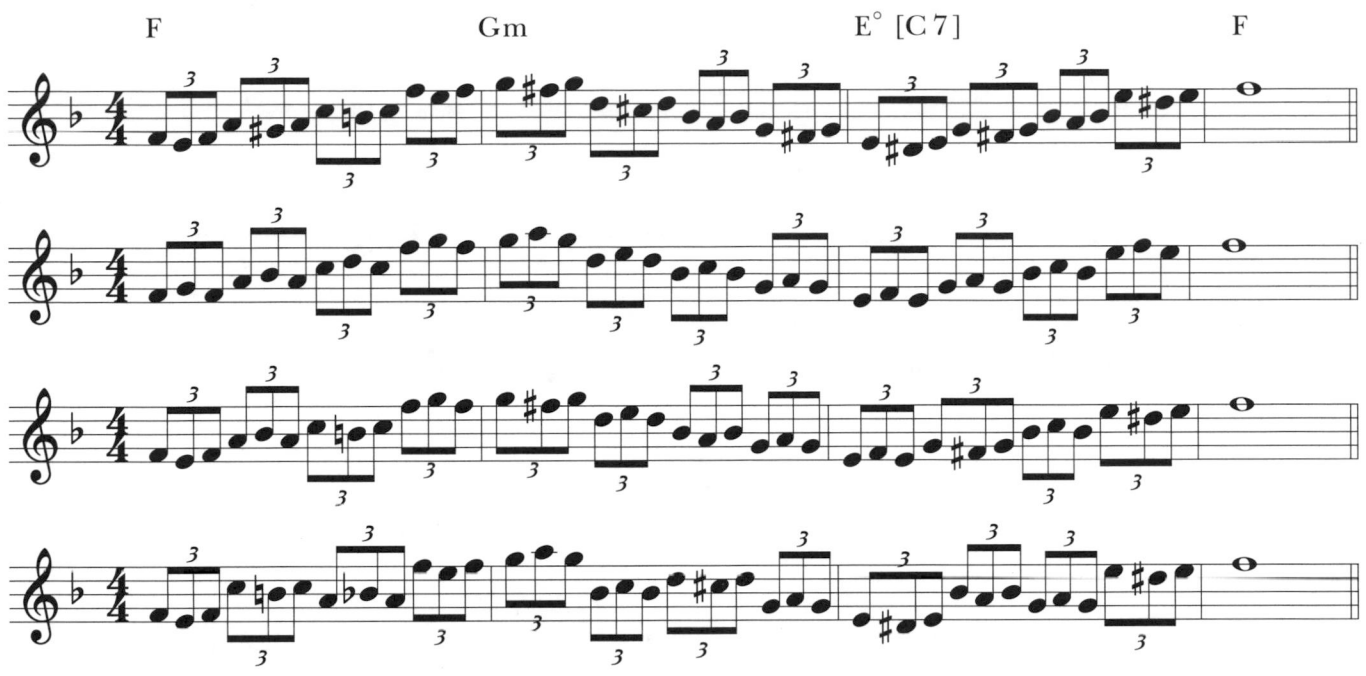

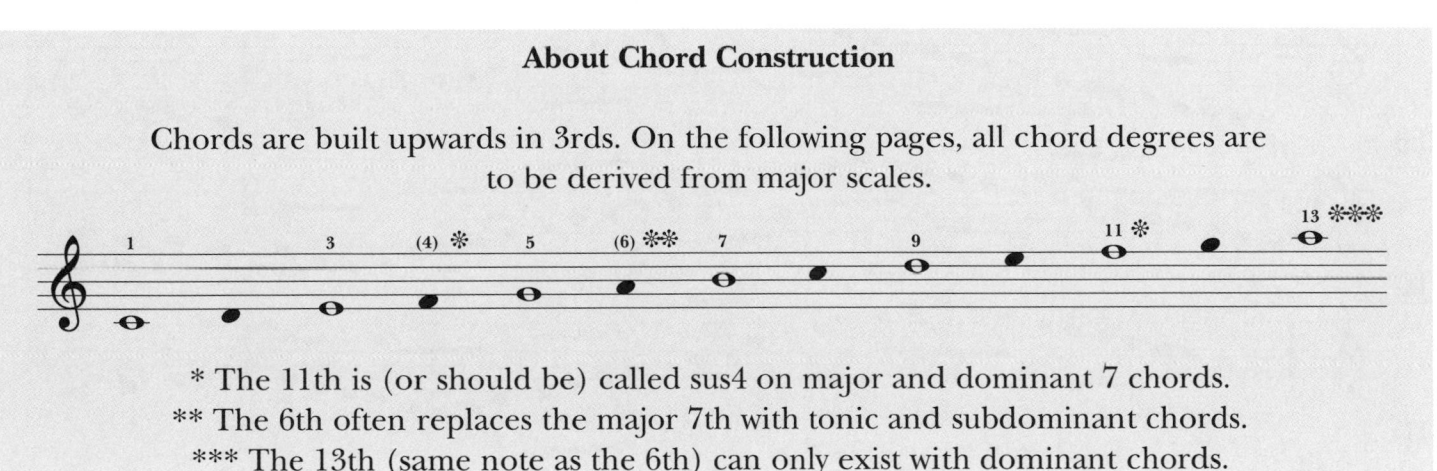

About Chord Construction

Chords are built upwards in 3rds. On the following pages, all chord degrees are to be derived from major scales.

* The 11th is (or should be) called sus4 on major and dominant 7 chords.

** The 6th often replaces the major 7th with tonic and subdominant chords.

*** The 13th (same note as the 6th) can only exist with dominant chords.

Major Scales—Position IV

Twelve keys—through cycle 5.

FINGERING TYPE

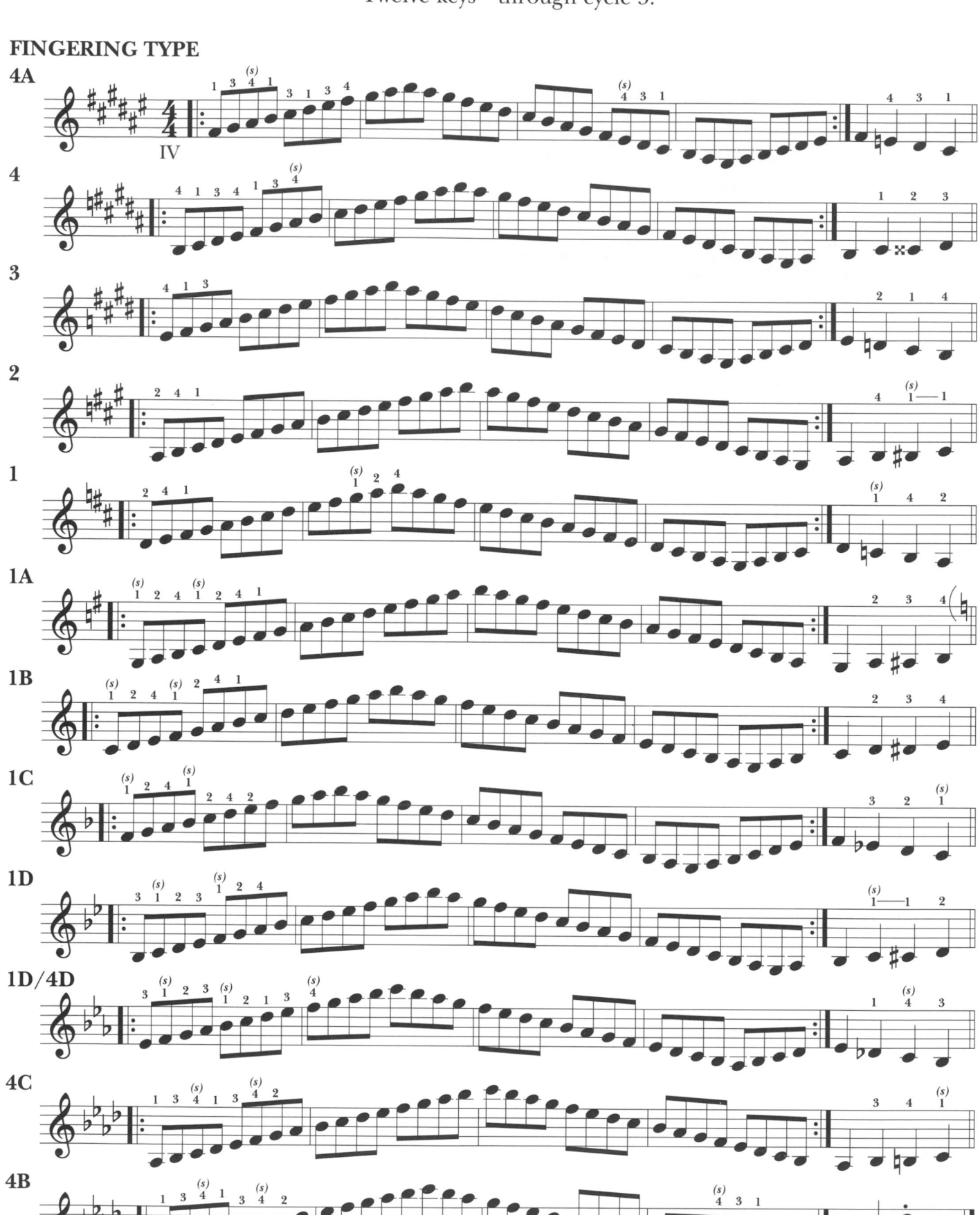

Principal Real Melodic Minor Scales—Position IV
Nine Practical Fingerings

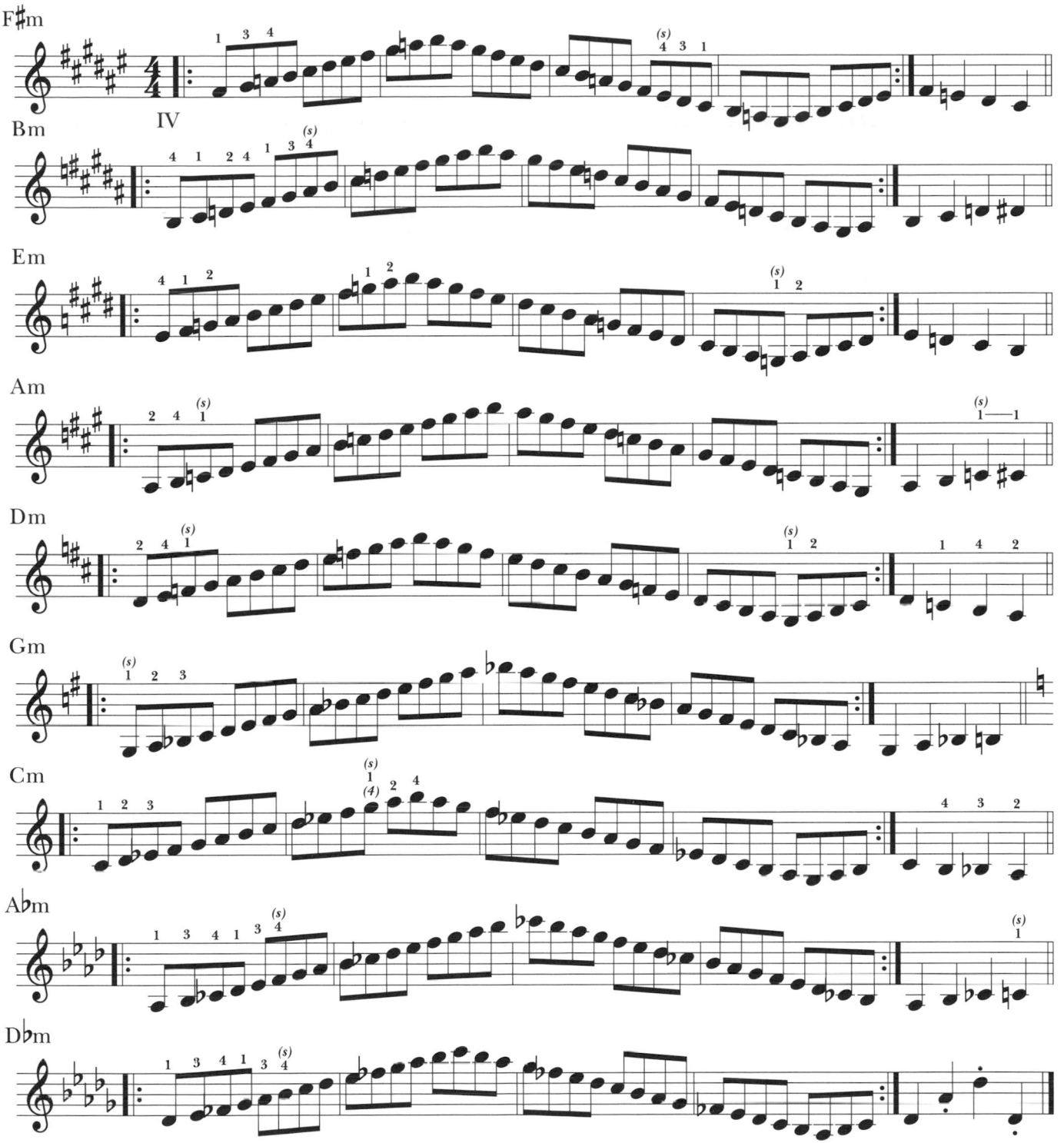

Chord Construction—Four-Part Harmony

All chords are constructed from major scale degrees.

■ Major scale degrees appear below each staff.

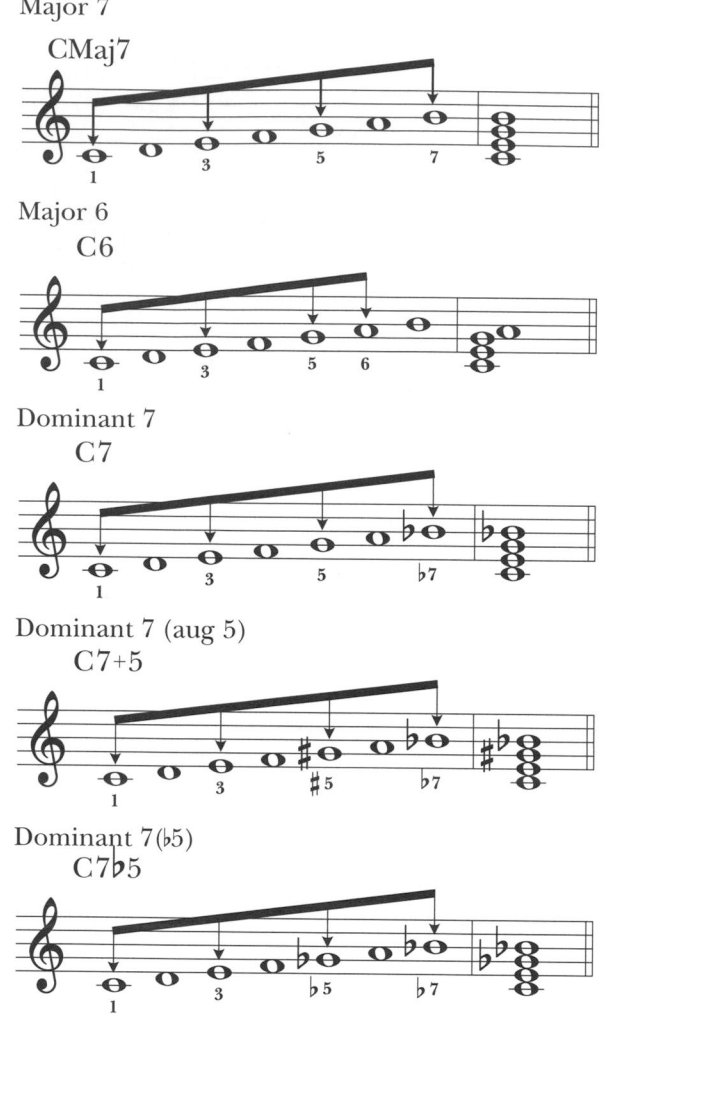
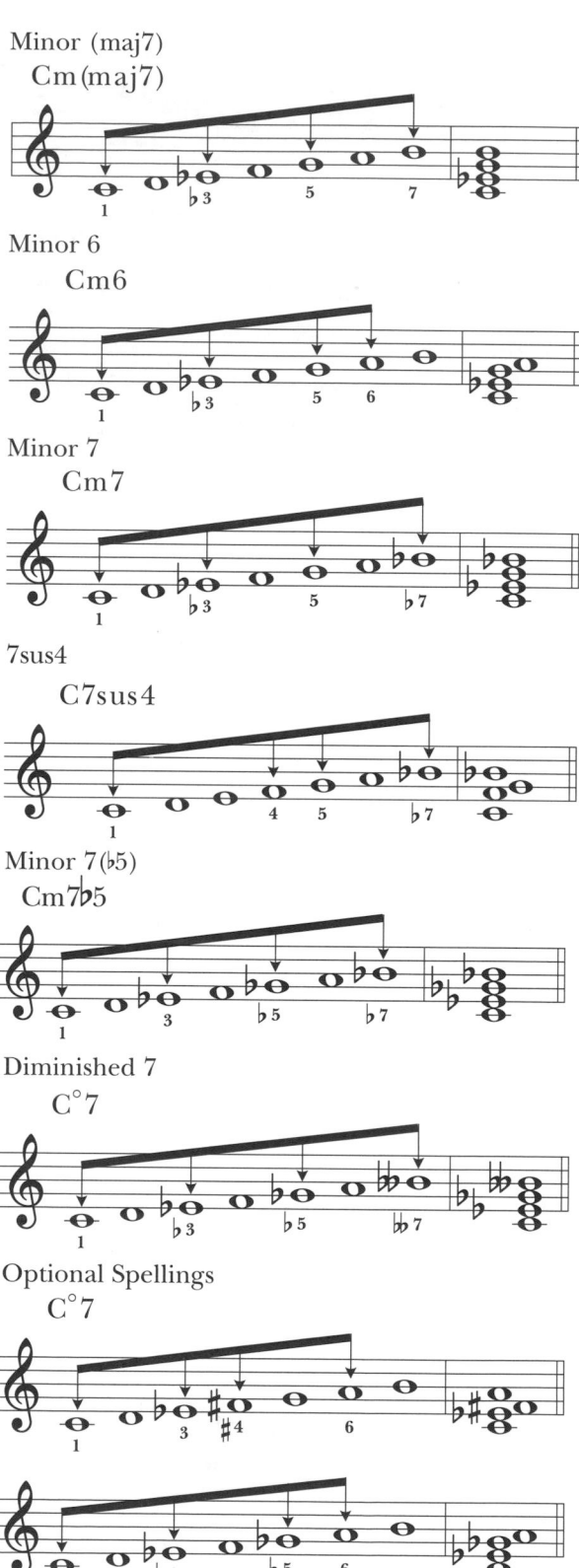

As it is impossible to play most close-voiced structures as chords, we must learn their spelling by practicing them as arpeggios. This must be done so thoroughly that chord spelling becomes automatic. Fingerings are derived from the twelve major scales. Practice them until they require very little, if any, conscious effort.

✷ See special pp. 96 and 97 for information on diminished 7 and dominant 7(♭5) chords.

Arpeggios—Four-Note C Chords
Chord Spelling

Fingering for all four-note chords is shown in the fifth position, with temporary changes to adjacent positions when necessary. After learning the spelling and fingering for each group of arpeggios as written, you must learn to spell and play all structures from all letter names existing from position II through position X. (I suggest doing this transposition on the guitar without writing it out.)

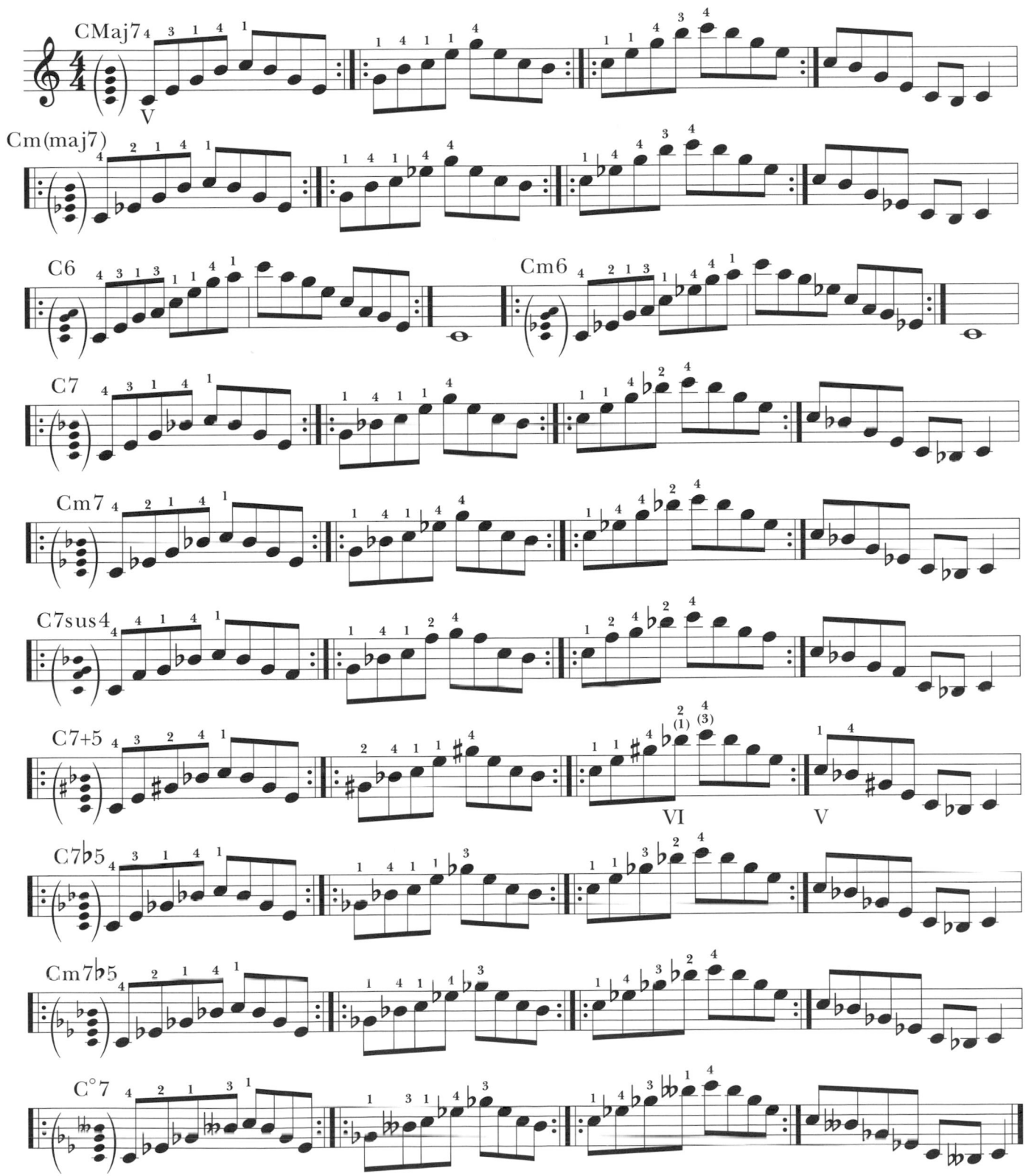

25

Rhythm Guitar—The Right Hand

Tango 1
Moderately slow
to slow

▮ Practice with each preceding tango beat.

▮ **EXERCISE**

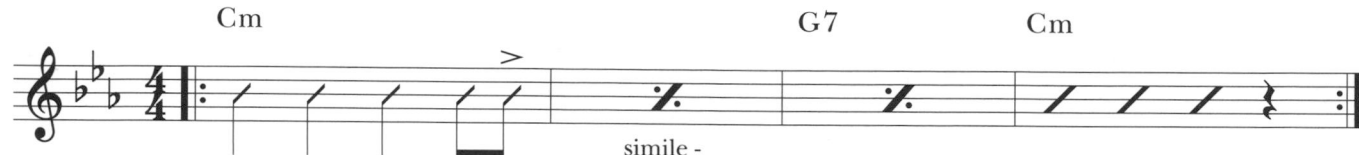

Tango 2

Tango 3

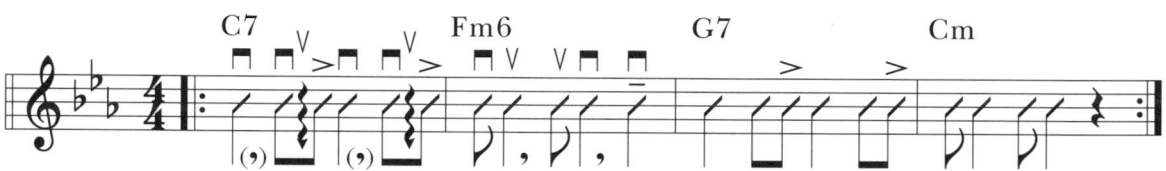

Merengue 1
Fast in 2

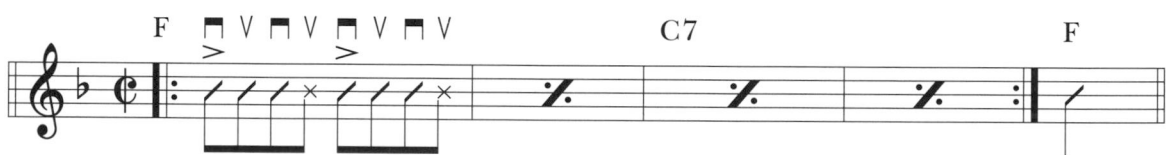

Merengue 2

Merengue 3

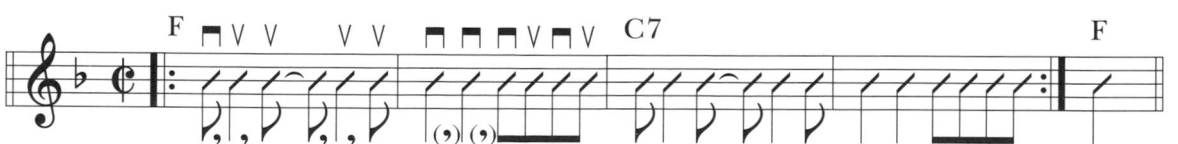

C Major Scale (Twelve Positions)

FINGERING TYPE

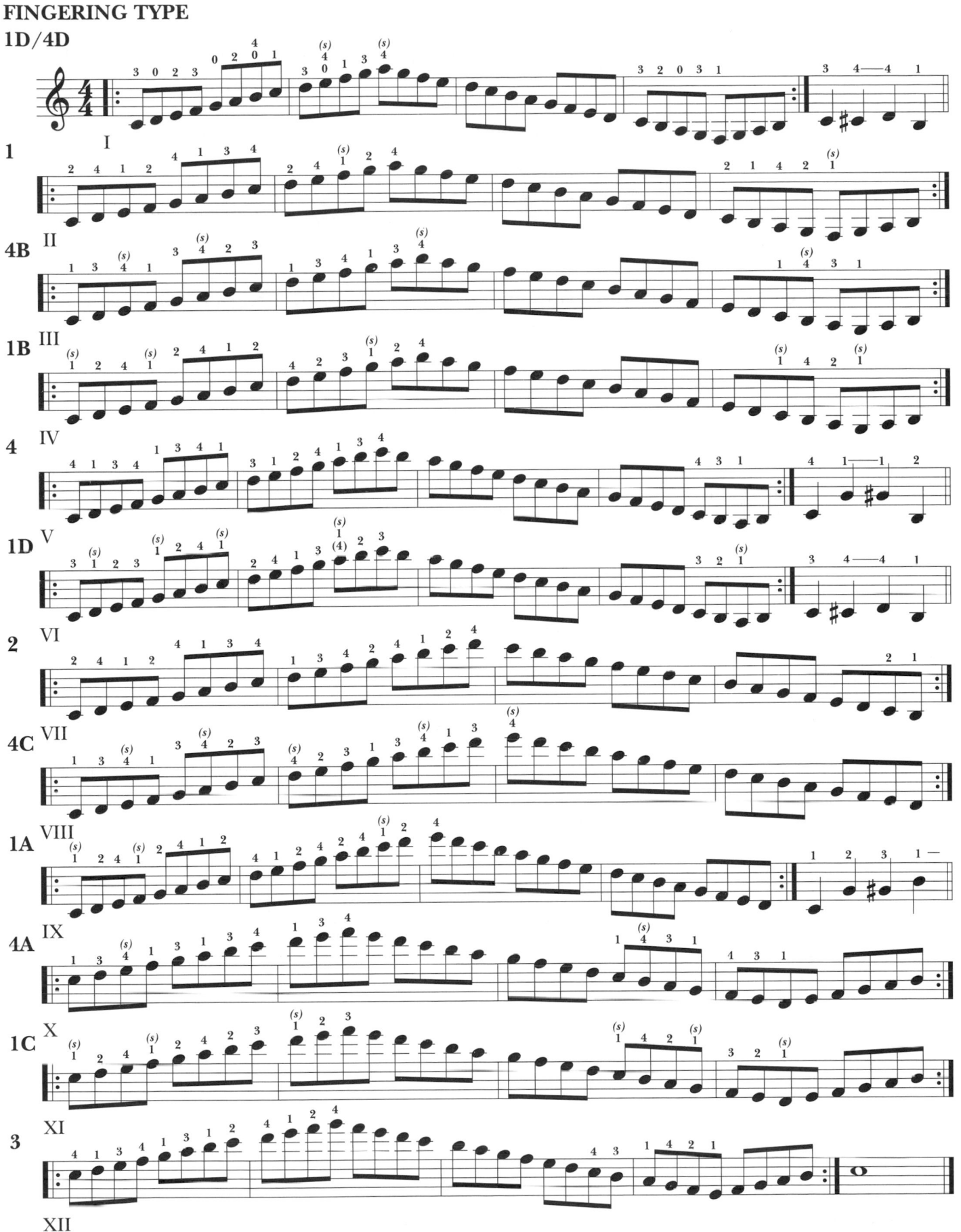

Natural Minor Scales

The natural minor scale has the same key signature and exactly the same notes as its relative major scale. Start on the 6th degree of any major scale. Consider this note as the tonic (1st degree) of the natural minor scale. Renumber the notes (1 through 7 upwards) from this new tonic for the degrees of the relative natural minor scale.

C Major Scale Degrees

A Natural Minor Scale Degrees

Another way to construct the relative natural minor scale is to take the notes a *diatonic 3rd below those of the major scale. (Take the note one line below a note on a line, or one space below a note on a space; do not change the key signature.)

✳ Diatonic: Confined to the notes of the key signature, with no alterations.

F Major

D Natural Minor

TABLE OF RELATIVE MAJOR-MINOR KEYS

| C major | D major | E♭ major | F major | G major | A major | B♭ major |
| A minor | B minor | C minor | D minor | E minor | F♯ minor | G minor |

When you play a major scale from its 6th degree, this is called the natural minor or Aeolian mode.

No exercises in natural minor scales are given here, as the fingerings are exactly the same as the relative major scales shown before. (For reading practice, see modal transposition in the next section.)

Modes: A Brief Discussion

There are names to indicate the playing of a scale from each of the seven notes. These are called modes. They are as follows:

Ionian   From the tonic (or 1)
Dorian   From the 2nd
Phrygian   From the 3rd
Lydian   From the 4th
Mixolydian   From the 5th
Aeolian   From the 6th
Locrian   From the 7th

There are two ways in which these modal terms are expressed:
- **Dorian mode, key of C**, means the C scale starting on the second note (D).
- **C Dorian**, means to start on the note C and play the scale of which C is the 2nd degree, i.e., the B♭ scale, starting from C.

To familiarize your ear with the sounds of these modes (and for extra reading practice from music you already own), refer to reading studies, speed studies, or any completely diatonic music in *Volumes I* and *II*. Transpose first into the Aeolian mode (add of three flats to the signature), as it has the most natural sound to our ears. Then later (in this order), transpose to Phrygian, Dorian, Lydian, Mixolydian, and Locrian modes.

Harmonic Minor Scales

The harmonic minor scale has the same key signature as its relative major scale and all notes but one are the same. Follow the same procedure as with natural minor, except raise the 7th degree one half step. This raised 7th degree becomes the leading tone of the harmonic minor scale.

EXAMPLE:

								1	2	3	4	5	6
Degrees	1	2	3	4	5	6	7	(8)	(9)	(10)	(11)	(12)	(13)
Major Scale ⟶	C	D	E	F	G	A	B	C	D	E	F	G	A
Minor Scale ⟶						a	b	c	d	e	f	g♯	a
					Degrees	1	2	3	4	5	6	7	(8)

The fingerings of a harmonic minor scale are easily mastered when you realize that it is nothing more than the relative major scale with one note raised. Therefore, all playing positions and fingering types coincide. Learn harmonic minor by converting from relative major to minor. Use any major scale fingering pattern. Sharp the 5th scale degree of the major, and you are playing the relative harmonic minor scale. Or, use the natural minor scale and give it a **leading tone** by raising its 7th degree.

C Major   A Harmonic Minor

F Major   D Harmonic Minor

Note: Harmonic minor is the only scale that contains an interval of an augmented 2nd. It occurs between the 6th and 7th scale degrees.

A Harmonic Minor (Nine Positions)

A Minor Etude (solo)

The guitar is a very difficult instrument on which to see exactly what you are playing. There are multiple choices for playing single notes and many chord voicings in the same octave. The strings are not tuned with constant intervals between them (like the violin, viola, or cello), so the relative location and fingering for the same group of notes varies from one set of strings to another.

The fact that the guitar is not a very visual instrument can prove to be quite a problem at times, especially when dealing with the study of harmony. Position marks are a great help, but they don't begin to clarify the layout of sounds like the physical appearance of the other harmonic instruments: the black and white keys of the piano, harpsichord, and accordion, the staggered bars of the xylophone and vibes, even the colored strings of the harp.

With regard to all this, and because I feel it is very important to be able to apply directly to the guitar (without any intermediate steps), in the following studies involving chord construction, melodic analysis, etc., we shall concentrate on three-note chord voicings.

Melodization of Triads

Melodization of triads is accomplished by replacing the top note of a triad (the root, 3rd, or 5th, depending on the inversion) with a higher degree of the scale from which the chord is formed. These notes (other than 1, 3, or 5) are referred to as tension notes, tensions, or high degrees.

MELODIC TENSIONS POSSIBLE FOR TONIC MAJOR CHORDS

Root Position			First Inversion		Second Inversion	
5	6	maj7	1	9	3	sus4*
3	3	3	5	5	1	1
1	1	1	3	3	5	5

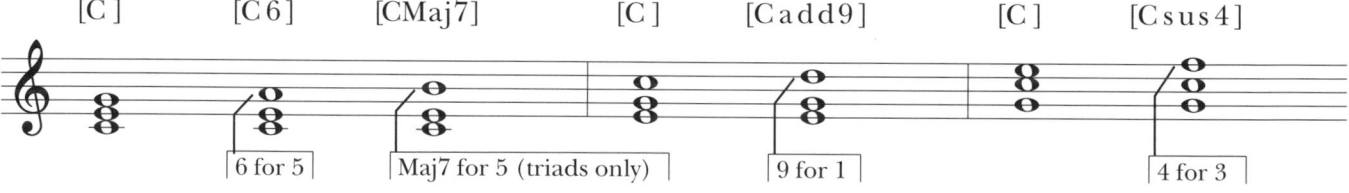

General Rule: A melodic tension replaces the first triadic tone directly below it in pitch (usually found on the same string).

Note that the 3rd is present in all voicings (except sus4*). The 3rd is the most important chord degree, as it alone indicates whether the chord structure is major or minor.

Tensions are also used as inside voices of chords, but because these are more difficult to "see," we shall not emphasize them until later.

*The symbol sus is an abbreviation for "suspension." It is a dissonant note that eventually resolves into the same chord. It usually moves downward to a lower chordal degree, or into a different chord that contains the same note.

Recognition of Melodic Degrees

✱ The 7th degree offers an exception to the general rule for tensions on three-part voicings, in that it may replace the 1st triadic tone above it, i.e., maj7 for 1 (usually located on the same string).

Melodic degrees:
♯5 to ♭5

The abbreviation alt (for altered), when used with chord symbols, means to chromatically raise and/or lower the indicated degree.

♯4 to 5 (♭5)

✱ ♯4 (like 7) may replace the first triadic tone above it, i.e. ♯4 for 5. This is because the ♯4 is the enharmonic equivalent of ♭5. (Enharmonic = two different letter or number designations for the same tone.)

Note: ♯4 is a diatonic tension on subdominant (IV) chords.

About Chord Voicings

On the guitar, it is usually impossible to play all notes in chords containing tensions or double alterations. The lack of mobility of five (or more) note structures and the sounding range involved in voicings with double alterations prohibits their use even when they are physically possible—which is seldom. However, any and all chord degrees that are present in a voicing must conform to the instructions contained in the chord symbol. Remember: Additions to chord structures are dangerous (major 7ths, 6ths, etc.), at least until after you have heard what is sounding around you. Alterations not indicated are madness; deletions are the norm, smart, sensible, and usually the most musical.

Because of all this, it is important to remember that the root and 5th are the most dispensable degrees of almost all types of chord structures. The 3rd is the most necessary. Like the frosting on a cake, more than one tension is nice if physically available, but it is certainly not a requisite.

Arpeggios—Four-Note F Chords
Chord Spelling

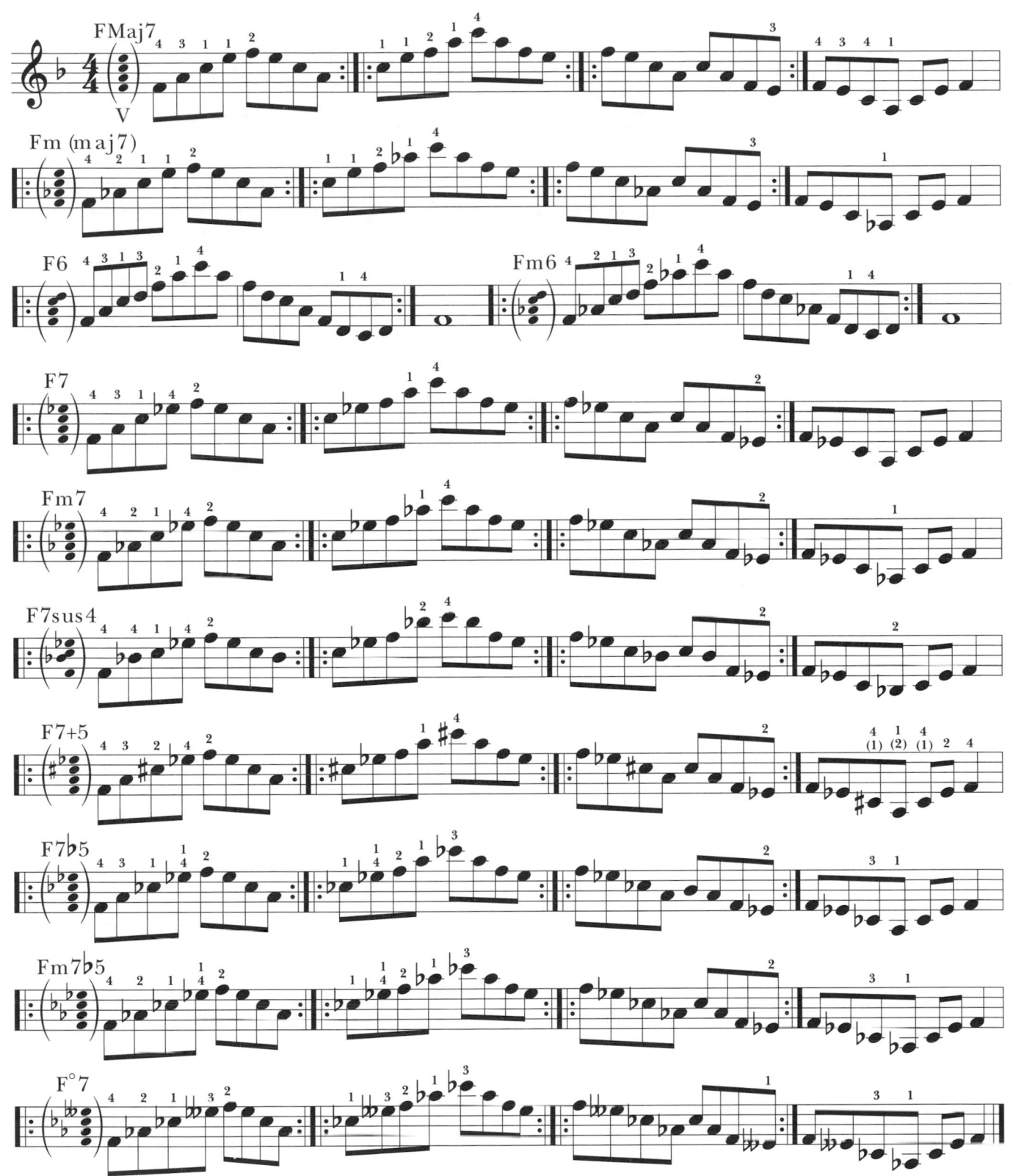

Arpeggios—Four-Note G Chords
Chord Spelling

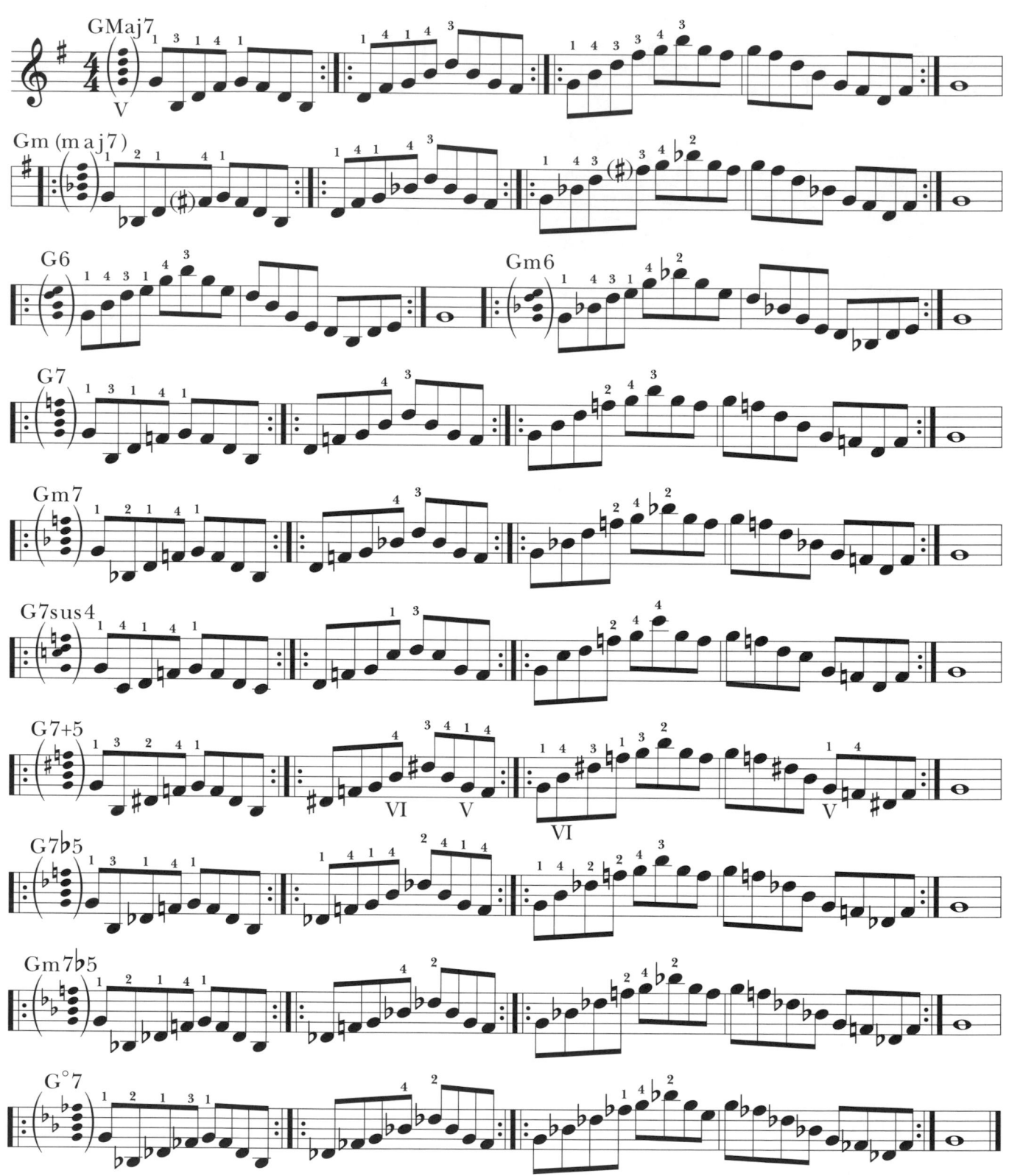

Chord-Scale Relationships—Dominant 7 Chords

For improvisation.

The Basic Idea: Chord-scale relationships are the result of alterations forced on the preceding scale sound by the actual construction of the chord itself.

An E7 chord occurring in the key of C major forces the G♮ to become G♯. Therefore until the occurrence of the next chord, you are functioning in the scale of A harmonic minor. An E7 chord occurring in F major alters the existing G♮ to G♯ and forces the B♭ to become B♮. Therefore, once again the scale for the duration of the E7 chord is A harmonic minor. An E7 chord occurring in the key of G raises the G to G♯, as in the previous examples, but when this G♯ is added to the F♯ that already exists in the scale, the sound that results is A real melodic minor.

EXAMPLES: Scales are named below each sequence of chords.

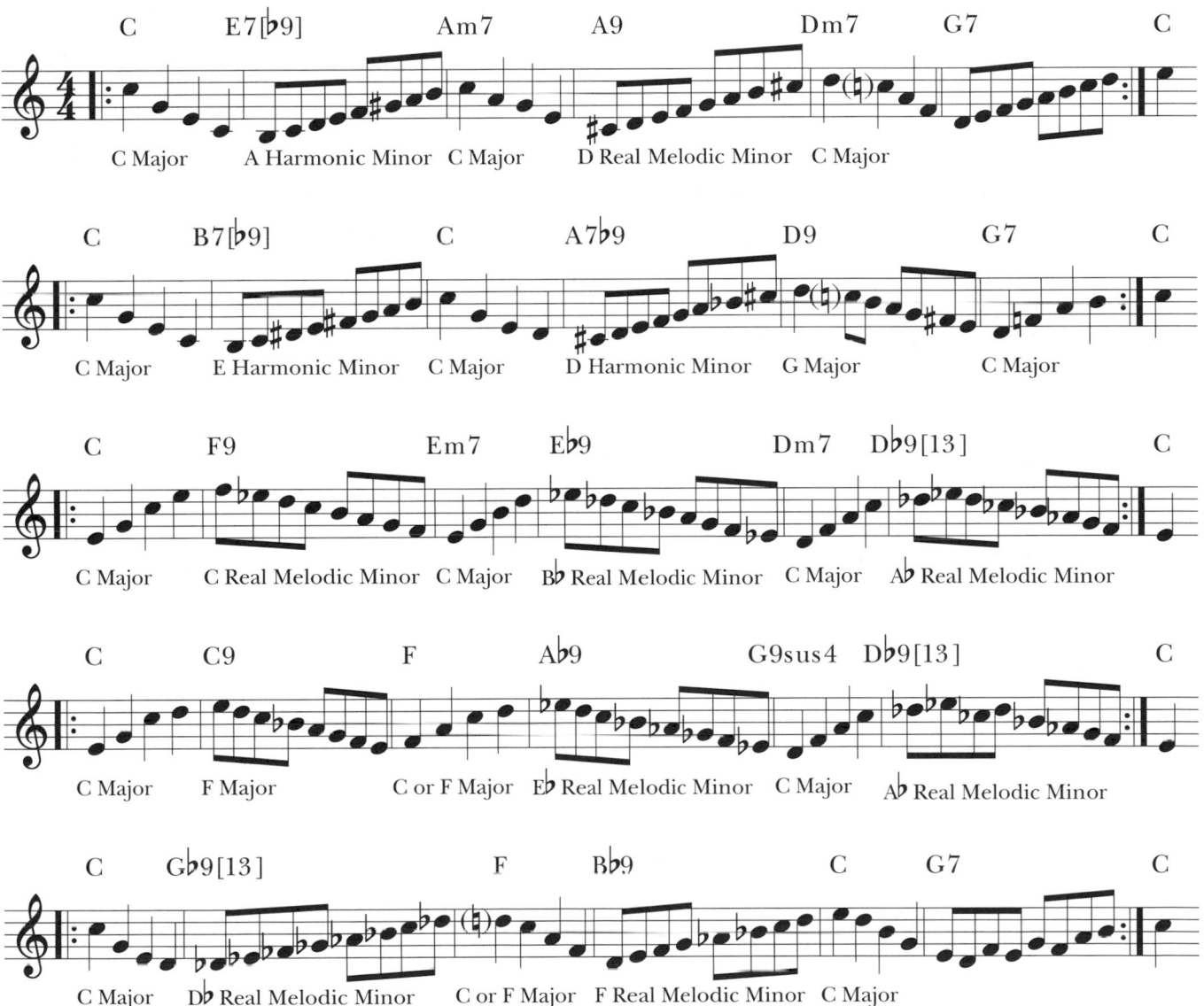

A more in-depth look at dominant 7 chord-scale relationships follows later.

Major Scales—Position V

Twelve keys—descending chromatically.

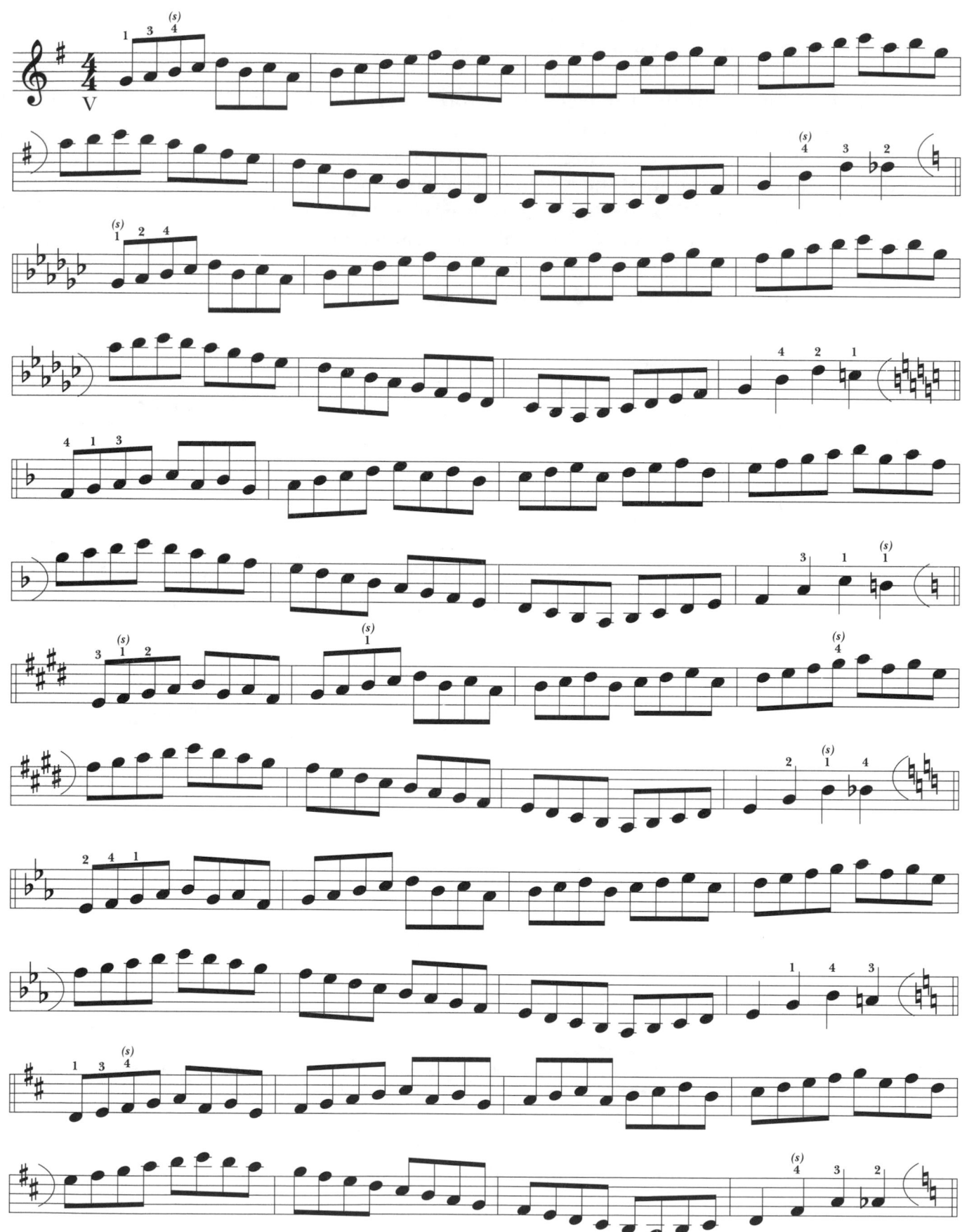

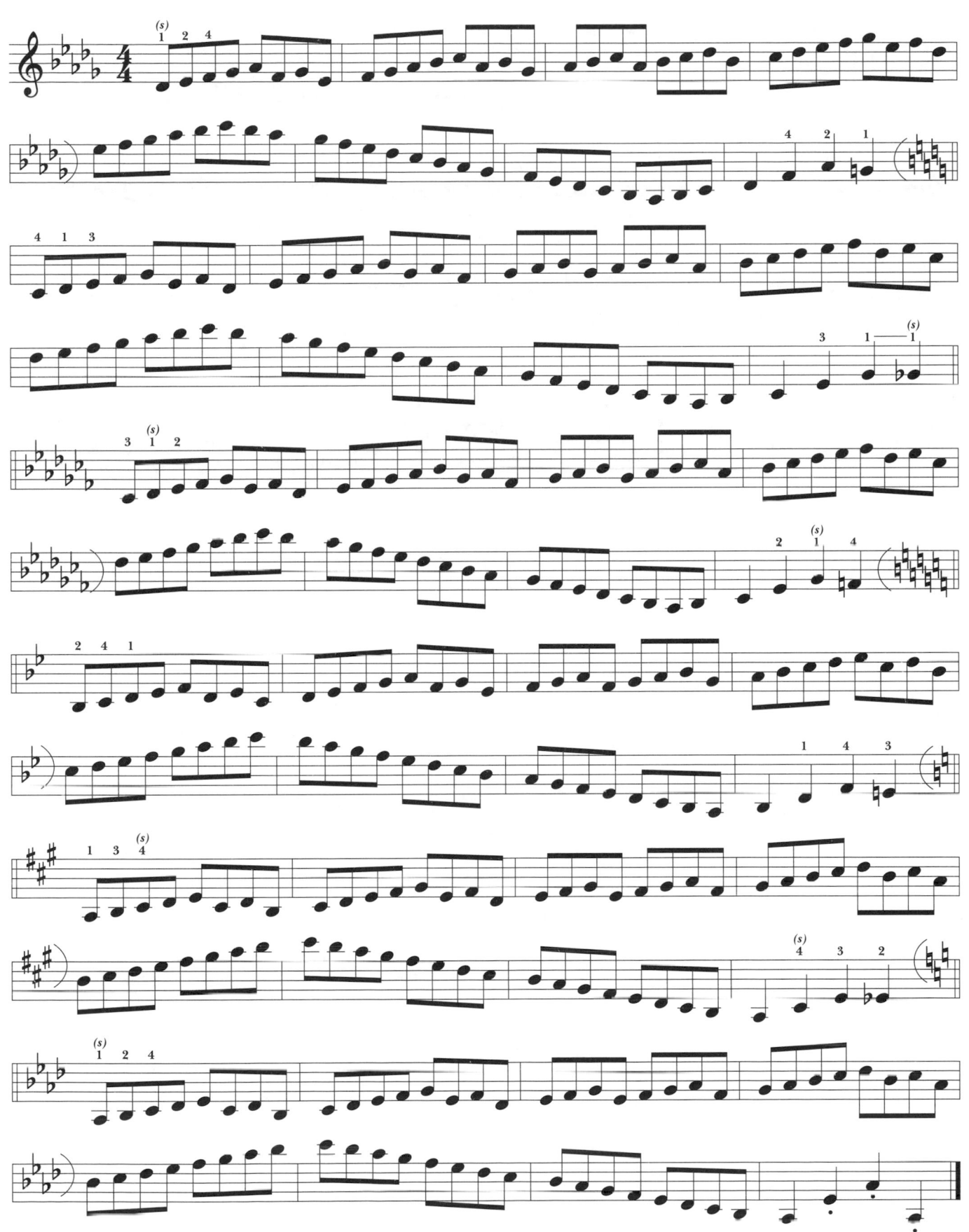

Principal Real Melodic Minor Scales—Position V

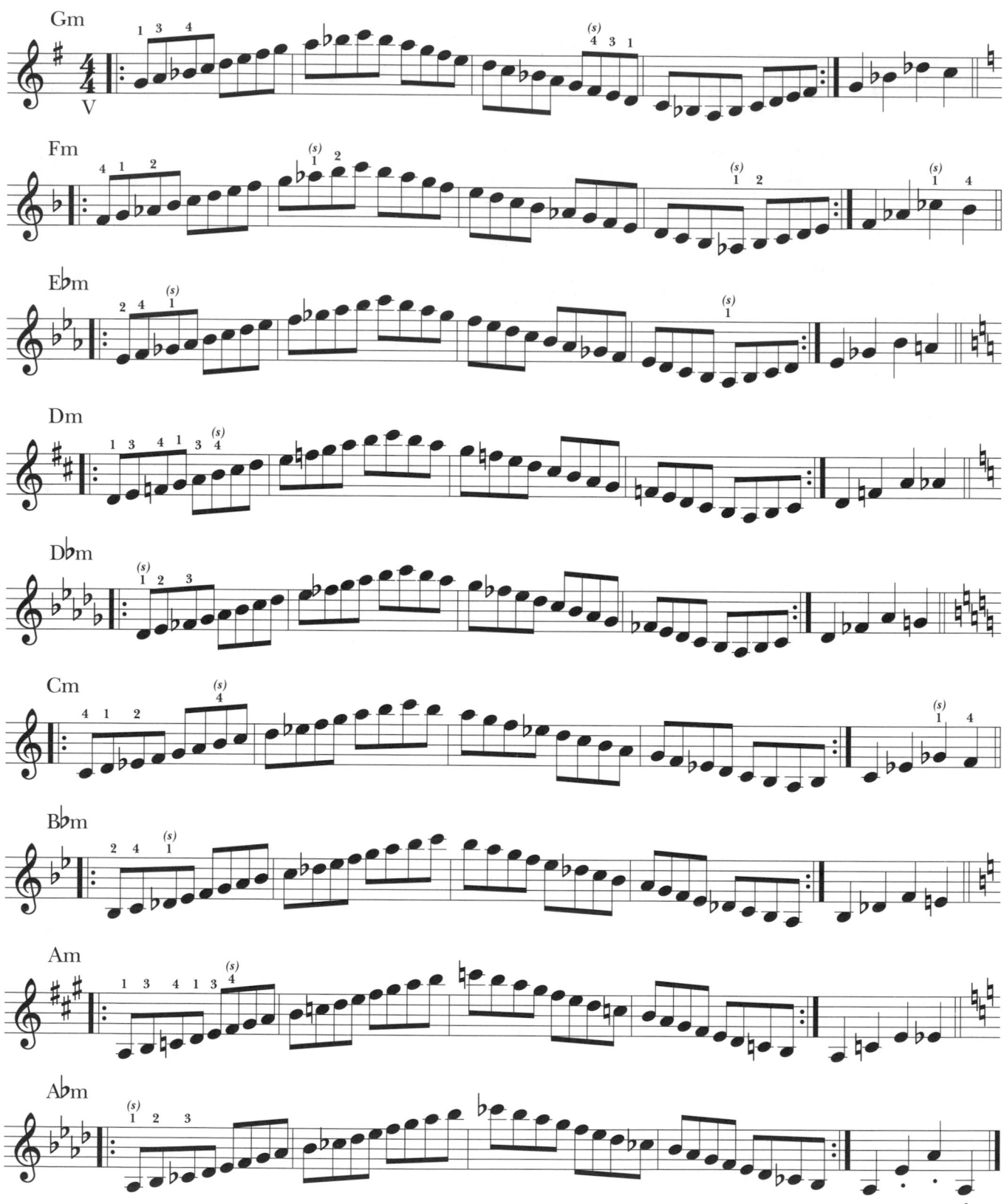

Chord Construction—Three-Note Voicings
Melodization of Tonic Major Chords

To melodize the above as subdominant (IV) chords, sharp the 4th degree.

Melodization of Tonic and Subdominant Minor 6 Chords

Diminished Scales—In Position

The diminished scale is made up of intervals 2, 1, 2, 1, 2, 1, 2, 1, 2, etc. Practice very carefully, as this uniformity produces a rather strange sound. Each fingering pattern contains at least one double stretch, indicated by $\overline{1\ 2\ 3\ 4}$ or $\overline{4\ 3\ 2\ 1}$. This extending of the 1st and 4th fingers may feel awkward at first, but it will prove very valuable for future scale situations. Remember: Stretch the fingers; don't move the hand.

The primary use of diminished scales in improvisation is over diminished 7 chords. When descending, it sounds better if you start on a high degree (or non-chord tone) of the diminished chord. When ascending, start from any note of the scale.

G°/Bb°/Db°/E°
FINGERING PATTERN 1

FINGERING PATTERN 2

Fingering pattern 2 employs the double stretch on strings 4 and 2.

G°/Bb°/Db°/E°

FINGERING PATTERN 3

Fingering pattern 3 employs the double stretch on strings 6 and 1.

G°/Bb°/Db°/E°

Memorize the fingering patterns. Practice all diminished scales, in all positions.

■ Practice as follows:

EXAMPLES OF APPLICATION FOR IMPROVISATION:

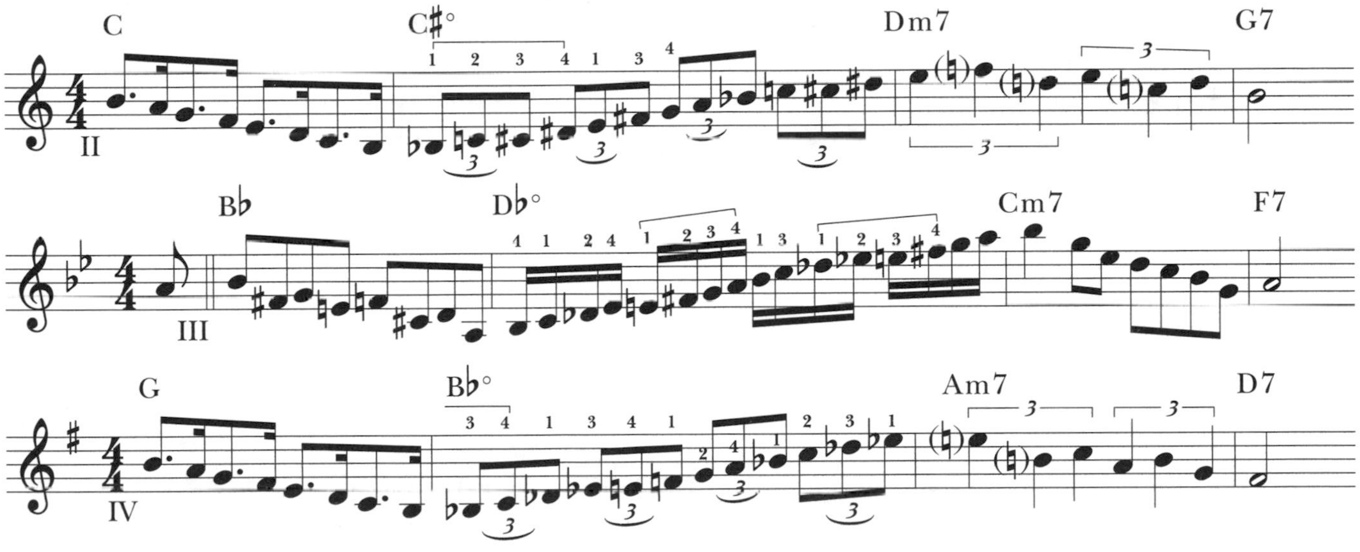

ANOTHER EXAMPLE OF APPLICATION:

❚ Treating (cycle 5) dominant 7 progressions like a chromatic sequence of diminished 7 chords.

ADDITIONAL DIMINISHED SCALE FINGERINGS:

❚ Constant fingering: one position change (two octaves; no stretches)

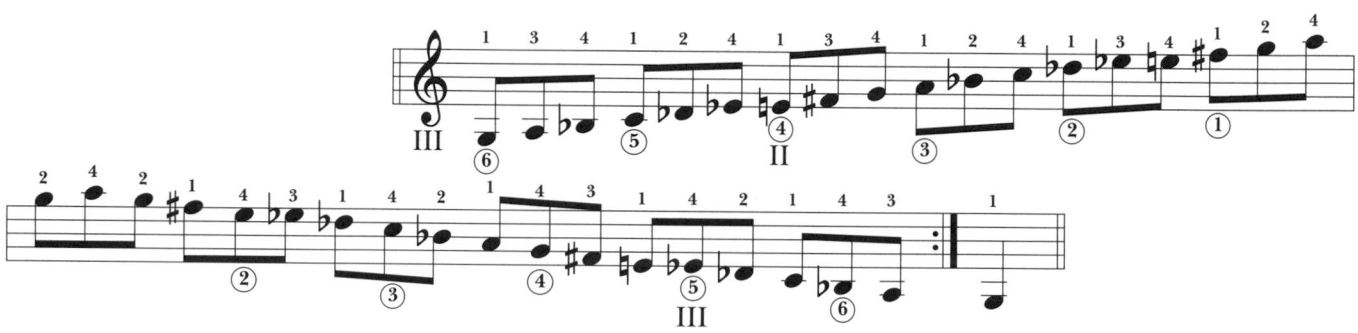

❚ Constant fingering: double stretch and position change on every string (three octaves)

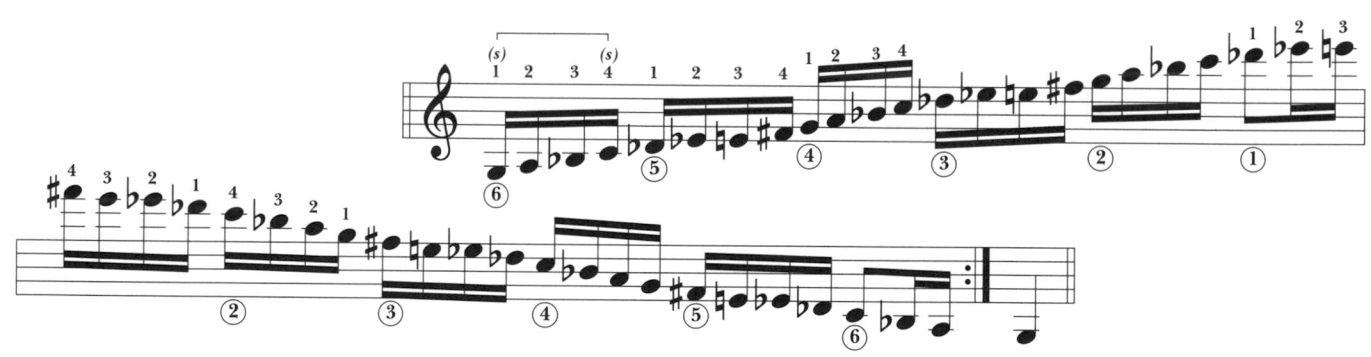

These additional fingerings are less practical for general use.

44

Chord Construction—Three-Note Voicings
Dominant 7 Chords

A complete dominant 7 chord contains four notes. To construct three-note voicings that accurately represent its sound, chord degrees 3 and ♭7 must be present. These two notes of the dominant 7 chord are called the tritone, as they are three whole steps apart. They form the unstable element that causes the restless sound and the need to resolve by moving on to another chord.

PREPARATION OF CLOSE VOICINGS

Recognition of Melodic Degrees—Dominant 7 Chords

Melodic degrees

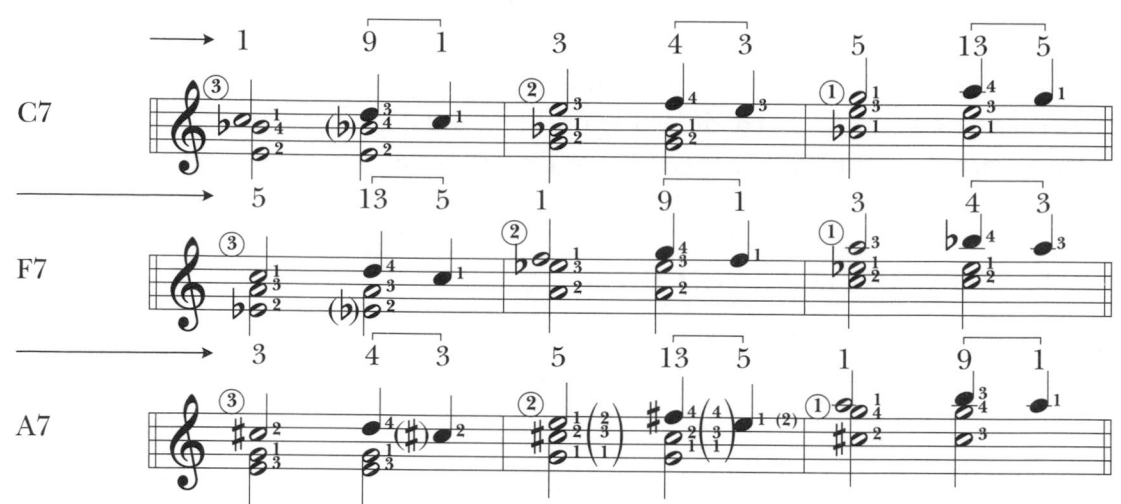

Speed Study

▌ Play thirteen times as written, but each time with a new key signature.

✱ Sequence of key signatures through cycle 5

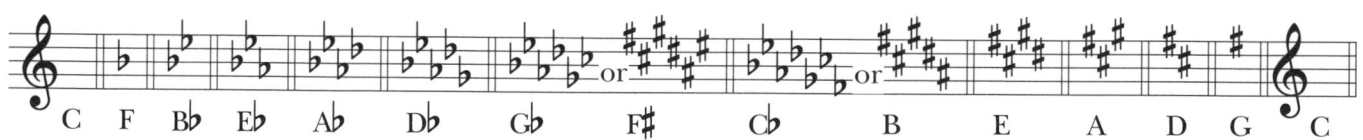

Also practice with minor scales. Nine of each are possible now, but all will be possible later.
Real Melodic Minor: Start with A (major with ♭3), then D, G, etc, through D♭.
Harmonic Minor: Start with (G major) E natural minor and add leading tone.

Arpeggios—Four-Note B♭ Chords
Chord Spelling

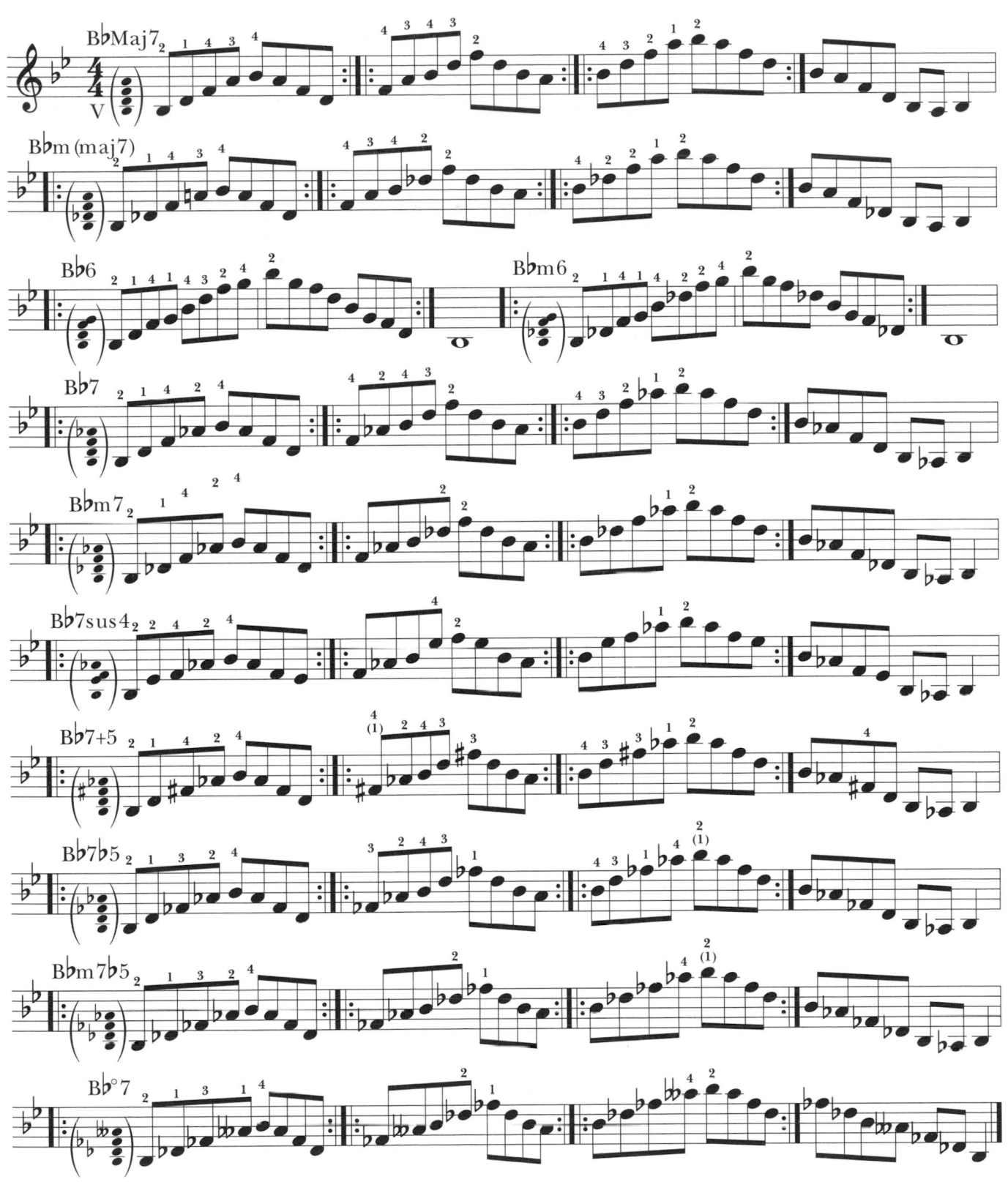

Arpeggios—Four-Note D Chords
Chord Spelling

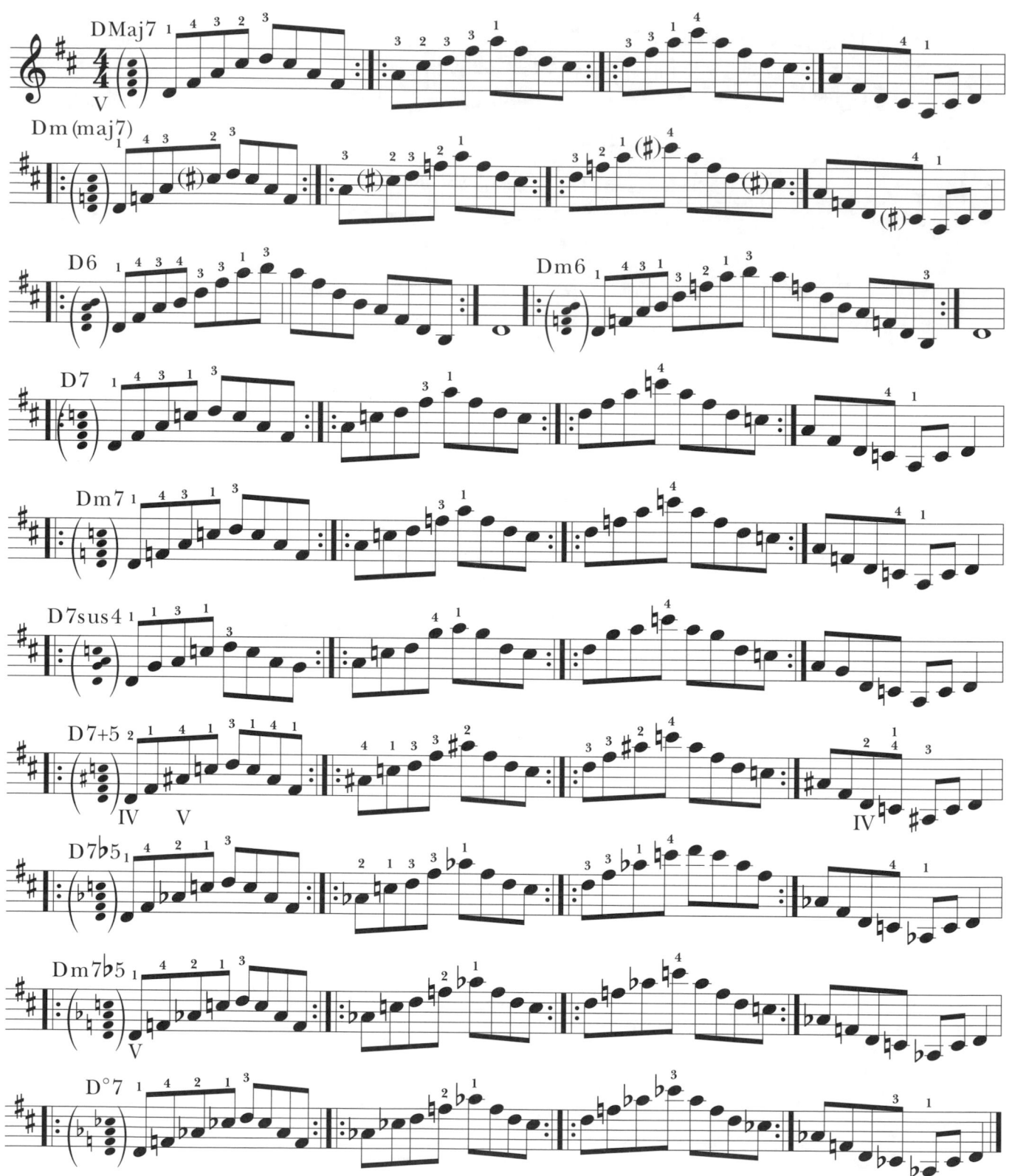

Melodic Rhythm Study No. 9 (duet)

Rhythm Guitar—The Right Hand

Mambo
Fast in 2

Variation

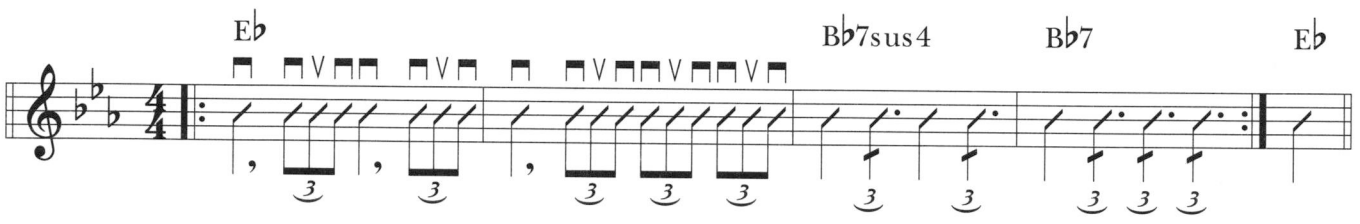

Bolero
Moderate 4

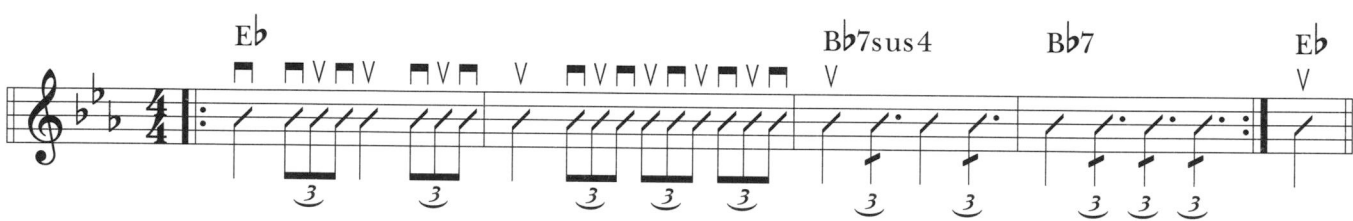

Fast 4

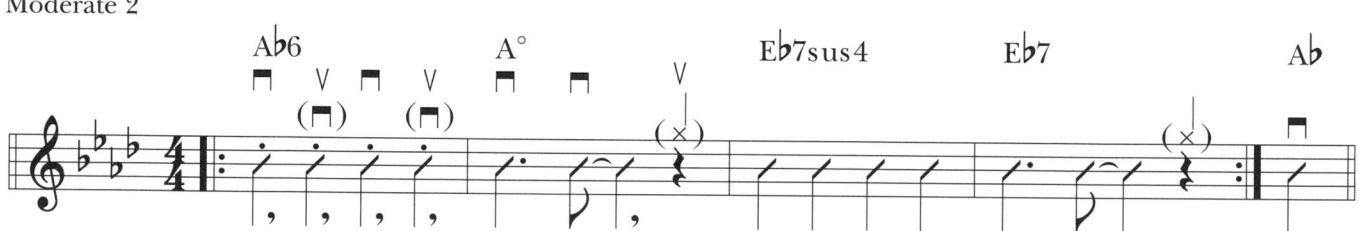

Conga
Moderate 2

Legato-Staccato
(Portato)
Long and short marks combined

(Written)   (Played)

F Major Scale (Twelve Positions)

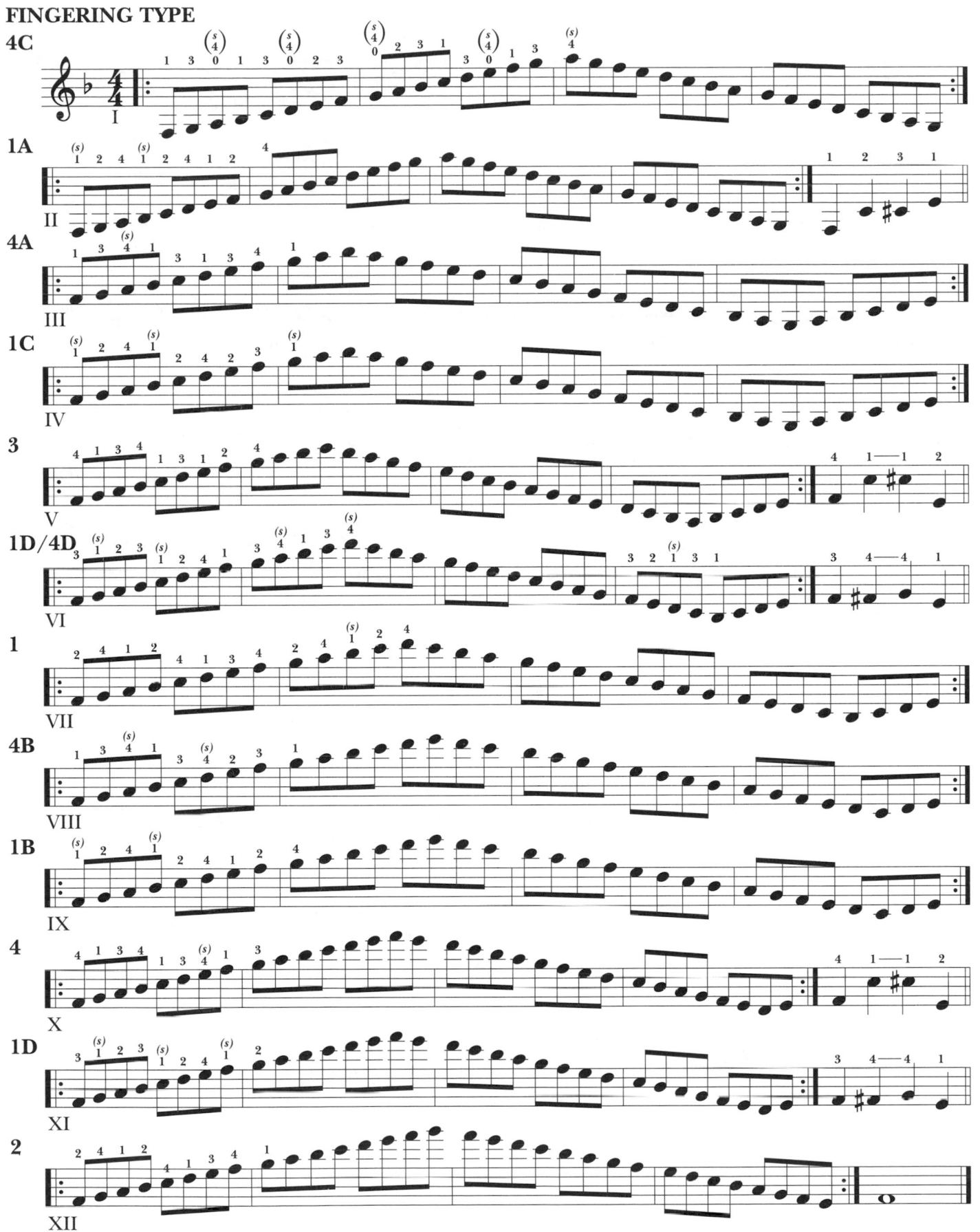

D Harmonic Minor (Nine Positions)

Etude in D Minor (solo)

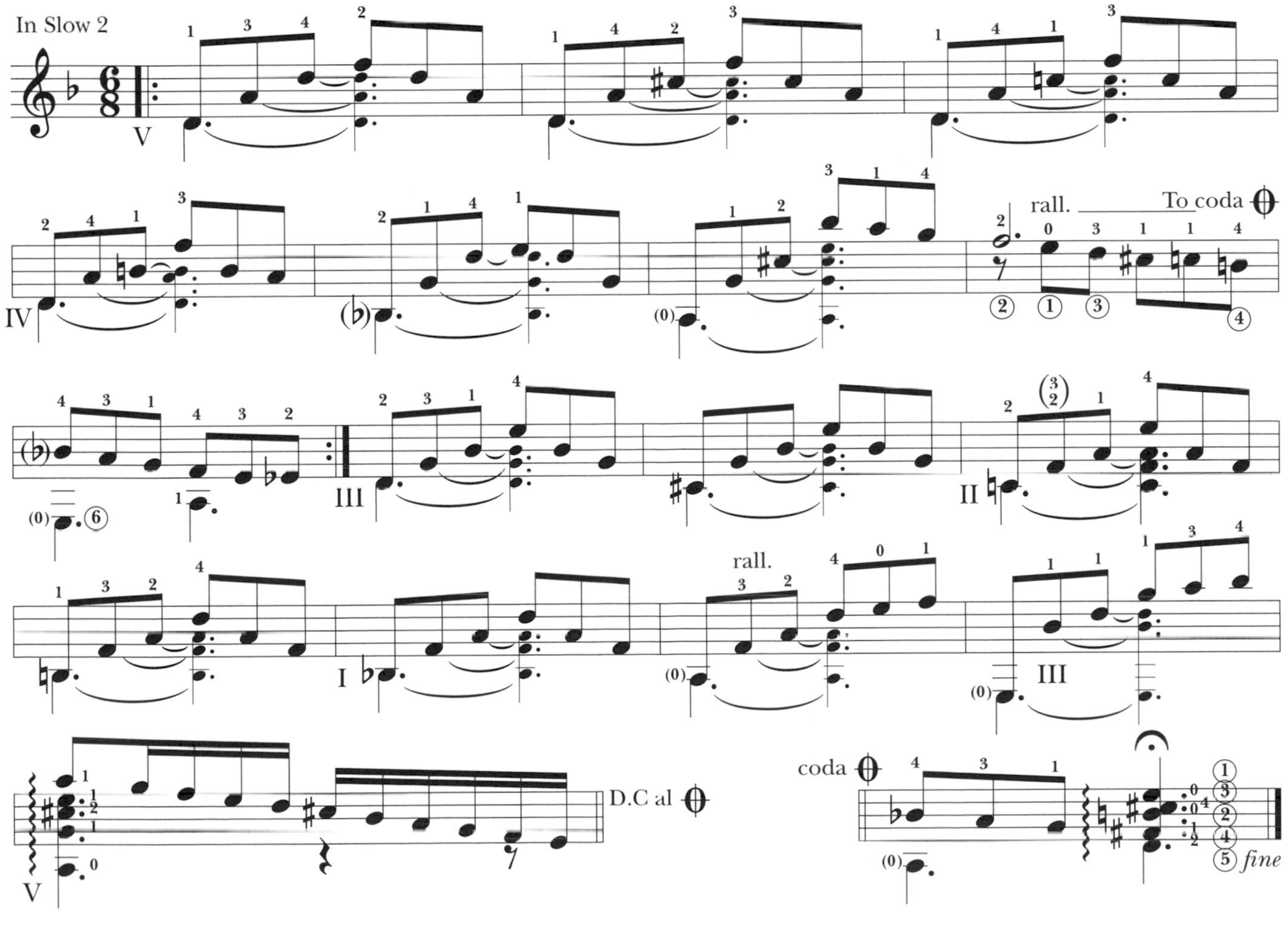

Melodic Embellishment

For improvisation.

The Appoggiatura: Temporary replacement of a note by a note that is one step above and/or below it.

The following exercises are based on three-note arpeggios. However, by extracting from all chords the smaller structures contained within them, the following has unlimited application.

Practice with all possible fingerings. Use finger stretches as often as possible; use slides only when absolutely necessary.

CHROMATIC APPROACH FROM BELOW: DIRECT RESOLUTION TO CHORD TONE

SCALE TONE APPROACH FROM ABOVE: DIRECT RESOLUTION TO CHORD TONE

INDIRECT CHROMATIC APPROACH: RESOLUTION DELAYED BY INSERTION OF SCALE TONE

INDIRECT SCALE TONE APPROACH: RESOLUTION DELAYED BY INSERTION OF CHROMATIC APPROACH

COMBINATION 1: ALTERNATING CHROMATIC AND SCALE TONE APPROACHES

COMBINATION 2: CHORD DEGREES NOT IN CONSECUTIVE ORDER; APPROACHES MIXED

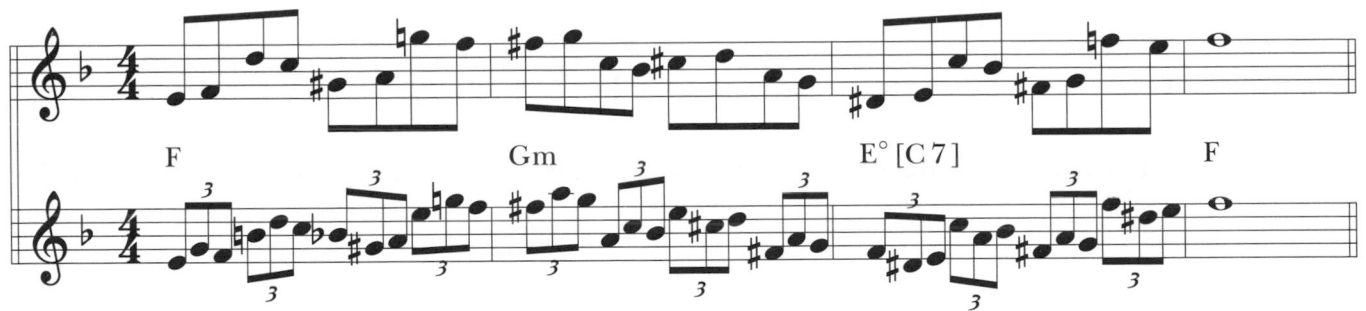

■ Many other combinations await your discovery.

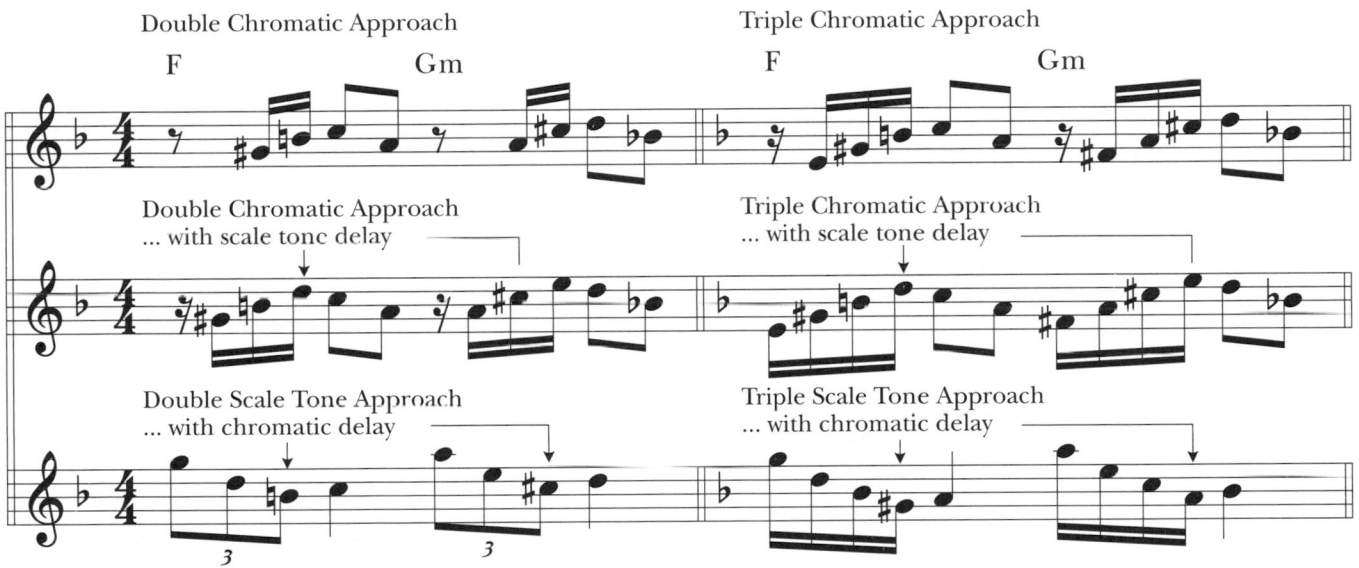

Rhythm Guitar—The Right Hand

5/4 Swing

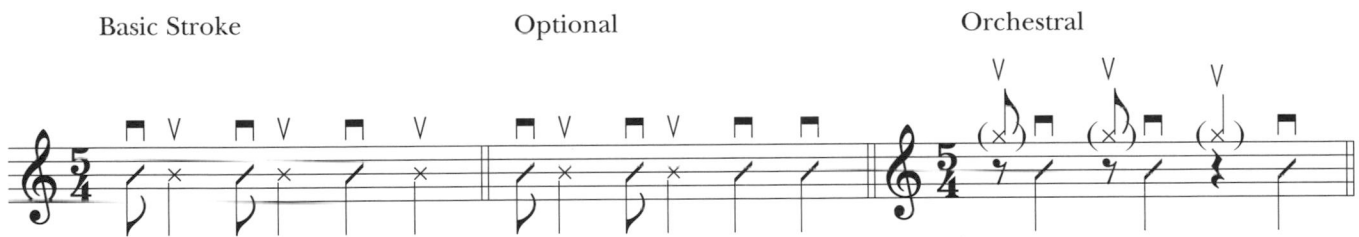

About Altered Chords and Chord Degrees

4th   The sus4 (suspended 4th) means that the 4th degree replaces the 3rd in all major and dominant 7 structures. The 3rd is available only as a melodic passing tone.

With minor chords, sus4 may replace or be used with ♭3. (See 11th, below.)

5th   When the 5th is specifically indicated as sharped or flatted on dominant 7 chords, you might assume that it is truly altered, but this is not so. Rather often the real meaning of a written ♭5 is +11, and ♯5 is ♭13. (See +11 and ♭13.)

When improvising, the player frequently can choose whether to raise or lower the 5th. Sometimes, it may be slightly imperfect theoretically, but ultimately it will be more musical.

For example, when the 5th is sharped, it may be treated melodically as a ♭13, and the normal 5th is used as a passing tone. When the 5th is flatted, it may be treated melodically as a +11, and the 5th is used as a passing tone.

With minor 7 chords, a specifically raised or lowered 5th does in fact represent a truly altered 5th degree.

9th   When the 9th is specifically flatted or sharped, it is truly altered harmonically and melodically. The ♯9 is sometimes melodically treated as ♭3. Alt 9 occurs with dominant 7 chords only.

11th   The 11th with dominant 7 structures is actually an enharmonically named sus4, but it indicates the possible presence of 9 and ♭7 in the voicing. An 11th chord therefore is a dominant 9sus4.

The 11th with minor chords represents the addition of another degree to the total structure, as it may be used with the ♭3 and/or 5, ♭7, and 9.

The augmented 11th (♯11, +11, 11+) exists only with major and dominant 7 chords. It is an added degree to the total structure (of 1, 3, 5, 7, 9) and is used with the 3rd. It does not necessarily replace any chord degree. It is often misleadingly called ♭5.

13th   The ♭13 is actually an enharmonically named ♯5. It cannot be used harmonically with a normal 5th, but it does not represent an altered 5th.

It is called ♭13 to indicate that the normal 5th is to be used as a melodic passing tone. The ♭13 is often misleadingly named ♯5. (13ths can occur only in dominant 7 chords.)

Whenever ♭13 seems to exist on a minor 7 chord, you are actually dealing with a I-for-IIIm7 situation. The appearance of an open voicing of the I chord with the 3rd in the bass, the root in the lead, and the sound brightened up with the 9th inside may mislead you into thinking otherwise. Probably the best name for this structure is minor 7 (add ♭6).

Arpeggios—Four-Note E♭ Chords
Chord Spelling

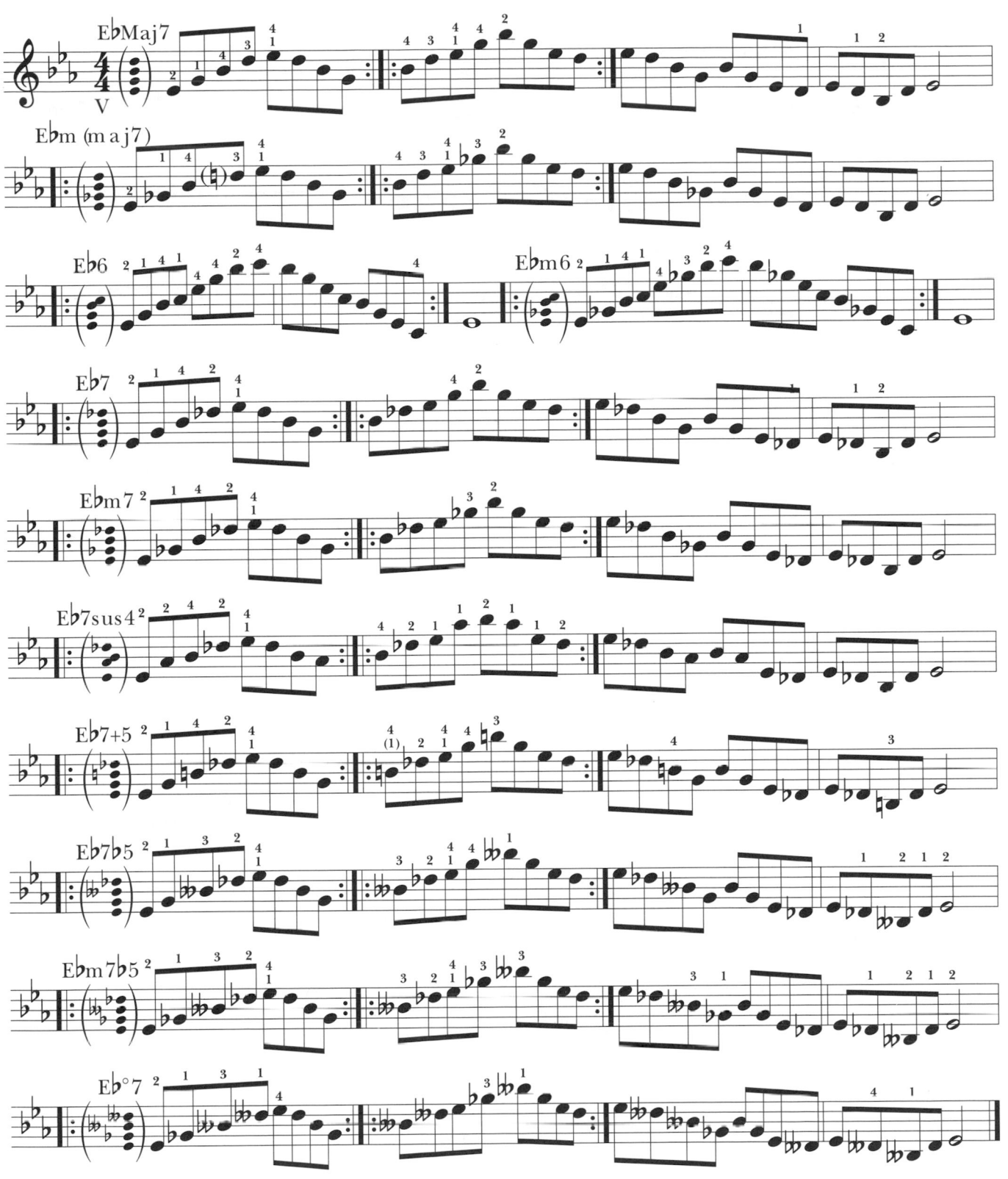

Arpeggios—Four-Note A Chords
Chord Spelling

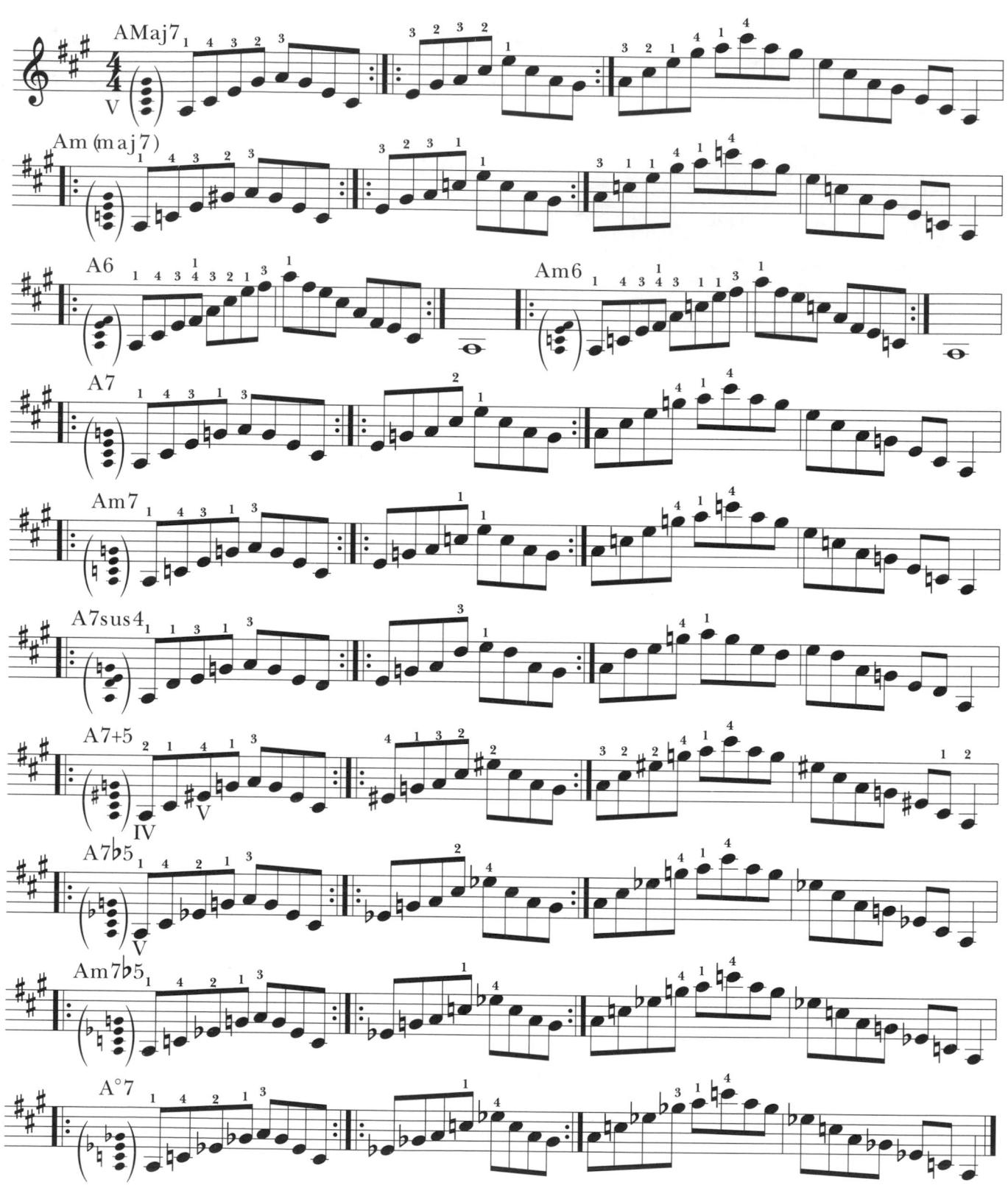

Chord Construction—Three-Note Voicings
Melodization of Dominant 7 Chords

Melodic degrees: Major scale from intended tonic

Once the dominant 7 sound has been established, voicings may be used in passing that do not contain the 3 and ♭7. The ear has a tendency to retain this sound.

Important: Because of their mobility, three-note voicings are very valuable for chord melody playing, for harmonized fills, and for comping. They are shown melodized according to chord-scale relationships and can really open up the harmonic/melodic potential of the guitar.

Melodization of I Minor Chords with Harmonic Minor Scale

Melodic degrees: Harmonic minor scale from chord name

Arpeggio Study—Seventh Chords

❚ Play from all fingers, but stay in position throughout the entire sequence.

✱ Also play the first chord of each measure as a minor 7 chord.

Melodic Rhythm Study No. 10 (duet)

Major Scales—Position VII

Twelve keys—ascending chromatically.

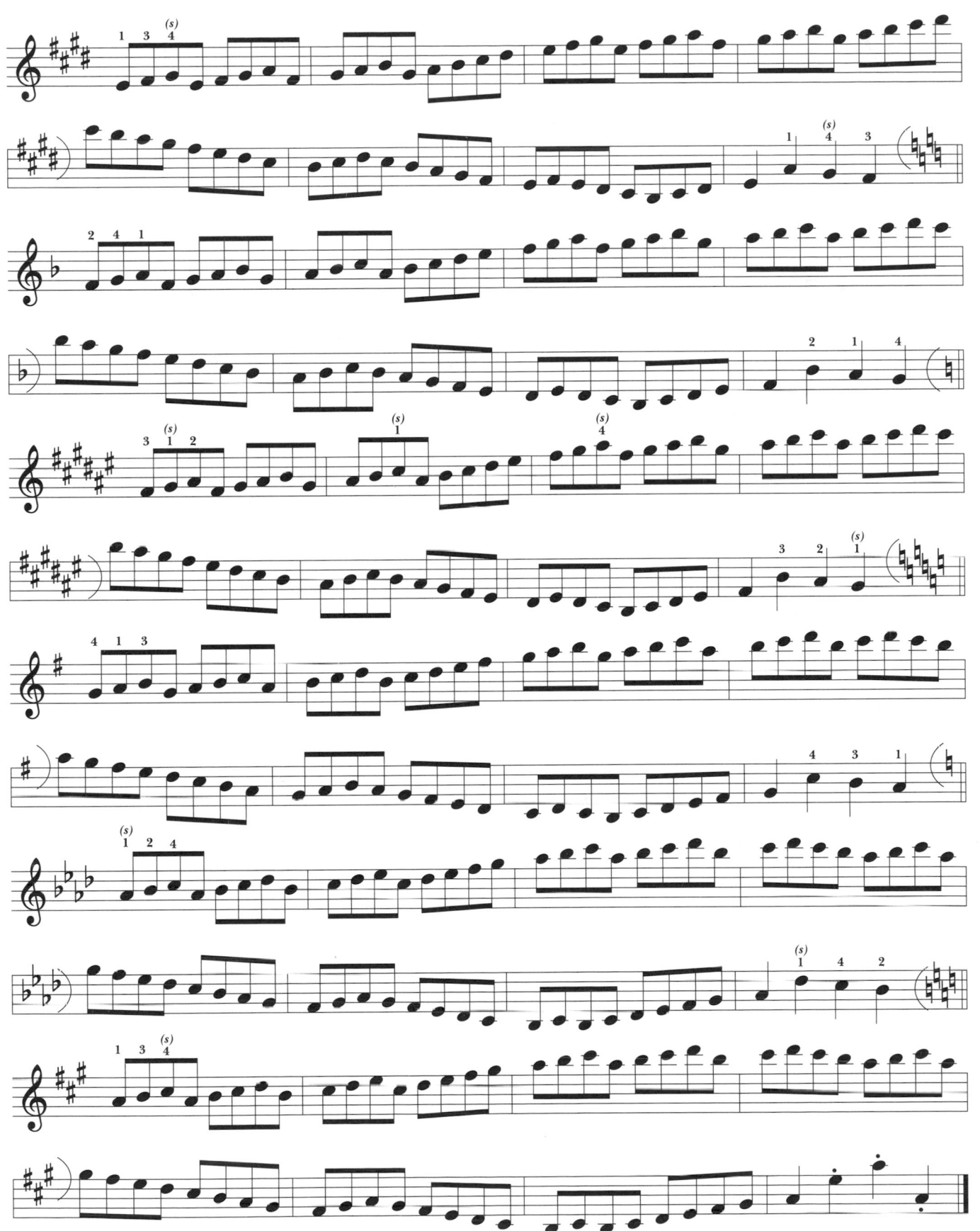

Principal Real Melodic Minor Scales—Position VII

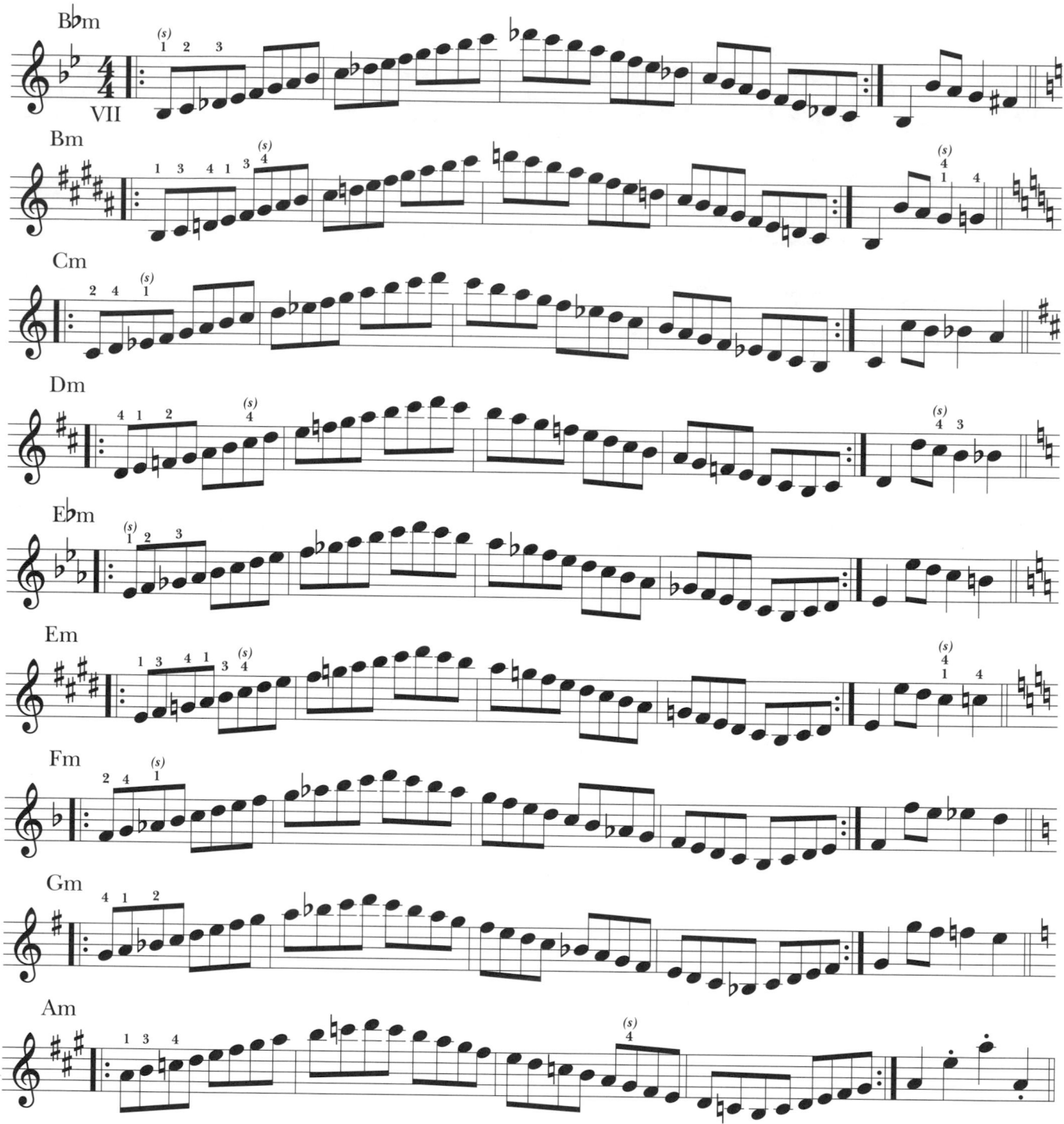

64

Chords—Three-Note Voicings
Melodization of Diminished Triads

Melodic degrees: Diminished scale (chord tones plus notes a whole step above and/or a half step below them).

▌ Fingering is constant if the sequence is played on the same set of strings.

Open Voicings

As we deal almost exclusively with diminished 7 chords, all of the preceding sequences may be played with any of the letter names that make up the four-note diminished 7 structure.

Melodization of Augmented Triads

Melodic degrees: Whole tone scale (chord tones plus notes a whole step above and/or below them—1, 9, 3, ♭5, +5, ♭7).

Fingering is constant if the sequence is played on the same set of strings. Note: These sequences also apply to dominant 7(+5) chords.

Open Voicings

As an augmented chord primarily represents the whole tone scale, the entire structure may move in whole steps.

Arpeggios—Four-Note A♭ Chords
Chord Spelling

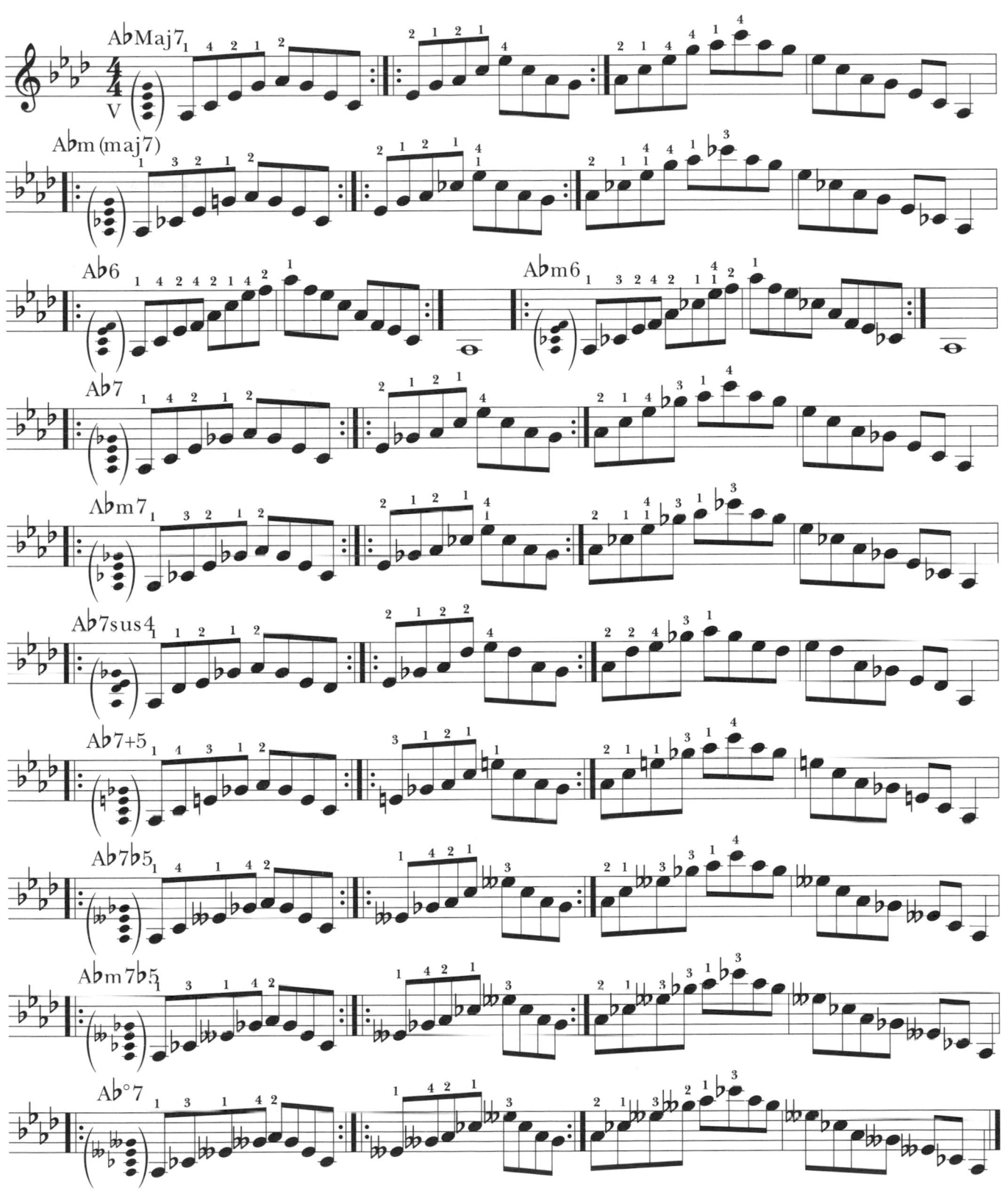

Arpeggios—Four-Note E Chords
Chord Spelling

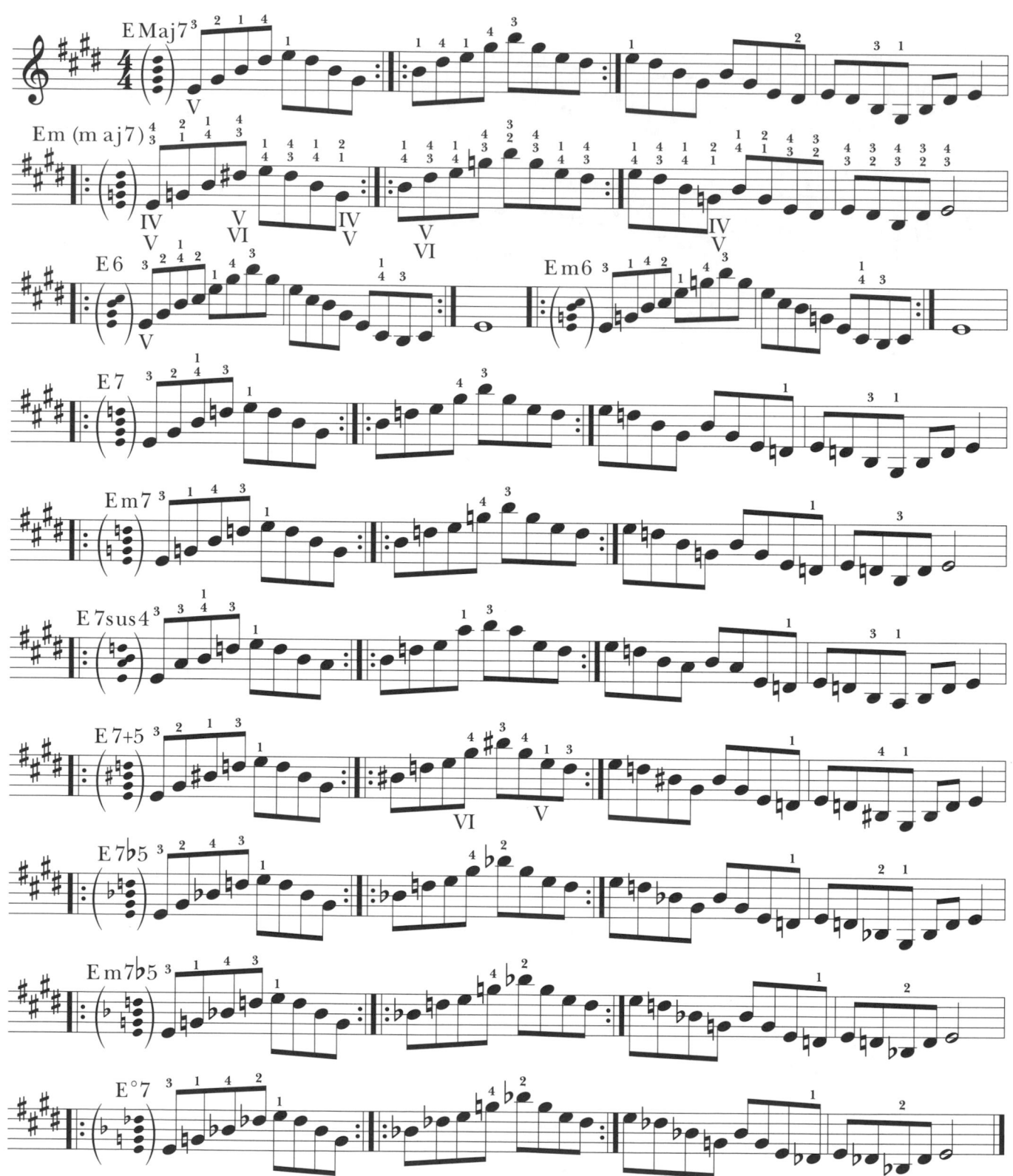

G Major Scale (Twelve Positions)

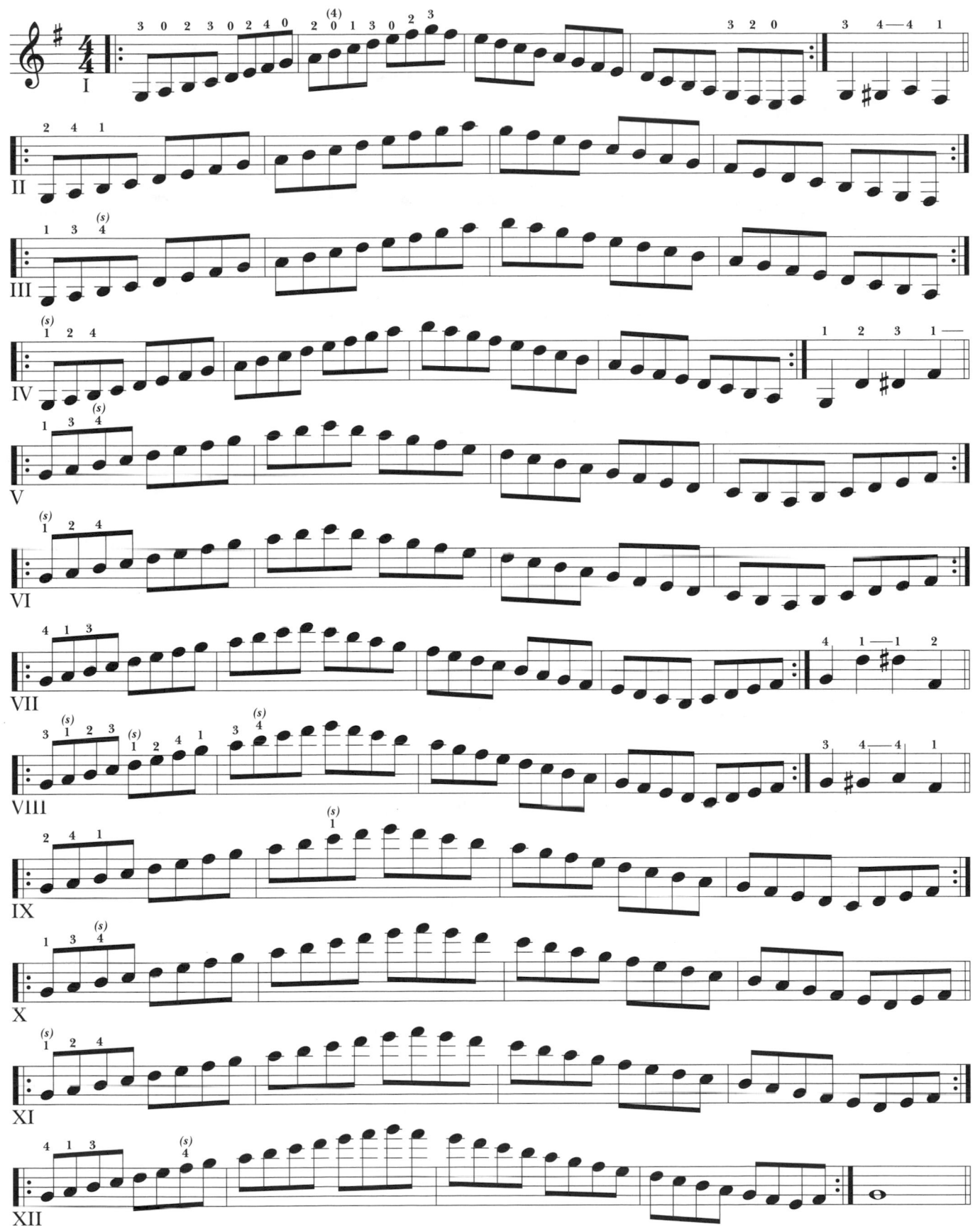

E Harmonic Minor—Nine Positions

E Minor Etude (solo)

Chord-Scale Relationships—Dominant 7 Chords

For improvisation.

1. The condition of the three highest degrees (tensions 9, 11, 13) on all dominant 7 chords with scale tone roots is controlled by the preceding scale.

I7, II7, V7	1, 3, 5, ♭7, 9, (11), 13	Major scale from intended tonic
VI7	1, 3, 5, ♭7, 9, (11), (♭13)	Real melodic minor scale from intended tonic
III7, VII7	1, 3, 5, ♭7, ♭9, (11), (♭13)	Harmonic minor scale from intended tonic
IV7	1, 3, 5, ♭7, 9, +11, 13	Real melodic minor scale from 5th

✱ The III7 and VII7 chords have a "built-in" ♭9. When the 9 is flatted, it is truly altered and ♯9 is compatible with it. By treating the ♯9 melodically as ♭3, the natural minor scale is the result. This is a second choice of related scale. All eight notes of the combined harmonic and natural minor scales are also used.

2. These three high degrees on all dominant 7 chords with non-scale tone roots are constant (9, +11, 13) and they are all treated the same as the IV7 chord.

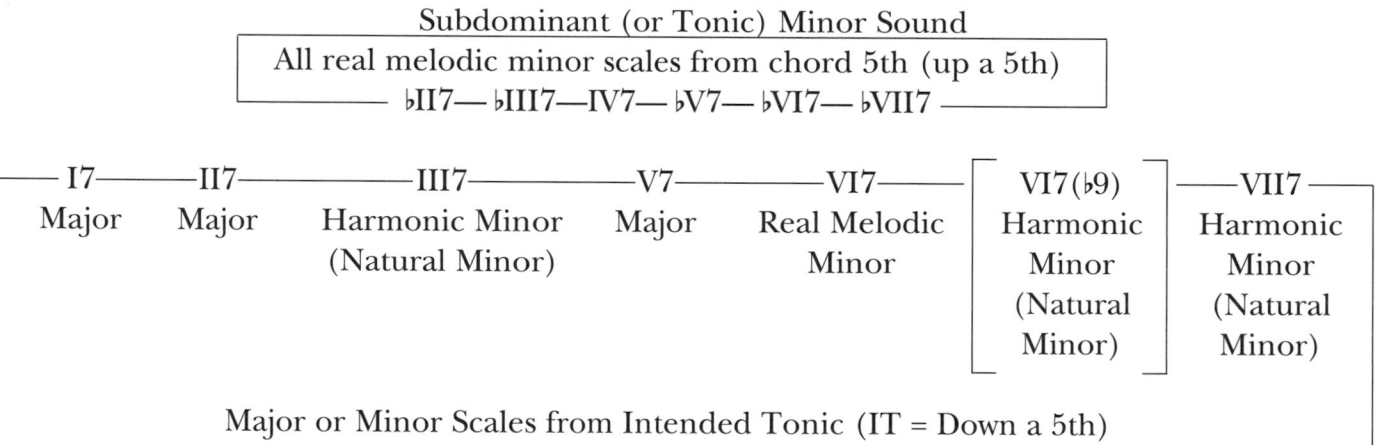

Dominant Sound

There is one structure containing an added alteration (not forced on it by the preceding scale sound): VI7(♭9). That chord has been included here, because it is encountered so often that we have become conditioned to hear it as the "norm." The VI7 with unaltered 9 is usually found only as a result of the melody being this note.

Have someone play the progressions for you (or use a tape recorder) and practice the proper scales over the following chord sequences.

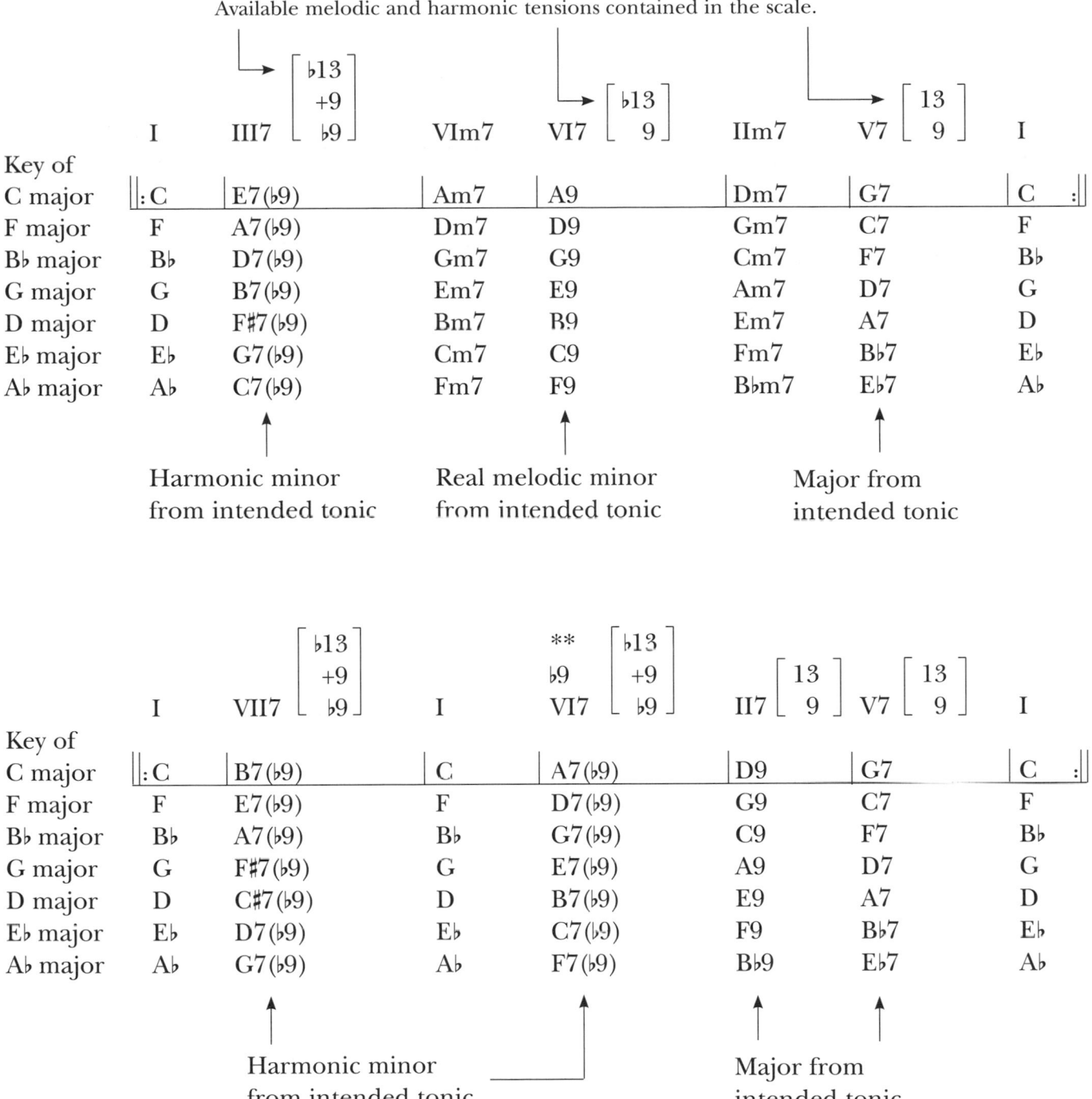

Available melodic and harmonic tensions contained in the scale.

	I	III7 [b13, +9, b9]	VIm7	VI7 [b13, 9]	IIm7	V7 [13, 9]	I
Key of C major	‖: C	E7(b9)	Am7	A9	Dm7	G7	C :‖
F major	F	A7(b9)	Dm7	D9	Gm7	C7	F
Bb major	Bb	D7(b9)	Gm7	G9	Cm7	F7	Bb
G major	G	B7(b9)	Em7	E9	Am7	D7	G
D major	D	F#7(b9)	Bm7	B9	Em7	A7	D
Eb major	Eb	G7(b9)	Cm7	C9	Fm7	Bb7	Eb
Ab major	Ab	C7(b9)	Fm7	F9	Bbm7	Eb7	Ab

Harmonic minor from intended tonic   Real melodic minor from intended tonic   Major from intended tonic

	I	VII7 [b13, +9, b9]	I	** VI7 [b13, +9, b9] b9	II7 [13, 9]	V7 [13, 9]	I
Key of C major	‖: C	B7(b9)	C	A7(b9)	D9	G7	C :‖
F major	F	E7(b9)	F	D7(b9)	G9	C7	F
Bb major	Bb	A7(b9)	Bb	G7(b9)	C9	F7	Bb
G major	G	F#7(b9)	G	E7(b9)	A9	D7	G
D major	D	C#7(b9)	D	B7(b9)	E9	A7	D
Eb major	Eb	D7(b9)	Eb	C7(b9)	F9	Bb7	Eb
Ab major	Ab	G7(b9)	Ab	F7(b9)	Bb9	Eb7	Ab

Harmonic minor from intended tonic   Major from intended tonic

	I	IV7 $\begin{bmatrix}13\\+11\\9\end{bmatrix}$	IIIm7	bIII7 $\begin{bmatrix}13\\+11\\9\end{bmatrix}$	IIm7	bII7 $\begin{bmatrix}13\\+11\\9\end{bmatrix}$	I
Key of C major	‖: C	F9	Em7	Eb9	Dm7	Db9	C :‖
F major	F	Bb9	Am7	Ab9	Gm7	Gb9	F
Bb major	Bb	Eb9	Dm7	Db9	Cm7	Cb9 (B9)	Bb
G major	G	C9	Bm7	Bb9	Am7	Ab9	G
D major	D	G9	F#m7	F9	Em7	Eb9	D
Eb major	Eb	Ab9	Gm7	Gb9	Fm7	Fb9 (E9)	Eb
Ab major	Ab	Db9	C7	Cb9 (B9)	Bbm7	A9	Ab

Real melodic minor from chord 5th

	I	I7 $\begin{bmatrix}13\\9\end{bmatrix}$	IV	bVI7 $\begin{bmatrix}13\\+11\\9\end{bmatrix}$	V7sus4 (IIm7)	bII7 $\begin{bmatrix}13\\+11\\9\end{bmatrix}$	I
Key of C major	‖: C	C9	F	Ab9	G9sus4 (Dm7)	Db9	C :‖
F major	F	F9	Bb	Db9	C9sus4 (Gm7)	Gb9	F
Bb major	Bb	Bb9	Eb	Gb9	F9sus4 (Cm7)	Cb9 (B9)	Bb
G major	G	G9	C	Eb9	D9sus4 (Am7)	Ab9	G
D major	D	D9	G	Bb9	A9sus4 (Em7)	Eb9	D
Eb major	Eb	Eb9	Ab	Cb9 (B9)	Bb9sus4 (Fm7)	Fb9 (E9)	Eb
Ab major	Ab	Ab9	Db7	Fb9	Eb9sus4 (Bbm7)	A9	Ab

Major from intended tonic

Real melodic minor from chord 5th

	I	♭V7 $\begin{bmatrix}13 \\ +11 \\ 9\end{bmatrix}$	IV	♭VII7 $\begin{bmatrix}13 \\ +11 \\ 9\end{bmatrix}$	I	B7$\begin{bmatrix}13 \\ 9\end{bmatrix}$	I
Key							
C major	‖: C	G♭9	F	B♭9	C	G7	C :‖
F major	F	C♭9	B♭	E♭9	F	C7	F
		(B9)					
B♭ major	B♭	F♭9	E♭	A♭9	B♭	F7	B♭
		(E9)					
G major	G	D♭9	C	F9	G	D7	G
D major	D	A♭9	G	C9	D	A7	D
E♭ major	E♭	A9	A♭	D♭9	E♭	B♭7	E♭
A♭ major	A♭	D9	D♭7	G♭9	A♭	E♭7	A♭

↑   ↑

Real melodic minor from chord 5th

It is necessary that you know (very well) the normal condition of tensions on all dominant 7 structures so you will instantly recognize any alterations that may be present. The effect of specially altered degrees on dominant 7 chord-scale relationships will be discussed later.

From this point on, all chord-scale pages consist of a great deal of information applicable to composition, spontaneous or otherwise, presented very concisely. As this concerted presentation can be confusing, the material must be worked out by the interested student very gradually over a considerable period of time.

Determine the scale for a chord by its effect on the scale preceding it.

Practical Fingerings for Moving from Position to Position

4-4 AND 1-1   FINGER SLIDES EMPLOYING THE HALF STEP

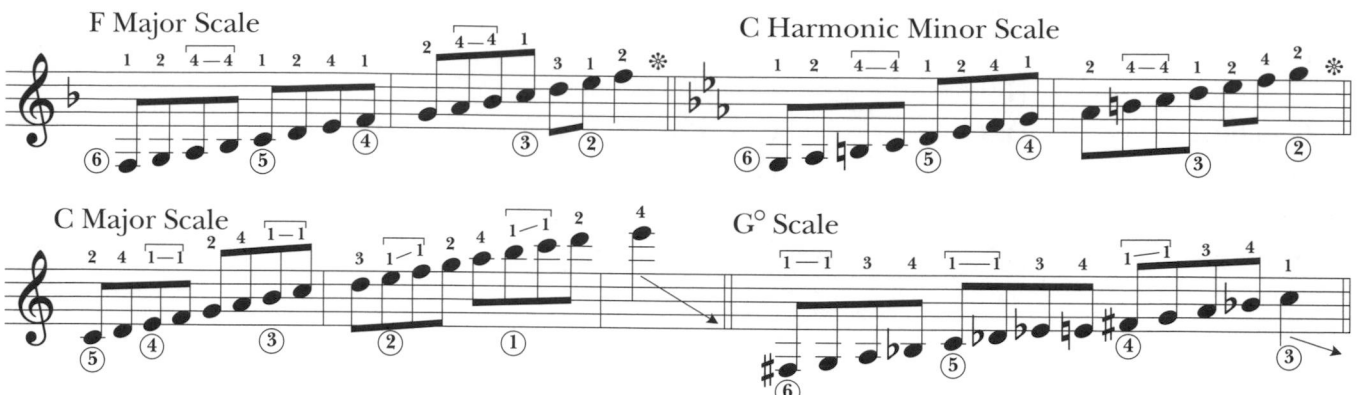

✻ No descent with 4th finger slides

The preceding 1st and 4th finger slides are also possible (and practical) for distances of from two to three frets.

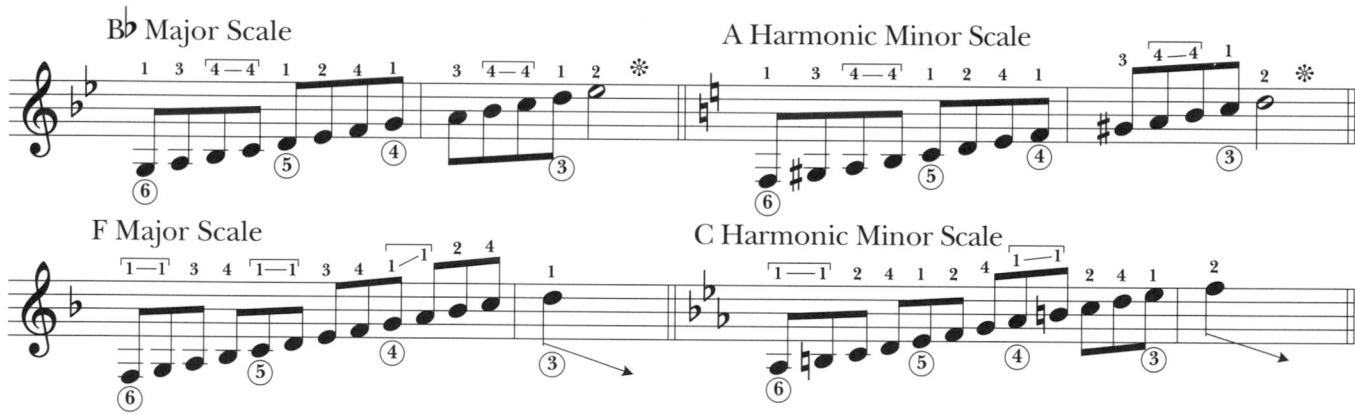

13-13 EMPLOYING THE HALF STEP 12-13, 12-12 VARIATIONS

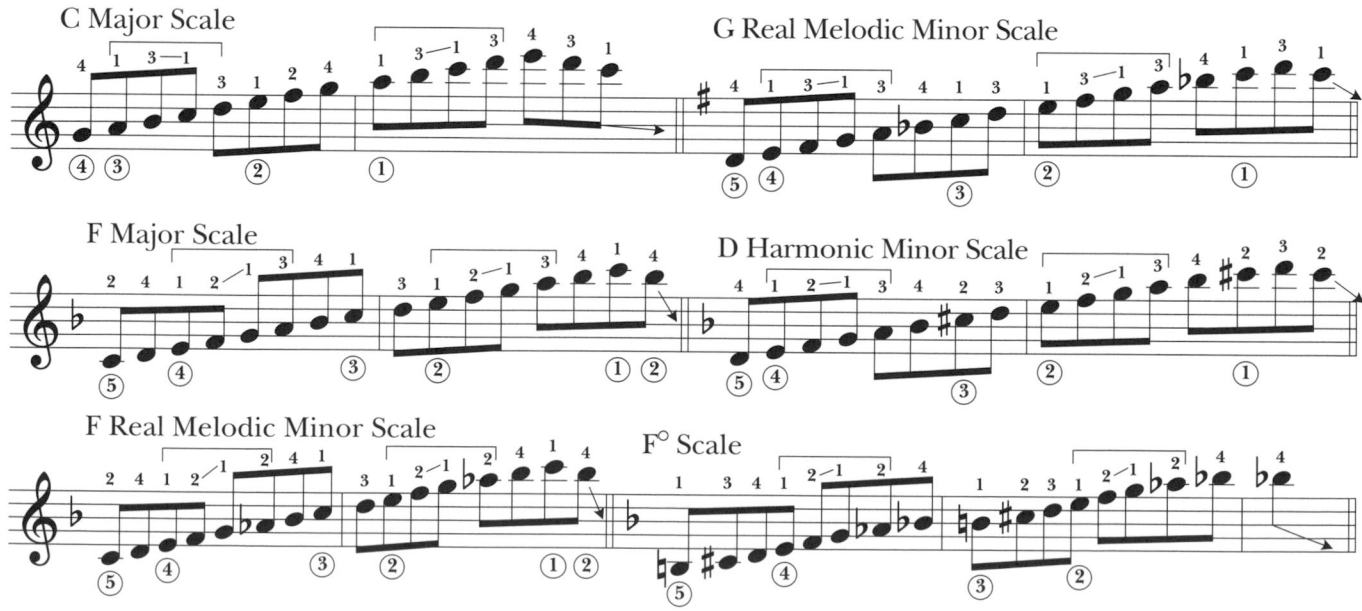

1-2 3-4 THE DOUBLE STRETCH—EMPLOYING THE HALF STEP

F Major Scale   G° Scale

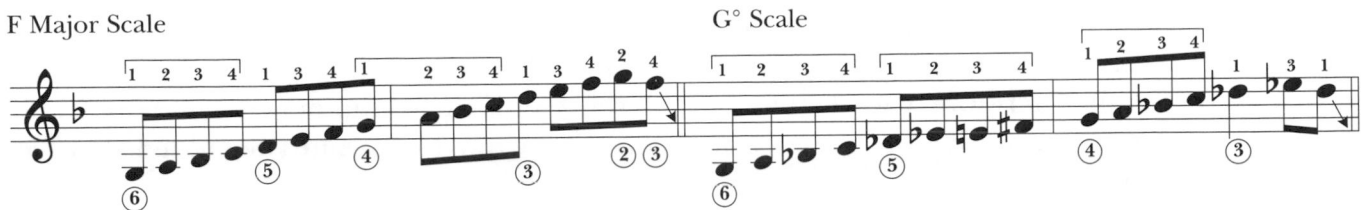

13-24 FINGER EXCHANGE—EMPLOYING THE HALF STEP

F Major Scale   D Major Scale

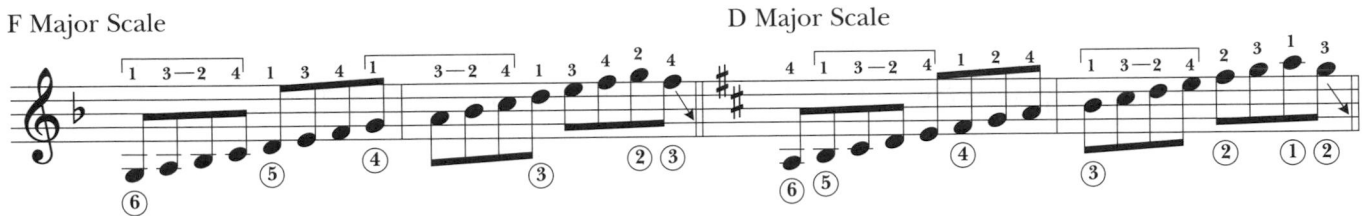

124-124 REPEATED FINGERING—SEPARATED BY A WHOLE STEP

G Major Scale   C Major Scale

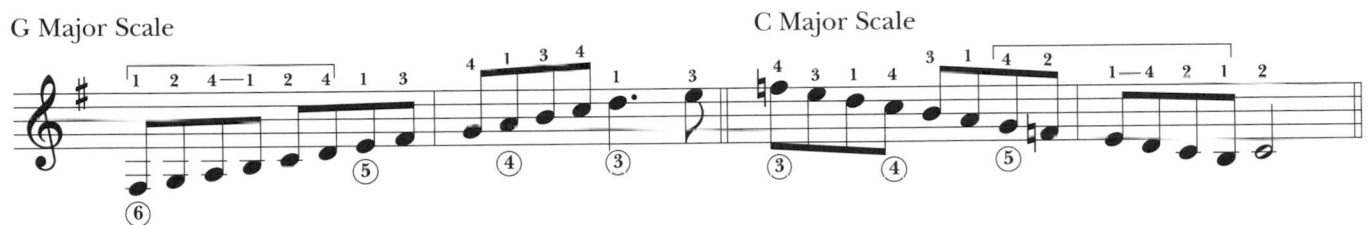

134-134 REPEATED FINGERING—SEPARATED BY A WHOLE STEP

G Major Scale   C Major Scale

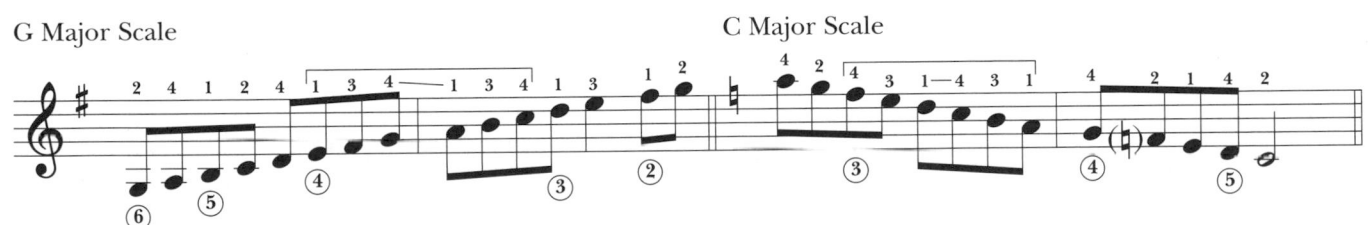

13-134 VARIATION OF ABOVE, 13-124 VARIATION OF ABOVE

C Real Melodic Minor Scale   F Major Scale

▌ Analyze the intervals involved in the preceding position-to-position fingerings. You will find many other possibilities for application, especially when used in combinations.

▌ All of the fingerings employing the half step are very reliable, as they do not require looking at the fingerboard. The others are sometimes dangerous when the music and/or conductor demand your full attention.

Chord Construction—Three-Note Voicings
Dominant 7 Chords—Preparation of Close and Open Voicings

The distance between the 3rd and ♭7th chord degrees of a dominant structure is called a tritone. This tritone interval (an augmented 4th or a diminished 5th) divides our twelve-tone (chromatic) scale exactly in half. Therefore, each tritone by itself represents the sound of two dominant 7 chords, their roots being separated by the same ♯4 or ♭5 interval. A third note must be added to a tritone to remove this ambiguity.

In a cycle 5 chord progression, tritones move chromatically downward. The ♭7 of the first chord moves to 3 of the next chord, which moves to ♭7 of the next, and so on.

In the following studies, root and 5th chordal degrees are added to chromatic tritone sequences (representing cycle 5 progressions as follows: 1) below, 2) above, and 3) between.

1

TRITONE ON 3RD AND 4TH STRINGS (♭7, 3 IN THE LEAD)

TRITONE ON 2ND AND 3RD STRINGS (♭7, 3 IN THE LEAD)

2

TRITONE ON 4TH AND 5TH STRINGS (ROOT, 5 IN THE LEAD)

TRITONE ON 3RD AND 4TH STRINGS (ROOT, 5 IN THE LEAD)

3

TRITONE ON 2ND AND 4TH STRINGS (3, ♭7 IN THE LEAD)

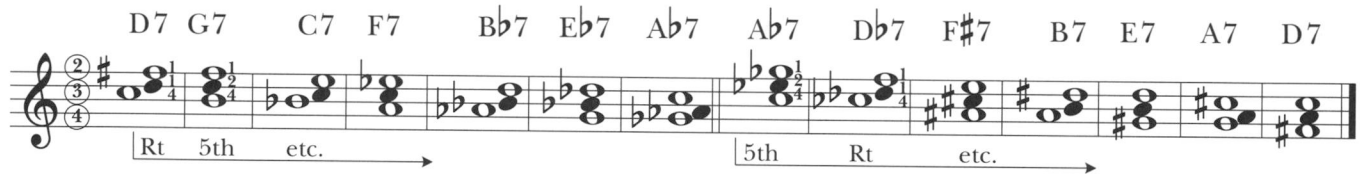

TRITONE ON 3RD AND 4TH STRINGS (3, ♭7 IN THE LEAD)

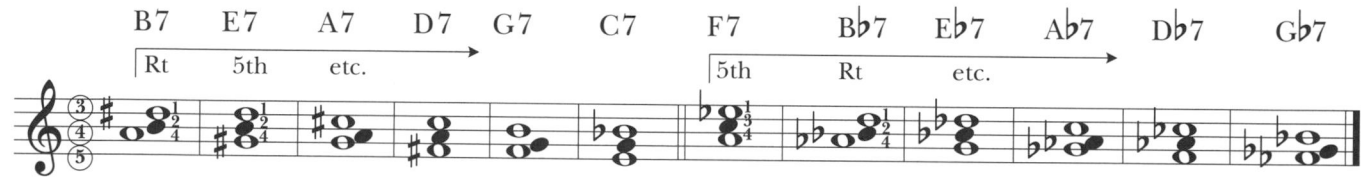

The tritone (interval of ♯4 or ♭5) should not be used below B–F as found on the 5th and 4th strings, respectively. The sound becomes cloudy in pitch from this point on down.

Arpeggios—Four-Note D♭ Chords
Chord Spelling

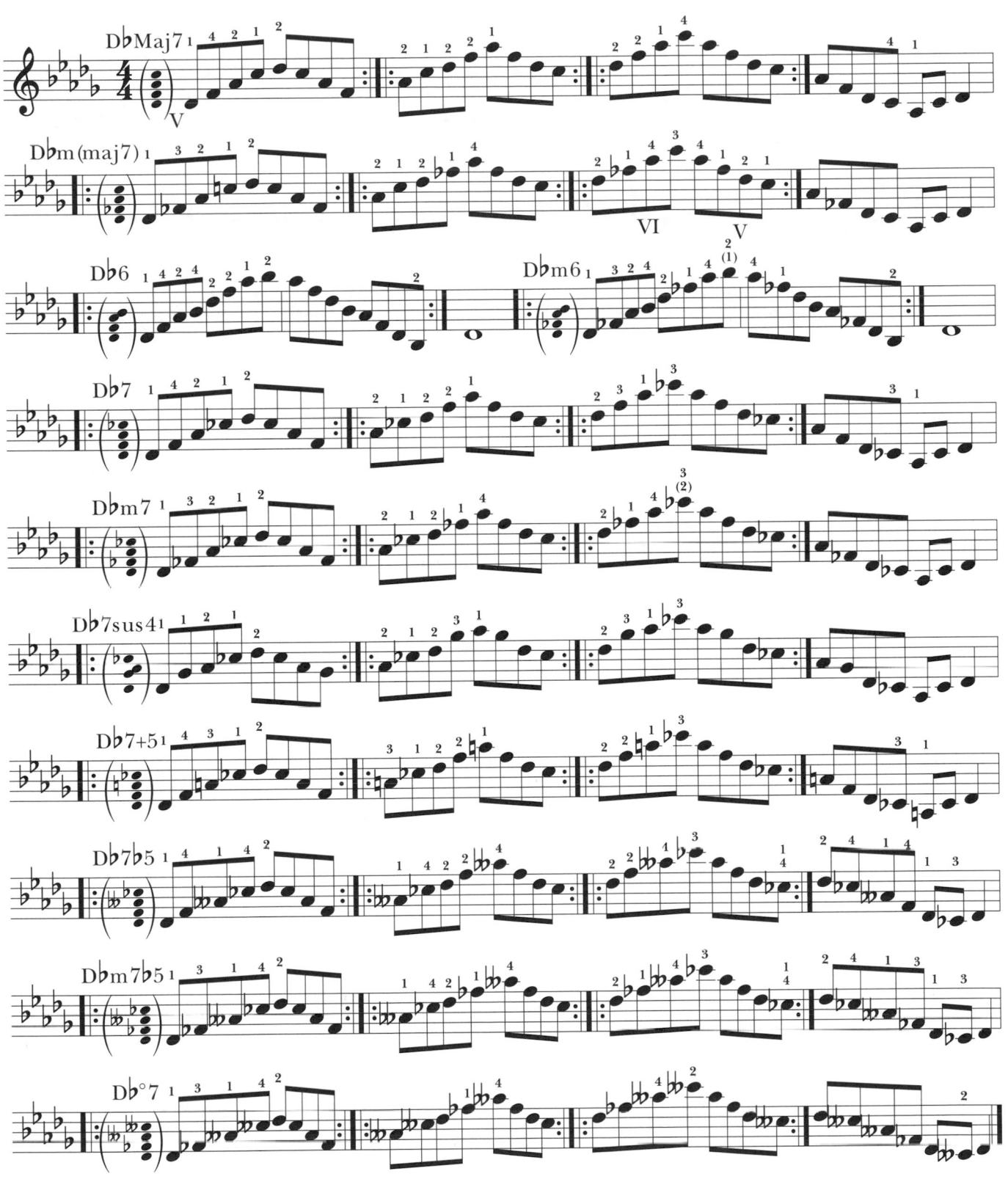

Arpeggios—Four-Note B Chords
Chord Spelling

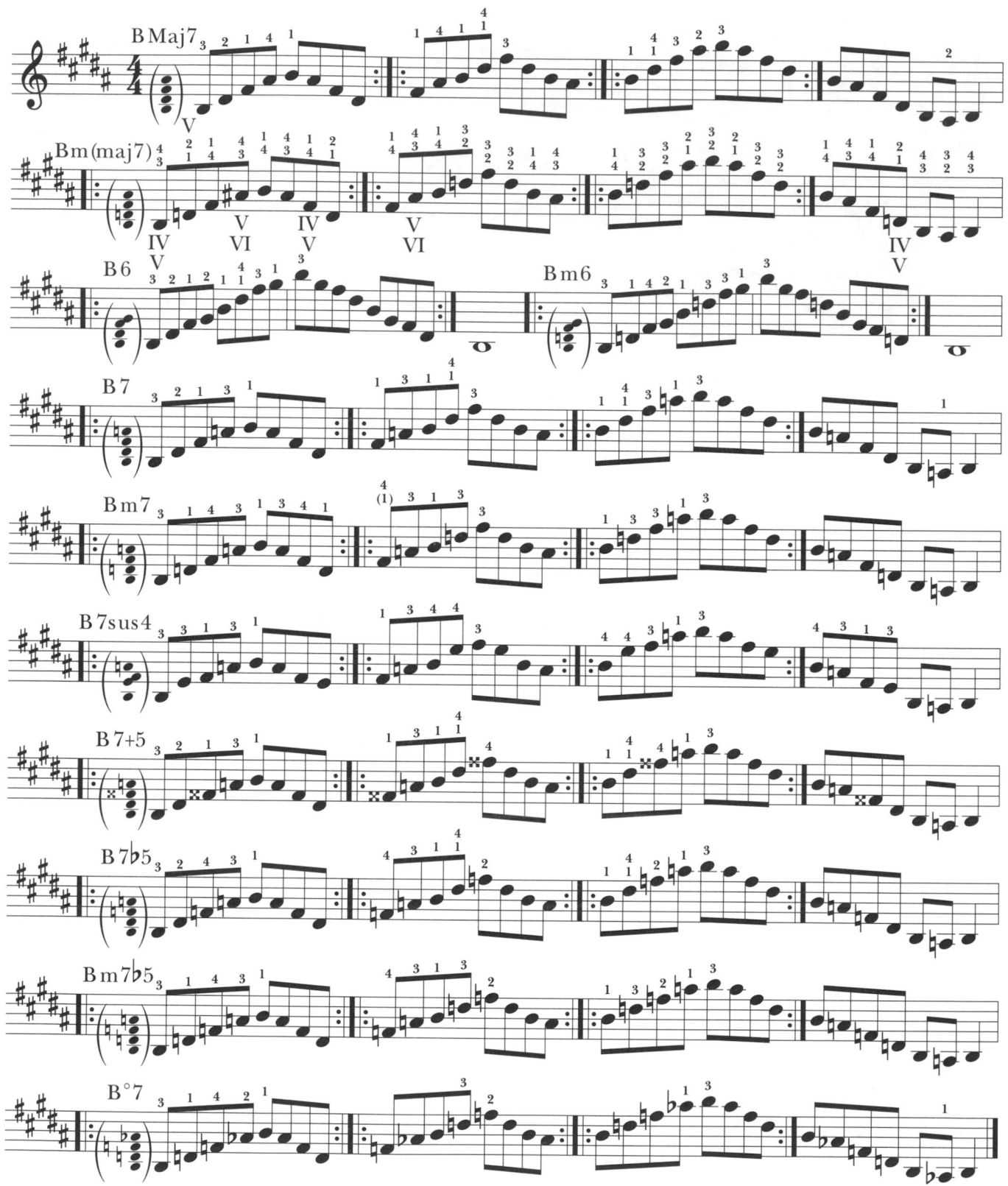

Rhythm Guitar—The Right Hand

Polka Dot (Polka-Duet)

Major Scales—Position VIII
Twelve Keys—Through Cycle 5

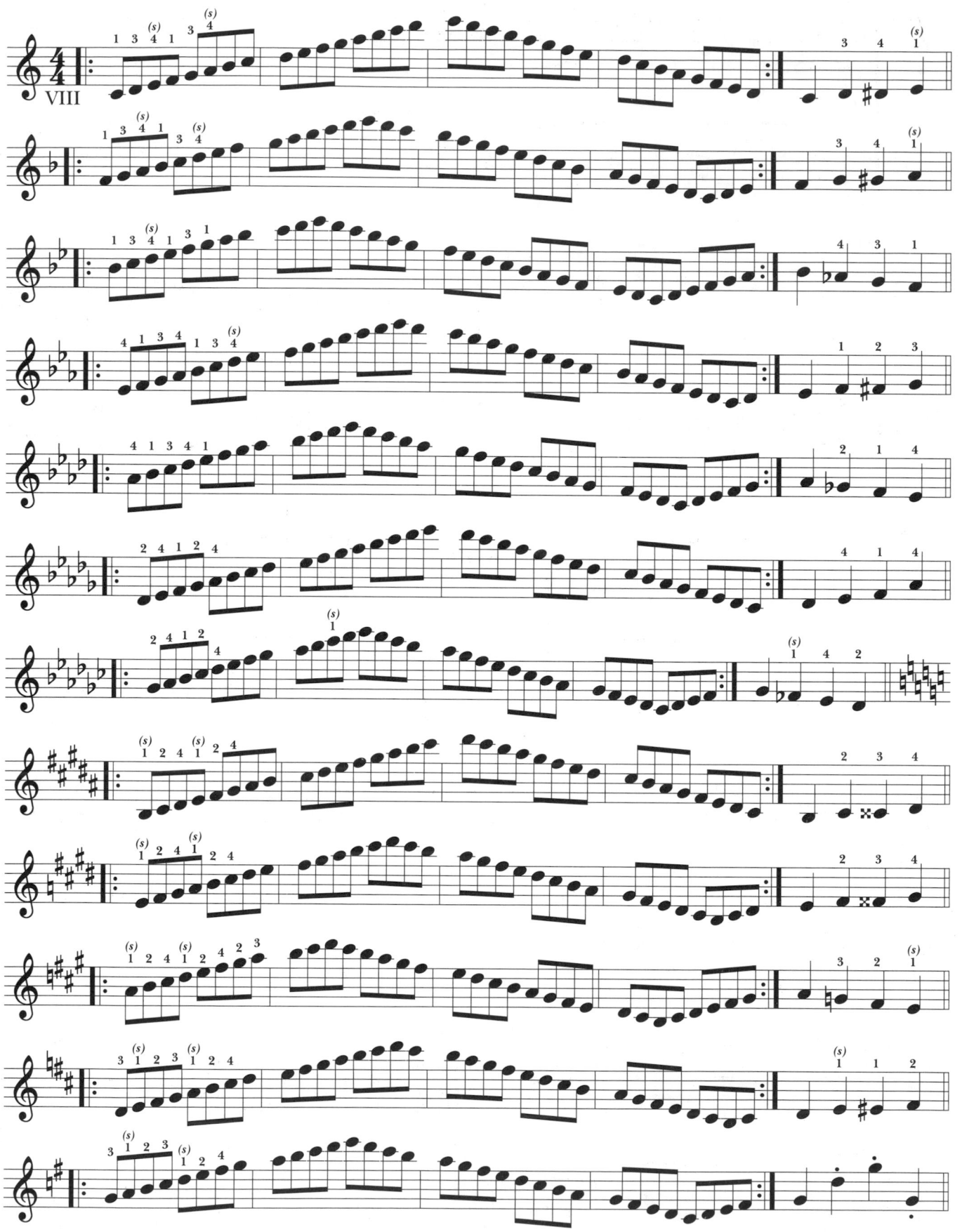

Chords—Three-Note Voicings
Melodization of Tonic Major Chords

Melodic degrees: Major scale from chord name

✳ 6th degree necessary as an undervoice

Melodization of Minor 7 Chords as VIm7

Melodic degrees: Major scale from ♭3 of chord

✳✳ Melodic degrees shown in parentheses must be used only in passing.

As the preceding I major and VIm7 chords produce the same tonic major sound, their voicings are interchangeable (C = Am7, F = Dm7, B♭ = Gm7). This is called **diatonic substitution**, or the replacement of one chord with another that represents the sound of the same scale and chord function (tonic, subdominant, and dominant) and whose chord tones are derived from higher or lower scale degrees.

Melodization of Subdominant Major Chords

Melodic degrees: Major scale from 5th of chord

✱ ♯4 is a diatonic tension on IV chords.

Melodization of Minor 7 Chords as IIm7

Melodic degrees: Major scale from ♭7 of chord

As the preceding IVmaj and IIm7 chords produce the same subdominant sound, their voicings are interchangeable (A♭ = Fm7, C = Am7, F = Dm7).

Arpeggios—Four-Note F# Chords
Chord Spelling

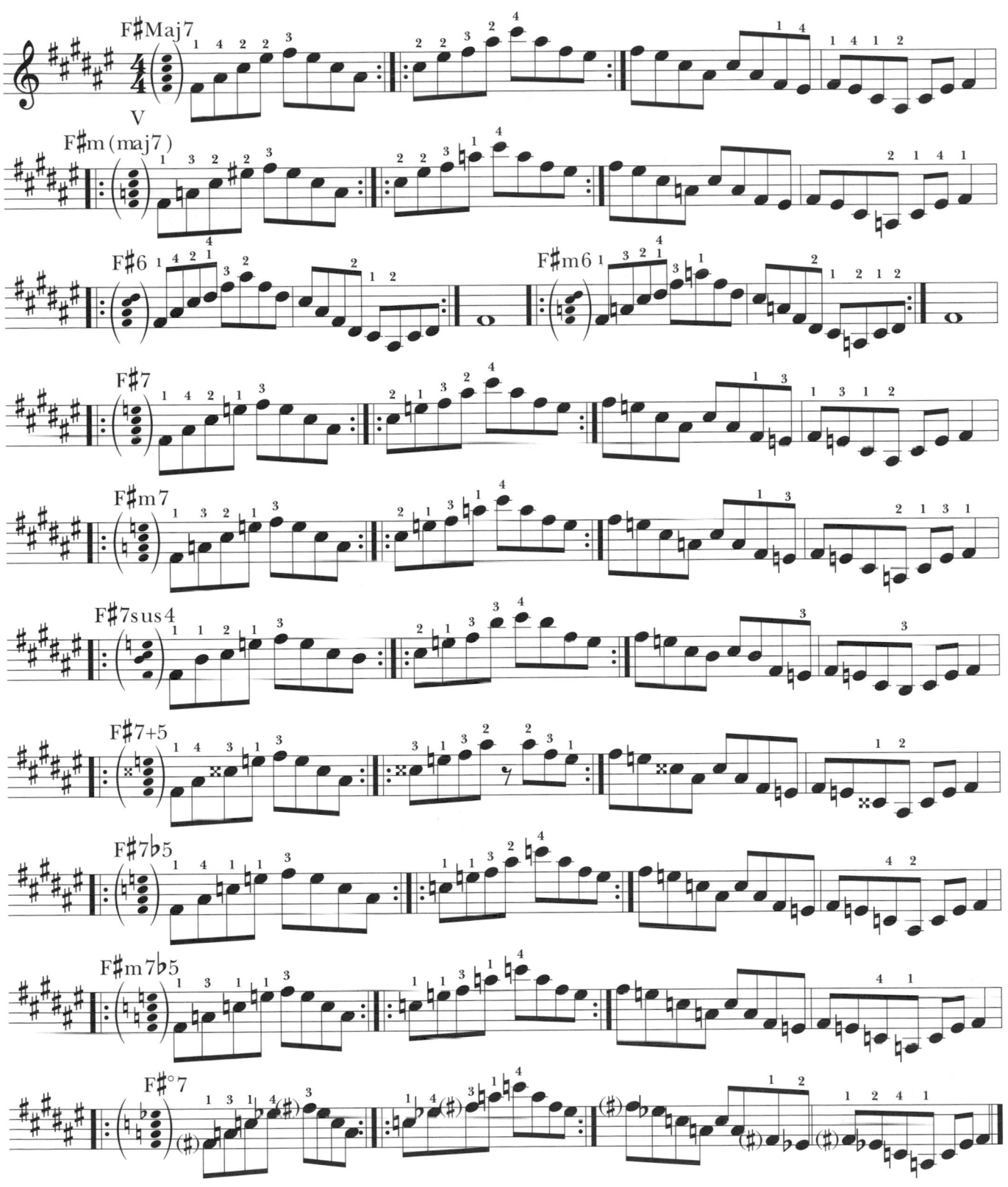

Arpeggios—Four-Note G♭ Chords

Chord Spelling

All fingering from preceding F♯ arpeggios.

Four-Note C♯ Chords

Fingering from preceding D♭ arpeggios.

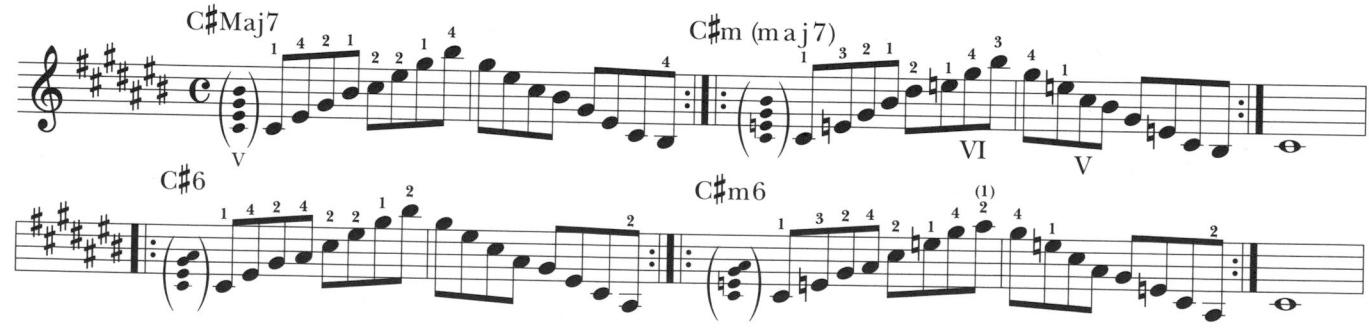

Four-Note C♭ Chords

Fingering from preceding B arpeggios.

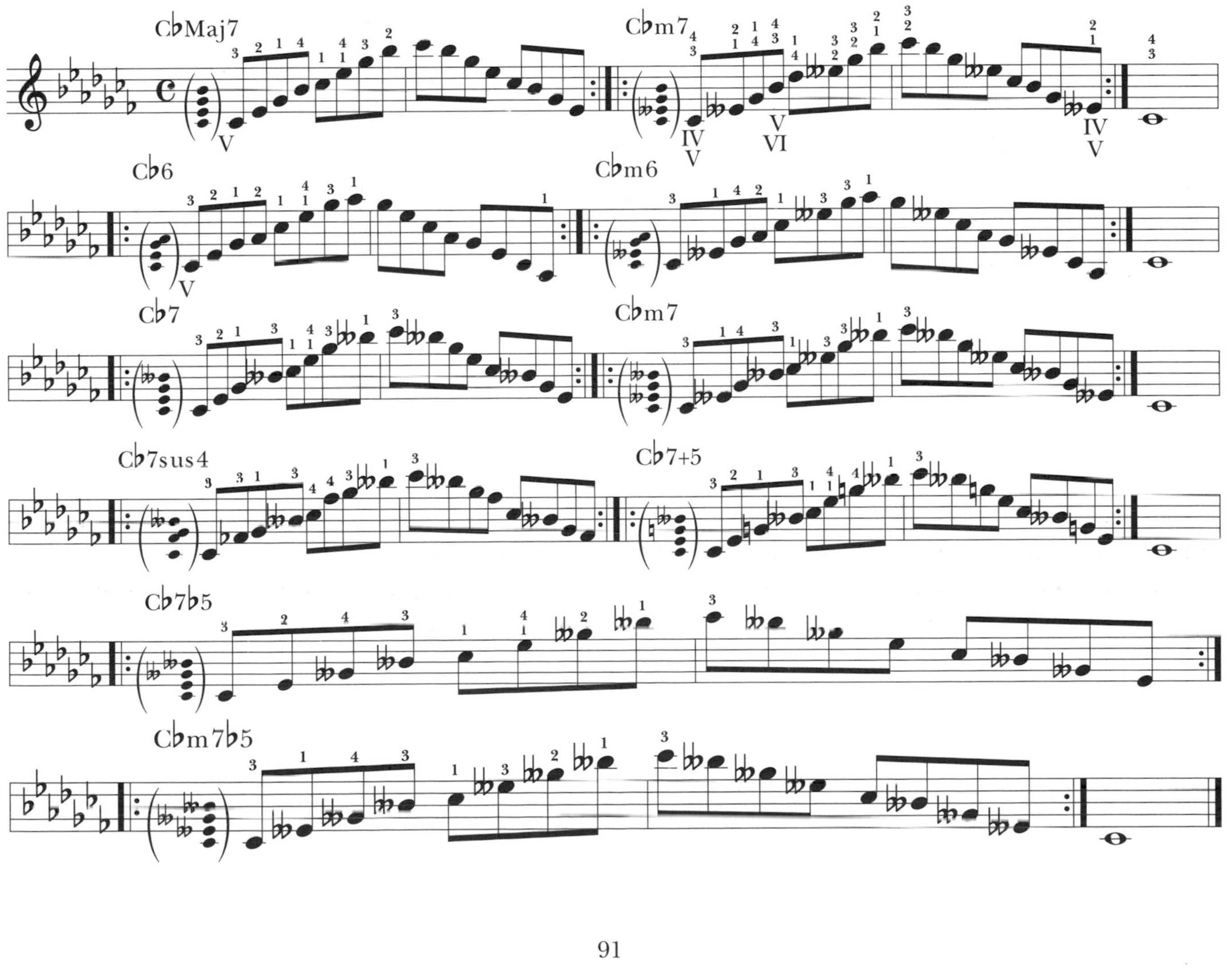

Chords—Three-Note Voicings
Dominant 7 Chords—Open Voicings, All Inversions

▌ Chord voicings notated here as (●) should be used only in passing for the following reasons:
  • Incomplete structure (indefinite sound)
  • Weak degree in the "bass"

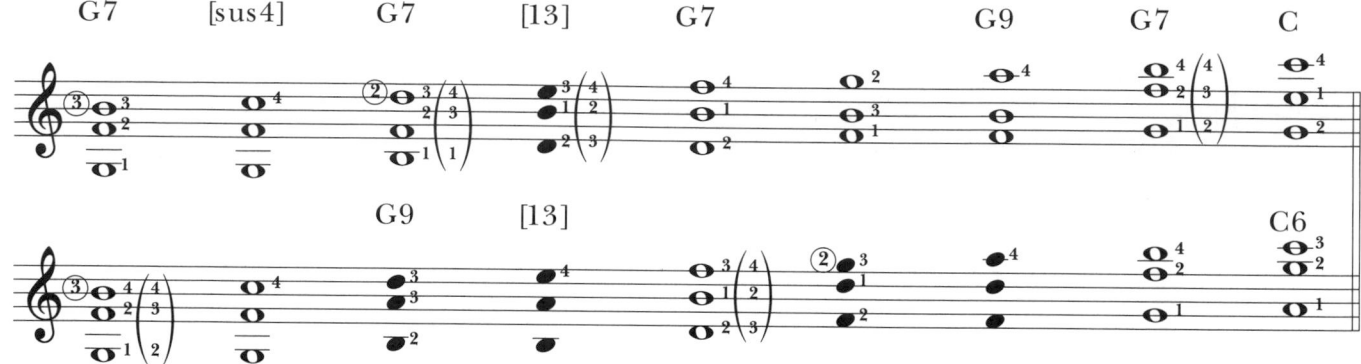

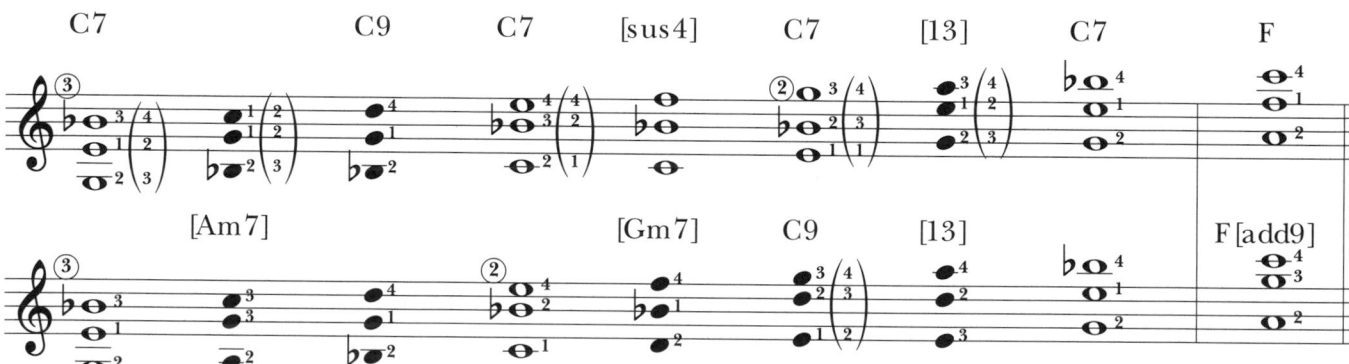

Most three-part chord voicings without the root do not have a well-defined sound, unless 1) they follow a strong voicing (including the root) of the same chord, or 2) they are the second chord of a strong cadence, closely voice led from the first chord (which has set the tonality), or 3) they are a spread voicing with the 5th degree on the bottom, sounding in the low register.

Bb Major Scale (Twelve Positions)

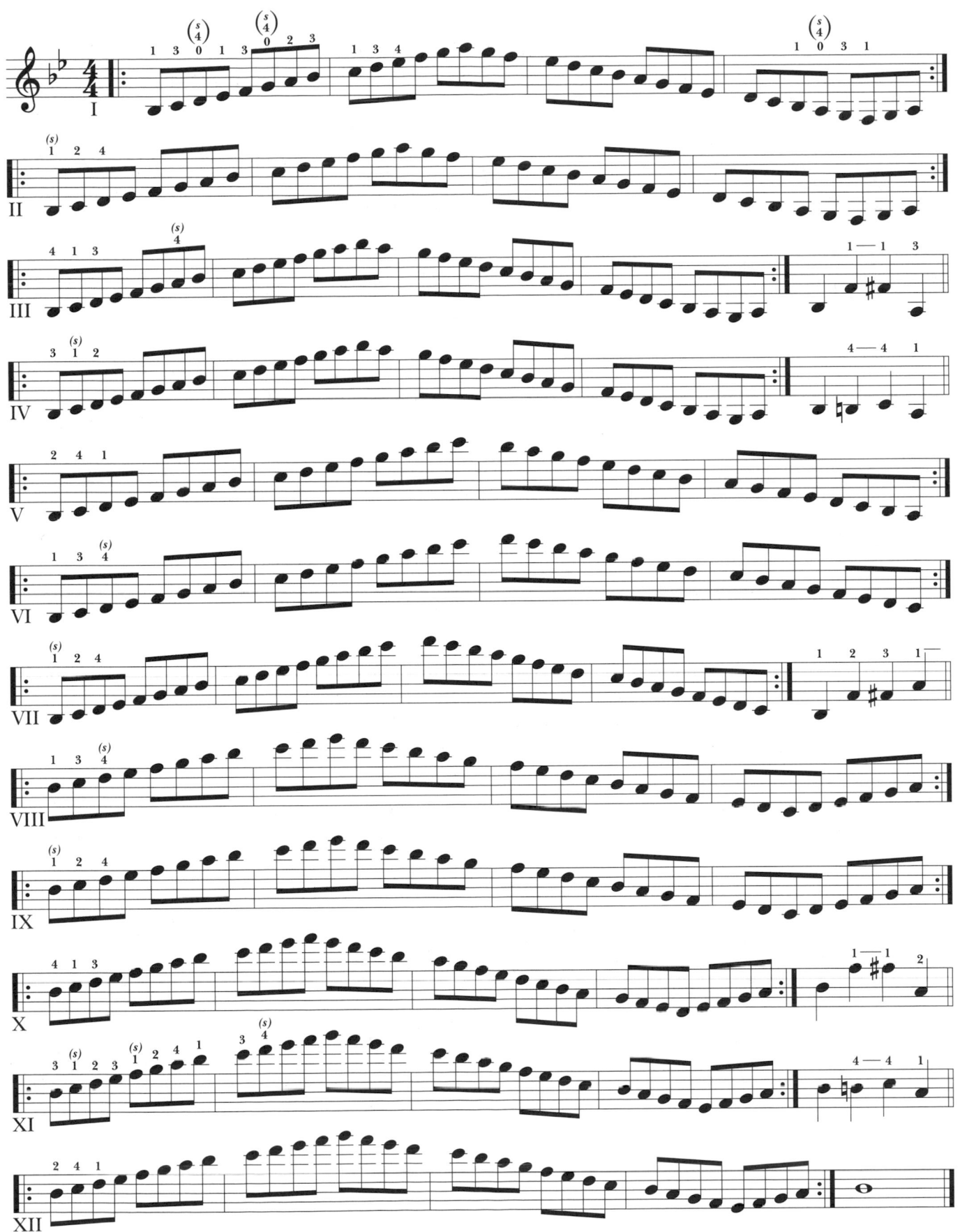

G Harmonic Minor (Nine Positions)

Etude in G Minor (solo)

Arpeggios—Diminished 7 Chords
Chord Spelling Most Used

Because the notes of the diminished 7 chord divide the chromatic scale into four equal parts (all minor 3rd intervals), any chord tone may be considered the root. To eliminate the use of double flats in notation, chord spelling varies. Diminished 7 chords are often notated as if they were constructed from major scale degrees 1, ♭3, ♭5/♯4, and 6, as well as 1, ♭3, ♭5, ♭♭7. The number 7 is not usually used with diminished chord symbols. The 7th chordal degree is always assumed (unless a three-note structure is specified by the word "triad").

Scale degrees from chord name.

Arpeggios—Dominant 7(♭5) Chords

Because the notes of the dominant 7(♭5) chord divide the chromatic scale into two like parts (each consisting of four half steps and two half steps), the structure can be named from the ♭5 as well as the root.

✴ Enharmonic spelling: same sound but different notation.

Theory: Diatonic 7th Chords—Harmonic Minor

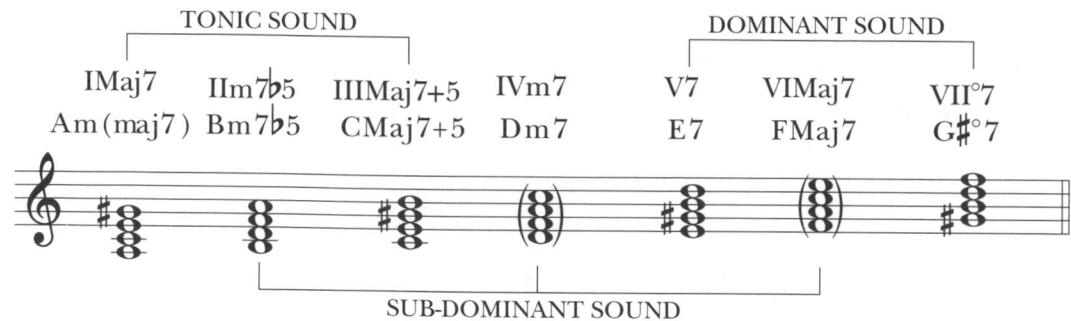

- Note the following:
  - The tonic chord is usually a minor triad. However it is sometimes found brightened up with the 6th degree borrowed from the melodic minor scale.
  - The II chord is always a minor 7(♭5).
  - IIm7(♭5) is often (and misleadingly) referred to as IVm6 (Bm7(♭5) = Dm6).
  - The 9th degree on V7 is always ♭9.
  - IVm7 and VIMaj7 usually occur as passing chords, for they tend to suggest the sound of the relative major (or natural minor).

Expect anything to happen in minor keys, from the most basic diatonic harmonic minor relationships to a conglomeration of (temporary) sounds borrowed from real or traditional melodic and natural minor scales.

Arpeggio and Scale Study

- Play in all possible areas of the fingerboard.

Play the entire sequence without changing position; don't "baby" your fingers.

Chords—Three-Note Voicings
Melodization of Subdominant Minor 6 Chords

Melodic degrees: Real melodic minor scale from chord name.

Melodization of Minor 7(♭5) as Altered IIm7 Chords

Melodic degrees: Real melodic minor scale from ♭3 of chord.

As the preceding IVm6 and IIm7(♭5) chords produce the same subdominant minor sound, their voicings are interchangeable (A♭m = Fm7(♭5), etc.).

Melodization of Dominant 7 Chords as IV7 and ♭VII7

Melodic degrees: Real melodic minor scale from 5th degree of chord.

Melodization of Dominant 7 Chords as VI7

Melodic degrees: Real melodic minor scale from intended tonic.

Major Scales—Position X

Twelve keys—through cycle 5.

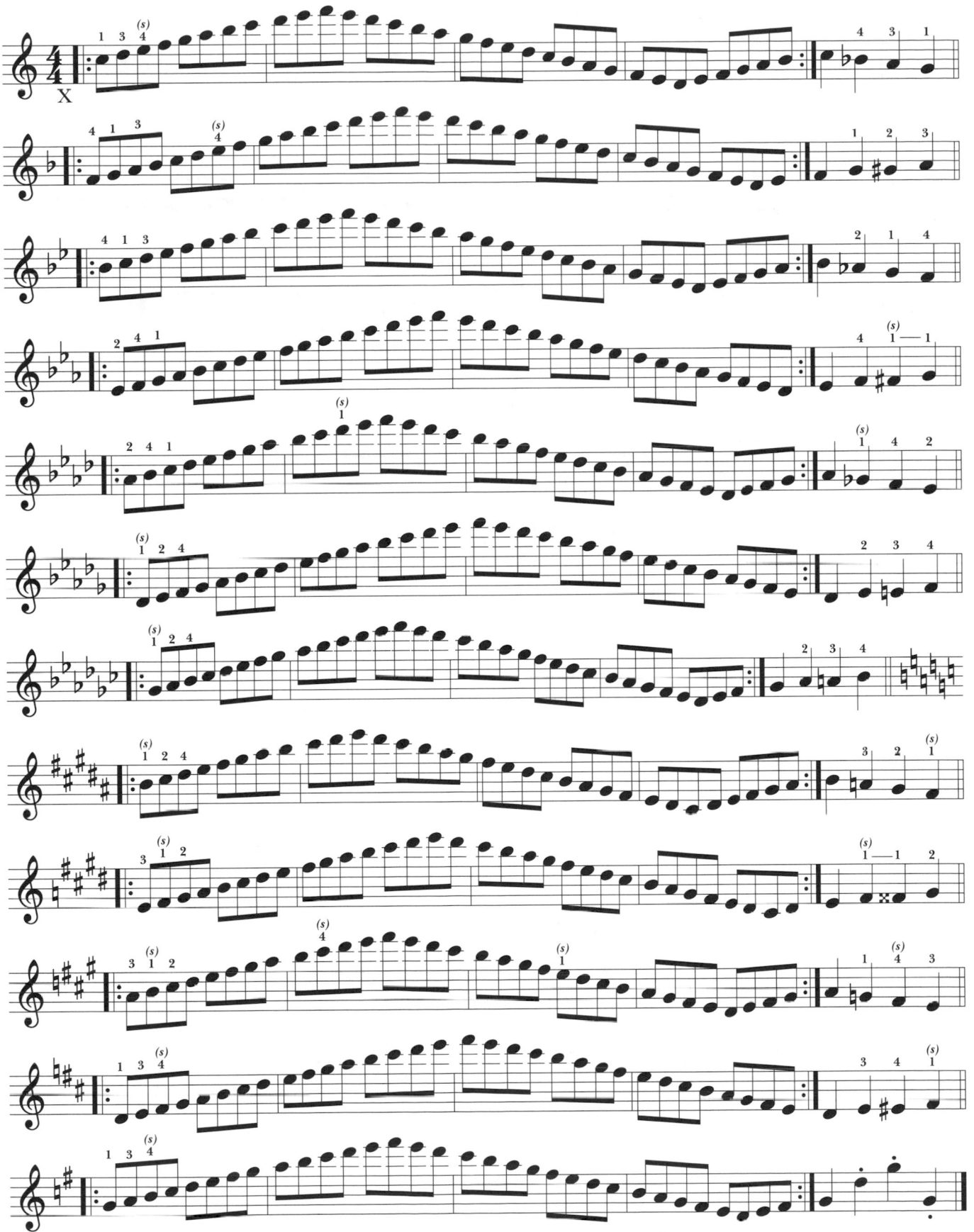

Principle Real Melodic Minor Scales—Position X

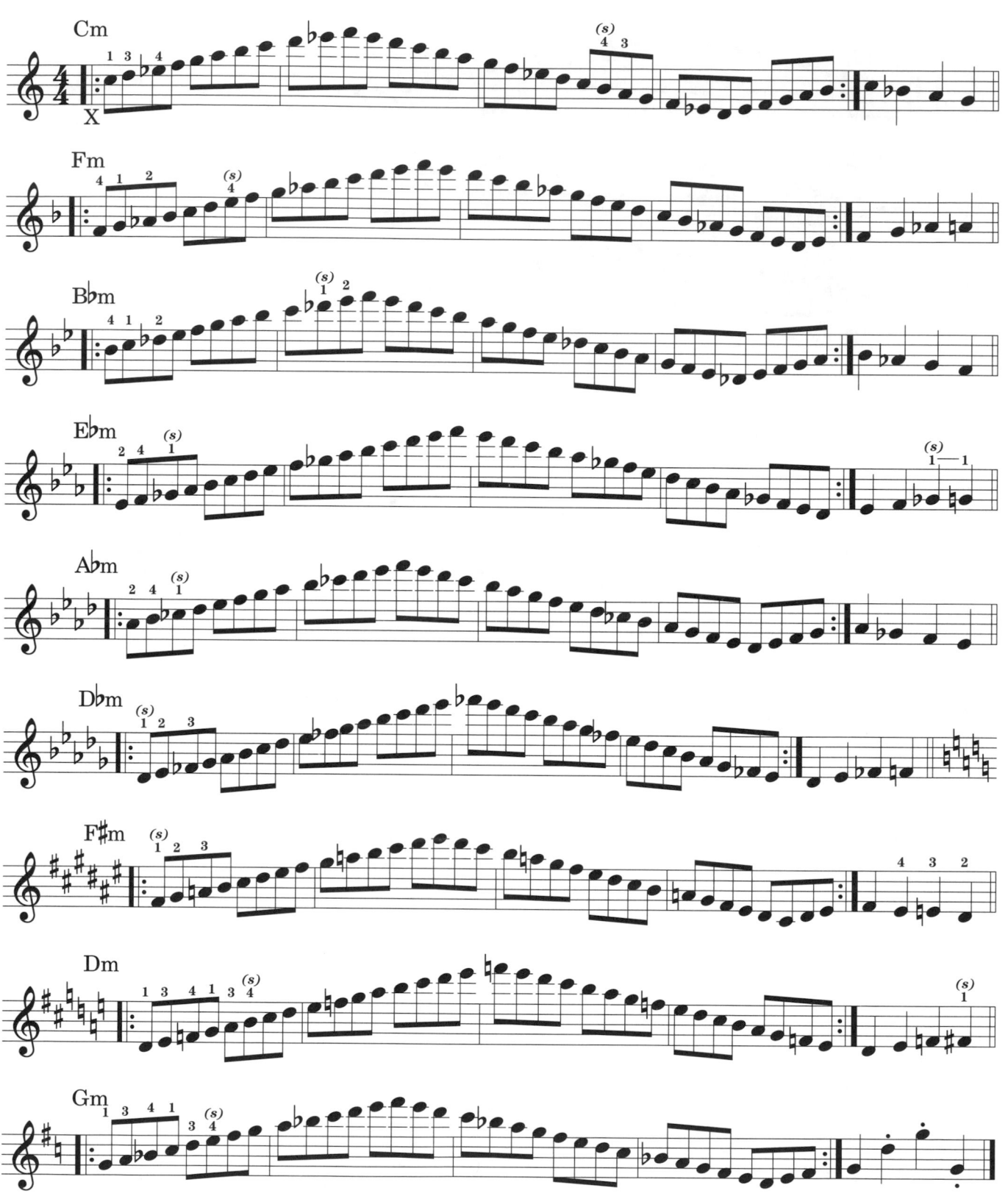

Chords—Three-Note Voicings
Major 6th Chords—Close and Open Voicings

6TH AND 3RD IN THE LEAD

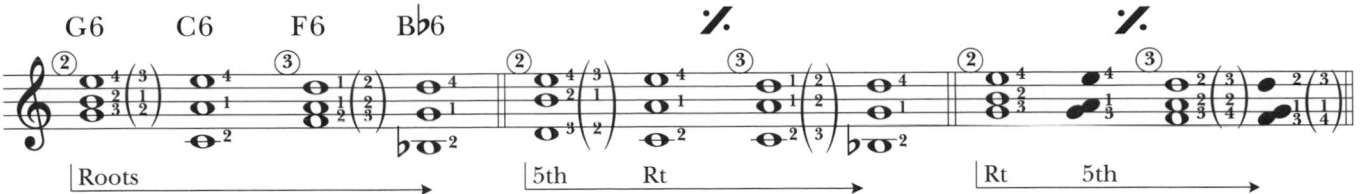

**3RD AND 6TH IN THE LEAD
(ROOT, 5TH, INSIDE VOICE)**

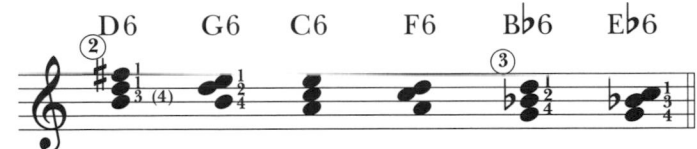

ROOT AND 5TH IN THE LEAD

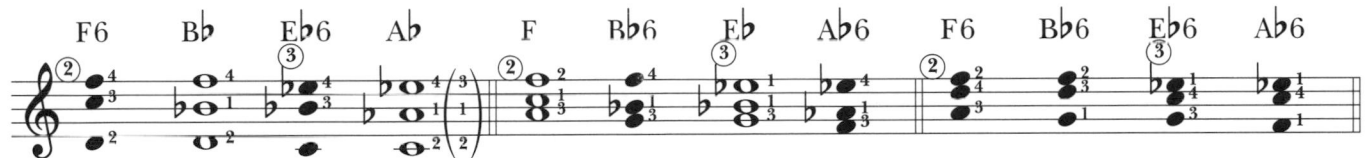

Major 7th Chords

3RD AND MAJOR 7TH IN THE LEAD

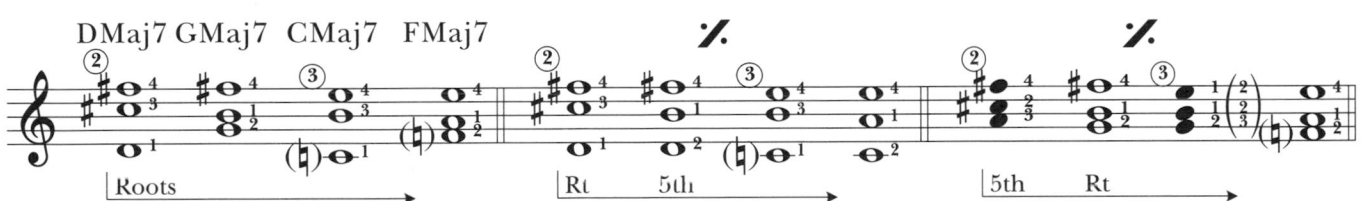

**3RD AND MAJOR 7TH IN THE LEAD
(ROOT, 5TH, INSIDE VOICE)**

NO ROOT, LEAD WITH MAJOR 7TH

5TH IN THE LEAD

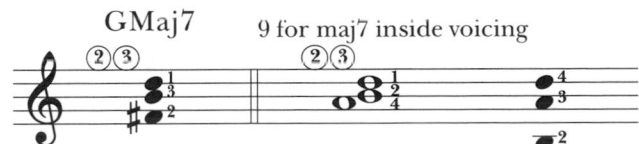

Major 6 and 7 Chords—Open Voicings, All Inversions

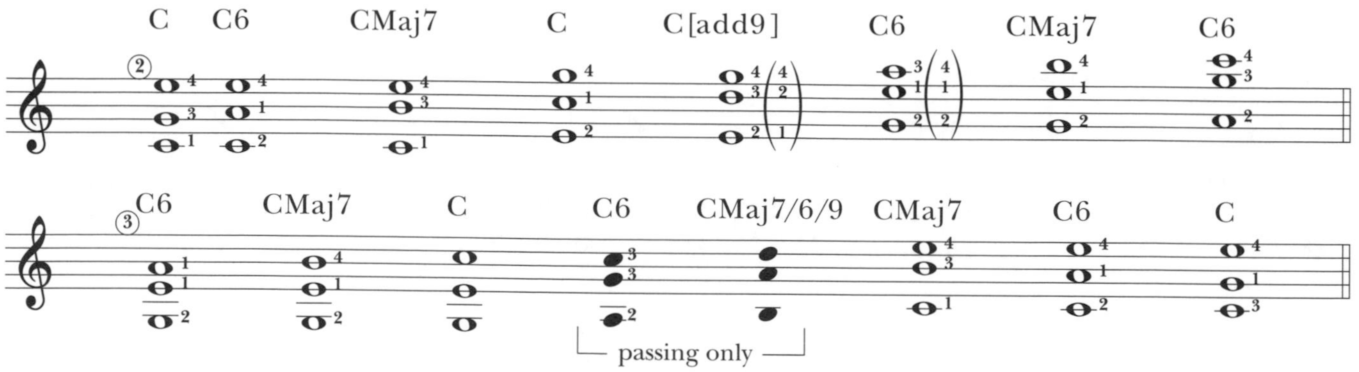

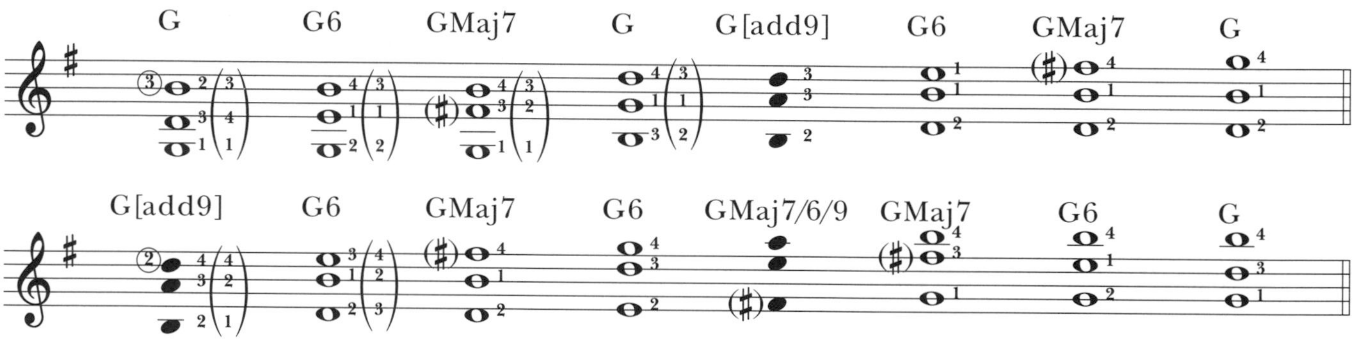

Chord Construction—Five-Part Harmony

■ A 9th chord (five notes) is built by adding another note a 3rd above the four-part structure.

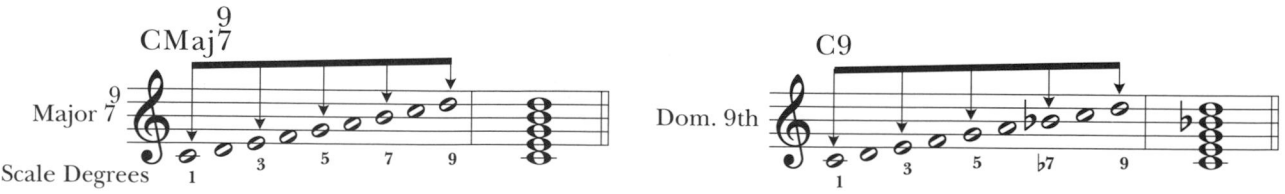

Only dominant 7 and sus4 chords will accept an alteration of a half step up or down to this added 9th,
i.e. C7(♭9), C7(♯9), or +9.

104

Five-Note Arpeggios
Major 7 (9) and Dominant 9 Chords—Chord Spelling

Fingering for all five-note chords is shown in the fifth position with temporary changes to adjacent positions when necessary. After learning as written, transpose and play all structures from all letter names existing from position II through position X.

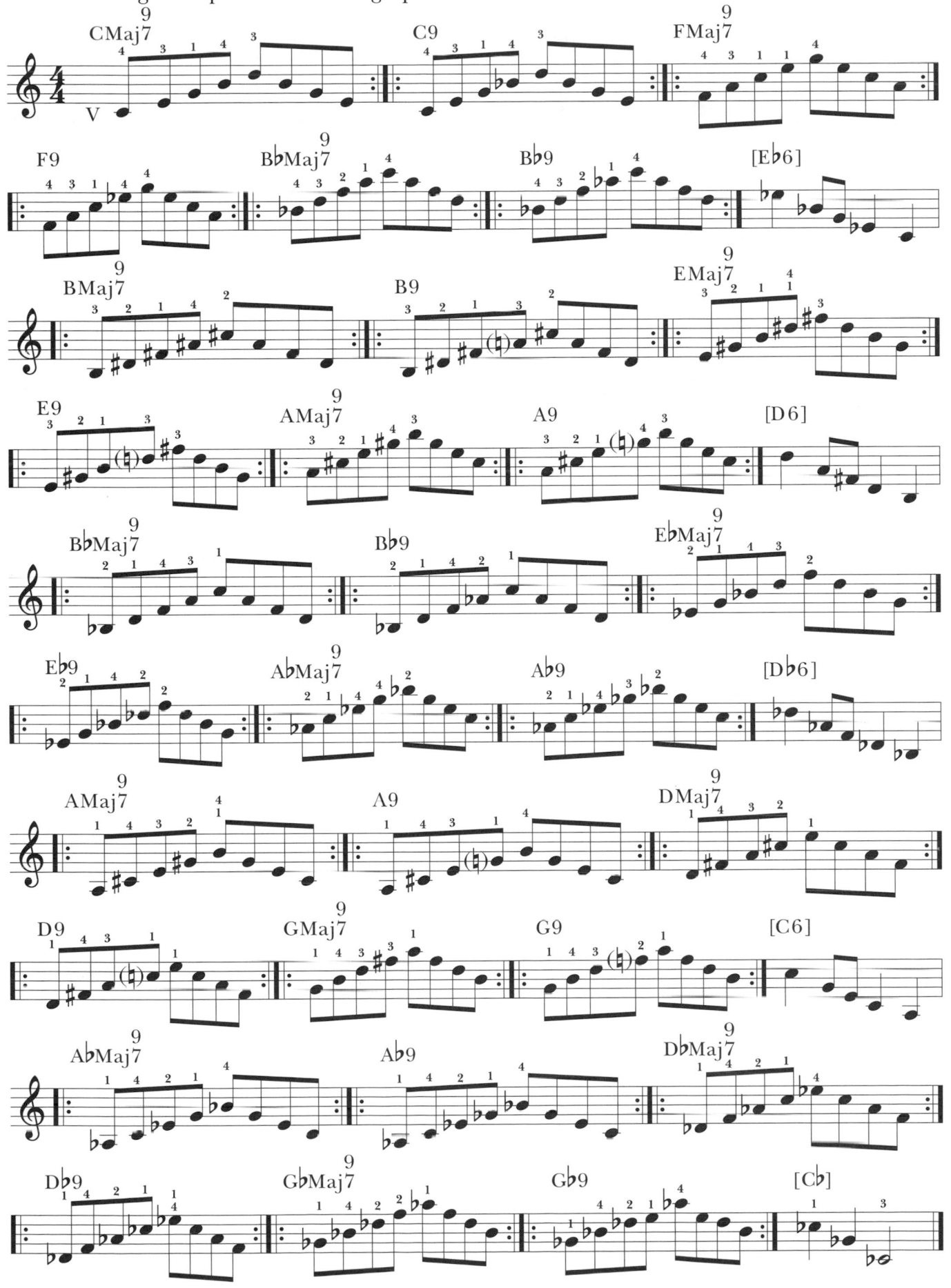

Daydreams (duet)

Slow 4

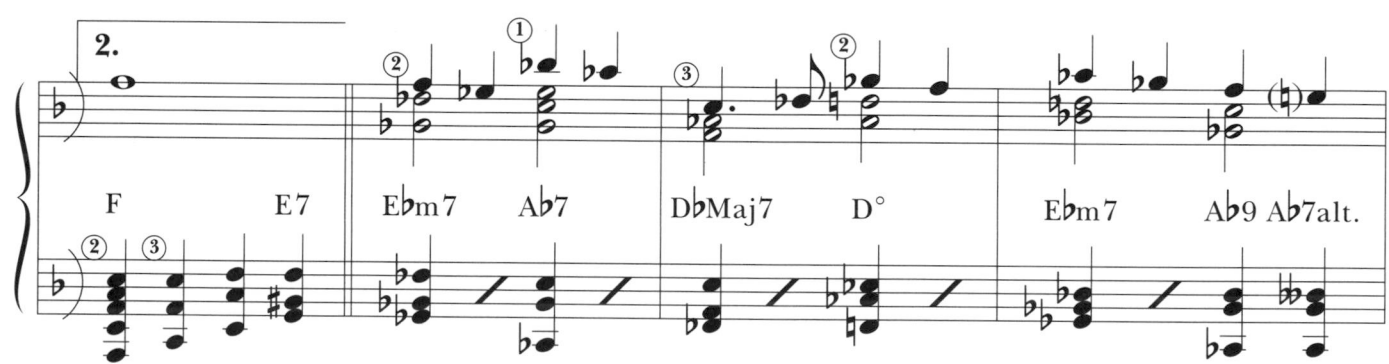

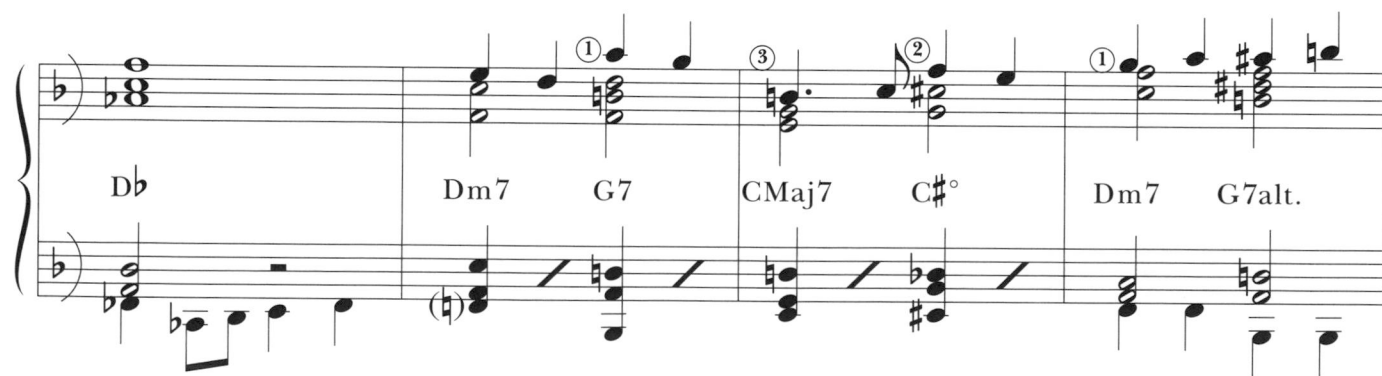

✳ Observe strings indicated for top note of chord voicings

Five-Note Arpeggios
Minor 9 and Diminished 9 Chords—Chord Spelling

The diminished 9 chord symbol used below does not indicate the lowering of the 9th chordal degree. Instead, it represents the four-part diminished 7 chord with the major 9th added. This is logical when you compare it with the meaning of minor 9 chord symbols, i.e. minor 7 with 9 added.

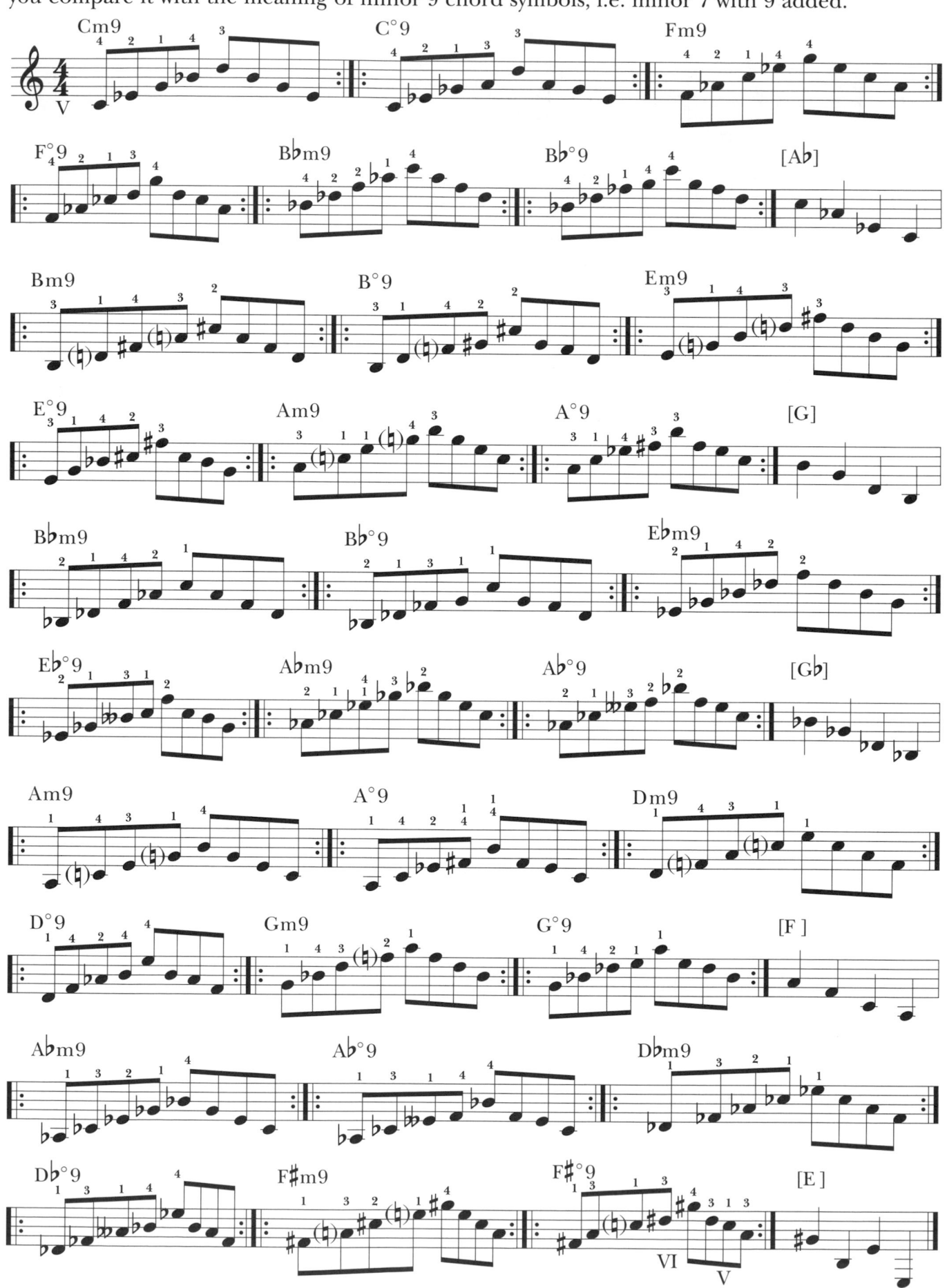

D Major Scale (Twelve Positions)

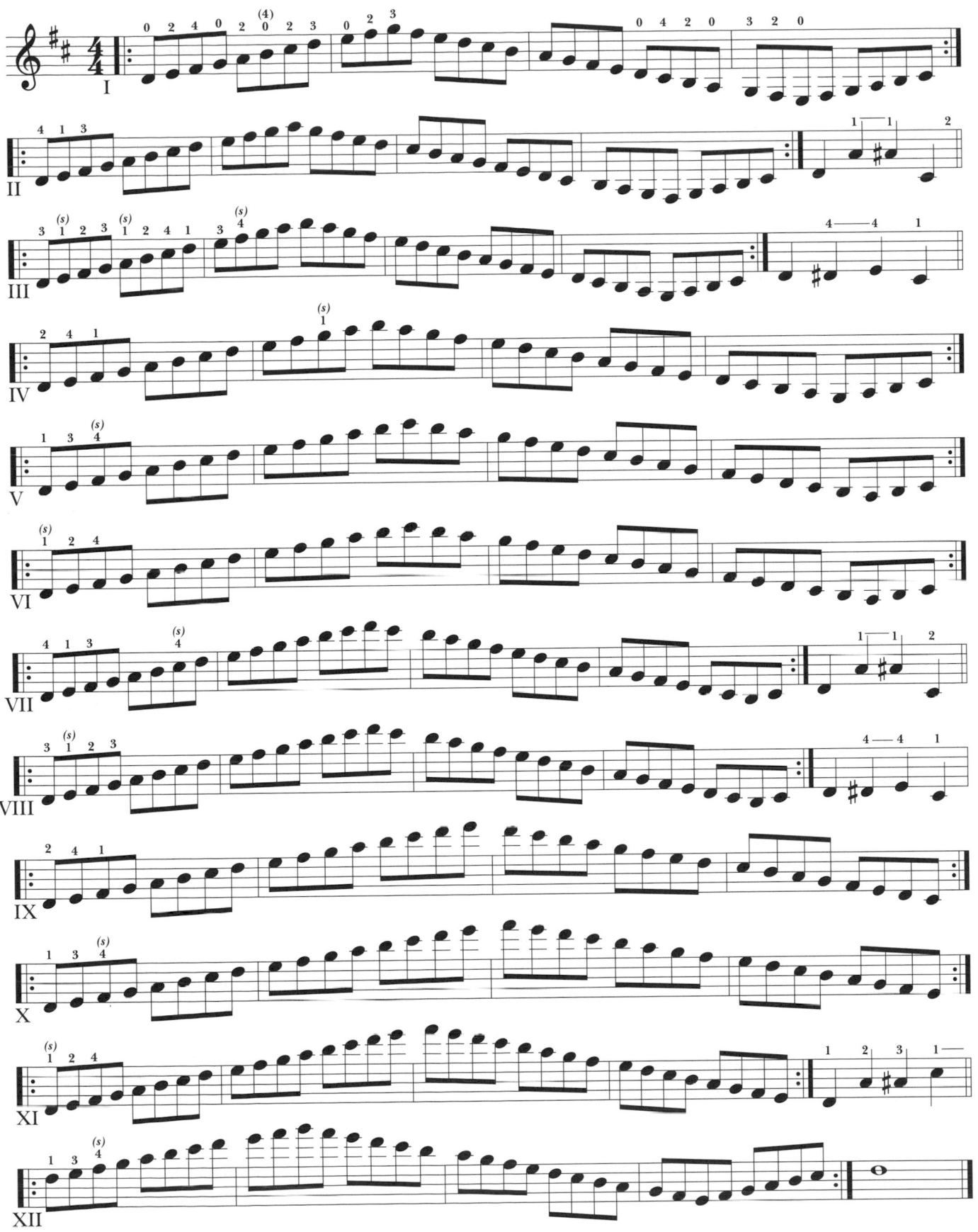

B Harmonic Minor (Nine Positions)

B Minor Etude (solo)

Rubato

Chords—Three-Note Voicings
Dominant 7 Chord Study with ♭5 (Chromatic Approach) in the Bass

Remember: ♭5 is a strong bass note.

About Chord Progressions (Cycle 5)

To aid in determining the true name of a chord structure, and therefore the related scale and function it represents, be aware that the strongest and most common chord movement is down a 5th (cycle 5). Investigate all possible names for the chord in question, and the one that makes the strongest cadence to the following chord will be the real name.

Examples:

Gm6 to F	= C9 to F	Gm6 to A7	= Em7(♭5) to A7
A° or F#° to Gm7	= D7(♭9) to Gm7	A7 to F6	= A7 to Dm7
G° or E° to F	= C7(♭9) to F	Gm7(♭5) or B♭m6 to D9	= *A7alt or E♭9 to D9

*When a dominant 7 chord is completely altered (both 9 and 5 chromatically raised and/or lowered), it takes on all the characteristics of the other dominant 7 chords containing the same tritone. This substitute dominant 7 (with tensions 9, +11, 13) is constructed from the ♭5 of the altered V7 chord. The chromatic approach (from above) created by this substitute dominant 7 constitutes a very strong progression, second only to cycle 5. To help in the investigation of multiple names for chord structures, study the information on the next page.

Note: Look ahead to the next chord to analyze a progression.
Look back to the preceding chord to determine the related scale.

Theory: Interchangeable Chord Structures

The following chord structures could be referred to as diatonic substitutions, in that they represent (in the proper setting) the exact same scale sound.

C6   Am7   FMaj7/9   Dm11
                              D9sus4

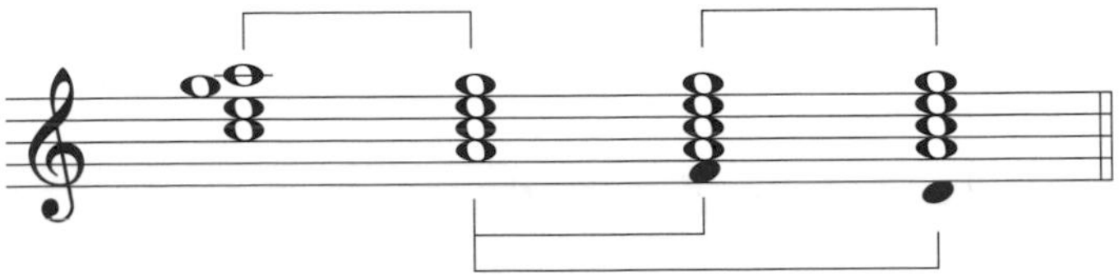

CMaj7   Am9✱   FMaj7/9/11+   D13sus4

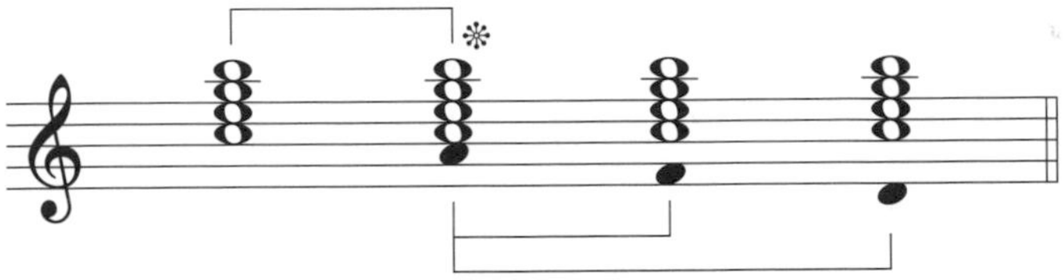

Cm6   Am7♭5   F9   D7♭9   sus4   B7alt[+5♭9]

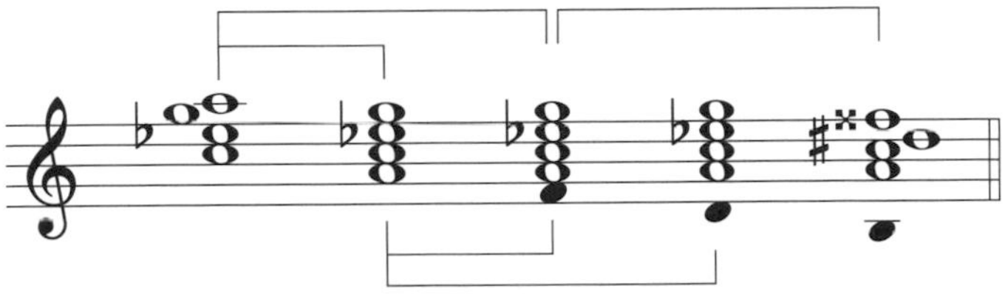

Cm(maj7)7   Am9♭5✱✱   F11+   D7♭9/13sus4

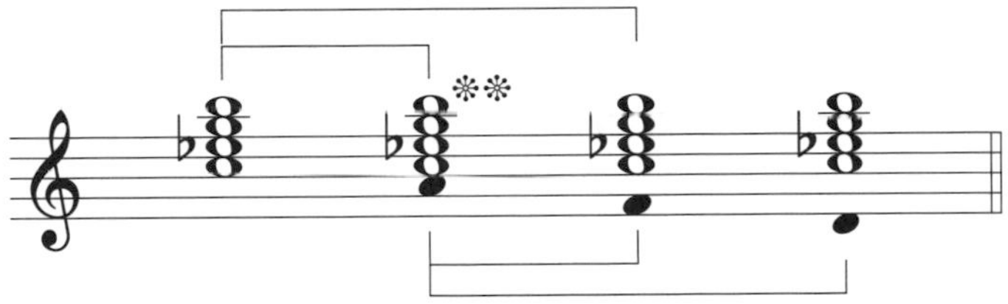

E♭Maj7+5   Cm(maj7)9   Am9/11/♭5   F11+/13

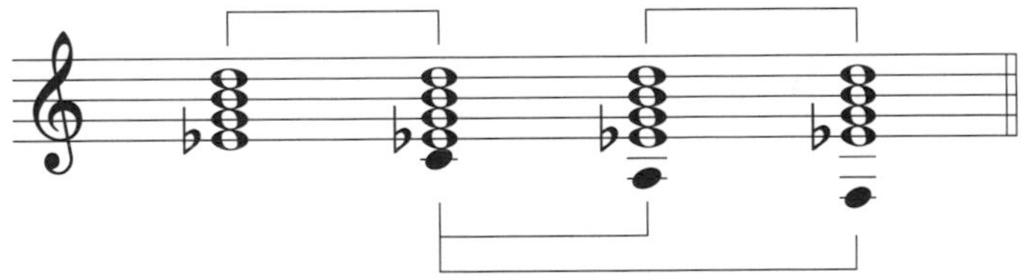

✱ Am9 can also be considered C6/7
✱✱ Am9(♭5) can be considered Cm6/7.

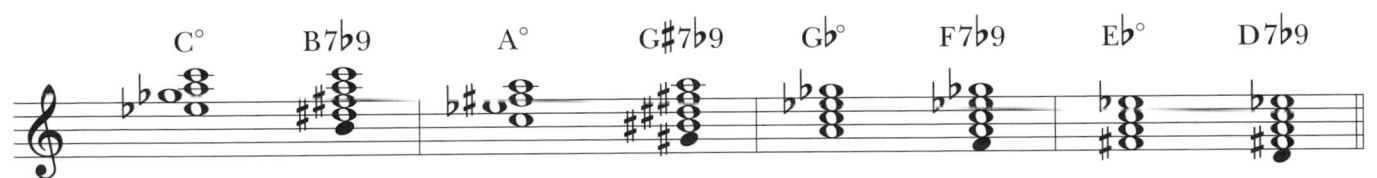

All four names of diminished 7 chords and their related dominant 7(♭9) chords are completely interchangeable.

113

Rhythm Guitar—The Right Hand

Joropo and Nanigo
Moderately fast to fast

Basic Strokes

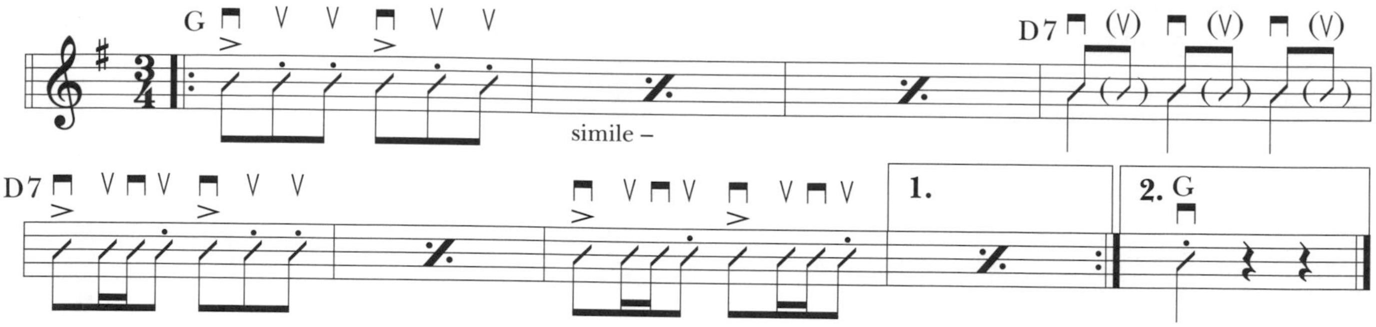

Basic and Orchestral

Orchestral

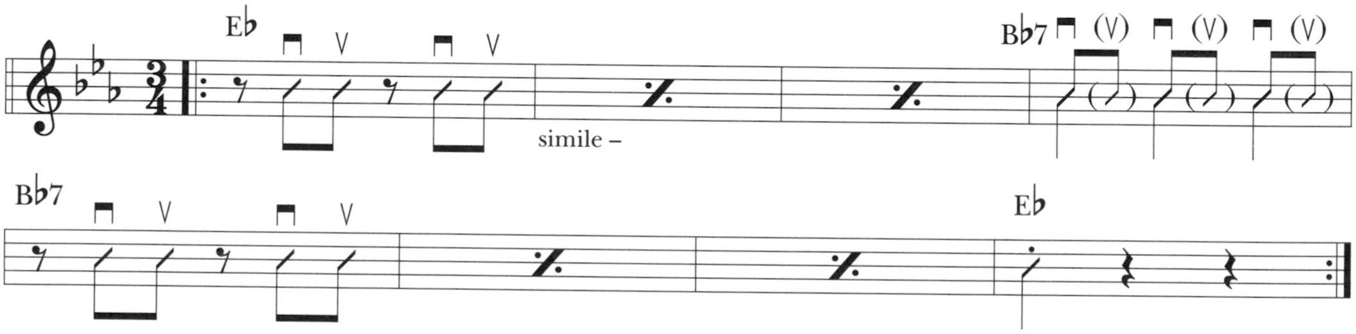

Arpeggio Study—7th Chords

■ Play from all fingers, but stay in position throughout the entire sequence.

＊ Also play first chord of each measure as a minor 9 and as a dominant 7(♭9).

Chords—Three-Note Voicings
Melodization of Minor 7 Chords as IIIm7

Melodic degrees: Major scale a 3rd below chord name.

✱ Passing tones only. (Note: ♭9 can be a chord tone of dominant 7 only.)

IIIm7 can be used as a diatonic substitution for I (Am7 = Fmaj7). But stay out of the low register when doing this. The 5th of the IIIm7 chord is the major 7 of the I chord, and the major 7th chord degree should not occur below the note D on the first space below the staff.

Melodization of Minor 7(♭5) Chords as VIIm7(♭5)

Melodic degrees: Major scale a half step above chord name.

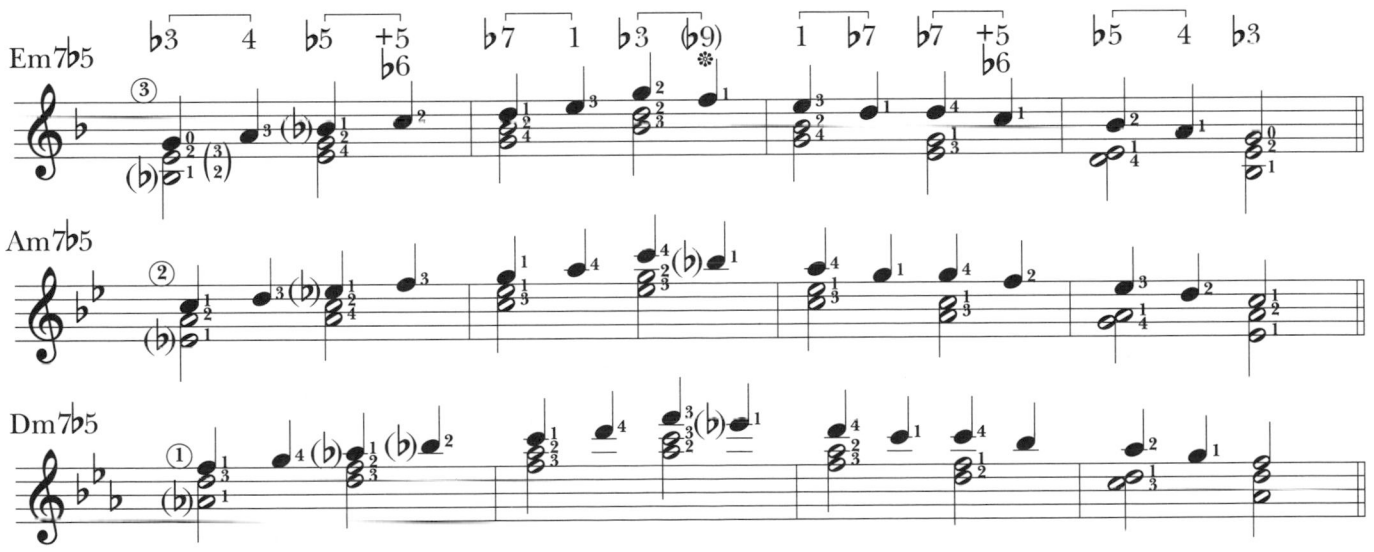

VIIm7(♭5) can be used as a diatonic substitution for V7 (Am7(♭5) = F9). But as with IIIm7 for I, this is not good in the low register.

Chromatic Melodization of Dominant 7 Chords

Eleven of the twelve chromatic tones can be considered chord degrees of a dominant 7 structure. The exception is the major 7.

CLOSE VOICINGS

✳ 9 for 1, inside voice.

Also possible with root, but somewhat more difficult physically.

EXAMPLE

OPEN VOICINGS

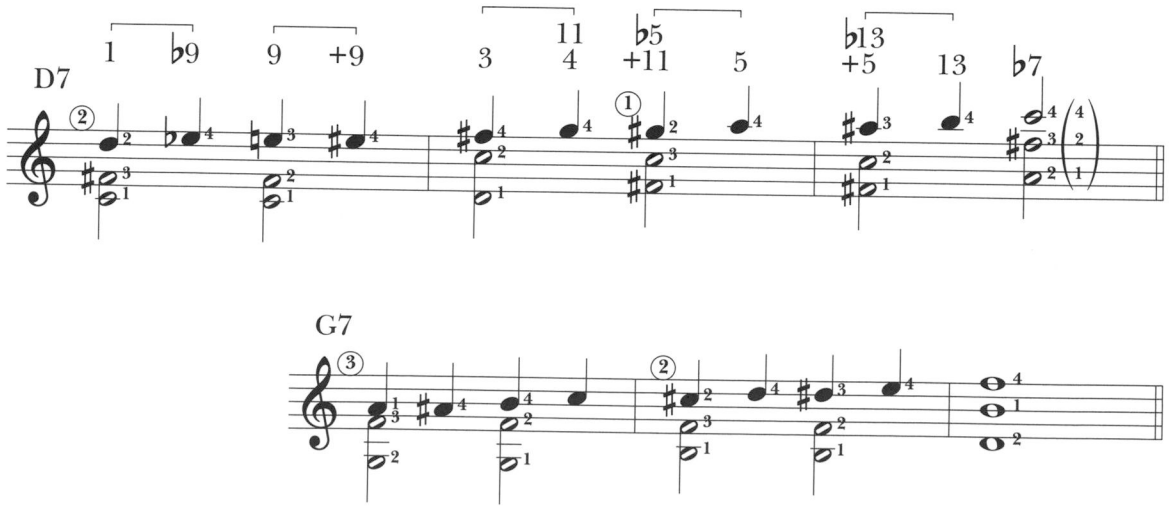

Melodic Rhythm Study No. 11

Five-Note Arpeggios

Minor (6, 9), Major (6, 9), Dominant 9sus4, and Dominant 7(♭9) Chords
Chord Spelling

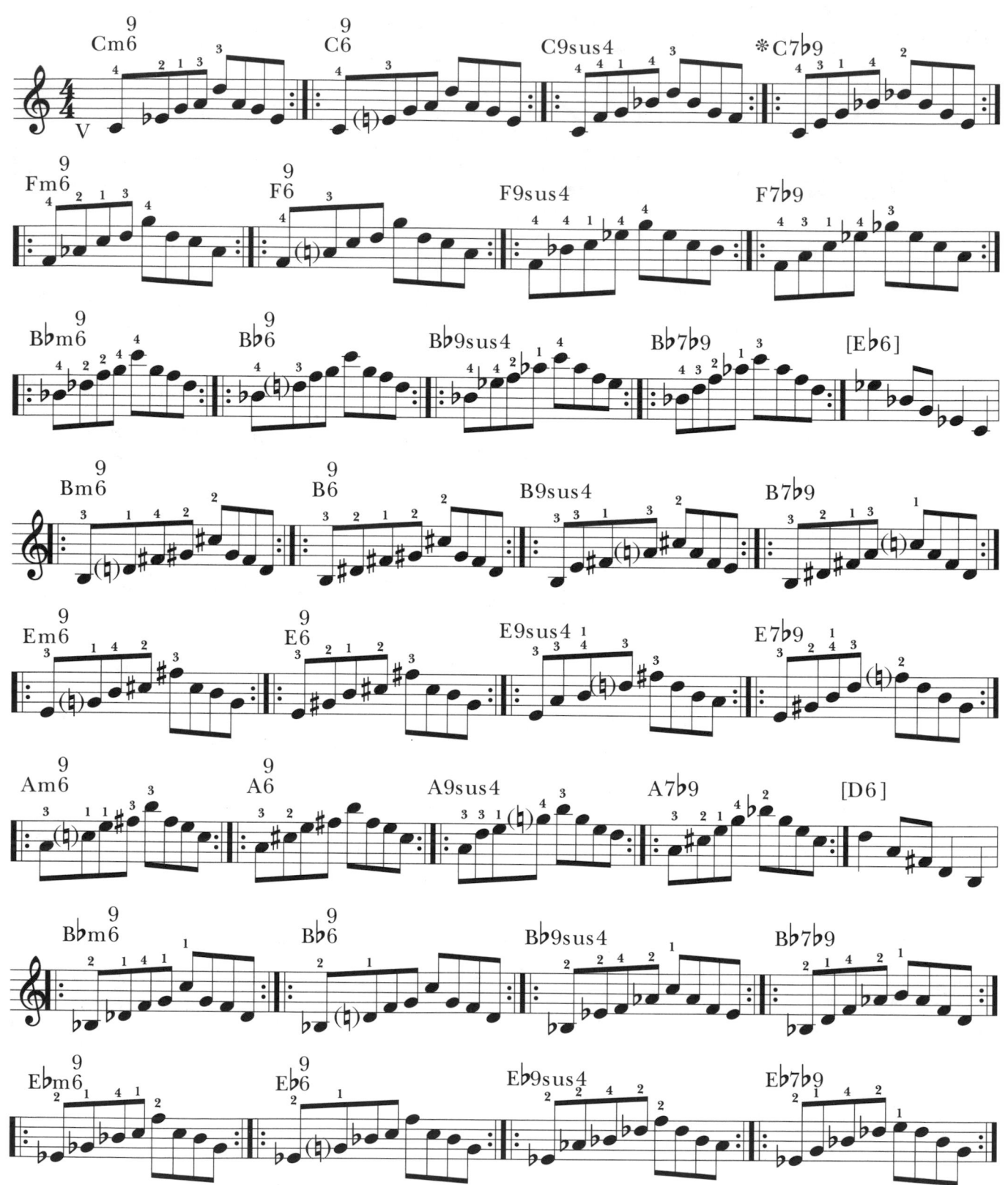

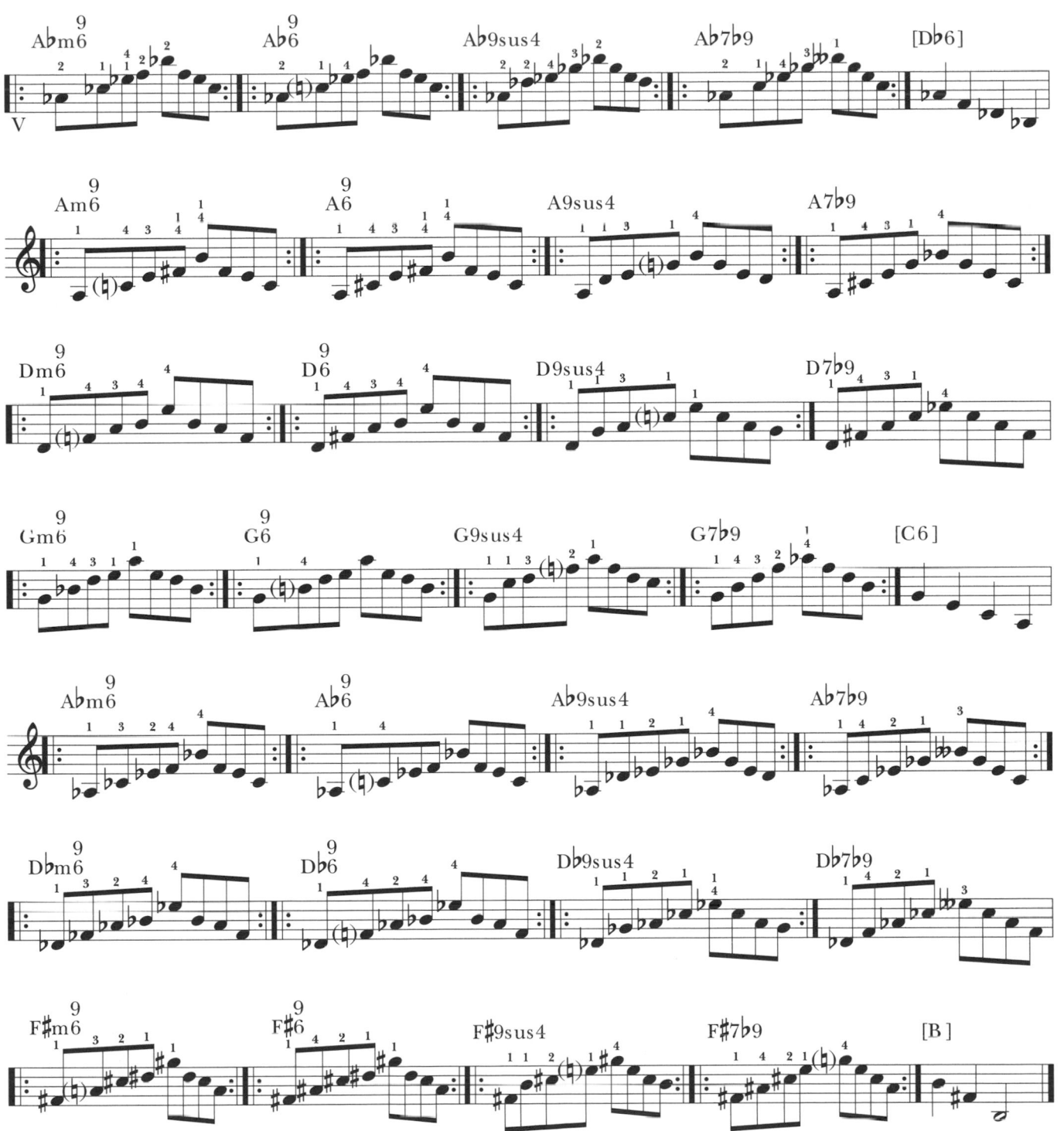

✳ Only dominant 7 and dominant 7sus4 chords will accept an alteration of a half step up or down to this added 9th chord degree.

Chords—Three-Note Voicings
Minor 7 Chords—Close and Open Voicings

♭3RD AND ♭7TH IN THE LEAD

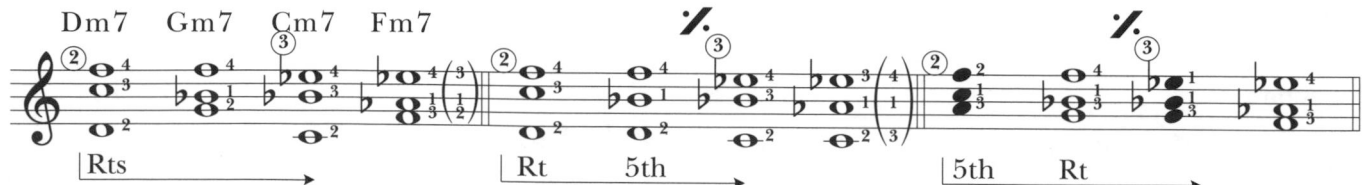

♭3RD AND ♭7TH IN THE LEAD
ROOT AND 5TH, INSIDE VOICE

ROOT AND 5TH IN THE LEAD

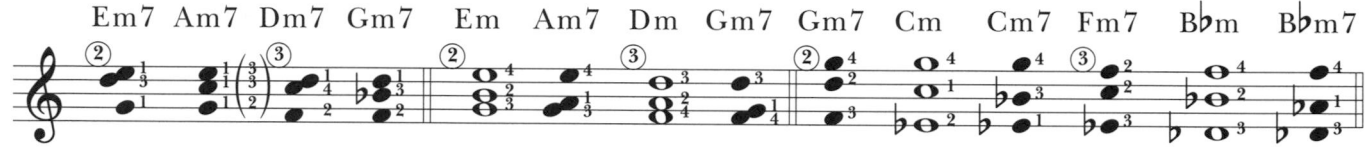

Minor 6 Chords

6TH AND ♭3RD IN THE LEAD

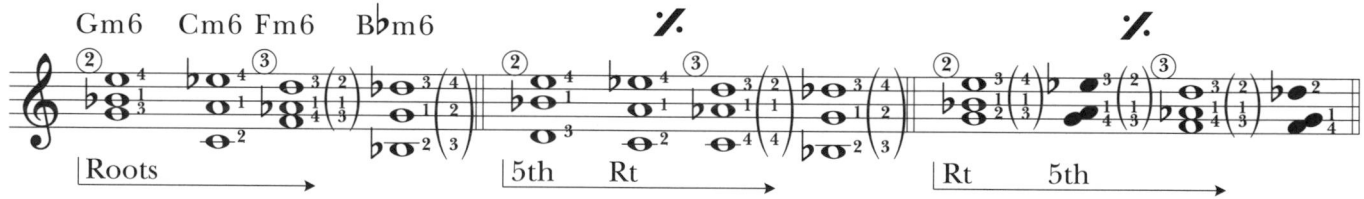

♭3RD AND 6TH IN THE LEAD
ROOT AND 5TH, INSIDE VOICE

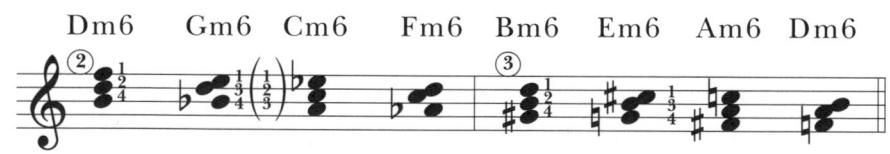

ROOT AND 5TH IN THE LEAD

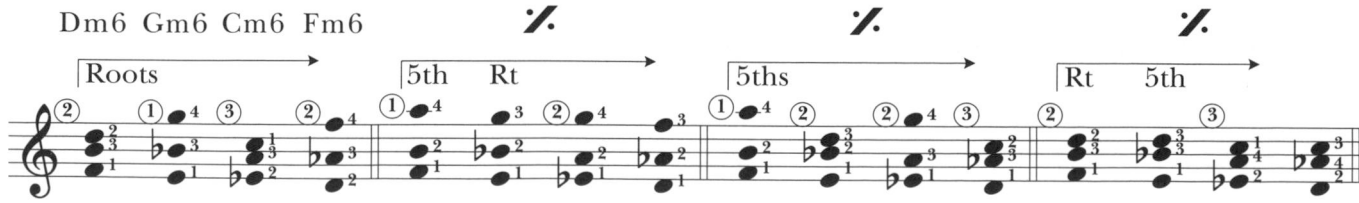

Minor 7 Chords—Open Voicings, All Inversions

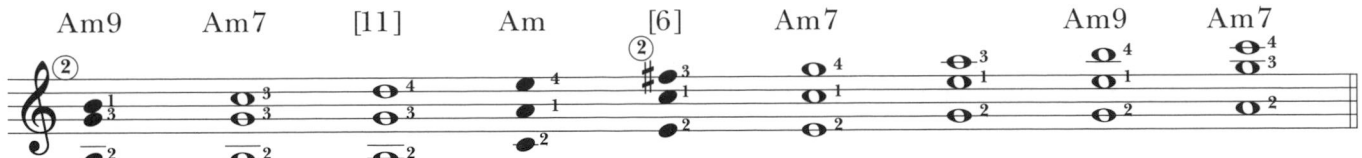

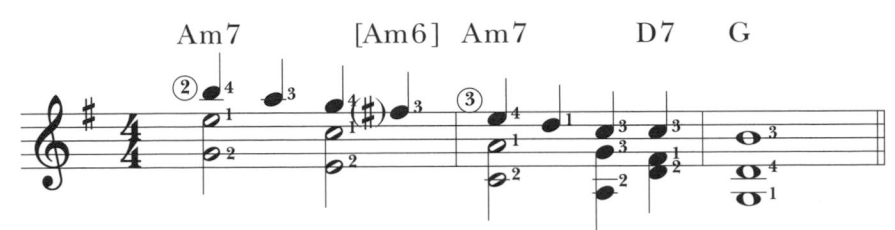

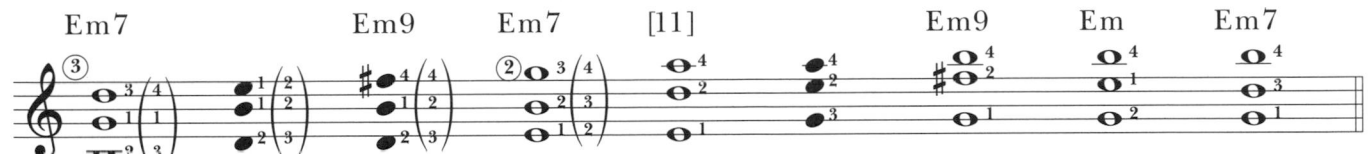

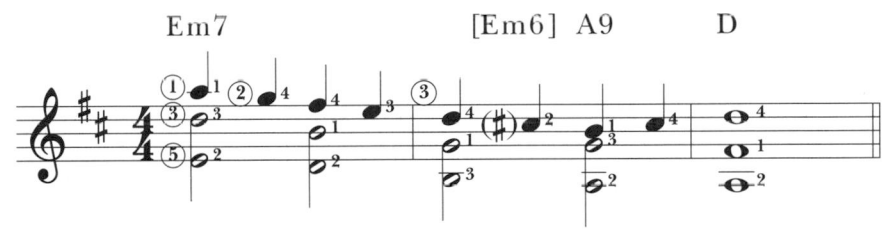

IIm7–V7–I Chord Study

121

Chord-Scale Relationships

For improvisation.

SPECIAL ALTERATIONS ON DOMINANT 7 CHORDS WITH SCALE TONE ROOTS (EXCEPT IV7)

sus4	The subdominant sound of IIm7 (or IV6); treat accordingly
sus4(alt 9)	Subdominant minor sound of IIm7(♭5) (IV6); treat accordingly
	Note: 3rd degree of sus4 chords must be a melodic passing tone only.
Alt 5	On dominant 7 chords that contain an unaltered 9th, I7, II7, V7, VI7 = Whole tone scale from any chord tone.
	The specified ♯5 can often be treated as ♭13, and specified ♭5 can be treated as +11.
	See below, ♭13 and +11.
Alt 5 and 9	Real melodic minor scale from ♭9 of chord.
	Sometimes the alt 9 is not specified and must be remembered as already being present. Ex: III7 and VII7.
	For optional melodic treatment of ♯5 (alt 9) see below, ♭13(alt 9).
	For optional melodic treatment of ♭5 (alt 9) see below, +11(alt 9).
Alt 9	On V7, II7, I7, use real melodic minor from ♭7 of chord. Or, major scale with ♭6 from intended tonic.
	Also, you may combine both scales, real melodic minor with added ♯4.
Alt 9 on VI7	Harmonic or natural minor from intended tonic.
Unaltered 9 on III7 and VII7	Real melodic minor from intended tonic.
11	Sus4 on dominant 7.
	(See sus4.)
Aug 11 ♯11, +11	On all dominant 7 chords, use real melodic minor from chord 5th.
	The 9th is considered unaltered with +11 unless specified alt.
+11(alt 9)	Diminished scale from chord degrees 3, 5, ♭7, ♭9.
13	On dominant 7 chords with scale tone roots (except IV7), use the major scale from the intended tonic.
	The 9th is considered unaltered and the 11th natural with these 13th chords unless otherwise specified.

13(alt 9)	Same as alt 9 on V7.
13(+11)	Same as augmented 11.
13 (+11, ♭9)	Same as +11(alt 9).
♭13	On dominant 7 chords with unaltered 9ths, I7, II7, V7, (VI7) use real melodic minor from intended tonic.
♭13 (alt 9)	Harmonic (or natural) minor from intended tonic.
	Remember ♭13 and alt 9 are already contained in III7 and VII7 and therefore do not constitute any alteration on them.

SPECIAL ALTERATIONS ON IV7 AND DOMINANT 7 CHORDS WITH NON-SCALE TONE ROOTS

sus4	The sound of IIm7 (or IV6); treat accordingly.
Alt 5	Whole tone scale from any chord tone.
♭5	No alteration; ♭5 is already present as +11.
♯5	Same as alt 5 because ♭5 is already present as +11.
Alt 9	Diminished scale from chord degrees 3, 5, ♭7, ♭9.
Alt 5 and 9	Real melodic minor scale from ♭9 of chord.
11	See sus4.
Aug 11	No alteration (already contained in chord).
+11(alt 9)	Same as alt 9.
13	No alteration.
13(alt 9)	Same as alt 9.
♭13	Same as alt 5 (♭13 must be considered ♯5 here).
♭13 (alt 9)	Same as alt 5 and 9.

Pretty Please (duet)

Five-Note Arpeggios

Dominant 7 (aug 9) Chords

Chord Spelling

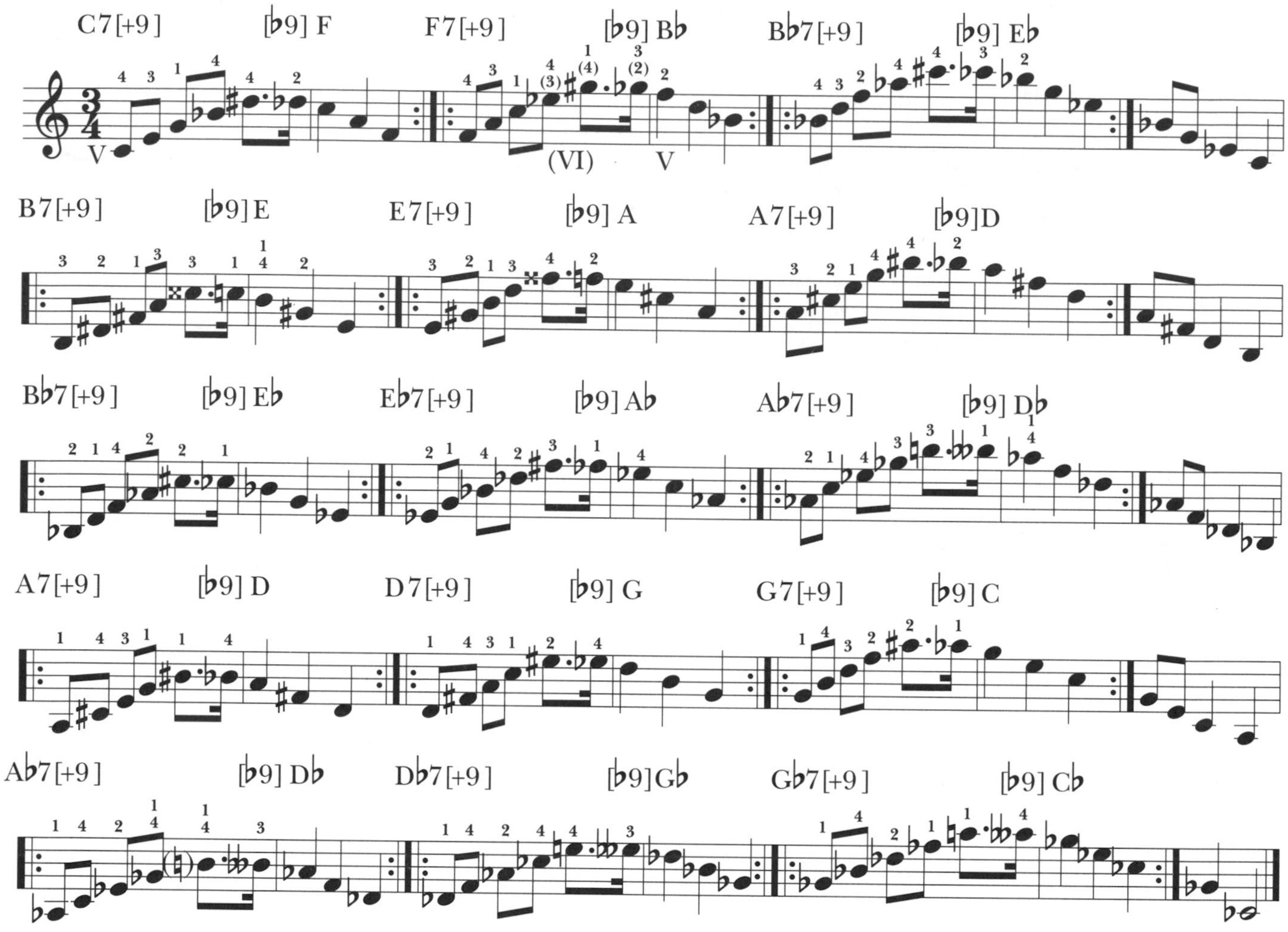
C7[+9]
[♭9] F
F7[+9]
[♭9] B♭
B♭7[+9]
[♭9] E♭
B7[+9]
[♭9] E
E7[+9]
[♭9] A
A7[+9]
[♭9]D
B♭7[+9]
[♭9] E♭
E♭7[+9]
[♭9] A♭
A♭7[+9]
[♭9] D♭
A7[+9]
[♭9] D
D7[+9]
[♭9] G
G7[+9]
[♭9] C
A♭7[+9]
[♭9] D♭
D♭7[+9]
[♭9]G♭
G♭7[+9]
[♭9] C♭

Harmonizing a Melody—From a Lead Sheet with Chords Indicated

Think of the melody as being written an octave higher. Add the most important chord tones under it that are physically available.

C
C7
F
Fm
[C] [C♯°]
Em7 A7 Dm7 G7+
C

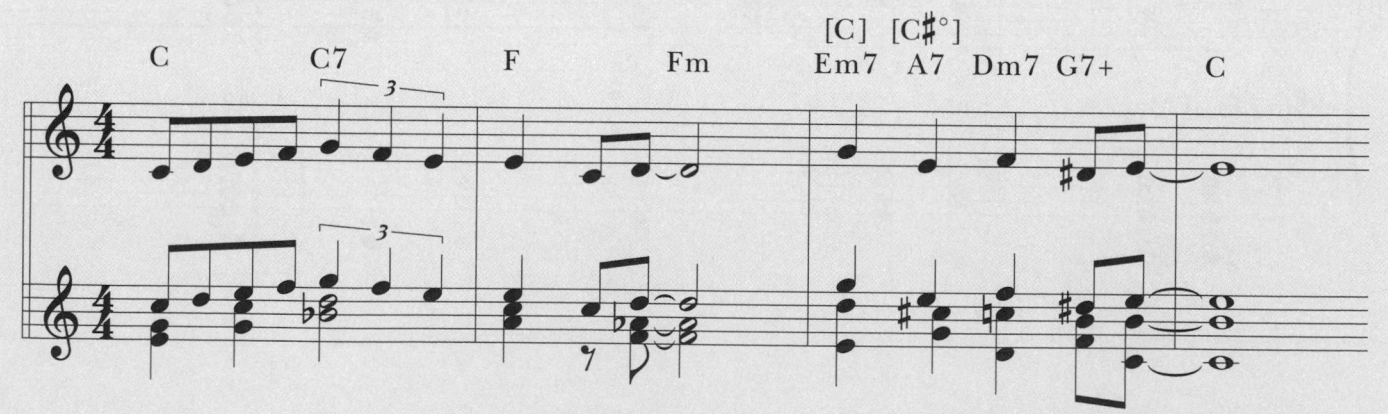

To attempt to play a chord for every melody note is impractical, and it denies you one of the most striking effects of guitar chord-melody playing—that of a moving melody over sustained chord tones.

E♭ Major Scale (Twelve Positions)

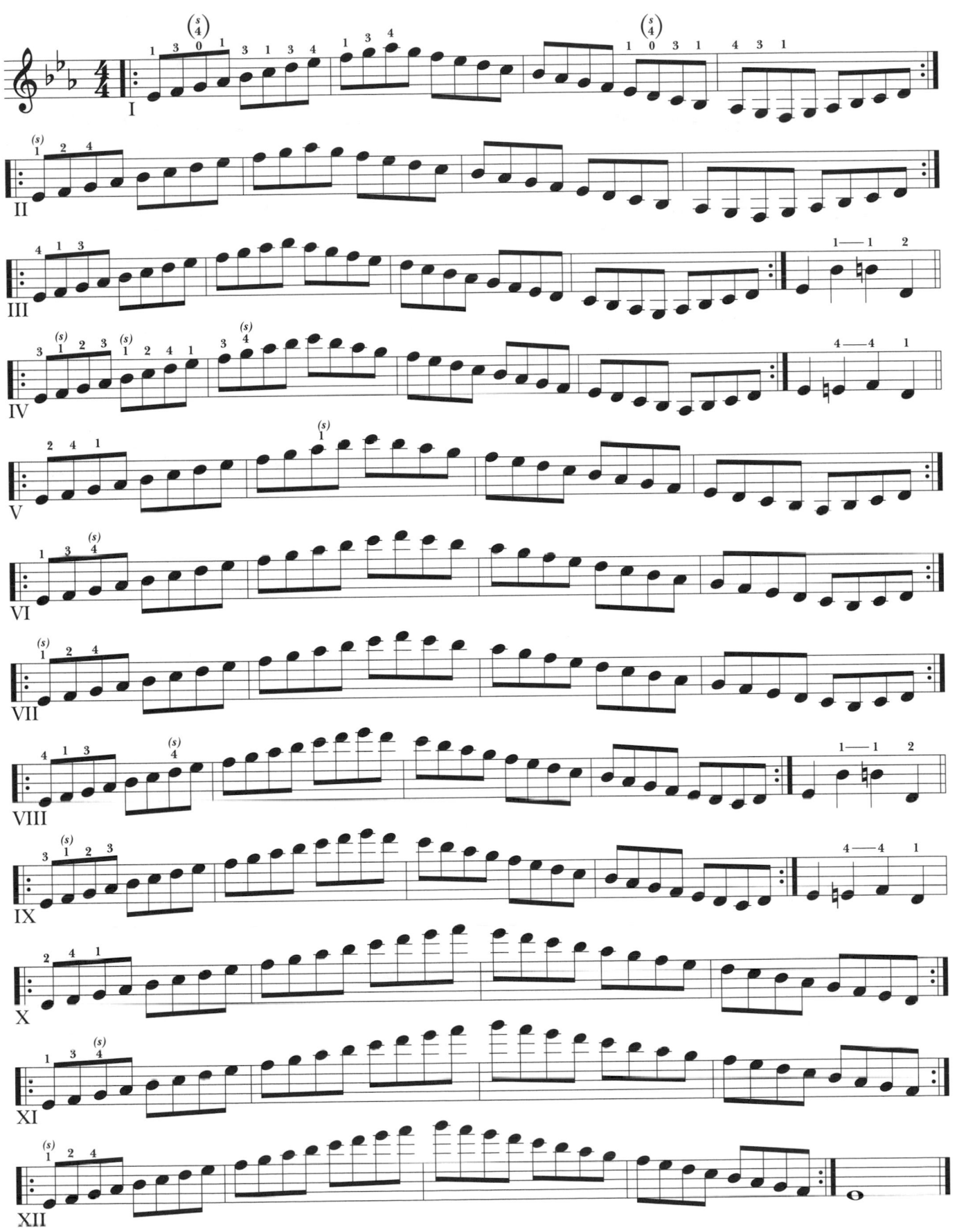

C Harmonic Minor (Nine Positions)

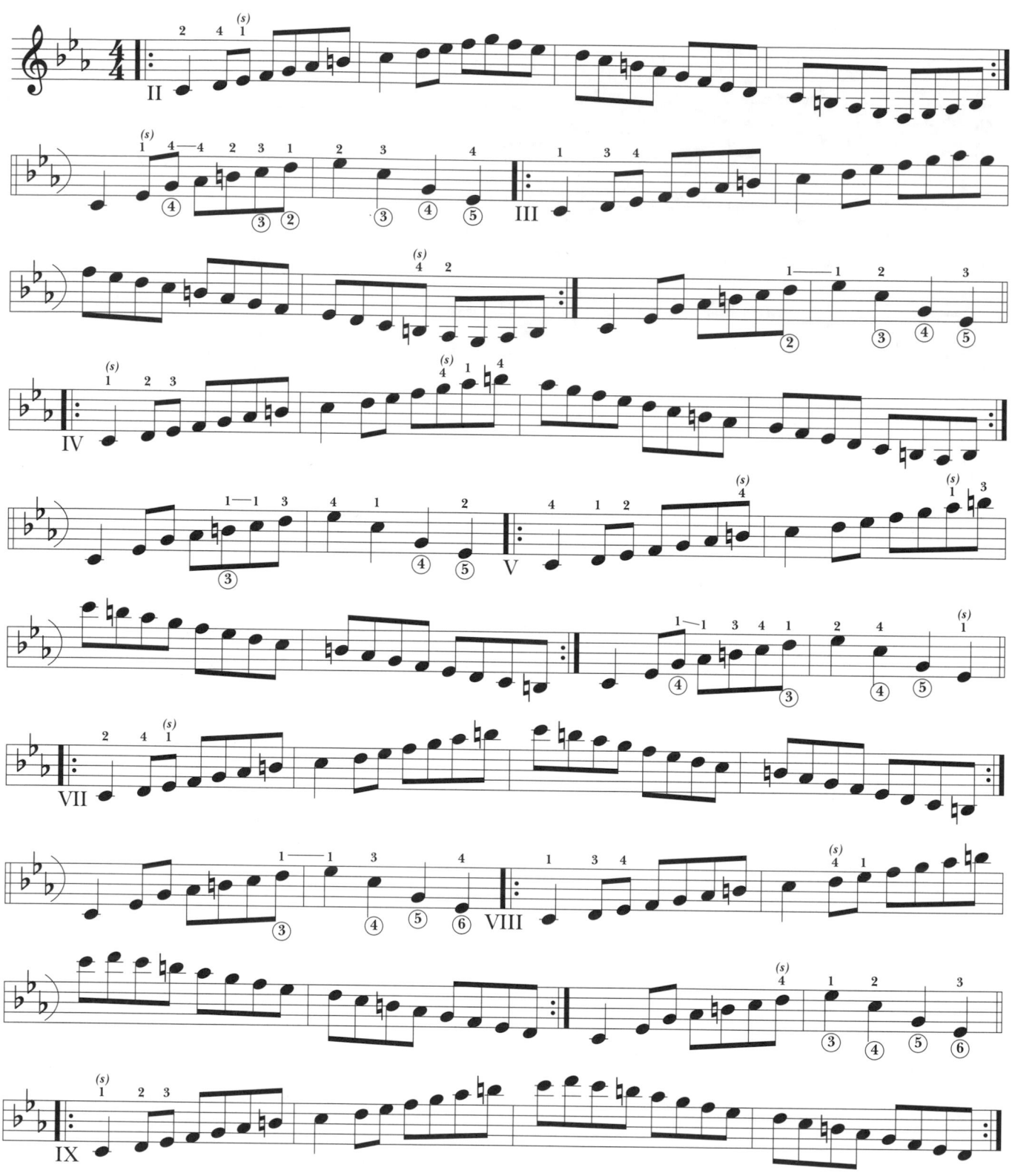

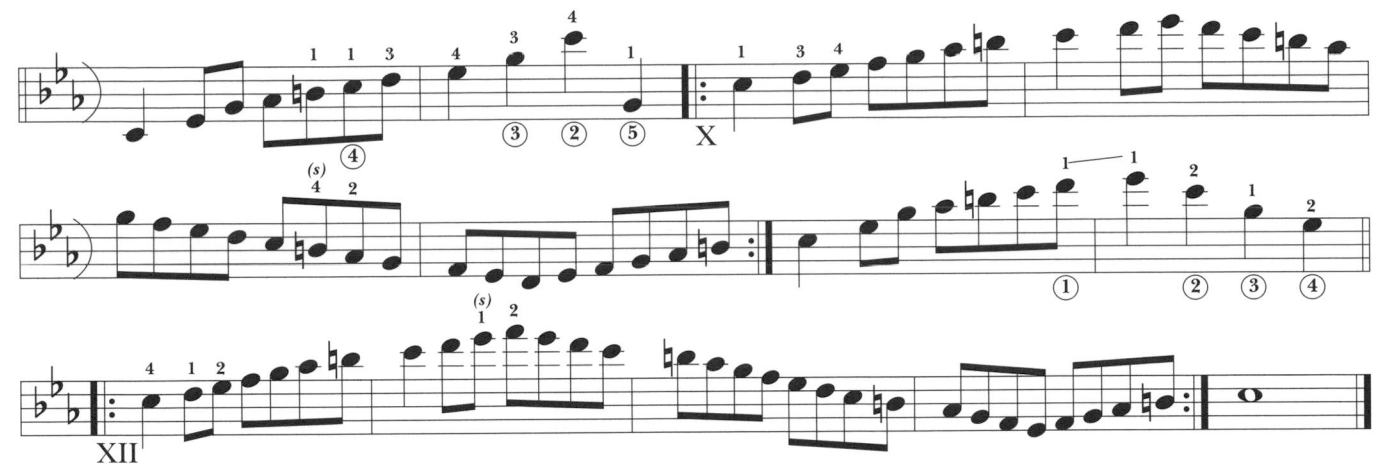

Etude in C Minor (solo)

Chord-Scale Relationships

For improvisation.

Remember: Look ahead to the next chord to analyze a progression. Look back to the preceding chord to determine the related scale.

Major Chords

Major chords with scale tone roots (except IV) represent a tonic sound. Scale = major from chord name.

The IV chord and all major structures with non-scale tone roots represent the subdominant sound. Scale= Major from 5th degree of chord.

All major chords will accept being melodized as IV chords. But realize that the +11 is being forced on those that normally represent the tonic sound.

Also be advised that very occasionally, a nondiatonic major chord with a scale tone root represents a modal sound. That is, the writer wants only the major triad harmonically, but the melodic tones are to be the same as those used with a dominant 7 structure of the same letter name.

Minor 7 Chords

All minor 7 chords represent the subdominant sound of IIm7 (for IV), except IIIm7, VIm7, and VIIm7, which represent tonic sounds. IIIm7 and VIm7 are diatonic substitutions for I. VIIm7 = IIIm7 for I (key of the dominant).

IIm7	Major scale from ♭7 of chord
IIIm7 and VIm7 (for I)	Major scale from name of tonic chord being replaced
VIIm7 (as IIIm7 for I)	Major scale from name of tonic chord being replaced

A comparison of minor 7 chords with their related major 6 chords (containing the same notes) will reveal some second choice VIm7-for-I relationships. Scale = Major from name of related major 6 chord.

Note: All second choice scale relationships must be handled with care.

Chords—Three-Note Voicings
Minor-Major 7 and 6 Chords—Close and Open Voicings

MAJOR 7TH AND 6TH IN THE LEAD

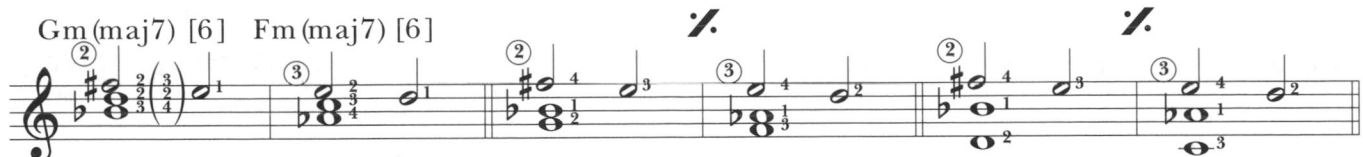

5TH IN THE LEAD   **♭3RD IN THE LEAD**

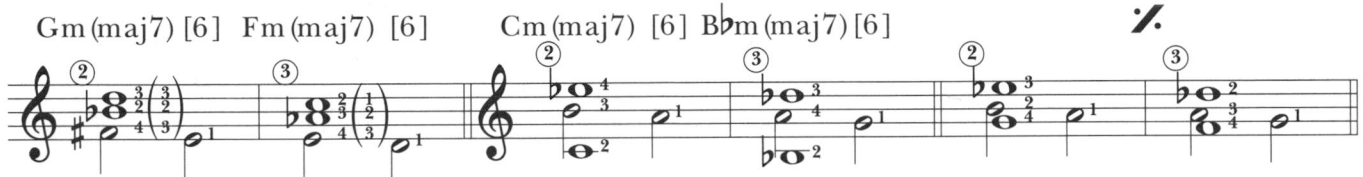

MAJOR 7TH AND 6TH COMBINED IN SAME VOICING

    MAJOR 7TH AND 6TH IN LEAD   **6TH AND 5TH IN LEAD**

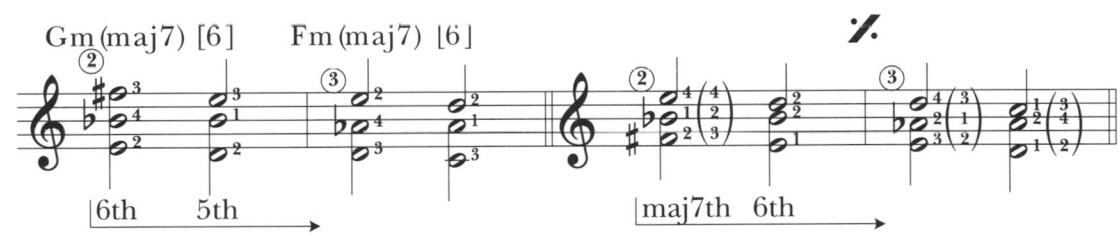

Minor 6, Minor-Major 7 Chords—Open Voicings, All Inversions

131

Five-Note Arpeggios

Dominant 7 (+5, +9), Dominant 9 (+5), and Dominant 7 (+5, ♭9) Chords
Chord Spelling

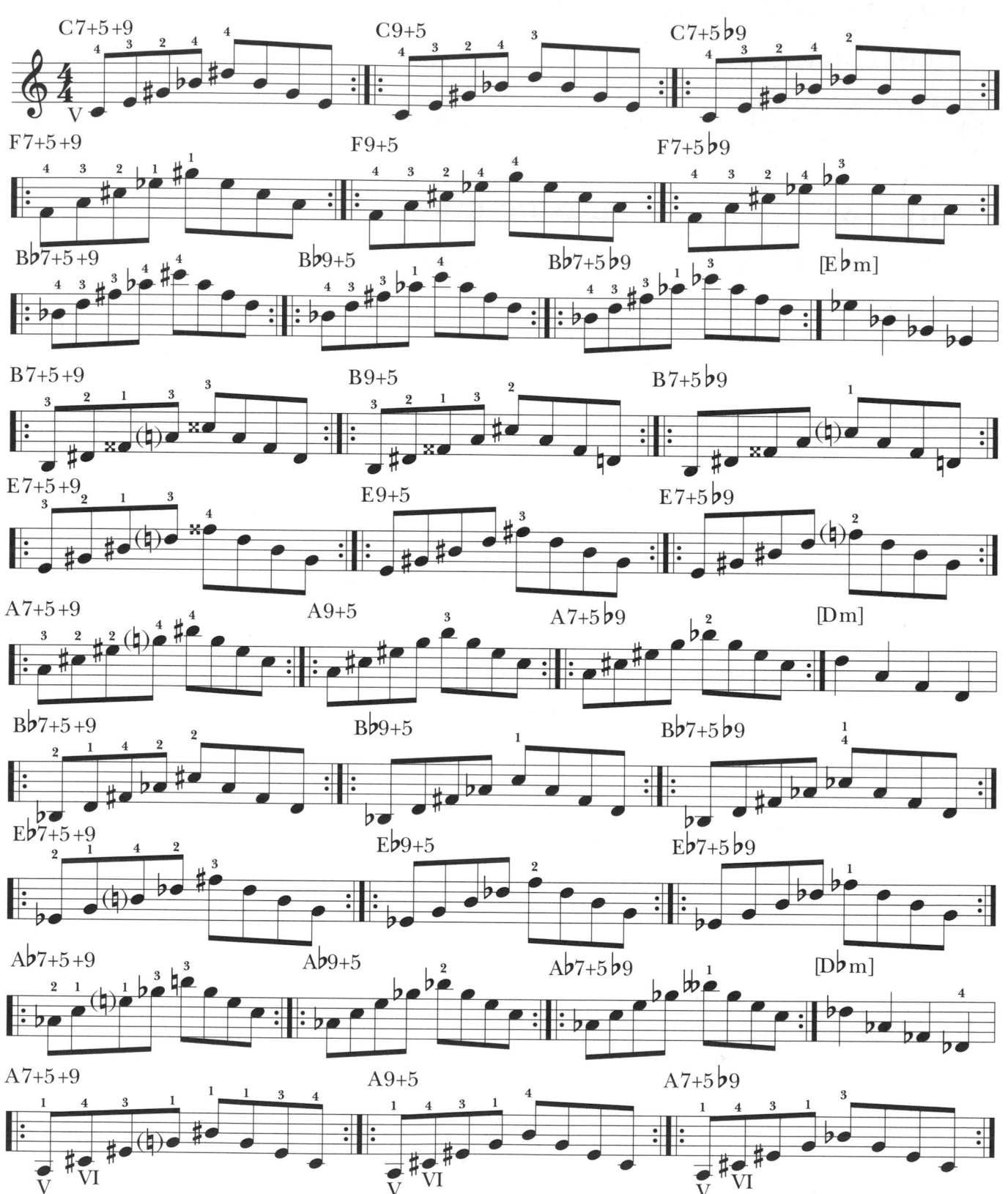

Chords—Three-Note Voicings
Melodization of I Minor Chords with Harmonic Minor Scale

Melodic degrees: Harmonic minor scale from chord name

✳ Major 7 necessary as an undervoice

Melodization of IIm7(♭5) Chords with Harmonic Minor Scale

Melodic degrees: Harmonic minor scale from ♭7 of chord

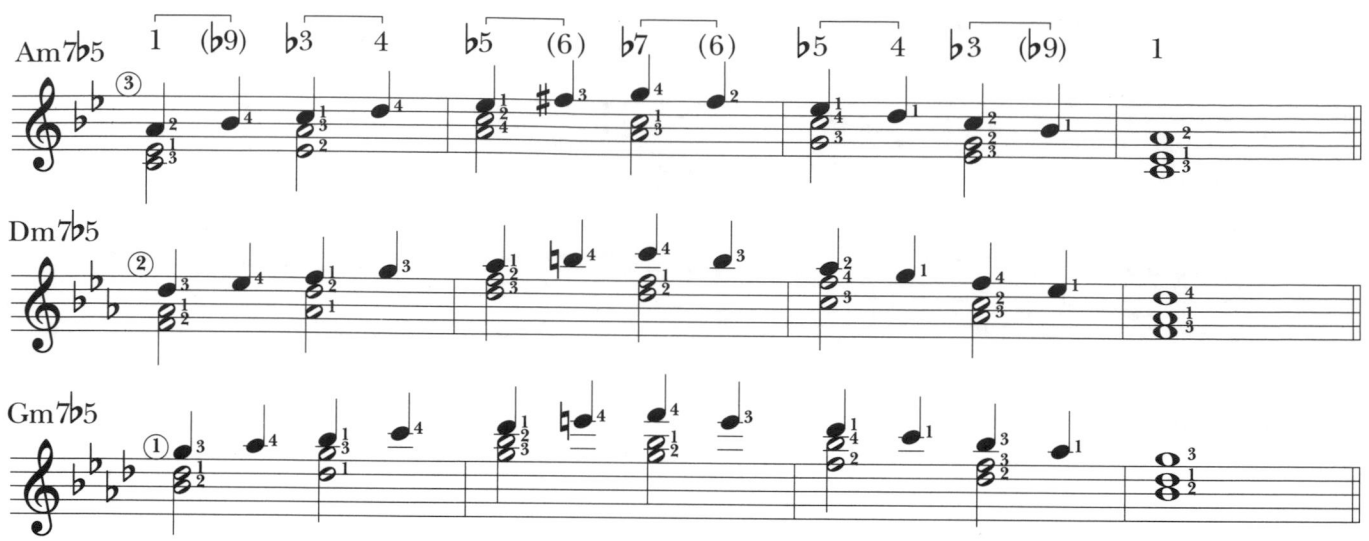

Melodization of Dominant 7 Chords with Harmonic Minor Scale

Melodic degrees: Harmonic minor scale from intended tonic

✳ = ♭9 for undervoice

Teeah-Wanna
Optional Duet with Rhythm Guitar

The notes contained in the bottom staves of the following study represent the chord-scale relationships. They are to be played with the rhythm guitar part (not the melody) to further acquaint the ear with these related sounds.

Moderate 4
Latin

To aid in the analysis of the preceding chord-scale relationships, observe the following numerical breakdown.

I	/	/	/	VII7	/	/	/	I	/	/	/	III7	/	/	/
VI9	/	/	/	**1.** II7	/	V7	/	IIIm7	/	bIII7	/	IIm7	/	V7b9	/ :||
2. II7	/	V7	/	I	/	/	/	I7	/	/	/ ||	IV7	/	/	/
IV7	/	/	/	I	/	IIm7	/	I	/	IIIm7	#I°	II7	/	/	/
II7	/	/	/	IIm7	/	/	/	V9+5	/	/	/ ||	I	/	/	/
VII7	/	/	/	I	/	/	/	III7	/	/	/	VI9	/	/	/
II7	/	V7	/	I	/	bVII7	/	I	/	/	≹	||			

137

Rhythm Guitar—The Right Hand

Moderately fast
Paso Doble 1

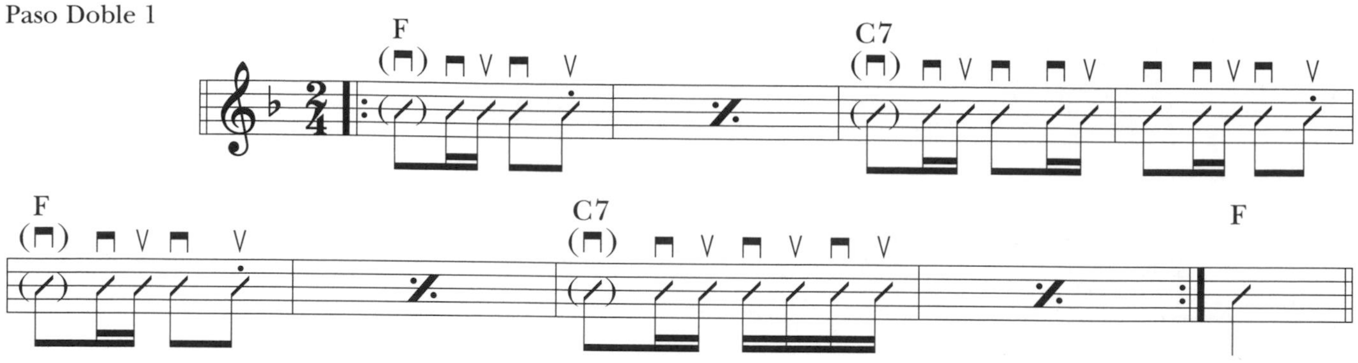

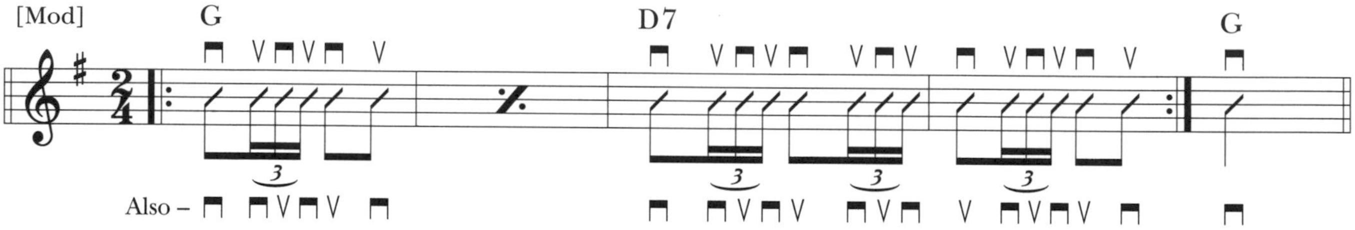

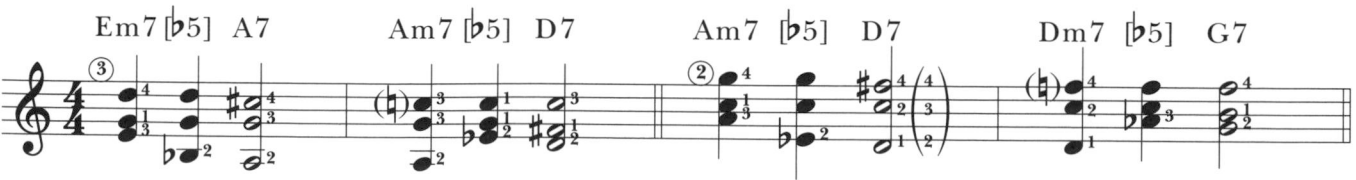

Moderately fast
Paso Doble 2

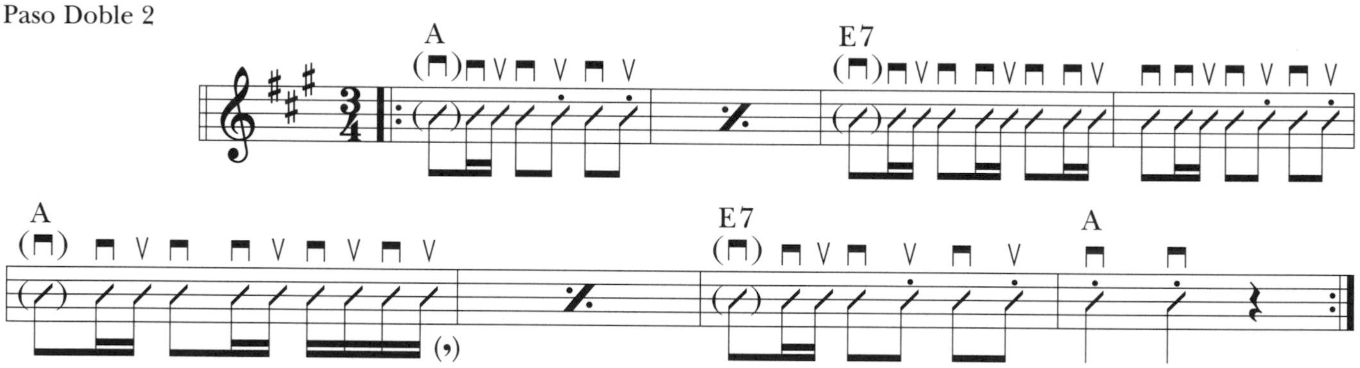

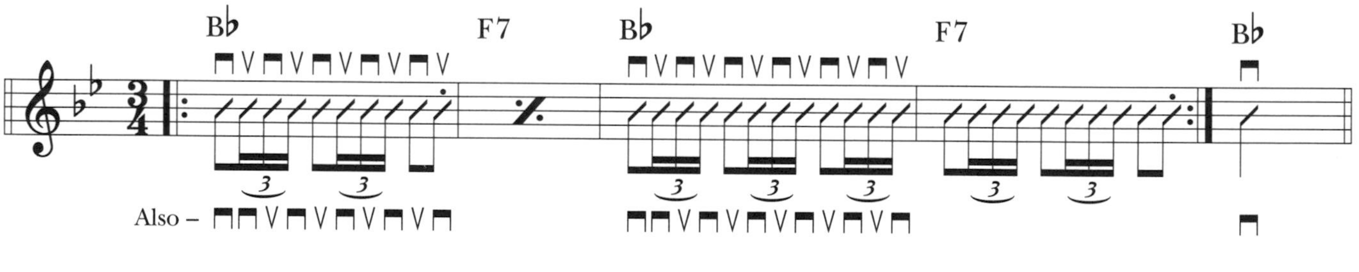

Chord Study—Minor 7 with ♭5 (Chromatic Approach) in the Bass

Five-Note Arpeggios
Minor (maj 7, 9) and Dominant (♭5, ♭9) Chords
Chord Spelling

139

Chord-Scale Relationships

For improvisation.

Minor 6 Chords

All minor 6 chords can be considered as representing the subdominant or tonic minor sound. Scale = Real melodic minor from chord name. (However, IIm6, Vm6, and VIm6 will sound slightly forced. See next relationship.)

IIm6, Vm6, VIm6 are best treated as representing the dominant sound of IIm6 for V9. Scale = major from a whole step below the minor 6 chord.

A comparison of other minor 6 chords with their related dominant 9 chords (containing the same notes) will reveal that IIIm6, #IVm6, VIIm6, and #Im6 can also be treated as IIm6 for V9. But the scale for this harmonic situation is real melodic minor from a whole step below the minor 6 chord.

Minor 7(♭5) Chords

Minor 7(♭5) chords most frequently represent the dominant sound of VIIm7(♭5) for V7. IIIm7(♭5), #IVm7(♭5), VIIm7(♭5) = Major scale from half step above chord.

All other minor 7(♭5) chords represent the subdominant or tonic minor sounds of IIm7(♭5) (for IVm6) or VIm7(♭5) (for Im6) = Real melodic minor from ♭3 of chord.

A comparison of minor 7(♭5) chords with non-scale tone roots (except #IV) with their related dominant 9 structures will reveal some second choice chord-scale relationships. Scale = Real melodic minor from half step above min 7(♭5).

(Ex: #Im7(♭5) = VI9, #IIm7(♭5) = VII9, #Vm7(♭5) = III9, #VIm7(♭5) = #IV9)

Also a minor 7(♭5) chord represents the II chord in a minor key. It is often treated as a "package deal" with the V7 of that minor key when it is the next chord. Example: Bm7(♭5) to E7 = The harmonic (or natural) minor scale for both chords. (It is always the option of the player to treat the chords in this situation as one unit or independently.)

A Major Scale (Twelve Positions)

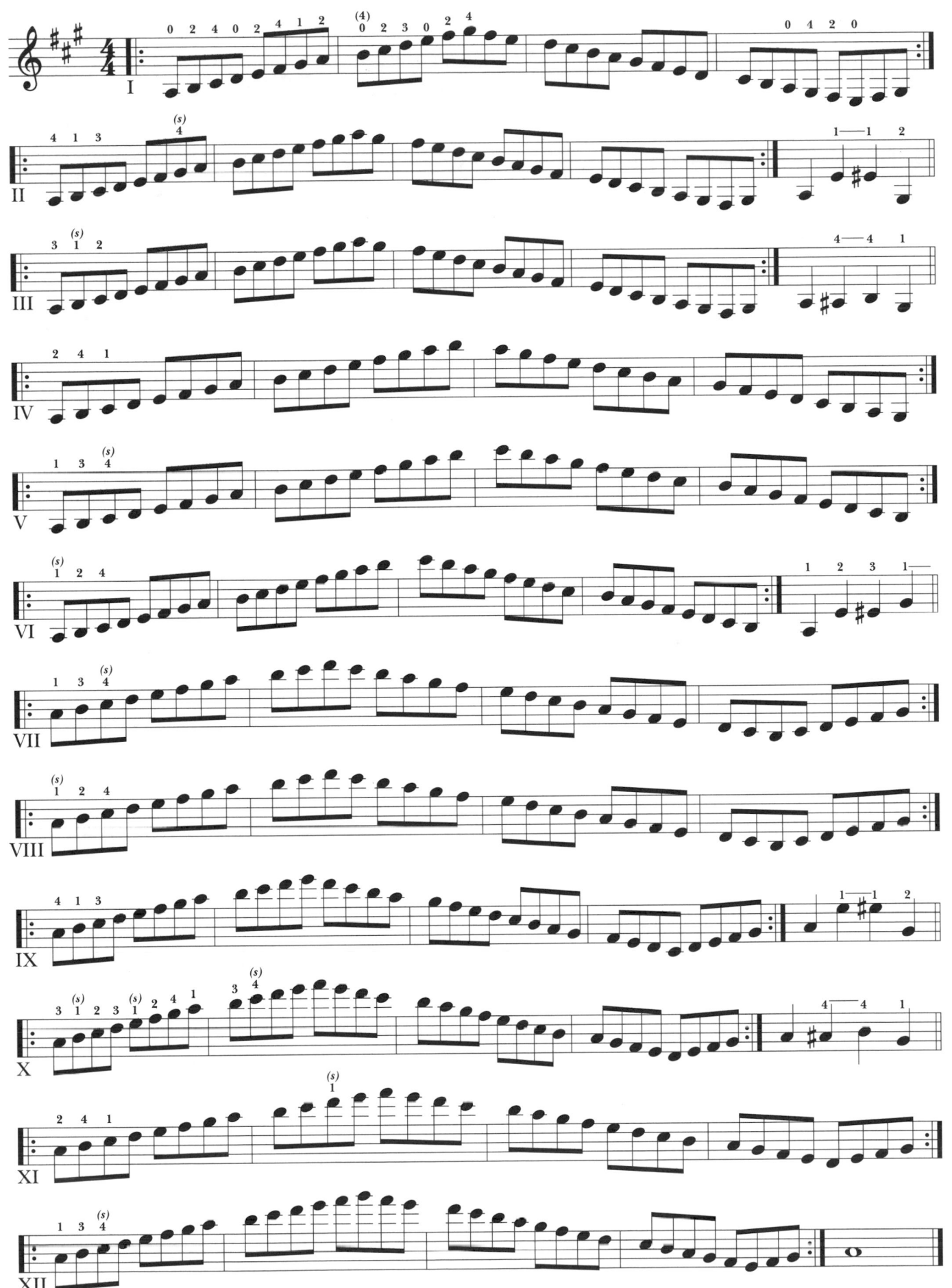

F# Harmonic Minor (Nine Positions)

F♯ Minor Etude (solo)

Five-Note Arpeggios
Dominant 7(♭9)sus4 and Dominant 9(♭5)
Chord Spelling

Chords—Three-Note Voicings
Study in F Major

Study in F Harmonic Minor

Five-Note Arpeggios
Dominant 7(+9)sus4 and Dominant 7(♭5, +9) Chords
Chord Spelling

✱ +9 on all sus4 chords enharmonically notated here as ♭3.

Additional Fingerings for Minor Scales

These fingerings are less practical for general use, as they will not accommodate as many interval combinations as those presented earlier.

The principal fingerings shown are a result of the alterations on the major scale fingering type from which the minor scale is derived.

The optional fingerings (shown in parentheses) suggest some of the combinations possible when fingering types are mixed. When all fingerings have been mastered by thorough and precise study, you can and will do this without conscious effort.

Real Melodic Minor

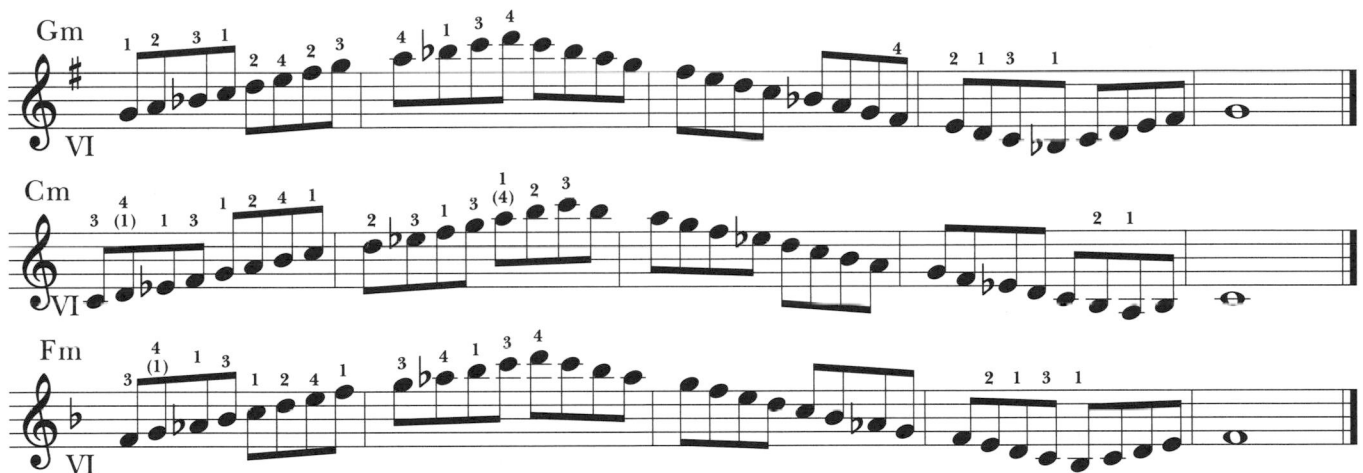

Harmonic Minor

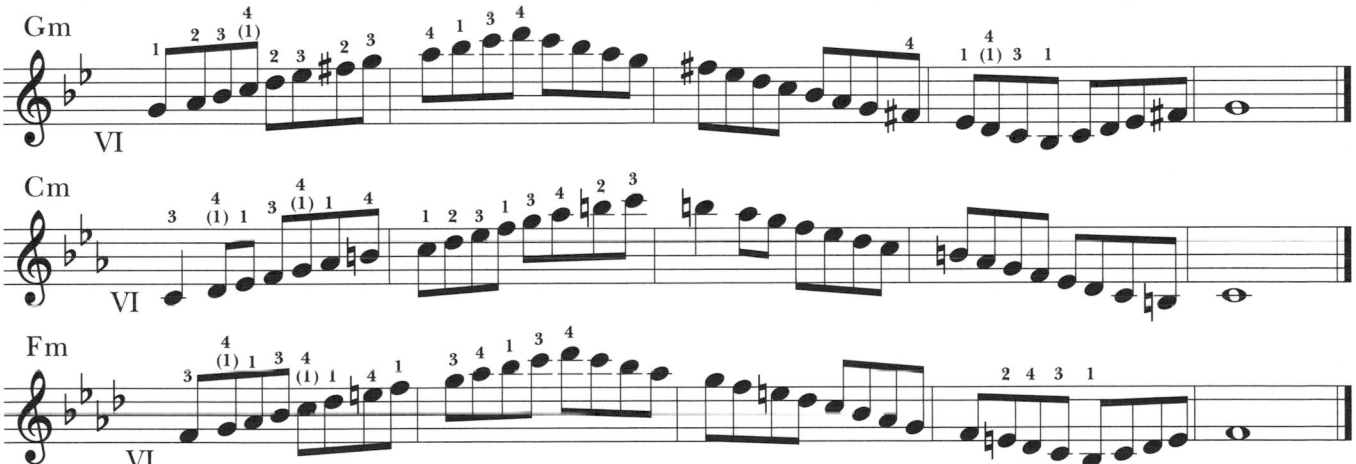

Chord-Scale Relationships

For improvisation.

Diminished 7 Chords

All diminished 7 chords will accept a diminished scale from any chord tone. (In most cases, these are not perfect relationships.)

Be advised that diminished 7 chord names are frequently misleading, in that most of the time they indicate only part of a larger harmonic structure. (The related scale remains hidden until the name of the complete chord is realized.)

The following will help in the proper treatment of diminished 7 chords.

Any diminished 7 chord that can be analyzed as:

♯I° almost always = VI7(♭9)	(Occasionally ♯I° = I7(♭9))
II° almost always = III7(♭9)	(Occasionally II° = V7(♭9))
I° is usually a true diminished 7, but is more musical when melodically treated as VII7(♭9).	(Occasionally I° = II7(♭9))

Also note: As all dominant 7(♭9) chords contain a diminished 7 built on 3, 5, ♭7, and ♭9 of the dominant 7, they will accept melodization with diminished scales from these notes. This chord-scale relationship is imperfect, but the uniformity of sound makes it work.

Augmented Triads

Augmented triads are primarily melodized with the whole tone scale from any chord tone (including 9).

As I and IV are the only scale degrees on which augmented structures could occur as strict triads, note the following relationships:

- I+ can be melodized with a harmonic or real melodic minor scale from a minor 3rd below the chord name. You may also use the real melodic minor from the intended tonic.

- IV+ can be melodized with the real melodic minor from a minor 3rd below.

Be advised that an augmented triad on anything other than I or IV is an incompletely named chord. Include the 7th in your analysis of these structures to determine the related scale.

Preparation of Four-Part Open Voicings
Adding the 5th Degree to Three-Part Open Voicings

Preparation of Four-Part Open Voicings
Adding the Root to Three-Part Open Voicings

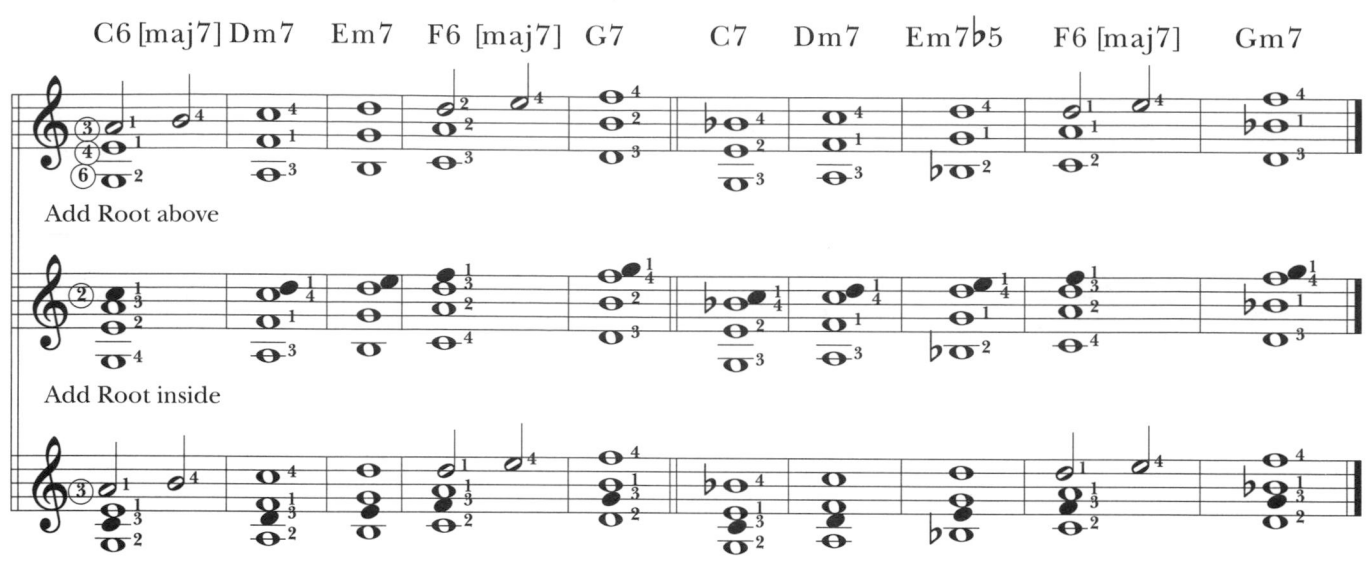

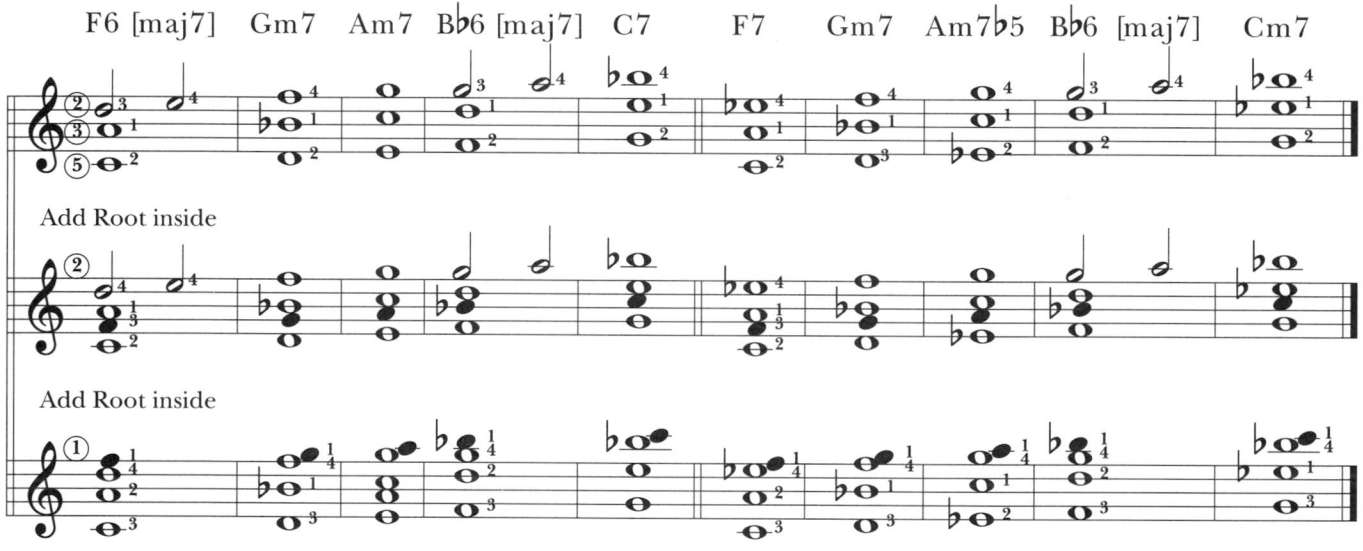

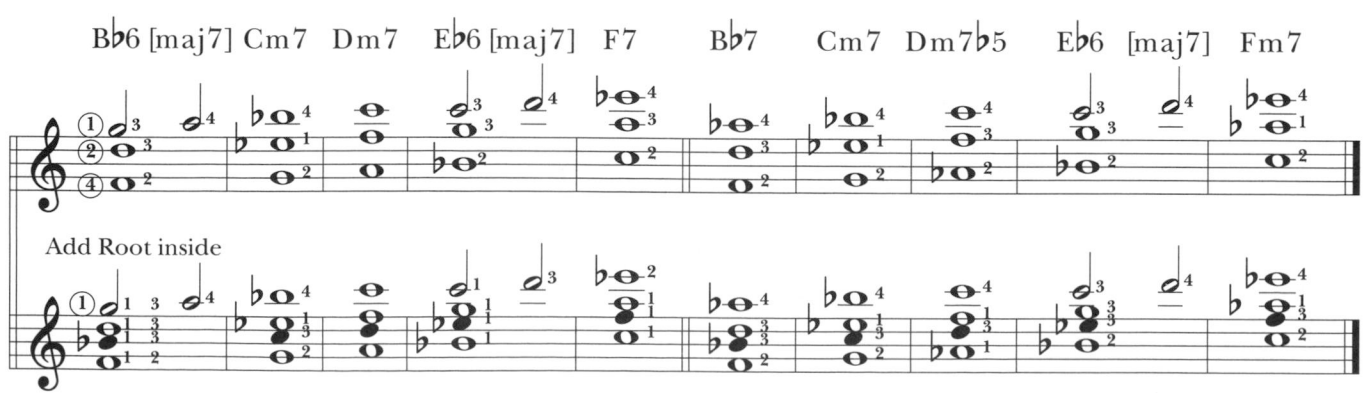

Five-Note Arpeggios
Minor 9(b5) Chords
Chord Spelling

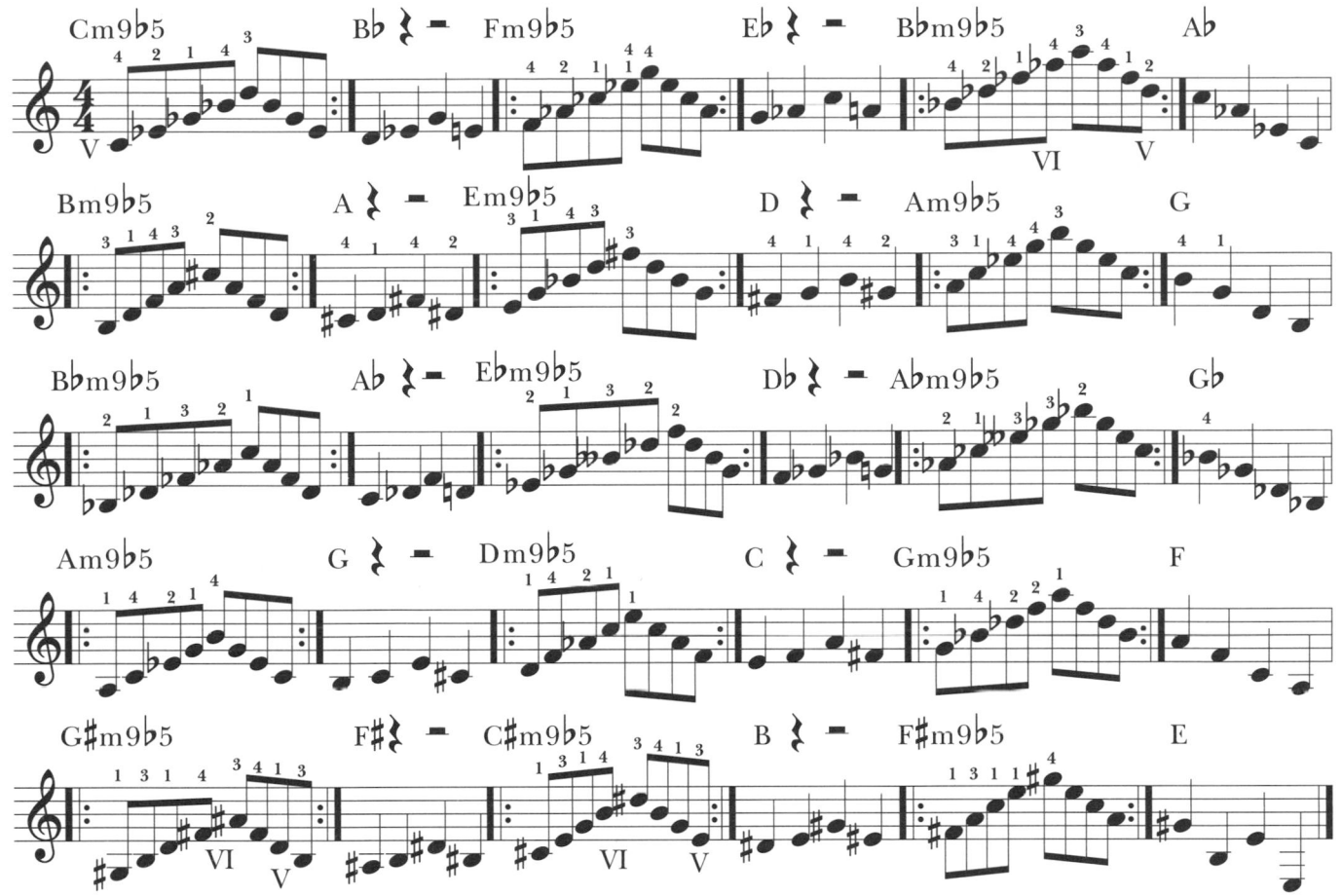

About Improvisation

Chord-scale relationships provide you with all the raw material (both melodic and harmonic) for any chord structure in any situation, but they will not make music for you.

In the final analysis, consideration must be given to each chord, for they contain a variety of sounds, such as the "warm" notes (3 and b7), the "bland" ones (1 and 5), and the various tensions and altered degrees that add the "sparkle" and/or the "buzz." Variety is certainly a factor in interesting music.

Also, very important are the "lines" that exist in a chord progression. These lines, resulting from the chromatic and scalewise movement of the inner voices of chords, form a solid basis for the creation of secondary melodies (especially valuable in comping.) Look for the chromatic motion that occurs between chords. Look for the tension and resolve possibilities available on each structure, for these are the pretty notes on which to build melodic ideas.

It's Late

Preparation of Four-Part Open Voicings
Adding the 3rd Degree to (Very Incomplete) Open Voicings

Because these voicings have the 7th (or 6th) degree as the bottom note, tonality must be established before using them.

No 3rd degree present—use with discretion:

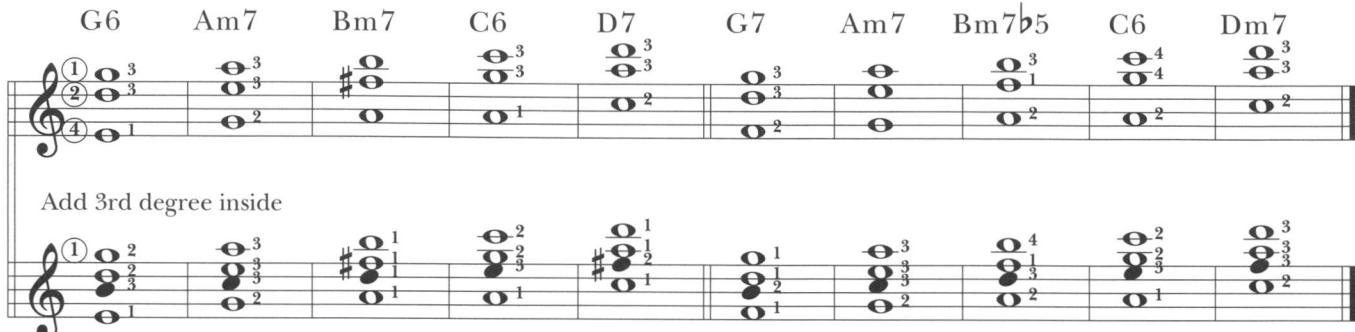

No 3rd degree present—use with discretion:

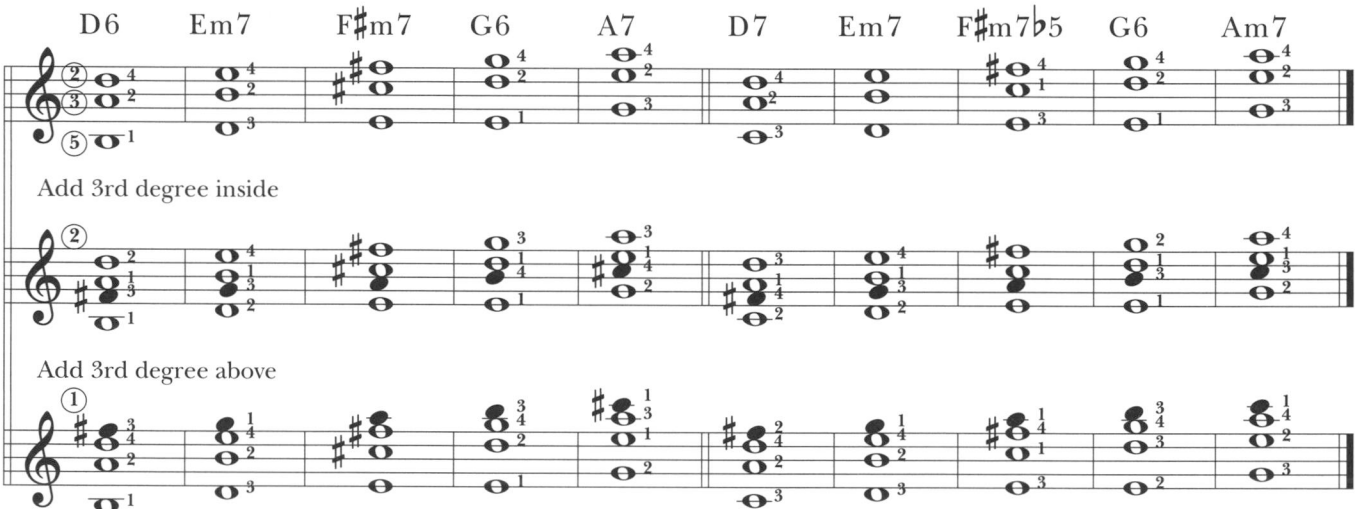

No 3rd degree present—use with discretion:

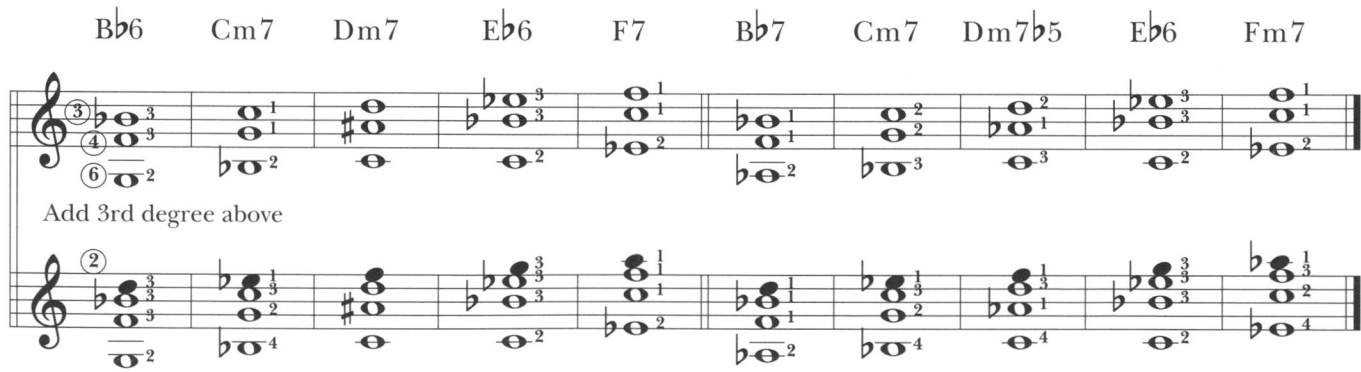

> Be especially careful of these 7ths (or 6ths) on the bottom in the low register. Observe rules for use.
> (See *Vol. II*, pg. 100.)

Preparation of Four-Part Open Voicings
Adding the 7th Degree to Three-Part Open Voicings

No 7th degrees present:

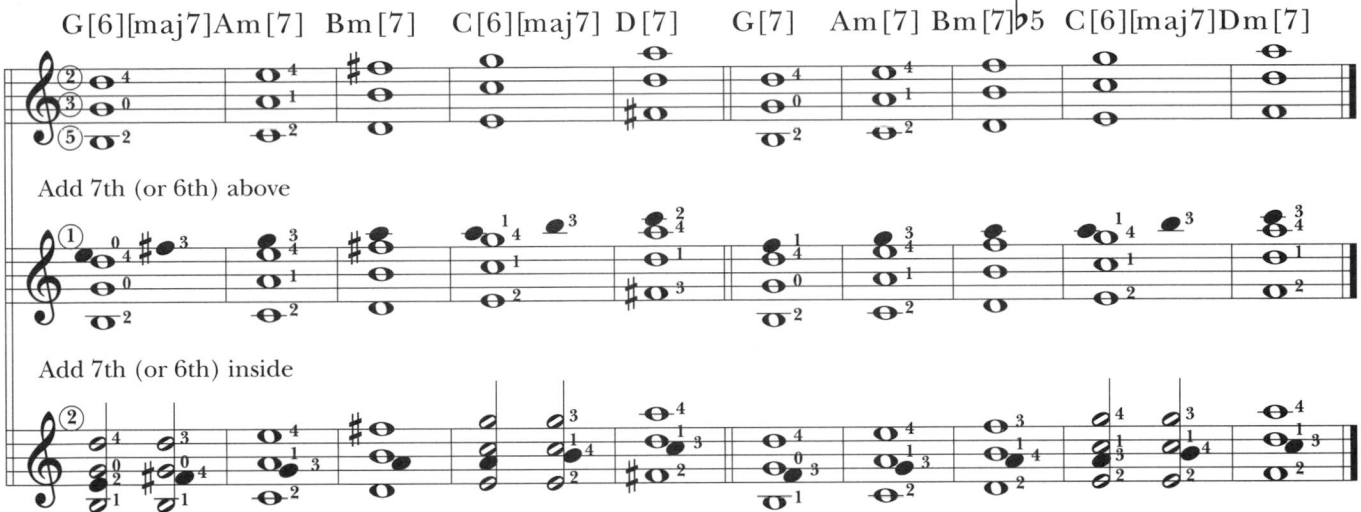

No 7th degrees present:

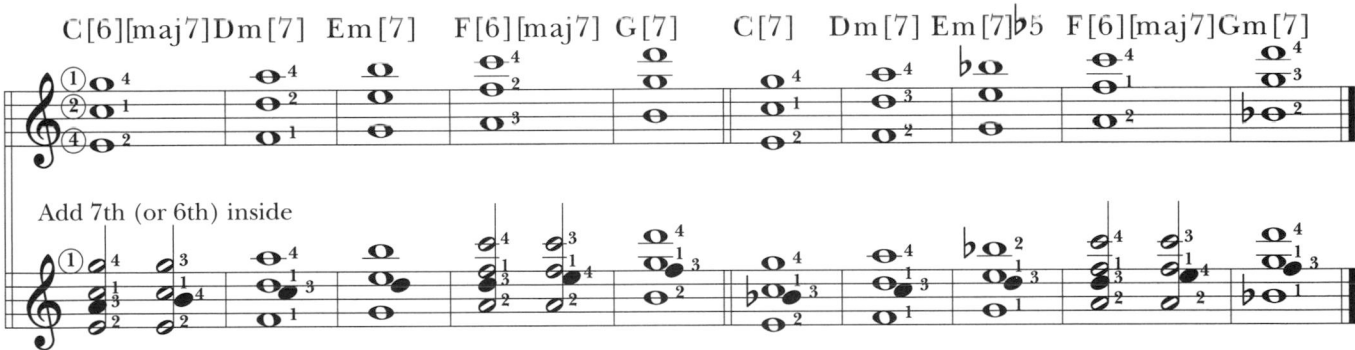

Scale-Chord Relationships
Major Scales

1. All diatonic structures in a major key.

2. All nondiatonic major chords with scale tone roots except IV, use the scale from chord name.

3. IV and all major chords with non-scale tone roots, use the scale from the 5th degree of the chord. (Note: This also includes all major chords with indicated +11 and is a second choice for the above-mentioned major chords with scale tone roots.)

4. All nondiatonic minor 7 chords (except VIIm7) usually function as IIm7. Use the scale from the ♭7 of the chord. (Note: VIIm7 is IIIm7 for I. Also note: A comparison of nondiatonic minor 7 structures with their related major 6 chords will reveal some second choice VIm7-for-I relationships.)

5. I7, II7, and all 13th chords with scale tone roots, except IV7, use the scale from the intended tonic.

6. III7, VI7(♭9), VII7 (second choice), use the scale from a major 3rd below the chord name. This can also be considered a natural minor scale from the intended tonic. (Note: The scale does not include the 3rd degree of the chord, and some melodic patterns may require the addition of this note.)

Harmonic Minor Scales

1. All diatonic structures in a minor key.

2. III7, VII7, VI7(♭9) (in a major key), use the scale from the intended tonic.

3. Diminished 7 chords that can be analyzed as I°, treat as VII7.
   Diminished 7 chords that can be analyzed as ♯I°, treat as VI7(♭9).
   Diminished 7 chords that can be analyzed as II°, treat as III7.

4. All dominant 7 chords with altered 9 and ♭13 (or ♯5 considered as ♭13), use the scale from the intended tonic.

5. I aug (I+) triad (second choice), use the scale from a minor 3rd below.

Real Melodic Minor Scales

1. IVm6 and Im6, use the scale from chord name.

2. IIm7(♭5) (occurring in a major tonality) treat as IVm6.

3. IV7 and all dominant 7 chords with non-scale tone roots, use the scale from the chord 5th. (Note: This is also a second-choice relationship for all dominant 7 chords, except IV7, with scale tone roots.)

4. All dominant 9 chords with specified +11 and 13, also I7(+11), II7(+11), V7(+11) (or ♭5 considered +11), scale from chord 5th.

5. All dominant 9ths with ♭13 (or ♯5 considered as ♭13), use the scale from the intended tonic. (Note: This includes VI7, III9, and VII9, each of which has a built-in ♭13.)

6. All completely altered dominant 7 chords (this means alt 9 and 5), use the scale from the ♭9 of the chord. This includes III7 (alt 5) and VII7 (alt 5), as they have a built-in altered 9. (Note: This can be a second choice of scale relationship for III7, VI7(♭9), and VII7 without the indicated alt 5 because of the built-in alterations of 9 and/or 13, and the fact that ♭13 can sometimes be treated as ♯5, or in this case, alt 5. However, the relationship is imperfect, so handle with care.)

7. I7(♭9), II7(♭9), V7(♭9), and all dominant 13(♭9) (or alt 9) chords with scale tone roots (except IV7), use the scale from the ♭7 of the chord. (Note: Scale does not contain the 3rd degree of the chord. Some melodic patterns may require the addition of this note.)

8. These do not occur very often. Use very cautiously:
Diminished 7 chords that can be analyzed as I°: treat as II7(♭9).
Diminished 7 chords that can be analyzed as ♯I°: treat as I7(♭9).
Diminished 7 chords that can be analyzed as II°: treat as V7(♭9).

9. I+ and IV+: Triads (second choice), use the scale from a minor 3rd below. I+ triad (second choice), use the scale from the intended tonic.

Whole Tone Scales

All major and dominant 7(♯5) (or alt 5), use the scale from any chord tone. (Note: The 9th must be unaltered in these structures. Whole tone scales are especially necessary for augmented dominant 7 chords with non-scale tone roots.

Diminished Scales

1. Diminished chords that can be analyzed as I°, use the scale from chord tones. (Note: This is theoretically more perfect than the previously mentioned treatment as VII7, but less musical.)

2. Diminished 7 chords that can be analyzed as ♯I° and II°, use the scale from chord tones. (Note: These are less perfect than the VI7(♭9) and III7 treatment, and less musical.)

3. IV7(♭9) and all dominant 7(♭9) chords with non-scale tone roots, use the scale from 3, 5, ♭7, ♭9 of chord. (Note: All dominant 7(♭9) chords may be treated in this manner with varying degrees of imperfection, however the consistent intervals of the scale will hold things together.)

4. All augmented 11(♭9) (or alt 9) chords, use the scale from 3, 5, ♭7, ♭9 of chord.

Remember: Look ahead to the next chord to analyze a progression.
Look back to the preceding chord to determine the related scale.

Index

More Fine Publications

Berklee Press

GUITAR

BEBOP GUITAR SOLOS
by Michael Kaplan
00121703 Book $16.99

BLUES GUITAR TECHNIQUE
by Michael Williams
50449623 Book/Online Audio...........$24.99

BERKLEE GUITAR CHORD DICTIONARY
by Rick Peckham
50449546 Jazz – Book.........................$12.99
50449596 Rock – Book........................$12.99

BERKLEE GUITAR STYLE STUDIES
by Jim Kelly
00200377 Book/Online Media $24.99

CLASSICAL TECHNIQUE FOR THE MODERN GUITARIST
by Kim Perlak
00148781 Book/Online Audio.............$19.99

CONTEMPORARY JAZZ GUITAR SOLOS
by Michael Kaplan
00143596 ....$16.99

CREATIVE CHORDAL HARMONY FOR GUITAR
by Mick Goodrick and Tim Miller
50449613 Book/Online Audio.............$19.99

FUNK/R&B GUITAR
by Thaddeus Hogarth
50449569 Book/Online Audio $19.99

GUITAR CHOP SHOP – BUILDING ROCK/METAL TECHNIQUE
by Joe Stump
50449601 Book/Online Audio $19.99

GUITAR SWEEP PICKING
by Joe Stump
00151223 Book/Online Audio.............$19.99

INTRODUCTION TO JAZZ GUITAR
by Jane Miller
00125041 Book/Online Audio $19.99

JAZZ GUITAR FRETBOARD NAVIGATION
by Mark White
00154107 Book/Online Audio $19.99

JAZZ SWING GUITAR
by Jon Wheatley
00139935 Book/Online Audio.............$19.99

A MODERN METHOD FOR GUITAR*
by William Leavitt
Volume 1: Beginner
00137387 Book/Online Video $24.99
*Other volumes, media options, and supporting songbooks available.

A MODERN METHOD FOR GUITAR SCALES
by Larry Baione
00199318 Book........................... $10.99

Berklee Press publications feature material developed at the Berklee College of Music.
To browse the complete Berklee Press Catalog, go to
www.berkleepress.com

BASS

BASS LINES
Fingerstyle Funk
by Joe Santerre
50449542 Book/Online Audio $19.95
Metal
by David Marvuglio
00122465 Book/Online Audio.............$19.99
Rock
by Joe Santerre
50449478 Book/CD $19.95

BERKLEE JAZZ BASS
by Rich Appleman, Whit Browne, and Bruce Gertz
50449636 Book/Online Audio $19.99

FUNK BASS FILLS
by Anthony Vitti
50449608 Book/Online Audio...........$19.99

INSTANT BASS
by Danny Morris
50449502 Book/CD $9.99

VOICE

BELTING
by Jeannie Gagné
00124984 Book/Online Media $19.99

THE CONTEMPORARY SINGER – 2ND ED.
by Anne Peckham
50449595 Book/Online Audio $24.99

JAZZ VOCAL IMPROVISATION
by Mili Bermejo
00159290 Book/Online Audio $19.99

TIPS FOR SINGERS
by Carolyn Wilkins
50449557 Book/CD............................$19.95

VOCAL TECHNIQUE
featuring Anne Peckham
50448038 DVD...$19.95

VOCAL WORKOUTS FOR THE CONTEMPORARY SINGER
by Anne Peckham
50448044 Book/Online Audio..........$24.99

YOUR SINGING VOICE
by Jeannie Gagné
50449619 Book/Online Audio $29.99

WOODWINDS/BRASS

TRUMPET SOUND EFFECTS
by Craig Pederson & Ueli Dörig
00121626 Book/Online Audio..............$14.99

SAXOPHONE SOUND EFFECTS
by Ueli Dörig
50449628 Book/Online Audio $15.99

THE TECHNIQUE OF THE FLUTE: CHORD STUDIES, RHYTHM STUDIES
by Joseph Viola
00214012 Book.......................................$19.99

PIANO/KEYBOARD

BERKLEE JAZZ KEYBOARD HARMONY
by Suzanna Sifter
00138874 Book/Online Audio............$24.99

BERKLEE JAZZ PIANO
by Ray Santisi
50448047 Book/Online Audio $19.99

BERKLEE JAZZ STANDARDS FOR SOLO PIANO
Arranged by Robert Christopherson, Hey Rim Jeon, Ross Ramsay, Tim Ray
00160482 Book/Online Audio............$19.99

CHORD-SCALE IMPROVISATION FOR KEYBOARD
by Ross Ramsay
50449597 Book/CD.............................$19.99

CONTEMPORARY PIANO TECHNIQUE
by Stephany Tiernan
50449545 Book/DVD $29.99

HAMMOND ORGAN COMPLETE
by Dave Limina
50449479 Book/CD $24.99

JAZZ PIANO COMPING
by Suzanne Davis
50449614 Book/Online Audio $19.99

LATIN JAZZ PIANO IMPROVISATION
by Rebecca Cline
50449649 Book/Online Audio $24.99

SOLO JAZZ PIANO – 2ND ED.
by Neil Olmstead
50449641 Book/Online Audio $39.99

DRUMS

BEGINNING DJEMBE
by Michael Markus & Joe Galeota
00148210 Book/Online Video $16.99

BERKLEE JAZZ DRUMS
by Casey Scheuerell
50449612 Book/Online Audio............$19.99

DRUM SET WARM-UPS
by Rod Morgenstein
50449465 Book.....................................$12.99

A MANUAL FOR THE MODERN DRUMMER
by Alan Dawson & Don DeMichael
50449560 Book.....................................$14.99

MASTERING THE ART OF BRUSHES – 2ND EDITION
by Jon Hazilla
50449459 Book/Online Audio $19.99

PHRASING: ADVANCED RUDIMENTS FOR CREATIVE DRUMMING
by Russ Gold
00120209 Book/Online Media............$19.99

WORLD JAZZ DRUMMING
by Mark Walker
50449568 Book/CD $22.99

STRINGS/ROOTS MUSIC

BERKLEE HARP
Chords, Styles, and Improvisation for Pedal and Lever Harp
by Felice Pomeranz
00144263 Book/Online Audio $19.99

BEYOND BLUEGRASS
Beyond Bluegrass Banjo
by Dave Hollander and Matt Glaser
50449610 Book/CD $19.99

Beyond Bluegrass Mandolin
by John McGann and Matt Glaser
50449609 Book/CD $19.99

Bluegrass Fiddle and Beyond
by Matt Glaser
50449602 Book/CD $19.99

EXPLORING CLASSICAL MANDOLIN
by August Watters
00125040 Book/Online Media $19.99

THE IRISH CELLO BOOK
by Liz Davis Maxfield
50449652 Book/Online Audio..........$24.99

JAZZ UKULELE
by Abe Lagrimas, Jr.
00121624 Book/Online Audio............$19.99

BERKLEE PRACTICE METHOD

GET YOUR BAND TOGETHER
With additional volumes for other instruments, plus a teacher's guide.

Bass
by Rich Appleman, John Repucci and the Berklee Faculty
50449427 Book/CD $16.99

Drum Set
by Ron Savage, Casey Scheuerell and the Berklee Faculty
50449429 Book/CD $14.95

Guitar
by Larry Baione and the Berklee Faculty
50449426 Book/CD $16.99

Keyboard
by Russell Hoffmann, Paul Schmeling and the Berklee Faculty
50449428 Book/Online Audio $14.99

WELLNESS

MANAGE YOUR STRESS AND PAIN THROUGH MUSIC
by Dr. Suzanne B. Hanser and Dr. Susan E. Mandel
50449592 Book/CD $29.99

MUSICIAN'S YOGA
by Mia Olson
50449587 Book $17.99

THE NEW MUSIC THERAPIST'S HANDBOOK – 3RD EDITION
by Dr. Suzanne B. Hanser
00279325 Book....................$29.99

AUTOBIOGRAPHY

LEARNING TO LISTEN: THE JAZZ JOURNEY OF GARY BURTON
by Gary Burton
00117798 Book $27.99

♭ HAL•LEONARD®

Prices subject to change without notice. Visit your local music dealer or bookstore, or go to **www.berkleepress.com**

MUSIC THEORY/EAR TRAINING/ IMPROVISATION

BEGINNING EAR TRAINING
by Gilson Schachnik
50449548 Book/Online Audio $16.99

THE BERKLEE BOOK OF JAZZ HARMONY
by Joe Mulholland & Tom Hojnacki
00113755 Book/Online Audio...........$27.50

BERKLEE MUSIC THEORY – 2ND ED.
by Paul Schmeling
Rhythm, Scales Intervals
50449615 Book/Online Audio $24.99
Harmony
50449616 Book/Online Audio $22.99

IMPROVISATION FOR CLASSICAL MUSICIANS
by Eugene Friesen with Wendy M. Friesen
50449637 Book/CD $24.99

REHARMONIZATION TECHNIQUES
by Randy Felts
50449496 Book....................$29.99

MUSIC BUSINESS

ENGAGING THE CONCERT AUDIENCE
by David Wallace
00244532 Book/Online Media..........$16.99

HOW TO GET A JOB IN THE MUSIC INDUSTRY – 3RD EDITION
by Keith Hatschek with Breanne Beseda
00130699 Book....................$27.99

MAKING MUSIC MAKE MONEY
by Eric Beall
50448009 Book $27.99

MUSIC LAW IN THE DIGITAL AGE – 2ND EDITION
by Allen Bargfrede
00148196 Book....................$19.99

MUSIC MARKETING
by Mike King
50449588 Book....................$24.99

PROJECT MANAGEMENT FOR MUSICIANS
by Jonathan Feist
50449659 Book....................$27.99

THE SELF-PROMOTING MUSICIAN – 3RD EDITION
by Peter Spellman
00119607 Book....................$24.99

MUSIC PRODUCTION & ENGINEERING

AUDIO MASTERING
by Jonathan Wyner
50449581 Book/CD....................$29.99

AUDIO POST PRODUCTION
by Mark Cross
50449627 Book $19.99

THE SINGER-SONGWRITER'S GUIDE TO RECORDING IN THE HOME STUDIO
by Shane Adams
00148211 Book $16.99

UNDERSTANDING AUDIO – 2ND EDITION
by Daniel M. Thompson
00148197 Book $24.99

SONGWRITING, COMPOSING, ARRANGING

ARRANGING FOR HORNS
by Jerry Gates
00121625 Book/Online Audio............$19.99

BEGINNING SONGWRITING
by Andrea Stolpe with Jan Stolpe
00138503 Book/Online Audio $19.99

BERKLEE CONTEMPORARY MUSIC NOTATION
by Jonathan Feist
00202547 Book....................$17.99

COMPLETE GUIDE TO FILM SCORING – 2ND ED.
by Richard Davis
50449607 ....$29.99

CONTEMPORARY COUNTERPOINT: THEORY & APPLICATION
by Beth Denisch
00147050 Book/Online Audio.........$22.99

THE CRAFT OF SONGWRITING
by Scarlet Keys
00159283 Book/Online Audio...........$19.99

JAZZ COMPOSITION
by Ted Pease
50448000 Book/Online Audio $39.99

MELODY IN SONGWRITING
by Jack Perricone
50449419 Book....................$24.99

MODERN JAZZ VOICINGS
by Ted Pease and Ken Pullig
50449485 Book/Online Audio $24.99

MUSIC COMPOSITION FOR FILM AND TELEVISION
by Lalo Schifrin
50449604 Book $34.99

MUSIC NOTATION
PREPARING SCORES AND PARTS
by Matthew Nicholl and Richard Grudzinski
50449540 Book....................$16.99

MUSIC NOTATION
THEORY AND TECHNIQUE FOR MUSIC NOTATION
by Mark McGrain
50449399 Book....................$24.95

POPULAR LYRIC WRITING
by Andrea Stolpe
50449553 Book $15.99

SONGWRITING: ESSENTIAL GUIDE
Lyric and Form Structure
by Pat Pattison
50481582 Book....................$16.99
Rhyming
by Pat Pattison
00124366 2nd Ed. Book $17.99

SONGWRITING IN PRACTICE
by Mark Simos
00244545 Book....................$16.99

SONGWRITING STRATEGIES
by Mark Simos
50449621 Book....................$24.99

THE SONGWRITER'S WORKSHOP
Harmony
by Jimmy Kachulis
50449519 Book/Online Audio $29.99
Melody
by Jimmy Kachulis
50449518 Book/Online Audio $24.99